StarBriefs 2001

StarBriefs 2001

A Dictionary
of Abbreviations, Acronyms and Symbols in Astronomy, Related Space Sciences and Other Related Fields

Compiled by

André Heck
Strasbourg Astronomical Observatory, France

KLUWER ACADEMIC PUBLISHERS
DORDRECHT / BOSTON / LONDON

A C.I.P. Catalogue record for this book is available from the Library of Congress.

ISBN 0-7923-6510-0 (StarBriefs 2001)

ISBN 0-7923-6509-7 (StarGuides 2001, companion volume to StarBriefs 2001)

ISBN 0-7923-6511-9 (2 Volume set)

Published by Kluwer Academic Publishers,
P.O. Box 17, 3300 AH Dordrecht, The Netherlands.

Sold and distributed in North, Central and South America
by Kluwer Academic Publishers,
101 Philip Drive, Norwell, MA 02061, U.S.A.

In all other countries, sold and distributed
by Kluwer Academic Publishers,
P.O. Box 322, 3300 AH Dordrecht, The Netherlands.

Printed on acid-free paper

Printed in the Netherlands.

Foreword

This compilation probably looks like one of the craziest things a human being could spend his or her time on. Yet nobody would wonder at someone taking a short walk every day – after twenty five years that person would have covered a surprisingly long distance. This is exactly the story behind this list, which appeared first as a few pages within the directory *StarGuides* (or whatever name it had at that time) and as a distinct sister publication since 1990.

The idea behind this dictionary is to offer astronomers and related space scientists practical assistance in decoding the numerous abbreviations, acronyms, contractions and symbols which they might encounter in all aspects of the vast range of their professional activities, including traveling. Perhaps it is a bit paradoxical, but if scientists quickly grasp the meaning of an acronym solely in their own specific discipline, they will probably encounter more difficulties when dealing with adjacent fields. It is for this purpose that this dictionary might be most often used. Scientists might also refer to this compilation in order to avoid identifying a project by an acronym which already has too many meanings or confused definitions.

Hence *StarBriefs 2001*, the enlarged and fully updated version of a series of dictionaries, gathers together more than 140,000 entries encountered in the literature of astronomy and space sciences, as well as of related fields such as aeronautics, aeronomy, astronautics, atmospheric sciences, chemistry, communications, computer sciences, data processing, education, electronics, energetics, engineering, environment, geodesy, geophysics, information handling, management, mathematics, meteorology, optics, physics, remote sensing, and so on.

Abbreviations, acronyms, contractions and symbols in common use and/or of general interest have also been included where appropriate. The traveling scientist has also not been forgotten, with the inclusion of the codes of locations, currencies and so on. Separate sections are devoted to Greek letters, mathematical symbols, special signs and characters, as well as to entries with a numerical beginning.

This compilation is essentially carried out in parallel with the permanent updating of the master files for *StarGuides*, a directory of astronomy, space sciences and related organizations of the world made available as a sister publication. In practice, all relevant abbreviations and acronyms encountered when scanning the literature received relating to the *StarGuides* directory are gathered here, the underlying principle being that they might also appear one day under the eyes of astronomers and related space scientists. We insist that these are ‘real’ abbreviations and acronyms, in other words that they are not constructed automatically from dictionaries as in some resources.

The master files for the present compilation of *StarBriefs 2001* have been made accessible as an on-line database at the *European Southern Observatory (ESO)* under the label *StarWords* (Albrecht & Heck 1994). *StarBriefs* files are reachable at the *Centre de Données astronomiques de Strasbourg (CDS)* as the database *StarBits* (Heck *et al.* 1994) via the URL http://vizier.u-strasbg.fr/starbits.html. All these resources belong to the *Star*s Family of Astronomy and Related Resources* (Heck 1997).

The associated directory *StarGuides 2001* (ISBN 0-7293-6509-7) and its derived databases gather together all practical data available on associations, societies, scientific committees, agencies, companies, institutions, universities, etc., and more generally organizations, involved in astronomy and space sciences. Many other types of entry have also been included such as academies, advisory and expert committees, bibliographic services, consultants, data and documentation centres, dealers, distributors, funding agencies and organizations, journals, manufacturers, meteorological services, museums, norms and standards offices, planetariums, public observatories, publishers, research institutions in related fields, software producers and distributors, etc.

StarBriefs is permanently updated and new versions will be released regularly in order to cope with the continuous development of new projects and new experiments. We also allowed an extensive number of technical acronyms to enter the list, because it is often difficult to find their explanations in other sources. Deviations from standard or 'official' acronyms are also listed if their frequency of occurrence in the literature justifies it – in such cases, the recommended explanation is flagged with an asterisk.

It should also be noted that the manner of abbreviation is not always carried out in the same way in all countries. The use of full points seems to be one of the features which varies the most (for instance, in some countries they might be used after contractions whilst this would not be the case in other countries).

The meaning of abbreviations, acronyms, contractions and symbols should always be explained at their first occurrence in any document. It would be even more practical and helpful if authors and editors systematically tabulated them, with their explanations, at the end of the papers, chapters, reports, proceedings, books, etc., they produce. Such a policy would at least remove any ambiguity, since, as shown many times in the following pages, a given group of characters may have numerous meanings.

Socio-linguists might be interested in the following table giving the sixty most frequent terms (stop words excluded) appearing in the acronym explanations out of a total of about 89,200 different words (at least two characters) for the whole dictionary. The combination of the most frequent term ('System') and of its plural ('Systems', ranking 46th) accounts for 9.53% of all words together (about 516,000). The last term of the table ('Review', ranking 60th) accounts for 0.67%.

Rank	Word	Freq.	Rank	Word	Freq.	Rank	Word	Freq.
1	System	7738	21	Technology	1201	41	Integrated	801
2	Data	3594	22	High	1129	42	Experiment	794
3	Information	2930	23	Engineering	1015	43	Automatic	785
4	Control	2859	24	Group	1008	44	Flight	775
5	Center/re	2471	25	Interface	972	45	New	773
6	Program/me	2050	26	Committee	951	46	Systems	766
7	Test	1893	27	Support	949	47	Technical	710
8	International	1844	28	Digital	940	48	Access	706
9	Research	1765	29	Satellite	911	49	Assembly	703
10	Unit	1609	30	Development	905	50	Software	698
11	Management	1607	31	Processing	883	51	Remote	696
12	Space	1516	32	Institute	883	52	University	692
13	Analysis	1486	33	Society	873	53	Energy	667
14	Science	1453	34	Facility	868	54	Optical	652
15	Association	1435	35	Language	860	55	Mission	646
16	Computer	1424	36	Operations	855	56	Standard	640
17	National	1385	37	Equipment	839	57	Design	640
18	Network	1364	38	Advanced	836	58	Power	619
19	Time	1256	39	Office	822	59	Laboratory	606
20	Service	1256	40	European	807	60	Review	602

A first conclusion from this pretty good picture of our everyday jargon is that all terms are English/American ones (in spite of the fact that the dictionary is multilingual). Whilst 'American' is not listed in the most frequent terms, 'European' does appear (ranking 40th), probably conditioned by the need to emphasize the differences with 'National' (ranking 17th) and 'International' (ranking 8th) projects and organizations. 'Space' (ranking 12th) is included in the top sixty most frequent terms, whilst 'Astronomical' is just after this leading group, in 65th position with 590 occurrences. The next directly astronomy-related term is 'Solar' in 71st position with 561 occurrences. 'Telescope' and 'Observatory' score respectively 549 and 538 appearances whilst 'Astronomy' is way behind with only 376 occurrences, a bit below 'Physics' with 412.

When compiling catalogues such as *StarGuides* and *StarBriefs*, one cannot help but be impressed by the very broad spectrum of disciplines to which astronomy and related space sciences are linked, and by the very rich variety of techniques applied in these fields.

Acknowledgements

The implementation, as databases, of the *Star*s Family* products by the European Space Agency, the European Southern Observatory and Strasbourg Astronomical Data Centre have been strong incentives to continue and always improve these time-consuming compilations for the benefit of the best possible communication within the astronomical community, as well as between it and the outside world.

June 2000.

REFERENCES

Albrecht, M.A. & Heck, A.: 1994, StarWords – An On-line Database of Acronyms, Abbreviations, and Contractions in Astronomy, Space Sciences, and Related Fields, *Astron. Astrophys. Suppl.* **103** 471.

Heck, A.: 1997, Electronic Yellow-Page Services: The Star*s Family as an Example of Diversified Publishing, in *Electronic Publishing for Physics and Astronomy*, Ed. A. Heck, Kluwer Acad. Publ., Dordrecht, 211-220.

Heck, A., Egret, D. & Ochsenbein, F.: 1994, StarWorlds – StarBits (announcement of two databases), *Astron. Astrophys. Suppl.* **108**, 447-448.

Author's address: **André HECK**
Observatoire Astronomique
11, rue de l'Université
F-67000 Strasbourg
France
Telephone: (+33)(0) 388 150 743
Telefax: (+33)(0) 388 491 255
Electronic mail: starpages@astro.u-strasbg.fr
WWW: http://vizier.u-strasbg.fr/~heck

Dictionary

The asterisks (*) indicate recommended explanations among several similar ones.

A

a: About • Absorptivity • Acceleration • Accepté • Accepted • Acre • Acreage • Acting • Active • Actual • Address • Adjective • Advance • Advanced • Afternoon • Age • Akzeptiert • Alto • Am • Amateur • Ampere • An • Année • Anno • Anonymous • Answer • Ante • Antes • Are • Arrive • Arriving • At • Atto • Auf • Deceleration • Electron Annihilation Operator • Lattice Space of Si • Linear Absorption Coefficient • Mean Sound Absorption • Radiation Density Constant • Semimajor Axis • Sonic Velocity • Specific Acidity • Specific Helmholtz Free Energy • Speed of Sound • Stefan-Boltzmann Radiation Constant • Surface Albedo • Thermal Diffusivity • Universal Affirmative • Year

Å: Ångström

A: Abell • Absolute • Absorbance • Absorption • Academician • Academy • Ace • Acre • Acreage • Activity • Adjutant • Adult • Advanced • Aerodynamic Reference Area • Affinity • Air • Albedo • Alp • Alpha (international alphabet) • Alt • Altesse • Alto • America • American • Ammeter • Ampere • Ampere-Turn • Amplitude • Analog • Anchor • Anemic • Anno • Annual • Anode • Answer • Ante • Apogee • Appendix • April • Area • Ariel • Argon • Arroyo • Art • Arterial • Assistant • Associate • Atom • Atomic • Atomic Weight • August • Austral (currency) • Australia • Australian • Austria (international vehicle registration code) • Austria (postal country code) • Azimuth • Cross Section • Distance Unit • Equivalent Absorption Area • Five Hundred (Roman numeral) • Helmholtz Free Energy • Helmholtz Function • Interstellar Extinction • Linear Current Density • Magnetic Vector Potential • Mass Number • Nuclear Number • Nucleon Number • Oort's Constant of Galactic Rotation • Optical Density • Permeance • Richardson Constant • Rotational Constant • Ten • Spectral Type • Standard Paper Size • Universal Affirmative • Work • 500 (Roman numeral)

aa: Acting Appointment • Afar (ISO language code) • After Arrival • Always Afloat • Años • Approximate Absolute • Area Administrator • Arithmetic Average • Attendance Allowance • Author's Alteration

A/A: Airplane Avionics • Air-to-Air • Analog-to-Analog • Angular Accelerometer

A&A: Additions & Amendments • Advertise & Award • Astronomy and Astrophysics (periodical code)

AA: Academic Affairs • Accelerated Assembly • Accelerometer Assembly • Access Authorization • Achieved Availability • Acte Authentique • Acting Appointment • Adjoint Administratif • Administrative Assistant • Advertise and Award • Affirmative Action • After Arrival • Age Allowance • Agent Administratif • Airplane Avionics • Air-to-Air • Aitreya Aranyaka • Alcoholics Anonymous • Altezas • Angular Accelerometer • Angular Aperture • Antiaircraft • Antiproton Accumulator • Approximate Absolute • Ardennes-Astronomie (France) • Area Administrator • Arithmetic Average • Artificial Agent • Assistant Administrator • Associate Administrator • Associate of Arts • Astrograph Astrometry • Astronautica Acta • Astronomy and Astrophysics • Atlas Agena • Atomic Absorption • Atomic Adsorption • Attendance Allowance • Auswärtiges Amt • Author's Alteration • Auto-Analysis • Auto-Answer • Automobile Association • Size of Battery • Size of Cell

AAA: Academic Assistance Access • Accélérométrie Astronomique Absolue • Advanced Aircraft Analysis • Agricultural Adjustment Administration • Agrupación Astronómica Aragonesa (Spain) • Agrupación Astrónomos Aficionados (Venezuela) • Agrupación de Aficionados a la Astronomía (Uruguay) • Airframe Airworthiness Assurance • Allocation-Assessment-Analysis • Amateur Astronomers Association (USA-NY) • Amateur Astronomers of Acadiana (USA-LA) • American Accounting Association (USA) • American Arbitration Association (USA) • American Automobile Association (USA) • Amor-Aten-Apollo • Anaa (Tuamotu Islands) (location identifier) • Anno Ante Christum • Apollo-Amor-Aten • Asociación Argentina de Astronomía (Argentina) • Assistant Associate Administrator • Association Astronomique d'Anjou (France) • Association Astronomique de l'Ain (France) • Association d'Astronomie de l'Artois (France) • Association of Amateur Astronomers (USA-TX) • Associazione Acquavivese Astrofili (Italy) • Astronaut-Actuated Abort • Astronomy and Astrophysics Abstracts • Aten-Apollo-Amor • Author's Argumentation Assistant • Automatic Amplitude Control

AAAA: Albany Area Amateur Astronomers (USA-NY) • American Association of Amateur Astronomers (USA-TX) • Ames Area Amateur Astronomers (USA-IA) • Ars Ad Astra Association (Switzerland) • Asociación Argentina Amigos de la Astronomía • Association des Astronomes Amateurs d'Auvergne (France) • Atlas Automatisé des Absorptions Atmosphériques • Automatized Atmospheric Absorption Atlas

AAAAA: Anne Arundel Association of Amateur Astronomers (USA-MD)

AAAAT: Association des Astronomes Amateurs Abitibi-Témiscamingue (Canada-QC)

AAACE: American Association for Adult and Continuing Education (USA)

AAAD: Amateur Astronomers Association of Delhi (India)

AAAE: Alliance of Associations for the Advacement of Education (USA) • American Association for Adult Education (USA) • American Association of Academic Editors (USA) • American Association of Airport Executives (USA)

AAAF: Association Aéronautique et Astronautique de France

AAAI: American Association for Artificial Intelligence (USA) • Association of Australian Aerospace Industries

AAAL: American Academy of Arts and Letters (USA)

AAAP: Amateur Astronomers Association of Pittsburgh (USA-PA) • Amateur Astronomers Association of Princeton (USA-NJ)

AAAPT: Asian-African Association for Plasma Training

AAAS: American Academy of Arts and Sciences (USA) • American Association for the Advancement of Science (USA) • Arizona Amateur Astronomy Society (USA-AZ) • Automated Airport Advisory Service

AAATE: Association for Advancement of Assistive Technology in Europe

AAB: Academic Advisory Board • Adaptive Abstract Builder • All-to-All Broadcast • Arrabury (Australia-QLD) (location identifier) • Association d'Astronomie El-Battani (Algeria) • Associazione Astrofili Bolognesi (Italy) • Astronomischer Arbeitskreis Bamberg (Germany)

AABNCP: Advanced Airborne National Command Post

AABV: Associazione Astrofili del Basso Vicentino (Italy)

AABW: Antarctic Bottom Water

AAC: Academic Advisory Council • Aeronautical Administrative Communication • Agrupación Astronómica Cántabra (Spain) • Agrupación Astronómica Complutense (Spain) • Agrupación Astronómica de Córdoba (Spain) • Alachua Astronomy Club (USA-FL) • Alamogordo Astronomy Club (USA-NM) • Al Arish (Egypt) (location identifier) • Alumni Advisory Council • Amateur Astronomy

Centre (UK) • Anglo-American Cataloguing (rules) • Angola Air Charter Ltd. • Apprentissage Anticipé de la Conduite • Asesores Astronómicos Casereños (Spain) • Asset Allocation Committee • Associate Analysis Center • Association des Archivistes Canadiens (Canada) • Association of American Colleges (USA) • Associazione Astronomica Cassino (Italy) • Associazione Astronomica Cortina (Italy) • Astronomisches Abenteuer Camp • Atlanta Astronomy Club (USA-GA) • ATM Access Concentrator • Atomic Absorption Coefficient • Attaché d'Administration Centrale • Automatic Aperture Control • Autonomous Activation Condition

AACB: Aeronautics and Astronautics Coordinating Board

AACC: Airport Associations Coordinating Council • American Association of Community Colleges (USA) • Anglo-American Cataloguing Committee

AACCM: Anglo-American Cataloguing Committee for Cartographic Materials

AACCS: Analytical Approach to Candidate Constellation Selection

AACE: Airworthiness Assurance Center of Excellence • Asociación Argentina de Ciencias Espaciales (Argentina) • Association for Adult and Continuing Education (UK) • Association for the Advancement of Computing in Education

AACGM: Altitude-Adjusted Corrected Geomagnetic

AACI: Amateur Astronomers of Central Iowa (USA-IA)

AACJC: American Association of Community and Junior Colleges (USA)

AACO: Arab Air Carriers Association • Association Astronomique de la Côte d'Or (France)

AACPSM: Actual Average Cost per Successful Mission

AACR: Anglo-American Cataloguing Rules

AACRAO: American Association of Collegiate Registrars and Admissions Officers (USA)

AACS: Advanced Automatic Compilation System • Attitude and Antenna Control System • Attitude and Articulation Control System • Automated-Access Control System • Auxiliary Attitude Control System

AACTE: American Association of Colleges for Teacher Education (USA) • Asociación para el Avance de la Ciencia y la Tecnología en España (Spain)

AAC&U: Association of American Colleges & Universities

AACU: Association of American Colleges and Universities

AACCUP: Accrediting Agency of Chartered Colleges and Universities of the Philippines (formerly PAASCU)

AACUP: Accrediting Agency of Colleges and Universities of the Philippines (formerly PAASCU)

AAD: Ad-Dabbah (Sudan) (location identifier) • Agrupación Astronómica de los Dolores (Spain) • Analog Alignment Diskette • Attitude Anomaly Detector • Australian-Antarctic Discordance

AADC: Alaska Aerospace Development Corporation (USA-AK) • Approach And Departure Control

AADEW: Attack Assessment for Directed-Energy Weapon

AADI: Asociación Argentina de Derecho Internacional (Argentina)

AADR: Applications Approval Design Review

AADS: Ascent Air Data System

AADSF: Advanced Automated Directional Solidification Furnace

AADZ: Astronomsko-Astronauticko Drustvo Zagreb (Croatia)

AAE: Abort Advisory Equipment • Aerospace Auxiliary Equipment • Agrupación Astronáutica Española (Spain) • American Association of Engineers (USA) • Annaba (Algeria) (location identifier) • Association for Adult Education (UK) • Association for Astronomy Education (UK)

AAEC: Attitude Axis Emergency Control • Australian Atomic Energy Commission

AAED: Advanced Airborne Expendable Decoy

AAEE: American Association of Electrical Engineers (USA) • Asuntos Exteriores

AAEO: Association of Adult Education Organizations

AAES: American Association of Engineering Societies (USA)

AAEU: Aquatic Animal Experiment Unit

AAF: Agrupación Astronómica de Fuerteventura (Canary Islands) • Annual Adjustment Factor • Antecedent Action Form • Apalachicola (USA-FL) (location identifier) • Association des Archivistes Français (France) • Associazione Astrofili Fiorentini (Italy) • Associazione Astronomica Frusinate (Italy)

AAFB: Alamogordo Air Force Base (USA-NM) (now HAFB)

AAFC: Association Astronomique de Franche-Comté (France)

AAFE: Advanced Applications Flight Equipment • Advanced Applications Flight Experiment (*) • Aeroassist Flight Experiment

AAFI: Association des Anciens Fonctionnaires Internationaux

AAG: American Association of Geographers (USA) • Asociación Astronómica de Guatemala • Association of American Geographers (USA) • Associazione Astrofili Garfagnana (Italy)

AAGC: Agrupación Astronómica de Gran Canaria (Canary Islands)

AAGG: Associazione Astrofili Galileo Galilei (Italy)

AAGS: American Association for Geodetic Surveying (USA) • Association of African Geological Surveys

AAH: Agrupación Astronómica de Huesca (Spain)

AAHam: Astronomische Abhandlungen Hamburger Sternwarte (periodical code)

AAHE: American Association for Higher Education (USA)

AAI: Amateur Astronomers Inc. (USA-NJ) • Angle of Approach Indicator • Application to Application Interface • Arraias (Brazil-TO) (location identifier) • Arrival Aircraft Interval • Association Astronomique de l'Indre (France) • Associazione Astrofili Imolesi (Italy) • Astronomischer Arbeitskreis Ingolstadt eV (Germany) • Astronomy Acknowledgement Index • Avionics-ATC Integration

AAIB: Air Accident Investigation Branch

AAIC: Aircraft Accident Investigation Committee

AAID: Arithmetic Array Identification

AAIE: American Association of Industrial Engineers (USA)

AAIL: Applied Artificial Intelligence Laboratory

AAIM: American Association of Industrial Management (USA)

AAIP: American Association of Independent Publishers (USA)

AAIR: Advanced Atmospheric Imaging Radiometer

AAIS: American Association for Information Science (USA) • Associate Administrator for Information Systems

AAIT: Air Accident Investigation Team

AAJ: Amateur Astronomers of Jackson (USA-MI) • Association of American Journalists (USA)

AAJA: Afro-Asian Journalists' Association

AAJC: American Association of Junior Colleges (USA)

AAK: Aranuka (Kiribati) (location identifier) • Astronomischer Arbeitskreis Kassel eV (Germany)

AAL: Above Aerodrome Level • Above Airport Level • Absolute Assembly Language • Acronym and Abbreviation List • Aero-Acoustic Levitation • Ålborg (Denmark) (location identifier) • Amateurs Astronomes du Luxembourg (Luxemburg) • Associated Absorption Line • Association of Assistant Librarians • Astronomische Arbeitsgruppe Laufen eV (Germany) • Atelier Astronomie de Limoges (France) • ATM Adaptation Layer

AALC: Association Astronomique du Loir-et-Cher (France)

AALL: American Association of Law Libraries (USA)

AAM: Absolute Angular Momentum • Advanced Access Mode • Advanced Analysis Module • Advanced Astro Master • Agrupación Astronómica de Madrid (Spain) • Agrupación Astronómica de Manresa (Spain) • American Association of Museums (USA) • Associazione Astrofili Mantovani (Italy) • Associazione Astronomica Milanese (Italy) • Atmospheric Angular Momentum • ATOS Application Module • Mala Mala (South Africa) (location identifier)

AAMC: Auto-Adaptive Multimedia Composer

AAMI: Advanced Active Microwave Instrument • Association for the Advancement of Medical Instrumentation

AAMP: Advanced Architecture Microprocessor

AAMS: Analytical Accounting Management System

AAMSI: American Association for Medical Systems Informatics (USA)

AAMT: Associazione Astronomica Monti della Tolfa (Italy)

AAMV: Ameteurastronomen Max Valier (Italy)

AAMVG: Astronomisches Arbeitskreis Merkur/Venus Göttingen (Germany)

AAN: Al Ain (UAE) (location identifier)

AANC: Astronomical Society of Northern California (USA-CA)

AANEAS: Anglo-Australian Near-Earth Asteroid Survey

AANT: Autonomous and Advanced Navigation Technique

AANZ: Astronautics Association of New Zealand

aaO: am angeführten Ort • am angegebenen Ort

AAO: Abastumani Astrophysical Observatory (Georgia) • Adjacent Arctic Ocean • Agrupación Astronómica d'Osona (Spain) • Anaco (Venezuela) (location identifier) • Anglo-Australian Observatory • Astronaut Activities Office (NASA)

AAOE: Airborne Antarctic Ozone Experiment

AAOEPP: Anglo-Australian Observatory - Epping (Australia-NSW)

AAofA: Amateur Astronomers of Acadiana (USA-LA)

AAOL: Association Against Obtrusive Lighting (Australia)

AAOPr: Anglo-Australian Observatory Preprint (periodical code)

AAOSPS: Anglo-Australian Old Stellar Populations Survey

AAP: Adjoint Administratif Principal • Advanced Automation Program • Affirmative Action Plan • Affirmative Action Program • Agrupación Astronómica de la Palma (Canary Islands) • Airlock Adapter Plate • American Association of Publishers (USA) • Analog Autopilot • Apollo Applications Program • Applications Access Point • Assemble At Pad • Association of American Publishers (USA) • Associative Array Processor • Associazione Amici dei Planetari (Italy) • Astronomische Arbeitsgemeinschaft Pfaffenwinkel (Germany) • Astronomischer Arbeitskreis Pforzheim (Germany) • Atmospheric Analysis and Prediction • Houston Andrau (USA-TX) (location identifier)

AAPA: Association of Asia Pacific Airlines

AAPG: American Association of Petroleum Geologists (USA)

AAPIER: American Association for Public Information, Education and Research (USA)

AAPM: American Association of Physicists in Medicine (USA)

AAPP: Administraciones Públicas

AAPPS: Association of Asian and Pacific Physical Societies

AAPR: Angle Averaged Partial Redistribution

AAPS: American Association for the Promotion of Science (USA)

AAPSS: American Academy of Political and Social Science (USA)

AAPT: American Association of Physics Teachers (USA)

AAQ: Anapa (Russia) (location identifier) • Association des Archivistes du Québec (Canada-QC)

A&AR: Astronomy and Astrophysics Review

AAR: After Action Report • Against All Risks • Airport Acceptance Rate • Århus (Denmark) (location identifier) • Association for Automated Reasoning • Association of American Railroads • Attaché d'Administration de la Recherche • Auto Analysis Results • Average Annual Rainfall

AARAO: Ålborg Amateur Radio Astronomy Observatory (Denmark)

AARC: Anglo-American Cataloging Rules • Arctic and Antarctic Research Group • Auroral Atmospheric Radiance Code

Aarg: Aargang

AARG: Astronomy and Astrophysics Research Group

AARI: Advanced Airborne Remote Instrumentation • Arctic and Antarctic Research Institute

AARL: Association for Academic and Research Libraries (USA)

AARN: Australian Academic Research Network

AARNet: Australian Academic and Research Network

AARNET: Australian Academic Research Network

AARP: American Association of Retired Persons (USA) • AppleTalk Address Resolution Protocol • Attaché d'Administration de la Recherche Principal

AARPS: Air-Augmented Rocket Propulsion System

AARS: Asian Association of Remote Sensing

AARU: Astronomy and Atmospheric Research Unit (USM)

A&ARv: Astronomy and Astrophysics Review (periodical code)

A&AS: Astronomy and Astrophysics Supplement Series (periodical code)

AAS: Abort Advisory System • Academy of Applied Science (USA) • Acta Astronomica Sinica • Advanced Antenna System • Advanced Automation System • African Academy of Sciences • Agrupació Astronòmica de Sabadell (Spain) • Airport Advisory Service • All-to-All Scatter • American Academy of Sciences (USA) • American Aerospace Society (USA) • American Astronautical Society (USA) • American Astronomical Society (USA) • Amphibious Assault Ship • Andersen Air Sampler • Apalapsili (Indonesia) (location identifier) • Archdale Astronomical Society (USA-NC) • Arithmetic Assignment Statement • Associate in Applied Science • Association Astronomique du Soissonnais (France) • Association des Archivistes Suisses (Switzerland) • Associazione Astrofili di Salò (Italy) • Associazione Astrofili Sardi (Italy) • Associazione Astrofili Segusini (Italy) • Associazione Astrofili Spezzini (Italy) • Astronomical Association - Sofia (Bulgaria) • Astronomischer Arbeitskreis Salzkammergut (Austria) • Athens Astronomical Society (USA-GA) • Atomic Absorption Spectrometer • Atomic Absorption Spectrometry • Atomic Absorption Spectrophotometry • Atomic Absorption Spectroscope • Atomic Absorption Spectroscopy • Auburn Astronomical Society (USA-AL) • Austin Astronomical Society (USA-TX) • Australian Academy of Sciences • Automated Accounting System • Automatic Addressing System • Ayrshire Astronomical Society (UK-Scotland) • Azimuth Alignment System

AASA: Association Astronomique Saint-Avertinoise (France)

AASANA: Administración de Aeropuertos y Servicios Auxiliares

AASAP: Applied Analysis Spectral Analysis Process

AASC: Astronomy and Astrophysics Survey Committee

AASCU: American Association of State Colleges and Universities (USA)

AASE: Airborne Arctic Stratospheric Expedition • Airborne Arctic Stratospheric Experiment

AASERT: Augmentation Award for Science and Engineering Research Training

AASES: ATA Aviation Safety Exchange System • Australian Amateur Space Engineering Society

AASG: Association of American State Geologists (USA)

AASI: Amateur Astronomical Society of Iran • Astronomical Association of Southern Illinois (USA-IL)

A'asia: Australasia

AASIR: Advanced Atmospheric Sounder and Imaging Radiometer • Advanced Atmospheric Sounding and Imaging Radiometer

AASM: Associated African States and Madagascar • Association of American Steel Manufacturers

AASP: American Association of Stratigraphic Palynologists (USA) • ASCII Asynchronous Support Package

AASR: Airport and Airways Surveillance Radar

A&ASS: Astronomy and Astrophysics Special Supplement Series (periodical code)

AASS: Astronomy and Astrophysics Supplement Series • Asuntos Sociales

AASSREC: Association of Asian Social Science Research Councils

AASTR: Advanced Along Track Scanning Radiometer

AASTRA: American Astronomical Society Teacher Resource Agent

AASU: Armstrong Atlantic State University (USA-GA) • Attaché d'Administration Scolaire et Universitaire (France)

AASW: American Association of Scientific Workers (USA)

AASZ: Astronomical and Astronautical Society of Zadar (Croatia)

A&AT: Astronomical and Astrophysical Transactions (periodical code)

AAT: Advanced Altimeter Tracker • Administration de l'Assistance Technique • Administrative Appeals Tribunal • Advanced Application of Technology • Agrupación Astronómica de Tenerife (Canary Islands) • Altay (China-PRC) (location identifier) • Amateur-Astronomie-Tagung • Analogic Alias Table • Anglo-Australian Telescope • Area Adaptive Thresholding • Art and Architecture Thesaurus • Associazione Astrofili Teatini (Italy) • Associazione Astrofili Thethys (Italy) • Associazione Astrofili Treutini (Italy) • Associazione Astrofili Triestini (Italy) • Attitude Acquisition Technique • Australian Antarctic Territory • Average Access Time • Avis d'Arrêt de Travail

AATB: Anglo-Australian Telescope Board

AATGS: AAT Guide Star

AATM: At All Times

AATO: Army Air Transport Organisation

AATS: AMOS Acquisition Telescope System

AATSE: Australian Academy of Technological Science and Engineering

AATSR: Advanced Along-Track Scanning Radiometer

AAU: Århus University (Denmark) • Asau (Samoa) (location identifier) • Association of African Universities • Association of American Universities (USA)

AAUA: Agrupación Astronómica Universitaria de Alicante (Spain)

aaub: Antwoord Alstublieft

AAUF: Associations Aéronautiques de l'Union Française

AAUI: Apple Attachment Unit Interface

AAUP: American Association of University Presses (USA) • American Association of University Professors (USA) • Association Astronomique Universitaire de Perpignan (France) • Association of American University Presses (USA) • Association of Australian University Presses

AAUW: American Association of University Women (USA)

AAV: Action-Angle Variable • Agrupación Astronómica Vizcaina (Spain) • Alah (Philippines) (location identifier) • Associazione Astrofili Valdinievole (Italy) • Associazione Astrofili Valtellineri (Italy) • Associazione Astrofili Veneziani (Italy)

AAVSO: American Association of Variable Stars Observers (USA-MA)

AAVV: Autores Varios

AAW: Abbottabad (Pakistan) (location identifier) • Aerospace and Astrophysics Working Group • Algemene Arbeidsongeschiktheidswet

AAWF: Auxiliary Aviation Weather Facility

AAWORD: Association of African Women for Research and Development

AAX: Araxá (Brazil-MG) (location identifier)

AAY: Al Ghaydah (Yemen) (location identifier)

AAZ: Airport Approach Zone

a/b: Airborne

ab: Abkhazian (ISO language code) • About • Abridgment

aB: Auf Bestellung

Ab: Astatine

A/B: Afterburner • Airborne • Aktiebolaget

AB: Aberdeen (UK postal code) • Able Seaman • Abstract • Abstracting Board • Acceleration Boundary • Adapter Booster • Advisory Board • Aft Bulkhead • Afterburner • Air Base • Air Bearing • Airborne • Aitreya Brāhmana • Alberta (Canada) • Algemeen Bestuur • Alliance Balkanique • Anchor Bolt • Answer Back • Antibody • Archive Branch • As Before • Assembly Bill • Assez Bien • Assyro-Babylonian • Atomic Bremsstrahlung • Autobeacon • Automated Bibliography • Avionics Bay • Bachelor of Arts

ABA: Abakan (Russia) (location identifier) • American Bankers' Association (USA) • American Bar Association (USA) • American Book Award • American Booksellers Association (USA) • Amplifier-Buffer-Attenuator • Asociación Boliviana de Astronomía (Bolivia) • Associate in Business Administration • Australian Bankers' Association • Australian Booksellers Association

AbAO: Abastumani Astrophysical Observatory (Georgia)

AbAOB: Byull. Abastumanskaya Astrofiz. Obs. (periodical code)

ABARS: Aggregate Backup And Recovery Support (*) • Application Backup And Recovery Support

ABATS: Automatic Bit Access Test System

Abb: Abbaye • Abbey • Abbildung • Abbuono

ABB: Abingdon (UK-England) (location identifier) • Asea Brown Boveri • Asociación Boliviana de Bibliotecarios (Bolivia) • Associaçāa Brasileira de Bibliotecários (Brazil)

ABBA: Automated Biomass Burning Algorithm

Abbr: Abbreviate • Abbreviated • Abbreviation

Abbrev: Abbreviate • Abbreviated • Abbreviation

ABBS: Astronomical Bulletin Board System

ABC: Academia Brasileira de Ciências (Brazil) • Active Broadcast Control • Activity-Based Costing • Ada Browse Capability • Advanced Biomedical Capsule • After Bottom Center • Airborne Control • Air Business Contact • Airway-Breathing-Circulation • Alcoholic Beverage Control • Allocations for Budgetary Control • America-Britain-Canada • American Bibliographical Center (USA) • American Born Chinese • American Broadcasting Co. • Animal Birth Control • Antwerp-Brussels-Charleroi (Belgium) • Apparel Business Control • Application Builder Class • Approach By Concept • Argentina-Brazil-Chile • Artifact-Based Collaboration • Aruba-Bonaire-Curaçao • Asociación de Bibliotecarios de Chile • Atanasoff-Berry Computer • Atomic-Biological-Chemical • Augmented Bibliographic

Citation • Australian Bibliographical Centre • Australian Broadcasting Commission • Australian Broadcasting Corporation • Automatic Binary Computer • Automatic Break Control • Automatic Brightness Control • Automation of Bibliography through Computerization • Axial Bay Camera

ABCA: Association of British Columbia Archivists (Canada-BC)

ABCAS: Algorithm Beam Center and Shape

ABCB: Air-Blast Circuit Breaker

ABCC: Association of British Chambers of Commerce

ABCCC: Airborne Command and Control Center

ABCD: Asymmetry-Border-Color-Diversity

ABCF: As-Built Configuration File

ABCG: Abell Galaxy Clusters

ABCL: As-Built Configuration List

ABCM: Associação Brasileira de Ciências Mecânicas (Brazil)

ABCP: Airborne Command Post

ABCR: As-Built Configuration Record

ABCS: Advanced Business Communication via Satellite

Abd: Abdicate • Abdicated • Abdomen • Abdominal

ABD: Abadan (Iran) (location identifier) • Aboard • Access Block Diagram • Administration de Banque de Données • Administration de Base de Données • Airborne Ballistics Division (NASA) • All But Dissertation • Association Belge de Documentation (Belgium) • Average Body Dose • Axial-Boom Drive

ABDA: American-British-Dutch-Australian • Analog-Based Domain Analysis

ABDC: After Bottom Dead Center

ABDL: Automatic Binary Data Link

ABDP: Association of British Directory Publishers (UK)

ABDPEA: Association Belge pour le Développement Pacifique de l'Énergie Atomique (Belgium)

ABE: Acetone-Butanol-Ethanol • Advisory Board on Energy • Agent Building Environment • Air-Based Electronics • Air-Breathing Engine • Allentown (USA-PA) (location identifier) • Arm-Based Electronics • Associação Brasileira de Educação (Brazil) • Bethlehem (USA-PA) (location identifier) • Easton (USA-PA) (location identifier)

ABEC: Aitken Bicentennial Exhibition Centre (Canada-New Brunswick)

ABEd: Bachelor of Arts in Education

ABEDA: Arab Bank for Economic Development in Africa

ABELF: Association des Éditeurs Belges de Langue Française (Belgium)

ABEND: Abnormal End

Aber: Aberdeen (UK-Scotland) • Aberdonian

Aberr: Aberration

ABERT: Auto Bit Error Rate Test

ABES: Air-Breathing Engine System

ABET: Accreditation Board for Engineering and Technology (*) • Accreditation Board for Engineering Technology

ABETS: Airborne Beacon Electronic Test Set

ab ex: Ab Extra

Abf: Abfahrt

ABF: Abaiang (Kiribati) (location identifier) • Application Binary Format • Association des Bibliothécaires Français (France) • Association des Bibliothèques de France • Average Branching Factor

ABFM: Airborne Field Mill Program

ABFN: Adaptive Beam-Forming Network • Adjustable Beam-Forming Network

Abg: Abgeordneter

ABG: Abingdon (Australia-QLD) (location identifier)

Abgk: Abgekürzt

Abh: Abhandlung • Abhandlungen

ABH: Alpha (Australia-QLD) (location identifier)

ABHRS: As Built Hardware Reporting System

ABI: Abilene (USA-TX) (location identifier) • Advance Book Information • Advanced Book Information • American Biographical Institute (USA-NC) • Application Binary Interface • Artificial Biosensing Interface • Association des Bibliothèques Internationales • Associazione di Bibliotecari Italiani (Italy) • Avionics Bus Interface

ABIDS: Automated Biometric Identification System

ab init: Ab Initio

ABIOS: Advanced Basic Input/Output System

ABIR: Agent-Based Information Retrieval

ABIRA: American Biographical Institute Research Association (USA)

ABIST: Automatic Built-In Self-Test

ABIT: Airborne Information Transmission

ABJ: Abidjan (Ivory Coast) (location identifier)

ABJPAA: Association Belge des Journalistes Professionnels de l'Aéronautique et de l'Astronautique (Belgium)

Abk: Abkürzung

ABK: Kabri Dar (Ethiopia) (location identifier)

Abl: Ablative • Amtsblatt

ABL: Adaptation Binary Load • Airborne Laser • Aircraft-Based Laser • Albedo • Ambler (USA-AK) (location identifier) • Astrometry-Based Luminosity • Atmospheric Boundary Layer

ABLE: Airborne Laser Evaluation • Amazon Boundary Layer Experiment • Arctic Boundary Layer Expedition • Arctic Boundary Layer Experiment (*) • Atmospheric Balloon-borne Lidar Experiment • Atmospheric Boundary Layer Expedition

ABLE ACE: Airborne Laser Extended Atmospheric Characterization Experiment

ABLISS: Association of British and Information Science Schools

ABLOS: Algorithm Beam Line Of Sight

ABLS: Association of British Library Schools • Bachelor of Arts in Library Science

Abm: Abeam

ABM: Acquisition Bus Monitor • Advanced Bill of Materials • Apogee Boost Motor • Associate in Business Management • Asynchronous Balanced Mode • Atmosphère Basse et Moyenne • Atomic Beam Method • Automatic Batch Mixing • Bamaga (Australia-QLD) (location identifier)

ABMCP: Airborne Military Command Post

aBN: Amorphous Boron Nitride

Abn: Aberdeen (UK-Scotland)

ABN: Airborne • Albina (Suriname) (location identifier) • African Biosciences Network • Agrupación de Bibliotecas Nacionales (Spain) • Algemeen Beschaafd Nederlands • Australian Bibliographic Network

ABNCP: Airborne National Command Post

ABNEI: Association Belge des Négociants Exportateurs et Importateurs (Belgium)

Abnl: Abnormal

Abnml: Abnormal

ABNT: Associação Brasileira de Normas Técnicas (Brazil)

ABO: Aboisso (Ivory Coast) (location identifier) • Annular Beam Oscillator

Abog: Abogado

ABOL: Advisor Business-Oriented Language

ABOMO: Airborne Optic and Magnetic Observatory

ABOP: Algemene Bond van Onderwijzend Personeel

ABOS: Algemeen Bestuur voor Ontwikkelingssamenwerking

ABp: Archbishop

ABP: Acoustic Backscatter Probe • Algemeen Burgerlijk Pensioenfonds • Arterial Blood Pressure • Associated Book Publishers • Association des Bibliothèques Pédagogiques • Atkamba (Papua New Guinea) (location identifier)

ABPA: American Book Producers Association (USA) • Australian Book Publishers' Association

ABPP: Arecibo-Berkeley Pulsar Processor

ABPS: Air-Breathing Propulsion System • Automatic Blood Pressure System

ABQ: Albuquerque (USA-NM) (location identifier)

Abr: Abrégé • Abridge • Abridged • Abridgment • Abril • Abrogé

ABR: Aberdeen (USA-SD) (location identifier) • Abril • Air Breathing Rocket • Asymmetric Bit Rate • Available Bit Rate

ABRACOS: Anglo-Brazilian Amazonian Climate Observational Study

ABRAMS: Automated Bidirectional Reflectance Measurement System

ABRC: Advisory Board for the Research Councils (UK) • Advisory Board to Research Council • Association des Bibliothèques de Recherche du Canada

ABRES: Antiballistic Reentry System

Abrev: Abreviación • Abreviado

Abrév: Abréviation

ABRIXAS: A Broadband Imaging X-ray All-sky Survey

ABRS: Automated Book Request System

ABRV: Advanced Ballistic Reentry Vehicle

Abs: Absatz • Absence • Absender • Absent • Absolu • Absolument • Absolute • Absolutely • Arsorbent • Abstract

ABS: Absolute • Abstract • Abus de Biens Sociaux • Abu Simbel (Egypt) (location identifier) • Acrylonitrile Butadiene Styrene • Activity-Based Statistics • Airborne Bathymetric Survey • Air-Break Switch • American Bureau of Shipping (USA) • American Business Systems Inc. • Ammonia Boiler System • Anti-Blockier-System • Anti-Blocking System • Antilock Braking System • Association des Bibliothécaires Suisses (Switzerland) • Atomic Beam Source • Australian Bureau of Statistics • Automatic Backup System • Automatic Bibliographic Service • Aux Bons Soins de

ABSC: Advanced Biological Safety Cabinet

Abschn: Abschnitt

abse. re.: Absente Reo

abse. feb.: Absente Febre

abs. feb.: Absente Febre

ABSMAG: Absolute Magnitude

Absol: Absolu • Absolument • Absolute • Absolutely

Absorp: Absorption

abs. re.: Ansente Reo

ABSS: Astronomical Bulletin Board System

ABST: Abstention

ABSTI: Advisory Board on Scientific and Technical Information

Abstr: Abstract • Abstracted

Abs. Visc.: Absolute Viscosity

ABSW: Association of British Science Writers (UK)

Abt: About • Abteilung

ABT: Abort • About • Air-Bearing Table • Al-Baha (Saudi Arabia) (location identifier) • Automatic Battery Test

ABTA: Association of Travel British Agents

ABTAC: Australian Book Trade Advisory Committee

ABTCU: Active Baseplate Thermal Control Unit

ABU: Asian Broadcasting Union • Asian-Pacific Broadcasting Union • Asociación de Bibliotecarios del Uruguay • Atambua (Indonesia) (location identifier)

ABUEN: Asociación de Bibliotecas Universitarias y Especializadas de Nicaragua

ABUP: Asociación de Bibliotecarios Universitarios de Paraguay

ABV: Above • Abuja (Nigeria) (location identifier)

ABVA: Algemene Bond Van Ambtenaren

ABVV: Algemeen Belgisch Vakverbond

ABW: Abau (Papua New Guinea) (location identifier) • Algemene Bijstandswet • Antarctic Bottom Water • Arctic Bottom Water • Aruba (ISO country code)

ABWR: Advanced Boiling Water Reactor

ABX: Albury (Australia-NSW) (location identifier)

ABY: Albany - Dougherty County (USA-GA) (location identifier)

Abys: Abyssinian

Abz: Abzüglich

ABZ: Aberdeen Dyce (UK-Scotland) (location identifier)

Abzw: Abzweigstelle

a/c: Account • Account Current • A Cuenta • Air Conditioning • Ao Cuidado de

àc: À Compte

ac: A Capo • Acre • Activity • Air Conditioning • Alternating Current • Anno Christi • Año Citada • Ante Christum • Ante Cibum • Anno Currente

aC: Antes de Cristo • Avanti Cristo

Ac: Acetate • Acetyl • Actinium • Altocumulus

A&C: Addenda & Corrigenda

A/C: Account • Account Current • Air Conditioning • Aircraft • Approach Control • Associate Contractor • Current Account

AC: Abstract Computer • Access Charge • Access Control • Accord Commercial • Accretion Column • Aceiro • Acre (Brazil) • Action Code • Activated Carbon • Active Centre • Active Component • Adjudication Committee • Administration Centrale • Advanced Camera • Advanced Composite • Advisory Circular • Advisory Committee • Advisory Council • Aerodynamic Center • Aerospace Center • Agent Comptable • Agent Consulaire • Agent Contractuel • Air Carrier • Air Conditioning • Air Control • Aircraft Carrier • Aire Corrigée • Allocation de Chômage • Alternating Current • Altocumulus • American Cartographer (journal) • Analog Computer • Analyse Chimique • Analysis Center • Angular Correlation • Année Civile • Année Courante • Anno Corrente • Annual Conference • Anomalous Component • Ante Christum • Ante Cibum • Anticenter • Antigalactic Center • Appeal Court • Appellation Contrôlée • Application Controller • Area Code • Army Corps • Arrêté de Compte • Asociación Civil • Assegno Circolare • Assemblée Constituante • Assembly Complete • Astrographic Catalog • Astronomical Constant • Astronomical Council • Atelier Central • Atlas-Centaur • Attitude Control • Audio Center • Audit Committee • Auracanian • Australia-California • Australian Capital Territory • Authorization Center • Authorization Code • Author's Correction • Autocollimation • Automatic Computer • Automobile Club • Auxiliary Computer • Avocat-Conseil • Axiom of Choice • Azimuth Center

AcA: Acta Astronomica (periodical code)

ACA: Academic Cooperation Association • Acapulco (Mexico) (location identifier) • Adjacent-Channel Attenuation

• Advanced Computer Application • Agency for cultural Affairs • Algorithme de Compression Avancé • Altitude Controller Assembly • American Cartographic Association (USA) • American Conservation Association (USA) • American Crystallographic Association (USA) • Annunciator Control Assembly • Applications Control Architecture • Aspect Camera Assembly • Associate Contractor Administration • Association Ciel d'Anjou (France) • Association of Consulting Actuaries • Associazione Cernuchese Astrofili (Italy) • Astro-Club Aubagnais (France) • Astronomie Centre Ardenne (Belgium) • Astronomy Club of Augusta (USA-NJ) • Attitude Controller Assembly • Author Co-citation Analysis • Waco (USA-TX) (location identifier)

ACAA: Asociación Cordobesa Amigos de la Astronomía (Argentina)

AcAAS: Acta Aeronautica et Astronautica Sinica (periodical code)

AcAau: Acta Astronautica (periodical code)

ACABQ: Advisory Committee on Administrative and Budgetary Questions

ACAC: Administration Centrale de l'Aviation Civile • African Civil Aviation Conference • Ancient City Astronomy Club (USA-FL)

ACACC: Association des Cartothèques et Archives Cartographiques du Canada

ACACE: Advisory Council for Adult and Continuing Education (UK)

ACACIA: A Consortium for the Application of Climate Impact Assessments

AcAcu: Acta Acustica (periodical code)

Acad: Academia • Academic • Academical • Academician • Académie • Académique • Academy

ACAD: Advanced Computer-Aided Design

Acad. Fran.: Académie Française

ACADS: Association for Computed-Aided Design

Acad. Sci.: Académie des Sciences

ACAE: Australian Commission for Advanced Education

AcAer: Acta Aerodynamica Sinica (periodical code)

ACAIS: Air Carrier Activity Information System

ACAL: Academia de Ciencias de América Latina • Advanced Computer Architecture Laboratory

ACAMPS: Automated Communications and Message Processing System

Acap: Acapulco (Mexico)

ACAP: Århus Centre for Advanced Physics (Denmark) • Advanced Computer for Array Processing • Application Configuration Access Protocol • Automatic Circuit Analysis Program

AcApS: Acta Astrophysica Sinica (periodical code)

ACaPS: Agile Cable Production Service

ACAPS: Automated Chart Production System

ACAQ: Advisory Committee on Air Quality

ACAR: Angular Correlation of Annihilation Radiation

ACARD: Advisory Council for Applied Research and Development (UK)

ACARS: Aircraft Communications Addressing and Reporting System • ARINC Communications, Addressing And Reporting System • ARINC Communications And Reporting System • Automatic Communications And Reporting System

ACARU: Association Canadienne des Administrateurs de Recherche Universitaire (Canada)

ACAS: Advisory Conciliation and Arbitration Service • Airborne Collision Avoidance System • Aircraft Collision Avoidance System • Application Control Architecture Services • Atlantic Canada Airborne Sensing

AcASn: Acta Astronomica Sinica (periodical code)

ACAST: Advisory Committee on the Application of Science and Technology

ACATA: Association for Computer-Assisted Text Analysis (USA)

ACATS: Advisory Committee for Advanced Television Service • Airborne Chromatograph for Atmospheric Trace Species

AcAuS: Acta Automatica Sinica (periodical code)

ACAVC: Advisory Committee on Agricultural and Veterinary Chemicals

ACB: Air Circuit Breaker • Air Cushion Barge • Application Control Block • Asociación Cubana de Bibliotecarios (Cuba) • Association Canadienne des Bibliothèques (Canada) • Astro Club de Beauce (France) • Bellaire - Antrim County (USA-MI) (location identifier)

ACBCT: Automatic Circuit Board Card Tester

ACBCU: Association Canadienne des Bibliothèques de Collèges et d'Universités (Canada)

ACBD: Association pour la Coopération entre Bibliothèques et Centre de Dpcumentation (France)

ACBE: Advisory Committee on Business and the Environment (UK)

ACBLF: Association Canadienne des Bibliothèques de Langue Française (Canada)

ACBSP: Association of Collegiate Business Schools and Programs

Acc: Accelerate • Accelerated • Acceleration • Accélération • Accélérer • Accent • Accept • Acceptance • Accepted • Accompanied • Accompaniment • Accord • Accordant • According • Account • Accountant • Accounted • Accumulate • Accumulation • Accumulator • Accumuler • Accusative

AcC: Acta Cosmologica (periodical code)

ACC: Academia de Ciencias de Cuba • Accra Kotoka (Ghana) (location identifier) • Accréditif • Accumulator • Acid Copper Chromate • Action Concertée Coordonnée • Active Clearance Control • Acute Cardiovascular Collapse • Adaptive Cruise Control • Administrative Committee of Coordination • Administrative Committee on Coordination • Administrative Coordinating Committee • Advanced Communications Course • Aft Cargo Carrier • Aft Cargo Compartment • Air Combat Command • Air Coordinating Committee • Altocumulus Castellanus • Annual Capital Charge • Antarctic Circumpolar Current • Antarctic Conservation Committee • Arab Cooperation Council • Area Control Center • Ateliers de Constructions du Centre (France) • Attitude Control Computer • Australian Chamber of Commerce • Australian Copyright Council • Automatic Control Console • Automation Classification Code • Auxiliary Crew Compartment • Aviation Civile et Commerciale

ACCA: Air Charter Carriers Association • Automatic Cloud Cover Assessment

Accad: Accademia

ACCAD: Advisory Committee on Climate Applications and Data

ACCBD: Advisory Committee on Conservation of Biological Diversity

ACCC: Area Control Computer Complex • Arthur C. Clarke Centre (Sri Lanka) • Australian Competition and Consumer Commission

ACCD: American Coalition of Citizens with Disabilities (USA) • Astrometric Charge-Coupled Device

Acce: Acceptance

ACCE: Advisory Committee on Chemicals in the Environment • Augmented Continuous Control Evaluator

Accel: Accelerate • Accelerated • Acceleration

Accél: Accélération • Accélérer

ACCES: Atelier de Capitalisation des Connaissances et d'Exploitation du Savoir-Faire

Access: Accessory

ACCESS: Advanced Cosmic-ray Composition Experiment for Space Station • All Children Can Explore the Solar System • Alliance for Collaborative Change in School Systems • Assembly Concept for Construction of Erectable Space Structures

Accid: Accident • Accidental

ACCIS: Advisory Committee for the Coordination of Information Systems (UNO)

ACCME: Accreditation Council for Continuing Medical Education

ACCMT: Arthur C. Clarke Centre for Modern Technologies (Sri Lanka)

ACCN: Audit Central Control Network

Accom: Accommodate • Accommodation

Accomp: Accompanied • Accompaniment • Accompany • Accomplishment

ACCORD: AURA Coordinating Committee of Observatory Research Directors

ACCP: Atlantic Climate Change Programme

Accred: Accredited

Accrd: Accrued

ACCS: Adaptive Communication Control System • Automatic Calling Card Service

ACCSIC: Atomic Collision Cross Sections Information Center

ACCSQ: ASEAN Consultative Committee for Standard and Quality

Acct: Account • Accountant • Accounting

ACCT: Account • Advisory Committee on Certification and Testing • Agence de Coopération Culturelle et Technique (France) • Ancillary Computer Compatible Tape

ACCU: Academisch Computer Centrum Utrecht (Netherlands) • Audio Central Control Unit

Accum: Accumulate • Accumulated • Accumulation • Accumulative

Accus: Accusative

Accy: Accountancy

ACCY: Accountancy

Acd: Accord

ACD: Acandi (Colombia) (location identifier) • Accuracy Control Document • Acrobat Capture Document • Adobe Capture Document • Aeronautical Charting Division • Allgemeine Computer-Datenbank • Analysis and Computation Division • Anode-Cathode Distance • Anticoincidence Detector • Architectural Control Document • Atmospheric Chemistry Division • Audio Control Panel • Autocommutateur Digital • Automated Call Distribution (*) • Automatic Call Directory • Automatic Call Distribution • Automatic Call Distributor

ACDA: Advisory Committee on Distinction Awards • Arms Control and Disarmament Agency (USA) • Asociación Columbiana De estudios Astronómicos (Columbia)

ACDAR: Acoustic Radar

ACDAS: Automatic Control and Data Acquisition System

AC/DC: Alternating Current/Direct Current

AC-DC: Alternating Current/Direct Current

ACDDS: Advanced Cartographic Data Digitizing System

ACDF: Access Control Decision Function

ACDI: Agence Canadienne de Développement International (Canada) • Asynchronous Communications Device Interface

ACDMS: Automated Control of a Document Management System

ACDNS: Attitude Control, Determination, and Navigation System

ACDO: Air Carrier District Office

ACDPI: Association for a Competitive Data Processing Industry (USA)

ACDS: Active Control Dataset • Archical Catalogue and Database System • Astronomical Catalogues Database System • Attitude Control and Determination System

Acdt: Accident

ACDT: Advanced Concept Technology Demonstrator

ACE: Acceptance Checkout Equipment • Acer Computer Explorer • Acidification in Clouds Experiment • Action Communautaire pour l'Environnement • Actuator Control Electronics • ADA Compilation Environment • Adaptive Communications Environment • Adaptive Contrast Enhancement • Admission and Credential Evaluators • Adobe Certified Expert • Advanced Cartographic Environment • Advanced Cartographic Equipment • Advanced Communication Engine • Advanced Communications Enhancement • Advanced Composition Explorer • Advanced Computing Environment • Advanced Computing Equipment • Advanced Cooled Engine • Adverse Channel Enhancement • Advisory Centre for Education (UK) • Advisory Committee on Environment • Advisory Committee on the Environment • Aerosol Characterization Experiment • Affinity Capillary Electrophoresis • Aids to Communication in Education • Airborne Command Element • Air Cargo Express Inc. • Alternating Conditional Expectation • American Council on Education (USA) • Americans Communicating Electronically • Analytical Computing Environment • Animateur de Club Expérimental • Application Construction Environment • Arctic Cloud Experiment • Army Corps of Engineers • Array Control Electronics • Assemblée Consultative Européenne • Association for the Conservation of Energy • Association of Conference Executives • Association of Consulting Engineers • Association of Cultural Exchange • Astronomical Consultants and Equipment Inc. (USA-AZ) • Atmospheric Chemistry Education • Atmospheric Compensation Experiment • Attitude Control Electronics • Australian College of Education • Automatic Calibration and Equalization • Automatic Calling Equipment • Automatic Centroid Extractor • Automatic Checkout Equipment • Automatic Computing Engine • Automatic Computing Engineering • Lanzarote (Canary Islands) (location identifier)

ACEA: Action Committee for European Aerospace • Association Canadienne pour l'Éducation des Adultes (Canada) • Association des Constructeurs Européens d'Automobiles

ACEB: Association Canadienne des Écoles de Bibliothécaires (Canada)

ACEC: Advisory Committee on Electromagnetic Compatibility • American Consulting Engineers Council • Area of Critical Environmental Concern • Asociación Canaria para la Enseñanza de las Ciencias (Spain) • Ateliers de Construction Électriques de Charleroi (Belgium)

ACEEE: American Council for an Energy-Efficient Economy (USA)

ACEID: Asian Center for Educational Innovation and Development

ACEL: Air Crew Equipment Laboratory

AcEle: Acta Electronica (periodical code)

ACELE: Accord Centreuropéen de Libre-Échange

ACELP: Algebraic Code-Excited Linear Prediction

AcElS: Acta Electronica Sinica (periodical code)

Acep: Acepción

ACEP: American College of Emergency Physicians (USA)

Acept: Aceptación

ACER: Advisory Committee on Environmental Resources • Australian Council for Educational Research

ACERSS: Advanced Civilian Earth Remote Sensing System

ACeS: Asia Cellular System

ACES: Acceptance Checkout and Evaluation System • Acceptance Control Equipment Section • Acoustic Containerless Experiment System • Active Control Equivalence Studies • Aerolineas Centrales de Colombia SA • Atomic Clocks Ensemble in Space • Automatic Checkout Equipment Sequencer

ACESJ: Applied Computational Electromagnetics Society Journal (periodical code)

Acet: Acetone

ACET: Advisory Committee on Electronics and Telecommunication

ACEU: Aerocontrol Electronics Unit

ACF: Académie Canadienne Française • Access Control Facility • Access Control Field • Advanced Common Flightdeck • Advanced Communications Facility • Advanced Communications Function • Angular Correlation Function • Applications Control Facility • Area Control Facility • Arizona Carbon Foil Co. Inc. (USA-AZ) • Astro-Club de France • Australian Conservation Foundation • Automated Control Function • Autocorrelation Function • Awareness Campaign Forum

ACFAS: Association Canadienne-Française pour l'Avancement des Sciences

ACFCI: Association des Chambres Françaises de Commerce et d'Industrie (France)

ACFG: Automatic Continuous Function Generation

ACFHE: Association of Colleges of Further and Higher Education

ACFM: Alternating-Current Field Measurement

ACFNET: Advanced Communications Function Network

ACFO: Aircraft Certification Field Office

AcFre: Acta Frequenza (periodical code)

ACFS: Advanced-Concept Flight Simulator

Acft: Aircraft

ACFT: Aircraft

ACG: Address Coding Guide • Archival Collection Guide • Arp's Compact Galaxy • Automatic Common Generator • Automatic Control Gear

AcGeP: Acta Geophysica Polonica (periodical code)

ACGI: Advanced Computer Generated Image • Asynchronous Common Gateway Interface

ACGIH: American Conference of Governmental Industrial Hygienists (USA)

ACGW: Analytical Construction by Generalized Work

ACH: Achat • Acheté • Air Changes per Hour • Altenrhein (Switzerland) (location identifier) • Association for Computer and the Humanities • Attempt per Circuit per Hour • Automated Clearinghouse

ACHA: American Clearinghouse Association (USA) • Applied and Computational Harmonic Analysis (journal)

ACHAYA: Asociación Chilena de Astronomía y Astronautica (Chile)

ACHE: Alabama Center for Higher Education (USA-AL)

ACHEX: Aerosol Characterization Experiment

Achiev: Achievement

Achievd: Achieved

AcHPh: Acta Helvetica Physica (periodical code)

ACHR: American Society of Human Rights (USA)

ACI: Age-Controlled Item • Airports Council International • Alderney - The Blaye (Channel Islands) (location identifier) • Alliance Coopérative Internationale • Allocated Configuration Identification • Association Cartographique Internationale • Aurigny - The Blaye (Channel Islands) (location identifier) • Authority to Change Information • Autonomous Citation Index • Aviation Civile Internationale

ACIA: Asynchronous Communications Interface Adapter

ACIAAT: Advisory Committee on Instrumentation for the Anglo-Australian Telescope

ACIAS: Automated Calibration Internal Analysis System

ACIBS: Architecture for Computer Integrated Business Systems

ACIC: Aeronautical Chart and Information Center • Australian Chemical Industry Council

ACiCh: Astron. Cirkular (periodical code)

ACID: Acidification Chemistry Information Database • Acoustic Imaging Development • Aircraft Identification • Atomic-Consistent-Isolated-Durable • Atomicity-Consistency-Isolation-Durability • Automatic Classification and Interpretation of Data

ACIGA: Australia Consortium for Interferometric Gravitational Astronomy

ACIL: Automatic Controlled Instrument Landing

ACIP: Aerodynamic Coefficient Instrumentation Package • Assembly Configuration and Integration Panel

ACIR: Advisory Commission on Intergovernmental Relations

ACIRI: AT&T Center for Internet Research at ICSI (USA-CA)

ACIS: Advanced CCD Imaging Spectrometer • American Committee for Interoperable Systems (USA) • American Council for International Studies (USA) • AXAF CCD Imaging Spectrometer

ACISP: Australian Conference on Information Security and Privacy

ACJ: Attitude Control Jet

Ack: Acknowledge • Acknowledgement

ACK: Acknowledge • Acknowledged • Acknowledgement • Amsterdam Compiler Kit • Nantucket (USA-MA) (location identifier)

Ackgt: Acknowledgement

ACL: Academia de Ciências de Lisboa (Portugal) • Access Control List • Advanced CMOS Logic • Advanced Computing Laboratory • Advisory Committee on Libraries • Agent Communication Language • Agent Communications Language • Aguaclara (Colombia) (location identifier) • Alowable Container Load • Alternate Concentration Level • An Chomhairle Leabharlanna (Ireland) • Applications Control Language • Ascent Closed Loop • Association for Computational Linguistics • Automatic Cartridge Library

ACLAE: Association Canadienne des Laboratoires d'Analyse Environnementale

ACLC: Adaptive Communication Live Controller

ACLD: Above Clouds • Advisory Council on Library Development

ACLL: Association for Literary and Linguistic Computing

ACLR: Access Control, Logging and Reporting

ACLS: American Council of Learned Societies (USA) • Automatic Carrier Landing System • Automatic Control and Landing System

ACLT: Actual Calculated Landing Time

ACLU: American Civil Liberties Union (USA)

ACM: Acquisition Control Module • Advanced Climate Model • Advanced Component Mounter • Agenda des Conférences Mathématiques (EMS) • Analyse des Correspondances Multiples • Arica (Colombia) (location identifier) •

Asbestos-Containing Material • Association for Computing Machinery • Astéroïdes-Comètes-Météores • Asteroids-Comets-Meteorites • Asteroids-Comets-Meteors • Atmospheric Control Module • Australian Chamber of Manufacturers • Authorized Controlled Material • Auto-Correlation Method • Auxiliary Core Memory • Average Cost per Mission

ACMA: Air Cargo Management Association

ACMAD: African Centre of Meteorological Applications for Development

ACMB: Applications Configuration Management Board • Attic Cycladic Metamorphic Belt

AcMCS: Acta Materiae Compositae Sinica (periodical code)

ACME: Advanced Computer for Medical Research • Advanced Cosmic Microwave Explorer • Association of Consulting Management Engineers • Attitude Control and Maneuvering Electronics

AcMec: Acta Mechanica (periodical code)

AcMeS: Acta Meteorologica Sinica (periodical code)

AcMet: Acta Metallurgica (periodical code)

ACMI: Air Combat Maneuvering Initiative

AcMik: Acta Mikrochimica (periodical code)

ACMIO: Arhangelsk Complex Magnetic Ionospheric Observatory (Russia)

ACML: Association of Canadian Map Libraries

ACMLA: Association of Canadian Map Libraries and Archives

ACM&M: Acta Metallurgica et Materialia (periodical code)

ACMMS: Acta Metallurgica et Materialia Supplement (periodical code)

ACMNS: Association for Computational Machinery Network Services

ACMO: Agent Chargé de la Mise en Oeuvre

Acmp: Accompany

ACMP: Advisory Committee on Marine Pollution • Applied and Computational Mathematics Program (ARPA)

ACMRR: Advisory Committee on Marine Resources Research

ACMS: Advanced Configuration Management System • Analytical Cost Management System • Application Control Management System • Automated Census Mapping System • Automated Configuration Management System • Avionics Control Management System

AcMSS: Acta Mechanica Solida Sinica (periodical code)

AcMSn: Acta Mechanica Sinica (periodical code)

ACN: Acceptance Change Notice • Accession Change Number • Advance Change Notice • Advisory Committee on Networking • Airborne Classification Number • Alkane Carbon Number • All Concerned Notified • Ante Christum Natum • Association Canadienne de Normalisation (Canada) • Authorized Code Number • Ciudad Acuna (Mexico) (location identifier)

ACNA: Arctic Institute of North America

ACNFP: Advisory Committee on Novel Foods and Processes

ACNN: Australian Conference on Neural Networks

ACNO: Advisory Committed of National Organizations (USA) • Assistant Chief of Naval Operations

AcNum: Acta Numerica (periodical code)

A&CO: Activation & Checkout

ACO: Abell-Corwin-Olowin • Acceptance Checkout • Activation and Checkout • Administrative Contracting Officer • Aircraft Certification Office • Ascona (Switzerland) (location identifier) • Assembly Checkout

ACOB: Altair Complex-Object Benchmark

ACOC: Associated Contractor Originated Change

ACOG: Atlanta Committee for the Olympic Games (USA-GA)

ACOH: Advisory Committee for Operational Hydrology

ACONET: Akademisches Computer Netz

ACONS: AMOS Control Software

AcOpS: Acta Optica Sinica (periodical code)

AcOpt: Acta Optica (periodical code)

ACORD: Advisory Council On Research and Development

ACORN: Acronym-Oriented Nut • Advanced Collaborative Open Resource Network • Associative Content Retrieval Network • Automated Conversion Routine • Automatic Checkout and Recording Network

ACOS: Advanced Computer-Oriented System • Advanced Coronal Observing System • Advanced Crystal Observation System • Advisory Committee on Safety • Arms Control Observation Satellite • Assistant Chief Of Staff • Attitude and Orbit Control System

ACOST: Advisory Council on Science and Technology

ACOT: Apple Classroom Of To-morrow

A&COTA: Assembly & Checkout Technical Analysis

ACOTA: Assembly and Checkout Technical Analysis

AcouL: Acoutics Letters (periodical code)

Acoust: Acoustique

ACP: Abstract Cauchy Problem • Acceptance Checkout Procedure • Access Control Protocol • Action Point • Active Control Point • Adaptive Control Process • Adjacent-Channel Power • Advanced Computational Processor • Advanced Computer Program • Advanced Computer Project • Advanced Cooperative Project • Aerosol Collector and Pyroliser • Aerosol Collector/Pyroliser • Affordable Composite for Propulsion • African-Caribbean-Pacific • Afrique-Caraïbes-Pacifique • Agence Comptable Principale • Algebra of Communicating Processes • Alternating Current Plasma • American Center for Physics (USA-MD) • Analyse en Composantes Principales • Analytical Computer Program • Ancillary Control Program • Aniline Cloud Point • Architecture and Capabilities Platform • Array Control Processor • Asia-Caribbean-Pacific • Aspen Center for Physics (USA-CO) • Assemblée Consultative Provisoire • Associate Computing Professional • Association Canadienne des Physiciens (Canada) • Association of Canadian Publishers (Canada) • Association of Contingency Planners • Astronaut Control Panel • Astronomical Computing Panel (UK) • Atmospheric Chemistry Program • Atypical Clustering Procedure • Audio Control Panel • Auxiliary Control Panel • Auxiliary Control Process • Azimuth Change Pulse

ACPA: Association of Computer Programmers and Analysts (USA) • Audio Capture and Playback Adapter

ACPAQ: Advisory Committee on Post-Adjustment Questions

ACPAR: Angular Correlation of Positron-Annihilation Radiation

ACPD: Alternating-Current Plasma Detector • Alternating-Current Plasma Difference

ACPDP: AC Plasma Display Panel

ACPE: Association for Continuing Professional Education (USA)

ACPG: Astronomical Community Participation Group

ACpH: Air Changes per Hour

ACPH: Air Changes Per Hour

AcPhA: Acta Physica Austriaca (periodical code)

AcPhH: Acta Physica Hungarica (periodical code)

AcPhy: Acta Physica (periodical code)

ACPI: Advanced Configuration and Power Interface

ACPL: Assembly Component Parts List

ACPM: Associate Contractor Program Manager • Attitude Control Propulsion Motor

ACPO: Associate Contractor Projects Office • Association of Chief Police Officers

ACPS: Attitude Control Propulsion System

AcPSl: Acta Physica Slovaca (periodical code)

AcPSn: Acta Physica Sinica (periodical code)

Acpt: Acceptance

ACPT: Accept

ACPU: Association Canadienne des Professeurs d'Université (Canada) • Auxiliary Computer Power Unit

ACPVI: Analyse en Composantes Principales avec Variables Instrumentales

ACPY: Accompany

Acq: Acquire • Acquisition

ACQ: Acquisition • Advisory Committee on Quality

ACQE: Acquisition and Control Query Executive

ACR: Access Rule Index • Acronyme • Active Cavity Radiometer • Actual Cost Report • Advanced Capabilities Radar • Air Carrier • Allowance Change Request • Allowed Cell Rate • Alternate CPU Recovery • Anomalous Cosmic Ray • Approach Control Radar • Araracuara (Colombia) (location identifier) • Astroclub Radebeul (Germany) • Astrometric Calibration Region • Attenuation-to-Crosstalk Ratio • Audio Cassette Recorder • Automatic Card Reader • Automatic Character Reader • Automatic Character Recognition • Automatic Current Regulation • Auxiliary Computer Room

ACRBC: Acceptance, Checkout, Retest, and Backout Criteria

ACRC: Advanced Computing Research Centre • Assured Crew Return Capability (now ACRV)

Acrd: Accrued

ACRE: Acquisition of Requirements • Advanced Clock/Ranging Experiment • Assessing Calculus Reform Efforts • Automatic Climatological Recording Equipment

ACRES: Australian Centre for Remote Sensing • Automated Change Request and Evaluation System

ACRF: Advanced Computing Research Facility • Atmospheric Cloud Radiative Forcing

Acrg: Acreage

AcRhe: Acta Rheologica (periodical code)

ACRI: Advanced Computer Research Institute • Advanced Computer Research International • Advanced Computing Research Institute (USA-NY)

ACRIM: Active Cavity Radiometer for Irradiance Monitor • Active Cavity Radiometer for Irradiance Monitoring • Active Cavity Radiometer Irradiance Monitor (*) • Active Cavity Radiometer Irradiance Monitoring

ACRL: Association of College and Research Libraries

Acron: Acronym

ACROSS: Automated Cargo Release and Operations Service System

ACRP: Advanced Concepts Research Project

ACRS: Accelerated Cost Recovery System • Across • Advanced Carbon Reactor Subassembly • Advisory Committee on Reactor Safeguards • African Council on Remote Sensing • Asian Conference on Remote Sensing • Astrographic Catalog of Reference Stars

ACRT: Accelerated Crucible Rotation Technique

ACRV: Advanced Crew Return Vehicle • Assured Crew Rescue Vehicle • Assured Crew Return Vehicle (formerly ACRC) (*) • Astronaut Crew Rescue Vehicle • Astronaut Crew Return Vehicle

a/cs: Accounts

ACS: Academic Computer Specialist • Acesso • Access • Access Control Set • Access Control System • Accounting Control System • Achinsk (Russia) (location identifier) • Active Cancellation System • Active Control System • Actuator Control System • Adaptive Code System • Adaptive Control System • Advanced Camera for Surveys • Advanced Communications System • Advanced Composite Structure • Advanced Computer Systems • Aft Crew Station • Air Charter Services • Air Conditioning System • Alternating Current Synchronous • Altitude Control System • American Chemical Society (USA) • Amphibious Command Ship • Analog Computer System • Analysis Computer System • Antarctic Conservation Strategy • Anti-Coincidence System • Asian Chemical Societies • Assembly Control System • Asynchronous Communication Server • ATM Circuit Steering • Atmosphere Control and Supply • Atmosphere Control System • Attitude Control and Stabilization • Attitude Control System • Australian Computer Society • Autocorrelation Spectrometer • Autograph Card Signed • Automated Cartridge System • Automated Commercial System • Automated Confirmation Service • Automated Control System • Automatic Class Selection • Automatic Climate Station • Automatic Control System • Autopilot Control System

ACSAD: Arab Centre for the Study of Arid Zones and Dry Lands

ACSB: American Ceramic Society Bulletin (periodical code)

ACSC: American Ceramic Society Communications (periodical code) • American Council on Schools and Colleges (USA) • Applied Communications System Center • Association of Casualty and Surety Companies

ACSE: Application Control Service Element • Arctic Coastal and Shelf Ecosystem • Association Control Service Element • Association Control Service Entity • Association of Consulting Structural Engineers

ACSET: Advisory Committee on the Supply and Training of Teachers

ACSF: Association des Centres Serveurs Français (France)

ACSI: Association Canadienne des Sciences de l'Information (Canada)

ACSIA: Aide à la Communication pour les Systèmes d'Information Automatisés

ACSIOM: Applied Computing System Institute of Massachusetts Inc. (USA-MA)

ACSIR: Advisory Council for Scientific and Industrial Research (UK)

ACSIS: Advisory Committee on Standards Information Services • Auto-Correlation Spectrometer Imaging System

ACSJ: American Ceramic Society Journal (periodical code)

ACSL: Advanced Continuous Simulation Language

ACSM: American Congress of Surveying and Mapping (USA)

ACSN: Advance Change Schedule Notice • Advance Change Study Notice • Advanced Change Study Notice

ACSNet: Australian Computer Science Network

ACSOE: Atmospheric Chemistry Studies in the Oceanic Environment

ACSP: Advanced Control Signal Processor • Advisory Council on Scientific Policy (UK)

ACSS: Analog Computer Subsystem • Automated Colour Separation System

Acst: Acoustic

ACST: Advisory Council on Science and Technology (Australia)

ACSTI: Advisory Committee for Scientific and Technical Information (UK)

ACSU: Alternating-Current Switching Unit

Acsy: Accessory

ACSyS: Arctic Climate System Study

Act: Actif • Acting • Action • Activates • Active • Activität • Activité • Activity • Actor • Actual • Actuary

ACT: Accredited Compaq Technician • Acquisition Coarse Tracker • Acquisition-Control-Test • Activated • Active • Active Control Technology • Active Cooling Test • Actor • Actriz • Actual • Adaptive Current Tomography • Advanced Composites Technology • Advanced Computer Technique • Advanced Corporation Tax • Advisory Council on Technology (UK) • Air-Čerenkov Telescope • Air Combat Training • Air-Cushion Technology • Alliance Commerciale Technologique • American College Test • Analogical-Circuit Technique • Analysis Complexity Tool • Analysis of Complexity Tool • Applied Coherent Technology • Association of Classroom Teachers • Association of Commercial Televisions • Atmospheric-Čerenkov Technique • Atmospheric-Čerenkov Telescope (*) • Audio Center Transmission • Australian Capital Territory • Automatic Code Translation • Azimuth Control Torquer • Waco (USA-TX) (location identifier)

a cta: A Cuenta

ACTA: Advanced Civilian Technology Agency (USA) • American Carriers Telecommunications Association (USA) (*) • America's Carriers Telecommunications Association (USA)

ACTAp: Automatic Control Theory Applications (periodical code)

ACTAS: Alliance of Computer-based Telephony Application Suppliers

AcTC: Acta Theoretica Chimica (periodical code)

ACTC: Associated Colleges of the Twin Cities (USA-MN)

ACTD: Advanced Concept Technology Demonstration • Attitude Control Torquing Device

ACTE: Approval Committee for Telecommunications Equipment • Automatic Checkout Test Equipment

AcTeC: Acta Technica (periodical code)

ACTFEL: AC Thin-Film Electroluminescent

ACTFL: American Council on Teaching Foreign Languages (USA)

Actg: Acting

ACTG: Acting

ACT GS: Actual Ground Speed

ACTI: Advisory Committee on Technology Innovation

ACTIM: Agence pour la Coopération Technique et Industrielle en Météorologie

ACTION: Assisting Carers using Telematics Interventions to meet Older person's Needs

Activ: Activement • Activité

ACTIVE: Advanced Control Technology for Integrated Vehicles

Actl: Actual

ACTM: Atmospheric Chemical Transport Model

Actnt: Accountant

ACTO: Association en Touraine et Centre-Ouest (France)

ACTP: Adobe Certified Training Provider • American College Testing Program

Actr: Actuator

ACTS: Acoustic Control and Telemetry System • Acquisition Coarse Tracking System • Advanced Civil Transport Simulator • Advanced Communications Technologies and Services • Advanced Communications Technology Satellite (*) • African Center for Technology Studies • Area Communications Terminal System • Attitude Control and Translation System • Automated Computer Time Service • Automatic Coin Telephone Service

ACTT: Advanced Communications and Timekeeping Technology

ACTU: Australian Council of Trade Unions

Actuat: Actuating

ACTV: Active

Actvt: Activate

Actvtd: Activated

Actvty: Activity

Acu: Acustica (periodical code)

AC&U: Association of Colleges & Universities

ACU: Acceleration Compensation Unit • Achutupo (Panama) (location identifier) • Address Calculation Unit • Address Control Unit • Alarm Control Unit • Alternating Current Control Unit • Altocumulus • Annunciator Control Unit • Antenna Control Unit • Arithmetic and Control Unit • Array Control Unit • Association of Colleges and Universities • Association of Commonwealth Universities • Association of Computer Users • Attitude Control Unit • Automatic Calling Unit • Automatic Call Unit • Automatic Client Upgrade • Avionics Cooling Unit

ACUC: Animal Care and Use Committee

ACUIM: Atmopsheric Correction Using Imbedded Models

ACUO: Avionics Cooling Unit Operator

ACUP: Association of Canadian University Presses

Acus: Acusativo

ACUS: Automated Conference of the United States

Acúst: Acústica

ACV: Actual Cash Value • Air-Cushion Vehicle • Algemeen Christelijk Vakverbond • Alternating Current Volt • Arcata (USA-CA) (location identifier) • Astro Club Voyager (Italy) • Eureka (USA-CA) (location identifier)

ACVC: ADA Compiler Validation Capability

ACVE: Accelerometer Calibration Vibration Exciter

ACVEN: Advisory Committee on Vehicle Emissions and Noise

ACW: Alternating Continuous Wave • Automatic Car Wash

ACWC: Advisory Committee on Weather Control

ACWP: Actual Cost of Work Performed

ACWS: Actual Cost of Work Scheduled • Aircraft Control and Warning System

ACX: Alpha Charge Exchange

ACY: Any Country • Atlantic City - Pomona (USA-NJ) (location identifier)

ACYC: Anticyclonic

a/d: A Dater de • A la Date de

ad: Adapt • Adaptation • Adapter • Adverb • After Date • an dem • an den • an der • Ante Diem • auf der • Autograph Document

aD: außer Dienst

Ad: Adiabatic • Admis • Admission • Advantage • Adverb • Advert • Advertisement

A&D: Aerospace & Defense

A/D: After Date • Analog-Digital • Analog to Digital • Assembly/Disassembly

AD: Abbreviated Dialing • Access Door • Accidental Damage • Accretion Disk • Accumulated Dose • Acquisition Director • Acte de Décès • Active Duty • Activity Description • Adjoint Administratif • Administrateur Délégué • Administration de Données • Administration Déléguée • Administration Départementale • Administrative Decision • Administrative Department • Administrative Directive • Advanced • Advanced Development • Advisory Direction • Aerodynamic Decelerator • Aeronautical Data • Aerospace and Defense • After Date • Air Density • Air Dried • Airworthiness Directive • Albrecht Dürer (1471-1528) • Algerian Dinar (currency) • Algorithmic Differentiation • Alternate Days • Altitude Deviation • Alzheimer's Disease • Ambipolar Diffusion • Amplifier Detector • Andorra (ISO country code) • Andorre (ISO country code) • Anno Domini • Anodal Deviation • Ante Diem • Antimatter Decelerator • Antiproton Decelerator • Aperture Door • Apparent Drift • Applicable Document • Application Development • Architectural Design • Area Director • Arithmetic Device • Art Director • As

Drawn • Assembly District • Assembly Drawing • Assistant Director • Associate Director • Asymmetric Drift • Atomic Drive • Attachment Device • Auditory Display • Auto Dial • Auto Doubler • Autograph Document • Autogroup Definition • Automatic Dictionary • Automatic Differentiation • Automatic Display • Autonomous Domain • Average Deviation • Avis de Débit • Ayant Droit

ADA: Acceleration-Deceleration-Acceleration • Action Data Automation • Adana (Turkey) (location identifier) • Adaptive Diagnostics Authoring • Administration Agent • Adobe Developers Association • Advanced Data Analysis • Advanced Disk Array • Aeronautical Development Agency • American with Disabilities Act (USA) • Amplifier Detector Assembly • Association Drômoise d'Astronomie (France) • Atomic Development Authority • Automatic Data Acquisition • Automatic Document Analysis • Azimuth Drive Assembly

ADAAC: Affiliated Distributed Active Archive Center

ADAAPS: Aircraft Data Acquisition Analysis and Presentation System

ADAB: Australian Development Aid Bureau

ADABAS: Adaptable Database System

ADAC: Astronomical Data Analysis Center • Attitude Determination And Control

ADACS: Attitude Determination And Control System • Automatic Data Acquisition and Control System

ADAD: Automatic-telephone Dialing-Announcing Device

ADADL: ADA Design And Documentation Language

ADAES: Advanced Data Assessment Expert System

ADAF: Advection-Dominated Accretion Flow

ADALT: Advanced Radar Altimeter

ADAM: A Dynamic Attribute Manager • Analog Data Acquisition Module • Art Design Architecture and Media • Astronomical Data Acquisition and Management • Astronomical Data Acquisition Monitor • Astronomical Data Analysis and Management • Automated Distributed Architecture Model • Automatic Data Acquisition Module

ADAMHA: Alcohol, Drug Abuse, and Mental Health Administration (now SAMHSA)

ADAMS: Automated Dynamic Analysis of Mechanical Systems

Adap: Adapted

ADAP: Advanced Data Analysis Program • Aerodynamic Data Analysis Program • Argos Data Acquisition Platform • Automated Data Analysis Program

AdApM: Advances in Applied Mathematics (periodical code)

ADAPS: Advanced Data Acquisition and Processing System • Atmospheric Data Acquisition and Processing System • AVHRR Data Acquisition and Processing System

ADAPSO: Association of Data Processing Organizations

ADAPT: Analog-Digital Automatic Program Tester

ADAPTS: Advanced Data And Picture Transformation System

ADAR: Airborne Data Acquisition and Registration

ADARTS: Ada-based Design Approach for Real-Time Systems

AdAs: Ad Astra (periodical code)

ADAS: Aerospace Data Acquisition System • Airborne Data Acquisition System • Aircraft Departure and Arrival Sequencing • Altrincham and District Astronomical Society (UK) • Astronomical Data Analysis Software • Astronomical Data Analysis System • Atmospheric Data Acquisition System • Atomic Data and Analysis Structure (package) • Automatic Data Acquisition System • Auxiliary Data Annotation Set

ADASS: Astronomical Data Analysis Software and Systems

ADATE: Automatic Digital Assembly Test Equipment

AdAtS: Advances in Atmospheric Sciences (periodical code)

ADATS: AMOS Data Transmission System

ADB: Accidental Death Benefit • African Development Bank • Apple Desktop Bus • Asian Development Bank • Astronomical Database • Izmir - Adnan Menderes (Turkey) (location identifier)

ADBIS: Advances in Databases and Information Systems (conference)

ADBK: Astronomical Database Kernel

ADBS: Association des Documentalistes et Bibliothécaires Spécialisés (Fran ce)

ADBU: Association des Directeurs des Bibliothèques Universitaires (France)

ADC: Accretion Disk Corona • Actual Direct Cost • Adaptive Data Compression • Add with Carry • Advanced Developing Country • Advanced Diamond Composite • Advice of Duration and Charge • Affiliated Data Center • Aide De Camp • American Digital Cartography • Analog-to-Digital Conversion • Analog-to-Digital Converter • Antenna Drive Cabinet • Apparent Diffusion Coefficient • Application Development Conference • Application Device Channel • Area Data Center • Astrometric Differential Coordinates • Astronomical Data Center • Astronomy Data Center (*) • Atlantic Data Coverage • Atmospheric Dispersion Compensator • Atmospheric Dispersion Corrector • Attitude Determination and Control • Automated Distribution Control • Automatic Data Collection • Automatic Detect Classification • Automatic Digital Calculator • Automatic Drift Control • Average Depth of Compensation

ADCBu: Astronomical Data Center Bulletin (periodical code)

ADCC: Advanced Data Communication Control Procedure • Automatic Data Cataloguing and Characterization

ADCCC: Astrophysics Data Center Coordinating Council

ADCCP: Advanced Data Communication Control Procedure

AD&CD: Advanced Displays & Controls Development

ADCD: Advanced Displays and Controls Development

ADCI: Automatic Display Call Indicator

ADCLS: Advanced Data Collection and Location System • Automated Data Collection and Location System

AdCM: Advanced Ceramic Materials (periodical code)

ADCN: Attitude Determination, Control, and Navigation

AdCoL: Advanced Composites Letters (periodical code)

ADCOM: Aerospace Defense Command (now AFSPC)

ADCOS: Advisory Committee for Space Sciences

ADCP: Acoustic Doppler Current Profiler

ADCR: Applicable Document Contractual Record

ADCS: Attitude Determination and Control System • Automated Document Conversion System • Automatic Data Collection System

ADCT: Adaptive Discrete Cosine Transform

add: Adidit

Add: Addenda • Addendum • Addition • Additional • Address • Addressed

ADD: Accessibility and Dissemination of Data • Adaptive Defect Detection • Adaptive Device Driver • Addendum • Addis Ababa (Ethiopia) (location identifier) • Analog Data Digitizer • Architectural Design Document • Architecture Definition Document • Arctic Data Directory • Assistive Devices Division • Attention Deficit Disorder • Automatic Document Detection

ADDA: American Design Drafting Association • Australian Database Development Association

ADDAS: Airborne Digital Data Acquisition System

ADDB: Antarctic Digital Database

ADDE: Abstract Data Distribution Environment

ADDEF: Association pour le Développement du Droit de l'Espace en France (France)

ADDF: Advanced Data Distribution Facility

Addit: Additional

Addl: Additional

Addn: Addition

ADDNET: Acid Deposition Data Network

Addnl: Additional

ADDP: Antarctic Digital Database Project

ADDS: Antarctic Data Directory System • Applications Developmental Data System • Automatic Document Distribution Service • Aviation Digital Data Service

Addsd: Addressed

Addv: Addvection

ADE: Actuator Drive Electronics • Aden (Yemen) (location identifier) • Aeronautical Development Establishment • Agency-wide Distributed Environment • Autodesk Data Extension • Automatic Directory Exchange • Avance et Développement Électronique

ADEA: Age Discrimination in Employment Act (USA)

ADEB: Association Des Éditeurs Belges (Belgium)

ADEC: Association pour le Développement des Entreprises et des Compétences

ADECAM: Association pour le Développement de la Communication Animée et Multimédia

ADEF: Action de Développement d'Évaluation et de Formation • Association de Défense des Économiquement Faibles

ad effect: Ad Effectum

ADEL: Antenne de Diffusion des Échanges du Livre (France)

ADELE: Articulation de Déploiement à Lames d'Enroulement

ADEME: Agence de l'Environnement et de la Maîtrise de l'Énergie (France)

ADENA: Asociación para la Defensa de la Naturaleza (Spain)

AdEOA: Advances in Earth-Oriented Applications (periodical code)

ADEOS: Advanced Earth Observation Satellite • Advanced Earth Observing Satellite (*) • Advanced Earth Observing Spacecraft

ADEPA: Association pour le Développement de la Production Automatisée

AdEPS: Advances in Earth and Planetary Sciences (periodical code)

ADEPT: Advanced Distributed Environment for Production Technology • Assisting in Deployment of Energy Practices and Technologies

ADEQUA: Aide à la Décision de la Qualité

ADERA: Association pour le Développement de l'Enseignement et de la Recherche en Aquitaine (France)

ADES: Association de Documentation Économique et Sociale

ADESS: Automated Data Editing and Switching System

ADEST: Association pour le Développement et l'Étude de la Scientométrie et de la Technométrie (France)

ADETEM: Association pour le Développement des Techniques d'Études du Marché (France)

ADETIM: Association pour le Développement des Techniques des Industries Mécaniques (France)

ad eund: Ad Eundem

ADF: Active Dark Filament • Adapter Description File • Algorithm Development Facility • Alliance for a Digital Future • Application Development Facility • Asian Development Fund • Astrophysics Data Facility • Automatically Defined Function • Automatic Direction Finder • Automatic Direction Finding • Automatic Display Finder • Automatic Document Feed • Automatic Document Feeder • Avionics Development Facility

ADFA: Australian Defence Force Academy

ad fin: Ad Finem

ADFRF: Ames-Dryden Flight Research Facility (NASA) (formerly DFRF)

ADFTO: Advanced Data Flow Technology Office

ADG: Adrian - Lenawee County (USA-MI) (location identifier) • Assistant Director General

AdGeo: Advances in Geophysics (periodical code)

ADH: Advanced Development Hardware • Aldan (Russia) (location identifier) • Alternative Decohering History • Antidiuretic Hormone • Automatic Data Handling

ADHD: Attention Deficit/Hyperactivity Disorder

ad h.l.: Ad Hunc Locum

ADHS: Analog Data Handling System

AdI: Agência de Inovação (Portugal)

ADI: Abstract Data Interface • Acceptable Daily Intake • Air Defense Initiative • Alternating Direction Implicit • Analytic and Design Investigation • Apollo Document Index (NASA) • Application Data Interchange • Application Data Interface • Arandis (Namibia) (location identifier) • Archivage et Documentation Intelligents • Area of Dominant Influence • Association de Droit International • Attitude Direction Indicator • Attitude Display Indicator • AutoCAD Device Interface • Autodesk Device Interface

ADIAF: Association pour le Développement de l'Informatique dans l'Administration Française (France)

ADIE: Acquisition Data Input Equipment

ADIF: Aide à la Définition d'Intercepteurs Futurs • Analog Data Interchange Format • Antarctic Directory Interchange Format

ADIFOR: Automatic Differentiation of Fortran

ADII: Agency for the Development of Innovations and Inventions

ADIL: Astronomy Digital Image Library

ad inf: Ad Infinitum

ad init: Ad Initium

ad int: Ad Interim

ADION: Association pour le Développement International de l'Observatoire de Nice (France)

ADIOS: Automatic Digital Input-Output System • Automatic Digitizing of Interpretation Overlay System

ADIRC: Association pour le Développement Informatique en Région Centre (France)

ADIRS: Air Data and Inertial Reference System • Air Data Inertial Reference System • Air Data/Inertial Reference System

ADIRU: Air Data/Inertial Reference Unit

ADIS: Arctic Data Interactive System • Association de Diffusion de l'Iconographie Scientifique • Association for the Development of Instructional Sytems (USA) • Automated Data Interchange System

ADIT: Agence pour la Diffusion de l'Information Technologique (France) • Analog-Digital Integrating Translator • Assistant Dean for Information Technology

ADIVA: Association des Distributeurs Informatiques à Valeur Ajoutée (France)

ADIZ: Air Defense Identification Zone

Adj: Adjacent • Adjective • Adjetivo • Adjoining • Adjoint • Adjourned • Adjudged • Adjunct • Adjust • Adjustable • Adjusted • Adjustment • Adjutant

ADJ: Accrétion-Disques-Jets • Adjacent • Adjustment • Amman (Jordan) (location identifier)

Adjt: Adjoint • Adjutant

ADK: Adak Island (USA-AK) (location identifier) • Application Development Kit

ADKMS: Advanced Data and Knowledge Management System

ADL: Acoustic Delay Line • Activity Definition Language • Ada Design Language • Address Data Latch • Adelaide (Australia-SA) (location identifier) • Advaned Development Laboratory • Advances in Digital Libraries (conference) • Aeronautical Data Link • Apollo Documentation List (NASA) • Après la Durée Légale • Architecture Description Language • Arthur D. Little (USA-MA) • Assertion Definition Language • Au-delà de la Durée Légale • Automated Data Library • Automatic Data Link • Avionics Development Laboratory

ADLAT: Adaptive Lattice

ad lib: Ad Libitum

ad loc: Ad Locum

ADLS: Airborne Doppler Lidar System

ADLY: Arrival Delay

Adm: Administración • Administratif • Administration • Administrative • Administrator • Admiraal • Admiral • Admiralty • Admission • Admitted

ADM: Add-Drop Multiplexer • Add-Drop Multiplexing • Administração • Administrador • Administratif • Administrative • Aeronautical Decision Making • Aggregate Data Manager • Air Data Management • Analysis/Design Model • Annual Delegate Meeting • Ardmore (USA-OK) (location identifier) • Automatically Defined Macro • Automatic Display Mode • Automatic Distribution of Microfiche • Average Daily Membership

ADMA: Aviation Distributors and Manufacturers Association

ADMATCH: Address Matching

ADMC: Agriresidue Dry Matter Content

AdMD: Administrative Management Domain

ADMD: Administrative Management Domain

ADMF: Asynchronous Data Mover Facility

Admin: Administratif • Administration • Administrative • Administrator

ADMIS: Automated Data Management Information System

ADMIT: African Drought Monitoring Integrative Technique • Automatic Drought Monitoring Integrative Technique

Admón: Administración

Admor: Administrador

ADMOS: Automatic Device for Mechanical Order Selection

admov: Admove • Admoveatur

Admr: Administrador

ADMS: Analog and Digital Monitoring System • Antarctic Data Management System • Automated Data Management System

ADMSR: Airborne Dual-frequency Microwave Radiometer-Scatterometer

ADMYTE: Archivo Digital de Manuscritos Y Textos Españoles (Spain)

ADN: Abbreviated Dialing Number • Acide Désoxyribonucléique • Advanced Digital Network • Allgemeiner Deutscher Nachrichtendienst • Andes (Colombia) (location identifier) • Anti-Dwarf Nova • Automatic Dobson Network • Yemen (international vehicle registration code)

ADNDT: Atomic Data and Nuclear Data Tables

ad neut: Ad Neutralizandum

ADNFI: Association des Nouvelles Formations d'Ingénieurs

ADO: Abstract Data Object • Active Data Object • ActiveX Data Object • Advanced Development Objective • Airport District Office • Andamooka (Australia-SA) (location identifier) • Assistant District Officer • Astronomy Department Observatory • Automotive Diesel Oil

ADONIS: Adaptive Optics Near-Infrared System • AMOS Daytime Optical Near-infrared Imaging System • Article Delivery Over Network Information Systems

ADOP: Advanced Distributed Onboard Processor

ADOR: Astrometry and Design Operations Review • Astronomy Design and Operations Review

ADOS: African Doppler Observation Survey

ADOT: Advanced Display Optimization Tool

AdP: Arbeitsgemeinschaft deutschsprachiger Planetarien

ADP: Acceptance Data Package • Acceptance Documentation Package • Administrative Data Processing • Advanced Data Processing • Advanced Design Program • Advanced Development Program • Advanced Ducted Prop • Advanced Ducted Pulsion • Aéroports de Paris (France) • Air Data Package • Air Data Probe • Anuradhapura (Sri Lanka) (location identifier) • Application Development Platform • Application Distribution Protocol • Arbeitsgemeinschaft deutschsprachiger Planetarien • Area Density of Particles • Astrophysics Data Program • Atmospheric Dynamics Program • Automated Data Processing • Automatic Data Processing • Peseta (Andorra) (ISO currency code)

ADPA: Air Data Probe Assembly

ADPC: Automated Data Processing Center

ADPCM: Adaptive Delta Pulse Code Modulation • Adaptive Difference Pulse Code Modulation • Adaptive Differential Pulse Code Modulation (*)

ADPE: Automated Data Processing Equipment • Automatic Data Processing Equipment • Auxiliary Data Processing Equipment

ADPF: Automated Data Processing Facility

ADPG: Advances in Planetary Geology (periodical code)

ADPL: As Designed Parts List

ADPLAN: Advanced Planning

ADPM: Automatic Data Processing Machine

ADPP: Automatic Data Processing Program

ADPS: Acceptance Data Package System • Automatic Data Processing System

ADPSO: Association of Data Processing Service Organizations

Adptr: Adapter

ADQ: Kodiak (USA-AK) (location identifier)

Adr: Adresse

ADR: Accident Data Recorder • Achievable Data Rate • Active Data Repository • Active Distributed Routing • Address • Adiabatic Demagnetization Refrigerator • Advisory Route • Aeroporti Di Roma (Italy) • Aggregate Data Rate • Alternative Dispute Resolution • Analog Digital Recorder • Andrews (USA-SC) (location identifier) • Architectural Design Review • Automated Demand Resolution

ADRIES: Advanced Digital Radar Imagery Exploitation System

ADRL: Automatic Drawing Release List

Adrs: Address

ADRS: A Departmental Reporting System • Astrometry Data Reduction System • Automated Data Retrieval System • Automatic Data Reporting System • Automatic Document Request Service

ADRST: Association pour le Développement de la Recherche Scientifique et Technique (France)

ad s: Ad Sectum

AdS: Académie des Sciences

ADS: Académie Des Sciences • Accelerator-Driven System • Accurately Defined System • Acid Deposition System • Action Démocratique et Sociale • ADF Data System • Advanced Data Streaming • Advanced Data System • Advanced Digital System • Advanced Dressing Station • Aerial Delivery System • Aide Décisive à la Sécurité • Airborne Data System • Air Data Sensor • Air Data System • Aitken Double Stars • American Dialect Society (USA) • Angular Displacement Sensor • Application Development Solution • Application Development System • Application-Driven Scalability • Ascent Design System • Aspect Determination System • Associate Director for Science • Astrophysical Data System • Astrophysics Data System • Attitude Determination System • Attitude Displacement System • Attitude Display System • Audio Distribution System • AutoCAD Development System • Autograph Document Signed • Automated Data Service • Automated Data System • Automated Dependent Surveillance • Automatic Data System • Automatic Dependent Surveillance • Automatic Depressurization System • Automatic Distribution System • Auxiliary Data System • Dallas Addison (USA-TX) (location identifier)

ADSA: Alternate Demonstration Structural Article • Angular Displacement Sensor Assembly

ad saec: Ad Saeculum

ADSAP: Advise As Soon As Possible

ad sat: Ad Saturandum

ADSC: Address Status Changed • Adobe Document Structuring Convention • Automatic Data Service Center

ADSEP: Advanced Separation Process • Automatic Data Set Editing Program

ADSF: Advanced Directional Solidification Furnace (*) • Automated Directional Solidification Furnace • Automatic Directional Solidification Furnace

ADSG: Agents Du Service Général

ADSI: Analog Directory Services Interface • Analog Display Services Interface

ADSL: ABR Digital Subscriber Line • Asymmetrical Digital Subscriber Line • Asymmetric Digital Subscriber Line (*) • Asymmetric Digital Subscriber Loop

ADSM: ADSTAR Distributed Storage Manager

ADSP: Advanced Digital SAR Processor • Aerially Deployed Soil Penetrator • AppleTalk Data Stream Protocol

AdSpR: Advances in Space Research (periodical code)

ADSR: Attach-Decay-Sustain-Release

ADSRN: ASLV Down Range Station

ADSS: Advanced Digital Stereophotogrammetric System • Analysis of Digitized Seismic Signals • Automatic Data Switching System

ADSTAR: Automatic Document Storage and Retrieval

adst. feb.: Adstante Febre

ADT: Abstract Data Type • Actual Departure Time • Ada (USA-OK) (location identifier) • Advanced Design Team • Advanced Development Technology • Advanced Display Technology • Air Data Transfer • Alaska Daylight Time • Anomalous Diffraction Theory • Application Data Type • Application Development Tool • Approved Departure Time • Astronomy Diagnostic Test • Atlantic Daylight Time • Automatic Data Transfer • Automatic Data Translator • Automatic Data Transmission • Automatic Data Transmitter • Automatic Detection and Tracking • Average Daily Traffic

ADTA: Air Data Transducer Assembly

ADTAC: Automatic Digital Tracking Analyzer Computer

ADTD: Association of Data Terminal Distributors

ADTN: Administrative Data Transmission Network

ADTP: Accelerated Development Test Program

ADTRS: Advanced Tracking and Data Relay Satellite

ADTS: Advanced Detection Technology Sensor • Automatic Data and Telecommunications Service

ADTU: Automatic Digital Test Unit • Auxiliary Data Translator Unit

ADU: Acceleration-Deceleration Unit • Alpha Demonstration Unit • Analog/Digital Unit • Analog Display Unit • Analog to Digital Unit • Annunciator Display Unit • Arbitrary Data Unit • Ardabil (Iran) • Ascent Driver Unit • Automatic Dialing Unit • Azimuth Drive Unit

ad us: Ad Usum

ad us. ext.: Ad Usum Externum

ad v: Ad Valorem

Adv: Advance • Advanced • Advancement • Advent • Adverb • Adverbial • Adverbially • Adverbio • Adversaire • Adverse • Adversus • Advertisement • Advertising • Advice • Advise • Adviser • Advisory • Advocate

ADV: Abstract Data View • Advise • Advogado • Air Distribution and Ventilation • Allgemeine Datenverarbeitung • Andover (UK-England) (location identifier) • Arbeitsgemeinschaft für Datenverarbeitung • Automatisierte Datenverarbeitung

ad val: Ad Valorem

Advb: Adverb

ADVCTN: Advection

Adven: Adventure

Advers: Adversativo

ADVISAR: All Digital Video Imaging System for Atmospheric Research

ADVISOR: Advanced Integrated Safety and Optimizing Computer

ADVN: Advance

Adv. Pmt.: Advance Payment

AdvSpaRes: Advances in Space Research

Advt: Advertisement

Advtg: Advertising

ADVY: Advisory

AdW: Akademie der Wissenschaften

ADW: Camp Springs (USA-MD) (location identifier)

ADWG: Astronomical Database Working Group

ADX: Automatic Data Exchange • Automatic Digital Exchange • Saint Andrews - Leuchars (UK-Scotland) (location identifier)

ADXPS: Angular Dependent X-ray Photoelectron Spectroscopy

ADY: Alldays (South Africa) (location identifier)

ADZ: Als Daar Zijn • Approach and Departure Zone • San Andrés Island (Colombia) (location identifier)

ae: Aetatis

A&E: Accident & Emergency • Airframe & Engine • Analysis & Evaluation • Architects & Engineers • Architectural & Engineering

AE: Above or Equal • Account Executive • Acid Ester • Acoustic Emission • Actual Evapotranspiration • Adjoint d'Enseignement • Adult Education • Aeronautical Engineer • Aeronautical Engineering • Affaires Étrangères • Afrique Équatoriale • Age Exemption • Aide-Éducateur • Airframe and Engine • All England • Almost Everywhere • American English • American Express • Analog Electronics • Application Entity • Application Environment • Application Execution • Applications Engineer • Applications Explorer • Architects and Engineers • Architectural and Engineering • Architectural Engineering • Associate Editor • Associate in Education • Associate in Engineering • Association Européenne • Astronomische Eenheden • Astronomische Einheit • Atmosphere Explorer • Atmospheric Explorer • Atomic Energy • Auroral Electrojet • Auto Estrada • Autoexposure • Automatic

Exposure • Autumn Equinox • Auxiliary Equipment • Avis d'Expédition • Émirats Arabes Unis (ISO country code) • United Arab Emirates (ISO country code)

AEA: Abemama Atoll (Kiribati) (location identifier) • Acoustic Emission Analysis • Aerial Experiment Association • Aft End Assembly • Agence Européenne pour l'Environnement • American Electronics Association (USA) • Antenna Elevation Angle • Association of European Airlines • Atomic Energy Act • Atomic Energy Authority (UK) • Automatic Error Analysis

AEAAC: Association Éducative des Amateurs d'Astronomie du Centre (France)

AeAm: Aerospace America (periodical code)

AEAP: Atmospheric Effects of Aviation Project

AEB: Aft Equipment Bay • Agência Espacial Brasileira (Brazil) • Analog Expansion Bus • Analysis and Evaluation Board • Area Electricity Board • Associated Examining Board • Astronomy Education Board (AAS)

AEBP: Atmospheric Equivalent Boiling point

AEBR: Association of European Border Regions

AEC: Acte d'État Civil • Additional Extended Coverage • Aft End Cone • Aft Events Controller • Altitude Engine Control • American Express Co. • Architectural and Engineering Construction • Architecture-Engineering-Construction • Archived Exposure Catalog • Archive Exposure Catalog • Association des Éditeurs Canadiens (Canada) • Association of Education Committees (UK) • Atomic Energy Commission • Atomic Energy Corporation • Atomic Energy Council • Atomic Energy of Canada • Australian Environmental Council • Automatic Experiment Container • Automatic Exposure Control • Automatic Exposures Catalog

AECB: Atomic Energy Control Board

AeCh: Aerospace China (periodical code)

AECI: Agencia Española de Cooperación Internacional (Spain)

AECIC: Affaires Étrangères et Commerce International du Canada

AECL: Atomic Energy of Canada Ltd.

AeCM: Aerospace Composites Materials (periodical code)

AECMA: Association Européenne de Constructeurs de Matériaux Aérospatiaux • Association Européenne de Constructeurs de Matériel Aéronautique • Association Européenne de Constructeurs de Matériel Aérospatial

AECS: Area Entry Control System • Atmospheric Effect Correction System • Automated Entry Control System

AECSI: ACM Electronic Community Services Initiative • ACM Electronic Community Strategic Initiative

AECT: Association of Educational Communications and Technology

AEd: Associate in Education

AED: Actual End Date • Advanced Electronics Design • ALGOL Extended for Design • Analog Event Distributor • Analysis and Evaluation Division • Association of Engineering Distributors • Atomic Emission Detector • Automated Engineering Design • Automatic External Defibrillator • Dirham (UAE) (ISO currency code)

AEDB: Apollo Engineering Documentation Board (NASA)

AEDC: Antarctic Environmental Data Centre • Arnold Engineering Development Center (USA-TN)

AEDD: Antarctic Environmental Data Directory

AEDDWG: Antarctic Environmental Data Directory Working Group

AEDE: Association Européenne des Enseignants

AEDI: Audiovisual Event Description Interface

AEDIMA: Asociación Española de Informática y Matemática Aplicada (Spai n)

AEDP: Astrometry and Engineering Data Processing

AEDS: Analog Event Distributor System • Arctic Environmental Data System

AEDT: Australian Eastern Daylight Time

AEDV: Adaptive Electronic Document Viewer

AeDy: Aerospace Dynamics (periodical code)

AeE: Aeronautical Engineer

AEE: Achieving Engineering Excellence • Administración Espacial Europea • Alkaline Earth Element • Application Execution Environment • Association for Environmental Education • Atomic Energy Establishment

AEEA: Association Européenne pour l'Enseignement de l'Astronomie

AEEC: Aeronautical Electronics Engineering Committee • Airline Electronic Engineering Committee

AEEI: Atmospheric Emitted Radiance Instrument • Autonomous Energy-Efficiency Index

AEELS: Automatic Electronic Emitter Locating System

AeEn: Aerospace Engineering (periodical code)

AEEN: Agence Européenne pour l'Énergie Nucléaire

AEENM: Association des Élèves de l'École Nationale de la Météorologie (France)

AEERL: Air and Energy Engineering Research Laboratory

AEES: Association des Élèves des Écoles Spéciales

AEESCA: Association for Engineering Education in South and Central Asia

AEESEAP: Association for Engineering Education in Southeast Asia and the Pacific

AEF: Aerospace Engineering Facility • Affaires Économiques et Financières • Air Expeditionary Force • American Expeditionary Force • Arbeitsgemeinschaft Extraterrestriche Forshung (Germany) • Average Emission Factor

AEFE: Agence pour l'Enseignement du Français à l'Étranger

AeFS: Aeronautical Fixed Service

aeg: Ad Eundem Gradum • All Page Edges Gilt

AEG: Aek Godan (Indonesia) (location identifier) • Aircraft Evaluation Group • Allgemeine Elektrizitäts-Gesellschaft (Germany) • Association of Engineering Geologists • Association of Exploration Geochemists • Asynchronous Exponential Growth

AEGE: ASEAN Experts Group on the Environment

AEGEE: Association des États Généraux des Étudiants d'Europe • Association des États Généraux des Étudiants Européens

AEGIS: An Existing General Information System • Australian Electronic Government Information Service

AEGS: Association of European Geological Societies

AEH: Abecher (Chad) (location identifier) • Antenna Effective Height

AEI: Aerial Exposure Index • Air Express International • Albert-Einstein-Institut (Germany) • Alternative Energy Institute • American Express International • Association des Écoles Internationales • Azimuth Error Indicator

AEIC: ASEAN Earthquake Information Center

AEIDC: Arctic Environmental Information Data Center

AEIE: Agrupación Europea de Interes Economico

AEIMEE: Antarctic Environmental Implications of Mineral Experiments

AEIMP: Apple Event Interprocess Messaging Protocol

AeJ: Aeronautical Journal (periodical code)

AEJ: Association of European Journalists

AEK: Aseki (Papua New Guinea) (location identifier)

AEL: Aeronautical Engine Laboratory • Aeronautics Education Laboratory (USA-MD) • Aircraft Engine Laboratory • Albert Lea (USA-MN) (location identifier) • Associated European Laboratories • Automation Engineering Laboratory

AELE: Association Européenne de Libre Échange

AELS: Airborne Electronic Library System

A&ME: Aeronautical & Mechanical Engineering

AEM: Acoustical Emission Monitoring • Aeronautical and Mechanical Engineering • Airborne Electromagnetic • Analytical Electron Microscope • Analytical Electron Microscopy • Animal Enclosure Module (*) • Animal Exposure Module • Application Explorer Mission • Arsenal Exchange Model • ASEAN Economic Ministers • Assembly and External Maintenance • Astronomical Equipment Maker • Astronomical Equipment Making • Astronomy Education Material • Atmospheric Explorer Mission

AeMat: Aerospace Materials (periodical code)

AEME: Agence de l'Environnement et de la Maîtrise de l'Énergie (France)

AeMed: Aerospace Medicine (periodical code)

AE&MP: Ambassador Extraordinary & Minister Plenipotentiary

AEMP: Association of European Management Publishers

AeMS: Aeronautical Mobile Service

AeMSS: Aeronautical Mobile Satellite Service

AEMU: Advanced Extravehicular Mobility Unit

aen: Aeneus

AEN: Agence pour l'Énergie Nucléaire

AEng: Associate in Engineering

AENOR: Asociación Española de Normalización y Certificación (Spain)

AENSB: Association de l'École Nationale Supérieure des Bibliothèques

AEO: Aioun - El Atrouss (Mauritania) (location identifier) • Alkaline Earth Oxide • All-European Observatory • An Element Of • Assistant Education Officer • Assistant Experimental Officer

AEOP: Adaptive Electro-Optical Processor

AEORMA: Asociación Española para la Ordenación del Medio Ambiente (Spain)

AEOS: Advanced Electro-Optical System

AE&P: Ambassador Extraordinary & Plenipotentiary

AEP: AppleTalk Echo Protocol • Application Environment Profile • Application Execution Platform • Aquatic Environmental Protection • Arbeitsgemeinschaft Extraterrestrische Physik (Germany) • Association Européenne de Productivité • Asymptotic Equipartition Property • Buenos Aires - Jorge Newbery (Argentina-BA) (location identifier)

AEPB: Astronomy Education Policy Board

AEPC: Asociación Española para el Progreso de las Ciencias (Spain)

AEPI: Atmospheric Emissions Photo Imaging • Atmospheric Emissions Photometric Imaging

AEPM: APACM Energy and Particle Monitor

AEPS: Airline Employment Placement Service • Arctic Environmental Protection Strategy • Arctic Environment Protection Strategy

AEPT: Asociacón Española de Prensa Técnica (Spain)

AEPUA: Association d'Éducation Permanente des Universités de l'Atlantique

AEPUC: Association pour l'Éducation Permanente dans les Universités du Canada

aeq: Aequalis

AeQ: Aeronautical Quarterly (periodical code)

AEQ: Active Equalizer • ADC External Query • Award Evaluation Questionnaire

Aér: Aéronautique

Aer: Aerodrome • Aeronáutica • Aeronautics • Aeroplane

AER: Adler Sochi (Russia) (location identifier) • Antenna Equipment Room • Association of European Radios • Astronomy Education Research • Atmospheric and Environmental Research Inc. (USA-MA) • Atomic Energy Research • Sochi (Russia) (location identifier)

Aera: Aeration

AERA: American Educational Research Association (USA) • Annus Erat Regni Augusti • Automated En Route ATC

AerAs: L'Aéronautique et l'Astronautique (periodical code)

AERE: Association of Environmental and Resource Economists • Atomic Energy Research Establishment (UK)

AEREA: Association of European Research Establishments in Aeronautics

AERI: Atmospheric Emitted Radiance Instrument • Atmospheric Emitted Radiance Interferometer (*)

AERIE: Airborne Equipment for Remote Imaging of the Environment

AERL: Association of European Research Libraries

AERN: Association of European Research Networks • Astronomy Education Resource Network

AeRNS: Aeronautical Radionavigation Service

AeRNSS: Aeronautical Radionavigation Satellite Service

Aero: Aerodrome • Aerodynamic • Aeronautic • Aeronautical • Aeronautics • Aeropuerto • Aerospace • Aerospace UK (periodical code)

AERODAS: Aeromagnetic Data System

Aerodyn: Aerodynamic

Aéron: Aeronomie

Aeron: Aeronautical • Aeronautics • Aeronomy

Aeronaut: Aeronautical • Aeronautics

Aéron: Aéronautique

AERONICA: Aerolineas Nicaraguenses SA

AEROS: Advanced Earth Resources Observational Satellite • Artificial Earth Research and Orbiting Satellite

Aerosp: Aerospace • Aerospatial

Aérosp: Aérospatial

AERP: Aquatic Effects Research Program

AerRv: Revue Aérospatiale (periodical code)

AERS: Advanced Earth Resource Satellite • Advanced ESA Resource Satellite • Association pour l'Expansion de la Recherche Scientifique

AerST: Aerosol Science Technology (periodical code)

AERT: Association for Education by Radiotelevision (USA)

AerTe: Aerojet Technology (periodical code)

AES: Accord Économique et Social • Acuerdo Económico y Social • Advanced Encryption Standard • Advanced Energy System • Advanced Expert System • Advanced Extravehicular Suit • Aerospace Electrical Society • Ålesund (Norway) (location identifier) • Ahmedabad Earth Station (India) • Airborne Emission Spectrometer • Air Evaporation System • Analog Event System • Analysis and Evaluation Staff • Application Environment Service • Application Environment Specification • Artificial Earth Satellite • Association des Écrivains Scientifiques (France) • Astronomical Explorer Satellite • ATM Edge Switch • ATM Enhancement Service • Atmosphere Exchange System • Atmospheric Environmental Service • Atmospheric Exchange System • Atomic Emission Spectrography • Atomic Emission Spectrometer • Atomic Emission Spectrometry • Atomic Emission Spectroscopy • Atomic Emission Spectrum • Audio Engineering Society • Auger Electron Spectroscopy • Auger Electron Spectrum

AESA: Active Electronically Scanned-array Antenna • Approximation Elimination Search Algorithm • ATM End-System Address • Atmospheric Effect of Stratospheric Aircraft

AESB: Avionics Engineering Subboard

AESC: American Engineering Standards Committee (USA)

AESE: Airborne Electrical Support Equipment • Association of Earth Science Editors

AESF: Association des Écrivains Scientifiques de France

AeSIJ: Aeronautical Society of India Journal (periodical code)

AESIS: Australian Earth Sciences Information System

AESL: Aerospace Energy Systems Laboratory

AESOP: Advanced End-to-end Simulation of Onboard Processing • Aircraft-Environment-Situation-Operation-Personnel • AMOS Ephemeris Satellite Orbit Predictor • An Evolutionary System for Online Processing • An Extensible Symbolic Optics Package • Anti-Electron Suborbital Payload • Automated Engineering and Scientific Optimization Program

AESP: Aerospace Education Services Project

AEST: Australian Eastern Standard Time

Aesth: Aesthetics

AESTM: Association Européenne des Sciences et Techniques de la Mer

aet: Aetatis

AET: Active Edge Table • Actual Elapsed Time • Actual Evapotranspiration • Actual Exposure Time • Algorithm Evaluation and Test • Allakaket (USA-AK) (location identifier) • Approximate Exposure Time

AETA: Association Européenne des Transporteurs Aériens

aetat: Aetatis

AETB: Advanced Expander Cycle Testbed • Alumina-Enhanced Thermal Barrier

AETC: Air Education and Training Command

AETL: Approved Engineering Test Laboratory

AetR: Aller et Retour

AETR: Advanced Engineering Test Reactor

AETT: Association for Educational and Training Technology (UK)

AEU: Altitude Encoder Unit • American Ethical Union (USA) • Archiv für Elektronik und Übertragungstechnik • Asian Electronics Union

AEV: Air Evacuation • Anti-Earth Viewing

AEW: Airborne Early Warning

AEW&C: Airborne Early Warning & Control

AEWC: Airborne Early Warning and Control

AEWG: Astronomy Education Working Group (AAS)

AEWS: Aircraft Early Warning System

AEX: Alexandria (USA-LA) (location identifier) • Automated Execution

AExpr: Astronomy Express (periodical code)

AEY: Akureyri (Iceland) (location identifier)

AEZ: Agro-Ecological Zone

a/f: A Favor

af: Ad Finem • Advanced Freight • A Favor • Afrikaans (ISO language code) • Anno Futuro • Audio Frequency

Af: Afghani (currency) • Africa • African • Afrikaans • Aruban Florin (currency) • Assez Faible

A/F: Airframe • Air-to-Fuel • As Found

AF: Académie Française • Active Filament • Adult Fiction • Affaires Foncières • Affiliation • Afghanistan (ISO country code) • Africa • Aft Fuselage • Agent de Fabrique • Air Force • Airframe • Alliance Française • Allocations Familiales • Alta Frequenza • Alternating Field • Ancien Franc (France) • Anglo-French • Année Fiscale • Antiferromagnet • Antiferromagnetic • Arrêté Fédéral • Assemblée Fédérale • Associate Fellow • Astronomical Foundation (USA-CA) • Attenuation Factor • Attitude File • Audio Fidelity • Audio Frequency • Autofocus • Automatic Filter • Auxiliary Flag • Award Fee • Axial Flow

AFA: Advanced Fuel Assembly • Afghani (Afghanistan) (ISO currency code) • Air Force Academy (USA) • Airframe Assembly • Analog Filter Assembly • Anonymous FTP Archive • Association des Fournisseurs d'Accès (France) • Association Française d'Astronomie (France) • Association of Flighht Attendents • Automatic Fire Alarm • San Rafael (Argentina-MD) (location identifier)

AFAC: Airborne Forward Air Controller • Asynchronous Fast Adaptive Composite

AFACTS: Automatic Facilities Test System

AFAIAA: Associate Fellow of the American Institute of Aeronautics and Astronautics

AFAIK: As Far As I Know

AFAL: Air Force Astronautical Laboratory

AFAM: Associazione Friulana di Astronomia e Meteorologia (Italy)

AFAQ: Association Française pour l'Assurance de la Qualité (France)

AFAS: Association Française pour l'Avancement des Sciences (France)

Afb: afbeelding

AFB: Air Force Base • Association Française des Banques (France)

AFBC: Atmospheric Fluidized-Bed Combustor

AFC: Active Flight Control • Administration and Finance Committee • Administrative and Finance Committee • Aerodynamic Flight Control • Aircraft Flight Correlator • Alkaline Fuel Cell • Analog to Frequency Converter • Analyse Factorielle des Correspondances • Area Forecast Center • Automatic Flight Control (*) • Automatic Flight Controller • Automatic Font Change • Automatic Frequency Control • Automatic Frequency Controller

AFCAC: African Civil Aviation Commission

AFCE: American Foundation for Continuing Education (USA) • Automatic Flight Control Equipment

AFCEE: Association Française du Commerce et des Échanges Électroniques (France) • Association Française pour le Commerce et les Échanges Électroniques (France)

AFCET: Association Française pour la Cybernétique Économique (France)

AFCF: Allocation Forfaitaire pour Charges de Famille

AFCO: Automatic Fuel Cutoff

AFCOHT: Association Française du Commerce Hors Taxes (France)

AFCOS: Association Française des Conseils en Organisation Scientifique (France)

AFCRL: Air Force Cambridge Research Laboratory (USAF)

AFCS: Adaptive Flight Control System • Automatic Flight Control System

AFCT: Affect

AFCVI: Analyse Factorielle des Correspondances avec Variables Instrumentales

Afd: Afdeling

AfD: Amt für Datenverarbeitung

AFD: Accelerated Freeze Dried • Accelerated Freeze Drying • Advance Function Printing • Aft Flight Deck • Agence Française de Développement (France) • Application Filing Date • Assistant Flight Director • Automated Flaw Detector • Automatic File Distribution • Port Alfred (South Africa) (location identifier)

AfDB: African Development Bank

AFDC: Aid for Families with Dependent Children • Aid to Families with Dependent Children

AFDE: Aft Flight Deck Equipment

AFDIE: Association Française pour le Développement de l'Intelligence Économique (France)

AFDIN: Association Française de Documentation et d'Information Nucléaires (France)

AFDK: After Dark

AFDM: Advanced Fluid-Dynamics Model • Ash Free Dry Mass

AFDO: Aft Flight Deck Operator • Assistant Flight Dynamics Officer

AFDR: Ascent Flight Design Review

AFDS: Autonomous Free-flight Dispenser System

AFDW: Ash-Free Dry Weight

AF&E: Airframe & Engine

AFE: Airframe and Engine • American Flight Echocardiograph • Apple File Exchange

AFEA: Association des Femmes et Étudiantes Africaines • Association Française pour l'Éducation des Adultes (France)

AFEAS: Alternative Fluorocarbons Environmental Acceptability Study

AFEB: Award Fee Evaluation Board

AFEC: Award Fee Evaluation Committee

AFECEI: Association Française des Établissements de Crédit et des Entreprises d'Investissements (France)

AFECTI: Association Française des Experts de la Coopération Technique Internationale (France)

AFEJ: Asian Forum of Environmental Journalists

AFEQ: Adiabatic Fast Equilibration

AFERO: Asia and the Far East Regional Office

AFES: Automated Feature Extraction System • Avionik-Flug-Erprobungs-System

AFESD: Arab Fund for Economic and Social Development

AFEUR: Association Française d'Étude pour l'Union Européenne (France)

AFEWES: Air Force Electronic Warfare Effectiveness Simulator

Aff: Affaire • Affairs • Affectionate • Affiliate • Affiliated • Affirmatif • Affirmative • Affix

AFF: Aberration-Free Focus • Acceptance and Ferry Flight • Autonomous Formation Flying • Colorado Springs (USA-CO) (location identifier)

AFFDL: Air Force Flight Dynamics Laboratory (USAF)

Aff.Etr.: Affaires Étrangères

AFFF: Aqueous Film Forming Foam

Affil: Affiliate • Affiliated

Affmo: Afectísimo

Afft: Affidavit

Afg: Afghanistan

AFG: Administrative and Finance Group • Advanced Frame Grabber • Afghanistan (international vehicle registration code) • Afghanistan (ISO country code) • Analog Function Generator • Antenna Field Gain • Arbitrary Function Generator

AFGE: American Federation of Government Employees (USA)

Afgh: Afghanistan

Afghan: Afghanistan

AFGL: Air Force Geophysical Laboratory • Air Force Geophysics Laboratory (*)

AFGWC: Air Force Global Weather Center (USAF)

AFH: Antenna Feed Horn

AFHB: Arbeitsgemeinschaft der Fachhochschulbibliotheken

AFI: Amalfi (Colombia) (location identifier) • American Film Institute (USA) • AppleTalk Filing Interface • Association Fiscale Internationale • Authorized Flight Instructor • Automatic Fault Isolation

AFIA: Association Française pour l'Intelligence Artificielle (France)

AFIAP: Artiste de la Fédération Internationale de l'Art Photographique

AFID: Association Française d'Information et de Documentation (France)

AFII: American Federation of International Institutes (USA) • Association for Font Information Interchange

AFIIM: Associate Fellow of the Institution of Industrial Managers

AFIJ: Association pour Faciliter l'Insertion des Jeunes (France)

AFIM: Antarctic Flight Information Manual

AFIMA: Associate Fellow of the Institute of Mathematics and its Applications

AFIP: Affichage des Fonctions d'Intérêt Prioritaire

AFIPS: American Federation of Information Processing Societies (USA)

AFIRO: Association Française d'Informatique et de Recherche Opérationnelle (France)

AFIS: Aerodrome Flight Information Service • Association Française pour l'Information Scientifique (France) • Automated Fingerprint Identification System • Automated Flight Inspection System • Automatic Fingerprint Identification System

AFISS: Airborne Flight Information Subsystem

AFIT: Air Force Institute of Technology (USAF)

AFITV: Air Force Instrumented Test Vehicle

AfK: Arbeitsgemeinschaft für Kommunikationsforschung

AFK: Away From Keyboard

Afl: Aflevering

AFL: Abone Field Level • Abstract Family of Languages • Alta Floresta (Brazil-MT) (location identifier) • American Federation of Labour (USA) • Automatic Fault Location

AFLA: Asian Federation of Library Associations

AFLC: Air Force Logistics Command

AFLCD: Antiferroelectric Light Emitting Diode

AFLEX: Automatic Flight Landing Experiment

AfRL: Akademie für Raumforschung and Landesplannung (Germany)

Aflt: Afloat

A/FM: Arm Firing Mechanism

AFM: Abrasive Flow Machining • Adobe Font Metrics • Aircraft Flight Manual • Airplane Flight Manual • Asociación Filosófica Mexicana (Mexico) • Atomic Force Microscope • Atomic Force Microscopy • Audio Frequency Modulation

AFMa: Air Force Magazine (periodical code)

AFMC: Air Force Material Command (USAF) • Asian Fluid Mechanics Congress

AFME: Agence Française pour la Maîtrise de l'Énergie (France)

Afmo: Afectísimo

AFMR: American Foundation for Management Research (USA)

AFMS: Automatische Fernmeldende Meteorologische Station

AFMSS: Air Force Mission Supprt System

AFN: Alaska Federation of Natives • Australian Fiducial Network • Average Failure Number • Jaffrey (USA-NH) (location identifier)

AFNIC: Association Française pour le Nommage Internet en Coopération (France)

AFNIL: Agence Francophone pour la Numérotation Internationale du Livre

AFNOR: Association Française de Normalisation (France)

AFO: Abort From Orbit • Afton (USA-WY) (location identifier) • Announced Flight Opportunity • Announcement of Flight Opportunity • Astronomischer Freundeskreis Ostsachsen (Germany)

AFOAT: Air Force Office of Atomic Test

AFOC: Advanced Fiber Optic Coupler

AFOE: Advanced Fiber Optic Echelle

AFOEV: Association Française des Observateurs d'Étoiles Variables (France)

AFOP: Association Française des Industries de l'Optique et de la Photonique (France)

AFOPT: Association Française des Opérateurs Privés en Télécommunications (France)

AFOS: Advanced Field Operations System • Automated Forecasting Operational System • Automation of Field Operations and Services

AFOSH: Air Force Occupational Safety and Health

AFOSR: Air Force Office of Scientific Research (USA-DC)

AFOTEC: Air Force Test and Evaluation Center (USA)

AFP: Adiabatic Fast passage • Advanced Flexible Processor • Advanced Function Presentation • Advanced Function Printing • Agence France-Presse • Air Force Program • Annual Financial Plan • AppleTalk File Protocol • AppleTalk Filing Protocol • Area Flight Plan • Ariane-Furnished Property • Attached Fortran Processor • Avis de Fortes Précipitations

AFPA: Association pour la Formation Professionnelle des Adultes (France)

AFPD: Authorization For Program Development

AFPDS: Advanced Function Presentation Data Stream

AFPI: Association Française des Professionnels de l'Internet (France)

AFPM: Advanced Fluid Physics Module

AFPRO: Air Force Plant Representative Office

AFPS: Advanced File and Print Server

AFPU: Association Française des Presses d'Université (France)

AFQA: Air Force Quality Assurance

Afr: Africa • African • Afrikaans • Afrique

A-Fr: Anglo-French

AFr: Anglo-French

AFR: Accident Frequency Rate • Actual Fuel Remaining • Afore (Papua New Guinea) (location identifier) • Africa • Air Force Regulation • Air-Fuel Ratio • Allocation Formation Reclassement • Annualized Failure Rate • Automatic Field Recognition • Automatic Fingerprint Recognition • Automatic Format Recognition • Available For Release

AFRA: Association Française de Régulation et d'Automatisme (France)

AFRAA: African Airlines Association

AFRALTI: African Advanced Level Telecommunications Institute (Kenya)

AFRAND: African Foundation for Research And Development

AFRASEC: Afro-Asian organization for Economic Cooperation

AFRC: Agricultural and Food Research Council (UK) • Agriculture and Food Research Council (UK)

AFRE: Alta Frequenza (periodical code)

AFREPEN: African Energy Policy Research Network

AFRG: Australian Fusion Research Group

AFRHC: Association Française de Recherche sur l'Histoire du Cinéma (France)

AFRI: Association Française de Robotique Industrielle (France)

Afrik: Afrikaans

AFRIMS: Armed Forces Research Institute of Medical Sciences (USA)

AFRL: Air Force Research Laboratory (USA)

AFRO: Africa Regional Office

AFRPL: Air Force Rocket Propulsion Laboratory (USAF)

AFRSI: Advanced Flexible Reusable Surface Insulation

AfS: Amt für Standardisierung

AFS: Air Force Station • AIRS Facility System • American Field Service • Andrew File System • Antenna Feeder System • Antenna Feed System • Arch Filament System • Ascent Flight System • Astronomisk Forening for Sydsjaelland (Denmark) • Atomic Fluorescence Spectrometry • Atomic Fluorescence Spectroscopy • Atomic Frequency Standard • Autogenic Feedback System • Automatic Flight System • Automatic Frequency Stabilization • Automation Feasibility Study • Axial Field Spectrometer

AFSA: American Foreign Service Association (USA)

AFSAC: Air Force Security Assistance Center (USA)

AFSATCOM: Air Force Satellite Communications (USAF)

AFSB: Air Force Studies Board (USAF)

AFSBO: American Federation of Small Business Organizations (USA)

AFSC: Aggregate Friction Seal Coat

AFSCF: Air Force Satellite Control Facility

AFSIG: Ascent Flight System Integration Group

AFSCME: American Federation of State, County, and Municipal Employees (USA)

AFSCN: Air Force Satellite Control Network (USAF)

Afsd: Aforesaid

AFSK: Automatic Frequency Shift Keying

AFSPACECOM: Air Force Space Command (Now AFSPC)

AFSPC: Air Force Space Command

AFSRR: Avionics Flight Systems Readiness Review

AFSRT: Avionics Flight Systems Readiness Test

AFSS: Automated Flight Service Station

AFSTB: Air Force Science and Technology Board (USAF)

AFSV: Automatische Fermelde-Speicher-Vermittlung

Aft: After • Afternoon

AFT: Active Fine Tracker • Aerodynamic Flight Test • After • Afutara (Solomon Islands) (location identifier) • Ahlfors Finiteness Theorem • Ambiguity Function Technique • American Federation of Teachers (USA) • Association Française de Topographie (France) • Association Française du Transport (France) • Atmospheric Flight Test • Automatic Flight Termination • Axial-Flow Turbine

AFTA: Acoustic Fatigue Test Article • Aft Frame Tilt Actuator

AFTAC: Air Force Technical Application Center

AFTE: American Federation of Technical Engineers (USA) • Autogenic Feedback Training Experiment

AFTEL: Association Française de Télématique (France)

AFTI: Advanced Fighter Technology Integration

Aftn: Afternoon

AFTN: Aeronautical Fixed Telecommunication Network • Afternoon • Automated Fixed Telecommunication Network

AFTP: Acceptance Functional Test Procedure

AFTRA: American Federation of Television and Radio Artists (USA)

AFTS: Aft Fuselage Thrust Structure

AFU: Advanced Flying Unit • All Fouled-Up • Analog Feedback Unit

AFUL: Association Française des Utilisateurs de Linux (France)

AFUTT: Association Française des Utilisateurs de Téléphone et de Télécommunications (France)

AFUU: Association Française des Utilisateurs d'Unix (France)

AFUSIE: Association Française des Utilisateurs de Systèmes d'Information Électroniques (France)

AFVG: Anglo-French Variable Geometry

AFVP: Association Française des Volontaires du Progrès (France)

AFW: Fort Worth - Alliance (USA-TX) (location identifier)

AFY: Afyon (Turkey) (location identifier)

Afz: Astrofizica (periodical code)

AFZ: Acronym-Free Zone • Astro Fax Zirkular

af2i: Association Française des Intermédiaires en Information

aG: auf Gegenseitigkeit

Ag: Acting • Agent • Agreement • Argent • August • Silber • Silver

A-G: Air-to-Ground

A/G: Air-to-Ground

A&G: Acquisition & Guidance • Acquisition & Guiding

AG: Accountant-General • Address Generation • Ad Gentes • Adjutant General • Administration Générale • Agent du Gouvernement • Agent-General • Agglomération • Aktiengesellschaft • Andean Group • And Gate • Annales Geophysicae • Antigravity • Antigua & Barbuda (ISO country code) • Arbeitsgemeinschaft • Arbeitsgruppe • Arbetsgrupp • Artificial Gravity • Assemblée Générale • Astronomische Gesellschaft (Germany) • Attitude Gyroscope • Attorney-General • Avocat Général • Auxiliary Gas • Barbuda (Antigua &_) (ISO country code)

AGA: Adaptive Genetic Algorithm • Agadir (Morocco) (location identifier) • Agent Administratif • American Gas Association (USA) • Arab Geologists Association • As Good As • Association of Government Accountants • Attitude Gyro Assembly

AGAA: Association Guadeloupéenne d'Astronomes Amateurs (France) • Attitude Gyroscope Accelerometer Assembly • Automatic Gain Adjusting Amplifier

AGAC: American Guild of Authors and Composers (USA) • Archivo General de la Administración Central

AGACS: Automatic Ground-to-Air Communications System

AGADES: Accès Général à la Documentation Électronique Spécialis ée

AGAGE: Advanced Global Atmospheric Gases Experiment

AGAL: Australian Government Analytical Laboratories

AGAP: Attitude Gyroscope Accelerometer Package

AGAPE: Andromeda Galaxy and Amplified Pixels Experiment (formerly) • Andromeda Gravitational Amplification Pixel Experiment (*)

AGARAS: Antiglare-Antireflective-Antistatic

AGARD: Advisory Group for Aerospace Research and Development (NATO)

AGASA: Akeno Giant Air Shower Array (Japan)

AGASC: AXAF Guide and Aspect Star Catalog

AGASP: Arctic Gas and Aerosol Sampling Programme

AGATA: Advanced Gas Turbine for Automobiles

AGATE: Advanced Gamma-ray Astronomy Telescope • Advanced Gamma-ray Astronomy Telescope Experiment • Advanced General Aviation Transport Experiment

AGB: Accessory Gearbox • A Goog Brand • Any Good Brand • Astronomische Gesellschaft Baden (Switzerland) • Astronomische Gesellschaft Bern (Switzerland) • Asymptotic Giant Branch • Augsburg (Germany) (location identifier)

AGBD: Above Ground Biomass Density

AGC: Abell Galaxy Cluster • Administration Générale Civile • Advanced Graduate Certificate • Association Géologique du Canada • Automatic Gain Control • Pittsburgh - Allegheny County (USA-PA) (location identifier)

AGCA: Automatic Ground-Controlled Approach

AGCD: Administration Générale de la Collaboration au Développement (*) • Association Générale de la Collaboration au Développement • Association Générale pour la Collaboration et le Développement

AGCFM: Advisory Group on Computational Fluid Mechanics

AGCI: Aspen Global Change Institute

AGCL: Answer Garden Command Language • Automatic Ground-Controlled Landing

AGCM: Atmospheric General Circulation Model

AGCP: Advanced Guidance and Control Processor

AGCPO: Arctic Global Change Programme Office

AGCS: Accord Général sur les Commerces de Services • Automatic Ground Control Station

AGCU: Air/Ground Cooling Unit • Attitude Gyro Coupling Unit

Agcy: Agency

AGCY: Agency

Agd: Agreed

AGD: Anggi (Indonesia) (location identifier) • Atelier de Génie Didacticiel • Axial Gear Differential

AGDA: Andøya Geophysical Data

AGDB: African Geodetic Databank • Australian Geographic Database

AGDS: Automatic Graphic Digitizing System

AGE: Aerospace Ground Equipment • Air/Ground Equipment • Angle of Greatest Elongation • Applied General Equilibrium • Associate in General Education • Association Générale des Étudiants • Automatic Ground Equipment • Automatic Guidance Electronics • Wangerooge (Germany) (location identifier)

AGED: Allocation de Garde d'Enfants à Domicile

AGEX: Address Generation/Execute Cycle

AGF: Agen (France) (location identifier) • Águas-Furtadas • Arbeitsgemeinschaft der Großforschungseinrichtungen (Germany) • Association des Géographes Français (France) • Assurances Générales de France • Automated Gradient Furnace

AGFA: Aktiengesellschaft für Anilinfabrikation (Germany)

ag. feb.: Aggrediente Febre

AGFS: Aviation Gridded Forecast System

Agg: Aggregate

AGG: Angoram (Papua New Guinea) (location identifier)

AGGG: Advisory Group on Greenhouse Gases

Aggl: Agglomération

Aggr: Aggregate

Agh: Afghani (currency)

AGH: Hälsingborg Ängelholm (Sweden) (location identifier) • Helsingborg Angelholm (Sweden) (location identifier)

AGHF: Advanced Gradient Heating Facility • Advanced Gradient Heating Furnace

AGI: Address Generation Interlock • Adjusted Gross Income • Advanced Ground Instructor • American Geographical Institute (USA) • American Geological Institute

(USA) • Analytical Graphics Inc. (USA-PA) • Année Géophysique Internationale • Annual General Inspection • Application Graphics Interface • Applied Geophysics Institute • Association for Geographic Information • Authorized Ground Instructor • Wageningen (Suriname) (location identifier)

AGICE: Automatic Graphic Identification of Constitutive Equations

AGID: Association of Geoscientists for International Development

AGIF: Arbeitsgemeinschaft für Innerbetriebliche Fortbildung (Germany)

AGIP: Agenzia Generale Italiana Petroli (Italy) • Azienda Generale Italiana Petroli (Italy)

AGIRC: Association Générale des Institutions de Retraite des Cadres (France)

AGIRE: Agence d'Imagerie pour la Recherche et l'Enseignement

AGIS: Arctic Geographic Information System • Auto-Guiding and Imaging System • Automated Geographic Information System

agit: Agitat • Agitatum

agit. ante sum.: Agita Ante Sumendum

AGJ: Aguni (Japan) (location identifier)

AGK: Astronomische Gesellschaft Katalog • Kagua (Papua New Guinea) (location identifier)

AGKR: AGK Reference Star

AGL: Above Ground Level • Above Ground Link • Absolute Ground Level • Allied Geophysical Laboratories • ASTRONET Graphics Library • Astronomische Gesellschaft Luzern (Switzerland) • Atelier de Génie Logiciel • Azores-Gibraltar Line • Wanigela (Papua New Guinea) (location identifier)

Aglm: Agglomerate

AGM: Advisory Group Meeting • Alternate Graphics Mode • Angmagssalik (Greenland) (location identifier) • Annual General Meeting • Arithmetic-Geometric Mean • Association General Meeting • Automatic Global Matcher

AGMS: Arbeitsgemeinschaft Massenspektrometrie (Germany)

Agn: Again

AGN: Active Galactic Nucleus • Additive Gaussian Noise • African Gravity Network • Again • Angoon (USA-AK) (location identifier) • Archives Générales de la Nation

AGN&C: Advanced Guidance, Navigation & Control

AGNC: Advanced Guidance, Navigation and Control

AGND: Analog Ground

AGNES: A Graphical Notation Editing System

Ago: Agosto

AGO: Angola (ISO country code) • Assemblée Générale Ordinaire • Astronomische Gesellschaft Oberwallis (Switzerland) • Automated Geophysical Observatory • Magnolia (USA-AR) (location identifier)

AGOALF: Agostoalferes

AGOCG: Advisory Group On Computer Graphics (UK)

AGOES: Advanced Geosynchronous Observation Environment Satellite

AGONET: Automatic Geophysical Observatory Network

AGOR: Accélérateur Groningen-Orsay • Auxiliary-General Oceanographic Research

AGOSS: Automated Ground Operations Scheduling System

AGP: Accelerated Graphics Port • Acide Gras Polyinsaturé • Advanced Geosynchronous Platform • Advanced Graphics Port • Astronomy, Geophysics, and Planetary sciences • Automatic Guidance Programming • Aviation General Policy • Malaga (Spain) (location identifier)

AGQ: Agrinion (Greece) (location identifier) • Arbeitsgemeinschaft Quantenoptik (Germany)

Agr: Agreement • Agrégé • Agricultura • Agricultural • Agriculture

AGR: Advanced Gas-cooled Reactor • Advanced Gas Reactor • Agra (India) (location identifier) • Agricultural • Agriculture • Apollo Gamma Ray • Archive Generation Request • Archives Générales du Royaume (Belgium)

AGRE: Active Geophysical Rocket Experiment

AGREE: Advisory Group on Reliability of Electronic Equipment

AGRHYMET: Agriculture-Hydrology-Meteorology Program

Agric: Agricultural • Agriculture

Agricult: Agricultural • Agriculture

AGRINET: Agricultural Network (UK)

AGRIS: Agricultural Information System

AgRISTARS: Agriculture and Resources Inventory Surveys Through Aerospace Remote Sensing

Agron: Agronomy

AGRS: Acid-Gas Removal System • Airborne Gamma-Ray Spectrometer

AGS: Abort Guidance Section • Abort Guidance System • Actual Ground Speed • Advance Guidance System • Aguascalientes (Mexico) • Aircraft General Standard • Alternating Gradient Synchrotron • American Geographical Society (USA) • Analog Guidance Sensor • Antigravity Suit • Artificial Gravity Structure • Assistant General Secretary • Associate in General Studies • Augusta Bush (USA-GA) (location identifier) • Automatic Gain Stabilization

AGSAC: Australian Gemini Science Advisory Committee

AGSC: Australian Gemini Steering Committee

AGSE: Astrophysical Grant in Support of Education • Astrophysics Grant Supplements for Education • Astrophysics Grant Supplement to Education

AGSO: Australian Geological Survey Organization

AGSS: American Geographical and Statistical Society (USA)

AGSSB: Atmospheric, Geophysical and Space Sciences Branch

Agst: Against

AGST: Applied General Systems Theory

Agt: Agent • Agreement

AGT: Adaptive Guidance Throttling • Advanced Gas Turbine • Agent • Agent Technique • Association fo Geology Teachers • Attractive-Game Theory • Average Generation Time • Ciudad del Este - Alejo Garcia (Paraguay) (location identifier)

AGTDC: Accord Général sur les Tarifs Douaniers et le Commerce

AGTM: Archive Generation Termination Message

AGTS: Automatic Gyro Test Set

AGU: Address-Generation Unit • Aerospace Ground Unit • Aguascalientes (Mexico) (location identifier) • American Geophysical Union (USA) • Automatic Ground Unit

AGV: Acarigua (Venezuela) (location identifier) • Assurance Générale de Vie • Automated Guided Vehicle • Automatic Guided Vehicle • Avion à Grande Vitesse

AGVA: Anglo-German Variable Geometry

AGVS: Automatically Guided Vehicle System

AGW: Acoustic-Gravity Wave • Actual Gross Weight • Agnew (Australia-QLD) (location identifier) • Atmospheric Gravity Wave

AGWP: Absolute Global Warming Potential

AGX: Agatti Island (India) (location identifier)

Agy: Agency

AGY: Argyle Downs (Australia-WA) (location identifier)

AGZ: Aggeneys (South Africa) (location identifier)

AGZO: Astronomische Gesellschaft Zürcher Oberland (Switzerland)

AGZU: Astronomische Gesellschaft Zürcher Unterland (Switzerland)

Ah: Ampere-Hour

A&H: Accident & Health

A/H: Altitude/Height

AH: Adresse Habituelle • After Hatch • Aft Hatch • Air Heater • Ampere-Hour • Angle Horaire • Angulo Horario • Anno Hebraico • Anno Hegirae • Aromatic Hydrocarbon • Artificial Horizon • AstroHubs • Attitude Hold • Authentication Header • Avant-Hier

AHA: American Historical Association (USA) • Arbeitsgemeinschaft Hildesheimer Amateursastronomen • Okinawa Naha (Japan) (location identifier)

AHAC: AEOS-Haleakala Atmospheric Characterization

AHAF: Advanced Hypervelocity Aerophysics Facility

AHAM: Adaptive Hypermedia Application Model • Association of Home Appliance Manufacturers

AHAS: Allegheny Highlands Astronomical Society (USA-PA)

AHB: Abha (Saudi Arabia) (location identifier)

AHBC: Apparent Horizon Boundary Conditions

AHC: Animal Health Committee • Herlong Amedee (USA-CA) (location identifier)

A&HCI: Arts & Humanities Citation Index

AHCI: Arts and Humanities Citation Index

AHCIET: Asociación Hispanoamericana de Centros de Investigación y Estudios de Telecomunicaciones

AHCPR: Agency for Health Care Policy and Research (USA)

AHCS: Advanced Helicopter Control System • Advanced Hybrid Computer System

Ahd: Ahead

AHD: Ahead • Ardmore Downtown (USA-OK) (location identifier)

AHDAS: Adaptable Hydrologic Data Acquisition System

AHDP: Advanced H/W Development Baseline

AHE: Adaptive Histogramme Equalization • Associate in Home Economics • Association for Higher Education

AHEAD: Animal Health/Emerging Animal Diseases

AHERA: Asbestos Hazard Emergency Response Act

AHF: American Heritage Foundation (USA) • Anhydrous Hydrogen Fluoride • Antihemophilic Factor • Arapahoe (USA-NE) (location identifier) • Attitude History File

AHFG: ATM-attached Host Functional Group

AHH: Amery (USA-NE) (location identifier)

AHI: Amahai (Indonesia) (location identifier) • American Health Institute (USA) • Artificial Horizon Indicator

ahl: Ad Hunc Locum

AHA: Aishalton (Guyana) (location identifier) • Angulo Horario Local

AHM: Amsterdam Hypermedia Model • Automatic Hold Mode

AH&MA: American Hotel & Motel Association (USA)

AHMA: American Hotel and Motel Association (USA)

AHN: Archivo Histórico Nacional • Athens (USA-GA) (location identifier)

A/H/O: Abort/Hold/Orbit

AHO: Alghero (Italy) (location identifier)

AHOF: Astronaut Hall Of Fame (USA-FL)

AHOI: Allgemeines Hochschul-Orientierungs- und Informationssystem

AHP: Absorption Heat Pump • Air Horsepower • Analytic Hierarchy Process

AHPCP: Affordable High-Performance Computing Prospect

AHPCRC: Army High-Performance Computation Research Center • Army High-Performance Computing Research Center

AHPL: A Hardware Programming Language

AHQ: Air Headquarters • Army Headquarters

AHR: Ablative Heat Rate • Acceptable Hazard rate

AHRB: Arts and Humanities Research Board

AHRIS: Advanced High-Resolution Imaging Spectrometer

AHRS: Attitude and Heading Reference System • Attitude, Heading and Reference Systems • Attitude Heading Reference System

AHRV: Advanced High-Resolution Visible (imager)

AHS: Adaptive Hypermedia System • Adult Health Study • Aft Heat Shield • Ahuas (Honduras) (location identifier) • Anno Humanae Salutis • Annual Housing Survey • Automated Highway System

AHSB: Authority, Health, and Safety Branch

AHSE: Assembly and Handling Support Equipment

AHSJ: American Helicopter Society Journal (periodical code)

AHT: Absorption Heat Transformer • Amchitka (USA-AK) (location identifier) • Attitude History Table

AHU: Air Handling Unit • Al Hoceima (Morocco) (location identifier)

ahv: Ad Hanc Vocem

AHW: Als Het Ware

AHX: Avionic Heat Exchanger

AHY: Ambatolahy (Madagascar) (location identifier)

AHZ: Alpe d'Huez (France) (location identifier)

ai: Ad Interim

A&I: Abstracting & Indexing • Accident & Indemnity

A/I: Actuator/Indicator

AI: Action Intégrale • Action Item • Activity Index • Adaptation Interdite • Adapter Interface • Admiralty Islands • Adresse Incomplète • Agenda Item • Airspeed Indicator • Altesse Impériale • Altimeter Indicator • Altitude Indicator • American Institute • Amnesty International • Analog Input • Anguilla (ISO country code) • Anno Inventionis • Ainti-Icing • Arbitrage International • Artificial Insemination • Artificial Intelligence • Assistant Ingénieur • Associate Investigator • Associative Ionization • Attitude Indicator • Audio Interface • Automatic Input • Avalanche Injection • Azimuth Indicator

AIA: Aerospace Industries Association (USA) • Alliance (USA-NE) (location identifier) • American Industries Association (USA) • American Institute of Aeronautics (USA) • American Insurance Association (USA) • Anguilla (ISO country code) • Applications Integration Architecture • Associazione Industrie Aerospaziali (Italy) • Automated Image Analysis • Automatic Information Acquisition

AIAA: Aircraft Industries Association of America • American Institute of Aeronautics and Astronautics (USA)

AIAAJ: American Institute of Aeronautics and Astronautics Journal (periodical code)

AIAB: Aerospace Industries Association of Brazil

AIAC: Aerospace Industries Association of Canada • Artificial Intelligence Advisory Committee

AIACP: Atelier Industriel de l'Aéronautique de Cuers-Pierrefeu (France)

AIAG: Automotive Industry Action Group

AIAI: Artificial Intelligence Applications Institute (UK-Scotland)

AIAO: Aerospace Industry Association of Oregon (USA-OR)

AIASJ: American Institute of Aeronautics and Astronautics Student Journal (periodical c ode)

AIB: Accidents Investigation Branch • Anita Bay (USA-AK) (location identifier) • Associazione Italiana Biblioteche (Italy)

AIBD: Asian Institute for Broadcasting Development

AIBS: American Institute of Biological Sciences (USA)

AIC: Academia de la Investigación Científica (Mexico) (now AMC) • Accretion-Induced Collapse • Acrobat Interapplication Communication • Airok (Marshall Islands) (location identifier) • Akaike Information Criterion • Alfén Ion Cyclotron • American Institute of Chemists (USA) • Artificial Intelligence Center • Astronomer In Charge • Automatic Information Center • Automatic Intercept Center

AICA: Associazione per l'Informatica ed il Calcolo Automatico

AICAS: Astronomical Institute of the Czech Academy of Sciences (Czech Republic)

AICB: Association Internationale Contre le Bruit

AICC: Action Item Control Card • Automatic Intercept Communications Controller

AICDT: Astronomer In Charge Discretionary Time

AICE: American Institute of Chemical Engineers (USA)

AIChE: American Institute of Chemical Engineers (USA)

AICMA: Association Internationale des Constructeurs de Matériel Aérospatial

AICP: Aerodynamic Instrumental Coefficient Package

AICRIP: All India Coordinated Programme for Rice Improvement

AICRO: Association of Independent Contract Research Organizations

AICS: Action Item Closeout Sheet • American Institute for Computer Science (USA) • Artificial Intelligence - Cognitive Science (conference) • Association Internationale du Cinéma Scientifique

AID: Abbreviated Item Description • Action Item Directive • Action Item Disposition • Acute Infection Disease • Aeronautical Inspection Directorate • Agence Internationale pour le Développement • Agency for International Development • Algebraic Interpretive Dialogue • Allgemeiner Informationsdienst • Analog Input Differential • Anderson (USA-IN) (location identifier) • Association Internationale de Développement • Association Internationale pour le Développement • Audit Item Disposition • Automated Interaction Detector • Automatic Indexing of Documents • Automatic Interaction Detection

AIDA: Advanced Integrated-circuit Design Aid • ALIS Internal Data Administration • Alternatives for International Document Availability • Arecibo Initiative on Dynamics of the Atmosphere • Associazione Italiana per la Documentazione Avanzata (Italy) • Astronomical Image Data Archive • Attention-Interest-Desire-Action

AIDAA: Associazione Italiana di Aeronautica e Astronautica (Italy)

AIDAB: Australian International Development Assistance Bureau

AIDAS: Advanced Instrumentation and Data Analysis System

AIDBP: Antarctic Ice Drifting Buoy Programme

AIDC: Aeronautic Industry Development Centre • Australian Industry Development Corporation

AIDDE: Ames' Interactive Dynamic Display Editor

AIDE: Aerospace Installation Diagnostic Equipment

AIDES: Automatic Image Data Extraction System

AIDOS: Automatisiertes Informations- und Dokumentationssystem

AIDJEX: Arctic Ice Deformation Joint Experiment • Arctic Ice Dynamics Joint Experiment (*)

AIDP: Associate of the Institute of Data Processing

AIDS: Abort Inertial Digital System • Accident Information Display System • Administrative Information Data System • Advanced Image Digitizing System • Advanced Integrated Data System • Advanced Integrated Display System • Airborne Integrated Data System • Aircraft Integrated Data System • Astronomical Information Distributed Service • Automated Information and Documentation System • Automated Information Data System • Automated Information Display System • Automated Information Dissemination System • Automatic Integrated Debugging System

AIE: Agence Internationale de l'Énergie • Aiome (Papua New Guinea) (location identifier) • Association for International Education • Associazione Italiana degli Editori (Italy) • Author Index Entry

AIEA: Agence Internationale de l'Énergie Atomique

AIEE: Associate of the Institution of Electrical Engineers • Association des Instituts d'Études Européens

AIEP: Additive Inverse Eigenvalue Problem

AIERI: Association Internationale des Études et Recherches sur l'Information

AIESEC: Association Internationale des Étudiants en Sciences Économiques et Commerciales

AIESI: Association Internationale des Écoles des Sciences de l'Information

AIEWS: Advanced Integrated Electronic Warfare System

AIF: Asociación Internacional de Fomento • Assis (Brazil-SP) (location identifier) • Associazione per l'Insegnamento della Fisica (Italy) • Astronomical Instrumentation Facility • Audio Interchange Format

AIFC: Audio Interchange File Compressed

AIFF: Audio Interchange File Format

AIFT: Association Interprofessionnelle pour la Formation Technique (France)

Aig: Aiguille

AIG: Address Indicating Group • Aerosol-Ionization Gas • Association Internationale de Géodésie • Astronomical Information Gateway • Astronomisches Institut Graz (Austria) • Astronomy Interest Group • Yalinga (Central African Rep.) (location identifier)

AIGA: American Institute of Graphic Arts (USA) • Association Internationale de Géomagnétisme et d'Aéronomie

AIGG: Association Internationale de Géodésie et de Géophysique

Aigle: Aiguille

AIGO: Australian Interferometric Gravitational-wave Observatory • Australian International Gravitational Observatory

AIH: Acoustical Imaging and Holography (periodical code) • All In Hands • American Institute of Hydrology

AIHA: American Industrial Hygiene Association (USA) • American International Health Alliance (USA)

AIHEP: Artificial Intelligence in High-Energy Physics

AIHS: Académie Internationale d'Histoire des Sciences • Association Internationale d'Hydrologie Scientifique

AII: Accord Inter-Institutionnel • Alisabieh (Djibouti) (location identifier) • Astronomical Information Infrastructure

AIIA: Australian Information Industry Association • Australian Institute of International Affairs

AIIC: Association Internationale des Interprètes de Conférences

AIIDS: Analyst Interpreter Interactive Display System

AIIE: American Institute of Industrial Engineers (USA)

AIIM: Association for Information and Image Management (USA)

AIIMS: All India Institute of Medical Sciences

AiInt: Air International (periodical code)

AIIP: Association of Independent Information Professionals

AIISA: Automated Ice Information System for Arctic

AIJ: Activities Implemented Jointly

AIK: Aiken (USA-SC) (location identifier) • Artificial Intelligence Knowledge

AIKR: Artificial Intelligence Knowledge Representation

Ail: Aileron

AIL: Ailigandi (Panama) (location identifier) • Année Internationale du Livre (1972) • Association of International Libraries • Authoring In the Large

AILAS: Automatic Instrument-Landing Approach System

AILIS: Association of International Librarians and Information Specialists

AILS: Advanced Integrated Landing System • Airborne Information for Lateral Spacing • ARC Imaging Library Server • Automatic Instrument Landing System

AIM: Acquisition Interface Module • Advanced Informatics in Medicine • Advanced Information Memory • Advanced Instrumentation and Measurements • Advanced Instrument Manufacturing • Aerospace Industries Malaysia • African Index Medicus • Ailuk Island (Marshall Islands) (location identifier) • Airman's Information Manual • Alternative Investment Market • American Institute of Management (USA) • Analytical Ion Microscope • Analytical Ion Microscopy • Apogee Injection Module • Application Interpreted Model • Approximate Intertial Manifold • Approximate Inverse Mapping • Artificial Ionospheric Mirror • Association Internationale de Météorologie • Associative Index Method • Astrometric Interferometry Mission • Astronomisches Institut Münster (Germany) • Astronomy Internet Master's (degree) • ATM Inverse Multiplexing • Atmosphere-Ionosphere-Magnetosphere • Attitude and Information Management • Australian Institute of Management • Auroral Ionospheric Mapper • Automated Information Management • Automated Internal Management • Automatic Immission Monitoring • Automatic Information Management

AIMA: As Interest May Appear

AIMAS: Associazione Italiana di Medicina Aeronautica e Spaziale (Italy)

AIMCC: Association of Ice Monitoring Contractors of Canada

AIMCS: Airborne Investigations of Mesoscale Convective Systems

AIME: American Institute of Mechanical Engineers (USA) • American Institute of Mining Engineers (USA) • Association for Information Media and Equipment • Association of the Institute of Mechanical Engineers • Autonomous Integrity Monitored Extrapolation

AIMIA: Australasian Interactive Multimedia Industry Association

AIML: Astronomical Instrument Markup Language

AIMPA: Association Internationale de Météorologie et de Physique de l'Atmosphère

AIMLIS: Automatic Intelligent Material Location and Identification System

AIMO: Association of Indian Meteor Observers (India)

AIMR: Airborne Imaging Microwave Radiometer

AImS: Acousto-optic Imaging Spectrometer

AIMS: Advanced Information Management System • Advanced Integrated Maintenance System • Advanced Integrated Millimetre-wave Subassembly • AFGL Interactive Meteorological System • Agile Infrastructure for Manufacturing Systems • Airborne Integrated Mapping System • Aircraft Integrated Monitoring System • Aircraft Information Management System • Analytical Imagery Matching System • Applied Information Management System • Australian Institute for Marine Sciences • Automated Image Measuring Software • Automated Image Measuring System • Automatic Image Matching System • Automated Industrial Management System

AIMT: Artificial-Intelligence-based Maintenance Tutor

AIMUX: ATM Inverse Multiplexing

AIN: Advanced Intelligent Network • Approved Item Name • Assembly Identification Number • Wainwright (USA-AK) (location identifier)

AINA: Arctic Institute of North America

AInd: Anglo-Indian

AINSE: Australian Institute for Nuclear Science and Engineering

AIO: Airborne Infrared Observatory • Assistent In Opleiding • Asynchronous Input/Output • Atlantic (USA-IA) (location identifier) • Auroral Imaging Observatory

AIOCS: Automated Industry and Occupation Coding System

AIOD: Automatic Identificated of Outward Dialing • Automatic Identificated Outward Dialing

AION: Alphabetical Index Of Names

AIOP: Analog Input/Output Package • Association Internationale d'Océanographie Physique • Association Internationale d'Orientation Professionnelle

AIP: Action Incitative Programmée (CNRS) • Advanced Information Processing • Aeronautical Information Publication • Agence Internationale de Presse • Ailinglapalap Island (Marshall Islands) (location identifier) • Aircraft Instrumented Payload • Airport Improvement Program • Algorithm Intercomparison Programme • Algorithm Intercomparison Project • American Institute of Physics (USA) • Archival Information Package • Association of Internet Professionals • Astronomical Image Processing • Astrophysikalisches Institut Potsdam (Germany) • Atmosphere-Independent Propulsion • Australian Institute of Physics • Automatic Information Processing • Avionics Integration Plan • Azimuthal Intensity Profile

AIPC: Asociación Iberoamericana de Periodistas Científicos • Association Internationale des Palais des Congrès

AIPCN: Association Internationale Permanente des Congrès de Navigation

AIPCR: Association Internationale Permanente des Congrès de la Route

AIPE: Apple Image Processing Educator

AIPEA: Association Internationale Pour l'Étude des Argiles

AIPG: American Institute of Professional Geologists (USA)

AIPLA: American Intellectual Property Law Association (USA)

AIPLF: Association Internationale des Parlementaires de Langue Française

AIPO: American Institute of Public Opinion (USA) • Australian Industrial Property Organization

AIPPI: Association Internationale pour la Protection de la Propriété Industrielle

AIPS: Association Internationale de Prévoyancce Sociale • Astronomical Image Processing System • Automated Information Processing System • Automated Intelligence Processing System • Automatic Indexing and Proofreading System

AIPU: Association Internationale des Professeurs d'Universités

AIPULF: Association Internationale des Presses Universitaires de Langue Française

AIR: Action Item Report • Adaptive Intercommunication Requirement • Advanced Information Retrieval • Aero

International Regional • After Initial Release • Airbornen Imaging Radar • All India Radio • American Indian Reservation • American Institute for Research (USA) • Annals of Improbable Research • Aripuanã (Brazil-MT) (location identifier) • Artist In Residence • Association Internationale des Registres • Automated Image Retrieval • Automatic Indexing and Retrieval

AIRAPT: Association Internationale pour la Recherche Avancée en Pneumatique Technique

AIRC: Association of Independent Radio Companies

AIRCAM: Aerospace Infrared Camera

AirCo: Air Conditioning • Air et Cosmos (periodical code)

AirE: Aircraft Engineering (periodical code)

AIREA: American Institute of Real Estate Appraisers (USA)

AIRES: Aerovías de Integración Regional SA • Airborne Infrared Echelle Spectrometer • Automatic Inverted Recording Echo Sounder

AIRG: Applied Imaging Research Group

AIRH: Association Internationale de Recherches Hydrauliques

AIRIS: Alaska Integrated Resource Inventory System

AIRO: Antarctic Infrared Observatory

AIROBICC: Airshower Observation By Integrating Čerenkov Counters

AIRP: Airborne Instrumentation Research Project • Association Internationale des Relations Publiques

Airpl: Airplane

AIRS: Acoustic-Imaging Recognition System • Advanced Infrared Radiation Sounder • Advanced Infrared Radiometer Sounder • Advanced Infrared Sounder • Aerometric Information Retrieval System • Aircrew Incident Reporting System • Antarctic International Radiometric Survey • Atmospheric Infrared Sounder • Auroral Imaging Remote Sensor • Auroral Ionospheric Remote Sensor • Automatic Image Retrieval System • Automatic Information Retrieval System

AIRSAR: Airborne Synthetic Aperture Radar

AirSp: Air and Space (periodical code)

AIRSS: Airborne Infrared Linescan System

AIRST: Advanced Infrared Search and Track

AIRT: Autonomous Infrared Technology

AIRTO: Association of Independent Research and Technology Organizations

AIS: Accident Investigation Site • Accounting Information System • Action Item Sheet • Active Interference Suppression • Active Isolation System • Address Information System • Administrative and Information Services • Administrator for Information Systems • Advanced Imaging System • Advanced Ionospheric Spectrometer • Advance In Schedule • Aeronautical Information Service • Agreed Industry Standard • Airborne Imaging Spectrometer • Alarm Indication Signal • American Information Systems Inc. (USA-IL) • Analog Instrumentation System • Analysis Instrument Set • Application Information System • Application Interface and Services • Arbeitsgruppe für Integrierte Schaltungen • Arizona Imager/Spectrometer • Arorae Island (Kiribati) (location identifier) • Association Internationale de Sociologie • Attitude Indicating System • Australian Information Service • Authoring In the Small • Automated Information Security • Automated Information System • Automated Instrumentation System • Automatic Intercept System

AISA: Association Internationale pour la Sécurité Aérienne • Astronomical Institute of the Slovak Academy

AISAO: Astrofiz. Issledovaniya (periodical code)

AISB: Association of Imaging Service Bureaus

AISC: Agence Internationale de Satellites de Contrôle • Airborne Imaging Spectrometer Campaign • Assessment and Information Service Centre

AISE: Association Internationale des Sciences Économiques

AISES: American Indian Science and Engineering Society (USA)

AISGC: Automatic Information System of Geodesy and Cartography

AISH: Association Internationale des Sciences Hydrologiques

AISI: American Iron and Steel Institute (USA)

AISIST: Australian Institute of Spatial Information Sciences and Technology

AISJ: Association Internationale des Sciences Juridiques

AIS&LUS: All India Soil & Land Use Survey

AISLUS: All India Soil and Land Use Survey

AISM: Association Internationale de Signalisation Maritime • Association Internationale des Sociétés de Microbiologie

AISP: Association Internationale des Sciences Politiques • Association of Information Systems Professionals

AISPIT: Association Internationale de Séismologie et de Physique de l'Intérieur de la Terre

AISR: Academic Information Services and Research • Applied Information Systems Research • Astrophysics Information Systems Research

AISRP: Applied Information Systems Research Program

AISS: Association Internationale de la Science du Sol • Association Internationale de la Sécurité Sociale • Australian International Space School

AIST: Abbreviated Integration System Test • Advanced Interferometric System Technology • Agency of Industrial Science and Technology • Astrometric Artificial Satellite Telescope

AISTR: Advanced Interferometric System Technology Requirement

AISTS: Associazione Italiana della Stampa Tecnica e Scientifica (Italy)

AI&T: Assembly, Integration & Testing

AIT: Acceptor Idle Table • Advanced Imaging Tracker • Advanced Information Technology • Advanced Intelligent Tape • Agency for Instructional Technology • Agency for Instructional Television • Aide Technique • Airborne Intelligence Technician • Aitutaki (Cook Islands) (location identifier) • Algorithmic Information Theory • American Institute in Taiwan • Analysis and Integration Team • Artificial Ionospheric Turbulence • Asian Institute of Technology (UNU) • Assembly, Integration and Test • Assembly-Integration-Testing • Associazione Italiana Telerilevamento (Italy) • Astrometric and Imaging Telescope • Astrometric Imaging Telescope (*) • Astronomisches Institut Tübingen (Germany) • Autoignition Temperature • Autoignition Test • Automated Information Transfer • Automatic Imaging Telescope

AITA: Association Internationale des Transports Aériens

AITAL: Asociación Internacional de Transporte Aéreo Latinoamericano

AITC: Airport Intermodal Transit Connector • Association Internationale des Traducteurs de Conférence • Association of Investment Trust Companies

AITMP: Arctic Ice Thickness Monitoring Project

AITP: Aerospace Industry Technology Program • Arctic Ice Thickness Project

AITS: Advanced Information Transmission System

AITT: Acadia Institute for Teaching and Technology

AIU: Abort Interface Unit • Association Internationale des Universités • Atiu Island (Cook Islands) (location identifier) • Auxiliary Interface Unit • Avionics Interface Unit

AIUB: Astronomisches Institut der Universität Bern (Switzerland)

AIUW: Astronomisches Institut der Universität Würzburg (Germany)

AIV: Aliceville - George Downer (USA-AL) (location identifier) • Assemblage-Intégration-Vérification • Assembly-Integration-Verification • Association Internationale de Volcanologie • Automatische Informationsverarbeitung

AIVS: Automatisertes Informationsverarbeitungssystem

AIW: APPN Implementer's Workshop • Arbeitskreis für Internationale Wissenschaftskommunikation • Astronomisches Institut Würzburg (Germany) • Atlantic Intercoastal Waterway

AIWEX: Arctic Internal Wave Experiment

AIWM: American Institute of Weights and Measures (USA)

AIX: Advanced Interactive Executive • Artificial Intelligence Executive

AIY: Atlantic City - Bader (USA-NJ) (location identifier)

AIZ: Kaiser - Lake of the Ozarks (USA-MO) (location identifier)

Aj: Ajouté

A/J: Antijam

AJ: Acte Judiciaire • Antijamming • Apojove • Assembly Jig • Astronomical Journal (periodical code) • Auberge de Jeunesse

AJA: Adjoint Administratif • Ajaccio - Campo dell Oro (France) (location identifier)

AJAP: Adjoint Administratif Principal

AJAS: Association des Jeunes Amis de la Science (France)

AJB: Astronomischer Jahresbericht

AJCL: American Journal of Computational Linguistics

AJCS: Automated Job Card System

AJF: Al-Jouf (Saudi Arabia) (location identifier) • Jouf (Saudi Arabia) (location identifier)

AJI: Association Juridique Internationale

AJJ: Akjoujt (Mauritania) (location identifier)

AJL: Aizawl (India) (location identifier)

AJN: Anjouan (Comoros) (location identifier)

AJO: Al-Jouf (Yemen) (location identifier)

AJP: American Journal of Physics

AJPAA: Association des Journalistes Professionnels de l'Aéronautique et de l'Astronautique (France)

AJPO: ADA Joint Program Office

AJR: Arvidsjaur (Sweden) (location identifier)

AJS: Abreojos (Mexico) (location identifier)

AJSE: Arabian Journal of Science Engineering (periodical code)

AJSPI: Association des Journalistes Scientifiques de la Presse d'Information (France)

AJT: Adjoint Technique

AJU: Aracaju (Brazil-SE) (location identifier)

AJY: Agades (Niger) (location identifier)

AK: Accadian • Alaska (USA) • Alphabetischer Katalog • Apogee Kick • Arbeitskreis

aka: Also Known As

AKA: Active Keel Actuator • Also Known As • Anisotropic Kinetic Alpha • Ankang (China-PRC) (location identifier) • Arbeitskreis Asteroiden (Germany) • Ausschuß zur Koordinierung der Auslandsarbeit

Akad: Akademie

AKAF: Arbeitskreis Astronomie Freiburg eV (Germany)

AKAM: Advanced Knowledge Area Module • Arbeitskreis Astrometrie

AKB: Atka (USA-AK) (location identifier)

AKCLIS: Australian Key Centre for Land Information Studies

AKD: Akola (India) (location identifier)

AKE: Akieni (Gabon) (location identifier) • Arbeitskreis Energie (Germany)

AKF: Arbeitskreis Festkörperphysik (Germany) • Kufrah (Libya) (location identifier)

AKG: Anguganak (Papua New Guinea) (location identifier)

AKI: Akiak (USA-AK) (location identifier)

AKJ: Asahikawa (Japan) (location identifier)

AKK: Akhiok (USA-AK) (location identifier)

AKL: Auckland (New Zealand) (location identifier)

AKM: Apogee Kick Motor • Arbeitskreis Meteore (Germany) • Zakouma (Chad) (location identifier)

AKN: King Salmon (USA-AK) (location identifier)

AKO: A Kind Of • Akron - Washington County (USA-CO) (location identifier)

AKP: Anaktuvuk Pass (USA-AK) (location identifier) • Arbeitskreis Planetenbeobachter (VdS)

AKPA: Association of Korean Physicists in America

AKQ: Astraksetra - Gunung Batin (Indonesia) (location identifier)

AKR: Akure (Nigeria) (location identifier) • Auroral Kilometric Radiation

AKS: Algorithmic Key Selection • Anormale Kosmische Strahlung • Auki (Solomon Islands) (location identifier)

AKSAS: Alaska State Accounting System

AKT: Akrotiri (Cyprus) (location identifier)

AKU: Aksu (China-PRC) (location identifier)

AKUP: Association of Korean University Presses (Korea-ROK)

AKV: Akulivik (Canada-QC) (location identifier)

AKW: Atomkraftwerk • Authors and Key Words

AKWIC: Altered Keyword In Context • Authors and Keywords In Context

AKX: Akt'ubinsk (Kazakhstan) (location identifier)

AKY: Akyab (Myanmar) (location identifier) • Sittwe (Myanmar) (location identifier)

al: Alia • Alii • Allée

Ál: Álava (Spain)

Al: Alemán • Alfvén Number • Alinéa • Aluminium

A/L: Airlock • Approach and Landing • Autoland

A&L: Approach & Landing

AL: Abraham Lincoln (1809-1865) • Absence Légale • Access Line • Acronym List • Administration Légale • Airlock • Alabama (USA) • Alagoas (Brazil) • Alameda • Albania (international vehicle registration code) • Albania (ISO country code) • Albanie (ISO country code) • Alberta (Canada) • Algeria • Algonquin • Altai • América Latina • Anglo-Latin • Année-Lumière • Anno Lucis • Antistokes Luminescence • Appleton Laboratory (UK) (see RAL) • Approach and Landing • Après Livraison • Arab League • Artificial Life • Aschgraues Licht • Ashen Light • Assemblée Législative • Assistance Légale • Astronomical League (USA) • Attachment Level • Attenuation Length • Attestation Légalisée • Autograph Letter • Autoland • Saint Albans (UK postal code)

Ala: Alabama (USA) • Alanine

ALA: All Letters Answered • Alma Ata (Kazakhstan) (location identifier) • Almaty (Kazakhstan) (location identifier) • American Librarians Association (USA) • American Library Association (USA) (*) • American Lung Association (USA) • Asia and Latin America • Associate in Liberal Arts • Association Lotoise d'Astronomie (France) • ATM Link Accelerator • Authorized Landing Area • Authors League of America

ALABOL: Algorithmic And Business-Oriented Language

ALACE: Autonomous Lagrangian Circulation Explorer

ALACT: Asociación Latinoamericana de Ciencias Fisiológicas

ALADA: Asociación Latinoamericana de Derecho Aeronáutico-espacial

ALADDIN: Atmospheric Layer And Density Distribution of Ions and Neutrals

ALADI: Asociación Latinoamericana De Integración

ALADIN: Atmospheric Laser Doppler Instrument

ALAE: Atmospheric Lyman Alpha Emission • Atmospheric Lyman Alpha Experiment

ALAGE: Asociación Latinoamericana de Geofísica Espacial

ALALC: Asociación Latinoamericana de Libre Comercio

ALAP: AppleTalk Link Access Protocol • As Late As Possible • As Low As Practicable • As Low As Practical

ALARA: As Low As Reasonably Achievable

ALARM: Airborne Lidar Agent Remote Monitor

ALARME: Alerte aux Risques Météorologiques Exceptionnels

Alas: Alaska (USA)

ALAS: Advanced Liquid Axial Stage • Approach/Landing Autopilot System • Atmospheric Laboratory for Application Science

ALASAT: Air-Launched Antisatellite

ALASKA: Arctic Lands And Shelves - Key Assessments

ALASS: Alternate Low-Altitude Surveillance System

ALAT: Aviation Légère de l'Armée de Terre (France)

alb: Albus

Alb: Albania • Albanian • Alberta (Canada) • Albion

ALB: Academic Libraries of Brooklyn (USA-NY) • Acceptance Log Book • Active Longitude Belt • Albania (ISO country code) • Albanie (ISO country code) • Albany (USA-NY) (location identifier)

Alba: Alberta (Canada)

Albac: Albacete (Spain)

Albq: Albuquerque (USA-NM)

ALBS: Air-Launched Balloon System

alc: À la Carte

Alc: Alcalde • Alcohol • Alcool

ALC: Aft Load Controller • Air Logistics Center • Alicante - El Altet (Spain) (location identifier) • Alphabet Code • Arithmetic and Logic Circuits • Artificial Luminous Cloud • Assembly Language Coding • Attached Laboratory Centre • Audio Load Compensator • Augmented Landing Site • Automatic Level Control • Automatic Light Control • Automatic Load Control

ALCA: Aft Load Controller Assembly • Automatic Level-Control Assembly

ALCAE: Association of Librarians of Colleges of Advanced Education (USA)

ALCATRAZ: A Large Catalogue of Atmospheric Transparencies for Relative Astronomical Zeroing

ALCC: Airborne Launch Control Center

Alch: Alchemy

AlCl: Aluminium Chloride

ALCOA: Aluminium Co. of America

ALCOM: ALGOL Compiler

ALCORSS: Australian Liaison Committee On Remote Sensing by Satellite

ALCS: Authors' Licensing and Collecting Society

ALCTS: Association for Library Collections and Technical Services

Ald: Alderney (Channel Islands)

ALD: Affection de Longue Durée • Alerta (Peru) (location identifier) • Analog Line Driver • Arc-Length Difference • Ariane Lift Derivative • Astronomy from Large Databases • Automatic Line Drawing

ALDC: Adaptive Lossless Data Compression • Automatic Load and Drive Control

ALDHU: Asociación Latino-americana para los Derechos Humanos

ALDO: Activity-Level-Dependent Operation

ALDOC: Arab League Documentation Centre

ALDP: Advanced Launch Deployment Program • Advanced Launcher Development Programme • Automatic Language Data Processing

ALDS: Automatic Lightning Detection System

ALDT: Aldeamento

ALE: Accord de Libre Échange • Additional Living Expense • Address Latch Enable • Agence Locale pour l'Emploi • Airborne Laser Experiment • Airport Lighting Equipment • Alpine (USA-TX) (location identifier) • Application Link Enabling • Arbitrary Lagrange Euler • Arbitrary Lagrangian Eulerian • Association de la Libre Entreprise • Association de Libre-Échange • ATM Link Enhancement • Atmospheric Lifetime Experiment • Atomic Layer Epitaxy • Automatic Link Establishment

ALEC: Alternative Local Exchange Carrier

ALECSO: Arab League Educational, Cultural and Scientific Organization

ALEGE: Asociación Latino-Americana de Geofisica Espacial

ALENA: Accord de Libre-Échange Nord-Américain

ALERT: Acute Launch Emergency Reliability Tip • Acute Launch Emergency Restraint Tip • Arctic Long-term Environmental Research Transect • Association pour L'Étude des Risques du Travail

ALES: Automated Land Evaluation System

ALEXIS: Array of Low-Energy X-ray Imaging Sensors • Atmospheric Lidar Experiment In Space

AlF: Aluminium Fluoride

ALF: Acoustic Levitating Furnace • Acoustic Levitation Furnace • Airborne Laser Fluoroscope • Aloft • Alta (Norway) (location identifier) • Automatic Letter Facer • Auxiliary Landing Field

ALFA: Adaptive Optics with Lasers for Astronomy • Aéroports de Langue Française Associés • Air-Lubricated Free Attitude • Astronomical Low-Frequency Array

ALFE: Advanced Liquid Feed Experiment • Analog Line Front End

a/Lfg: An Lieferung

ALFG: Additive Lagged-Fibonacci Generator

ALFI: Association des Laboratoires Français Indépendants (France)

AlFr: Alta Frequenza (periodical code)

ALFS: Airborne Low-Frequency SONAR

ALFT: Action-Level Fault Tolerance • Approach and Landing Flight Test

Álg: Álgebra

Alg: Algebra • Algebraic • Algêbre • Algébrique • Alger (Algeria) • Algeria • Algerian • Algiers (Algeria)

ALG: Alger (Algeria) (location identifier) • Along • Autonomous Landing Guidance

AlGaAs: Aluminium Gallium Arsenide

ALGEC: Algorithmic Language for Economic Calculations

Alger: Algeria • Algerian

ALGHNY: Allegheny

ALGI: Address-Linked Geological Inventory

ALGOL: Algorithmic-Oriented Language

ALH: Advanced Light Helicopter • Advanced Liquid Hydrogen • Alarm Handler • Albany (Australia-WA) (location identifier) • Allan Hills (Antarctica)

ALI: Accelerated Lambda Iteration • Alice (USA-TX) (location identifier) • American Law Institute (USA) • American Library Institute (USA) • Annual Limit on Intake • Approximate Lambda Iteration • Array Level Interface • Automatic Location Identification

ALIAS: Aircraft Laser Infrared Absorption Spectrometer

ALIBI: Adaptive Location of Internetworked Bases of Information

Alic: Alicante (Spain)

ALICE: Adventists Libraries Information Cooperative • A Large Ion Collider Experiment • A Large Isotopic Composition Experiment • Algorithme d'Injection de Courants Électriques • Analyse des Liquides Critiques dans l'Espace • Array-Limited Control Electronics • Array-Limited Instrument Control Environment

ALID: Automated Library Issue Document

Align: Alignment

ALIO: Activity-Level-Independent Operation

ALIS: Advanced Life Information System • Auroral Large Imaging System • Automated Land Information System • Automated Library and Information System

ALISE: Association of Library and Information Science Education

ALISSA: Atmosphère par Lidar Sur Saliout

ALIT: Automatic Line Insulation Testing

ALIVE: Artificial Life Interactive Video Entertainment

ALJ: Alexander Bay (South Africa) (location identifier) • Australian Law Journal

Alk: Alkali • Alkaline

ALK: Alkalinity • Asela (Ethiopia) (location identifier)

ALKS: Adaptive Least k-th Squares

All: Allegato

ALL: Airborne Laser Laboratory • Albenga (Italy) (location identifier) • Lek (Albania) (ISO currency code)

ALLC: Association for Literary and Linguistic Computing

Alld: Allemand

ALLEA: All European Academies

Alleg: Allegory

allg: allgemein

Allow: Allowance

ALLSTAR: Aeronautics Learning Laboratory for Science, Technology And Research

Alfz: Alférez

Alm: Almacén • Almanac • Almería (Spain)

ALM: Advanced Lightning Mapper • Alamogordo (USA-NM) (location identifier) • Almirante • AppWare Loadable Module • Asset and Liability Management • Assurance de Longue Maladie • Audiolingual Method

ALMA: Atacama Large Millimeter Array (Chile)

ALMDA: Airlock Multiple-Docking Adapter

ALMRS: Automated Land and Mineral Record System

ALMS: Airborne Laser Mapping System • Analytic Language Manipulation System • Automated Library Management System • Automated Logistics Management System

Almte: Almirante

AlN: Aluminium Nitride

ALN: Alton (USA-IL) (location identifier) • Asynchronous Learning Network

ALO: Advanced Lunar Operation • Alternate Launch Officer • Annonce Légale Obligatoire • Astronomical Latitude Observatory • Automatic Lock-On • Waterloo (USA-IA) (location identifier)

ALOE: AppleTalk Low Overhead Encapsulation

ALOFT: A Language Oriented for Flight Testing • AMOS Large Optical Test Facility

ALOHA: Aerial Location of Hazardous Atmospheres (model) • A Long-term Oligotrophic Habitat Assessment

Aloj: Alojamiento

ALOMAR: Arctic Lidar Observatory for Middle Atmosphere Research

ALOPE: Airborne Lidar Oceanographic Probing Experiment

ALOS: Advanced Land Observing Satellite

ALOT: Apple Library Of To-morrow (USA-CA)

Alp: Alpes • Alpin • Alpine

ALP: Accélération par Laser de Particules • Advanced Leadership Programme • Advanced Lunar Project • Advancement of Librarianship Programme (IFLA) • Aleppo Neirab (Syria) (location identifier) • Anomalous Luminous Phenomenon • Assembly Language Program • Automated Language Processing • Automated Learning Process • Automated Library Program

ALPA: Air Line Pilots Association (USA)

ALPAC: Automated Language Processing Advisory Committee • Automatic Language Processing Advisory Committee (*)

ALPAL: A Livermore Physics Applications Language

ALPEX: Alpine Experiment

Alph: Alphabetical • Alphabétique

Alpha: Alphabetical • Alphabétique

ALPHA: Airborne Lidar Plume and Haze Analyzer • Automatic Literature Processing, Handling and Analysis

ALPO: Association of Lunar and Planetary Observers (USA-CA)

ALPOJ: Association of Lunar and Planetary Observers Journal (periodical code)

ALPS: Accidental Launch Protection System • Addresses, Lists and Proposals System • Advanced Linear Programming System • Advanced Liquid Propulsion System • Airborne Laser Polarization Sensor • Alternative Launch-Point System • Approach and Landing Procedures Simulator • Automated Language Processing System • Automated Launch Processing System • Automatic Linear Positioning System

ALPSP: Association of Learned and Professional Society Publishers

Alq: Alquimia

ALQ: Alegrete (Brazil-RS) (location identifier) • Association des Librairies du Québec (Canada-QC)

ALQDS: All Quadrants

alr: Aliter

ALR: Advanced Logic Research Inc. • Alexandra (New Zealand) (location identifier) • Antenne Linéaire Remorquée • Art Lost Register

ALRI: Acute Lower Respiratory Infection

ALRS: Airborne Laser Ranging System

ALRU: Automatic Line Record Update

ALRUTS: Automated Line Replaceable Unit Tracking System

ALS: Absorption-Line System • Accident Localization System • Ada Language System • Adjacent Link Station • Administrative and Legal Service • Advanced Landsat System • Advanced Launching System • Advanced Launch System (*) • Advanced Light Source • Advanced Limb Scanner • Advanced Logistics System • Airborne Lidar System • Airlock System • Alamosa (USA-CO) (location identifier) • Alternate Launch System • Alternate Line Service • Alternative Landing Site • American Lunar Society (USA-PA) • Application Layer Structure • Approach/Landing System • Approach Lighting System • Attitude Limiting System • Australian Landsat Station

• Autograph Letter Signed • Automated Library System • Automated List Service • Automatic Landing System • Automatic Level Setting

ALSA: Astronaut Life Support Assembly

Alsat: Alsatian

ALSATER: Réseau Alsacien de Télécommunications de la Recherche (France)

ALSCC: Apollo Lunar Surface Closeup Camera • Ariane Launch Sales Control Committee

ALSE: Astronaut Life Support Equipment

ALSEC: All Sectors

al seg: Al Segno

ALSEP: Apollo Lunar Surface Experiment Package

ALSOR: Air-Launched Sounding Rocket

ALSP: Aggregate Level Simulation Protocol

ALSS: Advanced Life Support System • Airlock Support System

ALST: Alaska Standard Time

ALSTG: Altimeter Setting

Alt: Alter • Alteration • Alternate • Alternatif • Alternative • Alternator • Altesse • Altimeter • Altitude • Alto

ALT: Alenquer (Brazil-PA) (location identifier) • Alter • Alternate • Altimeter • Altitude • Approach and Landing Test • Automated Language Translation

Alta: Alberta (Canada)

ALTA: Active Long Term Archive • Alberta (Canada)

ALTAIR: Automatic Logical Translation And Information Retrieval

alt-az: Altazimuth

altaz: Altazimuth

Atld: Altitude

alt. dieb.: Alternis Diebus

Alte: Almirante

Altér: Altération

Alter: Alteration

Altern: Alternativement

alt. hor: Alternis Horis

Altm: Altimeter • Altmark

Altn: Alternate

ALTN: Alternate

ALTNLY: Alternately

alt.noct.: Alternis Noctibus

Altntr: Alternator

Altnv: Alternative

ALTR: Approach and Landing Test Requirement

Altrd: Altered

Altrn: Alternation

ALTS: Airborne Laser Terrain Scanner

ALTV: All-Light Television

ALU: Alula (Somalia) (location identifier) • Arithmetic and Logic Unit • Arithmetic/Logic Unit

ALUE: Admissible Linear Unbiased Estimator • Arizona Land Use Experiment

Alum: Aluminium • Alumna • Alumnae • Alumni • Alumnus

ALV: Advanced Launch Vehicle • Advanced Logistics Vehicle • Algemene Ledenvergadering • Andorra la Vella (Andorra) (location identifier) • Australian Launch Vehicle • Auswärtiger Leihverkehr • Autonomous Land Vehicle

ALVINN: Autonomous Land Vehicle In a Neural Net

ALW: Aard- en Levenwetenschappen • Walla Walla (USA-WA) (location identifier)

ALWF: Actual Wind Factor

ALWR: Advanced Light-Water Reactor

ALX: Alexander City - Thomas C. Russell (USA-AL) (location identifier)

Aly: Alley

ALY: Alexandria - El Nouzha (Egypt) (location identifier) • Alley

ALZ: Alitak (USA-AK) (location identifier)

a/m: Above Mentioned

am: Amber • Amharic (ISO language code) • Ante Meridiem • Attometre

aM: am Main

A/m: Ampere per Meter

Am: America • American • Americium • Amiral • Metallic-Line A-type Star

A&M: Agricultural & Mechanical

A/M: Area to Mass • Auto/Manual

AM: Active Matrix • Active Message • Actuator Mechanism • Adhesion Model • Adjacent Matrix • Administrative Memorandum • Administrative Module • Aeromedical Monitor • Agent de Maîtrise • Agile Mirror • Aide-Mémoire • Airline Mileage • Airlock Module • Air Mail • Allocation Mensuelle • Alpes-Maritimes (France) • Altazimuth Mount • Amazonas (Brazil) • American • Amplitude Modulated • Amplitude Modulation • Angular Momentum • Anno Mundi • Annus Mirabilis (1666) • Ante Meridiem • Antenna Management • Application Manager • Archive Manager • Area Manager • Armenia (ISO country code) • Arménie (ISO country code) • Arp-Madore • Arrêté Ministériel • Assertional Module • Assistant Manager • Associate Member • Associative Memory • Assurance Management • Assurance Mutuelle • Automated Mapping • Automatic Monitoring • Auxiliary Memory • Auxiliary Module • Avis de Mutation • Award of Merit • Master of Arts

AMA: Absolute Measurement Accuracy • Active Matrix Array • Against Medical Advice • Amarillo (USA-TX) (location identifier) • American Management Association (USA) • American Marketing Association (USA) • American Medical Association (USA) • American Microfilm Association (USA) • Area Minimum Altitude • Articulated Mirror Assembly (*) • Articulating Mirror Assembly • Association Malmédienne d'Astronomie (Belgium) • Associazione Marchigiana Astrofili (Italy) • Attitude Measurement Assembly • Australian Medical Association • Automatic Message Accounting • Automatic Memory Allocation

AMAD: Activity Median Aerodynamic Diameter • Airborne Missions and Applications Division • Aircraft-Mounted Accessory Drive • Auxiliary-Mounted Accessory Drive

AMAE: American Museum of Atomic Energy (USA)

AMAG: Austria Metal AG

Amal: Amalgamated

AMAL: Aeronautical Medical Acceleration Laboratory

AMANDA: Antarctic Muon And Neutrino Detector Array

AMANDDA: Automatic Messaging And Directory Assistance

AMAP: Academic Map Analysis Package • Arctic Monitoring and Assessment Programme

AMAPS: Apogee Motor Assembly with Paired Satellites

AMARC: Automatic Message Accounting Recording Center

AMAS: Advanced Millimeter-wave Atmospheric Sounder • Association Marseillaise d'Astronomie (France)

AMASE: Astrophysics Multispectral Archive Search Engine

AMASIC: Autonomous Mapping by Analytic Sensor and Intelligence Correlation

AMASS: Airport Movement Area Safety System • Archive Management And Storage System

Amat: Amateur

AMATF: Arctic Monitoring and Assessment Task Force

AMATIS: Automated Meteorological Terminal Information Service

Amb: Ambassade • Ambassadeur • Ambassador • Ambiguo • Ambulance

AMB: Ambilobe (Madagascar) (location identifier)

AMBAC: Asociación Mexicana de Bibliotecarios (Mexico)

Ambas: Ambassador

AMBD: Automated Maintenance Database

AMBER: Advanced Multiple Beam Equalisation Radiography • Astronomical Multiple Beam Recombiner

Ambig: Ambiguous

AMBIT: Algebraic Manipulation By Identity Translation • Acronym May Be Ignored Totally

AMC: Academia Mexicana de Ciencias (Mexico) (formerly AIC) • Adjoint Model Compiler • Aerospace Manufacturers' Council • Aerospace Manufacturing Center • Aft Motor Controller • Air Material Command • Air Mobility Command • All Major Cards • Alternate Master Clock • Am Timan (Chad) (location identifier) • Année Mondiale des Communications (1983) • Appalachian Mountain Club • Archives and Manuscripts Control • Area Management Committee • Associate Members' Committee • Association Minéralogique du Canada • Associative Memory Computer • Associative Memory Controller • Automatic Manual Control • Automatic Measuring Complex • Automatic Message Counting • Automatic Mission Control • Automatic Mixture Control • Axial Meridian Circle

AMCA: Aft Motor Controller Assembly

AMCD: Addressograph-Multigraph-Copier-Duplicator • Annular Momentum Control Device

AMCEN: African Ministerial Conference on the Environment

AMCM: Advanced Missions Cost Model

AMCR: Automated Master Charge Record • Automatic Magnetic Card Reader

AMCS: Attitude Measurement and Control System • Attitude Monitoring and Control System

Amd: Amend

AMD: Acid Mine Drainage • Active Matrix Display • Advanced Micro Devices Inc. • Aerospace Medical Division • Ahmedabad (India) (location identifier) • Air Movement Data • Amend • Amended • Amendment • Angular Momentum Deficit • Antarctic Master Directory • Application Missions Division • Applied Mathematics Division • Asset Management and Disposition • Auxiliary Memory Drum • Dram (Armenia) (ISO currency code)

AMDAPS: Automatic Meteorological Data Acquisition and Processing System

AMDAR: Aircraft Meteorological Data Relay

AMDCS: Assurance Management Document Control System

AMDEA: Association of Manufacturers of Domestic Electrical Appliances

AMDF: Average Magnitude Difference Function

AMDG: Ad Majorem Dei Gloria

AMDIS: Atomic and Molecular Data Interactive System

AMDL: Abstract Machine Description Language

AMDS: Advanced Missions Docking System • Automatic Message Distribution System

Amdt: Amendment

AME: Abort Maneuver Evaluator • Accord Monétaire Européen • Acuerdo Monetario Europeo • Adaptive Mixture of Experts • Advanced Master of Education • Alto Molocue (Mozambique) (location identifier) • Antarctic Marine Ecosystem • Application Model Editor • Astronaut Maneuvering Equipment • Attitude Measurement Electronics • AutoCAD Modelling Extension • Automatic Measuring Engine • Automatic Microfiche Editor

AMEC: Aft Master Events Controller • Arctic Military Environmental Cooperation • Association of Mining and Exploration Companies

AMeDAS: Automated Meteorological Data Acquisition System

AMEDAS: Automated Meteorological Data Acquisition System

AMEDE: Amélioration du Mécanisme de Déploiement

AMEGE: Antarctic Marinen Engineering Geological Expedition

AMEL: Active-Matrix Electroluminescent (display)

Am. Emb.: American Embassy

Amend: Amendment

Amendt: Amendment

Amér: América

Amer: America • American

Amér: Amérique

Amer. Ind.: American Indian

AmerInd: American Indian

Amer. Sp.: American Spanish

AmerSp: American Spanish

Amer. Std.: American Standard

AMES: Accent on Mathematics and Engineering Science • Automatic Message Entry System

AMET: Academy of Maritime Education and Training

AmEx: American Express

AMEX: American Express • Atlantic Meteorology Experiment • Australian Monsoon Experiment

AMF: Abort Motor Facility • Aide-Mémoire Fiscal • Airmass Factor • Ama (Papua New Guinea) (location identifier) • Apogee Motor Fire • Arab Monetary Fund • Area Maintenance Facility • Arrangement Multifibre • Astronaut Memorial Foundation • Automatic Micro Furnace • Automatic Mirror Furnace

AMFABS: Advanced Maintenance Free Aircraft Battery System

AMFM: Automated Manufacturing Facilities Management • Automated Mapping Facilities Management

Amg: Among

AMG: Activation Management Group • Algebraic Multigrid • Amboin (Papua New Guinea) (location identifier) • Amplitude Modulation Generator • Angle of Middle Gimbal • Applied Mathematics Group • Architecture Management Group • Australian Map Grid • Automatic Magnetic Guidance

AMGS: Acceleration Monitoring Guidance System • Advanced Meteorological Ground Station

Amh: Amharic

AMH: Arba Mintch (Ethiopia) (location identifier)

AMI: Accord Multilatéral sur les Investissements • Active Microwave Instrument • Active Microwave Instrumentation • Acuerdo Marco Interconfederal • Acute Myocardial Infarction • Advanced Microwave Instrument • Aerospace Materials Information • Agence Maritime Internationale • Air Mercury International • Air Mileage Indicator • Alternate Mark Inversion • American Megatrends Inc. • Analyses et Mesures Industrielles • Apogee Memory Image • Apogee Motor Igniter • Application of Metrics in Industry • Assurance Maladie-Invalidité • Astrophysique et Méthodes Interférométriques • Axial Meridian Circle • Mataram (Indonesia) (location identifier)

AMIA: Atelier Mobile d'Initiation à l'Astronomie (Canada) • Australian Map Industry Association

AMIAP: Associate Member of the Institution of Analysts and Programmers

AMIBE: Applications Multimodales pour Interfaces et Bornes Évoluées

AMIC: Acquisition Management Information Center • Aerospace Materials Information Center • Australian Mining Industry Council • Automatic Management Information Center

AMICA: Asteroid Multiband Imaging Camera • Automatic Multispectral Image Compaction Algorithm

AMICIST: Auroral Microphysics and Ion Conics: Investigation of Space and Time

AMICO: Art Museum Image Consortium • Asteroids, Meteorites, Impacts and their Consequences (conference)

AMIDE: Advanced Multimedia Information Dissemination Environment

AMIDS: Advanced Multispectral Image Descriptor System

AMIE: Airglow Measurements of Infrared Emissions • Assimilative Mapping of Ionospheric Electrodynamics • Automated Methane Instrument Evaluation

AMIERE: Associate Member of the Institute of Electronic and Radio Engineers

AMIF: Applied Mathematics for Industrial Flow • Automated Map Information File

AMIGAS: Advanced Meteorological Image and Graphics Analysis System

AMIGO: American Inter-hemispheric Geobiosphere Organization (USA)

AMIGOS: Access Method for Indexed Data Generalized for Operating System

Am. Ind.: American Indian

AMIP: Atmospheric Model Intercomparison Project

AMIR: Advanced Microwave Imaging Radiometer

AMIRA: Australian Mining Industry Research Association

AMIS: Accounting Management Information System • Advanced Management Information System • Advanced Microwave Imaging Sensor • Australian Municipal Information System • Automated Maintenance Information System • Automated Management Information System • Aviation Management Information System

AMISAR: Active Microwave Imager Synthetic Aperture Radar

AMIX: American Information Exchange

AMJ: Almenara (Brazil-MG) (location identifier) • April-May-June • Assemblée Mondiale de la Jeunesse

AmJC: American Journal of Cardiology (periodical code)

AmJPh: American Journal of Physics (periodical code)

AmJS: American Journal of Science (periodical code)

AMK: Aircraft Modification Kit • Durango Animas (USA-CO) (location identifier)

AML: Abandoned Mine Land • Acquisition Material List • Amplitude Modulated Link • Angular Momentum Loss • Applied Mathematics Laboratory • Approved Materials List • ARC Macro Language • Array Machine Language • Astronomical Markup Language • Asymmetric Maximum Likelihood • Puerto Armuellas (Panama) (location identifier)

AMLC: Asynchronous Multiline Controller

AMLCD: Active-Matrix Liquid Crystal Display

AMLI: Algebraic Multilevel Iteration

AMLLV: Advanced Multipurpose Large Launch Vehicle

AMLP: Amplitude Modulation Link Program

AMLR: Antarctic Marine Living Resources

AMLS: Advanced Manned Launch System • Master of Arts in Library Science

AMM: Adaptive Mathematical Model • Aircraft Maintenance Manual • Amman - Queen Alia (Jordan) (location identifier) • Analog Modem Module • Anomalous Magnetic Moment • Application Model Manager • ASEAN Ministerial Meeting • Association Médicale Mondiale • Auroral Multiscale Mission • Automatic Maintenance Monitor • Autorisation de Mise sur le Marché

AMMA: Advanced Memory Management Architecture

AMME: Automated Multimedia Exchange

AmMin: American Mineralogist (periodical code)

AMMIS: Automated Manpower Management Information System

AMML: Automated Microbial Metabolism Laboratory

AMMM: Autonomous Magnetic Momentum Management

AMMOS: Advanced Multimission Operations System

AMMP: Advanced Manned Missions Program • African Magnetic Mapping Project • Approved Modernization Maintenance Program

AMMR: Advanced Multichannel Microwave Radiometer • Aircraft Multichannel Microwave Radiometer • Algebraic Methods for Model Reduction (meeting)

AMMS: Advanced Microwave Moisture Sounder • Advanced Multichannel Microwave Sensor • Airborne Microwave Moisture Sounder • Automatic Multimode Mass Spectrometry

AMMT: Automated Multimedia Terminal

AMN: Alma - Gratiot Community (USA-MI) (location identifier) • Atomic Mass Number

AMNH: American Museum of Natural History (USA-DC)

AMO: Air Mass Zero • Alternate Molecular Orbital • Atomic-Molecular-Optical • Mao (Chad) (location identifier)

AMOB: Automatic Meteorological Oceanographic Buoy

AMOHAS: Aerospace Medicine and Occupational Health Advisory Subcommittee

AMOHD: Aerospace Medicine and Occupational Health Division

AMON: ATM Monitoring

AMOO: Aerospace Medical Operations Office

AMOOS: Advanced Maneuvering Orbit-to-Orbit Shuttle

AMOP: Arctic Marine Oil Program • Atomic, Molecular and Optical Physics

AMOR: Advanced Meteor Orbit Radar

AMORS: Australian Modular Optical Remote Sensor

AMORSA: Atmospheric and Meteorological Ocean Remote Sensing Assembly

Amort: Amortization

AMOS: Advanced Mechanical and Optical Systems (Belgium) • Aerodrome Meteorological Observing System • Aerospace Maintenance and Operational Status • Air Force Maui Optical Site (USA-HI) • Air Force Maui Optical Station (USA-HI) (*) • Antarctic Meteorological Observing System • Assembly Management Operating System • Associative Memory Organizing System • Australian Meteorological and Oceanographic Society • Automated Materials Ordering System • Automated Meteorological Observation Station • Automated Multiobject Spectrograph • Automated Multiobject Spectrometer • Automated Multiple Object Spectrograph • Automatic Magnetic Observatory System • Automatic Meteorological Observation Station • Automatic Meteorological Observation System • Automatic Meteorological Observing Station (*) • Automatic Meteorological Observing System

AMOSS: Airline Management and Operations Support System

AMOWG: Astronomy Management Operations Working Group • Astrophysics Management Operations Working Group

Amp: Amperage • Ampere • Amplified • Amplifier • Amplify • Amplitude

AMP: Accelerometer Monitoring Program • Access Module Processor • Adaptation Mathematical Processor • Advanced Management Program • Advanced Maneuvering Program • Advanced Maneuvers Package • Advanced Materials and Processes (periodical code) • Advanced Microprocessing • Advanced Microprocessor • Aerospace Media Publishing • Alarm Manager Process • Alliance for Minority Participation • Ampanihy (Madagascar) (location identifier) • Apparent Motion Parameter • ARGOS Meteorological Platform • Aseptic Maintenance by Pressurization • Association of Media Producers (USA) • Associative Memory Processor • Average Mean Pressure • Axially Magnetized Plasma

AMPAS: Academy of Motion Picture Arts and Sciences (USA-CA)

AMPC: Automatic Message Processing Center

AMPD: Applied Materials Physics Division

Ampl: Amplified • Amplifier • Amplify • Amplitude

AMPL: Advanced Microprocessor Programming Language • A Macro Programming Language • A Manuscript Preparation Language

AMPLTD: Amplitude

AMPO: Astronaut Memorial Planetarium and Observatory (USA-FL)

AMPP: Advanced Microprogrammable Processor • Arctic Meteorology Photographic Probe

AMPQC: Atomic and Molecular Physics and Quantum Chemistry

AMPR: Advanced Microwave Precipitation Radiometer • Aeronautical Manufacturer Planning Report • Aeronautical Manufacturer Progress Report

Amps: Amperes

AMPS: Active Microwave Pressure Sounder • Adaptive Mode Planning System • Advanced Maneuvering Propulsion System • Advanced Meteorological Processing System • Advanced Mobile Phone Service • Advanced Mobile Phone Standard • Advanced Mobile Phone System • Assembly of Mathematical and Physical Sciences • Association for Modelling, Planning, and Simulation • Atmospheric and Magnetospheric Plasma-in-Space (Shuttle) • Atmospheric and Magnetospheric Plasma System • Automated Mission Planning System • Automatic Mapping and Planning System • Automatic Material Processing System • Automatic Message Processing System

AMPT: Advanced Maneuvering Propulsion Technology

AMPTE: Active Magnetospheric Particle Tracer Explorer

AMQ: Ambon (Indonesia) (location identifier)

AMQUA: American Quaternary Association (USA)

AMR: Abnormal Mission Routine • Adaptive Mesh Refinement • Advanced Material Request • Advanced Microwave Radiometer • Advanced Micro Research • Aerospace Medical Research • Age-Metallicity Relation • Anisotropic Magnetoresistance • Area-to-Mass Ratio • Arno (Marshall Islands) (location identifier) • Automated Management Report • Automated Meter Reading • Automatic Membership Registration • Automatic Message Recording • Automatic Message Routing

AMRA: Adaptive Mesh Refinement for Astrophysics • Advanced Multimode Radar Altimeter

AMRC: Accès Multiple à Répétition de Codes • Antarctic Meteorological Research Center

AMRF: Accès Multiple par Répartition en Fréquence • Automated Manufacturing Research Facility

AMRIR: Advanced Medium Resolution Imagery Radiometer

Amrit: Amritsar (India)

AMRL: Aeronautical and Maritime Research Laboratory (Australia-VIC)

AMROC: American Rocket Co.

AMRPD: Applied Manufacturing Research and Process Development

AMRS: Airborne Microwave Radiometer System • Automated Mail Retrieval Service

AMRSA: Atmospheric and Meteorological Remote Sensing Assembly

AMRV: Astronaut Maneuvering Research Vehicle

AMRVI: Advanced Medium Resolution Visible Imager

AmS: Amateur Service

AMS: Aberdeen Mechanical Society (UK-Scotland) • Accelerator Mass Spectrometer • Accelerator Mass Spectrometry • Acceptable Means of Compliance • Access Method Service • Accident-Monitoring System • Acoustic Measurement System • Actuation Mechanism System • Acute Mountain Sickness • Advanced Manned Spacecraft • Advanced Mapping System • Advanced Memory System • Advanced Meteorological Satellite • Advanced Meteorological System • Advanced Mission Studies • Aeronautical Material Specification • Airborne Maritime Surveillance • Airborne Measurement System • Aircraft Multiband Scanner • Airlock Module Station • Airmass • Alpha Magnetic Spectrometer • Alternate Mission Scientist • Altitude Measurement System • American Mathematical Society (USA) • American Meteorological Society (USA) • American Meteor Society (USA) • Amsterdam Schipol (Netherlands) (location identifier) • Anchor Macroscopic Semantic • Anisotropy of Magnetic Susceptibility • Anomalous-Mass Star • Anomaly Management System • Apollo Mission Simulator • Applied Mathematical Sciences • Archive Management System • Arctic Marine Science • Army Map Service • Articulated Mirror System • Artificial Moon Satellite • Assistance Médico-Sociale • Associative Memory System • Attached Manipulator System • Attitude Measurement Sensor • Attitude Measurement System • Automated Meteorological Station • Automated Maintenance System • Automated Manifest System • Automated Mapping System • Automated Material System • Automated Measurement System • Automated Microbial System • Automatic Management Switch • Automatic Meteorological System • Auxiliary Memory Set

AMSA: Associazione Maremmana Studi Astronomici (Italy)

AMSAT: Amateur Satellite

AMSBu: American Meteorological Society Bulletin (periodical code)

AMSC: American Mobile Satellite Corporation • Attitude Measurement and Control System

AMSCAT: Airborne Microwave Scatterometer

AmSci: American Scientist (periodical code)

AMSD: Adjusted Mapping Support Data

AMSE: Asymptotic Mean Squarred Error

AMSEE: Association Méditerranéenne des Sciences de l'Environnement et de l'Espace (France)

AMSGA: Association of Manufacturers and Suppliers for the Graphic Arts (UK)

AMSIE: Annual Meeting and Science Innovation Exposition (AAAS)

AMSL: Above Mean Sea Level

AMSO: Arab Organization for Standardization and Metrology

AMSR: Advanced Mechanically Scanned Radiometer • Advanced Microwave Scanning Radiometer (*) • Advanced Microwave Sounding Radiometer

AmSS: Amateur Satellite Service

AMSS: Advanced Manned Space Simulator • Advanced Meteorological Sounding System • Aeronautical Mobile Satellite Service • Airborne Multispectral Scanner • Aircraft Multichannel Scanner System

Amst: Amsterdam (Netherlands)

AMSTS: Automatic Multiparameter Semiconductor Test Set

AMSU: Advanced Microwave Sounder Unit • Advanced Microwave Sounding Unit (*) • Advanced Microwave Sound Unit

Amt: Amount

AMT: Adaptive Matched Filter • Address Mapping Table • Air Mail Transfer • Alternative Minimum Tax • Amata (Australia-NT) (location identifier) • Amount • Amplitude-Modulated Transmitter • Angular Mapping Transformation • Apogee Motor Timer • Apple Media Tool • Area Management Team • Associate in Mechanical Technology • Astrograph Mean Time • Automatic Moon Tracking • Automated Monitoring Telescope • Automatic Motor Tester • Auxiliary Master Tape • Available Machine Time • Aviation Maintenance Technician

AMTA: Advanced Multicolor Tracker for AMOS • Association for Machine Translation in the Americas

AMTAS: Automatic Modal Tuning and Analysis System

AMTD: Automatic Magnetic-Tape Dissemination

AMTEC: Alkali Metal Thermal to Electric • Association des Média et de la Technologie en Éducation au Canada

AMTEX: Airmass Transformation Experiment • Airmass Transportation Experiment

AMTF: Acoustic Model Test Facility

AMTRAK: American Travel by Track

AMTRAN: Automatic Mathematical Translator

AMTS: Advanced Meteorological and Temperature Sounder • Advanced Moisture and Temperature Sounder (*) • Advanced Moisture Temperature Sounder • Association Mondiale des Travailleurs Scientifiques

AMU: Aligarh Muslim University (India) • Amanab (Papua New Guinea) (location identifier) • Arab Maghreb Union • Associated Midwestern Universities (USA) • Astronaut Maneuvering Unit • Atomic Mass Unit • Attitude Match Update • Audio Management Unit • Auxiliary Memory Unit • Average Monthly Usage

AMV: Amderma (Russia) (location identifier)

AMVER: Automated Mutual Vessel Rescue System

AMW: Ames (USA-IA) (location identifier) • Angular Momentum Wheel • Average Molecular Weight

AMWA: Area Microwave Assembly

AMX: Ammaroo (Australia-NT) (location identifier) • Automatic Message Exchange

AMY: Ambatomainty (Madagascar) (location identifier)

AMZ: Ardmore (New Zealand) (location identifier) • Astro Mail Zirkular

AM0: Air Mass Zero

AM2: Amperes per Meter Squared

an: Anno

An: Actinon • Annam • Annum • Anonymous • Answer

A/N: Account Name • Advice Note • Alphanumeric • As Needed

AN: Above Named • Accession Number • Acid Number • Alphanumeric • Ammonium Nitrate • Anglo-Norman • Angola • Annunciator • Antarctica • Antilles Néerlandaises (ISO country code) • Antilles (Netherlands_) (ISO country code) • Applications Notice • Associated Number • Astronautics Notice • Astronomische Nachrichten (periodical code) • Atomic Number • Audible Noise • Netherlands Antilles (ISO country code)

ANA: Abridged Nautical Almanac • Adaptive Null Antenna • Advanced Network Applications • American Newspaper Association (USA) • Anaheim (USA-CA) (location identifier) • Article Numbering Association • Association of National Advertisers • Automatic Number Analysis • Automatic Number Announcement

ANABAD: Asociación Nacional de Archiveros, Bibliotecarios, Antiquarios y Documentalistas (Spain)

ANAC: Automatic Number Announcement Circuit

AnaCh: Analytical Chemistry (periodical code)

ANADP: Association of North American Directory Publishers (USA)

ANAE: Académie Nationale de l'Air et de l'Espace (France)

Anag: Anagram

AnaIn: Analytical Instrumentation (periodical code)

Anal: Analogie • Analogique • Analogous • Analogy • Analyse • Analyser • Analysis • Analyst • Analytic • Analytical • Analytique • Analyze • Analyzer

ANAL: Advise-Negate-Arrange-Legislate

Analyt: Analytical

ANAM: Agence Nationale des Aérodromes et de la Météorologie

ANAO: Australian National Audit Office

AnAp: Annales d'Astrophysique (periodical code)

ANAP: Association Narbonnaise d'Astronomie Populaire (France)

ANARE: Australian National Antarctic Research Establishment • Australian National Antarctic Research Expeditions

ANARS: Autonomous Navigation and Attitude Reference System

ANAS: Adaptation in Natural and Artificial Systems

ANASA: Azerbaijanian National Aerospace Agency (Azerbaijan)

ANASE: Association des Nations de l'Asie du Sud-Est

Anat: Anatomia • Anatomical • Anatomie • Anatomique • Anatomist • Anatomy

ANB: Anniston (USA-AL) (location identifier)

ANBEF: Association Nationale des Bibliothèques d'Expression Française (Belgium)

ANBF: Australian National Beamline Facility

ANBRE: Analog Biomechanical Recorder

ANBS: Asian Network of Biological Sciences

Anc: Ancien • Ancient

ANC: Academia Nacional de Ciencias (Mexico) • Acid-Neutralizing Capacity • Active Noise Cancellation • Active Noise Control • Active Nutation Control • Advanced Nuclear Computer • All Number Calling • Alternate Network Connectivity • Anchorage (USA-AK) (location identifier) • Assistant Network Controller • Automatic Noise Control

ANCA: Australian Nature Conservation Agency

AnCap: Annals of Cape Observatory (periodical code)

ANCB: Academia Nacional de Ciencias de Bolivia

ANCCAC: Australian National Committee on Computation and Automatic Control

ANCE: Association Nationale du Commerce Extérieur

Anch: Anchorage • Anchored

ANCIC: Australian National Copyright Information Centre

ANCOSPAR: Australian National Committee on Space Research

ANCOVA: Analysis of Covariance

ANCSA: Alaska Native Claims Settlement Act

Anct: Ancient

and: Andante

And: Andalucía (Spain) • Andromeda-Andromedae (constellation)

AND: Alphanumeric Display • Anderson (USA-SC) (location identifier) • Andorra (international vehicle registration code) • Andorra (ISO country code) • Andorre (ISO country code)

ANDE: Alphanumeric Display Equipment

AnDea: Annals of Dearborn Observatory (periodical code)

ANDES: Aérolineas Nacionales Del Ecuador SA

ANDF: Architecture Neural Distribution Format • Architecture Neutral Development Format • Architecture Neutral Distribution Format • Application Neural Distribution Format

ANDI: Applied Neutron Diffraction for Industry • Automated Non-Destructive Inspector

ANDS: Advanced Navigation Display System

Andz: Anodize

ANE: Acuerdo Nacional de Empleo • Angers (France) (location identifier)

ANEC: American Nuclear Energy Council (USA) • Averaged Null Energy Condition

ANELE: Asociación Nacional de Editoriales de Libros de Enseñanza (Spain)

Aner: Aneroid

ANF: Adult Non-Fiction • Antofagasta - Cerro Moreno (Chile) (location identifier) • Automatic Number Forwarding

ANFC: Annual Natural Forest Cleared

ANFIAC: Association Nationale pour la Formation Artistique et Culturelle (France)

ANFIS: Adaptive Neuro-Fuzzy Inference System

ANFR: Agence Nationale des Fréquences Radioélectriques (France) • Algorithms with Neglecting Functional Residuals

Ang: Angle • Angular

ANG: Air National Guard • Angoulême (France) (location identifier) • Guilder (Netherlands Antilles) (ISO currency code)

ANGB: Air National Guard Base (USA)

AnGeo: Annales de Géophysique (periodical code)

Angew: Angewandt

Angl: Angle • Angleterre • Anglican • Anglicized

Anglr: Angular

ANGO: Adviescommissie Niet-Gebonden Onderwijsonderzoek

Anh: Anhalt • Anhang

ANH: Anuha Island (Solomon Islands) (location identifier)

AnHar: Annals of the Astronomical Observatory of Harvard College (periodical code)

ANH&MRC: Australian National Health & Medical Research Council

Anhyd: Anhydrous

ANI: Agência de Noticias de Informações • Agência de Noticias e Informações • Agencia Nacional de Informaciones • Agência Nacional de Informações • Aniak (USA-AK) (location identifier) • Authorized - Not Implemented • Automatic Number Identification

ANIBAL: Abaissement du Niveau de Bruit des Avions Légers

ANIF: Automatic Number Identification Failure

AnIHP: Annales de l'Institut Henri Poincaré (periodical code)

ANIL: Association Nationale d'Information sur le Logement (France)

ANIPLA: Associazione Nazionale Italiana per l'Automazione (Italy)

ANITA: A New Inspiration To Arithmetic

AnITPM: Antarctic Ice Thickness Measurement Programme

ANJ: Zanaga (Congo) (location identifier)

Ank: Ankunft

ANK: Alphanumeric Keyboard • Ankara (Turkey) (location identifier)

Anl: Anlage

ANL: Accademia Nazionale dei Lincei (Italy) • Active-Node List • Agence Nationale du Logiciel (France) • Andulo (Angola) (location identifier) • Argonne National Laboratory (USA-IL) • Australian National Library • Automatic Noise Limiter • Automatic Noise Limiting

ANLAR: Accademia Nazionale dei Lincei - Atti Rendiconti (periodical code)

AnLei: Annals of Leiden Observatory (periodical code)

Anlg: Analog

ANLP: Applied Natural Language Processing

Anlr: Annular

ANLYS: Analysis

Anm: Anmerkung

ANM: Académie Nationale de Médecine • Antalaha (Madagascar) (location identifier)

AnMet: Annalen Meteorologie (periodical code)

AnMN: Antarctic Meteorite Newsletter (periodical code)

ANMS: Automated Network Monitoring System • Automated Notice to Mariners System

Ann: Annales • Annals • Année • Annex • Annexe • Anni • Anno • Annonce • Annual • Annuel • Annuelle • Annuity • Annulation

ANN: Adaptive Non-linear Network • All-figures Number Now • Annette Island (USA-AK) (location identifier) • Annotations • Architecture for Neural Networks • Artificial Neural Network

ANNA: Air Force - NASA - Navy - Army

ANNE: Artificial Neural Network Emulator

Anniv: Anniversary

Annot: Annotate • Annotated • Annotation • Annotator

AnnRev A & A: Annual Review of Astronomy and Astrophysics

Annu: Annual • Annuel • Annuelle

Annul: Annulment

ANNX: Annex

ANO: Alphanumeric Output • Angoche (Mozambique) (location identifier) • Any Other

ANOC: Association des Nouveaux Opérateurs de Câble (France) • Authorized Notice Of Change

ANODE: Auroral Nitric Oxide Dayglow Experiment

ANOM: Authorized Network Operations Manager

Anon: Anonyme • Anonymous • Anonymously

AnORB: Annales de l'Observatoire Royal de Belgique (periodical code)

ANOV: Analysis Of Variance

ANOVA: Analysis Of Variance

ANP: Algemeen Nederlands Persbureau • Annapolis Lee (USA-MD) (location identifier)

ANPA: Agenzia Nazionale per la Protezione dell'Ambiente (Italy) • American Newspaper Publishers Association (USA)

AnPar: Annales de l'Observatoire de Paris (periodical code)

ANPC: Advanced Navigation and Positioning Corporation

AnPCS: Annales de Physique - Colloquium Supplement (periodical code)

ANPE: Agence Nationale Pour l'Emploi (France)

AnPh: Annales de Physique (periodical code)

AnPhy: Annales of Physics (periodical code)

ANPIT: Agency for the National Program on Information Technology

ANPRM: Advanced Notice of Proposed Rulemaking

ANQ: Angola - Tri-State Steuben County (USA-IN) (location identifier) • Archives Nationales du Québec (Canada-QC)

Anr: Another

ANR: Antwerpen Deurne (Belgium) (location identifier) • Automatic Network Routing

ANRA: Air Navigation Radio Aid

ANRAC: Aids Navigation Radio Control

ANRC: American National Red Cross (USA) • Australian National Research Council

AnRE: Annual Review of Energy (periodical code)

ANRE: Advanced Nuclear Rocket Engine • Agency of Natural Resources and Energy

AnRev: Annual Review

AnRFM: Annual Review of Fluid Mechanics (periodical code)

AnRMS: Annual Review of Materials Science (periodical code)

ANRT: Association Nationale de la Recherche Technique (France)

Ans: Answer

ANS: Advanced Network and Services Inc. • Advanced Neutron Source • Air Navigation System • American National Standard • American Nuclear Society (USA) • Andahuaylas (Peru) (location identifier) • Array Normal Spin • Artificial Nervous System • Artificial Neural System • Astronomical Netherlands Satellite • Astronomische Nederlandse Satelliet • Autograph Note Signed • Automatic Navigation System • Automatic Nervous System • Autonomous Navigation System

ANSA: Advanced Network System Architecture • Agenzia Nazionale Stampa Associata (Italy)

ANSARI: Astrophysical Network Space Archiving Research Institute

ANSBR: American National Standard for Bibliographic References

ANSC: American National Standards Committee (USA)

ANSCII: American National Standard Code for Information Interchange

ANSCR: Alphanumeric System for Classification of Recordings

ANSCSOCR: American National Standard Character Set for Optical Character Recognition

ANSE: Autonomous Nervous System Experiment

ANSEI: Asociación Nacional de la Seguridad de Entornos Informáticos (Spain)

ANSER: Actuated Nose Strakes for Enhanced Rolling

AnShO: Annals of Shanghai Observatory (periodical code)

ANSI: American National Standards Institute (USA)

ANSIC: Aerospace Nuclear Safety Information Center (USA)

ANSL: Australian National Standards Laboratory

ANSMET: Antarctic Search for Meteorites

ANSORC: Analog System for Optical Reading of Characters

ANSP: Academy of Natural Sciences of Philadelphia (USA-PA)

ANSR: Advanced Neutron Source Reactor

ANSS: Aeronautical Navigation Satellite System

ANSSR: Aerodynamically Neutral Spin-Stabilized Rocket

Anst: Anstalt

ANSTC: Australian Nuclear Science and Technology Council

ANSTI: African Network of Scientific and Technological Institutions

ANSTJ: Association Nationale Science Techniques Jeunesse (France)

ANSTO: Australian Nuclear Science and Technology Organization

ANSTP: Austrian National Space Technology Programme

Ant: Antarctica • Antenna • Antérieur • Anterior • Anticuado • Antigua • Antiguo • Antillas • Antlia-Antliae (constellation) • Antonym

ANT: Actor Network Theory • Antenna • Antenna Noise Temperature • Antilles Néérlandaises (ISO country code) • Antilles (Netherlands_) (ISO country code) • Articulated Navigation Testbed • Autonomous Navigation Technology • Netherlands Antilles (ISO country code) • Sankt Anton (Austria) (location identifier)

Antarc: Antarctica

ANTARES: Antarctic Research • Astronomy with a Neutrino Telescope and Abyss Environmental Research • Astrophysics Network for Training and Research

ANTAS: Automatische Naß-Trocken Ablagerungssammler

ANTC: Advanced Networking Test Center

AnTec: Annals Tectonicae (periodical code)

AnTel: Annales des Télécommunications (periodical code)

Anten: Antenna

Anthol: Anthology

Anthrop: Anthropological • Anthropology

Anthropol: Anthropological • Anthropology

ANTI: Automated Near-Term Improvement

Antífr: Antífrasis

Antig: Antigua

antilog: Antilogarithm

ANTIOPE: Acquisition Numérique et Télévisualisation d'Images Organisées en Pages d'Écriture

Antiq: Antiquarian • Antiquary • Antiquity

Antk: Antarktika (periodical code)

AnTok: Annals of Tokyo Astronomical Observatory (periodical code)

ANTOSTRAT: Antarctic Offshore Acoustic Stratigraphy

AnTou: Annales de l'Observatoire de Toulouse (periodical code)

Antrop: Antropología

ANTS: Automatic Nitrogen Transfer System

Antw: Antwerp (Belgium) • Antwort

Antwpn: Antwerpen (Belgium)

ANU: Antigua (Leeward Islands) (location identifier) • Australian National University (Australia-ACT)

ANUATC: Australian National University Astrophysical Theory Centre

ANULAS: Australian National University Landsat Analysis System

ANV: Anvik (USA-AK) (location identifier)

ANVAR: Agence Nationale de Valorisation de la Recherche (France) • Agence Nationale pour la Valorisation de la Recherche (France)

ANVR: Algemene Nederlandse Vereniging van Reisbureaus

ANW: Ainsworth (USA-NE) (location identifier)

AnWie: Annals of Vienna Observatory (periodical code)

ANWR: Arctic National Wildlife Refuge

Anx: Annex

ANX: Anexo • Annex • Automotive Network Exchange

ANY: Anthony (USA-KS) (location identifier)

Anz: Anzeiger

ANZ: Angus Downs (Australia-NT) (location identifier)

ANZAAS: Australian and New Zealand Association for the Advancement of Science

ANZEC: Australian and New Zealand Environment Council

ANZECC: Australian and New Zealand Environment and Conservation Council

ANZIAM: Australia and New Zealand Industrial and Applied Mathematics

ANZES: Australian and New Zealand Scientific Exploration Society

ANZLIC: Australia-New Zealand Land Information Council

ANZONE: Antarctic Zone

ANZUS: Australia - New Zealand - United States

a/o: Account of

a/O: An der Oder

aO: An der Oder

Ao: Arroyo

A/O: Account Of • Analog Output • And Others

AO: Abgabenordnung • Abnormal Occurrence • Abort Output • Abstract Object • Acousto-Optical • Accountant Officer • Account Of • Adaptive Optics • Additional Observation • Administrative Office • Administrative Officer • Administrative and Overhead • Among Others • Amplifier Output • Analog Output • And Others • Angola (ISO country code) • Annonce d'Opportunité • Anno Ordinis • Announcement of Opportunity • Appel d'Offre • Arctic Oscillation • Area Office • Area of Operations • Arecibo Occultation • Assembly Order • Assembly Outline • Atmosphere-Ocean • Atomic Orbital • Atomic Oxygen • Authorizing Official • Automated Operation • Autres Objets

AOA: Abort Once Around • Administration On Aging (USA) • Airborne Optical Adjunct • Airport Operating Area • Allen Osborne Associates • American Ordnance Association (USA) • Angle Of Arrival • Angle Of Attack • Aroa (Papua New Guinea) (location identifier) • Associazione Ogliastrina di Astronomia (Italy) • At Or Above

AOARD: Asian Office of Aerospace Research and Development

AOAS: Advanced Oceanic Automation System • Arkansas-Oklahoma Astronomical Society (USA)

AOB: Adaptive Optics Bonnette • Advanced Operational Base • Annanberg (Papua New Guinea) (location identifier) • Any Other Business • At Or Below

AOBP: Arctic Ocean Buoy Program

A&OC: Attitude & Orbit Control

AOC: Adaptive Optics Coronagraph • Advice Of Charge • Aeronautical Operational Control • Aircraft Operations Center • Airline Operations Center • Air Operations Center • Altenburg (Germany) (location identifier) • Annual Operating Cost • Anodic Oxide Coating • Appellation d'Origine Contrôlée • Apportionnement Of Credit • Array Operations Center (NRAO) • Astronomical Observatory Copenhagen (Denmark) • Attitude and Orbit Control • AURA Oversight Committee • Automatic Operation Control • Automatic Output Control • Auxiliary Output Chip • Average Operating Cost • Award Of Contract

AOCA: Amis de l'Observatoire des Creusets à Arbaz (Switzerland)

AOCB: Any Other Competent Business

AOCC: Astronomical Observatory of Campo catino (Italy)

AOCE: Apple Open Collaboration Environment • Apple Open Collaborative Environment (*) • Attitude and Orbit Control Electronics

AOCI: Airborne Ocean Color Imager Spectrometer

AOCM: Advanced Ocean Color Monitor

AOCMS: Attitude and Orbit Control and Measurement System

AOCP: AVCS Onboard Computer Program

AOCR: Advanced Optical Character Reader • Advanced Optical Character Recognition

AOCRD: Acceptance and Operational Checkout Requirements Document

AOCRS: African Organization for Cartography and Remote Sensing

AOCS: Active Optics Control System • Altitude and Orbit Control System • Arctic Ocean Climate Station • Attitude and Orbit Control System (*) • Attitude Orbit Control System • Automated Orbit Control System

AOCSN: Arctic Ocean Climate Station Network

AOD: Abou Deia (Chad) (location identifier) • Acousto-Optic Deflector • Acousto-Optic Device • Advanced Object Data • Aerosol Optical Depth • Analog Output Differential • Applied Optical Detection • As Of Date • Assistant Operations Director • Atmosphere and Ocean Dynamics

AODC: Acridine Orange Direct Count • Australian Oceanographic Data Centre

AODPA: Acquisition Orbit Determination Program Assembly

AOE: Airport Of Entry • Application Operating Environment • At Our Expenses

AOES: Advanced Orbit/Ephemeris System

AOET: Atomic Oxygen Exposure Tray

AOF: Afrique Occidentale Française • Anode Oxide Film • Archive Operations Facility

AOFEE: Advanced Orbiter Flight Evaluation Experiment

AOFP: Absolute Open-Flow Potential

AofS: Academy of Sciences

AOFSCE: Absolute Oil Fluorescence Spectral Conversion Efficiency

AOG: Abstract Object Generic • Aircraft On Ground • Anshan (China-PRC) (location identifier) • Auxiliary Output Generator

AOGC: Arctic Ocean Grand Challenge

AOGCM: Atmosphere-Ocean General Circulation Model

AOGO: Advanced Orbiting Geophysical Observatory

AOH: Ancient Ocean Hypothesis • Lima - Allen County (USA-OH) (location identifier)

AOI: Ancona (Italy) (location identifier) • Angle Of Incidence • Announcement of Opportunity Instruments • Arab Organization for Industrialization • Area Of Interest • Associazione Ottica Italiana (Italy) • Astrometric Optical Interferometer • Automated Optical Inspection

AOIP: Association des OUvriers en Instruments de Pression (France)

AOIPS: Atmospheric and Oceanographic Image Processing System

AOIS: Alps Observation and Information System • Automatic Optical Inspection System

AOIV: Automatically Operated Inlet Valve

AOJ: Aomori (Japan) (location identifier)

AOK: All Okay • Astrooptik Kohler (Switzerland) • Karpathos (Greece) (location identifier) • Kwanza (Angola) (ISO currency code)

AOL: Absent Over Leave • Airborne Oceanographic Lidar • Airborne Optical Laser • America On Line • Application-Oriented Language • Area Of Launch • Astronomy On Line • Paso de los Libres (Argentina-CR) (location identifier)

AOLO: Advanced Orbital Launch Operation

AOLR: Amplifier Open Loop Response

AOM: Acousto-Optic Modulator • Adaptive Optics Module • Amorphous Organic Matter

AOML: Atlantic Oceanographic and Meteorological Laboratory

AOMPS: Automatic Outgoing Message Processor System

AOMS: Advanced Optical Monitoring System

AON: All Optical Network • Arona (Papua New Guinea) (location identifier) • New Kwanza (Angola) (ISO currency code)

AONBS: Asian Oceania Network of Biological Sciences

AOO: Altoona (USA-PA) (location identifier) • American Oceanographic Organization (USA) • Analyse Orientée Objets • Any Other Object • Approche Orientée Objet • Area Of Operations

AOP: Advanced Operating Platform • Annual Operating Plan • Assembly Operations Plan • Association of Online Professionals • Astrophysikalisches Observatorium Potsdam (Germany)

AOPA: Aircraft Owners and Pilots Association (USA)

AOPL: Advance Order Parts List

AOPM: Airline Operations Planning Model

AOPP: Atmospheric, Oceanic, and Planetary Physics

a. o. Prof.: außerordentlicher Professor

AOPS: Airborne Operations

AOQ: Average Outgoing Quality

AOQL: Average Outgoing Quality Limit

a/or: And/Or

AOR: Advice Of Rights • Alor Setar (Malaysia) (location identifier) • Angle Of Reflection • Antenna Ohmic Resistance • Application Owning Region • Arctic and Offshore Research • Area Of Responsibility • Atlantic Ocean Region

AORA: Atlantic Ocean Recovery Area

AORD: Astronaut Operations Requirement Document

AORIS: Arctic and Offshore Research Information System

AORS: Abnormal Occurrences Reporting System • Acousto-Optical Radio Spectrograph

A&OS: Atmospheric & Oceanic Sciences

AOS: Academic Operating System • Acousto-Optical Spectrograph • Acousto-Optical Spectrometer • Acousto-Optical Spectroscopy • Acousto-Optic Spectrograph • Acousto-Optic Spectrometer • Acousto-Optic Spectroscopy • Acquisition Of Signal • Active Optical Sensor • Activity Operating Schedule • ACTS Optical System • Adaptive-Optics System • Add-On Stabilization • Add Or Subtract • Advanced Operating System • Advanced Orbiting System • Airborne Optical System • Alternate Operator Service • Aluminium Oxide Sphere • Amateur Observers' Society (USA-NY) • Amook (USA-AK) (location identifier) • Amount Of Something • Amplifier Output Stage • Astronomical Observatory Satellite • Astronomiczne Obserwatorium Szerokościowe (Poland) • Atmospheric and Oceanic Sciences • Australian Optical Society • Automated Office System • Automatic Observing System • Azimuth Orientation System

AOSB: Arctic Ocean Sciences Board

AOSD: Atomic Object State Diagram

AOSGE: Arctic Ocean System in the Global Environment

AOSIS: Association of Small Island States

AOSO: Advanced Orbiting Solar Observatory

AOSP: Active Optics Simulation Program • Advanced Onboard Signal Processor • Automatic Operating and Scheduling Program

AOSS: Active Optics Simulation System • Airborne Oil Surveillance System • Atmospheric, Oceanic and Space Sciences

AOST: Association des Opérateurs de Services de Télécommunications (Fran ce)

AOT: Actual Operating Time • Adaptive Optics Team • Aeroassisted Orbital Transfer • Aerosol Optical Thickness • Alignment Optical Telescope • Allowed Outage Time • Astronomical Observation Template • Astronomical Observing Template • Atmospheric Optical Thickness • Auxiliary Output Tester • Average Operation Time • Avionics Operating Time • Avionics Overall Test

AOTF: Acousto-Optical Tunable Filter (*) • Acousto-Optic Tunable Filter

AOTS: Advanced Orbital Test Satellite

AOTV: Aero-assisted Orbital Transfer Vehicle • Aero-assisted Orbit Transfer Vehicle

AOU: Astronomiska Observatoriet Uppsala (Sweden) • Attopeu (Laos) (location identifier)

AOV: Annonce d'Opportunité de Vol

AOVEF: Arecibo Observatory Visitor and Educational Facility (Puerto Rico)

AOVP: Arctic Ocean Variability Project

AOW: Algemene Ouderdomswet • Asian-Oceanic Workshop • Asia-Oceania Workshop

AOX: Adsorbable Organic Halogen

A°: Arroyo

Ap: Apartado • Aparte • Apelación • Apparent • Approach • April • Astrofizika (periodical code) • Astrophysics (periodical code) • Autopilot • Peculiar A-type Star

A&P: Airframe & Powerplant • Asia & Pacific • Attitude & Pointing

A/P: Account Paid • Accounts Payable • Additional Premium • Airport • Autopilot

AP: Above Proof • Absolute Parallelism • Academic Press (now HBJ) • Accelerometer Package • Access Panel • Access Point • Accident Professionnel • Accompte Provisionnel • Account Payable • Accreditation Panel • Acoustic Processor • Acquisition Plan • Active Prominence • Additional Premium • Adjective Phrase • Adjunct Processor • Administration Publique • Administrative Procedure • Advanced Placement • Advanced Post • Advance Pay • Aerosol Particle • Affranchissement Postal • After Peak • Agence de Placement • Agence de Presse • Agency Procedure • Aire Projetée • Air Pollution • Air Pressure • Alkaline Permanganate • All Points • All Purposes • Amapá (Brazil) • Ammonium Perchlorate • Analytical Plotter • Annales de Physique • Annual Production • Anomalous Propagation • Ante Prandium • Antipersonnel • Apartado Postal • Application Processor • Application Profile • Application Program • Application Project • Application Protocol • À Protester • Arecibo Pulsar • Arithmetic Progression • Array Processor • Arrêté Préfectoral • Asia-Pacific • As Prescribed • As Purchased • Ascent Phase • Assistance Publique • Associated Press • Association Professionnelle • Associative Processor • Assurance Produit • Astrometric Photometer • Atmospheric Physics • Atmospheric Pressure • Atomic Power • Atomic Powered • Attached Processor • Attendant Position • Attitude and Pointing • Authority to Pay • Authority to Purchase • Author's Proof • Automated Procedure • Automatic Programming • Autopilot • Avant-Projet • Avant-Propos • Average Payable • Average Price • Avis Préalable • Pakistan (international aircraft registration code)

APA: Abort Programmer Assembly • Adaptive Packet Assembly • Additional Personal Allowance • Administrative Procedure Act • Alkaline Phosphatase Activity • Allowance for Program Adjustment • All Points Addressable • American Physicists Association (ISA) • American Planning Association (ISA) • American Press Association (USA) • American Psychological Association (USA) • Annular Phased Array • Antarctic Protected Area • Antenna Pattern Analyzer • Antenna Phase Angle • Arithmetic Processing Accelerator • Arithmetic Processor Accelerator • Asociación Paraguaya de Autores (Paraguay) • Asociación Peruana de Astronomía (Peru) • Associate in Public Administration • Associate Principal Astronomer • Association of Patent Attorneys • Associative Principle for Addition • Augmented Predictive Analyzer • Australian Postgraduate Award • Austria Presse Agentur • Automatic Photographic Analysis • Denver - Arapahoe County (USA-CO) (location identifier)

APAA: Asian Patent Attorneys Association • Associação Portuguesa de Astrónomos Amadores (Portugal)

ApAc: Applied Acoustics (periodical code)

APAC: Aerial Photogrammetry, Analysis and Cartography • Antenna Pointing Angle Change • Asociación Peruanan para el Avance de la Ciencia (Peru) • Australia Pacific Airports Corporation

APACM: Atmospheric Physical and Chemical Monitor • Atmospheric Physics and Chemistry Monitor

APACS: Association for Payment Clearing Services

APAD: Application-Adaptable Database System

APAE: Attached Payload Accommodation Equipment

APAFO: Advanced Particles And Fields Observer

APAL: Array processor Assembly Language

APALS: Autonomous Precision Approach and Landing System

APAM: Array Processor Access Method

APAN: Asia-Pacific Advanced Network

APANA: Australian Public Access Network Association

APAP: Amis du Planétarium d'Aix-en-Provence (France)

APAPI: Abbreviated Precision Approach Path Indicator

APAR: Absorbed Photosynthetically Active Radiation • Authorized Program Analysis Report • Automatic Program Analysis Report • Automatic Programming and Recording

APAREN: Address Parity Enable

Apart: Apartment

APAS: ADABAS Performance Analysis System • Aerodynamic Performance and Sizing Program • Androgynous Peripheral Assembly System • Astrophysical, Planetary and Atmospheric Sciences • Attività de Preparazione, Accompagnamento e Sostegno • Automated Program for Aerospace-Vehicle Synthesis • Automated Programmable Assembly System • Automatic Performance Analysis System

APASEG: All-party Parliamentary, Astronomy, and Space Environment Group (UK)

APAST: Association Africaine pour l'Avancement des Sciences et des Techniques

APASU: Attaché Principal d'Administration Scolaire et Universitaire

APAT: Atmospheric Pressure and Ambient Temperature

APB: All-Points Bulletin • Apolo (Bolivia) (location identifier)

APBAD: Associação Portuguesa de Bibliotecários, Arquivistas e Documentalistas (Portugal)

Apble: Apreciable

APBO: Accumulated Postretirement Benefit Obligation

APC: Accelerometer Pulse Converter • Activity Processing Code • Adaptive Payload Carrier • Advance Procurement Change • Advance Procurement Check • Advanced Process Control • Advanced Programming Course • Advanced Propulsion Comparison • Advanced Propulsion Cooling • Aeronautical Public Correspondence • Aft Power Controller • Airborne Particle Counter • Aircraft-Pilot Coupling • Air Pollution Control • American Personal Communications • American Power Conversion • Antarctic Place-names Committee • Antenna Pattern Correction • Approach • Area Positive Control • Armored Personnel Carrier • Aspirin-Phenacetin-Caffeine • Association des Planétariums du Canada • Association for Progressive Communication • Association for Progressive Computing • Attitude Processing Center • Automated Process Software • Automatic Performance Control • Automatic Phase Control • Automatic Pitch Control • Automatic Power Control • Automatic Predicate Calculus • Automatic Public Convenience • Autonomous Payload Control • Autonomous Payload Controller • Napa County (USA-CA) (location identifier)

APCA: Aft Power Controller Assembly • Air Pollution Control Association (USA)

APCAJ: Air Pollution Control Association Journal (periodical code)

APCATS: Asian-Pacific Conference on Aerospace Technology and Science

APCB: Air Pollution Control Board

APCCE: Asian Pacific Confederation of Chemical Engineering

APCCIRN: Asian-Pacific Coordinating Committee for International Research

APCE: Agence Pour la Création d'Entreprises (France)

APCF: Advanced Protein Crystallization Facility (*) • Automated Protein Crystallization Facility

APCG: Advanced Protein Crystal Growth

APCH: Approach

APCHG: Approaching

APCI: Agence pour la Promotion et la Création Industrielle (France)

APCM: Aerial Photography Contract Management • Assemblée Permanente des Chambres de Métiers (France)

APCN: Anno Post Christum Natum • Asia-Pacific Cable Network

APCOM: Application Program to Communication

APCOPPLSRF: Analysis and Program for Calculation of Optimum Propellant Performance for Liquid and Solid Rocket Fuels

ApCpH: Attempt per Circuit per Hour

APCPH: Attempt Per Circuit Per Hour

APC-REX: Autonomous Payload Control Rocket Experiment

APCREX: Autonomous Payload Control Rocket Experiment

APCS: Active-Passive Correlation Spectroscopy • Aeronautical Production Control System • Alberta Photogrammetric Control System • Atmosphere Pressure Control System • Attitude and Pointing Control System • Automated Project Control System

APC-STEX: Autonomous Payload Control Satellite Experiment

APCSTEX: Autonomous Payload Control Satellite Experiment

APCUG: Association of Personal Computers Users Groups

APCVD: Atmospheric Pressure Chemical Vapour Deposition

Apd: Approved

APD: Absolute Pointing Drift • Acoustic-Phonetic Decoding • Adjustable Pitch Device • Advanced Planning and Design • Advanced Planning Document • Advanced Program Development • Aeronautical Propulsion Division •

Aide Publique au Développement • Amplitude Probability Distribution • Analog-to-Pulse Duration • Angular Position Digitizer • Area Passive Dosimeter • Area Postal Directory • Asian Pacific Division • Astronomical Pocket Diary • Automated Products Division • Auxiliary Power Distribution • Avalanche Photodiode • Avalanche Photodiode Detector • Avant-Projet Détaillé • Average Particle Diameter • Average Percentage Damage • Average Pore Diameter • Average Power Dissipation

APDA: Arctic Precipitation Data Archive

APDC: Association Pour le Développement des Communications (France)

APDE: Application Protocol Development Environment

APDF: Astronomical Data Processing Facility

APDL: Ada Programming Design Language • Algorithmic Processor Description Language

Apdo: Apartado

APDOOR: Aperture Door

APDS: Advanced Personnel Data System • Advanced Planning Data Sheet

APDU: Application Protocol Data Unit

APE: Absolute Pointing Error • Airborne Polar Experiment • Alfvén Propulsion Engine • Alpha-Proton-Electron • Arithmetic Processing Extension • Asemblea Parlamentaria Europea • Assemblée Parlementaire Européenne • Atmospheric Physics Experiment • Atomic Photoelectric Effect • Aurora Photography Experiment • Automatic Photomapping Equipment • Automatic Positioning Equipment • Available Potential Energy • Available Power Efficiency • San Juan Aposento (Peru) (location identifier)

APEA: Antenna Pattern Error Analysis

APEC: Advisory Panel on Environmental Change • Agence Pour l'Emploi des Cadres (France) • Asia-Pacific Economic Cooperation • Association Pour l'Emploi des Cadres (France) • Astrophysical Plasma Emission Code • Autonomous Robot Control and Simulation

APED: Astrophysical Plasma Emission Database

APEM: Association du Personnel de la Météorologie

ApEn: Applied Energy (periodical code) • Approximate Entropy

ApEnM: Applied Environmental Microbiology (periodical code)

Aper: Aperture

APERAL: Air Pollution Effects Research At Lancaster

APERT: Aerospace Programme for Education, Research and Technology

APES: Advanced Processor Emulator System

APEX: Active Plasma Experiment • Advanced Photovoltaic and Electronics Experiment • Advanced Purchase Excursion • Antiproton Experiment • Arctic Polynya Experiment • Ariane Payload Experiment

APF: Analyseur de Plasma Froid • Asia-Pacific Forum • Attitude Parameter File • Authorized Program Facility • Naples (USA-FL) (location identifier)

APFA: Association of Professional Flight Attendants (USA)

APFIM: Atom-Probe Field-Ion Microscope • Atom-Probe Field-Ion Microscopy

APFO: Aerial Photography Field Office

APFR: Automatic Programmable Film Reader

APFS: Apparent Places of Fundamental Stars

APFTI: Airline Passenger Fair Treatment Initiative

APG: Aberdeen Phillips (USA-MD) (location identifier) • Algorithmic Pattern Generator • All Paths Generator • Antenna Power Gain • Application Performance Group • Application Program Generator • Astrometric-Photometric Group • Atlas of Peculiar Galaxies • Azimuth Pulse Generator

APGC: Air Proving Ground Command

APGERN: Asian-Pacific Global Environmental Research Network

Aph: Aphetic • Aphorism

APh: Astroparticle Physics (periodical code)

APH: Attitude Pointing History • Bowling Green - Camp A.P. Hill (USA-VA) (location identifier)

APHA: American Public Health Association (USA)

APHELIE: Astronomie, Physique, Élaboration , Instrumentation et Observation

APHIS: Animal and Plant Health Inspection Service

API: Acoustic Properties Instrument • Advanced Procurement Information • Advance Publication Information • Aerial Photointerpretation • Aide aux Projets d'Innovation • Air Photo Interpretation • Air Pollution Index • Air Position Indicator • Alphabet Phonétique International • American Petroleum Institute (USA) • Angle Position Indicator • Antenna Position Indicator • Apiay (Colombia) (location identifier) • Application et Promotion de l'Innovation • Application Program Interface • Application Programmer Interface • Application Programmer's Interface • Application Programming Interface • A Priori Importance • Association Phonétique Internationale • Atmospheric Pressure Ionization • Automatic Priority Interrupt

APIB: Application Program Integration Board

APIC: Advanced Programmable Interface Controller • Advanced Programmable Interrupt Controller • Antecedent Precipitation Index Catchment • Automatic Power Input Controller • Automatic Programming Information Centre (UK)

APICS: American Production and Inventory Control Society (USA)

APID: Application Identifier • Applications Process Identification

APIF: Automated Process Information File

APIG: Applied Physics Interest Group

APII: Asia-Pacific Information Infrastructure

ApInv: Astrophysical Investigations (periodical code)

APIRD: Authorized Procurement Information Requirements Description

APIRL: Authorized Procurement Information Requirements List

APIS: Aerial Photography Information System • Antarctic Pack Ice Seals

ApJ: Astrophysical Journal (periodical code)

ApJS: Astrophysical Journal Supplement Series (periodical code)

APK: Apataki (Tuamotu Islands) (location identifier) • Astronaut Preference Kit

Apl: Aplicado • April

ApL: Astrophysical Letters (periodical code)

APL: Adult Performance Level • Advance Procurement List • Advanced Programming Language • Allowance Parts List • Alternating Polarization Laser • Applicable Price List • Applications Program Library • Applied Physics Laboratory • Applied Physics Letters • Approved Parts List • A Programming Language • Assembly Part List • Assembly Programming Language • Association for Programmed Learning • Associative Programming Language • Automated Programming Language • Average Picture Level • Nampula (Mozambique) (location identifier)

APLA: American Patent Law Association (USA)

APLAC: Analysis Program Linear Active Circuit • Asia Pacific Laboratory Accreditation Cooperation

ApLC: Astrophysical Letters Communications (periodical code)

APLCN: Appalachian

APLDI: APL Data Interface

APLF: Association des Planétariums de Langue Française

APLM: Applications Program Library Maintenance

APLMF: Asia-Pacific Legal Metrology Forum

APLO: Automation Planning and Liaison Office

APLSS: Attached-Payload Space Station • Attached-Payload Support System

APLU: Automatic Program Loading Unit

APM: Acquisition Project Manager • Advanced Power Management • Aerosol Physics Module • Airborne Particulate Matter • Air Pollution Meteorologist • Akustische Plattenmode • Alarm Panel Monitor • Antenna Pointing Mechanism • Antenna Positioning Mechanism • Appended Pressurized Module • Area Prediction Model • Ascent Particle Monitor • Assistant Paymaster • Assistant Project Manager • Associate Program Manager • Associative Principle for Multiplication • Attached Pressurized Module (now CAL) • Automated Plate Machine • Automatic Photographic Measuring • Automatic Plate Measurement • Automatic Plate Measuring • Automatic Printing Machine

ApMaC: Applied Mathematics Computation (periodical code)

APMAC: Asia-Pacific Market Awareness Committee

ApMaL: Applied Mathematics Letters (periodical code)

ApMaM: Applied Mathematics and Mechanics (periodical code)

ApMaO: Applied Mathematics and Optimization (periodical code)

ApMat: Applied Mathematics (periodical code)

APME: Associative Processor Microelectronic Element

APMG: Assistant Postmaster General

APMI: Aerial Photography Micrographic Index

ApMic: Applied Microbiology (periodical code)

APMIS: Automatic Project Management Information System

ApMM: Applied Mathematics and Mechanics (periodical code)

APMM: Association of Policy Market Makers

APMMRI: Automatic Point Marking, Measuring and Recording Instrument

APMP: Acid Precipitation Mitigation Programme • Asia-Pacific Metrology Programme

APMR: Adaptive Pointable Multibeam Radar

ApMRv: Applied Mechanics Reviews (periodical code)

APMS: Automated Performance Measurement System • Automated Performance Measuring System

APMSM: Assembly Phase Maintenance Simulation Model

Apmt: Appointment

ApMT: Applied Microgravity Technology (periodical code)

APMT: Advanced Planetary Mission Technology • Antenna Pattern Measurement Test • Asia Pacific Mobile Telecommunications

APN: All Pass Network • Alpena (USA-MI) (location identifier) • Asia-Pacific Network • Authorized Part Number

ApNM: Applied Numerical Mathematics (periodical code)

ApNr: Astrophys. Norvegica (periodical code)

Apo: Apogee

APO: Access Protection Option • Advanced Planner and Optimizer • Analysis Procedure Object • Apache Point Observatory (USA-NM) • Apartado (Colombia) (location identifier) • A Part Of • Army Post Office • Australian Patent Office • Austrian Patent Office • Automated Purchase Order

APOA: Arctic Petroleum Operators Association

Apóc: Apócope

APOD: Astronomy Picture Of the Day

Apog: Apogee

APOLLO: Article Procurement with Online Local Ordering

APOMA: American Photo-Optical Manufacturers Association (USA) • American Precision Optics Manufacturers Association (USA)

APOP: Authentication Post Office Protocol

Apops: Apostrophe

ApOpt: Applied Optics (periodical code)

APOR: Advisory Panel for Operations Research

APOS: Advanced Polar Orbiting Satellite

APOTA: Automatic Positioning Telemetering Antenna

App: Apparat • Apparatus • Appareil • Apparent • Apparently • Appeal • Appearance • Appendice • Appendix • Applicable • Application • Applied • Apply • Appointed • Appointment • Apprehend • Apprentice • Approach • Approval • Approve • Approved • Approximate • Approximately • Approximation

AP&P: Advanced Plans & Programs

APP: Advanced Planetary Probe • Advanced Plans and Programs • Advanced Procurement Package • Advanced Project Planning • Aerospace Propulsion Products • Agence pour la Protection des Programmes (France) • Amyloid Precursor Protein • Analytic Pointing Performance • Antenna Position Programmer • Apollo Applications Programme • Application • Applications Portability Profile • Approach • Asapa (Papua New Guinea) (location identifier) • Astrophysics Payload • Atmospheric Physics Program • Atmospheric Pressure Perturbation • Automated Photographic Patrol • Automatic Plate Processor • Automatic Position Planning • Auxiliary Power Package • Auxiliary Power Plant

APPAG: Association Pour un Planétarium dans l'Agglomération Grenobloise (France)

AppAn: Applicable Analysis (periodical code)

Appar: Apparatus • Apparent • Apparently

APPC: Advanced Peer-to-Peer Communication • Advanced Program-to-Program Communication

Appd: Approved

APPEAL: Academic Press Print and Electronic Access License • Asia-Pacific Programme of Education for All

APPECS: Adaptive Pattern Perceiving Electronic Computer System

Append: Appendice • Appendix

APPF: Automated Payload Processing Facility

ApPhC: Applied Physics Communications (periodical code)

ApPhL: Applied Physics Letters (periodical code)

ApPhy: Applied Physics (periodical code)

APPI: Advanced Peer-to-Peer Internetworking

Appl: Appeal • Applicable • Application • Applied • Apply

APPLE: Advanced Propulsion Payload Effects • Applied Parallel Programming Language Experiment • Ariane Passenger Payload Experiment • Associative Processor Programming Language Evaluation

Applic: Application

Appl.Op.: Applied Optics

Appl.Opt.: Applied Optics

APPM: Accident Prevention Program Manager • Avant-Projet de Plan de Masse

Appmt: Appointment

APPN: Advanced Peer-to-Peer Networking

ApPP: Astrophysics and Plasma Physics

Appr: Apprentice • Approved • Approximately

Appro: Approbation • Approval

Approp: Appropriate • Appropriation

Approx: Approximate • Approximately

Apprp: Appropriate

Apprx: Approximate

Apps: Appendices

APPS: Advanced Planning Program Scheduling • Aerosol Physical Properties of the Stratosphere • Analytical Photogrammetric Positioning System • Analytical Photogrammetric Processing System • Atmospheric Pressure Plasma System • Auxiliary Payload Power System

APPSR: Aerial Photography Progress Summary Report

Appt: Appartement • Appoint • Appointed • Appointment

Appd: Appointed

Appx: Appendix

APQ: Absolutely-Positively-Quality • Actual Pointing Quality • Arapiraca (Brazil-AL) (location identifier) • Associação Portuguesa para a Qualidade (Portugal)

APQCO: Asian-Pacific Quality Control Organization

APQF: Aerial Photography Quad File

Apr: April • Aprile

APR: Acoustic Paramagnetic Resonance • Active Prominence • Active Prominence Region • Adaptive Pointing Radar • Advance Parts Release • Agency Procurement Request • Agency Progress Report • Airborne Profile Recorder • All-Purpose Room • Alternate Path Retry • Annual Percentage Rate • Annual Planning Report • Annual Progress Report • Annual Purchase Rate • Antenna Position Recorder • April • April River (Papua New Guinea) (location identifier) • Area Planning Report • Atmospheric Processes Research • Automatic Pattern Recognition • Automatic Power Reserve • Automatic Programming and Recording

APRA: Australian Postgraduate Research Award

APRC: Anno Post Roman Conditam

Aprch: Approach

APRM: Asian-Pacific Regional Meeting • Average Power Range Monitor

APRO: Aerial Phenomena Research Organization

Aprox: Aproximadamente

APRP: Adaptive Pattern Recognition Processing

APRS: Acoustic Position Reference System • Applied Physics Research Section • Automatic Position Reference System

APRSS: Australian Photogrammetric and Remote Sensing Society

Aprt: Airport

APRT: Advanced Productivity Research and Technology

Aps: Apus-Apodis (constellation)

A&PS: Administration & Program Support

APS: Accelerometer Parameter Shift • Accessory Power Supply • Acquisition Processor Subsystem • Active Pixel Sensor • Administration and Program Support • Advanced Photon Source • Advanced Photosynthetic System • Advanced Photo System • Advanced Planning and Scheduling • Advanced Planning System • Advanced Propellant System • Aerodynamic Particle Sizer • Aft Propulsion System • Air Pollution Syndrome • Airport Planning Standard • Air Pressure Switch • Albuquerque Public Schools (USA-NM) • Alpha Particle Spectrometer • Altitude Proximity Sensor • American Philosophical Society (USA) • American Physical Society (USA) (*) • American Physics Society (USA) • American Polar Society (USA) • Amplifier Power Supply • Anápolis (Brazil-GO) (location identifier) • Ångström Pyrheliometric Scale • Angular Position Sensor • Antenna Pointing System • Appearance Potential Spectroscopy • Arc-Plasma Spraying • Area Parameter Specification • Ascent Propulsion System • Astronomy and Planetary Science • Astrophysical and Planetary Sciences • Asynchronous Production System • Asynchronous Protocol Specification • Atmospheric Pollution Sensor • Atmospheric Pressure Sensor • Attitude Propulsion System • Australasian Planetarium Society • Australian Public Service • Automated Planning System • Automated Plate Scanner • Automated Production System • Automatic Page Search • Automatic Pilot System • Automatic Processing System • Automatic Program System • Automatic Protection Switch • Auxiliary Power Supply • Auxiliary Power System • Auxiliary Program Storage • Auxiliary Propulsion System • Auxiliary Publication Service • Average Particle Size

APSAT: Asia-Pacific Satellite

APSB: Advanced Planning Systems Branch • Advanced Program Systems Branch • Astronomy and Planetary Science Board

APSC: Asia Pacific Space Centre Pty Ltd (Christmas Island)

APSCC: Asia-Pacific Satellite Communication Council

ApScR: Applied Scientific Research (periodical code)

APSE: ADA Programming Support Environment

ApSF: Amperes per Square Foot

APSF: Amperes Per Square Foot • Australian Pacific Science Foundation

APSG: Asia-Pacific Space Geodynamics

APSGP: Asia-Pacific Space Geodynamics Project

ApSI: Amperes Per Square Inch

APSI: Advanced Propulsion System Integration • Aircraft Propulsion System Integration • Amperes Per Square Inch

APSIN: Alaska Public Safety Information Network

APSO: Asian Physical Society • Audio Precision System One

APSP: Array Processor Subroutine Package

ApSpE: Applied Spectroscopy (periodical code)

APSR: Aerial Photography Summary Record

APSRS: Aerial Photography Summary Record System

Ap&SS: Astrophysics & Space Sciences (periodical code)

ApSS: Applied Surface Science (periodical code)

APSS: Advanced Planetary Spacecraft System • Applied Physics A - Solids and Surfaces (periodical code) • Atmospheric Pressure Supply System • Automated Program Support System • Automatic Planning and Scheduling System

APST: Association of Professional Scientists and Technologists

APSTM: Analytical Photon Scanning Tunneling Microscope • Analytical Photon Scanning Tunneling Microscopy

ApSup: Applied Superconductivity (periodical code)

Apt: Airport • Apartment

AP&T: Advanced Physics & Technology

APT: Address Pass Through • Address Path Through • Advanced Patent Technique • Advanced Passenger Train • Advanced Physics and Technology • Advanced Process Technology • Airborne Pointer Tracker • All-Picture Transfer • Analog Program Tape • Analyseur de Plasma Thermique • Apartamento • Apartment • Applatissement • Application Productivity Tool • Applied Potential Tomography • Architectural Power Tool • Asia-Pacific Telecommunity • Asia Pacific Telescope • Association of Professional Translators (USA) • Astronaut Preference Test • Astronomer's Proposal Tool • Automated Photometric Telescope • Automated Programming Tool • Automatically Programmed Tool • Automatic Photoelectric Telescope • Automatic Photometric Telescope • Automatic Picture Transmission (*) • Automatic Position Telemetering • Automatic Programmed Tool • Automatic Programming of Tools • Automation Planning and Technology • Autonomous Picture Transmission • Jasper - Marion County (USA-TN) (location identifier)

Aptdo: Apartado

APTE: Automatic Production Test Equipment

ApTI: Actions per Time Interval

APTI: Actions Per Time Interval

APTIC: Air Pollution Technical Information Center

Apto: Apartamento

APTP: Arithmetic Proficiency Training Program • Asia-Pacific Technology Program

Apts: Apartments

APTS: Aerial Profiling of Terrain System • Automatic Picture Transmission System

APTU: African Postal and Telecommunications Union

APU: Accessory Power Unit • Apucarana (Brazil-PR) (location identifier) • Arab Postal Union • Arithmetic Processing Unit • Assessment of Performance unit • Automated Program Understanding • Auxiliary Power Unit • Auxiliary Propulsion Unit

APUC: Association des Presses Universitaires Canadiennes (Canada)

APUHS: Automatic Program Unit High Speed

APULS: Automatic Program Unit Low Speed

APUPA: Alien-Penumbral-Umbral-Penumbral-Alien

APUS: Auxiliary Power Unit System

APUT: Auxiliary Power Unit Test

Apv: Approve

APV: Algemene Politieverordening • Apple Valley (USA-CA) (location identifier)

Apvd: Approved

Apvl: Approval

APW: Abel-Poisson Wavelet • Apia Faleolo (Samoa) (location identifier) • Apparent Polar Wander • Approximation of Plane Waves

APWA: American Public Welfare Association (USA)

APWL: Automatically Processed Wire List

APWP: Apparent Polar Wander Path

APWR: Advanced Pressurized Water Reactor

Apx: Appendix

APX: Alpha Proton X-ray • Arapongas (Brazil-PR) (location identifier)

APXS: Alpha Proton X-ray Spectrometer

APY: Alto Parnaíba (Brazil-MA) (location identifier)

APZ: Zapala (Argentina) (location identifier)

Aq: Aqua • Aqueous • Aquatic

AQ: Accomplishment Quotient • Accumulator-Quotient • Achievement Quotient • Air Quality • Antarctica (ISO country code) • Antarctique (ISO country code) • Assurance de Qualité • Astronomy Quarterly • Astrophysical Quantities • Astroqueyras (France) • Attainment Quotient • Authorized Qualifyer

AQA: Agence de la Qualité de l'Air • Air Quality Act (USA) • American Quaternary Association (USA) • Approved Quality Assurance • Araraquara (Brazil-SP) (location identifier) • Assessment and Qualifications Alliance (UK)

AQAB: Air Quality Advisory Board

AQAFO: Aeronautical Quality Assurance Field Office

AQAG: Air Quality Advisory Group

aq. bull.: Aqua Bulliens

AQC: Automatic Quality Control • Automatic Quench Calibration • Automatic Quench Correction • Automatic Quench Compensation

aq. cal: Aqua Calida

AQCD: Air Quality Criteria Document

AQCEN: Accès à la Qualité, à la Certification et à la Normalisation

AQCL: Analytical Quality Control Laboratory

aq. com.: Aqua Communis

aq. comm.: Aqua Communis

AQD: Air Quality Data

aq. dest.: Aqua Destillata

aq. ferv.: Aqua Fervens

aq. frig.: Aqua Frigida

AQDHS: Air Quality Data Handling System

AQE: Absolute Quantum Efficiency

AQG: Anqing (China-PRC) (location identifier)

AQI: Air Quality Index • Qaisumah (Saudi Arabia) (location identifier)

AQIS: Air Quality Information System

AQJ: Aqaba (Jordan) (location identifier)

Aql: Aquila-Aquilae (constellation)

AQL: Acceptable Quality Level • Astronomical Query Language • Astronomy Query Language • Average Quality Limit

AQM: Additive Quark Model • Air Quality Management • Air Quality Model • Ariquemes (Brazil-RO) (location identifier) • Atmospheric Quality and Modification

AQN: Azimuthal Quantum Number

AQP: Advanced Qualification Programme • Arequipa - Rodriguez Ballón (Peru) (location identifier)

AQPU: Association Québecoise des Presses Universitaires (Canada-QC)

aq. pur.: Aqua Pura

Aqr: Aquarius-Aquarii (constellation)

AQR: Acceptable Quality Rate

AQS: Air Quality Subsystem • Approximate Quadratic Search • Saqani (Fiji) (location identifier)

Aqstn: Acquisition

AQT: Acceptable Quality Test • Association Québecoise de Télédétection (Canada-QC) • Aviation Qualification Test

aq. tep.: Aqua Tepida

Aque: Aqueduct

Aqued: Aqueduct

AQY: Alyeska (USA-AK) (location identifier)

a/r: Accounts Receivable

ar: Arabic (ISO language code) • Argentina

aR: Am Rhein

Ár: Árabe

Ar: Arabia • Arabian • Arabic • Aragón (Spain) • Argon • Arrival • Arrive • Arrived • Arrives • Arroyo • Aryl Group

A/R: Accounts Receivable • Accusé de Réception • All Risks • Anno Regni • As Required • At the Rate of • Autoland/Rollout • Azimuth/Range

A&R: Artists & Repertory • Assault & Robbery • Assembly & Recycle • Automation & Robotics

AR: Acceptance Readiness • Acceptance Review • Accomplishment Ratio • Account Receivable • Accusé de Réception • Achievement Ratio • Acid Resisting • Acknowledgement of Receipt • Acquisition Radar • Acrylic Rubber • Action Register • Action Request • Active Radical • Active Region • Activity Report • Acute Rejection • Address Register • Administrative Ruling • Advanced Reactor • Advanced Robotics • Advice of Receipt • Advisory Report • Aerodynamic Report • Age Replacement • Aircraft Rocket • Alarm Report • Aller et Retour • Allocated Reserve • All Risks • Alternate Route • Altesse Royale • Anna Regina • Annual Register • Annual Report

• Annual Return • Annual Review • Annual Reviews Inc. • Anode Reaction • Anomaly Report • Anti-Reflection • Anti-Reflective • Application Region • Applied Research • Approved for Release • April • Arabia • Archival Research • Archival Researcher • Argentina (ISO country code) • Argentine (ISO country code) • Arkansas (USA) • Arrêté Royal • Arrival and Return • Arrives • Arriving • Ascención Recta • Aspect Ratio • As Required • Assembly and Recycle • Association Régionale • Astronomy-Related • Atmosphere Revitalization • Augmented Reality • Automation and Robotics • Autonomous Region • Autoregression • Autoregressive • Average Rating • Average Revenue • Avis de Réception • Avril

Ara: Ara-Arae (constellation)

ArA: Arkiv for Astronomi (periodical code)

ARA: Abbreviated Registered Address • Academic and Research Affairs • Accelerated Readiness Analysis • Address Register Area • Advanced Radar Altimeter • Advanced Rigid Array • Aerospace Research Association • Analyse du Rayonnement Atmosphérique • Angular Rate Assembly • Apple Remote Access • AppleTalk Remote Access • Associate of the Royal Academy • Association for Research Astronomy • Associazione Reggiana Astrofili (Italy) • Associazione Romana Astrofili (Italy) • Atmospheric Radiation Analysis • Attitude Reference Assembly • Austral (Argentina) (ISO currency code) • Auxiliary Recovery Antenna • Average Response Amplitude • New Iberia - Acadiana (USA-LA) (location identifier)

ARA&A: Annual Review of Astronomy & Astrophysics (periodical code)

ARAA: Annual Review of Astronomy and Astrophysics

Arab: Arabia • Arabian • Arabic

AraBN: Arab Biosciences Network

ARABSAT: Arab Satellite Communication Organization

ARAC: Aerospace Research Applications Center • Army Radar Approach Control • Atmospheric Release Advisory Capability • Aviation Rulemaking Advisory Committee

ARAD: Average Response Amplitude Data

Arag: Aragonés

ARAG: Antireflective-Antiglare

ARAL: Automatic Record Analysis Language

ARALL: Aramide Aluminium Laminate

ARALR: Association Régionale d'Astronomie pour le Languedoc-Roussillon (France)

ARAMCO: Arabian American Co.

ARAMIS: Adaptation du Radar À la Météorologie Infra-Synoptique • Antarctic Remote-sensing Aerial-photography and Mapping Information System • Antarctic Research in Applied Meteorology, Imaging and Sounding • Atmospheric Radiation And Minor Species

ARAMP: Arctic Remote Autonomous Measurement Platform

ARAP: Age-Rotation-Activity Paradigm • Association de Recherche et d'Action Pédagogique • Astronaut Rescue Air Pack • Automated Reliability Assessment Program

ArApM: Archive of Applied Mechanics (periodical code)

ARAR: Accident Risk Assessment Report • Applicable or Relevant and Appropriate Requirement • Associazione Ravennate Astrofili Rheyta (Italy)

ARARAT: Advanced Radiometric Ray Tracer

ARAS: Antireflective - Antistatic • Atomic Resonance Absorption Spectroscopy

ARAT: Avion de Recherche Atmosphérique et de Télédétection

Arauc: Araucano

Arb: Arbeit • Arbeiter • Arbitrer • Arbitrageur • Arbitraire • Arbitrary • Arbitration • Arbitrator

ARB: Adaptive Rate-Based • Address Recorder Buffer • Aerospace Research Bulletin (AIAA) • Air Registration Board • Ann Arbor (USA-MI) (location identifier)

ARBIRS: Advanced Region-Based Image Retrieval System

arc: Arcus

Arc: Arcade • Arcaico

A&RC: Application & Resource Control

ARC: Accelerating Rate Calorimeter • Action de Recherche Concertée • Active Radar Calibration • Active Radar Calibrator • Active Radical Combustion • Administrative Radio Conference • Advanced Radial Camera • Advanced Reentry Concept • Advanced Remote Control • Advanced RISC Computing • Aeronautical Research Council • Aerospace Remote Calculator • Aerospace Research Chamber • Aggregation of Red Cells • Alberta Research Council (Canada) • Altitude Rate Command • American Red Cross (USA) • Ames Research Center (NASA) • Analog Response Conditioner • Analyzer-Response-Controller • Antarctic Research Center • Antireflection Coating • Application and Resource Control • Applied Research Corporation • Arcade • Archive • Archived • Arctic Research Commission • Arctic Village (USA-AK) (location identifier) • Area Resources Center • Association pour la Recherche Cognitive (France) • Astrophysical Research Consortium (USA) • Astrophysics Research Center • Atelier de Recherche et de Création • Atlantic Research Corporation • Atmospheric Radiance Code • Attached Resources Computing • Australian Research Council • Automated Rendezvous and Capture • Automatic Rate Control • Automatic Ratio Control • Automatic Reconfiguration Criterion • Automatic Relay Calculator • Automatic Remote Control • Automatic Responsivity Correction • Auxiliary Roll Control • Average Resonance Capture • Average Response Computer

ARCA: Advanced RISC Computing Architecture

ARCAD: Arctic Auroral Density

ARCADE: Absolute Radiometer for Cosmology, Astrophysics, and Diffuse Emission • Ampere Remote Control Access Data Entry • Architectural Computer-Aided Design and Evaluation

ARCANE: Action de Recherche sur la Circulation en Atlantique Nord-Est

ARCB: Arcebispo

ARCC: Analysis of Recent Climate Change

ARCCOH: Asian Regional Coordinating Committee on Hydrology

ARCDR: Altimetry and Radiometry Composite Data Record

ARCFPCS: Algorithm with Residual Cancellation in Functionals and at Points of Control Switching

Arch: Archaic • Archaism • Archbishop • Archery • Archipel • Archipelago • Archipiélago • Architect • Architectural • Architecture • Archive • Archives • Archivo

Archbp: Archbishop

Archel: Archipel

ARCH: Argonne/Chicago Development Corporation • AutoRegressive Conditional Heteroscedasticity

Archaeol: Archaeological • Archaeology

Archbp: Archbishop

ARCHIMEDES: Advanced Research in Communications using Highly Inclined orbits for Mobile Experiments and Demonstration with European Satellites

ARCHV: Archive

ARCISS: Association of Research Centres in the Social Sciences (UK)

ARCL: Allowable Residual Contamination Level

arcmin: Minute of Arc

ARCN: Attached Resource Computer Network

ARCnet: Attached Resource Computing Network • Attached Resource Computer Network

arcos: Arc Cosine

ARCOS: Area Correlation for Determination of Object Surfaces

arcot: Arc Cotangent

ARCQUERY: Archive Query

ARCS: Admiralty Raster Chart Service • Aft Reaction Control System • Altitude Rate Command System • Arcminute Radio Cluster-lens Search • Associate of the Royal College of Science • Atmospheric Radiation and Cloud Station • Average Rapidness of Citation Successions

arcscs: Arc Cosecant

arcsec: Arc Second • Second of Arc

arcsin: Arc Sine

ARCSS: Arctic Systems Science

ARCST: Associate of the Royal College of Science and Technology

arctan: Arc Tangent

ARCU: American-to-Russian Converter Unit

ARCUS: Arctic Research Consortium of the United States

Ard: Ardennes (France)

ARD: Abort Region Determinator • Absolute Radiation Detector • Acute Radiation Disease • Acute Respiratory Disease • Aeronautical Research Division • Alor Island (Indonesia) (location identifier) • Analyse Relationnelle de Données • Arbeitsgemeinschaft der öffentlich-rechtlichen Rundfunkanstalten der Bundesrepublik Deutschland (Germany) • Arrière Droit • ASCII Region Definition • Atmospheric Recovery Demonstrator • Atmospheric Reentry Demonstrator (*) • Automatic Reentry Demonstrator • Average Response Data

ARDA: Analog Recording Dynamic Analyzer

ARDB: Arctic Runoff Database

ARDIS: Advanced Radio Data Information Service

ARDMS: Automated Route Design and Management System

ARDS: Advanved Remote Display Station

Ardt: Arrondissement

ARDU: Aircraft Research and Development Unit

ARE: Acoustic Radiation Element • Activated Reactive Evaporation • Admiralty Research Establishment (UK) • Algebraic Riccati Equation • Antenna Range Equipment • Arab Republic of Egypt • Arecibo (Puerto Rico) (location identifier) • Assemblée des Régions d'Europe • Asymptotic Relative Efficiency • ATM Routing Engine • Automatic Record Evaluation • Auxiliary Rocket Engine • Émirats Arabes Unis (ISO country code) • United Arab Emirates (ISO country code)

AREA: Angle Resolving Energy Analyzer

AREAL: Atmospheric Research and Exposure Assessment Laboratory

AREAS: Area Resource Analysis System

AREDOR: Association de Recherche et d'Exploitation de Diamant et d'Or

AREG: Antarctic Research Evaluation Group

ARELAS: Arequipa Laser Station (Peru)

ARELS: Association of Recognized English Language Services

AREN: Australian Regional Geodetic Network

ARENTS: ARPA Environmental Test Satellite

AREP: Atmospheric Research and Environment Programme

AREPS: Annual Review of Earth and Planetary Sciences (periodical code)

ARES: Advanced Railroad Electronic System • Advanced Research EMP Simulator • Architectural Reasoning for Embedded Software • Association pour la Recherche sur l'Environnement Spatial • Automated Runway Evaluation System • Automatic Record Evaluation System

ARESCE: Association Royale des Étudiants en Sciences Commerciales et Économiques

ArEVO: Arbeitsentgeltsverordnung

ARF: Abort-Retry-Fail • Acaricuara (Colombia) (location identifier) • Acute Respiratory Failure • Aeronautical Research Foundation • Aerospace Recovery Facility • Airport Reservation Function • Ancillary Response File • Ancillary Response Function • Anticipated Receive File • Aquatic Research Facility • Automatic Reporting Feature

ARFA: Allied Radio Frequency Agency

ARFDS: Automatic Reentry Flight Dynamics Simulator

ARFE: Association des Régions Frontalières d'Europe • Association des Régions Frontalières Européennes

ARFF: Airport Rescue and Firefighting

ARFM: Annual Review of Fluid Mechanics (periodical code)

Arg: Argent • Argentina • Argentine • Argentinian

Årg: Årgang

ARG: Air Rescue Group • Argentina (ISO country code) • Argentine (ISO country code) • Arrière Gauche • Attributed Relational Graph • Walnut Ridge (USA-AR) (location identifier)

Argen: Argentina • Argentine

Argent: Argentina • Argentine

ARGO: Astrophysical Radiation and Gamma Observatory

ARGOS: Advanced Remote Gaseous Oxides Sensor • Advanced Research and Global Observation Satellite

ARGOSI: Application-Related Graphics and OSI Standards Integration

ARGUS: Advanced Real-time Gaming Universal Simulation • Automatic Routine Generating and Updating System

a/Rh: Am Rhein

a. Rh.: Am Rhein

ARH: Arkhangelsk (Russia) (location identifier)

Ari: Aries-Arietis (constellation)

ARI: Academic Research Infrastructure • Accelerated Research Initiative • Acute Respiratory Infection • Advanced Research Initiative • Airborne Remote Instrumentation • Annual Recurrence Interval • Annual Reviews Inc. (USA-CA) • Arica Chacalutta (Chile) (location identifier) • Astronomisches Rechen-Institut (Germany) • Astronomy Research and Instrumentation • Automated Readibility Index • Automatic Return Item

ARIA: Advanced Range Information Aircraft • Advanced Range Instrumentation Aircraft (*) • Aeroflot Russia International Airlines • Aeroflot Russian International Airlines • Atmospheric Response In Aurora

ARIC: Acid Rain Information Centre

ARIDPCM: Adaptive Recursive Interpolated Differential Pulse Code Modulation

ARIES: Antarctic Reception Imagery for Environmental Studies • Applied Resource Image Exploitation System • Association de Recherche, d'Information et d'Échanges Européens • Astronomical Radio Interferometric Earth Survey • Astronomical Radio Interferometric Earth Surveying • Astronomy Resources for Intercurricular Elementary Science • Atmospheric Research Information Exchange Study • ATM Research and Industrial Enterprise Study • Australian Resources Information and Environment Satellite • Authentic Reproduction of an Independent Earth Satellite

ARIGC: Astronomisches Rechen-Institut General Catalogue

ARIMA: Autoregressive Integrated Moving Average

ARIN: Aerospace Research Information Network • American Registry of Internet Numbers

ARINC: Aeronautical Radio Inc.

ARIPO: African Regional Industrial Property Organization

ARIS: Advanced Range Instrumentation Ship • Aggregate Route-based IP Switching • Alabama Resource Information System • Arizona Resource Information System • Atomic Reactor In Space • Attitude and Rate Indicating System • Australian Resources Information System • Authority-Responsibility-Information-Support • Automatic Recording Infrared Spectrometer • Automatisches Retrieval Informationssystem

ARISAL: Association Régionale des Ingénieurs et des Scientifiques d'Alsace (France)

ARISE: Advanced Radio Interferometry between Space and Earth • Agricultural Resources Inventory and Survey Experiment • A Reusability Infrastructure for Software Engineering

Arist: Aristotle (384-322 BC)

ARIST: Agence Régionale d'Information Scientifique et Technique (France) • Annual Review of Information Science and Technology

Aristoph: Aristophanes (c. 445 - c. 380 BC)

ARISTOTE: Association de Réseaux Informatiques en Système Totalement Ouvert et Très Élaboré

ARISTOTELES: Application and Research Involving Space Techniques Observing The Earth fields from Low-Earth-orbit Spacecraft • Application and Research Involving Space Technologies Observing The Earth's field from Low Earth orbit Satellite

Arit: Aritmética

Arith: Arithmetic • Arithmetical • Arithmetician

Ariz: Arizona (USA)

ARJ: Arso (Indonesia) (location identifier)

Ark: Arkansas (USA)

ARK: Arusha (Tanzania) (location identifier)

ARL: Acceptable Reliability Level • Adaptive Representation through Learning • Adjusted Ring Length • Admiralty Research Laboratory • Aeronautical Research Laboratory • Aerospace Research Laboratory • Air Resources Laboratory • AKZO Research Laboratories • Applied Research Laboratory • Arctic Research Laboratory • Arly (Burkina Faso) (location identifier) • Army Research Laboratory (USA) • Association of Research Libraries • Astrophysics Research Laboratory • Australian Radiation Laboratory • Average Remaining Lifetime • Average Run Length

ARLA: Association of Residential Letting Agents

ARLC: Area Resource Learning Center

ARLL: Advanced Run Length Limited

ARLM: Airborne Reconnaissance Low-Multifunction

ARLTAB: Army Research Laboratory Technical Assessment Board

Arm: Armagh (UK-Northern Ireland) • Armenia • Armenian • Armorican

ArM: Master of Architecture

ARM: Aachen Robot Manipulator • Acorn RISC Machine • Adjustable-Rate Mortgage • Advanced RISC Machine • Airborne Remote Mapping • Algorithmic Remote Manipulation • All-Risk Management • Amateur Radio Monitor • Angular Resolution Measure • Anhysteretic Remanent Magnetization • Annotated Reference Manual • Application Reference Model • Application Response Measurement • Applied Research Management • Area Radiation Monitor • Armagh (Starlink internal code) • Armenia (ISO country code) • Arménie (ISO country code) • Armidale (Australia-NSW) (location identifier) • Asynchronous Response Mode • Atmospheric Radiation Measurement • Atmospheric Radiation Monitoring • Atomic-Resolution Microscope • Atomic-Resolution Microscopy • Automated Route Management • Automatic Restart Management • Automatic Restart Manager

ARMA: Accumulator-Reservoir Manifold Assembly • Alarmstelle für Radioaktivität an der Meteorologischen Anstalt • Association of Records Managers and Administrators • Autoregressive and Moving Average • Autoregressive Moving Average (*)

ARMADE: Automation and Robotics for Microgravity Applications Demonstrator

ARMAR: Airborne Rain Mapping Radar

ARMCAS: Arctic Radiation, Microphysics, Clouds and Aerosols Study

ARME: Automated Reseau Measuring Equipment

Armen: Armenia • Armenian

ARMM: Analysis and Research of Methods for Management • Automated Retroactive Minimal Moderation

ARMMS: Automatically Reconfigurable Modular Multiprocessing System • Automatically Reconfigurable Modular Multiprocessor System

A&RMOWG: Astronomy & Relativity Mission Operations Working Group

ARMS: Absolute Radiometric Measurements in Space • ADPE Resources Management System • Advanced Receiver Model System • Advanced Retrieval Mail System • Advanced Robotics Manipulation System • Aerial Radiological Measurements and Survey • Aerial Radiological Measuring System • Application of Remote Manipulators in Space • Automated Record Management Study • Automated Requirements Management System • Automated Retrieval Mail Service • Automated Retrieval Mail System • Automatic Radiation Monitoring System • Automatic Receiving and Measuring System • Automatic Remote Manned System • Aviation Reconfigurable Manned Simulator

ARMSAT: Atmospheric Radiation Measurement Satellite

ARMU: Addressable Remote Multiplexer Unit

ARMZ: Armazém

ARN: Access Remote Node • Acide Ribonucléique • Action and Reply Notice • Additional Reference Number • Alteration Request Number • Astronomical Research Network • Atmospheric Radio Noise • Stockholm Arlanda (Sweden) (location identifier)

ARND: Around

ARNEE: ASEAN Region Network on Environmental Education

ARNEWS: Acid Rain National Early Warning System

ARNFM: Annual Review of Numerical Fluid Mechanics (periodical code)

Arngmt: Arrangement

ARNICA: Arcetri Near-Infrared Camera

ARO: Advanced Research Objective • After Receipt of Order • Airport Reservation Office • Algonquin Radio Observatory (Canada) • Applied Research Objective • Arboletas (Colombia) (location identifier) • Asian Regional Organization • Automatic Range Only • Autonomous Robotic Observatory

AROC: Ariane Recoverable Orbital Carrier

AROD: Airborne Radar Orbital Determination • Airborne Ranging and Orbit Determination • Airborne Remote Operated Device

ARODS: Airborne Radar Orbital Determination System • Airborne Ranging and Orbit Determination System

AROF: Ambiguity Resolution On the Fly

AROM: Alterable Read-Only Memory

AROS: African Regional Organization for Standardization • Automated Regional Operations System

ARP: Address Resolution Protocol • Adjustable-Rate Preferred • Advanced Reentry Program • Advanced Research Project • Aerospace Reference Project • Airport Reference Point • Alternativa Racional a la Pseudociencias • Alternative Release Procedure • Antarctic Research Program • Antenna Radiation Pattern • Aragip (Papua New Guinea) (location identifier) • Arctic Research Plan • As-Run Procedure • Atmospheric Research Program • ATV Rendezvous Predevelopment • Peso (Argentina) (ISO currency code)

ARPA: Advanced Research Projects Agency (USA) (formerly DARPA) • Arctic Research and Policy Act • Automatic Radar Plotting Aid

ARPANet: Advanced Research Projects Agency Network

ARPAS: Automated Resource Planning and Analysis System

ARPCS: Atmospheric Revitalization Pressure Control System

ARPD: Advanced Research Program Directive

ARPDP: Association of Rehabilitation Programs in Data Processing

ARPDS: Auto-Regressive Power Density Spectrum

ARPEGE: Action de Recherche Petite Échelle et Grance Échelle

ARPEGES: Astronomie Radio Planétaire, Extragalactique, Galactique et Stellaire (Observatoire de Paris-Meudon)

ARPES: Angle-Resolved Photoelectric Spectroscopy • Angle-Resolved Photoelectron Spectroscopy

ARPESH: Accurate and Reliable Prototype Earth Sensor Head

ARPIC: Aerospace Radioisotope Power Information Center

Arpk: Airpark

ARPL: Adjust Requested Privilege Level

ARpM: Average Revenue per Message

ARPM: Average Revenue Per Message

ARPR: Annual Reviews Preprints and Reprints • Automatic Radar Pattern Recognition

ARPS: Advanced Radar Processing System

Arpt: Airport

Arq: Arquetipo • Arquitecto • Arquitectura

ARQ: Arauquipa (Colombia) (location identifier) • Arquitecto • Automatic Repeat-Request • Automatic Request

Arqto: Arquitecto

Arqueol: Arqueología

Arr: Arrange • Arranged • Arrangement • Arranger • Arrecife • Arrêté • Arrival • Arrive • Arrived • Arrivée • Arriver • Arrives • Arrondissement

ARR: Accounting Rate of Return • Advanced Requirements Review • Advanced Restricted Report • Advanced Robotics Research • Advance Release Record • Aerospace Rapid Response • Aerospace Rescue and Recovery • Alto Rio - Senguerr (Argentina) (location identifier) • Andøya Rocket Range (Norway) • Anno Regni Reginae • Anno Regni Regis • Antenna Radiation Resistance • Antenna Rotation Rate • Archive Readiness Review • Automatic Repeat Request

ARRC: Aerospace Rescue and Recovery Center

ARRCC: Analysis of Rapid and Recent Climate Change

Arrdt: Arrondissement

ARRE: Autoread Relocation

ARRECAD: Arrecadaçao

Arrest: Arresting

ARRF: Automatic Recording and Reduction Facility

Arrgt: Arrangement

ARRL: American Radio Relay League (USA)

ARRM: Arruamento

ArRMA: Archive of Rational Mechanics and Analysis (periodical code)

Arron: Arrondissement

Arrond: Arrondissement

ARRQ: Automatic Repeat Request

ARRS: Advanced Rescue and Recovery System • Aerospace Rescue and Recovery Service • Air Recovery and Rescue Service • American Roentgen-Ray Society (USA) • Attitude-Referenced Radiometer Study

ARRSTC: Asian Regional Remote Sensing Training Centre

ARRT: Absolute Reaction Rate Theory

ARS: Active Repeater Satellite • Acute Radiation Syndrome • Advanced Reconnaissance Satellite • Advanced Reconnaissance System • Advanced Record System • Advanced Reentry System • Advanced Rescue System • Aeronautical Research Scientist • Aerospace Research Satellite • Agricultural Research Service • Airborne Radiometer Scanner • Air Rescue Service • Air Revitalization System • Alpha Ray Spectrometer • Alternate Route Selection • Aménagement des Rythmes Scolaires • American Radium Society (USA) • American Rescue Service (USA) • American Rocket Society (USA) (now AIAA) • Analog Recording System • Angular Rate Sensor • Antenna Receiving Station • Aragarças (Brazil-GO) (location identifier) • Archive and Retrieval System • Area Scan • Assembleo Regionale Siciliana (Italy) • Atmosphere Revitalization Section • Atmosphere Revitalization System • Atmospheric Revitalization System • Atmospheric Radiance Study • Attitude Reference System • Audio Response System • Automated Retrieval System • Automatic Recording Spectrometer • Automatic Route Selection

ARSA: Airport Service Radar Area • Automated Radioxenon Sample Analyzer

ARSAS: Agricultural Remote Sensing Analyzing System

ARSC: African Remote Sensing Council • Alberta Remote Sensing Centre • Analog Rotation Speed Control • Association des Revues Scientifiques et Culturelles (Belgium)

ARSDA: Advanced Radiation Space Defense Application

ARSE: Alpha Ray Spectrometric Equipment

ARSEC: Association of Roman Catholics for Space Exploration and Colonization (USA-MA)

ARSFSS: Advanced Reduced Scale Fuel System Simulator

ARSL: Atmospheric Rendezvous Space Logistics

ARSLOE: Atlantic Remote Sensing Land Ocean Experiment

ARSM: Algebraic Reynolds Stress Model

ARSO: African Regional Standards Organization

ARSP: Aerospace Research Support Program • African Remote Sensing Programme • Analog Radar Signal Processor • Applied Remote Sensing Program

ARSR: Air Route Surveillance Radar

Art: Article • Artículo • Articulo • Artificial • Artikel • Artist

ART: Abteilung Reprotechnik • Adaptive Refinement Tree • Adaptive Resonance Theory • Admissible Rank Test • Advanced Research and Technology • Advanced Research Technologies • Advanced Resolution Technology • Airborne Radiation Thermometer • Algebraic Reconstruction Technique • Allgemeine Relativitätstheorie • Alpha Recoil Track • Animated Reconstruction of Telemetry • Arithmetic Reasoning Test • Article • Audible Ringing Tone • Automated Reasoning Tool • Automated Request Transmission • Automatic Radiotheodolite • Automatic Range Tracker • Automatic Range Tracking • Autorité de Régulation des Télécommunications (France) • Avionics Research Testbed • Watertown (USA-NY) (location identifier)

ARTA: Apple's Real-Time Architecture • Ariane Research and Technology Accompaniment • Association pour la Recherche et la Technologie en Alsace (France)

ARTC: Air Route Traffic Control • Annual Regional Telecommunications Conference

ARTCC: Air Route Traffic Control Centre

ARTE: ADA Run Time Environment

ARTEMIS: Advanced Relay and Technology Mission Satellite (*) • Advanced Relay Technology Mission • Advanced Research and Technology Mission • African Real-Time Environmental Monitoring Information System • African Real-Time Environmental Monitoring using Imaging Satellites • Antimatter Research Through the Earth Moon Ion Spectrometer

ARTEP: Ariane Technology Experiment Platform

ARTES: Advanced Research and Telecommunications Systems • Advanced Research in Telecommunications Systems (*)

ARTFL: American Research on the Treasury of French Language

ARTIC: A Real-Time Interface Coprocessor

Artif: Artificial • Artificiel

ARTIM: Automated Reasoning Tool for Information Management

ARTIS: Airborne Real-Time Imaging System

ARTIST: Astronomy-Related Teacher In-Service Training

ARTK: ADA Real-Time Kernel

ARTO: Advanced Radiation Technology Office

ARTOOS: Advanced Radiation Technology Office Optical Support

ARTS: Adapted Real-Time Strategy • Advanced Radiofrequency Tracking System • Advanced Remote Tracking Station • Annual Research Task Summary • Asynchronous Remote Takeover Server • Automated Radar Terminal System • Automated Resource Tracking System • Automated Restricted Tracking System • Automated Retrieval and Tracking System • Autonomous Range and Target Simulator

ArtSa: Artificial Satellites (periodical code)

ARTSE: Ada Run-Time Support Environment

ARTSIP: Adaptive Real-Time Strategies for Image Processing

ARTSS: Advanced Real-Time Simulation System

ARTT: Aménagement de la Réduction du Temps de Travail • Asynchronous Remote Takeover Terminal

ARTTel: Automatic Request Transmission by Telephone

ARU: Analog Remote Unit • Analog Response Unit • Araçatuba (Brazil-SP) (location identifier) • Attitude Reference Unit • Audio Response Unit

ARUPS: Angle-Resolved Ultraviolet Photoemission Study

ARV: Aeroballistic Reentry Vehicle • Aerospace Research Vehicle • American Revised Version • Arctic Research Vessel • Ascent and Reentry Vehicle • Minocqua (USA-WI) (location identifier)

ARVAD: Armagh (Starlink internal code)

ARVISA: Advanced Real-time Vision System and Architecture

ARW: Advanced Research Workshop • Arad (Romania) (location identifier) • Associated Random Walk • Atmospheric Radio Wave • Attitude Reaction Wheel

ARX: Asbury Park (USA-NJ) (location identifier) • Autoregressive with External input

ARXPS: Angle-Resolved X-ray Photoelectron Spectroscopy

ARY: Ararat (Australia) (location identifier)

ARZ: Autorestricted Zone • N'Zeto (Angola) (location identifier)

AR5I/F: Ariane 5 Interface

a/s: Au Sujet de • Aux Soins de

as: Aanstaande • Assamese (ISO language code)

As: Altostratus • Ampere Second • Arsen • Arsenic • Asia • Asian • Asiatic • Asociación • Astronomische Schemering • Asymmetrical

A/S: Account Sales • Acquisition Sensor • Advanced Supplementary • After Sight • Airspeed • Aktieselskab • Aktjeselskab • Alongside • Analysis/Synthesis • Arm/Safe • Ascent Stage • Auxiliary Stage

A&S: Accident & Sickness • Alignment & Stabilization

AS: Absorbing Substrate • Academy of Science • Accident de Service • Accumulator Sampler • Acquisition Section • Action Spécifique • Adapter Section • Additional Source • Additional Stars • Add-Subtract • Administrative Support • Admittance Spectroscopy • Advanced Schottky • Advanced System • Aerospace Standard • Aerospace Study • Aft Shroud • Agent de Service • Airspeed • Alignment and Stabilization • All Sections • Almost Surely • Altostratus • Aluminium Silicate • American Samoa (ISO country code) • American Standard • Ammunition Ship • Amorph Sport • Amorphous Sport • Analog Signal • Analog State • Angle Sidéral • Angulo Sidéreo • Anno Salutis • Anno Salvatoris • Anodic Stripping • Answer Supervision • Antenna System • Antisolar • Antispoof • Antispoofing • Anti-Sun • Aperture Synthesis • Apogee Stage • Applicability Statement • Applied Science • Applied Statistics • Area Source • Artificial Satellite • Ascent Stage • Asia • Assistant Secretary • Associate in Science • As Stated • Asymptotic Bias • Atmosphere and Space • At Sight • Attitude Set • Augmentation System • Australian Standard • Authentification Server • Automated Storage • Automatic Skin • Automatic Sprinkler • Autonomous System • Auxiliary Stage • Samoa (American_) (ISO country code) • Samoa Américaine (ISO country code)

ASA: Abort Sensor Assembly • Acetylsalicylic Acid • Acoustical Society of America • Adapter Service Area • Adjustable Shock Absorber • Advanced System Avionics • Aerospace States Association • Aerosurface Amplifier • Aerosurface Servoamplifier • Agency System Administrator • Air Starline AG • American Standards Association (USA) (now ANSI) • American Statistical Association (USA) • American Student Association (USA) • Amplifier and Switch Assembly • Antarctic Support Associates • Antenna Subassembly • Apprentissage Symbolique Automatique • Assab (Ethiopia) (location identifier) • Association Sammielloise d'Astronomie (France) • Association Scientifique d'Astronomie (Algeria) • Association Sologne Astronomie (France) • Astroclub Solaris Aarau (Switzerland) • Astronomical Society of Australia • Astronomical Society of the Atlantic (USA-FL) • Atmospheric Spectroscopy Application • Atomic Sphere Approximation • Austrian Space Agency • Automatic Steering Antenna • Automatic Systems Analysis • Azimuth Servo Assembly

ASAA: Association Saranaise des Astronomes Amateurs (France)

ASAAS: Asymmetric Stress Analysis of Asymmetric Solids

ASAC: Active Satellite Attitude Control • Aerodynamic Surfaces Assembly and Checkout • Altostratus and Altocumulus • Antarctic Science Advisory Council • Application-Specific Architecture Compilation • Australian Scientific Advisory Committee

AsAcS: Astronautica Acta Supplement (periodical code)

ASACS: Advanced Secure Access Control System

ASAE: American Society for Aerospace Education (USA) • American Society of Agricultural Engineers (USA)

AsAer: Astronautics and Aeronautics (periodical code)

ASAIHL: Association of Southeast Asian Institutions of Higher Learning

ASAJ: Acoustical Society of America Journal (periodical code)

ASAL: Arid and Semiarid Land

ASAMS: Automatic Structure Analysis of Mass Spectra

ASANCA: Advanced Study of Active Noise Control in Aircraft

ASAP: Active Scattering Aerosol Spectrometer • Advanced Scientific Array Processor • Advanced Sensor Analysis Program • Advanced System Analysis Package • Aerospace Safety Advisory Panel • AFGL Statistical Analysis Programs • Airborne Science and Applications Program • Amsterdam Special Action Programme • Analog System Assembly Pack • Any Service - Any Port • Any Service to Any Port • Ariane Structure for Auxiliary Payload • As Soon As Possible • Automated Shipboard Aerological Program • Automated Statistical Analysis Program • Automatic Self-Allocating Processor • Automatic Switching And Processing

ASAPTS: Automated Scheduling And Product Tracking System

ASAR: Advanced Synthetic Aperture Radar

ASARS: Advanced Synthetic Aperture Radar System • Airborne Search-And-Rescue System

ASAS: Active Scattering Aerosol Spectrometer • Advanced Solid Axial Stage • Advanced Solid-state Array Sensor • Advanced Solid-state Array Spectrometer • Advanced Solid-state Array Spectroradiometer • Aerodynamic Stability Augmentation System • All Source Analysis System • Australian Science and Application Spacecraft • Aviation Safety Analysis System

ASASAC: Asociación de Astrónomos Autodidactos de Colombia

ASASP: Active Scattering Aerosol Spectrometer Probe

ASAT: Anti-Satellite • Automatic Spares Analysis Technique

ASATS: Anti-Satellite System

Asb: Asbestos

ASB: Administration and Storage Building • Airlock Stowage Bag • Air Safety Board • Altitude Sensor Bypass • Analysis and Simulation Branch • Archenhold-Sternwarte Berlin (Germany) • Ashkhabad (Turkmenia) (location identifier) • Associated Standardisation Bodies • Associated Standardising Bodies • Asymmetrical Sideband • Australian Space Board • Automated Status Board

ASBC: Advanced Solid-Block Correlator

ASBCA: Armed Services Board of Contract Appeals

ASBL: Association Sans But Lucratif

ASBR: Academically-Separated Budgeted Research

ASBU: Arab States Broadcasting Union

Asc: Ascend • Ascenseur • Ascenso • Ascent

ASc: Associate in Science

ASC: Accelerometer Signal Conditioner • Accredited Standards Committee • ACE Science Center (USA-CA) • Adams State College (USA-CO) • Advanced Scientific Computer • Advanced Simulation Center • Advanced System Concept • Aerodynamic Surface Control • Aeronautical Systems Command • Aerospace Planning Chart • Aerospace Static Converter • Aerosurface Control • Agence Spatiale Canadienne (Canada) • Agnes Scott College (USA-GA) • Airlock Support Component • Albany State College • All-Sky Camera • American Satellite Co. • American Security Council (USA) • American Society of Cartographers (USA) • American Society of Cinematographers (USA) • American Society of Cybernetics (USA) • Analog Signal Converter • Analog Signal Correlator • Analog Strip Chart • Apache Survival Coalition (USA) • Applied Science Corporation • Applied Superconductivity Conference • Area Safety Coordinator • Arizona State College (USA-AZ) • Arlington State College • Array Stowage Container • Ascención (Bolivia) (location identifier) • ASCII • ASEAN Standing Committee • Assembly Shortage Control • Association Sportive et Culturelle • Associative Structure Computer • Astronautical Society of Canada • Astronautics Support Center • Astronomy Survey Committee • Astro Space Center (Russia) • Asynchronous Communication Procedure • Aughey Spark Chamber • Australian Seismological Centre • Australian Space Council • AUTODIN Switching Center • Automatic Scan Counter • Automatic Selectivity Control • Automatic Sensitivity Control • Automatic Switching Center • Automatic Synchronized Control • Automatic System Control • Auxiliary Switch Closed • AXAF Science Center (USA-MA)

AsCA: Asian Crystallographic Association

ASCA: Advanced Satellite for Cosmology and Astrophysics (Japan) (formerly Astro-D) • Asian Crystallographic Association • Association for Science Cooperation in Asia

ASCADS: Aft Station Command And Data System

ASCAN: Astronaut Candidate

ASCAP: American Society of Composers, Authors and Publishers (USA)

ASCAT: Advanced Scatterometer • Analog Self-Checking Automatic Tester

ASCATT: Advanced Scatterometer

ASCB: Address Space Control Block • Avionics Standard Communication Bus

ASCC: Academia Sinica Computing Center (China-ROC) • Aeronautical Satellite Communications Center (USA) • Asia-Pacific Satellite Communication Council • Astronomical Society of the Central Coast (USA-CA) • Automatic Sequence-Controlled Calculator

ASCD: ACTS Systems Concept Development • American Society of Computer Dealers (USA)

ASCDS: AXAF Science Center Data System

ASCE: Abrupt Space Charge Edge • Advanced Solar Coronal Explorer • Airlock Signal Conditioning Electronics • American Society of Civil Engineers (USA)

ASCEND: Agenda of Science for Environment and Development • Agenda on Science for Environment and Development

ASCHOT: Astrophysical Schmidt Orbiting Telescope

ASCENT: Assembly System for Central Processor

ASCF: Ames Satellite Communications Facility

ASCH: American Society for Circumpolar Health (USA)

ASChE: American Society of Chemical Engineers (USA)

ASCID: Altered State of Consciousness Induction Device

ASCII: American Standard Code for Information Interchange

ASCI: Accelerated Strategic Computing Initiative • Advanced Speech Cell Item • Association Canadienne des Sciences de l'Information (Canada)

ASCL: Astrophysics Source Code Library

ASCLA: Association of Specialized and Cooperative Library Agencies

ASCM: Advanced Spaceborne Computer Module

ASCNET: Australian Computer Science Network

ASCO: Aboard Sensing Control Unit • Asian Science Communicators' Organization

ASCOM: Astronomy Common Object Model (initiative)

ASCON: Automated Switched Communications Network

ASCORE: Automatic Shipboard Checkout and Readiness Equipment

ASCOT: Accelerated Simulation of Charge particle Orbits in a Tokamak • Atmospheric Studies in Complex Terrain • Automatic Scanning-system for Corpus-Oriented Tasks

ASCP: American Society of Consulting Planners • Attitude Set Control Panel • Aviation System Capacity Plan

ASCPL: Astro Scientific Centre Pte Ltd (Singapore)

ASCR: Analog Strip Chart Recorder • Asymmetrical Semiconductor Controlled Rectifier

ASCS: Advanced Stirling Conversion System • Aerospace Sector Certification Scheme • Aft Shroud Cooling System

• Association des Sociétés de Commercialisation de Services (France) • Atmosphere Storage and Control System • Atmospheric Storage and Control Section • Attitude and Spin Control System • Attitude Stabilization Control System • Automated Storage Control System • Automated Scan Control System • Automatic Stabilization and Control System

ASCSP: Australia Space Centre for Signal Processing

ASCU: Association of State Colleges and Universities (USA) (now AASCU) • Automatic Scanning Control Unit

AScW: Association of Scientific Workers (UK)

ASD: Accurate Space Derivative • Adjustable Speed Drive • Advanced Systems Development • Aerospace Services Division • Aircraft Situation Display • Air Support Division • After-Sales Department • Aircraft Situation Display • Allentown School District (USA-PA) • Alternate Source Development • Analysis and Simulation Division • Andros Town (Bahamas) (location identifier) • Application Support Division • Arbre Senonique de Décision • Arc Supérieur Droit • Assistant Secretary of Defense • Association for Software Design • Association Suisse de Documentation (Switzerland) • Astronomy Support Department • Astro Socio Dynamics • Astrostation Demmin (Germany) • Atmospheric Sciences Division • Automated Structural Design • Automatic Synchronized Discriminator

ASDA: Accelerate-Stop Distance Available • Atomic and Space Development Authority • Australian Standard Data Archive

ASDAC: Advanced Software Development And Commercialization

ASDAR: Aircraft-to-Satellite Data Relay

AsDB: Asian Development Bank

ASDC: Aeronomy and Space Data Center • Australian Space Development Conference

a/s de: Aux Soins de

ASDE: Aerosurface Driver Electronics • Airport Surface Detection Equipment

ASDEC: Automatic Selection of Digital Electronic Computers

ASDEM: Automatic Sample Detection and Efficiency Monitoring

ASDEX: Axially Symmetric Divertor Experiment

ASDG: Automated Systems Development Group

ASDI: Arctic Sustainable Development Initiative • Australian Space Data Infrastructure • Automatic Selective Dissemination of Information

ASDM: Aerosurface Driver/Monitor • Apollo-Soyuz Docking Module

ASDMS: Autonomous Spacecraft Data Management System

ASDO: Accredited Standards Developing Organization

ASDS: Astronomical Software and Documentation Service (*) • Astronomical Software Directory Service (*) • Astronomical Software Documentation Service

ASDSP: Application-Specific Digital Signal Processor

ASDT: Accumulated Standardized Data Tape

ASDTIC: Analog Signal to Discrete Time Interval Converter

ASDU: Aurora Software Development Unit

ASDWG: Astro Socio Dynamics Working Group

AS&E: American Science & Engineering • America Science & Engineering

ASE: Accredited Systems Engineer • Active Seismic Experiment • Advanced Software Engineering • Advanced Space Engine • Advanced Systems Engineering • Aerospace Support Equipment • Agence Spatiale Européenne • Airborne Support Environment • Airborne Support Equipment • Air-Sea Experiment • Allowable Steering Error • American Science and Engineering • American Society of Educators (USA) • American Society of Engineers (USA) • American Stock Exchange • Amplified Spontaneous Emission • Application Service Element • Application Software Entity • Application Specific Environment • Aspen (USA-CO) (location identifier) • Associate of the Society of Engineers • Association for Science Education • Association of Space Explorers • Astronomical Society of Edinburgh (UK-Scotland) • Atmospheric Structure Experiment • Attitude Sensor Electronics • Audio Support Equipment • Automatic Stabilization Equipment • Automatic Support Equipment • Automotive Stirling Engine • Avionics Simulated Environment • Snowmass (USA-CO) (location identifier)

ASEA: American Solar Energy Association (USA) • Association of South East Asia

ASEAMS: Association of Southeast Asian Marine Scientists

ASEAN: Association of Southeast Asian Nations

ASEAP: Association of Southeast Asian Publishers

ASEAUS: Association of South East Asian University Students

ASEB: Aeronautics and Space Engineering Board

ASEC: Applied Solar Energy Corporation

ASECNA: Agence pour la Sécurité de la Navigation Aérienne en Afrique et à Madagascar

ASED: Application Specific Environment Description • Avionics System Engineering Division

ASEDEM: Asociación Española de Empresas de Multimedia (Spain)

ASEE: American Society for Engineering Education (USA)

ASEER: Aerospace and Environmental Education Research

ASEL: Automatic Space Ecological Laboratory

ASEM: Advanced Space Environment Monitor • Application Specific Electronic Module • Assembly of Station by Extravehicular Methods

ASEP: Advanced Science Education Program • Advanced Scientific-computer Enhancement Project • Automated Signal Excess Prediction • Automatic Signal Execution and Processor

ASEPS: Astronomical Studies of Extrasolar Planetary Systems

ASER: Amplification by Stimulated Emission of Radiation

ASERL: Association of Southeastern Research Libraries (USA)

ASES: Aviation Safety Exchange System

ASESS: Aerospace Environment Simulation System

ASET: Advanced Sensor Evaluation and Test • Astronomical Society of East Texas (USA-TX) • Author System for Education and Training

ASF: Actively Star-Forming • Active Streaming Format • Additional Safety Factor • Additional Secondary Factor • Additional Selection Factor • Advanced Simulation Facility • Area Sampling Frame • Aerospace Security Force • Air Safety Foundation (USA) • Alaska SAR Facility • Amperes per Square Foot • Amplitude Spread Function • Angle Selection Function • Angular Selection Function • Area Sampling Frame • Arithmetic Statement Function • Astounding Science Fiction • Astrakhan (Russia) (location identifier) • Astronomical Society of Frankston (Australia-VIC) • Atmospheric Science Facility • Atmospheric Stabilization Framework • Atomic Scattering Factor • Atomic System Function • Automatic Store and Forward

ASFA: American Science Film Association (USA)

ASFE: Accelerometer Scale Factor Error

ASFG: Actively Star-Forming Galaxy • Atmospheric Sound-Focusing Gain

ASFIP: Accelerometer Scale Factor Input Panel

ASFIS: Aquatic Sciences and Fisheries Information System

ASFM: Astronomical Society of Fort McMurray (Canada-AB)

ASFS: Adelaide Space Frontier Society (Australia-SA)

ASFSE: American-Swiss Foundation for Scientific Exchange

ASFT: Abbreviated System Functional Test

ASFTS: Auxiliary System Function Test Stand

ASG: Acting Secretary-General • ADAM Support Group • Advanced Systems Group • Aeronautical Standards Group • Aggregate Specialization Graph • Antenna Steering Group • Ashburton (New Zealand) (location identifier) • Assessment Subgroup • Assistant Secretary-General • Astronomical Society of Georgia • Astronomical Society of Glasgow (UK-Scotland) • Australian Spaceport Group • Avionics Systems Group

ASGA: Advanced Solar GaAs Array • Association des Services Géologiques Africains • Australian Standard Geographical Classification

ASGB: Aeronautical Society of Great Britain

Asgd: Assigned

ASGH: Astronomical Society of Greater Hartford (USA-CT)

ASGI: Astronomical Science Group of Ireland (Northern Ireland)

ASGLS: Advanced Space-Ground Link System

Asgmt: Assignment

Asgn: Assign

ASGPD: Attitude Set and Gimbal Position Display

ASGRG: Australasian Society for General Relativity and Gravitation (Australia)

ASGSB: American Society for Gravitational and Space Biology (USA)

A&SH: Astronomie & Sciences Humaines

ASH: Action on Smoking and Health (USA) • Astronomical Society of Haringey (UK) • Astronomical Society of Harrisburg (USA-PA) • Astronomical Society of the Hunter (Australia) • Avion Spatial HERMES • Nashua Boire (USA-NH) (location identifier)

ASHE: A Simple Hypertext Editor

ASHOE: Airborne Southern Hemisphere Ozone Experiment

Ashp: Airship

AHRAE: American Society of Heating, Refrigerating and Air-conditioning Engineers (USA)

ASI: Advanced Science Instrument • Advanced Scientific Instrument • Advanced Studies Institute • Advanced Study Institute • Aerospace Static Inverter • Agenzia Spaziale Italiana (Italy) • Airspeed Indicator • Alphabetical Subject Indexing • Ajustement Structurel Intensif • Amended Shipping Instructions • American Safety Institute (USA) • American Society of Indexers (USA) • Analytic Subject Index • Artemis Society International (USA-CA) • Articulated Subject Index • Asian Statistical Institute • Astronomical Society of India • Atmospheric Structure Instrument • Atmospheric Structure Investigation • Augmented Spark Igniter (*) • Augmented Spark Indicator • Augmented System Igniter • Augmented System Ignition • Australian Science Index • Aviation Safety Inspector • Axial Scientific Instrument • Azimuth Speed Indicator • Georgetown Wideawake (Saint Helena) (location identifier)

ASIA: Action Spécifique d'Initiatives Académiques • Advanced Simulation Interoperable Architecture

ASIAA: Academia Sinica's Institute of Astronomy and Astrophysics (China-ROC)

ASIAC: Australian Surveying Industry Advisory Committee

ASIAS: Astronomy Science Information and Abstract Service

ASIC: Application-Specific Integrated Circuit • Australian Standard Industry Classification

ASICC: Australian Space Industry Chamber of Commerce

ASIDIC: Association for Scientific Information and Dissemination Centres • Association of Scientific Information Dissemination Centres

ASIDS: Australian Space Industry Development Strategy

ASIES: Academia Sinica's Institute of Earth Sciences (China-ROC)

ASII: Air-Sea-Ice Interaction • American Science Information Institute (USA)

ASIL: Application-Specific Integrated Lens

ASIP: All-Sky Imaging Photometer

ASIPI: Asociación Interamericana de la Propiedad Industrial

ASIS: Abort Sensing and Implementation System • Airborne Spectroradiometric Imagery System • American Society for Industrial Security (USA) • American Society for Information Science (USA) • Application Software Installation Server • Automated Sea Ice Segmentation

ASIT: Advanced Security and Identification Technology

ASIV: Automatisiertes System für Informationsverarbeitung

ASIZ: Advanced Sea Ice Zone

ASJ: Amami O-shima (Japan) (location identifier) • Astronomical Society of Japan

ASJC: American Society of Journalists and Composers (USA)

ASK: Acquisition of Strategic Knowledge • Actively Shared Knowledge • American School of Kuwait • Amplitude Shift Keying • Analog Select Keyboard • Anomalous State of Knowledge • Assisted Search for Knowledge • Astronaut Survival Kit • Yamoussoukro (Ivory Coast) (location identifier)

ASKA: Automatic System for Kinematic Analysis

ASKC: Astronomical Society of Kansas City (USA-MO)

ASKE: Association for Skeptical Enquiry (K)

ASKS: Automatic Station Keeping System

ASL: Above Sea Level • Adaptive Speed Levelling • Albuquerque Seismological Laboratory (USA-NM) • American Sign Language • Applied Science Laboratory • Approved Source List • Arbeitskreis Sternwarte Lübeck (Germany) • Arithmetic Shift Left • Association for Symbolic Logic • Astigmatic Spectral Line • Atmospheric Sciences Laboratory • Atmospheric Surface Layer • Automated Soft Lander • Automatic Systems Laboratories Ltd. (UK) • Average Staffing Level • Azimuth Steering Line • Marshall - Harrison County (USA-TX) (location identifier)

ASLC: Adaptive Sidelobe Canceller • Astronomical Society of Las Cruces (USA-NM)

ASLEP: Apollo Surface Lunar Experiments Package

ASLI: Astronomical Society of Long Island (USA-NY)

ASLK: Algemene Spaar- en Lijfrentekas

ASLM: Apple Shared Library Manager

ASLO: American Society of Limnology and Oceanography (USA) • Australian Scientific Liaison Office

ASLT: Advanced Solid Logic Technology

ASLU: Antenna Select Logic Unit

ASLV: Advanced Small Launch Vehicle • Augmented Satellite Launch Vehicle

Asm: Assemble

ASM: Accelerated Surface Mail • Administrator of Session Monitors • Advanced Servomanipulator • Aerospace Structural Material • Agrégats Stables Minimums • Algebraic Stress Model • Algemene Salarismaatregelen • Algorithmic State Machine • All-Sky Monitor • American Samoa (ISO country code) • American Society for Metals (USA) • Angular Second Moment • Angular Second

Momentum • Antenna Switching Matrix • Aperture Synthesis Map • Application Specific Module • Applied Simulation and Modelling • Applied Software Measurement • Asmara (Ethiopia) (location identifier) • Assembler • Assistant Sales Manager • Assistant Stage Manager • Assistant Station Master • Association for Systems Management • Association Set Manager • Astronomical Seeing Monitor • Astronomical Site Monitor • Audiovisual Multimedia Services • Austrian Society for Aerospace Medicine (Austria) • Automated Storage Management • Auxiliary Storage Manager • Available Seat Mile • Samoa (American_) (ISO country code) • Samoa Américaine (ISO country code)

ASMA: Aerospace Medical Association • Antarctic Specially Managed Area • Automatic Segment Matching Algorithm

ASMC: AUTODIN Station Maintenance Console

ASMD: Atmospheric Sciences Modelling Division

ASME: American Society of Magazine Editors (USA) • American Society of Mechanical Engineers (USA)

ASMES: Advanced Synchronous Meteorological Satellite

ASMI: Aircraft Surface Movement Indicator

ASMIEG: Advanced Structural Materials Information Exchange Group

ASMO: Arab Standards and Metrology Organization

ASMP: American Society of Magazine Photographers (USA) • American Society of Media Photographers (USA) • Asymmetric Multiprocessing

ASMS: Advanced Synchronous Meteorological Satellite • American Society for Mass Spectrometry (USA) • Atmosphere Sensing and Maintenance System • Automated Systems Management System

ASMSS: Airborne Stereo Multispectral Sensor

ASMU: Automatically Stabilized Maneuvering Unit

Asn: Asparagine

ASN: Abstract Syntactical Notation • Abstract Syntax Notation • Access Stack Node • Acknowledged Sequence Number • Advance Ship Notice • AeroSpace Network • Allotment Serial Number • AMOS Star Number • Army Service Number • Association Suisse de Normalisation (Switzerland) • Astronomical Society of Nevada (USA-NV) • Autonomous System Number • Average Sample Number • Talladega (USA-AL) (location identifier)

ASNAP: Automatic Steerable Null Antenna Processor

Asnau: Astronautik (periodical code)

ASNAV: Association Nationale pour l'Amélioration de la Vue (France)

ASNE: American Society of Newspaper Editors (USA)

ASNH: Astronomical Society of New Haven (USA-CT)

ASNNE: Astronomical Society of Northern New England (USA-ME)

ASNORA: Association Normande d'Astronomie (France)

ASNSW: Astronomical Society of New South Wales (Australia-NSW)

ASNT: American Society of Non-destructive Testing (USA) • Astronomical Society of North Texas (USA-TX)

ASNY: Astronomical Society of New York (USA-NY)

ASO: Administrative Service Office • Administrative Service Only • Advanced Solar Observatory • Ajustement Structurel Ordinaire • Algemeen Secundair Onderwijs • Amateur Satellite Observers (USA-AR) • Application Service Object • Area Safety Officer • Area Supply Officer • Asosa (Ethiopia) (location identifier) • Astronomía Sigma Octante (Bolivia) • August-September-October • Australian Space Office (now ASPU) • Australian Survey Office • Auxiliary Switch Open • Aviation Safety Office

Asoc: Asociación

ASOC: Air Sovereignty Operation Center • Antarctic and Southern Ocean Coalition

ASOCON: Asian Soil Conservation

ASOD: Astronomical Society of the Desert (USA-CA)

ASOP: Automatic Scheduling and Operating Program

ASOS: Automated Seismological Observation System • Automated Surface Observing System • Automatic Storm Observation Service • Automatic Surface Observing System • Auto Surface Observing System

ASOSV: Amateur Satellite Observers of Southeast Virginia (USA-VA)

ASOVAC: Asociación Venezolana para el Avance de la Ciencia (Venezuela)

Asp: Aspartic Acid

ASP: Accepté Sous Protêt • Active Server Page • Activity Scheduling Processor • Activity Scheduling Program • Advanced Signal Processor • Advanced Stokes Polarimeter • Advanced Study Program • Aerospace Plane • Aerosurface Position • Aggregated Switch Procurement • Airborne Science Program • Alice Springs (Australia-NT) (location identifier) • American Selling Price • American Society of Photogrammetry (USA) • Analog Signal Processor • Anti-Solar Point • AppleTalk Session Protocol • Application Service Provider • Arrival Sequencing Program • Aspirante • Association of Shareware Professionals • Association-Storing Processor • Associative Structures Package • Astronautics Standard Practice • Astronomical Society of the Pacific (USA-CA) • Astronomy Spacelab Payload • Atmosphere Sounding Projectile • Atmospheric Sensor Package • Attached Support Processor • Attitude Sensor Package • Attitude Sensor Payload (*) • Augmented Support Period • Augmented Support Plan • Auroral Scanning Photometer • Automated Schedule Procedure • Automated Sneak Program • Automated Spooling Priority • Automatic Sample Processor • Automatic Schedule Procedure • Automatic Servo Plotter • Automatic Switching Panel • Automatic Synthesis Program • Auto Sequence Program • Auto Sequence Programming • Auxiliary Spacecraft Power

ASPA: Antarctic Specially Protected Area • Association of South Pacific Airlines • Association pour la Surveillance de la Pollution Atmosphérique

ASPAC: Asian and Pacific Council

ASPB: Astronomical Society of the Palm Beaches (USA-FL)

ASPC: Accepté Sous Protêt pour Compte • Asia Pacific Space Centre (Christmas Island)

ASPCA: American Society for Prevention of Cruelty to Animals (USA)

ASPCS: Astronomical Society of the Pacific Conference Series

ASPE: Advances in Solar Physics Euroconference

ASPEI: Association of South Pacific Environmental Institutions

ASPEN: Asian Physics Education Network • Automatic System for Performance Evaluation of the Network

ASPERA: Automatic Solar-system Particle Experiment with Rotating Analyzer • Automatic Space Plasma Experiment with a Rotating Analyser

Asph: Asphalt

ASPH: Adaptive Smooth Particle Hydrodynamics

ASPI: Advanced SCSI Programming Interface • Advanced SCSI Protocol Interface • Aerosurface Position Indicator • Australian Space Policy Institute

ASPIC: Apparatus for Surface Physics and Interfaces at CERN

ASPIN: Arizona State Public Information Network

ASPIP: Arab Society for the Protection of Industrial Property

ASPJ: Airborne Self-Protection Jammer

ASPL: Advanced Space Propulsion Laboratory

ASPN: Algorithm for Synthesis of Polynomial Networks

ASPO: ATMOS Science Program Office

ASPOC: Active Satellite Potential Control • Active Spacecraft Potential Control (*)

ASPOL: A Simulation Process Oriented Language

ASPP: Advanced Satellite Products Project • American Society of Picture Professionals (USA) • American Society of Plant Physiologists (USA) • Antenna Solar Panel Positioner • Atmospheric and Space Plasma Physics

ASPRS: American Society of Photogrammetry and Remote Sensing (USA)

ASPRT: Adjusted Sequential Probability Ratio Test

ASPS: Action Spécifique de Physique Stellaire • Advanced Signal Processing System • Annular Suspension and Pointing System

ASpScROW: Astronomy, Space Sciences and Related Organizations of the World (now StarGuides)

Aspt: Aspirant

ASPU: Australia Space Policy Unit (formerly ASO)

ASQ: Austin (USA-NV) (location identifier) • Automated Software Quality

ASQC: American Society for Quality Control (USA)

ASQP: Airline Service Quality Performance

ASR: Acceptance Summary Report • Access Service Request • Accumulators Shift Right • Activation Status Register • Active Surge Region • Advanced Special Receiver • Advanced Systems Requirement • Advances in Space Research • Airport Surveillance Radar • Air-Sea Rescue • Alphabetisches Sachregister • Altimeter Setting Region • AMOS Spectral Radiometer • Analog Shift Register • Annual Summary Report • Approved System Requirement • Arithmetic Shift Right • Array Switching Regulator • Astronomy Space and Radio • Astrophysics and Space Research • Atmospheric Sounding Radiometer • Atmospheric Sound Refraction • Automatic Send and Receive • Automatic Send-Receive • Automatic Speech Recognition • Automatic Speech Recognizer • Automatic Step Regulator • Automatic Storage and Retrieval • Avionics System Review • Kayseri (Turkey) (location identifier)

AS&RA: Astrophysics Software & Research Aids

ASRA: ADP Systems Resources Analysis • Advanced Systems Research Aircraft • Applied Science and Research Application • Automatic Stereo Recording Amplifier

ASRAS: Astronomy Section of the Rochester Academy of Sciences (USA-NY)

ASRB: Alpine Surface Radiation Budget • Astronomy Space and Radio Board

ASRC: Astronomical Society of Rowan County (USA-NC) • Atmospheric Sciences Research Center

ASRE: American Society of Refrigerating Engineers (USA)

ASRI: Australian Space Research Initiative • Australian Space Research Institute

ASRL: Atmospheric Sciences Research Laboratory

ASRLO: Australian Scientific Research Liaison Office

ASRM: Abort Solid Rocket Motor • Advanced Solid Rocket Motor • Antenna System Readiness Monitor

ASRO: Astronaut Rover

ASRP: Australian Synchrotron Research programme

ASRS: Airborne Satellite Receiving Station • Automated Shuttle Requirements System • Automated Support Requirements System • Automatic Storage and Retrieval System • Auxiliary Support Reaction System • Aviation Safety Reporting System

ASRT: Academy of Scientific Research and Technology

Ass: Assegno • Assemblée • Assembly • Assistant • Association • Associazione • Assorted • Assurance

A&SS: Alignment & Stabilization System

ASS: Accessory Supply System • Acquisition Sun Sensor • Advanced Space Station • Aerosol Sampling System • Aerospace Surveillance System • Airlock Support System • Alignment and Stabilization System • Allocation Spécifique de Solidarité • Altitude Sensing System • Analog Simulation System • Analog Sun Sensor • Antenna Support Structure • Antenna Support System • Assistante Sociale • Automatic Space Station • Automatic Stabilization System • Automatic Switching System • Axisymmetric Spiral

ASSA: Astronomical Society of South Australia • Astronomical Society of Southern Africa • Austrian Solar and Space Agency (now ASA)

ASSAB: Académie Serbe des Sciences et des Arts - Bulletin (periodical code)

ASSAS: Advanced Solid-State Array Sensor

ASSC: Accounting Standards Steering Committee

ASSE: American Society of Sanitary Engineers (USA) • Automation System for Scientific Experiments

ASSEDIC: Association pour l'Emploi Dans l'Industrie et le Commerce (France)

ASSESS: Airborne Science Shuttle Experiments System Simulation • Airborne Science Spacelab Experiments System Simulation

ASSET: Abstracts of Selected Solar Energy Technology • Advanced Systems Synthesis and Evaluation Technique • American Society of Scientific and Engineering Translators (USA) • Asset Source for Software and Technology • Association for Schools Science, Engineering and Technology • Astronomical Society of SE Texas (USA-TX)

ASSH: Académie Suisse des Sciences Humaines (Switzerland) • Advanced Space Systems Hardening

ASSI: Active Stabilization in Stellar Interferometry • Applications Statistiques Scientifiques Informatiques • Atmospheric amd Solar (EUV) Spectrometer Instrument

ASSIA: Applied Social Science Index and Abstracts

Assigt: Assignment

Assim: Assimilate

ASSIST: Automatic System of Secondary Index Selection Technique

ASSL: Astrophysics and Space Science Library

ASSM: Académie Suisse des Sciences Médicales (Switzerland) • Automated Space Service Module

Assmt: Assessment

Assn: Association

ASSN: Académie Suisse des Sciences Naturelles (Switzerland) • Accademia Svizzera di Science Naturali (Switzerland)

ASSO: Assistant System Security Officer

Assoc: Associate • Associated • Association • Associé • Associer

Assocn: Association

Assoz: Associazione

ASSP: Aerosol Scattering Spectrometer Probe • Aerospace Systems Security Program

ASSR: Autonomous Soviet Socialist Republic (now obsolete)

ASSS: Aerospace Systems Safety Society

Ass/t: Assistant

Asst: Assistant

ASST: Académie Suisse des Sciences Techniques (Switzerland) • American Society for Steel Treating (USA)

Asstd: Assorted

ASSU: Ambulance de Secours et de Soins d'Urgence

ASSW: Astronomical Society of the South West (Australia)

Assy: Assembly

ASSY: Assembly

Ast: Asturiano • Asturias (Spain)

ASt: Außenstelle

AST: Abort-Scan Table • Absolute Space-Time • Abstract Syntax Tree • Active Solar-Type • Adaptive Structuration Theory • Add-Subtract Time • Advanced Solar Telescope • Advanced Switch Test • Advanced System Technology • Aerospace, Science and Technology • Aerospace Technologist • Agent de Service Technique • Agent des Services Techniques • Airborne Surveillance Testbed • Airbus Super Transporter • Airlock Systems Test • Alaskan Standard Time • Albert (Wong) - Safi (Qreshey) - Thomas (Yuen) • All-Systems Test • Antarctic Submillimeter Telescope • Apparent Sidereal Time • Assembly and Structure Test • Astoria (USA-OR) (location identifier) • Astrometry • Astrometry Science Team • Astronomical Society of Tasmania (Australia-TAS) • Astronomy Science Team • Astro Star Tracker • Asynchronous System Trap • Asynchronous Trap • Atlantic Standard Time • Atomized Suspension Technique • At Same Time • Automatic Spectroscopic Telescope • Average Sampling Time

AStA: Allgemeine Studentenausschuss (Germany)

ASTA: Advanced Software Technology and Algorithms • Aerospace Technologies of Australia • Airport Surface Traffic Automation • Allgemeiner Studenten-Ausschuß (Germany) • American Society of Travel Agents (USA) • Arab Scientists and Technologists Abroad • Association of Short-circuit Testing Authorities • Astronomical Time Series Analysis • Auckland Science Teacher's Association (New Zealand)

ASTAP: Advanced Statistical Analysis Program

Astab: Astable

ASTB: Astronomiska Sällskapet Tycho Brahe (Sweden)

ASTC: Advanced Satellite Tracking Center • Advanced Science and Technology Center • Advanced Science-Technology Center • Advanced Semiconductor Technology Center • AlliedSignal Technical Services Corporation (USA) • Association of Science and Technology Centers (USA-DC)

ASTD: Advanced Space Technology Division • Area Scale Temperature Display

ASTE: Action Spécifique "Télescopes Étrangers" (INSU) • Advanced Strategic/Tactical Expendable • Aerospace Systems Test Environment • American Society of Tool Engineers (USA)

ASTEC: Aerospace System Test and Evaluation Complex • Australian Science and Technology Council

ASTED: Avancement des Sciences et des Techniques de la Documentation

ASTEG: All Systems Test Equipment Group

ASTEP: Algorithm Simulation Test and Evaluation Program

ASTeR: Audio System for Technical Reading

ASTER: Advanced Spaceborne Thermal Emission and Reflection • Assessment Tools for the Evaluation of Risk • Atmosphere-Surface Turbulence Exchange Research • Audio System for Technical Readings

ASTERIX: Analyse Stéréoscopique par Radar à Impulsions X

ASTEX: Atlantic Stratocumulus Transition Experiment

ASTF: Aeropropulsion System Test Facility

ASTG: Aerospace Test Group

AstHe: Astronomical Herald (periodical code)

ASTI: Advanced Science and Technology Institute • Area of Special Tourist Interest

ASTIP: Army Scientific and Technical Information Program

ASTIS: Arctic Science and Technology Information Service • Australian Science and Technology Information Service

ASTL: Advanced Space Technology Laboratory • Automated SPARCS Command Link

ASTM: American Society for Testing and Materials (USA) • American Society for Testing Materials (USA) (*) • American Standards and Test Method

ASTMS: Association of Scientific, Technical and Managerial Staff • Association of Specialist Technical Organizations (UK)

ASTOR: Airborne Stand-Off Radar • Area Stand-Off Radar

ASTOVL: Advanced Short Takeoff and Vertical Landing

ASTP: Advanced Space Technology Program • Advanced Space Transportation Program • Advanced Systems and Technology Programme • Apollo-Soyuz Test Project

Astr: Astronomer • Astronomical • Astronomy

ASTR: Addition, Subtraction, Timing and Ratio • Aerospace Systems Test Reactor • Along-Track Scanning Radiometer • Astronome

ASTRA: Advanced Space Technologies for Robot Applications • Application of Space Techniques Relating to Aviation • Applied Space Technology Regional Advancement • Association in Scotland to Research into Astronautics • Associazione per lo Studio e la Ricerca Astronomica (Italy) • Astronome Adjoint • Astronomical and Space Techniques for Research on the Atmosphere • Astronomical Society of the Toms River Area (USA-NJ) • Astronomical Space Telescope Research Assembly • Automated Scale Transition • Automatic Satellite Tracking Antenna • Automatic Sorting, Testing, Recording Analysis

ASTRAL: Alliance for Strategic Token Ring Advancement and Leadership

Astr. Ap.: Astronomy and Astrophysics

ASTRE: Accéléromètre Sensible Triaxial Électrostatique • Astrophysiciens à la Recherche de l'Entreprise

ASTREX: Advanced Space Technology Research Experiment

ASTI: Analytical Stereorestitution Instrument

ASTRID: Advanced Stars: a Tool for Relating Information and Data • Association Scientifique et Technique pour la Recherche en Information Documentaire (Belgium)

ASTRIS: Astrofili della Società di Telecomunicazioni del Raggruppamento IRI-STET (Italy)

Astro: Astronaute • Astronautique • Astronomer • Astronomical • Astronomie • Astronomique • Astronomy

AST/RO: Antarctic Submillimeter Telescope/Remote Observatory

ASTRO: Air Space Travel Research Organization • Air Support To Regular Operations • Antarctic Submillimeter Telescope and Remote Observatory • Artificial Satellite Time and Ratio Orbit • Ascent Trajectory Optimization • Astronomers in the Schools Teaching Resource Organization

Astrol: Astrología • Astrologie • Astrologique • Astrology

Astron: Astronaute • Astronautique • Astronomer • Astronomía • Astronomical • Astronomie • Astronomique • Astronomy

ASTRON: Stichting Astronomisch Onderzoek in Nederland (Netherlands)

Astronaut: Astronautical • Astronautics • Astronautique

Astrophys: Astrophysical • Astrophysics • Astrophysique

ASTS: American Space Transport System • Associzaioine della Stampa Tecnica e Scientifica (Italy) • Avionics System Test Specification

Ast. T.: Astronomical Time

ASTU: Automatic Systems Test Unit

ASTUTE: Automated System for Thesaurus Updating, Testing and Editing

ASU: Administration Scolaire et Universitaire • Administrative Service Unit • Administrative Support Unit • Airborne Sensing Unit • Air Separation Unit • Allocation de Salaire Unique • American Students Union (USA) • Angelo State University (USA-TX) • Area Service Unit • Arizona State University (USA-AZ) • Arkansas State University (USA-AR) • Association pour la Statistique et ses Utilisations (France) • Astronomical Server URL • Astronomy Study Unit • Asunción - Silvio Pettirossi (Paraguay) (location identifier)

ASUP: AUTODIN Switch Upgrade Project

ASUR: Airborne Submillimeter Radiometer

ASUT: Association Suisse de Utilisateurs du Téléphone

ASV: Advanced Safety Vehicle • Aerospace Vehicle • Airline Schedule Vendor • Amboseli (Kenya) (location identifier) • American Standard Version • Anode Supply Voltage • Anodic Stripping Voltammetry • Arabic Vowelized Simplified • Arithmetic Simple Variable • Atlas Stellarum Variabilium • Automatic Self-Verification • Automatic Shuttle Valve • Autonomous Space Vehicle

ASVD: Analog Simultaneous Voice/Data

ASVIP: American Standard Vocabulary for Information Processing

ASVS: Advanced Space Vision System

ASVT: Address Space Vector Table • Applications Systems Verification and Transfer • Applications Systems Verification Test

ASW: Address and Synchronization Word • Amorphous Solid Water • Application Software • Association of Scientific Workers (UK) • Astronomical Society of Wakabadai (Japan) • Aswan (Egypt) (location identifier) • Augmented Spherical Wave

ASWA: Astronomical Society of Western Australia Inc. (Australia-WA)

ASWG: Ad hoc Science Working Group • American Steel Wire Gauge

ASX: Answer-Stock Exchange • Ashland (USA-WI) (location identifier)

ASY: Ashley (USA-ND) (location identifier)

Asym: Asymmetric • Asymmetry

Asymp: Asymptote

Async: Asynchronous

ASYNC: Asynchronous

AS3AP: ANSI SQL Standard Scalable And Portable

at: Austria

At: Ampere-turn • Astat • Astatine • À Temps • Atención • À Terme • Atmosphere • Atom • Atomic • Attitude • Attorney

A/T: American Terms

A&T: Acceptance & Transfer • Acquisition & Technology • Assembly & Test

AT: Absolute Threshold • Abstract Type • Accelerator Technology • Acceptance Test • Access Time • Access Tower • Accident de Trajet • Accident du Travail • Accidente de Trabajo • Achievement Test • Action Taken • Action Time • Admission Temporaire • Advanced Technology • Agent Technique • Aide Technicien • Air Temperature • Air Traffic • Aliveness Test • Allocation Temporaire • Alta Tensione • Alternative Technology • Ambient Temperature • Analysis Time • Anisotropy Telescope • Annual Training • Apparent Time • Appearance Time • Application Technology • Appropriate Technology • Aptitude Technique • Arrêt de Travail • Arrival Time • Assembly and Test • Assembly Truss • Astrometry Team • Astronomical Time • Atlantic Time • Atomic Time • Attainment Target • Attention • Attenuation • Auroral Time • Australian Telescope • Australia Telescope • Austria (ISO country code) • Automatic Telescope • Automatic Test • Automatic Translation • Autorisation Temporaire • Autriche (ISO country code) • Auxiliary Tank • Auxiliary Telescope • Auxiliary Timer

A&TA: Assembly & Test Area

ATA: Abort Time Assembly • Academic Travel Abroad • Actual Time of Arrival • Additional Training Assembly • Advanced Technologies and Applications • Advanced Technology Attachment • Advanced Test Accelerator • Aerocondor Transportes Aéreos Lda. • Aft Test Article • Airport Traffic Area • Air Transport Association • Air Transport Auxiliary • American Translators' Association (USA) • American Transportation Association (USA) • Ammonia Tank Assembly • Anta (Peru) (location identifier) • Antarctica (ISO country code) • Antarctique (ISO country code) • Assembly and Test Area • Assembly Test Article • Associate Technical Aide • Associazione Technica dell'Automobile • Associazione Tuscolana di Astronomia (Italy) • AT Attachment • Atlantic Treaty Association • Automatic Track Acquisition • Automatic Tracking Antenna • Automatic Trouble Analysis • Auxiliary Timer Assembly • Avionics Test Article • Azimuth Torquer Amplifier

ATAC: Advanced Technology Advisory Committee • Air Transport Advisory Council • Air Transport Association of Canada • Applied Technology Advanced Computer • Association des Transporteurs Aériens par Charter • Australian Time Assignment Committee (AAO)

ATACS: Advanced Telecommunications Abuse Control System

ATAI: Association de Transport Aérien International

ATALA: Association pour le Traitement Automatique des Langues

ATAM: Association for Teaching Aids in Mathematics

ATAO: Annals of Tokyo Astronomical Observatory

ATAP: Any Time - Any Place

ATAPI: AT Attachment Packet Interface

ATARR: Advanced Turbine Aerothermal Research Rig

ATARS: Advanced Tactical Air Reconnaissance System

ATAS: Advanced Technology Alert System • Airspace and Traffic Advisory Service • Automatic Test Analysis System • Automatic Three-Axis Stabilization • Automatic Tracking Antenna System

ATB: All Terrain Bike • All Trunks Busy • Association of Tropical Biology • Atbara (Sudan) (location identifier) • Automated Ticket and Boarding

ATBM: Average Time Between Maintenance

ATC: Ablative Thrust Chamber • Ablative Thrust Control • Absolute Time Command • Acoustical Test Chamber • Active Thermal Control • Active Transfer Command • Address Translation Cache • Adiabatic Toroidal Compressor • Advanced Technology Center • After Top Center • Aircraft Target Computer • Air Traffic Control • Air Training Command • Applied Technology Council • Arthur's Town (Bahamas) (location identifier) • Assistant Test Conductor • Astrophysical Theory Centre • Atmosphere Technology Division • ATM Transfer Capability • Australian Telecommunications Commission • Authorization To Copy • Authorized Training Centre • Automatic Target Cueing • Automatic Temperature Compensation • Automatic Temperature Control • Automatic Threshold Circuit • Automatic Tracking Control • Automatic Train Control • Automatic Transmission Control • Average Total Cost

ATCA: Attitude Translation Control Assembly • Australia Telescope Compact Array

ATCAA: Air Traffic Control Assigned Airspace

ATCBI: Air Traffic Control Beacon Interrogator

ATCC: Air Traffic Control Centre

ATCCC: Air Traffic Control Command Center

ATCE: Ablative Thrust Chamber Engine

Atch: Attach

Atchd: Attached

ATCM: Antarctic Treaty Consultative Meeting

ATCO: Air Taxi Commercial Operator • Air Traffic Control Officer • Ambient Temperature Catalytic Oxidizer • Astronomie en Touraine et Centre-Ouest (France)

ATCOS: Atmospheric Composition Satellite

ATCP: Absolute Time Command Processor • Antarctic Treaty Consultative Party • AppleTalk Control Protocol

ATCRB: Air Traffic Control Radar Beacon

ATCRBS: Air Traffic Control Radar Beacon System

ATCS: Active Thermal Control System • Air Traffic Control System • Auxiliary Telescope Control Software

ATCSCC: Air Traffic Control System Command Center

ATCSR: Alternate Transmitted Chirp Slope Reversal

ATCT: Airport Traffic Control Tower

ATD: Above Threshold Detachment • Actual Time of Departure • Advanced Technology Demonstration • Advanced Technology Development • Aerospace Technology Division • Alpha-Track Detector • Amphibious Transport Dock • Associative Template Dataflow • À Titre Définitif • Atmospheric Technology Division • Atoifi (Solomon Islands) (location identifier) • Automatic Tape Degausser • Automatic Teaching Device

ATDA: Alternate Target Docking Adapter • Augmented Target Docking Adapter

ATDB: Aerothermal Data Block • Aerothermodynamic Data Book

ATDBA: Association Tunisienne des Documentalistes, Bibliothécaires et Archivistes (Tunisia)

ATDC: After Top Dead Center

ATDD: Asynchronous Time Diversity Device • Atmospheric Turbulence and Diffusion Division

ATDE: Advanced Technology Demonstrator Engine

ATDF: Archival Tracking Data File • Automated Test Development Folder

ATDI: Appropriate Technology Development Institute

ATDL: Atmospheric Turbulence and Diffusion Laboratory

ATDLG: Advanced Technology Demonstration Laser Gyro

ATDM: Asynchronous Time Division Multiplexing

ATDN: Advanced Technology Demonstration Network

ATDnet: Advanced Technology Demonstration Network

ATDP: Acceptance Test Data Package • Attention Dial Pulse

ATDRS: Advanced Tracking and Data Relay Satellite

ATDRSS: Advanced Tracking and Data Relay Satellite System

ATDT: Attention Dial Tone

ATE: Accéléromètre Triaxial Électrostatique • Acceptance Test Equipment • Actual Time En route • Advanced Technological Education • Advanced Technology Engine • Aerospace Test Equipment • Aerospace Training Equipment • Airborne Test Equipment • Antlers (USA-OK) (location identifier) • Area Test Equipment • ASCII Table Extension • A Temporal Element • A Temporal Entity • Auslandstätigkeitserlaß • Automated Test Equipment (*) • Automatic Test Engineer • Automatic Test Equipment

ATEGG: Advanced Turbine Engine Gas Generator

ATEL: Atelier

ATEM: Analytical Transmission Electron Microscope • Analytical Transmission Electron Microscopy

AtEnv: Atmospheric Environment (periodical code)

ATEP: Advanced Technical Education Program

ATER: Attaché Temporaire d'Enseignement et de Recherche • Automatic Testing, Evaluation and Reporting

ATERP: Arctic Terrestrial Environmental Research Program

AtERv: Atomic Energy Review (periodical code)

ATES: Alcatel Thomson Espace (France)

ATEX: Atlantic Troposheric Experiment

ATF: Accelerator Test Facility • Acceptance Test Facility • Acoustic Test Facility • Advanced Tactical Fighter • Advanced Toroidal Facility • Aerodrome Traffic Frequency • Alcohol-Tobacco-Firearms • Algebraic Technological Function • Ambato Chachoan (Ecuador) (location identifier) • Anatomical Tranfer Function • Anstalt für Tieftemperaturforschung (Germany) • Antenna Test Facility • Astrometric Telescope Facility • Astronomical Test Facility • Atmospheric Transmission Factor • Attention To File • Automated Telescope Facility • Automatic Telescope Facility • Automatic Tracking Feature • Avionics Test Facility • French Southern Territories (ISO country code) • Southern Territories (French_) (ISO country code) • Territoire Français Austral (ISO country code)

ATFM: Air Traffic Flow Management

ATFOS: Alignment and Test Facility for Optical Systems

ATFS: Association of Track and Field Statisticians

ATFT: Affine Toda Field Theory

ATG: Address Translating Gateway • Advanced Technology Group • Aerospace Test Group • Antigua & Barbuda (ISO country code) • Automatic Test Grading • Automatic Timeline Generator • Barbuda (Antigua &_) (ISO country code)

ATGIA: Accurate Tidal Gravimetry In Antarctica

ATGS: Advanced Terminal Guidance System

Ath: Athlete • Athletic

ATH: Abbreviated Trouble History • Above The Horizon • Athens (Greece) (location identifier) • Athinai (Greece) (location identifier) • Attention Hang-up • Auxiliary Telescope Hall

ATHE: Association of Teachers in Higher Education

Athen: Athenian

ATHENA: Acquisition des données de Topographie de surface et d'Hydrologie dans l'Est Nord Atlantique

Athl: Athlete • Athletic

ATHL: Airborne Takeoff and Horizontal Landing

ATI: Above Threshold Ionization • Action Thématique Innovante (INSU) • Advanced Technical Information • Advanced Technologies and Instrumentation • Advanced Technology Insertion • Advanced Technology Instrumentation • Allocation Temporaire d'Invalidité • Apparent Thermal Inertia • Area Time Integral • Artigas (Uruguay) (location identifier) • Assistant Technique à l'Industrie • Association of Technical Institutions • Automatic Test Inhibit • Automation of Technical Information • Average Total Inspection

ATIA: Advanced Techniques for Information Analysis

ATIC: Access To Interpersonal Communication • Advanced Thin Ionization Calorimeter • Aerospace Technical Intelligence Center • Analog Telephony Interface Card • Avionics Test and Integration Complex

ATICA: ACIS Telemetry Interpreter for Calibration Analysis

ATICCA: Australian Tertiary Institution Consulting Companies Association

ATII: Advanced Technique for Imagery Interpretation

ATILO: Administration Tribunal of the International Labour Office

ATIP: Asian Technology Information Program

ATIRCM: Advanced Tactical Infrared Countermeasure (*) • Advanced Threat Infrared Countermeasure

ATIS: Agence de Traitement des Images Satellitaires • Alliance for Telecommunications Industry Solution (USA-DC) • Appropriate Technology Information Service • Atherton Tools Integration Service • Automated Terminal Information Service • Automatic Telescope Instruction Set • Automatic Telescope International Standard • Automatic Terminal Information Service • Automatic Terminal Information System • Automatic Transmitter Identification System

ATJ: Antsirabe (Madagascar) (location identifier)

ATJAM: ASME Transactions - Journal of Applied Mechanics (periodical code)

ATJDS: ASME Transactions - Journal of Dynamic Systems (periodical code)

ATJEG: ASME Transactions - Journal of Engineering, Gas Turbines and Power (periodical c ode)

ATJEM: ASME Transactions - Journal of Engineering Materials and Technology (periodical code)

ATJEP: ASME Transactions - Journal of Engineering Power (periodical code)

ATJFE: ASME Transactions - Journal of Fluids Engineering (periodical code)

ATJHT: ASME Transactions - Journal of Heat Transfer (periodical code)

ATJLT: ASME Transactions - Journal of Lubrication Technology (periodical code)

ATJSE: ASME Transactions - Journal of Solar Energy and Engineering (periodical code)

ATJTr: ASME Transactions - Journal of Tribology (periodical code)

ATJTu: ASME Transactions - Journal of Turbomachinery (periodical code)

ATJVA: ASME Transactions - Journal of Vibration Acoustics (periodical code)

ATK: Andrew Toolkit • Atqasuk (USA-AK) (location identifier) • Available Ton-Kilometer • Average Ton-Kilometer

Atl: Atlantic • Atlas

AT&L: Acceptance Test & Logistics

ATL: Acceptance Test and Logistics • Active Template Library • Active Time List • Actual Total Loss • Advanced Technology Laboratory • Analog Threshold Logic • Applications Terminal Language • Artificial Transmission Line • Atlanta Hartsfield (USA-GA) (location identifier) • Automated Tape Library • Automatic Tape Library • Automatic Test Line • Avionics Test Laboratory

ATLA: American Trial Lawyers Association (USA)

ATLAS: AAO Tunable Littrow Articulated Spectrograph • Abbreviated Test Language for All Systems • Abbreviated Test Language for Avionics Systems • Advanced Technology LADAR System • Airborne Tunable Laser Absorption Spectrometer • Analysis of Thrust Structure Loads And Stresses • Anglo-Australian Tunable Littrow Astronomical Spectrograph • Argonne Tandem-LINAC Accelerator System • Argonne Tandem-Linear Accelerator System (*) • Atmospheric Laboratory for Applications and Science (*) • Atmospheric Laboratory for Applications in Science • A Toroidal LHC Apparatus • Automatic Tape Label Assignment System • Automatic Thin-Layer Analytical System

ATLAST: Augment the Teaching of Linear Algebra through Software Tools

ATLC: Atlantic

ATLH: Atalho

ATLID: Atmospheric Lidar

ATLIS: Automated Technical Library Information System

ATLO: Acceptance Test and Launch Operations

ATLOS: Advanced Launcher Trajectory Optimization Software

at. m.: Atomic Mass

atm: Standard Atmosphere

At/m: Ampere-Turns per Meter

Atm: Atmosphere • Atmosphère • Atmospheric • Atmosphérique

ATM: Accroissement Total Moyen • Adaptive Triangular Mesh • Adobe Typeface Manager • Adobe Type Manager (*) • Advanced Telescope Mission • Aerodynamic Twisting Moment • Airborne Thematic Mapper • Airborne Topographic Mapper • Air Traffic Management • Altamira (Brazil-PA) (location identifier) • Amateur Telescope Maker • Analog-to-Time Module • Antenna Test Model • Apollo Telescope Mount • Approved Testing Material • Ascent Trajectory Model • Association of Teachers of Management • Association of Teachers of Mathematics • Asynchronous Time Mode • Asynchronous Transfer Mode (*) • Atmosphere • Atmospheric Transport Model • Automated Teller Machine • Automatic Teller Machine (*) • Auxiliary Tape Memory

ATMA: Accroissement Total Moyen Annuel • Amateur Telescope Maker's Association (USA-WA)

ATMAB: Association Technique Maritime et Aéronautique - Bulletin (periodical code)

ATMAP: Atmospheric Tides Middle Atmosphere Program

ATMARP: Asynchronous Transfer Mode Address Resolution Protocol

ATME: Atmospheric Transmission Measurement Equipment

AtmEn: Atmospheric Environment (periodical code)

ATMIS: Atmospheric and Meteorology Instrument System • Atmospheric Moisture Intercomparison Study

ATMJ: Amateur Telescope Making Journal

ATMOB: Amateur Telescope Makers Of Boston (USA-MA)

Atmos: Atmosphere • Atmosphere (periodical code)

ATMOS: Atmospheric Trace Molecular Spectroscopy • Atmospheric Trace Molecule Spectroscopy (*) • Atmospheric Trace Molecules Observed by Spectroscopy (*) • Atmospheric Trace Molecules Spectroscopy • Atmospheric Turbulence Measurement and Observation System • Automated Tactical Meteorological Observing System • Automatic Testing Multiple Operating System

Atmosph: Atmosphere • Atmosphère • Atmospheric • Atmosphérique

ATMOSPHERE: Advanced Techniques and Models of System Production in a Heterogeneous, Extensible and Rigorous Environment

ATMP: Ascend Tunnel Management Protocol

AtmRe: Atmospheric Research (periodical code)

AT&MS: Advanced Technology & Mission Studies (NASA)

ATMS: Advanced Traffic Management System • Assumption-based Thruth Maintenance System • Atmospheric Turbulence Measuring Set • Automated Technical Manual System • Automatic Transmission Measuring System

AtmTe: Atmospheric Technology (periodical code)

Atn: Atención

ATN: Actual Test Number • Aeronautical Telecommunications Network • Amateur Television Network (USA) • AMOS Test Number • Arc Tangent • Astrogeophysical Transmission Network • Attention • Augmented Transition Network • Namatanai (Papua New Guinea) (location identifier)

ATNC: Agent Transmissible Non Conventionnel

Atnd: Attended

Atndt: Attendant

ATNF: Australia Telescope National Facility

at. no.: Atomic Number

Ato: Atento

AtO: Atmosphere and Ocean (periodical code)

ATO: Abort To Orbit • Action Technical Order • Advanced Technologies Operation • Air Tasking Order • Assisted Takeoff • Athens - Ohio University (USA-OH) (location identifier) • Australian Tax Office • Automatic Turn-On

ATOC: Acoustic Thermometry of Ocean Climates • Average Total Operating Cost

ATOLL: Acceptance Test Of Launch Language • Analysis of the Tropical Oceanic Lower Layer

ATOM: Analog Tree-Organized Multiplexer • Astronomical Telescope Orientation Mount • Automatic Topographic Mapper

ATOMIC: Action of Transport On Minor Constituents • Astrophysical Transient Observatory Multiple Imaging Camera

ATOMS: Air Traffic Operations Management System

ATOP: Ada Technology Operational Planning

ATOS: Administratif, Technique, Ouvrier et de Service • Advanced Technology Operations System • Automated Technical Order System • Automatisation Totale de l'Observation de la Surface • Automatisation Totale de l'Observation de Surface

ATOVS: Advanced TIROS Operational Vertical Sounder

ATOX: Atomic Oxygen

ATP: Absolute Time Processor • Accelerated Training Program • Acceptance Test Plan • Acceptance Test Procedure (*) • Acceptance Test Program • Accroissement Total Périodique • Acquisition-Tracking-Pointing • Action Thématique Programmée (CNRS) • Activation Test Program • Adenosine Triphosphate • Adult Treatment Panel • Advanced Technical Payroll • Advanced Technology Park • Advanced Technology Program • Advanced Telescope Project • Advanced Turboprop • Aid Trade Provision • Airborne Tracking and Pointing • Aitape (Papua New Guinea) (location identifier) • All Tests Pass • AppleTalk Transaction Protocol • Array Transform Processor • Astronautics Test Procedure • Astrophysical Theory Program • Astrophysics Theory Program • Augmented Thrust Propulsion • Authority To Proceed • Authorization To Proceed • Author-Translation-Publisher • Automated Test Pattern • Automated Test Plan • Automated Test Procedure • Automated Tow Placement • Automatic Text Processing • Automatic Train Protection • Autorisation de Transfert Préalable • Available To Promise

ATPA: Accroissement Total Périodique Annuel • Alpha Temperature Probe Assembly

ATPDC: Atomic Transition Probabilities Data Center

ATPG: Automatic Test Path Generator • Automatic Test Pattern Generation • Automatic Test Program Generation • Automatic Test Program Generator

ATPS: Atmosphere Temperature Profile Sounder

ATPSE: Acquisition, Tracking, and Pointing System Emulator

ATQ: Amritsar (India) (location identifier)

ATR: Accès de Tiers au Réseau • Access Time of Reading • Accidental Trigger Rate • Advanced Technology Radar • Advanced Test Reactor • Advanced Thermal Reactor • Aided Target Recognition • Air Transport Rack • Air Transport Radio • Air Transport Rating • Air Turbo Rocket • Ambient Temperature Recorder • Analog Tape Recorder • Angle-Time-Range • Anti-Transmit Receive • Assistant Task Representative • Assistant Technique Régional • Atar (Mauritania) (location identifier) • Atlantic Tracking Range • Attenuated Total Reflection • Attribute • Australian Telecommunication Research (periodical code) • Automatic Target Recognition • Automatic Terminal Recognition • Automatic Thrust Restoration • Autonomous Target Recognition

ATRAN: Automatic Terrain Recognition And Navigation

ATRIMA: As They Respective Interests May Appear

ATRIP: Advancement of Teaching and Research in Intellectual Property

ATRRL: Advanced Technology Reusable Rocket Launcher

ATRS: Automated Training Records System • Automated Trouble Reporting System

ATRT: Atelier Thématique de Réadaptation par le Travail

ATRWG: Automatic Target Recognition Working Group

ATS: Absolute Temperature Scale • Acceptance Test Specification • Acquisition and Tracking System • Administrative Telephone System • Administrative Terminal System • Advanced Technology Satellite • Advanced Telecommunications Services • Advanced Terminal System • Air Traffic Service • American Trading Syndicate (USA-CA) • Angle Tracking System • Analog Tone Signal • Antarctic Treaty System • Antique Telescope Society (USA-PA) • Apple Terminal Services • Applications and Technology Satellite • Applications Technology Satellite • Artesia (USA-NM) (location identifier) • Assembly Trusses and Structure • Astronomical Time Switch • Asynchronous Task Storage • Atlantic Tracking Ship • Atlantic Tracking Station • Atmospheric Sounder • At The Suit of • Attitude Thrustor System • Attitude Transfer System • A Tutorial System • Automated Text System • Automated Tilting System • Automated Time Standard • Automated Trade System • Automated Tunning System • Automatic Telemetry System • Automatic Terminal System • Automatic Test Set • Automatic Test System • Automatic Transcribing System • Auxiliary Telescope System • Auxiliary Territorial Service • Aviation Transport Services • Ayudante Técnico Sanitario (Spain) • Schilling (Austria) (ISO currency code)

ATSA: Aerotransporte SA

AtSc: Atmospheric Science (periodical code)

ATSC: Advanced Television Systems Committee • AlliedSignal Technical Services Corporation (USA-MD)

ATSCM: Antarctic Treaty Special Consultative Meeting

ATsir: Astron. Tsirk. (periodical code)

ATSIS: Automated Technical-Specification Information System

ATSM: Automated Technique for Spacecraft Monitoring

ATSOCC: Applications Technology Satellite Operations Control Center

ATSP: Airborne Tracking Solar Photometer • Asymmetric Travelling Salesman Program

ATSR: Activity Time Status Report • Along-Track Scanning Radiometer

ATSS: Acquisition and Tracking Subsystem • Automatic Test Support System

AtST: Atomisation Spray Technology (periodical code)

Att: Attach • Attache • Attaché • Attached • Attempt • Attempted • Attend • Attendance • Attended • Attending • Attente • Attention • Attitude • Attorney

AT&T: American Telephone & Telegraph Co.

ATT: Accelerated Test Technology • Acceptance Thermal Testing • Advanced Technology Telescope • Advanced Transport Technology • Advanced Transport Telematics • All Thrust Termination • Aménagement du Temps de Travail • Astrometric Technology Telescope • Astronomical Tele-Telescope • Astronomischer Tausch- und Trödeltreff (Germany) • Asynchronous Time Trap • Atmautluak (USA-AK) (location identifier) • Atterrissage Tout Temps • Automatic Target Tracking • Average Task Time

Atta: Atenta

ATTA: Advanced Terrain Tracking Altimeter • Atom Trap Trace Analysis

ATTAC: Advanced Target/Threat Assessment Code

ATTAS: Advanced Technology Testing Aircraft System • Advanced Transport Technology Aircraft System • Advanced Transport Testing Aircraft System

ATTC: Automatic Transmission Test and Control

Atte: Atentamente

ATTE: Automatic Transistor Test Equipment

ATTIC: Alternative Technology Treatment Information Center

ATTIS: AT&T Information System

ATTM: AEG Telefunken Technische Mitteilungen (periodical code) • Authorization To Transfer Material

Attmte: Atentamente

Attn: Attention

ATTN: Attention

Atto: Atento

ATTOM: Acoustic Travel Time Ocean Current Monitor

ATTR: Average Time To Repair

ATTRAS: Automatic Telemetry Tracking Antenna System

Attrib: Attribute • Attributed to • Attribution • Attributive • Attributively

ATTRIB: Attribute

ATTS: Automated Test Technology Standard • Automatic Telemetry Tracking System

ATTTJ: ATT Technical Journal (periodical code)

Atty: Attorney

Atty. Gen.: Attorney General

ATU: Advanced Training Unit • American Technological University (USA) • Arab Telecommunication Union • Arkansas Tech University (USA-AR) • Attu Island - Casco Cove (USA-AK) (location identifier) • Audio Terminal Unit • Automatic Tracking Unit • Autonome Transfer Unit • Auxiliary Test Unit

ATUC: African Trade Union Confederation • Australia Telescope Users Committee • Average Total Unit Cost

ATUG: Australian Telecommunications Users Group

ATV: Advanced Television • Advanced Test Vehicle • All-Terrain Vehicle • Amateur Television • Arbeidstijdverkorting • Ariane Transfer Vehicle • Associated Television • Ati (Chad) (location identifier) • Atmospheric Test Vehicle • Automate à Transitions Valuées • Automated Transfer Vehicle • Automatic Threshold Variation

ATVA: Advanced Technology Visible Array

ATVC: Ascent Thrust Vector Control • Ascent Thrust Vector Controller • Automatic Thrust Vector Control

ATVCD: Ascent Thrust Vector Control Driver

ATVCS: Automatic Thrust Vector Control System

ATVT: Alignment/Thermal Verification Tool

ATW: Appleton (USA-WI) (location identifier) • Menasha (USA-WI) (location identifier) • Neenah (USA-WI) (location identifier)

ATWDDS: Automated Terminal Weather Dissemination Display System

At. Wt.: Atomic Weight

ATX: Atbasar (Kazakhstan) (location identifier)

ATY: Watertown (USA-SD) (location identifier)

ATZ: Airport Traffic Zone • Assiut (Egypt) (location identifier)

au: Australia

Au: Author • Gold • Or

AÜ: Automatische Übersetzung

AU: Accounting Unit • Alignment Unit • All Up • Alternate Use • American University (USA-DC) • Analyzable Unit • Ångstrom Unit • Août • Apollo Unit • Arbitrary Unit • Archival User • Arithmetic Unit • Astronomical Unit • Atomic Unit • Auburn University (USA-AL) • Audio • August • Australia (ISO country code) • Australie (ISO country code) • Author • Authorized User • Auxiliary Unit

AUA: Aruba - Reina Beatrix (Netherlands Antilles) (location identifier) • Associations des Universités Africaines • Associations des Universités Arabes • Associations des Universités de l'Atlantique

AUAGA: All-Union Astronomical-Geodetical Association (USSR)

AUASS: Arab Union for Astronomy and Space Sciences

AUB: Aft Utility Bridge • Alstublieft • American University of Beirut (Lebanon) • Itauba (Brazil-MT) (location identifier)

AUBC: Association of Universities of the British Commonwealth

AUC: Ab Urbe Condita • American University of the Caribbean • Anno Urbis Conditae • Arauca (Colombia) (location identifier) • Area Under Curve (*) • Area Under the Curve • Australian Universities Commission • Average Unit Cost

AUCBE: Advisory Unit for Computer-Based Education (UK)

AUCC: Association des Universités et Collèges du Canada • Association of Universities and Colleges of Canada

Auct: Auctorum

AUCV: Advanced Unmanned Cargo Vehicle

Aud: Audience • Audiencia • Audit • Auditor

AUD: Augustus Downs (Australia-QLD) (location identifier) • Automatic Data Processing • Dollar (Australia) (ISO currency code)

AUDACIOUS: Automatic Direct Access to Information with the Online UDC System

AUDE: Association des Utilisateurs de Détecteurs Électroniques (France)

AUDF: Automated Unit Development Folder

AUDIT: Automated Data Input Terminal

AUDIX: Audio Information Exchange

AUE: Abu Rudeis (Egypt) (location identifier) • Acta Unica Europea • Acte Unique Européen • Association des Universitaires d'Europe

AUEF: Association Universités-Entreprises pour la Formation

AUF: Agence Universitaire de la Francophonie • Authorized User File

Aufl: Auflage

Aug: Augmentation • Augmentative • Augmented • Augmenter • August

AUG: Arbeit-Umwelt-Gesundheit • August • Augusta (USA-ME) (location identifier)

Augm: Augmentative

AUH: Abu Dhabi (UAE) (location identifier)

AUI: Akhawayn University in Ifrane (Morocco) • Associated Universities Inc. (USA-DC) • Association des Utilisateurs d'Internet (France) • Attached Unit Interface • Attachment Unit Interface • Attachment Universal Interface • Aua Island (Papua New Guinea) (location identifier) • Auxiliary Unit Interface

Auj: Aujourd'hui

AUJ: Active Universal Joint • Ambunti (Papua New Guinea) (location identifier)

AuJPA: Australian Journal of Physics - Astrophysical Supplement (periodical code)

AuJPh: Australian Journal of Physics (periodical code)

AuJPS: Australian Journal of Physics - Supplement (periodical code)

AUK: Alakanuk (USA-AK) (location identifier)

AUL: Astronomy User's Library • Aur Island (Marshall Islands) (location identifier) • Average Useful Life

AULC: Arizona Universities Library Consortium (USA-AZ)

Aum: Aumentativo • Aumento

AUM: Adaptive User Model • Animal Unit Months • Association Universitaire de Mécanique • Austin (USA-MN) (location identifier)

AuMM: Australian Meteorological Magazine (periodical code)

AuMSJ: Australian Mathematical Society Journal (periodical code)

aun: Absque Ulla Nota

AUN: Auburn (USA-CA) (location identifier)

AUO: Administratively Uncontrollable Overtime • African Unity Organization • Auburn Opelika (USA-AL) (location identifier)

AUP: Aberdeen University Press • Acceptable Use Policy • Actual Unit Price • Agaun (Papua New Guinea) (location identifier) • American University of Paris (France) • Appropriate Use Policy • Astroparticle and Underground Physics

AUPELF: Association des Universités Partiellement ou Entièrement de Langue Française

AUPG: American University Publishers Group (USA)

AUQ: Atuona (Marquesas Islands) (location identifier)

Aur: Auriga-Aurigae (constellation)

AuR: Astronomie und Raumfahrt

AUR: All Up Around • Aurillac (France) (location identifier) • Automated Underreporter

AURA: Association of Universities for Research in Astronomy (USA-DC) • Atmospheric Ultraviolet Radiance Analyzer

AURBO: Aurora Borealis

AURELHY: Analyse Utilisant le Relief pour l'Hydrométéorologie

AURIO: Auroral Imaging Observatory

AUROC: Association des Utilisateurs du Réseau OCLC

AURP: AppleTalk Update-based Routing Protocol

AurPh: Auroral Physics (periodical code)

Aus: Australia • Australian • Austria • Austrian

AUS: Austin - Robert Mueller (USA-TX) (location identifier) • Australia (international vehicle registration code) • Australia (ISO country code) • Australie (ISO country code)

AUSAC: Arecibo Users and Science Advisory Council

AUSDEC: Australasian Spatial Data Exchange Centre

AUSEANET: Austroasian Network

Ausg: Ausgabe

AUSINET: Australian Information Network

AUSLAS: Austrian Laser Station

AUSLIG: Australian Survey and Land Information Group • Australian Surveying and Land Information Group (*)

AUSM: Advection Upstream Splitting Method

Aust: Australia • Australian • Austria • Austrian

Austr: Australia • Australian • Austria • Austrian

Austral: Australasia • Australasian • Australia • Australian

Aut: Auteur • Author • Autograph • Automatique • Automne • Automobile • Autonomy • Autorisation • Autriche • Autumn

AUT: Advanced User Terminal • Association of University Teachers (UK) • Atauro (Indonesia) (location identifier) • Austria (ISO country code) • Automated Ultrasonic Testing • Autriche (ISO country code)

Auth: Authentic • Author • Authoress • Authority • Authorization • Authorize • Authorized

Auto: Automatic • Automobile • Automotive

AUTO: Automatic

Autobiog: Autobiographical • Autobiography

AUTOBUS: Automated Budget System

AUTODIN: Automatic Digital Network

AUTOEXEC: Automatic Execution

AUTOFIB: Automatic Fibre

Autog: Autograph

Autogr: Autographe

Autom: Automatica (periodical code) • Automatique

Automov: Automovilismo

Auton: Autonomous

AUTOPIC: Automatic Personal Identification Code

AUTOSCRIPT: Automatic System for Composing, Revising, Illustrating, and Phototypesetting

AUTOSPEC: Automated Specification

Autostr: Autostrade

AUTOVON: Automatic Voice Network

Autp: Autopista

Autref: Autrefois

Autv: Autovía

AUU: Aurukun Mission (Australia-QLD) (location identifier)

AUUA: Americas UNIVAC Users Association

AUV: Auslandumzugskosten-Verordnung • Autonomous Underwater Vehicle

AUW: All-Up Weight • Wausau (USA-WI) (location identifier)

AUWS: Automatic Unmanned Weather Station

Aux: Auxiliary

AUX: Araguaína (Brazil-TO) (location identifier) • Auxiliary

AUXF: Auxiliary Frame

Auxil: Auxiliary

AUY: Aneityum (Vanuatu) (location identifier)

AUZ: Aurora (USA-IL) (location identifier)

av: Ad Valorem • Annos Vixit

Áv: Ávila (Spain)

Av: Avance • Avant • Avenida • Avenue • Average • Avocat • Avoirdupois • Avril

A&V: Assembly & Verification

A/V: Ad Valorem

AV: Abnormal Voltage • Acid Value • Acoustic Velocity • Actual Value • Acuité Visuelle • Ad Valorem • Antipodal Void • Aperture Value • Arteriovenous • Asset Value • Astronomicheskij Vestnik • Atharvaveda • Atrioventicular • Audiovisual • Augmented Virtuality • Authorized Version • Avenida • Avenue • Average Value • Average Variability • Avis de Versement • Avis de Virement

AVA: Aerodynamische Versuchanstalt • Asociación Valenciana de Astronomía (Spain) • Astronomische Vereinigung Aarau (Switzerland) • Astronomische Vereinigung Albstadt (Germany) • Astronomische Vereinigung Augsburg (Germany) • Audiovisual Authoring

AVADS: Automated Vacancy Announcement Distribution System

Avail: Availability • Available

AVASI: Abbreviated Visual Approach Slope Indicator

AVB: Analog Video Bandwidth • Aviano (Italy) (location identifier)

AVBATTS: Aviation Battery Standardization

Avbl: Available

AVBS: Absolute Value Bit Synchronizer

av. C.: Avanti Cristo

AVC: Additional Voluntary Contribution • Audio Visual Connection • Automatic Vibration Control • Automatic

Voltage Control • Automatic Volume Control • Average Variable Cost

AVCG: Automatic Vapor Crystal Growth

AVCS: Advanced Vidicon Camera System • Attitude and Velocity Control System

Avd: Avenida • Avoid

AVD: Acoustic Variable Density • Air Vehicle Detection • Alternate Voice Data • Anode Voltage Drop • Astronomischer Verein Dortmund (Germany) • Automatic Voice Data • Automatic Voltage Digitizer • Avant Droit

Avda: Avenida

AVDA: Associazione Vigevanese Divulgazione Astronomica (Italy)

avdp: Avoirdupois

AVDU: Audiovisual Display Unit

Ave: Avenue • Average

AVE: Aerospace Vehicle Electronics • Aerospace Vehicle Equipment • Airborne Vehicle Equipment • Alta Velocidad • Avenue • Aerospace Vehicle Equipment • Atmospheric Variability Experiment • AutoCAD Visual Extension • Automatic Volume Expansion

AVEN: Avenue

AVENSA: Aérovías Venezolanas SA

AVENU: Avenue

AVERT: Automatic Verification, Evaluation and Readiness Tester

AVESCA: Aerovías Especiales de Carga Ltda.

AVF: Astronomische Vereinigung Frauenfeld • Avoriaz (France) (location identifier)

Avg: Average

AVG: Allgemeine Verlagsgesellschaft • Amateurastronomische Vereinigung Göttingen (Germany) • American Volunteers Group • Approximation de la Vallée Généralisée • Audiovisual Group • Auvergne (Australia-NT) (location identifier) • Avant Gauche • Average • Avion à Grande Vitesse

AVGB: Astronomischer Verein der Grafschaft Bentheim (Germany)

Avge: Average

AVGR: Ames Vertical Gun Range

AVGSE: Average String Error

AVGWE: Average Word Error

AvH: Alexander von Humboldt (1769-1859)

AVH: Almenvitenskapelige Høyskole (Norway) • Astronomischer Verein Hoyerswerda eV (Germany)

AVHRR: Advanced Very High-Resolution Radiometer

AVI: Air Velocity Index • Arbitrary Variant of an Image • Audio Video Interlaced • Audio Video Interleave • Audio Video Interleaved • Audio Visual Interleaved • Automatic Vehicle Identification • Ciego de Avila (Cuba) (location identifier)

Aviac: Aviación

AVIACO: Aviación y Comercio SA

AVIANCA: Aérovías Nacionales de Colombia SA

AVIC: Aviation Industries of China (China-PRC)

AVICA: Association des Villes Câblées (France)

AVID: Advanced Visual Information Display • Automated Verification Information Database

AVIRG: Audiovisual Information Research Group

AVIRIS: Airborne Visible and Infrared Imaging Spectrometer • Airborne Visible/Infrared Imaging Spectrometer

AVIS: Advanced Visual Information System • Audiovisual Information System • Automatic Visual Inspection System

AVISO: Archivage, Validation et Interprétation des Satellites Océanographiques

AVISURS: Aerospace Vehicle Inventory, Status and Utilization Reporting System

av. J-C: Avant Jésus-Christ

av. JC: Avant Jésus-Christ

AVK: Arvaikheer (Mongolia) (location identifier) • Astronomische Vereiningung Karlsruhe (Germany) • Astronomische Vereiningung Kärtens (Austria) • Astronomische Vereiningung Kreuzlingen (Switzerland) • Audio Video Kernel

AVL: Address Validity • Adel'son-Vel'skii-Landis • Advanced Visualization Laboratory • Angle Versus Length • Asheville (USA-NC) (location identifier) • Automatic Vehicle Location • Avionics Verification Laboratory • Hendersonville (USA-NC) (location identifier)

Avlbl: Available

AVLIS: Atomic Vapor Laser Isotope Separation

AVLN: Automatic Vehicle Location and Navigation

AVLSI: Advanced Very-Large-Scale Integration

AVM: Accelerated Value Model • Audiovisual Modulator • Automatic Vehicle Monitoring

AVMI: Automated Video Maintenance Information

AVML: Asiatic Vegetarian Meal

AVMS: Automatic Vehicle Monitoring System

Avn: Aviation

AVN: Astronomische Vereinigung Nürtingen eV (Germany) • Avenue • Aviation • Avignon Caumont (France) (location identifier)

AVNA: Automatic Vector Network Analyzer

AVNIR: Advanced Visible and Near Infrared Radiometer

AVNL: Automatic Video Noise Limiter

AVNUE: Avenue

AVO: Algemeen Vormend Onderwijs • Astronomical Virtual Observatory • Astronomischer Verein Olten (Switzerland) • Astronomischer Verein Ortenau (Germany) • Average Vehicle Occupancy • A View Of • Avoid Verbal Orders • Avon Park (USA-FL) (location identifier)

AVOSS: Aircraft Vortex Spacing System

AVP: Adaptive Video Processor • Address Verification Pulse • Application Validation Project • Wilkes Barre (USA-PA) (location identifier) • Scranton (USA-PA) (location identifier)

AVPD: Antivirus Product Developers

AVPN: ATM Virtual Private Network

Avr: Avril

AVR: Arbeitskreis Volkssternwarte Recklinghausen (Germany) • Automatic Voice Relay • Automatic Voice Recognition • Automatic Voltage Regulation • Automatic Voltage Regulator • Automatic Volume Recognition • Avril • Axial Velocity Ratio

AVRA: Automatic Visual Range Assessor

AVRCS: Attitude-Velocity-Reaction Control System

AVRG: Automated Virtual Reality Generator

AVRP: Audiovisual Recording and Presentation

AVRS: Astronomischer Verein Remscheid eV (Germany)

AVS: Advanced Visualization System • Advanced Visual System • Aerospace Vehicle System • Altitude Vertical Scale • American Vacuum Society (USA) • Application Visualization Software • Application Visualization System • Artificial Visual System • Audiovisual System • Automated Verification System • Autonomous Vision System • Average Speed

AVSA: Associazione Valdostana Scienze Astronomiche (Italy)

AvSeM: Aviation, Space, and Environmental Medicine (periodical code)

AVSM: Auxiliary Video Switching Matrix • Auxiliary Video Switch Matrix

AVSO: Allgäuer Volkssternwarte Ottobeuren (Germany)

AvSP: Aviation Safety Programme

AVSS: Audio Video Support System

AVSSE: Atmospheric Variability Severe Storms Experiment

AVST: Advanced Video Simulation Technique

AVSTC: Apple Valley Science and Technology Center (USA-CA)

AVSTOVL: Advanced Short Takeoff and Vertical Landing

Avt: Avant

AVT: Acceptance Vibration Testing • Analog Voice Telephony • Applied Voice Technology • Astronomische Vereinigung Tübingen (Germany) • Audiovisual Tutorial

Avtn: Aviation

AVU: Artificial Vision Unit • Avu Avu (Solomon Islands) (location identifier)

AVUS: Automobilverkehrs- und Übungstraße

AVUSD: Apple Valley Unified School District (USA-CA)

AVV: Avalon (Australia-VIC) (location identifier)

AVVEU: Altitude and Vertical-Velocity Electronic Unit

AVVI: Altitude and Vertical-Velocity Indicator

AVVID: Architecture for Voice, Video and Integrated Data (Cisco)

AVW: Tucson - Avra Valley (USA-AZ) (location identifier)

A&VWG: Assembly & Verification Working Group

AVWG: Assembly and Verification Working Group

AVWM: Astronomische Vereinigung West-München (Germany)

AvWST: Aviation Week and Space Technology (periodical code)

AVX: Catalina Island - Avalon Bay (USA-CA) (location identifier)

AVZ: Analysen- und Vorhersagenzentrale • Astronomische Vereinigung Zürich (Switzerland)

A/W: Actual Weight

AW: Accompanying Wave • Actual Weight • Administrative Weight • Airlock Wall • Airworthiness • Alfén Wave • All Water • Armstrong-Whitworth • Aruba (ISO country code) • Assembly Workstand • Astronomical Watch • Atomic Weight • Axial Width • Schreiner Airways BV

AWA: Advise When Able • African West Air • All Wave Antenna • Argonne Wakefield Accelerator • Audio Wave Analyzer • Awassa (Ethiopia) (location identifier) • Away Without Authorization

AWAA: Attack Warning and Attack Assessment

AWAC: Airborne Warning And Control

AWACS: Airborne Warning And Control System

AWAD: Assembly Work Authorization Document

AWAGS: Australia-Wide Array of Geomagnetic Stations

AWAR: Area Weighted Average Resolution

AWAS: Automated Work Authorization System

AWB: Airway Bill • Awaba (Papua New Guinea) (location identifier)

AWBE: Automatic Weather Broadcast Equipment

AWC: Arbitrary Waveform Composer

AWCS: Agency-Wide Coding Structure • Automated Wafer Cartridge System • Automatic Warning and Control System

AWCU: Association of World Colleges and Universities (now AWE)

AWD: African Wave Disturbance • Aniwa (Vanuatu) (location identifier)

AWDAAS: Automated Weather Datat Acquisition and Archiving System

AWDL: Aviation Weather Development Laboratory

AWDS: Automated Weather Distribution System

AWE: Adaptable Windowing Environment • All-Weather Electronics • Alowe (Gabon) (location identifier) • Association for World Education (formerly AWCU) • Atomic Weapons Establishment (UK)

AWEA: American Wind Energy Association (USA)

AWES: Area Weapons Effects System

AWG: Activation Working Group • American Wire Gauge • Astronomy Working Group • Aviation Week Group • Guilder (Aruba) (ISO currency code)

AWGAA: Australian Working Group for Antarctic Astronomy

AWGN: Additive With Gaussian Noise

AWH: Alfvén Wave Heating • Awareh (Ethiopia) (location identifier)

AWHFY: Are We Having Fun Yet

AWI: Accommodation Weight Investigation • Alfred Wegener Institute • Assembly Work Instruction

AWIN: Aviation Weather Information

AWIPS: Advanced Weather Interactive Processing System

AWIS: Airport Weather Information System • Association for Women In Science (USA)

AWISE: Association for Women in Science and Engineering

AWJ: Abrasive Water Jet

AWK: Aho-Weinberger-Kernighan • Wake Island (US Minor Outlying Islands) (location identifier)

AWL: Absent With Leave • Absent Without Leave • All-Weather Landing • Application Wrapper Layer • Atmospheric Waves Lidar • Augmented Wire List • Automated Wire List • Average Work Load

AWLS: All-Weather Landing System

AWM: Association for Women in Mathematics (USA) • Automatic Writing Machine • West Memphis (USA-AR) (location identifier)

AWN: Alton Downs (Australia-SA) (location identifier) • Automated Weather Network

AWOL: Absent Without Leave

AWON: Stichting Aardwetenschappelijk Onderzoek (Netherlands)

AWOS: Absent Without Leave • Automated Weather Observation System

AWP: Annual Work Plan • At Work Printing • Austral Downs (Australia-NT) (location identifier) • Aviation Weather Processor

AWPA: American Word Processing Association (USA)

AWPG: Aviation Weather Products Generator

AWPSP: Association Wallonne du Personnel des Services Publics (Belgium)

AWPV: Automated Weld Process Verification

AWR: Advanced Weather Radar • Airborne Weather Radar • Automated Work Request • Awar (Papua New Guinea) (location identifier)

AWRA: American Water Resources Association (USA)

AWRE: Autowrite Relocation

AWS: Acoustic Well Sounding • Advanced Warning System • Advanced Workstations and Systems • Air Weather Service • Air Weather Station • All-Weather System • Alternating Work Schedule • Altitude Warning System • American Weather Service (USA) • Astronaut Work Station • Automated Wiring System • Automatic Warning System • Automatic Weather Station

AWSK: Amsterdaamse Weer- en Sterrenkundige Kring (Netherlands)

AWSO: All-Weather Surface Observations

AWSt: Automatische Wetterstation

AW&ST: Aviation Week & Space Technology

AWT: Abstract Windowing Toolkit (*) • Abstract Windows Toolkit • Actual Work Time • Advanced Windowing Toolkit • Average Working Time

AWTB: All-Weather Test Bed

AWU: Atomic Weight Unit

AWV: Atmospheric Wind Velocity

AWVVO: Außenwirtschaftsverordnung

AWY: Airway

AWZ: Ahwaz (Iran) (location identifier)

Ax: Axial • Axiom

AX: American Express • Architecture Extended • Automatic Transmission

AXA: Anguilla (Leeward Islands) (location identifier)

AXAF: Advanced X-ray Astrophysical Facility • Advanced X-ray Astrophysics Facility (now CXO) (*)

AXAF-I: AXAF Imaging

AXAFI: AXAF Imaging

AXAF-S: AXAF Spectroscopy

AXAFS: AXAF Spectroscopy

AXB: Alexandria Bay (USA-NY) (location identifier)

AXBT: Airborne Expendable Bathythermograph

AXC: Aramac (Australia-QLD) (location identifier)

AXCP: Airborne Expendible Current Profiler

AXCTD: Autonomous - Expendable Conductivity - Temperature Depth

AXD: Alexandroupolis (Greece) (location identifier)

Axe: Annexe

AXEL: Astrophysical Experimental Library

AXG: Algona (USA-IA) (location identifier)

AXI: Automated X-ray Inspection

AXIS: Atmospheric X-ray Imaging Spectrometer

AXK: Ataq (Yemen) (location identifier)

AXL: Alexandria (Australia-NT) (location identifier)

AXM: Armenia (Colombia) (location identifier)

AXN: Alexandria (USA-MN) (location identifier)

AXP: Anomalous X-ray Pulsar • Spring Point (Bahamas) (location identifier)

AXR: Arutua (Tuamotu Islands) (location identifier)

AXS: Alpha-backscatter/X-ray-fluorescence Spectrometer • Alpha-proton X-ray-fluorescence Spectrometer • Alpha-proton X-ray Spectrometer • Altus (USA-OK) (location identifier)

AXT: Akita (Japan) (location identifier) • Anomalous X-ray Pulsar

AXU: Axum (Ethiopia) (location identifier)

AXV: Wapakoneta - Neil Armstrong (USA-OH) (location identifier)

AXX: Angel Fire (USA-NM) (location identifier)

ay: Aymara (ISO language code)

AY: Alderney (Channel Islands) • Assyrian

AYA: Association of Yale Alumni (USA-CT) • Ayapel (Colombia) (location identifier)

AYC: Ayacucho (Colombia) (location identifier)

AYD: Alroy Downs (Australia-NT) (location identifier)

AYE: At Your Expenses • Fort Devens (USA-MA) (location identifier)

AYG: Yaguara (Colombia) (location identifier)

AYH: Alconbury (UK) (location identifier) • American Youth Hostels

AYHA: American Youth Hostels Association (USA)

AYI: Yari (Colombia) (location identifier)

AYK: Arkalyk (Kazakhstan) (location identifier)

AYL: Anthony Lagoon (Australia-NT) (location identifier)

AYLI: As You Like It

AYN: Anyang (China-PRC) (location identifier)

Ayo: Arroyo

AYO: Ayolas (Paraguay) (location identifier)

AYP: Ayacucho (Peru) (location identifier)

AYQ: Ayers Rock (Australia-NT) (location identifier)

Ayr: Ayrshire (UK)

AYR: Ayr (Australia-QLD) (location identifier)

AYS: Waycross - Ware County (USA-GA) (location identifier)

AYT: Antalya (Turkey) (location identifier) • Are You There?

Ayto: Ayuntamiento

AYU: Aiyura (Papua New Guinea) (location identifier)

AYW: Ayawasi (Indonesia) (location identifier)

AYZ: Amityville Zahns (USA-NY) (location identifier)

az: Azerbaijani (ISO language code) • Azimuth • Azure

aZ: auf Zeit

Az: Azimuth

AZ: Arizona (USA) • Azerbaijan (ISO country code) • Azimuth • Azinhaga • Aztec

AZB: Amazon Bay (Papua New Guinea) (location identifier)

AZD: Yazd (Iran) (location identifier)

AZE: Azerbaijan (ISO country code)

AZ-EL: Azimuth-Elevation

AZEL: Azimuth-Elevation

Azer: Azerbaijan

AZG: Apatzingan (Mexico) (location identifier)

AZh: Astronomicheskij Zhurnal (periodical code)

AZH: Astronomisches Zentrum Halberstadt (Germany)

AZI: Abu Dhabi - Bateen (UAE) (location identifier)

AZK: Alphabetischer Zentralkatalog

Azm: Azimuth

AZM: Astronomisches Zentrum Magdeburg (Germany) • Manat (Azerbaijan) (ISO currency code)

AZN: Andizhan (Uzbekistan) (location identifier)

Azo: Azores

AZO: Allgemeine Zollordnung • Arbeitszeitordnung • Kalamazoo (USA-MI) (location identifier)

AZOP: Antarctic Zone Of Peace

AZP: Mexico City - Atizapan (Mexico) (location identifier)

AZR: Adrar (Algeria) (location identifier)

AZS: Astronomisches Zentrum Schkeuditz (Germany) • Automatic Zero Set

AZT: Zapatoca (Colombia) (location identifier)

AZZ: Ambriz (Angola) (location identifier)

A0: Standard Paper Size

A1: Standard Paper Size

A10: Standard Paper Size

A2: Botswana (international aircraft registration code) • Standard Paper Size

A2LA: American Association for Laboratory Accreditation (USA)

A3: Astronomes Amateurs Aixois (France) • Standard Paper Size • Tonga (international aircraft registration code)

A3C: Action Concertée Chambre de Combustion

A3I: Accelerated Advance Acquisition Initiative

A3IE: Association des Ingénieurs de l'Institut d'Informatique d'Entreprise

A4: Standard Paper Size

A4O: Oman (international aircraft registration code)

A5: Bhutan (international aircraft registration code) • Standard Paper Size

A6: Standard Paper Size • UAE (international aircraft registration code) • United Arab Emirates (international aircraft registration code)

A7: Qatar (international aircraft registration code) • Standard Paper Size

A8: Standard Paper Size

A9: Standard Paper Size

A9C: Bahrain (international aircraft registration code)

B

b: Bachelor • Bajo • Bale • Bar • Barn • Bass • Battle • Bay • Beam • Bedroom • Before • bei • beim • Billion • Bin • Binding Factor • Bis • Bit • Black • Blend • Blended • Book • Born • Bottom • Bound • Breadth • Brother • Brotherhood • Burgers Vector • By • Collision Parameter • Ecliptic Latitude • Galactic Latitude • Heliocentric Latitude • Impact Parameter • Magnetic-Field Strength • Mobility Ratio • Phonon Annihilation Operator • Semiminor Axis • Wien Displacement Law Constant

B: Baai • Bachelor • Bacillus • Bad • Baht (currency) • Bahia • Baía • Baie • Baja • Balboa (currency) • Ban • Band • Banque • Barrage • Baryon Number • Base • Bass • Baud • Baumé • Bay • Beatus • Beau • Bel • Belgium (international vehicle registration code) • Belgium (postal country code) • Bien • Billion • Birmingham (UK postal code) • Bishop • Black • Block • Blue • Board • Bois • Bolívar (currency) • Boliviano • Book • Bor • Born • Boron • Boundary Conditions • Bowen Ratio • Bravo (international alphabet) • Breadth • Breakdown Potential • Brewster • Brief • British • Brother • Brotherhood • Bucht • Building • Bulb • Bulletin • Byte • China (PRC) (international aircraft registration code) • Ecliptic Latitude of the Sun • Eleven • Magnetic Flux Density • Magnetic Induction • Mass Defect • Oort's Constant of Galactic Rotation • PRC (international aircraft registration code) • Rotational Constant • Secondary Road • Spectral Type • Susceptance • Three Hundred (Roman numeral) • 300 (Roman numeral)

ba: Bashkir (ISO language code)

Ba: Bahía • Bath • Bathroom • Barium • Baryum

B/A: Bank Angle • Barometric Altimeter • Boost/Abort • Buffer Amplifier

BA: Bachelor of Arts • Bagages Accompagnés • Bahía (Brazil) • Balancing Allowance • Bangladesh • Bank Acceptance • Bank Angle • Bantu • Barometric Altimeter • Bas Âge • Base Assembly • Basic Agreement • Basic Authorization • Basses-Alpes (France) • Bath (UK postal code) • Behavior Aggregate • Bibliography of Astronomers • Bilan Annuel • Binary Add • Blind Approach • Blockausgleichung • Boolean Algebra • Bordereau d'Achats • Bosnia-Herzegovina (ISO country code) • Bosnie-Herzégovine (ISO country code) • Bourse d'Apprentissage • Breathing Air • Bridge Amplifier • British Academy • British America • British Association • Bronze Age • Buenos Aires (Argentina) • Buffer Amplifier • Bulletin d'Admission • Bureau Auxiliaire • Bureau d'Affrètement • Bureau d'Archives • Bus Automaton • Bus Available

BAA: Bachelor of Applied Arts • Bialla (Papua New Guinea) (location identifier) • British Airport Authority • British Astronomical Association • Broadband Active Analyzer • Buffalo Astronomical Association (USA-NY) • Burgenländische Amateurastronomen (Austria) • Business Area Analysis

BAAA: Boletín de la Asociación Argentina de Astronomía (periodical code)

BAAE: Bachelor of Aeronautical and Astronautical Engineering

BAAM: Basic Administration And Management

BAAPG: Bulletin of the American Association of Petroluem Geologists (periodical code)

BAAS: British Association for the Advancement of Science (UK) • Bulletin of the American Astronomical Society (periodical code)

Bab: Babylon

BAB: Booster Assembly Building • Budget Advisory Board • Bundesautobahn • Bureau pour l'Automatisation des Bibliothèques (France) • Marysville Beale (USA-CA) (location identifier)

BABAS: Bay Area Book Awareness Service (USA-CA)

BABEL: Broadcasting Across the Barriers of European Languages

BABOGAS: Balloon-Borne Gaschromatograph

BABS: Beam Approach Beacon System • Blind Approach Beacon System

BABT: British Approval Board for Telecommunications

Bac: Baccalauréat (France)

BAC: Barranca de Upia (Colombia) (location identifier) • Barometric Altitude Control • Basic Attitude Control • Beijing Astrophysics center (China-PRC) • Beleid Advies Commissie • Binary-Analog Conversion • Binary Asymmetric Channel • Blood-Alcohol Concentration • Boeing Aerospace Co. • Boeing Aircraft Co. • Booster Assembly Contractor • Bordeaux Astrographic Catalogue • Bristol Astronomy Club (UK) • British Aerospace Corporation • Broad Acoustic Category • Budget At Completion • Buffer Access Card • Bulletin d'Activité Cyclonique • Bureau d'Archives Collectives

BACA: Beleid Advies Commissie Aardwetenschappen (Netherlands)

BACE: Basic Automatic Checkout Equipment

Bach: Bachelor

BACH: Beautiful Aperiodic Crystal of Harmony

BACIS: Budget Accounting Information System

Backgrd: Background

BACL: Beleid Advies Commissie Levenwetenschappen (Netherlands)

BACODINE: BATSE Coordinates Distribution Network

BACP: Bandwidth Allocation Control Protocol

BACS: Bank Automated Clearing System (UK) • Bibliographic Access and Control System • Body Axis Coordinate System

Bact: Bacterial • Bacteriology • Bacterium

BACT: Best Available Control Technology

BACV: Budget At Completion Variance

Bad: Badajoz (Spain)

BAD: Banque Africaine de Développement • Broken As Designed • Shreveport Barksdale (USA-LA) (location identifier)

BADAS: Binary Automatic Data Annotation System • Blackpool And District Astronomical Society (UK)

BADC: Beijing Astronomical Data Centre (China-PRC) • Binary Asymmetric Dependent Channel • British Atmospheric Data Centre

BADEA: Banque Arabe de Développement Économique en Afrique

BAdW: Bayerische Akademie der Wissenschaften (Germany)

BAe: British Aerospace

BAE: Bachelor of Aeronautical Engineering • Barcelonnette (France) (location identifier) • Beacon Antenna Equipment • Biblioteca de Autores Españoles • Bureau of American Ethnology

BAeA: British Aerospace Australia

BAEF: Belgian-American Educational Foundation

BAeSC: British Aerospace Space and Communications

BAF: Bacterial Air Filter • Barnes Absorption Filter • Bâtiment d'Assemblage Final • Benefield Anechoic Facility (USA) • Biological Air Filter • Westfield Barnes (USA-MA) (location identifier)

BAFA: Brevet d'Aptitude aux Fonctions d'Animateur

BAFI: Base des Agents Financiers

BAFO: Best And Final Offer • British Air Forces of Occupation

Bag: Bagages • Baggage

BAG: Baguio (Philippines) (location identifier) • Branche Asymptotique des (étoiles) Géantes • Bundesamt für Gesundheitswesen (Switzerland)

BAGBI: Booksellers' Association of Great Britain and Ireland

BAGI: Backscatter Absorption Gas Imaging

Bah: Bahamas

BAH: Bahrain Muharraq (Bahrain) (location identifier) • Basic Adaptive Hardware

BAHC: Biospheric Aspects of the Hydrological Cycle

BAI: Barometric Altitude Indicator • Bearing Altitude Indicator • Beratende Ausschuß der Industriephysiker (Germany) • Budget Advisory Board • Buenos Aires (Costa Rica) (location identifier)

BAIC: Binary Asymmetric Independent Channel

BAICz: Bulletin of the Astronomical Institutes of Czechoslovakia (periodical code)

BAID: Boolean Array Identifier

BAIE: British Association of Industrial Editors (UK)

BAILER: British Association for Information and Library Education and Research (UK)

BAIR: Breathing Air

Ba. Is.: Bahamas Islands

BAIT: Berkeley Automated Imaging Telescope • Berkeley Automatic Imaging Telescope

BAJ: Bali (Papua New Guinea) (location identifier)

BAK: Backup • Baku (Azerbaijan) (location identifier) • Bildaufklärung • Binary Adaptation Kit

BAKSA: Bahagian Kajian Sains Angkasa (Malaysia)

Bal: Balance • Balanced • Baluchistan

BAL: Balance • Basic Assembly Language • Batman (Turkey) (location identifier) • Blood Alcohol Level • Boîte Aux Lettres • Broad Absorption Line

BALAST: Balloon Astronomy

BALI: Backweb Authoring Language Interface

BALIS: Bavarian Agriculture Information System (Germany)

Balk: Balkan

BALM: Block and List Manipulator

Baln: Balnéaire

BALO: Bulletin d'Annonces Légales Officielles

BAlp: Basses-Alpes (France)

BALQ: Broad Absorption Line Quasar

BALQSO: Broad Absorption Line Quasi-Stellar Object

BALR: Broad Absorption Line Region

BALSA: Brown University Algorithm Simulator and Animator

BALSAMINE: Ballons Surpressurisés pour l'Analyse de la Mousson Indienne d'Été

Balt: Baltic • Baltimore (USA-MD)

BaltA: Baltic Astronomy (periodical code)

BALUN: Balanced/Unbalanced

BAM: Bachelor of Applied Mathematics • Bacteriology and Applied Microbiology • Balloon Anisotropy Measurement • Basic Access Method • Basic Access Mode • Basic Acoustic Modelling • Battle Mountain - Lander County (USA-NV) (location identifier) • Bifunctional Adaptive Mesh • Bit Assignment Matrix • Boyan Action Module • Brackett Alpha Measurement • Bradley Aberration Method • Bundesanstalt für Materialforschung (Germany)

BAMBI: Better Access to Manuscripts and Browsing of Images

BAMF: Bundesamt für Militärflugplätze (Germany)

BAMIN: Biology/Applied Mathematics Instruction Model

BAMM: Balloon Altitude Mosaic Measurement

BAMN: By Any Means Necessary

BAMO: Polyazidomethyloxetane

BAN: Base Aéronavale • Basongo (Zaire) (location identifier) • Boundary-Access Node • Budget Allocation Notice • Bulletin of the Astronomical Institutes of the Netherlands (periodical code)

BANANA: Build Absolutely Nothing Anywhere Near Anyone

BANCS: Bell Application Network Control System

BAND: Boston Area NEXRAD Demonstration

Banl: Banlieue

Banq: Banque

BANR: Board on Agriculture and Natural Resources

BANS: Bulletin of the Astronomical Institutes of the Netherlands, Supplement Series (periodical code)

BANZARE: British, Australian, New Zealand Antarctic Research Expedition

BAO: Ban Mak Khaen - Udorn (Thailand) (location identifier) • Beijing Astronomical Observatory (China-PRC) • Bisei Astronomical Observatory (Japan) • Boulder Atmospheric Observatory (USA-CO) • Byurakan Astronomical Observatory (Armenia)

bàp: Billets à Payer

Bap: Baptist • Baptized

BAP: Baibara (Papua New Guinea) (location identifier) • Bandwidth Allocation Protocol • Basaltic Achondrite Parent • Basic Assembler Program • Best Adaptive Path • Biotechnology Action Programme • Bottom Agitation Process • Branche d'Activité Professionnelle • Brevet d'Aptitude Physique • British Association of Planetaria (*) • British Association of Planetariums • Bulletin d'Alerte aux Précipitations

BAPE: Balloon Atmospheric Propagation Experiment

BAPI: Business Applications Programming Interface

BAPMoN: Background Air Pollution Monitoring Network (*) • Baseline Air Pollution Monitoring Network

BAPS: Baseline Air Pollution Station • Berg Analysis and Prediction System

Bapt: Baptist • Baptized

BAPT: Bundesamt für Post und Telekommunikation (Germany)

BAPTA: Bearing And Power Transfer Assembly

BAPU: Bureau d'Aide Psychologique Universitaire

BAQ: Barranquilla - Ernesto Cortissos (Colombia) (location identifier)

bàr: Billets à Recevoir

Bar: Barometer • Barometric • Baron • Barones • Baronne • Barrel • Barrister

BAR: Baker Island (USA-AK) (location identifier) • Ballistocraft for Acceleration Research • Barão • Base Address Register • Bucket Activation Ratio • Buffer Address Register

Barb: Barbados

BARB: Ballast Aerating Retrieval Boom

Barc: Barcelona (Spain)

BARC: Bay Area Research Center (USA-CA) • Bhabha Atomic Research Centre (India) • Block Adaptive Rate Controlled

BARD: Bay Array Regional Deformation (USA-CA) • Binational Agricultural Research and Development

Barg: Bargain

BARN: Biological and Agricultural Reactor Netherlands

Baro: Barometer

BARO: Barometer

BARON: Baronesa

BARP: Bundesamt für Raumplanung (Switzerland)

Barr: Barrister

BARRNet: Bay Area Regional Research Network (USA-CA)

BARS: Backup Attitude Reference System • Barred And Ringed Spirals • Baseline Accounting and Reporting System • Bistatic Auroral Radar System • Budget Analysis Reporting System

BARSC: British Association of Remote Sensing Companies

BART: Bay Area Rapid Transit (USA-CA) • Billing Analysis Reporting Tool

BARTOC: Bus Advanced Real-Time Operational Control

BARTS: Bell Atlantic Regional Timesharing

BARWAN: Bay Area Research Wide Area Network

BAS: Bachelor of Applied Sciences • Balalae (Solomon Islands) (location identifier) • Ballarat Astronomical Society (Australia-VIC) • Baltimore Astronomical Society (USA-MD) • Barbados Astronomical Society • Barnard Astronomical Society (USA-TN) • BASIC • Bedford Astronomical Society (UK) • Birmingham Astronomical Society (UK) • Birmingham Astronomical Society (USA-AL) • Boise Astronomical Society (USA-ID) • Bolton Astronomical Society (UK) • Boolean Assignment Statement • Bradford Astronomical Society (UK) • Braintree Astronomical Society (UK) • Breathing Air System • Brevard Astronomical Society (USA-FL) • Bridgewater Astronomical Society (UK) • Bridport Astronomical Society (UK) • Brisbane Astronomical Society (Australia-QLD) • Bristol Astronomical Society (UK) • British Antarctic Survey • Broadcast Auxiliary Service • Budget Allocation Summary • Bulgarian Academy of Sciences • Bundesamt für Schiffsvermessung (Germany) (now BSH) • Bureau d'Aide Sociale

BASc: Bachelor of Applied Sciences

BASC: Board on Atmospheric Science and Climate • Brouwer Award Selection Committee (AAS)

BASD: Ball Aerospace Systems Division

BASE: Beaufort and Arctic Storms Experiment • Bipolar Advanced Silicon for Europe • British Aerospace Systems and Equipment

BASF: Badische Anilin- und Sodafabrik

BASG: Ball Aerospace Systems Group

BASH: Bird Aircraft Strike Hazard

BASI: Bulletin of the Astronomical Society of India (periodical code) • Bureau of Air Safety Investigation

BASIC: Basic Automatic Stored Instruction Computer • Beginner's All-purpose Symbolic Instruction Code • British-American Security Information Council

BASIL: Boeing Automated Systems Information and Logging

BASIN: Basic Studies on Air glow, Smog and Inversion

BASIS: Bibliothekarisch Analytisches System zur Informationsspeicherung • Budgetary And Scheduling Information System • Burst Arcsecond Imaging and Spectroscopy

BASJE: Bolivian Air Shower Joint Experiment

BASM: Built-in Assembler

BASS: Backup Avionics System Software • Base Solaire Sol • Broad Area Surveillance System • Broadband Array Spectrograph System • Broadband Array Spectrometer System (*) • Bulk Avalanche Semiconductor Switch

BAST: Best Available and Safest Technology • Board on Army Science and Technology

BASV: Baade-Swope Variable

BASW: Baade-Swope

Bât: Bâtiment

Bat: Batavia • Battery

BAT: Back Arc Tectonics • Barretos (Brazil-SP) (location identifier) • Basic Air Temperature • Batch • Battery • Beam Assembly Teleoperator • Best Available Technology • Biochemical Automated Technology • Blind Approach Training • British Antarctic Territory • Buffer Alias Table • Bureau d'Assistance Technique • Burst Alert Telescope

BATC: Ball Aerospace and Technologies Corporation (USA-CO) • Beijing-Arizona-Taipei-Connecticut

BATE: Base Activation Test Equipment

BATEA: Best Available Technology Economically Available

BATES: Battlefield Artillery Tactical Engagement System

BATF: Blind Approach Training Flight • Bureau of Alcohol, Tobacco, and Firearms (USA)

BATGE: Biosphere-Atmosphere Trace Gas Exchange • Biosphere-Atmosphere Tropical Global Exchange

Bath: Bathroom

BATH: Best Available True Heading

BATO: Balloon-Assisted Takeoff

Bâts: Bâtiments

BATS: Basic Additional Teleprocessing Support • Bayesian Analysis of Time Series • BDT Acquisition Telescope System • Biosphere-Atmosphere Transfer Scheme

BATSE: Burst And Transient Source Experiment

Batt: Battalion • Battery

BATT: Battery

Bau: Bureau

BAU: Baryon Asymmetry of the Universe • Bauru (Brazil-SP) (location identifier) • Business As Usual

BAUEM: Bundesamt für Übermittlungstruppen (Switzerland)

bàv: Bon à Vue

Bav: Bavaria (Germany) • Bavarian

BAV: Baotou (China-PRC) (location identifier) • Berliner Arbeitsgemeinschaft für Veränderliche Sterne eV (Germany) • Bundesdeutsche Arbeitsgemeinschaft für Veränderliche Sterne eV (Germany)

baw: Bis auf Weiteres

BAW: Bayerische Akademie der Wissenschaften (Germany) • Biawonque (Gabon) (location identifier)

BAX: Barnaul (Russia) (location identifier)

Bay: Bayern (Germany) • Bayonet

BAY: Baia Mare (Romania) (location identifier) • Bayer

Bayer: Bayerisch

BAYS: British Association Youth Section

BAZ: Barbelos (Brazil-AM) (location identifier) • Buchloer Astronomisches Zirkular

bb: Books

B&B: Bed & Breakfast

BB: Ball Bearing • Banc-Balance • Band-by-Band • Barbade (ISO country code) • Barbados (ISO country code) • Baseband • Bayesian Bound • Bed and Breakfast • Bénéfice Brut • Bescherming Burgerbevolking • Big Bang • Billet de Banque • Black Body • Blackburn (UK postal code) • Blattbereich • Blowback • Blue Book • Brassboard • Breadboard • Bright Band • Broad Band • Building Block • Bulletin Board • Bulletin de Bagages • Bureau de Bienfaisance • Bureau of Budget • Double Black

BBA: Bachelor in Business Administration • Balkan Bulgarian Airlines • Balmaceda (Chile) (location identifier) • Blue Book Abstract • Broadband Antenna • Bucket Brigade Algorithm

BBAA: Back Bay Amateur Astronomers (USA-VA)

BBAO: Black Birch Astronomical Observatory (New Zealand)

BBAS: Balloon-Borne Astronomical Studies • Big Bend Astronomical Society (USA-TX)

BBB: Beans-Bacon-Biscuits • Bed, Breakfast and Bath • Benson (USA-MN) (location identifier) • Treble Black

BBC: Base Band Convertor • Basic Bibliographic Citation • Bay City (USA-TX) (location identifier) • Beam-to-Beam Correlation • Before Bottom Center • Before Business Clearance • Big Bang Cosmology • Blockhouse Battery Charger • Breadboard Camera • British Broadcasting Corporation

BBCNC: British Broadcasting Corporation Networking Club

BBD: Black-Body Detection • Block Branch diagram • Brady Curtis (USA-TX) (location identifier) • Bucket Brigade Device • Dollar (Barbados) (ISO currency code)

BBDC: Before Bottom Dead Center

BBE: Big Bell (Australia-WA) (location identifier)

BBED: Broad-Band Energy Distribution

BBEM: Bed, Breakfast, and Evening Meal

BBF: Balloon-Borne Filter • Beijing Book Fair (China-PRC) • Berlin Brandenburg Flughafen (Germany) • Bulletin des Bibliothèques de France • Burlington (USA-MA) (location identifier)

BBFR: Balloon-Borne Filter Radiometer

BBG: Butaritari (Kiribati) (location identifier)

BBH: Barth (Germany) (location identifier)

BBHIR: Brandband Hemispheric Infrared Radiometer

BBHSR: Brandband Hemispheric Solar Radiometer

BBI: Berlin Brandenburg International (Germany) • Bhubaneswar (India) (location identifier) • Broadband Interferometer • Broadband Interferometry • Bureau des Brevets et Inventions

BBIAT: Be Back In A Tick

BBIR: Broadband Infrared

BBJ: Bitburg (Germany) (location identifier) • Boeing Business Jet

BBK: Breadboard Kit • Kasane (Botswana) (location identifier)

bbl: Barrel

BBL: Babolsar (Iran) (location identifier) • Barrel • Basic Business Language • Branch Back and Load

BBLT: Bus Block Transfer

BBM: Battambang (Cambodia) (location identifier) • Beam-Bending Magnet • Books By Mail • Breadboard Model • Break Before Make • Building Block Monochromator

BBMF: Balloon-Borne Magnet Facility

BBMIP: Branch-Bound Mixed Integer Programming

BBML: Baby Meal

BB&N: Bolt, Beranek & Newman Inc.

BBN: Bario (Malaysia) (location identifier) • Big Bang Nucleosynthesis • Bloomberg News Radio • Bolt, Beranek and Newman Inc.

BBNS: Broadband Network Services

BBO: Berbera (Somalia) (location identifier) • Beta Barium Borate

BBOL: Building-Block-Oriented Language

BBP: Bembridge (Isle of Wight) (location identifier) • Building Block Principle

BBPC: Broadband Polarizing Cube

BBPS: Behavior-Based Personal System

BBQ: Barbecue • Barbuda (Leeward Islands) (location identifier)

BBQC: British Board of Quality Control

BBR: Balloon-Borne Radio • Band-to-Band Registration • Basse Terre (Guadeloupe) (location identifier)

bbrg: Ball Bearing

BBRS: Balloon-Borne Radio System

BBS: Blackbush (UK) (location identifier) • Bright Star Spot • Bulletin Board Service • Bulletin Board System

BBSes: Bulletin Board Systems

BBSO: Big Bear Solar Observatory (USA-CA)

BBSP: Balloon-Borne Solar Pointer

BBSRC: Biotechnology and Biological Sciences Research Council (UK)

BBSS: Balloon-Borne Sounding System

BBT: Basal Body Temperature • Berberati (Central African Rep.) (location identifier) • Broadband Technology

BBU: Bit Buffer Unit • Bucureşti Baneasa (Romania) (location identifier)

BBV: Bereby (Ivory Coast) (location identifier)

BBW: Broken Bow (USA-NE) (location identifier) • Bundesamt für Bildung und Wissenschaft (Switzerland)

BBX: Blue Bell (USA-PA) (location identifier)

BBXRT: Broad Band X-Ray Telescope

BBY: Bambari (Central African Rep.) (location identifier)

BBZ: Zambezi (Zambia) (location identifier)

bc: Before Christ • Between Centers • Blind Copy

B/C: Bench Check • Birth Certificate

B&C: Boller & Chivens

BC: Bachelor of Chemistry • Bachelor of Commerce • Back Connected • Back Course • Bad Character • Bad Check • Bail Commercial • Baisse Comprise • Baisse des Cours • Baja California (Mexico) • Balance Commerciale • Baldone Catalogue • Ballistic Camera • Ballistic Centre • Base Construction • Baseline Configuration • Basic Control • Battery Charge • Battery Charger • Battery Conditioner • Beam Collimator • Bearing Change • Beat Cepheid • Beco • Before Christ • Before Computer • Bell Cord • Bench Check • Bénéfice Commercial • Between Centers • Bibliothekcentralen • Big Crunch • Billet Collectif • Binary Code • Binary Counter • Birth Certificate • Black Carbon • Blind Copy • Bloch Curve • Board of Control • Body-Centered • Bolometric Correction • Bolt Center • Bon de Caisse • Bon de Commande • Bordereau de Coupons • Borocarbon • Bose Condensation • Boston College (USA-MA) • Bottom Center • Boundary Condition • Bourse de Commerce • Brevet de Capacité • Brigade Criminelle • British Columbia (Canada) • British Commonwealth • British Council • Broadcast • Broadcast Control • Broad Classification • Bubble Chamber • Bubble Combustion • Budget Center • Budgeted Cost • Bulkhead Connector • Bulletin de Commande • Buoyancy Compensator • Bureau Central • Bureau de Chômage • Bureau de Contrôle • Bureau de Correspondance • Bureau of the Census (USA) • Bus Controller • Bus Coupler • Business Class • Business Component • Butyl Chloride

Bca: Boca

B/CA: Business and Commercial Aviation

BCA: Bachelor of Commerce and Administration • Baracoa (Cuba) (location identifier) • Benefit Cost Analysis • Bilderberg Continuum Atmosphere • Binary Collision Approximation • Booster Change Assembly • Bureau Commun d'Architecture • Business and Commercial Aviation • Business Council of Australia

BCAAS: Berks County Amateur Astronomical Society (USA-PA)

BCAC: Barker Coded Alternating Code • British Conference on Automation and Computation

BCAR: British Civil Airworthiness Requirement

BCB: Binary Code Box • Blacksburg - Virginia Tech (USA-VA) (location identifier) • Broadcast Band

BCBG: Bon Chic, Bon Genre

BCC: Basic Computer Concept • Beacon Control Console • Bear Creek (USA-AK) (location identifier) • Bellevue Community College (USA-WA) • Bibliographic Compilation Catalogue • Blind Carbon Copy • Blind Courtesy Copy • Block Check Character • Block Control Character • Blocked Call Cleared • Body-Centered Cubic • Brevard Community College (USA-FL) • British Copyright Council • Budget Classification Code • Bureau Central de Compensation • Bureau Central de Comptabilité • Business Cooperation Centre

BCCI: Bulletin de la Chambre de Commerce et d'Industrie

BCCT: Break Control Command Transducer

BCD: Bacolod (Philippines) (location identifier) • Bad Conduct Discharge • Baseline Concept Definition • Baseline Configuration Document • Behind Completion Date • Bibliographic Content Designator • Binary-Coded Data • Binary-Coded Decimal • Blocked Call Delay • Blue Compact Dwarf • Bottom Collider Detector • Budget Change Document

BCDC: Binary-Coded Decimal Counter

BCDG: Blue Compact Dwarf Galaxy

BCDI: Bureau Central de Documentation et d'Information

BCDIC: Binary-Coded Decimal Interchange Code

BCDS: Backup Control Data Set

Bce: Balance

BCE: Bachelor of Chemical Engineering • Bachelor of Civil Engineering • Backup Control Electronics • Banca Centrale Europea • Banque Centrale Européenne • Barium Cloud Experiment • Baseline Cost Estimate • Beam Collimation Error • Before Christian Era • Before Common Era • Biblioteca Central de Educação (Brazil) • British Commonwealth and Empire • Bryce (USA-UT) (location identifier) • Bubble Chamber Experiment • Burn Control Element • Bus Control Electronics • Bus Control Element

BCEAO: Banque Centrale des États de l'Afrique de l'Ouest

BCEE: Bureau Commun d'Études et d'Enquête

BCEN: Banque Commerciale pour l'Europe du Nord

BCEOM: Bureau Central pour les Équipements d'Outre-Mer

BCET: Balance Contrast Enhancement Technique

BCF: Bandpass Crystal Filter • Basic Calibration File • Basic Control Frequency • Beam Correction Factor • Biconcentration Factor • Billet de Chemin de Fer • Billion Cubic Feet • Bouca (Central African Rep.) (location identifier) • Bridge Control Facility

BCFHT: Bulletin of the Canada-France-Hawaii Telescope (periodical code)

BCFS: Backup Flight Control System

BCFSK: Binary Code Frequency Shift Keying

BCG: Bacille Calmette-Guérin • Bemichi (Guyana) (location identifier) • Biconjugate Gradient • Block-Connected Graph • Blue Compact Galaxy • Bright Cluster Galaxy • Brightest Cluster Galaxy

BCGM: Biconjugate Gradient Method

bch: Body-Centered Hypercubic

Bch: Branch • Bunch

BCh: Bachelor of Chemistry

BCH: Baker-Campbell-Hausdorff • Baucau - English Madeira (Indonesia) (location identifier) • Beach • Binary-Coded Hexadecimal • Binary-Coded Hollerith • Bose-Chandhuri-Hocquenheim • Bureau Central des Hôpitaux

Bche: Bouche

BChE: Bachelor of Chemical Engineering

BCI: Barcaldine (Australia-QLD) (location identifier) • Baseline Configuration Identification • Battery Condition Indicator • Binary-Coded Information • Broadcast Interference

BCID: Best Candidate Item Database

BCIE: Banco Centroamericano de Integración Economico

BCIR: Bureau Climatologique Interrégional

BCIS: Battlefield Combat ID System

BCI: Bureau Central de Séismologie

BCIU: Battery Control and Interface Unit • Bolometer Computer Interface Unit • Bus Control Interface Unit

BCJ: Baca Grande (USA-CO) (location identifier)

BCJS: Buffer Control Junction Switch

BCK: Bolwarra (Australia-QLD) (location identifier)

BCKG: Backing

Bcl: Bicyclette

BCL: Barra Colorado (Costa Rica) (location identifier) • Basic Contour Line • Batch Command Language • Burroughs Common Language

BCLA: Beam Combination Laboratory Area (VLT)

BCLF: Base Cardiac Life Support

BCM: Bacau (Romania) (location identifier) • Back Course Marker • Basic Control Monitor • Beam Change Monitor • Beam Charge Monitor • Become • Beyond Capacity of Maintenance • Block Coded Modulation • Bright Cluster Member • Brightest Cluster Member • Bureau Central Météorologique • Bureau de Contrôle Médical

BCMG: Becoming

BCMR: Boîtier de Commande des Mécanismes et Réchauffeurs

BCMT: Bureau Central de Magnétisme Terrestre

Bcn: Beacon

BCN: Backbone Concentrator Node • Barcelona - El Prat (Spain) (location identifier) • Beacon • Boarding Control Number • Breakdown Control Number • Broadband Communication Network

BCNet: Business Cooperation Network

BCNF: Boyce-Codd Normal Form • Bus Controller Navigation Filter

BCNU: Be Seeing You

BCNZ: Broadcasting Corporation of New Zealand

Bco: Banco

BCO: Binary-Coded Octal • Booster Cutoff • Jinka (Ethiopia) (location identifier)

BCOE: Bench Checkout Equipment

BCOI: British Central Office of Information (UK)

Bcp: Beaucoup

BCP: Base Condemnation Percent • Basic Calcium Phosphate • Basic Celestial Phenomenon • Benchmark Control Point • Bibliothèque Centrale de Prêt (France) • Book of Common Prayer • Bridging Control Protocol • Brilliant Computer Products • Budget Change Proposal • Bulk Copy Program • Byte Controlled Protocol • Byte Control Protocol

BCPL: Basic Combined Programming Language (*) • Basic Computer Programming Language

BCR: Bar Chart Report • Baseline Configuration Review • Battery Charge Regulator • Bôca do Acre (Brazil-AM) (location identifier) • Branch Constitutive Relation • Branched-Chain Reaction • Budget Change Request • Bureau Communautaire de Référence • Business Communications Review • Byte Count Register

BCRC: British Columbia Research Council (Canada) • Bureau Central de Recherches Criminelles

BCRCI: Brigade Centrale de Répression de la Criminalité Informatique (France)

BCRD: Basic Consolidated Requirements Document • Budget Civil de Recherche et de Développement (France)

BCRS: Beleidscommissie Remote Sensing (Netherlands)

BCRST: Bibliography on Cold Regions Science and Technology

BCRT: Binary-Coded Range Time • Binary Code Range Time (*) • Bright Cathode-Ray Tube

BCRTS: Binary-Coded Range Time Signal

BCS: Bachelor of Chemical Science • Bachelor of Commercial Science • Backscatter Cross Section • Backup Control System • Bar Code Sorter • Bardeen-Cooper-Schrieffer • Basic Catalog Structure • Basic Contract Specification • Basic Control System • Batch Change Supplement • Beam Control System • Belle Chasse (USA-LA) (location identifier) • Bent Crystal Spectrometer • Binary Compatibility Standard • Boeing Computer Services • Boîtier de Contrôle de Sauvegarde • Bragg Crystal Spectrometer • Bragg Curve Spectroscopy • Brightest Cluster Sample • British Computer Society • Bus Controller Simulator • Business Communications Services

BCSAIL: British Columbia Spatial Archive and Interchange Format

BCSD: Business Council for Sustainable Development

BCSCO: Bismuth-Calcium-Strontium-Copper Oxide

BCSD: Blank Corrected Sample Data

BCSF: Bureau Central Sismologique Français (France)

BCSS: Bociny Computer Support Services

BCSSS: British Columbia Space Sciences Society (Canada-BC)

Bcst: Broadcast

BCST: Board on Chemical Sciences and Technology

BCSVS: Bibliographic Catalogue of Suspected Variable Stars

BCT: Bandwidth Compression Technique • Barycentric Coordinated Time • Boca Raton (USA-FL) (location identifier) • Body Centered Tetragonal • Bureau Central de Transmission • Bus Configuration Table

BCU: Basic Computer Unit • Battery Control Unit • Bauchi (Nigeria) (location identifier) • Bench Checkout Unit • Bibliothèque Cantonale et Universitaire (Switzerland) • Branch Control Unit • Buffer Control Unit • Bus Control Unit • Bus Coupler Unit

BCV: Belmopan (Belize) (location identifier) • Brightness Contrast Value • Business Continuation Volume

BCVS: Bibliographic Catalogue of Variable Stars

BCW: Buffer Control Word

BCWP: Budget Costs for Work Performed • Budgeted Cost of Work Performed

BCWS: Budget Costs for Work Scheduled • Budgeted Cost of Work Scheduled

BCX: Beloreck (Russia) (location identifier)

BCY: Bulchi (Ethiopia) (location identifier)

BCZ: Bickerton Island (Australia-NT) (location identifier)

b/d: Barrels per Day • Brought Down

bd: Bis Die

bD: Bidécadaire

Bd: Band • Baud • Board • Bold • Bond • Boulevard • Bound • Broad • Bundle

B/D: Bank Draft • Bills Discounted • Binary to Decimal • Brought Down • Bulge-to-Disk

BD: Backward Difference • Bahrain Dinar (currency) • Banc de Dynalpie • Band • Bande Dessinée • Bangladesh (international vehicle registration code) • Bangladesh (ISO country code) • Bank Draft • Barrière Douanière • Base de Datos • Base de Données • Beam Deflection • Beam Diameter • Bidirectional • Bilan Définitif • Billet à Découvert • Bills Discounted • Binary Decoder • Binary Digit • Binary Discrete • Binary Divide • Bit Density • Blowing Dust • Board • Bon de Douane • Bonner Durchmusterung • Boot Descriptor • Booster Development • Bordereau Descriptif • Bradford (UK postal code) • Brief Description • Brown Dwarf • Buiten Dienst • Bus Director

BDA: Backup Drive Amplifier • Base de Données d'Amas • Battle Damage Assessment • Beacon Drive Amplifier • Bermuda • Bermuda Kindley (Bermuda) (location identifier) • Bermuda Tracking Station • Blast Danger Area • Block Decoder Assembly • Bomb Damage Assessment • Brazilian Decimetric Array • Business Data Analysis • Hamilton (Bermuda) (location identifier)

BDACS: BDM Data Acquisition and Conversion System

BDAM: Basic Direct-Access Method

BDAP: Banque de Données Analysées et Prévues

BDAS: Bendigo District Astronomical Society Inc. (Australia-VIC)

BDAT: Best Demonstrated Available Technology

BDB: Banque de Données Bibliographqiues • Bundaberg (Australia-QLD) (location identifier)

BDBS: Base de Données Bathymétriques du SHOM

BDC: Backgrounds Data Center (USA-DC) • Backup Digital Computer • Backup Domain Controller • Barra do Corda (Brazil-MA) (location identifier) • Beijing Data Center (China-PRC) • Bidirectional Converter • Binary Decimal Counter • Bottom Dead Center

BDCF: Baseline Data Collection Facility

BDCR: Baseline Document Change Request

Bdcst: Broadcast

BDCU: Base de Données Charge Utile

BdD: Banque de Données • Base de Données

BDD: Badu Island (Australia-QLD) (location identifier) • Banque De Données • Baseline Definition Document • Baseline Description Document • Binary-to-Decimal Decoder • Binary Digital Data • Burst Detection Dosimeter • Burst Detector Dosimeter

BDDE: Bulletin de la Direction Départementale de l'Équipement (France)

BDDN: Broadcast Data Dissemination Network

BDDPI: Banque de Données du Droit de la Propriété Industrielle (INPI)

BDDR: Banque de Données Documentaires Réparties

Bde: Bände • Brigade

BDE: Base de Données Extragalactiques • Baudette (USA-MN) (location identifier) • Bidirectional Electron-event • Borland Database Engine • Boundary of Diffuse Encroachment • Bureau des Élèves • Buyer-Designated Equipment

BDEAC: Banque de Développement des États de l'Afrique Centrale

BDEC: Bendix Field Engineering Corporation

BDES: Bibliographic Data Exchange System

BdF: Bibliothèque de France

BDF: Backward Difference Formula • Backward Differential Formula • Backward Differential Function • Bibliothèque De France • Binary Distribution Format • Bitmap Distribution Format • Block Data Format • Bradford Rinkenberger (USA-IL) (location identifier) • Bulk Data Frame

Bdg: Binding

BDG: Binding (USA-UT) (location identifier) • Blanding (USA-UT) (location identifier) • Blue Dwarf Galaxy • Bridge • Bureau de la Documentation et de Gestion

BDH: Bandar Lengeh (Iran) (location identifier)

BDI: Bande Diffuse Interstellaire • Base de Données Intelligente • Bearing Deviation Indicator • Binary Digital Image • Bird Island (Seychelles) (location identifier) • Both Dates Included • Both Days Included • Bourse de Doctorat pour Ingénieur • Budget Data and Integration • Buffered Direct Injection • Bundesverband der Deutschen Industrie (Germany) • Bureau de Documentation et d'Information • Burundi (ISO country code)

BDIN: Book Dealer Identification Number

BDIS: Boîtier de Distribution

BDJ: Banjarmasin (Indonesia) (location identifier) • Brans-Dicke-Jordan

BDK: Bondoukou (Ivory Coast) (location identifier)

Bdl: Bundle

BdL: Bureau des Longitudes (France)

BDL: Banque de Données Locales • Baseline Demonstrator Laser • Basic Design Language • Boundary Double Layer • Bureau Des Longitudes (France) • Hartford - Bradley (USA-CT) (location identifier) • Springfield - Bradley (USA-MA) (location identifier) • Windsor Locks - Bradley (USA-CT) (location identifier)

BDLC: Burroughs Data Link Control

Bdle: Bundle

BDLI: Bundesverband der Deutschen Luftfahrt-, Raumfahrt- und Ausrüstungindustrie eV (Germany)

BDM: Background Debug Mode • Bandirma (Turkey) (location identifier) • Baryonic Dark Matter • Base de Données Météorologiques • Binary Delta Modulation • Binary Digital Multiplier • Births-Deaths-Marriages • Bone Densitometer • Branch Delegates' Meeting • Bureau de Documentation Minière • Bureau Des Méthodes

BDMI: Bibliographic Dictionaries Master Index

BDMM: Basic Differential Multiplier Method

BDN: Badin (Pakistan) (location identifier) • Bell Data Network • Bright Dark Nebula

BDO: Bandung (Indonesia) (location identifier)

BDOS: Basic Disk Operating System

BDP: Bhadrapur (Nepal) (location identifier) • Bibliothèque Départementale de Prêt (France) • Bourse de Demi-Pension • Breakdown Pressure • Burst Detection Processor • Burst Detector Processor • Business Data Processing

BDPA: Black Data Processing Association • Bureau pour le Développement de la Production Agricole

BDPU: Bubble, Drop and Particle Unit (*) • Bubble Droplet Particle Unit

BDQ: Vadodara (India) (location identifier)

BDR: Baseline Design Review • Basic Data Record • Battery Discharge Regulator • Bidirectional Reflectance • Border • Bridgeport (USA-CT) (location identifier) • Build Design Review

BDRF: Bidirectional Reflectance Function

Bdrm: Bedroom

Bdry: Boundary

bds: Bis in Die Sumendus

Bds: Boards • Bundles

BDS: Barbados (international vehicle registration code) • Base de Données Scientifiques • Beijing Document Service (China-PRC) • Binary Digital Search • Biographical Dictionary of Scientists • Bioreactor Demonstration System • Bioreactor Development System • Brindisi (Italy) (location identifier) • Bulgarian Committee for Standardization and Metrology • Bulk Data System • Burnham's Double Stars

BDSF: Bidirectional Scattering Function

BDSP: Basic Data Set Project

BDSS: Base de Données Sédimentologiques du SHOM

BDST: British Double Summer Time

BDT: Barycentric Dynamical Time • Barycentric Dynamic Time • Basa de Dados Tropical • Base de Données Technologiques • Beam Deflection Tube • Beam Director/Tracker • Bidirectional Proton-event • Billing Data Transmitter • Binary Deck-to-Tape • Block Data Transfer • Bon Du Trésor • Bureau de Développement des Télécommunications • Gbadolite (Zaire) (location identifier) • Taka (Bangladesh) (ISO currency code)

BDTC: British Dependent Territory Citizen

BDU: Bardufoss (Norway) (location identifier) • Baryon-Dominated Universe • Basic Display Unit • Bundesverband Deutscher Unternehmensberater (Germany)

B-du-R: Bouches-du-Rhône (France)

BDV: Basis-Datenverarbeitung • Breakdown Voltage • Moba (Zaire) (location identifier)

BDW: Bedford Downs (Australia-WA) (location identifier)

Bdx: Bordeaux (France)

BDX: Broadus (USA-MT) (location identifier)

Bdy: Boundary

BDY: Bandon State (USA-OR) (location identifier) • Body

BDZ: Baindoung (Papua New Guinea) (location identifier)

be: Belgium • Bezüglich • Byelorussian (ISO language code)

Be: Beryllium

Bé: Baumé • Breveté

B&E: Beginning & Ending • Breaking & Entering

B/E: Bill of Exchange

BE: Bachelor of Education • Bachelor of Engineering • Back End • Bande Étroite • Bank of England • Banlieue Est • Barkhausen Emission • Beacon Explorer • Beam Expander • Belgique (ISO country code) • Belgium (ISO country code) • Bell End • Beloruss • Below or Equal • Benin • Best Estimate • Bharat Electronics • Bibliografía Española (Spain) • Bill of Exchange • Binary Encounter • Binding Edge • Binding Energy • Board of Education • Bon d'Électricité • Bon d'Épargne • Booster Engine • Bordereau d'Envoi • Bordereau d'Escompte • Bose-Einstein • Bourse d'Études • Bradford Engineering • Brevet Élémentaire • Brevet d'Études • British Element • British Embassy • Bureau d'Enregistrement • Bureau d'Études

BEA: Bereina (Papua New Guinea) (location identifier) • Binary Encounter Approximation • Bond Energy Analysis • Bright Earth Avoidance (region) • Bureau à l'Énergie Atomique • Bureau des Enquêtes et Accidents • Bureau of Economic Analysis

BEAC: Banque des États de l'Afrique Centrale

BEAD: Beginning Extended-Area Descriptor

BEAF: Beam Expander Alignment Function

BEAM: Brain Electrical Activity Mapping

BEAMS: Becoming Enthusiastic About Math and Science

BEAP: Banc d'Essai des Accélérateurs à Poudre

BEAR: Barents Euro-Arctic Region • Beacon and Auroral Research

bearb: Bearbeitet

BEARS: Berkeley EUV Airglow Rocket Spectrometer

BEAST: Background Emission Anisotropy Scanning Telescope

BEAT: Best Execution Analysis Tabulation

Beau: Bordereau

BEB: Benbecula (UK) (location identifier) • Bureau d'Études du Bâtiment

BEBA: Bolometer Electronics Box Assembly

BEBC: Big European Bubble Chamber

Bec: Because

BEC: Backup Event Controller • Base Extension Course • Bose-Einstein Condensate (*) • Bose-Einstein Condensation • Bose-Einstein Correlation • Brevet d'Études Comemrciales • Convertible Franc (Belgium) (ISO currency code) • Wichita Beech (USA-KS) (location identifier)

BECAT: Background Estimation Convolution And Threshold

BECAUSE: Benchmark of Concurrent Architectures for Use in Scientific Engineering

BECI: Bibliografía Española de Ciencias de la Información (Spain)

BECMG: Becoming

BECN: Backward Error-Congestion Notification • Backward Error-Correction Notification • Backward Explicit Congestion Notification

BECO: Booster Engine Cutoff

BECTA: Bureau d'Exploitation et de Coordination des Transports Aériens

BECU: Billion European Currency Units

Bed: Bedeutend

BEd: Bachelor of Education

BED: Bedford Hanscom (USA-MA) (location identifier) • Blue-Emitting Diode • Box External Data

BEDAC: Burst Error Detection And Correlation

BEDIS: Booktrade EDI Standards

BEDO: Burst Extended Data Out

Beds: Bedfordshire (UK-England) • Bedrooms

BEDT: Bisethylenedithia

BEE: Bachelor of Electrical Engineering • Beagle Bay (Australia-WA) (location identifier)

BEEd: Board on Engineering Education

BEEL: Biological Equivalent Exposure Limit

BEEM: Ballistic Electron Emission Spectroscopy

BEEP: Bureau Européen de l'Éducation Populaire

BEES: Board on Energy and Environmental Systems

BEET: Best Estimated Evaluation Trajectory

Bef: Before

BEF: Band Elimination Filter • Bluefields (Nicaragua) (location identifier) • Franc (Belgium) (ISO currency code)

BEFS: Berkeley Extreme and Far-ultraviolet Spectrometer

Beg: Beginning

BEG: Bach-Escher-Gödel • Beograd (Yugoslavia-Serbia) (location identifier) • Bureau d'Études Géologiques

Begr: Begründet

BEH: Benton Harbor (USA-MI) (location identifier)

BEI: Back-scattered Electron Imaging • Banque Européenne d'Investissement • Beica (Ethiopia) (location identifier) • Brevet d'Enseignement Industriel (France)

Beibl: Beiblatt

beif: Beifolgend

BeiGe: Beiträge zur Geophysik (periodical code)

beigeb: Beigebunden

beil: Beiliegend

BEIR: Biological Effects of Ionizing Radiation

Beitr: Beitrag • Beiträge

BEJ: Berau (Indonesia) (location identifier)

Bek: Bekanntmachung

BEK: Rae Bareli (India) (location identifier)

Bel: Belgian • Belgium • Belorussian

BEL: Basic Equipment List • Belém (Brazil-PA) (location identifier) • Belfast (Starlink internal code) • Belgique (ISO country code) • Belgium (ISO country code) • Bell • Broad Emission Line • Financial Franc (Belgium) (ISO currency code)

Belg: Belgian • Belgium

BELGOSPACE: Belgian Space Industry Association

BELINDIS: Belgian Information and Dissemination Service

BELLCORE: Bell Communications Research

BELR: Broad Emission Line Region

BELST: Belarussian Committee for Standardization, Metrology and Certification

Bem: Bemerkung

BEM: Bachelor of Engineering of Mines • Big Emerging Market • Biological Environment Monitor • Bossembele (Central African Rep.) (location identifier) • Boundary Element Method • Bug-Eyed Monster

BEMA: Biogenic Emissions in the Mediterranean Area

BEMS: Bioelectromagnetics Society

Ben: Bengali

BEN: Basic Encyclopedia Number • Benghazi Benina (Libya) (location identifier) • Benin (ISO country code) • Broadband Electronic Noise • Broadband Electrostatic Noise • Bureau d'Études Nucléaires

BENCHMARC: Biodegradation of Environmental Chemicals Modelled with Aquatic, Relative-rate Coefficients

Bend: Bendigo (Australia-VIC)

Benef: Benefice • Beneficiary

BENELUX: Belgium-Netherlands-Luxemburg

Benf: Beneficiario

Beng: Bengal • Bengali

BENS: Bounded Error Navigation System

BENT: Beginning Evening Nautical Twilight

BEO: Base Engineer Officer • Biologie-Energie-Ökologie • Boulder Extension Office (SWRI) • Newcastle Belmont (Australia-NSW) (location identifier)

Beob: Beobachter • Beobachtung

BEOP: Best Estimate of Orbital Parameters

BEP: Bachelor of Engineering Physics • Backward Error Propagation • Bellary (India) (location identifier) • Best Efficiency Point • Biomolecular Engineering Programme • Bit Error Probability • Bon d'Épargne Progressif • Brevet d'Études Professionnelles (France) • Bureau Exécutif Permanent

BEPC: Beijing Electron-Positron Collider (China-PRC) • Brevet d'Études du Premier Cycle

BEPD: Basic Entry Pay Date

BePhA: Beiträge zur Physik Atmosphäre (periodical code)

BEPI: Bureau d'Études et de Participation Industrielle

BePl: Beiträge Plasmaphysik (periodical code)

BEPP: Brevet Élémentaire de Pilote Privé

Beq: Bequeath

BEQ: Bury Saint Edmunds - Honington (UK-England) (location identifier)

Beqt: Bequest

Ber: Berber • Bericht • Berlin (Germany)

BER: Basic Encoding Rule • Berlin (Germany) (location identifier) • Binary Encoded Representation • Bit Error Rate

Berb: Berber • Berbère • Berberisco

BERD: Banque Européenne de Reconstruction et de Développement

BERDI: Balkan Environmental Research and Development Institute (Greece)

Bergstr: Bergstraße

BERI: Bureau d'Études et de Réalisations Industrielles

Ber. Is.: Bermuda Islands

Berks: Berkshire (UK-England)

Berl: Berlin (Germany)

Berm: Bermuda

BERNET: Berlin Netz (Germany)

BERT: Best Error Ratio Tester • Bit Error Rate Test • Bit Error Rate Tester

Bes: Besonders

BES: Bachelor of Engineering Science • Bachelor of Environmental Studies • Banc d'Essais Site • Basic Energy Sciences • Beam Emission Spectroscopy • Booster Exhaust Stream • Bose-Einstein Spectrum • Bose-Einstein Statistics • Bourse de l'Enseignement Secondaire • Brest Guipavas (France) (location identifier) • Brevet d'Enseignement Social • British Ecological Society • Business Expansion Scheme

BESEX: Bering Sea Experiment

BESI: Bus Electronic Scanning Indicator

BESR: Board on Earth Sciences and Resources

BESS: Balloon-Experiment with Superconducting Solenoid (*) • Balloon-Experiment with Superconducting Spectrometer • Biomedical Experiment Scientific Satellite • Biomedical Experiment Support Satellite • Bottom Environmental Sensing System

BESSY: Berlin Electron Storage Synchrotron • Berliner Elektronenspeicherringgesellschaft für Synchrotronstrahlung (Germany)

Best: Bestellung • Bestimmt

BEST: Bathymetric Estimator Search Technique • Battery Energy Storage Test • Bilan Énergétique du Système Tropical • Board on Environmental Sciences and Technology • Booster Exhaust Study Test • British Expertise in Science and Technology • Business Equipment Software Technique

Best.Nr.: Bestellnummer

Bet: Between

BET: Best Estimate of Trajectory • Best-Estimate Trajectory • Bethel (USA-AK) (location identifier) • Bit-Enhanced Technology • Black Entertainment Television • Budget Énergie Transport • Brunauer-Emmett-Teller

BETA: Ballistic Exploitation and Target Acquisition • Billion-channel Extraterrestrial Assay • Bureau d'Économie Théorique et Appliquée

BETEL: Broad Exchange over Trans-European Links

Betr: Betreff • Betreffend • Betreffs • Betrifft

Betrbf: Betriebsbahnhof

BETRS: Basic Exchange Telecommunications Radio Service

BETSCE: Brilliant Eyes Ten-Kelvin Sorption Cryocooler Experiment

Betw: Between

BEU: Bedourie (Australia-QLD) (location identifier) • BENELUX Economic Union • Bureau of Energy Utilization

BEUC: Bureau Européen des Unions de Consommateurs

Bev: Bevel • Bevölkerung

BeV: Billion Electronvolts

BEV: Beer Sheba (Israel) (location identifier) • Billion Electron Volts • Blacksburg Electronic Village • Bundesamt für Eich- und Vermessungswesen (Austria)

BEW: Beira (Mozambique) (location identifier)

BEX: Benson (UK-England) (location identifier)

BEY: Beirut (Lebanon) (location identifier)

bez: Bezahlt • Bezüglich

Bez: Bezeichnung • Bezirk

BEZ: Beru (Kiribati) (location identifier)

BEZS: Bandwidth Efficient Zero Support

bezw: Beziehungsweise

b/f: Bring Forward • Brought Forward

bf: Base Frequency • Boldface • Bona Fide

Bf: Bahnhof • Brief

B/F: Background/Foreground • Bring Forward • Brought Forward

BF: Back Feed • Back Focal • Background Field • Bad Flag • Bailleur de Fonds • Banque de France • Bassa Frequenza • Basse Fréquence • Beam Forming • Beat Frequency • Belgian Franc (currency) • Bellman-Ford • Bénéfice Forfaitaire • Black Forest (Germany) • Body Flap • Boyfriend • Breathing Frequency • Brevet Français • Brought Forward • Burkina Faso (ISO country code) • Franc (Belgium) (currency)

BFA: Burkina Faso (ISO country code)

BFAS: Basic File Access System

B/FAST: Breakfast

BFAST: Breakfast

BFB: Bang For the Buck • Blue Fox Bay (USA-AK) (location identifier) • Business Finance Branch

BFC: Backup Flight Control • Bloomfield (Australia-QLD) (location identifier) • Blue Featureless Continuum • Body Flap Control • Bohr Frequency Condition • Bold Face Capital

BFCE: Banque Française du Commerce Extérieur

BFCO: Band Filter Cutoff

BFCS: Backup Flight Control System

BFD: Back Focal Distance • Boolean Function Designator • Bradford (USA-PA) (location identifier)

BFDK: Before Dark

BFE: Base Flood Elevation • Beam Forming Electrode • Bielefeld (Germany) (location identifier) • Blue Fundamental Edge • Buyer-Furnished Equipment

BFEC: Bendix Field Engineering Corporation (USA-MD)

BFET: Ballistic Field-Effect Transistor

BFF: Buffered Flip-Flop • Bund Freischaffender Forodesigner (Germany) • Scottsbluff (USA-NE) (location identifier)

BfFS: Bundesanstalt für Flugsicherung (Germany)

BFG: Brute Force Gyro • Bullfrog Basin (USA-VT) (location identifier)

BfGR: Bundesanstalt für Geowissenschaften und Rohstoffe (Germany)

BFGS: Broyden-Fletcher-Goldfarb-Shanno

BFHZ: Bayerisch-Französisches Hochschulzentrum

BFI: British Film Institute • Brute Force and Ignorance • Seattle Boeing (USA-WA) (location identifier)

BFID: Boolean Function Identifier

BFII: Banking, Finance and Insurance Information

BFJ: Ba (Fiji) (location identifier)

BFK: Denver Buckley (USA-CO) (location identifier)

BFL: Back Focal Length • Bakersfield (USA-CA) (location identifier) • Back Focal Length • Bonded Film Lubrication • Bundesamt für Landestopographie (Switzerland)

BfLR: Bundesanstalt für Landeskunde und Raumentwicklung (Germany)

BFM: Blocking Flow Method • Boson-Fermion Model • Bowen Fluorescence Mechanism • Mobile (USA-AL) (location identifier)

Bfn: Bloemfontein (South Africa)

BFN: Beam-Forming Network • Bloemfontein (South Africa) (location identifier)

BFO: Beat Frequency Oscillator • Black Forest Observatory (USA-CO) • Buffalo Range (Zimbabwe) (location identifier)

BFOSC: Bologna Faint Object Spectrometer and Camera

BFP: Beaver Falls (USA-PA) (location identifier) • Buyer-Furnished Part

BFPOP: British Forces Post Office

Bfr: Before • Buffer

BFr: Belgian Franc (currency)

BFR: Bedford - Virgil I. Grissom (USA-IN) (location identifier) • Before • Belgian Franc (currency) • Biennial Flight Review • Big Fast Router • Block Format Recording • Franc Belge (currency)

BFRP: Boron Fiber-Reinforced Plastic

BfS: Bundesanstalt für Strahlenschutz (Germany)

BFS: Backup Flight Software • Backup Flight System • Band Filter Set • Belfast (UK-Northern Ireland) (location identifier) • Best Fit Sphere • Bureau Fédéral de Statistiques

BFSS: Backup Flight Software System

BFT: Beaufort County (USA-SC) (location identifier) • Binary File Transfer • Budget de Fonctionnement Total

BFU: Bengbu (China-PRC) (location identifier)

BFX: Bafoussam (Cameroon) (location identifier)

BFY: Budget Fiscal Year

b/g: Bonded Goods

bg: Bulgaria • Bulgarian (ISO language code)

Bg: Background • Bag • Berg • Bourg • Brug • Budget

BG: Back Gear • Background • Billet de Groupe • Block Group • Blood Group • Bologna Galactic • Bridgestone • Bulbe Galactique • Bulgaria (international vehicle registration code) • Bulgaria (ISO country code) • Bulgaria (postal country code) • Bulgarie (ISO country code) • Bureau de Garantie • Bureau Géographique • Burg

BGA: Ball Grid Array • British Geophysical Association (UK) • Bucaramanga (Colombia) (location identifier)

BGB: Booue (Gabon) (location identifier) • Bürgerliches Gesetzbuch

BGC: Bank Giro Credit • Biogeochemical Cycle • Bragança (Portugal) (location identifier) • Bright Galaxy Catalog • Burnham General Catalogue

BGD: Bangladesh (ISO country code) • Belgische Geologische Dienst (Belgium) • Borger (USA-TX) (location identifier)

Bge: Barrage • Beige

BGE: Bach-Gödel-Escher • Block Gaussian Elimination • Bologna Galactic Extended • Brainbridge - Decatur County (USA-GA) (location identifier) • Branch if Greater or Equal

BGeod: Bulletin Géodésique (periodical code)

BGF: Bangui (Central African Rep.) (location identifier)

BGG: Bongouanou (Ivory Coast) (location identifier) • Booster Gas Generator

BGGC: Beijing Group of Galaxy and Cosmology (China-PRC)

BGGM: BGS Global Geomagnetic Model

BGH: Boghe Abbaye (Mauritania) (location identifier) • Bundesgerichtshof (Germany)

BGHT: Bight

BGI: Barbados (location identifier) • Bibliographie Géographique Internationale • Borland Graphics Interface • Bureau Gravimétrique International

BGJ: Borgarfjördur Eystri (Iceland) (location identifier)

BGK: Berliner Gesamtkatalog • Bhatnagar-Gross-Krook • Big Creek (Belize) (location identifier)

BGL: Baglung (Nepal) (location identifier) • Below Ground Level • Lev (Bulgaria) (ISO currency code)

BGLS: Bragg Grating Laser Sensor

BGM: Binghamton (USA-KY) (location identifier) • Endicott (USA-KY) (location identifier) • Johnson City (USA-KY) (location identifier)

BGN: Began • Begin • Board on Geographic Names (USA) • Brüggen (Germany) (location identifier)

BGO: Bergen (Norway) (location identifier) • Bismuth Germanate • Bismuth Germanate Oxide • Bismuth Germanium Oxide • Bismuth Germinate • Burke-Gaffney Observatory (Canada-NS)

BGP: Bongo (Gabon) (location identifier) • Border Gateway Protocol • Borland Gateway Protocol • Bulletin Grand Public • Byzantine Generals Problem

BGQ: Big Lake (USA-AK) (location identifier)

BGR: Bangor (USA-ME) (location identifier) • Bulgaria (ISO country code) • Bulgarie (ISO country code) • Bundesanstalt für Geowissenschaften und Rohstoffe (Germany)

BGS: Backup Guidance System • Big Spring - Webb (USA-TX) (location identifier) • British Geological Survey (UK) • Burgs

BGSU: Bowling Green State University (USA-OH)

Bgt: Bight • Bought

BGT: Bagdad (USA-AZ) (location identifier) • Biorack Ground Team • Branch if Greater Than

BGU: Bangassou (Central African Rep.) (location identifier) • Ben Gurion University (Israel)

BGV: Bento Gonçalves (Brazil) (location identifier)

Bgw: Bergwerk

BGW: Baghdad (Iraq) (location identifier) • Bienaymé-Galton-Watson

BGX: Bagé (Brazil) (location identifier)

BGY: Bergamo (Italy) (location identifier) • Milano - Orio al Serio (Italy) (location identifier)

BGZ: Braga (Portugal) (location identifier)

bh: Bihari (ISO language code)

BH: Bahrain (ISO country code) • Bahrein (ISO country code) • Bâtiment Hydrographique • Belize (international vehicle registration code) • Black Hole • Block Handler • Bournemouth (UK postal code) • Breather Hose • Burlington House • Bushongo

BHA: Bahia de Caraquez (Ecuador) (location identifier) • Base Helix Angle • Bibliographie de l'Histoire de l'Art • Butt-Head Astronomer

B'ham: Birmingham (UK-England)

BHarO: Bulletin of Harvard Observatory (periodical code)

BHB: Bar Harbor (USA-ME) (location identifier) • Black Hole Binary • Blue Horizontal Branch

BHC: Beam-Heated Cathode • Black Hole Candidate • Born-Haber Cycle • Bullhead City (USA-AZ) (location identifier) • Busy Hour Call • Laughlin (USA-NV) (location identifier)

Bhd: Berhad • Bulkhead

BHD: Belfast City (Northern Ireland) (location identifier) • Dinar (Bahrain) (ISO currency code)

BHE: Biharmonic Equation • Blenheim (New Zealand) (location identifier)

B'head: Birkenhead (UK-England)

Bhf: Bahnhof

BHF: Background Heat Flux • Bahia Cupica (Colombia) (location identifier)

BHG: Blue Hertzsprung Gap • Brus Laguna (Honduras) (location identifier) • Bureau de l'Hygiène Générale

BHH: Bisha (Saudi Arabia) (location identifier)

BHI: Bahia Blanca - Commandante Espora (Argentina-BA) (location identifier) • Balbus-Hawley Instability • Branch if Higher • British Humanities Index • Bureau Hydrographique International

BHIS: Branch if Higher or Same

BHJ: Bhuj (India) (location identifier)

BHK: Bukhara (Uzbekistan) (location identifier)

BHL: Bahia Angeles (Mexico) (location identifier) • Bursty High Latitude

B-HLI: Broadband Higher-Layer Information

BHLI: Broadband Higher-Layer Information

Bhm: Birmingham (UK-England)

BHM: Birmingham (USA-AL) (location identifier)

BHMO: Blue Hill Meteorological Observatory (USA)

BHN: Beihan (Yemen) (location identifier) • Brinell Hardness Number

BHNC: Bridge-function Hypernetted Chain

BHND: Behind

BHO: Bâtiment Hydro-Océanographique • Bhopal (India) (location identifier) • Black Hole Observatory (Germany)

bhp: Brake Horsepower

BHP: Basic Hydrogen Peroxide • Bhojpur (Nepal) (location identifier) • Brake Horsepower

BHQ: Broken Hill (Australia-NSW) (location identifier)

BHR: Bahrain (ISO country code) • Bahrein (ISO country code) • Bharatpur (Nepal) (location identifier) • Block Handler Routine

BHRA: British Hydraulics Research Association

BHRC: Budget and Human Resources Committee

BHS: Bahamas (ISO country code) • Base Heat Shield • Bathurst Raglan (Australia-NSW) (location identifier) • Black Hole System • Burke High School (USA-NE)

BHT: Bipolar Heterojunction Transistor • Bottom-Hole Temperature • Branch History Table • Brighton Downs (Australia-QLD) (location identifier) • Butylated Hydroxytoluene

Bhu: Bhutan

BHU: Bhavnagar (India) (location identifier)

BHV: Bahawalpur (Pakistan) (location identifier)

BHVAD: Birmingham (Starlink internal code)

BHvaO: Hvar Observatory Bulletin (periodical code)

BHW: Sargodha Bhagatanwala (Pakistan) (location identifier)

BHX: Birmingham (UK-England) (location identifier)

BHXN: Black Hole X-ray Nova

BHXRT: Black Hole X-Ray Transient

BHXT: Black Hole X-ray Transient

BHY: Beihai (China-PRC) (location identifier)

BHZ: Belo Horizonte - Pamphula (Brazil-MG) (location identifier)

BH1: Bâtiment Hydrographique de Première Classe

BH2: Bâtiment Hydrographique de Deuxième Classe

bi: Bislama (ISO language code)

Bi: Bismuth • Wismuth

BI: Background Information • Back-Illuminated • Back-side Illuminated • Basic Index • Binary Input • Base d'Imposition • Basic Instinct • Bénéfice d'Inventaire • Biens Immobiliers • Big Island • Billet Individuel • Binary • Bordereau d'Inscription • Brevet d'Invention • Brief Introduction • Buffer Index • Built In • Bulk Issue • Bulletin d'Information • Bureau d'Instruction • Burundi (ISO country code) • Bus Interface

BIA: Bastia Poretta (France) (location identifier) • Blue Ice Area • Bomb Impact Assessment • Boost, Insertion and Abort • Booster Interstage Assembly • British Island Airways Ltd. • Bureau of Indian Affairs (USA) • Business Impact Analysis • Bus Interface Adapter

BIAC: Business Industry Advisory Committee

BIAPS: Battery Inverter Accessory Power Supply

BIAS: Bakosurtanal Image Analysis System • Better International Air Service

BIATEX: Biosphere-Atmosphere Exchange

bib: Bibe

Bib: Bible • Biblical • Bibliothèque

BIB: Baidoa (Somalia) (location identifier) • Balanced Incomplete Block • Bibliography • Biographical Information Bank • Blocked Impurity Band • Board for International Broadcasting • Boundary First Born • Bulletin d'Information des Bibliothèques

BIBB: Bundesinstitut für Berufsbildung

BIBD: Balanced Incomplete Block Design

BIBEX: Biomass Burning Experiment

BIBIB: Back-Illuminated Blocked Impurity Band

BIBID: Bibliographic Identifier

Bibl: Biblical • Bibliographer • Bibliographical • Bibliographie • Bibliographique • Bibliography • Biblioteca • Bibliothèque

Biblio: Biblioteca

Bibliogr: Bibliografía • Bibliographer • Bibliographical • Bibliographie • Bibliographique • Bibliography

BIBLIS: Bibliotheeks Informatie Systeem

BIBNET: Bibliographic Network

BIBSYS: Bibliotek System

BIC: Barium Ion Cloud • Battery Interconnecting Cable • Bayes Information Criterion • Bénéfices Industriels et Commerciaux • Biased Ion Collector • Big Creek (USA-AK) (location identifier) • Book Industry Communications • Bulletin d'Information Cyclonique • Bureau International du Cinéma • Business Information Conference • Bus Interface Circuit • Bus Interface Controller • Byte Input Control

BICA: Bureau International de Chimie Analytique

BICAA: Bénéfices Industriels, Commerciaux, Agricoles et Artisanaux

Bicarb: Bicarbonate

BICC: Berne International Copyright Convention • Biomedical Information Communication Center • Bureau International des Chambres de Commerce

BICDS: Bulletin d'Information du Centre de Données de Strasbourg (periodical code)

BICE: Board on Infrastructure and Constructed Environment

BICEP: Basic Image Classification and Enhancement Package

BICEPT: Book Indexing with Context and Entry Points from Text

BICER: Baikal International Centre for Ecological Research

BICFET: Bipolar Inversion-Channel Field-Effect Transistor

BiCG: Biconjugate Gradient

BICI: Book Item and Contribution Identifier • Bureau International du Commerce et de l'Industrie

BICM: Bilan Individuel de Compétences Managériales

BiCMOS: Bipolar Complementary Metal Oxide Silicon

BICS: Burroughs Inventory Control System

BICSE: Board on International Comparative Studies in Education

bid: Bis in Die

BID: Bachelor of Industrial Design • Banco Interamericano de Desarollo • Banque Interaméricaine de Développement • Banque Islamique de Développement • Basic Iterative

Deconvolution • Bis In Die • Blind Iterative Deconvolution • Block Island (USA-RI) (location identifier) • Böblingen Instruments Division (HP) • Business Improvement District

BIDA: Bureau International des Droits d'Auteurs

BIDC: Barbados Investment and Development Corporation

BIDE: Bulletin International de Découvertes et d'Explorations

BIDF: Bureau International de Documentation Fiscale

BiDi: Bidirectional

BIDKA: Bibliotheek-, Informatie- en Documentatiecentrum van de Kristelijke Arbeidersbeweging (Belgium)

BIDM: Basic Interoperability Data Model

BIDP: Bunyip Internet Directory Project

BIDR: Basic Image Data Record

BIDS: Bangladesh Institute of Development Studies • Bath Information and Data Service (UK) • Bath ISI Data Service (UK) • Bendix Integrated Data System • Broadband Integrated Distributed Star

Bie: Bergerie

BIE: Bachelor of Industrial Engineering • Beatrice (USA-NE) (location identifier) • Boundary Integral Equation • Bureau International d'Électrothermie • Bureau International de l'Éducation • Bureau International des Expositions • Bureau of Industry Economics

BIEF: Banque d'Information sur les États Francophones

Bien: Biennial

BIET: British Institute of Engineering Technology • Bureau International de l'Enseignement Technique

BIF: Banded Iron Formation • Beta-Induced Fluorescence • Bus Interface Function • El Paso - Biggs (USA-TX) (location identifier) • Franc (Burundi) (ISO currency code)

BIFADEC: Board for International Food and Agricultural Development and Economic Cooperation

BIFET: Bipolar Field-Effect Transistor

BIFR: Board for Industrial and Financial Reconstruction

BIG: Beliefs in Government • Big Delta (USA-AK) (location identifier) • Brazilian Institute of Geography and Statistics

BIGFET: Bipolar Insulated Gate Field-Effect Transistor

BIGIS: Bendix Interactive Geographic Information System

BIGRAT: Bicentennial Gamma-Ray Telescope

BIGS: Booster Inertial Guidance System

BIH: Bishop (USA-CA) (location identifier) • Board on International Health • Bosnia-Hercegovina (ISO country code) • Bosnie-Hercégovine (ISO country code) • Built-In Hold • Bulletin d'Information Hebdomadaire • Bureau International de l'Heure (France)

BII: Background Illumination Intensity • Bikini Atoll (Marshall Islands) (location identifier) • Binary Independence Indexing

BIICC: Bureau International d'Information des Chambres de Commerce

BIIEA: Bureau International d'Information sur les Éphémérides Astronomiques

Bij: Bijouterie

BIJ: Biliau (Papua New Guinea) (location identifier)

Bijl: Bijlage

Bijv: Bijvoorbeeld

Bijz: Bijzonder

BIK: Biak (Indonesia) (location identifier)

BIL: Band Interleaved by Line • Bâtiment d'Intégration Lanceur (CSG) • Billings (USA-MT) (location identifier) • Block Input Length • Braille Institute Library

Bilat: Bilatéral

BILBO: Built-In Logic Block Observer

BILD: Bureau International de Liaison et de Documentation

BILE: Balanced Inductor Logical Element

BiLuf: Bildmessung und Luftbildwesen (periodical code)

Bim: Bimestrale • Bimestre

BIM: Balanced Implicit Multiway • Beginning of Information Marker • Best In Match • Bimestriel • Bimini (Bahamas) (location identifier) • Binary Ionic Mixture • Bit Image Memory • Boundary Integral Method • Bulletin Individuel de Mouvement

BIMA: Berkeley-Illinois-Maryland Array • Berkeley-Illinois-Maryland Association (*) • British Interactive Multimedia Association

Bimest: Bimestriel

BIML: Bureau International de Métrologie Légale

BIMS: Browse Image Management System • Bus Ion Mass Spectrometer

Bin: Binary

BIN: Bamiyan (Afghanistan) (location identifier) • Barrier Intrinsic-N+ • Basic Information Network • Bell Information Network • Billing Information Number • Binary • Bureau International de Normalisation • Business Information Network

BINAC: Binary Automatic Computer

BINAS: Biosafety Information Network and Advisory Servers

Bind: Binding

BIND: Berkeley Internet Name Daemon (*) • Berkeley Internet Name Domain

BINR: Basic Intrinsic Noise Ratio

BINS: Boreal Institute of Northern Studies

BIO: Bedford Institute of Oceanography (Canada-NS) • Bilbao Sondica (Spain) (location identifier)

BIOC: Break Into Other Computers

Biochem: Biochemical • Biochemist • Biochemistry

Biogr: Biographer • Biographical • Biographie • Biographique • Biography

Biol: Biología • Biological • Biologist • Biology

BIOM: Buffer Input-Output Memory

BIOMASS: Biological Investigation of Marine Antarctic Systems and Stocks

BIOME: Biological Imaging and Observational Mission to Earth

Bion: Bionika (periodical code)

BION: Believe It Or Not

BIOP: Blatant Interruption of Operational Potentialities • Bulletin Intérieur de l'Observatoire de Paris

BIOPAR: Bifurcating Optical Pattern Recognizer

Bioquím: Bioquímica

BIOS: Basic Input/Output System • Biological Investigations Of Space • Biological Satellite

BioSc: BioScience (periodical code)

BIOT: British Indian Ocean Territory

BIP: Balloon Interrogation Package • Band Interleaved by Pixel • Bâtiment d'Intégration des Propulseurs (CSG) • Biens et Intérêts Privés • Bruttoinlandsprodukt • Bulimba (Australia-QLD) (location identifier) • Bureau d'Information et de Propagande • Bureau International de Presse

BIPAR: Bistatic Parasitic Radar

BIPM: Bureau International des Poids et Mesures (France)

BIpS: Billion Instructions per Second

BIPS: Billion Instructions Per Second • Bulletin of the Israel Physical Society (periodical code)

BIQ: Biarritz Anglet Bayonne (France) (location identifier)

BIQUUE: Best Invariant Quadratic Uniformly Unbiased Estimate

BIR: Band Irradiance Response • Binary Independence Retrieval • Biratnagar (Nepal) (location identifier) • Birmingham (Starlink internal code) • Browser Information Request • Bulletin d'Informations Rapides • Bureau of Immigration Research • Bureau of Industry Research

BIRA: Belgische Instituut voor Ruimte Aëronomie (Belgium)

BIRAMIS: Bilan Radiatif de l'Atmosphère par Micro-accélérométrie Spatia le

BIRAP: Balloon Infrared Astronomy Platform

BIRCH: Balanced Iterative Reducing and Clustering using Hierarchies

BIRD: Banco Internacional para la Reconstrucción y el Desarollo • Banque Internationale pour la Reconstruction et le Développement • Binational Industrial Research and Development • Browsing Interface for the Retrieval of Documents

BIRDE: Banque Internationale pour la Reconstruction et le Développement Économique

BIRF: Banco Internacional para la Reconstrucción y Fomento

Birm: Birmingham (UK-England)

BIRM: Beijing Institute of Radiometrology and Measurement (China-PRC)

BIRMD: Balanced Incomplete Repeated Measures Design

BIRS: Basic Indexing and Retrieval System • Basic Information Retrieval System • Binary Infrared Stars • Bright Infrared Stars

Bis: Bissextile

BIS: Bachelor of Science in Information Systems • Bande Intermédiaire Satellite • Bank for International Settlements • Basic Information Sheet • Beam-Induced Stripping • Bibliotheks-Informations-System • Bien de Interés Cultural • Biocide Injection System • Bio-geographic Information System • Bismarck (USA-ND) (location identifier) • Bremsstrahlung Isochromat Spectroscopy • British Interplanetary Society • Building Information Society • Bureau of Indian Standards • Bureau of Information Science • Business Information System

BISA: Belgian Institute for Space Aeronomy (Belgium)

BISAC: Book Industry System Advisory Committee (USA)

BISAM: Basic Indexed Sequential Access Method

BISAR: Bistatic Array

BISCA: Bolzano International Schools on Cognitive Analysis (Italy)

B-ISDN: Broadband Integrated Services Digital Network

BISDN: Broadband Integrated Services Digital Network

BISE: Best Index Slope Extraction

BISHOP: Better Image Software with Help On-line for Programs

BiSON: Birmingham Solar Oscillation Network (*)

BISON: Birmingham Solar Oscillation Network

BISP: Business Information System Program

BISS: Berkeley Interactive Statistical System

BISSCAN: Binary Identification System from S-Curve Analysis

BIST: Baseline Integrated Satellite Test • Built-In Self-Test

BISY: British Industrial Space Year • British Industry in Space Year

BiSync: Binary Synchronous Communication

BISYNC: Binary Synchronous Communication

BIT: Bachelor of Industrial Technology • Baitadi (Nepal) (location identifier) • Behavior and Information Technology • Bilateral Investment Treaty • Binary Digit • Binary Information Transfer • Born-Infeld Theory • Browser Information Transfer • Built-In Test • Bureau International du Travail (Switzerland)

BitBlT: Bit Block Transfer

BITBLT: Bit Block Transfer

BITE: Built-In Test Equipment

BITFTP: BITNET FTP

BITNET: "Because It's Time" Network

BITNIC: BITNET Network Information Center

BITon: Bol. Inst. Tonantzintla y Tacubaya (periodical code)

BiTraC: Bidirectional Transfer Channel

BITS: Book Integrity Token System • Boot Integrity Token System • Budget Impact Tracking System • Building Integrated Timing Supply

BIU: Basic Information Unit • Basic Interface Unit • Bibliothèque Interuniversitaire • Bildudalur (Iceland) (location identifier) • Buffer Interface Unit • Bureau International des Universités • Bus Interface Unit

BIV: Bria (Central African Rep.) (location identifier) • Built-In Variance

BIW: Billiluna (Australia-WA) (location identifier)

BIX: Biloxi Keesler (USA-MS) (location identifier) • Binary Information Exchange • Byte Information Exchange

BIY: Bisho (South Africa) (location identifier)

BIZ: Bimin (Papua New Guinea) (location identifier)

Bi2S3: Bismuth Sulfide

BJ: Bachelor of Journalism • Benin (ISO country code) • Bubble-Jet

BJA: Bejaia (Algeria) (location identifier)

BJC: Broomfield Jeffco (USA-CO) (location identifier) • Bubble Jet Color

BJCEB: British Joint Communications Electronics Board

BJD: Bakkafjördur (Iceland) (location identifier)

BJF: Batsfjord (Norway) (location identifier)

BJG: Bolaang (Indonesia) (location identifier)

BJH: Bajhang (Nepal) (location identifier)

BJI: Bemidji (USA-MN) (location identifier)

BJJ: Wooster - Wayne County (USA-OH) (location identifier)

BJK: Benjina (Indonesia) (location identifier)

BJL: Banjul (Gambia) (location identifier)

BJM: Bujumbura (Burundi) (location identifier)

BJN: Bajone (Mozambique) (location identifier)

BJO: Bermejo (Bolivia) (location identifier)

Bjr: Bonjour

BJR: Bahar Dar (Ethiopia) (location identifier) • Blockhouse Junction Rack

BJS: Beijing (China-PRC) (location identifier) • Peking (China-PRC) (location identifier)

BJT: Bipolar Junction Transistor

BJU: Bajura (Nepal) (location identifier) • Bob Jones University (USA-SC)

BJW: Bajawa (Indonesia) (location identifier)

BJX: León-Guanajuato - Del Bajio (Mexico) (location identifier)

BJY: Beograd Batajnica (Yugoslavia-Serbia) (location identifier)

BJZ: Badajoz - Talavera la Real (Spain) (location identifier)

Bk: Back • Backwardation • Bank • Barrack • Berkelium • Black • Block • Book • Break

BK: Backup • Bundeskanzler

BKA: Bundeskriminalamt (Germany) • Moscow Bykovo (Russia) (location identifier)

BKB: Bikaner (India) (location identifier)

Bkble: Bookable

Bkbndr: Bookbinder

BKC: Benzalkonium Chloride • Buckland (USA-AK) (location identifier)

Bkcy: Bankruptcy

Bkd: Blackboard • Booked

BKD: Breckenridge - Stephens County (USA-TX) (location identifier)

Bkdn: Breakdown

BKE: Baker (USA-OR) (location identifier)

BKF: Brooks Lake (USA-AK) (location identifier)

Bkfst: Breakfast

Bkg: Banking • Booking • Bookkeeping • Breakage

BKG: Background • Bundesamt für Kartographie und Geodäsie (Germany)

Bkgd: Background

BKH: Kekaha - Barking Sands (USA-HI) (location identifier)

BKI: Kota Kinabalu (Malaysia) (location identifier)

BKJ: Boke (Guinea) (location identifier)

BKK: Bangkok Donmuang (Thailand) (location identifier)

Bkkg: Book-Keeping

BKL: Cleveland - Burke Lakefront (USA-OH) (location identifier)

bklr: Black Letter

Bklt: Booklet

Bklyn: Brooklyn (USA-NY)

BKM: Bakalalan (Malaysia) (location identifier)

BKN: Birni Nkoni (Niger) (location identifier) • Broken

BKNO3: Boron-Potassium Nitrate

BKO: Bamako (Mali) (location identifier) • Barkhausen-Kurz Oscillator

bKOM: Broadband Kilometric Radiation

BKP: Backup • Barkly Downs (Australia-QLD) (location identifier)

Bkpg: Bookkeeping

Bkpr: Bookkeeper

Bkpt: Bankrupt

BKQ: Blackall (Australia-QLD) (location identifier)

BKR: Bokoro (Chad) (location identifier)

Bkrpt: Bankrupt

Bks: Banks • Barracks • Books

BKS: Basic Kernel System • Bengkulu (Indonesia) (location identifier)

BKSP: Backspace

Bkt: Basket • Bracket • Bucket

BKT: Blackstone (USA-VA) (location identifier) • Camp Pickett - Blackstone (USA-VA) (location identifier)

BKU: Betioky (Madagascar) (location identifier)

BKW: Beckley (USA-WV) (location identifier)

BKX: Brookings (USA-SD) (location identifier)

Bky: Berkeley (USA-CA)

BKY: Berkeley (USA-CA) • Bukavu (Zaire) (location identifier)

BKZ: Bukoba (Tanzania) (location identifier)

b/l: Bill of Lading

bl: Barrel • Bel

Bl: Bale • Black • Blatt • Block • Blue • Boulevard

B-L: Bill of Lading

B/L: Baseline • Bill of Lading

BL: Backlit • Bande Large • Baseline • Between Layers • Billet de Logement • BioLuminescence • Black Letter • Blank Line • Bloch Line • Bloco • Blowing • Blue Lightning • Bolton (UK postal code) • Bon de Livraison • Boundary Layer • Bowling Lane • Boy's Life • British Library • Bulleted List • Bureau de Location • Busy Line • By-Laws

BLA: Barcelona (Venezuela) (location identifier) • Base de Lancement Ariane (CSG) • BL Lacertae (object) • Puerto de la Cruz (Venezuela) (location identifier)

BLACS: Basic Linear Algebra Communication Subprogram • Basic Linear Algebra Communication Subroutine

BLADE: Basic Level Automation of Data through Electronics • Bell Laboratories Automatic Device

BLADES: Bell Laboratories Automatic Device System

Bladz: Bladzijde

BLAI: Brown Leaf Area Index

BLAISE: British Library Automated Information Service

Blas: Blasón

BLAS: Basic Linear Algebra Subprogram • Basic Linear Algebra Subroutine • Blackburn Leisure Astronomy Section (UK)

BLAST: Basic Local Alignment Search Tool • Bates Large Acceptance Spectrometer Toroid • Battlefield Laser Acquisition Sensor Test • Blocked Asynchronous Transmission • Burst Locations with an Arcsecond Telescope

BLB: Balboa (Panama) (location identifier)

BLBD: Binary Light Beam Deflector

BLC: Bali (Cameroon) (location identifier) • Baseline Configuration • Boundary Layer Control • British Library Catalogue • Broad Line Cloud • Broad Line Component

BLCE: Baseline Calibration Equipment

BLCIP: British Library Cataloguing In Publication

BLCMP: Birmingham Libraries Cooperative Mechanisation Project

BLCTD: British Library Computing and Telecommunications Department

Bld: Boldface • Boulevard

BLD: Below Limit of Detection • Bload • Boulder City (USA-NV) (location identifier) • Build

BLDB: Bilingual Lexical Database

BLDC: Brushless Digital Converter

Bldg: Building

BLDG: Building

Bldgs: Buildings

Bldr: Builder

BLDSC: British Library Document Supply Centre

BLDUP: Buildup

BLE: Bahadur-Lazersfeld Expansion • Binary Logic Element • BorlÄnge (Sweden) (location identifier) • Borlange (Sweden) (location identifier) • Branch if Less or Equal

BLEDE: Backscattered Laser Energy Digitizing Equipment

BLER: Block Error Rate

BLEU: Belgo-Luxembourg Economic Union

BLF: Bluefield (USA-WV) (location identifier) • Bluff • Boundary Layer Flow • Busy Line Field • Princeton (USA-WV) (location identifier)

BLFS: Bluffs

Blg: Belge

BLG: Belaga (Malaysia) (location identifier) • Binary Line Generalization

BLH: Band-Limited Hiss • Blythe (USA-CA) (location identifier)

BLH&SS: British Library Humanities & Social Sciences

BLI: Bellingham (USA-WA) (location identifier)

BLIMP: Balloon-Launched Integral Model Payload • Boundary Layer Integral Matrix Procedure

BLING: Bladed Ring

BLIP: Background-Limited Infrared Photoconductor (*) • Background-Limited Infrared Photodetector • Background-Limited Infrared Photography • Background Limiting Performance • Basic Landsat Image Processing • Boundary Layer Instrumentation Package

BLIS: Bell Laboratories Interpretive System • Bibliographic Information System • Boundary Layer Instrumentation System

BLISK: Bladed Disk

BLISS: Balloon-borne Laser In Situ Sensor • Basic Language for the Implementation of System Software • Bibliographic and Library Information Search Service • British Library Information Sciences Service

BLIT: Block Image Transfer

BLJ: Batna (Algeria) (location identifier)

BlJMH: Bulgarian Journal of Meteorology and Hydrology (periodical code)

BlJPh: Bulgarian Journal of Physics (periodical code)

Blk: Black • Blank • Block • Bulk

BLK: Blackpool (UK-England) (location identifier) • Block

Blkd: Bulkhead

Blkg: Blanking • Blocking

Blks: Blocks

BLKT: Blanket

BLL: Billund (Denmark) (location identifier)

BLLE: Balanced Line Logical Element

BLLS: Boundary Layer Lidar System

BLM: Basic Language Machine • Belmar - Monmouth County (USA-NJ) (location identifier) • Bi-Linear Modelling • Boundary Layer Model • Bureau of Land Management (USA)

BLMC: Buried Logic Macrocell

BLML: Bland Meal

Bln: Balloon

BLN: Backbone Link Node • Benalla (Australia-VIC) (location identifier)

BLNAF: British Library Name Authority File

BLNBS: British Library National Bibliographic Service

BLNSA: British Library National Sound Archive

BLNT: Broadband Local Network Technology

BLO: Below • Below Clouds • BL Lacertae Object • Blönduós (Iceland) (location identifier)

BLOb: Binary Large Object

BLOB: Binary Large Object

BLOS: Branch if Lower Or Same

BLOW: Booster Lift-Off Weight

BLP: Basic Launch Plan • Bellavista (Peru) (location identifier) • Best Linear Predictor • Boundary Layer Physics • Boundary Layer Profile • Bypass Label Processing

BLPG: Buried Long-Period Grating

BLPU: Basic Land and Property Unit

BLQ: Bologna (Italy) (location identifier)

BLR: Bangalore (India) (location identifier) • Belarus (ISO country code) • Biélorussie (ISO country code) • Boiler • Boucle Locale Radio • Boundary Layer Radar • Broad Line Region

BLR&DD: British Library Research & Development Department

BLRDD: British Library Research and Development Department

BLRG: Broad-Line Radio Galaxy

BLRIC: British Library Research and Innovation Centre (UK)

BLS: Balanced Line System • Balloon Launching Station • Band-Limited Signal • Baseline Spectroscopy • Basic Linked System • Bell Log System • Bern-Lötschberg-Simplon • Bidimensional L System • Block Level Sharing • Bollon (Australia-QLD) (location identifier) • Botswana-Lesotho-Swaziland • Bottom Left Side • Boundary Layer Separation • Brillouin Light Scattering • Bundle of Lines of Sight • Bureau of Labor Statistics

BLSN: Blowing Snow

BLSR: Bidirectional Line-Switched Ring

BLSRIS: British Library Science Reference and Information Service

Blt: Built

BLT: Beacon-Lettuce-Tomato • Blackwater (Australia-QLD) (location identifier) • Block Transfer • Branch Less Than • Bundesamt für Landestopographie (Switzerland)

Bltn: Bulletin

BLU: Ben Lee User • Blue Canyon (USA-CA) (location identifier)

BLUE: Best Linear Unbiased Estimator

BLUP: Best Linear Unbiased Prediction

BLUUE: Best Linear Uniformly Unbiased Estimate

BLV: Belleville Scott (USA-IL) (location identifier) • Bleed Valve • Busy Line Verification

BLVA: Bayerisches Landesvermessungsamt (Germany)

Blvd: Boulevard

BLVD: Boulevard

Blvrd: Boulevard

Blw: Below

BLW: Below • Waimanalo Bellows (USA-HI) (location identifier)

BLWN: Band-Limited White Noise

BLWR: Blower

BLX: Belluno (Italy) (location identifier)

BLY: Belmullet (Ireland) (location identifier)

Blz: Bladzijde

BLZ: Belize (ISO country code) • Blantyre Chileka (Malawi) (location identifier) • Boundary Layer Zone

Blzd: Blizzard

BLZD: Blizzard

B/M: Bench Maintenance • Buffer/Multiplexer • Bulletin Maritime

BM: Banque Mondiale • Basal Metabolism • Battle Management • Beam Monitor • Bench Maintenance • Bench Mark • Bermuda (ISO country code) • Bermudes (ISO country code) • Bibliothèque Municipale • Bimensuel • Binary Multiply • Black Marker • Body-Mounted • Boltzmann Machine • Bon Marché • Boyer-Moore • Branch Manager • Branch Memorandum • British Museum (UK) • Brownian Motion • Bubble Memory • Buffer Manager • Buffer Module • Bulletin Mensuel • Bulletin Météorologique • Business Manager

BMA: Balanced Magnetic Amplifier • Body-Mounted Accelerometer • British Medical Association • Bulletin Météorologique d'Alerte • Bulletin Météorologique d'Avertissement • Stockholm Bromma (Sweden) (location identifier)

BMaA: Bundesministerium für auswärtige Angelegenheiten

BMAA: Bays Mountain Amateur Astronomers (USA-TN) • Bucks-Mont Astronomical Association (USA-PA)

BMAED: Board on Manufacturing and Engineering Design

BMAG: Body-Mounted Attitude Gyro

BMAN: Broadband Metropolitan-Area Network

BMAP: Background Measurement and Analysis Program • Biological Monitoring and Abatement Program • Buffer Map

BMAS: Body-Mounted Accelerometer System

BMath: Bachelor of Mathematics

BMATT: Briefcase Multimission Advanced Tactical Terminal

BMB: Base Maintenance Building • Bench Maintenance Building • Biomedical Belt • Bumba (Zaire) (location identifier)

BMBW: Bundesministerium für Bildung und Wissenschaft (Germany)

BMC: Battelle Monte Carlo • Binary Magnetic Core • Block Multiplexor Channel • Brigham City (USA-UT) (location identifier) • British Museum Catalogue • Bulk Motion Comptonization

BMCD: Blind Mate Connector Bracket

BMC3: Battle Management Command, Control, and Communications

BMD: Bangladesh Meteorological Department • Bearths-Marriages-Deaths • Belo (Madagascar) (location identifier) • Biomedical • Business Management Department (STScI) • Dollar (Bermuda) (ISO currency code)

BMDADS: Biomedical Data Analysis and Display System

BMDP: Biomathematical Data Processing • Biomedical Package

BME: Bachelor of Mechanical Engineering • Bachelor of Mining Engineering • Bench Maintenance Equipment • Bench-Mounted Echelle • Bipolar Mode Ejection • Born-Mayer Equation • Broome (Australia-WA) (location identifier)

Bment: Bâtiment

BMEP: Brake Mean Effective Pressure

BMF: Background Mean Field • Bakouma (Central African Rep.) (location identifier) • Basic Main Frame • Beryllium Machining Facility • Boron Metal Fiber • Bundesministerium der Finanzen • Business Master File

BMFT: Bundesministerium für Forschung und Technologie (Germany)

BMG: Badan Meteorologi dan Geofisika (Indonesia) • Bloomington (USA-IN) (location identifier) • Bundesministerium für Gesundheit

BMH: Bomai (Papua New Guinea) (location identifier)

BMI: Backplane Message Interface • Backplane Message Interfacer • Baltimore Museum of Industry (USA-MD) • Battelle Memorial Institute • Bismaleimide • Bloomington Normal (USA-IL) (location identifier) • Body-Mass Index • Branch if Minus • Bravais-Miller Index • Bundesministerium des Innern (Germany)

BMIC: Bus Master Interface Controller

BMIS: Basic Management Information System

BMJ: Baramita (Guyana) (location identifier)

BMK: Borkum (Germany) (location identifier)

BML: Basic Machine Language • Berlin (USA-NH) (location identifier) • British Museum Library

BMLS: Balloon Microwave Limb Sounder

BMM: Berührungslose Mittendickenmeßgerät • Biological Macromolecule • Bitam (Gabon) (location identifier) • Bohr and Mottleson Model

BMMD: Body Mass Measuring Device

BMN: Bamerny (Iraq) (location identifier) • Bulletin Météorologique National

BMNH: British Museum of Natural History (UK)

BMNT: Beginning Morning Nautical Twilight

BMO: Bhamo (Myanmar) (location identifier) • Black Moshannon Observatory (USA-PA) • Bond Molecular Orbital • British Meteorological Office (UK) • Business Machine Operator

BMOC: Big Machine On Campus • Big Man On Campus

B'mouth: Bornemouth (UK-England)

BMP: Bad Manufacturing Practice • Basic Multilingual Plane • Batch Message Processing • Binary Millisecond Pulsar • Bitmap Picture • Blue Metal-Poor • Break Mean Power • Brampton Island (Australia-QLD) (location identifier) • Burst Mode Protocol

BMPA: Broadband Microwave Power Amplifier

BMPR: Bimonthly Progress Report

BMPT: Bundesministerium für Post und Telekommunikation (Germany) • Buren-Mosset-Parmentier-Toroni

BMQ: Bamburi (Kenya) (location identifier)

BMR: Baltrum (Germany) (location identifier) • Bulletin Météorologique Régional • Bureau of Mineral Resources

BMRC: Bureau of Meteorology Research Centre (Australia)

BMS: Background Mapping Sensor • Background Measurement Satellite • Basic Mapping Support • Basic Meteorological Service • Battle Management Simulation • Below Minimum Standards • Bias Management System • Biomedical Monitoring System • Board on Mathematical Sciences • Bondi-Metzner-Sachs • Boston Museum of Science (USA-MA) • Broadcast Message Server • Brumado (Brazil-BA) (location identifier) • Building Management Science • Bulletin Météorologique Spécial • Burst Measuring System • Business Management System • Business Modelling System • Bypass Monochrome Signal

BMT: Beaumont (USA-TX) (location identifier) • Beginning of Magnetic Tape • British Mean Time • Brooklyn-Manhattan Transit (New York)

BMTC: Bureau of Meteorology Training Centre

BMTE: Bright Magnitude Theoretical Excess

BMTS: Bench Maintenance Test Set

BMU: Battery Management Unit • Bermuda (ISO country code) • Bermudes (ISO country code) • Bima (Indonesia) (location identifier) • Bus Monitor Unit

BMV: Ban Me Thuot - Phung-Duc (Viet Nam) (location identifier) • Bulletin Météorologique de Vigilance • Bundesministerium für Verkehr

BMVg: Bundesministerium für Verteidigung

BMW: Bayerische Motoren-Werke • Bordj Badji Mokhtar (Algeria) (location identifier)

BMwA: Bundesministerium für wirtschaftliche Angelegenheiten

BMWF: Bundesministerium für Wissenschaft und Forschung

BMWi: Bundesministerium für Wirtschaft

BMWV: Bundesministerium für Wissenschaft und Verkehr (Austria)

BMX: Big Mountain (USA-AK) (location identifier)

BMY: Belep Island (New Caledonia) (location identifier)

BMZ: Bamu (Papua New Guinea) (location identifier) • Bundesministerium für wissenschaftliche Zusammenarbeit (Germany)

bn: Bangla (ISO language code) • Bengali (ISO language code)

Bn: Baan • Basin • Billion • Born • Brown

BN: Ballistic Number • Bank Note • Becklin-Neugebauer • Biblioteca Nacional • Bibliothèque Nationale • Binary Number • Boron Nitride • Brighton (UK postal code) • Brunei (ISO country code) • Bull Nose • Bureau de Normalisation • Bureau National • But Not

BNA: Base-Neutralized Acid • Block Numbering Area • British North America • Burroughs Network Architecture • Nashville (USA-TN) (location identifier)

BNAE: Bureau de Normalisation de l'Aéronautique et de l'Espace (France)

BNAP: Basic Hydrological Network Assessment Project

BnB: Branch and Bound

BNB: Boende (Zaire) (location identifier) • British National Bibliography

BNC: Baby N Connector • Base-Neutralizing Capacity • Bayonet Collector • Bayonet Nut Collector • Bénéfice Non-Commercial • Beni (Zaire) (location identifier) • Boisson Non Comprise • British National Connector

BNCAR: British National Committee for Antarctic Research

BNCC: British National Committee for Chemistry

BNCE: Banque Nationale pour le Commerce Extérieur

BNCF: Biblioteca Nazionale Centrale di Firenze (Italy)

BNCI: Banque Nationale pour le Commerce et l'Industrie

BNCIHP: Belgisch Nationaal Comité voor het Internationaal Hydrologisch Programma

BNCOD: British National Conference On Databases

BNCSR: British National Committee on Space Research

BNCT: Boron Neutron Capture Therapy

BND: Bandar Abbas (Iran) (location identifier) • Beam Normalization Detector • Bend • Bénéfices Non Déclarés • Bound • Bundesnachrichtendienst (Germany) • Dollar (Brunei) (ISO currency code)

BNDRY: Boundary

BNDT: Base Nationale des Données Topographiques (Canada)

Bne: Borne

BNE: Branch if Not Equal • Brisbane (Australia-QLD) (location identifier) • But Not Exceeding

BNEC: British Nuclear Energy Conference

BNES: British Nuclear Energy Society

BNF: Backus-Naur Form • Backus Normal Form • Baranof (USA-AK) (location identifier) • Bibliographie Nationale Française • Bibliothèque Nationale de France (*) • Big-Name Fan • Boolean Normal Form • Boron Nitride Fiber

BNFL: British Nuclear Fuels Ltd.

BNG: Banning (USA-CA) (location identifier) • Blue Nucleated Galaxy • Branch No Group

BNH: Hartford Barnes (USA-CT) (location identifier)

BNI: Benin City (Nigeria) (location identifier) • Bibliografia Nazionale Italiana

BNIAB: British National Institute of Agricultural Botany (UK)

BNIST: Bureau National de l'Information Scientifique et Technique

BNJ: Bonn (Germany) (location identifier)

BNK: Ballina (Australia-NSW) (location identifier) • Bank

Bnkg: Banking

BNL: Barnwell County (USA-SC) (location identifier) • Bibliothèque Nationale du Luxembourg • Brookhaven National Laboratory (USA-NY)

Bnls: Boneless

BNM: Bibliotheca Neerlandica Manuscripta (Netherlands) • Bodinumu (Papua New Guinea) (location identifier) • Bureau National de Métrologie (France)

BNMS: Bus Neutral Mass Spectrometer

BNN: Boundary Network Node • Brønnøysund (Norway) (location identifier)

BNO: Baksan Neutrino Observatory (Russia) • Burns (USA-OR) (location identifier)

BNP: Bannu (Pakistan) (location identifier) • British National Bibliography • Bruto Nationaal Produkt

BNPC: Bathymetric Navigation Planning Chart

BNPD: Bivariate Normal Probability Density

BNQ: Baganga (Philippines) (location identifier) • Bibliothèque Nationale du Québec (Canada-QC) • Bureau de Normalisation du Québec (Canada-QC)

Bnr: Burner

BNR: Banfora (Burkina Faso) (location identifier) • Bloomberg News Radio

BNRSC: British National Remote Sensing Centre

BNS: Bachelor of Natural Science • Backbone Network Services • Barinas (Venezuela) (location identifier) • Bathymetric Navigation System • Beacon Numerical Service • Bibliothèque Nationale Suisse (Switzerland) • Billed Number Screening • Binary Number System • Bomb Navigation System • Braille 'N Speak • Buyer No Seller

BNSC: British National Space Centre

Bnsh: Burnish

BNSI: Barbados National Standards Institution

BNT: Bundi (Papua New Guinea) (location identifier)

BNTH: Beneath

BNU: Basic Networking Utilities • Bibliothèque Nationale et Universitaire (France) • Blumenau (Brazil-SC) (location identifier)

BNUS: Bibliothèque Nationale et Universitaire de Strasbourg (France)

BNV: Belgische Natuurkundige Vereniging (Belgium) • Boana (Papua New Guinea) (location identifier)

BNW: Boone (USA-IA) (location identifier)

BNX: Banja Luka (Bosnia-Hercegovina) (location identifier)

BNY: Bellona (Solomon Islands) (location identifier)

Bnz: Bronze

BNZ: Banz (Papua New Guinea) (location identifier)

b/o: Brought Over

bo: Bolivia • Tibetan (ISO language code)

Bo: Barrio

B/O: Booster/Orbiter • Breakout • Brought Over • Burnout • Buyer's Option

BO: Back Order • Bâtiment Océanographique • Beating Oscillator • Beat Oscillator • Bench Order • Billet à Ordre • Binary Output • Binary to Octal • Blackout • Blocking Oscillation • Body Odor • Boletín Oficial • Bolivia (ISO country code) • Bolivie (ISO country code) • Booster Orbiter • Born-Oppenheimer • Box Office • Braeside Observatory (USA-AZ) • Branch Office • Breakout • Broker's Order • Bulletin Officiel • Burnout • Business Object • Buyer's Option

BOA: Basic Object Adapter • Basic Ordering Agreement • Big Optical Array • Big Orbital Array • Bipolar Operational Amplifier • Board Of Advisors • Boma (Zaire) (location identifier) • British Optical Association • Broad Ocean Area • Building Omission Area

BOAD: Banque Ouest-Africaine de Développement

BOAO: Bohyunsan Optical Astronomy Observatory (Korea-ROK)

BoB: Break-out Box

BOB: Beginning Of Business • Best on Best • Bijzondere Opsporingsbrigade • Boliviano (Bolivia) (ISO currency code) • Bolometer Box • Bora Bora (Society Islands) (location identifier) • Break-Out Box • Broker of Observation Blocks • Bureau Of the Budget

BOBCATSSS: Barcelona-Oslo-Budapest-Copenhagen-Amsterdam-Tampere-Sheffield- Szombathely-Stuttgart

BOC: Base Operations Contract • Beginning Of Cycle • Bell Operating Co. • Beoordelingscommissie (Netherlands) • Bimodal Optical Computer • Block-Oriented Computer • Blowout Coil • Bocas del Toro (Panama) (location identifier) • Boîte d'Outils de Communication • Bureau Of Census • Byte Output Control

BOCA: Borland Object Component Architecture

BOCES: Board Of Cooperative Educational Services

BOCOL: Basic Operating Consumer-Oriented Language

BOCS: Business Object Component Set

Bod: Bodega

BOD: Base Operations Division • Basic Operational Data • Beneficial Occupancy Date • Bid Opening Date • Biochemical Oxygen Demand • Board Of Directors • Bordeaux Mérignac (France) (location identifier) • Bright-Object Detector • Buy-Off Date

BODACC: Bulletin Officiel des Annonces Civiles et Commerciales

BODC: British Oceanographic Data Centre

BOE: Board Of Education • Boletín Oficial del Estado • Bottom Of Edge • Boundji (Congo) (location identifier) • Break Of Entry

BOEN: Bulletin Officiel de l'Éducation Nationale (France)

BOF: Basic Oxygen Furnace • Beginning Of File • Beurre-Oeufs-Fromage • Bias Oscillator Frequency • Binary Oxide Film • Bird Of a Feather • Washington Bolling (USA-DC) (location identifier)

BofB: Back of Board

BofE: Bank of England

BOFROST: Bonner Frozen Spin Target

BOFS: Biogeochemical Ocean Flux Study • Bird-Of-a-Feather Session

BOG: Bogotá Eldorado (Colombia) (location identifier)

BOGSAAT: Bunch Of Guys Sitting Around A Table

Boh: Bohemia • Bohemian

BOH: Bournemouth (UK-England) (location identifier)

BOI: Boise (USA-ID) (location identifier) • Branch Output Interrupt • Break Of Integrity • Business Object Interface

BOJ: Bourgas (Bulgaria) (location identifier)

BOK: Brookings State (USA-OR) (location identifier)

bol: Bolus

Bol: Boletín • Bolivia • Bolivian

BOL: Bally Kelly (UK) (location identifier) • Beginning Of Life • Beginning Of Line • Bolivia (ISO country code) • Bolivie (ISO country code) • Business-On-Line

BOLD: Bibliographic Online Display • Blood Oxygenation Level Dependent • Boeing On-Line Data

Boll: Bollettino

BoLMe: Boubdary-Layer Meteorology (periodical code)

BOLT: Broadband Observatory for the Localization of Transients

BOLTS: Broad Area Optical Lightning Telescope Sensor

Bom: Bombay (India)

BoM: Bureau of Meteorology

BOM: Base Operation Manager • Basic Operating Monitor • Beginning Of Message • Beginning Of Mission • Beginning Of Month • Bill Of Material • Binary Order of Magnitude • Bombay (India) (location identifier) • Born-Oppenheimer Method • Bureau Of Meteorology • Bureau Of Mines • By Other Means

Bomb: Bombay (India)

BOMB: Bombeiros

BOMEX: Barbados Oceanographic and Meteorological Experiment

BOMS: Bollettino delle Opere Moderne Straniere

BON: Better Object Notation • Bonaire Flamingo (Netherlands Antilles) (location identifier)

BonD: Bonner Durchmusterung (periodical code)

BOND: Bandwidth On Demand

BONTT: Bergen-Oslo Nuclear Theory Team (Norway)

Boo: Bootes-Bootis (constellation)

BOO: Bodø (Norway) (location identifier)

BOOK: Bibliographic Online Organized Knowledge • Built-in Orderly Organized Knowledge

BOOM: Binocular Omni-Orientation Monitor • Binocular Omni-Oriented Monitor

BOOMERANG: Balloon Observations of Millimetric Extragalactic Radiation and Geophysics

BOOST: Broadband Object-Oriented Service Technology

BOOTP: Bootstrap Protocol

BOP: Baseline Operations Plan • Basic Overall Polarity • Basic Oxygen Process • Beginning Of Period • Binary Output Program • Biological Oxygen Demand • Bouar (Central African Rep.) (location identifier) • Building Optimization Program • Bureau Of Printing

BOPACE: Boeing Plastic Analysis Capability for Engines

BOPI: Bulletin Officiel de la Propriété Industrielle

BOpS: Billion Operations per Second

BOPS: Billion Operations Per Second

BOPul: Bulletin of Pulkovo Observatory (periodical code)

BOQ: Boku (Papua New Guinea) (location identifier)

BOQT: Bit Optimized Query Technology

Bor: Borough

BOR: Basic Output Record • Belfort Fontaine (France) (location identifier) • Body Of Revolution • Bureau Of Reclamation

BORAM: Block-Oriented Random Access Memory

BOREAS: Boreal Ecosystem and Atmosphere Study (*) • Boreal Ecosystem Atmospheric Study

BORIS: BOREAS Information System

Boro: Borough

BORPQU: Borland Pro Quattro

BORQU: Borland Quattro

Borr: Borrador

BORSCHT: Battery-Overvoltage-Ringing-Supervision-Coding-Hybrid-Testing

BOS: Backup Operating System • Basic Operating System • Book Order and Selection • Boston Logan (USA-MA) (location identifier) • Bright-Object Sensor • Business Office Supervisor

BOSC: Base Operations Support Contract

BOSEX: Baltic Open Sea Experiment

BOSO: Ballistic Orbital Support Operations

BOSP: Bulletin Officiel du Service des Prix

BOSS: Behavior of Offshore Structures • Big Occulting Steerable Satellite • Billing and Order Support System • Block-Oriented System Simulation • Business-Oriented Software System

BOST: Baselined Operational System Test

Bot: Botanic • Botánica • Botanical • Botanist • Botany • Bottle • Bottom • Bought

BOT: Beam On Target • Beginning Of Table • Beginning Of Tape • Beginning Of Track • Board Of Trade • Board Of Trustees • Boset (Papua New Guinea) (location identifier) • Bottom • Build-Operate-Transfer • Bureau des Objets Trouvés • Bureau d'Organisation Technique • Bureau d'Orientation Technique • Bureau Of Trade

BOTA: Board On Testing and Assessment

BOTT: Bol. Obs. Tonantzintla y Tacubaya (periodical code)

BOTTM: Bottom

BOU: Bourges (France) (location identifier)

Boul: Boulevard

BOUL: Boulevard

BOULV: Boulevard

Bourg: Bourgmestre

BOV: Bias Output Voltage • Boang (Papua New Guinea) (location identifier)

BOW: Bartow (USA-FL) (location identifier) • Basic Operating Weight • Bill Of Work • Bright Object Warning

BOX: Borroloola (Australia-NT) (location identifier)

BOY: Bobo Dioulasso (Burkina Faso) (location identifier)

BOZ: Bozoum (Central African Rep.) (location identifier)

b/p: Blueprint • Bills Payable

Bp: Beaucoup • Bon pour • Peculiar B-type Star

B&P: Bid & Proposal • Budgetary & Planning

B/P: Bills Payable • Blood Pressure • Blueprint

BP: Back Pressure • Back Projection • Back Propagation • Bandpass • Barber Pole • Barometric Pressure • Base Pair • Base Plate • Base Pointer • Basic Protocol • Basis Point • Basse Pression • Battery Package • Before the Present • Below Proof • Between Perpendiculars • Bid and Proposal • Biens Propres • Big Point • Bills Payable • Birth Place • Bit Pattern • Blood Pressure • Blueprint • Boiler Plate • Boiling Point • Boîte Postale • Bon à Payer • Bonum Publicum • Boulevard Périphérique • Brennpunkt • Brevet Professionnel • Bright Point • British Patent • British Petroleum • Budgetary and Planning • Budget Program • Bulletin de Paye • Bureau de Placement • Bureau de Poste • Bureau du Personnel • Buttock Plane • Bypass

BPA: Bahnpostamt • Balanced Parametric Amplifier • Bandpass Amplifier • Basic Pressure Altitude • Beam Plasma Amplification • Bethpage Grumman (USA-NY) (location identifier) • Bioshield Power Assembly • BNL Plant Analyzer • Board on Physics and Astronomy • Brookhaven Plant Analyzer

BPAA: Battle Point Astronomical Association (USA-WA)

BPAM: Basic Partitioned Access Method

BPANZ: Book Publishers Association of New Zealand

bpb: Bank Post Bills

BPB: BIOS Parameter Block • Boridi (Papua New Guinea) (location identifier)

Bpc: Bishopric

BPC: Back Pressure Control • Bamenda (Cameroon) (location identifier) • Biomass Production Chamber • Blood Pressure Control • Book Prices Current • Boost Protective Cover • Brevet Professionnel Commercial • Bureau de Pilotage et de Coordination • Business Practices Course • Butyl Pyridium Chloride

BPCA: Bioshield Pyrotechnic Control Assembly

BPCD: Barrels Per Calendar Day

BPCF: Bandpass Crystal Filter

BPD: Bapi (Papua New Guinea) (location identifier) • Barrels Per Day • Baseline Program Document • Baseline Program Documentation • Beam-Plasma Discharge • Beam-Pulse Detector

BPDF: Bidirectional Polarization Distribution Function

BPDU: Bridge Protocol Data Unit

BPE: Best Preliminary Estimate • Bit-Plane Encoding • Boiling Point Elevation

BPED: Basic Pay Entry Date

BPF: Bandpass Filter • Batuna (Solomon Islands) • Berkeley Packet Filter • Bon Pour Francs • Bottom Pressure Fluctuation

BPG: Barra do Garças (Brazil-MT) (location identifier) • Break Pulse Generator

BPGC: Beam-Plasma-Gas Coupling

BPh: Bachelor of Philosophy

BPH: Bislig (Philippines) (location identifier)

bpi: Bits per Inch

BPI: Bibliothèque Publique d'Information (France) • Big Piney Marbleton (USA-WY) (location identifier) • Boost Phase Intercept • Bureau de Presse et d'Information • Bureau of Public Inquiries • Bytes Per Inch

BPJ: Balanced Pressure Joint

BPKM: Billion Passenger Kilometre

BPKT: Basic Programming Knowledge Test

Bpl: Birthplace

B/PL: Book Plate

BPL: Branch if Plus

BPLA: Beijing Proton Linear Accelerator (China-PRC)

bpm: Beats per Minute • Bits per Minute

BPM: Barrels Per Minute • Basic Parameters Module • Batch Processing Monitor • Beam Positioning Magnet • Beam Profile Monitor • Beam Propagation Method • Beats Per Minute • Bits Per Minute • Bruce Proper Motion • Business Performance Management

BPMS: Blood Pressure Measuring System

BPN: Balikpapan (Indonesia) (location identifier) • Bibliography and Program Notes

BPNN: Back Propagation Neural Network

BPO: Base Post Office • Blanket Purchase Order • Borland Pascal with Objects • Business Process Outsourcing

bpp: Bits per Pixel

BPP: Bon Pour Pouvoir • Bounded-error Probabilistic Polynomial • Brevet de Pilote Professionnel • Bubble Pulse Period

BPR: Battery Powered Recorder • Béton de Poudres Réactives • Big Pulkovo Radiotelescope (Russia) • Bimonthly Progress Report • Bottom Pressure Recorder • Bureau of Public Roads • Business Process Redesign • Business Process Reengineering (*) • Bypass Ratio

BPRC: Beard Polar Research Center

BPRM: Breit-Pauli R Matrix

bps: Bits per Second

BpS: Bytes per Second

BPS: Banque de Programmes et de Services • Beats Per Second • Belarusian Physical Society • Belgian Physical Society • Boost Propulsion System • Boundary Plasma Sheet • Bulk Processing System • Bureau of Product Standards (Philippines) • Bureau of Professional Standards • Bytes Per Second • Porto Seguro (Brazil-BA) (location identifier)

bpsi: Bit per Square Inch

BPSK: Binary Phase Shift Keying (*) • Bi-Phase Shift Key • Bi-Phase Shift Keying

BPSN: Budget Project Symbol Number

BPSS: Basic Packet-Switching Service • Biopack Subsystem

bpt: Bits per Track • Boiling Point

BPT: Bandpass Transformer • Beaumont - Jefferson County (USA-TX) (location identifier) • Best Practice Technology • Body Point • Port Arthur (USA-TX) (location identifier)

BPU: Basic Processing Unit • Beppu (Japan) (location identifier) • Branch Prediction Unit • Branch Processing Unit

BPV: Bipropellant Valve

BPX: Burning Plasma Experiment

BPY: Besalampy (Madagascar) (location identifier)

Bq: Becquerel

BQ: Billet de Quai

BQA: Baler (Philippines) (location identifier)

BQE: Bubaque (Guinea-Bissau) (location identifier)

BQH: Biggin Hill (UK) (location identifier)

BQK: Brunswick Glynco (USA-GA) (location identifier)

BQL: Basic Query Language • Boulia (Australia-QLD) (location identifier)

BQN: Aguadilla (Puerto Rico) (location identifier)

BQO: Bouna (Ivory Coast) (location identifier)

BQP: Bounded-error Quantum Polynomial

BQQ: Barra (Brazil-BA) (location identifier)

BQR: Bonus Qualité Recherche • Boqueirão

BQS: Berkeley Quality Software • Blagoveschensk (Russia) (location identifier) • Bright Quasar Survey

BQT: Brest (Belarus) (location identifier)

BQU: Port Elizabeth (Saint Vincent & Grenadines) (location identifier)

Bque: Banque

Bquin: Bouquin

BQV: Gustavus Bartlett (USA-AK) (location identifier)

BQW: Balgo Hills (Australia-WA) (location identifier)

b/r: Bills Receivable

br: Brazil • breit • Breton (ISO language code)

Br: Bearing • Bedroom • Birr (currency) • Branch • Brass • Brazil • Breite • Breton • Brevet • Breveté • Brick • Bridge • Brief • Brig • Britain • British • Broché • Brom • Brome • Bromine • Bronze • Brother • Brown • Brücke • Brume • Brut • Bruto

B-R: Beobachtung-Rechnung

B/R: Backup/Restore • Bills Receivable

BR: Bacteriorhodopsin • Bad Register • Bairro • Bank Rate • Basic Research • Basic RPC • Bedroom • Begin Row • Bibliothèque Royale (Belgium) • Bills Receivable • Binary Relation • Biorack • Biosphere Reserve • Bit Rate • Bit Reverse • Blind Reflex • Block Release • Block Replacement • Book of Reference • Book Review • Booster Regulator • Boundary Representation • Branch • Branching Ratio • Branch Report • Brazil (international vehicle registration code) • Brazil (ISO country code) • Break Request • Bremsstrahlung Radiation • Brésil (ISO country code) • Briefing Room • Bright Region • British Rail • Bromley (UK postal code) • Browse • Buffer Register • Bureau of Reclamation • Buriat

BRA: Barreiras (Brazil-BA) (location identifier) • Basic Rate Access • Bee Research Association • Bench Replaceable Assembly • Bradford (Starlink internal code) • Brazil (ISO country code) • Brésil (ISO country code)

BRAC: Base Realignment And Closure • Basic Rest-Activity Cycle

Brad: Bradford (UK-England)

BRAD: British Rates And Data

BRAG: Brandeis Radio Astronomy Group (USA-MA)

BRAHMS: Broad Range Hadron Magnetic Spectrometer

BRAIN: Basic Research in Adaptive Intelligence and Neurocomputing • Basic Research in Aircraft Interior Noise

BRAINIAC: Brain-Imitating Almost-automatic Computer

Br.Am.: British America

BRAM: Bulletin Régional d'Alerte Météorologique • Bulletin Régional d'Avertissement Météorologique

BRAO: Bundesrechtsanwaltsordnung

Bras: Brasil

BRAS: Baton Rouge Astronomical Society (USA-LA) • Black River Astronomical Society (USA-OH)

BRASS: Berkeley Reconfigurable Architecture, Systems and Software

BRAT: Basic RISC Architecture Timer

BRAVO: Back-end Redesign for Added Value for Observers

BRAVOS: Business Risk And Value of Operations in Space

Braz: Brazil • Brazilian

Brazz: Brazzaville (Congo)

BRB: Barbade (ISO country code) • Barbados (ISO country code) • Barreirinhas (Brazil-MA) (location identifier) • Be Right Back • Broad Band

Br.C.: British Columbia (Canada)

BRC: Below Regulatory Concern • Below Regulatory Control • Budget Review Committee • Business Reply Card • San Carlos de Bariloche (Argentina-RN) (location identifier)

Brch: Branch

Brck: Brücke

BRCS: Basic Reference Coordinate System

BRD: Biological Resources Division (USGS) • Blank Recording Disc • Booster Requirements Document • Brainerd (USA-MN) (location identifier) • Bundesrepublik Deutschland (Germany)

Brdcst: Broadcast

BRDF: Bidirectional Reflectance Distribution Function (*) • Bidirectional Reflective Distribution Function

Brdg: Bridge

BRDGE: Bridge

BRDT: Bayesian Reliability Demonstration Test

BRE: Bistatic Radar Equation • Bremen Neuerland (Germany) (location identifier) • Bridge Relay Encapsulation • British Research Establishment • Bureau de Rapprochement des Entreprises (France) • Cruzeiro (Brazil) (ISO currency code)

BREBS: Bowen Ratio Energy Balance System

BREC: Battery Regulator and External Converter

BREM: Boeing Radiation Exposure Model

BRER: Board on Radiation Effects Research

BRESEX: Brazilian Remote Sensing Experiment

BRESM: Bureau de Recherches et d'Études SHOM/Météo-France

Brev: Brevet • Breveté • Brevetto

BREW: Biospherics Research on Emissions from Wetlands

Brf: Brief

BRF: Band Rejection Filter • Basic Research Fund • Bidirectional Reflectance Factor • Blind Relevance Feedback • Bradford (UK-England) (location identifier) • Bridge Relay Function • Brief

BRFI: Bidirectional Reflectance Field Instrument

Brg: Bearing

BRG: Bearing • Bridge • Bright Red Galaxy • Business Research Group • Whitesburg (USA-KY) (location identifier)

BRGC: Binary Reflected Gray Code

BRGG: Bureau de Recherches Géologiques et Géophysiques

BRGGM: Bureau de Recherches Géologiques, Géophysiques et Minières

BRGM: Bureau de Recherches Géologiques et Minières (France)

BRH: Brahman (Papua New Guinea) (location identifier)

BRHR: Basic Research and Human Resources

bRi: Bulk Richardson Number

Br. I.: British India • British Isles

BRI: Banque des Règlements Internationaux • Bari (Italy) (location identifier) • Bartol Research Institute (USA-DE) • Basic Rate Interface (*) • Basic Rate Interference • Basic Rate ISDN • Brain Response Interface • Bristol (Starlink internal code) • Bureau de Renseignements Internationaux

BRIAN: Barrier Reef Image Analysis

BRIC: Biological Research In Canisters • Bureau de Recherche pour l'Innovation et la Convergence (France) • Bureau de Recherches Industrielles et Commerciales

BRICS: Basic Research In Computer Sciences

BRIDGE: Biotechnology, Research for Innovation, Development and Growth in Europe

BRIEF: Basic Reconfigurable Interactive Editing Facility

Brig: Brigade • Brigadier

BRIG: Brigadeiro

BRIL: Budget Receipt/Issuance Log

BRIM: Biosphere Reserve Integrated Monitoring

Brisb: Brisbane (Australia-QLD)

BRISCA: Bristol Ice and Snow Cover Algorithm

Brist: Bristol (UK-England)

Brit: Britain • Britannia • British • Briton

BRITE: Basic Research in Industrial Technologies for Europe • Bright Radar Indicator Terminal Equipment

Brit.Mus.: British Museum

Brit.Pat.: British Patent

BRJ: Bright (Australia-VIC) (location identifier)

Brk: Brake • Brick

BRK: Bourke (Australia-NSW) (location identifier) • Brake • Break • Brook

Brkg: Breaking

BRKN: Broken

Brkr: Breaker

BRKS: Breaks • Brooks

Brkt: Bracket

BRKT: Bracket

Brl: Barrel

BRL: Ballistics Research Laboratory • Bendix Research Laboratories • Beyond Richardson-Lucy • Building Restriction Line • Burlington (USA-IA) (location identifier) • Real (Brazil) (ISO currency code)

BRLO: Broad Recombination of Line Objects

BRM: Barometer • Barquisimeto (Venezuela) (location identifier) • Baseline Reference Mission • Basic Reference Model • Basic Remote Module • Binary Rate Multiplier • Biological Research Module • Bio Research Module

BRMIC: Behavior Research, Methods, Instruments and Computers (periodical code)

BRMP: Background Radiation Monitoring Programme

Br.n.: Broché Neuf

Brn: Brown

BRN: Bahrain (international vehicle registration code) • Bern Belp (Switzerland) (location identifier) • Berne Belp (Switzerland) (location identifier) • Brazilian New Cruzado (ISO currency code) • Brunei (ISO country code)

BRNCH: Branch

Brng: Bearing • Burning

Bro: Brother • Brotherhood

BRO: Breault Research Organization Inc. (USA-AZ) • Brownsville (USA-TX) (location identifier)

Broc: Brochure

Brom: Bromide

BROOM: Ballistic Recovery Of Orbiting Man

Bros: Brothers

Brosch: Broschiert

BROTS: Beneficial Rays Of The Sun

BRP: Biaru (Papua New Guinea) (location identifier) • Biological Reclamation Process • Bulletin Régulier de Précipitations

BRPM: Breath Rate Per Minute

BRPRS: Blue Ridge Parkway Regional Stargaze (USA-NC)

BRQ: Brno (Czech Rep.) (location identifier)

BRR: Barra (UK) (location identifier)

BRRS: Banana River Repeater Station

BRRU: Business Regulation Review Unit

BRS: Balloon Radio System • Bandwidth Reservation System • Basic Research Symposium • Bibliographic Reference Service • Bibliographic Research Service • Bibliographic Retrieval Service (*) • Bibliographic Retrieval System • Biennial Reporting System • Big Red Switch • Binary Relational Store • Binary Ring Sequence • Block Recorder System • Bottom Right Side • Break Request Signal • Bristol Lulsgate (UK-England) (location identifier) • Bureau of Resource Sciences • Buried Ridge Structure

BRSC: Beijing Remote Sensing Corporation (China-PRC)

BRSL: Balasore Rocket Launching Station (India)

Brt: Bright • Brightness • Brut

BrT: Brevet Technique

BRT: Base Resistance Transistor • Bathurst Island (Australia-NT) (location identifier) • Belgische Radio en Televisie (Belgium) • Binary Run Tape • Bright • Bruttoregistertonne

BRU: Battery Regulator Unit • Branch Registration Unit • Brunei (international vehicle registration code) • Brussel Zaventem (Belgium) (location identifier) • Brussels Zaventem (Belgium) (location identifier) • Bruxelles Zaventem (Belgium) (location identifier) • Bureau de Renseignements Universitaires

Brunsw: Brunswick

Brux: Bruxelles (Belgium)

BRV: Backup Recovery Vector • Ballistic Reentry Vehicle • Bremerhaven (Germany) (location identifier)

BRW: Barrow Wbas (USA-AK) (location identifier)

BrWK: Brennstoff Wärme Kraft (periodical code)

BRWM: Board on Radioactive Waste Management

Brx: Brussels (Belgium)

BRX: Barahona (Dominican Rep.) (location identifier)

BRY: Bardstown Samuels (USA-KY) (location identifier)

Brz: Brazil • Bronze

BRZ: Borotou (Ivory Coast) (location identifier)

BRZN: Bibliotheksrechenzentrum Niedersachsen (Germany)

b/s: Bill of Sale

Bs: Bags • Bales • Bas • Bolivars (currency) • Boliviano (currency)

B/S: Balance Sheet • Behind Schedule • Bill of Sale • Bits per Second

BS: Bachelor of Science • Backspace • Backup System • Bahamas (international vehicle registration code) • Bahamas (ISO country code) • Balance Sheet • Base Station • Bashkir • Battery Simulator • Baudhāyana Sulbasūtra • Below Specification • Be Specific • Beam Splitter • Beam Steering • Beam Stop • Bibliothèque Spécialisée • Bill of Sale • Binary Scale • Binary Star • Binary State • Bit-Sliced • Bit Sync • Block Specification • Blowing Snow • Blue Straggler • Bon de Sortie • Bordereau de Situation • Bow Shock • Brevet de Spécialité • Brevet Supérieur • Bright Spot • Bright Stars (catalogue) • Bristol (UK postal code) • British Standard • Broadcasting Satellite (Japan) • Broadcasting Service • Broadcast Satellite (Japan) • Broadcast Station • Broad Structure • Browse Service • Building Section • Bulletin de Salaire • Bureau of Standards • Bureau of Statistics • Burgerlijke Stand • Bus Switch

BSA: Backwards Stepping Aggregation • Battery Solar Array • Bibliographical Society of America (USA) • Bit Sync Acquisition • Board of Scientific Affairs • Bohr-Sommerfeld Atom • Bon de Souscription d'Actions • Bossaso (Somalia) (location identifier) • Boy Scouts of America • Brookhaven Science Associates • Burst Spectrum Analyzer • Business Software Alliance • Bus Solar Array

BSAA: Bachelor of Science in Applied Arts

BSAdv: Bachelor of Science in Advertising

BSAE: Bachelor of Science in Aeronautical Engineering

BSAF: Bids Sollicited As Follows

BSAM: Basic Sequential Access Method

BSAP: Base Station Application Port

Bs. As.: Buenos Aires (Argentina)

BSAS: Barnard-Seyfert Astronomical Society (USA-TN)

Bsb: Brisbane (Australia-QLD)

BSB: Brain-State-in-the-Box • Brasília (Brazil-DF) (location identifier) • British Satellite Broadcasting Ltd.

BSBA: Bachelor of Science in Business Administration

BSBB: Baryon-Symmetric Big bang

BSBD: Biblio Service Data Bank

BSc: Bachelor of Science

B/SC: Brake Skid Control

BSC: Bahia Solano (Colombia) (location identifier) • Base Station Controller • Beam Splitter Chamber • Beam Steering Computer • Bergen Scientific Centre (IBM) • Binary Symmetric Channel • Binary Synchronous Communication • Bi-Synchronous Communication • Bit Scan Command • Boyan Script • Brake Skid Control • Brake Skid Controller • Brevet Supérieur Commercial • Bright Stars Catalogue • British Steel Corporation • Business Service Center

BSCA: Binary Synchronous Communications Adapter • Boresight Camera Assembly

BSCCO: Bismuth Strontium Calcium Copper Oxide

BSCE: Bachelor of Science in Civil Engineering

BSChE: Bachelor of Science in Chemical Engineering

BSCM: Bisynchronous Communications Module

BSCom: Bachelor of Science in Communications

BSCP: British Standard Code of Practice

BSCS: Bachelor of Science in Computer Science

BSCW: Basic Support for Cooperative Work

BSD: Baoshan (China-PRC) (location identifier) • Ballistic Systems Division • Beam Steering Device • Bergedorfer Spektral Durchmusterung • Berkeley Software Design Inc. (USA-CO) • Berkeley Software Distribution • Berkeley Standard Distribution • Berkeley System Distribution • Bit Storage Density • Board of Sustainable Development • Bright Surge on Disk • British Space Development • Business System Development • Dollar (Bahamas) (ISO currency code)

BSDAM: Basic Sequential Data Access Method

BSDC: Binary Symmetric Dependent Channel

BSDF: Bidirectional Scatter Distribution Function

BSDI: Berkeley Software Design Inc. (USA-CO) • Berkeley Software Design International (USA-CO) (*)

BSDP: Booster Stage Discharge Pressure

BSDS: Biological Standoff Detection System • Boot Strap Data Set

BSE: Bachelor of Science in Education • Bachelor of Science in Engineering • Back-Scattered Electron • Base Station Equipment • Basic Service Element • Bâtiment de Stockage des EAP (CSG) • Bâtiment de Stockage des Étages (CSG) • Bethe-Salpeter Equation • Booster Systems Engineer • Sematan (Malaysia) (location identifier)

B/sec: Bits per Second

BSEc: Bachelor of Science in Economics

BSEC: Black Sea Economic Cooperation • Brevet Supérieur d'Études Commerciales

BSEd: Bachelor of Science in Education

BSED: Ball Systems Engineering Division

BSEE: Bachelor of Science in Electrical Engineering

BSEEng: Bachelor of Science in Electrical Engineering

BSElE: Bachelor of Science in Electronic Engineering

BSEng: Bachelor of Science in Engineering

BSEP: Bachelor of Science in Engineering Physics

BSES: Bachelor of Science in Engineering Sciences

BSF: Backspace File • Back Surface Field • Berliner Sternfreunde eV (Germany) • Binational Science Foundation • Bit Scan Forward • Pohakuloa Bradshaw (USA-HI) (location identifier)

BSFC: Brake Specific Fuel Consumption

BSFR: Back Surface Field Reflector (*) • Back Surface Field with Reflection • Back Surface Filled Reflector

BSG: Bata (Equatorial Guinea) (location identifier) • Bénéficiaire de Soins Gratuits • Bit Sync Generator • Blue Supergiant • British Standard Gauge

BSGDG: Breveté sans Garantie du Gouvernement

BSGE: Bachelor of Science in General Engineering

BSH: Brighton (UK-England) (location identifier) • Bundesamt für Seeschiffahrt und Hydrographie (Germany)

BSHM: Bâtiment de Soutien de Haute Mer • British Society for the History of Mathematics (UK)

BSHS: British Society for the History of Science

BSI: Basic Shipping Information • Basic Shipping Instruction • Battery Status Indicator • Béton Spécial Industriel • Bibliographical Star Index • Blairsville (USA-PA) (location identifier) • Booster Situation Indicator • Branch and Store Instruction • British Standards Institution • Bureau de Statistique Industrielle

BSIC: Binary Symmetric Independent Channel

BSIR: Bachelor of Science in Industrial Relations

BSIRA: British Scientific Instrument Research Association

BSIT: Bachelor of Science in Industrial Technology

BSJ: Bachelor of Science in Journalism • Bairnsdale (Australia-VIC) (location identifier)

Bsk: Basket

BSK: Biostack • Biskra (Algeria) (location identifier)

Bskt: Basket

Bs/L: Bills of Lading

BSL: Bachelor of Science in Law • Bachelor of Science in Linguistics • Bâle Mulhouse (Switzerland) (location identifier) • Basel Mulhouse (Switzerland) (location identifier) • Basle Mulhouse (Switzerland) (location identifier) • Biot-Savart Law • Blackwell Science Ltd. • Bright Source List • Bright Surge on Limb • Mulhouse Bâle (France) (location identifier)

BSLS: Bachelor of Science in Library Science

BSM: Austin Bergstrom (USA-TX) (location identifier) • Basic Storage Module • Basic System Memory • Basic System Module • Beam-Steering Mechanism • Boîtier de Servitude Moteur • Booster Separation Motor • Branch Sales Manager

BSMD: Bulk Store Memory Device

BSME: Bachelor of Science in Mechanical Engineering

BSMF: Bit Sync Matched Filter

Bsmnt: Basement

B/SMPL: Bits per Sample

BSMPL: Bits per Sample

Bsmt: Basement

BSMT: Basement

BSN: Basis Systeme Netzwerk • Bossangoa (Central African Rep.) (location identifier)

BSNE: Bachelor of Science in Nuclear Engineering

BSNI: Barbados National Standards Institution

Bsns: Business

BSO: Bacteria-Shaped Object • Ballon Stratosphérique Ouvert • Basco (Philippines) (location identifier) • Beroeps Secundair Onderwijs • Blue Stellar Object • Bright Stellar Object • Broad System of Ordering • Buro voor Systeemontwikkeling BV (Netherlands)

Bson: Buisson

BSP: Bensbach (Papua New Guinea) (location identifier) • Bibliographical Service Provider • Binary Space Partition • Biodiversity Support Program • Biospecimen Sharing Program • Blackwell Scientific Publications • Boolean Satisfiability Problem • Bulk Synchronous Parallel • Burroughs Scientific Processor • Business Systems Planning

BSPA: Bachelor of Science in Public Administration

BSQ: Band Sequential • Bisbee (USA-AZ) (location identifier)

BSR: Back Surface Reflector • Basic System Release • Basra (Iraq) (location identifier) • Bâtiment de Soutien de Région • Bistatic Spaceborne Radar • Bit Scan Reverse • Bit Slippage Rate • Blood Sedimentation Rate • Board of Standard Review • Bragg Scattering Region • Brevet de Sécurité Routière • Bulk Shielding Reactor • Byte Status Register

BSRAM: Burst Static Random Access Memory

BSRC: Behavioral Science Research Centre

BSRF: Biological Space Research Facility

BSRM: Booster Solid-Rocket Motor

BSRN: Baseline Surface Radiation Network

BSROC: Belgium Space Remote Operations Center

BSRSL: Bulletin de la Société Royale des Sciences de Liège (periodical code)

BSRT: Bachelor of Science in Radiological Technology

BSS: Bachelor of Secretarial Science • Bachelor of Social Science • Backup System Service • Balanced Salt Solution • Balloon-borne Superconducting Selenoid • Balsas (Brazil-MA) (location identifier) • Base Station System • Basic Safety Standard • Beam Steering System • Bigelow Sky Survey • Bisymmetric Spiral • Biurakan Second Survey • Block Started by Symbol • Blue Straggler Star • Borehole Seismic System • Bottom-Side Sinusoidal • Breakthrough Science Society (India) • Bright Star Spot • British Standards Specification • British Sundial Society (UK) • Broadcasting Satellite Service • Broadcast Satellite Service • Business Support System

BSSA: Bachelor of Science in Secretarial Administration • Bulletin of Seismological Society of America (periodical code)

BSSC: Bit Synchronising and Signal Conditioner • Board for Software Standardization and Control • Board of Software Standard and Control

Bsse: Basse

BSSE: Bachelor of Science in Secondary Education • Basic Set Superposition Error

BSSS: Bachelor of Science in Secretarial Studies • Bachelor of Science in Social Science

BST: Barium Strontium Titanate • Basic Services Terminal • Beam Steering Transducer • Beam-Switching Tube • Belgian Space Training (Belgium) • BioServe Space Technologies (USA-CO) • Bost (Afghanistan) (location identifier) • Brevet Supérieur Technique • British Standard Time • British Summer Time • Byggstandardiseringen i Sverige (Sweden)

Bstb: Bistable

BSTC: Biotechnology Specimen Temperature Controller

Bstg: Bastogne (Belgium)

BSTI: Bangladesh Standards and Testing Institution

BST&IE: Bachelor of Science in Trade & Industrial Education

Bstr: Booster

BSTR: Booster

BSTS: Boost Surveillance and Tracking System • Boost Surveillance Tracking System • Broadband Series Test System

BSU: Ball State University (USA-IN) • Basankusu (Zaire) (location identifier) • Base Spatial Unit • Basic Sounding Unit • Basic Structural Unit • Beam Steering Unit • Bench Scale Unit

BSV: Boot Sector Virus • Bosset (Papua New Guinea) (location identifier)

BSW: Basic Software • Boswell Bay (USA-AK) (location identifier) • Business Softwar

BSX: Bassein (Myanmar) (location identifier)

BSY: Bardera (Somalia) (location identifier) • Busy

BSYNC: Binary Synchronous Communication

BSZ: Bartletts (USA-AK) (location identifier)

b/t: Between

Bt: Bâtiment • Beat • Benefit • Bent • Billet • Boat • Bought • Brevet • Brut • Bukit

BT: Bachelor of Teaching • Background Test • Ballistic Transport • Bassa Tensione • Basse Tension • Beijing Time • Belfast (UK postal code) • Bit Test • Board of Trade • Body Text • Bon à Tirer • Botswana • Boutan (ISO country code) • Brevet de Technicien • Brevet Technique • Brightness Temperature • British Telecom • Broader Term • Bulletin Technique • Bureau de Transport • Bureau Technique • Burn Time • Burst Tolerance • Business Telecommunications • Bus Terminator • Buthan (ISO country code)

BTA: Ballistic Trajectory Aggregate • Basic Trading Area • Bertoua (Cameroon) (location identifier) • Better Than Average • Binary Tables Agreement (FITS) • Bolshoi Teleskop Azimutal'nij (Russia)

BTAC: Binary Tree Algebraic Computation • Branch Target Address Cache • Branch Target Address Code

BTAIM: Be That As It May

BTAM: Basic Tape Access Method • Basic Telecommunications Access Message • Basic Telecommunications Access Method • Basic Terminal Access Method

BTB: Betou (Congo) (location identifier) • Block Transfer Bus • Branch Target Buffer • Burst Time Plan

BTC: Batticaloa (Sri Lanka) (location identifier) • Before Top Center • Bench Test Console • Big Throughput Camera • Binary Time Code • Bit Test and Complement • Bit Time Counter • Block Truncating Coding • Breakthrough Curve • Bus Tie Contractor

Btca: Biblioteca

BTCAM: Big Throughput Camera

BTD: Balanced Tape Drive • Binary To Decimal • Boston Transportation Department (USA-MA) • Breath Testing Device • Brief Task Description • Brightness Temperature Difference • Brunette Downs (Australia-NT) (location identifier) • Burn To Depletion

BTDAM: Basic Telecommunications Data Access Method

BTDC: Back of Top Dead Center • Before Top Dead Center

BTDL: Basic Transient Diode Logic

bte: Binte

Bté: Breveté

Bte: Boîte

BTE: Bench Test Equipment • Better Than Expected • Boltzmann Transport Equation • Bonthe (Sierra Leone) (location identifier) • British Telecommunications Engineering (periodical code) • Bulk Tape Eraser

BTECC: Book Trade Electronic Communications Committee

BTESS: Brevet Technique d'Études Supérieures Spécialisées

BTEX: Benzene-Toluene-Ethylbenzene-Xylene

BTF: Battery Test Facility • Beam Transfer Function • Bench Test Fixture • Biotechnology Facility • Bountiful - Salt Lake (USA-UT) (location identifier)

BTG: Batangafo (Central African Rep.) (location identifier) • Bon de Transport Gratuit

Bth: Bath • Bathroom • Berth

BTH: Batu Besar (Indonesia) (location identifier) • Beyond The Horizon

BTHP: Béton à Très Hautes Performances

BThU: British Thermal Unit

BTI: Barter Island (USA-AK) (location identifier)

BTIC: Business and Technological Information Centre

BTJ: Banda Aceh (Indonesia) (location identifier)

Btk: Buttock

BTK: Bijzonder Tijdelijk Kader • Bratsk (Russia) (location identifier)

Btl: Bottle

BTL: Battle • Battle Creek (USA-MI) (location identifier) • Beacon Tracking Level • Bell Telephone Laboratories • Below The Line • Between Layers • Bread-Lettuce-Tomato

BTLB: Block Translator Look-aside Buffer

BTLC: Bureau Technique de Liaison et de Coordination

BTLS: Basic Tape Library Support

Btm: Bottom

BTM: Bench Test Model • Bottom • Bromotrifluoromethane • Butte (USA-MT) (location identifier)

BTMS: Body Temperature Measuring System

Btn: Baton • Button

BTN: Bennettsville (USA-SC) (location identifier) • Between • Bhutan (ISO country code) • Boutan (ISO country code) • Ngultrum (Bhutan) (ISO currency code)

BTO: Botopasie (Suriname) (location identifier) • Brief Task Outline

BtoB: Back to Back

BTOBD: Be There Or Be Dead

BTOBS: Be There Or Be Square

BToIT: Bulletin of the Tokyo Institute of Technology (periodical code)

BTok: Bulletin of the Tokyo Astronomical Observatory (periodical code)

BTP: Back-Traffic Protection • Batch Transfer Program • Bâtiment et Travaux Publics • Broader Term Partitive • Butler Graham (USA-PA) (location identifier)

BTQ: Butare (Rwanda) (location identifier)

BTR: Baton Rouge - Ryan (USA-LA) (location identifier) • Better • Bimetric Theory of Relativity • Bit Test and Reset • Block Tape Recorder • Boolean Topic Representation • Burn Time Remaining • Bus Transfer

BTRL: British Telecom Research Laboratory

BTS: Balloon Transport System • Base Transceiver Station • Batch Terminal Simulator • Battery Test Set • Beacon Tracking System • Beam Transport System • Bench Test Specification • BIH Terrestrial System • Bit Test and Set • Boolean Time Sequence • Bratislava - M.R. Stefanik (Slovakia) (location identifier) • Brevet de Technicien Supérieur (France) • Bureau of Transportation Statistics

BTSA: Brevet de Technicien Supérieur Agricole (France)

BTSCC: Block Transfer Serial Crate Controller

BTSF: Biotechnology Strategic Forum

BTSHD: Block Transfer Serial Highway Driver

BTSS: Basic Time-Sharing System

BTT: Beginning-to-Tape Test • Bettles (USA-AK) (location identifier)

BTTF: Blue Taurus Tunable Filter

Btu: British Thermal Unit

BTU: Basic Transmission Unit • Bintulu (Malaysia) (location identifier) • British Thermal Unit • Bus Terminal Unit

BTUM: Braunschweig Technische Universität - Mitteilungen (periodical code)

BTV: Bulletin Télévisé • Burlington (USA-VT) (location identifier)

BTW: Batu Licin (Indonesia) (location identifier) • Belasting over de Toegevoegde Waarde • Between • By The Way

BTWC: Biological and Toxic Weapons Convention

Btwn: Between

BTWN: Between

BTX: Benzene-Toluene-Xylene • Betoota (Australia-QLD) (location identifier) • Bildschirmtext

Bty: Battery

BTY: Beatty (USA-NV) (location identifier)

BTZ: Bursa (Turkey) (location identifier)

Bu: Bureau • Bushel • Butyl Group

B/U: Backup

BU: Backup • Base Unit • Bibliothèque Universitaire • Birmingham University (UK-England) • Bollettino Ufficiale • Boston University (USA-MA) • Bottom Up • Break-Up • Branch Unit • Brihadāranyaka Upanishad • Brown University (USA-RI) • Buildup • Bulgaria • Burma (ISO country code) • Bushman

BUA: Booster Umbilical Assembly • Buka (Papua New Guinea) (location identifier)

BUAAJ: Beijing University Aeronautics and Astronautics Journal (periodical code)

BUAF: Big Ugly ASCII Font

BUAG: Big Ugly ASCII Graphic

BuAIC: Bulletin of the Astronomical Institutes of Czechoslovakia (periodical code)

BUAN: Baseline Upper Air Network

BuARB: Bulletin de l'Académie Royale de Belgique (periodical code)

BUB: Burwell (USA-NE) (location identifier)

BUBL: Bulletin Board for Libraries

BUC: Backup Computer • Bangor University College (UK-Wales) • Burketown (Australia-QLD) (location identifier)

Bucks: Buckinghamshire (UK-England)

BUCOE: Boulder Upslope Cloud Observation Experiment

Bud: Budget

BUD: Beneficial Use Date • Budapest Ferihegy (Hungary) (location identifier) • Budget

BUDC: Backup Digital Computer

Budpst: Budapest (Hungary)

BUDS: Backup Digital System

BUE: Buenos Aires (Argentina-BA) (location identifier)

BUEC: Backup Emergency Communications

BUECE: Backup Emergency Communications Equipment

BUES: Budapest University of Economic Sciences (Hungary)

BUF: Buffalo (USA-NY) (location identifier) • Buffer

BUFC: Backup Flight Controller

BUFF: Buffer

BUFORA: British Unidentified Flying Objects Research Association

BUFR: Binary Universal Form for Representation

BUG: Benguela (Angola) (location identifier)

BUGS: Backup Guidance System • Bayesian (interface) Using Gibbs Sampling

BUH: Bucharest (Romania) (location identifier) • Bucureşti (Romania) (location identifier)

BUI: Bokondini (Indonesia) (location identifier)

BUIC: Backup Interceptor Control • Bureau Universitaire d'Information sur les Carrières

BuIPS: Bulletin of the Israel Physical Society (periodical code)

BUJ: Boussaada - Ain Eddis (Algeria) (location identifier)

BUK: Albuq (Yemen) (location identifier) • Burman Kyat (ISO currency code)

BUKS: Bibliotheeks Uitleen- en Katalogus Systeem (Netherlands)

Bul: Bulgarian • Bulletin

BUL: Bulolo (Papua New Guinea) (location identifier)

Bulg: Bulgaria • Bulgarian

Bull: Bulletin

BUM: Broad Upshifted Maximum • Butler (USA-MO) (location identifier)

BUMA: Bureau voor Muziek-Auteursrecht

BuMat: Bulletin Mathématique (periodical code)

BUN: Buenaventura (Colombia) (location identifier) • Bgultrum (Bhutan) (ISO currency code)

BUNCH: Burrougs - Univac - NCR - Control Data - Honeywell

BUO: Burao (Somalia) (location identifier)

BUOU: Backup Optical Unit

BUP: Bachillerato Unificado Polivalente (Spain) • Backup Plate • Bathinda (India) (location identifier) • British United Press

BUPT: Beijing University of Posts and Telecommunications (China-PRC)

BUQ: Bulawayo (Zimbabwe) (location identifier)

Bur: Bureau • Buried • Burma • Burmese

BUR: Backup Rate • Burbank (USA-CA) (location identifier) • Burma (international vehicle registration code)

Burg: Burgalés • Burgess • Burgemeester • Burgomaster • Burgos

Burl: Burlesque

Burm: Burma • Burmese

BURP: Backup Rate of Pitch

BURR: Backup Rate of Roll

BURU: Bottom Up Review Update

BURY: Backup Rate of Yaw

Bus: Bushel • Business

BUS: Backup Structure • Batumi (Georgia) (location identifier) • Beilstein Unique Sequence • Broadcast and Unknown Server • Bureau Universitaire de Statistiques

BUSCA: Bonn University Simultaneous Camera

BUSD: Blow Up/Settle Down

BUSDSP: Bureau Universitaire de Statistiques et de Documentation Scolaires et Professionnelles

Bush: Bushel • Bushing

BUSS: Balloon-borne Ultraviolet Stellar Spectrometer

BUST: Baksan Underground Scintillation Telescope (Russia)

But: Butter • Button

BUT: Brevet Universitaire de Technologie • Burtonwood (UK-England) (location identifier)

BUTE: Butte

BUTEC: Bureau Technique

BUU: Boltzmann-Ühling-Uhlenbeck • Buyo (Ivory Coast) (location identifier)

BUV: Backscatter Ultraviolet • Bella Union (Uruguay) (location identifier)

BUVS: Backscatter Ultraviolet Spectrometer

BUW: Baubau (Indonesia) (location identifier)

BUX: Bunia (Zaire) (location identifier)

Buy: Buyer • Buying

BUY: Bunbury (Australia-WA) (location identifier)

BUZ: Bushehr (Iran) (location identifier)

BuZa: Buiten Zaken

bv: Bijvoorbeeld

B/V: Bleed Valve

BV: Back View • Balanced Voltage • Bamberg Variable • Basic Version • Bene Vale • Berechnungsverordnung • Besloten Vennootschap • Bijvoorbeeld • Bleed Valve • Boîte avec Valeur (déclarée) • Bonnet Valve • Book Value • Booster Vehicle • Boost Vehicle • Bouvet Island (ISO country code) • Brightness Value • Bulletin de Visite • Bulletin de Vote • Île Bouvet (ISO country code)

BVA: Beauvais (France) (location identifier) • Billing Validation Application • Bureau des Vins et Alcools • Bureau de Voyages Aériens

BVAC: Beaver Valley Astronomy Club (USA-PA)

BVAS: Boston University Astronomical Society (USA-MA)

BVAT: Before Value-Added Tax

BVB: Boa Vista (Brazil-RR) (location identifier)

BVBA: Besloten Vennootschap met Beperkte Aansprakelijkheid

BVC: Boa Vista (Cape Verde) (location identifier) • Billing Validation Center • Boost Voltage Converter

BVD: Beacon Video Digitizer • Beaver Inlet (USA-AK) (location identifier) • Binnenlandse Veiligheidsdienst

BVE: Barotropic Vorticity Equation • Binocular Visual Efficiency • Bivariate Exponential • Breadboard Verification Equipment • Brive-la-Gaillarde (France) (location identifier) • Bureau de la Vie Étudiante

BVF: Bua Dama (Fiji) (location identifier)

Bvg: Beverage

BVG: Berlevåg (Norway) (location identifier) • Berliner Verkehrsgesellschaft (Germany)

BVH: Vilhena (Brazil-RO) (location identifier)

BVI: Birdsville (Australia-QLD) (location identifier) • British Virgin Islands

BVLS: Battery-Voltage Limit System

BVK: Bundesverdienstkreuz

BVM: Bayerische Volkssternwarte München eV (Germany) • Belmonte (Brazil-BA) (location identifier) • Boundary Value Method

BVO: Bartlesville (USA-OK) (location identifier) • Bundesverdienstorden

BVol: Bulletin of Volcanology (periodical code)

BVP: Beacon Video Processor • Bolovip (Papua New Guinea) (location identifier) • Booster Vacuum Pump • Boundary Value Problem • British Visitor's Passport • Bureau de Vérification de la Publicité

BVPA: Boundary Value Problem Approach

BVPN: Broadband Virtual Private Network

BVPS: Beacon Video Processing System • Booster Vacuum Pump System

BVQI: Bureau Veritas Quality International

BVR: Beyond Visual Range • Brava (Cape Verde) (location identifier)

BVRO: Belgische Vereniging voor Raketonderzoek (Belgium)

BVRS: Breadboard Visual Reference System

BVS: Basic Variable Set • Bergische Volkssternwarte Wuppertal eV (Germany) • Best Value Selection • Bibliothekverbundsystem • Branch VME System • Breves (Brazil-PA) (location identifier) • Broadcast Video Server • Buoyant Venus Station

Bvt: Brevet • Brevetted

BVT: Basic Vehicle Timing • Bouvet Island (ISO country code) • Île Bouvet

BVU: Brightness Value Unit

BVW: Batavia Downs (Australia-QLD) (location identifier)

BVX: Batesville (USA -AR) (location identifier)

BVY: Beverly (USA-MA) (location identifier)

BVZ: Beverly Springs (Australia-WA) (location identifier)

b/w: Black and White

B-W: Baade-Wesselink

B&W: Black & White

B/W: Black and White • Black to White

BW: Baade's Window • Baade-Wesselink • Bandwidth • Beam Width • Bitte Wenden! • Black and White • Body Weight • Botswana (ISO country code) • Bridge Wire • Bulk Water • Burgerlijk Wetboek • Business Week

BWA: Backward Wave Amplifier • Bent Wire Antenna • Bhairawa (Nepal) (location identifier) • Botswana (ISO country code)

BWB: Barrow Island (Australia-WA) (location identifier) • Blended Wing Body

BWC: Backward Wave Converter • Bandwidth Compression • Basic Weight Controller • Biological Weapons Convention (1972) • Brawley (USA-CA) (location identifier) • Broadband Waveguide Circulator • Bureau Weather Control

Bwd: Backward

BWD: Backward • Brownwood (USA-TX) (location identifier)

BWE: Bandwidth Expansion • Billet de Weekend • Braunschweig (Germany) (location identifier)

BWER: Bounded Weak Echo Region

BWF: Barrow-in-Furness (UK-England) (location identifier)

BWFR: Black and White Film Recorder

BWG: Birmingham Wire Gauge • Bowling Green - Warren County (USA-KY) (location identifier)

BWH: Butterworth (Malaysia) (location identifier)

BWI: Baltimore Washington International (USA-MD) (location identifier) • British West Indies

BWIA: British West Indian Airways

BWiN: Breitbauwissenschaftsnetz

BWJ: Bawan (Papua New Guinea) (location identifier)

BWK: Bol (Croatia) (location identifier)

BWL: Blackwell (USA-OK) (location identifier)

BWM: Backward Wave Magnetron • Block-Write Mode • Bowman (USA-ND) (location identifier) • Broadcast Warning Message

BWN: Bandar - Seri Begawan (Brunei) (location identifier) • Bergmann-Wagoner-Nordtvedt

BWO: Backup While Open • Backward Wave Oscillator • Balakovo (Russia) (location identifier)

BWOS: Backward Wave Oscillator Synchronizer

BWOT: Backward Wave Oscillator Tube

BWP: Bewani (Papua New Guinea) (location identifier) • Pula (Botswana) (ISO currency code)

BWPA: Backward Wave Power Amplifier

BWQ: Brewarrina (Australia-NSW) (location identifier) • Buzz Word Quotient

BWR: Bandwidth Ratio • Boiling Water Reactor

BWS: Base Weather Station • Beaufort Wind Scale • Black-White Symmetry • Blaine (USA-WA) (location identifier)

BWT: Backward Wave Tube • Bohr-Wheeler Theory • Broadcast Warning TWX • Burnie Wynyard (Australia-TAS) (location identifier)

BWTP: Basalt Waste Isolation Project

BWTS: Bandwidth Test Set

BWU: Bankstown (Australia-NSW) (location identifier)

BWV: Bach Werke-Verzeichnis • Back Water Valve

BWY: Woodbridge (UK-England) (location identifier)

Bx: Box • Bordeaux (France)

BX: Base Exchange • Box

BXA: Bogalusa - George R. Carr (USA-LA) (location identifier)

BXB: Babo (Indonesia) (location identifier)

BXC: Boxborough (USA-MA) (location identifier)

BXD: Bade (Indonesia) (location identifier)

BXE: Bakel (Senegal) (location identifier)

BXI: Boundiali (Ivory Coast) (location identifier)

BXK: Buckeye (USA-AZ) (location identifier)

Bxl: Bruxelles

BXL: Blue Lagoon (Fiji) (location identifier)

BXM: Batom (Indonesia) (location identifier)

BXN: Bodrum Imsik (Turkey) (location identifier)

BXO: Bissau (Guinea-Bissau) (location identifier)

BXR: Bureau for External Relations

BXS: Borrego Springs (USA-CA) (location identifier)

BXT: Bontang (Indonesia) (location identifier)

BXU: Butuan (Philippines) (location identifier)

BXV: Breiddalsvík (Iceland) (location identifier)

BXX: Borama (Somalia) (location identifier)

By: Barony • Byte

BY: Babylon • Bayern (Germany) • Belarus (ISO country code) • Biélorussie (ISO country code) • Billion Years • Byelorussia (ISO country code) • Budget Year

BYA: Billion Years Age • Boundary (USA-AK) (location identifier)

BYB: Baya (Oman) (location identifier) • Dibaa (Oman) (location identifier)

BYC: Yacuiba (Bolivia) (location identifier)

BYD: Beidah (Yemen) (location identifier) • Beyond

Bye: Byelorussia • Byelorussian

BYG: Buffalo (USA-WY) (location identifier)

BYH: Blytheville (USA-AR) (location identifier)

BYI: Burley (USA-ID) (location identifier) • Rupert (USA-ID) (location identifier)

BYK: Bouake (Ivory Coast) (location identifier)

BYL: Bella Yella (Liberia) (location identifier)

BYM: Bayamo (Cuba) (location identifier)

BYN: Bayankhongor (Mongolia) (location identifier)

Bynd: Beyond

Byo: Bulawayo (Zimbabwe)

BYO: Bring Your Own • Build Your Own

BYOB: Bring Your Own Beer • Bring Your Own Booze • Bring Your Own Bottle

BYOP: Bring Your Own Popcorn

BYOS: Bring Your Own Snacks

Byp: Bypass

BYP: Bypass

BYPA: Bypass

BYPS: Bypass

BYQ: Bunyu (Indonesia) (location identifier)

BYr: Billion Years

BYR: Rouble (Belarus) (ISO currency code)

BYS: Fort Irwin - Bicycle Lake (USA-CA) (location identifier) • Biélorussie (ISO country code) • Byelorussia (ISO country code)

BYT: Bantry (Ireland) (location identifier)

BYU: Bayoo • Bayou • Bayreuth (Germany) (location identifier) • Brigham Young University (USA-UT)

BYUP: Brigham Young University Press

BYW: Blakely Island (USA-WA) (location identifier)

BYX: Baniyala (Australia-NT) (location identifier)

Byz: Byzantine

Bz: Benzene • Benzoyl • Bezirk

B-Z: Belousov-Zhabotinskii

BZ: Belize (ISO country code) • Belousov-Zhabotinskii • Blandford-Znajek • British Zone • Byzantine

BZA: Bonanza - San Pedro (Nicaragua) (location identifier) • Breathing Zone Air

BZC: Búzios (Brazil-RJ) (location identifier)

BZD: Balranald (Australia-NSW) (location identifier) • Dollar (Belize) (ISO currency code)

BZE: Belize City - Goldston (Belize) (location identifier)

BZG: Bydgoszcz (Poland) (location identifier)

Bzgl: Bezüglich

BZI: Balikesir (Turkey) (location identifier)

BZK: Briansk (Russia) (location identifier)

BZL: Barisal (Bangladesh) (location identifier)

BZM: Bergen op Zoom - Woensdrecht (Netherlands) (location identifier)

BZN: Bozeman Gallatin (USA-MT) (location identifier)

BZO: Bolzano (Italy) (location identifier)

BZP: Bizant (Australia-QLD) (location identifier)

Bzr: Buzzer

BZR: Béziers (France) (location identifier)

BZS: Basic Zirconium Sulphate • Washington - Buzzards Point (USA-DC) (location identifier)

BZT: Nrazoria - Hinkles Ferry (USA-TX) (location identifier)

BZU: Buta (Zaire) (location identifier)

BZV: Brazzaville (Congo) (location identifier)

Bzw: Beziehungsweise

BZY: Bel'cy (Moldova) (location identifier)

BZZ: Brize Norton (UK) (location identifier)

B0: Paper Size

B1: First Bologna Survey • Paper Size

B2: Paper Size • Second Bologna Survey

B2B: Business to Business

B2C: Business to Consumer

B2X: Binary to Hexadecimal

B3: Paper Size

B4: Paper Size

B5: Paper Size

B6: Paper Size

B7: Paper Size

B8: Paper Size

B9: Paper Size

B10: Paper Size

C

c/: Calle • Cargo • Cuenta

c: About • Calorie • Candle • Capacity • Caput • Carbohydrate • Carbon • Carton • Case • Cathode • Cent • Centavo (currency) • Centi • Centigrade • Centime (currency) • Centimeter • Centre • Century • Chairman • Chairperson • Chapter • Characteristic Velocity • Charm • Chief • Child • Church • Cibus • Circa • Circiter • Circum • Cirrus • City • Cloudy • Cognate • Cold • Colour • Como • Compact • Compound • Concentration • Congius • Constant • Consul • Contra • Contrast • Convection • Copper • Copy • Copyright • Corps • Court • Cube • Cubic • Cum • Curie • Currency • Cycle • Exhaust Velocity • Molecular Velocity Vector • Specific Heat Capacity • Speed of Light in Vacuum • Velocity of Light • Velocity of Sound

C/: Calle • Cargo • Cuenta

C: Acceleration Correction • Cabo • Caldo • Caliente • Calle • Calorie • Cambrian • Campo • Canada (international aircraft registration code) • Candle • Cap • Capacitance • Cape • Capo • Carbon • Carbone • Carboniferous • Cargo • Cathode • Celestron • Celsius • Cent • Centavo (currency) • Center • Centesimo • Centigrade • Centime • Centre • Centrifugal Force • Centro • Centum • Century • Chapter • Charge Conjugation • Charlie (international alphabet) • Charm Quantum Number • Chaud • Chief • Circling • Ciudad • Closed • Cloudy • Coated • Code • Coefficient • Cold • Collector • College • Colon • Color Index • Common • Commonwealth • Complement • Complete • Compliance • Conceptual • Conductivity • Congress • Conservative • Consonant • Constant • Convective Transport • Core • Cosine • Costa • Coulomb • Council • Count • Country • County • Cuba (international vehicle registration code) • Cycle • Cytidine • Euler Number • Heat Capacity • Hundred (Roman numeral) • Kohlenstoff • Molecular Concentration • Parallactic Angle • Programming Language • Quantity of Electricity • Rotational Constant • Set of Complex Numbers • Spectral Type • Viscous Damping • 100 (Roman numeral)

ca: Canada • Catalan (ISO language code) • Cathode • Centare • Centiare • Circa

Ca: Cable • Calcium • Canada • Canadian • Carretera • Compagnia • Companhia • Compañia

C/A: Capital Account • Cash Account • Coarse/Acquisition (*) • Course/Acquisition • Credit Account • Cuenta de Ahorros • Current Account

C&A: Calibration & Analysis

CA: Cabin Attendant • Calar Alto (Spain) • Calculateur Analogique • Calculateur Arithmétique • Calibrated Altitude • Calibration Adjust • California (USA) • Canada (ISO country code) • Capital Account • Capital Asset • Carlisle (UK postal code) • Carrier Aircraft • Catamarca (Argentina) • Cellular Automata • Cellular Automaton • Central America • Centrifugal Atomization • Certificat Administratif • Certificate Authority • Certification Authority • Chargé d'Affaires • Chartered Accountant • Chemical Abstracts • Chief Accountant • Chiffre d'Affaires • Chromospherically Active • Chronological Age • City Attorney • Civil Aviation • Clear and Add • Clear Aperture • Clerical Assistant • Closest Approach • Closing Angle • Cloud Analysis • Coarse Acquisition • Coarse Alignment • Collision Avoidance • Commercial Agent • Completion of Assembly • Compressed Air • Comptroller of Accounts • Computer Animation • Computer Assembly • Cone Angle • Conseil d'Administration • Constant Amplitude • Construction Analysis • Consular Agent • Continental Arctic • Contingency Abort • Continuous Averaging • Contour Anomaly • Contract Advice • Contract Authorization • Contract Award • Contractor Assisted • Control Area • Control Assembly • Control Authority • Controlled Atmosphere • Controller of Accounts • Controlling Authority • Controlled Atmosphere • Convening Authority • Cooperative Agreement • Cornerstone Agent • Coronagraphic Apodizer • Correct or Amplify • Corrective Action • Correspondence Analysis • Corriente Alterna • Cost Account • Coulomb Approximation • Courant Alternatif • Cours d'Appel • Course Alignment • Court of Appeal • Court of Auditors • Cover Aft • Credit Account • Critical Analysis • Criticality Analysis • Current Account • Current Assets • Current Awareness • Cylindrical Anode

CAA: Cambridge Astronomical Association (UK-England) • Canadian Authors' Association • Canadian Aviation Authority • Career Advancement Award • Catacamas (Honduras) (location identifier) • Cedar Amateur Astronomers (USA-IA) • Central Arid Asia • Centre à Autonomie d'Acheminement • Chinese Arctic Administration • Chinese Astronomy Abstracts • Circular Aperture Antenna • Civil Aeronautics Administration • Civil Aeronautics Authority • Civil Aviation Agency • Civil Aviation Authority • Clean Air Act (USA) • Committee for Astronomy and Astrophysics • Committee on Astronomy and Astrophysics • Composite Audio Amplifier • Computational Aeroacoustics • Computer Amplifier Alarm • Computer-Assisted Accounting • Cost Account Authorization • Cuyahoga Astronomical Association (USA-OH)

CAAA: Clean Air Act Amendment (*) • Clean Air Amendment Act • Composers, Authors, and Artists of America

CAAB: Club d'Astronomes Amateurs de Bienne (Switzerland)

CAAC: Capital Area Astronomy Club (USA-MI) • Charlotte Amateur Astronomers Club (USA-NC) • Civil Aviation Administration of China (China-PRC) • Club Astrónomos Aficionados del Cerro (Uruguay) • Club d'Astronomie Alpha Centaure (France)

CAACTD: Comité Asesor sobre la Aplicación de la Ciencia y la Tecnología al Desarollo (ONU)

CAADRP: Civil Aircraft Airworthiness Data Recording Program

CAAE: Canadian Association for Adult Education • Certificat d'Aptitude à l'Administration des Entreprises

CaAer: Canadian Aeronautics and Space Journal (periodical code)

CAAIS: Computer-Assisted Action Information System

CAAL: Club des Astronomes Amateurs de Laval (Canada-QC) • Club des Astronomes Amateurs de Longueuil (Canada-QC)

CAALIS: Computer-Aided Assembly-Line Inspection System

CAAO: Center for Astronomical Adaptive Optics

CAAP: Composition and Acidity of Asian Precipitation

CAAR: Compressed Air Accumulator Rocket

CAAS: Central Arkansas Astronomical Society (USA-KS) • Chinese Association for the Advancement of Science • Civil Aviation Authority of Singapore • Computer-Assisted Acquisition System • Club des Astronomes Amateurs de Sherbrooke (Canada-QC) • Crimean Amateur Astronomical Society (Ukraine)

CAAT: Centro Astronómico del Alto Turia (Spain) • Computer-Aided Assembly Tool • Coordinamento Associazioni Astrofili Toscane (Italy)

CAATS: Canadian Automated Air Traffic System • Compact Airborne Automatic Tracking System

Cab: Cabin • Cabinet • Cable

CAB: Cabin • Cabinda (Angola) (location identifier) • Cabinet • Calculateur Arithmétique Binaire • Centro Atomico Bariloche (Argentina) • Cercle Astronomique de Bruxelles (Belgium) • Change Analysis Board • Change Appeals Board • Change Authorization Board • Circolo Astrofili Bergamaschi (Italy) • Circolo Astrofili Bolzano (Italy) • Circuit Assurance Block • Citizens Advice Bureau • Civil Aeronautics Board • Civil Aviation Board • Climate Applications Branch • Commonwealth Agricultural Bureau • Communication Accelerator Board • Computer Address Bus • Corrective Action Board • Cost Audit Board

CABEI: Central American Bank for Economic Integration

CABS: Carrier Access Billing System • Chromospherically Active Binary Star • Committee on Automated Bibliographic Services (UK) • Computerized Annotated Bibliography System

CABSCAT: Catalog of Chromospherically Active Binary Stars

Các: Cáceres (Spain)

CAC: Cable Access Cover • Call Admission Control • Carrier Access Code • Cascavel (Brazil-PR) (location identifier) • Center for Advanced Computation • Central Air Conditioning • Centro Atomico Constituyentes (Argentina) • Chemical Abstracts Condensates • Circuit Administration Center • Clear All Channels • Climate Analysis Center • COBE Analysis Center • Colonial Astronomy Club (USA-GA) • Command And Control • Committee on Atmospheric Chemistry • Community Access Center • Community Advisory Committee • Complete Address Constant • Compressed Aeronautical Chart • Computer Advisory Committee • Computer-Aided Classification • Concordia Astronomy Club (USA-NJ) • Connection Admission Control • Consumer Advisory Council • Contract Award Committee • Conversational Associative Category • Cooperation And Coordination • Correction Action Committee • Coupling Among Classes • Crossroads Astronomy Club (USA-TX) • Cumberland Astronomy Club (USA-MD) • Curriculum Advisory Committee

CACA: Computer-Aided Circuit Analysis

CACAE: Commonwealth Advisory Committee on Advanced Education (Australia)

CACB: Center Aisle Connector Bracket

CACC: Communications And Configuration Console • Computer Application Control Code

CACD: Collins Avionics and Communications Division • Computer-Aided Circuit Design

CACDS: Centre for Advanced Computer and Decision Support

CACE: Central Advisory Council for Education (UK) • Computer-Aided Control Engineering

CACGP: Commission on Atmospheric Chemistry and Global Pollution

CACHE: Computer Aids for Chemical Engineering Education

CACID: Computer-Aided Concurrent Integral Design

CACM: Central Advisory Council for the Ministry • Central American Common Market • Communications of the Association for Computing Machinery

CACMI: Comité Africain pour la Coordination des Moyens d'Information

CACOFD: Caribbean Congress of Fluid Dynamics

CACON: Cargo Container

CACP: Central Arbitration Control Point

CACS: Center for Advanced Computer Studies • Computer-Aided Communications System

CACSD: Computer-Aided Control System Design

CACST: Central Advisory Council for Science and Technology (UK)

CACT: Carrier-Aided Code Tracking

CACTUS: Capteur Accélérométrique Capacitif Triaxial Ultra-Sensible

CACUL: Canadian Association of College and University Libraries

CAC&W: Continental Aircraft Control & Warning

c.-à-d.: C'est-à-Dire

càd: C'est-à-Dire

Cád: Cádiz (Spain)

Cad: Cadauno • Cadmium

CAD: Cadillac (USA-MI) (location identifier) • Cash Against Documents • Catalogue of Astrometric Data • Center for Astronomical Data • Civil Aviation Department • Collecte Automatique des Données • Comité d'Aide au Développement • Comité d'Assistance au Développement • Command Action Directive • Commercial Aircraft Development • Computation and Analysis Division • Computer Access Device • Computer Adaptor Display • Computer-Aided Design • Computer-Aided Design (periodical code) • Computer-Aided Detection • Computer-Aided Dispatching • Computer-Aided Drafting • Computer-Aided Drawing • Computer-Assisted Device • Computer-Assisted Diagnosis • Computer-Assisted Dialog • Computer-Associated Device • Consultatiebureau voor Alcohol en Drugs • Contract Award Date • Coronary Artery Disease • Dollar (Canada) (ISO currency code)

CADA: Commission d'Accès aux Documents Administratifs • Computer-Aided Design and Analysis • Computer-Assisted Distribution and Assignment

CADAF: Colorado Astronomical Data Analysis Facility (USA-CO)

CADAM: Computer-Augmented Design And Manufacturing

CADAP: Computer-Aided Dentrite Analysis Program

CADARS: Catalogue of Apparent Diameters and Absolute Radii of Stars

CaDAS: Cleveland and Darlington Astronomical Society (UK)

CADAS: Comité des Applications De l'Académie des Sciences

CADAT: Computer-Aided Design And Test

CADC: Canadian Astronomical Data Centre • Canadian Astronomy Data Centre (*) • Canadian Astrophysical Data Centre • Central Air-Data Computer • Colour Analysis Display Computer

CADD: Computer-Aided Design and Development • Computer-Aided Design and Drafting • Computer-Aided Drafting and Design

CADDAC: Central Analog Data Distributing and Computing System

CADDE: Central Automatic Digital Data Encoder

CADDET: Centre for Analysis and Dissemination of Demonstrated Energy Technologies

CADDMAS: Computer-Assisted Dynamics Data Monitoring and Analysis System

CADDR: Change Address

CADDS: Computer-Aided Design and Drafting System • Computer-Aided Design Drafting System

CADE: Computer-Aided Design Engineering • Computer-Aided Design and Evaluation • Computer-Aided Document Engineering • Computer-Assisted Data Evaluation • Controller/Attitude-Direct Electronics

CADES: Computer-Assisted Data Entry System

CADET: Computer-Aided Design and Electrical Test • Computer-Aided Design Experiment Translator

CADF: Commutated Antenna Direction Finder

CADFISS: Computation And Data Flow Integration Subsystem

CADI: Central Apollo Data Index (NASA) • Centre Algérien d'Information et de Documentation (Algeria) • Certificat d'Aptitude à la Documentation Informatisée • Computer Access Device Input

CADIAP: Comment And Data Integration And Printing

CADIC: Computer-Aided Design of Integrated Circuits

CADIS: Calar Alto Deep Imaging Survey • Computer-Aided Design Information System

CADIST: Centre d'Acquisition De l'Information Scientifique et Technique • Centre d'Acquisition et de Diffusion de l'Information Scientifique et Technique (France) • Centre d'Acquisition et D'Information Scientifique et Technique

CADITS: Coupling And Dynamics of the Ionosphere-Thermosphere System

CADIZ: Canadian Air Defense Identification Zone

CADL: Communications And Data Link • Critical Application Distribution List

CADM: Configuration Administration and Data Management

CADMAP: Computer-Aided Drafting, Mapping, and Photogrammetry

CADMAT: Computer-Aided Design, Manufacture And Test • Computer-Aided Design, Manufacturing, And Testing

CADMOS: Centre d'Aide au Développement de la Microgravité et aux Opérations Spatiales

CADMP: Computer-Aided Data Management Procedure

CADO: Computer Access Device Output

CADPO: Communications And Data Processing Operation

CADPOS: Communications And Data Processing Operation System

CADRE: Centre d'Animation, de Développement et de Recherche en Éducation • Computer-Aided Data Retrieval and Evaluation • Coupling And Dynamics of Regions Equatorial • Current Awareness and Document Retrieval for Engineers

CADS: Central Air Data System • Command And Data Simulator • Computer-Aided Debriefing System • Computer-Aided Design System • Computer-Assisted Dispatching System • Control-Accessibility-Disposal-Storage • Control Authority Data Structure

CADSI: Communications and Data Systems Integration

CADSS: Combined Analog-Digital Systems Simulator

CADU: Control And Display Unit • Controller And Data Unit (*)

CADV: Combined Alternate Data/Voice

Cae: Caelum-Caeli (constellation)

CAE: Calculateur Arithmétique Électronique • Córbrese Al Entregar • College of Advanced Education • Columbia (USA-SC) (location identifier) • Committee on Advanced Education • Common Applications Environment • Compositeur-Auteur-Éditeur • Computer-Aided Education • Computer-Aided Engineering • Computer-Assisted Entry • Computer-Assisted Engineering • Computer-Assisted Estimating • Conseil des Affaires d'État • Control Actuation Electronics • Corrective Action Engineer

CAEA: Climate and Atmospheric Environment Activity

CAEAL: Canadian Association for Environmental Analytical Laboratories

CAECAL: Computer-Aided Engineering for Cargo Accommodation and Location

CAEDE: Computer-Aided Engineering and Design for Electronics

CAEG: Computer-Aided Engineering Graphics

CAEI: Certificat d'Aptitude à l'Enseignement des Inadaptés (France)

CAELEP: Certificat d'Aptitude à l'Enseignement dans les Lycées Professionnels (France)

CAeM: Commission for Aeronautical Meteorology (WMO)

CAEM: Comité d'Assistance Économique Mutuelle • Conseil d'Assistance Économique Mutuelle

CAEO: Coalition of Adult Education Organizations

CAEPE: Centre d'Achèvement et d'Essais des Propulseurs et Engins

Caer: Caernarvonshire (UK-England)

CAER: Concours d'Accès aux Échelles de Rémunération

Caern: Caernarvonshire (UK-England)

Caerns: Caernarvonshire (UK-England)

CAERPA: Concours d'Accès aux Échelles de Rémunération des Professeurs Agrégés

Caes: Caesar

CAES: Comité d'Action et d'Entraide Sociales • Compressed-Air Energy Storage

CAESAR: CCD Airborne Experimental Scanner for Applications in Remote Sensing

CAESR: Cold Atmospheric Emission Spectral Radiometer

CAET: Computer-Assisted Electrical Test

CAETI: Computer-Assisted Education and Training Initiative

CAETS: Council of Academies of Engineering and Technological Sciences (USA-DC)

CAEU: Council of Arab Economic Unity

CAF: Caisse d'Allocations Familiales (France) • Call Attach Facility • Carauari (Brazil-AM) (location identifier) • Central African Federation • Central African Rep. (ISO country code) • Chemical Analysis Facility • Clean Air Facility • Clean Assembly Facility • Cleared As Filed • Cost And Freight • Cost-Assurance-Freight • Coût-Assurance-Fret • République Centrafricaine (ISO country code)

CAFB: Certificat d'Aptitude aux Fonctions de Bibliothécaire (France)

CAFBC: Circulating Atmospheric Fluidized Bed Combustion

CAFE: Computer-Aided Film Editor • Conditional Access For Europe • Corporate Average Fuel Economy

CAFF: Conservation of Arctic Flora and Fauna

CAFFE: Computer-Aided Fast Fabrication Exploration

CAFIE: Centre d'Autoformation et d'Ingénierie Éducative (France)

CAFIT: Computer-Assisted Fault Isolation Test

CAFM: Commercial Air Freight Movement

CAFO: Central Analysis and Forecast Office

CAFOC: Centre Académique à la Formation Continue (France)

CAFOS: Calar Alto Faint Object Spectrograph

CAFS: Content-Addressable File Store

CAFTA: Concours Annuel des Fabricants de Télescopes Amateurs

CAG: Cagliari (Italy) (location identifier) • Canadian Association of Geographers • Canopus Acquisition Gate • Change Analysis Group • Column Address Generator • Commercial Arbitration Group • Competitiveness Advisory Group • Comptroller and Auditor General • Computer-Aided Gauging • Contrôle Automatique de Gain

CAGC: Coded Automatic Gain Control

CAGD: Computer-Aided Geometric Design • Computer-Assisted Geometric Design

CAGE: Central Arctic Geological Expedition • Composition of Aleatorically Generated Elements • Computerized Aerospace Ground Equipment

CAgM: Commission for Agricultural Meteorology (WMO)

CAGR: Civil Advanced Gas-Cooled Reactor • Commercial Advanced Gas-Cooled Reactor

CAGS: Canadian Association of Graduate Schools • Certificate of Advanced Graduate Study

Cah: Cahier • Cahiers

CAH: Va Mau (Viet Nam) (location identifier)

CAHA: Centro Astronómico Hispano-Alemán (Spain)

CAHD: Computer-Aided Hypermedia Development

CAHDE: Computer-Aided Hypermedia Development Environment

C/AHRS: Compass/Attitude and Heading Reference System

CAHRS: Compass/Attitude and Heading Reference System

CAHS: Comprehensive Automation of the Hydrometeorological Service

CAHT: Chiffre d'Affaires Hors Taxes

Cai: Caithness (UK-Scotland)

CAI: Cairo (Egypt) (location identifier) • Calcium-Aluminium-rich Inclusion • Call Assembly Index • Canadian Aeronautical Institute • Centre d'Analyse de l'Information • Centre d'Analyse des Images (France) • Coded-Aperture Imaging • Color Alteration Index • Comisión Asesora para la Investigación • Comité Arctique International • Compression After Impact • Compression of Astronomical Images • Computer-Administered Instruction • Computer-Aided Instruction • Computer-Assisted Instruction • Computer-Assisted Interviewing • Computer Associates International • Computer-Analog Input • Confederation of Australian Industry • Configuration Acceptance Instruction • Current-Awareness Information

CAIA: Conference on Artificial Intelligence for Applications

CAIC: Computer-Assisted Indexing and Classification • Computer-Assisted Instruction Center

CAICYT: Centro Argentino de Información Científica y Técnologica (Argentina) • Comisión Asesora de Investigación Científica y Técnica (Spain)

CAID: Computer-Aided Industrial Design • Control Authority Identification • Control Authority Identifier

CAIDA: Cooperative Association for Internet Data Analysis (USA-CA)

CAIG: Cost Analysis Improvement Group

CAIL: Conditional Attributed Interaction Lyndenmaier System

CAILE: Comité sur l'Accès à l'Information et la Liberté d'Expression (Canada)

CAIM: Computer-Assisted Instruction Management

CAIMS: Canadian Applied and Industrial Mathematics Society

CAIN: Camara Infraroja • Computer Algebra Information Network • Conflict Archive on the Internet

CAIP: Computer Analyses of Images and Patterns (conference) • Computer-Assisted Indexing Program • Consortium of Affiliates for International Programs

CAIRP: Centro Argentino para la Información y Refutación de la Pseudociencia (Argentina)

CAIR: Committed Aggregate Information Rate

CAIRS: Central Automated Inventory and Referral System • Computer-Assisted Information Retrieval System • Computer-Assisted Interactive Resources Scheduling • Computer-based Automatic Information Retrieval System

CAIS: Canadian Association for Information Science • Central Abstracting and Indexing Services • Central Abstracting and Information Services • Centre for Advanced Instrumentation Systems • Collocated Automatic Intercept System • Common Airborne Integrated System • Common APSE Interface Set • Common APSE Interface Specification • Computer-assisted Action Information Service

CAISc: Canadian Association fo rInformation Science

CAISMI: Centro per l'Astronomia Infrarossa e lo Studio del Mezzo Interstellare (Italy)

CAISMS: Computer-Assisted Instruction Study Management System

CAISSE: Computer-Aided Information System on Solar Energy

CAIT: Center for Advanced Information Technology • Computer-Aided Instruction Terminal • Computer-Aided Instruction Trainer

Caith: Caithness (UK-Scotland)

CAITS: Computerized Automatic Inertial Test Set

CAIVRS: Credit Alert Interactive Voice Response System

CAJ: Canaima (Venezuela) (location identifier)

CaJCh: Canadian Journal of Chemistry (periodical code)

CaJES: Canadian Journal of Earth Sciences (periodical code)

CaJPh: Canadian Journal of Physics (periodical code)

CaJRS: Canadian Journal of Remote Sensing (periodical code)

CAJUN: CD-ROM Acrobat Journals Using Networks

CAK: Akron Canton (USA-OH) (location identifier) • Castor-Abbott-Klein

cal: Calando • Calorie • Calvus

Cal: Calcutta (India) • Calendar • Calibrate • Calibrated • Calibration • Calibre • California (USA) • Called • Kilocalorie

CAL: Calculated Average Life • Calendar • Calibrate • Calibration • Calibration Abstract Layer • Campbeltown (UK-England) (location identifier) • Canadian Astronautics Ltd. • Canister Airlock • Central Association of Libraries (USA-CA) • Client Access License • Columbia Astrophysics Laboratory • Columbus Attached Laboratory (formerly APM) • Command Authorization List • Common Assembler Language • Component Action List • Computer-Aided Learning • Computer-Aided Logistics • Computer-Assisted Learning • Computer Animation Language • Contract Advice Letter • Conversational Algebraic Language • Cornell Aeronautical Laboratory • Creative Applications Laboratory

CALA: Club d'Astronomie de Lyon-Ampère (France) • Chinese American Librarians Association • Computer-Aided Loads Analysis

CALAS: Computer-Assisted Language Analysis System

Calç: Calçada

Calc: Calculate • Calculated • Calculation • Calculus • Calcutta (India)

CALC: Calculate • Calculus As a Laboratory Course

CALCE: Computer-Aided Life Cycle Engineering

CALCRE: Comité des Auteurs en Lutte Contre le Racket de l'Édition (France)

Cald: Calculated

CALD: Caldeira

CalDB: Calibration Database

CALEEP: California Earthquake Education Program

Calg: Calculating • Calgary (Canada-AB)

CALGO: Collected Algorithms (ACM)

Calib: Calibrate • Calibrated • Calibration

Calibr: Calibrate • Calibrated • Calibration

CALIBRE: Common Access to Libraries in Europe

Calif: California (USA)

Caligr: Caligrafía

CALIM: Consortium of Academic Libraries in Manchester (UK)

CALIMS: Computer-Assisted Library and Information System

CALL: Computer-Aided Language Learning • Computer-Aided Learning Laboratory • Computer-Assisted Language Learning • Computer-Assisted Learning Laboratory • Computer-Augmented Loft Lines

CALM: COBOL Automatic Language Modifier • Collected Algorithm for Learning Machines • Committee for Administrative and Legal Matters • Computer-Assisted Library Mechanization • Conservation And Land Management • Contra-Almirante

Caln: Calculation

CALN: Computer-Assisted Learning Network

CALP: Computer-Assisted Learning Package

CALPERS: California Public Employees Retirement System (USA-CA)

CALRS: Centralized Automatic Loop Reporting System

CALS: Cloud and Aerosol Lidar System • Comprehensive Automated Learning System • Computer-aided Acquisition and Logistics Support • Computer-aided Acquisition Logistics Support • Computer-Aided Logistics and Support • Computer-Aided Logistics Support • Computer-Aided Logistics System • Continuous Acquisition and Life-cycle Support • Current Awareness Literature Search • Current Awareness Literature Service

CALST: Cornell Aeronautical Laboratory Shock Tunnel

CALTECH: California Institute of Technology (USA-CA)

CALTF: Combined Air Lift Task Force

CALTRANS: California Department of Transportation (USA-CA)

CALUTRON: California University Cyclotron

Calv: Calvinist

Cam: Cambodia • Cambodian • Cambridge • Camelopardalis (constellation) • Camera • Cameroon • Camino

CAM: Cambridge (Starlink internal code) • Camera • Caminho • Camiri (Bolivia) (location identifier) • Cancellation of Amplitude Modulation • Carrier Aircraft Modification • Carte À Mémoire • Cascade Access Method • Cell Adherence Molecule • Cellular Adhesion Molecule • Cellular Automata Machine • Cellular Automata Model • Central Address Memory • Centre d'Aviation Météorologique • Centrifuged Accommodation Module • Cercle Astronomique Mosan (Belgium) • Channel Access Method • Checkout and Automatic Monitoring • Circolo Astrofili di Milano (Italy) • Circular Area Method • Civil Aeronautics Manual • Cockpit Area Microphone • Coded Aperture Mask • Common Access Method • Communication Access Method • Communication-Advertising-Marketing • Computer Achievement Monitoring • Computer Address Matrix • Computer-Aided Machining • Computer-Aided Makeup • Computer-Aided Management • Computer-Aided Manufacture • Computer-Aided Manufacturing • Computer-Aided Mapping • Computer Annunciation Matrix • Computer-Assisted Makeup • Computer-Assisted Maintenance • Computer-Assisted Manufacture • Computer-Assisted Manufacturing • Computerized Anatomical Man • Content-Addressable Memory • Content-Addressed Memory • Contingency Analysis Model • Continuous Air Monitor • Control Access Manager • Coronagraphic Apodizer Mechanism • Coronagraphic Apodizer Mechanizer • Cost Account Manager • Coupled-Angular Mode • Customer Asset Management

CAMA: Central Arctic Management Area • Centralized Automatic Message Accounting • Civil Aerospace Medical Association

CAMAC: Computer-Aided Measurement And Control • Computer-Automated Measurement And Control

CAMARC: Computer-Aided Movement Analysis in a Rehabilitation Context

CAMAS: Central Automatic Message Accounting System

Camb: Cambridge

Cambs: Cambridgeshire (UK-England)

CAMC: Carlsberg Automatic Meridian Circle

CAMCOS: Computer-Assisted Maintenance and Control System

CAMD: Center for Advanced Microstructures and Devices • Computer-Aided Mechanical Drawing • Computer-Aided Molecular Design

CAME: Consejo de Ayuda Mutua Económica

CAMEO: Computer-Aided Management of Emergency Operations • Computer-Assisted Management of Emergency Operations

CAMERA: Control And Monitoring Equipment Resource Allocation

CAMEX: Convection and Moisture Experiment

CAMI: Civil Aeromedical Institute

CaMILE: Collaborative and Multimedia Interactive Learning Environment

CAMILE: Concerted Action on Management Information for Libraries in Europe

CAMIS: Center for Advanced Medical Informatics at Stanford

CAML: Consequential Angular Momentum Loss

CAMM: Computer-Aided Maintenance Management • Computer-Aided Manufacturing Management • Computer-Aided Molecular Modelling • Computer-Assisted Maintenance Management • Computer-Assisted Manufacturing Management • Computer-Assisted Molecular Modelling

CAMMICE: Charge And Mass Magnetospheric Ion Composition Experiment

CAMMU: Cache/Memory Management Unit

CAMOS: Committee on Atomic, Molecular, and Optical Sciences

CAMP: Campeche (Mexico) • Campo • Code Applications and Maintenance Program • Cold Arctic Mesopause Project • Compiler for Automatic Machine Programming • Computer-Aided Message Processing • Computer-Assisted Mathematics Program

Camr: Camera

CAMR: Camera • Conférence Administrative Mondiale des Radiocommunications • Conferencia Administrativa Mundial de Telecomunicaciones

CAMRSS: Center for Autonomous and Man-controlled Robotic and Sensing Systems

CAMS: Cabin Atmosphere Monitoring System • Calibrated Airborne Multispectral Scanner • Caminhos • Canadian Applied Mathematics Society • Collision Activation Mass Spectrum • Computer-Aided Manipulation System • Control And Monitoring System

CAMSP: Center for Atomic, Molecular and Surface Physics (UWA)

CAMST: Community Action on Magnetic-Storage Technology

Can: Canada • Canadian • Canal • Canarias • Canberra (Australia-ACT) • Cancel • Cancelled • Cañon • Canton • Canyon

CA/N: Convertisseur Analogique/Numérique

CAN: Campus Area Network • Canada (ISO country code) • Cancel • Certification Analysis Network • Committee of an Advisory Nature • Computer-Aided Navigation • Computer Algebra Nederland • Computer Algebra Netherlands (*) • Controller Area Network • Convertisseur Analogique/Numérique • Cooperative Agreement Notice • Cost Account Number • Customer Access Network • Guangzhou (China-PRC) (location identifier)

CANA: Centro Astronomico Neil Armstrong (Italy)

Canad: Canada • Canadian

CANAMEX: Canadian-American-Mexican

CANARI: Code pour l'Analyse par Navier-Stokes de l'Aérodynamique d'une Roue Isolée

CANARIE: Canadian Network for the Advancement of Research, Industry and Education

Canc: Cancel • Cancellation • Cancelled

CANC: Cancelled

Cand: Candelabra • Candidate

CANDI: Combined Analogue Digital Integration

CANDU: Canadian Deuterium-Uranium

CANEL: Connecticut Advanced Nuclear Engineering Laboratory (USA-CT)

CANEX: Canadian Experiment

CanF: Canadian French

Can. Fr.: Canadian French

CANGAROO: Collaboration between Australia and Nippon for a Gamma-Ray Observatory in Outback

Can. I.: Canary Islands (Spain)

CANM: Communications in Applied Numerical Mathods (periodical code)

CANOPUS: Canadian Auroral Network for the Origin of Plasma in Earth Neighborhood Program Unified Study

Can. Pac.: Canadian Pacific

Cant: Canterbury (UK-England) • Cantería • Cantilever • Contonese

Cantil: Cantilever

CANTO: Caribbean Association of National Telecommunication Organizations

CANTRAN: Cancel in Transmission

CANUS: Canada - United States

Canv: Canvas

CANZ: Canada - Australia - New Zealand

CAO: Cagliari Astronomical Observatory (Italy) • Career Access Opportunity • Central Aerological Observatory • Central Astronomy Observatory (Russia) • Centro Astronomico Orione (Italy) • Chief Accountant Officer • Chief Administrative Officer • Circuit Activation Order • Classification Advisory Officer • Clayton (USA-NM) (location identifier) • Cold Air Outbreak • Commissie Antarctisch Onderzoek (Netherlands) • Commission for Asia and Oceania • Computer-Aided Optimization • Conception Assistée par Ordinateur • Control Authority Office • Cost Administration Office • County Advisory Officer • Coupled Atmosphere-Ocean • Crimean Astrophysical Observatory

CAOI: Catalogue of Astronomical Object Identifications

CAOIS: Control Authority Office Information System

CAOQ: Center for Astronomical Observing Quality (USA-CO)

CAOS: Comité Académique des Oeuvres Sociales • Computer-Aided Observation System • Computer-Aided Organic Synthesis

cap: Capillatus

Cap: Capacitance • Capacitor • Capacity • Capiat • Capital • Capitale • Capitalize • Capitán • Capitol • Capitolo • Capitúlo • Capitulum • Capricornus-Capricorni (constellation) • Captain • Caput

CAP: Canadian Association of Physicists • Capability Assurance Program • Capacitor • Capacity • Cap Haitien (Haiti) (location identifier) • Capitão • Capture • Card Assembly Program • Carrier Amplitude Phase • Carrierless Amplitude and Phase • CCMS Application Program • Center for Astronomy and Physics • Central Arbitration Point • Centre d'Analyse et de Programmation • Centre d'Archivage et de Prétraitement • Certificat d'Aptitude Pédagogique • Certificat d'Aptitude Professionnelle (France) • Civil Air Patrol • Cloud Atmospheric Parameter • Club Astronómico del Paraguay • Codice di Avviamento Postale • Collective Action Plan • Combat Air Patrol • Command Acceptance Pattern • Commission Administrative Paritaire • Committee on Applied Physics • Common Agricultural Policy • Communications Access Processor • Communications Alternative Provider • Communications Application Platform • Competitive-Access Provider • Complex Autonomous Payload • Composer-Author-Publisher • Computer Address Panel • Computer-Aided Planning • Computer-Aided Production • Computer-Aided Publishing • Computer Application Program • Computer-Assisted Printing • Computer-Assisted Production • Computer-Assisted Publishing • Concurrent Algorithmic Programming • Conservation Assessment Program • Contamination Analysis Program • Contractor-Acquired Property • Contrat d'Achat Privilégié • Control Assembly Program • Control and Authorization Process • Convertible Adjustable Preferred • Cooperative Agreement Program • Corporate Affiliates Program • Cost Account Package • Cost Account Plan • Cost Analysis Plan • Council on Advanced Programming • Crew Activity Plan • Cryogenic Associative Processor • Current Assessment Plan

CAPA: Commission Administrative Paritaire Académique (France) • Computer-Aided Plastic Application

CAPAC: Composers', Authors', and Publishers' Association of Canada

CAPAL: Computer and Photographic Assisted Learning

CAPAV: Committee on Atmospheric Problems of Aerospace Vehicles

CAPCA: Canadian Advanced Protein Crystallization Apparatus

CAPCHE: Component Auto-Programmed Checkout Equipment

CAPCOM: Capsule Communicator

CAPD: Commission Administrative Paritaire Départementale (France) • Computer Assistance to Persons with Disabilities • Computing to Assist Persons with Disabilities

CAPE: Centre d'Analyse de Politique Économique • Certificat d'Aptitude au Professorat des Écoles (France) • Communication Automatic Processing Equipment • Computer-Aided Production Engine • Computer Applications in Production and Engineering • Connecting Asia-Pacific to Europe • Consortium for the Advancement of Physics Education • Convective Available Potential Energy

CAPEB: Confédération de l'Artisanat et des Petites Entreprises du Bâtiment (France)

CAPEPS: Certificat d'Aptitude au Professorat d'Éducation Physique et Sportive (France)

CAPER: Cleft Accelerated Plasma Experimental Rocket • Collaborative Applications for Project-based Education • Computer-Aided Pattern Evaluation and Recognition

CAPES: Certificat d'Aptitude au Professorat de l'Enseignement Secondaire (France) • Certificat d'Aptitude Pédagogique à l'Enseignement Secondaire • Certificat d'Aptitude Préparatoire à l'Enseignement Secondaire • Certificat d'Aptitude Professionnelle l'Enseignement Secondaire • Coordenação de Aperfeiçoamento de Pessoal de Nível Superior (Br azil)

CAPET: Certificat d'Aptitude au Professorat de l'Enseignement Technique (France) • Certificat d'Aptitude Professionnelle l'Enseignement Technique

CAPI: Common Application Program Interface • Common Application Programmers Interface • Computer-Assisted Personal Interviewing • Computer-Assisted Photo-Interpretation • Cryptographic Application Program Interface

Capit: Capitalist

CAPL: Capillary Pump Loop • Coastal Anti-Pollution League

CAPLP: Certificat d'Aptitude au Professorat de Lycée Professionnel (France)

CAPM: Capital Asset Pricing Model

Capn: Captain

CAPO: Canadian Army Post Office

CAPP: Canadian Association of Professional Physicists • Center for Applied Parallel Processing • Committee on Astronomy and Public Policy (AAS) • Computer-Aided Process Planning • Computer-Aided Production Planning • Computer-Assisted Process Planning • Computer-Assisted Production Planning • Content-Addressable Parallel Processor

CAPPI: Constant Altitude Plan Position Indicator

CAPRI: Capacitance Rate Integrator • Captive Reset Ignitor • Card And Printer Remote Interface • Computational Analysis Programming Interface • Computer-Aided Personal Reference Index

Capric: Capricorn

CAPRICE: Cosmic Antiparticle Ring-Imaging Čerenkov Experiment

Caps: Capitals • Capsule

CAPS: Cabin Air Pressure System • Capitals • Captive Animals Protection Society • Cassette Programming System • Cassini Plasma Spectrometer • Catalèg Automatitzat de Publicacions en Série • Center for Analysis and prediction of Storms • Center for Analysis of Particle Scattering • Common Attitude Pointing System • Crew Activity Planning System • Crew Altitude Protection System

CAPSAIS: Certificat d'Aptitude au Professorat Spécialisé d'Adaptation et d'Intégration Scolaires (France)

CAPSIM: Captive Simulation

CAPSS: Computer-Aided Planning and Scheduling System

Capt: Captain • Caption

CAPT: Computer-Assisted Procedures Trainer

CAPTAC: Conférence des Administrations des Postes et Télécommunications de l'Afrique Centrale

CAPTAIN: Computer-Aided Processing and Terminal Access Information • Covariance Analysis Program for the Study of Augmented Inertial Navigators

CAPU: Crew Altitude Protection Unit

Capy: Capacity

CAQ: Caucasia (Colombia) (location identifier)

CAQA: Computer-Aided Quality Assurance

Car: Carat • Cargo • Carina-Carinae (constellation)

CAR: Capital Authorization Request • Cardiff (Starlink internal code) • Caribou (USA-ME) (location identifier) • Center for Aerospace Research • Central African Rep. • Certification Analysis Requirement • Certification Approval Request • Channel Address Register • Chargé d'Administration de la Recherche • Civil Airworthiness Regulation • Cloud Absorption Radiometer • Cloudtop Altitude Radiometer • Collectieve Arbeidsvoorwaardenregeling • Collisionally Activated Reaction • Command Action Response • Committed Access Rate • Compounded Annual Rate • Compounded Annual Return • Computer-Assisted Reproduction • Computer-Assisted Research • Computer-Assisted Retrieval • Conditional Autoregression • Configuration And Acceptance Review • Configuration Audit Review • Content of Address Register • Contract Authorization Request • Contractual Action Request • Corporation for Astronomical Research • Corrective Action Record • Corrective Action Reply • Corrective Action Report • Corrective Action Request • Customer Acceptance Review

CARA: California Association for Research in Astronomy (USA-CA) • Center for Astrophysical Research in Antarctica • Centre for Astrophysical Research in the Antarctic • Chemical Assessments and Related Activities • Current Aerospace Research Activities

CARAFE: Cloud, Aerosol, and Radiation Arctic Field Experiment

CARAIB: Carbon Assimilation In the Biosphere

CARAM: Content-Addressable Random Access Memory

CARAN: Centre d'Accueil et de Recherche des Archives Nationales (France)

Carb: Carbon • Carbonate • Carburetor

CARB: Central Acquisition Review Board

Card: Cardinal

CARD: Cartographic Representation of Data • Comet and Asteroid Rendezvous Docking • Compact Automatic Retrieval Device • Compact Automatic Retrieval Display • Computer-Aided Research and Development • Constraints And Restriction Document

CARDA: Computer-Aided Reliability Data Analysis

CARDIF: Cardiff (Starlink internal code)

CARDME: Concerted Action for Research on Demand Management in Europe

Cards: Cardiganshire (UK-Wales)

CARDS: Climatological Aerological Reference Data Set • Computer-Aided Reliability Data Systems

CARE: Career Awareness Resource Education • Center of Applied Research in Education • Computed-Aided Requirements Engineering • Computed-Aided Risk Evaluation • Cooperative for American Relief Everywhere • Customers Are Really Everything

CAREM: Centre Armoricain de Recherche en Environnement (France)

CARES: Cartographic Retrieval Service • Computer-Assisted Retrieval Service

CARF: Central Altitude Reservation Facility • Central Altitude Reservation Function

CARIAN: Centre for Advanced Research in International Affairs in the Netherlands

Carib: Caribbean

CARIB: Caribbean

CARICOM: Caribbean Community

CARID: Customer Acceptance Report Item Disposition • Customer Acceptance Review Item Disposition

CARIFTA: Caribbean Free Trade Area

CARIN: Center for Applied Research on Information Networking

CARIRI: Caribbean Research Institute (Trinidad & Tobago)

CARIS: Computer-Aided Resource Information System • Computerized Agricultural Research Information Service

CARISMA: Corrections to Applied Research Laboratories Ion-Sputtering Mass Analyzers

CARJ: Clube de Astronomia do Rio de Janeiro (Brazil-RJ)

CARL: California Academic and Research Libraries (USA-CA) • Canadian Association of Research Libraries • Capacité d'Absorption des Radicaux Libres • Chemical Algorithm for Reticulation Linearization • Club Astronomique de la Région Lilloise (France) • Colorado Alliance of Research Libraries (*) • Colorado Association of Research Libraries • Compressor Aero Research Laboratory

CARLA: Code Actuated Random Load Apparatus

CARM: Customer Accommodation Requirement Model

CARMA: Combined Array for Research in Millimeter-wave Astronomy • Controlled Auto-Regressive Moving Average

Carms: Carmarthenshire (UK-Wales)

CARN: Competitive Applied Research Network

Carns: Caernarvonshire (UK-Wales)

Carol: Carolingian

Carp: Carpathian • Carpenter • Carpentry • Carpintería

CARP: Cache Array Routing Protocol • Computer-Assisted Regional Planning

Carr: Carrera • Carretera • Carriage • Carrier

CARR: Customer Acceptance Readiness Review • Customer Accommodation Requirement Representative

Carr. Fwd.: Carriage Forward

CARRS: Coherent Anti-Stokes Resonance Raman Scattering

CARS: Center for Applications of Remote Sensing • Climate Applications Referral Service • Climate Applications Referral System • Coherent Anti-Stokes Raman Scattering (*) • Coherent Anti-Stokes Raman Spectrum • Coherent Anti-Stokes Raman Spectroscopy • Combined Altimeter Radiometer Sensor • Community Antenna Relay System • Computer-Aided Reference Service • Computer-Aided Routing System • Computer-Assisted Reference Selection • Computer Audit Retrieval System • Configuration Accounting and Reporting System • Control And Release System • Corrective Action Reporting System

CARSO: Carnegie Southern Observatory (now Las Campanas Observatory) (Chile)

CARSS: Center for Atmospheric and Remote Sensing Studies

Cart: Cartonné

CART: Center for Advanced Research and Telecommunications • Centralized Automatic Recorder and Tester • Classification And Regression Tree • Cloud and Atmospheric Radiation Testbed • Cloud And Radiation Testbed • Complete And Ready for Test • Computerized Automatic Rating Technique • Custom Asynchronous Receiver/Transmitter

CARTE: Center for Advanced Research and Technology for Education • Centre d'Applications et de Recherches en Télédétection • Collaboration on Aeronautical Research and Technology in Europe

Carth: Carthage

Cartog: Cartographer • Cartography

cas: Castellanus

Cas: Cassegrain • Cassiopeia-Cassiopeiae (constellation) • Castle • Casual • Casualty

CAS: Calculated Airspeed • Calibrated Airspeed • California Academy of Sciences (USA-CA) • Calumet Astronomical Society (USA-IN) • Canadian Astronautical Society • Canberra Astronomical Society (Australia-ACT) • Canterbury Astronomical Society (New Zealand) • Cardiff Astronomical Society (UK) • Casablanca (Morocco) (location identifier) • Center for Aerospace Structures • Center for Astrophysical Sciences • Central Administrative Services • Certificate of Advanced Studies • Change Analysis Section • Channel-Associated Signaling • Charlottesville Astronomy Society (USA-VA) • Chemical Abstracts Service • Chesmont Astronomical Society (USA-PA) • Chester Astronomical Society (UK) • Cheyenne Astronomical Society (USA-WY) • Chicago Academy of Sciences (USA-IL) • Chicago Air Shower Array (USA-IL) • Chicago Astronomical Society (USA-IL) • Chinese Academy of Sciences (China-PRC) • Chinese Astronomical Society (China-PRC) • Cincinnati Astronomical Society (USA-OH) • Coarse Alignment Servo • Cockermouth Astronomical Society (UK) • Cognizant Area Supervisor • Collision Avoidance System • Columbus Astronomical Society (USA-OH) • Column Address Select • Column Address Strobe • Command Augmentation System • Commission for Atmospheric Sciences (WMO) • Communicating Applications Specification • Communications Analysis Section • Communications Antenna Sleeve • Communications Applications Specification • Communications Applications Standard • Complex Adaptive System • Computational Aeroscience • Computed Airspeed • Computer-Aided Styling • Computer-Aided System • Computer Algebra System • Computer-Assisted Search • Computerized Acquisition System • Computerized Autodial System • Conception de l'Architecture du Système • Configuration Accounting System • Control Augmentation System • Cooperative Application Satellite • Corps Administratif Supérieur • Cosmo Astronomical Society (USA-FL) • Cost Accounting Standard • Cotswold Astronomical Society (UK) • Council on Atmospheric Studies • Crew Accommodation System • Criticality Alarm System • Croydon Astronomical Society (UK) • Cumberland Astronomical Society (USA-TN) • Current Awareness Services • Current Awareness Situation

CASA: Center for Advanced Spatial Analysis • Center for Astrophysics and Space Astronomy • Centi Astro-Space Activities (USA-PA) • Central And South America • Centre of Advanced Study in Astronomy (India) • Chef Adjoint de Service Administratif • Chicago Air Shower Array • Construcciones Aeronáuticas SA • Controller Automation Spacing Aid

CASAC: Clean Air Scientific Advisory Committee

CASAFA: Commission on the Application of Science to Agriculture, Forestry and Aquaculture

CAsB: Clube de Astronomia de Brasília (Brazil)

CASB: Catalogs And Surveys Branch • Cost Accounting Standards Board

CASBAA: Cable And Satellite Broadcast Association of Asia

CASC: China Aerospace Corporation (China-PRC)

CASCA: Canadian Astronomical Society - Société Canadienne d'Astronomie

CASCADE: Computer-Aided SONAR Classification And Data Extraction

CASCADES: Calculs et Analyses de Seuils Capables d'Annonces des Débordements Soudains • Calculs et Analyses de Signaux Capables d'Annonces des Débordements Soudains

CASCDS: Cascades

CASCO: Conformity Assessment Committee

CASD: Chemical Analysis Solutions Division (HP) • Computer-Aided Structural Design • Computer-Aided System Design

CASE: Care Application Service Element • Center for Advanced Studies in the Environment • Circum Arctic Structural Events • Committee on Academic Science and Engineering (USA) • Common Access Switching Equipment • Common Application Service Element • Computer-Aided Software Engineering • Computer-Aided Support Equipment • Computer-Aided Systems Engineering • Computer-Aided Systems Evaluation • Computer-Assisted Software Engineering • Computer-Assisted Systems Engineering • Computer-Assisted Systems Evaluation • Computer-Automated Support Equipment • Confederation for the Advancement of State Education • Cooperative Award in Science and Engineering • Coordinating Agency for Supplier Evaluation • Correlation of Alpine Structural Events • Council for Advancement and Support of Education • Council for Advancement of Secondary Education (US)

CASEA: Computer-Assisted System Engineering and Analysis

CASF: China Association of Science Foundations (China-PRC)

Cash: Cashier

CASH: Catalogue of Available and Standard Hardware

CASI: Canadian Aeronautics and Space Institute • Center for Aerospace Information • Compact Airborne Spectrographic Imager (*) • Compact Airborne Spectrographic Imagery • Compact Airborne Spectrographic Instrument • Computer-Assisted Self-Interviewing

CASIA: Center for Analysis of Satellite Interference with Astronomy

CASIC: Computer-Assisted Survey Information Collection

CASID: Committee for Aviation and Space Industry Development (China-ROC)

CASIMIR: Catalyst Studies for Industry through Microgravity Research

CASIN: Centre for Applied Studies in International Negotiations

CASIS: Center for the Advanced Study of Information Systems

CASJ: Canadian Aeronautics and Space Journal (periodical code)

CASJQ: Canadian Aeronautics and Space Journal Quarterly (periodical code)

CASL: Calibration Service Laboratory • California Association of School Libraries (USA-CA) • Crosstalk Application Scripting Language

CASLEO: Complejo Astronómico El Leoncito (Argentina-SJ)

CASLIN: Czech And Slovak Library Information Network

CASLIS: Canadian Association of Special Libraries and Information Services

CASO: Cancellation Addendum Sales Order

CASOAR: Chambre Adaptée à la Simulation à l'Oxygène Atomique pour les Revêtements

CASP: Cambridge Arctic Shelf Programme • Capability Support Plan • Center for Advanced Space Propulsion • Center for Applied Solar Physics • Center for Astronomy and Space Physics • Change Assessment Summary Package • Computer-Assisted Search Planning • Crew Activities Scheduling Program

CASPAR: Cambridge Analog Simulator for Predicting Atomic Reactions

CASPEC: Cassegrain Echelle Spectrograph

CASPIA: Computer-Assisted System for the Processing of International Applications

CASPIR: Cryogenic Array Spectrometer for the Infrared • Cryogenic Array Spectrometer/Imager

CASR: Cost and Schedule Reporting System

CASRS: Countdown and Status Receiving Station • Countdown and Status Receiving System

CASS: Center for Advanced Space Studies • Center for Astronomy and Space Sciences • Center for Astrophysics and Space Sciences • Center for Atmospheric and Space Sciences • Center for Automated Space Science • Chinese Academy of Social Sciences (China-PRC) • Coarse Analog Sun Sensor • Computer-Assisted Search Service • Computerized Algorithmic Satellite Scheduler • Conditional Access Subsystem • Conférence des Académies Scientifiques Suisses (Switzerland) • Consolidated Automated Support System • Copyright Accounting and Settlement System

CASSAD: Centre for African Settlement Studies and Development

CASSI: Chemical Abstracts Service Source Index

CASSIOPE: Computer-Aided System for Scheduling, Information and Operation of Public transport in Europe

CASSIS: Classification and Search Support Information System

CASSM: Content Addressable Segment Sequential Memory

CASST: Commercial Aviation Safety Strategy Team

CASSY: Cockpit Assistant System

Cast: Castellano • Castle • Castilian • Castilla (Spain)

CAST: Castelo • Casting And Solidification Technology • Center for Application of Sciences and Technology • Centre d'Actualisation Scientifique et Technique (France) • China Association for Science and Technology • Chinese Academy of Space Technology • Common Access Security Terminal • Computer-Aided Software Testing • Computer-Aided Structural Technology • Computer-Assisted Scanning Technique • Computer-Assisted Self-Training • Computer-Assisted Software Testing • Computerized Automatic Systems Tester • Cooperation in Applied Science and Technology

CASTD: Committee on Application of Science and Technology to Development

CASTI: Canter for the Analysis of Science and Technical Information

CASTLe: CfA-Arizona Space Telescope Lens (survey)

CASTLES: CfA-Arizona Space Telescope Lens Survey

CASTNET: Clear Air Status and Trends Network

CASTOR: Chaîne d'Acquisition, de Surveillance et de Traitement des Observations Radar

CASTS: Countdown And Status Transmission System

CASU: Conseiller d'Administration Scolaire et Universitaire (France) • Consolidated Administrative Support Unit

CASW: Council for the Advancement of Scientific Writing

Cat: Catalán • Catalog • Catalogue • Cataluña • Cataract • Catégorie • Category • Catena • Catenate

CAT: Calling Attention To • Carburetor Air Temperature • Catalogue • Category • Category Alias Table • Cat Island (Bahamas) (location identifier) • Celestial Atomic Trajectory • Centralized Automatic Testing • Centre for Advanced Technology • Centre for Alternative Technology • Centro de Astrofísica Teórica • Čerenkov Array at Themis • Čerenkov Atmospheric Telescope (*) • Civil Air Transport • Classical Analytic Technique • Cleanup And Treatment • Clear-Air Turbulence • Close Approach Tool • Club Astronomique Tarbais (France) • College of Advanced Technology • Columbus Automation Testbed • Comisión para Asignación de Tiempo • Comité de Asignación de Tiempos • Comité para la Asignación del Tiempo • Common Assessment Task • Common Authentication Technology • Communications And Tracking • Compile And Test • Complementary Analysis Team • Composition Automatique de Textes • Comprehensive Analysis Tool • Computed Axial Tomography • Computer-Aided Teaching • Computer-Aided Technology • Computer-Aided Telescope • Computer-Aided Test • Computer-Aided Testing • Computer-Aided Tomography • Computer-Aided Trading • Computer-Aided Training • Computer-Aided Transcription • Computer-Aided Translation • Computer-Aided Typesetting • Computer-Assisted Teaching • Computer-Assisted Technology • Computer-Assisted Telescope • Computer-Assisted Testing • Computer-Assisted Thermography • Computer-Assisted Tomography • Computer-Assisted Trading • Computer-Assisted Training • Computer-Assisted Transcription • Computer-Assisted Translation • Computer-Assisted Typesetting • Computerized Axial Tomography (*) • Computer of Average Transients • Concatenate • Consolidated Atomic Time • Control and Assessment Team • Control Attenuator Timer • Coordinated Analysis of the Thermosphere • Cosmic Anisotropy Telescope • Coudé Auxiliary Telescope • Craft Access Terminal • Crisis Action Team

Catal: Catalan

CATAM: Computer-Aided Teaching of Applied Mathematics

CATCH: Couplage avec l'Atmosphère en Conditions Hivernales

CATE: Computer-Aided Test Equipment • Contrat d'Aménagement du Temps de l'Enfant

CATEG: Category

Cath: Cathedral • Cathode

CATI: Colorado Advanced Technical Institute (USA-CO) • Colorado Advanced Technology Institute (USA-CO) • Computer-Assisted Telephone Interviewing

CATIA: Cabine Téléphonique Intégrée à l'Architecture • Computer-Aided Three-Dimensional Interactive Application

CATIE: Centro Agronomico Tropical de Investigación y Enseñanza

CATIES: Common Aperture Technique for Imaging Electro-Optical Sensors

CATIS: Common Applications and Tools Integration Services

CATLAS: Centralized Automatic Trouble Locating and Analysis System

CATM: Cashless Automatic Teller Machine

CATNIP: Computer-Assisted Technique for Numerical Indexing Purposes

Catóp: Catóptrica

Catópr: Catóptrica

CATP: Computer-Aided Typesetting Process

CATR: Compact Antenna Test Range

CATS: Change Action Tracking System • Cheap Access To Space • Chicago/Argonne Terabyte System • Committee on Applied and Theoretical Statistics • Communications And Tracking System • Comprehensive Analytical Test System • Computer-Accessed Telemetry System • Computer-Aided Teaching System • Computer-Aided Telemetry System • Computer-Aided Theatrical Store • Computer-Aided Troubleshooting • Computer-Assisted Test Shop • Computer-Assisted Test System • Computer-Assisted Trading System • Computer-Assisted Training System • Computer-Assisted Troubleshooting System • Computer-Automated Test System • Computing And Telecommunications Services • Conference And Travel Services • Conformance Abstract Testing Suite • Correspondence Action Tracking System • Credit Accumulation Transfer Scheme

CATSAT: Cooperative Astrophysics and Technology Satellite

CATT: Center for Advanced Technology in Telecommunications • Centre d'Accueil et de Traitement Téléphonique • Combined Arms Tactical Trainer

CATV: Cable And Television • Cable Television • Common Antenna Television • Community Access Television • Community Antenna Television

CATVS: Community Antenna Television System

CATWG: Calibration Target Working Group

CAU: Calculatrice Arithmétique Universelle • Caruaru (Brazil-PE) (location identifier) • Coarse Alignment Unit • Command Acquisition Unit • Command Activation Unit • CORDIC Angle Unit • Command Acquisition Unit • Command Activation Unit • Complex Arithmetic Unit • Control Arithmetic Unit • Controlled Access Unit • Converter Amplifier Unit • CPU Access Unit • Cryptographic Ancillary Unit • Customer Acquisition Unit

CAUB: Common Area User Block

CAUCA: Código Aduanero Uniforme Centroamericano

CAUCE: Canadian Association for University Continuing Education • Coalition Against Unsollicited Commercial E-mail

CAUL: Council of Australian University Libraries

CAUP: Canadian Association of University Presses • Centro de Astrofísica da Universidade do Porto (Portugal)

CAURA: Canadian Association of University Research Administrators

Caus: Causation • Causative

CAUSE: College And University System Exchange

CAUT: Canadian Association of University Teachers • Caution

CAUV: Centro de Astronomia Ulisses Villela (Brazil)

Cav: Cavalier • Cavaliere • Caveat • Cavity

CAV: Cazombo (Angola) (location identifier) • Circolo Astrofili Veronesi (Italy) • Composite Analog Video • Computer-Aided Verification • Constant Air Volume • Constant Angular Velocity

CAVAD: Cambridge (Starlink internal code)

CAVAL: Cavaleiro

CAVE: CAVE Automatic Virtual Environment • Compañía Aérea de Viajes Expresos SA • Computer Access for Vision and Employment • Computer-Augmented Video Education • Computer-Automated Virtual Environment

CAVERN: Computer Access for Vision and Employment Research Network

CAVI: Centre Audiovisuel International

CAVIAR: Computer-Aided Visitor Information and Retrieval

CAVM: Circumpolar Arctic Vegetation Mapping

CAVOK: Cloud And Visibility OK

CAVORT: Coherent Acceleration and Velocity Observations in Real Time

CAVT: Constant Absolute Vorticity Trajectory

CAVU: Ceiling And Visibility Unlimited • Clear And Visibility Unlimited

CAW: Campos - Bartolomeu Lisandro (Brazil-RJ) (location identifier) • Channel Address Word • Club Astro de Wittelsheim (France)

CAWG: Coaxial Adapter Waveguide • Customer Accommodation Working Group

CAWS: Commercial Analyst Workstation • Common Aviation Weather System

CAX: Carlisle (UK-England) (location identifier) • Community Automatic Exchange

Cay: Cayenne (French Guyana) • Cayman Islands

CAY: Cayenne (French Guyana) (location identifier) • Centre Astronomique des Yvelines (France) • Centro Astronómico de Yebes (Spain)

CAZ: Cobar (Australia-NSW) (location identifier)

Cb: Columbium • Cumulonimbus

C/B: Circuit Breaker

CB: Call Back • Cambridge (UK postal code) • Cantilever Beam • Cape Breton (Canada) • Carte Bancaire • Carte Blanche • Carte Bleue • Cash Book • Cataclysmic Binary • Census Bureau • Center of Buoyancy • Central Bureau • Change Board • Change Bulletin • Character-Based • Chemical and Biological • Chlorobromomethane • Chubut (Argentina) • Circuit Breaker • Citizens' Band • Column Builder • Common Base • Common Battery • Common Bench • Comparator Buffer • Compass Bearing • Component Broker • Condition Bit • Conduction Band • Configuration Baseline • Constant Bandwidth • Continental Breakfast • Continuous Breakdown • Control Board • Control Break • Cost Benefit • County Borough • Covalent Bonding • Crystal Ball • Currency Bond

CBA: Center for Backyard Astrophysics • Center for Basement Astrophysics • Central Baffle Assembly • Collective Bargaining Agreement • Colliding Beam Accelerator • Commonwealth Broadcasting Association • Corner Bay (USA-AK) (location identifier) • Cosmic Background Anisotropy • Cosmological Baryon Asymmetry • Cost Benefit Analysis

CBAA: Comité Belge des Astronomes Amateurs (Belgium)

CBAIPI: CMM-Based Appraisal for Internet Process Improvement

Cbal: Counterbalance

CBAM: Content-Based Access to Multimedia

CBARS: Classified Baseline Accounting and Reporting System

CBAS: Cape Breton Astronomical Society (Canada-NS)

CBASSE: Commission on Behavioral and Social Sciences and Education

CBAT: Central Bureau for Astronomical Telegrams (USA-MA) • College Board Achievement Test

CBB: Cluster-Based Browsing • Cochabamba (Bolivia) (location identifier) • Commission Belge de Bibliographie

(Belgium) • Consumer Buying Behavior • Contract Budget Baseline

CBBS: Computer-Based Bibliographic Search • Computerized Bulletin Board System

CBC: Canadian Broadcasting Corporation • Caribbean Broadcasting Co. • Change Board Commitment • Cherrabun (Australia-WA) (location identifier) • Cipher Block Chaining • Closed Baryton Cycle • Complete Blood Count • County Borough Council • Current Book Contents

CBCR: Channel Byte Count Register

CBCS: Commonwealth Bureau of Census and Statistics • Computer-Based Conference System

C-BD: C-Band

CBD: Car Nicobar (India) (location identifier) • Cash Before Delivery • C-Band • Center of Business District • Central Business District • Centrale Bibliotheekdienst (Netherlands) • Centre de Bibliographie et Documentation • Chemical Bath Deposition • Chemical-Biological Defense • Climatologie - Base de Données • Commerce Business Daily • Compton Backscatter Densitometer • Constant Bit Density • Convention on Biological Diversity • Crossed-Beam Display • Cypher Block Chaining

CBDE: Carbonated Beverage Dispenser Evaluation

CBDS: Calibration Database System • Centre Belge de Documentation Scientifique (Belgium) • Connectionless Broadband Data Service

CBE: Chemical Beam Epitaxy • Chemical-Biological-Environmental • Coherent Backscatter Enhancement • Collisionless Boltzmann Equation • Companion of the British Empire • Computer-Based Education • Connector Bracket Experiment • Convention sur le Brevet Européen • Costs, Budgeting and Economics • Council for Basic Education • Cumberland (USA-MD) (location identifier)

CBEC: Canadian Book Exchange Centre

CBECI: Centre Belge des Échanges Culturels Internationaux (Belgium)

CБED: Convergent-Beam Electron Diffraction

CBEL: Cambridge Bibliography of English Literature

CBEM: Computer-Based Educational Material

CBEMA: Computer and Business Equipment Manufacturers Association

CBEP: Community-Based Environmental Protection

CBER: Center for Biologics Evaluation and Research

CBERS: China-Brazil Earth Resources Satellite • Chinese-Brazilian Earth Resource Satellite

CBET: Computer-Based Education and Training

CBEVE: Central Bureau for Educational Visits and Exchanges

CBF: Central Board of Finance • Cerebral Blood Flow • Cold Brittle Point • Council Bluffs (USA-IA) (location identifier) • Current Basis Function

CBG: Cambridge (UK-England) (location identifier)

CBGA: Carpathian Balkan Geological Association

CBH: Bechar (Algeria) (location identifier)

CBI: Cape Breton Island • Caribbean Basic Initiative • Center for Biological Informatics • Center for Business Information • Central Bureau of Identification • Commission Baleinière Internationale • Competency-Based Instruction • Complementary Binary • Complete Background Investigation • Compound Batch Identification • Computer-Based Information • Computer-Based Instruction • Confederation of British Industries • Confidential Business Information • Cosmic Background Imager • Cumulative Book Index

CBIL: Common and Bulk Items List

CBIOS: Compatibility Basic Input/Output System

CBIPO: Custom Build Installation Productivity Option

CBIQS: Content-Based Indexing and Querying System

CBIR: Conten-Based Information Retrieval

CBIRF: Chemical and Biological Incident Response Force

CBIS: Central Bureau Information System • Centre de Rectification des Images Spatiales • Computer-Based Information System

CBIV: Computer-Based Interactive Videodisk

CBJ: Cabo Rojo (Dominican Rep.) (location identifier)

CBK: Checkbook • Colby (USA-KS) (location identifier)

Cbl: Cable

CBL: Calculator-Based Laboratory • Ciudad Bolivar (Venezuela) (location identifier) • Cloud Boundary Layer • COBOL • Commercial Bill of Lading • Computer-Based Learning • Convective Boundary Layer

CBLT: Character Block Transfer

CBM: Centre for Bibliographic Management • Coalbed Methane • Columbus (USA-MS) (location identifier) • Common Bathing Mechanism • Communications Buffer Memory • Condition-Based Maintenance • Conduction Band Mission • Confidence Building Measure

CBMIS: Computer-Based Management Information System

CBMS: Computer-Based Mail System • Computer-Based Medical System • Computer-Based Message System • Conference Board of Mathematical Sciences

CBMU: Current Bit Monitor Unit

cBN: Cubic Boron Nitride

CBN: Centre Bibliographique National • Cirebon (Indonesia) (location identifier) • Computerized Bibliographic Network • Content-Based Navigation • Cooperative Base Network • Cubic Boron Nitride

CBNM: Central Bureau for Nuclear Measurements

Cbo: Colombo (Sri Lanka)

CBO: Community-Based Organization • Computer Burst Order • Congressional Budget Office • Cooperative Business Object • Cotabato (Philippines) (location identifier)

CBOE: Chicago Board of Options Exchange (USA-IL)

CB-OHS: Component-Based Open Hypermedia System

CBOHS: Component-Based Open Hypermedia System

CBOT: Chicago Board Of Trade (USA-IL)

CBP: Coimbra (Portugal) (location identifier) • Commission for Bibliography and Publications • Connector Bracket Power • Convergent Billing Platform • Cosmic Background Polarization

CBQ: Calabar (Nigeria) (location identifier) • Class-Based Queuing

CBR: Canberra (Australia-ACT) (location identifier) • Case-Based Reasoning • Centraal Bureau voor de Rijvaardigheidsbewijzen (Netherlands) • Change Board Register • Chemical, Bacteriological, and Radiation • Cloud Base Recorder • Code Coverage Analysis • Constant Bit Rate • Content-Based Retrieval • Cosmic Background Radiation • Crude Birth Rate

CBRS: Coherent Bremsstrahlung Radiation Spectrum • Computer-Based Reference Service

CBS: Cabimas (Venezuela) (location identifier) • Case Blue Stars • Cauchy-Buniakowskii-Schwarz • Centraal Bureau voor de Statistiek • Central Bibliographic Service • Central Bibliographic System • Central Broadcasting System • Central Bureau of Statistics • Close Binary System • Cluster-Based Search • Coherent Bremsstrahlung-radiation Spectrum • Columbia Broadcasting System • Commission for Basic Systems (WMO) • Common Basic Services • Community Base Station • Connector Bracket Signal • Contract and Business Services

CBSA: Cargo Bay Stowage Assembly

CBSS: Close-Binary Supersoft Source • Council of the Baltic Sea States

CBST: Current Bibliography on Science and Technology

CBT: Case Builder Tool • Catumbela (Angola) (location identifier) • Cesium Beam Tube • Computer-Based Training • Conduction-Band Tail

CBTR: Carrier Bit Timing Recovery

CBTS: Cesium Beam Time Standard

CBU: Cluster Bomb Unit • Completely Built-Up

CBV: Cabin Bleed Valve • Cerebral Blood Volume • Coban (Guatemala) (location identifier)

CBW: Chemical and Biological Warfare • Constant Bandwidth • Convert Byte to Word • Critical Bandwidth

CBX: Common Branch Extender • Company Branch Exchange • Computer-controlled Branch Exchange • Computerized Branch Exchange • Condobolin (Australia-NSW) (location identifier)

CBY: Canobie (Australia-QLD) (location identifier)

CBZ: Cabin Creek (USA-AK) (location identifier) • Coastal Boundary Zone

c/c: Compte Courant • Cuenta Corriente

cc: Carbon Copy • Centuries • Chapters • Cicero • Copies • Crosscouple • Cubic Centimeter

Cc: Cirrocumulus

C&C: Candidate & Calendar • Command & Control • Communication & Control • Computer & Communications • Computing & Communications

C-C: Carbon-Carbon • Center-to-Center • Computer-to-Computer

C/C: Carbon/Carbon • Combustion Chamber • Compte Courant • Configuration Control • Conto Corrente

CC: Calçada • Calibration Curve • Carbon-Carbon • Carbon Copy • Carriage Control • Cartographic Catalog • Casa de Campo • Cascade Correlation • Cash Credit • Cashier's Check • Casilla de Correo • Caucasus • Central Commission • Central Committee • Central Computer • Central Console • Central Control • Centre de Calcul • Cepstrum Coefficient • Chamber of Commerce • Change Course • Channel Controller • Charged Current • Chart Convergency • Cheap Chip • Chief Clark • Chile (international aircraft registration code) • Chronometer Correction • Chonometric Cosmology • Circuit Court • Circulating Copy • City Centre • City Council • City Councilor • Civil Court • Classification Code • Close Control • Close-Coupled • Close Coupling • Closed Circuit • Closing Coil • Cloud to Cloud • Cluster Controller • Cocos (Keeling) Islands (ISO country code) • Code Civil • Code Control • Code Converter • Código Civil • Coefficient of Contingency • Coefficient of Correlation • Collaborative Coefficient • Collision Course • Color Center • Color Code • Color Compensation • Color Contrast • Color Conversion • Color Correction • Column Chromatography • Combined Cycle • Combustion Chamber • Comité Central • Comité de Coordination • Command and Control • Command Computer • Command Conference • Commercial Control • Common Collector • Common Control • Communication and Control • Communication Center • Communications Control • Community Council • Compact Companion • Compass Course • Compatibility Change • Compensating Current • Compilation Catalogue • Complete Controllability • Compte Courant • Computation Center • Computer Calculator • Computer Code • Computer Complex • Concave • Conceptual Clustering • Concurrency Control • Condition Code • Configuration Control • Constant Current • Consular Corps • Contamination Control • Continuation Clause • Continuous Current • Conto Corrente • Contra Credit • Contract Completion • Control Center • Control Computer • Control Console • Conversion Complete • Coordinate Converter • Coordinating Center • Coordination Center • Copie Conforme • Corner Cube • Corps Consulaire • Correction Coil • Corriente Continua • Cost Center • County Clerk • County Commissioner • County Council • County Court • Country Code • Country to Country • Coupled Channel • Courant Continu • Courtesy Car • Credit Card • Crew Certified • Crew Compartment • Cross Correlation • Current Contents (periodical) • Current Cost • Cursor Control • Cushion Craft • Cusp Cap • Îles Cocos Keeling (ISO country code) • Keeling Islands (Cocos_) (ISO country code)

CCA: Cadre Communautaire d'Appui • Call Control Agent • Canadian Cartographic Association • Canonical Circulation Analysis • CENELEC Certification Agreement • Center for Computing and Automation • Centre Culturel de l'Astronomie (France) • Centre de Calcul Analogique • Chromated Copper Arsenate • Circuit Card Assembly • Circuit Court of Appeals • Cloud Chamber Analysis • Cluster-Cluster Agglomeration • Collaborative Compression Architecture • Comité de Coordination des Actions • Commission on Crystallographic Apparatus • Common Cryptographic Architecture • Communications Carrier Assembly • Composite Cell Address • Computer Corporation of America • Configuration Control Article • Consumers' Cooperative Association • Continental Control Area • Continuum Computer Architecture • Contract Change Authority • Contract Change Authorization • Coolant Control Assembly • County Court of Appeals • Cryogenic Cooler Assembly • Current Cost Accounting • Customer Control Access • Fort Chaffee (USA-AR) (location identifier)

CCAA: Center for Computer-Aided Analysis • Common Cabin Air Assembly

CCAB: Canadian Circulations Audit Board • Consultative Committee of Accounting Bodies

CCAD: Central American Commission on the Environment and Development

CCAF: Circolo Culturale di Farra d'Isonzo (Italy)

CCAFS: Cape Canaveral Air Force Station (USA-FL)

CCAM: Canadian Congress of Applied Mechanics • Commission for Climatology and Applications of Meteorology • Conversational Communication Access Method

CCAMLR: Commission for Conservation of Antarctic Marine Living Resources • Convention for Conservation of Antarctic Marine Living Resources

CCAP: Cable Coupling Analysis Program • Coastal Change Analysis Program • Commercial Complex Autonomous Payload • Competitive Cooperative Agreement Program

CCAPS: Circuit Card Assembly and Processing System

CCAQ: Consultative Committee on Administrative Questions

CCAS: Cape Canaveral Air Station (USA-FL) • Cape Cod Astronomical Society (USA-MA) • Center of Curvature Alignment Sensor • Chester County Astronomical Society (USA-PA) • Cleveland County Astronomical Society (USA-NC) • Convention for the Conservation of Arctic Seals

CCAT: Centre de Coordination et d'Appui Technique

CCATS: Command, Communications And Telemetry Systems • Communications, Command And Telemetry Systems

CCASTD: Comité Consultatif sur l'Application de la Science et de la Technique au Développement

CCAT: Comité de Coordination de l'Assistance Technique (ONU)

CCB: Catalogue de Compilation Bibliographique • Change Control Board • Civil Cooperation Bureau • Clock Control Board • Clock Correction Board • Collective Catalogue Belgium • Command Control Block • Configuration Change Board • Configuration Control Board • Contract Change Board • Upland Cable (USA-CA) (location identifier)

CCBD: Change Control Board Directive • Configuration Control Board Directive

CCBS: Change Control Board Summary • Committee on Capacity Building in Science

CCBT: Centre de Coordination Bibliographique et Technique

CCBX: Computer-Controlled Branch Exchange

CCC: Canadian Chamber of Commerce • Canadian Climate Centre • Canadian Committee on Cataloging • Cape Communications Control (NASA) • Carriage Control Character • Catalogue de Compilation Critique • Cayo Coco (Cuba) (location identifier) • CCATS Command Controller • CEN Certification Committee • Central Classification Committee • Central Computer Complex • Central Control Commission • Central Control Complex • Change and Configuration Control • Change Current Controller • Chemical Contamination Control • Chronological Coverage Code • Civilian Conservation Corps (USA) • Cold and Climate Change • Comité de Certification du CEN • Command Control Console • Communication Center Console • Communications Command Center • Communications Control Center • Community of Common Concern • Complex Control Center • Computer Checkout Console • Computer Command Control • Computer Communications Console • Computer Communications Converter • Computer Control Center • Computer Control Communication • Computer Control Complex • Concourse Computer Center • Conseil Consultatif Conjoint • Conseil de Coopération Culturelle • Contaminant Control Cartridge • Controller Checkout Console • Coordinate Conversion Computer • Copyright Clearance Center (USA) • Correlator Control Computer • Countercurrent Chromatography • Crew Coordination Concept • Critical Compilation Catalogue • Cryogenic Current Comparator • Cusp Cap Collar • Customs Cooperation Council • Cyclic Check Character

CCCB: Configuration Change Control Board

CCCC: Computerized Conferencing and Communications Center (USA-NJ)

CCCE: Closed-Cycle Cryogenic Equipment

CCCD: Crew Compartment Configuration Drawing

CCCF: Central Committee on Communications Facilities • Chaos Computer Club de France

CCCG: Canadian Conference on Computational Geometry

CCCGH: Centro Científico y Cultural de Geografía e Historia (Ecuador) (now CCIGM)

CCCH: Central Computer Complex Host

CCCI: Command-Control-Communications-Intelligence

CCCII: Chinese Character Code For Information Exchange

CCCL: Complementary Constant Current Logic

CCCM: Canadian Climate Cenre Model • Centro de Control de Calidad y Metrología • Council Committee on Consumer Matters

CCCN: CDIAC Computing System Network

CCCNT: Conseil Consultatif Canadien sur les Normes de Télécommunications

CCCO: Committee on Climate Changes and the Ocean • Committee on Climatic Changes and the Ocean (*) • Corning Community College Observatory

CCCR: Communication and Control Command Requirement

CCCS: Central Control Computer System • Common Core Character Set • Conseil Canadien des Centres des Sciences

CCCU: Crew Compartment Cooling Unit

CCCUNY: City College of City University of New York (USA-NY)

CC-D: Current Contents on Diskette

CCD: Calcite Compensation Depth • Calibration Curve Data • Cambridge Crystallographic Database (UK) • Camouflage, Concealment and Deception • Census County Division • Central Command Decoder • Central Computing Department • Charge-Coupled Detector • Charge-Coupled Device (*) • Checkout Command Decoder • Cold-Cloud Duration • Color-Color Diagram • Computer Center Division • Computer-Controlled Display • Configuration Change Directive • Conseil de Coopération Douanière • Constant Change Display • Construction Completion Date • Consumer Computer Device • Control Center DORIS • Current Contents on Diskette • Los Angeles - Century City (USA-CA) (location identifier)

CCDA: Centre Canadien de Données Astronomiques (Canada-BC) • Commercial Chemical Development Association • Commission pour la Coordination de la Documentation Administrative (France)

CCDB: Carbon Content of Dead Biomass • Committee on Chemical Databases • Common Core Database in Biotechnology

CCDC: Cambridge Crystallographic Data Centre • Central Control and Display Console

CCDF: Complementary Cumulative Distribution Function

CCDIS: Crustal Dynamics Data Information Station

CCDM: Catalogue des Composantes de Doubles et Multiples • Comité Consultatif pour la Définition du Mètre • Consultative Committee on the Definition of the Meter

CCDP: Climate Change Detection Project

CCDR: Contractor Cost Data Reporting • Contractor Crtical Design Review

CCDS: Center for the Commercial Development of Space • Comité Consultatif pour la Définition de la Seconde • Consultative Committee for the Definition of the Second

CCE: Centre de Calcul Électronique • Centro de Calculo Electrónico • Centro de Computação Electrônica • Change Compositive Explorer • Charge Collection Efficiency • Charge Composition Explorer • Chartered Civil Engineer • Chesler-Cram Equation • Comité Central d'Entreprise • Comité Consultatif d'Électricité • Command Control Equipment • Commission des Communautés Européennes • Communications Control Equipment • Compound-Cycle Engine • Computer Command Engineer • Conseil des Communes d'Europe • Continuous Control Evaluator • Corporate Component Engineering • Cryocooler Electronics • Current-Carrying Element • Saint-Martin - Grand Case (French Antilles) (location identifier)

CCEE: Countries of Central and Eastern Europe

CCEI: Composite Cost Effectiveness Index • Coordinating Committee on Engineering Information (USA)

CCEL: Centre Canadian de l'Échange du Livre (Canada)

CCEMRI: Comité Consultatif pour les Étalons de Mesure des Rayonnements Ionisants

CCES: Cold-Cathode Emission Spectroscopy

CCETT: Centre Commun d'Études de Télédiffusion et de Télécommunicat ions (France)

CCEU: Council on Continuing Education Units (USA)

ccf: Hundred Cubic Feet

CCF: Carcassonne Salvaza (France) (location identifier) • Catalogue Collectif de France • Central Computer Facility • Central Computing Facility • Central Control Facility • Chondrite Carbonée Fission • Common Cause Failure • Common Communication Format • Configuration Control Form • Converter Compressor Facility • Cross-Correlation Function • Cumulative Count Function • Current Calibration File • Custom Control Factory

CCFA: Capillary and Connective Flow Apparatus

CCFD: Complementary Cumulative Frequency Distribution

CCFE: Coordinating Committee on Fusion Energy

CCFF: Cape Canaveral Forecast Facility (USA-FL) • Crew Compartment Fit and Function

CCFL: Counter-Current Flow Limit

CCFM: Centre Canadien de Fusion Magnétique (Canada) • Cryogenic Continuous Film Memory

CCFP: Consultative Committee for the Fusion Programme

CCFT: Captive Carry Flight Test • Cold-Cathode Fluorescent Tube • Controlled Current Feedback Transformer

CCG: Carl-Cranz-Gesellschaft (Germany) • Cartesian Coordinate Grid • Catalogue de Compilation Général • Central Cluster Galaxy • Centre Canadien de Géomatique (Canada) • Certification Consultative Group • Change Control Group • Computer Coordination Group • Constant Current Generator • Country Commercial Guide • Crane (USA-TX) (location identifier)

CCGA: Communications Control Group Assembly

CCGCR: Centre for Climate and Global Change Research (Canada)

CCGE: Canadian Council on Geographic Education

CCGS: Canadian Coast Guard College (Canada-NS)

CCGT: Combined Cycle Gas Turbine

CCGTM: Commonwealth Consultative Group on Technology Management

CCH: Chile Chico (Chile) (location identifier) • Colegio de Ciencias y Humanidades • Commercial Clearinghouse • Connections per Circuit per Hour • Cubic Capacity of Holds

CCHP: Constant-Conductance Heat Pipe

CCHS: Crew Compartment Heat Shield

CCI: Canadian Copyright Institute • Card Computer Interface • Centre de Calcul Interfacultaire • Centro di Calcolo Interfacolta • Chambre de Commerce et d'Industrie • Chambre de Commerce Internationale • Charge Conjugation Invariance • Charge-Coupled Imager • Client Communication Interface • Comissão Central de Informática • Comité Científico Internacional • Commercial and Customer Intelligence • Common Client Interface • Component Cost Index • Computer Carrier Interrupt • Concôrdia (Brazil-SC) (location identifier) • Contract Change Identification • Cost Control Item

CCIA: Comité de Coordination pour l'Information et l'Automatisme • Computer and Communications Industry Association (USA) • Console Computer Interface Adapter

CCIC: Cold Climatic Information Center (USA) • Committee on Computing, Information and Communications

CCID: Comité Consultatif pour l'Information et la Documentation • Communities Committee for Information and Documentation (EU)

CCIF: Centre de Calcul Interfacultaire

CCIFQ: Centre de Coopération Interuniversitaire Franco-Québecoise

CCIG: Centre Canadien d'Information sur les Géosciences • Cold Cathode Ion Gauge

CCIGM: Centro Cultural del Instituto Geográfico Militar (Ecuador) (formerly CCCGH)

CCIM: Command Computer Input Multiplexer

CCINU: Comité Commun de l'Information des Nations Unies (ONU)

CCIP: Chambre de Commerce et d'Industrie de Paris (France)

CCIR: Center for Communication and Information Research • Comité Consultatif International des Radiocommunications • Consultative Committee on International Radiocommunications

CCIRN: Coordinating Committee for Intercontinental Research Networking • Coordinating Committee for Intercontinental Research Networks (*)

CCIS: Center for Computer and Information Services • Command and Control Information System • Command, Control, and Information System • Command Control Information System • Common Channel Interoffice Signaling • Computer-Controlled Interconnect System • Coordinating Committee for Information Services

CCISD: Centre de Coopération Internationale en Santé et en Développement

CCIST: Commission Consultative sur l'Information Scientifique et Technologique (Canada)

CCIT: Cable Communications for Information Transfer

CCITT: Comité Consultatif International Télégraphique et Téléphoniq ue (now ITU) • Consultative Committee for International Telegraph and Telephone • Consultative Committee on International Telegraph and Telephone (*) • Consultative Committee on International Telegraphy and Telephony

CCIW: Canadian Centre for Inland Waters

CCJ: Calicut (India) (location identifier)

CCK: Cell Culture Kit • Cocos Islands (location identifier) • Cocos (Keeling) Islands (ISO country code) • Communications Commission of Kenya • Îles Cocos Keeling (ISO country code) • Keeling Islands (Cocos_) (ISO country code)

CCl: Commission for Climatology (WMO)

CCL: Calibration Control Language • Cambridge Consultants Ltd. • Canadian Circumpolar Library • Chinchilla (Australia-QLD) (location identifier) • Closed-Circuit Loop • Closed Conference Loop • Command Connection Language • Commonality Candidate List • Common Command Language • Communications Command Language • Communications Control Language • Composite Cell Logic • Computer Control Loading • Configuration Control Logic • Connection Control Language • Constraint in Computational Logics • Control Center Line • Council for the Central Laboratory • Coupled Cavity LINAC • Critical Characteristics List • Critical Components List

CCLB: Carbon Content of Living Biomass

CCLI: Course, Curriculum and Laboratory Improvement

CCLKWS: Counterclockwise

CCLN: Council for Computerized Library Networks (USA)

CCLO: Coordinating Council of Library Organizations (USA)

CCLRC: Council for the Central Laboratory of the Research Councils (UK)

CCls: Court of Claims

CCLS: Computer-Controlled Launch Set

CCM: Cell Culture Module • Central Customer Manager • Centre de Contrôle et de Mission • Centrum voor Constructie en Mechatronica (Netherlands) • Chromatographie en Couches Minces • Climate Community Model • Comité Consultatif pour la Masse • Commission Centrale des Marchés • Committee on Challenges of a Modern Society • Communications Controller Multichannel • Community Climate Model • Company-Conformed Name • Computer Control Module • Consejo Cultural Mundial • Constant Current Modulation • Continuous Core Measurements • Controlled Carrier Modulation • Cost Centre Manager • Counter-Countermeasure • County College of Morris (USA-NJ) • Crew Cargo Module • Crisciúma (Brazil-SC) (location identifier) • Current Confinement Model

CCMA: Comité de Compradores de Material Aeronáutico • Converging Computing Methodologies in Astronomy (ESF)

CCMC: Community Coordinated Modelling Center

CCMD: Canadian Centre for Management Development • Continuous-Current Modulating Device

CCME: Contract Change Mass Estimate

CCMI: Comité Consultatif sur les Météorites et les Impacts (Canada)

CCMM: Centre Contre les Manipulations Mentales (France)

CCMMP: Committee on Condensed-Matter and Materials Physics (BPA)

CCMP: Coordinating Committee on the Moon and Planets

CCMS: Capacity-Constrained Machine Scheduling • Checkout Control and Monitor System • Command Control and Monitor System • Commission on the Challenges of Modern Society • Committee on the Challenges of Modern

Society (*) • Committee on the Challenges to Modern Society • Computer Center Management System

CCMSS: Computer-Controlled Microfilm Search System

CCMU: Computer Controller Multiplexer Unit

CCN: Campus Computing Network • Catalogue Collectif National (France) • Center for Civic Networking • Central Command Network • Centre Climatologique National • Chakcharan (Afghanistan) (location identifier) • Cisco Communications Networks (now AVVID) • Classification Change Notice • Cloud Condensation Nucleus • Command Control Number • Common Communications Network • Community Care Network • Configuration Change Negotiation • Configuration Change Notice (*) • Configuration Control Number • Conseil Canadien des Normes • Contract Change Negotiation • Contract Change Notice • Contract Change Notification

CCNAA: Coordination Council for North American Affairs

CCNC: Common Channel Network Controller • Computer-Communications Network Center

CCNDT: Canadian Council for Non-Destructive Technology

CCNH: Calçadinha

CCNNU: Comité de Coordination et de Normalisation des Nations Unies (ONU)

CCNPS: Catalogue Collectif National des Publications en Série

CCNR: Consultative Committee for Nuclear Research

CCNS: Computer-Controlled Navigation System

CCNT: Centre de Contrôle Nacelle-Télescope

CCNU: Chloroethylcyclohexylnitrosourea

CCNUMA: Cache Coherent Non-Uniform Memory Access

CCNY: City College of New York (USA-NY)

CC&O: Customer Cost & Obligation

CCO: Campus Computing Organization • Carimagua (Colombia) (location identifier) • Catalogue Collectif des Ouvrages • Central Coding Office • Computational Cosmology Observatory • Configuration Change Order • Contract Change Order • Contrôle-Commande Opérationnelle • Current-Controlled Oscillator • Customer Contracting Officer • Customer Cost and Obligation • Crystal-Controlled Oscillator

CCOD: Current Contents On Disquette

CCOE: Catalogue Collectif des Ouvrages Étrangers

CCOF: Complementary Columbus Orbital Facility

CCOH: Corrosive Contaminants, Oxygen, and Humidity

CCOHS: Canadian Centre for Occupational Health and Safety

CCOL: Coordinating Committee on the Ozone Layer

CCOM: Centro Comercial

CCOP: Carnot Coefficient Of Performance

CCORE: Centre for Cold Ocean Resources Engineering

CCOSM: Chemistry and Circulation Occultation Spectroscopy Mission

CCP: Canadian Climate Programme • Catalogue Collectif des Périodiques • Center Console Panel • Central Cluster Panel • Certificate in Computer Programming • Certified Computing Professional • Change Control Package • Change Control Panel • Charge Capacitance Probe • Civil Code of Procedure • Code of Civil Procedure • Collaborative Computational Project • Comisión del Canal de Panamá • Command Control Panel • Command Control Post • Commercial Change Proposal • Commission on Comparative Planetology • Communications Control Program • Communications Control Package • Communications Control Panel • Communications Control Processor • Community College of Philadelphia (USA-PA) • Compression Control Protocol • Compte Chèque Postal • Compte Courant Postal • Computer Control Panel • Concepción - Carriel Sur (Chile) (location identifier) • Conference on Computational Physics • Configuration Change Plan • Configuration Change Point • Configuration Control Panel • Configuration Control Phase • Configuration Control Point • Console Command Processor • Contamination Control Plan • Continuous Correlation Processing • Conto Corrente Postale • Contract Change Proposal • Control Computerizado de Procesos • Controlled Components Program • Coordinated Commentary Programming • Corrosion Control Plan • Cost Control Program • Court of Common Pleas • Cross-Check Procedure • Credit Card Purchase • Critical Compression Pressure • Cubic-Close Packed

CCPA: Cloud Chamber Photographic Analysis

CCPB: Catálogo Colectivo del Patrimonio Bibliográfico

CCPC: Communication Computer Programming Center • Control Center Programming Center

CCPDSR: Common Center Processing and Display System Replacement • Communications Center Processing and Display System Replacement

CCPE: Canadian Council of Professional Engineers

CCPEP: Center for Chemistry and Physics of the Earth and Planets (formerly CEPI)

CCPI: Consultative Committee on Public Information

CCPIT: China Council for the Promotion of International Trade (China-PRC)

CCPL: Computer Center Program Library

CCPOQ: Consultative Committee on Programme and Operational Questions

CCPP: Catálogo Colectivo de Publicações Periódicas

CCPR: Comité Consultatif de Photométrie et de Radiométrie

CCPS: Command/Control Pressure Suit

Ccpto: Concepto

CCQ: Cachoeira (Brazil-BA) (location identifier) • Cross-Crew Qualification

CCQJC: Christopher Columbus Quincentenary Jubilee Commission

CCQM: Comité Consultatif pour la Quantité de Matière

CCR: Cargo Compatibility Review • Cassette Camera Recorder • Center for Cell Research (USA-PA) • Central Control Room • Centre Commun de Recherche • Centre de Contrôle Régional • Centro Comune de Ricerche • Closed-Circuit Radio • Commission of Civil Rights • Commission on Civil Rights • Commitment, Concurrency and Recovery • Common Center of Research • Communicating Cells Ratio • Complex Chemical Reaction • Complex Control Room • Computer and Computation Research • Computer Character Recognition • Computer-Controlled Retrieval • Concord (USA-CA) (location identifier) • Configuration Change Request • Configuration Control Record • Console Card Reader • Consumer-Controlled Reconfiguration • Continuous Catalyst Regeneration • Contract Change Record • Contract Change Release • Contract Change Request • Contractor Change Request • Contractor Change Review • Contractor Cost Reduction • Control Center Rack • Conversation Control Record • Cube-Corner Reflector • Current Cell Rate

CCRA: Cape Canaveral Reference Atmosphere (USA-FL)

CCRB: Configuration Control Review Board

CCRE: Conseil des Communes et Régions d'Europe

CCRF: Consolidated Communication Recording Facility

CCRI: Comité Consultatif de Recherches en Informatique

CCRIA: Comité Consultatif de Recherches en Informatique et Automatique

CCRRDT: Comité Consultatif Régional de la Recherche et du Développement Technologique

CCRS: Canada Centre for Remote Sensing • Canadian Centre for Remote Sensing • Centre Canadien de Ressources Spatiales (Canada)

CCRST: Comité Consultatif de la Recherche Scientifique et Technique (France)

C&CS: Command & Control System

CCS: Calculus for Communicating Systems • Calculus of Communicating Systems • Camera Control System • Caracas - Simon Bolivar (Venezuela) (location identifier) • Central Checkout System • Central Command Schedule • Central Computer and Sequencer • Central Computer Services • Central Control Section • Central Control Software • Central Control Station • Central Control System • Centre de Calcul de Strasbourg (France) • Centre de Contrôle des Satellites • Centre de Contrôle Spécialisé • Centre du Cadran Solaire (France) • Centro de Control de Satélites • Charge Control System • Classified Catalogue System • Clock Comparator System • Collective Call Sign • Command and Communications System • Command and Control Segment • Command and Control System • Command Control System • Common-Channel Signaling • Common Command Set • Common Communications Support • Common Control Software • Complex Control Set • Computer Command System • Computer Consulting Service • Computer Core Segment • Computing Classification System • Computing Consulting Service • Concurrency Control Service • Confidential Cover Sheet • Configuration Control Secretariat • Console Communication System • Contamination Control System • Continuous Commercial Service • Continuous Composite Servo • Contract Control System • Contrôle, Commande et Servitude • Controlled Combustion System • Controlled Component Synthesis • Cool Carbon Stars • Cooperative Credit Society • Coördinatiecommissie Synchrotronstraling • Cryocooler System

CCSA: Centre de Calcul Scientifique de l'Armement • Common Control Switching Arrangement

CCSAIT: Colorado College Student Astronomy Instrument Team

CCSC: Canadian Council of Science Centres • Central Connecticut State College (USA-CT) • Coordinating Committee for Satellite Communications • Customer Support Call Center

CCSD: Command Communications Service Designator • Couple-Cluster Single and Double

CCSDS: Consultative Committee for Space Data Systems • Consultative Committee for Standard Data Services

CCSE: Comité pour la Compétitivité et la Sécurité Économique

CCSEM: Centro di Cultura Scientifica Ettore Majorana (Italy) • Computer-Controlled Scanning Electron Microscope • Computer-Controlled Scanning Electron Microscopy

CCSF: CALTECH Concurrent Supercomputing Facility (USA-CA) • City College of San Francisco (USA-CA)

CCSG: Computer Components and System Group

CCSI: California Commercial Spaceport Inc. (USA-CA)

CCSID: Coded Character Set Identification • Coded Character Set Identifier

CCSM: Cross Correlogram Spectral Matching

CCSN: Community College of Southern Nevada (USA-NV)

CCSO: Computing and Communicating Services Office

CCSP: Communications Concentrator Software Package

CCSQ: Consultative Committee on Substantive Questions

CCSR: Carbon-Carbon Survivable Radiator • Crew Compartment Stowage Review

CCSS: Control Center Standard Software • Coordination and Contract Summary Sheet

CCSSC: Coca-Cola Space Science Center (USA-GA)

CCST: Center for Computer Sciences and Technology • Centre Culturel Scientifique et Technique • Centre de Culture Scientifique et Technique • Coordination Committee for Science and Technology

CCSTG: Computer-Controlled Small Tool Polisher

CCSTI: Centre de Culture Scientifique, Technique et Industrielle

CCSU: Configuration Control and Sensing Unit

CCSV: Comité de Coordination du Service Volontaire (France)

CCT: Cauchy Convergence Test • Central Control Terminal • Centre Canadien de Télédétection (Canada) • Cerebral Circulation Time • Clean Coal Technology • Colonia Catriel (Argentina-RN) (location identifier) • Comité Consultatif de Thermométrie • Command/Control Transmitter • Common Customs Tariff • Complex Coordination Test • Computer-Compatible Tape • Computer-Controlled Typesetting • Constant Current Tokamak • Constant Current Transformer • Consultative Committee on Thermometry • Contact Charge Transfer • Continuous Cooling Transformation • Continuous-Current Tokamak • Crew Compartment Trainer • Crystal-Controlled Transmitter • Current Completion Time

CCTA: Central Computer and Telecommunications Agency • Commission de Coopération Technique pour l'Afrique • Computer Communications and Telecomms Agency • Coordinating Committee of Technical Assistance

CCTAC: Computer Communications Trouble Analysis Center

CCTCF: Communication Circuit Technical Control Facility

CCTF: Command and Control Test Facility

CCTG: Constant Concentration Tracer Gas

CCTI: Centre de Calcul et de Traitement de l'Information

CCTR: Cassette Compact Toroid Reactor • Chromosphere-Corona Transition Region

CCTS: Comité de Coordination des Télécommunications par Satellite

CCTV: Closed-Circuit Television

CCU: Calcutta - Dum Dum (India) (location identifier) • Camera Control Unit • Cell Culture Unit • Central Computer Unit • Central Control Unit • Charge Control Unit • Chart Comparison Unit • Cluster Control Unit • Comité Consultatif des Unités • Comité Consultatif des Universités • Communication and Consultation Unit • Communication Control Unit • Communications Carrier Umbilical • Communications Carrier Unit • Communications Coupling Unit • Computer Control Unit • Conversion Computer Unit • Convolutional Coding Unit • Coronary-Care Unit • Coupling Control Unit • Crewman Communications Umbilical • Crew Communications Unit • Crew Compartment Umbilical

CCUFB: Centre de Coopération Universitaire Franco-Bavarois

CCUMC: Consortium of College and University Media Centers

CCUS: Chamber of Commerce of the United States

CCV: Chamber Coolant Valve • Control-Configured Vehicle • Coolant Control Valve • Craig Cove (Vanuatu) (location identifier)

CCVA: Chamber Coolant Valve Actuator

CCVD: Combustion Chemical Vapor Deposition

CCVR: Centre de Calcul Vectoriel pour la Recherche

CCVS: COBOL Compiler Validation System

CCVT: Coupling Capacitor Voltage Transformer

CCW: Carbon Copy for Windows • Channel Command Word • Channel Control Word • Component Cooling Water • Continuous Composite WORM • Counterclockwise • Cowell (Australia-SA) (location identifier)

CCWCP: Coordinating Committee for the World Climate Programme

CCWS: Component Cooling Water System

CCX: Cáceres (Brazil-MT) (location identifier)

CCY: Charles City (USA-IA) (location identifier)

CCZ: Chub Bay (Bahamas) (location identifier) • Command and Control Zone

c/d: Carried Down • Certificate of Deposit

cd: Candela • Cum Dividendo

Cd: Cadmium • Card • Ciudad • Cord • Could

C&D: Cease & desist • Communication & Data • Control & Display

C/D: Certificate of Deposit • Code/Debug • Command/Data select • Construction/Development • Consular Declaration • Control/Data • Countdown • Cup/Disc • Customs Declaration

CD: Calendar Day • Calibration Device • Call Director • Canadian Dollar • Capacitive Discharge • Capacitor Diode • Card • Carried Down • Carrier Density • Carrier Detect • Carrier Detection • Cash Discount • Center Director • Center Distance • Central Detector • Central District • Certificate of Deposit • Change Directive • Change Directory • Chapter Director • Chart Distance • Chemiluminescence Detection • Chemistry Division • Chilldown • Chromatic Dispersion • Circuit Description • Circuit Diagram • Circular Dichroism • Circular Dispersion • Civil Defense • Classification Décimale • Classification of Defects • Clearance Delivery • Climatological Data • Clock Driver • Closing Date • Cloud Detection • Cluster-Directed • Coefficient of Drag • Cold Drawn • Collaborative Document • College Diploma • Collision Detection • Color Display • Comité de Dirección • Command Decoder • Committee Draft • Common Doigitizer • Communication and Data • Community Development • Compact Disk • Compound Document • Computational Differentiation • Concept Definition • Concept Development • Confidential Document • Congressional District • Consiglio Direttivo • Content Dependent • Contour Dynamics • Contract Definition • Contract Design • Contrast Detail • Control and Display • Convection-Diffusion • Convective-Dispersive • Converging-Diverging • Converted Distance • Coordination Document • Coordinated Development • Córdoba (Argentina) • Córdoba Durchmusterung • Core Damage • Corporate Development • Corps Diplomatique • Così Detto • Countdown • Critical Dimension • Criticality Dosimeter • Cross Disperser • Crystal Diode • Crystal Driver • Current Density • Current-Driven • Current Driver

CDA: Call Data Accumulator • Catalogue des Données Astrométriques • Central Design Authority • Centro de Datos Astronómicos • Centro de Divulgação da Astronomia (Brazil) • Circuit Distribution Assembly • Columbus Design Authority • Command and Data Acquisition • Common Data Access • Communication Decency Act (USA) • Compound Document Architecture • Computer Design and Applications • Conceptual Design Activity • Convolutional Decoder Assembly • Cooinda (Australia-NT) (location identifier) • Coordinated Data Analysis • Cosmic Dust Analyzer • Craft Digital Assistant • Critical Design Audit

CDAA: Central Data Analysis Area

CDAC: Cosmology Data Analysis Center

Cdad: Ciudad

CDAG: Computational Directed Acyclic Graph

CDAI: Centre de Documentation Aéronautique Internationale (France)

CDAP: Climatic Data Analysis Program

CDAS: Central Data Acquisition System • Cleethorpes and District Astronomical Society (UK) • Climate Data Assimilation System • Command and Data Acquisition Station • Configuration Design Analysis Simulation • Configuration Development Analysis Simulation • Core Design Analysis Simulation

CDAW: Coordinated Data Analysis Workshop

CDB: Calibration Database • Caribbean Development Bank • Central Data Bank • Central Data Buffer • Central Database • Cold Bay (USA-AK) (location identifier) • Command Database • Common Database • Common Data Bus • Commonwealth Development Bank • Comprehensive Database • Computer Data Bus • Congested District Board (Ireland) • Cosmic Diffuse Background • Current Data Bit

Cdbd: Cardboard

CDBFR: Common Data Buffer

CDBN: Column-Digit Binary Network

CDBS: Calibration Database Software • Calibration Database System

CdC: Carte du Ciel

CDC: Capsule Drive Core • Cedar City (USA-UT) (location identifier) • Center for Disease Control • Central Digital Computer • Central Document Control • Central Drift Chamber • Centrale De Commutation • Centre De Contrôle • Centre de Documentation Chimique • Centro de Documentação Cientifica • Centro de Documentacíín Científica • Ceramic Disk Capacitor • Chemical Data Center • Classified Document Control • Climate Diagnostics Center (NOAA) • Cloud Droplet Concentration • Coalition to Diversify Computing (ACM) • Coaxial Directional Coupler • Code Décimal Cyclique • Code Directing Character • Cold-Drawn Copper • Comisión de Documentación Científica • Command and Datahandling console • Command Decoder Coaxial • Common Development Cycle • Computer Development Center • Computer Display Channel • Configuration Data Control • Confined Detonating Cord • Context-Dependent Clustering • Contract Data Coordination • Control Data Corporation • Control Distribution Center • Convective Day Category • Cost Determination Committee • Count - Double Count • Countdown Clock • Critical Density of Current • Cryogenics Data Center • Crystal Data Center • Crystallographic Data Center

CDCA: Centre de Dépouillement des Clichés Astronomiques

CD&CC: Central Data & Cataloguing Center

CDCC: Central Data and Cataloguing Center • Centro de Divulgação Cientifica e Cultural

CDCD: Cargo Design Control Drawing

CDCF: Cahier Des Charges Fonctionnel • Cosmic Dust Collection Facility • Cosmic Dust Collector Facility

CDCI: Continuous Duration and Constant Impact

CDCL: Collective Distributive Constraint Language

CDCM: Carbon Dioxide Concentrating Module

CDCN: Contract Data Change Notice • Contract Document Change Notice

CDCP: Center for Disease Control and Prevention • Command Display and Control Processor

CDCR: Center for Communication and Documentation Research (USA-OH) • Conceptual Design and Cost Review

CDCS: Central Data Collection System • Centre for Development Cooperation Services

CDCT: Centro de Documentação Científica e Técnica • Centro de Documentación Científica y Tecnológica

CDCTE: Centro de Documentación Científica, Técnica y Económica

CDCU: Communications Digital Control Unit

CDD: Central Data Display • Centre De Documentation • Certificate of Disability for Discharge • Change Drive and Directory • Charge Density Distribution • Charge/Discharge Device • Charging/Discharging Device • Chart Distribution Data • Circumplanetary Dust Dynamics • Classification Décimale de Dewey • Coded Decimal Digit • Color Data Display • Come Dovevasi Dimostrare • Command Decoding and Distribution • Common Data Directory • Compact-Disk Drive • Computer-Directed Drawing • Contrat à Durée Déterminée • Cosmic Dust Detector

CD-DA: Compact Disk - Digital Audio

CDDA: Compact Disk - Digital Audio

CDDB: Common Diagnostic Database

CDDC: Center Data Descriptions Catalog

CDDF: Central Data Distribution Facility

CDDI: Computer-Directed Drawing Instrument • Copper-Distributed Data Interface

CDDIS: Crustal Dynamics Data Information Station • Crustal Dynamics Data Information System (*)

CDDN: China Digital Data Network (China-PRC)

CDDP: Centre Départemental de Documentation Pédagogique (France)

CDDS: CLUSTER Data Disposition System

CDDT: Countdown Demonstration Test

Cde: Conde

CDE: Caledonia (Panama) (location identifier) • Centre de Documentation Européenne • Commercial Derivative Engine • Commission for Development and Exchange • Commission on Development and Exchange • Commission pour le Développement et les Échanges • Common Desktop Environment • Compact Disk Erasable • Complex Data Entry • Concept Demonstrator Engine • Conde • Content Directory Entry • Continuous Distribution of Ellipsoids • Convection-Dispersion Equation

CEDC: Commission Départementale d'Équipement Commercial (France)

CDEDS: Collection and Distribution of Ecological Data by Satellite

C. de E.: Consejo de Estado

C. de F.: Chemin de Fer

CDEFI: Conférence des Directeurs d'Écoles et Formations d'Ingénieurs (France)

CDEK: Computer Data Entry Keyboard

CDEN: Conseil Départemental de l'Éducation Nationale (France)

CDEP: Concept Development and Educational Programme

CDER: Centre de Développement des Énergies Renouvelables

CDES: Computer Data Entry System • Conversational Data Entry System • Crossed-Dispersed Echelle Spectrograph

CdF: Collège de France

CD&F: Class Determination & Finding

CDF: Cable Distribution Frame • California Department of Forestry (USA-CA) • Catalog Data File • Central Database Facility • Central Data Facility • Channel Definition Format • Charte des Droits Fondamentaux • Circuit Design Fabrication • Class Determination and Finding • Collider Detector at Fermilab • Collider Detector Facility • Combined Data File • Combined Distributing Frame • Combined Distribution Function • Comma-Delimited Format • Command Display Facility • Common Data Form • Common Data Format • Community Development Fund • Complete Disjunctive Form • Confined Detonating Fuse • Conventional Data Format • Cool-Down Facility • Core-Damage Frequency • Cortina d'Ampezzo - Fiames (Italy) (location identifier) • Cumulative Density Function • Cumulative Distribution Function • Custom-Defined Function

CDFCHB: Command Data Format Control Handbook

CDFNT: Cold Front

CDFS: CD-ROM File System • Compact Disk File System • Core Driven Fan Stage

CDF&TDS: Circuit Design, Fabrication & Test Data Systems

CDFTDS: Circuit Design, Fabrication, and Test Data Systems

cd. fwd.: Carried Forward

Cdg: Cardiganshire (UK-Wales) • Coding

CDG: Capacitance Diaphragm Gauge • Capacitor Diode Gate • Central Design Group • Central Dominant Galaxy • Centrally Dominant Galaxy • Circular Diffraction Grating • Compact Dwarf Galaxy • Concept Development Group • Cosmic Diffuse Gamma-ray • Paris - Charles de Gaulle (France) (location identifier)

C&DH: Command & Data Handling • Control & Data Handling

CDH: Camden (USA-AR) (location identifier) • Command and Data Handling • Communications and Data Handling • Constant Delta Height • Constant Differential Height • Control and Data Handling

CDHC: Command and Data Handling Console

CDHF: Central Data Handling Facility • Cluster Data Handling Facility • Coordinated Data Handling Facility

CDHS: Command and Data Handling System • Comprehensive Data Handling System

CD-I: Common Disk Interactive • Compact Disk Interactive

CDI: Centre de Documentation et d'Information • Centre des Impôts (France) • Centre for the Development of Industry • Centre pour le Développement Industriel • Circle Digit Identification • Classification-Discrimination-Identification • Coloured Digital Image • Command Data Interface • Command Detector Interface • Command Display Indicator • Commercial Delta Inc. • Common Disk Interactive • Compact Disk Interactive • Computer-Directed Instrument • Constrained Data Item • Contract Definition Instruction • Contrat à Durée Indéterminée • Core Data Item • Course Deviation Indicator • Critical Decision Issue

CDIA: Centre de Documentation et d'Information de l'Assurance • Certified Document Imaging Architect

CDIAC: Carbon Dioxide Information Analysis Center

CDIC: Carbon Dioxide Information Center

CDID: Contract Data Item Description

CDIF: Compound Document Interchange Format

CDIS: Comité Directeur de l'Information Statistique

CDIT: Centro de Documentación e Información Tecnológica

c. div.: Cum Dividend

CDIVS: Core Data Item Verification System

CDJ: Conceição do Araguaia (Brazil-PA) (location identifier)

CDK: Cedar Key - Lewis (USA-FL) (location identifier) • Control Development Kit • Custom Development Kit • Cyberspace Developer Kit

CDL: Cable Delay Line • California Digital Library (USA-CA) • Candle (USA-AK) (location identifier) • Carbon Dioxide Laser • Case Definition Language • Central Development Laboratory • Client Display Library • Cloud Detection Lidar • Coaxial Diode Limiter • Comité De Liaison • Command Downlink • Common Diode Limiter • Component Description Language • Computer Description Language • Computer Design Language • Control Development Language • Course Description Language • Current-carrying Double Layer • Cyberspace Development Kit

CDLB: Carbon Dioxide Laser Beam

CDLI: Common Data Link Interface

CDM: Category Discrimination Method • C Data Manager • Central Data Management • Centre Départemental de la Météorologie • Centre Départemental Météorologique • Chemistry Data Management • Clean Development Mechanism • Code Division Multiplexing • Cold Dark Matter • Collider Dipole Magnet • Commission on Dynamic Meteorology • Communications Data Manager • Conceptual Data Model • Configuration Design Module • Contrat de Durée Maximale • Control Development Methodology

• Córdoba Durchmusterung • Core Discharge Monitor • Critical Damping Mode

CDMA: Code Division Multiple Access

CDMF: Collaborative Document Management Facility • Commercial Data Masking Facility

CD-MO: Compact Disk - Magneto-Optical

CDMO: Compact Disk Magneto-Optical

CDMR: Cyclic Data Management Routine

CDMS: Central Data Management System • Coherent Doppler Measurement System • Command and Data Management System • Command Data Management System • Communications and Data Management System • Communications Data Management System • Configuration Definition Management System • Contract Data Management System • Contractor Data Management System • Control and Data Management System • Cryogenic Dark Matter Search • Cryospheric Data Management System • Cyberspace Data Monitoring System

CDMU: Central Data Management Unit • Command and Data Management Unit

Cdn: Canadian

CDN: Camden Woodward (USA-SC) (location identifier) • Canada (international vehicle registration code) • CDR Discrepancy Notice • Centre de Documentation Nationale • Chicago Daily News • Coded Decimal Notation • Comparison Dark Nebula • Conceptual Design Review • Content-Delivery Network • Coordinations-Definitions-Notations

CDNC: Cloud Drop Number Concentration

Cdnes: Condiciones

CDNNET: Canadian Network

CDNR: CDR Discrepancy Notice Record

CDNU: Control Display Navigation Unit

Cdo: Commando • Consulado

CDO: Central Dense Overcast • Centre de Documentation Ophtalmologique (France) • Collaborative Data Object • Comité des Directeurs d'Organismes • Continuous Distribution of Oscillators • Continuous Duty Overnight • Cradock (South Africa) (location identifier)

CDOM: Comité Directeur de l'Océanographie Militaire

CDOS: Chief Data and Operations Scientist • Comité Départemental des Oeuvres Sociales • Concurrent Disk Operating System • Customer Data and Operations System (*) • Customer Data Operations System

CDOT: Colloidal Disorder-Order Transition

CD-P: Compact Disk - Photographic

C&DP: Control & Data Processor

CDP: Census Designated Place • Central Data Processing • Central Data Processor • Central Distribution Panel • Centre de Données Planétaires (France) • Certification Data Processing • Certified Data Processor • Cisco Discovery Protocol • Closest Discrete Pairwise • Color Display Processor • Command and Data Processor • Command Data Processor • Command, Data and Processing • Common Depth Point • Common Document Pool • Communications Data Processor • Compact Disk - Photographic • Compressor Discharge Pressure • Computational Data Processing • Computer Data Processing • Concept Definition Proposal • Concept Development Plan • Condensed Dispersed Phase • Confidence Development Program • Constant Deviation Prism • Control Data Panel • Control and Data Processor • Control and Display Panel • Cooperative Development Program • Correlated Data Processor • Cost Data Plan • Critical Decision Point • Critical Design Parameter • Crustal Dynamics Project • Cuddapah (India) (location identifier)

CDPA: California Desert Protection Act (USA) • Classified Data preparation Area

CDPC: Central Data Processing Computer • Computation and Data Processing Center

CDPD: Cellular Digital Packet Data

CDPF: Central Data Processing Facility

CDPI: Command, Data Processing and Instrumentation

CDPIC: Caisson à Dépression de la Partie Inférieure du Corps

CDPIE: Command Data Processing Interface Equipment

CDPIS: Command, Data Processing and Instrumentation System

C/DPM: Cost/Design Performance Management

CDPM: Cost-Design Performance Management

CDPP: Centre de Données de la Physique des Plasmas (France)

CDPS: Communications Data Processing System • Computing and Data Processing Society (Canada)

CDPU: Central Data Processing Unit

CDQ: Core-Dominated QSO • Croydon (Australia-QLD) (location identifier)

CDQR: Critical Design and Qualification Review

Cdr: Commander

CD-R: Compact Disk Recordable

CDR: Call Detail Record • Call Detail Recording • Centrale Des Revues • Chadron (USA-NE) (location identifier) • Cleaning Decontamination Request • Column Density Ratio • Command Destruct Receiver • Commander • Common Data Record • Common Data Representation • Compact Disk Recordable • Conceptual Design Requirement • Conceptual Design Review • Configuration Design Review • Consortium De Réalisation • Constant Dose Range • Content of Decrement Register • Content of the Decrement of a Register • Continuous Data Replication • Contract Data Requirement • Cooperative Development Research • Coordinator of Data Requirements • Cost Detail Report • Countdown Deviation Request • Critical Design Review • Crude Death Rate • Cumulative Data Report • Current Design Reusable

CDRA: Carbon Dioxide Removal Assembly

CDRAM: Cached Dynamic Random Access Memory

CDRB: Critical Design Review Baseline

CDRD: Carbon Dioxide Research Division • Computational and Data Reduction Division

CDRH: Center of Devices and Radiological Health

CDRL: Contract Data Requirements List (*) • Contractor Data Requirements List

CD-ROM: Compact-Disk - Read-Only Memory

CDROM: Compact-Disk Read-Only Memory

CDRP: Carbon Dioxide Research Program • Certification for Disaster Recovery Planners • Coherent Dispersion Removal Processor

CDRR: Contract Documentation Requirements Record

CDRS: Computer Data Recording System • Contour Drawing Rating Scale • Contract Data Requirements Schedule • Control and Data Retrieval System

CD-RTOS: Compact-Disk - Real-Time Operating System

CDRTOS: Compact-Disk Real-Time Operating System

CdS: Cadmium Sulfide

C&DS: Command & Data Simulator

CDS: Cataloguing Distribution Service • Cathode Dark Space • Cavity Digitizing System • Cell Directory Service • Central Data Station • Central Data System • Centralized Data System • Centre de Données de Strasbourg (France) (*) • Centre de Données Stellaires (France) • Centre for Development Studies • Centre for Disaster Studies (China-PRC) • Characteristic Damage State • Chemical Discriminator System • Childress (USA-TX) (location identifier) • Chromatography Data System • Circuit Data Sheet •

Circumstellar Dust Shell • Climate Data Set • Climate Data Service • Closeout Door System • Cockpit Development Station • Coded Doppler Sonography • Cold-Drawn Steel • Collision Detector System • Color Data System • Color Difference Signal • Comité Directeur Scientifique • Command and Data Simulator • Command and Data System • Communications and Data System • Compatible Duplex System • Complex Data System • Compliance Data System • Comprehensive Display System • Compressed Data Storage • Computer Data Switchboard • Computer Data System • Computer Design and Simulation • Computer-Drive System • Computer Duplex System • Computerized Dispersive Spectroscopy • Computerized Documentation System • Conceptual Design Study • Configuration Design System • Configuration Development System • Consolidated Direct Support • Container Delivery System • Continuous Dynamical System • Control and Data System • Control and Display System • Control and Dynamical System • Control Data Set • Control Data System • Control Distribution System • Controller Select Direction • Coronal Diagnostic Spectrometer • Correlated Double Sampler • Correlated Double Sampling • Cortex Dealer System • Coupling Display System • Cryogenic Dilution System • Current Directory Structure

CDSA: Common Data Security Architecture • Condessa

CDSB: Concise Dictionary of Scientific Biography

CD&SC: Communications Distribution & Switching Center

CDSC: Communications Distribution and Switching Center

CDSCC: Canberra Deep Space Communications Complex (Australia-ACT)

CDSE: Computer-Driven Simulation Environment

CDSF: Central Data Services Facility • Commercially-Developed Space Facility

CDSI: Continuous Duration and Sporadic Impact

CDSL: Computer Data System Laboratory • Consumer Digital Subscriber Line

CDSLR: Crustal Dynamics Satellite Laser Ranging

CDSPH: Chemo-Dynamical Smoothed Particle Hydrodynamics

CDSR: COBE Science Data Room • Contract Data Status Report • Cost Data Summary Report

C&DSS: Communication & Data Subsystem

CDSS: Communication and Data Subsystem • Compressed Data Storage System • Customer Decision Support System

CDST: Central Daylight Saving Time • Centre de Documentation Scientifique et Technique (CNRS)

Cdt: Commandant • Cordialement

CdT: Cadmium Telluride

CDT: Call Data Transmitter • Catalogues and Databases Team • Center for Democracy and Technology • Central Daylight Time • Centre de Documentation Technique • Centro de Documentación Técnica • Class Descriptor Table • Command Descriptor Table • Compressed Data Tape • Compressor Discharge Temperature • Computer Data Transmission • Computer Data Transmitter • Computer Display Terminal • Concrete Data Type • Configuration Data Table • Connaissance des Temps (periodical code) • Constrained Delaunay Triangulation • Control Data Terminal • Control Departure Time • Control Differential Transformer • Controlled Departure Time • Coordinate Data Transmission • Countdown Demonstration Test • Countdown Time • Craft-Design-Technology

CDTA: Capital District Transportation Authority

CDTCP: Communication and Data Transmission Control Program

CdTe: Cadmium Telluride

CDTED: Compressed Digital Terrain Elevation Data

CDTI: Centro para el Desarollo Tecnológico e Industrial • Centro para el Desarollo Tecnológico Industrial (*) • Cockpit Display of Traffic Information

CDTL: Common Data Translation Language

CDTP: Centre de Documentation Technique et Pédagogique (Belgium)

CDTS: Computer-Directed Training System

CD-TV: Commodore Dynamic - Total Vision • Compact Disk - Television

CDTV: Commodore Dynamic - Total Vision • Compact Disk - Television

CDU: Calculatrice Digitale Universelle • Camden (Australia-NSW) (location identifier) • Central Display Unit • Centre de Documentation Universitaire • Clasificación Decimal Universal • Classification Décimale Universelle • Command Data Unit • Command Decoder Unit • Command Detector Unit • Computer Display Unit • Control Data Unit • Control and Diagnostic Unit • Control and Display Unit • Control-Display Unit • Counter Display Unit • Coupling Data Unit • Coupling Display Unit • CRT Display Unit • Current Data Unit

CDUA: Conservation District Usage Application

CD-V: Compact Disk Video

CD/V: Concept Definition/Validation

CDV: Carte De Visite • Cartographic Data Visualizer • Cell Delay Variation • Compact Disk - Video • Cordova - Mile 13 (USA-AK) (location identifier) • Current Domestic Value

CDVT: Cell Delay Variation Tolerance

CDW: Caldwell Wright (USA-NJ) (location identifier) • Car Damage Waiver • Centrifugally Driven Wind • Charge Density Wave • Chilled Drinking Water • Circumpolar Deep Water • Cold Drinking Water • Collision Damage Waiver • Command Data Word • Command Word • Computer Data Word • Continuum Distorted Wave • Control Data Word

CDWG: Countdown Working Group

CD-WO: Compact-Disk - Write Once

CDWO: Compact-Disk - Write Once

CDWR: Chilled Drinking Water Return

CDX: Compound Index • Control Differential Transmitter

CD-XA: Compact-Disk - Extended Architecture

CDXA: Compact-Disk Extended Architecture

CDY: Cagayan de Sulu (Philippines) (location identifier)

Cdz: Cádiz (Spain)

CdZnTe: Cadmium Zinc Telluride

ce: Caveat Emptor • Correo Electrónico

Ce: Cer • Cerium • Cérium

C&E: Ciel et Espace (periodical code)

C/E: Cost Effectiveness

CE: Cache Enable • Calculateur Électronique • Capillary Electrophoresis • Carbon Equivalent • Cargo Element • Ceará (Brazil) • Celestial Equator • Cellular Encoding • Celtic • Centre of Effort • Change Engineer • Change Evaluation • Chapman-Enskog • Charge Exchange • Chemical Engineer • Chief Engineer • Chip Enable • Christian Era • Chromospheric Emission • Church of England • Circular Error • Civil Engineer • Civil Engineering • Clear End • Clear Entry • Clinical Evaluation • Close Encounter • Coarse Erection • Coil Experiment • Collision Elimination • Combustion Efficiency • Comitato Ejecutivo • Comité d'Entreprise • Commercial Engineer • Common Emitter • Common Entrance • Common Envelope • Common Era • Communauté Européenne (now UE) • Communications Electronics • Commutator End • Compact Elliptical • Compass Error • Complement Edge • Compression Engine • Computer Entry • Computing

Efficiency • Concurrent Engineering • Conformité Européenne • Conseil de l'Europe • Conseil d'Établissement • Conseil Économique • Consiglio d'Europa • Consumer Electronics • Contamination Experiment • Continuing Education • Contrast-Enhanced • Control Element • Control Engineer • Conversion Efficiency • Convert Enable • Corps of Engineers • Correlation Energy • Correlation Environment • Cost Effectiveness • Cost Element • Coulomb Explosion • Council of Europe • Cours Élémentaire • Crew Equipment • Critical Examination • Cryocooler Electronics • Current Efficiency • Current Expendable • Cusp Extension • Customer Engineer

CEA: Canadian Exporters' Association • Cause-and-Effect Analysis • Center for EUV Astrophysics (USA-CA) • Central Electricity Authority • Central Electronics Assembly • Chemical Engineering Abstract • Circular Error Average • Clube Estudiantil de Astronomia (Brazil) • Commissariat à l'Énergie Atomique (France) • Commission Économique pour l'Afrique • Comité Européen des Assurances • Conference of Educational Associations • Control Electronics Assembly • Control Element Assembly • Consorzio Engineering Aeroportuale • Cost Effective Analysis • Cost Effectiveness Analysis (*) • Council of Economic Advisers • Council of Educational Advance • Wichita Cessna (USA-KS) (location identifier)

CEAA: Center for Editions of American Authors • Council of European-American Associations

CEAAL: Centro de Estudios Astronômicos de Alagoas (Brazil)

CEABA: Chemical Engineering and Biotechnology Abstracts

CEAC: Commission Européenne de l'Aviation Civile

CEADEN: Centro de Estudios Aplicadas para el Desarollo Nuclear

CEAEO: Commission Économique pour l'Asie et l'Extrême-Orient

CEAL: Civil Engineering Automation Library

CEAM: Center for Exposure Assessment Modelling • Coarse Earth Acquisition Mode • Concerted European Action on Magnets • Cost Effectiveness Analysis Methodology

CEAMIG: Centro de Estudios Astronômicos de Minas Gerais (Brazil-MG)

CEAO: Communauté Économique de l'Afrique de l'Ouest • Communauté Économique pour l'Afrique de l'Ouest

CEAREX: Coordinated Eastern Arctic Experiment • Coordinated Eastern Arctic Regional Experiment

CEAS: Computer Engineering and Application Society • Confederation of European Aerospace Societies

CEASE: Campaign for the Elimination of Acronyms in Scientific Exchange

CEB: Cebu (Philippines) (location identifier) • Central Electricity Board • Centre Électronique de Brest (France) • Communications Electronics Board

CEBAF: Continuous Electron Beam Accelerator Facility (now TJNAF)

CEBAS: Closed Equilibrated Biological Aquatic System

CEBC: China-EC Biotechnology Center

CEBEDOC: Centre Belge de Documentation (Belgium)

CeBIT: Centrum für Büro- und Informationstechnik (Germany)

CEBM: Centre Européen de Biologie Moléculaire • Conférence Européenne de Biologie Moléculaire (*)

CEBO: Coordinated EISCAT and Balloon Observations

CEC: California Energy Commission • Canadian Electrical Code • Capillary Electrochromatography • Capsule End Cover • Cation Exchange Capacity • Central European Country • Centre d'Expertise Clientèle • Centre Éducatif et Culturel • Civil Engineering Corps • Commission of the European Communities • Commonwealth Education Conference • Complex Equipment Contract • Compound Elliptical Concentrator • Computer Electronic Complex • Confédération des Étudiants Chercheurs • Control Electronics Container • Control Encoder Coupler • Controlled Element Computer • Cooperative Engagement Capability • Coordinating European Council • Crescent City (USA-CA) (location identifier) • Cryogenic Engineering Conference

CECA: Communauté Européenne du Charbon et de l'Acier

CECAM: Centre Européen de Calcul Atomique et Moléculaire

CECC: CENELEC Electronic Components Committee • Central European Candidate Country (EU)

CECED: Collaborative Environment for Concurrent Engineering Design • Conseil Européen de la Construction Électrodomestique

CECL: Cascade Rmitter-Coupled Logic

CECLES: Centre Européen de Construction de Lanceurs d'Engins Spatiaux • Conseil Européen pour la Construction de Lanceurs d'Engins Spatiaux (now merged with CERS into ASE)

CECM: Composite Engineering Change Memo

CECO: Center Engine Cutoff

CECP: Compatibility Engineering Change Proposal • Country Extended Code Pages

CECS: Centro de Estudios Científicos de Santiago (Chile) • Civil Engineering Computing System

CECSD: Citizens for Energy Conservation and Solar Development

CED: Capacitance Electronic Disc • Ceduna (Australia-SA) (location identifier) • Centre d'Édition et de Documentation • Centre Européen de Documentation • Cohesive Energy Density • Collins English Dictionary • Communauté Européenne de Défense • Computer Entry Device • Cost Estimate Dispersion • Council for Economic Development

CEDA: Centro de Estudios de Astronomía (Venezuela)

CeDAR: Center for Database Access Research

CEDAR: Center of Excellence for Document Analysis and Recognition • Central European Environmental Data Request • Coupling, Energetics and Dynamics of Atmospheric Regions

CEDAS: Cluster EFW Data Analysis

CEDC: Centre d'Études, de Documentation et de Coordination • Cyclic Error Detection Code

CEDD: Committee on Exchange of Digital Data

CEDDA: Center for Experiment Design and Data Analysis

CEDE: Centre d'Études et de Documentation Européennes • Centro de Estudios sobre el Desarollo Económico

CEDEA: Communauté Économique Des États d'Afrique

CEDEFOP: Centre Européen pour le Développement de la Formation Professionnelle

CEDEP: Centre Européen D'Éducation Permanente

CEDEX: Courrier d'Entreprise à Distribution Exceptionnelle

CEDF: Conception-Étude-Développement-Fabrication

CEDH: Convention Européenne des Droits de l'Homme

CEDI: Centre Européen de Documentation et d'Information

CEDISA: Centro de Diseño de Sistemas Automatizados

CEDO: Centre for Educational Development Overseas

CEDRE: Centre de Documentation, de Recherche et d'-Expérimentation • Centre d'Études et de Recherches • Contamination Et Dégradation des Revêtements dans l'Espace

CEDT: Centre for Electronics Design and Technology

C&EE: Central & Eastern Europe

CEE: Center for Excellence in Education • Central and Eastern European • Centro de Estudios Espaciales • Certificate of Extended Education • Cherepovets (Russia) (location identifier) • Civil and Environmental Engineering • Comité Estatal de Estatisticas • Commission Économique pour l'Europe • Common Entrance Examination • Communauté Économique Européenne (now UE) • Comprehensive Environmental Evaluation • Computer Ergonomics Evaluation • Computers and Electrical Engineering (periodical code) • Comunità Económica Europea • Contrat d'Échange Express • Council of Environment Education

CEEA: Communauté Européenne de l'Énergie Atomique

CEEAC: Communauté Économique des États de l'Afrique Centrale

CEEB: College Entrance Examination Board

CEEC: Central and Eastern European Countries • Council for European Economic Education

CEED: Centre for Economic and Environmental Development • Centre for Economic and Environment Development

CEEF: Controlled Ecological Experiment Facility

CEEI: Center for Energy and Environmental Information • Centre Européen d'Études et d'Information • Comité Européen d'Entreprise et d'Innovation

CEEJ: Canadian Electrical Engineering Journal

CEEL: Core Electron Energy Loss

CEEM: Centre for Energy and Environmental Management

CEEP: Centre Européen des Entreprises Publiques • Centre Européen d'Études de Population • Comité Européen de l'Entreprise Publique • Composite Entry Evaluation Program

CEES: Committee on Earth and Environmental Sciences

CEESAR: Centre Européen d'Études Socio-économiques et Accidentologiques des Risques

CEESIM: Combat Electromagnetic Environment Simulator

CEESP: Conselho Estadual de Educação de São Paulo (Brazil-SP)

CEF: Cable Entrance Facility • Chicopee (USA-MA) (location identifier) • Computer Execute Function • Crystalline Electric Field • Springfield Westover (USA-MA) (location identifier)

CEFI: Centre d'Économie et de Finance Internationale • Centre d'Études sur les Formations d'Ingénieurs • Comité d'Études sur les Formations d'Ingénieurs

CEFIC: Conseil Européen des Fédérations de l'Industrie Chimique

CEFOAM: Checkout Equipment For Onboard Automatic Maintenance

CEFTA: Central European Free Trade Agreement

CEG: Chester (UK-England) (location identifier) • Collège d'Enseignement Général • Collins English-German • Combined Environment Group • Computer Education Group • Continuous Edge Graphics • Council for Excellence in Government

CEGB: Central Electricity Generating Board

CEGEP: Collège d'Enseignement Général et Professionnel (Canada)

CEGOS: Commission d'Études Générales de l'Organisation Scientifique

CEGP: Close-in Extrasolar Giant Planet

CEGRN: Central Europe GPS Reference Network

CEHT: Camera Electrical High Tension

CEI: Center for Environmental Information • Central European Initiative • Centre d'Études Industrielles • Cercle Europe Ingénieurs • Chiang Rai (Thailand) (location identifier) • Cloudtop Entrainment Instability • Commercial-Economic Information • Commission Électrotechnique Internationale • Committee for Engineering Information • Communauté des États Indépendants • Communications-Electronics Instruction • Comparably Efficient Interconnection • Computer-Extended Instruction • Conducted Electromagnetic Interference • Configuration End Item • Connected Element Interferometer • Contract End Item • Cost Effectiveness Index • Coulomb Explosion Imaging • Council of Engineering Institutions (UK)

CEIA: Canadian Environmental Industry Association

CEIAC: Coastal Engineering Information Analysis Center

CEIB: Computer Equipment Information Bureau (USA)

CEIC: Caribbean Environmental Information Center • Chemical Industry Education Centre • Commission on Electronic Information and Communications

CEICN: Centre Européen d'Information pour la Conservation de la Nature

CEIF: Council of European Industrial Federations

CEIE: Centre d'Études et d'Information sur l'Enseignement

Ceil: Ceiling

CEIN: Contract End Item Number

CEIPI: Centre d'Études Internationales sur la Propriété Industrielle

CEIR: Corporation for Economic and Industrial Research

CEIS: Combined Elements and Integrated Systems • Contract End Item Specification

CEIST: Centre Européen d'Informations Scientifiques et Techniques

CEIT: Cargo Equipment Integrated Test • Crew Equipment Interface Test • Crew Equipment Integration Test

CEJ: Chernigov (Ukraine) (location identifier) • Counter Electrojet • Cour Européenne de Justice

CEK: Chelyabinsk (Russia) (location identifier) • Computer Entry Keyboard

Cel: Celebrate • Celebrated • Celebration • Celsius • Celestial

CEL: Cape Eleuthera (Bahamas) (location identifier) • Carbon Equilibrium Loop • Celescope • Célula • Certificat Européen de Langue • Contrat d'Emploi Local • Conversational Extensible Language • Coronal Emission Line • Cosine Emission Law • Critical Experiment Laboratory • Cryogenic Engineering Laboratory

CELC: Commonwealth Education Liaison Committee

CELENV: Cellule d'Environnement

CELESCOPE: Celestial Telescope

CELESTE: Čerenkov Low-Energy Sampling and Timing Experiment

CELI: Computer-Enhanced Learning Initiative

CELIA: Centre Lasers Intenses et Applications (France) (formerly PMOH)

CELIAS: Charges, Elements and Isotope Analysis System (*) • Charges, Elements and Isotopes Analysis for SOHO

CELJ: Conference of Editors of Learned Journals

CELLS: California Educational Linkage in the Life Sciences

Cel.Mech.: Celestial Mechanics

CELP: Code-Excited Linear Prediction

CELPA: Centro de Experimentación y Lanzamiento de Proyectiles Autopropulsados

Cels: Celsius

CELS: Controlled Ecological Life Support

CELSS: Closed Ecological Life Support Systems • Controlled Ecological Life Support Systems (*)

Célt: Céltico

Celt: Celtic • Celtique

CELT: California Extremely Large Telescope (USA-CA) • Classified Entries in Lateral Transposition

Celtolat: Celtolatino

CELV: Commercial Expandable Launch Vehicle • Complementary Expandable Launch Vehicle

Cem: Cement • Cemetery

CEM: Camera Electronics Module • Cascade Exiton Model • Central (USA-AK) (location identifier) • Centro Español de Metrología (Spain) • Channel Electron Multiplier • Chromospheric Evaporation Model • Circular Electric Mode • Color Evaporation Model • Column Emission Measure • Communications-Electronics-Meteorological • Compatibilité Électromagnétique • Complex Electromechanical • Computational Electromagnetics • Configuration Element Management • Contact Electronic Mailbox • Continuous Emission Measure • Continuous Emission Monitoring • Conventional Enhancement Modification • Cost and Effectiveness Method • Cost Estimating Method • Coulomb Explosion Method • Crew Escape Module • Cumulus Ensemble Model

CEMA: Consumer Electronics Manufacturers Association • Council for Economic Mutual Assistance

CEMAC: Committee of European Associations of Manufacturers of Active Electronic Components

CEMAGREF: Centre National du Machinisme Agricole, du Génie Rural et des Eaux et Forêts (France)

CEMAP: Centre Européen de Mathématique Appliquée

CEMC: Continental Electronics Manufacturing Co.

CeMDA: Celestial Mechanics and Dynamical Astronomy (periodical code)

CeMec: Celestial Mechanics (periodical code)

CEMF: Counter-Electromotive Force

CeMiSS: Centro Militare di Studi Strategici (Italy)

CEMM: Compaq Expanded Memory Manager

CEMOS: Central European Motivation Seminar

CEMP: CCAMLR Ecosystem Monitoring Program

CEMR: Conférence Européenne des Ministres de la Recherche • Council of European Municipalities and Regions

CEMS: Central Electronic Management System • Community of European Management Schools • Conversion Electron Mössbauer Spectroscopy • Conversions-Elektronen-Mössbauer Spektroskopie

CEMT: Conférence Européenne des Ministres des Transports

Cen: Centaurus-Centauri (constellation) • Central • Centre • Century

CEN: Center • Center for Electronic Materials • Centre d'Étude de la Neige • Centre d'Études Nucléaires • Centre Européen de Normalisation • Ciências Exactas e Naturais • Ciudad Obregón (Mexico) (location identifier) • Comité Estatal de Normalización • Comité Européen de Normalisation • Comité Européen des Normes • Cygnus Egg Nebula • RAL (Starlink internal code)

CENA: Centre d'Études de la Navigation Aérienne • Coopération d'Étalonnage Nord-Américaine

CENAM: Centro Nacional de Metrología

CENBG: Centre d'Études Nucléaires de Bordeaux Gradignan (France)

CENCER: CEN Certification

CEND: Compton-Emission Neutron Detector

CENDIC: Centro Nacional de Documentación e Información Científica

CENDL: Chinese Evaluated Nuclear Data Library (China-PRC)

CENELEC: Comité Européen de Normalisation Électrotechnique • Comité Européen des Normes Électrotechniques

CENFAR: Centre d'Études Nucléaires de Fontenay-aux-Roses (CEA)

CENG: Centre d'Études Nucléaires de Grenoble (CEA)

CENIC: Centro Nacional de Investigaciones Científicas

CENIDE: Centro Nacional de Investigaciones para el Desarollo de la Educación (Spain)

CENL: Council of European National Libraries

CENNA: Convention on Early Notification of a Nuclear Accident

CENR: Committee on Environmental and Natural Resources

Cens: Censor • Censorship

CENS: Centre d'Études Nucléaires de Saclay (CEA)

CenSIS: Centre for Spatial Information Studies

Cent: Cantavo (currency) • Center • Centesimo (currency) • Centigrade • Centigrado • Centime • Central • Centre • Centrifugal • Centro • Centum • Century

CENT: Center

CENTAR: Code for Extended Non-linear Transient Analysis of Reactors

Centig: Centigrade

CENTO: Central Treaty Organization

CENTORA: Centre des Technologies Optiques en Rhône-Alpes (France)

CENTR: Center • Central

CENTX: Central Telephone Exchange

CEO: Catalyzed Electrochemical Oxidation • Center for Earth Observation • Centre Européen de l'Orgue • Chief Education Officer • Chief Engineer's Office • Chief Executive Officer • Combination of Expert Opinions • Comité Européen de l'Outillage • Confederation of Employee Organizations • Waco Kungo (Angola) (location identifier)

CEOS: Committee on Earth Observation from Satellites • Committee on Earth Observation Satellites (*) • Committee on Earth Observing Satellites

CEOSR: Center for Earth Observing abd Space Research (GMU)

Cep: Cepheus-Cephei (constellation)

CEP: Camera Evaluation Program • Capability Evaluation Plan • CEA Event Pipeline • Celestial Ephemeris Pole • Central East Pacific • Circle End Point • Circle of Equal Probability • Circular Error Probability • Circular Error Probable (*) • Codigo de Endereço Postal • Color Evaluation Program • Comié Ecumenico de Proyectos • Committee for Environmental Protection • Committee on Environmental Protection • Common Electronics Parts • Community Education Programme • Component Error Propagation • Comprehensive Event Pipeline • Computed Ephemeris Position • Computer Entry Punch • Concepción (Bolivia) (location identifier) • Conduction Electron Polarization • Configuration Evaluation Plan • Consejo Electoral Provisional • Constraint Exploiter Process • Continuous Estimation Program • Cylindrical Electrostatic Probe

CEPA: Canadian Environmental Protection Act • Commonwealth Environment Protection Agency • Coupled Electron Pair Approximation

CEPAC: COSTEP/ERNE Particle Analyser Collaboration

CEPAL: Comisión Económica (de las Naciones Unidas) Para América Latina (y el Caribe)

CEPD: Committee on Education and Professional Development

CEPE: Compagnie d'Électricité et de Piézo-Électricité • Cylindrical Electrostatic Probe Experiment

CEPES: Centre Européen pour l'Enseignement Supérieur • Comité Européen pour le Progrès Économique et Social

CEPEX: Central Equatorial Pacific Experiment

CEPGL: Communauté Économique des Pays des Grands Lacs

CEPH: Centre d'Étude du Polymorphisme Humain (France)

CEPHAG: Centre d'Études des Phénomènes Aléatoires Géophysiques

CEPI: Center for Earth and Planetary Interiors

CEPIS: Council of European Professional Informatics Societies

CEPMART: Centre for Promotion of Mathematical Research in Thailand

CEPMMT: Centre Européen de Prévision Météorologique à Moyen Terme

CEPO: Center Education Program Officer

CEPP: Committee on Elementary Particle Physics

CEPPAD: Comprehensive Energetic Particle and Pitch Angle Distribution • Comprehensive Energetic Particle Pitch Angle Distribution

CEPr: Centre d'Essai des Propulseurs

CEPR: Centre for Economic Policy Research

CEPS: Center for Earth and Planetary Studies (USA-DC) • Centre d'Étude et de Prospective Stratégique (France)

CEPT: Conférence Européenne des Postes et Télécommunications

CEPUP: Chemical Education for Public Understanding Program

CEQ: Cannes (France) (location identifier) • Council on Environmental Quality

CEQA: California Environmental Quality Act

Cer: Ceramic

CER: Cell Error Ratio • Centre des Énergies Renouvelables • Ceramic Electrochemical Reactor • Certification Evaluation Review • Charge Exchange Recombination • Cherbourg (France) (location identifier) • Closer Economic Relation • Command and Expenditure Report • Component Engineering Request • Consultative Environmental Review • Contactless Electroreflectance • Corotation Eccentricity Resonance • Cost Estimating Relationship

CERA: Centre d'Étude et de Recherche pour l'Automobile (France)

CERAG: Centre d'Études et de Recherches Appliquées à la Gestion (France)

CERAP: Central Radar Approach

CERC: Center for Energy Research Computation • Central Engine Room Control

CERCA: Centre de Recherche en Calcul Appliqué

CERCG: Centre d'Études et de Réalisations Cartographiques et Géographiques (France)

CERCLA: Comprehensive Environmental Response, Compensation, and Liability Act

CERCLIS: Comprehensive Environmental Response, Compensation and Liability Information System

CERCO: Comité Européen de Responsables de la Cartographie Officielle

CERCOA: Centre d'Étude et de Recherche en Chimie Organique Appliquée

CERD: Catalog of Environmental Resource Data

CERDU: Central European Research and Development Unit

CERE: Computer Entry and Readout Equipment

CEREG: Centre d'Études et de Recherches Éco-Géographiques (France)

CEREGE: Centre Européen de Recherche et d'Enseignement des Géosciences de l'Environnement (France)

CEREM: Centre d'Études et de Recherches sur les Matériaux (France)

CEReS: Center for Environmental Remote Sensing

CERES: California Environmental Resources Evaluation System • Center for Educational Resources • Cloud and Earth Radiant Energy System • Clouds Earth Radiant and Energy System (*) • Crop Environment Resource Synthesis

CERESI: Cellule de Recherche en Sciences de l'Information

CERESIS: Centro Regional de Seismología

CERET: Centre Européen de Réflexion et d'Étude en Thermodynamique

CERFACS: Centre Européen de Recherche et de Formation Avancée en Calcul Scientifique (France)

CERFnet: California Education and Research Federation Network

CERGA: Centre d'Études et de Recherches en Géodynamique et Astrométrie (France) (*) • Centre d'Études et de Recherches Géodynamiques et Astronomiques (Franc e)

CERGOP: Central Europe Regional Geodynamics Project

CERI: Center for Educational Research and Innovation • Center for Environmental Research Information • Centre d'Enseignement et de Recherche en Informatique (France) • Centre d'Études des Relations Internationales

CERIF: Common European Research Information Format

CERII: Centre d'Études et de Recherche sur l'Information et l'Innovation

CERIS: Centre d'Étude des Relations Internationales et Stratégiques (Belgium)

CERIST: Centre d'Études et de Recherche d'Information Scientifique et Technique

CERL: Computer-based Education Research Laboratory • Consortium of European Research Libraries • Corvallis Environmental Research Laboratory

CERM: Coordinated Emergency Response Measures

CERMA: Centre d'Études et de Recherches en Médecine Aérospatiale (France)

CERMAV: Centre de Recherche sur les Macromolécules Végétales (France)

CERN: Centre Européen pour la Recherche Nucléaire (*) • Conseil Européen pour la Recherche Nucléaire

CERP: Centre Européen des Relations Publiques • Confédération Européenne des Relations Publiques • Cooperative Ecological Research Project

CERPET: Centre de Recherches Pédagogiques pour les Enseignements Technologiques (France)

CERR: Centre d'Écologie des Ressources Renouvelables (France)

Cerraj: Cerrajería

CERS: Centre Européen de Recherches Spatiales (now merged with CECLES into ASE) • Cost-Estimating Relationship

CERSI: Centre d'Études et de Recherche en Sciences de l'Information (France)

Cert: Certificate • Certificated • Certification • Certified • Certify

CERT: Centre d'Études et de Recherche de Toulouse (ONERA) • Centre d'Études et de Recherche Technique • Certification • Committee on Energy Research and Technology • Computer Emergency Response Team

CERTI: Centro Regional de Tecnologia em Informatica

Certif: Certificate • Certificated

Cert. Inv.: Certified Invoice

CES: Center for Energy Studies • Central Electronics System • Centre d'Études Sociologiques • Certificat d'Études Spéciales • Certificat d'Études Supérieures • Cessnock (Australia-NSW) (location identifier) • Chemically Enhanced Self-sputtering • Circuit Emulation Service • Closed Ecological System • Coast Earth Station • Collège d'Enseignement Secondaire • Combined Earth Sensor •

Comité Économique et Social • Committee on Earth Sciences • Committee on Earth Studies • Commonwealth Employment Service • Community Energy System • Computational Engineering and Science • Computer Engineering Service • Concept Evaluation Study • Confédération Européenne des Syndicats • Conference for European Statisticians • Conference for European Statistics • Conical Earth Sensor • Conseil Économique et Social • Constant Elasticity of Substitution • Consumers Electronic Show • Contrat Emploi-Solidarité • Control Electronics Section • Control Electronics System • Controlled Environmental System • Cooperative Extension Service • Coordinated Evaluation System • Corpus Encoding Standard • Cost Effectiveness Study • Cost-Estimating System • Coudé Echelle Spectrograph • Coudé Echelle Spectrometer • Crew Escape System

CESA: Central Electronics Subassembly • Commission des Études Spatiales Amont

CESAP: Commission Économique et Sociale pour l'Asie et le Pacifique (ONU)

CESAR: Canadian Expedition to Study Alpha Ridge • Capsule Escape and Survival Applied Research • Center for Engineering Systems Advanced Research • Central European Satellite for Advanced Research

CESB: Centre d'Études Supérieures de la Banque (France)

CESBIO: Centre d'Études Spatiales de la Biosphère (France)

CESC: Conferencia Europea de Seguridad y Cooperación

CESDIS: Center of Excellence for Space Data and Information Sciences (USA-MD)

CESE: Center for Earth Science and Engineering • Coalition for Earth Science Education • Coalition of Earth Science Education

CESEPUP: Committee on Science, Engineering and Public Policy

CESG: Cryogenic Electrically Suspended Gyroscope

CESGA: Centro de Supercomputación de Galicia (Spain)

CESHR: Center for Electrochemical Systems and Hydrogen Research

CESID: Centro Superior de Información para la Defensa (Spain)

CESLS: Constant Energy Synchronous Luminescence Spectrometry

CESN: Canadian Electronic Scholarly Network

CESO: Council of Engineering and Scientific Organizations

CESP: Centre de Supercomputació de Catalunya (Spain) • Centre d'Étude des Supports Publicitaires (France)

CESQG: Conditionally Exempt Small Quantity Generator

CESR: Center for Earth System Research • Centre d'Étude Spatiale des Rayonnements (France) • Conduction Electron Spin Resonance • Cornell Electron-positron Storage Ring (USA-NY)

CESRA: Committee of European Solar Radio Astronomers • Community of European Solar Radio Astronomers (*)

CESSDA: Council of European Social Science Data Archives

CEST: Center for Exploitation of Science and Technology • Central European Saving Time • Central European Summer Time

Cet: Cetus-Ceti (constellation)

CET: Capsule Elapsed Time • Central Earth Terminal • Central European Time • Cholet - Le Pontreau (France) (location identifier) • Clutter Equivalent Target • Collège d'Enseignement Technique • Combined Element Test • Commission on Education and Training • Common External Tariff • Concern for Europe's Tomorrow • Constant Elasticity of Transformation • Council for Educational Technology • Creative and Emerging Technology • Cumulative Elapsed Time

CETA: Carbon Emissions Trajectory Assessment • Centre d'Étalonnage Agréé • Comprehensive Education and Training Administration • Comprehensive Employment and Training Act • Corrosion Evaluation and Test Area • Crew and Equipment Translation Aid (*) • Crew Equipment Translation Aid

CETAF: Consortium of European Taxonomic Facilities

CETD: Cross Enterprise Technology Development

CETDP: Cross-Enterprise Technology Development Program

CETEL: Cycle d'Enseignement de la Télédétection

CETEX: Ceramic Tile Experiment • Committee on Extraterrestrial Exploration • Contamination by Extraterrestrial Exploration • Contamination of Extraterrestrial Exploration

CETF: CALTECH Environment Task Force • Critical Evaluation Task Force

CETH: Center for the Electronic Texts in the Humanities

CETI: Communication with Extraterrestrial Intelligence

CETIM: Centre Technique des Industries Mécaniques (France)

CETIS: Centre Européen de Traitement de l'Information Scientifique

CETM: Customized Electronic Technology Magazine

CETO: Centro de Ciências e Tecnologias Ópticas

cet. par.: Ceteris Paribus

Cetr: Cetrería

CETR: Center for Explosives Technology Research

CETS: Chemical Energy Transmission System • Commission on Engineering and Technical Systems • Conference Européenne des Télécommunications par Satellite

CEU: Camera Electronics Unit • Clemson - Oconee County (USA-SC) (location identifier) • Command Execute Unit • Continuing Education Unit • Control Electronics Unit • Coupler Electronics Unit

CEUS: Central and Eastern United States

CEV: Connersville Mettle (USA-IN) (location identifier) • Controlled Environment Vault • Cryogenic Explosive Valve

CEVAT: Combined Environmental, Vibration, Acceleration, Temperature

CEVI: Common Equipment Voltage Indicator

CEVT: Contingency Extravehicular Transfer

CEVU: Conseil des Études et de la Vie Universitaire

CEW: Crestview - Bob Sikes (USA-FL) (location identifier)

CEWES: Corps of Engineers Waterways Experiment Station

CEX: Chena Hot Springs (USA-AK) (location identifier)

Cey: Ceylon

CEY: Murray - Calloway County (USA-KY) (location identifier)

Ceyl: Ceylon

CEZ: Cortez (USA-CO) (location identifier)

CE3: Close Encounters of the Third Kind

c/f: Carried Forward • Cost and Freight

cf: Confer • Cubic Foot • cum figuriis

Cf: Californium • Confer

C&F: Circulation & Fulfillment • Cost & Freight

C/F: Center Frequency

CF: Calibration Factor • Canada (international aircraft registration code) • Cannot Find • Capacity Factor • Cape Faint • Carbon Film • Carbon Filtered • Carbon Fraction • Carboxylfluorescine • Cardiff (UK postal code) • Carriage Forward • Carried Forward • Carrier-Free • Carrier Frequency • Carry Flag • Carry Forward • Catalogue of

Faint • Cathode Follower • Center Focus • Center Forward • Center Frequency • Center of Flotation • Central African Republic (ISO country code) • Central File • Central Formation • Centrally Funded • Centrifugal Force • Centripetal Force • Certainty Factor • Change in Formula • Change to Frequency • Characteristic Function • Chemin de Fer • Circuit Finder • Circulation and Fulfillment • Cloud Fraction • Cold Front • Color Filter • Communication Factor • Communications Facility • Commutation Factor • Comorian Franc (currency) • Compensation Fee • Completion Force • Computers and Fluids (periodical code) • Concentration Factor • Concept Feasibility • Concept Formulation • Concrete Floor • Consequence of Failure • Constant Frequency • Content Filtering • Context Free • Continuous Flow • Contract Formulation • Controlled Facility • Conversion Factor • Cooling Flow • Corner Fitting • Correction Factor • Correlation Factor • Corresponding Fellow • Cost and Freight • Count Forward • Coupling Facility • Crowding Factor • Crystal Field • Cubic Feet • Cumulative Frequency • République Centrafricaine (ISO country code)

CfA: Center for Astrophysics

CFA: Centre de Formation d'Apprentis (France) • Centrifugal Fast Analyzer • Chartered Financial Analyst • Cloud Fractional Absorptance • Coffee Point (USA-AK) (location identifier) • Communauté Financière Africaine • Communauté Française d'Afrique • Compact Flash Association • Concept Feasibility Analysis • Confirmatory Factor Analysis • Continuous Flow Analysis • Controlled Firing Area • Correspondence Factor Analysis • Correspondence Factorial Analysis • Cross-Field Amplifier

CFAD: Contoured Frequency by Altitude Diagrams

CFADC: Controlled-Fusion Atomic Data Center

CFAE: Contractor-Furnished Airborne Equipment • Contractor-Furnished Aircraft Equipment

CFAM: Cogeneration Feasibility Analysis Model

CFAO: Conception de la Fabrication Assistée par Ordinateur • Conception et Fabrication Assistées par Ordinateur (*) • Contrôle de la Fabrication Assistée par Ordinateur

CFAR: Constant False Alarm Rate

CFAS: Central Florida Astronomical Society (USA-FL)

CFAT: Carnegie Foundation for the Advancement of Teaching

CFB: Carey Foster Bridge • Change Franc Pacifique • Cipher Feedback • Circulating Fluid Bed

CFBC: Circulating Fluid Bed Combustion

CFBS: Canadian Federation of Biological Societies

CFC: Carbon Fiber Composite • Carbon Fiber Compound • Central Fire Control • Central Flow Control • Centre Français du Copyright (France) • Chlorinated Fluoro Carbon • Chlorofluorocarbon • Circulating Fluidized Combustion • Comité Français de Cartographie • Complex Facility Console • Congés de Fin de Carrière • Conseiller en Formation Continue (France) • Cube à Faces Centrées

CFCA: Communications Fraud Control Association

CFCB: Computer Format Control Buffer

CFCC: Census Feature Class Code • Coupling Facility Control Code

CFCE: Centre Français du Commerce Extérieur (France)

CFCF: Central Flow Control Facility

CFCS: Central Flow Control Service

cfd: Cubic Feet per Day

CFD: Bryan Coulter (USA-TX) (location identifier) • Centre de Fabrication et de Distribution • Computational Fluid Dynamics • Continuous Flow Diffusion • Control Flow Diagram • Cumulative Frequency Distribution

CFDA: Catalog of Federal Domestic Assistance • Computational Fluid Dynamics Analysis

CFDCC: Computational Fluid Dynamics Community Club

CFDP: Calibration File Derivation Procedure • Coherent File Distribution Protocol

CfDS: Campaign for Dark Skies

CFDS: Centralized Fault Display System

CFDT: Cartesian Full Disk Telescope

CFE: Cathode Flicker Effect • Central Framework Executive • Clermont-Ferrand Aulnat (France) (location identifier) • College of Further Education • Consolidated Front End • Contractor-Furnished Equipment • Control Fuel Element • Control Function Equipment • Controlled Flash Evaporation • Critical Function Equipment • Customer-Furnished Equipment

CFEM: Combined Fates and Effects Model

CFEPP: Comité Français pour l'Étude des Phénomènes Paranormaux (France)

CFES: Continuous Flow Electrophoresis System

CFF: Canfunfo (Angola) (location identifier) • Carbon Fraction of Fuel • Cartographic Feature File • Chemins de Fer Fédéraux • Columbus Free Flyer • Commission on Fossil Fuels • Compressible Flow Facility • Conical Flow Field • Counter Flip-Flop • Critical Fusion Frequency

CFFT: Columbus Free Flying Laboratory

CFFTP: Canadian Fusion Fuels Technology Project

CFG: Caine Farber and Gordon • Certificat de Formation Générale • Cienfuegos (Cuba) (location identifier) • Configuration • Constant-Frequency Generator • Context-Free Grammar

CFGT: Comité Français des Grands Télescopes (France) (*) • Conseil Français des Grands Télescopes (France)

CFGU: Common Frequency Generation Unit

cfh: Cubic Feet per Hour

CFH: Clifton Hills (Australia-SA) (location identifier) • Cost-Freight-Insurance

CFHT: Canada-France-Hawaii Telescope

CFI: Camfield (Australia-NT) (location identifier) • Card Format Identifier • Certified Flight Instructor • Cloud-Free Interval • Comprehensive Flare Index • Corporación Financiera Internacional • Cost-Freight-Insurance • Court of First Instance

CFIA: Component Failure Impact Analysis

CFISU: Canadian Foundation for the International Space University

CFIT: Controlled Flight Into Terrain

CFK: Carbonfaserverstärkte Kunststoffe • Chloorfluorkoolwaterstof

CFL: Calibrated Focal Length • Close Focus Lens • Compact Fluorescence Lamp • Context-Free Language • Courant-Friedrichs-Levy

CFLOS: Cloud-Free Line Of Sight

CFLSE: Critical Fluid Light Scattering Experiment

CFLSI: Critical Fluid Light Scattering Instrument

cfm: Confirm • Cubic Feet per Minute

Cfm: Confirm • Confirmation

CFM: Candle Flames in Microgravity • Cathode Follower Mixer • Center Frequency Modulation • Chlorofluoromethane • Closed Flux Memory • Code Fragment Manager • Computer Facilities Management • Configuration Management • Conformally Flat Model • Contingency For Movement • Contractor-Furnished Material • Crystal Frequency Multiplier • Customer-Furnished Material

CFMA: Centre de Formation National d'Avignon (France)

CFMF: Crip Flow Management Facility

CFMS: Central Financial Management System • Chained File Management System

CFMU: Central Flow Management Unit

CFN: Carrickfinn (Ireland) (location identifier) • Confine • Donegal (Ireland) (location identifier)

CFO: Calling For Orders • Central Facilities Operations • Central Forecasting Office • Ceramics Fiber Optics • Channel For Orders • Chief Financial Officer • Comité Français d'Organisation • Commissioning and Fitting Out • Complex Facility Operator • Confreza (Brazil-MT) (location identifier)

CFONB: Comité Français d'Organisation et de Normalisation Bancaire (France)

CFOP: Club des Fibres Optiques Plastiques

CFOR: Conversational Fortran

cFP: Core Fundamental Plane

CFP: Call For Papers • Call For Participation • Carpentaria Downs (Australia-QLD) (location identifier) • Carrier Frequency Pulse • Center of Filtering and Plotting • Centre de Formation Professionnelle • Certified Financial Planner • Code Federal Regulation • Cold Frontal Passage • Cold Front Passage • Combined Filter and Plot • Communauté Française du Pacifique • Communauté Financière du Pacifique • Computer-Freedom-Privacy • Conceptual Flight Profile • Contractor-Furnished Property • Franc (French Pacific Islands) (ISO currency code)

CfPA: Center for Particle Astrophysics (USA-CA)

cfpd: Cubic Feet per Day

cfph: Cubic Feet per Hour

CFPL: Computational Fluid Physics Laboratory

CFpM: Cubic Feet per Minute

CFPO: Canadian Forces Post Office

CFPP: Centre de Formation Professionnelle et de Perfectionnement

CFPPA: Comité Français de Pilotage du Plan d'Action

cfps: Cubic Feet per Second

CFPs: Calls For Papers

Cfr: Chauffeur • Confronta

CFR: Caen Carpiquet (France) (location identifier) • Call For Release • Carbon-Film Resistor • Carbon Formation Reactor • Catastrophic Failure Rate • Centre des Faibles Radioactivités (France) • Centre for Field Research • Code of Federal Regulations (USA) • Commercial Fast Reactor • Computerized Facial Recognition • Constant Frame Rate • Coupling Facility Receiver • Cumulative Failure Rate

CFRD: Confidential, Formerly Restricted Data

CRFE: Contract Financial Requirements Estimate

CFRM: Coupling Facility Resource Management • Coupling Facility Resource Manager

CFRO: Comité Français de Recherche Opérationnelle

CFRP: Carbon Fibre Reinforced Plastic • Conceptual Framework for Reuse Processes

CFRS: Canada France Redshift Survey

cfs: Coste-Flete-Seguro • Cubic Feet per Second

CFS: Carrier Frequency Shift • Cassegrain Feed System • Center For Standards • Center Frequency Stabilization • Central Frequency Synthesizer • Cesium Feed System • Chronic Fatigue Syndrome • Coffs Harbour (Australia-NSW) (location identifier) • Coherent Frequency Synthesizer • Cold-Finished Steel • Combination Frequency Signal • Combined File Search • Committee on Fundamental Science • Common File System • Completely Finished Set • Component Failure Summary • Composite Feed System • Contract Financial Status • Coupling Facility Sender • Cryogenic Fluid Storage • Cubic Feet per Second

CFSC: Cryogenic Fluid Storage Container

CFSD: Citizens For Space Demilitarization • Critical Function Sequence Diagram

CFSE: Crystal-Field Stabilization Energy

CSFK: Coherent Frequency Shift Keying

CSFM: Communication Finite State Machine

CFSP: Common Foreign and Security Policy (EU)

CFSR: Center For Solar Research (NJIT) • Contract Fund Status Report • Contract Fund Status Review

CFSS: Combined File Search Strategy

CFST: Context-Free Syntactical Translator

CFSTI: Clearinghouse for Federal Scientific and Technical Information

Cft: Craft

CFT: Chirp Fourier Transform • Clifton Morenci (USA-AZ) (location identifier) • Cluster Fault Tolerance • Common Facilities Test • Companion Fault Tolerance • Complement Fixation Test • Complex Fourier Transform • Conformal Field Theory • Continuous Fourier Transport • Cray Fortran Compiler • Cross File Transfer • Cross-Flow Turbine • Cross-Functional Team • Crystal Field Theory

CFTA: Centre de Formation Technique Agréé

CFTC: Commodity Futures Trading Commission • Commonwealth Fund for Technical Cooperation

CFTE: Critical Fluid Thermal Equilibration

Cftmn: Craftsman

CFU: Call Forwarding Unconditional • Control Functional Unit • Corfu (Greece) (location identifier) • Kerkyra (Greece) (location identifier)

CFUVM: Call Forwarding Unconditional to Voice Mailbox • Club Français des Utilisateurs de VM

CFV: Call For Votes • Coffeyville (USA-KS) (location identifier) • Continuous Flow Ventilation

CFVME: Critical Fluid Viscosity Measurement Experiment

CFWP: Central Flow Weather Processor

CFWSU: Central Flow Weather Service Unit

CFWU: Central Flow Weather Unit

CFY: Company Fiscal Year • Current Fiscal Year (*)

CFZF: Commercial Float Zone Facility • Commercial Float Zone Furnace

cg: Centigram

Cg: Congress

C&G: Control & Guidance

CG: Case Galaxy • Center of Gravity • Character Generator • Cloud to Ground • Coast Guard • Cometary Globule • Compact Galaxy • Compact Group • Comptroller General • Computer-Generated • Computer Graphics • Conceptual Graph • Congo (ISO country code) • Conjugate Gradient • Consul General • Contadora Group • Contrainte Grammaticale • Control Group • Coolant Gas • Coordonnées Généralisées • Coulomb-Glauber

CGA: Call Global Area • Canadian Gas Association • Canonical Genetic Algorithm • Color Graphics Adapter • Compressed Gas Association • Craig (USA-AK) (location identifier)

CGAC: China General Aviation Corporation

CGAM: Centre for Global Atmospheric Modelling

CGAS: Coast Guard Air Station • Cooled Grating Array Spectrometer

CGAU: Cabin Gas Analysis Unit

CGB: Convert Gray to Binary • Cosmic Gamma-ray Background • Cuiabá - M. Rondon (Brazil-MT) (location identifier)

CGBA: Commercial Generic Bioprocessing Apparatus

CGC: Canadian Geoscience Council • Cape Gloucester (Papua New Guinea) (location identifier) • Clebsch-Gordan Coefficient • Coast Guard College • Color Graphics Converter • Comité de Gestion et de Coordination • Command Guidance Computer • Commission Géologique du

Canada • Confédération Générale des Cadres • Coût Global Client

CGCD: Crossed Grid Charge Detector

CGCG: Catalogue of Galaxies and Clusters of Galaxies

CGCM: Coupled General Circulation Model

CGCP: Canadian Global Change Programme

CGCT: Compagnie Générale de Constructions Téléphoniques

CGCU: Centre de Gestion de la Charge Utile

CGD: Changde (China-PRC) (location identifier) • Climate and Global Dynamics • Committee for Geophysical Data • Coplanar Grid Detector

CGDB: Central Geographic Database

CGDI: Canadian Geospatial Data Infrastructure

CGDS: Computer Graphics Display System

Cge: Carriage • Charge

CGE: Cambridge (USA-MD) (location identifier) • Capillary Gel Electrophoresis • Chadwick-Goldhaber Effect • Compagnie Générale d'Électricité • Compagnie Générale des Eaux • Computable General Equilibrium • Conférence des Grandes Écoles • Conseil Général des Études

CGED: Committee on Geophysical Environmental Data

CGER: Caisse Générale d'Épargne et de Retraite • Commission on Geosciences, Environment and Resources

CGES: Center for Global Environmental Studies

CGF: Cleveland - Cuyahoga County (USA-OH) (location identifier) • Color Guard Fiber • Command Generation Function • Computer-Generated Force • Crystal Growth Facility • Crystal Growth Furnace • Czech Government Flying Service

CGG: Casiguran (Philippines) (location identifier) • Centro Galileo Galilei (CNAA) • Commission on Global Governance • Compagnie Générale de Géophysique • Compensated Geothermal Gradient • Contingency Gravity Gradient

CGGTTS: CCDC Group on GPS Time Transfer Standards

CGH: Cape of Good Hope (South Africa) • CHON Grain Halo • Computer-Generated Hologram • Computer-Generated Holographic • Concept Generalization Hierarchy • São Paulo - Congonhas (Brazil-SP) (location identifier)

CGHA: Castel Gondolfo H-Alpha

CGI: Calibration Group Identifier • Canadian Glacier Inventory • Cape Girardeau (USA-MO) (location identifier) • Common Gateway Interface • Computer-Generated Image • Computer-Generated Imagery • Computer-Generated Imaging • Computer Graphics Interface • Corrugated Galvanized Iron

CGIA: Centre for Geographical Information and Analysis

CGIAR: Consultative Group on International Agricultural Research

CGIC: Canadian Geoscience Information Centre

CGIP: Computer Graphics Image Processing (periodical code)

CGIS: Canadian Geographical Information System • Continental Geographical Information System

CGJ: Chingola (Zambia) (location identifier)

CGK: Jakarta Soekarno-Hatta (Indonesia) (location identifier)

CGL: Charge Generation Layer • Circling Guidance Light • Commercial General Liability • Complex Ginzburg-Landau • Comprehensive General Liability • Continuous Gas Laser • Corrected Geomagnetic Latitude

CGLE: Complex Ginzburg-Landau Equation

CGLM: Classical General Linear Model

cgm: centigram

CGM: Camiguin Mambajao (Philippines) (location identifier) • Computer Graphics Metafile • Conjugate Gradient Method • Corrected Geomagnetic

CGMO: Coast and Geomagnetic Observatory

CGMRF: Compound Gaussian Markov Random Field

CGMS: Coordination Group on Meteorological Satellites • Coordination on Geostationary Meteorological Satellites (*)

CGMW: Commission for the Geological Map of the World

CGN: CCSDS Ground Network • Centre de Gravité National • Cologne Bonn (Germany) (location identifier) • Concentrator Group Number • Köln Bonn (Germany) (location identifier)

Cgo: Cargo

CGO: Cruz Gonzales O-stars) • Zhengzhou (China-PRC) (location identifier)

CGOM: Consultative Group on Ocean Mapping

CGP: Canonical Genetic Programming • Central Grounding Point • Chittagong (Bangladesh) (location identifier) • Command Generation Program • Committee on Gravitational Physics • Control Ground Point

CGPA2: Second General Catalogue of stars observed with Photoelectric Astrolabes

CGPL: Conversational Graphical Programming Language

CGPM: Conférence Générale des Poids et Mesures

CGPME: Confédération Générale des Petites et Moyennes Entreprises (Fran ce)

CGPO: Canadian Gemini Project Office

CGPS: Canadian Galactic Plane Survey • Compiling Genetic Programming System

CGQ: Changchun (China-PRC) (location identifier)

CGR: Campo Grande (Brazil-MS) (location identifier) • Captured Gamma Ray • Carbon Glass Resistor

CGRI: Commissariat Général aux Relations Internationales

CGRM: Computer Graphics Reference Model

CGRO: (Arthur Holley) Compton Gamma-Ray Observatory

CGROSSC: (Arthur Holley) Compton Gamma-Ray Observatory Science Support Center

CGRS: Central Gyro Reference System • Compact Gamma-Ray Spectrometer

CGRT: Carbon-Glass Resistance Thermometer

cgs: Centimeter-Gram-Second

C&GS: Coast & Geodetic Survey (USA) • Commercial & Government Systems

CGS: Center for Government Studies (USA-CA) • Central Geological Survey • Centre de Gestion des Servitudes • Centre de Gestion Scientifique • Centro di Geodesia Spaziale • Civil GPS Service • Coast and Geodetic Survey (USA) • Coast Guard Station • College Park (USA-MD) (location identifier) • Conjugate Gradient Squared • Continuous Grain Silicon • Cooled Grating Spectrometer • Cryogenic Grating Spectrometer

CGSA: Computer Graphics Structural Analysis

CGSB: Canadian General Standards Board • Canadian Government Specifications Board

CGSC: Chinese Geodetical Star Catalogue

CGSE: Common Ground Support Equipment

CGSIC: Civil GPS Service Interface Committee

CGSPO: Combined Gas Steam Plant Optimization

CGSS: Cryogenic Gas Storage System

CGT: Capital Gains Tax • Ceramic Gas Turbine • Chinguitti (Mauritania) (location identifier) • Closed Gas Turbine • Compagnie Générale Transatlantique • Compressed-Gas Turbomotor • Corrected Geomagnetic Time

CGTE: Global Changes in Terrestrial Ecosystems

CGTI: Conseil Général des Technologies de l'Information (France)

CGTO: Contracted Gaussian-Type Orbital

CGU: Canadian Geophysical Union • Ciudad Guayana (Venezuela) (location identifier)

CGV: Caiguna (Australia-WA) (location identifier) • Conditions Générales de Vente • Critical Grid Voltage

CGW: Computer Graphics World • Corning Glass Works

CGWIC: China Great Wall Industry Corporation (China-PRC)

CGWT: Cylindrically Guided Wave Technique

CGX: Chicago - Meiggs Field (USA-IL) (location identifier)

CGY: Cagayan de Oro (Philippines) (location identifier)

CGZ: Casa Grande (USA-AZ) (location identifier)

ch/: Cheque

ch: Candle-Hour • Choke • Cosinus Hyperbolique • Switzerland

Ch: Chain • Chair • Chairman • Chaldron • Chamber • Chambre • Chancellor • Channel • Chapel • Chapelle • Chapitre • Chaplain • Chapter • Charge • Charges • Chart • Château • Chaussée • Check • Chemical • Chemistry • Cheval-vapeur • Chief • Child • Children • Chile • Chilean • China • Chinese • Choir • Choke • Christ • Church • Methyladyne

C&H: Cosmic & Heliospheric

C/H: Camera Head • Central Heating

CH: Candle-Hour • Carnegie Hall • Case Hardening • Central Heating • Centre d'Hydrographie • Certified Hardware • Chaco (Argentina) • Channel • Chapel • Character Height • Chester (UK postal code) • Chickasaw • Chief Hydrologist • Chile • China • Church • Chuvash • Clearinghouse • Collision Hypothesis • Compass Heading • Conductor Head • Confederatio Helvetica • Confédération Helvétique • Control Heading • Control House • Coronal Hole • Court House • Critical Height • Cursor Horizontal • Customs House • Methylidyne • Suisse (ISO country code) • Switzerland (international vehicle registration code) • Switzerland (ISO country code) • Switzerland (postal country code)

Cha: Chameleon-Chameleonis (constellation)

CHA: Chattanooga (USA-TN) (location identifier)

ChA&A: Chinese Astronomy & Astrophysics (periodical code)

CHAC: Computer History Association of California (USA-CA)

CHAID: CHI-squared Automatic Interaction Detector

ChaIRN: Chamaeleon Infrared Nebula

Chal: Chaldron • Chaleur • Challenge

CHAM: Chemical Abstract Machine

CHAMADE: Change Management in Distributed Environment

Chamb: Chambers

CHAMMP: Computer Hardware, Advanced Mathematics and Model Physics

ChaMP: Chandra Multiwavelength Project

CHAMP: Comet Halley Active Monitoring Program

CHAMPs: Charge Massive Particles

CHAMPS: Configurable Hardware for Multidisciplinary Projects in Space

Chan: Chancellor • Chancery • Channel

CHAN: Channel

Chanc: Chancellor • Chancery

ChAOS: Chicago Adaptive Optics System

CHAOS: Canadian High Acceptance Orbit Spectrometer • Chapel Hill Astronomical and Observational Society (USA-NC)

Chap: Chapitre • Chaplain • Chapter

CHAP: Challenge Authentication Protocol • Challenge Handshake Authentication Protocol • Channel Processor

CHAPS: Clearinghouse Automatic Payments System

Chapt: Chapter

Char: Character • Characteristic • Characterize • Charter

CHAR: Character

CHARA: Center for High-Angular Resolution Astronomy (USA-GA)

Charact: Characterize

CHARC: Characteristics

CHARGE: Chemical Analysis of Released Gas Experiment

CHARM: Camera High-Angular-Resolution Module • Composite High-Altitude Radiation Model

CHARON: Calern High-Angular-Resolution Optical Network (France)

chart: Charta

chart. cerat.: Charta Cerata

Chas: Chassis

CHAS: Chagrin Valley Astronomical Society (USA-OH)

CHASE: Chamber Association of Small Enterprises • Coronal Helium Abundance Skylab Experiment

Chât: Château

CHAT: Conversational Hypertext Access Technology

Chb: Cherbourg (France)

CHB: Chilas (Pakistan) (location identifier)

Chbre: Chambre

Chbrs: Chambers

ChC: Choke Coil

CHC: Chance • Christchurch (New Zealand) (location identifier)

CHCK: Channel Check

CHCP: Change Code Page

CHCS: Cabin Humidity Control System

Chd: Chord

ChD: Doctor of Chemistry

CHD: Chandler (USA-AZ) (location identifier) • Coronary Heart Disease

Ch. de F.: Chemin de Fer

chdir: Change Directory

CHDL: Computer Hardware Description Language

CHDM: Cold and Hot Dark Matter • Cold plus Hot Dark Matter

CHDU: Current Header Data Unit

ChE: Chemical Engineer • Chief Engineer

CHE: Caherciveen Reenroe (Ireland) (location identifier) • Suisse (ISO country code) • Switzerland (ISO country code)

CHEAF: Centrum voor Hoge-Energie Astrofysica (Netherlands)

CHEASE: Cubic Hermite Element Axisymmetric Static Equilibrium

Chem: Chemical • Chemically • Chemin • Chemist • Chemistry

CHEM: Chemicals-Health-Environment-Me (*) • Chemicals-Health-Environment-Medicine

ChemE: Chemical Engineer

CHEMFET: Chemical-sensing Field Effect Transistor

CHEMRAWN: Chemical Research Applied to World Needs

CHEMS: Charge Energy Mass Spectrometer

ChEnC: Chemical Engineering Communications (periodical code)

ChEnS: Chemical Engineering Science (periodical code)

CHEOPS: Chemical Operations System • Chemistry of Ozone in the Polar Stratosphere • Chemistry Of the Polar Stratosphere

CHEP: Computing in High-Energy Physics

CHEPS: Center for Higher Education Policy Studies

CHERS: Charge Exchange Recombination Spectroscopy

Ches: Cheshire (UK-England)

CHESS: Cornell High-Energy Synchrotron Source (USA-NY)

CHEST: Combined Higher Education Software Team

CHeX: Confined Helium Experiment

CHEX: Committee on Human Exploration

Chf: Chief

CHF: Calibration History File • Chemical Heritage Foundation • Chinhae (Korea-ROK) (location identifier) • Continuous Heating Furnace • Coupled Hartree-Fock • Franc (Liechtenstein & Switzerland) (ISO currency code)

Chg: Change • Charge

CHG: Centre Hospitalier Général • Change • Chaoyang (China-PRC) (location identifier) • Charge

Chgd: Charged

Chge: Change • Charge

ChGeo: Chemical Geology (periodical code)

Chgov: Changeover

CHGR: Charger

CHGRP: Change Group

ChGSE: Chiral Gaussian Symplectic Ensemble

CHH: Chachapoyas (Peru) (location identifier)

Chi: Chicago (USA-IL) • China • Chinese

CHI: Catalogue Handling Interface • Chicago (USA-IL) (location identifier) • Computer Horizon Inc. • Computer-Human Interaction • Computer-Human Interface • Computer-Human Interfacing

CHIA: Cumulative Hydrologic Impact Assessment

Chic: Chicago (USA-IL)

CHIC: Community Health Information Classification

Chich: Chichester

CHIEF: Combined Helmholtz Integral Equation Formulation

ChIEJ: Chinese Institute of Engineers Journal (periodical code)

CHIH: Chihuahua (Mexico)

Chil: Chilean

Chim: Chimie • Chimique

Chin: China • Chinese

CHIN: Canadian Heritage Information Network • Community Health Information Network

CHIO: Cultural Heritage Information Online

CHIP: Character-Handling Interface Processor • Chromospheric Helium Imaging Photometer • Chromospheric HI Imaging Photometer • Computerized Homework In Physics • Constraint-Handler In PROLOG • Constraint-Handling In PROLOG

CHIPS: Clearinghouse Interbank Payments System • Cosmic Hot Interstellar Plasma Spectrometer

CHIRLAM: Chemical High-Resolution Limited Area Model

CHIRP: Confidential Hazardous Incident Reporting Programme • Confidential Human Incidence Reporting Programme • Confidential Human Incidence Reporting Project

CHIS: Chiapas (Mexico) • Computerized Hospital Information System

ChJ: Chief Justice

CHJ: Chipinge (Zimbabwe) (location identifier)

ChJA: Chinese Journal of Aeronautics (periodical code)

ChJIR: Chinese Journal of Infrared Research (periodical code)

ChJL: Chinese Journal of Lasers (periodical code)

ChJNP: Chinese Journal of Nuclear Physics (periodical code)

ChJS: Chinese Journal of Semiconductors (periodical code)

ChJSS: Chinese Journal of Space Science (periodical code)

Chk: Check

CHK: Chickasha (USA-OK) (location identifier)

ChkB: Check Bit

chkdsk: Check Disk

Chkr: Checker

Chl: Chloride • Chloroform

Ch-L: Chef-Lieu

CHL: Certified Hardware List • Challis (USA-ID) (location identifier) • Channel • Chef-Lieu • Chile (ISO country code) • Chili (ISO country code)

Chld: Chilled

Chlo: Chloride • Chloroform

CHLW: Commercial High-Level Waste

Chm: Chairman

CHM: Capital Himano y Movilidad • Chimbote (Peru) (location identifier) • Constructable Habitat Module

Chmbr: Chamber

CHMBR: Chamber

CHMI: Czech Hydrometeorological Institute

CHML: Child Meal

CHMM: Continuous Hidden Markov Model

Chmn: Chairman

CHMMN: Commission Hydrographique de la Méditerranée et de la Mer Noire

Chmn: Chairman

chmod: Change Mode

Chn: Channel

CHN: China (PRC) (ISO country code) • Chine (RPC) (ISO country code) • Chonju (Korea-ROK) (location identifier) • Hydrogen Cyanide • PRC (ISO country code) • République Populaire de Chine (ISO country code)

Chng: Change

CHNO: Hydrogen Isocyanate

CHNS: Hydrogen Isothiocyanate

CHO: Aldehyde • Charlottesville (USA-VA) (location identifier)

Choc: Chocolate

CHOICE: Chemical Online Information Center

CHOL: Common High Order Language

CHON: Carbon-Hydrogen-Oxygen- Nitrogen

CHOPS: Chief of Operations

CHORUS: CERN Hybrid Oscillation Research Apparatus

CHOWN: Change Owner

Chp: Chairperson • Chopper

C&HP: Cosmic & Heliospheric Physics

CH&P: Crew Habitability & Protection

CHP: California Highway Patrol • Catalogue Handling Plus • Center for the History of Physics (USA-MD) • Channel Path • Chapter • Chemical Heat Pump • Circle Hot Springs (USA-AK) (location identifier) • Combined Heat and Power

CHPC: Center for High-Performance Computing

Ch. Pd.: Charges Paid

ChPhL: Chinese Physics Lasers (periodical code)

ChPhy: Chinese Physics (periodical code)

CHPI: Chung Hua Polytechnic Institute (China-ROC)

Ch. Ppd.: Charges Prepaid

Chq: Cheque • Chèque

CHQ: Chania (Greece) (location identifier)

CHR: Centre Hospitalier Régional • Character • Character Handling Routine • Chateauroux (France) (location identifier) • Committee on Human Rights • Cooper-Harper Rating

CHRD: Channel Ready

Chrg: Charge

Chrgr: Charger

CHRIS: Chemical Hazard Response Information System • Compact High-Resolution Imaging Spectrometer

CHRISP: Coronagraphic High-Resolution Infrared Spectrograph

CHRISTA: Cryogenic Infrared Spectrometers and Telescopes for the Atmosphere

Chrm: Chairman

CHRNS: Center for High-Resolution Neutron Scattering

CHROMEX: Chromosomes Experiment

Chron: Chronicle • Chronique • Chronograph • Chronological • Chronologically • Chronologique • Chronology • Chronometer • Chronomètre • Chronometry

Chronol: Chronological • Chronologically • Chronologique • Chronology

CHRP: Calcul et Réseaux à Haute Performance • Common Hardware Reference Platform

CHRU: Centre Hospitalier Régional Universitaire

Chrs: Chambers

Chrst: Characteristic

Chrt: Chart

Chs: Chapters

CHS: Calibration Heat Source • Canadian Hydrographic Service • Carlsbad High School (USA-NM) • Change Sign • Charleston (USA-SC) (location identifier) • Comité d'Hygiène et de Sécurité • Community Health Service • Compact Helical System • Cooperative Hypermedia Server • Cooperative Hypertext Server • Cross Hole Seismic • Cylinder Head Sector

ChSAJ: Chinese Society of Astronautics Journal (periodical code)

ChSBu: Chinese Science Bulletin (periodical code)

CHSCT: Comité d'Hygiène, de Sécurité et des Conditions de Travail (Fran ce)

CHSF: Cargo Hydrazine Service Facility

ChSMJ: Chinese Society of Mechanical Engineers Journal (periodical code)

CHSSI: Common HPC Software Support Initiative

ChSST: Chinese Space Science Technology (periodical code)

CHT: Cathode Heating Time • Chatham Island (New Zealand) (location identifier) • Cylinder Head Temperature

Chtg: Charting

CHU: Centigrade Heat Unit • Centre Hospitalier Universitaire • Chuathbaluk (USA-AK) (location identifier) • Current Header Unit

ChUI: Character User Interface

CHUNET: Swiss University Network

CHUNNEL: Channel Tunnel

CHUR: Chondritic Uniform Reservoir

CHV: Chaves (Portugal) (location identifier)

CHW: Command Header Word • Constant Hot Water • Jiuquan (China-PRC) (location identifier)

CHWS: Chilled Water System

CHX: Cabin Heat Exchanger • Changuinola (Panama) (location identifier) • Condensing Heat Exchanger

CHW: Chilled Water • Cold Heading Wire • Cold and Hot Water

CHX: Condensing Heat Exchanger

CHy: Commission for Hydrology (WMO)

CHY: Choiseul Bay (Solomon Islands) (location identifier) • Columbia-Harvard-Yale

CHZ: Chiloquin (USA-OR) (location identifier) • Continuously Habitable Zone

CH3CN: Methyl Cyanide

CH4: Methane

c/i: Certificate of Insurance

Ci: Cirrus • Curie

C&I: Commerce & Industry • Commercial & Industrial • Communication & Instrumentation • Cost & Insurance

C/I: Carrier-to-Interference • Carrier-to-Intermodulation • Certificate of Insurance

CI: Call Indicator • Canary Islands • Card Input • Caroline Islands • Cascade Ionization • Cast Iron • Catalog Interoperability • Cédula de Identidad • Center of Impact • Central Interval • Centre d'Information • Certification Inspection • Chain Index • Channel Islands (*) • Chemical Inspectorate • Chemical Ionization • Chief Inspector • Chief Instructor • Chimalateco • Christmas Island • Circuit Intégré • Circuit Interrupter • Close-In • Code Instrument • Coefficient of Intelligence • Collisional Ionization • Color Index • Color Interior • Combustion Index • Commonwealth Institute • Communication and Instrumentation • Compass and Inclinometer • Competitive Intelligence • Component Interface • Compound Interferometer • Compression-Ignition • Computer Indicator • Confidence Interval • Configuration Information • Configuration Inspection • Configuration Interaction • Configuration Item • Congiunzione Inferiore • Consular Invoice • Context Independent • Continuous Improvement • Continuous Interlock • Contract Initiation • Contract Item • Contrast Index • Control Interval • Converging Inflow • Coordinate Index • Coronal Index • Coronagraphic Imager • Corporate Identity • Corrosion Inhibitor • Cost Index • Côte d'Ivoire (international vehicle registration code) • Côte d'Ivoire (ISO country code) • Courrier International • Course Interpreter • Course Invalid • Crew Interface • Cubic Inch • Cumulative Index • Curriculum and Instruction • Ivory Coast (international vehicle registration code) • Ivory Coast (ISO country code)

Cia: Compagnia • Companhia • Compañia

CIA: Cash In Advance • Centrale Informatie Afdeling (Netherlands) • Central Intelligence Agency (USA) • Centre d'Intelligence Artificielle • Chemical Industries Association • Chemical Institute of Australia • Collision-Induced Absorption • Communications Interface Assembly • Compagnía Aeronautica Italiana • Competitive Intelligent Agent • Computer Industry Association (USA) • Computer Interface Adaptor • CONE Interface Adapter • Conference Interface Agent • Configuration Identification Accounting • Conseil International des Archives • Control Interface Assembly • Controller Interface Assembly • Cooperative Information Agent • Cooperative Information Agents (workshop) • Coronographe Interférentiel Achromatique • Roma Ciampino (Italy) (location identifier)

CIAA: Centro de Investigación Astronómico de Alicante (Spain) • Coordinator of Inter-American Affairs

CIAB: Conseil International des Agences Bénévoles

CIAC: Centro de Información Ambiental del Caribe • Computer Incident Advisory Capability

CIAD: Climate Impact Assessment Division

CIAE: Canadian Institute for Adult Invitation

CIAHG: Copyright Issues Ad Hoc Group

CIAI: Commerce and Industry Association Institute

Cial: Commercial

CIALS: Confédération Internationale des Arts, des Lettres et des Sciences

CIAM: Cooperative Institute for Applied Meteorology

CIAME: Comité Interprofessionnel pour l'Automatisation et la Mesure Électronique (France)

CIAO: Conception Industrielle Assistée par Ordinateur • Coronagraph Imager with Adaptive Optics

CIAP: Climatic Impact Assessment Program

CIAR: Canadian Institute for Advanced Research (Canada-ON) • Center for Inter-American Relations

CIArb: Chartered Institute of Arbitrators

CIAS: Conference of Independent African States

CIASE: China Institute of Aeronautic System Engineering • Computer Institute for Applications in Science and Engineering

CIAT: Contrat Interministériel d'Aménagement du Territoire

CIATF: Comité International des Associations Techniques de Fonderie

CIAX: Cassegrain Instrument Auto Exchanger

cib: Cibus

CIB: Catalina Island (USA-CA) (location identifier) • Central Intelligence Board • Centre d'Information Bibliographique • Centre International Bibliographique • Centrum voor Informatiebeleid (Netherlands) • Change Impact Board • Change Implementation Board • Classification Internationale des Brevets • COBOL Information Bulletin • Command Input Block • Control Interface Block • Controlled Image Base • Cosmic Infrared Background

CIBA: Control Instrumentation for Balloon

CIBR: Cosmic Infrared Background Radiation

CIBTC: China International Book Trading Corporation

Cic: Cicero (106-43 BC)

C-I-C: Cloud-In-Cell

CIC: Cable In Conduit • Canadian Institute of Chemistry • Cargo Integration Contractor • Carrier Identification Code • Centro de Información Científica • Centro Interdipartimentale della Communicazione • Chemical Information Center • Chemical Institute of Canada • Chico (USA-CA) (location identifier) • China Internet Corporation (China-PRC) • CIGNET Information Center (USA-MD) • Cloud-In-Cell • Combined Intelligence Center • Comet Impact Campaign • Comisión de Investigaciones Científicas • Comité International de Coordination • Commander In Chief • Command Input Coupler • Command Interface Control • Committee on Information and Communications • Committee on Institutional Cooperation • Commonwealth Information Centre • Communication Intelligence Corporation • Communications Instructor Console • Compensated Ion Chamber • Complex Integrated Circuit • Computer Instruments Corporation • Computer Interface Control • Computerized Information Center • Connector Integrated Cell • Consumer Information Catalog • Consumers Interest Council • Contamination Instrument Complement • Control and Information Center • Control Inteligente Computerizado • Controller In Charge • Core Image Converter • Coronado Instrument Group Inc. (USA-AZ) • Coronagraphic Imaging Camera • Creative Incentive Coalition • Crew Interface Coordinator • Critical Item Code • Critical Item Consolidation • Customer Identification Code

CICA: Center for Innovative Computer Applications • Center for Innovative Computing Applications • Council of International Civil Aviation • Customer Identification Control and Accounting

CICADA: Central Instrumentation Control And Data Acquisition

CICAP: Comitato Italiano per il Controllo delle Affermazioni sul Paranormale (Italy)

CICAR: Cooperative Investigations of the Caribbean and Adjacent Regions

CICATS: Computer Industry Coalition for Advanced Television Service

CICC: Cable-In-Conduit Conductor • Cargo Integration Control Center • Center of the International Cooperation for Computerization

CICCR: Configuartion Information Control Center Requirement

CICE: Centre International de Calcul Électronique (ONU)

CICESE: Centro de Investigación Científica y de Educación Superior de Ensenada (Mexico)

CICG: Centre International des Conférences de Genève (Switzerland) • Centre Interuniversitaire de Calcul de Grenoble (France)

CICH: Centro de Información Científica y Humanística (UNAM)

CICHE: Committee for International Cooperation in Higher Education

CICI: Confederation of Information Communication Industries

CICIAM: Committee for the International Conferences on Industrial and Applied Mathematics

CICIS: Chemicals In Commerce Information System

CICP: Committee to Investigate Copyright Problems

CICR: Comité International de la Croix-Rouge

CICRIS: Cooperative Industrial and Commercial Reference and Information Service

CICRP: Centre Interuniversitaire de Calcul de la Région Parisienne (France)

CICS: Center for Intelligent Control Systems • Central Integrated Checkout System • Core Instrument Control System • Customer Information Control System • Custom Information Control System

CICT: Centro de Información Científica-Ténica • Conseil International du Cinéma et de la Télévision

CICU: Central Interface Converter Unit • Commission Internationale de la Classification Universelle • Communications Interface Control Unit • Computer Interface Conditioning Unit • Computer Interface Control Unit

CICYT: Comisión Interministerial de Ciencia y Tecnología

CID: Cable Interconnection Diagram • Cabling Interface Drawing • Cedar Rapids (USA-IA) (location identifier) • Center for Information and Documentation • Centre d'Information et de Documentation • Centre d'Informatique Documentaire • Centre International de Documentation • Centro de Investigación Documentaria • Centro de Investigación y Desarollo • Change In Design • Channel Identification • Characteristic Item Description • Charge Injection Device • Chemical Information Division • Cidade • Circular Intensity Difference • Civil Investigative Demand • Collection of Indexable Data • Collisionally-Induced Dissociation • Collision-Induced Dissociation • Commercial Item Description • Commercial Item Drawing • Component Identification • Computer-assisted Imaging Device • Computer-Integrated Design • Computer-Integrated Documentation • Computer Interface Device • Configuration-Installation-Distribution • Converted In Distance • Criminal Investigation Department • Cubic Inch Displacement • Current-Image Diffraction • Customer Interaction Data • Iowa City (USA-IA) (location identifier)

CIDA: Canadian International Development Agency • Centre d'Informatique et de Documentation Automatique • Centro de Investigaciones de Astronomía (Venezuela) • Centro de Investigaciones y Desarollo de la Armada (Spain) • Cometary and Interstellar Dust Analyzer • Comité Intergouvernemental du Droit d'Auteur

CIDA-E: Centro de Investigación y Difusión Aeronáutico-Espacial (Uruguay)

CIDAE: Centro de Investigación y Difusión Aeronáutico-Espacial (Uruguay)

CIDAS: Centro Italiano Divulgazione Astronomica e Spaziale (Italy)

CIDAV: Centro Internacional de Documentación Audiovisual

CIDE: Centro de Investigación y Desarollo Espacial • Centro de Información y Documentación de Empresas • Centro de Investigación y Desarollo de la Educación • Comisión de Inversión y Desarollo Económico

CIDEP: Chemically Induced Dynamic Electron Polarization

CIDESA: Centre International de Documentation Économique et Sociale Africaine

CIDEX: Courrier Individuel à Distribution Exceptionnelle

CIDF: Control Interval Definition Field

CIDIE: Committee of International Development Institutions for the Environment

CIDIM: Cooperating Intelligent System for Distributed Management

CIDL: Canadian Initiative on Digital Libraries • Configuration Item Data List • Control Interface Document List

CIDN: Change In Drawing Notice

CIDNP: Chemically Induced Dynamic Nuclear Polarization

CIDOC: Center for Intercultural Documentation

CIDP: Centre International de Documentation Parlementaire (Switzerland) • Control Instrument Data Processor

CIDR: Classless Interdomain Routing • Classless Internet Domain Routing • Critical Intermediate Design Review

CIDRE: Circulation Intelligente de Dossiers Répartis

CIDS: CEPEX Integrated Data System • Circular Intensity Differential Scattering • Circularity and Inner Diameter Station • Collateral Information Display System

CIDSC: Contractor In-house Data Support Capability

CIDSP: Centre d'Information des Données Socio-Politiques

CIDST: Comité d'Information et de Documentation Scientifique et Technique

Cie: Compagnie

CIE: Cargo Integration Equipment • Centre for International Economics • Cesium Ion Emission • Chinese Institute of Electronics • Chinese Institute of Engineers • Ciclo de Información Espacial • Collie (Australia-WA) (location identifier) • Collisional Ionization Equilibrium • Comité Inter-Entreprises • Commissariat International de l'Éclairage • Commission Internationale de l'Éclairage • Common Ion Effect • Communications Interface Equipment • Compagnie Internationale de l'Éclairage • Computer-Integrated Engineering • Confédération Internationale des Étudiants • Contrat Initiative Emploi (France) • Controlled Internal Extension

CIEA: Conseil International d'Éducation des Adultes • Committee on International Environmental Activities

CIEC: Citizens Internet Empowerment Coalition • Commission Internationale de l'État Civil • Conseil International des Études Canadiennes (Canada)

CIECC: Consejo Interamericano para Educación, Ciencia y Cultura

CIEE: Council of International Educational Exchange

CIEF: Capillary Isoelectric Focusing

CIEM: Comité International pour l'Exploration de la Mer • Computer-Integrated Engineering and Manufacturing • Conseil International pour l'Exploration de la Mer

CIEMAT: Centro de Investigaciones Energeticas Medioambientales y Tecnológicas (Spain)

CIEMS: Catalog for Information Exchange and Message Standards

Cien: Ciencia • Científico

CIEPS: Centre International d'Enregistrement des Publications en Série

CIES: Centre International des Étudiants et Stagiaires (*) • Centre International des Étudiants Stagiaires

CIESIN: Consortium for International Earth Science Information Network

CIESM: Commission Internationale pour l'Exploration Scientifique de la Mer Méditerranée

CIESP: Centro de Estudos de Informática e Educação de São Paulo (Brazil-SP)

CIETT: Confédération Internationale des Entreprises de Travail Temporaire

CIF: Capacitor Input Filter • Cell In Frames • Central Index File • Central Information Facility • Central Information File • Central Instrumentation Facility • Central Integration Facility • Centro Internacional de Física • Chifeng (China-PRC) (location identifier) • Cloning Inhibiting Factor • Código de Identificación Fiscal (Spain) • Cohesive Intermolecular Force • Color Interior Film • Common Interchange Format • Common Intermediate Format • Computer Integrated Factory • Computer Interface Function • Congé Individuel de Formation • Conseil International des Femmes • Contrast Improvement Factor • Corrected Impact Factor • Cost-Insurance-Freight • Crystallographic Information File • Crystallographic Input Format • Customer Information File

CIFC: Cost-Insurance-Freight-Commission

CIFCI: Cost-Insurance-Freight-Commission-Interest

CIFE: Colleges and Institutes for Further Education • Conseil International du Film d'Enseignement • Cost-Insurance-Freight-Exchange

CIFEG: Centre International pour la Formation et les Études Géologiques

CIFI: Cost-Insurance-Freight-Interest

CIFJ: Centre International du Film pour la Jeunesse

CIFKU: Conditional Instability of the First Kind Upside-down

CIFMA: Centro Internacional de Física y Matemáticas Aplicadas (Mexico)

CIFOR: Center for International Forestry Research

CIFRE: Convention Industrielle de Formation par la Recherche

CIFS: Committee for the International Freedom of Scientists (APS) • Common Internet File System • Copenhagen Institute for Future Studies (Denmark)

CIG: Cable Integrity Group • Canadian Institute of Geomatics • Centro de Investigaciones Geoténicas • Chemical Ion Generator • Climate Interest Group • Clumpy Irregular Galaxy • Comité International de Géophysique • Communications and Interface Group • Compound Interlace Grating • Computer Image Generator • Computerized Interactive Graphics • Conférence Intergouvernementale • Connector Integration Group • Connector Integrity Group • Coordinate Indexing Group • Craig Moffat (USA-CO) (location identifier) • Customer Interface Group

CIGAR: Channel Inspection and Gauging Apparatus of Reactors

CIGB: Commission Internationale des Grands Barrages

CIGIS: Comprehensive Institute of Geotechnical Investigations and Surveying

CIGMA: Cooperative Institute for Geoscience Management and Applications

CIGNET: Cooperative International GPS Network

CIGR: Commission Internationale du Génie Rural

CIGRE: Conférence Internationale des Grands Réseaux Électriques

CIGREF: Club Informatique des Grandes Entreprises Françaises (France)

CIGS: Cooled Infrared Grating Spectrometer • Copper Indium Gallium Diselenide

CIGSS: Common Imagery Ground/Surface System

CIH: Changzhi (China-PRC) (location identifier) • Coronographe Interférentiel Hybride

CIHS: Conseil de l'Informatique Hospitalière et de Santé

CII: Center for Industrial Innovation • Change Installation Instruction • Computer-Integrated Instruction • Configuration Identification Index • Configuration Item Identifier

CIIA: Canadian Institute of International Affairs

CIIBA: Comité Interministériel pour l'Informatique et la Bureautique dans l'Administration (France)

CIID: Centro Internacional de Investigaciones para el Desarollo • Commission Internationale des Irrigations et du Drainage

CIIL: Configured Items Identification List • Control Interface Intermediate Language

CIIR: Center for Intelligent Information Retrieval (USA-MA) • Central Institute for Industrial Research

CIISTU: Commissióne Italiana per l'Insegnamento della Scienze della Terra e dell'Universo (Italy)

CIJ: Cobija (Bolivia) (location identifier)

CIJB: Cradle Interface Junction Box

CIJE: Current Index to Journals in Education

CIJEDAE: Comisión Interamericana de Juristas Expertos en Derecho Aéreo y Espacial

CIK: Chalkyitsik (USA-AK) (location identifier) • Crypto Ignition Key

CIKM: Conference on Information and Knowledge Management

CIL: Common Interface Language • Complex Intermediate Language • Component Integration Laboratories • Computer Interpreter Language • Configuration Inspection Log • Contract Inspection Log • Contrat d'Initiative Locale • Council (USA-AK) (location identifier) • Critical Items List

CILB: Commission Internationale de Lutte Biologique

CILET: Circular Letter

CILF: Conseil International de la Langue Française

CI/LI: Corrosion Inhibitor/Lubricity Improver

CILI: Corrosion Inhibitor/Lubricity Improver

CILN: Condensation de l'Information en Langue Naturelle

CILS: Computer Information Library Service

CILT: Center for Innovative Learning Technologies (USA)

CIM: Cable Interface Module • Cimitarra (Colombia) (location identifier) • Collection Insertion Map • Command Interface Module • Common Information Model • Common Interface Model • Communications Interface Module • CompuServe Information Manager • Computer Input from Microfilm • Computer Input Matrix • Computer Input Media • Computer Input Microfilming • Computer Input Multiplexer • Computer Input on Microfilm • Computer-Integrated Manufacturing • Conférence Islamique Mondiale • Council on Information Management • Cryogenic Transfer Module • Crystal Impedance Meter

CIMASI: Conférence Internationale sur les Mathématiques Appliquées et les Sciences de l'Ingébieur

CIME: Centro Internazionale Matematico Estivo • Comité Intergouvernemental pour les Migrations Européennes • Computer Integrated Manufacturing and Engineering • Conseil International des Moyens d'Enseignement • Consejo Intergubernamental para las Migraciones Europeas • Council of Industry for Management Education • Cursus Intégré pour la Mobilité des Étudiants

CIMH: Caribbean Institute for Meteorology and Hydrology (formerly CMI) • Commission Internationale pour la Météorologie Historique

CIMI: Computer Interchange of Museum Information • Consortium for the Interchange of Museum Information

CIMIC: Center for Information Management and Connectivity (USA-NJ)

CIMM: Canadian Institute of Mining and Metallurgy

CIMMS: Cooperative Institute for Mesoscale Meteorological Studies

CIML: Comité International de Métrologie Légale • Contract Item Material List

CIMNE: Centro Internacional de Métodos Numéricos en Ingeniería (Spain)

CIMO: Commission des Instruments et des Méthodes d'Observation • Commission des Instruments et Méthodes d'Observation • Commission for Instruments and Methods of Observation (WMO)

CIMP: Climatic Index Monitoring Program • Commission on Igneous and Metamorphic Petrogenesis • Common Information Management Protocol

CIMS: Chemical Information Management System • Chemical Ionization Mass Spectrometry • Consolidated Information Management System

CIMSS: Cooperative Institute for Meteorological Satellite Studies

CIMU: Compatibility Integration Mock-Up

CIN: Carroll (USA-IA) (location identifier) • Center Information Network • Change Identification Number • Change Identifier Number • Change Incorporation Notice • Change Instrumentation Notice • Chemical Industry Notes • Code Identification Number • Comité International de Normalisation • Contract Item Number • Cooperative Information Network • Corporate Internet

CINA: Comisión Internacional de Navegación Aérea

CINAA: Cyclic Instrumental Neutron Activation Analysis

CinC: Commander in Chief

CINC: Combined Intelligence Center • Commander in Chief

CIND: Computer Index of Neutron Data

CINDA: Chrysler Integrating Numerical Difference Analyzer

CINDAS: Center for Information and Numerical Data Analysis and Synthesis (USA-IN)

CINE: Comet Impact Network Experiment

Cinem: Cinematografía

Cinn: Cincinnati (USA-OH)

CINOF: Campaign for Integrated Observations of Solar Flares

CINR: Central Institute for Nuclear Research

CINS: Cryogenic Inertial Navigation System

CINTEL: Computer Interface for Television

CINU: Centre d'Information des Nations Unies (ONU)

CINTEX: Catalog Interoperability Experiment

CINVESTAV: Centro de Investigación y de Estudios Avanzados (Mexico)

CIO: Calibrated Individual Observation • Catalog of Infrared Observations • Central Imagery Office (USA) • Central Information Office • Centre d'Information et d'Orientation • Centro de Investigaciones en Optica • Chief Information Officer • Chief Investment Officer • Comité International d'Organisation • Comité International Olympique • Commission International d'Optique • Common Item Order • Compagnie des Indes Orientales • Congress of Industrial Organizations • Conventional International Origin

CIOB: Central Indian Ocean Basin

CIOCS: Communications Input and Output Control System • Communications Input/Output Control System

CIOMS: Council for International Organizations of Medical Sciences

CION: Centre Informatique de l'Observatoire de Nice

CIOP: CAMAC Input-Output Processor • Centro de Investigaciones Opticas • Committee on International Organizations and Programs

CIOS: Conseil International pour l'Organisation Scientifique

CiP: Computers in Physics

CIP: Capital Investment Plan • Capsule Internal Programmer • Cast Iron Pipe • Catalog Interoperability Protocol • Cataloguing In Publication • Cellule d'Information Préventive • Cellule Information Prévention • Central Information Processor • Centralized Information Processing • Centre d'Information de la Presse • Change Implementation Plan • Change Improvement Plan • Channel Interface Processor • Chipata (Zambia) (location identifier) • Classroom Instruction Program • Clustering Initial Pattern • Cold Isostatic Pressing • Combined Instrument Panel • Command Interface Port • Commercially Important Person • Committee on International Programs • Common Industrial Policy • Common Input Processor • Common Integrated Processor • Communication Improvement Program • Communications Interface Processor • Component Improvement Program • Compressor Inlet Pressure • Computer Image Processing • Computer Information Processing • Computer-Investitions-Programm • Computers In Physics • Constant Injection Pressure • Contour Interpolation Program • Contract Information Processor • Control and Interface Panel • Convergence and Implementation Plan • Coordinated Instrument Package • Coronal/Interplanetary Physics • Cost Improvement Plan • Cryogenic Insulation Program • Current Injection Probe • Current In the Plasma

CIPA: California Institute for Physics and Astrophysics (USA-CA) • Chartered Institute of Patent Agents • Contributions for International Peacekeeping Activities

CIPE: Center for Image Processing in Education (USA-AZ) • Central Institute for Physics of the Earth • Committee on Computers In Physics Education • Concurrent Image Processing Executive • Conseil Interministériel de Planification Économique

CIPI: Canadian Intellectual Property Institute

CIPM: Comité International des Poids et Mesures

CIPO: Canadian Intellectual Property Office • Comité International pour la Protection des Oiseaux (now BirdLife)

CIPP: Context-Input-Process-Product

CIpR: Cubic Inches Per Revolution

CIPR: Commission Internationale de Protection contre les Radiations • Cubic Inches Per Revolution

CIPRA: Commission Internationale pour la Protection des Alpes

CIPS: Canadian Information Processing Society • Centralized Information Processing System • Cesium Ion Propulsion System • C Image Processing System • Continuous Ice Particle Sampler

CIPSH: Conseil International de Philosophie et de Sciences Humaines • Conseil International pour la Philosophie et les Sciences Humaines (*)

CIPSO: Commercial INTERNET Protocol Security Option

CIPW: Cross-Iddings-Pirsson-Washington

CIQ: Chiquimula (Guatemala) (location identifier)

CIQM: Centre for Information Quality Management

Cir: Circa • Circinus-Circini (constellation) • Circle • Circonscription • Circuit • Circular • Circulation • Circumference • Circus • Cirugía

CIR: Cairo (USA-IL) (location identifier) • Cargo Integration Review • Central Intensity Ratio • Circle • Circulate • Color Infrared • Comité Interrégional • Commission on Industrial Relations • Committed Information Rate • Compagnie Industrielle Radioélectrique • Computerized Information Retrieval • Computerized Inspection Report • Configuration Inspection Report • Configuration Investigation Review • Conversational Information Retrieval • Corotating-Interacting Region • Corotating Interaction Region • Corotation-Inclination Resonance • Corotation Inclined Resonance • Cost Information Report • Council on Industrial Relations • Critical Item Report • Current Instruction Register

CIRA: Center for Information Revolution Analyses • Centre Interministériel de Renseignements Administratifs (France) • Centro Italiano Ricerche Aerospaziale (Italy) • Conference on Industrial Research Associations • COSPAR International Reference Atmosphere

CIRAC: Canadian Institute for Research in Atmospheric Chemistry

CIRB: Cosmic Infrared Background • Create Interrupt Request Block

Circ: Circa • Circle • Circuit • Circular • Circulation • Circumference • Circumstance • Circus

CIRC: Central Information Reference and Control • Circle • Circular • Circular Reference • Circulation Input Recording Center • Cross-Interleaved Reed-Solomon Code

CIRCE: Centre Interrégional de Calcul Électronique • Complex Instruments Ranking with a new Computational Environment

CIRCL: Circle

CIRCLE: Champollion Infrared Comet Lander Experiment

Circum: Circumference

CIRCUM: Circumference

CIRD: Cargo Integration Review Dry (run)

CIRE: Collaborative to Integrate Research and Education

CIRED: Centre International de Recherche sur l'Environnement et le Développement

CIRES: Cooperative Institute for Research in Environmental Sciences (USA-CO)

CIRHS: Critical Items and Residual Hazards List

CIRIS: Central Inertial Reference Instrumentation System • Completely Integrated Range Instrumentation System • Cryogenic Infrared Radiance Instrumentation for Shuttle

CIRLA: Cheasapeake Information Research Library Alliance

CIRM: Celestial Infrared Mapping • Celestial Infrared Measurement • Comité International Radio-Maritime

CIRMES: Centre for Integrated Resource Management and Environmental Science

CIRMM: Centre Interuniversitaire de Rencontres Mathématiques de Marseille (France)

CIRMS: Celestial Infrared Measurement System

CIRP: Conference on Infrared Physics

CIRPAS: Center for Interdisciplinary Remotely-Piloted Aircraft Studies

CIRPASS: Cambridge Infrared Panoramic Survey Spectrograph

CIRRIS: Cryogenic Infrared Radiance Instrumentation for Shuttle • Cryogenic Infrared Radiance Instrument for Shuttle

CIRS: Center for Information and Research Services • Circles • Composite Infrared Spectrometer • Computerized Information Retrieval Service

CIRSI: Cambridge Infrared Survey Instrument

CIRST: Centre Interministériel de la Recherche Scientifique et Technologique (France) • Centre Interuniversitaire de Recherche sur la Science et la Technologie

CIRTV: Centre International de Radio et Télévision

CIRUISEF: Conférence Internationale des Responsables d'Universités et Instituts Scientifiques d'Expression Française

CIRUS: Consortium for Industrial Research in the Use of Space

CIS: Canadian Institute of Surveying • Canton Island (Kiribati) (location identifier) • Capillary Ion Source • Card Information Structure • Cargo Integration Schedule • CASE Integration Service • Cataloguing In Source • CDIAC Information System • Cells In Space • Center for Interdisciplinary Studies (UK) • Center for International Studies • Centre for Information Science • Central Integration Site • Cesium Ion Source • Change Impact Summary • Chemical Information Services • Chemical Information System • CLUSTER Ion Spectrometer • CLUSTER Ion Spectrometry • CLUSTER Ion Spectroscopy • Commentator Information System • Common Information System • Commonwealth of Independent States • Communication and Instrumentation System • Communications and Information System • Communications Information System • Communications Interface System • Community Information System • Compact Image Sensor • Compact Interactive Standard • Compensated Image System • Compensated Imaging System • Component Identification Sheet • Complementary Information System • CompuServe Information Service • Computer Information System • Configuration Information System • Congressional Information Service • Contact Image Sensor • Contractor's Information Submittal • Contrat d'Intervention sur Site • Conventional Inertial System • Cooperative Information System • Coordinate Indexing System • Copper Indium Diselenide • Corporate Information System • Cost Information System • Countryside Information System • Cryogenic Instrumentation System • Cryogenic Interferometer/Spectrometer • Customer Information System • Customer Interaction Software • Customs Information System • Cycles In Service

CISA: Circular Index Sequential Access

CISAC: Committee on International Security and Arms Control (NAS) • Confédération Internationale des Sociétés d'Auteurs et de Compos iteurs

CISAM: Compressed Index Sequential Access Method

CISAS: Centro Interdipartimentale di Studi e Attivitá Spaziali (Italy)

CISC: Complex Instruction Set Computer • Complex Instruction Set Computing

CISCO: Cooled Infrared Spectrograph and Camera for OHS

CISDA: Centre for Information Society Development in Africa

CiSE: Computer in Science and Engineering (periodical)

CISE: Centro de Investigaciones y Servicios Educativos • Computer and Information Science and Engineering

CISET: Committee on International Science, Engineering and Technology

CISG: Communauté Industrielle Spatiale de Guyane (French Guyana)

CISIG: Geographic Information Systems Interest Group

CISIR: Ceylon Institute of Scientific and Industrial Research

CISK: Conditional Instability of the Second Kind (*) • Convective Instability of the Second Kind

CISM: Canadian Institute of Surveying and Mapping • Centre International pour les Sciences Mécaniques

CISP: Cooperative Interactive Storyboarding Prototyping

CISPR: Comité International Spécial des Perturbations Radioélectriques

CISR: Centre for Information Science Research • Commercial International Security Requirements • Configuration Item and Status Report

CISS: Centaur Integrated Support Structure • Comprehensive Integrated Security System • Cooperation of ISO Serendipity Survey

CIST: Centre d'Information sur les Sciences de la Terre • Centre International pour la Science et la Technologie • Computer-Aided Surgery and Therapy

CISTI: Canadian Institute for Scientific and Technical Information

CISTOD: Confederation of International Scientific and Technological Organizations for Development

Cit: Citation • Cited • Citizen • Citrate

CIT: Calibration Implementation Team • California Institute of Technology (USA-CA) • Carnegie Institute of Technology • Case Institute of Technology • Center for Innovative Technology • Central Institute of Technology • Centro de Información Técnica • Chemical Instrumentation Test and Evaluation • Chimkent (Kazakhstan) (location identifier) • Circumstellar Imaging Telescope • Combustor Inlet Temperature • Comité International de Télévision • Command Interface Test • Communication and Information Technology • Compact Ignition Tokamak • Compression In Transit • Compressor Inlet Temperature • Computer-Illustrated Text • Computer Interface Terminal • Contact Ion Thruster • Counselor In Training • Cranfield Institute of Technology (UK) • Critical Item Tag

CITA: Canadian Institute for Theoretical Astrophysics

CITAC: Cooperation on International Traceability in Analytical Chemistry

CITADEL: Citations and Delivery

CITC: Computing Information Technology Center

CITCE: Comité International de Thermodynamique et de Cinétique Électrochimique

CITE: Capsule Integrated Test Equipment • Cargo Integration Test Equipment • Centre for International Technology and Education (UK-England) • College of Information Technology and Engineering (USA-WV) • Computer Integration Test Equipment • Controller Input Test Equipment • Council on Technology Education

CITED: Copyright In Transmitted Electronic Documents

CITEL: Committee for Inter-American Telecommunications • Conferencia Interamericana de Telecomunicaciones

CITEP: Colorado Io Torus Emissions Package

CITERE: Convention Industrielle et Technique d'Études et de Recherche à l'Étranger

CITES: Convention on International Trade of Endangered Species

CITI: Centre Interuniversitaire de Traitement de l'Information • Columbia Institute for Teleinformation • Common Internetworking Topology Initiative • Computer Internet Telephony Integration • Confédération Internationale des Travailleurs Intellectuels

CITIC: China International Trust and Investment Corporation

CITIS: Centralized Integrated Technical Information Service • Centralized Integrated Technical Information System • Contractor Integrated Technical Information Service

CITO: Centraal Instituut voor Toetsontwikkeling

CITP: Capillary Isotachophoresis

CITRAL: Commission on International Trade Law

CITS: Center for Information Technology Services • Central Integrated Test System

CIU: Central Information Unit • Central Interpretation Unit • Command Interface Unit • Communication Interface Unit • Computer Interface Unit • Controller Interface Unit • Sault Sainte Marie - Chippewa County (USA-MI) (location identifier)

CIUI: Crustal Intelligent User Interface

CIUP: Centro de Informática da Universidade do Porto (Portugal)

CIUS: Conseil International des Unions Scientifiques

CIUT: Columbus IVA Underwater Testing

Civ: Civil • Civilian • Civilization

CIV: Centrum voor Informatieverwerking • Chomley (USA-AK) (location identifier) • Colegio de Ingenieros de Venezuela • Commission Internationale du Verre • Côte d'Ivoire (ISO country code) • Critical Ionization Velocity • Ivory Coast (ISO country code)

CIVR: Computer and Interactive Voice Response

CIVT: Cargo Interface Verification Test

CIVU: Cargo Integration Verification Unit

CIW: Canouan Island (Grenadine Islands) (location identifier) • Carnegie Institution of Washington (USA-DC) • Ceramic Insulated Wire

CIWG: Cargo Integration Working Group • Catalog Interoperability Working Group

CIX: Chiclayo (Peru) (location identifier) • Commercial Internet Exchange • Compulink Information Exchange • Computer Information Exchange

CIY: Comiso (Italy) (location identifier)

CIZ: Coari (Brazil-AM) (location identifier)

cj: Conjectural

CJ: CalTech - Jodrell Bank • Chief Justice • Cour de Justice • Court of Justice

CJA: Cajamarca (Peru) (location identifier)

CJB: Coimbatore (India) (location identifier)

CJC: Calama (Chile) (location identifier) • Clean Japan Centre

CJChE: Canadian Journal of Chemical Engineering (periodical code)

CJD: Candilejas (Colombia) (location identifier)

Cje: Corretaje

CJE: Cour de Justice Européenne

CJECE: Canadian Journal of Electrical Computer Engineering (periodical code)

CJH: Chiko Lake (Canada-BC) (location identifier)

CJI: Crafton Island (USA-AK) (location identifier)

CJK: Chinese-Japanese-Korean

CJL: Chitral (Pakistan) (location identifier)

CJLI: Command Job Language Interpreter

CJLTP: Chinese Journal of Low-Temperature Physics (periodical code)

CJMA: Communications Junction Module Assembly

CJN: El Cajon (USA-CA) (location identifier)

CJP: Club Jean Perrin (France) • Cornu-Jellet Prism • Cultureel Jongerenpaspoort

CJPE: Certified Java Programmer Examination

CJS: Ciudad Juarez (Mexico) (location identifier)

CJTCS: Chicago Journal of Theoretical Computer Science

CJU: Cheju (Korea-ROK) (location identifier)

Ck: Cask • Check • Cometary Knot • Cook • Cook Islands (ISO country code) • Cork • Creek • Îles Cook (ISO country code)

CKA: Cherokee Kegelman (USA-OK) (location identifier)

CKB: Clarksburg (USA-WV) (location identifier)

Ckbd: Corkboard

CKC: Cherkassy (Ukraine) (location identifier)

CKD: Complete Knocked Down • Completely Knocked Down • Count-Key-Data • Crooked Creek (USA-AK) (location identifier)

CKDC: Chemical Kinetics Data Center

CKE: Clear Lake (USA-CA) (location identifier) • Cumulus Kinetic Energy

CKG: Chongqing (China-PRC) (location identifier)

CKH: Chokurdah (Russia) (location identifier)

CKI: Croker Island (Australia-NT) (location identifier)

CKIP: Chinese Knowledge Information Processing

CKIPG: Chinese Knowledge Information Processing Group (China-PRC)

CKK: Ckerokee (USA-AR) (location identifier)

CKM: Cabibbo-Kobayashi-Masakawa (matrix) • Clarksdale Fletcher (USA-MS) (location identifier)

CKN: Crookston (USA-MN) (location identifier)

CKO: Centrum voor Klimaatonderzoek (Netherlands) • Chief Knowledge Officer • Cornélio Procópio (Brazil-PR) (location identifier)

CKP: Cayley-Klein Parameter

Ckpt: Cockpit

CKPT: Checkpoint

CKR: Commercial Key Recovery • Crane Island (USA-WA) (location identifier)

CKRA: Cape Kennedy Reference Atmosphere

CKS: Carajás (Brazil-PA) (location identifier) • Checks

CKSM: Check Sum

Ckt: Circuit

CKT: Circuit

CKU: Cordova (USA-AK) (location identifier)

CKV: Clarksville (USA-TN) (location identifier)

ckw: Clockwise

CKW: Chlorkohlwasserstoff

CKX: Chicken (USA-AK) (location identifier)

CKY: Conakry Gbessia (Guinea) (location identifier)

CKZ: Canakkale (Turkey) (location identifier)

c/l: Cash Letter

cl: Centiliter • Chile • Cum Laude

Cl: Calle • Canal • Chlor • Chlore • Chlorine • Claim • Clarinet • Class • Classe • Classification • Clause • Clearance • Clerk • Climb • Clos • Close • Closet • Closing • Closure • Cloth • Cluster • Clutch • Council

C/L: Cash Letter • Checklist

C&L: Commerce & Labor

CL: Calculateur Logique • Calendar Line • Car Load • Carnegie Library (USA) • Cathode Luminescence • Cathodoluminescence • Centerline • Centerline Light • Central Laboratory • Chain Length • Checklist • Chemiluminescence • Child-Langmuir • Chile (ISO country code) • Chili (ISO country code) • Circular Letter • Civil Law • Clark Lake • Clear • Clearance Limit • Close • Closed • Closed Loop • Cluster • Coding Language • Coefficient of Lift • Command Language • Common Law • Common LISP • Competitive Learning • Compiler Language • Complete Linearization • Computational Linguistics • Condition aux Limites • Condition Lever • Condition Limite • Confidence Level • Connectionless • Control Language • Control Leader • Control Logic • Convertible Lens • Cristal Líquido • Critical List • Current Loop • Cygnus Loop • Sri Lanka (international vehicle registration code)

CLA: Camera Lens Adaptor • Canadian Library Association • Carry Lookahead Adder • Catch Limit Algorithm • Center for Laser Applications (UTSI) • Cloud Analysis • Comilla (Bangladesh) (location identifier) • Commission for Latin America • Conjugated Linoleic Acid • Control Logic Array • Copyright Licensing Agency (UK) • Couche Limite Atmosphérique • Cover Layer Assembly

CLAES: Cryogenic Limb Array Etalon Spectrometer

CLAIMS: Class Assignee Index Method Search

CLAIR: Center for Library And Information Resources • Clevers Leaf Area Index by Reflectance

CLAIRE: Clear Air Infrared Radiation Experiment

CLAM: Computational Linear Algebra Machine

CLAMS: Classification Automatique de Données Multispectrales

CLANS: Computerized Link Analysis System

CLAR: Channel Local Address Register

CLARA: Computer Load And Resources Analysis

CLARCS: Copyright Licensing Agency's Rapid Clearance Service (*) • Copyright Licensing Agency's Rapid Clearance System

CLARET: Cloud Lidar And Radar Exploratory Test

CLAROM: Club pour les Actions de Recherche sur les Ouvrages en Mer

CLAS: Calibration Laboratory Assessment Service • California Learning Assessment System • CEBAF Large Acceptance Spectrometer • China Lake Astronomical Society (USA-CA) • Classification • Communications Link Analyzer System • Computerized Loads Analysis System

CLASP: Comprehensive Logistics Automated Support Program • Computer Language for Aeronautics and Space Programming • Computer Launch And Separation Problem

Class: Classic • Classical • Classification • Classified • Classify

CLASS: Classification Literature Automated Search Service • Client Access to Systems and Services • Coherent Lidar Airborne Shear Sensor • College Library Access and Storage System • Computer-based Laboratory for Automated School Systems • Computerized Literature Access Search Service • Cooperative Library Agency for Systems and Services • Cosmic Lens All-Sky Survey • Cross-chain Loran Atmospheric Sounding System • Cryogenic Liquid Acquisition, Storage and Supply • Cumulative and Linguistic Advanced Search System • Custom Local-Area Signaling Services

CLASSI: Classification System for Information

CLASSMATE: Computer Language to Aid and Stimulate Scientific, Mathematical, and Technical Education

CLAWS: Classify, Locate and Avoid Wind Shear • Coherent Launch Atmospheric Wind Sounder

CLB: Cape Lookout Bight (USA-NC) • Castlebar (Ireland) (location identifier) • Club • Community Land Bank • Constant Level Balloon

cLBV: Candidate Luminous Blue Variable

CLC: Change Letter Control • Clean Lakes Clearinghouse • Clear Carry • Clearlake (USA-TX) (location identifier) • Combinatorial Logic Circuit • Command Line Controller • Commonwealth Liaison Committee • Computer Liaison Committee • Control Launch Center • Couche Limite Convective • Course Line Computer • Critical Load Cycle

CLCC: Closed-Loop Continuity Check

CLCE: Communications Link Characterization Equipment

Cl-Cl: Cluster-Cluster

ClCl: Cluster-Cluster

CLCS: Current-Logic-Current-Switching

Cld: Called • Cancelled • Cleared • Cloud • Colored • Cooled • Could

CLD: Carlsbad (USA-CA) (location identifier) • Central Line Data • Clear Direction • Cloud • Colour-Luminosity Diagram • Command Language Directive • Computer Logic Demonstrator • Congé de Longue Durée • Cost Laid Down • Cumulative Luminosity Distribution • Current-Limiting Device

CLDA: Control Logic and Drive Assembly

CLDB: Common Land Database

CLDC: Centre de Loisirs et de Diffusion Culturelle

CLDP: Convertible Laser Designation Pod

Clds: Clouds

CLDS: Canada Land Data System • Clouds

CLDST: Closed Loop Dynamic Stability Test

Cldy: Cloudy

ClDy: Climate Dynamics (periodical code)

CLE: Centre de Loisirs Éducatifs • Cleveland Hopkins (USA-OH) (location identifier) • Communication-Lecture-Écriture • Continuous Legal Education • Control Law Electronics • Council of Legal Education • Customer-Located Equipment

CLEA: Comité de Liaison Enseignants Astronomes (France) • Contemporary Laboratory Experiences in Astronomy • Council of Local Education Authorities

CLEAPSE: Consortium of Local Education Authorities for the Provision of Science Equipment

CLEAR: Closed-Loop Evaluation And Reporting • Compiler, Executive Program, Assembler Routines • Components Life Evaluation And Reliability • Coronagraph and Low-Emissivity Astronomical Reflector • Coronagraphic and Low-Emissivity Astronomical Reflector (*)

CLEARS: Cornell Laboratory for Environmental Applications of Remote Sensing

CLEC: Competitive Local-Exchange Carrier

CLEF: Comité du Livre Électronique Français • Cross-Language Evaluation Forum

CLEI: Common-Language Equipment Identification

CLEM: Cargo Lunar Excursion Module • Closed Loop Eigenvector Matrix

CLEO: Conference on Lasers and Electro-Optics

CLEOPATRA: Cloud Experiment Oberpfaffenhofen And Transports

CLF: Center-to-Limb Function • Central Laser Facility • Chilean Investment Unit (ISO currency code) • Clear Lake Facility • Cliff • Control and Limit File • Control Laser Facility • Controller - Load Forward • Control Lyapunov Function • Critical Link Factor

CLFIC: Center Launch and Flight Instrumentation Center

CLFS: Cliffs

CLFST: Cambridge Low-Frequency Synthesis Telescope

CLFT: Classical Local Field Theories

Clg: Ceiling • Cooling

CLG: Ceiling • Central Laboratory for Geodesy • Coalinga (USA-CA) (location identifier) • Company Limited by Guarantee • Compile, Load and Go

CLGEE: Comité de Liaison des Géomètres-Experts Européens

CLH: Coolah (Australia-NSW) (location identifier)

CLI: Call-Level Interface • Canada Land Inventory • Cellular Line Interface • Certified Lotus Instructor • C Language Interface • Clear Interrupt • Clintonville (USA-WI) (location identifier) • Coefficient of Luminous Intensity • Coherent Laser Illumination • Command-Language Interpreter • Command-Line Interface • Command Line Interpreter • Commission Locale d'Information • Compare Logical Immediate • Contractor Line Item • Cost-of-Living Index • Cumulative Leakage Index

CLIB: Clementine Lunar Image Browser

CLIBS: Colinear Laser Ion Beam Spectroscopy

CLIC: Collisionneur Linéaire du CERN • Command Language for Interrogating Computers • Compact Linear Collider

CLICOM: Climate Computing

CLICS: Categorical Logic In Computer Science

CLID: Calling Line Identification

CLIF: Computational Linguistics in Flanders (Belgium) • Concept Lattice Interchange Format

CLIIP: Common Law Institute of Intellectual Property

Clim: Climate • Climatic

CLIM: Common LISP Interface Manager

CLIMAP: Climate Applications Project • Climate: Long-range Interpretation, Mapping, And Prediction

CLIMEX: Climatic Extremes

Clin: Clinic • Clinical

CLIN: Contract Line Item Number

CLINS: Climatic Laboratory Instrumentation System

CLIODN: Committee for Liaison with International Organizations and Developing Nations

CLIP: Calling-Line Identification Parameter • Calling-Line Identification Presentation • Cancel Launch In Progress • Cirrus Lidar Probe • Combined Laser Instrumentation Package • Compiler Language for Information Processing • Computer Launch Interference Problem • Crew Loads Instrument Pallet • Croydon Library Internet Project (UK)

CLIPS: C Language Integrated Production System

CLIR: Cross-Language Information Retrieval

CLIS: College of Library and Information Services • Contract Line Item Status

CLISP: Conversational List Processing

CLIST: Command List

CLIVAR: Climate Variability

CLJ: Cluj (Romania) (location identifier)

Cljo: Complejo

Clk: Clerk • Clock

CLK: Chek Lap Kok (Hong Kong) • Clinton (USA-OK) (location identifier) • Clock

Clkw: Clockwise

Clkws: Clockwise

CLL: Bryan (USA-TX) (location identifier) • Chronic Lymphocytic Leukaemia • Collaborative Learning Laboratory • College Station (USA-TX) (location identifier) • Corrected Line Length • Critical Load Level

CLLI: Common-Language Location Identification

Cllj: Callejón

CLLM: Consolidated Link Layer Management

Cllr: Councillor

Clm: Column

CLM: Career Limiting Move • Care Logic Module • Cell-Loss Margin • Central Meridian Longitude • Circumlunar Mission • Congé de Longue Maladie • Conversational Line Multiplexer • Port Angeles (USA-WA) (location identifier)

CLMC: Central Logistics Management Center • Chemical Laser Mode Control

Cln: Clean

CLN: Carolina (Brazil-MA) (location identifier) • Central Library Network

Clnc: Clearance

CLNP: Connectionless-mode Network Protocol • Connectionless Network Protocol

CLNR: Cleaner

CLNS: Connectionless-mode Network Service • Connectionless Network Service • Cornell Laboratory of Nuclear Studies (USA-NY)

Clnt: Coolant

Clo: Closet • Clothing

ClO: Chlorine Monoxide

CLO: Cali (Colombia) (location identifier) • Chapter Liaison Officer • Chief Liaison Officer • Club Liaison Office • Computer Lock-On • Concentric Line Oscillator • Congressional Liaison Office • Consular Liaison Officer • Contrat Logiciel • Crystal Lake Observatory (USA)

CLONES: Central Location On-line Entry System

CLORA: Club des Organismes de Recherche et Associés

Clos: Closet

CLOS: Common LISP Object System • Coordination of Land Observing Satellites

CLOT: Combined Loads Orbiter Test

CLOUD: Cosmics Leaving Outdoor Droplets (experiment)

CLOUDS: Cloud Logic to Optimize Use of Defense Systems

CLP: Cell Loss Priority • Clarks Point (USA-AK) (location identifier) • Clipboard • Club Laser de Puissance (France) • Command Language Processor • Concurrent Logic Programming • Constraint Logic Programming • Continuous Line Plotter • Contract Laboratory Program • Cornell List Processor • Couche Limite Planétaire • Peso (Chile) (ISO currency code)

Clpbd: Clapboard

Clpr: Clapper

CLPS: Calibration Lamp Power Supply

CLQ: Colima (Mexico) (location identifier)

Clr: Clear • Color • Cooler • Couleur • Councilor

CLR: Calipatria (USA-CA) (location identifier) • Cell Loss Ratio • Circuit Layout Record • Clear • Combined Line and Recording • Command-Leadership-Resource • Computer Language Recorder • Coronal-Line Region • Council on Library Resources • Councilor • Current-Limiter Resistor

CLRB: Cost Limit Review Board

CLRC: Central Laboratory of the Research Councils (UK) • Circuit Layout Record Card

Clrg: Clearing

CLRG: Clearing

Clrm: Classroom

CLRO: Clark Lake Radio Observatory

CLRP: Central Laboratory for Radiological Protection

Clrs: Colors • Couleurs

CLRS: Clear and Smooth

CLRTAP: Convention on Long-Range Transportation of Anthropogenic Pollutants

cls: Clear Screen

CLS: Case Late Stars • Center for Language Studies • Certificate in Library Science • Chehalis Centralia (USA-WA) (location identifier) • Cislunar Space • Classical Least Squares • Close Lunar Satellite • Cloud Lidar System • Collecte et Localisation par Satellite • Command and Launch System • Commercial Launch Services • Committee on Life Sciences • Common Language System • Concept Learning System • Constrained Least Squares • Contingency Landing Site

CLSA: Canada Land Use Monitoring Program

ClSCC: Cleveland State Community College (USA-TN)

Clsd: Closed

CLSES: Closed Loop Solvent Extraction System

Clsg: Closing

CLSI: Computer Library Service International

CLSID: Class Identifier

CLSM: Confocal Laser Scanning Microscope • Crew Life Support Monitor

CLSO: Contingency Landing Support Officer

CLSP: Center for Language and Speech Processing

CLSR: Cooled Limb Sounding Radiometer

CLSS: Computerized Literature Searching Service

Clsy: Classify

CLT: Calculated Landing Time • Cell Loss Priority • Central Limit Theorem • Charlotte (USA-NC) (location identifier) • Closed-Loop Telemetry • Closed-Loop Test • Communications Line Terminal • Computer Language Translator • Culture

Clthg: Clothing

CLTP: Connectionless Transport Protocol

CLTR: Cross-Language Text Retrieval

CLTS: Clear Task Switch • Cryogenic Linear Temperature Sensor

CLTV: Closed-Loop Television

CLU: California Lutheran University (USA-CA) • Central Logic Unit • Columbus (USA-IN) (location identifier) • Combination Logic Unit • Command Logic Unit • Cryptographic Logic Unit

CLUD: Conceptual Clustering with Usage Dependency

CLUE: Čerenkov Light Ultraviolet Experiment • Clustering Environment • Common LISP User-interface Environment • Compensated Laser Ultrasonic Evaluation

CLUI: Command Language User Interface

CLUMP: Canada Land Use Monitoring Program

CLUSIF: Club de la Sécurité Informatique Français (France)

CLUT: Color Look-Up Table

CLV: Caldas Novas (Brazil-GO) (location identifier) • Center-Limb Variation • Center-to-Limb Variation • Constant Linear Vector • Constant Linear Velocity • Corrected Loss Variance

CLW: Clearwater (USA-FL) (location identifier) • Cloud Liquid Water • Correction Location Word

Clws: Clockwise

CLX: Clorinda (Argentina-FO) (location identifier) • Common LISP X-interface

CLY: Calvi Sainte-Catherine (France) (location identifier)

CLZ: Calabozo (Venezuela) (location identifier)

cm: Carat Métrique • Causa Mortis • Centimeter • Centimetric

c/m: Call of More

Cm: Curium

C-M: Color-Magnitude

C&M: Control & Monitoring

CM: Calibrated Magnification • Calibration Module • Cambodia • Camera Module • Cameroon (ISO country code) • Cameroun (ISO country code) • Cargo Manager • Carnegie Monitor • Cartographic Material • Celestial Mechanics • Center of Mass • Center of Momentum • Centimeter • Central Maine (USA-ME) • Central Memory • Central Meridian • Certificated Master • Certified Master • Change Management • Chef de Mission • Chelmsford (UK postal code) • Chrome-Molybdenum • Circular Measure • Circulation Manager • Classification Model • Classified Message • Color-Magnitude • Combustion Modification • Combustion Module • Command Module • Commercial Middeck • Common Market • Common Meter • Communications Multiplexer • Complex Mixture • Composite Material • Computer Module • Computing Media • Condition Monitoring • Confidential Memorandum • Configuration Management • Configuration Manager • Configuration Monitor • Congé de Maladie • Connection Machine • Connectionnist Model • Construction and Machinery • Construction Management • Consumable Management • Contamination Mode • Contraction Mapping • Contract Management • Contract Modification • Contrast Medium • Control and Monitoring • Contrôle Médical • Control Mark • Control Mode • Control Monitor • Core Meltdown • Core Memory • Corporate Membership • Correction Memo • Corrective Maintenance • Corresponding Member • Countermeasure • Cours Moyen • Credit Memo • Crew Module • Critical Mass • Cruise Mode • Cursor Movement

Cma: Cima

CMa: Canis Major - Canis Majoris (constellation)

CMA: Calcium Magnesium Acetate • Canadian Medical Association • Canadian Micrographic Association • Canadian Museums Association • Cash Management Account • Cell Management Agent • Census Metropolitan Area • Central Management Agency • Central Mapping Agency • Central Mapping Authority • Centre de Mathématiques Appliquées • Centre for Mathematical Analysis • Centro Meteorológico Antártico • Cercle Méridien Automatique • Certificate of Management Accounting • Channel Multiplier Array • Chef de Mission Adjoint • Chemical Manufacturers' Association • Chemical Manufacturing Association • China Meteorological Administration (China-PRC) • Círculo Meridiano Automático • Classified Mail Address • Command Modulator Assembly • Committee on Minorities in Astronomy • Communications Managers Association • Complex Modulus Apparatus • Computer Management Association (USA) • Computer-Marked Assignment • Concert Multithread Architecture • Configuration Management Accounting • Conical Monopole Antenna • Conseil Mondial de l'Alimentation • Consistent Multifluid Advection • Contrat Matériel • Cunnamulla (Australia-QLD) (location identifier) • Cylindrical Mirror Analyser

CMAA: Central Missouri Astronomical Association (USA-MO)

CMAC: Centralized Maintenance and Administration Center • Cerebellar Model Articulation Controller

CMACS: Central Monitor And Control System

CMAK: Conical Monopole Antenna Kit

CMAL: Controlled Multiple-Address Letter

CMAM: Commercial Middeck Augmentation Module

CMAME: Computer Methods in Applied Mechanics and Engineering (periodical code)

CMAO: Calcul Mathématique Assisté par Ordinateur • Contract Management Assistance Officer

CMAP: Climate Modelling, Analysis and Prediction • Climatic Modelling and Applications Program

CMAR: Configuration Management Action Request

CMAS: Computational Methods in Applied Sciences • Confédération Mondiale des Activités Subaquatiques • Configuration Management Applications Software • Council for Meteorology and Atmospheric Sciences (India)

CMAT: Compatible Materials

CMB: Caractères Magnétiques en Bâtonnets • Chemical Mass Balance • Colombo Katunayake (Sri Lanka) (location identifier) • Configuration Management Board • Configuration Management Branch • Configuration Management Bulletin • Core-Mantle Boundary • Corrective Maintenance Burden • Cosmic Microwave Background

CMBR: Cosmic Microwave Background Radiation

CMBT: Cosmic Microwave Background Telescope

CMC: Camocim (Brazil-CE) (location identifier) • Canadian Meteorological Centre • Carlsberg Meridian Catalogue • Carlsberg Meridian Circle • Center for Macromolecular Crystallography • Central Maintenance Computer • Central Methodist College (USA-MO) • Ceramic Matrix Composite • Certificate Management Consultant • Code for Magnetic Characters • Coherent Multichannel • Combined Meteorological Committee • Command Module Computer • Common Mail Call • Common Messaging Call • Communications Management Configuration • Communications Mode Control • Compact Molecular Cloud • Complement Carry • Complex Monitor and Control • Composite à Matrice Céramique • Computer-Mediated Communication • Conservation Monitoring Centre • Constant Mean Curvature • Construction Maintenance Center • Convolve-Multiply-Convolve • Crew Module Computer • Current Mode Coupler • Customized Multimedia Connection • Cylindrical Mesh Cathode

CMCC: Central Mission Control Centre

CMCD: Cylindrical Mesh Cathode Device

CMCI: CompuMath Citation Index

CMCIRS: Computerized Map Cataloging and Information Retrieval System

CMCK: Centre Médico-Chirurgical de Kourou (French Guyana)

CMCL: Command Management Control List

CMCS: Centralized Maintenance Computer System • Communications Monitoring and Control System • COMSAT Maritime Communications Satellite • Construction Management Control System

Cmd: Command

CMD: Catalogued Metadata • Centralized Message Distribution • Central Meridian Distance • Centre Mondial de Données • Chargé de Mission Diplomatique • Color-Magnitude Diagram • Command • Compatibility Mode Driver • Cootamundra (Australia-NSW) (location identifier) • Core Memory Drive • Core Memory Driver • Cryogenic Magnetic Detector • Current Meter Digitizer

CMDAC: Current-Mode Digital-to-Analog Converter

CMDB: Composite Modified Double Base • Configuration Management Database

CMDF: Combined Main Distribution Frame • Command Masking Data Facility

Cmdg: Commanding

CMDL: Climate Monitoring and Diagnostics Laboratory (*) • Climate Monitoring Diagnostics Laboratory

CMDM: Control Multiplexer/Demutiplexer

Cmdo: Comodoro

Cmdr: Commander

Cmdre: Commodore

CMDS: Centralized Message Data System • Configuration Management Data System • Consortium for Materials Development in Space • Computer Misuse Detection System

Cmdt: Comandante

Cmdte: Comandante

Cme: Centime

CME: Camera Module Electronics • Center for Media Education • Chemically Modified Electrode • Circuit Multiplication Equipment • Ciudad del Carmen (Mexico) (location identifier) • Commercial Measuring Equipment • Common Mode Error • Community Modelling Effort • Computer Memory Element • Conférence Mondiale de l'Énergie • Conseil Mondial de l'Énergie • Continuing Medical Education • Coronal Mass Ejection

CMEA: Cellular Message Encryption Algorithm • Council for Mutual Economic Assistance

CMEO: Coronal Mass Ejection Onset

CMER: Center for Management and Economic Research

CMERI: Central mechanical Research Institute

CMES: Common Modular Elint System

CMF: Cancelling Magnetic Feature • Carbon Monofluoride • Cartesian Mapping Function • Central Maintenance Facility • Central Microstructure Facility • Chambéry (France) (location identifier) • Chip-Matched Filter • CICS Monitor Facility • Cockpit Motion Facility • Coherent Memory Filter • Comoving Frame • Complex Maintenance Facility • Comprehensive Management Facility • Compressed Magnetic Field • Configuration Management Facility • Conseil des Marchés Financiers • Consolidated Maintenance Facility • Core-Mantle Friction • Cross-Modulation Factor • Cylindrical Magnetic Film

CMFD: Computational Magneto-Fluid Dynamics

CMG: Commercial Middeck Galley • Commission for Marine Geology • Computer Management Group • Computer Measurement Group • Control Moment Gyro • Control Monitor Group • Corumbá (Brazil-MS) (location identifier)

CMGA: Control Moment Gyro Assembly

CMGD: Centre of Marine Geological Data

CMGEA: Control Moment Gyro Electrical Assembly

CMGIA: Control Moment Gyro Inverter Assembly

CMGR: Committee on Microgravity Research

CMGS: Control Moment Gyro System

CMH: Columbus - Port Columbus (USA-OH) (location identifier)

CMHAS: Crayford Manor House Astronomical Society (UK)

CMHMM: Continuous Mixture Hidden Markov Model

CMi: Canis Minor - Canis Minoris (constellation)

CMI: Cable Microcell Integrator • Caribbean Meteorological Institute (now CIMH) • Certain Meteoritical Inclusion • Champaign (USA-IL) (location identifier) • Chr. Michelsen Institute (Norway) • Comité Maritime International • Comité Météorologique International • Computer-Managed Instruction • Continuous Measurable Improvement • Copyright Management Information • Correlative Measurement Investigator • Crop Moisture Index • Cryomagnetics Inc. • Cyclotron Maser Instability • Urbana (USA-IL) (location identifier)

CMIB: Controlled Multispectral Image Base

CMIC: Computer Microfilm International Corporation

CMIF: Common Module Integration Facility • Core Module Integration Facility • CWI Multimedia Interchange Format

C/Min: Counts per Minute • Cycles per Minute

CMIP: Carrier Management Information Protocol • Common Management Information Protocol • Cost Management Improvement Program

CMIQ: Centro Méxicano de Información Química (Mexico)

CMIR: Centre Météorologique Interrégional • Corotating Merged Interaction Region

CMIS: Committee Members Information System • Common Management Information Service • Common Management Information System • Common Module Integration Simulator • Computer Management Information System • Computer Microimage System • Computer-orinted Management Information System • Configuration Management Information System • Connection Machine Instruction Set • Control Monitor Isolation System • Corporate Management Information System

CMITU: Centre Mondial d'Informations Techniques et d'Urbanisme

CMIX: Commercial Materials ITA Experiment

CMJ: Chi Mei (China-ROC) (location identifier)

CMJN: Cyan-Magenta-Jaune-Noir

CMK: Club Makokola (Malawi) (location identifier)

Cml: Commercial

CML: Camooweal (Australia-QLD) (location identifier) • Central Meridian Longitude • Chemical Markup Language • Collimated Monochromatic Light • Combined Membership List • Commands and Measurement List • Common Machine Language • Computer-Managed Laboratory • Computer-Managed Learning • Computer-Management Learning • Conceptual Modelling Language • Conference Mailing List • Conversational Modelling Language • Coupled Map Lattice • Current-Mode Logic

CMLS: Computer Multiple Listing Service

CMM: Capability Maturity Model (*) • Carmelita (Guatemala) (location identifier) • Centre de Météorologie Marine (France) • Centre de Météorologie Maritime (*) • Coherent Microwave Memory • Command Memory Management • Commission for Marine Meteorology (WMO) • Commission for Maritime Meteorology • Compact Master Manipulator • Condition-Monitored Maintenance • Coordinate Measurement Machine • Coordinate Measuring Machine

CMMACS: CSIR Centre for Mathematical Modelling and Computer Simulation

CMMDA: Command Module Multiple-Docking Assembly

CMME: Compton's Multimedia Encyclopaedia

CMMP: Condensed Matter and Materials Physics

CMMR: Common Mode Rejection Ratio

CMMS: Computerized Maintenance Management Software

CMMU: Cache/Memory Management Unit

Cmn: Commission

CMN: Casablanca - Mohamed V (Morocco) (location identifier) • Centre Météorologique National • Cerium Magnesium Nitrate • Cerous Magnesium Nitrate • Common • Constructions Mécaniques de Normandie (France) • Contract Management Network • Contract Master Network • Crewman

CMC: Commence

CMNH: Cleveland Museum of Natural History (USA-OH)

CMNPS: Canadian Minimum Navigation Performance Specification

Cmnr: Commissionner

Cmnt: Comment

CMO: Canada Map Office • Caribbean Meteorological Organization • Central Management Office • Central Meteorological Observatory • Centre Militaire d'Océanographie • Chief Meteorological Officer • Chief of Mission Operations • Class Mix Option • Collateralized Mortgage Obligation • Common Main Objective • Compact Massive Object • Configuration Management Office • Configuration Management Officer • Configuration Management Organization • Contract Management Office • Contract Management Officer • Contracts Management Office • Coordinating and Monitoring Office • Coordination and Monitoring Office • Obbia (Somalia) (location identifier)

CMOD: Centre Mondial d'Orientation Documentaire

CMOPE: Confédération Mondiale des Organisations de la Profession Enseignante

CMOS: Canadian Meteorological and Oceanographic Society • Carbon Molybdenum Steel • Complementary Metal-Oxide Semiconductor • Complementary Metal-Oxide Silicon (*)

CMOSM: Configuration Management Operating Systems Manual

CMOT: CMIP Over TCP

CMOV: Conditional Move

Cmp: Compromise

CMP: Camp • Celestial Mapping Program • Central Meridian Passage • Central Modelling Programme • Centre de Maintien à Poste • Centre de Mise à Poste • Chemical Mechanical Polishing • Chemical Mechanical Processing • Christa McAuliffe Planetarium (USA-NH) • Climate Modelling Programme • Coastal Mapping Program • Command Module Pilot • Common Mid-Point • Communications and Media Policy • Communiqué Météorologique de Presse • Compare • Computer Modelling Package • Computer • Configuration Management Panel • Configuration Management Plan • Configuration Management Procedure • Configuration Management Program • Console Message Processor • Contrast Mode Photometer • Correlation of Molecular Pairs • Cost of Maintaining Project • Coût Moyen Pondéré • Custom Mass Production • Santana do Araguaia - Campo Alegre (Brazil-PA) (location identifier)

Cmpd: Compound

CmpEn: Composites Engineering (periodical code)

CMPh: Condensed Matter Physics

CMPL: Complete

Cmplmt: Complement

Cmplte: Complete

Cmpltn: Completion

CmpMa: Composites Manufacturing (periodical code)

Cmpn: Companion

Cmpns: Compensate

Cmpnt: Component

CMPP: Configuration Management Program Plan

Cmpr: Compare

CMPR: Compare

CMPRT: Compartment

CMPS: Command Module Procedures Simulator • Compare String • Computer, Mathematical, and Physical Sciences

Cmpsg: Compensating

Cmpsn: Compensation

Cmpst: Composite

CmpSt: Composite Structures (periodical code)

CmpST: Composites Science Technology (periodical code)

Cmpt: Compute

Cmptd: Computed

CMPTD: Computed

Cmptg: Computing

Cmptr: Computer

CMPT: Computer

CMQ: Center for Measurement Quality • Clermont (Australia-QLD) (location identifier)

CMQA: Configuration Management/Quality Assurance

CMR: Cameroon (ISO country code) • Cameroun (ISO country code) • Catalytic Membrane Reactor • Cell Misinterpretation Rate • Cellule Multifonctionnelle Régionale • Center for Marine Resources • Center Materials Representative • Central Meter Reading • Chargé de Mission de la Recherche • Cockpit Management Resources • Colmar (France) (location identifier) • Color-Magnitude Relation • Colossal Magnetoresistance • Commercial Mail Relay • Common Mode Rejection • Communications Monitoring Report • Communications Moon Relay • Computer Modern Roman • Configuration Management Representative • Configuration Management Review • Contract Management Review • Contractor Management Review • Countdown Modification Request

CMRA: Chemical Marketing and Resources Association

CMRB: Contractor Material Review Board

CMRDI: Central Metallurgical Research and Development Institute

CMRN: Cooperative Meteorological Rocket Network

CMRP: Configuration Management Requirements Plan

CMRR: Common Mode Rejection Ratio

CMRS: Calibration/Measurement Requirements Summary • Centre Météorologique Régional Spécialisé • Configuration Management Requirements Standard

cms: Cras Mane Sumendus

CM&S: Communications Maintenance & Storage

CMS: Cable Management System • Call Management System • Canadian Mathematical Society • Card Management Service • Center for Measurement Standards • Central Maintenance System • Central Materials Supply • Centre de Mécanique Spatiale • Centre de Météorologie Spatiale (France) • Centre Médico-Social • Cepstral Mean Subtraction (method) • Change Management System • Circuit Maintenance System • Code Management System • Color Management System • Command Management System • Command Module Simulator • Common Mode Signal • Communications Maintenance and Storage • Communications Management System • Communiqué Météorologique Spécial • Compact Muon Solenoid • Compiler Monitor System • Component Model Synthesis • Component Mode Synthesis • Computer Management System • Configuration Management System • Contact

Management Software • Contract Master Schedule • Conventional Monitoring System • Conversational Monitoring System • Conversational Monitor System • Conversation Monitoring System • Conversation Monitor System • Core Module Simulator • Corrective Maintenance System • Cost Measurement System • Current-Mode Switching • Customer Management System • Scusciuban (Somalia) (location identifier)

CMSA: Consolidated Metropolitan Statistical Area

Cmsd: Commissioned

CMSE: Candidate Materials Space Exposure

Cmshft: Camshaft

CMSI: Command Monitor System Interface

CMSM: Committee on Marine Surveying and Mapping

Cmsn: Commission

Cmsnd: Commissioned

Cmsr: Commissioner

CMSS: Commission on Molecular Structure and Spectroscopy

CMT: Cadmium Mercury Telluride • Cametá (Brazil-PA) (location identifier) • Carlsberg Meridian Telescope (Canary Islands) • Cellular Mobile Telephone • Centroid-Moment-Tensor • Code Matching Technique • Computer Memory Tester • Connection Management • Continuous Media Toolkit • Continuum Mechanics and Thermodynamics (periodical code) • Corrected Mean Temperature

CMTA: Centre Multifonction de Télécommunications Avancées

Cmte: Comandante • Committee

CMTI: Celestial Moving Target Indicator

CMTS: Computerized Maintenance Test System • Computer Maintenance Test Set

CMU: Carnegie Mellon University (USA-PA) • Central Michigan University (USA-MI) • Chiang Mai University (Thailand) • Command and Monitoring Unit • Communications Management Unit • Construction and Maintenance Unit • Control and Monitor Unit • Core Memory Unit • Cotisation Maladie Universelle • Couverture Maladie Universelle • Kundiawa (Papua New Guinea) (location identifier)

CMV: Cloud Motion Vector • Common Mode Voltage • Coromandel (New Zealand) (location identifier)

CMVA: Coordinated Millimeter VLBI Array

CMVAS: Configuration Management Verification Accounting System

CMVC: Cooperative Model-View Controller

CMW: Camaguey (Cuba) (location identifier) • Circular Magnetic Wave • Cloud-Motion Wind • Compartmented Mode Workstation

CMwA: Computers and Mathematics with Applications (periodical code)

CMX: Hancock (USA-MI) (location identifier)

CMY: Cyan-Magenta-Yellow • Sparta McCoy (USA-WI) (location identifier)

CMYB: Cyan-Magenta-Yellow-Black

CMYK: Cyan-Magenta-Yellow-Black

CMZ: Caia (Mozambique) (location identifier) • Central Molecular Zone

cn: China (PRC) • Cras Nocte • PRC

C/N: Carbon/Nitrogen • Carrier/Noise • Circular Note • Consignment Note • Contract Note • Cover Note • Credit Note

CN: Canadian National • Carbon-Nitrogen • Change Notice • China (PRC) (ISO country code) • Chine (RPC) (ISO country code) • Circular Note • Classical Nova • Code Napoléon • Column • Combined Nomenclature • Cometary Nebula • Comité de Normalización • Comité National • Commande Numérique • Common Name • Common Network • Communication Network • Compound Nucleus • Condensation Nucleus • Contract Number • Control Number • Coordination Number • Correction Notice • Current Node • Cyanogen Radical • Morocco (international aircraft registration code) • PRC (ISO country code) • République Populaire de Chine (ISO country code)

CN/A: Convertisseur Numérique/Analogique

CNA: Canadian National Archives • Cananea (Mexico) (location identifier) • Centre National de l'Automatisation (France) • Certified NetWare Administrator • Certified Network Administrator • Certified NetWare Analyst • Chemical Notation Association • Code Not Allocated • Comisión Nacional de Astronomía • Communications Network Application • Convertisseur Numérique/Analogique • Copper-Nickel Alloy • Cosmic Noise Absorption • Customer Name and Address

CNAA: Consorzio Nazionale per l'Astronomia e l'Astrofisica (Italy) • Council for National Academic Awards (UK)

CNAC: Chinese National Aviation Corporation

CNAE: Comisión Nacional de Actividades Espaciales (Argentina)

CNAM: Caisse Nationale d'Assurance Maladie (France) • Conservatoire National des Arts et Métiers (France)

CNAMTS: Caisse Nationale d'Assurance Maladie des Travailleurs Salariés (France)

CNAP: Conseil National des Astronomes et Physiciens (France)

CNAPS: Connected Network of Adaptive Processors System

CNAR: Compound Net Annual Rate

CNAV: Caisse Nationale d'Assurance Vieillesse (France)

CNB: Comité National Belge (Belgium) • Coonamble (Australia-NSW) (location identifier)

CNBOS: Comité National Belge de l'Organisation Scientifique (Belgium)

CNBPHI: Comité National Belge pour le Programme Hydrologique International

Cnc: Cancer-Cancri (constellation)

CNC: Central Navigation Computer • Centre National de la Cinématographie (France) • Centre National de la Consommation (France) • Coconut Island (Australia-QLD) (location identifier) • Computerized Numerical Control • Computerized Numerical Controlled • Computerized Numerically Controlled • Computer Numerical Control • Computer Numerical Controller • Computer Numerically Controlled (*) • Computer Numeric Control • Condensation Nucleus Counter • Configurable Network Computing • Contrôle Numérique par Ordinateur

CNCB: Computer Networks and Communications Branch

CNCC: Customer Network Control Center

CNCIM: Centre National de Coordination et de l'Indexation Matières

CNCIS: Commission Nationale de Contrôle des Interceptions de Sécurité

Cncl: Conceal • Council

CNCL: Cancel • Common Network Command Language

Cncld: Cancelled • Concealed

CNCL: Commission Nationale de la Communication et des Libertés

Cnclr: Councillor

CNCPI: Chambre Nationale des Conseils en Propriété Industrielle (France)

CNCPRST: Centre National de Coordination et de Planification de la Recherche Scientifique et Technique (Morocco)

Cncr: Concurrent

CNCR: Characterization of Neurospora Circadian Rythms

CNCT: Consejo nacional de Ciencia y Tecnología (Mexico)

Cnctrc: Concentric

Cncv: Concave

Cñd: Cañadón

Cnd: Conduit

CND: Cannot Duplicate • Centre National de Documentation • Circumnuclear Disk • Condition • Constanta (Romania) (location identifier) • Contrôle Non-Destructif

Cñda: Cañada

Cndct: Conduct

CNDCT: Centro Nacional de Documentación Científica y Tecnológica (Bolivia)

CNDCTE: Centro Nacional de Documentación Científica, Ténica y Económica (Uruguay)

Cndctr: Conductor

Cndcty: Conductivity

CNDEP: Centre National de Documentation sur l'Enseignement Privé (France)

CNDL: Centre National de Dépôt et de Labellisation (France)

CNDN: Canadian

CNDO: Complete Neglect of Differential Overlap

CNDP: Centre National de Documentation Pédagogique (France)

Cnds: Condensate

CNDS: Centro Nazionale di Documentazione Scientifica (Italy)

CNDST: Centre National de Documentation Scientifique et Technique (Belgium) • Centro Nazionale di Documentazione Scientifico-Tecnica (Italy) • Conseil National de la Documentation Scientifique et Technique

CNDT: Centro Nazionale di Documentazione Tecnica (Italy)

Cndtn: Condition

CNE: Canon City (USA-CO) (location identifier) • Certified NetWare Engineer • Certified Network Engineer • Certified Novell Engineer • Comisión Nacional de Energía • Comité National d'Évaluation • Concentrateur Numérique Éloigné

CNEA: Comisión Nacional de Energía Atomica (Argentina)

CNEAF: Coal, Nuclear, Electric, and Alternate Fuels

CNED: Center for Non-conventional Energy Development • Centre National d'Éducation à Distance (France) • Centre National d'Enseignement à Distance (France) (*)

Cnel: Coronel

CNELG: Compact Narrow Emission-Line Galaxy

CNEN: Comissão Nacional de Energia Nuclear (Portugal) • Comitato Nazionale per l'Energia Nucleare (Italy) • Comité National d'Études Nucléaires

CNEP: Centre National d'Études à Distance (France)

CNER: Comité National d'Évaluation de la Recherche (France)

CNES: Centre for Neural and Evolutionary Systems (UK) • Centre National d'Études Spatiales (France)

CNESER: Conseil National de l'Enseignement Supérieur Et de la Recherche (France)

CNESR: Conseil National de l'Enseignement Supérieur et de la Recherche (France)

CNET: Centre National d'Études des Télécommunications (France)

CNEXO: Centre National pour l'Exploitation des Océans (France)

CNF: Belo Horizonte - Confins (Brazil-MG) (location identifier) • Belo Horizonte - Tancredo Neves (Brazil-MG) (location identifier) • Canadian Neutron Facility • Configuration • Conjunctive Normal Form • Coordinator Nationale Faciliteit (Netherlands)

CNFA: Comité National Français d'Astronomie (France)

CNFL: Confluence

CNFGG: Comité National Français de Géodésie et de Géophysique (France) (*) • Conseil National Français de Géodésie et de Géophysique (France)

Cnfig: Configuration

CNFPT: Centre National de Formation Publique Territoriale (France)

CNFRS: Comité National Français de Radioélectricité Scientifique (Franc e) • Compact Nuclear Flat-Spectrum Radio Source

CNFWMP: Canadian Nuclear Fuel Waste Management Program

CNG: Calling • Charge Number Grouping • Cognac Parvaud (France) (location identifier) • Compressed Natural Gas

CNH: Claremont (USA-NH) (location identifier)

CNI: Calcul Numérique Intensif • Carte Nationale d'Identité • Centre National de l'Informatique • Changhai (China-PRC) (location identifier) • Coalition for Networked Information • Commissariat National à l'Informatique • Committee for Nuclear Information (USA) • Common Network Interface • Communications-Navigation-Identification • Consejo Nacional de Investigación • Coulomb-Nuclear Interference

CNIB: Centro Nazionale di Informazioni Bibliografique (Italy)

CNIC: Centre National de l'Informatique Chimique (France) • Centro Nacional de Investigaciones Científicas • Computer Network and Information Center • Computer Network Information Center

CNICT: Centro Nacional de Información Científica y Ténica (Venezuela) • Consejo Nacional de Investigaciones Científicas y Ténicas (Argentin a)

CNIDC: Centro Nacional de Información y Documentación Científica (Spain)

CNIDCYT: Centro Nacional de Información y Documentación Científica Y Tecnológica (Peru)

CNIDE: Centre National d'Information et de Documentation Économiques (Algeria)

CNIDR: Clearinghouse for Networked Information Discovery and Retrieval

CNIDS: CommerceNet/Nielsen Internet Demographic Survey

CNIDST: Comité National de l'Information et de la Documentation Scientifique et Technique (France)

CNIE: Comisión Nacional de Investigaciones Espaciales (Argentina)

CNIES: Centro Nacional de Información Económica y Social (Mexico)

CNIG: Centre National de l'Information Géographique (France) • Centro Nacional de Informação Geográfica (Portugal) • Conseil National de l'Information Géographique (France)

CNIL: Commission Nationale de l'Informatique et des Libertés (France) (*)

CNIN: Centraal Normalisatie Instituut in Nederland (Netherlands)

CNIP: Center for Networked Information and Publishing

CNIPA: Committee of National Institutes of Patent Agents

CNIPTG: Communications, Networks and Information Processing Theory Group

CNIR: Centre National d'Information Routière (France)

CNISF: Conseil National des Ingénieurs et Scientifiques de France

CNIT: Centre National des Industries et des Techniques (France)

CNJ: Cloncurry (Australia-QLD) (location identifier)

CNK: Coomon Noun Keyword • Concordia Blosser (USA-KS) (location identifier)

Cnl: Cancel

CNL: Canadian National Library • Carrier Noise Level • Central National Library • Circuit Net Loss • Committee on National Libraries (UK) • Commonwealth National Library (Australia) • Computer Newsletter • Sindal (Denmark) (location identifier)

CNLA: Council of National Library Associations (USA)

CNLD: Cancelled

CNLH: Canelha

CNLIA: Council of National Library and Information Associations (USA)

CNLLS: Complex Non-Linear Least Squares

CNLS: Complex Non-linear Least Squares

CNM: Carlsbad (USA-NM) (location identifier) • Cold Nuclear Medium • Communication Network Management • Customer Network Management

CNMA: Communication Network for Manufacturing Applications

CNME: Centre National des Mouvements d'Énergie (France) • Communications in Numerical Methods in Engineering (periodical code)

CNMS: Canadian National Measurement System

CNMV: Comisión Nacional del Mercado de Valores (Spain)

CNN: Cable News Network • Cellular Neural Network • Conference on Neural Networks

Cno: Corno

CNO: Carbon-Nytrogen-Oxygen • Chief of Naval Operations • Chino (USA-CA) (location identifier)

CNOC: Canadian Network for Observational Cosmology • Commission on Nomenclature in Organic Chemistry (IUPAC)

CNOFS: Communication Navigation Outgae Forecast System

CNOUS: Centre National des Oeuvres Universitaires et Scolaires (France)

CNP: Centre National de Prêt • Committee on Nuclear Physics • Connexion Non-Pertinente • Conseil National des Programmes • Customized Networking Platform • Eastgreenland (Greenland) (location identifier)

CNPCN: Comité National pour la Protection du Ciel Nocturne (France)

CNPF: Conseil National du Patronat Français (France)

CNPq: Conselho Nacional de Pesquisas (Brazil)

CNPR: Centre National de Promotion Rurale (France)

CNPS: Centre National des Professions de Santé (France) • Compact Nuclear Power Source • Conseil National de la Politique Scientifique (Belgium)

CNQ: Conselho Nacional da Qualidade (Portugal) • Corrientes - Camba Punta (Argentina-CR) (location identifier)

Cnr: Corner

CNR: Carrier-to-Noise Ratio • Chanaral (Chile) (location identifier) • Change Notice Request • Circumnuclear Ring • Committee on Natural Resources • Conseil National de la Recherche • Consiglio Nazionale delle Ricerche (Italy) • Corner • Cosmic-Ray Nucleus • Council of National Representatives

CNRC: Conseil National des Recherches du Canada

CNRET: Centre for Natural Resources Energy and Transport

CNRF: Cold Neutron Research Facility

CNRI: Corporation for National Research and Initiatives • Corporation for National Research Initiatives (*) • Corporation of National Research Initiatives

CNRL: Communications and Navigation Research Laboratory

CNRM: Centre National de Recherche Météorologique (France) • Centre National de Recherches Météorologiques (France) (*) • Comité National de Formation en Microélectronique (France)

CNRS: Centralized Network Reference Service • Centre National de la Recherche Scientifique (France)

CNRT: Centre National de Recherche Technologique

CNS: Cairns (Australia-QLD) (location identifier) • Catalogue of Nearby Stars • Center for Nonlinear Science (UNT) • Central Navigation Support • Central Nervous System • Certified Novell Salesperson • Cognitive and Neural System • Cold Neutron Source • Comet News Service • Common Number System • Communications, Navigation and Surveillance • Complimentary Network Service • Composite Nuclear Spectrum • Computer Network System • Computing and Network Services • Context Name Service • Continuous

CNSA: Centro Nacional de Sanidad Ambiente (Spain) • China National Space Administration (Chine-ROC)

CNSC: Center for Non-Linear Studies

CNSF: Coalition for National Science Funding • Cornell National Supercomputer Facility (USA-NY)

CNSG: Catalogue of Nearby Stars from the Ground

CNSH: Catalogue of Nearby Stars from Hipparcos

CNSL: Console

Cnsld: Consolidate

Cnsltnt: Consultant

CNSNS: Comisión Nacional de Seguridad Nuclear y Salvaguardias (Mexico)

Cnsp: Conspicuously

CNSR: Canadian Network for Space Research • Comet Nucleus Sample Return

CNSS: Core Nodal Switching System

CNSTAT: Committee on National Statistics

Cnstr: Canister

Cnt: Count

CNT: Canadian National Telegraph • Celestial Navigation Trainer • Centre National de Télédétection • Charata (Argentina-CH) (location identifier) • Computer Network Technology

Cntd: Contained • Continued

CNTE: Centre National de Télé-Enseignement (France)

CNTER: Center

CNTI: Centre National de Tranfert de l'Information

CNTL: Control

Cntn: Contain

Cntor: Contactor

Cntr: Container • Contribute • Contribution

CNTR: Center • Counter

CNTL: Central

CNTRL: Control • Controller

Cntrln: Lenterline

CNTRS: Centers

CNTS: Centre National des Techniques Spatiales

CNU: Chanute - Martin Johnson (USA-KS) (location identifier) • Chungnam National University (Korea-ROK) • Conseil National des Universités (France) • CSM Navigation Update

CNUCE: Centro Nazionale Universitario di Calcolo Elettronico (Italy)

CNUCED: Conférence des Nations Unies sur le Commerce Et le Développement

CNUED: Conférence des Nations Unies sur l'Environnement et le Développement

CNUL: Council on National University Libraries

CNUSC: Centre National Universitaire Sud de Calcul (France)

CNUST: Conférence des Nations Unies pour la Science et la Technique

CNV: Canavieiras (Brazil-BA) (location identifier) • Centre National du Volontariat (France) • Conventional

Cnvc: Convenience

Cnvr: Conveyor

Cnvt: Convert

CNVT: Convert

Cnvtb: Convertible

Cnvtd: Converted

Cnvtr: Converter

Cnvx: Convex

CNWRA: Centre for Nuclear Waste Regulatory Analyses

CNX: Certified Network Expert • Chiang Mai (Thailand) (location identifier)

CNY: Moab (USA-UT) (location identifier) • Yuang Ren-Min-Bi (China-PRC) (ISO currency code)

CNYN: Canyon

CNYTDO: Central New York Technology Development Organization (USA-NY)

CNZ: Cangamba (Angola) (location identifier)

c/o: Care Of • Carried Over • Cash Order • Change Over • Consist of

co: Corsican (ISO language code)

C/o: Care Of • Carried Over

Co: Camino • Cerro • Cobalt • Colombia • Colorado (USA) • Company • Compagnie • Continuity • Continuous • Corsican (ISO language code) • County • Cowling Number • Kobalt

C/O: Carbon to Oxygen Ratio • Cash Order • Certificate of Origin • Changeout • Checkout • Contamination/Overpressure • Cutoff

CO: Cabinet Office • Carbon Monoxide • Cardiac Output • Carried Over • Cash Order • Catalogue d'Observation • Central Office • Change Order • Channel Overflow • Checkout • Chief Officer • Chief Operator • Clerical Officer • Clock Oscillator • Colchester (UK postal code) • Colombia (international vehicle registration code) • Colombia (ISO country code) • Colombie (ISO country code) • Color • Colorado (USA) • Colour • Column • Comité d'Organisation • Commanding Officer • Command Output • Complete Observability • Compteur ordinal • Compte Ouvert • Computer Operator • Connection-Oriented • Conscientious Objector • Conseiller d'Orientation • Console Operator • Contracting Officer • Convert Out • Co-Occurrence • Coordinated Organizations • Correction Officer • Coupled Oscillator • Crystal Oscillator • Current Oscillation • Cutoff • Cut Out

CoA: Comments on Astrophysics

COA: Center Operations Area • Certificate Of Authority • Certificate of Operating Authority • Change Of Address • Change Of Assignment • Coherent Optical Array • Columbia (USA-CA) (location identifier) • Commission On Accreditation • Condition On Admission • Confirmation Of Arrival • Constant Output Amplifier • Course Of Action

COAA: Centro de Observação Astronômica do Algarve (Portugal)

COACH: Cognitive Adaptive Computer Help

COADC: Committee for Oceanographic Advice to Developing Countries

COADS: Comprehensive Ocean-Atmosphere Data Set

COAH: Coahuila (Mexico)

COALA: Calibration for Ozone by Atmopsheric Link Acquisitions

CoAp: Comments on Astrophysics (periodical code)

COAP: Center for Ocean Analysis and Prediction

COAPI: Coordinamento degli Osservatori Astronomici Popolari Italiani (Italy)

COARDS: Cooperative Ocean-Atmosphere Research Data Service

COARE: Coupled Ocean-Atmosphere Response Experiment

COAS: Coarse Optical Alignment Sight • Council of the Organization of American States • Crewman Optical Alignment Sight • Crew Optical Alignment Sight (*)

CoAsi: Contr. Oss. Astrofis. Padova in Asiago (periodical code)

COASST: Committee On the Application of Satellite and Space Technology

COAST: Cache On A Stick • Cambridge Optical Aperture Synthesis Telescope • Computer Operations, Audit and Security Technology

COAT: Coherent Optical Adaptive Technique

Coax: Coaxial

COB: Change Order Board • Chip On Board • Clip Onboard • Close Of Business • Columbus Operations Board • Commission des Opérations de Bourse (France) • Communications Office Building • Contenu Optimal des Bases • Coolibah (Australia-NT) (location identifier) • Cryogenic Optical Bench

COBAE: Comissão Brasileira de Atividades Espaciais (Brazil)

COBALT: Collocation-Based Language Translator

COBE: Cosmic Background Explorer

COBIS: Computer-Based Instruction System

COBOL: Common Business-Oriented Language

CoBos: Contributions of Bosscha Observatory (periodical code)

cobq: Cum Omnibus Bonis Quiescat

CoBRA: Computerised Bibliographic Record Actions

COBRA: Compatible Onboard Ranging • Complete Oil Breakdown Rate Analyzer • Computerised Bibliographic Record Actions • Computerized Boolean Reliability Analysis • Cooperation in Bureau, Research and Administration • Cosmic Background Radiation Anisotropy

COBRAS: Cosmic Background Radiation Anisotropy Satellite

COBY: Current Operating Budget Year

COC: Certificate Of Completion • Chamber Of Commerce • Circuit Order Control • Close-Open-Close • Coded Optical Character • Computer Output Committee • Concordia (Argentina-ER) (location identifier) • Cone Of Confusion

COCA: Clearinghouse On Computer Accommodation

CoCam: Contributions of Cambridge Observatory (periodical code)

COCAST: Council for Overseas Colleges of Arts, Science, and Technology

COCESNA: Central American Corporation for Air Navigation Services

COCF: Centre for Our Common Future

coch: Cochlear

COCHISE: Cold Chemi-excited Infrared Simulation Experiment

COCI: Committee On Chemistry and Industry • Consortium On Chemical Information

COClA: Catalogue of Open-Cluster Ages

CoCo: Configuration Control

COCO: Coaxial-Collinear • Collaborative Computing • Coordinator Of Chain Operations

CoCom: Control and Computers (periodical code)

COCOM: Coordination Committee

CoCoMo: Constructive Cost Model

COCOMO: Constructive Cost Model

COCONA: Conseil de Coopération Nord-Atlantique

COCONUDS: Coordinated Constellation of User-Defined Satellites

COCS: Camera Override Control System

Cód: Código

Cod: Codex • Codicil • Codification

CoD: Córdoba Durchmusterung

COD: Cash On Delivery • Center Operations Directorate • Centre for Orbit Determination • Chamber Of Deputies • Code • Cody (USA-WY) (location identifier) • Coherent Optical Device • Collect On Delivery • Concise Oxford Dictionary • Confirmation Of Delivery • Converted Out Distance • Correction Of Deficiency • Crack Opening Displacement

CODA: Chemical Online Data Analyzer • Conservation Option for Decision Analysis

CODAB: CODATA Bulletin (periodical code)

CODAG: Combined Diesel And Gas • Cosmic Dust Aggregation

CODAN: Carrier-Operated Device Anti-Noise

CoDAO: Contributions of the Dominion Astrophysical Observatory (periodical code)

CODAP: Client-Oriented Data Acquisition Program • Computer Documentation of Anomalous Phenomena • Control Data Assembly Program

CODAR: Coastal Ocean Dynamics Application Radar • Correlation Detection And Ranging • Correlation Display Analyzing And Recording

CoDAS: Condensed Matter Directed Alerting Service

CODAS: Customer-Oriented Data System

CODASYL: Conference on Data System Languages

CODATA: Committee on Data for Science and Technology • Confederation of Design and Technology Associations

CODC: Canadian Oceanographic Data Centre

CODCF: Central Office Data Connecting Facility

Codd: Codices

CODD: Central Online Data Directory

CoDDO: Contributions of the David Dunlap Observatory (periodical code)

CODE: Canadian Organization for Development through Education • Centre for Orbit Determination in Europe • Client/Server Open Development Environment • Coastal Ocean Dynamics Experiment • Cooperative Olympus Data Experiment

CoDec: Coder/Decoder • Compression/Decompression

CODEC: Coder/Decoder • Compression/Decompression

CODED: Computer-Oriented Design of Electronic Devices

CODER: Composite Document Effective Retrieval • Composite Document Expert Retrieval • Composite Document Extended Retrieval

CODEST: Comité de Développement Européen pour la Science et la Technologie • Comité pour le Développement Européen de la Science et de la Technol ogie • Committee for the Development of European Science and Technology

CoDEX: Chemo-Dynamical Evolution of Galaxies

CODF: Crystallite Orientational Distribution Function

CoFl: Combustion and Flame (periodical code)

CODI: Cloud Ozone Dust Imager

CODIAC: Centralized Or Distributed Integrated Access Control

CoDIAK: Concurrent Development, Integration, and Application of Knowledge

CODIF: Composition and Distribution Function

CODISC: Centre Opérationnel de la Direction de la Sécurité Civile (France)

CODMAC: Committee on Data Management And Computation (*) • Committee on Data Management And Computing • Committee On Data Management, Archiving and Computing

CODN: Component Operational Data Notice

CODOC: Code Documentaire

CODOG: Combined Diesel Or Gas

CODOT: Classification of Occupations and Directory of Occupational Titles

CODP: Coherent Optical Dynamic Processor

CoDR: Conceptual Design Review

CoE: Council of Europe

COE: Center Of Excellence • Central Office Equipment • Certificate Of Entitlement • Chief Operating Engineer • Code Of Ethics • Coeur d'Alene (USA-ID) (location identifier) • Cognizant Operations Engineer • Commission On Ecology • Committee On ECDIS • Committee On Education • Common Operating Environment • Complete Operating Equipment • Corps Of Engineers • Cost Of Electricity • Cost Of Energy

COEA: Cost and Operational Effectiveness Analysis • Cost Operational Effectiveness Analysis

Co-Ed: Co-Editor

CoEd: Co-Editor

COED: Computer-Operated Electronic Display • Computer-Oriented Electronic Display • Concise Oxford English Dictionary

COEES: Central Office Equipment Engineering System

Coef: Coefficient

Coeff: Coefficient

Coéff: Coéfficient

COEL: Chef des Opérations de l'Ensemble de Lancement

COEM: Commercial Original Equipment Manufacturer • Component Environmental Model

CoEnCo: Committee for Environmental Conservation

COER: Central Office Equipment Report

COES: Committee for Operations and EGSE Standardisation • Committee for Operations and EGSE Standards

COESA: Committee On Extension for the Standard Atmosphere

COF: Cocoa Patrick (USA-FL) (location identifier) • Columbus Orbital Facility • Computer Operations Facility • Computer-Optimized Fabrication • Construction Of Facilities • Cost Of Facilities

COFAC: Construction Facility

CofC: Chamber of Commerce

COFDM: Coded Orthogonal Frequency Division Multiplex

COFECUB: Comité Français d'Évaluation de la Coopération Universitaire avec le Brésil (France)

COFF: Common Object File Format

CofG: Center of Gravity

COFI: Checkout and Fault Isolation

COFIDA: Comité des Fonds Internationaux de Droits d'Auteurs

COFR: Certificate Of Flight Readiness • Certification Of Flight Readiness

CoFra: Contributions of the CNR in Frascati (periodical code)

COFRAC: Comité Français d'Accréditation (France)

COFRAD: Central-Office Frame Relay Access Device

COFRS: Chinese Operational Earth Resources Satellite

CofS: Chief of Staff

COFS: Coastal and Ocean Flux Study • Coastal Ocean Forecasting System

COFUSI: Comité Français des Unions Scientifiques Internationales (France)

COFW: Certificate Of Flight Worthiness

Cog: Cognate • Cognizant

CoG: Centre of Gravity

COG: Centre of Gravity • Cluster Of Galaxies • Collaborative Open Group • Computer Operations Group • Condoto (Colombia) (location identifier) • Congo (ISO country code) • Council Of Governments • Curve Of Growth

COGB: Certified Official Government Business

COGENE: Committee on Genetic Experimentation

COGEODATA: Committee on Geological Data

COGEODOC: Commission on Geological Documentation

COGEP: Confined One Gluon Exchange Potential

COGI: Centre d'Organisation Générale pour l'Informatique (France)

CoGIS: Cooperative Geographic Information System

COGO: Coordinate Geometry

COGS: Combined Operations Ground System • Continuous Orbital Guidance System

COGSIS: Committee on Government Statistics and Information Services (USA)

COH: Cash On Hand • Cooch Behar (India) (location identifier)

COHC: China Ocean Helicopter Corporation

COHMAP: Cooperative Holocene Mapping Project

COHO: Coherent Oscillator • Coordinated Heliospheric Observations

COHRA: Cambridge Observing Handbooks for Research Astronomers (CUP)

COHSI: Cambridge OH Suppression Instrument

Co-I: Co-Investigator

CoI: Co-Investigator

COI: Central Office of Information • Certificate of Origin and Interest • Coast Orbital Insertion • Cocoa - Merritt Island (USA-FL) (location identifier) • Co-Investigator • Comité Olympique International • Commission Océanographique Intergouvernementale • Commission Océanographique Internationale • Composite Optics Inc. (USA-CA) • Conflict Of Interest • Contingency Orbit Insertion • Cost Of Illness

CoIAP: Contributions de l'Institut d'Astrophysique de Paris (periodical code)

COIC: Canadian Oceanographic Identification Centre • Careers and Occupational Information Center

COIF: Charities Official Investment Fund

COIL: Chemical Oxygen Iodine Laser

COIM: Checkout Interpreter Module

COIN: Committee On Information Needs • Common Operators Interface Nucleus

COINS: Committee On Improvement of National Statistics

COIR: Centent-Oriented Information Retrieval

COIS: Canals Otoliths Integration Studies • Computation-Oriented Information System

COISTD: Confédération des Organisations Internationales Scientifiques et Techniques pour le Développement

COJ: Coonabarabran (Australia-NSW) (location identifier)

COK: Cochin (India) (location identifier) • Cook Islands (ISO country code) • Îles Cook (ISO country code)

CoKon: Communications of Konkoly Observatory (periodical code)

CoKwa: Contr. Astrophys. Kwasan Observ. Univ. Kyoto (periodical code)

Col: Cola • Colección • Colegio • Collect • Collected • Collection • Collector • College • Collegiate • Colombia • Colombian • Colon • Colonel • Colonia • Colonial • Colony • Color • Colorado (USA) • Colored • Columba-Columbae (constellation) • Column • Columna • Coronel

COL: Carry-On Laboratory • Checkout Language • Coherent Optical Laser • Colima (Mexico) • Collision • Coll Island (UK) (location identifier) • Colombia (ISO country code) • Colombie (ISO country code) • Communications-Oriented Language • Computer-Oriented Language • Cosmos-On-Line • Cost Of Living

COLA: Center for Ocean-Land-Atmosphere (USA-MD) • Collision Avoidance • Committee On Library Automation • Compact Objects in Low-powered AGNs • Constant-Output Level Adapter • Cost Of Living Adjustment • Cost Of Living Allowance

COLAGE: Conferencia Latinoamericana de Geofísica Espacial

colat: Colatus

COLD: Computer Output to Laser Disk

COLDSAT: Cryogenic On-orbit Liquid Depot Storage, Acquisition and Transfer

Colec: Colección

Colect: Colectivo

colent: Colentur

colet: Coletur

COLI: Company Owned Life Insurance

CoLic: Contributions of Lick Observatory (periodical code)

COLICO: Committee On Library Cooperation (Ireland)

COLIDAR: Coherent Light Detection And Ranging

CoLIS: Conceptions of Library and Information Science

COLIS: Council On Library and Information Services

Coll: Collateral • Colleague • Collect • Collected • Collection • Collective • Collector • College • Collège • Collegiate • Collègue • Collaboration • Colloque • Colloquial • Colloquium

Collab: Collaborate • Collaboration • Collaborator

COLLAGE: Collaborative Analysis and Graphics Environment • Collision and Aggregation Experiment

Collat: Collateral • Collaterally

Collect: Collective • Collectively

COLLIDE: Collision Into Dust Experiment

COLLOG: Columbus Logbook

Colloq: Colloque • Colloquial • Colloquialism • Colloquially • Colloquium

Collr: Collector

collun: Collunarium

collut: Collutorium

collyr: Collyrium

COLM: Column

Colo: Colorado (USA)

colog: Cologarithm

Colon: Colonial

CoLon: Communications of the University of London Observatory (periodical code)

Coloq: Coloquial

color: Coloretur

COLOS: Columbus Operating System

CoLPL: Communications of the Lunar and Planetary Laboratory (periodical code)

Cols: Columns

COLS: Communications for Online Systems

COLT: Central Office Line Tester • Computerized Online Testing

COLTRIMS: Cold Target Recoil - Ion Momentum Spectroscopy

Com: Coma Berenices - Comae Berenices (constellation) • Comedy • Comercio • Comisión • Comité • Comma • Command • Commander • Commentary • Commerce • Commercial • Commission • Commissioner • Committee • Commodore • Common • Commonality • Commonly • Commonwealth • Commune • Communicate • Communicated • Communication • Community • Común • Comunidad

CoM: Center of Mass • Chief of Mission

COM: Change Order Modification • Checkout Mode • Checkout Operations Manual • Chief Of Mission • Coleman (USA-TX) (location identifier) • Color Octet Model • Command • Commercial • Common Object Model • Common Organization of the Market • Comores (ISO country code) • Comoros (ISO country code) • Complex Organic Mixture • Component Object Model • Computerized Operations Management • Computer Output Management • Computer Output Microfilm • Computer Output Microfiche • Computer Output Microform • Computer Output on Microfilm • Computer Output on Microfiche • Computer Output on Microform • Cyclic Object Methodology

COMA: Conserving Momentum Approximation

COMAC: Continuous Multiple-Access Collator

COMADAB: Columbus Marketing Database

COMANDOS: Construction and Management of Distributed Open Systems

COMAP: Consortium for Mathematics and its Applications

COMAS: Central Office Maintenance and Administration System • Combined Orbital Maneuvering and Abort System

Comb: Combination • Combine • Combined • Combining • Combustible • Combustion

COMB: Combine • Combined • Combiner • Combustion • Console-Oriented Model Building

Combl: Combustible

COMBO: Correlation Of Miss Between Orbits

ComBr: Communication Broadcasting (periodical code)

Combu: Combustion

ComCon: Command Console

Comd: Command • Commander

COMD: Component Modeling

COMDEX: Communications Development Exposition • Computer Dealers Exposition

Comdg: Commanding

Comdr: Commander

Comdt: Commandant

COMEDI: Commerce Electronic Data Interchange

COMEND: Comendador

COMES: Comité Mexicano para el Eclipse Solar (Mexico) • Commissariat à l'Énergie Solaire

COMESA: Common Market for Eastern and Southern Africa

COMET: Coherent Electromagnetic-Energy Transmission • Command Executive Translator • Commercial Experiment Transporter • Committee for Middle East Trade • Computer-Operated Management Evaluation Technique • Coordinated Multimedia Explanation Testbed • Cornell Macintosh Terminal Emulator

COMETE: Commission de Mémoire Technique

COMETS: Career-Oriented Module to Explore Topics in Science • Commission d'Éthique pour les Sciences • Computer-Operated Multifunction Electronic Test System

COMETT: Community in Education and Training for Technology

COMEX: Commodity Exchange

COMI: Comité d'Orientation des Moyens Informatiques

COMICS: Cooled Mid-Infrared Camera and Spectrograph

CoMil: Contributions of Milano Observatory (periodical code)

COMIND: Cognitive Match Interface Design

COMIREX: Committee on Imagery Requirements and Exploitation

COMIS: Cartographic Operations Management Information System • Coded Mask Imaging Spectrometer • Committee Meeting Information System

COMIX: Cooperative Multiuser Interface to X-window

Coml: Commercial

COMLO: Compass Locator

Comm: Command • Commander • Comment • Commentary • Commerce • Commercial • Commercially • Commission • Commitment • Committee • Common • Commonwealth • Communicate • Communication • Community • Commutator

COMMANDOS: Construction and Management of Distributed Open Systems

Commd: Commissioned

Commn: Commission

Commnd: Commissioned

ComMP: Communications in Mathematical Physics (periodical code)

COMMS: Communications

COMMSEC: Communications Security

Commun: Communication • Community

Commy: Commissary

COMn: n-th Serial Port

COMNAP: Council of Managers of National Antarctic Programmes

ComNet: Communications Network

COMOD: Comodoro

Comp: Compañía • Companion • Comparative • Comparativo • Comparator • Compare • Comparing • Comparison • Compass • Compatible • Compensate • Compensating • Compensation • Compete • Competitive • Competitor • Compilación • Compilador • Compilation • Compile • Compiled • Compiler • Compiling • Complement • Complete • Component • Compose • Composed • Composer • Composite • Composition • Compositor • Compound • Compounded • Comprehensive • Compression • Comprising • Compromise • Compromiso • Computation • Compute • Computer • Computer (periodical code) • Computing

CoMP: Contributions to Mineralogy and Petrology (periodical code)

COMP: Compare • Compass • Compensate • Component • Computer • Conceptually Oriented Mathematics Program • Controlled Optics Modelling Package

COMPACT: Commercial Product Acquisition Team

Compar: Comparative • Comparison

COMPARE: Console for Optical Measurement and Precise Analysis of Radiation from Electronics

COMPAS: Coastal Ocean Management, Planning and Assessment System • Computer-Oriented Metering Planning and Advisory System

COMPASS: Common Muon and Proton Apparatus for Structure and Spectroscopy (CERN) • COMPTEL Processing and Analysis Software System • Computer-Aided Subject Search

Compd: Compound

COMPEN: Compensator

ComPh: Computers and Physics (periodical code)

Compl: Complement • Complément • Complete • Completed • Compliment • Complimentary

COMPL: Completed

COMPLEX: Committee on Planetary and Lunar Exploration (NAS)

COMPLIS: Collaboration for Measurements using a Pulsed Laser Ion Source

Complt: Complaint

CompM: Computational Mechanics (periodical code)

Compn: Compensate

Compo: Composites (periodical code) • Composition

Compr: Compression • Compressive

COMPR: Compress • Compressor

Comps: Compliments

Compt: Compartment • Comptroller

COMPT: Compartment • Controlled Optics Modelling Package

COMPTEL: Compton Telescope

Comptr: Comparator • Comptroller

Compu: Computing (periodical code)

COMPUSEC: Computer Security

Comr: Commissioner

COMS: Computer-based Operations Management System

COMSAT: Communications Satellite

COMSCEE: Committee on Science in Central and Eastern Europe

ComSE: Computing Systems in Engineering (periodical code)

COMSEC: Communications Security

COMSER: Commission on Marine Science and Engineering Research

COMSL: Communication Simulation Language

Comsnd: Commissioned

ComSo: Communications Society (periodical code)

COMSOFT: Consortium for the Management of Software Technologies

ComSp: Commercial Space (periodical code)

COMSS: Coastal Oceans Monitoring Satellite System (now ERS)

COMSTAC: Commercial Space Transportation Advisory Committee

CoMStOC: Coronal Magnetic-Structures Observing Campaign

COMTR: COMSAT Technical Review (periodical code)

Comunic: Comunicación

COM1: First Serial Port

COM2: Second Serial Port

COM3: Third Serial Port

COM4: Fourth Serial Port

con: Congestus

Con: Concentration • Concerning • Concerto • Conclusion • Confidence • Conjunx • Connection • Conservative • Consolidate • Consolidated • Consol • Consul • Continue • Continued • Contra • Convenience • Conversation

CON: Concord (USA-NH) (location identifier) • Condensation • Cónego • Console • Contact • Continuous Orthonormalization

CONACYT: Consejo Nacional de Ciencia y Tecnología (Mexico)

CONAE: Comisión Nacional de Actividades Espaciales

CONAFE: Consejo Nacional de Fomento Educativo

Conbd: Contributed

Conc: Concentrate • Concentrated • Concentrating • Concentration • Concentric • Concerning • Concesivo • Concrete

Concd: Concentrated

CONCERT: Conformity Certification

Concg: Concentrating

CONCISE: COSINE Network's Central Information Service for Europe

Concl: Conclude • Concluded • Concluding • Concludingly • Conclusion

Concn: Concentration

CONCN: Concentration

CONCORDE: Connected Car-Oriented Entity

Concr: Concrete

CONCYTEC: Consejo Nacional de Ciencia y Tecnología (Peru)

Cond: Condense • Condensed • Condenser • Condensing • Condicional • Condition • Conditional • Conditioning • Conditioner • Conduct • Conducted • Conducting • Conductivity • Conductor

COND: Condenser • Condition • Conditioner

CONDORS: Convective Diffusion Observed by Remote Sensors

Condr: Conductor

CONDR: Conditioner

CONE: Combined Neutral and Electron • Common OSI Networking Environment • Cryogenic Orbital Nitrogen Experiment

CONEE: Comisión Nacional del Espacio Exterior

CONES: Consejo Nacional de Educación Superior (Bolivia)

Conf: Confédération • Confer • Conference • Confidential • Configuration • Confirmación • Confirmed • Conformance

CONF: Conference

CONFDC: Confidence

Confed: Confederacy • Confederate • Confederation

Confid: Confidential

Config: Configuration

CONFIG: Configuration

CONFU: Conference on Fair Use

cong: Congius

Cong: Congress • Congressional

Congr: Congruent

Coni: Conical

CONICA: Coudé Near-Infrared Camera

CONICET: Consejo Nacional de Investigaciones Científicas y Técnicas (Argentina)

CONICIT: Consejo Nacional de Investigaciones Científicas y Técnologicas (Venezuela) • Consejo Nacional para Investigaciones Científicas y Técnologicas (Costa Rica)

CONICOR: Consejo de Investigaciones de Córdoba (Argentina-CO)

CONICYT: Comisión Nacional de Investigación Científica y Técnologica (Ch ile) • Consejo Nacional de Investigaciones Científicas y Técnicas (Uruguay) • Consejo Nacional de Investigaciones Científicas y Técnologicas (Venezuela)

CONIDA: Comisión Nacional de Investigaciones y Desarollo Aerospacial (Peru)

CONIE: Comisión Nacional de Investigaciones del Espacio (Spain)

CONIRIS: Calibrated Optical and Near Infrared Imaging System

Conj: Conjugation • Conjunción • Conjunction • Conjunctive • Conjunto

CoNLL: Conference on Natural Language Learning

Conn: Connect • Connecté • Connected • Connecter • Connecticut (USA) • Connecting • Connection • Connector • Connexion • Connotation

CONN: Connect • Connector

CONNEX: Connection Matrix

CONOPS: Concept of Operations

CONP: Connection-Oriented Network Protocol

ConPh: Contemporary Physics (periodical code)

CONPS: Carbon-Oxygen-Nitrogen-Phosphorus-Sulfur

CONQUEST: Concurrent Querying Space and Time

CONRAIL: Consolidated Rail Corporation (USA)

CONRIM: Council on Northern Resources Information Management

Cons: Consecrated • Consecutive • Consequence • Conséquence • Conservation • Conservative • Conservatoire • Conservatorium • Conservatory • Consider • Consideration • Considération • Considérer • Considering • Consist • Consistance • Consistency • Consisting • Consolidate • Consolidated • Consonant • Constable • Constant • Constante • Constitution • Constitutional • Construction • Consul • Consult • Consultant • Consulter • Consulting

CONS: Connection-Oriented Networking Session • Connection-Oriented Network Service • Conselheiro • Console

CONSAL: Congress of Southeast Asian Libraries

CONSCAN: Conical Scanning

Consec: Consecutive

Conseq: Consequence

Conséq: Conséquence

Conserv: Conservatoire • Conservatorium • Conservatory

Consid: Consideration

CONSOIL: Conservation Of Natural Resources through Sharing Of Information Layers

Consol: Consolidated

Conspec: Construction Specification

Const: Constable • Constant • Constante • Constitución • Constitution • Constitutional • Construction • Constructor

Constl: Constitutional

Constr: Constraint • Construct • Construction

CONSTR: Construct • Construction

CONSULT: Consultório

Cont: Contabilidad • Contact • Contador • Contain • Container • Containing • Content • Contents • Continent • Continental • Continuación • Continue • Continued • Continuous • Continuum • Contour • Contra • Contract • Contraction • Control • Controller

CONT: Container • Contingency • Continue • Continued • Continuous

CONTAC: Central Office Network Access

CONTAM: Contaminate

Contbg: Contributing

Cont'd: Continued

Contd: Contained • Continued

CONTDVD: Continental Divide

Contemp: Contemporary

Contg: Containing

CONTG: Continuing

CONTIG: Contiguous

Contin: Continued • Continuous

CONTLY: Continuously

Contn: Continuation

CONTONE: Continuous Tone

CONTOUR: Comet Nucleus Tour

Contr: Contracción • Contract • Contracted • Contraction • Contrary • Contrast • Contrasted • Contribution • Control • Controller

CONTR: Control • Controller

Contracc: Contracción

CONTRAIL: Condensation Trail

Contrib: Contributed • Contributing • Contribution • Contributor

CONTU: Commission on New Technological Uses

CONUS: Continental United States

Conv: Convection • Convective • Convenient • Convention • Conventional • Conversation • Conversion • Convert • Converter • Converting • Convertible • Convocation

CONV: Converter

Convn: Convection

COO: Centre d'Orbitographie Opérationnelle • Centre Opérationnel d'Orbitographie • Chief Operating Officer • Computerondersteund Onderwijs • Cotonou (Benin) (location identifier)

COOL: Chorus Object-Oriented Layer • Common Object-Oriented LISP • Concurrent Object-Oriented Language

COOLLAR: Cryogenic On-Orbit Long-Life Active Refrigerator

Coop: Cooperativa • Coopérative • Cooperative

COOP: Concurrent Object-Oriented Programming • Cooperative

COOR: Centre for Organizational and Operational Research

Coord: Coordinador • Coordinate • Coordinating • Coordination • Coordinator

COORD: Coordinate

COOS: Chemical Orbit-to-Orbit Shuttle • Coastal Ocean Observing System

COOSRA: Canadian Offshore Oil Spill Research Association

COOTS: Conference on Object-Oriented Technologies and Systems

Cop: Copernican • Copia • Copper • Copyright • Copyrighted

COP: Celescope Optical Package • Center of Pressure • Central Operations Program • Chane-Over Point • Coefficient Of Performance • Coherent Optical Processor • Combinatorial Optimization Problem • Combined Operations Plan • Committee for Operations and Plans • Common Online Package • Communication Output Printer • Community-Oriented Policing • Composite Operations Plan • Computer Optimization Package • Computer Optimization Program • Conference Of the Parties • Conseiller d'Orientation - Psychologue • Constabulary On Patrol • Constituent Object Parser • Contingency Operations Plan • Continuous Operation Program • Co-Orbiting Platform • Copperstown (USA-NY) (location identifier) • Cross-Order Profile • Current Operating Procedure • Customer-Orienting Program • Customer-Owned Property • Peso (Colombia) (ISO currency code)

COPA: Child Online Protection Act (USA) • Columbus Payload • Compañía Panameña de Aviación SA • Coordinated Observations of Polar Electrodynamics • Council On Postsecondary Accreditation

COPAC: Continuous Operation Production Allocation and Control

COPANIT: Comisión Panameña de Normas Industriales y Técnicas

COPANT: Comisión Panamericana de Normas Técnicas

COPARSS: Committee On Practical Applications of Remote Sensing from Space

COPAT: Comisión Panamericana Técnica

COPD: Chronic Obstructive Pulmonary Disease

COPE: Carbon Monoxide Pollution Experiment • Communications-Oriented Processing Equipment • Co-Orbiting Platform Elements

COPEARMS: Coordinating Project for Electronic Author's Rights Management Systems

COPEM: Columbus Precursor Environment Monitor

COPEPP: Committee on Public Education and Public Policy

CoPer: Contributions to Perkins Observatory (periodical code)

COPES: Committee on Program Evaluation and Support • Conceptually Oriented Program in Elementary Science • Control Program for Engineering Synthesis

CoPhC: Computer Physics Communications (periodical code)

COPI: Computer-Oriented Programmed Instruction

COPICAT: Copyright Ownership Protection for Computer-Assisted Training

COPICS: Copyright Office Publication and Interactive Cataloging System

COPIDAB: Columbus Pictures Data Bank

COPL: Centre d'Optique, Photonique et Laser • Core Persistent Object Library

Copol: Copolarized

COPOLCO: Consumer Policy Committee

CoPP: Contributions to Plasma Physics (periodical code)

COPPUL: Council of Prairie and Pacific University Libraries (USA)

Copr: Copyright

COPS: Canadian Optics and Photonics Society • Chrysler Optical Processing Scanner • Circuit Order Preparation System • Circuit Order Processing System • Coastal Ocean Prediction Systems • Coherent Optical Processing System • Common Open Policy Service • Computerized Oracle and Password System • Computer Oracle and Password System • Computer-Oriented Partial Sum • Concept-Oriented Programming System

COPT: Convection Profonde Tropicale

Coptr: Copartner

COPTRAN: Communication Optimization Program Translator

Copul: Copulativo

COPUOS: Committee on the Peaceful Uses of Outer Space (UNO)

COPUS: Committee On the Public Understanding of Science (UK)

COPV: Composite Overwrapped Pressure Vessel

COPYME: Confederación de la Pequeña y Mediana Empresa (Spain)

COQ: Certificate Of Qualification • Certification Of Qualification • Choibalsan (Mongolia) (location identifier)

Cor: Corner • Corollaire • Corpus • Correct • Corrected • Correction • Correlation • Correlative • Correspondence • Correspondent • Corresponding

CoR: Committee of Region

COR: Center for Operations Research • Centrale Ondernemingsraad • Centre d'Opérations Réseau • Circular Of Requirements • Coherent Optical Receiver • Cold Optics Radiometer • Commande Optimale du Rotor • Committee Of Region • Committee Of the Regions (*) • Component Overhaul and Repair • Concentric Orbit Rendezvous • Continental Operating Range • Contracting Officer's Representative • Córdoba - Pajas Blancas (Argentina-CD) (location identifier) • Corner • Coronel • Correction • Corrigendum • Cut-Off Rigidity

CORA: Coherent Optical Radar Amplifier

CORAD: Correlation Radiometer

CORAL: Class-Oriented Ring-Associative Language • Computer Online Real-time Applications Language

CORAM: Centre d'Observation et de Recherche Astronomique Mandeville (Belgium)

CORAVEL: Correlation Radial Velocities

CORB: Communications de l'Observatoire Royal de Belgique (periodical code)

CORBA: Common Object Request Broker Architecture

Córd: Córdoba

Cord: Cordillera

CorD: Córdoba Durchmusterung (periodical code)

CORD: Computer with Online Remote Devices

CORDIC: Coordinate Rotation Digital Computer

CORDIS: Community Research and Development Information Service

CORDS: Center for Orbital and Reentry Debris Studies • Copyright Office Registration and Deposit System

CORE: Central Operation of Resources for Educators • Chemical Online Retrieval Experiment • Coherent Optical Receiver Engineering • Cold Ocean Resources Engineering • Common Operational Research Equipment • Complex Operating Rule Engine • Continuous Observation of the Rotation of the Earth • Contractor Renormalization • Controlled Requirement Expression • Correlation-Regression-Effectiveness • Council of Registrars

CORECT: Collaborative Requirements Capture • Committee on Renewable Energy, Commerce and Trade

CoREN: Corporation for Research and Enterprise Network

COREPER: Comité des Représentants Permanents

CORES: Computer-assisted Order Routing and Execution System

CORETA: Centre d'Organisation de Recherches et d'Expertises en Technologies Avancées

CORF: Committee on Radio Frequencies

CORI: Celestial Observers of Rhode Island (USA-RI)

CORIA: Complexe de Recherche Interprofessionnelle en Aérothermisme

CORINE: Coordination of Information on the Environment

CORKS: Computer-Oriented Record Keeping System

CORLAT: Collection of Reference Laterite

Corn: Cornish • Cornwall (UK)

CORN: Computer Resources Nucleus

Corol: Corollary

Coroll: Corollary

Corp: Corporal • Corporate • Corporation

Corpn: Corporation

CORPS: Computing, Organizations, Policy and Society

Corr: Correct • Corrected • Correctement • Correction • Corrective • Correctly • Correlate • Correlating • Correlation • Correlative • Correlator • Correlation • Correlative • Corrente • Correspond • Correspondence • Correspondent • Corresponding • Corridor • Corrigenda • Corriger • Corrugated • Corrupt • Corrupted • Corruption

CoRR: Computing Research Repository

Correg: Corregidor

Correl: Correlative

Corresp: Correspondence • Corresponding

Corrup: Corrupción

Cors: Corsica • Corsican

CORS: Canadian Operational Research Society • Catalytic Oxygen Removal System • Continuously Operating Reference Station • Corners

CORSA: Cosmic Ray Satellite

CORSAIR: Coastal Optical Remote Sensing Airborne Radiometer • Computer-Oriented Reference System for Automatic Information Retrieval

CoRSE: Communications of the Royal Society of Edinburgh (periodical code)

CORSE: Chabot Observatory Remote Sensing Experiment • Conference on Remote Sensing Education

CORSPERS: Committee on Remote Sensing Programs for Earth Resource Survey

Corspnd: Correspond

CORSS: Coordination of Ocean Remote Sensing Satellites

CORSSAC: Civil Operational Remote Sensing Satellite Advisory Committee

CORTA: Cellule Opérationnelle de Régulation du Trafic Aérien

CORU: Candidate Orbital Replaceable Unit

cos: cosine

Cos: Companies • Counties

COS: Carbonyl Sulfide • Carry-on Oxygen System • Cash On Shipment • Cassette Operating System • Central Office of Statistics • Changeover Switch • Checkout Station • Chief Of Section • Chief Of Staff • Chief Of State • Chief Of Station • Class Of Service • Cloud Observing System • Coefficient d'Occupation des Sols • Colorado Springs - Peterson (USA-CO) (location identifier) • Combinatorial Organic Synthesis • Communication Operation Station • Communications Operating System • Compatibility Operating System • Compatible Operating System • Conceptual Operational System • Console Operating System • Corporation for Open Systems • Cosine • Cosmic Origins Spectrograph • Cray Operating System • Crystal Observation System

co. sa.: Come Sopra

COSA: Colorado Office of Space Advocacy (USA-CO)

COSAC: Cometary Sampling And Composition • Communications Sans Connexions

COSAG: Combined Steam And Gas

COSAMC: Commission for Special Applications of Meteorology and Climatology

COSAR: Compression Scanning Array Radar

COSATI: Committee On Scientific And Technical Information

COS-B: Cosmic gamma-ray Satellite B

COSB: Cosmic gamma-ray Satellite B

COSC: Chabot Observatory and Science Center (USA-CA)

COSCAT: Coherent Scatter

COSCL: Common Operating System Control Language

COSCQ: Central Organization for Standardization and Quality Control (Iraq)

COSD: Composite Object State Diagram

COSE: Combined Office Standard Environment • Committee on Optical Science and Engineering • Committee On Science Education • Common Open Software Environment • Common Open Systems Environment • Common Operating Software Environment • Cooperative Operating System Environment

cosec: Cosecant

cosech: Hyperbolic Cosecant

COSEPP: Committee on Science, Engineering and Public Policy

COSEPUP: Committee on Science, Engineering and Public Policy

COSETI: Columbus Optical Search for Extraterrestrial Intelligence

CoSGI: Contributions of the Slovak Geophysical Institute (periodical code)

cosh: Hyperbolic Cosine

COSI: Center of Science and Industry • Closeout System Installation • Combiner Simulator • Comet Simulation

COSIMA: Cometary Secondary Ion Mass Analyzer • Crystallization of Organic Substances In Microgravity for Applications

COSINE: Cooperation for OSI Networking in Europe

COSIT: Conference On Spatial Information Theory

CoSka: Contributions of Skalnate Pleso Observatory (periodical code)

COSM: Checkout Servicing and Maintenance

COSMIC: Carnegie Observatories Spectroscopic Multislit and Imaging Camera • Common System Main Interconnecting • Computer Software Management and Information Center • Constellation Observing System for Meteorology, Ionosphere, and Climate

CoSMO: Consolidated Supercomputing Management Office

COSMOSOMAS: Cosmological Structures On Medium Angular Scales

Cosmog: Cosmogony • Cosmographical • Cosmography

Cosmogr: Cosmografía

COSMOS: Computer-Oriented System for Management Order Synthesis • Computer System for Mainframe Operations • Console-Oriented Statistical Matrix Operator System • Coordinates, Size, Magnitude, Orientation and Shapes

COSMT: Czech Office for Standards, Metrology and Testing

COSNA: Composite Observing System for the North Atlantic

co. so.: Come Sopra

COSOD: Conference on Scientific Ocean Drilling

COSOMA: Copyright Society of Malawi

COSP: Centre d'Orientation Scolaire et Professionnelle

COSPAC: Cost/Schedule Planning And Control

COSPAR: Committee on Space Research

COSPEC: Correlation Spectrometer

COSPIN: Cosmic-ray and Solar-charged Particles Investigation

COSPUP: Committee on Science and Public Policy

COSQ: Cost On Software Quality

COSRAD: Coastal Ocean Surface Radar

CosRe: Cosmic Research (periodical code)

COSRIMS: Committee on Support of Research in the Mathematical Sciences

Coss: Consules

COSS: Common Object System Specification • Customer-Oriented Strategic System

COSSA: CSIRO Office of Space Science and Applications

COSSAC: Chief Of Staff to Supreme Allied Commander

COSSC: Compton Observatory Science Support Center

CosSe: Cosmic Search (periodical code)

Cost: Costante

COST: Committee for Overseas Science and Technology • Committee on Science and Technology • Coopération en Science et Technologie • Cooperation in Science and Technical Research • Cooperation On Science and Technology

• Coopération Scientifique et Technique • Cost-Oriented Systems Technique

Costa: Costanilla

COSTA: Cost Accounting

COSTAR: Corrective Optics Space Telescope Axial Replacement (*)

CoSte: Contr. Univ. Arizona, Steward Obs. (periodical code)

COSTED: Committee on Science and Technology in Developing Countries

COSTEP: Comprehensive measurements of the Supra-Thermal and Energetic Particle populations • Comprehensive Supra-Thermal and Energetic Particles

CoStr: Computers and Structures (periodical code)

COSTR: Collaborative Solar-Terrestrial Research

COSY: Checkout Operating System

cot: Cotangent

CoT: College of Technology

COT: Central Office Terminal • Change Of Temperature • Checkout Time • Columbus Operations Team • Committee On Toxicology • Computer-Output-Typesetting • Consistent Orbit Tracing • Cotulla (USA-TX) (location identifier) • Critical Optical Technology • Customer-Operated Terminal • Cyclooctatetraene • Cyclooctatetraenyl

cotan: Cotangent

COTASP: Council of Technology and Allied Sciences of the Philippines

COTE: Checkout Terminal Equipment

coth: Hyperbolic Cotangent

COTIF: CEDAR Optical Tomographic Imaging Facility • CEDAR Optical Tomography of the Ionosphere Facility

COTM: Communications On The Move

CoTol: Contributions of Cerro Tololo Observatory (periodical code)

COTR: Contracting Officer Technical Representative

COTS: Checkout Test Set • Commercial Off-The-Shelf

Cott: Cottage

COTT: Central Organization for Technical Training

Cou: Coudé

COU: Columbia (USA-MO) (location identifier) • Concept of Operation and Utilization • Curso de Orientación Universitaria (Spain)

Coun: Council • Councillor • Counsellor

COUNT: Computer-Operated Universal Test

COUP: Consolidated Operations and Utilisation Plan

Cour: Courant

Cov: Covariance • Coventry (UK-England) • Cover • Covered • Covering

COV: Calculus Of Variation • Coefficient Of Variation • Composé Organique Volatile • Concept-Oriented View • Corona Onset Voltage • Covariance • Covilha (Portugal) (location identifier) • Cutoff Value • Cutoff Valve • Cutoff Voltage

COVAM: Capture Orbit Vehicle Assembly Mode

COVE: Coronal Velocity Explorer

COVENIN: Comisión Venezolana de Normas Industriales (Venezuela)

COVIRA: Computer Vision in Radiology

CoVis: Collaborative Visualization

CoVVO: Contributions of Van Vleck Observatory (periodical code)

COW: Cell-site On Wheel • Cluster Of Workstations • Committee on Water • Copy On Write

COWAR: Committee on Water Research

CoWat: Contributions of the University of Waterloo Observatory (periodical code)

COWAT: Coaxial Waveguide Transformer

COWI: Commissie Wetenschappelijke Informatieverzorging (Netherlands)

COx: Carbon Oxide

COX: Congo Town (Bahamas) (location identifier)

COY: Coolawanyah (Australia-WA) (location identifier)

COYOTES: Coordinated Observations of Young Objects from Earthbound Sites

COZ: Costanza (Dominican Rep.) (location identifier)

COZAC: Conservation Zone Advisory Committee

CO2: Carbon Dioxide

CO2EL: Carbon Dioxide Electrolysis

C°: Cerro

cP: Centipoise

Cp: Camp • Candlepower • Carriage Paid • Compare • Copy

C&P: Capabilities & Procedures • Contracts & Pricing

C/P: Charter Party • Cold Plate

CP: Bolivia (international aircraft registration code) • Caixa Postal • Calibration Procedure • Call for Proposals • Cambridge Pulsar • Camp • Canadian Pacific • Canadian Press • Candlepower • Cape • Cape Photographic • Cape Province (South Africa) • Cardinal Point • Card Punch • Cargo Projects • Carriage Paid • Carte Perforée • Casella Postale • Case Postale • Case Processor • Center of Pressure • Center Point • Central Peak • Central Point • Central Processor • Chamber Pressure • Change Package • Change Point • Change Proposal • Character Printer • Charge Parity • Checkpoint • Chemically Peculiar • Chemically Pure • Chemische Physik • Circuit Package • Circular Pitch • Circular Polarization • Circular Pure • Civil Procedure • Clock Phase • Clock Pulse • Code of Practice • Code of Procedure • Code Postal • Codice Penale • Código Postal • Coherent Processor • Código Postal • Cold Plate • Colegio Público • Colis Postal • College Park (USA-MD) • Color Printing • Command Processor • Command Post • Command Pulse • Common Point • Common Program • Communications Processor • Communications Program • Community Program • Completed Path • Compteur de Programme • Computer Program • Concentration Polarization • Concurrent PROLOG • Condensation Product • Conditional Processor • Conditional Proof • Conference Paper • Conference Proceedings • Conference Publication • Congés Payés • Congressional Presentation • Conjugation Parity • Connector Panel • Connexion Pertinente • Conservation Park • Console Processor • Constant Potential • Constant Pressure • Constrained Procedure • Contact Point • Content Particle • Continuous Path • Contracting Party • Contracts and Pricing • Contributed Presentation • Control Panel • Control Point • Control Procedure • Control Processor • Control Program • Coordinated Procurement • Copilot • Copy • Copy Protected • Core Program • Cosmological Principle • Cours Préparatoire • Cox Process • Crack Propagation • Crossover Point • Cross Polarization • Cross Product • Current Paper • Cyberspace Protocol • Cyclic Protocol • Phosphorus Carbide

CPA: Cape Palmas (Liberia) (location identifier) • Cardiopulmonary Arrest • Center for Particle Astrophysics • Central Planning Area • Centro de Pesquisas Antárticas • Certified Public Accountant • Cessation Progressive d'Activités • Change Proposal Analyst • Chartered Public Accountant • Chirped Pulse Amplification • Circularly Polarized Antenna • Circularly Polarized Array • Classification of Products by Activity • Closest Point of Approach • Coherent Potential Approximation • Cold-Plasma Analyzer • Commission on Preservation and Access • Commutative Principle for Addition • Conjugation-Parity Asymmetry • Constant Potential Accelerator • Contingency

Planning Aid • Contract Price Adjustment • Cost Planning and Appraisal • Critical Path Analysis

CPAA: Charge Particle Activation Analysis • China Patent Agents Association (China-PRC)

CPAC: Cambridge Planck Analysis Center (USA-MA) • Castle Point Astronomy Club (UK-England)

CPAF: Cost Plus Award Fee

CPAG: Commission on Preservation and Access Glossary

CPAL: Computer Processing on Asian Languages

CPAM: Caisse Primaire d'Assurance Maladie (France) • Columbus Payload Accommodation Handbook • Continental Polar Air Mass

CPAN: Circumpolar Protected Area Network • Comprehensive Perl Archive Network

CPAO: Conception de Programme Assistée par Ordinateur

CPAP: Chef de Projet Ariane Production • Control Parameter Assembly Program

CPAS: Centre Public d'Aide Sociale

CPB: Capurgana (Colombia) (location identifier) • Centraal Planbureau • Central Planning Bureau • Channel Program Block • Control Processor Board • Corporation for Public Broadcasting

CPC: Canadian Polar Commission • Calculatrice à Programmes par Cartes • Cape Photographic Catalogue • Card Programmed Calculator • Career Planning Center • Cargo Processing Contract • Center for Personal Communication • Center for Plant Conservation • Central Pennsylvania Consortium (USA-PA) • Central Planning Center • Central Processing Complex • Central Processor Complex • Central Product Classification • Centrifugal Partition Chromatography • Characteristics Properties Code • Chiron Perihelion Campaign • Chopped Photometric Channel • Climate Prediction Center • Clock Pulsed Control • Coastal Patrol Craft • Coated Paper Copier • Color Pack Camera • Column Position Counter • Comité du Programme et de la Coordination • Commission Professionnelle Consultative • Common Product Classification • Communications Processing Center • Complementary Protection Certificate • Compound Parabolic Concentrator • Computer Physics Communications • Computer Print Control • Computer Process Control • Computer Program Command • Computer Program Component • Computer Programming Concept • Condensation Particle Counter • Conference Planning Committee • Constant Pressure Chart • Contact Postcode • Contract Progress Control • Control Pressure Conditioner • Control Processing Complex • Corps Propulseur Chargé • Creative Programming Consultants (USA-TX) • Critical Propulsion Component • Cyclic Permutation Code • Cyclic Program Control • San Martin de los Andes - Chapelco (Argentina-NE) (location identifier)

CPCB: Crew Procedures Control Board

CPCC: Communications Processor Conversion Center

CPCEI: Computer Program Contract End Item

CPCG: Commercial Protein Crystal Growth

CPCI: Commission Permanente de Concertation pour l'Industrie (France) • Computer Program Change Instruction • Computer Program Configuration Item • Computer Program Configured Item • Cross-Pointer Course Indicator

CPCL: Computer Program Change Library • Computer Program Control Library

CPCP: Center for Planetary Chemistry and Physics • Center for the Physics and Chemistry of the Planets

CpCpH: Connections per Circuit per Hour

CPCPH: Connections Per Circuit Per Hour

CPCR: Computer Program Change Request • Crew Procedures Change Request

CPCS: Check Processing Control System • Communication Payload Checkout System • Conversion Process Controller System • Cost Planning and Control System

CPCT: Center for Photoinduced Charge Transfer

CPCU: Computer Power Conditioning Unit

CPC2: Second Cape Photographic Catalogue

Cpd: Compound

CpD: Cards per Day • Cycles per Day

CPD: Cape Photographic Durchmusterung • Cards Per Day • Center Program Director • Change Plot Device • Chemical Property Data • Chemical Propulsion Division • Circuit Protection Device • Coaxial Power Divider • Commercial Product Development • Commercial Program Development • Common Position Date • Communications Planning and Development • Contact Potential Difference • Continuing and Professional Development • Continuing Professional Development • Coober Pedy (Australia-SA) (location identifier) • Cosmic-Particle Detector • Course Pointing Device • Crew Passive Dosimeter • Crew Procedures Division • Cycles Per Day

CPDC: Centre Pédagogique de Documentation et de Communication • Command Processor Distributor Control • Committee on the Participation of Developing Countries

CPDD: Conceptual Project Design Description

CPDDS: Computer Program Detail Design Specification

CPDF: Central Personnel Data File

CPDLC: Controller-to-Pilot Data Link Communication

CPDP: Computer Program Development Plan • Computer Program Development Program • Continuing Professional Development Plan

CPDR: Contractor's Preliminary Design Review

Cpds: Compounds

CPDS: Computer Program Design Specification • Computer Program Detailed Specification • Computer Program Development Specification • Crew Procedures Documentation System

Cpdt: Cependant

CPDU: Central Power Distribution Unit • Command Pulse Distribution Unit

CPE: Campeche (Mexico) (location identifier) • Cape • Cathodic Protection Equipment • Centrally Planned Economy • Central Processing Element • Central Processing Equipment • Central Programmer and Evaluator • Centre de Prospective et d'Évaluation • Certified Processing Element • Chief Program Engineer • Chief Project Engineer • Circadian Periodicity Experiment • Circular Probable Error • Cloud Processing Equipment • Commission Paritaire d'Établissement (France) • Communauté Politique Européenne • Computer Performance Evaluation • Computer Peripheral Equipment • Constant Phase Element • Continuing Professional Education • Contractor Performance Evaluation • Coopération Politique Européenne • Cosmic-Particle Event • Customer Premises Equipment • Customer-Provided Equipment

CPEG: Certificat Préparatoire aux Études de Gestion

CPEI: Computer Program End Item

CPEM: Conference on Precision Electromagnetic Measurements

CPEN: Centre Permanent d'Étude la Nature (Belgium)

CPEP: Contemporary Physics Education Project

CPER: Contrat de Plan État-Région (France)

CPERT: Continuous Professional Education Round Table

CPES: Chemistry, Physics, and Environmental Sciences • Contractor Performance Evaluation System • Crew Procedures Evaluation Simulator • Crew Procedures Evaluator Simulator

CPET: Charged Particle Electrostatic Thruster

CPEUG: Computer Performance Evaluation Users Group

CPEV: Customer Premises Equipment Voice

CpF: Cost per Flight

CPF: Calibration Product File • Cargo Processing Facility • Central Processing Facility • Central Provident Fund • Cepu (Indonesia) (location identifier) • Change Parameter File • Command Propagation Facility • Common Publishing Format • Context Passing Format • Control Program Facility • Cost Per Flight • Critical Point Facility • Cumulative Percentage Frequency

CPFF: Cost Plus Fixed Fee

CPG: Calibration Proposal Generation • Candidate Pass Generator • Carmen de Patagones (Argentina-BA) (location identifier) • Catalog of Paired Galaxies • Catalogue of Pairs of Galaxies • Central Processing Group • Change Planning Group • Clock Pulse Generator • Coarse-Pointing Gimbal

CPGA: Catalogue of Peculiar Galaxies and Associations

CPGE: Classes Préparatoires aux Grandes Écoles (France)

CPGS: Compiling Genetic Programming System

cph: Cards per Hour • Characters per Hour • Counts per Hour • Cycles per Hour

Cph: Cost Per Hour

CPH: Central Powerhouse • Copenhagen (Denmark) (location identifier) • København (Denmark) (location identifier)

CPhD: Cape Photographic Durchmusterung (periodical code)

CPHE: Comets in the Post-Halley Era

cpi: Characters per Inch • Clock per Instruction • Cycles per Instruction

CpI: Cost per Instruction

CPI: Carbon Preference Index • Center of Programmed Instruction • Central Patent Index • Centro Público de Investigación • Channel Port Index • Charge Particle Instrument • Chemical Patent Index • Clock Pulse Interval • Code de la Propriété Intellectuelle (France) • Code-Page Information • Code-Page Interface • Coherent Processing Interval • Committee on Public Information • Common-Path Interferometry • Common Programming Interface • Comprehensive Plasma Investigation • Conference Papers Index • Consumer Price Index • Continuous Process Improvement (*) • Continuous Progress Improvement • Cost Performance Index • Cost Per Instruction • Cost Plus Incentive • Crédits de Politique Industrielle • Cross Pointer Indicator • Current Physics Index

CPIA: Close-Pair Interstitial Atom

CPIC: Charged Particles Information Center (USA) • China Patent Information Center (China-PRC) • Common Programming Interface for Communications • Computer Program Integration Contract • Computer Program Integration Contractor

CPIF: Character Position In Frame • Cost Plus Incentive Fee

CPIN: Computer Program Identification Number

CPIP: Club des Professionnels de l'Industrie du Photon (France) • Computer Program Implementation Process

CPIPI: Commission pour la Protection Internationale de la Propriété Industrielle

CPIRSS: Catalog of Positions of Infrared Stellar Sources

CPIS: Canadian Polar Information System • Center for Planning and Information Systems • Comprehensive Planning Information System

CPJI: Cour Permanente de Justice Internationale

CPK: Corey-Pauling-Koltun

Cpl: Corporal • Couple

CpL: Characters per Line

CPL: Callable Personal Librarian • Capillary Pumped Loop • Central Public Library • Chaparral (Colombia) (location identifier) • Characters Per Line • Chemical Physics Letters (periodical code) • Combined Programming Language • Common Program Language • Common Programming Language • Common Pulse Line • Computer Program Library • Consolidated Parts List • Contractor Procurement List • Conversationnal Planning Language • Conversationnal Programming Language • Couple • Critical Path Length • Current Privilege Level

CPLC: Commission du Personnel des Laboratoires Conventionnés

Cpld: Coupled

CPLEE: Charged Particle Lunar Environment Experiment

CPLF: Certificat Pratique de Langue Française

Cplg: Coupling

CPLOM: Center for Production, Logistics and Operations Management

Cplr: Coupler

Cplry: Capillary

CPLT: Complete

cpm: Cards per Minute • Characters per Minute • Counts per Minute • Cycles per Minute

CpM: Cost per Minute

CP/M: Control Program for Microcomputers • Control Program Monitor

CPM: Capsule Positioning Mechanism • Cargo Propulsion Module • Cathode Pulse Modulation • Cell Performance Monitor • Center Program Manager • Central Path Method • Central Processing Module • Central Processor Module • Change Proposal Manager • Characters Per Minute • Combinatorial Pattern Matching • Command Processor Module • Common Particular Meter • Common Proper Motion • Compton (USA-CA) (location identifier) • Computer Program Module • Conference Preparatory Meeting • Continuous Phase Modulation • Control Program for Microcomputers • Control Program Monitor • Conversational Program Module • Cost and Performance Measurement • Cost Per Minute • Cost Per Thousand • Coût Par Millier • Critical Path Method

CPMA: Centre de Projets en Microélectronique Avancée (France)

CPMAS: Cross Polarization Magic Angle Spinning

CPMB: Common Proper Motion Binary

CPME: Charged Particle Measurement Experiment

CPMF: Cost Plus Management Fee

CPMIS: Consolidated Personnel Management Information System

Cpmm: Character per Millimeter

CPMM: Character Per Millimeter

CPMOH: Centre de Physique Moléculaire Optique et Hertzienne (France) (now CELIA)

CPMP: Carrier Performance Measurement Plan • Circum-Pacific Man Project • Crew Procedures Management Plan • Crew Products Management Plan

CPMS: Climatological Prediction by Model Statistics • Computer-assisted Photogrammetric Mapping System • Contractor Property Management System

CPMT: Components, Packaging and Manufacturing Technology

Cpn: Copenhagen (Denmark) • Coupon

CPN: Cape Rodney (Papua New Guinea) (location identifier) • Causal Probabilistic Network • CCSDS Principal Network • Corps de Propulseur Nu • Customer Premise Network

CPND: Charged Particle Nuclear Data

Cpnhgn: Copenhagen (Denmark)

CPNI: Consumer Proprietary Network Information

Cpntr: Carpenter

Cpo: Campo

CPO: Canberra Planetarium and Observatory (Australia-ACT) • Cancel Previous Order • Capilla Peak Observatory (USA-NM) • Catalytic Partial Oxidation • Central Pennsylvania Observers (USA-PA) • Central Planning Office • Central Post Office • Central Project Office • Charged Particle Optics • Chief Petty Officer • Chinese Patent Office (China-PRC) • Circular Parking Orbit • Cloud Physics Observatory • Command Pay Office • Compulsory Purchase Order • Controlled Precision Oscillator • Copiapo (Chile) (location identifier) • Core Project Office

CPODA: Contention Priority-Oriented Demand Assignment

CPOL: Communications Procedure-Oriented Language

CPOZ: Compressed Profile Ozone

CPP: Calling Party Pays • Cancellation Protection Plan • Capacitive Phase Probe • Caractères Par Pouce • Card Punching Printer • Chemical Processing Plant • Cloud Photopolarimeter • Columbus Polar Platform • Controllable Pitch Propeller • Corps Propulseur Protégé • Critical Path Planning • Current Perpendicular to the Plane • Current Purchasing Power

CPPA: Canadian Periodical Publishers Association • Centralized Parts Procurement Agency • Cosmological Pressureless Parabolic Advection

CPPC: Core Project Planning Committee

CPPL: Command Processor Parameter List

CPPM: Centre de Physique des Particules de Marseille (France)

CPPS: Commission Permanente pour le Pacifique Sud • Computer Programming Performance Specification • Computer Program Product Specification • Critical Path Planning and Scheduling

Cpq: Compaq

CPQ: Campinas (Brazil-SP) (location identifier) • Comissaõ Permanente para a Qualificaçao

cpr: Cycles per Revolution

CPR: Canadian Pacific Railway • Cardiopulmonary Rescucitation • Card Puncher and Reader • Casper (USA-WY) (location identifier) • Center for Public Resources Inc. • Centre de Programmation Récréative • Clock Pulse Repeater • Cloud Physics Radiometer • Committee of Permanent Representatives • Common Position Reference • Computer-based Patient Record • Consolidated Progress Report • Continuous Plankton Recorder • Contract Procurement Request • Cost Performance Report • Critical Problem Report • Customer Payload Requirement • Cut Paraboloidal Reflector • Cycle Pressure Ratio

CPRA: Compressed Pulse Radar Altimeter

CPRC: Central Price Regulation Committee

CPRDB: Comprehensive Pacific Rainfall Database

CPRI: Computerized Patient Record Institute

Cpris: Compris

Cprs: Compress

CPRS: Central Policy Review Staff • Cloud Profiling Radar System

Cprsd: Compressed

Cprsn: Compression

Cprsr: Compressor

cps: Cards per Second • Centipoise • Characters per Second • Counts per Second • Cycles per Second

CpS: Connections per Second

CPS: Camera Power Supply • Caractères Par Seconde • Carte de Professionnel de Santé • Cartographic Production System • Cathode Potential Stabilized • Center for Policy Studies • Central Pain Syndrome • Central Plasma Sheet • Central Power Supply • Central Power System • Central Processing System • Centrum voor Plasmafysica en Stralingstechnologie • Certificate Practice Statement • Certification Practice Statement • Certified Professional Secretary • Chicago Public Schools (USA-IL) • Close-Packed Sphere • Coarse Pointing System • Color Picture Signal • Commercial Processing System • Commission Permanente de Standardisation (France) • Commonwealth Public Service • Communications Processing System • Complementary Polar Satellite • Computer Power Supply • Computer Programming Service • Computer Program Specification • Computer Program System • Connections Per Second • Contour Plotting System • Control Power Supply • Control Pressure System • Conversational Programming System • Conversation Program System • Copy Processing System • Crew Procedures Simulator • Critical Path Schedule • Critical Path Scheduling • Critical Path System • Current Population Survey • Saint Louis - Bi-State Parks (USA-MO) (location identifier)

CPSC: Consumer Product Safety Commission

CPSD: Corrected Pattern Standard Deviation • Cross-Power Spectral Density

Cpse: Counterpoise

CPSE: Common Payload Support Equipment

CPSG: Common Power Supply Group

CPSK: Cathodic Protection Survey Kit • Coherent Phase Shift Keyed

CPSMA: Commission on Physical Sciences, Mathematics and Applications

CPSO: Cataloging Policy and Support Office • Cumberland Plateau Seismological Observatory

CPSR: Computer Professionals for Social Responsibility • Contractor's Procurement System Review

CPSS: Central Processing Subsystem • Cold Plate Support Structure • Common Program Support System • Cooperative Program in Space Science • Critical Phase System Software

CPSST: Cross-Product Sea Surface Temperature

CPST: Commission on Professionals in Science and Technology • Committee to Promote Science and Technology

CPSU: Central Processor Subunit

Cpt: Capitain • Cockpit • Comptant • Counterpoint

CpT: Cost per Thousand

CPT: Cape Town - D.F. Malan (South Africa) (location identifier) • Captain • Cargo Processing Technician • CCI-PS-TRI • Central Planning Team • Centre de Physique Théorique • Charged-Particle Telescope • Charge-Parity-Time • Clock, Programming and Timing • Combination Photoelectric Telescope • Command Pass Through • Comprehensive Performance Test • Computer Program Tape • Critical Path Technique • Cryogenic Pressure Transducer • Crystal Pressure Transducer

CPTA: Comité Technique Paritaire Académique (France) • Core Propulsion Test Article

Cptán: Capitán

CPTC: Central Processor Test Console

CPTD: Comité Technique Paritaire Départemental (France)

Cpte: Capitaine

CPTEC: Centro de Previsão de Tempo e Estudos Climáticos

CPTP: Coordinated Project Technology Programme

CPTR: Compact Payload Test Range

CPTYC: Committee on Physics in Two-Year Colleges

CpU: Cost per Unit

CPU: Central Packaging Unit • Central Processing Unit (*) • Central Processor Unit • Collective Protection Unit • Command Processor Unit • Communications Processing Unit • Computer Printer Unit • Computer Processing Unit • Computer Program Update • Conférence des Présidents d'Universités • Control Processing Unit • Cururupu (Brazil-MA) (location identifier)

CPUAX: Central Processing Unit - Arithmetic Extended

Cpv: Capoverso

CPV: Campina Grande - João Suassuna (Brazil-PB) (location identifier) • Cape Verde (ISO country code) • Cap Vert (ISO country code) • Centre de Programmation Végétation • Chlorure de Polyvinyle • Common Pressure Vessel • Common Procurement Vocabulary

CPX: Culebra (Puerto Rico) (location identifier)

cq: Casu Quo

CQ: Cercle de Qualité • Conditionally Qualified

CQAO: Contrôle de Qualité Assisté par Ordinateur

CQC: Critères-Qualité-Contrôles

CQCM: Cryogenically-cooled Quartz Crystal Microbalance

CQCT: Centre Québécois de Coordination de la Télédétection

CQDR: Critical Qualification Design Review

CQF: Calais (France) (location identifier)

CQFD: Ce Qu'il Fallait Démontrer

CQFP: Ceramic Quad-Flatpack

CQG: Classical and Quantum Gravity

CQGra: Classical Quantum Gravity (periodical code)

CQI: Continuous Quality Improvement

CQIM: Configuration and Quality Information Management

CQL: Computers and the Quality of Life

CQM: Colloquium • Control Quality Monitor

CQP: Cape Flattery (Australia-DLD) (location identifier)

CQS: Costa Marques (Brazil-RO) (location identifier)

CQSSP: Completed Quasar, Satellite and Star Positioning • Coupled Quasar, Satellite and Star Positioning (*)

CQT: Capacitor Qualification Test • Caquetania (Colombia) (location identifier)

CQTP: Capacitor Qualification Test Program

c/r: Company's Risk

cr: Con Riserva • Costa Rica • Crux • Cum Rights

Cr: Chrom • Chrome • Chromium • Commander • Councillor • Create • Created • Creation • Credit • Creditor • Creek • Crown • Cruise • Cruiser

C/R: Commutation Rate

CR: Calculated Risk • Calendrier Républicain • Camera Ready • Cameroon • Card Reader • Carriage Return • Carrier's Risk • Carrington Rotation • Cash Receipt • Cathode Ray • Cathode Reaction • Central Region • Central Registry • Čerenkov Radiation • Certification Request • Certification Requirement • Challenge Ring • Change Record • Change Recommendation • Change Request • Chargé de Recherche • Civis Romanus • Classification Research • Comb Response • Commutation Rate • Company's Risk • Complete Redistribution • Complete Regression • Compression Ratio • Compte-Rendu • Comptes-Rendus • Comptes-Rendus de l'Académie des Sciences (periodical code) • Computing Reviews • Concentration Ratio • Conception Review • Conditioned Reflex • Conditioned Response • Conference Report • Configuration Review • Confirm Response • Conflict Resolution • Congo Red • Consciousness-Raising • Conservation Reserve • Consultant Report • Contractor's Response • Controlled Rectifier • Control Relay • Control Room • Coordinate Reader • Coordinate Registration • Corner Reflector • Corotation Resonance • Correlation Radiometer • Corrientes (Argentina) • Cosmic Ray • Costa Rica (international vehicle registration code) • Costa Rica (ISO country code) • Cost Reimbursement • Counter-Rotating • Count Reverse • Credit • Credit Rating • Credit Report • Creek • Crew • Critical Ratio • Critical Resource • Croatia • Cross Reference • Croydon (UK postal code) • Crystal Rectifier • Current Rate • Current Regulator • Current Relay • Curvature Radiation • Customer's Request • Cyclic Reduction • Cyclotron Resonance • Portugal (international aircraft registration code)

CrA: Corona Australis - Coronae Australis (constellation) • Corona Austrina - Coronae Austrinae (constellation)

CRA: Carbon Rod Atomizer • Cassegrain Reflector Antenna • Catalog Recovery Area • Centre de Recherches Atmosphériques (France) • Centro di Ricerche Aerospaziale • Composite Research Aircraft • Computer Research Association • Computing Research Association • Conflict Resolution Advisory • Consiglio per le Ricerche Astronomiche • Coordinated Regional Assessment • Cooperative Research Agreement • Corrosion-Resistant Alloy • Cosmic-Ray Altimeter • Council for Research in Astronomy • Craiova (Romania) (location identifier) • Crew Reception Area

CRAA: Corporation for Research Amateur Astronomy (USA-CA)

CRAAE: Centro de Radio-Astronomia e Aplicações Espaciais (Brazil)

CRAAG: Centre de Recherches en Astronomie, Astrophysique et Géophysique (Algeria)

CRAAM: Centro de Rádio-Astronomia e Astrofísica Mackenzie (Brazil-SP) (formerly GRAM)

CRAB: Compte-Rendu d'Activité des Bâtiments

CRAC: Career Research Advisory Centre (UK)

CRADA: Cooperative Research And Development Agreement • Cooperative Research And Development Award

CRADO: Collaborative Research And Development Operation

CRAF: Comet Rendezvous and Asteroid Flyby • Committee on Radio Astronomy Frequencies

CRAFT: Computerized Relative Allocation of Facilities Technique • Cooperative Research Action For Technology • Cray Research Adaptive Fortran

CRAL: Centre de Recherche Astronomique de Lyon (France)

CRAM: Card Random Access Memory • Challenge-Response Authentication Mechanism • Communication Routing And Metering • Compte-Rendu Automatisé Mensuel • Computerized Reliability Analysis Method

CRAND: Cosmic-Ray Albedo Neutron Decay

CRANE: Cosmic-Ray Nuclear Experiment

CrAO: Crimean Astrophysical Observatory (Ukraine)

CRAP: Cleveland Regional Association of Planetariums (USA-OH)

CRAPE: Central Register of Aerial Photography for England

CRAR: Critical Reliability Action Report

CRAS: CIMSS Regional Assimilation System • Comet Rapid Announcement Service • Comptes-Rendus de l'Académie des Sciences (France) • Cost Reduction Alternative Study

CRAST: Coordinate Refinement Analytical Strip Triangulation

CRATE: Centimetric Radiometer Absolute Temperature

CRAWS: Cold Region Automatic Weather Station

CRAZI: Count Routine Applied to Zero Input

CrB: Corona Borealis - Coronae Borealis (constellation)

CRB: Central Radio Bureau • Change Review Board • Collarenebri (Australia-NSW) (location identifier) • Concept Review Board • Configuration Review Board • Correspondence Review Board • Cramer-Rao Bound

CRBE: Conversational Remote Batch Entry

CRBP: Cold Regions Bibliography Project

CRC: Camera-Ready Copy • Cancer Research Campaign • Candidate Review Committee • Carriage Return Contact • Cartago (Colombia) (location identifier) • Catalogiung Rules Committee • Centre Régional de Coordination • Centre Régional de la Consommation (France) • Centre Réseau Communication • Centro de Rastreiro e Controle • Chargé de Recherche Contractuel • Civil Rights Commission • Class-Responsibility-Collaborator • Climate Research Committee • Code de Redondance Cyclique • Colón (Costa Rica) (ISO currency code) • Color

Ratio Composite • Communications Research Centre • Compact Radio Core • Computer Resources Center • Contractor-Recommended Code • Cooperative Research Centre • Coordinating Research Council • Cost Reduction Curve • Cour Régionale des Comptes • Cyclical Redundancy Character • Cyclic Redundancy Character • Cyclic Redundancy Check • Cyclic Redundancy Checking • Cyclic Redundancy Code

CRCC: Center for Research on Concepts and Cognition

CRCD: Coordinated Research and Curriculum Development

CRCG: Center for Research in Computer Graphics

CRCH: Central Register and Clearinghouse • Central Reprint Clearinghouse

CRCI: Chambre Régionale de Commerce et d'Industrie

CRCL: Circle

CRCLE: Circle

Crclt: Circulate

Crcltd: Circulated

CRCM: Conference on Recent Crustal Movements • Commission on Recent Crustal Movements

Crcmf: Circumference

CRCS: Cassegrain Rotator Control System

CRCSS: Cooperative Research Centre for Satellite Systems (QUT)

CRCT: Centre de Recherche sur la Combustion Turbulente • Centre de Recherche sur la Culture Technique • Congé pour Recherches ou Conversion Thématique

Crctn: Correction

CRCV: Circunvalação

CRCW: Concurrent-Read - Concurrent-Write

CRD: Capacitor-Resistor Diode • Cardfile • Centre de Recherche Documentaire • Change Request Disposition • CODATA Referral Database • Comodoro Rivadavia (Argentina-CB) (location identifier) • Computer Readout Device • Confidential Restricted Data • Congressional Reference Division • Cosmic-Ray Detector • Cost Reporting Document

CRDA: Converging Runway Display Aid • Cooperative Research and Development Agreement

CRDD: Calibrated Raw Detector Data • Calibrated Reconstructed Detector Data (*)

CRDDP: Centre Régional de Documentation et de Diffusion Pédagogique

CRDEC: Chemical Research, Development, and Engineering Center

CRDF: Cathode-Ray Direction Finding • Civilian Research and Development Foundation

CRDI: Centre de Recherches pour le Dévelopement International

CRDM: Control Rod Drive Mechanism

CRDP: Centre Régional de Documentation Pédagogique (France) • Certification Review Data Package • Computer Resources Development Plan

CRDS: Calligraphic Raster Display System • Contribution au Remboursement de la Dette Sociale (France)

CRE: Cation-Responsive Electrode • Cauchy-Riemann Equation • Center for Renewable Energy • Cold Readout Electronics • Conférence des Recteurs Européens • Cosmic-Ray Electron • Cosmic-Ray Exposure • Cross-Range Error • Cryogenic Readout Electronics • Cumulative Radiation Effect • Myrtle Beach - Grand Strand (USA-SC) (location identifier)

CREA: Cosmic-Ray Exposure Age

CREAM: Cosmic-Radiation Effect and Activation Monitor (*) • Cosmic-Ray Effect and Activation Monitor • Cosmic-Ray Effect and Activation Monitoring

Crec: Creciente

CREDAS: Centro Regional de Datos Satelitales

CREDIF: Centre de Recherche et d'Étude pour la Diffusion du Français

CREDO: Cosmic-Ray Effects and Dosimetry

CREFAL: Centro Regional de Educación Fundamental para la América Latina

CREI: Centro Regional de Educación en Informática • Centro Regional para la Ensañanza de la Informática

CREID: Conversational Response for ETL Image Database

CREIM: Canadian Rockies Ecosystem Interagency Management

CREIT: Centre de Recherches et d'Études en Informatique et Télématique

CREME: Cosmi-Ray Effects on Microelectronics

CREN: Chemical Resources and Environmental Needs • Computer Research Education Network • Corporation for Research and Educational Networking

CREOL: Center for Research and Education in Optics and Lasers • Center for Research in Electro-Optics and Lasers

CREPE: Cosmic-Ray Emulsion Plastic Equipment

CREPUQ: Conférence des Recteurs Et Principaux des Universités du Québec (Canada-QC)

CRERI: Centre of Empirical Research in International Relations

CRERS: Crew Rescue and Equipment Retrieval System

Cres: Crescent

CRES: Centre for Research and Environmental Studies • Centre for Research on Environmental Systems • Centre Régional d'Énergie Solaire • Corrosion-Resistant Steel • Crescent

CRESALC: Centro Regional para la Educación Superior de América Latina y el Caribe

CRESP: Consortium for Risk Evaluation with Stakeholder Participation • Coordinated Research and Environmental Surveillance Project

CRESS: Center for Research in Experimental Space Science

CRESST: Cryogenic Rare Event Search with Superconducting Thermometers

CREST: Cenozoic Reef Evolution in Space and Time • Centre de Recherche en Évaluation Sociale des Technologies • Comité de la Recherche Scientifique et Technique • Core Research for Evolutional Science and Technology • Creativity in Science and Technology • Curriculum Resources in Energy Science and Technology

CRESTech: Centre for Research in Earth and Space Technology (Canada-ON) (merger of ISTS and WCGR)

CRESTEL: Club de Réflexion et d'Études sur les Systèmes de Télécommunications (France)

CRESU: Cinétique de Réaction en Écoulement Supersonique Uniforme

CRET: Centre de Recherches et d'Échanges Technologiques • Cathode-Ray Electron Tube

CRF: Cable Retransmission Facility • Calibration Resource File • Capacitor Resonance Frequency • Carnot (Central African Rep.) (location identifier) • Cathode-Ray Furnace • Celestial Receiver Frame • Celestial Reference Frame • Cell Relaying Function • Cloud Radiation Forcing • Cloud Radiative Forcing • Combustion Research Facility • Compressor Research Facility • Computer Readable Form • Concentrator Relay Function • Consolidated Revenue Fund • Content Retrieval Filter • Corotating Frame • Corticotropin Releasing Factor • Cosmic-Ray Flux • Cross-Reference File

Crg: Carriage

CRFP: Centre Régional de Formation Professionnelle

CRG: Centre de Recherche en gestion • Centre de Recherches Géophysiques • Change Review Group • Collaborative

Research Grant • Collaborative Research Group • Control Rate Gyro • Correspondence Review Group • Cosmic-Ray Gas • Cost Report Generator • Jacksonville Craig (USA-FL) (location identifier)

CRGP: Converse Reduced Game Property

CRH: Cherribah (Australia-QLD) (location identifier)

CRHI: Centre de Recherche en Histoire de l'Innovation

CRHST: Centre de Recherche en Histoire des Sciences et des Techniques

CRI: Cambridge Research and Instrumentation (USA-MA) • Centraal Rekeninstituut (Netherlands) • Centrale de Règlements Interbancaires (France) • Centre de Réception des Images • Centre de Recherche en Informatique • Centre de Ressources Informatiques • Centre Relai Innovation • Certification Requirements Index • Climate Research Institute • Color Reversal Intermediate • Comité de Recherches en Informatique • Contrôle des Règlements Interbancaires • Corporate Research Institute • Costa Rica (ISO country code) • Cray Research Inc. • Criterion-Referenced Instruction • Crooked Island (Bahamas) (location identifier) • Cross-Reference Index • Cyclotron Resonance Feature

CRIA: Center for Research on Information Access

CRIBABB: Centro Regional de Investigaciones Basicas y Applicadas Bahia Blanca (Argentina)

CRIB: Computerized Resources Information Bank

CRIC: Centre de Recherche Industrielle du Québec (Canada-QC) • Centre for Research on Innovation and Competition (UK) • Copyright Research and Information Center

CRICCS: Center for Real-time, Intelligent, Complex Computer System

CRICR: Centre Régional d'Information et de Coordination Routière (France)

CRICyT: Centro Regional de Investigaciones Científicas y Técnologicas (Argentina-MD)

CRID: Centre de Recherche et d'Information pour le Développement

CRIE: Cosmic-Ray Isotope Experiment

CRIEPI: Central Research Institute of Electric Power Industry (Japan)

CRIES: Comprehensive Resource Inventory and Evaluation System

CRIG: Capacitor Rate-Integrating Gyro

CRIHAN: Centre de Ressources Informatique de Haute-Normandie (France)

Crim: Criminal

CRIM: Centre de Recherche Informatique de Montréal (Canada-QC) • Commercial Refrigerator Incubator Module

Criminol: Criminologist • Criminology

CRIN: Centre de Recherches en Informatique de Nancy (France) • Club Recherche et Industrie

CRIPTE: Calcul sur Réseaux des Interactions Perturbatrices en Topologie Électromagnétique

CRIRES: Cryogenic Infrared Echelle Spectrograph • Cryogenic Infrared Echelle Spectrometer (*)

CRIS: Calibration Recall and Information System • Catania Relativistic Ion Studies • Center for Research on the Information Society • Centre de Rectification des Images Spatiales (CNES) • Clinical Radiology Imaging System • Command Retrieval Information System • Computerized River Information System • Connector Removal and Installation Sheet • Cosmic-Ray Isotope Spectrometer • Current Research Information System

CRISAP: Centro de Referencia y Información de Suelos de Amazonia Peruana (Peru)

CRISD: Computer Resource Integrated-Support Document • Computer Resource Integration Support Document

CRISP: Centre for Remote Imaging, Sensing and Processing • C-machine Reduced Instruction Set Processor • Computer Resources Integrated Support Plan • Counter Rotating Integrated Shrouded Propfan • Critical Surface Spectrometer • Cryptographic Reduced Instruction Set Processor

CRISTA: Cryogenic Infrared Spectrometers and Telescopes for the Atmosphere

Crit: Criterion • Critic • Critical • Critically • Criticism • Critique • Criticized

CRIT: Critical

CRITT: Centre Régional d'Innovation et de Transfert de Technologie • Centre Régional d'Innovation et de Transfert Technologique (France)

CRJ: Canadair Regional Jet • Coorabie (Australia-SA) (location identifier)

Crk: Cork (Ireland) • Crank

CRK: Creek • Luzon Island - Clark (Philippines) (location identifier)

CRL: Cambridge Research Laboratories • Cartographic Research Laboratory • Cathode-Ray Lamp • Central Reference Library • Central Regional Laboratory • Central Research Library • Certificate Revocation List • Certified Record Librarian • Certified Reference Librarian • Charleroi Gosselies (Belgium) (location identifier) • Communications Research Laboratory • Complex Record Language • Computing Research Laboratory • Contrat de Recherche Libre • Customer Requirements List

CRLCF: Circulation

CRLF: Carriage Return + Line Feed

CRLIS: Current Research in Library and Information Science

CRLJ: Communications Research Laboratory Journal (periodical code)

CRLRv: Communications Research Laboratory Review (periodical code)

CRM: Catarman (Philippines) (location identifier) • Central Rocky Mountains (USA) • Centre de Recherche des Macromolécules • Certified Reference Material • Change Request Material • Chemical Release Module • Chemical Remanent Magnetization • Cloud Resolving Model • Cockpit Resource Management • Conceptual Reference Mission • Count Rate Meter • Crew Resource Management • Criterion-Referenced Measurement • Customer Relationship Management • Cyclotron Resonance Maser

CRMC: Communication Resources Management Centre

CRMD: Centre de Recherche en Météorologie Dynamique • Computer Resources Management Data

CRMP: Coastal Resources Management Project

Crn: Crane • Crown

CRN: Cable Routing Notation • Canadian Road Network • Centre de Recherche Nucléaire • Change Review Notice • Conflict Resolution Network • Contract Revision Number • Coronal Rain • Cosmic-Ray Nucleus • Cromarty (UK-Scotland) (location identifier)

CRNL: Chalk River Nuclear Laboratories

Crnmtr: Chronometer

Cro: Croatian

CRO: Cathode-Ray Oscilloscope • Cathode-Ray Oscillograph • Chemical Release Observation • Chief Recruiting Officer • Community Relations Officer • Component Repair and Overhaul • Compulsory Rights Order • Contract Research Organization • Corcoran (USA-CA) (location identifier)

CROAP: Conception et Réalisation d'Outils d'Aide à la Programmation

Croat: Croatia • Croatian

CROC: Computer Review and Orientation Course

Crom: Cromarty (UK-Scotland)

CROM: Control Read-Only Memory

Cronol: Cronología

CROP: Customized Reduction Of Products

CROS: Computerized Reliability Organization System • Crystalline Operating System

CROSS: Cross Support Service

CROSSPATE: Coordinative Retrieval of Selectively Sorted Permuted Analogue-Title Entries

CROUS: Centre Régional des Oeuvres Universitaires et Scolaires (France)

CRP: Card Reader Program • Center for Responsive Politics • Change Review Panel • Climate Research Program • Compulsory Reporting Point • Computer Reset Pulse • Configuration Requirements Processing • Coordinated Research Programme • Corpus Christi (USA-TX) (location identifier) • Cosmic-Ray Particle • Cost Reduction Program • Cross-Reference Project • Current Research Profile

CRPA: Centre de Recherches en Physique de l'Atmosphère • Controlled Radiation Patterned Antenna • Controlled-Reception Pattern Antenna • Controlled-Reception Patterned Antenna • Croatian Radio Protection Association

CRPD: City Real Property Database

CRPE: Centre de Recherches en Physique de l'Environnement (France) • Concours de Recrutement des Professeurs des Écoles

CRPG: Centre de Recherches Pétrographiques et Géochimiques (France)

CRPHT: Centre de Recherches sur la Physique des Hautes Températures

CRPL: Central Radio Propagation Laboratory (USA) • Cosmic-Ray Physics Laboratory

CRPP: Center for Research in Particle Physics • Centre de Recherches en Physique des Plasmas

CRPPH: Committee on Radiation Protection and Public Health

crPSO: Crystallographic Point Symmetry Operation

Crpt: Carpet

CRPZ: Centre Régional de Prévisions de Zone (France)

CRQ: Caravelas (Brazil-BA) (location identifier) • Command Response Queue • Compte-Rendu Quotidien • Corequisite

CRR: Cargo Readiness Review • Ceres (Argentina-SF) (location identifier) • Commissioning Results Review • Computer Run Report • Critical Requirements Review

CRRE: Combined Radiation and Release Experiment

CRREL: Cold Regions Research and Engineering Laboratory

CRRERIS: Commonwealth Regional Renewable Energy Resources Information System

CRRES: Combined Release and Radiation Effects Satellite

Crs: Course

CRS: Calibration Requirements Summary • Candidate Review Subcommittee • Cavity-Ringdown Spectrometer • Center for Remote Sensing • Certified Repair Station • Character Recognition System • Child Restraint System • Coastal Radio Station • Cold-Rolled Steel • Command Relay Section • Command Retrieval System • Committee of Remote Sensing • Computerized Reservation System • Computerized Retrieval Service • Computer Resources Support • Congressional Research Service (USA) • Control and Reporting Service • Controlled Random Search • Corsicana (USA-TX) (location identifier) • Cosmic Radio Source • Cosmic-Ray Shower • Cosmic-Ray System • Crew Rescue System • Cross • Course

CRSC: Center for Remote Sensing and Cartography • Center for Research in Scientific Communication • Center for Research in Scientific Computation

Crscnt: Crescent

CRSCNT: Crescent

CRSDP: Codar Remote Sensing Demonstration Programme

CRSE: Centre de Recherche des Sciences Économiques • Course

CRSENT: Crescent

CRSEST: Crescent

CRSFD: Crossfeed

CRSI: Ceramic Reusable Surface Insulation

Crsn: Corrosion

CRSN: Centre de Recherche en Sciences Naturelles

CRSNG: Conseil de Recherche en Sciences Naturelles et Génie

CRSNT: Crescent

CRSP: Commercial Remote Sensing Program • Cooperative Research Support Project • Cryogenic Spectrograph • Cryogenic Spectrometer

CRSR: Center for Radiophysics and Space Research (USA-NY)

CRSS: Cambridge-ROSAT Serendipity Survey • Canadian Remote Sensing Society • Chemically Rigidized Space Structure • Commercial Remote Sensing Satellite • Commercial Remote Sensing System • Critical Resolved Shear Stress

Crssng: Crossing

CRSSNG: Crossing

Crst: Crescent

CRST: Crescent • Crest

CRSTI: Canadian Remote Sensing Training Institute

Crsv: Corrosive

Crsvr: Crossover

CRSVR: Crossover

Crt: Courant • Court • Crater-Crateris (constellation)

CRT: Cathode-Ray Tube • Centre Régional de Télécommunications • Chinese Remainder Theorem • Complex Ridge Terrain • Composite Rate Tax • Composite Readiness Test • Computer Research Technology • Computer Response Time • Cosmic-Ray Telescope • Cosmic-Ray Tracking • Counter Recovery Time • Court • Criterion-Referenced Test • Criterion-Referenced Testing • Crossett (USA-AR) (location identifier)

CRTAM: Computer Real-Time Access Method

CRTBT: Centre de Recherches sur les Très Basses Températures (France)

CRTC: Canadian Radio-television and Telecommunications Commission • Cathode-Ray Tube Controller

Crtg: Cartridge • Crating

CRTG: Cartridge

Crtkr: Caretaker

Crtl: Courtil

CRTL: C Run-Time Library

CRTO: Cathode-Ray Tube Oscillograph

CRTP: Compressed Real-Time Protocol

CRTR: Crater

Crts: Courtes

CRTS: Centre Royal de Télédétection Spatiale (Morocco) • Cathode-Ray Tube Shield

CRTST: Centre de Recherches Transdisciplinaires sur les Sciences et les Techniques

CRTT: Cathode-Ray Tube Tester

CRTU: Combined Receiving and Transmitting Unit • Cryogenic Radiator Test Unit

CRTV: Composite Reentry Test Vehicle

Cru: Crux-Crucis (constellation)

CRU: Card Reader Unit • Carriacou Island (Windward Islands) (location identifier) • Central Registration Unit • Climatic Research Unit • Comité Réseau des Universités • Command Relay Unit • Communications Register Unit • Computer Resource Unit • Control Register User

CRUC: Catalèg de Revistes de les Universitats Catalanes

CRUD: Create-Retrieve-Update-Delete

CRUDDES: Crustal Dynamics Database Expert System

CRUME: Committee for Research in Undergraduate Mathematics Education

CRUSK: Center for Research on Utilization of Scientific Knowledge

CRUSO: Coordinated Request and User Support Office

CRUX: Cosmic-Ray Upset Experiment

Crv: Corvus-Corvi (constellation) • Curve

CRV: Cargo Return Vehicle • Centre de Recherche en Volcanologie (France) • Compte-Rendu du Vent • Computer-Readable Version • Crew Rescue Vehicle • Crew Return Vehicle • Crotone (Italy) (location identifier)

CRW: Charleston (USA-WV) (location identifier) • Control Read/Write • Crew

CRWI: Coalition for Responsible Waste Incineration

CRX: Corinth - Roscoe Turner (USA-MS) (location identifier)

CRY: Carlton Hill (Australia-WA) (location identifier)

Cryo: Cryogenic • Cryogenics (periodical code)

CRYO: Cryogenic

CRYOFD: Cryogenic Flexible Diode

Crypto: Cryptographic • Cryptopgraphy

CRYSP: Crystal Sample Package

CRYSYS: Cryospheric System • Cryospheric System Study

Cryst: Crystal • Crystalline • Crystallized • Crystallography

Crystallog: Crystallography

Crystd: Crystallized

Crystn: Crystallization

CRZ: Chardzhou (Turkmenia) (location identifier)

CRZT: Cruzamento

c/s: Cases • Cycles per Second

cs: Case • Census • Come Supra • Consul • Czech (ISO language code)

cS: Centistoke

Cs: Cäsium • Cesium • Césium • Cirrostratus

C&S: Calendaring & Scheduling

C/S: Client/Server • Concentrator/Spectrometer • Cost/Schedule • Counts per Second • Cycles per Second

C-S: Client-Server • Cox-Stewart

CS: Calcium Silicate • Calendaring and Scheduling • Calibration System • Camera Site • Canadian Standard • Canal Sémaphore • Capital Stock • Captain Starlink • Carbon Monosulfide • Carbon Steel • Card Station • Cast Steel • Catalog Subgroup • Center Section • Centraal Station • Central Secretariat • Central Station • Central Storage • Central Supply • Certificate in Statistics • Change of Status • Change Status • Charge Shuffling • Chartered Surveyor • Checkout Station • Chemical Society • Chief of Staff • Chief Scientist • Chief Secretary • Chip Select • Circumstellar • Circumstellar Shell • Cirrostratus • City Surveyor • Civil Servant • Civil Service • Classifier Syntax • Classifier System • Clear and Subtract • Clear to Send • Clock Synchronization • Close Support • Code Segment • Cognitive System • College of Science • Color Scanner • Color Specification • Colour Start • Comité Scientifique • Command Set • Command System • Commercial Service • Common Set • Common Subsequence • Communications Satellite • Communications Station • Communications Switcher • Communications System • Comparison Side • Component Specification • Composite Superconductor • Compton Scattering • Computer Science • Computer Section • Computer Simulation • Computer Software • Computer System • Computing Science • Conditioned Stimulus • Confederate States • Confinement System • Congiunzione Superiore • Conseil Scientifique • Consumables Status • Contact Scientist • Continental Shelf • Contingency Sample • Contractor Share • Control Scanner • Control Segment • Control Set • Control Signal • Control Slip • Control Star • Control Station • Control Strobe • Control Switch • Convergent Stereoscopic • Coolant System • Cooperative Society • Core Segment • Core Shift • Corporate Source • Correct Selection • Correlador Solar • Cosine-Sine • Coupled States • Cox-Stewart • Crew Simulator • Crew Station • Crew System • Critical Sensitive • Critical Surface • Cross Section • Cross-Strapping • Cubic Spline • Current Sensor • Current Source • Current Strength • Cusp Shortening • Cycle Sequence • Czechoslovakia (ISO country code) (now obsolete) • Czech Republic (international vehicle registration code) • Portugal (international aircraft registration code) • Slovakia (international vehicle registration code)

CSA: Calendaring and Scheduling API • Cambridge Scientific Abstracts • Canadian Space Agency • Canadian Standards Association (*) • Canadian Standards Authority • Character Scan or Alternate • Charge-Sensitive Amplifier • Chinese Society of Astronautics • Classical Simulated Annealing • Client Services Approach • Cloud Size Analyzer • Colonsay Island (UK) (location identifier) • Commission Spécialisée d'Astronomie • Common Storage Area • Common System Area • Communications Service Authorization • Compression Service Adapter • Computer Science Association • Computer Sciences of Australia • Computer System Analyst • Computing Society of America • Confederate States of America • Conical Scan Antenna • Conseil Supérieur de l'Audiovisuel (France) • Contextual Sequences Analysis • Contractor Support Area • Contract Specification Amendment • Contrat Support Application • Control Switching Assembly • Covariance Structure Analysis • Cryogenic Society of America (USA) • Current Source Amplifier • Cyclic Strain Attenuator

CSAA: Committee on Space Astronomy and Astrophysics

CSAB: Computing Sciences Accreditation Board

CSAC: Census Statistical Area Committee (USA) • Computer Science Accreditation Commission

CSAD: Capsule Systems Advanced Development

CSAGI: Comité Scientifique de l'Année Géophysique Internationale • Comité Spécial de l'Année Géophysique Internationale (*)

CSAIN: Canadian Substance Abuse Information Network

CSAIO: Chef du Service Académique d'Information et d'Orientation

CSAM: Circular Sequential Access Memory • Coarse Sun Acquisition Mode • College of Science And Mathematics • Crinkled Single Aluminized Mylar

CSAO: CAS Shaanxi Astronomical Observatory (China-PRC)

CSAP: Control System Analysis Program

CSAPI: Common Speller Application Programming Interface

CSAR: Canadian Space Agency Rocket • Center for Simulation for Advanced Rockets • Channel System Address Register • Communications Satellite Advanced Research • Custom Scanned Absolute Response

CSAS: Colorado Springs Astronomical Society (USA-CO) • Computerized Status Accounting System

CSAT: Combined Systems Acceptance Test

CSATI: Centro Studi ed Applicazione sulle Tecnologie dell'Informazione

CSB: Calcium Silicate Vrick • Caransebes (Romania) (location identifier) • Change Schedule Board • Closely Spaced Basing • Complex Support Building • Conseil Supérieur des Bibliothèques

CSBM: Committee on Space Biology and Medicine

CSBTS: China State Bureau of Technical Supervision (China-PRC)

c&sc: Capitals & Small Capitals

csc: Cosecant

CSc: Candidate of Sciences

C/SC: Cost/Schedule Control

CSC: Canadian Society for Chemistry • Canas (Costa Rica) (location identifier) • Carnegie Science Center • Carrier-Smoothed Code • Cathod Strip Chamber • Center for Scientific Computing • Center for the Study of Communications • Central Switching Concept • Civil Service Commission • Coastal Service Center (NOAA) • Commercial Space Center • Common Signaling Channel • Commonwealth Science Council • Communications Switchboard Console • Communications Systems Center • Communications Systems Control • Compact Spherical Coil • Complex Support Controller • Comprehensive Self-Check • Computer Science Conference • Computer Science Corporation • Computer Service Center • Computer Software Component • Computer Software Configuration • Computer System Controller • Conical-Shaped Charge • Conseil Scientifique Consultatif • Contingency Support Center • Convention for Safe Containers • Cost-Schedule-Control

CSCA: Catalogue of Star Clusters and Associations

CSCB: Contractor's Summary Cost Breakdown

CSCC: Cleveland State Community College (USA-TN) • Concurrent Supercomputing Consortium

CSCD: Character Set Computer Development • Control System Command Decoder

CSCE: Conference on Security and Cooperation in Europe • Conférence sur la Sécurité et la Coopération en Europe

csch: Hyperbolic Cosecant

CSCI: Computer Software Configuration Item • Current Source Current Intensive

CSCL: Computer-Supported Collaborative Learning • Computer-Supported Cooperative Learning

CSCR: Classification System for Computing Reviews

C/SCS: Cost/Schedule Control System

CSCS: Central Security Control System • Centro Svizzero di Calcolo Scientifico (Switzerland)

CSCSAT: Commercial Synchronous Communication Satellite

C/SCSC: Cost/Schedule Control System Criteria

CSCSI: Canadian Society for Computational Studies of Intelligence

CSCU: Conseil Supérieur des Corps Universitaires

CSCW: Computer-Supported Collaborative Work • Computer-Supported Cooperative Work (*) • Contract Start Date • Cresswell Downs (Australia-NT) (location identifier) • Crystal Size Distribution • Cumulative Sum Diagram

CSD: Chemical Systems Division • CICS System Definition • Circuit Switched Data • Civil Service Department • Commission on Sustainable Development • Common Subdivision • Comparative Soot Diagnostics (*) • Computer Security Day • Computer System Director • Conference on Sustainable Development • Configuration Standardization Document • Constant Speed Drive • Contract Start Date • Control System Development • Cooperative Soot Diagnostics • Core-Shift Drive • Corrective Service Disk • Corrective Service Diskette • Crew Systems Division • Current Source Density

CSDA: Computational Statistics and Data Analysis • Continuous Slowing Down Approximation

CSDB: Central Specialized Design Bureau (Russia)

CSDC: Circuit Switched Digital Capability • Cluster Science Data Centre • Commonwealth Spatial Data Committee

CSDD: Control Systems Development Division

CSDF: Central Source Data File

CSDG: Common Software Development Group

CSDI: Cross Spectral Dispersion Imaging

CSDL: Charles Stark Draper Laboratory (MIT) • Conceptual Scheme Definition Language

CSDM: Computer System Diagnostic Manual

CSDN: Circuit Switched Data Network

CSDR: COBE Science Data Room

CSDRA: Centre for Spatial Data Research and Applications

CSDRBL: Considerable

CSDS: Cluster Science Data System • Critical Space Debris Sphere

CSDT: Computer Software Data Tape

Cse: Course

CS&E: Computational Science & Engineering (*) • Computer Science & Engineering

CSE: Camel Service Environment • Center for Science and Engineering • Center for Statistical Education • Centraal Schriftelijk Eindexamen • Central Signal Establishments • Centre de Suivi Écologique • Certificate of Secondary Education • Certified System Engineer • Circumstellar Envelope • Cognitive Systems Engineering • Cold Start Entry • Commission des Spécialistes d'Établissement • Common Support Equipment • Commonwealth Service of Employment (Australia) • Communications Systems Engineer • Computational Science and Engineering • Computer Science and Engineering • Computer Support Equipment • Conférence Spatiale Européenne • Configuration Switching Equipment • Conseil Supérieur de l'Éducation • Contrat de Suivi d'Exploitation • Crested Butte (USA-CO) (location identifier) • Cryostat Servicing Equipment

CSEA: Civil Service Employees Association

CSEC: Centre for the Study of Environmental Change

CSECT: Control Section

CSEE: Compagnie de Signaux et d'Équipements Électroniques (France) • Computer Science and Electrical Engineering

CSELI: Centro Studi E Laboratori Telecomunicazioni (Italy)

CSEM: Centre Sismologique Euro-Méditerranéen • Centre Suisse d'Électronique et de Microélectronique (Switzerland)

CSEN: Conseil Supérieur de l'Éducation Nationale

CSEOL: Center for the Study of Evolution and Origin of Life (IGPP)

CSER: Centre for Satellite Engineering Research

CSERGE: Centre for Social and Economic Research on the Global Environment

CSES: Center for the Study of Earth from Space

CSET: College of Science, Engineering and Technology

CSETI: Center for the Study of Extraterrestrial Intelligence

CSF: Central Serials File • Central Simulation Facility • Central Simulation Form • Central Supply Facility • Central Switching Facility • Cerebrospinal Fluid • Chaos-Solitons-Fractals (periodical code) • Character Scan or Fail • Chi-Squared Function • Coding Sequence Finder • Columbus Simulation Facility • Community Support Framework • Configuration State Function • Continuous Spontaneous Localisation • Contractor Support Facility •

Contract Supply Facility • Contrast Sensitivity Function • Cost Sensitivity Factor • Creil (France) (location identifier) • Crew Storage Facility • Critical Success Factor

CSFA: Canadian Scientific Film Association

CSFC: Cruise Specific Fuel Consumption

CSFD: Canadian Symposium on Fluid Dynamics

CSFR: Color Strip Film Recorder • Committee on Scientific Freedom and Responsibility • Continuous Scan Film Rescorder

CSFS: Committee on Sciences for Food Security

CSFT: Combined System Functional Test

Csg: Casing

CSG: Calibration Signal Generator • Centre Spatial Guyanais (French Guyana) • Clock Sequence Generator • Columbus (USA-GA) (location identifier) • Concept Synthesis Group • Constructive Solid Geometry • Consulting Services Group • Consumer Support Group • Contribution Sociale Généralisée

CSGD: Chinese Solar-Geophysical Data

CSGE: Common Software Engineering Environment

CSH: Calcium Silicate Hydrate • Cape Sarichef (USA-AK) (location identifier) • Communications Soft Hat • Convective Stratiform Heating

CSHARP: Cooperative Student High-Altitude Rocket Payload

CSHELL: Cryogenic Echelle Spectrometer

CsI: Caesium Iodide

CSI: Carrier-Scale Internetworking • Cartographie par Satellite et Informatique • Casino (Australia-NSW) (location identifier) • Catalogue of Stellar Identifications • Centre de Sociologie de l'Innovation • Cercle Scientifique Interfacultaire • Channel Status Information • Code Structure Identifier • Coelliptic Sequence Initiation • College of Staten Island (USA-NY) • Commercial Sites Index • Commercial Systems Integration • Common Software Interface • Common System Interface • Compact Source Iodide • Computational Sciences and Informatics • Computer Science Institute • Computer Security Institute • Computer Society of India • Concentric Sequence Initiation • Consolidated Software Inventory • Consorzio Servizi Informatici • Control and Structural Integrity • Control Servo Input • Control Structure Interaction • Cost Since Inception • Cost Since Inspection • Crew Software Interface • Critical Success Index • Customer Satisfaction Index • Customer Source Inspection

CSIAM: Chinese Society for Industrial and Applied Mathematics

C/SiC: Carbon/Silicium Carbide

CSIC: Consejo Superior de Investigaciones Científicas (Spain)

CSICOP: Committee for the Scientific Investigations of Claims of the Paranormal (USA-NY)

CSID: Calling Station Identification • Convergence Source-Image Distortion

CSIDA: CSIRO System for Interactive Data Analysis

CSIESR: Comité des Services Informatiques de l'Enseignement Supérieur et de la Recherche

CSII: Configuration Similarity Iterative Improvement

CSIIS: Canadian Sea Ice Information System

CSILE: Computer-Supported Intentional Learning Environment (*) • Computer-Supported Intentional Learning Experiment

CSIM: Cosmic Simulator • Custom Spectral Irradiance Measurement

CSIN: Computer Systems Information Network

CSIP: Committee for the Scientific Investigations of the Paranormal • Continental Scale International project

CSIR: Commonwealth Scientific and Industrial Research (South Africa) • Computer Software Integration Review • Computer Systems Integration Review • Council for Scientific and Industrial Research (South Africa) • Council of Scientific and Industrial Research (India) • Crew Software Interface Review

CSIRO: Commonwealth Scientific and Industrial Research Organization (Australia)

CSIRONET: CSIRO Network

CSIS: Center for Secure Information Systems (CMU) • Center for Strategic and International Studies • Centralized Storm Information System • Centre for Spatial Information Systems

CSISO: Conditioned Single Input - Single Output

CSISRS: Cross-Section Information Storage and Retrieval System

CSIST: Chung Shan Institute of Science and Technology (Taiwan)

CSIT: Customer Service Improvement Team

CSIU: Core Segment Interface Unit

CSJ: Cape Saint Jacques (Viet Nam) (location identifier)

Csk: Cask

CSK: Cap Skirring (Senegal) (location identifier) • Cathodic Survey Kit • Committee of Standardization of the DPRK

CSL: California State Library (USA-CA) • Casal • Centre Spatial de Liège (Belgium) • Clock Sync Logic • Code Selection Language • Command Sequence Language • Command Signal Limiter • Common Specification Language • Compact Shuttle Lidar • Computer Science Laboratory • Computer Sensitive Language • Computer Simulation Language • Computer Status Light • Computer Structure Language • Computer System Language • Console • Context Sensitive Language • Context Switch Latency • Control and Simulation Language • Control and Status Logic • Crew Systems Laboratory • Critical Sequence List • San Luis Obispo (USA-CA) (location identifier)

CSLA: China Satellite Launch Agent

CSLG: Campus-wide Software License Grants

CSLIN: Contract Subline Item Number

CSLIP: Compressed Serial Line Interface Protocol

CSLM: Client Service Level Monitor • Confocal Scanning Laser Microscope

CSLO: Canadian Scientific Liaison Office

CSLP: Cooperative Satellite Learning Program

CSLR: Continuous Spoken Language Recognition

CSM: Camera-Speaker-Microphone • Cargo System Manual • Circumstellar Matter • Circumstellar Medium • Climate System Model • Climate System Modelling • Climate System Monitoring • Clinton (USA-OK) (location identifier) • Coaxial Switching Matrix • Color-Singlet Model • Combined Symbol Matching • Command and Service Module • Command Service Module • Commission for the Synoptic Meteorology • Common Support Module • Communications Services Manager • Communications Storage Manager • Component and Supplier Management • Computer Status Matrix • Computer System Manual • Conseil Supérieur de la Météorologie • Continuous Sheet Memory • Convexity-Symmetry-Maximum • Cumberland Science Museum (USA-TN) • Cyclotron Scattering Model

CSMA: Carrier-Sense Multiple Access • Computational Structural Mechanics and Applications (periodical code)

CSME: Canadian Society of Mechanical Engineering

CSMEE: Committee for Support of Mathematicians from Eastern Europe • Committee on Science, Mathematics and Engineering Education

CSMF: Communications Satellite Monitoring Facility

CSMM: Critical Stream Meteoroid Mass

CSMNR: Complex Studies and Mapping of Natural Resources

CSMOL: Control Station Manual Operating Level

CSMP: Climate System Modelling Program • Continuous System Modelling Program • Control System Modelling Program

CSMPS: Computerized Scientific Management Planning System

CSMR: Course Spatial-resolution Multispectral Radiometer

CSMT: Carlos Sanchez Magro Telescope • Center for Space Microelectronics Technology • Circuit Switching Magnetic Tape

CS-MUX: Circuit-Switched Multiplexer

CSMUX: Circuit-Switched Multiplexer

CSN: Card Select Number • Caribbean Satellite Network • Carson City (USA-NV) (location identifier) • Central Satellite Numérique • Centro di Scienze Naturali • Circuit Switched Network • Circuit Switching Network • Comité Supérieur de Normalisation • Concentrateur Satellite Numérique • Consejo de Seguridad Nuclear • Control Symbol Number • Coopérant dy Service National (France) • Customer Source Number • Cycles Since New

CSNA: Classification Society of North America (USA-IL)

CSNE: Coopérant du Service National en Entreprise (France)

CSNet: Computer Science Network

CSNF: Common Source Noise Figure

CSNH: Casalinho

CSNI: Committee on the Safety of Nuclear Installations • Communications Systems Network Interoperability

CSNLP: Cognitive Science of Natural Language Processing

CSNR: Comet Nucleus Sample Return

CSNSM: Centre de Spectroscopie Nucléaire et de Spectrométrie de Masse (France)

CSNW: Client Service for NetWare

Cso: Corso

CSO: CALTECH Submillimeter Observatory (USA-HI) • Case Stellar Object • Central Selling Organization • Central Services Organization • Central Statistical Office • Chained Sequential Operation • Chief Scientific Officer • Chief Signal Officer • Chief Staff Officer • Closely-Spaced Object • Color Separation Overlay • Commonwealth Scientific Office • Community Service Order • Compact Symmetric Object • Complex Safety Office • Computer Service Office • Computer Systems Officer • Consolidated Space Operations • Customer Service Office

CSOAR: Colorado Student Ozone Atmospheric Rocket

CSOB: Calibration Science and Operations Branch • Crew System and Operations Branch

CSOC: Consolidated Space Operations Center • Consolidated Space Operator's Center • Consolidated Support Operations Center

CSOM: Computer System Operator's Manual • Confocal Scanning Optical Microscope

CSOP: Crew Systems Operating Procedures

C/S/P: Cost/Schedule/Performance

CSP: Canada Scholarship Program • Cape Spencer (USA-AK) (location identifier) • Carbon Sheet Pumping • Center for Space Physics • Center for Space Power (USA-TX) • Central Signal Processor • Centro di Supercalcolo Piemonte (Italy) • Certified Special Processor • Coder Sequential Pulse • Coherent Signal Processor • Commercial Subroutine Package • Commercial Support Program • Communicating Sequential Process • Communication Sequential Process • Communications Satellite Program • Component Service Provider • Composants et Systèmes de Précision (France) • Composite Stellar Population • CompuCom Speed Protocol • Computer Simulation Program • Computer Support Program • Conformal Space Projection • Conseil Scientifique et Pédagogique • Constraint Satisfaction Problem • Control Signal Processor • Control Switching Point • Convention de Site et de la Profession • Cooperative Statistical Program • Council for Science Policy • Council for Scientific Policy • Cryptographic Support Processor • Cutting Stock Problem

CSPAA: Conférence de Solidarité des Pays Afro-Asiatiques

CSPAAR: Center for Space Plasma, Aeronomy and Astrophysics Research (UAH)

CSPAR: Center for Space Plasma and Aeronomic Research (USA-AL)

CSPC: Conditional Stored Program Command • Cost and Schedule Planning and Control

CSPDN: Circuit Switched Public Data Network

CSPF: Critical Single-Point Failure

CSPG: Common Source Power Gain

CSPI: Coherent State Path Integral

CSPISE: Center for Stochastic Processes in Science and Engineering

CSPL: Certified Special Processor List

CSPM: Cercle des Sciences Physiques et Mathématiques

CSPN: Central Star of Planetary Nebula

CSPO: Communications Satellite Project Office • Control Systems Procurement Office

CSPP: Collaborative Special Project and Programme • Computer Systems Policy Project

CSPSW: Center for Solar Physics and Space Weather (IACS)

CSPU: California State Polytechnic University (USA-CA) • Core Segment Processing Unit

CSPW: Computer-Supported Personal Work

CSQ: Creston (USA-IA) (location identifier)

C/SR: Cost/Schedule Report

CSR: Casuarito (Colombia) (location identifier) • Cell Sequenced Ratio • Cell Switch Router • Center for Space Research • Center for Subatomic Research (Canada) • Central Serials Record • Certification Status Report • Certified Shorthand Reporter • Check Signal Return • Check Status Register • Check Status Reply • Check Status Report • Clementine Spectral Reflectance • Command Status Register • Communications Satellite Relay • Communications Service Request • Compact Stellar Remnant • Complete Spatial Randomness • Comprehensive Spending Review • Conical Scan Radiometer • Continuous Speech Recognition • Contract Status Report • Control and Status Register • Control Status Register • Cost-Schedule-Risk • Crew Station Review • Customer Service Representative • Customer Signature Required • Customer Support Room

CSRA: Centro de Sensoriamento Remoto da Amazonia (Brazil) • Civil Service Reform Act • Comité Scientifique pour les Recherches Antarctiques

CSRAM: Custom Static Random Access Memory

CSRC: Canadian Space Resource Centre • Crash-Safety Research Centre

CSRG: Catalogue of Southern Ringed Galaxies • Computer Science Research Group • Computer Systems Research Group

CSRI: Chemistry and Soil Research Institute

CSRL: Conceptual Structures Representation Language

CSRP: Computers and Software Review Panel

CSRS: Canadian Spatial Reference System • Centre Suisse de Recherches Scientifiques (Switzerland) • Civil Service Retirement System • Committee Services and Research Section • Continuous Speech Recognition System • Cooperative State Research Service

CSRSR: Center for Space and Remote Sensing Research

CSRT: Conseil Supérieur de la Recherche et de la Technologie

CSS: Cargo Storage and Securing • Cascading Style Sheet • Cassilândia (Brazil-MS) (location identifier) • Catalogue of S Stars • Center for Space Science • Central Site System • Certificat de Sécurité-Sauvetage • Character String Scanner • Circuit Switching Station • Channel Signaling System • Channel Subsystem • Clock Subsystem • Coarse Sun Sensor • Cockpit System Simulator • Cognitive Science Society • Combat Stores Ship • Commonwealth Superannuation Scheme • Communications Security System • Compact Steep Spectrum • Complete Statistical System • Computational Support Section • Computer Search Services • Computer Service System • Computer Subsystem • Computer System Simulator • Connecticut Space Center (USA-CT) • Continuous System Simulator • Control Stick Steering • Control Subsystem • Conversational Software System • Coordination de la Sécurité au Sol • Core Segment Simulator • Council for Science and Society • Crew Safety System

CSSA: Center for Space Science and Astrophysics • Coarse Sun Sensor Assembly • Computer Society of South Africa • Configuration Similarity Simulated Annealing

CSSAR: Centre for Space Science and Applied Research

CSSC: Canadian Strategic Software Consortium

CSSD: Communications System Status Display

CSSE: Center for Space Science and Exploration (USA-NM) • Control System Simulation Equipment

CSSEC: Computer Systems Support and Evaluation Command

CSSH: Congress of the Social Sciences and Humanities

CSSIN: Conseil Supérieur de la Sûreté des Installations Nucléaires (France)

CSSIP: Centre for Sensor Signal and Information Processing

CSSL: Communications and Space Science Laboratory • Continuous Systems Simulation Language

CSSO: Computer System Security Officer

CSSP: Circuits, Systems, and Signal processing (periodical code) • Committee on Solar and Space Physics • Council of Scientific Society Presidents • Cutting and Skiving Stock Problem

CSSPPT: Commission Supérieure du Service Public des Postes et Télécommunications (France)

C/SSR: Cost/Schedule Status Report

CSSR: Center for Solar and Space Research • Cost Schedule Status Report

CSST: China Space Solar Telescope • Commission Spécialisée des Sciences de la Terre

CSSTE: Centre for Space Science and Technology Education

CSSU: Converter Simulator Signal Unit

Cst: Coast • Constant

CST: Cadre Spécial Temporaire • Capsule Systems Test • Carlos Sánchez Telescope • Castaway (Fiji) (location identifier) • Caustic Surface Trace • Central Standard Time • Centre Spatial de Toulouse (France) • Certified Survey Technician • Channel Status Indicator • Character Set Technology • Classification on Science and Technology • Coast • College of Science and Technology • Combined Station and Tower • Combined Systems Test • Combustion Science and Technology (periodical code) • Comité Supérieur de la Télématique • Compact Standard Transponder • Complex Safety Technician • Computational Structure Technology • Computer Systems and Telematics • Constant • Contract Supplemental Tooling • Control System Test • Convective-Stratiform Technique • Convulsive Shock Therapy • Council for Science and Technology • Crew Software Trainer • Crew Station Trainer • Crew Systems Trainer • Culture Scientifique et Technique

CSTA: Canterbury Science Teachers' Association • Computer Sciences Technicolor Associates • Computer-Supported Telecommunications Applications • Computer-Supported Telephony Applications • Crew Software Training Aid

CSTAG: Centre for the Study of Telematics and Governance (UK-Scotland)

CSTAR: Center for Space Transportation and Applied Research • Classified Scientific and Technical Aerospace Report • Continuous Stellar-Tracking Attitude Reference

CSTB: Centre Scientifique et Technique du Bâtiment • Computer Science and Telecommunications Board

CSTC: Computer Science Teaching Center • Consolidated Satellite Test Center • Consolidated Space Test Center

CSTD: Centre of Science and Technology for Development (*) • Committee on Science and Technology for Development

Cste: Constante

CsTe: Cesium Tellurium

CSTE: Certified Software Test Engineer

CSTEA: Center for the Studies of Terrestrial and Extraterrestrial Atmospheres

Cstg: Casting

CSTG: Computer Systems Technical Group • Coordination of Space Techniques for Geodesy and Geodynamics

CSTGAG: Coordination of Space Techniques for Geodesy And Geophysics

CSTI: Center for Scientific and Technological Information • Civil Space Technology Initiative • Committee for Scientific and Technical Information • Council of Science and Technology Institutes • Council of Scientific and Technical Institutes

CSTIN: China Science and Technology Information Network (China-PRC)

CSTL: Coastal

CSTM: Custom Spectral Transmission Measurement

CSTP: Columbus Solar-Terrestrial Physics • Committee for Science and Technology Policy • Committee on Science and Technology Policy

CSTR: Canister • Committee on Solar-Terrestrial Research • Computer Science Technical Report

CSTS: Centro di Studio per le Telecomunicazioni Spaziali (Italy) • Commercial Space Transportation Study • Cryogenic Storage and Transfer System

CSTU: Combined Systems Test Unit

CSU: California State University (USA-CA) • Central Statistical Unit • Central Switching Unit • Channel Service Unit • Channel Synchronizer Unit • Circuit Switching Unit • Colorado State University (USA-CO) • Composante Sol-Utilisateur • Computer Software Unit • Conseil Supérieur des Universités • Constant Speed Unit • Conventional Speed Unit • Customer Set-Up • Santa Cruz do Sul (Brazil-RS) (location identifier)

CSUCA: Consejo Supérior de Universidades de Centroamérica

CSUDH: California State University at Dominguez Hills (USA-CA)

CSUF: California State University at Fullerton (USA-CA)

CSUL: Colorado State University Libraries (USA-CO)

CSULB: California State University at Long Beach (USA-CA)

CSUN: California State University at Northridge (USA-CA) • Commission for Symbols, Units and Nomenclature (IUPAC)

CSUS: California State University at Sacramento (USA-CA)

CSV: Catalogue of Suspected Variables • Cathodic Stripping Voltammetry • Characteristic Statistical Value • Circuit-Switched Voice • Comma-Separated Values • Comma-Separated Variables • Common Services Verb • Crossville (USA-TN) (location identifier)

CSVC: Core Sample Vacuum Container

CSVS: Canadian Space Vision System • Catalogue of Suspected Variable Stars (periodical code)

CSW: Certified Social Worker • Channel Status Word • Continuous Seismic Wave

CSWA: Committee on the Status of Women in Astronomy (AAS)

CSWG: Circumpolar Seabird Working Group

CSWP: Committee on the Status of Women in Physics (ASP) • Common Software Processing

Cswy: Causeway

CSWY: Causeway

CSX: Cayman Stock Exchange (Cayman Islands) • Changsha (China-PRC) (location identifier)

CSY: Cheboksary (Russia) (location identifier)

CS2: Carbon Disulfide

ct: Carat • Centum

Ct: Canton (China-PRC) • Caught • Cent • Certificate • Circuit • Connecticut (USA) • Content • Count • County • Courant • Court • Credit • Current

C&T: Chips & Technologies • Ciel et Terre (periodical code) • Ciência & Tecnologia • Communications & Tracking

C/T: Conference Terms • Control and Timing

CT: Cable Transfer • Canal de Transit • Canterbury (UK postal code) • Cape Town (South Africa) • Captive Test • Cassette Tape • Center Tap • Center Thickness • Central Time • Črenkov Telescope • Certificated Teacher • Certified Teacher • Chef Technicien • Chemical Transfer • Circuit Theory • Clock Time • Cognitive Technology • College of Technology • Colour Type • Comissaõ Técnica • Commercial Traveller • Commission Technique • Compact Tension • Component Test • Compressed Term • Computational Technique • Computed Tomography • Computed Tomography • Computerized Tomography • Computer Technology • Computer Telephony • Connecticut (USA) • Connectivity Table • Contrôle Technique • Controlled Term • Control Transformer • Copyright Treaty • Cordless Telephone • Correlation Tracker • Counter Timer • Court • Crawler-Transporter • Critical Temperature • Cryogenic Trapping • Cryptological Technician • Cryptology Technician • Current Transformer • Cursor Table • Cycle Time • Cyclic Time

cta: Cum Testamento Annexo

Cta: Caleta • Cuenta • Cuesta

CTA: Canadian Target Assembly • Caribbean Technical Assistance • Catania (Italy) (location identifier) • Centro Técnico Aeroespacial • Chief Technical Adviser • Cognitive Task Analysis • Comité de la Télématique Anonyme • Command and Telemetry Assembly • Commission for the Teaching of Astronomy (IAU) • Commodities Trading Adviser • Compatibility Test Area • Component Test Area • Computational Technology Area • Computer Technology Associates • Computer Traders Alliance • Contractor Technical Assistance • Contrat de Télé-Assistance • Controlled Thrust Assembly • Controlled Time of Arrival

CTAC: Central Technology Advisory Committee • Cetyltrimethylammonium Chloride • Computer Tester Analyzer Controller

CTAD: Conference on Trade And Development

CTAF: Common Traffic Advisory Frequency

CTAG: Canadian Time Allocation Group

CTAM: Continental Tropical Air Mass

CTAMN: Centre de Télédétection et d'Analyse des Milieux Naturels

CTAN: Comprehensive TEX Archive Network

CTAS: Center/Tracon Automation System • Central Texas Airborne Systems (USA-TX)

CTB: Calibration Test Box • Centraal Technische Bibliotheek (Netherlands) • Central Terminal Building • Cipher Type Byte • Commonwealth Telecommunications Board • Communications Toolbox • Complex-to-Basic • Comprehensive Test Ban • Control Test Bed • Cut Bank (USA-MT) (location identifier)

CTBA: Centre Technique du Bois et de l'Ameublement

CTBL: Cloud-Topped Boundary Layer

CTBT: Comprehensive Test Ban Treaty

CTBTO: Comprehensive Test Ban Treaty Organization

CTC: Camera, Timing and Control • Capsule Test Conductor • Carbon Tetrachloride • Catamarca (Argentina-CA) (location identifier) • CCATS Telemetry Controller • Center for Technology Commercialization • Centralized Traffic Control • Central Technical Catalogue • Central Test Center • Chief Test Conductor • City Technology College • Climate Test Chamber • Closed Timelike Curve • Committee for Technical Cooperation • Communications Technology Center • Community Technology Center • Compañía de Teléfonos de Chile • Compatibility Test Capsule • Computer Technology Center • Constant Torque Compensation • Cooke Transit Circle • Cornell Theory Center (USA-NY) • Cost To Completion • Counter/Timer Circuit • Counter-Timer Control • Crew Training Centre • Crew Training Complex

CTCA: Channel To Channel Adaptor

Ctcas: Características

CTCI: Classification Type pour le Commerce International • Contract Technical Compliance Inspection

CTCP: Configuartion Test Change Period • Contract Task Change Proposal

CTCPEC: Canadian Trusted Computer Product Evaluation Criteria

CRCS: Central Thermal Control System

C&TCS: Communications & Tracking Checkout System

CTCS: Communications and Tracking Checkout System

CTCSS: Continuous Tone Coded Squelch System

Ct. Ct.: Compte Courant

CTCU: Correlation Tracker and Control Unit

Ctd: Coated • Continued • Created

CTD: Cell Transfer Delay • Centrale Technische Dienst • Central Tracking Detector • Centre for Telecommunications Development • Certified Test Data • Charge-Transfer Device • Chemical Technology Division • Classified Telephone Directory • Conductivity-Temperature-Depth • Consecutive Triple Difference • Cumulative Trauma Disorder • Current-Temperature-Density

CTDC: Chemical Thermodynamics Data Center

CTDIF: Cambridge Tabular Data Interchange Format

CTDP: Centre de Traitement Doris/Poséidon

CTDU: Cross-link Transponder Data Unit

Cte: Commandante • Corriente

CTE: Carti (Panama) (location identifier) • Center for Teaching Excellence • Central Timing Equipment • Certification Test and Evaluation • Charge Transfer Effect • Charge Transfer Efficiency (*) • Ciencias de la Tierra y del Espacio (periodical code) • Coefficient of Thermal Expansion • Commercial Test Equipment • Component Test Equipment • Concurrent Test Environment • Content-Transfer-Encoding • Contractor Test and Evaluation • Cross-Track Error • Cruise Thermal Efficiency

CTEA: Cost and Training Effectiveness Analysis

CTEC: Certified Technical Education Center

CTEI: Centre de Traitement et d'Exploitation des Images • Cloud-Top Entrainment Instability

CTEM: Conventional-Transmission Electron Microscope • Conventional-Transmission Electron Microscopy

CTEP: Camera Tube Evaluation Program

Cter: Center

CTER: Command Terminal

CTERM: Communications Terminal

CTERMP: Communications Terminal Protocol

Ctf: Certificate • Certify

CTF: Central Training Facility • CLIC Test Facility (CERN) • Combined Test Force • Commit To Flight • Common Test Facility • Communauté des Télévisions Francophones • Community Task Force • Contrast Transfer Function • Core Test Facility • Correction To Follow • Counting Test Facility

CTFC: Central Time and Frequency Control

CTFE: Chlorotrifluoroethylene

CTFM: Continuous-Transmission Frequency Modulated

Ctg: Cartridge • Coating • Cutting

CTG: Cartagena (Colombia) (location identifier) • Center for Technology in Government (USA) • Central Training Group • Centrum voor Technische Geowetenschappen (Netherlands) • Chemical Topological Graph • Commonality Task Group

Ctge: Cartage • Cartridge • Cottage

CTGY: Category

CTH: Cloud Top Height • Coatesville - Chester County Carlson (USA-PA) (location identifier)

CTI: Camera Timing Indicator • CCD Transit Instrument • Centre des Technologies de l'Information • Centre des Télécommunications et de l'Informatique • Centre de Traitement de l'Information • Centre de Traitement Informatique • Centre Technique Industriel • Charge Transfer Inefficiency • Climate Technology Initiative • Coherent Toroidal Interconnect • Commission des Titres d'Ingénieur (France) • Compile Time Instrumentation • Computational Testbed for Industry • Computer Telephone Integration • Computer Telephony Integration • Computer-Telescope Interface • Consumer Technology Index • Couplage Téléphonie-Informatique • Critical Technology Institute • Critical Transportation Item • Cuito Canavale (Angola) (location identifier) • Current Technology Index

CTIA: Capacitative Trans-Impedance Amplifier • Cellular Technology Industry Association (USA-DC) (*) • Cellular Telecommunications Industry Association (USA-DC)

CTIF: Cellule de Traitement des Informations Financières

CTIM: Coupled Thermosphere-Ionosphere Model

CTIO: Cerro Tololo Interamerican Observatory (Chile) • Coatings Technology Integration Office (USAF)

CTIR: Computation Theory and Information Retrieval

CTIS: Center for Trade and Investment Services • Comprehensive Telemetry Indexing System • Crawler-Transporter Intercom System

CTIV: Centre de Traitement des Images de Végétation

CTK: Canton (USA-SD) (location identifier)

Ctl: Central • Control

CTL: Calibration Task List • Canoga Test Laboratory • Central • Central Technical Library • Charleville (Australia-QLD) (location identifier) • Checkout Test Language • Complementary Transistor Logic • Component Test Laboratory • Constructive Total Loss • Control • Controller Tape Location • Core Transistor Logic • Critical To Launch

Ctlry: Cutlery

Ctlst: Catalyst

CTLV: Center-To-Limb Variation

CTLZ: Control Zone

ct/m: Counts per Minute

CTM: Cell Transfer Matrix • Centrifugal Twisting Moment • Chemical Tracer Model • Chemical Transport Model • Chemistry/Transport Model • Chetumal (Mexico) (location identifier) • Ciências e Tecnologias dos Materiais • Collapsible Tube Mast • Communications Terminal Module • Community Trademark (EU) • Computerized Text Management • Continuum Storm • Contract Technical Manager • Cordless Terminal Mobility • Critical Thermal Maximum • Crystalline Transitional Material

CTMA: Center Traffic Management Advisor

CTMS: Carrier Transmission Measuring System

ctn: Cotangent

Ctn: Carton

CTN: Cable Termination Network • Canadian Technology Network • Centre des Technologies Nouvelles (Belgium) • Certification Test Network • Comité Technique de Normalisation • Cooktown (Australia-QLD) (location identifier)

CTNE: Compañía Telefónica Nacional de España (Spain)

Ctnr: Container

CTNS: Center for Theology and the Natural Sciences (USA-CA)

Cto: Conto

CTO: Call Transfer Outside • Calverton - Peconic River (USA-NY) (location identifier) • Cancelled To Order • Centre de Topographie des Océans (France) • Chief Technical Officer • Circular Terminal Orbit

CTOL: Conventional Takeoff and Landing

CTOS: Cassette Tape Operating System • Computerized Tomography Operating System • Converging Technologies Operating System

CTP: Capacitor Test Program • Carutapera (Brazil-MA) (location identifier) • Center for Theoretical Physics • Charge-Transfer Photography • Christmas Tree Pattern • Closed-Time Path • Clustering Terminal Pattern • Coded Telemetry Processor • Comité Technique Paritaire • Command Translator and Programmer • Common Transport Policy • Communications Timing Procedure • Compiled Translation Process • Confidence Test Program • Conseil Technique Paritaire • Construction Test Procedure • Conventional Terrestrial Pole • Coordinated Test Program • Cyclic Time Processor

CTPA: Coax-to-Twisted-Pair Adapter

CTPC: Comité Technique Paritaire Central

CTPD: Crew Training and Procedures Division

CTQ: Santa Vitória do Palmar (Brazil-RS) (location identifier)

Ctr: Center • Contour • Contribution • Contributor • Culture • Cutter

CTR: Cattle Creek (Australia-NT) (location identifier) • Center • Centre de Traitement Régional • Centre de Traitements Régionaux • Certification Test Requirement • Certification Test Review • Certified Test Record • Certified Test Result • Civil Tiltrotor • Classical Transition Region • Common Technical Regulation • Computer Technology Research • Consolidate Time Rate • Contractor Task Report • Contract Technical Representative • Contractual Technical Report • Control • Controlled Thermonuclear Reaction • Controlled Thermonuclear Research • Controller • Cooperative Threat Reduction • Coordinate Transformation Restoration • Core Transistor Register

CTRC: Colorado Technical Reference Center (USA-CO)

Ctre: Centre

CTRF: Conventional Terrestrial Reference Frame

Ctrfgl: Centrifugal

Ctrg: Centering

Ctrl: Central

CTRL: Control

Ctrlr: Controller

Ctro: Centro

CTRR: Component Test Results Review

CTRS: Centers • Component Test Requirements Specification • Conventional Terrestrial Reference System

Ctrst: Contrast

CTRW: Continuous-Time Random Walk

ct/s: Counts per Second

Cts: Centavos • Centimes • Cents • Certificates • Counts • Crates

C&TS: Communications & Tracking System

CTS: Cable Telemetry System • Cable To Shore • Canadian Technology Satellite • Captive Trajectory System • Cassegrain Twin Spectrograph • Center for Theoretical Study • Centralized Translation System • Central Timing System • Centro Turistico Studentesco • Certification Test Specification • Cesium Time Standard • Change Tracking System • Characteristic Time Scale • Charge-Transfer Spectrum • Chirp Transform Spectrometer • Circuit Test Set • Clear To Send • Cloud Top Scanner • Coherent Thought Sequencing • Command and Telemetry System • Committee on the Teaching of Science • Common Transport Semantics • Communications and Tracking System • Communications Technology Satellite • Communications Terminal Synchronous • Communications Test Station • Compatible Time Sharing • Component Test Set • Computerized Tomographic Scanner • Computerized Training System • Computer Test Set • Computer Transition Systems (Australia-VIC) • Computer Typesetting • Computer Typing System • Conformance Testing Service • Conformance Test Service • Console Test Set • Context Thesaurus System • Contingency Transfer System • Continuous Tracking Survey • Contractor Technical Service • Contractor Technical Support • Contrat de Télé-Services • Conventional Terrestrial System • Conversational Terminal System • Conversational Time Sharing • Coordinate Transformation System • Cosmic Top Secret • Counts • Courts • Cryogenic Temperature Sensor • Cryogenic Temperature Source • Sapporo Chitose (Japan) (location identifier)

CTSD: Computer Test Sequences Document

C&TSS: Communications & Tracking Subsystem

CTSS: Communications & Tracking Subsystem • Compatible Time-Sharing System • Conversational Time-Sharing System • Cray Time-Sharing System

CTT: Card-to-Tape Tape • Cartridge Tape Transport • Centre for Technology Transfer • Chemical Thermodynamic Table • Classical T Tauri • Cloud Top Temperature • Command Tranfer Tape • Crew Transfer Tunnel • Le Castellet (France) (location identifier)

CTTB: Checkout Techniques Testbed

CTTC: Cartridge Tape Transport Controller • Copper To The Curb

Cttee: Committee

CTTH: Copper To The Home

CTTMP: Center for Transport Theory and Mathematical Physics

cTTS: Classical T Tauri Star

CTTY: Console Teletype

CTU: Cabine Telecommunications Unit • Capsule Test Unit • Captive Test Unit • Cargo Transport Unit • Caribbean Telecommunication Union • Cartridge Tape Unit • Centigrade Thermal Unit • Central Telemetry Unit • Central Terminal Unit • Central Thermal Unit • Central Timing Unit • Chengdu (China-PRC) (location identifier) • Command and Telemetry Unit • Command Telemetry Unit • Compatibility Test Unit • Component Test Unit • Custom, Tradition and Usage

CTV: Cable Television • Channel Television • Color Television • Commercial Television • Compatibility Test Van • Computerisierte Textverarbeitung • Constant Tangential Velocity • Control Test Vehicle • Crew Tranfer Vehicle • Crew Transport Vehicle (*)

CTVA: Canadian Transportable VLBI Antenna

CTW: Cottonwood (USA-AZ) (location identifier)

CTWG: Calibration Targets Working Group

Ctwlk: Catwalk

Ctwt: Counterweight

CTWT: Counterweight

CTX: Cortland (USA-NY) (location identifier)

Cty: County

CTY: Center for Talented Youth • County • Cross City (USA-FL) (location identifier)

CTZ: Clinton - Sampson County (USA-NC) (location identifier) • Control Traffic Zone

c/u: Cada Uno

cu: Basque (ISO language code) • Cuba • Cubic

Cu: Copper • Cuivre • Cumulus • Kupfer

CU: Calibration Unit • Calibration Uplinks • Cambridge University (UK-England) • Camera Unit • Central Unit • Centroiding Unit • Certificat d'Université • Charge Utile • Chinese University • Ciudad Universitaria • Close-Up • Coefficient of Utilization • Colorado University (USA-CO) • Columbia University (USA-NY) • Combining Unit • Command Unit • Communications Unit • Computer Unit • Computing Unit • Construction Unit • Control Unit • Cornell University (USA-NY) • Crystal Unit • Cuba (international aircraft registration code) • Cuba (ISO country code)

CUA: Catholic University of America (USA-DC) • Ciudad Constitución (Mexico) (location identifier) • Commercial Utilization Area • Common User Access • Computer Users Association • Corporate Unit Agreement

CUAS: Cambridge University Astronomical Society (UK-England) • Champaign-Urbana University Astronomical Society (USA-IL)

Cub: Cubic • Cubicle

CUB: Columbia Owens (USA-SC) (location identifier) • Commonality Usage Board • Cuba (ISO country code) • Cursor Backward

CUBE: Concertation Unit for Biotechnology in Europe • Cosmic Ultraviolet Background Experiment

CUBIC: Caribbean Uniform Building Code • Cosmic Unresolved Background Imaging Camera • Cosmic Unresolved Background Instrument using CCDs

CUBRC: Calspan-University of Buffalo Research Center (USA-NY)

CUBS: Coronal Ultraviolet Berkeley Spectrometer

CUC: Computer Usage Control • Crystal Unit Cell • Cucuta (Colombia) (location identifier)

CUCAP: Cambridge University Committee on Aerial Photography

CUCCA: Columbia University Center for Computing Activities (USA-NY)

CUCI: Clasificación Uniforme para el Comercia Internacional

CUCRS: Commonwealth Users Committee on Remote Sensing

CUD: Caloundra (Australia-QLD) (location identifier) • Computer Underground Digest • Cursor Down

CUDI: Centro Universitario de Documentación e Información

CUDIXS: Common User Digital Information Exchange System

CUDN: Common User Data Network

CUDSS: Canada-UK Deep SCUBA Survey

CUE: Cognizant User Engineer • Collaborative Ukrainian Experiment • Command Uplink Electronics • Command Use Equipment • Common Usage Equipment • Communication Unit Executor • Computer Update Equipment • Cuenca (Ecuador) (location identifier)

CUED: Cambridge University Engineering Department

CUEJ: Centre Universitaire d'Enseignement et de Journalisme (France)

Cuen: Cuenca (Spain)

CUF: Capacity Utilization Factor • Command Unit Function • Common University Fund • Cross-Utilization File • Cuneo (Italy) (location identifier) • Cursor Forward

CUFOS: Center for Unidentified Flying Object Studies

cu. ft.: Cubic Foot

CUG: Closed User Group • Cudal (Australia-NSW) (location identifier) • C Users' Group • Orange Cudal (Australia-NSW) (location identifier)

CUH: Cushing (USA-OK) (location identifier)

CUHK: Chinese University of Hong Kong

CUI: Centre Universitaire d'Informatique • Change User Interface • Character-oriented User Interface • Character User Interface • Close-Up Imager • Common Usage Item • Common User Interface • Currillo (Colombia) (location identifier)

CUIL: Common Usage Item List

cu. in.: Cubic Inch

CUIO: Centre Universitaire d'Information et d'Orientation

CUIP: CPS UC Internet Project

Cuis: Cuisine

CUIS: Columbus Users Information System • Columbus Utilisation Information System

CUISP: Columbus Users Information System Prototype

cuj: Cujus

CUJ: Culion (Philippines) (location identifier)

CUK: Caye Caulker (Belize) (location identifier)

CUL: Cambridge University Library • Command Uplink • Common Use List • Cornell University Library • Culiacan (Mexico) (location identifier) • See You Later

CULER: Cryogenic Upper (atmosphere) Limb Emission Radiometer

CULR: Common Upper Layer Requirements

Cult: Cultivé • Cultural • Culture • Culturel

Cum: Cumulative

CUM: Central-Unit Memory • Cumana (Venezuela) (location identifier) • Cumulative

Cumb: Cumberland (UK-England)

CUMD: Continuous Update Memory Display

CUMP: Central-Unit Memory Programmer

CUN: Cancún (Mexico) (location identifier)

CUNUMBR: Control Unit Number

CUNY: City University of New York (USA-NY)

CUO: Caruru (Colombia) (location identifier) • Copenhagen University Observatory (Denmark)

CUP: Cambridge University Press • Canadian University Press • Carupano (Venezuela) (location identifier) • Columbia University Press (USA-NY) • Commonality Usage Proposal • Communications User Program • Cursor Position • Peso (Cuba) (ISO currency code)

CUPE: Cranfield Unit for Precision Engineering

CUPID: Completely Universal Processor I/O Design • Conversation-mode Utilities for Processing Image Data • Cosmic Ultraviolet Polarimetric Imaging Device

CUPLE: Comprehensive Unified Physics Learning Environment

CUPR: Catholic University of Puerto Rico

CUPRI: Cornell University Portable Radar Interferometer

CUpS: Connection Updates per Second

CUPS: Connection Updates Per Second • Consolidated Uniform Payroll System • Consortium for Upper-level Physics Software

CUQ: Coen (Australia-QLD) (location identifier)

Cur: Currency • Current

CUR: Council on Undergraduate Research • Curaçao Hato (Netherlands Antilles) (location identifier) • Cursor

CURDIF: Centre Universitaire de Recherche et Développement Informatique

CURDS: Centre for Urban and Regional Development Studies (UK)

CUREA: Consortium for Undergraduate Research and Education in Astronomy

CURI: Comité d'Utilisateurs des Ressources Informatiques

CURL: Consortium of University Research Libraries

Curr: Currency • Current • Curriculum

CURRI: Centre Universitaire Régional de Ressources Informatiques

Currt: Current

Curt: Current

CURV: Commandable Underwater Recovery Vehicle • Curve

CUS: Catalog Update Server • Central User Service • Columbus (USA-NM) (location identifier) • Conférence Universitaire Suisse (Switzerland) • Continental US

CUSA: Centrifugal Urine Separator Assembly

CUSBL: Cumulus Under Stratocumulus Boundary Layer

CUSE: Committee on Undergraduate Science Education

CUSH: Comité des Utilisateurs du SHOM (France)

CUSI: Club des Utilisateurs de Systèmes Informatiques (France) • Configurable Unified Search Index • Configurable Unified Search Interface (*)

CUSP: Compton Upscattering Process • Cyclotron Upscattering Process

Cust: Custodian • Custody • Customer • Customs

Custod: Custodian

CUT: Canonical Unit of Time • Class Under Test • Control Unit Terminal • Control Unit Tester • Coordinated Universal Time • Cutral (Argentina-NE) (location identifier)

CUTE: Clarkson University Terminal Emulator • Common Use Terminal Equipment • Constrained and Unconstrained Test Environment • Constrained and Unconstrained Testing Environment

CUTLASS: Collaborative UK Twin Located Auroral Sounding System

CUTS: Computer Users' Tape System • Customer Usage Tracking System

CUU: Chihuahua (Mexico) (location identifier) • Computerunterstützter Unterrricht • Cursor Up

CUV: Casigua (Venezuela) (location identifier)

CUVRA: Characteristics of Ultraviolet in the Alps

CUW: Cube Cove (USA-AK) (location identifier)

CUX: Corpus Christi - Cuddihy (USA-TX) (location identifier)

CUY: Cue (Australia-WA) (location identifier)

CUZ: Cuzco (Peru) (location identifier)

cv: Cras Vespere

C-V: Capacitance-Voltage

CV: Calibrated Volume • Calorific Value • Cape Verde (ISO country code) • Cap Vert (ISO country code) • Cardiovascular • Carrier Vehicle • Cataclysmic Variable • Cavallo Vapore • Cavalos Vapor • Cave • Central Version • Check Valve • Chèque de Voyage • Cheval-Vapeur • Chief Value • Circuit Virtuel • Classification Value • Coefficient of Variation • Collective Vector • Command Vehicle • Command Verification • Common View • Complex Visibility • Computer Vision • Configuration Verification • Constant Voltage • Constant Volume • Continuously Variable • Conventional Vehicle • Convertible • Córdoba Variables • Core vehicle • Cost Variance • Counter Voltage • Cove

• Coventry (UK postal code) • Crecimiento Vegetativo • Cruise Vehicle • Cryptovariable • Crystal Violet • Curriculum Vitae • Cyclic Voltammetry

CVA: Central Valley Astronomers (USA-CA) • Cerebrovascular Accident • Pittsburgh Civic (USA-PA) (location identifier)

CVAC: Catawba Valley Astronomy Club (USA-NC) • Comox Valley Astronomy Club

CVAF: Common VTOC Access Facility

CVAP: Cataclysmic Variables Alert Programme

CVAS: Chagrin Valley Astronomical Society (USA-OH) • Chattaboochee Valley Astronomical Society (USA) • Chippewa Valley Astronomical Society (USA-OH) • Configuration Verification and Accounting System

CVB: Chungribu (Papua New Guinea) (location identifier)

CVC: Capacitance Voltage Characteristics • Card Value Center • Cesium Vapor Cathode • Cleve (Australia-SA) (location identifier) • Current-to-Voltage Converter • Current Voltage Characteristics

CVCF: Constant Voltage and Constant Frequency

CVCM: Collected Vacuum Condensible Materials • Collected Volatile Condensible Materials

CVCP: Committee of Vice-Chancellors and Principals

CVD: Cash Versus Documents • Cave Direito • Central Velocity Dispersion • Čerenkov Viewing Device • Chemical Vapor Deposition • Chromatic Variation in Dispersion • Come Volevasi Dimostrare • Congressional Visit Day (USA)

CVDA: Commercial Vapor Diffusion Apparatus

CVDR: Cockpit Video Data Recorder

CVE: Cave Esquerdo • Centrale Verwerkingseenheid • Collaborative Virtual Environment • Complete Vehicle Erector • Cooperative Virtual Environment • Covenas (Colombia) (location identifier) • Escudo (Cape Verde) (ISO currency code)

CVEI: Convertisseur de l'Électronique Image

CVEsc: Escudo of Cape Verde (currency)

CVF: Central Visual Field • Circular Variable Filter • Command Verified File • Compressed Volume File • Continuously Variable Filter • Courchevel (France) (location identifier)

CVFR: Controlled Visual Flight Rules

CVFS: Cesium Vapor Feed System • Circular Variable Filter Spectrometer

CVG: Cincinnati - Northern Kentucky (USA-OH) (location identifier)

CVGA: Color Video Graphics Adapter

CVGIP: Computer Vision, Graphics, and Image Processing (periodical code)

CVH: Caviahue (Argentina-NE) (location identifier) • Common VTOC Handler

CVI: Caleta Olivia (Argentina-SC) (location identifier) • Carbon Vapor Impregnation • Centrum voor Informatieverwerking (Netherlands) • Chemical Vapor Infiltration • Configuration Verification Index • Configuration Verification Inspection • Counterflow Virtual Impactor • Current Variable Inductor

CVIA: Computer Virus Industry Association

CVIM: Collaboration, Visualisation and Information Management • Configurable Vision Input Module

CVJ: Constant Velocity Joint • Cuernavaca (Mexico) (location identifier)

CVL: Cape Vogel (Papua New Guinea) (location identifier) • Configuration Verification List

CVLP: Chinese Visiting Lecturer Programme

CVLS: Central Vacuum Loading System

CvM: Cramer - von Mises (test)

CVM: Cash Value Machine • Centre de Veille Météorologique • Chromospheric Variability Monitor • Ciudad Victoria (Mexico) (location identifier) • Cluster Variation Method • Control Valve Module

CVn: Canes Venatici - Canum Venaticorum (constellation)

CVN: Clovis (USA-NM) (location identifier)

Cvntl: Conventional

CVO: Albany (USA-OR) (location identifier) • Chief Veterinary Officer • Corvallis (USA-OR) (location identifier)

CVOS: C Virtual Operating System

CVP: Circuit Virtuel Permanent • Commissioning and Verification Phase • Content Vectoring Protocol

CVPP: Command and Validation Peripheral Processor

CVPR: Computer Vision and Pattern Recognition

CVQ: Carnarvon (Australia-WA) (location identifier)

Cvr: Cover

CVR: Change Verification Recort • Change Verification Report • Cockpit Voice Recorder • Command Voltage Regulator • Computer Voice Response • Configuration Verification Review • Constant Velocity Recording • Constant Voltage Reference • Cover • Culver City - Hughes (USA-CA) (location identifier) • Crystal Video Receiver

CVRR: Computer Vision and Robotics Research

CVRS: Cockpit Voice Recording System

Cvrsn: Conversion

CVS: Clovis Cannon (USA-NM) (location identifier) • Comisión Vela Solar • Common Video System • Concurrent Version System • Coves

CVSD: Continuous Variable Slope Data • Continuous Variable Slope Delta

CVSE: Conferentie voor Veiligheid en Samenwerking in Europa

CVSRF: Crew-Vehicle Systems Research Facility

CVSS: Cuyhoga Valley Space Society (USA-OH)

Cvt: Convertible

CVT: Canonical Variational Theory • Chemical Vapor Transport • Color Video Tape • Command Vetting Translator • Communications Vector Table • Concept Verification Test • Concept Verification Testing • Configuration Verification Test • Constant Variable Transmission • Constant Voltage Transformer • Continuing Vocational Training • Continuously Variable Transmission • Convert • Cosmic Virial Theorem • Coventry (UK-England) (location identifier) • Cray Visualization Toolkit • Current Value Table

CVTE: Chemical Vapor Transport Experiment • Crystals by Vapor Transport Experiment • Crystals of Vapor Transport Experiment

CVTS: Compressed Video Transmission Service

CVU: Calibration Verification Unit • Constant Voltage Unit • Corvo Island (Portugal) (location identifier)

CVV: Cryostat Vacuum Vessel • Cryo Vacuum Vessel

CVW: CodeView for Windows

CVX: Critical Velocity Experiment • Critical Viscosity of Xenon

CVZ: Continuous Viewing Zone

c/w: Complete with

C&W: Cable & Wireless • Caution & Warning • Control & Monitoring

C/W: Carrier Wave • Caution & Warning

CW: Carrier Wave • Caution and Warning • Character Width • Chemical Weapon • Chi-Writer • Clockwise • Close Workstation • Cockcroft-Walton • Colliding Wind • Command Word • Commercial Weight • Community Workshop • Complete With • Computer Warfare • Computer Week • Computing Week • Continuous Wave • Continuous Waveform • Control Word • Conventional Wisdom • Crewe (UK postal code) • Curie-Weiss

CWA: Caution and Warning Annunciator • Center Weather Advisory • CEN Workshop Agreement • Chemical Warfare Agent • Civil Works Administration • Clean Water Act (USA) • Clean Work Area • Closed World Assumption • Collective Worling Agreement • Controlled Work Area • Mosinee (USA-WI) (location identifier) • Stevens Point (USA-WI) (location identifier) • Wausau - Central Wisconsin (USA-WI) (location identifier)

CWAIP: Centre for Wavelets, Approximation and Information Processing (Singapore)

CWAR: Continuous Wave Acquisition Radar

CWARC: Canadian Workplace Automation Research Centre

CWAS: Central Wyoming Astronomical Society (USA-WY)

CWB: Central Weather Bureau • Curitiba - Afonso Pena (Brazil-PR) (location identifier)

CWBS: Contract Work Breakdown Structure

CWC: Central Water Commission • Centre for Wireless Communications (Singapore) • Chemical Weapons Convention (1993) • Chernovtsy (Ukraine) (location identifier) • Condensed-Water Content • Cyclone Warning Centre

CWCS: Celestial World Coordinate System

CWD: Cable Wiring Diagram • Change Working Directory • Convert Word to Double • Cyclotron Wave Device

CWDB: Close White Dwarf Binary

CWDE: Center for World Development Education

CWDS: Cyclone Warning Dissemination System

CWDW: Convert Word to Double Word

CWE: Caution and Warning Electronics • Caution and Warning Equipment

CWEA: Caution and Warning Electronics Assembly

C'wealth: Commonwealth

Cwealth: Commonwealth

CWEAR: Critical-Work Element Audit Review

CWES: Cloud Wind Estimation System

CWEU: Caution and Warning Electronics Unit

CWF: Composite Wave Filter • Cornwell-Weisskopf Formula

CWFM: Continuous Wave Frequency-Modulated

CWFS: Calibration Wave-Front Sensor

CWG: Callaway Gardens (USA-GA) (location identifier) • Closed Waveguide • Commonality Working Group • Communications Working Group • Constant Wear Garment • Coordination Working Group

CWI: Centrum voor Wiskunde en Informatica (Netherlands) • Clinton (USA-IA) (location identifier) • Collection Wide Information • Continuous Wave Illuminator

CWIC: Classification by Weighted Iterative Clustering

CWIF: Continuous Wave Intermediate Frequency

CWIFS: Compact Wide Field Sensor

CWIK: Chemical World Index Key

CWIN: Cockpit Weather Information Needs

CWIP: Camera with Workstation for Image Processing • Construction Work In Progress

CWIS: Campus-Wide Information Service • Campus-Wide Information System • Current Weather Instrument System

CWISP: Campus-Wide Information System Protocol

CWIT: CDF Windowing Image Tool

CWL: Cardiff Wales (UK-Wales) (location identifier) • Center Wavelength • Continuous Wave Laser

CWLM: Caution and Warning Limit Module

CWLS: Conventional Weighted Least Squares

Cwlth: Commonwealth

CWN: Contract Work Notification

CWO: Cash With Order • Chief Warrant Officer • Continuous-Wave Oscillator • Mineral Wells - Fort Wolter (USA-TX) (location identifier)

CWP: Campbellpore (Pakistan) (location identifier) • Central Weather Processor • Circulating Water Pump • Cumulative Weight Percent • Current Workspace Pointer

CWpM: Correct Words per Minute

CWPM: Correct Words Per Minute

CWPS: Council on Wage and Price Stability

CWR: Continuous-Wave Radar • Cowarie (Australia-SA) (location identifier)

CWRS: Core Water-Recycling System

CWRU: Case Western Reserve University (USA-OH)

CWS: Cartographic Workstation • Caution and Warning Status • Caution and Warning System • Center Island (USA-WA) (location identifier) • Center Work System • Chilled Water Supply • Clinician Workstation • Continuous Wave Spectroscopy • Crew Workstation • Customer Warning System • Cyclone Warning System

CWSF: Canada-Wide Science Fair • Chemical Waste Storage Fcaility

CWSU: Caution and Warning Status Unit • Center Weather Service Unit • Central Weather Service Unit • Condensate Water Servicing Unit

cwt: Hundredweight

CWT: Call Waiting • Center Wing Tank • Code Walk-Through • Continuous Wavelet Transform • Cowra (Australia-NSW) (location identifier)

CWTS: Centrum voor Wetenschaps- en Technologiestudies (Netherlands)

CWVC: Campus-Wide Virtual Computer

CWVW: Cloud and Water Vapor Wind

CWW: Corowa (Australia-NSW) (location identifier)

CWY: Clearway

Cx: Complex • Convex

CX: Central Exchange • Christmas Island (ISO country code) • Color Exterior (film) • Control Transmitter • Cycle Crossover • Île de Noël (ISO country code) • Uruguay (international aircraft registration code)

CXA: Caicara (Venezuela) (location identifier)

CXB: Cosmic X-ray Background • Cox's Bazar (Bangladesh) (location identifier)

CXBR: Cosmic X-ray Background Radiation

CXC: Chitina (USA-AK) (location identifier)

CXF: Coldfoot (USA-AK) (location identifier)

CXH: Vancouver - Coal Harbour (Canada-BC) (location identifier)

CXI: Christmas Island (Kiribati) (location identifier) • Common X-windows Interface

CXJ: Caxias do Sul - Campo dos Bugres (Brazil-RS) (location identifier)

CXL: Calexico (USA-CA) (location identifier)

CXMS: Conversion X-ray Mössbauer Spectroscopy

CXN: Candala (Somalia) (location identifier)

CXO: Chandra X-ray Observatory (formerly AXAF) • Conroe - Montgomery County (USA-TX) (location identifier)

CXP: Cilacap (Indonesia) (location identifier)

CXQ: Christmas Creek (Australia-WA) (location identifier)

CXR: Christmas Island (ISO country code) • Île de Noël (ISO country code)

CXS: Charge Exchange Spectroscopy

CXT: Charters Towers (Australia-QLD) (location identifier)

CXY: Cat Cay (Bahamas) (location identifier)

cy: Welsh (ISO language code)

Cy: Capacity • Century • City • County • Currency • Cyanide • Cycle

CY: Calendar Year • Chypre (ISO country code) • Current Year • Cycle • Cyprus (international vehicle registration code) • Cyprus (ISO country code)

CYA: Les Cayes (Haiti) (location identifier)

CYB: Cayman Brac (Cayman Islands) (location identifier)

Cyber: Cybernetics

Cyc: Cycle • Cycling

CYC: Caye Chapel (Belize) (location identifier) • Cyclonic

CYE: Crystal Lake (USA-PA) (location identifier)

CYF: Chefornak (USA-AK) (location identifier)

Cyg: Cygnus-Cygni (constellation)

CYG: Corryong (China-ROC) (location identifier)

CYI: Chiayi (China-ROC) (location identifier)

Cyl: Cylinder • Cylindrical

CYL: Coyoles (Honduras) (location identifier) • Cylinder

CYLINA: Cyberspace Leveraged Intelligent Agent

CYM: Cayman Islands (ISO country code) • Chatham (USA-AK) (location identifier) • Îles Cayman (ISO country code)

CYMK: Cyan-Yellow-Magenta-Black

Cyn: Canyon

CYN: Canyon

CYO: Cayo Largo del Sur (Cuba) (location identifier)

Cyp: Cyprian • Cyprus

CYP: Calbayog (Philippines) (location identifier) • Cape York Peninsula (Australia-QLD) • Chypre (ISO country code) • Cyprus (ISO country code) • Pound (Cyprus) (ISO currency code)

CYR: Colonia (Uruguay) (location identifier)

CYS: Cheyenne (USA-WY) (location identifier)

CYSA: Cape York Space Agency Pty. Ltd. (Australia)

CYT: Yakataga (USA-AK) (location identifier)

CYTED: Ciencia Y Tecnología para el Desarollo

CYU: Cuyo (Philippines) (location identifier)

CYX: Cherskiy (Russia) (location identifier)

CYZ: Cauayan (Philippines) (location identifier)

cz: Czech Republic

Cz: Cenozoic • Czech

CZ: Canal Zone • Conservation Zone • Control Zone • Czechoslovakia (now obsolete) • Czech Republic (ISO country code) • République Tchèque (ISO country code)

CZA: Chichén Itzá (Mexico) (location identifier)

CZB: Cruz Alta - Carlos Ruhl (Brazil-RS) (location identifier)

CZC: Copper Center (USA-AK) (location identifier)

CZCS: Coastal Zone Color Scanner • Coastal Zone Color Scanning

Cze: Czech

CZE: Capillary Zone Electrophoresis • Coro (Venezuela) (location identifier) • Czech Republic (ISO country code) • République Tchèque (ISO country code)

CZF: Cape Romanzof (USA-AK) (location identifier)

CZH: Corozal (Belize) (location identifier)

CZJ: Corazón de Jesús (Panama) (location identifier)

CzJPh: Czechoslovak Journal of Physics (periodical code)

CZK: Cascade Locks (USA-OR) (location identifier) • Koruna (Czech Republic) (ISO currency code)

CZL: Constantine (Algeria) (location identifier)

CZM: Coastal Zone Management • Cozumel (Mexico) (location identifier)

CZMA: Coastal Zone Management Act (USA)

CZMG: Coastal Zone Management Group

CzMJ: Czechoslovak Mathematical Journal (periodical code)

CZMS: Coastal Zone Management Subgroup

CZN: Chisana (USA-AK) (location identifier)

Czo: Cozzo

CZO: Chistochina (USA-AK) (location identifier)

CZP: Cape Pole (USA-AK) (location identifier)

CZS: Cruzeiro do Sul (Brazil-AC) (location identifier)

CZT: Cadmium Zinc Telluride • Carrizo Springs (USA-TX) (location identifier)

CZU: Corozal (Colombia) (location identifier)

CZW: Czestochowa (Poland) (location identifier)

CZX: Changzhou (China-PRC) (location identifier)

CZY: Cluny (Australia-QLD) (location identifier)

CZZ: Campo (USA-CA) (location identifier)

C2: Command and Control • Nauru (international aircraft registration code)

C2D: Character to Decimal

C2GCR: Centre for Climate and Global Change Research

C2I: Command, Control, and Intelligence

C2P: Centre de Compétence en Projection Plasma

C2S: Command and Control Subsystem

C2VR: Centre de Calcul Vectoriel pour la Recherche

C2W: Command and Control Warfare

C2X: Character to Hexadecimal

C3: Andorra (international aircraft registration code) • Command, Control and Communications

C3A: Club Ajaccien des Amateurs d'Astronomie (France)

C3AD: Computer-Aided Aerospace Analysis and Design

C3I: Command, Control, Communication and Information • Command, Control, Communication and Intelligence • Command, Control, Computing and Intelligence

C4: Center for Clouds, Chemistry and Climate • Comet Coma Chemical Composition

C4I: Command-Control-Communications-Computers-Intelligence

C4ISR: Command-Control-Communications-Computer-Intelligence-Surveillance Reconnaissance

C5: Gambia (international aircraft registration code)

C6: Bahamas (international aircraft registration code)

C9: Mozambique (international aircraft registration code)

c/-: Care of • Currency

D

d: Apparent Diameter • British Penny • Da • Date • Daughter • Day • Dead • Deceased • Deci • Decision • Decree • Deep • Degeneracy of Vibrational Mode • Degree • Delete • Deliver • Delivery • Delta • Dem • Denarius • Density • Depart • Depth • Deputy • Der • Desert • Dextro • Deuteron • Día • Dialect • Dialectal • Diameter • Died • Dime • Diminutif • Diminutive • Diminutivo • Dinar (currency) • Discharge • Distance • Distance Modulus • Dividend • Dollar • Dose • Down • Drachma (currency) • Droit • Droite • Lattice Plane Spacing • Pence • Penny • Relative Density • Thickness

D: Absorbed Dose • Da • Dag • Daily • Dalasi (currency) • Damen • Damping Constant • Darcy • Day • Debye-Waller Factor • December • Deck • Declination • Deep • Deliver • Delta (international alphabet) • Democrat • Democratic • Demy • Densité • Density • Department • Depression • Depth • Deputy • Destroyer • Detectivity • Deus • Deuterium • Deutschland • Devonian • Diameter • Diffusion Coefficient • Diffusive Transport • Digital • Dimension • Dimensional • Dinar (currency) • Dioptre • Director • Diretto • Discrete • Dispersion • Dissociation Energy • Distinguished • Divorced • Doctor • Dogana • Dom • Don • Dong (currency) • DOS • Dose • Douane • Double • Down • Drag • Droit • Droite • Drop Diameter • Dutch • Dynamic • Electric Displacement • Electric Flux Density • First Derivative • Five Hundred (Roman numeral) • Fractal Dimension • Germany (international aircraft registration code) • Germany (international vehicle registration code) • Germany (postal country code) • Linear Diameter • Mass Diffusion Coefficient • Mean Elongation • Optical Density • Route Départementale • Telescope Aperture • 500 (Roman numeral)

d/a: Days after Acceptance • Documents against Acceptance

d-a: Digital to Analog

da: Danish (ISO language code) • Déca • Deka

d/A: Documentos contra Aceptación

Da: Danish • Daughter • Day

D/A: Days after Acceptance • Decoder Activate • Delivery on Acceptance • Deposit Account • Digital to Analog • Documents against Acceptance • Documents for Acceptance

DA: Algerian Dinar (currency) • Dakota • Dartford (UK postal code) • Data Acquisition • Data Adapter • Data Analysis • Data Automation • Data Available • Decimal Add • Decimal-to-Analog • Decision Altitude • Decoder Activate • Deflection Amplifier • Delayed Action • Density Altitude • Département d'Astrophysique • Department of the Army • Deployed Assembly • Deployment Assembly • Deposit Account • Deputy Administrator • Deputy Assistant • Derecho Administrativo • Descent Advisor • Descriptor Area • Design Authorization • Design Automation • Desk Accessory • Destination Address • Destructive Analysis • Detector Assembly • Development Assistance • Dictionary of Americanisms • Differential Algebra • Dinar (currency) • Dip Angle • Direct Access • Direct Action • Direct Ascent • Direction of Administration • Directory Assistance • Discrete Address • Dissociative Attachment • Distribution Assembly • District Administrator • District Attorney • Divergence Angle • Docking Adapter • Documents against Acceptance • Documents Attached • Documents for Acceptance • Doesn't Answer • Dona • Don't Answer • Double Aisle • Double Amplitude • Drift Angle • Drive Actuator • Drive Assembly • Dummy Antenna • Dutch Antilles

DAA: Data Access Arrangement • Data Analysis Assistant • Data Analysis in Astronomy • Data Automation Activity • Decimal Adjust for Addition • Designated Approval Authority • Devon Astronomical Association (UK) • Digital Automatic Acquisition • Division Administrative Assistant • Fort Belvoir - Davison (USA-VA) (location identifier)

DAAC: Data Active Archive Center • Distributed Active Archive Center (*) • Division des Archives Audiovisuelles et Cartographiques

DAACS: Dump Analysis And Consulting System

DAAD: Deutscher Akademischer Austauschdienst (Germany)

DAAK: Deutsch-Amerikanisches Akademisches Konzil

DAAM: Dynamic Associative Access Memory

DAAS: Data Acquisition and Analysis System • Department of Astronomy and Atmospheric Sciences

DAAT: Distributed Architecture Analysis Tool

DAB: Data Acquisition Bus • Data Announcement Bulletin • Daytona Beach (USA-FL) (location identifier) • Dictionary of American Biography • Diffuse And Bright • Diffusion Absorption Band • Digital Audio Broadcast • Digital Audio Broadcasting • Display Attention Bit • Distributeur Automatique de Billets • Docking And Berthing

DABBLE: Definitive Atmospheric Buoyant Boundary Layer Experiment

DABRK: Day Break

DABS: Discrete Address Beacon System

DABU: Data Bus Unit

DAC: Data Acquisition and Control • Data Acquisition Camera • Data Acquisition Chassis • Data Acquisition Computer • Data Acquisition and Control • Data Acquisition Controller • Data Analysis and Control • Data Analysis Center • Data Analysis Console • Data Analysis Control • Data Archiving Centre • Data Assembly Centre • Days After Contract • Deductible Average Clause • Delayed Action Command • Design Analysis Cycle • Design Augmented by Computer • Design Automation Conference • Dhaka (Bangladesh) (location identifier) • Diamond Anvil Cell • Digital-to-Analog Converter • Digital-to-Analytical Conversion • Digital-to-Audio Converter • Digital Arithmetic Center • Direct-Access Computing • Direct Air Cycle • Discrete Absorption Component • Discretionary Access Control • Display Analysis Console • Distance Amplitude Correction • District Access Coordinator • Document Availability Code

DACA: Division des Archives Cartographiques et Audiovisuelles

DACB: Data Acquisition and Control Buffer

DACBU: Data Acquisition and Control Buffer Unit • Digital Acquisition and Control Buffer Unit

DACC: Direct-Access Communications Channels

DACCS: Digital Access Cross Connect System

DACD: Digital Audio Campact Disk

DACE: Design and Analysis of Computer Experiments

DACEMS: Data Communications Equipment Monitoring and Switching

DACG: Discriminant Analysis with Constellation Graph

DACI: Direct Adjacent Channel Interference

DACP: Data Acquisition and Control Processor • Data Acquisition, Control, and Processing

DACS: Data Acquisition and Control System • Data and Analysis Center for Software • Dedicated Address Computation Stage • Digital Access Cross-connect Switch • Digital Access Cross-connect System • Digital Acquisition and Control System • Digital Auto-Correlation Spectroscopy • Directory Assistance Charging System • Discrete-Address

Communication System • Divert and Attitude Control System

DACST: Department of Art, Culture, Science and Technology (South Africa)

DACT: Disposable Absorption Collection Trunk

DACU: Data Acquisition and Control Unit • Digital-to-Analog Converter Unit

D&AD: Design & Art Direction

DA&D: Data Acquisition & Distribution

DAD: Da Nang (Viet Nam) (location identifier) • Data Access Diagram • Data Acquisition and Distribution • Data Acquisition Division • Deputy Assistant Director • Design and Art Direction • Design Approval Data • Desktop Application Director • Detector Array Development • Diode-Array Detector • Double Amplitude Displacement • Draft Addendum • Driveability Analysis Data • Dual Air Density • Due At Department

DADA: Distributed Astronomical Data Archive

DADAS: Directory Assistance Direct Access Service

DADB: Data Analysis Database

DADC: Digital Audio Disk Corporation

DADE: Data Acquisition and Decommutation Equipment • Data Acquisition and Documentation Equipment • Dual Air Density Explorer

DADEE: Dynamic Analog Differential Equation Equalizer

DADP: Devil's Advocate Decision Program

DADS: Data Acquisition and Distribution System • Data Archive and Distribution Service • Data Archive and Distribution System (*) • Digital Assisted Database System • Document Availability and Distribution Service • Dual Air Density Satellite • Dynamic Analysis and Design of Systems • Dynamic Analysis and Design System

DADSM: Direct-Access Device Space Management • Direct-Access Device Space Manager

DAdW: Deutsche Akademie der Wissenschaften (Germany)

DA&E: Data Analysis & Evaluation

DAE: Daparizo (India) (location identifier) • Data Acquisition Equipment • Data Analysis and Evaluation • Data Automation Equipment • Department of Atomic Energy • Dictionary of American English • Differential Algebraic Equation (*) • Differential and Algebraic Equation • Distributed Application Environment • Dump Analysis and Elimination • Dynamic Asian Economies

DAEC: Département d'Astrophysique Extragalactique et de Cosmologie (Observatoire de Paris-Meudon)

DAEF: Data Analysis and Evaluation Facility

DAEMON: Disk And Execution Monitor

DAET: Délégué Académique à l'Enseignement Technique (France)

DAEU: Diplôme d'Accès aux Études Universitaires

DAF: Data Acquisition Facility • Data Analysis Facility • Daup (Papua New Guinea) (location identifier) • Described As Follows • Discard At Failure • Dissolved Air Flotation • Document Absolute Frequency

DAFC: Digital Automatic Frequency Control

DAFCO: Délégation Académique à la Formation Continue (France)

DAFI: Documents And Forms Inventory

DAFNE: Double Annular Factory for Nice Experiments

DAFT: Data Acquisition Frequency Table • Digital-to-Analog Function Table

dag: Decagram

DAG: Barstow Daggett (USA-CA) (location identifier) • Daggett Barstow (USA-CA) (location identifier) • Development Assistance Group • Direct Acyclic Graph • Directed Acyclic Graph • Documentation Advisory Group

DAGC: Delayed Automatic Gain Control • Digital Automatic Gain Control

DAGH: Distributed Adaptive Grid Hierarchy

DAGS: Dartmouth Institute for Advanced Graduate Studies

DAH: Dathina (Yemen) (location identifier)

DAHAP: Digital Archive of Historical Astronomy Pictures

DAHNI: Demonstrator on the Access Hypermedia Non-visual Interface

DAI: Darjeeling (India) (location identifier) • Datastream Access Interface • Deferred Acceptance Item • Discrete-Activity Indicator • Dissertation Abstracts International • Distributed Artificial Intelligence

DAIC: Data Acquisition and Instrument Control

DAID: Data Applications and Interactive Display

DAIM: Data Analysis Information Memo • Dynamic Active Index Matrix

DAIR: Direct Altitude and Identity Readout • Driver Aid, Information and Routing • Dynamic Allocation Interface Routine

DAIRS: Dial Access Information Retrieval System • Distributed Aperture Infrared System

DAIS: Data Avionics Information System • Digital Airborne Imaging Spectrometer • Distributed Applications and Interoperable Systems • Distributed Automatic Intercept System • Document Analysis and Information System

DAISY: Data Acquisition and Interpretation System • Data Acquisition Integrated System

DAIU: Digital-to-Analog Interface Unit

DAJ: Dauan Island (Australia-QLD) (location identifier)

Dak: Dakota (USA)

DAK: Double-Adding KNMI

dal: Decaliter

DAL: Dallas - Fort Worth - Love (USA-TX) (location identifier) • Data Abstraction Layer • Data Access Language • Data Access Layer • Data Access Line • Data Accession List • Data Acquisition Language • Data Acquisition List • Data-Aided Loop • Data Analysis Laboratory • Data Analysis Language • Database Access Language • Dedicated Access Line • Disk Access Lockout • Distribution Authority List • Drawing Assembly List

DALC: Divided Access Line Circuit

DALF: Diplôme Approfondi de Langue Française

DALGT: Daylight

DALI: Dispositif d'Accès et de Lecture de l'Inventaire • Document And Library Integration

DALO: Disconnect At Lift-Off

DALR: Dry Adiabatic Lapse Rate

DALS: Data Acquisition Logging System

DALY: Disability-Adjusted Life Year

dam: Decameter

DAM: Damascus (Syria) (location identifier) • Data Access Method • Data Acquisition and Monitoring • Data Addressed Memory • Data Analysis Management • Data Association Message • Decameter • Decametric • Decametric Radiation • Definition, Analysis and Mechanization • Descriptor Attribute Matrix • Detection And Mapping • Deutsche Akademie für Metrologie (Germany) • Digitized Analog Magnetogram • Direct-Access Memory • Direct-Access Method • Distributed Associative Memory • Double Alumunized Mylar • Driver Amplifier Module • Dynamic Address Mapping

DAMA: Dark Matter (search) • Data Administration Management Association • Demand-Assigned Multiple Access • Demand Assignment Multiple Access

DAMAp: Département Atomes et Molécules en Astrophysique (Observatoire de Paris-Meudon)

DAME: Data Acquisition and Monitoring Equipment

DAMI: Designated Airworthiness Maintenance Inspection

DAMMPS: Damping And Metal Matrix for Precision Structure

DAMOP: Division of Atomic, Molecular and Optical Physics

DAMOYR: Day-Month-Year

DAMPS: Digital Advanced Mobile Phone Service

DAMPT: Department of Applied Mathematics and Theoretical Physics

DAMR: Dynamic Adaptive Mesh Refinement

DAMS: Direct-Access Management System

DAMT: Distributed Applications Monitor Tool

DAMTP: Department of Applied Mathematics and Theoretical Physics

DAMUS: Data Management and User Services

Dan: Danish

DAN: Danville (USA-VA) (location identifier) • Data Analyzer • Data Analysis Network • Deutsches Astronetz (Germany) • Deutsches Astronomisches Netz (Germany)

DANEOPS: DLR - Archenhold Near-Earth Objects Prediscovery Survey

DANIDA: Danish International Development Agency

DANOF: Département d'Astronomie Fondamentale (Observatoire de Paris-Meudon)

DAO: Data Access Objects • Data Acquisition Object • Data Assimilation Office • Defense Acquisition Organization • Department Administrative Order • Dessin Assisté par Ordinateur • Dial Assist Operator • District Advisory Officer • Documentation Assistée par Ordinateur • Dominion Astrophysical Observatory (Canada-BC) • Duly Authorized Officer

DAOG: Data Archive Operations Group

DAOPHOT: DAO Photometric Package

DAP: Daily Activity Plan • Darchula (Nepal) (location identifier) • Data Access Protocol • Data Acquisition and Processing • Data Acquisition Plan • Data Acquisition Platform • Data Acquisition Programme • Data Analysis Program • Data Associated Parameter • Data Automation Proposal • Delayed Alpha Particle • Demande d'Admission Préalable • Descent Analysis Program • Design Analysis Phase • Detail Assembly Panel • Developer Assistance Program • Diastolic Arterial Pressure • Differential Accelerometer Package • Digital Area Photometer • Digital Autopilot • Direct Access Program • Directed Audit Program • Direction de l'Administration et du Personnel • Director of Administrative Planning • Directory Access Protocol • Distributed Array Processor • Division Advisory Panel • Do All Possible • Do Anything Possible • Documents Against Payment • Donor-Acceptor Pair • Double-Amplitude Peak • Draw A Person • Dynamic Adaptive Modelling

DAPAF: Data Processing And Analysis Facility

DAPF: Data Analysis and Processing Facility

DAPFS: Direct-Ascent Powered-Flight Simulation

DAPHNE: Document Application Processing in a Heterogeneous Network Environment

DAPN: Directional Antenna Phasing Network

DAPNIA: Département d'Astrophysique, de Physique Nucléaire et d'Instrumentation Associée

DAPP: Data Acquisition and Processing Program

DAppSc: Doctor of Applied Science

DAPS: DCS Automated Processing System • Decommutation And Processing System • Direct-Access Programming System

DAPU: Data Acquisition and Processing Unit

DAQ: Data Acquisition

DAR: Daily Activity Report • Dar-Es-Salaam (Tanzania) (location identifier) • Data Access Register • Data Acquisition Rack • Data Acquisition Recorder • Data-Aided Receiver • Data Automation Requirement • Daughters of the American Revolution • Defense Acquisition Regulation • Designated Agency Representative • Designated Airworthiness Representative • Deutscher Akkreditierungsrat (Germany) • Deviation Approval Request • Deviation Authorization Request • Differentiation with Asymmetrical Reinforcement • Digital Audio Recording • Digital Autopilot Requirement • Direct Absorption Receiver • Direct-Access Recorder • Division of Atmospheric Research • Doppler Acoustic Radar • Drawing Analysis Record • Driver-Augmented Readout

DARA: Deutsche Agentur für Raumfahrtsangelegenheiten GmbH (Germany)

DARAI: Deutsche Arbeitsgemeinschaft für Raumfahrt und Astronomie Information (Germany)

DARC: Delayed-Action Radar Calibrator • Département d'Astrophysique Relativiste et de Cosmologie (Observatoire de Paris-Meudon) • Description-Acquisition-Restitution-Correlation • Direct Access Radar Channel • Direct Access Radar Control • Document Archive Controller

DARD: Data Acquisition Requirements Document

DARDC: Device for Automatic Remote Data Collection

DARE: Data Archival and Retrieval Enhancement • Data Automatic Reduction Equipment • Data Automation Research and Experimentation • Data Rescue • Demand And Resource Evaluation • Dictionary of American Regional English • Documentation Automated Retrieval Equipment • Drug Abuse Resistance Education

DARHF: Dual Axis Radiographic Hydrodynamic Facility

DARI: Database Application Remote Interface

DARIS: Diffusion et Accès Rapide à l'Information Scientifique

DARM: Data Acquisition Room

DARMIS: Data Requirements Management Information System

DARMS: Drifting Automatic Radiometeorological Station

DARN: Dual Auroral Radar Network

DARO: Defense Airborne Reconnaissance Office

DARP: Direct Access RTCS Pointer

DARPA: Defense Advanced Research Projects Agency (USA) (now ARPA)

DARR: Delft Atmospheric Research Radar (Netherlands)

DARS: Data Accumulating and Reporting Sheet • Data Acquisition Recording System • Data Acquisition and Reduction System • Decommutation And Readout System • Digital Adaptive Recording System • Digital Attitude Reference System • Digital Audio Radio Service

DART: Daily Automatic Rescheduling Technique • Data Analysis Recording Tape • Data Retrieval Tool • Decision-Aided Rescheduling Tool • Decomposed Ammonia Radioisotope Thruster • Deutsche Amateur-Radioastronomische Tagung (Germany) • Dexterous Anthropomorphic Robotic Testbed • Dial-A-Ride-Transportation • Directional Automatic Realignment of Trajectory • Director And Response Tester • Display for Advanced Research Technology • Dual Axis Rate Transducer • Dublin Area Rapid Transit (Ireland) • Dust Accumulation and Removal Tool

DARTE: Discrete Angle Radiative Transfer Equation

DARTFire: Diffusive and Radiative Transport in Fire

DARTS: Digital Automated Radar Tracking System • Digital Azimuth Range Tracking System • Dynamics Algorithms for Real-Time Simulation

DARWIN: Design Assessment of Rotors With Inspection

DAS: Dairo Air Services • Dakota Astronomical Society (USA-ND) • Data Access Security • Data Acquisition Station • Data Acquisition System • Data Analysis Software • Data Analysis Station • Data Analysis System • Data Assimilation System • Data Automation System • Dataset Attribute Structure • Decimal Adjust for Subtraction • Deflection Amplifier System • Delaware Astronomical Society (USA-DE) • Denver Astronomical Society (USA-CO) • Department of Administrative Services • Delivered Alongside Ship • Design, Analysis, and Simulation • Development Advisory Service • Dial Access System • Differential Absorption and Scattering • Digital Address System • Digital Analog Simulator • Digital Angiography Subtraction • Digital Autocorrelation Spectrometer • Digital Avionics System • Dipole Antenna System • Direct Access System • Directory Assistance System • Distributed Authentication Service • Distributor And Scanner • DME-based Azimuth System • Document Analysis System • Documentation Accountability Sheet • Doppler Acoustic Sounder • Double Algebraic Sum • Dual-Attached Station • Dual-Attach Station • Dual Attachment Station • Dundee Astronomical Society (UK-Scotland) • Dwingeloo Autocorrelation Spectrometer • Dynamically Assigned Socket

DASA: Deutsche Aerospace AG (now part of EADS) • Dual Aerospace Servo Amplifier • Dual Aerosurface Servo Amplifier

DASC: Deep Atmosphere Spectral Camera

DASD: Data Acquisition Support Document • Direct-Access Storage Device

DASDL: Data And Structure Definition Language

DASE: Differential Absorption and Scattered Energy • Differential Absorption and Scattering Equipment

DASF: Direct-Access Storage Facility

DASGal: Département d'Astronomie Stellaire et Galactique (Observatoire de Paris-Meudon)

DASH: Digital Audio-recorder with Stationary Head • Display And Sight Helmet • Distributed Analysis Software Hierarchy

DASI: Degree Angular Scale Interferometer • Digital All-Sky Imager • Digital Array Scanned Interferometer

DASL: Data Access System Language • Data Approximation Subroutine Library

DA&SM: Data Acquisition & Systems Maintenance

DASM: Data Acquisition System Manager

DASO: Demonstration And Shakedown Operations

DASOC: Disturbance Accommodation Standard-Deviation Optimal Controller (HST)

DASOP: Département d'Astronomie Solaire et Planétaire (Observatoire de Paris-Meudon)

DASR: Data Acquisition Statistical Recorder • Digital Airport Surveillance Radar

DASS: Data Analysis Support Staff • Data Archive and Science Support • Defensive Aids Subsystem • Direct Access Secondary Storage

Dat: Date • Datif • Dative • Dativo • Datum

DAT: Data • Data Acceptance Test • Data Acquisition Terminal • Data Acquisition Test • Datong (China-PRC) (location identifier) • Design Acceptance Test • Design Approval Test • Design, Assembly and Test • Desktop Analysis Tool • Detail Assembly Template • Digital Archival Tape • Digital Audio Tape • Direct-Ascent Threat • Disk Array Technology • Docking Alignment Target • Dummy Alias Table • Dynamic-Address Translation

DATA: Direct-Access Terminal Application • Distribution and Automation Technology Advancement

DATACOM: Data Communications

DATAR: Délégation à l'Aménagement du Territoire et à l'Action Régionale (France)

DADTE: Direction de l'Aménagement du Territoire et du Développement Économique (France)

DATE: Dynamics, Acoustics, and Thermal Environment

DAtF: Deutsches Atomforum (Germany)

DATIS: Digital Airborne Topographic Imaging System

DATOS: Detection And Tracking Of Satellites

DATP: Detailed Acceptance Test Procedure

DATPE: Direction de l'Aménagement du Territoire et de la Protection de l'Environnement

DATR: Design Acceptance Test Report • Design Approval Test Report

DATS: Data Acquisition and Transmission System • Detail Acceptance Test Specification • Dynamic Accuracy Test Set • Dynamic Accuracy Test System

DATTS: Data Acquisition, Telecommand and Tracking Station • Data Acquisition, Tracking and Telecommand Station

DATU: Direct Access Testing Unit

DATV: Digitally-Assisted Television

Dau: Daughter

DAU: Daru (Papua New Guinea) (location identifier) • Data Acquisition Unit • Data Adapter Unit • Data Arithmetic Unit • Digital Adapter Unit • Document Administratif Unique • Dümmster Anzunehmender User

DAUF: Degenhardt Amateur User Facility (USA-TN)

DAV: Data Above Voice • Data Available • Data Valid • David (Panama) (location identifier) • Deutscher Autorenverband (Germany) • Digital Audio/Video

DAVE: Do Anything Very Easily

DAVIC: Digital Audio Video Council • Digital Audiovisual Council

DAVID: Digital Audio/Video Interactive Decoder • Distributed Access View Integrated Database

DAW: Data Analysis Workshop • Deutsche Akademie der Wissenschaften (Germany) • Digital-Analog-Wandler • Dry Active Waste

DAWG: Data Acquisition Working Group

DAWID: Device for Automatic Word Identification and Discrimination

DAWMS: Deep Attack Weapons Mix Study

DAWN: Digital Automatic Weather Network

DAX: Daxian (China-PRC) (location identifier) • Deutsche Aktienindex

DAY: Dayton - James M. Cox (USA-OH) (location identifier)

DAZ: Darwaz (Afghanistan) (location identifier)

DAZD: Double Anode Zener Diode

dB: Decibel

Db: Dobra (currency)

D/B: Daybook • Decimal to Binary • Disk to Bulge • Docking/Berthing

DB: Dagelijks Bestuur • Daily Bulletin • Daimler-Benz • Dark Band • Databank • Database • Database Browser • Data Block • Data Buffer • Data Bus • Daybook • Deadband • Decibel • Delayed Broadcast • Design Baseline • Design Bureau • Detailed Balance • Deutsche Bibliothek (Germany) • Deutsche Bundesbahn (Germany) • Deutsche Bundesbank (Germany) • Diffusion-Bonded • Diffusion Bonding • Digital Block • Diode Bus • Direct Beam • Direct Broadcast • Dirty Beam • Disparition Brusque • Display Buffer • Disposition Board • Distribution Box • Document Base • Double Base • Double Bed • Double Biased • Double Break • Drifting Buoy • Dunkle Bänder • Dry Bulb • Dynamic Breaking

d/b/a: Doing Business As

dBa: Decibel Adjusted

DBA: Adjusted Decibel • Dalbandin (Pakistan) (location identifier) • Dassault Aviation Belgique (Belgium) • Database Accelerator • Database Access • Database Accessor • Database Administration • Database Administrator • Diode Box Assembly • Doctor of Business Administration • Doing Business As • Doppelbesteuerungsabkommen • Drum Buffer Assembly

DBAAM: Disk Buffer Area Access Method

DBAC: Database Administration Center

DBAF: Database Access Facility

DBAM: Database Access Method • Database Access Module

DBAS: Database Administration System

DBAWG: Database Administration Working Group

DBB: Dynamic Bass Boost

DBC: Database Computer • Data Buffer Coupler • Data Bus Control • Data Bus Coupler • Diameter Bolt Circle • Differential Box Counting • Digital-to-Binary Converter • Distributed Batch Controller

DBCC: Database Consistency Check

DBCL: Database Command Language

DBCO: Digital Block Clock Oscillator

DBCP: Drifting Buoy Cooperation Panel

DBCR: Database Change Request

DBCS: Double-Byte Character Set

DBCU: Data Bus Control Unit

DBd: Disparition Brusque Dynamique

DBD: Database Description • Database Descriptor • Database Driver • Dead Band • Dhanbad (India) (location identifier) • Double-Base Diode • Double Beta Decay

DBDA: Database Design Aid

DBDB: Digital Bathymetric Database

DBDD: Database Design Document

DBDF: Data Block Definition File

DBDL: Database Definition Language

DBDQ: Distributed Queue Dual Bus

DBDU: Digital Bargraph Display Unit

DBE: Database Element • Data Bus Element • Data Bus Enable • Deep Basin Experiment • Design-Basis Event • Digital Based Exciter • Disadvantaged Business Enterprise

DBEATS: Dispatch Payable Both Ends All Time Saved

DBELTS: Dispatch Payable Both Ends on Lay Time Saved

DBES: Database and Expert System Program

DBF: Database File • dBase Format • Digital Beam Forming

DBFF: Digital Block Flip-Flop

DBFN: Data Base File Number • Data Bus File Number

DBFW: Digital Fly-By-Wire

DBG: Data Bus Group • Deutsche Bunsengesellschaft (Germany)

DBGMP: Data Bus Generation and Maintenance Package

DBH: Database Handler • Diameter at Breast Height

DBHS: Database Handling System

DBI: Database Interface • Data Bus Interface • Decomposition-Based Induction • Deutsches Bibliotheksinstitut (Germany) • Digital Bus Interface • Drag Balance Instrument

DBIA: Data Bus Interface Adapter • Data Bus Isolation Amplifier • Digital Block Inverter Amplifier

DBIF: Database Interface

DBIL: Database Input Language

DBIMA: Database of Images

DBIN: Description Bibliographique Internationale Normalisée

DBIO: Database Input/Output

DBIU: Data Bus Interface Unit

Dbk: Drawback

DBK: Database Kernel • Dominant Battlespace Knowledge

Dbl: Double

DBL: Damage Before Launch • Database Language • Database List • Database Load

Dblr: Doubler

dBM: Decibels above one Milliwatt

D/BM: Docking/Berthing Mechanism

DBM: Database Management • Database Manager • Data Buffer Memory • Data Bus Monitor • Debra Marcos (Ethiopia) (location identifier) • Decibel Meter • Décision Budgétaire Modificative • Detailed Balancing Model • Diploma in Business Management • Docking Berthing Mechanism

DBME: Database Management Environment

DBMIST: Direction des Bibliothèques, des Musées et de l'Information Scientifique et Technique (France)

DBML: Database Management Language • Diabetic Meal

DBMS: Database Management Software • Database Management System

DBMV: Digital Block Multivibrator

Dbn: Durban (South Africa)

DBN: Data Bus Network • Design By Numbers • Dublin (USA-GA) (location identifier)

DBO: Dual Beam Oscilloscope • Dubbo (Australia-NSW) (location identifier)

DBOMP: Database Organization and Maintenance Processor

DBOS: Disk-Based Operating System

dBP: Decibels above one Picowatt

DBP: Debepare (Papua New Guinea) (location identifier) • Design Baseline Program • Deutsche Bundespost (Germany) • Division of Biological Physics

DBPC: Database Processing Center

DBPL: Database Programming Language

DBpSM: Decibels per Square Meter

DBPSM: Decibels Per Square Meter

DBQ: Database Query • Dubuque (USA-IA) (location identifier)

DBR: Database Retrieval • Development Baseline Review • Distributed Bragg Reflector • DOS Boot Record

DBRAD: Database Relational Application Directory

DBRC: Database Recover Control

DBRD: Dépense Brute de Recherche et de Développement

Dbre: Diciembre

DBRITE: Digital Bright Radar Indicator Tower Equipment

dBRN: Decibels above Reference Noise

DBRN: Databank Release Notice

DBRTD: Double Barrier Resonance Tunnelling Diode

DBS: Database Server • Database Specification • Database Storage • Database System • Data Bus Simulator • Définition des Besoins du Système • Demand Broadcast System • Dibromostyrene • Diffraction Back-Scattering • Digital Book System • Digital Broadcasting by Satellite • Direct-Broadcasting Satellite • Direct-Broadcasting Service • Direct-Broadcast Satellite • Direct-Broadcast Service • Direct-Broadcast System • Docking Berthing System • Doppler Beam Swinging • Double-Beam Spectrophotometer • Dubois (USA-ID) (location identifier) • Duplex Bus Selector

DBSA: Database and System Administration

DBSC: Digital Block Slave Clock

DBSCAN: Density-Based Spatial Clustering of Applications with Noise

DBSM: Decibels per Square Meter

DBST: Digital Block Schmitt Trigger • Double British Summer Time

DBt: Disparition Brusque Thermale

DBT: Debra Tabor (Ethiopia) (location identifier) • Design Built Team • Deutsche Bundespost Telekom (Germany) • Disparition Brusque Thermale • Double-Base Transistor • Dual-Buffer Tritation • Dynamique et Bilan de la Terre

DBTF: Database Task Force

DBTG: Database Task Group

dBU: Decibel Unit

DBU: Data Bus Unit • Deutsche Bundesstiftung Umwelt (Germany) • Digital Bus Unit • Disk Buffer Utilization

DBUR: Databank Update Request

DBUT: Database Update Time

dBV: Decibels above one Volt • Deutscher Bibliothekverbund (Germany) • Dubrovnik Cilipi (Croatia) (location identifier)

DBV: Database Variable

dBW: Decibel reference to one Watt

DBW: Data Bus Wire • Design Bandwidth • Differential Ballistic Wind

DBWA: Design By Walking Around

DBWR: Database Writer

DBY: Dalby (Australia-QLD) (location identifier)

dC: Después de Cristo • dopo Cristo • Dwarf Carbon

D&C: Display & Console • Display & Control • Distribution & Control

D/C: Displays/Controls • Down-Converter

DC: Da Capo • Dairy Cattle • Data Camera • Data Cell • Data Center • Data Channel • Data Check • Data Code • Data Collection • Data Communication • Data Concentrator • Data Control • Data Conversion • Data Coordinator • Dead Center • Dearborn Catalogue • Death Certificate • Decimal Classification • Decimal Currency • Decreto • Default Colour • Define Constant • Definitive Contract • Degré de Condensation • Degree of Condensation • Delete Character • Delta Clipper • Deputy Chief • Deputy Commissioner • Deputy Consul • Deputy Counsel • Derecho Civil • Design Change • Design Contractor • Detector Controller • Developed Country • Developing Country • Development Center • Development Characteristic • Development Cost • Device Control • Device Coordinates • Diagnostic Centre • Dialogue Controller • Dictionary of Canadianisms • Dienstcommissie • Differential Calculus • Differential Correction • Digital Clock • Digital Code • Digital Comparator • Digital Computer • Diners Club • Diplomatic Corps • Direct Command • Direct Current • Direct Cycle • Directional Control • Display and Console • Display and Control • Display Computer • Display Console • Display Coupler • Displays/Controls • Disque Compact • Dissimilarity Coefficient • District Chief • District Commissioner • District Council • District of Columbia (USA) • Division Chief • Double Column • Double Compact • Double Contact • Double Coronagraph • Drift Chamber • Drift Correction • Dry Chemical

DC&A: Data Collection & Analysis

DCA: Data Capture Agent • Data Correction Amplifier • Debit-Credit Account • Defense Communications Agency (USA) • Deflection Coil Amplifier • Department of Civil Aviation • Design Change Authorization • Detailed Configuration Accounting • Detector Control Assembly • Detrended Correspondence Analysis • Digital Command Assembly • Digital Communications Associates Inc. (USA-GA) • Direct-Current Amplifier • Distributed Communication Architecture • Distribution Control Assembly • Document Content Architecture • Double Conversion Adapter • Double Coulomb Approximation • Double Cyclotron Absorption • Drift Correction Angle • Droplet Combustion Apparatus • Ronald Reagan Washington National (USA-DC) (location identifier)

DCAA: Defense Contracting Audit Agency (USA) • Dual Automatic Answer Device

DCAC: Defense Communication Agency Circular • Define and Control Airplane Configuration

DCAF: Distributed Console Access Facility

DCAI: Direct-Current Analog Input

DCAJ: Direction des Contrats et des Affaires Juridiques

DCAM: Data Communications Access Method • Department of Civil Aviation and Meteorology (Qatar)

DCAMIP: Data Center for Atomic and Molecular Ionization Processes

DCAO: Digital Card And-Or Gate

DCAP: Data Capture And Presentation • Dynamic Controls and Analysis Package

DCAR: Danish Centre for Atmospheric Research • Design Corrective Action Report • Discrepancy and Corrective Action Report

DCARS: Data Collection And Reduction System

DCAS: Data Collection and Analysis System • Defense Contract Administration Service

DCASMA: Defense Contract Administration Service Management Area

DCASR: Defense Contract Administration Service Region

DCASS: Digital Cartographic Software System

DCAT: Developmental Comparative Active Telescope

DCATT: Data Certification And Technology Transfer

DCB: Damage Control Book • Data Control Block • Data Control Board • Device Control Block • Design Change Board • Design Control Board • Digital Continuum Backend • Disk Coprocessor Board • Drawing Configuration Baseline

DCBS: Doubly-Compressed Bit Slices

DCC: DAQ and Control Computer • Data Collection Center • Data Communications Channel • Data Compression Conference • Data Computation Complex • Data Condition Code • Data Country Code • Dedicated Control Centre • Descriptive Cataloging Committee • Design Change Control • Design Concept Change • Development Control Center • Digital Command Control • Digital Communications and Control • Digital Compact Cassette • Digital Control Computer • Direct Cable Connection • Discrete Cell Control • Discrimination and Control Computer • Display Channel Complex • Display Combination Code • Display Computer Control • Display Control Console • Distributed Call Center • Distribution Control Center • Documentation Coordinating Committee • Document Control Center

DCCF: Distance Classifier Correlation Filter

DCCH: Digital Control Channel

DCCO: Defense Commercial Communications Office • Digital Card Clock Oscillator

DCCP: Design Change Control Program • Digital Computer Control Panel • Distributed Call Control Protocol

DCCR: Display Channel Complex Rehost

DCCS: Digital Camera Control System • Digital Command Communications System • Digital Comparator Correlator System • Discontiguous Shared Segments

DCCU: Data Communications Control Unit • Data Correlation Control Unit • Decommutator Conditioning Unit • Digital Cluster Control Unit • Digital Command and Control Unit • Digital Communications and Control Unit

Dcd: Decode

DCD: Data Carrier Detect • Data Compression Device • Decimal Classification Division • Deflection Coil Drive • Design Change Document • Design Control Document • Differential Current Density • Dimensional Control Drawing • Diode-Capacitor-Diode • Double Channel Duplex • Double-Crystal Diffraction • Drain-Current Drift • Dual Command Decoder

DCDB: Digital Cadastral Database • Digital Cartographic Database

DCDD: Design Characteristics Definition Document

DCDG: Diode-Capacitor-Diode Gate

DCDL: Digital Control Design Language

DCDOS: Deputy Chief Data and Operations Scientist

Dcdr: Decoder

DCDS: Digital Cartographic Data Standard • Digital Countdown Display System • Distributed Computing Design System

DCDSTF: Digital Cartographic Data Standards Task Force

DCE: Data and Control Electronics • Data Circuit-terminating Equipment • Data Circuit-termination Equipment • Data Collection Event • Data Communications Equipment • Data Compression Equipment • Data Conversion Equipment • Department of Conservation and Environment • Deployable Cargo Element • Deployment Control Electronics • Despin Control Electronics • Detector Control Electronics • Digital Communication Experiment • Diploma in Chemical Engineering • Distributed Computer Environment • Distributed Computing Environment (*) • Distributed Computing Equipment • Doctor of Civil Engineering • Domestic Credit Expansion • Drive Control Equipment • Droplet Combustion Experiment

DCEA: Digital Control Electronics Assembly

DCED: Département de la Coordination, de l'Exploitation et du Développement

DCEP: Data Collection and Evaluation Plan

DCES: Data Collection and Evaluation System • DSS Communications Equipment System

DCESS: Danish Center for Earth System Science (Denmark)

DCF: Data Capture Facility • Data Channel Filter • Data Communication Facility • Data Compression Facility • Data Control Facility • Data Correlation Facility • Data Count Field • Differential Counts Function • Digital Cartographic File • Discounted Cash Flow • Discrete Correlation Function • Distributed Computing Facility • Distributed Coordination Function • Document Composition Facility • Dominica Cane Field (Dominican Rep.) (location identifier) • Driver Configuration File • Dynamic Coercive Force

DCFEM: Dynamic Crossed-Field Electron Multiplication

DCFF: Digital Card Flip-Flop

DCFG: Direct-Current Free Gyro

DCFL: Direct Coupled FET Logic

DCFP: Dynamic Crossed-Field Photomultiplier

DCG: Data Control Group • Data Coordination Group • Definite Clause Grammar • Design Control Group • Dichromated Gelatin • Diode-Capacitor Gate • Direct-Current Generator • Double-Current Generator • Dwarsverband Commissie Gewasbescherming

DCGT: Direction et Contrôle des Grands Travaux

DCH: D Channel Handler

DChE: Doctor of Chemical Engineering

Dcha: Derecha

Dcho: Derecho

DCI: Data Capture Interface • Data Collection Initiative • Decimomannu Rafsu (Italy) (location identifier) • Definite Clause Grammar • Definite Gaussian Grammar • Deliverable Contract Item • Desorption Chemical Ionization • Dielectric Constant Indicator • Digital Clock Indicator • Direct Computer Input • Director of Central Intelligence • Display Control Interface • Documentation Change Instruction • Documentation, Control, and Integration • Document de Contrôle d'Interface • Double Column Inch • Dwarsverbandcommissie Informatica (Netherlands) • Rafsu Decimomannu (Italy) (location identifier)

DCIA: Digital Card Inverting Amplifier

DCIB: Data Communication Input Buffer

DCILM: Direct Computer Input Load Module

DCIP: Data Correction Indicator Panel

DCIS: Distribution Construction Information System • Downrange Computer Input System

DCIST: Directory of Computerized Information in Science and Technology

DCIU: Digital Control and Interface Unit

DCK: Dahl Creek (USA-AK) (location identifier)

Dckg: Docking

DCKNG: Docking

Dcl: Declaration

DCL: Data Checklist • DEC Command Language • DEC Control Language • Declaration • Delphi Common LISP • Depôt Central du Livre (France) • Designate Command Line • Designer Choice Logic • Detailed Checklist • Detector Characterization Laboratory • Device Clear • Dickens-Cuppie-Lucey • Digital Command Language • Digital Computer Laboratory • Digital Control Language • Digital Control Logic • Direct Communications Link • Doctor of Civil Law • Document Change List • Dynamic Characteristic Load

DCLC: Drift Cyclotron Loss Cone

DCld: Dark Cloud

DCLI: Direct Connection Line Interface

DCLIR: Dead Cat Lying In the Road

DCLM: Department of Conservation and Land Management

Dclr: Decelerate

DCLS: Data Collection and Location System • Dual Channel Line Scanner

DCLU: Digital Carrier Line Unit

DCM: Castres (France) (location identifier) • Data and Configuration Management • Data and Control Management • Data Circulation Model • Data Communication Mission • Data Communications Manager • Data Communications Module • Data Control Module • Decommutator Control Memory • Deputy Chief of Mission • Device Communication Manager • Diagnostic-Controlled Modem • Diagramme Couleur-Magnitude • Differential Code Modulation • Digital Carrier Module • Digital Cartographic Model • Digital Circuit Module • Digital City Map • Dimension Constraint Manager • Direction Cosine Matrix • Display and Control Module • Distributed Computer Model • Distributed Computing Manager • Double Crystal Monochromator • Dwarsverbandcommissie Materialonderzoek (Netherlands)

DCMB: Development Configuration Management Board

DCMD: Dose Calculation for Multiple Data

DCME: Digital Circuit Multiplication Equipment

DC-MEF: Disque Compact à Mémoire Fixée

DCMF: Data Capture and Management Facility

DCMP: Division of Condensed Matter Physics

DCMS: Data Communication Management System • Data Control Multiplex System • Digital Capacitance Measuring System • Distributed Call Measurement System

Dcmsn: Decommission

Dcmsnd: Decommissioned

DCMT: Dabish-Chinese Meridian Telescope

DCMU: Digital Concentrator Measurement Unit

DCMV: Digital Card Multivibrator

DCN: Data Center Network • Data Change Notice • Design Change Notice • Development Change Notice • Direction des Constructions Navales (France) • Discontinued Component Notice • Distributed Computer Network • Document Change Notice • Document Control Number • Draft Change Notice • Drawing Change Notice

DCNA: Data Communication Network Architecture

DCNP: Document Change Notice Proposal

DCO: Data Collection Order • Data Control Office • Detailed Checkout • Digital Central Office • Drawing Change Order

d. col: Double Column

DCOM: Distributed Component Object Model

DCOMP: Division of Computational Physics

DCOP: Detailed Checkout Procedure • Displays, Controls and Operation Procedure

DCOS: Data Communication Output Selector • Downrange Computer Output System

DCP: Data Collecting Platform • Data Collection Platform • Data Collection Program • Data and Command Processor • Data Communication Processor • Delayed Command Processor • Depot Condemnation Percentage • Design Change Proposal • Development Concept Paper • Development Cost Plan • Diffuse Cloud Phase • Digital Cartographic Production • Digital Clock Pulse • Digital Computer Programming • Direct Coupled Plasma • Direct Coupled Pump • Direct-Coupling Photometer • Direct-Current Plasma • Display Control Panel • Display Control Program • Distributed Communications Processor • Division of Chemical Physics • Drug Control Programme • DTS Calibration Parameter • Duplex Central Processor

DCPEI: DEU Control Program End Item • Display Control Program End Item

DCPG: Digital Clock Pulse Generator

DCPL: Distributed Control Programming Language

DCPLS: Data Collection Platform Location System

DCPRS: Data Collecting Platform Retransmission System

DCPS: Data Communication Protocol Standard • Data Compression Processing System • Digital Cartographic Production Segment • Dynamic Crew Procedures Simulator

DCPSK: Differential Coherent Phase Shift Keyed

DCPSP: Direct-Current Power Supply Panel

DCQM: Digital Circuit Quality Monitor

DCR: Daily Communication Report • Data Change Request • Data Conversion Receiver • Data Coordinator and Retriever • Decatur Hi-Way (USA-IN) (location identifier) • Decision Circuit Reception • Decrease • Dedicated Control Room • Design Certification Review • Design Change Record • Design Change Request • Design Characteristic Review • Design Clarification Review • Design Conception Review • Design Concept Report • Design Concern Report • Detector Count Rate • Differential Colour Refraction • Differential Correlation Radiometer • Digital Continuum Receiver • Digital Conversion Receiver • Disposition of Contract Request • Distributed Computing Revolution • Document Change Record • Document Change Request • Document Change Review • Dual Channel Radiometer • Dynamic Color Rendition

DCRB: Design Change Review Board

DCRE: Degenerate Configuration of Reflective Equilibrium

DCRG: Decreasing

DCRR: Design Change Requirement Record

DCRS: Data Collection and Reduction System • Danish Centre for Remote Sensing

DCRT: Division of Computer Research and Technology

DC&S: Display & Control System

DCS: DASD Calculation Services • Data Classification System • Data Collecting System • Data Collection System • Data Communications System • Data Conditioning System • Data Control Sequencer • Data Control Services • Data Control System • Data Conversion System • Defense Communications System • Deflection Coil Set • Delayed Coincidence Spectroscopy • Department of Computer Science • Design Change Schedule • Design Change Summary • Design Communication System • Design Control Specification • Design Criteria Specification • Desktop Colour Separation • Despin Control System • Detailed Checkout Specification • Detector Controller Simulator • Detector Control System • Development of Communications Systems • Differential Cross Section • Digital Camera System • Digital Cartography Section • Digital Cellular System • Digital Color Separate • Digital Command System • Digital Communications Services • Digital Communication System • Digital Control Signal • Digital Control Station • Digital Control System • Digital Cross-connect System • Direct-Current Sensor • Dispatch Critical System • Display and Control System • Distributed Computer System • Distributed Computing System • Distributed Control System • Diversified Computer Systems Inc. (USA-CO) • Doctor of Commercial Science • Document Control Sheet • Document Control Software • Document Control System • Doncaster Finningley (UK-England) (location identifier) • Double Compton Scattering • Dual Checkout Station

DCSA: Direct-Current Servo Amplifier

DCSC: Digital Card Slave Clock

DCSI: Data and Control Signal Interface

DCSP: Digital Control Signal Processor

DCST: Data Collection System Tape • Data Collection System Transponder • Digital Card Schmitt Trigger • Direction de la Coopération Scientifique et Technique

DCSTS: Data Collection Storage and Transmission System

DCSU: Differential Corrected Spectral Unit • Digital Computer Switching Unit

Dct: Direct • Document

DCT: Database Configuration Tool • Data Communications Terminal • Data Congestion Threshold • Data Conversion Transmitter • Decimal Code Translator • Dedicated Carbon Tax • Destination Control Table • Dictionary • Digital Carrier Trunk • Direct Cosine Transform • Discipline Coordination Team • Discrete Cosine Transform • Dissector Camera Tube • Distributed Computer Telephony • Divisão de Ciências de Terra • Document Control Ticket • DSS Communications Terminal • Duncan Town (Bahamas) (location identifier) • Dynamic Circuit Timing

DCTI: Direction Centrale du Traitement de l'Information

DCTL: Direct-Coupled Transistor Logic

DCTM: Direct-Current Torque Motor

DCTN: Defense Commercial Communications Network

DCTS: Digital Coordinate Transformation System • Dimension Custom Telephone Service

DCTV: Digital Color Television

DCU: Data-Cache Unit • Data Collection Unit • Data Communications Unit • Data Command Unit • Data Control Unit • Date Conversion Utility • Decade Counting Unit • Decatur Pyor (USA-AL) (location identifier) • Decimal Counting Unit • Detection and Control Unit • Detector Control Unit • Digital Coefficient Unit • Digital Computer Unit • Digital Control Unit • Digital Counting Unit • Display and Control Unit • Distribution Control Unit • Drum Control Unit • Dublin City University (Ireland)

DCUPS: Densely-Centered Uniform-P Search

DCUTL: Direct-Coupled Unipolar Transistor Logic

DCV: Demain Chez Vous • Direct-Current Volt • Directional Control Valve • Document Cut-off Value

DCVR: Direct-Current Voltage Reference • Direct-Current Voltage Regulator

DCVZ: Denver Convergence Vorticity Zone

DCW: Digital Chart of the World

DCWS: Debris Collision Warning System

DCWV: Direct-Current Working Volt

DC-X: Delta Clipper Experimental

DCX: Delta Clipper Experimental • Double Charge Exchange

DC-XA: Delta Clipper Experimental Advanced

DCXA: Delta Clipper Experimental Advanced

DCXRD: Double Crystal X-Ray Diffraction

dd: De Dato • Días • Dono Dedit

d/d: Delivered

D/d: Days after Date • Delivered

Dd: Dated • Dedicated • Delivered

D/D: Days after Date

D&D: Decontamination & Decommissioning • Design & Development • Drug & Disease free

DD: Damage Done • Data Definition • Data Depository • Data Description • Data Descriptor • Data Dictionary • Data Discrepancy • Data Display • Days • Day's Date • Days after Date • Days after Delivery • Deadline Date • Decimal Divide • Decoder Driver • De Dato • Dedicated Display • Deep Down (*) • Deep-Drawn • Degree-Day • Delayed Delivery • Delete Description • Demand Draft • Departure Date • Deputy Director • Deputy Directorate • Description Description • Design and Development • Design Deviation • Destructive Dilemna • Detailed Design • Dewey Decimal • Digital Data • Digital Display • Directives Documentation • Director's Discretionary • Disc Diameter • Disconnecting Device • Dishonorable Discharge • Disk Drive • Doctor of Divinity • Document Delivery • Document Director • Domain Decomposition • Donnée de Définition • Dossier de Définition • Double Deck • Double Degenerate • Double Demy • Double Density • Dry Dock • Due Date • Dundee (UK postal code)

DDA: Data Descriptive Area • Data Distribution Assembly • Dedicated Digital Access • Demand Deposit Application • Development Display Assembly • Digital Differential Analyzer • Digital Directory Assistance • Discrete Dipole Approximation • Division on Dynamical Astronomy • Domain-Defined Attribute • Drawing Departure Authorization

DDAE: Dossier de Demande d'Autorisation d'Exploiter (France)

DDAS: Digital Data Acquisition System • Digital Data Archives System

DDB: Data Display Board • Data Display Buffer • Design Data Block • Device-Dependent Bitmap • Digital Data Buffer • Distributed Database • Double Declining Balance • Dynamic Database

DDBMS: Distributed Database Management System

DDBS: Descriptor Database System

DD&C: Data Dictionary & Catalogue

DDC: Data Display Central • Data Dissemination Center • Data Distribution Center • Data Documentation Cost • Defense Documentation Center • Dewey Decimal Classification • Diffraction Dilution Coefficient • Digital Data Cell • Digital Data Channel • Digital Data Control • Digital Data Conversion • Digital Data Converter • Digital Display Converter • Direct Data Channel • Direct Department Calling • Direct Digital Control • Direction du Développement et de la Coopération (France) • Discipline Data Center • Display Data Channel • Disruptive Discharge Cleaning • District Data Center • District Development Community • Dodge City (USA-KS) (location identifier) • Double-Doped Crystal • Dual Diversity Comparator

DDCE: Digital Data Conversion Equipment

DDCM: Differential Decay Curve Method

DDCMP: Digital Data Communications Message Protocol

DDCP: Draft Development Concept Paper

DD&CS: Dedicated Display & Control System

DDCS: Dedicated Display and Control System • Digital Database Connection Service • Digital Data Calibration System • Direct Digital Control System • Distributed Data Connection Services • Double Differential Cross Section

DDCU: DC-to-DC Converter Unit

DDD: Data Definition Document • Dat, Dicat, Dedicat • Date Deficiency Discovered • Date Discrepancy Discovered • Deadline Delivery Date • Désétatisation-Défiscalisation-Dérégulation • Design Definition Document • Desired Delivery Date • Detailed Data Display • Detailed Design Document • Deutscher Depeschen-Dienst (Germany) • Digital Daemon Disciple • Digital Data Distributor • Digital Design Document • Digital Diagnostic Diskette • Digital Display Driver • Direct Distance Dial • Direct Distance Dialing • Display Decoder Drive • Dono Dedit Dedicavit • Drug Delivery Device • Duplexed Display Distributor

DDDA: Decimal Digital Differential Analyzer

DDDL: Digital Data Downlink

DDDS: Digital Data Display System • Digital Data Distribution System

DDDU: Digital Decoder Driver Unit

DDE: Data Dictionary Entry • Direct Data Entry • Direct Digital Encoder • Direction Départementale de l'Équipement (*) • Division Départementale de l'Équipement • Dwight David Eisenhower (1890-1969) • Dynamic Data Exchange

DDEML: Dynamic Data Exchange Management Library • Dynamic Data Exchange Manager Library

DDEN: Délégué Départemental de l'Éducation Nationale

DDES: Data Display Expert System • Digital Data Editing System • Direct Data Entry Station

DDF: Data Description File • Data Descriptive File • Data Distribution Facility • Descriptive Data File • Dielectric Dissipation Factor • Director's Discretionary Fund • Dynamic Data File • Dynamic Data Format • Dynamic Data Formatting

DDFT: Design-Development-Fabrication-Testing

DDG: Dandong (China-PRC) (location identifier) • Data Display Generator • Deputy Director-General • Digital Data Group • Digital Display Generator

DDGE: Digital Display Generator Element

DDH: Digital Display Handling

DDHA: Digital Data Handling Assembly

DDHS: Digital Data Handling System

DDHU: Digital Data Handling Unit

DDI: Data Display Indicator • Daydream Island (Australia-QLD) (location identifier) • Dedicated Display Indicator • Device Driver Interface • Digital Display Indicator • Diplôme de Docteur Ingénieur • Direct Dial In • Direct Dial Inwards • Directed Drawing Instrument • Discrete Data Input • Discrete Digital Input • Dynamic Data Interchange

DDIF: Digital Document Interchange Format

d.d. in d.: De Die in Diem

DDIP: Division de Documentation et d'Information Pédagogique

DDIS: Data Depository Index System • Depository Distribution Information System • Document Data Indexing Set

DDJ: Digital Differencing Junction

DDK: Device Development Kit • Device Driver Kit • Disk Driver Kit • Driver Development Kit

DDL: Database Description Language • Data Definition Language • Data Description Language • Data Descriptive Language • Data Distribution Laboratory • Data Distribution List • Data Downlink • Data Drawing List • Delay-Locked Loop • Dialogue Description Language • Differential Delay Line • Differential Distribution Law • Digital Data Link • Digital Data Logger • Digital Design Language • Diode-Diode Logic • Direct Downlink • Dispersive Delay Line • Display Description Language • Distributed Design Language • Documentation Distribution List • Document Description Language

DDLT: Diagnostic Decision Logic Table

DDM: Data Display Module • Data Display Monitoring • Decaying Dark Matter • Délégué Départemental de Météo France • Design Decision Memo • Desktop Document Manager • Difference in Depth of Modulation • Digital Display Machine • Digital Display Makeup • Discrete Data Management • Distributed Data Management • Distributed Data Manager • Distributed Design Methodology • Dodoima (Papua New Guinea) (location identifier) • DOME Data Manager

DDMA: Disk Direct Memory Access

DDMan: Data Display Module Alphanumeric

DDMg: Data Display Module Graphical

DDMP: Domain Decomposition Message Passing

DDMS: Deputy Daily Mission Specialist • Digital Data Measuring System • Digital Desertification Monitoring System

DDMWG: Data Distribution And Marketing Working Group

DDM3d: Data Display Module 3D Graphical

DDN: Data Dissemination Network • Defense Data Network • Delta Downs (Australia-QLD) (location identifier) • Digital Data Network • Documentation Development Notification

DDNA: Divided Difference Noise Analysis

DDNC: Direct Digital Numerical Controller

DDNMC: Data Dissemination Network Management Centre

DDNS: DEC Distributed Name Service • Dynamic Domain Naming Service

DDO: Data-Driven Object • David Dunlap Observatory (Canada-ON) • Director's Discretionary Observation • Discrete Data Output • Discrete Digital Output • Dynamic Data Object

DDOCE: Digital Data Output Conversion Equipment

DDOP: Differential Dilution Of Precision

DDoS: Distributed Denial of Service

DDP: Data Distribution Panel • Data Distribution Plan • Datagram Delivery Protocol • Deep Discount Price • Design Data Package • Design Development Plan • Digital Data Processing • Digital Data Processor • Director's Discretionary Program • Distributed Data Processing • Dorado (Puerto Rico) (location identifier)

DDPC: Digital Data Processing Center • Digital Data Processing Complex

DDPE: Digital Data Processing Equipment

DDPF: Dedicated Display Processing Function

DDPO: Defense Dissemination Program Office

DDPS: Data-Directed Programming System • Digital Data Processing System • Discrimination Data Processing System

DD&R: Data Display & Review

DDR: Daily Demand Rate • Data Definition Record • Data Description Record • Data Descriptive Record • Data Direction Register • Data Discrepancy Report • Data Display and Review • Data Display Room • Design Development Record • Detail Design Review • Deutsche Demokratische Republik (now merged into BRD) • Development Discrepancy Report • Digital Data Receiver • Dynamic Device Reconfiguration

DDR&E: Deputy Director for Research & Engineering

DDRG: Double-Double Radio Galaxy

DDRH: Digital Data Recording Head

DDRI: Design Drafting Reference Information

DDRP: Direct/Delayed Response Project

DD&RR: Data Display & Review Room

DDRS: Digital Data Recording System

DDRT: Data Distribution Review Team

Dds: Dedans

DDS: Data Design Specification • Data Dialog System • Data Display Set • Data Display System • Data Disposition System • Data Dissemination System • Data Distribution System • Deputy Directorate of Science • Descent Design System • Design Data Sheet • Detailed Design Specification • Development Data Sheet • Dewey Decimal System • Digital Dataphone Service • Digital Data Service • Digital Data Servo • Digital Data Storage • Digital Data System • Digital Display Scope • Digital Drafting System • Direct Digital Synthesis • Discipline Data System • Division of Digital Systems • Doctor of Dental Science • Documentation Distribution System • Document Delivery Service • Drug Delivery System • Dust Detection System • Dust Detector System

DDSM: Digital Data Switching Matrix

DDSOT: Digital Daily System Operability Test

DDSR: Deleted Data Space Release

DDST: Direction Départementale des Services Techniques (France)

DDSU: Digital Data Storage Unit

DD&T: Design, Development & Test

DDT: Data Description Table • Debye Dipole Theory • Deflagration-Detonation-Transition • Department of Development and Tourism • Design, Development and Test • Diagnostic Decision Table • Dichloro-Diphenyl-Trichlorethane • Digital Data Terminal • Digital Data Transmitter • Digital Debugging Tape • Digital Demodulation Technique • Director's Discretionary Time • Domain Definition Table • Don't Do That • Doppler Data Translator • Double Deflection Tube • Dynamic Debugging Technique • Dynamic Display Tester

DDTC: Director's Discretionary Time Committee

DDT&E: Design, Development, Testing & Engineering • Design, Development, Testing & Evaluation

DDTE: Design-Development-Test-Engineering • Design-Development-Test-Evaluation • Digital Data Terminal Equipment • Direction Départementale du Travail et de l'Emploi (France)

DDTEFP: Direction Départementale du Travail, de l'Emploi et de la Formation Professionnelle (France)

DDTESM: Digital Data Terminal Equipment Service Module

DDTF: Docking Dynamics Test Facility • Dynamic Docking Test Facility

DDTP: Director's Discretionary Time Proposal

DDTS: Digital Data Transmission System • Direct Dial Telephone System • Dynamic Docking Test System

DDU: Dadu (Pakistan) (location identifier) • Data Display Unit • Decommutator Distribution Unit • Dedicated Display Unit • Diagnostic Display Unit • Digital Display Unit • Digital Distributing Unit • Disc Data Unit • Display and Debug Unit • Display Driver Unit

DDV: Dark Dense Vegetation • Datendirektverbindung

DDWG: Digital Data Working Group

DDX: Digital Data Exchange • Distributed Data Exchange • Double Differential Cross-Section

de: German (ISO language code) • Germany

dE: Diffuse Elliptical • Dwarf Elliptical

DE: Allemagne (ISO country code) • Dassault Électronique • Data Element • Data Encoder • Data Extraction • Decay • December • Décembre • Decision Element • Decision Error • Deflection Error • Delaware (USA) • Denmark • Derby (UK postal code) • Descent Engine • Design Engineering • Design Evaluation • Detection Efficiency • Detector Efficiency • Développement Exploratoire • Diesel Engine • Digital Electronics • Digital Encoder • Diplôme d'État • Direct Electrolysis • Direct Encounter • Direct Excitation • Directory Enquiry • Directory Entry • Discard Eligibility • Discard Eligible • Disconnection Event • Display Electronics • Distant Element • Doctor of Engineering • Dokumentationseinheit • Doppler Extractor • DOS Extender • Double Elephant • Double End • Double Entry • Double-Excited • Draw End • Driver Education • Dynamics Explorer • Germany (ISO country code)

Dea: Dean

DEA: Data Encryption Algorithm • Data Exchange Agreement • Data Exchange Annex • Defense Exchange Agreement • Délégation aux Études et aux Audits • Deployed Electronics Assembly • Detector Electronics Assembly • Dielectric Analysis • Diffusione Edizioni Anglo-americane (Italy) • Digital Electronics Assembly • Diplôme d'Études Approfondies (France) • Directory-Enabled Application • Display Electronics Assembly • Divisão de Electrônica Aeroespacial • Division des Études Avancées • Drug Enforcement Agency (USA) • Dynamo Electric Amplifier

DEACT: Deactivate

DEAG: Data Extraction and Analysis Group

DEAL: Data Entry Application Language • Decision Evaluation and Logic

DEAR: Department of Energy Acquisition Regulation

DEB: Data Extent Block • Debrecen (Hungary) (location identifier) • Déclaration d'Échange de Biens • Detached Eclipsing Binary • Detector Electronics Box • Digital Electronics Box

DEBIS: Distributed Expert-Based Information System

DEBITS: Deposition of Biogeochemically Important Trace Species

DEBS: Digital Electron Beam Scanner

DEBRA: Diffuse Extragalactic Background Radiation

dec: Decena • Decimeter • Declination

Dec: Deceased • Deceleration • December • Decimal • Decimeter • Decision • Declaration • Declare • Declared • Declination • Decode • Decoded • Decoder • Decoding • Decompose • Decomposition • Decrease • Decreased • Decreasing • Decrement • Decrescendo

Déc: Décembre

DEC: Decatur (USA-IL) (location identifier) • December • Décembre • Declination • Decoder • Decrement • Department of Environmental Conservation • Device Clear • Digital Equipment Corporation • Digital Evaluation Computer • Division des Examens et Concours

DECA: Descent Engine Control Assembly

Decaf: Decaffeinated

DECAL: DEC Author Language

DECB: Data Event Control Block

DECC: Digital European Cordless Communications

Decd: Deceased

DECDEE: Designing Centers for Distance Education in Estonia

DECIMAL: Decision Making in Libraries

DECIMALS: Dynamical, Electrodynamical and Chemical Interactions near the Mesopause at Arctic Latitudes during Summer

dECL: Differential Emitter-Coupled Logic

DECL: Deliverable Equipment Checklist • Direct Energy Conversion Laboratory

Decn: Decision

DECNET: DEC Network

DECNET E-net: DEC Engineering Network

DECOM: Decommutate • Decommutator

DECOMATE: Delivery of Copyright Materials to End-users

Decomp: Decompose • Decomposed • Decomposition

DECOMP: Decompiler

Decompn: Decompression

Decontn: Decontamination

Decr: Decrease

DECR: Decrease

DECS: Department of Education, Culture and Sports

DECSAI: Departamento de Ciencias de la Computación e Inteligencia Artificial (Spain)

DECT: Digital Enhanced Cordless Telecommunications • Digital European Cordless Telecommunications (*)

DECU: Data Exchange Control Unit • Development and Educational Communication Unit

DECUS: Digital Equipment Computer Users' Society

Ded: Dedicate • Dedicated • Deduce • Deduct • Deduction

DEd: Doctor of Education

DED: Dark-Emitting Diode • Data Element Dictionary • Data Element Directory • Dedicate • Dedicated • Dehra Dun (India) (location identifier)

DEDA: Data Entry and Display Assembly

DEDB: Data Entry Database

de d. in d.: De Die in Diem

DEDL: Data Element Description List

DEDP: Data Entry and Display Channel • Document d'Examen des Droits à la Pension • Dossier d'Examen des Droits à la Pension

DEDS: Data Entry and Display System • Digital Error Detection System

DEDSL: Data Entity Dictionary Specification Language

DEE: Dexterous End Effector • Digital Evaluation Equipment • Digital Events Evaluator • Diploma in Electrical Engineering

DEEC: Digital Electronic Engine Control

DEEP: Deep Extragalactic Evolutionary Probe • Directly Elected European Parliament

DEES: Dynamic Electromagnetic Environment Simulator

DEET: Department of Employment, Education and Training

DEETYA: Department of Employment, Education, Training and Youth of Australia

Def: Default • Defect • Defectif • Defective • Defectivo • Defense • Deferred • Deficiency • Deficient • Deficit • Define • Defined • Definite • Definition • Definitive • Deflate • Deflation • Deflect • Deflected • Deflection • Defunct • Defunctus

Déf: Définition

DEF: Default • Definition • Desktop Functional Equivalent • Development and Evaluation Facility

DEFA: Digital Event Filter and Address

DEFCAL: Deflection Calibration

DefEl: Defense Electronics (periodical code)

Defek: Defektoskopiia (periodical code)

DEFI: Documentation Économique et Financière Informatisée

Defl: Deflate • Deflation • Deflect • Deflected • Deflection • Deflector

DEFL: Deflect

DEFM: Demographic and Economic Forecasting Model

DEFRAG: Defragment

DEFT: Direct Electronic Fourier Transform • Discrete Electronic Fourier Transform • Dynamic Error-Free Transmission

Deg: Degré • Degree

DEG: Degenerate Electron Gas • Degree • Directorate of Environmental Geology • Dynamique des Espaces Géographiques

DEGA: Deutsche Gesellschaft für Akustik (Germany)

de g. à dr.: De Gauche à Droite

DEH: Decorah (USA-IA) (location identifier)

DeHyZ: Deutsche Hydrographische Zeitschrift (periodical code)

DEI: Deliverable End Item • Denis Island (Seychelles) (location identifier) • Descent Imager • Design, Engineering, and Integration • Design Engineering Identification • Design Engineering Inspection • Dispositif d'Étalonnage Interne • Division des Études et de l'Information • Dutch East Indies

DEIC: Differential Equations in Industry and Commerce

DEIMOS: Deep Extragalactic Imaging Multiobject Spectrograph

DEIS: Design Engineering Inspection Simulation • Design Engineering Inspection Simulator • Design Evaluation and Inspection Simulator • Draft Environmental Impact Statement

Dek: Deken

DEK: Dansk Elektroteknisk Komite (Denmark) • Data Encryption Key • Donald E. Knuth

del: Delineator • Delineavit

Del: Delaware (USA) • Delegación • Delegate • Delegated • Delegation • Delete • Deleted • Deletion • Delhi (India) • Delineation • Deliver • Delivered • Delivery • Delphinus-Delphini (constellation)

DEL: Data Entry Language • Delete • Delhi - Indira Gandhi (India) (location identifier) • Deliver • Delivery • Deorbit-Entry-Landing • Diode Électroluminescente

Deld: Delivered

Dele: Delete

Deleg: Delegación • Delegate • Delegation

DELHP: Division on Education, Labor, and Human Performance

DELICAT: Data Enhancement of Library Catalogues

delin: Delineavit

DELM: Department of the Environment and Land Management

DELO: Diagonal Element Lambda Operator

DELOG: Design Engineering Logistics

DELP: Department of the Environment, Land and Planning

DELPHI: Detector with Lepton Photon and Hadron Identification

DELSTR: Delete String

Delt: Delineator

DELT: Dynamic Environmental Laboratory Test

DELTA: Description Language for Taxonomy • Detailed Labor and Time Analysis • Determination Effective Levels of Task Automation • Development of European Learning through Technological Advance

Delv: Deliver • Delivered

Dely: Delivery

Dem: Demand • Demanded • Democrat • Democratic • Demodulate • Demodulated • Demodulator • Demonstrate • Demonstrated • Demonstration • Demonstrative

DEM: Dembidollo (Ethiopia) (location identifier) • Demonstration • Department of Environmental Management • Differential Emission Measure • Digital Electronics Module • Digital Elevation Map • Digital Elevation Mapping • Digital Elevation Matrix • Digital Elevation Model • Dynamic Enterprise Modeller • Mark (Germany) (ISO currency code)

DEMA: Direct E-mail Marketing Association

DEMAT: Department of Educational Media and Technology

DEMIRM: Département de Matière Interstellaire et Radioastronomie Millimétrique (Observatoire de Paris-Meudon)

DEMKO: Danmarks Elektriske Materielkontrol (Denmark)

DEMM: Division of Educational Media Management

Demo: Demonstration

Demod: Demodulator

Demon: Demonstrate • Demonstrative

DEMON: Diminishing Error Method of Optimization for Networks

DEMR: Department of Energy, Mines, and Resources (Canada)

DEMS: Digital Error Monitoring System

DEMUX: Demultiplexer

Den: Denmark • Denomination • Denotation • Denote • Denoted • Density • Denver (USA-CO)

DEN: Data Element Number • Denver Stapleton (USA-CO) (location identifier) • Directory-Enabled Network • Document Enabled Networking

DEng: Doctor in Engineering

DEngS: Doctor in Engineering Science

DENI: Department of Education of Northern Ireland

DENIS: Deep Near-Infrared Survey

DENR: Department of Environment and Natural Resources (Philippines)

Dens: Density

Dent: Dental • Dentist • Dentista • Dentiste • Dentistry

DEO: Dearborn - Hyatt Regency (USA-MI) (location identifier)

DEOF: Double End Of File

D/EOM: Directorate of Earth Observation and Microgravity • Director of Earth Observation and Microgravity

DEORB: Deorbit

DEOS: Deeply Embedded Outburst Star • Delft Institute for Earth-Oriented Space Research (Netherlands) • Digital Engine Operating System

DEOSS: Demonstration of Electro-Optical Surveillance Sensor

Dep: Depart • Departs • Department • Departure • Dependency • Dependent • Deponent • Deportes • Depose • Deposed • Deposit • Depósito • Depot • Deputy

Dép: Départ • Département • Député

DEP: Data Entry Panel • Data Exchange Program • Dedicated Experiment Processor • Deep • Deflection Error Probable • Delft Electronic Products • Deparizo (India)

(location identifier) • Department • Département de l'Économie Publique • Department of Environmental Protection • Department of Environment and Planning • Departure • Depth • Descanse En Paz • Design Engineering Program • Direction de l'Évaluation et de la Prospective • Display Electronics Package • Division of Educational Programs

DEPA: Danish Environmental Protection Agency • Defense Evaluation and Research Agency

DEPART: Departamento

DEPI: Differential Equations Pseudocode Interpreter

Depl: Deplete • Depleted • Depletion

DEPL: Deploy

DepLIS: Department of Library and Information Services

DEPOS: Depósito

Depr: Depreciation • Depression

DEPRESS: Depressurize

DEPS: Dernier Entré - Premier Sorti • Differential Empirical Population Synthesis

DEPSK: Differential Encoding Phase Shift Keying

Dept: Department • Deponent • Deputy

Dépt: Département

DEPT: Department

Depto: Departamento

Depts: Departments

DEPULP: Département d'Éducation Permanente de l'Université Louis Pasteur (France)

DEQ: Department of Environmental Quality

Dér: Dérivatif • Dérivé

Der: Derecha • Derecho • Derivado • Derivative • Derive • Derived

DER: Data Entity Relationship • Declining Error Rate • Defective Equipment Review • Derim (Papua New Guinea) (location identifier) • Designated Engineering Representation • Designated Engineering Representative • Design Error Report • Direction des Études et des Recherches • Document Error Report • Drawing Error Report

DERA: Defence Evaluation and Research Agency (UK)

DERAD: Département de Radioastronomie Décimétrique (Observatoire de Paris-Meudon)

DERAT: Département d'Études et de Recherches en Aérothermodynamique

Derby: Derbyshire (UK-England)

DERF: Dynamical Extended Range Forecasting

DERI: Départment d'Études et de Recherche en Informatique

Deriv: Derivation • Derivative • Derive • Derived

DERL: Derived Emergency Reference Level

DERM: Digital Elevation and Radiometry Model

DERO: Département d'Études et de Recherches en Optique (ONERA)

DERP: Defective Equipment Replacement Program

DERRA: Department of Extragalactic Research and Relativistic Astrophysics

DERSI: Développement des Relations Spatiales Industrielles

DERV: Diesel-Engine Road Vehicle

Des: Desert • Desierto • Design • Designate • Designated • Designation • Designer

DES: Data Encryption Standard (*) • Data Encryption System • Data Entry Sheet • Data Exchange System • Data Extension Segment • Data Extraction Segment • Delhi Earth Station (India) • Department for Education and Science • Department of Education and Science • Descend • Descent • Description • Design Expansion System • Desroches (Seychelles) (location identifier) • Dessinateur • Deutsches Elektronen Synchrotron • Development Enquiry Service • Differential Energy Spectrum • Differential Equation Solver • Digital Editing Station • Digital Expansion System • Diplôme d'Études Spécialisées • Diplôme d'Études Supérieures • Discrete Event System • Documentation Économique et Sociale

DESA: Defense Evaluation Support Agency

Desap: Desaparecida

DESAT: Desaturated • Desaturation

Desc: Descarga • Descend • Descendant • Descenso • Descent • Describe • Described • Description

DESC: Defense Electronics Supply Center • Descent • Display ESR SAR Coverage

DeScE: Defense Science Electronics (periodical code)

Descr: Describe • Description

Descto: Descuento

DESCW: Display ERS Swath Coverage for Windows

DESD: Department for Economics and Social Development

Desgl: Desgleichen

DESHE: Distributed Expert System Hosting Environment

Desid: Desideratum

Desig: Designate

DESIG: Designate

DESIRE: Design Information Recovery Environment

Desp: Dispatch • Dispatched

DESPA: Département de Recherche Spatiale (Observatoire de Paris-Meudon)

Desp: Despectivo

DESPOT: Design Performance Optimization

DèsS: Docteur ès Sciences

DESS: Diplôme d'Enseignement Supérieur Spécialisé • Diplôme d'Études Supérieures Spécialisées (France) (*)

DESSA: Distressed and Expatriate Scientists and Scholars from Africa

DèsSc: Docteur ès Sciences

Dest: Destroyed

DEST: Destination • Diplôme d'Études Supérieures Techniques • Direction de l'Enseignement Supérieur des Télécommunications (France)

DESTEC: Département des Sciences et Techniques de l'-Expression et de la Communication

Destn: Destination

DESTU: Diplôme d'Études Scientifiques et Techniques Universitaires

Desus: Desusado

DESY: Deutsches Elektronen-Synchrotron (Germany)

Det: Detach • Detached • Detachment • Detail • Detailed • Detect • Detected • Detection • Detector • Determination • Determine • Determined • Determiner

Dét: Détaché • Détachement • Détacher • Détail • Détaillé • Détailler • Détecté • Détecter • Détection • Détermination • Déterminé • Déterminer

DET: Data Entry Terminal • Data Evaluation Terminal • Decision Error Tradeoff • Déclaration d'Entrée sur le Territoire • Density Equilibration Timescale • Design Evaluation Test • Detail • Detector • Detent • Detroit City (USA-MI) (location identifier) • Device Execute Trigger • Digital Event Timer • Direct Energy Transfer • Division of Educational Technology

DETA: Dielectric Thermal Analysis • Diethylene Triamine

DETEC: Defensive Technology Evaluation Code

Deter: Determinado

DETIR: Defensive Technology Information Repository

Detn: Determination

DETN: Detection

DETR: Department of the Environment, Transport and the Regions (UK) • Detector

DETS: Digital Element Test Set

DETYA: Department of Education, Training and Youth Affairs (Australia)

DEU: Allemagne (ISO country code) • Data Encoder Unit • Data Entry Unit • Data Exchange Unit • Digital Electronics Unit • Digital Evaluation Unit • Display Electronics Unit • Germany (ISO country code)

DEUG: Diplôme d'Enseignement Universitaire Général • Diplôme d'Études Universitaires Générales (France)

DEUP: Diplôme d'Études Universitaires Professionnalisées (France)

DEUST: Diplôme d'Études Universitaires en Sciences et Techniques (France) • Diplôme d'Études Universitaires Scientifiques et Techniques

DEUT: Data Encoder Unit Transmitter

Dev: Develop • Developed • Developer • Developing • Development • Developmental • Deviate • Deviated • Deviation • Device • Devon (UK-England) • Devonshire (UK-England)

Dév: Développé • Développer • Développement

DEV: Develop • Development • Deviation • Device

DEVCO: Development Committee

Devel: Development

DEVN: Deviation

Devon: Devonshire (UK-England)

Devp: Develop

Devpt: Development

DEVS: Driver's Enhanced Vision System

DEVSIS: Development Science Information System

DEW: Directed Energy Weapon • Distant Early Warning

DEWIZ: Distant Early Warning Identification Zone

dex: Power of Ten

DEX: Data Exchange

Dez: Dezember • Dezembro

DEZ: Deirezzor (Syria) (location identifier) • Deutsches Energiezentrum (Germany) • Dezembro

d/f: Días Fecha

D/F: Direction Finder • Direction Finding

DF: Damage Free • Data Facility • Data Field • Data Flow • Data Fusion • Dead Freight • Dean of Faculty • Decimal Fraction • Decontamination Factor • Degree of Freedom • Density Functional • Derechos Fundamentales • Describing Function • Destination Field • Detection Fraction • Development Flight • Development Forward • Device Flag • Dielektrische Festkörper • Dirac-Fock • Direct Flight • Direct Flow • Direction Finder • Direction Finding • Disappearing Filament • Disassembly Facility • Discrimination Filter • Disk File • Display Function • Dissipation Factor • Distribution Factor • Distribution Function • Distrito Federal • Djibouti Franc (currency) • Don't Fragment • Double Feeder • Double Flag • Double Focussing • Double Foolscap • Draft • Drawing File • Droits Fondamentaux • Duty Factor • Duty Free

DFA: Department of Foreign Affairs • Describing Function Analyzer • Design For Automation • Deterministic Finite Automaton • Detrended Fluctuation Analysis • Development Fund for Africa • Digital Fault Analysis • Digital Frequency Analysis • Direction Finding Antenna • Discriminant Function Analysis • Dominant Feature Analysis

DFACS: Drag-Free and Attitude Control System

DFAD: Digital Feature Analysis Data

DFAE: Département Fédéral des Affaires Étrangères (Switzerland)

DFAIT: Department of Foreign Affairs and International Trade

DFAT: Department of Foreign Affairs and Trade

DFAX: Digital Facsimile

DFB: Distributed Feedback (*) • Distribution Feedback • Driftwood Bay (USA-AK) (location identifier)

DFBLD: Distribution Feedback Laser Diode

DFBSC-LD: Dispersion Feedback Semiconductor Laser Device

DFBSCLD: Dispersion Feedback Semiconductor Laser Device

DFC: Data Feed Client • Data Field Channel • Data Filter Controller • Data Flow Control • Data Format Converter • Data Format Converter • Département des Finances et de la Comptabilité • Disk File Control • Division Financière et Comptable • Dossier de Fabrication et de Contrôle • Dust-Free Chamber

DFCG: Directeur Financier et de Contrôle de Gestion

DFCS: Digital Flight Control Software • Digital Flight Control System (*)

DFD: Data Flow Diagram • Data Function Diagram • Deutsches Fernerkundungsdatenzentrum (Germany) • Digital Flight Display • Digital Frequency Display • Direct Fringe Detection • Displaced Frame Difference • Division of Fluid Dynamics

DFDC: Disk File Descriptor Control

DFDL: Dual-Frequency Doppler Lidar • Dual-Frequency Dye Laser

DFDR: Digital Flight Data Recorder

DFDSS: Data Facility Dataset Services

DFDT: Dynamic Fault Diagnosis Technique

DfE: Department for Education

DFE: Dark Frame Error • Data Flow Editor • Data Flow Engineer • Debye-Falkenhagen Effect • Decision Feedback Equalizer • Department For Education • Designated Federal Entity • Direction-Finding Equipment • Distributed Framework Executive

DfEE: Department for Education and Employment (UK)

DFG: Data Flow Graph • Deutsche Forschungsgemeinschaft (Germany) • Digital Function Generator • Diode Function Generator • Discrete Frequency Generator • Display Format Generator • Distinguished Foreign Guest

DFGI: Deutsches Geodätisches Forschungsinstitut (Germany)

DFGII: Deutsches Geodätisches Forschungsinstitut - Abteilung I (Germany)

DFH: Dong Fang Hong

DFHSM: Data Facility Hierarchical Storage Manager

DFI: Dark Field Illumination • Defiance (USA-OH) (location identifier) • Developmental Flight Instrument • Developmental Flight Instrumentation • Development Flight Instrumentation • Development Flight Integration • Digital Facility Interface • Direct Foreign Investment • Direction des Finances • Direction Finding Indicator • Double Fourier Interferometry

DFIB: Data Function Information Block

DFI PLT: Development Flight Instrumentation Pallet

DFIR: Data Flow Instrument Responsible

DFIS: Digital Facsimile Interface System • Dual Filament Ion Source

Dfl: Deflating

DFL: David Florida Laboratory (Canada) • Decommutator Format Load • Deterministic Fragment Loss • Display Formatting Language

DFLD: Distribution-Free Logic Design

DFM: Data Filter Manager • Data Flow Management • Data Flow Manager • Design For Manufacturability • Design For Manufacture • Design For Manufacturing • Digital Frequency Meter • Dual-Frequency Method

DFMA: Design For Manufacturing and Assembly

DFMR: Dual Frequency Microwave Radar

DFMS: Digital Facility Management System

DFN: Deutsches Forschungsnetz (Germany)

DFNT: Dipartimento di Fisica Nucleare e Teorica

DFO: Data Flow Operations • Department of Fisheries and Oceans • Disk File Optimizer • Dynamic Feedback Optimization • Dynamics Functional Objective

DFOS: Danish Faint Object Spectrograph

DFP: Data Facility Product • Davidon-Fletcher-Powell • Diode Flat Pack • Dipole Flat Plate • Drumduff (Australia-QLD) (location identifier) • Dry Film Processor • Dry Filter Processing • Dynamic Flow Parameter

DFPL: Data Flow Programming Language

Dfr: Defrost

DFR: Decreasing Failure Rate • Defrost • Document Filing and Retrieval • Drag-Free Reference • Dual-Frequency Receiver • Dust-Free Room

DFRC: Dryden Flight Research Center (USA-CA) (formerly DFRF)

DFRF: Dryden Flight Research Facility (NASA) (now DFRC)

DFRNL: Differential

DFS: Daily Flow System • Dansk Fysisk Selskab (Danemark) • Data Flow System • Decision Fusion Simulator • Demodulation and Frame Synchronizer • Deutscher Fernmelde Satellit • Digital Field System • Digital Frequency Synthetizer • Direct Flow Sampler • Directional Finding System • Display Formatting System • Distributed File Service • Distributed File System • Document Finding System • Donnerstag-Freitag-Samstag • Doppler Frequency Shift • Dynamic Flight Simulator

DFSMS: Data Facility Storage Management System

DFSS: Debt-for-Science Swap • Depth From Shading and Stereo

Dft: Draft • Drift

DFT: Density Functional Theory • Design Feasibility Test • Design For Test • Design For Testability • Diagnostic Function Test • Digital Filtering Technique • Discrete Fourier Transform • Distributed Function Terminal • Document File Transfer • Down Flow Tent • Dual-Flow Turbine

Dftg: Drafting

DFTNIC: Data Flow Technology Network Information Center

Dftr: Deflector

DFTS: Dispersive Fourier Transform Spectrometer • Dispersive Fourier Transform Spectroscopy

DFTUZ: Departamento de Física Teórica de la Universidad de Zaragoza (Spain)

DFÜ: Datenfernübertragung

DFU: Data File Utility

DFUS: Diffuse

DFV: Datenfernverarbeitung

DFVLR: Deutsche Forschungs- und Versuchanstalt für Luft- und Raumfahrt (Germany) (now DLR)

DFW: Dallas - Fort Worth (USA-TX) (location identifier)

DFWG: Data Flow Working Group

DFWM: Degenerate Four Wave Mixing

DFWMAC: Distributed Foundation Wireless Media Access Control

DFyT: Derecho Financiero y Tributario

dg: Decigram • Dei Gratia

D/G: Dust/Gas Ratio

DG: Data General • Data Good • Data Group • Dei Gratia • Density Gradient • Deo Gratias • Dependence Graph • Deutsche Grammophon • Differential Generator • Differential Geometry • Diode Gate • Directeur Général • Directional Gyroscope • Direction Générale • Directorate General • Director General • Direttore Generale • Direzione Generale • Displacement Gyro • Display Generator • Double Groove • Drafting Group • Dumfries (UK postal code)

DGA: Dangriga (Belize) (location identifier) • Délégation Générale pour l'Armement (France) • Demi-Grand Axe • Direcção Geral do Ambiente (Portugal) • Distributed Genetic Algorithm

DGAC: Direction Générale de l'Aviation Civile (France)

DGACM: Direction Générale de l'Aviation Civile et de la Météorologie (Central African Rep.)

DGAFP: Direction de l'Administration Générale et de la Fonction Publique (France)

DGaO: Deutsche Gesellschaft für angewandte Optik (Germany)

DGB: Danger Bay (USA-AK) (location identifier) • Disk Gap Band

DGC: Data Gathering Document • Data General Corporation • Degahbur (Ethiopia) (location identifier) • Deutsche Gesellschaft für Chronometrie (Germany) • Direct Geodetic Constraint

DGCA: Directorate General of Civil Aviation

DGCCRF: Direction Générale de la Concurrence, de la Consommation et de Répression des Fraudes (France)

DGCL: Direction Générale des Collectivités Locales (France)

DGCM: Direct Geodetic Constraint Method • Dynamic Global Vegetation Model

DGCR: Direcção Geral de Desenvolvimento Regional (Portugal)

DGD: Dynamic Gravity Detector

DGDAS: Davenport Group Draft Advisory Standard

DGDF: Digital Geospatial Data File

DGE: Davisson-Germer Experiment • Délégation Générale à l'Espace • Deutsche Gesellschaft für Ernährung (Germany) • Direcção Geral de Energia (Portugal) • Doppler Gravity Experiment • Mudgee (Australia-NSW) (location identifier)

DGED: Direction Générale des Études et de la Documentation

DGEMP: Direction Générale de l'Énergie et des Matières Premières (F rance)

DGF: Dotation Globale de Fonctionnement • Douglas Lake (Canada-BC) (location identifier)

DGfB: Deutsche Gesellschaft für Biophysik (Germany)

DGFI: Deutsches Geodätisches Forschungsinstitut (Germany)

DGfK: Deutsche Gesellschaft für Kartographie (Germany)

DGfS: Deutsche Gesellschaft für Sprachwissenschaft (Germany)

DGG: Daugo (Papua New Guinea) (location identifier) • Department of Geology and Geophysics • Deutsche Geologische Gesellschaft (Germany) • Deutsche Geophysikalische Gesellschaft (Germany) • Deutsche Glastechnische Gesellschaft (Germany) • Dynamic Gravity Generator • Dysprosium Gallium Garnet

DGGMNT: Deutsche Gesellschaft für Geschichte der Medizin, Naturwissenschaft und Technik (Germany)

DGGMR: Directorate General of Geology and Mineral Resources

DGI: Department of Geographic Information • Directional Gyroscopic Indicator • Direction Générale des Impôts

DGICYT: Dirección General de Investigación Científica y Técnica (Spain)

DGIS: Direct Graphics Interface Standard • Directoraat Generaal Internationale Samenwerking (Netherlands)

DGIWG: Digital Geographic Information Working Group

D-GIX: Distributed Global Internet Exchange

DGIX: Distributed Global Internet Exchange

DGK: Deutsche Geodätische Kommission (Germany) • Deutsche Gesellschaft für Kristallographie (Germany)

Dgl: Dergelijke • Dergleichen • Desgleichen

DGL: Diffuse Galactic Light • Doped Glass Laser • Douglas (USA-AZ) (location identifier)

DGLR: Deutsche Gesellschaft für Luft- und Raumfahrt (Germany)

DGM: Deputy General Manager • Deutsche Gesellschaft für Materialkunde (Germany) • Digital Ground Modelling • Digital Group Multiplexer • Directional Gyro Mode

DGMP: Deutsche Gesellschaft für Medizinische Physik (Germany)

DGN: Dahlgren (USA-VA) (location identifier) • Dirección General de Normas (Mexico) • Domestic Geographic Name

DGO: Directional Gyro Operation • Director General's Office • Durango (Mexico) • Durango - Guadalupe Victoria (Mexico) (location identifier)

DGON: Deutsche Gesellschaft für Ortung und Navigation (Germany)

DGOR: Deutsche Gesellschaft für Operations Research eV (Germany)

DGP: Data Generating Program • Daugavpils (Latvia) (location identifier) • Digital Graphic Product • Dirección General de la Policía • Direction Générale de la Production • Distributed Genetic Programming

DGPA: Department of General Physics and Astronomy

DGPC: Dave's Genetic Programming Code

DGPF: Deutsche Gesellschaft für Photogrammetrie und Fernerkundung

D-GPS: Differential Global Positioning System

DGPS: Differential Global Positioning System

DGPT: Direction Générale des Postes et Télécommunications

DGQ: Deutsche Gesellschaft für Qualität

Dgr: Degrease

DGR: Dargaville (New Zealand) (location identifier) • Dwarf-to-Giant Ratio

DGRB: Diffuse Gamma-Ray Background

DGRCST: Direction Générale des Relations Culturelles, Scientifiques et Techniques

DGRST: Direction Générale à la Recherche Scientifique et Technique (France) • Direction Générale de la Recherche Scientifique et Technologique

Dgs: Degaussing

DGS: Data Gathering System • Data Ground Station • Density Gradient Sendimentation • Deployable Ground Station • Deployable Ground System • Deutsche Gesellschaft für Standardisierung (Germany) • Digital Ground Station • Digital Graduate Student • Dirección General de Seguridad • Display Generation System • Distributed Graphics System • Distributed Graph Storage

DGSE: Direction Générale de la Sécurité Extérieure (France)

DGSRD: Directorate-General of Scientific Research and Development

Dgt: Digit

DGT: Deutsche Große Teleskop (Germany) • Directional Gamma-ray Telescope • Direction Générale des Télécommunications • Direction Générale Technique • Directorate General of Telecommunications • Dumaguete (Philippines) (location identifier)

Dgtl: Digital

DGU: Dedougou (Burkina Faso) (location identifier) • Digital Graphics Unit • Directional Gyro Unit • Display Generator Unit

DGV: Dark Gray Varnish

DGVM: Dynamic Global Vegetation Model

DGW: Douglas - Converse County (USA-WY) (location identifier)

DGWK: Deutsche Gesellschaft für Warenkennzeichnung (Germany)

DGZ: Desired Ground Zero

DGZfP: Deutsche Gesellschaft für Zerstörungsfreie Prüfung (Germany)

dh: das heißt • Dirham (currency)

D/H: Deuterium/Hydrogen Ratio

DH: Dahomean • Damage History • Data Handling • Dead Heat • Decision Height • Decoherent History • Default Hierarchy • De Havilland • Derechos Humanos • Design History • Direction Indicator • Dirham (currency) • Double Heterostructure • Double Hydrant • Drill Hole • Droits de l'Homme • Durham (UK postal code) • Dynamically Hot • Moroccan Dirham (currency)

DHA: Department of Humanitarian Affairs • Design Hazards Analysis • Destination Hardware Address • Dhahran (Saudi Arabia) (location identifier)

DHAE: Department of Home Affairs and Environment

Dhc: Doctor Honoris Causa

DHC: Dark-Haloed Crater • Data Handling Center

DHCN: Daily Historical Climate Network

DHCP: Dynamic Host Configuration Program • Dynamic Host Configuration Protocol (*)

DHCU: Data Handling and Control Unit

DHD: Durham Downs (Australia-QLD) (location identifier)

DHDC: Don Harrington Discovery Center (USA-TX)

DHDD: Digital High-Definition Display

DHDS: Data Handling and Display System

DHE: Data Handling Equipment • Debye-Hueckel Equipment • Double Hop Experiment

DHF: Abu Dhabi - Al Dhafra (UAE) (location identifier) • Data Handling Facility

DHG: Double Henyey-Greenstein (function)

DHHCS: Department of Health, Housing and Community Services

DHHS: Department of Health and Human Services (USA)

DHI: Deutsches Hydrographisches Institut (Germany) (now BSH) • Dhangarhi (Nepal) (location identifier) • Disk-Halo Interaction

DHL: Désastre Humanitaire pour les Localités • Dhala (Yemen) (location identifier) • Dutch Holographic Laboratories

DHM: DEVISE Hypermedia • Dharamshala Gaggal (India) (location identifier)

DHMM: Discrete Hidden Markov Model

DHMO: Dihydrogen Monoxide

Dhmy: Dehumidify

DHN: Dothan (USA-AL) (location identifier)

DHO: Damped Harmonic Oscillator

DHP: Developed Horsepower

DHPE: Data Hardware Project Engineer

DHR: Decay Heat Removal • Den Helder - De Kooy (Netherlands) (location identifier) • Diagramme Hertzsprung-Russell

DHRD: Department of Housing and Regional Development

DHRM: Dexter Hypermedia Reference Model • Dexter Hypertext Reference Model

DHS: Data Handling System • Department of Human Services • Discrete Horizon Sensor • Division of History of Science

DHSH: Department of Human Services and Health

DHSS: Data Handling Subsystem

DHT: Dalhart (USA-TX) (location identifier) • Discrete Hartley Transform

DHTE: Dynamic-Mapping Hypertext Engine

dHTML: Dynamic Hypertext Markup Language

DHU: Detector Head Unit

DHV: Deutscher Hochschulverband (Germany)

DHW: Dengle-Hart-Weinberg • Domestic Hot Water • Double-Hung Window

Dhyr: Dehydrator

di: das ist • Dat is • De-ice • Dienstags • Dinsdag • Dwarf Irregular

Di: Didymium • Diameter • Dienstag

DI: Daily Inspection • Data In • Data Integration • Data Integrator • Data Interface • Data Item • Dat Is • Daily Inspection • Decrement Interrupt • Délégation à l'Informatique (France) • Department of the Interior • Descent Image • Descent Imager • Desk Instruction • Destination Index • Development Integration • Diameter • Dienstag • Dienstags • Digital Image • Digital Input • Diode for Isolation • Diplomatic Immunity • Diplomingenieur • Directeur Informatique • Direct-Injected • Direct Injection • Direction Informatique • Dirty Image • Discrete Input • Disk Instability • Display Interface • Distance Indicator • District Inspector • Divisional Inspector • Document Identifier • Document Imaging • Doppler Imaging • Double Imperial • Drill Instructor • Driving Issue • Dwarf Irregular

Dia: Diagram • Dialect • Diameter

DIA: Darmstadt Indexing Approach • Date Impact Analysis • Date of Initial Appointment • Defense Intelligence Agency • Denver International Airport (USA-CO) • Diameter • Difference Image Analysis • Digital Interface Adapter • Direct Interaction Approximation • Document Indexing Approach • Document Interchange Architecture

DIAD: Drum Information Assembler/Dispatcher

DIADS: Digital Image Analysis and Display System • Digital Image Analysis Display System • Digital Integrated Air Defense System

Diag: Diagnose • Diagnosis • Diagonal • Diagram

DIAG: Diagonal

DIAGD: Diagnostic Daemon

DIAGN: Diagnose

Dial: Dialect • Dialecte • Dialectal • Dialectic • Dialéctica • Dialectical • Dialectique

DIAL: Department of Information on the Astronomical Life • Differential-Absorption Lidar • Digital-Image Analysis Laboratory • Direct Information Access Line • Disability Information Awareness Line

Dialect: Dialectal

Diam: Diameter • Diamètre

DIAM: Data Independent Accessing Model • Data Independent Access Model • Defense Intelligence Agency Manual

DIANA: Data and Information Available Now in Africa • Digital Integrated Access Network Arrangement • Direct Information Access Network for Africa

DIANE: Direct Information Access Network for Europe

Diaph: Diaphragm

DIARAD: Dual Irradiance Absolute Radiometer

DIAS: Digital Image Analysis System • Distributed Intelligent Actuator and Sensor • Dublin Institute for Advanced Studies (Ireland) • Dynamic Inventory Analysis System

DIB: Data Input Bus • Déchet Industriel Banal • Device-Independent Bitmap • Dibrugarh (India) (location identifier) • Dictionary of International Biography • Diffuse Interstellar Band • Direct Inward Dialing • Directory Information Base • Dual Independent Bus

DIBI: Device-Independent Backup Interface

DIBOC: Distributed Input Buffering with Output Control

DIBOL: DEC's Business-Oriented Language

Dic: Dicembre • Diciembre

DIC: Data Information Center • Data Input Clerk • Data Input Console • Data Insertion Converter • Data Item Category • Defence Industries Council (UK) • Departamento de Investigaciones Científicas • Dictionary • Differential Interference Contrast • Digital Input Computer • Digital Integrated Circuit • Digital Integrating Computer • Digital Interface Controller • Dili (Zaire) (location identifier) • Dissolved Inorganic Carbon • Documentation and Information Centre • Document Identifier Code • Double Ionization Cell • Double Ionization Chamber

DICAP: Direct-Current Circuit Analysis Program

DICAPS: Document Imaging and Computer-Assisted Publication System

DICB: Data Interface Control Board

Dicc: Diccionario

DICC: Digital Interface Code Converter

DICD: Draft Interface Central Document

DICE: DAO Instrument Control Environment • DARPA Initiative in Concurrent Engineering • Defense Initiative for Concurrent Engineering • Digital Integrated Circuit Element • Digital Intercontinental Conversion Equipment • Direct Intercommunication Experiment • Direct Interestablishment Communication Experiment • Direct Interestablishment Communication in Europe • Distributed Interactive Collaboration Environment • Dual Imaging Čerenkov Experiment • Dynamic Integrated Climate Economy

DICIFER: Digital Interactive Complex for Image Feature Extraction and Recognition

DICS: Digital Image Correction System

Dict: Dictation • Dictator • Dictionary

DICT: Dirección de la Información Científica y Técnica

DICU: Data Interface Control Unit • Digital Interface and Control Unit

DID: Damage-Injuries-Death • Data Information Delivery • Data Information Description • Data Input Display • Data Item Description • Destination Identification • Digital Image Data • Digital Image Document • Digital Information Display • Digital Interactive Decoder • Direct Inbound Dialing • Direct Injection Diesel • Direct Inward Dial • Direct Inward Dialing • Double Isotope Derivative • Drum Information Display • Dust Impact Detection

DIDA: Dynamic Instrumentation Digital Analyzer

DIDAC: Digital Data Acquisition

DIDAK: Digital Image Data Analysis Karlsruhe

DIDC: Data Input Display Console

DIDD: Dynamic Integrated Data Display

DIDDS: Dynamic Integrated Data Display System

DIDF: Dual Input Describing Function

DIDL: Digital Integrated Design Language

DI/DO: Data Input / Data Output

DIDOT: Digitising and Designing Of Type

DIDS: Decision Information Display System • Defense Integrated Data System • Digital Information and Display System • Digital Information Display System • Distributed Information Delivery System • Domestic Information Display System • Donor-Impurity Density of States

DIDSY: Dust Impact Detection System

DIE: Antsiranana (Madagascar) (location identifier) • Diploma in Industrial Engineering

dieb. alt.: Diebus Alternus

DIEBRA: Digital Image Enlarging Balanced Reconstruction Algorithm

dieb. secund.: Diebus Secundis

dieb. tert.: Diebus Tertius

Diel: Dielectric

DIES: Data Interpretation and Evaluation System • Digital Image Enhancement System

DIESA: Department of International and Social Affairs

DIEU: Digital Interface Electronics Unit

Dif: Differ • Difference • Different • Differential

DIF: Data Integration Function • Data Interchange Format • Data Interface Facility • Decay In Flight • Descriptive Item File • Difference • Diffuse • Digital Interface Frame • Directory Interchange Format • Discrete Increment Filter • Display Information Facility • Document Interchange Format

DIFAS: Digital Ice Forecast and Analysis System

Diff: Differ • Difference • Different • Differential • Difficult • Difficulty • Diffraction

DIFF: Differential

Dig: Digest • Digit • Digital • Digitalize • Digitalized

DIG: Deputy Inspector-General • Diffuse Ionized Gas • Direction Informatique Généralisée • Dwarf Irregular Galaxy

DIGACC: Digital Guidance And Control Computer

DIGBE: Diffuse Galactic Background Experiment

DIGENOR: Dirección General de Normas y Sistemas de Calidad

DIGEST: Digital Geographic Exchange Standard

DIGI: Digital

DIGITEF: Diffusion par l'Informatique de Graphiques, d'Images et de Textes pour l'Enseignement du Futur

DIGRAPH: Directed Graph

DIGS: Delta Inertial Guidance System • Digital Inertial Guidance System

DIH: Discrete Input High

DII: Diode Ion Injector • Dynamic Invocation Interface

DIID: Data Independent Interface for Databases

DIIG: Digital Information Infrastructure Guide

DIIP: Direct Interrupt Identification Port

DIIPS: Digital Interactive Image Processing System

DIIR: Digital Image-Intensifier Radiography

Dij: Dijon (France)

DIJ: Dijon (France) (location identifier)

DIK: Dickinson (USA-ND) (location identifier)

Dil: Dilute • Diluted • Dilution

DIL: Deliverable Items List • Dili (Indonesia) (location identifier) • Discrete Input Low • Dual In-Line • Dynamic Imaging Library

DILGEA: Department of Immigration, Local Government and Ethnic Affairs

DILIB: Document and Information Library

Diln: Dilution

DILSA: Distributed Integrated Logistics Support Analysis

dim: Dimidius

Dim: Dimanche • Dimension • Diminish • Diminuendo • Diminutive

DIM: Data In Memory • Data Input Module • Data Interface Module • Data In the Middle • Department of Immigration and Multicultural Affairs (Australia) • Description, Installation and Maintenance • Design Impact Meeting • Design Information Manual • Design Interface Manual • Design Interface Meeting • Differential Inverse Method • Diffuse Ionized Medium • Digital Image Matching • Digital Input Module • Dimbokro (Ivory Coast) (location identifier) • Disk Instability Model • Distributed Information Management • Document Image Management • Dust Impact Monitor

DIMA: Direct Imaging Mass Analyzer

DIMACS: Discrete Mathematics and Theoretical Computer Science

DIMAP: Distributed Image Management And Projection

DIMAPS: Digital Image Manipulation Analysis and Processing System • Digital Image Manipulation And Processing System

DIMD: Dichroic Medium Dispersion Mode

DIME: Development of Integrated Monetary Electronics • Dual Independent Map Encoding

Dimen: Dimensions (periodical code)

DIMEN: Dimension

DIMES: Delfts Interfacultair Instituut voor Micro-Elektronica en Submicrontechnologie (Netherlands) • Diffuse Microwave Emission Survey • Digital Image Manipulation and Enhancement System

DIMIAS: Digital Interactive Multi-Image Analysis System

Dimin: Diminish • Diminuendo • Diminutive

DIMM: Differential Image Motion Monitor • Dual In-line Memory Module

DIMS: Data and Information Management System • Data Information and Manufacturing System • Data Input Management System

DIMT: Diagonal Intrinsically Multiply Transitive

DImTe: Display Imaging Technology (periodical code)

DIMWG: Data and Information Management Working Group

Din: Dinar (currency) • Dining • Dinner • Dinsdag

DIN: Data Identification Number • Deutsche Industrie-Normen (Germany) • Deutsches Institut für Normung (Germany) • Dien Bien Phu (Viet Nam) (location identifier) • Digital Imaging Network • Direct Injection Nebulizer • Document Identification Number

DINA: Detector of Ions and Neutral Atoms • Distributed Information Network Architecture

DINAMEL: Dirección Nacional de Metrologia Legal (Uruguay)

DIng: Doctor of Engineering

DINS: Digital Inertial Navigational System

Dio: Diode

DIO: Data Input/Output • Device Input/Output • Digital Input/Output • Diomede Island (USA-AK) (location identifier)

DIOB: Digital Input/Output Buffer

DIODE: Détermination Immédiate d'Orbites DORIS Embarquée

DIOI: Digital Input/Output Interface

DION: Documentair Informatiesysteem voor Onderwijsliteratuur in Nederland

Dióp: Dióptrica

DIOP: Digital Input/Output Package

Diopt: Diopter

Dióptr: Dióptrica

DIOS: Direct Inout/Output System

Dip: Diploma • Diputación

DIP: Data In-line Package • Data Input Processor • Data Integrated Plan • Data Integration Program • Debtor In

Possession • Deep Inelastic Process • Designated Inspection Point • Design Improvement Program • Detailed Integration Plan • Diapaga (Burkina Faso) (location identifier) • Digital Image Processing • Digital Image Processor • Digital Incremental Plotter • Digital Instrumentation Programmer • Dimensional Inspection Planner • Direct Imaging Plotter • Display Information Processor • Display Input Processor • Display Interface Processor • Distributed Information Processing • Document Image Processing • Document Improvement Programme • Drop and Insert Point • Dual In-line Package • Dual In-line Pin

DipAM: Diploma in Applied Mechanics

DipAppSc: Diploma of Applied Science

DipCE: Diploma in Civil Engineering

DipChemEng: Diploma in Chemical Engineering

DipEd: Diploma in Education

DipEl: Diploma in Electronics

DipEng: Diploma in Engineering

Dipl: Diplexer • Diploma • Diplomat • Diplomatic

DIPL: Display Initial Program Load

DipLib: Diploma of Librarianship

Dipl.-Ing.: Diplomingenieur

DipLSc: Diploma of Library Science

Diplxr: Diplexer

DipMechE: Diploma of Mechanical Engineering

DIPORS: Digital Image Processing Of Remotely Sensed Data

DipPA: Diploma in Public Administration

DIPPA: Digital Parallel Processing Array

DIPPR: Design Institute for Physical Property Data (USA)

DIPS: Development Information Processing System • Digital Image Processing System • Digital Information Processing System • Display Image Processing System • Distributed Internet Patent Service • Dynamic Isotope Power System

DipT: Diploma in Teaching

DipTech: Diploma in Technology

DIQ: Delivered Image Quality • Divinópolis (Brazil-MG) (location identifier)

dir: Dirham (currency)

Dir: Dirección • Direct • Directed • Directeur • Direction • Directional • Directive • Director • Directorate • Directory • Direttore • Direzione

DIR: Data Item Requirement • Daytime Infrared • Department of Industrial Relations • Developer Inhibitor Release • Direct • Direct Information Reading • Direct Information Recording • Direction Interrégionale • Directory • Dire Dawa (Ethiopia) (location identifier) • Discipline-oriented Information Retrieval • Document Information Retrieval • Documentation Information Record • Document Information Record

DIRA: Data Information Retrieval in Astronomy • Distribution Information Retrieval in Astronomy

DIRBE: Diffuse Infrared Background Experiment

DIRC: Detection of Internally Reflected Čerenkov (light)

DIRCM: Directed Infrared Countermeasure

DIRCN: Décennie Internationale pour la Réduction des (effets des) Catastrophes Naturelles

DIRECT: Digital Rectifier

Dir.Gen.: Director General

Dir.-Genl.: Director-General

dIrr: Dwarf Irregular (galaxy)

DIRS: Dialog-Information Retrieval System • Digital Image Rectification System • Digital Image Registration System • Distributed Information Retrieval System • Dynamic Information Retrieval System

DIRT: Dusty Infrared Test

DIRTS: Data and Information Related To Seismology

Dis: Discharge • Disciple • Discipline • Disconnect • Discontinue • Discount • Dispense • Distance • Distant • Distribute

DIS: Data and Information System • Data Information System • Data Input System • Data Interpretation System • Deep Inelastic Scattering • Defense Industrial Security • Defense Investigative Service • Design Improvement Study • Development Information System • Device Interface System • Digital Image Stabilizing • Digital Imaging Spectrophotometry • Digital Imaging Spectroscopy • Digital Imaging Subset • Digital Instrumentation System • Diplôme Interuniversitaire de Spécialisation • Disagree • Distributed Information System • Distributed Interactive Simulation • Divisional Information System • Documentation Index System • Document Imaging System • Document Inquiry and Storage • Document Inquiry System • Doppler Inertial System • Double Imaging Spectrograph • Draft International Standard • Dynamic Impedance Stabilization • Loubomo (Congo) (location identifier)

DISA: Data Interchange Standards Association • Defense Information Systems Agency • Direct Inward System Access

Disabld: Disabled

DISAM: Direct and Index Sequential Access Method

DISAP: Disapproved

Disassm: Disassemble

Disassy: Disassembly

DISASSY: Disassembly

Disc: Disciple • Discipline • Disconnect • Disconnected • Discontinue • Discontinued • Discount • Discover • Discover (periodical code) • Discovered • Discoverer • Discovery • Discrete

DISC: Differential Scattering • Digital Information System CD-ROM • Digital Stereo Correlation System • Discone • Disconnect • Discrete

Disch: Discharge • Discharged

DISCH: Discharge

DISCO: Direct-Image Stabilized Camera Option • Direct SCF Optimization • Dual Spectral Irradiance and Solar Constant Orbiter

DISCOS: Database and Information System Characterising Objects in Space • Distributed Computing Solution • Disturbance Compensation System

Discr: Discriminate • Discriminated • Discrimination

Disct: Discount

DISCUS: Delaware Innovative Science/Math Collaborative for Undergratuate Success • Digital Self-healing Control for Upgraded Safety

DISD: Data and Information Systems Division

DISENG: Disengage

DISIMP: Device Independent Software for Image Processing

DISN: Defense Information Systems Network

DISOSS: Distributed Office Support System • Distributed Office Systems Support

Disp: Dispatcher • Dispense • Dispenser • Disperse • Dispersion • Display • Displacement • Disposición

DISP: Directory Information Shadowing Protocol • Dispenser • Display • Displacement

DISPAC: Domestic and International Scientific Planning and Cooperation

Displ: Displaced • Displacement • Displays (periodical code)

DISR: Descent Imager/Spectral Radiometer

Diss: Dissipation

DISSPLA: Display Integrated Software System and Plotting Language

Dist: Distance • Distant • Distinguish • Distinguished • Distort • Distorted • Distortion • Distribute • Distributed • Distribution • District

DIST: Daewoo Institute of Science and Technology (Korea-ROK) • Délégation à l'Information Scientifique et Technique (France) • Department of Industry, Science and Technology (Australia) • Direction de l'Information Scientifique et Technique • Distance • Distribution • District • Documentation et Information Scientifique et Technique

DISTB: Direction de l'Information Scientifique et des Bibliothèques

Distn: Distortion

DISTN: Direction à l'Information Scientifique et aux Technologies Nouvelles (France)

DISTNB: Direction de l'Information Scientifique, des Technologies Nouvelles et des Bibliothèques (France)

Distr: Distribución • Distribute • Distributed • Distribution • Distributivo • Distributor • District • Distrikt

DISTR: Distributor

Dists: Districts

Disyunt: Disyuntivo

DIT: Data Inquiry Terminal • Depth in Inheritance Tree • Detector Integration Time • Detroit Institute of Technology (USA-IL) • Directory Information Tree • Display Index Table • Divisão de Informação Técnica • Division Informatique et Telecommunications • Division of Information Technology • Documentation Implementation Team • Documentation Industrielle et Technique • Double Income Tax • Dynamic Integrated Test

DITAC: Department of Industry, Technology And Commerce

DITDOS: Distributed Inventory Tracking and Data Ordering Specifications

DITEN: Direction de l'Information et des Technologies de l'Éducation Nationale (France)

Dithm: Dithmarschen (Germany)

DITMO: Development, Integration, Test, and Mission Operations

DITR: Department of Industry, Technology and Resources • Deutsches Informationszentrum für Technische Regeln (Germany)

DITS: Distributed Instrumentation Tasking System

DITU: Digital Interface Test Unit

DIU: Data Interface Unit • Data Information Unit • Digital Input Unit • Digital Insertion Unit • Digital Interface Unit • Diu (India) (location identifier)

Div: Divergence • Diverse • Diversion • Diverter • Divide • Dividend • Divided • Divine • Divinity • Division • Divisione • Divisor • Divorced

DIV: Data In Virtual • Distance d'Itakura Vectorielle • Divide • Division • Divo (Ivory Coast) (location identifier) • Dynamic Imagery Viewer

DIVA: Département d'Information sur la Vie Astronomique • Deutsches Interferometer für Vielkanalphotometrie und Astrometrie (Germany) • Dialogprogramm zur Versuchsdaten-Analyse • Dicyanovinyl Anisole • Double Interferometer for Visual Astrometry

DIVE: Distributed Interactive Virtual Environment

DIVIMP: Divertor Impurity

Divn: Division

Divnl: Divisional

DIVOLI: Diviseur Optique de Ligne

DIVOTS: Data Input Voice Output Telephone System

DIW: Defense Information Warfare • Department of Industrial Works • Deutsche Institut für Wirtschaftsforschung (Germany) • D-Inside Wire

DIXRD: Delft Instruments X-Ray Diffraction (Netherlands)

DIY: Diyarbakir (Turkey) (location identifier) • Do It Yourself

dJ: der Jüngere • dieses Jahres

Dj: Djebel

DJ: Deceptive Jamming • Dinner Jacket • Disk Jockey • District Judge • Djibouti (ISO country code) • Doctor Juris • Doctor of Law • Dust Jacket

DJA: Djougou (Benin) (location identifier) • Dose Journalière Autorisée

DJB: Jambi (Indonesia) (location identifier)

DJDE: Dynamic Job Definition Entry

DJE: Deflected Jet Exhaust • Djerba (Tunisia) (location identifier)

DJF: December-January-February • Franc (Djibouti) (ISO currency code)

DJG: Djanet (Algeria) (location identifier)

DJI: Djibouti (ISO country code)

DJIA: Dow Jones Industrial Average

DJJ: Jayapura (Indonesia) (location identifier)

DJM: Djambala (Congo) (location identifier)

DJN: Delta Junction (USA-AK) (location identifier)

DJNR: Dow Jones News Retrieval

DJO: Daloa (Ivory Coast) (location identifier)

DJU: Degré-Jour Unifié • Djúpivogur (Iceland) (location identifier)

dk: Denmark

Dk: Dark • Deck • Dock

DK: Danemark (ISO country code) • Dark • Data and Knowledge • Deck • Denmark (international vehicle registration code) • Denmark (ISO country code) • Denmark (postal country code) • Dezimal-Klassifikation • Display Keyboard • Docking

DKB: Decimal Keyboard • Distributed Knowledge Base

DKBMS: Data and Knowledge Base Management System

DKC: Design Knowledge Capture

DKD: Deutsche Kalibrierdienst (Germany) • Deutsches Komitee für Dokumentation (Germany)

dkg: Dekagram

DKG: Deutsche Keramische Gesellschaft (Germany)

DKI: Deutsche Kommission für Ingenieur-Ausbildung (Germany) • Dunk Island (Australia-QLD) (location identifier)

DKK: Dunkirk (USA-NY) (location identifier) • Krone (Denmark) (ISO currency code)

dkl: Dekaliter

dkm: Dekameter

DKM: Dreikoordinatenmeßgerät

DKMM: Deutsches Komitee für Meeresforschung und Meerestechnik (Germany)

Dkr: Dansk Krone (currency)

DKR: Dakar (Senegal) (location identifier) • Dansk Krone (currency)

DKRZ: Deutsches Klimarechenzentrum (Germany)

dks: Decastere

DKS: Dikson (Russia) (location identifier) • Direct Keying System • Dote-Kivelson-Schwartz

DKT: Dahl-Kirkam Telescope

DKUUG: Danish Unix-systems User Group (Denmark)

DKV: Docker River (Australia-NT) (location identifier)

dl: Deciliter

d/L: Decreto Ley

D/L: Data Link • Demand Loan • Deorbit/Landing • Down Link • Down List • Down Load

DL: Dale • Daresbury Laboratory (UK-England) • Darlington (UK postal code) • Data Language • Data Link • Data Logging • Daylight • Deadline • Dead Load • Decreto-Lei • Definition List • Delay Line • Deorbit/Landing • Depósito Legal • Dépôt Légal • Description List • Descriptive List • Detection Limit • Developed Length • Development Left • Diesel • Diffraction Limited • Digital Library • Diode Logic • Directive de Lancement • Direct Line • Disk Light • Display Level • Distinguished Lecture • Distinguished Lecturer • Distributed List • Distribution Line • Dividing Line • Doctor of Laws • Document Librarian • Dose Level • Double Layer • Double Ledger • Double Lobe • Downlink • Download • Drawing List • Driving Licence

DLA: Damped Ly-alpha Absorber • Damped Ly-alpha Absorption • Damped Lyman Alpha • Data Link Address • Declination of Launch Asymptote • Defense Logistics Agency (USA) • Delay • Delay Line Assembly • Dérivé Léger Ariane • Diffuse Limited Aggregation • Diffusion-Limited Aggregation • Division de la Logistique Administrative • Division of Library Automation • Documentation-Libraries-Archives • Douala (Cameroon) (location identifier) • Double Logarithmic Approximation

DLAI: Documentation, Libraries and Archives Infrastructure

DLAS: Damped Lyman-alpha Absorption System

DLAT: Destructive Lot Acceptance Testing

DLB: Dalbertis (Papua New Guinea) (location identifier) • Digital Line Backend • Dynamic Load Balance

DLBI: Differential Long-Baseline Interferometer • Differential Long-Baseline Interferometry

DLC: Dalian (China-PRC) (location identifier) • Data Link Control • Delay Line Control • Diamond-Like Carbon • Diamond-Like Coating • Digital Library Coordinator • Digital Logic Circuit • Digital Loop Carrier • Direction des Lycées et des Collèges • Direct Lift Control • Distributed Loop Carrier • District Library Center • Dual-Layer Capacitor

DLCF: Data Link Control Field

DLCI: Data Link Connection Identifier

DLCP: Data Link Control Panel

DLCQ: Discretized Light-Cone Quantization

DLCS: Data Line Concentration System • Delay Line Control Software

DLCU: Digital Line Carrier Unit

Dld: Delivered

DLD: Data Link Decoder • Deadline Date • Geilo Dagali (Norway) (location identifier)

DLE: Data Link Equipment • Data Link Escape • Density and Luminosity Evolution • Digital Live Editor • Dôle Travaux (France) (location identifier)

DLEC: Digital Library Education Collaboratory

DLED: Dedicated Loop Encryption Device

DLEP: Distributed List Expansion Point

DLF: Data List File • Del Rio - Laughlin (USA-TX) (location identifier) • Diffraction Limited Focusing • Dopolavoro Ferroviario

DLG: Digital Line Graph • Dillingham (USA-AK) (location identifier)

DLGE: Digital Line Graph Enhanced

DLH: Direct Labour Hour • Dulluth Superior (USA-MN) (location identifier)

DLI: Dalat Lienkhang (Viet Nam) (location identifier) • Digital Library Initiative • Dilute Lens Imager • Direct Liquid Introduction • Division de la Logistique Immobilière

DLIRIM: Diffraction-Limited Infrared Imager

DLIS: Department of Library and Information Science • Department of Library and Information Studies

DLK: Dulkaninna (Australia-SA) (location identifier)

DLL: Data Linked Library • Data Link Layer • Data Link Library • Delay Lock Loop • Dillon (USA-SC) (location identifier) • Direction des Livres et de la Lecture • Dynamically Loadable Library • Dynamic Link Library

DLM: Dalaman (Turkey) (location identifier) • Deep-Layer Mean • Delay Line Memory • Digital Landscape Model • Digital Logic Module • Dimanche-Lundi-Mardi • Distributed List Member • Distributed Lock Manager • Dynamic Link Module

DLMPS: Division of Logic, Methodology and Philosophy of Science

DLMS: Digital Landmass System

DLN: Dillon (USA-MT) (location identifier) • Double Length Number

DLNR: Department of Land and Natural Resources

DLO: Dead Letter Office • Dienst Landbouwkundig Onderzoek (Netherlands) • Director of Launch Operations • Distribution List Owner • Division Liaison Officer • Document-Like Object • Dolomi (USA-AK) (location identifier)

DLOGLR: Deutsche Lilienthal Oberth Gesellschaft für Luft- und Raumfahrt (Germany)

DLOS: Double Line Of Sight

DLP: Data Link Processor • Digital Light Processing • Digital Light Processor

DLPI: Data Link Provider Interface

DLPR: Dual Linear Polarization Radar

DLPS: Data Link Processing System

DLPU: Data Link Processor Unit

Dlr: Dealer • Dollar

DLR: Data Link Receiver • Deutsche Forschungsanstalt für Luft- und Raumfahrt eV (Germany) • Deutsches Zentrum für Luft- und Raumfahrt eV (Germany) (*) • Docklands Light Railway (UK-England) • DOS LAN Requester

DLRL: Diffraction-Limited Raman Laser

DLRNa: DLR Nachrichten (periodical code)

DLRP: Data Link Reference Point

Dlrs: Dollars

DLS: Damped Lyman-alpha System • Data Link Set • Data Link Simulator • Data Link Switching • Data Loader Simulator • Data Logging System • Department of Library Services • Department of Library Studies • Differential Light Scattering • Digital Link Service • Digital Logic System • Dirty Little Secret • Distributed Link Service • DME-based Landing System • Dock Landing Ship • Doctor of Library Science • Domestic Substances List • Doppler Lidar Sounder • Dynamic Limb Sounder • Dynamic Load Simulator • The Dalles (USA-OR) (location identifier)

DLSc: Doctor of Library Science

DLSC: Daily Living Science Center (USA-LA) • Data Link to a Satellite Channel • Defense Logistics Service Center

DLSM: Data Link Summary Message

DLSP: Distribution List Submit Permission

DLSw: Data Link Switching

Dlt: Delete

DLT: Data Link Terminal • Data Link Translator • Data Loop Transceiver • Digital Library Technology • Digital Linear Tape • Direct Linear Transform • Distributed Language Translator • Dual Language Translation

Dltd: Deleted

DLTM: Data Link Test Message

DLTR: Data Link Terminal Repeater • Data Link Transmission Repeater

DLTS: Deep Level Transient Spectroscopy

DLTU: Digital Line/Trunk Unit

DLU: Dependent Logical Unit • Development Laboratory Unit • Digitizer Logic Unit • Display Logic Unit

DLUR: Dependent Logical Unit Requester

DLUS: Dependent Logical Unit Server

DLV: Delissaville (Australia-NT) (location identifier)

DLVA: Detector Logarithmic Video Amplifier

Dlvd: Delivered

DLVQ: Dynamic Learning Vector Quantization

Dlvy: Delivery

Dlx: Deluxe

Dly: Daily • Delay

DLY: Delay • Dillons Bay (Vanuatu) (location identifier)

DLZ: Dalanzadgad (Mongolia) (location identifier)

dm: Decimeter • Decimetric

dM: dieses Monats

D/M: Demodulate/Modulate • Demodulation/Modulation • Demodulator/Modulator

DM: Dam • Dark Marking • Dark Matter • Data Management • Data Manager • Data Manual • Data Mark • Data Master • Data Mining • Debugging Mode • Decimal Multiply • Decision Maker • Deep Monitoring • Definitional Module • Deformable Mirror • Delta Modulation • Demand Meter • Demonstration Model • Density Meter • Descriptive Method • Design Manual • Design Margin • Design Memorandum • Detached Module • Detector Module • Deutsche Mark (currency) • Development and Maintenance • Development Milestone • Development Model • Development Motor • Diagonal Mirror • Dialectic Method • Dialog Manager • Dialogue Mode • Dipole Moment • Direct Mail • Disassembly Manual • Discard Message • Dispersion Measure • Dispersion Measurement • Display Manager • Distributed Management • Distributed Memory • District Manager • Docking Mechanism • Docking Module • Docteur en Médecine • Doctor of Mathematics • Domain Model • Dominica (ISO country code) • Dormant Mode • Double Module • Downshifted Maximum • Drafting Manual • Durchmusterung • Dynamical Meteorology

DMA: Data Management Administrator • Data Management Analysis • Data Management Assembly • Data Management Association • Decision-Making Aid • Deep Map Assistant • Defense Mapping Agency (USA) • Designated Market Area • Differential Mobility Analyzer • Direct Marketing Association • Direct Memory Access • Direct Memory Address • Direct Memory Addressing • Document Management Alliance • Document Management Attribute • Dominica (ISO country code) • Drive Motor Assembly • Drum Memory Assembly • Dynamic Mechanical Analysis • Dynamic Memory Access • Tucson - Davis Monthan (USA-AZ) (location identifier)

DMAAC: Defense Mapping Agency Aerospace Center

DMAC: Data Management Analysis Center • Data Management and Analysis Center • Direct Memory Access Controller

DMAHTC: Defense Mapping Agency Hydrographic and Topographic Center

DMAP: Distributor Management Application Package

DMAPS: Design, Manufacturing And Producibility Simulation

DMAS: Data Management and Archive System • Des Moines Astronomical Society (USA-IA)

DMAT: Data Management and Archive Team • Digital Module Automatic Tester

DMATC: Defense Mapping Agency Topographic Center

DMath: Doctor of Mathematics

DMB: Database Management Branch • Data Management Branch • Dzhambul (Kazakhstan) (location identifier)

DMC: Dark Molecular Cloud • Database Maintenance Component • Data Management Center • Data Management Channel • Data Management Component • Data Management Computer • Digital Microcircuit • Digital Monitor Computer • Direct Maintenance Costs • Direct Manufacturing Costs • Direct Monte Carlo • Direct Multiplexor Channel • Discrete Memoryless Channel • Disk Memory Controller • Divisão de Mecânica e Controle • Drought Monitor Centre • DSIF Monitor and Control • Dwarf Molecular Cloud

DMCA: Digital Millennium Copyright Act (USA)

DMCF: Deservicing, Maintenance, and Checkout Facility

DMCL: Device Media Control Language

DMCM: Double-density Modular Core Memory

DMCS: Data Management and Control System

DMD: Data Management Division • Data Monitor and Display • Demodulate • Département Maintenance et Documentation • Dienstag-Mittwoch-Donnerstag • Digital Methods Division • Digital Micromirror Device • Doomadgee Mission (Australia-QLD) (location identifier) • Dual Mode Display

DMDCP: Dual-Mass Dynamic Cone Penetrometer

DMDF: Digital Map Data Format

DMDP: Data Maintenance Diagnostic Program

DMDS: Dimethyl Disulfide

DME: Department of Mines and Energy • Diagnostic Monitor Executive • Diploma in Mechanical Engineering • Direction de la Maîtrise d'Étude • Direct Measurements Explorer • Distance Measurement Equipment • Distance-Measuring Equipment • Distributed Management Environment • Dropping Mercury Electrode • Dynamic Mission Equipment • Moscow Domodedovo (Russia) (location identifier)

DMEA: Damage Modes and Effects Analysis

DMED: Digital Message Entry Device

DMEMP: Distributed Memory Explicit Message Passing

DMES: Digital Message Entry System • Dimethylethoxis

DMF: Data Management Facility • Depot Maintenance Facility • Digital Multiplexing and Formatting • Dimethylformamide • Distribution Media Floppy • Drug Master File • DSIF Maintenance Facility

Dmg: Damage

DMG: Data Management Group • Deutsche Meteorologische Gesellschaft (Germany) • Digital Media Group

DMGD: Dark Moon Geophysical Day

DMH: Dual Mode Hydrazine

DMI: Danish Maritime Institute • Danish Meteorological Institute • Danmarks Meteorologiske Institut (Denmark) • Deferred Maintenance Item • Desktop Management Information • Desktop Management Interface • Diagnostic Mathematics Inventory • Digital Multiplexed Interface • Direct Manipulation Interface • Direct Memory Interface • Distance Measuring Instrument

DMIA: Dual Multiplexer Interface Adapter

DMICB: Daily Mission Integration Control Board

DMILL: Durci Mixte sur Isolant Logico-Linéaire

DMIR: Designated Manufacturing Inspection Representative

DMIS: Data Management Information System • Dynamic-Mapping Information System

DMJ: Defense Management Journal (periodical code)

DMK: Dynamic Management Kit

DML: Database Management Language • Data Macro Language • Data Management Language • Data Manipulation Language • Data Manipulation Logic • Digitized Message Link • Dronning Maud Land (Antarctica) • Dual Mode Laser

DMLA: Dégénérescence Macilaire Liée à l'Âge

DMLCC: Design and Manufacturing of Low-Cost Composites

DMLE: Distributed Multimedia Learning Environment

DMLS: Dissecting Microscope Lighting System • Doppler Microwave Landing System • Dynamical Method of Least Squares

DMM: Digital Multimeter • Disaster Monitoring and Mitigation

DMMG: Displacement Method Matrix Generator

DMMH: Direct Maintenance Man-Hours

DMMM: Division of Materials, Mechanics, and Manufacturing

DMMP: Distributed-Memory Microprocessor

DMMS: Dynamic Memory Management System

Dmn: Dimension • Dimensional

DMN: Data Multiplexing Network • Deming (USA-NM) (location identifier) • Dimethylnitrosamine • Direction de la Météorologie Nationale

DMNA: Dimethylnitrosamine

DMNH: Denver Museum of Natural History (USA-CO)

DM&O: Data Management & Operations

DMO: Data Management Office • Data Management Officer • Diffuse Massive Object • Direction de la Maîtrise d'Ouvrage • Direct Marketing Organization • Distributed Management Object • Sedalia (USA-MO) (location identifier)

DMOC: DSPSE Mission Operations Center (USA-VA)

DMON: Discrete Monitoring

DMOS: Diffusion Metal Oxide Semiconductor • Diffusive Mixing of Organic Solutions • Double-diffused Metal Oxide Semiconductor

DMP: Daily Management Plan • Data Management Plan • Data Management Pointer • Data Management Processor • Data Merger Processor • Deployable Maintenance Platform • Deployment and Maintenance Platform • Direct Manipulation • Division of Materials Physics • Documentation Material Processed • Dot Matrix Printer • Dump

DMPC: Distributed Memory Parallel Computer

DMPI: Desired Mean Point of Impact

Dmpr: Damper

DMQS: Display Mode Query and Set

Dmr: Dimmer

DMR: Data Management Routine • December • Differential Microwave Radiometer • Ohamar (Yemen) (location identifier)

DMRC: Directorate of Meteorology Research Committee

DMRG: Density-Matrix Renormalization Group

DMRS: Data Management and Retrieval System

DMRT: Dense Medium Radiative Transfer

DMS: Daily Mission Scientist • Database Management System • Data Management Software • Data Management System • Data Measuring System • Data Memory Set • Data Monitoring System • Data Multiplex System • Daily Mission Schedule • Decision-Making System • Deep Multicolor Survey • Deployed Maintenance System • Deputy Mission Scientist • Design Maintenance System • Design Management System • Desktop Mapping System • Development Management System • Differential Maneuvering Simulator • Digital Management System • Digital Modem Server • Digital Multiplexed System • Digital Multiplexing Synchronizer • Digital Multiplex Switch • Digital Multiplex System • Dimethyl Sulfide • Directeur de la Mission Satellite • Disk Monitor System • Display Management System • Distance Measuring System • Distribution Management System • Division of Mathematical Sciences • Docking Mechanism System • Docking Module System • Documentation Management System • Documentation of Molecular Spectroscopy • Document Management Service • Document Management System • Doppler Measuring System • Drawing Management System • Drift Magnetic Separator • Drum Memory System • Dutch Meteor Society (Netherlands) • Dynamic Micromagnetic Structure • Dynamic Motion Simulator • Dynamic Motion System

DMSA: Double and Multiple Systems Annex

DMSB: Disposable Molecular Sieve Bed

DMSD: Digital Multistandard Decoding

DMSF: Data Management Systems Facility

DMSH: Diminish

DMSI: Data Management Segment Interface

DMSO: Dimethylsulphoxide

DMSP: Data Management Summary Processor • Defense Meteorological Satellite Program • Dichroic Microspectrophotometer • Distributed Mail Service Protocol

DMSS: Data Management System Simulator • Data Multiplex Subsystem • Dedicated Mission Support System • Defense Meteorological Satellite System • Distributed Mission Support System

Dmst: Demonstration

DMST: Division of Military Science and Technology

Dmstn: Demonstration

Dmstr: Demonstrator

DMT: Dark Matter Telescope • Desired Merge Total • Deutsche Meteorologentagung (Germany) • Diagonal Multiply Transitive • Diamantino (Brazil-MT) (location identifier) • Digital Message Terminal • Digital Multitone • Direct Modulation Technique • Discrete Multitone • Distributed Mission Training

DMTF: Desktop Management Task Force • Distributed Management Task Force

DMTG: Data Manipulation Task Group

DMTH: Data Manager Threshold

DMTPS: Digital Magnetic Tape Plotting System

DMTS: Digital Module Test Set

DMTU: Digital Magnetic Tape Unit

DMU: Data Management Unit • Data Manipulation Unit • Data Measurement Unit • Data Multiplexer Unit • Data Multiplex Unit • Decision-Making Unit • Digital Management Unit • Digital Message Unit • Digital Monitor Unit • Dimapur (India) (location identifier) • Directly Managed Unit • Dual Maneuvering Unit

dmv: Door Middel van

DMV: Department of Motor Vehicles (USA) • Deutsche Matematiker Vereinigung eV (Germany) • Digital Motion Video

DMWG: Data Management Working Group

DMX: Data Multiplex

DMY: Day-Month-Year

DMZ: Demilitarized Zone

Dn: Don • Down • Dozen • Dyne

D/N: Debit Note

DN: Data Name • Data Number • Date Number • Day and Night • Decimal Number • Delayed Neutron • De Novo • Despacho Normativo • Digital Number • Directorate Notice • Directory Number • Discrepancy Notice • Dispatch

Note • Distinguished Name • Distrito Nacional • Document Number • Domain Name • Doncaster (UK postal code) • Down • Dwarf Nova

Dna: Doña

DNA: DEC Network Architecture • Defense Nuclear Agency • Designatic National Agency • Desoxyribonucleic Acid • Did Not Attend • Digital Network Architecture • Directional Neighbourhoods Approach • Directorio Nacional Argentino • Division de la Navigation Aérienne • Does Not Answer • Does Not Apply • Okinawa Kadena (Japan) (location identifier)

DNAG: Decade of North American Geology

DNB: Deutsches Nachrichtenbüro (Germany) • Dictionary of National Biography (UK) • Dunbar (Australia-QLD) (location identifier)

DNC: Danane (Ivory Coast) (location identifier) • Delayed-Neutron Counting • Depreciation of National Capital • Did Not Come • Digital Nautical Chart • Direct Numerical Control • Direct Numerically Controlled • Distributed Numerical Control • Dynamic Network Controller

DNCRI: Division of Networking and Communications Research and Infrastructure

DNCS: Day/Night Camera System

DNCT: Direction Nationale de la Cartographie et de la Topographie

DND: Department of National Defence • Department of National Defense • Direct Numerical Diagonalization • Do Not Duplicate • Dundee (UK-Scotland) (location identifier)

Dne: Douane

DNe: Dwarf Novae

DNE: Dallas North (USA-TX) (location identifier) • Digital News Editor • Does Not Exist • Dynamic Node Exchange

DNEL: Disconnect Non-Essential Loads

DNES: Department of Non-conventional Energy Sources

DNESYS: Drainage Network Extraction System

DNF: Derna (Libya) (location identifier) • Did Not Finish • Disjunctive Normal Form

DNH: Department of National Heritage (UK) • Dunhuang (China-PRC) (location identifier)

DNHR: Dynamic Non-Hierarchical Routing

DNI: Desktop Network Interface • Detailed Network Interface • Documento Nacional de Identidad • Do Not Invite • Wad Medani (Sudan) (location identifier)

DNIC: Data Network Identification Code

DNIS: Dialed Number Identification Service

DNK: Danemark (ISO country code) • Denmark (ISO country code) • Deutscher National Komitee (Germany) • Dnepropetrovsk (Ukraine) (location identifier)

DNL: Augusta Daniel (USA-GA) (location identifier) • Differential Non-Linearity • Do Not List

DNLB: Danmarks Natur- og Laegevidenskabelige Bibliotek

DNLK: Downlink

DNLS: Derivative Non-Linear Schrödinger

DNLT: Downlist

DNM: Damped Newton Method • DEC Network Manager • Delayed Neutron Monitor • Denham (Australia-WA) (location identifier) • Dirección Nacional de la Meteorología

DNMR: Dynamic Nuclear Magnetic Resonance

DNN: Dalton (USA-GA) (location identifier)

DNO: Descending-Node Orbit • Dianópolis (Brazil-TO) (location identifier)

DNOM: Direction of Naval Oceanography and Meteorology

DNOS: Distributed Network Queuing System

DNP: Dang (Nepal) (location identifier) • Document Non Pertinent • Do Not Publish • Dynamic Nuclear Polarization

DNQ: Deniliquin (Australia-NSW) (location identifier) • Did Not Qualify

DNR: Department of Natural Resources • Détection par Neutrons Rapides • Dialed Number Recorder • Digital Noise Reduction • Dinard Pleurtuit (France) (location identifier) • Do Not Rescucitate • Do Not Return

DNRC: Department of Natural Resources and Conservation

Dns: Downs

DNS: Decimal Number System • Denison (USA-IA) (location identifier) • Dense • Departure Nuclear Stage • Did Not Show up • Direct Numerical Simulation • Discrete Network Simulation • Domain Name Server • Domain Name Service • Domain Name System • Domain Naming Service • Domain Naming System

DNSC: Data Network Service Center

DNSLP: Downslope

DNSSEC: Domain Name System Security

DNSTRM: Downstream

DNT: Desired Non-merge Total • Détection par Neutrons Thermiques • Santa Ana Downtown (USA-CA) (location identifier)

DNU: Dinangat (Papua New Guinea) (location identifier) • Do Not Use

DNV: Danville (USA-IL) (location identifier)

DNW: Dedicated Network • Deutsch-Niederländischer Windkanal • Duits-Nederlandse Windtunnel

DNX: Dinder Galegu (Sudan) (location identifier) • Dynamic Network X-connect

DNZ: Denizli Cardak (Turkey) (location identifier)

d/o: Daughter of

do: Ditto • Dominican Rep. • donnerstags

dO: Der Obige

Do: Ditto • Donnerstag

D&O: Description & Operations

D-O: Germany (formerly East Germany) (postal country code)

D/O: Delivery Order

DO: Data Organizer • Data Out • Dearborn Observatory (USA-IL) • Delivery Order • Dépense Ordinaire • Description and Operations • Digital Ortho • Digital Output • Directeur des Opérations • Direct Object • Direct Optimized • Director of Operations • Director's Office • Discrete Output • Disponibilité d'Office • Dissolved Oxygen • District Office • District Officer • Ditto • Divisional Office • Divisional Officer • Document-Ordered • Dominican Rep. (ISO country code) • Donnerstag • Donnerstags • Drawing Office • République Dominicaine (ISO country code)

DOA: Date Of Availability • Dead On Arrival • Department Of Aviation • Display-Oriented Application • Dissolved-Oxygen Analyzer • Doany (Madagascar) (location identifier) • Dutch Occultation Association (Netherlands)

DOAMS: Distant-Object Attitude-Measuring System

DOAS: Differential Optical Absorption Spectrometer • Differential Optical Absorption Spectroscopy (*) • Differentielle Optische Absorptionspektroskopie • Dual Optical Absorption Spectroscopy

DOB: Date Of Birth • Deployable Optical Bench • Dobo (Indonesia) (location identifier)

DOBIS: Dortmunder Bibliotheks Informations System

Doc: Doctor • Document • Documentation • Documento

DoC: Department of Commerce

DOC: Daily Operating Cost • Data, Operations and Control • Data Operations Center • Data Operations Console

• Data Operations Control • Data Operations Controller • Data Optimizing Computer • Data Output Channel • Date Of Change • Decimal to Octal Conversion • Deep-sky Observer's Companion • Department Of Commerce • Department Of Communications • Department Of Conservation • Designated Operational Coverage • Design Operation Capability • Direct Operating Cost • Discipline Operations Center • Disque Optique Compact • Dissolved Organic Carbon • Document • Documentation • Document-Oriented Computing • Dornoch (UK-England) (location identifier) • Drift-Orbit Correction • Dynamic Overload Control

DOCA: Date Of Current Appointment • Distributed Object Computing Architecture

DOCASE: Distribution and Objects in CASE

DOCDEL: Document Delivery

DOCED: Document Edition

DocEng: Doctor of Engineering

DOCKET: Document and Code Knowledge Elicitation Toolset

DOCP: Degree Of Circular Polarization

DOCS: Display Operator Console System • Document Organization and Control System

DOCSIS: Data Over Cable System Interface Specification

DOCSV: Data Over Circuit Switched Voice

Docu: Document • Documentary • Documentation

DOCUS: Display-Oriented Computer Usage System

DoD: Department of Defense • Depth of Discharge

DOD: Date Of Death • Department Of Defense • Department of Ocean Development • Depth Of Discharge • Deutsches Ozeanographisches Datenzentrum (Germany) • Development Operations Division • Digital Object Delivery • Direct Outward Dialing • Division Organisation et Développement • Dodoma (Tanzania) (location identifier) • Drops On Demand

DODAS: Digital Oceanographic Data Acquisition System

DoDISS: Department of Defense Index of Specifications and Standards

DODISS: Department Of Defense Index Specifications and Standards

DODL: DOCASE Design Language

DO-DO: Drive-On - Drive-Off

DODR: Digital Original Data Record

DODS: Definitive Orbit Determination System • Different Orbitals for Different Spins • Distributed Oceanographic Data System

DODT: Design Option Decision Tree

DoE: Department of Energy

DOE: Department Of Electronics • Department Of Energy • Department Of the Environment • Depending On Experience • Depends On Experience • Design Of Experiments • Diffractive Optical Element • Distributed Objects Everywhere • Djoemoe (Suriname) (location identifier)

DOEM: Designated Official for Environmental Matters

DOEn: Department Of Environment

DOES: Disk-Oriented Engineering System

DOF: Degree Of Freedom • Department Of Finance • Direction Of Flight • Dora Bay (USA-AK) (location identifier)

DoG: Difference of Gaussians

DOG: Difference Of Gaussian • Dongola (Sudan) (location identifier) • Dot On the Ground

DOGE: Doris Orbitography and Geopotential Evaluation

DOGS: Dwingeloo Obscured Galaxy Survey

DOH: Department Of Health (USA) • Discrete Output High • Doha (Qatar) (location identifier)

DOHC: Double Overhead Camshaft

DoI: Department of Interior

DOI: Degree Of Interest • Department of Industry • Department of the Interior • Descent Orbit Insertion • Differential Orbit Improvement • Digital Object Identifier • Direction des Opérations Informatiques • Direct Orbit Insertion • Director Of Information • Division Of Infrastructure • Document Object Identifier • Document-Oriented Interface • Doini (Papua New Guinea) (location identifier)

DO-IT: Disabilities, Opportunities, Internetworking and Technology

DOIT: Disabilities, Opportunities, Internetworking and Technology

DoJ: Department of Justice

DOJ: Department Of Justice

DOK: Deutsches Optisches Komitee (Germany) • Donetsk (Ukraine) (location identifier)

DOKM: Diagonal Optokinetic Nystagmus

Dol: Dollar

DOL: Daily Operating Log • Day Of Launch • Department Of Labor • Deauville Saint-Gatien (France) (location identifier) • Discrete Output Low • Display-Oriented Language • Dynamic Object Library

DOLA: Department Of Land Information

DOLARS: Digital Off-Line Automatic Recording System

DOLP: Degree of Linear Polarization

Dols: Dollars

Dom: Domain • Domenica • Domestic • Domicile • Dominant • Domingo • Dominica • Dominican • Dominion • Dominus

DOM: Data Organization and Management • Data Output Multiplexer • Day Of Month • Deo Optimo Maximo • Département d'Outre-Mer (France) • Description, Operation and Maintenance • Development Operations Manager • Digital Ohmmeter • Dirty Old Man • Discrete Ordinate Method • Dispersed Organic Material • Dissolved Organic Material • Dissolved Organic Matter • Distributed Object Management • Document Object Model • Dominica - Melville Hall (Dominica) (location identifier) • Dominican Rep. (international vehicle registration code) • Dominican Rep. (ISO country code) • République Dominicaine (ISO country code)

DOMAIN: Distributed Operating Multiaccess Interactive Network

DOMAINS: Distributed Open Management Architecture In Networked Systems

DOME: Distributed Object Management Environment

DOMF: Distributed Object Management Facility

Domin: Dominican

DOMINO: Demonstrationszentrum für Optische Meßtechnik und Integrierte Optik (Germany)

DOMIS: Directory of Materials Information Sources

DOMM: Distributed Object Management Model

DOMS: Distributed Object Management System

DOMSAT: Domestic Satellite

don: Donec

DON: Department Of Navy • Disque Optique Numérique • Dissolved Organic Nitrogen • Dos Lagunas (Guatemala) (location identifier)

DONA: Decentralized Open Network Architecture

DONUT: Direct Observation of the Nu Tau • Documentation Offered with No Usability Testing

DOO: Dorobisoro (Papua New Guinea) (location identifier)

DOOD: Deductive and Object-Oriented Database

DOOL: Distributed Object-Oriented Language

doorg: Doorgang

DOORS: Dynamic Object-Oriented Requirements System

DOP: Daily Observing Plan • Daily Operating Plan • Degree Of Polarization • Dermo-Optical Perception • Detailed Operating Procedure • Developing-Out Paper • Dilution Of Precision • Directory Operational Protocol • Dolpa (Nepal) (location identifier) • Driver-Operated Plug • Peso (Dominican Rep.) (ISO currency code)

DOPIE: Department Of Primary Industries and Energy

DOPLID: Doppler Lidar

DOPLIGHT: Doppler Lightning

D/OPS: Directorate of Operations

DOQ: Digital Orthophoto Quadrangle • Digital Orthoquad

DOQQ: Digital Orthophoto Quarter Quadrangle

Dor: Dorado-Doradus (constellation) • Dorian • Doric

DOR: Daily Operational Report • Date Of Request • Design Objective Reliability • Differenced One-way Range • Digital Optical Recording • Director of Operational Requirements • Direct Oxide Reduction • Dori (Burkina Faso) (location identifier)

DORAN: Doppler Range And Navigation

DORAS: Development of Operational Retrieval Algorithm from Spectra

DORCA: Dynamic Operational Requirements and Cost Analysis

DORIS: Détermination d'Orbite et Radiopositionnement Intégré par Satellite • Development Of Reasoning In Science • Doppler Orbit determination and Radiopositioning Integrated on Satellite • Doppler Orbitography and Radiopositioning Integrated by Satellite • Doppler Orbitography and Radiopositioning Integrated from Space

DorPo: Dornier Post (periodical code)

Dors: Dorset (UK-England)

DORS: Departementaal Overleg Remote Sensing (Netherlands)

dor. s. tr.: Doré sur Tranche

dOS: Direct Opacity Sampling

DoS: Denial of Service • Department of State

D.O/S: Distributed Operating System

DOS: Density Of States • Department Of Space • Department Of State • Digital Operation System • Digital Orthophoto System • Dios (Papua New Guinea) (location identifier) • Discrete Ordinates System • Discrete Orthonormal Sequence • Disk Operating System • Division of Optics and Spectroscopy

DOSE: Dictionary Of Substances and their Effects • Dynamics Of the Solid Earth

DOSEM: Disk Operating System Emulation

DOSES: Development of Statistical Expert Systems

DOSIM: Dosimeter

DOSLI: Department Of Survey and Land Information

DOSS: Decision-Oriented Scheduling System • Department Of State Services • Dexterous Orbiter Servicing System

DOST: Department Of Science and Technology

DOSV: Deep Ocean Survey Vehicle

DOSX: DOS Extender

Dot: Dotation

DoT: Department of Tourism • Department of Transport

DOT: Data and Operations Team • Data Output Tape • Deep Ocean Transponder • Department Of Overseas Trade • Department Of Telecommunications • Department Of Transportation • Deployment Operations Team • Designating Optical Tracker • Design Optimization Tool • Digital Optical Tape • Digital Output Timer • Directly Observed Therapy • Dutch Open Telescope (formerly OTT)

DOTA: Deployable Optical Telescope Assembly

DOTAC: Department Of Transport And Communications

DOTC: Dansk Organisation for Test og Certificering (Denmark) • Data Observing Testing Console

DOT&E: Director, Operational Test & Verification

DOTIS: Département d'Orientation Technique et d'Information Scientifique

DOTL: DOT Leader

DOTS: Digital Office Timing Supply • Dynamic Ocean Tracking System • Dynamic Ocean Track System

Dott: Dottore

DOTT: Detection Of Tornadic Thunderstorms

DOU: Dourados (Brazil-MS) (location identifier)

Doubl: Doublet

DOV: Data Over Voice • Director of Orbital Verification • Dover Cheswold (USA-DE) (location identifier)

DOVAP: Doppler Velocity And Position

DOVF: Dependent Overflow

DOW: Day Of Week • Direct Overwrite

DOWS: Directional Ocean Wave Spectrum

DOX: Dongara (Australia-WA) (location identifier)

DOY: Day Of Year

Doz: Dozen

d/P: Documentos contra Pago

Dp: Deep • Department • Departamento

D&P: Development & Production • Drain & Purge

D/P: Data Pack • Delivery against Payment • Delivery on Payment • Documents against Payment

DP: Damp-Proof • Damp-Proofing • Data Pack • Data Policy • Data Printer • Data Processing • Data Processor • Data Product • Data Production • Davis-Putnam • Deep • Deep Penetration • Definitive Parameter • Deflection Plate • Degree of Polimerization • Delayed Procurement • Delivery Point • Delta Pressure • Demarcation Point • Departure Point • Deployable Payload • Depreciation Percentage • Depth • Description Pattern • Design Phase • Design Proof • Development and Production • Development Phase • Development Plan • Development Prototype • Dew Point • Dial Pulse • Dial Pulsing • Diametral Pitch • Differential Pressure • Digital Photogrammetry • Directione Propria • Direct Playback • Disabled Person • Displaced Person • Display Panel • Display PostScript • Distributed Publishing • Distribution Point • Distrito Postal • Division President • Doctor of Philosophy • Document Pertinent • Domus Procerum • Double Paper • Double Play • Double Pole • Double Precision • Drafting Panel • Draft Proposal • Drain and Purge • Dual Purpose • Durable Press • Dynamic Programming

D&PA: Design & Product Assurance

DP&A: Data Processing & Analysis

DPA: Chicago - Du Page County (USA-IL) (location identifier) • Data Processing Activity • Data Processing Agency • Data Processing Algorithm • Data Processing Area • Data Processing Assembly • Deferred Payment Account • Delegation of Procurement Authority • Demand Protocol Architecture • Demi-Petit Axe • Departamento de Pesquisa e Aplicações • Department of Political Affairs • Design and Product Assurance • Destructive Physical Analysis • Detector Position Angle • Deutsches Patentamt (Germany) • Deutsche Press-Agentur (Germany) • Diphenylamine • Diploma in Public Administration • Division Planning Authorization • Doctor of Public Administration • Double-Precision Arithmetic

DPAC: Dedicated Plant Assignment Center

DPAIS: Data Processing Advanced Information System

DPAM: Demand Priority Access Method

DPANS: Draft Proposed American National Standard

DPAREN: Data Parity Enable

DPARS: Data Processing Automatic Record Standardization

DPAS: Door Peninsula Astronomical Society (USA-WI)

DPASV: Differential Pulse Anodic Stripping Voltammetry

DPB: Data Plotting Board • Deposit Pass Book • Dewpoint Temperature • Document Processing Branch • Drive Parameter Block

DPC: Database Promotion Centre (Japan) • Data Processing Center • Data Processing Central • Data Processing Control • Dedicated Personal Communicator • Destination Point Code • Differential Pressure Control • Digital Phase Comparator • Digital Pressure Converter • Digital Process Controller • Direct Program Control • Display Power Control • Domestic Policy Council • Dual Payload Carrier

DPCA: Data Processing Control Area • Displaced Phase Centre Antenna

DPCC: Data Processing Control Center • Data Processing Control Console • Differential Pulse Code Modulation

DPCI: Domain PC Interconnect

DPCL: Dedicated Payload Communication Link

DPCM: Delta Pulse Code Modulation • Differential Pulse Code Modulation

DPCS: Distributed Process Construction System

DPCSD: Department for Policy Coordination and Sustainable Development

DPCTE: Data Processor and Computer Test Equipment

DPCU: Digital Processing and Control Unit

DPD: Data Processing Department • Data Processing Division • Data Procurement Document • Data Project Directive • Data Protection Directive • Debrecen Photoheliographic Data • Digital Phase Difference • Digital Product Definition • Discrete Probability Distribution

DPDM: Differential Pulse Duration Modulation • Double Pulse Duration Modulation

DPDS: Data Processing and Display System • Data Processing Design System

DPDT: Double-Pole Double-Throw • Dual-Pole Dual-Throw

DPDU: Direction de la Programmation et du Développement Universitaire

DPE: Data Parity Error • Data Path Extender • Data Processing Electronics • Data Processing Environment • Data Processing Equipment • Delayed Pulse Extraction • Development Project Engineer • Dieppe (France) (location identifier) • Digital Processing Electronics

DPESR: Direction des Personnels de l'Enseignement Supérieur et de la Recherche

DPF: Data Private Facility • Data Processing Facility • Dense Plasma Focus • Differential Pressure Feedback • Directional Potential Function • Dynamic Pressure Feedback

Dpg: Damping

DPG: Dedicated Packet Group • Deutsche Physikalische Gesellschaft (Germany) • Deutsche Plannungsgesellschaft (Germany) • Digital Pattern Generator • Dual-Purpose Gyroscope • Dugway Michael (USA-UT) (location identifier)

DPGS: Data Product Generation System

DPh: Doctor of Philosophy

DPH: Department of Public Health • Director of Public Health

DPhG: Département de Physique Générale

DPhil: Doctor of Philosophy

dpi: Dots per Inch

DPI: Data Processing Installation • Department of Primary Industries • Department of Public Information • Derivative Port Identifier • Detail Program Interrelationship • Deuterium Pellet Injector • Digital Products Inc. • Divisão de Processamento de Imagens • Dot Pitch Integer • Double Pendulum Interferometry

DPIE: Department of Primary Industries and Energy

DPIR: Data Processing and Information Retrieval

DPIV: Digital Particle Image Velocimeter

DPK: Deer Park (USA-NY) (location identifier) • Deutsche Patentklassifikation (DPA)

DPKO: Department of Peacekeeping Operations

Dpl: Diplomat • Diplomatic • Duplex

DPL: Data Processing Laboratory • Data Processing Language • Data Programming Language • Davis-Putnam-Loveland • Declared Parts List • Descriptor Privilege Level • Diplomat • Dipolog (Philippines) (location identifier) • Distributed Processing Library • Distributed Program Link • Document Processing Language • Dual Polarization Lidar • Dublin Public Libraries (Ireland)

DPLE: Declarative Programming Languages in Education

DPLL: Digital Phase-Locked Loop

DPLOA: Draft Proposed Letter Of Agreement

DPLR: Doppler

DPLY: Deployment

DPM: Data Processing Machine • Data Processing Management • Data Processing Manager • Data Processing Modelling • Deflectable Photomultiplier • Deputy Prime Minister • Deputy Program Manager • Deputy Project Manager • Development Planning Memo • Dichroic Parametric Mirror • Digital Panel Meter • Digital Prominence Monitor • Domaine Public Maritime • Double Projection Model • Drop Physics Module (*) • Drop Physics Molecule • Dual Parton Model • Dual-Port Memory

DPMA: Data Processing Management Association • Deutsches Patent und Markenamt (Germany) • Distributive Principle of Multiplication over Addition

DPM&C: Department of Prime Minister & Cabinet

DPMI: DOS Protected Mode Interface

DPMIS: Data processing Management Information System

DPMMS: Department of Pure Mathematics and Mathematical Statistics

DPMR: Deputy Project Manager Resources

DPMS: Dahlgren Polar Monitoring Service (USA) • Display Power Management Signaling • Display Power Management System • DOS Protected Mode Services

DPMT: Deputy Project Manager Technical

DPN: Data Packet Network • Deepen

DPNG: Deepening

DPNH: Dinitrophenylhydrazine

DPO: Data Phase Optimization • Data Processing Operation • Devonport (Australia-TAS) (location identifier) • Digital Phosphor Oscilloscope • Digital Processing Oscilloscope • Direction Par Objectifs • Distinct Point of Origin • Distributing Post Office • District Pay Office

DPOB: Date and Place Of Birth

DPODP: Double-Precision Orbit Determination Program

DPOSS: Digital Palomar Observatory Sky Survey • Digitized Palomar Observatory Sky Survey

DPP: Decision Process Pattern • Deferred Payment Plan • Deployment/Pointing Panel • Detailed Project Plan • Development Program Plan • Differential Pulse Polarography • Digital Preprocessor • Director of Public Prosecutions • Direct Product Profitability

DPPA: Double Pumped Parametric Amplifier

DPPE: Data Processing Project Engineer

DPPR: Direction de la Prévention de la Pollution et des Risques (France)

DPQ: Double-Precision Quantity

DpR: Degrees per Revolution

DPR: Data Processing Request • Data Protection Register • Debrecen Photoheliographic Results • Definition Phase Review • Degrees Per Revolution • Democratic People's Republic • Department of Professional Regulation • Development Planning Review • Direction of Public Relations • Document Problem Report • Dual Precipitation Radar

DPRK: Democratic People's Republic of Korea (North Korea)

DPRS: Docking Probe Retraction System

DpS: Degrees per Second • Dividend per Share

DP&S: Data Processing & Software

DPS: Data Processing and Software • Data Processing Standard • Data Processing System • Degrees Per Second • Denpasar-Bali - Ngurah Rai (Indonesia) (location identifier) • Descent Power System • Descent Propulsion System • Design and Performance Specification • Design and Procedure Standard • Development Processing System • Diagnostic Problem Solver • Differential Power Switch • Digisonde Portable Sounder • Digital Phase Shifter • Digital Photogrammetric Station • Digital Photogrammetric System • Digital Photogrammetry System • Digital Plotter System • Digital Power Supply • Digital Processing System • Diode Phase Shifter • Director of Personal Services • Director of Postal Services • Disk Programming System • Display PostScript • Display Power Supply • Distributed Processing System • Disturbance Propagation System • Dividend Per Share • Division of Planetary Science • Doctor of Public Service • Document Processing System • Droit Préférentiel de Souscription • Dust Protection System

DPSC: Data Processing Service Center

DPSCA: Department of Political and Security Council Affairs

DPSD: Data Processing Systems Division

DPSK: Differential Phase Shift Keying

DPSM: Diode Phase Shifter Module • Double Passage Soustractif Multicanal

DPSN: Department of Physics of Stars and Nebulae

DPSPAR: Digital Processing Systems Personal Animation Recorder

DPSR: Data Processing Service Request

DP&SS: Data Processing & Software System

DPSS: Data Processing and Services System • Data Processing and Software System • Data Processing Software System • Data Processing Subsystem • Data Processing Switching System • Data Product Screening System

DPSST: Data Product Screening Software Team

DPST: Double-Pole Single-Throw • Dual-Pole Single-Throw

Dpstk: Dipstick

DPSU: Downstream Processing Unit

Dpt: Département • Department • Deponent • Deposit • Depot

DPT: Data Punched Tape • Delayed Picture Transmission • Descent Performance Test • Design Proof Test • Diphteria-Pertussis-Tetanus • Distributed Processing Technology • Division of Production Technology • Drive Parameter Table • Drive Parameter Tracking • Dynamic Pressurization Test

DPTH: Depth

Dpto: Departamento

DPtoTP: Display Coordinates to Tablet Coordinates

Dpty: Deputy

DPU: Data Parallel Unit • Data Processing Unit • Digital Processing Unit • Distributed Processing Unit • Dumpu (Papua New Guinea) (location identifier)

DPV: Differential Pulse Voltammetry

DPW: Department of Public Works • Digital Photogrammetric Workstation

DPWB: Diensten voor de Programmatie van het Wetenschapsbeleid

DPWR: Data Process Work Request

DPWS: Digital Photogrammetric Workstation

Dpx: Duplex

DPX: Duplex • Duplexer

DPY: Deploy

DQ: Data Quality • Definite Quantity • Dernier Quartier • Design Qualification • Destination Queue • Direct Question • Dispersion Quotient • Disqualified • Disqualify • Duque • Dynamic Query • Fiji (international aircraft registration code)

DQA: Data Quality Assessment • Diagonal Quadratic Approximation

DQC: Data Quality Control • Definite Quantity Control • Dynamic Quality Control

DQCM: Data Quality Control Monitor

DQDB: Distributed-Queue Dual Bus

DQE: Detective Quantum Efficiency • Detector Quantum Efficiency (*)

DQF: Data Quality File

DQI: Digital Quartz IMU

DQL: Database Query Language

DQM: Digital Quality Monitor • Distributed Queue Management

DQMCS: Data Query and Manipulation Common Service

DQO: Data Quality Objective

DQP: Dès Que Possible • Diode Qualification Program

DQPSK: Differential Quadratic Phase Shift Keying • Differential Quaternary Phase Shift Keying

DQQ: Digital Quarter Quadrangle

DQR: Design Qualification Requirement

DQS: Distributed Queuing System • Duques

DQSA: Duquesa

DQT: Diode Qualification Test

DQTP: Design Qualification Test Plan • Diode Qualification Test Program

Dque: Duque

dr: Drachma (currency)

Dr: Debtor • Directeur • Director • Docteur • Doctor • Dokter • Doktor • Dollar • Door • Drachma (currency) • Drain • Dram • Drama • Draw • Drawer • Drawn • Dreef • Drève • Drill • Drive • Driver • Droit • Druk • Drum

D/R: Deposit Receipt • Digital to Resolver • Direct or Reverse

DR: Daily Report • Daily Review • Data Ready • Data Receive • Data Received • Data Receiver • Data Record • Data Recorder • Data Reduction • Data Register • Data Report • Data Requester • Data Requirement • Dead Reckoning • Decay Ratio • Decommissioning Review • Decreto Regulamentar • Deficiency Report • Definitively Relevant • Délégation Régionale • Demi-Reliure • Designated Router • Design Requirement • Design Review • Detailed Report • Development Report • Development Right • Deviation Range • Deviation Ratio • Dielectronic Recombination • Digital Research • Differential Rate • Differential Rotation • Dilution Refrigerator • Dining Room • Directeur de Recherche • Direction Régionale • Discount Rate • Discrepancy Record • Discrepancy Report • Dispatch Reliability • Dispersion Relation • Disposition Record • Disposition Report • Dissociative Recombination • Distinguished Reviewer • Distributive Recursion • District Railway • District Registry • Documentation Requirement • Document Report • Document Retrieval •

Dominican Republic • Double Royal • Doutor • Drachma (currency) • Drafting Request • Draft Recommendation • Drifting • Drift Rate • Drive • Drizzle and Rain • Dutch Reformed • Dynamic Range • Dynamic Relaxation

Dra: Draco-Draconis (constellation)

DR&A: Data Reduction & Analysis • Data Requirements & Analysis

DRA: Data Reduction Agent • Data Reduction and Analysis • Data Reformatter Assembly • Data Release Authorization • Data Requirements and Analysis • Data Resource Adapter • Defence Research Agency • Deployment and Retraction Assembly • Design Review Agreement • Digital Read-in Assembly • Digital Recorder Analyzer • Discontinuous Reinforced Aluminium • Document Release Authorization • Doutora • Drum Read Amplifier • Mercury - Desert Rock (USA-NV) (location identifier)

DRAC: Direction Régionale de l'Aviation Civile • Distributed Read Address Counter

DRACO: Dynamics, Reconnection And Configuration Observatories (mission)

D/RADEX: Digitized Radar Experiment

DRAE: Diccionario de la Real Academia de la Lengua Española

DRAGN: Double Radiosource Associated with Galactic Nuclei

DRAGNs: Double Radiosource in Active Galactic Nuclei

DRAL: Daresbury and Rutherford Appleton Laboratories (UK) (now CCLRC)

DRAM: Detection-Radar Automatic Monitoring • Digital Random Access Memory • Dynamic Random Access Memory (*)

DRAMS: Digital Recording And Measuring System

DRANS: Data Reduction and Analysis System

DRAO: Daeduk Radio Astronomy Observatory (Korea-ROK) (now TRAO) • Dominion Radio Astrophysical Observatory (Canada-BC)

dr. ap.: Dran (Apothecaries)

DRAPE: Data Recording And Processing Equipment • Digital Recording And Playback Equipment

DRASC: Diffusion Raman Anti-Stokes Cohérente

DRAT: Data Reduction Analysis Tape • Data Requirements Analysis Tool • Demonstration Reliability Acceptance Test

dr. av.: Dran (Avoirdupois)

DRAW: Direct Read After Write

DRB: Decimal Register Binary • Derby (Australia-WA) (location identifier) • Design Requirements Baseline • Design Review Board

DRBTM: Dynamic Radiation Belt Test Model

DRC: DAMA Remote Controller • Data Recording Camera • Data Recording Control • Data Recording Controller • Data Recovery Component • Data Reduction Center • Data Reduction Compiler • Data Reduction Complex • Data Return Capsule • Data Return Container • Design Rule Check • Directeur de Recherche Contractuel • Direct-Reading Compass • Dirico (Angola) (location identifier) • Discrete Rate Command

DRCCE: Directeur de Recherche Contractuel de Classe Exceptionnelle

DRCE: Directeur de Recherche de Classe Exceptionnelle • Direction Régionale du Commerce Extérieur (France)

DRCG: Dressler Cluster of Galaxies

Drch: Drachma (currency)

DRCS: Data Receiver and Command Sender • Digital Radio Concentrator System

Drct: Direct

DRD: Data Reading Device • Data Recording Device • Data Requirements Description • Data Requirements Document • Design Requirements Document • Detailed Requirements Document • Document Requirement Description • Dorunda Station (Australia-QLD) (location identifier) • Drum Read Driver

DRDA: Distributed Relational Data Architecture • Distributed Relational Database Access • Distributed Relational Database Architecture (*)

DRDBMS: Distributed Relational Database Management System

DRDC: Detector Research and Development Committee

DRDM: Distributed Relational Data Manager

DRDS: Dynamic Reconfiguration Demonstration System

DRDW: Direct Read During Write

DRE: Data Reduction Equipment • Department of Resources and Energy • Differential Riccati Equation • Direction Régionale de l'Équipement • Downrange Error • Drummond Island (USA-MI) (location identifier)

DREAM: Definitive Realization of Energy and Matter

DRED: Data Routing and Error Detecting • Direction de la Recherche et des Enseignements Doctoraux • Direction de la Recherche et des Études Doctorales

DREE: Direction des Relations Économiques Extérieures

DREN: Defense Research and Education Network

DrEng: Doctor of Engineering

DREO: Defence Research Establishment Ottawa (Canada-ON)

DRES: Defence Research Establishment Suffield (Canada-ON) • Direction des Relations avec l'Enseignement Supérieur • Diversity Resources and Educational Services

DRET: Direction de la Recherche et de la Technologie • Direction des Recherches et Études Techniques

DREWS: Direct Readout Equatorial Weather Satellite

DRF: Data Recovery Field • Data Reporting Form • Data Request Form • Data Requirement Form • Deutsche Rettungsflugmacht (Germany) • Digital Raster File • Digital Radio Frequency • Discrepancy Report Form • Doctoral Records File • Documentation Requisition Form • Document Relative Frequency • Drift River (USA-AK) (location identifier)

DRFM: Digital Radio Frequency Memory

DRFT: Drift

Drftd: Drifted

DRG: Data Record Generator • Data Reduction Group • Deering (USA-AK) (location identifier) • Defense Research Group • Deutsche Röntgengesellschaft (Germany) • Digital Random Generator • Digital Ranging Generator • Digital Raster Graphic

DRGP: Digital Representation of Graphic Products

DRGS: Direct Readout Ground Station

DRH: Département des Ressources Humaines • Digital Readout Head • Directeur des Ressources Humaines • Direction des Ressources Humaines

DRHAF: Direction des Ressources Humaines et des Affaires Financières

Dr. hc: Doctor Honoris Causa

DRI: Data Rate Indicator • Data Recording Interface • Data Reduction Interpreter • Data Representation Interface • Dead-Reckoning Indicator • Defense Reinvestment Initiative • Defense Research Internet • De Ridder - Beauregard Parish (USA-LA) (location identifier) • Desert Research Institute (USA-NV) • Design Research Institute (USA-NY) • Differential Refraction Instrument • Diffuse-Reflectance Infrared Fourier Transform • Digital Research Inc. • Direction des Relations Internationales • Directly Riemann Integrable • Distributed Recovery Intelligence

DRID: Direct Readout Image Dissector

DRIG: Digital Rate-Integrating Gyro

Dr.-Ing.: Doktor der Ingenieurwissenschaft

DRIP: Data Reduction Input Program

DRIPS: Dynamic Real-Time Information Processing System

DRIR: Direction Régionale de l'Industrie et de la Recherche • Direct Readout Infrared Radiometer

DRIRE: Direction Régionale de l'Industrie, de la Recherche et de l'Environnement (France)

DRIRU: Dry Rotor Inertial Reference Unit

DRISS: Digital Read-In Subsystem

DRIV: Drive

DRIVE: Dedicated Road Infrastructure for Vehicles in Europe

DRK: Display Request Keyboard

DRL: Data Reduction Laboratory • Data Requirements Language • Data Requirements List • Data Retrieval Language • Daytime Running Lamp • Decision Representation Language • Design Review List • Digital Resource Librarian • Document Requirements List • Double-Rack Laboratory

DRLH: Directed Recursive Labelnode Hypergraph

DRLMS: Digital Radar Landmass Simulation

DRM: Data Records Management • Data Relay Mode • Département des Recherches Météorologiques • Design Reference Mission • Design Reference Model • Design Requirements Manual • Design Review Meeting • Detector Response Matrix • Detrital Remanent Magnetization • Digital Range Machine • Digital Road Map • Diploma in Resource Management • Direction des Renseignements Militaires • Discipline Requirements Manager • Distinct Region Model • Division du Renseignement Militaire • Drafting Room Manual • Drama (Greece) (location identifier) • Drawing Requirements Manual

DRMI: DOS Routed Mode Interface

DRMO: Defense Reutilization Marketing Organization

DRMS: Data Resource Management System • Distance Root Mean Square

DRMU: Digital Remote Measurement Unit

Drn: Drawn

DRN: Data Record Number • Data Reference Number • Dewan Riset Nasional (Indonesia) • Dirranbandi (Australia-QLD) (location identifier) • Document Release Notice • Document Revision Notice • Drain

DRNN: Diagonal Recurrent Neural Network • Dynamic and Recurrent Neural Network • Dynamic Recurrent Neural Network

DRO: Daily Routine Order • Data Request Output • Datenverarbeitung-Rationalisierung-Organisation • Destructive Readout • Digital Readout • Digital Revision Overlay • Direct Readout • Document Release Order • Durango - La Plata (USA-CO) (location identifier) • Purgatory (USA-CO) (location identifier)

DROG: Drogue

DROID: Distributed Readout Imaging Device

DROM: Decoder Read-Only Memory

DROO: Digital Readout Oscilloscope

DROP: Data Printout Program

DROPPS: Distribution and Role Of Particles in the Polar Summer

DROS: Direct Readout Satellite

DROWS: Direct Readout Weather Satellite

DRP: Data Reduction Procedure • Data Reduction Program • Data Restoration Program • Densest Random Packing • Digital Recording Process • Distribution and Replication Protocol • Dividend Reinvestment Plan

DRPP: Data Relay Preparatory Programme

DRR: Data Recorder/Reproducer • Data Redundancy Removal • Design Readiness Review • Design Release Review • Design Request Review • Design Requirements Review • Digital Radio-Relay • Document Release Record • Durrie (Australia-QLD) (location identifier)

DRRD: Design Requirements Review Disposition

Dr. rer. nat.: Doctor of Natural Science

DRRT: Délégation Régionale à la Recherche et à la Technologie (Fra nce)

DRS: Data Rate Selector • Data Reaction System • Data Receiving Station • Data Recording Set • Data Recording System • Data Reduction Software • Data Reduction System • Data Relay Satellite • Data Relay Station • Data Relay System • Data Requirement Specification • Data Retrieval System • DCP Retransmission System • Digital Range Safety • Digital Readout System • Digital Receiver Station • Digital Recording System • Direct Reception System • Disassembly/Reassembly Station • Discrepancy Reporting System • Documentation Retrieval System • Document Retrieval System • Donning and Recharge System • Dresden Klotzsche (Germany) (location identifier) • Driver Resource • Drives • Dynamic Reflectance Spectroscopy

DRSA: Data Recording System Analyst

DRSAM: Dynamic Random-Sequential Access Method

DRSCS: Digital Range Safety Command System

Drsg: Dressing

DRSG: Digital Recorder Signal Generator

DRSN: Drifting Snow

DRSS: Data Relay Satellite System • Data Relay System Satellite

DRT: Data Receive Terminal • Data Relay Transponder • Data Review Technician • Decision Response Time • Del Rio (USA-TX) (location identifier) • Deployment Readiness Test • Design Reference Timeline • Design Reference Trajectory • Digital Readout Timer • Direction de la Recherche et de la Technologie (France) • Discourse Representation Theory

DRTE: Direction Régionale du Travail et de l'Emploi

DRTL: Diode Resistor Transistor Logic

DRTM: Data Relay and Technology Mission (*) • Data Relay Technology Mission

DRTS: Data Relay and Tracking Satellite • Data Relay Test Satellite

DRTSS: Data Relay and Tracking Satellite System

DRU: Data Referenc eUnit • Digital Range Unit • Digital Recovery Unit • Digital Register Unit • Digital Remote Unit • Digital Retiming Unit • Drummond (USA-MT) (location identifier)

DRUB: Digital Remote Unit Buffer

DRUC: Disposition Record Unsatisfactory Condition

DRUID: Dynamic Rules for User Interface Design

DRUMS: Defeasible Reasoning and Uncertainty Management System

Dr. und Vrl.: Druck und Verlag

Dr. u. Vrl.: Druck und Verlag

DRV: Daily Reference Value • Dark Red Varnish • Democratic Republic of Vietnam • Drive • Driver

DRVID: Differenced Range Versus Integrated Doppler

Drvr: Driver

DRW: Darwin (Australia-NT) (location identifier) • Draw • Drawing

DRWG: Data Reduction Working Group

drx: Drachma (currency)

Drzl: Drizzle

DRZL: Drizzle

d/s: Days' Sight

ds: Dominus

Ds: Dans • Decistere • Destro • Dysprosium • Element of Path

dS: Dwarf Spheroidal

D&S: Davis & Sanford Co. (USA-NY) • Demand & Supply

D/S: Days after Sight • Deploy/Stow • Descent Stage • Digital to Synchro

DS: Danish Standards • Dansk Standard (Denmark) • Dansk Standardiseringsråd (Denmark) • Dark Spot • Data Scanning • Data Segment • Data Send • Data Server • Data Set • Data Sheet • Data Storage • Data Strobe • Data Synchronization • Data System • Date of Service • Daylight Saving • Decimal Subtract • Decision Sheet • Deep Space • Deep Survey • Defect Size • Define Storage • Delivery Schedule • Demonstration Science • Dempster-Schafer • Department of State • Department Scientifique • Depleted Sample • Deputy Secretary • Descent Stage • Descent System • Design Sheet • Design Specification • Design Standard • Detached Subgiant • Detail Specification • Diagnostic Software • Dictionary of Scientists • Diesel Ship • Diffractive System • Diffuse Surface • Digital Selection • Digital Signal • Diode for Suppression • Diode Switch • Diplomatic Security • Directing Staff • Directory Service • Direct Sequence • Direct Signal • Discipline Scientist • Discover • Disk Storage • Docking System • Doctor of Science • Document Signed • Domestic Service • Dornier System • Double-Sided • Double Star • Draw Start • Driver Seated • Drive System • Dünne Schicht • Dust Storm • Duty Scientist • Dwarf Spheroidal

DSA: Data Service Adapter • Data Systems Administration • Defense Supply Agency • Deployable Solar Array • Demonstration Structural Article • Design Schedule Analysis • Destination Software address • Dial System Assistance • Digital Signature Algorithm • Digital Spectrum Analyzer • Dimensionally Stabilized Anode • Directory System Agent • Directory System Alert • Discrete Sample Analyzer • Display System Activity • Distributed System Architecture • Doppler Spectrum Analyzer • Dynamic Scalable Architecture • Dynamic Signal Analyzer

DSAC: Data Synthetic Analysis Center

DSAD: Data Systems and Analysis Directorate

DSAM: Direct Sequential Access Method

DSAO: Data Sources in Asian-Oceanic Countries

DSAP: Data Systems Automation Program • Destination Service Access Point

DSAR: Data Sampling Automatic Receiver

DSAS: DAMUS Satellite Archive System • Density Scalable Atomic Sphere • Digital Sun Angle Sensor • Distributed Systems Application Services

DSASS: Digital Sun Angle Sensor System

DSAZ: Deutsche-Spanische Astronomiches Zentrum (Spain)

DSB: Defense Science Board • Device Status Byte • Dictionary of Scientific Biography • Direct-Sound Broadcast • Direct-Sounder Broadcast • Direct-Sounding Broadcast • Document Status Bulletin • Double Sideband • Double Strand Break

DSBAM: Double-Sideband Amplitude Module

DSBSC: Double Sideband Suppressed Carrier

Dsbl: Disable

Dsbld: Disabled

DSBS: Digital Sound Broadcasting Satellite

DSc: Doctor of Science

DSC: Albanian Direction of Standards • Data Separator Card • Data Services Center • Data Synchronizer Channel • Debye-Sears Cell • Dedicated Signal Conditioner • Delaware State College (USA-DE) • Differential Scanning Calorimeter • Differential Scanning Calorimetry • Differential Signal Control • Digital Scan Converter • Digital Setting Circle • Digital-to-Synchro Converter • Direction de la Sécurité Civile (France) • Direct Satellite Communications • Distant Station Connected • Documentation Standards Committee • Document Service Center • Document Structure Convention • Document Supply Center • Drift Scan Camera • Dschang (Cameroon) (location identifier) • Dynamic Sequential Control • Dynamic SQL Caching • Dynamic Standby Computer

DScA: Docteur en Sciences Appliquées

DSCB: Data Set Control Block

DSCC: Deep Space Communication Complex

DScEng: Doctor of Science in Engineering

D/SCI: Directorate of Science • Director of Science

DSCIM: Display System Computer Input Multiplexer

DSCN: Design and Simulator Computation Network

Dscont: Discontinue

Dscontd: Discontinued

DSCR: Direction de la Sécurité et de la Circulation Routières (France)

Dscrm: Discriminate

Dscrmd: Discriminated

Dscrmr: Discriminator

DSCS: Defense Satellite Communications System • Desk Side Computer System • Digital Satellite Compression Service • Digital Simulator Computer System

DSCT: Directional Solidification of Cadmium Telluride

DScTech: Doctor of Technical Science

DSCU: Digital Computer Switching Unit

DSD: Dark Surge on Disk • Data Status Display • Data Storage Device • Data Structure Diagram • Data Systems Division • Disk Storage Device • Dual-Scan Display • La Désirade (Guadeloupe) (location identifier)

DSDC: Direct Service Dial Capability

DSDD: Double-Side - Double Density (*) • Double-Sided - Double Density

DSDIC: Design Support for Distributed Industrial Control

DSDL: Data Storage Description Language

DSDM: Delamination Simulation by Damage Mechanism • Dynamic System Development Method

DSDP: Deep-Sea Drilling Project

DSDS: Data Storage and Distribution System • Data Systems Dynamic Simulator • Digital Synchro Data Source • Document Survey Data Sheet

DSDT: Discrete Space and Discrete Time

DSDU: Data Storage Distribution Unit

DSE: Data Set Experiment • Data Set Extension • Data Storage Encoder • Data Storage Equipment • Data Structure Editor • Data System Equipment • Data Systems Engineering • Debye-Sears Effect • Derecho del Trabajo • Dessie (Ethiopia) (location identifier) • Development Support Equipment • Dirección de Seguridad del Estado • Direction pour les Systèmes et l'Espace (ONERA) • Distributed Systems Environment • Division of Solar Energy • Division Support et Environnement

DSEA: Data Storage Electronics Assembly • Display Station Emulation Adapter

DSECT: Dummy Control Section

DSEE: Domain Software Engineering Environment

DSES: Division of Social and Economic Studies

DSF: Data Scanning and Formatting • Data Stream Formation • Delivery Sequence File • Design Safety Factor • Deputy Secretary General • Digital Signal Generator • Directional Solidification Furnace • Disappearing Solar Filament • Disk Storage Facility

DSFG: Deep-Sky Field Guide

DSFT: Distributed System Fault Tolerance

DSG: Datenschutzgesetz (Austria) • Defense Systems Group • Dilasag (Philippines) (location identifier) • Dynamique des Systèmes Gravitationnels

Dsgn: Design • Designer

DSHD: Double-Sided - High Density • Double-Side - High Density (*)

DSI: Data System Integration • Délégation aux Systèmes d'Information • Destin (USA-FL) (location identifier) • Detailed Spectrum Investigation • Development Sciences Inc. • Differential Speckle Interferometry • Diffusion Sélective de l'Information • Digital Speech Interpolation • Directeur des Systèmes d'Information • Direction des Services Informatiques • Direction des Systèmes d'Information • Directorate of Scientific Information • Discrete Smooth Interpolation • Discrete Surface Integral • Diseminación Selectiva de Información • Distribution System Integration • Division des Services Informatiques • Division of Science Information • Dynamic Skeleton Interface

DSIC: Departamento de Sistemas Informáticos y Computación • Documentation and Scientific Information Centre

DSID: Data Set Identification • Divergence Source-Image Distortion

DSIF: Deep Space Instrumentation Facility

DSIM: Distributed Shared Image

DSIN: Direction de la Sûreté et des Installations Nucléaires (France)

DSIO: Directeur des Systèmes d'Information et d'Organisation

DSIP: Development Support and Integration Program

DSIPT: Dissipate

DSIPTG: Dissipating

DSIR: Department of Scientific and Industrial Research

DSIRnet: New Zealand Government Network

DSIS: Deep-Structure Indexing System • Distributed Statistical Information System • Distributed Support Information Standard

DSJ: Defense Science Journal (periodical code)

DSK: Dera - Ismail Khan (Pakistan) (location identifier)

Dsl: Diesel

DSL: Dark Side Limb • Daru (Sergio Leone) (location identifier) • Data Set Label • Data Structures Language • Design Specification List • Digital Simulation Language • Digital Subscriber Line • Domain-Specific Language • Dynamic Simulation Language

DSLAM: Digital Subscriber Line Access Multiplexer

DSLS: Division of Space Life Sciences

Dslv: Dissolve

Dslvd: Dissolved

DSM: Data Status Message • Data Storage Memory • Data Systems Modernization • Deep Space Measurement • Demand-Side Management • Des Moines (USA-IA) (location identifier) • Detector Structural Model • Diagnostic and Statistical Manual • Digital Select Matrix • Digital Select Module • Digital Simulation Model • Direction of Systems Management • Direct-Solution Method • Directorate of Servicing and Maintenance • Distributed Shared Memory • Distributed System Management • Distribution System Management

DSMA: Digital Sense Mutiple Access

DSMAC: Digital Scene Matching Area Correlation

DSMC: Direct Simulation Monte Carlo

DSMCC: Digital Storage Media Command and Control

DSMCS: Document Structure Management Common Service

Dsmd: Dismissed

DSML: Directory Service Markup Language

DSMP: Defense Meteorological Satellite System

DSMRD: Distribution of Sporadic Meteor Radiants Density

DSN: Data Set Name • Data Steering Network • Deep-Space Network (NASA) • Delta Scuti Network • Derived Services Network • Dewan Standardisari Nasional (Indonesia) • Digital Switching Network • Diminishing Source Notice • Distributed System Network • Dongsheng (China-PRC) (location identifier) • Standardization Council of Indonesia

DSNA: Dictionary Society of North America

DSNT: Distant

DSO: Dark-Sky Observer • Data Systems Office • Deep-Sky Object • Deep-Sky Observer • Deputy Safety Officer • Designated Senior Official • Design Secondary Objective • Detailed Secondary Objective • Detailed Supplementary Objective • Digital Storage Oscilloscope • Direct System Output • Display Switching Oscilloscope • Distributed Shared Object • District Staff Officer

DSOB: Data Systems Operations Branch

DSOC: Discipline Science Operations Center

DSOF: Dispersion-Shifted Optical Fiber

DSOM: Distributed System Object Model

DSOP: Draft Statement of Principles

dsp: Decessit Sine Prole

DSp: Datenspeicher

DSP: Data Source Panel • Data Source Provider • Deep Space Probe • Defense Support Program • Densité Spectrale de Puissance • Department of State Planning • Departure Sequencing Program • Deployable Solar Panel • Desktop Storage Pack • Despacho • Differential Scanning Polarimetry • Digital Signal Processing • Digital Signal Processor • Digital Strip Printer • Direction de la Stratégie et des Programmes • Directory System Protocol • Display Simulation Program • Distributed System Program • Doppler Spectrum Processor • Dynamic Site Publisher • Dynamic Support Program

DSPE: Digital Signal Processing Experiment • Division of Scientific Personnel and Education

DSpec: Design Specification

Dspl: Display • Disposal

Dspld: Displayed

dSph: Dwarf Spheroidal

Dsplcd: Displaced

DSPM: Designated System Project Manager

Dspn: Disposition

Dspo: Disposal

DSPO: Defense Support Project Office

DSPRM: Digital Signal Processing Resource Manager • Digital Signal Processor Resource Manager

DSPRMI: Digital Signal Processing Resource Manager Interface • Digital Signal Processor Resource Manager Interface

DSPS: Digital Signal Processing System

DSPSE: Deep Space Program Science Experiment

DSPT: Département Scientifique, Pédagogique et Technique

DSPU: Down Stream Physical Unit

DSQ: Développement Social des Quartiers

DSQD: Double-Sided - Quad Density

DS&R: Data Storage & Retrieval

DSR: Daily Status Report • Dansk Selskab for Rumfartsforskning (Denmark) • Data Scanning and Routing • Data Set Ready • Data Specification Request • Data Storage and Retrieval • Data Survey Report • Device Status Register • Device Status Report • Digital Shift Register •

Digit Storage Relay • Direct Server Return • Direction of Scientific Research • Director of Scientific Research • Direct Storage Recorder • Display System Replacement • Division of Solar Research • Dynamic Service Register • Dynamic Spatial Reconstruction • Dynamic Spatial Reconstructor • Dynamic Storage Region

DSRC: Dedicated Short Range Communications

Dsrd: Desired

DSREDS: Digital Storage and Retrieval Engineering Data System

DSRI: Danish Space Research Institute • Digital Standard Relational Interface

DSRMC: Defense Systems Review Military Communications (periodical code)

DSRR: Digital Short-Range Radio

DSRT: Deep Sea Reversing Thermometer

DSS: Daily Science Schedule • Dakota Space Society (USA-ND) • Dark-Sky Switzerland • Data Server System • Data Set Service • Data Simulation System • Data Station Selector • Data Storage System • Data Support Section • Data Support System • Decision Support System • Deductive Subsystem • Deep Seismic Sounding • Deep Space Station • Deep Survey Spectrometer • Delft Sensor System • Deliverable Survival Shelter • Department of Social Security • Department of Supply and Services • Department Summary Schedule • Deployed Supply System • Deutsche Satellitensysteme GmbH (Germany) • Diagnostic Simulation System • Digital Satellite System • Digital Signal Synchronizer • Digital Signature Standard • Digital Simulator System • Digital Sky Survey • Digital Storage System • Digital Subset • Digital Subsystem • Digitized Sky Survey • Directory and Security Services • Direct Station Selection • Discrete-Source Scan • Distributed Snuffer System • Documentation Support Services • Document Search System • Document Status Sheet • Document Storage System • Dornier Satellitensysteme • Double Spot System • Drum Storage System • Dynamic Simulation System • Dynamic Steady State • Dynamic Support System

DSSA: Domain-Specific Software Architecture

DSSB: Double Single-Sideband

DSSC: Distributed Services Steering Committee

DSSD: Double Sided - Single Density • Double Side - Single Density (*)

DSSM: Dedicated Solar Sortie Mission • Dynamic Subsystem Management

DSSMOF: Dispersion-Shifted Single Mode Optical Filter

D/SSP: Directorate of Space Station and Platforms • Director of Space Station and Platforms

DSSR: Deep-Space Surveillance Radar

DSSS: Direct-Sequence Spread Spectrum

DSSSD: Double-Sided Si Strip Detector

DSSSL: Document Style Semantics and Specification Language

Dst: Disturbance Short Time

DST: Dark Streak • Data Source Terminal • Data System Test • Daylight Saving Time • Deep-Space Transponder • Department of Science and Technology • Detailed System Test • Developer Solutions Team • Digital Storage Tape • Dimensional Special Tooling • Direction de la Surveillance du Territoire (France) • Director of Supplies and Transport • Direct-Sounding Transmission • Discrete Sine Transform • Discrete Symmetry Transform • Distance • Distribution of Sign Times • Documentation Scientifique et Technique • Domeless Solar Tower • Double Solar Telescope • Double Summer Time • Dual Standard Transponder • Dunkle Striche • Dunn Solar Telescope (USA-NM) • Dynamic Surface Tension

DSTF: Delta Spin Test Facility

DSTI: Division of Scientific and Technical Information

Dstlt: Distillate

Dstltd: Distillated

Dstn: Destination

DSTN: Double Supertwisted Nematic • Dual-Scan Supertwisted Nematic

Dstng: Distinguish

Dstngd: Distinguished

DSTO: Defence Science and Technology Organization (Australia)

DSTP: Data Self-Test Program

D/STS: Directorate of Space Transportation Systems • Director of Space Transportation Systems

DSTS: Deep-Space Tracking System

DSTU: State Committee of Ukraine for Standardization, Metrology and Certification

DSU: Data Service Unit • Data Storage Unit • Data Synchronization Unit • Decoder Switching Unit • DEU Switching Unit • Digital Service Unit • Digital Standard Update • Digital Storage Unit • Disk Storage Unit • Drum Storage Unit

DSUP: Defense System Upgrade Program (*) • Defensive System Upgrade Program

DSUWG: Data Systems Users Working Group

DSV: Dansville (USA-NY) (location identifier) • Design-Specification-Verification • Diameter Spherical Volume • Digital System Validation

DSVD: Digital Simultaneous Voice/Data

DSVM: Distributed Shared Virtual Memory

DSVT: Digital Subscriber Voice Terminal

DSW: Data Status Word • Device Status Word

DSX: Digital System X-connect

DS2: Second Dark Spot (Neptune)

dt: Deutsch

dT: Thymidine

Dt: Doit

D&T: Development & Technology

DT: Daily Telegraph • Damage-Tolerant • Dark Time • Data Technician • Data Terminal • Data Transfer • Data Transmission • Data Transmit • Daylight Time • Decay Time • Decision Table • Deduction Theorem • Definition Term • Delayed Time • Delay Time • Delete Translation • Delta Time • Demi-Traitement • Department of Transportation • Department of Treasury • Description Title • Destructive Testing • Detective • Deuterium-Tritium • Deutsche Telekom (Germany) • Developmental Test • Development and Technology • Development Test • Development Testing • Device Type • Digital Technique • Digital Tracker • Dispersion Time • Displaced Threshold • Disposición Transitoria • Distance Test • Division Tehnique • Document Type • Dorchester (UK postal code) • Double Throw • Double Time • Downtime • Downtown • Drift Tube • Drop Tank • Dwell Tine • Dynamic Tester • Dynamic Time

DTA: Damage Tolerance Analysis • Data • Daily Testability Analyzer • Delta (USA-UT) (location identifier) • Deutsche Technische Akademie (Germany) • Development Test Article • Differential Thermal Analysis • Disk Transfer Area • Distributing Terminal Assembly

DTAD: Digital Terrain Analysis Data

DTAP: Direct Transfer Application Port

DTAS: Data Transmission And Switching • Digital Test Access System

DTB: Danmarks Tekniske Bibliotek • Decimal To Binary

DTBA: Date To Be Advised

DT&C: Department of Transport & Communications

DTC: Data Terminal Controller • Data Transmission Center • Dead Time Correction • Decision Threshold Computer • Defense Trade Control • Department of Transport and Communications • Design-Time Control • Design To Cost • Desk Top Computer • Differential Thermal Calorimetry • Digital Tape Conversion • Digital Trunk Controller • Display Test Chamber • Display Timing Control • Dispositif de Transfert de Charges • Distributed Terminal Controller • Distributed Transaction Coordinator • Domain Technology Committee • DSIF Telemetry and Command • Dynamic Traffic Control

DTCD: Department of Technical Cooperation for Development

DTCR: Data Transfer and Certification Record

DTCS: Data Transmission and Control System • Digital Test Command System

DTCU: Data Transmission Control Unit

DTCW: Data Transfer Command Word

dtd: Dentur Tales Doses

Dtd: Dated

DTD: Datadawai (Indonesia) (location identifier) • Data Type Definition • Detailed Test Description • Digital Terrain Data • Directorate of Technical Development • Director of Technical Development • Document Type Declaration • Document Type Definition

DTDP: Different Time - Different Place

DT&E: Demonstration, Testing & Evaluation • Development Test & Evaluation

DTE: Daet - Camarines Norte (Philippines) (location identifier) • Data Terminal Equipment • Data Terminating Equipment (*) • Data Transmission Equipment • Data Transmission Exchange • Data Transmitting Equipment • Deliverable Test Equipment • Development-Test-Evaluation • Digital Television Equipment • Digital Tone Extractor • Direct To Earth • Distributed Transaction Equipment • Down To Earth • Dumb Terminal Emulator

DTED: Digital Terrain Elevation Data (*) • Digital Terrain Evaluation Data

DTel: Deutsche Telekom (Germany)

D/TEL: Directorate of Telecommunications • Director of Telecommunications

DTEM: Digital Terrain Elevation Mapping

DTEMS: Digital Terrain Elevation Mapping System

DTEO: Defence Test and Evaluation Organisation

DTF: Daily Transaction File • Data Transmission Factor • Data Transmission Feature • Data Transmission Function • Data Transmittal Form • Define The File • Demonstration Test Flight • Development Test Facility • Dial Tone First • Digital Tape Format • Distributed Transaction Facility • Domain Task Force

Dtg: Dienstag

DTG: Digital Tape Generation • Direct Trunk Group • Display Transmission Generator • Dry Tuned Gyro

DTH: Death Valley (USA-CA) (location identifier) • Direct To Home

DThPT: Diploma in Theory and Practice of Teaching

DTI: Danish Technological Institute • Department of Trade and Industry • Development Test Instrumentation • Direction des Transmissions et de l'Informatique • Display Technologies Inc.

DTIC: Defense Technical Information Center

DTIF: Digital Transmission Interface Frame

DTK: Desired Track

Dtl: Detail

DTL: Dead Time Log • Decomposition Type List • Desktop Library • Detroit Lakes (USA-MN) (location identifier) • Detroit Testing Laboratory (USA-IL) • Dictionary Term Lookup • Diode-Transistor Logic

DTLCC: Design To Life Cycle Cost

Dtld: Detailed

DTLN: Dateline

DTLS: Descriptive Top-Level Specification

DTM: Data Transfer Mechanism • DEC Test Manager • Department of Terrestrial Magnetism • Design Test Model • Desktop Manufacturing • Developmental Test Model • Diamond Turning Machine • Digitaal Terrein Model • Digital Terrain Map • Digital Terrain Model • Digital Terrain Modelling • Dortmund (Germany) (location identifier) • Dynamic Test Model • Dynamic Transfer Mode

DTMAG: Dual Technique Magnetometer

DTMF: Data Tone Multiple Frequency • Dial Tone Modulated Frequency • Dial Tone Multifrequency • Dual Tone Multifrequency (*)

DTMO: Design Test and Mission Operation • Development, Test and Mission Operation

DTMS: Development, Test, and Mission Support • Digital Test Measurement System • Digital Test Monitor System

DTN: Dataflow Technology Nederland BV • Designated Transport Node • Shreveport Downtown (USA-LA) (location identifier)

DTNI: Direction Technique Nationale Informatique

Dto: Descuento • Distrito

DT&O: Data, Test & Operation

DTO: Data, Test and Operation • Descuento • Design Test Objective • Detailed Test Objective • Developmental Test Objective • Development Test Objective • Direito • Documentation Technique Officielle

DTOL: Digital Test-Oriented Language

Dtor: Director

DTP: Design To Price • Desktop Publishing • Detailed Test Plan • Detailed Test Procedure • Detail Test Plan • Development Test Plan • Diode Test Program • Diphteria-Tetanus-Pertussis • Directory Tape Processor • Distributed Transactional Processing • Dolph-Tchebyscheff Pattern • Dynamic Test Panel • Dynamic Testing Program

DTPM: Distributed Transaction Processing Middleware • Distributed Transaction Processing Monitor

DTQE: Detection Quantum Efficiency

DTR: Daily Total Rate • Data Terminal Ready • Data Transfer Rate • Decatur Island (USA-WA) (location identifier) • Dedicated Token Ring • Delay-Throughput-Reliability • Diffusion-Transfer Reversal • Digital Tape Recorder • Distribution Tape Reel • Diurnal Temperature Rate • Draft Technical Report • Dynamic Transaction Routing

DTRA: Defense Threat Reduction Agency (USA)

Dtrbd: Daughterboard

DTRD: Development Test Requirement Document

DTRE: Data Type Refinement Environment

DTREM: Dust, Thermal and Radiation Engineering Measurement

DTRF: Data Transmittal and Routing Form

DTRIC: Dielectric Totally Internally Reflecting Concentrator

DTRM: Development and Technology Review Mission • Development Technology Review Mission

Dtrs: Distress

DTRS: Dryden Technical Report Server

DTRT: Deteriorate

DTS: Data Test Station • Data Transfer System • Data Transmission System • Data Transport System • Dedicated Transmission Service • Defense Technology Study • Detailed Test Specification • Development Test Satellite • Development Test Sensor • Digital Telemetry System •

Digital Terrain System • Digital Tracking System • Digital Transmission System • Distributed Time Service • Droits de Tirages Spéciaux • Dynamic Test System

Dtsch: Deutsch

DTSL: Data Transfer Systems Laboratory

DTSS: Dartmouth Time-Sharing System • Digital Topographic Support System

DTST: Defense Technology Study Team

DTT: Data Throttling Time • Data Transfer Timing • Design Test Tool • Detroit (USA-MI) (location identifier) • Direction des Transports Terrestres • Double Taxation Treaty

dTTL: Differential Transistor-Transistor Logic

DTTU: Data Transmission Terminal Unit

DTTV: Digital Terrestrial Television

DTU: Data Transfer Unit • Data Transmission Unit • Development Test Unit • Digital Telemetry Unit • Digital Test Unit • Direct To User

D'ture: Departure

DTUTF: Digital Tape Unit Test Facility

DTV: Data Translation and Verification • Desktop Video • Development Test Vehicle • Digital Television • Driver Thermal Viewer • Dynamic Test Vehicle

DTVC: Data Transmission and Verification Converter • Desktop Video Conferencing • Digital Transmission and Verification Converter

DTV&I: Development, Test, Verification, & Integration

DTVI: Development, Test, Verification, and Integration

DTW: Destop Workstation • Detroit - Wayne County (USA-MI) (location identifier) • Dynamic Time Warping

Dtzd: Dutzend

du: Duplicatus

dU: der Unterzeichnete

Du: Duke • Dutch

DU: Decision Unit • Depleted Uranium • Detector Unit • Digital Unit • Diplôme d'Université • Diplôme Universitaire • Disk Usage • Display Unit • Distribution Unit • Dobson Unit • Doctor of the University • Documentation Unit • Do Until • Dust • Dutch

DUA: Demande d'Utilisation d'Ariane • Deterministic Uncertainty Analysis • Digital Uplink Assembly • Directory User Agent • Durant Eaker (USA-OK) (location identifier)

DUAL: Dynamic Universal Assembly Language

DUAP: Dual Use Application Program

DUAT: Direct User Access Terminal

DUATS: Direct User Access Terminal Service

Dub: Dubious • Dubitans • Dublin (Ireland)

DUB: Dublin (Ireland) (location identifier)

DUBBLE: Dutch-Belgian Beamline Experiment

DUBS: Dual-bus Unidirectional Broadcast System

DUC: Dense Upper Cloud • Digital Uplink Command • Display Use Chain • Duncan Halliburton (USA-OK) (location identifier) • Dutch Utilization Centre

DUD: Dunedin (New Zealand) (location identifier)

DUE: Datenübertragungseinrichtungen • Division of Undergraduate Education • Dundo (Angola) (location identifier)

DUEL: Data Update Edit Language

DUES: Diplôme Universitaires d'Études Scientifiques

DUF: Duck (USA-NC) (location identifier)

DUG: Douglas (USA-AZ) (location identifier)

DUI: Data Use Identifier • DELPHI User Interface • Directory User Interface • Driving Under Influence • Driving Under the Influence

DUJ: Dubois (USA-PA) (location identifier)

DUK: Dukuduk (South Africa) (location identifier)

Dulc: Dulcis

DULCE: Demonstration Unit for Low-order Cryogenic Etalon

DULCO: Diplôme Unilingue des Langues et Civilisations Orientales

DUM: Dumai (Indonesia) (location identifier) • Dummy

DUMAND: Deep Underwater Muon And Neutrino Detection • Deep Underwater Muon And Neutrino Detector

DUMC: Dutch Union Map Catalogue

Dun: Dundee (UK-Scotland)

DUN: Dial-Up Networking • Dundas (Greenland) (location identifier)

DUNDEE: Down Under Doppler and Electricity Experiment

DUniv: Doctor of the University

DUNS: Data Universal Numbering System

Duo: Duodecimo

DUO: Disk Unseen Object

Dup: Duplicate

DUP: Data User Programme • Data Utilisation Programme • Déclaration d'Utilité Publique • Décret d'Utilité Publique • Disk Utility Program • Duplicate

Dupdo: Duplicado

Dupl: Duplicado • Duplicate

Dupld: Duplicated

DUPLEX: Dust-Prominent Longitudinally Extended (PPN)

Duplx: Duplex

Duplxr: Duplexer

DUPS: Datenübertragungsprogrammsystem

DUQ: Duncan Quam (Canada-BC) (location identifier)

Dur: Duration • Durban (South Africa) • Durham (UK-England)

DUR: Duration • Durban - Louis Botha (South Africa) (location identifier) • Durham (Starlink internal code)

DURC: During Climb

DURD: During Descent

DURG: During

Durn: Duration

DUS: Data Utilization Station • Diplôme Universitaire Scientifique • Düsseldorf Lohausen (Germany) (location identifier) • Düsseldorf Rhein-Rhur (Germany) (location identifier)

DUSO: Dutch User Support Organization

Dut: Dutch

DUT: Delft University of Technology (Netherlands) • Device Under Test • Diplôme Universitaire de Technologie (France) • Dutch Harbor (USA-AK) (location identifier)

DUTA: Display Unit Test Assembly

DUTE: Digital Universal Test Equipment

DUV: Damaging Ultraviolet • Data Under Voice • Deep Ultraviolet

DUVAD: Durham (Starlink internal code)

DUVE: Diffuse Ultraviolet Experiment

DUVS: Deep Ultraviolet Survey

DUZ: Deutsche Universitätszeitung

d/v: Días Vista

dv: Deo Volente

DV: Daily Value • Datenverarbeitung • Defective Vision • Deo Volente • Designated Verification • Designee for Verification • De Vaucouleurs • Differential Velocity • Digital Video • Direct Vision • Discrete Variational (method) •

Discrete Velocity • Discrimination Value • Divide • Double Vision • Driver

DVA: Datenverarbeitungsanlage • Deutsche Verlagsanstalt (Germany) • Deva (Romania) (location identifier) • Differential Velocity Aberration • Differential Voltage Amplifier • Directed Vector Autocorrelation • Direct Vector Autocorrelation • Direct Video Access • Diverse Vector Area • Document Validation Audit

DVAA: Delaware Valley Amateur Astronomers (USA-PA)

DVB: Digital Video Broadcast • Digital Video Broadcasting

DVC: Desktop Video Conferencing • Device • Digital Video Cassette • Digital Video Compression • Digital Video Conferencing • Digital Voice Communication

D/VD: Data/Voice Data

DVD: Data/Voice Data • Datenverarbeitungsdienst • Delta Velocity Display • Digital Versatile Disk • Digital Video Disk • Direct Versatile Drive • Divide

DVDP: Dry Valley Drilling Program

DVDS: Digital Video Display System

DVE: Datenverarbeitungseinrichtung • Design Verification Experiment • Differential Vector Equation • Digital Video Effect • Distributed Virtual Environment

DVF: Digital Visual Fortran • Displacement Vector Field

DVFR: Day Visual Flight Rule • Defense Visual Flight Rule

DVG: Datenverarbeitungsgesellschaft • Deutsche Vakuum Gesellschaft (Germany) • Digital Video Generator

DVGW: Deutscher Verein des Gas- und Wasserfaches (Germany)

DVI: Data/Voice Integration • Datenverarbeitungsinstitut • Device Independent • Difference Vegetation Index • Differential Vegetation Index • Digital Video Imaging • Digital Video Interaction • Digital Video Interactive • Direct Voice Input • Disque Vidéo Interactif

DVIP: Digital Video Integrator and Processor

DvJ: Dag van het Jaar

Dvl: Develop

DVL: Devils Lake (USA-ND) (location identifier) • Direct Voice Line

DVLP: Develop

Dvlpd: Developed

DvM: Dag van de Maand

DVM: Digital Velocity Meter • Digital Video Microscope • Digital Voltmeter • Discontinuous Variational Method • Displayed Virtual Machine

DVMA: Direct Virtual Memory Access

DVMRP: Distance Vector Multicast Routing Protocol

DVN: Davenport (USA-IA) (location identifier)

DVO: Davao Mati (Philippines) (location identifier) • Decimal Voltage Output • Delta Velocity Off • Delta Velocity On • Digital-Video Object • Durchführungsverordnung

DVOR: Doppler VHF Omnidirectional Range

DVP: Data Validation Program • Davenport Downs (Australia-QLD) (location identifier) • Design Verification Program • Digital Video Plotter • Digital Video Processing • Digital Video Producer • Dynamic Virus Protection

DVPDF: Dry Vacuum Pump Discharge Filter

DVPF: Dry Vacuum Pump Filter

Dvr: Driver

DVR: Daly River (Australia-NT) (location identifier) • Direct Voltage Readout • Discrete-Variable Representation • Driver

DVS: Datenverarbeitungssystem • Daten-Video-Sprache • Daily Visits Statistics • Delta Velocity System • Design Verification Specification • Digital Voice System • Digital Voltage Source • Distributed Virtual System • Document Vector Space

DVSZ: Datenverarbeitungs-Servicezentrum (Germany)

Dvt: Deviate

DVT: Design Verification Test • Design Verification Testing • Deutsche Verband Technisch-Wissenschaftlicher Vereine (Germany) • Digital Video Tape • Phoenix - Deer Valley (USA-AZ) (location identifier)

Dvtd: Deviated

DVTR: Digital Video Tape Recorder

DVU: Delta Velocity Ullage

DVV: Downward Vertical Velocity

DVVP: Development, Verification and Validation Plan

DvW: Dag van de Week

DVWG: Deutsche Verkehrswissenschaftliche Gesellschaft (Germany)

DVX: Digital Voice Exchange

DVZ: Datenverarbeitungszentrum

d/w: Dust Wrapper

dw: Deelswoord • Dienstwillig

Dw: Dwarf

D-W: Germany (formerly West Germany) (postal country code)

D/W: Dock Warrant

DW: Data Warehouse • Data Warehousing • Data Word • Dead Weight • Delivery Weight • Developed With • Diabatic Warming • Direct Weighted • Dishwasher • Distilled Water • Double Weight • Drum Write • Dwingeloo

DWA: Dwangwa (Malawi) (location identifier)

DWAC: Distributed Write Address Counter

DWARF: Dwingeloo/Westerbork Astronomical Reduction Facility (Netherlands)

DWB: Data Word Buffer • Soalala (Madagascar) (location identifier)

DWBA: Distorted Wave Born Approximation

DWC: Digital Wireless Communications

DWCD: Drinking Water Criteria Document

DWD: Deutscher Wetterdienst (Germany) • Dynamic Weather Display

DWDM: Dense Wavelength Division Multiplexer (*) • Dense Wavelength Division Multiplexing (*) • Digital Wave Division Multiplexer • Digital Wave Division Multiplexing

DWE: Delivery With Equipment • Doppler Wind Experiment

DWF: Dayton Wright (USA-OH) (location identifier)

Dwg: Drawing

DWG: Digital Waveform Generator • Discipline Working Group • Drawing

DWH: Data Warehousing • Houston - David Wayne Hooks (USA-TX) (location identifier)

DWI: Data Word In • Descriptive-Word Index • Descriptor Word Index • Driving While Intoxicated • Dutch West Indies (now Netherlands Antilles)

DWIA: Distorted Wave Impulse Approximation

DWIM: Diffuse Warm Ionized Medium • Do What I Mean

DWIPS: Digital Weather Imaging Processing System

DWIS: Do What I Say

DWIT: Do What I Think

DWL: Doppler Wind Lidar

Dwn: Drawn

DWN: Oklahoma City Downtown (USA-OK) (location identifier)

DWNDFTS: Downdrafts

DWO: Development Work Order • Direct Writing Oscillograph

DWP: Department of Water and Power • Digital Wave Processor

DWPNT: Dew Point

DWQT: Deferred Write Queue Threshold

Dwr: Drawer

DWR: Development Work Request • Digital Wired Recorder • Disaster Warning Receiver • Doppler Weather Radar

DWS: Demineralized Water System • Design Work Study • Detailed Work Statement • Detector Workstation • Development Work Statement • Disaster Warning Satellite • Disaster Warning System (*) • Doppler Wind Sensor • Doppler Wind Sounder • Orlando - Walt Disney World (USA-FL) (location identifier)

dwt: Pennyweight

DWT: Dead Weight Tonnage • Dead Weight Tons • Discrete Wavelet Transform • Dog Wags Tail

DWV: Dielectric Withstanding Voltage

DWZ: Dat Wil Zeggen • Directie Wetenschap en Technologie

Dx: Diagnosis

DX: Daily Exposure • Data Explorer • Distance • Long Distance

DXA: Deux-Alpes (France) (location identifier)

DXB: Diffuse X-ray Background • Drawing Interchange Binary • Dubai (UAE) (location identifier)

DXC: Data Exchange Control • Drawing Exchange Control

DXD: Dixie (Australia-QLD) (location identifier)

DXF: Data Exchange Format • Digital Exchange Format • Drawing Exchange Format • Drawing Interchange File • Drawing Interchange Format

DXI: Data Exchange Interface

DXP: Data Exchange Protocol • Digital Experiment Processor

DXR: Danbury (USA-CT) (location identifier) • Deep X-Ray

DXRB: Diffuse X-Ray Background

DXRBS: Deep X-ray Radio Blazar Survey

DXS: Data Exchange Standard • Data Exchange System • Diagnostic X-ray Spectrometry • Diffuse X-ray Spectrometer • Diffuse X-ray Spectrometry

DXT: Data Extract

dy: Demy

d/y: Delivery

Dy: Delivery • Deputy • Dysprosium

DY: Benin (international vehicle registration code) • Benin (ISO country code) • Dahomey (ISO country code) • Day • Dockyard • Dudley (UK postal code)

DYA: Dysart (Australia-QLD) (location identifier)

DYANA: Dynamics-Adapted Network for Atmosphere • Dynamics-Adapted Network for the Atmosphere • Dynamik Adaptiertes Netz von Atmosphärer Netzstationen

DyAtO: Dynamics of Atmospheres and Oceans (periodical code)

DYB: Dynamic Breaking

Dyd: Dockyard

DYF: Do Yourself In

DYG: Dayong (China-PRC) (location identifier)

DYL: Doylestown (USA-PA) (location identifier)

DYM: Diamantina Lakes (Australia-QLD) (location identifier)

dyn: Dyne

Dyn: Dynamic • Dynamics • Dynamo • Dynamometer • Dynastic • Dyne

DyN: Día y Noche

DYN: Dynamic

DYNAPAC: Dynamic Plasma Analysis Code

DynCo: Dynamics and Control (periodical code)

DYR: Anadyr (Russia) (location identifier)

DYS: Abilene Dyess (USA-TX) (location identifier)

DYSIM: Dynamic Simulator

DYTEC: Dynamique de la Terre er Évolution du Climat

DYU: Dushanbe (Tajikistan) (location identifier)

DYW: Daly Waters (Australia-NT) (location identifier)

dz: Buthani (ISO language code)

Dz: Dozen • Drizzle

DZ: Algeria (international vehicle registration code) • Algeria (ISO country code) • Algérie (ISO country code) • Doppelzentner • Dozen • Drizzle

DZA: Algérie (ISO country code) • Algeria (ISO country code) • Dzaoudzi (Comoros) (location identifier)

DZB: Deutsche Zentralbibliothek (Germany)

DZC: Dual-Zone Chamber

DZCh: Deutscher Zentralausschuß für Chemie (Germany)

DZD: Dinar (Algeria) (ISO currency code)

DZI: Codazzi (Colombia) (location identifier)

DZN: Dzezkazgan (Kazakhstan) (location identifier)

DZNM: Croatian State Office for Standardization and Metrology

DZO: Durazno (Uruguay) (location identifier)

DZU: Dazu (China-PRC) (location identifier)

D-Zug: Durchgangszug

DZug: Durchgangszug

D0: Day Zero

D2: Angola (international aircraft registration code) • Degrade and Destroy

D2C: Decimal to Character

D2T2: Dye Diffusion Thermal Transfer

D2X: Decimal to Hexadecimal

D3: Demand Driven Dispatch

D4: Cape Verde (international aircraft registration code)

D6: Comoros Islands (international aircraft registration code)

D8: Group of Eight Developing Countries

E

e/: Entre • Envío

e: Base of Natural Logarithms • Eccentricity • Edition • Educated • Elder • Eldest • Electric • Electricity • Electromotive • Electron • Elementary Charge • End • Engineer • Engineering • Entrance • Época • Equatorial Conformation • Erlang • Error • Euler Number • Excellence • Excellent • Internal Energy • Polarization Vector • Positron • Relative Elongation • Specific Internal Energy • Transcendental Number • Unit Coordinate Vector • Wet Air

E: Color Excess • Earl • Earth • East • Easter • Eastern • East London (UK postal code) • Eccentric Anomaly • Echo (international alphabet) • Edinburgh (UK-Scotland) • Efficiency • Egypt • Egyptian • Einsteinium • Einwohner • Electric Field • Electric Field Strength • Electrode Potential • Electrometric Effect • Electromotive Force • Electronic • Elevation Angle • Elevator • Elimination Reaction • Elliptical Galaxy • Elongation • Emalangeni (currency) • Eminence • Éminence • Ended • Energy • Engineer • Engineering • England (UK) • English • E-number • Eötvös • Equation of Time • Equator • Equatorial • Erdgeschoß • Erg • Error • España • Est • Estacionamiento • Este • Estimated • Evening • Exa • Excellent • Expected • Expressway • Illuminance • Internal Energy • Irradiance • Kinetic Energy of an Electron • London (East_) (UK postal code) • Modulus of Elasticity • Radiative Flux • Spain (international vehicle registration code) • Spain (postal country code) • Two Hundred Fifty (Roman numeral) • Unit Matrix • Universal Negative Categorical Proposition • Young Modulus • 250 (Roman numeral)

ea: Each • En Andere • En Anderen

Ea: Estancia

E/A: Ephemeris/Attitude

EA: Each • East Africa • Eastern Arctic • Economic Adviser • Educational Affairs • Effective Action • Effective Agent • Effective Address • Effective Area • Electronics Assembly • Elemental Analyzer • Elevation Angle • Emergency Area • En Andere • En Anderen • Ente Autonomo • Enumeration Area • Environment Agency • Environmental Agent • Environmental Assessment • Erreur d'Alimentation • Ethylacrylate • Event Analyzer • Evolution Artificielle • Evolutionary Algorithm • Executing Authority • Executive Assistant • Experimental Area • Experiment Assembly • Expert Agent • Expert Assistant • Exposure Adjust • Exposure Age • Extended Area • Extended Attribute

EAA: Eagle (USA-AK) (location identifier) • Encyclopaedia of Astronomy and Astrophysics • Engineer in Aeronautics and Astronautics • Ethylene Acrylic Acid • Excitatory Amino-Acid • Experimental Aircraft Association

EAAA: Escambia Amateur Astronomers Association (USA-FL) • European Association of Advertising Agencies

EAAE: European Association for Astronomy Education (Germany)

EAAI: European Association of Aerospace Industries

EAAM: Effective Atmospheric Angular Momentum • European Association of Aerospace Manufacturers

EAAP: European Association for Animal Production

EAAS: Euro-Asian Astronomical Society (Russia)

EAB: Abbse (Yemen) (location identifier) • Editorial Advisory Board • Extra-Anatomic Bypass

EAC: Early Analysis Capability • East African Community • Educational Advisory Committee • Employee Action Circle • Energy-Absorbing Capacity • Engineering Accreditation Commission • Engineering Advisory Committee • Equipment Activity Controller • Estimate At Completion • European Accreditation of Certification • European Aerospace Conference • European Astronauts Centre (ESA) • European Atomic Commission • Expect Approach Clearance • Expect Approach Control • Experimental Apparatus Canister • Experiment Apparatus Container • Extended Attitude Control

EACE: Early Analysis Capability Enhanced • European Association of Cognitive Ergonomics

EACH: European Academies Clearinghouse

EACL: Énergie Atomique du Canada Limitée (Canada) • European Association for Computational Linguistics

EACN: Equivalent Alkane Carbon Number

EACRO: European Association of Contract Research Organisations

ead: Eadem

EAD: Electrically Alterable Device • Encoded Archival Description • Enseignement À Distance • European Aerospace Database

EADAS: Engineering and Administrative Data Acquisition System

EADB: East African Development Bank

EADI: Electronic Attitude Direction Indicator (*) • Electronic Attitude Director Indicator • Electronic Attitude Display Indicator • European Association of Development Institutes

EADN: European Astrophysical Doctoral Network

EADS: Engineering Analysis and Data System • Environmental Assessment Data System • European Aeronautic, Defense and Space Co. (merging of DASA and Aerospatiale-Matra)

EADT: Experiment Agency Data Tape

EADTU: European Association of Distance Teaching Universities

EAE: Emae (Vanuatu) (location identifier) • Enquête Annuelle d'Entreprise

EAEC: European Atomic Energy Community

EAEG: European Association of Exploration Geophysicists • Extended Asynchronous Exponential Growth

EAEMP: European Agency for the Evaluation of Medical Products

EAEO: Equal Access End Office

EAERE: European Association of Environmental and Resource Economists

EAES: European Atomic Energy Society

EaEvS: Earth and Evolution Sciences (periodical code)

EAF: Electric-Arc Furnace • Emergency Action File • Engineering Analysis Facility • Engineering Analysis Function

EAFB: Edwards Air Force Base (USA-CA)

EAFS: Effective Aerial Film Speed

EAFT: European Association For Terminology

EAG: Experimental Astrophysics Group • External Advisory Group

EAGA: East Asian Growth Accord

EAGB: Early Asymptotic Giant Branch (star) • Expéditions Astrophysiques et Géophysiques Belges (Belgium)

EAGE: European Association of Geoscientists and Engineers

EAGIL: Expert Associates Guide to Industrial Learning

EAGLE: EOS Atmospheric Global Lidar Experiment • European Association for Grey Literature Exploitation

EAGLES: Expert Advisory Group for Language Engineering Standards (*) • Expert Advisory Group on Language Engineering Standards

EAGS: European Association of Exploration Geophysicists

EAHIL: European Association for Health Information and Libraries

EAHP: École d'Application des Hauts Polymères

EAI: Enseignement Animé Interactif • Enterprise Application Integration • External Authoring Interface

EAIC: European Aerospace Industry Council

EAID: Engine Auxiliary Inlet Door

EAIE: European Association for International Education

EAIS: European Astronomical Information Service

EAITC: External Affairs and International Trade Canada

EAITech: Euro-American Institute of Technology

EAK: Kenya (international vehicle registration code)

EAL: Échelle Atomique Libre • Electromagnetic Amplifying Lens • Equipment Airlock • ESA Attached Laboratory • European Accreditation of Laboratories • European Astronomical Landscape

EAM: Earth Acquisition Mode • Effective Angular Momentum • Electrical Accounting Machine • Electronic Accounting Machine • Embedded-Atom Method • Emergency Action Message • Nejran (Saudi Arabia) (location identifier)

EAMAC: École Africaine pour la Météorologie et l'Aviation Civile

EAMEDPM: Electronic Accounting Machine and Electronic Data Processing Machine

EAMPAG: European Astronaut Medical Psychological Advisory Group

EAMPS: Extended Advanced Mobile Phone Service

EAMREA: Environmental Impact of Mineral Exploration-Exploitation in Antarctica

EAN: Effective Atomic Number • Equivalent Atomic Number • European Academic Network • European Article Number • Wheatland Phifer (USA-WY) (location identifier)

EANDC: European-American Nuclear Data Committee

EANS: European Article Numbering System

EAO: Engelhardt Astronomical Observatory (Kazakhstan) • Enseignement Assisté par Ordinateur • Étude Assistée par Ordinateur

EAON: European Asteroidal Occultation Network • Except As Otherwise Noted

EAOS: Easy Access Ordering System • Environment Australia Online Service

EAP: Basel Mulhouse (Switzerland) (location identifier) • Educational Awareness Program • Effective Air Path • Employee Assistance Program • Enterprise Architecture Planning • Environmental Affairs Program • Environmental Assessment Panel • Environment Assistance Programme • Equivalent Air Pressure • Étage d'Accélération à Poudre • Experimental Aircraft Project • Mulhouse Bâle (France) (location identifier)

EAPA: European Alliance of Press Agencies

EAPE: Eddy Available Potential Energy

EAPEE: European Association for Political and Evolutionary Economy

EAPG: European Association of Petroleum Geoscientists

EAPGE: European Association of Petroleum Geoscientists and Engineers

EApJL: Electronic Astrophysical Journal Letters

EAPOE: Extensible Authentication Protocol Over Ethernet

EAPROM: Electrically Alterable Read-Only Memory

EAPS: Earth, Atmospheric and Planetary Sciences • Euro-Asian Physical Society

EAR: Electronically Agile Radar • Employee Attitude Research • Energy-Absorbing Resin • Engineering Action Request • Engineering Analysis Report • Entity-Attribute-Relationship • Environmental Analysis Report • Kearney (USA-NE) (location identifier)

EARA: European Association for Research in Astronomy

EARCOM: East Africa Regional Remote Sensing Management Committee

EARG: Environmental Adaptation Research Group • Estación Astronómica de Río Grande (Argentina-TF)

EARL: Easy Access Report Language • Electronic Access to Resources in Libraries • Electronically Accessible Russian Lexicon

EARMA: European Association for Research Managers and Administrators

EARN: European Academic and Research Network • European Academic Research Network • European Asteroid Research Node

EAROM: Electrically Alterable Read-Only Memory

EAROS: Electrically Alterable Read-Only Store

EARP: Environmental Assessment Review Process

EARS: Electronic Access to Reference Services • Environmental Analysis and Remote Sensing

EARSC: European Association of Remote Sensing Companies

EARSEC: European Airborne Remote Sensing Capabilities

EARSeL: European Association of Remote Sensing Laboratories

EARSEL: European Association of Remote Sensing Laboratories

EARSF: European Airborne Remote Sensing Facility

EARSO: European Airborne Remote Sensing Consortium

EARTO: European Association of Research and Technology Associations (Belgium)

EARTS: En-route Automated Radar Tracking System

EAS: Earth's Artificial Satellite • Eastbay Astronomical Society (USA-CA) • Electronic Animation System • Engine Actuation System • Engineering Administrative System • Engineering Analysis System • Environmental Analyses Section • Equivalent Airspeed • Error Analysis Study • European Astronomical Society • Evolutionary and Adaptive System • Exeter Astronomical Society (UK) • Extended Announcement System • Extended Area Service • Extensive Air Shower • Extensive Atmospheric Shower • San Sebastian (Spain) (location identifier)

EASA: European Aviation Safety Authority

EASB: European Astronauts Selection Board

EASD: Equal Access Service Date

EASE: East Auroral Snapshot Experiment • Easy Access Screen Entry • Electric Automatic Support Equipment • Empirical Assessment in Software Engineering • Engineering Applications Support Programme • European Association of Science Editors • Experimental Assembly of Structures in EVA • Expert Activity Scheduling Environment

EASEP: Early Apollo Scientific Experiments Package

EASF: Enhanced All-Source Fusion

EASI: European Academic Supercomputing Initiative • Expert Analysis System Interface

EASIT: European Association for Software Access and Information Transfer

EASIZ: Ecology of the Antarctic Sea Ice Zone

EASL: Engineering Analysis and Simulation Language

EASLS: East African Scientific Literature Service

EASOE: European Arctic Stratospheric Ozone Experiment

EASS: European Astronomy and Space Service

EASSIS: European Astronomy and Space Sciences Service

EASST: European Association for the Study of Science and Technology

East: Eastern

EAST: East Australian Standard Time • Euro-African Satellite Telecommunications • European Assistance for Science and Technology Ltd. Co. (UK)

EASTT: Environmental Advisory Service for Transfer of Technology

EASy: Evolutionary and Adaptive System

EASY: Electronic Application System • Evolutionary and Adaptive System

EAT: Earliest Arrival Time • Encoder Address Translator • Environmental Acceptance Test • Environnement, Aménagement et Télédétection • Expected Approach Time • Extended Alias Table • Tanzania (international vehicle registration code) • Wenatchee (USA-WA) (location identifier)

EATA: Enhanced AT-bus Attachment

EATCHIP: European Air Traffic Control Harmonization and Integration Programme • European ATC Convergence Harmonisation and Implementation Programme

EATCS: European Association for Theoretical Computer Science • European Association for Theoretical Computing Science • European Association on Theoretical Computer Science • External Active Thermal Control System

EATF: Externally Applied Thermal Field

EATMS: European Air Traffic Management System

EATR: Extravehicular Assist Teleoperator Robot

EAU: Eau Claire (USA-WI) (location identifier) • Economic Analysis Unit • Émirats Arabes Unis • Erase All Unprotected • Uganda (international vehicle registration code)

EAV: Effective Angular Velocity • End of Active Video • Engine Assembly Vehicle

EAW: Equivalent Average Word

EAWEP: East Asia and Western Pacific

EAWS: Engineering Analysis Workstation

EAX: Electronic Automatic Exchange

EAZ: Tanzania (international vehicle registration code)

EAZO: Energy Active Zone of the Ocean

E/B: Equipment Bay

EB: Eastbound • Eclipsing Binary • Economy and Business • Electricity Board • Electron Beam • Electron Bombarded • Electron Bombardment • Electronic Beam • Electronic Book • Electronics Bay • Ellerman Bomb • Emergency Box • Empirical Bayes • Encyclopaedia Britannica • Equatorial Band • Equipment Bay • Executive Board • Executive Body • Executive Bulletin • Expendable Booster

EBA: Elba Island (Italy) (location identifier) • Electronics Bay Assembly • Exponentiated Born Approximation

EBAA: European Business Aviation Association

EBAD: École des Bibliothécaires, Archivistes et Documentalistes

EBAE: European Bureau for Adult Education

EBAR: Edited Beyond All Recognition

EBB: Electronic Bulletin Board • Elegant Breadboard • Entebbe Kampala (Uganda) (location identifier)

EBBR: Energy-Balance Bowen Ratio

EBBT: Equivalent Black-Body Temperature

EBC: EISA Bus Controller • Electron Beam Coating • Electron Beam Control • Electronics Begin Countdown • Emulated Buffer Computer • European Bibliographical Center

EBCCD: Electron-Bombarded Charge-Coupled Device

EBCD: Extended Binary-Coded Decimal

EBCDIC: Extended Binary Coded Decimal Interchange Character • Extended Binary Coded Decimal Interchange Code

EBCG: European Biotechnology Coordination Group

ebd: Ebenda

EBD: El Obeid (Sudan) (location identifier) • Equivalent Binary Digit

EBE: Excédent Brut d'Exploitation • Except British Commonwealth • Extraterrestrial Biological Entity

e-beam: Electron Beam

EBEB: Europäisches Büro für Erwachsenenbildung

EBER: Electron-Beam Electroreflectance

EBES: European Board for Edifact Standards

EBF: Electronic Book Format • Emergency Bug Fix

EBG: El Bagre (Colombia) (location identifier) • Electron Beam Generator • Electron Beam Gun • Escher-Bach-Gödel

EBI: Equivalent Background Input • European Bioinformatics Institute • Extended Background Investigation

EBIC: Electron Beam Induced Current • Electron Beam Induction Current • European Business Information Centre

EBICON: Electron Bombardment Induced Conductivity

EBIN: Eindhovens Bibliotheek Informatie Netwerk

EBIP: European Biotechnology Information Project

EBIR: Electron Beam Image Recorder

EBIS: Electron Beam Ion Source • ESO Budget Information System • European Biotechnology Information Service

EBIT: Electron Beam Ion Trap • European Broadband Interconnection Trial

EBJ: Esbjerg (Denmark) (location identifier)

EBK: Einstein-Brilloin-Keller

EBL: Electron Beam Lithography • Explanation-Based Learning • Extragalactic Background Light

EBLIDA: European Bureau of Library, Information and Documentation Associations

EBLP: Estimated Best Linear Predictor

EBM: El Borma (Tunisia) (location identifier) • Electrical Bridgman Method • Electro-Beam Machine • Electro-Beam Machining • Energy Balance Model • Estimation Before Modelling • Evidence-Based Medicine • Extended Boolean Model

EbN: East by North

EBN: Ebon (Marshall Islands) (location identifier) • Extended Border Node

EBND: Eastbound

EBNF: Extended Backus-Naur Form (*) • Extended Backus Normal Form

EBNN: Explanation-Based Neural Network

EBO: Ebon (Marshall Islands) (location identifier)

EBOF: Enterprise Business Objects Facility

EBONE: European Backbone

EBOP: Eclipsing Binary Orbit Program • Enhanced BATSE Occultation Package

EBP: Electron Beam Process • Electron Beam Pumped • Electronic Book Player • Electronics Base Plate • European Books in Print • EUV Bright Point

EBPP: Electronic Bill Payment and Presentment

EBr: Encyclopaedia Britannica

EBR: Baton Rouge Downtown (USA-LA) (location identifier) • Electron Beam Recorder • Electron Beam Recording • Electron Beam Regulator • Experimental Breeder Reactor • Extended Boot Record • Extended-partition Boot Record

EBRD: European Bank for Reconstruction and Development

EBRF: Extended Bidirectional Reflectance Factor

EbS: East by South

EBS: Economics and Business Studies • Electron Beam System • Electron-Bombarded Semiconductor • Electron-Bombarded Silicon • Electronically-controlled Braking System • Electronic Band Spectra • Electronic Brain Storming • Emergency Breathing System • Entrée Banalisée Synchrone • Equipment Breakdown Structure • Executive Briefing System • Webster City (USA-IA) (location identifier)

EBSA: European Biophysical Societies Association

EBSI: École de Bibliothéconomie et des Sciences de l'Information (Canada-QC) (*) • École des Bibliothèques et des Sciences de l'Information (Canada-QC)

EBT: Electron Beam Technique • Electron Beam Test • Electronic Benefits Transfer • Electronic Book Technologies • Electronic Book Technology • Elmo Bumpy Torus

EBTE: Electronics Bay Test Equipment

EBTR: Electronic Bearing-Time Recorder

EBU: Eddy Break-Up • Electronics Bay Unit • European Broadcasting Union • Saint-Étienne Bouthéon (France) (location identifier)

EBV: Electron Beam Vaporization

EBW: Ebolowa (Cameroon) (location identifier) • Effective Bandwidth • Electron Beam Welding • Exploding Bridge Wire • Explosive Bridge Wire

EBWR: Experimental Boiling-Water Reactor

e/c: En Cuenta

ec: Ecuador • Exempli Causa

Ec: Ecuador

E/C: Encoder/Coupler

E&C: Engineering & Commissioning

EC: Earth Convergency • East Caribbean • East Central London (UK postal code) • East Coast • Economy Class • Ecuador (international vehicle registration code) • Ecuador (ISO country code) • Eddy Correlation • Eddy Current • Edinburgh-Cape (survey) • Educational Committee • Education Committee • Effective Concentration • Elastic Center • Electoral Committee • Electrochemical • Electrolytic Corrosion • Electron Capture • Electron Cyclotron • Electronic Combat • Electronic Computer • Elemental Carbon • Element Contractor • Emergency Commission • Enabling Command • Encoder-Coupler • Engine Control • Engine Cutoff • Engineering Change • Engineering Corps • Environmental Capacity • Environmental Compliance • Environmental Control • Environment Canada • Équateur (ISO country code) • Equipment Change • Erase Character • Error Correction • Essentiallity Code • Ethylene Carbonate • Eurocheque • European Commission • European Communities • European Community • Europese Commissie • Events Controller • Events Coupler • Evolutionary Computation • Exchange Carrier • Executive Committee • Executive Council • Expansive Classification • Experiment Canister • Experiment Computer • Extended Control • Extended Coverage • Extension Course • London (East Central_) (UK postal code) • Spain (international aircraft registration code)

ECA: Earth-Crossing Asteroid • East Tawas - Emmet County (USA-MI) (location identifier) • Economic Commission for Africa • Economic Cooperation Administration • Economic Cooperation Agreement • Eigenvalue Change Analysis • Electron-Cyclotron Absorption • Electronic Computer Assembly • Electronic Control Amplifier • Electronics Control Assembly • Engine Control Assembly • Engineering Change Analysis • Epoxy Curing Agent • European Common Allocation • European Court of Auditors • European Crystallographic Association • Europe Continental Airlines

ECAC: Earth-Crossing Asteroids and Comets • Electromagnetic Compatibility Analysis Center (USA) • European Civil Aviation Conference

ECACC: European Centre for Animals Cell Cultures

ECAD: Electronic Computer-Aided Design • Electronics and Computer Applications Division • Electronics Computer-Aided Design

ECAFE: Economic Commission for Asia and the Far East (now ESCAP)

ECAI: European Conference on Artificial Intelligence

Ecal: Electronic Calibration

ECAL: East Coast Alert Landing • Electronic Channel Alignment • Enjoy Computing And Learn

ECALPO: European Computer-Aided Learning Program

ECAM: Electronic Centralized Aircraft Monitor • Electronic Centralized Aircraft Monitoring • Electronic Centralized Aircraft Monitoring • European Conference on Application of Meteorology • Extended Communications Access Method

ECAMP: Environmental Compliance Assessment and Management Programme • European Conference on Atomic and Molecular Physics

ECAN: Electronic Calibration And Normalization

ECAO: Environmental Criteria and Assessment Office

ECAP: Electronic Circuit Analysis Program • Electronic Customer Access Program

ECARP: European Computational Aerodynamic Research

ECAS: Eight-Color Asteroid Survey • European Courses in Advanced Statistics • Experiment Computer Application Software

ECASS: Electronically-Controlled Automatic-Switching System • Export Control Automated Support System

ECAT: European Centre for Automatic Translation

ECB: Electronic Code Book • Electronic Components Board • Engineering Change Board • Entry Control Block • Environment Coordination Board • European Central Bank • Event Control Block • Event Control Buffer • Event Controlled Block

ECBC: Extended Cross Border Coverage

Ecc: Eccentric • Eccetera

ECC: Earth Continuity Conductor • Electrochemical Concentration Cell • Electron Capture to the Continuum • Electronic Calibration Center • Electronic Command Center • Electronic Computer Center • Electronic Cuff Checklist • Element Control Center • Emergency Core Coolant • Emergency Core Cooling • Engineering Change Control • Engineering Critical Component • Enter Cable Change • Error Check and Correcting • Error Check Code • Error Checking and Correcting • Error Checking and Correction • Error Code Correction • Error-Correcting Code • Error Correction Code • Estimated Correction Cost • EUROCONTROL Experimental Centre • European Coordinating Committee • European Coordination Committee • European Crystallographic Committee • Experimental Computer Complex

ECCA: Electronic Component Checkout Area

ECCAI: European Coordinating Committee on Artificial Intelligence

ECCAP: Electromagnetic Compatibility Circuit Analysis Program • Electromagnetic Compatibility Computer Analysis Proposal • Electromagnetic Compatibility Computerized Analysis Program

ECCB: Engineering Change Control Board

ECCC: Electronic Community Coordinating Committee (ACM)

ECCCS: Emergency Command Control Communications System

ECCD: Electron-Cyclotron Current Drive

Eccl: Ecclesiastic

Eccles: Ecclesiastic

ECCM: Electronic Counter-Counter Measure • Equilibrium Cloud Condensation Model

ECCN: European Climate Computing Network

ECCO: European Communities Communications Office • Extremely-heavy Cosmic-ray Composition Observatory

ECCOMAS: European Community on Computational Methods in Applied Sciences

ECCP: Experiment Configuration Control Panel

ECCS: Emergency Core-Cooling System

ECCV: European Conference on Computer Vision

ECD: Early Closing Day • Effective Cutoff Diameter • Electric Control Drive • Electrochromic Device • Electron-Capture Detection • Electron-Capture Detector • Electronic Cash Disbursement • Electronic Control Display • Electrostatic Charge Decay • Engineering Control Drawing • Enhanced Color Display • Entry Corridor Display • ESA Computer Department • Estimated Completion Date • Extragalactic Cepheid Database

ECDA: Engine Control Development Area • Experiments on Cosmic Dust Analogues • Extensible Compound Document Architecture

ECDB: Engineering Chemistry Database

ECDC: Economic Cooperation among Developing Countries • Engineering Configuration Data Control • External Countdown Clock

ECDIN: Environmental Chemical Data and Information Network

ECDIS: Electronic Chart Display and Information System

ECDL: European Computer Driving Licence • European Conference on Digital Libraries

ECDR: Electrostatic Corona Discharge Reactor • Experiment Conceptional Design Review

ECDU: Electrical Coupling Display Unit • Electronic Coupling Display Unit

ECE: Eastern and Central Europe • Economic Commission for Europe • Electrical and Computer Engineering • Electrical Checkout Equipment • Electron-Cyclotron Emission • Element Characteristics Equation • Experiment Checkout Equipment • External Consultant Examiner

ECEF: Earth-Centered Earth-Fixed

ECEH: European Centre for Environment and Health

ECF: Echo Frame • Effective Cloud Fraction • Effective Cutoff Frequency • Electron Cyclotron Frequency • Energy Conversion Factor • Engine Calibration Facility • Enhanced Connectivity Facility • Equipment Calibration Facility • Equivalency Capability File • European Coordinating Facility • Experimental Calibration Facility • Extended-Care Facility • Externally Caused Failure

ECFA: European Committee for Future Accelerators

ECFY: Engineering Careers Foundation Year

ECG: Ecosystem Conservation Group • Electrocardiogram • Electrocardiograph • Electrocardiography • Electrocatalytic Gas • Electroepitaxial Crystal Growth • Elizabeth City (USA-NC) (location identifier)

ECGC: Electron Capture Gas Chromatograph

ECGI: Error-Correcting Grammatical Inference

Ech: Échelon

ECH: Echuca (Australia-VIC) (location identifier) • Electron Cyclotron Heating

ECHAM: European Centre Hamburg Model

ECHIVAL: European project on Climatic and Hydrological Interactions between Vegetation, Atmosphere and Land

ECHO: European Commissions Host Organisation • European Community Humanitarian Office • Extraction and Classification of Homogeneous Objects

ECHP: European Community Household Panel

ECHR: European Convention on Human Rights

ECHT: European Conference on Hypertext

ECI: Earth-Centered Inertia • Earth-Centered Inertial (*) • Eau Chaude Industrielle • Electrical Circuit Interrupter • Electronic Control Instrumentation • Employment Cost Index • Energy-Cost Indicator • European Corpus Initiative • External Call Interface

ECIA: European Council of Information Associations

ECICS: European Customs Inventory of Chemical Substances

ECID: En Route Computer Identification

ECIDC: Educational Cooperation between Industrialized and Developing Countries

ECIE: Executive Council on Integrity and Efficiency

ECIL: Electronic Corporation of India Ltd.

ECIP: Electronic Cataloguing in Publications • European Community Investment Partner

ECIR: Étoile Carbonnée Infrarouge

ECIS: Earth-Centered Inertial System • Error Correction Information System • European Committee for Interoperable Systems • European Community Information Service • European Conference on Information Systems • Extragalactic Card Index System

ECISS: European Committee for Iron and Steel Standardization

ECIT: Electronic Combat Integrated Test

ECITC: European Committee for Information Technology Certification

ECJ: European Court of Justice

ECK: Emergency Communication Key

ECKD: Extended-Count-Key-Data

ECL: École Centrale de Lyon (France) • Effluent Concentration Limit • Electronic Circuit Logic • Emergency Cooling Limit • Emitter Control Logic • Emitter-Coupled Logic • Existing Chemicals List

ECLA: Economic Commission for Latin America • European Community Library Association

ECLAC: Economic Commission for Latin America and the Caribbean

ECLAIR: European Collaborative Linkage of Agriculture and Industry through Research • Évaluation de Critères Liés pour l'Automatisation des Interrogations et des Recherches • Extensible Class Library for Information Retrieval

Ecli: Eclipse • Ecliptic

ECLIPS: Experimental Cloud Lidar Pilot Study (*) • Experimental Cloud Lidar Project Study

ECLIS: Economic and Cultural Landscape Information System

ECLO: Emergency Locust Operations Center • Emitter-Coupled Logic Operator

ECLRv: Electrical Communication Laboratories Review (periodical code)

ECLS: Environmental Control and Life Support • Environmental Control and Life-support System

ECLSS: Environmental Control and Life Support System

ECM: Electric Coding Machine • Electrochemical Machining • Electromagnetic Countermeasure • Electromechanical Counting Mechanism • Electron-Cyclotron Maser • Electronic Control Module • Electronic Countermeasure • Ellipsoid Collector Mirror • Emulation Control Module • Energy Conservation Measure • Energy Conversion

Management (periodical code) • Engine Condition Monitoring • Engineering Change Management • Environmental Corrosion Monitor • Error Correcting Memory • Error Correction Mode (*) • European Common Market • European Crystallographic Meeting • Extended Core Memory • Extra-Cloud Medium • Extra Crew Member

ECMA: East Coast Magnetic Anomaly • European Computer Manufacturers' Association

ECMB: Energy Center Management Board

ECMEA: European Conference of Meteorological Experts for Aeronautics

ECMI: European Consortium for Mathematics in Industry

ECML: European Conference on Machine Learning

ECMOG: Erie County Mobile Observing Group (USA-PA)

ECMP: Equal Cost Multipath

ECMR: Electronic Calibration Maintenance Record

ECMS: Electronic Copyright Management System • Engine Condition Monitoring System

ECMU: Extended Core Memory Unit

ECMWF: European Centre for Medium-range Weather Forecasting • European Centre for Medium-range Weather Forecasts (UK) (*)

ECN: Energieonderzoek Centrum Nederland • Engineering Change Notice • Environmental Change Network • Environmental Communications Network • Epoxy-Cresol-Novolak • Ercan (Cyprus) (location identifier) • European Centres and Networks • European Consulting Network • European Cosmological Network • European Cosmology Network (*) • Explicit Congestion Notification

ECNE: Enterprise-Certified NetWare Engineering

ECNOD: ESA Computer and Network Operations Department

Eco: Ecological • Ecology

ECO: Earth-Crossing Object • Ecological • Ecology • Economic Cooperation Organization • Effective Competitive Opportunity • Electron-Coupled Oscillator • Electronic Central Office • El Encanto (Colombia) (location identifier) • Energy Conservation Opportunity • Engine Combustion • Engine Cutoff • Engineering Change Order • Etscorn Campus Observatory (USA-NM) • Expected Cost of Observing

ECOA: Equal Credit Opportunity Act (USA)

ECOD: European Consortium for Ocean Drilling

ECODU: European Control Data Users

Ecol: Ecología • Ecological • Ecology

ECOLE: European Collaborative Learning Environment

E-COM: Electronic Computer-Originated Mail

ECOM: Electronic Computer-Originated Mail • En-route Communications • ESA Costing Model

ECOMARGE: Ecology of the Continental Margins

ECOMET: European Cooperation in Meteorology

Econ: Economia • Economic • Economics • Economist • Economy

ECOOP: European Conference on Object-Oriented Programming

ECOPS: European Committee for Ocean and Polar Sciences • European Committee on Ocean and Polar Sciences

ECOR: Engineering Commission on Ocean Resources • Engineering Committee on Oceanic Resources (*)

ECORS: Étude des Continents et Océans par Réflexion Sismique

ECOS: ESA Costing Software • Experiment Computer Operating System

ECOSAT: Étude du Climat par Observation Spatiale de l'Atmosphère de la Terre

ECOSOC: Economic and Social Council

ECOSSE: Educational Channel Out of Scotland Servicing Europe

ECOWARM: European Committee for Water Resources Management

ECOWAS: Economic Community of West African States

ECP: Écart Circulaire Probable • École Centrale de Paris (France) • Effective Core Potential • Encryption Control Protocol • Engineering Change Proposal • Engineering Control Program • Enhanced Capabilities Port • Environmental Choice Program • European Common Position • European Common Proposal • Executive Control Program • Explicitly-Coded Program • Extended Capabilities Port

ECPA: Electronic Communications Privacy Act (USA) • European Commission on Preservation and Access

ECPF: Electromagnetic Containerless Processing Facility

ECPIU: Electronic Circuit Plug-In Unit

ECPM: École de Chimie, Polymères et Matériaux

ECPR: Electrically Calibrated Pyroelectric Radiometer • European Consortium for Political Research

ECQAC: Electronic Components Quality Assurance Committee

Écr: Écrire • Écrit • Écriture

ECR: Earth Centered Rotating • Efficient Customer Response • El Charco (Colombia) (location identifier) • Electron-Cyclation Radiation • Electron-Cyclotron Resonance • Electron-Cyclotron Response • Electronic Cash Register • Engineering Change Request • Equipment Change Record • Executive Control Routine • Expected Cost of Repair • Experiment Change Record • Experiment Concept Review • Extragalactic Cosmic Ray

ECRA: Engineering Change Request Authorization • Environmental Cleanup Responsibility Act (USA)

ECRC: Engineering Contract Requirements Committee • European Computer Research Centre

ECRH: Electron-Cyclotron Resonance Heating

ECRQP: Equality Constraints Recursive Quadratic Programming

ECRU: Environmental Consulting and Research Unit (CSIRO)

ECRV: Exponentially Corrected Random Variable

ECS: Eau Chause Sanitaire • Edgeworth-Cramer Series • Effective Cross Section • Electrical Control Station • Electron-Capture Spectroscopy • Electronically Collimated System • Electronic Charting System • Electronic Chart System • Electronic Composing System • Electronic Cross-connect System • Electronics Control System • Emergency Coolant System • Emergency Cooling System • Enable Control System • Enclosure Control System • Energy Conversion System • Energy Corrected Sudden • Engine Control System • Enterprise Computer Solutions • Envelope Circumstellaire • Environmental Control System • EOSDIS Core System • European Committee for Standardization • European Communications Satellite • Exact Cubic Search • Executive Communication Systems • Executive Control System • Experimental Communications Satellite • Extended Core Storage • Newcastle Mondell (USA-WY) (location identifier) • Sucre (Ecuador) (ISO currency code)

ECSA: European Computing Services Association • Exchange Carrier Standards Association

ECSC: European Coal and Steel Community • European Conference on Satellite Communications

ECSCW: European Conference on Computer-Supported Cooperative Work

ECSE: Experimental Computer Science and Engineering

EcSJ: Electrochemical Society Journal (periodical code)

ECSL: European Centre for Space Law

ECSN: European Climate Support Network

ECSO: European Council of Skeptical Organizations

ECSS: European Cooperation for Space Standardisation • Extendable Computer System Simulator

ECSTASY: Economical Storage and Access System

ECSU: Eastern Connecticut State University (USA-CT)

ECT: Économie des Changements Techniques • Emergency Communication Terminal • Emission-Computed Tomography • Emission-Computerized Tomography • Emulsion Chamber Technology • Environment Control Table • Equatorial Crossing Time • Estimated Completion Time • Experiment Command Tape • Exposure Control Technique

ECTA: Electronics Component Test Area • European Communities Trademark Association

ECTF: Edinburgh Centre for Tropical Forests • Enterprise Computer Telephony Forum • European Community Telework Forum

ECTL: Electronic Communal Temporal Lobe

ECTP: European Conference on Thermophysical Properties

ECTRA: European Committee for Telecommunications Regulatory Affairs

ECTS: European Computer Trade Show • European Conference on Telecommunications by Satellite • European Credit Transfer System

ECU: Earth-Centered Universe • Ecuador (ISO country code) • EISA Configuration Utility • Electrical Control Unit • Electrical Conversion Unit • Electronic Control Unit • Electronic Coupling Unit • Electronics Control Unit • Engine Control Unit • Environmental Change Unit • Environmental Control Unit • Équateur (ISO country code) • European Currency Unit • Experiment Control Unit • Extreme Close-Up

Ecua: Ecuador

Ecuad: Ecuador

ECUP: European Copyright User Platform

ECV: Electric Cargo Vehicle • Enhanced Customer View

ECVAM: European Centre for Validation of Alternate Methods

ECVFP: Expanded Charted Visual Flight Procedures

ECWA: Economic Commission for Western Asia

ECWMT: Europese Centrum voor Weersverwachtingen op Middellange Termijn

ECWS: European Centre for Work and Society

ed: En Dergelijk • Ex Divident

Éd: Éditer • Éditeur • Édition

Ed: Edición • Edile • Edilizia • Edinburgh (UK-Scotland) • Edit • Edited • Editing • Editio • Edition • Editor • Educated • Education

E/D: Edge Distance • Enable/Disable

E&D: Engineering & Development • Evaluation & Development

ED: Earth Distance • Eastern District • Edge Distance • Education Department • Effective Dose • Election District • Electromagnetic Dissociation • Electron Device • Electron Diffraction • Elevation Data • End Delimiter • Energy Division • Engineering and Development • Engineering Data • Engineering Directive • Engineering Directorate • Engineering Division • Entry Date • Enumeration District • Environmental Damage • Equivalent Dose • Erase Display • Error Detecting • Estimated Date • European Date • European Datum • Evaluation and Development • Event Designator • Events Dispatcher • Evolutionary Design • Ex Dividend • Executive Director • Experimental Detector • Explosive Device • External Device • Extra Dividend • Extra-high Density • Extra-low Dispersion

EDA: Early Departure Authorized • Economic Development Administration • Edge Damage Analysis • Edna Bay (USA-AK) (location identifier) • Electronic Design Automation • Electronic Display Assembly • Elevation Difference Accuracy • Elevation Drive Assembly • Embedded Document Architecture • Engineering Design Activity • Environmental Damage and its Assessment • Error Detector Assembly • Error and Dispersion Analysis • Estimated Date of Arrival • Estimated Date of Availability • Exploratory Data Analysis • Explorative Datenanalyse

EDAC: Electromechanical Digital Adapter Circuit • Error Detecting And Correcting • Error Detection And Correction

EDAP: Environmental Decision Alignment Process

EDAPS: Electronic Data Processing System

EDARC: Enhanced Direct Access Radar Channel

EDAS: Enhanced Data Acquisition System

EDASRES: Engineering Drawings Automated Storage and Retrieval System

EDAX: Energy-Dispersive Analysis of X-rays

EdB: Bachelor of Education

EDB: Economic Development Board • Eldebba (Sudan) (location identifier) • Electronic Development Board • Elektronische Datenverarbeitung • Embrittlement Database • End of Data Block • Engineering Database • Entry Database • Environmental Data Book • Environmental Data Buoy • Ethylene Dibromide • Extended Dynamic Backtracking • Extensional Database • External Database • External Data Bus

EDBD: Environmental Database Directory

EDBMS: EXOSAT Database Management System

EDBS: Expert Database System

EDBT: Extending Database Technology

EDC: Edinburgh District Council (UK-Scotland) • Education in Developing Countries • Effective Date of Change • Effective Dielectric Constant • Einstein Data Center • Electrochemical Depolarized Cell • Electron Distribution Curve • Electronic Data Communications • Electronic Digital Computer • Electronic Documentation Project • Engineering Data Center • Engineering Design Centre • Engineering Design Change • Enhanced Data Correction • Environmental Data Center • EOS Data Center • EROS Data Center • Error Detecting Code • Error Detection and Correction • Error Detection Code • Estimated Date of Completion • Euroavia Design Contest • EUROLAS Data Centre • European Data Center • European Defense Community • European Dialing Code • European Documentation Centre • External Device Code • Extra Dark Colour

EDCARS: Engineering Data Computer-Assisted Retrieval System

EDCAS: Equipment Designer's Cost Analysis System

EDCC: Edinburgh-Durham Cluster Catalogue • Electronic Digital Control Computer • Engineering Drawing Control Center

EDCD: Electronic Digital Control Device

EDCP: Engineering Design Change Proposal

EDCPF: Environmental Data Collection and Processing Facility

EDCS: Electrical Distribution and Control System • EURECA Dedicated Control System

EDCT: Expected Departure Clearance Time

EDCW: External Device Control Word

Edd: Ediderunt • Editiones

EdD: Doctor of Education

EDD: Electronic Document Delivery • Environmental Data Directory • Erldunda (Australia-NT) (location identifier) • Estimated Delivery Date • Exchange of Digital Data • Expected Date of Delivery

EDDC: Extended Distance Data Cable

EDDD: Electronic Document Development and Delivery

EDDF: Error Detection and Decision Feedback

EDDI: Extended Disk Driver Interface

EDDR: Electron Dipole-Dipole Reservoir

EDDS: Electronic Document Delivery Service • Experiment Data Distribution System

EDE: Edenton (USA-NC) (location identifier) • Effective Dose Equivalent • Electrical Design Engineer • Elliptic Differential Equation • Exact Differential Equation

EDED: Error Detection Encoder-Decoder

EDELWEISS: Expérience pour Détecter Les WIMPs en Site Souterrain

EDEN: Environmental Data European Network

EDEX: Electrodynamics Explorer

EDF: Anchorage Elmendorf (USA-AK) (location identifier) • Earliest Deadline First • Edifício • Électricité de France • Empirical Distribution Function • Engineering Data File • European Development Fund • Event Data Format • Event Definition File • Experiment Data Facility • Extensible Description Formalism • External Delay Factor

EDFA: Erbium-Doped Fibre Amplifier

EDFE: EVA Development Flight Experiment

EDFL: École Polytechnique Fédérale de Lausanne (Switzerland)

EDFT: EVA Demonstration Flight Test

EDG: Edgewood Weide (USA-MD) (location identifier) • Educational Development Group • Emergency Diesel Generator

EDGAR: Electronic Data Gathering, Analysis and Retrieval

EDGE: Electronic Document Gathering Environment • Enhanced Data rates for Global Evolution • Enhanced Data rates for GSM Evolution • Enhancing Diversity in Graduate Education

EDGT: Elevation Data Editing Terminal

EDH: EPIC Data Handling • Error Detection and Handling

EDHE: Experimental Data Handling Equipment

EDI: Échange de Données Informatisé • Échange de Données Informatisées (*) • Edinburgh (Starlink internal code) • Edinburgh Turnhouse (UK-Scotland) (location identifier) • Eidgenössiches Departement des Innern (Switzerland) • Electron Drift Instrument • Electronic Data Interchange • Electronic Data Interfacing • Electronic Document Interchange • Engineering Data Interface • Environnement de Développement Intégré • Equivalent-Damage Index • ESA-Developed Instrument • Explicitely Defined Intron

EDIC: Engineering Data Identification and Control

EDID: Extended Display Identification

EDIES: EDC Digital Image Enhancement System • EROS Digital Image Enhancement System

Edif: Edificio

EDIF: Electronic Data Interchange Format • Electronic Design Interchange Format

EDIFACT: Electronic Data Interchange For Administration, Commerce and Transport

EDIL: Electronic Data Interchange between Libraries • Electronic Document Interchange between Libraries

EDILIBE: Electronic Data Interchange for Libraries and Booksellers in Europe

Edin: Edinburgh (UK-Scotland)

Ed. in Ch.: Editor in Chief

EDIPAS: Engineering Data Interactive Presentation and Analysis System

EDIPS: EDC Digital Image Processing System • EROS Digital Image Processing System

EDIS: Educational Documentation and Information System • Electrical Distribution Information System • Electronic Document Information System • Engineering Data Information System • Environmental Data and Information Service (now merged with NESS into NESDIS) • Environmental Data Information Service • Erderkundungs- Daten und Informations System (Germany)

Edit: Edited • Éditer • Éditeur • Editing • Édition • Editor • Editore • Editorial

EDIT: Electronic Dissemination of Information Technology • Engineering Development Integration Test • Error Deletion by Iterative Transmission

EDITOR: ERTS Data Interpreter and Tenex Operations Recorder

EDIV: Engineering Design Intent Verification

EDK: El Dorado (USA-KS) (location identifier)

EdL: Édition de Luxe

EDL: Edit Decision List • Eldoret (Kenya) (location identifier) • Electrical Double Layer • Electrodeless Discharge Lamp • Engineering Development Laboratory • Entry-Descent-Landing • Equatorial Dust Lane • Extended Disk Life • Extensor Digitorum Longus

EDLC: Ethernet Data Link Control

EDLN: Engineering Development Logic Network

EDLS: Engineering Drawing Library System

EDLU: Encoder and Decoder Lander Unit

EdM: Master of Education

EDM: Electrical Discharge Machined • Electrical Discharge Machining • Electric Dipole Moment • Electric Discharge Machining • Electrodischarge Machining • Electronic Distance Measurement • Electronic Distance Measuring • Electronic Distance Meter • Electronic Document Management • Electronic Drafting Machine • Embryo Development Module • Energy-Density Model • Engineering Data Management • Engineering Development Mission • Engineering Development Model • Enterprise-wide Data Model • Environmental Descriptor Manager • La Roche sur Yon (France) (location identifier)

EDMA: Extended Direct Memory Access

EDMAS: Environmental Dose Modelling and Analysis System

EDMF: Electric Dipole Moment Function • Extended Data Management Facility

EDMI: Electronic Distance Measuring Instrument

EDMICS: Engineering Data Management Information and Control System

Edmn: Edmonton (Canada-AB)

EDMP: Engineering Drawing Management Program

EDMS: Electronic Data Management System • Electronic Document Management System • Engine Distress Monitoring System • Engineering Data Management System • Engineering Document Management System • Expert Dialogue Management System • Extended Data Management System

Edn: Edition

EDN: Electronic Design Notebook • Electronic Document and Notebook • Environmental Data Network • European Distribution Network

EDNA: European Data Network Agency

Ednbhg: Edinburgh (UK-Scotland)

EDNS: Expected Demand Not Supplied

Edo: Estado

EDO: Extended Data Out • Extended-Duration Orbiter

EDOC: Effective Date Of Change

EDOF: Electronic Document Order Form

EDOMP: Extended Duration Orbiter Medical Project

EDORAM: Extended Data Out Random Access Memory

EDOS: Effective Date Of Supply • Enhanced Disk Operating System • EOS Data and Operations System • Extended Disk Operating System

EDP: Earth Dynamics Program • Educational Data Processing • Electricidade De Portugal • Electron Density Profile • Electron Diffraction Pattern • Electronic Data Processing • Electronic Display Panel • Embedded Data Processor • Engineering Design Proposal • Engineering Development Phase • Entry Date Patent • Environmental Definition Program • Estimated Date of Publication • Executive Development Programme • Expedite Departure Path

EDPC: Electronic Data Processing Center

EDPE: Electronic Data Processing Equipment

EDPICS: Experience-Diagram-Principles-Investigate-Check-Summarize

EDPIS: Electronic Data Processing and Information System

EDPITAF: Educational Development Projects Implementing Task Force

EDPL: Engineering Development Processor Laboratory

EDPM: Electronic Data Processing Machine • Every Damm Problem Multiplied

EDPO: Einstein Data Products Office

EDPRT: Expert Diagnostic, Predictive, and Resolution Tools

EDPS: Electronic Data Processing System

EDPT: Electronic Data Processing Test • Enhanced Drive Parameter Table

EDQ: Erandique (Honduras) (location identifier)

EDR: Edward River (Australia-QLD) (location identifier) • Effective Data Rate • Elastic Recoil Detection • Electronic Data Reader • Electronic Data Reading • Electronic Dictionary Research • Electronic Document Report • Engineering Data Record • Engineering Design Review • Environmental Data Record • Experiment Data Record

EDRAM: Enhanced Dynamic Random Access Memory • Extended Dynamic Random Access Memory

EDRC: Engineering Design Research Center

EDRS: Enforcement Document Retrieval System • Engineering Data Retrieval System • European Data Relay Satellite

EDRSAM: Extended DRSAM

EDRSS: European Data Relay Satellite System

EDRT: Effective Date of Release from Training

Eds: Editions • Editors

EdS: Educational Specialist

EDS: Earth Data System • Electrical Distribution System • Electron Devices Society • Electronic Data System • Electronic Document Storage • Electronic Drive System • Emergency Deorbit System • Emergency Detection System • Energy-Dispersive Spectrometer • Energy-Dispersive Spectrometry • Energy-Dispersive Spectroscopy • Energy-Dispersive Spectrum • Engine Diagnostic System • Engineering Data System • Engineering Design System • Engineering Development Schedule • Entry Data System • Environmental Data Service • Environmental Data System • Error Detection System • Estimated Date of Separation • Exchangeable Disk Store • Executive Database System • EXOSAT Database System • Experiment Display System • Explosive Detection System • Explosive Device System • Express Delivery Service

EDSAC: Electronic Data Storage Automatic Computer • Electronic Discrete Sequential Automatic Computer

EDSAT: Educational Satellite

EDSC: Early Deployment System Concept

EDSG: Energy Demand Steering Group

EDSGC: Edinburgh/Durham Southern Galaxy Catalogue

EDSI: Enhanced Small Device Interface

EdSIG: Education Special Interest Group

EDSR: Electronic Document Storage and Retrieval • Engineering Data Storage and Retrieval

EDSS: Environment Decision-making Support System • Expert Decision Support System

EDST: Eastern Daylight Saving Time

EDSX: Electronic Digital Signal X-connect

EDT: Eastern Daylight-saving Time • Eastern Daylight Time • Effective Date of Training • Electronic Data Transmission • Electronic Theodolite System • Eligible Device Table • Energy Design Technique • Estimated Delivery Time • Estimated Departure Time • Extra Dispersion Triplet

EDTA: Ethylene Diamine Tetraacetic Acid

EDTCC: Electronic Data Transmission Communications Central

Edtn: Edition

Edtr: Editor

EDTR: Experimental, Developmental, Test and Research

EDTS: Electronic Data Transfer System

EDTV: Extended Definition Television

Edu: Education • Educational

EDU: Education • EPIC Data Unit • Event Detection Unit • Exchange Data Unit • Experimental Development Unit

Educ: Educate • Educated • Education • Educational

EDUSAT: Educational Satellite

EDV: Electronic Depressurization Valve • Electronic Depressurizating Valve • Elektronische Datenverarbeitung

EDVAC: Electronic Discrete Variable Automatic Computer

EDVM: Elektronische Datenverarbeitungsmaschine

EDVS: Elektronische Datenverarbeitungssystem

EDVZ: Elektronische Datenverarbeitungszentrum

EDW: Edwards (USA-CA) (location identifier) • Elementary Data Word • Engineering Data Writer

EDX: Energy Dispersive X-ray • Enterprise Data Exchange

EDXA: Energy Dispersive X-ray Analysis

EDXD: Energy Dispersive X-ray Diffraction

EDXF: Energy Dispersive X-ray Fluorescence

EDXRD: Energy Dispersive X-Ray Diffraction

EDXRFA: Energy Dispersive X-Ray Fluorescence Analysis

EDXS: Energy Dispersive X-ray Spectrometer • Energy Dispersive X-ray Spectroscopy

EDZ: Europäisches Dokumentationszentrum

e&e: Each & Every

ee: Estonia

E-E: End to End

E/E: Electrical to Electrical

E&E: Electrical & Electronic

EE: Early English • Edge to Edge • Electrical Engineer • Electrical Engineering • Electrodynamic Explorer • Electron Emission • Electronic Editing • Electronics Enclosure • Electronics Engineer • Electronics Engineering • Emergency Engineering • Employment Exchange • End Effector • End to End • Environmental Education • Environmental Engineering • Errors Excepted • Estonia (ISO country code) • Estonia (postal country code) • Estonie (ISO country code) • Experimental Evaluation • Extended Edition • Eye and Ear

EEA: Electromagnetic-Environment Analysis • Electronic Engineering Association • Électronique et Électrotechnique Automatique • Equal Employment Act (USA) • Essential Elements of Analysis • European Economic Area • European Environment Agency

EEB: Eccentric Eclipsing Binary • European Environmental Bureau • European Environment Bureau

EEC: East European Country • East European Craton • Electronic Engine Control • Energy-Energy Correlation • European Economic Community (now EU) • European Extragalactic Collaboration • Explosive-Emission Cathode • Extended Error Correction

EECA: Electrostatic Energy-Charge Analyzer

EECF: Earthnet ERS-1 Central Facility • ESRIN ERS Central Facility ((*))

EECOM: Electrical, Environmental, and Communications

EECR: Electronic Entities Cataloguing Rules

EECS: Electrical-Energy Conservation System • Electrical Engineering and Computer Science

EECT: End-to-End Call Trace

EED: Effective Equivalent Dose • Electroexplosive Device • Extracted Engineering Data • Needles (USA-CA) (location identifier)

EEE: Electrical-Electronic-Electromechanical • Électrique-Électromécanique-Électronique • Electromagnetic-Environment Experiment • Electronic-Electrical-Electromechanical • Ephemeris Enhancement Endeavor • Espace Économique Européen

EEEA: Électriciens-Électrotechniciens-Électroniciens-Automaticiens

EEED: Electronic and Electrical Engineering Division

EEEE: Electronic Encyclopaedia of Examples and Exercises

EEEU: End Effector Electronics Unit

EEF: Encircled Energy Function • Engineering Employers Federation • Enhanced Emission Feature • Equivalent-Energy Function

EEFSU: Eastern Europe and Former Soviet Union

EEG: Electroencephalogram • Electroencephalograph • Electroencephalographic • Electroencephalography • Étiquette Électronique de Gondole • Europese Economische Gemeenschap

EEK: Eek (USA-AK) (location identifier) • Kroon (Estonia) (ISO currency code)

EEI: Edison Electric Institute • Energy and Environment Information • Equipment-to-Equipment Interface • Essential Elements of Information • External Environment Interface

EEIC: Environmental Education and Information Committee

EEIG: European Economic Interest Group (*) • European Economic Interest Grouping

EEIS: End-to-End Information System

EEJ: Equatorial Electrojet

EEL: Electrical Equipment List • Electron Energy Loss • Epsilon Extension Language

EELG: Extended Emission Line Gas

EELR: Extended Emission Line Region (*) • Extensive Emission Line Region

EELS: Electron Energy Loss Spectroscopy • European Evaluation of Language Systems

EELV: Evolutionary Expendable Launch Vehicle • Evolved Expendable Launch Vehicle (*)

EEM: Earth Entry Module

EEMA: European Electronic Messaging Association

EEMIN: European Environment Monitoring and Information Network

EE&MP: Envoy Extraordinary & Minister Plenipotentiary

EEMS: Electrostatic Engine Monitoring System • Enhanced Expanded Memory Specification • European Environment Monitoring Satellite • Evaluation of Expanded Memory System

EEN: Brattleboro (USA-VT) (location identifier) • Keene (USA-NH) (location identifier)

EENR: Economic Evaluation of Natural Resources

EENT: Eye-Ear-Nose-Throat

EEO: Elliptical Earth Orbit • End-to-End Operation • Equal Employment Office • Equal Employment Officer • Equal Employment Opportunity • European Express Organization

EEOC: Equal Employment Opportunity Commission

EEOS: European Earth Observation System • European Earth Observing System

EEP: Earth Equatorial Plane • East Equatorial Pacific • East European Programme • Einstein Equivalence Principle • Electron-Excited Particle • Electronic Equipment Package

EEPC: European Educational Programme Coordinator

EEPG: European Engineering Planning Group

EEPROM: Electrically Erasable Programmable Read-Only Memory

EER: Energy Efficiency Ratio • Engineering and Economic Research • Enhanced Early Retirement • Enhanced Edge Recognition • Equal Error Rate • Equilibrium Equivalent Radon • Estonian Crown (currency) • Extended Entity Relationship • External Engineering Review

EERI: Earthquake Engineering Research Institute

EERIE: École pour les Études et la Recherche en Informatique et Électronique (France)

EERM: Établissement d'Études et de Recherches Météorologiques (Franc e)

EERO: European Environmental Research Organization

EEROM: Electrically Erasable Read-Only Memory

Eerw: Eerwaarde

EES: Earth and Environmental Sciences • Earth Elevation Sensor • Earth Exploration Satellite • Ejection Escape Suit • Electron Energy Spectroscopy • Electron Energy Spectrum • Emergency Ejection Suit • End-to-End System • Engineering Education Scheme • Environment Effects Statement • Escape Ejection Suit • Escrowed Encryption Standard • Escrow Encryption Standard • European Economic Space • European Economic Sphere • European Exchange System • Expendable Explorer Spacecraft

EESA: Electron Electrostatic Analyzer

EESAT: Environmental Education, Science and Technology

EESD: Earth Environmental Science Database

EESI: Earth Environment Satellite Initiative • Earth Environment Space Initiative

EESS: Earth Exploration Satellite Service • Emergency Egress Severance System • Emergency Egress Slide System • Engineering Education Science Scheme

EEST: Enhanced Ethernet Serial Transceiver • ESAnet Engineering Support Team

EET: Eastern European Time • Edge Enhancement Technology • Electronic Engine Thrust • Engineering Evaluation Test • Entry Elapsed Time • Equator Earth Terminal • Equivalent Exposure Time • ESA Earth Terminal • Estimated Elapsed Time • Estimated Exposure Time

EETB: Electronic/Electrical Termination Building

EETE: European Training Technology Event

EEU: Experiment Environmental Unit • Experiment Environment Unit • Extravehicular Excursion Unit

EEUU: Estados Unidos

EEV: English Electric Valve

EEVL: Edinburgh Engineering Virtual Library

EEVP: End-to-End Virtual Path

EEVPN: End-to-End Virtual Private Network

EEVT: Electrophoresis Equipment Verification Test

EEWS: Emergency Egress Window System

EEX: Electronic Egg Exchange

EExSc: Earth and Extraterrestrial Sciences (periodical code)

EEZ: Exclusive Economic Zone

Ef: Efectos

E&F: Equipment & Facilities

EF: Edge to Face • Eisenbahnfähre • Electronic Field • Electronic Frontier • Élement Fini • Elevator Effectiveness • Emergency Fund • Emission Factor • Emitter Follower • Empty Field • Enrichment Factor • Equilibrium Flow • Event Foreground • Experimental Flight • Explicit Forwarding • Exposed Facility • Extra Fine

EFA: Engine Fault Analysis • Eurofighter Aircraft • European Fighter Aircraft • Experiment Flight Application • Extended File Attribute

EFAM: Extreme Floods Alarm Mission

EFAS: En-route Flight Advisory Service • Experimental Fiction Analysis System

EFB: Eight Fathom Bight (USA-TX) (location identifier) • European Federation of Biotechnology

EFC: Electronic Fee Collection • Electronic Filing Cabinet • Estimated Final Cost • Expect Further Clearance

EFCC: Externally Fired Combined Cycle

EFCE: European Federation of Chemical Engineering

EFCI: Explicit Forward Congestion Indication • Explicit Forward Congestion Indicator

EFCS: Earth-Fixed Coordinate System • Electronic Flight Control System

EFD: Earliest Finish Date • Early Fault Detection • Electric Fields Detector • Electronic Forms Designer • Energy Flux Density • External Floppy Disk • External Floppy Drive • Houston Ellington (USA-TX) (location identifier)

EFDA: European Federation of Data processing Associations

EFDP: European Federation of Data Processing

EFE: Electronic Field Experiment

EFEI: Escola Federal de Engenharia de Itajuba (Brazil)

EFET: École Française d'Enseignement Technique (France)

Eff: Effect • Effective • Effetto • Efficace • Efficiency

EFF: Electronic Frontier Foundation • Experimental Forecast Facility

EFFECT: Exchange Format for Electronic Components and Texts

EFFF: Electric Field Flow Fractionation

Effy: Efficiency

EFG: Efogi (Papua New Guinea) (location identifier) • Electric-Field Gradient

EFGF: Epitaxial Ferrite Garnet Film

EFH: Earth Far Horizon

EFI: Electromechanical Frequency Interference • Electronic Fuel Injection • Electronics For Imaging • Enrico Fermi Institute (USA-IL) • Exploding Fail Initiator

EFIC: European Fuel Information Centre

EFIE: Electric Field Integral Equation

EFILA: European Forum for Implementors of Library Applications

EFIN: Environmental Financing Information Network

EFIS: Electronic Flight Information System (*) • Electronic Flight Installation System • Electronic Flight Instrumentation System • Electronic Flight Instrument System

EFK: Newport (USA-VT) (location identifier)

Efl: Effluent

EFL: Effective Focal Length • Emitter Follower Logic • English as a Foreign Language • Equivalent Focal Length • Extended Fortran Language • External Financial Limit • Kefallinia (Greece) (location identifier)

EFLA: Education Film Library Association (USA) • European Feeder Link Antenna

EFLC: European Foundation for Library Cooperation

EFM: Eight-to-Fourteen Modulation • Extended Functionality Module

EFMD: European Foundation for Management Development

EFMS: Experimental Flight Management System

EFO: East Fork (USA-AK) (location identifier) • Error-Freak-Oddity

EFOSC: ESO Faint Object Spectrograph and Camera

EFOV: Effective Field Of View

EFP: Éducation et Formation Professionnelle • Einstein-Fokker-Planck • Electronic Field Production • Emulating Floating Point • Error-Free Performance • ESA-Furnished Property • European Fusion Programme

EFPA: Educational Film Producers' Association (UK)

EFPD: Effective Full-Power Day

EFPH: Effective Full-Power Hour • Equivalent Full-Power Hour

EFPISS: Extrinsic Fabry-Pérot Interferometric Strain Sensor

EFPOP: Electrical Field Poling by Optical Polarimetry

EFPY: Effective Full-Power Year

EFQ: Estimated Fuel Quantity

EFQM: European Foundation for Quality Management

EFR: Emerging Flux Region • Enhanced Full Rate • Estimated Fuel Remaining • European Fast Reactor • Experimental Fast Reactor

EFRAP: Electronic Feeder Route Analysis Program

EFRAPS: Extracted Feature Rectification and Processing System

EFS: Earth-Fixed System • Eesti Füüsika Selts (Estonia) • Electronic Filing System • Epoch Folding Search • European Forward Study • Extensible File System

EFSM: Ethernet/FDDI Switching Module • Extended Finite State Machine

EFSN: European Federation of Scientific Networks

EFSPI: European Federation of Statisticians in Pharmaceutical Industry

EFSSS: Engine Failure Sensing and Shutdown System

EFSY: ESA Financial System

Eft: Efecto

EFT: Early Finish Time • Electronic Field Trip • Electronic Fund Transfer

EFTA: European Free Trade Association

EFTO: Encrypted For Transmission Only

EFTPoS: Electronic Funds Transfer at Point of Sale

EFTPOS: Electronic Funds Transfer at Point Of Sale

EFTS: Electronic Funds Transfer System

EFTSU: Equivalent Full Time Student Unit

EFU: Energetic Feed Unit

EFUV: Extreme and Far Ultraviolet

Efvo: Efectivo

EFW: Electric Fields and Waves (experiment)i (CLUSTER) • Energy From Waste • Jefferson (USA-IA) (location identifier)

EFY: End of Fiscal Year

eg: Egypt • Exempli Gratia • Ex Grege • For Example

eG: Eingetragene Genossenschaft

Eg: Egypt • Egyptian

EG: Eggen • Egypt (ISO country code) • Égypte (ISO country code) • Electronographic • Encyclopaedia Galactica •

Europäische Gemeinschaft • Europeiska Gemenskaperna • Europese Gemeenschap • Expert Group • Exponentiated Gradient • External Gradient

EGA: Earth Gravity Assist • Earth-Grazing Asteroid • Electron Generator Assembly • Engati (Papua New Guinea) (location identifier) • Enhanced Graphics Adapter • Evolved Gas Analyser • Evolved Gas Analysis • Extended Graphics Adapter • Extended Graphics Array

EGAL: European Group of Astronomy Librarians

EGARD: Environmental Gamma-ray And Radon Detector

EGAS: European Group on Atomic Spectroscopy

EGB: Educación General Básica • Electronic Gimbal Box • Escher-Gödel-Bach

EGC: Bergerac (France) (location identifier) • Education in Global Change • Engine Gimbal Control • Enhanced Group Call • Experiment Ground Computer

EGCAT: Extragalactic Catalog

EGCM: Eddy-resolving General Circulation Model

EGD: Electrogravimmetric Device • Evolved Gas Detection

EGE: Enhanced Greenhouse Effect • Vail - Eagle County (USA-CO) (location identifier)

EGEAS: Electric Generating Expansion Analysis System

EGF: Electrical Grapple Fixture • Epidermal Growth Factor

EG&G: Edgerton, Germershausen & Grier

EGG: Evaporating Gaseous Globule

EGGF: European Guidance and Guarantee Fund

EGH: Electrogastrogram • Europäischer Gerichtshof

EGI: Embedded GPS/Inertial • Exhaust Gas Ignition • Extended Gaussian Image • Valparaiso Duke (USA-FL) (location identifier)

EGIC: Éditeur Graphique Interactif de Calculs

EGIL: Electrical General Instrumentation and Lighting

EGIS: European Geographical Information System

EGKS: Europese Gemeenschap voor Kolen en Staal

EGL: Extragalactic Light • Neghelli (Ethiopia) (location identifier)

EGLE: Extreme Gravitational Lensing Event

EGLS: Estimated Generalized Least Squares

EGM: Eggen Proper Motion • Extraordinary General Meeting • Sege (Solomon Islands) (location identifier)

eGmbH: Eingetragene Genossenschaft mit Beschränkter Haftpflicht

EGMS: Escuela de Geología, Minas y Geofísica (UCV)

EGMT: Elapsed Greenwich Mean Time

EGN: Geneina (Sudan) (location identifier)

EGNOS: European Geostationary Navigation Overlay Service • European Geostationary Navigation Overlay System

EGO: Belgorod (Russia) (location identifier) • Eccentric Geophysical Observatory • Économie-Gouvernements-Organisations • Edit Graphics Object • EUVE Guest Observer • Experimental Geophysical Orbiting • Extragalactic Object

EGOC: EUVE Guest Observer Center

EGOCS: EUVE Guest Observer Center Software

EGORD: Experiment Ground Operations Requirements Documennt

EGOS: European Group on Ocean Stations

EGOWG: EUVE Guest Observer Working Group

EGP: Eagle Pass - Maverick County (USA-TX) (location identifier) • European Glaciological Programme • Experimental Geodetic Payload • Exterior Gateway Protocol (*) • External Gateway Protocol • Extra-solar Giant Planet • Pound (Egypt) (ISO currency code)

EGPWS: Enhanced Ground Proximity Warning System

Egr: Egregio

EGR: Earned Growth Rate • Exhaust Gas Recirculation

EGRET: Energetic Gamma-Ray Experiment Telescope

EGRS: Extragalactic Radio Source

EGS: Earth Gateway Station • Egilsstadir (Iceland) (location identifier) • Electron and Gamma-ray Shower • Electron Gamma-ray Shower • European Geophysical Society • Expeditionary Ground Station • Experimental Geodetic Satellite

EGSE: Electrical Ground Support Equipment

EGSMA: Egyptian Geological Survey and Mining Authority

EGSP: Electronics Glossary and Symbol Panel

EGT: Einstein-invariant Gauge Theory • Elapsed Ground Time • Estimated Ground Time • European Geotraverse • Exhaust Gas Temperature

EGV: Eagle River (USA-WI) (location identifier) • External Gelation of Uranium

EGX: Egegik (USA-AK) (location identifier)

EGY: Egypt (ISO country code) • Égypte (ISO country code)

Egypt: Egyptian

EH: Edinburgh (UK postal code) • Ehrenhalber • English Heritage • Era Hispánica • Event Header • Evolvable Hardware • Sahara Occidental (ISO country code) • Sahara (Western_) (ISO country code) • Western Sahara (ISO country code)

EHA: Earth Horizon Angle • Electro-Hydraulic Actuator • Electro-Hydrostatic Actuator • ESO Historical Archive

EHAPE: Environmental Health Action Plan for Europe

EHAZ: Equatorward-moving High-latitude Activity Zone

EHB: Extended Horizontal Branch • Extreme Horizontal Branch (*)

EHBO: Eerste Hulp Bij Ongelukken

EHC: Effective Heat Capacity • Electrical Heating Control

EHD: Electrohydrodynamic • Electrohydrodynamical • Electrohydrodynamically • Electrohydrodynamics

EHe: Extreme Helium (star)

EHE: Extended HII Envelope • Extraordinary Hall Effect • Extremely High Energy

Ehem: Ehemalig • Ehemals

EHEP: Experimental High-Energy Physics

E-HEPNET: European High-Energy Physics Network

EHEPNET: European High-Energy Physics Network

EHESS: École des Hautes Études en Sciences Sociales

EHF: European Headache Federation • Extremely High Frequency

EHIA: Environmentally High Impact Area

EHL: Effective Half-Life • Elastohydrodynamic Lubrication • El Bolsón (Argentina-RN) (location identifier) • Electron-Hole Liquid

EHLLAPI: Emulator High-Level Language Applications Programming Interface • Enhanced High-Level Language Applications Programming Interface

EHM: Cape Newenham (USA-AK) (location identifier) • Electron Holographic Microscope • Extremely Hazardous Material

EHMO: Extended Hückel Molecular Orbital

EHO: Environmental Health Officer

EHOT: External Hydrogen/Oxygen Tank

EHP: Effective Horsepower • Electric and Hybrid Propulsion • Electric Horsepower • Electron-Hole Pair • Electron-Hole Plasma • Equivalent Horsepower • Estimated Horsepower

EHR: Education and Human Resources

EHS: Earth Horizon Scanner • Earth Horizon Sensor • Environmental Health Service • Environmental Health Standard • Euro-Hermespace • European Home System • European Hybrid Spectrometer • Extra-High Strength

EHSD: Existing Hardware/Software Database

EHSI: Electronic Horizontal Situation Indicator

EHT: East Hartford - Rentschler (USA-CT) (location identifier) • Electrical High Tension • Extra-High Tension

EHTO: European Health Telematics Observatory

EHTS: Emacs-based Hypertext System

EHU: Environmentally Homogeneous Unit • Extension Header Unit

EHV: Electric and Hybrid Vehicle • Extra-High Voltage • Extremely High Velocity

EHW: Evolvable Hardware

EHX: Experiment Heat Exchanger

EHz: Exahertz

EI: Earth Interface • East Indian • East Indies • Electrical Insulation • Electromagnetic Interaction • Electromagnetic Interference • Electronics Interface • Electron Impact • Electron Ionization • Emission Index • End Item • Energy Intake • Engineering Index • Engineering Instruction • Enterprise Integration • Entry Interface • Environmental Impact • Exposure Index • External Input • External Interrupt • External Irradiation • Ireland (international aircraft registration code)

EIA: Economic Impact Assessment • Education Initiative in Astronomy • Eia (Papua New Guinea) (location identifier) • Electrical Industries Association • Electronic Industries Association (USA) • Energy Information Administration • Engineering In Astronomy • Environmental Impact Analysis • Environmental Impact Assessment • Environment Institute of Australia • Equatorial Ionization Anomaly • Equatorial Ionospheric Anomaly • European Information Association • Experimental and Integrative Activities

EIAC: Environmental Information and Analysis Center (USA)

EIAF: Expanded Inward Access Feature

EIAJ: Electronics Industry Association of Japan

EIARD: European Initiative for Agricultural Research for Development

EIASN: End Item Assembly Sequence Number

EIAT: End Item Acceptance Test

EIB: Electrical Interface Box • Engineering Information Bureau • Europäische Investitionsbank • European Investment Bank • Evaporating Infalling Body

EIC: Editor-In-Chief • Educational Information Center • Electrostatic Ion Cyclotron • End Item Contract • Engineering Information Center • Engineering Institute of Canada • Environment Information Centre • Equatorial Infrared Catalogue • Euro Info Centre • Experimental Intercom

EICAS: Engine Indicating and Crew Alerting System • Engine Indicating and Crew Alert System • Engine Indication and Crew Alerting System • Engine Indication and Crew Alert System • Engine Instrument and Crew Alerting System • Engine Instrument and Crew Alert System (*)

EICC: Euro Info Correspondance Centre

EICD: Experiment Interface Control Document

EICIT: Electrical Interface Continuity/Isolation Test

EICM: Electronic Information and Communication Module • Employer's Inventory of Critical Manpower

EICOS: European Initiative for Communicators in Science

EICS: Energetic Ion Composition Spectrometer

EICW: Electrostatic Ion Cyclotron Wave

EID: Electron-Induced Disordering • End Item Delivery • End Item Description • End Item Documentation • Entry Identification Number • Evolution of Interstellar Dust • Experiment Interface Document

EIDE: Enhanced Integrated Drive Electronics

EIDERS: Externally Ionized Disks in the Environs of Radio Sources

EIDES: ETL Image Database for Experimental Studies

Eidg: Eidgenössisch

EIDOSTAS: European Infrastructure for Document Supply in Technology and Applied Sciences

EIDP: End Item Data Package

EIDR: External Intermediate Design Review

EIDS: Electronic Information Display System • Electronic Integrated Display System • ESA IUE Data Server

EIE: Electron Impact Excitation • Eniseysk (Russia) (location identifier) • Equilibrium Isotope Effect • European Industrial Engineering

EIED: Engine Instrument Electronics Display

EIES: Electronic Information Exchange System

EIF: Electronic Interface Format • European Investment Fund • External Interface • External Interface File

EIFA: Element Interface Functional Analysis

EIFOV: Effective Instantaneous Field Of View

EIFS: Equalized Integral Field Spectroscopy

Eig: Eigenlijk

EIG: Exchange Information Group

EIGRP: Enhanced Interior Gateway Routing Protocol • Enhanced Internet Gateway Routing Protocol

Eigtl: Eigentlich

EIH: Einasleigh (Australia-QLD) (location identifier) • Electronic Information Handling

EII: Electron Impact Ionization

EIIA: European Information Industry Association

EIII: Electrical Industry Information Institute

EIISO: Edison International Infrared Space Observatory (ESA)

Eil: Eiland • Eilanden

EIL: Fairbanks Eielson (USA-AK) (location identifier)

EILIS: Electronic Integrated Library and Information Service

EIM: Electronic Image Management • Engineering Information Management • Environmental Information Management

EIMB: Engelhardt Institute for Molecular Biology (Russia)

EIMI: Environmental Interactions Measurement Investigation

EIMS: Employment Information in the Mathematical Sciences

EIMU: Environmental Information Management Unit

EIN: Educational Information Network • Eindhoven (Netherlands) (location identifier) • Electronic Information Network • Employer Identification Number • Enterprise Information Network • Environmental Information Network • Equivalent Input Noise • European Informatics Network

E-in-C: Engineer in Chief

EinC: Engineer in Chief

E. Ind.: East Indian • East Indies

EINECS: European Inventory of Existing Commercial Chemical Substances

Einl: Einleitung

EINS: European Information Network Service

Einschl: Einschließlich

Einw: Einwohner

EIO: Extended Interaction Oscillator

EIOL: European Infrastructure for Open Learning

EIONET: Environmental Information and Observation Network

EIOOTS: European In-Orbit Operations Technology Symposium

EIP: Electronic Information Publishers • End Item Parameter • Experiment Implementation Plan

EIPA: European Institute of Public Administration

EIPL: Environnement Intégré de Production de Logiciel

EIPO: European IGAC Project Office

EIPPCB: European Integrated Pollution and Prevention and Control Bureau

EIPS: Electronic Information Products and Services • Electron Impact Point Source

EIR: Earned Income Relief • End Item Requirement • Environmental Impact Report • Excess Information Rate • Experiment Integration Review • Extended Information Rate • Extreme Infrared • Ireland (international vehicle registration code)

EIRENE: European Integrated Radio Enhanced Network

EIRI: Energy Information Resource Inventory (DoE)

EIRMA: European Industrial Research Management Association • European Industrial Research Manager Association

EIRO: European Industrial Relations Observatory

EIRP: Effective Isotropically Radiated Power • Effective Isotropic Radiated Power • Equivalent Isotropically Radiated Power • Equivalent Isotropic Radiated Power

EIRR: External Independent Readiness Review

EIRSG: European Information Retrieval Specialist Group

Eis: Eisern

EIS: Beef Island (British Virgin Island) (location identifier) • Earth Imaging System • Economic Information Service • Economic Information System • Effluent Information System • Electrical Integration System • Electronic Information Service • Electronic Instrument System • Ellipsoïde Isotherme Singulier • E-mail Interface Server • End Item Specification • Energy Information System • Engineering Information Service • Engineering of Intelligent Systems • Enterprise Information System • Environmental Imaging Spectrometer • Environmental Impact Statement • Environmental Impact Study • Environmental Information Service • Environmental Information System • Environment-Induced Scattering • ESO Imaging Survey • EUV Imaging Spectrometer • Executive Information System • Expanded Inband Signaling • Explosives Inspection Section • Tortola (British Virgin Island) (location identifier) • Virgin Gorda (British Virgin Island) (location identifier)

EISA: Extended Industry Standard Architecture

EISAC: European Imaging Spectrometer Aircraft Campaign

EISCAT: European Incoherent Scatter

EISG: Experimental Investigation of Spacecraft Glow

EISMINT: European Ice Sheet Modelling Initiative

EISN: Environmental Information and Support Network

EISO: Edison Infrared Space Observatory • Environmental Information System Office (ORNL)

EISS: Economic Impact Study System

EIT: Electrical Impedance Tomography • Electromagnetically-Induced Transparency • Électronique-Informatique-Télécommunications • Enterprise Integration Technologies • Entry Interface Time • Equatorial Ionosphere-Thermosphere • EUV Imager Telescope • EUV Imaging Telescope (*)

EITO: European Information Technology Observatory

EIU: Economic Intelligence Unit • Economist Intelligence Unit • Engine Interface Unit • Ethernet Interface Unit • Event Identification Unit • (The) Economist Intelligence Unit

EIUF: European ISDN User Forum

EIV: Error-In-Variable

EIVM: Elektronische Informationsverarbeitung Maschine

EIVT: Electrical and Instrumentation Verification Test • Electrical Interface Verification Test • Electronic Installation Verification Test

EIWB: End Item Workbook

EIWG: Earth Imaging Working Group

EIY: Ein Yahav (Israel) (location identifier)

EJ: Electronic Journal • Exajoule • Ireland (international aircraft registration code)

EJA: Barrancabermeja (Colombia) (location identifier)

EJAM: European Journal of Applied Mathematics

EJASA: Electronic Journal of the Astronomical Society of the Atlantic

EJB: Enterprise Java Bean

EJC: Engineering Joint Council

EJCSC: European Joint Committee of Scientific Cooperation

Ejem: Ejemplar • Ejemplo

EJH: Wedjh (Saudi Arabia) (location identifier)

EJIC: Education and Job Information Center

EJMF: European Journal of Mechanics - Fluids (periodical code)

EJMS: European Journal of Mechanics - Solids (periodical code)

Ejn: Ejection

EJN: Electronic Journal Navigator (Blackwell)

EJO: Electronic Journals Online

EJP: Effacement des Jours de Pointe

EJPh: European Journal of Physics (periodical code)

EJRT: Engineering Job Release Ticket

EJS: Electric Jet System

Ejtr: Ejector

Ejud: Ejusdem

EK: Armenia (international aircraft registration code) • Eastman Kodak • Eerste Kwartier • Eerstkomend • E-Kernel • Eskimo • Event Clock • Event Kernel

EKA: Eureka Murray (USA-CA) (location identifier)

EKB: Edgeworth-Kuiper Belt • Ekibastuz (Kazakhstan) (location identifier)

EKC: Eastman Kodak Co.

EKD: Elkedra (Australia-NT) (location identifier) • Evangelische Kirche in Deutschland

EKE: Eddy Kinetic Energy • Ekereku (Guyana) (location identifier) • Equatorial Kerr Effect

EKF: Extended Kalman Filter

EKG: Electrocardiogramme • Elektrokardiogramme

EKI: Elkhart (USA-WI) (location identifier)

EKN: Elkins (USA-WV) (location identifier)

EKO: Edgeworth-Kuiper Object • Elko (USA-NV) (location identifier)

EKS: Electrokinetic Surveying • Electronic Keyboard System • Energy Kernel System • Europäische Kulturstiftung

EKT: Eskilstuna (Sweden) (location identifier)

EKTS: Electronic Key Telephone Set

EKU: Eastern Kentucky University (USA-KY)

EKV: Exoatmospheric Kill Vehicle

EKW: Equivalent Kilowatt

EKX: Elizabethtown (USA-KY) (location identifier)

EKZ: Europäisches Kulturzentrum (Switzerland)

el: Greek (ISO language code)

El: Elected • Electo • Electric • Electrical • Électricité • Electricity • Électrique • Electroluminescent • Electron • Électron • Electronic • Électronique • Element • Elevated • Elevation • Elongation

E/L: Entry/Landing

EL: Elastic Limit • Electrical Laboratory • Electroluminescence • Electroluminescent • Electronics Laboratory • Electron Lattice • Element Layer • Elevation • Epitaxial Layer • Equipment Laboratory • Erase Line • Europhysics Letters (periodical code) • Liberia (international aircraft registration code)

ELA: Eagle Lake (USA-TX) (location identifier) • Eight-Letter Acronym • Electronic Learning Aid • Electron Linear Accelerator • Ensemble de Lancement Ariane

ELAB: Electronic Laboratory

ELAC: Economically Less Advantaged Country • European Lidar Airborne Campaign

ELACS: Extended Life Attitude Control System

ELAG: European Library Automation Group

ELAIS: European Large Area Infrared Survey • European Large Area ISO Survey

ELAN: Emulated Local Area Network • Environment in Latin-America Network

ELAP: EtherTalk Link Access Protocol

Elas: Elasticity

Elast: Elastic (periodical code)

Elb: Elbow

ELB: El Blanco (Colombia) (location identifier) • Electrical Load Bank • Emergency Locator Beacon • Environment Liaison Board

ElBah: Elektrische Bahnen (periodical code)

ELBBO: Extended Life Balloon-Born Observatory

ELBL: Extended Line-By-Line

ELC: Earth Limb Clutter • Elcho Island (Australia-NT) (location identifier) • Electron Loss to Continuum • Emission-Line Cloud • End of Line Code • Environmental Law Centre • Environment Law Centre • Environment Liaison Centre

ELCA: Earth Control Landing Assembly

ELCD: Electrolytic Conductivity Detector

ELCI: Environment Liaison Centre International

ELCID: European Living Conditions Information Directory

ELCO: Environnement Limité Concentrique Ordonné

ElCom: Electronics Communications of Japan (periodical code)

ELCP: Emission-Line Coronal Photometer

Elctd: Electrode

Elctlt: Electrolyte

Elctrn: Electron

Elctrochem: Electrochemical

Eld: Elder • Eldest

ELD: Earth Launch Date • Economic Load Distribution • El Dorado (USA-AR) (location identifier) • Electron Landau Damping • Extra-Long Distance

ELDA: European Language resources Distribution Agency

ELDO: European Launcher Development Organization (merged now with ESRO into ESA)

ELDORA: Electra Doppler Radar

ELDP: European Lake Drilling Programme

ELDS: Editorial Layout Display System

ELDV: Electrically-Operated Depressurization Valve

Ele: Electo

ELE: Earth Limb Experiment • El Real (Panama) (location identifier) • Extinction Level Event

Élec: Électricité • Électrique • Électron • Électronique

Elec: Election • Elector • Electoral • Electric • Electrical • Electrically • Electrician • Electron • Electronic

Elecm: Electromagnetics (periodical code)

Élect: Élection • Électoral • Électricité • Électrique • Électron • Électronique

Elect: Election • Electoral • Electric • Electrical • Electrician • Electricity

Électr: Électricité • Électrique • Électron • Électronique

Electr: Electric • Electrical • Electricidad • Electricity • Electronical • Electronics

Electromagn: Electromagnetismo

Electrón: Electrónica

Électron: Électronique

Electron: Electronical • Electronics

ELEGANT: Electron Gamma-ray Neutrino Telescope

Elek: Elektronika (periodical code)

Elekt: Elektrisch • Elektrizität

Elektr: Elektrisch • Elektrizität • Elektronisch • Elektronik

Elektron: Electronisch • Elektronik

Elem: Elektromekhanika (periodical code) • Element • Elementary

Elev: Elevate • Elevation • Elevator

ELF: Early Lunar Flare • El Fasher (Sudan) (location identifier) • European Laser Facility • Extension Language Facility • Extra Low Frequency • Extremely Low Frequency

ELFIN: European Laminar Flow Investigation

ELG: Education Librarians Group • El Golea (Algeria) (location identifier) • Emission-Line Galaxy • Engineering Log Generator • European Launching Group • European Library Group

ELGRA: European Low-Gravity Association

ELH: North Eleuthera (Bahamas) (location identifier)

ELI: Elim (USA-AK) (location identifier) • English Language Interpreter • Environmental Law Institute • Eurocell Library and Interfaces

ELIA: European League of Institutes of Arts

ELIAS: Earth Limb Infrared Atmospheric Structure • Entry Level Interactive Applications System

Elim: Eliminate

ELINCS: European List of Notified Chemical Substances

ELINOR: Electronic Information Online Retrieval

ELINT: Electronic Intelligence

ELISE: Electronic Library Image Service for Europe

ELITE: Electron Isotope Telescope • European Laboratory of Intelligent Techniques Engineering

ELJ: El Recreo (Colombia) (location identifier)

ELK: Elk City (USA-OK) (location identifier)

ElL: Electronics Letters (periodical code)

ELL: Ellisras (South Africa) (location identifier)

ElLC: Electrotechnical Laboratory Circulars (periodical code)

Ellip: Elliptical

ElLR: Electrotechnical Laboratory Researches (periodical code)

ELM: Edge-Localized Mode • Elaboration Likelihood Model • Electronic Mail • Elmira Corning (USA-NY) (location identifier) • Experimental Logistics Module (*) • Experiment Logistics Module • Extended Lunar Mission

ELMC: Electrical Load Management Center

ELMF: European Large Magnetic-field Facility

ELMO: Electronic Map Ordering

ELMS: Earth Limb Measurement Satellite • Elastic-Loop Mobility System • Electrical Load Management System • Engineering Lunar Model Surface • EOS Microwave Limb Sounder • Experimental Library Management System

ELN: Ellensburg Bowers (USA-WA) (location identifier)

ELNET: European Library Network

Elng: Elongate

ELNGT: Elongate

Elo: Elocution • Eloquence

ELO: Eldorado (Argentina-MI) (location identifier) • Emerging Local Operator • Emission-Line Object • European Limitless Office

ELOISE: European Land-Ocean Interaction Studies • European Large Orbiting Instrumentation for Solar Experiments

Elong: Elongate • Elongated • Elongation

E. Long.: East Longitude

ELOR: Extended Lunar Orbital Rendezvous

ELOS: Experimental Land Observation System • Extended Line Of Sight

ELOT: Hellenic Organization for Standardization (Greece)

Elp: Eliptical

ElP: Electronics Power (periodical code)

ELP: El Paso (USA-TX) (location identifier) • English Language Program • Equivalent Local Potential • European Librarians and Publishers

ElPro: Electronic Progress (periodical code)

ELQ: Gassim (Saudi Arabia) (location identifier)

ELQP: Extended Linear-Quadratic Program

ELR: Emission-Line Region • Environmental Lapse Rate • Environment Lapse Rate • Environment Lapse Rate • Exceptional Leave to Remain • Extended-Line Region

ELRA: European Language Resources Association

ELRAD: Earth Limb Radiance

ELS: Earth Landing System • Eastern Launch Site • East London (South Africa) (location identifier) • Electronic Library System • Electron Lodgement Service • Elevon Load System • Emission-Line Star • Energy-Loss Spectroscopy • Entry Level System • Environment Labelling Scheme • External Lighting System

ELSA: Electronic Library SGML Application • Environmental Life Support Assembly • European Laboratory for Structural Assessment • European Law Students' Association

ELSC: Earth Landing Sequence Controller

ELSET: Element Set

Elsewh: Elsewhere

ELSI: Enhanced Longwave Spectrometer/Imager

ELSIE: Electronic Speech Information Equipment

ELSPA: European Leisure Software Publishers Association

ELSS: Emergency Life-Support System • EVA Life-Support System

ELSW: Elsewhere

ELSy: Experimental-Laserkreisel-System

ELT: Effective Lagrangian Theory • Electrical Impedance Tomography • El Tor (Egypt) (location identifier) • Emergency Locating Transmitter • Emergency Locator Transmitter (*) • Emission Line Tool • Extrapolated Launch Time • Extremely Large Telescope

ELTIS: European Local Transport Information Service

ELU: El Oued (Algeria) (location identifier)

ELV: Earth Launch Vehicle • Elfin Cove (USA-AK) (location identifier) • Expendable Launch Vehicle • Extra-Low Voltage

ELVIL: European Legislative Virtual Library

ELVIRA: Electronic Library and Visual Information Research Association

ELVIS: Early Light Vision Isomorphic Singular (system) • Experimenter's Laboratory for Visualized Interactive Science

Elvn: Elevation

ElW: Elektrizitätswerk

ELW: Earth Launch Window • Ellamar (USA-AK) (location identifier)

ELWRT: Electrowriter

ELWW: European Laboratory Without Walls

Ely: Easterly

ELY: Ely (USA-NV) (location identifier)

Elz: Elzevieren

ELZ: Wellsville (USA-NY) (location identifier)

Em: Emanation • Eminent

E&M: Electrical & Mechanical • Elektrotechnik und Maschinenbau (periodical code)

E/M: Engineering Model • Escape Motor

E-M: Expectation-Maximization (algorithm)

EM: Earth-Moon • Effective Mass • Electrical and Mechanical • Electrical Model • Electromagnetic • Electromotive • Electromodulation • Electronic Mail • Electron Microscope • Electron Microscopy • Electron Model • Emergency Maintenance • Emission Measure • End Matched • Engineering Management • Engineering Memorandum • Engineering Model • Engineer of Mines • Enhanced Monitoring • En Mano • Enterprise Manager • Environmental Modelling • Environmental Module • Equatorial Mount • Equipment Module • Escape Motor • Estrada Municipal • Evaluation Model • Exception Monitor • Expanded Memory • Expectation-Maximization (algorithm) • Experienced Member • Experiment Module • External Memorandum • External Memory

EMA: East Midlands (UK-England) (location identifier) • École des Mines d'Alès (France) • Effective Medium Approximation • Egyptian Meteorological Authority • Electromechanical Actuation • Electromechanical Actuator • Electromechanic Analysis • Electronic Mail Association • Electronic Messaging Association (*) • Elektronische Mitteilungen zur Astronomiegeschichte • Engineering and Manufacturing Aid • Enhanced Monitoring Array • Entreprise Management Architecture • Environmental Management Agency • European Monetary Agreement • Exponential Moving Average • Extended Memory Addressing

EMAC: Europe Market Awareness Committee

EMACH: Extended Maximum Average Correlation Height

EMACS: Editing Macros • Editor Macros • Engine Monitoring And Control System

EMAD: Engine Maintenance And Disassembly • Equatorial Middle Atmosphere Dynamics

EMAE: EPIC MOS Analogue Electronics

e-mail: Electronic Mail

EMail: Electronic Mail

EMALS: Electromagnetic Aircraft Launch System

EMAP: Environmental Monitoring and Analysis Program • Environmental Monitoring and Assessment Program

EMaPS: European (boards) for Marine and Polar Science

EMARAC: Electromagnetic Memory And Research Arithmetical Calculator

EMAS: Eco-Management and Audit Scheme • Edinburgh Multiaccess System

EMATS: Equipment Manipulation And Transportation System • Equipment Manipulation And Transport System • Experiment Manipulation And Transportation System • Experiment Manipulation And Transport System

Emb: Embajada • Embajador • Embankment • Embassy

EMB: Electromagnetic Bias • Embaixador • Environmental Management Bureau • Extended Memory Block • San Francisco - Embarkadero (USA-CA) (location identifier)

EMBARC: Electronic Mail Broadcast to A Roaming Computer

EMBD: Embedded

EMBDD: Embedded

EMBL: European Molecular Biology Laboratory

EMBnet: European Molecular Biology Network

EMBO: European Molecular Biology Organization

Embriol: Embriología

EMC: École des Métiers de la Communication • Educational Media Council (USA) • EISCAT Magnetospheric Cross • Electromagnetic Capability • Electromagnetic Cleanliness • Electromagnetic Compatibility (*) • Electromagnetic Control • Electronic-Mail Connection • E-Mail Connection • Energy Management Centre • Energy Management Company • Enhanced Machine Controller • Enhanced Memory Chip • Enhanced Multicoating • Environmental Module Computer • Environmental Monitoring and Compliance • Environment Management Committee • Equivalent Mission Cycle • European Muon Collaboration • Extended Math Coprocessor

EMCC: Eight-Man Crew Capability • Electromagnetic Code Consortium • Emergency Mission Control Center

EMCD: Electromechanical Control Diagram

Emcee: Master of Ceremonies

EMCF: European Monetary Cooperation Fund

EMCFA: Electromagnetic Compatibility Frequency Analysis

EMCH: EPIC MOS Camera Head

EMCI: Engineering Model Configuration Inspection

EMCOF: European Monetary Cooperation Fund

EMCON: Emission Control

EMCR: EPIC MOS Control and Recognition

EMCS: Energy Monitoring and Control System • EPIC MOS Camera System • Event Management Control System

EMCSC: Ettore Majorana Centre for Scientific Culture

EMCT: Extended-duration Metabolic Control Test

EMCU: EPIC MOS Camera Unit

E&MD: Engineering & Manufacturing Development

EMD: Electrolytic Manganese Dioxide • Elliptical Mass Distribution • Emerald (Australia-QLD) (location identifier) • Engineering and Manufacturing Development • Engineering-Manufacturing-Development • Entry Monitor Display • Exploitation Maximale des Données

EMDB: Equipment Maintenance Database

EMDM: Enhanced Multiplexer/Demultiplexer

EMDR: Eye Movement Desensitization and Reprocessing

EMDS: EURECA Mimic Diagram System

EMDSO: Eastern Missouri Dark-Sky Observers (USA-MO)

EMDU: EPIC MOS Data Unit

EME: Earth-Mars-Earth • Earth-Moon-Earth • Electromagnetic Environment • Emden (Germany) (location identifier)

EMEA: Europe - Middle East - Africa

EMEC: Electromagnetic Effects Capability • Electromagnetic Effects Compatibility • Enhanced Master Events Controller

EMEP: European Monitoring and Evaluation Programme

Emer: Emergency • Emeritus

EMERAUDE: Élaboration de Modèles d'Étude, de Recherche et d'Application, Utilisant des Domaines Étendus

Emerg: Emergency

EMES: Earth Monitoring Educational System • Electrical, Mechanical, and Environmental Systems

EMEX: Equatorial Mesoscale Experiment

EMF: Earth Magnetic Field • Electromagnetic Field • Electromagnetic Force • Electromotive Force • Energy Modelling Forum • Enhanced Metafile Format • Evaporated Metal Films Corporation (USA-NY)

EMG: Electromagnetogram • Electromyogram • Electromyograph • Electromyography • Empangeni (South Africa) (location identifier) • Executive Management Group

EMH: Efficient Market Hypothesis

EMHD: Electron Magnetohydrodynamics

EMI: Earthquakes and Megacities Initiative • Électrodynamique des Milieux Ionisés • Electromagnetic Interface • Electromagnetic Interference (*) • Electromotive Interference • Electronic Money Issuer • Emergency Management Institute • Emirau (Papua New Guinea) (location identifier) • Ernst-Mach-Institut (Germany) • European Monetary Institute • External Muon Identifier

EM&IAS: Electromechanical & Integrated Avionics System

EMIAS: Electromechanical and Integrated Avionics System

EMIC: Emission Microscope • Emission Microscopy

EMIF: ESCON Multi-Image Facility

EMIM: Expected Mutual Information Measure

EMIP: European Market for Infrastructural Projects

EMIR: Espectrógrafo Multi-objeto Infrarojo • European Microgravity Research • European Multilingual Information Retrieval

EMIRS: Electrochemically Modulated Infrared Spectroscopy

EMIS: Earth Moon Ion Spectrometer • Educational Management Information System • European Mathematical Information Service (EMS)

EMISM: Electromagnetic Interference Safety Margin

EMITS: Electronic Mail for Invitations to Tender Software (*) • Electronic Mailing of Invitations to Tender Software

EMK: Emergency Medical Kit • Emmonak (USA-AK) (location identifier)

EML: Electromagnetic Launcher • Electronic Mail • Element Management Layer • Expected Measured Loss

eMLPP: Enhanced Multilend Precedence and Preemption

EMM: Electromechanical Mockup • Enterprise Mail Manager • État-Major de la Marine • Expanded Memory Manager • Extended Memory Management • Kemerer (USA-WY) (location identifier)

EMMCVPR: Energy Minimization Methods in Computer Vision and Pattern Recognition (workshop)

EMMI: ESO Multimode Instrument

EMML: Expectation Maximization Maximum Likelihood

EMMS: Electronic Mail and Messaging System • Engine Maintenance and Management Service

EMN: Engineering Management Network • Nema (Mauritania) (location identifier)

EMNLP: Empirical Methods in Natural Language Processing

EMO: Eagle Mountain Observatory (USA-NY) • Edgeworth-Kuiper Object • Electronic Microorganism • Emo (Papua New Guinea) (location identifier)

EMOC: EOS Mission Operations Center

EMOD: Erasable Memory Octal Dump

EMON: Exception Monitoring

EMOP: Équipe Mobile d'Ouvriers Professionnels

EMOS: Earth Mean Orbital Speed • ECMWF Meteorological Operational System

EMOT: European Management and Organisations in Transition

emp: Emplastrum • Ex Modo Praescripto

Emp: Empennage • Emperor • Empire • Empress

EMP: Earth, Moon and Planets (periodical code) • Earth-Moon-Planets • Ecological Monitoring Programme • Electromagnetic Pulse • Electron Microprobe • Emporia (USA-KS) (location identifier) • End of Month Payment • Engineering Master Plan • Enhanced Monitoring Period • Environmental Management Plan • Environmental Management Programme • Environmental Monitoring Plan • Environmental Monitoring Programme • Environment Management Plan • Environment Management Program • Ephemerides of Minor Planets • Equipment Mounting Plate • Erasable Memory Program • Event Manager Process • Extremely Metal-Poor

EMPA: Eidgenössicher Materialprüfungs- und Versuchsanstalt (Switzerland)

EMPACT: Electrons Muons Partons Air Core Toroid

EMPAR: European Multipurpose Phased Array Radar

EMPB: European Multiprotocol Backbone

EMPC: Electrostatic Molecular Potential Contour

EMPDB: Electronic Materials Properties Data Bank

Empf: Empfohlen

EMPGS: Electromechanical Power Generation System

EMPIRE: Early Manned Planetary-Interplanetary Round Trip Experiment

EMPIS: Engineering Materials and Processes Information Service • Environmental Management and Planning Information System

Empl: Employee

Empld: Employed

EMPR: Empreendimento

EMPS: Elevation Matrix Processing System • En-route Maintenance Processor System

EMPST: Energy-Matter-Personality-Space-Time

EMQ: Event Manager Queue

EMR: Electromagnetic Radiation • Electromagnetic Response • Electronic Medical Record • Energy, Mines and Resources (Canada) • Engineering Management Review • Engine Mixed Ratio • Engine Mixture Ratio • Enhanced Metafile Record • Executive Management Review • Expanding Molecular Ring

EMRC: European Medical Research Council

EMRL: Equipment Maintenance Requirements List

EMRTC: Energetics Materials Research and Testing Center (USA-NM)

EMS: Edinburgh Mathematical Society (UK-Scotland) • Effluent Management System • Electromagnetic Spectrum • Electromagnetic Surveillance • Electromagnetic Susceptibility • Electro-Mechanical Systems • Electronic Mail Service • Electronic Mail System • Electronic Manuscript Standard • Electronic Manuscript Submission • Electronic Meeting System • Electronic Message Service • Electronic Message System • Electron Multiplex Switch • Electronic Mail System • Electronic Management System • Element Management System • Embessa (Papua New Guinea) (location identifier) • Emergency Management System • Emergency Medical Service • Energy Management System • Engineering Master Schedule • Enregistreur Magnétique Spatial • Enterprise Messaging Server • Entry Monitoring System • Entry Monitor System • Environmental Management System • Environmental Monitoring Satellite • ENVISAT Mission System • Error Mean Square • Error Message Service • European Mathematical Society • European Meteorological Service • European Mobile System • European Monetary System • Europees Monetair Systeem • Evolving Magnetic Structure • Expanded Memory Specification • Expanded Memory Support • Expanded Memory System • Expanding Molecular Shell • Experimental Monitoring Satellite • Experiment Mounting Structure • Exposure Management System • Extended Memory Specification

EMSA: Electron Microscopy Society of America

EMSAW: En-route Minimum Safe Altitude Warning

EMSC: European Mediterranean Seismological Centre

EMSI: Electromagnetic Sensing and Interpretation • European Manned Space Infrastructure

Emsl: Emsland (Germany)

EMSL: Environmental Molecular Sciences Laboratory • Environmental Monitoring Systems Laboratory • European Microwave Signature Laboratory

Emsn: Emission

EMSP: Enhanced Modular Signal Processor

EMSPS: European Mobility Scheme for Physics Students

EMSS: Einstein Medium-Sensitivity Survey (*) • Emulated Multispectral Scanner • Extended Medium-Sensitivity Survey

EMST: Extended Magic-Set Transformation

EMSU: Environmental Meteorological Support Unit

EMT: Effective-Mass Theory • Effective Medium Theory • Electrical Metallic Tubing • El Monte (USA-CA) (location identifier) • Emergency Medical Technician • End of Magnetic Tape • Energy Momentum Tensor • Équivalent Mégatonnique • Éducation Manuelle et Technique • European Mathematical Trust

EMTA: Engineering Marine Training Authority

EMTE: Electromagnetic Test Environment

EMTED: Electromagnetic Test and Evaluation Data System

Emtr: Emitter

EMU: Eastern Michigan University (USA-MI) • Economic and Monetary Union • Electrical Multiple Unit • Electromagnetic Unit • Engineering Model Unit • Engine Management Unit • European Monetary Union • European Monetary Unit • EVA Maneuvering Unit • Extended Memory Unit • Extravehicular Manoeuvering Unit • Extravehicular Mobility Unit (*)

emuls: Emulsio

EMV: Europay-Mastercard-Visa

EMVC: EPIC MOS Voltage Converter

EMVICT: Engineering Model Verification and Interface Compatibility Test

EMWA: Engineering Minor Work Authorization

EMWAC: European Microsoft Windows Academic Centre

EMX: El Maiten (Argentina-CB) (location identifier) • Enterprise Mail Exchange

EMY: El Minya (Egypt) (location identifier)

En: Energy • Enero • Engineer • English (ISO language code)

EN: École Normale • Économie Nationale • Éducation Nationale • El Niño • Emissora Nacional • Enfield (UK postal code) • England • Estrada Nacional • Europastandard • European Network • European Norm • Exception Noted • External Network • Extrapolation Number

ENA: Eastern North America • École Nationale d'Administration (France) (*) • École Nationale d'Affichage • Energetic Neutral Atom • European Neuroscience Association • Extended Network Architecture • Kenai (USA-AK) (location identifier)

ENAA: Epithermal Neutron Activation Analysis

ENAC: École Nationale de l'Aviation Civile (France) • Energetic Neutral Atom Camera

ENACEOS: Energetic Neutral Atom Camera for Earth-Orbiting Satellite

ENACS: ESO Nearby Abell Cluster Survey

Enam: Enamel

ENAM: European and National Astronomy Meeting • European North Atlantic Margin • Exotic Nuclei and Atomic Masses

ENAN: Education Native American Network (USA)

ENAP: Energetic Neutral Atom Precipitation

ENAV: En-route Navigational Aid

ENB: Eneabba West (Australia-WA) (location identifier) • Escuela Nacional de Bibliotecarios (Argentina)

ENBA: Escuela Nacional de Biblioeconomía y Archivonomía (Mexico)

Enbl: Enable

ENBW: Equivalent Noise Bandwidth

Enc: Enclosed • Enclosure • Encoder • Encuadernación • Encyclopaedia

EnC: Energy Conversion (periodical code)

ENC: École Nationale des Chartes (France) • Eisenhower National Clearinghouse (USA-OH) • Electronic Navigational Chart • Encoded • Equivalent Noise Charge • Nancy (France) (location identifier)

ENCA: Enterprise Network Computing Architecture

Encap: Encapsulate • Encapsulated

Enc. Brit: Encyclopaedia Britannica

ENCC: Environmental Noise Control Committee

Encd: Encased • Encode

ENCD: Electronic Navigation Chart Data

Encdd: Encoded

Encl: Enclosed • Enclosure

ENCORE: Evolutionary Computations Repository

ENCS: Existing and New Chemical Substance

Encuad: Encuadernación

Ency: Encyclopaedia • Encyclopaedic

Ency.Brit: Encyclopaedia Britannica

Encyc: Encyclopaedia • Encyclopaedic

Encyc.Brit: Encyclopaedia Britannica

Encycl: Encyclopaedia • Encyclopaedic

End: Endorsed

eND: Enhanced Navigation Display

END: Enid Vance (USA-OK) (location identifier) • Expert National Détaché

ENDEX: Environmental Data Index (NOAA)

ENDF: Evaluated Neutron Data File • Evaluated Nuclear Data File

ENDG: Ending

ENDIF: Enterprise Network Data Interconnectivity Family

ENDOC: Environmental Documentation (EU)

ENDOR: Electron Nuclear Double Resonance

Endow: Endowment

ENDR: Entretien Non-Directif de Recherche

ENDS: European Nuclear Documentation System

Endvr: Endeavour (periodical code)

Ene: Enero

ENE: East-Northeast • East-Northeastern • Ende (Indonesia) (location identifier) • Est-Nord-Est

ENEA: Energia Nucleare e Energie Alternative (Italy) • European Nuclear Energy Agency

ENEC: European Nuclear Energy Community

EnEng: Environmental Engineering (periodical code)

ENEP: Escuela Nacional de Estudios Profesionales (Mexico)

Energ: Energetika (periodical code)

ENERN: East-Northeastern

ENEWD: East-Northeastward

ENF: Enfermeira • Enontekio (Finland) (location identifier) • Européen Non-Français

ENFI: Electronic Networks For Interaction

ENFIA: Exchange Network Facility for Interstate Access

ENFJ: Extraverted Intuitive Feeling Judging

EnFM: Engineering Fracture Mechanics (periodical code)

ENFP: Extraverted Intuitive Feeling Perceiving

eNFS: Enhanced Network File System

Eng: Engine • Engineer • Engineering • England (UK) • English • Engrave • Engraver • Engraving

ENG: Electronic News Gathering • Electronystagmogram • Engenheiro • England

Enga: Engage

EngAn: Engineering Analysis (periodical code)

EngCo: Engineering Computations (periodical code)

Engg: Engineering

Engin: Engineer • Engineering

Engl: England (UK) • English

ENGO: Environmental Non-Governmental Organization

Engr: Engineer • Engineering • Engraved • Engraver • Engraving

Engrg: Engineering

Engrv: Engrave

Engy: Energy

ENH: Earth Near Horizon • Enshi (China-PRC) (location identifier)

ENHA: Electronic Newsletter for the History of Astronomy

ENI: École Nationale d'Ingénieurs (France) • El Nido (Philippines) (location identifier)

ENIAC: Electronic Numerical Integrator Analizer and Computer • Electronic Numerical Integrator And Calculator (*) • Electronic Numerical Integrator And Computer

ENIC: École Nouvelle d'Ingénieurs en Communications (France)

ENID: Endnote Identification

ENIM: École Nationale d'Ingénieurs de Monastir (Tunisia)

ENIS: Electronic Networks and Information Sharing • European Nuclear Information Service

Enj: Enjuiciamiento

ENJ: El Naranjo (Guatemala) (location identifier)

Enk: Enkelvoud

ENK: Enniskillen - Saint Angelo (Northern Ireland) (location identifier)

Enl: Enlarge • Enlarged • Enlargment • Enlisted

ENL: Centralia (USA-IL) (location identifier)

Enlg: Enlarge

ENLR: Extended Narrow-Line Region

ENM: École Nationale de la Météorologie (France) • Enterprise Network Management

ENN: Nenana (USA-AK) (location identifier)

ENNA: Établissement National de Navigation Aérienne

ENNS: European Neural Network Society

ENO: Encarnación - Teniente Prim Alarcón (Paraguay) (location identifier) • European Northern Observatory (Canary Islands)

ENP: École Nationale de la Photographie (France) • Escuela Nacional Preparatoria

ENPA: Energetic Neutral Particle Analyzer

ENPI: Energetic Neutral Particle Imager

ENPC: École Nationale des Ponts et Chaussées (France)

ENPT: École Nationale des Postes et Télécommunications (France)

ENQ: Coronel E. Soto Cano (Honduras) (location identifier) • Enquire • Enquiry

EnQL: English-like Query Language

ENR: Enterprise Network Roundtable • Equivalent Noise Ratio • Excess Noise Ratio

Enrgz: Energize

ENRIC: Environment and Natural Resources Information Center

ENRICH: European Network for Research In global Change • European Network for Research on global Change

ENRM: Environment and Natural Resources Management

Ens: Ensign

ENS: École Normale Supérieure (France) • Enschede Twente (Netherlands) (location identifier) • Enterprise Network Services • Enterprising Messaging System • European Nervous System

ENSA: European Neutron Scattering Association

ENSAE: École Nationale Supérieure de l'Aéronautique et de l'Espace (France)

ENSAI: École Nationale de la Statistique et de l'Analyse de l'Information (France)

ENSAIS: École Nationale Supérieure des Arts et Industries de Strasbourg (France)

ENSAM: École Nationale Supérieure des Arts et Métiers (France)

ENSAR: Établissement National des Sciences Agricoles et Rurales (France)

ENSB: École Nationale Supérieure des Bibliothécaires (France) (now ENSSIB) (*) • École Nationale Supérieure des Bibliothèques (France) (now ENSSIB) (*)

ENSCP: École Nationale Supérieure de Chimie de Paris (France)

ENSDF: Evaluated Nuclear Structured Data File

ENSE: Enterprise Network Strategies Europe

ENSEEIHT: École Nationale Supérieure d'Électrotechnique, d'Électronique, d'Informatique et d'Hydraulique de Toulouse (France)

ENSERB: École Nationale Supérieure d'Électricité et de Radio de Bordeaux (France)

ENSERG: École Nationale Supérieure d'Électronique de Grenoble (France)

ENSET: École Nationale Supérieure de l'Enseignement Technique (France)

ENSG: École Nationale des Sciences Géographiques (France)

ENSI: École Nationale Supérieure d'Ingénieurs (France) • Equivalent Noise Sideband Input

ENSICA: École Nationale Supérieure d'Ingénieurs de Constructions Aéronautiques (France)

ENSIETA: École Nationale Supérieure des Ingénieurs des Études et Techniques d'Armenent

ENSIMEV: École Nationale Supérieure d'Ingénieurs en Mécanique et Énergétique de Valenciennes (France)

ENSITM: École Nationale Supérieure des Industries Textiles de Mulhouse (France)

ENSJF: École Normale Supérieure de Jeunes Filles (France)

ENSL: École Normale Supérieure de Lyon (France)

ENSMA: École Normale Supérieure de Mécanique et d'Aéronautique (France)

ENSMP: École Normale Supérieure des Mines de Paris (France)

ENSO: El Niño Southern Oscillation

ENSPS: École Nationale Supérieure de Physique de Strasbourg (France)

ENSPTT: École Nationale Supérieure des Postes et Télécommunications (France)

ENSS: European Navigation Satellite System • Exterior Nodal Switching System

ENSSIB: École Nationale Supérieure des Sciences de l'Information et des Bibliothèques (France)

EnST: Environmental Science Technology (periodical code)

ENSTA: École Nationale Supérieure des Techniques Avancées (France)

Ent: Entrance • Entry

ENT: Ear-Nose-Throat • Eniwetok Island (Marshall Islands) (location identifier) • Entrada • Entropy • European Negotiating Team

ENTEL: Empresa Nacional de Telecomunicaciones (Chile)

ENTICE: Energetic Trans-Iron Composition Experiment

ENTJ: Extraverted Intuitive Thinking Judging

Entlo: Entresuelo

ENTP: Extraverted Intuitive Thinking Perceiving

Entom: Entomología

Entr: Enter • Entrance

EnTR: Energy Technology Review (periodical code)

ENTR: Entire • Entrance

Entro: Entropie (periodical code)

Entw: Entweder

ENU: Enugu (Nigeria) (location identifier)

Enum: Enumerate • Enumeration

ENUR: Enterprise Network Users Roundtable

ENUWAR: Environmental consequences of Nuclear War

Env: Envelope • Environ • Envoi • Envoyer

ENV: Europäische Vornorm • Wendover (USA-UT) (location identifier)

ENVI: Environment for Visualizing Images

Environ: Environment • Environmental • Environnement

ENVISAT: Environmental Satellite

Envr: Environment • Environmental

ENW: Kenosha (USA-WI) (location identifier)

ENY: Yan'an (China-PRC) (location identifier)

Enz: Enzovoort • Enzovoorts

Enzkr: Enzkreis (Germany)

eo: Esperanto (ISO language code) • Ex Officio

Eö: Eötvös

E&O: Engineering & Operations • Errors & Omissions

E/O: Electrical to Optical • Electro-Optical • Engineering/Operations

E-O: Electro-Optical

EO: Earth Observation • Earth Observing • Earth Orbit • Earth Orbiter • Édition Originale • Education Officer • Einstein Observatory • Electronic Operation • Electro-Optic • Electro-Optical • Electro-Optics • Elliptical Orbit • Emergency Operation • End Office • Engineering and Operations • Engineering Order • Engine-Out • Equal Opportunity • Erasable Optical • Escape Orbit • Executive Officer • Executive Order • Ex Officio • External Output

EOA: End Of Address • Examination-Opinion-Advice

EOAC: Earth Observation Advisory Committee

EOAE: Earth-Orientated Applications Experiment

EOAP: Earth Observations Aircraft Program

EOARD: European Office of Aerospace Research and Development

EOAS: Engineering Order Accountability System

EOAST: Earth-Oriented Applications and Space Technology (periodical code)

EOB: Electronic Order of Battle • Electro-Optical Breadboard • End Of Block • Executive Office Building • Extensible Optical Bench

EOBT: Estimated Off-Block Time

Eoc: Eocene

EOC: Early Operational Capability • Earth Observation Center • Edge Of Coverage • Electron Optical Camera • Electro-Optical Center • Electro-Optics Camera • Emergency Operating Center • Emergency Operations Center • End Of Carrier • End Of Charge • End Of Contract • End Of Conversation • End Of Conversion • End Of Cycle • Engine Order Capability • Engine-Out Capability • Equal Opportunities Commission • ESPRIT Optical Computing • Executive Oversight Committee

EOCAP: Earth Observations Commercial Applications Program (*) • Earth Observations Commercialization Application Program • Earth Observing Commercialization and Applications Program

E/OCM: Electromagnetic/Optical Countermeasure

EOCM: Electro-Optical Countermeasure

EOCR: Engineering Order Control Record

EOCY: End Of Calendar Year

Eod: Eodem

EoD: End of Discharge

EOD: End Of Data • End Of Day • Engineering Operations Division • Entry On Duty • Erasable Optical Disk • Estimated On Dock • Estimated Operational Date • Every Other Day • Expected Occupancy Date • Explosive Ordnance Disposal

EODB: End Of Data Block

EODC: Earth Observation Data Centre

Eod.Loc.: Eodem Loco

EODMS: Earth Observation Data Management System

Eod.Op.: Eodem Opere

EODRS: Earth Observation Data Retrieval System

EODS: Electro-Optic Direction Sensor

E&OE: Errors & Omissions Excepted

EOE: Earth Observation Experiment • Earth Orbit Ejection • Earth Orbit Equipment • Electronic Order Exchange • Equal Opportunity Employer • Errors and Omissions Excepted

EOED: Earth Orbit Escape Device

EOEP: Earth Observation Experimental Programme

EOF: Earth Orbital Flight • Emergency Operating Facility • Empirical Orthogonal Function • End Of File • Enterprise Object Framework • Expected Operations Forecast • Experiment Operations Facility

EOFS: Erasable Optical File System

EOFY: End Of Fiscal Year

EOG: Electrooculogram • Electrooculograph • Electrooculographic

EOH: Engineering Open House • Medellín - Enrique Olaya Herrera (Colombia) (location identifier)

EOHP: Except Otherwise Herein Provided

EOHT: External Oxygen and Hydrogen Tank

EOI: Earth Orbit Insertion • Eday (UK) (location identifier) • Electro-Optical Industries Inc. (USA-AZ) • End Of Integration • End Of Interrupt • End Or Identify • Expression Of Interest

EOIM: Evaluation of Oxygen Interaction with Materials

EOIS: Earth Observing Information System • Electro-Optical Imaging System

EOJ: End Of Job

EOK: Keokuk (USA-IA) (location identifier)

EOL: Earth Observation Laboratory • Earth Orbit Launch • End Of Life • End Of Line • End Of List

EOLM: Electro-Optical Light Modulator

EOLS: EINSTEIN Online Service

EOLT: End Of Logical Tape

EOM: Earth Observation Mission (now ATLAS) • Earth Orbital Mission • End Of Medium • End Of Message • End Of Mission • End Of Month • End of Observation Marker • Engineering Operations Manual • Equation Of Motion • European Optical Module • Every Other Month • External Optical Modulator • Extractable Organic Material • Extractable Organic Matter

EOMF: End Of Minor Frame

EON: EOSAT Online Network

EONR: European Organization for Nuclear Research

EONS: Environments Of Nearby Stars

EOOE: Erreur Ou Omission Exceptée

EOP: Earth Observations Program • Earth Orbit Plane • Earth Orientation Parameter • Emergency Operating Procedure • Emergency Oxygen Pack • End Of Part • End Of Period • End Of Program • Equal Opportunity Program • Executive Office of the President • Experiment of Opportunity Program

EOPAG: ERS-1 Operations Planning Advisory Group

EOPAP: Earth and Ocean Physics Applications Program

EOPF: End Of Powered Flight

EOPGS: École et Observatoire de Physique du Globe de Strasbourg (France)

EOPM: Electro-Optic Phase Modulator

EOPP: Earth Observation Preparatory Programme

EOPR: Engineering Order Purchase Request • Engineering Order Purchase Requisition

EOQ: Economic Order Quantity • End Of Quarter • European Organization for Quality

EOQC: European Organization for Quality Control

EOQL: End Of Qualified Life

EOR: Earth Observing Radar • Earth Orbit Rendezvous • El Dorado (Venezuela) (location identifier) • Electro-Optical Research • End Of Record • End Of Refresh • Exclusive Or

EORC: Earth Observation Research Center

EORCU: European Ozone Research Coordinating Unit

EORM: Enhanced Object Relationship Model

EORQ: Engineering Order Request for Quotation

EORSAT: Electronic Ocean Reconnaissance Satellite • ELINT Ocean Reconnaissance Satellite

EORTC: European Organisation for Research and Treatment of Cancer

EOS: Earth Observation Satellite • Earth Observatory Satellite • Earth Observation System • Earth Observing System • Earth Orbital Shuttle • Egyptian Organization for Standardization (Egypt) • Electrical Overstress • Electro Optical Systems • Electro Optic Systems • Electrophoresis Operations in Space • Electrostatic Overstress • Ellery Open System • Emergency Oxygen System • End Of Service • End Of String • Equation Of State • Erasable Optical Storage • ERIN Online Service • Estación de Observación Solar (Mexico-SON) • European Optical Society • Extended Operating System • Neosho (USA-MO) (location identifier)

EOSAT: Earth Observation Satellite Corporation

EOSC: Earth Observations Science Committee

EOSCAT: Einstein Observatory Source Catalog

EOSD: Electro-Optical Systems Design (periodical code)

EOSDIS: Earth Observing System Data and Information System • Earth Observing System Data Information System (*)

EOSH: Electro-Optical Sensor Hardening

EOSP: Earth Observing Scanning Polarimeter

EOSRAM: Earth Observing System Regional Amazon Model

EOSS: Earth Observation Satellite System • Earth Orbital Space Station • Electro-Optic Sensor System

EOST: Observatoire et École des Sciences de la Terre (France)

EOSTAG: Earth Observation Scientific and Technical Advisory Group

EOT: Earth-Observed Time • End Of Table • End Of Tape • End Of Task • End Of Test • End Of Text • End Of Track • End Of Train • End Of Transfer • End Of Transmission

EOTC: European Organization for Testing and Certification

EOTF: Engineering Order To Follow

EOTS: Electro-Optical Threat Sensor • Electro-Optical Tracking System

EOU: End Of User

EOV: End Of Volume • Explanation Of Votes

EOW: Empty Operating Weight • End Of Word • Energy Over Weight • Every Other Word

EOWG: Earth Observation Working Group • Engineering and Operations Working Group

EOY: End Of Year

EOZ: Elorza (Venezuela) (location identifier)

ep: Editio Princeps

E/P: Electron/Proton

EP: Earth Probe • Early Picture • Easy Projection • Edge Preserving • Editio Princeps • Effective Pitch • Elbow Pitch • Electrically-conducting Polymer • Electrically Polarized • Electrical Power • Electric Propulsion • Electronic Publishing • Electronic Pulse • Electrophotographic • Electrostatic Precipitator • Element Probability • Empty Pattern • Emulation Program • Endpaper • End Point • Enfermedad Profesional • Engineering Personnel • Entrepiso • Entreplanta • Entry Point • Environmental Park • Environmental Protection • Environment Protection • Equivalence Principle • Equivalent Person • Equivalent Personnel • Estimated Position • European Parliament • European Plan • Evoked Potential • Evolutionary Programming • Exit Pupil • Expanded Polystyrene • Experience Point • Experiment Planning • Extended Play • Extraction Procedure • Extraterrestriche Physik • Extreme Pressure • Iran (international aircraft registration code)

EPA: Earth Polar Axis • Economic Planning Agency • Edge-Preserving Adaptive • Educational and Public Affairs • Educational Paperback Association (USA) • Educational Priority Area • Electron-Positron Accumulator • Electron-Positron Annihilation • Electron Probe Analyzer • Electro-optical Polarization Analyzer • Employment Protection Act • Energetic Particles Analyzer • Energy-Performance Assessment • Engineering Personnel Authorization • Engineering Project Authorization • Environmental Protection Agency (USA) • Environmental Protection Authority • Environment Planning Authority • Environment Protection Agency (*) • Equatorial Pitch Angle • Equilibrium Point Analysis • Estimated Plan of Action • Établissement Public à caractère Administratif • Établissement Public d'Aménagement • Europäischer Patentamt • European Photochemistry Association • Exoatmospheric Penetration Aid • Explicitely Parallel Algorithm

EPAC: Economic Planning Advisory Group • Energetic-Particle Anisotropy and Composition (experiment) (*) • Energetic-Particle Composition • European Particle Accelerator Conference

EPACML: Environmental Protection Agency Composite Model for Landfills

EPACT: Energetic Particle Acceleration, Composition and Transport

EPAD: Electrically Powered Actuation Design • Electrically Powered Actuator Design

EPAIRS: Environmental Protection Agency Information Retrieval System (USA)

EPALL: Emergency Preparedness At Local Level

EPAM: Electron, Proton, and Alpha Monitor

EPAO: Educational and Public Affairs Office

EPAQ: Electronic Parts of Assessed Quality

EPAS: Energetic-Particle Anisotropy Spectrometer

EPB: Electronic Picture Book • Equivalent Passband • Equivalent Pension Benefit • ESS Parallel Benchmark • Eucrite Parent Body • Europäischer Beschäftigungspakt

EPC: Earth Prelaunch Calibration • Easy Processing Channel • Edge Punched Card • Edinburgh Portable Compiler • Editorial Policy Committee • Editorial Processing Center • Educational Publishing Council • Electrical Power Conditioner • Electronic Parts Control • Electronic Power Conditioner • Environmental Protection Committee • Error Protection Code • Étage à Propulsion Cryotechnique • Étage Principal Cryotechnique • Étude Personnelle Chiffrée • European Paleoclimate • European Patent Convention • Event-driven Process Chain • Executive Policy Committee • External Power Contractor

EPCA: Energy Policy and Conservation Act • Extended Principle Component Analysis

EPCC: Environment Policy Coordinating Committee

EPCDC: Electrical Power Conditioning, Distribution, and Control

EPCE: EPIC p-n Control Electronics

EPCH: EPIC p-n Camera Head

EPCO: Emergency Power Cutoff • European Planetary and Cometary Observers

EPCP: Étude du Plateau Continental Polaire

EPCRA: Emergency Planning and Community Right-to-know Act

EPCS: EPIC p-n Camera System • Experiment Physics Control System • Experiment Point Control System

EPCSCP: Établissement Public à Caractère Scientifique, Culturel et Professionnel (France)

EPCU: Édifice de Préparation de la Charge Utile • Ensemble de Préparation des Charges Utiles (CSG) • EPIC p-n Camera Unit

EPD: Earliest Possible Date • Earliest Practical Date • Early Packet Discard • Electrical Power Distribution • Emergency Preparedness Division • Emergency Procedures Document • Energetic-Particle Detector (*) • Energy-charged Particle Detector • Engineering Product Definition • Environmental Protection Division • ESA Publications Division • Estimated Propagation Delay • Etch-Pit Density • European Pollen Database

EPDA: Education Professions Development Act (USA)

EPDB: Environmental Protection Database • Experiment Power Distribution Box

EPD&C: Electrical Power Distribution & Control

EPDC: Electrical Power Distribution and Control

EPD&CS: Electrical Power Distribution & Control System

EPDCS: Electrical Power Distribution and Control System

EPDH: EPIC p-n Data Handling

EPDM: Ethylene-Polypropylene-Diene-Monomer

EPDS: Electrical Power Distribution System • ERS-1 Products Distribution Service

EPDU: Electrical Power Distribution Unit • EPIC p-n Data Unit

EPE: Editorial Project for Education • Emergency Power Engineering • Energetic Particles Experiment • Energetic Particles Explorer • Environmental Protection Expenditure • European Partnership for Education • European Power Electronics

EPEA: EPIC p-n Event Analyser • Experiment Pointing Electronic Assembly

EPER: Environmental Processes and Effects Research

EPEX: Extreme Phenomena Explorer

EPF: Emission-Phase Function

EPFL: École Polytechnique Fédérale de Lausanne (Switzerland)

EPFS: Equipment Pool for Field Spectroscopy

EPFZ: École Polytechnique Fédérale de Zurich (Switzerland)

EPG: Electrical Power Generator • Electronic Program Guide • Electronic Proving Ground • Emergency Procedure Guideline • Engineering Proposal Generation • European Paleoclimatology Glaciology • Weeping Water - Browns (USA-NE) (location identifier)

EPGC: Earth Processes and Global Changes

EPGCS: Electric Power Generation and Control System

EPGS: École de Physique du Globe de Strasbourg (France) • Electric Power Generating System • Electric Power Generation System

EPGWS: Enhanced Proximity Ground Warning System

Eph: Ephemeris

EPH: Ephrata (USA-WA) (location identifier)

EPHC: Eastern Pacific Hurricane Center

EPHIN: Electron-Photon-Helium Instrument • Electron Proton Helium Instrument

EPHOS: European Procurement Handbook for Open Systems

EPI: Eagle Picher Industries • Earth Path Indicator • Echo Planar Imaging • Effective-Pair Interaction • Electrical Patents Index • Electronic Position Indicator • Energetic Particles Investigation • Environmental Productivity Index • Environnement de Programmation Intégré • Epi (Vanuatu) (location identifier) • Équipement de Protection Individuelle • ESOC Precipitation Index • Evolutionary Program Induction • External Presentation Interface • External Protocol Interface

EPIC: Earth Pointing Instrument Carrier • Earth Polychromatic Imaging Camera • Educational Policy Information Center • Efficient Pyramid Image Coder • Electronic Privacy Information Center (USA) • Energetic Particle and Ion Composition • Engineering Performance Information Center • Engineering Physics Information Center • Equatorial Processes Including Coupling • Établissement Public Industriel et Commercial • European Photon Imaging Camera • Explicit Planetary Isentropic Coordinate

EPICA: European Project for Ice Coring in Antarctica

EPICS: Electrolysis Performance Improvement Concept Study • Experimental Physics and Industrial Control System

EPIE: Educational Products Information Exchange

Epil: Epilogue

EPIOTIC: ESPRIT Projects In Optical Techniques of Interface Characterization

EPIRB: Emergency Position-Indicating Radio Beacon (*) • Emergency Position-Indicator Radio Beacon

EPIRSS: Extended Performance Infrared Search Set

EPISD: El Paso Independent School District (USA-TX)

Epit: Epitaph • Epitome

EPITA: École Pour l'Informatique et les Techniques Avancées (France)

EPJ: European Physical Journal

EPK: Episkopi (Cyprus) (location identifier)

EPL: Early Programming Language • Effective Privilege Level • Electrical Power Level • Electronic Program Language • Emergency Power Level • Encoder Programming Language • Épinal (France) (location identifier) • Eruptive Prominence on Limb • Evaluated Products List • Exact Power Law • Expected Power Level • Extensible Programming Language

EPLD: Electrically Programmable Logic Device • Erasable Programmable Logic Device (*)

EPLE: Établissement Public Local d'Enseignement

EPLRS: Enhanced Position Location Reporting System

EPLV: Enseignement Précoce des Langues Vivantes

epm: En Propria Mano

EPM: Earth Pointing Mode • Electric Printing Machine • Electron-Probe Microanalysis • Energetic Particle Monitor • Engineering Project Manager • Enhanced Presentation Manager • Enterprise Process Management • Environmental Planning and Management • European Physiology Module • Expanding Photosphere Method • Expanding Photosphere Model • Extent of Presence Metaphor

EPMA: Electron Probe Microanalyser • Electron-Probe Microanalysis

EPMC: Estimación de Parámetros por Mínimos Cuadrados

EPMR: Electric Power Margin Report

EPMS: Engineering Performance Management System

EPN: Elemental Point Number • Epena (Congo) (location identifier) • European Planetarium Network • European Pulsar Network

EPNdB: Effective Perceived Noise Decibel

EPNI: European Premier Network Integrator

E&PO: Education & Public Outreach

EPO: Earthnet Programme Office • Earth Parking Orbit • Education and Public Outreach • Electronic Publishing Only • Element Project Office • Engineering Placement Office • Enseignement Programmé par Ordinateur • Europäische Patentorganisation • European Patent Office (*) • European Patent Organisation

EPOC: Educational Point Of Contact • Employee Participation in Organisational Change • External Payload Operations Center

EPOCH: European Programme On Climatology and Hazards

EPOCS: Eastern Pacific Ocean Climate Study • Equatorial Pacific Ocean Climate Studies

EPONA: Energetic Particle Onset Admonitor

EPOP: Electric Propulsion Orbital Platform • European Polar Orbiting Platform • European Polar Platform

EPOPE: Étude de Processus dans l'Ocean Pacifique Équatorial

EPOQUE: European Parliament On-line Query • European Patent Office Query (system) (*)

EPOS: Earth Parameter and Orbit System • Earthquake Phenomena Observation System • Electronic Point of Sale • Engineering and Project-management Oriented Support System • Enhanced Preferred Observation Schedule • European Polarstern Study • European Proximity Operations Simulator

EPP: Emergency Preparedness Plan • Energetic Particle Package • Enhanced Parallel Port • Enterprise Portal Plan • Environment Protection Programme • European Polar Platform • Experimental Physics Project

EPPI: Electronic Programmed Procurement Information

EPPL: Environmental Planning and Programming Language

EPPP: Electronic Publishing Promotion Project

EP&PS: Electrical Power & Pyrotechnics System

EPPS: Electrical Power and Pyrotechnics System • Energetic Particles and Plasma Stability

EPPSS: Electrical Power/Pyro Sequential System

EPPT: European Printer Performance Test

EPR: East Pacific Rise • Education and Public Relations • Einstein-Podolsky-Rosen • Electric Rocket Engine • Electronic Patient Record • Electronic Patient Report • Electronic Proposal Review • Electronic Publishing Resource • Electron Paramagnetic Resonance • Engineering Purchase Request • Engine Pressure Ratio • Enseignement Primaire Rénové (Belgium) • Esperance (Australia-WA) (location identifier) • Estimated Probability of Relevance • EUVE Problem Reporting • Exhaust-Pressure Ratio • Explorer Platform Retrieval

EPRB: Einstein-Podalsky-Rosen-Bohm

EPRCA: Enhanced Proportional Rate Control Algorithm

EPRG: Engineering Psychology Research Group

EPRI: Electric Power Research Institute (USA-CA) • Electron Paramagnetic Resonance Imaging

EPRN: Emergency Program Release Notice

EPROM: Electrically Programmable Read-Only Memory • Erasable Programmable Read-Only Memory • Eraseable and Programmable Read-Only Memory

EpS: Earnings per Share

EPS: Earnings Per Share • Éducation Physique et Sportive • Electrical Power System • Electronic Proposal Submission • Electronic Publishing System • Electrostatic Plotting Software • Electrostatic Positioning System • Emergency Power System • Encapsulated PostScript • Engineered Performance Standard • Enhanced Press-Schechter (theory) • Ensemble Prediction Scheme • Ensemble Propulsif Stockable • Entreprise de Portage Salarial • Établissement Psychiâtrique Spécialisé • Étage à Propergols Stockables (*) • Étage Propulsif Stockable • EUMETSAT Polar System • Euregio Publiekscentrum voor Sterrenkunde (Netherlands) • European Physical Society • European Pointing System • European Polar Satellite • European Polar System • Experimental Power Supply • Experimental Publication System • Experiment Pointing System • Extensible Programming System

EPSAT: Estimation des Précipitations par Satellite • Étude des Précipitations par Satellite

EPSCOR: Experimental Program to Stimulate Competitive Research

EPSEX: Electric-Propulsion Space Experiment

EPSF: Encapsulated PostScript File • Encapsulated PostScript Format

EPSHOM: Établissement Principal du Service Hydrographique et Océanographique de la Marine (France)

EPSIG: Electronic Publishing Special Interest Group

E&PSL: Earth and Planetary Science Letters (periodical code)

EPSL: Earth and Planetary Science Letters

EPSP: Experiment Power Switching Panel

EPSRC: Engineering and Physical Sciences Research Council (UK)

EPSS: Electrical Power Subsystem • Electrical Power Supervision System • Electronic Performance Support System • Enhanced Packet Switched Service • Experimental Packet-Switching System

EPST: Établissement Public Scientifique et Technologique

EPSTF: Electrical Power System Test Facility

EPT: Effective Performance Time • Eliptamin (Papua New Guinea) (location identifier) • Emergency Procedure Trainer • Environmental Proof Test • Environment Policy and Technology • Ethylene-Polyprepylene-Terpolymer • Excess-Profits Tax • Experiment Parameter Table • External Pipe Thread

EPTCE: Electrical Power Thermal Control Electronics

EPTCS: Electrical Power Thermal Control System

EpTU: Events per Time Unit

EPTU: Events Per Time Unit

EPU: Emergency Power Unit • Energy Process Unit • Europäisches Patentübereinkommen • European Picture Union

EPÜ: Europäisches Patentübereinkommen

EPUSP: Escola Politécnica da Universidade de São Paulo (Brazil)

EPV: Ensemble de Prise de Vues

EPVAC: Electronic Publishing Volunteer Advisory Committee

EPVC: EPIC p-n Voltage Converter

EPWAT: Equatorial Plasma Wave Turbulence

EPWG: Experiment Planning Working Group

EPZ: Export Processing Zone

Eq: Equal • Equality • Equate • Equation • Équation • Equator • Equatorial • Équatorial • Équipement • Equipment • Equitable • Équitable • Equity • Equivalence • Équivalence • Equivalent • Équivalent

EQ: Earthquake • Educational Quotient • Emotional Quotient • Equal • Equality • Equalization • Equalizer • Equatorial • Équatorial • Equipment Qualification • Equivalence • Équivalence • External Inquiry

EQA: European Quality Alliance • European Quality Association • European Quality Award

EQC: External Quality Control

EQE: Event Queue Element

EQF: Embedded Query Facility • Embedded Query Function

EQI: Environmental Quality Index • Exhaust Quality Index

EQIMG: Equipment Industrial Management Group

EQIP: Environmental Quality Information Panel

Eql: Equal • Equally

EQLIPSE: Evaluation and Quality in Library Performance Systems

Eqlz: Equalize • Equalizer

EQM: Electrical Qualification Model • Engineering Qualification Model • Evaluation and Qualification Model

Eqn: Equation

EQN: Equation • Équation

EQNET: European Quality Network

Eqns: Equations

EQO: Environment Quality Objective

Eqpt: Equipment

EQS: Edinburgh Quasar Survey • Environmental Quality Standard • Esquel (Argentina-CB) (location identifier) • European Quality System

Equ: Equuleus-Equulei (constellation)

Equat: Equatorial

Equil: Equilibrium

Equip: Equipment

Équip: Équipement

EQUIP: Equipment Usage Information Programme

EQUIS: Equatorial Ionospheric Studies

Equit: Equitación

Equiv: Equivalent

Équiv: Équivalent

EQUS: Electronic Questionnaires System

EQV: Equivalent

er: Elder

Er: Erbium

E-R: Entity-Relationship

ER: Earth Resources • Earth Rotation • Eastern Range • East Reading (UK-England) • East River (USA-NY) • Echo Ranging • Edge Recombination • Effectiveness Report • Efficiency Report • Elastic Reality • Electron Reflectometer • Electroreflectance • Elongation Rate • Emergency Request • Emergency Rescue • Emergency Room • Emission Region • End Row • Energy Research • Engine Room • Enhanced Reactivation • Entity-Relation • Entity-Relationship • Entre Rios (Argentina) • Environmental Requirement • Equatorial Red (survey) • Équilibre Radiatif • Équipe de Recherche • Equipment Room • Equivalence Ratio • Eritrea (ISO country code) • Error Register • Érythrée (ISO country code) • Established Reliability • Evaporation Residue • Expanded Response • Experiment Representative • Explanation Report • Explicit Rate • Extended Range • Extended Reliability • Moldova (international aircraft registration code)

ERA: Early Retirement Adjustment • Earned Run Average • Electrical Replaceable Assembly • Electronic Reading Automation • Element-Relationship-Attribute • Emergency Relief Administration • Entités-Relations-Attributs • Entity-Relationship-Attribute • Environmental Resources Analysis • Equal Rights Amendment (USA) • Équipe de Recherche Associée • Erigavo (Somalia) (location identifier) • European Research Area • European Robot Arm • European Robotic Arm (*) • Exobiology and Radiation Assembly • Exobiology Radiation Assembly • External Robotic Arm

ERAA: European Regional Airline Association

ERAB: Energy Research Advisory Board

ERAD: En-route Radar

ERAF: Earth Resources Aircraft Facility

ERAM: Earth Resources Applications Mission • European Regional Astronomy Meeting

ERAMS: Environmental Radiation Ambients Monitoring System

ERAP: Earth Resources Aircraft Program • Earth Resources Aircraft Project

ERAR: Error Return Address Register

Eras: Erasmus (1466-1536)

ERAS: Electronic Routing and Approval System

ERASME: Étude Radar des Sols et de la Mer

ERASMUS: European Action Scheme for the Mobility of University Students • European Robotics, Automation, Simulation and Mission Utilization Support

ERAST: Environmental Research Aircraft and Sensor Technology

ERATO: Exploratory Research for Advanced Technology

ERAU: Embry-Riddle Aeronautical University

ERB: Earth Radiation Budget • Editorial Review Board (*) • Engineering Reference Branch • Engineering Review Board (*) • Engineering Review Body • Ernabella (Australia-SA) (location identifier)

ERBD: Équipe de Recherche sur les Bases de Données

ERBE: Earth Radiation Backscatter Experiment • Earth Radiation Budget Experiment (*)

ERBES: Earth Radiation Budget Explorer Satellite

ERBF: Earth Resources Browse Facility

ERBI: Earth Radiation Budget Instrument

ERBS: Earth Radiation Budget Satellite

ERBSS: Earth Radiation Budget Satellite System

ErbStDV: Erbschaftsteuer-Durchführungsverordnung

ERC: Earth Rate Compensation • Earth Re-entry Capsule • Earth Return Capsule • Economic Research Council • Educational Reference Center • Educational Resource Center • Electrical Rule Checker • Electronics Research Council • Embedded Real-time Computer • Employee Recognition Center • Employment Rehabilitation Centre • Energy Research Corporation • Engineering Research Center • Erzincan (Turkey) (location identifier) • European Radiocommunications Committee • European Research Conference • Evaluation Review Committee • Excessive Requirement Cost • Expérience de Repérage de Champ

ERCA: European Research Courses on Atmosphere

ERCB: Energy Resources Control Board

Ercg: Erecting

ERCH: Electron Resonance Cyclotron Heating

ERCIM: European Research Consortium for Informatics and Mathematics

ERCM: Engineering Requirements and Change Management

ERCO: European Radio Communications Office

ERCOS: Expérience Rayonnements Cosmiques

ERCS: Emergency Response Computer System

Erct: Erection

ERD: Berdyansk (Ukraine) (location identifier) • Ecological Research Division • Edited Raw Data • Elastic Recoil Detection • Electronic Reference Document • Electronic Report Distribution • Emergency Response Division • Emergency Return Device • Entity-Relation Diagram • Entity-Relationship Diagram • Environmental Radiation Data • Equivalent Residual Dose • Error Recording Device • Estimated Receival Date • Estimated Release Date • Extrapolated Response Dose

ERDA: Elastic Recoil Detection Analysis • Energy Research and Development Administration • Exchange Regional Development Assistant

ERDAS: Earth Research Data Analysis System • Earth Resources Data Analysis System (*) • Earth Resources Digital Analysis Software • Earth Resources Digital Analysis System

ERDC: Earth Resources Data Center • Energy Research and Development Centre

ERDE: Engineering Research and Development Establishment

ERDF: European Regional Development Fund

ERDIC: Energy Research, Development, and Information Centre

ERDL: Engineering Research and Development Laboratory

ERDOS: Expérience Radiations Dosimétrie

ERDS: Emergency Response Data System

ERE: Effective Resolution Element • Émission Rouge Étendue • Entity Relationship Editor • Erave (Papua New Guinea) (location identifier) • Extended Red Emission • Extent of Reaction Effect

EREA: École Régionale d'Enseignement Adapté

Erec: Erection

EREP: Earth Resources Experimental Package • Environmental Recording Editing and Printing

ERF: Edge Response Function • Energy Rejection Filter • Erfurt (Germany) (location identifier) • European Research Foundation

ERFA: European Radio Frequency Agency

ERFDP: Earth Resources Flight Data Processor

ERFEN: Estudio Regional del Fenomeno El Niño

ERFIS: Emergency Response Facility Information System

ERFU: Environmental Restoration and Facilities Upgrade

Erg: Ergänze!

ERG: Electrical Resistance Gauge • Electroretinogram • Electroretinograph • Electroretinography

ERGO: European Research Gateway Online

ERGOM: European Research Group On Management

Ergon: Ergonomics

ERH: Errachidia (Morocco) (location identifier) • Europäischer Rechnungshof

Eri: Eridanus-Eridani (constellation)

ER&I: Engine Receiving & Inspection

ERI: Earthquake Research Institute (Japan) • Educational Research Information • Engine Receiving and Inspection • Erie (USA-PA) (location identifier) • Eritrea (ISO country code) • Érythrée (ISO country code) • Extravehicular Reference Information

ERIA: Estudios y Realizaciones en Informatica Aplicada

ERIC: Educational Research and Information Center • Educational Resources Information Center • Educational Resources Information Clearinghouse • Energy Rate Input Controller • Environmental Research and Information Centre

ERICA: Experiment on Rapidly Intensifying Cyclones over the Atlantic

ERIM: Earth Resource Information Management • Environmental Research Institute of Michigan (USA-MI)

ERIN: Earth Resources Information Network • Environmental Resources Information Network

ERIPS: Earth Resources Interactive Processing System

ERIS: Earth Resources Information System • Earth Resources Inventory System • Emergency Response Information System • Energy Research Information System • Environmental Resources Information System • Exoatmospheric Reentry System

ERISA: Employee Retirement Income Security Act (USA) • Employment Retirement Income Security Act (USA) (*)

ERISTAR: Earth Resources Information Storage, Transformation, Analysis, and Retrieval

Erit: Eritrea

ERKE: Excess Rotational Kinetic Energy

Erl: Erläuterung

ERL: Earth Resources Laboratory • Echo Return Loss • Electronic Reference Library • Energy Research Laboratory • Environmental Research Laboratory • Event Record Log

ERLAP: European Reference Laboratory for Air Pollution

ERLAS: Earth Resources Laboratory Application Software

ERLL: Enhanced Run Length Limited

ERM: Earth Resource Mapping • Earth Return Module • École Royale Militaire • Electronic Rights Management • Empirical Risk Minimization • Enterprise Resource Management • Entity-Relationship Model • EPIC Radiation Monitor • Erechim - Comandante Kraemer (Brazil-RS) (location identifier) • Exact Repeat Mision • Exchange Rate Mechanism • Executive Review Meeting • Extended Relational Model • External Resource Manager

ERMD: EPIC Radiation Monitor Detector

ERME: EPIC Radiation Monitor Electronics

ERMES: European Radio Messaging System

ERMP: Environment Management and Review Program

ERMS: Environmental Resources Mapping System • EPIC Radiation Monitor System

ERN: Eirunepé (Brazil-AM) (location identifier)

ERNE: Energetic and Relativistic Nuclei and Electrons

ERNIE: Electronic Random Number Indicator Equipment

ERNN: External Recurrent Neural Network

ERNS: Emergency Response Notification System

ERO: Early Release Observation • Eldred Rock (USA-AK) (location identifier) • Engineering Release Operation • Enhanced Reliability Option • European Radiocommunications Office • European Regional Organization • Executive Records Office • Extremely Red Object

EROC: Ecological Rate Of Change

EROM: Erasable Read-Only Memory

EROMM: European Register of Microform Masters

EROPS: Extended-Range Operations

EROS: Earth Resources Observation Satellite • Earth Resources Observation System • Earth Resources Observing System • EISCAT Real-time Operating System • European Rivers Outflow Studies • Expérience de Recherche d'Objets Sombres • Experimental Reflection Orbital Shot

ERP: Earth Resources Program • Earth Rotation Parameter • Educational Reimbursement Program • Education and Public Relations • Effective Radiated Power • Elevator Reference Plane • Energy Radiation Pattern • Enterprise Resource Planning • Environmental Research Park • Episodic Response Project • Eppawala Rock Phosphate • Error Recovery Package • Error Recovery Procedure • European Recovery Programme • Event-Related Potential • Extensional Recursive Processing • Eye Reference Point

ERPS: Experimental Rocket Propulsion Society (USA-CA)

ERR: Electrical Resistance Requirement • Energy Release Rate • Engineering Release Record • Errol (USA-NH) (location identifier) • Error

ErrC: Error Correction

ERRC: Expendability, Recoverability, and Repair Capability

Erron: Erroneous • Erroneously

ERRP: Equipment Replacement and Refurbishment Plan

ERRRIS: E-Region Rocket/Radar Instability Study

ERRSAC: Eastern Regional Remote Sensing Applications Center

ERRT: Error Rate Relative to Truncation

Ers: Erase • Erased

ERS: Earth Recovery System • Earth Remote Sensing • Earth Resources Satellite • Earth Resources Survey • Economic Research Service • Electric Resource System • Element Rotation Stand • Emergency Radio Service • Emergency Response System • Engineering Release System • Enterprise Router Switch • Entrée Réservée Synchrone • Entry and Recovery Simulation • Environmental Research Satellite • Error Reporting System • Escadron de Reconnaissance Stratégique • Estimated Release Schedule • European Remote Sense • European Remote Sensing • Extended Review Session • Extragalactic Radio Source • Windhoek Eros (Namibia) (location identifier)

ERSAC: Environmental Remote Sensing Applications Centre

ERSAF: Environment Remote Sensing Analysis Facility • European Remote Sensing Aircraft Facility

ERSAL: Environmental Remote Sensing Applications Laboratory

ERSAR: Earth Resources Synthetic Aperture Radar

ERSC: Environmental Remote Sensing Centr • Eurimage, Radarsat International, Spot Image Consortium • European Research Standing Committee

ERSCOS: European Remote-sensing Satellite Central User Service

ERSDAC: Earth Remote Sensing Data Analysis Center • Earth Resources Satellite Data Analysis Center

ERSDIS: Earth Resources Satellite Data and Information Center

ERSI: Elastomeric Reusable Surface Insulation • Environmental and Remote Sensing Institute

ERSIS: Earth Resources Spectral Information System

ERSISS: European Remote-sensing Satellite Interface Subset

ERSOS: Earth Resources Survey Operational System

ERSP: Earth Resources Survey Program • Electronic Reservation Services Provider • Epipolar Rectified Stereo Pair

ERSS: Earth Resources Survey Satellite

ERSSP: European Remote Sensing Satellite Programme

ERST: European Report on Science and Technology (periodical code)

ERSTI: European Report on Science and Technology Indicators

ERSUN: Environmental Remote Sensing Unit

ERT: Earth Received Time • Earth Resources Technology • Education Round Table • Environmental Response Team • Environment Round Table • Erdenet (Mongolia) (location identifier) • European Round Table • Extended Range Transmitter

ERTBP: Elliptic Restricted Three-Body Problem

ERTCS: Emergency Recirculation and Tritium Cleanup System

ERTE: Exact Radiative Transfer Equation

ERTG: Economic Radioisotope Thermoelectric Generator

ERTMS: European Railway Traffic Management System

ERTS: Earth Resources Technology Satellite (later called Landsat)

ERTU: Egyptian Radio and Television Union

ERU: Earth Rate Unit • Emergency Recovery Unit • Energy Research Unit • Error Return Update • Erume (Papua New Guinea) (location identifier)

ERUS: ERS User Services

ERV: Earth Return Vehicle • English Revised Version • Experimental Reentry Vehicle • Expiratory Reserve Volume • Kerrville (USA-TX) (location identifier)

Erw: Erweitert

Erweit: Erweitert

ERY: Early

ERZ: Erzurum (Turkey) (location identifier)

Erzgeb: Erzgebirge (Germany)

es: Spain • Spanish (ISO language code)

Es: Einsteinium • Esempio

E/S: Engine Speed • Entrées/Sorties • Equipment Section

E&S: Earth & Space (periodical code) • Engineering & Science • Evans & Sutherland Computer Corporation (USA-UT)

ES: Earth Sensor • Education Specialist • Éducation Surveillée • Elastic Scattering • Eldest Son • Electrical Sounding • Electric Starting • Electrochemical Society • Electronic Structure • Electronic Support • Electronic Switch • Électronique Séquentielle • Electron Synchrotron • Electrostatic • El Salvador (international vehicle registration code) • Employment Strategy • End System • Energy Spectrum • Engineering and Science • Engineering Specification • Engine Speed • Enriched Sample • Enseignement Supérieur • Enter Shadow • Equal Section • Equipment Section • Equipment Shelf • Escape System • Eskimo • Espagne (ISO country code) • Espíritu Santo (Brazil) • Estonia (international aircraft registration code) • Étoile Symbiotique • Euler Sum • Evolution Strategy • Executive Secretary • Exercise Server • Expert System • Expanded Storage • Experiment Segment • Expert System • Exposed Section • Extra Segment • Spain (ISO country code)

ESA: Ecological Society of America • EFTA Surveillance Authority • Electronically Scanned Array • Electronically Scanning Antenna • Electronically Steerable Antenna • Electrostatic Analyzer • Electrostatic Sector Analyzer • Electrostatic Spherical Analyzer • Endangered Species Act (USA) • Energy Sensor Array • Energy System Analysis • Engineering Supply Area • Engineering Support Area • Engineering System Assembly • Enterprise Systems Architecture • Entomological Society of America • Environmentally Sensitive Area • Equatorial South America • Esa'Ala (Papua New Guinea) (location identifier) • European Sociological Association • European Space Agency • European Symposium on Algorithms • European System of Accounts • Evening Sector Arc • Excited State Absorption • Explosive Safe Area • Extended Service Area

ESAAB: Energy System Acquisition Advisory Board

ESABu: European Space Agency Bulletin (periodical code)

ESAC: Electronic Systems Assistance Center • Enterprise-wide Security and Audit Conference

ESAD: Earth Science and Applications Division

ESADS: Earth Science and Applications Data Systems

ESAGB: Early Super Asymptotic Giant Branch

ESAJ: European Space Agency Journal (periodical code)

ESAM: Earth Sensor Assembly Module • Energy Saving Assessment Model • Environmental Science And Management

ESANET: European Space Agency Network

ESANN: European Symposium on Artificial Neural Networks

ESAP: ECLS Systems Assessment Program

ESAR: Electronically-Steerable Array Radar • Exercice-Symbole-Assemblage-Règle

ESARS: Expert System for Accessing Remotely Sensed data

ESAS: Enhanced Situational Awareness System • Extended Source Analysis Software

ESAST: European Space Agency Scientific Technical Review (periodical code)

ESAT: ERB Solar Analysis Tape

ESATAN: ESA Thermal Analysis Network

ESB: Ankara Esenboga (Turkey) (location identifier) • B-Type Spanish Peseta (ISO currency code) • Electrical Stimulation of the Brain • Electricity Supply Board • Electric Storage Battery • Empire State Building (USA-NY) (*) • Empty State Building (USA-NY) (earlier nickname for (*)) • Engineering Standards Board • Engineering Support Branch • Essential Switching Box • Executive Safety Board

ESBS: École Supérieure de Biotechnologie de Strasbourg (France)

ESBT: Experimental S-Band Terminal

Esc: Escalera • Escape • Escompte • Escrow • Escudo (currency) • Escultura

ESC: École Supérieure de la Communication • Economic and Social Committee • Economic and Social Council • Educational Scientific Computer • Electronic Security Command • Electronic Still Camera • Electronic Structural Correlator • Electronic Systems Center • Engineering Service Circuit • Engineering Support Center • EPIC Science Console • Escada • Escanaba (USA-MI) (location identifier) • Escape • Escudo (currency) • ESIS Steering Committee • Étage Supérieur Cryogénique • European Seismological Commission • European Space Conference

• Euro Space Center (Belgium) • Expedited Site Characterization • Extended Core Storage

ESCA: Electron Spectroscopy for Chemical Analysis • Engenharia de Systemas de Controle e Automação • European Speech Communication Association

Escand: Escandinavo

ESCAP: Economic and Social Commission for Asia and the Pacific (UN)

ESCAPE: Experiment of the Sun for Complementing the ATLAS Payload for Education

Escarp: Escarpment

ESCB: European System of Central Banks

ESCC: Experiment Scoring and Control Centre • External Stress Corrosion Cracking

ESCCD: Electrical Short-Circuit Current Decay

ESCDR: Escadaria

ESCES: Experimental Satellite Communication Earth Station

ESCI: Environmental Science Citation Index

ESCIN: European Science Communications and Information Network

Escl: Escalator • Escalinata

ESCM: Extended Services Communications Manager

ESCNH: Escadinha

ESCO: Educational, Scientific, and Cultural Organization • European Satellite Consulting Organization

ESCON: Enterprise Systems Connection

ESCPI: École Supérieure de Chimie et de Physique Industrielle

ESCR: Escritório • External Standard Channel Ratio

ESCS: Electronic Spacecraft Simulator

ESCULT: Escultor

ESCWA: Economic and Social Commission for Western Asia

ESD: Earliest Start Date • Eastsound (USA-WA) (location identifier) • Echo-Sounding Device • Ecologically Sustainable Development • Educational School District • Educational Service District • Educational Systems Division • Electromagnetic Shower Detector • Electronic Software Distribution • Electronic Spectroscopic Diffraction • Electronic Systems Division • Electronics Systems Division • Électronique Serge Dassault • Electron Stimulated Desorption • Electrostatic Discharge • Electrostatic Sensitive Device • Electrostatic Spray Deposition • Emergency Shutdown • Emission Standards Division • Engineering Services Department • Environmental Sciences Division • Estimated Shipping Date • European Standard Drawer • Experiment Systems Division • Exploitation Support Data • External Symbol Dictionary

ESDA: Earth Sciences Data Standards • Electrostatic Deposition Analysis

ESDB: Endangered Species Database • Extended Source Database

ESDD: Earth Science Data Directory

ESDI: Enhanced Small Device Interface (*) • Enhanced Standard Device Interface

ESDIM: Earth System Data and Information Management

ESDIS: Earth Science Data and Information System

ESDOC: European Science Data and Operations Centre

ESDP: External Sensor Data Processing

ESDS: Entry Sequenced Dataset

ESDU: Engineering Sciences Data Unit • Event Storage and Distribution Unit

E&E: Electrical & Systems Engineering

ESE: Earth Science Enterprise • East-Southeast • East-Southeastern • Electrical and Systems Engineering • Electrical Support Equipment • Electronic Support Equipment • Electron Spin Echo • Engineering Support Equipment • Ensenada (Mexico) (location identifier) • Est-Sud-Est • Extreme Scattering Event

ESEA: Elementary and Secondary Education Act (USA)

ESEC: European Software Engineering Conference

ESECA: Energy Supply and Environmental Coordination Act

ESEE: Exhaustive Set of Exclusive Events

ESEM: Evaluation of Space Environment with Materials

ESEO: École Supérieure d'Électronique de l'Ouest (France)

ESEP: Earth System Evolution Program • European Science Exchange Programme

ESERN: Eastern

ESES: Environmentally Sound Energy System • ESA Software Engineering Standards

ESEWD: East-Southeastward

ESEX: Electric Space Experiment

ESF: Alexandria Esler (USA-LA) (location identifier) • Economic Support Fund • Edge Spread Function • Equatorial Spread-F • Eureka Software Factory • European Science Foundation • European Social Fund • Evans Solar Facility (NSO) • Explosive Safe Facility • Extended Service Frame • Extended Superframe • Extended Superframe Format

ESFC: Equivalent Specific Fuel Consumption

ESFJ: Extraverted Sensing Feeling Judging

ESFON: European Star Formation Network

ESFP: Extraverted Sensing Feeling Perceiving

ESG: Education Support Grant • Empirical Study Group • Enseignement Secondaire Général • Mariscal Estigarribia (Paraguay) (location identifier)

ESGEO: ESRO Geostationary Earth Orbiting Satellite

ESGP: Earth Science Geostationary Platform

Esgr: Esgrima

ES&H: Environmental Safety & Health

ESH: Environmental Safety and Health • Equivalent Solar Hour • Equivalent Standard Hour • Sahara Occidental (ISO country code) • Sahara (Western_) (ISO country code) • Shoreham by Sea (UK-England) (location identifier) • Western Sahara (ISO country code)

ESHP: Equivalent Shaft Horsepower

ESI: Earth Sciences Information • Earth Space Institute (France) • Earth Systems Institute • École des Sciences de l'Information • Electrical System Integration • Electronic Share Information • Electrospray Ionization • Ellery Systems Inc. (USA-CO) • Enhanced Serial Interface • Enseignement Secondaire Inférieur • Environmental Sensitivity Index • Environment Sensitivity Index • Espinosa (Brazil-MG) (location identifier) • European Satellite Industry • Experimental Systemic Initiative • Exploration Science Institute (USA-TX) • Externally Specified Index • External System Interface

ESIAC: Electronic Satellite Imaging Analysis Console

ESIC: Earth Science Information Center • Environmental Science Information Center

ESIE: Elementary, Secondary, and Informal Education

ESIEA: École Supérieure d'Informatique Électronique Automatique (France)

ESIEE: École Supérieure d'Ingénieurs en Électrotechnique et Électronique

ESIG: Environmental and Society Impact Group

ESIM: Earth Science Information Manager

ESIMUD: European Space Industry Multimedia Database

ESIN: Earth Science Information Network

ESIO: Earth Science Information Office

ESIP: Earth Science Information Partner

ESIPP: Earth Science Imaging Processing Package

ES-IS: End System to Intermediate System

ESIS: End System to Intermediate System • ESCAP Statistical Information System • European Social-science Information System • European Space Information System • European Survey of Information Society

ESIST: European Society and its Interactions with Science and Technology

ESJ: École Supérieure de Journalisme (France)

ESJWG: Earth Sciences Joint Working Group

Esk: Eskimo

ESK: Eskisehir (Turkey) (location identifier)

ESL: Electronic Software Licensing • Elista (Russia) (location identifier) • Energy Sciences Laboratories • Engineering Societies Library • English as a Second Language • ESA Simulation Language • European Small Launcher • European System Language • Expected Search Length • Expected Significance Level

ESLAB: European Space Laboratory

ESLIT: Entrance Slit

ESLSCA: École Supérieure Libre des Sciences Commerciales Appliquées (France)

ESM: Earth Search Mode • École Supérieure de Métrologie • Electronic Support Measure • Energy Storage Module • En Sus Manos • Enterprise Storage Management • Environmental Systems Monitor • Esmeraldas (Ecuador) (location identifier) • Executive Secretaries Meeting • Experiment Support Module • Experiment Systems Monitor • Extended Sum-rule Model • External Storage Module

ESMAP: Energy Sector Management Assistance Programme

ESMDM: Enhanced Space-Station Multiplexer/Demultiplexer

ESME: École Supérieure de Mécanique et d'Électricité

ESMF: Efficient Separable Median Filter

ESMN: European Solar Magnetograph Network

ESMOS: European Stratospheric Monitoring Station

ESMR: Electrically Scanned Microwave Radiometer • Electrically-Scanning Microwave Radiometer • Enhanced Specialized Mobile Radio

ESMS: Environmental and Special Measurements System

Esn: Esential

ESN: Earth Science Network • Easton (USA-MD) (location identifier) • Educationally Subnormal • Electronic Security Number • Electronic Serial Number • Electronic Switched Network • Enhanced Service Node • European Science Notes (periodical code)

Esntl: Esential

ESNU: Enseignement Supérieur Non-Universitaire (Belgium)

ESO: Emergency Services Officer • Enseñanza Secundaria Obligatoria • Entry Server Offering • Espanola (USA-NM) (location identifier) • European Southern Observatory • Event Sequence Override • Expanded Storage Only

ESOB: European Southern Observatory Bulletin (periodical code)

ESOC: European Space Operations Centre (ESA) • EUVE Science Operations Center (USA-CA) • Extrasolar Oort Cloud

ESOL: English for Speakers of Other Languages

ESOMAR: European Society for Opinion and Marketing Research

ESOMe: European Southern Observatory Messenger (periodical code)

ESOP: Employee Share Ownership Plan • Employee Stock Ownership Plan • European Symposium on Occultation Projects

ESOSP: European Southern Observatory Scientific Preprints (periodical code)

ESOSR: European Southern Observatory Scientific Reports (periodical code)

ESOW: Engineering Statement Of Work

Esp: Espace • Espagne • España • Español • Especial • Especially • Especialmente • Esperanto • Esplanade

ESP: Early Support Program • East Stroudsburg - Birchwood-Pocono (USA-PA) (location identifier) • Economic Simulation Program • Electric Submersible Pump • Electronic Schematic Processor • Electron Spin Polarization • Electron Stream Potential • Electronic Stability Program • Electronic Standard Procedure • Electroshock Protection • Elsevier Science Publishers • Empirical Studies of Programers • Employee Suggestion Program • Emulation Sensing Processor • Encapsulating Security Payload (*) • Encapsulating Security Protocol • Encapsulation Security Payload • Energetic Solar Particle • Energy Savings Performance • Energy Simulation Package • Energy Sum Processor • Engineering and Scientific Processor • Engineering Simulation Program • English for Special Purposes • English for Specific Purposes • Enhanced Scalar Processing • Enhanced Serial Port • Enhanced Service Provider • Enhanced Support Package • En-route Spacing Program • Enseignement Secondaire Professionnel • Enterprise Storage Platform • Environmental Sensing Platform • ESO Slice Project • Esoteric Scholarly Publication • Espagne (ISO country code) • Exchangeable Sodium Percentage • Executive System Planning • Expandable Stored Program • Expanding Space Paradigm • Experiment Sensing Platform • Extended Self-contained PROLOG • Extended Surface Photometry • Extended Support Plan • Extrasensory Perception • Extrasolar Planet • Extravehicular Support Pack • Peseta (Spain) (ISO currency code) • Spain (ISO country code)

ESPACE: European System for Particle and Astrophysics Cybereducation • Évaluation et Suivi de la Production Agricole en fonction du Suivi et de l'Environnement

E-SPAN: European SPAN

ESPAN: European SPAN

ESPAS: ESO Search for Potential Astronomical Sites

ESPC: European Space Power Conference

ESPE: Escuela Politécnica del Ejercito

Espec: Especial • Especially

ESPF: Experimental SAR Processing Facility

ESPG: ESR Software Project Group

ESPI: Electronic Speckle Pattern Interferometry

ESPIN: European Space Information Network

Espl: Esplanada

ESPL: Extensible Structure Processing Language

ESPM: Escola Superior de Propaganda e Marketing

ESPN: Entertainment Sports Network

ESPOL: Executive System Problem-Oriented Language

ESPP: Employee Share Purchase Plan

ESPRIT: European Strategic Programme for Research in Information Technology • European Strategic Programme for R&D in Information Technology (*)

ESPS: Entropic Signal Processing System • Experiment Segment Pallet Simulator

Esq: Esquina • Esquire

ESQ: Esquerdo

Esq°: Esquerdo

Esqr: Esquire

ESR: Earth Science Review • Earth Space Review • Effective Sunrise • EISCAT Svalbard Radar • Electrical Substitution Radiometer • Electron Spin Resonance • El Salvador (Chile) (location identifier) • Emergency Sun Reacquisition • Engineering Support Request • Enseignement Secondaire Rénové (Belgium) • Equipment Status Report • Equivalent Series Resistance • Experimental Storage Ring

ESRA: European Safety and Reliability Association

ESRANGE: European Sounding Rocket Launching Range • European Space Range

ESRC: Economic and Social Research Council (UK) • Environmental Science Research Center • European Science Research Council

ESRF: Environmental Studies Revolving Funds • European Synchrotron Radiation Facility

ESRI: Earth Sciences and Resources Institute • Environmental Systems Research Institute

ESRIN: European Space Research Institute

ESRM: Enterprise Storage Resource Management

ESRO: European Space Research Organization (merged now with ELDO into ESA)

ESRP: European Supersonic Research Program

ESRU: Earth Satellite Research Unit

Ess: Essence • Essential • Essentially • Essex (UK-England)

ESS: Earth and Space Sciences • Earth System Science • Effective Sunset • Efficient Sensing System • Electronic Signalling System • Electronic Switching System • Elementary Science Study • Emergency Survival System • Energy Storage System • Engineering Support System • Enseignement Secondaire Supérieur • Enterprise Storage Server • Environmental Support System • Environment, Science and Society • Equipment Support Section • ESO Sculptor Survey • Essen (Germany) (location identifier) • Ethernet Segment Switch • European Social Survey • European Spallation Source • European Statistical System • Executive Support System • Expendable Second Stage • Experiment Support System • Experiment System Simulator • Expert Support System • Expert System Shell • Extended Self-Similarity • Extended Service Set

ESSA: Electrically Steerable Spherical Array • Electronic Switching Spherical Array • Environmental Science Services Administration (USA) • Environmental Science Services Agency • Environmental Survey Satellite

ESSAT: European Society for the Study of Science and Theology

ESSC: Earth System Science Center • Earth System Sciences Committee • Edmonton Space and Science Centre (Canada-AB) • Engineering Support Services Contract • European Space Science Committee

ESSCO: Electronics Space Systems Co.

ESSD: Electronic Sensors and Systems Division • Environmentally Sound and Sustainable Development • European Space Science Directory

ESSDE: European Space Software Development Environment

ESSE: Earth System Science Education

ESSEC: École Supérieure des Sciences Économiques et Commerciales (France)

ESSENTIAL: European Systems Strategy for the Evolution of New Technology In Advanced Learning

ESSEX: Experimental Solid-State Exchange

ESSI: European Software and Systems Initiative • European Software Systems Initiative

ESSII: Expert System for Satellite Image Interpretation

ESSIR: European Summer School on Information Retrieval

ESSIS: Earth and Space Science Information System • Earth System Science Information System

ESSL: Engineering and Scientific Subroutine Library

ESSM: Expert System Simulation Model

ESSN: Earth System Science Network

ESSO: Earth Science Support Office

ESSOPE: Expert System for Spacecraft Operations Planning and Execution

ESSP: Earth System Science Pathfinder • Educational Space Simulations Project • Enchanted Skies Star Party (USA)

ESSS: External Stores Support System

ESST: European Studies in Science and Technology

ESSX: Electronic Switching System Exchange

Est: Established • Establishment • Estate • Estatuto • Estimate • Estimated • Estimation • Estimator • Estonia • Estonian • Estrecho • Estuary • Estudio

ESt: Eisenbahn Station

ES&T: Environmental Science & Technology • Environment, Science & Technology

EST: Earliest Start Time • Earth-Satellite-Target • Eastern Standard Time (*) • Eastern Summer Time • Electroshock Therapy • Electroshock Treatment • Ellipsoidal Shell Tokamak • Engineering Support and Technologies • Enseignement Secondaire Technique • Environment, Science and Technology • Estate • Estherville (USA-IA) (location identifier) • Estimate • Estonia (ISO country code) • Estonie (ISO country code) • Expert System Tools • Expressed Sequence Tag

ESTA: Escape System test Article • European Science and Technology Assembly

Estab: Establish • Established • Establishment

ESTAB: Estabelecimento

ESTAC: Estação

Estad: Estadística

ESTAR: Electronically Scanned Thinned Array Radiometer (*) • Electronically Steered Thinned Array Radiometer • External Servicing Technology Testbed for Automation and Robotics

Estát: Estática

EStDV: Einkommensteuer-Durchführungsverordnung

ESTE: Electrical Special Test Equipment

ESTEC: European Space Research and Technology Centre (ESA)

ESTEEM: Earth Science Teachers Exploring Exemplary Materials

ESTF: Electronic System Test Facility

Estg: Estimating

ESTJ: Extraverted Sensing Thinking Judging

ESTL: Electronic Systems Test Laboratory • European Space Tribology Laboratory

Estn: Estimation

ESTP: Extraverted Sensing Thinking Perceiving

EStR: Einkommensteuerrichtlinien

ESTR: Engineering and Science Tape Recorder • Environment, Science and Technology Resources • Estrada

ESTRACK: European Space Tracking (ESA)

ESTS: Estates

ESTU: Extended Stretched Trans-Uranium (WNSL)

ESU: Electrostatic Unit • Emporia State University (USA-KS) • Equal Size Use • Eyepiece Storage Unit

ESV: Earth Satellite Vehicle • Emergency Shutdown Valve • Emergency Shutoff Valve • Erstschichtversagen • Experimental Safety Vehicle

ESVA: Earth Station Verification and Acceptance

ESVS: Escape Suit Ventilation System

ESW: Easton State (USA-WA) (location identifier) • East-Southwest • Engine Status Word • Error Status Word

ESWG: Electronic Services Working Group (ACM)

ESYS: Expert System

ESZB: Europäisches System der Zentralbanken

Ét: Étage

et: Estonian (ISO language code) • Ethiopia

Et: Ethil

E&T: Education & Training

ET: Eastern Time • Edge Thickness • Educational Television • Education and Training • Effectiveness Team • Effective Temperature • Egypt (international vehicle registration code) • Eigenvalue Transform • Elapsed Time • Electrical Transcription • Electric Telegraph • Electric Typewriter • Electron Transfer • Electron Transition • Element Test • Elevated Temperature • Elongated Tokamak • Emerging Technology • Emptying Time • Enabling Technology • Enhanced Telephone • Ephemeris Time • Equation of Time • Equivalent Term • Escape Tower • Est • Estimated Time • Ethiopia (international aircraft registration code) • Ethiopia (ISO country code) • Éthiopie (ISO country code) • European Technology • Evapotranspiration • Event Timer • Excitation Transport • Execution Territory • Extended Time • External Tank • Extraterrestrial

ETA: Educational Television Association (UK) • Electronic Travel Aid • Emanation Thermal Analysis • Energetic Transient Array • Engineering Test Article • Estação Terrena de Alcântara (Brazil) • Estimated Time of Acquisition • Estimated Time of Appearance • Estimated Time of Arrival • European Teachers' Association • Expected Time of Arrival • Explosive Transfer Assembly • External Tank Attachment

Étab: Établissement

ETAC: Environment Technical Advisory Committee • Environment Technical Applications Center

ETACS: Extended Total Access Communication System

ETAGO: End Tag Open

ETAI: Electronic Transactions on Artificial Intelligence

et al: Et Alia • Et Alibi • Et Alii

ETAM: Entry Telecommunications Access Method

ETAN: European Technology Assessment Network

ETANN: Electrically Trainable Analog Neural Network

ETAP: Expanded Technical Assistance Program • Experimental Technical Assessment Panel

ETAS: Effective True Airspeed

ETB: Birr (Ethiopia) (ISO currency code) • End of Transmission Block • Equipment Transfer Bag • West Bend (USA-WI) (location identifier)

ETBE: Ethyl Tertio Butyl Ether

etc: Et Caetera

ETC: Earth Terrain Camera • Electronic Test Centre • Energy Technology Center • Engineering and Training Center • Enhanced Throughput Cellular • Estação Terrena de Cuiabá (Brazil) • Estimated Time of Completion • Estimate To Complete • Et Caetera • European Test Conference • European Translation Centre • European Travel Commission • Explosive Transient Camera • Exposure Time Calculator

ETCA: Établissement Technique Central de l'Armement (France) • Études Techniques et Constructions Aérospatiales (Belgium)

ETCD: Équipement de Terminaison de Circuit de Données

ETCO: Equipment Transfer/Change Order • External Tanks Corporation

ETCS: European Train Control System

ETD: Electrical Terminal Distributor • Electronic Theses and Dissertations • End-Transaction Descriptor • Entwicklung Technische Dokumentation • Estimated Time of Departure • Estimated Turnover Date • Etadunna (Australia-SA) (location identifier) • Evolutionary Terminal Defense • Expected Time of Departure • Extension Trunk Dialing • External Tank Door

ETDC: East Tennessee Discovery Center (USA-TN)

ETDE: Energy Technology Data Exchange

ETDs: Electronic Theses and Dissertations

ETE: Éditions Techniques Européennes (France) • End-To-End • Engenharia e Tecnologia Espacial • Estimated Time En-route • Evacuation Time Estimate • Genda Wuha (Ethiopia) (location identifier)

ETEDB: External Thermal Environment Database

ETEEM: End-To-End Engagement Model • End-To-End Engineering Model

ETES: Éditions Techniques, Économiques et Sociales

ETEX: European Tracer Experiment

ETF: Educational Trust Fund • Electrical Test Facility • Electrothermal Feedback • Elgin Test Facility (NASA) • Engineered Test Facility • Environmental Task Force • Estimated Time of Flight • European Technology Facility • European Training Foundation • European Transfer Format

ETFE: Ethylene-Tetrafluoroethylene

ETFRN: European Tropical Forest Research Network

Étg: Étang

ETG: Enhanced Target Generator • European Tripartite Group

ETGTS: Electronic Text and Graphics Transfer System

Eth: Ether • Ethical • Ethics • Ethiopia • Ethiopian

ETH: Eidgenössische Technische Hochschule (Switzerland) • Elat (Israel) (location identifier) • Ethiopia (international vehicle registration code) • Extraterrestrial Hypothesis • Falkland Islands (ISO country code) • Îles Malouines (ISO country code) • Malvinas (ISO country code)

ETHICS: Effective Technical and Human Implementation of Computer-based Systems

Ethnog: Ethnography

Ethnol: Ethnological • Ethnology

ETHP: Equivalent Thrust Horsepower

ETHZ: Eidgenössische Technische Hochschule Zürich (Switzerland)

ETI: Economic and Technological Intelligence • Economic Thickness of Insulation • Economic Transition and Integration • Éditions Techniques et Industrielles • Elapsed-Time Indicator • Environmental Technology Initiative • Epner Technologies Inc. (USA-NY) • Estimated Time of Interception • Expertcenter for Taxonomic Identification • Extraterrestrial Intelligence

ETIC: Environmental Technical Information Center • Estimated Time In Commission

Etim: Etimología

ETIS: Environmental Technical Information System • ESA Tele-Invoicing System

ETKM: Every Test Known to Man

ETL: Educational Technology Language • Electrotechnical Laboratory • Ending Time Label • Endorsed Tools List • Environmental Technology Laboratory • Environmental Test Laboratory • Équilibre Thermodynamique Local • Extraction-Transformation-Loading • Extract-Transform-Load

ETLA: Equivalent Two-Level Atom • Extended Three-Letter Acronym • Extended Time Limited Area

ETLOW: External Tank Lift-Off Weight

ETLS: Executable TEX Language Specification

ETM: Elapsed-Time Meter • Electrical Thermal Model • Electronically Transmitted Message • Electron-Trapping Material • Engineering Test Model • Enhanced Thematic Mapper • Évapotranspiration Maximale • Exonération du Ticket Modérateur

ETMA: Efficient Turbulence Model for Aeronautics

ETMD: Extended Thematic Mapper Derivative

ETMS: Enhanced Traffic Management System

ETN: Eastland (USA-TX) (location identifier) • École Technique Normale • Educational Teleconferencing Network • Electronic Tandem Network • Electronic Telecommunications Network

ETNO: European Telecommunications Network Operators

Etnogr: Etnografía

Etnol: Etnología

ETO: Earth-To-Orbit • Electronic Trading Opportunity • Engineering Task Order • Espace de la Transformation en Ondelettes • Estimated Takeoff • Estimated Time Off • Estimated Time of Operations • Ethylene Oxide • European Transport Organization • Extended Time Observation

ETOC: Estimated Time Of Correction

EtoE: End to End

ETOL: ESA Test and Operations Language • European Test and Operations Language • European Test Operations Language

ETOM: Electron-Trapping Optical Memory

ETOP: Education and Training in Optics and Photonics

ETOPS: Extended Twin-Engine Operations

ETOS: Earthquake and Tsunami Observation System • Extended Tape Operating System

ETP: Engineering Test Phase • Engineering Test Program • Environmental Test Plan • Equal Time Point • Equal Treatment Property • Equipment Test Plan • Équivalent à Temps Plein • Estimated Turnaround Point • Estimated Turning Point • European Training Programme • Évapotranspiration Potentielle • Exploitants de Télécommunications Publiques • Exploration Technology Program • Extra-Thin Pellicle

ETPL: Endorsed Tempest Products List

E. Tpo.: Ecuación de Tiempo

ETR: Early Token Release • Eastern Test Range • Educational Technology Review (AACE) • Engineering Tape Recorder • Engineering Tape Recording Data • Engineering Test Reactor • Engineering Test Request • Enhanced Temperature Region • Estimated Time of Return • Évapotranspiration Réelle • Experimental Test Reactor • Experimental Thermonuclear Reactor

ETRD: Engineering Tape Recorded Data

ETRF: European Travel Research Foundation

ETRI: Electronics and Telecommunications Research Institute

ETOD: Eastern Test Range Operations Directive

ETRP: Education and Training Programme

ETRR: Environmental Test Readiness Review

ETRS: European Terrestrial Reference System

Éts: Établissements

ETS: Earth Thermal Storage • Econometric Time Series • Educational Testing Service • Edwards Test Station (NASA) • Electrical Test Set • Electrical Test Setup • Electronic Tandem Switching • Electronic Translation System • Electronic Translator System • Energy Transfer System • Engineering and Technical Services • Engineering Test Satellite • Enterprise (USA-AL) (location identifier) • Equinox Transition Study • Estimated Time of Separation • European Telecommunication Standard • European Teleprocessing System • Experimental Technical Satellite • Experimental Test Site • Expertise Transfer System • External Tank System

ETSA: Electronic Tandem Switching Administration • Electron Transport System Activity

ETSACI: Electronic Tandem Switching Administration Channel Interface

ETSC: European Transport Safety Council

ETSEP: External Tank Separation

ETSF: Éditions Techniques et Scientifiques Françaises (France)

et seq: Et Sequens • Et Sequentia

et seqq: Et Sequentes • Et Sequentia

ETSI: European Telecommunications Standards Institute

et sqq: Et Sequentes • Et Sequentia

ETSS: Entry Time-Sharing System • Experimental Time-Sharing System • External Tank Separation System

ETSSP: Electronic Tandem Switching Status Panel

ETSU: East Tennessee State University (USA-TN)

ETT: Education and Technology Transfer • Entreprise de Travail Temporaire • ESTEC Telescience Testbed • Estimated Time of Track

ETTC: European Test and Telemetry Conference

ETTD: Équipement de Terminaison de Traitement de Données • Équipement Terminal de Transmission de Données

ETTF: Extreme Temperature Transition Furnace

ETTRT: European Transactions on Telecommunications-Related Technologies (periodical cod e)

ETU: Easy To Use • Enhanced Terminology Upgrade • Experimental Test Unit

ETUA: European Teleconferencing User Association

ETUC: European Trade Union Confederation

ETUD: External Tank Umbilical Disconnect

et ux: Et Uxor

ETV: Educational Television • Electrical Test Vehicle • Electronic Tape Vaulting • Electrothermal Vaporization • Electrothermal Vaporizer • Engine Test Vehicle • Extraterrestrial Visitation

ETVA: External Tank Vent Arm

ETVM: Electrostatic Transistorized Voltmeter

ETVX: Entry-Task-Validation-Exit

ETW: End of Tape Warning • Error Time Word • European Transsonic Windtunnel

ETX: End of Text

Ety: Etymology

Etym: Etymological • Etymology

Etymol: Etymological • Etymology

ETZ: Metz-Nancy-Lorraine (France) (location identifier)

ETZL: European Time Zone Logistics

eu: Basque (ISO language code)

Eu: Euler Number • Europium

E-U: États-Unis

EU: Economic Updating • Einstein Universe • Electrical Unit • Emotional Universe • Encyclopaedia Universalis • Engineering Unit • États-Unis • Europäische Union (formerly EWG) • Europe • European Union (formerly EEC) • European University • Europese Unie • Europe Unie • Execution Unit • Expander Unit • Experimental Unit

EUA: États-Unis d'Amérique • Eua (Tonga) (location identifier) • External User Access

EUATMS: European Unified Air Traffic Management System

Euc: Euclid (c. 330 - c. 275 BC)

EUC: End-User Computing • Engineering Unit Conversion • Eucla (Australia-WA) (location identifier) • Extended Unix Code

EUCLID: European Conference of Library and Information Educational Research • European Cooperation for the Long term In Defence

EUCMG: End-User Computing Management Group

EUCOS: European Composite Observing System

EUCREM: European Cloud Resolving Modelling

EUCREX: European Cirrus Radiation Experiment • European Cloud Radiation Experiment

EUDISED: European Documentation and Information System for Education

EUE: Eureka (USA-NV) (location identifier)

EUF: Eufaula Weedon (USA-AL) (location identifier)

EUG: Eugene (USA-OR) (location identifier) • European Union of Geosciences

EuGH: Europäischer Gerichtshof

EUI: Easy User Interface • Energy Utilization Index • Energy Utilization Intensity • European University Institute

EUIT: Educational Uses of Information Technology

EULA: End User Licensing Agreement

EUM: Neumünster (Germany) (location identifier)

EUMAC: European Modelling of Atmospheric Constituents

EUMETNET: European Meteorological Network

EUMETSAT: European Meteorological Satellite Organization

EUN: Egyptian Universities Network • Laayoune (Morocco) (location identifier)

EUNEASO: European Near-Asteroid Search Observatories

EUnet: European Unix Network

EUO: Paratebueno (Colombia) (location identifier)

EUP: Edinboro University Planetarium (USA-PA) • English Universities Press (UK)

EUPEN: European Physics Education Network

EUPRO: European Union of Physics Research Organisations

Eur: Europe • European • Européen

EUR: Euro (ISO currency code) • Europe

EURACA: European Air carriers Association

EURACS: European Radar Cross Section

EURAD: European Acid Deposition

EURADOS: European Radiation Dosimetry

EURAM: European Research on Advanced Materials

EURASEP: European Association of Satellite Experiments on Pollution

EURASIP: European Association for Signal Processing

EURECA: European Retrievable Carrier

EUREF: European Reference Frame

EUREKA: European Research Coordination Agency

EURESCO: European Research Conference

EURET: European Research for Transport

EURILIA: European Initiative in Library and Information in Aerospace

EURIPA: European Information Providers Association

EURIS: European Information Service

EURISY: European Association for the International Space Year • Europe in Space Year

EURO: European Regional Office

EUROCAE: European Organization for Civil Aviation Electronics

EUROCASE: European Council of Applied Sciences and Engineering

EUROCEAN: European Oceanographic Association

EUROGI: European Umbrella Organization for Geographic Information

EUROHEPNET: European High-Energy Physics Network

EuroHORC: European Heads Of Research Councils

EUROHORCS: European Union Research Organizations Heads Of Research Councils

EUROLAS: European Laser Consortium

EUROMECH: European Mechanics Committee

EUROMET: European Meteorological Education and Training • European Metrology

EuroPlaNet: European Planetarium Network

EUROS: European Retrievable Orbiting System

EUROSEP: European Association of Scientists in Environmental Pollution

EUROSIM: European Real-time Operations Simulator

EUROSPACE: European industrial Space study group

EUROVISION: European Television

EUS: Eastern United States

EUSC: European Utility Support Centre

EUSJA: European Union of Science Journalists Associations

EUSS: Extended Update and Support Service

EUT: End-User Training

EUTELSAT: European Telecommunications Satellite Organization

EUUG: European Unix User Group

EUV: Extreme Ultraviolet

EUVE: Extreme Ultraviolet Explorer

EUVL: Extreme Ultraviolet Light

EUVIP: Extreme Ultraviolet Imaging Photometer

EUVITA: Extreme Ultraviolet Imaging Telescope Array

EUVS: Extreme Ultraviolet Spectrograph

EUVSAT: Extreme Ultraviolet Satellite

EUVT: Extreme Ultraviolet Telescope

EUX: Saint Eustatius (Netherlands Antilles) (location identifier)

ev: En Volgende

eV: eingetragener Verein • Electronvolt • Elektronvolt

EV: Earned Value • Earth Viewing • Eerstvolgend • Electric Vehicle • Electronic Vision • English Version • Enhanced Version • Entry Vehicle • En Ville • Era Volgare • Étoile Variable • Expandable Vehicle • Expanded Volume • Experiment Version • Exposure Value • Extravehicular

EVA: Earned-Value Analysis • Economic Value Added • Electronic Valve Actuator • Europese Vrijhandelsassociatie • Evadale (USA-TX) (location identifier) • Extravehicular Activity

Evac: Evacuate • Evacuated • Evacuation

Évac: Évacuation • Évacué • Évacuer

EVAC: East Valley Astronomy Club (USA-AZ) • Extravehicular Activity Communications

Eval: Evaluate • Evaluated • Evaluation

Éval: Évaluation • Évalué • Évaluer

EVAL: Earth-Viewing Application Laboratory

Evap: Evaporate • Evaporated • Evaporation • Evaporator

Évap: Évaporateur • Évaporation • Évaporé • Évaporer

Evapn: Evaporation

EVAS: Enhanced Vortex Advisory System • Extravehicular Activity System

EVATA: Extravehicular Activity Translational Aid

EVBA: Equivalent Vector Boson Approximation

EVC: Engine Valve Control • Extravehicular Communications

EVCA: European Venture Capital Association • Evolving Cellular Automata

Evce: Evidence

EVCF: Eastern Vehicle Checkout Facility

EVCS: Emergency Voice Communications System • Extravehicular Communications System

EVCSD: Electron Vision Corporation of San Diego (USA-CA)

EVCU: Extravehicular Communications Umbilical

EVD: Economische Voorlichtingsdienst • Eva Downs (Australia-NT) (location identifier) • External Visual Display

EVDE: External Visual Display Equipment

Eve: Evening

EVE: Espace Vidéo Européen • European Videotelephony Experiment • Evenes (Norway) (location identifier) • Evening • Extensible VAX Editor

EVES: Emergency Voice Evacuation System

EVF: Equipment Visibility File

EVFU: Electronic Vertical Format Unit

Evg: Evening

EVG: Sveg (Sweden) (location identifier)

EVGA: Extended Video Graphics Adapter • Extended Video Graphics Array

EVH: Evans Head (Australia-NSW) (location identifier)

EVI: Economic Value of Information

Evid: Evidence • Evident

Évid: Évidence • Évident

E-ViEWS: Emergency Vehicle Early Warning and Safety (system)

EViEWS: Emergency Vehicle Early Warning and Safety (system)

EVIL: Extended Video Interactive Language

EVINCO: Evaluation and Interviewing Committee

EVIRI: Enhanced Visible and Infrared Imager

EVL: Electronic Visualization Laboratory

EVLDF: Evaluation Data File

EVLSS: Extravehicular Life Support System

EVM: Earth-Viewing Module • Electronic Voltmeter • Entry Vocabulary Module • Error Vector Magnitude • Eveleth (USA-MN) (location identifier)

EVMU: Extravehicular Mobility Unit

EVN: European VLBI Network • Yerevan (Armenia) (location identifier)

Evng: Evening

EVO: Engineering Verification Order • Extravehicular Operation

Evol: Evolution • Evolutionary

EVOM: Electronic Voltohmmeter

EVOP: European Volcanological Project • European Volcanology Project (*)

EVP: Elastoviscoplastic • Entrer dans la Vie Professionnelle • Executive Vice President • Expert Visits Programme • External Viewing Platform

EVPA: Early VLBI Polarization Axis • Electric Vector Position Angle • Extragalactic Variable Position Angle

EVPI: Expected Value of Perfect Information

EVR: Electronic Video Recording • Electrovideo Recorder

EVRIS: Étude de la Variabilité, de la Rotation et des Intérieurs Stellaires

EVS: Enhanced Vision System • Equipment Visibility System • European Voluntary Service • External Visibility System • Extravehicular Suit • National Standards Board of Estonia

EVSC: Extravehicular Suit Communications

EVSM: Electrostatic Voltmeter System

EVSS: Extravehicular Space Suit

EVSTC: Extravehicular Suit Telecommunications

EVSU: Extravehicular Space Unit

EVT: ESO Video Team • Experiment Verification Test • Extravehicular Transfer

Evtl: Eventuell

EVU: Energieverwertungsunternehmen

EVUA: European VPN User Association

EVV: Evansville (USA-IN) (location identifier)

EVW: Evanston (USA-WY) (location identifier)

EVX: Electronic Voice Exchange • Évreux (France) (location identifier)

Evy: Every

Ew: Einwohner

E/W: Energy/Weight

EW: Belarus (international aircraft registration code) • Early Warning • Edwards-Wilkinson • Electronic Warfare • Entrance Window • Equivalent Width • Estonia (international vehicle registration code) • Evanescent Wave

EWA: End Warning Area • Estimated Warehouse Arrival

EWAN: Emulator Without A Name • Enterprise Wide-Area Network

EWAT: Electronic Warfare Advanced Technology

EWB: Fall River (USA-MA) (location identifier) • New Bedford (USA-MA) (location identifier)

EWC: East West Center • Equatorial Wave Campaign • European Works Council • European Writers Congress

EWD: Eastward • Extended Warnier Diagram

EWDT: Early Warning Data Transmission

EWE: Emergency Window Escape • Ewer (Indonesia) (location identifier) • Extreme Weather Event

EWEB: Eugene Water and Electricity Board (USA-OR)

EWF: Extrazellulare Waschflüßigkeit

EWG: Europäische Wirtschaftsgemeinschaft (now EU) • Executive Working Group

EWI: Electronic Work Instruction • Enarotali (Indonesia) (location identifier) • Europäisches Währungsinstitut • Europäisches Wirtschaftinstitut

eWiC: Electronic Workshop in Computing

EWL: Evaporated Water Loss

EWK: Extent of World Knowledge • Newton (USA-KS) (location identifier)

EWM: Engineering Wringout Meeting

EWN: Enterprise-Wide Network • New Bern (USA-NC) (location identifier)

EWO: Electronic Warfare Officer • Engineering Work Order • Essential Work Order • Ewo (Congo) (location identifier)

EWON: European Work Organization Network

EWOS: European Workshop for Open Systems

EWR: Early-Warning Radar • Engineering Work Request • Europäischer Wirtschaftsraum • New York - Newark (USA-NJ) (location identifier)

EWS: Early Warning System • Emergency Water Supply • Employee-Written Software • Engineering Workstation • Europäisches Währungssystem • European Warranty System • European Workgroup on SGML • European Working-group on SGML • European Workstation • Expert Workbench System

EWSK: East-West Station Keeping

EWSTC: European Week for Scientific and Technological Culture

EWTA: Enterprise-Wide Technical Architecture

EWU: Eastern Washington University (USA-WA)

EWW: Extended Workweek

EWY: Newbury Greenham (UK-England) (location identifier)

e. Wz.: eingetragenes Warenzeichen

ex: Exit

Ex: Examination • Examine • Examined • Examiner • Example • Excellent • Except • Excepted • Exception • Excess • Exchange • Exclude • Excluding • Exclusive • Excursion • Execute • Executed • Executive • Exemple • Exempt • Exeter (UK-England) • Export • Express • Extra

EX: Execute • Exeter (UK postal code) • Extract • Kyrgyzstan (international aircraft registration code)

ExA: Experimental Astronomy (periodical code)

EXAFS: Extended X-ray Absorption Fine Structure

Exag: Exaggerate • Exaggeration

Exam: Examination • Examine • Examined • Examinee • Examiner

Examd: Examined

Examg: Examining

Examn: Examination

EXAMS: Exposure Analysis Modelling System

EXAP: Extended Arithmetic Processor

ex aq.: Ex Aqua

EXAS: Experiment on Autonomous SAR

ex b.: Ex Bonus

exc: Excudit

Exc: Excellence • Excellency • Excellent • Excellentie • Except • Excepted • Exception • Excess • Exchange • Excursion

EXC: Experiment Computer

ExCA: Exchangeable Card Architecture

EXCA: Exchangeable Card Architecture

EXCEDE: Excitation by Electron Deposition

Exch: Exchange • Exchequer

Excl: Exclamation • Exclamativo • Exclude • Excluded • Excluding • Exclusif • Exclusive

Exclam: Exclamación • Exclamation • Exclamatory

ExCo: Executive Committee

EXCO: Executive Council

EXCOM: Executive Committee

EXCP: Except • Executive Channel Program

Excsv: Excessive

Exctr: Exciter

excud: Excudit

Exd: Examined

EXDB: Experimental Database

Ex. Doc.: Executive Document

EXE: Executable

Exec: Execute • Execution • Executive • Executor

EXELFS: Electron Energy Loss Fine-Structure Spectroscopy

Exempl: Exemplaire

EXEMSI: Experimental European Manned Space Infrastructure

Exer: Exercise

Exes: Expenses

EXES: Echelle-Cross-Echelle Spectrograph • Echelon-Cross-Echelle Spectrograph

ExFl: Experiments in Fluids (periodical code)

ex. g.: Exempli Gratia

Exgl: Extragalactic

ex. gr.: Exempli Gratia

Exh: Exhaust • Exhibit • Exhibition

EXH: Exchange Hardware

Exhib: Exibit • Exhibition

Exhbn: Exhibition

EXI: Excursion (USA-AK) (location identifier)

ExInEd: Exploration In Education

ex int.: Ex Interest

EXIST: Energetic X-ray Imaging Survey Telescope

EXITE: Energetic X-ray Imaging Telescope Experiment

Exkl: Exklusive

ex lib.: Ex Libris

ExM: Experimental Mechanics (periodical code)

EXM: Exmouth Gulf (Australia-WA) (location identifier)

EXNOR: Exclusive NOR

ExNPS: Exploration of Neighbouring Planetary Systems

EXNPS: Exploration of Neighbouring Planetary Systems

EXO: Executive Office • Experiment Operator

ex off.: Ex Officio

Exor: Executor

EXOR: Exclusive OR

EXOS: Exosphere Satellite

EXOSA: EXOSAT Express (periodical code)

EXOSAT: European X-ray Observatory Satellite

ex p.: Ex Parte

Exp: Expand • Expanded • Expansion • Expect • Expected • Expecting • Expedition • Expenses • Experience • Experiment • Experimental • Experimented • Experimenting • Expert • Expiration • Expire • Expired • Expires • Explanation • Explain • Explained • Explaining • Expletivo • Exponential • Export • Exportation • Exported • Exporter • Expose • Exposed • Exposing • Exposure • Express • Expression • Expulsion

EXP: Expect • Expected • Experiment • Exponent • Expressway

EXPC: Expect

Expdn: Expedition

Exped: Expedite

Expen: Expendable

EXPEN: Expendable

Exper: Experiment • Experimental

Expl: Exemplar • Explain • Explanation • Explanatory • Exploration • Explore • Explored • Explorer • Explosion • Explosive

Expld: Explode

Expln: Explosion

EXPLODE: Expanding List in the OSI Directory Environment

Exploit: Exploitation

Expn: Exposition

EXPN: Expand

Expnt: Exponent • Exponential

Expo: Exposition

Exp. O.: Experimental Order

Expr: Expresión • Express

EXPR: Expressway

ExpRes: Experiment Results

EXPRES: Experimental Research in Electronic Submission

EXPRESS: Expedited Processing of Experiments to Space Station • Experimental Reentry Space System • Experiment Reusable Space System

EXPRESSO: Experiment for Regional Sources and Sinks of Oxidants

EXPRIM: Expert pour la Recherche d'Images

Expsr: Exposure

Expt: Experiment

Expte: Expediente

Exptl: Experimental

EXPW: Expressway

Expy: Expressway

EXPY: Expressway

Exr: Executor

Exs: Expenses

EXSAS: Extended Scientific Analysis System

ExSEL: Excellence in Science and Engineering Leadership

Exstg: Existing

Ex Stre.: Ex Store

Ext: Extend • Extended • Extending • Extension • Extent • Exterior • External • Externally • Extinct • Extinguish • Extra • Extract • Extraction • Extreme

ExT: Experimental Techniques (periodical code)

EXT: Exeter (UK-England) (location identifier) • Experiment Terminal • Extension • External

Extd: Extend • Extended

EXTD: Extend

EXTEC: Exploration science and Technology

Exten: Extended • Extension

EXTER: External

ExTFS: Experimental Thermal Fluid Science (periodical code)

EXTGH: Extinguish

Exting: Extinguishing

Extm: Extreme

Extn: Extension • External

EXTN: Extension

Extnr: Extinguisher

Extr: Exterior • Extract • Extraction (periodical code) • Extracto • Extrait • Extraordinary • Extremadura (Spain)

EXTRAP: Extrapolate

ExTRAS: Experiment on Timing Ranging and Atmospheric Soundings

Extraterr: Extraterrestre • Extraterrestrial • Extraterrestrische • Extraterritorial

Extraterrit: Extraterritorial

EXTRM: Extreme

EXTRN: External Reference

Extns: Extensions

EXTNSN: Extension

EXTS: Extensions

EXTSV: Extensive

EXTV: European Experimental Test Vehicle

Exx: Examples

EXU: Execution Unit

EXUG: European X User Group

Ex Whf.: Ex Wharf

Ex Whse.: Ex Warehouse

Exz: Exzellenz

EY: Tajikistan (international aircraft registration code)

EYL: Yelimane (Mali) (location identifier)

Eylt: Eyelet

EYP: El Yopal (Colombia) (location identifier)

Eypc: Eyepiece

EYR: Yerington (USA-NV) (location identifier)

EYS: Eliye Springs (Kenya) (location identifier)

EYW: Key West (USA-FL) (location identifier)

EZ: Easy • Economische Zaken • Empowerment Zone • Entrainment Zone • Equatorial Zone • Ephemeridenzeit • Exclusion Zone • Turkmenistan (international aircraft registration code)

EZB: Europäische Zentralbank • Exclusive Zone Boundary

EZE: Buenos Aires - Ezeiza (Argentina-BA) (location identifier) • Buenos Aires - Ministro Pistarini (Argentina-BA) (location identifier)

eZIF: Electrically-activated Zero Insertion Force

EZMW: Europäisches Zentrum für Mittelfristige Wettervorhersagen

EZS: Elazig (Turkey) (location identifier)

EZU: Europäische Zahlungsunion

EZW: Embedded Zero-tree Wavelet

E2PROM: Electrically Erasable Programmable Read-Only Memory

E2S: Expert Software Systems (Belgium)

E3: Electromagnetic Environmental Effect • Electronic Entertainment Exhibition (USA-CA)

E3I: École d'Ingénieur en Informatique pour l'Industrie (France)

F

f: Distribution Function • Earth Flattening • Fac • Facing • Faible • Fair • Farad • Farthing • Father • Fathom • Feet • Female • Féminin • Feminine • Feminino • Femto • Feuillet • Fiant • Fiat • Fiction • Field • Fine • Firm • Flat • Flattening • Fluid • Focal Length • Folgende • Foliación • Folio • Follow • Following • Follow Up • Foot • Fondé • For • Force • Form • Formula • Founded • Fraction • Franc • Frequency • From • Fugacity • Function of • Guilder (currency) • Mixture Fraction • Packing Fraction • Specific Helmholtz Function • Terrestrial Flattening • Thermal Utilization Factor • True Anomaly

F: Electric Field Strength • Faculté • Faculty • Fahrenheit • Failure • Fair • False • Family • Farad • Faraday Constant • Farthing • Fast • Father • Fathom • February • Federation • Fellow • Female • Féminin • Feminine • Fen • Fermé • Fermi • Ferrovia • Field • Filial • Filter • Final • Fine • Firm • Firth • Flat • Fleet • Florijn • Fluor • Fluorine • Flux • Flux Density • Focal Length • Fog • Folio • Follow • Following • Follow Up • Force • Forint (currency) • Fort • Forte • Forty (Roman numeral) • Forward • Foxtrot (international alphabet) • Franc (currency) • France (international aircraft registration code) • France (international vehicle registration code) • France (postal country code) • Frauen • Freddo • Free Energy • Freighter • French • Frequency • Frère • Friday • Frigate • Frio • Frío • Froid • Front • Fuel • Function of • Helmholtz Function • Hyperfine Structure Quantum Number • Kinematic Flux • Magnetomotive Force • Nodal Forces • Quantum Number of Total Angular Momentum • Radiant Flux • Rotational Term • Spectral Type • Thrust • 40 (Roman numeral)

f&a: Fore & Aft

fa: Farsi (ISO language code) • Persian (ISO language code)

Fa: Factura • Firma • Florida (USA)

F/A: Fabrication and Assembly • Failure Analysis • Fuel/Air

F&A: Fabrication & Assembly

FA: Fabrication and Assembly • Factor Analysis • Faculty Advisor • Failure Analysis • Family Allowance • Fatty Acid • Feedback Analysis • Field Activities • Field Allowance • Filter Array • Filtered Air • Final Approach • Final Assembly • Finance Act • Financial Adviser • Fine Art • Fire Alarm • First Aid • Fixed Asset • Flight Acceptance • Flight Aft • Flight Attendant • Fluoranthene • Focus Amplifier • Forced Air • Foreign Agent • Free Agent • Free Aperture • Free Atmosphere • Fuel Assembly • Full Adder • Fully Automatic • Functional Analysis • Functional Area • Fuse Alarm

FAA: Fachmesse für Amateur-Astronomen (Germany) • False Alarm Avoidance • Faranah (Guinea) (location identifier) • Federal Aviation Agency • Fellow of the Australian Academy • Foreign Assistance Act • Föreningen för Astronomi och Astronautik (Sweden) • Formalin Acetic Alcohol • Forum der Amateur-Astronomie • Free of All Average • Furnace Atomic Absorption

FAAAS: Fellow of the American Academy of Arts and Sciences • Fellow of the American Association for the Advancement of Science

FAAC: Ford Amateur Astronomy Club (USA-MI)

FAAO: Fellow of the American Academy of Optometry

FAAP: Federation of Accrediting Agencies of the Philippines

FAAQ: Fédération des Astronomes Amateurs du Québec (Canada-QC)

FAAS: Flame Atomic Absorption Spectroscopy • Foreign Affairs Administrative Support

FAAT: Fully Analytical Aerotriangulation

fàb: Franco à Bord

Fab: Fabric • Fabricate • Fabricated • Fabrication

FÀB: Franco À Bord

FAB: Fabrica • Fabricate • Fabrication • Fast Atom Bombardment • Finance And Budget • First-Aid Box • Flight Assignment Baseline • Fraction Actually Burned • French-American-British

Fabbr: Fabbrica

FABF: Fraction of Agriresidue Burned in Fields

FABITH: Förderverband Astronomische Bildung in Thüringen (Germany)

FABIUS: Fabrication Automatique de Bibliographies et d'Index Utilisant des Syntagmes

FABMS: Fast Atom Bombardment Mass Spectroscopy

FABP: Family of Alternative Bandit Processes

Fabr: Fabricate • Fabrication

FABU: Fuel Additive Blender Unit

FABx: Fire Alarm Box

Fac: Facade • Facial • Facility • Facsimile • Factor • Factory • Facultad • Faculté • Faculty

FAC: Facilities • Facility • Failure Analysis Committee • Fast Adaptive Composite • Fast As Can • Feature Analysis Code • Federal Affairs Committee • Field-Aligned Current • File Access Code • Final Approach Course • Fonds d'Aide et de Coopération • Forecast At Completion • Forward Air Controller • Fresneau Astrographic Catalogue

FACA: Federal Advisory Committee Act (USA)

FACC: Feature and Attribute Coding Catalog • Ford Aerospace Communications Corporation • Fundo de Apoio Comunidade Científica

FACE: Feature Analysis Comparison and Evaluation • Field Ancillary Computer Effort • Florida Area Cumulus Experiment • Forests Absorbing Carbon Dioxide Emission • Framework for Academic Cooperation in Europe • Free-Air CO Enrichment

FACELIF: Flowing Atmospheric Chemistry Experiment with Laser-Induced Fluorescence

FACES: Fortran Automatic Code Evaluation System

FACET: Future Automated Commercial Environment Team

FACI: First Article Configuration Inspection

Facil: Facility

FACIT: Fast Automatic Conversion with Integrated Tools • Formal Approaches to Computing and Information Technology

FACM: Florida Association of Cadastral Mappers (USA-FL) • Functiones et Approximatio Comentarii Mathematici (periodical code)

FACO: Fabrication and Acceptance Checkout • Factory Acceptance Checkout • Factory Assembly and Checkout • Final Assembly Checkout

FACP: Fellow of the American College of Physicians • Fully Automated Computer Program

FACR: First-Article Configuration Review • Fourier Analysis and Cyclic Reduction

Facs: Facsimile

FACS: Facilities Assignment and Control System • Feature Attribute Coding Standard • Feature Attribute Coding System • Federation of Asian Chemical Societies •

Fellow of the American College of Surgeons • Finance And Control System • Fine Attitude Control System • Fluorescence-Activated Cell Sorter • Fully Automatic Compiling System

FACSFAC: Fleet Area Control and Surveillance Facility

FACSI: Federal Advisory Council on Scientific Information (USA)

Facsim: Facsimile

Fact: Factura • Facture

FACT: Factor Analysis Chart Technique • Federation Against Copyright Theft • First-Address Control Table • Flexible Automatic Circuit Tester • Flight Acceptance Composite Test • Foundation for Advanced Computer Technology • Fully Automated Cataloging Technique • Fully Automatic Compiler-Translator • Fully Automatic Computer-Translator • Fully Automatic Compiling Technique

FACTS: FDF Automated Control and Tracking System • Fingerprint Analysis and Criminal Tracing System • Flexible Astronomical Catalogue Transport System • Florida Atlantic Coast Transport Study • Fortran Analytical Cross Reference Tabulation System

FAD: Fast Automatic Differentiation • Fernsprechansagedienst • Fiber Actuator Device • Filter Analog-to-Digital • Final Approach Display • Final Archive Definition • Flight Assessment Delta • Fraction Actually Degraded • Free Air Delivered

FADABEF: Fédération des Associations de Bibliothécaires, Archivistes et Documentalistes des États Francophones

FADC: Fast Analog Digital Converter • Flash Analog Digital Converter • Final Archive Definition Committee

FADEC: Full-Authority Digital Electronic Control • Full-Authority Digital Electronic Control • Full-Authority Digital Engine Control (*) • Full-Authority Digital Engine Controller

FADS: Fast Area Digitizing Scanner

FADU: File Access Data Unit

FAE: Faroe Islands (location identifier) • Faeroese • Final Approach Equipment

FAEP: Fédération des Associations d'Éditeurs de Publications

FAES: Flame Atomic Emission Spectroscopy

FAF: Final Approach Fix • Final Assembly Fixture • First Aerodynamic Flight • Fort Eustis (USA-VA) (location identifier)

FAFS: Falcon Air Force Station • Flame Atomic Fluorescence Spectroscopy

FAFTT: Fonds d'Assurance Formation du Travail Temporaire

FAG: Fagurholmsmyri (Iceland) (location identifier)

FAGS: Federation of Astronomical and Geophysical Services • Fellow of the American Geographical Society

Fah: Fahrenheit

FAH: Farah (Afghanistan) (location identifier) • Fournisseur d'Applications Hébergées

FAHQMT: Fully-Automatic High-Quality Machine Translation

FAHQT: Fully-Automatic High-Quality Translation

Fahr: Fahrenheit

Fahrb: Fahrbahn

FAI: Fairbanks (USA-AK) (location identifier) • Fédération Aéronautique Internationale • Field-Aligned Irregularity • Fonds Autoroute de l'Information (France) • Fournisseur d'Accès à Internet • FREJA Auroral Imager • Fresh-Air Inlet

FAIAA: Fellow of the American Institute of Aeronautics and Astronautics

FAIFE: Free Access to Information and Freedom of Expression

FAIM: Fast Aerial Image Model • Fully Analytic Ionospheric Model

FAIP: Fellow of the Australian Institute of Physics

FAIR: Fabrication, Assembly and Inspection Record • Failure Analysis Information Retrieval • Fairness and Accuracy In Reporting • Fast Access Information Retrieval • Federal Agricultural Improvement and Reform (USA) • Fly-Along Infrared • Fully-Automatic Information Retrieval

FAIRS: Federal Aviation Information Retrieval System (USA) • Fully Automatic Information Retrieval System

FAISAT: Final Analysis Inc. Satellite

FAITC: Foreign Affairs and International Trade Canada

FAJ: Fajardo (Puerto Rico) (location identifier) • Filtered Accumulated Jitter

Fak: Faktura • Fakultät

FAK: Fachnormenausschuss Klassifikation • False Island (USA-AK) (location identifier) • File Access Key

FAKYR: Fachbereich-Kybernetik-Retrieval-system

FAL: Facility Area Layout • Finite Automation Language • First Approach and Landing • Roma - Falcon State (USA-TX) (location identifier)

FALCON: Fission-Activated Laser Concept

Falk.I.: Falkland Islands

Falk.Is.: Falkland Islands

Fam: Familia • Familiar • Family

FAM: Faculdade Anhembi Morumbi (Brazil) • Familiarization • Farmington (USA-MO) (location identifier) • Fast Access Memory • File Access Method • Five-Axis Mount • Full-Aperture Metrology • Fully Adaptive Method • Fully Adaptive Multigrid

FAMAF: Faculdad de Matemáticas, Astronomía y Física

FAMAS: Flutter And Matrix Algebra System

FAME: Fizeau Astrometric Mapping Explorer • Fluids And Materials Experiment • Forest Assessment and Monitoring Environment • Formatting And Multiplexing Equipment • Full-sky Astrometric Mapping Explorer

FAMIS: Financial And Management Information System • Fulfillment And Member Information System

FAMOS: Flexible Automatisierte Montagesysteme • Flight Acceleration Monitor Only System • Future of Atomic, Molecular, and Optical Sciences

FAMOUS: Flexible Automation Monitoring and Operation User Station

FAMS: FDF Automated Monitoring System • File Attribute Management System

FAMSCO: First Article Master Scheduling Conference

FAMT: Fully Automatic Machine Translation

FAMU: Fuel Additive Moisture Unit

FAN: Family Area Network • Farsund (Norway) (location identifier) • Federal Aviation News

FANE: Fonds d'Action Nationale Européen

FANGIO: Feedback Analysis for GCM Intercomparison and Observations

FANS: Fully Automatic Navier-Stokes • Future Air Navigation System

FAO: Fabrication Assistée par Ordinateur • Faro (Portugal) (location identifier) • Finish All Over • Flight Activities Officer • Food and Agriculture Organization (UNO) • For the Attention Of • Foyer d'Accueil et d'Orientation

FAOB: Federation of Asian and Oceanian Biochemists

FAP: False-Alarm Probability • Fast Atmospheric Pulsation • File Access Protocol • Filtre À Particules • Final Approach Plane • Final Approach Point • First Aid Post

• Food and Agriculture Project • Fortran Assembly Program • Franc d'Avaries Particulières • Freedom Attached Payload

FAPAM: Fédération d'Astronomie Populaire du Midi (France)

FAPE: Fund for Assistance to Private Education

FAPESP: Fundação de Amparo à Pesquisa do Estado de São Paulo (Brazil)

FAPI: Family Application Program Interface

FAPIN: Fabry-Pérot Interferometer

FAPS: Fault Analysis Projection System • Filtered Accumulated Phase Slope

FAQ: Fair Average Quality • Foire Aux Questions • Frequently Asked Question

FAQL: Frequently Asked Question List

FAR: Failure Analysis Report • Fargo (USA-ND) (location identifier) • Faster-than-Average Rotation • Federal Acquisition Regulation (USA) • Federal Aviation Regulation (USA) • Final Acceptance Review • Final Assessment Report • Fine Angular Resolution • Flight Acceptance Review • Flight Analysis Review • Force d'Action Rapide • Fuelflow to Airflow Ratio • Fund for Astrophysical Research

FARC: Federal Archives and Record Center (USA)

FARE: Fluid Acquisition and Resupply Experiment

Farm: Farmacia

FARM: Farmaceútica • Fast Archive Recovery Method

FARNET: Federation of Academic Research Networks • Federation of Advanced Research Networks • Federation of American Research Networks (*)

FARR: Frequency Agile Rain Radar

FARS: Facility for Atmospheric Remote Sensing • Fatal Accident Reporting System

FAS: Facilité d'Ajustement Structurel • Failure Analysis Section • Family of Alternative Superprocesses • Faskrudsfjördur (Iceland) (location identifier) • Feature Analysis Section • Federation of American Scientists (USA) • Federation of Astronomical Societies (UK) • Final Approach Segment • Flame Absorption Spectroscopy • Flexible Assembly System • Flight Analysis System • Flight Application Software • Fonds d'Action Sociale • Foothills Astronomical Society (USA-SC) • Forsyth Astronomical Society (USA-NC) • Forward Acquisition Sensor • Frame Alignment Signal • Free Alongside Ship • Full Approximation Scheme • Full Authorization System

FASAB: Federal Accounting Standards Advisory Board (USA)

FASAS: Federation of Asian Scientific Academies and Societies

FASB: Financial Accounting Standards Board

Fasc: Fascicolo • Fascicule • Fascículo • Fasciculus

FASc: Fellow of the Indian Academy of Sciences

FASCOS: Flight Acceleration Safety Cutoff System

FASE: Federation of Acoustical Societies in Europe

FASEB: Federation of American Societies for Experimental Biology

FASED: Facilitation Awards for Scientists and Engineers with Disabilities

FASINEX: Frontal Air-Sea Interaction Experiment

FASIT: Fully Automatic Syntactically-based Indexing of Text

FASO: Forward Airfield Supply Organisation

FASP: Faculdades Associadas de São Paulo (Brazil) • Fly-Away Sensor Package

FASR: Frequency-Agile Solar Radiotelescope

FASRC: Federation of Arab Scientific Research Councils

FASS: Frequency Agile Signal Simulation

FASST: Flexible Architecture Standard System Technology

FAST: Facility for Automatic Sorting and Testing • Factor Analysis System • Failure Analysis by Statistical Techniques • Fast Acquisition Search and Track • Fast Algorithm for the Satisfaction Test • Fast All Sky Telescope • Fast Auroral Snapshot • Fast Automatic Shuttle Transfer • Federation Against Software Theft (UK) • Federazione delle Assoziazioni Scientifiche e Tecniche (Italy) • File Analysis and Selection Technique • Final Approach Spacing Tool • First Application System Test • Flexible Algebraic Scientific Translator • Flow Analysis Software Toolkit • Forecasting and Assessment in Science and Technology • Formula And Statement Translator • Fourier Amplitude-Sensitivity Test • Functional Analysis System Technique • Fundamental Astronomy by Space Techniques

FASTEX: Frontal Atlantic Storm Track Experiment

FASTI: Fast Access to Systems Technical Information

Fastnr: Fastener

FASTR: Fast Term Recognizer

FASTS: Federation of Australian Scientific and Technical Societies • Federation of Australian Scientific and Technological Societies (*)

FASTT: Flagstaff Astrometric Scanning Transit Telescope

FASU: Filter And Shutter Unit

FAT: Factory Acceptance Test • Fast Automatic Transfer • File Allocation Table • File Attribute Table • Final Acceptance Test • Final Acceptance Testing • Final Assembly Test • Flight Attitude Table • Foreign Alias Table • Foreign Area Translation • Formula Assembler Translator • Fresno (USA-CA) (location identifier) • Functional Area Training

FATAL: Fully Automatic Test Algebraic Language

FATCAT: Frequency And Time Circuit Analysis Technique

FATE: Formation of Aerosols and their Transformation over Europe

Fath: Fathom

FATP: Future Astronaut Training Program

FATS: Factory Acceptance Test Specification • Faint Asteroid Taxonomy Survey • Fault Tracking System • Fortran Automatic Timing System

Fatt: Fattura

Faub: Faubourg

FAUST: Far Ultraviolet Space Telescope

Fav: Favour • Favourable • Favourite

FAV: Fakarava (Tuamotu Islands) (location identifier)

FAWG: Flight Assignment Working Group

FAWN: Florida Automated Weather Network

FAWS: Flight Advisory Weather Service

Fax: Facsimile

FAX: Facsimile

FAY: Fayetteville (USA-NC) (location identifier)

FAZ: Frankfurter Allgemeine Zeitung (Germany)

Fb: Fabrik

FB: Fachbereich • Faint Blue • Flare Brightness • Franc Belge • Feedback • Field Bulletin • Film Badge • Final Braking • Flaring Branch • Flat Bar • Fleet Broadcast • Forschungsbericht • Forward-Backward • Fraction Burned • Freight Bill

FBA: Fellow of the British Academy • Fixed Block Architecture

FBAC: Fort Bend Astronomy Club (USA-TX)

FBAS: Fixed Base Aft Station

FBB: Flyback Booster

FBC: Fallen Building Clause • Fluidized-Bed Combustion

FBCG: Faint Blue Compact Galaxy

FBCR: Faster, Better, Cheaper Report (NASA)

FBCS: Fellow of the British Computer Society • Fixed Base Crew Station

FBCW: Fallen Building Clause Waiver

Fbd: Freeboard

FBD: Faizabad (Afghanistan) (location identifier)

FBE: Fédération Bancaire Européenne • Francisco Beltrão (Brazil-PR) (location identifier) • Free Buffer Enquiry • Fundamental Blue Edge

FBF: Frankfurt Book Fair (Germany)

Fbg: Faubourg

FBG: Faint Blue Galaxy • Fiber Bragg Grating • Fort Bragg (USA-NC) (location identifier)

FBI: Federal Bureau of Investigations (USA)

F-BIDR: Full-resolution Basic Image Data Record

FBIDR: Full-resolution Basic Image Data Record

FBIS: Foreign Broadcast Information Service

Fbk: Firebrick

FBK: Fairbanks - Fort Wainwright (USA-AK) (location identifier)

FBL: Faribault (USA-MN) (location identifier) • Fly-By-Light

FBM: Foot Board Measure • Frame Buffer Module • Fuzzy Bag Model • Fuzzy Bitmap • Lubumbashi (Zaire) (location identifier)

FBML: Francis Bitter Magnet Laboratory

FBN: Federal Base Network (USA) • Feedback Network

FBNCC: Fallback Network Control Center

F/B/O: For the Benefit Of

FBO: Fixed-Base Operator • Flaring Branch Oscillation • For the Benefit Of

FBOA: Fellow of the British Optical Association

FBP: Filtered Back-Projection • Final Boiling Point • Free Boundary Problem

FBPS: Four-Band Photometric System

FBQS: FIRST Bright Quasar Survey

Fbr: Fiber • Fibre

FBR: Fast Breeder Reactor • Fluidized-Bed Reactor • Fort Bridger (USA-WY) (location identifier)

Fbrbd: Fireboard

Fbro: Febrero

FBS: Fall Back Switch • Feedback Signal • Feedback System • Firefighter Breathing System • First Byurakan Survey • Flare Build-up Study • Fluidized Bed Combustor • Forward-Based System • Frontal Boundary Study

FBSR: Feedback Shift Register

FBT: Fringe Benefit Tax

FBu: Burundi Franc (currency)

FBU: Oslo Fornebu (Norway) (location identifier)

FBV: Field Base Visit • Fuel Bleed Valve

FBW: Fly-By-Wire • Fractional Bandwidth

FBY: Fairbury (USA-NE) (location identifier) • Future Budget Year

fc: Fin Courant

F&C: Full & Change

F/C: Fire Control • Fit Check • Flight Control • Flight Controller • Flow Control • Fuel Cell

FC: Fachgruppe • Facilitated Communication • Facility Control • Facultad de Ciencias • False Color • Feature Count • Featureless Continuum • Ferrite Core • Fibre Channel • Fidei Commissum • Field Camera • Field Completion • Field Component • Field-Cooled • Fieri Curavit • Fifth Column • File Cabinet • File Compare • File Conversion • File Copy • Financial Consultant • Finding Chart • Fire Control • First Class • Fit Check • Flight Computer • Flight Control • Flight Controller • Flight Critical • Follow Copy • Foot-Candle • For Cash • Forward Checking • Fragment Code • Frame Control • Fuel Cell • Fuel Consumption • Full Course • Fully Coated • Functional Component • Functional Core • Function Code • Funnel Cloud

Fca: Fábrica

FCA: Fibre Channel Association • Field Change Analysis • Figure Control Actuator • Flexible Ceramic Armor • Flight Critical Aft • Flight Control Actuator • Flow Control Assembly • Fluids Control Assembly • Forward Controller Assembly • Frequency Control Analysis • Frequency Control and Analysis • Fuel Control Assembly • Full Cost Accounting • Functional Compatibility Analysis • Functional Configuration Analysis • Functional Configuration Assessment • Functional Configuration Audit • Functional Core Adapter • Kalispell (USA-MT) (location identifier)

FCAD: Five College Astronomy Department (USA-MA)

FCAF: Flight Crew Accommodation Facility

FCAG: Facultad de Ciencias Astronómicas y Geofísicas

FCAGLP: Facultad de Ciencias Astronómicas y Geofísicas de La Plata (Argentina-BA)

FCAL: Fibre Channel - Arbitrated Loop

FCAP: Facility Capacity • Flight Control Applications Program

FCAPT: Four College Automated Photoelectric Telescope

FCAR: Formation de Chercheurs et Aide à la Recherche • Free of Claim for Accident Reported

FCAS: Fairfield County Astronomical Society (USA-CT)

FCASI: Fellow of the Canadian Aeronautics and Space Institute

FCAST: Forecast

FCB: Faster-Cheaper-Better • Ficksburg (South Africa) (location identifier) • File Control Block • Forms Control Buffer • Fracture Control Board • Fresh Cut Basalt

fcc: Face-Centered Cubic

FCC: Face-Centered Cubic • False Colour Composite • Federal Communications Commission (USA) • Federal Computer Conference • File Carbon Copy • Fits-Class Certificate • Flat Conductor Cable • Flight Control Computer • Flight Crew Compartment • Fluid Catalytic Cracking • Fornax Cluster Catalog • Forward Command Channel • Four College Consortium (USA)

FCCA: Floating Current-to-Current Amplifier

FCCAPT: Four College Consortium Automatic Photoelectric Telescope

FCCC: Framework Convention on Climate Change

FCCE: Fondos de Cohesión de las Comunidades Europeas

FCCN: Fundação para a Commputação Científica Nacional (Portu gal)

FCCP: Firm Contract Cost Proposal

FCCSET: Federal Coordinating Council for Science, Engineering and Technology (USA) (now NSTC)

FCCU: Fluid Catalytic Cracker Unit

FCD: Fecal Collection Device • Flight Control Division

FCDB: Flight Control Data Bus

FCDC: Flexible Confined Detonating Cord • Flight Control Data Concentrator • Fundamental Constants Data Center

FCDR: Final Critical Design Review

FCDS: Fleet Consolidated Data Set

FCE: Fast-Cycle Executive • Flight Control Equipment • Flight Crew Equipment • Fondo de Cultura Económica • Forward Control Element

FCEF: Flight Crew Equipment Facility

FCEI: Facility Contract End Item • Facility Contractor End Item

FCEL: Fibre Channel - Enhanced Loop • Flight Crew Equipment Laboratory

FCELP: Facultad de Ciencias Exactas de la Plata (Argentina-BA)

fcém: Force Contre-Électromotrice

FCF: First Captive Flight • Flight Critical Forward • Fluids Combustion Facility • Functional Check Flight • Fusion Critical Frequency

FCFM: Flight Combustion Facility Monitor

FCFOV: Full Coded Field Of View

FCFS: First Come - First Served

Fcg: Facing

FCG: Fundaçao Calouste Gulbenkian (Portugal)

FCH: Flight Controllers Handbook • Fresno Chandler (USA-CA) (location identifier)

FCHL: Flight Control Hydraulics Laboratory

FCI: Flight Control Indicator • Flight Crew Integration • Flight Critical Item • Flux Changes per Inch • Force Concept Inventory • Fractally Compressed Image • Functional Configuration Identification • Functional Configuration Inspection

FCIC: Federal Crop Insurance Co. • Fellow of the Chemical Institute of Canada

FCIF: Formal Context Interchange Format • Full Common Intermediate Format

FCIM: Flight Control Interface Module

FCIP: Flight Cargo Implementation Plan

FCIR: False Colour Infrared

FCIT: First Computer Interface Tester

FCKW: Fluorchlorkohlenwasserstoffe

FCL: F-Center Laser • Feeder Control Logic • Flight Crew Licensing • Fluid Coolant Loop • Freon Coolant Line • Freon Coolant Loop

FCLA: Florida Center for Library Automation (USA-FL)

FCLC: Fibre Channel Loop Community

FCLT: Freeze Calculated Landing Time

FCM: Fast Component Mounter • Feature Change Map • Flight Contamination Monitor • Flow Control Module • Fond Cosmologique en Micro-ondes • Fondo Cósmico de Microondas • Fuzzy Cognitive Map • Minneapolis - Flying Cloud (USA-MN) (location identifier)

FCMC: Facility Control and Monitor Console

FCMS: Flight Crew Mission Simulator • Fuel Control and Monitoring System

FCMU: Foot-Controlled Maneuvering Unit

FCN: Free Core Nutation • Function

f. co.: Fair Copy

Fco: Franco

FCO: Field Change Order • Final Checkout • Flight Crew Operations • Foreign and Commonwealth Office (UK) • Fraction of Carbon Oxidized • Full Cutoff • Functional Checkout • Rome - Leonardo da Vinci - Fiumicino (Italy) (location identifier)

FCOB: Flight Control Operations Branch • Flight Crew Operations Branch

FCOD: Flight Control Operations Directorate • Flight Crew Operations Directorate

FCOH: Flight Controllers Operational Handbook

FCOL: Faculty of Communications and Open Learning

FCOS: Flight Computer Operating System • Flight Control Operating System • Flight Control Operational Software

Fcp: Foolscap

FCP: Failure Correction Panel • Fatigue-Crack Propagation • File Control Processor • Final Control Point • Firm Cost Proposal • Flat Concurrent PROLOG • Flight Control Procedure • Flight Correction Proposal • Fonds Commun de Placement • Forest Canopy Profile • Franck-Condon Principle • Franck-Condon Pumping • Fuel Cell Plant • Fuel Cell Power • Fuel Cell Propellant • Function Control Package

FCPA: Federal Chamber of Patent Attorneys • Fractional Cutting-Plane Algorithm

FCPC: Federal Computer Products Center • Field Coil Power Conversion • Flight Control Primary Computer

FCPGA: Flip Chip Pin Grid Array

FCPh: Fundamentals of Cosmic Physics (periodical code)

FCpI: Flux Changes per Inch

FCPI: Flux Changes Per Inch • Fonds Commun de Placement pour l'Innovation (France)

FCPLR: Formation Complémentaire Par La Recherche

FCPS: Fairfax County Public Schools (USA-VA) • Fuel Cell Power System

FCPU: Facility Central Processing Unit

FCR: Facility Change Request • Final Configuration Review • First Customer Release • Flight Configuration Review • Flight Control Room • France Câbles et Radio • Fuse Current Rating

FCRA: Fecal Collection Receptable Assembly

FCRAO: Five College Radio Astronomy Observatory (USA-MA)

FCRC: Federated Computing Research Conference

FCRSWG: Forestry Canada Remote Sensing Working Group

FCRT: Flight Cathode Ray Tube

Fcs: Francs

FCS: Colorado Springs - Butts (USA-CO) (location identifier) • Fast Control System • Fecal Containment System • Federal Communications System (USA) • Federation of Communication Services • Feedback Control System • Fermion-Chern-Simons (theory) • Field Citation Score • File Compression Software • Finally Came to my Senses • Fine Calibration Source • Finite Coordinate Space • Fire Control System • Flat Crystal Spectrometer • Flight Control Simulator • Flight Control System • Flight Crew System • Foreign Commercial Service • Frame Check Sequence • Front Crew System

FCSC: Flight Control Secondary Computer

FCSD: Flight Control Systems Department • Flight Crew Support Division

Fcsg: Focusing

Fcsle: Forecastle

FCSM: Flight Combustion Stability Monitor

FCSS: Fast Combat Support Ship • Fuel Cell Servicing System

FCST: Federal Council for Science and Technology (USA) • Forecast

FCSTF: Fuel Cell System Test Facility

FCT: Fast Current Transformer • Filament Center Tap • File Control Table • Film-Cooled Tailpipe • First Coding Theorem • Fixed Conversion Time • Flight Controller Trainer • Flight Control Team • Flight Crew Trainer • Flux-Corrected Transport • Fuel Cell Test • Fuel Consumption Theorem • Function • Yakima (USA-WA) (location identifier)

FCTB: Flight Crew Training Building

FCTF: Fuel Cell Test Facility

FCTR: Flight Controller Training Room

FCTS: Flight Crew Trainer Simulator

Fctn: Function

Fctnl: Functional

Fcty: Factory

FCU: Fluid Checkout Unit • Frequency Control Unit • Fuel Control Unit

FCV: Flow Control Valve • Fuel Cell Vehicle

FCW: Federal Computer Week (journal) • Format Control Word

Fcy: Fancy

FCY: Forrest City (USA-AR) (location identifier)

f&d: Fill & Drain

f/d: Focal Distance

Fd: Feed • Filed • Fiord • Fjord • Flight Deck • Focal Distance • Ford • Forward • Found • Founded • Free Delivery • Free Discharge • Free Dispatch • Fund

F&D: Fill & Drain

F/D: Face of Drawing • Field of Drawing • Fill/Drain

FD: Fast Delivery • Fault Detection • Federal District • Feed • Fermi-Dirac • Fictitious Domain • Filament Darkness • File Description • File Descriptor • Fill and Drain • Fill Draw • Finite Difference • Fire Department • Flight Day • Flight Deck • Flight Director • Flight Dynamics • Floppy Disk • Floppy Drive • Focal Distance • Forbush Decrease • Fraction Dead • Free Delivery • Freeze Dried • Full Disk • Full Duplex • Fully Digital • Functional Dependency • Functional Description • Functional Designator • Function Designator

FDA: Facility Disposal Area • Fault Detection and Annunciation • Flight Deck Assembly • Food and Drug Administration (USA)

FDAC: Flight Demonstration of ASTA Concepts

FDAI: Flight Director Attitude Indicator

FDAO: Flight Dynamics Analysis Operation

FDAU: Flight Data Acquisition Unit

FDB: Facility Database • Fahrenheit Dry Bulb • Field Dynamic Braking • Fit Database

Fdbk: Feedback

FDBS: Federated Database System

FdC: Fin de Citation

FDC: First-Day Cover • Flight Data Center • Floppy-Disk Controller • Fluorosensor Data Correlator

FDCD: Fluorescence-Detected Circular Dichroism

FDCF: Flight Design Control Facility

FDCI: Fixed Duration and Constant Impact

FDCT: Forward Discrete Cosine Transform

FDD: Floppy-Disk Drive • Formation De Développement • Frequency Division Duplex

FDDA: Four-Dimensional Data Assimilation

FDDB: Function Designator Database

FDDF: Flux Decrement Distribution Function

FDDI: Fiber Digital Data Interface • Fiber Digital Device Interface • Fiber Distributed Data Interface (*)

FDDL: Field Data Description Language • Frequency-Division Data Link

FDE: Fault Detection Electronics • Finite Difference Equation • Flight Data Entry • Fluid Dynamics Experiment • Førde (Norway) (location identifier) • Functional Differential Equation

FDEP: Flight Data Entry and Printout • Formatted Data Entry Program

FDF: First Drop Flight • Flight Data File • Flight Dynamics Facility • Form Definition File • Form Description File • FORS Deep Field • Fort de France (Martinique) (location identifier)

FDFU: Federation of Documentary Film Units

Fdg: Funding

FDH: Fixed Dynamical Heating • Friedrichshafen (Germany) (location identifier)

FDHD: Floppy Disk High-Density • Floppy Drive High-Density

FDHS: Fast Data Handling Simulator

FDI: Failure Detection and Isolation • Failure Detector Indicator • Fast Delivery Image • Fault Detection and Identification • Fault Detection and Isolation • Field Data Investor • Flight Director Instrument • Foreign Direct Investment • Frequency Domain Interferometry

FDIC: Federal Deposit Insurance Corporation

FDIIR: Fault Detection, Isolation, Identification, and Recompensation

FDIO: Flight Data Input/Output

FDIOC: Flight Data Input/Output Center

FDIOR: Flight Data Input/Output Remote

FDIR: Failure Detection, Isolation, and Recovery • Fault Detection, Identification, and Recovery • Fault Diagnosis, Isolation, and Recovery

FDISK: Fixed Disk • Format Disk

FDK: Flight Display Keyboard • Frederick (USA-MD) (location identifier)

FDL: Fast Delay Line • File Definition Language • Fixed Document Language • Flight Director Loop • Flight Dynamics Laboratory • Formatted Definition Language • Form Definition Language

FDLC: Final Design Load Cycle

FDLN: Feed Line

FDLP: Federal Depository Library Program (GPO)

FDM: Feasibility Demonstration Model • Finite Difference Method • Flight Data Manager • Flight Design Manager • Formal Development Methodology • Frequency Data Multiplexer • Frequency Division Multiplex • Frequency Division Multiplexer • Frequency Division Multiplexing • Frequency Domain Multiplex • Frequency Domain Multiplexer • Frequency Domain Multiplexing • Full Descriptive Method

FDMA: Frequency Demand Multiple Access • Frequency-Division Multiple Access (*)

Fdn: Foundation

FDN: Fixed Dialling Number

FDNU: Format-Dependent Nonuniformity

Fdo: Firmado

FDO: Fee Determination Official • Flight Director's Office • Flight Dynamics Officer • For Declaration Only

FDOA: Frequency Difference Of Arrival

FDOC: Fraction of Degradable Organic Carbon

FDOR: Flight Design Operations Review

FDP: Fast Data Processor • Fast Delivery Processor • Fast Delivery Product • Fast Digital Processor • Feasibility Demonstration Program • Federal Demonstration Project • Field Development Program • Field Distortion Pattern • Flight Data Processing • Flight Design Panel • Formation De Perfectionnement • Fourniture de Documents Primaires • Frequency-Dependent Polarization • Full Disk Patrol • Full Dog Point

FDPA: Frequency-Dependent Position Angle

FDPC: Federal Data Processing Center

FDPP: Framework Demonstration Projects Program

FDPR: Flight Data Processor Replacement

Fdr: Feeder • Finder • Founder

FDr: Fire Door

FDR: Failure Detection and Recovery • Fast Dump Restore • Final Data Report • Final Design Review • Fit Data Record • Flight Data Recorder • Flight Definition Requirement • Flight Dynamics Room • Franklin Delano Roosevelt (1882-1945) • Frederick (USA-OK) (location identifier) • Functional Design Review

FDRD: Flight Definition and Requirements Directive • Flight Definition and Requirements Document • Flight Determination and Requirements Directive

FDRI: Flight Director Rate Indicator

FDRS: Flight Data Recording System

FDRT: Flexible Digital Receiving Terminal

Fdry: Foundry

FDS: Failure Detection Electronics • Failure Dynamics Software • Faraday Dark Space • Fast Diode Switch • Fermi-Dirac Statistics • Financial Data System • Fire Detection and Suppression • Fire Detection System • Flare Detection System • Flexible Display System • Flight Data System • Flight Design and Scheduling • Flight Design System • Flight Director System • Flight Display System • Flight Dynamics Simulator • Flight Dynamics Software • Flight Dynamics Support • Flight Dynamics System • Fluid Distribution System • Flux Difference Splitting • Formal Data Structure • Frame Difference Signal • Full Development System • Functional Design Specification

FDSC: Flight Dynamics Simulation Complex (*) • Flight Dynamics Situation Complex

FDSCU: Fluid Distribution System Control Unit

FDSMM: Fluid Dynamics and Solidification of Metallic Melts

FDSN: Federation of Digital Seismographic Networks

FDSR: Final Design Status Review

FDSWG: Future Data Systems Working Group

FDT: Failure Diagnostic Team • Fault Detection System • Flux-limited Diffusion Theory • Formal Description Technique • Full Duplex Teletype • Fully Developed Turbulent

FDTD: Finite-Difference Time Domain

FDTE: Final Development Test and Evaluation

FDU: Bandundu (Zaire) (location identifier) • Flight Divider Unit • Fluid Distribution Unit • Frequency Determining Unit • Frequency Doubling Unit

FDV: Fill and Drain Valve

FDx: Full Duplex

FDX: Full Duplex

FDY: Findlay (USA-OH) (location identifier)

fe: Fecit

Fe: Eisen • Fer • Iron

F&E: Facilities & Equipment • Facility & Equipment • Facility & Environment • Forschung & Entwicklung

F/E: Forschung/Entwicklung • Full/Empty

FE: Facility and Environment • Failure Equation • Farbexzeß • Far East • Far Encounter • Feature Extraction • February • Février • Field Elevation • Field Emission • Field Engineer • Fill End • Fine Erection • Finite Element • Fire Extinguisher • First Edition • Flight Engineer • Flight Engineering • Flight Execution • Flight Executive • Flight Experiment • Force Execution • Foreign Editor • For Example • Format Effector • Forschung und Entwicklung • Fossil Energy • Front End • Further Education

FEA: Failure Effects Analysis • Federal Editors Association (USA) • Fetlar (UK) (location identifier) • Finite-Element Analysis • Flight Experiment Apparatus • Fluid Experiment Apparatus • Fluid Experiment Assembly • Front End Amplifier • Front End Analysis

FEAAS: Federation of East Anglian Astronomical Societies (UK-England)

FEAC: Far-End Alarm and Control

FEACO: Fédération Européenne des Associations de Conseils en Organisation

FEANI: Fédération Européenne d'Associations Nationales d'Ingénieurs

FEANTSA: Fédération Européenne des Associations Nationales Travaillant avec les Sans-Abri

FEAP: Finite-Element Analysis Program

FEARO: Federal Environmental Assessment Review Office

FEAST: Fast Ethernet using Advanced SMC Technology • Finite Element Analysis for Structural Test • Finite Element Analysis of Structures

FEAT: Final Engineering Acceptance Test • Fine Error Averaging Test • Formal Evaluation Acceptance Test

Feb: Febbraio • Febrero • February

FEB: Fair Employment Board • Falling Evaporating Body • February • Fédération des Éditeurs Belges (Belgium) • Fédération des Entreprises Belges (Belgium) • Filter Electronics Box • Flight Evaluation Board • Fluidized Ejecta Blanket • Forward Equipment Bay • Fourth Entry Battery • Franklin Electronic Book • Functional Electronic Block • Functional Energy Block • Sanfebagar (Nepal) (location identifier)

Febb: Febbraio

FEBEL: Faint Extended Broad Emission Line

FEBS: Federation of European Biochemical Societies

Feby: February

fec: Fecit

FEC: Federal Election Commission (USA) • Field Engineering Change • Fine Erection Complete • Fondation Européenne de la Culture • Foreign Exchange Carrier • Foreign Exchange Certificate • Foreign Exchange Cost • Forward Error Correction • Forward Events Controller • Forwarding Equivalent Class • Front-End Computer • Front-End Controller

FECB: File Extended Control Block • Foreign Exchange Control Board

FECM: Fonds Européen de Coopération Monétaire

FECOM: Fonds Européen de Coopération Monétaire

FECN: Forward Error Code Notification • Forward Explicit Congestion Notification

FECR: Final Experiment Concept Review

FECS: Federation of European Chemical Societies

Fed: Federal • Federalist • Federate • Federated • Federating • Federation

Féd: Fédéral • Fédéraliste • Fédération • Fédéré • Fédérer

FED: Far End Data • Federal • Field-Emission Display (*) • Field-Emitter Display • Field-Emittor Display • Final Estimation of Data • Flight Events Demonstration • Fondo Europeo de Desarollo • Fonds Européen de Développement • Front End Driver

FEDAC: Forward Error Detection And Correction

FEDB: Flight Execution Database

FEDC: Fusion Energy Design Center

FEDCAC: Federal Computer Acquisition Center

FEDER: Fonds Européen de Développement Régional

FEDGE: Finite Element Data Generation

FEDIA: Fédération d'Études des Infrastructures Alternatives (France)

Fedn: Federation

FEDP: Facility and Equipment Design Plan

FEE: Failure Effects Evaluation • Feature Extraction Environment • Fédération des Étudiants Étrangers • French Echocardiograph Equipment • Front End Electronics • Full Exit Entry

FEED: Field Exploitation of Elevation Data

FEEP: Field Emission Electric Propulsion

FEER: Fast Eigensolution Extraction Routine • Fundamental Equilibrium Exchange Rate

FEF: Fédération des Étudiants Francophones • Flight Engineering Facility • Fondation Européenne pour la Formation

FEFI: Flight Environmental Fault Isolation

FEFO: First Ended - First Out

FEG: Fergana (Uzbekistan) (location identifier)

FEI: Faculdade de Engenharia Industrial • Federation of the Electronics Industry • Financial Executive Institute • Flexible External Insulation • Fonds Européen d'Investissement

FEID: Flight Equipment Interface Design • Flight Equipment Interface Device • Functional Engineering Interface Device

FEIN: Federal Employer Identification Number (USA)

FEIS: Final Environment Impact Statement

FEISEAP: Federation of Engineering Institutions of Southeast Asia and the Pacific

FEIT: Fujitsi Enhanced Imaging Technology

FEK: Ferkessedougou (Ivory Coast) (location identifier) • Forskningsbibliotekernes Edb-Kontor (Denmark)

FEL: First-Element Launch • Free-Electron Laser • Fürstenfeldbruck (Germany) (location identifier) • Fysich Elektronisch Laboratorium • Fysich en Elektronisch Laboratorium

FELASI: Federación Latin-Americana de Sociedades de Física

FELASOFI: Federación Latin-Americana de Sociedades de Física

FELI: Free Electron Laser Institute

FeLiX: Fe-Line region X-ray

FELIX: Free Electron Laser for Infrared Experiment • Front End of Laser Interceptor Transmitter

Fell: Fellow • Fellowship

Felr: Feeler

fém: Force Électromotrice

Fem: Female • Féminin • Feminine

FEM: Field Electron Microscope • Field Electron Microscopy • Field Emission Microscope • Field Emission Microscopy • Field Engineering Maintenance • Field Evaluation Model • Field Evaluation Modelling • Finite-Element Method • Finite-Element Model • Finite-Element Modelling • Flight Engineering Model • Flight Engineering Modelling • Flyable Engineering Model • Flyable Engineering Modelling • Fonds pour l'Environnement Mondial • Force Électromotrice • Free Electron Maser • Front End Module • Functional Electrical Model • Functional Electrical Modelling

FEMA: Federal Emergency Management Agency (USA) • Federation of European Meteor Observers

FEMCPL: Facilities and Environmental Measurement Component Parts List

FEMF: Foreign Electromotive Force

FEMIS: Fondation Européenne des Métiers de l'Image et du Son

FEMMS: Frontiers of Electron Microscopy in Materials Science

FEMO: Finite Element Modelling Optimization • Free-Electron Molecular Orbital

FEMP: Federal Energy Management Program • Free Energy Minimization Procedure

FEMS: Federation of European Microbiological Societies

FEN: Fédération de l'Éducation Nationale • Fernando de Noronha (Brazil-FN) (location identifier)

FEO: Federal Environment Office • Few Entry Object • Field Engineering Order

FEORP: Federal Equal Opportunity Recruitment Program (USA)

FEP: Faculté d'Éducation Permanente • Fast Evolutionary Programming • Feature Extraction Processor • Federation of European Publishers • Finite Element Program • Floral Ethel Propane • Flow Enhancing Protocol • Fluorinated Ethene Propene • Fluorinated Ethylene Propylene (*) • Fonction d'Étalon de Point • Forecast Expenditure Plan • Formal Evaluation Phase • Format Evolution Process • Free Endpaper • Freeport Albertus (USA-IL) (location identifier) • Front-End Processor

FEPA: Fair Employment Practices Act (USA)

FEPC: Fair Employment Practices Commission • Fair Employment Practices Committee • Flight Equipment Processing Contractor • Format Evolution Process Committee

FEPCA: Federal Employees Pay Comparability Act (USA)

FEPG: Federal Engineering Planning Group

FEPS: Facility and Equipment Planning System • Far Encounter Planet Sensor

FEPT: Fédération Eurpéenne de la Presse et de la Télévision

FER: Fergusons Gulf (Kenya) (location identifier) • Final Evaluation Report

FERA: Federal Emergency Relief Administration (USA)

FERC: Federal Energy Regulatory Commission (USA)

FERD: Facility and Equipment Requirements Document • Forum Européen de la Recherche et du Développement

FERF: Far-End Remote Failure • Far-End Reporting Failure

FERIAS: Forest Environment and Resource Information Analysis System

FERMI: Front-End Readout Microsystem

FERN: Forest Ecosystems Research Network • Further Education Research Network

FEROS: Fiber-fed Extended Range Optical Spectrograph

Ferr: Ferrocarriles

FERRO: Far Eastern Regional Research Office

FERS: Federal Employees Retirement System (USA) • Fédération Européenne des Réseaux Scientifiques

FERT: Fortitudo Eius Rhodum Tenuit

Frev: Fervens

FES: Fast Erect System • Feature Extraction Segment • Fédération des Éditeurs Suisses (Switzerland) • Field Emission Spectroscopy • Final Environmental Statement • Fine Error Sensor • Fine Error Signal • Flame Emission Spectrometry • Flame Emission Spectroscopy • Flash Evaporator System • Flight Element Set • Fluids Experiments System • Fondation Européenne de la Science • Functional Electrical Simulation • Fuzzy Expert System

FESA: Finite Element Structural Analysis

FESC: First EUVE Source Catalog

FESEM: Field Emission Scanning Electron Microscope • Field Emission Scanning Electron Microscopy

FESIP: Fast External Satellite Image Processing

FESL: Failure Effects Summary List

FESP: Flight Execution Support Plan

FESR: Finite Energy Sum Rules

FESS: Facilities and Engineering Services Section (Fermilab) • Flight Experiment Shielding Satellite

Fest: Festival • Festivo

FESTIP: Future European Space Transportation Investigation Programme

FESWMS: Finite Element Surface Water Model System

FESWS: Feature Extraction Workstation

FET: Federal Estate Tax • Federal Excise Tax • Field-Effect Transistor (*) • Field Electron Transistor • Flight Elapsed Time • Fossil-Energy Technology • Fremont (USA-NE) (location identifier)

Dear Sir / Madam,

Thank you for purchasing StarGuides 2001 and / or StarBriefs 2001. Each of these volumes are valuable reference works which will continue to be updated in the future.

It is the intention of the publisher and compiler to issue new editions of both of these books (although the frequency of these updates has not yet been determined). If you would like to be informed when future editions of either of these books are released then please complete and return the attached reply slip.

Thank you,

The Publisher

StarGuides 2001 / StarBriefs 2001

I would like to be informed when future editions of StarGuides / StarBriefs (please delete where applicable) are released.

Name: ____

Position: ____

____

Address: ____

____

____

Telephone / Fax / E-mail (optional): ____________________________

____

Signed / Dated: _______________________________________

Please return to:
Ms. Livia Iebba, Reference Work Unit, Kluwer Academic Publishers, Spuiboulevard 50, PO Box 17, 3300 AA Dordrecht, The Netherlands. Fax: +31 78 6392388

FETEM: Field Emission Transmission Electron Microscope • Field Emission Transmission Electron Microscopy

FETV: Full Envelope Thrust Vectoring

FeU: Fern Universität (Germany)

FEU: Flight Equivalency Unit • Further Education Unit

Fév: Février

Fev: Fevereiro

FEV: Far End Voice • Fevereiro • Février • Fisheye View • Forced Expiratory Volume • Front Expansion Velocity

FEVARO: Friedrich Ebert Visual and Radiotelescope Observatory (Germany)

FEW: Cheyenne Warren (USA-WY) (location identifier)

FEWG: Flight Evaluation Working Group

FEWS: Famine Early Warning System • Follow-on Early Warning System • Follow-on-Earth Warning System

FExt: Fire Extinguisher

FEXT: Far-End Crosstalk

FEZ: Fez (Morocco) (location identifier) • Freizeit- und Erholungszentrum (Germany)

ff: Faisant Fonction • Fecerunt • Feuillets • Folgende • Folios • Following • Forms • Fortissimo • Thick Fog

F-F: Face to Face • Flip-Flop

F/F: Fill/Full • Full/Fill

FF: Failure Factor • Feed Forward • Fellows • Filling Factor • Find File • Fixed Focus • Flat Field • Flight Forward • Flip-Flop • Ford Foundation • Forecasting Funding • Form Feed • Fortsetzung Folgt • Forward Fuselage • Free Fall • Free Flight • Free Flyer • Free Format • French Franc • Frozen Flow • Fuel Flow • Full Field • Full Functional

FFA: Fast Fourier Analyzer • Federal Facilities Agreement • Flygtekniska Försörksanstalten (Sweden) • For Further Assignment • Free Fall Acceleration • Free from Alongside • Kill Devil Hills (USA-NC) (location identifier)

FFAST: Forest Fire Advanced System Technology

FFBD: Functional Flow Block Diagram

FFC: Father Function Coefficient • Federal Facilities Council • Final Flight Certification • First-Flight Capability • Flagstaff Field Center (NIST) • Flip-Flop Complimentary • Foreign Funds Control • Free-Flyer Centre • Free From Chlorine

FFCB: Fédération Française de Coopération entre Bibliothèques (F rance)

FFCC: Ferrocarriles • Free-Flyer Control Centre

FFD: Fairford (UK-England) (location identifier) • Fixed-Format Display • Functional Flow Diagram • Function Flow Diagram

FFDF: Fluid Physics and Dynamics Facility

FFDP: Far Field Diffraction Pattern

FFE: Fluorescence par Faisceau d'Électrons • Forced Fault Entry

FFEL: Förderung der Forschung in Entwicklungsländern

FFEM: Freeze-Fracture Electron Microscope • Freeze-Fracture Electron Microscopy

FFES: Fossil Free Energy System

FFEU: Free-Flow Electrophoresis Unit

fff: Fortississimo

FFF: Field Flow Fractionation • Final Factory Functional • Fission-Fusion-Fission • Forschungsförderungsfonds (Austria)

FFFF: Flow Field Flow Fractionation

FFFP: Film Forming Fluoroprotein

FFFT: Forced Flow Flamespreading Test

FFGF: Fill Frame Generation Function

FFHR: Fusion-Fission Hybrid Reactor

FFI: Fine Field Imager • For Further Information • Forsvarets Forskningsinstitutt (Norway) • Fraunhofer Filling In • Free From Infection • Full Field Investigation

FFIS: Federal Facilities Information System

FFL: Fairfield (USA-IA) (location identifier) • Fédération Française du Livre (France) • Flip-Flop Latch

FFLOP: Field Fresnel Lens Optical Platform

Ffly: Faithfully

FFM: Fergus Falls (USA-MN) (location identifier) • Free-Flying Module

FFN: Feedforward Neural

ffo: Folios

FfO: Forschungsinstitut für Optik

FFO: Dayton Patterson (USA-OH) (location identifier) • Flux-Flow Oscillator

FFOL: FDDI Follow-On LAN

FFon: Faisant Fonction

FFP: Fabrication-Falsification-Plagiarism • Falsification - Fabrication of Results - Plagiarism • Faser-Fary-Pérot • Fast Fluorescence Photometry • Fédération de la Formation Professionnelle (France) • Firmed Fixed Price • Firm Fixed Price • Frequent Flyer Programme

FFPE: Front Focal Plane Enclosure

FFPI: Flip-Flop Position Indicator

FFQS: FIRST Faint Quasar Survey

FFr: French Franc (currency)

FFR: Factory Flight Readiness • Fission-Fusion Ratio • Fixed Frequency Receiver • Flight Feasibility Review • Frame Forwarding Rate

FFRD: Flip-Flop Relay Driver

FFR&DC: Federally-Funded Research & Development Center

FFRDC: Federally-Funded Research and Development Center

FFRR: First Flight Readiness Review • Full-Frequency Range Recorder • Full-Frequency Range Recording

FFRRV: Free-Flight Rotorcraft Research Vehicle

FFS: Fast File System • Fat Fingered Syndrome • Federal Financial System • Flame Fluorescence Spectroscopy • Force-Free Surface • Formatted File System • Free Fall Sensor

FFSA: Fédération Française des Sociétés d'Assurance (France)

FFSAC: Field Functional Systems Assembly and Checkout

FFSD: Failed Fuel Surveillance and Diagnosis

FFSM: Fast Fourier Synoptic Mapping

FSDPS: Flight Service Data Processing System

FFSS: Full-Frequency Stereophonic Sound

FFST: First Failure Support Technology

FFT: Factory Functional Test • Fast Fourier Transform • Final Form Text • For Further Transfer • Frame Format Table • Frankfort - Capital City (USA-KY) (location identifier) • Full Fuselage Trainer

FFTB: Final Focus Test Beam

FFTD: Finite Fourier Transform Definition

FFTF: Fast Flux Test Facility • Final Focus Test Facility

FFTH: Fiber To The Home

FFTO: Free-Flying Teleoperator

FFU: Formatted Field Unit • Futaleufu (Chile) (location identifier)

FFV: Fédération Française de Voile (France) • Focus Field Variation

FFwd: Fast Forward

FFWF: Fonds zur Förderung der Wissenschaftlichen Forschung

Ffy: Faithfully

Fg: Faubourg • Fine Grain • Fog • Fully Good

FG: Fachgremium • Faktengruppe • Federal Government • Fibonacci Generator • Fifth Generation • File Gap • Filter Gate • Filtergewinde • Fixed Gear • Floated Gyro • Fluorescence Gated • Flux Gradient • Fog • Forward Gate • Frais Généraux • Frame Ground • French Guiana • Fully Good

FGA: Foreign General Average • Free of General Average

FGAN: Forschungsgesellschaft für Angewandte Naturwissenschaften

FGB: Functional Energy Block (in Russian)

FGBA: Fluids Generic Bioprocessing Apparatus

FGC: Fifth Generation Computer • Fixed Gain Control

FGCC: Federal Geodetic Control Committee (USA)

FGCS: Federal Geodetic Control Subcommittee (USA) • Fifth Generation Computer Systems

Fgd: Forged

FGD: F Derik (Mauritania) (location identifier) • Flux Gas Desulphurisation

FGDC: Federal Geographic Data Commission (USA) • Federal Geographic Data Committee (USA) (*) • Federal Geospatial Data Clearinghouse

FGE: Fine Guidance Electronics

FGEF: Federal Geographic Exchange Format (USA)

FGFC: Fixed Gas-Filled Capacitor

FGGE: First GARP Global Experiment

FGH: Flexible Gyro Header

FGHA: Flexible Gyro Header Assembly

FGHC: Flat Guarded Horn Clause

FGI: Apia Fagali'i (Samoa) (location identifier) • Fédération Graphique Internationale • Financial Guarantee Insurance • Finnish Geodetic Institute

FGJ: Forschungsgesellschaft Joanneum (Germany)

FGK: Forschungsgesellschaft Kunstoffe (Germany)

FGKL: Fifth Generation Kernel Language

FGKRL: Fifth Generation Knowledge Representation Language

FGL: Fox Glacier (New Zealand) (location identifier)

FGM: Fluxgate Magnetometer

Fgn: Foreign

FGN: Flying Ghost Nebula • Fundamental Geodetic Network

FGNP: Field Generated Normal Point

FGO: Fderik (Mauritania) (location identifier) • Finanzgerichtsordnung

FGP: Faktengruppenprofil • Fetch-Generalize-Project

FGS: Fellow of the Geological Society • Fine Guidance Sensor • Fine Guidance System • Florida Geological Survey (USA-FL) • Forschungsgruppe Satellitengeodäsie

FGSAA: Fine Guidance Sensor Access Area

FGSMRAO: France-Germany-Spain Millimeter Radio Astronomy Observatory

FGSS: Fine Guidance Sensor System • Flexible Guidance Software System

Fgt: Freight

FGT: Federal Geographic Technology (USA) • Federal Gift Tax • Fundaçao Gomes Teixeira (Portugal)

FGU: Fangatau (Tuamotu Islands) (location identifier)

FGV: Ferry à Grance Vitesse

FGW: Filter Grating Wheel

FGWE: First Global Weather Experiment

Fgy: Foggy

fh: Fiat Haustus

F/H: Freehold

FH: Field Headquarters • Fire Hose • Fire Hydrant • Flex Hose • Flexible Hose • Flight Hour • Fog Horn • Frequency Hopping • Fore Hatch • Forward Hatch

FHA: Federal Highway Administration (USA) • Federal Housing Administration • Flexible Header Assembly

FHB: Family Hold Back • Field Horizontal Branch

FHBD: Fachhochschule für Bibliothekswesen und Dokumentation (Germany)

FHC: Flight Half Coupling

FHD: Farben-Helligskeits-Diagramm • First-Hand Distribution • Fixed-Head Disk • Flame Hydrolysis Deposition

FHF: Flächenhelligskeit-Fluktuation • First Horizontal Flight

FhG: Fraunhofer Gesellschaft (Germany)

FHI: Fuji Heavy Industries

FHIL: Forbidden High-Ionization Line

FHLC: Faint High-Latitude Carbon

Fhld: Freehold

FHLMC: Federal Home Loan Mortgage Corporation

FHMA: Frequency Hopping Multiple Access

FHNC: Fermi Hypernetted Chain Theory

FHO: Fachhochschule Offenburg (Germany) • Failed Handover • Fenton Hill Observatory (USA-NM)

F'hold: Freehold

fhp: Friction Horsepower

FHP: Forum on the History of Physics • Fuel High Pressure

FHS: Fachhochschule • Foran High School (USA-CT) • Forward Heat Shield • Frequency Hopping Signal • Fuel Handling System

FHSF: Fixed-Head Storage Facility

FHSS: Forward Heat-Shield Separation • Frequency-Hopping Spread Spectrum

FHST: Fixed-Head Star Tracker

FHT: Fast Hartley Transform

FHTC: Fixed High-Temperature Capacitor

FHTS: Fluid Heat Transport System

FHU: Fort Huachuca (USA-AZ) (location identifier) • Sierra Vista (USA-AZ) (location identifier)

FHVC: Fixed High-Volt Capacitor

FHW: Fachhochschule Wiesbaden (Germany)

FHWA: Federal Highway Administration (USA)

FHy: Fire Hydrant

FHZ: Fakahina (Tuamotu Islands) (location identifier)

fi: Finland • Finnish (ISO language code)

F&I: Forschung im Ingenieurwesen (periodical code)

FI: Fachinformation • Faculty Index • Faeroe Islands • Falkland Islands • Fault Identification • Fault Isolation • Field Intensity • Field Ionization • Fiji Islands • Final Issue • Finland (ISO country code) • Finlande (ISO country code) • Fire Insurance • Fixed Interval • Flare Index • Flight Instrumentation • Flow Indicator • Flow Injection • Forecast International (USA-CT) • For Instance • Formal Inspection • Free In • Fréquence Intermédiaire • Front-Illuminated • Front-side Illuminated • Fuel Injected • Full Internet

FIA: Fault Isolation Analysis • Federal Insurance Administration • Fédération Internationale d'Automatique • Flame Ionization Analyser • Flight Integration Assessment • Flow Injection Analysis • Flow Injection Analyzer • Freedom of Information Act (USA) • Free In Air • Free Interstitial Atom • Full Interest Admitted • Fully Illuminated Area • Future Imagery Architecture

FIAB: Fédération Internationale des Associations de Bibliothécaires • Flight Integration Assessment Board

FIAC: Foire Internationale d'Art Contemporain • Foire Internationale des Arts Contemporains

FIAF: Federación Internacional de Archivos de Film

FIAI: Fédération Internationale des Associations d'Institu-teurs

FIAMS: Flinders Institute for Atmospheric and Marine Studies

FIANN: Fédération Internationale des Associations Nationales de Normalisation

FIAP: Fellow of the Institution of Analysts and Programmers • Forum on Industrial and Applied Physics

FIAR: Fabbrica Italiana Apparecchiature Radioelettiche • Failure Investigation Action Report

FIAS: Formation Internationale Aéronautique et Spatiale

FIASc: Fellow of the Indian Academy of Sciences

FIAT: Federación Internacional de Archivos de Televisión

fib: Fibratus

FIB: Focused Ion Beam

FIBEX: First International BIOMASS Experiment

FIBRE: Friendly Inputting of Bibliographic Records

Fic: Fiction • Fictional • Fictitious

FIC: Fault Isolation Code • Federal Information Center • Federal Information Commission • Field Installed Connector • Fire Cove (USA-AK) (location identifier) • Fleet Installed Cost • Flight Information Center • Flight Information Controller • Fogarty International Center • Fonds Interministériel de Coopération • Frequency Interference Control

FICA: Federal Insurance Contributions Act (USA)

FICCDC: Federal Inter-Agency Coordinating Committee on Digital Cartography (USA)

FICCI: Federation of Indian Chambers of Commerce and Industry

FICOME: Fédération Interprofessionnelle de la Communication d'Entreprise (France)

FICS: Fault Isolation Checkout System • Financial Information Control System • Forecasting and Inventory Control System

Fict: Fiction • Fictitious

FICU: Fonds International de la Coopération Universitaire

Fid: Fidelity • Fiduciary

FID: Failure Identification • Fault Isolation Detection • Fédération Internationale d'Information et de Documentation • Final Instrument Definition • Fishers Island (USA-NY) (location identifier) • Flame Ionization Detector • Flammen Ionisationsdetektor • Flight Implementation Directive • Free Induction Decay

FIDAC: Film Input to Digital Automatic Computer • Fireball Data Center (IMO)

FIDAE: Feria Internacional del Aire y del Espacio (Chile)

FIDDO: Focused Investigation of Document Delivery Options

FIDE: Fully Integrated Data Environment • Fuzzy Inference Development Environment

FIDEA: Federación Interamericana de Educación de Adultos

FIDES: Forecaster's Intelligent Discussion Expert System

FIDEX: Floating Ice Detection Experiment

FIDIT: Fonds d'Intervention pour le Développement Industriel et Technologique

FIDLE: FIFE Doppler Lidar Experiment

FIDNet: Federal Instrusion Detection Network (USA)

FIDO: Facility for Integrated Data Organizations • Filing, Imaging and Distributed Objects • Flight Dynamics Officer • Flight Inspection District Office • Fog, Intense, Dispersion Of • Fog Investigation Dispersal Operations

FIDOM: Fonds d'Investissement dans les Départements d'Outre-Mer (France)

FIDP: Fellow of the Institute of Data Processing

F&IE: Facilities & Industrial Engineering

FIE: Facilities and Industrial Engineering • Fair Isle (UK) (location identifier) • Flight Instrumentation Engineer • Fourier Integral Estimate • Ferric Ion Free • Formaldehyde-Induced Fluorescence

FIEC: Formation des Investigateurs aux Essais Cliniques

FIED: Fédération Interuniversitaire de l'Enseignement à Distance

FIEE: Fédération des Industries Électriques et Électroniques • Fellow of the Institution of Electrical Engineers

FIEEC: Fédération des Industries Électriques, Électroniques et de Communication (France)

FIEEE: Fellow of the Institute of Electrical and Electronics Engineers

FIEI: Fellow of the Institution of Engineers in Ireland

FIEJ: Fédération Internationale des Éditeurs de Journaux

FIELD: Foundation for International Environmental Law and Development

FIEP: Federación Internacional de Editores de Periódicos

F&IER: Forschung im Ingenieurwesen - Engineering Research (periodical code)

FIERA: Fast Imager Electronic Readout Assembly

FIES: Fiber-coupled Echelle Spectrograph

FIF: Failure Indicating Fuse • Fiber Instrument Feed • Fractal Image Format

FIFCLC: Fédération Internationale des Femmes de Carrières Libérales et Commerciales

FIFDU: Fédération Internationale des Femmes Diplômées des Université s

FIFE: First ISLSCP Field Experiment

FIFI: Far Infrared Fabry-Pérot Interferometer • Field Imaging Far-Infrared • Flexible Ideal Format for Information

FIFI LS: Field Imaging Far-Infrared Line Spectrometer

FIFILS: Field Imaging Far-Infrared Line Spectrometer

FIFO: First In - First Out • Flight Inspection Field Office • Floating Input - Floating Output

FIFRA: Federal Insecticide, Fungicide, and Rodenticide Act (USA)

fig: Figura

Fig: Figurado • Figurative • Figuratively • Figuré • Figure

FIG: Fédération Internationale des Géomètres • Figure • Flight Inspection Group • Floating Integrating Gyro • FORTH Interest Group • Fria (Guinea) (location identifier) • Friends of Internet in Greece

FIGARO: French Italian Gamma-Ray Observatory

FIGS: Fabry-Pérot Infrared Grating Spectrometer

FIH: Kinshasa N'Djili (Zaire) (location identifier)

FII: Federal Item Identification • Fédérations Internationale des Interprètes

FIICPI: Fédération Internationale des Ingénieurs-Conseillers en Propriété Industrielle

FIIG: Federal Item Identification Guide

FIIGMO: Forget It, I've Got My Orders

FIIM: Fédération Internationale des Ingénieurs Municipaux • Fellow of the Institution of Industrial Managers

FIIN: Federal Item Identification Number

FIIWG: Facility Integration and Installation Working Group

FIJPA: Fédération Internationale des Journalistes Professionnels de l'Aéronautique

FIJPAA: Fédération Internationale des Journalistes Professionnels de l'Aéronautique et de l'Astronautique

FIK: Field Ionization Kinetics • Finke (Australia-NT) (location identifier)

Fil: Filament • Fillet • Fillister • Filosofia • Filter • Filtrate

FIL: Filament • Fillmore (USA-UT) (location identifier) • Foire Internationale du Livre • Formation-Information-Liaison • Fuel Injection Line

FiLA: Five-Letter Acronym

FILAO: Fiber Link to Adaptive Optics

Filat: Filatelia

FiLCHeRS: Falsifiability - Logic - Comprehensiveness - Honesty - Replicability - Sufficiency

FILE: Feature Identification and Location Equipment • Feature Identification and Location Experiment • Future Identification and Location Experiment

FILER: File Information Language Executive Routine

FILG: Filing

Filip: Filipinas

FILM: Far Infrared Line Mapper • Foire Internationale du Livre de Montréal (Canada-QC)

FILO: First In - Last Out

Filol: Filología

FILOS: Far-Infrared Limb Observing Spectrometer

FILS: Federal Information Locator System (USA)

Filt: Filter

FILX: File Index

FIM: Facing Identification Mark • Failure Indication Module • Fault Isolation Meter • Fédération des Industries Mécaniques (France) • Field Instruction Memorandum • Field Intensity Meter • Field Ion Microscope • Field Ion Microscopy • Finnish Mark (ISO currency code) • Flight Information Manual • Full Individual Membership • Markka (Finland) (ISO currency code)

FIMA: Fault Isolation Maintainability Analysis • Fellow of the Institute of Mathematics and its Applications

FIMATE: Factory-Installed Maintenance Automatic Test Equipment

FIML: Full Information Maximum Likelihood

FIMR: Finnish Institute of Marine Research

FIMS: Facility Information Management System • Far-ultraviolet Imaging Spectrograph • Fast Ion Mass Spectrometer • Fault Isolation and Monitoring System • Fellow of the Institute of Mathematical Statistics • Field Intensity Measuring System • Field Ionization Mass Spectrometry • Field-Ion Mass Spectrometer • Field-Ion Mass Spectrometry • Financial Information Management System • Fine-split Infrared Multispectral Scanner • Forms Interface Management System • FORS Instrumental Mask Simulator • Framework for Integrated Manufacturing System • FUV Imaging Spectrograph

Fin: Final • Finance • Finances • Financial • Financier • Finis • Finish • Finland • Finnish

FIN: Federal Item Name • Finland (international vehicle registration code) • Finland (ISO country code) • Finland (postal country code) (formerly SF) • Finlande (ISO country code) • Finschhafen (Papua New Guinea) (location identifier) • Fish Information Network • Focused Information Network • Future International Network

FINA: Following Items Not Available

Finan: Financial

FIND: Fault Isolation by Nodal Dependence • Federal Item Name Directory

FINDAS: FIRST Integrated Network and Data Archive System

FINDS: Foundation for the International Non-governmental Development of Space

FINE: First In the Nation in Education

FINGIS: Finnish Geographical Information System

FINNARP: Finnish Antarctic Research Program

FINNIDA: Finnish International Development Agency

FINO: Flight Inspection National Office

FINRACS: File of Normalized Radar Cross Sections

Fin.Sec.: Financial Secretary

FIO: File Input/Output • For Information Only • Fortran Input/Output • Fourier Integral Operator • Free In and Out • Frequency In and Out • Furnished and Installed by Others

FIOC: Final Initial Operational Capability • Frame Input/Output Controller

FIOpt: Fiber Integrated Optics (periodical code)

FIOS: Free In and Out and Stowed

FIOT: Free In and Out and Trimmed

FIP: Facility Interface Processor • Fair Information Practice • Fairly Important Person • Far-Infrared Photometer • File Processor Buffering • First Ionization Potential • Forum on International Physics • Free Instrument Package • Fuel Injection Pump • Future Impact Point

FIPA: Foundation for Intelligent and Physical Agents

FIPEX: Flux Probe Experiment

FIPLF: Fédération Internationale de la Presse de Langue Française

FIPP: Fédération Internationale de la Presse Périodique

FIPRAP: Federal Information Processing Resource Acquisition Plan

FIPS: Federal Information Processing Standard (USA) • Federal Information Publication Standard (USA) • Fisheries Image Processing System • Flagstaff Image Processing System • Flight Inspection Positioning System • FOC Image Processing System • Freiburg Image processing System

FIPSE: Fund for the Improvement of Postsecondary Education

FIQ: Fast Interrupt Request

FIR: Facility Installation Review • Facility Interference Review • Far Infrared • Far-Infrared Radiation • Fast Infrared • Fault Isolation Routine • Fern Infrarot • Field Intensity Receiver • Finite Impulse Response • Finite Input Response • Flight Information Region • Flight Information Report • Flight Information Requirement • Flight Inspection Report • Fonds d'Intervention pour les Rapaces (France) • Fuel Indicator Reading • Functional Input Report • Future Issue Requirement

FIRAS: Far-Infrared Absolute Spectrophotometer

FIRB: Far-Infrared Background

FIRBE: Far-Infrared Balloon-borne Experiment

FIRD: Far-Infrared Detector • Fault Isolation Requirement Document

FIRE: Far Infrared Experiment • Fast Imaging Ranicon Experiment • Film Image Recorder • Fire In Resource and Environment • First ISCCP Regional Experiment • Flare Infrared Experiment • Flight Investigation of the Reentry Environment

FIREC: Fédération Internationale des Rédacteurs en Chef

FIREX: Free-flying Imaging Radar Experiment

FIRG: Far-Infrared Galaxy

FIRGA: Far Infrared GaAs Array • Far-Infrared Gallium Arsenide (detector)

FIRLS: Fast Iteratively Reweighted Least Squares

FIRM: Factory Information Resource Management

FIRMPC: Federal Information Resources Management Policy Council

FIRMR: Federal Information Resource Management Regulation

FIRMS: Forest Information Resource Management System • Fourier Ion Resonance Mass Spectrometer

FIRO: Far-Infrared Observation

FIRP: Far-Infrared Photometer • Far-Infrared Pointer

FIRPO: Fast Infrared Photometer for Occultations

FIRPP: Far-Infrared Pointer Package

FIRS: Far-Infrared Spectrometer • Far-Infrared Survey • File Interrogation and Reporting System • Fourier Transform Infrared Sounder • Free-text Information Retrieval System • Future Information Retrieval System

FIRSA: Far-Infrared Stressed Array

FIRSDIMS: Flexible Interactive Remore Sensing Data Information and Management System

FIRSSE: Far Infrared Sky Survey Experiment

FIRST: Faint Images of the Radio-Sky at Twenty-cm • Far-Infrared and Submillimeter Telescope • Far-Infrared Search and Track • Far-Infrared Space Telescope • Fast Interactive Retrieval System Technology • First Image of the Radio Sky Telescope • Flexible Information Retrieval System for Text (USA) • For Inspiration and Recognition of Science and Technology • Forum of Incident Response and Security Team

FIRTECH: Formation des Ingénieurs par la Recherche Technique

Fís: Física

FIS: Fachinformationssystem • Facilities Inventory Study • Facility Interface Sheet • Fahrtgastinformationssystem • Far-Infrared Spectrometer • Far-Infrared Surveyer • Fault Isolation System • Federal Information System • Field Information System • Field Installation Simulator • Filipino Inventors Society • Flight Information Service • Fondation Internationale pour la Science • Forest Information System • Free In Store

FISA: Facility Inspection Shakedown Acceptance • Foreign Intelligence Surveillance Act (USA)

Fisc: Fiscal • Fiscality

FISC: Foundation for International Scientific Coordination

FISCH: Fibre-Schmidt

FISDW: Field-Induced Spin-Density Wave

FISH: First In - Still Here • Fluorescence In Situ Hybridization

Fisiol: Fisiología

FISLIB: Fortran Interactive Subroutine Library

FISS: Fédération Internationale des Sociétés Scientifiques

FIST: Fault Injection Security Tool • Fault Isolation by Semiautomatic Techniques • Final Integrated System Test • First In - Still There • First Integrated System Test • Fully Integrated System Test • Functional Integrated System Test

FISTA: Flying Infrared Signatures Technology Aircraft

FISTNI: Fédération Internationale des Sociétés pour le Traitement Numérique de l'Information

FIT: Fabrication, Integration and Test • Fault Isolation Test • Federal Income Tax • Federal Insurance Tax • Fédération Internationale des Traducteurs • Field Investigation team • First Indication of Trouble • Fixed Interval Timer • Florida Institute of Technology (USA-FL) • Forschungsgesellschaft für Informationstechnik (Germany) • Foundation for International Training • Free and Independent Traveller • Free of Income Tax • Frequent Independent Traveller • Front Intertropical • Functional Integration Test

FITA: Fault Isolation Test Adapter

FITCE: Fédération des Ingénieurs des Télécommunications de la Communauté Européenne

FITH: Fire-In-The-Hole

FITI: Fédération Internationale pour le Traitement de l'Information

FITL: Fiber In The Loop

FITNR: Fixed In The Next Release

FITS: File Image Transfer System • Flexible Image Transport System (*) • Flexible Interchange Transport System • Functional Interpolating Transformational System • Functional Interpolating Transform System

FITSIO: Flexible Image Transport System Input/Output

FITT: Full Immersion Telepresence Testbed

FITW: Federal Income Tax Withholding

FIU: Florida International University (USA-FL)

FIV: Five Finger (USA-AK) (location identifier) • Fuel Isolation Valve

FIUO: For Internal Use Only

FIX: Federal Interagency Exchange • Federal Interconnection Exchange

Fiz: Fizika (periodical code)

FIZ: Fachinformationszentrum • Fitzroy Crossing (Australia-WA) (location identifier)

FizAt: Fiziki Atmosfery (periodical code)

FI3G: Forum International de l'Instrumentation et de l'Information • Géographiques

fj: Fiji (ISO language code)

Fj: Fjord

FJ: Fidji (ISO country code) • Fiji (ISO country code)

FJCC: Fall Joint Computer Conference

Fjd: Fjord

FJD: Dollar (Fiji) (ISO currency code)

FJI: Fidji (ISO country code) • Fiji (international vehicle registration code) • Fiji (ISO country code)

FJR: Al-Fujairah (UAE) (location identifier)

Fk: Fork

FK: Falkirk (UK postal code) • Falkland Islands (ISO country code) • Fundamental Katalog • Îles Malouines (ISO country code) • Malvinas (ISO country code)

FKB: Flight Display Keyboard

FKFO: Fonds voor Kollektief Fundamenteel Onderzoek (Belgium)

FKH: Fakenham Sculthorp (UK-England) (location identifier)

FKI: Kisangani (Zaire) (location identifier)

FKJ: Fukui (Japan) (location identifier)

FKL: Franklin - Chess Lamberton (USA-PA) (location identifier)

FKN: Franklin (USA-VA) (location identifier)

FKO: Franz-Keldysh Oscillation

FKP: Förderverein Kieler Planetarium eV (Germany) • Pound (Falkland Islands) (ISO currency code)

FKQ: Fak Fak (Indonesia) (location identifier)

FKr: Faroese Krone (currency)

FKS: Franz-Kroller-Sternwarte (Austria) • Fukushima (Japan) (location identifier)

FK3: Fundamental Katalog Nr. 3

FK4: Fundamental Katalog Nr. 4

FK5: Fundamental Katalog Nr. 5

fl: Florin (currency) • Floruit • Guilder (currency)

fL: Foot-Lambert

Fl: Flanders • Flashing • Flat • Flemish • Fleuve • Floor • Florijn • Fluid • Flush • Guilder

FL: Fail • Far Left • Feed Line • Flemish • Flight Leg • Flight Level • Floor • Florida (USA) • Flow Line • Flying Logo • Focal Length • Fonction de Luminosité • Foreign Language • Forward Link • Fraction Living • Franc Luxembourgeois (currency) • Full Lift • Full Load •

Function Library • Funzione di Luminosità • Fuzzy Logic • Liechtenstein (international vehicle registration code) • Liechtenstein (postal country code)

Fla: Florida (USA)

FLA: Fiducial Light Assembly • Firn Line Altitude • Florencia (Colombia) (location identifier) • Four-Letter Abbreviation • Four-Letter Acronym • Future Large Aircraft

FLAC: Fast Lagrangian Analysis of Continua • Field Level Access Control

FLACC: Full Level ALGOL Checkout Compiler

FLAD: Fundaçao Luse-Americana para o Desenvolvimento (Portugal)

FLAG: Fiberoptic Link Around the Globe • Finite-Length-scale Anti-Gravity • Flemish Aerospace Group (Belgium) • Flexible Assignment Generator

FLAGE: Flexible Lightweight Agile Guided Experiment

FLAIR: Fibre-Linked Array-Image Reformatter • Fibre-Linked Astronomical-Image Reformatter • Food-Linked Agro-Industrial Research • Functional Language Articulated Interactive Resource

Flam: Flamenco

FLAM: Fault Location And Monitoring

FLAME: Fire Logistics Airborne Mapping Equipment

FLAMES: Fibre Large Area Multi-Element Spectrograph

FLAP: Fatal Light Awareness Program • First Level Adaptive Program

FLAPIR: FLIR and Lidar Atmospheric Propagation in the Infrared

FLAPS: Flight Application Software

FLAPW: Full-potential Augmented Plane Wave

FLAR: Fault Location And Repair • Forward-Looking Airborne Radar

FLASH: Fast Low-Angle Shot • Fiber-Linked Astronomical Spectrograph of Heidelberg

FLAT: Families for Learning Accurate Theories (USA)

flav: Flavus

FLB: Floriano Cangapara (Brazil-PI) (location identifier)

FLC: Falls Creek (Australia-VIC) (location identifier) • Federal Laboratory Consortium • Federal Library Committee • Ferro-electric Liquid Crystal • Forward Load Control • Frequency and Load Controller • Friedmann-Lemaire Cosmology • Fuzzy Logic Controller

FLCA: Forward Load Control Assembly • Forward Load Controller Assembly

FLCB: Frequency and Load Controller Box

FLCD: Ferroelectric LCD

Fld: Field • Fluid

FLD: Fault Logic Diagram • Field • Flux Limited Diffusion • Fond du Lac (USA-WI) (location identifier) • Fraunhofer Line Discriminator • Functional Line Diagram

FLDC: Fixed-Length Don't Care

Fldg: Folding

fl. dr.: Fluidram

FLDS: Fields

Fldt: Floodlight

FlDyR: Fluid Dynamics Research (periodical code)

FlDyT: Fluid Dynamics Transactions (periodical code)

Fle: Flemish

FLE: Petersburg - Fort Lee (USA-VA) (location identifier)

FLEA: Flux Logic Element Array

FLECSE: Flexible Environment for Collaborative Software Engineering

FLEEP: Flying Lunar Excursion Experimental Platform

Flem: Flemish

FLEP: Fédération Laïque d'Éducation Populaire (France)

FLES: Français Langue Étrangère et Seconde

FLETC: Federal Law Enforcement Training Center

FLEVO: Facility for Liquid Experimentation and Verification in Orbit

Flex: Flexible

FLEX: Flexible

FLF: Fixed-Length Field • Flensburg Schaferhaus (Germany) (location identifier) • Foul Language Filter

FLFS: Fast and Large File System

Flg: Flagging • Flashing • Flooring • Flying • Following

FLG: Falling • Flag • Flagstaff (USA-AZ) (location identifier)

FLH: Flotta (UK) (location identifier) • Fluorescence Line Height

FLHO: Federal Lands Highway Office (USA)

FLI: Fault Location Indicator • Flateyri (Iceland) (location identifier) • Fluorescence Line Imager • Free Language Indexing

FLIC: Fault Location Indicating Console • Florida Imaging Center (USA-FL)

FLIDAR: Fluorescence Lidar

FLIER: Fast Low-Ionization Emission Region • Fast Low-Ionization Emitting Region

FLIH: First-Level Interrupt Handler

FLIHTT: Flexible Intermodal Horizontal Transhipment Technique

FliIn: Flight International (periodical code)

FLILO: Flying Infrared for Low-level Operations

FLIM: Forest-Light Interaction Model

FLIMAP: Fast Laser Imaging Mapping And Profiling

FLINN: Fiducial Laboratories for an International Natural-sciences Network • Fiduciary Laboratory for an International Natural-sciences Network

FLIP: Flexible Image Processing • Floating Indexed Point • Floating Instrument Platform • Floating Point Interpretive Program • Flow Interferogram Processing

FLIPS: Future Language Information Processing System

FLIR: Forward-Looking Infrared

FLIRT: Fortran Logical Information Retrieval Technique

FLITE: Free-flyer Laser Interferometer Technology Experiment

FLIZ: F-Layer Irregularity Zone

FLJ: Falls Bay (USA-AK) (location identifier)

FLK: Falkland Islands (ISO country code) • Îles Malouines (ISO country code) • Malvinas (ISO country code)

FLL: Fixed Loss Loop • Fort Lauderdale (USA-FL) (location identifier) • Frequency-Locked Loop • Frequency Lock-Loop • Fresnel-Like Lens • Hollywood (USA-FL) (location identifier)

FLLG: Frustum Lifting Lug Kit

Flli: Fratelli

FLLK: Frustum-Lifting Lug Kit

Flm: Flame

FLM: Federal Land Manager • Filadelfia (Paraguay) (location identifier) • Flight-weighted Laser Module • Fluorescent Light Module • Functional Level Management

Flmb: Flammable

FlMSR: Fluid Mechanics Soviet Rosearch (periodical code)

FLN: Florianópolis - Hercilio Luz (Brazil-SC) (location identifier) • Fuel Line

flo: Floccus

FLO: Fault Location Oscillator • First Lunar Outpost • Florence (USA-SC) (location identifier) • Functional Line Organization

FLOH: Flugzeuglidar für Ozeanographie und Hydrologie

FLOP: Floating-Point Operation

FLOpS: Floating-Point Operations per Second

FLOPS: Floating-Point Operations Per Second (*)

FloorCo: Floor Cooling

flor: Floruit

Flor: Florence (Italy) • Florentine • Florida (USA)

FLOR: Future Lunar Objectives and Requirements

FLORENCE: Flux Océaniques Restitués par bilan Énergétique des Couches limites à moyenne Échelle

FLOS: Fixed Line Of Sight

FLOSIAN: Flow Simulation Analysis

Flot: Flotation

FLOT: Flotation

FLOTUS: First Lady Of The United States

fl. oz.: Fluid Ounce

FLP: Facility Layout Problem • Fault Location Panel • First-Level Processor • Flag Link Protocol • Flap • Flippin (USA-AR) (location identifier) • Fuzzy Logic Pack

FLPS: Flight Load Planning System • Flight Load Preparation System

Flr: Failure • Filler • Floor • Florijn

FLR: Failure • Field Line Resonance • Finite Larmor Radius • Firenze - Amerigo Vespucci (Italy) (location identifier) • Firenze Peretola (Italy) (location identifier) • Flare • Flight Log Recorder • Florence - Amerigo Vespucci (Italy) (location identifier) • Florence Peretola (Italy) (location identifier)

FLRA: Federal Labor Relations Authority (USA)

fl. rt.: Flow Rate

FLRW: Friedmann-Lemaître-Robertson-Walker

FLRY: Flurry

FLS: Falls • Feeder Link Station • Finance Ledger System • First-Look Survey • Flight Surgeon • Flinders Island (Australia-TAS) (location identifier) • Front Line State • Future Launching System • Off-Line System

FLSA: Fair Labor Standards Act

FLSC: Flexible Linear Shaped Charge • Flexible Linear Shaped Cord

Flt: Flashlight • Flight • Float

FLT: Fermat's Last Theorem • First Level Trigger • Flat • Flat (USA-AK) (location identifier) • Flight • Floating

Fltg: Floating

FLTP: Future Launcher Technology Programme

Fltr: Filter

FLTR: Filter

FLTS: Flats

FLTSATCOM: Fleet Satellite Communications (USN)

FLU: Fault Location Unit • New York - Flushing (USA-NY) (location identifier)

Fluc: Fluctuant • Fluctuate • Fluctuated • Fluctuating • Fluctuation

FLUC: Fluctuating

Fluor: Fluorescent • Fluoridation • Fluoride

FLUOR: Fluorescent

FluQ: Fluids Quarterly (periodical code)

Fluid: Fluidika (periodical code)

FLUREX: Fluorescence Experiment

FLV: Fort Leavenworth (USA-KS) (location identifier)

FLW: Fault Location Word • Fellow • Flores Island (Azores) (location identifier) • Follow

FLWO: Fred Lawrence Whipple Observatory (USA-AZ)

Flwp: Followup

FLX: Fallon (USA-NV) (location identifier)

FLY: Finley (Australia-NSW) (location identifier)

fm: Fathom • Femtometre • Fermi • Festmeter • Fiat Mistura

Fm: Farm • Fermium • From • Metallic-Line F-type Star

F&M: Feinwerktechnik und Meßtechnik (periodical code) • Force & Moment

FM: Facility Maintenance • Facilities Management • Facility Module • Factory Management • Fax Modem • Ferritic and Martensitic • Field Magnet • Field Manual • Field Marshall • File Maintenance • Financial Management • Financial Manager • Fine Measurement • Fine Mese • First Mold • Flight Manual • Flight Model • Floor Manager • Flow Meter • Fluid Management • Fluid Module • Focusing Mount • Folding Mirror • Force and Moment • Foreign Mission • Formerly Married • Forms Management • Forward Motion • Foundation Manager • Foundation Member • Fracture Mechanism • Fréquence Modulée • Frequency Meter • Frequency Modulated • Frequency Modulation • Frequency Multiplex • From • Full Match • Full Moon • Functional Manager • Functional Module • Funzione di Massa • Micronesia (ISO country code) • Micronésie (ISO country code)

FMA: Failure Mode Analysis • February-March-April • Final Monitor Aid • Flight Margin Assessment • Formosa - El Pucu (Argentina-FO) (location identifier) • Forward Maintenance Area • Fragment Mass Analyzer • Fuzzy Medial Axis

FMAC: Facility Maintenance And Control • Fargo-Moorhead Astronomical Club (USA-MN)

FMAD: Fluid Management And Management

FMAG: Fluxgate Magnetometer

FMAHTS: Flight Manifest And Hardware Tracking System

Fman: Foreman

FMAPE: Fast Maximum A Posteriori with Entropy

FMAR: Final Mission Analysis Review

FMB: Federal Maritime Board

FMC: Federal Maritime Commission (USA) • Ferrite Memory Core • First Magnitude Corporation (USA-WY) • Five Mile (USA-AK) (location identifier) • Flexible Monte Carlo • Flight Management Computer • Food Machinery Corporation • Formation et Mobilité des Chercheurs • Forward Motion Compensation • Forward Motor Controller • Frequency-Modulated Cyclotron • Fuel Management Computer • Fully Mission Capable • Fully Multi-coated

FMCA: Forward Motor Controller Assembly

FMCF: First Manned Captive Flight

FMCS: Factory Management and Control System • Federal Mediation and Conciliation Service • Flight Management Computer System

FMCT: Fissile Material Cutoff Treaty

FMCW: Frequency Modulation Continuous Wave

Fmd: Formed

FMD: Ferrous Metal Detector • Fields Metadata • Fixed-Module Drive • Forward Multiplicity Detector • Frequency Modulation Discriminator

FMD&C: Flight Mechanics, Dynamics & Control

FMDC: Flight Mechanics, Dynamics, and Control

FMDM: Flexible Multiplexer/Demultiplexer • Frequency Modulation Deviation Meter

FMDS: Flight Model Discharge System

FMDT: Frequency Modulated Differential Transducer

FME: Fort Meade - Tipton (USA-MD) (location identifier) • Frequency-Measuring Equipment

FMEA: Failure Mode and Effects Analysis • Failure Mode Effects Analysis (*)

FMEC: Forward Master Events Controller

FMECA: Failure Mode and Effects Criticality Analysis • Failure Mode Effects and Criticality Analysis

FMES: Failure Modes Effects and Analysis • Full Mission Engineering Simulation

FMF: Facility Master File • Free Molecular Flow • Fuzzy Membership Function

FMFB: Frequency Modulation with Feedback

FMFIA: Federal Manager's Financial Integrity Act

FMFR: Four-Minus-Five Restoral

FMFT: Four-Minus-Five Test

FMG: Ferrovia Monte Generoso (Switzerland) • Flamingo (Costa Rica) (location identifier) • Franc Malgache (currency) • Frequency Modulation Generator

FMGEC: Flight Management Guidance and Envelope Computer • Flight Management Guidance Envelope Computer

FMGIS: Flight Management Geographic Information System

FMGS: Flight Management and Guidance System • Flight Management Guidance System

FMH: Falmouth Otis (USA-MA) (location identifier)

FMI: Failure Mode Indicator • Fermilab Main Injector • Finnish Meteorological Institute • Flow Measurement and Indication • Fluorescent Microthermographic Imaging • Focused Michelson Interferometer • Fonction de Masse Initiale • Fondo Monetario Internacional • Fonds Monétaire International • Forest Management Institute • Free Motion Impedance • Funzione di Massa Iniziale • Kalemie (Zaire) (location identifier)

FMIC: Frequency Modulation and Interference Control

FMICW: Frequency-Modulated Interrupted Continuous Wave

F-MIDR: Full-resolution Mosaicked Image Data Record

FMIDR: Full-resolution Mosaicked Image Data Record

FMIM: Foreign Mass Injection Method

FMIP: Financial Management Improvement Program

Fmk: Finnish Mark (currency)

FMk: Finnish Mark (currency)

FMK: Fast Moving Knot • Field Modification Kit

FMKs: Fast Moving Knots

Fml: Formal

FML: Fault Message Line • File Manipulation Language • Final Materials List • Flamsteed

Fmla: Formula

FMLIS: Federal Mineral Lands Information System (USA) • Fuzzy Multispectral Landcover Information System

FMM: Fast Multipole Method • Financial Management Manual

FMMC: Fixed Mylar Metallized Capacitor

Fmn: Formation

FMN: Farmington (USA-NM) (location identifier)

FMNBNA: Frequency Modulation and Narrowband Noise Analyzer

FMO: Fast Moving Object • Financial Management Office • Flatland Meteorological Observatory • Flight Medical Officer • Flux de Matière dans l'Océan • Frequency Management Office • Münster (Germany) (location identifier)

FMOF: First Manned Operational Flight • First Manned Orbital Flight

FMOFEVA: First Manned Orbital Flight with Extravehicular Activity

FMOFPL: First Manned Orbital Flight with Payload

FMOI: Fédération Mondiale des Organisations d'Ingénieurs

FMOS: Fiber Multiobject Spectrograph

FMP: Facility Management Plan • Finance and Management Policy • Financial Management Plan • Functional Maintenance Procedure

FMPAC: Fixed Multipurpose Application Console

FMPM: Family-Manned Planetary Mission

FMPS: Fast Multidimensional Processing System • Functional Mathematical Programming System

FMPT: First Materials Processing Test

FMQ: Frequency-Modulated Quartz

Fmr: Former

FMR: Failure and Malfunction Report • Fair Market Rent • False Match Rate • Fast Magnetic Rotator • Ferromagnetic Resonance • Field Modification Request • Frequency-Modulated Receiver • Function Maximum Rate

FMRB: Forschungs- und Meßreaktor Braunschweig (Germany)

FMRC: Factory Mutual Research Corporation

FMRD: Flight Mission Rules Document

FMRI: Functional Magnetic Resonance Imaging

FMRL: Functional Machine Representation Language

Fmrly: Formerly

FMS: Facilities Management System • False Memory Syndrome • Fast Matrix Solver • Fault Management System • Fecal Management System • File Maintenance System • File Management System • Filter Management System • Final Monthly Schedule • Final Multiple Score • Financial Management Service • Financial Management System • Flexible Manufacturing System • Flexible Measuring System • Flight Management System (*) • Flight Managing System • Flow Measuring System • Fluid Management System • Flux Monitoring System • Food Management System • Foreign Military Sale • Forms Management System • Fort Madison (USA-IA) (location identifier) • Frequency-Mixer Stage • Frequency Management System • Frequency Multiplier Storer • Functional Master Schedule

FMSF: False Memory Syndrome Foundation

FMSP: Frequency Modulation Signal Processor

FMST: Frequency Mass Spectrometer Tube

FMT: Flight Management Team • Format • Frequency-Modulated Transmitter • Friction Measurement Test • Full Machine Translation • Full Mission Trainer

FMTA: Flash Mass Thermal Analysis

FMTP: File Management Transaction Processor • First Materials Processing Test

FMU: File Migration Utility • Financial Management Unit • Flight Monitor Unit • Force Measurement Unit • Francis Marion University (USA-SC) • Functional Mock-Up

FMV: Full Motion Video

Fmw: Firmware

FMX: Frequency Modulation Transmitter

FMY: Fort Myers (USA-FL) (location identifier)

FN: False Negative • Fernando de Noronha (Brazil) • Financial Network • First Name • Flux Number • Footnote • Forêt Noire

FNA: Fast Neutral-particle Analyzer • For Necessary Action • Freetown Lungi (Sierra Leone) (location identifier) • Frequency Network Analyzer

FNAC: Future Needs in Astronomical Computing

FNAL: Fermi National Accelerator Laboratory (USA-IL)

FNAP: Fortran Network Analysis Program

FNATH: Fédération Nationale des Accidentés du Travail et des Handicapés

FNAUT: Fédération Nationale des Associations d'Usagers de Transports (France)

FNB: Food and Nutrition Board

FNBR: Fast Neutron Breeder Reactor

FNC: Fast Neutron Cavity • Federal Networking Council • Fiber Nickel Cadmium • Forward Neutron Calorimeter • Funchal (Madeira Island) (location identifier)

FNCAC: Federal Networking Council Advisory Committee (USA)

Fnct: Function

Fnd: Found

FND: Facility Need Date • Fast Neutron Dose

FNDAE: Fonds National pour le Développement des Adductions d'Eau (France)

Fndd: Founded

Fndr: Founder

FNDVA: Fonds National du Développement de la Vie Associative (France)

Fne: Fine

FNE: Fane (Papua New Guinea) (location identifier) • Fonds National de l'Emploi (France)

FNEP: Fondation Nationale Entreprise et Performance (France)

FNET: French UUCP Network

FNF: First Normal Form

FNG: Fada Ngourma (Burkina Faso) (location identifier)

FNI: Nîmes Arles Camargue (France) (location identifier)

FNJ: Pyongyang (Korea-DPRK) (location identifier)

FNK: Fin Creek (USA-AK) (location identifier)

FNL: Final • Fort Collins - Loveland (USA-CO) (location identifier)

FNLN: Fine Line

FNM: Fast Multipole Method • Financial Network Manager

FNMA: Federal National Mortgage Association

FNMT: Fábrica Nacional de Moneda y Timbre (Spain)

FNNPE: Federation of Nature and National Parks of Europe

FNOC: First Numerical Oceanography Center • Fleet Numerical Oceanographic Center (USN) (*) • Fleet Numeric Oceanography Center

FNOES: Fondation Nationale pour l'Organisation d'Expéditions Scientifiques (Belgium)

FNPA: Foreign Numbering Plan Area

FNPO: For NASA Personnel Only

FNR: False Nonmatch Rate • Funter Bay (USA-AK) (location identifier)

FNRI: Food and Nutrition Research Institute

FNRS: Fonds National de la Recherche Scientifique (Belgium)

FNS: Failure Notification Sheet • Fonds National Suisse (Switzerland) • Food and Nutrition Service • Fortified Navier-Stokes

Fnsh: Finish

FNSU: Fédération Nationale des Sports Universitaires (France)

Fnt: Front

FNT: Failure Notification Telex • Fermat Number Transform • Flint (USA-MI) (location identifier) • Font • Front

FNTGNS: Frontogenesis

FNTLYS: Frontolysis

FNV: Federatie Nederlandse Vakverenigingen (Netherlands)

FNWC: Fleet Numerical Weather Center (USN)

f/o: Full Out

fo: Faeroese (ISO language code) • Folio

Fo: Fourier Number

F/O: Fuel/Oxidizer

FO: Fail Operating • Fail Operation • Fail Operational • Falsified Observation • Faroe Islands (ISO country code) • Fast Operate • Fast Operating • Federal Official • Fellowship Office • Fiber Optics • Field Office • Field Officer • Filter Output • Firm Offer • Firm Order • First Officer • First Overtone • Flat Oval • Flight Operation • Foldout • Foreign Object • Foreign Office • Formatted Object • Formatting Object • Formosa (Argentina) • For Orders • Forward Observer • Functional Objective • Furnace Oil • Îles Faroë (ISO country code)

FOA: Figl-Observatorium für Astrophysik (Austria) • Foula (Shetland Islands) (location identifier) • Future Offensive Aircraft

FOAC: Federal Office Automation Center

FOAF: Friend Of A Friend

FOAS: Future Offensive Air System

FOB: Fiber Optics Board • Flight Operations Building • Fort Bragg (USA-CA) (location identifier) • Fractional Orbit Bombardment • Free On Board • Friend Of Bill • Fresh Off the Boat

FOBE: First-Overtone Blue Edge

FOBS: Fractional Orbit Bombing System

Foc: Focal • Focus

FOC: Faint Object Camera • Fiber Optic Communication • Fleet Operational Capability • Flight Operating Costs • Flight Operations Center • Focus • Follow-On Contract • Foreign and Overseas Commission (UK) • Form Of Occurrence • Free Of Charge • Full Operational Capability (*) • Full Operations Capability • Fuzhou (China-PRC) (location identifier)

FOCA: Focal Objective Corrective Astigmatism

FOCAL: Facility for Optical Calibration At Liège (Belgium) • Forum Of Continuing Astronomical Liaison

FOCAM: Faint Object Camera

FOCAP: Fibre-Optic Coupled Aperture Plate

FOCAS: Faint Object Camera And Spectrograph • Faint Object Classification and Analysis System • Fiber Optic Communications for Aerospace Systems

FOCES: Fibre-Optic-coupled Cassegrain Echelle Spectrograph

FOCI: Fisheries Oceanography Coordinated Investigations

FOCIS: Forest Classification and Inventory System

FOCL: Fibre-Optic Communications Line

FOCpS: Force Calculations per Second

FOCPS: Force Calculations Per Second

FOCR: Final Operational Concept Review

FOCS: Fiber Optic Chemical Sensor • Fiber-Optic Current Sensor • Force Calculations per Second • Foundations Of Computer Science (symposium)

FOCSI: Fiber-Optic Control System Integration

FOCUS: Fast Orbit Computation Utility Software • Federation on Computing in the United States • Forum Of Control Data Users • Fotoproduction of Charm: Upgraded Spectrometer

FOD: Field Operations Division • Finger Of Death • First Order Design • Flight Operations Directorate • Flight Operations Division • Foreign-Object Damage • Fort Dodge (USA-IA) (location identifier) • Free Of Damage

FODS: Faint Object Detection System • Fiber-Optic Demonstration System

FOE: Fiber Optic Echelle • Flight Operations Engineer • Follow-On Evaluation • Friends Of the Earth • Topeka Forbes (USA-KS) (location identifier)

FOEB: Flight Operations Evaluation Board

FOF: Field Observing Facility • First Orbital Flight • First Operational Flight • Flight Operations Facility • Friends Of Friends

FOFCH: First Order Flux Conservative Hyperbolic

FOFE: Fiber Optic Front End

FOFP: Fiber-Optic Face Plate

FOG: Fiber Optic Gyroscope • First Osborne Group • Foggia (Italy) (location identifier) • Forecast Generator • Fourth Order Gravity • Frequency Offset Generator • Fiber Optics Guidance

FOGA: Foundations of Genetic Algorithms

FOGN: First-Order Gravity Net

FOGT: First-Order Gradient Technique

FOH: First-Order hold

FOI: Field Operations Installation • Freedom Of Information

FOIA: Freedom Of Information Act (USA)

FOID: Fibre Optic Image Dissector

FOIH: Flight Operations Integration Handbook

FOIL: File-Oriented Interpretive Language • First-Outer-Inner-Last • Freedom Of Information Legislation

FOIM: Field Office Information Management

FOIRL: Fiber-Optic Interrepeater Link

FOK: Westhampton - Suffolk County (USA-NY) (location identifier)

fol: Folio • Folium

Fol: Folio • Folium • Follow • Followed • Following

FOL: Fiber Optics Laser • Fiber Optics Light • Friends Of LISA

FoLA: Four-Letter Acronym

FOLACL: Federation of Local Authority Chief Librarians

FOLD: Federally-Owned Landsat Data (USA)

Folg: Following

FOLG: Fiber Optics Laser Gyro

Foll: Follow • Followed • Following

fols: Folios

FOM: Fault Of Management • Federation Object Model • Field Offset Mirror • Field Operations Manual • Field Operations Memorandum • Figure Of Merit • Foumban (Cameroon) (location identifier) • Stichting voor Fundamenteel Onderzoek der Materie (Netherlands)

FOMM: Functionally-Oriented Maintenance Manual

FOMOS: Fiber-Optics Multiobject Spectrograph

FOMR: Flight Operations Management Room

FOMTE: Forme, Orientation et Mouvement de la Terre dans l'Espace

Fon: Fonética • Fonología

FON: Fiber Optic Network • Font • Phone

Fond: Fondamenta • Fondamental • Fondation • Fondazione

FONEMA: Fast Omnidirectional Non-scanning Energy Mass Analyser

FONSI: Finding Of No Significant Impact

FONTX: Forbid One-Node Tree Crossover

FOO: Numfoor (Indonesia) (location identifier)

FOOS: Function-Oriented Organizational Structure

FOOT: Follow-On Operational Test

FOP: Facility Operating Procedure • Facility Operations Plan • Fiber Optics Probe • Financial Operating Plan • Flight Operations Panel • Flight Operations Plan • Follow-On Production • Forest Park - Morris (USA-GA) (location identifier)

FOPC: First-Order Predicate Calculation

FOPG: Flight Operations Planning Group

FoPh: Found. Phys. (periodical code)

FOPI: First-Order Polynomial Interpolator

FOPP: First-Order Polynomial Predictor • Follow-On Parts Production

FOPPR: Flight Operations Plans, Procedures and Rules

FOPS: Flight Operations Planning Schedule

FOPSS: Flight Operation Product Scheduling System

FOQA: Flight Operations Quality Assurance

For: Foreign • Foreigner • Fornax-Fornacis (constellation)

FOR: Field Of Regard • File Owning Region • Flight Operations Review • Force • Fortaleza - Pinto Martins (Brazil-CE) (location identifier) • Fortran • Free On Rails

Forc: Forcing

FORD: Flight Operations Requirement Document

Foregd: Foreground

FOREGS: Forum of European Geological Surveys

Form: Formation • Formula • Formulate • Formulated • Formulation • Formule • Formuler

Formn: Foreman • Formation

FORNN: Forenoon

FORS: Faint Object Red Spectrograph • Fiber-Optic Rate Sensor • Focal Reducer/Spectrograph

Forsch: Forschung

Fort: Fortification • Fortified

FORT: Foyer Optimisé pour le Radiotélescope

FORTE: Fast On-orbit Recording and Transient Events • Fast On-orbit Recording of Transient Events (*)

FORTH: Foundation for Research and Technology - Hellas (Greece)

FORTLP: Fortran Linear Programming

FORTRAN: Formula Translator

FORTRESS: Framework for Operating, Recommending and Testing a Range of European Security Services

Forts: Fortsetzung

FORUM: Formula for Optimizing through Real-Time Utilization of Multiprogramming

FOS: Faint Object Spectrograph • Field Of Science • File Organization System • Flight Operations Security • Flight Operations Segment • Flight Operations Support • Forrest (Australia-WA) (location identifier) • Fortran Operating System • Forward-looking Optronic System • Free On Station • Free On Steamer • Full Operational Status

FOSA: Flight Operations Support Annex

FOSD: Functional Operational Sequence Diagram

FOSDIC: Film Optical Sensing Device for Input to Computers

FOSE: Federal Office Systems Exposition

FOSF: Field Observing Support Facility

FOSI: Formatting Output Specification Instance • Formatted Output Specification Instance • Frontiers Of Space Imaging

FOSMS: Facility Operations Scheduling and Monitoring System

FOSO: Flight Operations Scheduling Office

FOSOL: Florian's Own Statistically Oriented Language

FOSP: Flight Operations Support Personnel

FOSR: Flexible Optical Solar Reflector

FOSS: Fore Optic Support Structure • Full Option Science System • Future of Space Sciences

FOSSIL: Fide/Opus/Seadog Standard Interface Layer

FOST: Flight Operations Support Team

FOSTER: Flight Opportunities for Science Teacher Enrichment • Flight Opportunities for Science Teacher Environment

FOT: Faint-Object Telescope • Fifth-Order Theory • Final Observation Tape • Fleet Ocean Tug • Flight Operations Team • Flight Operations Trainer • Forster (Australia-NSW) (location identifier) • Fourier Optical Transform • Fraction Of Time • Free Of Tax • Free On Truck • Frequency of Optimal Traffic • Fundamental Overtone

FOTD: Fiber-Optic Towed Decoy

FotO: Friends of the Observatory

FOTRA: Free Of Tax to Residents Abroad

FOTS: Fiber Optic Transmission System

FOU: Fougamou (Gabon) (location identifier)

FOUDRE: Fourniture de Documents sur Réseau Électronique

Found: Foundation • Founded

FOUO: For Official Use Only

FOUPA: Fiber-Optic Universal Port Adapter

FOURBI: Fast Object/User Retrieval and Bibliographic Index • Follow-Up Relational Base for Images

FoV: Field of View

FOV: Face Of Venus • Field Of View • Field Of Vision • First Orbital Vehicle

FOVIA: Fotómetro Visible Adaptado

FOWG: Flight Operations Working Group

FOX: Field Operational X.500 • Fox (USA-AK) (location identifier)

FOY: Foya (Liberia) (location identifier)

fp: Forte-Piano

Fp: Finite Part • Frontispiece

F-P: Fabry-Pérot • Finson-Probstein

F&P: Fluids & Propulsion

FP: Fabry-Pérot • Faceplate • False Positive • Fast Processor • Federal Parliament • File Processor • Film Pack • Filter Paper • Final Point • Financial Plan • Fine Paper • Fine Pointer • Fine Pointing • Fire Plug • First Performance • First Performed • Fission Product • Fixed Point • Fixed Price • Flash Point • Flat Panel • Flexible Processor • Flight Plan • Floating Point • Fluids and Propulsion • Fluorescent Particle • Focal Plane • Focal Point • Fokker-Planck (approximation) • Fonction Publique • Foolscap • Footpath • Foot-Pound • Forest Park • Formación Profesional • Formation Permanente • Formation Professionnelle • Forte-Piano • Forward Pass • Framework Programme • Freezing Point • Fresh Paragraph • Frontispiece • Front Page • Fuel Pressure • Full Period • Full Point • Full Power • Full Propagation • Fully Paid • Functional Path • Functional Programming • Function Path • Function Point • Fundamental Plane • Fusing Point

FPA: Failure Probability Analysis • Final Power Amplifier • First Point of Aries • Flat-Plate Antenna • Flat-Plate Array • Flexible Production Automation • Flight Path Analysis • Flight Path Angle • Floating-Point Accelerator • Floating-Point Arithmetic • Fluid Processing Apparatus • Focal-Plane Array • Focal-Plane Assembly • Foreign Press Association • Formation Professionnelle des Adultes • Fractal Pixon Axis • Free of Particular Average

FPAA: Field Programmable Analog Array • Flat-Plate Array Antenna • Flight Path Analysis Area

FPAC: Flight Path Analysis and Command

FPAG: Fundamental Physics Advisory Group (ESA)

FPAS: Focal Plane Assembly Shroud

FPAU: Fringe Phase Alignment Unit

FPB: Fractal Pixon Basis • Fuel Preburner

FPBOV: Fuel Preburner and Oxidizer Valve

F&PC: Finance & Program Control

FPC: Facility Power Control • Federal Power Commission (USA) • Federal Publishers Committee (USA) • Final Processing Center • Finance and Program Control • Fixed-Point Calculation • Fixed-Precision Capacitor • Fixed-Price Call • Fixed-Price Contract • Fixed-Price Contractor • Fixed-Program Computer • Flexible Printed Circuit • Flight Path Control • Flight Processor Card • Floating-Point Calculation • Flow Proportional Counter • Fluids Pressure Control • Formation Professionnelle Continue • For Private Circulation • Forward Power Controller • Front Panel Control

FPCA: Forward Power Control Assembly • Forward Power Controller Assembly

FPCC: Fixed Polycarbonate Capacitor • Fixed Printed Circuit Capacitor

FPCE: Floating-Point C Extension

FPCS: Fast Profiling Current Source • Focal Plane Crystal Spectrometer • Freezing-Point Calibration Standard • Full-Page Composition System

FPCT: FITS Product Conformance Tester

FPD: Flame Photometric Detector • Flat-Panel Display • Flight Projects Directorate • Full-Page Display

FPDA: Focal Plane Deck Assembly

FPDE: Focal Plane Drive Electronics

FPDIS: Fabry-Pérot Doppler Imaging System

FPE: Factory Process Engine • Final Prediction Error • Floating-Point Engine • Fonds de Propriété d'État (France) • French Postural Experiment • Functional Program Element

FPEG: Fast Pulse Electron Gun • Fundamental Plane of Elliptical Galaxies

FPF: Fine Pointing Facility • Fluid Physics Facility • Full Power Frequency

FPG: Förderkreis Planetarium Göttingen eV (Germany)

FPGA: Field Programmable Gate Array

FPH: Flight Procedures Handbook • Fondation pour le Progrès de l'Homme

FPHA: Federal Public Housing Authority

FPHB: Flight Procedures Handbook

FPI: Fabry-Pérot Interferometer • Fabry-Pérot Interferometry • First Periodic Inspection • Fixed-Price Incentive • Floating-Point Interface • Fluorescent Penetrant Inspection • Focussed Process Improvement • Formación de Personal Investigador • Formation Professionnelle Initiale • Fuel Pressure Indicator

FPIF: Fixed Price Incentive Fee

FPIII: Third Framework Programme

FPIP: Focal Plane Instrument Package

FPIS: Forward Propagation by Ionospheric Scatter

FPIT: Field Preinstallation Test

FPITS: Fabry-Pérot Inverse Transform Spectrometer

FPIV: Fourth Framework Programme

FPK: Fixed-Position Keyboard

Fpl: Fireplace

FPL: Federal Public Library (USA) • Ferry-Porter Law • Fetch Pointer List • Final Parts List • Flight Plan • Flight Propulsion Laboratory • Frequency Phase Lock • Full-Performance Level • Full Power Level • Full Power Load • Functional Programming Language • Fuzzy Programming Language

FPLA: Field-Programmable Logic Array

FPLC: Franklin Pierce Law Center

FPLMTS: Future Public Land Mobile Telecommunication System

FPLP: Fortran List-Processing Language

fpm: Feet per Minute

FPM: Facility Power Monitor • Faculté Polytechnique de Mons (Belgium) • Fast-Page Mode • Federal Personnel Manual • Feet Per Minute • File Protect Memory • Finite Population Mean • First Polar Mission • Flashes Per Minute • Fluid Physics Module • Flux Particle Monitor • Folding Platform Mechanism

FPMIS: Federal Personnel Management Information System (USA)

FPMs: Faculté Polytechnique de Mons (Belgium)

FPN: Fixed Pattern Noise

FPNNC: Fédération des Parcs Naturels et Nationaux d'Europe

FPO: Field Post Office • Fleet Post Office • For-Profit Organization • Freeport (Bahamas) (location identifier) • Future Projects Office

FPOE: First Port Of Entry

FPOV: Fuel Preburner Oxidizer Valve

FPP: Facility Power Panel • Family Planning Programme • Fantasy-Prone Personality • Fixed Path Protocol • Fixed Point Program • Floating-Point Processor • Focal Plane Package • Focal Plane Polarimeter • Freon Pumped Package • Freon Pump Package

FPPA: Field Programmable Processor Array

FPPO: Fox Park Public Observatory (USA-MI)

FPPR: Fixed Price with Price Revision

FPPS: Full Page Phototypesetting System

FPPU: Floating-Point Processor Unit

FpR: Feet per Revolution

FPR: Failure/Problem Report • Fan Pressure Ratio • Federal Procurement Regulation • Feet Per Revolution • Final Progress Report • Fixed-Point Representation • Flat-Plate Radiometer • Flight Performance Reserve • Floating-Point Register • Floating-Point Routine • Forest-Penetrating Radar • Fort Pierce (USA-FL) (location identifier) • Full Power Response • Full Propellant Requirement

Fprf: Fireproof

FPRS: Federal Property Resources Service

FPRV: Flight Product Requirements Verification

fps: Feet per Second • Foot-Pound-Second • Frames per Second

FPS: Fabry-Pérot Spectrometer • Favorite Picture Selection • Financial Planning System • Fine Pointing Simulation • Fine Pointing Simulator • Flight Per Second • Floating-Point System • Focal Plane Structure • Forum on Physics and Society • Forward Power Supply • Full-Pressure Suit

FPSA: Focal Plane Structure Assembly

FPSNW: File and Print Server for NetWare

FPSO: Flight Projects Support Office

FPSP: Focal Plane Signal Processor

FPSR: Flight Planning and Storage Review

fpsps: Feet per Second per Second

FPSS: Fine-Pointing Sun Sensor • Focal-Plane Support Structure

FPSTU: Full-Pressure Suit Training Unit

FPS2: Feet Per Second Squared

FPT: Female Pipe Thread • Finite Population Total • Fixed Price Tenders • Forced Perfect Termination • Full Production Training • Functional Performance Test

FPTE: Facility Portable Test Equipment

FPTS: Forward Propagation by Tropospheric Scatter

FPU: First Production Unit • Floating-Point Unit • Focal Plane Unit • Fuel Purification Unit

FPV: Fifth Framework Programme • Floating-Point Validating • Flow Proportioning Valve

FPWG: FSD Planning Working Group

FPY: Perry Foley (USA-FL) (location identifier)

FPYD: Fission Product Yield Data

FQ: Facteur de Qualité • Fiscal Quarter • Flight Qualification • Formal Qualification • Fused Quartz

FQAS: Flexible Query Answering System

FQDN: Fully Qualified Domain Name

FQF: First Qualification Flight

FQHE: Fractional Quantum Hall Effect

FQHN: Fully Qualified Host name

FQI: Federal Quality Institute • Flight Qualification Instrumentation • Fuel Quantity Indicator

FQIL: Fuel Quartz Incandescent Lamp

FQIS: Fuel Quantity Indicating System

FQL: Functional Query Language

FQLFE: Functional Query Language Front End

FQM: Four-Quadrant Multiplier

FQP: Fundamental Questions Program

FQPA: Flight Quality Photomultiplier Assembly

FQR: Flight Qualification Recorder (*) • Flight Qualification Review (*) • Formal Qualification Review

FQT: Formal Qualification Test • Frequent

FQTR: Flight Qualification Tape Recorder

FQTV: Frequent Traveller

fr: Folio Recto • Franc (currency) • Frei • French (ISO language code)

Fr: Father • Fragment • Frame • Franc (currency) • Français • France • Francium • Franco • Frase • Fratelli • Frater • Frau • Free • Freitag • Freitags • French • Frequency • Frequent • Frère • Friar • Friday • From • Front • Froude Number

F/R: Flared Rudder • Folio Reference

FR: Failure Rate • Failure Report • Falling Rapidly • Fanaroff-Riley • Faraday Rotation • Faroe Islands (international vehicle registration code) • Faroe Islands (postal country code) • Far Right • Fast Release • Federal Register • Federal Republic • Field Reversing • File Received • File Register • Final Release • Final Report • Final Review • Firing Room • Fixed Ratio • Flat Rate • Fletcher-Reeves • Flight Readiness • Flight Reliability • Flight Rule • Focal Reducer • Force-Reflective • Foreign Reimbursement • Frame • Frame Rate • Frame Relay • France (ISO country code) • Frei • Frequency Range • Frequency Rate • Frequency Response • Froude Number • Full Range • Full Rate • Fully Reusable • Functional Requirement • Funding Request • Fusion Reactor

fra: Factura • Fractus

FRA: Federal Railroad Administration (USA) • Federal Records Act • Formation de Recherche Associée • Forward Rate Agreement • France (ISO country code) • Francfort (Germany) (location identifier) • Frankfurt (Germany) (location identifier) • Full Range Analyzer

Frac: Fractional

FRAC: Flint River Astronomy Club • Fonds Régional d'Art Contemporain • Fracção • Fractionator Reflux Analog Computer

FRACAS: Failure Reporting Analysis and Corrective Action System • Filter Response Analysis for Continuously Accelerating Spacecraft

FRACTAL: Fractional Dimensional

FRAD: Frame Relay Access Device

Frag: Fragile • Fragment • Fragmentation • Fragmento

FRAG: Fragile • Fragment • Fragmentation

FRAGNET: Fragmented Network

FRAM: Ferroelectric Random Access Memory • Ferroelectronic Random Access Memory • Fine-Resolution Antarctic Model

FRANEC: Frascati Raphson Newton Evolutionary Code

Fränk: Fränkische

FRAS: Fellow of the Royal Astronomical Society

FRASC: Front Range Astronomical Super-Cluster (USA-CO)

FRATT: Fonds Régional d'Aide au Transfert de Technologie

Fraud: Fraudulent

FRB: Failure Review Board • Fast Rise Balloon • Federal reserve Bank • Federal reserve Board • Forbes (Australia-NSW) (location identifier)

Frbd: Freeboard

FRC: Facility Review Committee • Federal Radio Commission (USA) • Federal Records Center (USA) • Fiber-Reinforced Composite • Field-Reversed Configuration • Fixed Radio Control • Flight Readiness Certification • Flight Research Center • Focal Reducing Camera • França (Brazil-SP) (location identifier) • Fresnel Reflection Coefficient • Front Range Consortium (USA-CO) • Full Route Clearance • Functional Redundancy Check • Functional Residual Capacity • Fusion Research Center

FRCB: Flight Rules Change Board • Flight Rules Control Board

FRCD: Floating-Rate Certificate of Deposit

FRCI: Fibrous Refractory Composite Insulation

FRCN: Fundamental Reference and Calibration Network

FRCP: Facility Remote Control Panel • Fellow of the Royal College of Physicians

FRCS: Fellow of the Royal College of Surgeons • Forward Reaction Control System

Frd: Friend

FRD: Facilities Requirements Document • Failure Rate Data • Flight Requirements Document • Focal Ratio Degradation • Ford • Foundation for Research Development (South Africa) • Friday Harbor (USA-WA) (location identifier) • Functional Requirements Document

FRDB: Failure Rate Database

FRDI: Flight Research and Development Instrumentation

FRDM: Finite Range Droplet Model

FRDR: Flight Readiness Design Review

FRDS: Federal Reporting Data System • Fords

Fre: Fracture • Freitag • French

FRE: Fera Island (Solomon Islands) (location identifier) • Flight-Related Element • Format Request Element • Fundamental Red Edge

FRECOPA: French Cooperative Package • French Cooperative Payload

FRed: Focal Reducer

FRED: Figure Reading Electronic Device • Frame Editor • Front-End to Disk

Free: Freeway

Freew: Freeway

freiw: Freiwillige

Fréq: Fréquence

Freq: Frequency • Frequent • Frequentative • Frequently • Frequenz (periodical code)

FREQ: Frequency

FRESA: Frame Relay End-System Address

FRESCO: Full Resolution Coadder • Full Resolution Survey Coaddition

FRESIP: Frequency of Earth-Sized Inner Planets

FRESS: File Retrieval and Editing System

FREX: Fonds Régional à l'Exportation • Fridays Excepted

FRF: First Reduction File • Flight Readiness Firing • Frame Relay Forum • Franc (France) (ISO currency code) • Frankfurt Rhein-Main (Germany) (location identifier) • Frequency Response Function • Frequency Reuse Factor

FRFC: Fonds de la Recherche Fondamentale Collective (Belgium)

FRFCIM: Fonds de la Recherche Fondamentale Collective d'Initiative Ministérielle (Belgium)

FRFI: Flug Revue Flugwelt International (periodical code)

FRFT: File Ready For Transfer • Flight Readiness Firing Test • Fractional Fourier Transform

FRG: Farmingdale (USA-NY) (location identifier) • Federal Republic of Germany • Floated-Rate Gyro • Forge

FRGS: Fellow of the Royal Geographical Society • Forges

Frgt: Freight

FRGU: Frequency Reference Generator Unit

FRH: Fly Runway Heading • French Lick (USA-IN) (location identifier)

Fri: Fribourg (Switzerland) • Friday

FRI: Faranoff-Riley class I • Forest Research Institute • Forest Resource Inventory • Fort Riley (USA-KS) (location identifier) • Friday

FRIA: Fellow of the Royal Irish Academy

Fric: Friction • Frictional

FRICC: Federal Research Internet Coordinating Committee

Frict: Friction • Frictional

Friesl: Friesland

FRII: Faranoff-Riley class II

FRINGE: File and Report Information Processing Generator

FrInJ: Franklin Institute Journal (periodical code)

FRIS: Forest Resource Information System

Frisco: San Francisco (USA-CA)

FRISPI: Fast Recording Imager and Spectrophotometric Instrument

FRJ: Fréjus Saint-Raphaël (France) (location identifier)

FRJD: Forward Reaction Jet Driver

Frk: Frøken • Fröken

FRK: Fork • Fregate Island (Seychelles) (location identifier)

FRKS: Forks

Frl: Fräulein

FRL: Forli (Italy) (location identifier) • Frame Reference Line • Frame Representation Language • Fuselage Reference Line

Frld: Foreland • Freehold

Frm: Frame • From

FRM: Facility Requirements Manager • Fairmont (USA-MN) (location identifier) • Farm • Fast Resource Management • Fast Rotating Model • Fiber-Reinforced Material • Film-Reading Machine • Foucault Rotating Mirror • Functional Reference Model

FRMG: Forming

FRMT: Format

FRN: Floating-Rate Note • Forskningsrådsnämmden • Fort Richardson (USA-AK) (location identifier)

Frnc: Furnace

Frng: Fringe

FRNT: Front

FRO: Faroe Islands (ISO country code) • Fire Risk Only • Fleurs Radioastronomical Observatory (Australia-NSW) • Floro (Norway) (location identifier) • Îles Faroë (ISO country code)

FROG: Fédération Orcéenne en Géosciences (France) • Free-flight Rocket Over Ground • Frequency Resolved Optical Gating

Front: Frontispiece

FRONT: Fronteira • Frontière

FRONTIERS: Forecasting Rain Optimized using New Techniques of Interactively Enhanced Radar and Satellite

Frontis: Frontispiece

FROST: First Regional Observing Study of the Troposphere

FRP: Fast Reservation Protocol • Feature Recognition Processor • Federal Radionavigation Plan (USA) • Fiber-Reinforced Plastic • Filament-Reinforced Plastic • Flag Register Processing • Fresh Water Bay (USA-AK) (location identifier) • Fuel-Reprocessing Plant • Fuselage Reference Plane

Frpf: Fireproof

FRPG: Fantasy Role Playing Game

FRpI: Flux Reversals per Inch

FRPI: Flux Reversals Per Inch

Frpl: Fireplace

FRQ: Feramin (Papua New Guinea) (location identifier) • Frequent

FRR: Facility Readiness Review • Failure and Rejection Report • Failure and Replacement Report • Failure Reporting Review • Fast Repetition Rate • Federal Reuse Repository • Federatie van Russische Republieken • Fédération des Républiques Russes • Federation of the Russian Republics • Flight Readiness Review • Front Royal - Warren County (USA-VA) (location identifier) • Full Reimbursement Rate • Functional Recovery Routine

FRRB: Fast Rise Reflective Balloon

FRRID: Flight Readiness Review Item Description • Flight Readiness Review Item Disposition

FRRPA: Forest and Rangeland Renewable Planning Act (USA)

Frs: Francs (currency)

FRS: Faint Reference Star • Fast Retrieval Storage • Federal Reserve System • Fellow of the Royal Society • Figure Rating Scale • Filtered Rayleigh Scattering • First Readiness State • Flexible Route Selection • Flores (Guatemala) (location identifier) • Functional Requirement Specification • Function Repository Server

FRSAS: Fast-Response Solar-Array Simulator

FRSI: Felt Reusable Surface Insulation • Flexible Reusable Surface Insulation

FRSIS: Fuzzy Relational Soil Information System

FRSL: Fellow of the Royal Society of Literature

FRSQ: Flat Radio Spectrum Quasar

FRSS: Federal Register Search System • Fellow of the Royal Statistical Society • Flight Reference Stabilization System

FRST: Forest • Frost

Frt: Freight • Front

FRT: Fast Rotating Target • File Received Timeout • Film Register Treaty • Flight Readiness Test • Flight Readiness Training • Fonds de la Recherche et de la Technologie • Fonds de la Recherche Technologique • Frequency Response Test

FrTel: France Telecom

FRTS: Frame Relay Traffic Shaping

FRU: Bishkek (Kyrgyzstan) (location identifier) • Field-Replaceable Unit

FRUSA: Flexible Rolled-Up Solar Array

FRW: Francistown (Botswana) (location identifier) • Friedmann-Robertson-Walker

FRWAY: Freeway

FRWF: Forecast Wind Factor

Frwk: Framework

Frwy: Freeway

FRWY: Freeway

Fry: Freeway

FRY: Ferry • Fryeburg (USA-ME) (location identifier)

FRRY: Ferry

Frz: Freeze

FRZ: Freeze • Fritzlar (Germany) (location identifier)

FRZN: Frozen

Frzr: Freezer

FRZR: Freezer

fs: Facsimile • Foot-Second

FS: Fachschule • Factor of Safety • Factor Space • Fail Safe • Fail to Sync • Faire Suivre • Far Side • Fast Screening • Fast Slew • Fault Summary • Fax Software • Feasibility Study • Federal Specification • Federal Standard • Federal Survey • Fermi Surface • Ferrovie dello Stato (Italy) • Field Security • Field Separator • Field Service • Field Stop • File Segment • File Separator • File Start • File System • Fill Start • Final Settlement • Financial Secretary • Financial Statement • Fine Structure • Fin Stabilized • Fire Station • Fire Suppression • First Stage • First Step • Fixed Service • Fixed Silo • Flight Safety • Flight Simulator • Flight Spare • Flight Standard • Flight System • Floating Sign • Float Switch • Fluid System • Flying Saucer • Flying Status • Foreign Service • Forest Service • Forward Shell • Frame Scan • Frame Status • Franc Suisse • Freon Servicer • Frequency Shift • Frequency Stability • Frequency Standard • Frequency Switching • Frontal Strength • Full Scale • Full Size • Functional Schedule • Functional Schematic • Functional Selector • Function Set • Fuselage Station • Future System

FSA: Farm Services Agency • Fast Simulated Annealing • Federal Security Agency • Final Site Acceptance • Final Stage Acceptance • Firewall Switching Agent • Finite State Automation • Flexible Solar Array • Fonction de Sélection Angulaire • Foreign Service Allowance • Formal Safety Assessment • Free Support Area • Frequency Stability Analyzer • Frequency Standard Assembly • Fuel Storage Area

FSAA: Flight Simulator for Advanced Aircraft

FSAM: Factor Space Access Method • Fine Sun Acquisition Mode • Flexible Solar-Array Mechanism

FSAR: Final Safety Analysis Report

FSAS: Flight Service Automation System

FSB: Frequency Selective Bolometer

Fsbl: Feasible • Fusible

FSc: Full Scale

FSC: Faint-Source Catalogue • Fairchild Space Co. • Fault Simulation Comparator • Federal Stock Classification • Federal Supreme Court • Fernbank Science Center (USA-GA) • Figari (France) (location identifier) • Final Systems Check • Finance Subcommittee • Financial Supervisory Commission • FIRST Science Centre • First-Stage Conduit • Fixed Silicon Capacitor • Flat-Spectrum Component • Florida Southern College (USA-FL) • Fluid Storage Container • Forest Stewardship Council • Frequency Shift Converter • Fuel Sulfur Content • Fuel Systems Capability • Full Systems Capable

FSCC: Fastbus Smart Crate Controller • First-Stage Conduit Container

FSCI: Frequency Space Characteristic Impedance

FSCL: Frequency Sensitive Competitive Learning

FSCM: Federal Supply Classification Manufacturer • Federal Supply Code for Manufacturers

FSCP: Feasibility Study Change Proposal • Fire Sensor Control Panel • Flight Support Computer Program

FSCR: Faint Source Catalogue Reject (file)

FSCS: Flight Simulation Computing System

FSCU: Frequency Select Control Unit

FSD: File System Descriptor • File System Driver • Financial Services Department • Fixed Shutter Dome • Flight Simulation Division • Full-Scale Deflection • Full-Scale Development • Functional Sequence Diagram • Sioux Falls (USA-SD) (location identifier)

FSDA: Fastbus Silicon Data Accumulator

FSDB: Faint-Source Database

FSDC: Fiber-Supported Droplet Combustion

FSDKO: Formalisation of Satellite Design Knowledge for Operations

FSDO: Flight Standards District Office

FSD&P: Flight Systems Design & Performance

FSDP: Flight Systems Design and Performance

FSDS: File System Descriptor Set

FSDU: Frame Synchronization and Decommutation Unit • Frequency Synthesizer and Distribution Unit

FSE: Facilities System Engineer • Facility Support Equipment • Factory Support Equipment • Feuille de Soins Électronique • Field Support Equipment • Fill Start Entry • Final State Effect • Flight Simulation Engineer • Flight Support Equipment • Flight Systems Engineering • Fonds Social Européen • Full-Screen Edit • Full-Screen Editor

FSEA: Fellow of the Society of Engineering Associates

FSEC: Fairchild Space and Electronics Co. • Federal Software Exchange Catalog • Federal Software Exchange Center (USA)

FSECO: First-Stage Engine Cutoff

FSED: Full-Scale Engineering Development

FSET: FIRE Science Experiment Team

FSF: Filtered Science File • First Static Firing • Fixed Sequence Format • Fragmented Signature File • Free Software Foundation • Front Surface Flat • Fundamental System of Functions

FSFDP: Free Standing and Film Deposited Photodiode

FSFM: Full Screen Full Motion

FSG: Flexible Space Garment • Forschungseinrichtung Satellitengeodäsie

FSGC: Florida Space Grant Consortium (USA-FL)

FSGH: Fused Sphere Guided Homotopy

FSGS: Flare/Shallow Glide Slope

FSH: First-Stage Hydraulics • Fixed Superhard

FSI: Facility-class Science Instrument (SOFIA) • Final Systems Installation • Flight Software Integration • Focus Software Inc. (USA-AZ) • Foreign Service Institute • Fort Sill - Henry Post (USA-OK) (location identifier) • Forum sur la Société de l'Information

FSIC: Feature Space Iterative Clustering

FSII: Fuel System Icing Inhibitor

FSIM: Functional Simulator

FSIOP: File Server Input/Output Processor

FSIRR: Filter Scanning Infrared Radiometer

FSIS: First-Stage Ignition System

FSIWG: Flight System Interface Working Group

FSJ: Feedback Summing Junction

FSK: Fort Scott (USA-KS) (location identifier) • Frequency Shift Key • Frequency Shift Keying

FSKM: Fuzzy Segmental K-Means

FSL: Flight Simulation Laboratory • Flight Systems Laboratory • Fluid Science Laboratory • Forecast Systems Laboratory • Foreign Science Library • Fossil Downs (Australia-WA) (location identifier) • Fundamental Straight Line

FSLP: First Spacelab Payload

FSLT: First Sea Level Test

FSM: Fast-Steering Mirror • Fédération Syndicale Mondiale • Field-Strength Meter • Finite-State Machine • Firmware Support Manual • First Servicing Mission • First-Stage Motor • First Surface Mirror • Flight Schedule Monitor • Flight System Mockup • Fluid Science Laboratory • Flying-Spot Microscope • Flying-Spot Microscopy • Formal Specification Maturity • Fort Smith (USA-AR) (location identifier) • Forward Scatter Meter • Free Speech Movement • Frequency Shift Modulation • Fuel Supply Module • Micronesia (ISO country code) • Micronésie (ISO country code)

FSMC: First-Stage Motor Container

FSMMS: Frequency Stability Module Management System

FSMR: Fixed Simultaneous Multibeam Radar

FSMWI: Free Space Microwave Interferometer

FSN: Federal Stock Number • Federation Stock Number • File System Navigator • Foreign Service National • Fort Sheridan - Haley (USA-IL) (location identifier) • Full Service Network

FSO: Facility Security Office • Facility Security Officer • Field Support Office • Flight Safety Officer • Foreign Service Officer • Frequency Sweep Oscillator • Full-Scale Output • Functional Supplementary Objective

FSOC: Fairchild Satellite Operation Center

FSOH: Flight Support Operations Handbook

FSP: Fault Summary Page • Flat Salary Payroll • Flight Safety Plan • Flight Scheduling Precedence • Flight Strip Printer • Food Stamp Program • Foreign Service Pay • Forschungschwerpunkt • Frequency Shift Pulsing • Full-Scale Production • Saint-Pierre (Saint-Pierre & Miquelon) (location identifier)

FSPC: Federal Science Policy Council (USA)

FSPD: Freeze Speed Parameter

FSPL: Free Space Path Loss

FSQL: Fuzzy Structured Query Language

FSR: Faint Source Reject • Fast Slew Rate • Feedback Shift Register • Fermi Selection Rule • Field Service Regulation • Final Software Review • Final System Release • Final System Review • Final System Run • Fin Stabilized Rocket • Fixed Sample Rate • Flight Safety Research • Flight Simulation Report • Flight Solar Reflectometer • Flight Specific Requirements • Flight Summary Report • Flight Support Request • Flight Support Room • Flux Sensitive Resistor • Folio Search and Replace • Force Sensitive Resistor • Free Spectral Range • Free System Resources • Frequency Shift Receiver • Frequency Shift Reflector • Full-Scale Record

FSRB: Flight Safety Review Board

FSRD: Federal Support of Research and Development (USA)

FSRM: First-Stage Rocket Motor

FSRQ: Flat-Spectrum Radio Quasar

FSRR: Flight Software Readiness Review • Flight System Readiness Review • Flight System Recording System

FSRS: Flat Spectrum Radio Source • Flight System Recording System

FSRT: Flight Systems Redundancy Test

FS&S: Flight Systems & Servicing

FSS: Facilities and Support Services • Faint-Source Survey • Feature Subset Selection • Federal Security Service (Russia) • Federal Supply Service • Field Spectrometer System • Fine Sun Sensor • Fire Suppression System • First-Stage Separation • Fixed Satellite Service • Fixed Service Structure • Flight Service Station • Flight Servicing System • Flight Support Station • Flight Support Structure • Flight Support System • Flight Systems Simulator • FOC Safing Sequence • Fokker Space and Systems • Font Selection Scheme • Forres Kinloss (UK-Scotland) (location identifier) • Frame Storage System • Freitag-Samstag-Sonntag • Frequency Selective Surface • Friday-Saturday-Sunday • Full-Scale Station • Full-text Storage System

FSSC: Frequency Shifted Subcarrier

FSSD: First-Stage Separation Device

FSSDE: Formalisation of Standards for Satellite Data Exchange

FSSE: Flight Sensor Stimuli Equipment

FSSM: Flexible Second Surface Mirror • Formal Specifications Strategies Maturity

Fssn: Fission

FSSO: Flight Science Support Office

FSSP: Forward Scattering Spectrometer Probe

FSSR: Flight System Software Request • Flight System Software Requirement • Functional Subsystem Software Requirement

FSSS: Future Strategic Systems Study

FSSTD: Financing System for Science and Technology for Development (UNO)

FS&T: Federal Science & Technology

FST: Facility Science Team • Field Support Terminal • File Status Table • Fine Star Tracker • Finite Sampling Time • Finite State Technology • FIRE Science Team • Fixed Service Tower • Flat Square Tube • Fort Stockton - Pecos County (USA-TX) (location identifier) • Free Space Transfer • French Summer Time • Frequency Shift Transmission • Full-Scale Tunnel • Full-Scale Turbulence • Full Spectrum Theory • Functional Simulator and Translator

FSTC: Financial Services Technology Consortium • Foreign Science and Technology Center (USA)

FSTD: Flight Simulation Test Data

FSTE: Factory Special Test Equipment

FSTI: Federal Scientific and Technical Information (USA)

FSTJ: Fujitsu Scientific Technical Journal (periodical code)

Fstnr: Fastener

FSTS: Federal Secure Telephone Service (USA) • Frequency Standard Test Set • Future Space Transportation System

FSTV: Full-Scale Test Vehicle

FSU: Fcator of Safety Ultimate • Fast Service Upgrade • FLIR Sensor Unit • Florida State University (USA-FL) • Fonds de Solidarité Universitaire • Former Soviet Union • Fort Sumner (USA-NM) (location identifier) • Freja System Unit • Freon Servicing Unit • Frequency Synthesizer Unit • Friedrich-Schiller-Universität (Germany) • Fringe Sensing Unit • Fringe Sensor Unit • Frostburg State University (USA-MD) • Full-Scale Unit • Functional Storage Unit

FSV: Final Stage Vehicle • Flight Space Vehicle

FSW: Flight Software

FSY: Factor of Safety Yield

FSz: Full Size

FS3: Future Security Strategy Study

ft: Feet • Fiat • Foot

Ft: Forint (currency) • Fort

F/T: Full Time

F&T: Fire & Theft

FT: Factory Test • Failure Tolerance • Fast Track • Fast Train • Fault Tolerance • Feed Through • Feet • Field Terminator • File Transfer • File Transferred • Filing Time • Financial Times • Fine Tracker • Fischer-Tropsch • Flight Team • Flight Test • Flow Tube • Foot • Formal Training • Fort • Fourier Tachometer • Fourier Transform • France Telecom • Free Troposphere • Frente • Frequent Traveller • Full Terms • Full Text • Full Time • Fully Tracked • Functional Test

FTA: Failure To Appear • Fast Time Analysis • Fatigue Test Article • Fault Tree Analysis • Flight Test Article • Free-Trade Agreement • Futuna Island (Vanuatu) (location identifier)

FTAM: File Transfer, Access and Management (*) • File Transfer and Access Method • File Transfer And Management

FTAS: Fast Time Analyzer System

FTB: Film Transfer Boom • FITS Table Browser • Flying Testbed • Frequency Time Base

FT&C: Formal Training & Certification

FTC: Fast Time Constant • Federal Trade Commission (USA) • Fiber to the Curb • Flight Test Center • Flight Test Conductor • Florida Test Center • Formal Training and Certification • Fourier Transform Collimator • Fractal Transform Compression • Frequency Time Control

FTCCD: Frame Transfer Charge Coupled Device

FTCP: Flight Test Change Proposal

FTCR: Functional Test Change Request

FTCS: Fault-Tolerant Computing System • Forward Time Centered Space

Ftd: Fitted

FTD: Flight Test Direction • Flight Test Directive • Flight Test Director • Flight Training Device • Foreign Technology Division • Formal Technical Document • Foundational Technical Development • Functional Test Data • Functional Track Directory

FTDR: Flight Test Data Recorder

FTDS: Formal Training Data System

Fte: Fonte

FTE: Facility Training Equipment • Factory Test Equipment • Flight Test Engineer • Flight Test Equipment • Flight Test Evaluation • Flux Transfer Event • Fonte • Forced Test End • Forschung und Technologische Entwicklung • Full-Time Employee • Full-Time Equivalent • Functional Test Equipment

FTEE: Full-Time Equivalent Employee

FTF: File Transfer Facility • Floating Track Following • Functional Test Flight

FTFD: Field Test Force Director

Ftg: Fitting • Footing

FTG: Final Trunk Group

FTGE: Frontage

fth: Fathom

FTH: Fourier Transform Heterodyne

fthm: Fathom

FTHR: Farther • Further

FTI: Fitiuta (American Samoa) (location identifier) • France Télécom Interactive • Frequency Time Indicator • Frequency Time Intensity

FTIR: Fourier Transform Interferometer • Fourier Transform Interferometer Radiometer • Fourier Transform in the Infrared

FTIS: Flight Test Instrumentation System • Fourier Transform Infrared Spectrometer • Fourier Transform Infrared Spectroscopy • Fourier Transform Interferometer Spectrometer • Function Test Information Sheet

FTK: Fort Knox - Godman (USA-KY) (location identifier) • Freight Ton Kilometer

FTL: Farran Technology Ltd. (Ireland) • Faster Than Light • Food-Transportation-Lodging • Fortuna Ledge (USA-AK) (location identifier)

FTLRS: French Transportable Laser Ranging System

FTLS: Formal Top-Level Specification

FTM: Flat Technology Monitor • Flat Tension Mask • Flight Test Manual • Flight Training Mission • Force/Torque Module • Freight Ton-Mile • Frequency Time Modulation

FTMC: Fault-Tolerant Memory Control • Frequency and Time Measurement Counter

FTMP: Fault-Tolerant Multiprocessor

FTMS: Flexible Test and Measurement Systems

FTO: Fellow Traveller Orbit • Flight Test Objective • Flight Test Operation • Foreign Technology Office • Fourier Transform Operator • Functional Test Objective

FTOC: Future Table Of Contents

FTOH: Flight Team Operations Handbook • Flight Test Operations Handbook

FTP: Federal Test Procedure • Field Task Proposal • File Transfer Protocol (*) • File Transport Protocol • Fixed-Term Plan • Fixed Throttle Point • Flight Test Plan • Flight Test Procedure • Flight Test Program • Foiled Twisted Pair • Full Throttle Position • Full-Time Permanent • Full-Time Personnel • Functional Test Procedure • Functional Test Program • Functional Test Progress • Function Test Progress

FTPD: File Transfer Protocol Daemon

FTPF: Flight and Training Planning Facility

FTR: Fermi-surface Traversal Resonance • Final Technical Report • Final Test Rack • Flight Test Report • Flight Test Requirement • Flow-Tube Reactor • Formal Technical Review • Functional Test Report • Functional Test Requirement

FTRB: Flight Test Review Board

FTRD: Flight Test Requirements Document • Functional Test Requirements Document

FTRM: Flight Test Request Memorandum

ft/s: Feet per Second

FTS: Far-Term Study • Far-Term System • Fault Tolerance System • Federal Telecommunications System (USA) • Federal Telephone System • Fiducial Transfer System • File Transfer System • Fine Track Sensor • Flight Telemetry Subsystem • Flight Telerobotic Servicer • Flight Termination System • Flight Test Station • Flight Test Support • Flight Test System • Fourier Transform Spectrograph • Fourier Transform Spectrometer • Fourier Transform Spectrometry • Fourier Transform Spectroscopy • Fourier Transform System • Free Text Search • French Technology Survey • Frequency and Timing Subsystem • Frequency Time Schedule • Frequency Time Standard • Full Turbulence Simulation • Functional Test Specification

FTSA: Fault-Tolerant System Architecture

FTSDBM: Fourier Transform Spectrometer Database Management

FTSDTF: Flight Telerobotic Servicer Demonstration Test Flight

FTSE: Finite Temperature Schrödinger Equation

FTT: Field Test Telescope • File Transfer Timeout • Fischer-Tropsch Type • Fractal Transform Template • Full-Time Temporary • Fully Threaded Tree

FTTA: Federal Technology Transfer Act

FTTC: Fibre To The Curb

FTTH: Fibre To The Home

FTTO: Fibre To The Office

FTTP: Full-Time Temporary Personnel

Fttr: Fitter

FTU: Flight Test Unit • Florida Technological University (USA-FL) • Formal Training Unit • Fort Dauphin (Madagascar) (location identifier) • Freeman Time Unit • Frequency Transfer Unit • Fuel Transfer Unit

FTV: Flight Television • Flight Test Vehicle • Masvingo (Zimbabwe) (location identifier)

FTVE: Functional Technology Validation Experiment

FTW: Fort Worth - Meacham (USA-TX) (location identifier) • Forward Travelling Wave

FTX: Owando (Congo) (location identifier)

Fty: Factory

FTY: Atlanta - Fulton County (USA-GA) (location identifier)

FTZ: Fernmeldetechnisches Zentralamt (Germany) • Forschungs- und Technologiezentrum • Free Trade Zone

Fu: Fuel Mass • Fuse

FU: Field Unit • Flight Unit • Follow-Up • Fouled Up • Freie Universität • Fuel

FUA: Follow-Up Amplifier • Fuel Use Act

FUACE: Fédération Universelle des Associations Chrétiennes d'Étudiants

FUB: Forward Utility Bridge • Fulleborn (Papua New Guinea) (location identifier)

FUBAR: Fouled Up Beyond All Recognition

FUCAM: Faculté Universitaire Catholique de Mons (Belgium)

FUD: Fear-Uncertainty-Doubt • First-Use Date • Follow-Up Detector

FUDD: Follow-Up Detection Device

FUE: First Unit Equipped • Fuerteventura (Canary Islands) (location identifier) • Puerto del Rosario (Canary Islands) (location identifier)

FUEGOS: Fibre Unit for European General Optical Spectroscopy

FUES: Follow-up and Evaluation Section

FUFOR: Fund for Unidentified Flying Object Research

FUG: Fuyang (China-PRC) (location identifier)

FUGI: Future of Global Interdependence

Fuhld: Fuseholder

FUI: Federal Unemployment Insurance • File Update Information

FUJ: Fujairah Aviation Centre • Fukue (Japan) (location identifier)

FUK: Fukuoka (Japan) (location identifier)

Ful: Fulcrum

FUL: Fondation Universitaire Luxembourgeoise (Belgium) • Fullerton (USA-CA) (location identifier)

FULBI: Fédération des Utilisateurs de Logiciels pour Bibliothèques

FUM: File Utilities Module • Fuma (Papua New Guinea) (location identifier)

FUMO: Functional Model

FUN: Funafuti (Tuvalu) (location identifier)

FUNC: Function

FUNCT: Functional

Fund: Fundación • Fundador • Fundamental

FUND: Fundação

FUNDAP: Fundação do Desenvolvimento Administrativo (Brazil)

FUNDP: Facultés Universitaires Notre-Dame de la Paix (Belgium)

FUNET: Finnish University Network

FUNI: Frame-based User-to-Network Interface

Funl: Funnel

FUNY: Free University of New York (USA-NY)

FUO: Follow-Up Output • Fuoshan (China-PRC) (location identifier)

FUORS: FU Orionis Star

FUP: Facility Utilization Plan

FUPAZ: Frente Unido Pro Astronomía Zuliana (Venezuela)

Fur: Furlong • Further

FUR: Follow-Up Report

Furl: Furlough

Furn: Furnace • Furnish • Furnished • Furniture

FURS: Follow-Up Reporting System

Fus: Fuselage

FUS: Fuselage

FUSE: Far-Ultraviolet Spectroscopic Explorer

FUSL: Facultés Universitaires Saint-Louis (Belgium)

Fuslg: Fuselage

FUSLG: Fuselage

FUSRAP: Formerly Utilized Sites Remedial Action Program

Fut: Future • Futuro

FUT: Flight Unit Test • Futuna (Wallis & Futuna Islands) (location identifier)

FUV: Far Ultraviolet

FUVITA: Far Ultraviolet Imaging Telescope Array

FUVOP: Far Ultraviolet Objective Prism

fv: Folio Verso

FV: Fachverband • Flight Vehicle • Flight Version • Flow Valve • Flux Valve • Front View • Full Voltage

FVAS: Fox Valley Astronomical Society (USA-IL) • Fraser Valley Astronomical Society (Canada-BC)

FVC: Forced Vital Capacity • Frequency-to-Voltage Converter

FVDR: Flightdeck Video Data Recorder

FVDS: Fuel Vent and Drain System

FVE: Finite-Volume Element

FVF: First Vertical Flight

FVIPS: First Virtual Internet Payment System

FVK: Faserverstärkte Kunststoffe

FVM: Facility Verification Module • Finite-Volume Method

FVMS: Fluid Volume Measurement System

FVP: Feasibility Validation Program • Feedback Verification Procedure • Fluid Velocity Potential

FVQ: Fuzzy Vector Quantization

FVS: Flight Vehicle Structure • Flux Vector Splitting

FVSEP: Formal Verification System Evaluation Program

FVSL: Full Viscous Shock Layer

FVSS: Fuel Vapor Scrubber System

FVSW: Fox Valley Skywatchers (USA-IL)

FVT: Flash Vacuum Thermolysis • Full Video Translation

FVU: Functional Verification Unit

FVV: Facility Verification Vehicle • Fill and Vent Valve

FVW: Fortran Visual Workbench

Fw: Fußweg

F/W: Filter Wheel • Firmware

FW: Falle Wave • Filter Wheel • Firewall • Flag Word • Formula Weight • Forschung im Weltraum • Fourier-Wiener • Fresh Water • Fuel Wasting • Full Wave

FWA: Federal Works Agency • First Word Address • Fort Wayne - Baer (USA-IN) (location identifier) • Forward Wave Amplifier • Full-Wave Amplifier

FWAS: Fort Wayne Astronomical Society (USA-IN) • Fort Worth Astronomical Society (USA-TX)

F'way: Farmway

FWB: Farenheit Wet Bulb • Four Wheel Brake • Four Wheel Braking

FWBA: Full-Wave Balanced Amplifier

FWBR: Full-Wave Bridge rectifier

FWC: Filament-Wound Case • Full Well Capacity

FWCA: Fish and Wildlife Coordination Act (USA)

FWCS: Flight Watch Control Station

Fwd: Foreword • Forward

FWD: Forward • Four-Wheel Drive • Free World Dialup • Front-Wheel Drive

FWDD: Forwarded

Fwdg: Forwarding

FWF: Förderung der Wissenschaftlichen Forschung • Free Widget Foundation

FWG: Facility Working Group • FITS Working Group • Flexible Waveguide

FWGA: Folded Waveguide Antenna

FWH: Flexible Wire Harness • Flexible Working Hours • Fort Worth - Carswell (USA-TX) (location identifier)

FWHH: Full Width Half Height

FWHIP: Full Width between Half-Intensity Points

FWHM: Full Width at Half Maximum • Full Width at Half Meridian

FWHP: Full Width at Half Peak

FWI: Federation of West Indies • French West Indies

FWIE: Fish and Wildlife Information Exchange

FWIW: For What It's Worth

FWL: Farewell (USA-AK) (location identifier) • Fixed Word Length

FWM: Filter Wheel Mechanism • Food Web Model • Fort William (UK-Scotland) (location identifier) • Four-Wave Mixing

FWOP: First Window Of Protection

FWP: Field Work Proposal

FWPCA: Federal Water Pollution Control Act (USA) • Federal Water Pollution Control Administration (USA)

FWPD: Firewall Product Developers

FWR: Full-Wave Rectification • Full-Wave Rectifier

FWS: Filter Wedge Spectrometer • Final Work Statement • Fish and Wildlife Service

FWT: Forward Wave Tube • Free Water Tritium • Full Wavelet Transform

FWV: Fixed-Wing Vehicle • Full Wave

Fww: Flugwetterwarte

FWW: Flugwetterwarte • Food, Water and Waste

FWWM: Food, Water and Waste Management

FWWMS: Food, Water and Waste Management System

Fwy: Freeway

FWY: Freeway

FWZI: Full Width at Zero Intensity

Fx: Fracture • Fractured

FX: Facsimile • Foreign Exchange • France (Metropolitan_) (ISO country code) • Metropolitan France (ISO country code)

FXCH: Floating-point Exchange

Fxd: Fixed

FXD: Fixed

FXE: Fort Lauderdale - Executive (USA-FL) (location identifier)

Fxg: Fixing

FXM: Flaxman Island (USA-AK) (location identifier)

FXO: Cuamba (Mozambique) (location identifier) • Foreign Exchange Office

FXP: Flaring X-ray Pulsar

FXR: Flash X-Ray

FXS: Foreign Exchange Station

FXT: Foreign Exchange Service

Fxtr: Fixture

FXU: Fixed-point Unit

FXX: France (Metropolitan_) (ISO country code) • Metropolitan France (ISO country code)

FXY: Forest City (USA-IA) (location identifier)

fy: Frisian (ISO language code)

Fy: Ferry

FY: Blackpool (UK postal code) • Financial Year • Fiscal Year • Full Year

FYA: For Your Attention

FYDO: Fiscal Year Design Objective

FYDP: Fiscal Year Development Plan

FYE: Full-Year Equivalent

FYEO: For Your Eyes Only

FYI: For Your Information

FYIG: For Your Information and Guidance

FYM: Fayetteville (USA-TN) (location identifier)

FYN: Fuyun (China-PRC) (location identifier)

FYO: Fiscal Year Option

FYP: Five-Year Plan

FYR: Former Yugoslav Republic

FYROM: Former Yugoslav Republic Of Macedonia

Fys: Fysica

FYT: Faya (Chad) (location identifier)

FYTQ: Fiscal Year Transition Quarter

FYU: Fort Yukon (USA-AK) (location identifier)

FYV: Fayetteville (USA-AR) (location identifier)

FZ: Float Zone • Fracture Zone • Franc Zone • Freezing • Free Zone • French Zone • Fuze

FZDZ: Freezing Drizzle

FZFG: Freezing Fog

FzH: Forschungszentrum Karlsruhe (Germany)

FZO: Filton (UK-England) (location identifier)

FZP: Fresnel Zone Plate

FZRA: Freezing Rain

F-Zug: Fernschnellzug

FZug: Fernschnellzug

G

g: Acceleration of Free Fall • Degeneracy • Density of Vibrational Modes • Gale • Gallon • Garade • Gas • Gaseous • Gauche • Gauge • Gender • General • Generally • Genitive • Geographical • Gerade • g-factor • Gluon • Gold • Good • Government • Grade • Grain • Gram • Grand • Grav • Gravitational Acceleration • Gravity • Gravity Constant • Great • Green • Grid • Gros • Groschen (currency) • Grosse • Guide • Guinea • Gun • Osmotic Coefficient • Specific Gibbs Free Energy • Statistical Weight • Surface Gravity • Variance of Mixture Fraction • Vibrational Term

G/: Giro

G: Conductance • Flexibility • Four Hundred (Roman numeral) • Gain • Garage • Gauche • Gauss • Geld • General • Generator • Geometry • German • Germany • Gibbs Function • Giclas (catalogue) • Giga • Giornale • Giorno • Glasgow (UK postal code) • Golf (international alphabet) • Golfe • Golfo • Good • Gora • Gourde (currency) • Grand • Gravitational Constant • Gravity • Green • Ground • Growth Factor • Guarani (currency) • Guernsey • Gulden (currency) • Gulf • Gunong • Gunung • Gust • Modulus of Rigidity • Newtonian Constant of Gravitation • Phase Factor • Reciprocal Lattice Vector • Shear Modulus • Spectral Type • Storm • United Kingdom (international aircraft registration code) • United Kingdom (international vehicle registration code) • Weight • 400 (Roman numeral)

ga: Gaelic (ISO language code) • Irish (ISO language code)

Ga: Billion Years • Gallium • Georgia (USA)

G&A: General & Accounting • General & Administration • General & Administrative

G/A: General Average • Ground to Air

G-A: Ground to Air

GA: Gabon (ISO country code) • Gambia • Gate Array • Gated Attenuation • Gear Assembly • General Accounting • General Agent • General Agreement • General Assembly • General Assignment • General Availability • General Average • General Aviation • Generating Application • Génération d'Adresse • Generic Algorithm • Genetic Algorithm • Geographic Access • Geographical Association • Geographic Analysis • Georgia (USA) • Glide Angle • Global Analysis • Global Array • Go Ahead • Go Around • Government Agency • Grammaire d'Attributs • Gran Atractor • Grand Attracteur • Great Attractor • Ground Antenna • Growing Amplitude • Gyro Assembly

GAA: Gain Adjuster Adapter • General Aviation Activity • Générateur Atomique Alpha • Gruppo Astrofili Astigiani (Italy) • Guamal (Colombia) (location identifier)

GAAA: General Aviation Activity and Avionics

GAAC: German-American Academic Council

GAAE: Grupo Astrofili Arthur Eddington (Italy)

GAANN: Graduate Assistance in Areas of National Need

GAAP: General Aviation Action Plan • Generally Accepted Accounting Principles • Geological Analysis Aid Package

GAAREX: Global Atmospheric Aerosol Radiation Experiment

GAARS: Global Atmospheric and Aerosol Radiation Study

GAART: Government Astronomy Administration Round Table

GaAs: Gallium Arsenide

Gab: Gabon

GAB: Gabbs (USA-NV) (location identifier) • Gabinete • Gabon (ISO country code) • Générateur Atomique Beta • Gesellschaft für Astronomische Bildung • Gewestelijk Arbeidsbureau • Group Asynchronous Browsing • Guichet Automatique Bancaire

GABA: Gamma-Aminobutyric Acid

GAC: General Area Coverage • Geographical Area Code • Geological Association of Canada • Global Accessibility Cone • Global Area Coverage • Goddard Astronomy Club (USA-MD) • Gracias (Honduras) (location identifier) • Greenbelt Astronomy Club (USA-MD) • Ground Attitude Control • Group-Average Clustering • Grumman Aerospace Corporation (USA-NY) • Gruppo Amici del Cielo (Italy) • Gustavus Adolphus College (USA-MN)

GACCOS: Galactic Anticenter CO Survey

GACE: Grupo de Astronomia y Ciencias del Espacio

GACEP: Global Aerosol Climatology and Effects Program

GA&CS: Ground Acquisition & Command Station

GACS: Ground Acquisition and Command Station

GACU: Ground Air Conditioning Unit • Ground Avionics Cooling Unit

GAD: Gadsden (USA-AL) (location identifier) • Générateur Atomique Delta • Geocentric Axial Dipole • German Astronomical Directory • Gruppo Astronomia Digitale

GADB: General Attribute Database

GADG: Gravity or Acceleration Due to Gravity

GADL: Ground-to-Air Data Link

GADO: General Aviation District Office

GADS: Geodata Analysis and Display System • Geographic Analysis and Display System

Gae: Gaelic

G&AE: General & Administrative Expense

GAE: Gabes (Tunisia) (location identifier) • General American English • Gibbs Adsorption Equation

GAEC: Grumman Aircraft Engineering Corporation (USA-NY)

Gaél: Gaélico

GAF: Gafsa (Tunisia) (location identifier) • Generic Application Frame • Gesellschaft für Angewandte Fernerkundung (Germany)

GAFD: Geophysical and Astrophysical Fluid Dynamics

GAFIA: Getting Away From It All

GAFOR: General Aviation Forecast

GAFPIC: Goddard Astronomical Fabry-Pérot Imaging Camera

G/A/G: Ground to Air to Ground

GAG: Gage (USA-OK) (location identifier) • Gemini Acquisition and Guiding • Générateur Atomique Gamma • Genetic Algorithm Group • Genetic Algorithms Group • Georg-Agricola-Gesellschaft (Germany) • Gruppo Astrofili Genovesi (Italy)

GAGE: Global Atmosphere Gases Experiment

GAH: Gayndah (Australia-QLD) (location identifier)

GAHC: Group-average Agglomerative Hierarchical Clustering

GAHQ: Groupe Astronomique Hague Querqueville (France) (formerly GAQ)

GAI: Gaithersburg - Montgomery County (USA-MD) (location identifier) • Gemeenschappelijk Antenne-Inrichting • General Accounting Instruction • Gesellschaft für Angewandte Informatik • Gibbs Adsorption Isotherm • Guaranteed Annual Income

GAIA: Geographic Access Information and Analysis • Global Astrometric Interferometer for Astrophysics • Graphical Astronomy and Image Analysis

GAIIA: Global Alliance of Information Industry Associations

GAIM: Global Analysis, Interpretation and Modelling

GAIN: Global Analysis and Information Network • Graphic Aids for Investigating Networks

GAIS: Global Area Information System • Global Area Intelligent Search

GAJ: Yamagata (Japan) (location identifier)

GAK: Gakona (USA-AK) (location identifier) • Gemeenschappelijk Administratiekantoor • Générateur Atomique Kappa • Göteborgs Astronomiska Klubb (Sweden)

Gal: Galerie • Galerij • Galicia (Spain) • Gallon

GAL: Galena (USA-AK) (location identifier) • Galeria • Gas Analysis Laboratory • Générateur Atomique Lambda • Generic Array Logic • Grow And Learn

GALAXY: General Automatic Luminosity And XY (measuring machine)

GALCIT: Guggenheim Aeronautical Laboratory - California Institute of Technology (USA-CA)

GALD: Greatest Axial Linear Dimension

GALE: Genesis of Atlantic Lows Experiment

GALEX: Galaxy Evolution Experiment • Galaxy Evolution Explorer (*)

Gall: Gallego • Gallery • Gallon • Galloway (UK-Scotland)

GALLEX: Gallium Experiment

GALP: Good Automated Laboratory Practice

gals: Gallons

Gals: Gallons

Galv: Galvanic • Galvanize • Galvanized • Galvanometer

Galvnm: Galvanometer

Galy: Galley

Gam: Gambia

GAM: Gambell (USA-AK) (location identifier) • Geographical Analysis Machine • Geostationary Airglow Monitor • GPS-Aided Munition • Graphics Access Method • Groupe d'Activation Moteur • Groupe d'Assistance Mutuelle • Gruppo Astrofili Massesi (Italy) • Gruppo Astrofili Menkalinan (Italy) • Guaranteed Annual Minimum • Guide Acquire Module

GAMA: General Aircraft Manufacturers Association • General Aviation Manufacturers Association (*) • General Aviation Manufacturing Association • Group of Amateur Astronomers (Czech Rep.)

GAME: Generic Applications and Monitors Environment • GEWEX Asian Monsoon Experiment • Groupe d'Étude de l'Atmosphère Météorologique

GAMES: Gravity And Magnetic Earthprobe Studies • Gravity And Magnetic Earth Surveyor • Gravity And Magnetic Experiment Satellite

GAMETAG: Global Atmospheric Measurements Experiment on Tropospheric Aerosols and Gases

GAMM: Gesellschaft für Angewandte Mathematik und Mechanik (Germany)

GAMMA: General Abstract Model for Multiset Manipulation • Génération Automatique des Maillages et Méthodes d'Adaptation • GSSS Automated Measuring Machine • Guide-star Automatic Measuring Machine

GAMP: Global Atmospheric Measurement Program

GAMS: General Algebraic Modelling System • Guide for Available Mathematical Software • Guide to Available Mathematical Software (*)

GaMuM: Gesellschaft angewandte Mathematik und Mechanik (periodical code)

GaN: Gallium Nitride

GAN: Gan Island (Maldives) (location identifier) • Global Area Network

GANE: GPS Attitude and Navigation Experiment

GANIL: Grand Accélérateur National d'Ions Lourds (France)

GAO: General Accounting Office • Générateur Atomique Oméga • Gesellschaft für Angewandte Organisation • Gothard Astrophysical Observatory (Hungary) • Guantanamo - Los Canos (Cuba) (location identifier)

GAP: Gap Analysis Program • Gemeinsame Agrarpolitik • General Assembly Programme • General Aviation Propulsion • Geographic Application Program • Geostationary Atmospheric Profiler • Giant Aperiodic Crystal • Global Activity Plan • Glycidil Acide Polymer • GOAL Automatic Procedure • Grant Assistance Program • Group Activation Pack • Gruppo Astrofili di Palermo (Italy) • Gruppo Astrofili Pavese (Italy) • Gruppo Astrofili Peloritani (Italy) • Gruppo Astrofili Pesarese (Italy) • Gruppo Astrofili Piceni (Italy) • Guidance-Acquisition Package • Guildhall Automation Project (UK) • Gusap (Papua New Guinea) (location identifier)

GAPC: Global Alternative Propulsion Center (GM)

GAPE: Great Alaska and Pacific (VLBI) Experiment

GAPEX: Ground-based Atmospheric Profiling Experiment

GApFD: Geophysical and Astrophysical Fluid Dynamics (periodical code)

GAPHYOR: Gas Physics Orsay (database)

GAPI: Gateway Application Programming Interface • Gestion Automatisée des Personnels Intérimaires

GAPL: Group Assembly Parts List

GAPM: Global Atmospheric Prediction Model

GAPP: Geometrical Arithmetic Parallel Processor • Global Alliance Partnership Program

GAPRA: Groupement Astronomique Populaire de la Région d'Antibes (France)

GApS: Giga-Actions per Second • Billions of Actions per Second

GAPS: Giga-Actions Per Second • Billions of Actions Per Second

GAQ: Gao (Mali) (location identifier) • Groupe Astronomique de Querqueville (France) (now GAHQ)

Gar: Garage

GAR: Garagem • Garaina (Papua New Guinea) (location identifier) • Groupe d'Astrophysique Relativiste • Growth Analysis and Review • Gruppo Astrofili di Rozzano (Italy) • Gruppo Astrofili Romani (Italy)

GARAG: Genetic Algorithms Research and Application Group

GARAGe: Genetic Algorithms Research and Application Group

Gard: Garden

GARD: General Address Reading Device • Group for Advanced Receiver Development

GARP: Genetic Algorithm for Rule Production • Global Atmosphere Research Program • Global Atmospheric Research Program

GARR: Gruppo Armonizzazione delle Reti per la Ricerca (Italy)

GARS: Geological Applications of Remote Sensing • Geostationary Archival and Retrieval System • GOES Archival and Retrieval System

GARSOM: Garchinger Sonnenmodell

GARTEUR: Group for Aeronautical Research and Technology in Europe

Gas: Gasoline

GAS: Gallium Arsenide • Garissa (Kenya) (location identifier) • Gas Acquisition System • Gas Analyzer System • General Aviation Simulator • Generalized Applications

System • Générateur Atomique Sigma • Genesee Astronomical Society (USA-MI) • Getaway Special • Giant Air Shower • Gibraltar Astronomical Society • Gisborne Astronomical Society (New Zealand) • Gislaved Astronomiska Sällskap (Sweden) • Groupe Astronomie de Spa (Belgium)

GASAT: Gender And Science And Technology

GASB: Governmental Accounting Standards Board

Gasc: Gascoigne

GASC: Graphics and Software Communication System

GASCAN: Getaway Special Canister

GASER: Gamma-ray Laser

GASIL: General Aviation Safety Information Leaflet

GASL: General Activity Simulation Language

GASP: Galaxy Surface Photometry • Gammas At South Pole • Gemeinsame Außen- und Sicherheitspolitik • General Activity Simulation Program • Gettysburg Automated Stellar Photometer • Global Air Safety Program • Global Air Sampling Program • Global Atmospheric Sampling Program • Graded Assessment in Science Project • Grand Accelerated Space Platform • Graphic Applications Subroutine Package • Gravity-Assisted Space Probe • GSSS Archive Support Project • GSSS Astrometric Support Package • Guide-star Astrometric Support Program • Guide-star system Astrometric Support Package

GASPAR: Groupe Académique de Soutien et de Prévention pour les Adolescents à Risques (France)

GASPI: Gimbal Angle, Surface Position Indicator

GASPRA: Garden State Planetarium Resource Association

GASS: Ground Analysis Subsystem

GASSATA: Genetic Algorithm for Simplified Security Audit Trail Analysis

GAST: Globally Averaged Surface Temperature • Greenwich Apparent Sidereal Time

GAT: Gap Tallard (France) (location identifier) • General Alias Table • General Analysis Technique • General Application Template • Generalized Algebraic Translator • Gifted And Talented • Greenwich Apparent Time • Ground-to-Air Transmitter • Groupe d'Activation Tuyère

GATAM: General Automation Telecommunication Access Method

GATC: Graphic Arts Technical Committee

GATD: Graphical Analysis of Three-dimensional Data

GATE: Galileo Telescope Environment • GARP Atlantic Tropical Experiment • Generic Automatic Test Equipment • Global Atmospheric Tropical Experiment

GATES: Global Access Telehealth and Educational System • Global Access Telehealth and Education System

GATF: Graphics Art Tcchnical Foundation

GATIQ: Groupe pour l'Avancement Technologique et Industriel du Québec (Canada-QC)

GATP: Ground Acceptance Test Procedure • Ground Article Test Procedure

GATS: General Agreement on Trade in Services • Guidance Acceptance Test Set

GATSS: Global Air Transportation Systems and Services

GATT: Gate-Assisted Turnoff Thyristor • General Agreement on Tariffs and Trade

GATU: Geophysical Automatic Tracker Unit

GAU: Gauhati (India) (location identifier) • General Accounting Unit • Größterer Anzunehmender Unfall • Guwahati (India) (location identifier)

GAUCHO: General Architecture for Unified Communication in Heterogeneous OSI environments

GAUSS: Galactic Ultrawide-angle Schmidt System • Galaktisches Ultraweitwinkel-Schmidt-System

GAV: Gag Island (Indonesia) (location identifier) • Gaveto • Gross Annual Value • Gruppo Astrofili Vesuviano (Italy) • Gruppo Astrofili Vicentini (Italy)

GAVRT: Goldstone Apple Valley Radio Telescope (USA-CA)

GAW: Gangaw (Myanmar) (location identifier) • Global Atmosphere Watch • Global Atmospheric Watch • Guaranteed Annual Wage

GAWA: German-American Writers' Association

GAWV: Gunstig Antwoord Wordt Verwacht

GAX: Gamba (Gabon) (location identifier)

GAY: Gaya (India) (location identifier)

Gaz: Gazetta • Gazette • Gazetteer

GAZ: Guasopa (Papua New Guinea) (location identifier)

Gb: Gigabit • Gilbert

G&B: Gordon & Breach

GB: Gain Bandwidth • Gamma-ray Burst • Gauss-Bonnet • Gebiedsbestuur • General Background • Geologische Bundesanstalt (Austria) • Gigabyte • Good Bye • Grande Bretagne • Great Britain • Great Burst • Green Bank • Groot-Brittannië • Guanabara (Brazil) • Guide Book • Royaume-Uni (ISO country code) • United Kingdom (international vehicle registration code) • United Kingdom (ISO country code)

GBA: Alderney (international vehicle registration code) • Big Bay (Vanuatu) (location identifier) • GAS Bridge Assembly • Global Biodiversity Assessment

GBB: Gara Djelibet (Algeria) (location identifier) • Generic Blackboard

GBC: Gasuke (Papua New Guinea) (location identifier) • Ground-Based Computer

GBCL: Gravitationally-Bound Current Loop

GBD: Global Burst Detector • Government Business Division • Great Bend (USA-KS) (location identifier)

GBDF: Generalized Backward Differentiation Formula

GBE: Gaborone (Botswana) (location identifier) • Gödel-Escher-Bach • Gun-Barrel Equation

GBERL: Gulf Breeze Environmental Research Laboratory

GBf: Guterbahnhof

GBF: Geographic Base File • Negarbo (Papua New Guinea) (location identifier)

Gbg: Garbage

GBG: Galesburg (USA-IL) (location identifier) • Guernsey (international vehicle registration code)

GBH: Galactic Black Hole • Galbraith Lake (USA-AK) (location identifier)

GBHC: Galactic Black Hole Candidate

GBHRIL: Ground-Based High-Resolution Imaging Laboratory

GBI: Goodwin Biotechnology Inc. • Grand Bahama (Bahamas) (location identifier) • Green Bank Interferometer • Green Biomass Index • Ground-Based Interceptor

GBIC: Gigabit Interface Connector • Gigabit Interface Converter

GBII: Ground-Based Infrared Instrumentation

gBIS: Graphical Issue Based Information System

GBIS: Geo-Based Information System • Geographic Base Information System

GBIIS: Ground-Based Infrared Instrumentation System

GBJ: Jersey (international vehicle registration code) • Marie Galante (Guadeloupe) (location identifier)

GBK: Gbangbatok (Sierra Leone) (location identifier)

GBL: Goulburn Island (Australia-NT) (location identifier) • Government Bill of Lading • Great Belt Link • Ground-Based Laser

GBM: Gamma-ray Burst Monitor • Garbaharcy (Somalia) (location identifier) • Global Book Market • Isle of Man (international vehicle registration code)

GBML: Genetic-Based Machine Learning • Genetics-Based Machine Learning

GBN: Global Business Network • Graphite Boron Nitride

Gbo: Grabador

GBO: Baltimore Greenbelt (USA-MD) (location identifier) • Goods in Bad Order • Ground-Based Observatory • Grundbuchordnung

Gbor: Gobernador

GBP: Gamboola (Australia-QLD) (location identifier) • Group Buffer Pool • Pound Sterling (UK) (ISO currency code)

GBPC: Ground-Based Programme Committee

G/bps: Gigabits per Second

GBR: Glass Bead Rating • Governed Burst Rate • Great Barrier Reef • Great Barrington (USA-MA) (location identifier) • Green-Blue-Red • Ground-Based Radar • Ground-Balloon-Rocket • Royaume-Uni (ISO country code) • United Kingdom (ISO country code)

Gb/s: Gigabits per Second

GBS: Gamma Burst Source • Geographic Base System • Global Broadcast Service • Global Broadcast System • Goal-Based Scenario • Government Bureau of Standards • Ground-Based Scanner • Ground-Based Software • Ground-Beacon System • Port Fitzroy (New Zealand) (location identifier)

GBSRN: Global Baseline Surface Radiation Network

GBSS: Ground-Based Surveillance System

GBT: Green Bank Telescope • Gridded Brightness Temperature • Ground-Based Telemetry

GBTS: Ground-Based Training System

GBU: Khashm El Girba (Sudan) (location identifier)

GBUV: Geostationary Backscatter Ultraviolet

GBV: Gate Breakdown Voltage • Gemeinsamer Bibliotheksverband (Germany) • Gibb River (Australia-WA) (location identifier) • Graphischer Bildungsverband

GBW: Guild of Book Workers (USA)

GBX: Glovebox

Gbyte: Gigabyte

GBZ: Gibraltar (international vehicle registration code) • Great Barrier Island (New Zealand) (location identifier)

GBzG: Gerlands Beiträge zur Geophysik (periodical code)

GB2: Green Bank Second Survey

Gc: Gigacycle

G&C: Guidance & Control

GC: Gain Control • Galactic Centre • Garbage Collect • Gas Chromatograph • Gas Chromatographic • Gas Chromatography • General Catalog • General Comment • General Corpus • Generic Code • Gigacycle • Gimbal Case • Globular Cluster • Gold Coast • Good Condition • Good Conduct • Governing Council • Grande Coupole • Graphics Context • Great Circle • Green Bank C • Ground Computer • Ground Contract • Ground Control • Ground Controller • Ground Cover • Groupe de Coordination • Guidance and Control • Guidance Computer • Guidance Control • Gyro Compass • Gyro Control • Gyrodine Complex

GCA: Gain Control Amplifier • Galactic Center Arm • Gas Control Assembly • Geophysics Corporation of America • Gesellschaft für Computeranwendungen • Global Change Atlas • Global Coalition for Africa • Graphics Communication Association (USA) • Ground Control Approach • Ground-Controlled Approach • Group Capacity Assessment • Guacamaya (Colombia) (location identifier) • Guatemala (international vehicle registration code) • Guidance and Control Assembly • Guidance Coupler Assembly • Gyro Control Assembly

GCAG: Grand Challenge Application Group

GCAS: Ground Collision Avoidance System

GCAT: Gestion Commerciale des Activités Techniques

GCC: GAME Command Chain • General Compilation Catalogue • General Conference Control • Gillette (USA-WY) (location identifier) • Global Change Category • Global Change Coalition • Global Change Consortium • Global Climatic Change • Globular Cluster Candidate • Gnu C Compiler • Goddard Communications Center • Goddard Computing Center • Governmental Coordinating Committee • Ground Communications Controller • Ground Communications Coordinator • Ground Computer Controller • Ground Control Center • Guidance Checkout Computer • Guidance and Control Computer • Gulf Cooperation Council

GCCA: Graphic Communications Computer Association

GCCC: Ground Control Computer Center

GCCD: Global Climate Change Digest

GCCE: Graphics Correctness Criteria Editor

GCCHK: Global Cross Check

GCCO: Ground Control Checkout

GCCS: Global Command and Control System

GCD: Gain Control Driver • Geometric Correction Data • Gnu C Debugger • Great Circle Distance • Greatest Common Denominator • Greatest Common Divisor • Ground Controlled Descent

GCDA: Geocentric Datum of Australia

GCDB: Geographic Coordinate Database

GCDC: Global Change Data Center • Ground Checkout Display and Control

GCDIS: Global Change Data and Information System

GCDMWG: Global Change Data Management Working Group

GCDP: Global Change Database Project

GCDU: Graphical Control Display Unit

GCE: General Certificate of Education • Global Change Encyclopaedia • GPS Cargo Element • Ground-based Cloud Experiment • Ground Checkout Equipment • Ground Communications Equipment • Ground Control Equipment • Groupe de Coordination Espace

GCEM: Geographic Correlator Exploration Machine • Goddard Cumulus Ensemble Model

G&CEP: Guidance & Control Equipment Performance

GCEP: Global Change Education Program • Guidance and Control Equipment Performance

GCERF: Governor's Council on Education Reform and Funding

GCEV: Grupo Canario de Estrellas Variables (Spain)

GCF: Gestion Comptable et Financière • Globular Cluster Feature • Greatest Common Factor • Ground Calibration File • Ground Command Facility • Ground Communications Facility

GCFAP: Guidance and Control Flight Analysis Program

GCFP: Globular Cluster Fundamental Plane

GCG: Generalized Conjugate Gradient • Gravity-Controlled Gyro • Ground Command Guidance

GCGS: Gravity-Controlled Gyro System

GCGSS: General Catalogue of Galactic S Stars

GCH: Generalized Continuum Hypothesis

GCHC: Gulf Coast Hydroscience Center

GCHQ: Government Communications Headquarters (UK)

GCHX: Ground Cooling Heat Exchanger

G&CI: Guidance & Control Integration

GCI: Generalized Communications Interface • Geocentric Inertial • Ground-Controlled Interception • Group Complexity Index • Guernesey - La Villiaze (Channel Islands) (location identifier) • Guernsey - La Villiaze (Channel Islands) (location identifier) • Guest Computational Investigator • Guidance and Control Integration

GCID: Ground Computer Interface Device

GCIL: Ground Command Interface Logic • Ground Control Interface Logic

GCILC: Ground Command Interface Logic Controller

GCILU: Ground Command Interface Logic Unit

GCIP: GEWEX Continental International Project

GCIR: Goodwin Institute for Cancer Research

GCJ: Johannesburg - Grand Central (South Africa) (location identifier)

GCK: Garden City (USA-KS) (location identifier)

GCl: Globular Cluster

GCL: Galactic Center Lobe • Gestion de Configuration Logicielle • Global Commerce Link • Graphics Card Language • Graphics Command Language • Ground Control Landing • Ground Coolant Loop

GCLF: Globular-Cluster Luminosity Function

GCM: General Circulation Model • General Collective Model • General Control Module • Global Circulation Model • Global Climate Model • Grand Cayman (Cayman Islands) (location identifier) • Graph-Cache Manager • Greatest Common Measure • Greatest Common Multiple • Ground Control Message • Groupe de Concertation de Mécanique • Groupe de Contrôle Modifié

GCMAM: Global Climate/Middle Atmosphere Model

GCMC: Grand Canonical Monte Carlo

GCMD: Global Change Master Directory

GCMR: Ground Control Message Request

GCMRC: Grand Canyon Monitoring and Research Center (USA-AZ)

GC/MS: Gas Chromatograph/Mass Spectrometer

GCMS: Gas Chromatograph/Mass Spectrometer • Gas Chromatograph Mass Spectrometry

GCN: GIRD Change Notice • Grand Canyon National Park (USA-AZ) (location identifier) • GRB Coordinates Network • Ground Communications Network • Ground Control Network

GCNMS: Gas Chromatograph and Neutral Mass Spectrometer

GCNS: Glasgow College of Nautical Studies (UK-Scotland)

GCO: Gemeenschappelijk Centrum voor Onderzoek • Generation Certification Official • Ground Checkout • Ground Control Officer • Grove Creek Observatory (Australia-NSW)

GCOM: Global Change Observation Mission

GCOS: General Comprehensive Operating System • General Computer Operating System • General Computer Operational System • Global Change Observing System • Global Climate Observation System • Global Climate Observing System (*) • Ground Computer Operating System

GCP: Generalized Computer Program • Geodetic Control Point • Gestion du Cycle de Projet • Global Change Program • GLO/CRYOHP Payload • Glow and Cryogenic Payload • Ground Command Post • Ground Control Point • Group Command Post • Group Control Point • Guidance Control Package

GCPA: General Catalogue of stars with the system of Photoelectric Astrolabes

GCPB: General Catalogue of Printed Books

GCPD: General Catalogue of Photometric Data

GCPO: Gemini Canadian Project Office

GCpS: Gigacycles per Second

GCPS: Gigacycles Per Second

GCQMS: Groupe Conseil pour la Qualité des Matériaux Spatiaux

GCR: Galactic Cosmic Radiation • Galactic Cosmic Ray • Gamma Cosmic Ray • Gas-Cooled Reactor • Grand Cercle de Référence • Gravitation et Cosmologie Relativistes • Gray Component Replacement • Ground-Controlled Radar • Group-Coded Recording • Group-Coded Request • Group-Code Recording • Group-Code Request

GCRA: Generic Cell Rate Algorithm

GCRIO: Global Change Research Information Office

GCRP: Galactic Cosmic Ray Particle • Global Change Research Plan • Global Change Research Programme

GCRV: General Catalog of Radial Velocities • Generalized Conjugate Residual

Gc/s: Gigacycles Per Second

GCS: Gate-Controlled Switch • Gate Controller Switch • General Classification Scheme • General Communication System • Generator-Controlled Switch • Geo-Common System • Geostationary Communications Satellite • Globular-Cluster System • Ground Central Site • Ground Command System • Ground Communications System • Ground Control Station • Group Control System • Guidance and Control System • Guidance Cutoff Signal • Gyroless Control System

GCSA: Genetic Configuration Similarity Algorithm

GCSC: Guidance Control and Sequencing Computer

GCSE: General Certificate of Secondary Education (UK)

Gc/sec: Gigacycles per Second

GCSP: Grand Canyon Star Party (USA)

GCSS: GEWEX Cloud System Study • Global Cloud System Study • Global Communications Satellite System • Ground-Controlled Space System

GCT: Geocentric Coordinated Time • Greenwich Civil Time (*) • Greenwich Conservatory Time • Gregory-Coudé Telescope • Ground Checkout and Test • Guidance Command Test • Guidance Computer Test

GCTA: Ground-Commanded Television Assembly

GCTC: Gagarin Cosmonaut Training Center (Russia)

GCTE: Global Change and Terrestrial Ecosystems (*) • Global Change in Terrestrial Ecosystems • Guidance Computer Test Equipment

GCTM: Global Chemical Transport Model

GCTP: General Coordinate Transformation Package • Geographic Coordinate Transformation Package

GCTS: Gas Component Test Stand • Ground Communications Tracking System

GCTSP: General Catalogue of Trigonometric Stellar Parallaxes

GCU: General Control Unit • Generator Control Unit • Glasgow Caledonian University (UK-Scotland) • GNC Computer Unit • Ground Checkout Unit • Ground Control Unit • Ground Cooling Unit • Guidance and Control Unit • Guidance Coupler Unit • Gyro Coupling Unit

GCUpS: Giga Connection Updates per Second

GCUPS: Giga Connection Updates Per Second

GCV: Generalized Cross-Validation • Global Change Video • Gross Calorific Value

GCVS: General Catalogue of Variable Stars

GCWD: Globular Cluster White Dwarf

GCX: Galactic Center X-ray

GCXSE: GSFC Cosmic X-ray Spectroscopy Experiment

GCY: Greenville (USA-TN) (location identifier)

GC3: Grand Challenge Cosmology Consortium • Great Circle Cell Count

gd: Scots Gaelic (ISO language code)

Gd: Gadolinium • Good • Grand • Ground • Guard

GD: Gap Detector • Garbage Disposal • Gate Driver • General Delivery • General Design • General Development • General Diagram • General Duties • General Duty • General Dynamics • Generic Decryption • Geologic Division • Geospatial Data • Giclas Dwarfs • Glow Discharge • Good Delivery • Gradient Descent • Grand Duchess • Grand Duke • Graphics Digitization • Graphics Disable • Gravimetric Density • Grenada (ISO country code) • Grenade (ISO country code) • Ground Detector • Guarda

GDA: Galvo-Drive Amplifier • GEBCO Digital Atlas • Geographic Data Analysis • Gimbal Drive Actuator • Gimbal Drive Assembly • Global Data Area • Global Directory Agent • Gounda (Central African Rep.) (location identifier)

GDAAC: Goddard Distributed Active Archive Center

GDAM: General Data Access Method

GDAP: GEOS Data Adjustment Program

GDAS: Global Data Assimilation System • Ground Data Acquisition System

GDAU: General Data Acquisition Unit

GDB: Genome Database • Geodetic Data Beacon • Gesellschaft für Dokumentation und Bibliographie (Austria) • Global Database • Gnu Debugger • Graphical Database

GDBMS: Generalized Database Management System

GDBS: Generalized Database System • Grenada Bureau of Standards

GDC: General Design Criteria • General Digital Computer • General Dynamics Corporation • Geocentric Dust Cloud • Geodetic Data Center • Geographic Data Council • Geological Data Center • Geomagnetic-Data Center • Geophysical Data Center • Gesellschaft für Datensysteme und Computer • Global Data Centre • Global Data Control • Glow Discharge Cleaning • Greenville - Donaldson Center (USA-SC) (location identifier) • Guidance Data Converter • Gyro Display Coupler

GDCC: Guarded Definite Clause with Constraints

GDCD: Glow Discharge Cleaning Device

GDCF: Generic Data Capture Facility

GDCh: Gesellschaft Deutscher Chemiker (Germany)

GDCP: Geostationary Data Collection Platform

GDCS: Government Documents Catalog Service

GDD: Gesellschaft für Datenschutz und Datensicherung (Hermany) • Global Data Disable • Gordon Downs (Australia-WA) (location identifier) • Graphics Data Display

GDDL: Graphical Data Definition Language

GDDM: Graphical Data Display Manager • Graphics Data Display Manager

GDDSYN: Geodesic Distribution Binary Synthesis

Gde: Gourde (currency) • Grande

GDE: Generalized Data Entry • Gode (Ethiopia) (location identifier) • Ground Data Equipment • Group for Development and Environment

GDEM: Global Digital Elevation Model

GDEW: Global Delivery Express Worldwide

GDF: Gaz De France • General Data File • Generalized Data Format • Geographic Data File • Geophysical Data Facility • Grammar Description File

GDFCF: Gross Domestic Fixed Capital Formation

GDG: Generation Data Group • Generation Dataset Group • Magdagachi (Russia) (location identifier)

GDH: Golden Horn (USA-AK) (location identifier) • Ground Data Handling

GDHS: Ground Data Handling System

GDI: Gasoline Direct Injection • General Dissemination of Information • General Drafting Interface • Gordil (Central African Rep.) (location identifier) • Graphical Device Interface • Graphics Device Interface (*) • Grupo de Desarollo Instrumental

GDJ: Gandajika (Zaire) (location identifier)

Gdk: Gdansk (Poland)

GDL: Gas Dynamics Laboratory • Global Data Link • Glow Discharge Lamp • Grammar Description Language • Guadalajara - Miguel Hidalgo (Mexico) (location identifier)

GDLI: Global Distance Learning Initiative

GDM: Gardner (USA-MA) (location identifier) • Generalization Data Model • Geodetic Distance Measurement • Geospatial Data Management

GDMO: Global Definition for Managed Objects • Guidelines for the Definition of Managed Objects

GDMS: Generalized Data Management System • Generic Data Management System • Glow Discharge Mass Spectroscopy • Ground Data Management System

GDMT: Global Desired Merge Total

Gdn: Garden • Guardian

GDN: Garden • Gdansk (Poland) (location identifier) • Goal Dependency Network • Government Data Network

GDNÄ: Gesellschaft Deutscher Naturforscher und Ärzte (Germany)

Gdnc: Guidance

Gdns: Gardens

GDNS: Gardens

GDNT: Global Desired Non-merge Total

GDO: Guasdualito (Venezuela) (location identifier)

GDOP: Geometric Dilution Of Precision

Gdor: Gobernador

GDP: Gás de Portugal • General Development Plan • Generalized Distributor Program • Generalized Documentation Processor • GOME Data Processor • Grammar Description Programme • Graphic Display Processor • Graphic Draw Primitive • Gross Domestic Product • Ground Data Processing • Guadalupe (Brazil-PI) (location identifier)

GDPS: Geographically-Dispersed Parallax Sysplex • Global Data Processing System • Government Document Publishing Service • Ground-based Data Processing System

GDQ: Gondar (Ethiopia) (location identifier)

GdR: Groupe de Recherche • Groupement de Recherche

GD&R: Grinning, Ducking & Running

GDR: Gas-to-Dust Ratio • Geodetic Data Reduction • Geophysical Data Record • German Democratic Republic (now merged into FRG) • Giant Dipole Resonance • Global Depository Receipt • Grinning, Ducking and Running • Groupement De Recherche

GDRC: Gyro Drift Rate Compensation

Gds: Goods

GDS: General Data Stream • Generation Dataset • Geodetic Data Site • Geographic Data Service • Global Directory Service • Global Distribution System • Goldstone • Graphic Data System • Great Dark Spot (Neptune) • GRETACODER Data Systems AG (Switzerland) (*) • Gretag Data Systems AG (Switzerland) • Ground Data System • Ground Display System • Guarded Discrete Stochastic • Guide and Directory Services

GDSD: Ground Data Systems Division

GDSIDB: Global Digital Sea-Ice Databank

GDSM: Ground Data Systems Manager

GDSMQ: General Department for Standardization Metrology and Quality Control

GDSN: Global Digital Seismic Network • Global Digital Seismograph Network

GDSO: Ground Data Systems Officer

GDSS: Group Decision Support System

GdT: Gemeinschaftsausschuß der Technik

GD&T: Geometric Dimensioning & Tolerancing

GDT: Geometric Dimensioning and Tolerancing • Gestion de Données Techniques • Global Descriptor Table • Grand Turk (Turks and Caicos Islands) (location identifier) • Graphic Display Terminal • Graphics Development Tool • Graphics Development Toolkit

GDTA: Groupement pour le Développement de la Télédétection Aérospatiale

GDU: Graphic Display Unit

GDV: Gesellschaft für Datenverarbeitung • Glendive (USA-MT) (location identifier)

GDW: Gladwin (USA-MI) (location identifier)

GDX: Magadan (Russia) (location identifier)

GDZ: Gelendzik (Russia) (location identifier)

ge: Gilt Edges

Ge: Germanium

G/E: Graphite/Epoxy

GE: Gas Exchange • General Election • General Electric • Geocentric Gravitational Constant • Geographic Equivalent • Georgia (ISO country code) • Géorgie (ISO country code) • Germany • Grande École • Graphics Enable • Graphite Epoxy • Greater than or Equal • Greatest Elongation • Greedy Execution • Ground Equipment • Group of Experts • Gyro Error

GEA: Generalized Extensions Agreement (FITS) • Geometric Element Analysis • Gigabit Ethernet Alliance • Gimbal Electronics Assembly • Global Environment Assessment • Grup d'Estudis Astronomics (Spain) • Grupo de Estudos de Astronomia (Brazil) • Nouméa Magenta (New Caledonia) (location identifier)

GEAP: Grupo de Estudios Astronómicos de Puertollano (Spain)

GEAUDE: Gestion Automatisée Décentralisée

geb: geboren • gebunden

Geb: Gebergte • Gebirge

GEB: Gebe (Indonesia) (location identifier) • Gödel-Escher-Bach • Gemeentelijk Energiebedrijf

GEBA: Gebäudeautomation • Global Energy Balance Archive

GEBCO: General Bathymetric Chart of the Oceans

GEBECOMA: Groupement Belge des Constructeurs de Matériel Aérospatial (Belgium)

Gebr: Gebroeders • Gebrüder

GEBR: Galactic and Extragalactic Background Radiation

GEC: Gecitkale (Cyprus) (location identifier) • General Education Curriculum • General Electric Co. • Genetic and Evolutionary Computation • Genetic and Evolutionary Computing • Geospace Electrodynamics Connections (mission) • Global Electrodynamic Monitor • Global Environmental Change • Ground Environment Complex

GeCAS: Geochimica et Cosmochimica Acta Supplement (periodical code)

GECAS: General Electric Capital Aviation Services

GECCO: Genetic and Evolutionary Computation Conference

GECHS: Global Environmental Change and Human Security

GECMA: Global Electrodynamics and Coupling through the Middle Atmosphere

GeCoA: Geochimica et Cosmochimica Acta (periodical code)

GECOS: General Electric Comprehensive Operating System

GECR: Global Environment Change Report

GECS: Geostationary Earth Climate Sensor

GED: General Educational Development (*) • General Education Development • General Equivalency Diploma • Georgetown - Sussex County (USA-DE) (location identifier) • Gestión Electrónica de Documentos • Gestion Électronique de la Documentation • Gestion Électronique de Documents • Global Electrodynamics • Graduate Equivalent Diploma

GEDEMON: Geometrics, Derivatives, Moments and Noise

GEDEONEX: Générateur de Dossiers d'Environnement d'Océanographie Numérique Expérimental

GEDES: Groupement Étudiant de l'Enseignement Supérieur

GEDEX: Geodetic Demonstration Experiment • Greenhouse Effect Detection Experiment

GEDI: Groupement Europages de Diffusion d'Information • Group on Electronic Document Interchange

GEDIS: Geological, Exploration and Development Information System

GEDR: Global Emissivity Data Record

GEE: Galaxy Evolution Explorer • Generalized Estimating Equation (*) • Generalized Estimation Equation • George Town (Australia-TAS) (location identifier) • Group of Economic Experts

GEEP: Group of Experts on Environmental Pollutants

GEERS: Groupe d'Études Européennes pour la Recherche Spatiale

GEF: Gauss Error Function • Gestion Électronique des Factures • Geva (Solomon Islands) (location identifier) • Global Environmental Facility • Global Environment Facility (*) • Global Environment Fund • Ground Equipment Failure • Groupement d'Exercice Fonctionnel

GEFD: Geophysical and Environmental Fluid Dynamics

GEFE: Grupo de Estudios de Formación Estelar

GEFES: Grupo Especializado de Física del Estado Solido (Spain)

gefl: gefälligst

GEG: Spokane (USA-WA) (location identifier)

GeGe: Geophysica et Geodaetica (periodical code)

gegr: gegründet

geh: geheftet

GEHN: Groupe d'Experts de Haut Niveau

GEHR: Giant Extragalactic HII Region

GEI: Gesellschaft für Elektronische Informationsverarbeitung mbH • Gestion Électronique de l'Information • Graphics Engine Interface • Green Islands (Papua New Guinea) (location identifier)

GEIA: Global Emission Inventory Activity

GEID: Gestion Électronique de l'Information et de Documents

GEIE: Groupement Européen d'Intérêt Économique

GEIP: Greenhouse Education and Information Programme

GEIS: Gebäudeinformationssystem • General Edited Information Set • General Electric Information Service • General Electric Information System • Generic Edited Information Set • Geographical Environmental Information System • German Encyclopaedic Internet Service • Groupe d'Étude sur l'Information Scientifique

GEISA: Gestion et Études des Informations Spectroscopiques Atmosphériques

GEISCO: General Electric Information Services Co.

GEISHA: Gestion, Extraction, Interrogation d'une Structure Hiérarchisée Arborescente • Groningen Exportable Infrared Survey High-Resolution Analysis

GEK: Ganes Creek (USA-AK) (location identifier) • Lari (Georgia) (ISO currency code)

GEL: Santo Angelo - Sepe Tiaraju (Brazil-RS) (location identifier)

Gem: Gemeinde • Gemeinschaft • Gemini-Geminorum (constellation)

GEM: Galactic Emission Mapping • Galaxy Emission Mapping • Galileo Europa Mission • Gamma-Electron-Muon • Gas Electron Multiplier • Gaseous Electron Multiplier • Generalized Ensemble Mothod • Generic Electronics Module • Generic Engineering Model • Geometric Element Modelling • Geospace Environmental Modelling • Geospace Environment Modelling • German Equatorial Mount • Giant Earth Mover • Giotto Extended Mission • Gobal Electrodynamics Monitor • Global Enterprise Management • Global Enterprise Manager • Goddard Earth Model • Goddard Environment Model • Graphical Environment Manager • Graphical Execution Manager • Graphic Engine Monitor • Graphics Environment Manager • Graphite Epoxy Motor • Ground Effect Machine • Ground Electronics Maintenance • Ground Environmental Model • Ground Equipment Model

GEMCOS: Generalized Message Control System

GEMINI: General Excitation Mechanisms in Nightglow

GEMMS: Geophysical Exploration Manned Mobile Submersible

GEMO: Ground Electronic Maintenance Officer

GEMPAK: General Meteorology Package

GEMPS: Groupe d'Études en Médecine et Physiologie Spatiales

GEMS: Gas Exchange Monitoring System • General Electric Multispectral System • General Energy Management System • Generalized Evaluation Model Simulator • Geographic and Environmental Modelling System • Geographic Monitoring System • Geological Exploration and Mapping System • Geostationary European Meteorological Satellite • Glasses with Embedded Metals and Sulfides • Global Environmental Monitoring System (*) • Global Environment Monitoring System • Goodyear Electronic Mapping System • Graphical Exposure Modelling System • Great Explorations in Math and Science

GEMU: Global Electronic Multimedia University

GemVO: Gemeinnützigkeitsverordnung

gen: genannt

Gen: Gender • General • Generally • Generator • Generic • Genitive

Gén: Général • Género

Gen: Generaal • General • Generate • Generated • Generation • Generator • Genesis • Geneva (Switzerland) • Genève (Switzerland) • Gennaio • Genossenschaft • Genova (Italy)

GEN: General • Global Engineering Network • Global European Network • Oslo Gardermoen (Norway) (location identifier)

Gend: Gendarme

Geneal: Genealogía

GENESIS: Global Ecosystem Numerical Estimation Using Satellite Imaging System

GENIAC: Genius Almost Automatic Computer

GENIE: General Electric Network for Information Exchange • Global Environmental Network for Information Exchange

GENISIS: General Imaging Spectrometry Interpretation System

Génit: Génitif

Genit: Genitive • Genitivo

GENIUS: Germanium in liquid Nitrogen Underground Setup • Global Environment Network and Information User System

Genl: General

GENLY: Generally

Gen.Mgr.: General Manager

Genn: Gennaio

Gent: Gentleman

Geo: Geographic • Geology (periodical code) • George

GEO: Georgetown Timehri (Guyana) (location identifier) • Georgia (ISO country code) • Géorgie (ISO country code) • Geostationary Earth Observatory • Geostationary Earth Orbit • Geosynchronous Earth Orbit • German-England Observatory • Global Environment Outlook • Gravitational European Observatory • Grupo Especial de Operaciones

Geoch: Geochemistry (periodical code)

GeocJ: Geochemical Journal (periodical code)

Géod: Géodésie • Géodésique

Geod: Geodesia • Geodesy • Geodetic • Geodetical

GEODAS: Geophysical Data System

GEODE: Geodynamics and Ore Deposit Evolution

GEODES: Geographic Decision Expert System

GEODIS: Geographic Design and Implementation System

GEODOM: Geographic Data, Overlay and Manipulation

GEODSS: Ground-based Electro-Optical Deep-Space Surveillance (*) • Ground-based Electro-Optical Deep-space Surveillance Station

Geofís: Geofísica

Geog: Geography

Géogr: Géographie • Géographique

Geogr: Geografía • Geographer • Geographic • Geographical • Geography

GeoIn: Geocarto International (periodical code)

GeoJ: Geophysical Journal (periodical code)

GeoJI: Geophysical Journal International (periodical code)

Geol: Geology

Géol: Géologie • Géologique

Geol: Geología • Geologic • Geological • Geologist • Geology

GEOLAN: Geological Local Area Network

Geom: Geometría • Geometric • Geometrical • Geometry

GeoM: Geological Magazine (periodical code)

Géomag: Géomagnétique • Géomagnétisme

Geomag: Geomagnetic • Geomagnetism

GEOMER: Geostationary Multispectral Electroscanning Radiometer

Geomorf: Geomorfología

GEOMPAK: Geometric Manipulation Package

GEOMULDAT: Geochemical Multielement Database

GEON: Gyro-Erected Optical Navigation

GEONESS: Geostationary Earth Orbiting Nadir Etalon Sounding Spectrometer

GeoNr: Geophysica Norvegica (periodical code)

GeoO: Georgetown Observatory Monograph (periodical code)

Geoph: Geophysica (periodical code)

Geophys: Geophysical • Geophysics

Géophys: Géophysique

Geopol: Geopolitics

GEOREF: Geological Reference

GeoRL: Geophysical Research Letters (periodical code)

GEOS: Geodetic Earth-Orbiting Satellite • Geodynamic Experimental Ocean Satellite • Geostationary Earth-Orbiting Satellite • Graphical Environment Object System • Graphic Environment Operating System • Groupement Européen d'Observation Stellaire • Grupo Europeo de Observaciones Estelares • Gruppo Europeo di Osservazione Stellare

GEOSAR: Geologic Synthetic Aperture Radar

GEOSAT: Geodesy Satellite • Geodetic Satellite

GeoSCIA: Geostationary Scanning Imaging Absorption Spectrometer

GEOSEPS: Geosynchronous Solar Electric Propulsion Stage

GeoSJ: Geological Society Journal (periodical code)

GeoSu: Geophysical Surveys (periodical code)

Geot: Geotimes (periodical code)

Geoth: Geothermics (periodical code)

GEP: Goddard Experiment Package • Gross Energy Product

GEPAN: Groupe d'Études sur les Phénomènes Aérospatiaux Non-identifié s (France)

GEPATA: Groupe d'Étude des Problèmes de l'Automatisation des Travaux Administratifs (France)

GEPP: Gaussian Elimination with Partial Pivoting

GEPS: Geostationary Earth Processes Spectrometer

GEQI: Groupe d'Expertise de la Qualité Image

Ger: German • Germany • Gerundio • Gerundive

GER: Gerência • Global Environmental Research • Gross Energy Requirement • Nueva Gerona (Cuba) (location identifier)

GERAILP: Groupe d'Étude et de Recherches pour les Applications Industrielles des Lasers de Puissance (France)

GERB: Geostationary Earth Radiation Budget

GERD: Graduate Education and Research Development

GERDIEN: General European Road Data Information Exchange Network

GERDSM: Groupe d'Études et de Recherches en Détection Sous-Marine

GERE: Génie Électrique et Radioélectricité

GEREMI: Groupe de Réflexion sur la Maintenance et l'Instrumentation

GERES: Geometric Rectification Expert System

GERG: Global Environment Research Group

GERIS: Geophysical and Environmental Research Imaging Spectrometer • Group on Evaluation of Research in Information Science

Germ: Germanía • Germanic • Germánico

GERMEA: Groupe d'Entraînement et de Recherche pour les Méthodes d'Éducation Active (France)

GERSULP: Groupe d'Étude et de Recherche sur les Sciences de l'Université Louis Pasteur (France)

GERT: Giant Equatorial Radiotelescope • Graphical Evaluation and Review Technique

GERTS: General Electric Range Tracking System

GERV: General Electric Reentry Vehicle

Ges: Gesamt- • Gesellschaft • Gesetz

GES: Gaz à Effet de Serre • General Santos - Buayan (Philippines) (location identifier) • Generic External Shell • Geographic Entry System • Gesellschaft für Elektronische Systemforschung (Germany) • Gift and Exchange Section • Global Enterprise Services • Ground Earth Station • Ground Electronic System • Ground Equipment System • Ground Exploitation System

GESAMP: Group of Experts on Scientific Aspects of Marine Pollution

Gesch: Geschichte

Geschw: Geschwindigkeit • Geschwister

Gesell: Gesellschaft

GESF: General Equation Specification File

GESMAR: Geodetic Survey Marks Register

GesmbH: Gesellschaft mit beschränkter Hafting

GESOC: General Electric Satellite Orbit Control

GESPL: Generalized Edit System Programming Language

GESS: Geostationary Environmental Satellite System • Goddard Experiment Support System • Grating Element Support Structure

GESSY: Groupe d'Études de Signaux et Systèmes

gest: gestorben

Get: Getekend

GET: Generalized Einstein Theory • Geraldton (Australia-WA) (location identifier) • Get-Execute-Trigger • Ground Elapsed Time • Groupe d'Études Techniques • Group Execute Trigger

GETA: Government Employees Training Act • Groupe d'Études pour la Traduction Automatique (France)

GETF: Global Environment Trust Fund

GETI: Ground Elapsed Time of Ignition • Groupe d'Études Techniques et Informatiques (France)

GETL: Ground Elapsed Time of Landing

GETMS: Geostationary Experimental Temperature and Moisture Sounder

GETO: Ground Equipment Turn Off

GETS: General Environmental Test Specification • Ground Equipment Test Set

GETSCO: GE Technical Services Co.

GEU: GAS Electronics Unit

GeV: Giga Electronvolt

GEV: Gällivare (Sweden) (location identifier) • Gallivare (Sweden) (location identifier)

GEVS: General Environmental Verification Specification

GEW: Gewoya (Papua New Guinea) (location identifier)

GewB: Gewerkschaftsbibliothek

GEWEX: Global Energy and Water-cycle Experiment (*) • Global Energy Water-cycle Experiment

GewO: Gewerbeordnung

GewStDV: Gewerbesteuer-Durchführungsverordnung

GewStR: Gewerbesteuer-Richtlinien

GEX: Gas Exchange • Gas Exchange Experiment • Gaussian Extraction • Geelong (Australia-VIC) (location identifier)

GEY: Greybull - South Big Horn County (USA-WY) (location identifier)

gez: gezeichnet

GF: French Guyana (ISO country code) • Gain Factor • Gauge Factor • General Form • General Format • General Formula • Generator Field • Generic Failure • Girlfriend • Glass Fiber • Globular Filament • Government Form • Government-Furnished • Gradient Freezing • Green Function • Ground Fog • Growth Factor • Guinean Franc (currency) • Guyana (French_) (ISO country code) • Guyane Française (ISO country code)

GFA: Good Fair Average • Great Falls - Malmstrom (USA-MT) (location identifier) • Groupe Fermé d'Abonnés

GFAAS: Graphite Furnace Atomic Absorption Spectroscopy

GFAE: Government-Furnished Aeronautical Equipment • Government-Furnished Aircraft Equipment

GFAI: Gesellschaft für Förderung der Angewandten Informationsforschung (Germany)

GFB: Go For Broke • Togiak Fish (USA-AK) (location identifier)

GFC: Gas Filter Correlation • Generic Flow Control • Gestion Financière et Comptable

GFCC: Ground Facilities Control Center

GFCF: Gross Fixed Capital Formation

GFCI: Ground-Fault Circuit Interruptor

GFCM: General Fisheries Council for the Mediterranean

GFCR: Gas Filter Correlation Radiometer

GfD: Gesellschaft für Datenverarbeitung

GFD: General Forecasting Division • Geophysical Fluid Dynamics • Gone For the Day • Government-Furnished Data • Government-Furnished Documentation • Greenfield Pope (USA-IN) (location identifier)

GFDL: Geophysical Fluid Dynamics Laboratory

GFE: Government-Furnished Equipment • Graphical Front End • Grenfell (Australia-NSW) (location identifier)

GFEL: Government-Furnished Equipment List

GFERD: Government-Furnished Equipment Requirement Description

GFF: Government-Furnished Facility • Gradient Freeze Furnace • Griffith (Australia-NSW) (location identifier)

GFFC: Geophysical Fluid Flow Cell

GFFIL: Groupement Français des Fournisseurs d'Information en Ligne (France)

GFH: Global Failure Handler

GFI: Gas Flow Indicator • Government Free Issue • Government-Furnished Information • Ground-Fault Interrupt • Ground-Fault Interruptor

GFII: Groupement Français de l'Industrie de l'Information (France) • Groupement Français de l'Industrie de l'Informatique (France)

GFIT: Glass Fiber Insulation Tubing

GfK: Gesellschaft für Klassifikation (Germany)

GFK: Grand Forks (USA-ND) (location identifier)

GFL: Gas Film Levitation • Gas Film Levitator • Glens Falls - Warren County (USA-NY) (location identifier)

GFlops: Billion Floating-Point Operations per Second

GFLOPS: Billion Floating-Point Operations Per Second • GigaFLOPS

GFM: Generic Function Manager • Government-Furnished Material

GFML: Gluten Free Meal

GFMS: Gaseous Flow Measuring System • Generalized File Management System • Generic Function Management System

GFN: Grafton (Australia-NSW) (location identifier)

GFO: Bartica (Guyana) (location identifier) • General Fiber Optics • GEOSAT Follow-On

GFOES: Groupement Français pour l'Observation et l'Étude du Soleil (France)

GFOS: Ground Facilities and Operations Section • Ground Facilities Operations Section

GF&P: Gases, Fluids & Propellants

GFP: General Forecasting Program • Generalized Filter Program • Government Full Period • Government-Funded Procurement • Government-Funded Program • Government-Furnished Property • Graphical Form Painter

GFPBBD: Groupement Français des Producteurs de Bases et Banques de Données (France)

GFPI: Goddard Fabry-Pérot Imager

GFPT: Green-Function Perturbation Technique

GFQ: Gradient Facility with Quenching • Gradient Furnace with Quenching

GFR: German Federal Republic • Glass Fiber Reinforced • Government Facilities Request • Granville (France) (location identifier) • Grim File Reaper • Guaranteed Frame Rate

GFRP: Glass-Fiber-Reinforced Plastic (*) • Graphite-Fiber-Reinforced Plastic

GFRS: Global Foundation for Research and Scholarship (Japan)

GFS: Gas Flow Shielded • Generalized Finite Sequence • Government-Furnished Service • Government-Furnished Software • Grandfather-Father-Son

GFSK: Gaussian Filtered Shift Keying • Gaussian Frequency Shift Keying

GFTM: Generic Fluids Transfer Model

GFTU: General Federation of Trade Unions

GFU: Groupe Fermé d'Usagers • Groupe Fermé d'Utilisateurs

GFURD: Ground Facilities User Requirements Document

GFY: Government Fiscal Year • Grootfontein (Namibia) (location identifier)

GFZ: Geodätisches Forschungszentrum • Geoforschungszentrum

G&G: Geology & Geophysics

G-G: Ground to Ground

G/G: Ground to Ground

GG: Gas Generator • Golden Gate • Governor General • Gravity Gradient • Gravity Gradiometer • Great Gross • Ground Guidance • Grundgesetz

GGAO: Goddard Geodetic and Astronomy Observatory (formerly GORF) (USA-MD) • Goddard Geophysical and Astronomical Observatory (USA-MD)

GGAV: Gefahrgutausnahmeverordnung

GGB: Golden Gate Bridge (USA-CA)

GGC: Galactic Globular Cluster • Gemeenschappelijke Geautomatiseerde Catalogssystemen • Generalized Genetic Code • Ground Guidance Computer • Lumbala (Angola) (location identifier)

GGCM: Geospace General Circulation Model

GGD: Gemeentelijke Geneeskundige Dienst • Gregory Downs (Australia-QLD) (location identifier)

GGDPAC: Government Geoscience Database Policy Advisory Committee

GGE: General Graphical Editing • Georgetown (USA-SC) (location identifier) • Ground Guidance Equipment

ggf: gegebenfalls

GGFC: Global Geophysical Fluids Center

GGFM: Generalized Green's Function Method

GGFR: Geosynchronous Gas Filter Radiometer

GGG: Gadolinium Gallium Garnet • Glacial Geology and Geomorphology • Kilgore (USA-TX) (location identifier) • Longview - Gregg County (USA-TX) (location identifier)

GGGS: Galileo Geometry and Graphics Software

GGI: GPS Geoscience Instrument • Greenhouse Gas Index

GGL: Gilgal (Colombia) (location identifier)

GGM: Gravity Gradient Mode • Gravity Gradiometer Mission

GGN: Gagnoa (Ivory Coast) (location identifier)

GGO: Guiglo (Ivory Coast) (location identifier)

GGP: Gateway-to-Gateway Protocol • General Global Parameter • Generic Genetic Programming • Global Geodynamics Project • GPS Guidance Package • Gross Global Product

GGPD: General Global Parameter Definition • Generalized Gaussian-Poisson Distribution

GGPP: Giant Gaseous Protoplanet

G. gr.: Great Gross

GGR: Garoe (Somalia) (location identifier)

GGS: Geometry and Graphics Software • Global Geoscience • Global Geospace • Global Geospace Science • Global Geospace System • Gobernador Gregores (Argentina-SC) (location identifier) • Goddard GRO Simulator • Graphic Generator System • Gravity Gradient Satellite • Gravity Gradient Sensor • Ground Guidance System

GGSE: Gravity Gradient Stabilization Experiment

GGSN: Gateway GPRS Support Node

GGSS: Gravity Gradiometer Survey System • Grid Giant Star Survey

GGSSS: Global Geospace Surveillance Satellite System

GGT: George Town - Exuma (Bahamas) (location identifier) • Global Geoscience Transects • Gravity Gradient Torque

GGTS: Gravity Gradient Test Satellite

GGVM: Gas Generator Valve Module

GGVS: Gefahrgutverordnung Straße

GGW: Glasgow (USA-MT) (location identifier)

GH: Gaseous Hydrogen • Gas Heater • Ghana (international vehicle registration code) • Ghana (ISO country code) • Giclas Hyades • Global Header • Good Health • Ground Handling • Gaseous Hydrogen

GHA: General Housekeeping Area • Ghana (ISO country code) • Ghardaia (Algeria) (location identifier) • Greenwich Hour Angle • Gyro Header Assembly

GHAAS: Greater Hazleton Area Astronomical Society (USA-PA)

GHAT: Ground Handling And Transportation

GHB: Gesamthochschulbibliothek • Governor's Harbour (Bahamas) (location identifier)

GHC: Cedi (Ghana) (ISO currency code) • Gating Half-Cycle • Generalized Hyperbolic Class • Giant Hydrogen Cloud • Great Harbour Cay (Bahamas) (location identifier) • Ground Half-Coupling • Group Handling Council • Guarded Horn Clause

GHCD: Growth Hormone Concentration and Distribution

GHCN: Global Historical Climatology Network

GHDBS: Generic Hierarchical Database System

GHe: Gaseous Helium

GHE: Garachine (Panama) (location identifier) • Gauss Hypergeometric Equation • Gibbs-Helmholtz Equation • Ground Handling Equipment

GHF: Gauss Hypergeometric Function • Gradient Heating Facility • Gradient Heating Furnace

GHG: Greenhouse Gas

GHIS: GOES High-resolution Interferometer Sounder

GHK: Gush Katif (Israel) (location identifier)

GHL: Garden Hill (Canada) (location identifier)

GHM: Centerville (USA-TN) (location identifier) • Generalized Hall Model • Global History Manager • Golla-Hughes-McTavish

GHMA: Greek Hellenic Mapping Agency

GHMI: Ground Human Machine Interface

GHN: Guanghan (China-PRC) (location identifier)

GHOST: Global Horizontal Sounding Technique

GHP: Gas High Pressure • Geroch-Held-Penrose • Grand Handicapé Physique

GHQ: General Headquarters

GHR: Giant HII Region

GHRIL: Ground-based High-Resolution Imaging Laboratory

GHRS: Goddard High-Resolution Spectrograph

GHS: Gas Handling System • Ground Handling System

GHSM: Groupe Hydraulique Servomoteur

GHT: Gas Holding Tank • Ghat (Libya) (location identifier)

GHU: Gualeguaychu (Argentina-ER) (location identifier)

GHW: Guaranteed Hourly Wage

GHX: Ground Heat Exchanger

GHz: Gigahertz

GHZ: Greenberger-Horne-Zeilinger (experiment)

gi: Gill

Gi: Gilbert

GI: Galvanized Iron • Gated Integrator • General Index • General Information • General Issue • Generalized Instance • Generic Issue • Geographic Information • Geological Institute • Geophysical Institute • Gesellschaft für Informatik (Germany) • Gibraltar (ISO country code) • Government Issue • Grammatical Inference • Guest Investigator

GIA: General-purpose Interface Adapter • Geographical Information Analysis • Geographic Information and Analysis • Gestionnaire Informatique d'Applications

GIAC: Geomatics Industry Association of Canada

GIADA: Grain Impact Analyzer and Dust Accumulator

GIAG: Government-Industry Advisory Group

GIAL: Geographic Information and Analysis Laboratory

GIAM: Global Implications of Applied Microbiology

GIANS: German Integrated Air Navigation System

GIANT: General Information and Analysis Tool • General Integrated Analytical Triangulation

GIAO: Gauge-Invariant Atomic Orbital

GIAPP: Geological Image Analysis Program Package

GIAS: Global Integration And Synthesis

Gib: Gibraltar

GIB: General Instruction Book • Generierung Intelligent Bedienoberflächen • Geographic Information Board • Gibraltar (ISO country code) • Gibraltar (location identifier) • Guy In the Back

GIBIS: Graphical Issue-Based Information System

GIBS: Générateur d'Interface Bord-Sol

GIC: Boigu Island (Australia-QLD) (location identifier) • Geocarto International Centre • Geomagnetically-Induced Current • GPC Interface Console • Graphic Intercalation Compound • Graphite Intercalation Compound • Ground-Induced Current

GICC: Geographic Information Coordinating Committee

GICEL: Groupement des Industries de la Construction Électrique (France)

GICLA: Grupo de Implantação do Centro de Lançamento de Alcântara (Brazil-MA)

GICS: Geocoded Image Correction System • Geographic Identification Code Scheme • Grants Information and Control System

GID: Gesellschaft für Information und Dokumentation (Germany) • Gitega (Burundi) (location identifier) • Grazing Incidence Diffraction • Group Identification • Group Identificator

GIDAP: Guidance Inertial Data Analysis Program

GIDB: Geographic Information Database

GIDD: GSE Interface Definition Document

GIDEP: Government and Industry Data Exchange Program • Government-Industry Data Exchange Program (*) • Government Inter-agency Data Exchange Program

GIDPS: Ground Image Data Processing Software

GIDS: General-purpose Interactive Display System • Generic Intelligent Driver Support • Generic Intelligent Driver System

GIDSC: GLAST Instrument Development Steering Committee

GIDW: General Inverse-Distance Weighted (model)

GIE: Groupe d'Intérêt Économique • Groupement d'Intérêt Économique • Ground Instrumentation Equipment

GIEC: Groupe Intergouvernemental d'Évolution du Climat • Groupe Intergouvernemental pour l'Évolution du Climat • Groupe Intergouvernemental sur l'Évolution du Climat (*)

GIEE: Groupe d'Intérêt Économique Européen • Groupement d'Intérêt Économique Européen

GIEWS: Global Information and Early Warning System

GIF: GAME Interface • General Image Format • General Information File • Graphical Interchange Format • Graphics Interchange Format • Gravito-Inertial Force • Ground Interface Facility • Gulf It to Fortran • Winter Haven - Gilbert (USA-FL) (location identifier)

GIFAD: Graphical Interface For Astronomical Data • Graphical Interface For Astronomical Databases

GIFAS: Groupement des Industries Françaises Aéronautiques et Spatiales (France)

GIFO: Groupement des Industries Françaises de l'Optique (France)

GIFOV: Ground Instantaneous Field Of View

GIFT: Gas-Insulated Flow Tube • General Internet File Transfer • Grupo Interuniversitario de Física Teórica (Spain)

GIG: Geometric Interpolation Grid • GPS IERS and Geodynamics (experiment) • Group Infrastructure Grants • Rio de Janeiro (Brazil-RJ) (location identifier)

GIGA: Grupo de Informatica Gráfica Avanzada

GIGO: Garbage In - Garbage Out

GIGP: Generalized Inverse Gaussian-Poisson

GIGT: Groupe d'Instrumentation des Grands Télescopes (France)

GIGUE: Groupement pour l'Informatique de Gestion des Universités et des Établissements

GIHS: Generalized Information Handling System

GII: Global Information Infrastructure • Siguiri (Guinea) (location identifier)

GIIC: Global Information Infrastructure Commission

GIIS: Groupement International d'Informatique de Santé

GIIT: Groupe Institutionnel de Terminologie et de Documentation

GIITS: General Imagery Intelligence Training System

GIIV: Gated-Image Intensifier Viewer

GIK: Glucose-Insulin-Potassium

GIL: Gaseous Ion Laser • Gilgit (Pakistan) (location identifier) • Graphical Intermediate Language • Green Indicating Lamp

GILC: Global Internet Liberty Campaign

GILD: Gas-Immersed Laser Doping

GILDAS: Grenoble Image and Line Data Analysis Software

GILS: Global Information Locator Service • Government Information Locator Service

GIM: Generalized Information Management • Geomagnetically Induced Motion • Geomagnetic Image Motion • Miele Mimbale (Gabon) (location identifier)

GIMADS: Generic Integrated Maintenance and Diagnostic System • Generic Integrated Maintenance Diagnostic System

GIMEX: Greenland Ice Margin Experiment

GIMI: Gestion Intégrée de Maintenance Informatisée • Global Imaging Monitor of the Ionosphere

GIMM: Grand Interféromètre Millimétrique

GIMMS: Geographic Information Mapping and Management System • Global Inventory Modelling and Monitoring Study • Global Inventory Monitoring and Modelling Study • Global Inventory Mounting Modelling and Simulation

GIMP: Geomagnetically-induced Image Motion Problem (*) • Geomagnetically-Induced Motion Problem • GNU Image Manipulator Program

GIMS: Generalized Information Management System • Geographic Information and Modelling System • Ground Identification of Missions in Space • Group on Instrument and Measurement Science

GIN: Geomagnetic Information Node • Global Information Network • Graphics Input • Green Index Number • Guinea (ISO country code) • Guinée (ISO country code)

GINA: GPS Inertial Navigation Assembly

GINAS: GPS Integrated Navigation and Attitude System

GINO: Graphical Input Output

GINOF: Graphical Input and Output in Fortran

Gio: Giovedi

GIO: Geographic Investigations Office • Government Insurance Office (Australia-NSW) • Graphics Input/Output • Great Island Observatory (USA-RI) • Gurushikhar Infrared Observatory (India)

GIOP: General Inter-ORB Protocol

GIOSP: Groupement International d'Observation des Surfaces Planétaires

GIP: Gaussian Image Point • General Information Programme • Genetic Improvement Program • Gileppe (Belgium) • Glazed Imitation Parchment • Global Inventory Project • Gross Internal Product • Groupement d'Intérêt Public • Guest Investigator Program • Pound (Gibraltar) (ISO currency code)

GIPE: Generation of Interactive Programming Environments

GIPME: Global Investigation of Pollution in the Marine Environment

GIPR: Gigabit Internet Protocol Router

GIpS: Giga-Instructions per Second

GIPS: Generalized Image Processing System • Geographical Information Processing System • Giga-Instructions Per Second • Ground Information Processing System

GIPSY: General Image Processing System • General Information Processing System • Groningen Image Processing System

GIR: Generalized Information Retrieval • Girardot (Colombia) (location identifier) • Groupe Informatique Régional

GIRAFFE: Graphic Interface For Finite Elements

GIRAS: Geographic Information Retrieval and Analysis System

GIRD: General Incentive for Research and Development (Canada) • Ground Integration Requirements Document • Group for the Study of Reaction Motion

GIREP: Groupe International de Recherches en Éducation de la Physique

GIRL: Generalized Information Retrieval Language • German Infrared Laboratory • Graph Information Retrieval Language

GIRLS: General Indexing in Reciprocal Lattice Space • Generalized Information Retrieval and Listing System • Global Interrogation Recording and Location System

GIRMOS: Gemini Infrared Multiobject Spectrograph

GIROS: Graphic Interactive Raster-Oriented System

GIRS: General Information Retrieval System • Graph Information Retrieval System

GIRSS: General Information Retrieval System Simulation

GIRTA: Gestion Intégrée des Rechanges par Traitement Informatisé

GIS: Gas Imaging Spectrometer • Gas-scintillation Imaging Spectrometer • Gemini Interlock System • General Information System • Generalized Information System • Generalized Inquiry System • Generalized Instance Set • Geografisch Informatie Systeem • Geographical Information System • Geographic Information System • Geohydrologic Information Service • Geological Information Service • Geoloscience Information Society • Geospectral Imaging Services (Australia-WA) • Gisborne (New Zealand) (location identifier) • Global Indexing System • Global Information Society • Global Ionospheric Studies • Government Information Services (USA) • Grant Information System • Graphic Information System • Graphic Input System • Groupement d'Intérêt Scientifique • Guidance Information System

GISA: Geochemical Interactive Systems Analysis

GISAT: Ground Identification of Satellites

GISB: Gas Industry Standards Board

GISC: Government Information Services Committee • Guest Investigator Selection Committee

GISE: Generalized Integrated Square Error

GISIG: Geographic Information System Interest Group • Geographic Information Systems International Group

GISMO: General Interpretative System for Matrix Operations

GISMOS: Global Ionospheric Simultaneous Measurements Of Substorms

GISO: General Industry Safety Order

GISP: General Information System for Planning • Geophysical In Situ Probe • Greenland Ice Sheet Project • Group Independent Study Project

GISS: Goddard Institute for Space Studies (USA-NY)

GISSAT: Geographic Information System and Spatial Analysis Technologies

GISSMO: Gas Ionization Solar Spectral Monitor

GIST: Geographic Information System Toolkit • Geographic Information System for Transportation • Geographic Information System Tutorial • Global Information System Technology • Global Information System Test • Government Information Systems Technology • Groupe Informatique Scientifique et Technique

GISW: Generalized Interacting Stellar Wind

GIT: Gastrointestinal Tract • Geita (Tanzania) (location identifier) • General Information Test • Georgia Institute of Technology (USA-GA) • Graph Isomorphism Tester

GITA: Geospatial Information and Technology Association • Groupe d'Informatique et de Traitement Automatique

GITCAD: Global Ionosphere-Thermosphere Coupling And Dynamics • Global Ionospheric Thermospheric Coupling And Dynamics

GITG: Ground Interface Technical Group

GITP: Groupement des Industries de Télécommunication et d'Électronique Professionnelle (France) • Ground Integration Test Program

GITS: Government Information Technology Services

Giu: Giugno

GIU: Guidance Integration Unit

GIUB: Geodetic Institute of the University of Bonn (Germany)

GIUK: Greenland - Iceland - United Kingdom

GIV: Geologic Information Visualization

GIWG: Ground Interface Working Group

GIWTLAF: Globally Integrated Workbench for Teaching and Learning Accounting and Finance

GIX: Global Internet Exchange

GIXS: Grazing-Incidence X-ray Scattering

GIY: Giyani (South Africa) (location identifier)

GIZ: Gizan (Saudi Arabia) (location identifier)

GI2T: Grand Interféromètre à Deux Télescopes

GI3T: Grand Interferometer of Three Telescopes • Grand Interféromètre à Trois Télescopes

GJ: Gigajoule • Gliese-Jahreiß (catalogue) • Grapefruit Juice

GJA: Guanaja (Honduras) (location identifier)

GJI: Geophysical Journal International

GJL: Jijel (Algeria) (location identifier)

GJM: Guajará-Mirim (Brazil-RO) (location identifier)

GJP: Galactic Jupiter Probe • Graphic Job Processor

GJR: Gjogur (Iceland) (location identifier)

GJRO: German-Japanese (Earth) Rotation Observations

GJT: Grand Junction (USA-CO) (location identifier)

Gk: Greek

GKA: Goroka (Papua New Guinea) (location identifier)

GKD: Gemeinsame Körperschaftsdatei • Gordon-Kendall-Davison (notation)

GKE: Geilenkirchen (Germany) (location identifier)

GKH: Gorkha (Nepal) (location identifier)

GKL: Gacs-Kurdyumov-Levin • Great Keppel Island (Australia-QLD) (location identifier)

GKN: Gulkana (USA-AK) (location identifier)

GKO: Kongoboumba (Gabon) (location identifier)

GKS: Graphical Kernel System (*) • Graphics Kernel System

g/l: Grams per Liter

gl: Galician (ISO language code) • Gill • Greenland

Gl: Glaciar • Glacier • Glas • Gleichung • Gleis • Gloss • Glucinum • Gold

G/L: General Ledger

GL: Garage Level • Gas Laser • Gauge Length • General Ledger • General Lexicon • General Linear • Geophysics Laboratory • Geosciences Laboratory • Gliese (catalogue) • Gloucester (UK postal code) • Gothic Letter • Government Laboratory • Grande Ligne • Graphical Link • Graphics Library • Gravitational Lens • Gravitational Lensing • Gray Level • Greatest Length • Greenland (ISO country code) • Green Light • Groenland (ISO country code) • Ground Level

GLA: Gamma-Linolenic Acid • General Ledger Account • Glasgow (Starlink internal code) • Glasgow (UK-Scotland) (location identifier) • Ground Long Activity

GLACS: Ground Loop Attitude Control System

GLADIS: Global Astronomical Data and Information Server

GLADYS: Geophysics Laboratory Array Detector Spectrometer

GLAI: Green Leaf Area Index

Glam: Glamorgan (UK-Wales) • Glamorganshire (UK-Wales)

GLARS: Geoscience Laser Altimetry Ranging System

Glas: Glasgow (UK-Scotland)

GLAS: Geodesic Laser Altimeter System • Geoscience Laser Altimeter System • Goddard Laboratory for Atmospheric Sciences • Global Land AVHRR Sampled

GLASOD: Global Assessment of Soil Degradation

GLASS: Generalized Language and Speech System

GLAST: Gamma Large-Area Silicon Telescope • Gamma-ray Large-Area Space Telescope (*)

GLAT: Geomagnetic Latitude • Groupe de Linguistique Appliquée des Télécommunications

GLB: Glas Lubricated Bearing • Global Location Broker • Grease Lubricated Bearing • Greatest Lower Bound • Ground Location Beacon

GLBC: Great Lakes Basin Commission

GLC: Gas-Liquid Chromatography • Geladi (Ethiopia) (location identifier) • Generator Line Contractor • Gray-Level Cooccurrence • Greater London Council • Ground-Level Concentration

GLCM: Gray-Level Cooccurrence Matrix

Gld: Guilder • Gulden

GLD: Gas Leak Detector • Generalized Levenshtein Distance • Goodland (USA-KS) (location identifier) • Gross Logical Design • Ground Laser Designator • Groupe de Légitime Défense

GLDH: Gray Level Difference Histogram

GLDT: Gas Laser Discharge Tube

GLDV: Gesellschaft für Linguistiche Datenverarbeitung (Germany) • Gray Level Difference Vector

GLE: Gainesville (USA-TX) (location identifier) • Generalized Luminosity Evolution • Ground Level Effect • Ground Level Enhancement • Ground Level Event

GLEAM: Graphic Layout and Engineering Aid Method • Ground-based Lidar Experiment for Atmospheric Measurement

GLEP: Group for Lunar Exploration and Planning

GLERL: Great Lakes Environmental Research Laboratory

GLF: Gaussian Lens Formula • Generalized Lambda Family • Golfito (Costa Rica) (location identifier)

GLG: Glengyle (Australia-QLD) (location identifier)

GLH: Greenville (USA-MS) (location identifier)

GLI: Glen Innes (Australia-NSW) (location identifier) • Global Imager

GLIDS: Grey Level Indicating Display System

GLIM: Generalized Linear Interactive Modelling • Generalized Linear Model

GLIMPCE: Great Lakes International Multidisciplinary Program on Crustal Evolution

GLIMPSE: Global Implicit Search

GLIN: Global Legal Information Network • Grijze Literatuur in Nederland

GLINT: Ground-based Lidar Investigation Transportable

GLIPAR: Guidelines for Investigation, Planning and Research

GLIS: GEAC Library Information System • Geographic and Land Information Society • Global Land Information System

GLK: Galcaio (Somalia) (location identifier)

GLL: Gol Klanten (Norway) (location identifier)

GLLD: Ground Locator Laser Designator

GLM: Generalized Linear Model • General Land Office (USA) • General Linear Model • Gigabaud Link Module • Gigabit Link Module • Glenormiston (Australia-QLD) (location identifier) • Government-Loaned Material • Graduated-Length Method • Grand Livre du Mois

GLN: Glen • Goulimime (Morocco) (location identifier)

GLNS: Glens

GLO: Get the Lead Out • Gloucester (UK-England) (location identifier) • Goddard Launch Operations • Ground Logistics Operation

GLOB: Galaxy-Like Object

GLOBE: Global Backscatter Experiment • Global Land One-kilometer Base Elevation • Global Observations to Benefit the Environment • Global Learning and Observations to Benefit the Environment (*) • Global Learning by Observations to Benefit the Environment

GLOBEC: Global Ocean Ecosystem

GLOBUS: Global Budget of the Stratosphere

GLOCHANT: Global Change Antarctica

GloED: Global Emissions Database

GLOM: Gross Lift-Off Mass

GLOMAC: Global Modelling of Atmospheric Constituents • Global Modelling of Atmospheric Chemistry • Global Modelling of the Atmosphere Chemistry

GLOMEX: Global Meteorological Experiment (now GARP)

GLOMR: Global Low-Orbiting Message Relay • Global Low-Orbit Message Relay (*)

GLONASS: Global Navigation Satellite System

GLOPAC: Gyroscopic Low-Power Attitude Control

GLORI: Global and Local Radar Imager

GLORIA: Geological Long-Range Inclined Asdic

Glos: Glossary • Gloucestershire (UK-England)

Gloss: Glossary

GLOSS: Global Observation Surveillance System • Global Sea Level Observing System • Global Sounding System

GLOW: Gross Lift-Off Weight (*) • Ground Lift-Off Weight

GLP: General Layout Plan • General Linguistic Processor • Good Laboratory Practice • Gross Lawyer Product • Guadeloupe (ISO country code) • Gulgubip (Papua New Guinea) (location identifier)

GLPA: Great Lakes Planetarium Association (USA)

Glpg: Glowplug

GLPS: Genetic Logic Programming System

GLQ: Glennallen (USA-AK) (location identifier)

GLR: Gaylord - Otsego County (USA-MI) (location identifier) • Government Land Register • Graphic Level Recorder

GLRAL: Great Lakes Region of the Astronomical League (USA)

GLRC: Gas-Liquid Radiochromatography • Great Lakes Research Center

GLRD: Great Lakes Research Division

GLRS: Geodynamics Laser Ranging System • Geoscience Laser Ranging System

GLS: Galveston Scholes (USA-TX) (location identifier) • Generalized Least Squares • Generic Learning Support • Global Learning Scheme • Green Laser System • Ground Launch Sequencer

Glsgw: Glasgow (UK-Scotland)

GLSR: Geodynamics Laser Ranging System

Glsry: Glossary

GLT: Gas Laser Tube • Gladstone (Australia-QLD) (location identifier) • Ground Laser Tracking

GLTN: Goddard Laser Tracking Network (USA-MD)

GLTP: General Language Transformation Process

GLU: General Linking Utility • Générateur de Liens Uniformes (*) • Générateur de Liens Universels • Global Land Use

GlV: Globe Valve

GLV: Ginzburg-Landau Vortex • Golovin (USA-AK) (location identifier)

GLW: Glasgow (USA-KY) (location identifier)

GLX: Galela (Indonesia) (location identifier)

GLY: Glossary • Goldsworthy (Australia-WA) (location identifier)

Gly: Glycine

Glyc: Glycerite

Glycn: Glycerine

GLZ: Breda Gilze-Rijen (Netherlands) (location identifier)

gm: Gigameter • Gram

Gm: Gigameter

G&M: Geography & Map

GM: Gambia (ISO country code) • Gambie (ISO country code) • Gaseous Mixture • Gauss-Markov • Geiger-Müller

• Generalized Method • General Maintenance • General Manager • General Meeting • General Member • General Merchandise • General Mortgage • General Motors Corporation • Genetically Modified • Geometric Mean • Grand Master • Gravitational Mass • Greenwich Meridian • Grid Modulation • Ground Malfunction • Ground Mobile • Ground Mode • Group Mark • Group Mobile • Guam

GM&A: General Management & Administration

GMA: Gemena (Zaire) (location identifier) • General Management and Administration • General Mental Ability • Giant Molecular Association • Government Modification Authorization • Graphe des Marquages Accessibles • Gruppo Marsicano Astrofili (Italy) • Guest Observer

GMAB: Give Me A Break

GMAG: Geophysics Magnetometer

GMAN: Global Microlensing Alert Network

GMAO: Gestion de Maintenance Assistée par Ordinateur

GMAP: Geometric Modelling Applications Programme

GMAPS: General Map Analysis Planning System

GMAT: Graduate Management Admission Test • Greenwich Mean Astronomical Time

GMAW: Gas Metal Arc Welding

GMB: Gambela (Ethiopia) (location identifier) • Gambia (ISO country code) • Gambie (ISO country code) • Gran Mancha Blanca (Saturno)

GmbH: Gesellschaft mit beschränkter Hafting

Gmc: Germanic

GMC: General Management Committee • Giant Molecular Cloud • Glinka Memorial Collection • Guerima (Colombia) (location identifier)

GMCA: Genetic Maximum Clique Algorithm

GMCC: General Monte Carlo Code • Geophysical Monitoring for Climate Change • Ground Mobile Control Centr

GMCF: Goddard Mission Control Facility

GMCL: Ground Measurements Command List

GMCS: Giant Molecular Cloud Structure

GMD: Dalasi (Gambia) (ISO currency code) • Gel Microdrop • General Management Directive • Geometric Mean Distance • Gesellschaft für Mathematik und Datenverarbeitung (Germany) • Ground Meteorological Detector • Ground Meteorological Device

GMDA: Group Method of Determining Arguments

GMDB: Grid Meta-Database

GMDES: Gridded Meteorological Data Extraction System

GMDH: Group Method of Data Handling

GMDL: GEC Marconi Dynamics Ltd.

GMDSS: Global Maritime Distress and Safety System

GME: Gimbal Module Electronics • Goddard Medium Energy • Gomel (Belarus) (location identifier) • Grupo de Mecánica Espacial

GMES: Global Monitoring for Environment and Security

GMF: Ground Module Force

GMI: Gasmata (Papua New Guinea) (location identifier) • Goddard Management Instruction

GMIP: Government Mandatory Inspection Point

GMIR: Global Merged Interaction Region

GMIS: Generalized Management Information System • GSSC Management Information System

GMK: Gyromagnetic Kompass

GML: Generalized Markup Language • Generalized Maximum Likelihood • General Markup Language • General Measurement Loop • Graphic Machine Language

GMLCC: Ground Mobile Launch Control Center

GMLS: Geometrical Method of Least Squares

GMM: Gamboma (Congo) (location identifier) • Gaussian Mixture Model • Generalized Mixed Model • Geometrical Mathematical Model

GMMSS: Global Multimedia Mobile Satellite System

GMN: Greymouth (New Zealand) (location identifier)

GMO: Genetically-Manipulated Organism • Genetically-Modified Organism

GMOS: Gemini Multiobject Spectrograph

GMP: Genetically Modified Plant • Good Manufacturing Practice

GMPCS: Global Mobile Personal Communications by Satellite

GMPI: General and Mechanical Patents Index

GMPL: Global Managed Private Line

GMPLS: Global Multimode Precision Landing System

GMPR: Geostationary Microwave Precipitation Radiometer

GMQ: Good Merchantable Quality

GMQS: Gruppo Milanese Quadranti Solari (Italy)

GMR: Gambier Island (location identifier) • Geometric Mean Radius • Giant Magneto-Resistance • Giant Magneto-Resistive • GOES-Meteosat Relay • Ground Mapping Radar

GMRES: Generalized Minimal Residual • Generalized Minimum Residual

GMRF: Gaussian Markov Random Field

GMRP: GARP Multicast Registration Protocol

GMRT: Giant Meterwave Radio Telescope (India)

GM&S: General, Medical & Surgical

GMS: General Maintenance System • Geophysical Monitoring Satellite • Geostationary Meteorological Satellite • Global Mapping Satellite • Global Messaging Service • Grant-Maintained Status • Gravitational Mass Sensor • Gravity Measuring System • Ground Maintenance Support • Guidance Monitor Set • Guimarães (Brazil-MA) (location identifier)

GMSC: Geologic Map Standards Committee

GMSD: Gas Microstructure Detector

GMSK: Gaussian Minimum-Shift Keying

GMSS: Geostationary Meteorological Satellite System • Global Meteorological Satellite System

GMST: Greenwich Mean Sidereal Time

GMT: Geiger-Mueller Tube • General Machine Test • Geomarine Technology • Granite Mountain (USA-AK) (location identifier) • Greenwich Mean Time

GMTA: Great Minds' Think Alike

GMTF: Geometric Modulation Transfer Function

Gmtry: Geometry

GMU: George Mason University (USA-VA) • GPS Monitoring Unit • Greenville Downtown (USA-SC) (location identifier)

GMV: Grupo de Mecánica de Vuelos SA (Spain) • Guaranteed Minimum Value

GMW: Gram Molecular Weight • Große Magellansche Wolke

GMY: Rheindahlen (Germany) (location identifier)

gn: Green • Guarani (ISO language code) • Guinea

G&N: Guidance & Navigation

G/N: General Notes

GN: Gauss-Newton • General Notes • Graduate Nurse • Ground Network • Guidance and Navigation • Guinea (ISO country code) • Guinée (ISO country code)

GNA: Gaseous Nitric Acid • Globewide Network Academy (USA-TX) • Grodna (Belarus) (location identifier)

GNAAC: Global Network Associated Analysis Center

GNAS: General National Airspace System

GNAT: General Numerical Analysis of Transport • Global Network of Astronomical Telescopes • Global Network of Automatic Telescopes

GNATS: General Non-linear Analysis of Two-Dimensional Structures

GNB: Grenoble Saint-Geoirs (France) (location identifier) • Guinea-Bissau (ISO country code) • Guinée-Bissau (ISO country code)

GNBS: Gyuana National Bureau of Standards

GN&C: Guidance, Navigation & Control

GNC: Geologic Names Committee • Global Navigation Chart • Goddard Network Control • Graduated Non-Convexity • Graphical Numerical Control • Graphic Numerical Control • Guidance and Navigation Computer • Guidance and Navigation Control • Guidance, Navigation and Control (*)

GNCIS: Guidance, Navigation and Control Integration Simulator

GN&CS: Guidance, Navigation & Control System

GNCS: Guidance and Navigation Control System • Guidance, Navigation and Control System

Gnd: Ground

GND: Grenada (Windward Islands) (location identifier) • Ground

GNDFG: Ground Fog

GNE: Gent (Belgium) (location identifier) • Guidance and Navigational Electronics • Guidance and Navigation Electronics • Guidance and Navigation Equipment

GNF: Franc (Guinea) (ISO currency code)

GNFMS: Gaseous Nitrogen Flow Measuring System

GNG: Gaussian Noise Generator • Gooding (USA-ID) (location identifier)

GNI: Get Net Info • Global Network Infrastructure • Green Island (China-ROC) (location identifier) • Gross National Income • Gross National Investment

GNIR: Get Net Info Reply

GNIRS: Gemini Near-Infrared Spectrograph

GNIS: Geographic Names Information System

GNL: Gaz Naturel Liquéfié

GNLS: GPS Navigation and Landing System

Gnltd: Granulated

GNM: Grand Nuage de Magellan • Guanambi (Brazil-BA) (location identifier)

GNMS: Gaseous Nitrogen Measuring System

GNN: Ghinnir (Ethiopia) (location identifier) • Global Network Navigator

GNO: Gallium Neutrino Observatory • Gross Product Originating

GNOL: Good Neighbor Lighting

GNOLE: Good Neighbor Lighting Experiment

Gnom: Gnomónica

GNOMES: Generalized NRAO Observe, Multi-telescope Experiment Scheduler

GNOS: Goddard Network Operations Support

GNOSIS: Gesellschaft für Normierte Organisations-Software und Informationsstandards

GNP: Graphics Nesting Program • Gross National Product

GNQ: Equatorial Guinea (ISO country code) • Guinea (Equatorial_) (ISO country code) • Guinée Équatoriale (ISO country code)

GNR: Gaseous Nuclear Rocket • General Roca (Argentina-RN) (location identifier) • Genetics-Nanotechnology-Robotics • Geographical Names Register

GNRT: Generalized Near-Real Time

gns: Guineas • Guinées

G&NS: Guidance & Navigation System

GNS: Franc (Guinea) (ISO currency code) • Global Network Service • GOMR Nadir Sounder • Graphics Name Service • Guidance and Navigation Simulator • Guidance and Navigation System • Guidance Navigation System • Gunungsitoli (Indonesia) (location identifier)

GNSO: Goddard Network Support Operations

GNSS: Global Navigation Satellite System

GNT: Grants Milan (USA-NM) (location identifier)

GNU: General Neural Unit • Geologic Names Unit • GNU is Not Unix • Goodnews Bay (USA-AK) (location identifier)

GNULEX: Geologic Names Unit Lexicon

GNUT: Goddard-Navy Universal Time

GNV: Gainesville (USA-FL) (location identifier) • Gaz Naturel pour Véhicule

GNVQ: General National Vocational Qualification

GNY: Granby - Grand County (USA-CO) (location identifier)

GNZ: Ghanzi (Botswana) (location identifier)

GO: Garantie d'Origine • Gas Operated • Gaussian Orbital • Gebruikerondersteuning • General Observer • General Office • General Order • General Organization • Generalized Operation • Geometrical Optics • Geometry-Optimized • Global Optimization • Goiás (Brazil) • Government-Operated • Government-Owned • Grandes Ondes • Graphitic Oxide • Guest Observer

GOA: Generic Open Architecture • Gênes (Italy) (location identifier) • Genoa (Italy) (location identifier) • Geologisch-Oceanografisch-Atmosferisch • Gyro-Optics Assembly • Gyro Output Amplifier

GOAA: Greater Orlando Airport Authority (USA-FL)

GOAL: Galactic Origin and the Acceleration Limit • General Orientation Analysis Language • Ground Operations Aerospace Language

GOALI: Grant Opportunities for Academic Liaison with Industry

GOALS: Global Ocean-Atmosphere-Land System

GOAS: Guidance Optical Alignment Shelter

GOAT: Gerber Oscillogram Amplitude Translator • Goes Over All Terrain

GOB: Goba (Ethiopia) (location identifier)

GOBELIN: GOME Breadboard Model Experimental Link

GOBEP: Generalized One-Boson Exchange Potential

GOC: General Optical Council • Gora (Papua New Guinea) (location identifier) • Greek Orthodox Church • Ground Operations Coordinator • Ground Operations Center

GOCA: Ground Operations Control Area

GOCAP: Graphic Output Circuit Analysis Program

GOCC: GATE Operational Control Center • Geodetic Operations Control Center • Global Operations Control Center

GOCE: Gravity and Ocean Circulation Explorer

GOCO: Government-Owned - Contractor-Operated

GOCP: Groupe des Observateurs du Ciel Profond (France)

GOD: Giant Optical Device

GODAE: Global Ocean Data Assimilation Experiment

GODAS: Graphically-Oriented Design and Analysis System

GODC: Geosciences Operational Data Center

GODIVA: GOME Data Interpretation, Validation and Application

GODSEP: Guidance and Orbit Determination for Solar Electric Propulsion

GOE: Gaussian Orthogonal Ensemble • General Operating Expense • Gonalia (Papua New Guinea) (location identifier) • Government-Owned Equipment • Ground Operating Equipment • Ground Operational Equipment • Ground Outfitting Equipment

GOED: Global Oceans Ecosystems Dynamics

GOES: Geostationary Operational Environmental Satellite (*) • Geostationary Orbiting Environmental Satellite • Geosynchronous Operational Environmental Satellite • Geosynchronous Orbiting Environmental Satellite • Global Omnibus Environmental Survey

GOEZS: Global Ocean Euphotic Zone Study

GOF: Global Ocean Flux • Goodness Of Fit • Government-Owned Facility • Guest Observer Facility • San Angelo - Goodfellow (USA-TX) (location identifier)

GOFS: Global Ocean Flux Study

GOGO: Government-Owned - Government-Operated

GOH: Nuuk (Greenland) (location identifier)

GOHREM: Geosynchronous Orbit High-Resolution Earth Monitoring

GOHSS: Galileo OH Subtracted Spectrometer

GOI: Geostationary Operational Imager • Goa Dabolim (India) (location identifier) • Government-Owned Installation

GOIN: Global Observation Information Network

GOJ: Nizhnij Novgorod (Russia) (location identifier)

GOK: Guthrie (USA-OK) (location identifier)

GOL: General Operating Language • Gold Beach State (USA-OR) (location identifier)

GOLD: General Organization of Large Databases • Geophysical Online Data • Global Online Data

GOLDIS: Geophysics Online Data and Information System

GOLEM: Großpeicher Orientierte Listenorganisierte Ermittlungsmethode

GOLF: Global Oscillations at Low Frequencies (*) • Global Oscillations Low-Frequency

GOLISS: Galileo Online Search System

GOLS: German Oceanic Lidar System

GOM: Global Ocean Monitoring • Goma (Zaire) (location identifier) • Government-Owned Material • Grand Old Man

GOME: Global Ozone Monitoring Experiment

GOMI: Global Ozone Monitoring Instrument

GOMMS: Ground Operations and Material Management System

GOMOS: Global Ozone Monitoring by Occultation of Stars

GOMPS: Global Ocean Monitoring Payload Study

GOMR: Global Ozone Monitoring Radiometer

GOMS: Geostationary Orbiting Meteorological Satellite • Global Ocean Monitoring Satellite • Global Ozone Monitoring Satellite • Goals-Operators-Methods-Selection • Ground Operations Management System • Geostationary Operational Meteorological Satellite

GOMY: Get Out of My Way!

GON: Groton (USA-CT) (location identifier) • New London (USA-CT) (location identifier)

GONG: Global Oscillations Network Group

GOO: Generalized Overhauser Orbital • Goondiwindi (Australia-QLD) (location identifier) • Ground Observer Organization • Ground Operation Order

GOOB: Going Out Of Business

GOOS: Global Ocean Observing System • Global Ozone Observing System

GOP: Geographical Observatories Programme • Good Old Paper • Gorakhpur (India) (location identifier) • Government-Owned Property • Grand Old Party (USA) • Ground Operations Panel • Ground Operations Plan • Group Of Pictures • Guest Observer Program

GOPE: Government-Owned Plant Equipment

GOPEX: Galileo Optical Experiment

GOPG: Ground Operations Planning Group

GOpS: Gigabyte Operations per Second • Giga Operations per Second

GOPS: Gigabyte Operations Per Second • Giga Operations Per Second

GOQ: Golmud (China-PRC) (location identifier)

GOR: General Operating Room • Gore (Ethiopia) (location identifier) • Ground Operations Review

GORBI: Graphical Object Retrieval and Browsing Interface

GORD: Ground Operations Requirements Document

GORDES: Groupe Opérationnel de Recherche, de Documentation et d'Étude sur la Science (Belgium)

GORF: Goddard Optical Research Facility (now GGAO)

GORMP: Global Ozone Research and Monitoring Program • Global Ozone Research and Monitoring Project

GORP: Ground Operational Requirements Plan • Ground Operations Requirements Plan • Ground Operations Review Panel

GORR: Ground Operations Readiness Review

GORS: General Online Retrieval System • General Organization Of Remote Sensing

GOS: Gemenebest van Onafhenkelijke Staten • Generalized Oscillator Strength • Geodetic Optical System • Geomagnetic Observing System • Geostationary Operational Sounder • Global Observing System • Global Ozone Sensor • Gosford (Australia-NSW) (location identifier) • Grade Of Service • Gross Operating Surplus • Ground Operations System • Ground Optical Station • Guest Operating System

GOSAM: Gelatin of Sols: Applied Microgravity Research

GOSB: General Observer Support Branch

GOSEAC: Group of Specialists on Environmental Affairs and Conservation

GOSIP: Government Open Systems Interconnection Profile

GOSP: Goldendale Observatory State Park (USA-WA) • Ground Operational Support Plan

GOSRP: Ground Operational Support Requirements Plan

GOSS: GOES Operational Support System • Ground Operational Support System • Ground Operations Support System • Groupe des Opérateurs de Stations SPOT

GOSSP: Global Observing Systems Space Panel

GOSST: Global Ocean Sea Surface Temperature

GOST: Gravito-Optical Surface Trap

Gót: Gótico

GOT: General Observing Time • Göteborg Landvetter (Sweden) (location identifier) • Gothenburg Landvetter (Sweden) (location identifier) • Government-Owned Terminal

GOTF: Goddard Optical Test Facility

Goth: Gothic

GOTS: Governmental Off-The-Shelf • Government Off-The-Shelf • Gravity-Oriented Test Satellite

Gou: Gourde (currency)

GOU: Garoua (Cameroon) (location identifier)

Gouv: Gouvernement • Gouvernemental

Gov: Government • Governmental • Governor

GOV: Generator Output Voltage • Gove Nhulunbuy (Australia-NT) (location identifier) • Government

Gov. Gen.: Governor General

Govt: Government

GOW: Garrow Lake (Canada-NT) (location identifier)

GOWG: Ground Operations Working Group

GOWI: Get On With It

GOX: Gaseous Oxygen

GOY: Gal Oya - Amparai (Sri Lanka) (location identifier)

GOYA: Get Off Your After-end

GOZ: Gorna Orjahovica (Bulgaria) (location identifier)

GO3OS: Global Ozone Observing System

g/p: Giro Postal

gp: Galley Proofs

Gp: Group

GP: Galactic Probe • Gallup Poll • Gas-Permeable • Gas Plasma • Gas Pressure • General Paralysis • General Pause • General Practitioner • General Processing • General Product • General Protection • General Provision • General Publication • General Purpose • Generalized Programming • Genetic Programming • Geographical Pole • Geographical Position • Geographic Point • Geometric Pitch • Geometric Progression • Geschichte der Physik • Glidepath • Gloria Patri • Government Property • Graduated Pension • Grand Prix • Grand Public • Graphic Production • Graphics Processor • Gratuity Passenger • Gravity Probe • Great Primer • Group Permit • Group Processor • Guadeloupe (ISO country code) • Guidance Package • Guinea Pig • Gunn-Peterson • Gun Position • Gyro Package

GPa: Gigapascal

GPA: General-Purpose Amplifier • Gestion Partagée des Approvisionnements • Grade Point Average • Gravity Probe A • Great Problem of Astrophysics • Ground Plane Antenna • Grounded Plate Amplifier • Groupe Patrimoine Astronomique (France) • Guidance Positioning Assembly • Patras Araxos (Greece) (location identifier)

GPAA: General Physics Advanced Abstracts

GPAC: General-Purpose Analog Computer

gpad: Gallons per Acre per Day

GPADS: Guided Parafoil Airborne Aerial Delivery System • Guided Precision Aerial Delivery System

GPAMS: General-Purpose Automated Maintenance System

GPAO: Gestion de Production Assistée par Ordinateur

GPAPD: Gallons Per Acre Per Day

GPAS: General-Purpose Airborne Simulator

GPB: General-Purpose Buffer • Gravity Probe B • Group Parameter Block • Guarapuava - Tancredo Thomaz Faria (Brazil-PR) (location identifier)

GPBIM: General-Purpose Buffer Interface Module

GP&C: Global Positioning & Communication

GPC: Gauge Pressure Control • Gel Permeation Chromatograph • Gel Permeation Chromatography • General Precision Connector • General-Purpose Committee • General-Purpose Computer • Global Planetary Change (periodical code) • Global-Processing Center • Granville-Phillips Co. • Graphic Performance Characterization • Graphics Performance Committee • Gyro Pulse Count

GPCC: Global Precipitation Climatology Centre

gpcd: Gallons per Capita per Day

GPCIPL: General-Purpose Computer Initial Program Load

GPCOE: General-Purpose Checkout Equipment

GPCP: Global Precipitation Climatology Project

GPCPD: Gallons Per Capita Per Day

gpd: Gallons per Day

GPD: Gas Pressure Dominated • Generalized Probabilistic Descent • General Presentation Descriptor • General Publishing Division • General-Purpose Data • Gibbs Probability Distribution • Gimbal Position Display

GPDC: General-Purpose Digital Computer

GPDL: Graphical Picture Drawing Language

GPDS: General-Purpose Display System

GPE: Galactic Plane Enhancement • Gas Power Exchange • General-Purpose Equipment

GPEC: Gestion Prévisionnelle des Emplois et des Compétences

GPF: General Protection Fault • Geospatial Prototype Facility

GPFS: General-Purpose Flight Simulator

GPG: Grant Proposal Guide • Graphic Product Generation

GPGL: Graphical-Purpose Graphic Language

gph: Gallons per Hour

Gph: Graphite

GPHS: General-Purpose Heat Source

GPI: General Physics Institute • General-Purpose Interface • General-Purpose Inverter • Genuine Progress Indicator • Gimbal Position Indicator • GOES Precipitation Index • Grand Projet Innovant • Graphics Programming Interface • Ground Point of Interception • Ground Position Indicator • Guapi (Colombia) (location identifier)

GPIB: General-Purpose Interface Bus

GPID: Goals, Processes and Indicators of Development

GPIO: General-Purpose Input/Output

GPIP: General-Purpose Image Processor

GPIRS: GSFC Photographic Image Retrieval System

GPKD: General-Purpose Keyboard and Display

GPL: Gaz de Pétrole Liquéfié • General Public Licence • General-Purpose Laboratory • General-Purpose Language • Generalized Programming Language • Geometric Processing Library • GNU Public License • Graphic Programming Language • Guapiles (Costa Rica) (location identifier)

GPLD: Government Property Lost or Damaged • Government Property Lost or Destroyed

GPLI: Groupement Permanent de Lutte contre l'Illettrisme (France)

gpm: Gallons per Minute

GpM: Gallons per Mile • Groups per Minute • Groups per Message

GPM: Gaussian-Peak Method • Genotype-Phonetype Mapping • Gestalt Photomapper • Graduated Payment Mortgage • Ground Potential Model • Groups Per Minute • Groups Per Message

GPME: General-Purpose Mission Equipment

GPMS: Galileo Probe Mass Spectrometer • Global Position Management System

GPN: Garden Point (Australia-NT) (location identifier) • General Performance Number

GPNDB: Galactic Planetary Nebulae Database

GPNDS: Global Positioning and Nuclear Detection System

Gpo: Grupo

GPO: Gemini Project Office • General Pico (Argentina-LP) (location identifier) • General Post Office • General-Purpose Oscilloscope • General-Purpose Output • Government Printing Office • Grand Prisme-Objectif

GPOS: Government Printing Office Style (manual)

GPP: Gamma-Poisson Process • General Plant Projects • GLOW/PADSE Payload • Gross Primary Production • Gyro Pitch Position

GPPA: Great Plains Planetarium Association (USA)

GPPE: General Purpose Processing Equipment

GPPF: Giant Planet Probe Facility

GPPL: General-Purpose Programming Language

gppm: Graphics Pages per Minute

GPR: General-Purpose Radiometer • General-Purpose Receiver • General-Purpose Register • General-Purpose Relay • General-Purpose Representative • Genio Populi Romani • Government Property Register • Graphical Patient Record • Greenwich Photoheliographic Result • Ground-Penetrating Radar

GPRA: Government Performance and Results Act (USA)

GPRC: Geophysical and Polar Research Center

GPRDR: Global Power Reflection Data Record

GPRN: GOAL Procedure Release Notice

GPRP: Government Production Research Property

GPRR: General-Purpose Radio Receiver

GPRS: General Packet Radio Service

GPRT: General-Purpose Radio Transmitter

gps: Gallons per Second

GPS: Galactic Plane Survey • Galapagos Island (Ecuador) (location identifier) • Gauge Pressure Switch • Generality and Problem Solving • Generalized Parabolic Sheave • General Problem Solver • General Processing System • General-Purpose Shelter • General-Purpose Simulator • Genetic Programming System • Geophysical Processor System • Géopositionnement Par Satellite • GHz Peaked Spectrum • Gigahertz Peaked Spectrum • Global Positioning Satellite • Global Positioning System (*) • Global Position Sensor • Global Position System • Government Paper Specification • Government Paper Standard • Graduated Pension Scheme • Ground-based Planetary Spectroscopy • Ground Power Supply • Ground Processing Simulation • Groups of Pulses per Second • Guidance Power Supply • Gyroscope Parameter Shift

GPSCO: Global Position System Consortium

GPSDIC: General-Purpose Scientific Document Image Code

GPSDR: Global Positioning System Demonstration Receiver

GPSDW: General-Purpose Scientific Document Writer

GPSE: General-Purpose Simulation Environment

GPSI: General-Purpose Serial Interface

GPSIC: Global Positioning System Information Center

GPSIM: Global Position System Integrity Monitoring

GPSP: General-Purpose Signal Processor • General-Purpose Software Program • Great Plains Star Party (USA)

GPSR: Global Positioning System Receiver

GPSS: General-Purpose Simulation System • General-Purpose Simulator System • General-Purpose Systems Simulator

GPSSP: General-Purpose Simulation System Package

GPST: Global Positioning System Time

GPSU: Ground Power Supply Unit

GPSW: Global Positioning System World (periodical code)

Gpt: Groupement

GPT: Biloxi (USA-MS) (location identifier) • General-Purpose Technology • General-Purpose Tool • Geometric and Positional Tolerance • Guidance Position Tracking • Gulfport (USA-MS) (location identifier)

GPTA: Guangdong Post and Telecommunications Administration (China-PRC)

GPTE: General-Purpose Test Equipment

GPTS: Geomagnetic Polarity Time Scale

GPU: Gas Power Unit • General Postal Union • Generating Power Unit • Graphics Processing Unit • Ground Power Unit

GPV: General-Public Virus • General-Purpose Vehicle • Gyroscope Pickoff Voltage

GPVI: Graphic Processor Video Interface

GPVS: General-Purpose Visual Simulator

GPWS: Ground Proximity Warning System

GPX: Galactic-Plane X-ray • Generalized Programming Extended

GPZ: Grand Rapids (USA-MN) (location identifier)

GQ: Equatorial Guinea (ISO country code) • General Quarters • Geologic Quadrangle • Guinea (Equatorial_) (ISO country code) • Guinée Équatoriale (ISO country code)

GQA: Get Quick Answer • Give Quick Answer

GQED: Gravitational Quasi-Equilibrium Distribution

GQG: Grand Quartier Général

GQJ: Machrihanish (UK) (location identifier)

GQL: Graphical Query Language

GQM: Global Query Manager • Goal-Question-Metric

GQQ: Galion (USA-OH) (location identifier)

GQR: Gauss Quadrature Rule • Giant Quadrupole Resonance

gr: Gram • Greece • gross • Grote

Gr: Gear • Graad • Graben • Grade • Grain • Grammar • Grand • Grashof Number for Mass Transfer • Gravity • Great • Greece • Greek • Grey • Griego • Groschen (currency) • Ground • Group • Grove • Gründung

GR: Game Reserve • Gamma Ray • Gas Retention • Géante Rouge • General Reconnaissance • General Relativistic • General Relativity • General Research • General Reserve • Generic Relation • Giclas Red • Glide Ratio • Government Regulation • Government Report • Grande Randonnée • Grande Remise • Grand Répulseur • Gravitational Radiation • Great Repulsor • Grèce (ISO country code) • Greece (international vehicle registration code) • Greece (ISO country code) • Greece (postal country code) • Greedy Refinement (algorithm) • Greek • Ground Rule • Groupe de Recherche • Growth Rate

GRA: Gamarra (Colombia) (location identifier) • Gamma-Ray Amplification • Gamma-Ray Astronomy • Gamma Resonance Absorption • Graduate Research Assistant • Graduate Research Assistantship • Graduate Research Associate • Gyro Reference Assembly

GRAAA: Grand Rapids Amateur Astronomical Association (USA-MI)

GRAAL: Graph Algorithmic Language • Groupe de Recherche en Astronomie et Astrophysique du Languedoc (France)

Grab: Grabado

GRACE: Gamma-Ray Astrophysics Coordinated Experiments • Gravity Recovery And Climate Experiment • Ground Radar Calibrating Equipment

Grad: Gradient • Grading • Gradual • Graduate • Graduated

GRAD: Gamma-Ray Advanced Detector • Generalized Remote-Access Database • Gradient

GRADE: Graphical Airspace Design Environment

GrADS: Grid Analysis And Data System

GRADS: Grid Analysis And Data System • Grid Analysis And Display System

GrAeH: Grumman Aerospace Horizons (periodical code)

GRAF: Graphic Addition to Fortran

GRAFH: Groupe de Radioastronomie Franco-Helvétique

GRAID: Graphical Aid

GRAIL: General Real-time AIDA Interface Language

GRAIN: Graphics-oriented Relational Algebraic Interpreter

Gral: General

Gram: Gramática • Grammar • Grammarian • Grammatical

GRAM: Generalized Rank Annihilation Method • General Reference Atmospheric Model • Generic Radio Access Network • Grupo de Rádio-Astronomia Mackenzie (Brazil-SP) (now CRAAM)

GRAMME: Groupement pour la Recherche et les Applications de la Micropesanteur aux Matériaux de l'Espace

GRAMS: Ground Receiving And Monitoring System

Gran: Granada (Spain) • Granite • Granular • Granulated

GRAND: Gamma-Ray Astrophysics at Notre-Dame

GRANDE: Gamma-Ray And Neutrino Detection Experiment • Gamma-Ray And Neutrino Detector

GRANTECAN: Gran Telescopio de Canarias

GRAPE: Gamma-Ray Attenuation Porosity Evaluator • Graphical Analysis of Program Execution • Gravitational Pipe • Gravitational Pipeline • Gravity Pipe • Gravity Pipeline

GRAPES: Group for the Advancement of Positional and Evolutionary Studies

GRAPHIC: Group And Phase Ionosphere Calibration

GRAPWG: Gamma-Ray Astronomy Program Working Group

GRARD: Goddard Range And Range Data

GRARR: Goddard Range And Range Rate

GRAS: Generally Recognized As Safe • Generally Regarded As Safe • George R. Agassiz Station (USA-AZ) • Glonass Receiver for Atmospheric Sounding • GNSS Receiver for Atmospheric Sounding

GRASP: Gamma-Ray Astronomy with Spectroscopy and Positioning • General-purpose Relativistic Atomic Structure Programme • General Resource Allocation and Selection Program • Generalized Reentry Application Simulation Program • Generalized Retrieval and Storage Program • Generic Retrieve/Archive Services Protocol • GONG Reduction and Analysis Software Package • Government Relations And Science Policy

GRASS: Geographical Resources Analysis Support System

GRATE: Generic Rules and Agent model Testbed Environment

GRATIS: Gamma-Ray Arcminute Telescope and Imaging System • Gamma-Ray Arcminute Telescope Imaging System (*)

Grav: Gravitational • Gravity • Gravure

GRAZAR: Gamma-Ray BLAZAR

GRB: Gamma-Ray Burst • Gamma-Ray Burster • Green Bay (USA-WI) (location identifier)

GRBM: Gamma-Ray Burst Monitor

Gr. Br.: Great Britain

Gr. Brit.: Great Britain

Grbx: Gearbox

GRC: Generation of Reproducible Copy • Gestion de la Relation Client • Global Reference Code • Grand Cess (Liberia) (location identifier) • Grêce (ISO country code) • Greece (ISO country code) • Ground Resolution Cell

GRCA: Ground Reference Coverage Area

Gr. Cap.: Grande Capitale

GRCDA: Governmental Refuse Collection and Disposal Association

Grd: Grand • Grind • Guard

GRD: Drachma (Greece) (ISO currency code) • General Requirements Document • Genetic Reconfiguration of DSPs • Graphics Display • Greatest Response Data • Greenwood (USA-SC) (location identifier) • Grenada (ISO country code) • Grenade (ISO country code) • Ground • Ground-Resolved Distance

GRDB: Geoscientific Resource Database

GRDC: Global Runoff Data Centre

GRDEN: Garden

GRDL: Geodetic Research and Development Laboratory • Geosciences Research and Development Laboratory • Gradual

GRDN: Garden

GRDNS: Gardens

Grdtn: Graduation

Gre: Greek

GRE: Galactic Ridge Emission • Gamma-Ray Experiment • Gamma-Ray Explorer • Giotto Radio-science Experiment • Graduate Record Examination • Grant-Related Expenditure • Graphics Engine • Greenville (USA-IL) (location identifier) • Ground Reconstruction Electronics • Ground Resolution Element

GREAS: Generic Resource Event and Activity Scheduler

GREB: Galactic Radiation Experiment Background

GREC: Groupement de Recherche sur les Émetteurs de Chaleur

GRECE: Groupement de Recherches et d'Études sur la Civilisation Européenne

GRECO: Groupement de Recherches Coordonnées

Grecolat: Grecolatino

GRED: Generalized Random Extract Device • General Relativistic Electrodynamics

GREDB: Gamma-Ray Event Database

GREDR: Global Reflectivity Data Record

GREEN: Global Data Resource for Earth and the Environment

GREENS: Global Redevelopment with Energy Environment Sustainability

GREG: Groupe de Relativité Générale

GReGr: General Relativity and Gravitation (periodical code)

GREOS: Groupe de Recherches et d'Études en Océanographie Spatiale • Groupement pour la Recherche En Océanographie Spatiale

GREP: Global Regular Expression Print

GRESIC: Groupe de Recherche Expérimentale sur les Systèmes Informatisés de Communication

GRESS: Gradient Enhanced Software System

GRETA: Groupe d'Établissements

GREX: Gamma-Ray Experiment

GRF: Gaussian Random Field • Gibbs Random Field • Gradual Rise and Fall • Graph • Gravity Research Foundation • Großer Roter Fleck (Jupiter) • Tacoma Gray (USA-WA) (location identifier)

GRFL: Groundwater Remediation Field Laboratory

GRG: Gardez (Afghanistan) (location identifier) • General Relativity and Gravitation • Giant Radio Galaxy

GRGS: Groupe de Recherches de Géodésie Spatiale • Groupe de Recherches en Géodésie Spatiale

GRGT: Guam Remote Ground Terminal

GRH: Garuahi (Papua New Guinea) (location identifier) • Gestion des Ressources Humaines • Gramm-Rudman-Hollings

GRI: Government Reports Index • Grand Island (USA-NE) (location identifier) • Group Repetition Interval

GRID: Gamma-Ray Imaging Device • Global Resources Information Database • Gradient Imaging Display Software • Graphic Remote Integrated Display • Graphic Retrieval and Information Display

GRIDS: Geographic Resources Information and Data System • Gridded Resource Inventory Data System • Grid-Referenced Information Display System

GRIN: Graded Index • Gradient Index • Gradient Radial Index

GRIND: Graphic Relation Index

GriO: Griffith Observatory (periodical code)

GrIP: Gravis Interface Protocol

GRIP: Gamma-Ray Imaging Payload • General Retrieval of Information Program • Greenland Icecore Project

GRIPS: General Relation-Based Information Processing System • Government Raster Image Processing Software • Groupe Informel de Problèmes Stratégiques

GRIS: Gamma-Ray Imaging Spectrometer • Global Resources Information System • Grating Infrared Spectrometer

GRISM: Grazing Incidence Prism

GRIST: Grazing Incidence Solar Telescope

GRITS: Gamma-Ray Imaging Telescope System • Goddard Range Instrumentation Tracking System

GRJ: George (South Africa) (location identifier)

GRK: Killeen Gray (USA-TX) (location identifier)

GRL: Gamma-Ray Line • Garasa (Papua New Guinea) (location identifier) • Greenland (ISO country code) • Groenland (ISO country code)

GRM: Geopotential Research Mission • Grand Marais - Devils Track (USA-MN) (location identifier) • Growing Rigour Method • Guidance Rate Measurement

GRMHD: General Relativistic Magnetohydrodynamics

Grn: Green

GRN: Good Receipt Note • Gordon (USA-NE) (location identifier) • Green • Ground

GRNC: Group Not Counted

Grnd: Ground

GRNN: General Regression Neural Network

GRNS: Greens

gro: gross

Gro: Gros • Grove

GRO: Gamma-Ray Observatory • General Register Office • Gerona (Spain) (location identifier) • Government Research Organization • Gravitational Radiation Observatory • Guerrero (Mexico)

GROCSE: Gammay-Ray Optical Counterpart Search Experiment

GROSSC: GRO Science Support Center (USA-MD)

GROT: Grota

GROV: Grove

Grp: Group

GRP: Gamma-Ray Projector • Gaussian Random Process • Glass-Reinforced Plastic (*) • Glass-Reinforced Polyester • Ground Relay Panel • Group • Groupement pour la Recherche en Productique • Gurupi (Brazil-TO) (location identifier)

GRPA: Government Performance and Results Act

Grph: Graphic

GRQ: Groningen (Netherlands) (location identifier)

GRR: Grand Rapids (USA-MI) (location identifier) • Graphical Rewrite Rules • Guidance Reference Release

GRRH: Gestion des Ressources et des Relations Humaines

grs: gross

Grs: Grains • Grease • Gros • Gross

GRS: Galactic Radio Sources • Galactic Ring Survey • Gamma-Radiation Source • Gamma-Radiation Spectrometer • Gamma-Ray Spectrometer • Gamma-Ray Spectrum • General Retrieval System • Generic Retrieval System • Geodetic Reference System • Global Receiving System • Global Reference System • Global Resource Serialization • Graph-Rewriting System • Gravity Reference Signal • Gravity Reference System • Great Red Spot (Jupiter) • Grid Reference System • Grille de Référence SPOT • Grosseto (Italy) (location identifier) • Ground Receiving Station • Gyro Reference System

GRSE: Gamma-Ray Spectrometric Equipment

GRSFE: Geologic Remote Sensing Field Experiment • Global Remote Sensing Field Experiment

GRSG: Geological Remote Sensing Group

GRSP: General Range Safety Plan

GRSS: Geoscience and Remote Sensing Society

GRSST: Gamma-Ray Silicon Space Telescope

GrStR: Grundsteuer-Richtlinien

GRSU: Geography Remote Sensing Unit

Grt: Gracht • Great

GRT: Gamma-Ray Telescope • Gamma-Ray Tube • General Relativity Theory • Generic Run Time • Germanium Resistance Thermometer • Great • Gross Registered Tonnage • Gross Registered Tonnes • Ground Radio Telescope • Ground-Received Time • Gujrat (Pakistan) (location identifier)

GRTAD: Ground-based Real-Time Attitude Determination

Grtg: Grating

GRTLKS: Great Lakes

GRTLS: Glide Return To Launch Site

Grtr: Greater

GRTR: Groupement de Recherche en Télédétection et Radiométrie

GRTS: Goddard Real-Time System

Gru: Grus-Gruis (constellation)

GRU: Ground Replaceable Unit • Guidance Regulator Unit • Gyroscope Reference Unit • São Paulo - Guarulhos (Brazil-SP) (location identifier)

Grv: Groove • Grove

GRV: Grove • Groznyj (Russia) (location identifier)

Grvd: Grooved

Grvl: Gravel

GRVS: Groves

GRW: Galactic Radio Wave • Graciosa Island (Azores) (location identifier)

gr. wt.: Gross Weight

GRX: Granada (Spain) (location identifier)

GRY: Grímsey (Iceland) (location identifier) • Gross Redemption Yield

GRZ: Graz (Austria) (location identifier)

g/s: Grant Supported

gs: Ghostscript • Groundspeed • Ground State

Gs: Gauss

G/S: Glide Slope • Ground Speed • Ground Station

GS: Galactic Shell • Gas Servicer • Generalized System • General Schedule • General Secretary • General Service • General Staff • Geographical Survey • Geological Survey • Geomagnetic Storm • Georgia and South Sandwich Islands (South_) (ISO country code) • Geprüfte Sicherheit • Giclas Shell • Glide Slope • Global Symmetry • Ghostscript • Grand Schmidt • Gray Scale • Ground Segment • Ground Speed • Ground Stabilized • Ground State • Ground Station • Ground System • Groupement Scientifique • Groupe Spécialisé • Group Separator • Guidance Station • Guidance System • Guide Star • Heliocentric Gravitational Constant • Sandwich Islands (South Georgia and South_) (ISO country code) • South Georgia and South Sandwich Islands (ISO country code) • South Sandwich Islands (South Georgia &_) (ISO country code)

G&SA: Goods & Service Allowance

GSA: Gaseous Sulfuric Acid • Generalized Semi Analytical • Generalized Similarity Analysis • General Services Administration (USA) • Geological Society of Africa • Geological Society of America (USA-CO) • Geological Society of Australia • Geologic Spatial Analysis • GNC Simulation Assembly • Gnomonicae Societas Austriaca (Austria) • Goods and Service Allowance • Government Service Administration • Guidance System Astrometer • Long Pasia (Malaysia) (location identifier)

GSAC: Graduate Student Advisory Committee

GSAM: Generalized Sequential Access Method • Generalized Standard Addition Method

GSAMB: Geological Society of America Bulletin (periodical code)

GSAMm: Geological Society of America Memoir (periodical code)

GSAR: Generalized Synthetic Aperture Radar • Guide Star Availability Request

GSARP: Global Scale Atmospheric Research Program

GSAS: Graduate School in Arts and Sciences

GSAT: Geosynhcronous Satellite • Ground System Acceptance Test

GSB: Gamma-Strahl-Blitze • Goldsboro - Seymour Johnson (USA-NC) (location identifier) • Guandong Seismological Bureau (China-PRC)

GSBCA: General Services Board and Contract Appeals

GSC: Galileo Science Catalog • Gascoyne Junction (Australia-WA) (location identifier) • Gas-Solid Chromatography • Gas-Star Complex • Gemini Science Committee • Geological Survey of Canada • Guidance System Console • Guide Star Catalog (*) • Guide Star Coordinate

GSCU: Ground Service Cooling Unit

GSD: General Supply Depot • Geodetic Survey Division • Geographic Situation Display • Geological Survey Division • Greelay Storage Division • Ground Sample Distance • Guide Star Data • Guide Star Determination

GSDC: Gas Scintillation Drift Chamber

GSDF: Global Sustainable Development Facility

GSDI: Global Spatial Data Initiative

GSDL: Ground Software Development Laboratory

GSDM: Global Spatial Data Model

GSDR: Global Slope Data Record

GSDS: Global Spatial Data System

GSE: Geocentric Solar Ecliptic • Gesellschaft für Software-Engineering • Government Sponsored Enterprise • Government Support Equipment • Göteborg Saeve (Sweden) (location identifier) • Gothenburg Saeve (Sweden) (location identifier) • Grad-Shafranov Equation • Graduate School of Education • Graphic Screen Editor • Ground Service Equipment • Ground Servicing Equipment • Ground Support Equipment • Ground Systems Engineering

GSED: Ground Systems Engineering Department

GSE&I: General Systems Engineering & Integration

GSEI: General Systems Engineering and Integration

GSEL: Ground Support Equipment List

GSERD: Ground Support Equipment Recommendation Data

GSESP: Graduate in Science and Engineering Survey Programme

GSF: Gesellschaft für Strahlen- und Umweltforschung mbH (Germany) • Global Status Flag • Goldreich-Schubert-Fricke • Graduate Student Council • Gyro Scale Factor

GSFC: Goddard Space Flight Center (USA-MD)

GSG: General Support Group

GSGE: Computational Science Graduate Fellowship

GSGP: Global Sedimentary Geology Program

GSH: Global Schoolhouse • Goshen (USA-IN) (location identifier)

GSHAP: Global Seismic Hazard Assessment Program (*) • Global Seismic Hazard Assessment Project

GSI: Generalized Scale Invariance • Geographical Survey Institute (Japan) • Gesellschaft für Schwerionenforschung (Germany) • Giga-Scale Integration • Glide Slope Indicator • Government Secure Intranet • Government Source Inspection • Ground Segment Infrastructure • Gruppo Sole Italia (Italy) • Guadalcanal (Solomon Islands) (location identifier)

GSIG: Ground Systems Integration Group

GSIM: Global Satellite Image Mapping

GSIP: Glideslope Intercept Point

GSIR: Guide Star Inventory Request

GSIS: Geographic Snow Information System

GSIU: Ground Standard Interface Unit

GSJ: Geological Survey of Japan

GSK: General Semantic Knowledge

GSKM: Galactic Structure and Kinematic Model

GSL: Generalized Simulation Language • Graphics Software Laboratories • Guaranteed Student Loan

GSLIS: Graduate School of Library and Information Science (UCLA)

GSLMP: Global Sea Level Monitoring Programme

GSLS: Ground Station Link Survivability

GSLV: Geostationary Launch Vehicle • Geostationary Satellite Launch Vehicle • Geostationary Space Launch Vehicle • Geosynchronous Satellite Launch Vehicle

gsm: Gram per Square Meter

GSM: General Sales Manager • General Systems Model • Geocentric Solar Magnetospheric • Global Satellite Messaging • Global Spectral Model • Global Standard for Mobiles • Global State Manager • Global System for Mobiles • Ground Segment Manager • Ground Station Module • Groupe Spécial Mobile

GSME: Ground Support Maintenance Equipment

GSMN: Gerencia del Servicio Meteorológico Nacional

GSMNP: Great Smoky Mountains National Park

GSMO: Ground Systems Management Office

GSMP: Generalized Semi-Markov Process • General Switch Management Protocol

GSMQ: Global Services Management Queue

GSN: Global Seismic Network • Global Seismographic Network (*) • Mount Gunson (Australia-SA) (location identifier)

GSNF: Growing Season Net Flux

GSNP: Group on Statistical and Non-linear Physics

GSNW: Gateway Service for NetWare

GSO: General Spin Orbital • General Staff Officer • Geostationary Orbit • Geosynchronous Orbit • Global Seismic Oscillation • Graduate School of Oceanography • Greensboro (USA-NC) (location identifier) • Ground Support Office • Ground Support Operations • Ground Systems Operations • GSFC Security Office • Gyro Storage Oven • High Point (USA-NC) (location identifier) • Winston-Salem (USA-NC) (location identifier)

GSOC: German Space Operations Center

GSOP: Guidance Systems Operation Plan

GSOV: Ground Segment Overall Verification

GSOWG: Ground Systems Operations Working Group

GSOWM: Global Spectral Ocean Wave Model

GSP: Generalized Scheme of Preferences • Generalized System of Preferences • General Simulation Program • General System of Preferences • Geodetic Satellite Program • Geological Survey of Pakistan • Glassfibre-Strengthened Polyester • Government-Specified Purchase • Graphics System Processor • Graphic Subroutine Package • Greenland Sea Project • Greenville (USA-SC) (location identifier) • Greer (USA-SC) (location identifier) • Gross Social Product • Gross State Product • Ground Safety Plan • Guidance Signal Processor • Spartanburg (USA-SC) (location identifier)

GSPB: Geodetic Satellite Policy Board

GSPC: Gas Scintillation Proportional Counter • Guide Star Photometric Catalogue

GSPD: Ground Speed

GSPDC: Geostationary Satellite Precipitation Data Centre

GSPF: Ground Software Production Facility

GSPH: Gridded Smooth Particle Hydrodynamics

GSPL: Global Services Priority List

GSPN: Generalized Stochastic Petri Net

GSPO: Ground Systems Project Office

GSPR: Groupement Spécial de Protection Rapprochée • Guidance Signal Processor-Repeater • Guide Star Parallel Request

GSQ: Generalized Sinusoidal Quantity

GSQL: Gateway to Structured Query Language

GSR: Galactic Standard of Rest • Galvanic Skin Reflex • Galvanic Skin Response • Gardo (Somalia) (location identifier) • Générateur Solaire Rigide • Generic Record Syntax • Glide Slope Receiver • Global Self-Regulation

GSRC: Geological Survey Research Committee

GSRF: Ground Software Reference Facility

GSRN: Global Surface Radiation Network

GSRP: Graduate Student Researchers Program

GSRR: Ground Segment Requirements Review

GSS: Generic Security Service • Geodetic Survey Squadron • Geographic Support System • Geostationary Satellite • Geosynchronous Space Station • Global Space Station • Global Surveillance System • Government Statistical Service • Government Systems Sector • Grad-Schlüter-Shafranov (equation) • Ground Support Software • Ground Support System • Grouped Sweeping Scheduling • Growth Space Station • Grumman Sensor Systems Inc. • Guidance System Simulator • Guided Self-Scheduling • Guide Star Selection • Sabi Sabi (South Africa) (location identifier)

GSSA: Ground Support Systems Activation

GSSAPI: Generic Security Service Application Program Interface

GSSC: General Systems Simulation Center • Ground Support Simulation Computer • Ground Support Systems Contractor

GSSDR: Global Surface Slope Data Record

GSSF: General Services Stock Fund

GSSI: Ground Support System Integration

GSSI&C: Ground Support Systems Installation & Checkout

GSSIC: Ground Support Systems Installation and Checkout

GSSP: Global Stratotype Section and Point

GSSR: Goldstone Solar System Radar • Ground Support System Review • Guide Star Selection Request

GSSS: Ground Support System Specification • Guide Star Selection System • Guide Star Support System

GSST: Global Sea Surface Temperature

GST: Goods and Services Tax • GPS Science and Technology • Greenwich Sidereal Time • Greenwich Standard Time • Ground System Test • Groupe de Servitude Table • Groupe de Soutien Technique • Gust • Gustavus (USA-AK) (location identifier)

GSTDN: Ground Spacecraft Tracking and Data Network • Ground Spaceflight Tracking and Data Network

GSTE: Geometrical Similarity as Topological Equivalence • Guidance System Test Equipment

GSTF: Ground Systems Test Flow

GSTP: General Supporting Technology Programme (*) • General Support Technology Programme • Ground System Test Procedure

GSTS: Ground-based Surveillance and Tracking System • Groupe Scientifique de Télédétection Spatiale • Gusts

GSTU: Guidance System Test Unit

GSTV: Ground System Thermal Vacuum

GSTY: Gusty

GSU: Gas Servicer Unit • Gedaref (Sudan) (location identifier) • Georgia Southern University (USA-GA) • Georgia State University (USA-GA)

GSUF: Gesellschaft für Strahlen- und Umweltforschung mbH (Germany)

GSV: Geostationary Service Vehicle • Glove Stop Valve • Guided Space Vehicle

GSVD: Generalized Singular Value Decomposition

GSVOO: Ground Station Visibility of Orbiting Objects

GSVP: Ground Support Verification Plan

GSW: Ground Saucer Watch

GSWM: Global-Scale Wave Model

Gsy: Guernsey (Channel Islands)

GSY: Grimsby Binbrook (UK-England) (location identifier)

gt: Gilt • Guatemala

Gt: Gent (Belgium) • Gigatonne • Great • Groot

G/T: Facteur de Mérite • Gain/Temperature

GT: Game Theory • Gamow-Teller • Gas Tank • Gas Turbine • Gauge Theory • General Tool • Generic Task • Geographic Term • Geometry and Topology • Gigaton • Gravitational Telescope • Greater Than • Greenwich Time • Gross Ton • Gross Tonnage • Ground Team • Ground Test • Ground Transmit • Ground Truth • Group Technology • Guaranteed Time • Guatemala (ISO country code) • Guidance Trajectory • Guidance Transmitter • Gyro Torque

GT&A: Ground Test & Acceptance

GTA: Gas-Tungsten Arc • Gatokae (Solomon Islands) (location identifier) • Gestionnaire des Titres d'Accès • Graduate Teaching Assistantship • Graduate Teaching Associate • Ground Test Access • Ground Test and Acceptance • Ground Test Article • Ground Test Assembly • Ground Torquing Assembly • Groupe de Télédétection Aquatique

GTAS: Grand Traverse Astronomical Society (USA-MI)

GTAW: Gas Tungsten Arc Welding

GTB: Genting (Malaysia) (location identifier)

Gt. Br.: Great Britain

Gt. Brit.: Great Britain

GTC: General Telephone Co. • Global Theme Conference • Glycol Trim Console • Good Till Cancelled • Grand Tour Cluster • Gran Telescopio de las Canarias • Green Turtle (Bahamas) (location identifier) • Ground Test Conductor

GTCC: Gas Turbine Combined Cycle

GTCE: Global Tropospheric Chemistry Experiment

GTCP: Global Tropospheric Chemistry Program

GTCU: Ground Thermal Conditioning Unit

Gtd: Guaranteed

GTD: Generalized Tracking Data • Geometrical Theory of Diffraction • Geothermodynamic Data • Groupe de Travail Disciplinaire • Groupe Technique Disciplinaire

GTDAR: GEC-Thompson-DASA Airborne Radar

GTDM: Generic Time Division Multiplex

GTDR: Global Topographic Data Record

GTDS: Goddard Trajectory Determination System

GTE: General Telephone and Electric • General Telephone Electronics • Global Troposphere Experiment • Global Tropospheric Experiment • Greater Than or Equal • Groote Eylandt - Alyangula (Australia-NT) (location identifier) • Ground Test Equipment • Guidance Test Equipment

Gtee: Guarantee

GTF: Generalized Transformation Function • Gestion Transaction Financement • Government Test Facility • Great Falls (USA-MT) (location identifier) • Ground Test Facility

GTG: Grantsburg (USA-WI) (location identifier) • Ground-To-Ground • Groupe de Tomographie Géophysique

GTHRES: Gravity Threshold

GTI: Gesellschaft fü Technische Informationssysteme • Good Time Interval • Grand Turk Island • Graphiques-Textes-Images • Grote Technologische Instituten • Ground Test Instrumentation • Guettin (Germany) (location identifier)

GTIC: Groupe Technique Interministériel des Composants

GTIM: Groupe de Travail Inter-Ministériel

GTIS: Geographical Technical Information System • Government Technology Information System

GTK: Gross Tonne Kilometer • Sungei Tekai (Malaysia) (location identifier)

GTL: Gas To Liquid • Gunning Transceiver Logic

gTLD: Generic Top Level Domain • Global Top Level Domain

GTLD: Generic Top Level Domain

GTLV: Government-furnished Target Launch Vehicle

GTM: Geometry Technology Module • Global Telecom Market • Global Transaction Manager • Good This Month • Gran Telescopio Milimétrico • Ground Team Manager • Ground Test Motor • Guatemala (ISO country code)

GTMS: Graphical Text Management System

GTN: Glasgow Telecolleges Network (UK-Scotland) • Mount Cook - Glentanner (New Zealand) (location identifier)

Gto: Guanajuato (Mexico)

GTO: Gate Turn Off • Gaussian-Type Orbital • Generic Technical Order • Geostationary Transfer Orbit • Geosynchronous Transfer Orbit • Gorontalo (Indonesia) (location identifier) • Guanajuato (Mexico) • Guaranteed Time Observation • Guaranteed-Time Observer

Gtos: Gastos

GTOS: Global Terrestrial Observing System

GTOW: Gross Takeoff Weight

GTOWG: Guaranteed Time Observer Working Group

GTP: General Test Plan • Geophysical Turbulence Program • Global Theme Program

GTPA: Gestion Technologique des Passagers Aériens

GTQ: Quetzal (Guatemala) (ISO currency code)

Gtr: Greater

GTR: Columbus - Golden Triangle (USA-MS) (location identifier) • Garantie de Temps de Réparation • General Theory of Relativity • Geoid/Topography Ratio • Ground Truth Radiometer • Starkville (USA-MS) (location identifier) • West Point (USA-MS) (location identifier)

GTRE: Gas Turbine Research Establishment

GTREP: Georgia Tech Regional Engineering Program (USA-GA)

GTRF: GPS Terrestrial Reference Frame

GTRI: Georgia Tech Research Institute (USA-GA)

GTS: Gas Turbine Ship • General-Technical-Security • General Technology Systems • General Test Support • Geostationary Technology Satellite • Global Telecommunications Society • Global Telecommunications System • Global Translation System • Goldstone Tracking Station (NASA) • Go To Standby • Granites (Australia-NT) (location identifier) • Graphics Terminal Scheduler • Graphics Terminal System • Great Trigonometric Survey • Greenwich Time Signal • Ground Telemetry System • Ground Terminal System • Ground Test Station • Guam Tracking Station (NASA) • Guidance Test Set • Guidance Test Station • Gyro Tilt Signal

GTSN: Global Telemetered Seismograph Network

GTSS: General Time-Sharing System

gtt: Guttae

GTT: Geiger Tube Telescope • Georgetown (Australia-QLD) (location identifier) • Global Title Transmission

GTTS: Gyro Transfer Table System

GTU: Ground Test Unit • Guidance Test Unit

GTUB: Geographic Tabulation Unit Base

GTV: Ground Test Vehicle • Ground Transport Vehicle • Guided Test Vehicle

GTW: Gesellschaft für Terminologie und Wissenstransfer • Good This Week • Gross Takeoff Weight • Zlin Holesov (Czech Rep.) (location identifier)

GTWAY: Gateway

GTWY: Gateway

GTY: Gettysburg (USA-PA) (location identifier)

GTZ: Gesellschaft für Technische Zusammenarbeit (Germany)

gu: Guam • Guinea • Gujarati (ISO language code)

Gu: Guest • Guinea

GU: Georgetown University • Government Unit • Guam (ISO country code) • Guardia Urbana • Guatemala • Guidance Unit • Guilford (UK postal code) • Guinea

GUA: Gestionnaire Utilisateur d'Applications • Grupo Universitario de Astronomía (Spain) • Guatemala City - La Aurora (Guatemala) (location identifier) • Guidance Unit Assembly

Guad: Guadalajara (Spain)

Guadal: Guadalajara (Spain)

GUAM: Georgia-Ukraine-Azerbaijan-Moldova

Guar: Guarantee • Guaranteed

Guat: Guatemala

Guay: Guyaquil

GUB: Generalized Upper Bounding • Greatest Upper Bound • Guerrero Negro (Mexico) (location identifier)

GUC: Crested Butte (USA-CO) (location identifier) • Gunnison (USA-CO) (location identifier)

GUCP: Ground Umbilical Carrier Plate

GUD: Goundam (Mali) (location identifier)

GUE: Gaussian Unitary Ensemble • Guriaso (Papua New Guinea) (location identifier)

GUESS: General Purpose Expert System Shell

GUF: French Guiana (ISO country code) • General University Funds • Guiana (French_) (ISO country code) • Guyane Française (ISO country code) • Gulf Shores - Edwards (USA-AL) (location identifier)

GUG: Guari (Papua New Guinea) (location identifier)

GUH: Gunnedah (Australia-NSW) (location identifier)

Gui: Guiana

GUI: Give Us Incentives • Graphical User Interface • Graphic User Interface • Guiria (Venezuela) (location identifier)

GUIB: Graphical User Interface for the Blind

GUID: Globally Unique Identifier • Global Unique Identifier • Guidance

GUIDE: Graphical User Interface for a Data Environment

GUIDO: Guidance Officer

Guil: Guilder (currency)

Guin: Guinea

Guip: Guipúzcoa (Spain)

GUIRR: Government-University-Industry Research Roundtable

GUISDAP: Grand Unified Incoherent Scatter Design and Analysis Package • Grand Unified Incoherent Scatter Design and Analysis Program (*)

Guj: Gujarat

GUJ: Guaratinguetá (Brazil-SP) (location identifier)

GUL: Goulburn (Australia-NSW) (location identifier)

GULP: General Utility Language Processor • General Utility Library Program • Global Upgrade LAN Product • Grotesque Large Useless Project

GULS: Generic Upper Layer Security

GUM: Gamma Ursa Majoris • Guam Agana (Guam) (location identifier) • Guatemala (ISO country code)

GUN: Montgomery Gunter (USA-AL) (location identifier)

GUO: Gualaco (Honduras) (location identifier)

GUP: Gallup (USA-NM) (location identifier) • Gas Under Pressure

GUPS: Grand Unified Photon Spectrum

GUPT: Grand Unified Phase Transition

GUQ: Guanare (Venezuela) (location identifier)

GUR: Alotau (Papua New Guinea) (location identifier)

GUREN: Groupe Universitaire de Recherche sur l'Environnement

GUS: Gemeinschaft der Unabhängigen Staten • Peru Grissom (USA-IN) (location identifier)

GUT: Grand Unified Theory • Grand Unified Transition • Great Unification Theory • Groupe francophone des Utilisateurs de TEX • Gütersloh (Germany) (location identifier)

GUU: Grundarfjördur (Iceland) (location identifier)

GUUS: Greatest Useful Unfounded Set

guv: gerecht und vollkommen

GUV: Mougulu (Papua New Guinea) (location identifier)

GUW: Guryev (Kazakhstan) (location identifier)

GUX: Guna (India) (location identifier)

GUY: Guyana (international vehicle registration code) • Guyana (ISO country code) • Guyane (ISO country code) • Guymon (USA-OK) (location identifier)

G-V: Gravity-Velocity

GV: Generalversammlung • Giant Void • Gigavolt • Goed Voor • Grande Vitesse • Gran Velocidad • Gravimetric Volume • Gross Valuation • Ground Visibility

GvA: Gesellschaft für volkstümliche Astronomie (Germany)

GVA: Geneva Cointrin (Switzerland) (location identifier) • Genève Cointrin (Switzerland) (location identifier) • Gyroscope Vibration Absorber

GVB: Generalized Valence Bond

GVDS: Global Vegetation Data Set

GVE: Gordonsville (USA-VA) (location identifier)

GVF: Goodness of Variance Fit

GVHRR: Geosynchronous Very-High-Resolution Radiometer

GVI: Global Vegetation Index (*) • Green Vegetation Index • Green River (Papua New Guinea) (location identifier)

GVIS: Geographic Visualization

Gvl: Gravel

GVL: Gainesville - Lee Gilmer (USA-GA) (location identifier) • Graphics and Visualization Laboratory • Graphics Visualization Laboratory • Ground Vibration Laboratory

GVLS: Global Verification and Location System

GVN: Global Volcanism Network

GVP: Global Volcanism Program (Smithsonian Institution) • Greenvale (Australia-QLD) (location identifier) • Gross Value of Production

GVR: Governador Valadares (Brazil-MG) (location identifier)

GVSM: Generalized Vector Space Model

GVSP: Goddard Visiting Scientist Program

Gvt: Government

GVT: Greenville Majors (USA-TX) (location identifier) • Ground Vibration Test

GVTA: Ground Vibration Test Article

GVTF: Ground Vibration Transfer Function

Gvtl: Governmental

GVU: Graphics, Visualization and Usability

GVW: Grandview - Richards Gebaur (USA-MO) (location identifier) • Great Virtual Wall • Gross Vehicle Weight • Gross Vehicular Weight

GVX: Gävle (Sweden) (location identifier) • Gavle (Sweden) (location identifier)

Gw: Gigawatt

G/W: Gateway

GW: Gas Well • Gateway • Gee Whiz • Gierer-Witz • Gigawatt • Global Warming • Gravitational Wave • Gravity Wave • Gross Weight • Ground Water • Guinea Bissau (ISO country code) • Guinée Bissau (ISO country code)

GWA: General Work Area • Global Atmosphere Watch • Gwa (Myanmar) (location identifier)

GWAAT: Global Wide-Area Applications Testbed

GWADIR: Gravitational-Wave Antenna Detector Interferometer

GWC: Generic WIYN Client • Global Weather Center

GWD: Gwadar (Pakistan) (location identifier)

GWDG: Gesellschaft für Wissenschaftliche Datenverarbeitung Göttingen (Germany)

GWE: Global Weather Experiment • Gravitational-Wave Experiment • Gweru (Zimbabwe) (location identifier)

GWEC: Global Wireless Education Continuum

GWEN: Ground Wave Emergency Network

GWG: Gesellschaft für Wirtschaftsförderung Göttingen (Germany)

GWh: Gigawatt-Hour

GWHIS: Global-Wide Help and Information System

GWIB: Gravity-Wave-Induced Bias

GWIGWO: Good Will In - Good Will Out

GWL: Gwalior (India) (location identifier)

GWM: Generic Window Manager • Global Wind Measurement • Graphics Windows Manager • Guaranteed Weekly Minimum

GWN: Gnarowein (Papua New Guinea) (location identifier) • Graphics Window Manager

GWO: Greenwood Leflore (USA-MS) (location identifier)

GWP: Generalized Waring Process • Gloval Warning Potential • Global Water Partnership • Government White Paper • Greenhouse Warming Potential • Gross World Product • Peso (Guinea-Bissau) (ISO currency code)

GWP2: Guelph-Waterloo Programme for Graduate Work in Physics

GWS: General Working Scheme • Glashow-Weinberg-Salam • Glenwood Springs (USA-CO) (location identifier) • Global Wave Statistics • Graphical Windowing System • Great White Spot (Saturn)

GWSI: Ground Water Site Inventory

GWT: Griffith-Woodley Technique • Gross Weight • Ground Winds Tower • Westerland (Germany) (location identifier)

GWTR: Ground Water Treatment Rule

GWU: Gardner-Webb University (USA-NC) • George Washington University (USA-DC)

GWUI: Ground Water Under the Influence

GWUP: Gesellschaft zur Wissenschaftlichen Untersuchung von Parawissenschaften eV (Germany)

GWV: Glendale (USA-WV) (location identifier)

GWW: Berlin Gatow (Germany) (location identifier) • Gauss-Weierstraß Wavelet • Guaranteed Weekly Wage

GWY: Galway (Ireland) (location identifier)

GX: Galactic X-source • Gateway Accelerated

GXF: Seiyun (Yemen) (location identifier)

GXG: Negage (Angola) (location identifier)

GXH: Mildenhall (UK-England) (location identifier)

GXQ: Coyhaique (Chile) (location identifier)

GXUVDC: German XUV Data Centre

GXX: Yagoua (Cameroon) (location identifier)

GXY: Greeley (USA-CO) (location identifier)

Gy: Gray

GY: Good Year • Guernsey (Channel Islands) • Guyana (ISO country code) • Guyane (ISO country code) • Gypsy

GYA: Guayaramerin (Bolivia) (location identifier)

GYD: Dollar (Guyana) (ISO currency code)

GYE: Guayaquil - Simon Bolivar (Ecuador) (location identifier)

GYI: Gisenyi (Rwanda) (location identifier)

GYL: Argyle (Australia-WA) (location identifier)

GYM: Guaymas (Mexico) (location identifier)

GYN: Goiãnia - Santa Genoveva (Brazil-GO) (location identifier)

GYP: Gympie (Australia-QLD) (location identifier) • Gyro Yaw Position

Gyr: Gigayear • Gyroscope

GYR: Goodyear Litchfield (USA-AZ) (location identifier)

Gyro: Gyroscope

Gyrs: Gigayears

GYY: Chicago Gary (USA-IN) (location identifier)

G-Z: Giacobini-Zinner (comet)

GZ: Giacobini-Zinner (comet) • Goldene Zahl • Ground Zero

GZI: Ghazni (Afghanistan) (location identifier)

GZK: Greisen-Zatsepin-Kuzmin (cutoff)

GZM: Gozo (Malta) (location identifier)

GZO: Gizo (Solomon Islands) (location identifier)

GZT: Gaziantep (Turkey) (location identifier) • Greenwich Zone Time

G2RA: Groupement Régional de Recherche en Acoustique

G7: Groupe des Sept • Group of Seven Industrial Nations (now G8) • Grupo de los Siete

G8: Groupe des Huit • Group of Eight Industrial Nations (formerly G7) • Grupo de los Ocho

H

h: Altitude • Elevation • Enthalpy • Habitant • Habitante • Hacia • Harbour • Hard • Hardness • Hart • Heat • Heat Transfer Coefficient • Heavy Sea • Hecto • Height • Heiß • Hekto • Hell • Help • Hence • Heure • High • Hinten • Hoch • Homme • Hoogte • Hora • Horizontal • Horse • Hot • Hour • House • Hundred • Husband • Hybrid • Miller Index • Planck Constant • Specific Enthalpy • Step Size • Thickness

H: Absolute Helligkeit • Altitude • Angular Impulse • Boltzmann Function • Dose Equivalent • Dynamic Amplification Factor • Eccentric Anomaly of Hyperbolic Orbit • Elevation • Enthalpy • Haben • Hafen • Haft • Haftung • Hail • Haler • Half • Halt • Haltestelle • Hamiltonian • Handel • Handschrift • Harbour • Hard • Hardness • Haupt • Haus • Hauteur • Hazardous • Haze • Head • Header • Heat Capacity • Heating Value of the Fuel • Heavy • Heck • Heft • Height • Heilige • Heim • Heimat • Heiß • Heizung • Helligkeit • Help • Hengst • Henry • Hermano • Hersteller • Hessen (Germany) • High • Highest • High-Pressure Center • Hill • Hinfahrt • Hired • Hitze • Hoch • Hochdruck • Hochschule • Höhe • Hoheit • Hora • Horizontal Component of the Magnetic Vector • Horn • Hospital • Hot • Hotel (international alphabet) • Hour • Hour Angle • Hub • Hubble Constant • Hund • Hundert • Hungary (international vehicle registration code) • Hungary (postal country code) • Hydrant • Hydrogen • Hydrogène • Hydrogenium • Light Exposure • Magnetic Field Strength • Magnetic Intensity • Momentum • Thickness of Lithosphere • Two Hundred (Roman numeral) • Wasserstoff • 200 (Roman numeral)

ha: hauptamtlich • Hausa (ISO language code) • Hectare • Hektar • Hoc Anno • Huius Anni

hA: herrschende Ansicht

Ha: Hafen • Hahnium • Haiti • Handel • Hartmann Number • Hawaii

H&A: Health & Accident

H/A: Hazardous Area

HA: Apogee Altitude • Haftanstalt • Haiti • Handausgabe • Harrow (UK postal code) • Hauptabteilung • Hauptamt • Hautes Alpes (France) • Hawaii • Hazard Analysis • Hazardous Area • Health Authority • Height of Apogee • High Altitude • High Angle • High Aperture • High Availability • Higher Authority • Highlights of Astronomy • Hochamt • Homogeneous Arc • Host Agent • Host Astronomer • Hour Angle • Hungary (international aircraft registration code) • Hydrologic Atlas

HAA: Hamilton Amateur Astronomers (Canada-ON) • Hasvík (Norway) (location identifier) • Hawkesbury Astronomical Association (Australia-NSW) • Height Above Airport • High-Altitude Abort

HAADS: High-Altitude Attitude Determination System

HAAP: High-Altitude Aerial Photograph

HAARP: High-Frequency Active Auroral Research Program

HAARS: High-Altitude Attitude Reference System

HAAS: High Altitude Air Sampler

hab: habilitieren • habilitiert

Hab: Habilitand • Habilitation • Habitant • Habitat • Habitation

HAB: Habitaçao • Habitation • Hamilton - Marion County (USA-AL) (location identifier) • Hammelburg (Germany) • Harmful Algal Bloom • High-Altitude Burst • Horizontal Assembly Building • Horizontal Axis Bearing

HABCIT: High-Altitude Balloon Circumstellar Imaging Telescope

Hab.Corporation: Habeas Corpus

habil: Habilitatus

Habil: Habilitand • Habilitation

HABP: Hypersonic Arbitrary Body Program

Habt: Habeat

HAC: Hachijojima (Japan) (location identifier) • Haleakala Atmospheric Characterization • Harbor and Approach Chart • Hardware Access Layer • Heading Alignment Circle • Heading Alignment Cone • Heading Alignment Cylinder • Head of Astronomical Computing • Hereinafter Described • House Appropriations Committee • Hughes Aircraft Co. • Hydrogenated Amorphous Carbon

HACBSS: Homestead And Community Broadcasting Satellite Service

HACCP: Hazard Analysis and Critical Control Point • Hazard Analysis Critical Control Point

HACE: Hubble Advanced Camera for Exploration

HACMP: High-Availability Cluster Multiprocessing

HACP: HTML Astronomy Classroom Project

HAD: Halmstad (Sweden) (location identifier) • Hauptabteilungsdirektor • Health Assessment Document • Heat Actuated Device • Helium Abundance Detector • High-Accuracy Data • Historical Astronomy Division • Hospitalisation À Domicile

HADS: Hierarchical Applications Data Structure • High-Amplitude Delta Scuti (star)

HADTS: High-Accuracy Data Transmission System

HAdW: Heidelberger Akademie der Wissenschaften (Germany)

HAE: Havasupai (USA-AZ) (location identifier) • Height Above Ellipsoid • Helicity-induced Alfvén Eigenmode • High-Altitude Endurance • High-Amplification Event

HAEBE: Herbig Ae/Be (star)

HAEC: Hughes Aerospace and Electronics Co.

Haf: Hafen

HAF: Half Moon (USA-CA) (location identifier) • Height Above Field

HAFB: Holloman Air Force Base (USA-NM) (formerly AAFB)

HaFraBa: Hamburg-Frankfurt-Basel (Autobahn)

Haft: Haftung

Haftb: haftbar

Haftg: Haftung

HAG: Den Haag (Netherlands) (location identifier) • Hardware Advisory Group • Help Authoring Guide • Hochschularbeitsgemeinschaft

HAGE: Human Activity and Global Environment

HAH: Moroni Hahaya (Comoros) (location identifier)

HAI: Hellenic Aerospace Industry (Greece) • Three Rivers - Dr. Haines (USA-MI) (location identifier)

HAIC: Hetero-Atom In Context

HAIL: High Altitude Inversion Layer

HAINS: High-Accuracy Inertial Navigation System

HAIRDS: High-Altitude Infrared Detecting Set

HAIS: Hughes Applied Information Systems

HAIZY: Hamburger Isochron-Zyklotron (Germany)

HAJ: Hanover Langenhagen (Germany) (location identifier)

HAK: Haikou (China-PRC) (location identifier) • Horizontal Access Kit

Hal: Halogen

HAL: Halali (Namibia) (location identifier) • Hard Array Logic • Hardware Abstraction Layer • Hardware Array Logic • Height Above Landing • Heuristically Programmed Algorithmic (2001) • Highly Automated Logic • High-order Articulated Language • High-order Assembly Language • High-temperature Acoustic Levitator • Hindustan Aeronautics Ltd. (India) • House-programmed Array Logic • Houston Aerospace Language

halbj: halbjährig

Halbj: Halbjahr

halbjhg: halbjährig

HALCA: High Altitude Laboratory for Communications and Astrophysics • Highly Advanced Laboratory for Communications and Astronomy (formerly VSOP) (*)

HALDB: High-Availability Large Database

HALE: High-Altitude Long-Endurance

HALLPASS: Hadamard Transform Laser Long-Path Absorption Spectrometer System

HALO: High-Altitude Large Optics • High-Altitude Lift Off • High-Altitude Long Operation • Hughes Automated Lunar Observer

HALOE: Halogen Occultation Experiment

HALS: Hydrographic Airborne Laser Sounder

HALSS: High-Altitude Lidar Sensing Station

Ham: Hamburg (Germany)

HAM: Hamburg Fuhlsbüttel (Germany) (location identifier) • Hardware Associative Memory • Height Adjustment Maneuver • Hierarchical Access Method • Hold And Modify • Hopfield Associative Memory • Horizontal Access Module • Hypertext Abstract Machine

HAMM: Hypertext Abstract Machine Management

HAMP: Handling And Moving Procedure • High-Altitude Measurement Probe

HAMS: Half-Age Main Sequence

HAMT: Human-Aided Machine Translation

Han: Hanover (Germany)

HAN: Hanoi (Viet Nam) (location identifier) • Home Area Network • Home ATM Network

HANA: Haiti National Airlines

HANAIR: Haiti National Airlines

HAND: Have A Nice Day

Handb: Handbook • Handbuch

handgeschr: handgeschrieben

handl: handlich

Handl: Handling • Handlung

HANDS: High-Altitude Nuclear Detection Studies

handschr: handschriftlich

Handschr: Handschreiben • Handschrift

Handw: Handwerk

Handwb: Handwörterbuch

HANE: High-Altitude Nuclear Explosion

HANG: Hawaiian Air National Guard

HANOMAG: Hannoversche Maschinenbau-Aktiengesellschaft (Germany)

HANOVA: High-dimensional Analysis Of Variance

HANS: High-Altitude Navigation System

Hants: Hampshire (UK-England)

HANUL: High-energy Astrophysics Neutrino Laboratory (Korea-ROK)

HAO: Hamilton (USA-OH) (location identifier) • High-Altitude Observatory (USA-CO) • Hydrogenated Anthracene Oil

HAP: Handelsabgabepreis • Hardware Allocation Panel • Hazardous Air Pollutant • High-Altitude Platform • High-Altitude Probe • High-speed Arithmetic Processor • Honeywell Array Processor • Horizontal Axis Pivot • Kontiki Resort - Whitsunday - One Hundred (Australia-QLD) (location identifier) • Long Island (Australia-QLD) (location identifier)

HAPAG: Hamburg-Amerikanische Packetfahrt-Aktiengesellschaft

HAPDAR: Hard Point Demonstration Array Radar

HAPEX: Hydrological Atmospheric Pilot Experiment

HAPI: Harris Application Program Interface • High-Altitude Plasma Instrument

HAPP: High-Altitude Powered Platform

HAPPE: High-Altitude Particle Program Experiment

HAPPEX: Hall A Proton Parity Experiment

HAPRS: High-Accuracy Positioning Reference System

HAPS: Huddersfield Astronomical and Philosophical Society (UK) • Hybrid Approximated Solution • Hydrazine Auxiliary Propulsion System

HAQ: Hanimaadhoo (Maldives) (location identifier)

Har: Harbour • Harmonizaçao • Harmonizada • Harmonisation • Harmonisé • Harmonization • Harmonized

HAR: Harrisburg (USA-PA) (location identifier) • Hazard Analysis Report • Heterodyne Array Receiver • High-Altitude Research

HARA: High-Altitude Radar Altimeter

HarCi: Harvard Circular (periodical code)

HARB: Harbor • Harbour

HARC: Houston Advanced Research Centre (USA-TX)

HARD: High-Accuracy Reflector Development

HARDI: High Angular Resolution Deployable Interferometer

HARDTS: High-Accuracy Radar Data Transmission System

HAREM: Halieutic Radar Experiment in the Mediterranean

harm: Harmonisch

Harm: Harmonic • Harmonie • Harmonique • Harmonium • Harmony

HARM: Hazard Assessment Rating Methodology

harml: Harmlos

HARN: High-Accuracy Reference Network

HARNET: Hong-Kong Academic and Research Network

HaRP: Hawaiian Rainband Project

HARP: Heterodyne Array Receiver Program • High-gain Avalanche Rushing Photoconductor • HST Aberration Recovery Program

HARPI: Height Azimuth Range Position Indicator

HARS: Heading And Reference System • High-Altitude Route System

HART: Harmonic Aerodynamic Rotor Test • Hayden Analysis and Reporting Tool • Heartland Astronomical Research Team (USA-KS) • Height Area Rain Threshold • Hypermedia Authoring Research Toolkit

HartRAO: Hartebeesthoek Radio Astronomy Observatory (South Africa)

HARTRAO: Hartebeesthoek Radio Astronomy Observatory (South Africa)

HARTS: Hydrometeorological Automatic Recording and Telemetering System

Harv: Harvard

HARV: High-Alpha Research Vehicle • High-Angle-of-attack Research Vehicle • High-Aspect Ratio Vessel • Hybrid Augmented Reusable Vehicle

HAS: Hail (Saudi Arabia) (location identifier) • Halmstads Astronomiska Sällskap (Sweden) • Hamilton Astronomical Society (New Zealand) • Harrogate Astronomical Society (UK) • Hawaiian Astronomical Society (USA-HI) • He Atom Scattering • Helicopter Air Service • High-Altitude Sensor • Holddown Alignment Support • Houston Astronomical Society (USA-TX) • Hungarian Academy of Sciences • Hydraulic Actuation System • Hydrogen Actuation System

HASAWA: Health And Safety At Work Act

HASC: High-Altitude Surveillance Concept

HASE: High-Assurance Systems Engineering

HASI: Huygens Atmospheric Structure Instrument

HASL: Health And Safety Laboratory • Hot Air Solder Levelling

HASP: High-Accuracy Satellite Position • High-Altitude Sampling Program • High-Altitude Space Platform • Houston Automatic Spooling Priority

HAST: High-Altitude Supersonic Target

HASVR: High-Altitude Space Velocity Radar

HASYLAB: Hamburg Synchrotron Laboratory (Germany)

HAT: Handling and Archiving Tool • Hatfield (Starlink internal code) • Heathlands (Australia-QLD) (location identifier) • Height Above Touchdown • Hierarchical Alias Table • High-Altitude Test • High-Altitude Testing • Hot-Anode Thruster

HATM: Home Automatic Teller Machine

HATVAD: Hatfield (Starlink internal code)

HAU: Haugesund (Norway) (location identifier)

Haub: Haubitze

haupts: hauptsächlich

haus: hausieren

Hausbes: Hausbesetzer • Hausbesitzer • Haubesorger • Hausbesuch

Hausgeh: Hausgehilfe

Haush: Haushalt • Haushaltung • Hausherr

Hausm: Hausmacht • Hausmädchen • Hausmarke • Hausmeister

Hausr: Hausrat • Hausrecht

haust: Haustus

Hausverw: Hausverwalter • Hausverwaltung

hausw: hauswirtschaftlich

Hausw: Hauswirtschaft

Hav: Havanna

HAV: Havana - José Marti (Cuba) (location identifier) • La Habana - José Marti (Cuba) (location identifier)

HAVAGO: Horizontal Adjustment and Vertical Adjustment of Geodetic Observations

HAVO: Hoger Algemeen Voortgezet Onderwijs

Haw: Hawaii (USA)

HAW: Haverfordwest (UK-England) (location identifier) • Hawaii (STDN) • Heidelberger Akademie der Wissenschaften (Germany)

HAWAII: HgCdTe Astronomical Wide-Area Infrared Imager

HAWC: High-resolution Airborne Wideband Camera

HAWIE: Hamburg-Wechsler-Intelligenztest für Erwachsene

HAWIK: Hamburg-Wechsler-Intelligenztest für Kinder

HAWK: Have Alimony - Will Keep

HAWT: Horizontal-Axis Wind Turbine

HAX: Muskogee Hatbox (USA-OK) (location identifier)

HAY: Haycock (USA-AK) (location identifier)

Haz: Hazard • Hazardous

HAZ: Hatzfeldthaven (Papua New Guinea) (location identifier) • Heat-Affected Zone • Heiße Abfall-Zellenanlage • High-latitude Activity Zone

HAZMAT: Hazardous Material

HAZW: Hazardous Waste

h/b: Hardbound

hb: Halb • Heilbar • Hemoglobin

Hb: Handbedienung • Handbetrieb • Handbuch • Hausbriefkasten • Herberge

H/B: Half Board • Hyperbaric

HB: Half-Breadth • Handbibliothek • Handbook • Handbücherei • Handelsbetrieb • Hardback • Hard Black • Hauptbibliothek • Hauptbuch • Hauptbüro • Heat Budget • Height Balanced • Helvetisches Bekenntnis • Herringbone • Hexachlorbutadien • Hierarchical Basis • High Bay • Hochbau • Hochschulbibliothek • Hofbibliothek • Hollow Beam • Homogeneous Band • Horizontal Branch • Horizontal Bridgman • Human Being • Hyperbaric • Liechtenstein (international aircraft registration code) • Switzerland (international aircraft registration code)

HBA: Hobart (Australia-TAS) (location identifier) • Hochbauamt • Host Bus Adapter

HBB: Handbook Bulletin • Hobbs - Industrial Park (USA-NM) (location identifier) • Horizontal Blue Branch • Hot Bottom Burning • Hot Big Bang

HBBC: Hot Big Bang Cosmology

HBBM: Hot Big Bang Model

HBBW: Half-Beam Bandwidth

HBC: Handlebae Control • Herbig and Bell Catalogue • Hierarchical Bayesian Clustering • High Breaking Capacity

HBCSE: Homi Bhabha Centre for Science Education (India)

HBCU: Historically Black Colleges and Universities

Hbd: Halbband • Hochbedarf

HBD: Had Been Drinking • Has Been Drinking

HBDS: Hierarchical Binary Digital Search

HBE: First-overtone Blue Edge

Hbf: Hauptbahnhof • Heimatbahnhof

HBF: Horizon Break File

hbfl: hauptberuflich

Hbg: Hamburg (Germany)

HBG: Hattiesburg (USA-MS) (location identifier)

HBH: Hauptbuchhalter • Hobart Bay (USA-AK) (location identifier)

Hbhf: Hauptbahnhof

HBI: Harbour Island (Bahamas) (location identifier)

Hbj: Halbjahr

HBJ: Harcourt Brace Jovanovich

Hbl: Halbblut

HBl: Handelsblatt • Hauptblatt

HBL: Bebelegi (South Africa) (location identifier) • High-energy-peaked BL Lacertae (object)

HBLL: Hydrogen Burning Limiting Layer

HBM: Halo Beaming Model • Her Britannic Majesty • His Britannic Majesty • Hydrologic Benchmark

HBMM: Hydrogen-Burning Minimum Mass

HBMN: Hydrologic Benchmark Network

HBMS: Hyperbase Management System

HBN: Phu-bon (Viet Nam) (location identifier)

HBO: HB Oscillation • Hoger Beroeps Onderwijs • Home Box Office • Horizontal-Branch Oscillation • Humboldt (USA-NE) (location identifier)

HBOSC: Howard B. Owens Science Center (USA-MD)

HBP: High Blood Pressure

HBPIC: High Blood Pressure Information Center

HBQS: Homogeneous Bright Quasar Survey

Hbr: Harbour

HBR: Harbor • Harvard Business Review • Has Been Reviewed • High Bit Rate • High Burst Rate • Hobart (USA-OK) (location identifier)

HBRA: High Background Radiation Area

HBRACW: Has Been Reviewed And Concurred With

HBRS: Harbors

HBS: Harvard Business School (USA-MA) • Hogere Burger School • Hydrostatic Bearing System

Hbschr: Hubschrauber

HBSN: Humboldt Bay Seismic Network

HBSS: High-Band SIGINT System • High-Band Subsystem

hbstl: halbstündlich

Hbstr: Herbiester (Belgium)

Hbt: Hobart (Australia-TAS)

HBT: Hafr Albatin (Saudi Arabia) (location identifier) • Hanbury-Brown Twiss • Heflex Bioengineering Test • Heterojunction Bipolar Transistor

HBTD: High Bit Density Tape

HBTM: Hybrid Bispectral Threshold Method

HBV: Hamburg-Bergedorf Variables

HBW: Half Bandwidth • Hot Bridge Wire

HBX: Hubli (India) (location identifier)

Hby: Hereby

HBZ: Hochschulbibliothekszentrum

h/c: Held Covered

hc: Honoris Causa

H/C: Head Count • Held Covered

H&C: Hot & Cold

HC: Ecuador (international aircraft registration code) • (The) Hague Convention • Hand Control • Hand Controller • Handling Capacity • Hard Copy • Hazardous Cargo • Head Count • Health Certificate • Heat Content • Heater Controller • Heterogeneous Computing • High Capacity • High Carbon • High Court • High Current • High-speed CMOS • Historical Commission • Hollow Cathode • Hors Commerce • Hors Concours • House of Commons • Humidity Control • Hybrid Computer • Hydrocarbon • Hydrocarbure

HCA: Big Spring - Howard County (USA-TX) (location identifier) • Hierarchical Cluster Analysis

Hcap: Handicap

HCAS: Harford County Astronomical Society (USA-MD)

HCB: Hexachlorobenzene • House of Commons Bill • Shoal Cove (USA-AK) (location identifier)

HCBP: Hexachlorobiphenol

HCC: Han Characters Collection • Hexachlorocyclohexan • Hibbing Community College (USA-MN) • Highly Competitive Carrier • Hollow Cathode Contactor • Hudson - Columbia County (USA-NY) (location identifier) • Hydraulic Cement Concrete

HCCB: Hardware Configuration Control Board

HCCCHO: Propynal

HCCD: Historical Canadian Climate Database • High-performance Charge-Coupled Device

HCCI: Homogeneous Charge Compression Ignition

HCCUP: Health Care Cost and Utilization Project

HCD: Hardware Configuration Definition • Hardware Configuration Dialog • High-Current Density • High-Current Diode • Hollow Cathode Discharge

HCDB: Historical Cost Database

HCDD: Hexachlorodibenzodioxin

HCDM: Hot Cold Dark Matter

HCDN: Hydroclimate Data Network

HCE: Hic Conditus Est • High-Cycle Fatigue • Human-Caused Error

HCED: Hand Controller Engage Driver

HCF: Halt and Catch Fire • Heat Control Filter • Height Correction Factor • High-Calorific Fuel • High-Cycle Fatigue • Highest Common Factor • Highest Confidence First • Honeycomb Foundation • Hundred Cubic Feet

HCFC: Hydrochlorofluorocarbon

HCFO: Hispanic Community Fund Office

HCG: Hickson Compact Group • Hierarchical Concept Graph

HCH: Hechingen (Germany) • Hexachlorocyclohexane

HCHO: Formaldehyde

HCHy: Hovering Craft and Hydrofoil (periodical code)

HCI: Hardness-Critical Item • Hierarchically Classified Index • High-Current Inductor • Highly Charged Ion • Host Controller Interface • Hughes Communications Inc. • Human-Computer Interaction • Human-Computer Interface • Human-Computer Interfacing

HCIL: Hague Conference on International Law • Human Computer Interaction Laboratory

HCIRV: Human Computer Interface Research Centre

HCITE: Horizontal Cargo Integration Test Equipment

HCJ: High Court of Justice

HCl: Hydrogen Chloride

HCL: Hardware Compatibility List • Harvard College Librray (USA-MA) • Helium Cadmium Laser • High Cost of Living • Hollow Cathode Lamp • Horizontal Center Line • Horizontal Component of Lift • Hummingbird Communications Ltd. (Canada) • Hydrochloric Acid

HCM: Eil (Somalia) (location identifier) • Hard Copy Module • Heat Capacity Mapper • Her Catholic Majesty • His Catholic Majesty • Human Capital and Mobility • Hybrid Coupled Model

HCMC: Ho Chi Minh City (Viet Nam)

HCMM: Heat Capacity Mapping Mission

HCMOS: High-performance Complementary Metal Oxyde Semiconductor

HCMP: Hardware Configuration and Maintenance Plan

HCMR: Heat Capacity Mapping Radiometer

HCMT: HOSS Communications Model Toolkit

HCN: Historical Climatology Network • Hydrocyanic Acid • Hydrogen Cyanide

HCO: Formyl Radical • Harper College Observatory (USA-IL) • Harvard College Observatory (USA-MA) • Hastings College Observatory (USA-NH)

HCON: Hour-Confirmed

HCOOH: Formic Acid

Hcp: Handicap

HCP: Habitat Conservation Plan • Hexachlorocyclopentadiene • Hexachlorophene • Hexagonal Close-Packed • House of Commons Paper • Hybrid Combustion Process • Hydraulic Circulation Pump

Hcptr: Helicopter

HCQ: Halls Creek (Australia-WA) (location identifier)

HCR: High Cross Range • Highway Contract Route • Holy Cross (USA-AK) (location identifier) • Horizon Crossing Sensor

HCRB: Hardware Configuration Review Board

HCRD: Human-Carrying Rotation Device

HCRO: Hat Creek Radio Observatory (USA-CA) • High Cross-Range Orbiter

HCRS: Heritage Conservation and Recreation Service (USA)

HCRT: Hanna City Robotic Telescope

HCRW: Hot and Cold Running Water

HCS: Heliospheric Current Sheet • High-Carbon Steel • Homogeneous Computer System • Johannesburg Randburg (South Africa) (location identifier)

HCSD: High-Speed Circuit Switched Data

HCSDS: High-Capacity Satellite Digital Service

HCSHT: High-Carbon Steel Heat-Treated

HSCI: Hughes Communications Services Inc.

HCSL: Hybrid Computation and Simulation Laboratory

HCSS: High-Capacity Storage System

hct: Hora Cum Tempore

HCT: Hardware Compatibility Test • High Critical Temperature • Hollow Cathode Tube • Hypertext Compiling Tools

HCTDS: High-Capacity Terrestrial Digital Service

HCTS: Hydraulic Component Test Stand

HCU: Heater Control Unit • Helium Charging Unit • Host Connection Unit • Hydraulic Charging Unit • Hydraulic Control Unit • Hydraulic Cycling Unit

HCUP: Health Cost and Utilization Project

HCV: Helium Check Valve • Hydrogen Check Valve

HCW: Cheraw (USA-SC) (location identifier) • Heterogeneous Computing Workshop • Horizontal Component of Weight

HCZ: Hydrogen Convection Zone

hd: Hand • Hard • Head • Hora Decubitus • Hundert

Hd: Hand • Handel • Hard • Head • Hemd • Hund • Hundert

H&D: Hardened & Dispersed

HD: Hafendienst • Half Duplex • Hand-Drawn • Hard Disk • Hard Drive • Hardened and Dispersed • Harmonization Document • Hauptdeck • Hausdorff Distance • Heavy Distillate • Heavy Duty • Heißdampf • Helium Dewar • Henry Draper (catalogue) • Hierarchical Diagnosis • High Density • High Definition • High Dose • Highly Desirable • Hochdecker • Hochdruck • Holddown • Horizontal Drain • Horizontal Drive • Huddersfield (UK postal code) • Hurter-Driffield • Hydrodynamical • Hydrodynamics • Hydrogen Deuteride • Hydrogen-Deuterium • Hydrogen Drain • Hydrographic Department

Hda: Hacienda

HDA: Hard-Disk Attack • Hardware Disposition Area • Head Disk Assembly • Hexanediamine • Hexamethylenediamine • Hidden Falls (USA-AK) (location identifier) • High-Duty Alloy • Housekeeping Data Acquisition • HST Data Archive • Hubble Data Archive (HST)

HDAL: Hexadecenal • Hexadecimal

HDAM: Hierarchical Direct-Access Method

HDAS: Hinckley and District Astronomical Society (UK) • Hydrographic Data Acquisition System

HDATZ: High-Density Air-Traffic Zone

Hdb: Handbuch

HdB: Hydrogen-deficient Binary (star)

HDB: Haut Débit Binaire • HEASARC Database • Heidelberg (Germany) (location identifier) • Hierarchical Database • Housing and Development Board • Housing Development Board • Hypersonic Database

Hdbk: Handbook

hdbr: handbreit

Hdbr: Handbremse

HdC: Hydrogen-deficient Carbon-rich

HDC: Hard Disk Controller • Hasselblad Data Camera • Hawaii DUMAND Center (USA-HI) • Henry Draper Catalogue • Hierarchical Distributed Control • Highly Developed Country • Hungarian Data Center • Hybrid Device Controller • Hydrodynamic Chromatography

HDCD: High-Definition Compatible Digital

HDCL: High-level Data Link Control

Hdcp: Handicap

HDCS: HIPPARCOS Dedicated Computer System

HDD: Hard-Disk Drive • Hardware Design Document • Head-Down Display • High-Density Data • Horizontal Directional Drilling • Hyderabad (Pakistan) (location identifier)

HDDA: Hexanediol Diacrylate

HDDBMS: Heterogeneous Distributed Database Management System

HDDL: High-Definition Distance Learning

HDDMC: High-Definition Display Manufacturing Consortium

HDDR: High-Density Digital Recorder • High-Density Digital Recording

HDDS: High-Density Data System

HDDT: High-Density Digital Tape

HDDTR: HHDT Recorder

HDE: Hauptgemeinschaft des Deutschen Einzelhandels • Head-Drop-Einheit • Henry Draper Extension (catalogue) • Holdrege Brewster (USA-NE) (location identifier) • Homogeneous Differential Equation

HDEP: High-Density Electronic Packaging

HDF: Hardware Display Function • Hartmann Dispersion Formula • Hierarchical Data Format • High-frequency Direction Finder • Historical Data File • Hubble Deep Field

HDF-N: Hubble Deep Field North

HDFN: Hubble Deep Field North

HDF-S: Hubble Deep Field South

HDFS: Hubble Deep Field South

Hdg: Heading

HDG: Heading

HDGCP: Human Dimension of Global Change Programme • Human Dimension of Global Change Project

HDGEC: Human Dimensions of Global Environmental Change • Human Dimensions of Global Environment Change

hdgest: handgesteuert

hdgew: handgewebt

hdgm: handgemacht

hdgr: handgroß

Hdgr: Handgriff

HDGRS: High-Density Ground Recording System

HDH: Oahu Dillingham (USA-HI) (location identifier)

Hdhbg: Handhabung

HDHN: High-Degree Helioseismology Network

HDI: Haftpflichtverband der Deutschen Industrie • Head-to-Disk Interface • Head-to-Disk Interference • High-Definition Imaging • High-Density Interconnect • Human Development Index • Human Development Institute

HDIC: Helium Double Ionization Cell

HdJ: Haus der Jugend

HDJ: Hydrographic Department of Japan

HdK: Handkasse • Haus der Kultur

Hdkf: Handkerchief

hdl: handlich

Hdl: Handel • Handle • Handlung

HDL: Hardware Description Language • High-Density Lipoprotein

Hdlbg: Heidelberg (Germany)

HDLC: High-level Data Link Control

HDLD: Headland

Hdlg: Handling • Handlung

HDM: Hamadan (Iran) (location identifier) • Hardware Device Module • Hierarchical Particle Mesh • High-Duty Metal • Hot Dark Matter • Hypermedia Design Model • Hypermedia Document Modelling • Hypertext Design Model • Hypertext Document Model

HDMI: High-Density Multichip Interconnect

HDML: Handheld Device Markup Language

HDMS: Heidelberg Dark Matter Search • Hermes Documentation Management System • High-Density Memory System

Hdn: Harden

HDN: Hayden - Yampa Valley (USA-CO) (location identifier) • Steamboat Springs (USA-CO) (location identifier)

Hdns: Hardness

HDO: Eau Lourde • Heavy Water

HDOP: Horizontal Dilution Of Precision

HDOS: Hughes Danbury Optical Systems (USA-CT)

HDP: High-Density Polyethylene • Holddown Post • Horizontal Data Processing • Human-Dimension Programme • Hydroxydimethylpyrimidine

HDPE: High-Density Polyethylene

HDPS: High-Density Power Supply

HDQ: Headquarters

Hdqrs: Headquarters

Hdr: Header

HDR: Header • Health Data Recorder • Heißdampfreaktor • High Data Rate • High Data Register • High-Density Recorder • High-Density Rule • High-Dose Rate • Hold Down and Release • Hot Dry Rock

HDRA: High Data Rate Assembly

HDRL: High-Data-Rate Laser • High-Dose Reference Laboratory

HDRM: High Data Rate Multiplexer

HDRR: High Data Rate Recorder

HDRS: High-Data-Rate System

HDRSS: High-Data-Rate Storage System

HDRT: High Data Rate Terminal

HDS: Hardware Description Sheet • Hawaii Deep Survey • Hierarchical Data Structure • Hierarchical Data System • High-Density Sample • High-Dispersion Spectrograph • Hitachi Data Systems • Hydrogen Disposal System

HDSBUV: High-Density SBUV

HDSC: High-Density Signal Carrier

hdschr: handschriftlich

Hdschr: Handschreiben • Handschrift

Hdshk: Handshake

HDSL: High-bit-rate Digital Subscriber Line (*) • High-data-rate Digital Subscriber Line • High-rate Digital Subscriber Line • High-rate Digital Subscriber Link

Hdst: Headset

HDST: Headset

hdt: hundert

Hdt: Hundert

HdT: Haus der Technik

HDT: Hard-Disk Toolkit • Head-Down Tilt • High-Density Tape • Hochdruckturbine • Host Data Terminal • Host Digital Terminal

HDTOMS: High-Density TOMS

HDTP: Handheld Device Transport Protocol • Hoofddirectie Telecommunicatie en Post (Netherlands)

HDTR: High-Density Tape Recorder

HDTS: High-Density Transcription System

HDTV: High-Definition Television

HDU: Head-Drop-Unit • Header and Data Unit • Header Data Unit

HdV: Heure de Vérité

HDVD: High-Definition Video Disk • High-Definition Video Display • High-Definition Volumetric Display

Hdw: Handwagen • Handwerk • Hardware

Hdwb: Handwörterbuch

HDW: Hardware

Hdwd: Hardwood • Headword

Hdwe: Hardware

HDWE: Hardware

Hdwr: Hardware

HDx: Half Duplex

HDX: Half Duplex

HDY: Hat Yai (Thailand) (location identifier)

Hdz: Handzeichen

HDZ: Hochdruckzylinder

he: Hic Est • Hoc Est

He: Helium • Hélium

H/E: Heat Exchanger

HE: Half Effect • Hardware Evaluation • Hardware Evaluator • Hauptentschädigung • Heat Engine • Heat Exchange • Heat Exchanger • Heavy Equipment • Hebrew • Henize • Her Eminence • Her Excellency • High Eccentricity • High Efficiency • High End • High Energy • Higher Education • High Explosive • Highly Explosive • His Eminence • His Excellency • Horizontal Equivalent • Hub End • Human Engineering • Hydroelectric • Hydrogen Embrittlement • Hyperbolic Equation

HEA: Health Effects Assessment • Herat (Afghanistan) (location identifier) • High-Energy Approximation • High-Energy Astrophysics • Higher Education Act (USA) • Higher Education Authority

HEAD: High-Energy Astrophysics Division

HEADS: High-Energy Astrophysics Database Service

HEAL: Human Exposure Assessment Location

HEAMOWG: High-Energy Astrophysics Management Operations Working Group

HEAMWG: High-Energy Astrophysics Management Working Group

HEANet: Higher Education Authority Network

HEAO: High-Energy Astronomical Observatory • High-Energy Astronomy Observatory • High-Energy Astrophysical Observatory (*) • High-Energy Astrophysics Observatory • Hoger Economisch en Administratief Onderwijs

HEAP: High-Energy Aim Point • High-Energy Astrophysics • High-Energy Astrophysics Program

HeaPh: Health Physics (periodical code)

HEAR: High-Efficiency Anti-Reflecting • High-Efficiency Anti-Reflection • Human-Error Action Report

HEASARC: High-Energy Astrophysics Science Archive Research Center

HEASRC: High-Energy Astrophysics Science Research Center

HEAT: High-Energy Antimatter Telescope • High-lift Engine Aeronautics Technology

Heb: Hebrew

HEB: Handheld Electronic Book • Henzada (Myanmar) (location identifier) • Hersbruck (Germany) • High-Energy Buncher • High-Excitation Blob • Hollow Electron Beam • Hot Electron Bolometer

Hebd: Hebdomadaire

HEBIS: Hessiches Bibliotheks-Informations-System

HEBM: Hot Electron Bolometer Mixer

Hebr: Hebrew • Hebrides

HEC: Hardened Electronic Component • Hasselblad Electric Camera • Hautes Études Commerciales • Header Error Check • Header Error Control • Heure de l'Europe Centrale • High-Emission Cathode • High-Energy Chemistry • Hollerith Electronic Computer • Hydroelectricity Commission

HECaR: High-sensitivity Electrically Calibrated Radiometer

HECC: Higher Education Coordinating Cuncil • High-End Computing and Computation

HECK: Header Error Check • High-Energy Chemical Kernel

HECP: Herbaceous Energy Crops Program

HECR: High-Energy Cosmic-Ray

HECRE: High-Energy Cosmic-Ray Experiment

HECSE: Higher Education Consortium for Special Education (USA)

HeCV: Helium Cataclysmic Variable

HECV: Helium Check Valve

HED: Herendeen (USA-AK) (location identifier) • High-Energy Detector • Howardite-Eucrite-Diogenite • Hypohidrotic Ectodermal Dysplasia

HEDC: Hasselblad Electric Data Camera • HESSI European Data Center

HEDCC: Human-Error Data Control Center

HEDCO: Higher Education for Development Cooperation

HEDM: High-Energy Density Matter

HEDS: Human Exploration and Development of Space

HEE: Helena - Thompson-Robbins (USA-AR) (location identifier) • Human Energy Expenditure

HEEC: Heure d'Été de l'Europe Centrale

HEED: Health and Environmental Effects Document • High-Energy Electron Diffraction

HEEEL: High-Energy Electronically Excited Laser

HEEN: Human Evolution Education Network

HEEO: Highly Elliptical Earth Orbit • Highly Elliptical Equatorial Orbit

HEEP: Health and Environmental Effects Profile

HEES: Hypertext Evaluation Expert System

HEF: Hochschulkonferenz für Erziehungswissenschaft und Fachdidaktik

HEFA: Higher Education Facilities Act (USA)

HEFC: Higher Education Funding Council

HEFCE: Higher Education Funding Council for England (UK)

HEFCS: Higher Education Funding Council for Scotland (UK)

HEFCW: Higher Education Funding Council for Wales (UK)

HEFT: High-Energy Focussing Telescope

HeG: Helium Gauge

HEG: Helium Gauge • High-Energy Gamma-ray • High-Energy Grating • High-Enthalpy (facility at) Göttingen (Germany)

HEGCS: High-Energy Gas Čerenkov Spectrometer

HEGR: High-Energy Gamma Ray

HEGRA: High-Energy Gamma-Ray Array (*) • High-Energy Gamma-Ray Astronomy

HeGV: Helium Gauge Valve

HEGV: Helium Gauge Valve

HEH: Heho (Myanmar) (location identifier)

HEHI: High-Energy Heavy Ion

HEI: Heide Büsum (Germany) (location identifier) • Higher Education Institute • Higher Education Institution

HEIDI: High-Energy Imaging Device

Heim: Heimat

heimatl: heimatlich • heimatlos

HeiTS: Heidelberg Transport System

HEK: Hanseatische Ersatzkasse • Heihe (China-PRC) (location identifier)

Hel: Helicopter

HEL: Hardware Emulation Layer • Helsinki (Finland) (location identifier) • High-Energy Laser

HELAS: Hellenic Astronomical Society (Greece)

HELB: High-Energy Laser Beam

HELCUSS: High-Energy Laser Computerized Unmanned Space Synthesis

HELF: High-Energy Laser Fluence

Helg: Helgoland

Heli: Heliport

Hélico: Hélicopter

Heliogr: Heliograph • Heliographie

HELIO: Heliocentric

HELIOS: Handicapped European Living In Open Society

hell: hellenisch

HELLO: High-Energy Laser Light Opportunity

HELO: High Elliptical Lunar Orbit

HELOS: Highly Eccentric Lunar Occultation Satellite

HELP: Highly Extendable Language Processor

HELS: High-Energy Laser System

HELSTF: High-Energy Laser System Test Facility

helv: helvetisch

Helv: Helvetia • Helvetica

HEM: Harmonization of Environmental Measurement • Helsinki Malmi (Finland) (location identifier) • High-Energy Monitor • Hitchhike Experiment Module • Hybrid Electromagnetic Wave

HEMA: High-Energy Neutral Atom

HEMI: Hemispherical

HEMIS: Harmonization of Environmental Measurement Information System

HEMM: Heavy Earth-Moving Machinery

HEMP: High-altitude Electromagnetic Pulse

HEMS: High-level Entity Management System

HEMT: High Electron Mobility Transistor (*) • High Electron Movability Transistor

HeMV: Helium Manual Valve

HEMV: Helium Manual Valve

HEN: Hendon (UK-England) (location identifier)

HENA: High-Energy Neutral Atom

HENP: High Energy and Nuclear Physics

HENSA: Higher Education National Software Archive • Higher Education National Software Authorities

HEO: Haelogo (Papua New Guinea) (location identifier) • High Earth Orbit • High Earth Orbiter • High-Energy Orbit • High Eccentric Orbit • High Elliptical Orbit • Higher Executive Officer • Highly Eccentric Orbit • Highly Elliptical Orbit

HEOO: Honeywell Electro-Optical Operations

HEOR: Historical Event Operation Record

HEOS: High-Eccentricity Earth-Orbiting Satellite • Highly Eccentric Orbit Satellite (*) • Highly Elliptical Orbit Satellite

HEP: Heterogeneous Element Processor • High-Energy Particle • High-Energy Physics • High-Energy Pulse • Human Error Probability • Hydroelectric Power

HEPA: High-Efficiency Particle Absolute • High-Efficiency Particle Accumulator • High-Efficiency Particulate Air • High-Efficiency Particulate Atmosphere

HEPAD: High-Energy Proton and Alpha Detector

HEPAP: High-Energy Physics Advisory Panel

HEPC: High-Energy Proportional Counter

HEPI: High Earch Processes Imager

HEPIC: High-Energy Physics Information Center

HEPL: High-Energy Physics Laboratory

HEPNet: High-Energy Physics Network

HEPP: High Energy and Particle Physics

HEPS: High-Energy Particle Spectrometer

Her: Herald • Hercules-Herculis (constellation) • Hertog

HER: Heraklion (Greece) (location identifier) • High-Energy Resolution • Human-Error Rate

HeRA: Hermes Robot Arm • Hermes Robotic Arm

HERA: Hadron Electron Research Accelerator • Hadron Electron Ring Accelerator • Hadron Elektron Ring Anlage • Hermes Robot Arm • Hermes Robotic Arm

HERAP: Human-Error Research and Analysis Program

HERAS: Hull and East Riding Astronomical Society (UK)

HERB: High-Energy Resolution Burst

HERC: Hercules • High-Energy Cosmic Ray

HERCULES: Hand-held, Earth-oriented, Real-time, Cooperative, User-friendly, Location-targeting and Environmental System

HERD: Herdade • High Education and Research Development

Héréd: Héréditaire

Hered: Hereditary

HERF: High-Energy Radiated Field • High-Energy Radio Frequency

herg: hergestellt

Herg: Hergabe • Hergang

hergest: hergestellt

HERL: Health Effects Research Laboratory

HERMES: High-resolution Evaluation of Radiances from Meteorological Satellites

HERO: Hazards of Electromagnetic Radiation to Ordnance • High-Energy Replicated Optics

HEROS: Heidelberg Extended Range Optical Spectrograph

HERP: Hydrological Emscher Radar Project

HERS: High-Energy Range Spectrometer • Higher Education Resource Services

HERSCP: Hazardous Exposure Reduction and Safety Criteria Plan

Herst: Hersteller • Herstellung

HERT: Headquarters Emergency Relocation Team

Herts: Hertfordshire (UK-England)

HERU: Higher Education Research Unit

HES: Hamburg/ESO Survey • Hermiston State (USA-OR) (location identifier) • Hypertext Editing System

HESAC: Hermes Safety Advisory Committee

HESF: Hermes Engineering Support Facility

HESP: High-Energy Solar Physics

HESPE: High-Energy Solar-Physics Explorer

hess: hessisch

Hess: Hessen (Germany)

HESS: High-Energy Squib Simulator • High-Energy Stereoscopic System • Hydrology and Earth System Sciences (journal)

HESSE: Hard Embedded Software Support

HESSI: High-Energy Solar Spectrographic Imager • High-Energy Solar Spectroscopic Imager (*)

HESYRL: Hefei National Synchrotron Radiation Laboratory (China-PRC)

Het: Heterodyne

HET: Hall Effect Thruster • Heavy-Equipment Transporter • High-Energy Telescope • High-Energy Transfer • Hobby-Eberly Telescope • Hohhot (China-PRC) (location identifier)

HETC: High-Energy Transport Code

HETE: High-Energy Transient Experiment (*) • High-Energy Transient Explorer

HETG: High-Energy Transmission Grating

HETGS: High-Energy Transmission Grating Spectrometer

HETM: Hybrid Engineering Test Model

HETP: Healthcare Engineering and Technology Policy • Height Equivalent per Theoretical Plate • Height Equivalent to a Theoretical Plate • Hexaethyl Tetraphosphate

HETS: High-Energy Telescope System

HEU: Highly Enriched Uranium

Heur: Heuristic

HEUS: High-Energy Upper Stage

HEV: Hybrid Electric Vehicle

HEW: Half Energy Width • Hamburgische Elektricitätswerke • Health-Education-Welfare

HeWD: Helium White Dwarf

HEWS: Humanitarian Early Warning System

Hex: Hexadecimal • Hexagon • Hexagonal

HEX: Heat Exchanger • Hexadecimal • Hexagon • Santo Domingo - Herrera (Dominican Rep.) (location identifier)

HEXAGONE: High-Energy X- And Gamma-ray Observatory of Nuclear Emission

HEXE: High-Energy X-Ray Experiment

HEXIS: High-Energy X-ray Imaging Spectrometer

HEXOS: Humidity Exchange Over the Sea

HEXTE: High-Energy X-ray Timing Experiment

HEY: Ozark Hanchey (USA-AL) (location identifier)

HEZ: Natchez - Hardy-Anders (USA-MS) (location identifier)

hf: Half

Hf: Hafen • Hafnium • Handelsform • Hauptform • Hypothèse Faible

H/F: Homme/Femme • Human Factor

HF: Halbe Fahrt • Hard Failure • Hard Firm • Hartree-Fock • Hauptfunktion • Haute Fréquence • Haze Filter • Hickey-Frieden • High Fidelity • High Frequency • Hilfsfunktion • Hochfrequenz • Horizontal Flight • Hors Fonction • Hot Firing • Human Factor • Hydrofluoric Acid • Hydrogen Fill • Hydrogen Fluoride • Hyperfiltration • Hypothèse Forte

Hfa: Haifa (Israel)

HFA: Haifa (Israel) (location identifier) • High-Frequency Accelerometer • Housing Finance Agency • Hydrofluoroalkane

HFAARP: High-Frequency Active Auroral Research Program

HFAS: High-Frequency Acoustic Sounder • High-Frequency Antenna System

HFB: Hartree-Fock-Bogolioubov • Heptafluorobutyrate

HFBA: Heptafluorobutyric Acid

HFBC: High-Frequency Broadcasting

hf. bd.: Half Binding • Half Bound

Hfbf: Hafenbahnhof

HFBI: Heptafluorobutyrylimidazole

HfBK: Hochschule für Bildende Künste

HFBR: High-Flux Beam Reactor

HFC: Heat Flow and Convection • Helsinki Fair Centre (Finland) • High-Frequency Correction • High-Frequency Current • Hybrid Fiber and Coax • Hybrid Fiber Coax • Hybrid Fiber Coaxial • Hydraulic Flight Control • Hydrofluorocarbon

HFCC: Hermes Flight Control Centre

HFCL: Hydraulic Flight Control Laboratory

HFCT: Hydraulic Flight Control Test

HFCV: Helium Flow Control Valve

HFCW: Hydrogen Fluoride Continuous Wave

HFD: Hartford Brainard (USA-CT) (location identifier) • Heat Flow Density • High-Frequency Domain

HFDF: High-Frequency Direction Finder

HFDL: High-Frequency Data Link • Host Forms Description Language

HFDP: Human Factors Development Plan

HFDS: Head-up Flight Display System • Hydrogen Fluid Distribution System

HFDU: High-Frequency Data Unit

HFE: Hefei (China-PRC) (location identifier) • High-Frequency Excess • High-Frequency Executive • Historia de Formación Estelar • Human Factors and Ergonomics • Human-Factors Engineering

HFES: Human Factor and Ergonomics Society

HFET: Hellmann-Feynmann Electrostatic Theorem

HFF: High-Frequency Filter • High-Frequency Furnace • Hochschule für Fernsehen und Film • Hoffman Mackall (USA-NC) (location identifier) • Hubble Flanking Field

HfG: Hochschule für Gestaltung

HFG: Harmonic Frequency Generator • High-Frequency Generator • Hochfrequenzgesetz

HFI: High-Frequency Instrument

HFIP: Hexafluoroisopropanol • Hexafluoroisopropyl

HFIS: History and Foundations of Information Science

HFL: Dutch Guilder (currency) • Hollandse Florijn (currency)

HFM: Hazardous Fluids Module • Health and Fault Management • Hollow Fiber Membrane • Human-Factor Modelling

HFML: High Fibre Meal • High Field Magnet Laboratory

HFMP: Human Factors Management Plan

HFMS: High Fidelity Mechanical Simulator

HFMU: High-Fidelity Mock-Up • High-Frequency Mock-Up

Hfn: Hafen • Hafner

HFN: High-Frequency Noise • Höfn (Iceland) (location identifier) • Hornafjördur (Iceland) (location identifier)

HFO: Heavy Fuel Oil • High-Frequency Oscillation • High-Frequency Oscillator

HFOVR: Hemispherical Field-Of-View Radiometer

HFP: Hardened Focal Plane • Hexafluoropropylene

HFPL: HF-induced Plasma Line

HFPO: Hexafluoropropylene Oxide

HFR: Hartree-Fock Relativistic • High-Flux Reactor • High Focal Ratio • High-Frequency Range • High-Frequency Receiver • High-Frequency Resistor • Hochflußreaktor • Human and Family Resources

HFRA: High-Frequency Recovery Antenna

HFRT: High-Frequency Radio Transmitter

HFRW: High-Frequency Resistance Welding

HFS: Hagfors (Sweden) (location identifier) • Heliogeophysical Forecasting Service • Hierarchical File System (*) • Hierarchical Filing System • High-Fidelity Simulator • Horizontal Flight Simulator • Hyperfine Structure • Hypothetical Future Sample

HFSD: Heating and Fueling Systems Division

HFSE: High Field Strength Element

HFSJG: Hochalpine Forschungsstationen Jungfraujoch und Gornergrat (Switzerland)

HFSN: High-Frequency Seismic Noise

HFSP: Hierarchical and Functional Software Process • Human Frontier Science Program

HFSS: High-Frequency Structure Simulation • Hochschule für Sozialpädagogik und Sozialökonomie

Hft: Haftung

HFT: Hammerfest (Norway) (location identifier) • Heating Facility for Turbines • High-Flux Telescope • Horizontal Flight Test • Human factors in Telecommunications

HFTE: Human Factors Test and Engineering

HFTF: Horizontal Flight Test Facility

Hftg: Haftung

HFTS: Horizontal Flight Test Simulator

HFW: Henschel Flugzeug-Werke

HFWA: High-Frequency Wave Analyzer

HFX: High-Frequency Transceiver • High-Frequency Transducer

hg: Hectogram • herausgegeben

Hg: Hang • Herausgeber • Hergang • Hofgebäude • Hügel • Mercure • Mercury • Quecksilber

HG: Handelsgenossenschaft • Handelsgericht • Hendelsgesellschaft • Harmonic Generator • Harrogate (UK postal code) • Haute-Garonne (France) • Henyey-Greenstein • Her Grace • High Gain • High German • His Grace • Hyades Giclas • Hydride Generation

HGA: Hardware Graphics Accelerator • Hargeisa (Somalia) (location identifier) • Hercules Graphics Adapter • High-Gain Antenna • Human-Genome Analysis • Hybrid Genetic Algorithm

HGAC: High-Gain Antenna Controller

HGAS: High-Gain Antenna System

hgb: herausgegeben

Hgb: Herausgeber

Hgbf: Hauptgüterbahnhof

HGC: Hercules Graphics Card

HGCC: Hot Gas Clean Up

HgCdTe: Mercury-Cadmium Telluride

HGCP: Hercules Graphics Card Plus

HGD: Hughenden (Australia-QLD) (location identifier)

HGDS: Hazardous Gas Detection System

HGE: Handling Ground Equipment

HGF: Hermann-von-Helmholtz-Gemeinschaft deutscher Forschungzentren (Germany)

hgg: herausgegeben

HGH: Hangzhou (China-PRC) (location identifier) • Human Growth Hormone

Hgl: Hügel

HGL: Helgoland (Germany) (location identifier) • High-Gain Link • High Galactic Latitude

hgm: hausgemacht

HGM: Hot-Gas Manifold

HGMF: High-Gradient Magnetic Filtration

HGMS: High-Gradient Magnetic Separation

HGN: Mae Hong Son (Thailand) (location identifier)

HGO: Haleakala Gamma Observatory (USA-HI) • Henry Grieb Observatory (USA-CA) • Hermes Global Orbiter • Hidalgo (Mexico) • Korhogo (Ivory Coast) (location identifier)

h. g. pt.: Hard Gloss Paint

Hgr: Hangar • Hanger

HGR: Hagerstown (USA-MD) (location identifier) • Hangar • High Growth Rate

HGS: Freetown Hastings (Sierra Leone) (location identifier) • Head-up Guidance System • Human Genome Sciences • Hydrogen Gas Saver • Hydrogenous Gas System

HGSC: Hubble Guide Star Catalogue

Hgsk: Hogskole

HGSt: Hauptgeschäfstelle

Hgt: Height

HGT: Height • Jolon Hunter (USA-CA) (location identifier)

Hgts: Heights

HGU: Mount Hagen (Papua New Guinea) (location identifier)

HGV: Heavy Goods Vehicle • Hydrogen Gas Valve

HGVT: Horizontal Ground Vibration Test

HGW: Herbert George Wells • Higher Gross Weight • Hochgeschwindigkeitswolken

Hgwy: Highway

Hgy: Highway

HGZ: Hogatza (USA-AK) (location identifier)

hh: Hands

HH: Double Hard • Haiti (international aircraft registration code) • Handelshochschule • Hand-Held • Handhold • Hansen-Hurwitz • Hare Hoogheid • Heavy Hydrogen • Herbig-Haro • Heren • Her Highness • Hermanos • Herr • His Highness • His Holiness • Hitchhiker • Hochschulen • Horizontal-Horizontal • Hour • Hydrometeor Heating

HHA: Huanghua (China-PRC) (location identifier)

HHB: Hot Horizontal Branch

HHC: Hammer Head Crane • Handheld Computer • Higher Harmonic Control • Hydrohalocarbon

HHD: Hypertensive Heart Disease

HHE: Hachinohe (Japan) (location identifier)

HHeF: Hot Helium Flasher

HHF: Höchstfrequenz • Hyper-High Frequency

HHFA: Housing and House Finance Agency

HHG: High Harmonic Generation

HHH: Hilton Head Island (USA-SC) (location identifier) • Treble Hard

HHI: Heinrich-Hertz-Institut (Germany) • Wahiawa Wheeler (USA-HI) (location identifier)

HHIP: Handheld Information Processor

HHMI: Howard Hughes Medical Institute (USA)

HHMS: Hydrostatic Head Monitoring Station

HHMMSS: Hours-Minutes-Seconds

HHMU: Handheld Maneuvering Unit

HHN: Hahn (Germany) (location identifier)

HHO: Hypertext Hands-On

HHOK: Ha Ha Only Kidding

HHOS: Ha Ha Only Serious

HHP: Hong Kong (China-PRC) (location identifier)

HHPA: Hexahydrophthalic Anhydride

HHQ: Höchste Quantität • Hua Hin (Thailand) (location identifier)

HHR: Hawthorne (USA-CA) (location identifier)

HHRR: Handheld Ratioing Radiometer

HHS: Handelshochschule • Health and Human Services • Heterodyned Holographic Spectrometer

HHT: Hand-Held Tester • Heinrich Hertz Telescope • High-Temperature Tetragonal

HHV: Higher Heating Value

HHW: Höchstes Hochwasser

Hhz: Hochdruckdampfheizung

HHZ: Hikueru (French Polynesia) (location identifier)

hi: Hindi (ISO language code)

Hi: Hindi

HI: Dominican Rep. (international aircraft registration code) • Halbinsel • Handelsinstitut • Handling Instruction • Hawaii (USA) • Hawaiian Islands (USA) • Health Insurance • Hearing Impaired • Heat Index • Height Indicator • Height of Instrument • Hic Iacet • High • High Intensity • Honeywell Inc. • Horizontal Interval • Host Integrator • Human Interaction • Human Interface • Humidity Index • Hydriodic Acid • Hydrodynamic Interaction • Hydrogen Iodide • Hydrographisches Institut • Neutral Hydrogen

HiA: Highlights in Astronomy (periodical code)

HIA: Herzberg Institute of Astrophysics (Canada) • Hot Ion Analyzer

HIAA: Hydroxyindoleacetic Acid

Hi.Ac.: High Accuracy

HIAD: Handbook of Information for Aircraft Designers

HIALS: High-Intensity Approach Light System

HIAS: Human Intellect Augmentation System

HI/AYH: Hostelling International/American Youth Hostels

HIB: Chisholm (USA-MN) (location identifier) • Heavy-Ion Beam • Hibbing-Chisholm (USA-MN) (location identifier)

HIBAL: High-altitude Balloon

HIBE: Hildescheimer Borde Experiment

HIBEX: High Acceleration Booster Experiment

HiBFiN: High-Brightness Filamentary Network

HIBP: Heavy Ion Beam Probe

HIBREL: High-Brightness Relay

HIC: Habitat International Coalition • Harsware Integration Contractor • Headend Interface Converter • Heavy Ion Counter • HIPPARCOS Input Catalogue • Human Interaction Component • Humidity Indicator Controller • Hybrid Integrated Circuit • Pretoria Iscor (South Africa) (location identifier)

HICAMP: Highly Calibrated Airborne Measurements Program

HICAP: Histogram Cluster Analysis System

HICAT: High-Altitude Clear Air Turbulence

HICDRP: Hybrid Incoherent Coherent Dispersion Removal Processor

HICIP: Hybrid Image Classification Instruction Package

HICOG: High Commissioner for Germany

HICOM: High Commission for Germany

HICSS: Hawaii International Conference on System Science

HID: Hardware Interface Device • Heavy-Ion Discontinuity • High-Intensity Discharge • Horn Island (Australia-QLD) (location identifier)

HIDAM: Hierarchical Indexed Direct-Access Method

HiDAS: High Desert Astronomical Society (USA-CA)

HIDAS: Helicopter Integrated Defensive Aids System

HiDEC: Highly-integrated Digital Electronic Control

HIDM: High Information Delta Modulation

Hidrául: Hidráulica

Hidrom: Hidrometría

HIE: Whitefield (USA-NH) (location identifier)

HIER: Higher

HIES: Hydrographic Image Exploitation System

HIF: Hand-edited Information File • Host Interface • Human-Initiated Failure • Hydrologic Instrumentation Facility • Hyperfile Interface Language • Hyper-G Interchange Format • Ogden Hill (USA-UT) (location identifier)

HIFAM: High-Fidelity Amplitude Modulation

HIFD: High-Density Floppy Disk

HIFE: Height Interpolation by Finite Elements

Hi-Fi: High Fidelity

HIFI: Hawaii Imaging Fabry-Pérot Interferometer • Heterodyne Instrument for FIRST • High Fidelity • Hypertext Interface For Information

HIFOGS: High-efficiency Infrared Faint-Object Grating Spectrometer (*) • High-resolution Faint-Object Grating Spectrometer

HIFR: Hover-In-Flight Refueling

Hig: Higiene

HIG: Hawaii Institute of Geophysics (USA-HI) • Highbury (Australia-QLD) (location identifier) • High-Integrating Gyroscope • Honeywell Integrating Gyro

HIGED: Handbook of Instructions for Ground Equipment Designers

HIH: Her Imperial Highness • His Imperial Highness • Hook Island (Australia-QLD) (location identifier)

HII: High Input Impedance • Human-Information Interaction • Ionized Hydrogen • Lake Havasu City (USA-AZ) (location identifier)

HIIP: Harvard Information Infrastructure Project

HIIPS: Home Institute Image Processing System • Home Institution Image Processing Subsystem (*) • Home Institution Image Processing System • Hud Integrated Information Processing Services

HIIS: Heavy Ions In Space

HIJ: Hiroshima (Japan) (location identifier)

HIK: Honolulu Hickam (USA-HI) (location identifier)

HIL: Hardware In Loop • Hazardous Immiscible Liquid • High-Ionization Line • High-Intensity Light • Human Interface Link • Human In the Loop • Shillavo (Ethiopia) (location identifier)

HILAC: Heavy-Ion Linear Accelerator

HILAT: High-Latitude

HILS: Hardware-In-the-Loop Simulation

HILT: Heavy Ion Large Telescope

HIM: Hardware Interface Module • Highly Ionized Medium • Hot Intercloud Medium • Hypertext - Information Retrieval - Multimedia

HIMAC: Heavy-Ion Medical Accelerator

HiMAT: Highly Maneuverable Aircraft Technology

HIMAT: Instituto Colombiano de Hidrología, Meteorología y Adecuación de Tierras (Colombia)

HiMaTE: High Mach Turbine Engine

HIMES: Highly Maneuverable Experimental Space-vehicle

HIMSIS: Hymalayan Snowcover Information System

HIMSPEC: High- and Mid-latitude Speciation

HIMSS: Healthcare Information and Management System Society (USA) • High-Resolution Microwave Spectrometer Sounder

Hin: Hindi

HIN: Chinju (Korea-ROK) (location identifier) • Health Information Network • Hors Instituts Nationaux • Hybrid Integrated Network

Hind: Hindi • Hindu • Hindustan • Hindustani

HINDU: Histogram Inspired Neighborhood Discerning Unsupervised

HINIL: High-Noise-Immunity Logic

hint: hinten • hinter

HInt: High Intensity

HINT: Hanoi Institute of Nuclear Technology (Viet Nam)

Hinw: Hinweg • Hinweis • Hinwendung

HIO: Hillsboro Portland (USA-OR) (location identifier)

HIORP: HST Independent Optical Review Panel

HIOS: High-performance Input/Output System

HIP: Hardware Interface Panel • Hardware Interface Processor • Hardware Interface Program • Headingly (Australia-QLD) (location identifier) • Health Information Provider • Health Insurance Plan • Hierarchical Information Processor • High-Impact Pressure • Highly Ionized Plasma • Hilpoltstein (Germany) • Host Interface Protocol • Hot Isostatically Pressed • Hot Isostatic Press • Hot Isostatic Pressing

HIPA: Health Information Portability Act (USA)

HIPAA: Health Insurance Portability and Accountability Act (USA)

HIPAR: High-Power Acquisition Radar

HIPAS: Hunting Image Processing and Analysing System

HIPASS: HI Parkes All Sky Survey

HIPC: Heavily Indebted Poor Country • High Chamber Pressure • Highly Indebted Poor Country

HIPDA: Hydroxyisophthalyl Dihydroxamic Acid

HIPERLAN: High-Performance Local Area Network

HIPO: Hierarchical Input-Process-Output • Hierarchy plus Input-Process-Output

HiPot: High Potential

Hipp: Hippocrates

HIPPARCOS: High-Precision Parallax Collecting Satellite

HIPPI: High-Performance Parallel Interface

HIPPO: Hawaii Imaging Polarimeter for Proto Objects

HIPPS: Highly Integrated Pluto Payload System

HIPR: Hypermedia Image Processing Reference

HIPS: Harwell Image Processing System • High-Impact Polystyrene

HIPSS: Hub Integrated Power Switching System

HIPWAC: Heterodyne Instrument for Planetary Wind and Composition

HIR: Higher • Honiara (Solomon Islands) (location identifier)

HIRAAS: High-Resolution Airglow/Aurora Spectroscopy

HIRAN: High-Range Navigation

HiRAP: High-Resolution Accelerometer Package

HiRDES: High-Resolution Double Echelle Spectrograph

HIRDES: High-Resolution Double Echelle Spectrograph

HiRDLS: High-Resolution Dynamics Limb Sounder

HIRDLS: High-Resolution Dynamics Limb Sounder

HIREFS: High-Resolution Erect Field Spectrometer

HIREGS: High-Resolution Gamma-ray Spectrometer

HiRel: High Reliability

HIRELS: High-Resolution Emission-Line Spectrometer

HIRENS: High-Resolution Narrow Swath

HiRes: High Resolution

HIRES: Highly Integrated Raster-based Exploration System

HIRF: High-Intensity Radiated Field • High-Intensity Radiation Field • High-Intensity Radio Frequency

HIRG: Hyperluminous Infrared Galaxy

HIRIS: High-Resolution Imaging Spectrometer

HIRL: High-Intensity Runway Light

HIRLAM: High-Resolution Limited Area Model (*) • High-Resolution Local Area Model

HIRO: HiScat International Radio Observatory

HIRRLS: High-Resolution Research Limb Sounder

HIRS: High-Impulse Retrorocket System • High-resolution Infrared Radiation Sounder (*) • High-resolution Infrared Sounder • Holographic Information Retrieval System

His: Histoire • Historia

HIS: Hauptinformationsstelle • Hayman Island (Australia-QLD) (location identifier) • Heavy-Ion Source • Hic Iacet Sepultus • Hic Iacet Sepulta • Hierarchical Intensive Search • High-Intensity Spectrometer • High-resolution Infrared Sounder • High-resolution Interferometer Sounder • Hochschul-Informations-System • Hue-Intensity-Saturation • Hybrid Infrared Source • Hydrologic Information System • Hypertext Information System

HISA: HI Self-Absorption

HISAM: Hardware-Initiated Standalone Memory • Hierarchical Indexed Sequential Access Method

HISAR: Hughes Integrated Synthetic Aperture Radar

HISARS: Hydrologic Information Storage and Retrieval System

HISDAM: Hierarchical Indexed Sequential Direct Access Method

Hisp: Hispanic • Hispánico

HISP: Hybrid Image and Signal Processing

HISPARS: High-Speed Pattern Analysis and Representation System

HISS: Heavy Ion Superconducting Spectrometer • Heritage Information Statistical System • Holographic Ice Surveying System

HISSR: Hydrogeological Information System with Spatial Reference

HiSSS: High Speed Strike System

Hist: Histoire • Histology • Historia • Historian • Historical • Historique • History

HIST: High-Sensitivity Telescope

HISTEP: High-speed Integrated Space Transportation Evaluation Program

HISTORIA: Heraldic Images Storing Applications

HIT: Hierarchical Incremental Testing • Human Interface Technology

HiTec: High Technology (periodical code)

HITL: Hardware In The Loop • Human Interface Technology Laboratory

HITLS: Hardware In The Loop Simulation

HiTOP: High Integration of Technology, Organization and People

HITS: Hierarchical Integrated Test Simulator • High-speed Integrated Test System

Hitt: Hittite

HIU: Headset Interface Unit • Host Interface Unit • Hydrologic Information Unit

HIUS: Hispanic Institute of the United States

HIV: Helium Isolation Valve

HIVE: Highly-parallel Integrated Virtual Environment

HIVOS: High-Vacuum Orbital Simulator

HIWAS: Hazardous In-flight Weather Advisory Service

HIWY: Highway

HIX: Hiva Oa (French Polynesia) (location identifier)

HiZSST: High Z Supernova Search Team

hj: halbjährig • halbjährlich • Hic Jacet

Hj: Haji • Halbjahr • Hilfsjahr

HJ: Halt and Jump • Hamilton-Jacobi (equation) • Hic Jacet

HJB: Halbjahresbeitrag • Hamilton-Jacobi-Bellman

HJBI: Hamilton-Jacobi-Bellman-Isaacs (equation)

HJBT: Heterojunction Bipolar Transistor

HjD: Halbjahresdurchschnitt

HJD: Heliocentric Julian Date

HJR: Khajuraho (India) (location identifier)

hjs: Hic Jacet Sepultus • Hic Jacet Sepulta

HJS: Hic Jacet Sepultus • Hic Jacet Sepulta

HJT: Khujirt (Mongolia) (location identifier)

h/k: Housekeeping

hk: Hoek • Hong Kong (China-PRC) • Housekeeping

H&K: Homing & Kill

HK: Colombia (international aircraft registration code) • Handelskammer • Hauptkasse • Hauptkommission • Hefner-Kerze • Heilkosten • Hochspannungskabel • Hohe Kommission • Hoger Kommissar • Homing and Kill • Hong Kong (international vehicle registration code) (obsolete) • Hong Kong (ISO country code) (obsolete) • Housekeeping

HKA: Blytheville (USA-AR) (location identifier)

HKAAS: Kong Kong Amateur Astronomical Society

HKAS: Kong Kong Astronomical Society

HKB: Healy Lake (USA-AK) (location identifier)

HKCS: Hong Kong Computer Society

HKD: Dollar (Hong Kong) (ISO currency code) (obsolete) • Hakodate (Japan) (location identifier) • Housekeeping Data

HKDB: Housekeeping Database

Hkf: Handkerchief

HKF: Housekeeping File

HKFRD: Hoher Kommissar der Französischen Republik in Deutschland

HKG: Hong Kong (ISO country code) (obsolete) • Hong Kong - Kai Tak (China-PRC) (location identifier)

HKH: Hare Koninklijke Hoogheid

HKIEd: Hong Kong Institute of Education (China-PRC)

HKIGS: Hong Kong Internet and Gateway Services

HKIX: Hong Kong Internet Exchange

HKJ: Jordan (international vehicle registration code)

HKK: Hokitika (New Zealand) (location identifier)

HKl: Handelsklasse

HKN: Hoskins (Papua New Guinea) (location identifier)

HKO: Hong Kong Observatory (China-PRC) (formerly RHKO)

HKP: Housekeeping • Hubble Key Program

HKPG: Hubble Key Program Group

HKS: Jackson Hawkins (USA-MS) (location identifier)

HKSAR: Hong Kong Special Administration Region (China-PRC)

Hksw: Hookswitch

HKT: Hollow Kathode Tube • Hot Kathode Tube • Phuket (Thailand) (location identifier)

HKTA: Hong Kong Tourism Association (China-PRC)

HKTDC: Hong Kong Trade Development Council

HKU: Hong Kong University

HKUST: Hong Kong University of Science and Technology

HKV: Hit-to-Kill Vehicle • Kaskovo (Bulgaria) (location identifier)

HKW: Heizkraftwerk

HKY: Hickory (USA-NC) (location identifier)

HKZ: Heizkostenzuschluß

hl: Haul • Hectoliter • Hectolitre • Heilig • Hell • Hoc Loco

Hl: Heilige • Heiliger

H/L: Hard Line

HL: Half Life • Hard Lander • Hard Line • Harmonic Line • Haute-Loire (France) • Heel Line • High Level • Hill • Hinge Line • Hochleistung • Hora Local • House of Lords • Hydrogen Line • Korea (ROK) (international aircraft registration code) • ROK (international aircraft registration code)

HLA: High-Level Analog • High-Level Architecture • Lanseria (South Africa) (location identifier)

HLAL: High-Level Assembler Language

H-LAM/T: Human using Language, Artifacts and Methodology in which he is Trained

HLAMT: Human using Language, Artifacts and Methodology in which he is Trained

HLASM: High-Level Assembler

HLB: Batesville Hillenbrand (USA-IN) (location identifier) • Harvard Library Bulletin • Hydrophilic-Lipophilic Balance

HLC: Heavy-Lift Capability • High-Latitude Cloud • High-Level Committee • Hill City (USA-KS) (location identifier) • Hybrid Load Controller

HLCF: Hardened Launch Control Facility

Hlcl: Helical

HLCL: High-Level Command Language

HLCM: High-Level Control Module

HLCO: Hard Labor Creek Observatory (USA-GA) • High-Low-Close-Open

Hlcptr: Helicopter

HLD: Hailar (China-PRC) (location identifier) • Head-Level Display • High-Level Discrete • Hold

Hldg: Holding

Hldn: Holddown

Hldr: Holder

HLDS: Hydrogen Leak Detection System

HLE: Heure Limite d'Enregistrement • High-Level Exposure • Homogeneous Linear Estimator • Hydrogen Line Emission

HLF: Half • High-Latitude Field • High-Level Formatting • Horizontal Laminar Flow • Hultsfred (Sweden) (location identifier)

HLFC: Hybrid Laminar Flow Control

hlg: heilig • hochliegend

Hlg: Halogen • Heilige

HLG: Helpline Group • High-Level Group • Wheeling - Ohio County (USA-WV) (location identifier)

HLH: High-L Helioseismometer • Ulanhot (China-PRC) (location identifier)

HLI: Higher-Layer Information • High-Level Interface • Hollister (USA-CA) (location identifier)

HLIPS: High-Level Image Processing System

HLIV: High-Level Input Voltage

Hlk: Heilkunde

HLK: Handlochkarte

Hll: Heilige

HLL: Halbleiterlabor • High-Level Language • Hill • Hillside (Australia-WA) (location identifier)

HLLAPI: High-Level Language Application Programming Interface

HLLS: Hills

HLLV: Heavy-Lift Launch Vehicle

HLLW: Hollow

Hlm: Hefner-Lumen

HLM: Habitation à Loyer Modéré • High-Level Meeting • Holland Park Township (USA-MI) (location identifier)

HLMT: Helmet

HLN: Helena (USA-MT) (location identifier)

HLNW: High-Level Nuclear Waste

HLOV: High-Level Output Voltage

HLP: Hardy-Littlewood-Pólya • Help • Jakarta - Halim Perdana Kusuma (Indonesia) (location identifier)

Hlpr: Helper

HLPS: Hot Liquid Process Simulator

HLQ: High-Level Qualifier • High-Luminosity Quasar • Highly Luminous Quasar (*)

HLR: High-Level Radiation • High-Level Radioactivity • Home Location Register • Home Location Regulator • Killeen - Fort Hood (USA-TX) (location identifier)

HLRM: High-Level Radio Modulator

HLRW: High-Level Radioactive Waste

HLRZ: Höchstleistungsrechenzentrum

HLS: Harvard Law School (USA-MA) • Hills • Hogere Landbouwschool • Hue-Luminance-Saturation • Saint Helens (Australia-TAS) (location identifier)

HLSP: Heitler-London-Slater-Pauling

HLSTO: Hailstones

HLT: Halt • Hamilton (Australia-VIC) (location identifier) • Helium II Telescope • Human-Language Technology

HLTA: Halt Acknowledge

Hlth: Health

HLTL: High-Level Test Language • High-Level Transistor Logic

HLTP: Hilltop

HLTTL: High-Level Transistor-Transistor Logic (*) • High-Level Transistor-Translator Logic

HLU: Houailou (New Caledonia) (location identifier)

HLUE: Homogeneous Linear Unbiased Estimator

HLV: Heavy-Lift Vehicle • Helenvale (Australia-QLD) (location identifier)

HLW: High-Level Waste • Hluhluwe (South Africa) (location identifier)

Hlx: Hefner-Lux

HLY: Holyhead (UK-England) (location identifier)

HLZ: Hamilton (New Zealand) (location identifier)

hm: Hallmark • Hectometer • Hectometric • Hoc Mense • Huius Mensis • Hydroxymethyl

hM: Harmonisches Mittel

H/m: Henry per Meter

Hm: Halbmesser • Hausmeister • Heckium • Heimat • Tiempo Solar Medio

HM: Hallmark • Handelsmarine • Handelsministerium • Hare Majesteit • Harmonic Mean • Haute-Marne (France) • Hazardous Material • Headmaster • Head Motion • Heard and McDonald Islands (ISO country code) • Heavy Maintenance • Heavy Metal • Her Majesty • Hertzsprung Method • Hinge Mount • His Majesty • Hochleistungsmotor • Homogenization Medium • Hungarian Method • Hydrogen Maser • Îles Heard et McDonald • McDonald Islands (Heard and_) (ISO country code)

HMA: High-Memory Area • Hub Management Architecture • Hybrid Mirror Antenna • Hypergolic Maintenance Area • Hypergol Maintenance Area • Malmö City (Sweden) (location identifier)

HMAC: Hashed Message Authentication Code

HMAS: Her Majesty's Australian Ship • His Majesty's Australian Ship

Hmb: Hamburg (Germany)

HMB: Hard Mobile Basing • Hexamethylbenzene

HMBA: Hydroxymethylbenzanthracene

HMBP: High-Mass Binary PULSAR

Hmc: Harmonic

HMC: Halley Multicolor Camera • Hardware Management Console • Harvey Mudd College (USA-CA) • Horizontal Meridian Circle • Horizontal Microcode • Hybrid Microcircuit • Hydrogen Maser Clock • Hypergol Maintenance and Checkout

HMCA: Hardware Management Console Application

HMCC: Houston Mission Control Center (NASA) • Hypergolic Maintenance and Checkout Cell • Hypergol Maintenance and Checkout Cell

HMCF: Hypergol Maintenance and Checkout Facility

HMCK: Hummock

HMCRI: Hazardous Materials Control Research Institute

HMCS: Helmet-Mounted Cuing System • Her Majesty's Canadian Ship • His Majesty's Canadian Ship

Hmd: Humid • Humidity

HMD: Handbuch der Maschinelle Datenverarbeitung • Head-Mounted Device • Head-Mounted Display • Heard and McDonald Islands (ISO country code) • Helmet-Mounted Display • Heterodyne Matrix Detector • Hydraulic Mean Depth • Hydrostatic Motor Driven • Îles Heard et McDonald • McDonald Islands (Heard and_) (ISO country code)

HMDS: Hawaii Medium Deep Survey

HMDE: Hanging Mercury Drop Electrode

Hme: Homme

HME: Handheld Microgravity Experiment • Hassi Messaoud (Algeria) (location identifier) • High Magnification Event • Hough Mode Extension

HMF: Health Maintenance Facility • Heavy-Metal Fluoride • Heliospheric Magnetic Field • High Mach Flow • High Magnetic Field • Horizontal Mating Facility • Hypergolic Maintenance Facility • Hypergol Maintenance Facility

HMG: Her Majesty's Government • Hermannsburg (Australia-NT) (location identifier) • Hydraulic Motor Generator

HMGF: High Modulus Glass Fiber

HMI: Hahn-Meitner Institute (Germany) • Hami (China-PRC) (location identifier) • Handbook of Maintenance Instructions • Hardware Monitor Interface • Hub Management Interface • Human-Machine Interface

HMINET: Hahn-Meitner-Institut Netz (Germany)

HMIS: Hazardous Materials Information System

HMJ: Khmelnitskiy (Ukraine) (location identifier)

HML: Hardened Mobile Launcher • Hard Mobile Launcher • Heure Moyenne Locale • Hierarchical Model Learning

HMLE: Hypermedia-based Learning Environment

HMM: Hidden Markov Model • Hidden Markov Modelling

HMMP: Hypermedia Management Protocol

HMMR: High-resolution Multifrequency Microwave Radiometer

HMMS: Hypermedia Management Schema • Hypermedia Management Scheme

HMMU: Hardware Memory Management Unit

HMMWV: High-Mobility Multipurpose Wheeled Vehicle • High-Mobility Multiwheeled Vehicle • Humvee

HMN: Alamogordo Holloman (USA-NM) (location identifier)

HMNAO: Her Majesty's Nautical Almanac Office (UK)

HMNS: Houston Museum of Natural Sciences (USA-TX)

HMO: Health Maintenance Organization • Hermosillo (Mexico) (location identifier) • Honolulu Magnetic Observatory (USA-HI) • Hueckel Molecular Orbital

HMOM: Hypermedia Object Manager

HMOS: High-density Metal Oxide Semiconductor • High-performance Metal Oxide Semiconductor • High-speed Metal Oxide Semiconductor

HMP: Hand-Made Paper • Hexasodium Metaphosphate • High Melting Point • Hoc Monumentum Posuit • Host Monitoring Protocol • Humidity Monitoring Panel • Hydraulic Maintenance Panel • Hydrazine Monopropellant • Hydrozene Monopropellant • Latvian Meteorological Agency

HMPA: Hexamethylphosphoramide

HMPAC: Helmet-mounted Multi-Purpose Applications Control

HMPT: Hexamethylphosphorous Triamide

Hmr: Hammer

HMR: Hamar (Norway) (location identifier) • Hermes Resource Module

HMRT: Harmonic Mean Response Time

HMS: Hardened Memory System • Harmonic Multiplier Source • Hazard Monitor System • Hazards Monitoring System • Health Monitoring System • Heavy Media Separation • Helmet-Mounted Sight • Her Majesty's Service • Her Majesty's Ship • Hermes • Hessen-Mittelrhein-Saar (Germany) • Hierarchical Memory Storage • High-Mass Star • High-Momentum Spectrometer • His Majesty's Service • His Majesty's Ship • History Memory System • Homeshore (USA-AK) (location identifier) • Hours-Minutes-Seconds • Huge Magnetic Field • Hungarian Meteorological Service • Hypermedia Management System

HMSO: Her Majesty Stationary Office • His Majesty Stationary Office • Honolulu Magnetic and Seismological Observatory (USA-HI)

Hmstd: Homestead

HMSU: High-resolution Microwave Sounding Unit

HMT: Hemet Ryan (USA-CA) (location identifier)

HMTA: Hazardous Materials Transportation Act (USA)

HMTF: Hydrodynamic Modulation Transfer Function

HMU: Hardware Mockup • Hydraulic Mockup • Hydromechanical Unit

HMUX: Hybrid Multiplexer

HMV: Heavy Maintenance Visit • High Magnification Viewer • His Master's Voice • Hydrodynamically Modulated Voltammetry • Hydrogen Manual Valve

HMW: High Molecular Weight

HMWS: Hawaii Medium Wide Survey

HMX: Cyclotetramethylene Tetranitramine

HMXB: High-Mass X-ray Binary

HMXRB: High-Mass X-Ray Binary

hn: Hac Nocte • Honduras

Hn: Horn

HN: Honduras (ISO country code) • Host to Network • Hotel Night • Hypernova

Hna: Hermana

HNA: High Nickel Alloy • Morioka (Japan) (location identifier)

Hnas: Hermanas

HNB: Huntingburg (USA-IN) (location identifier)

HNC: Hatteras (USA-NC) (location identifier) • Heavy Nucleus Collector • Higher Education Certificate • Hybrid Newton-Raphson • Hydrogen Isocyanide • Isocyanure d'Hydrogène

HNCO: High-precision Numerical-Controlled Oscillator • Isocyanic Acid

HNCS: Isothiocyanic Acid

HND: Higher National Diploma • Honduras (ISO country code) • Hundred • Tokyo Haneda (Japan) (location identifier)

Hndbk: Handbook

HNDLER: Handler

Hndrl: Handrail

Hndst: Handset

Hndwl: Handwheel

HNE: Heavy Nuclei Experiment • Heure Normale de l'Est • Tahneta Pass Lodge (USA-AK) (location identifier)

HNF: Hydrazinium Nitroformate

Hng: Hinge

HNG: Hienghene (New Caledonia) (location identifier)

HNGL: Helium Neon Gas Laser

HNGR: Hangar • Hanger

HNH: Hoonah (USA-AK) (location identifier)

HNI: Heiweni (Papua New Guinea) (location identifier)

HNIL: High-Noise-Immunity Logic

HNK: Hinchinbrooke Island (Australia-QLD) (location identifier)

HNL: Helium Neon Laser • Honolulu (USA-HI) (location identifier) • Lempira (Honduras) (ISO currency code)

HNM: Hana (USA-HI) (location identifier)

HNML: Hindu Meal

HNN: Honinabi (Papua New Guinea) (location identifier)

HNNC: Hecht-Nielsen Neurocomputer Corporation

Hno: Hermano

HNO: Hercegnovi (Yugoslavia-Montenegro) (location identifier) • Nitroxyl Radical

Hnos: Hermanos

HNRC: HOMS National Reference Centre

Hnrs: Honours

HNS: Haines (USA-AK) (location identifier) • Hexanitrostilbene • High Nitrogen Steel • Hughes Network Systems • Hybrid Neural System

HNSC: House National Security Committee

Hntg: Hunting

HNX: Hanna (USA-WY) (location identifier)

HNY: Happy New Year • Hengyang (China-PRC) (location identifier)

Ho: Holmium • Hotel • House

H/O: Handover

HO: Habitual Offender • Hale Observatories (USA-CA) • Handelsorganisation • Handover • Harmonic Oscillator • Haystack Observatory (USA-MA) • Head Office • High Output • Hochofen • Hoger Onderwijs • Hold Over • Home Office • Honduras • Hopi • Hours of Operation • House Officer • Hydrogen-Oxygen • Hydrographic Office • Hydroxyl Radical

HOA: Hands-Off Automatic • High-Order Algebra • Hola (Kenya) (location identifier)

HOB: Harvest Object Browser • High-Order Bit • High-Order Byte • Hobbs - Lea County (USA-NM) (location identifier)

HOBU: Hoger Onderwijs Buiten de Universiteit

HoC: House of Commons

HOC: Halogenated Organic Compound • Hands-On Component • Heavy Organic Chemical • Held On Charge • Higher-Order Crossing • High Output Current • Hourly Operating Cost • Hydrophobic Organic Compound • Komako (Papua New Guinea) (location identifier)

hochd: hochdeutsch

Hochf: Hochformat

Hochsch: Hochschule

Hochschwarzw: Hochschwarzwald (Germany)

höchst: höchstens

hochw: hochwertig • hochwürdig

Hochw: Hochwasser

HOCl: Hypochlorous Acid

HOCUS: Hand Or Computer Universal Simulation

HoD: Head of Department

HÖD: Handbuch des öffentlichen Dienstes

HOD: Hodeidah (Yemen) (location identifier)

HOE: Holographic Optical Element • Homing Optical Experiment • Homing Overlay Experiment • Houeisay (Laos) (location identifier) • Hydraulically Operated Equipment

HOEAS: Heart of England Astronomical Society (UK)

HOED: Hypermedia Online Educational Database

HOEM: Heliosynchronous Out-of-Ecliptic Mission

HOF: Hall Of Fame • Hofuf (Saudi Arabia) (location identifier)

H. of C.: House of Commons

HOFCO: Horizontal Functional Checkout

HOFD: Heterogeneous Opposed Flow Diffusion

H. of L.: House of Lords

höfl: höflich

Höfl: Höflichkeit

H. of R.: House of Representatives

HOFS: Hydrogen-Oxygen Fuel System

HOFT: High-Order Formal Tool

HOG: Hermann-Oberth Gesellschaft (Germany) • High-Order Godunov • Hofgeismar (Germany) • Holguin (Cuba) (location identifier)

Hoh: Hoheit

HOH: Hash Of Hashes • Hofheim (Germany) • Hohenems (Austria) (location identifier) • Water

Hohenz: Hohenzollern

HOHI: High Order Harmonic Interferometer

HOI: Hao Island (Tuamotu Islands) (location identifier) • Homogeneous Object Identifier

HOK: Hooker Creek (Australia-NT) (location identifier)

Hol: Holandés • Holiday • Holland • Hollandais • Hollande • Holocene

HOL: Head Of Line • Higher Order Language • High-Order Language • Holikachu (USA-AK) (location identifier)

HOLAB: Holographic Laboratory Brassboard

HOLC: High-Order Language Computer

HOLIS: Harvard Online Information Server

Holl: Holland • Hollandais • Hollande

HOLM: Higher Order Language Machine

HOLOP: Holographic Optics • Holographische Optiklabor

Holst: Holstein (Germany)

HOLW: Hollow

HOLWG: High-Order Language Working Group

HOM: Hectometric Radiation • Homer (USA-AK) (location identifier) • Homogeneity • Hotine Oblique Mercator

HOME: Highly Optimized Microscope Experiment

HOMES: Huron-Ontario-Michigan-Erie-Superior

HOMO: Highest Occupied Molecular Orbit • Highest Occupied Molecular Orbital

HOMS: Hydrological Operational Multipurpose Subprogramme • Hydrological Operational Multipurpose System (*)

Hon: Honor • Honorable • Honorably • Honorary

HON: Huron (USA-SD) (location identifier)

Hond: Honduras

Hono: Honolulu (USA-HI)

Hons: Honors

Hon.Sec.: Honorary Secretary

HOO: Quanduc - Nhon Co (Viet Nam) (location identifier)

HOOD: Hierarchical Object-Oriented Design

HOOP: Handbook Of Operating Procedures • Hoger Onderwijs- en Onderzoekplan

HOP: Harvest Object Protocol • Hierarchical Objective Procedure • Hierarchical Observing Protocol • High-Order Partition • High Output Paraffin • High-Oxygen Pressure • Hopkinsville Campbell (USA-KY) (location identifier) • Hourly Operating Plan

HOPE: Halley Optical Probe Experiment • H-II Orbiting Plane • Hydrogen-Oxygen Primary Extraterrestrial • H-Z Orbital Plane

HOPFI: High-Order Fabry-Pérot Interferometer

HOPG: Highly Oriented Pyrolitical Graphite

HOPI: High-speed Occultation Photometer and Imager

HOPL: History Of Programming Languages

HOPLA: Harmonized Optical Programmable Logic Array

HOPR: Hubble Observation Problem Report

HOPS: Haystack Observatory Post-processing Software

HOQ: Hof (Germany) (location identifier)

Hor: Horizon • Horizontal • Horologium-Horologii (constellation) • Horology

HOR: Heliocentric Orbit Rendezvous • Horb (Germany) • Horta (Azores) (location identifier)

hor. decub.: Hora Decubitus

HoReCa: Hotels-Restaurants-Cafés

Horiz: Horizon • Horizontal

HORIZ: Horizon • Horizontal

HORM: Hybrid Orbital Rehybridization Method

Horol: Horological • Horology

HORSE: Horizon Science Experiment

Horz: Horizontal

HÖS: Höchstadt (Germany)

HOS: Chos Malal (Argentina-NE) (location identifier) • Harvest Object System • Heckscher-Ohlin-Samuelson • High-level Observing Software • High-level Operating Software • High-level Operation Software • High-Order Software • Hybrid Operating System

HOSC: Huntsville Operations Support Center (USA-AL)

Hosp: Hopices • Hospital

HOSS: Halo Orbit Space Station • Hamilton Outer Space Services

Host: Hostería

HOST: Hardened Optical Sensor Testbed • Hubble Orbital System Test

Hot: Hotel

HOT: Hands-On Technology • Home Order Television • Hot Springs (USA-AR) (location identifier)

HOTAS: Hands On Throttle And Stick

HOTAY: Hands-On-Throttles-And-Yokes

HOTLIPS: Home and Office Techniques for a Local Image Processing System

HOTOL: Horizontal Takeoff and Landing

HOTS: Holographic Optical Telescope and Scanner

HOTT: Hot Of The Tree

Hotzenw: Hotzenwald (Germany)

HOU: Hands On Universe • Houston Hobby (USA-TX) (location identifier)

HOV: High-Occupancy Vehicle • Homogeneity Of Variance • Ørsta-Volda (Norway) (location identifier)

HOW: Fort Kobbe - Howard (Panama) (location identifier) • Handover Word • Handover Work

HOX: Homalin (Myanmar) (location identifier)

HOY: Hoy Island (UK) (location identifier)

Hp: Haltepunkt • Hilfspunkt

H/P: Heliport

HP: Halbpension • Half Pay • Half Plate • Haltepunkt • Handicapped Person • Handling and Propulsion • Handling Procedure • Hard Point • Harmonic Progression • Harvard Pulsar • Hauptpreis • Hauptpunkte • Haute Pression • Haute Provence • Hautes Pyrénées (France) • Haut-Parleur • Health Physicist • Health Physics • Heated Pitot • Heating Plant • Heat Pipe • Height of Perigee • Hemel Hempstead (UK postal code) • Herbstprüfung • Hewlett-Packard • Higher Power • High Pass • High Performance • High Power • High Pressure • High Priority • Himachal Pradesh (India) • Hiper Pool • Hire Purchase • Holding Pattern • Homepage • Horizontally Polarized • Horizontal Parallax • Horizontal Parallelism • Horizontal Polarization • Horsepower • Hot Pressed • Hot Prop • Human Performance • Hydrazinium Perchlorate • Hydrogen Purge • Hydrostatic Pressure • Panama (international aircraft registration code)

hPa: Hecto-Pascal

HPA: Ha'Apai (Tonga) (location identifier) • Handling and Positioning Aid • Heuristically-Programmed Algorithm • Hidrocarburo Policíclico Aromático • High-Power Amplifier • High-Power Antenna

HPAL: House-Programmed Array Logic

HPAU: High-Power Amplifier Unit

HPB: Hooper Bay (USA-AK) (location identifier)

Hpbf: Hauptpersonenbahnhof

HPBRM: High Payload Baseline Reference Mission

HPBW: Half-Power Bandwidth • Half-Power Beam Width

HPC: Handheld Personal Computer • High-density Projection Chamber • High-Performance Computing • High-Pressure Compressor • Hollerith Punched Card

HPCA: High-Performance Computer Architecture

HPCB: High-Performance Computing Branch

HPCC: High-Performance Computing and Communications • High-Performance Control Center

HPCCD: High-Performance Charge-Coupled Device

HPCCIT: High-Performance Computing, Communications and Information Technology

HPCCP: High-Performance Computing and Communications Program

HPCCV: High-Performance Computing, Communications and Visualization

HPCE: High-Performance Capillary Electrophoresis

HPCF: High-Performance Computing Facility

HPCG: Handheld Protein Crystal Growth

HPCI: High-Performance Computing Initiative

HPCMP: High-Performance Computing Modernization Program

HPCN: High-Performance Computing and Networking

HPCR: High-Pressure Compressor Rotor

HPCRC: High-Performance Computing Research Center

HPCS: High-Performance Computing System

HPCU: High-Performance Computing Unit

HPD: Half-Power Diameter • Highest Posterior Density • High Performance Demodulator • Hourly Precipitation Data • Hybrid Photocathode Detector • Hybrid Photodetector • Hybrid Photon Detector • Hydrometeorological Processes Division

HPDC: High-Performance Distributed Computing

HPDD: High-Performance Device Driver • High-Performance Disk Driver

HDPE: High-Density Polyethylene

HPDF: High-Performance Demonstration Facility

HPDJ: Hewlett-Packard DeskJet

HPDT: High-Pressure Drift Tube

HPE: Heat Pipe Experiment • High-Power Effect • Hope Vale (Australia-QLD) (location identifier)

HPER: Health - Physical Education - Recreation

HPF: Hazardous Processing Facility • High-Pass Filter • High-Performance Fortran • High-Power Field • Highest Portable Frequency • Highest Possible Frequency • Hochpass-Filter • Horizontal Processing Facility

Hpfl: Halbpflicht

HPFS: High-Performance File System • High-Pressure Fire Service

HPFT: High-Pressure Fuel Turbopump

HPFTP: High-Pressure Fuel Turbopump

HPG: Hewlett-Packard Graphics • High-Performance Group • High-Power Generator • High-Power Group • High-Pressure Gas • Homopolar Generator

HPGe: High-Purity Germanium

HPGL: Hewlett-Packard Graphics Language

HPGN: High-Precision Geodetic Network

HPGS: High-Pressure Gas System

HPGSPC: High-Pressure Gas Scintillation Proportional Counter

HPH: Haiphong Catbi (Viet Nam) (location identifier) • Horsepower-Hour

HPHD: High-Pressure High-Density

HPHS: Hartford Public High School (USA-CT)

HPI: High-Performance Insulation

HPIB: Hewlett-Packard Interface Bus

HPICS: Heat Pipe Instrumentation and Control System

HPIL: Hewlett-Packard Interface Loop

HPL: High Polar Latitude • High-Power Laser • High-Productivity Language • Hybrid Programming Language • Hypergolic Plume Impingement

HPLC: High-Performance Liquid Chromatograph (*) • High-Performance Liquid Chromatography (*) • High-Precision Liquid Chromatography • High-Pressure Liquid Chromatograph • High-Pressure Liquid Chromatography

HPLJ: Hewlett-Packard LaserJet

HPM: Harvard Project Manager • Hierarchical Particle Mesh • High-Pass Mode • High-Performance Motor • High-Power Microwave • Holdup Processing Mode • Holdup Process Mode • Hyperpage Mode

HPMA: High-Power Microwave Assembly

HPMT: HOSS Process Model Toolkit

HPN: High-Pass Network • Westchester County (USA-NY) (location identifier) • White Plains (USA-NY) (location identifier)

HPNS: High-Pressure Nervous Syndrome

HPO: Head Post Office • High-Performance Option • High-Pressure Oxygen

HPOC: Huygens Payload Operations Centre • Huygens Probe Operations Center

HPOL: Halfwave Polarimeter • High-Power Optics Laboratory

HPOP: High-Precision Orbit Propagator • High-Pressure Oxidizer Pump

HPot: Helical Potentiometer

HPOT: High-Pressure Oxidizer Turbopump

HPOTP: High-Pressure Oxidizer Turbopump

HPP: Half Page Printer • Heat Pipe Performance • High-Precision Photogrammetry • High-Pressure Polyethylene • Holding under Promise of Payment • Homogeneous Poisson Process

HPPA: Hewlett-Packard Precision Architecture

HPPF: Horizontal Payloads Processing Facility

HPPI: High-Performance Peripheral Interface • High-Precision Parallel Interface

HPPN: Hydrogen-Poor Planetary Nebula

HPQ: Highly Polarized Quasar • High-Polarization Quasar

HPR: Hardware Problem Report • High-Performance Router • High-Performance Routing • High-Power Receiver • High-Pressure Regulator • Horizontal Pressure Regulator • Human Performance Reliability • Hydrogen Pressure Regulator • Pretoria Central (South Africa) (location identifier)

HPRF: High Pulse Recurrence Frequency

HPRV: Hydrogen Pressure Relief Valve

HPS: Health Physics Society (USA-MD) • High Performance Systems Inc. (USA-NH) • High-Precision Subset • High-Pressure Sodium • High-Pressure Steam • High-Protein Supplement • Hybrid Propulsion System • Hydraulic Power System

HPSEQT: Hiper Pool Sequential Threshold

HPSIR: Hewlett-Packard Serial Infrared

HPSLT: High Power Semiconductor Laser Technology

HPSN: High-Performance Scalable Networking

HPSP: High-Precision Scanning Platform • High-Precision Scan Platform

HPT: Hampton (USA-IA) (location identifier) • Hexa-Pod Telescope • Hexapod-Teleskop • High-Potential Test • High-Precision Thermostat • High-Pressure Test • High-Pressure Tin • High-Pressure Turbine • High-Productivity Tool • Horizontal Plot Table

Hptb: Hauptband • Hauptberuf • Hauptbuch

Hptbd: Hauptband

HPTE: High-Precision Tracking Experiment

Hptl: Hauptlehrer

HPTLC: High-Performance Thin-Layer Chromatography

Hptm: Hauptmann

Hpts: Hauptsache

HPTS: High-Performance Transaction System

Hptsch: Hauptschule

Hptst: Hauptstadt • Hauptstation

hptw: hauptwörtlich

Hptw: Hauptwache • Hauptwerk • Hauptwort

HPU: High-Power Unit • High-Pressure Unit • Hydraulic-Power Unit • Hydraulic-Propulsion Unit

HPUX: Hewlett-Packard Unix

HPV: Helium Pressure Vessel • High-Performance Video • High-Pressure Valve • Human-Powered Vehicle • Kauai Island - Princeville (USA-HI) (location identifier)

HPVL: High-Performance Visualization Laboratory

HPVM: High-Precision Visibility Mode

HPW: Half Power Width • High-Performance Workstation

HPWM: HP Window Manager

HPY: Baytown (USA-TX) (location identifier)

HPZ: Helicopter Protection Zone

hq: Hoc Quaere

Hq: Headquarters

HQ: Headquarters • Hydroquinone • Highly Qualified • High Quality

HQDM: Headquarters Data Manager

HQF: Homogeneous Quadratic Form

HQM: Aberdeen (USA-WA) (location identifier) • Hamburg Quasar Monitoring • Hoquiam Bowerman (USA-WA) (location identifier)

HQMD: Headquarters Management Directive

HQP: Hybrid Query Processor

HQR: Handling Quality Rating

HQS: Hamburg Quasar Survey

hr: Croatia • Croatian (ISO language code) • Hear • Here • Hour

Hr: Heer • Herr • Horn • Hour

H-R: Hertzsprung-Russell

H/R: High Rate

HR: Croatia (ISO country code) • Croatie (ISO country code) • Hail and Rain • Handelsrat • Handelsrechnung • Handelsrecht • Handelsregister • Handling Room • Hardness Ratio • Harvard Revised • Hardware Reliability • Hauptredaktion • Hauptreferent • Heart Rate • Heat-Resistant • Henry Russell • Hereford (UK postal code) • Hertzsprung-Russell • High Range • High Rate • High Reduction • High Resolution • Historical Record • Holiday Rambler • Honduras (international aircraft registration code) • Horizontal Rule • Hour • House of Representatives • Human Reliability • Human Resource • Humidité Relative • Hydrogen Relief

HRA: Haute Résolution Angulaire • Health Resource Administration • Hemispherical Reflective Antenna • High-Rate Acquisition • High-Resolution Apodizer • Horizontal Resolution Advisory • Mansehra (Pakistan) (location identifier)

HRAA: High-Rate Acquisition Assembly

HRAO: Hartebeesthoek Radio Astronomy Observatory (South Africa)

HRAP: Hydrologic Rainfall Analysis Project

HRAS: Harvard Radio Astronomy Station

HRAV: High-Resolution Airborne Video

HRB: Handelsregisterblatt • Harbin (China-PRC) (location identifier) • Highway Research Board

HRC: Hasselblad Reflex Camera • Herbig-Rao Catalog • Highly Reflective Cloud • High-Resolution Camera • High-Resolution CCD • High-Resolution Channel • High-Resolution Chromatography • High-Resolution Converter • Human Resources Center • Human Rights Commission • Hybrid Receiver Circuit

HRCam: High-Resolution Camera

HRCGC: High-Resolution Capillary Gas Chromatography

HRCT: High-Resolution Computed Tomography • High-Resolution Computerized Tomography

HRD: Hertzsprung-Russell Diagram • High-Rate Detector • High-Rate Discharge • High-Rate Dositometer • High-Resolution Data • High-Resolution Diagnostic • High-Resolution Diskette • High-Resolution Display • Human Resource Development • Human Resources Data • Hurricane Research Division (NOAA)

HRDA: High-Rate Data Assembly

HRDB: High-Resolution Database

HRDC: Human Resources Development Canada

HRDI: High-Resolution Doppler Imager

HRDL: High-Resolution Doppler Lidar

HRDM: High-Rate Demultiplexer

HRDP: Hypothetical Reference Digital Path

HRDR: High-Rate Digital Recorder

HRDS: High-Rate Data Station

HRDTM: High-Resolution Digital Terrain Model

HRE: Harare (Zimbabwe) (location identifier) • Hardware Reuse Evaluation • Hardware Routing Engine • Highly Relativistic Electron • Holy Roman Emperor • Holy Roman Empire • Homogeneous Reactor Experiment • Hydrazine Rocket Engine

HREE: Heavy Rare Earth Element

HREELS: High-Resolution Electron Energy Loss Spectrometer • High-Resolution Electron Energy Loss Spectroscopy

HREF: Hypertext Reference

HReg: Handelsregister

HREM: High-Resolution Electron Microscope • High-Resolution Electron Microscopy

HREP: Highly Relativistic Electron Precipitation

HRES: High-Resolution Electronic System

HRESC: High-Resolution Still Camera

HRF: High-Resolution Facsimile • Human Research Facility

HRFD: High-Resolution Fourier Diffractometer

HRFM: High-Rate Frame Multiplexer

HRFPI: High-Resolution Fabry-Pérot Imaging

HRFR: High-Resolution Film Recorder

HRFRS: High-Resolution Film Recorder System

hrg: herausgegeben

Hrg: Hearing • Herausgeber

HRG: Hemispherical Resonating Gyro • Hemispherical Resonator Gyro (*) • High-Resolution Geometric • High-Resolution Graphics • Hochschulrahmengesetz • Hurghada (Egypt) (location identifier)

HRGC: High-Resolution Gas Chromatography

HRH: Her Royal Highness • His Royal Highness

HRHR: High-Resolution Hamilton Retrofit

HRI: High-Resolution Image • High-Resolution Imager • High-Resolution Imaging • High-Resolution Interferometer

HRII: High-Resolution Imaging Interferometer

HRIIS: High-Resolution Infrared Imaging Spectrometer

HRIP: Hic Requiescit In Pace

HRIR: High-Resolution Infrared Radiometer

HRIRS: High-Resolution Infrared Radiation Sounder

HRIS: High-Resolution Imaging Spectrometer • High-Resolution Infrared Sounder • Human-Resource Information System

HRISI: High-Resolution Imaging Spectrometer Industrial

HRJ: Chaurjhari (Nepal) (location identifier)

HRK: Hochschulrektorenkonferenz (Germany) • Kharkov (Ukraine) (location identifier) • Kuna (Croatia) (ISO currency code)

HRL: Harlingen (USA-TX) (location identifier) • High-Refraction Layer • High-Repetition Laser • Horizontal Reference Line • Hughes Research Laboratories • Hypermedia Research Laboratory (USA-TX) • San Padre Island (USA-TX) (location identifier)

HRLD: High-Resolution Laser Diffraction

HRM: Handheld Radiation Meter • Hassi R'Mel - Tilrempt (Algeria) (location identifier) • High-Rate Multiplexer • High-speed Road Monitor • Hopwise Reliable Multicast • Hudson River Museum (USA-NY) • Human Resource Management

HRMA: High-Resolution Mirror Assembly

HRMI: High-Resolution Microwave Imager

HRMP: Harvard Radio Meteor Project • Haui-Recurrent Markov Process

HRMS: High-Resolution Mass Spectrometry • High-Resolution Mass Spectroscopy • High-Resolution Microwave Search • High-Resolution Microwave Survey (*) • High-Resolution Multispectral Scanner • Human-Resource Management System

HRMSI: High-Resolution Multispectral Stereo Imager

HRMTI: High-Resolution Multispectral Thermal Imager

HRMVS: High-Resolution Multispectral Video System

Hrn: Herren • Herrn

HRN: Heron Island (Australia-QLD) (location identifier)

HRNTWT: High Reynolds Number Transonic Wind Tunnel

HRO: Harmonic Reference Oscillator • Harrison (USA-AR) (location identifier)

HROI: High-Resolution Optical Imager • High-Resolution Optical Instrument

HROM: High-Resolution Ozone Mapping

HROS: High-Resolution Optical Spectrograph

HRP: High-Resolution Profile • Horizontal Radiation Pattern • Human Reliability Program

HRPI: High-Resolution Pointable Imager

HRPS: Hazard Reduction Precedence Sequence

HRPT: High-Resolution Picture Transmission

HRR: Herrera (Colombia) (location identifier) • Higher Reduced Rate • High-Range Resolution • High-Resolution Radar • High-Resolution Radiometer

HRRC: Human Resources Research Center

HRRDA: Human Resources Regional Development Assistant

HRRE: Hydrogen Recombination Radiation Experiment

HRRI: Human Resources Research Institute

HRRL: Human Resources Research Laboratory

HRRO: Human Resources Research Office

HRRR: High-Range Resolution Recognition

HRRS: High-Range Resolution Signature

Hrs: Heures • Horas • Hours

HRS: Harrismith (South Africa) (location identifier) • Haute Résolution Spectrale • Hazard Ranking System • High-Rate Science • High-Rate Switch • High-Resolution Spectrograph • High-Resolution Spectrum • High-Resolution System • Hours

HRSC: High-Resolution Stereo Camera • Human Resources Steering Committee

HRSCMR: High-Resolution Surface Composition Mapping Radiometer

hrsg: herausgegeben

Hrsg: Herausgeber

HRSG: Heat Recovery Steam Generator

HRSGS: High-Resolution Shuttle Glow Spectroscopy

HRSI: High-Temperature Reusable Surface Insulation

HRSMR: High-Resolution Soil Moisture Radiometer

HRSO: High-Resolution Solar Observatory

HRSR: High-Resolution Scanning Radiometer

HRSS: High-Resolution Stereo Camera

Hrst: Hersteller • Herstellung

HR&T: Heat Rejection & Transport

HRT: Harrogate Linton-on-Ouse (UK-England) (location identifier) • Heat Rejection and Transport • Helmholtz Reciprocal Theorem • High-Rate Telemetry • High-Resolution Telescope • High-Resolution Thermometer • High-Resolution Tracker • High Res Technologies (USA-NY)

HRTEM: High-Resolution Transmission Electron Microscope • High-Resolution Transmission Electron Microscopy

HRTF: Head-Related Transfer Function

HRTI: High-Resolution Thermal Imager

HRTIR: High-Resolution Thermal Infrared • High-Resolution Thermal Infrared Radiometer

HRTS: High-Rate Telemetry System • High-Resolution Telescope and Spectrograph (*) • High-Resolution Telescope Spectrograph • High-Resolution Transition Region

HRV: Croatia (ISO country code) • Croatie (ISO country code) • Handelsregisterverfügung • Haute Résolution dans le Visible • Haute Résolution Visible • Heart Rate Variability • Heat Rate Variability • High Resolution in the Visible • High Resolution Video • High Resolution Visible

HRVIR: Haute Résolution Visible-Infrarouge • High-Resolution Visible-Infrared

HRW: Human Right Watch

HRWFS: High-Resolution Wavefront Sensor

HRX: Hypothetical Reference Connection

HRY: Henbury (Australia-NT) (location identifier)

Hrzg: Herzog

Hrzgt: Herzogtum

Hrzn: Horizon

HRZN: Horizon

hs: handschriftlich • Hic Situs • Hoc Sensu • Hora Sommi

Hs: Handschrift • Haus • Hours

H/S: Heat Shield

H-S: Hamilton Standard • Horizon Scanner • Horizon Sensor

HS: Habitability System • Hail and Snow • Haltestelle • Hamilton Standard • Handschrift • Handset • Hard Silo • Harmonized System • Haute Saône (France) • Hawker Siddeley • Header Size • Heat Shield • Heat Suppression • Heaviside • Heck-Sause • Helical Scan • Heliosphere • Heliospheric Shock • Hic Sepulta • Hic Sepultus • Hic Situs • Hierarchically Structured • Hierarchical Signature • Highest Score • High School • High Session • High Speed • High State • Hochschule • Home Secretary • Hora Sidérea • Horizon Scanner • Horizon Sensor • Hors Service • Host Scientist • Hot Short • Hot Spot • Hydrogen Sulfide • Hypersensibilisation • Hypersensibilisierung • Hypersensitization • Hyperstructural • Hyperstructure • Thailand (international aircraft registration code)

HSA: Hardware System Area • Heat Shield Abort • Heure Supplémentaire-Année • High-Speed Arithmetic • Historic Sites Act (USA) • Hören-Sprechen-Aufnehmen • Hollandse Signaalapparaten BV • Horizon Sensor Assembly

HSAC: High-Speed Analog Computer

HSAM: Hierarchical Sequential Access Method

HSAP: Honeycomb Sandwich Aluminium Panel

HSAR: Holographic Synthetic Aperture Radar

H-Sat: Heavy Satellite

HSAT: Heavy Satellite

Hsb: Hefner-Stilb

HSB: Harrisburg Raleigh (USA-IL) (location identifier) • Heat-Shield Boost • High-Speed Bus • High Surface Brightness • Hochleistungsschnellbahn • Hochschulbibliothek • Hue-Saturation-Brightness • Hungarian Space Board

HSBG: High Surface Brightness Galaxy

HSBIP: Hsinchu Science-Based Industrial Park (Taiwan)

HSC: Hardware/Software Coordination • Harmon Science Center (USA-OK) • Headwaters Science Center (USA-MN) • Hierarchical Storage Controller • Higher School Certificate • High-Speed Channel • High-Speed Craft • Honeycomb Strip Chamber • Host Software Component • House Spacecraft • Shaoguan (China-PRC) (location identifier)

HSCA: Harvard-Smithsonian Center for Astrophysics (USA-MA)

Hsch: Hochschule

hschr: handschriftlich

Hschr: Handschreiben • Handschrift

HSCL: High-Speed Command Link

HSCP: High-Speed Command Punch

HSCR: High-Speed Card Reader • High-Speed Command Reader

HSCS: High-Speed Communications System

HSCSD: High-Speed Circuit-Switched Data

HSCT: High-Speed Civil Transport (*) • High-Speed Commercial Transport • High-Speed Compound Terminal • Hypersonic Civil Transport

HSCU: Hydraulic Supply and Checkout Unit

HSD: Hamilton Standard Division • Hautschichtdick • Health and Safety Directive • Heat Storage Device • High-Speed Data • High-Speed Detector • High-Speed Device • High-Speed Diesel • Horizontal Situation Display

HSDA: High-Speed Data Acquisition • High-Speed Data Assembly

HSDARS: High-Speed Data Acquisition and Reduction System

HSDF: High-Speed Digital Filter

HSDG: Hamburg-Südamerikanische Dampfschiffahrts-Gesellschaft

HSDL: High-Speed Data Line • High-Speed Digital Line • Hypertext Structure Description Language

HSDMC: High-Speed Direct Multiplex Control

HSDS: High-Speed Detector Survivability

HSDT: Hypersonic Small-Disturbance Theory

Hse: House

H&SE: Health & Safety Executive

HSE: Halley-centered Solar Ecliptic • Health and Safety Executive • Heat Shield Entry • Heure Supplémentaire d'Enseignement • Hic Sepulta Est • Hic Sepultus Est • Hic Situs Est

Hsekpr: Housekeeper

HSEP: High-Speed Electrostatic Printer

HSES: High-Speed Experimentation and Simulation • Hughes Satellite Earth Station

HSF: HERA Simulator Facility • Hypergolic Servicing Facility • Hypergol Servicing Facility

HSFC: High-Speed Framing Camera

HSFD: High-Speed Flight Demonstration

HSF-P: HERA Simulator Facility Prototype

HSFP: HERA Simulator Facility Prototype

HSFS: High Sierra File System • Hobart Space Frontier Society (Australia-TAS)

Hsg: Housing

HSG: Hierarchical Symbolic Grouping • High Sierra Group • High Speed Gear • High Sustained Gs (acceleration) • Hochschulgemeinschaft

HSGC: Hawaii Space Grant College (USA-HI)

Hsh: Haushalt • Haushaltung

HSH: Her Serene Highness • His Serene Highness • Las Vegas - Henderson (USA-NV) (location identifier)

HSHC: High-Speed H-alpha Camera

Hshld: Household

HSHT: Helen Sawyer Hogg Telescope

HSI: Habitat Suitability Index • Handling Speed Intelligently • Hastings (USA-NE) (location identifier) • Heat Stress Index • High Solar Intensity • High-Speed Imager • High-Speed Interface • High-Speed Interferometer • High-speed Serial Interface • Horizontal-Situation Indicator • Host System Interface • Hot Section Inspection • Hue-Saturation-Intensity • Human-System Interaction • Human-System Interface • Human-Systems Integration • Hyperspectral Imager • Hyperspectral Instrument

HSIC: High-Speed Integrated Circuit

HSIG: Hermetically Sealed Integrating Gyroscope

H/SIR: Hardware/Software Integration Review

HSIR: Hardware and Software Integration Review • High-Speed Information Retrieval

HSJ: Honeycombed Sandwich Joint

HSK: Horizontal Sling Kit

HSKP: Housekeeping

hsl: häuslich

HSL: Hardware Simulation Laboratory • Hazardous Substance List • High-Speed Launch • High-Speed Logic • Hora Sidérea Local • Hue-Saturation-Lightness • Hue-Saturation-Luminance • Huslia (USA-AK) (location identifier)

HSLT: High-Speed Telemetry Link

HSM: Her Serene Majesty • Hierarchical Storage Management • Hierarchical Storage Manager • High-Speed Memory • His Serene Majesty • Horsham (Australia-VIC) (location identifier) • Hyperbase Session Manager

HSMA: Hotel Sales and Marketing Association

HSMB: Hybrid Superconductor Magnet Bearing

HSMD: High-Speed Multiplexer/Demultiplexer

HSMIMP: High-Speed Modular Interface Message Processor

HSMS: High-Speed Microwave Switch

HSN: Home Shopping Network

HSNG: Housing

HSO: Habitation and Space Operations • Hoger Secundair Onderwijs • Hungarian Space Office

Hsp: Hörspiel • Hospital

HSP: Health Stabilization Program • Hermes Spaceplane • Hydrological Simulation Programme • High-Speed Photometer • High-Speed Printer • High-Speed Pulse • Hot Springs (USA-VA) (location identifier)

HSPA: High-Speed Parallel Adder

HSPD: Hybrid Superconducting Pixel Detector

HSPTAL: High-Speed Paper Tape Absolute Loader

HSPTP: High-Speed Paper Tape Punch

HSPTR: High-Speed Paper Tape Reader

HSQ: Heat Shield Qualification

HSR: Hardware Status Register • Harvest Server Registry • Heat Shield Recovery • High-Speed Reader • High-Speed Research • Hochschulring

HSRA: Harvard-Smithsonian Reference Atmosphere • High-Speed Regeneration Assembly

HSRD: Hermes System Requirements Document

HSRL: High Spectral Resolution Lidar

HSRM: High Spectral Resolution Mode

HSRMWR: High Spatial Resolution Millimeter Wave Radiometer

HSRN: High Spectral Resolution Sounder

HSRO: High-Speed Repetitive Operation

HSRP: High-Speed Research Program • High-Speed Rotary Prism • Hot Standby Router Protocol

HSRS: High Spectral Resolution Sounder

HSRV: Heat Source Reentry Vehicle

Hss: Handschriften • Hessen

HSS: Hamburg/SAO Survey • Hic Sepulti Sunt • Hic Siti Sunt • High-School Student • High-Speed Scientific • High-Speed Service • High-Speed Steel • High-Speed Stream • High-Speed System • High Stress Strain • Hissar (India) (location identifier) • History of Science Society • Hybrid Simulation System • Hydraulic System Simulator

HSSDS: High-Speed Switched Digital Service

HSSI: High-Speed Serial Interface

HSSM: Hardened Space Sensor Model • Hydrocarbon Spill Screening Model

HSSP: High-Speed Switch Port

HSSS: Highly Structured Stochastic System

HSSU: Housing and Supporting Structure Unit

hst: Hora Sine Tempore

Hst: Haltestelle • Handsteuerung • Hauptstadt • Hauptstelle • Hilfstelle

HST: Hawaiian Standard Time • Highest Spring Time • High-Speed Taxiway • High-Speed Technology • High-Speed Telemetry • High-Speed Train • High-Speed Transit • High-Speed Transport • History • Hoge-Snelheids Tunnel • Homestead (USA-FL) (location identifier) • Horizontal Solar Telescope • Host • Hubble Space Telescope • Hypersonic Transport

HSTCT: Hubble Space Telescope Comet Team

HSTCXO: High-Stability Temperature-Compensated Crystal Oscillator

HSTF: Hypergolic System Test Facility

HSTJ: HST Superconducting Tunnel Junction

HSTP: Hard Stop

HSTR: High-Speed Token Ring

HSTRA: High-Speed Token Ring Alliance

Hstrx: Hestreux (Belgium)

HSTS: High-Speed Text System • Horizontal Solar Telescope with Spectrograph

HSTTL: High-Speed Transistor-Transistor Logic

HSU: Heater Switching Unit • Helium Service Unit • Helium Servicing Unit

HSV: Decatur (USA-AL) (location identifier) • Hermes Space Vehicle • Hue-Saturation-Value • Huntsville - Madison County (USA-AL) (location identifier) • Hypersonic Vehicle

HSW: Hauptschlagwort • High Temperature Winkler

HSWA: Hazardous and Solid Waste Act

HSWG: Helioseismology Working Group

HSWO: Hierarchical Stepwise Optimization

HSWS: Hartmann-Schack Wavefront Sensor

HSWT: Huygens Science Working Team

HSYNC: Horizontal Synchronization

ht: Hoc Tempore • Hoc Titulo

Ht: Haut • Heat • Height

H&T: Handling & Transportation

H/T: Heat Threat

HT: Hadamard Transform • Haiti (ISO country code) • Halftime • Half Title • Halftone • Halt and Transfer • Handling Time • Hankel Transform • Haute Tension • Hawaiian Time • Head-Tail • Head-to-Tail • Heat Threat • Heat Transfer • Heat Treatment • Heavy Tank • Height • High Technology • High Temperature • High Tensile • High Tension • High Tide • Hilbert Transformation • History Tape • Hochtemperatur • Hoc Tempore • Hoc Titulo • Horizontal Tab • Hors Taxes • Hors Texte • Hough Transform • Hubble Type • Hypertext

Hta: Huerta

HTA: Chita (Russia) (location identifier) • Hungarian Trademark Association • Hypertension Artérielle

HTB: Haupttagebuch • High-Tension Battery • Terre de Bas (Guadeloupe) (location identifier)

HTC: Halley-Type Comet • Heat Transfer Coefficient • Hermes Training Complex • High-Tech Curve • High-Technology Center • High-Temperature Carbon • Hybrid Technology Computer • Hydrographic and Topographic Center • Hydrothermical Coefficient

Htd: Heated

HTDP: Horizontal Time-Dependent Positioning

Hte: Haute

HTE: Hazardous Top-level Event • High Temperature Elongation • Horwitz-Thompson Estimator • Hypertext Engine

H. Tens.: Haute Tension

Hter: Hinter

HTF: Hatfield (UK-England) (location identifier) • Hyper-G Text Format • Hypertext Functionality

HTFA: Host Transfer File Access

Htg: Heating

HTG: Gourde (Haiti) (ISO currency code) • Hatanga (Russia) (location identifier) • Heating

HTGCR: High-Temperature Gas-Cooled Reactor

HTGR: High-Temperature Gas-cooled Reactor • High-Temperature Gas Reactor

HTH: Hawthorne (USA-NV) (location identifier) • Helix-Turn-Helix

HTHL: Horizontal Takeoff and Horizontal Landing

HTHP: High Temperatures and High Pressures (periodical code)

HTHY: Hope This Helps You

HTI: Haiti (ISO country code) • Hamilton Island (Australia-QLD) (location identifier) • Handwerkstechnisches Institut • Horizons Technology Inc.

HTIS: Hadamard Transform Imaging Spectroscopy

HTJR: Heat Transfer Japanese Research (periodical code)

HTK: HMM Toolkit

Htl: Hotel

HTL: High Threshold Logic • Höhere Technische Lehranstalt • Houghton - Roscommon County (USA-MI) (location identifier)

HTLL: High Test Level Language

HTLRS: Hydrographic Transportable Laser Ranging System

HTM: Hard To Match • Hard Tube Modulator • Heat Transfer Medium • Hierarchical Triangular Mesh • High-Temperature Material • High-Throughput Mission

HTMA: High-Throughput Mirror Assembly

HTML: Hypertext Markup Language

HTMPL: High-Temperature Materials Processing Laboratory

HTMS: High-Temperature Mass Spectrometry • High-Temperature Measurement System

HTN: Hotan (China-PRC) (location identifier)

HTO: East Hampton (USA-NY) (location identifier) • Horizontal Takeoff • Hydrous Titanium Oxide • Hypertext Optimization

HTOC: High-Throughput Organic Chemistry

HTOHL: Horizontal Takeoff - Horizontal Landing

HTOL: Horizontal Takeoff and Landing

Htp: Haltepunkt

HTP: Hermes Technology Programme • High Temperature-Pressure • High Test Peroxide • Hochtemperaturpyrolyse • Hydroxyl Terminated Propellant

HTPB: Highly Terminated Polybutadiene • Hydroxyl-Terminated Polybutadiene (*)

HTQL: Hypertext Query Language

Htr: Hauteur • Heater

HTR: Hateruma (Japan) (location identifier) • Heater • High-Temperature Reactor • High-Temperature Region • High-Temperature Resistor • High Time Resolution

HTRM: High Time Resolution Mode

HTRP: High Time Resolution Processor

HTRQ: Hypertext Request

Hts: Heights

HTS: Ashland (USA-KY) (location identifier) • Harm Targeting System • Hawaii Tracking Station (USA-HI) • Heat Transfer Section • Heat Transfer System • Heat Transport Section • Heat Transport System • Heights • Higher Technical School • High-Technology System • High-Temperature Superconducting • High-Temperature Superconductivity • High-Temperature Superconductor • Hogere Technische School • Huntington (USA-WV) (location identifier) • Hybrid Test Set • Hydrodynamic Test System • Hypertext System

HTSC: High-Temperature Superconducting • High-Temperature Superconductor

HTSR: Heat Transfer Soviet Research (periodical code)

HTemS: High Temperature Science (periodical code)

HTrEn: Heat Transfer Engineering (periodical code)

Htt: Hütte

HTTB: High-Technology Testbed

HTTC: Higher Teacher Training College

HTTP: Hypertext Transfer Protocol (*) • Hypertext Transport Protocol

HTTPD: Hypertext Transport Protocol Daemon

HTTPS: Hypertext Transport Protocol Server

HTTR: High-Temperature Test Reactor

HTU: Heat Transfer Unit • Height of Transfer Unit • Hopetoun (Australia-VIC) (location identifier)

HTV: H-II Transfer Vehicle • Hope Transfer Vehicle • Huntsville (USA-TX) (location identifier)

HTVA: Hors Taxe sur la Valeur Ajoutée

HTVL: Horizontal Takeoff - Vertical Landing

HTW: Chesapeake - Huntington County (USA-OH) (location identifier)

HTXS: High-Throughput X-ray Spectroscopy

HTZ: Hato Corozal (Colombia) (location identifier) • Helicopter Taxi Zone

hu: Hungarian (ISO language code) • Hungary

HU: Harvard University (USA-MA) • Hauptuntersuchung • High Up • High Usage • Histogram Updating • Hongrie (ISO country code) • Hull (UK postal code) • Hungary (ISO country code)

HUA: Huntsville Redstone (USA-AL) (location identifier) • Hydrologic Unit Area

HUAC: House Un-American Activities Committee

HUB: Humbert River (Australia-NT) (location identifier) • Humboldt-Universitätsbibliothek (Germany) • Humboldt-Universität zu Berlin (Germany)

HUBE: Hopkins Ultraviolet Background Explorer • Hot Universe Background Explorer

Hubschr: Hubschrauber

HUC: Hydrologic Unit Code

HUCR: Harvard University Character Recognition • Harvard University Character Recognizer • Highest Useful Compression Ratio • Humacao (Puerto Rico) (location identifier)

HUD: Header and Data Unit • Head-Up Display • Housing and Urban Development • Humboldt (USA-IA) (location identifier)

HUDE: Head-Up Display Electronics

HUDOM: Heads of University Departments of Mathematics

HUE: Halo Ultraviolet Explorer • Humera (Ethiopia) (location identifier)

HUF: Forint (Hungary) (ISO currency code) • Terre Haute (USA-IN) (location identifier)

HUFEB: Hokkaido University Faculty Engineering Bulletin (periodical code)

HUFEM: Hokkaido University Faculty Engineering Memoirs (periodical code)

HUFS: Histogram Updating with Full Sorting

HUG: HEASARC Users Group • Huehuetenango (Guatemala) (location identifier)

HUGO: Highly Unusual Geophysical Operation

HUH: Huahine Island (Society Islands) (location identifier)

HUI: Hue - Phu Bai (Viet Nam) (location identifier)

HUJ: Hugo (USA-OK) (location identifier)

HUK: Haftpflicht-, Unfall- und Kraftverkehrsversicherung • Hukuntsi (Botswana) (location identifier)

HUL: Hardware Utilization List • Harvard University Libraries (USA-MA) • Houlton (USA-ME) (location identifier)

HULL: Hardware Utilization and Location List

hum: Humilis

Hum: Human • Humane • Humanities • Humanity • Humble • Humidity • Humorous

HUM: Highly Unstable Mother • Hilbert Uniqueness Method • Houma Terrebonne (USA-LA) (location identifier) • Humidity

HUMINT: Human Intelligence

HUMS: Health and Usage Monitoring System

HÜN: Hünfeld (Germany)

Hun: Hungarian

HUN: Hongrie (ISO country code) • Hualien (China-ROC) (location identifier) • Hungary (ISO country code)

Hund: Hundred

Hung: Hungarian • Hungary

HUP: Harvard University Press (USA-MA) • Horizon Ultraviolet Program

HUQ: Houn (Libya) (location identifier)

Hur: Hurricane

HUR: Hardware Usage Report

HURA: Hexagonal Uniformly Redundant Array

HURCN: Hurricane

HURD: Heads Up Range Device

Huricn: Hurricane

HUS: Hughes (USA-AK) (location identifier) • Husum (Germany)

Husb: Husbandry

HUT: Hard Upper Torso • Helsinki University of Technology (Finland) • Hopkins Ultraviolet Telescope • Hutchinson (USA-KS) (location identifier)

HUTSPA: Hopkins Ultraviolet Telescope Spartan

HUU: Huanuco (Peru) (location identifier)

HUV: Hudiksvall (Sweden) (location identifier)

HUX: Huatulco (Mexico) (location identifier) • Santa Cruz - Huatulco (Mexico) (location identifier)

HUY: Humberside (UK-England) (location identifier)

HUZ: Huizhou (China-PRC) (location identifier)

hv: Hoc Verbum

H/V: Horizontal/Vertical

HV: Handelsverbindung • Handelsverkehr • Handelsvertrag • Handelsvertreter • Handelsvertretung • Harvard Variable • Haulage Vehicle • Hauptvortrag • Hausverwalter • Hausverwaltung • Haute-Volta (ISO country code) • High Vacuum • High Velocity • High Voltage • High Volume • Hydrogen Vent • Hyperventilation • Upper Volta (ISO country code) • Vatican (international aircraft registration code)

HVA: Analalava (Madagascar) (location identifier) • Hauptverwaltung Aufklärung • High-Velocity Absorption • High-Voltage Amplifier

HVAC: Heating - Ventilating - Air Conditioning • High-Voltage Alternating Current

HVAG: Halls Valley Astronomical Group (USA-CA)

HvaOB: Hvar Observatory Bulletin (periodical code)

HVB: Hervey Bay (Australia-QLD) (location identifier) • High-Voltage Bias

HVC: Hardened Voice Channel • Hardened Voice Circuit • High-Velocity Cloud • High-Velocity Complex • High-Voltage Connector • High-Voltage Control • Hovercraft

HVD: Humanistischer Verband Deutschlands (Germany) • Khovd (Mongolia) (location identifier)

HVDC: High-Voltage Direct Current

HVDS: Hazardous Vapor Detection System • Hypergol Vapor Detection System

HVE: Hanksville (USA-UT) (location identifier) • High-Vacuum Environment • High-Vacuum Evaporation • High-Voltage Electronics

HVEC: High Vacuum Equipment Corporation

HVEM: High-Voltage Electron Microscope • High-Voltage Electron Microscopy

HVES: High-Vacuum Evaporation System

HVF: Harmonically Varying Field • Hi-res Variable Frequency

HVG: High-Voltage Gradient • Honningsvåg (Norway) (location identifier)

HVI: High-Value Item • High-Velocity Impact • High-Volume Impactor

HVK: Hólmavík (Iceland) (location identifier)

HVL: Half-Value Layer • Hora Verdadera Local

HVM: Helium Vector Magnetometer • Hvammstangi (Iceland) (location identifier) • Hydraulic Valve Module

HVN: Haven • New Haven (USA-CT) (location identifier)

HVNS: High-Velocity Neutron Star

HVO: Hawaiian Volcano Observatory (USA-HI)

HVOA: High-Voltage Operational Amplifier

HVoDS: Hunstville Voice Distribution System (USA-AL)

HVOF: High-Velocity Oxygen Fuel

HVP: Half-Value Period • High-Vacuum Pump • High Video Pass • High-Voltage Pump • Horizontal and Vertical Position • Horizontal and Vertical Projection

HVPF: Hauptverwaltung für das Post- und Fernmeldewesen

HVPS: High-Voltage Power Supply • High-Volume Printing System

HVPVE: High-Voltage Photovoltaic Effect

HVR: Havre (USA-MT) (location identifier) • Heavier • High-Vacuum Rectifier • High-Voltage Rectifier • High-Voltage Regulator • High-Voltage Relay • High-Voltage Resistor • Hover

HVS: Hartsville (USA-SC) (location identifier) • High-Velocity System • High-Voltage Supply • High-Voltage Switch • High-Volume Sampler

HVSA: High-Voltage Solar Array

HVSE: High-Voltage Solar Experiment

HVSF: Honeywell Verification Simulation Facility

HVSL: Holidays - Vacation - Sick Leave

HVSP: High-Voltage Solar Panel

HVST: High-Voltage Switching Transistor

HVT: Half-Value Thickness

HVU: High-Voltage Unit

HVW: High-Voltage Wire • Hochschule für Verwaltungswissenschaften

Hvy: Heavy

HVY: Heavy

HVYR: Heavier

h/w: Herewith • Husband and Wife

Hw: Handwagen • Hauptwagen • Hauptwert • Hauswart

H/W: Hardware • Head Wind

HW: Half Wave • Half Word • Halle/Westfalen (Germany) • Handelsware • Handelswert • Handelswissenschaft • Handwerk • Handwörterbuch • Hardware • Hardwood • Hazardous Waste • Head Wind • Heavy Weight • High Water • Hochwasser • Hot Water • Hot Wire

HWA: Hawabango (Papua New Guinea) (location identifier)

HWAY: Highway

Hwb: Handwörterbuch

HWB: Halbwertbreite

HWCA: Human Waste Collection Assembly

HWCI: Hardware Configuration Item

HWCP: Hardware Code Page

HWD: Hayward (USA-CA) (location identifier) • Height-Width-Depth • Horizontal Weather Depiction

HWDMS: Hazardous Waste Data Management System

Hwg: Handwagen • Hochwohlgeboren

HWH: Hochwasserhöhe

HWI: Hawk Inlet (USA-AK) (location identifier) • Hochwasserintervall

HWIL: Hardware In Loop

HWK: Handwerkskammer • Hawker - Wilpena Pound (Australia-SA) (location identifier)

HWL: High-Water Line

HWM: High-Water Mark • Horizontal-Wind Model

HWMC: Harware Management Console

HWMCA: Harware Management Console Application

HWMS: Handbook of Writing for the Mathematical Sciences

HWN: Hwange National Park (Zimbabwe) (location identifier)

HwO: Handwerksordnung

HWO: Hollywood - North Perry (USA-FL) (location identifier)

HWP: Half-Wave Plate • Hochschule für Wirtschaft und Politik

Hwr: However

HWR: Heavy-Water Reactor

HWRP: Hydrology and Water Resources Programme

HWS: Hawaii Wide Survey • Heat Wedge Spectrometer • Hurricane Warning System

HWSFC: Hazardous Waste Superfund Collection

HWSFD: Hazardous Waste Superfund Database

HWT: Heavyweight Tank • Hypersonic Wind Tunnel

Hwvr: However

HWW: Hot-and-Warm Worked

HWWI: Halley Worldwide Investigations

HWWRS: Hyperfiltration Wash Water Recovery System

HWWS: Hyperfiltration Wash Water (recovery) System

Hwy: Highway

HWY: Highway

HWZ: Hochwasserzeit • Hochwasserzeit

HWZI: Half-Width at Zero Intensity

H/X: Heat Exchanger

HX: Halifax (UK postal code) • Heat Exchanger

HXD: Hard X-ray Detector

HXDA: HRMA X-ray Detection Assembly

HXDS: HRMA X-ray Detection System (*) • HRMA X-ray Detector System

HXIS: Hard X-ray Imaging Spectrometer (*) • Hard X-ray Imaging Spectrometry

HXR: Hard X-Ray

HXRBS: Hard X-Ray Burst Spectrometer

HXS: Hard X-ray Spectrometer

HXT: Hard X-ray Telescope • High-energy X-ray Telescope

HXTS: High Throughput X-ray Spectroscopy

HXX: Hay (Australia-NSW) (location identifier)

hy: Armenian (ISO language code)

Hy: Heavy • Highway • Hydraulique • Hydrazine

Hya: Hydra-Hydrae (constellation)

HYA: Hyannis (USA-MA) (location identifier)

HYADES: Hydroxyl Atmospheric Detection System

Hyb: Hybrid

Hybr: Hybrid

HYC: High Wycombe (UK-England) (location identifier)

Hyd: Hydrant • Hydrate • Hydraulic • Hydraulics • Hydraulique • Hydrographic • Hydrostatic • Hydrostatics

HYD: Hyderabad (India) (location identifier) • Hydraulics • Hydrology

HYDAC: Hybrid Digital-Analog Computer

Hydm: Hydrometer

Hydr: Hydrant • Hydrate • Hydraulic • Hydraulics • Hydraulique • Hydrazin • Hydrographic • Hydrographique • Hydrography • Hydrostatic • Hydrostatics

Hydraul: Hydraulic • Hydraulics

Hydrelc: Hydroelectric

Hydrelcty: Hydroelectricity

Hydrogr: Hydrographic • Hydrographique • Hydrography

Hydros: Hydrostatic • Hydrostatics

Hydt: Hydrant

HYF: Hayfields (Papua New Guinea) (location identifier)

HYFA: Hydrometeor Free Atmosphere

HyFaCC: Hypergravity Facility for Cell Culture

HYFLEX: Hypersonic Flight Experiment

Hyg: Hygiène • Hygiene • Hygiénique

HYG: Hydaburg (USA-AK) (location identifier) • Hypergol • Hypergolic

HYGMA: Hygromagmaphile Mantle Array

Hygr: Hygromètre • Hygrométrie

Hyi: Hydrus-Hydri (constellation)

HyInt: Hyperfine Interactions (periodical code)

HYL: Hollis (USA-AK) (location identifier)

HYN: Huangyan (China-PRC) (location identifier)

Hyp: Hypodermic • Hypophyse • Hypotenuse • Hypothèque • Hypothèse • Hypothesis • Hypothetical

HYP: Hyphenation

HYPACE: Hybrid Programmable Attitude Control Electronics

Hyper: Hyperbole • Hyperbolic • Hyperbolique

HyperTIES: Hypertext based on The Interactive Encyclopaedia System

Hypn: Hypnose • Hypnotic • Hypnotique

Hypot: Hypotenuse

Hypoth: Hypothèse • Hypothesis • Hypothetical • Hypothétique

HYR: Hayward (USA-WI) (location identifier) • Higher

HYS: Hays (USA-KS) (location identifier)

HYSAT: Hybrid Satellite • Hybrid Sensor Experimental Satellite • Hyperspectral Satellite

HySTP: Hypersonic System Technology Program

HYTEA: Hypertext Authoring

HyTech: Hypersonic Technology

HYV: Hyvinkää (Finland) (location identifier)

Hz: Handzeichen • Heizung • Hertz • Hochzeit

HZ: Dust Haze • Habitable Zone • Haftungszeit • Harrison-Zel'dovich • Haze • Hertz • Saudi Arabia (international aircraft registration code)

hzb: heizbar

HZB: Hazebrouck Merville-Calonne (France) (location identifier)

HZD: Hydrated Zirconium Dioxide

HZE: High Atomic Number and Energy

Hzg: Heizung • Herzog

HZG: Hanzhong (China-PRC) (location identifier)

Hzgin: Herzogin

Hzgt: Herzogtum

Hzgtm: Herzogtum

HZK: Hessicher Zentralkatalog • Húsavík (Iceland) (location identifier)

HZL: Hazleton (USA-PA) (location identifier)

HZO: Hydrated Zirconium Oxide

HZRG: High-Redshift Radio Galaxy

HZSS: High-Z Supernova Search

Hzt: Herzogtum

Hztm: Herzogtum

Hzy: Hazy

H0: Hubble Constant

H2: Hydrogen

H2O: Water

H2S: Hydrogen Sulfide

H4: Solomon Islands (international aircraft registration code)

I

i: Bahnneigung • Im • Imaginary Number • Impulse • In • Inclination • Indicate • Information • Innen • Innerhalb • Innerlich • Instantaneous Current • Interest • Intransitive • Island • Isle • Orbital Inclination • Phase Angle • Pitch Angle • Tourist Information • Unit Coordinate Vector • van't Hoff Factor

I: Current • Electric Current • Île • Ilha • Imperator • Imperium • Impulse • Inch • Inclination • Independent • India (international alphabet) • Indian • Inductive Effect • Inductometric Effect • Industrie • Inertia • Injected • Input • Insel • Inspection • Inspector • Institute • Instructor • Intelligence • Intensity • International • Interpreter • Interstate • Inti (currency) • Invar • Iode • Iodine • Ionic Strength • Ionization Potential • Ireland • Irish • Irradiated • Isla • Island • Isle • Isola • Isoleucine • Isospin Quantum Number • Isotopic Spin • Issue • Italian • Italy (international aircraft registration code) • Italy (international vehicle registration code) • Italy (postal country code) • Jod • Luminous Intensity • Moment of Inertia • Momentum • Nuclear Spin Quantum Number • One (Roman numeral) • Particular Affirmative • Point-Spread Function • Radiant Intensity • Sound Intensity • Supergiant Star • Unit Matrix • 1 (Roman numeral)

ia: im allgemeinen • In Absentia • Inter Alia • Interlingua (ISO language code)

iA: Im Aufbau • Im Auftrag • In Abwesenheit • In Ausbildung

Ia: Iowa (USA) • Supergiant Star of High Luminosity • Super-Supergiant

I&A: Identification & Authorization • Indexing & Abstracting

IA: Ice Age • Illuminated Atmosphere • Im Allgemeinen • Immediately Available • Impedance Angle • Implementation Agency • Impulse Analysis • In Absentia • Incremental Analysis • Indicated Airspeed • Indicated Altitude • Indirect Access • Indirect Addressing • Information Architect • Ingénieur de l'Armement • Initial Allowance • Initial Appearance • Input Axis • Inspecteur d'Académie • Inspection Académique • Inspection Authorization • Institute of Astronomy • Instrumentation Amplifier • Intel Architecture • Intelligence Artificielle • Intelligenzalter • Interactive • Interactive Analysis • Inter-American • Intercept Angle • Interface Agreement • Interface Amplifier • Interim Action • Interim Archive • Internal Audit • Inverter Assembly • Iowa (USA) • Iron Age • Isochron Age • Issuing Agency

IA&A: Ingenieria Aeronautica y Astronautica (periodical code)

IAA: Igarka (Russia) (location identifier) • Immediate Action Authority • Institut d'Astronomie et d'Astrophysique • Institute of Applied Astronomy • Instituto Antarctica Argentina • Instituto de Astrofísica de Andalucia (Spain) • Instrumental Activation Analysis • Interactive Achievement Award • International Academy of Astronautics • International Advertising Association • International Aerospace Abstracts • International Astronautical Academy • Internationales Arbeitsamt • Irish Astronomical Association (Ireland) • Ithaca Amateur Astronomers (USA-NY)

IAAA: I Am An Accountant • International Association for Astronomical Arts (USA-CA) • International Association for Astronomical Artists (USA-CA) (*)

IAAB: Interamerican Association of Broadcasters

IAAC: Interamerican Accreditation Cooperation • International Association for Analog Computing

IAAE: Institute of Automotive and Aeronautical Engineers

IAAER: International Association for the Advancement of Educational Research

IAAF: International Amateur Athletic Federation

IAAG: Institut d'Astronomie, d'Astrophysique et de Géophysique • ISEE Archiving Advisory Group

IAAI: Innovative Applications of Artificial Intelligence (conference)

IAAIL: International Association for Artificial Intelligence and Law

IAAL: I Am A Lawyer

IAAMBA: I Am A Master in Business Administration

IAAP: Institute of Astrophysics and Atmospheric Physics • International Association of Applied Psychology

IAARC: International Administrative Aeronautical Radio Conference • International Association for Automation and Robotics in Construction

IAASM: International Academy of Aviation and Space Medicine

IAAT: Institut für Astronomie und Astrophysik Tübingen (Germany)

IAB: ICSU Abstracting Board • Industrial Advisory Board • Industrial Arbitration Board • Industry Advisory Board • Industry Arbitration Board • Information Access Branch • Inter-American Bank • Interessengemeinschaft Astrofotografie Bochum (Germany) • International Abstracting Board • International Association of Broadcasting • Internet Activities Board (*) • Internet Architecture Board • IUS Assembly Building • Wichita McConnell (USA-KS) (location identifier)

IABA: International Aircraft Broker's Association • International Amateur Boxing Association

IABBE: International Association for Better Basic Education

IABC: International Association of Business Communicators

IABE: Ibero-American Bureau of Education

IABG: Industrie Anlagen-Betriebs Gesellschaft mbH (Germany)

IABLA: Inter-American Bibliographical and Library Association

IABO: International Association for Biological Oceanography

IABP: International Arctic Buoy Programme

IABS: Integrated Apogee Booster System • International Association of Biological Standardization

IABSE: International Association for Bridge and Structural Engineering

IABV: Integriertes Automatisches Besteuerungsverfahren

IAC: Import Advisory Committee • Improved Anode Catalyst • In Any Case • Industrial Advisory Council • Industrial Application Center • Industrial Assessment Center • Industry Advisory Commission • Information Analysis Center • Ingénieurs Assimilés et Cadres • Instituto de Astrofísica de Canarias (Spain) • Instruction Atomicity Controller • Instrument Approach Chart • Instrument Array Cable • Instrument-on-A-Card • Integrating Assembly Contractor • Integrating Associate Contractor • Integration, Assembly and Checkout • Interagency Assessment Committee • Inter-American Conference • Inter-Application Communication • Interim Archive Catalogue • International Activities Committee • International Advisory Committee • International Agricultural Centre (Netherlands) • International Air Convention • International

Algebraic Compiler • International Analysis Code • International Astronautical Congress • Internet Application Component • Internet As Control • Interstate Aviation Committee

IACA: Independent Air Carriers' Association • Inter-American Cultural Association • International Air Carriers' Association • International Air Charters' Association

IACB: International Advisory Committee on Bibliography

IACBDT: International Advisory Committee on Bibliography, Documentation and Terminology

IACC: Industrial Analysis and Control Council • Integrating Assembly and Checkout Contractor • International Advanced Correlator Consortium • International Air Cargo Corporation

IACCA: Inter-Agency Committee on Climate Change

IACCP: Inter-American Council of Commerce and Production

IACDLA: International Advisory Committee on Documentation, Libraries, and Archives

IACED: Inter-Agency Committee on Environment and Development

IACET: International Association for Continuing Education and Training

IACG: Interagency Committee on Geomatics • Interagency Consultative Group

IACGEC: Interagency Committee on Global Environmental Change

IACHEI: International Association of Consultants in Higher Education Institutes

IACHR: Inter-American Commission on Human Rights

IACI: Inter-American Copyright Institute • International Association of Conference Interpreters

IACIS: Intelligent Area Communication and Information System

IACIT: International Association of Conference Interpreters and Translators

IACKL: Interrupt Acknowledgement Latency

IACM: International Association for Computational Mechanics

IACO: Installation, Assembly, and Checkout

IACODLA: International Advisory Committee on Documentation, Libraries and Archives

IACOMS: International Advisory Committee on Marine Sciences

IACP: International Aerosol Climatology Project • International Association of Computer Programmers

IACR: International Association for Cryptologic Research

IACRS: Inter-Agency Committee on Remote Sensing

IACS: Indian Association for the Cultivation of Science • Inertial Attitude Control System • Institute for Astronomy and Computational Sciences • Institute for Astrophysics and Computational Sciences (CUA) • Interactive Computer System • International Association of Classification Societies

IACSS: Inter-American Conference of Social Security

IACT: Imaging Atmospheric Čerenkov Telescope

IACWD: Interagency Advisory Committee on Water Data

IaD: Imad-ad-Dean (USA-MD)

IAD: Immediate Action Directive • Indemnité Annuelle de Départ • Information And Documentation • Inspecteur Adjoint au Directeur • Installation, Assembly or Detail • Institut d'Assistance au Développement • Integrated Access Device • Integrated Automatic Documentation • Intelligence Artificielle Distribuée • Interactive Debug • Interface Agreement Document • Interface Analysis Document • International Astrophysical Decade • Internationaler Austauschdienst • Internet Addiction Disorder • Inventory Adjustment Document • Ion-Assisted Deposition • Washington Dulles (USA-DC) (location identifier)

IADB: Inter-American Development Bank

IADC: Inter-Agency Debris Coordination • Inter-Departmental Advisory and Development Committee

IADD: Index to American Doctoral Dissertations

IADIWU: International Association for the Development of International and World Universities

IADL: International Association of Democratic Lawyers

IADLA: International Association for the Development of Libraries in Africa

IADP: INTELSAT Assistance and Development Programme

IADPC: Interagency Data Processing Committee

IADPG: Intelligence Automatic Data Processing Group

IAE: Inflatable Antenna Experiment • Information Analysis and Evaluation • Infrared Auroral Emission • Institut d'Administration des Entreprises • Institute for the Advancement of Engineering • Institute of Atomic Energy • Instituto de Aeronaática e Espaço (Brazil) • Integral of Absolute Error • International Aero Engines

IAEA: Inter-American Education Association • International Atomic Energy Agency (Austria) (*) • International Atomic Energy Authority

IAEAA: International Antitrust Enforcement Assistance Act

IAEC: International Atomic Energy Committee

IAECOSOC: Inter-American Economic and Social Council

IAEE: International Association for Earthquake Engineering • International Association of Energy Economists

IAEEA: International Association for the Evaluation of Educational Achievement

IAEEL: International Association for Energy Efficient Lighting

IAEG: International Association for Engineering Geology • International Association of Engineering Geologists • International Association of Engineering Geology (*)

IAEO: International Atomic Energy Organization • Internationale Atomenergie-Organisation

IAEP: International Assessment of Educational Progress

IAES: Intelligent Array Expansion System • International Association of Exchange Students

IAESM: IEEE Aerospace Electronic Systems Magazine (periodical code)

IAESR: Institute of Applied Economic and Social Research

IAESTE: International Association for the Exchange of Students for Technical Experience

IAET: International Association of Educational Television

IAEVI: International Association for Educational and Vocational Information

IAF: Image Analysis Facility • Imageur d'Aurores pour FREJA • Indian Air Force • Initial Approach Fix • Interactive Facility • Inter-American Foundation • International Accreditation Forum • International Aeronautical Federation • International Astronautical Federation • International Automobile Federation • Istituto de Análisi Numerica

IAFA: Internet Anonymous FTP Archive

IAFC: Instantaneous Automatic Frequency Control

IAFE: Instituto de Astronomía y Física del Espacio (Argentina)

IAG: Instituto Astronômico e Geofísico • Instituto de Astronomía y Geodesia • Instrument Advisory Group • Interagency Group • International Association of Geodesy (*) • International Association of Geology • International Association of Geomorphologists (*) • Niagara Falls (USA-NY) (location identifier)

IAGA: International Association of Geomagnetism and Aeronomy

IAGB: Industrieanlagen Betriebsgesellschaft

IAGBN: International Absolute Gravity Base Network

IAGC: Instantaneous Automatic Gain Control • International Association on Geochemistry and Cosmochemistry • International Association of Geophysical Contractors

IAGIS: International Association of Geodesy Information Service

IAGL: Institut d'Astrophysique et de Géophysique de Liège (Belgium) (formerly IAL)

IAGO: Interaction Atmosphère Glace Océan

IAGOD: International Association on the Genesis of Ore Deposits

IAGRG: Indian Association for General Relativity and Gravitation

IAGS: Inter-American Geodetic Survey

IAH: Houston (USA-TX) (location identifier) • Interface Attachment Hardware • International Association of Hydrogeologists (*) • International Association of Hydrogeoly

IAHB: International Association of Human Biologists

IAHC: International Ad Hoc Committee • Internet Ad Hoc Committee

IAHF: International Aerospace Hall of Fame

IAHR: International Association for History of Religion • International Association for Hydraulic Research

IAHS: International Association of Hydrological Sciences

IAI: Institut Africain d'Informatique • Instituto de Automática Industrial • International Acquisition and Interpretation • International African Institute • International Alliance for Interoperability • International Anthropological Institute • International Association for Identification • International Automotive Institute • Ion-Acoustic Instability

IAIA: International Association for Impact Assessment

IAIS: Industrial Aerodynamics Information Service • Institute for Advanced International Studies

IAIWC: International Association of Independent Web Consultants

IAJ: Irish Astronomical Journal

IAK: Information-Anchor-Knowledge • Internationale Arbeitskonferenz

IAKV: Interessengemeinschaft Autoren, Komponisten und Musikverleger (Austria)

IAL: Immediate Action Letter • Information Access Laboratory • Immediate Action Letter • Institut d'Astrophysique de Liège (Belgium) (now IAGL) • Instrument Analysis Laboratory • Instrument Approach and Landing • International Algebraic Language • International Algorithmic Language • International Association of Limnologists • International Association of Linguistics • International Astronomical Landscape • Irish Academy of Letters (Ireland)

IALA: International Association of Lighthouse Authorities • International Auxiliary Language Association

IALBA: Inter-American Biographical and Library Association

IALD: International Association of Lighting Designers

i. allg.: im allgemeinen

IALPG: Interprétation et Adaptation Locale des Prévisions Générales

IALS: International Association of Legal Science

IAM: Iberian Atlantic Margin • ILA Associative Memory • Improved Acoustic Modelling • Impuls-Amplituden Modulation • In Amenas (Algeria) (location identifier) • Information Asset Management • Institute for Applied Mechanics • Institute of Administrative Management • Institute for Advanced Materials • Interactive Algebraic Manipulation • International Academy of Management • International Association of Machinists • International Association of Meteorology • Internationale Arbeitsgemeinschaft für Müllforschung

IAMAP: International Association of Meteorology and Atmospheric Physics (now IAMAS)

IAMAS: International Association of Meteorology and Atmospheric Sciences (formerly IAMAP)

IAMAW: International Association of Machinists and Aerospace Workers

IAMCR: International Association for Mass Communication Research

IAMG: International Association for Mathematical Geology

IAMP: Industrial Applied Mathematics Program

IAMS: Initial Attack Management System

IAMT: International Association for Machine Translation

IAN: Institutional Area Network • International Astronomical Network • Internationale Anatomische Nomenklatur • Kiana (USA-AK) (location identifier)

IANA: Internet Assigned Numbers Authority

IANAA: I Am Not An Accountant

IANAL: I Am Not A Lawyer

IANAMBA: I Am Not A Master in Business Administration

IANC: International Airline Navigators Council • International Anatomical Nomenclature Committee

IANEC: Inter-American Nuclear Energy Commission

IANI: Intelligent Access to Nordic Information

IANW: International Academic Networkshop

IAO: In And Out of Cloud • Indexation Assistée par Ordinateur • Ingénierie Assistée par Ordinateur • Institut für Arbeitswirtschaft und Organisation • Interdepartementaal Antarctica Overleg (Netherlands) • Internationale Arbeitsorganisation

i. ao. D.: im außerordentlichen Dienst

IAOE: International Arctic Ocean Expedition

IAOO: International Association of Assessing Officers

IAOPA: International Council of Aircraft Owners and Pilot Associations

IAOS: International Association for Official Statistics

IAP: Image Analysis Processor • Imaging Atom Probe • Improved Accuracy Program • Indiscrétions Astronomiques Publiables • Institut d'Astrophysique de Paris (France) • Institut für Atmosphärenphysik (Germany) • Institute of Applied Physics • Institute of Atmospheric Physics • Instrument Approach Procedure • Integrated Action Plan • Inter-Academy Panel • Inter-Array Processor • International Academy of Pathology • International Association of Photoplatemakers • International Association of Planetology • International Association of Psychotechnics • Internet Access Provider • Internet Application Platform

IAPA: Instrument Approach Procedure Automation • Inter-American Press Association • International Airline Passengers Association

IAPCO: International Association of Professional Congress Organizers

IAPG: Institut für Astronomische und Physikalische Geodäsie (Germany) • International Association of Physical Geography

IAPH: International Association of Ports and Harbours

IAPIP: International Association for the Protection of Industrial Property

IAPM: IEEE Antennas Propagation Magazine (periodical code)

IAPO: International Association of Physical Oceanography

IAPP: Inter-Access Point Protocol • International Association of Plant Physiology

IAPPP: International Amateur-Professional Photoelectric Photometry (USA-TN)

IAPR: International Association for Pattern Recognition

IAPS: Integrated Apogee Boost System • Integrated Avionics Processing System • Interim Antenna Pointing System • International Academy of Political Science • International Association for the Properties of Steam • International Association of Physics Students

IAPSO: International Association for the Physical Sciences of the Ocean

IAPT: International Association for Plant Taxonomy

IAPU: Improved Auxiliary Power Unit

IAPW: In A Perfect World

IAQ: Bahregan (Iraq) (location identifier) • Indoor Air Quality

IAR: Independent Annual Review • Institute for Aerospace Research • Institute for Astrophysical Research • Institut für Atmosphärische Radioaktivität • Instituto Argentino de Radioastronomía (Argentina-BA) • Institut pour les Applications de la Recherche • Instruction Address Register • Yaroslavl (Russia) (location identifier)

IARA: Inter-Allied Reparations Agency

IARAS: Institute of Astronomy of the Russian Academy of Sciences

IARB: Inspection Analysis Review Board • Inter-American Review of Bibliography

IARC: International Agency for Research on Cancer

IARD: Information Analysis and Retrieval Division

IARES: Illustrateur Autonome de Robotique pour l'Exploration Spatiale

IARIW: International Association for Research in Income and Wealth

IARR: Internal Average Relative Reflectance

IARS: Independent Air Revitalization System • Integrated Airborne Reconnaissance Strategy

IARU: International Amateur Radio Union

IARUS: International Association for Regional and Urban Statistics

IAS: Iasi (Romania) (location identifier) • Image Acquisitions Systems • Image Analysis System • Immediate Access Storage • Immediate Access Store • Impact Assessment Sheet • Impact Assessment Study • Indiana Astronomical Society (USA-IN) • Indicated Airspeed • Individual Article Supply • Information Acquisition System • Information Adaptive System • Infrared Absorbed Spectroscopy • Institut d'Astrophysique Spatiale (France) • Institute for Advanced Studies • Institute for Advanced Study (USA-NJ) • Institute for Atmospheric Sciences • Institute of Aeronautical Sciences • Institute of Applied Systems • Institute of the Aerospace Sciences • Institut für Angewandte Systemtechnik • Instrument Adaptor System • Instrument Approach System • Integrated Avionics Suite • Integrated Avionics System • Intelligent Array Subsystem • Intelligent Autonomous System • Interactive Analysis System • International Aircraft Standards • International Association of Sedimentologists • International Association of Seismology • Internationale Arbeitsgemeinschaft Ski • Internationaler Antwortschein • Iowa Academy of Science (USA-IA) • Irish Astronomical Society (Ireland) • Istituto di Astrofisica Spaziale (Italy)

IASA: Improving America's Schools Act (USA) • International Air Safety Association • International Association of Sound Archives • International Aviation Safety Assessment

IASAI: Image Analysis Systems and Artificial Intelligence

IASB: Institut d'Aéronomie Spatiale de Belgique (Belgium) • International Aircraft Standards Bureau

IASc: Indian Academy of Sciences

IASC: Indexing and Abstracting Society of Canada • Institut Aéronautique et Spatial Canadien • Interagency Steering Committee • International Arctic Science Committee • International Association for Statistical Computing (Netherlands)

IASCC: Irradiation-Assisted Stress Corrosion Cracking

IASE: International Association for Statistical Education

IASH: International Association of Scientific Hydrology

IASI: Improved Atmospheric Sounder in the Infrared • Improved Atmospheric Sounding Interferometer • Infrared Atmospheric Sounding Instrument • Infrared Atmospheric Sounding Interferometer (*) • Integral Applied Studies Institute (Moldova) • Inter-American Statistical Institute

IASL: Institute of Air and Space Law • Inter-American School of Librarianship (Venezuela) • International Association of School Librarianship

IASLS: Institute for Advanced Studies in Life Support

IASM: Institute of Aerospace Safety and Management

IASOR: Ice And Snow On Runway

IASP: Interactive Automatic System for Photointerpretation • International Association for Social Progress • International Association of Scholarly Publishers

IASPEI: International Association of Seismology and Physics of the Earth's Interior

IASPM: Infrared Atmospheric and Signature Prediction Model

IASPS: International Association for Statistics in Physical Sciences

IASR: Intermediate Altitude Sounding Rocket

IASS: International Association of Survey Statisticians • Inverter/ATCS Support Structure

IASSIST: International Association for the Social Sciences Information Services and Technology

IASTE: International Association for the Exchange of Students for Technical Experience

IASTED: International Association of Science and Technology for Development

IASTG: International Association of Structural/Tectonic Geologists

IASTP: Inter-Agency Solar-Terrestrial Physics

IASY: International Active Sun Year

IA&T: Integration, Assembly & Test

IAT: Image Annotation Tape • Import Address Table • Individual Acceptance Test • Information Assessment Team • Informazioni e Assistenza Turistica (Italy) • Inside Air Temperature • Integrated Avionics Test • International Alias Table • International Atomic Time • Inverse Abel Transform • Ion Acoustic Turbulence

IATA: International Air Transport Association • Is Amended To Add

IATCA: International Auditor and Training Certification Association

IATD: Is Amended To Delete

IATM: International Association for Testing Materials

IATME: International Association of Terrestrial Magnetism and Electricity

IATOS: Ingénieurs, Administratifs, Techniques et Ouvriers de Service • Ingénieurs, Administratifs, Techniques, Ouvriers et de Service (*)

IATR: Is Amended To Read

IATU: Inter-American Telecommunications Union

IATUL: International Association of Technical University Libraries • International Association of Technological University Libraries

IATV: Interactive Alphanumeric Television

IAU: IAU AG URL • Iaura (Papua New Guinea) (location identifier) • Image Alignment Unit • Incremental Analysis Update • Internal Alignment Unit • International Academic Union • International Association of Universities • International Astronomical Union

IAUC: International Astronomical Union Circular (periodical code)

IAUCo: International Astronomical Union Colloquium (periodical code)

iAufR: im Auftrage und für Rechnung

IAUFWG: International Astronomical Union FITS Working Group

iAuiV: im Auftrag und in Vertretung

IAUL: IDL Astronomy User's Library

IAUP: International Association of University Presidents • Internet User Account Provider

IAUPL: International Association of University Professors and Lecturers

IAUS: International Astronomical Union Symposium (periodical code)

IAV: International Association of Vulcanology • Internationales Aisschreibungsverfahren • Inventory Adjustment Voucher

IAVC: Instantaneous Automatic Volume Control

IAVCEI: International Association of Volcanology and Chemistry of the Earth's Interior

IAVG: International Association for Vocational Guidance

IAVSD: International Association for Vehicle System Dynamics

iaW: in anderen Worten

IAW: In Accordance With • International AGN Watch • Isotopic Atomic Weight

IAWA: International Association of Wood Anatomists

IAWG: International Affiliation of Writer's Guild

IAWPRC: International Association on Water Pollution Research and Control (now IAWQ)

IAWQ: International Association on Water Quality (formerly IAWPRC)

IAWR: Internationale Arbeitsgemeinschaft im Rheineinzugsgebiet

IAX: Image Application Executor

IAYC: International Astronomical Youth Camp

ib: Ibidem • im besonderem

iB: im Bau • im Bayern • im Betrieb • in Buchstaben

Ib: Ibidem • Supergiant Star of Low Luminosity

IB: Inboard • In Bond • Inbound • Incentive-Based • Industrial Business • Industriebank • Industriebau • Inert Building • Information Bulletin • Information Bureau • Informationsbank • Informationsblatt • Informationsbüro • Ingenieurbau • Ingenieurbüro • Inner Button • Inspection Bulletin • Instruction Book • Instruction Buffer • Interface Bus • Internal Bremsstrahlung • International Baccalaureate • International Bank • Internationales Büro • Interrogation Bus • Inverted Barometer • Invoice Book • Ionization-Bounded • Isobar

ibA: in besonderem Auftrag

IBA: Ibadan (Nigeria) (location identifier) • Igniter Booster Assembly • Independent Broadcasting Authority • Inner Buffer Assembly • Inspection By Attribute • INTEGRAL Burst Alert • International Bar Association • International Biographical Archive • International Book Association • Internationale Buchausstellung • Internationales Biographisches Archiv • Ion Beam Analysis

IBAC: International Business Aviation Council

IBAD: Ion-Beam-Assisted Deposition

IBAME: International Board on the Applications of the Mössbauer Effect

IBB: International Broadcasting Bureau

IBBD: Instituto Brasileiro de Bibliografia e Documentação (Brazil)

IBBH: Internationaler Bund der Bau- und Holzarbeiter

IBBN: Inhomogeneous Big Bang Nucleosynthesis • Inhomogeneous Black Body Nucleosynthesis

IBBR: Interbank Bid Rate

IBBY: Internationale Board on Books for the Youth

IBC: Impurity Band Conduction • Impurity Band Conductor • Information-Based Complexity • Inside Back Cover • Instrument Bus Computer • Integrated Broadband Communications • Intermediate Bulk Container • International Biographical Centre (UK) • International Boundary Commission • International Brightness Coefficient • International Broadcasting Corporation • International Business Co.

IBCA: International Book Collectors Association

IBCC: International Building Classification Committee • International Bureau of Chambers of Commerce

IBCCA: International Bathymetric Chart of the Caribbean

IBCCEA: International Bathymetric Chart of the Central Eastern Atlantic

IBCE: International Bureau for Cultural Exchange

IBCM: International Bathymetric Chart of the Mediterranean

IBCN: Integrated Broadband Communications Network • Integrated Broad Channel Network

IBCS: Inflight Blood Collection System • Intel Binary Compatibility Standard

IBCWIO: International Bathymetric Chart of the Western Indian Ocean

ibd: Ibidem

IBD: Ion Beam Deposition • Iterative Blind Deconvolution

IBE: Ibague (Colombia) (location identifier) • Inner Back End • Institute of British Engineers • International Bureau for Education (*) • International Bureau of Education

IBEC: International Bank for Economic Cooperation • Irish Business and Employers Confederation

IBECO: Inboard Booster Engine Cutoff

IBEP: Instituto Brasileiro de Edições Pedagógicas (Brazil) • International Border Environmental Plan

Ibér: Ibérico

i. bes.: im besonderes

i. bes. A.: im besonderen Auftrag

IBEW: International Brotherhood of Electrical Workers

IBEX: International Burst Explorer • International Business Exchange • Iodine, Betas and X-rays

IBF: Internally Blown Flap • International Book Festival

IBFF: Impulse Base Flow Facility

IBFM: Interacting Boson-Fermion Model

IBFN: Integriertes Breitbandfernmeldenetz

IBG: Institute of British Geographers • Interblock Gap

IBGE: Instituto Brasileiro de Geografia e Estatística (Brazil)

IBGIS: Image-Based Geographic Information System

IBGM: Internationales Büro für Gewichte und Maße

IBH: Instituto Bibliográfico Hispánico (Spain)

IBHB: Intermediate Blue Horizontal Branch

IBI: Iboki (Papua New Guinea) (location identifier) • International Bureau for Informatics • Internationales Burgenforschungsinstitut • Invoice Book Inwards

ibid: Ibidem

IBICT: Instituto Brasileiro de Informação em Ciência e Tecnologia (Brazil)

IBID: International Bibliographical Description

IBIP: International Bureau of Intellectual Property

IBIR: Internal Band Irradiance Response

IBIS: Image-Based Information System • Image on-Board INTEGRAL Spacecraft • Imager on-Board INTEGRAL Spacecraft (*) • Instant Business Information System • Instrument de Biologie Spatiale • Instrument for Biological Investigations in Space • Integrated Bibliographic Information System • Integrated Business Information System • Interagency Border Inspection System • Interaktives Bildinterpretationssystem • Interferometer-Based Information System • Intranet Business Information System • Issue-Based Information System

IBIT: Intergovernmental Bureau for Information Technology

IBJ: Internationales Buchjahr

IBK: Internationale Beleuchtungskommission

IBL: Inter-Bibliothecair Leenverkeer • Intermediate BL Lacertae (object) • Intraband Luminescence

IBM: Inferior But Marketable • Insidious Black Magic • Integrated Background Monitoring • Interacting Boson Model • International Business Machines Corporation • It's Better Manually

IBMAL: Ion Beam Modification and Analysis Laboratory (UNT)

IBMJ: International Business Machines Journal (periodical code)

IBMP: Institute for Biomedical Problems

IBMBR: Interbank Market Bid Rate

IBMC: Institut de Biologie Moléculaire et Cellulaire

IBMG: Internationales Büro für Maße und Gewichte

IBMP: Institut de Biologie Moléculaire des Plantes • Institute for Biomedical Problems

IBMR: International Bureau for Mechanical Reproduction

IBMS: Ion Beam Mass Spectrometer

IBMtr: Inboard Motor

IBN: Identification Beacon • Indexed By Name • Institut Belge de Normalisation (Belgium) • Integrated Business Network • International Biosciences Network

Ibnd: Inbound

IBNORCA: Instituto Boliviano de Normalización y Calidad (Bolivia)

IBO: Ibo (Mozambique) (location identifier) • Information Bearing Object • International Baccalaureate Organization • International Broadcasting Organization • Invoice Book Outwards

IBOC: In-Band On-Channel

IBOL: Interactive Business-Oriented Language

IBOY: International Biodiversity Observation Year

IBP: Iberia (Peru) (location identifier) • Initial Boiling Point • Institut für Bauphysik (Germany) • Institute of British Photographers • International Biological Programme • International Boundary Point • Ion Beam Projector • Itty-Bitty Processor

IBPA: International Book Printers Association • Irish Book Publishers Association

IBPGR: International Board for Plant Genetic Resources

IBPIO: International Buoy Programme for the Indian Ocean

i. Br.: im Breisgau

IBR: Implementation Baseline Review • Incorporate By Reference • Intermediate Bit Rate

IBRD: International Bank for Reconstruction and Development

IBRO: International Brain Research Organization

IBRS: International Business Reply Service

IBS: Impulse Balance System • Information Brokering System • INTELSAT Business Service • Interafrican Bureau of Soils • International Business Service • Intrabeam Scattering • Ion Beam Scanning • Ion Beam Sputtering • Ion Beam Synthesis

IBSAT: Indexing By Statistical Analysis Techniques

IBSE: Initial Blood Storage Equipment

IBSH: Informational Bulletin of the Southern Hemisphere (periodical code)

IBSL: Internationales Berufssekretariat für Lehrer

IBSR: Interactive Bibliographic Search and Retrieval

IBSRAM: International Board for Soil Research And Management

IBSS: Infrared Background Signature Survey • Institute of Biology of the Southern Seas

IBT: International Brotherhood of Teamsters • International Bureau of Time • Ion Beam Technology

IBTE: International Bureau for Technical Education

IBTT: International Bureau for Technical Training

IBTTA: International Bridge, Tunnel and Turnpike Association

IBU: Interference Blanker Unit • International Broadcasting Union

IBVPA: Inverse Boundary-Value Problems in Aerodynamics

IBVPH: Inverse Boundary-Value Problems in Hydrodynamics

IBVS: Information Bulletin on Variable Stars (periodical code)

IBW: Instantaneous Bandwidth • Intelligent Bibliographic Workstation • Ion Bernstein Wave

IBWC: International Boundary Water Commission

IBWD: Interacting Binary White Dwarf

IBWM: Intelligent Bandwidth Manager • International Boundary and Water Commission • International Bureau of Weights and Measures

IBWZ: Internationale Bank für Wirtschaftliche Zusammenarbeit

IBY: International Book Year (1972)

IBZ: Ibiza (Spain) (location identifier)

i/c: Icing Condition • In Charge of

I&C: Installation & Checkout • Instrumentation & Communications

Ic: Icarus

I/C: Intercom

I&C: Installation & Checkout • Instrumentation & Communication • Integration & Checkout

IC: Ice Crystal • Iceland • Identification Card • Identity Card • Image Capture • Image Compression • Imperial College • Implementation and Conversion • Important Customer • Impulse Conductor • Incomplete Contract • Incremental Cost • Independent Carrier • Index Catalogue • Index Correction • Indicator and Control • Inductance-Capacitance • Inductive Coupling • Industrial Court • Industrialized Country • Industry Commission • Inertial Component • Informal Communication • Information and Communication • Information Carrier • Information Center • Information Circular • Information Content • Ingénieur Commercial • Ingénieur en Chef • Initial Condition • Inner Cluster • Inorganic Carbon • Input Circuit • Input Current • Inspection Card • Inspection Committee • Inspection General • Inspector General • Instabilité de Courant • Installation and Checkout • Instruction Card • Instruction Code • Instruction Counter • Instrumentation and Communication • Instrument Controller • Instrument Correction • Integrated Circuit • Integration and Checkout • Integration Cell • Intelligence Community • Intensive Care • Interaction Coefficient •

InterCity • Intercom • Intercommutation • Intercomparison • Intercomputer • Intercomputer Communication • Interface Control • Interface Controller • Interface Coordinator • Intergovernmental Council • Interim Change • Intermittent Combustion • Internal Combustion • Internal Communication • Internal Conversion • International Conference • International Congress • International Cooperation • Interrupt Controller • Inverse Check • Inverse Compton • Ion Chromatography • Ionization Chamber • Iron Column • Itinerário Complementar • Ivory Coast

ICA: Icabaru (Venezuela) (location identifier) • Ignition Control Additive • Immediate Constituent Analyzer • Independent Computing Architecture • Independent Cost Assessment • Industrial Communications Association • Instrument Coordination Analyzer • Integrated Chameleon Architecture • Integrated Circuit Array • Integrated Communications Adapter • Integrated Control Architecture • Intelligent Communication Agent • Intelligent Console Architecture • Interim Change Analysis • International Cartographic Association • International Commission on Acoustics • International Communication Agency • International Communication Association • International Congress on Acoustics • International Cooperation Administration • International Cooperative Alliance • International Council on Archives • International Court of Arbitration • Intra-application Communication Area • Intracavity Assembly • Invalid Care Allowance • Ion Composition Analyser • Item Change Analysis • Item Control Area

ICAA: International Civil Airports Association

ICAAD: Interactive Computer-Aided Analysis and Design

ICAAP: International Consortium for Alternative Academic Publications

ICAC: Independent Commission Against Corruption • Information Centre Advisory Committee • Instituto de Contabilidad y Auditoria de Cuentas

ICACGP: International Commission on Atmospheric Chemistry and Global Pollution

ICACS: International Conference on Atomic Collisions and Solids

ICAD: Integrated Computer-Aided Design • Integrated Control And Display • International Committee for Accessible Document Design • International Community for Auditory Display

ICADD: International Committee for Accessible Document Design

ICADE: Interactive Computer-Aided Design Evaluation

ICAE: Integrated Computer-Aided Engineering • International Commission on Agricultural Engineering • International Commission on Atmospheric Electricity • International Conference of Agricultural Economists • International Council for Adult Education

ICAENA: Ion Composition Analyser and Energetic Neutral Atom (imager)

ICAES: International Congress of Anthropological and Ethnological Sciences

ICAG: International Comparison of Absolute Gravimeters

ICAI: Image Classification Artificial Intelligence • Institute for Computer-Assisted Information • Intelligent Computer-Assisted Instruction • International Commission for Aid to Intellectuals

ICAIL: International Conference on Artificial Intelligence and Law

ICAIR: International Centre for Antarctic Information and Research

ICAITI: Instituto Centroamericano de Investigación y Tecnología Industrial

ICAL: Image and Chemical Analysis Laboratory

ICALP: International Colloquium on Automata, Languages and Programming

ICALPE: International Centre for Alpine Environment

ICAM: Integrated Communications Access Method • Integrated Computer-Aided Manufacturing • Interactive Cartographic Automated Mapping

ICAMD: International Commission on Antigens and Molecular Diagnostics

ICAME: International Conference on the Applications of the Mössbauer Effect

ICAN: Individual Circuit Analysis • International Commission for Air Navigation

ICANN: International Conference on Artificial Neural Networks • Internet Corporation for Assigned Names and Numbers

ICAO: Institute of Chartered Accountants of Ontario (Canada-ON) • International Civil Aviation Organization (Canada-QC)

ICAOB: International Civil Aviation Organization Bulletin (periodical code)

ICAP: Improved Capability Program • International Conference on Atomic Physics • International Crashworthiness Analysis Program • Internet Calendar Access Protocol

ICAPC: Increased Capability Adaptive Payload Carrier

Icar: Icarus (periodical code)

ICAR: Indian Council of Agricultural Research • Initiated Corrective Action Report • Integrated Camera And Radiometer • Interface Control Action Request • Investigation and Correction Action Report • Investigation and Corrective Action Report (*)

ICARDA: International Centre for Agricultural Research in Dry Areas

ICARE: Image Cartography Expert • International Center for the Advancement of Research and Education

ICARUS: Image Correction And Registration Utility System • Imaging Cosmic and Rare Underground Signal

ICAS: Image Capture and Analysis System • Instrument Crew Alerting System • Integrated Computer-Assisted Surveillance • Integrated Cost And Schedule • Intel Communicating Applications Specification • Interdepartmental Committee for Atmospheric Sciences • Interdepartmental Committee on Aerial Surveys • Intermittent Commercial and Amateur Service • International Committee for Atmospheric Sciences • International Computer Access System • International Council of Aeronautical Sciences (*) • International Council of Aerospace Sciences

ICASE: Institute for Computer Applications in Science and Engineering • Integrated Computer-Aided Sustaining Engineering • International Council of Associations for Science Education

ICASL: International Center for the Advancement of Scientific Literacy (USA-IL)

ICASSP: International Conference on Acoustics, Speech and Signal Processing

ICAST: International Conference on Adaptive Structures and Technologies

ICASVR: International Committee on Atmosphere-Soil-Vegetation Relations

ICAT: Institut Canadien d'Astrophysique Théorique (Canada-ON)

ICATI: Instituto Centramericano de Tecnología Industrial

ICATU: International Council of Arab Trade Unions

ICB: Inner Core Boundary • Instituto Catalán de Bibliografía • Instrument Control Bus • Interface Change Board • Interface Control Board • Interim Change Bulletin • Interior Control Board • International Competitive Bid • International Cooperation Board • Interrupt Control Block • Inventions and Contributions Board

ICBC: IMAX Cargo Bay Camera • International Cataloguing and Bibliographic Control

ICBL: International Campaign to Ban Landmines

ICBM: Intercontinental Ballistic Missile

ICBN: Initiative Canadienne sur les Bibliothèques Numériques (Canada) • International Code of Botanical Nomenclature

ICBP: International Committee for Bird Preservation (now BirdLife)

ICC: Ice Crystal Cloud • Image Converter Camera • Import Commodity Clearance • Incident Command Center • Industrial Communications Council • Information Center Complex • Information Coding Classification • INSAT Coordination Committee • Inspectorate Coordinating Committee • Installation, Calibration and Checkout • Instrument Competency Check • Instrument Control Center • Instrument Control Computer • Instrumentation Checkout Complex • Instrumentation Control Center • Integrated Cargo Carrier • Integrated Chip Circuit • Integrated Circuit Card • Integrated Communications Control • Intelligent Call Center • Intercompany Correspondence • Intercomputer Channel • Intercomputer Communications Corporation • Interchannel Communicator • Intercomputer Channel • Intercomputer Communication • Interface Control Chart • Interface Control Contractor • Intergovernmental Consultative Committee • Intergovernmental Copyright Committee (UNESCO) • Interim Consultative Committee • International Cartographic Conference • International Chamber of Commerce • International Climatological Commission • International Color Consortium • International Communications Committee • International Communications Conference • International Computation Centre • International Computing Centre • International Conference on Communications • International Congregational Council • International Congress Center • International Consultative Committee • International Control Commission • International Convention Centre • International Coordinating Committee • International Copyright Convention • International Copyright Convention • International Cultural Center • Interstate Commerce Commission (USA)

ICCA: Interagency Coordinating Committee for Astronomy • International Conference on Computer Applications • International Congress and Convention Association • International Council of Chemical Associations

ICCAIA: International Coordination Council of Aerospace Industries Associations

ICCAL: International Conference on Computers And Learning

ICCB: Integrated Change Control Board • Intergovernmental Consultation and Coordination Board • Internet Configuration Control Board

ICCBR: International Conference on Case-Based Reasoning

ICCC: Interclient Communication Convention • Intercouncil Coordination Committee • International Computers Communications Conference • International Conference on Computers and Communication • International Council for Computer Communication

ICCD: Intensified Charge-Coupled Device

ICCE: International Commission on Continental Erosion • International Conference on Computers in Education • International Council of Commerce Employers

ICCEE: International Classification Commission for Electrical Engineering

ICCG: Incomplete Choleski Conjugate Gradient • Interagency Coordinating Committee for Geophysics

ICCL: Interface Control Configuration List • International Commission on Climate • International Committee on Computational Linguistics

ICCN: Intercenter Council on Computer Networking

ICCP: Institute for Certification of Computer Professionals (*) • Institute for the Certification of Computer Professionals • Institute of Certified Computer Professionals • Institute of Certified Computing Professionals • Interclient Communications Protocol • Interface Coordination and Control Procedure • International Cloud Climatology Program • International Commission on Cloud Physics • International Commission on Clouds and Precipitation • International Conference on Cataloguing Principles

ICCPR: International Covenant on Civil and Political Rights

ICCR: Institute for Cosmic-Ray Research (Japan)

ICCS: Ice Centre Communications System • Integrated Communications Cabling System • Intercomputer Communication System • International Conference in Computer Science • International Conference on Conceptual Structures • International Council for Canadian Studies

ICCSTI: Interdepartmental Coordinating Committee on Scientific and Technical Information (UK)

ICCTA: International Committee for the Coordination of Terminological Activities • International Consultative Council of Travel Agents

ICCTI: Instituto de Cooperação Científica et Tecnológica Internacional (Portugal)

ICCU: Istituto Centrale per il Catalogo Unico (Italy)

ICD: Industry Council for Development • Installation Completion Date • Instrumentation Control Document • Interactive Call Distribution • Interface Control Document • Interface Control Documentation • Interface Control Drawing • International Center for Documentation • International Classification of Diseases • International Climatic Decade • International Code Designation • International Code Designator

ICDA: Integrated Cached Disk Array

ICDAR: International Conference on Document Analysis and Recognition

ICDB: Ice Core Database

ICDD: International Centre for Diffraction Data

ICDE: International Conference on Data Engineering • International Conference on Distance Education

ICDG: Institute of Chemistry and Dynamics of the Geosphere

ICDL: International Centre for Distance Learning

ICDO: International Civil Defense Organization

ICDP: International Conference on Distributed Platforms

ICDR: Incremental Critical Design Review

ICDS: Interface Control Document Schedule

ICDU: Inertial Coupling Data Unit • Inertial Coupling Display Unit

Ice: Iceland • Icelandic

ICE: Improved Cost Estimate • In-Circuit Emulation • In-Circuit Emulator • Independent Cost Estimate • Information Center for the Environment • Information Center on Education • Initial Cooling Experiment • Innovative Control Effector • Input-Checking Equipment • Input Control Element • Institut du Cinéma Éducatif (France) • Institution of Civil Engineers • Instituto de Ciencias de la Educación (Spain) • Instrumentation/Communication Equipment • Instrument Checkout Equipment • Instrument/Communication Equipment • Integrated Communications Environment • Intelligent Concept Extraction • Interactive Computer Ephemeris • Interactive Conference Environment • Interactive Control Experience • InterCity Express • Interface Configuration Equipment • Interface Configuration Experiment • Interferometry Center of Excellence • Internal Combustion Engine • International Cirrus Experiment • International Cometary Explorer (formerly ISEE-3) • International Cultural Exchange • Intrusion Countermeasure Electronics • IRAF Control Environment • Isolated and Confined Environment

ICEAM: International Committee on Economic and Applied Microbiology

ICEAR: International Centre for Equatorial Atmospheric Research

ICECC: Information Centre for European Culture Collections

ICED: International Council for Educational Development • Intraprofessional Council on Environmental Design • Ionospheric Conductivity and Electron Density

ICEDIS: International Committee on Electronic Data Interchange for Serials

ICEF: International Children's Emergency Fund • International Council for Educational Film

ICEHL: International Conference on English Historical Linguistics

ICEI: Institution of Civil Engineers of Ireland

ICEIS: International Conference on Enterprise Information Systems

Icel: Iceland • Icelandic

ICEM: Incremental Cost Effectiveness Model • Independent Cluster Emission Model • Intergovernmental Committee for European Migration • International Conference on Electronic Materials • International Council for Educational Media

ICEMAP: Interagency Committee for Extramural Mathematics Program

ICEPRO: Icelandic Procedures

ICER: Information Centre of the European Railways

ICES: Initiative pour la Coopération dans l'Europe du Sud-Est • Internal-Conversion Electron Spectroscopy • International Committee on Erosion and Sedimentation • International Conference on Environment Systems • International Conference on Evolvable Systems • International Council for the Exploration of the Seas • International Council of Environmental Sciences

ICESA: International Commission for Earth Sciences in Africa

ICESAT: Ice, Cloud and land Evaluation Satellite

ICESCR: International Covenant on Economic, Social and Cultural Rights

ICET: International Centre for Earth Tides • International Centre of Earth Tides (*) • International Council for Engineering and Technology • International Council on Education for Teaching

ICETT: Industrial Council for Educational and Training Technology

ICEUM: International Conference on Exploration and Utilization of the Moon

ICEX: Ice and Climate Experiment

ICF: Inertial Confinement Fusion • Information and Content Exchange • Inertial Confinement Fission • Initial Confinement Fission • Innovation Capital Funds • Institut Cinématographique de France • Instrument Control Facility • Integrated Catalog Facility • Interact Communication Facility • Interface Control Function • International Computer Forum • International Congress on Fracture • Intrinsic Correlation Function • Ionization Correction Factor • Isothermal Casting Furnace

ICFA: International Committee for Future Accelerators

ICFAC: Inertial Confinement Fusion Advisory Committee

ICFES: Instituto Colombiano para el Fomento de la Educación Superior (Colombia)

ICFG: International Commission on Fungal Genetics

ICFM: International Commission on Food Mycology

ICFMH: International Committee on Food Microbiology and Hygiene

ICFPM: International Centre for Fundamental Physics in Moscow (Russia)

ICFT: Inverse Chip Fourier Transform

ICFTU: International Confederation of Free Trade Unions

ICG: Icing • In-flight Coverall Garment • Intellectual Capital Group • Interactive Computer Graphics • International Commission on Glass • International Coordinating Group • International Crisis Group • Internet Commerce Group • Interunion Commission on Geodynamics

ICGA: IEEE Computer Graphics Applications (periodical code) • International Conference on Genetic Algorithms

ICGC: International Conference on Gravitation and Cosmology

ICGEB: International Centre for Genetic Engineering and Biotechnology

ICGI: International Conference on Grammar Inference

ICGRG: International Commission on General Relativity and Gravitation

ICGS: Interactive Computer-Graphics System

ICGTMP: International Colloquium on Group Theoretical Methods in Physics

ICGW: International Commission on Ground Water

ICH: Interchanger • International Conference on Harmonization • Ion Cyclotron Heating

ICHA: Integrated Cargo Hazard Assessment

ICHB: International Comet Handbook

ICHEO: Inter-University Council for Higher Education Overseas

ICHEP: International Conference on High-Energy Physics

ICHM: Intelligent Component Health Monitor • International Commission on the History of Mathematics

ICHMT: International Centre for Heat and Mass Transfer • International Communications in Heat and Mass Transfer (periodical code)

ICI: Cicia (Fiji) (location identifier) • Image Component Information • Imagery/Communication Interface • Imperial Chemical Industries • Instituto de Cooperación Iberoamericana • Instituto de Cooperación Internacional • Intelligent Communication Interface • Interim Configuration Inspection • International Commission on Illumination • International Copyright Institute • Internet Computer Index

ICIA: International Communications Industries Association

ICIAM: International Conference on Industrial and Applied Mathematics

ICIAP: International Conference on Image Analysis and Processing

ICIC: Intercalibrations-Intercomparisons • International Copyright Information Center

ICICIS: International Conference on Interoperability in Cooperative Information Systems

ICID: Intensified Charge Injection Device • International Commission of Irrigation and Drainage

ICIDR: International Centers for Infectious Diseases Research

ICIMOD: International Centre for Integrated Mountain Development (Nepal)

ICIO: Interim Cargo Integration Operations

ICIP: International Conference on Image Processing

ICIPE: International Centre for Insect Physiology and Ecology

ICIS: Integrated Cockpit Information System • Inter-Campus Information System

ICiSM: IEEE Circuits Systems Magazine (periodical code)

ICIST: Institut Canadien de l'Information Scientifique et Technique (Canada-ON)

ICIT: Instituto Cubano de Investigaciones Tecnológicas (Cuba)

ICITV: Institut de la Carte Internationale du Tapis Végétal

ICIV: Institut de la Carte Internationale de la Végétation

ICJ: International Court of Justice

ICK: Nieuw Nickerie (Suriname) (location identifier)

ICL: Clarinda (USA-IA) (location identifier) • Instrument Calibration Laboratory • Integrated Cloud Liquid • Interface Clear • Interface Control Layout • International Computers Ltd. • International Confederation of Labour • Interpretive Coding Language • Interunion Commission on the Lithosphere

ICLA: International Commission on the Low Atmosphere

ICLAS: International Council on Laboratory Animal Science

ICLD: International Commission on Large Dams

ICLI: Institute for Logic, Cognition, Language and Information • Integrated Call Level Interface

ICLID: Incoming-Caller Identification • Incoming-Call Line Identification

Iclnd: Iceland

ICLP: International Conference on Logic Programming

ICLS: Intercavity Laser Spectroscopy • Irish Central Library for Students

ICM: Image Color Matching • Increased Coherence Model • Installable Compression Manager • Instantaneous Center of Motion • Institute of Computer Management • Instrumentation and Communication Monitor • Integrated Call Management • Integrated Circuit Mask • Integrated Cost Module • Integrated Coverage Measurement • Interactive Condition Mode • Interbus Caching Module • Intercloud Medium • Intercluster Medium • Intercommunication • Interface Coordination Memo • Interference Control Monitor • Intergovernmental Committee for Migrations • Interim Control Module • Intermediate Coupled Model • International Conference on Magnetism • International Conference on Materials • International Congress of Mathematicians • Intra-Cluster Medium • Inventory Control Manager • Iterated Conditional Mode • Iterated Conditioned Mode • Iterative Conditional Mode

ICMA: International Card Manufacturers Association • International Commission on the Middle Atmosphere

ICMC: International Cryogenic Materials Conference

ICMCTF: International Conference on Metallurgical Coatings and Thin Films

ICME: International Conference on Mathematical Education • International Congress on Mathematical Education • Interplanetary Coronal Mass Ejection

ICMES: International Cooperation on Marine Engineering Systems

ICMF: Integrated Coupling Migration Facility

ICMG: International Commission for Microbial Genetics

ICMH: Institut Canadien de Microreproductions Historiques (Canada)

ICMI: International Commission for Mathematical Instruction

ICMID: International Committee for Microbiological and Immunological Documentation

ICMIS: Integrated Computerized Management Information System

ICML: Instrument Control Markup Language • International Conference on Machine Learning

ICMO: Integrated Configuration Management Office

ICMOD: International Conference on Management of Data

ICMOVPE: International Conference on Metal Organic Vapor Phase Epitaxy

ICMP: Interchannel Master Pulse • Internet Control and Message Protocol • Internet Control Message Protocol (*)

ICMR: Instrument Calibration and Maintenance Record

ICMS: Indirect-Cost Management System • Integrated Circuit and Message Switch • International Centre for Mathematical Sciences

ICMSF: International Commission on Microbiological Specifications for Foods

ICMSSR: Interagency Committee on Meteorological Services and Supporting Research

ICMST: International Conference on Machine Searching and Translation

ICMT: Intercontract Material Transfer • International Commission on Mycotoxicology

ICMUA: International Commission on Meteorology of the Upper Atmosphere

ICN: Ice Condensation Nucleus • Instrumentation and Calibration Network • Integrated Computer Network • Inter-Computer Network • Interface Change Notice • International Communications Network • Iodine Cyanide

ICNAF: International Council for Northwest Atlantic Fisheries

ICNI: Integrated Communications Navigation Identification

ICNIA: Integrated Communications, Navigation, Identification Analysis

ICNN: International Conference on Neural Networks

ICNP: International Committee for Nuclear Physics

I&C/O: Integration & Checkout

ICO: Icon • Idle Cutoff • In Case Of • Input Current Offset • Instrumentation Control Officer • Integrated Checkout • Integrated-Communication Officer • Integration and Checkout • Intermediate Circular Orbit • International Carbohydrate Organization • International Commission for Optics • International Crew Office • Internet Connectivity Option • Islamic Conference Organization • Sicogon Island (Philippines) (location identifier)

ICOD: International Centre for Ocean Development • International Conference On Databases

ICODP: International Conference on Open Distributed Processing

ICOHTEC: International Committee for Cooperation in the History of Technology

ICOLC: International Coalition Of Library Consortia

ICOLD: International Commission On Large Dams

ICOLE: Indiana Council on Outdoor Lighting Education (USA-IN)

ICOLP: Industry Cooperative for Ozone Layer Protection

ICOM: International Council Of Museums

ICOME: International Commission on Microbial Ecology

ICOMIDC: International Council On Mathematics In Developing Countries

IComM: IEEE Communications Magazine (periodical code)

ICOMOS: International Council Of Monuments and Sites

iCOMP: Intel Comparative Microprocessor

ICONE: Index Comparatif Entre Normes Nationales et Européennes

ICONO: International Conference on Coherent and Nonlinear Optics

ICONTEC: Instituto Colombiano de Normas Técnicas y Certificación (Colombia)

ICOR: Intergovernmental Conference on Oceanic Research

ICORP: International Cooperative Research Project

ICOS: Improved Crew Optical Sight • Integrated Checkout System

ICoSM: IEEE Control Systems Magazine (periodical code)

ICOT: Isothermal Corrosion Oxidation Test

ICOTS: International Conference On the Teaching of Statistics

ICOU: Inertial Coupling Data Unit

ICP: Ignition Control Programmer • IMA Control Protocol • Image Coprocessor • Imperial College Press (UK) • Independent Content Provider • Indicator Control Panel • Inductively-Coupled Plasma • Inherently Conducting Polymer • Institut de la Communication Parlée • Institute of Chemistry Physics (Russia) • Instituto das Comunicações de Portugal • Instrument Calibration Procedure • Integrated Channel Processor • Integrated Communications Provider • Intensive Course Programme • Interface Change Proposal • Interface Control Panel • International Classification of Patents • International Commission on Photobiology • International Comparison Programme • International Computer Programs • International Congress of Publishers • International Cooperative Program • Internet Cache Protocol • Internet Caching Protocol • Intra-Cluster Pattern • Intracranial Pressure • Inventory Control Point • Item Control Point

ICPAE: International Commission on Planetary Atmospheres and their Evolution

ICPAES: Inductively Coupled Plasma Atomic Emission Spectrometer

ICPAM: International Centre for Pure and Applied Mathematics

ICPD: International Conference on Population and Development

ICPIC: International Cleaner Production Information Clearinghouse

ICPL: Istituto Centrale per la Patologia del Libro (Italy)

ICPM: International Commission for Polar Meteorology

ICP/MS: Inductive Coupled Plasma/Mass Spectrometry

ICPMS: Inductively Coupled Plasma Mass Spectrometer

ICPO: International Criminel Police Organization

ICPP: Interactive Computer Presentation Panel • International Congress on Plasma Physics

ICPR: International Conference on Pattern Recognition

ICPS: Image Collecting and Processing System • Informatique et Calcul Parallèle de Strasbourg • International Conference of Physics Students • International Conference on Physics of Semiconductors

ICPSR: Inter-university Consortium for Political and Social Research

ICQ: International Comet Quarterly

ICR: Ice Crystal Replicator • Indirect Collective Referencing • Inductance-Capacitance-Resistance • Information Collection Request • Input and Compare Register • Institute for Cooperative research • Institute for Creation Research • Institute of Cancer Research • Instruction Change Request • Intelligent Character Reader • Intelligent Character Recognition • Interdepartementale Commissie Ruimtevaart (Netherlands) • Interface Compatibility Record • International Commission on Rheology • Interrupt Control Register • Ion Cyclotron Resonance • Ionic Cyclotronic Resonance • Nicaro (Cuba) (location identifier)

ICRA: International Center for Relativistic Astrophysics (Italy)

ICRAD: Interactive Color Radar Display

ICRAF: International Council for Research in Agroforestry

ICRB: Interface Change Review Board

ICRC: Interagency Classification Review Committee (USA) • International Committee of the Red Cross • International Cosmic-Ray Conference

ICRCCM: Intercomparison of Radiation Codes in Climate Models

ICRCM: International Centre for Recent Crustal Movements • International Centre on Recent Crustal Movements

ICRD: International Cooperative on Research and Development

ICRF: IERS Celestial Reference Frame • International Celestial Reference Frame • Ion Cyclotron Radio Frequency • Ion Cyclotron Range of Frequencies

ICRH: Ion Cyclotron Resonance Heating

ICRIS: Integrated Customer Record Information System

ICRISAT: International Crops Research Institute for Semi-Arid Tropics

ICRM: International Committee on radionuclide Meteorology

ICRO: International Cell Research Organization

ICRP: International Climatic Research Program • International Commission on Radiation Protection • International Commission on Radiological Protection (*)

ICRR: Institute for Cosmic-Ray Research

ICRS: Imprimatur Common Reference Set • International Celestial Reference System

ICRSDT: International Committee on Remote Sensing and Data Transmission

ICRU: International Commission on Radiation Units

ICS: Incident Command System • Industrial Chemistry Society • Information and Computer Science • Information and Computer Systems • Information Control System • Infrared Calibration System • Infrared Camera System • Infrared Communications System • Input Contactor Switch • Insert Card Section • Institute of Computer Sciences • Instrumentation Checkout Station • Instrumentation and Communications System • Instrumentation and Control System • Instrument Command Sequence • Instrument Control Software • Instrument Control System • Integrated Checkout System • Integrated Command System • Integrated Control Storage • Integration Control Sequencer • Interactive Chatting Service • Intercommunication System • Interdepartemental Committee on Space • Interface and Control Systems Inc. (USA) • Interface Control Specification • Interferometer Control Software • International Centre for Science • International Chamber of Shipping • International Classification of Standards • International Commission on Stratigraphy • International Consultancy Service • Internet Client Station • Interphone Control Station • Interpretive Computer Simulator • Intuitive Command Structure • Inventory Control System • Inverse Compton Scattering • Istituto Centrale di Statistica

ICSA: International Computer Security Association

ICSB: International Committee on Systematic Bacteriology

ICSC: International Civil Service Commission

ICSD: International Continental Scientific Drilling

ICSE: Impact of Computing in Science and Technology (periodical code) • International Certificate of Secondary Education • International Conference on Software Engineering

ICSEI: International Cooperation in the Search and the Examination of Inventions

ICSG: International Civil Service Commission

ICSH: Interstitial Cell Stimulating Hormone

ICSI: International Commission on Snow and Ice • International Computer Science Institute (USA-CA) • International Conference on Scientific Information

ICSID: International Center for the Settlement of Investment Disputes • International Council of Societies of Industrial Design

ICSL: Interactive Continuous Simulation Language

ICSLS: International Convention for Safety of Life at Sea

ICSM: IEEE Conference on Software Maintenance • Intergovernmental Committee on Surveying and Mapping • International Committee on Scientific Management

ICSO: International Conference on Space Optics

ICSOC: International Conventions and Scientific Organizations

ICSOS: International Conference on Structure of Surfaces

ICSPRO: Intersecretariat Committee on Scientific Programmes Relating to Oceanography

ICSPS: International Council for Science Policy Studies

ICSRD: Interdepartmental Committee on Scientific Research and Development

ICSS: IBM Continuous Speech Series • Intelligent Customer Support System • International Communications Switching Systems • Internet Connection Secure Server

ICST: Imperial College of Science and Technology (UK) • Institute for Computer Sciences and Technology • Integrated Combined System Test • International Centre for Science and Technology (Russia) • International Cooperation in Science and Technology

ICSTI: International Council for Scientific and Technical Information

ICSTM: Imperial College of Science, Technology and Medicine (UK)

ICSU: International Council of Scientific Unions

ICSW: International Commission on Surface Water

I&CT: Information & Communication Technology

ICT: Image Converter Tube • Influence Coefficient Test • Información Científica y Técnica • Information and Communication Technologies • Information Centre for Technology • Ingénieur Civil Technico-commercial • Institute for Chemical Technology • Institute for Computer Typesetting • Institute of Computer Technology • Instrument Command Template • Instrument Control Terminal • Integer Cosine Transform • Integrated Circuit Tester • Integrated Computer Telemetry • Interactive Command Test • Interactive Coordinate Transformation • Interface Control Tooling • International Call for Tenders • Wichita Mid-Continent (USA-KS) (location identifier)

ICTA: International Centre for Tropical Agriculture • International Council of Travel Agents

ICTAF: Interdisciplinary Center for Technological Analysis and Forecasting

ICTB: International Customs Tariffs Bureau

ICTC: Inertial Component Temperature Controller

ICTD: Interchannel Time Displacement

ICTDR: International Center for Tropical Disease Research

ICTE: Inertial Component Test Equipment

ICTF: International Commission on the Taxonomy of Fungi • Inverse Collection Term Frequency

ICTL: International Conference on Temporal Logic

ICTO: Interkerkelijk Comité voor Tweezijdige Ontwapening

ICTP: International Centre for Theoretical Physics (Italy)

ICTS: Integrated Circuit Test Set

ICTV: International Commission on the Taxonomy of Viruses

ICU: Image Control Unit • Indicator Control Unit • Inflight Calibration Unit • Instruction Cache Unit • Instruction Control Unit • Instrumentation Control Unit • Instrument Controller Unit • Instrument Control Unit • Intelligent Control Unit • Intensive Care Unit • Interactive Chart Utility • Interface Control Unit • International Code Use • ISA Configuration Utility

IC&V: Intelligent Collaboration & Visualization

ICV: Intelligent Collaboration and Visualization • Internal Combustion Vehicle

ICVA: International Council of Voluntary Agencies

ICVAD: Imperial College of Science, Technology and Medicine (Starlink internal code)

ICVS: IRAS Catalogue of Variable Sources • IRAS Catalogue of Variable Stars

ICW: In Compliance With • In Connection With • Initialization Command Word • Input Control Word • Integrated Cooperative Workspace • Interrupted Continuous Wave • Ion-Cyclotron Wave

ICWA: Institute of Current World Affairs

ICWE: International Conference on Water and the Environment

ICWED: Informal Conference on Western European Directors

ICWG: Interface Control Working Group • International Coordination Working Group

ICWM: International Committee on Weights and Measures

ICWQ: International Commission on Water Quality

ICWRS: International Commission on Water Resources Systems

ICY: Icy Bay (USA-AK) (location identifier)

ICYYLM: International Commission on Yeasts and Yeast-Like Microorganisms

ICZ: International Climate Zone • Intertropical Convergence Zone

ICZN: International Commission on Zoological Nomenclature

id: Idem • in dem • in den • in der • Indonesia

Id: Idaho (USA) • Idem • Identification • Identity

I&D: Information & Documentation • Investigação & Desenvolvimento

ID: Idaho (USA) • Identification • Identification Data • Identification Dissector • Identifier • Identity Document • Image Dissector • Immediate Delivery • Incremental Delivery • Indefinite Delivery • Identification • Identity • Index of Discrimination • Index of Dissimilarity • Indicating Device • Individual Development • Indonesia (ISO country code) • Indonésie (ISO country code) • Industrial Democracy • Infectious Desease • Information Department • Information Documentaire • Informationsdienst • Inner Detector • Inside Diameter • Inside Dimensions • Installation Data • Institut für Datenverarbeitung • Instrument-on-a-Disk • Integrated Demonstration • Integrated Diagnostics • Interactive Discrimination • Interface Device • Interface Document • Interférométrie Différentielle • Internal Development • Internal Diameter • Intrusion Detection • Iraqi Dinar (currency) • Issue Date • Item Description • Iterative Deconvolution

Ida: Idaho (USA)

IDA: Idaho Falls (USA-ID) (location identifier) • Image Dissector Assembly • Industrial Development Agency • Information-Decision-Action • Infrared Detection Array • Initial Data Analysis • Initial Design Activity • Input Data Assembler • Institut de l'Audiovisuel • Institute for Defense Analyses • Integrated Digital Access • Integrated Diode Array • Integro-Differential Analyzer • Intelligent Data Analysis • Intelligent Disk Array • Intelligent Drive Array • Interactive Data Archive • Interactive Differential Analyzer • Interactive Digital Archive • Interchange of Data between Administrations • Intercommunication Data Area • Interface Definition Agreement • Interface Display Assembly • Intermediate Data Array • International Dark-sky Association (USA-AZ) • International Datacasting Association • International Deployment of Accelerometers • International Development Agency • International Development Association • International Documentary Association • Internet Developers Association • Isotope Dilution Analysis

IDAAS: International Directory of Astronomical Associations and Societies (merged with IDPAI into ASpScROW) (now StarGuides)

IDAC: Instant Data Access Control • Integrated Data Acquisition and Control • Interconnecting Digital-Analog Converter

IDACA: Integrated Data Acquisition and Control Algorithm

IDAM: Image Data Access Method • Indexed Direct Access Method

IDAMAP: Intelligent Data Analysis in Medicine and Pharmacology

IDAMIS: Integrated Data Analysis and Management Information System

IDAMS: Image Display And Manipulation System • Integrated Data Analysis and Management System • Internationally developed Data Analysis and Management Software

IDAPI: Independent Database Application Programming Interface • Integrated Database Application Programming Interface • Integrated Data Application Programming Interface

IDAPS: Image Data Processing System

IDAR: Iterim Data Assessment Review

IDAS: Ilkeston and District Astronomical Society (UK) • Image Data Analysis System • Integrated Data Acquisition System • Integrated Data Analysis System • Integrated Design Automation System • Integrated Development Approval System • International Database Access Service

IDASS: Integrated Data Assimilation and Sounding System • Intelligence Data Analysis System for Spacecraft

IDAST: Interpolated Data and Speech Transmission

IDATE: Institut de l'Audiovisuel et des Télécommunications en Europe

IDATS: International Data Acquisition and Transmission System

IDB: Idre (Sweden) (location identifier) • Image Database • Industrial Development Board • Industrial Development Bond • Inertial Data Box • Information Database • In-suit Drink Bag • Integrated Database • Intentional Database • Interamerican Development Bank • Internal Database • Internal Data Bus • International Database • Islamic Development Bank

IDBM: Intelligent Database Management

IDBMS: Image Database Management System • Integrated Database Management System

ID&C: Integrated Display & Control

IDC: Image Dissector Camera • Indirect Cost • Industrial Development Certificate • Information and Documentation Center • Infrared Detector Cryostat • Input Display Console • Inspection Data Card • Insulation Displacement Connection • Insulation Displacement Connector • Integrated Disk Channel • Integrated Display and Control • Integrated Display Assembly • Inter-Departmental Committee • Interface Document Control • International Data Center • International Data Corporation • Internet Database Connector • Interactive Devices Control • Interface Document Control • Inverse Dual Converter • Item Design Change

ID&CA: Inverter Distribution & Control Assembly

IDCA: Inverter Distribution and Control Assembly

IDCC: International Data Comunications Centre • International Data Coordination Centre

IDCCC: Interior Department Cartographic Coordinating Committee (USA)

IDCE: Institut pour le Développement du Conseil d'Entreprise

IDCMA: Independent Data Communications Manufacturers Association

IDCP: Interdivision Change Proposal

IDCS: Image Dissector Camera System • Integrated Data Coding System • Intelligent Data Collection System • International Data Collecting System • International Data Collection System • ISO Dedicated Computer System

IDCT: Inverse Discrete Cosine Transform

IDCU: Integrated Data Communications Utility

ID&D: Inspection Damage & Discrepancy

IDD: Image Definition Device • Indirect by Direct • Information and Data Dissemination • Information Delivery Device • Inspection Damage and Discrepancy • Instrument Deployment Device • Integrated Data Dictionary • Interface Definition Document • Interface Description Document • Interface Design Document • Interface Development Document • Integrated Data Denial • International Direct Dial • International Direct Dialing • Internet Data Distribution

IDDD: International Direct Distance Dialing

IDDE: Integrated Development and Debugging Equipment • Interactive Development and Debugging Environment

IDDL: Integrated Data Description Language

IDDS: Improved Data Display System • Instrumentation Data Distribution System • Integrated Data Display System

I&DE: Investigação & Desenvolvimento Experimental

IDE: Imbedded Drive Electronics • Information-Documentation-Evaluation • Initial Design Evaluation • Institute of Developing Economies • Instructional Design Environment • Integrated Design Environment • Integrated Development Environment • Integrated Device Electronics • Integrated Disk Electronics • Integrated Drive Electronics • Intelligent Drive Electronics • Interactive Design and Engineering • Interactive Development Environment • Interface Design Enhancement • Interplanetary Dust Experiment • Investissement Direct Étranger

IDEA: Improved Data Effectiveness and Availability • Inductive Data Exploration and Analysis • Initiative to Develop Education through Astronomy • Innovation Development and Engineering Applications • Instituto Internacional de Estudios Avanzados • Instrument Driver for Experimental Applications • Integrated Database Environment for Assessment • Integrated Data for Enforcement Analysis • Intelligence-Diagnostic-Expertise-Administration • Interact Decision Evolution Aid • Interactive Data Entry Access • Interactive Data Exploration and Analysis • Interactive Deconvolution with Error Analysis • Interface and Display Electronics Assembly • International Data Encryption Algorithm • International Digital Electronic Access Library

IDEAL: Interactive Databased Environment for Accelerated Learning • Interactive Dialogues for Explanation And Learning • International Decade for the East African Lakes • International Digital Electronic Access Library

IDEALS: Information Directory of European Automated Library Services • Intelligent Direct European Access to Library Systems

IDEAM: Instituto de Estudios del Ambiente

IDEAS: Information Dissemination, Editing and Searching System • Initial Design and Evaluation Analysis System • Initiative to Develop Education through Astronomy and Space Science • Interactive Design and Evaluation of Advanced Spacecraft • Interactive Document Easy Access System • Interior Department's Electronic Acquisition System • International Database Engineering and Applications Symposium

IDECM: Integrated Defensive Electronic Countermeasures

IDECS: Image Discrimination, Enhancement and Combination System

IDEF: Integrated Data Exchange Facility

iDEN: Integrated Digital Enhanced Network

Iden: Identification • Identify

IDEN: Interactive Data Entry Network

Ident: Identical • Identification • Identifier • Identify

IDEP: Idustry Data Exchange Program • Interagency Data Exchange Program • International Data Exchange Program • Interservice Data Exchange Program

IDES: Image Detail Enhancement System • Image Dissector Echelle Spectrograph • Information and Data Exchange System • Interactive Data Entry System • Interactive Doppler Editing Software • International Diffuse EUV Spectrometer

IDEX: Image-Data Exchange

IDF: Idiofa (Zaire) (location identifier) • IDL Data Format • Image Description File • Image Display Format • Industrial Diesel Fuel • Information Design Facility • Institut für Diagnostikforschung • Integrated Data File • Interactive Dialogue Facility • Interactive Distant Learning • Interface Definition File • Intermediate Distributing Frame (*) • Intermediate Distribution Frame • Internal Delay Factor • International Dairy Federation • Inverse Document Frequency • Isotropic Distribution Function

IDFA: International Duty Free Association

IDFT: Inverse Discrete Fourier Transform

idFv: in der Fassung von

IDG: Ida Grove (USA-IA) (location identifier) • Institute of Dynamics of Geospheres • Integrated Drive Generator • Inter-Dialog Gap • International Data Group

IDGC: Institutional Dimensions of Global Change

IDGE: Isothermal Dendritic Growth Experiment

IDGS: Isotope-Dilution Gamma Spectrometry

IDHAF: One-Dimensional Heisenberg Antiferromagnet

IDHS: Information Data Handling System • Integrated Data Handling System

IDHT: Instrument Data Handling and Transmission

IDHTS: Instrument Data Handling and Transmission System

IDI: Image Display Interface • Imaging Doppler Interferometer • Improved Data Interchange • Indiana (USA-PA) (location identifier) • Information Dynamics Inc. (USA-DC) • Institut de Droit International • Instrumentation Data Item • Ion Dipole Interaction • Investissement Direct International

IDIAS: Ice Data Integration and Analysis System

IDICT: Instituto de Documentación e Información Científica y Técnica

IDID: Integrated Dual Image Detector

IDIDAS: Interactive Digital Image Display and Analysis System

IDIMS: Interactive Data Integration and Management System • Interactive Digital Image Manipulation System

IDIOT: Ice Depth Instrument Operator Transportable

ID/IQ: Indefinite Delivery/Indefinite Quantity

IDIQ: Indefinite Delivery/Indefinite Quantity

IDIS: Integrated Data Information System • International Directory Inquiry System

IDIV: Integer Divide

IDK: I Don't Know • Indulkana (Australia-SA) (location identifier)

IDL: Image Display Language • Indentured Drawing List • Information Description Language • Instruction Definition Language • Intelligent Database Language • Interactive Data Language • Interface Definition Language (*) • Interface Description Language • Intermediate-Density Lipoprotein • International Date Line

IDLC: Integrated Digital Logic Circuit

IDLE: Integrated Distributed Learning Environment

IDLH: Immediately Dangerous to Life and Health

IDLI: International Development Law Institute

IDM: Identification of Dark Matter • Improved Data Modem • Induced Dipole Moment • Information and Data Management • Information Distribution Management • Instrument Demonstration Module • Intelligent Database Machine • Intelligent Data Management • Intelligent Document Management • Interactive Dynamic Map • Inter-Departmental Meeting • Ion Drift Meter

IDMM: Intermediate and Depot Maintenance Manual

IDMRD: Intermediate and Depot Requirements Document

IDMRS: Intermediate and Depot Requirements Specification

IDMS: Image Database Management System • Information and Data Management System • Ingest Debris Monitoring System • Integrated Database Management System • Integrated Data Management System • Intelligent Data Management System • Interactive Distributed Multimedia Systems • Isotope Dilution Mass Spectrometry

IDMTF: Interim Data Management and Transfer Facility

IDN: Indagen (Papua New Guinea) (location identifier) • In Dei Nomine • Indonesia (ISO country code) • Indonésie (ISO country code) • Integrated Data Network • Integrated Digital Network • Intelligent Data Network • Interface Definition Notation • International Directory Network

IDNDR: International Decade for Natural Disaster Reduction

IDO: Integrated Development Operation • Interdivisional Operation • Interface Definition Object • Santa Isabel do Morro (Brazil-TO) (location identifier)

IDOC: Inner Diameter of Outer Conductor

IDOE: International Decade for Ocean Exploration

IDOH: Instrument Data Operations Handbook

IDOS: Integrated Display Operating System

IDOT: Instrument Digital Online Transcriber

IDP: Image Data Processing • Independence (USA-KS) (location identifier) • Industrial Data Processing • Information Data Processing • Input Data Processor • Institute of Data Processing • Instrument Development Program • Integrated Data Package • Integrated Data Presentation • Integrated Data Processing • Integrated Data Processor • Integrated Development Programme • Integrated Display Processor • Interface Digital Processor • Interim Digital Processor • International Driving Permit • Internet Datagram Protocol • Internetwork Datagram Protocol • Interplanetary Dust Particle • Interstellar Dust Particle

IDPAI: International Directory of Professional Astronomical Institutions (merged with IDAAS into ASpScROW) (now StarGuides)

IDPC: Integrated Data Processing Center

IDPD: ISO Data Product Document

IDPF: Integrated Digital Photogrammetric Facility

IDPGF: Institutional Data Products Generation Facility

IDPI: Isophoromdiisocyanate

IDPM: Institute of Data Processing Management (UK)

IDPMA: International Data Processing Management Association

IDPMX: Interim Digital Processor with Multiple Execution

IDPR: Illinois Department of Professional Regulation (USA-IL) • Interdomain Policy Routing

IDPRS: Interactive Data Processing and Research System

IDPS: Image Data Production System • Instrument Data Processing System • INTEGRAL Data Processing System • Integrated Data Processing System • Integrated Design and Production System • Integrated Distributed Power System

IDPT: Image Dissector Photomultiplier Tube

IDPU: Information, Documentation and Publication Unit • Instrument Data Processing Unit • Integrated Data Processing Unit

idR: in der Regel

IDR: Image Data Resampler • Incremental Design Review • Indore (India) (location identifier) • Information Dissemination and Retrieval • Initial Design Review • Input Data Request • Inspection Discrepancy Report • Installation Data Record • Instantaneous Data Rate • Institute for Dream Research • Instrumentation Development Request • Interface Design Review • Interim Data Review • Interim Design Review • Interim Discrepancy Report • Intermediate Data Rate • Intermediate Data Record • Intermediate Design Review • Internal Development Report • International Drawing Rights • Rupiah (Indonesia) (ISO currency code)

IDRA: Interdivision Revision Authorization • Istituto per la Didattica e la Ricerca Applicata

IDRC: Improved Data Recording Capability • International Development Research Centre

IDRD: Information Definition Requirements Document • Internal Data Requirements Document

IDRIS: Institut de Développement et des Ressources en Informatique Scientifique

IDRP: Interdomain Routing Protocol

IDRS: Integrated Data Retrieval System

IDS: Image Display System • Image Dissection System • Image Dissector Scanner • Income Data Services • Index of Double Stars • Inertial Data System • Inertial Doppler System • Information and Documentation Service • Information and Decision System • Information Data System • Information Display System • Information Dissemination Service • Information Dissemination System • Information Distribution System • Infrared Detection System • Initiative de Défense Stratégique • Input-Data Strobe • Institut de Documentation Scientifique • Institute of Development Studies • Instrument Data System • Instrument Development Section • Integrated Data Storage • Integrated Data Store • Integrated Data System • Integrated Design Support • Integrated Design System • Integrated Diagnostic System • Integrated Display System • Interactive Data System • Interactive Development System • Interdepartmental Committee on Space • Interdisciplinary Science • Interdisciplinary Scientist • Interdisciplinary Study • Interface Data Sheet • Intermediate-Dispersion Spectrograph • Internal Directory System • Interstellar Dust System • Intrusion Detection System • Item Description Sheet

IDSC: Information and Decision Support Centre

IDSCS: International Decade of Solar Cycle Studies

IDSD: Institutional Data System Division (JSC)

IDSIPS: Interactive Digital Satellite Image Processing System

IDSL: ISDN Digital Subscriber Line

IDSO: Interdivisional Sales Order

IDSR: Initial Design Status Review • Inspecteur Départemental de Sécurité Routière • Interdivisional Support Request

IDSS: Image Data System Simulation • Integrated Data Sounding System • Integrated Design Support System • Integrated Diagnostic Support System • Intelligent Decision Support System

IDST: Integrated Digital Services Terminal

IDT: Idêntica • Image Dissector Tube • Industrial Design Technology • Information-Documentation-Transfert • Ingénieur Divisionnaire des Travaux • Instrument Dedicated Team • Instrument Definition Team • Instrument Development Team • Interactive Development Tool • Interactive Display Terminal • Interdigital Transducer • Interdisciplinary Team • Interférométrie Différentielle des Tavelures • International Discount Telecommunication • Interrupt Descriptor Table • Investigação e Desenvolvimento Tecnológico • Investigation Definition Team

IDTA: Interdivisional Technical Agreement

IDTC: IEEE Design Test Computers (periodical code)

IDTP: Integrated Data Transmittal Package

IDTS: Improved Doppler Tracking System • Instrumentation Data Test Station • Instrumentation Data Transmission System • Instrument Data Telemetry System

IDU: Industrial Documentation Unit • Instrument Development Unit • Integrated Display Unit • Interface Demonstration Unit

IDUM: ISO Data User's Manual

i. Durchschn.: im Durchschnitt

IDV: Institut für Datenverarbeitung • Integrating Digital Voltmeter • Integrierte Datenverarbeitung • Interlibrational Derived Vehicle • Intra-Day Variability • Intra-Day Variable

IDVC: Integrated Data/Voice Channel

IDVS: Integriertes Datenverarbeitungssytem

IDW: Informations- und Dokumentationswissenschaft • Institut für Dokumentationswesen • IRAF Developers' Workshop

IDWA: Interdivisional Work Authorization

IDWG: Inter-Departmental Working Group • IRAF Developers' Workshop

Idx: Index

IDX: Index

IDY: Identify • Île d'Yeu (France) (location identifier)

IDYL: Image et Dynamique de la Litosphère • Imagerie et Dynamique de la Litosphère (*)

IDZ: Informations- und Dokumentationszentrum

ie: Id Est • Interlingue (ISO language code) • Ireland

IE: Imaging Element • Index Error • Indo-European • Industrial Engineer • Industrial Engineering • In Excess • Information Element • Information Engineer • Information Engineering • Information Environment • Information Exchange • Information Extraction • Ingénieur d'Études • Inside Edge • Integrated Environment • Intelligence Économique • Interface Engineer • Intermediate Equipment • Intermediate Erection • Internet Explorer • Ion Exchange • Ionization Energy • Ionospheric Explorer • Ireland (ISO country code) • Irlande (ISO country code)

IEA: Image Enhancement Algorithm • Industrial and Engineering Application • Industrial Editors Association • Information Exchange Agreement • Institute of Energy Analysis • Institution of Engineers - Australia • Instituto de Estudos Avançados • Integrated Electronics Assembly • International Energy Agency

IEAS: Ice Elevation Altimeter System

IEAUST: Institution of Engineers, Australia

IEB: Instituto de Estudos Brasileiros • Instruction Engineering Bit

IEBS: Institute of Environmental and Biological Sciences

IEC: Industrial Energy Conservation • Information Exchange Center • Integrated Electronic Component • Integrated Electronic Control • Integrated Environmental Control • Intelligence Économique et Concurrentielle • Interexchange Carrier • Interim Executive Committee • International Electrotechnical Commission (*) • International Electrotechnical Committee • International Engineering Consortium • International Executive Committee • Ion Exchange Chromatography • Ionospheric Electron Content • Item Entry Control

IECA: Independent Election Corporation of America

IECC: International Express Carriers Conference

IECG: Io-Europa-Ganymede-Callisto

IECM: Induced Environment Contamination Monitor

IECO: Inboard Engine Cutoff

IECQ: International Electrotechnical Common Quality

IECS: Igloo Environment Control System • Institut d'Études Commerciales de Strasbourg (France)

IED: Information Engineering Directorate • Institution of Engineering Designers • Interacting Equipment Document • Intercambio Electronico de Datos • Ionospheric Electron Density

IEDAL: International Electronic Digital Access Library

IEDL: IEEE Electron Device Letters (periodical code)

IEDM: International Electron Devices Meeting • International Electronic Devices Meeting

IEDS: International Environment and Development Service

IEE: Induced Electrical Effect • Induced Electron Emission • Initial Environmental Evaluation • Institution of Electrical Engineers • Instituto de Electrotécnica e Energia

IEEE: Institute of Electrical and Electronic Engineering • Institute of Electrical and Electronics Engineers (*)

IEEEN: Institute of Electrical and Electronics Engineers Network (periodical code)

IEEEP: Institute of Electrical and Electronics Engineers Proceedings (periodical code)

IEEES: Institute of Electrical and Electronics Engineers Spectrum (periodical code)

IEEP: Incapacitated Emergency Egress Practice • Institute for European Environment Policy • Institution of Electrical Engineers Proceedings (periodical code) • International Environment Education Program

IEERv: Institution of Electrical Engineers Reviews (periodical code)

IEF: Information Engineering Facility • Informatique Électronique Française (France) • Infrared Emission Feature • International Environment Facility • International Environment Forum • Inventory Exchange Format • Isoelectric Focusing

IEFT: Integrated Experiment Functional Test

IEG: Information Exchange Group • Institut d'Électronique de Grenoble (France) • Zielona Gora (Poland) (location identifier)

IEGI: Institut Européen pour la Gestion de l'Information

IEGP: Information Exchange Group Program

IEH: International EUV/FUV Hitchhiker

IEHT: Intensifier Electrical High Tension

IEI: Industrial Engineering Institute • Institution of Engineers of Ireland

IEIC: International Education Information Center

IEICE: Institute of Electronics, Information and Communication Engineers

IEIWG: International Element Interface Working Group

IEJ: Iejima (Japan) (location identifier)

IELS: Isotope Exciter Light Source

IELV: Intermediate Expandable Launch Vehicle

IEM: Impulsion Électromagnétique • Institute of Experimental Meteorology • Instituto de Enseñanza Media • Integral Equation Model • Integrated Electronics Module • Interférence Électromagnétique • Institution of Engineers - Malaysia

IEMC: Intrasystem Electromagnetic Compatibility

IEMCCAP: Intrasystem Electromagnetic Compatibility Computer Analys Program

IEMP: Induced Electromagnetic Pulse

IEMSI: Interactive Electronic Mail Standard Identification

IEN: Individualized Electronic Newspaper • Integrated Enterprise Network • Integrated Enterprise Node • Interactive Employment Network • Internet Experiment Note • Istituto Elettrotecnico Nazionale (Italy)

IENM: Ingénieur de l'École Nationale de la Météorologie

IENR: Investigation of the Earth Natural Resources

IEO: Incoherent Electronic Oscillator • International Exchange Office

IEOS: International Earth Observing System

IEOSC: International Earth Observation Satellite Committee

IEP: Independent Electric Plant • Individual Educational Plan • Individualized Educational Programme • Information Exchange Program • Information Exchange Project • Inner Equipment Platform • Instantaneous Effective Photocathode • Institut d'Études Politiques • Instituto Electrotécnico Português (Portugal) • Integrated Engineering Program • International Executive Program • Inverse Eigenvalue Problem • Isoelectric Point • Punt (Ireland) (ISO currency code)

IEPA: Independent Electron Pair Approximation

IEPG: Independent European Programme Group • Intercontinental Engineering Planning Group

IEPRC: International Electronic Publishing Research Centre

IER: Independent Evaluation Report • Individual Evaluation Report • Installation Enhancement Release • Institute of Engineering Research • Institute of Environmental Research • Interface Evaluation Report • Internationale Elektronische Rundschau (periodical code)

IERE: Institution of Electronic and Radio Engineers

IEREJ: Institution of Electronic and Radio Engineers Journal (periodical code)

IERS: International Earth Reference System • International Earth Rotation Service (*) • International Ellipsoid Reference System

ieS: im engeren Sinn

I&ES: Information & Electronics System

IES: Illuminating Engineering Society • Imaging Electron Spectrometer • Information and Electronics System • Information Exchange System • Information Extraction System • Institute of Environmental Sciences • Instituto de Educación Secundaria • Integral Energy Spectrum • Intelligent Energy Saver • Interactive Exploration System • International Exchange Service • Ion and Electron Sensor • Ion and Electron Spectrometer • Ion Electron Spectrometer • Ion Energy Selector • Ion Engine Simulator • Ion Engine System • IRAS Explanatory Supplement • Isolated Emission Source

IEASANZ: Illuminating Engineering Society of Australia and New Zealand

IESC: Information Exchange Steering Committee

IESD: Instrumentation and Electronic Systems Division

IESG: Internet Engineering Steering Group

IESI: Institut Européen des Sciences Informatiques

IESNA: Illuminating Engineering Society of North America (USA)

IESP: Integrated Electronic Signal Processor

IESS: Internet Entrepreneurs Support Server

IESSD: Institute for Environmental Studies for Sustainable Development

IEST: Institut Européen des Standards de Télécommunications

IESTO: Institut d'Études Supérieures des Techniques d'Organisation

IET: Impact Excited Transmitter • Initial Engine Test • Initial Entry Training • Interest Equalization Tax

IETA: Ingénieur des Études et Techniques d'Armement

IETE: Institution of Electronics Telecommunication Engineers

IETF: Internet Engineering Task Force

IETM: Interactive Electronic Technical Manual

IETS: Inelastic Electron Tunnelling Spectroscopy

IEU: Input Expansion Unit • Instrument Electronics Unit • Interface Electronics Unit • Ion Exchange Unit

IEV: International Electrotechnical Vocabulary • Kiev Zhulyany (Ukraine) (location identifier)

IEVD: Integrated Electronic Vertical Display

IEW: In Eén Woord • International Energy Workshop

IEWCS: Intelligence and Electronic Warfare Common System

IExp: Institute of Electrical and Electronics Engineers Expert (periodical code)

IEY: International Education Year (1970)

if: Ipse Fecit

I/F: Interface

IF: Image Fusion • Impact Factor • Importance Factor • Individual Focus • Inertial Fusion • Information Feedback • Information Fetch • Information Filtering • Inhibiting Factor • Initialization File • Inside Frosted • Institut de France • Integration Facility • Interactive Facsimile • Interface • Intermediate Fix • Intermediate Frequency • Internal Fault • Interrupt Flag • Inverted File • Investissement Formation

IfA: Institute for Astronomy

IFA: Igniter-Fuel Assembly • Immunofluorescence Assay • Independent Financial Adviser • Industry Film Association • In-Flight Abort • In-Flight Analysis • In-Flight Anomaly • Information Flow in Astronomy • Institute For Astronomy • Institut for Fysik og Astronomi • Instrumented Fuel Assembly • Integrated Feed Antenna • Integrated File Adapter • Integrated Financial Adviser • Intelligent Filtering Agent • Intensive Flux Array • Interface Functional Analysis • Intermediate Frequency Amplifier • Intermediate Frequency Attenuator • Iowa Falls (USA-IA) (location identifier) • Istituto di Fisica dell'Atmosfera

IFAA: International Federation of Advertising Agencies

IFAC: International Federation of Automatic Control

IFAD: International Fund for Agricultural Development

IFAE: Interamerican Federation for Adult Education

IfAG: Institut für Angewandte Geodäsie

IFAG: Institut Für Angewandte Geodäsie

IFALPA: International Federation of Airline Pilots' Associations

IFAM: Inverted File Access Method

IFAORS: Institute for Atmospheric Optics and Remote Sensing

IFAPA: International Foundation of Airline Passengers Associations

IFARE: Institut Franco-Allemand de Recherches sur l'Environnement

IFAS: Idaho Falls Astronomical Society (USA-ID) • Institute of Food and Agricultural Sciences

IFASC: Integrated Functions Assessment Steering Committee

IFAST: Integrated Facility for Avionic Systems Test

IFATCA: International Federation of Air Traffic Controllers Associations

IFATT: Integrated Feature And Terrain Triangulation (algorithm)

IFB: Invitation For Bid

IFC: Industry Foundation Class • In-Flight Calibration • Inside Front Cover • Intelligent Forms Language • Intensive Field Campaign • Interface Clear • Interface Controller • International Finance Corporation

IFCAI: International Federation of Commercial Arbitration Institutions • Istituto di Fisica Cosmica ed Applicazioni dell'Informatica (Italy)

IFCE: In-Flight Contamination Experiment

IFCF: Intermediate Frequency Crystal Filter

IFCG: Interface Control Group

IFCIS: International Foundation on Cooperative Information Systems

IFCN: Interfacility Communication Network

IFCP: Interfacility Communications Processor

IFCPAR: Indo-French Centre for the Promotion of Advanced Research

IFCPLV: Institut Français de la Communication et de la Publicité sur le Lieu de Vente (France)

IFCS: International Federation for Computer Sciences • International Federation of Classification Societies • Integrated Flight Control System • Irish Federation of Computer Societies

IFCTR: Istituto per Ricerche in Fisica Cosmica e Tecnologie Relative (Italy)

IFCU: International Federation of Catholic Universities

IfD: Institut für Dokumentation • Institut für Dokumentationswesen

IFD: Image File Directory • Industrial Facilities Discharge • Inflatable Floating Device • In-Flight Deployment • Information Flow Diagram

IFDA: International Federation for Data Processing

IFDC: International Fertilizer Development Center

IFDL: Interfacility Data Link

IFDM: Integrated Finite-Difference Method

IFDO: International Federation for Data Organizations

IFDS: Inertial Flight Data System • Interfacility Data System

IfE: Institut für Erdmessung (Germany)

IFE: Industry-Furnished Equipment • In-Flight Entertainment • Inner Front End • Intelligent Front End • Isoelectric Focussing Experiment

IFEA: In-Flight Emergency Assistance • Integrated Furnace Experiment Assembly

IFEL: Induction Free-Electron Laser

IfEN: Institut für Erdmessung und Navigation (Germany)

IFEN: Intercompany File Exchange Network

IFEOS: International Forum for Earth Observations from Space

IFERS: International Flat Earth Research Society

IFEU: Institut für Energie- unf Umweltforschung GmbH (Germany)

iff: If and Only If

IFF: Identification Friend-Foe • Iffley (Australia-QLD) (location identifier) • Independent Fourier Frequency • Induced Fluid Flow • Institute For the Future • Interchangeable File Format • Interchange File Format • Ionized Flow Field

IFFA: Interactive Flash Food Analyzer • International Federation of Film Archives

IFFET: Institut Français du Film Éducatif et Technique (France)

IFFN: Identification Friend-Foe-Neutral

IFFP: Institut Français de Formation Permanente (France)

IFFRO: International Federation of Reproduction Rights Organizations

IFFS: International Federation of Film Societies

IFFT: Inverse Fast Fourier Transform

IFG: Incoming Fax Gateway • Information File Generator • Intellectual Freedom Group • InterFrame Gap

IFGL: Initial File Generation Language

IFH: Image File Header • In-Flight Helium

IFI: Independent Fourier Interval • International Film Institute • International Fund for Ireland

IFIA: International Federation of Inventors' Associations

IFIAS: International Federation of Institutes for Advanced Study (Canada) • International Federation of Institutes of Advanced Studies

IFICS: Inflight Interceptor Communications System

IFID: International Federation for Information and Documentation

IFIG: Intellectual Freedom Interest Group

IFIM: International Flight Information Manual

IFIP: International Federation for Information Processing (*) • International Federation of Information Processors

IFIPS: International Federation of Information Processing Societies (now IFIP)

IFIR: Instituto de Física de Rosario (Argentina-SF)

IFIRS: Integral Field Infrared Spectrograph • Integral Field Infrared Spectrometer (*)

IFIT: International Forest Investigation Team

IFJ: International Federation of Journalists • Isafjördur (Iceland) (location identifier)

IFK: Integral of First Kind

IFL: Innisfail (Australia-QLD) (location identifier) • Intelligent Forms Language • International Frequency List

IFLA: International Federation of Library Associations

IFLC: International Frequency List Committee

IFLOT: Intermediate Focal Length Optical Tracker

IFLOWS: Integrated Flood Observation and Warning System

IFLTT: Intermediate Focal Length Tracking Telescope

IFM: In-Flight Maintenance • Instantaneous Frequency Measurement • Integrated Financial Management • Intelligent Flow Management • Intermediate Frame Memory • Io Flow Model

IFMBE: International Federation for Medical and Biological Engineering

IFMIS: Integrated Facilities Management Information System

IFMP: Ipsilon Flow Management Protocol

IFMS: Impact Force Measuring System • In-Flight Management System • Integrated Financial Management System

IFMU: Integrated Flight Management Unit

IFN: Ice-Forming Nucleus • Ice-Freezing Nucleus • Institut Français de Navigation (France) • Interferon • Isfahan (Iran) (location identifier)

IFNP: International Federation of Newspaper Publishers

IFO: Improved Fiber Optics • Information Systems Office • Intensive Field Observation • Intensive Field Operation • International Field Office • Ivano-Frankovsk (Ukraine) (location identifier)

IFOBS: International Forum on Open Bibliographic Systems

IFODA: Institut Français d'Organisation Documentaire et Administrative (France)

IFOG: Interferometric Fiberoptic Gyro

IFOP: Institut Français d'Opinion Publique (France) • Instrument Flight Operation Plan • Instrument Flight Operation Procedure

IFOR: Interactive Fortran

IFORS: International Federation of Operational Research Societies

IFOT: In-Flight Operations and Training

IFOV: Individual Field Of View • Instantaneous Field Of View • Instrument Field Of View

IfP: Institut für Planetologie (Germany)

IFP: Bullhead City (USA-AZ) (location identifier) • In-Flight Performance • Information Flow Process • Institut Français du Pétrole (France) • Intensive Field Program • Interface Processor • Interferometry Panel • Laughlin Bullhead (USA-AZ) (location identifier)

IFPA: Industrial Film Producers Association (USA) • Information Film Producers of America • Instituto de Física de Porto Alegre (Brazil-RS)

IFPC: International Fund for the Promotion of Culture (UNESCO)

IFPI: Imaging Fabry-Pérot Interferometer • International Federation of Photographic Industries • International Federation of the Phonographic Industry (*)

IFPM: In-Flight Performance Monitor

IFPP: International Federation of the Periodical Press

IFPS: Imaging Fabry-Pérot Spectrograph • Imaging Fabry-Pérot Spectrometer • International Federation of Palynological Societies • Intra-Formation Positioning System

IFPSI: Institut Français de Presse et des Sciences de l'Information (France)

IFPUG: International Function Point Users Group

IFQ: Invitation For Quote

IFR: Image-to-Frame Ratio • Infrared Filter Radiometer • Initial Flight Requirement • Instrument Flight Recovery • Instrument Flight Rules • Instrument Flying Regulation • Integral Fast Reactor • Interface Repository

IFRB: International Frequency Registration Board

IFRC: International Fusion Research Council

IFREMER: Institut Français de Recherche pour l'Exploitation de la Mer (France) (*) • Institut Français de Recherche pour l'Exploitation des Mers (France)

IFRR: Institut für Raumflug- und Reaktortechnik (Germany)

IFRRO: International Federation of Reproduction Rights Organizations

IFRT: Intellectual Freedom Round Table

IFRTP: Institut Français pour la Recherche et la Technologie Polaires (France)

IFRU: In-Flight Replaceable Unit • In-Flight Replacement Unit

IFS: Impact and Flotation System • In-Flight Safety • Information Flow Structure • Information Flow System • Installable File System • Institute for Fusion Studies (USA-TX) • Institute for Future Studies • Institute of Fundamental Studies (Sri Lanka) • Integral Field Spectrography • Integrated Flight Simulator • Integrated Forest Study • Interchange File Separator • Interface Specification • Interface System • Integrated Forecasting System • Internal Field Separator • International Foundation for Science • Iterated Function System

IFSA: International Fuzzy Systems Association

IfSAR: Interferometric Synthetic Aperture Radar

IFSARE: Interferometric Synthetic Aperture Radar for Elevation

IFSC: Instituto de Física de São Carlos (Brazil-SP)

IFSD: In-Flight Shutdown

IFSE: International Federation of Scientific Editors

IFSEA: International Federation of Scientific Editors' Association

IFSEG: International Federation of Societies of Economic Geologists

IFSEM: International Federation of Societies for Electron Microscopy

IFSI: Istituto di Fisica dello Spazio Interplanetario (Italy)

IFSMgr: Installable File System Manager

IFSS: Index of Federal Specifications and Standards (USA) • Inspection Field Support System • International Flight Service Station

IFSTD: Interim Fund for Science and Technology for Development

IFT: In-Flight Test • Ingénieur Divisionnaire des Travaux • Input Frequency Tolerance • Instituto de Física Teórica • Interactive Flight Technology • Interface Tool • Intermediate Frequency Transformer • International Federation of Translators • International Frequency Table • Inverse Fourier Transform • Io Flux Tube • Ion Focusing Technique

IFTA: International Federation of Teachers' Association • International Federation of Television Archives • International Federation of Travel Agencies • International Free Trade Area

IFTC: International Film and Television Council

IFTD: Interface Terminal Distributor

IFTE: Intermediate Forward Test Equipment

IFTF: Institute For The Future

IFTM: In-Flight Test and Maintenance • In-Flight Test Maintenance

IFToMM: International Federation for the Theory of Machines and Mechanisms

IFTPP: International Federation of the Technical and Periodical Press

IFTR: Ion Flow Tube Reactor

IFTS: Imaging Fourier Transform Spectrograph (*) • Imaging Fourier Transform Spectrometer • In-Flight Test System

IFTU: International Federation of Teachers' Unions • International Federation of Trade Unions

I/FU: Interface Unit

IFU: Imprimé Fiscal Unique • Institut Für Umweltforschung • Institut Für Umweltwissenschaften (Germany) • Integral Field Unit • Integrated Field Unit • Interface Unit

IFV: Institut de Force de Vente

IFWEA: International Federation of Workers' Education Association

IFWG: Interface Working Group

IFYCSEM: Integrated First-Year Curriculum in Science, Engineering, and Mathematics

IFYGL: International Field Year for the Great Lakes

Ig: Ignition • Ignitor

IG: Igloo • Ignitor • Ilford (UK postal code) • Indo-Germanic • Industrial Group • Industriegewerkschaft • Inertial Guidance • Information Gateway • Infrastructure Group • Ingénieur Général • Inner Gimbal • Inspector General • Institute of Geophysics • Instruction Guide • Instrumentation Group • Instrument Ground • Interacting Galaxies • Interagency Group • Interessengemeinschaft • Internal Guidance • Inverse Gain • Ionization gauge

IGA: Inagua (Bahamas) (location identifier) • Inner Gimbal Angle • Instituto de Geofísica y Astronomía • Integrated Graphics Array • Integrating Gyro Accelerometer • Intergimbal Assembly • Inter-Governmental Agreement • International Geographic Association • Ion Gun Assembly

IGAC: Instituto Geografico Augustin Codazzi (Colombia) • Interessengemeinschaft Astronomie Crimmitschau (Germany) • International Global Atmospheric Chemistry (program) (*)

IGACEM: Inspection Générale de l'Aviation Civile et de la Météorologie

IGACS: Integrated Guidance And Control System

IGACSM: Inter-Governmental Advisory Committee on Surveying and Mapping

IGADD: Inter-Governmental Authority on Drought and Development

IGAE: Inter-Governmental Agreement on the Environment

IGAP: International Global Aerosol Program

IGARSS: International Geoscience and Remote Sensing Society • International Geoscience and Remote Sensing Symposium

IGAX: Inner Gimbal Axis

IGB: Ingeniero Jacobacci (Argentina-RN) (location identifier) • International Gravity Bureau

IGBMC: Institut de Génétique et de Biologie Moléculaire et Cellulaire

IGBP: International Geosphere-Biosphere Program

IGBT: Insulated Gate Bipolar Transistor

IGC: Institute for Global Communications (USA-CA) • Institutional Grants Committee • Integrated Graphical Controller • Integrated Graphics Controller • Intelligent Graphics Controller • Intergovernmental Committee • Intergovernmental Conference • Intergovernmental Council • International Geological Congress • International Geophysical Committee • International Geophysical Cooperation • International Gravity Commission • Internet Governance Committee • Inverse Gas Chromatography • Ion Gun Collector

IGCC: Integrated Gasification Combined Cycle • Intergovernmental Copyright Committee

IGCP: Instituto Geografico e Cadastral do Portugal • International Geological Correlation Program

IGCRE: Institute for Global Change Research and Education

IGCS: Integrated Guidance and Control System

IGD: Institut für Graphische Datenverarbeitung • Integrated Genomic Database • Interaction Graphics Display • Interactive Graphical Document • Interactive Graphical Documentation • Interessengemeinschaft Dokumentation • Inverse Gaussian Distribution

IGDB: Instrument Group Database

IGDC: Interagency Geographic Data Committee

IGDR: Interim Geophysical Data Record

IGDS: Interactive Graphics Design System • Iodine Generating and Dispensing System

IGE: Iguela (Gambia) (location identifier) • Image Generating Equipment • In-Ground Effect • Instrumentation Ground Equipment • International Geophysical Extension

IGEB: Interagency GPS Executive Board

IGeC: International Geoid Commission

IGERT: Integrative Graduate Education and Research Training

IGeS: International Geoid Service

IGES: Initial Graphical Exchange Specification • Initial Graphics Exchange Specification • Initial Graphics Exchange Standard • Institute of Global Environment and Society (USA-MD) • Intermediate Graphics Exchange Standard • International Graphics Exchange System

IGEX: International GLONASS Experiment

IGF: Inset Graphics Format • Insulin-like Growth Factor • International Graphical Federation

IGFA: International Group of Funding Agencies

IGFES: Interactive Graphics Finite Element System

IGFET: Insulated-Gate Field-Effect Transistor

IGFM: Internal Gamma Flux Monitor

IGFN: Italian GPS Fiducial Network

IGFOV: Instantaneous Geometric Field Of View

IGG: Igiugig (USA-AK) (location identifier)

IGH: Ingham (Australia-QLD) (location identifier) • Intelligent Gas Handling • Intervening Galaxies Hypothesis

IGHC: International Group Handling Council

IGI: Imaging Grism Instrument • Informatique de Gestion pour l'Imprimerie • Instrument Ground Instructor • Interactive Graphics Interpreter

IGIA: Industrial Guest Investigator Agreement

IGIF: International Geographic Information Foundation

IGIPAC: Integrated GPS Inertial Position and Attitude Control

IGIPAS: Interagency Group on International Programs in Atmospheric Sciences

IGIS: Integrated Geographical Information System • Intelligent Geographical Information System

IGL: Information Grouping Logic • Interactive Graphics Library • Ionized Gas Laser • Izmir Cigli (Turkey) (location identifier)

IGLD: International Great Lakes Datum

IGLO: Informal Group of Liaison Offices

IGM: Inertial Guidance Mode • Instituto Geográfico Militar • Interactive Guidance Mode • Intergalactic Medium • Inter-Governmental Meeting • Intragroup Medium • Iterative Guidance Mode • Kingman (USA-AZ) (location identifier)

IGMA: Instituto Geográfico Militar de Argentina

IGMF: Intergalactic Magnetic Field

IGMP: Internet Group Management Protocol (*) • Internet Group Membership Protocol • Internet Group Message Protocol • Internet Group Multicast Protocol

ign: Ignotus

Ign: Ignite • Ignition • Ignotus

IGN: IBM Global Network • Iligan - Maria Cristina (Philippines) (location identifier) • Institut Géographique National • Instituto Geográfico Nacional

IGN&C: Integrated Guidance, Navigation & Control

IGNC: Integrated Guidance, Navigation and Control

IGNOU: Indira Gandhi National Open University (India)

IGNS: Institute of Geological and Nuclear Sciences Inc. (New Zealand)

IGO: Chigorodo (Colombia) (location identifier) • Inter-Governmental Organization • International Gamsberg Observatory (Namibia) • International Gemini Observatory • International Gemini Office • Stichting voor Isotopen-Geologisch Onderzoek

IGOE: Inputs-Guides-Outputs-Enablers

IGOR: Intelligent Graphics Object Recognition • Intercept Ground Optical Recorder

IGORE: Integrated Gamma-Ray Observatory Reduction Environment • Interactive GRO/OSSE data Reduction Environment

IGORTT: Intercept Ground Optical Recorder Tracking Telescope

IGOS: Integrated Global Observing Strategy • Interactive Graphic Orbit System

IGOSS: Integrated Global Ocean Services System • Integrated Global Ocean Station System (*)

IGP: Inertial Guidance Package • Inertial Guidance Platform • Information Generation and Publishing • Institut für Geodäsie und Photogrammetrie • Institute of General Physics • Instituto Geofísico del Perú • Interior Gateway Protocol • International Gemini Project • International Geodynamics Project • Internet Gate Protocol • Internet Gateway Protocol • Ion Getter Pumping

IGPCF: Inert Gas Phase Condensation Facility

IGPD: Interdivisional Group on Physics for Development

IGPO: International Gemini Project Office • International GEWEX Project Office

IGPP: Institute of Geophysics and Planetary Physics (USA-CA) • Interactive Graphics Packaging System

IGPRAD: Inter-Governmental Panel on Radioactive waste

IGPS: Intelligent Gateway Processing System

IGQMS: Ion Guide Quadruple Mass Spectrometer

IGR: Igreja • Iguassu Falls (Argentina-MI) (location identifier) • Iguazu Cataratas (Argentina-MI) (location identifier) • Index to Government Regulation (USA)

IGRAP: Institut Gassendi pour la Recherche Astronomique en Provence (France)

IGRF: International Geomagnetic Reference Field (*) • International Geomagnetic Reference File

IGRP: Interior Gateway Routing Protocol (*) • Internet Gateway Routing Protocol

IGRSS: International geoscience and Remote Sensing Society

IGS: Improved Gray Scale • Inertial Guidance System • Inspection Générale des Services • Instrumentation Ground System • Instrument Guidance Ground System • Integrated Graphics System • Integrated Ground Segment • Interactive Graphics System • Interaktives Graphisches System • Interchange Group Separator • Interconnection Ground Subnetwork • International Geodynamics Service • International Geophysics Series (periodical code) • International Glaciological Society • International GPS Service • Internet Go Server

IGSC: International Group for Scientific Coordination

IGSCB: International GPS Service Central Bureau

IGSE: Instrument Ground Support Equipment

IGSMA: Inertial Guidance System Maintenance Area

IGSPC: Imaging Gas Scintillation Proportional Counter

IG StaR: Interessen Gemeinschaft Sternfreunde an der Ruhr (Germany)

Igt: Ingot

IGT: Institute of Gas Technology • Interactive Graphics Terminal

IGU: Iguassu Falls (Brazil-PR) (location identifier) • International Gas Union • International Geographical Union • Saltos do Iguaçu (Brazil) (location identifier)

IGV: Inlet Guide Vane

IGW: Increased Gross Weight • Internal Gravity Wave

IGWMC: International Ground Water Modelling Center

IGY: International Geophysical Year

IH: Iacet Hic • Information Handling • Information Hub • Installation Handbook • Integrated Hardware • Intercept Heading • Interferometric Hartmann • Internet Hub • Interrupt Handler

iha: In Het Algemeen

IHA: Indian Housing Authority • Information Handling in Astronomy • Institut Herzberg d'Astrophysique (Canada) • Interface Hazard Analysis • Niihama (Japan) (location identifier)

IHADSS: Integrated Helmet and Display Sight System

IHAP: Image Handling And Processing

IHAS: Integrated Hazard Avoidance System

ihb: In Het Bijzonder

IHB: Intermediate Horizontal Branxh • International Hydrographic Bureau

IHC: Intellectually Handicapped Child • Ionic Heated Cathode

IHCL: International Conference on Historical Linguistics

IHCP: Inverse Heat Conduction Problem

IHD: International Hydrological Decade

IHDP: International Human Dimension Program

IHDS: Infrared High-Dispersion Spectrograph

IHE: Infrastructural-Hydraulic-Environmental • Institut pour l'Hygiène et l'Épidémiologie • Instuut voor Hygiëne en

Epidemiologie • Intermediate High Energy • Internal Hydrogen Embrittlement

IHEA: Irish Higher Education Authority

IHEDN: Institut des Hautes Études de Défense Nationale (France)

IHEG: Institute of Hydrogeology and Engineering Geology

IHEP: Institute of High-Energy Physics

IHES: Institut des Hautes Études Scientifiques

IHEU: International Humanist and Ethical Union

IHF: Inverse Hyperbolic Function • Isothermal Heating Facility

IHFAS: Integrated High-Frequency Antenna System

IHFF: Inhibit Halt Flip-Flop

IHFR: Institut Hydrométéorologique de Formation et de Recherche (Algeria)

IHFT: Institut für Hochfrequenztechnik (Germany)

IHI: Improved Holographic Image • Ishikawajima-Harima Industries

IHIS: Interoperability of Heterogeneous Information Systems

IHK: Industrie- und Handelskammer

ihl: In Hoc Loco

IHL: Internet Header Length

IHM: Interface Homme-Machine

IHMC: Institute for Human and Machine Cognition (USA-FL)

IHN: Qishn (Yemen) (location identifier)

IHO: Ihosy (Madagascar) (location identifier) • International Hydrographic Organization

IHP: Indicated Horsepower • Information Handling Project • Institut Henri Poincaré (France) • Internationaal Hydrologisch Programma • International Hydrological Programme

IHPAn: Annales de l'Institut Henri Poincaré (periodical code)

IHPEG: Information Highway Parental Empowerment Group

IHPP: Institute of High-Pressure Physics

IHPRPT: Integrated High Payoff Rocket Propulsion Technology

IHPTET: Integrated High-Performance Turbine Engine Technology

IHR: Infrared Heterodyne Radiometer • Institute of Historical Research • Integrate-Hold-Reset

IHS: Information Handling Service • Information Handling System • Intensity-Hue-Saturation • International Heliospheric Study • Internet Host Server • Isotope Heat Source

IHSA: Interrupt Handler Save Area

IHSDS: Interconnections within High-Speed Digital Systems

IHSM: Institute of Health Services Management

IH/SR: Integrated Harware/Software Review

IHSR: Integrated Harware - Software Review

IHT: Inheritance Tax • International Herald Tribune

IHTEP: Integrated High-Performance Turbine Engine Technology

IHTV: Interim Hypersonic Test Vehicle

IHU: Ihu (Papua New Guinea) (location identifier) • Image Handling Utility

IHV: Independent Hardware Vendor

IHX: Interloop Heat Exchanger

IHW: International Halley Watch

IHwy: Info Highway • Information Highway

II: Bright Giant Star • Igniter Initiator • Image Intensifier • Imagery Interpretation • Implementing Instruction • Information for Industry • Information Index • Initial Issue • Input Impedance • Instituto de Informática • Interdisciplinary Investigator • Ion Implantation • Item Identification • Luminous Giant Star

IIa: Bright Giant Star

IIA: Image Intensifier Assembly • Indian Institute of Astrophysics • Indicator of Industrial Activities • Inertial Instrument Assembly • Information Industry Association (USA) • Information Interchange Architecture • Inishmaan (Ireland) (location identifier) • Institute of Industrial Administration • Instituto de Investigaciones Antropológicas (Mexico) • Instrument Interface Agreement • Intelligent Industrial Automation • Intelligent Information Agent

IIab: Bright Giant Star

IIACE: Instituto de Investigaciones Aplicadas de Ciencias Espaciales (Argentina)

IIAF: Idaho Institute Analysis Facility (USA-ID)

IIAL: International Institute of Arts and Letters

IIAP: Institut International d'Administration Publique

IIApN: Indian Institute of Astrophysics Newsletter (periodical code)

IIAS: International Institute for Administrative Sciences • International Institute for Advanced Studies

IIASA: International Institute for Applied Systems Analysis

IIb: Bright Giant Star

IIB: Institut International de la Bibliographie • Institut International des Brevets • International Investment Bank

IIBR: Inheritance Inference-Based Refinement

IIC: Ignitor Initiator Cartridge • Ignitor Initiator Container • Image Interpretation Cell • Independent Inspection Co. • Interactive Information Centre • Investigation In Charge • Ion-Ion Collision • Item Identification Code

IICE: Interactive Image Client Environment

IICIS: Institute for International Commerce and Information Sciences

IICO: Ion Implantation Corporation

IICT: Instituto de Investigação Científica et Tropical (Portugal)

IID: Image Intensifier Device • Impact Ionization Diode • Independent and Identically Distributed • Independent Identically Distributed • Instantiation Identifier • Interface Identifier • Ion Implantation Doping • Ionospheric Ion Density

IIDP: Integrated Instrument Development Program

IIDS: Integrated Instrumentation Display System • Integrated Instrument Display System

IIDT: Ion Implantation Doping Technique

IIE: Initial Ion Event • Institute for International Education • Institute of Industrial Engineers • Institute of International Education

IIED: International Institute for Environment and Development

IIEE: Ion-Induced Electron Emission

IIEF: Institut International d'Études Françaises

IIELR: International Institute for Electronic Library Research (UK)

IIEP: International Institute of Educational Planning

IIF: Image Information File • Information Item File • Institute of International Finance (USA-DC) • Instituto de Investigaciones Filosóficas (Mexico) • Intelligent Information Fusion

IIFS: Intelligent Information Fusion System

IIG: Indian Institute of Geomagetism • Interrange Instrumentation Group

IIGDT: Interactive Intelligent Image Graphics Display Terminal

IIGM: Indian Institute of Geomagnetism

IIGS: Initial Image Generating System

IIH: Instituto de Investigaciones Históricas (Mexico) • Integrated Interactive Home

III: Giant Star • Improved Initialization Inversion • Intelligent Information Interface

IIIa: Giant Star

IIIA: Internet Information Infrastructure Architecture

IIIab: Giant Star

IIIb: Giant Star

IIID: International Institute of Information Design

IIIT: Indian Institute of Information Technology (India)

IIJ: Instituto de Investigaciones Jurídicas (Mexico)

IIK: Imagery Interpretation Key

IIL: Induction Ion Laser • Institute of International Law • Integrated Injection Logic • Intermediate Inversion Layer

IILR: Inner Inner Lindblad Resonance

IiM: Integration in Manufacturing

IIM: Institute for Industrial Mathematics (Israel) • Institute for Information Management • Institution of Industrial Managers

IIMAS: Instituto de Investigaciones en Matemáticas Aplicadas y en Sistemas

IIMI: International Irrigation Management Institute (Sri Lanka)

IIMS: Imaging Ion Mass Spectrometer • Integrated Information Management System

IIN: Instituto Isaac Newton (Chile) • Integrated Information Network • Item Identification Number • Nishinoomote (Japan) (location identifier)

IINA: Istituto Italiano di Navigazioni - Atti (periodical code)

IInfSc: Institute of Information Scientists

IINREN: Interagency Interim National Research and Education Network

IIO: Information Item Only

IIOC: Intelligent Input/Output Controller

IIOE: International Indian Ocean Expedition

IIOP: Internet Inter-ORB Protocol (*) • Interoperable Inter-ORB Protocol

IIP: Institute for Image Processing • Institute of Intellectual Property • Institut International de la Presse • Instrument Information Profile • Instrument Interface Proposal • International Informatics Programme

IIPA: International Intellectual Property Alliance

IIPAH: International Institute for Problems of the Asteroid Hazard

IIPE: Investigation of the Ionospheric Plasma Escape

IIPH: ISO Instantaneous Pointing History

IIPL: Interactive Image Processing Laboratory

IIPS: Interactive Image Processing System

IIPTI: International Intellectual Property Training Institute

IIR: Imaging Infrared Sensor • Immediate Impulse Response • Infinite Impulse Response • Institute for International Research • Intelligent Information Retrieval • Intermediate Infrared

IIRG: Istituto Internazionale per le Ricerche Geotermiche (Italy)

IIRS: Indian Institute of Remote Sensing • Institute for Industrial research and Standards • Intelligent Information Retrieval System

IIRV: Improved Interrange Vector

IIRW: Intelligent Information Retrieval Workstation

IIS: Image Intensified System • Implant Ion Sensor • Information and Intelligent Systems • Inspection Item Sheet • Institute of Information Scientists • Institut International de la Soudure • Institut International de Statistique • Integrated Information System • Integrated Instrument Simulator • Intelligent Information System • Interactive Instructional System • Interface Identification Sheet • Internal Integral Starter • Internet Information Server • Internet Information Services • Ion Implantation Study • Nissan Island (Papua New Guinea) (location identifier)

IISA: Institut International des Sciences Administratives

IISc: Indian Institute of Science (India)

IIS/G: Internal Integral Starter/Generator

IISG: Interferometric Imaging Science Group • Interferometric Instrumentation Science Group • Internal Integral Starter/Generator

IISI: International Iron and Steel Institute

IISL: International Institute of Space Law (France)

IISME: Industry Initiatives for Science and Math Education

IISN: Institut Interuniversitaire des Sciences Nucléaires

IISP: Information Infrastructure Standards Panel • Interim Interface Switch Protocol • Interim Inter-Switch Protocol • Interim Inter-Switch Signalling Protocol

IISR: Institute of Ionospheric and Space Research

IISS: Image Inventory Search and Summary • Integrated Information Support System • International Institute for Strategic Studies

IIST: International Institute for Software Technology

IISU: ISRO Inertial System Unit

IIT: Illinois Institute of Technology (USA-IL) • Image Intensification Tube • Image Intensifier Tube • Indian Institute of Technology • Industrial Information Transfer • Institute for Information Technology • Integrated Information Technology • Israel Institute of Technology

IITA: Information Infrastructure Technologies and Applications (*) • Information Infrastructure Technology Applications • International Institute of Tropical Agriculture (Nigeria)

IITAP: International Institute of Theoretical and Applied Physics

IITCS: Igloo Internal Thermal Control Section

IITF: Information Infrastructure Task Force

IITM: Indian Institute of Tropical Meteorology

IITS: Igniter Initiator Test Set

IIU: Input Interface Unit • International Islamic University

IIV: Image Intensifier Viewer

IIVC: Institut National de l'Ingénierie de la Vision et du Contrôle (France)

IIVD: Image Intensifier Viewing Device

IIVRS: International Institute for Vital Registration and Statistics

IIW: International Institute of Welding

IIWPA: International Information/Word Processing Association

iJ: im Jahre

IJ: IJssel • Indemnité Journalière

IJAA: International Journal of Adhesion and Adhesives (periodical code)

IJACS: International Journal of Adaptive Control and Signal Processing (periodical code)

IJAEM: International Journal of Analytical and Experimental Model Analysis (periodical code)

IJAIS: International Journal of Accounting Information Systems

IJAP: International Journal of Aviation Psychology (periodical code)

IJBC: International Journal on Bifurcation and Chaos

IJC: International Joint Commission • International Journal of Control (periodical code)

IJCAI: International Joint Conference on Artificial Intelligence

IJCAII: International Joint Conferences on Artificial Intelligence Inc.

IJCM: International Journal of Computer Mathematics (periodical code)

IJCPR: International Joint Conference on Pattern Recognition

IJDM: International Journal of Damage Mechanics (periodical code)

IJE: International Journal of Electronics (periodical code) • Inverse Joule Effect

IJEFM: International Journal of Engineering and Fluid Mechanics (periodical code)

IJEM: International Journal of Electronic Markets

IJER: International Journal of Energy Research (periodical code)

IJES: International Journal of Engineering Science (periodical code)

IJET: International Journal of Educational Telecommunications (AACE)

IJFa: International Journal of Fatigue (periodical code)

IJFE: International Journal of Fusion Energy (periodical code)

IJFr: International Journal of Fracture (periodical code)

IJGIS: International Journal of Geographical Information Systems

IJHE: International Journal of Hydrogen Energy (periodical code)

IJHFF: International Journal of Heat and Fluid Flow (periodical code)

IJHM: International Journal of Hybrid Microelectronics (periodical code)

IJHMT: International Journal of Heat and Mass Transfer (periodical code)

IJICS: International Journal of Intelligent Control and Systems

IJIE: International Journal of Impact Engineering (periodical code)

IJIMW: International Journal of Infrared Millimeter Wave (periodical code)

IJIST: International Journal of Imaging Systems Technology (periodical code)

IJK: Izhevsk (Russia) (location identifier)

IJLT: IEEE Journal of Lightwave Technology (periodical code)

IJMeS: International Journal of Mechanical Sciences (periodical code)

IJMF: International Journal of Multiphase Flow (periodical code)

IJMHG: Indian Journal of Meteorology, Hydrology, and Geophysics (periodical code)

IJMM: International Journal of Mini and Microcomputers (periodical code)

IJMoS: International Journal of Modelling and Simulation (periodical code)

IJMSI: Intrnational Journal of Mass Spectrometry and Ion Processes (periodical code)

IJNA: IMA Journal on Numerical Analysis (periodical code)

IJNLM: International Journal of Non-Linear Mechanics (periodical code)

IJNME: International Journal for Numerical Methods in Engineering (periodical code)

IJNMF: International Journal for Numerical Methods in Fluids (periodical code)

IJOE: IEEE Journal of Oceanic Engineering (periodical code)

IJP: Ink Jet Printing • International Journal of Plasticity (periodical code)

IJQC: International Journal of Quantum Chemistry (periodical code)

IJQE: IEEE Journal of Quantum Electronics (periodical code)

IJRA: IEEE Journal of Robotics Automation (periodical code)

IJRAI: International Journal of Radiation Applications and Instrumentation (periodical code)

IJRR: International Journal of Robotics Research (periodical code)

IJRS: International Journal of Remote Sesing (periodical code)

IJRSP: Indian Journal of Radio and Space Physics (periodical code)

IJSA: International Journal of Supercomputer Applications (periodical code)

IJSAC: IEEE Journal on Selected Areas in Communications (periodical code)

IJSC: International Journal of Satellite Communications (periodical code)

IJScA: International Journal of Supercomputer Applications (periodical code)

IJSS: International Journal of Solids and Structures (periodical code)

IJSSC: IEEE Journal of Solid-State Circuits (periodical code)

IJTJE: International Journal of Turbo Jet Engines (periodical code)

IJTP: International Journal of Theoretical Physics (periodical code)

IJU: Ijuí - João Batista Bos Filho (Brazil-RS) (location identifier)

IJVTOP: International Jupiter Voyager Telescope Observations Program • International Jupiter Voyager Telescope Observing Program

IJW: International Jupiter Watch

IJX: Jacksonville (USA-IL) (location identifier)

ik: Inupiak (ISO language code)

IK: Image Kernel • Indemnité Kilométrique • Inner Keel • Institut für Kernphysik

IKAT: Interactive Keyboard And Terminal

IKB: International Klein Blue • Wilkesboro - Wilkes County (USA-NC) (location identifier)

IKBS: Intelligent Knowledge-Based System

IKE: Internet Key Exchange

IKES: Ion Kinetic Energy Spectrometry

IKET: Individual Knowledge Evaluation Test

IKF: Institut für Kosmoforschung

IKFIA: Institute of Cosmophysical Research and Aeronomy (Russia)

IKH: Instabilité de Kelvin-Helmholtz

IKI: Iki (Japan) (location identifier) • Institut Kosmitcheskij Isledovanja (Russia) • Space Research Institute (Russia)

IKIP: Institut Keguruan dan Ilmu Pendidikan (Indonesia)

IKK: Kankakee - Greater Kankakee (USA-IL) (location identifier)

IKL: Ikela (Zaire) (location identifier)

IKMP: Internet Key Management Protocol

IKO: Nikolski (USA-AK) (location identifier)

IKON: Interkerkelijke Omroep Nederland

IKP: Inkerman (Australia-QLD) (location identifier)

IKS: Image Kernel System • Integrated Key Set • Inverse Kinetics Simulator • Tiksi (Russia) (location identifier)

I&KT: Informations- & Kommunikationstechnologien

IKT: Informations- und Kommunikationstechnologien • Irkutsk (Russia) (location identifier)

IKU: Interface Keying Unit

IKV: Interkerkelijk Vredesberaad

IKZ: Internationales Kongresszentrum

il: Israel

Il: Illustrated • Illustration

I/L: Import Licence

IL: Illinois (USA) • Ilyushin • Imaging Library • Important Lady • Important Launch • Incoming Letter • Inertial Lab • Informative Language • Insertion Loss • Inside Leg • Instruction Leaflet • Instrumentation Laboratory • Instrument Landing • Interaction List • Interactive Lyndenmaier • Intermediary Letter • Intermediate Language • Intermediate Loop • Internal Letter • International Light (USA-MA) • Interpretive Language • Investigation Leader • Ionization Loss • Israel (international vehicle registration code) • Israel (ISO country code)

ILA: Illaga (Indonesia) (location identifier) • Induction Linear Accelerator • Initial Line Address • Instruction Look Ahead • Instrument Landing Approach • International Law Association • Internet Learning Agent

ILAC: International Laboratory Accreditation Committee • International Laboratory Accreditation Conference

ILAMS: Infrared Laser Atmospheric Monitoring System

ILANA: Inertial Land Navigation

ILAP: Integrated Local Area Planning • Interactive Lively Application Project

ILAS: Improved Limb Atmospheric Spectrometer • Infrared Limb Atmospheric Spectrometer • Instrument Landing Approach System

Ilat: Ilativo

ILB: Interactive Linear Browser

ILBM: Image Lithologic Boundary Manifestation

ILBS: Image Library and Browse System

ILC: Industrial Liaison Committee • Initial Launch Capability • Instruction Length Code • Integrated Launch Complex • Integrated Logic Circuit • Interim Library Catalog • Interim Load Cycle • Intermediate Load Cycle • International Linear Collider

ILCA: International Livestock Centre for Africa

ILCBT: Instructor-Led Computer-Based Training

ILCC: Initial Launch Capability Complex

ILCCS: Integrated Launch Control and Checkout System

ILCE: Integrated Logistics and Concurrent Engineering

ILCO: Instantaneous Launch Control Officer

iLCS: Integrated Laser Control System

ILCSO: Illinois Libraries Computer System Organization

ILD: Injection Laser Diode • Interaural Level Difference • International Lunar Dateline

ILDA: Intensified Linear Diode Array • International Laser Display Association (USA-FL)

ILDE: Initiative Locale de Développement et d'Emploi

ILDP: Instrument Library Developer Program

ILDS: Integrated Logistics Data System

Ile: Ilustre

ILE: Fort Hood (USA-TX) (location identifier) • Initiative Locale pour l'Emploi • Interactive Learning Environment • Interbase Local Engine • Killeen (USA-TX) (location identifier)

ILEC: Incumbent Local-Exchange Carrier • Independent Local-Exchange Carrier (*)

ILES: Indian Landsat Earth Station

ILEWG: International Lunar Exploration Working Group

ILF: Ilford (Canada-MB) (location identifier) • Intermediate Latitude Field • Internal Logical File

ILFC: International Lease Financial Corporation

ILFOSS: International Lunar Far-side Observatory and Science Station

ILG: Wilmington (USA-DE) (location identifier)

ILH: Illisheim (Germany) (location identifier)

ILHP: Integrated Liquid Hydrogen Plant

ILI: Iliamna (USA-AK) (location identifier) • Instant Lunar Ionosphere • Institute for Land Information • Instrumentation and Laboratory Improvement • Intelligent Link Interface • International Law Institute

ILIAD: Internet Library Information Assembly Database

ILIERS: Integrated Library Information Education and Retrieval System

ILIM: Institute for Land Information Management

ILISS: Integrated Logistics Information Support System

ILK: Ilaka (Madagascar) (location identifier)

ill: Illustrissimus

Ill: Illinois (USA) • Illustrate • Illustrated • Illustration • Illustrator • Illustré • Illustrissimus

ILL: Input Logic Level • Institut Max von Laue - Paul Langevin (France) • Interlibrary Loan • Willmar (USA-MN) (location identifier)

ILLIAC: Illinois Integrator and Automatic Computer

IlliGAL: Illinois Genetic Algorithms Laboratory (USA-IL)

ILLINET: Illinois Library and Information Network (USA-IL)

Illit: Illiterate

Ills: Illustrations

ILLS: Interlibrary Loan System

ILLSS: Identification List of Lines in Stellar Spectra

Illum: Illuminate • Illuminated

ILLUM: Illuminate

Illus: Illustrate • Illustrated • Illustration • Illustrator

Illust: Illustrate • Illustrated • Illustration • Illustrator

ILM: Immeuble à Loyer Moyen • Independent Landing Monitor • Integrated Layer Management Interface • Interface Logiciel-Matériel • Wilmington (USA-NC) (location identifier)

ILMF: Intermediate-Level Maintenance Facility

ILMI: Interim Local Management Interface

Ilma: Ilustrísima

Ilmo: Ilustrísimo

ILMP: Integrated Logistics Management Program

ILMR: International Laboratory of Marine Radioactivity

ILMS: Integrated Library Management System

ILN: Immeuble à Loyer Normal • Wilmington Clinton (USA-OH) (location identifier)

ILO: Iloilo (Philippines) (location identifier) • In Lieu Of • Integrated Logistics Operations • International Labour Office • International Labour Organization (*)

ILOC: International Lunar Occultation Centre

ILOM: International Latitude Observatory of Mizusawa (Japan)

ILP: Île des Pins (New Caledonia) (location identifier) • Incapacidad Laboral Permanente • Inductive Logic Programming • In-Line Printer • Instruction Level Parallelism • Integrated Logistics Panel • Intermediate Language Processor • International Lithosphere Program

ILPC: Intermediate Long-Period Comet

ILPF: Ideal Low-Pass Filter • Internet Law and Policy Forum

ILPO: International Lunar Planning Office

ILPS: International Logic Programming Symposium

ILQ: Ilo (Peru) (location identifier) • International Lunar Quinquennium

ILR: Ilorin (Nigeria) (location identifier) • Ingoing Long-wave Radiation • Inner Lindblad Resonance • Ingenieurbüro für Luft- und Raumfahrt

ILRAD: International Laboratory for Research on Animal Diseases

ILRI: Institute for Land Reclamation and Improvement (Netherlands)

ILRS: International Laser Ranging Service

ILRV: Integral Launch and Recovery Vehicle • Integral Launch and Reentry Vehicle • Integrated Launch and Recovery Vehicle • Integrated Launch and Reentry Vehicle

ILRVS: Integral Launch and Recovery Vehicle System • Integral Launch and Reentry Vehicle System • Integrated Launch and Recovery Vehicle System • Integrated Launch and Reentry Vehicle System

ILS: Illumination System • Instrument Landing System • Instrument Line of Sight • Integrated Light Spectrum • Integrated Linearization System • Integrated Logistics Support • Integrated Logistics System • Interactive Laboratory System • Interferometric Laser Source • International Latitude Service • International Launch Services • International Lunar Society • Internet Locator Server • Intracavity Laser Spectroscopy • Inverse Least Squares • Ionization Loss Spectroscopy • Iterative Least-Squares • Shekel (Israel) (ISO currency code)

ILSA: Improved Lightweight Solar Array • Integrated Logistics Support Analysis

ILSC: Integrated Logistics Support Concept

ILSCM: Integrated Logistics Support Control Manual

ILSCP: Integrated Logistics Support Concept Plan

ILSDF: Integrated Logistics Support Data File

ILSES: Integrated Library and Survey Extraction Service

ILSI: International Life Sciences Institute

ILSMT: Integrated Logistics Support Management Team

ILSP: Institute for Language and Speech Processing (Greece) • Integrated Logistics Support Plan

ILSS: Integrated Life Support System

ILSSE: Integrated Life Science Shuttle Experiment

ILST: Indonesian Low-Speed Tunnel

ILSW: Interrupt Level Status Word

ILSWG: Integrated Logistics Support Working Group

ILT: Imprecisely Located Target • Incapacidad Laboral Transitoria • Installation Lead Time • Institute for Learning Technologies • Institut für Lasertechnik (Germany)

ILTF: Internet Law Task Force

Iltma: Ilustrísima

Iltmo: Ilustrísimo

ILTP: Integrated Long-Term Programme

Iltre: Ilustre

ILU: Independent Logical Unit • Interlanguage Unification • Kilaguni (Kenya) (location identifier)

ILUG: Innovative Low Users Group

ILV: Industrial Launch Vehicle • Internationaler Leihverkehr

ILW: Immediate-Level Waste • Intermediate-Level Waste (*) • International Leonid Watch

ILWAS: Integrated Lake-Watershed Acidification Study

ILWIS: Integrated Land and Water Information System • Integrated Land and Watershed Information System

ILX: Ileg (Papua New Guinea) (location identifier)

ILY: I Love You • Islay (UK) (location identifier)

ILZ: Zilina (Slovakia) (location identifier)

iM: im Mittel

I&M: Inspection & Maintenance • Installation & Maintenance

IM: Imaging Mode • Impulse Modulation • Information Memory • In Maintenance • Industrial Manager • Informal Memo • Information Management • Information Manager • Information Modelling • Infrastructure Module • Ingénieur de la Météorologie • Inner Marker • Inspection Manual • Inspection Memorandum • Instruction Management • Instruction Manual • Instrumentation Manager • Instrumentation and Measurement • Integration Management • Intensity Modulation • Interface Module • Interim Memo • Intermediate Maintenance • Intermediate Model • Intermediate Modulation • Intermodulation • Internal Memo • Internal Message • Intra Muros • Intra-Muscular • Inventory Management • Inverted Microscope • Inverted Microscopy • Irregular Marking • Isle of Man • Item Management • Item Mark

IMA: Iamalele (Papua New Guinea) (location identifier) • Ice Motion Algorithm • Impedance Matching Attenuator • Information Management and Archives • Information Médicale Automatisée • Institute for Mathematics and its Applications • Institute of Management Accountants • Institute of Mathematics and its Applications • Integrated Modular Avionics • Interactive Measurement and Analysis • Interactive Multimedia Association • Intermediate Maintenance Activity • International Mineralogical Association • Inverse Multiplexing for ATM • Ion Microprobe Analyzer

IMAC: Integrated Microwave Amplifier Converter • Isochronous Media Access Control

IMACS: Inamori Magellan Areal Camera and Spectrograph • Integrated Manufacturing and Accounting System • Integrated Manufacturing Control System • International association for Mathematics And Computers in Simulation

IMADS: Information Management Assimilation and Delivery System

Imag: Imaginary • Imagination • Imagine

IMAG: Information Management Advisory Group • Informatique et Mathématiques Appliquées de Grenoble (France)

IMAGE: Imager for Magnetopause-to-Aurora Global Exploration • Integrated Mapping And Geographic Encoding • Integrated Model for the Assessment of the Greenhouse Effect • Integrated Model to Assess the Greenhouse Effect • Investigative Materials About Global Environment

IMAGES: Integrated Mapping And Geographic Encoding System • International Marine Global Change Study

IMAGINE: Integrated Multiagent Interactive Environment

IMALS: Integrated Maintenance of Assets and Logistics Support

IMAO: Ingénierie Mécanique Assistée par Ordinateur

IMAP: Imager for the Magnetosphere, Aurora and Plasmasphere • Indian Middle Atmosphere Programme • Input-Management-Analysis-Presentation • Interactive Mail Access Protocol • Internet Message Access Protocol (*) • Internet Messaging Access Protocol

IMAPS: Interactive Microcomputer Array Processing System • Interstellar Medium Absorption Profile Spectrograph

IMAR: Institute for Manufacturing and Automation Research

IMAS: Institutional Management Accounting System • Integrated Mass Announcement System • Intelligent Mobile Autonomous System

IMASP: Instantiable MASP • Instantiated MASP

IMASS: Intelligent Multiple Access Spectrum Sharing

IMAU: Instituut voor Marien en Atmosferisch Onderzoek Utrecht (Netherlands)

IMAX: Image Maximum • Isotope Matter Antimatter Experiment

IMB: Imbaimadai (Guyana) (location identifier) • Indice Moyen Budgétaire • Information Management Branch • Input Memory Buffer • Irvine-Michigan-Brookhaven

IMBLMS: Integrated Medical Behavioral Laboratory Measurement System

IMC: Ice Mass Content • Image Motion Compensation • Image Motion Compensator • Image Motion Configuration • Information Management Center • Information Management Conference • Information Modelling by Composition • Institute of Management Consultants • Institute of Measurement and Control • Instrument Meteorological Conditions • Integrated Maintenance Chart • Integrated Maintenance Concept • Integrated Microelectronic Circuitry • Integrated Microwave Circuit • Integrated Monolithic Circuit • International Machine Consortium • International Maintenance Control • International Management Council • International Maritime Committee • International Meteor Conference • International Meteorological Centre • International Meteorological Committee • Internet Microcode Load

IMCC: Image Motion Compensation and Calibration • Integrated Mission Control Center • Item Management Control Code

IMCCE: Institut de Mécanique Céleste et de Calcul des Éphémérides (France)

IMCD: Input Marginal Checking and Distribution

IMCE: Inter-Ministerial Committee for Environment

IMCO: Intergovernmental Maritime Consultative Organization (now IMO)

IMCoS: International Map Collectors Society

IMCP: Integrated Monitor and Control Panel

IMCR: Integrated Master Control Record

IMCS: Integrated Machinery Control System

IMD: Imonda (Papua New Guinea) (location identifier) • India Meteorological Department • Information Media Department • Institute for Management Development (*) • Institute for Managing Development • Intermediate • Intermittent Motion Driver • Intermodulation Distortion

IMDA: International Map Dealers Association

IMDACS: Integrated Multivariate Data Analysis and Classification System

IMDAS: Interactive Multispectral Data Analysis System

IMDB: Instruction Management Database • Integrated Maintenance Database

IMDC: Internal Message Distribution Center

IMDG: International Maritime Dangerous Goods

IMDPS: INSAT Meteorological Data Processing System

IMDS: Image Data Stream • Integrated Monitoring and Diagnostics System • International MD Service

IMDT: Intermediate

IME: Indirect Manufacturing Expense • Information Management and Engineering • Input Method Editor • Institution of Mechanical Engineers • Institut Monétaire Européen • International Magnetospheric Explorer • Interplanetary Meteoroid Experiment

IMEC: Interuniversitair Micro-Electronica Centrum • Interuniversity Micro-Electronics Centre

IMechE: Institute of Mechanical Engineers

IMEKO: International Measurement Confederation

IMEO: Initial Mass in Earth Orbit

IMEP: Indicated Mean Effective Pressure • Institution of Mechanical Engineers - Proceedings (periodical code) • International Measurement Evaluation Programme

IMERSE: Indonesian Marine Environment Remote Sensing Experiment

IMeRv: International Metals Review (periodical code)

IMEWG: International Mars Exploration Working Group

IMEX: Inner Magnetosphere Explorer

IMF: Ice Mass Flux • Image Furnace • Imphal (India) (location identifier) • Individual Master File • Initial Mass Function • Instituteur-Maître-Formateur • Intense Magnetic Field • Intermediate Maintenance Facility • Intermediate Mass Fragment • Internal Magnetic Focus • International Monetary Fund • Interplanetary Magnetic Field • Interstellar Magnetic Field • Inventory Master File

IMFK: Integrated Multifunction Key

IMFL: Institut de Mécanique des Fluides de Lille (France)

IMFM: Institut de Mécanique des Fluides de Marseille (France)

IMFP: Inelastic Mean Free Path • Interaction Mean Free Path

IMFT: Institut de Mécanique des Fluides de Toulouse (France) • Interferometric Moving Target Focusing

IMG: Image • Information Management Group • Inhaminga (Mozambique) (location identifier) • Institut de Mécanique de Grenoble (France) • Institut für Meteorologie und Geophysik • Inverted Magnetron Gauge

IMGC: Istituto di Metrologia G. Colonnetti (Italy)

IMGG: Institute of Meteorology and Geophysics of Graz (Austria)

IMGWL: IEEE Microwave and Guided Wave Letters (periodical code)

IMH: Institute of Meteorology and Hydrology

IMHE: Institutional Management of Higher Education

IMHO: In My Humble Opinion

IMI: Ine Island (Marshall Islands) (location identifier) • Inner Magnetosphere Imager • Inner Magnetosphere Imaging • Installation and Maintenance Instruction • Interactive Menu Interface • Internal Multilayer Insulation • Internal Multiscreen Insulator • International Meteorological Institute (Sweden) • Ion Mass Imager

IMIC: Integrated Management Information Center • Integrated Management Information Computer

IMICS: Integrated Management Information Control System

IMINT: Imagery Intelligence

IMIO: Image Input/Output

IMIP: Industrial Modernization Incentive Program

IMIS: Integrated Maintenance Information System • Integrated Management Information System • Ion Mass Imaging Spectrograph

Imit: Imitate • Imitation • Imitative

IMK: Institut für Meteorologie und Klimaforschung • Simikot (Nepal) (location identifier)

IML: Imperial (USA-NE) (location identifier) • Initial Measurement List • Inner Mold Line • Inside Mold Line • Instrument Markup Language • Interactive Maintenance Language • Intermediate Maintenance Level • International Microgravity Laboratory

IMLI: International Maritime Law Institute

IMLR: Iron Mass to Light Ratio

Imm: Immeuble

IMM: Immokalee (USA-FL) (location identifier) • Independent Manned Manipulator • Input Message Manual • Institute of Molecular Medicine • Integrated Maintenance

Management • Integrated Maintenance Manual • Intelligent Memory Manager • Interactive Multimedia • International Mars Mission • International Monetary Market • International Money Market

IMMA: Ion Microprobe Mass Analyzer

IMME: Isobaric Multiplet Mass Equation

Immed: Immediate

IMMPS: Intelligent Multimedia Presentation System

Immun: Immunity • Immunization

IMN: Imane (Papua New Guinea) (location identifier)

IMNHO: In My Not (so) Humble Opinion

IMNSHO: In My Not So Humble Opinion

IMO: Information Market Observatory • Initiatives de Microprocesseurs Ouverts • In My Opinion • Integration Management Office • Interband Magneto-Optic • Interferometer Module • Internationale Meteorologische Organisatie • International MARC Office • International Maritime Organization • International Mathematical Olympiad • International Meteorological Organization (now WMO) • International Meteor Organization • Zemio (Central African Rep.) (location identifier)

IMOP: Instruments and Methods of Observation Programme

IMOS: Integrated Modelling of Optical Systems

Imp: Impact • Impedance • Imperative • Imperator • Imperatrix • Imperfect • Imperial • Impersonal • Implement • Implementation • Implement • Implemented • Important • Import • Important • Imported • Importer • Imprenta • Impression • Imprimatur • Imprimé • Imprimeur • Imprint • Improper • Improve • Improved • Improvement • Impulse

IMP: Ice Mass Path • Imager for Mars Pathfinder • Impasse • Imperatriz (Brazil-MA) (location identifier) • Imperial College of Science, Technology and Medicine (Starlink internal code) • Importance • Important • Improvement Maintenance Program • Improving Management Performance • Indeterminate Mass Particle • Inflatable Micrometeoroid Paraglider • Information Management Plan • Information Message Processor • Initial Memory Protection • Input Message Processor • Integrated Maintenance Plan • Integrated Maintenance Procedure • Integrated Master Plan • Integrated Memory Processor • Integrated Monitoring Panel • Integrated Multiprotocol Processor • Interface Message Processor • Interim Monitoring Program • Internal Macro-instruction Procedure • International Mars Patrol • International Match Point • Interplanetary and Magnetospheric Physics • Interplanetary Magnetometer Probe • Interplanetary Measurement Probe • Interplanetary Monitoring Platform • Interplanetary Monitoring Probe • Interprocess Message Parsing • Inventory Management Plan

IMPA: Intelligent Multiport Adapter • International Master Printers' Association

IMPAC: International Merchant Purchase Authorization Card

IMPACT: Illinois Microarchitecture Project utilizing Advanced Computer Technology • Implementation, Planning, And Control Technique • Information Market Action Programme • Innovative Microelectronic Pixellated and Advanced CCD Technology • Integrated Manufacturing, Planning and Control Technology • Integrated Material Productivity Action • International Monitoring Programs for Asteroid and Comet Threat • Inventory Management Program And Control Technique • Investigations of Magnetospheric Particle Acceleration and Turbulence

IMPAE: Institute of Meteorology and Physics of the Atmospheric Environment

IMPATT: Impact Ionization Avalanche Transit Point

Impce: Importance

Impd: Impedance

IMPD: Impedance

Impér: Impératif

Imper: Imperative • Imperativo

Impérat: Impératif

Imperat: Imperative • Imperativo

Imperf: Imperfect

Impers: Impersonal

Impf: Imperfect

Impft: Imperfect

IMPI: Internal Microprogrammed Interface

Impl: Implement • Implementation • Implemented

IMPLS: Impulse

IMPO: Industrial Maintenance and Plant Operation

IMPOL: Imaging Polarimeter

Imposs: Impossible

IMPOV: In My Point Of View

IMPP: Imaging Photometer and Polarimeter

IMPPACT: Integrated Modelling of Products and Processes using Advanced Computer Technologies

Impr: Imprenta • Impreso

IMPR: Improving

IMPReSS: Image Perimeters of Sky Surveys

IMPRIMATUR: Intellectual Multimedia Property Rights Model And Terminology for Universal Reference

IMPRINT: Imagery Prototype Initiative

Imprl: Imperial

Improv: Improvement • Improvisation

Imprsn: Impression

IMPS: Institut de Médecine et Physiologie Spatiales • Integrated Master Programming and Scheduling • Interaction Measurement Payloads • IRAS Minor Planet Survey

Impt: Important

IMPT: Important

Imptr: Importer

Impv: Imperative

IMR: Imaging Microwave Radiometer • Indice Moyen Réel • Individual Medical Report • Informal Memorandum Report • Insect Monitoring Radar • Internet Monthly Report

IMRADS: Information Management, Retrieval And Dissemination System

IMRB: Interim Mission Reconfiguration Nase

IMRC: Imaging Media Research Center (KITST) • International Mission Review Committee

IMRD: Initial Mass Ratio Distribution

IMRMS: Ion Molecule Reaction Mass Spectrometry

IMRO: Inspection, Modification, and Repair Order

IMRR: Institute for Reference Materials and Measurements

IMRT: Indo-Mauritian Radiotelescope • Intensity-Modulated Radiation Therapy

IMRv: International Materials Review (periodical code)

IMS: Image Manipulation System • Image Motion Simulator • Imaging Spectrometer • Indirect Measuring System • Industrial Mathematics Society (USA-MI) • Industrial-Medical-Scientific • Inertial Measuring Set • In-flight Management System • Information Management Service • Information Management System • Information Manufacturing System • Institute for Molecular Science • Institute of Management Services • Institute of Mathematical Statistics • Institute of Museum Services • Institut Suisse de Météorologie (Switzerland) • Integrated Maintenance System • Integrated Mapping System • Intelligent Manufacturing System • Interim Meteorological Satellite

• Interlock Management System • Intermediate Maintenance Standard • Intermediate Map Scale • Intermediate-Mass Star • International Magnetosphere Study • International Magnetospheric Study (*) • International Metallographic Society • International Metric System • Interoperability in Multidatabase Systems (workshop) • Interplanetary Mission Support • Inventory Management System • Ion Mass Spectrometer • Ion Mobility Spectrometer • Ion Mobility Spectrometry • Ion Mobility Spectroscopy • Ionization and Momentum Sensor • Irradiance Measuring System

IMSA: Illinois Mathematics and Science Academy (USA-IL)

IMSAT: Imaging Satellite

IMSc: Institute of Mathematical Sciences

IMSC: Integrated Media Systems Center • International Mobile Satellite Conference

IMSCIE: International Magnetosphere Study Central Information Exchange

IMSD: Information Management and Services Division • Integrated Mission for Sustainable Development

IMSE: Integrated Mean Square Error • Integrated Methods in Science and Engineering (conference) • Integrated Modelling Support Environment

IMSI: Information Management System Interface • International Mobile Subscriber Identity

IMSL: International Mathematical and Statistical Library • International Mathematical Subroutine Library

IMSP: Institut de Mathématiques et de Sciences Physiques • Intense Morning Sector Precipitation • Internet Message Support Protocol

IMSR: In-flight Medical Support System • Interplanetary Mission Support Requirement

IMSS: In-flight Medical Supply System • Istituto e Museo di Storia della Scienza

IMST: Integrated Mission Simulation Test

IMT: Immediately • Inductive Modelling Technology • Industrial and Materials Technologies • Industrial Matters and Technology • Information Management Technology • Instantiated Mission Timeline • Interactive Modelling Technology • International Mobile Telecommunications (I-TU) • Internet Media Type • Iron Mountain (USA-MI) (location identifier)

IMTA: Intensive Military Training Area • International Map Trade Association

IMTC: International Multimedia Teleconferencing Consortium

IMTDS: Introduction to the Modern Theory of Dynamical Systems

IMTF: Integrated Manufacturing Test Facility

IMTS: Improved Mobile Telephone Service • Improved Mobile Telephone System

IMTV: Interactive Multimedia Television

IMU: Impedance Matching Unit • Income Maintenance Unit • Inertial Measurement Unit • Inertial Measuring Unit • Interface Master Unit • Internal Measurement Unit • International Mathematical Union

IMUIMG: ISDN Memorandum of Understanding Implementation Management Group

IMUL: Integer Multiply

IMUPLM: Inertial Measurement Unit Prelaunch Monitor

IMUX: Inverse Multiplexing

IMV: Intermodule Ventilation

IMW: Instituut voor Milieu Wetenschappen • Intelligent Machining Workstation • Intelligent Manufacturing Workstation • Intermediate Molecular Weight • International Map of the World

IMX: International Monetary Exchange

IMXB: Intermediate-Mass X-ray Binary

IMZ: Nimroz (Afghanistan) (location identifier)

in: Inch • India • Indonesian (ISO language code) • Intortus

In: India • Indian • Indium • Inertia • Inertial • Inlet • Inner

IN: Ice Nucleus • Inca • Inde (ISO country code) • India (ISO country code) • Indiana (USA) • Individual Network • Inexact Newton (algorithm) • Information Need • Inlet • Input • Institut National • Instrumentation Notice • Instrument Note • Intelligent Network • Intelligent Networking • Internal Note • International Notice

INA: Information National Association • Information Networking Alliance • Information Networking Architecture • Institut National Agronomique (France) • Institut National de l'Audiovisuel (France) • Inta (Russia) (location identifier) • Interessengemeinschaft Niederrheinischer Astronomievereine (Germany) • International Normal Atmosphere • Iron Nickel Alloy

INAA: Instrumental Neutron Activation Analysis • Integrated Neutron Activation Analysis

INADS: Initialization and Administration System

INAF: International Name Authority File

INAG: Institut National d'Astronomie et de Géophysique (France) (now INSU)

INALCO: Institut National des Langues et Civilisations Orientales

INAMHI: Instituto Nacional de Meteorología e Hidrología (Ecuador)

INAOE: Instituto Nacional de Astrofísica, Optica y Electrónica (Mexico)

INAP: Integrated Neutron Activation Prediction • Intelligent Network Application Part

INAPI: Institut Algérien de Normalisation et de Propriété Industrielle (Algeria)

InARP: Inverse Address Resolution protocol

INAS: Indexing and Abstracting Service • Institutional Need Analysis System • International Near-Earth Asteroid Survey

InASP: Indian Academy of Sciences Proceedings (periodical code)

INASP: International Network for the Availability of Scientific Publications

InAsSb: Indium-Arsenide-Antimonide

Inaug: Inaugurate • Inaugurated

INB: Independence (Belize) (location identifier) • Internal Network Bus

Inbd: Inboard • Inbound

INBD: Inbound

Inbegr: Inbegriffen

INBIT: Input Bit

inc: Incisus • Incus

Inc: Include • Included • Including • Inclusive • Income • Incomplete • Incorporate • Incorporated • Incorporation • Income • Incoming • Incomplete • Inconclusive • Increase • Increased • Increasing • Increment • Incumbent

INC: In Cloud • Increment • Inertial Navigation Computer • In Nomine Christi • Installation Notice Card • Installation Notification Certificate • Institut National de Cartographie • Intergovernmental Negotiating Committee • International Carrier • International Negotiating Committee • Intranuclear Cascade • Yinchuan (China-PRC) (location identifier)

INCA: Indian National Cartographic Association • Input Catalogue • Intelligence Communications Architecture • Ion and Neutral Camera (Cassini)

Incand: Incandescent

INCB: International Narcotics Control Board

Ince: Insurance

InCh: Inorganic Chemistry (periodical code)

INCH: Intelligent Chart

Incid: Incidental

Incin: Incinerator

Incl: Inclination • Incline • Inclined • Include • Included • Including • Inclusion • Inclusive • Incoming • Incomplete • Inconclusive

INCL: Inclination

INCO: Instrumentation Communications Officer • Integrated Communications Officer

Incoat: Incoativo

Incog: Incognito

Incoh: Incoherent

Incompl: Incomplete

Incor: Incorporated • Incorporation • Incorrect

Incorp: Incorporated • Incorporation

Incorr: Incorrect

INCOSPAR: Indian National Committee for Space Research

Incr: Increase • Increased • Increasing • Increment

INCR: Increase

Incrt: Increment

INCS: Institut National de Cinématographie Scientifique

in d.: In Dies

ind: In Nomine Dei

Ind: Independence • Independent • Index • India • Indian • Indiana (USA) • Indicate • Indicated • Indication • Indicative • Indicator • Indices • Indies • Indigo • Indirect • Indirectly • Individual • Indonesia • Indonesian • Inductance • Induction • Indus-Indi (constellation) • Industrial • Industrie • Industriel • Industry

IND: India (international vehicle registration code) • Inde (ISO country code) • India (ISO country code) • Indianapolis (USA-IN) (location identifier) • Indústria • In Nomine Dei • Investigative New Drug

INDAC: Inductor Accelerator

INDAE: Instituto Nacional de Derecho Aeronáutico y Espacial (Argentina)

INDAT: Incoming Data

INDB: Integrated Network Database

INDC: Indicate • International Nuclear Data Committee (IAEA)

IndE: Industrial Engineer

INDECOPI: Instituto Nacional de Defensa de la Competencia y de la Protección de la Propiedad Intelectual (Peru)

Indéf: Indéfini

Indef: Indefinido • Indefinite

INDEF: Indefinite

INDEFLY: Indefinitely

Indep: Independent

INDES: Integrated Nuclear Data Evaluation System

Indét: Indéterminé

Indet: Indeterminado

INDFT: Indefinite

INDI: Ion-Neutral Dynamics Investigation

Indic: Indicating • Indication • Indicative • Indicativo • Indicator

Indiv: Individual

Individ: Individual

Indl: Industrial

Indn: Induction

INDO: Intermediate Neglect of Differential Overlap

Indo-Eur: Indo-European

INDOEX: Indian Ocean Experiment

Indo-Ger: Indo-German • Indo-Germanic

Indon: Indonesia

INDOTEC: Instituto Dominicano de Tecnologia Industrial (Dominican Rep.)

Indt: Indent

Indtry: Industry

Induc: Induction

Indum: Indumentaria

Indus: Industrial • Industrious • Industry

Indust: Industrial • Industrious • Industry

Indv: Individual

INE: Chinde (Mozambique) (location identifier) • Inertial Navigation Equipment • Instituto Nacional de Estatística

INEAS: International Near-Earth Asteroid Survey

Ined: Ineditus

INED: Institut National d'Études Démographiques (France) • International Network for Educational (information)

INEGI: Instituto de Engenharia Mecânica e Gestão Industrial (Portugal) • Instituto Nacional de Estatistica, Geografía e Informatica (Mexico)

INEL: Idaho National Engineering Laboratory (USA-ID) • International Network of Emerging Libraries

INEM: Instituto Nacional de Empleo (Spain)

INEMET: Instituto Nacional de Meteorologia (Brazil-DF)

INEN: Instituto Ecuatoriano de Normalización (Ecuador)

InEPS: Indian Earth and Planetary Sciences (periodical code)

INER: Institute of Nuclear Energy Research

INES: International Network of Engineers and Scientists • International Nuclear Event Scale • IUE Newly Extracted Spectra

INESC: Instituto Nacional de Engenharia de Sistemas e Computadores (Portugal)

INET: Institute of Nuclear Energy Technology

INETER: Instituto Nicaraguense de Estudios Territoriales (Nicaragua)

INEWS: Integrated Electronic Warfare System

in ex.: In Extenso

in f.: In Fine

inf: Infinum • Infunde

Inf: Inferior • Infinite • Infinitive • Infinity • Infinum • Infirmier • Infirmière • Influence • Inform • Informal • Information • Informed • Infra

INF: Information • In Guezzam (Algeria) (location identifier)

INFA: Infanta

InfCo: Information Control (periodical code)

INFCO: Information Committee

InfD: Information Display (periodical code)

INFE: Infante

Infin: Infinitive

Infinit: Infinitivo

INFIS: International Federation of Information Services

INFJ: Introverted Intuitive Feeling Judging

Infl: Inflammable • Inflated • Influence • Influenced • Influencia • Influido

Infm: Information

INFM: Istituto Nazionale di Fisica della Materia (Italy)

INFN: Istituto Nazionale di Fisica Nucleare (Italy)

INFNET: Instituto Nazionale di Fisica Nucleare Network (Italy)

Info: Information

INFO: Information

In-Fo: In-Folio

InFo: In-Folio

INFO: Information Network and File Organization • Information Network For Operations • Integrated Network Fiber Optics

INFOL: Information-Oriented Language

Inform: Informática • Information

INFORMS: Institute For Operations Research and the Management Sciences (merging of ORSA and TIMS) • Institute For Operations Research and the Management Society

INFOSEC: Information Security

INFP: Introverted Intuitive Feeling Perceiving

InfPh: Infrared Physics (periodical code)

INFSWG: Information Services Working Group

INFU: Institut für Umweltforschung (Germany)

InFuJ: Institute of Fuel Journal (periodical code)

Ing: Ingeniero • Ingenieur • Ingénieur • Inglaterra • Inglés

ING: Inertial Navigation and Guidance • Inertial Navigation Gyro • Intense Neutron Generator • Isaac Newton Group (Canary Islands) • Istituto Nazionale di Geofisica (Italy) • Lago Argentino (Argentina-SC) (location identifier)

InGaAs: Indium Gallium Arsenide

IngAr: Ingenieur Archiv (periodical code)

INGE: Intelligent Generator

Ingeg: Ingegneria (periodical code)

Ingen: Ingeniería

InGeo: International Geophysics (periodical code)

Ingl: Inghilterra

INGO: International Non-Governmental Organization

INGRES: Institut für Gewerblichen Rechtschutz (Switzerland) • Interactive Graphics and Retrieval System

Inh: Inhaber • Inhabitant • Inhalt

INH: Inhambane (Mozambique) (location identifier)

Inhab: Inhabitant

Inher: Inheritance

INI: Information Network Interface • Initialize • In Nomine Iesu • Nis (Yugoslavia-Serbia) (location identifier)

INIB: Internationale Normeninformationsbank

INIC: Instituto Nacional de Investigação Científica (Portugal)

INIS: International Nuclear Information System (IAEA)

InISJ: Indian Institute of Science Journal (periodical code)

INISSE: International Institute of Space Sciences and Electronics

INIST: Institut National de l'Information Scientifique et Technique (France)

Init: Initial • Initialization • Initialize • Initially • Initio

INIT: Initialization • Initialize

INITCO: Initialization and Checkout

Inj: Injection • Injury

INJ: Injune (Australia-QLD) (location identifier) • In Nomine Jesu

InJAE: Indian Journal of Aerospace Engineering (periodical code)

inject: Injectio

Inject: Injection

InJET: Indian Journal of Electronics Telecommunication (periodical code)

InJME: Indoan Journal of Mechanical Engineering (periodical code)

InJP: Indian Journal of Physics (periodical code)

InJRSP: Indian Journal of Radio and Space Physics (periodical code)

InJTP: Indian Journal of Theoretical Physics (periodical code)

INK: Information-Networks-Knowledge • Wink (USA-TX) (location identifier)

inkl: Inklusive

Inl: Inlet

INL: Integral Non-Linearity • International Falls (USA-MN) (location identifier) • Internode Link

INLD: Inland

in lim.: In Limine

in loc.: In Loco

in loc. cit.: In Loco Citato

INLR: Item No Longer Required

INLT: Inlet

INM: Innamincka (Australia-SA) (location identifier) • Institut National de Métrologie • Instituto Nacional de Meteorología (Spain) • Integrated Network Management • International Nautical Mile

INMARSAT: International Maritime Satellite Organization

in mem.: In Memoriam

INMETRO: Instituto Nacional de Metrologica, Normalização e Qualidade Industrial

INMG: Instituto Nacional de Meteorologia e Geofísica (Portugal)

INMH: Instituto Nacional de Meteorología e Hydrología (Ecuador)

INMS: Ion and Neutral Mass Spectrometer

Inn: Inning

INN: Innsbruck Kranebitten (Austria) (location identifier) • Instituto Nacional de Normalización (Chile) • International Negotiation Network • Internode Network

INNERTAP: Information Network on New and Renewable Energy Technologies for Asia and the Pacific

INNORPI: Institut National de la Normalisation et de la Propriété Industrielle (Tunisia)

INNS: International Neural Network Society

INO: Inflow - No Outflow • Inongo (Zaire) (location identifier) • Iraqi National Observatory (Iraq) • Istituto Nazionale di Ottica (Italy) • Iterative Natural Orbital

Inop: Inoperative

INOPV: Inoperative

Inorg: Inorganic

INOSHAC: Indian Ocean and Southern Hemisphere Analysis Centre

Inp: Input

InP: Indium Phosphide

INP: If Not Possible • Initial Program Load • Institute of Nuclear Physics • Institut National Polytechnique • Interdepartemental Natuurbeleids Plan • Intranode Policing

INPA: Institute for Nuclear and Particle Astrophysics

INPAC: Institute for Nuclear and Particle Astrophysics and Cosmology (USA)

INPACK: Input Acknowledge

INPADOC: International Patent Documentation Center

INPC: International Nuclear Physics Conference

INPE: Instituto Nacional de Pesquisas Espaciais (Brazil)

INPG: Institut National Polytechnique de Grenoble (France)

INPI: Institut National de la Propriété Industrielle (France) (*) • Institut National de la Propriété Intellectuelle (France) • Ion and Neutral Particle Imager

INPL: Institut de Physique Nucléaire de Lyon (France) • Institut National Polytechnique de Lorraine (France) • Israel National Physical Laboratory

INPO: Institute of Nuclear Power Operations

in pr.: In Principio

INPRES: Instituto Nacional de Prevención Sísmica (Argentina)

In Pro: In Proportion

INPS: Institut National Pour la Simplification

INPT: Institut National Polytechnique de Toulouse (France)

INPUT: Induced Pulse Transient

Inq: Inquiry • Inquisition

INQ: Inisheer (Ireland) (location identifier)

INQUA: International Union for Quaternary Research

Inr: Inner

INR: Impôt de Non-Résident • Inference-to-Noise Ratio • Rupee (India) (ISO currency code) • Sault Sainte Marie - Kincheloe (USA-MI) (location identifier)

INRA: Institute for Natural Resources in Africa • Institut National de Recherche Agronomique (France)

INRE: Institute of Natural Resources and Environment

INREQ: Information Request

INRENA: Instituto Nacional de Recursos Naturales (Peru)

INRETS: Institut National de Recherche sur les Transports et leur Sécurité (France)

INRIA: Institut National de Recherche en Informatique et en Automatique (France)

INRP: Institut National de Recherche Pédagogique (France)

INRPVC: Institut National de Recherche sur la Prévention du Vieillissement Cérébral (France)

INRS: Institut National de la Recherche Scientifique

in s.: In Situ

ins: Inches

Ins: Inscribe • Inscribed • Inscription • Insert • Inserted • Insertion • Inserto • Inside • Inspector • Insular • Insulate • Insulated • Insulation • Insurance • Insure

INS: Immigration and Naturalization Service (USA) • Improved Navigational Satellite • Indian Springs (USA-NV) (location identifier) • Inelastic Neutron Scattering • Inertial Navigational System • Inertial Navigation Sensor • Inertial Navigation System (*) • Input String • Insulation • Integrated Navigation System • Integrated Network Server • International Navigation System • International Network Service • Ion-Neutralization Spectroscopy • Isolated Neutron Star

INSA: Indian National Science Academy • Institut National des Sciences Appliquées (France)

INSAG: International Nuclear Safety Advisory Group

INSAP: Indian National Science Academy Proceedings (periodical code) • Inspiration of Astronomical Phenomena

InSAR: Interferometry using Synthetic Aperture Radar

INSAR: Interferometric Synthetic Aperture Radar

INSAT: Indian National Satellite

InSb: Indium Antimonide

Insc: Inscribe • Inscribed • Inscripción • Inscription

Insce: Insurance

Inscr: Inscribe • Inscription

INSDOC: Indian National Scientific Documentation Centre

INSEE: Institut National de la Statistique et des Études Économiques (France)

Insep: Inseparable

Insép: Inséparable

INSERM: Institut National de la Santé et de la Recherche Médicale (France) (*) • Institut National Supérieur des Études et de la Recherche Médicale (France)

INSIDE: Interior Structure and Internal Dynamical Evolution

InSIGHT: Investigative Stimuli for Intuitive Growth using High Technology

INSITE: Institutional Space Inventory Technique • Integrated Sensor Interpretation Technique • Interactive Nonlinear Systems Investigative Toolkit for Everyone

INSIVUMEH: Instituto Nacional de Sismología, Vulcanología, Meteorología e Hidrología (Guatemala)

INSM: Integrated Network and System Management

Insol: Insoluble

Insolv: Insolvent

Insp: Inspect • Inspected • Inspection • Inspector

INSP: Integrated National Space Programme

INSPEC: Information Services for the Physics and Engineering Communities

INSPIRE: Integrated Spatial Information Research and Engineering • Interactive NASA Space Physics Ionosphere Radio Experiment

Inst: Install • Installation • Installed • Instance • Instant • Instantaneous • Institución • Institut • Institute • Institution • Institutional • Institutionnel • Instituto • Instruct • Instruction • Instructional • Instructor • Instrument • Instrumental • Instrumentation • Instrumented

INSTA: Internordisk Standardisering

INSTAAR: Institute for Arctic and Alpine Research

INSTAR: Inertialess Scanning, Tracking And Ranging

INSTARS: Information Storage And Retrieval System

INSTBY: Instability

IN-STEP: In-Space Technology Experiment Program

INSTEP: In-Space Technology Experiment Program

Instl: Install • Installation

Instld: Installed

Instm: Instrumentation

Instn: Institution

INSTN: Institut National des Sciences et Techniques Nucléaires

Instr: Instruction • Instructor • Instrument • Instrumentation • Instrumental

INSTR: Instruct • Instrument • Instrumentation

Instrum: Instrumental • Instrumental • Instrumentation • Instrumented

INSU: Institut National des Sciences de l'Univers (France)

Insuf: Insufficient

Insul: Insulate • Insulation

Insuld: Insulated

Int: Integer • Integral • Integrated • Intelligence • Intendente • Intensity • Interest • Intérêt • Interim • Interior • Intérieur • Interjection • Intermediate • Intermittent • Internal • International • Interpret • Interpretation • Interpreter • Interrogate • Interrupt • Intersection • Interval • Interview • Intransitive

INT: Institut National des Télécommunications (France) • Integer • Intendente • Intensity • Internal • International • Interrupt • Isaac Newton Telescope (Canary Islands) • Winston-Salem (USA-NC) (location identifier)

INTA: Instituto Nacional de Técnica Aerospacial (Spain) • Instituto Nacional de Tecnologia Agropecuaria (Argentina) • International Trademark Association (formerly USTA) • Interrupt Acknowledge

int. al.: Inter Alia

INTAP: Interoperability Technology Association for Processing

Intchg: Interchangeable

INTCHG: Interchange

Intcp: Intercept

INTCP: Intercept

INTD: Institut National des Techniques de la Documentation

INTEC: Institut National des Techniques Économiques et Comptables

INTECH: Institute for New Technologies

INTECO: Instituto de Normas Técnicas de Costa Rica

INTECOL: International Association for Ecology

Integ: Integral • Integrate • Integration

INTEGR: Integrate • Integration

INTEGRAL: International Gamma-Ray Astrophysics Laboratory

Integrg: Integrating

Intel: Intelligence • Intelligent

Intell: Intelligence • Intelligent

INTELSAT: International Telecommunication Satellite Consortium

Intens: Intensifier • Intensify • Intensive • Intensivo

Inter: Interavia (periodical code) • Interim • Interior • Intermediate • International • Interrogation • Interrogative

INTER: Intermittent

Interj: Interjección • Interjection • Interjectiva

INTERMAPS: Interactive Multimedia Access Publishing services

Intern: International

Internac: Internacional

Internat: International

InterNIC: Internet Network Information Center

Interp: Interpreter

Interplanet: Interplanetary

Interrog: Interrogate • Interrogation • Interrogative • Interrogativo

INTERROG: Interrogate

INTERTACOS: Interpretation, Treatment, Analysis and Correlation Of Spectra

INTESA: Iniziative Telematiche per Servizi Applicativi

INTF: Interface • Interim National Test Facility

INTG: Integration

INTI: Instituto Nacional de Tecnologia Industrial

INTJ: Introverted Intuitive Thinking Judging

Intk: Intake

Int'l: International

Intl: Internal • International

INTL: International

Intlk: Interlock

INTLK: Interlock

Intlz: Initialize

Intmd: Intermediate

Intmt: Intermittent

INTMT: Intermittent

INTN: Instituto Nacional de Tecnología y Normalización

INTO: Interrupt if Overflow

INTOR: International Tokamak Reactor

INTP: Introverted Intuitive Thinking Perceiving

Intpr: Interpret

Intr: Interior • Intransitif • Intransitive • Intransitivo • Introduce • Introduced • Introducing • Introduction • Introductory

INTR: Interior

INTRA: Intelligent Tracker

Intrans: Intransitive

In Trans: In Transit

Int. Rev.: Internal Revenue

Intrf: Interference

Intrg: Interrogate

INTRG: Interrogate

INTRLVR: Interleaver

Intro: Introduce • Introduced • Introduction • Introduire • Introduit

Introd: Introducción • Introduce • Introduced • Introduction • Introduire • Introduit

Intrpl: Interpolate

Intrpt: Interrupt

INTRPT: Interrupt

INTS: Intense

Intsct: Intersect

INTSF: Intensify • Intensifying

INTSFY: Intensify

IntSM: Interavia Space Markets (periodical code)

INTST: Intensity

Intstg: Interstage

INTUG: International Telecommunications Users Group

Intvl: Interval

INTVL: Interval

INTVLM: Intervalometer

INTX: Interrupt and Timing Extension

INU: Inertial Navigation Unit • Ion Nucleation • Nauru Island (location identifier)

Inus: Inusitado • Inusité

inv: Invenit

Inv: Invenit • Invent • Invented • Inverter • Invention • Inventor • Inventory • Inverse • Inversed • Inversion • Invert • Inverted • Investigate • Investigated • Investigation • Investment • Invitation • Invite • Invited • Invoice • Invoiced

INV: Inductive Null Voltage • Inverness (UK-Scotland) (location identifier) • Inverter

Invar: Invariable

Invest: Investigate • Investigated • Investigation • Investigational

inv. et del.: Invenit et Delineavit

INVOF: In Vicinity Of

InvPr: Inverse Problems (periodical code)

INVRN: Inversion

Invs: Inverse

Invt: Inventory • Invert

Invtr: Inverter

INW: Institut für Nachrichtentechnik und Wellenausbreitung • Winslow (USA-AZ) (location identifier)

INWATS: Inward Wide Area Telephone Service

INWG: International Network Working Group • Internetwork Working Group

INX: Inanwatan (Indonesia) (location identifier)

INY: Inyati (South Africa) (location identifier)

INZ: In Salah (Algeria) (location identifier)

IN2P3: Institut National de Physique Nucléaire et de Physique des Particules (France)

In2S3: Indium Sulfide

IN3: Internet Interdisciplinary Institute

Io: Iowa (USA)

I/O: Inboard/Outboard • Input/Output • Inspecting Order

I&O: Inlet & Outlet

IO: British Territory of Indian Ocean (ISO country code) • Idea Organizer • Inflow-Outflow • Image Orthicon • Indian Ocean (British Territory of_) • Indirect Object • Industrial Operations • Information Operation • Inlet and Outlet • In Order • Information Officer • Initial Only • Input/Output • Inspection Order • Inspection Outline • Integrated Optics • Interior Orientation • International Organization • Inward Operator • Ion Engine • Iterative Operation • Territoire Britannique de l'Océan Indien (ISO country code)

IoA: Institute of Astronomy

IOA: Input-Output Adapter • Input-Output Address • Input-Output Assembly • Inquiry-Ordering-Accounting • Instrumentation Operating Area • Instrument Operating Assembly • Ioannina (Greece) (location identifier)

IOAA: Independent Office Appropriations Act

IOAT: Indicated Outside Air Temperature • In-Orbit Automation Technology

I/OB: Input/Output Bus

IOB: Ink On Bitmap • Input-Output Block • Input-Output Box • Input-Output Buffer • Input-Output Bus

IOBC: International Organization for Biological Control

IOBM: In Ocean By Mistake • Input/Output Bitmap

IOBPS: Input/Output Box and Peripheral Simulator

I/OC: Input/Output Controller

IOC: Image Orthicon Camera • Image Orthicon Control • Indirect Operating Cost • Initial Operating Configuration • Initial Operational Capability • In-Orbit Calibration • In-Orbit Checkout • In-Orbit Communication • Inorganic Chemical • In-Out Converter • Input Offset Current • Input-Output Card • Input-Output Channel • Input-Output Comparator • Input-Output Controller • Input-Output Converter • Integrated Operations Configuration • Integrated Optical Circuit • Intergovernmental Oceanographic Commission • Interim Operational Capability • International Observations Campaign • International Oceanographic Commission • International Olympic Committee • International Operating Center • International Operating Center • International Ozone Commission • Interorbit Communication • ISDN Ordering Code • Item Of Concern • Iterative Orbit Calculator

IOCC: Input-Output Control Center • Input-Output Control Command

IOCD: International Organization for Chemistry in Development

IOCG: International Organization on Crystal Growth

IOCM: Interim Operational Contamination Monitor

IOCRD: In-Orbit Calibration Requirements Document

IOCS: Input-Output Control System

IOCTL: Input-Output Control

IOCU: Input-Output Control Unit • International Organization of Consumers Unions

IOD: Identified Outward Dialing • Initial Orbit Determination • Input/Output Daemon • Integrated Optical Density

IODA: Input-Output Drive Assembly

IODB: Integrated Object Database

IODC: Input-Output Data Channel • Input-Output Delay Counter

IODE: International Oceanographic Data and Exchange • International Oceanographic Data Exchange

IODS: Improved Optical Detection System

IoE: Institute Of the Environment

IOE: Initial Operating Experience • Input/Output Error • Institute Of the Environment

IOEM: Integrated Opto-Electronics Module

IOF: Initial Operational Flight • International Oceanographic Foundation

IofM: Isle of Man (UK)

IOG: Input-Output Gate

IOGE: Input-Output Operational Ground Equipment

IOH: Integrated Open Hypermedia

IOHA: Integrated Operating Hazard Analysis

IOHR: Imagerie Optique à Haute Résolution

IOI: Imaging Optical Interferometer • In-Orbit Infrastructure • Interactive Observing Infrastructure • Interim Operating Instruction • Internal Operating Instruction • International Ocean Institute • International Olympiad in Informatics • International Ozone Institute

IOIS: Integrated Office Information System

IOJ: International Organization of Journalists

IOK: Iokea (Papua New Guinea) (location identifier)

IOL: Initial Outfitting List • Intermediate Objective Lens • Interoffice Letter • Intraocular Lens • Ireland On Line

IoM: Institute of Meteoritics

IOM: Input-Output Module • Input-Output Multiplexer • Institute Of Meteoritics • International Organization for Migration • Interoffice Memorandum • Isle of Man (UK) (location identifier)

IOMAC: Indian Ocean Marine Affairs Cooperation

IOMAS: Ion Mass Spectrometer

I/OMI: Integration, Operations, and Maintenance Instruction

IOMP: International Organization for Medical Physics

Ion: Ionian

IoN: Institute of Navigation

ION: Impfondo (Congo) (location identifier) • Institute Of Navigation • Interlending OSI Network • International Ocean Network

IONA: Input/Output Network Adapter

IONDS: Integrated Operational Nuclear Detection System • Integrated Operational Nuclear Detonation System

IONIC: Integrated Optics Near-infrared Interferometric Camera

Iono: Ionosphere (periodical code)

Ionos: Ionosphere • Ionospheric

IONL: Internal Organization of the Network Layers

IOOFC: Integrated Optics and Optical Fiber Communication

IoP: Institute of Physics

I/OP: Inboard/Outboard Profile

I&OP: In and Out Processing

IOP: Increment Operation Plan • Indian Ocean Programme • Initial Operational System • In-Orbit Plane • Input-Output Package • Input-Output Parallelism • Input-Output Part • Input-Output Processor • Input-Output Pulse • Installation Operating Program • Institute of Ocean Sciences • Institute Of Physics • Instituto de la Opinión Física (Spain) • Intensive Observing Period • Interim Operating Procedure • Internal Ocular Pressure • Internal Operating Procedure • International Opacity Project • Interoperability • Intraocular Pressure • Ioma (Papua New Guinea) (location identifier)

IOPAB: International Organization for Pure and Applied Biophysics

IOPC: International Oil Pollution Compensation

IOPG: Institut et Observatoire de Physique du Globe

IOPL: Input-Output Privilege Level • Integrated Open Problem List

IoPP: Institute of Physics Publishing

IOPP: Institute Of Physics Publishing

IOPS: Input-Output Programming System

IOPT: Information Operations Planning Tool • Input/Output Parameter Table

IOQE: Input/Output Queue Element

IOR: Index Of Refraction • Indian Ocean Region • Industrially-Oriented Research • Inishmore (Ireland) (location identifier) • Input-Output Register • Interoperable Object Reference • Interorder Overlap Region • Iterative Orthophoto Refinements

IORD: Input/Output Read

IORL: Input/Output Requirements Language

IORTD: Input/Output Relationships and Timing Diagram

IOS: Ilhéus - Eduardo Gomes (Brazil-BA) (location identifier) • Image Optical Scanner • Image Orthicon System • Initial Operational System • In-Orbit Servicing • Input-Output Sense • Input-Output Skip • Input-Output System • Input-Output Supervision • Input-Output Supervisor • Input-Output Switch • Inspection Operation Sheet • Inspection Operation System • Institute of Oceanographic Sciences • Instructor Operator Station • Instrumentation Operation Station • INTEGRAL Overall Simulator • Integrated Office System • Integrated Operations Schedule • Interim Operational System • Internetworking Operating System • Interorganizational System

IOSB: Input/Output Standard Block

IOSC: Integrated Operations Support Center • Integrated Operations Support Complex

IOSF: In-Orbit Simulation Facility

IOSGA: Input-Output Support Gate Array

IOSH: Institution of Occupational Safety and Health

IOST: ISEMSI Operations Support Team

IOSTE: International Organization for Science and Technology Education

I/OT: Input/Output Test

IOT: British Territory of Indian Ocean (ISO country code) • Indian Ocean Territory • Initial Orbit Time • In-Orbit Testing • In-Out_thru • Input-Output Termination • Input-Output Transfer • Instrument Operations Team • Territoire Britannique de l'Océan Indien (ISO country code)

IOTA: Infrared and Optical Telescope Array • Infrared-Optical Telescope Array (*) • Institut d'Optique Théorique et Appliquée (France) • International Occultation Timing Association

IOTA-ES: International Occultation Timing Association - European Section

IOTAES: International Occultation Timing Association - European Section

IOTD: Image Of The Day

IOT&E: Initial Operational Test & Evaluation • Integration Operational Test & Evaluation

IOTE: Initial Operational Test and Evaluation • Integration Operational Test and Evaluation

IOTG: Input-Output Task Group

IOTM: Image Of The Month

IOTR: Item Operation Trouble Report

IOTW: Image Of The Week

IOTY: Image Of The Year

I/OU: Input/Output Unit

IOU: Île Ouen (New Caledonia) (location identifier) • Input-Output Unit • Input-Output Utility • I Owe You

IOUBC: International Office for Universal Bibliographic Control

IOV: Input Offset Voltage

IOVF: Independent Overflow

IOW: Inflow-Outflow Wind • In Other Words • Iowa City (USA-IA) (location identifier) • Isle Of Wight (UK)

IOWR: Input/Output Write

I/P: Impact Predictor • In Progress • Input

IP: Ice Pellet • Identification Point • Identification of Position • Igloo Pallet • Image Point • Image Processing • Imaginary Part • Impact Point • Impact Prediction • Impact Predictor • Impingement Point • Implementation Plan • Important Paper • Important Passenger • Important Person • Improvement Program • Incentive Pay • Increment Plan • Indexed and Paged • Index of Performance • Indicator Panel • Indirect Proof • Individual Permit • Induced Polarization • Industrial Preparedness • Inertial Platform • Inertial Processing • Information Processing • Information Product • Information Provider • Ingénieur Principal • Initial Phase • Initial Plan • Initial Point • Initial Position • In Plant • Input Power • Input Primary • Input Processor • Installation Procedure • Instalment Plan • Instruction Pamphlet • Instruction Pointer • Instructor Pilot • Instrumentation Payload • Instrumentation Plan • Instrumentation Power • Instrument Payload • Instrument Processor • Intellectual Property • Interaction Point • Interested Parties • Interface Processor • Interfacing Processor • Interior Point • Intermediate Pallet • Intermediate Point • Intermediate Polar • Intermediate Pressure • Internet Protocol • Internetwork Protocol • Interpedia • Interplanetary • Interrupt Process • Invited Paper • Invited Presentation • Ionic Propulsion • Ionization Potential • Ion Pairing • Ipswich (UK postal code) • Iron Pipe • Issuing Point • Item Processing • Itinerário Principal

IPA: Imagery Product Archive • Indice Ponctuel d'Abondance • Industrial Preparedness Action • Infinitesimal Perturbation Analysis • Information Process Analysis • Information Processing Association • Institute for the Physics of the Atmosphere • Institute of Public Affairs • Institut für Physik der Atmosphäre • Intergovernmental Personnel Act (USA) • Intergovernmental Personnel Agreement • Intermediate-Power Amplifier • International Palaeontological Association • International Permafrost Association • International Pharmaceutical Abstracts • International Phonetic Alphabet (*) • International Phonetic Association • International Press Association • International Priority Airmail • International Publishers Association • Intrinsic Polarization Angle • Ipota (Vanuatu) (location identifier) • Isopropyl Alcohol

IPAA: Instrumental Photon Activation Analysis

IPAC: Industrial Programme Advisory Committee (PPARC) • Information Processing And Control • Infrared Processing and Analysis Center (USA-CA) (*) • IRAS Processing and Analysis Center

IPACS: Integrated Power and Attitude Control System

IPAD: Integrated Program for Aerospace Design

IPADS: Interactive Processing And Display System

IPAG: Information Planning and Analysis Group

IPAL: Image Processing Algorithms Library

IPAM: Institute for Pure and Applied Mathematics

IPAR: Initial Product Assessment Report • Instantiated Parameterized Action Representation • Intrapulse Polarization Agile Radar • Irregular Parts Authorization Request

IPAS: Image Processing and Analyzing System • Integrated Problem Assessment System

IPAT: Impact-Population-Affluence-Technology

IPB: Illuminated Push Button • Illustrated Parts Breakdown • Inert Processing Building • Information Policy Board • Institut de Physique Biologique • Institut für Pflanzenernährung und Bodenkunde (Germany)

IpC: Instructions per Clock • Instructions per Cycle

IPC: Easter Island (Chile) (location identifier) • Illustrated Parts Catalogue • Image Processing Center • Imaging Proportional Counter • Impact Programme Committee • Indicative Planning Council • Industrial Policy Committee • Information and Publicity Committee • Information Policy Committee • Information Processing Center • Information Processing Code • Information Product Characteristics • Inner Tracking Chamber • Institute for Interconnecting and Packaging • Instructions Per Clock • Instructions Per Cycle • Integrated Pollution Control • Integrated Procedures Control • Integrated Processor Controller • Integrated Procurement Control • Intellectual Property Council • Intermediate Processing Center • Intermittent Positive Control • International Patent Classification • International Polar Commission • International Privacy Coalition • International Program Committee • International Pyrheliometer Comparison • Internet Proxy Cache • Interplanetary Communication • Interprocess Communication • Interprocessor Communication • Isla de Pascua (Chile) (location identifier)

IPCB: ISO Parts Coordination Board

IPCC: Intergovernmental Panel on Climate Change • International Property Cooperation Center • International Publishers Copyright Council • Internet Protocol Contact Center (Cisco)

IPCD: Image Photon Counting Detector

IPCDT: IEE Proceedings on Computers and Digital Techniques (periodical code)

IPCE: International Phone Card Exchange

IPCES: Intelligent Process Control by Expert Systems

IPCL: Instrumentation Program and Component List • Instrument Program and Component List

IPCMS: Institut de Physique et Chimie des Matériaux de Strasbourg (France)

IPCP: Internet Protocol Control Protocol

IPCRP: IEE Proceedings on Communications and Radar Processing (periodical code)

IPCRS: IEE Proceedings on Communications, Radar and Signal Processing (periodical code)

IPCS: Image Photon Counting System • Image Production Control System • Integrated Propulsion Control System • Intelligent Process-Control System • Interactive Problem Control System • International Programme on Chemical Safety

IPCSV: IEE Proceedings on Communications, Speech, and Vision (periodical code)

IPCT: Integrated Parametric Current Transformer

IPCTA: IEE Proceedings on Control Theory and Applications (periodical code)

IPCU: Interface Power Control Unit

IPD: Image Plane Detector • Imaging Photon Detector • Individual Package Delivery • Individual Potential Demand • Information Processing Division • Insertion Phase Delay • Instructional Program Development • Integrated Pin Diode • Integrated Powerhead Demonstration • Integrated Product Development • Interplanetary Dust • Isotope-Powered Device • Iterated Prisoner's Dilemna

IPDA: Intrapulse Demodulation Analysis

IPDC: International Programme for the Development of Communication • Internet Protocol Device Control

IPDD: Initial Project Design Description

IPDE: Integrated Product Development Environment

IPDM: Isolated Power and Data Module

IPDN: International Paleoclimatic Data Network

IPDP: Interval of Pulsations of Diminishing Period

IPDPG: Image Processing Data Products Group

IPDR: Incremental Preliminary Design Review

IPDS: IBM Personal Dictation System • Integrated Product Design System • Intelligent Printer Data Stream

IPDT: Integrated Product Development Team

IPDU: Instantaneous Panoramic Display Unit

IPE: Image Pac Extension • Image Processing Engine • Immeuble du Parlement Européen • Improved Performance Engine • Incentive PERT Event • Industrial Plant Equipment • Information Processing Equipment • Integrated Programming Environment • Interpret Parity Error • Inverse Photoelectric Effect • Ipil (Philippines) (location identifier)

IPEA: International Preliminary Examining Authority

IPEC: IVS Proposal Evaluation Committee

IPECS: IEE Proceedings on Electronics, Circuits, and Systems (periodical code) • Integrated Power and Environmental Control System

IPEE: Inclination of a Plane to the Earth's Equator

IPEEFI: Identification de Projets Énergétiques Éligibles à des Financements Internationaux

IPEI: Ionospheric Plasma and Electrodynamics Instrument

IPEPA: IEE Proceedings on Electric Power and Applications (periodical code)

IPERS: International Peer Review System

IPES: Inorganic Properties Estimation System • Inverse Photoemission Study

IPEX: Interferometry Program Experiment

IPF: Image Processing Facility • Immediate Parameter File • Imposición a Plazo Fijo • Information Presentation Facility • Instrument Pointing File • Integrated Processing Facility

IPFC: Indo-Pacific Fisheries Council

IPFR: Institute of Plasma and Fusion Research

IPG: Image Processing Group • Institut de Physique du Globe • Institut für Physikalische Geodäsie (Germany) • Institut für Planetarische Geodäsie (Germany) • Instrument Projects Group • Interference Protection Group • Ipiranga (Brazil-AM) (location identifier) • Isotope Power Generator

IPGP: Institut de Physique du Globe de Paris (France)

IPGR: Iterative Parallel Region Growing

IPGS: Institut de Physique du Globe de Strasbourg (France)

IPGTD: IEE Proceedings on Generation, Transmission, and Distribution (periodical code)

iph: Impressions per Hour • Inches per Hour

IPH: Ipoh (Malaysia) (location identifier) • Instantaneous Pointing History

IPHAC: Ion Produced Hydrogenated Amorphous Carbon

IPHIR: Interplanetary Helioseismology by Irradiance

IPI: Institute of Production Innovation • Instrument Principal Investigator • Intelligent Peripheral Interface • International Patent Institute • International Press Institute • Interpulse Interval • Ipiales (Colombia) (location identifier)

IPIC: Intellectual Property of Integrated Circuits

IPIP: Information Processing Improvement Program

IPIPS: Interactive Planetary Image Processing System

IPIR: Internal Peak Irradiance Response

IPIX: Intelligent Pixel-processing

IPK: Internationale Patentklassifikation

IPL: El Centro - Imperial County (USA-CA) (location identifier) • Identified Parts List • Illustrated Parts List • Image Processing Laboratory • Image Processing Log •

Imperial County (USA-CA) (location identifier) • Indentured Parts List • Information Processing Laboratory • Information Processing Language • Information Programming Language • Initial Program Load • Initial Program Loader • Instrumentation Program List • Internet Public Library • Interrupt Priority Level • Ion Projection Lithography

IPLAN: Integrated Planning and Analysis

IPLB: Internet Protocol Load Balancing

IPLC: Initial Program Load Complete

IPLL: Instituto Português do Livro e da Leitura (Portugal)

IPLS: Initial Program Load Start

IPLV: Intermediate Payload Launch Vehicle

ipm: Inches per Minute

IpM: Impulses per Minute • Interruptions per Minute

IP&M: Information Processing & Management

IPM: Illustrated Parts Manual • Impulses Per Minute • Inches Per Minute • Incidental Phase Modulation • Incidental Pulse Modulation • Information Processing and Management • Inspection Procedure Manual • Institut für Physikalische Meßtechnik • Integration Program Manager • Interference Prediction Model • International Prototype Meter • Inter-Personal Message • Inter-Personal Messaging • Interruptions Per Minute • Italian Panchromatic Monochromator • Italian Panoramic Monochromator (*)

IPMAF: IRAS Post Mission Analysis Facility

IPMAP: IEE Proceedings on Microwaves, Antennas, and Propagation (periodical code)

IPMI: Intelligent Platform Management Interface • Internet Provider Multicast Initiative

IPMOA: IEE Proceedings on Microwaves, Optics, and Antennas (periodical code)

IPMP: Investigation into Polymer Membranes Processing

IPMS: Impact Predictor Monitor Set • Initial Performance Measurement System • Institution of Professionals, Managers and Specialists • International Polar Motion Service (now IERS) • Interpersonal Message Service • Impact Predictor Monitor Set

IPN: Institut de Physique Nucléaire • Instituto Politécnico Nacional (Mexico) • International Product News • Internet Protocol Number • Interplanetary Network • Interprocessor Node • Intracluster Planetary Nebula • Ipatinga Usiminas (Brazil-MG) (location identifier) • Instituto Politécnico Nacional

IPNE: Institut de Physique Nucléaire Expérimentale

IPng: Internet Protocol Next Generation

IPNG: Internet Protocol Next Generation

IPNNI: Integrated Private Network-to-Network Interface

IPNO: Institut de Physique Nucléaire d'Orsay (France)

IPNS: Intense Pulsed Neutron Source

IPN1: First Interplanetary Network

IPN2: Second Interplanetary Network

IPN3: Third Interplanetary Network

IPO: IEEE/IEI Publication On-disk • Initial Public Offering • Input-Process-Output • Instantaneous Power Output • Intellectual Property Owners Inc. • International Planning Office • International Procurement Office • International Program Office

IPoATM: Internet Protocol over Asynchronous Transfer Mode

iPOC: Interim Policy Oversight Committee

IPOC: International Partner Operations Center

IPOMS: International Polar Orbiting Meteorological Satellite

IPOP: International Polar Orbiting Platform

IPOpt: IEE Proceedings on Optoelectronics (periodical code)

iPP: Isostatic Polypropylene

IPP: Imaging Photopolarimeter • Imaging Polarimeter and Photopolarimeter • Impact Prediction Point • Individual Program Plan • Industrial Partnering Program • Industrial Preparedness Plan • Information Processing Profile • Information Production Process • Inhomogeneous Poisson Process • Initiative for Proliferation Prevention • Input Processor Package • Institute of Physical Problems • Institute of Plasma Physics • Instrument Programmable Parameter • Integrated Plotting Package • Integrated Product Plan • Intermittent Positive Pressure • Inverse Polarity Protection • Ionospheric Propagation Path

IPPAP: Institute for Particle Physics and Astrophysics

IPPB: Illustrated Provisioning Parts Breakdown

IPPC: International Policies and Plans Committee

IPPD: Integrated Product and Process Development

IPPE: Institute for Physics and Power Engineering (Russia)

IPPEX: Internet Plasma Physics Education Experience

IPPM: Internet Provider Performance Metrics

IPPNW: International Physicians for the Prevention of Nuclear War

IPPR: Internal Peak Power Response

IPPS: Interactive Picture Processing System

IPPSM: IEE Proceedings on Physical Science and Measurements (periodical code)

IPPV: Intermittent Positive Pressure Ventilation

IPQ: Information Processing Quotient • Information Processing Quality • Information Product Quality • Instituto Português da Qualidade (Portugal)

IPQC: In-Process Quality Control

ipr: Inches per Revolution

IPR: Ice-Point Reference • In-Process Review • Informal Progress Report • Inspecteur Pédagogique Régional • Institute for Plasma Research • Institute of Plasma Research • Institute of Public Relations • Instrument Performance Report • Intellectual Property Right • Intellectual Proprietary Right • Interim Problem Report • Interim Progress Report • Internal Progress Report • Ion Production Rate • Irish Publishing Record

IPRA: International Public Relations Association • Internet PCA Registration Authority • Internet Policy Registration Authority

IPRF: Impage Processing Research Facility • Impuesto sobre la Renta de Personas Físicas

IPRG: Institute for Petroleum Research and Geophysics (Israel)

IPRO: International Patent Research Office

IPRS: Information and Pubic Relations Service • International Postgraduate Research Scholarship

IPRSP: IEE Proceedings on Radar and Signal Processing (periodical code)

ips: Inches per Second • Instructions per Second • Interconnections per Second • Interruptions per Second

Ips: Ipswich (UK)

IPS: Image Processing Software • Image Processing Specialist • Image Processing System • Impact Predictor System • Improved Processing System • Inches Per Second • Inertial Pointing System • Inertial Positioning System • Information and Photographic Service • Information Processing Standard • Information Processing System • Institute of Polar Studies • Institut Paul Scherrer (Switzerland) • Instrumentation Power Supply • Instrumentation Power System • Instrument Pointing System • Integrated Power Source • Integrated Power System • Integrated Product Service • Integrated Propulsion System • Intelligent Page Store • Intelligent Power System • Interconnections Per Second • Interface Problem Sheet • Interim Policy Statement • International Pipe Standard • International Planetarium Society (USA-UT) • Interplanetary Scattering • Interplanetary Scintillation • Interpretative Programming

System • Interstellar Propulsion Society (USA-CA) • Inverter Power Supply • Invoice Processing System • Ionospheric Prediction Service (Australia) • Ion Propulsion System • Israel Physical Society • Item Processing System

IPSA: Interplanetary Space Agency (UK) • Irish Physics Students' Association

IPSC: Information Processing Standards for Computers • International Program for Scientific Cooperation

IPSD: Image Processing Systems Division

IPSE: Integrated Programming Support Environment • Integrated Project Support Environment • Intelligent Program Support Environment • Italian Package for Science Experiments (MSR)

IPSEC: Internet Protocol Security

IPSGSM: Instituto Politécnico Superior General San Martín (Argentina-SF)

IPSI: Institut de Promotion des Sciences de l'Ingénieur • Integrated Publication Systems Institute

IPSJ: Information Processing Society of Japan

IPSL: Institut Pierre-Simon Laplace (France) • Interface Problem Status Log

IPSN: Institut de Protection et de Sûreté Nucléaire

IPSO: Institut pour la Promotion des Systèmes Ouverts

IPSP: Internet Protocol Security Protocol

IPSS: Information Processing System Simulator • Institute of Planetary and Space Science • Integrated Parts Statusing System • Integrated Project Scheduling System • International Packet Switched Service

IPSSE: IEE Proceedings on Solid-State Electron Devices (periodical code)

IPST: In-Process Self Test • Institute for Physical Science and Technology • Institut Professionnel des Sciences et Technologies

IPT: Image Perspective Transform • Improved Programming Technology • Incremental Proof Testing • Indexed-Paged-Titled • Informatique Pour Tous (France) • In-Plant Test • In-Plant Transporter • In-Process Test • Instituto de Pesquisas Tecnológicas • Insurance Premium Tax • Integrated Product Team • Intelligent Printhead Technology • Intermediate Pressure Turbine • Internal Parameter Table • International Pipe Thread • Interplanetary Travel • Interprocess Communications • Io Plasma Torus • Williamsport (USA-PA) (location identifier)

IPTCS: Igloo Passive Thermal Control System

IPTF: Infrastructure Protection Task Force

IPTL: IEEE Photonics Technology Letters (periodical code)

IPTO: Independent Power Takeoff • Information Processing Techniques Office

IPTR: In-Process Technical Review

IPTS: Institute for Prospective Technological Studies • Institut für Praxis und Theorie der Schule (Germany) • International Practical Temperature Scale

IPU: Image Processing Unit • Information Processing Utility • Instruction Processing Unit • Integrated Plot Utility • Integrated Power Unit • Intelligent Processing Unit • Interface and Priority Unit • Inter-Parliamentary Union • Interphase Unit • Ipiaú (Brazil-BA) (location identifier) • Isotope Power Unit

i. Pl. v.: In Plaats Van

IPV: Improve • Individual Pressure Vessel • In Plaats Van • Isentropic Potential Velocity • Isentropic Potential Vorticity

IPVS: Ion Pump Vacuum System

IPW: Ipswich (UK-England) (location identifier)

IPWG: Integrated Project Working Group • Internet Privacy Working Group

IPX: Integrated Packet Exchange • Internet Packet Exchange • Internet Protocol Exchange • Internetwork Packet Exchange (*)

ipy: Inches per Year

IPY: International Polar Year • Ion Pair Yield

IP2D: Integrated Product and Process Development

iq: Idem Quod

I-Q: In-Quadrature

I/Q: Information/Query • In-Quadrature

IQ: Image Quality • Image Query • Indefinite Quantity • Information Quality • In Quadrature • Intelligence Quotient • Intelligent Queuing • International Quota • Intrinsic Quality • Iraq (ISO country code)

IQA: Institute for Quality Assurance • Irish Quality Association

IQAS: International Quality Assurance System

IQD: Dinar (Irak) (ISO currency code) • Image Quality Detector • International Quiet Day • Interquartile Difference

IQE: Image Quality Estimator • Interactive Query Expansion

IQEC: International Quantum Electronics Conference

IQI: Image Quality Indicator

IQL: Information Query Language • Intelligent Query Language • Interactive Query Language • Intermediate Query Language

IQM: Institute for Quality and Management • Qiemo (China-PRC) (location identifier)

IQN: Inner Quantum Number • Qingyang (China-PRC) (location identifier)

IQQ: Iquique - D. Aracena (Chile) (location identifier)

IQRP: Interactive Query and Report Processor

IQSY: International Quiet Sun Year

IQT: Initial Qualification Training • Iquitos (Peru) (location identifier)

ir: Iran

iR: im Ruhestand

Ir: Ingenieur • Ingénieur • Ireland • Iridium • Irish

I&R: Inspect & Review • Interchangeability & Replaceability

IR: Ice on Runway • Imaging Radiometer • Immediate Reserve • Impulse Response • Incidence Rate • Index Register • Individual Rational • Informal Report • Information Report • Information Request • Information Retrieval • Infrared • Infra-Rot • Infra-Rouge • Ingénieur de Recherche • Initial Release • Inland Revenue • Inner Roll • Inside Radius • Inspect and Review • Installation Report • Institutional Representative • Instruction Register • Instrumentation Report • Instrument Reading • Instrument Reference • Insulation Resistance • Intelligence Ratio • Interchangeability and Replaceability • Interim Report • Intermediate Review • Internal Reliability • Internal Report • Internal Resistance • Internal Revenue • International Registration • International Rendezvous • Internet Registry • Interpretation Report • Interrogation Report • Interrogator-Responder • Ionizing Radiation • Iran (international vehicle registration code) • Iran (ISO country code) • Iranian Rial (currency) • Ireland • Iroquois • Irreductible Representation • Isoprene Rubber

IRA: Individual Retirement Account • Infrared Astronomy (periodical code) • Institutional Research Award • Institut Régional d'Administration • Istituto di Radioastronomia (Italy) • Kirakira (Solomon Islands) (location identifier)

IRAA: Indoor Radon Abatement Act

IRAC: Infrared Array Camera • Interdepartmental Radio Advisory Committee • International Radio Advisory Committee (*) • International Relations Advisory Committee

IRACE: Infrared Array Control Electronics

IRAD: Independent Research And Development • Infrared Radiometer

IRAF: Image Reduction and Analysis Facility

IRAIT: Italian Robotic Antarctic Infrared Telescope

IrAJ: Irish Astronomical Journal (periodical code)

IRAM: Implicitly-Restarted Arnoldi Method • Improved Reliability And Maintainability • Improved Repairable Asset Management • Indexed Random Access Method • Input Random Access Memory • Institut de Radioastronomie Millimétrique • Instituto de Racionalización de Materiales (Argentina) • Instituto de Radioastronomía Milimetrica

IRAMMP: Infrared Analysis Measurement and Modeling Program

IRAMS: Inertial Reference Alignment Monitor System

Iran: Iranian

IRAN: Inspect and Repair As Necessary • Inspection and Repair As Necessary

IRAND: Interactive RAND

IRAO: Itapetinga Radio Astronomy Observatory (Brazil)

IRAP: ISLSCP Retrospective Analysis Project (now IDSLSC)

IRAPS: Image Recording And Processing System

IRAR: Infrared Airborne Radar

IRAS: Infrared Astronomical Satellite • Istituto Spezzino Ricerche Astronomiche (Italy)

IRATS: Infrared Ranging And Tracking System

IRAW: International Registration of Audiovisual Works

IRB: Industrial Relations Bureau • Industrial Revenue Bond • Infrared Background • Inspection Review Board • Institutional Review Board • Instruction Recorder Buffer • Instruction Reorder Buffer (*) • Inter-Repeater Bus • Iraan (USA-TX) (location identifier)

IRBM: Intermediate Range Ballistic Missile

IRBS: Infrared Background Sensor • Intermediate-Range Ballistic System • Intermediate-Range Booster System

IRc: Infrared Compact

IRC: Circle (USA-AK) (location identifier) • Imaging Research Center • Individual Request for Classification • Information Research Center • Information Resource Catalogue • Information Resources Centre • Information Retrieval Center • Infrared Catalogue • Infrared Companion • Innovation Relay Centre • Instrument Remote Control • Interactive Rule Checking • Interdisciplinary Research Center • Internal Revenue Code • International Radiation Commission • International Record Carrier • International Red Cross • International Reference Centre • International Research Council (now ICSU) • International Response Coupon • Internet Relay Chat • Interregion Communication • Interrupt Request Controller • Ionosphere Research Committee • Ion Recombination Chamber

IRCAM: Infrared Camera • Institut de Recherche et Coordination Acoustique-Musique

IRCC: Indian River Community College (USA-FL) • International Radio Consultative Committee

IRCCD: Infrared Charge Coupled Device

IRCCM: Infrared Counter-Countermeasure

IRCD: International Research Centers Directory

IRCFE: Infrared Communications Flight Experiment

IRCM: Infrared Countermeasures

IRCS: Inertial Reference and Control System • Infared Camera and Spectrograph • Institute for Research in Cognitive Science • Integrated Radiocommunications System • Integrated Resource Control System

IRCTTR: International Research Centre for Telecommunications, Transmission and Radar

IR&D: Independent Research & Development

IRD: Independent Research and Development • Information Requirements Description • Information Requirements Document • Information Resources Division • Infrared Data • Infrared Detector • Infrared Display • Infrared Divergence • Initiating Reference Document • Institut de Recherche pour le Développement • Integrated Receiver/Decoder • Integrated Receiver/Descrambler • Integration Requirements Document • Interface Requirements Document • International Relations Division • International Research and Development • Internet Resource Database • Ishurdi (Bangladesh) (location identifier)

IrDA: Infrared Data Association

IRDA: Industrial Research and Development Authority • Infrared Data Association • Infrared Detection Array • Infrared Developers Association • Infrared Device Association

IRDAC: Industrial Research and Development Advisory Committee

IRDB: Integrated Regional Database

IRDC: Infrared Dark Cloud

IRDL: Information Retrieval and Display Language

IRDM: International Rendezvous and Docking Mission

IRDNET: Ice Reconnaissance Data Network

IRDS: Information Resource Dictionary Standard • Information Resource Dictionary System • Infrared Detection Set

IRDU: Internet Research and Development Unit

Ire: Ireland

IRe: Infrared Extended

IRE: Information Resource Engineering • Infrared Emission • Institute for Radioelectronics and Engineering (Russia) • Institute of Radio Engineers • Institute of Radioengineering and Electronics • Irecê (Brazil-BA) (location identifier)

IREB: Intense Relativistic Electron Beam

IRED: Infrared Emitting Diode

IREE: Institute for Radio Engineering and Electronics • Institution of Radio and Electronics Engineers (Russia)

IREM: Institut de Recherche et d'Enseignement des Mathématiques

IRENA: Instituto Nicaragüense de Recursos Naturales y del Ambiente (Nicaragua)

IREP: Interdisciplinary Research Equipment Program

IREPA: Institut Régional de Promotion de la Recherche Appliquée

IReS: Institut de Recherches Subatomiques

IRES: Infrared Earth Sensor

IRESTE: Institut de Recherche et d'Enseignement Supérieur aux Techniques de l'Électronique (France)

IRET: Interrupt Return

IREX: International Research and Exchange

IRF: Information Retrieval Facility • Inheritance Rights Filter • Inherited Rights Filter • Instantaneous Radiative Flux • Institut de Recherche Fondamentale • Institutet för Rymdfysik (Sweden) • Internal Raster File • Internal Raster Format

IRFDA: Infrared Focal-Plane Assembly

IRFEST: Institut de Recherche sur les Fondements et les Enjeux des Sciences et des Techniques

IRFK: Institutet för Rymdfysik, Kiruna (Sweden)

IRFN: Inhibited Red Fuming Nitric

IRFNA: Inhibited Red Fuming Nitric Acid

IRFU: Institutet för Rymdfysik, Uppsala (Sweden)

IRFUm: Institutet för Rymdfysik, Umeå (Sweden)

IRG: Inertial Rate Gyro • Information Research Group • Infrared Galaxy • Infrared Generator • Inner Roll Gimbal

• Internet Research Group • Interrecord Gap • Lockhart River (Australia-QLD) (location identifier)

IRGA: Infrared Gas Analyzer

IRH: Informatique des Ressources Humaines • Initial Radiohologram

IRHC: Ingénieur de Recherche Hors Classe

IRHS: Infrared Heterodyne Spectrometer

Iri: Irish

IRI: Industrial Research Institute • Infrared Imagery • Infrared Instrumentation • Interagency Research Internet • International Reference Ionosphere • Ionospheric Research Instrument • Iringa (Tanzania) (location identifier) • Istituto per la Recostruzione Industriale (Italy)

IRIA: Infrared Information Analysis

IRIC: Infrared Image Converter

IRICC: Interagency Resource Information Coordination Council

IRICE: Instituto Rosario de Investigación en Ciencias de la Educación (Argentina-SF)

IRID: Infrared Image Display

IRIE: Infrared Imaging Equipment • Infrared Imaging Experiment • Infrared Information Exchange

IRIG: ILL Protocol Implementation Group • Inertial Rate Integrating Gyroscope • Inertial Reference Integrating Gyroscope • Interrange Instrumentation Group

IRIM: Infrared Imager

IRIMO: Islamic Rep. of Iran Meteorological Organization

IRIS: Image Reading Instrument System • Images, Reflets, Initiation Scientifique (France) • Incorporated Research Institutions for Seismology • Increased Readiness Information System • Information, Robotics and Intelligent Systems • Infrared Image Scanner • Infrared Imager Spectrograph • Infrared Imaging Spectrometer • Infrared Information Symposium • Infrared Instrumentation System • Infrared Intelligent Spectroradiometer • Infrared Interferometer Spectrometer • Infrared Intruder Sensor • Initiative for Research in Informatics applied to Society • Institute for Research in Information and Scholarships • Institute for Robotics and Intelligent Systems • Integrated Radar Imaging System • Integrated Raster Imaging System • Integrated Remote Imaging System • Integrated Resource Information System • Integrated Risk Information System • Integriertes Realtime Informationssystem • Interactive Radar Information System • Interactive Research Imaging System • Interactive Retrieval of Information System • Interconexión de Recursos Informáticos • International Radiation Investigation Satellite • International Radio Interferometric Surveying (*) • International Radio Interferometry Survey • International Relations Information Service • International Research and Information Service • International Research on the Interior of the Sun • Italian Research Interim Stage

IRISA: Institut de Recherche en Informatique et Systèmes Aléatoires

IRISI: Inter-Regional Information Society Initiative

IRISP: Infrared Imaging and Spectro-Polarimeter

IRIT: Institut de Recherche en Informatique de Toulouse (France)

IRJ: La Rioja (Argentina-LR) (location identifier)

IRJE: Internet Remote Job Entry

IrJST: Iranian Journal of Science Technology (periodical code)

IRK: Kirksville (USA-MO) (location identifier) • Implicit Runge-Kutta

Irl: Irlandés

IRL: Indexed Retrieval Language • Industrial Robot Language • Information Retrieval Language • Infrared Lamp • Infrared Light • Infrared Lens • Initiating Reference Letter • Initiating Reference List • In Real Life • Institute for Research on Learning • Interactive Reader Language • Interface Requirements List • Interrepeater Link • Ireland (international vehicle registration code) • Ireland (ISO country code) • Irlande (ISO country code) • Irish Pound (currency)

IRLAP: Infrared Link Access Protocol

IRLDA: Infrared Linear Detector Array

IRLED: Infrared Light Emitting Diode

IRLM: IMS Resource Lock Manager • Internal Resource Lock Manager

IRLS: Infrared Line Scanner • Integrated Remote Imaging System • Interrogation, Recording and Location System • Interrogation Recording Location System • Iteratively Reweighted Least Squares

IRM: IERS Reference Meridian • Image Refresh Memory • Imagerie par Résonnance Magnétique • Induced Remanent Magnetization • Information and Records Management • Information Resources Management • Infrared Mapper • Infrared Measurement • Infrared Montgolfier • Inherent Rights Mask • Initial Release Memorandum • Institut Royal Météorologique • Integrated Range Mission • Interconnect Resource Manager • Interim Research Memo • Intermediate Range Monitor • International Relations Manager • Ion Release Module • Isothermal Remanent Magnetization

IRMA: Information Régionale Météorologique Automatisée • Information Revision and Manuscript Assembly • Infrared Michelson Array • Institut de Recherche Mathématique Avancée

IRMB: Institut Royal Météorologique de Belgique (Belgium)

IRMBP: Institut Royal Météorologique de Belgique - Publications (periodical cod e)

IRME: Initiator Resistance Measuring Equipment

IRML: Interactive Media Research Laboratory

IRMP: Information Resources Management Plan • Infrared Measurement Program

IRMS: Infrared Mapping System • Information Resources Management System • Information Retrieval and Management System • Isotope Ratio Mass Spectrometry

IRN: Interface Revision Note • Interface Revision Notice • Interim Revision Notice • Internal Reference Number • Invoice Register Number • Iran (ISO country code) • Iriona (Honduras) (location identifier)

IRO: Birao (Central African Rep.) (location identifier) • Industrial Relations Officer • Inland Revenue Office • Institut für Rationalisierung und Organisation • Internal Representation Optimization • International Refugee Organization • International Relief Organization • Iowa Robotic Observatory (USA-IA)

IROAN: Inspect and Repair Only As Necessary

IROD: Instantaneous Readout Detector

IROMM: International Register of Microform Masters

Irón: Irónico

IROR: Internal Rate Of Return

IROS: Increase Reliability of Operational Systems • Instant Order Response System

IRP: IERS Reference Pole • Image Restoration Project • Independent Research Project • Individualized Reading Program • Inertial Reference Package • Infrared Passive • Initial Receiving Point • Innovative Research Program • Installation Restoration Program • Institut de Recherches et d'Études Politiques (France) • Integrated Resource Planning • Interest Rate Parity • Invited Review Paper • I/O Request Packet • Isiro (Zaire) (location identifier)

IRPA: Intensification of Research in Priority Areas • International Radiation Protection Agency • International Radiation Protection Association

IRPS: Image Retrieval and Processing System • Infrared Photometer Spectrometer • Infrared Photometer System

IRPTC: International Register of Potentially Toxic Chemicals

IRQ: Interrupt Request • Iraq (international vehicle registration code) • Iraq (ISO country code)

IRQC: Infrared Quantum Counter

Irr: Irregular

IRR: Independent Readiness Review • Information Research and Retrieval • Infrared Radiation • Infrared Radiometer • Infrared Receiver • Infrared Reflectography • Initial Requirements Review • Inspection Rejection Report • Installation and Removal Record • Integral Rocket Tamjet • Integrated Radio Room • Integrated Readiness Review • Interface Requirements Review • Interim Release Request • Internal Rate of Return • Rial (Iran) (ISO currency code)

IRRD: International Road Research Documentation

Irreg: Irregular • Irregularly

Irrglr: Irregular

IRRI: International Rice Research Institute

IRRL: Information Retrieval Research Laboratory

IRRMP: Infrared Radar Measurement Program

IRRS: Infrared Reflection Spectroscopy

IRRV: Institute of Revenues, Rating, and Valuation

I&RS: Information & Referral Service • Instrumentation & Range Safety

IRS: Indian Remote-sensing Satellite • Indian Resources Satellite • Inertial Reference System • Information and Referral Service • Information and Research Service • Information Recovery System • Information Retransmission System • Information Retrieval Service • Information Retrieval System • Informationsrecherchesystem • Infrared Scanner • Infrared Sounder • Infrared Source • Infrared Spectrograph • Infrared Spectrometer • Infrared Spectroscopy • Institut de Recherche Scientifique • Institut für Raumfahrtsysteme • Instrumentation and Range Safety • Integrated Rate System • Integrated Record System • Intensified Reticon Scanner • Interchange Record Separator • Interface Requirements Specification • Intermediate Reference Structure • Intermediate Resolution Spectrograph • Internal Reflection Spectroscopy • Internal Revenue Service (USA) • International Radiation Symposium • International Radio Silence • International Reference Stars • International Referral System • Ionospheric Radio Signal • Romanian Institute of Normalization • Sturgis Kirsch (USA-MI) (location identifier)

IRSA: Infrared Science Archive • Institute for Remote Sensing Applications

IRSAP: Interactive Remote Sensing Applications Package

IRSBF: Infrared Surface Brightness Fluctuation

IRSC: International Relations Subcommittee • Iranian Remote Sensing Center

IRSCS: Interregional Subject Covering Scheme

IRSD: Information Retrieval, Storage and Dissemination

IRSG: Information Retrieval Specialist Group • Internet Research Steering Group

IRSHELL: Infrared Echelle Spectrometer

IRSI: Infrared Space Interferometer • Infrared Space Interferometry

IRSIA: Institut pour l'encouragement de la Recherche Scientifique dans l'Industrie et l'Agriculture (Belgium)

IRSN: Institut Royal des Sciences Naturelles (Belgium)

IRSNB: Institut Royal des Sciences Naturelles de Belgique (Belgium)

IRSOL: Istituto Ricerche Solari Locarno (Switzerland)

IRSp: Informationsrecherchesprache

IRSPEC: Infrared Spectrometer

IRSR: Infrared Scanning Radiometer

IRSS: Indian Remote Sensing Satellite • Inertial Reference Stabilization System • Infrared Sensor System

IRSSP: Interactive Remote Sensing Software Package

IRS&T: Infrared Search & Track

IRST: Infrared Search and Track • Institut de Recherches Scientifiques et Techniques

IRSTS: Infrared Search and Track System

IRSU: Infrared Radio Scientific Union

IRSV: Infrared Survey of Valinhos

IRT: Icing Research Tunnel • Individual Reliability Test • Information Retrieval Technique • Informationsrecherchetechnologie • Infrared Radiation Thermometer • Infrared Telescope • Infrared Thermometer • Infrared Tracker • Infrared Tube • Initialize Reset Tape • In Reference To • In Regard To • In Reply To • In Response To • Instantaneous Radiative Transfer • Institut für Rundfunkteknik • Integrated Radar Target • Integrated Readiness Testing • Interborough Rapid Transit (USA-NY) • IUE Review Team

IRTAD: International Road and Traffic Accident Database

IRTC: Information Technology Research Centre

IRTCM: Integrated Real-Time Contamination Monitor

IRTE: Integrated Road Telematics Environment

IRTF: Infrared Telescope Facility (USA-HI) • Infrared Tully-Fisher • Internet Research Task Force

IRTFRD: Internet Research Task Force on Resource Discovery

IRTM: Infrared Thermal Mapper • Infrared Thermal Mapping • Infrared Transparent Material

IRTOS: Intelligent Real-Time Operating System

IRTP: Integrated Reliability Test Program

IRTS: Infrared Telescope in Space • Infrared Temperature Sounder

IRU: Indefatigable Right of Use • Industrial Rehabilitation Unit • Inertial Reference Unit • Information Retrieval Unit • Instrument Reference Unit • Intermediate Registration Unit

IRUS: Irvine Research Unit in Software (USA-CA)

IRV: Interange Vector • Isoradial Velocity • Isotope Reentry Vehicle

IRVR: Instrument Runway Visual Range

IRW: Indirect Reference Word • Infrared Window

IRWC: International Registry of World Citizens

IRWG: Image Reconstruction Working Group • Image Restoration Working Group

IRWL: Interchangeability and Replaceability Wording List

IRX: Information Retrieval Experiment

IRZ: Informations- und Recherchezentrum

is: Iceland • Icelandic (ISO language code)

iS: im Sinn

Is: Îles • Îlets • Ilhas • Islam • Islamic • Island • Islands • Islas • Isle • Isles • Islets • Isolas • Israel • Israeli

I&S: Interchangeability & Substitutability

I/S: Installation of Systems

IS: Iceland (international vehicle registration code) • Iceland (ISO country code) • Iceland (postal country code) • Image Selection • Image Sharpening • Image Subtraction • Incomplete Sequence • Indexing in Source • Indian Standard • Industrial Society • Information Science • Information Separator • Information Series • Information Service • Informationsstelle • Informationssystem • Information System • Initialize • Input Secondary • Inquiry Server • Installation Support • Instruction Sheet • Instrumentation Summary • Instrumentation System • Instrument

Scientist • Instrument Section • Instrument Station • Integrated Satellite • Integrated Software • Interchangeability and Substitutability • Interconnecting Station • Intermediate School • Intermediate System • Internal Shield • International Staff • International Standard • Interplanetary Scintillation • Interrupt Set • Interrupt Status • Interstellaire • Interval Signal • Intrinsically Safe • Intrinsic Safety • Invited Speaker • Island • Islande (ISO country code) • Isomer Shift • Israel

ISA: Ignition and Separation Assembly • Imprimé Sans Adresise • Industrial Standard Architecture • Industry Standard Architecture (*) • Inertial Sensor Assembly • Information Science Abstracts • Initial Sample Address • Initial Sun Acquisition • Institute of Space Aeronomy • Instructional Systems Approach • Instruction Set Architecture • Instrumentation Society of America (*) • Instrument Society of America • Instruments SA Inc. • Insulated Structure Assembly • Integrated Services Architecture • Integrated Support Area • Intelligent Software Assistant • Interactive Services Association • Interface Switching Assembly • Interim Stowage Assembly • Intermediate Specific Activity • Internal Security Act • International Searching Authority • International Sociological Association • International Standard Atmosphere • Internet Security Association • Interstage Adapter • Israel Space Agency • Mount Isa (Australia-QLD) (location identifier)

ISAA: Israel Society of Aeronautics and Astronautics

ISAAC: Infrared Spectrograph And Array Camera • Infrared Spectrometer And Array Camera (*) • International Symposium on Algorithms And Computation • Internet Security, Applications, Authentication and Cryptography

ISAAG: Imaging Spectrometer Science Advisory Group

ISAB: International Scientific Advisory Board • International Society for Adaptive Behavior

ISABP: International South Atlantic Buoy Programme

ISAC: Information Society Activity Center • INTELSAT Solar Array Coupon • Interferometry Science Advisory Committee • ISRO Satellite Centre (India)

ISAD: Information Society And Development

ISADPM: International Society for the Abolition of Data Processing Machines

ISADS: International Symposium on Autonomous Decentralized Systems

ISAGEX: International Satellite Geodesy Experiment

ISAIA: Interoperable Systems for Archival Information Access

ISAKMP: Internet Security Association and Key Management Protocol

ISAL: Information Service Access Line • Information System Access Line • International Surface Air Lift • Investigation of STS Atmospheric Luminosities

ISAM: Indexed Sequential-Access Management • Indexed Sequential-Access Method • Integrated Switching And Multiplexing

ISAMS: Improved Stratosphere And Mesosphere Sounder • Improved Stratospheric And Mesospheric Sounder (*)

ISAN: International Standard Agreement Number • International Standard Astronomical Number • International Standard Audiovisual Number

ISAO: Istituto di Scienze dell'Atmosfera e dell'Oceano (Italy)

ISAP: Index to South African Periodicals • Information Sort And Predict • Integrated System for Agriculture Productivity • ISO Spectral Analysis Package

ISAPI: Information Server Application Programming Interface • Internet Server Application Programming Interface (*) • Internet Service Application Programming Interface

ISAR: Information Storage And Retrieval • Intelligent Synthetic Aperture Radar • Interferometric Synthetic Aperture Radar • Inverse Synthetic Aperture Radar

ISARA: Interferometric Synthetic Aperture Radar Altimeter

ISARD: Integrated Support for Aeronautical Research and Development

ISAS: Imager Spectrometer and Atmospheric Sounder • Institute of Space and Astronautical Science (Japan) • INTEGRAL Scientific Analysis System • International School for Advanced Studies (Italy) • Ipswich School Astronomical Society (UK)

ISATS: Integration Spacecraft Automated Test System

ISAV: Instituto de Sistemas Audiovisuales

ISB: Idependent Sideband • Instrument Support Branch • Internal Serial Bus • International Society of Biometeorology • Islamabad (Pakistan) (location identifier)

ISBA: Interactions Surface-Biosphère-Atmosphère • International Seabed Authority • International Society for Bayesian Analysis

ISBD: International Standard Bibliographic Description

ISBF: Interactive Search of Bibliographic Files

ISBL: Information System Base Language • Institution Sans But Lucratif

ISBN: Integrated Satellite Business Network • International Standard Book Number

ISBRv: International Space Business Review (periodical code)

ISC: Index of Status Characteristics • Industrial Support Contractor • Inertial Start Command • Information Service Center • Information Society of Canada • Information Systems Center • Information Systems Committee • Initial Software Configuration • Institute for Scientific Computation • Institut Supérieur du Commerce (France) • Instruction Set Computer • Integrated Storage Control • Intelligent Subject Code • Interdisciplinary Scientific Commission • Interferometry Science Center • Interim Steering Committee • International Seismological Centre • International Signal and Control • International Space Corporation • International Statistical Classification • International Statistical Congress • International Steering Committee • International Student Conference • International Student Council • International Switching Center • Inter-State Commission • Interstellar Cloud • Interstellar Communication • Isles of Scilly - Saint Mary's (UK) (location identifier) • Item Status Code • IVS Steering Committee

ISCA: Information Systems Consultants Association • International Speech Communication Association • International Symposium on Computer Architecture • Ionization Spectroscopy for Chemical Analysis

ISCAN: Inertialess Steerable Communications Antenna

ISCC: International Standard Commodity Classification

ISCCB: Internal Software Change Control Board

ISCCP: International Satellite Cloud Climatology Project

ISCD: Interface Specification Control Document

ISCE: Integrated Scientific Computing Environment • International Staff Committee Europe (ESO)

ISCED: International Standard Classification of Education

ISCGM: International Steering Committee for Global Mapping

ISCJS: ISRO Satellite Centre Journal of Spacecraft Technology (periodical code)

ISCM: Initial Software Configuration Map

ISCMS: Integrated Supply Chain Management System

ISCN: International Software Consulting Network • International Standard Cartographic Number

ISCO: Initial Systems Checkout • International Soil Conservation Organisation (India) • International Standard Classification of Occupations

ISCP: Integrated Services Control Point

ISCR: Institute for Scientific Computing Research

ISCS: Information Service Computer System • Intermediate Science Curriculum Study • International Solar Cycle Studies

ISD: Image Section Descriptor • Independent School District • Information Systems Development • Information Systems Division • Inspection Supervision Disposition • Instructional Systems Design • Instructional Systems Development • Instructional Systems Device • Instruction Sequence Diagram • Integrated Symbolic Debugger • International Standard Depth • International Subscriber Dialing • Interstellar Dust • Iscuande (Colombia) (location identifier)

ISDA: Imaging Spectrometer Data Analyzer • INTEGRAL Spectrometer Data Analysis Group • Interactive Software Developers Association • Invariant Subspace Decomposition Algorithm

ISDB: Information Systems Development Branch • IRF Scientific Database

ISDC: INTEGRAL Science Data Centre • International Space Development Conference

ISDE: Integrated Software Development Environment

ISDEIV: International Symposium on Discharges and Electrical Insulation in Vacuum

ISDF: Institut de Sûreté De Fonctionnement

ISDN: Integrated Services Data Network • Integrated Services Digital Network (*) • Integrated Switch Data Network • International Standard Decimal Number

ISDS: Illini Space Development Society (USA-IL) • Inadvertent Separation Destruct System • International Serials Data System

ISDSS: International Society for Decision Support Systems

ISDT: Integrated Systems Development Tool

ISDWG: Information Systems Development Working Group

ISE: Image Symbol Editor • Information and Software Engineering • Information Science Education • Information Systems Engineering • Information Systems Executive • Installation Support and Evaluation • Institute for Software Engineering (USA) • Instrumentation Suitability Evaluation • Integral Squared Error • Interact Simulation Environment • International Space Enterprises (USA-CA) • International Space Explorer • International Stock Exchange • Interpret Sign Error • Ion Selective Electrode

ISEA: In-Service Engineering Agent • International Symposium on Electronic Art

ISEC: Information Security Evaluation Criteria

ISECCo: International Space Exploration and Colonization Company (USA-AK)

ISEE: International Sun-Earth Explorer

ISEF: International Science and Engineering Fair

ISELA: International Space Enterprises - Lavochkin Associates

ISEM: Information Systems, Engineering and Management • International Society for Ecological Modelling

ISEMSI: Isolation Study for the European Manned Space Infrastructure

ISEN: Interactive Satellite Education Network • International Standard Exploration Number

ISENA: Imaging Spectrometer for Energetic Neutral Atom

ISEP: Institute of Solid-Earth Physics (Norway) • Instructional Scientific Equipment Program • Isolated/Stabilized Exercise Platform

ISEPS: International Sun-Earth Physics Satellite

ISES: Information Sciences Experiment System • Information-Society-Eenrgy-System • International Solar Energy Society • International Space Environment Service

ISESCO: Islamic Educational, Scientific and Cultural Organization

ISF: Impôt de Solidarité sur la Fortune (France) • Industrial Space Facility • Industrial Support Fund • Information System Factory • Infrasonic Frequency • Installation de Simulation Fonctionnelle • Integrated Space Factory • Integrated Subject File • Integrated Systems Facility • International Science Foundation • Israel Science Foundation

ISFA: International Scientific Film Association

ISFET: Ion-Selective Field Effect Transistor • Ion-Sensitive Field Effect Transistor (*)

ISFJ: Introverted Sensing Feeling Judging

ISFL: International Scientific Film Library

ISFMS: Indexed Sequential File Management System

ISFN: International Standard File Number

ISFP: Introverted Sensing Feeling Perceiving

ISFRA: Institut Supérieur de Formation et de Recherche Appliquée

ISFUG: Integrated Software Federal User Group

ISG: Integrated Survey Grid • Ishigaki (Japan) (location identifier)

ISGA: International Society for Genetic Algorithms (now ISGEC)

ISGEC: International Society for Genetic and Evolutionary Computation

ISGI: International Service of Geomagnetic Indices

Isgn: Insignia

ISGN: International Space Geodetic and Gravimetric Network • International Space Geodetic Network • International Standard Graphical Number

ISGRG: International Society on General Relativity and Gravitation

ISGS: INTEGRAL Science Ground Segment

ISH: Information Superhighway • Ischia (Italy) (location identifier)

ISHA: Integrated System Hazard Analysis

ISHAM: International Society for Human and Animal Mycology

ISHS: International Society of Horticultural Science

ISI: Image System Interface • Indian Standards Institution (India) • Indian Statistical Institute • Induced Spatial Incoherence • Information Sciences Institute • Information Society Initiative • Information Systems Inventory • Infrared Spatial Interferometer • Infrared Spectroscopic Imaging • Initial Systems Installation • Institute for Scientific Information • Institut Supérieur d'Informatique • Instrumentation Support Instruction • Integrated Solutions Inc. • Intelligent Serial Interface • Internally Specified Index • International Statistical Information • International Statistical Institute • International Systems Institute • Ion Source Injector • Isisford (Australia-QLD) (location identifier)

ISIC: Information Seeking In Context • International Standard Industrial Classification • International Student Identity Card

ISID: International Symposium on Inverse Problems

ISIDE: Interactive System for Image Data Elaboration

ISIF: Intermediate Standard Transformation Format

ISIL: Interim Support Items List

ISIM: Instrument Simulator • Integrated Science Instrument Module

ISIMM: International Society for the Interaction of Mechanics and Mathematics

ISIN: International Standard Image Number

ISIP: International Science and Its Partners

ISIR: Improved SGM Imaging Radiometer • Infrared Spectral Imaging Radiometer • International Space Industry Report

ISIRA: International Science Initiative in the Russian Arctic

ISIRI: Institute of Standards and Industrial Research of Iran

IS-IS: Intermediate System to Intermediate System

ISIS: Inflatable Sunshield In Space • Institute for Science and International Security (USA-DC) • Integrated Safety Information System • Integrated Scientific Information System • Integrated Software for Image Spectrometry • Integrated Systems and Information Services • Integrated Standards Information System • Interactive Sharable Image System • Interferometric Stellar Imaging System • Intermediate-dispersion Spectrograph and Imaging System • Intermediate System to Intermediate System • Internally Switched Interface System • International Satellite for Ionosphere Studies • International Satellite for Ionospheric Studies (*) • International Space Information Service • Investigative Support Information System • ISRIC Soil Information System • Item Standardization Information System

ISITE: Integrated Self-Instructive Test Equipment • Intensified Silicon Intensifier Target

ISITEM: Institut des Sciences de l'Ingénieur en Thermique-Énergétique des Matériaux (France)

ISJ: Isla Mujeres (Mexico) (location identifier)

IsJT: Israel Journal of Technology (periodical code)

ISK: Icelandic Króna (currency) • Ion Source Kit • Krona (Iceland) (ISO currency code) • Nasik (India) (location identifier)

ISKN: International Standard Key Number

ISKO: International Society for Knowledge Organization

Isl: Islamic • Island • Isle

ISL: Iceland (ISO country code) • Inertial Systems Laboratory • Information Search Language • Information System Language • Initial System Loading • Institute of Space Law • Instrument Standards Laboratory • Interactive Simulation Language • Interactive System Language • Internetworking Switch Link • Intersatellite Link • Inverse Square Law • Isabel Pass (USA-AK) (location identifier) • Island • Islande (ISO country code) • Item Selection List

ISLA: Information System for Los Angeles (USA-CA) • International Stratospheric Laboratory for Astrophysics

ISLC: International Semiconductor Laser Conference

Isld: Island

Islds: Islands

ISLM: Integrated Services Line Module • Integration Shop and Lab Manager

Isln: Isolation

ISLR: Integrated Sidelobe Ratio

Isls: Islands

ISLS: Islands

ISLSCP: International Satellite Land Surface Climate Project • International Satellite Land Surface Climatology Program • International Satellite Land Surface Climatology Project (*)

ISLSSPWG: International Space Life Sciences Strategic Planning Working Group

ISLU: Integrated Services Line Unit

ISLV: Institut Supérieur des Langues Vivantes

ISM: Image Sensor Module • Image Steering Mirror • Image Storage Module • Industrial-Scientific-Medical • Industrial Security Manual • Information Sources Map • Information System Manager • Infrared Sky Monitor • Infrared Spectrometer • Institut Supérieur des Métiers (France) • Institut Suisse de Météorologie (Switzerland) • Institut Svizzer da Meteorologia (Switzerland) • Integrated System Management • Interactive Surface Modelling • Intermediate Service Module • International Safety Management • International Solar Month • International Staff Member • Internet Service Manager • Interpretive Structural Modelling • Interstellar Matter • Interstellar Media (USA-CA) • Interstellar Medium • Istituto Svizzero di Meteorologie (Switzerland) • Kissimmee (USA-FL) (location identifier)

ISMA: International Satellite Monitoring Agency

ISMARSC: Information Systems Mapping And Remote Sensing Consultancy

ISMB: Information Systems Management Board • Intelligent Systems for Molecular Biology

ISMC: Inflight Services Management Council • International Seminars on Mathematical Cosmology

ISMCS: International Station Meteorological Climate Summary

ISME: Instrument Software Maintenance Environment

ISMF: Interactive Storage Management Facility

ISMIS: International Symposium on Methodologies for Intelligent Systems

ISMLS: Interim Standard Microwave Landing System

ISMM: INTEGRAL Software Management Meeting

ISMN: International Standard Music Number

ISMRB: Information Systems Management Review Board

ISMS: Image Store Management System • Interactive Supply Management System • International Society for Mine Surveying • International Society for Mushroom Science

ISN: Information Systems Network • Initial Sequence Number • Institut des Sciences Nucléaires • Integrated Systems Network • Internal Sequence Number • Internal Statement Number • International Standard Nomenclature • International Supernovae Network • Internet Shopping Network • Williston (USA-ND) (location identifier)

ISNAR: International Service for National Agricultural Research (Netherlands)

ISND: IGN Seismic Network Databank

ISNS: International Satellite Navigation System

ISNU: Illinois State Normal University (USA-IL)

ISO: Imaging Spectrometer Observatory • Imaging Spectrometric Observatory (*) • Imaging Spectroscopic Observatory • Incentive Stock Option • Individual System Operation • Information Security Office • Information Systems Office • Infrared Space Observatory • In Search Of • Institute for Space Observations (USA-NY) • International Organization for Standardization • International Science Organization • International Standards Organization • Interoperable System Office • Isolation • ISRO Safety Office • Kinston (USA-MO) (location identifier)

ISOC: INTEGRAL Science Operations Centre • Integrated Science Operations Center • Internet Society • IVS Science Operations

ISOCAM: Infrared Satellite Observatory Camera

ISOCLS: Interactive Self-Organizing Cluster Program

ISODE: International Standards Organization Development Environment

ISOE: Integrated Sequence Of Events

ISO-IP: International Standards Organization Internetworking Protocol

Isol: Isolate • Isolated • Isolation

ISOL: Isolated • Isotope Separator Outline

ISOLD: Isolated

ISOM: Information Systems and Operations Management

ISOMAX: Isotope Magnet Experiment

ISON: International Standard Organization Number

ISONET: ISO Information Network

ISOO: Information Security Oversight Office

ISOON: Improved Solar Observing Optical Network

ISOP: Integrated Spacecraft Operations Plan • International Seismological Observation Period

ISOPAR: Improved Symbolic Optimizing Assembly Routine

ISOS: International Southern Ocean Studies

ISOSDC: Infrared Space Observatory Spectrometer Data Center

ISO-SP: International Standards Organization Session Protocol

ISOT: International Scientific Oversighht Team

ISOTOF: Isochronous Time-Of-Flight

ISO-TP: International Standards Organization Transport Protocol

ISOWAKE: International Standardized Oil Wake Experiment

ISOWG: Information Systems and Operations Working Group

ISP: Independent Study Program • Independent Study Project • Indexed Sequential Processor • Informatie Stimuleringsplan • Information Systems Plan • Information Systems Program • Initial Specific Impulse • Institute of Solid-state Physics • Instruction Set Processor • Instructional System Package • Instrumentation Support Plan • Integrated Space Plan • Integrated Steel Plant • Integrated Support Plan • Intelligent Services Peripheral • Interim Support Plan • Internally Stored Program • International Scientific Product • International Scientific Publication • International Society for Photogrammetry • International Standardized Profile • International Study Program • Internet Service Provider • Interrupt Stack Pointer • Interrupt Status Port • Inverse Sampling Procedure • Inverse Scattering Problem • Islip - Long Island (USA-NY) (location identifier) • Isotope Separation Power • Italian Society of Physics • Long Island - Macarthur (USA-NY) (location identifier)

ISPA: International Software Producers Association • Internet Service Providers Association • Inverted Socket Process Architecture

ISPAE: Institute for Space Physics, Astrophysics, and Education

ISPBX: Integrated Services PBX

ISPC: International Statistical Program Center • ISI Studies and Publishing Center

ISPCS: Italian Solar Physics Community Server

ISPDR: Integrated System Preliminary Design Review

ISPEC: International Space Physics Educational Consortium (USA-CA)

ISPF: Interactive System Productivity Facility • Interactive System Programming Facility

ISPG: Institutional Support Planning Group

ISPI: Intermediate-Scale Product Inventory • Internationales System für Patentinformation

ISPL: Incremental System Processing Language

ISPM: IEEE Signal Processing Magazine (periodical code) • International Solar Polar Mission (now Ulysses)

ISPN: International Standard Picture Number • International Standard Program Number • International Standard Protocol Number

ISPO: Information Society Project Office

ISPP: Industry Standard Plotting Package • Information System for Policy Planning • Institute for Solid-State Physics

ISPR: Institute for Survey and Policy Research • International Society for Paranormal Research • International Society for Psychotherapy Research • International Standard Payload Rack • Inter Services Public Relations

ISPRS: International Society for Photogrammetry and Remote Sensing

ISPS: Instruction Set Processor Specification • International Standard Paper Size

ISPT: Inverse Scattering Perturbation Theory

ISQ: In Status Quo • Manistique - Schoolcraft County (USA-MI) (location identifier)

ISQAP: International Standards and Quality Assurance Programme

ISQL: Interactive Structured Query Language

Isr: Israel • Israeli

IS&R: Information Storage & Retrieval

ISR: Image Storage/Retrieval • Incoherent Scatter Radar • Independent Science Review • Index to Scientific Reviews • Industrial Security Regulation • Information Storage and Retrieval • Information Systems Research • Informatique de Systèmes et Réseaux • Infrared Scanning Radiometer • Infrared, Submillimeter and Radio • Ingoing Shortwave Radiation • Initial System Release • Institute for Space Research • Institute of Social Research • Instrument Science Report • Integrated Safety Review • Integrated Switch Router • Intelligence-Surveillance-Reconnaissance • Interim Scientific Report • Interim Shop Release • Interim System Review • Intermediate Session Routing • Internal Scientific Report • International Simple Resale • International Society for Radiology • Interrupt Service Routine • Intersecting Storage Ring • Intersecting Storage Road • Inventory Status Report • Israel (ISO country code)

ISRA: Intelligente Systeme Roboter und Automatisierung

ISRAD: Information Storage, Retrieval And Dissemination

ISRAM: International Symposium on Robotics and Automated Manufacturing

ISRAMAP: Information System for Regional Automated Mapping Analysis and Planning

ISRB: Integrated System Review Board

ISRC: International Standard Recording Code

ISR&D: Information Storage, Retrieval & Dissemination

ISRD: Information Storage, Retrieval and Dissemination

ISRF: Interstellar Radiation Field

ISRG: Independent Space Research Group • Inferred Stochastic Regular Grammar

ISRI: Information Science Research Institute

ISRIC: International Soil Reference and Information Centre (Netherlands)

ISRM: Information System Resource Manager • International Society of Rock Mechanics

ISRN: International Standard Recording Number • International Standard Reporting Number • International Standard Report Number

ISRO: Indian Space Research Organization • Institute for Space Research Observatory

ISRS: Inflatable Space-Rigidized Structure • Information Search and Report System • Information Search and Retrieval System • Information Storage and Retrieval System • Integrated Status Reporting System

ISRU: In-Situ Resource Utilization

ISRv: Interdisciplinary Science Reviews (periodical code)

Iss: Issue

ISS: Ice Status System • Ignition Shielding System • Image Sensor System • Image Stabilization System • Imaging Science System • Industry Standard Specification • Inertial Sensor System • Inertial Subsystem • Information Search System • Information Seeking Situation • Information Sharing System • Information Subsystem • Information Support System • Initial Space Station • Input-State-Stability • In Seam Seismic • Installation Support Services

• Institute for Space Sciences • Institute for Space Studies • Institute for Systems Sciences • Institute of Soil Science • Instruction Set Simulator • Instruction Summary Sheet • Instrumentation Support Service • Instrument Servo System • Instrument Software System • Instrument Support Structure • Integrated Satellite System • Integrated Scheduling System • Integrated Sensor System • Integrated Sounding System • Integrated Storage System • Integrated Sunlight Spectrometer • Integrated Switching System • Integrated System Schematic • Integration Support Service • Interface Subset • Interface Subsystem • Interferometer Supervisor Software (VLTI) • Interim Stowage Shelf • Internal Switching System • International Scientific Secretariat • International Space Station • International Summer School • International Switching System • Intersatellite Service • Interstellar Scattering • Interstellar Scintillation • Ionosphere Sounding Satellite • Ion-Scattering Spectrometer • Ion-Scattering Spectrometry • Ion-Scattering Spectroscopy • Islands • Isotope Separation System • Wiscasset (USA-ME) (location identifier)

ISSA: Information Systems Security Association • Infrared Sky Survey Atlas • Institute of Space Sciences and Astronomy • Integrated Software Safety Analysis • International Space Station Alpha • IRAS Sky Survey Atlas

ISSAC: International Symposium on Symbolic and Algebraic Computation

ISSAG: Imaging Spectrometer Science Advisory Group

ISSAT: Institut des Sciences Spatiales et Applications de Toulouse (France)

ISSB: Information Systems Standards Board

ISSC: Integrated Systems Solution Corporation • International Social Science Council

ISSCC: International Solid-State Circuit Conference • International Space Station Coordination Committee

ISSCL: Integrated Services with Specific Link Layers (IETF)

ISSCO: Integrated Software Systems Corporation

ISSE: International Secondary School in Eindhoven (Netherlands)

ISSEP: International Soros Science Educational Programme

ISSI: International Space Science Institute (Switzerland)

ISSL: Initial Spares Support List

ISSM: Inertial Surveying System Mining

ISSMFE: International Society of Soil Mechanics and Foundation Engineering

ISSN: Integrated Special Services Network • International Standard Serial Number

ISSO: Information Systems Security Office • International Small Satellite Organization

ISSOF: International Solar System Observer's Fund

ISSOL: International Society for the Study of the Origin of Life

ISSP: ICSU Solar-System Panal • Information Systems Strategic Plan • International Solar Systems Program

ISSPP: Information Systems Strategic Planning Project

ISSR: Information Storage, Selection and Retrieval

ISSRE: International Symposium on Software Reliability Engineering

ISSS: IBM Speech Server Series • Information Society Standardization System (CEN) • Initial Sector Suite System • Intelligent Signals, Sensors and Surveillance • International School of Space Simulations • International Society for the Systems Sciences • International Society of Soil Science • International Student Space Simulations

ISSSR: International Symposium on Spectral Sensing Research

ISST: Institute for Space Science and Technology (USA-FL)

ISSTI: Information System for Scientific and Technical Information

ISSW: Instrument Specific Software

ISSWG: International ScaRab Science Working Group

Ist: Istituto

IS&T: Information Science & Technology • Information Systems & Technologies

IST: Icelandic Standard • Indian Standard Time • Information Science and Technology • Information Sciences Technology • Information Scientifique et Technique • Information Set Trailer • Information Society Technologies • Initial Service Test • Initial System Test • In-service School Teacher • Institut für Software-Technologie (Germany) • Institute of Science and Technology • Institute of Science Technology • Institute of Software Technology • Instituto Superior Técnico • Instrumentation Support Team • Instrument Science Team • Instrument Support Terminal • Integrated Systems Test • Integrierte Systemtest • International Space Technology Inc. • International Standard Thread • International Standard Time • Interstellar Travel • ISO Science Team • Istanbul Atatürk (Turkey) (location identifier)

ISTA: International Seed Testing Association • Intertank Structural Test Assembly

ISTAP: International Space Technology Assessment Program

ISTAR: Imagerie Stéréo Appliquée au Relief • Image Storage Translation And Reproduction • Information Storage Translation And Reproduction

ISTB: Information Sciences and Technology Branch • Integrated Subsystem Test Bed

ISTC: Incunabula Short Title Catalogue • Institute of Scientific and Technical Communicators • International Science and Technology Centre • International Student Travel Confederation

ISTD: In Situ Thermal Desorption • Integrated Space Technology Demonstration

ISTE: Information Scientifique, Technique et Économique • International Society for Technology in Education

ISTeC: International Small Telescope Cooperative (formerly NASTeC)

ISTEC: International Superconductivity Technology Center (Japan)

ISTEEM: Institut des Sciences de la Terre, de l'Eau et de l'Espace de Montpellier (France)

ISTEI: Institute for Scientific, Technical and Economic Information

ISTF: Integrated System Test Flow

ISTG: Institut des Sciences et Techniques de Grenoble (France)

ISTGG: Interferometric Spectrometer for Temperature and Greenhouse Gases

Isth: Isthmus

ISTI: Institut des Sciences et Techniques de l'Information • Institute of Scientific and Technical Information

ISTIC: Institute of Scientific and Technical Information of China

ISTJ: Introverted Sensing Thinking Judging

ISTL: Issues in Science and Technology Librarianship

ISTM: Information System Management Team

ISTN: Institute Sains and Technology Nasional (Indonesia)

ISTO: Institut des Sciences de la Terre d'Orléans (France)

ISTP: Index of Scientific and Technical Proceedings • Index of Scientific and Technical Publications • Integrated Systems Test Procedure • International Solar-Terrestrial Physics (*) • International Solar-Terrestrial Program • Introverted Sensing Thinking Perceiving

ISTR: Indexed Sequential Table Retrieval

ISTRA: Inter-Satellite Tracking

ISTRAC: ISRO Telemetry, Tracking And Command

ISTS: Institute for Space and Terrestrial Science (Canada-ON) (now CRESTech) • International Symposium on Space Technology and Science • Intersite Transmission System

ISU: Idaho State University (USA-ID) • Indiana State University (USA-IN) • Infrared Sounding Unit • Iowa State University (USA-IA) • Inertial Sensing Unit • Instrument Support Unit • International Scientific Union • International Space University • International System of Units • Iowa State University (USA-IA)

ISUP: Institut de Statistique de l'Université de Paris (France) • Integrated Services User Part

ISURSL: Indiana State University Remote Sensing Laboratory

ISUS: Imaging Stratospheric Ultraviolet Spectrometer • Information Services and User Support

ISV: Independent Software Vendor • Input Signal Voltage • International Scientific Vocabulary

ISVA: International Satellite Verification Agency

ISVL: Internet Software Visualization Laboratory

ISVR: Institute of Sound and Vibration Research

ISVTOP: International Saturn Voyager Telescope Observing Program

ISW: Industrial SPARC Workstation • Integrated Sachs-Wolfe • Interacting Stellar Wind • Wisconsin Rapids - Alexander (USA-WI) (location identifier)

ISWA: International Science Writers Association (USA-MA)

ISWC: International Standard Work Code

ISWE: Integrated Sachs-Wolfe Effect

ISWG: Imperial Standard Wire Gauge • Integrated Support Working Group

ISWN: International Standard Work Number

ISWT: INTEGRAL Science Working Team

ISX: Impurity Studies Experiment

ISY: International Space Year

ISYA: International School for Young Astronomers

ISZ: Iskustvenij Sputnik Zemli

it: Italian (ISO language code) • Italy

It: Italia • Italian • Italiano • Italie • Italien • Italy

I&T: Integration & Test

I/T: Intertank

IT: Identification Transponder • Ignition Temperature • Important Traveller • Importation Temporaire • Inclusive Tour • Income Tax • Index Term • Individual Training • Industrial Technology • Informatie Technologie • Information Technology • Information Theory • Ingénieur des Travaux • Injection Tank • Input Translator • Inspection Tag • Installation Test • Instrument Test • Integrated Technology • Integration and Test • Integration Test • Intelligent Terminal • Interline Transfer • Internal Thread • International Table • Intersection Table • Interval Timer • In Transit • Italie (ISO country code) • Italy (ISO country code)

Ita: Italian

ITA: Independent Television Authority • Industrial and Technical Assistance • Industrial Telecommunication Association (USA) • Ingénieurs-Techniciens-Administratifs • Initial Teaching Alphabet • Institut du Transport Aérien • Institute for Theoretical Astronomy • Institut für Theoretische Astrophysik • Instrumentation Technology Associates • Integrated Test Area • Integrated Test Article • Integrated Thruster Assembly • Integrated Truss Assembly • Ionization Test Apparatus • Itacoatiara (Brazil-AM) (location identifier) • Italie (ISO country code) • Italy (ISO country code) • Iterative Transfer Algorithm

ITAA: Information Technology Association of America

ITAC: IEEE Transactions on Automatic Control (periodical code) • International Target Audience Code • International Time Allocation Committee

ITAE: Integrated Time and Absolute Error

ITAES: IEEE Transactions on Aerospace Electronic Systems (periodical code)

Ital: Italian • Italic • Italicized • Italique • Italy

ITALY: I Trust And Love You

ITAM: Instituto Tecnológico Autónomo de México

ITAMP: Institute for Theoretical Atomic and Molecular Physics

ITAP: IEEE Transactions on Antennas and Propagation (periodical code) • Integrated Technical Assessment Panel

ITAR: International Traffic in Arms Regulations

ITARF: Ingénieurs, Techniciens et Administratifs de Recherche et de Formation

ITAS: IEEE Transactions on Applied Superconductivity (periodical code)

ITASE: International Trans-Antarctic Scientific Expedition

ITASS: IEEE Transactions on Acoustics, Speech, and Signal Processing (periodical code)

ITAV: Independent Test and Verification Plan

ITB: IEEE Transactions on Broadcasting (periodical code) • Industry Training Board • Information Technology Branch • Institut Teknologi Bandung (Indonesia) • Integral Terminal Block • Intermediate Text Block • Intermediate Transmission Block • International Time Bureau • Invitation To Bid • Ion Thruster Beam • Itaituba (Brazil-PA) (location identifier)

ITBD: Intraplate Tectonics and Basin Dynamics

ITBE: IEEE Transactions on Biomedical Engineering (periodical code)

ITC: Igloo Thermal Control • Independent Telephone Co. • Independent Television Commission • Industrial Training Council • Information Technology Center • Information Technology Committee • Ingénieur Technico-Commercial • Installation Time and Cost • Instructional Telecommunications Council • Integrated Telemetry Complex • Integrated Test Configuration • Integrated Trajectory Computation • Integration Time Calculator • International Telephone Code • International Teletraffic Congress • International Television Commission (UK) • International Television Committee • International Test Conference • International Toki Conference (Japan) • International Trade Center • International Trade Commission • International Traffic Congress • International Training Center • International Translation Centre • International Typeface Corporation • Intertask Communication • Intertropical Confluence • In-Trail Climb • Investment Tax Credit • Ionic Thermoconductivity

ITCA: Intrnational Topographic Composition Association

ITCAD: IEEE Transactions on Computer-Aided Design (periodical code)

ITCAN: Inspect, Test and Correct As Necessary

ITCF: Institut Technique des Céréales et des Fourrages (France

ITCG: Information Technology Coordinating Group

ITCHM: IEEE Transactions on Components, Hybrids and Manufacturing Techniques (periodica l code)

ITCmp: IEEE Transactions on Computers (periodical code)

ITCom: IEEE Transactions on Communications (periodical code)

ITCS: IEEE Transactions on Circuits and Systems (periodical code) • Internal Thermal Control System

ITCZ: Intertropical Convergence Zone

ITD: Ibrahim's Time Domain • Informal Test Description • Information Technology Division • Inner-Truncated Disk • Institute for Technology Development • Institute of Theoretical Dynamics • Integrated Technology Demonstration

• Integrated Test Document • Integrated Test Documentation • Interaural Time Difference • Interim Terrain Data • Intertropical Discontinuity • In-Trail Descent

ITDB: Inorganic Thermochemistry Database

ITDC: Instructional Technology Development Consortium

ITDI: Industrial Technology Development Institute

ITDM: Intelligent Time-Division Multiplexer

ITDP: International Telecom Development Corporation

ITDR: Integrated Technology and Design Review

ITDS: Improved Technical Data System • Integrated Technical Data System

IT&E: Information Technology & Engineering

ITE: Information Technology and Engineering • Information Technology Equipment • Input Test Equipment • Institute of Terrestrial Ecology • Instrumentation Test Equipment • Instrument Test Equipment • Instrument Test Engineer • Integral Throat Entrance • Integration Test Equipment • International Tunguska Expedition • Intersite Transportation Equipment • Inverse Time Element • Ituberá (Brazil-BA) (location identifier)

ITEC: IEEE Transactions on Electromagnetic Compatibility (periodical code) • Information Technology, Electronics and Communications • Information Technology Executive Committee

ITECS: Integrated Terminal Emulation Software

ITED: IEEE Transactions on Electronic Devices (periodical code)

ITEdu: IEEE Transactions on Education (periodical code)

ITEI: IEEE Transactions on Electrical Insulation (periodical code)

ITElC: IEEE Transactions on Electromagnetic Compatibility (periodical code)

ITEM: IEEE Transactions on Engineering Management (periodical code) • Information Technology in Educational Management

ITEME: Institution of Technician Engineers in Mechanical Engineering

ITEnC: IEEE Transactions on Energy Conversion (periodical code)

ITEP: Institute for Theoretical and Experimental Physics • Integrated Test and Evaluation Program • International Traineeship Exchange Program

ITEPC: Inter-Mars Tissue Equivalent Proportional Counter

ITER: International Thermonuclear Experimental Reactor • International Tokamak Engineering Reactor

Iterat: Iterativo

ITeSRE: Istituto di Tecnologie e Studio delle Radiazioni Extraterrestri (Italy)

ITF: Impulse Transfer Function • Integrated Thermal Flux • Intensity Transfer Function • International Training Forum • International Transfer Format • Intertropical Front • Inverse Trigonometric Function

ITFA: In The Final Analysis

ITFE: Inductive Terrain Feature Extraction

ITFS: IERS Terrestrial Reference Frame • Instructional Television Fixed Service

ITG: Independent Test Group • Informationstechnische Gesellschaft (Germany) • Information Technology Group • Integrated Terminal Guidance • Interactive Test Generator • International Technology Group • Internet Translation Gateway

ITGA: Isothermogravimetric Analysis

ITGB: Interféromètre à Très Grande Base

ITGE: IEEE Transactions on Geoscience Electronics (periodical code) • Instituto Tecnológico Geominero de España (Spain)

ITGRS: IEEE Transactions on Geosciences and Remote Sensing (periodical code)

ITH: Ithaca (USA-NY) (location identifier)

ITI: Indian Telephone Industries • Industrial Technology Institute • Information Technology Integration • Inspection and Test Instruction • Interactive Terminal Interface • International Technologies Inc. • International Thermal Instrument Co. (USA-CA) • International Training Institute • Itambacuri (Brazil-MG) (location identifier)

ITIA: IEEE Transactions on Industry Applications (periodical code)

ITIC: International Tsunami Information Centre • Internet Telephony Interoperability Consortium

ITIE: IEEE Transactions on Industrial Electronics (periodical code)

ITIEC: IEEE Transactions on Industrial Electronics and Control (periodical code)

ITIM: IEEE Transactions on Instrumentation Measurement (periodical code)

ITIMS: In-service Transmission Impairment Measurement Set

Itin: Itinerary

ITIP: IEEE Transactions on Image Processing (periodical code)

ITIR: Infrared Thermal Imaging Radiometer • Intermediate Thermal Infrared Radiometer

ITIS: Industrial Technical Information Service • Interagency Taxonomic Information System • Internal Translation Information System

ITIT: IEEE Transactions on Information Theory (periodical code)

ITK: Instrument Time Key • Itokama (Papua New Guinea) (location identifier)

Itl: Italian

ITL: Integrated Teaching and Learning • Integrated Transport Launcher • Integrate-Transfer-Launch • Integration, Test and Launch • Intent To Launch • Inverse Time Limit • Lira (Italy) (ISO currency code)

ITLB: Instruction Translation Lookaside Buffer

ITLD: International Top-Level Domain

ITLRS: Italian Transportable Laser Ranging System

ITLSA: Integrated Torso Limb Suit Assembly

ITM: IEEE Transactions on Magnetics (periodical code) • Information Tools and Methods • Ingénieur des Travaux de la Météorologie • Installation Team Meeting • Interim Technical Memorandum • Ionospheric-Thermospheric-Mesospheric • Osaka Itami (Japan) (location identifier)

ITMA: Institute of Trade Mark Agents

ITMag: IEEE Transactions on Magnetics (periodical code)

ITMG: Integrated Thermal Micrometeoroid Garment

ITMS: Intelligent Text Management System • Ion Trap Mass Spectrometry

ITMTT: IEEE Transactions on Microwave Theory Techniques (periodical code)

ITN: Identification Tasking and Networking • Independent Television News • Instituto Tecnológico e Nuclear (Portugal) • Interim Technical Note • Itabuna (Brazil-BA) (location identifier)

ITNN: IEEE Transactions on Neural Networks (periodical code)

ITNS: IEEE Transactions on Nuclear Science (periodical code)

ITO: Hilo (USA-HI) (location identifier) • Impulse Transfer Orbit • Indium Tin Oxide • Information Technology Office • Initial Timeout • Integrated Technical Order •

Integration and Test Order • Interface Technology Operation • International Test Organization • International Trade Organization

IToBu: Instituto de Tonantzintla Boletín (periodical code)

ITOCP: International Target-Oriented Complex Project

ITOF: Isochronous Time-Of-Flight

ITOG: Instrument Test and Operations Group

ITOpS: Information Technology Operation Services

ITOS: Improved TIROS Operational Satellite • Interactive Terminal Operating System

ITOY: International Tropospheric Ozone Year

ITP: Incidental Take Permit • Informal Test Plan • Informal Test Procedure • Informatic Technology Programme • Information Transformation Point • Input Translator Program • Inspection Test Procedure • Installation Test Program • Institute of Theoretical Physics • Instruction To Proceed • Integral Tile Part • Integrated Test Plan • Integrated Test Program • Integrated Training Plan • International Thomson Publishing • International Time Programme • International Time Project • International Training Programme • Internet Transaction Processing

ITPA: Institute for Terrestrial and Planetary Atmospheres • Institute for Theoretical Physics and Astronomy

ITPAMI: IEEE Transactions on Pattern Analysis and Machine Intelligence (periodical code)

ITPAS: IEEE Transactions on Power Apparatus Systems (periodical code)

ITPB: Integrated Test Program Board

ITPE: IEEE Transactions on Power Electronics (periodical code)

ITPO: International TOGA Project Office

ITPHP: IEEE Transactions on Parts, Hybrids, and Packaging (periodical code)

ITPP: International TOVS Processing Package

ITPR: Infrared Temperature Profile Radiometer

ITPS: IEEE Transactions on Plasma Science (periodical code)

ITQ: Individual Transferable Quota • Itaqui (Brazil-RS) (location identifier)

ITR: IEEE Transactions on Reliability (periodical code) • Incremental Tape Recorder • Informal Test Result • Initial Training Requirement • Inspection Test Report • Institute for Telecommunications Research (Australia-SA) • Instrumentation Tape Recorder • Integrated Technology Review • Integrated Test Requirement • Intense Thermal Radiation • Interfacial Thermal Resistance • Interim Technical Report • Interim Test Report • Internal Technical Report • Internal Throughput Rate • In-Transit Rendezvous

ITRA: IEEE Transactions on Robotics Automation (periodical code) • Intercomparison of Transmittance and Radiation Algorithms • International Telescope for Radio Astronomy

ITRB: Integrated Technical Review Board

ITRC: Information Technology Research Center

ITRD: Integrated Test Requirements Document

ITRDB: International Tree-Ring Databank

ITRF: IERS Terrestrial Reference Frame • International Terrestrial Reference Frame

ITRI: Industrial Technology Research Institute (USA-OH)

ITRM: Information Technology Resources Management

ITRR: Integrated Technology Requirements Review • Internal Throughput Rate Ratio

ITRS: International Terrestrial Reference System

ITS: Ignition Test Simulator • Incompatible Time-sharing System • Industrial Technology Services • Industrial Television Services Inc. • Infinite Time Span • Information and Technology Services • Informationstekniska Standardiseringen • Information Technology Security • Information Technology Standards • Information Transfer Satellite • Information Transfer System • Information Transmission Satellite • Information Transmission System • Inner Tracking System • Institute for Telecommunications Sciences • Institute of Telecommunication Science • Institute of Transportation Studies (USA) • Instrumentation Telemetry Station • Instrumentation Terrestre et Spatiale • In-Tank Solidification • Integrated Trajectory System • Integrated Transfer System • Integrated Truss Segment • Integrated Truss Structure • Intelligent Transportation System • Intelligent Transport System • Intelligent Tutoring System • Interactive Tutoring System • Intercept Trajectory Simulation • Interim Teleprinter System • Intermediate Tape Store • International Television System • International Temperature Scale • Internet Telephony Server • Internet Terminal Server • Ion Thrust System

ITSC: Intelligent Transportation System Council (IEEE) • International Telecommunications Satellite Consortium (now INTELSAT) • International TOVS Study Conference

ITSCG: Information Technology Strategy Cooperation Group

ITSE: IEEE Transactions on Software Engineering (periodical code)

ITSEC: Information Technology Security Evaluation Criteria

ITSIE: Intelligent Training System in Industrial Environments

ITSIG: Information Technology Strategies Implementation Group

ITSM: Instrument Team Software Manager

ITSMC: IEEE Transactions on Systems, Man, and Cybernetics (periodical code)

ITSN: Italian Telemetered Seismic Network

ITSP: IEEE Transactions on Signal Processing (periodical code) • Internet Telephony System Provider

ITSS: Incompatible Time-Sharing Systems • Information Technology and Scientific Services • Integrated Tactical Surveillance System

ITST: International Tsunami Survey Team

ITSTC: Information Technology Steering Committee

ITSU: IEEE Transactions on Sonics and Ultrasonics (periodical code) • Information Technology Standard Unit

itt: In Tegenstelling Tot

IT&T: Information Technologies & Telecommunications

ITT: Image Transformation Table • Integration and Test Team • Intensity Transfer Table • Internationales Teleskoptreffen • International Telephone and Telegraph • International Teleskop Treffen • Interstage Turbine Temperature • Inter-Turbine Temperature • Invitation To Tender • Invitation To Transmit

ITTA: Information Technology Training Association

ITTF: Information Technologies Task Force (*) • Information Technology Task Force

ITTI: Initiation à la Télédétection et au Traitement de l'Image

ITTO: International Tropical Timber Organization

ITTS: Interactive Text Transmission System

ITU: Image Transfer Unit • Infrared Telescope Unit • Institute for Trans-Uranium • Integrated Terrain Unit • Intensive Therapy Unit • International Technological University • International Telecommunication Union (formerly CCITT) • International Telegraphic Union • International Typographical Union • Istanbul Technical University (Turkey)

ITUFF: IEEE Transactions on Ultrasonics, Ferroelectrics, and Frequency (periodical code)

ITUG: International Telecommunications User Group

ITUTJ: International Telecommunication Union Telecommunication Journal (periodical code)

IT&V: Integrated Test & Verification

ITV: Independent Television • Industrial Television • Inspección Técnica de Vehículos • Inspection Technique des Véhicules • Instructional Television • Instrumented Test Vehicle • Integrated Technology Validation • Integrated Test and Verification • Integrated Test Vehicle • Interactive Television

ITVE: Integrated Technology Validation Experiment

ITVT: IEEE Transactions on Vehicular Technology (periodical code)

ITW: International TOAD Watch

ITWG: Interface Technical Working Group

ITWS: Integrated Terminal Weather System • International Tsunami Warning System

ITX: Information Technology Exchange • Intermediate Text

I/U: Instrumentation Unit

IU: Immunizing Unit • Impedance Unit • Indiana University (USA-IN) • Information Unit • Instrument Unit • Integer Unit • Interface Unit • Interface Utilisateur • International Unit • Internet User • Inter-Union

IUA: Inertial Unit Assembly • Interface Unit Adapter • International Union of Academies • International University of America

IUAA: International Union of Amateur Astronomers

IUAES: International Union of Anthropological and Ethnological Sciences

IUAP: Internet User Account Provider

IUAPPA: International Union of Air Pollution Prevention Associations

IUB: International Union of Biochemistry • International Universities Bureau

IUBMB: International Union of Biochemistry and Molecular Biology

IUBS: International Union of Biological Sciences

IUC: Information Unit for Conventions • International Union of Crystallography • International Users' Committee

IUCAA: Inter-University Centre for Astronomy and Astrophysics (India)

IUCAF: Interunion Commission on Frequency Allocations for Radio Astronomy and Space Science

IUCC: Information Unit on Climate Change • Interuniversity Committee on Computing

IUCF: Indiana University Cyclotron Facility (USA-IN)

IUCI: Interunion Committee on the Ionosphere

IUCN: International Union for the Conservation of Nature

IUCr: International Union of Crystallography

IUCRC: Industry-University Cooperation Research Centre

IUCRM: Interunion Commission on Radio Meteorology

IUCS: Instrumentation Update Command System • Interunion Commission on Spectroscopy

IUCSTP: Interunion Commission on Solar-Terrestrial Physics

IuD: Information und Dokumentation

IUD: Implementation Use Descriptor • Intrauterine Device

IUDS: Integrated Utilities Distribution System

IUE: Institut Universitaire Européen • Interface Unit Error • International Ultraviolet Explorer • Niue (location identifier)

IUEAC: International Ultraviolet Explorer Allocation Committee

IUEAIMS: International Ultraviolet Explorer Automated Information Management System

IUED: Institut Universitaire d'Études du Développement

IUEDAC: International Ultraviolet Explorer Data Analysis Center

IUEEN: International Ultraviolet Explorer ESA Newsletter (periodical code)

IUEF: International University Exchange Fund

IUEFA: International Ultraviolet Explorer Final Archive

IUEFPA: International Ultraviolet Explorer FES Photometric Archive

IUELDA: International Ultraviolet Explorer Low-Dispersion Atlas

IUEML: International Ultraviolet Explorer Merged Log

IUEOCC: International Ultraviolet Explorer Operations Control Center

IUESIPS: International Ultraviolet Explorer Spectral Image Processing System • International Ultraviolet Explorer Standard Image Processing Software

IUESOC: International Ultraviolet Explorer Science Operations Center

IUF: Institut Universitaire de France

IUFM: Institut Universitaire de Formation des Maîtres (France)

IUFOMC: International Unidentified Flying Object Monitoring Center

IUFoST: International Union of Food Science and Technology

IUFRO: International Union of Forestry Research Organizations

IUGG: International Union of Geodesy and Geophysics

IUGS: International Union of Geological Sciences

IUHEI: Institut Universitaire des Hautes Études Internationales (Switzerland)

IUHPS: International Union of the History and Philosophy of Science

IUI: Integrated User Interface • Intelligent User Interface • Interactive User Interface • Interim Use Item • International Union of Interpreters

IUIC: Interuniversity Information Committee

IUIS: International Union of Immunological Societies

IUL: Ilu (Indonesia) (location identifier)

IULA: International Union of Local Authorities

IULC: Interuniversity Library Council

IUM: Interim Use Material • Summit Lake (USA-CA) (location identifier)

IUMRS: International Union of Materials Research Societies

IUMA: Interim Use Material Authorization

IUMJ: Indiana University Mathematics Journal (periodical code)

IUMP: International Upper Mantle Program

IUMS: International Union of Microbiological Sciences • International Union of Microbiological Societies (*)

IUNS: International Union of Nutritional Sciences

IUNT: Instituto Uruguayo de Normas Técnicas (Uruguay)

IUO: Interunit Order

IUP: Indiana University of Pennsylvania (USA-PA) • Indiana University Press (USA-IN) • Installed User Program • Institut Universitaire Professionalisé (*) • Institut Universitaire Professionnel • Irish University Press • IRS Utilisation Programme • Israel Universities Press

IUPAB: International Union of Pure and Applied Biophysics

IUPAC: International Union of Pure and Applied Chemistry

IUPAP: International Union of Pure and Applied Physics • International User Resource Allocation Panel

IUPESM: International Union for Physical and Engineering Sciences in Medicine

IUPHAR: International Union of Pharmacology

IUPHY: Inter-University Programme in Hydrology

IUPP: Introductory University Physics Project

IUPS: International Union of Physiological Sciences

IUPsyS: International Union of Psychological Sciences

IUPUI: Indiana University - Purdue University at Indianapolis (USA-IN)

IUR: Interuser Reliability

IUS: Inertial Upper Stage • Interchange Unit Separator • Interim Use Sheet • International Union of Students • Inus (Papua New Guinea) (location identifier)

IUSB: Indiana University in South Bend (USA-IN)

IUSM: International Union for Surveys and Mapping

IUT: Institut Universitaire de Technologie (France)

IUTAM: International Union of Theoretical and Applied Mechanics

IUTE: Industry University Technology Experiment

IUTOX: International Union of Toxicology

IUV: Interlibrational Utility Vehicle

IUVSTA: International Union for Vacuum Science, Technique and Applications

IUWDS: International URSIGRAM and World Days Service

iV: in Vertretung • im Vorjahre • in Vorbereitung

I/V: Ida/Vuelta • Inlet Valve

IV: Increased Value • Informationsvermittlung • Information Visualization • Initialization Variable • Initialization Vector • Initial Velocity • Input Voltage • Insignificant Vector • Integrated Vehicle • Intensifier Vidicon • Interactive Video • Interface Volume • Intermediate Valency • Intermediate Voltage • Intravehicular • Intravenous • Intravenously • Inverness (UK postal code) • Invoice Value • Subgiant Star

IVa: Subgiant Star

IVA: Ambanja (Madagascar) (location identifier) • Impuesto al Valor Agregado • Impuesto sobre el Valor Añadido • Individual Voluntary Arrangement • Ingenjörsvetenskapsakademien (Sweden) • Instrument de Vision Artificielle • International Voice Association • Intravehicular Activity

IVab: Subgiant Star

iVAD: Integrated Voice And Data

IVAR: Insertion Velocity Adjust Routine • Internal Variable

IVb: Subgiant Star

IVB: Intermediate Vector Boson

IVBC: Integrated Vehicle Baseline Characteristics • Integrated Vehicle Baseline Configuration

IVC: Inner Vacuum Center • Institute Visiting Committee • Integrated Vacuum Circuit • Intermediate-Velocity Cloud • International Visiting Committee • Intervehicular Communications • Invercargill (New Zealand) (location identifier)

IVCE: International Video and Communications Exhibition

IVCM: Integration Vegetation-Complex Map

IVD: Information Viewing Device • Interactive Visual Display

IVE: Interface Verification Equipment

IVF: International Video Federation • In Vitro Fertilization

IVG: Ivangrad (Yugoslavia-Montenegro) (location identifier)

IVH: Ivishak (USA-AK) (location identifier)

IVHS: Intelligent Vehicle Highway System

IVI: Incremental Velocity Indicator • Interchangeable Virtual Instrument • Ion Velocity Instrument

IVIC: Instituto Venezolano de Investigaciones Científicas (Venezuela)

IVICS: Interactive Visual Image Classification System

IVIS: Interactive Velocity Interpretation System • Interactive Video Information System

IVL: Institutet för Vatten och Luftvård (Sweden) • Intel Verification Laboratory • Ivalo (Finland) (location identifier)

IVLN: Integrated Vehicle Logic Network

ivm: In Verband Met

IvM: Instituut voor Milieuvraagstukken (Netherlands)

IVM: Imaging Vector Magnetograph • Informationsverarbeitungsmaschine • Interactive Volume Modelling

IVMMS: Integrated Vehicle Mission Management System

IVN: Interactive Video Network

IVO: Chivolo (Colombia) (location identifier) • Individueel Voortgezet Onderwijs

IVP: Initial-Value Problem • Installation Verification Procedure • Installation Verification Program • Integrity Verification Procedure • Interface Verification Plan • Ion Vacuum Pump

IVPA: Initial Value Problem Approach

IVPS: Integrated Vacuum Processing System

IVR: Inner Vertical Resonance • Interactive Voice Response • International Vehicle Registration • Intramolecular Vibrational Relaxation • Inverell (Australia-NSW) (location identifier)

IVRS: Interim Voice Response System

IVS: Informationsverarbeitungssytem • Informationsvermittlungsstelle • Informationsvermittlungsystem • Input Voltage Supply • Interactive Videodisk System • Interface Verification Summary • International VLBI Satellite • International VLBI Service (*)

IVSI: Instantaneous Vertical Speed Indicator

IVT: Initial Vocational Training • Integrated Verification Test • Integration-Verification-Training • Interface Verification Test • Interface Verification Testing • Interrupt Vector Table • Intervehicular Transfer

IVTS: International Video Teleconferencing Service

IVU: Intravehicular Umbilical

IV&V: Independent Validation and Verification • Independent Verification & Validation

IVV: Independent Validation and Verification • Independent Verification and Validation • Informations-Vermittlung und Verarbeitung • Instantaneous Vertical Velocity

IVVI: Instantaneous Vertical Velocity Indicator

IVVS: Instantaneous Vertical Velocity Sensor

IVVV: Internationaal Verbond van Vrije Vakverenigingen

IVW: Interferometrically Verified Window • International Venus Watch • Inverway (Australia-NT) (location identifier)

iw: Hebrew (ISO language code)

I/E: In Work

IW: Ice Water • Indirect Waste • Information Warfare • Inside Width • Inside Wiring • Internet World • Interstellar Wind • Intrawing • In Work • Isle of Wight (UK) • Isolated Word • Isotopic Weight

IWA: IAYC Workshop for Astronomy • Institute of World Affairs • Ivanova (Russia) (location identifier)

IWAA: Interdisziplinäres Workshop Astronomie und Astrophysik

IWALC: International Water Associations Liaison Committee

I-WAY: Information Highway

IWAY: Information Highway • Information Wide Area Year

IWBNI: It Would Be Nice If

IWBS: Indirect Work Breakdown Structure

IWC: Ice Water Column (*) • Ice Water Content • International Whaling Commission • Iowa Wesleyan College (USA-IA)

IWCA: International Workshop on Cometary Astronomy

IWCASE: International Workshop on Computer-Aided Software Engineering

IMCO: International World Commission on the Oceans

IWCV: International Workshop on Climate Variability

IWD: Inland Waters Directorate (Canada) • Integral Wind Defect • Ironwood (USA-MI) (location identifier)

IWDS: International World Days Service • Interwork Distributed System

IWEM: Institute of Water and Environmental Management

IWF: Institut für den Wissenschaftlichen Film (Germany) • Institut für Weltraumforschung (Austria) • Internationaler Währungsfonds

IWFS: Integrated Waste Fluid System

IWG: Impact Working Group • Implementation Working Group • Instrumentation Working Group • Integration Working Group • Interagency Working Group • Intercommission Working Group • Interface Working Group • Intersessional Working Group • Investigating Working Group • Investigator Working Group

IWGATWR: International Working Group on Advanced Technologies for Water-Cooled Reactors

IWGDMCG: Interagency Working Group on Data Management for Global Change

IWGFPT: International Working Group on Fuel Performance and Technology

IWGFR: International Working Group on Fast Reactors

IWGGCR: International Working Group on Gas-Cooled Reactors

IWGGE: Interdepartmental Working Group on the Greenhouse Effect

IWIS: International Water Information System

IWJ: Iwami (Japan) (location identifier)

IWO: Inter-White-Oval (region) (Jupiter) • Iwo Jima (Japan) (location identifier)

IWP: Ice Water Path • Independent Weather Productions • Indirect Work Package • Interim Working Party • Internal Working Paper

IWPA: International Word Processing Association

IWPC: Institute of Water Pollution Control

IWR: Information World Review • Internet Weather Report

IWRA: International Water Resources Association

iwS: im weiteren Sinn

IWS: Instrument Workstation • Integrated Water System • Integrated Wiring System • Integrated Work Statement • Intelligent Workstation • West Houston (USA-TX) (location identifier)

IWSA: International Water Supply Association

IWT: Integral Wavelet Transform • Inverse Wavelet Transform

IWTC: International World Trade Center

IWU: Illinois Wesleyan University (USA-IL) • Intermediate Working Unit

IWWG: Internetwork Working Group

IW3C1: First International WWW Conference

IW3C2: Second International WWW Conference • International WWW Conference Committee

IW3C3: Third International WWW Conference

IW3C4: Fourth International WWW Conference

IX: Interactive Executive • Inverted Index • Ion Exchange

IXA: Agartala (India) (location identifier)

IXAS: Indian X-ray Astronomy Satellite

IXB: Bagdogra (India) (location identifier)

IXC: Chandigarh (India) (location identifier) • Interexchange Carrier

IXD: Allahabad (India) (location identifier)

IXE: Mangalore (India) (location identifier)

IXG: Belgaum (India) (location identifier)

IXH: Kailashahar (India) (location identifier)

IXI: Interim X.25 Infrastructure • International X.25 Infrastructure • International X.25 Interconnect • Lilabari (India) (location identifier)

IXJ: Jammu (India) (location identifier)

IXK: Keshod (India) (location identifier)

IXL: Leh (India) (location identifier)

IXM: Madurai (India) (location identifier)

IXN: Khowai (India) (location identifier)

IXP: Pathankot (India) (location identifier)

IXQ: Kamalpur (India) (location identifier)

IXR: Ranchi (India) (location identifier)

IXS: Silchar (India) (location identifier)

IXT: Pasighat (India) (location identifier)

IXU: Aurangabad (India) (location identifier)

IXV: Along (India) (location identifier)

IXW: Jamshedpur (India) (location identifier)

IXY: Kandla (India) (location identifier)

IXZ: Port Blair (Andaman Islands) (location identifier)

IYDK: If You Didn't Know

IYFEG: Insert Your Favorite Ethnic Group

IYGTFUI: If You Get The Font, Use It

IYIS: Internal Photopic Illuminance Response

IYK: Inyokern (USA-CA) (location identifier) • Ridgecrest (USA-CA) (location identifier)

IYL: International Youth Library

IYQS: International Year of the Quiet Sun

IYSWIM: If You See What I Mean

IZ: Informationszentrum

Izda: Izquierda

Izdo: Izquierdo

IzKry: Izv. Krym. Astrofiz. Obs. (periodical code)

IZM: Izmir (Turkey) (location identifier)

IZMIRAN: Institute of Terrestrial Magnetism, Ionosphere and Radio Wave Propagation (Russia)

IZO: Izumo (Japan) (location identifier)

IzPul: Izv. Glav. Astron. Obs. (periodical code)

Izq: Izquierda • Izquierdo

Izqda: Izquierda

Izqdo: Izquierdo

IZS: Informationszentrum Socialwissenschaften

IZT: Ixtepec (Mexico) (location identifier)

I2ICE: Integrated Instrumentation and In-Circuit Emulation

I2L: Integrated Injection Logic

I2T: Interférométrie à 2 Télescopes

i2u: Italian Unix Systems User Group

I3: Intelligent Information Interface

I3A: Informatique et Astronomie pour Astronomes Amateurs (France)

I3L: Isoplanar Integrated Injection Logic

I3R: Intelligent Interface for Information Retrieval

I4T: Interférométrie à 4 Télescopes

J

j: Current Density • Imaginary Number • Jährlich • Jemand • Jetzt • Jour • Jung • Unit Coordinate Vector

J: Action Integral • Advance Ratio • Angular Momentum • Coupling Constant • Current Density • Electric Current Density • Exchange Integral • Flux Density • Jack • Jacobian Determinant • Jahr • Jahrgang • January • Japan (international vehicle registration code) • Jet • Jod • Joule • Journal • Judge • Jugend • Jugendlich • Juliet (international alphabet) • July • June • Junge • Jupiter • Jurassic • Justice • Magnetic Polarization • Massieu Function • Moment of Inertia • Nuclear Spin Quantum Number • One (Roman numeral) • Performance Criterion • Quantum Number of Total Angular Momentum • Rotational Quantum Number • Sound Intensity • Total Angular Momentum Quantum Number • 1 (Roman numeral)

ja: Japanese (ISO language code)

Ja: January

J/A: Joint Account

JA: Jahresabschluß • Jahresausgleich • Jamaica (international vehicle registration code) • January • Janvier • Japan (international aircraft registration code) • Joint Account • Joint Agent • Jugendamt • Jump Address • Jump if Above • Junior Achievement

JA&A: Journal of Astrophysics & Astronomy

JAA: Jackson Astronomers Association (USA-GA) • Jackson Astronomical Association (USA-MS) • Jalalabad (Afghanistan) (location identifier) • Joint Airworthiness Authorities • Joint Aviation Authorities

Jaarb: Jaarboek

JAAS: Jewish Academy of Arts and Sciences

JAB: Jabiru (Australia-NT) (location identifier) • Jahresabschlußbericht

JAC: Jackson Hole (USA-WY) (location identifier) • Joint Astronomy Centre (USA-HI) • Jongerenadviescentrum

JACARA: Joint Australian Centre for Astrophysical Research in Antarctica

JACEE: Japanese-American Cooperative Emulsion Experiment

JACERS: Japanese Committee of Environmental Remote Sensing

JACH: Joint Astronomy Centre in Hawaii

JAChS: American Chemical Society Journal (periodical code)

JACKPHY: Japanese-Arabic-Chinese-Korean-Persian-Hebrew-Yiddish

JACM: Journal of the Association for Computing Machinery

JAco: Journal d'Acoustique (periodical code)

JACS: Jet Attitude Control System • Journal Article Copying Service • Journal of the American Ceramic Society (periodical code) • Journal of the American Chemistry Society

JAD: Jandakot (Australia-WA) (location identifier) • Joint Application Design • Joint Application Development • Joint Astrophysics Division • Julian Astronomical Day

JADC: Japan Aircraft Development Co.

JADT: Journée d'Analyse des Données Textuelles

JAE: Atlanta - Technology Park (USA-GA) (location identifier) • Journal of Acoustic Emission (periodical code) • Jump if Above or Equal

JAEC: Japan Aero Engine Corporation • Japan Atomic Energy Commission • Joint Atomic Energy Committee

JAEI: Japan Aviation Electronics Industry Ltd.

JAEIC: Joint Atomic Energy Intelligence Committee

JAEL: JSC Avionics Evolution Laboratory

JAEMA: Journal of Analytical and Experimental Modal Analysis (periodical code)

JAerE: Journal of Aerospace Engineering (periodical code)

JAERI: Japan Atomic Energy Research Institute

JAerP: Journal of Aerospace Power (periodical code)

JAerS: Journal of Aerosol Science (periodical code)

JAES: Journal of Audio Engineering Society (periodical code)

JAF: Jaffna Kankesanturai (Sri Lanka) (location identifier) • Journal des Astronomes Français (periodical code)

JAFCO: Just Another Foolish Console Operator

JAfES: Journal of African Earth Sciences (periodical code)

JAFSA: Japan Foundation for Shipbuilding Advancement

JAG: Jacobabad (Pakistan) (location identifier) • Jaggies Are Gone • Joint Association of Geophysics (now BGA) • Jugendarbeitsgemeinschaft

JAH: Aubagne Agora (France) (location identifier)

Jahrb: Jahrbuch

Jahresber: Jahresbericht

Jahrg: Jahrgang

Jahrh: Jahrhundert

Jahrt: Jahrtausend

Jahrtsd: Jahrtausend

JAH2: Journal of Astronomical History amd Heritage

JAI: Jaipur (India) (location identifier) • Japanese Aerospace Industry

JAICI: Japan Association for International Chemical Information

JAIED: Journal of Artificial Intelligence in Education (AACE)

JAIEG: Joint Atomic Information Exchange Group

JaIMJ: Japan Institute of Metals Journal (periodical code)

JAIPA: Japan Association for Information Processing in Astronomy

JAir: Journal of Aircraft (periodical code)

JAJ: Atlanta - Perimeter Mall (USA-GA) (location identifier)

JaJAP: Japanese Journal of Applied Physics (periodical code)

JAK: Jacmel (Haiti) (location identifier)

JAKIS: Japan Keyword Indexing Simulator

Jal: Jalisco (Mexico)

JAL: Jalapa (Mexico) (location identifier) • Jalisco (Mexico) • Jet Approach and Landing • Journal of Academic Librarianship

JAllC: Journal of Alloys and Compounds (periodical code)

JALN: Journal of Asynchronous Learning Networks

JALPG: Joint Automatic Language Processing Group

Jam: Jamaica • Jamaican

JAM: Jamaica (ISO country code) • Jamaique (ISO country code) • Jambol (Bulgaria) (location identifier) • Job Analysis Memorandum • Job Assignment Memorandum • Journal of Applied Meteorology • Jugend-Astrocamp Mühlhausen (Germany) • JYACC Application Manager

JAMA: Journal of the American Medical Association

JAMAC: Job Analysis Memo Activity Chart

JAMAS: Japan Medical Abstracts Society

JAMES: Joint Automated Message Editing System

JAMIC: Japanese Microgravity Center

JAMIE: Joint Analogue Microsystems Initiative of Europe

JAMP: Joint Automated Mapping Project

JAMSTEC: Japan Marine Science and Technology Center

Jan: Janeiro • Janitor • Januar • January

JAN: Jackson Thompson (USA-MS) (location identifier) • Janeiro • January • Janvier • Joint Army-Navy

JANA: Jamahiriya Arab News Agency (Libya)

JANAP: Joint Army-Navy-Air Force Publication

JANet: Joint Academic Network (UK)

JANIP: Joint Army Navy Instrumentation Program

JANNAF: Joint Army, Navy, NASA, and Air Force

JAnSc: Journal of Astronautical Sciences (periodical code)

JANUS: Joint Academic Network Using Satellites

Janv: Janvier

JANZOS: Japan-Australia-New-Zealand Observation of Supernovae

JAO: Atlanta - Beaver Ruin (USA-GA) (location identifier)

JAOS: Junior Amateur Observers' Society (USA-NY)

JAOT: Journal of Atmospheric and Oceanic Technology

Jap: Japan • Japanese • Japon • Japonais • Japonés

JAP: Jahresarbeitsprogramm • Journal of Applied Physics (periodical code) • Jupiter Atmospheric Probe

JApA: Journal of Astrophysics and Astronomy (periodical code)

JAPACS: Japanese Pacific Climate Studies

JAPATIC: Japan Patent Information Center

JAPE: Journal of Applied Photographic Engineering (periodical code)

JApEl: Journal of Applied Electrochemistry (periodical code)

JAPh: Journal of Applied Physiology (periodical code)

JAPIO: Japan Patent Information Organization

JApMa: Journal of Applied Mathematics (periodical code)

JApMe: Journal of Applied Meteorology (periodical code)

JApMw: Journal of Applied Metalworking (periodical code)

Japn: Japan • Japanese

JAPOA: Japan Amateur Photoelectric Observers Association

JAQ: Jacquinot Bay (Papua New Guinea) (location identifier) • Job Activities Questionnaire

Jar: Jargon

JAR: Arcadia - Civic Center (USA-CA) (location identifier) • Joint Airworthiness Regulations • Joint Airworthiness Requirement • Joint Aviation Requirement • Jovian Atmospheric Reference • Jupe Arrière

JARE: Japanese Antarctic Research Expedition

JARECT: Japan Annual Reviews in Electronic Computers and Telecommunications

JARING: Joint Advanced Research Integrated Networking

JAROS: Japan Resources Observation System

JAS: Japanese Awareness Service • Jasper County (USA-TX) (location identifier) • Job Accounting System • Job Analysis System • Job Applicant System • Joint Administration Services • Jordanian Astronomical Society (Jordan) • Journal of Aerospace Science • Journal of the Atmospheric Sciences • July-August-September • Junior Astronomical Society (UK)

JASA: Japan Astronomical Study Association

JAsAs: Journal of Astrophysics and Astronomy (periodical code)

JASCE: Journal of the American Society of Civil Engineers (periodical code)

JASCOME: Japan Society for Computational Methods and Engineering

JASE: Journal of Applied Science and Engineering (periodical code)

JASEg: Journal of the Astronomical Society of Egypt (periodical code)

JASG: Joint Advanced Study Group

JASIN: Joint Air-Sea Interaction Experiment

JASIS: Journal of the American Society for Information Science

JASJa: Journal of the Acoustical Society of Japan (periodical code)

JASRAC: Japanese Society of Rights of Authors and Composers

JASS: Journal of Applied Systems Studies • Journal of Astronomy and Space Sciences

JAtC: Journal of Atmospheric Chemistry (periodical code)

JASTIS: Japanese Science and Technology Information System

JAT: Jabat (Marshall Islands) (location identifier) • Japanese Alias Table

JATC: Joint Atomic Time Commission

Jato: Jahrestonne

JATO: Jet-Assisted Takeoff

JAtOT: Journal of Atmospheric and Oceanic Technology (periodical code)

JATP: Journal of Atmospheric and Terrestrial Physics (periodical code)

JAtS: Journal of Atmospheric Sciences (periodical code)

JAU: Jauja (Peru) (location identifier)

JAuGG: Journal of Australian Geology and Geophysics (periodical code)

Jav: Java • Javanese

JAV: Jahresarbeitsverdienst • Jakobshavn (Greenland) (location identifier) • Jupe Avant

JAVA: James A. Van Allen (1914-)

JAVIC: Japan Audio-Visual Information Center

JAVSO: Journal of the American Association of Variable Star Observers (periodical code)

JAWOP: Joint Automated Weather Observing Program

JAWS: Joint Airport Weather Study • Junk Acronyms When Speaking

JAX: Jacksonville (USA-FL) (location identifier)

JAZ: Jahresarbeitszeit

Jb: Jahrbuch • Jahresbericht

JB: Jahresbeitrag • Joint Board • Jugendbewegung • Jugendbund • Jukebox • Jump if Below • Junction Box

JBAA: Journal of the British Astronomical Association

JBall: Journal of Ballistics (periodical code)

JBAS: Journal of the British Astronomical Society (periodical code)

Jbb: Jahrbücher

Jber: Jahresbericht

Jberr: Jahresberichte

JBC: Boston City (USA-MA) (location identifier) • Japan Broadcasting Corporation • Joint Board on Communication

JBD: Jodan-Brans-Dicke

JBE: Jump if Below or Equal

JBEM: Junction Box Environmental Module

JBEMC: Junction Box Environmental Module Controller

Jber: Jahresbericht

JBES: Jodrell Bank Experimental Station (UK)

JBIG: Joint Bilevel Image-experts Group • Joint Bilevel Image Group • Joint Binary Image Group

JBiom: Journal of Biomechanics (periodical code)

JBIS: Journal of the British Interplanetary Society (periodical code)

JBK: Berkeley (USA-CA) (location identifier)

JBM: Judgment Base Module

JBOD: Just a Bunch Of Disks

JBOSC: Joint Base Operations Support Contract

JBOX: Junction Box

JBP: Los Angeles - Commerce Business Plaza (USA-CA) (location identifier)

JBPA: Japan Book Publishers Association

JBR: Jonesboro (USA-AR) (location identifier)

JBS: Jamaica Bureau of Standards • Jamaican Bureau of Standards • Joint Broadcast System • Joly Black Screen • Pleasanton - Hacienda Business Park (USA-CA) (location identifier)

JBT: Bethel City Landing (USA-AK) (location identifier)

JBU: John Brown University (USA)

JBVAD: Jodrell Bank (Starlink internal code)

JBW: Jugendbildungswerk

JC: Job Card • Joint Committee • Joule Cycle • Julius Caesar • Jump if Carry • Jump on Condition • Junior College

JCA: Cannes Croisette (France) (location identifier) • Jettison Controller Assembly • Jupiter Closest Approach

JCAB: Japanese Civil Aviation Bureau

JCAC: Joint Civil Affairs Committee

JCAE: Joint Committee on Atomic Energy

JCALS: Joint Computer-aided Acquisition and Logistics Support

JCAM: Journal of Climate and Applied Meteorology (periodical code)

JCAST: Joint Committee on the Archives of Science and Technology

JCB: Job Control Block • Jocaba (Brazil-SC) (location identifier) • Joint Board on Communication • Juris Canonici Baccalaureatus • Juris Civilis Baccalaureatus

JCC: Job Control Card • Joint Computer Conference • Joint Consultative Committee • Joint Consultative Council • San Francisco - China Basin (USA-CA) (location identifier)

JCCDS: Joint Committee on Cartographic Data Standardization (USA)

JCCE: Journal of Computing in Childhood Education (AACE)

JCCOMNET: Joint Coordination Center Communications Network

JCCS: Job Card Control System

JCD: Jeune Cadre Dynamique • Juri Canonici Doctor • Juri Civilis Doctor • Sainte Croix Island (US Virgin Islands) (location identifier) • Ste. Croix Island (US Virgin Islands) (location identifier)

JCE: Java Cryptographic Extension • Oakland - Convention Center (USA-CA) (location identifier)

JCEE: Joint Council on Economic Education

JCET: Joint Council on Educational Telecommunications • Joint Council on Educational Television

JCG: Joint Consultative Group

JCh: Journal of Chromatography (periodical code)

JCH: Christianshåb (Greenland) (location identifier)

JChPh: Journal of Chemical Physics (periodical code)

JChS: Journal of Chromatographic Science (periodical code)

JCI: Kansas City - Johnson Industrial Estate (USA-MO) (location identifier)

JCIS: Journal of Colloid and Interface Science (periodical code)

JCIST: Japan Information Center of Science and Technology

JCJ: Cheju - Chu Ja (Korea-ROK) (location identifier)

JCK: Jacobus Cornelius Kapteyn (1851-1922) • Java Compatibility Kit • Julia Creek (Australia-QLD) (location identifier)

JCL: Jeans' Color Law • Job Command Language • Job Control Language (*)

JCLC: Joint Committee on the Library of Congress

JCLDS: Joint Committee for Library and Documentation Standardization

JCLE: Joint Committee on Library Education

JCli: Journal of Climate (periodical code)

JClim: Journal of Climatology (periodical code)

JCLOT: Joint Closed-Loop Operations Test

JClP: Journal of Clinical Pharmacology (periodical code)

JCLWC: Joint Committee on Library Works as a Career

JCM: Jacobina (Brazil-BA) (location identifier) • Joint Commission Meeting • Joint Committee on Microcards • Joint Consultative Machinery

JCMS: Johnson County Middle School (USA-KY)

JCMST: Journal of Computers in Mathematics and Science Teaching (AACE)

JCMT: James Clerk Maxwell Telescope (USA-HI)

JCMTB: James Clerk Maxwell Telescope Board

JCN: Incheon (Korea-ROK) (location identifier) • Job Change Notice • Job Control Number • Junction

JCO: Comino (Malta) (location identifier)

JCoAM: Journal of Computational and Applied Mathematics (periodical code)

JCoCh: Journal of Computational Chemistry (periodical code)

JCom: Journal of Communications (periodical code)

JCoM: Journal of Computational Mathematics (periodical code)

JCoMa: Journal of Composite Materials (periodical code)

JCoPh: Journal of Computational Physics (periodical code)

JCP: Joint Committee on Printing • Journal of Chemical Physics

JCPCUS: Joint Committee on Printing of the Congress of the United States

JCPDS: Joint Committee on Powder Diffraction Standards

JCPI: Jpont Committee on Paleontologic Information

JCR: Jacareacanga (Brazil-PA) (location identifier) • Journal Citation to Reports

JCRF: Joint Climate Research Fund

JCrGr: Journal of Crystal Growth (periodical code)

JCS: Joint Chiefs of Staff • Journal of the Chemical Society

JCSAT: Japan Communication Satellite (now merged with SAJAC into JSS)

JCSC: Joint Computer Security Conference

JCSS: Journal of the Chinese Silicate Society (periodical code)

JCSTC: Joint Council for Scientific and Technical Information

Jct: Junction

JCT: Japanese Cordless Telephone • Joint Central Team • Joint Committee on Technology • Junction • Junction - Kimble County (USA-TX) (location identifier)

JCTION: Junction

Jctn: Junction

JCTN: Junction

JCTNS: Junctions

JCTR: Journal of Composites and Technology Research (periodical code)

JCTS: Junctions

JCU: James Cook University (Australia-QLD) • John Caroll University (USA)

JCUDI: Japan Computer Usage Development Institute

JCULS: Joint Committee on the Union List of Serials

JCV: Joule-Clausius Velocity

JCX: Los Angeles - Citicorp Plaza (USA-CA) (location identifier)

JCY: Johnson (USA-TX) (location identifier)

Jd: Joined

JD: Doctor of Jurisprudence • Doctor of Laws • Jahresdurchschnitt • Joint Discussion • Jordan Dinar (currency) • Julian Date • Julian Day • Julianisches Datum • Juris Doctor • Justice Department

JDA: Japanese Dark-sky Association • Japanese Defense Agency • Japan's Defense Agency • John Day (USA-OR) (location identifier) • Joint Development Agreement

JDAM: Joint Direct Attack Munition

JDB: Dallas Downtown (USA-TX) (location identifier)

JDBC: Java Database Connect • Java Database Connectivity (*)

JDC: Japan Documentation Center • Japanese Digital Cellular • Job Description Card

JDD: JAICI Document Delivery

JDE: Journal of Differential Equations (periodical code)

JDF: Juiz de Fora - Francisco de Assis (Brazil-MG) (location identifier)

JDH: Jodhpur (India) (location identifier)

JDIMP: Joint Data and Information Management Panel

JDK: Java Development Kit

JDL: Job Description Language • Job Description Library • Job Descriptor Language • Joint Directors of Laboratories

JDM: Miami Downtown (USA-FL) (location identifier)

JDN: Jordan (USA-MT) (location identifier) • Julian Day Number

JDO: Juàzeiro do Norte - Cariri (Brazil-CE) (location identifier)

JDOP: Joint Doppler Operational Project

JDP: Héliport de Paris (France) (location identifier)

JDPA: Japan Design Protection Association

JDS: Japan Documentation Society • Job Data Sheet

JDT: Minneapolis Downtown (USA-MN) (location identifier)

JDX: Houston - Central Business District (USA-TX) (location identifier)

JDY: Downey (USA-CA) (location identifier)

JDZ: Jingdezhen (China-PRC) (location identifier)

Je: June

JE: Japanese Ephemeris • Joshi Effect • Joule Effect • Jump if Equal

JEA: Japan Environmental Agency • Joint Endeavor Agreement • Journal of Experimental Algorithmics

JEAS: Journal of Engineering and Applied Sciences (periodical code)

Jeb: Jebel

JEC: Joint Economic Committee • Joint European Conference • Joint Evaluation Committee • Junta Electoral Central

JECC: Japan Electronic Computer Corporation

JECF: Java Electronic Commerce Framework

JECFA: Joint Expert Committee on Food Additives

JECL: Job Entry Control Language

JECS: Journal of Electronic Circuits and Systems (periodical code)

JECSIM: Joint Electronic Combat Simulation

JED: Jeddah - King Abdul Aziz (Saudi Arabia) (location identifier) • Joint Environmental Data Center • Julian Ephemeris Data • Julian Ephemeris Date (*)

JEDEC: Joint Electronic Devices Engineering Committee • Joint Electronic Devices Engineering Council (*)

JEdI: Joint Education Initiative

JEDI: Joint Education Initiative

JEE: Jeremie (Haiti) (location identifier) • Journal of Earthquake Engineering

JEEEA: Journal of Electrical and Electronic Engineering - Australia (periodical code)

JEEM: Journal of Environmental and Economic Management (periodical code)

JEF: Jacobi Elliptic Function • Jefferson City (USA-MO) (location identifier) • Jeunesses Européennes Fédéralistes

JEG: Egedesminde (Greenland) (location identifier)

JEHLT: Jet Engine of High Limited Thrust

JEI: Joint Education Initiative (formerly JEdI)

JEIA: Joint Electronic Information Agency • Joint Export-Import Agency

JEIDA: Japan Electronics Industry Development Association • Japanese Electronics Industry Development Association (*)

JEIPAC: JICST Electronic Information Processing Automatic Computer

JEJ: Jeh (Marshall Islands) (location identifier)

JElas: Journal of Elasticity (periodical code)

JElS: Journal of the Electrochemical Society (periodical code)

jem: jemals • jemand

JEM: Emeryville (USA-CA) (location identifier) • Japanese Experimental Module • Japanese Experiment Module (*) • Japan Experimental Module • Japan Experiment Module • Joint European Monitor

JEMat: Journal of Electronic Materials (periodical code)

JEMD: Japan English Magazine Directory

JEMH: Journal of Educational Multimedia and Hypermedia (AACE)

JEMP: Japanese Ephemerides of Minor Planets

JEMRMS: Japanese Experimental Module Remote Manipulator System

JEMT: Journal of Electron Microscopy Technique (periodical code)

JENAM: Joint European and National Astronomy Meeting

JENC: Joint European Networking Conference

JEner: Journal of Energy (periodical code)

JEngM: Journal of Engineering and Mechanics (periodical code)

JEngS: Journal of Engineering Sciences (periodical code)

JEnMa: Journal of Engineering and Mathematics (periodical code)

JenRv: Jena Review (periodical code)

jens: jenseitig • jenseits

JEnvS: Journal of Environmental Sciences (periodical code)

JEOS: Janus Earth Observation Satellite • Japanese Earth Observation System • Japanese Earth Observing Satellite • Japanese Earth Observing System • Journal of the European Optical Society

JEP: Joint European Project • Joint Experiment Programme • Jupiter Entry Probe

JEPI: Joint Electronic Payments Initiative

JEPS: Joint European Printer Server

JEQ: Jequié (Brazil-BA) (location identifier)

Jer: Jersey (Channel Islands)

JER: Jersey (Channel Islands) (location identifier)

JERC: Joint Electronic Research Committee

JERICO: Joint European Robotics Interactive and Calibration Operations

JERS: Japanese Earth Remote Sensing • Japanese Earth Resources Satellite • Japan's Earth Resources Satellite (*)

Jes: Jesuit

JES: Japanese Engineering Standard • Job Entry System • Joint Environment Simulator

JESA: Japan Engineering Standards Association

JESAUG: Journal of the Environmental Satellite Amateur Users' Group

JESC: Japan Engineering Standards Committee

JESCT: Job Entry System Communication Table

JESRP: Journal of Electron Spectroscopy and Related Phenomena (periodical code)

JESSI: Joint European Submicron Silicon Initiative • Joint European Submicron Silicon (*)

JEST: Joint Entrance Screening Test

JET: Japan Exchange and Teaching • Job Element Text • Job Euro Transfer • Joint Estimate Team • Joint European Telescope • Joint European Torus • Joint European Transport

JETDLAG: Joint European development of Tunable Diode Laser Absorption spectroscopy for measurement of Atmospheric trace Gases

JETEC: Joint Expendable Turbine Engine Concepts

JETP: Journal of Experimental and Theoretical Physics • Joint European Torus Project

JETRO: Japanese External Trade Organization

JETS: Joint Electronic Type System

JETh: Journal of Engineering and Thermophysics (periodical code)

JETT: Jettison

Jeu: Jeudi

JEV: Évry (France) (location identifier) • Jever

jew: jeweilig • jeweils

JEWA: Journal of Electromagnetic Waves and Applications (periodical code)

JEWM: Joint Expert Working Meeting

JEXAM: Japanese Experiment on Asia Monsoon

JEXM: Joint European X-ray Monitor

J/F: Journal Folio

JF: Junior Fiction

JFACT: Joint Flight Acceptance Composite Test

JFAA: Journal of Fourier Analysis and Applications

JFAP: Joint Faculty Appointment Program • Joint Frequency Allocation Panel

JFAS: Jet Fuel Accounting System

JFC: Java Foundation Class • Journal of Fluid Control (periodical code) • Jupiter Family Comet

JFCB: Job File Control Block

JFCL: Jump if Flag set and Clear it

JFD: Japanese Flight Demonstration • Japanese Flight Demonstrator

JFET: Junction Field Effect Transistor

JFF: Journal of Fire and Flammability (periodical code)

JFI: James Frank Institute (USA-IL)

JFIF: JPEG File Interchange Format

JFIT: Joint Framework for Information Technology

JFK: John Fitzgerald Kennedy (USA-NY) • New York - John Fitzgerald Kennedy (USA-NY) (location identifier)

JFKL: John Fitzgerald Kennedy Library

JFM: Fremantle (Australia-WA) (location identifier) • January-February-March • Job Function Manual • Journal of Fluid Mechanics (periodical code) • Jupiter Flyby Mission

JFMIP: Joint Financial Management Improvement Program

JFN: Jefferson Ashtabula (USA-OH) (location identifier)

JFPCC: Joint Fusion Power Coordination Committee

JFR: Frederikshåb (Greenland) (location identifier)

JFS: Journaled File System • Journal of Fluid Structures (periodical code)

JFT: Joint Firing Test

JFTOT: Jet Fuel Thermal Oxidation Tester

JFV: Jupiter Flyby Vehicle

jg: jung

Jg: Jaargang • Jahrgang

JG: Journal of Geology (periodical code) • Jugendgericht • Jump if Greater • Junior Grade • Junta General

JGA: Jamnagar (India) (location identifier)

JGB: Jagdalpur (India) (location identifier)

JG&C: Joint Guidance & Control

JGC: Grand Canyon (USA-AZ) (location identifier)

JGCD: Journal of Guidance Control and Dynamics (periodical code)

Jgd: Jugend

Jgdl: Jugendlich

JGE: Geoje (Korea-ROK) (location identifier) • Jump if Greater or Equal

JGeo: Journal of Geophysics (periodical code)

Jgg: Jahrgänge

JGG: Journal of Geomagnetism and Geoelectricity (periodical code)

JGGS: Journal of Geomagnetism and Geoelectricity - Supplements (periodical code)

JGI: Jolly Good Idea

JGL: Atlanta Galleria (USA-GA) (location identifier)

JGla: Journal of Glaciology (periodical code)

JGM: Joint Gravity Model

JGN: Jiayuguan (China-PRC) (location identifier)

JGO: Godhavn (Greenland) (location identifier)

JGOFS: Joint Global Ocean Flux Study

JGP: Houston - Greenway Plaza (USA-TX) (location identifier) • Journal of Geometry and Physics

JGQ: Houston - Transco Tower Galleria (USA-TX) (location identifier)

JGR: Gronnedal (Greenland) (location identifier) • Journal of Geophysical Research (periodical code)

JGRS: Journal of Geophysical Research - Supplements (periodical code)

JGuC: Journal of Guidance Control (periodical code)

JGX: Glendale (USA-CA) (location identifier)

JGZG: Journal of Geophysics - Zeitschrift für Geophysik (periodical code)

Jh: Jahresheft • Jahrhundert

JH: Jagerhaus

JHA: Job Hazard Analysis • Journal for the History of Astronomy (periodical code) • Justice and Home Affairs

JHATD: Johns Hopkins APL Technical Digest (periodical code)

Jhb: Jahrbuch

JHB: Johannesburg (South Africa • Johor Bahru (Malaysia) (location identifier)

JHC: Garden City (USA-NY) (location identifier)

JHD: Japan Hydrographic Department

Jhdt: Jahrhundert

JHE: Hälsingborg (Sweden) (location identifier) • Helsingborg (Sweden) (location identifier)

JHENP: Journal of High-Energy and Nuclear Physics (periodical code)

JHEP: Journal of High-Energy Physics

Jhg: Jahrgang

JHG: Jinghong Gasa (China-PRC) (location identifier)

Jhgg: Jahrgänge

Jhh: Jahreshefte

J. Hist. A.: Journal for the History of Astronomy

JHM: Kapalua (USA-HI) (location identifier)

JHMCS: Joint Helmet-Mounted Cueing System

JHO: Jeremiah Horrocks Observatory (UK-England)

JhPCh: Journal of Physical Chemistry (periodical code)

JHQ: Shute Harbour (Australia-QLD) (location identifier)

Jhrb: Jahrbuch

Jhrg: Jahrgang

JHS: Holsteinborg (Greenland) (location identifier) • Jesus Hominum Salvator • Junior High School

Jhtsd: Jahrtausend

Jhtsde: Jahrtausende

JHU: Johns Hopkins University (USA-MD)

JHV: Jahreshauptversammlung

JHW: Jamestown (USA-NY) (location identifier) • Jugendhilfswerk

Jhy: Jalhay (Belgium)

JHY: Cambridge - Hyatt Regency (USA-MA) (location identifier)

JHyd: Journal of Hydronautics (periodical code)

ji: Yiddish (ISO language code)

JI: Io (satellite of Jupiter) • Job Instruction • Joint Implementation • Junction Isolation • Jupiter Inlet

JIA: Juiná (Brazil-MT) (location identifier)

JIAFS: Joint Institute for the Advancement of Flight Sciences

JIAPS: Journal of the International Association of Physics Students

JIAWG: Joint Integrated Avionics Working Group

JIB: Djibouti (location identifier) • Job Information Block

JIC: Job Instruction Card • Joint Ice Center • Joint Implementation Committee • Joint Industrial Council • Journal of Internet Cataloging • Just In Case

JICA: Japan International Cooperation Agency

JICST: Japan Information Centre of Science and Technology

JID: Los Angeles - City of Industry (USA-CA) (location identifier)

JIE: Journal of Industrial Ecology

JIECE: Journal of the Institute of Electronic Communication Engineers (periodical code)

JIEO: Joint Interoperability and Engineering Organization

JIEq: Journal of Integral Equations (periodical code)

JIETE: Journal of the Institution of Electronics and Telecommunications (periodical cod e)

JIF: Joint Infrastructure Fund (UK)

JIG: Jane's Information Group

JIGI: Java Interface for Geospatial Information

JIHOP: J.I. Holcomb Observatory and Planetarium (USA-IN)

JII: Europa (satellite of Jupiter)

JIII: Ganymedes (satellite of Jupiter) • Japanese Institute of Invention and Innovation (Japan)

JIIM: Journal of Information and Image Management

JIK: Ikaria Island (Greece) (location identifier)

JIL: Jilin (China-PRC) (location identifier)

JILA: Joint Institute for Laboratory Astrophysics (USA-CO)

JILR: Journal of Interactive Learning Research

JIM: Jimma (Ethiopia) (location identifier) • Job Information Memorandum • Job Instruction Manual • Jovian Ionospheric Model

JIMA: Joint Institute for Millimeter Astronomy (now IRAM)

JIMAR: Joint Institute for Marine and Atmospheric Research

JIMIA: Journal of the Institute of Mathematics and Its Applications (periodical code)

JIMS: Jukebox Interface Management System

JIMSS: Journal of Intelligent Material Systems and Structures (periodical code)

JIMW: Journal of Infrared and Millimeter Wave (periodical code)

JIN: Jinja (Uganda) (location identifier) • Joint Implementation Network

JINR: Joint Institute for Nuclear Research (Russia)

JINS: Japan Institute of Nuclear Safety

JIO: Joint Intelligence Organization • Ontario (USA-CA) (location identifier)

JIP: Jipijapa (Ecuador) (location identifier) • Job Improvement Plan • Job Installation Plan

JIPA: Java Image Preview Application

JIPDEC: Japan Information Processing Development Centre

JIPID: Japanese International Protein Information Database

JIPNET: Japan Information Processing Network

JIPS: JANET IP service

JIR: Jiri (Nepal) (location identifier) • Job Improvement Request • Journal of Irreproducible Results

JIRA: Japanese Robotic Industries Association

JIRAD: Joint Independent Research And Development

JIRC: Journal of International Research Communications

JIRSG: Joint Interservice Regional Support Group

JIRW: Journal of Indexing and Reference Work

JIS: Jamaica Information Service • Japanese Industrial Standard • Japan Industrial Standard • Joint Integrated Simulation • Joint Integration Simulation • Journal of Information Science

JISAO: Joint Institute for the Study of Atmosphere and Ocean (USA-WA)

JISC: Japanese Industrial Standards Commission • Japanese Industrial Standards Committee (*) • Joint Information Services Committee • Joint Information Systems Committee

J/IST: JSF/Integrated Subsystem Technology

JIST: Journal of Imaging Science and Technology (periodical code) • JSF Integrated Subsystem Technology

JISWG: Joint Integrated Simulation Working Group • Joint IRAS Science Working Group

JISYA: Japan ISY Association

JIT: Just In Time

JITC: Joint Interoperability Test Center

JITR: Japan Industry Technical Reports

JIU: Jiujiang (China-PRC) (location identifier) • Joint Inspection Unit

JIV: Callisto (satellite of Jupiter)

JIVE: Joint Institute for VLBI in Europe

JIW: Jiwani (Pakistan) (location identifier)

JIWP: Joint Interim Working Party

JIX: Sinope (satellite of Jupiter)

jJ: jedes Jahr

J/J: Januar/Juli

JJ: Josephson Junction • Judges • Justices

JJA: June-July-August

JJAPL: Japanese Journal of Applied Physics - Letters (periodical code)

JJAPS: Japanese Journal of Applied Physics - Supplement (periodical code)

JJC: Joliet Junior College (USA-IL)

JJI: Juanjui (Peru) (location identifier)

JJN: Jinjiang (China-PRC) (location identifier)

JJPPAA: Jornadas de Puertas Abiertas

JJT: Jumbo Jet Transport

JJU: Julianehåb (Greenland) (location identifier)

J/K: Joule per Kelvin

Jk: Jack

JK: Jugendkreis • Julianischer Kalender

JKC: Johnson-Kron-Cousins

JKG: Jönköping (Sweden) (location identifier) • Jönköping (Sweden) (location identifier)

JKH: Chios (Greece) (location identifier)

JKR: Janakpur (Nepal) (location identifier)

Jkt: Jacket

JKT: Jacobus Kapteyn Telescope (Canary Islands) • Jakarta (Indonesia) (location identifier) • Job Knowledge Test

JKU: Johannes-Kepler-Universität (Austria)

JKV: Jacksonville (USA-TX) (location identifier)

jl: Jongsleden

Jl: Journal • July

JL: Joule's Law • Juillet • July • Jump if Less

JLA: Cooper Lodge - Quartz Creek (USA-AK) (location identifier) • Jamaica Library Association • Japan Library Association

JLB: Long Beach (USA-CA) (location identifier)

JLCM: Journal of Less Common Metals (periodical code)

JLD: Hälsingborg Landskrona (Sweden) (location identifier) • Helsingborg Landskrona (Sweden) (location identifier)

JLE: Jump if Less than or Equal

JLH: Arlington Heights (USA-IL) (location identifier)

Jln: Jalan

JLN: Joplin (USA-MO) (location identifier)

JLNT: Japanese Large National Telescope

JLO: Jesolo (Italy) (location identifier)

JLP: Juan-les-Pins (France) (location identifier)

JLR: Jabalpur (India) (location identifier)

JLS: Jales (Brazil-SP) (location identifier) • Jamaica Library Service • Joint Least Squares

JLTP: Journal of Low Temperature Physics (periodical code)

JLwT: Journal of Lightwave Technology (periodical code)

JLX: Los Angeles - Union (USA-CA) (location identifier)

Jly: July

jm: Jamaica

jM: jeder Monat

J/M: Jettison Motor

JM: Jamaica (ISO country code) • Jamaique (ISO country code) • Jettison Motor • Jolly Mad

JMA: Houston Marriot (USA-TX) (location identifier) • Japan Meteorological Agency • Joint Mission Analysis

JMAA: Journal of Mathematical Analysis Applications (periodical code)

JMAC: Japan Market Awareness Committee

JMagR: Journal of Magnetic Resonance (periodical code)

JMAPI: Java Management Application Programming Interface

JMASS: Joint Modelling And Simulation System

JMatR: Journal of Materials Research (periodical code)

JMatS: Journal of Materials Science (periodical code)

JMB: Jamba (Angola) (location identifier) • Joint Meteorological Board

JMBM: Journal of Mechanical Behavior and Materials (periodical code)

JMC: Japan Map Center • Sausalito - Marin County (USA-CA) (location identifier)

jmd: jemand

JMD: Dallas - Market Center (USA-TX) (location identifier) • Dollar (Jamaica) (ISO currency code)

JME: Jones Matrix Eigenanalysis

JMec: Journal de Mécanique (periodical code)

JMecA: Journal de Mécanique Appliquée (periodical code)

JMecE: Journal of Mechanical and Engineering Science (periodical code)

JMecT: Journal de Mécanique Théorique et Appliquée (periodical code)

JMED: Jungle Message Encoder-Decoder

JMemS: Journal of Microelectromechanical Systems (periodical code)

JMES: Journal of Materials and Energy Systems (periodical code)

JMet: Journal of Metals (periodical code)

JMH: Schaumburg Marriott (USA-IL) (location identifier)

JMK: Mykonos (Greece) (location identifier)

JMLS: John Menzies Library Service (UK)

JMM: Jacobi Matrix Method • Malmö Harbour (Sweden) (location identifier)

JMMM: Journal of Magnetism and Magnetic Materials (periodical code)

JMN: Mankato (USA-MN) (location identifier)

JMO: Jomsom (Nepal) (location identifier)

JMOA: Journal of Microwave, Optics, and Acoustics (periodical code)

JMoEl: Journal of Molecular Electronics (periodical code)

JMolE: Journal of Molecular Evolution (periodical code)

JMOp: Journal of Modern Optics (periodical code)

JMoSp: Journal of Molecular Spectroscopy (periodical code)

JMoSt: Journal of Molecular Structure (periodical code)

JMP: Joint Monitoring Programme • Journal of Mathematical Physics (periodical code) • Jump

JMPC: Joint Mapping and Photography Committee (USA)

JMPMS: Journal of Materials, Processing, and Manufacturing Science (periodical code)

JMPS: Journal of Mathematical and Physical Sciences (periodical code)

JMPSo: Journal of Mechanics, Physics, and Solids (periodical code)

Jms: Jamais

JMS: Jamestown (USA-ND) (location identifier) • Java Message Service

JMSAC: Joint Meteorological Satellite Advisory Committee

JMSEC: Journal of Mathematical Systems, Estimation, and Control

JMSJa: Journal of the Meteorological Society of Japan (periodical code)

JMSL: Journal of Materials Sciences - Letters (periodical code)

JMSME: Journal of Materials Science and Materials Electronics (periodical code)

JMTAS: Journal de Mécanique Théorique et Appliquée - Suppléments (perio dical code)

JMTK: Joint Mapping Tool Kit

JMTSS: Joint Multichannel Trunking and Switching System

JMU: James Madison University (USA-VA) • Jiamusi (China-PRC) (location identifier) • John Moores University (UK)

JMV: Jahresmitgliederversammlung

JMW: Journal of Marginal Worthiness

JMY: Freetown - Mammy Yoko (Sierra Leone) (location identifier)

Jn: Join • Junction • Junior

JN: Jahresnachweis • Juin • June

JNA: Januária (Brazil-MG) (location identifier) • Journée Nationale de l'Astronomie • Jump if Not Above

JNAE: Jump if Not Above or Equal

JNav: Journal of Navigation (periodical code)

JNB: Johannesburg - Jan Smuts (South Africa) (location identifier) • Jump if Not Below

JNBE: Jump if Not Below or Equal

Jnc: Junction

JNC: Jet Navigational Chart • Joint Negotiating Committee

JNCA: Journal of Network and Computer Applications

JNCASR: Jawaharlal Nehru Centre for Advanced Scientific Research (India)

JNCS: Journal of Non-Crystalline Solids (periodical code)

JND: Just Noticeable Difference

JNDI: Java Naming and Directory Interface • Java Naming Directory Interface • Java Network Directory Interface

Jne: June

JNE: Journal of Nondestructive Evaluation (periodical code)

JNET: Journal of Non-Equilibrium Thermodynamics (periodical code)

JNF: Junior Non-Fiction

JNG: Jining (China-PRC) (location identifier) • Jump if Not Greater

JNGE: Jump if Not Greater or Equal

JNH: Dallas - North Park Inn (USA-TX) (location identifier)

JNI: Java Native Interface • Junin (Argentina-BA) (location identifier)

JNICT: Junta Nacional de Investigação Científica e Tecnológica (Portugal)

JNK: Junk

Jnl: Journal

JNL: Jump if Not Less

JNLE: Jump if Not Less or Equal

JNLM: Journal of Non-Linear Mechanics (periodical code)

Jnlst: Journalist

JNLT: Japanese National Large Telescope • Japan National Large Telescope

JNN: Nanortalik (Greenland) (location identifier)

JNND: Just Not Noticeable Difference

JNNIE: Joint NSF-NASA Initiative in Evaluation

JNNS: Japanese Neural Network Society

JNO: Jump if No Overflow

JNP: Jump if No Parity • Newport Beach (USA-CA) (location identifier)

JNPEA: Japan Newspaper Publishers and Editors Association

Jnr: Junior

JNr: Jahresnummer

JNR: Japan National Railways

JNS: Jump if No Sign • Just Noticeable Shift • Narsaq (Greenland) (location identifier)

Jnt: Joint

JNT: Joint Network Team

JNTF: Joint National Testing Facility

JNU: Juneau (USA-AK) (location identifier)

JNuM: Journal of Nuclear Materials (periodical code)

JNWPU: Joint Numerical Weather Prediction Unit

JNX: Naxos Island (Greece) (location identifier)

JNZ: Jump if Not Zero

JO: Jagdordnung • Jeux Olympiques • Job Order • Jordan (ISO country code) • Jordanie (ISO country code) • Journal des Observateurs (periodical code) • Journal Officiel • Journal of Optics (periodical code) • Junta Ordinaria • Jupiter Orbiter

JOA: Joint Operating Agreement

JOBRESA: Joint Board of Remote Sensing Activities

Jo'burg: Johannesburg (South Africa)

JOBS: Job Opportunity in Basic Skills

JOC: Joint Operations Center • Joint Organizing Committee • Journal of Optical Communications (periodical code) • Santa Ana (USA-CA) (location identifier)

JOCC: Joint Operations Control Center

JOCE: Journal Officiel des Communautés Européennes

JOCR: Joint Observatory for Cometary Research (USA-NM)

JOD: Dinar (Jordan) (ISO currency code) • Jodrell Bank (Starlink internal code) • Joint Occupancy Date • Joint Occupancy Data

JODC: Japan Ocean Data Center • Japan Oceanographic Data Center

JODL: Journal Of Digital Libraries

JOE: Java Objects Everywhere • Joensuu (Finland) (location identifier)

JOG: Joint Operations Graphic • Yogyakarta (Indonesia) (location identifier)

JOH: Port Saint Johns (South Africa) (location identifier)

Johan: Johannesburg (South Africa)

JOI: Joint Oceanographic Institutions • Joinville Cubatão (Brazil-SC) (location identifier) • Jupiter Orbital Insertion • Jupiter Orbit Insertion

JOIDES: Joint Oceanographic Institutions for Deep Earth Sampling

JOINT-net: Japanese Open Integration Network

JOINTnet: Japanese Open Integration Network

JOIP: Joint Operations Interface Plan • Joint Operations Interface Procedure

JOIS: JICST Online Information System

JOK: Joshkar-Ola (Russia) (location identifier)

JOL: Job Organisation Language • Jolo (Philippines) (location identifier)

JOLA: Journal Of Library Automation

JOLIS: Joint Library and Information System

JOM: Japanese Optical Module • Job Operation Manual • Njombe (Tanzania) (location identifier)

JOMACHEM: Journal of Mathematical Chemistry

JOMAR: Joint Office for Mapping and Research (USGS)

JON: Job Order Number • Johnston Island (US Minor Outlying Islands) (location identifier)

JONSWAP: Joint North Sea Wave Analysis Project

JOP: Joint Observing Programme • Joint Operating Plan • Joint Operating Procedure • Joint Operation Procedure • Josephstaal (Papua New Guinea) (location identifier) • Jupiter Orbiter Probe (now Galileo)

JOPAL: Journal of Patent-Associated Literature

JOptT: Journal of Optical Technology (periodical code)

JOR: Job Operations Report • Job Order Request • Jordan (ISO country code) • Jordanie (ISO country code) • Orange (USA-CA) (location identifier)

Jord: Jordan • Jordanian

JOS: Job Order Supplement • Jos Heipang (Nigeria) (location identifier)

JOSE: Jindalee Ocean Surveillance System • Jose Observatory Site Evaluation

JOSO: Joint Organization for Solar Observations

JOSS: Joint Office for Science Support

JOT: Joint Observer Team • Joliet (USA-IL) (location identifier)

JOTA: Journal of Optimization Theory and Applications (periodical code)

JOULE: Joint Opportunities for Unconventional Long-term Energy • Joint Opportunities for Unconventional or Long-term Energy supply (*)

Jour: Journal • Journalism • Journalist • Journey • Journeyman

Journ: Journal • Journalism • Journalist • Journey • Journeyman

JOV: Joint Orbital Verification • Joint Orbit Verification

JOVE: Joint Venture

JOVIAL: Joule's Own Version of the International Algebraic Language • Jule's Own Version of the International Algorithmic Language

JOWIP: Joint Ocean Wave Investigation Project

JOY: Job Opportunity for Youth

jp: Japan

jP: juristische Person

JP: Jahresprämie • Japan (ISO country code) • Japon (ISO country code) • Jet Pilot • Jet Propellant • Jet-Propelled • Jet Propulsion • Julian Period • Justice de Paix • Justice of the Peace

JPA: Jamaica Press Association • Japan Patent Association • Japan Printers Association • João Pessoa - Castro Pinto (Brazil-PB) (location identifier) • Job Package Authorization • Johnstone Plasma Analyzer

JPal: Journal of Paleontology (periodical code)

JPALS: Joint Precision Approach/Landing System

JPApS: Journal of Pure and Applied Sciences (periodical code)

JPAS: Journal of Practical Applications in Space (periodical code)

JPASO: Joint Pacific Area Scheduling Office

JPATS: Joint Primary Aircraft Training System

JPB: Jahrprüfungsbericht • Joint Planning Board

JPBM: Joint Policy Board for Mathematics

JPC: Japan Patent Center • Joint Planning Council • Joint Power Conditioner • Joint Power Conditions • Joint Production Council • Joint Propulsion Conference

JPCM: Journal of the Physics of Condensed Matter (periodical code)

JPCRD: Journal of Physical and Chemical Reference Data (periodical code)

JPD: Pasadena (USA-CA) (location identifier)

JPDC: Journal of Parallel Distributed Computing (periodical code)

JPDRD: Joint Programme Definition and Requirements Document

JPE: Jump if Parity Even

JPEG: Joint Photographic Experts Group (*) • Joint Photographic-image Experts Group • Joint Picture Experts Group

JPER: Journal of Physics Education Research

JPet: Journal of Petrology (periodical code)

JPF: Job Planning Form

JPGM: J. Paul Getty Museum (USA-CA)

JPI: Jensen and Partners International

JPIC: Joint Programme Integration Committee

JPhA: Journal of Physics A (periodical code)

JPhB: Journal of Physics B (periodical code)

JPhC: Journal of Physics C (periodical code)

JPhCh: Journal of Physical Chemistry (periodical code)

JPhD: Journal of Physics D (periodical code)

JPhE: Journal of Physics E (periodical code)

JPhF: Journal of Physics F (periodical code)

JPhG: Journal of Physics G (periodical code)

JPhys: Journal de Physique (periodical code)

JPJ: Paterson (USA-NJ) (location identifier)

JPL: Jet Propulsion Laboratory (USA-CA)

JPlPh: Journal of Plasma Physics (periodical code)

JPM: Job Performance Measure

Jpn: Japan • Japanese

JPN: Japan (ISO country code) • Japon (ISO country code) • Washington Pentagon (USA-DC) (location identifier)

JPO: Japanese Patent Office • Joint Program Office • Journal of Physical Oceanography (periodical code) • Journée Portes Ouvertes • Jump if Parity Odd • Junior Professional Officer • Jupiter Polar Orbiter

JPOC: JSC Payload Operations Center

JPOP: Japanese Polar Orbiting Platform

JPoSc: Journal of Polymer Science (periodical code)

JPP: Joint Plenary Presentation • Joint Program Plan • Journal of Propulsion Power (periodical code)

JPR: Jiparaná (Brazil-RO) (location identifier)

JPRD: Joint Program Requirements Document

JPRS: Joint Publications Research Service • Journal of Photogrammetry and Remote Sensing (periodical code)

JPS: Jan Paagman Sterrenwacht (Netherlands) • Japan Physical Society • Jet-Propulsion System • JICST Photoduplication Service • Joint Planning Staff • Journal of Power Sources (periodical code)

JPSG: Japanese Phrase Structure Grammar

JPT: Houston - Park Ten (USA-TX) (location identifier) • Jet Pipe Temperature • Journal of Propulsion Technology (periodical code)

JPTO: Jet-Propelled Takeoff

JPTS: Jet Propellant Thermally Stable

JPU: Job Processing Unit • Paris - La Défense (France) (location identifier)

JPVRC: Japan Pressure Vessel Research Council

JPW: Job Processing Word

JPY: Yen (Japan) (ISO currency code)

JQ: Job Questionnaire

JQE: Jaque (Panama) (location identifier) • Job Queue Entry

JQRST: Journal of Quantitative Spectroscopy and Radiative Transfer (periodical code)

JQT: Journal of Quality Technology (periodical code)

Jr: Jahr • Januar • Jour • Journal • Junior

JR: Jahresrechnung • Joanneum Research • Joint Resolution

JRA: New York - West 30th Street (USA-NY) (location identifier)

JRASA: Journal of the Royal Aeronautical Society of London (periodical code)

JRASC: Journal of the Royal Astronomical Society of Canada (periodical code)

J. RAS Canada: Journal of the Royal Astronomical Society of Canada

JRB: Joint Review Board • New York - Manhattan (USA-NY) (location identifier)

JRC: Jet Reaction Control • Joint Research Centre • Rochester (USA-MN) (location identifier)

JRD: Jardim • Riverside (USA-CA) (location identifier)

JRDC: Japan Research Development Corporation

JRE: New York - East 60th Street (USA-NY) (location identifier)

JReAt: Journal de Recherches Atmosphériques (periodical code)

JREF: James Randi Educational Foundation (USA-FL)

JREI: Joint Research Equipment Initiative

JRes: Journal of Research (periodical code)

JResB: Journal of Research B (periodical code)

JRH: Jorhat (India) (location identifier)

JRheo: Journal of Rheology (periodical code)

JRI: Journaliste Reporter d'Images

JRK: Arsuk (Greenland) (location identifier)

JRMS: Japanese Remote Manipulator System

JRN: Juruena (Brazil-MT) (location identifier)

JRNBS: Journal of Research of the National Bureau of Standards (periodical code)

JRO: Jicamarca Radio Observatory (Peru) • Kilimanjaro (Tanzania) (location identifier)

JROC: Joint Requirements Oversight Committee • Joint Requirements Oversight Council

JRoS: Journal of Robotic Systems (periodical code)

JRP: Joint Requirement Planning

JRPC: Journal of Reinforced Plastics and Composites (periodical code)

JRRC: Japanese Reprographic Rights Center

JRS: Jerusalem (location identifier)

JRSp: Journal of Raman Spectroscopy (periodical code)

J/S: Jammer/Signal

JS: Job Specification • Jump if Sign

JSA: Jaisalmer (India) (location identifier) • Japanese Standards Association

JSAI: Japan Society for Artificial Intelligence

JSAP: Japan Society of Applied Physics

JSAPJ: Japan Society of Air Pollution Journal (periodical code)

JSAPS: Japanese Journal of Applied Physics - Supplement (periodical code)

JSAR: Journal of Space Astronomy Research (periodical code)

JSASJ: Japan Society of Aeronautical Space Sciences Journal (periodical code)

JSASL: Journal of Southern Academic and Special Librarianship

JSASS: Japan Society for Aeronautical and Space Sciences

JSAST: Japan Society of Aeronautical Space Sciences Transactions (periodical code)

JSAT: Joint Structural Assessment Team

JSBC: Japan Satellite Broadcasting Co.

JSBD: Japanese Standard Bibliographic Description

JSBSS: Japan Society for Biological Sciences in Space

JSC: Japanese Science Council • Japan Scientific Committee • Johnson Space Center (USA-TX) • Joint Scientific Committee • Joint Steering Committee

JSCAS: Johnson Space Center Astronomical Society (USA-TX)

JSCM: Japanese Science Council Manual

JSCMJ: Japan Society of Composite Materials Journal (periodical code)

JSCNOET: Joint Standing Committee on Nuclear and Other Energy Technologies

JSCom: Journal of Scientific Computing (periodical code)

JSD: Jackson System Development • Stratford Sikorsky (USA-CT) (location identifier)

JSDAP: Jupiter System Data Analysis Program

JSDF: Japanese Self-Defence Forces

JSDT: Java Shared Data Toolkit

JSE: Jamieson Research and Engineering (USA-CA) (now VRI) • Johannesburg Stock Exchange (South Africa) • Journal of Scientific Exploration • Journal of Statistics Education • Juneau (USA-AK) (location identifier)

JSEC: Joint Scientific and Educational Center

Jsey: Jersey (Channel Islands)

JSF: Japan Science Foundation • Japan Space Forum • Joint Strike Fighter

JSG: Journal of Structural Geology (periodical code) • San Rafael (USA-CA) (location identifier)

JSGeo: Journal of Series Geophysics (periodical code)

JSH: Sitia (Greece) (location identifier)

JShR: Journal of Ship Research (periodical code)

JSI: Skiathos (Greece) (location identifier)

JSIA: Japan Software Industry Association

JSIAM: Japan Society for Industrial and Applied Mathematics

JSIC: Joint Space Intelligence Center

JSIK: Japan Society of Information Knowledge

JSIS: Japan Science Information Service

JSIT: Journal of Systems and Information Technology

JSK: Saint Cloud (USA-MN) (location identifier)

JSL: Atlantic City - Steel Pier (USA-NJ) (location identifier) • Jet Select Logic • Job Specification Language

JSLA: Japann Special Library Association

JSLC: Jinquan Satellite Launch Facility (China-PRC) • Joint Standing Committee on Library Cooperation

JSLEJ: Japan Society of Lubrication Engineers Journal (periodical code)

JSLR: Journal of Soviet Laser Research (periodical code)

JSLS: Japan Society of Library Science

JSLWG: Joint Spacelab Working Group

JSM: José de San Martin (Argentina-CB) (location identifier) • Journal of Structural Mechanics (periodical code)

JSMAJ: Japan Society of Microgravity Applications Journal (periodical code)

JSMSJ: Japan Society of Materials Science Journal (periodical code)

JSN: Junction Switch Number • Los Angeles - Sherman Oaks (USA-CA) (location identifier)

JSO: Sodertalje (Sweden) (location identifier) • Södertälje (Sweden) (location identifier)

JSOC: Joint Science Operations Centre

JSORD: Joint Systems Operational Requirement Document

JSOW: Joint Stand Off Weapon

JSP: Cheju Sogwipo (Korea-ROK) (location identifier) • Jackson Structured Programming • Java Server Page • Joint Seismic Program • Journal of Statistical Physics (periodical code)

JSPC: Joint Seismic Program Center

JSpRo: Journal of Spacecraft and Rockets (periodical code)

JSPRS: Japan Society of Photogrammetry and Remote Sensing

JSPS: Japan Society for the Promotion of Science

JSpT: Journal of Spacecraft Technology (periodical code)

JSR: Jessore (Bangladesh) (location identifier)

JSRA: Joint Sponsored Research Agreement

JSRC: Joint Services Review Committee

JSS: Japan Satellite Systems (merger of JCSAT & SAJAC) • JavaScript Style Sheet • Joint Services Standard • Joint Surveillance System • Spetsai Island (Greece) (location identifier)

JSSED: Journal of Solid-State Electron Device (periodical code)

JST: Japanese Standard Time • Japan Society of Translators • Japan Standard Time • Johnstown (USA-PA) (location identifier) • Joint Systems Test

JStA: Journal of Strain Analysis (periodical code)

JSTA: Joint Science and Technology Alliance

JSTARS: Joint Surveillance Target Attack Radar System

JSTC: Joint Scientific and Technical Committee

JSTF: Joint Society Task Force

JSTPS: Joint Strategic Targeting Planning Staff

JSU: Sukkertoppen (Greenland) (location identifier)

JSup: Journal of Superconductivity (periodical code)

JSUS: Joint Science Utilisation Study

JSV: Journal of Sound Vibration (periodical code)

JSWG: Joint Science Working Group

JSY: Syros Island (Greece) (location identifier)

JSZ: Saint-Tropez (France) (location identifier)

Jt: Jahrestonne • Jahrtausend • Joint

J-T: Joule-Thompson

JT: Joule-Thompson • Journal Télévisé

JTA: Japan Trademark Association • Job Table Area • Job Task Analysis

JTAC: Joint Technical Advisory Committee

JTACS: Japanese Total Access Communication System

JTAG: Joint Task Action Group

JTAGG: Joint Turbine Advanced Gas Generator

JTATE: Journal of Technology and Teacher Education (AACE)

JTC: Joint Technical Committee • Joint Trade Committee • Joint Transfer Correlation • Joint Transfer Correlator • Joint Tripartite Committee • Joule-Thomson Coefficient • Joule-Thomson Cooler

JTC3A: Joint Tactical Command, Control, and Communications Agency

JTDE: Joint Technology Demonstrator Engine

JTE: Joule-Thomson Effect • Journal of Technology Education

JTech: Journal of Technology (periodical code)

JTeEv: Journal of Testing Evaluation (periodical code)

JTePh: Journal of Technical Physics (periodical code)

JTF: Joint Task Force • Joule-Thomson Flow

JTFA: Joint Time-Frequency Analysis

JTFEX: Joint Task Force Exercise

Jtg: Jahrestag

JTG: Joint Training Group

JTGCM: Jovian Thermospheric General Circulation Model

JThCM: Journal of Thermoplastic Composite Materials (periodical code)

JThSt: Journal of Thermal Stresses (periodical code)

JTHT: Journal of Thermophysics and Heat Transfer (periodical code)

JTI: Jatai (Brazil-GO) (location identifier)

JTIDS: Joint Tactical Information Distribution System (*) • Joint Tactical Integrated Display System • Joint Technical Information Distribution System

JTIS: Japanese Technical Information Service

JTJ: Josephson Tunnel Junction

Jtly: Jointly

JTMP: Job Transfer and Manipulation Protocol

JTMS: Justification-based Truth Maintenance System

JTO: Jump Takeoff • Thousand Oaks (USA-CA) (location identifier)

JTP: Job Training Package

JTPA: Job Training Partnership Act

JTPC: Joint Technical Planning Committee • Joint Technical Programming Committee

JTPT: Job Task Performance Test

JTR: Thira (Greece) (location identifier)

JTS: Java Transaction Services • Job Training Standards

JTSC: Joint Test Support Center

Jtsd: Jahrtausend

JTST: Jet Stream • Journal of Thermal Spray Technology (periodical code)

JTSTR: Jet Stream

JTTAS: Japan Technology Transfer Association

JTU: Jackson Turbidity Unit

JTWC: Joint Typhoon Warning Center

JTY: Astypalaia Island (Greece) (location identifier)

Ju: June

JU: Joint Use • Junge Unternehmer

JUA: Juara (Brazil-MT) (location identifier)

Jub: Jubilar • Jubiläum

JUB: Juba (Sudan) (location identifier)

JUC: Los Angeles - Universal City (USA-CA) (location identifier)

JUCS: Journal of Universal Computer Science

Jud: Judge • Judgment • Judicial • Judiciary

JUD: Juris Utriusque Doctor

Judgt: Judgment

Juev: Jueves

JUG: Joint User Group

JUGFET: Junction-Gate Field-Effect Transistor

JUGL: JANET User Group for Libraries

JUI: Juist (Germany) (location identifier)

Juil: Juillet

Juill: Juillet

JUJ: Jujuy - El Cadillal (Argentina-PJ) (location identifier)

Jul: Julho • Julio • July

JÜL: Jülich (Germany)

JUL: Julho • Juliaca (Peru) (location identifier) • July

JUM: Jumla (Nepal) (location identifier)

Jun: June • Junio • Junior • Junta

JUN: Jundah (Australia-QLD) (location identifier) • June • Junho

Junc: Junction

JUNet: Japanese University Network • Japanese Unix Network • Japanese UUCP Network

JUNIC: Joint United Nations Information Committee

JUNO: Joint Ultraviolet Night-sky Observer

Junr: Junior

JUO: Jurado (Colombia) (location identifier)

Jup: Jupiter

JUP: Upland Cable (USA-CA) (location identifier)

JUR: Jurien Bay (Australia-WA) (location identifier)

JURG: Joint Users Requirements Group

Juris: Jurisprudence

JURIS: Justice Retrieval and Enquiry System

Jurisp: Jurisprudence

JUS: Japanese UNIX Society

JUSE: Japanese Union of Scientists and Engineers

Just: Justice • Justicia • Justification

JUST: Japan Universal System Transport

JUT: Juticalpa (Honduras) (location identifier)

Juv: Juvenile

JUV: Upernavik (Greenland) (location identifier)

Juw: Juwelen • Juwelier

Jux: Juxtaposition

JUZ: Juzhou (China-PRC) (location identifier)

Jv: Java • Javanese

JV: Amalthea (satellite of Jupiter) • Jahresversammlung • Joint Venture • Jugendverband • Jugendvertreter • Jugendvertretung

JVA: Ankavandra (Madagascar) (location identifier) • Jahresverbrauchsabrechnung

JVAS: Jodrell Bank - VLA Astrometric Survey

JVCIR: Journal of Visual Communication and Image Representation (periodical code)

JVE: Joint Verification Experiment • Journal of Vibration Engineering (periodical code)

JVerz: Jahresverzeichnis

JVGR: Journal of Volcanology and Geothermal Research (periodical code)

JVI: Himalia (satellite of Jupiter) • Manville Kupper (USA-NJ) (location identifier)

JVII: Elara (satellite of Jupiter)

JVIII: Pasiphae (satellite of Jupiter)

JVIRD: Joint Verification and Integration Requirements Document

JVL: Janesville (USA-WI) (location identifier)

JVM: Java Virtual Machine

JvNC: John von Neumann Center

JvNCC: John von Neumann Computer Center

JVP: Joint Verification Plan

JvS: Jongeren Vereniging voor Sterrenkunde (Belgium)

JVS: Jeudi-Vendredi-Samedi

JVST: Journal of Vacuum Science and Technology (periodical code)

jw: Javanese (ISO language code)

JW: Jahreswagen • Jewish

JWA: Jwaneng (Botswana) (location identifier)

JWC: Joint Wet Case • Los Angeles - Warner Center Business Plaza (USA-CA) (location identifier)

JWG: Joint Working Group

JWH: Houston - Westchase Hilton (USA-TX) (location identifier)

JWL: Houston Woodlawns (USA-TX) (location identifier)

Jwlr: Jeweler

JWO: Job Work Order

JWT: Job Wait Time

JX: Lysithea (satellite of Jupiter)

JXI: Carme (satellite of Jupiter)

JXII: Ananke (satellite of Jupiter)

JXIII: Leda (satellite of Jupiter)

JXIV: Thebe (satellite of Jupiter)

JXN: Jackson (USA-MI) (location identifier)

JXST: Journal of X-ray Science and Technology (periodical code)

JXV: Adrastea (satellite of Jupiter)

JXVI: Metis (satellite of Jupiter)

Jy: Jansky • July

JY: Jersey (Channel Islands) • Jordan (international aircraft registration code)

JYP: John Young Planetarium (USA-FL)

JYV: Jyväskylä (Finland) (location identifier)

JZ: Jahreszahl • Jodzahl • Jump if Zero

J2: Djibouti (international aircraft registration code)

J3: Grenada (international aircraft registration code)

J5: Guinea Bissau (international aircraft registration code)

J6: Saint Lucia (international aircraft registration code)

J7: Dominica (international aircraft registration code)

J8: Saint Vincent and the Grenadines (international aircraft registration code)

K

k: Absorption Coefficient • Boltzmann Constant • Capacity • Circular Wave Number • Circular Wave Vector • Cold • Constant • Coupling Coefficient • Curvature • Effective Thermal Conductivity • Extinction Coefficient • Gaussian Gravitational Constant • Grid Index • Heat Conductivity • Height Index • Kaiserlich • Kalt • Karat • Katholisch • Kill • Kilo • Kilogram • Kindergarten • King • Knight • Knot • Kopeck (currency) • Mass Transfer Coefficient • Miller Index • Molecular Conductivity • Multiplication Factor • Phase Factor • Propagation Vector • Radius of Gyration • Rate Coefficient • Rate Constant • Stiffness • Thermal Conductivity • Unit Coordinate Vector • von Kármán Constant

K: Aberration Constant • Bulk Modulus • Cambodia (international vehicle registration code) • Capacity • Carat • Cathode • Compression Modulus • Cretaceous • Eddy Viscosity • Equilibrium Constant • Heat Transfer Coefficient • Inertia • Kaap • Kabel • Kaiser • Kalium • Kalorie • Kalt • Kälte • Kammer • Kanal • Kaon • Kap • Kapitel • Kapp • Karat • Kärnten (Austria) • Kasse • Kathode • Kelvin • Kerma • Kette • Kill • Kilo (international alphabet) • Kilocycle • Kilogram • Kina (currency) • Kindergarten • Kinetic Energy • King • Kip (currency) • Kirche • Kiste • Knight • Knot • Köchel • Kommandant • Kommissar • König • Konsonant • Konsul • Konto • Konzentration • Koordinaten • Kopf • Korps • Kote • Krona • Krone (currency) • Kwacha (currency) • Kyat (currency) • Lift-Dependent Drag • Luminous Efficacy • Moment of Inertia • One Thousand • Permeability • Potassium • Quantum Number of Component of Angular Momentum along Axis • Rotational Quantum Number • Solar Absorption Index • Solar Constant • Spectral Type • Stiffness • Thousand • Zero Point

ka: Georgian (ISO language code)

kA: Kiloampere

Kä: Kärnten (Austria)

Ka: Auroral Absorption Index • Thousands of Years

KA: Karelia • Kassenanwalt • Keep Alive • Kill Assessment • Kilmarnock (UK postal code) • Kilo-Ampere • Kleine Anfrage • Klimaanlage • Knotenamt • Knowledge Acquisition • Kontoauszug • Kulturamt • Kunstakademie • Kurzarbeit

KAA: Kasama (Zambia) (location identifier)

KAAU: King Abdul Aziz University (Saudi Arabia)

Kab: Kabel • Kabine • Kabinett

KAB: Kariba (Zimbabwe) (location identifier)

KABLE: KSC Atmospheric Boundary Layer Experiment

KAC: Kameshli (Syria) (location identifier)

KACST: King Abdulaziz City for Science and Technology (Saudi Arabia)

Kad: Kadenz • Kader • Kadett

KAD: Kaduna (Nigeria) (location identifier)

KaDeWe: Kaufhaus des Westens (Berlin)

KADC: Korean Aircraft Development Consortium

KADOC: Katholiek Archief en Documentatiecentrum

KADS: Knowledge Acquisition and Design Support • Knowledge Acquisition and Documentation System

KAE: Kake (USA-AK) (location identifier)

KAERI: Korean Atomic Energy Research Institute

KAF: Karato (Papua New Guinea) (location identifier) • Karlskrona Astronomi Förening (Sweden) • Københavns Astronomiske Forening (Denmark) • Kost-Assekuranz-Fracht

KAG: Kangnung (Korea-ROK) (location identifier) • Kapitalanlagegesellschaft • Kreisarbeitsgemeinschaft • Künstlerische Arbeitsgruppe

Kah: Kahoolawe (USA-HI)

KAH: Melbourne City (Australia-VIC) (location identifier)

KAI: Kaieteur (Guyana) (location identifier)

KAIS: Korea Advanced Institute of Science (Korea-ROK)

KAIST: Korean Advanced Institute of Science and Technology (Korea-ROK)

KAIT: Katzmann Automatic Imaging Telescope

Kaj: Kajüte

KAJ: Kajaani (Finland) (location identifier)

KAK: Kar (Papua New Guinea) (location identifier) • Key-Auto-Key

kal: Grammkalorie (now cal.)

Kal: Kalendae • Kalendas • Kalender • Kaliber • Kilokalorie (now Kcal.)

KAL: Kaltag (USA-AK) (location identifier) • Keep-Alive Loop

Kalf: Kalfactor

KALI: Korea Academy of Library and Information Science (Korea-ROK)

Kam: Kamerad

KAM: Kamaran Island (Yemen) (location identifier) • Kinematic Analysis Method • Knowledge Acquired by Media • Knowledge Acquisition Model • Knowledge Area Module • Kolmogorov-Arnold-Moser

KAMBIS: KARMAC Modulair Bibliotheek Informatie Systeem

KaMI: Kaons at the Main Injector

KAMS: Kawasaki Automatic Machining System

KAMSC: Kalamazoo Area Math and Science Center (USA-MI)

Kan: Kanaal • Kanada • Kanadier • Kanal • Kamister • Kansas (USA)

KAN: Kano (Nigeria) (location identifier) • Knowledge And Naming • Kriminalaktennachweis

Kand: Kandidat

KANDIDATS: Kansas Digital Image Data System

Kans: Kansas (USA)

KANSA: Kansas Association for Networked Supercomputer Applications (USA-KS)

kant: kantonal

Kant: Kanton

KAO: Kansas Astronomical Observers (USA-KS) • Korea Astronomy Observatory (Korea-ROK) • Kuiper Airborne Observatory (USA-CA) • Kuusamo (Finland) (location identifier)

KAOS: Kansas Astrophotographers and Observers Society (USA-KS) • Kiepenheuer-Institut Adaptives Optiksystem

Kap: Kapazität • Kapelle • Kapillare • Kapital • Kapitän • Kapitel

KAP: Kapanga (Zaire) (location identifier) • Kluwer Academic Publishers

KAPIS: Kluwer Academic Publishers Information Service

KAPL: KSC Approved Parts List

KAPN: KSC Automated Payloads Notice

KAPPA: Kernel Application Package

KAPPS: KSC Automated Payloads Project Specification

Kapt: Kapitän

KAPTUR: Knowledge Acquisition for Preservation of Tradeoffs and Underlying Rationales

KAQ: Kamulai (Papua New Guinea) (location identifier)

Kar: Karabiner • Karachi (Pakistan) • Karaffe • Karat • Karosse

KAR: Kamarang (Guyana) (location identifier) • Karlstadt (Germany) • Kodak Automated Retrieval

KARC: Kansai Advanced Research Center (Japan)

Kard: Kardinal • Kardiogramm

Karel: Karelia • Karelian

KARI: Korea Aerospace Research Institute (Korea-ROK) (*) • Korean Aerospace Research Institute (Korea-ROK)

KARMA: Knowledge-based Augmented Reality for Maintenance Assistance

KARMEN: Karlsruhe Rutherford interMediate Energy Neutrino • Karlsruhe Rutherford Medium-Energy Neutrino (experiment) (*)

KARP: Knowledge Acquisition, Representation and Processing

KARRN: Kyushu Area Research Regional Network

KARS: Kansas Applied Remote Sensing Program • Kilometer-square Area Radio Synthesis Telescope

kart: kartoniert

Kart: Kartell • Kartoffel • Kartograph • Karton

KAS: Kalamazoo Astronomical Society (USA-MI) • Karasburg (Namibia) (location identifier) • Kern Astronomical Society (USA-CA) • Knowledge Availability System • Korean Astronomical Society (Korea-ROK)

KASC: Knowledge Availability System Center

KASCADE: Karlsruhe Shower Core and Array Detector

Kash: Kashmir

KASL: Kansas Association of School Librarians (USA-KS)

KASLIB: Kanagawa Association of Special Libraries and Information Bureaus

Kass: Kassation • Kasserolle • Kassette • Kassier

KASSPR: Knowledge Augmented Severe Storms Predictor

Kat: Katalog • Katalysator • Katamaran • Katastrophe • Kategorie

KAT: Kaitaia (New Zealand) (location identifier) • Key Alias Table • Koordinierungsausschuß Terminologie

KATE: Knowledge-based Autonomous Test Engineer

Kath: Kathedral

KATS: Kennedy Avionics Test Set

KAU: Kauhava (Finland) (location identifier) • Keystone Adapter Unit • Kilo Accounting Units

KAUDYTE: Knowledge Acquisition and Use in Dynamic Task Environments

Kaufh: Kaufhalle • Kaufhaus

Kaufm: Kaufmann

KAUS: Knowledge Acquisition and Utilization System

KAV: Kavanayen (Venezuela) (location identifier) • Kindergeldauszahlungs-Verordnung

KAVAS: Knowledge Acquisition, Visualization and Assessment Study

KAvB: Koninklijke Academie van Belgien (Belgium)

KAW: Kawthaung (Myanmar) (location identifier)

KAX: Kalbarri (Australia-WA) (location identifier)

KAY: Wakaya Island (Fiji) (location identifier)

KAZ: Kau (Indonesia) (location identifier) • Kazakhstan (ISO country code)

Kazakh: Kazakhstan

KAZMENST: Committee for Standardization, Metrology and Certifications of Kazakhstan

kb: Kilobar • Kilobit

K&B: Kitchen & Bathroom

KB: Kantonsbibliothek • Kartenbeilage • Kassenbericht • Kaushītaki Brāhmana • Key Based • Keyboard • Keyboard Button • Kilobyte • Knowledge Base • Knowledge-Based • Knowledge Building • Kommissionsberatung • Kommissionsbericht • Kongelige Bibliotek (Denmark) • Königliche Bibliothek • Koninklijk Besluit • Koninklijke Bibliotheek • Koninkrijk België • Kontaktbereich • Kostenbereich • Kriegsbeschädigte • Kuiper Belt • Kulturbund • Kungliga Biblioteket • Kurbestimmungen • Kursbuch • Kurzbericht

KBA: Kabala (Sierra Leone) (location identifier) • Knowledge-Based Accounting • Knowledge-Based Analysis

KBAC: Kennedy Booster Assembly Contractor

KBAD: Knowledge-Based Analysis and Design

KBAIS: Knowledge-Based Action Information System

kbar: Kilobar

KBB: Kirkimbie (Australia-NT) (location identifier) • Knowledge-Based Bibliography

KBBE: Knowledge-Based Business Engineering

KBC: Birch Creek (USA-AK) (location identifier) • Kapton Beta Cloth

KBCA: Knowledge-Based Circuit Analysis

KBCD: Knowledge-Based Circuit Design

KBCE: Knowledge-Based Control Engineering

KBCID: Knowledge-Based Concurrent Integral Design

Kbcm: Kubikzentimeter

KBCSD: Knowledge-Based Control System Design

KBD: Keyboard • Kimberley Downs (Australia-WA) (location identifier) • Knowledge-Based Database • Knowledge-Based Design • Knowledge-Based Detection • Knowledge-Based Development • Knowledge-Based Device • Knowledge-Based Diagnosis • Knowledge-Based Directory • Knowledge-Based Dispatching • Knowledge-Based Drafting

KBDA: Knowledge-Based Design and Analysis • Knowledge-Based Distribution and Assignment

KBDAT: Knowledge-Based Design And Test

KBDB: Knowledge-Based Database

KBDBMS: Knowledge-Based Database Management System

KBDCE: Knowledge-Based Distributed Computing Environment

KBDD: Knowledge-Based Design and Drafting • Knowledge-Based Drafting and Design

KBDDS: Knowledge-Based Design Drafting System

KBDE: Knowledge-Based Design and Evaluation • Knowledge-Based Design Engineering • Knowledge-Based Document Engineering

KBDET: Knowledge-Based Design and Electrical Test • Knowledge-Based Design Experiment Translator

KBDIC: Knowledge-Based Design of Electrical Circuits

KBDIS: Knowledge-Based Design Information System

Kbdm: Kubikdezimeter

KBDMP: Knowledge-Based Data Management Procedure

KBDMS: Knowledge-Based Database Management System

KBDS: Knowledge-Based Design System • Knowledge-Based Dispatching System

KBDSS: Knowledge-Based Decision Support System

KBE: Bell Island - Hot Springs (USA-AK) (location identifier) • Keyboard Encoder • Keyboard Entry • Knowledge-Based Education • Knowledge-Based Engineering • Knowledge-Based Entry • Knowledge-Based Environment • Knowledge-Based Estimating • Köln-Bonner-Eisenbahnen (Germany)

KBEDE: Knowledge-Based Engineering and Design for Electronics

KBEG: Knowledge-Based Engineering Graphics

KBES: Knowledge-Based Expert System

KBET: Knowledge-Based Electrical Test

KBF: Karubaga (Indonesia) (location identifier) • Koordinationsstelle für Forschungsprojekten

KBFE: Knowledge-Based Film Editor • Knowledge-Based Front End

KBFF: Knowledge-Based Feature Finder

KBFIT: Knowledge-Based Fault Isolation Test

KBFS: Knowledge-Based File System

KBG: Kabalega Falls (Uganda) (location identifier) • Knowledge-Based Gauging • Knowledge-Based Graphics

KBGD: Knowledge-Based Geometric Design

KBGIS: Knowledge-Based Geographical Information System

KBH: Kalat (Pakistan) (location identifier)

KBHD: Knowledge-Based Hypermedia Development

KBHDE: Knowledge-Based Hypermedia Development Environment

KBHDM: Knowledge-Based Hypermedia Design Method

Kbhvn: København (Denmark)

KBI: Knowledge-Based Instruction • Kreisbrandinspektor • Kribi (Cameroon) (location identifier)

KBIC: Key Base In Context

KBID: Knowledge-Based Imaging Device

KBIM: Keyboard Interface Module

KBIN: Koninklijk Belgisch Instituut voor Natuurwetenschappen (Belgium)

KBJ: Kings Canyon (Australia-NT) (location identifier)

KBK: Klag Bay (USA-AK) (location identifier)

KBL: Kabul (Afghanistan) (location identifier) • Knowledge-Based Language • Knowledge-Based Learning • Knowledge-Based Library • Knowledge-Based Listing

KBLA: Knowledge-Based Loads Analysis

KBLCE: Knowledge-Based Life Cycle Analysis

Kblg: Kabellänge

KBLL: Knowledge-Based Learning Analysis

KBLS: Knowledge-Based Logistics and Support • Knowledge-Based Logistics Support • Knowledge-Based Logistics System

Kbm: Kubikmeter

KBM: Kabwum (Papua New Guinea) (location identifier) • Keyboard Monitor • Kinetic Ballooning Mode • Knowledge-Based Maintenance • Knowledge-Based Management • Knowledge-Based Manufacture • Knowledge-Based Manufacturing • Knowledge-Based Mapping

KBMAC: Knowledge-Based Measurement And Control

KBMAT: Knowledge-Based Design, Manufacture, And Test • Knowledge-Based Design, Manufacturing, And Testing

KBMCS: Knowledge-Based Maintenance and Control System

KBMD: Knowledge-Based Mechanical Drawing

KBMM: Knowledge-Based Maintenance Management • Knowledge-Based Manufacturing Management

KBMP: Knowledge-Based Management Planning

KBMS: Knowledge-Based Management System • Knowledge-Based Manipulation System • Knowledge Base Management System

KBMU: Knowledge-Based Makeup

KBN: Kabinda (Zaire) (location identifier) • Knowledge-Based Navigation • Knowledge-Based Numerics

KBO: Kabalo (Zaire) (location identifier) • Kuiper Belt Object

KBOS: Knowledge-Based Observation System • Knowledge-Based Operating System

KBP: Keyboard Process • Kiev Borispol (Ukraine) (location identifier) • Knowledge-Based Pattern • Knowledge-Based Planning • Knowledge-Based Printing • Knowledge-Based Production • Knowledge-Based Programming • Knowledge-Based Publishing

kBpI: Kilobits per Inch

KBPL: Knowledge-Based Programming Language

KBPP: Knowledge-Based Process Planning • Knowledge-Based Production Planning

kbps: Kilobits per Second

KBPS: Kilobytes Per Second • Knowledge-Based Programming and Systems • Knowledge-Based Programming System

KBQA: Knowledge-Based Quality Assurance

KBR: Knowledge-Based Retrieval • Kota Bharu (Malaysia) (location identifier)

KBRA: Knowledge-Based Research Analysis

KBRD: Knowledge-Based Research and Development

KBRIS: Knowledge-Based Resource Information System

KBRP: Knowledge Base Research Program

KBRS: Knowledge-Based Reference Service • Knowledge-Based Routing System

kb/s: Kilobits per Second

KBS: Bo (Sierra Leone) (location identifier) • Ka-Band System • Knowledge-Based Software • Knowledge-Based Styling • Knowledge-Based System • Knowledge Base of Stars • Knowledge-Base Server • Korean Bibliographical Society (Korea-ROK) • Korean Bureau of Standards (Korea-ROK)

KBSA: Kenya Booksellers ans Stationers Association • Knowledge-Based Software Assistant • Knowledge-Based System Analysis

KBSD: Knowledge-Based Software Design • Knowledge-Based Structural Design • Knowledge-Based System Design

KBSE: Knowledge-Based Support Equipment • Knowledge-Based Systems Engineering • Knowledge-Based Systems Evaluation

KBST: Knowledge-Based Software Testing • Knowledge-Based Structural Technology

KBT: Kaben (Marshall Islands) (location identifier) • Knowledge-Based Technology • Knowledge-Based Tomography • Knowledge-Based Trading • Knowledge-Based Transcription • Knowledge-Based Typesetting

KBTE: Knowledge-Based Test Equipment

KBTM: Knowledge-Based Testing and Maintenance

KBTP: Knowledge-Based Typesetting Process

KBTS: Knowledge-Based Telemetry System • Knowledge-Based Test Shop • Knowledge-Based Test System • Knowledge-Based Trading System • Knowledge-Based Troubleshooting

KBU: Keyboard Unit • Kotabaru (Indonesia) (location identifier)

KBUI: Knowledge-Based User Interface

kbV: keine besonderen Vorkommnisse

KBV: Kleinbetragsverordnung • Knowledge-Based Verification • Knowledge-Based View

KBW: Chignik Bay (USA-AK) (location identifier)

KBX: Kambuaya (Indonesia) (location identifier)

KBY: Streaky Bay (Australia-SA) (location identifier)

Kbyte: Kilobyte

KBZ: Kaiboura (New Zealand) (location identifier)

kc: Kilocurie • Kilocycle

KC: Kansas City (USA) • Kyle Classification

KCA: Ku-band Communications Adapter • Kuqa (China-PRC) (location identifier)

kcal: Große Kalorie • Kilocalorie • Kylocycle

KCAO: Kumamoto Civil Astronomical Observatory (Japan)

KCAS: Knots Calibrated Airspeed

KCC: Coffman Cove (USA-AK) (location identifier) • Keyboard Common Contact

KCDC: Korea Commercial Aircraft Development Consortium

KCE: Collinsville (Australia-QLD) (location identifier)

KCG: Chignik Fisheries (USA-AK) (location identifier)

KCH: Kuching (Malaysia) (location identifier)

kCi: Kilocurie

KCI: Kenya Computer Institute • Kon (Indonesia) (location identifier)

KCK: Kansas City - Fairfax (USA-KS) (location identifier)

KCl: Potassium Chloride

KCL: Chignik Lagoon (USA-AK) (location identifier) • Kirchoff Current Law • Kyoto Common LISP

KCLN: Kansas City Library Network

KCMS: Kodak Color Management System

KCMSD: Kansas City Missouri School District (USA-MO)

KCN: Chernofski (USA-AK) (location identifier)

KCNA: Korean Central News Agency (Korea-ROK)

KCO: Kappler Crystal Optics (USA-MA) • Keystone College Observatory (USA-PA)

KCP: Kamenets-Podolskiy (Ukraine) (location identifier) • Kheda Communications Project

KCQ: Chignik Lake (USA-AK) (location identifier)

KCR: Colorado Creek (USA-AK) (location identifier) • KSC Change Request

KCRT: Keyboard Cathode Ray Tube

kc/s: Kilocycles per Second

kcs: Kilocycles per Second

Kčs: Koruna (currency)

KCS: Keyboard Configuration Studies • Key Configuration Study • Kings Creek Station (Australia-NT) (location identifier)

KCSC: Kansas Cosmosphere and Space Center (USA-KS)

kc/sec: Kilocycles per Second

KCU: Masindi (Uganda) (location identifier)

KCV: KKV Carrier Vehicle

KCZ: Kochi (Japan) (location identifier)

KD: Knock Down • Knocked Down • Kongeriget Danmark • Kontrolldienst • Kreisdirektor • Kundendienst • Kursor Down • Kuwaiti Dinar (currency)

KDA: Kolda (Senegal) (location identifier)

KDAAS: Kettering and District Amateur Astronomers Society (UK) (now NAA)

KDAD: Knowledge Discovery from Advanced Databases

KDB: Kambalda (Australia-WA) (location identifier) • Kernel Debugger • Kinematically Decoupled Bulge • Koordinaten-Datenbanksystem

KDC: Kandi (Benin) (location identifier) • Kinematically Decoupled Core • Knocked-Down Condition • Korean Decimal Classification

KDD: Khuzdar (Pakistan) (location identifier) • Knowledge Discovery in Databases • Kokusai Denshin Denwa Co.

Kde: Kunde

KDE: Koroba (Papua New Guinea) (location identifier)

KdF: Kraft durch Freude

KDF: Key Distribution Facility • Kouba (Algeria) (location identifier)

Kdg: Kündigung

KDG: Kardjali (Bulgaria) (location identifier)

Kdgg: Kündigung

KDH: Kandahar (Afghanistan) (location identifier) • Key Distribution Host

KDI: Kendari (Indonesia) (location identifier) • Knowledge and Distributed Intelligence • Korea Development Institute (Korea-ROK)

KDJ: N'Djole (Gambia) (location identifier)

KDK: Kodiak (USA-AK) (location identifier)

Kdm: Kingdom

KDM: Kaadedhdloo (Maldives) (location identifier) • Kongelige Danske Marine (Denmark)

KDMS: Kennedy Data Management System

KDN: Kinetically Designed Nozzle • N'Dende (Gabon) (location identifier)

Kdo: Kommando

KDO: Kadhdhoo (Maldives) (location identifier)

KDP: Kandep (Papua New Guinea) (location identifier) • Potassium Dihydrogen Phosphate

KDPN: KSC DOD Payloads Notice

KDPSN: KSC DOD Payloads Project Specification

Kdr: Kalender • Kinder • Kommandeur

KDR: Kandrian (Papua New Guinea) (location identifier) • Keyboard Data Recorder

Kdre: Kommodore

KDS: Kamaran Downs (Australia-QLD) (location identifier) • Kathode Dark Space • Key Display System • Knowledge Data System • Knowledge Dictionary System • KSC-Developed Software

Kdt: Kommandant

KDT: Keyboard and Display Test • Keyboard Display Terminal • Key Definition Table • Kinetic Desintegration Technology

KDU: Keyboard and Display Unit • Skardu (Pakistan) (location identifier)

KDV: Kandavu (Fiji) (location identifier) • KKV Deployment Vehicle

KDVZ: Kommunale Datenverarbeitungszentrale

Ke: Kessel • Kette

KE: Karteneinheit • Kenya (ISO country code) • Keypad End • Kinetic Energy • Kinetic Equation • Knowledge Engineering • Knowledge Exchange

KEA: Keisah (Indonesia) (location identifier)

KEALSE: Key-search Algorithm Selection

KEAS: Knots Equivalent Airspeed • Knots Estimated Airspeed

KEB: English Bay (USA-AK) (location identifier)

KEBS: Kenya Bureau of Standards

KEC: Kasenga (Zaire) (location identifier) • Key Encryption Component

KECNET: Kentucky Educational Computr Network

KECU: Kilo European Currency Unit • Thousand European Currency Units

KED: Kaedi (Mauritania) (location identifier)

KEDDS: Kansas Educational Dissemination/Diffusion System

KEE: Keele (Starlink internal code) • Kelle (Congo) (location identifier) • Knowledge Engineering Environment

KEEP: Kestrel Educational Exchange Program

KEEPS: Kodak Ektaprint Electronic Publishing System

KEF: Reykjavík Keflavík (Iceland) (location identifier)

KEG: Keglsugl (Papua New Guinea) (location identifier) • Kommission der Europäischen Gemeinschaft

KEGS: KAO Echelle Grating Spectrometer • Kuiper Echelle Grating Spectrometer

KEH: Kenmore (USA-WA) (location identifier)

KEI: Kepi (Indonesia) (location identifier)

KEIS: Kentucky Economic Information System

KEJ: Kemerovo (Russia) (location identifier)

KEK: Ekwok (USA-AK) (location identifier) • Kommission fßr Erschliessung and Katalogmanagement • Konferenz der Europäischen Kirchen • Koo Energy Ken

KEL: Kehl (Germany) • Kiel (Germany) (location identifier)

KELOG: Kiso Emission-Line Objects and Galaxies

KELV: KKV/ESM Launch Vehicle

KEM: Kemi Tornio (Finland) (location identifier) • Kemnath (Germany)

KEMAR: Knowles Electronics Manakin for Acoustic Research

Ken: Kensington • Kentucky (USA) • Kenya

KEN: Kenema (Sierra Leone) (location identifier) • Kent (Starlink internal code) • Kenya (ISO country code)

Kennz: Kennzahl • Kennzeichen • Kennziffer

KENVAD: Kent (Starlink internal code)

KEO: Odienne (Ivory Coast) (location identifier)

Kep: Kepulauan

KEP: Nepalganj (Nepal) (location identifier)

KEQ: Kebar (Indonesia) (location identifier)

KER: Kerman (Iran) (location identifier)

KERA: Kentucky Education Reform Act (USA-KY)

KERMIT: Kl-10 Error-free Reciprocal Micro Interconnect Over Tty lines

KES: Kesley (Canada) (location identifier) • Knowledge-based Engineering System • Knowledge Engineering Software • Knowledge Engineering System • Shilling (Kenya) (ISO currency code)

KET: Kiel Electron Telescope • Keng Tung (Myanmar) (location identifier)

keV: Kiloelectronvolt

KEV: Kinetic-Energy Vehicle • Kuorevesi Halli (Finland) (location identifier)

KEW: Kinetic-Energy Weapon

KEX: Kanabea (Papua New Guinea) (location identifier)

KEY: Keep Extending Yourself • Kericho (Kenya) (location identifier)

KEYB: Keyboard

KEYBBE: Keyboard for Belgium

KEYBBR: Keyboard for Brazil

KEYBCF: Keyboard for Canada-France

KEYBCZ: Keyboard for Czech Republic

KEYBDK: Keyboard for Denmark

KEYBFR: Keyboard for France

KEYBGR: Keyboard for Germany

KEYBHU: Keyboard for Hungary

KEYBIT: Keyboard for Italy

KEYBLA: Keyboard for Latin America

KEYBNL: Keyboard for Netherlands

KEYBNO: Keyboard for Norway

KEYBPL: Keyboard for Poland

KEYBPO: Keyboard for Portugal

KEYBSF: Keyboard for Switzerland-France

KEYBSG: Keyboard for Switzerland-Germany

KEYBSL: Keyboard for Slovakia

KEYBSP: Keyboard for Spain

KEYBSU: Keyboard for Finland

KEYBSV: Keyboard for Sweden

KEYBUK: Keyboard for United Kingdom

KEYBUS: Keyboard for United States

KEYBYU: Keyboard for Yugoslavia

kF: Kilofrancs

Kf: Kauf • Kurzfassung

KF: Kaltfront • Kilo Francs • Kleine Fahrt • Knowledge Finder • Kümmerly and Frey AG

KfA: Kommission für Astronomie

KFA: Kernforschungsanlage • Kiffa (Mauritania) (location identifier) • Kosten-Fracht-Assekuranz

KFAS: Keyed File Access System • Kuwait Foundation for the Advancement of Science

KFDD: Kronecker Functional Decision Diagram

Kffr: Kauffrau

KFG: Kalkurang (Australia-NT) (location identifier)

Kfh: Kaufhaus

KFK: Kernforschungszentrum Karlsruhe (Germany)

KFL: Kinderferienlager

Kfm: Kaufmann • Kaufmännisch

KFNT: Kinematika i Fisika Nebesbykh Te.

KFP: False Pass (USA-AK) (location identifier) • Konferenz der Fachbereiche Physik (Germany)

KFT: Kilofeet

Kfz: Kraftfahrzeug • Kubisch-flächenzentrierte

kg: Kilogram

kG: Kilogauss

Kg: Kampong • Kegel • Kenntnis genommen • König

KG: Kammergericht • Kantonsgericht • Kassationsgericht • Kaufmannsgericht • Key Generator • Kindergarten • Klein-Gordon • Kommanditgesellschaft • Korte Golf • Kreisgericht • Kyrgyzstan (ISO country code)

KGA: Kananga (Zaire) (location identifier)

KGB: Konge (Papua New Guinea) (location identifier)

KGC: Kingscote (Australia-SA) (location identifier) • Kurt Gödel Colloquium

KGD: Kaliningrad (Russia) (location identifier) • Known Good Die

KGer: Kammergericht

kgf: Kilogram-Force

Kgf: Kriegsgefangene

KGF: Karaganda (Kazakhstan) (location identifier)

KGG: Kedougou (Senegal) (location identifier)

KGH: Yongai (Papua New Guinea) (location identifier)

KGI: Kalgoorlie (Australia-WA) (location identifier)

KGJ: Karonga (Malawi) (location identifier)

KGK: New Koliganek (USA-AK) (location identifier)

kgl: königlich

KGL: Kigali (Rwanda) (location identifier)

kgm: Kilogram-Meter

KGM: Known Good Module • Kungum (Papua New Guinea) (location identifier)

Kgn: Königin

KGN: Kasongo Lunda (Zaire) (location identifier)

KGO: Kirovograd (Ukraine) (location identifier)

kgps: Kilograms per Second

KGPS: Kilograms Per Second • Kinematic Global Positioning System

Kgr: Kongreß • Königreich

KGR: Kammergerichtsrat • Kulgera (Australia-NT) (location identifier)

kgs: Kilograms

KGS: Known Geological Structure • Kos (Greece) (location identifier) • Kurt Gödel Society (Austria) • Som (Kyrgyzstan) (ISO currency code)

KGU: Keningau (Malaysia) (location identifier)

KGV: K Giant Variable • Kleinste Gemene Veelvoud

KGW: Kagi (Papua New Guinea) (location identifier)

KGX: Grayling (USA-AK) (location identifier)

KGY: Kingaroy (Australia-QLD) (location identifier)

KGZ: Glacier Creek (USA-AK) (location identifier) • Kyrgyzstan (ISO country code)

Kh: Krankenhaus

K-H: Kelvin-Helmholtz

KH: Cambodge (ISO country code) • Cambodia (ISO country code) • Kelvin-Helmholtz • Keyhole • Khmer Republic (ISO country code) • Kinderheim • Kulturhaus

KHA: Khaneh (Iran) (location identifier)

KHB: KSC Handbook

KHBV: Krankenhaus-Buchführungsverordnung

KHC: Kerch (Ukraine) (location identifier)

KHD: Katastrophenhilfsdienst • Klöckner-Humboldt-Deutz AG (Germany)

Khdw: Kunsthandwerk

KHE: Kherson (Ukraine) (location identifier)

Khf: Kirchhof

KHG: Kashi (China-PRC) (location identifier)

KHH: Kaohsiung (China-ROC) (location identifier)

KHI: Karachi - Quaid-e-Azam (Pakistan) (location identifier) • Kawasaki Heavy Industries • Kelvin-Helmholtz Instability

KHIL: KKV Hardware In the Loop

KHILS: KKV Hardware In the Loop Simulator

KHIT: KKV Hovering Interceptor Test

KHJ: Kauhajoki (Finland) (location identifier)

KHK: Khark Island (Iran) (location identifier)

KHL: Khulna (Bangladesh) (location identifier)

KHM: Cambodge (ISO country code) • Cambodia (ISO country code) • Khamti (Myanmar) (location identifier)

KHN: Nanchang (China-PRC) (location identifier)

KHO: Khoka Moya (South Africa) (location identifier)

KHR: Kharkhorin (Mongolia) (location identifier) • Riel (Cambodia) (ISO currency code)

KHS: Khasab (Oman) (location identifier)

KHT: Khost (Afghanistan) (location identifier)

KHU: Konferenz der Hessischen Universitäten • Kremenchug (Ukraine) (location identifier)

KHV: Khabarovsk (Russia) (location identifier)

KHW: Khwai River Lodge (Botswana) (location identifier)

kHz: Kilohertz

Ki: Kilo • Kind • Kinder • Kirche

KI: Kiribati (ISO country code) • Kiwanis International • Known Icing • Kosmicheskie Issledovaniya • Kreisinspektor • Künstliche Intelligenz

KIA: Kaiapit (Papua New Guinea) (location identifier) • Keck Interferometric Array

KIAM: Keldysh Institute of Applied Mathematics (Russia)

KIAS: Knots Indicated Airspeed

KIB: Ivanof Bay (USA-AK) (location identifier) • Kiso Information Bulletin

KIBO: Knowledge In - Bullshit Out

KIC: King City - Mesa del Rey (USA-CA) (location identifier)

KICH: Kyoto International Conference Hall (Japan)

KID: Kristianstad (Sweden) (location identifier)

KIDS: Kestrel Interactive Development System

KIE: Kieta (Papua New Guinea) (location identifier) • Knowledge Integration Environment

KIER: Korea Institute of Energy and Resources (Korea-ROK)

KIET: Korea Institution for Economics and Technology (Korea-ROK)

KIF: Kingfisher Lake (Canada-ON) (location identifier) • Knowledge Interchange Format

KIFU: Koordinationsstelle Für Umweltforschung

KIG: Koingnaas (South Africa) (location identifier)

KIH: Kish Island (Iran) (location identifier)

KIJ: Niigata (Japan) (location identifier)

KIK: Kirkuk (Iraq) (location identifier)

Kil: Kilometer

KIL: Kilwa (Zaire) (location identifier)

Kild: Kildare (Ireland)

Kilk: Kilkenny

KIM: Kalev Iteration Method • Keyboard Input Matrix • Kimberley (South Africa) (location identifier) • Kinetic Isotope Method • Knowledge-based Information Management

KIMMJ: Korean Institute of Metals and Materials Journal (periodical code)

KIMS: Kodak Image Management System • KSC Inventory Management System

KIN: Kingston - Norman Manley (Jamaica) (location identifier)

Kind: Kindergarten

Kinderg: Kindergarten

King: Kingdom

Kingd: Kingdom

KINITI: Korea Institute of Industry and Technology Information

KIO: Kili Island (Marshall Islands) (location identifier)

KIP: Keyboard Input Processor • Key Information Provider • Knowledge Information Processing • Thousand Pounds • Wichita Falls - Kickapoo (USA-TX) (location identifier)

KIPO: Korea Industrial Property Office (Korea-ROK)

KIpS: Thousands of Instructions per Second

KIPS: KAIST Image Processing System • Knowledge Information Processing System • Thousands of Instructions Per Second

KIQ: Kira (Papua New Guinea) (location identifier)

KIR: Kerry County (Ireland) (location identifier) • Kiribati (ISO country code) • Kyocera Image Refinement

KIRDI: Kenya Industrial Research and Development Institute

KIS: Kiepenheuer-Institut für Sonnenphysik (Germany) • Kisumu (Kenya) (location identifier) • Kitting Instruction

Sheet • Knowbot Information Service • Knowledge-based Information System • Kodak Infrared Scope • Ku-band Interface System

KISMIF: Keep It Simple, Make It Fun

KISR: Kuwait Institute for Scientific Research

KISS: Keep It Simple, Sir • Keep It Simple, Stupid • Keep It Straight and Simple • Keyed Indexed Sequential Search • Kinomatic Surveying System • Knowledge-based Interactive Signal-monitoring System • Korea Information Science Society • KPNO International Spectroscopic Survey • Kurs Information Service System

KIST: Korea Institute of Science and Technology (Korea-ROK)

KIT: Kernel for Intelligent Terminals • Key Issue Tracking • Kithira (Greece) (location identifier) • KWIC Interactive Tracker

KITE: Kuiper Infrared Technology Experiment

KITSAT: Korea Institute of Technology Satellite

KIU: Kiunga (Kenya) (location identifier)

KIUSN: KSC Inertial Upper Stage Notice

KIUPSN: KSC Inertial Upper Stage Project Specification

KIV: Kishinev (Moldova) (location identifier)

KIW: Kitwe (Zambia) (location identifier)

KIX: Osaka Kansai (Japan) (location identifier)

KIY: Kilwa (Tanzania) (location identifier)

KIZ: Karlsruhe Isochron Zyklotron (Germany) • Kikinonda (Papua New Guinea) (location identifier)

kJ: Kilojoule • Kommenden Jahres • Künftigen Jahres

Kj: Kalenderjahr • Konjunktion • Konjunktiv

KJA: Krasnoyarsk (Russia) (location identifier)

KJK: Kortrijk (Belgium) (location identifier)

KJP: Kerama (Japan) (location identifier)

KJS: Kenya Journal of Sciences

KJU: Kamiraba (Papua New Guinea) (location identifier)

kk: Kazakh (ISO language code)

KK: Kabushiki Kaisha • Kaiserlich-Königlich • Kaluza-Klein • Kempen-Krefeld (Germany) • Kernkraft • Kleinkaliber • Kompaßkurs • Kontaktkreis • Kreiskarte

KKA: Koyuk (USA-AK) (location identifier)

KKB: Kitoi Bay (USA-AK) (location identifier)

KKC: Khon Kaen (Thailand) (location identifier)

KKD: Kokoda (Papua New Guinea) (location identifier)

KKE: Bay of Islands (New Zealand) (location identifier) • Kerikeri (New Zealand) (location identifier)

KKG: Konawaruk (Guyana) (location identifier) • Kundenkreditgemeinschaft

KKH: Kongiganak (USA-AK) (location identifier)

KKI: Akiachak (USA-AK) (location identifier)

KKJ: Kita Kyushu - Kokura (Japan) (location identifier)

KKK: Kalakaket (USA-AK) (location identifier) • Kreiskontrollkommission

KKL: Karluk Lake (USA-AK) (location identifier)

KKM: Lop Buri (Thailand) (location identifier)

KKN: Kirkenes (Norway) (location identifier)

KKO: Kaikohe (New Zealand) (location identifier)

KKP: Koolburra (Australia-QLD) (location identifier)

KKR: Kaukura Atoll (Tuamotu Islands) (location identifier) • Koringa-Kohn-Rostocker

KKT: Karush-Kuhn-Tucker • Kentland (USA-IN) (location identifier)

KKU: Ekuk (USA-AK) (location identifier)

KKV: Kinetic Kill Vehicle

KKVWS: Kinetic Kill Vehicle Weapon System

KKW: Kernkraftwerk • Kikwit (Zaire) (location identifier) • Kinetic Kill Weapon

KKX: Kikaiga Shima (Japan) (location identifier)

KKY: Kilkenny (Ireland) (location identifier)

KKZ: Koh Kong (Cambodia) (location identifier)

kl: Greenlandic (ISO language code) • Kiloliter • klagen • klar • klären • klassisch • kleben • kleiden • klein • kleine • klingen • klingeln • klug

Kl: Klage • Klang • Klappe • Klasse • Klassifikation • Klassifizierung • Klassik • Klavier • Kleid • Kleider • Kleidung • Klein • Kleine • Kleiner • Klemme • Klient • Klinge • Klingel • Klinik • Klinker • Klippe • Kloster • Klub

KL: Kabellänge • Kalmyk • Karhunen-Loeve (transform) • Kartenleser • Keller's Language • Kelvin Law • Kernel Language • Key Length • Kleinman-Low • Knowledge Level • Kondo Lattice • Kontrollampe • Kuala Lumpur (Malaysia) • Kursor Left

KLA: Kansas Library Association • Karachi Library Association • Kentucky Library Association • Kenya Library Association • Korean Library Association

klass: klassisch

Klass: Klassifikation • Klassifizierung • Klassiker

Klassif: Klassifikation • Klassifizierung

K'lautern: Kaiserslautern

Klautern: Kaiserslautern

Klav: Klavier

Klb: Thousands of Pounds

KLB: Kalabo (Zambia) (location identifier)

KlBer: Kleinheubacher Berichte (periodical code)

Klbf: Kleinbahnhof

KlC: Kleincomputer

KLC: Kaolack (Senegal) (location identifier) • Karl Lambrecht Corporation (USA-IL)

KLD: Kalinin (Russia) (location identifier)

KLE: Kaele (Cameroon) (location identifier)

Kleinb: Kleinbahn

Kleinbf: Kleinbahnhof

klf: klassifiziert

KLF: Kaluga (Russia) (location identifier)

KLG: Kalskag (USA-AK) (location identifier)

KLH: Keyhole Limpet • Kolhapur (India) (location identifier)

KLI: Kota Koli (Zaire) (location identifier)

KLIC: KBS Life Cycle • Key-Letter-In-Context

KLJ: Klaipeda (Lithuania) (location identifier)

KLK: Kalokol (Kenya) (location identifier)

KLL: Levelock (USA-AK) (location identifier)

KLM: Kilometer • Knowledge-Level Model

KLN: Larsen Bay (USA-AK) (location identifier)

KLNN: Kullback-Leibler Nearest Neighbour

KLO: Kalibo (Philippines) (location identifier)

KLOC: Thousand Lines Of Code

KLOE: k Long Experiment

KLP: Kalp Bay (USA-AK) (location identifier)

KLQ: Keluang (Indonesia) (location identifier)

KLR: Kalmar (Sweden) (location identifier)

Kls: Kloster

KLS: Kelso Longview (USA-WA) (location identifier)

KLSS: Korean Library Science Society

KLT: Karhunen-Loéve Transform

KLu: Koninklijke Luchtmacht

KLU: Klagenfurt (Austria) (location identifier)

KLUN: Kinematics of the Local Universe

KLUT: Knowledge Look-Up Table

KLV: Karlovy Vary (Czech Rep.) (location identifier)

KLVAD: Keele (Starlink internal code)

KLW: Klawock (USA-AK) (location identifier)

KLX: Kalamata (Greece) (location identifier)

KLY: Kalima (Zaire) (location identifier)

KLZ: Kleinzee (South Africa) (location identifier) • Kopier- und Lieferzentrale

km: Cambodian (ISO language code) • Kilometer • Kilometric • Kilómetro

Km: Kingdom

kM: Künftigen Monats

K-M: Kaplan-Meier

KM: Comores (ISO country code) • Comoros (ISO country code) • Kaplan-Meier • Knowledge Management • Koninklijke Marine

KMA: Kerema (Papua New Guinea) (location identifier) • Koninklijke Militaire Academie • Korean Meteorological Administration • Koronaler Massenauswurf • Koronaler Materieauswurf

KMB: Koinambe (Papua New Guinea) (location identifier)

KMBAA: Knowledge and Model-Based Algorithm Adoption

kMc: Kilomegacycle

KMC: King Model Cluster • McCord (USA-AK) (location identifier)

KMD: Mandji (Gabon) (location identifier)

KME: Kamembe (Rwanda) (location identifier)

KMF: Franc (Comoros) (ISO currency code) • Kamina (Papua New Guinea) (location identifier)

KMG: Kunming (China-PRC) (location identifier)

km/h: Kilometers per Hour

KMH: Kilometers per Hour • Kosmischer Mikrowellenhintergrund • Kuruman (South Africa) (location identifier) • Speed

KMI: Key Management Infrastructure • Kleine en Middelgrote Industrieën • Knowledge Media Institute (UK) • Koninklijk Meteorologisch Instituut • KSC Management Instruction • Miyazaki (Japan) (location identifier)

KMIB: Koninklijk Meteorologisch Instituut van België (Belgium)

KMIO: Kaliningrad Magnetic Ionospheric Observatory (Russia)

KMIPS: KAIST Map and Image Processing Station

KMIT: King Mongkut's Institute of Technology (Thailand)

KMJ: Kumamoto (Japan) (location identifier)

KMK: Kultusministerkonferenz (Germany) • Makabana (Congo) (location identifier)

KML: Kamileroi (Australia-QLD) (location identifier)

KMM: Kimam (Indonesia) (location identifier)

KMN: Kamina (Zaire) (location identifier)

KMO: Kleine en Middelgrote Ondernemingen • Manokotak (USA-AK) (location identifier)

KMP: Keetmanshoop (Namibia) (location identifier) • Key Management Protocol • Knuth-Morris-Pratt

KMPA: Kenya Master Printers Association

kmph: Kilometers per Hour

kmps: Kilometers per Second

KMPS: Kernel Multiple-Processing System

KMQ: Komatsu (Japan) (location identifier)

KMR: Karimui (Papua New Guinea) (location identifier) • Key Milestone Review

km/s: Kilometers per Second

KMS: Key Management System • Knowledge Management System • Koninklijke Militaire School (Belgium) • Kubo-Martin-Schwinger • Kumasi (Ghana) (location identifier)

km/sec: Kilometers per Second

km/st: Stundenkilometer

km/t: Kilometer/Time

KMT: Kampot (Cambodia) (location identifier) • Kreiss Matrix Theorem

km/u: Kilometer per Uur

KMU: Kismayu (Somalia) (location identifier) • Kleine und Mittlere Unternehmen

KMUL: Karl-Marx-Universität Leipzig (former GDR)

KMV: Kalemyo (Myanmar) (location identifier)

KMW: Kleine Magellansche Wolke • Kostroma (Russia) (location identifier)

KMX: Khamis Mushait (Saudi Arabia) (location identifier)

KMY: Moser Bay (USA-AK) (location identifier)

KMZ: Kaoma (Zambia) (location identifier)

kn: Kannada (ISO language code) • Krona (currency) • Krone (currency)

Kn: Croatian Kuna (currency) • Knot • Knudsen Number

KN: Îles Saint Kitts et Nevis (ISO country code) • Kip (currency) • Knowledge Networking • Kongeriket Norge • KSC Notice • Nevis (Saint Kitts and_) (ISO country code) • Saint Kitts and Nevis (ISO country code)

KNA: Kenya News Agency • Saint Kitts and Nevis (ISO country code) • Viña de Mar (Chile) (location identifier)

KNAAAS: Kenya National Academy for Advancement of Arts and Sciences

KNAC: Keck Northeast Astronomy Consortium • Koninklijke Nederlandse Automobielclub (Netherlands)

KNAS: Kenya National Academy of Sciences

KNAW: Koninklijke Nederlandse Akademie van Wetenschappen (Netherlands)

KNB: Kanab (USA-UT) (location identifier)

KNC: Ji'An (China-PRC) (location identifier)

KNCSS: Thousand of Non-Comment Source Statements

KND: Kindu (Zaire) (location identifier) • Kirchlicher Nachrichtendienst (Germany)

KNE: Kanainj (Papua New Guinea) (location identifier)

KNET: Kangaroo Network

KNF: Kings Lynn - Marham (UK-England) (location identifier) • Klein-Nishina Formula

KNG: Kaimana (Indonesia) (location identifier)

KNH: Kinmen (China-ROC) (location identifier)

KNI: Katanning (Australia-WA) (location identifier)

KNITQ: Korean National Institute of Technology and Quality (Korea-ROK)

KNJ: Kindamba (Congo) (location identifier)

KNJK: Koninklijke Nederlandse Journalistenkring (Netherlands)

KNK: Kakhonak (USA-AK) (location identifier)

KNL: Kelanoa (Papua New Guinea) (location identifier) • Knoll

KNLS: Knolls

KNM: Kaniama (Zaire) (location identifier) • Kongelige Norske Marine (Norway)

KNMI: Koninklijk Nederlands Meteorologisch Instituut (Nethe lands)

k-NN: k Nearest Neighbours

kNN: k Nearest Neighbours

KNN: Kankan (Guinea) (location identifier)

KNO: Knokke - Het Zoute (Belgium) (location identifier)

KNOB: Knowledge Object

KNP: Capanda (Angola) (location identifier)

KNQ: Kone (New Caledonia) (location identifier)

KNS: Kerridge Knowledge System • King Island (Australia-TAS) (location identifier)

Knt: Knight

KNT: Kennett (USA-MO) (location identifier)

Knts: Knots

KnU: Knowledge Utility

KNU: Kanpur (India) (location identifier) • Kyungpook National University (Korea-ROK)

KNUB: Koninklijke Nederlandse Uitgeversbond (Netherlands)

KNV: Knights Inlet (Canada-BC) (location identifier)

KNVD: Koninklijk Nederlands Verbond van Drukkerijen (Netherlands)

KNVvL: Koninklijke Nederlandse Vereniging voor Luchtvaart (Netherlands)

KNW: New Stuyahok (USA-AK) (location identifier)

KNX: Kununurra (Australia-WA) (location identifier)

KNY: Kinoosao (Canada-SK) (location identifier)

KNZ: Kenieba (Mali) (location identifier)

ko: Knockout • Korean (ISO language code)

Ko: Kilo-octet

KO: Keep Off • Keep Out • Kilo-Octet • Knockout • Konkursordnung

KOA: Kailua (USA-HI) (location identifier) • Kampgrounds Of America • Kona - Ke-Ahole (USA-HI) (location identifier)

KOB: Koutaba (Cameroon) (location identifier)

KOC: Koumac (New Caledonia) (location identifier)

KOD: Karaoke On Demand • Kick-Off Drift • Knowledge-Oriented Design • Kotabangun (Indonesia) (location identifier)

KodOB: Kodaikanal Observatory Bulletin (periodical code)

KOE: Kupang (Indonesia) (location identifier)

KOG: Khong (Laos) (location identifier)

KOH: Koolatah (Australia-QLD) (location identifier) • Potassium Hydroxide

KOI: Kirkwall (UK-Scotland) (location identifier) • KSC Operation Instruction • Orkney Island (UK) (location identifier)

KOJ: Kagoshima (Japan) (location identifier)

KOK: Kokkola-Pietarsaari-Kruunupyy (Finland) (location identifier)

KOL: Koumala (Central African Rep.) (location identifier)

Kolloq: Kolloquium

KOM: Komo-Manda (Papua New Guinea) (location identifier) • KSC Organizational Manual

KOMA: Koordinaten-Meßapparat

KomMe: Komety i Meteory (periodical code)

Komm: Kommission

Komp: Kompanie

Kon: Koninklijk

KON: Kontum (Viet Nam) (location identifier)

Konf: Konferenz

KONIC: Kiso Observatory Near-Infrared Camera

KOO: Kongolo (Zaire) (location identifier)

kop: Kopeck

KOP: Nakhon Phanom (Thailand) (location identifier)

KOpS: Thousands of Operations per Second

KOPS: Thousands of Operations Per Second (*)

Kor: Korea • Korean

KOR: Corée du Sud (ISO country code) • Kokoro (Papua New Guinea) (location identifier) • Korea (ROK) (ISO country code) • ROK (ISO country code)

KORDI: Korea Ocean Research and Development Institute

KORG: Knowledge in Organizations Research Group

KOS: Kompongsom (Cambodia) (location identifier) • Sihanoukville (Cambodia) (location identifier)

KOSC: KSC Operations Support Center

KOSEF: Korean Science and Engineering Foundation (Korea-ROK)

KOSI: Kaiser OPtical Systems Inc • Knowledge-Oriented Statistical Inference • Kodak Optical Storage Interface

KOSMA: Kölner Observatorium für Submillimeter Astronomie (Switzerland) • Kõlner Observatorium für Submillimeter und Millimeter Astronomie (Switzerland)

KOSS: Kirschner-Oemler-Shechter-Shectman

KostO: Kostenordnung

KOT: Kotlik (USA-AK) (location identifier)

KOU: Koulamoutou (Gabon) (location identifier)

KOV: Kokchetav (Kazakhstan) (location identifier)

KOW: Ganzhou (China-PRC) (location identifier)

KOX: Kokonao (Indonesia) (location identifier)

KOY: Olga Bay (USA-AK) (location identifier)

KOZ: Ouzinkie (USA-AK) (location identifier)

kp: Kilopond

KP: Corée du Nord (ISO country code) • DPRK (ISO country code) • Keypad • Key Parameter • Key Person • Key Personnel • Key Point • Key Programme • Key Project • Key Pulse • Key Punch • Kitt Peak • Klein Paradox • Korea (North_) (ISO country code) • Kukarkin-Parenago • Kurie Plot

kPa: Kilopascal

KPA: Kenya Publishers Association • Key Process Area • Klystron Power Amplifier • Kopiago (Papua New Guinea) (location identifier) • Kraft Paper Association

KPB: Point Baker (USA-AK) (location identifier)

kpc: Kiloparsec

KPC: Kakuda Propulsion Center (Japan) • Keyboard/Printer Control • Keypunch Cabinet • Port Clarence (USA-AK) (location identifier)

KPCMS: Kodak Precision Colour Management System

KPD: King of Prussia (USA-PA) (location identifier) • Kitt Peak Downes (survey) • KSC Program Directive

KPE: Yapsiei (Papua New Guinea) (location identifier)

KPG: Kurupung (Guyana) (location identifier)

KPGO: Kokee Park Geophysical Observatory (USA-HI)

kph: Kilometers per Hour

KPH: Pauloff Harbor (USA-AK) (location identifier)

KPI: Kapit (Malaysia) (location identifier) • Kernel Programming Interface • Korean Periodical Index

KPIC: Key Phrase In Context

KPK: Parks (USA-AK) (location identifier)

KPM: Kathode Pulse Modulation • Kometen-Planetoiden-Meteore • Kompiam (Papua New Guinea) (location identifier)

KPN: Kipnuk (USA-AK) (location identifier)

KPNO: Kitt Peak National Observatory (USA-AZ)

KPO: Keypunch Operator • P'Ohang (Korea-ROK) (location identifier)

KPP: Kalpowar (Australia-QLD) (location identifier) • Key Programme Publication • Korean Pacific Press

kpps: Kilopulses per Second

Kpr: Keeper

KpR: Knots per Revolution

KPR: Knots Per Revolution • Kodak Photo Resist • Port Williams (USA-AK) (location identifier)

KPRD: KSC Program Requirements Document

KPRNA: Konza Prairie Research Natural Area (USA-KS)

KpS: Kilometers per Second

KPS: Kempsey (Australia-NSW) (location identifier) • Kilometers Per Second • Klystron Power Supply • Koch Process Systems Inc.

kpsf: KIPs per Square Foot

kpsi: KIPs per Square Inch • One Thousand Pounds per Square Inch

KPSM: Klystron Power Supply Modulator

KPT: Jackpot (USA-NV) (location identifier) • Kai's Power Tool

KPTC: Kenya Post and Telecommunications Corporation

KPV: Perryville (USA-AK) (location identifier)

KPVT: Kitt Peak Vacuum Telescope (USA-CA)

KPW: Won (Korea-DPRK) (ISO currency code)

KPY: Port Bailey (USA-AK) (location identifier)

KPZ: Kardar-Parisi-Zhang

KQA: Akutan (USA-AK) (location identifier)

KQB: Koonibba (Australia-SA) (location identifier)

KQL: Kol (Papua New Guinea) (location identifier)

KQML: Knowledge Query and Manipulation Language

kr: Korea (South_) • Kreutzer • Króna (currency) • Krona (currency) • Krone (currency) • Kroner (currency) • Kronor (currency) • Krónur (currency) • Kursor Right • ROK

Kr: Kreis • Krona (currency) • Króna (currency) • Krone (currency) • Kroner (currency) • Kronor (currency) • Krónur (currency) • Krypton

K&R: Kernighan & Ritchie

KR: Corée du Sud (ISO country code) • Knowledge Representation • Korea (South_) (ISO country code) • ROK (ISO country code)

KRA: Kerang (Australia-VIC) (location identifier) • Key Recovery Agent

KRB: Karumba (Australia-QLD) (location identifier)

KRC: Kerinci (Indonesia) (location identifier) • Key Recovery Center • Kodak Reflex Camera

KRD: Kurundi (Australia-NT) (location identifier)

KRDL: Kent Ridge Digital Laboratories (Singapore)

KRE: Key Recovery Encryption • Kirundo (Burundi) (location identifier)

KREEP: Potassium, Rare-Earth Elements and Phosphorus

KREMU: Kenya Rangeland Ecological Monitoring Unit

KRF: Korea Research Foundation (Korea-ROK) • Kramfors (Sweden) (location identifier)

KRG: Kantonale und Regionale Gesellschaften • Karasabai (Guyana) (location identifier)

KRI: Kikori (Papua New Guinea) (location identifier)

KRIMO: Kenya Rift Monitoring

KRIS: Korea Research Institute of Standards (Korea-ROK)

KRISS: Korea Research Institute of Standards and Science (Korea-ROK) (Formerly KSRI)

KRJ: Karawari (Papua New Guinea) (location identifier)

KRK: Cracow (Poland) (location identifier) • Kraków (Poland) (location identifier)

KRL: Kirchhoff Radiation Law • Knowledge Representation Language • Korla (China-PRC) (location identifier)

KRM: Karanambo (Guyana) (location identifier) • Klein-Rydberg Method

KRMS: Ka-band Radiometric Mapping System

KRN: Kiruna (Sweden) (location identifier)

KRO: Kathode Ray Oscilloscope • Kurgan (Russia) (location identifier)

KRP: Karup (Denmark) (location identifier) • Kodak Relief Plate

KRQ: Kramatorsk (Ukraine) (location identifier)

KRR: Knowledge Representation and Reasoning • Krasnodar (Russia) (location identifier)

krs: Kurus

Krs: Kreis

KRS: Knowledge Retrieval System • Kristiansand (Norway) (location identifier)

Krsn: Kerosene

KRT: Kathode Ray Tube • Key Renewal Theorem • Keyword Review Team • Khartoum (Sudan) (location identifier)

KRTO: Kathode Ray Tube Oscillograph

KRTS: Kathode Ray Tube Shield

KRTT: Kathode Ray Tube Tester

KRU: Kerau (Papua New Guinea) (location identifier)

KRV: Kerio Valley (Kenya) (location identifier)

KRW: Krasnovodsk (Turkmenia) (location identifier) • Won (Korea-ROK) (ISO currency code)

KRX: Kar Kar (Papua New Guinea) (location identifier)

KRY: Karamay (China-PRC) (location identifier)

Krz: Kubisch-raumzentrierte

KRZ: Kiri (Zaire) (location identifier)

ks: Kashmiri (ISO language code)

K-S: Kolmogorov-Smirnov

K/S: Kick Stage

KS: Kansas (USA) • Kantowski-Sachs • Katroptic System • Keypad Start • Kick Stage • Knowledge Sharing • Knowledge Source • Knowledge Space • Kodak Standard • Kolmogorov-Sinai • Kolmogorov-Smirnov • Konungariket Sverige • Koordinatensystem • Kosmische Strahlung

KSA: K-band Single Access • Key Search Algorithm • Knowledge-Skills-Abilities • Kosrae (Caroline Islands) (location identifier) • Ku-band Single Access

KSAM: Keyed Sequential Access Mode

KSB: Kasanombe (Papua New Guinea) (location identifier) • Koninklijk Sterrenwacht van België (Belgium)

KSC: Kagoshima Space Center (Japan) • Kearney State College (USA-NE) • Kennedy Space Center (USA-FL) • Kosice (Slovakia) (location identifier) • Krasnoyarsk Scientific Center (Russia)

KSCAP: Kennedy Space Center Area Permit

KSD: Karlstad (Sweden) (location identifier)

KSE: Kasese (Uganda) (location identifier)

KSEC: Kopernik Space Education Center (USA-NY)

KSF: Kassel Calden (Germany) (location identifier)

KSG: Kisengan (Papua New Guinea) (location identifier) • Kolen- en Staalgemeenschap

KSGA: Koninklijk Sterrenkundig Genootschap van Antwerpen (Belgium)

KSGC: Kentucky Space Grant Consortium (USA-KY)

KSh: Kenya Shilling (currency)

KSH: Bakhtaran (Iran) (location identifier) • Kermanshah (Iran) (location identifier) • Korn Shell • Kugelsternhaufen

KSI: Kissidougou (Guinea) (location identifier) • Knowledge Source Index • Knowledge Systems Inc.

KSIE: Kinetic Solvent Isotope Effect

KSJ: Kasos Island (Greece) (location identifier)

KSK: Karlskoga (Sweden) (location identifier)

KSL: Kassala (Sudan) (location identifier)

KSLOC: Thousand Source Lines Of Code

KSM: Kungliga Svenska Marinen (Sweden) • Saint Mary's (USA-AK) (location identifier)

KSMP: Kosher Meal

KSN: Kustanay (Kazakhstan) (location identifier)

KSNM: Thousand of Square Nautical Miles

KSO: Karl-Schwarzschild-Observatorium (Germany) • Kastoria (Greece) (location identifier)

KSOC: Key Symbol Out of Context

KSP: Key Science Programme • Keystone Project • Kodak Special Plate • Kosipe (Papua New Guinea) (location identifier)

KSPEC: K-band Spectrograph

KSpH: Keystrokes per Hour

KSPH: Keystrokes Per Hour

KSQ: Karshi (Uzbekistan) (location identifier)

KSR: Keyboard Send-Receive • Kopf-Schwanz-Radioquelle • Sandy River (USA-AK) (location identifier)

KSRC: Kashima Space Research Centre (Japan)

KSRI: Korea Standards Research Institute (Korea-ROK) (now KRISS)

KSS: Kandy School of Sciences (India) • Kenya Soil Survey • Sikasso (Mali) (location identifier)

KSSS: Korean Space Science Society (Korea-ROK) • KSC Station Set Specification

KSSU: KLM-SAS-Swissair-UTA

KST: Kiruna-Sodankylä-Tromsø • Knowledge Space Theory • Korean Standard Time • Kosti (Sudan) (location identifier)

KSU: Kansas State University (USA-KS) • Kennesaw State University (USA-GA) • Kent State University (USA-OH) • Key Service Unit • Key System Unit • King Saud University (Saudi Arabia) • Kristiansund (Norway) (location identifier)

KSV: Katalog Sluzhbij Vremeni • Springvale (Australia-WA) (location identifier)

KSVH: Kungl. Svenska Vetensk. Hand. (periodical code)

KSW: Kapteyn Sterrenwacht Werkgroep (Netherlands) • Kiryat Shmona (Israel) (location identifier) • Kuffner-Sternwarte Wien (Austria)

KSX: Yasuru (Papua New Guinea) (location identifier)

KSY: Kars (Turkey) (location identifier)

KSZ: Kotlas (Russia) (location identifier)

KSZE: Konferenz über Sicherheit und Zusammenarbeit in Europa

kt: Karat • Kilotonne • Knot

Kt: Kanton • Knight

K/T: Cretaceous/Tertiary

KT: Cretaceous-Tertiary • Kayser-Threde GmbH (Germany) • Kingston-upon-Thames (UK postal code) • Kit • Knot • Kohler Tube • Kuhn-Tucker

KTA: Karratha (Australia-WA) (location identifier)

KTAS: Knots True Airspeed

KTB: Cretaceous-Tertiary Boundary • Losterlitz-Thouless-Berezinski • Thorne Bay (USA-AK) (location identifier)

KTC: Katiola (Ivory Coast) (location identifier)

KTD: Kitadaito (Japan) (location identifier)

KTDS: Key To Disc Software

kte: Korte

KTE: Kerteh (Malaysia) (location identifier)

KTeV: Kaons at the Tevatron

KTF: Kaua'i Test Facility • Takaka (New Zealand) (location identifier)

KTFR: Kodak Thin-Film Resist

KTG: Kerntechnische Gesellschaft (Germany) • Ketapang (Indonesia) (location identifier)

KTH: Kungliga Tekniska Hoegskolan (Sweden) • Tikchik (USA-AK) (location identifier)

KTHNY: Kosterlitz-Thouless-Halperin-Nelson-Young

KTI: Kratie (Cambodia) (location identifier)

KTK: Kanua (Papua New Guinea) (location identifier)

KTL: Keck Task Library • Kitale (Kenya) (location identifier)

KTM: Kathmandu Tribhuvan (Nepal) (location identifier) • Keyword Translation Module

KTN: Ketchikan (USA-AK) (location identifier)

KTNF: Kodak Timing Negative Film

Kto: Konto

KTO: Kato (Guyana) (location identifier)

KTP: Kingston Tinson (Jamaica) (location identifier) • Potassium Titanyl Phosphate

KTQ: Kitee (Finland) (location identifier)

KTR: Katherine Tindal (Australia-NT) (location identifier)

kts: Knots

KTS: Kelvin Temperature Scale • Key Telephone Set • Key Telephone System • Knots • Komitee für Terminologie und Sprachfragen • Teller Mission (USA-AK) (location identifier)

KTT: Kittilä (Finland) (location identifier)

KTU: Key Telephone Unit • Kota (India) (location identifier)

KTV: Kamarata (Venezuela) (location identifier) • Kongress für Textverarbeitung

KTW: Katowice (Poland) (location identifier)

KTX: Koutiala (Mali) (location identifier)

KTY: Terror Bay (USA-AK) (location identifier)

ku: Kurdish (ISO language code)

Ku: Kurchatovium

KU: Keyboard Unit • Knowledge Unit • Kursor Up • Kuwait

KUA: Kuantan (Malaysia) (location identifier)

KUB: Katholieke Universiteit Brabant • Kommission für Umweltbeobachtung

KUC: Kuria (Kiribati) (location identifier)

KUD: Kudat (Malaysia) (location identifier)

KUE: Kukundu (Solomon Islands) (location identifier)

KUED: Kodak Unitized Engineering Data

KUF: Samara (Russia) (location identifier)

KUFEM: Kyushu University Faculty Engineering Memoirs (periodical code)

KUG: Katholieke Universiteit Brabant (Netherlands) • Kiso UV-excess Galaxies (catalogue) • Kubin Island (Australia-QLD) (location identifier)

KUH: Kushiro (Japan) (location identifier)

KUI: Kawau Island (New Zealand) (location identifier) • Keyboard User Interface

KUJ: Kushimoto (Japan) (location identifier)

KUK: Kasigluk (USA-AK) (location identifier)

KUL: Katholieke Universiteit Leuven (Belgium) • Kuala Lumpur (Malaysia) (location identifier) • Kuwait University Library

KUM: Yakushima (Japan) (location identifier)

KUN: Katholieke Universiteit Nijmegen (Netherlands) • Kaunas (Lithuania) (location identifier)

KUO: Kuopio (Finland) (location identifier)

KUP: Kupiano (Papua New Guinea) (location identifier)

KUQ: Kuri (Papua New Guinea) (location identifier)

Kur: Kurdish

KUR: Kuran-o-Munjan (Afghanistan) (location identifier)

KUS: Kulusuk Island (Greenland) (location identifier)

KUSP: Ku-band Signal Processor (*) • Ku-band Single Processor

KUT: Kutaisi (Georgia) (location identifier)

KUTD: Keep Up To Date

KUTGW: Keep Up the Good Work

KUU: Kulu (India) (location identifier)

KUV: Kiso UV-excess (object) • Kunsan (Korea-ROK) (location identifier)

Kuw: Kuwait

KUY: Uyak (USA-AK) (location identifier)

KUZ: Kusan (Korea-ROK) (location identifier)

kV: Kilovolt

K&V: Korteweg-de Vries

KV: Kennis Verrijkt • Kerr Vector • Kill Vehicle • Köchel Verzeichnis

kVA: Kilo-Volt-Ampere

KVA: Kavala (Greece) (location identifier) • Kungliga Vetenskaps Akademien (Sweden)

KVACC: KHILS Vacuum Cold Chamber

kVAhm: Kilo-Volt-Ampere Hour Meter

kVAr: Kilovar

KVAS: Kanawha Valley Astronomical Society (USA-WV)

KVB: Skövde (Sweden) (location identifier) • Skovde (Sweden) (location identifier)

KVC: King Cove (USA-AK) (location identifier)

KVCC: Kalamazoo Valley Community College (USA-MI)

KVDT: Keyboard Video Display Terminal

KVE: Kitava (Papua New Guinea) (location identifier)

KVG: Kavieng (Papua New Guinea) (location identifier)

KVI: Kernfysisch Versnellerinstituut (Netherlands)

KVIV: Koninklijk Vlaamse Ingenieurs Vereniging (Belgium)

KVK: Kirovsk (Russia) (location identifier)

KVL: Kirchoff's Voltage Law • Kivalina (USA-AK) (location identifier)

KVM: Kalamazoo Valley Museum (USA-MI)

KVO: Keyword Visual Organizer • Kraftverkehrsordnung

kVp: Kilovolts at Peak

KVP: Kodak Vacuum Probe • Kodak Versamat Processor • Kontinuierlicher Verbesserungsprozess

KVT: Kanzelhöhe Vacuum Telescope (Austria)

KVU: Korolevu (Fiji) (location identifier)

KVX: Kirov (Russia) (location identifier)

kw: Kuwait

kW: Kilowatt

KW: Keyword • Kirkwall (UK postal code) • Koweit (ISO country code) • Kuwait (ISO country code)

KWA: Kwajalein (Marshall Islands) (location identifier)

KWAC: Keyword And Context

KWADE: Keyword As a Dictionary Entry

KWC: Kentucky Wesleyan College (USA-KY)

KWD: Dinar (Kuwait) (ISO currency code) • Kawadjia (Central African Rep.) (location identifier)

KWE: Guiyang (China-PRC) (location identifier)

KWEST: Keyword Extracted as a String of Terms

KWF: Keyword File • Waterfall (USA-AK) (location identifier)

KWG: Krivoy Rog (Ukraine) (location identifier)

kWh: Kilowatt-Hour

KWH: Khwahan (Afghanistan) (location identifier)

kWhr: Kilowatt-Hour

KWI: Kuwait (location identifier)

KWIC: KAO Wide-field Infrared Camera • Keyword In Context • Kuiper Wide-field Infrared Camera

KWICK: Knowledge Workers Intelligently Collecting, Coordinating, Creating Consulting Knowledge

KWIDR: Keyword In Defined Rotation

KWIM: Know What I Mean

KWIOC: Keyword In/Out of Context

KWIP: Keyword In Permutation

KWIT: Keyword In Title

KWJ: Kwangju (Korea-ROK) (location identifier)

KWK: Kwigillingok (USA-AK) (location identifier)

KWL: Guilin (China-PRC) (location identifier)

KWLF: Kodak Wratten Light Filter

KWM: Kowanyama (Australia-QLD) (location identifier)

KWN: Quinhagak (USA-AK) (location identifier)

KWO: Kawito (Papua New Guinea) (location identifier)

KWOC: Keyword Out of Context

KWOT: Keyword Out of Title • Kilometer-Wave Orbiting Telescope

KWP: West Point (USA-AK) (location identifier)

KWPSI: Keyword Phrase Subject Index

KWPT: Keyword Plus Title

KWR: Kwai Harbour (Solomon Islands) (location identifier)

KWS: Kwailabesi (Solomon Islands) (location identifier)

KWT: Koweit (ISO country code) • Kuwait (international vehicle registration code) • Kuwait (ISO country code) • Kwethluk (USA-AK) (location identifier)

KWU: Kansas Wesleyan University (USA-KS) • Mansion House (New Zealand) (location identifier)

KWUC: Keyword and UDC out of Context

KWV: Kurwina (Papua New Guinea) (location identifier)

KWX: Kiwai Island (Papua New Guinea) (location identifier)

kwy: Keyway

KWY: Kiwayu (Kenya) (location identifier)

KWZ: Kolwezi (Zaire) (location identifier)

KX: Quit

KXA: Kasaan (USA-AK) (location identifier)

KXE: Klerksdorp (South Africa) (location identifier)

KXF: Kodak X-ray Film • Koro Island (Fiji) (location identifier)

KXK: Komsomolsk na Amure (Russia) (location identifier)

KXR: Karoola (Papua New Guinea) (location identifier)

ky: Kirghiz (ISO language code) • Kyat (currency) • Thousand Years

Ky: Kentucky (USA)

KY: Cayman Islands (ISO country code) • Îles Cayman (ISO country code) • Kentucky (USA) • Kenya • Key • Kirkcaldy (UK postal code)

KYA: Konya (Turkey) (location identifier)

KYB: Yangoonabie (Australia) (location identifier)

Kybd: Keyboard

KYBD: Keyboard

KYD: Dollar (Cayman Islands) (ISO currency code) • Orchid Island (China-ROC) (location identifier)

KYE: Tripoli (Lebanon) (location identifier)

KYF: Yeelirrie (Australia-WA) (location identifier)

Kyffh: Kyffhäuser

KYI: Yalata Mission (Australia-SA) (location identifier)

KyITB: Kyushu Institute Technology Bulletin (periodical code)

KYK: Karluk (USA-AK) (location identifier)

KYL: Key Largo - Port Largo (USA-FL) (location identifier)

KYN: Milton Keynes (UK) (location identifier)

KYO: Tampa - Top of Tampa (USA-FL) (location identifier)

KYP: Kyaukpyu (Myanmar) (location identifier)

Kypd: Keypad

kyr: Thousands of Years

Kyrgyz: Kyrgyzstan

KYS: Kayes (Mali) (location identifier) • Keys

KYT: Kyaukpyu (Myanmar) (location identifier)

KYU: Koyukuk (USA-AK) (location identifier)

KyUTR: Kyushu University Technology Reports (periodical code)

KYX: Yalumet (Papua New Guinea) (location identifier)

KYZ: Kyzyl (Russia) (location identifier)

kz: Duststorm • Kazakhstan • Sandstorm

KZ: Kazakhstan (ISO country code) • Killing Zone

KZB: Zachar Bay (USA-AK) (location identifier)

KZC: Kompong-Chhna (Cambodia) (location identifier)

KZD: Krakor (Cambodia) (location identifier)

KZF: Kaintiba (Papua New Guinea) (location identifier)

KZH: Kizhuyak (USA-AK) (location identifier)

KZI: Kozani (Greece) (location identifier)

KZK: Kompong Thom (Cambodia) (location identifier)

KZN: Kazan (Russia) (location identifier)

KZP: Katalog Zvezd Zapodozrennykh v Peremennosti

KZS: Kastelorizo (Greece) (location identifier)

KZT: Tenge (Kazakhstan) (ISO currency code)

K2K: KEK to Kamioka

K7: Cassette

K9: Skyward Aviation

L

l: Direct Supply • Ecliptic Longitude • Galactic Longitude • Geometrical Path Length • Heliocentric Longitude • Lake • Lambda • Land • Lang • Large • Late • Lateral • Latitude • Laut • Law • Leaf • League • Left • Legal • Legitimate • Leicht • Length • Length Scale • Lepton Number • Leu (currency) • Lies! • Light • Lightning • Line • Link • Links • Lira • Line • Links • Liquid • Liter • Literate • Litre • Little • Loch • Long • Lost • Low • Mean Anomaly of the Moon • Mean Free Path • Miller Index • Mixing Length • Orbital Angular Momentum Quantum Number • Pendulum Length • Pound (currency) • Quantum Number of Rotation • Quantum Number of Vibrational Angular Momentum • Reduced Pendulum Length • Relaxation Length • Thickness

L: Angular Momentum • Avogadro Constant • Characteristic Length • Conservation Equation • Convective Operator • Diffusion Length • Fifty (Roman numeral) • Geographical Longitude • Inductance • Inductor • Lac • Ladung • Lady • Lage • Lago • Lagóa • Lagoon • Lagrangian • Laguna • Lahn-Dill-Kreis (Germany) • Lake • Lambert • Land • Language • Large • Largeur • Last • Latent Heat • Latin • Latitude • Launch • Lautsprecher • Law • League • Learner • Left • Lei (currency) • Leistung • Lempira (currency) • Length • Leu (currency) • Lev (currency) • Leva (currency) • Level • Lex • Ley • Liber • Liberal • Libra • Licenciate • Licht • Lied • Lieferung • Lift • Liga • Light • Lima (international alphabet) • Line • Linear • Linear Energy Transfer • Linear Integral Operator • Link • Linking Number • Lira (currency) • Lire (currency) • Liste • Liter • Litre • Live • Liverpool (UK postal code) • Lobby • Locator • Loch • Locus • Lodge • London (UK-England) • Long • Longitude • Longueur • Lord • Lorenz Coefficient • Lough • Low • Lowest • Low-Pressure Center • Lumen • Luminance • Luminosity • Luxemburg (international vehicle registration code) • Luxemburg (postal country code) • Luyten (catalogue) • Kinetic Potential • Mean Longitude of a Planet • Orbital Angular Momentum Quantum Number • Pound (currency) • Radiance • Self-Inductance • Solar Luminosity • Sound Intensity • 50 (Roman numeral)

l/a: Letter of Authority

la: Lacunosus • Latin (ISO language code) • Lege Artis

lA: laut Akte • laut Angabe • laut Auftrag

La: Lane • Lanthan • Lanthane • Lanthanum • Louisiana (USA)

LA: Laboratoire d'Aérologie • Lancaster (UK postal code) • Land Access • Landesamt • Landesarchiv (Germany) • Lane • Language Availability • Lao PDR (ISO country code) • Laos (ISO country code) • Large Aircraft • Large Aperture • Large Array • Latin • Latin America • Latin American • Launch Abort • Launch Aft • Launch and Ascent • Launch Area • Launch Azimuth • Law Agent • Learning Agreement • Leave Allowance • Legal Adviser • Legislative Assembly • Leseranalyse • Letter of Authority • Library Association (UK) • Library Automation • Licensing Act • Licensing Authority • Lightning Arrester • Limited Area • Linear Accelerator • Link Assurance • List Attributes • Local Agent • Long Arc • Longitudinal Acoustic • Los Angeles (USA-CA) • Louisiana (USA) • Low Altitude • Low Attitude

LAA: Lamar (USA-CO) (location identifier) • Landesarbeitsamt • Laser Attenuator Assembly • Launch Area Antenna • Library Association of Australia • Local Airport Advisory • Los Alamos Astrophysics (USA-NM) • Louisville Amateur Astronomers (USA-KY)

LAAB: Laboratoire d'Astronomie et d'Astrophysique de Bordeaux (France)

LAAM: Laboratoire d'Astrophysique, Atomes et Molécules

LAAME: Large Antenna Assembly and Measurement Experiment

LAAOL: Los Alamos Astrophysical Opacity Library

LAAPD: Large-Area Avalanche Photodiode

LAAR: Liquid-Air Accumulator Rocket

LAAS: Laboratoire d'Analyse et d'Architecture des Systèmes • Laboratoire d'Automatique et d'Analyse des Systèmes (France) • Local Area Augmentation System • Los Angeles Astronomical Society (USA-CA) • Low-Altitude Alert System

Lab: Label • Labial • Laborant • Laboratoire • Laboratorium • Laboratory • Labour • Labourer • Labrador

LAB: Lablab (Papua New Guinea) (location identifier) • Laboratory • Latvijas Astronomijas Biedriba (Latvia) • League of American Bicyclists (USA) • Legal Advisory Board • Linear Alkyl Benzene • Lithosphere-Asthenosphere Boundary • Lloyd Aéreo Boliviano SA • Local Activities Board

LABATYD: Life's A Bitch And Then You Die

LABCEDE: Laboratory Cryogenic Electron-Dependent Emissions

lac: Locus Ante Citatus

Lac: Lacerta-Lacertae (constellation)

LAC: Laboratório Asociado de Computação • Lake City (Papua New Guinea) (location identifier) • Large Area Counter • Laser Amplifier Chain • Latin America and the Caribbean • LEISA Atmospheric Corrector • Levelized Annual Cost • Library Advisory Committee • Library Advisory Council • Linear Absorption Coefficient • Linear Accelerator Center • Lineas Aéreas del Caribe • Local Area Coverage • Long-run Average Cost • Loop Assignment Center • Lunar Astronautical Chart

LACAC: Latin American Civil Aviation Conference

LACATE: Lower Atmosphere Composition And Temperature Experiment

LACB: Landing Aids Control Building

LACCD: Large-Area Charge-Coupled Device

LACE: Liquid-Air Cycle Engine • Limited Area of Central Europe • Local Automatic Circuit Exchange • Los Angeles Contemporary Exhibition • Low-power Atmospheric Compensation Experiment • Lunar Atmospheric Composition Experiment

LACES: Low-Altitude Conical Earth Sensor

LASI: Laser Autostigmatic Cube Interferometer

LACIE: Large-Area Crop Inventory Experiment • Laser Altimeter Camera Irim Experiment

LACIP: Large-Area Crop Inventory Program

LACMA: Los Angeles County Museum of Arts (USA-CA)

LACN: Local Area Communications Network

LACONIQ: Laboratory Computer Online Inquiry

LACOS: Large-scale Correct System

LACS: Laboratory Acquisition and Control System • Light Aircraft Camera System

LACSA: Lineas Aéreas Costaricenses SA

LACVD: Laser-Activated Chemical Vapor Deposition

lad: laden

Lad: Ladung

LAD: Laboratory of Advanced Design • Ladeira • Language Acquisition Device • Large-Area Detector • Large-Area

Display • Lecture Automatique de Documents • Logical Analysis Device • Logistics Approval Data • Low-Accuracy Data • Low-Accuracy Designation • Luanda (Angola) (location identifier) • Lunar Atmosphere Detector

LADA: Langage de Description d'Architectures • Local Area Data Access

LADAR: Laser Detection And Ranging • Laser Radar • Light Amplification for Detection And Ranging

LADD: Lens Antenna Deployment Demonstration

LADDR: Layered Device Driver

LADE: Lineas Aéreas del Estado

LADECO: Linea Aérea Del Cobre

Ladest: Ladestelle

LADGPS: Local Area Differential Global Positioning System

LADIES: Lapan Digital Image Evaluation System • Los Alamos Digital Image Enhancement System

LADIMAS: Latitudinal Middle-Atmosphere Structure

LADISIS: Laboratoire des Dimensions Sociales et Incidences Subjectives du Langage

LADL: Laboratoire d'Automatique Documentaire et Linguistique (France)

Ladp: Ladyship

LADS: Laser Airborne Depth Sounder • Local-Area Data Service • Logistics Automated Data System

LADT: Local Access Data Transport

LAE: Lae (Papua New Guinea) (location identifier) • Linear Algebraic Equation • Liquid Apogee Engine

LAEFF: Laboratorio de Astrofísica Espacial y Física Fundamental (Spain)

LaEO: Laser ElektroOptik (periodical code)

LAER: Lowest Achievable Emission Rate

LAERTE: Laboratoire des Écoulements Réactifs et de leurs Techniques d'Études

LAES: Latin American Economic System • Letters in Applied and Engineering Sciences (periodical code)

LAET: Limiting Actual Exposure Time

Laev: Laevus

LaF: Louisiana French

LAF: Lafayette (USA-IN) (location identifier) • Local Adaptive Filter • Lower Atmospheric Facility

LaFoc: Laser Focus (periodical code)

LAFORIA: Laboratoire des Formes et d'Intelligence Artificielle

LAFTA: Latin-American Free Trade Association

Lag: Lager • Lagerung • Lagging • Lagoon • Laguna • Lagune

LAG: Laboratoire d'Astrophysique et de Géophysique • Lagrangian • La Guaria (Venezuela) (location identifier) • Library Action Group • Linzer Astronomische Gesellschaft (Austria) • Lovers of Active Galaxies

LAGB: Late Asymptotic Giant Branch

LAGCM: Loa Alamos General Circulation Model

LAGEOS: Laser Geodetic Satellite • Laser Geodynamic Earth-Orbiting Satellite • Laser Geodynamics Satellite (*)

LAGOS: Laser Gravitational-wave Observatory in Space • Laser-interferometer Gravitational-wave Observatory in Space

LAGS: Laser-Activated Geodetic Satellite • Launch Abort Guide Simulation

Lah: Lahore (Pakistan)

LAH: Labuha (Indonesia) (location identifier)

LAHS: Low Altitude - High Speed

LAHSO: Land And Hold Short Operations

LAI: Laboratory for Astronomical Imaging • Lannion (France) (location identifier) • Leaf Area Index • Library Association of Ireland • Logistics Artificial Intelligence

LAIA: Latin American Integration Association

LAICA: Laboratorio di Informatica e Calcolo

LAID: Land-Atmosphere Interaction Dynamics

LAIR: Laboratory for Astronomical Image Reduction

LAIRT: Large Aperture Infrared Telescope

LAIRTS: Large Aperture Infrared Telescope System

LAIS: Local Automatic Intercept System

LAIU: Launch Abort Interface Unit

LaJ: Lapidary Journal (periodical code)

LAJ: Lages (Brazil-SC) (location identifier)

LAK: Aklavik (Canada-NT) (location identifier) • Kip (Laos) (ISO currency code)

LAL: Laboratoire de l'Accélérateur Linéaire • Lakeland (USA-FL) (location identifier)

LALA: Large Area Lyman Alpha (survey)

LALD: Laboratory of Astronomical Light Detectors

LALL: Longest Allowed Lobe Length

LALP: Laboratoire d'Application des Lasers de Puissance (France)

LALR: Look-Ahead Left-Recursive

LALSR: Low-Altitude Large-Scale Reconnaissance

Lam: Laminated

LAM: Laboratoire d'Astrophysique de Meudon (Observatoire de Paris-Meudon) • Lameira • Lane Adaptor Module • Liberalium Artium Magister • Liquid Apogee Motor • Limited Area Model • Line Access Module • Local Area Model • Local Area Multicomputer • Long-Axis Mode • Los Alamos (USA-NM) (location identifier) • Low-Altitude Mission • Lyman Alpha Monitor

LAMA: Laboratoire de la Montagne Alpine (France) • Large Active Mirror for Astronomy • Large Active Mirror in Aluminium • Laser Machine • Library Administration and Management Association • Local Automatic Message Accounting

LAMAR: Large Area Modular Array of Reflectors

LAMARS: Large-Amplitude Multimode Aerospace Research Simulator

LAMAS: Laboratoire de Médecine Aérospatiale

LAMBDA: Laboratory Measurements and Borehole Data Analysis

LAMMA: Laser Microprobe Mass Analyzer

LAMMR: Large Antenna Multichannel Microwave Radiometer • Large Antenna Multifrequency Microwave Radiometer

LAMOD: Laser Optical Motion Detector

LAMOST: Large-Area Multiobject Spectroscopic Telescope (*) • Large Multiobject Spectroscopic Telescope

LAMP: Laboratoire Associé de Météorologie Physique • Large Advanced Mirror Program • Laser, Atomic and Molecular Physics • Local Area Multiprocessor • Low-Altitude Manned Penetration • Low-Altitude Mapping Photogrammetry • Lunar Analysis and Mapping Program

LAMPE: Langage Algorithmique de Mise en Page et d'Édition

LAMPF: Los Alamos Meson Physics Facility (USA-NM)

LAMPS: Laser-scan Automated Map Production System • Light Airborne Multipurpose System • Limited Area Mesoscale Prediction System • Linear and Mathematical Programming System

LAMRDT: Large Aperture Microwave Radiometer Development Test

LAMS: Large-Area Mosaicking System • Large-Area Mosaic System • Launch Acoustic Measuring System

Lan: Langley

LAN: Inland • Landau/Isar (Germany) • Lansing (USA-MI) (location identifier) • Lembaga Administrasi Negara (Indonesia) • Lineas Aéreas Nacionales • Local Apparent Noon • Local-Area Network • Longitude of the Ascending Node

LANACS: Local Area Network Asynchronous Connection Server

Lanc: Lancaster (UK-England)

Lancs: Lancashire (UK-England)

LANCSE: Los Alamos Neutron Science Center (USA-NM)

LAND: Lead Astrophysical Neutrino Detector

Landk: Landkarte

Landkr: Landkreis

Landr: Landrat

LANDS: Land Analysis and Decision Support

LANDSAT: Land Satellite

LANDSCAPE: Large Astrophysical Network Detecting Showers from Cosmic Accelerated Particles to the Earth

Landsch: Landschaft • Landschutz

Landstr: Landstraße

Landw: Landwehr • Landwirt • Landwirtschaft

LANE: Local Area Network Emulation • Local Area Network Extension

Lang: Language • Languedoc (France)

LANICA: Lineas Aéreas de Nicaragua

LANL: Los Alamos National Laboratory (USA-NM)

LANRES: Local Area Network Resource Extension and Services

LANSCE: Los Alamos Neutron Science Center (USA-NM)

LANTIRN: Low-Altitude Navigation and Targeting Infrared (system) for Night

LANTR: LOX-Augmented Nuclear Thermal Rocket

Lao: Laotian

LAO: Laoag (Philippines) (location identifier) • Laos (international vehicle registration code) • Laos (ISO country code) • Large Assembly Order • Legal Assistance Officer

LAOG: Laboratoire d'Astrophysique de l'Observatoire de Grenoble (France)

LAOL: Los Alamos Opacity Library

Lap: Lapland • Lappish

LAP: Laboratório Asociado de Plasma • Langmuir Probe on Rosetta • Language Analysis Package • La Paz (Mexico) (location identifier) • Large Area Panel • L-Arginine Phosphate • Library Access Program • Linear Accelerometer Package • Linear Array of Processors • Link Access Procedure • Link Access Protocol • List Assembly Programming • Local Access Point • Low-Altitude Photography

LAPA: Lineas Aéreas Privadas Argentinas

LaPaB: Laser Particle Beams (periodical code)

LAPAN: National Institute of Aeronautics and Space (Indonesia)

LAPB: Link Access Procedure Balanced • Link Access Protocol Balanced

LAPD: Laser-Aided Plasma Diagnostics • Link Access Procedure Direct

LAPH: Laboratoire d'Annecy-le-Vieux de Physique Théorique (France)

LAPI: Low-Altitude Plasma Instrument

LAPM: Link Access Procedure for Modems • Link Access Protocol Modem

LAPO: Lake Afton Public Observatory (USA-KS)

LAPP: Laboratoire d'Annecy-le-Vieux de Physique des Particules (France)

LAPR: Linear Array Pushbroom Radiometer

LAPRIDI: Laboratorio de Procesamiento de Imagenes Digitalizadas

LAPS: LAN Adapter and Protocol Support • LAN Adapter and Protocol Support • Left Aft Propulsion System • Local Analysis and Prediction System

LAPSS: Large Area Pulsed Solar Simulator

LAPST: Lunar And Planetary Sample Team

LAPW: Linear Augmented Plane Wave • Linearized Augmented Plane Wave

LAQ: Beida (Libya) (location identifier)

LAR: La Alternativa Racional (periodical) • Laramie (USA-WY) (location identifier) • Large Adaptive Reflector • Leaf Area Ratio • Ley de Arrendamientos Rústicos • Libya (international vehicle registration code) • Limit Address Register • Linear Associative Retrieval • Linhas Aéreas Regionais • Liniile Aeriene Romane • Load Access Right • Low-Angle Reentry

LARA: Laminar Flow Research Action

LARAM: Line Addressable Random Access Memory

LaRC: Langley Research Center (NASA)

LARC: Langley Research Center (NASA) • Large Automatic Research Computer • Livermore Automatic Research Computer • Local Accessibility Rank Condition • Los Alamos Radiometry Code

LARCA: Laboratoire de Réalisation de Composants pour l'Astronomie

LARE: Local Asymptotic Relative Efficiency

Larg: Largamente • Largeur • Largo

LARGE: Laboratoire de Recherche en Gestion

LARGOS: Laser-Activated Reflecting Geodetic Optical Satellite

LARI: Los Alamos Radiometric Instrument

LARIE: Laser-Activated Reactive Ion Etching

LAROE: Large Area Operational Experiment

LARP: Local And Remote Printing • Local Approvals Review Program

LARPS: Local And Remote Printing Station

LARS: Laboratory for Applications of Remote Sensing • Laminar Angular Rate Sensor • Launch And Recovery System • Load Alleviation and Ride Smoothing • Lower-Atmosphere Research Satellite

LARST: Land Applications of Remote Sensing Techniques • Local Applications of Remote Sensing Techniques

LAS: Laboratoire d'Astronomie Spatiale • Laboratório de Sensores e Materiais • LAGEOS Apogee Stage • Land Analysis Software • Land Analysis System • Landsat Assessment System • Large Angular Size • Large Astronomical Satellite • Laser Absorption Spectrometer • Las Vegas - McCarran (USA-NV) (location identifier) • Launch Area Supervisor • Launch Auxiliary System • League of Arab States • Lebanese Astronomical Society (Lebanon) • Leeds Astronomical Society (UK) • Leicester Astronomical Society (UK) • Lethbridge Astronomy Society (UK) • Lettre Autographe Signée • Level of Archive Services • Leven Astronomical Society (UK) • Library Access System • Library Association of Singapore • Light-Activated Switch • Ligue of Arab States • Limb Atmospheric Spectrometer • Linear Array Scanner • Liverpool Astronomical Society (UK) • Local Area Screening • Longmont Astronomical Society (USA-CO) • Louisville Astronomical Society (USA-KY) • Low-Altitude Satellite • Low-Altitude Sensor • Low-Angle Scattering • Lunar Attitude System

LASA: Landesagentur für Struktur und Arbeit (Germany) • Large Aperture Seismic Array • Large Area Solar Array • Lidar Atmospheric Sounder and Altimeter

LASC: Library Automation Steering Committee

LASCAR: Large-Scale Reprocessing

LASCO: Large-Angle Spectrometric Coronagraph (*) • Light And Spectrometric Coronagraph

LASCR: Light-Activated Silicon-Controlled Rectifier

LASD: Large Angle Silicon Detector

LASE: Large Aperture Seismic Experiment • Large Aperture Speckle Experiment • Lidar Atmospheric Sensing Experiment • Lidar Atmospheric Sounder Experiment

LASER: Light Amplification by Stimulated Emission of Radiation • Lincoln Advanced Science and Engineering Reinforcement • London And Southeastern Region (UK-England)

LASIC: Laboratoire Associé des Sciences de l'Information et de la Communication (France)

LASII: Laser Stabilized Imaging Interferometer

LasJ: Laser Journal (periodical code)

LASL: Los Alamos Scientific Laboratory

LASP: Laboratory for Astronomy and Solar Physics • Laboratory for Atmospheric and Solar Physics • Laboratory for Atmospheric and Space Physics • Leaflet of the Astronomical Society of the Pacific (periodical code) • Low-Altitude Space Platform

LASR: Laboratory of Automation of Scientific Research • Library Access, Search and Retrieval

LASRE: Linear Aerospike SR-71 Experiment

LASS: Land Application Satellite System • Large-Amplitude Space Simulator • Large Area Sky Survey • Library Automated Service System • Local Area Signalling Service • Local Area Sounding System • Low-Altitude Surveillance System

LASSII: Low-Altitude Satellite Study of Ionospheric Irregularities

LASSO: Laramie Astronomical Society and Space Observers (USA-WY) • Laser Synchronization from Satellite Orbit • Laser Synchronization from Stationary Orbit (*)

LAST: Local Area Storage Transport

LASTIR: Laser Sensor Thermal Infrared

LASTP: Linhas Aéreas de São Tome e Principe

LASTport: Local Area Storage Transport

LAstr: L'Astronomie (periodical code)

LASV: Low-Altitude Supersonic Vehicle

Lat: Latent • Lateral • Latin • Latín • Latitude • Latus • Latvia • Latvian

LAT: Laboratoire d'Astrophysique de Toulouse (France) • Laboratoire d'Astrophysique Théorique • Language Actions Table • Large Amateur Telescope • Large Angle Torque • Large Area Triangulation • Latch • Lateral • Latitude • La Uribe (Colombia) (location identifier) • Lauterbach (Germany) • Local Access Terminal • Local Address Table • Local Apparent Time • Local Area Terminal • Local Area Transport • Logistics Asset Tracking • Long Alias Table • Lot Acceptance Test

LATA: Local Access and Transport Area • Local Access Transport Area

LATAS: Laser True Airspeed System

LATIS: Loop Activity Tracking Information System

Latl: Lateral

LATM: Local Asynchronous Transport Mode

Latr: Latrine

LATRS: Laser Acquisition and Tracking System

LATTE: Lesson and Technical Training Environment

Latv: Latvia

Latvn: Latvian

Lau: Laundry

LAU: Lamu (Kenya) (location identifier) • Ley de Autonomía Universitaria (Spain)

LAUC: Librarians Association of the University of California (USA-CA)

Lauenb: Lauenburg (Germany)

Laun: Launch • Launched

Lav: Lavatory

LAV: Lalomalava (Samoa) (location identifier) • Lineas Aéreas Venezolanas • Load Average

LAVA: Look-Ahead Vector Analysis

LAVC: Local Area VAX Cluster • Los Angeles Valley College (USA-CA)

LAVE: Laser-Activated Vacuum Evaporation

LAVIP: Land-Atmosphere-Vegetation Interaction Project

LAVO: Lager Algemeen Voortgezet Onderwijs

LAVS: Lavatories

Law: Lawine • Lawyer

LAW: Fort Sill (USA-OK) (location identifier) • Lawton (USA-OK) (location identifier) • Low-Altitude Weapon

LAWD: Luyten Atlas of White Dwarfs

LAWN: Local Area Wireless Network

LAWP: Lower Atmosphere Wind Profiler

LAWRS: Limited Aviation Weather Reporting System

LAWS: Laser Atmospheric Wind Sounder • Low-Altitude Wind Shear

LAX: Los Angeles (USA-CA) (location identifier)

LAXS: Large Area X-ray Spectroscopy

LAY: Ladysmith (South Africa) (location identifier)

Laz: Lazarett

LAZ: Bom Jesús da Lapa (Brazil-BA) (location identifier) • Lebensarbeitszeit

lb: Binary Logarithm • Libra • Liebe • Lieber • Pound

lB: laut Bericht

Lb: Leben • Lehrbuch • Leib • Libelle • Lichtbild • Luftbild

LB: Landbund • Landesbezirk • Landesbibliothek • Langmuir-Blodgett • Launch Boost • Launch Bus • Lebanon (ISO country code) • Lectori Benevolo • Lecture Bottle • Left Button • Lesebuch • Letter Box • Liban (ISO country code) • Liberia (international vehicle registration code) • Libya • Limb Brightening • Line Buffer • Load Bank • Local Battery • Local Board • Logbook • Long Baseline • Low Bay • Lower Brace • Luyten Blue

LBA: Learning By Analogy • Leeds Bradford (UK-England) (location identifier) • Linear Block Array • Logical Block Address • Logical Block Addressing • Long-Baseline Array • Luftfahrt-Bundesamt

lb. ap.: Pound (Apothecaries)

lb. av.: Pound (Avoirdupois)

LBB: Leak Before Break • Little Brown Bird • Lubbock (USA-TX) (location identifier)

LBBY: Lobby

Lbc: Lübeck (Germany)

LBC: Liaison Banque-Client • Lübeck (Germany) (location identifier)

LBCR: Logbook Change Request

lbd: lebend

LBD: Laser Beam Deflection • Laser Beam Detector • Learning By Demonstration • Light Beam Deflection

lbdg: lebendig

LBDS: Leiden-Berkeley Deep Survey

LBE: Latrobe (USA-PA) (location identifier) • Lattice Boltzmann Equation • Learning By Example • Location-Based Entertainment • Low Business Ethics

LBET: Large Beam Expander Technology

lbf: Pound-Force

LBF: North Platte (USA-NE) (location identifier)

LBG: Lyman Break Galaxy • Paris - Le Bourget (France) (location identifier)

lf-ft: Pound-Foot

LBH: Length-Breadth-Height • Lyman-Birge-Hopfield • Sydney - Palm Beach (Australia-NSW) (location identifier)

LBI: Albi (France) (location identifier) • Long-Baseline Interferometer • Long-Baseline Interferometry

LBIR: Laser Beam Image Recorder • Low-Background Infrared • Low-Background Infrared Radiometer

LBJ: Labuan Bajo (Indonesia) (location identifier) • Lyndon Baines Johnson (1908-1973)

LBJL: Lyndon B .Johnson Library (USA)

LBK: Left Bank • Liboi (Kenya) (location identifier)

Lbl: Label

LBL: Label • Laminar Boundary Effect • Land Between the Lakes (USA-KY) • Lawrence Berkeley Laboratory (USA-CA) • Left Buttock Line • Liberal (USA-KS) (location identifier) • Line By Line • Low-energy-peaked BL Lacertae (object)

LBLRR: Lawrence Berkeley Laboratory Research Review (periodical code)

LBLRTM: Line-By-Line Radiative Transfer Model

LBLS: Laminar Boundary-Layer Separation

LBM: Lattice Boltzmann Method • Liquid Boost Module • Load Buffer Memory • Locator Back Marker • Logic Bus Monitor • Luabo (Mozambique) (location identifier)

LBMR: L-Band Microwave Radiometer

LBN: Lake Baringo (Kenya) (location identifier) • Latvia (ISO country code) • Lettonie (ISO country code) • Logical Bibliographic Network • Logical Block Number • Lynds Bright Nebulae (catalogue)

LBNL: (E.O.) Lawrence Berkeley National Laboratory (USA-CA)

LBNP: Lower-Body Negative Pressure

LBNPD: Lower-Body Negative Pressure Device

LBO: Las Brisas Observatory (USA-CO) • Laser Blow-Off • Lean Blow Out • Leveraged Buyout • Line Buildout • Lusambo (Zaire) (location identifier)

LBP: Length Between Perpendiculars • Logarithmic Barrier Problem • Long Banga (Malaysia) (location identifier) • Pound (Lebanon) (ISO currency code)

LBQ: Lambarene (Gabon) (location identifier)

LBQS: Large Bright QSO Survey • Large Bright Quasar Survey

Lbr: Labour • Lumber

LBR: Lábrea (Brazil-AM) (location identifier) • Laser Beam Recorder • Laser Beam Recording • L-Band Radiometer • Liberia (ISO country code) • Librarian • Library • Low Bit Rate

Lbry: Library

lbs: Pounds

LBS: Labasa (Fiji) (location identifier) • L-Band System • Lectori Benevolo Salutem! • Line Buffer System • Load Balance System • Locaal Bibliotheek System • London Business School • Low-Brightness Source

LBSAAS: Life and Biomedical Sciences and Applications Advisory Subcommittee

LBSAD: Life and Biomedical Sciences and Applications Division

LBSD: Low-Barrier Schottky Diode

LBSS: Low-Band SIGINT System • Low-Band Subsystem

lb. t.: Pound (Troy)

LBT: Large Binocular Telescope • Listen Before Talk • Long-Based Transponder • Long-Baseline Transponder • Lumberton (USA-NC) (location identifier)

LBTB: Large Binocular Telescope Beteiligungsgesellschaft (Germany)

LBTI: Large Binocular Telescope Interferometer

LBTM: Layer Bispectral Threshold Method

LBTP: Launch Base Test Plan • Launch Base Test Procedure

LBU: Labuan (Malaysia) (location identifier) • Launcher Booster Unit

LBV: Leuchkräftige Blaue Veränderliche (Stern) • Libreville (Gabon) (location identifier) • Local Bus Video • Low Bandwidth Video • Luminous Blue Variable

LBVN: Luminous Blue Variable Nebula

LBW: Live Body Weight • Long Bawan (Indonesia) (location identifier)

LBX: Laser Beam Expander • Local Bandwidth X • Lubang (Philippines) (location identifier)

LBY: La Baule - Montoir (France) (location identifier) • Libya (ISO country code) • Libye (ISO country code)

l/c: Letter of Credit

lc: Loco Citato

L/C: Land Cover • Letter of Credit

L&C: Laboratory & Checkout

LC: Inductance-Capacitance • Laboratory and Checkout • Land Cover • Landing Chart • Launch Complex • Launch Cost • Launch Countdown • Land Cover • Large Coronagraph • Launch Complex • Launch Conference • Launch Control • Launch Coordinator • Launch Count • Launch Countdown • Launch Critical • Left Center • Legal Currency • Letter of Credit • Lettres et Cartes • Level Control • Level Crossing • Library of Congress • Light Curve • Ligne Centrale • Limit Circle • Linear Chain • Linear Communication • Linear Contact • Line Contractor • Link-Cell • Liquid Chromatography • Liquid Crystal • Load Carrier • Local Call • Local Channel • Local Chebychev • Local Committee • Local Control • Lower California • Lower Canada • Luminosity Class • Lumped Constant • Saint Lucia (ISO country code)

LCA: Laboratory for Computational Astrophysics • Land Cover Analysis • Larnaca (Cyprus) (location identifier) • Launch Control Amplifier • Launch Control Analyst • Light Combat Aircraft • Light-Curve Analysis • Line Check Airman • Load Control Assembly • Load Controller Assembly • Local Communications Area • Local Component Analysis • Local Context Analysis • London City Airport (UK-England) • Longitudinal Chromatic Aberration • Low-Cost Automation • Lower Case Alphabet • Lufthansa Cargo Airlines • Nicosia Larnaca (Cyprus) (location identifier) • Saint Lucia (ISO country code)

LCAAS: La Crosse Area Astronomical Society (USA-WI)

LCAM: Laboratoire des Collisions Atomiques et Moléculaires (France)

LCAMOS: Loop Cable Maintenance Operation System

LCAO: Linear Combination of Atomic Orbitals

LCAP: Loosely Coupled Array of Processors

LCAR: Launch Complex Assessment Report

LCAS: Lake County Astronomical Society (USA-IL)

LCB: Launch Control Building • Least Common Bit • Line Control Block • Line of Constant Bearing • Low-Cost Booster • Lower Crustal Block

LCC: Lambert Conformal Conical (projection) • Land Cover Change • Landing Control Center • Langage Commun de Commande • Langage de Commande de Communications • Langley Complex Coordination • Language for

Conversational Computing • Launch Command and Control • Launch Commit Criteria • Launch Control Center • Launch Control Complex • Launch Control Console • Leadless Chip Carrier • Lecce Galatina (Italy) (location identifier) • Leeward Community College (USA-HI) • Levelized Compiled Code • Library Coordination Committee • Library of Congress Classification • Life-Cycle Cost • Life-Cycle Costing • Load-Carrying Capability • Local Communications Complex • Local Communications Console • Local Control Common • Lockheed-Martin California Co. • Loran C navigational Chart

LCCA: Laboratório de Computação Científica Avançada • Logical Configuration Communication Area

LCCD: Launch Commit Criteria Document

LCCIS: Local Common Channel Interoffice Signaling

LCCMS: Launch Control Center Measuring Station

LCCN: Library of Congress Card Number

LC&CO: Launch Control & Checkout

LCCO: Launch Control and Checkout

LCCS: Large Capacity Core Storage

LCD: Laboratoire de Combustion et de Détonique (France) • Laboratory for Computational Dynamics • Language for Computer Design • Launch Control Design • Launch Countdown • Learner-Centered Design • Least Common Denominator • Liquid Crystal Display • List of Chosen Descriptors • Local Change Directory • Local Climatological Data • Local Curvature Diminishing • Loss-of-Cell Delineation • Louis Trichardt (South Africa) (location identifier) • Lowest Common Denominator

LCDA: Local Committee Development Assistant

LCDB: Land Cover Database

LCDIA: Low-Cost Data Image Analyzer

LCDM: Low-density Cold Dark Matter

LCDN: Last Called Directory Number

LCDP: Lateral Control Departure Parameter

LCDRS: Las Campanas Deep Redshift Survey

LCE: La Ceiba (Honduras) (location identifier) • La Commande Électronique (France) • Latest Cost Estimate • Launch Complex Engineer • Launch Complex Equipment • Launch Control Equipment • Launch Countdown Exercise • Limited Career Extension • Lyapunov Characteristic Exponent

LCEB: Launch Control Equipment Building

LCEC: Local Committee Exchange Controller

LCEF: Launch Centered Earth-Fixed

LCELSS: Lunar CELSS

LCF: Large Cooling Flow • Launch Control Facility • Least Common Factor • Level Control Function • Liquid Crystal Filter • Local Control Facility • Logical for Computable Functions • Low-Cost Fiber • Low-Cycle Fatigue • Lowest Common Factor

LCFL: Last Closed Field Line

LCFS: Last Closed Flux Surface • Last Come - First Served

LCG: La Coruña (Spain) (location identifier) • Linear Congruence Generator • Linear Congruential Generator (*) • Liquid Cooled Garment • Liquid Cooling Garment • Load Classification Group • Longitudinal Centre of Gravity

LCH: Lake Charles (USA-LA) (location identifier) • Latch • Launch

LCHTF: Low Cycle High Temperature Fatigue

LCH4: Liquid Methane

LCI: Laconia (USA-NH) (location identifier) • Langage Commun de l'Information • Lyapunov Characteristic Indicator

LCIE: Laboratoire Central des Industries Électriques

LCK: Columbus Rickenbacker (USA-OH) (location identifier) • Library Construction Kit • Lock

LCKS: Locks

Lcl: Local

LCL: La Coloma (Cuba) (location identifier) • Laser Centerline Localizer • Latching Current Limiter • Less Car Load • Lifting Condensation Level • Linkage Control Language • Local • Lower Control Limit • Lunar Cargo Lander

LCLLY: Locally

LCLOC: Line Card Location

LCLS: Low-Cost Launch System

LCLV: Liquid-Crystal Light Wave • Low-Cost Launch Vehicle

LCLY: Locally

LCM: Laboratório Central de Metrologia • La Cumbre (Argentina-CD) (location identifier) • Large-Core Memory • Launch Control Monitor • Launch Crew Member • LCD Clock Module • Least Common Multiple • Life Cycle Management • Life Cycle Manager • Line Center Magnetogram (technique) • Liquid Crystal Meter • Local Committee Meeting • Local Content Manager • Local Customer Manager • Longitude of Central Meridian • Lowest Common Multiple

LCMI: Laboratoire des Champs Magnétiques Intenses

LCML: Low Calorie Meal

LCMN: Lauthanum Cerium Magnesium Nitrate

LC-MS: Liquid Chromatography - Mass Spectrometry

LCMS: Launch Control and Monitoring System • Liquid Chromatography - Mass Spectrometry • Low-Cost Modular Spacecraft

LCMT: Lfe-Cycle Mutation Testing

LCN: Balcanoona (Australia-SA) (location identifier) • Load Classification Number • Local Civil Noon • Local Communications Network • Logical Channel Number • Logic Change Notice • Logistics Control Number • Loosely Coupled Network • Lyapunov Characteristic Number

LCNO: Linear Combination of Orbitals

LCNP: Long Capitalized Noun Phrases

LCO: Laboratoire de Chimie Océanographique • Lague (Congo) (location identifier) • Las Campanas Observatory (Chile) • Launch Control Office • Launch Control Officer • Launch Control Operation • Limit-Cycle Oscillation • Limiting Condition for Operation

LCOG: Local Committee Operations Guide

LCOM: Lack of Cohesion in Methods • Local Committee Operations Manual • Logic Control Output Module

Lconc: Local Concavity

LCOSE: Launch Complex Operational Support Equipment

LCP: Language Conversion Program • Large Combustion Plant • Last Complete Program • Launch Control Panel • Least-Cost Planning • Left Circularly-Polarized • Left Circular Polarization • Light Charged Particle • Linear Complementarity Problem (*) • Linear Complementary Problem • Link Configuration Protocol • Link Control Protocol • Liquid-Crystal Polymer • Local Calibration Procedure • Local Committee President • Low-Cost Production • Lower Cost Processor

LCPE: Laboratoire de Chimie et Physique de l'Environnement

LCPL: Left Circularly Polarized Light

LCPM: Local Committee Presidents' Meeting

LCQ: Light-Cone Quantization • Logical Channel Queue

l/cr: Lettre de Crédit

LCR: Inductance-Capacitance-Resistance • Laboratoire Central de Recherches • La Chorrera (Colombia) (location identifier) • Least Cost Routing • Low Cross Range

LCRS: Las Campanas Redshift Survey

LCS: Laboratory for Computer Science • Landsat Customer Services • Large Capacity Storage • Large Cassegrain

Spectrograph • Large Core Storage • Las Cañas (Costa Rica) (location identifier) • Laser Crosslink System • Launch Control Sequence • Launch Control Simulation • Launch Control Station • Launch Control System • LCU Common Software • Learning Classifier System • Library Circulation System • Library Computer System • Library Control System • Lightcurve Shape • Limiting Control Setting • Line Coding Storage • Liquid-Crystal Shutter • List of Commands Signal • Literature Classification System • Local Common Software • Load Carrying Structure • Longest Common Subsequence • Lotus Communication Server

LCSA: Load Carrying Structure Assembly

LCS&E: Laboratory for Computational Science & Engineering

LCSE: Laboratory for Computational Science and Engineering • Laser Communication Satellite Experiment

LCSH: Library of Congress Subject Heading

LCSI: Launch Critical Support Item

LCSO: Launch Control Safety Officer

LCSQA: Laboratoire Central de Surveillance de la Qualité de l'Air

LCSS: Launch Control and Sequencer System

LCST: Laboratoire des Composants et Systèmes pour Télécommunications

LCT: Laboratoire Central de Télécommunications • Laplace-Carson Transform • Local Civil Time • Local Correction Tracking • Local Correlation Technique • Local Correlation Tracking • Low-Cost Terminal • Low Critical Temperature

Lctd: Located

LCTGM: Library of Congress Thesaurus for Graphic Materials

LCTMP: Little Change in Temperature

Lctr: Locator

LCTV: Liquid-Crystal Television

LCTY: Locality

LCU: Large Close-Up • Last Cluster Used • Launch Control Unit • Library Control Unit • Lightweight Computer Unit • Line Control Unit • Line Coupling Unit • Load Control Unit • Local Control Unit

LCUG: Liquid Cooling Undergarment

LCUP: Least-Cost Utility Planning

LCV: Light Commercial Vehicle • Logistics Cargo Vehicle • Lucca (Italy) (location identifier)

LCVG: Liquid Cooling and Ventilation Equipment • Liquid Cooling/Ventilation Equipment

LCVR: Liquid-Crystal Variable Retarder

LCW: Low-Conductivity Water

LCY: London City (UK-England) (location identifier)

ld: ledig • leitend • leuchtend

lD: Lichter Durchmesser

Ld: Ladung • Land • Landung • Lead • Leading • Limited • Load • Lord

L/D: Length to Diameter • Letter of Deposit • Lift to Drag

L&D: Landing & Deceleration • Loss & Damage

LD: Landesdienst • Landing and Deceleration • Laser Desorption • Laser Diode • Lateral Direction • Launch Director • Learning Disability • Learning-Disabled • Lethal Dose • Level Discriminator • Libyan Dinar (currency) • Linear Dimension • Line Drawing • Line Driver • Linguistics in Documentation • List Dump • Llandrindod Wells (UK postal code) • Loading Dock • Logic Driver • Long Distance • Low Density • Low Dutch

Lda: Licenciada • Limitada

LDA: Landing Directional Aid • Landing Distance Available • Laser Diffraction Analyzer • Laser Doppler Anemometer • Laser Doppler Anemometry • Latsimus Dorsi Anterior • Leading Dimension of the Array • Linear Diode Array • Linear Discriminant Analysis • Line Driving Amplifier • Local Database Agent • Local Density Approximation • Localizer Directional Aid • Logical Device Address • Malda (India) (location identifier)

LDAC: Leiden Data Analysis Center (Netherlands)

LDAP: Lightweight Directory Access Protocol

LDAR: Lightning Detection and Ranging System

LDAS: Land Data Access Subsystem • Letchworth and District Astronomical Society (UK)

LDB: Large Database • Launch Data Bus • Lexical Database • Light Distribution Box • Limited Data Block • Local Data Block • Logistics Databank • Londrina (Brazil-PR) (location identifier) • Long-Duration Balloon • Long-Duration Ballooning

LDBD: Logical Database Design

LDBF: Long-Duration Balloon Flight

LDBS: Land Database System

LDC: Least-Developed Country • Less-Developed Country • Level Decision Circuit • Lindeman Island (Australia-QLD) (location identifier) • Linguistic Data Consortium • Local Data Center • Local Distribution Company • Logistics Data Communications • Long-Distance Call • Long-Distance Carrier • Long-Distance Communication • Lot Date Code • Lotus Development Corporation • Low-Developed Country

LDCE: Limited Duration Candidate Exposure

LD-CELP: Low-Delay Code Excited Linear Prediction

LDCELP: Low-Delay Code Excited Linear Prediction

LDD: La Définition des Données • Lightly Doped Drain • Logical Database Design • Low-Density Data • Lunar Dust Detector

LDDI: Local Distributed Data Interface

LDDL: Logical Data Definition Language

LDDMWG: LGSOWG Data Distribution and Marketing Working Group

LDDS: Langage de Description des Données Statistiques • Low-Density Data System

LDE: Lagrangian Differential Equation • Linear Differential Equation • Long-Delayed Echo • Long-Duration Event • Lourdes Tarbes Ossun (France) (location identifier)

LDEC: Lunar Docking Events Controller

LDEF: Land Data Entry Form • Long-Duration Exposure Facility

LDEO: Lamont-Doherty Earth Observatory (USA-NY)

Lderry: Londonderry (Northern Ireland)

LDEV: Logical Device

LDF: Laboratoire David Florida (Canada-ON) • Large Deployable Reflector • Liquid Drop Facility • Log-normal Distribution Function

Ldg: Ladung • Landing • Landung • Leading • Loading • Lodging

LDG: Landing • Lodge

LDGE: Lodge

LDGO: Lamont-Doherty Geological Observatory

LDH: Large Dynamic Hyperbase • Lord Howe Island (Australia-NSW) (location identifier)

LDI: Lactate Dehydrogenase • Landing Direction Indicator • Landsat Digital Image Analysis System • Lean Direct Injection • Life Detection Instrument • Lindi (Tanzania) (location identifier) • Linear Differential Inclusion • Linear Displacement Indicator

LDIF: LDAP Data Interchange Format

LDIM: Lunar Digital Image Model

LDIS: Land Data Ingest System • Lightning Data and Information System

LDIU: Launch Data Interface Unit

LDJ: Linden (USA-NJ) (location identifier)

Ldk: Landkarte

LDK: Lidköping (Sweden) (location identifier) • Lower Deck

LDL: Language Description Language • Logical Data Language • Logic-based Data Language • Long Delay Line • Low-Density Lopoprotein

LDLP: Low-Density Lopoprotein

LDM: Last Day of the Month • Lens Data Manager • Library Data Management • Liquid-Drop Model • Local Data Manager • Logical Data Model • Long-Distance Modem • Ludington - Mason County (USA-MI) (location identifier)

LDMA: Laboratoire de Dynamique Moléculaire et Atomique

LDMD: Localized Dark Mantle Deposit

Ldmk: Landmark

LDMK: Landmark

LDMP: Library Data Management Project

LDMS: Laboratory Data Management System • Laser Distance Measuring System • Laser Desorption Mass Spectrometry • Lunar Distance Measuring System

LDMX: Local Digital Message Exchange

Ldn: London

LDN: Lamidanda (Nepal) (location identifier) • Less-Developed Nation • Level Day/Night • Local Data Network • Lynds Dark Nebulae

LDNG: Landing

Ldo: Licenciado

LDO: Ladouanie (Suriname) (location identifier) • Linear Distribution of Objects • Logically Deleted Object • Lunar Discovery Orbiter

LDOCE: Longmans Dictionary of Contemporary English

LDOS: Local Density of Occupied States • Local Density Of States

Ldp: Ladyship • Lordship

LDP: Label Switching Router • Language Data Processing • Library Development Program • Linux Documentation Project • Local Data Package • Local Data Processor • Logistics Data Package • Long-Distance Path • Lorentz Doppler Profile • Low-Density Polyethylene

LDPE: Low-Density Polyethylene

LDQ: Lobe-Dominant QSO • Lobe-Dominated QSO

Ldr: Ladder • Leader • Ledger • Lodger

LDR: Large Deployable Reflector • Leader • Level Distribution Recorder • Light-to-Dark Ratio • Linear Depolarization Ratio • Linear Dynamic Range • Liquid-Droplet Radiator • Lodar (Yemen) (location identifier) • Lorentz Double Refraction • Low Data Rate • Low Data Register • Low Dose Rate

LDRC: Logic Design Rules Checker

LDRD: Laboratory Directed Research and Development

LDRP: Logistics Depot Repair Procedure

Ldry: Laundry

lds: Laus Deo Semper

Lds: Loads

LDS: Landing/Deceleration System • Landing-Deservicing-Safing • Laser Docking Sensor • Latter Day Saints • Launch Data System • Launch Detection Satellite • Layered Defense System • Leadership Development Seminar • Leicester Database System • Linear Decision Surface • Linear Dynamic System • Local Distribution System • Logical Data Structure • Low-Density Sample • Low-Dispersion Spectrograph • Luyten Double Stars

LDSL: LBR Digital Subscriber Line

LDSS: Lunar Deep Seismic Sounding • Low-Dispersion Spectroscopic Survey • Low-Dispersion Survey Spectrograph

LDT: Laser Display Technology • Level Delay Time • Library Development Team • Linear Differential Transformer • Linear Displacement Transducer • Local Descriptor Table • Logic Design Translator • Long Distance Transmission

LDTP: Laboratoire de Dynamique Terrestre et Planétaire (OMP)

LDU: Lahad Datu (Malaysia) (location identifier) • Light Distribution Unit • Line Driver Unit

LDV: Laser Doppler Velocimeter • Laser Doppler Velocimetry • Laser Doppler Velocity • Laser Doppler Vibrometer • Leonardo Da Vinci • Linear Differential Vector • Linguistische Datenverarbeitung

LDVE: Linear Differential Vector Equation

LDW: Lansdowne (Australia-WA) (location identifier)

LDX: Long-Distance Extender • Long-Distance Xerography • Saint-Laurent-du-Maroni (French Guiana) (location identifier)

LDY: Londonderry (Northern Ireland) (location identifier)

LDZ: Lithium-Depleted Zone • Londolozi (South Africa) (location identifier)

Le: Lebanese • Lebanon • Leone (currency) • Lewis Number

L/E: Left Engine

LE: Egyptian Pound (currency) • LAN Emulation • Language Engineering • Language Environment • Large Enterprise • Launch Escape • Launching Equipment • Leading Edge • Left End • Left Engine • Left Eye • Leicester (UK postal code) • Lesotho • Less than or Equal to • Library Edition • Light Equipment • Limited Edition • Lobster Eye • Low End • Low Energy • Low Excitation • Luminosity Evaluation • Lyapunov Exponent

Lea: League • Leave

LEA: Laboratoire Européen Associé • Laboratorium voor Elektromagnetisme en Acustica • Laboratory for Experimental Astrophysics • Late Embryonic Abundance • Launch Escape Assembly • Learmonth (Australia-WA) (location identifier) • Light-Emitting Array • Ligue des États Arabes • Load Effective Address • Local Education Agency • Local Education Authority • Logistics Engineering Analysis

LEAD: Leadership for Environment And Development

LEAF: Land Ecosystem Atmosphere Feedback • Large European Acoustic Facility • Laser Environmental Active Fluorosensor • Laser Excited Automatic Fluorometer • Law Enforcement Access Field

LEAFS: Laser Excited Atomic Fluorescence Spectrometry

LEANDRE: Lidar embarqué pour l'Étude des Aérosols, des Nuages, de la Dynamique, du Rayonnement et des Espèces Minoritaires

LEAO: Lager Economisch en Administratief Onderwijs

LEAP: Lasers and Electro-optic Applications Program • Light-weight Exo-Atmospheric Projectile • Low-Energy Antiproton • Lunar Environment Assessment Package

LEAPS: Low Electron Attachment Potential Species

LEAR: Low-Energy Antiproton Ring

LEARP: LAN Emulation Address Resolution Protocol

LEAST: Learning Systems Standardization

Leb: Lebanese • Lebanon

LEB: Hanover (USA-NH) (location identifier) • Lebanon (USA-NH) (location identifier) • Low-Energy Booster • Low-Energy Buncher • Lower Equipment Bay • White River (USA-VT) (location identifier)

LEBS: Low-Emission Boiler System

LEC: LAN Emulation Client • Launch Escape Control • Light-emitting Electrochemical Cell • Liquid Encapsulated Czochralski • Local Exchange Carrier

LECP: Low-Energy Charged Particle

LEC/S: Low-Energy Concentrator/Spectrometer

LECS: Land Evaluation Computer System • LAN Emulation Configuration Server • Launching Equipment Checkout Set • Liquid Effluent Collection System • Low-Energy Concentration Spectrometer • Low-Energy Concentrator-Spectrometer (*)

Lect: Lectio • Lecture • Lecturer

LECT: Liquid Effluent Collection Tank

Lectr: Lecturer

Led: Ledig

LED: Language Editor • Light-Emitting Diode • Low-Energy Detector • Low-Energy Diffraction • Low-Energy Dislocation • Saint Petersburg - Pulkovo (Russia) (location identifier)

LEDA: Library of Efficient Data and Algorithms • Local Employment Development Action • Lunar European Demonstration Approach • Lyon-Meudon Extragalactic Database • Online Earthnet Data Availability

LEDAS: Leicester Database and Archive Service • Leicester Database Archive Service

LEDC: Low-Energy Detonating Charge

LEDCOP: Light Element, Detailed Configuration Opacity Code

LEDD: Light-Emitting Diode Display

LEDR: Laboratory for Environmental Data Research

LEDSS: Laboratoire d'Études Dynamiques et Structurales de la Sélectivité (France)

LEE: Leeds (Starlink internal code) • Leesburg (USA-FL) (location identifier)

LEED: Low-Energy Electron Diffraction

LEEM: Low-Energy Electron Microscope

LEET: Limiting Equivalent Exposure Time

LEF: Leading Edge Flap • Lebakeng (Lesotho) (location identifier) • Linear Electric Field • Low-frequency Electric Field

LEFI: Local Electric Field Instrument

LEFM: Linear Elasticity Fracture Mechanics

Leg: Legajo • Legal • Legislation • Legislative • Legislature

LEG: Aleg (Mauritania) (location identifier) • Légua • Low-Energy Gamma-ray • Low-Energy Germanium

LEGG: Longevity Evaluation Government Group

LEGI: Laboratoire des Écoulements Géophysiques et Industrieles (France)

Legis: Legislation • Legislative • Legislature

LEGM: Low-Energy Gamma Monitor

LEGOS: Laboratoire d'Études en Géophysique et Océanographie Spatiale (OMP)

LEGR: Low-Energy Gamma-Ray

LEGRI: Low-Energy Gamma-Ray Imager

LEGS: Liquid Encapsulated Gallium-arsenide Solidification

LEGSPC: Low-Energy Gas Scintillation Proportional Counter

LEGT: Lycée d'Enseignement Général et Technologique (France)

LEH: Launch/Entry Helmet • Le Havre - Octeville (France) (location identifier)

LEI: Almeria (Spain) (location identifier) • Laser-Enhanced Ionization • Leicester (Starlink internal code) • Legal Education Institute • Local Employment Initiative

LEICA: Low-Energy Ion Composition Analyzer

Leics: Leicestershire (UK-England)

LEID: Low-Energy Ion Detector

LEIDS: Logistics Electronic Information Delivery System

Leip: Leipzig (Germany)

LEIR: Low-Energy Ion Ring

LEISA: Linear Etalon Imaging Spectral Array

LEIT: Low-Energy Imaging Telescope

LEJ: Leipzig (Germany) (location identifier)

LEK: Labe (Guinea) (location identifier) • Liquid Encapsulated Kyropoulos

LEL: Laser à Électrons Libres • Lake Evella (Australia-NT) (location identifier) • Linear Energy Loss • Link-Edit Language • Link, Embed and Launch-to-edit • Lower Explosive Limit • Lower Exposure Limit

LELU: Launch Enable Logic Unit

LEM: Laboratoire d'Études des Microstructures • Laboratory Environment Model • Landform Elevation Model • Language Extension Module • Languages in Educational Mobility • Launch Escape Motor • Lecteur Électronique de Microfiches • Lemmon (USA-SD) (location identifier) • Low-Energy Monitor • Lunar Excursion Module • Lunar Exploratory Module

LEMA: Long-term Ecological Modelling Activity

LEMAC: Leading Edge Mean Aerodynamic Chord

LEMD: Laboratoire d'Électrostatique et de Matériaux Diélectriques (Fra nce)

LEMDE: Lunar Excursion Module Descent Engine

LEMMS: Low-Energy Magnetospheric Measurement System

LEMO: Laboratoire d'Électromagnétisme Microondes et Optoélectronique (Fr ance)

LEMPA: Low-Energy Magnetospheric Particle Analyzer

LEMS: Lunar Excursion Module Simulator

LEMT: Low-Energy Matrix Telescope

LEMTG: Laboratoire d'Étude des Mécanismes de Transfert en Géologie (OMP)

LEMZ: Liquid Encapsulated Melt Zone

len: Lenticaris

LEN: León (Mexico) (location identifier) • Line Equipment Number • Local Electrical Neutrality • Low Entry Networking

LENA: Low-Energy Neutral Astrophysics • Low-Energy Neutral Atom (*)

LEND: Large External Network Database

LENSC: Lake Erie Nature and Science Center (USA-OH)

LENST: Large Energy National Shock Tunnel

LENTIC: Laboratoire d'Étude des Nouvelles Technologies de l'Information et de la Communication (Belgium)

Leo: Leo-Leonis (constellation)

LEO: Leconi (Gabon) (location identifier) • Liaison Engineering Order • Librating Equidistant Observer • Linear Electro-Optic • Link Everything Online • Low Earth Orbit • Low Earth Orbiter

LEOMA: Laser and Electro-Optics Manufacturers Association

Leon: Leonés

LEONESS: Low Earth Orbiting Nadir Etalon Sounding Spectrometer

LEOP: Launch and Early Operation Phase • Launch and Early Orbit Phase (*) • Low Earth Operational (orbit) • Low Earth Orbit Phase

LEOPR: Laboratoire d'Électrochimie Organique et Photochimie Redox (France)

LEOS: Laboratory for Electro-Optic Sensors • Low Earth Orbiting Satellite

LEOSAR: Low Earth Orbit Synthetic Aperture Radar

Lep: Lepus-Leporis (constellation)

LEP: Laboratoire d'Électronique Philips • Laboratory for Extraterrestrial Physics • Large Electron-Positron (collider) • Large External Pressurized (tank) • Least Energy Principle • Leopoldina (Brazil-MG) (location identifier) • Light-Emitting Polymer • Limited English Proficiency • Linear Error Probable • Low-Energy Particle • Lycée d'Enseignement Professionnel (France) (now LP)

LEPC: Local Emergency Planning Committee • Low-Energy Proportional Counter

LEPEDEA: Low-Energy Proton and Electron Differential Energy Analyzer

LEPES: Laboratorie d'Études des Propriétés Électroniques des Solides (France)

LEPI: Laboratoire d'Évaluation et de Prospective Internationales

LEPMI: Laboratoire d'Électrochimie et de Physicochimie des Matériaux et Interfaces (France)

LEPOR: Long-term and Expanded Program of Oceanic Research

LEPORE: Long-term and Expanded Program of Oceanic Research and Exploration

LEPS: Launch Escape Propulsion System • Lidar Echo Processing System

LEPSI: Laboratoire d'Électronique et de Physique des Systèmes Instrumentaux

LEPT: Low-Energy Particule Telescope

LEQ: Lands End (UK-England) (location identifier)

LER: Leinster (Australia-WA) (location identifier) • Light-Emitting Resistor • Line-Emitting Region • Long Eye Relief (optics) • Lunar Eyepiece Receptacle

LERAM: Littoral Ecosystem Risk Assessment Model

LeRC: Lewis Research Center (NASA) (now John H. Glenn Research Center)

LERec: L'Écho des Recherches (periodical code)

LERS: Learning from Examples based on Rough Sets • Lewis Research Center (NASA)

LERTS: Laboratoire d'Études et de Recherches en Télédétection Spatial e

LERX: Leading Edge Root Extension

LES: Land Earth Station • LAN Emulation Server • Large Eddy Simulation • Large Eddy Simulator • Launch and Entry Suit • Launch Enabling System • Launch/Entry Suit • Launch Environmental Simulator • Launch Equipment Shop • Launch Escape System • Leading Edge System • Lesobeng (Lesotho) (location identifier) • Licensing Executives Society • Lincoln Experimental Satellite • Loi sur l'Enseignement Supérieur (France) • Lunar Escape System

LESA: Land Evaluation and Site Assessment • Legal Expert Systems Association

LESC: Life and Environmental Sciences • Life and Environmental Sciences Committee • Life and Environment Standing Committee

LESIP: Little Experimental System for Image Processing

LESS: Lake Effect Snow Studies • Laser Excited Sholskiij Spectroscopy • Leading Edge Structural System • Leading Edge Structure System • Leading Edge Subsystem • Leascost Estimating and Scheduling System (IBM) • Least Squares Solution

L. ès Sc.: Licencié ès Sciences

LEST: Laboratoire d'Économie et de Sociologie du Travail (France) • Large Earth Survey Telescope • Large Earth-based Solar Telescope (*) • Large Earth-borne Solar Telescope • Large Earthbound Solar Telescope • Large Earth Survey Telescope • Large European Solar Telescope

Let: Lettish

LET: Laboratoire d'Essais des Télécommunications (France) • Launch Escape Tower • Leticia (Colombia) (location identifier) • Letter • Limiter Experiment Testband • Linear Energy Transfer • Linearer Energietransfer • Link Evaluation Terminal • Local Essential Tree • Low-Energy Telescope

LETAM: Low End Telecommunications Access Method

LETF: Launch Equipment Test Facility

LETG: Low-Energy Transmission Grating

LETGS: Low-Energy Transmission Grating Spectrometer

LETI: Laboratoire d'Électronique et des Techniques de l'Informatique • Laboratoire d'Électronique et de Technologie de l'Informatique

LETS: Local Employment Trading System • Local Exchange Trading System • Low-Energy Telescope System

Lett: Letters • Lettres

LEU: Large External Unpressurized (tank) • Launch Enable Unit • Launcher Electronic Unit • Low-Enriched Uranium • Seo de Urgel (Spain) (location identifier)

Lev: Levante

LEV: Bureta Levuka (Fiji) (location identifier) • Launch Escape Vehicle • Levuka (Fiji) (location identifier) • Lunar Excursion Vehicle

LEW: Auburn (USA-ME) (location identifier) • Lewiston Auburn (USA-ME) (location identifier)

LEWEX: Labrador Extreme-Wave Experiment

LEWP: Line Echo Wave Pattern

Lex: Lexical • Lexicon • Lexikon

LEX: Leading Edge Extension • Lexicon • Lexington - Blue Grass (USA-KY) (location identifier)

LExEn: Life in Extreme Environments

LExSWG: Lunar Exploration Science Working Group (NASA)

LEZ: La Esperanza (Honduras) (location identifier) • Lunar Equatorial Zone

L/F: Load Factor

LF: Franc (Luxemburg) (currency) • Large Field • Large Format • Launch Facility • Launch Forward • Left Forward • Leuchtkraftfunktion • Light Face • Linear File • Line Feed • Line Finder • Load Factor • Loaf • Local Fluff • Locator Field • Log File • Logical File • Low Frequency • Luminosity Function

LfA: Laboratory for Astrophysics • Linux for Astronomy

LFA: Locally Flat Approximation • Logistics Functional Analysis • Long-Focus Astrograph • Low Frequency Active • Low-Frequency Astronomy

LFAF: Low-Frequency Accelerometer Flutter

LFAM: Low-Frequency Accelerometer Mode

LFAP: Low-Frequency Accelerometer Pogo

LFC: Large-Format Camera • Level of Free Convection • Limit of Free Convection • Lit Fluidisé Circulant • Loss Free Counting • Low-Frequency Current

LFCTR: Laboratorio di Fisica Cosmica e Tecnologie Relative (CNR)

lfd: laufend

LFD: Linear Finite Difference • Live Functional Diagram • Local Fractional Derivative • Low-Fat Diet • Low-Frequency Domain

LFDLTS: Low-Frequency Deep-Level Transient Spectroscopy

lfd. M.: laufenden Monats

LFE: Laboratory For Electronics

LFF: Logical Feature Filter • Lower Forward Fuselage

LFFT: Long Form Functional Test

Lfg: Lieferung

LFG: Lexical Functional Grammar

LFGM: Low-Frequency Gravitomagnetic

LFH: Local Failure Handler • Lunar Far Horizon

LFI: Hampton Langley (USA-VA) (location identifier) • Laser Fantasy International (USA-WA) • Last File Indicator • Link Fragmentation and Interleaving • Low-Frequency Imager • Low-Frequency Instrument (*) • Lost and Found International

LFK: Log File Key • Lufkin - Angelina County (USA-TX) (location identifier)

LFL: Linear Field Line • Lower Flammability Limit

LFM: Lateral Force Microscope • Lateral Force Microscopy • Lateral Force Mode • Limited-area Fine Mesh • Limited Fine Mesh • Limited Fine-mesh Model • Linear Field per Minute • Linear Frequency Modulated • Local Focus Mechanism • Long Focus Mechanism • Low-Frequency Modulation • Low-powered Fan Marker

LFML: Low Fat Meal

LFMR: Low-Frequency Microwave Radiometer

LFN: Louisburg Franklin (USA-NC) (location identifier) • Low-Frequency Noise

LFO: Low-Frequency Oscillator • Kelafo Callaf (Ethiopia) (location identifier)

LFOE: Large Format Optical Enlarger

LFOP: Landing and Ferry Operations Panel

LFOS: Launch and Flight Operations System

LFP: Lakefield (Australia-QLD) (location identifier)

LFPCA: Large-Format Photon Counting Array

LFPL: Lewis Flight Propulsion Laboratory (NASA)

LFpM: Linear Field per Minute

LFPM: Linear Field Per Minute

LFPRL: Lewis Flight Propulsion Research Laboratory (NASA)

LFR: Franc Luxembourgeois (currency) • La Fría (Venezuela) (location identifier) • Liechtenstein Förderkreis für Raumfahrt • Low-Frequency Range • Low-Frequency Receiver • Luxemburg Franc (currency)

LFRA: Low-Frequency Radio Astronomy

LFRIG: Luminous Far-Infrared Galaxy

LFRR: Low-Frequency Radio Range

LFS: Labour Force Survey • Laser Fluorescent Spectroscopy • Laser Fluorosensor • Laser-induced Fluorescent Spectroscopy • Life From the Stratosphere • Live From the Stratosphere (*) • Logic Fault Simulator • Loop Feedback Signal

LFSO: Lunar Far Side Observatory

LFSR: Linear Feedback Shift Register

LFSS: Launch Facility Security System

LFST: Largest Feasible Steerable Telescope

Lft: Leaflet

LFT: Lafayette (USA-LA) (location identifier) • Least Frequent Trigram • Lift • Luyten Five Tenths • New Iberia - Lafayette (USA-LA) (location identifier)

LfU: Landesamt für Umweltschutz

LFU: Laser Firing Unit • Least Frequently Used • Lunar Flying Unit

LFV: Low-Frequency Variability • Low-Frequency Variable • Low-Frequency Vibration • Lunar Flying Vehicle

LFW: Lome (Togo) (location identifier)

LFZ: Laminar Flow Zone

lg: Common Logarithm • lang

Lg: Lagoon • Lange • Large • Length • Liège (Belgium) • Long

LG: Landing Gear • Largo • Length • Line Generator • Local Government • Local Group • Locomotive Generator • Low German

LGA: Lattice Gas Automaton • Local Government Agency • Local Government Area • Local Government Authority • Low-Gain Antenna • New York - La Guardia (USA-NY) (location identifier)

LGB: Laser-Guided Bomb • Lateral Geniculate Body • Long Beach (USA-CA) (location identifier)

Lgc: Logic

LGC: La Grange - Calloway (USA-GA) (location identifier) • Launch Guidance Computer • Lunar Gas Chromatograph • Lunar Geological Camera

LGCA: Laboratoire de Géodynamique des Chaînes Alpines (France)

LGCAI: Latin-Greek-Cyrillic-African-IPA

LGCP: Lexical-Graphical Composer Printer

LGCR: Laboratoire de Gravitation et Cosmologie Relativistes (France)

LGD: La Grande (USA-OR) (location identifier)

LGDT: Load Global Descriptor Table • Local Global Descriptor Table • Local Government Development Program

Lge: Large • Liège (Belgium)

LGE: Logic Gate Expander • Lunar Geological Equipment

LGEIES: Local Government Environmental Information Exchange Scheme

LGF: Yuma Laguna (USA-AZ) (location identifier)

LGG: Liège Bierset (Belgium) (location identifier)

LGGE: Laboratoire de Glaciologie et de Géophysique de l'Environnement (France)

LGGST: Laboratoire de Géologie-Géomorphologie Structurale et de Télédétection

LGH: Leigh Creek (Australia-SA) (location identifier)

LGI: Deadmans Cay (Bahamas) (location identifier) • Long Island (Bahamas) (location identifier)

LGIO: Local Government Information Office

LGIS: Local Government Information Service

LGIT: Laboratoire de Géophysique Interne et Tectonophysique (France)

LGIU: Laser Gyro Interface Unit

LGI2P: Laboratoire de Génie Informatique et de l'Ingénerie de Production (France)

LGK: Langkawi (Malaysia) (location identifier)

LGL: Legal • Long Lellang (Malaysia) (location identifier) • Low Galactic Latitude

LGM: Laboratoire de Géologie et Mécanique (France) • Laiagam (Papua New Guinea) (location identifier) • Last Glacial Maximum • Link Gateway Module • Little Green Man

LGN: Lagoon • Landelijke Grondgebruiksclassificatie van Nederland • Line Gate Number • Linga Linga (Papua New Guinea) (location identifier) • Logical Group Node

LGO: Langeoog (Germany) (location identifier) • Low Gravity Orbit

LGP: Legaspi (Philippines) (location identifier) • Length Growth Period • Limit Generation Program

LGPL: Library General Public License

LGQ: Lago Agrio (Ecuador) (location identifier)

Lgr: Longueur

LGR: Cochrane (Chile) (location identifier)

LGRNG: Long Range

LGS: Landing Gear System • Laser Guide Star • Local Group Suspected • Malargue (Argentina-MD) (location identifier)

LGSF: Laser Guide Star Facility

Lgsltd: Legislated

Lgsltr: Legislature

LGSOWG: Landsat Ground Station Operation Working Group • Landsat Ground Station Operators Working Group

Lgstcs: Logistics

LGT: Las Gaviotas (Colombia) (location identifier) • Light • Liquid-Gauging Technology • Local Geomagnetic Time

LGTA: Landing Gear Test Article

Lgtd: Lighted

Lgth: Length

Lgtps: Longtemps

Lgts: Lights

LGTS: Lights

LGU: LOcal Government Unit • Logan Cache (USA-UT) (location identifier)

LGW: London Gatwick (UK-England) (location identifier) • Low Gross Weight

LGWR: Log Writer

LGX: Lugh Ganana (Somalia) (location identifier)

LGY: Lagunillas (Venezuela) (location identifier)

LGZ: Leguizamo (Colombia) (location identifier)

lH: lichte Hohe

L/H: Local/Horizontal

LH: Labour Hour • Left Half • Left Hand • Left Handed • Lick H-alpha (catalogue) • Liquid Hydrogen • Load High • Local/Horizontal • Lower Half • Lower Hybrid • Lucke-Hodge

LHA: Laboratoire de l'Horloge Atomique (France) • Lahr (Germany) (location identifier) • Lamont-Hussey H-Alfa • Landshauptarchiv • Local Hour Angle • Lower Hour Angle

LHATS: Local Hour Angle of True Sun

LHB: Landes- und Hochschulbibliothek • Late Heavy Bombardment • Line History Block • Local Hot Bubble • Lokale Heiße Blase • Lost Harbor (USA-AK) (location identifier)

LHBl: La Houille Blanche (periodical code)

LHC: Large Hadron Collider • Left-Hand Circular

LHCC: Large Hadron Collider Committee

LHCP: Left-Hand Circularly Polarized • Left-Hand Circular Polarization

LHD: Large Helical Device • Left-Hand Drive

LHDr: Left-Hand Drive

LHe: Liquid Helium

LHE: Lahore (Pakistan) (location identifier)

LHEA: Laboratory for High-Energy Astrophysics

LHFEB: Left-Hand Forward Equipment Bay

LHG: Lightning Ridge (Australia-NSW) (location identifier)

LHI: Lereh (Indonesia) (location identifier)

LHK: Guanghua (China-PRC) (location identifier)

LHM: Learning in Humans and Machines

LHMS: Lithuanian Hydrometeorological Service

LHMT: Letters on Heat Mass Transfer (periodical code)

LHN: Lishan (China-ROC) (location identifier)

LHNO: Lager Huishoud- en Nijverheidsonderwijs

LHOP: Laser-Hardened Optical Payload

LHP: Left-Hand Panel • Lehu (Papua New Guinea) (location identifier) • Loop Heat Pipe

lhr: Lumen-Hour

LHR: London Heathrow (UK-England) (location identifier) • Lower Hybrid Resonance

LHRCS: Left-Hand Reaction Control System

LHS: Laser Heterodyne Spectrometer • Las Heras (Argentina-SC) (location identifier) • Lawrence Hall of Science (USA-CA) • Layered Half Space • Left-Hand Side • Library of Health Sciences • Lunar Horizon Sensor • Luyten Half-Second (catalogue)

LHSC: Left-Hand Side Console • Liquid Hydrogen System Complex

LHSS: Linear Hashing Superimposed Signature

LHSSC: Left-Hand Side Storage Container

LHSTF: Liquid Hydrogen Structural Test Facility

LHSV: Liquid Hourly Space Velocity

LHT: Layman's Hypertext • Local Heaven Time

LHU: Lake Havasu (USA-AZ) (location identifier)

LHUP: Lock Haven University of Pennsylvania (USA-PA)

LHV: Lock Haven - W.T. Piper (USA-PA) (location identifier) • Lower Heating Value

LHW: Langzhou (China-PRC) (location identifier)

LHX: Liquid Hydrogen

LH2: Liquid Hydrogen

Li: Link • Linker • Lithium

LI: Language-Independent • Laser Illuminator • Laser Ionization • Lead Investigator • Leeward Islands • Left Intermediate • Letter of Intent • Letter of Introduction • Level Indicator • Liberia • Liechtenstein (ISO country code) • Linearized Image • Line Item • List • List Item • Lithuania • Load Immediate • Long Island (USA-NY) • Longitudinal Interval • Lower Intercept • Low Intensity • Lyons Instruments Ltd. (UK)

LIA: Laboratoire d'Informatique d'Avignon (France) • Laboratory of Image Analysis • Land Information and Analysis • Laser Industry Association • Laser Institute of America (USA-FL) • Liangping (China-PRC) (location identifier) • Limiting Interval Availability • Linear-Induction Accelerator • Little Ice Age

LIAC: Liège International Astrophysical Colloquium

LIACC: Laboratorio de Inteligência Artificial e Ciência de Computadores

LIADA: Liga Ibero-Americana de Astronomía

LIAISE: Local Interactive In-school Science and Engineering

LIAM: Local Infrastructure and Access Maintenance

LIAS: Library Information Access System

lib: Liber

Lib: Liber • Liberal • Liberia • Liberty • Libra-Librae (constellation) • Librarian • Library • Libretto

LIB: Left Inboard • Liberdade • Library • Light-Ion Beam • Limbunya (Australia-NT) (location identifier)

LIBAL: Low-Ionization Broad Absorption Line

LIBER: Ligue des Bibliothèques Européennes de Recherche

LIBG: Luminous Infrared Bright Galaxy

LIBIS: Leuvens Integraal Bibliotheek- en Informatiesysteem

Libr: Libration

LIBRA: Logiciel Intégré pour les Bibliothèques en Réseau Automatisé

LIBRIME: Library and Information Management in Europe

LIBRIS: Land Image-Based Resource Information System

LIBS: Library Information and Bibliographic System

Lic: Licenciado • License • Licensed • Licenciado • Licenciate

LIC: Land Information Center • Land Information Council • Licensed Internal Code • Ligne Inférieure de Contrôle • Limon (USA-CO) (location identifier) • Linear Integrated Circuit • Load Interface Circuit • Local Interstellar Cloud

LICA: Low-energy Ion Composition Analyzer

Licdo: Licenciado

LICIL: Long Island Consortium for Interconnected Learning (USA)

LicOB: Lick Observatory Bulletin (periodical code)

LICQ: Laboratório Industrial de Controlo da Qualidade (Portugal)

LiCr: Liquid Crystals (periodical code)

LICS: Library Information Control System • Laser-Induced Continuum Structure • Lotus International Character Set

LID: Land Information Division • Laser-Induced Damage • Leadless Inverted Device • Leiden Valkenburg (Netherlands) (location identifier) • Library Issue Document • Light-Induced Drift • Line Interface Device • Line Item Description • Linear Imaging Device • Local Island Diversity • Logistics Identification Document

LIDAR: Laser Image Detection And Ranging • Laser Imaging Detector And Ranger • Light Detection And Ranging

LIDB: Line Information Database

LIDQA: Landsat Image Data Quality Analysis • Landsat Image Data Quality and Analysis (*)

LIDS: Logistics Integrated Data System • Loral Information Display System

LIDSE: Laboratoire Interdisciplinaire de Didactique des Sciences Expérimentales (France)

LIDT: Load Interrupt Descriptor Table

LIE: Left Inboard Elevon • Libenge (Zaire) (location identifier) • Liechtenstein (ISO country code) • Lightning-Induced Electron • Limited Information Estimation • Linear Incremental Encoder • Localized Induction Equation

Liech: Liechtenstein

LIEF: Launch Information Exchange Facility

LIEP: Large Internet Exchange Packet

LIESIB: Laboratoire d'Informatique, d'Électronique, du Signal, et de l'Imagerie de Bourgogne (France)

LIET: Local Inclusive Electron Trigger

Lieut: Lieutenant

LiF: Laser-induzierte Fluoreszenz • Lithium Fluoride

LIF: Large Isothermal Furnace • Laser-Induced Fluorescence • Laser Interference Filter • Lifou (Loyalty Islands) (location identifier) • Low Insertion Force

LIFD: Laser-Induced Fluorescence Detection

LiFE: Learning is a Family Experience

LIFE: Laboratory for International Fuzzy Engineering • Language Industry For Europe • Laser-Induced Fluorescence Emission • Laser Infrared Flyout Experiment

LIFES: Laser-Induced Fluorescence and Environmental Sensing

LIFFE: London International Finance Futures Exchange

LIFIA: Laboratoire d'Informatique Fondamentale et d'Intelligence Artificielle

LIFO: Last In - First Out

LIFS: Laser-Induced Fluorescence Spectroscopy

LIFT: Laser-Induced Fluorescence of Trees • Laser-Induced Forward Transfer • Life science Technology • Linking Innovation, Finance and Technology (EU) • Logically Integrated Fortran Translator

Lig: Liguria

LIG: Limoges Bellegarde (France) (location identifier) • Luminous Infrared Galaxy

LIGHT: Life-cycle Global Hypertext

LIGM: Luminous Infrared Galaxy Merger

LIGMALS: Landsat Interactive Grey Map And Level Slice

LIGO: Laser Interferometer Gravitational-wave Observatory (*) • Laser Interferometer Gravitational Observatory • Laser Interferometric Gravitational-wave Observatory

LIH: Kauai Island (USA-HI) (location identifier) • Laser Interferometric Holography • Light Intensity High • Lihue (USA-HI) (location identifier)

LII: Mulia (Indonesia) (location identifier)

LIIP: Laser-Induced Infrared Photochemistry

LIJ: Long Island (USA-AK) (location identifier)

LIK: Likiep Island (Marshall Islands) (location identifier)

LIL: Large Immersion Lens • Light Intensity Low • Lille Lesquin (France) (location identifier) • Linear Injector for LEP (CERN) • Local Interaction Language • Low-Ionization Line

LILE: Large-Ion Lithophile Element

LILO: Last In - Last Out

LILT: Low-Intensity Low-Temperature

Lim: Limerick (Ireland) • Limit • Limitation • Limité • Limited • Limiting

LIM: Land Identification Method • Land Information Management • Land Inventory and Management • Lima (Peru) (location identifier) • Limit • Limiter Impurity • Linear Induction Motor • Line Interface Module • Lotus-Intel-Microsoft

LIMA: Laboratoire d'Instrumentation Météorologique Aéroportée (Franc e) • Laser Ionization Mass Analyzer • Lotus-Intel-Microsoft-AST

LIMAC: Large Integrated Monolithic Array Computer

LIMAP: Landsat Imagery Analysis Package

LIMARS: Laser Imaging and Ranging System

LIMD: Large-scale In-Memory Database

LIME: Laboratorie d'Instrumentation en Micro-informatique et Électronique (France)

LIMIS: Liquid Metal Ion Source

LIMOS: Laser Intensity Modulation System

LIMP: Lunar-anchored Interplanetary Monitoring Platform

LIMRC: LRU Identification and Maintenance Requirements Catalog

LIMS: Laboratory Information Management Software • Laboratory Information Management System • Land Information Management System • Landsat Image Management System • Laser Ionization Mass Spectroscopy • Library Information Management System • Lightning Mapper Sensor • Limb Infrared Monitoring System • Limb Infrared Monitor of Stratosphere • Logistics Information Management System • Logistics Inventory Management System

LIMSI: Laboratoire d'Informatique pour la Mécanique et les Sciences de l'Ingénieur (France)

Lin: Lineal • Linear

LIN: Less Industrialized Nation • Linear • Line Item Number • Logical Infostructure Name • Milano - Forlanini Linate (Italy) (location identifier)

LINAC: Linear Accelerator

LINC: Laboratory Instrument Computer • LBT Interferometric Camera • Logic and Information Network Compiler

LINCOTT: Liaison, Interface, Coupling, Technology Transfer

Lincs: Lincolnshire (UK-England)

LINDA: Langmuir Interferometry and Density for Astrid 2 • Line Drawing Analyzer

LINE: Library Information Network Exchange

LINEAR: Lincoln Near-Earth Asteroid Research

LINER: Low-Ionization Nuclear Emission-line Region • Low-Ionization Nuclear Emission Region

Ling: Lingüística

LINJET: Liquid Injection Electric Thruster

LINK: Live and Interactive Network of Knowledge

lin-log: Linear-Logarithmic

LINS: Laser Inertial Navigation System

LINUS: Laser-Induced Nuclear Simulation

LINX: London Internet Exchange

LINZ: Land Information of New Zealand

LIO: Land Information Office • Limon (Costa Rica) (location identifier)

LIOCS: Logical Input/Output Control System

LiOH: Lithium Hydroxide

LION: Lunar International Observer Network

LIP: Laboratoire d'Information du Parallélisme • Laboratory for Image Processing • Land Impact Program • Large Installation Program • Large Internet Packet • Laser-Induced Plasma • Lightning Instrument Package • Limited Installation Program • Lins (Brazil-SP) (location identifier) • Local Image Processing • Local Interested Party • Loop Initialization Process • Lunar Impact Probe

LIPI: Indonesian Institute of Sciences • Lembaga Ilmu Pengetahuan Indonesia

LIPL: Linear Information Processing Language

LIpS: Logical Instructions per Second • Logical Interferences per Second

LIPS: Laboratory Information Processing System • Laser Image Processing Scanner • Library of Image Processing Software • Logical Instructions Per Second • Logical Interferences Per Second

LIPX: Large Internetwork Package Exchange

Liq: Liquid • Liquor

LIQ: Liquid • Lisala (Zaire) (location identifier)

LIR: Liberia (Costa Rica) (location identifier)

LIRA: Library and Information Retrieval Application • Linear Recognition and Analysis • Lunar Infrared Array

LIRACS: Library Information Resources Automated Control System

LIRAD: Lidar-Radiometer

LIRES: Literature Retrieval System

LIRF: Low-Intensity Reciprocity Failure

LIRG: Luminous Infrared Galaxy

LIRIS: Long-slit Intermediate Resolution Infrared Spectrograph

LIRS: Laser Imaging Radar Sensor • Library Information Retrieval Service • Library Information Retrieval System • Limb Infrared Sounder • Logical Information Retrieval System

Lis: Lisboa (Portugal)

LIS: Laboratory Information System • Land Information System • Landsat Information System • Language-Independent Specification • Language-Independent Standard • Laser Interferometer System • Laser Isotope Separation • Launch Instant Selector • Learning and Intelligent System • Library and Information Science • Library and Information Service • Library and Instructional Services • Library Information System • Lightning Imaging Sensor • Lisboa Portela (Portugal) (location identifier) • Lisbon Portela (Portugal) (location identifier) • Lithium Ion Storage • Logical IP Subnetwork • Logistics and Inventory System • Logistics Information System • Loop Input Signal

LISA: Laboratoire Interuniversitaire des Systèmes Atmosphériques (France) • Land Information System for Agriculture • Large Interferometer for Submillimeter Astronomy (*) • Large Installation System's Administrator • Large Isotope Spectrometer for Astromag • Laser Interferometer Space Antenna • Laser Interferometric Space Antenna (*) • Library and Information Science Abstracts • Library and Information Services in Astronomy • Lichthinder Informatie Site Associatie (Netherlands) • Limb Imaging Spectrograph for Airglow • Linked Indexed Sequential Method • Local Integrated Software Architecture

LISAA: Low-cost Imaging System for Advanced Applications

LISAG: Laser Interferometry Satellite Array for Gravimetry

LISAN: Libraries on the Information Superhighway Advocacy Network

LISAR: LaRC Image Scanning, Archival and Retrieval

LISB: Lithium Ion Storage Battery

LISC: Land Information System Committee • Lockheed-Martin Integrated Solutions Co.

LISI: Laser-Induced Surface Improvement

LISITT: Laboratorio Integrado de Sistemas Inteligentes y Tecnologias en Trafico

LISKA: Lithium Ion Source Karlsruhe (Germany)

LISM: Local Interstellar Medium

LISN: Line Impedance Stabilization Network

LiSOCl2: Lithium-Thionyl Chloride

LISP: List Processing • List Processor

LISPEC: LLNL Immersion Spectrograph

LISR: Line Information Storage and Retrieval

LISS: Linear Imaging Self-Scanner • Linear Imaging Self-Scanning • Linear Imaging Self-scanning Sensor • Lockheed Intermediate Sun Sensor

LISSA: Life Support Systems Analysis

LISSP: Laboratory for Image Signal and Statistical Processing

LIST: Library Index Search and Transcribe • Lunar Interferometry Study Team

LISTED: Library Integrated System for Telematics-based Education

Lit: Literal • Literalmente • Literally • Literary • Literatur • Literatura • Literature • Lithuanian • Litre • Litter • Little

LIt: Lira Italiana (currency)

LIT: Italian Lira (currency) • Laboratório de Integração e Testes • Large Image Terminal • Linear Ion Trap • Line Identifier Tag • Lira Italiana (currency) • Literature Interpretation Theory • Little Rock (USA-AR) (location identifier) • Lizensintorg • Local Implementation Team • Low-Impedance Transmission

LITA: Library and Information Technology Association (*) • Library Information and Technology Association

LITC: Library Information Technology Centre

LITE: Large Imagery Telescope • Large Imaging Telescope • Laser In-space Technology Experiment • Lidar In-space Technology Experiment • Light-Time Effect • Links and Interfaces for Tool-data Exchange • Lower Ionosphere Thermosphere Explorer

Lith: Lithograph • Lithographer • Lithographic • Lithography • Lithuania • Lithuanian

Litho: Lithos (periodical code)

Lithol: Lithology

LITS: Linear Ion Trap Standard

Litt: Littéraire • Littera • Littéral • Littérature

LITU: Life In The Universe

LIU: Line Interface Unit • Long Island University (USA-NY)

LIUAPV: Laboratoire d'Informatique de l'Université d'Avignon et des Pays de Vaucluse (France)

LIUM: Laboratoire d'Informatique de l'Université de Mons (Belgium)

Liv: Liverpool (UK-England) • Livonian • Livre

LIV: Line Item Value • Livengood (USA-AK) (location identifier) • Low-Input Voltage • Lunar and Interplanetary Vehicle

LIVA: Liga Venezolana de Astronomía (Venezuela) (formerly LIVAA)

LIVAA: Liga Venezolana de Aficionados a la Astronomía (Venezuela) (now LIVA)

LIVC: Low-Input Voltage Converter

LIVCR: Low-Input Voltage Conversion and Regulation

LIVE: Large Image Visualization Environment

LIVES: Landsat Image Verification and Extraction System

LIVR: Low-Input Voltage Regulation

LIW: Loikaw (Myanmar) (location identifier) • Long Instruction Word

LIY: Hinesville Wright (USA-GA) (location identifier)

LIZ: Limestone Loring (USA-ME) (location identifier)

lJ: laufenden Jahres

Lj: Lichtjahr

LJ: Loja

LJA: Lodja (Zaire) (location identifier)

LJC: Louisville (USA-KY) (location identifier)

LJJ: Long Josephson Junction

LJLV: Little Joe Launch Vehicle

LJM: Liverpool (Starlink internal code)

LJMU: Liverpool John Moores University (UK-England)

LJN: Lake Jackson - Brazoria County (USA-TX) (location identifier)

LJO: Licksele Ionospheric Observatory (Sweden)

LJU: Ljubljana - Brniki (Slovenia) (location identifier)

LK: Lake • Leak • Lochkarte • Lock • Sri Lanka (ISO country code)

LKA: Landeskriminalamt (Germany) • Larantuka (Indonesia) (location identifier) • Sri Lanka (ISO country code)

LKB: Lakeba (Fiji) (location identifier) • Large Knowledge Base • Lochkartenanlage

LKC: Lekana (Congo) (location identifier)

Lkd: Locked

LKD: Lakeland Downs (Australia-QLD) (location identifier)

LKE: Seattle - Lake Union (USA-WA) (location identifier)

LKEI: Lockheed Martin - Khrunichov - Energia International

LKF: Leuchtkraftfunktion

Lkg: Locking

Lkge: Linkage

LKHA: Lick H-Alpha

LKI: Duluth Lakeside (USA-MN) (location identifier)

LKK: Kulik Lake (USA-AK) (location identifier)

LKL: Lakselv (Norway) (location identifier) • Lochkartenleser

LKLY: Likely

LKM: Lochkartenmaschine

LKN: Leknes (Norway) (location identifier) • Lembaga Kimia Nasional (Indonesia)

LKO: Lucknow (India) (location identifier)

LKP: Lake Placid (USA-NY) (location identifier)

Lkr: Locker

LKR: Las Khoreh (Somalia) (location identifier) • Rupee (Sri Lanka) (ISO currency code)

LKS: Lakes • Lakeside (USA-TX) (location identifier)

LKT: Lakota (Ivory Coast) (location identifier)

LKU: Lake Rudolf (Kenya) (location identifier)

LKV: Lakeview - Lake County (USA-OR) (location identifier) • Lochkartenverfahren

LKW: Lastkraftwagen

LKY: Lake Manyara (Tanzania) (location identifier)

LKZ: Brandon Lakenheath (UK-England) (location identifier)

ll: Loco Laudado

L-L: Line-to-Line

L&L: Launch & Landing

LL: Laatstleden • Landau-Lifshitz • Late Latin • Launch and Landing • Launch Left • Laws • Leased Line • Lebanese Pound (currency) • Lever Lock • Ligne Louée • Limiting Level • Limited Liability • Lincoln Laboratory (MIT) • Line-by-Line • Linear-Linear • Line Length • Lines • Live Load • Llandudno (UK postal code) • Lobby Level • Local Language • Long Lead • Long Line • Loose Leaf • Lords • Lower Left • Lower Level • Lower Limit • Lower Lobby • Low Lead • Low Level • Lunar Landing

LLA: Luleå (Sweden) (location identifier)

LLabJ: Lincoln Laboratory Journal (periodical code)

LLAGN: Low-Luminosity Active Galactic Nucleus

LLB: Bachelor of Laws • Laboratoire Léon Brillouin (France) • Legum Baccalaureus

LLBL: Low-Latitude Boundary Layer

LLC: Limited Liable Company • Line Length Code • Liquid-Liquid Chromatography • Logical Link Control • Lower Left Centre

LLCF: Launch and Landing Computational Facilities

LLCS: Length of Longest Common Subsequence

LLCSC: Low-Level Computer Software Component

LLD: Location de Longue Durée • Logic Level Driver • Lower Level Discriminator • Lower Level of Detection • Lower Limit of Detection • Low Level Discrete

LLDC: Least Developed Country

LLDPE: Linear Low-Density Polyethylene

LLDT: Load Local Descriptor Table

LLE: Laboratory for Laser Energetics • Liquid-Liquid Extraction • Load List Element • Loss of Life Expectancy • Low-Level Exposure • Low-Luminosity Elliptical • Malelane (South Africa) (location identifier)

LLF: Low Level Format • Low Level Formatting

LLFS: Lyman-Limit Flat Spectrum

LLG: Chillagoe (Australia-QLD) (location identifier)

LLGCX: Low-Luminosity Globular-Cluster X-ray (source)

LLH: Las Limas (Honduras) (location identifier)

LLI: Lalibela (Ethiopia) (location identifier) • Latitude and Longitude Indicator • Limited-Life Item • Local Lorentz Invariance

LLIA: Local Language Inference Algorithm

LLIC: Low Level Interaction Component

LLIL: Long Leadtime Items List

LLIV: Low-Level Input Voltage

LLL: Learning Language in Logic (workshop) • Limited-Life List • Lissadell (Australia-WA) (location identifier) • Lowest Landau Level • Low-Level Logic • Low Light Level

LLLTV: Low Light Level Television

LLLW: Liquid Low-Level Waste

LLM: L-band Land Mobile • Long Lama (Malaysia) (location identifier) • Lunar Landing Mission • Lunar Landing Module

LLMC: Low-Latitude Meridian Circle

LLMM: Leurs Majestés

LL.MM.: Leurs Majestés

LLMS: Little LEO Messaging System

LLN: Kelila (Indonesia) (location identifier) • Line Link Network • Lire La Nature (Canada-QC) • Louisiana Library Network (USA-LA)

LLNL: Lawrence Livermore National Laboratory (USA-CA)

LLNW: Low-Level Nuclear Waste

LLO: Low Lunar Orbit

LLOS: Landmark Line Of Sight

LLOV: Low-Level Output Voltage

LLOX: Lunar Liquid Oxygen

LLO2: Lunar Liquid Oxygen

LLP: Launch and Landing Project • Liaison Logique Permanente • Lightning Location and Prediction • Linda Downs (Australia-QLD) (location identifier) • Line Link Pulsing • Lunar Landing Program • Lyman Line Pumping

LLPN: Lumped, Linear, Parametric Network

LLPO: Launch and Landing Project Office

LLPR: Lattimer, Lamb, Pethick and Ravenhall

LLR: Load-Limiting Resistor • Lunar Laser Ranging

LLRA: LapLink Remote Access

LLRC: Library Learning Resource Center (USA) • Luneberg Lens Rapid Commutator

LLRF: Lunar Landing Research Facility

LLRS: Library and Learning Resource System • Luminous Linear Radio Source

LLRV: Lunar Landing Research Vehicle

LLRW: Low-Level Radioactive Noise

L&LS: Launch & Landing Site

LLS: Laser Light Scattering • Laser Light Source • Laser Line Scanner • Las Lomitas (Argentina-FO) (location identifier) • Launch and Landing Site • Lunar Landing Simulator • Lunar Logistics System • Lyman Limit System

LLSA: Literary Legacy of Space Age

LLSF: Linear Least Squares Fit

LLSI: Laser Light Scattering Instrument

LLSN: Lawrence Livermore Seismic Network

LLST: Linear Least Squares Transformation

LLSV: Lunar Logistics System Vehicle

LLT: Lander Local Time • Long Lead Time • Low-Level Terminal

LLTI: Long Lead-Time Item

LLTIL: Long Lead Time Item List

LLTV: Low-Light Television • Lunar Landing Training Vehicle

LLU: Lending Library Unit

LLV: Lockheed-Martin Launch Vehicle • Lunar Landing Vehicle • Lunar Logistics Vehicle

LLVPG: Large Launch Vehicle Planning Group

LLW: Language Learning Workbench • Lilongwe Kamuzu (Malawi) (location identifier) • Low-Level Waste

LLWAS: Low-Level Windshear Alert System

LLWS: Low-Level Wind Shear

LLWSAS: Low-Level Wind Shear Alert System

LLX: Lyndonville (USA-VT) (location identifier)

LLY: Mount Holly - Burlington County (USA-NJ) (location identifier)

Llz: Localizer

lm: Lumen

lM: laufenden Monats

Lm: Maltese Lira (currency)

L/M: Leadership/Management • List of Materials

L-M: Levenberg-Marquardt

L&M: Light & Medium

LM: Laboratory Module • Land Mobile • Langage de Machine • Large Model • Launch Mode • Launch Mount • Leadership/Management • Left Mid • Light and Medium • Light Microscope • Light Microscopy • Lightning Mapper • Limiting Magnitude • Line Monitor • Liquid Metal • Liquid Mirror • List of Material • Livello del Mare • Load Management • Logic Module • Logistics Manager • Logistics Manual • London Museum • Long Module • Lord Mayor • Low Mass • Lunar Module

LMA: Laboratory for Millimeter Astronomy • Laboratory for Millimeter-wave Astronomy • Lake Minchumina (USA-AK) (location identifier) • Laser Microprobe Analyser • Lockheed Martin Astronautics • Lunar Meteoroid Analyzer • Lunar Module Adapter

LMAE: Lunar Module Ascent Engine

LMAN: Last-Minute Astrometry Network

LMARS: Library Management and Retrieval System

LMB: Left Mouse Button • Le Micro-Bulletin (CNRS) • Local Message Box • Low-Mass Binary • Salima (Malawi) (location identifier)

LMBCS: Lotus Multibyte Character Set

LMBP: Low-Mass Binary Pulsar

LMC: Laboratoire de Modélisation et Calcul (France) • La Macarena (Colombia) (location identifier) • Large Magellanic Cloud • Laser Mirror Coating • Link Monitor and Control • Liquid-Metal Coolant • Local Management Committee • Low-Mass Companion

LMCA: Land Management Cooperative Agreement

LMCS: Local Multipoint Communications System

LMD: Laboratoire de Météorologie Dynamique • Laboratory of Meteorological Dynamics • La Manipulation des Données • Large Mirror Development • Local Medical Doctor • Los Menucos (Argentina-RN) (location identifier) • Lunar Meteoroid Detector

LMDA: Lunar Meteoroid Detector-Analyzer

LMDE: Lunar Module Descent Engine

LMDS: Local Multipoint Distribution Service (*) • Local Multipoint Distribution System • Logistics Management Data System

LME: Launch Monitor Equipment • Layer Management Entity • Le Mans - Arnage (France) (location identifier) • Lewis-Michaels Engineering (USA-NY) • Liquid Motion Experiment • Lower Mid Fuselage • Lunar Module Engine

LMER: Land Margins Ecosystem Research

LMES: Laboratory for Meteorology and Earth Sciences

L/MF: Low/Medium Frequency

LMF: Light Manufacturing Facility • Load Module Format • Low and Medium Frequency • Lower Mid-Fuselage • Low/Medium Frequency

LMFBR: Liquid-Metal-cooled Fast Breeder Reactor

LMFC: Limited Field Moisture Capacity

LMG: Lamassa (Papua New Guinea) (location identifier) • Left Main Gear • Light Machine Gun

LMGC: Lunar Module Guidance Computer

LMH: Limón (Honduras) (location identifier)

lm-hr: Lumen-Hour

LMi: Leo Minor - Leonis Minoris (constellation)

LMI: Les Marques Internationales • Linear Matrix Inequality • Link Management Interface • Local Management Interface • Lockheed-Martin Intersputnik • Lumi (Papua New Guinea) (location identifier)

LMIC: Land Management Information Center

LMIF: Lower Mechanical Integration Frame

LMIS: Liquid-Metal Ion Source • Logistics Management Information System

LMK: Landmark

LML: Lae Island (Marshall Islands) (location identifier)

LMLE: Local Maximum Likelihood Estimate

LMLG: Left Main Landing Gear

LMLR: Load Memory Lockout Register

LMLV: Lockheed Martin Launch Vehicle

LMM: Le Matériel Magnétique • Length-Measuring Machine • Linear Multistep Method • Locator Middle Marker • Logical Memory Manager • Los Mochis (Mexico) (location identifier) • Lunar Mapping Module • Lundi-Mardi-Mercredi

LMMS: Laser Microprobe Mass Spectrometry • Local Message Metering System • Lockheed Martin Missiles & Space

LMN: Library Management Network • Limbang (Malaysia) (location identifier)

LMO: Learning Material Object • Lens-Modulated Oscillator • Lossiemouth (UK-Scotland) (location identifier) • Low Mars Orbit • Low-Mass Object • Lucile Miller Observatory (USA-NC)

LMOS: Line Maintenance Operating System • Line Maintenance Operations System

LMP: Lampedusa (Italy) (location identifier) • Land management Planning • Link Management Protocol • List of Measurement Points • Locally Most Powerful • Lunar Module Pilot

LMPRT: Locally Most Powerful Rank Test

LMQ: Marsa Brega (Libya) (location identifier)

LMR: Large Microwave Radiometer • Laser Magnetic Resonance • Launch Mission Rule • Launch Monitor Room • Lime Acres (South Africa) (location identifier) • Liquid-Metal Reactor • Lowest Maximum Range

LMRD: Launch Mission Rules Document

LMREP: Lower Mississippi River Environmental Program

LMRT: Logistics Management Responsibility Transfer

LMS: Land Mobile Service • Landsat Mapping System • Large-scale Mapping System • Laser Mass Spectrometer • Laser Mass Spectrometry • Lead Mission Scientist • Least Mean Square • Least Median of Squares • Library Management Software • Library Management System • Life and Microgravity Sciences • Life and Microgravity Spacelab • Lightning Mapper Sensor • Linear Measuring Set • Load Measurement System • Logistics Master Schedule • Loss of Memory Syndrome • Lotus Messaging Link • Louisville - Winston County (USA-MS) (location identifier) • Low-Mass Star • Lunar Mass Spectrometer • Lunar Measuring System • Lunar Module Simulator

LMSA: Large Millimeter and Submillimeter Array

LMSAAC: Life and Microgravity Sciences and Applications Advisory Committee

LMSAL: Lockheed-Martin Solar and Astrophysics Laboratory

LMSC: Lockheed-Martin Missiles and Space Co.

LMSS: Land Mobile Satellite Service • Land Mobile Satellite System • Lunar Mapping and Survey System

LMST: Lightweight Multiband Satellite Terminal

LMSW: Load Machine Status Word

LMT: Klamath Falls (USA-OR) (location identifier) • Large Millimeter Telescope • Large Millimeter-wave Telescope • Launch Motor Test • Length-Mass-Time • Limit • Liquid-Mirror Telescope • Load Module Table • Local Mean Time • Local Minima Table • Logistics Management Transfer • Luchtvaartmaatschappij Twente (Netherlands)

LMTA: Librray and Media Technical Assistant

LMTAS: Lockheed Martin Tactical Aircraft Systems

LMTE: Liquid Metal Thermal Experiment

LMTL: Long-term Mission Time Line

LMTO: Linear Muffin-Thin Orbital (method)

Lmtr: Limiter

LMU: Lincoln Memorial University (USA) • Line Monitor Unit • Logical Memory Unit • Loyola Marymount University (USA) • Ludwig-Maximilians-Universität (Germany)

LMV: Laboratoriet för Mekaniska Vågor (Sweden)

lm/W: Luman/Watt

LMW: Low Molecular Weight

LM/X: LAN Manager for Unix

LMX: LAN Manager for Unix • Lopez de Micay (Colombia) (location identifier)

LMXB: Low-Mass X-Ray Binary

LMXRB: Low-Mass X-Ray Binary

LMY: Lake Murray (Papua New Guinea) (location identifier)

LMZ: Palma (Mozambique) (location identifier)

ln: Lingala (ISO language code) • Natural Logarithm

Ln: Laan • Lane • Link

LN: Lane • League of Nations • Line • Link • Liquid Nitrogen • Lincoln (UK postal code) • Load Number • Localization • Logarithm (Natural_) • Luna Nuova • Natural Logarithm • Norway (international aircraft registration code)

LNA: Laboratório Nacional de Astrofísica • Layered Network Algorithm • Low-Noise Amplifier • Low-Noise Antenna • West Palm Beach - Palm Beach County (USA-FL) (location identifier)

LNAAC: Local Network Associate Analysis Center

LNAI: Lecture Notes on Artificial Intelligence (Springer)

LNB: Lamen Bay (Vanuatu) (location identifier) • Low-Noise Block

LNC: Lengbati (Papua New Guinea) (location identifier) • Low-Noise Cable • Low-Noise Converter

LNCC: Ligue Nationale Contre le Cancer (France)

LNCH: Launch

LNCS: Lecture Notes in Computer Science (Springer)

LND: Lander Hunt (USA-WY) (location identifier) • Limiting Noise Dive

LNDF: Log Normal Distribution Function

Lndg: Landing

LNDG: Landing

LN:DI: Lotus Notes Document Imaging

LNDI: Lotus Notes Document Imaging

LNDNG: Landing

LNE: Laboratoire National d'Essais (France) • Lonorore (Vanuatu) (location identifier)

LNES: Lecture Notes in Earth Sciences (Springer) (periodical code)

LNF: Laboratori Nazionali di Frascati (Italy) • Low-Noise Feed • Munbil (Papua New Guinea) (location identifier)

LNEC: Laboratório Nacional de Engenharia Civil (Portugal)

LNEMS: Lecture Notes in Economic and Mathematical Systems (Springer)

LNEn: Lecture Notes in Engineering (Springer) (periodical code)

LNETI: Laboratório Nacional de Engenharia e Tecnologia Industrial (Portugal)

Lng: Landing

LNG: Lese (Papua New Guinea) (location identifier) • Liquefied Natural Gas • Liquid Natural Gas

LNGS: Laboratori Nazionali del Gran Sasso (Italy)

Lnh: Lunch

LNH: Lake Nash (Australia-NT) (location identifier) • Lunar Near Horizon

LNHM: Lafayette Natural History Museum (USA-LA)

LNI: Local Network Interconnect • Lonely Dew (USA-AK) (location identifier)

LNK: Lincoln (USA-NE) (location identifier)

LNM: Langimar (Papua New Guinea) (location identifier) • Lecture Notes in Mathematics (Springer) (periodical code)

LNN: Linear Neural Network • Willoughby - Lost Nation (USA-OH) (location identifier)

LNNI: LAN Network Node Interface • LAN Network-to-Network Interface

LNO: Leonora (Australia-WA) (location identifier)

LNP: Lecture Notes in Physics (Springer) (periodical code) • Liquid Nitrogen Processing • Loch Ness Productions (USA-MA) • Local Number Portability • Lunar Neutron Probe • Wise (USA-VA) (location identifier)

LNPAM: Lecture Notes in Pure and Applied Mathematics (Springer) (periodical code)

LNPE: Lunar Neutron Probe Experiment

LnPT: Level n Page Table

LNQ: Loani (Papua New Guinea) (location identifier)

LNR: Lone Rock - Tri County (USA-WI) (location identifier) • Low-Noise Receiver

LNRF: Locally Non-Rotating Frame

Lnrty: Linearity

LNS: Laboratory for Nuclear Science • Lancaster (USA-PA) (location identifier) • Land Navigation System • Lines

LNT: Liquid-Nitrogen Temperature

LNTS: Liquid-Nitrogen Transfer System

LNUG: Library Network User Group

LNTWA: Low-Noise Travelling Wave Amplifier

LNV: Landbouw, Natuurbeheer en Visserij (Netherlands) • Londolovit (Papua New Guinea) (location identifier)

LNX: Smolensk (Russia) (location identifier)

LNY: Lanai City (USA-HI) (location identifier)

LNZ: Linz (Austria) (location identifier)

LN2: Liquid Nitrogen

lo: Laothian (ISO language code)

Lo: Lodge • Lord

L/O: Lift Off

LO: Lager Onderwijs • Launch Operations • Launch Operator • Leading Order • Letter of Offer • Level Off • Liaison Office • Liaison Officer • Lichtenknecker Optics (Belgium) • Lift-Off • Linear Optimization • Line Occupancy • Local Oscillator • Located Object • Lock-On • Lock-Out • Logical Operation • Low Observable • Lowest Order • Low Observable • Low Orbit • Lubricating Oil • Lunar Observer • Lunar Orbiter • Lund Observatory (Sweden)

LOA: Laboratoire d'Optique Appliquée • Laboratoire d'Optique Atmosphérique • Landing Operations Area • Launch Operations Agency • Launch Operations Area • Leave Of Absence • Length Over All • Letter Of Agreement • Letter Of Authorization • Lorraine (Australia-QLD) (location identifier)

LOAL: Lock-On After Launch

LOAN: Library Online Automated Network • Low Observable Axisymmetric Nozzle

LOAP: List Of Applicable Publications

LOAS: Lift-Off Acquisition System

LOB: Laboratory Office Building • Launch Operations Branch • Launch Operations Building • Left Outboard • Line Of Balance • Line Of Business • Low-Order Bit • Low-Order Byte

LOBES: Low-Bidget Extraterrestrial-intelligence Search

LOBO: Low-frequency Background Observatory

LOBSTER: Large Angle Observatory for Burst, Supernova and Transient Event Research

Loc: Local • Localizer • Located • Location • Locative • Locator • Locución

LoC: Library of Congress

LOC: Language-Oriented Computer • Large Optical Cavity • Launch Operations Center • Launch Operations Complex • Launch Operations Contractor • Launch Operator's Console • Letter Of Comment • Letter Of Credit • Level Of Conflict • Library Of Congress (USA) • Limited Operational Capability • Limit Of Convection • Line Of Code • Line Of Communication • Local • Localidade • Localizer • Locally • Local Operating Co. • Local Optimally-emitting Cloud • Local Organizing Committee • Location • Lock (Australia-SA) (location identifier) • Logic Optimizing Compiler • Loop Online Control • Loss Of Control

LOCA: Loss Of Coolant Accident

LOCAAS: Low-Cost Autonomous Attack System

LOCAL: Laboratory Of Computer-Assisted Learning

LOCAP: Low Capacitance

LOCB: Launch Operations Control Building

LOCC: Launch Operations Control Center • Logistical Operations Control Center

LOCF: Last Observation Carried Forward

loc. cit.: Loco Citato

LOCID: Location Identifier

LOCIS: Library Of Congress Information System

LOCK: Logical Co-processing Kernel

LockH: Lockheed Horizons (periodical code)

loc. laud.: Loco Laudato

Locn: Location

loc. primo cit.: Loco Primo Citato

LOCS: Land-Ocean Contamination Study • Local Optical Correction System

LOCSS: Land-Ocean Climate Satellite System

LOD: Launch Operations Directive • Launch Operations Director • Launch Operations Directorate • Launch Operations Division • Launch Order Document • Leading Ones Detector • Legion Of Doom • Length Of Day • Level Of Detail • Limit Of Detection • Locally One-Dimensional • Longana (Vanuatu) (location identifier)

LODA: Large Object Database

LODCS: Lunar Orbiter Data Conversion System

LODE: Large Optics Demonstration Equipment • Large Optics Demonstration Experiment • Ley Orgánica del Derecho a la Educación (Spain)

LODESTAR: Logically Organized Data, Entry, Storage And Recording

LODF: Line Opacity Distribution Function

LODI: List Of Deleted Items

LODIA: Low-Cost Digital Image Analyzer

LODP: Lunar Orbiter Data Printer

LODR: Length Of Day - Regularized

LODSB: Load String Byte

LODUS: Low Data Rate UHF Satellite

LODYC: Laboratoire d'Océanographie Dynamique et de Climatologie (*) • Laboratoire d'Océanographie et de Dynamique du Climat

LOE: Left Outboard Elevon • Level Of Effort • Line Of Effort • Loei (Thailand) (location identifier)

LOEAT: Lowest Temperature Equaled for All Time

LOEFM: Lowest Temperature Equaled For the Month

LOEle: L'Onde Électrique (periodical code)

LOEPS: Land-Ocean Evolutionary Perspective Study

LOESE: Lowest Temperature Equaled So Early

LOF: Local Oscillator Frequency • Lock Off-line • Loen (Marshall Islands) (location identifier) • Loss Of Flow • Loss Of Function

LOFAR: Low-Frequency Acquisition And Ranging • Low-Frequency Array

Lo-Fi: Low Fidelity

LoFi: Low Fidelity

LoFLYTE: Low-Observable Flight Test Experiment

LOFO: Low-Frequency Oscillation

LOFPI: Low-Order Fabry-Pérot Interferometer

LOFT: Launch Operations Flight Test • Line-Oriented Flight Training • Low-Frequency Radio Telescope

log: Logarithm

Lóg: Lógica

Log: Logic • Lógica • Logical • Logistics

LoG: Laplacian of Gaussian

LOG: Large Optical Generator • Logarithm • Login • Longview (USA-WA) (location identifier)

LOGACS: Low-G Accelerometer Calibration System

LOGAL: Logical Algorithmic Language

LOGEL: Logic Generating Language

LOGIPAC: Logical Processor And Computer

LOGIT: Logical Inference Tester

LOGMOD: Logic Model

LOGO: Limit Of Government Obligation

LOGR: Logradouro

LOGSE: Ley de Ordenación General del Sistema Educativo (Spain)

LOH: Loja (Ecuador) (location identifier)

LoI: Letter of Instruction • Letter of Intent • Letter of Interest • List of Items

LOI: Letter Of Instruction • Letter Of Intent • Letter Of Interest • List Of Items • Loss On Ignition • Low-resolution Oscillation Imager • Luminosity Oscillation Imager • Lunar Orbital Insertion • Lunar Orbit Insertion

LOICZ: Land-Ocean Interactions in the Coastal Zone

LOID: Location Identifier

LOIS: Land Oceans Interactions Study • Library Orders Information System • Logistics Information Simulator • Lowell Observatory Instrumentation System

LOK: Lodwar (Kenya) (location identifier)

LOL: Laughing Out Loud • Limited Operating Life • Limit Of Liability • Lovelock Derby (USA-NV) (location identifier)

LOLA: Library Online Acquisition • Lunar Orbit and Lunar Approach • Lunar Orbit Landing Approach

LOLI: Limited-Operational-Life Item

LOLITA: Language for the Online Investigation and Transformation of Abstractions • Library Online Information and Text Access

LOM: Lagos de Moreno (Mexico) (location identifier) • Laser Optical Modulator • Launch Operations Manager • Life Of Mission • List Of Modifications • Locator at Outer Marker • Low-Order Memory • Lunar Orbital Mission

LOMA: Letters Of Map Amendment • Life Office Management Association

Lomb: Lombard • Lombardy

LOMC: Letters Of Map Correction

LOME: Low-rate Minimum Energy

LOMR: Letters Of Map Revision

Lon: Longitude

LON: Launch On Need • Letter Of Notification • Line Of Nodes • Lock Online • London Metropolitan (UK-England) (location identifier) • Longitude

Lond: London (UK-England) • Londonderry (Northern Ireland)

LONEOS: Lowell Observatory's Near-Earth Object Search (*) • Lowell Observatory's Near-Earth Object Survey

Long: Longitude • Longitudinal • Longueur

LONG: Longitude

Longl: Longitudinal

LonGSp: Long-slit Gornergrat Spectrometer

LONGSTOP: Long-term Gravitational Stability Test for the Outer Planets

LONS: Local On-line Network System

LOO: Laghouat L'Mekrareg (Algeria) (location identifier) • Launch Operations Order • Log Of Operation

LOOM: Laboratoire d'Optique de l'Observatoire de Marseille (France) • Low Overhead Object Management

LOOMS: Laser Optics Operations and Maintenance Support

LLOP: Low Overhead Object Protocol

LOOP: Library-Operated Outreach Program

LOOPE: Loop while Equal

LOOPNE: Loop while Not Equal

LOOPNZ: Loop while Not Zero

LOOPS: LISP Object-Oriented Programming System

LOOPZ: Loop while Zero

LOP: Last Operation Completed • Launch Operations Panel • Launch Operator Panel • Life Of Program • Line Of Position • Low-Order Partition

LOPAR: Low-Power Acquisition Radar

LOPC: Lunar Orbit Plane Change

LOPLAS: Letters on Programming Language and Systems

LOPP: Lunar Orbiter Photographic Project

LOPS: Launch Operations • Lunar Orbiting Photographic System

loq: Loquitur

LOQ: Lobatse (Botswana) (location identifier)

Lor: Lorong

LOR: Large Optical Reflector • Light Output Ratio • Line Outage Recorder • Loss Of Reference • Lunar Orbital Rendezvous • Lunar Orbit Rendezvous • Ozark Lowe (USA-AL) (location identifier)

LORA: Lecturer-Oriented Response Analysis • Level-Of-Repair Analysis

LORAC: Long-Range Accuracy

LORAD: Long-Range Active Detection

LORAN: Long-Range Aid to Navigation • Long-Range Navigation (*)

LORCS: League of the Red Cross and Red Crescent Societies

LORD: Local Optical Reference Device

LORE: Langley Online Research Explorer (NASA) • Line-Oriented Editor

LOREORS: Long-Range Electr-Optical Reconnaissance System

LoRes: Low Resolution

LORL: Large Orbital Research Laboratory

LOROP: Long-Range Oblique Photography

LOROPS: Long-Range Oblique Photography System

LORV: Low Observable Reentry Vehicle

LOS: Lagos (Nigeria) (location identifier) • Land Observation Satellite • Launch Operations System • Length Of

Service • Length Of Stay • Liaison Office Support • Lift-Off Simulator • Line Of Sight • Line Of Supply • Line Operational Simulation • Loop Output Signal • Loss Of Sight • Loss Of Signal • Loss Of Synchronization • Lunar Orbiting Satellite

LOSF: Lunar Orbit Staging Facility

LOSHEA: Line-Of-Sight Hard Earth Angle

LOSIC: Logiciel de Simulation de Couverture

LOSM: Launch Operations Simulation Model

LOSP: Landing Operations Support Panel

LOSR: Line-Of-Sight Range

LOSS: Leuschner Observatory Supernova Search • Lunar Orbital Space Station • Lunar Orbital Survey System • Lunar Orbit Space Station

LOSV: Line-Of-Sight Velocity

LOSVD: Line-Of-Sight Velocity Distribution

Lot: Lotion

LOT: Large Orbital Telescope • Large Orbiting Telescope • Ley de Ordenación de las Telecommunicaciones (Spain) • Lift-Off Time • Lockport Lewis (USA-IL) (location identifier)

LOTAWS: Laser Obstacle Terrain Avoidance Warning System

LOTEAM: Loteamento

LOTIS: Livermore Optical Transient Imaging System

LOTMOP: Lick Observatory Torus Monitoring Program

LOTOS: Language Of Temporal Ordering Specification • Long-Term Ozone Simulation

LOTREX: Longitudinal Traverse Experiment

LOTS: Low-Overhead Time-Sharing System

LOTV: Launch Operations and Test Vehicle

Lou: Louisiana (USA)

LOU: Louisville Bowman (USA-KY) (location identifier)

LOUISA: Lunar-Optical-Ultraviolet-Infrared Synthesis Array

LOV: Limit Of Visibility • Loss Of Vehicle • Loss Of View • Loss Of Visibility • Monclova (Mexico) (location identifier)

LOVLA: Lunar Optical Very Large Array

LOWESS: Locally Weighted Least Squares

LOWG: Landing Operations Working Group

LowOB: Lowell Observatory Bulletin (periodical code)

LOWR: Lower

LOX: Liquid Oxygen

LOXT: Large Orbiting X-ray Telescope

LOY: Loyengalani (Kenya) (location identifier)

LOZ: Corbin London (USA-KY) (location identifier)

LO2: Liquid Oxygen

Lp: Ladyship • Lordship

L-P: Low Pressure

LP: Laboratoire Propre • Land Process • Language Planning • La Pampa (Argentina) • Lapp • Large Paper • Large Particle • Large Photometer • Large Post • Last Post • Latent Period • Launch Pad • Launch Platform • Launch Processing • Legacy Project • Legal Procurator • Life Policy • Limit Point • Linear Phase • Linear Polarization • Linear Potential • Linear Prediction • Linear Program • Linear Programming • Line Printer • Linguistic Processor • Linkage Programme • Liquid Petroleum • Liquid Propellant • List Processing • List Processor • Logic Programming • Lonely Planet • Long Period • Long Play • Long Player • Long Playing • Long Primer • Loop • Lower Panel • Low Pass • Low Power • Low Pressure • Luna Piena • Luyten Palomar (catalogue) • Lycée Professionnel (France) (formerly LEP)

LPA: Gran Canaria (Canary Islands) (location identifier) • Launch Phase Analyst • Least Perturbative Action • Light Pulser Array • Linear Power Amplifier • Local Potential Approximation • Log Periodic Antenna • Low-Power Amplifier

LPAA: Log Periodic Array Antenna

LPAC: Launching Programme Advisory Committee

LPAI: Logic Programming and Artificial Intelligence

LPAP: Laboratoire de Physique Atmosphérique et Planétaire

LPAR: Logical Partition • Logical Partitioning • Logic Programming and Automated Reasoning

LPARL: Lockheed-Martin Palo Alto Research Laboratory (USA-CA)

LPATS: Lightning Position And Tracking System

LPB: La Paz - El Alto (Bolivia) (location identifier)

LPBBA: Log Periodic Broadband Antenna

LPC: Laboratoire de Physique Corpusculaire • Laboratory Precision Connector • Laser Processing Consortium • Leading Particle Cascade • Linear Plate Constant • Linear Prediction Coding • Linear Prediction Coefficient • Linear Predictive Coding • Link Priority Change • Local Procedure Call • Lompoc (USA-CA) (location identifier) • Long-Period Comet • Long Periodic Comet • Low-Pressure Chamber

LPCA: Lunar Pyrotechnic Control Assembly

LPCDF: Low-Profile Combined Distributing Frame

LPCE: Laboratoire de Physico-Chimie de l'Environnement • Laboratoire de Physique et Chimie de l'Environnement • Laboratoire de Physique et de Chimie de l'Environnement

LPCVD: Low-Pressure Chemical Vapor Deposition

LPD: Landing Point Designator • Language Processing and Debugging • La Pedrera (Colombia) (location identifier) • Laser Polarization Detector • Launch Point Determination • Launch Preparation Document • Launch Procedure Document • Line Printer Daemon • List of Physical Dimensions • Log Periodic Dipole • Low-Power Device • Low Probability of Detection • Luyten Palomar Double

LPDA: Laboratory Program Development Activity • Laser-Phased Doppler Anemometer • Linear Photodiode Array • Link Problem Determination Aid • Log Periodic Dipole Array

LPDM: Lagrange Particle Dispersion Model • Lagrangian Particle Dispersion Model • List of Physical Dimensions

LPDT: Low-Power Distress Transmitter

LPE: Laboratoire de Physique de l'Environnement • Lambda Point Experiment • La Primavera (Colombia) (location identifier) • Launch Preparation Equipment • Liquid-Phase Epitaxial • Liquid-Phase Epitaxy • Loi sur la Protection de l'Environnement (France) • Look Point Equation • Lunar and Planetary Ephemerides

LPEE: Liquid-Phase Electro-Epitaxial

LPES: Launch Preparation Equipment Set

LPETS: Lunar Prospector Electrical Test Set

LPEV: Launch Preparation Equipment Vault

LPF: Lambda Point Fridge • League for Programming Freedom • Library Processing Facility • Low-Pass Filter • Low-Power Field

LPFT: Low-Pressure Fuel Turbopump

LPFTP: Low-Pressure Fuel Turbopump

LPG: Langage de Programmation Numérique • La Plata (Argentina-BA) (location identifier) • Last Page Generator • Launch Preparations Group • Liquefied-Petroleum Gas • Liquid-Petroleum Gas • Liquid-Propane Gas • Liquid-Propellant Gun • Long-Path Gas • Low-Pressure Gas

LPGE: LEM Partial Guidance Equipment

LPGP: Laboratoire de Physique des Gaz et des Plasmas

LpH: Lines per Hour

LPH: Lines Per Hour • Lochgilphead (UK-Scotland) (location identifier)

LPHA: Laboratoire de Physique des Hautes Atmosphères

LPHS: Lunar and Planetary Horizon Scanner

lpi: Lines per Inch

LPI: Latent Photographic Image • Launching Position Indicator • Linköping (Sweden) (location identifier) • Lithium Pellet Injector • Local Position Invariance • Lotus Programming Language • Low-Power Illuminator • Low-Pressure Impactor • Low Probability of Intercept • Lunar and Planetary Institute (USA-TX)

LPIB: Laser Pulse Induced Breakdown • Lunar and Planetary Information Bulletin (*) • Lunar and Planetary Institute Bulletin

LPID: Logical Page Identifier

LPIF: Launch Package Integration Facility

LPJ: Pijiguaos (Venezuela) (location identifier)

LPK: Lipetsk (Russia) (location identifier)

LPL: Laser-Pumped-Laser • Linear Programming Language • List Processing Language • List Programming Language • Liverpool (UK-England) (location identifier) • Long Pulse Laser • Lotus Programming Language • Low Polar Latitude • Lunar and Planetary Laboratory • Lunar Personnel Lander • Lunar Projects Laboratory

LPLC: Low-Pressure Liquid Chromatography

LPLWS: Launch Pad Lightning Warning System

lpm: Lines per Millimeter • Lines per Minute • Liters Per Minute

LPM: Laboratorie de Physique Mathématique • Lamap (Vanuatu) (location identifier) • Landsat Program Management • Linkable Protection Module • Load Point Marker • Logical Process Manager • Lunar Payload Module • Lunar Portable Magnetometer • Luyten Proper Motion

LPMA: Laboratoire de Physique Moléculaire Atmosphérique • Laboratoire de Physique Moléculaire et d'Applications • Limb Profile Monitor of the Atmosphere

LpMin: Lines per Minute

LPML: Low Protein Meal

lpmm: Lines per Millimeter

LPN: Leron Plains (Papua New Guinea) (location identifier) • Logical Page Number • Low-Pass Network

LPNHE: Laboratoire de Physique Nucléaire des Hautes Énergies (France)

LPO: La Palma Observatory (Canary Islands) • Laporte (USA-IN) (location identifier) • Low-Power Output • Lunar Parking Orbit • Lunar Polar Orbiter

L'pool: Liverpool (UK-England)

LPOP: Low-Pressure Oxidizer Pump

LPOT: Low-Pressure Oxidizer Turbopump

LPOTP: Low-Pressure Oxidizer Turbopump

LPOX: Low-Pressure Oxygen

LPP: Lappeenranta (Finland) (location identifier) • Lean Premixed Prevaporized • Licensed Program Product • Lignes Par Pouce • Long Periodic Perturbation • Lunar Precepts Positioner

LPPM: Laboratoire de Photophysique Moléculaire

LPQ: Low-Polarization Quasar • Luang Prabang (Laos) (location identifier)

LPR: Large Pulkovo Radiotelescope • Least-Positive Residue • Light Pollution Reduction • Light Pollution Rejection • Line Printer • Liquid Propellant Rocket

LPRB: Loaded Program Request Block

LPRC: Launch Pitch Rate Control

LPRD: Light Pollution Reduction Device

LPRL: Lockheed-Martin Palo Alto Research Laboratory (USA-CA)

LPRT: Low-Power Radio Transmitter

lps: Liters per Second

LPS: Large Poincaré System • Large Pointing System • Laser Particulate Spectrometer • Laser Power Supply • Launch Phase Simulator • Launch Processing System • Linear Programming System • Liquid-Phase Sintering • Lithuanian Physical Society • Loop Prominence System • Loops • Lopez Island (USA-WA) (location identifier) • Low-Power Schottky • Low-Power Sodium • Low-Pressure Sodium • Lunar Penetrometer System • Lunar Pilotage System

LPSA: Lunar Periodic Scattering Array

LPSC: Liquid Propulsion System Centre • Lunar and Planetary Science Conference

LPSH: Laboratoire de Physique Solaire et de l'Héliosphère

LPSIRS: Linear Potential-Sweep Infrared Reflectance Spectroscopy

LPSP: Laboratoire de Physique Stellaire et Planétaire • La Palma Stokes Polarimeter

LPT: Lampang (Thailand) (location identifier) • Laplacian Pyramid Transform • Lewthwaite-Pickover-Truchet • Line Printer • Local Printer Terminal • Long-Term Prediction • Low-Power Test • Low-Power Transmitter • Low-Pressure Test • Low-Pressure Turbine • Lunar Propellant Tanker

LPTF: Laboratoire Primaire du Temps et des Fréquences (France)

LPTHE: Laboratoire de Physique Théorique et des Hautes Énergies

LPTL: Laboratoire de Physique Théorique des Liquides (France)

LPTO: Long-Term Programme Office

LPTTL: Low-Pressure Transistor-Transistor Logic

LPTV: Low-Power Television

LPT1: First Parallel Printer Port

LPT2: Second Parallel Printer Port

LPT3: Third Parallel Printer Port

LPU: Long Apung (Indonesia) (location identifier) • Low Pay Unit

LPUU: Linear Programming Under Uncertainty

LPV: Launching Point Vertical • Line Profile Variability • Line Profile Variable • Line Profile Variation • Long-Period Variable • Lunar Piloted Vehicle

LPVAD: Preston (Starlink internal code)

lpW: Lumens per Watt

LPW: Linear Polarized Wave • Little Port Walter (USA-AK) (location identifier)

LPWA: Lucent Personalized Web Assistant

LPX: Liepaya (Latvia) (location identifier)

LPY: Le Puy (France) (location identifier)

Lpz: Leipzig (Germany)

lq: Lege Quaeso

LQ: Argentina (international aircraft registration code) • Laser Quality • Lege Quaeso • Leptoquark • Letter Quality • Life Quality • Linear Quadratic • Lower Quartile

LQG: Large Quantity Generator • Linear Quadratic Gaussian

LQK: Pickens (USA-SC) (location identifier)

LQM: Link Quality Monitoring

LQN: Qala-Nau (Afghanistan) (location identifier)

LQP: Letter-Quality Printer

LQR: Linear Quadratic Regulator • Link Quality Report

LQS: Local Query Server

LQSS: Loral Qualcomm Satellite Services Inc.

lr: Laar

lR: laufende Rechnung

Lr: Lawrencium • Lower

L/R: Locus of Radius

LR: Labelled Release • Laboratory Report • Laboratory Request • Landing Radar • Landing Report • Language Resource • Lapse Rate • La Rioja (Argentina) • Laser Reflectometer • Latent Root • Launch Right • Lead Radial • Left Rear • Left Rudder • Letter Report • Letter Requirement • Lettre Recommandée • Liaison Report • Liaison Request • Liberia (ISO country code) • Library Representation • Ligne Réseau • Linear Representation • Line Receiver • Liquid Rocket • Living Room • Load Register • Logistics Regression • Log Run • Long Range • Long Run • Loop Resistance • Lorentz Resonance • Loss of Retranslation • Lower Right • Lucy-Richardson • Luft und Raumfahrt (periodical code)

LRA: Large Reflector Assembly • Larisa (Greece) (location identifier) • Laser Reflector Array • Laser Retroreflector Array • Local Register Allocation

LRAP: Long Route Analysis Program

LRB: Laboratoire René Bernas (France) • Leribe (Lesotho) (location identifier) • Level Reference Base • Liquid Rocket Booster

LRBA: Laboratoire de Recherches Ballistiques et Aérodynamiques (France)

LRC: Laarbruch (Germany) (location identifier) • Langley Research Center (NASA) • Learning Resources Center • Lewis Research Center (NASA) • Library Research Committee • Library Resource Center • Local Register Cache • Longitudinal Redundancy Character • Longitudinal Redundancy Check • Longitudinal Redundancy Checking • Luneberg Rapid Commutator

LRCA: Long-Range Combat Aircraft

LRCCP: Laboratoire de Recherche et de Contrôle du Caoutchouc et des Plastiques (France)

LRCO: Limited Remote Communications Outlet

LRD: Dollar (Liberia) (ISO currency code) • Laredo (USA-TX) (location identifier) • Launch Readiness Date • Launch Recovery Division • Launch Requirements Document • Liberian Dollar (ISO currency code) • Lightning and Radioemission Detector • Long-Range Detection • Low-Resolution Data

LRDA: Land Resources Data Analysis

LRDB: Low-Resolution Database

LRDC: Land Resources Development Centre

LRDG: :earning Resource Development Group

LRDP: Long-Range Development Plan

LRDTF: Low-Rate Data Transcription Facility

LRE: Language Research and Engineering • Latest Revised Estimate • Linguistic Research and Engineering • Linguistic Research Engineering • Liquid Rocket Engine • Longreach (Australia-QLD) (location identifier) • Lunar Retrograde Engine

LREC: Language Resources and Evaluation Conference

LRECL: Logical Record Length

LREE: Light Rare-Earth Element

LRES: Liquid Rocket Engine Stability

LRF: Jacksonville - Little Rock (USA-AR) (location identifier) • Large-scale Research Facility • Linear Ramp Filter • Liquid Rocket Fuel • Local Rest Frame • Long-Range Fuel

Lrg: Large

LRG: Library Research Group • Lora Lai (Pakistan) (location identifier)

LRH: La Rochelle (France) (location identifier)

LRI: Least Recently Loaded • Low-Resolution Image • Low-Resolution Imager • Low-Resolution Imaging • Lorica (Colombia) (location identifier)

LRIC: Long Run Incremental Cost

LRIDP: Land Resource Inventory Demonstration Project

LRIP: Low-Rate Initial Production

LRIR: Limb Radiance Inversion Radiometer • Low-Resolution Infrared Radiometer

LRIS: Land Resources Information System • Low-Resolution and Imaging Spectrograph • Low-Resolution Imaging Spectrograph (*) • Low-Resolution Imaging Spectrometer

LRIT: Low-Rate Image Transmission

LRITF: Land Resources Inventory Task Force

LRJ: Lemars (USA-IA) (location identifier)

LRK: Lincoln Rock (USA-AK) (location identifier)

LRL: Lama Kara (Togo) (location identifier) • Lawrence Radiation Laboratory • Learning Resource Laboratory • Least Recently Loaded • Lectronic Research Laboratories (USA-NJ) • Lunar Receiving Laboratory • Niamtougou (Togo) (location identifier)

LRLS: Long Range Lineup System

LRM: Casa de Campo (Dominican Rep.) (location identifier) • Laboratório Regional de Metrologia • Language Reference Manual • La Romana (Dominican Rep.) (location identifier) • Laser Raman Microprobe • Likelihood Ratio Method • Linear Referencing Method • Line-Reflect-Match • Line-Replaceable Module • Liquid Rocket Motor • Long Radius Meridian • Lunar Reconnaissance Mission • Lunar Reconnaissance Module

LRN: Local Resource Name • Location Routing Number • Moses Lake - Larson (USA-WA) (location identifier)

LRNAV: Long-Range Navigation

LRNR: Low-Resolution Non-scanning Radiometer

LRO: Large Radio Observatory • Lathrop Sharpe (USA-CA) (location identifier) • Local and Remote Operations

LRONIS: Low-Resolution Optical/Near-Infrared Spectrophotometer

LROR: Low-Resolution Omnidirectional Radiometer

LROSS: Long-Range Optical System Study

LRP: Lander Robotic Payload • Launching Reference Point • Long-Range Plan • Long-Range Planning • Low-Resolution Profile

LRPA: Liferaft Portside Aft • Long-Range Patrol Aircraft

LRPF: Liferaft Portside Forward

LRPM: Liferaft Portside Middeck

LRPP: Long-Range Planning Panel

LRPS: Long-Range Precision Strike

LRPT: Low-Rate Picture Transmission • Low-Resolution Picture Transmission

LRQ: Laurie River (Canada-MB) (location identifier)

LRR: Laser Retro-Reflector • Launch Readiness Report • Launch Readiness Review • Long-Range Radar

LRRR: Laser Ranging Retro-Reflectometer

LRRS: Library and Learning Resource Service

LRS: Laboratoire de Recherches Spatiales (Observatoire de Paris-Meudon) • Lander Radio System • Language Resource • Language Resource System • Laser Raman Spectroscopy • Laser Ranging Station • Leros (Greece) (location identifier) • Linear Referencing System • Line Repeater Station • Locally Rotationally Symmetric • Low-Rate Science • Low-Rate Switch • Low-Resolution Spectrograph • Low-Resolution Spectrometer • Low-Resolution Spectrum

LRSA: Land Remote Sensing Assembly • Liferaft Starboard Aft

LRSF: Liferaft Starboard Forward

LRSI: Low-temperature Reusable Surface Insulation

LRSM: Liferaft Starboard Middeck

LRSN: Log Record Sequence Number

LRSP: Low-Rate Science Processor

LRT: Launch, Recovery and Transport • Lehrstuhl für Raumfahrtechnik • Light Rail Train • Light Rail Transport • Local Rapid Transport • Loi de Réglementation des Télécommunications (France) • London Regional Transport (UK-England) • Long-Range Transport • Lorient Lann-Bihoué (France) (location identifier)

LRTAP: Long-Range Transport of Air Pollutants

LRTF: Linear Radial Transmission Filter

LRU: Las Cruces (USA-NM) (location identifier) • Last Register Used • Least Recently Used • Ley de Reforma Universitaria (Spain) • Linear Replaceable Unit • Line Replaceable Unit • Line Replacement Unit

LRV: Launch Readiness Verification • Lifting Reentry Vehicle • Long-Range Video • Los Roques (Venezuela) (location identifier) • Lunar Rover Vehicle • Lunar Roving Vehicle

LRW: Left Rendezvous Window

LRZ: Leibniz-Rechenzentrum (Germany)

ls: Locus Sigilli

L/S: Landing Site • Load System

L&S: Land & Sea • Lange & Springer • Launch & Steering • Logistics & Support

LS: Land and Sea • Landing Site • Launch and Streering • Launching System • Launch Sequence • Launch Service • Launch Simulator • Launch Site • Launch Station • Least Squares • Lectori Salutem • Leeds (UK postal code) • Left Side • Legal Seal • Lesotho (international vehicle registration code) • Lesotho (ISO country code) • Letter Service • Letter Signed • Lettre Signée • Liaison Scientist • Liaison Spécialisée • Library Science • Library Search • Licensed Surveyor • Licensing Server • Lichtsäule • Life Science • Life Support • Life System • Lighting System • Light Shield • Light Source • Ligne Spécialisée • Limit Switch • Lindsey-Smith • List Status • Livret Suiveur • Load System • Local Source • Local Store • Local Sunset • Local Symmetry • Locus Sigilli • Logistics and Support • Long Scarf • Long Sight • Loop System • Loudspeaker • Low Session • Low Speed • Low State • Luminous Stars • Lump Sum • Lunar Surface • Syrian Pound (currency)

LSA: Labor Surplus Area • Large Scale Area • Large-scale Sounding Array • Large Science Aperture • Large Southern Array • Large Space Antenna • Large Spherical Array • Laser with Saturable Absorber • Latent Semantic Analysis • Lateral Spherical Aberration • Launch Services Agreement • Library Services Act (USA) • Line Sharing Adapter • Linguistic Society of America • Local Security Authority • Logistics Support Analysis • Logistics Support Area • Longitudinal Spherical Aberration • Losuia (Papua New Guinea) (location identifier)

LSAA: Library Services Authority Act (USA)

LSACN: Logistics Support Analysis Control Number

LSAH: Launch Site Accommodation Handbook

LSALT: Lowest Safe Altitude

LSAN: LWS Automatic Analysis Result

LSAP: Launch Sequence Applications Program • Linear Systems Analysis Program • Link Service Access Point • Logistics Support Analysis Plan • Low Sun Angle Photography

LSAPI: License Service Applications Programme Interface

LSAPT: Lunar Sample Analysis Planning Team

LSAR: Logistics Support Analysis Report

LSAS: Logistics Support Analysis Strategy

LSAT: Large Scale Applications Test

LSB: Launcher Support Building • Launch Service Building • Launch Support Building • Least Significant Bit • Least Significant Byte • Line Segment Block • Linux Standard Base • Logistics Support Base • Lordsburg (USA-NM) (location identifier) • Lower Sideband • Low Surface Brightness • Lunar Surface Base

LSBD: Low Surface Brightness Dwarf

LSBG: Low Surface Brightness Galaxy

LSBY: Least Significant Byte

lsc: Loco Supra Citato

LSc: Licenciate in Sciences

LSC: Land Surface Climatology • Large-Scale Computer • La Serena (Chile) (location identifier) • Laser Spectral Control • Launch Sequence Control • Launch Services Contract • Launch Services Contractor • Launch Services Corporation • Launch Support Contractor • Launch Support Controller • Least Significant Character • Least-Square Circle • Least-Squares Collocation • Lens Sign Convention • Liberty Science Center (USA-NJ) • Light-speed c • Ligne Supérieure de Contrôle • Linear Sequential Circuit • Linear-Shaped Charge • Line Scan Camera • Liquid Scintillation Counter • Liquid Scintillation Counting • Liquid-Solid Chromatography • LISA Science Centre • Local Staff Committee • Loco Supra Citato • Louisville Science Center (USA-KY)

LSCA: Library Services Construction Act • Logistics Support Cost Analysis

LSCE: Launch Sequence and Control Equipment

LSCF: Least Squares Curve Fitting

LSCM: Laser Scan Confocal Microscope • Laser Scan Confocal Microscopy • Least-Squares Collocation Method

LSCO: Lanthanum Strontium Copper Oxide

LSCP: Low-Speed Card Punch

LSCT: Laser Spectral Control Technique

lsd: Librae Solidi Denarii

LSd: Sudanese Pound (currency)

LSD: Lake Shore Drive • Landing Ship Dock • Landing Site Determination • Land Surface Datum • Language for Systems Development • Large-Scale Domain • Large Screen Display • Launch Support Division • Launch Systems Data • Least Significant Difference • Least Significant Digit (*) • Least-Squares Deconvolution • Lexington Creech (USA-KY) (location identifier) • Librae-Solidi-Denarii • Local Spin Density • Logarithmic Series Distribution • Low-Speed Data • Lunar Surface Drill • Pounds-Shillings-Pence

LSDA: Life Sciences Data Archive

LSDS: Large-Screen Display System • Less Structured Dataset • Low-Speed Digital System

LSE: Laboratory Support Equipment • La Crosse (USA-WI) (location identifier) • Language Sensitive Editor • Launch Sequencer Equipment • Launch Support Equipment • Learning Support Equipment • Life Support Equipment • Limited Signed Edition • Liquid-Solid Extraction • Liquid-Solvent Extraction • Local Stellar Environment • London School of Economics (UK-England) • London Stock Exchange (UK-England) • Luminous Stars Extension • Lunar Surface Experiment

LSEAT: Launch Systems Evaluation Advisory Team

LSEC: Logistical Support Element Communications

LSEET: Laboratoire de Sondages Électromagnétiques de l'Environnement Terrestre

LSEP: Lunar Surface Experiment Package

LSEV: Lunar Surface Exploration Vehicle

LSF: Columbus Lawson (USA-GA) (location identifier) • Fort Benning - Lawson (USA-GA) (location identifier) • Laboratory Simulation Facility • Laser Support Facility • Launch Support Facility • Least Squares Filter • Least Squares Fit • Linear Swept Frequency • Line Spectral-pair Frequency • Line Spread Function • Liquid Structure Facility • Log-Structured File • Lunar Scientific Facility

LSFE: Life Sciences Flight Experiment

LSFR: Launch Site Flow Review

LSG: Liaison Scientists Group • Linear Sequence Generator • Local Sponsor Group • Local Studies Group • Lunar Surface Gravimeter

LSH: Laboratoire de Spectroscopie Hertzienne • Lashio (Myanmar) (location identifier)

LSHTIPMP: Laughing So Hard That I Pee My Pants

LSI: Laboratoire de Systèmes Intégrés • Laboratório de Sistemas Integráveis • Laboratory for Scientific Interrogation (USA-AZ) • Large-Scale Integrated • Large-Scale Integration • Latent Semantic Indexing • Launch Success Indicator • Lead System Integrator • Life Space Interviewing • Linear Shift Invariant • Liquid Silicon Infiltration • Lunar Surface Instrument • Shetland Islands - Sumburgh (UK) (location identifier)

LSIC: Large-Scale Integrated Circuit • Launch Support Integration Contractor • Launch System Integration Contractor

LSID: Launch Sequence and Interlock Document • Lunar Sequence and Interlock Document

LSIEP: Least Squares Inverse Eigenvalue Problem

LSIIT: Laboratoire des Sciences de l'Image de l'Informatique et de la Télédétection

LSIN: Logistics System Information Network

LSJ: Long Island (Papua New Guinea) (location identifier)

LSK: Lochstreifenkarte • Lusk (USA-WY) (location identifier)

LSKBS: Link-Structured Knowledge-Based System

LSL: Landing Ship Logistic • Larch Shared Language • Link Support Layer • Load Segment Limit • Los Chiles (Costa Rica) (location identifier) • Loti (Lesotho) (ISO currency code) • Low-Speed Logic

LSLE: Life Sciences Laboratory Equipment

lsm: Litera Scripta Manet

LSM: Laboratoire Souterrain de Modane (France) • Large Solid Motor • Laser Scan Microscope • Laser Scan Microscopy • Laser Scanning Microscope • Laser Scanning Microscopy • Launcher Status Multiplexer • Launching System Module • Layered Synthetic Microstructure • Least Squares Mean • Least Squares Method • Life Science Module • Linear Synchronous Motor • Link Services Manager • Linux Software Map • Local State Manager • Logistics Support Manager • Longitudinal Solar Magnetograph • Long Semado (Malaysia) (location identifier) • Lunar Surface Magnetometer

LSMC: Life Sciences Master Catalog

LSMDB: Logistics Simulation Models and Databases

LSMF: Large-Scale Magnetic Field

LSMI: Logistics Support Management Information

LSML: Low Salt Meal

LSMQ: Local Services Management Queue

LSMS: LNP Service Management System

LSN: Large-Scale Networking • Linear Scaling Number • Log Sequence Number • Los Banos (USA-CA) (location identifier) • Luminous Stars Northern

LSND: Liquid Scintillator Neutrino Detector

LSO: Lager Secundair Onderwijs • Landing Support Officer • Large Solar Observatory • Large Solar Orbiter • Launch Safety Officer • Launch Support Operations • Lesotho (ISO country code) • Les Sables - Talmont (France) (location identifier) • Life Systems Officer

LSOC: Launch Support Operations Contractor

LSOCE: Linear Stochastic Optimal Control and Estimation

LSOP: Lunar Surface Operations Planning

LSP: Laboratoire de Spectrométrie Physique (France) • Large-Scale Phenomenon • Large Space Photography • Las Piedras (Venezuela) (location identifier) • Launcher Status Panel • Least Significant Portion • Least-Squares Polynomial • Libray Software Package • Lightest Supersymmetric Particle (*) • Lightest Supersymmetric Partner • Light Scattering Photometer • Lightspeed Pascal • Linear-Spectrum Pair • Linguistic String Processor • Link State Packet • Logical Support Program • Logistics Support Plan • Low-Speed Printer • Lunar Solar Power • Lunar Spectral Photometrics • Lunar Surface Probe

LSPC: Local Serial Port Controller • Long-term Space Policy Committee • Lunar Solar Power Coalition

LSPDF: Life Science Payload Development Facility

LSPE: Lunar Seismic Profiling Experiment

LSPET: Lunar Sample Preliminary Examination Team

LSPL: Local Services Priority List

LSPO: Large-Scale Parameter Optimization • Lunar Surface Project Office

LSPPS: Logistics Support Plan for Preoperational Support

LSPTP: Low-Speed Paper Tape Punch

LSQ: Line Squall • Los Angeles (Chile) (location identifier)

LSQA: Local System Queue Area

LSR: Label Switching Router • Land-Sea Rescue • Lanthanide Shift Reagent • Laser Scanning Radar • Launch Signal Responder • Launch Site Recovery • Launch Support Request • Launch Support Requirement • Locally Shared Resource • Local Shared Resource • Local Space Rectangular • Local Standard of Rest • Local Sunrise • Logiciel, Systèmes et Réseaux • Logistics Shop Release • Logistics Success Rate • Logistics Support Requirement • Lost River (USA-AK) (location identifier) • Low Storage Requirement

LSRA: Landing Systems Research Aircraft

LSRM: Life Science Research Module

LSRO: Life Sciences Research Office

LSRP: Laser Printer

LSRPS: Line-Scan Receiving and Processing System

LSRR: Loose Source and Record Route

LSRV: Lunar Surface Rover Vehicle • Lunar Surface Roving Vehicle

LSS: Landing and Separation Simulator • Landing Sensor System • Langley Software Server (NASA) • Large-Scale Structure • Large-Scale Survey • Large-Scale System • Large Space Simulator • Large Space Structure • Large Space System • Last Scattering Surface • Launch Sequency Simulator • Launch Site Support • Launch Status Summarizer • Launch Support Section • Launch Support Services • Launch Support Structure • Launch Support System • Licensing Support System • Life-Saving Station • Life-Support System • Link Server System • Literature Search Service • Load Sensing System • Local Speed of Sound • Logistics Support Services • Long-Slit Spectrography • Long-Slit Spectroscopy • Loop Switching System • Low-Speed Stream • Luminous Stars Southern • Lunar Surveying System • Terre-de-Haut (Guadeloupe) (location identifier)

LSSC: Large Space Simulation Chamber • Launch Site Support Contractor

LSSD: Level-Sensitive Scan Design • Level-Sensitive Scan Detection • Lunar Surface Sampling Device

LSSE: Laser Sensor Support Equipment

LSSF: Life Sciences Support Facility

LSSG: Lunar Study Steering Group

LSSL: Life Sciences Space Laboratory

LSSM: Launch Site Support Manager

LSSP: Launch Site Support Plan • Lunar Surveying System Program

LSST: Large-Scale Seismic Test • Large Space System Technology • Launch Site Support Team

LSSW: Launch Site Software

Lst: Ladestelle

LST: Lage-Snelheids Tunnel • Land Surface Temperature • Laplace-Stieltjes Transform • Large Simple Trial • Large Space Telescope • Large Stellar Telescope • Laser Scattering Tomograph • Launceston (Australia-TAS) (location identifier) • Launch Support Team • Light Scattering Temperature • Limited Streamer Tube • Liquid Storage Tank • List • Lithuanian Standardization Office • Local Sidereal Time • Local Solar Time • Local Standard Time • Lunar Surface Telescope • Lunar Surface Transponder

L*STAR: Lockheed-Martin Satellite Telemetry Analysis in Real Time

LSTD: Laser System Technology Demonstration • Lunar Satellite Tracking Data

LStDV: Lohnsteuer-Durchführungsverordnung

LSTE: Launch Site Transportation Equipment

LSTH: Liu-Siebahn-Truhlow-Horowitz

LSTM: Lander Static Test Model

LStR: Lohnsteuer-Richtlinien

LSTS: Launch Site Test Set • Lunar Surface Thermal Simulator

LSTSRFA: Launch Station Test Set Radio Frequency Adapter

LSTWA: List of Serial Title Word Abbreviations

LSU: Launcher Selector Unit • Library Storage Unit • Life Support Umbilical • Life Support Unit • Lochschriftübersetzer • Long Sukang (Malaysia) (location identifier) • Louisiana State University (USA-LA)

LSUR: Launch Site Utilization Report

LSV: Landelijk Samenwerkende Volkssterrenwachten (Netherlands) • Las Vegas - Nellis (USA-NV) (location identifier) • Letztschichtversagen • Line Status Verifier • Low-Speed Vehicle • Lunar Shuttle Vehicle • Lunar Surface Vehicle

LSVE: Large-Scale Virtual Environment

LSVI: Large-Size Visual Interface

LSW: Landessternwarte • Lhok Seumawe (Indonesia) (location identifier)

LSWG: Life Science Working Group

LSVT: Large Solar Vacuum Telescope

LSX: Lhok Sukon (Indonesia) (location identifier)

LSY: Lismore (Australia-NSW) (location identifier)

LSZ: Limited-Speed Zone • Low-Speed Zone • Mali Losinj (Croatia) (location identifier)

lt: Lithuania • Lithuanian (ISO language code)

Lt: Laut • Lieutenant • Light • Luitenant

L&T: Laboratory & Test

L/T: Load Test

LT: Laboratory and Test • Language Translation • Laplace Transform • Lead Time • Left • Left Turn • Less Than • Light • Linear Triangulation • Line Termination • List Translation • Lithuania (international vehicle registration code) • Lithuania (ISO country code) • Lithuanie (ISO country code) • Liverpool Telescope (JMU) • Live Theatre • Local Time • Locum Tenens • Logic Theory • Logic Tree • London Transport (UK-England) • Long Ton • Lorentz Transformation • Lote • Lot Traceability • Low Temperature • Low Tension • Turkish Lira (currency)

LTA: Lead Time Analysis • Library Technical Assistant • Lighter Than Air • Lower Torso Assembly • Low-Temperature Adsorber • Tzaneen (South Africa) (location identifier)

LTAB: Line Test Access Bus

LTAS: Laboratoire des Techniques Aéronautiques et Spatiales

LTB: Latrobe - Westmoreland County (Australia-TAS) (location identifier) • Line Term Buffer • Low-Tension Battery

LTC: Lai (Chad) (location identifier) • Language Translation Computer • Library Technology Center • Light Transfer Characteristics • Live Time Correction • Local Test Cabinet • Local Time Clock • Longitudinal Time Code • Long Time Coming • Low-Temperature Carbonization • Lunar Terrain Camera • Lunar Topographic Camera

LTCC: Low-Temperature Cofired Ceramics

LTCM: Long-Term Capital Management

LTCS: Lower Thermosphere Coupling Study • Lower Thermospheric Coupling Study

Ltd: Limited

LTD: Ghadames (Libya) (location identifier) • LaTeX Tree Drawer • Launch Test Directive • Lift-To-Drag • Limited • Local Test Desk • Long-Term Depression • Long-Term Disability • Low-Temperature Detector

Ltda: Limitada

LTDS: Launch Tracking Data System • Launch Trajectory Data System • Linear Time Delay System

LTDT: Langley Transonic Dynamics Tunnel (NASA)

LTE: Large Table Electroplotter • Less Than or Equal • Letter To the Editor • Limited Test Equipment • Linear Threshold Element • Local Thermal Equilibrium • Local Thermodynamic Equilibrium (*) • Long-Term Effect • Low-Thrust Engine

Ltée: Limitée

LTER: Long-Term Ecological Research

LTerm: La Termotécnica (periodical code)

LTF: Leitre (Papua New Guinea) (location identifier) • Lightwave Terminating Frame • Line Trunk Frame

LTFD: Logic and Test Function Drawer

Ltg: Lettering • Lighting

LTG: Langtang (Nepal) (location identifier) • Lightning • Linear Tangent Guidance • Line Trunk Group • Low-Temperature Growth • Lunar Traverse Gravimeter

LTH: Lathrop Wells (USA-NV) (location identifier) • Lethality and Target Hardening

LTHG: Laboratoire d'Étude des Transferts en Hydrologie et Environnement (France)

LTHIS: Laboratory Test Hardware Inventory Status

LtHo: Light House

Lthr: Leather

LtHse: Light House

LTI: Altai (Mongolia) (location identifier) • Linear Time-Invariant

LTIRF: Laboratoire de Traitement d'Images et Reconnaissance de Formes (France)

LTIV: Lunar Trajectory Injection Vehicle

LTK: Latakia (Syria) (location identifier)

LTL: Lastourville (Gabon) (location identifier) • Line-To-Line • Lit (Lithuania) (ISO currency code) • Little • Long-term Timeline

LTLCG: Little Change

LTM: Lead Time Matrix • Lethem (Guyana) (location identifier) • Life Test Model • Linear Tangent Model • Load Transformation Matrix • Long-Term Memory • Long-Term Monitoring

LTMH: Largeur Totale à Mi-Hauteur

LTMS: Local Time Management System • Lunar Terrain Measuring System

LTN: London Luton (UK-England) (location identifier)

LTNG: Lightning

LTNR: Local Thermonuclear Runaway

LTO: Loreto (Mexico) (location identifier) • Low-Temperature Orthorhombic • Lunar Transfer Orbit

LTOP: Long-Term Operation Plan

LTP: Laboratory for Terrestrial Physics • Learning Technologies Project • Long-Term Planning • Long-Term Potentiation • Low-Temperature Phosphorescence • Low-Temperature Phosphorimetry • Low-Temperature Physics • Low-Temperature Platform • Lunar Transient Phenomenon • Lyndhurst (Australia-QLD) (location identifier)

LTPG: Long-Term-Planning Group

LTPMS: Lidar Temperature/Pressure/Moisture Sounder

LTPO: Long-Term Program Office

LTPS: Low-Temperature Polysilicon

LTPV: Long-Term Photometric Variables • Long-Term Photometry of Variables

LTQ: Le Touquet (France) (location identifier)

Ltr: Letter • Lighter

LTR: Laboratory Test Report • Lander Trajectory Reconstruction • Later • Left To Right • Letter • Letterkenny (Ireland) (location identifier) • Load Task Register • Long-Term Relationship • Long-Term Research • Loop Transfer Recovery • Low-Temperature Region

LTREB: Long-Term Research in Environmental Biology

LTRF: LLR Terrestrial Reference France

LTRMP: Long-Term Resource Monitoring Program

LTRR: Laboratory for Tree-Ring Research

LTRRS: Long-Term Regional Research Site

LTRS: Langley Technical Report Server (NASA)

Lts: Lights

LTS: Altus (USA-OK) (location identifier) • Labelled Transition System • Lagere Technische School • Language Translation System • Large Truss Structure • Launch Telemetry Station • Launch Telemetry System • Launch Tracking Station • Launch Tracking System • Lexicon/Thesaurus System • Library Technical Service • Light Tactical Satellite • Linearity Test Set • Long-Term Schedule • Long-Term Scheduler • Long-Term Scheduling • Long-Term Storage • Loss Test Set • Low-Temperature Solidification • Low-Temperature Superconductor

LTSA: Long-Term Space Astrophysics

LTSAB: Long-Term Stability

LTSEM: Low Temperature Scanning Electron Microscope

LTSG: Laser-Triggered Spark Gap

LTSP: Long-Term Space Plan

LTSPC: Long-Term Space Policy Committee

LTT: Language Teacher and Tester • Learning-Teaching-Technology • Low-Temperature Test • Lunar Test Table • Lunar Transfer Trajectory • Lunar Transit Telescope • Luyten Two Tenths (catalogue) • Saint-Tropez (France) (location identifier)

LTTL: Long-Term Time Line

LTTM: Long-Term Terrain Model

LTTR: Long-Term Technology Review

LTU: Linear Threshold Unit • Line Terminating Unit • Line Termination Unit • Line Test Unit • Lithuania (ISO country code) • Lithuanie (ISO country code) • Lufttransport-Unternehmen GmbH & Co. KG • Luleå Tekniska Universitet (Sweden)

LTV: Launch Test Vehicle • Logistics Transfer Vehicle • Lotusvale (Australia-QLD) (location identifier) • Lunar Transfer Vehicle

LTVAD: Leicester (Starlink internal code)

LTVC: Linear Terminal Velocity Constraint

LTVGF: Low-Temperature Vapour Growth Facility

LTW: Leonardtown - Saint Marys County (USA-MD) (location identifier) • Localized Travelling Wave

LTWG: Landsat Technical Working Group • Large Telescope Working Group • Launch Test Working Group

LTWTA: Linearized Travelling-Wave Tube Amplifier

LTXDB: Leicester (Starlink internal code)

lt-yr: Light-Year

lu: Luxembourg

Lu: Lucerne (Switzerland) • Luglio • Luglio • Lutetium • Lutétium • Luzern (Switzerland)

L/U: Land Use

LU: Land Unit • Leicester University (UK-England) • Litterae Uraniae • Logical Unit • Logistical Unit • Loudness Unit • Luminosity Uniformity • Luton (UK postal code) • Luxembourg (ISO country code) • Luxemburg (ISO country code)

LUA: Logical Unit Application • Lukla (Nepal) (location identifier)

lub: Least Upper Bound

Lub: Lubricant • Lubricate • Lubricating • Lubrication

LUB: Least Upper Bound • Logical Unit Block • Lubricate • Lumid Pau (Guyana) (location identifier) • Lunds Universitetsbibliotek (Sweden)

LubEn: Lubrication Engineering (periodical code)

Lubt: Lubricant

LUC: Convertible Luxemburg Franc (ISO currency code) • Laucala Island (Fiji) (location identifier) • Loyola University Chicago (USA-IL)

LUCC: Land Use and Cover Change

LUCI: Leading Us in Commerce Industry

LUCID: Language for Utility Checkout and Instrumentation Development

LUCS: Library Updating and Compiling System

LUD: Lower-Upper Decomposition • Luderitz (Namibia) (location identifier)

LUDA: Land Use Data Analysis • Land Use Development Assistance

LUE: Linear Unbiased Estimator • Lucenec (Slovakia) (location identifier)

LUEMA: Land Use and Environmental Management Authority

LUF: Franc (Luxemburg) (ISO currency code) • Lowest Usable Frequency • Phoenix Luke (USA-AZ) (location identifier)

Lug: Luggage • Luglio

LUG: Linux Users Group • Local Users Group • Lugano (Switzerland) (location identifier) • Lugar

LUGAM: Lunar Gamma Spectrometer

LUH: Ludhiana (India) (location identifier)

LUHG: Lowest Useful High Frequency

LUI: La Unión (Honduras) (location identifier) • Load Upper Immediate • Local User Input

LUIS: Landes-Umwelt-Informations-System

LUJ: Lusikisiki (South Africa) (location identifier)

LUK: Cincinnati (USA-OH) (location identifier)

LUKAS: Lowell-UK Asteroid Survey

LUL: Financial Luxemburg Franc (ISO currency code) • Laurel (USA-MS) (location identifier)

LULI: Laboratoire d'Utilisation des Lasers Intenses (France)

Lum: Luminosity • Luminous

LuM: Luchtmacht

LUM: Living Utility Module • Luxi Mangshi (China-PRC) (location identifier) • Lynx - Ursa Major

LUMAS: Lunar Mapping System

LUMI: Lunar Microwave Instrument

LUMIS: Land Use Management Information System

LUMO: Lowest Unoccupied Molecular Orbital

LUMS: Land Use Management System

Lun: Lundi • Lunedi • Lunes

LUN: Logical Unit Number • Lusaka (Zambia) (location identifier)

LUNI: LAN Emulation User Network Interface

LUNIS: Loyola University Nuclear Information System

LUNO: Loyola University New Orleans (USA-LA)

LUNOX: Lunar Oxygen

LUNR: Land Use and Natural Resources

LUO: Luena (Angola) (location identifier)

Lup: Lupus-Lupi (constellation)

LUP: Kalaupapa (USA-HI) (location identifier)

LUPF: Linear Utility Prediction Function

LUPI: Laser Unequal-Path Interferometer

LUPIN: Land Use Planning Information Network

LUPIS: Land Use Planning Information System

LUPWT: Langley Unitary Plan Wind Tunnel (NASA)

LUQ: San Luis (Argentina-SL) (location identifier)

LUR: Cape Lisburne (USA-AK) (location identifier)

LuRaQ: Luft und Raumfahrt Quarter (periodical code)

LURE: Laboratoire pour l'Utilisation du Rayonnement Électromagnétique (France) • Lunar Ranging Experiment

LUS: Liquid Upper Stage • Lusanga (Zaire) (location identifier)

LUSE: Land Use Suitability Evaluation

LUSEX: Lunar Surface Explorer

LUSI: Lunar Surface Inspection

LUSIS: Land Use and Suitability Information System

LUSS: Lunar Subsatellite

LUST: Leaking Underground Storage Tank

LUSTEC: Lunar Stereo Camera

LUT: Land Utilization Type • Launch Umbilical Tower • Laura Station (Australia-QLD) (location identifier) • Local User Terminal • Look-Up Table

LUTE: Lunar Ultraviolet Telescope Experiment

Luth: Lutheran

LUTT: Launcher Umbilical Tower Transporter

LUU: Laura (Australia-QLD) (location identifier)

LUV: Langgur (Indonesia) (location identifier) • Lunar Utility Vehicle

LUVIMS: Lunar UV-Visible-Infrared Mapping Spectrometer

LUVO: Lunar Ultraviolet Observatory

LUW: Luwuk (Indonesia) (location identifier)

Lux: Luxembourg • Luxemburg

LUX: Luxembourg (ISO country code) • Luxembourg (Luxemburg) (location identifier) • Luxemburg (ISO country code)

Luxmbrg: Luxemburg

LuxF: Luxembourg Franc (currency)

LUY: Lushoto (Tanzania) (location identifier)

LUZ: Lushan (China-PRC) (location identifier)

lv: Latvia • Latvian (ISO language code) • Lettish (ISO language code)

Lv: Leave • Lev (currency) • Livre

L/V: Launch Vehicle

LV: Argentina (international aircraft registration code) • Langue Vivante • Large Volume • Lasting Value • Latching Valve • Latvia (international vehicle registration code) • Latvia (ISO country code) • Launch Vehicle • Launch Vehicle • Leaves • Leaving • Lettonie (ISO country code) • Lev (currency) • Leva (currency) • Levenberg-Marquardt • Lift Vector • Light and Variable • Limit Value • Linear Velocity • Loading Valve • Load Vertical • Local Vertical • Local Void • Logiciel Vol • Long Vector • Loss Variance • Low Velocity • Low Voltage • Luncheon Voucher

LVA: Latvia (ISO country code) • Launch Vehicle Adaptor • Launch Vehicle Availability • Laval Entrammes (France) (location identifier) • Lettonie (ISO country code)

LVAAS: Lehigh Valley Amateur Astronomical Society (USA-PA)

LVAR: Launch Vehicle Assessment Report • Launch Vehicle Assessment Review

LVAS: Las Vegas Astronomical Society (USA-NV)

LVB: Livramento - Dos Galpões (Brazil-RS) (location identifier) • Luchtverkeersbeveiliging

LVC: Low-Velocity Cloud

LVCSR: Large-Vocabulary Continuous-Speech Recognition

LVD: Large Volume Detector • Lime Village (USA-AK) (location identifier) • Low Voltage Differential

LVDA: Launch Vehicle Data Adapter • Launch Vehicle Deployment Assembly

LVDC: Launch Vehicle Data Center • Launch Vehicle Digital Computer

LVDE: Low-Vibration Drive Electronics

LVDT: Linear Variable Differential Transformer • Linear Variable Differential Transducer (*) • Linear Velocity Displacement Transducer • Linear Voltage Differential Transformer • Linear Voltage Displacement Transducer

LVE: Linear Variable Etalon • Linear Variable Equation • Low-Voltage Electronics

LVES: Low-Vision Enhancement System

LVET: Left Ventricular Ejection Time

LVF: Linear Variable Filter • Linear Vector Function

LVFC: Launch Vehicle Flight Control

LVFCS: Launch Vehicle Flight Control System

LVFRR: Launch Vehicle Flight Readiness Review

LVG: Large Velocity Gradient • Low-Viscosity Gyro

LVGC: Launch Vehicle Guidance Computer

LVGSE: Launch Vehicle Ground Support Equipment

LVI: Livingstone (Zambia) (location identifier) • Low Viscosity Index

LVIP: Launch Vehicle Integration Plan

LVIS: Launch Vehicle Instrumentation System • Live Video Insertion System • Logistics Verification Information System

LVK: Livermore (USA-CA) (location identifier)

Lvl: Level

LVL: Lat (Latvia) (ISO currency code) • Lawrenceville (USA-VA) (location identifier) • Level

LVLASO: Low-Visibility Landing And Surface Operations

LVLH: Local Vertical/Local Horizontal

LVM: Liceo Víctor Mercante (Argentina-BA) • Livingston - Mission Field (USA-MT) (location identifier) • Logical Volume Manager

LVMD: Limb Volume Measurement Device

LVN: Library Video Network

LVO: Label-Value Object • Launch Vehicle Operation • Laverton (Australia-WA) (location identifier)

LVP: Lagrangian Variational Principle • Lavan (Iran) (location identifier) • Low-Voltage Pressure • Low-Voltage Protection

LVPD: Launch Vehicle Pressure Display

LVPG: Launch vehicle Planning Group

LVPP: Launch Vehicle Propulsion Program

LVPS: Low-Voltage Power Supply

LVQ: Learning Vector Quantification • Learning Vector Quantization (*) • Linear Vector Quantization

LVR: Large Vocabulary Recognition • Laser Videodisk Recorder • Local Vertical Reference • Low-Voltage Release

LVS: Large Volume Sampling • Las Vegas (USA-NM) (location identifier) • Launch Vehicle Simulator • Launch Vehicle System • Limburgse Volkssterrenwacht (Belgium) • Low-Voltage Supply • Low-Volume Sampling

LVSEM: Low-Voltage Scanning Electron Microscope • Low-Voltage Scanning Electron Microscopy

LVSG: Launch Vehicle Study Group

LVSSTS: Launch Vehicle Safety System Test Set

LVTC: Launch Vehicle Test Conductor

LVU: Liaison Virtuelle Unitaire

l/W: Luman/Watt

lW: lichte Weite

Lw: Lawrencium (now Lr)

LW: Launch Window • Left Wing • Light Weight • Line Weight • Liquid Water • Longwave • Longwavelength • Low Water • Low Weight

LWA: Local Wavefront Analyzer • Long-Wavelength Array • Long Wire Antenna • Lwbak (Philippines) (location identifier)

LWB: Lewisburg (USA-WV) (location identifier)

LWC: Lawrence (USA-KS) (location identifier) • Lightweight Coated Paper • Liquid-Water Concentration • Liquid-Water Content

LWCF: Land and Water Conservation Fund

LWCI: Leaf Water Content Index

LWD: Launch Window Display • Left Wing Down • Logistics Working Group • Loomis-Wood Diagram • Luyten White Dwarf

LWE: Lewoleba (Indonesia) (location identifier)

LWEST: Low-Water Equinoctial Spring Tide

LWFA: Laser Wakefield Acceleration

LWG: Landelijke Werkgemeenschap

LWH: Lawn Hill (Australia-QLD) (location identifier)

LWHS: Lightweight Headset

LWI: Laser Weather Identifier • Lowai (Papua New Guinea) (location identifier)

LWIC: Long-Wavelength Infrared Camera

LWIR: Long-Wave Infrared • Long-Wavelength Infrared

LWIRC: Long-Wavelength Infrared Camera

LWK: Shetland Islands - Lerwick Tingwall (UK) (location identifier)

LWL: Lichtwellenleiter • Wells Harriet (USA-NV) (location identifier)

LWLA: Long Wavelength Large Aperture

LWM: Lawrence (USA-MA) (location identifier) • Local Window Manager • Low Water Mark

LWN: Leninakan (Armenia) (location identifier)

LWO: L'viv (Ukraine) (location identifier) • L'vov (Ukraine) (location identifier)

LWOP: Leave Without Pay

LWP: Leave With Pay • Lightweight Process • Liquid Water Path • Long-Wavelength Primary (camera) • Long-Wavelength Prime (camera)

Lwr: Lower

LWR: Leeuwarden (Netherlands) (location identifier) • Light-Water Reactor • Long-Wavelength Radar • Long-Wavelength Radiation • Long-Wavelength Redundant (camera) • Lower

LWRM: Longwave Radiation Model

LWS: Lewiston (USA-ID) (location identifier) • Lightning Warning System • Linear White Space • Loathy Worm of Sendmail • Long-Wavelength Spectrograph • Long-Wavelength Spectrometer

LWSA: Long Wavelength Small Aperture

LWT: Lewistown (USA-MT) (location identifier) • Lightweight Tank

LWV: Lawrenceville (USA-IL) (location identifier)

LWW: Launch Window Width

LWY: Lawas (Malaysia) (location identifier)

Lwyr: Lawyer

lx: Lux • Meter-Candle

LX: Low Index • Luxemburg (international aircraft registration code)

LXA: Lhasa (China-PRC) (location identifier)

LXe: Liquid Xenon

LXE: Lightguide Express Entry

LXeGRIT: Liquid Xenon Gamma-Ray Imaging Telescope

LXeTPC: Liquid Xenon Time projection Chamber

LXG: Luang Namtha (Laos) (location identifier)

LXI: Linxi (China-PRC) (location identifier)

Lxmbrg: Luxembourg

LXN: Lexington (USA-NE) (location identifier)

LXR: Luxor (Egypt) (location identifier)

LXS: Lemnos (Greece) (location identifier)

LXU: Lukulu (Zambia) (location identifier)

LXV: Leadville (USA-CO) (location identifier)

ly: Light Year

Ly: Lyman

LY: Libya (ISO country code) • Libye (ISO country code) • Light Year • Lithuania (international aircraft registration code)

LYA: Luoyang (China-PRC) (location identifier)

LYB: Little Cayman (Cayman Islands) (location identifier)

LYC: Lycksele (Sweden) (location identifier)

LYD: Dinar (Libya) (ISO currency code)

LYE: Lyneham (UK-England) (location identifier)

LYG: Lianyun'Gang (China-PRC) (location identifier)

LYH: Lynchburg (USA-VA) (location identifier)

LYI: Linyi (China-PRC) (location identifier)

LYK: Lunyuk (Indonesia) (location identifier)

Lyn: Lynx-Lyncis (constellation)

LYN: Lyon Bron (France) (location identifier)

LYO: Lyons - Rice County (USA-KS) (location identifier)

LYP: Faisalabad (Pakistan) (location identifier)

Lyr: Layer • Lyra-Lyrae (constellation)

LYR: Layer • Layered • Longyearbyen (Norway) (location identifier)

LYRD: Layered

LYRIC: Language for Your Remote Instruction by Computer

LYS: Lyon Satolas (France) (location identifier)

Lyt: Layout

LYT: Lady Elliot Island (Australia-QLD) (location identifier)

LYU: Ely (USA-MN) (location identifier)

LYX: Lydd (UK-England) (location identifier)

LZ: Bulgaria (international aircraft registration code) • Landing Zone • Lazy Write • Lempel-Ziv • Luftschiff Zeppelin

LZA: Landau-Zener Approximation • Luiza (Zaire) (location identifier)

LZC: Lazaro Cardenas (Mexico) (location identifier)

LZD: Lanzhou (China-PRC) (location identifier) • Launch Zone Display

LZEEBE: Long-term Zonal Earth Energy Budget Experiment

LZF: Launch Zone Flag

LZGF: Lewis Zero Gravity Facility (NASA)

LZH: Liuzhou (China-PRC) (location identifier)

LZI: Luozi (Zaire) (location identifier)

LZO: Launch Zone Override • Luzhou (China-PRC) (location identifier)

LZS: Lempel-Ziv-Stacker

LZR: Lizard Island (Australia-QLD) (location identifier)

LZT: Lead Zirconate-Titanate • Local Zone Time

LZW: Lempel-Ziv-Walsh • Lempel-Ziv-Welch

L1: First Language • Lagrange Point # 1

L1PT: Level 1 Page Table

L2: Lagrange Point # 2 • Second Language

L2B: Linienzugbeeinflussung

L2F: Layer 2 Forwarding

L2FET: Logic Level Field Effect Transistor

L2M: Laboratoire de Microstructures et de Microélectronique

L2PT: Level 2 Page Table

L2TP: Layer 2 Tunnelling Protocol

L3: Lagrange Point # 3 • Latitude/Longitude Locator • Third Language

L3AB: Laboratoire d'Astrodynamique, d'Astrophysique et d'Aéronomie de Bordeaux (France)

L3C: Laboratoire Capteurs-Caractérisation-Contrôle

L3E: Laboratoire d'Études Expérimentales en Électromagnétisme (France)

L3G: Language de Troisième Génération

L4: Lagrange Point # 4

L4G: Language de Quatrième Génération

L5: Lagrange Point # 5

M

m: Apparent Magnitude • Electromagnetic Moment • Electron Mass • Inertial Constant • Mach Number • Magnetic Pole Strength • Magnetic Quantum Number • Magnetic Moment • Magnitude • Magnitude Apparente • Male • Manipulus • Mare • Mark (currency) • Married • Masculine • Masculino • Mass • Mass of Electron • Matured • Medieval • Medium • Member • Membre • Memorandum • Meridian • Meridies • Meridional • Meta • Meter • Mètre • Metric • Midday • Middle • Mile • Mill (currency) • Mille • Milli • Million • Minim • Minor • Minute • Misce • Mist • Mittlere • Moderate • Modification • Modulus • Mois • Molar • Molarity • Molecular Mass • Mole Mass • Month • Moon • Morning • Mort • Mountain • Mouth • Noon • Number of Phases • Particle Mass • Phase Number • Scheinbare Helligkeit

M: Absolute Helligkeit • Absolute Magnitude • Angular Momentum • Anomalie Moyenne • Em • General Precession in Right Ascension • Loti (currency) • Luminous Exitance • Mach Number • Macintosh • Magister • Magnetic Moment • Magnetic Polarization • Magnetic Quantum Number • Magnetization • Magnitude • Magnitude Absolue • Maintenance • Majesty • Major • Male • Maloti (currency) • Malta (international vehicle registration code) • Manchester (UK postal code) • Manitoba (Canada) • March • Mark (currency) • Markka (currency) • Marquis • Married • Masculine • Mass • Master • May • Me • Mean • Mean Anomaly • Mean Molar Mass • Measure • Medal • Medicine • Medieval • Medium • Meer • Mega • Member • Mer • Meridian • Meridies • Mesomeric Effect • Messier (catalogue) • Metal • Metropolitan • Mezzo • Middle • Migration Length • Mike (international alphabet) • Mile • Million • Minim • Ministère • Ministerio • Minus • Mitte • Mittlere Anomalie • Modal • Modifier • Molar Mass • Moment of Force • Monday • Monsieur • Monte • Mother • Motorway • Mountain • Moyen • Moyenne • Mud • Mutual Inductance • Noon • One Thousand (Roman numeral) • Radiant Exitance • Sea Mile • Spectral Type • Thermal Insulance • Thermal Insulation Coefficient • Thousand (Roman numeral) • Wind Speed • 1000 (Roman numeral)

m/a: My Account

ma: Myria

mA: Milliampere

Ma: Mach Number • Mars • Master • Million Years

M&A: Maintenance & Assembly • Management & Administration • Monitor & Alarm

M/A: Memorandum of Association

MA: Macedonian • Magnetic Amplifier • Magneto-Acoustics • Mai • Maintenance Ability • Maintenance Advisor • Maintenance Administrator • Maintenance Analysis • Maître Auxiliaire • Malaysia • Management Adviser • Mandan • Manufacturing Assembly • Map Analysis • Mapper Application Interface • Maranhão (Brazil) • Maritime Administration • Marketing Assistance • Maroc (ISO country code) • Marquesas Islands • Massachusetts (USA) • Master Alarm • Master of Arts • Material Authorization • May • Mean Annual • Medium Angle • Memory Address • Mental Age • Mercury-Atlas • Message Assembler • Metabolic Analyzer • Meter Amplifier • Middeck Act • Middle Age • Mike Amplifier • Mirror Assembly • Missed Approach • Mission Analysis • Mission Authority • Mobile Agent • Mobility Allowance • Model Age • Modified Achromat • Modulation d'Amplitude • Monitor and Alarm • Morocco (international vehicle registration code) • Morocco (ISO country code) • Moving Average • Moyen-Age • Multiple Access • Multiplex Advantage

MAA: Madras Meenambakkam (India) (location identifier) • Madras Astronomical Association (India) • Martz Astronomical Association (USA-NY) • Mathematical Association of America (USA) • Maximum Authorized Altitude • Mechanical Arm Assembly • Mémoire à Accès Aléatoire • Message Authenticator Algorithm • Metropolitan Area Acquisition • Moist Adiabatic Adjustment

MAAA: Mankato Area Astronomy Association (USA-MN)

MAAB: Materials Application Advisory Board

MAACS: Multiaddress Asynchronous Communication System

MAAP: Maintenance And Administration Panel

Maar: Maarschalk

MAAR: Monthly Associate Administrator's Review

MAARC: Magnetic Annular Arc

MAAS: Mission Area Analysis System

MAB: Magnetic-Amplifier Bridge • Malfunction Analysis Branch • Management Advisory Board • Man And the Biosphere • Marabá (Brazil-PA) (location identifier) • Maschinelles Austauschformat für Bibliotheken • Master Acquisition Bus • Materials Applications Board • Materials Advisory Board • Mechanical Automation Breadboard • Mise À Bord

MABIE: Mauna Loa Aerosol Backscatter Intercomparison Experiment

MABL: Marine Atmospheric Boundary Layer • Mass Addition Boundary Layer

MABM: Multilayer Absorbing Bottom Layer

MABP: Mean Arterial Blood Pressure

MABV: Makle- und Bauträgerverordnung

Mac: Macintosh

MAc: Master of Accountancy

MAC: Macau (ISO country code) • Machine-Aided Cognition • Macon Smart (USA-GA) (location identifier) • Macro • Magnetic Attitude Control • Magnetic Automatic Calculator • Maintenance Advisory Committee • Maintenance Aid Computer • Maintenance Allocation Chart • Maintenance Analysis Center • Major Activity Center • Management Advisory Committee • Man And Computer • Mandatory Access Control • Maneuver Analysis and Command • Mapping Advisory Committee • Mapping Applications Center • Mass Absorption Coefficient • Massive Algebraic Computation • Material Availability Commitment • Maximum Allowable Concentration • Maximum Allowable Cost • Maximum Atmospheric Concentration • Mean Aerodynamic Chord • Mechanical Analog Computer • Media Access Control • Medium Access Control • Meeting of Aid Coordinators • Memory-Assisted Cognition • Meridian Axial Circle • Message Authentication Code • Microgravity Advisory Committee • Middle-Atmosphere Cooperation • Midland's Astronomy Club (USA-SC) • Military Airlift Command • Millimeter Autocorrelator • Millstream Astronomy Club (USA-OH) • Mineralogical Association of Canada • Miniature Advanced Coronagraph • Mirror-Advanced Coronagraph • Money Access Center • Monitoring and Assessment Centre • Months After Contract • Motion Analysis Camera • Mouse-Activated Computer • Multiaccess Computer • Multiinstrument Airborne Campaign • Multimission Application Concept • Multiphase Atmospheric Chemistry • Multiple-Access Computer • Multiple-Access Computing • Multiple-Access Control • Multiple-Address Code • Multiplexed Analog Component • Multiplicator/Accumulator • Multiply/Accumulate • Multisensory Airborne Campaign • Multiservice Access Concentrator

MACABRE: Material Ablation with Chemically Active Boundary Layer in Reentry

MACAS: Magnetic Capability and Safety System

MACBM: Multiaccess Cable Billing System

MACCS: Magnetometer Array for Cusp and Cleft Studies

MACDIF: Map And Chart Data Interchange Format • Mapping And Charting Data Interchange Format

MACDS: Monitor And Control Display System

MACE: Machine-Aided Composition and Editing • Macintosh Audio Compression and Expansion • Maintenance Analysis Checkout Equipment • Material Ablation Conduction Erosion • Mechanical Antenna Control Electronics • Middeck Active Control Experiment • Minimum Average Correlation Energy

Maced: Macedonia • Macedonian

Mach: Machine • Machinery • Machinist

MACH: Machine • Maximum Average Correlation Height • Mesothermal Auroral Charging Code • Multilayer Actuator Head

MACHO: Massive Compact Halo Object (*) • Massive Condensed Halo Object

MACI: Monitor Access and Control Interface

MACIS: Management and Contracts Information Service

MACMATC: Mid-Atlantic Consortium for Mathematics and its Applications Throughout the Curriculum

MACMIS: Maintenance And Construction Management Information System

MACO: Major Assembly Checkout

MaCom: Mathematics of Computation (periodical code)

MACOS: Modelling and Analysis for Controlled Optical Systems

MACP: Mission Analysis Computer Program

MACRES: Malaysia Centre for Remote Sensing • Malaysian Centre for Remote Sensing

MACRIIM: Marine And Coastal Resources Integrated Information and Management

MACRO: Merge And Correlate Recorded Output • Monopoles, Astrophysics and Cosmic-Ray Observatory • Muon And Cosmic-Ray Observatory

MACS: Magellanic Catalogue of Stars • Maintenance Assistance Capability for Software • Management And Control System • Medium-Altitude Communications Satellite • Microwave Attitude Control Sensor • Modular Attitude Control System • Modulator Attitude Control System • Multiaccess Conversional System • Multifunction Automated Crawling System • Multiproject Automated Control System

MACSATS: Multiple-Access Communications Satellite

MACSE: Minority Access to Careers in Science and Engineering • Module Accélérateur à Cavités Supraconductrices pour Electrons

MACSS: Medium-Altitude Communications Satellite System

MACSYMA: Massive Algebraic Computation Symbolic Manipulation • Massive Algebraic Computation Symbolic Manipulator

MACWS: Machine Area Cooling Water System

Mad: Madame • Madeira (Azores)

MA&D: Mission Analysis & Design

MAD: Dirham (Morocco) (ISO currency code) • Machine Analysis Display • Madrid Barajas (Spain) (location identifier) • Magnetic-Anomaly Detection • Magnetic-Anomaly Detector • Magnetic-Anomaly Director • Magnetic-Azimuth Detector • Main Assembly Drawing • Maintenance Analysis Data • Maintenance-Assembly-Disassembly • Management Analysis Division • Manufacturing Assembly Drawing • Material Analysis Data • Material Availability Date • Mean Absolute Difference • Median Absolute Deviation • Méthode d'Analyse de Données • Michigan Algorithmic Decoder • Mission Analysis Division • Momentum Accumulation and Dumping • Monitor Address Data • Motor Assembly and Disassembly • Multiple-Access Device • Multiple-Aperture Device • Multiply And Add • Multispectral Analyzer Display • Multiwavelength Anomalous Diffraction • Mutual Assured Destruction

Madag: Madagascar

MADAM: Mean And Dispersion Additive Model • Moderately Advanced Data Management • Multipurpose Automatic Data Analysis Machine

MADAR: Malfunction Analysis Detection And Recording

MADARS: Maintenance Analysis, Detection and Reporting System • Malfunction Analysis Detection And Recording System

MADC: Machine-Assisted Detection and Classification • Multiplexer Analog to Digital Converter

MADCAP: Mammoth Decimal Arithmetic Program • Mosaic Array Data Compression And Processing

MADCS: Modular Attitude Determination and Control System

MADD: Mirror Added Dual Doppler • Module for Automatic Dock and Detumble • Mothers Against Drunk Driving

MADE: Manufacturing and Automated Design Engineering • Manufacturing Automation and Design Engineering • Micron Accuracy Deployment Experiment

MaDes: Machine Design (periodical code)

MADEX: Magnetic-Anomaly Detection Exercise

MADL: Maryland Advanced Development Laboratory (USA-MD)

Madm: Madame

MADM: Maintenance Automated Data Management • Multi-Attributive Decision Making • Multiple-Attributes Decision Making

MADN: Multiple Access Directory Number

MADO: Mulliken Approximation for Differential Overlap

Madr: Madras (India) • Madrid (Spain)

MADRAF: Midwestern Astronomical Data Reduction and Analysis Facility

MADRE: Manufacturing Data Retrieval

MADS: Management And Distribution System • Mars Atmosphere Density Sensor • Meteorological Airborne Data System • Modular Attitude Control System • Modular Auxiliary Data System • Multiple Area Dataset

MADT: Mean Administrative Delay Time • Modified Anomalous Diffraction Theory

MADYMO: Mathematical Dynamic Model

MaE: Master in Engineering

MAE: Machine Arithmétique Électronique • Macintosh Application Environment • Madera (USA-CA) (location identifier) • Management Application Entity • Master of Aeronautical Engineering • Matrix Arithmetic Expression • Mean Absolute Error • Mean Area of Effectiveness • Metropolitan Area Exchange • Ministère des Affaires Étrangères • Mission Accomplishment Estimate • Mission Analysis and Engineering • Multiaccess Enclosure • Musée de l'Air et de l'Espace (France)

MAEB: Macintosh Application Environment

MAECI: Ministère des Affaires Étrangères et du Commerce International

MAED: Middle Atmosphere Electrodynamics

MAEDS: Meteosat Argos Extended Dissemination Service • Multisatellite Applications Extended Dissemination Service

MAEG: Multiple Asynchronous Exponential Growth

MAEIF: Mean Absolute Error Improvement Factor

MAEL: Mobile Aeronautics Education Laboratory

MAEP: Minimum Autoland Entry Point

MAES: Management Assistant Expert System • Mexican American Engineers and Scientists

MAESA: Measurements for Assessing the Effects of Stratospheric Aerosols

Maest: Maestro

MAESTRO: Machine-Assisted Educational System for Teaching by Remote Operation • Méthodes d'Administration, d'Exploitation de Systèmes et de Télédistribution en Réseau Ouvert • Mission Analysis Evaluation and Space Trajectory Operations • Multiple Airborne Experiments Towards Radar Observation

MAET: Microwave Amplifier Electron Tube

MAF: Magnetic-Anisotropy Field • Main Assembly Fixture • Maintenance Action Form • Manpower Authorization File • Master Address File • Maximum Amine Fuel • Maximum Amplitude Filter • Michoud Assembly Facility (NASA) • Midland (USA-TX) (location identifier) • Million Acre Feet • Mission Analysis File • Mixed Amine Fuel • Multiple-Access Facility • Odessa (USA-TX) (location identifier)

MAFD: Minimum Acquisition Flux Density

MAFET: Microwave and Analog Front-End Technology

MAFF: Ministry of Agriculture, Fisheries and Food

MAFPEN: Mission Académique de la Formation des Personnels de l'Éducation Nationale (France)

mag: Magnitude

Mag: Magasin • Magazine • Maggio • Magnesium • Magnet • Magnetic • Magnetism • Magneto • Magnetometer • Magnification • Magnitude • Magnus • Magyar

MAG: Madang (Papua New Guinea) (location identifier) • Magazine • Magnetic • Magnetometer • Magnitude • Mission Advisory Group • MPP Advisory Group

MAGA: Medium-Accuracy Gyro Assembly

MAGE: Map Authoring and Generalization Expert • Marine Aerosol and Gas Exchange

MAGELLAN: Magasin Électronique d'Archivage Numérique

MAG/ER: Magnetometer/Electron Reflectometer (MO)

MAGER: Magnetometer/Electron Reflectometer (MO)

MagFE: Magnetic Fusion Energy (periodical code)

Magg: Maggio • Maggiore

MAGGE: Medium-Altitude Gravity Gradient Experiment

MAGIA: Méthodes Avancées de Gestion de l'Information en Astronomie

MAGIC: Machine for Automatic Graphics Interface to a Computer • Magnetic And Germanium Integer Calculator • Manipulator And Gantry Intelligent Control • Matrix Analysis via Generative and Interpretive Computations • Model of Acidification of Groundwater In Catchment • Modern Analytical Generator of Improved Circuits • Modified Action Generated Input Control • Modular Architecture for Graphics Image Control • MPI for Astronomy General-purpose Infrared Camera • Multidimensional Applications and Gigabit Internetwork • Multimission Advanced Ground Intelligent Control Consortium • Multipurpose And Generalized Interface to COBOL

MAGICA: Multimedia Agent-based Interactive Catalogues

MAGICS: Meteorological Application Graphics Integrated Colour System

MAGLEV: Magnetic Levitation

MAGMA: Magnetic fields near Mars

Magn: Magnet • Magnetic • Magnetismo • Magnetron

MAGN: Mildly Active Galactic Nucleus

MAGNUM: Multicolor Active Galactic Nuclei Monitoring

MAGO: Martian Atmospheric Grain Observer

MAGPIE: Mega Ampere Generator for Plasma Implosion Experiments

MAGPIES: Magnetospheric Plasma Imaging Explorer Satellite

MAGR: Miniature Advanced GPS Receiver • Miniature Airborne GPS Receiver • Miniaturized Airborne GPS Receiver

MAGRET: Modèle Agrométéorologique d'Évaporation et de Température

MAGSAT: Magnetic-field Satellite • Magnetometer Satellite

MAGSS: Medium-Altitude Global Satellite System

Magtd: Magnitude

mAH: Milliampere-Hour

Mah: Maharashtra (India) • Mahogany

MAH: Menorca (Baleares) (location identifier) • Multilingual Aligned Hypertext

MAHIS: Meridian Automatic Horizontal Instrument

MAHLOVS: Middle And High Latitudes Ocean Variability Study

MAHRSI: Middle-Atmosphere High-Resolution Spectrograph Instrument • Middle-Atmosphere High-Resolution Spectrograph Investigation

MAHT: Machine-Assisted Human Translation

Mai: Maio

MAI: Machine-Aided Indexing • Mail • Mangochi (Malawi) (location identifier) • Manufacturer Access Interface • Mean Annual Increment • Multilateral Agreement on Investments • Multiple-Address Instruction • Myanmar Airways International

MAIAA: Member of the American Institute of Aeronautics and Astronautics

MAIB: Marine Accident Investigation Branch (UK)

MAICO: Manufacturing-Assembly-Integration-Checkout

MAID: Maintenance Automatic Integration Director • Market Analysis and Information Database • Merger Acquisition Improved Decision • Multimedia Assets for Industrial Design • Multipurpose Automatic Inspection and Diagnostic • Münchener Arbeitskreis für Information und Dokumentation

MAIDS: Machine-Aided Information and Dissemination System • Management Automated Information Display System • Multipurpose Automatic Inspection and Diagnostic System

MAIL: Mockup And Integration Laboratory

MAIN: Maintenance • Material Automated Inventory Network • Multiple-Access Internal Network

MAINE: Maintaining Availability in Europe

Maint: Maintain • Maintenance

MAINT: Maintenance

MAIP: Member of the Australian Institute of Physics

MAIPL: Mid-America Image Processing Laboratory (USA)

MAIR: Manufacturing And Inspection Record • Multisensor Autonomous Image Registration

MAIS: Module Airborne Imaging Spectrometer • MOTIF Advanced Imaging System • Multicolor Airglow Imaging System • Multicolor All-sky Imaging System • Multiresource Analysis and Information System

MAIT: Management Automation Information Technology • Manufacturers' Association for Information Technology

Maj: Majoor • Major • Majority

MAJ: Major • Majuro (Marshall Islands) (location identifier) • Mission Astronomie Jeunesse

MAJA: Mise À Jour Automatique

MAK: Malakal (Sudan) (location identifier) • Mariestads Astronomiska Klubb (Sweden)

Mál: Málaga (Spain)

Mal: Malay • Malayan • Malaysia • Malaysian • Malta • Maréchal

MAL: Macro-Assembly Language • Malaysia (international vehicle registration code) • Malfunction • Material Allowance List • Mangole (Indonesia) (location identifier) • Member-At-Large • Meta-Assembly Language • Mobile Airlock • Model Assembly List

MALCD: Matrix-Addressed Liquid Crystal Display

MALD: Master of Arts in Law and Diplomacy • Miniature Air-Launched Decoy

MALDI: Matrix-Assisted Laser Desorption Ionization

MALE: Multiaperture Logic Element

Malf: Malfunction

MALF: Mobile Aerobee Launch Facility

MALI: Mémoire Adaptée aux Langages Indéterministes • Multilevel Accelerated Lambda Iteration

MALITAS: Mali Tinbouctou Air Service

MALL: Malleable

MALS: Master of Arts in Liberal Studies • Master of Arts in Library Science • Medium-intensity Approach Lighting System

MALT: Model Assembly List Transmittal

MALTED: Mesosphere and Lower Thermosphere Equatorial Dynamics

MAM: Management and Administration Manual • March-April-May • Matamoros (Mexico) (location identifier) • Measurement of Atoms and Molecules • Microarcsecond Metrology • Middle Atmosphere Model • Mirror Attenuation Mosaic • Mot-À-Mot • Multifunctional Autocorrelation Method • Multiple Allocation Memory • Musée des Arts et Métiers

MAMA: Machine à Mesurer Automatique • Machine Automatisée à Mesurer pour l'Astronomie (*) • Machine Automatique à Mesurer pour l'Astronomie • Manual-Automatic Multipoint Apparatus • Mobile Automated Metabolic Analyzer • Multianode Microchannel Array • Multianode Multiarray

MAMBO: Mediterranean Association for Marine Biology and Oceanography

MAMI: Machine-Aided Manufacturing Information

MAMIE: Minimum Automatic Machine for Interpolation and Extrapolation

MAMMAX: Machine-Made and Machine-Aided Index

MAMOS: Marine Automatic Meteorological Observing Station

MAMS: Model Attitude Measuring System • Multispectral Atmospheric Mapping Sensor • Multiple Access to Memory System

MAMT: Mean Active Maintenance Time

mAn: Meiner Ansicht nach

Man: Management • Manager • Manchester (UK-England) • Manila (Philippines) • Manitoba (Canada) • Manual • Manually • Manuel • Manufacture • Manufactured • Manufacturer • Manufacturing

MAN: Manchester (Starlink internal code) • Manchester (UK-England) (location identifier) • Manitoba (Canada) • Manual • Maschinenfabrik Augsburg-Nürnberg (Germany) • Medium Area Network • Metropolitan Area Network • Microwave Aerospace Navigation

Manch: Manchester (UK-England) • Manchuria

Man. Dir.: Managing Director

MANDOP: Multiple Analytical Doppler

Man. Ed.: Managing Editor

Manf: Manifold • Manufacturer

MANF: Manifold • May-August-November-February

ManGe: Manuscripta Geodaetica (periodical code)

MANIA: Multichannel Analysis of Nanosecond Intensity Alterations • Multichannel Analyzer of Nanosecond Intensity Alterations

MANIAC: Mathematical Analyzer, Numerical Integrator, and Computer • Mechanical And Numerical Integrator And Computer • Mid- And Near-Infrared Array Camera

MANIP: Manual Input

Manit: Manitoba (Canada)

MANOLIA: Matière Noire et Laboratoires Interactifs

Man.Op.: Manually Operated

MANOVA: Multivariate Analysis Of Variance

Manp: Manpower

MANPAGE: Manual Page

man. pr.: Mane Primo

Mans: Mansion

MANS: Mansarda • Microcosm Autonomous Navigation System

MANSAT: Manned Satellite

Manuf: Manufacture • Manufactured • Manufacturer • Manufacturing

Manufac: Manufacture • Manufactured • Manufacturer • Manufacturing

Manz: Manzana

MAO: Magnetic-Amplifier Output • Main Astronomical Observatory • Manaus - Eduardo Gomes (Brazil-AM) (location identifier) • Modulateur Acousto-Optique • Moerser Astronomische Organisation (Germany) • Musique Assistée par Ordinateur

MAOA: Meteorological Aspects of Ocean Affairs

MAOS: Magnetic-Amplifier Output Stage

MAOT: Maximum Allowable Operating Time

MA&P: Maintenance Analysis & Planning

Map: Mapping

MAP: Macro Arithmetic Processor • Macroassembly Program • Maintenance Analysis and Planning • Maintenance Analysis Procedure • Maintenance Analysis Program • Maintenance and Administration Position • Maintenance Automation Program • Major Air Pollutant • Mamai (Papua New Guinea) (location identifier) • Managed Application Provider • Management Analysis Program • Management Assessment Program • Manifold Absolute Pressure • Manufacturing Automation Protocol • Map Analysis Package • Margins of Active Plants • Mars Atmosphere Probe • Master Activity Programming • Master Automation Plan • Mathematical Analysis without Programming • Maximum A Posteriori • Maximum Average Price • Mean Arterial Pressure • Mean Average Precision • Measurement Allocation Plan • Media Access Project • Medical Aid Post • Mediterranean Action Plan • Memory Allocation Map • Mesoscale Alpine Programme • Message Acceptance Pulse • Microabrasion Package • Microwave Anisotropy Probe • Middle-Atmosphere Program • Military Airport Plan • Minimum Ascending Path • Missed Approach Point • Mission Application Program • Maintenance Analysis and Planning • Mobile Anisotropy Telescope • Mobile Application Part • Model Accreditation Plan • Modified Access Pricing • Modified American Plan • Modular Application System • Multiannual Programme • Multichannel Aperture Photometer • Multichannel Astrometric Photometer • Multifoil Abrasion Package • Multinational Account Program • Multiple-Allocation Procedure • Multiple-Array Processor • Multiple-Avoidance Program

MAPAQ: Ministère de l'Agriculture, des Pêcheries et de l'Alimentation du Québec (Canada-QC)

MAPI: Mail Application Program Interface • Mail Application Programming Interface • Messaging Application Program Interface • Messaging Application Programming Interface (*) • Microsoft Application Programming Interface (*)

MAPICS: Manufacturing, Accounting, and Production Information Control System • Morphological Analysis with Parameters Information Coding Scheme

MAPIPS: MacAulay Automated Photogrammetric and Interactive Processing System

MAPK: Mitogen-Activating Protein Kinase

MAPL: Manufacturing Assembly Parts List

MAPM: Minimum-cost Augmented Path Method

MAPMT: Multianode Photomultiplier Tube

MAPOS: Multiple Access Protocol Over SONET

MAPP: Mission Analysis and Performance Program

MAPPIT: Masked Aperture-Plane Interference Telescope

MAppSc: Master of Applied Science

MaPr: Mathematical Proceedings (periodical code)

MAPR: Marine Airborne Patrol Radar

MAPRP: Mesoscale Atmospheric Processes Research Program

MAPS: Madrid Agreement and Protocol System • Maintenance Analysis and Procedures System • Malabar Advanced Photometric System • Management Analysis and Planning System • Map Analysis Package System • Measurement of Air Pollution from Satellite • Measurement of Atmospheric Pollution from Satellites • Mesoscale Analysis and Prediction System • Meteorological Application and Presentation Software • Meteorological Application and Presentation System • Microcomputer-based Automated Projection System • Microwave Active and Passive Spectrometer • Middle Atlantic Planetarium Society (USA-PA) • Minnesota Automated Plate Scanner • Mitsubishi Array Processing System • Mobile Atmospheric Profiling System • Modular Azimuth Position System • Monopropellant Accessory Power Supply • Moteur Atomique de Propulsion Spatiale • Multichannel Astrometric Photometer with Spectrograph • Multidimensional Analysis Processing Scheme • Multiple-Address Processing System • Multiple Aircraft Positioning System • Multisatellite Attitude Program System • Multiservice Access Platforms • Multispectral Active and Passive System • Multispectral Active/Passive Scanner • Multivariate Analysis and Prediction of Schedules • Multivariate Analysis, Participation and Structure

MAPSAC: Machine-Aided Programming, Scheduling And Control

MAPSAT: Mapping Satellite

MAPSEP: Mission Analysis Program for Solar Electric Propulsion

MAPSS: Maintenance and Analysis Plan for Special Services

MAPSTAR: Middle Atmosphere Periodic Structured Associated Radiance

MAPTIP: Marine Aerosol Properties and Thermal Imager Performance

MAPTIS: Materials And Processes Technical Information System

MAPUG: Meade Advanced Products User Group

MAP3S: Multistate Atmospheric Power Production Pollution Study

MAQ: Mae Sot (Thailand) (location identifier) • Maximum Acceptance Quantity

Mar: March • Maréchal • Marina • Marine • Maritime • Marriage • Married • Martedi • Marzo

MAR: Maatschappelijke Advies Raad • Magnetic-Amplifier Relay • Maintenance Action Report • Maintenance And Refurbishment • Management Assessment Review • Maracaibo - La Chinita (Venezuela) (location identifier) • March • Março • Maroc (ISO country code) • Mars • Marzo • Material Availability Report • Mean Annual Runoff • Memory Address Register • Microprogram Address Register • Mid-Atlantic Ridge • Middeck Accommodations Rack • Minimal Angle Resolution • Minimum Acceptable Reliability • Mission Analysis Representative • Missing At Random • Morocco (ISO country code) • Multiple-Access Relay

MARA: Multimode Airborne Radar Altimeter

MARBI: Machine-Readable Bibliographic Information

MaRBu: Materials Research Bulletin (periodical code)

Marc: Marcato

MARC: Machine-Readable Cataloguing (*) • Machine-Readable Catalogue • Microgravity Advanced Research Centre • Mid-America Remote Sensing Center (USA-KY) • Middle Atmosphere Response to Change • Modern Asia Research Centre • Monitoring and Assessment Research Centre • Multiaccess Research Catalog

March: Marechal

MARCI: Mars Color Imager

MARCIA: Mathematical Analysis of Requirements for Career Information Appraisal

MARCS: Multispectral Airborne Reference-Aided Calibrated Scanner

marcsec: Milliarcsecond

MARDI: Malaysian Agricultural Research and Development Institute

MARE: Marine Activities, Resources and Education

MARECS: Maritime Communications Satellite

MARF: Meteorological Archive and Retrieval Facility (Germany)

Marg: Margin • Marginal

MarGe: Marine Geodesy (periodical code)

MARGIE: Minute-of-Arc Resolution Gamma-ray Imaging Experiment

MarGR: Marine Geophysical Researches (periodical code)

MARIA: Methods for Assessing the Radiological Impact of Accidents

MARIAN: Multiple Access and Retrieval of Information with Annotations

MARICA: Mariage de l'Image et de la Carte

MARIE: Mobile Autonomous Robot in an Industrial Environment

MARIS: Mississippi Automated Resource Information System

Marit: Maritime

Markgräflerl: Markgräflerland (Germany)

MARNR: Ministerio del Ambiente y de los Recursos Naturales Renovables (Venezuela)

MAROSIM: Multiarm Robot Simulator

MAROTS: Maritime Orbital Test Satellite

MARP: Markov Analysis and Reliability Prediction

MarPG: Marine Petroleum Geology (periodical code)

Marq: Marquês • Marquess • Marquis

Marqa: Marquesa

MARRS: Mission Analysis and Reconfiguration Requirements System

MARS: Machine À Réduction Symbolique • Machine À Résolution Symbolique • Maintenance Activities and Resources Simulation • Malmö Astronomi- & Rymdfarts Sällskap (Sweden) • Management Analysis Reporting System • Manned Aerodynamic Reusable Spaceship • Manned Astronautical Research Station • Maritime Access and Retrieval System • Marketing and Advertising Reference Service • Master Attitude Reference System • Materials Analysis Research Station • Meeting of Astronomy Research Students • Membrane with Active Rapid Segment • Memory Address Register Storage • Message Automatic

Reporting System • Meteorological and Aeronautical Retrieval System • Meteorological Automatic Reporting Station • Microcomputer-Assisted Rating System • Microdynamics Archiving and Retrieval System • Microgravity Advanced Research and Support • Microwave Atmospheric Remote Sensing • Miniature Attitude Reference System • Mission Maintenance and Reliability Simulation • Mobile Agricultural Radar Scatterometer • Mobile Access and Retrieval System • Mobile Access for Resources on Space • Mobile Atmospheric Remote Sensing • Modular Airborne Recorder System • Monitoring Agriculture with Remote Sensing • Monograph Acquisitions and Record System • Multiaccess Radio System • Multiactivity and Resources Scheduling • Multiaperture Reluctance Switch • Multicast Address Resolution Server • Multiple-Access Ranging System • Multiple-Access Retrieval System • Multiuser Archival and Retrieval System • Multivariate Adaptive Regression Splines • Multivariate Analysis, Retrieval and Storage

MARSIS: Marine Remote Sensing Information System

MARSNET: Mars Network

MARSS: Microwave Airborne Radiometer and Scanning System • Multiactivity and Resources Scheduling System

MARSSS: Modular Airborne Remote Sampling and Sensing System

MARSYAS: Marshall System for Aerospace Simulation (NASA)

MART: Maintenance Analysis Review Technique

MARTINI: Multiple Aperture Real-Time Image Normalisation Instrument

MARV: Maneuverable Reentry Vehicle

MARVEL: Machine-Assisted Realization of the Virtual Electronic Library

mas: Milliarcsecond

mAS: Milliampere-Second

Mas: Masculine

MAS: Macintosh Application Services • Madison Astronomical Society (USA-WI) • Malta Astronomical Society • Manipulateur Agile Spécialisé • Manuscrit Autographe Signé • Manus Island (Papua New Guinea) (location identifier) • Marquette Astronomical Society (USA-MI) • Mars Approach Sensor • Mass Announcement System • Master Analysis Scheme • Master of Accounting Science • Master of Applied Science • Mattatuck Astronomical Society (USA-CT) • Measurement Approval Sheet • Medium-Altitude Sensor • Memory and Auxiliary Storage • Metastable Atomic State • Meteorological Aids Service • Michiana Astronomical Society (USA-IN) • Microwave Active Spectrometer • Microwave Active System • Microwave Atmospheric Sounder • Millimeter-wave Atmospheric Sounder • Milwaukee Astronomical Society (USA-WI) • Minnesota Academy of Science (USA-MN) • Minnesota Astronomical Society (USA-MN) • Mobile Access Structure • Mobile Astronomical Society (USA-AL) • Modélisation Aléatoire et Statistique • Modis Airborne Simulator • Mountain Astronomical Station • Multiaccess Spool • Murdoch Astronomical Society (Australia) • Muskegon Astronomical Society (USA-MI)

MASA: Massachusetts Astronomy and Space Association (USA-MA) • Member of the Astronomical Society of Australia • Modular Avionics Systems Architecture

MASAR: Mid-Atlantic Slope And Rise

Masc: Masculine

MASc: Master of Applied Science

MASC: Methylaluminium Sesquichloride • Mountain Administration Support Center • Multiple Award Schedule Contract • Multiscale Alternating Shadow Collimator

MASCAM: Masking-pattern Adaptive Sub-band Coding And Multiplexing

MASCON: Mass Concentration

MASCOT: Multiaccess System Control Terminal

MASCOTTE: Montage Autonome Simplifié pour la Cryocombustion dans l'Oxygène et Toutes Techniques Expérimentales

MASDC: Military Aircraft Storage and Disposition Center

MA&SE: Mission Analysis & System Engineering

MASE: Mission Analysis and System Engineering

MASER: Material Science Experiment Rocket • Microwave Amplification by Stimulated Emission of Radiation • Multispectral Airborne System for Environmental Research

MASERATI: Middle Atmosphere Spectrometric Experiment on Rockets for Analysis of Trace-gas Influences • Middle Atmosphere Spectroscopic Experiment on Rockets for Analysis of Trace-gas Influences

MASEX: Mesoscale Air-Sea Exchange

MASH: Multimedia Architecture that Scales across Heterogeneity

MASL: Multimission Analysis Subroutine Library

MASM: Macro Assembler • Motorized Antenna Switching Matrix

MASN: Moving-Actor Subnetwork

MASP: Mid-Atlantic Star Party • Model for Assisted Software Process • Modular Atmosphere Simulation Program

MASPnet: Multiple Access-type Satellite Packet Network

MASR: Microwave Atmospheric Sounding Radiometer

Mass: Massachusetts (USA)

MASS: Magnetically Active Stellar System • Manned Activity Scheduling System • Matter Antimatter Superconducting Spectrometer • Maximum Availability and Support System • Mesoscale Analysis and Space Sensor • Mesoscale Atmospheric Simulation Sensor • Miniature Acquisition Sun Sensor • Model Atmospheres and Spectrum Synthesis • Monitor And Assembly System • Multiaccess Storage System • Multiple-Access Sequential Selection

Mass. Ave.: Massachusetts Avenue

MASSDAR: Modular Analysis, Speedup, Sampling and Data Reduction

MASSPO: Manned Space Flight Support Project Office

MAST: Maintenance Standardization Team • Marine Action in Science and Technology • Marine Science and Technology • Marine Sciences and Technologies • Mass Spectrometer Telescope • Measurement And Stimulus • Military Application of Ship Tracks • Model Assembly Sterilizer for Testing • Multiapplication SONAR Trainer • Multimission Archive at STScI • Multiple-Applications Storage Tube

MASTER: Matching Available Student Time to Educational Resources • Meteoroid And Space Debris Terrestrial Environment Reference • Multiple-Access Shared-Time Executive Routine

MASTEX: Mediterranean Aircraft and Ship Transmission Experiment

MASTIC: Malaysian Science and Technology Information Center

MASTIF: Multiaxis Space Test Inertia Facility (*) • Multiaxis Spin Test Inertia Facility

MASTIFF: Middle Atmosphere Studies of Instabilities, Fluctuations and Fields

MASTIR: Microfilmed Abstract System for Technical Information Retrieval

Mat: Matemáticas • Material • Matériel • Matière • Matinée • Matrix • Maturity

MA&T: Manufacturing Assembly & Test

MAT: Machine-Aided Translation • Machine-Assisted Translation • Machine Available Time • Maintainability Analysis Tool • Maintenance Access Terminal • Manufacturing Assembly and Test • Marconi Applied Technologies • Marine Air Temperature • Marine-Aviation-Transport

• Master Alias Table • Master Archive Tape • Master of Arts in Teaching • Matadi (Zaire) (location identifier) • Mathematical Astronomical Text • Mathematical Automata Theory • Matrix • Measurement of Atmospheric Turbulence • Mechanical Aptitude Test • Medial Axis Transform • Meteorological Atmospheric Turbulence • Microwave Antenna Tower • Mobile Aeronautical Telemetry • Mobile Anisotropy Telescope • Molecular Analysis Team • Mount Alignment Tool • Multilingual Aligned Text • Multiple-Access Test • Multiple Actuator Test • Multiple-Address Telegram • Multiple-Attribute Tree

MATA: Multiple Answering Teaching Aid

MATC: Mathematics and their Applications Throughout the Curriculum • Middle Atmosphere and Tropospheric Climate

MATCH: Mesoscale Atmospheric Transport and Chemistry

MATCHES: MAPS-Assisted Translation and Classification Help for Examiners System

MATCO: Materials Analysis, Tracking and Control

MATE: Mars Array Technology Experiment • Mission Analysis Technique for Experiments • Mobile Analysis and Training Extension • Modular Approach to Technology Education • Modular Automated Test Equipment • Modular Automatic Test Equipment • Multiaircraft Training Environment • Multiband Automatic Test Equipment • Multipurpose Automatic Test Equipment • Multisystem Automatic Test Equipment

Mater: Materialprüfung (periodical code)

MATES: Multiband Antiship Tactical EW System

MatEv: Materials Evaluation (periodical code)

MatG: Mathematical Geology (periodical code)

Math: Mathematical • Mathematically • Mathematician • Mathematics • Mathematika (periodical code) • Mathématique • Mathématiques

Maths: Mathematics

MATIAS: Mediciones Atmosféricas, Telecomunicaciones e Ingeniería y Aplicaciones de los Satélites

MATIC: Multiple-Area Technical Information Center

MATICO: Machine Applications to Technical Information Center Operations

MATIF: Marché À Terme International de France

MATIP: Mapping of Airline Traffic over Internet Protocol

Matl: Material

MatL: Materials Letters (periodical code)

MATL: Material

MatM: Mathematical Modelling (periodical code)

MATOGW: Maximum Allowable Takeoff Gross Weight

MATPS: Machine-Aided Technical Processing System

MATS: Maintenance Analysis Test Set • Major Assembly Tracking System • Manufacturing And Test Schedule • Mesoscale Atmospheric Transport Study • Military Air Transport Service • Mission Analysis and Trajectory Simulation • Monitoring And Test System • MOTIF Acquisition Telescope System • Museum Advisory and Training Service

MATSS: Mobile Aerial Target Support System

MatST: Materials Science Technology (periodical code)

MATT: Multimission Advanced Tactical Terminal

MatTe: Matériaux et Techniques (periodical code)

MATU: Micro-Accelerometer Test Unit

MATV: Master Antenna Television • Multiaxis Thrust Vectoring

MAU: Maintenance Analysis Unit • Math Acceleration Unit • Maupiti Island (Society Islands) (location identifier) • Media Access Unit • Medium Access Unit • Medium Attachment Unit • Mega Accounting Units • Million Accounting Units • Modular Avionics Unit • Mount Allison University (Canada-NB) • Multipoint Access Unit • Multistation Access Unit

Maur: Mauritius

Mau Re: Mauritian Rupee (currency)

MAUS: Materialwissenschaften Autonome (Experiment) Unter Schwerelosigkeit

MAUTO: Miranda-Ariel-Umbriel-Titania-Oberon • Mispronunciation Afflicts Uranus Too Often • My Angel Uriel Takes Opium

mA/V: Milliamperes per Volt

MAV: Maloelap Island (Marshall Island) (location identifier) • Mars Ascendant Vehicle • Mars Ascent Vehicle • Maturitätsanerkennungs-Verordnung • Mean Absolute Value • Micro Air Vehicle

MAVAD: Manchester (Starlink internal code)

MAVDM: Multiple Application Virtual DOS Machine

MAVES: Mars And Venus Exploration Studies

MAVIS: Multiaxis Vibration System

MAVIX: Multimedia Audiovisual Information Exchange

MAVO: Middelbaar Algemeen Voortgezet Onderwijs

MAVS: Mean Absolute Value Slope • Multispectral Airborne Video System

MAVSS: Monostatic Acoustic Vortex Sensing System

maW: mit anderen Worten

MAW: Malden (USA-MO) (location identifier) • Mathematics Awareness Week • Met Andere Woorden • Metropolitan Area Window • Microsoft At Work • Ministry of Agriculture and Water • Modular Abstandwaffe

MAWD: Mars Atmospheric Water Detection • Mars Atmospheric Water Detector

MAWG: Maintenance Analyzer Working Group

Max: Maximum

MAX: Matam (Senegal) (location identifier) • Maximum • Metropolitan Area Exchange • Metropolitan Area Express • Microwave Anisotropy Experiment • Mid-infrared Array Expandible • Millimeter-wave Anisotropy Experiment • Millimeter Anisotropy Experiment • Mountain Air Express

MAXENT: Maximum Entropy

MAXIE: Magnetospheric Atmospheric X-ray Imaging Experiment

MAXIM: Micro-Arcsecond X-ray Interferometer Mission

MAXIMA: Millimeter Anisotropy Experiment Imaging Array

MAY: Mangrove Cay (Bahamas) (location identifier) • Mayo

MAZ: Main Activity Zone • Mayaguez - Eugenio M. de Hostos (Puerto Rico) (location identifier)

mb: Millibar • Millibarn

Mb: Megabit • Millibar

M&B: Mild & Bitter

MB: Maandelijks Bulletin • Magnetband • Magnetic Bearing • Magnetic Breaking • Main Battery • Main Bus • Management Baseline • Manitoba (Canada) • Manned Base • Marine Board • Marker Beacon • Master Beacon • Matter-Bounded • Mbundu • Megabaud • Megabyte • Memory Buffer • Mid-Body • Middle Button • Millibar • Monthly Bulletin • Moored Buoy • Multiband

MBa: Mombasa (Kenya)

MBA: Main Baffle Assembly • Main-Belt Asteroid • Massbus Adapter • Master in Business and Administration • Master of Business Administration (*) • Milne Bay Air • Mombasa (Kenya) (location identifier) • Multibeam Antenna • Multiple Beam Antenna • Multiple Berthing Adapter

MBAC: Marshall Booster Assembly Contractor • Master of the British Association of Chemists

MBAL: Minimum Biologically Acceptable Level

mbar: Millibar

MBAR: Marshall Baseline Accounting and Reporting (system)

MBARI: Monterey Bay Aquarium Research Institute (USA-CA)

MBAS: Methylene Blue Active Substance

MBASIC: Microsoft BASIC • Model-Based Analysis Synthesis Image Coding

MBB: Make-Before-Break • Marble Bar (Australia-WA) (location identifier) • Messerschmitt-Bölkow-Blohm GmbH (Germany)

MBC: Maximum Breathing Capacity • M'Bigou (Gabon) (location identifier) • Methyl Benzimidazolecarbamate

MBCP: Mission Buget Capture Program • Multibyte Code Page

MBCS: Meteor Burst Communication System • Motion-Base Crew Station • Multibyte Character Set

mbd: Million Barrels per Day

MBD: Mission Baseline Description • Mmabatho (South Africa) (location identifier) • More-Better-Different • Multibody Dynamics

MBDB: Microsecond Burst Database

MBDS: Minimum Basic Dataset

MBE: Modified Bernouilli Equation • Molecular Beam Epitaxy • Monbetsu (Japan) (location identifier)

MBF: Molecular Beam Facility • Mount Buffalo (Australia-VIC) (location identifier)

MBFP: Manufacturing Build and Flow Plan

MBG: Mobridge (USA-SD) (location identifier) • Montréal Blue Galaxy

mbH: mit beschränkter Haftung

MBH: Maryborough (Australia-QLD) (location identifier) • Massive Black Hole

MBI: May Be Issued • Mbeya (Tanzania) (location identifier) • Molecular Biology Institute • Multibus Interface

Mbit: Megabit

MBIU: Multibus Interface Unit

MBJ: Montego Bay (Jamaica) (location identifier)

MBK: Matupa (Brazil-MT) (location identifier) • Medications and Bandage Kit

Mbl: Mobile

MBL: Manistee (USA-MI) (location identifier) • Marine Boundary Layer • Martian Boundary Layer • Memory-Based Learning • Microcomputer-Based Laboratory • Mobile

MBLCC: Marine Boundary Layer Convective Complex

MBLV: Multiple-Beam Laser Velocimeter

MBM: Magnani-Blitz-Mundy • Magnetic Bubble Memory • Mkambati (South Africa) (location identifier) • Model Buffer Manager

MBMM: Massively Parallel Mesoscale Model

MBMS: Model Base Management System • Molecular-Beam Mass Spectrometry

MBN: Mount Barnett (Australia-WA) (location identifier)

MBO: Mamburao (Philippines) (location identifier) • Management By Objectives • Middelbaar Beroeps-Onderwijs • Modine-Benstead Observatory (USA-WI) • Monostable Blocking Oscillator

MBOH: Minimum Break Off Height

MBONE: Multicast Backbone

MBOR: Management By Objectives and Results

MBox: Mail Box

MBP: Mean Blood Pressure • Moyobamba (Peru) (location identifier) • Multibody Platform

mbpd: Million Barrels per Day

MBPN: Multilayer Back Propagation Network

MBPR: Memory-Based Program Response

MBPRE: Multitype Branching Process in a Random Environment

Mbps: Megabits per Second • Millions Bits per Second

MBps: Megabytes per Second

MbpS: Megabits per Second

MBPS: Main Bus Power Supply • Megabytes Per Second (*) • Millions Bits Per Second

MBPT: Many-Body Perturbation Theory

MBQ: Mbarara (Uganda) (location identifier)

MBQS: Medium-Bright Quasar Sample • Medium-Bright Quasar Survey

Mbr: Member

MBR: Management By Results • Marker Beacon Receiver • Mars Balloon Relay • Master Boot Record • Mbout (Mauritania) (location identifier) • Mechanical Buffer Register • Medium Bit Rate • Memory-Based Reasoning • Memory Buffer Register • Microwave Background Radiation • Minimum Bounding Rectangle • Mission Briefing Room • Model-Based Reasoning

MBRA: Multibeam Radiometer Antenna

MBRV: Maneuverable Ballistic Reentry Vehicle

Mb/s: Megabits per Second

MBS: Bay City (USA-MI) (location identifier) • Magnetron Beam Switching • Malta Board of Standards (now MSA) • Master Boot Sector • Maximum Burst Size • Midland (USA-MI) (location identifier) • Minimum Bias Set • Mjølner BETA System • Mobile Base Support • Mobile Broadband Service • Mobile Broadband System • Mobile Broadcast Service • Mobile Broadcast System • MRS Base System • Multibody System • Saginaw (USA-MI) (location identifier)

MBSc: Master of Business Science

MBSL: Multiple Bubble-Sonoluminescence

MBST: Multiple-Switching Tube

MBSU: Main Bus Switching Unit

mbt: Met Betrekking Tot

MBT: Masbate (Philippines) (location identifier) • Maximum Burst Tolerance • Mean Body Temperature • Mechanical Bathythermograph • Meyer Binocular Telescope

MBTA: Massachusetts Bay Transportation Authority (USA-MA)

MBTI: Myers-Briggs Temperament Indicator • Myers-Briggs Type Indicator

MBU: Mbambanakira (Solomon Islands) (location identifier) • Multiple Bit Upset

mbv: Met Behulp Van

MBV: Main Base Visit • Masa (Papua New Guinea) (location identifier)

MBW: Moorabbin (Australia-VIC) (location identifier)

MBWA: Management By Wandering Around

MBX: Mailbox • Maribor (Slovenia) (location identifier)

MBY: Moberly (USA-MO) (location identifier)

Mbyte: Megabyte

MBZ: Mandatory Broadcast Zone • Maués (Brazil-AM) (location identifier) • Must Be Zero

m/c: Mi Cuenta • Motorcycle

mc: Millicurie

mC: Millicoulomb

Mc: Megacurie • Megacycle

M&C: Maintenance & Calibration • Maintenance & Checkout • Monitor & Control • Monitoring & Control

M/C: Main Chamber • Manchester (UK-England)

MC: Machine Check • Machine Console • Magellanic Cloud • Magnetic Core • Magnetic Course • Main Chamber • Main Chute • Major Component • Management Committee • Mantle Cavity • Manual Control • Mapping Camera • Mapping Center • Mare Crisium (Moon) • Marginal Check • Marginal Cost • Mars Chart • Mastercard • Master Clock • Master Control • Master of Ceremonies • Material Code • Material Control • Materiel Control • Maximum Clique • Medical Center • Medical Certificate • Medical Corps • Medium Coeli • Member of Congress • Member of Council • Memorandum of Conditions • Memory Clear • Memory Configuration • Memory Consumption • Memory Control • Méridien Central • Message Composer • Meter-Candle • Metric Camera • Metropolitan Counties (UK) • Micronesia • Midcourse • Midcourse Correction • Military Characteristic • Military Cross • Mills Cross • Mini-Cartridge • Mínimos Cuadrados • Mission Capability • Mission Completion • Mission Computer • Mission Continuation • Mission Control • Mission Controller • Mission-Critical • Mnemonic Code • Mode Change • Modular Computer • Modulation Collimator • Moindres Carrés • Mois Courant • Molecular Cloud • Molecular Cluster • Molecular Computing • Molonglo Cross • Momentary Contact • Monaco (international vehicle registration code) • Monaco (ISO country code) • Monaco (postal country code) • Monte Carlo • Monte-Charge • Morse Code • Moyens Communs • Multicarrier • Multicoated • Multiconfiguration • Multiple Choice • Multiple Copy • Multiply-Convolve • Mushroom-shaped Current

MCA: Macenta (Guinea) (location identifier) • Machine à Calculer Analogique • Machine à Calculer Arithmétique • Main Console Assembly • Main Control Assembly • Management Control Authority • Master Clock Assembly • Master Control Assembly • Maximum Crossing Altitude • Median Cut Algorithm • Message Channel Agent • Microchannel Architecture • Microwave Control Assembly • Minimum Controllable Airspeed • Minimum Crossing Altitude • Mise en Configuration d'Arrêt • Monetary Compensatory Amount • Monitoring and Control Assembly • Monitor Controller Assembly • Motor Control Assembly • Motoring and Control Assembly • Multichannel Analyzer • Multichip Assembly • Multicriteria Analysis • Multiple Channel Analyzer • Multiple-Classification Analysis • Multiple Correspondence Analysis

MCAC: McLennan County Astronomy Club (USA-TX)

MCAD: Mechanical Computer-Aided Design

MCAE: Mechanical Computer-Aided Engineering

MCAI: Microcomputer-Assisted Instruction

MCAIERS: Microcomputer Aerial Image Enhancement Reconnaissance System

MCAO: Mount Cuba Astronomical Observatory (USA-DE) • Mt. Cuba Astronomical Observatory (USA-DE) • Multiconjugate Adaptive Optics

MCAP: Multiple-Channel Analysis Program

MCAR: Missing Completely At Random

MCAS: Marine Corps Air Station • Medina County Astronomical Society (USA-OH)

MCAT: Manufacturing Cell Action Team • Medical College Admission Test

MCATU: Master Code Acquisition and Tracking Unit

MCAV: Modified Constant Angular velocity

MCB: McComb (USA-MS) (location identifier) • Memory Control Block • Message Control Block • Miniature Circuit Breaker • Module Control Block

MCBF: Mean Cycle Between Failures

MCBIAS: Microcomputer Based Image Analysis System

McC: McCormick

MC&C: Measurement, Command & Control

MCC: Main Combustion Chamber • Main Communications Center • Main Control Console • Maintenance Control Center • Man-Computer Communication • Master Control Center • Master Control Computer • Master Control Console • Material Control Code • Measurement, Command and Control • Mechanical-Chemical Code • Medical Care Centre • Mesoscale Cellular Convection • Mesoscale Convection Complex • Mesoscale Convective Complex • Micro Compact Car • Microelectronics and Computer Corporation • Midcourse Correction • Middlesex Community College • Ministère de la Culture et de la Communication (France) • Mission Control Center • Mobile Command Center • Mosaic Communications Corporation (USA-IL) • Motor Control Center • Multichannel Communications Controller • Multiple-Chip Carrier • Multiple-Communications Control • Multiple-Computer Complex • Museo de la Ciencia y del Cosmos (Spain) • Sacramento McClellan (USA-CA) (location identifier)

MCCA: Manual Cloud Cover Assessment • Marché Commun Centre-Américain • Media Conversion Computer Assembly

MCCC: Mission Control and Computing Center

MCCD: Mees Charge-Coupled Device • Monthly Cold Cloud Duration

MCCEM: Multichamber Concentration and Exposure Model

MCCG: Microgravity Combustion Coordination Group • Mission Command and Control Group

MCCI: Monte Carlo Configuration Interaction

MCCL: Maintainability Criteria Checklist • Maintenance Criteria Checklist • Microcomputer Compiler Language

MCCP: Maintenance Control Console Panel • Mission Contamination Control Plan • Mission Control Computer Program • Multichannel Colour Photometer

MCCR: Medical Care Cost Recovery • Mission-Critical Computer Resource

MCCS: Mission Control Center Simulation • Mission Control Center System • Mission-Critical Computer System

MCCT: Materials Center for Combustion Turbines

MCCU: Multiple-Communications Control Unit

McD: McDonald

M&CD: Metals & Ceramics Division (ORNL)

MCD: Mackinac Island (USA-MI) (location identifier) • Magnetic Circular Dichroism • Major Civil Division • Massimo Comune Divisore • Master Clerical Data • Mathematics and Computer Division • Maximum-likelihood Convolutional Decoder • Mean Cell Diameter • Mean Corpuscular Diameter • Memory Control Data • Minimum Cost Design • Minor Civil Division • Mission Control Directorate • Monitor Criteria Data • Multichannel Detector • Multiconcentration Diffusion • Multimedia Cartridge Drive

MCDF: Multiconfiguration Dirac-Fock

MCDM: Multi-Criteria Decision Making • Multiple Criteria Decision Making

MCDN: Micro Cellular Digital Network

MCDP: Microprogrammed Communication Data Processor

MCDR: Mission Critical Design Review

MCDS: Maintenance Control and Display System • Malfunction CRT Display System • Management Communication and Data System • Mission Control Data Set • Monitor and Control Display System • Multicommand Data System • Multifunction CRT Display System

MCDU: Multifunction Control Display Unit • Multifunction CRT Display Unit • Multipurpose Computer Display Unit

MCDV: Maximum Cell Delay Variance

MCDW: Monthly Climatic Data for the World (*) • Monthly Climatic Data of the World

MCE: Machine à Calculer Électronique • Magnetic Cumulative Explosion • Management Center Europe • Manpower and Cost Estimate • Marché Commun Européen • Marginal Cost Efficiency • Master of Civil Engineering • Media Conversion Equipment • Mercato Comune Europeo • Merced (USA-CA) (location identifier) • Mini-Creation Event • Mission Control Element • Mission Control Equipment • Model Control Element • Modular Camera Electronics • Monitor and Control Electronics • Multiple Coulomb Excitation

MCET: Massachusetts Corporation for Educational Telecommunications (USA-MA)

Mcf: Thousands Cubic Feet

MCF: Magnetic Confinement Fusion • Maintenance and Checkout Facility • Master Calibration File • Master Code File • Master Control Facility • Master Control File • Methane Conversion Factor • Mission Control Facility • Moderate Cooling Flow • Modular Combustion Facility • Modulation Contrast Function • Multicomplementary Filter • Multiple-Cassegrain Feed • Multiple-Code Field • Multiple-Cost Factor • Mutual Coherence Function • Tampa - Mac Dill (USA-FL) (location identifier) • Thousands Cubic Feet

MCFC: Molten Carbonate Fuel Cell

Mcfd: Thousands Cubic Feet per Day

MC&G: Mapping, Charting & Geodesy

MCG: Magnetic-Cumulative Generator • Mapping, Charting and Geodesy • McGrath (USA-AK) (location identifier) • Modèle de Circulation Générale • Morphological Catalogue of Galaxies

MCGA: Modèle de Circulation Générale de l'Atmosphère • Monochrome/Color Graphics Adapter • Multicolor Graphics Adapter • Multicolor Graphics Array (*)

m/cgo: Mi Cargo

McGU: McGill University (Canada-QC)

MCH: Machala (Ecuador) (location identifier) • Maintenance Channel • Methylcyclohexane • Monitor and Command Handler • Monitor and Control Handler

MCHB: Maintenance Channel Buffer

MChE: Master of Chemical Engineering

MCHF: Multiconfiguration Hartree-Fock

Mcht: Merchant

Mchry: Machinery

Mchy: Machinery

mCi: Millicurie

MCI: Kansas City (USA-MO) (location identifier) • Massive Cluster Ionization • Master Configuration Index • Matsushita Communication Industries (Japan) • Media Command Interface • Media Control Interface • Meridian Control Integrator • Microwave Communications Inc. • Midcourse Interceptor • Modular Common Inlet • Multiport Communications Interface

MCIAS: Multichannel Intelligent Announcement System • Multichannel Intercept Announcement System

MCID: Modular Common Inlet Device

McIDAS: Man-computer Interactive Data Access System (*) • Map-computer Interactive Data Access System

MCIDAS: Man-Computer Interactive Data Access System

MCIF: Miniature Card Implementers Forum • Molecular Contamination Investigation Facility

MCIR: Mission Configuration and Implementation Requirements

MCIS: Master of Communication and Information Studies • Material Control and Information System

MCIU: Manipulator Controller Interface Unit • Master Control and Interface Unit • Mission Control and Interface Unit • Master Control Interface Unit

MCJ: Maicao (Colombia) (location identifier)

MCK: McCook (USA-NE) (location identifier)

MCL: Denali National Park (USA-AK) (location identifier) • Maison de la Culture et des Loisirs • Mass Cargo Lander • Mass Change Log • Master Configuration List • Maximum Contaminant Level • Maximum Contamination Level • Message Control Language • Microfax Command Language • Microprogram Control Logic • Microsoft Compatibility Laboratories • MIDAS Command Language • Middle Compass Locator • Minimal Computer Load • Mount McKinley (USA-AK) (location identifier) • Mt. McKinley (USA-AK) (location identifier) • Multicolor Laser

MCLA: Massachusetts College of Liberal Arts (USA-MA)

MCld: Molecular Cloud

MCLD: Molecular Cloud • Multicolor Laser Display

MCLG: Maximum Contaminant Level Goal

MCLR: Maximum Cell Loss Ratio

MCLS: Multicolor Lightcurve Shape (method)

MCLT: Maximum Cruise Level Thrust

mcm: Minimo Comune Multiplo

MCM: MADS Control Module • Magnetic-Core Memory • Main Control Module • Manned Circumlunar Mission • Martian Climate Model • Mathematical Contest in Modelling • Maximum-Correlation Method • Megawatt Cassegrain Monopulse • Meson Cloud Model • Microcircuit Module • Microcontaminometer • Million Cubic Meters • Ministerial Council Meeting • Montants Compensatoires Monétaires • Monte Carlo (Monaco) (location identifier) • Monte Carlo Method • Multicarrier Modulation • Multicarrier Modulator • Multichip Module • Multicloud Model • Multiconference Management • Multiply-Convolve-Multiply

MCMC: Markov Chain Monte Carlo

MCMG: Man-Carrying Motion Generator

MCMM: Multichip Memory Module

MCMN: Marché Commun de la Mer Noire

MCMS: Multichannel Memory System

MCN: Macon - Lewis B. Wilson (USA-GA) (location identifier) • Maintenance Change Notice • Maintenance Control Number • Management Change Notice • Manual Control Number • Manufacturing Change Notice • Marégraphe Côtier Numérique • Master Change Notice • Master Control Number • Material Change Notice • Mathieu Characteristic Number • Minimum-Cost Network

MCNC: Microelectronics Center of North Carolina

MCNF: Minimum Cost Network Flow

MCNP: Monte Carlo Neutron Photon

MCNS: Multimedia Cable Network System

Mco: Morocco

M&CO: Maintenance & Checkout

MCO: Maintien en Condition Opérationnelle • Mars Climate Orbiter • Miscellaneous Charges Order • Mission Control Operation • Monaco (ISO country code) • Multiple-Channel Oscilloscope • Orlando (USA-FL) (location identifier)

MCOMIB: Mars Climate Orbiter Mishap Investigation Board (NASA)

MComM: Mathematical and Computer Modelling (periodical code)

MCOP: Mission Control Operations Panel

MCOR: Member of Committee of Region

MCOV: Main Chamber Oxidizer Valve

MCP: Macapá (Brazil-AP) (location identifier) • Main Control Panel • Maintenance Control Point • Management Control Plan • Maneuver and Constraints Program • Manual Control Panel • Master Change Package • Master

Change Proposal • Master Computer Program • Master Control Program • Master of City Planning • Materials Control Plan • Maximum Clique Problem • Maximum Continuous Power • Measurements Control Procedure • Memory-Centered Processing • Memory-Centered Processor • Merrillville Community Planetarium (USA-IN) • Message Control Program • Metal-Coated Particle • Meteorological Communications Package • Microchannel Plate • Microsoft Certified Professional • Minimum Convex Polygon • Minor Cycle Period • Mission Communications Payload • Mission Concept Paper • Mission Control Program • Mission Control Programmer • Mode Control Panel • Molecular Cloud Phase • Monitoring and Control Panel • Multibeam Communications Package • Multi-Channel Plate • Multicomponent Plasma • Multiple-Chip Package • Multiple-Comparison Procedure • Multiprotocol Communications Processor

MCPA: Memory Clock Pulse Amplifier

MCpB: Multiple Channel per Burst

MCPB: Multiple Channel Per Burst

MCpC: Multiple Channel per Carrier

MCPC: Manipulator Controller Power Conditioner • Multiple Channel Per Carrier

MCPF: Modular Containerless Processing Facility

MCpH: Maintenance Cost per Hour

MCPH: Maintenance Cost Per Hour

MCPP: Mission Configuration Product Plan

MCPPM: Materiel Control Policies and Procedure Manual

MCP-PMT: Microchannel Plate Photomultiplier

MCPPMT: Microchannel Plate Photomultiplier

MCPS: Major Cost Proposal System • Mechanical Copyright Protection Society • Microsoft Certified Product Specialist

MCPT: Microchannel Plate Tube

MCQ: Miskolc (Hungary) (location identifier)

MCR: Magnetic Card Reader • Magnetic Character Reading • Magnetic Character Recognition • Main Control Room • Maintenance Control Report • Manpower Control Report • Manufacturing Change Request • Master Change Record • Master Change Request • Master Clock Receiver • Master Control Room • Master Control Routine • Maximum Continuous Rating • Melchor de Menco (Guatemala) (location identifier) • Middle Common Room • Minimum Cell Rate • Mission Concept Review • Mission Control Room • Mission Control Routine • Mobile Control Room • Multichannel Radiometer • Multichannel Receiver • Multispectral Cloud Radiometer

MCRDF: Mission Configuration Requirements Data File

MCRF: Mission Configuration Requirements File

MCRG: Mission Configuration Requirements Group

MCRR: Machine Check Recording and Recovery

MCRS: Maintenance Computing and Recording System • Malaysian Centre for Remote Sensing

MCRT: Multichannel Rotary Transformer

MCS: Maintenance and Checkout Station • Maintenance and Control System • Maintenance Cost System • Major Component Schedule • Management Control System • Maneuver Control System • Manned Core Station • Manufacturing Control System • Mapping Camera System • Maritime Communication System • Master Circuit System • Master Control Station • Mathematical Code System • Mathematics and Computer in Simulation • Matrix Common Substructure • Matrix Common Structure • Mean Crew Size • Measurements Calibration System • Media Compensation System • Mercalli-Cancani-Sieberg • Meridian Control Signal • Mesoscale Cloud System • Mesoscale Convective System • Message Control System • Michigan Curtis Schmidt • Microcomputer System • Microcryogenic System • Microsoft Consulting Services • Microwave Carrier Supply • Microwave Communication System • Minimum Chi-Square • Miscellaneous Command Ship • Mission Control Station • Mission Control System • Model-Controlled System • Module Computer System • Monitor and Control System • Monitor Control System • Monitoring and Control System • Monte Carlo Simulation • Monte Caseros (Argentina-CR) (location identifier) • Monthly Cost Summary • Motor Circuit Switch • Mount Control System • Multichannel Scaler • Multichannel Scaling • Multichannel Survey • Multiple Character Set • Multiple Chemical Sensitivity • Multiple-Computer System • Multiple-Console Support • Multivendor Customer Service

MCSA: Mir Cooperative Solar Array • Multichannel Spectrum Analyser

MCSC: Microcathod Strip Chamber

MCSCF: Multiconfiguration Self-Consistent Field

MCSE: Microsoft Certified Systems Engineer

MCSEE: Member of the Canadian Society of Electrical Engineers

MCSS: Mask Cassette Support Structure • Mathematics of Control, Signal, and Systems • Monitor and Control Subsystem • Multiformat Communications Signal Simulator

MCSSCC: Multicolor Spin-Scan Cloud Camera

MCSSP: Multicomponent Spatial Search Program

MCSST: Multichannel Sea Surface Temperature

MCST: Materials Information for Science and Technology

MCSV: Mars Crew Sortie Vehicle

MCT: Magnetic Character Typewriter • Mainstream Corporation Tax • Master Control Terminal • Maximum Climb Thrust • Maximum Continuous Thrust • Memory Cycle Time • Mercury Cadmium Telluride • Metabolic Control Test • Methodology Coordination Team • Microwave Ceramic Triode • Minimum Competency Test • Minimum Connecting Time • Minimum Connection Time • Ministério de Ciência e Tecnologia (Brazil) • Mission Control Table • Mission Control Team • Montréal-Cambridge-Tololo (survey) • MOS-Controlled Thyristor • MOS-Controlled Transistor • Multistrip Cesium Thrustor • Muscat Seeb (Oman) (location identifier)

m/cta: Mi Cuenta

MCTEX: Marine Continental Thunderstorm Experiment

MCTF: Microcomputing Task Force

MCTRAP: Mechanized Customer Trouble Report Analysis Plan

MCTS: Master Central Timing System

MCTT: Metal-Ceramic Transmitting Tube

M&CU: Monitor & Control Unit

MCU: Machine Control Unit • Main Control Unit • Manual Control Unit • Master Clock Unit • Master Control Unit • Mechanism Control Unit • Medium Close-Up • Message Construction Unit • Message Control Unit • Microcomputer Unit • Minimum Coded Unit • Mission Computer Upgrade • Mission Control Unit • Montluçon (France) (location identifier) • Multichip Unit • Multipoint Control Unit

MCUIS: Master Control and User Interface Software

MCUOX: Multi-Component Uniform Order-based Crossover

mCV: Magnetic Cataclysmic Variable (star)

MCV: McArthur River (Australia-NT) (location identifier) • Mean Cell Volume • Microbial Check Valve

MCVF: Multichannel Voice Frequency

MCVP: Materials Control and Verification Program

MC&W: Master Caution & Warning

MCW: Mason City (USA-IA) (location identifier) • Master Caution and Warning • Modulated Continuous Wave

MCX: Makhachkala (Russia) (location identifier)

MCY: Maroochydore (Australia-QLD) (location identifier) • Sunshine Coast - Maroochydore (Australia-QLD) (location identifier)

MCZ: Maceio Palmares (Brazil-AL) (location identifier)

MC1: Molonglo Catalogue 1

MC2: Measurement-Computing-Communications • Molonglo Catalogue 2

MC2R: Mobile Computing and Communications Review

MC3: Molonglo Catalogue 3

MC4: Molonglo Catalogue 4

MC5: Molonglo Catalogue 5

m/d: Months after Date

Md: Maryland (USA) • Mendelevium • Mendélévium • Milliarde

M/D: Man/Day

M&D: Modelling & Design

MD: Doctor of Medicine • Machine Direction • Madagascar • Magnetic Deflection • Magnetic Disk • Magnetic Drum • Main Droite • Make Directory • Malfunction Detection • Malicious Damage • Management Data • Management Directive • Management Domain • Managing Director • Manual Direct • Manual Disconnect • Map Distance • Market Day • Marketing Direct • Maryland (USA) • Master Diagram • Master Dimension • Master Directory • Matter-Dominated • Mean Defect • Mean Deviation • Medical Department • Medicinae Doctor • Memorandum of Deposit • Mendoza (Argentina) • Mentally Deficient • Message Data • Message Digest • Metadata • Metallicity Distribution • Meteorology Department • Microdot • Middeck • Middle Dutch • Millenium Dome • Millidiopter • Mini Disk • Miscellaneous Document • Mission Day • Mission Director • Mobile Disk • Modèle Dynamique • Modular Design • Modular Dewar • Moldavie (ISO country code) • Moldova (ISO country code) • Molecular Dynamics • Monitor Display • Monochrome Display • Monotically Decreasing • Months after Date • Monsoon Depression • Motor Drive • Multidirectional • Multidomain • Multinomial Distribution

MDA: MacDonald, Dettwiler and Associates Ltd. (Canada) • Magnetic Deflection Amplifier • Mail Delivery Agent • Main Distribution Assembly • Maintainability Design Approach • Maintenance Data Analysis • Manufacturing Design Analysis • Mars Dust Analyzer • Materials Dispersion Apparatus • McDonnell Douglas Aerospace • Measurement, Decision and Actuation • Mechanically Despun Antenna • Metal Deactivator • Meteoroid Detector-Analyzer • Methodology of Decision Analysis • Middeck Assembly • Minimum Descent Altitude • Minimum Detectable Activity • Minimum Detectable Amount • Mission Data Analysis • Modular Digital Architecture • Moldavie (ISO country code) • Moldova (ISO country code) • Monochrome Display Adapter • Motor Drive Amplifier • Motorized Door Assembly • Multidimensional Access • Multidimensional Analysis • Multiple Diode Array • Multiple-Discriminant Analysis • Multiple-Docking Adapter • Multivariate Data Analysis • Munich Dust Analyzer • San Antonio - Martindale (USA-TX) (location identifier)

MDAC: McDonnell Douglas Astronautics Co. • Multiplying Digital to Analog Converter

MDACS: Modular Digital Access Control System

MDAIS: McDonnell Douglas Aerospace Information Services

MDAM: Multidimensional Access Memory

MDAP: Mutual Defense Assistance Programme

MDAR: Malfunction Detection, Analysis, and Recording

MDAS: Magnetic Drum Auxiliary Sender • Meteorological Data Acquisition System • Mission Data Acquisition System • Mount Diablo Astronomical Society (USA-CA) • Multispectral Data Analysis System

MDASER: Multispectral Data Analysis System for Earth Resources

MDAU: Maintenance Data Acquisition Unit

MdB: Mitglied des Bundestages

MDB: Master Database • Melinda (Belize) (location identifier) • Memory Databank • Meteorological Database • Mission Database • Mission Data Book • Mission Display Board • Movie Database • Multilateral Development Bank • Multiple-Drive Block • Multiplexed Data Bus • Museum Databank

MDBL: Maintainability Data Baseline • Maintainability Design Baseline

MDBST: Multidimensional Binary Search Tree • Multidimensional Binary Spanning Tree

MDC: Main Display Console • Manado (Indonesia) (location identifier) • Marker Distributor Control • Mars Dust Counter • Master DAMA Controller • Master Data Center • McDonnell Douglas Corporation • Meridian Digital Centrex • Message Distribution Center • Meteorological Data Collection • Metropolitan District Commission • Metropolitan District Council • Midcourse Data Center • Middle Developed Country • Minimum Detectable Concentration • Minimum Detectable Contrast • Minimum Distance Classification • Mission Director Center • Mission Duty Cycle • MMA Development Consortium • Modification and Design Control • More Developed Country • Moscow Data Center (Russia) • Motor Driver Chassis • Multidevice Controller • Munich Dust Counter (HITEN)

MDCA: Main Distribution Control Assembly

MDCC: Master Data Control Console

MDCS: Maintenance Data Collection System • Malfunction Display and Control System • Master Digital Command System • Material Data Collection System • Mission Data Collection Sheet

MDD: Machine-Dependent Data • Maintenance Due Date • Mate/Demate Device • Meteorological Data Dissemination • Meteorological Data Distribution • Midland (USA-TX) (location identifier) • Minimum Detectable Dose • Mission Data Display • Model-Dependent Data • Multidimensional Database

MDDB: Multidimensional Database

MDDBMS: Multidimensional Database Management System

MDDS: Material Directory Data Sheet • Meteorological Data Dissemination Service • Modular Design Display System • Multimedia Digital Distribution System

Mddx: Middlesex (UK-England)

MDE: Magnetic Decision Element • Matrix Difference Equation • Matter-Dominated Era • Mechanical Design Engineer • Mechanical Drive Electronics • Mechanism Drive Electronics • Medellín - José Maria Cordova (Colombia) (location identifier) • Medellín - Rio Negro (Colombia) (location identifier) • Meteoroid Detection Experiment • Mission-Dependent Element • Mission-Dependent Equipment • Mission-Dependent Experiment • Modular Design Environment • Modular Design of Electronics • Modular Display Electronics • Multimedia Document Environment

MDES: Multiple Data Entry System

MDF: Macroscopic Degree of Freedom • Magnetic Direction Finding • Main Distributing Frame • Main Distribution Frame (*) • Manipulator Development Facility • Master Document File • Mate/Demate Facility • Mating/Demating Facility • Maxwellian Distribution Function • Mean Daily Frequency • Medford (USA-WI) (location identifier) • Median Destructive Field • Medium-Density Fiberboard • Medium-Density Fibre • Menu Definition File • Metallicity Distribution Function • Mild Detonating Fuse • Minimum-Dose Focusing • Mittwoch-Donnerstag-Freitag

MDFP: MAST Data Format Processing • Mission Data Format Processing

MDFRR: Mission Directors Flight Readiness Review

MDG: Madagascar (ISO country code) • Mission Definition Group • Mudanjiang (China-PRC) (location identifier) • Multiplier Decoder Gate • Multipurpose Display Group

MDGC: Multidimensional Gas Chromatography

MDGD: Mercury-Doped Germanium Detector

MDH: Carbondale - Southern Illinois (USA-IL) (location identifier) • Magnetic Drum Head • Mechanical Data Handling • Minimum Descent Height

MDHMS: McDonnell Douglas Human Modelling System

MDHS: Meteorological Data Handling System

MDI: Magnetic Direction Indicator • Makurdi (Nigeria) (location identifier) • Management Data and Information • Manual Data Input • Market Data Input • Master Direction Indicator • McDonnell Douglas Inc. • Medium-Dependent Interface • Metered Dose Inhaler • Michelson Doppler Imager • Micro Design International • Minimum Discrimination Information • Mission Data Input • Model Deployment Initiative • Multidocument Interface • Multiple-Display Indicator • Multiple-Document Interface

MDIC: Manchester Decoder and Interface Chip

MDIE: Mother-Daughter Ionosphere Experiment

MDIF: Manual Data Input Function • Map Data Interchange Format

MDIM: Mars Digital Image Map

MDIO: Multidimensional Input/Output

MDIS: Manual Data Input Section • Metadata Interchange Specification

Mdise: Merchandise

MDIU: Manned Data Insertion Unit • Manual Data Input Unit

MDJ: Madras (USA-OR) (location identifier)

MDK: Mbandaka (Zaire) (location identifier) • Multimedia Developers Kit • Multimedia Development Kit

Mdl: Middle • Model • Module

MdL: Mitglied des Landtages

MDL: Lei (Moldova) (ISO currency code) • Machine-Dependent Library • Macro Description Language • Magnetic Delay Line • Magnetic Double Layer • Maintenance Diagnostic Logic • Management Data List • Mandalay (Myanmar) (location identifier) • Maquette de Déploiement Lanceur • Master Data Library • Microdevices Laboratory (JPL) • Middeck Locker • Minimum Description Length • Minimum Detectable Level • Minimum Detectable Limit • Mirrored Data Link • Modular Design Language • Modular Dummy Load • Module Design Language • Momentum Dump Logic

Mdlle: Mademoiselle

MDLP: Minimum Description Length Principle • Mobile Data Link Protocol

MDLS: Master's Degree in Library Science

Mdm: Madam • Medium

MDM: Magnetic Disc Memory • Manipulator Deployment Mechanism • Materials Database Management • Materials Division Manual • Maximum Detectable Momentum • Media Device Manager • Medical Decision Making • Medical Monitor • Mega DM • MetaData Manager • Michigan-Dartmouth-MIT • Midnight Density Maximum • Mirror Demonstration Model • Mirror Development Model • Mixed Dark Matter • Module Design Methodology • Monolithic Diode Matrix • Montag-Dienstag-Mittwoch • Mount Diablo Meridian • Multiplexer-Demultiplexer • Munduku (Papua New Guinea) (location identifier)

Mdme: Madam

MDMI: Monodisperse Microparticulate Injection

MDMO: Michigan-Dartmouth-MIT Observatory (USA)

MDMPI: Monodisperse Microparticulate Injection

MDMS: Multiple Database Management System

Mdn: Median

MDN: Madison Jefferson (USA-IN) (location identifier) • Mobile Directory Number

Mdnt: Midnight

Mdnz: Modernize

Mdnzd: Modernized

M&DO: Mission & Data Operations

MDO: Massive Dark Object • Middleton Island (USA-AK) (location identifier) • Multidisciplinary Design Optimization (*) • Multidisciplinary Optimization

MDOA: Mount Diablo Observatory Association (USA-CA)

M&DOD: Mission & Data Operations Directorate

MDOD: Mission and Data Operations Directorate

MDOP: Maximum Design Operating Pressure

MDP: Main Display Panel • Maintenance Display Panel • Malfunction Detection Package • Management Development Programme • Markovian Decision Process • Master Data Processor • Master Display Panel • Meteorid Debris Protection • Meteorological Datum Plane • Mindiptana (Indonesia) (location identifier) • Minimum Deceptive Problem • Minimum Distance Probability • Mission Data Pressure • Mission Data Processing • Mission Data Processor • Modified Dynamic Programming • Mound Prominence • Multidivisional Programme

MDPG: Magnetic Digital-Pulse Generator

MDPH: Maquette de Déploiement Partie Haute

MDPS: Micrometeorid and Debris Protection Shield • Micrometeorid Debris Protection Shield • MOPS Data Processing System

MDQ: Mar del Plata (Argentina-BA) (location identifier) • Minimum Detectable Quantity

MDQL: Multidimensional Query Language

MDQS: Management Data Query System

MDR: Magnetic Dipole Radiation • Magnetic Disc Recorder • Magnetic Document Reader • Magnetic Drum Recorder • Maintenance Demand Rate • Maintenance Design Requirement • Major Design Review • Manufacturing Design Review • Master Data Record • Master Discrepancy Report • Materials Disposition Review • Mechanical Design Review • Medfra (USA-AK) (location identifier) • Medium Data Rate • Memory Data Register • Midterm Design Review • Minimum Daily Requirement • Minimum Data Recorder • Minimum Design Requirement • Minimum Detectable Radiance • Minor Discrepancy Repair • Missing Data Report • Mission Data Reduction • Mission Definition Review • Mock-up Discrepancy Report • Module Description Resource • Monthly Director's Review • Morphology-Density Relation • Motion Detection Radar • Multichannel Data Recorder • Multidisk Reader • Multidrug-Resistant • Multifunction Dual-band Radar

MDRAM: Multiband Dynamic Random Access Memory

MDRC: McDonnell Douglas Research Center

MDRD: Mission Data Requirements Document

MDRE: Mass Driver Reaction Engine

MDRI: Mission Des Relations Internationales • Multipurpose Display Repeater Indicator

MDRS: Management Data Reporting System • Manufacturing Data Retrieval System • Mission Data Retrieval System

MDRU: Mineral Deposit Research Unit

MDS: Macintosh Development System • Magnetic Disk System • Magnetic Drum System • Maintenance Data System • Maintenance Documentation System • Malfunction Detection System • Management Data System • Mass Digital Storage • Master Delivery Schedule • Master Development Schedule • Mécanique Des Structures • Mechanized Documentation System • Medium-Deep Survey •

Message Distribution System • Metadata Standards • Meteoroid Detection Satellite • Microprocessor Development System • Microsoft Directory Services • Microwave Design System • Middle Caicos (Turks & Caicos Islands) (location identifier) • Minimum Discernible Signal • Minimum Discernible System • Mission Development Simulator • Mobile Data Service • Modulate-Demodulate System • Motion Detection System • Motivational Development Seminar • Multidimensional Scaling • Multidisciplinary Simulation • Multiple-Dataset System • Multiple-input Delivery Service • Multipoint Distribution Service

MDSC: Management Data Service Center • Modular Digital Scan Converter

MDSCC: Madrid Deep Space Communications Complex

MDSD: Magnetic Disk Storage Device • Mate/Demate Stiffleg Derrick • Monodigitizer/Stereodigitizer

MDSDS: Mass Data Storage and Delivery System

Mdse: Merchandise

MDSE: Manufacturing and Development Support Equipment

MdsF: Milliards de Francs

MDSF: Mass Data Storage Facility

MDSL: MBR Digital Subscriber Line

MDSP: Multichannel Subtractive Double Pass

MDSRS: Morphology-Dependent Stimulated Raman Scattering

MDSS: Magnetic Drum Storage System • Mass Digital Storage System • Meteorological Data Sounding System • Mission Data Staging Service • Mission Data Staging System (*) • Mission Data Support Service (*) • Mission Data Support System

MDSSC: McDonnell Douglas Space Systems Corporation

MDST: Mission Director Support Team

MDSV: Manned Deep Space Vehicle

MDT: Harrisburg (USA-PA) (location identifier) • Maintenance Data Terminal • Maintenance Demand Time • Mal Des Transports • Management Discovery Tool • Mean Detonating Time • Mean Downtime • Measurement Descriptor Table • Middleton - Olmsted State (USA-PA) (location identifier) • Minimum Detectable Temperature • Mobile Data Terminal • Moderate • Modified Data Tag • Modified Distance Transform • Modular Differentiation Technique • Monitored Drift Tube • Mountain Daylight Time

MDTF: Mixing Due to Tidal Forces

MDTS: Megabit Digital Troposcatter System • Mobile Doppler Tracking Station • Modular Data Transport System

MDTSCO: McDonnell Douglas Technical Services Co.

MDU: Marker Decoder Unit • Market Development Unit • Master Drive Unit • Mendi (Papua New Guinea) (location identifier) • Message Decoder Unit • Mission Data Unit • Multidimensional Unfolding • Multidisplay Unit • Multifunction Display Unit • Multiplexer/Demultiplexer Unit

MdUC: Milliards d'Unités de Compte

MDUC: Meteorological Data Utilisation Centre

MDUS: Machine Drawing Understanding System • Medium-Scale Data Utilization Station • Meteorological Data Utilization Station

MDV: Maldives (ISO country code) • Map and Data Viewer • Medouneu (Gabon) (location identifier)

MDW: Chicago Midway (USA-IL) (location identifier) • Mars Departure Window • Meadow

MDWS: Meadows

MDX: Mercedes (Argentina-CR) (location identifier) • Modular Digital Exchange

MDY: Midway Island (US Minor Outlying Islands) (location identifier) • Month-Day-Year

MDZ: Mendoza - El Plumerillo (Argentina-MD) (location identifier)

mE: meines Erachtens

Me: Maine (USA) • Maître • Mercury • Methyl

M&E: Mechanical/Electrical • Monitoring & Enforcement

ME: Magnitude Estimation • Maine (USA) • Main Engine • Main Entry • Management Engineering • Managing Editor • Manpower Estimate • Manufacturing Engineer • Marine Engineer • Marine Engineering • Mark End • Marriage Encounter • Maser Emission • Master of Education • Master of Engineering • Maximum Effort • Maximum Energy • Maximum Entropy • Mean Error • Mechanical Efficiency • Mechanical/Electrical • Mechanical Energy • Mechanical Engineer • Mechanical Engineering • Medical Examiner • Medio Evo • Medium Energy • Medway (UK postal code) • Melanesia • Metabolizable Energy • Metal Evaporated • Metallurgical Engineer • Methodist Episcopal • Mexico • Microelectronics • Middeleeuwen • Middle East • Middle Eastern • Middle English • Military Engineer • Milne-Eddington • Minimum Energy • Minimum Entropy • Mining Engineer • Mining Engineering • Miscellaneous Equipment • Mission Envelope • Mission Evaluation • Mobiele Eenheid • Mobile Equipment • Mobility Equipment • Modulation Efficiency • Molecular Electronics • Moneda Extranjera • Monitoring and Enforcement • Monoscopic Extraction • Mössbauer Effect • Most Excellent • Motion Estimator • Mouvement Européen • Movimiento Europeo • Multiengine • Multiple Element • Multiprotocol Encapsulation

MEA: Macaé (Brazil-RJ) (location identifier) • Main Electronics Assembly • Main Error Amplifier • Maintenance Engineering Analysis • Material Experiment Assembly • Mean En-route Altitude • Medium-Energy Accelerator • Member of the European Assembly • Minimum Energy Absorbed • Minimum En-route Altitude • Monoethanolamine • More Electric Aircraft • Multimode Error Analysis

MEAD: Maintenance Engineering Analysis Data

MEADS: Maintenance Engineering Analysis Data System • Medium Extended Air Defense System

MEAO: Middelbaar Economisch en Administratief Onderwijs

MEAR: Maintenance Engineering Analysis Request • Maintenance Engineering Analysis Record • Manufacturing Engineering Analysis Report

MeARB: Mémoires de l'Académie Royale de Belgique (periodical code)

Meas: Measurable • Measure • Measurement • Measurements (periodical code)

MEAS: Measurement • Measuring

MEASAT: Malaysia East Asia Satellite

MEB: Main Electronics Box • Master Electronics Board • Maximum Entropy Bubble • Medium-Energy Burst • Melbourne Essendon (Australia-VIC) (location identifier) • Memory Expansion Board • Microscope Électronique à Balayage

MEBO: Main Engine Burnout

Méc: Mécanique

Mec: Mecánica

MEC: Main Engine Console • Main Engine Controller • Manta (Ecuador) (location identifier) • Manual Emergency Control • Master Event Controller • Master Executive Council • Master of Engineering Chemistry • Maximum Endurable Concentration • Member Education Cycle • Member of the Executive Council • Meson Exchange Current • Microbiological Education Committee • Ministerio de Educação e Cultura (Brazil) • Ministerio de Educación y Ciencia (Spain) • Mission Events Controller • Mobile Equipment Console • Multiplex Exchange Channel

MECA: Main Engine Controller Assembly • Maintainable Electronics Component Assembly • Malfunctioned Equipment Corrective Action • Mars Environmental Compatibility Assessment • Mercury Evaporation and Condensation Analysis • Multielement Component Array

MECAS: Mission Equipment Configuration Accounting System

MECB: Missão Espacial Completa Brasileira

Mecc: Meccanica (periodical code)

MECC: Minnesota Educational Computing Consortium (USA-MN)

MECCA: Master Electrical Common Connector Assembly • Model Evaluation Consortium for Climate Assessment

MECE: Mutually Exclusive and Collectively Exhaustive

MecEn: Mechanical Engineering (periodical code)

MECF: Main Engine Computational Facility

Mech: Mechanic • Mechanical • Mechanically • Mechanics • Mechanism

Meckl: Mecklenburg (Germany)

MECL: Motorola Emitter Coupled Logic

MECLSI: Mission Equipment Configuration Launch Site Installation

MECO: Main Engine Cutoff • Manual Equipment Checkout

MECR: Maintenance Engineering Change Request

MEC/S: Medium-Energy Concentrator/Spectrometer

MECS: Mechanical Elevator Control System • Media Equipment Control System • Medium-Energy Concentrator/Spectrometer

MECSLSI: Mission Equipment Cargo Support Launch Site Installation

MECT: Mission Endurance Cycle Test

MECU: Million ECUs

med: Mediocris

Méd: Médico

Med: Median • Medical • Médical • Medicina • Médecine • Medicine • Medieval • Mediterranean • Medium

MEd: Master of Education

ME&D: Manufacturing Engineering & Development

MED: Madinah - Prince Mohammad Bin Abdulaziz (Saudi Arabia) (location identifier) • Manual Electron Device • Manual Entry Device • Manufacturing Engineering and Development • Maximum Equivalent Dose • Mechanical Equipment Design • Median • Medical • Medina (Saudi Arabia) (location identifier) • Medium • Message Entry Device • Microelectronic Device • Minimum Effective Dose • Molecular Electronic Device • Momentum Exchange Device • Monitor Execution and Dump

MEDA: Multiplex Electronic Doppler Analyzer

MEDAC: Meteor Echo Detection And Collection

MEDAD: Mean Effective Depth of the Atmosphere Duct

MEDALUS: Mediterranean Desertification And Land Use

MEDAS: Meteorological Data Acquisition System

MEDC: Mössbauer Effect Data Center

Medd: Meddedeling

MEDEA: Materialwissenschaftliche Experimente Doppelrack für Einzelexperimente und Anlagen • Measurements of Earth Data for Environmental Analysis • Multidiscipline Engineering Design, Evaluation and Analysis

MEDEVAC: Medical Evacuation

MEDF: Maximum Energy Distribution Function

MEDH: Maintainability Design Handbook

MEDI: Marine Environmental Data Information • Mössbauer Effect Data Index

MEDIA: Modèle Eulérien de Dispersion Atmosphérique

MEDICS: Medical Information Computer System

Medit: Mediterranean

MEDLAS: Mediterranean Laser (campaign)

MEDO: Multipole Expansion of Diatomic Overlap

MeDOC: Multimedia Electronic Documents

MEDOC: Multiexperiment Data and Operations Centre • Multiexperiment Data Operations Centre

MedPh: Medical Physics (periodical code)

MEDS: Marine Environmental Data Service • Mechanical Engineering Design System • Minimum Essential Dataset • Multifunctional Electronic Display System (*) • Multifunction Electronic Display System

MEDUSA: Miniaturized Electrostatic Dual-top-hat Spherical Analyzer

MEDY: Mesure Dynamique

MEE: Mare (Loyalty Islands) (location identifier) • Mass Energy Equivalent • Master of Electrical Engineering • Mechanical Evaluation Equipment • Minimum Essential Equipment • Mission-Essential Equipment • Multimedia Enabled Ethernet • Multiple Effect Evaporator

MEECN: Minimum Essential Emergency Communication Network

MEED: Microbial Ecology Evaluation Device

MEEGA: Mars-Earth-Earth Gravity Assist

MEEL: Mission Equipment Essentiallity List

MEEP: Management and Equipment Evaluation Program • Mir Environmental Effects Payload

MEEPS: Mir Environmental Effects Payloads

MEER: Mechanical-Electrical Equipment Room

MEES: Multipurpose Electromagnetic Environment Simulator

MEETAT: Maximum Improvement in Electronics Through Advanced Techniques

MEF: Master E-mail Facility • Melfi (Chad) (location identifier) • Mémoire Fixée • Microwave Emission Fluctuation • Microwave Emission Flux • Multiextension FITS (*) • Multiextension File • Multiextension Format • Multiple Extension FITS

MEFE: Mission Essential Function Equipment

MEFISTO: Mise En Forme de l'Information Scientifique et Technique

MEFOS: Meudon-ESO Fibre Optics Spectrograph

MEFTA: Middle-East Free-Trade Agreement

MEG: Magnetoencephalogramme • Magnetoencephalography • Malange (Angola) (location identifier) • Medium-Energy Grating • Megabyte • Message Entry Generator • Miniature Electrostatic Gyro

MEGA: Mars-Earth Gravity Assist • Meridian of Galaxy

MEGASTAR: Meaning of Energy Growth: an Assessment of Systems, Technologies And Requirements

MEGG: Merging

MEGO: My Eyes Glaze Over

MEGRIN: Multipurpose European Ground-Related Information Network

MEH: Mehamm (Norway) (location identifier)

MEHALICE: Methane and Halocarbons Intercalibration Experiment

MEHQ: Monomethyl Ether Hydroquinone

Mehrw.St.: Mehrwertsteuer

MEHT: Minimum Eye Height over Threshold

MEI: Main Economic Indicator • Main Engine Ignition • Maintenance Effectiveness Inspection • Maintenance Engineering Inspection • Maintenance Evaluation Inspection • Master End Item • Mathematical Engineering Industry (periodical code) • Meridian (USA-MS) (location identifier) • Ministry of Electronic Industries • Modified-Exclusive-Invalid

MEIC: Member of the Engineering Institute of Canada • Multichannel Electronic Image Classifier

MEIG: Main Engine Ignition

MEIMN: Multiend Item Modification Notice

MEIN: Medium-Energy Intense Neutron

MEIRS: Machinery and Electric Information Retrieval Service

MEIS: Medium Energy Ion Scattering • Mineral Exploration Information System • Multispectral Electro-optical Imaging Spectrometer

MEIT: Multiple Element Integration Test

MEIU: Main Engine Interface Unit • Management Education Information Unit

Méj: Méjico

Mej: Mejuffrouw

MEJ: Meadville (USA-PA) (location identifier)

Mejic: Mejicano

MEK: Meknes (Morocco) (location identifier) • Methyl Ethyl Ketone • Mission Equipment Kit

MEKASPEK: Mehrkanal-Spektral-Photometer

MEKC: Micellar Electrokinetic Chromatography

MEKF: Mission Equipment Kit Facility

MEL: Many-Element Laser • Master Equipment List • Material Engineering Laboratory • Mechanical Engineering Laboratory • Medium Expandable Launcher • Melbourne Tullamarine (Australia-VIC) (location identifier) • Minimum Equipment List • Mobile Electronics Laboratory • Modular Electromagnetic Levitator

Melan: Melanesia • Melanesian

Melb: Melbourne (Australia-VIC)

MELCO: Mitsubishi Electric Corporation

MELI: Master Equipment List Index • Minimum Equipment List Index

MELISSA: Micro-Ecological Life-Support System Alternative

MELT: Mantle Electromagnetic and Tomography (experiment)

MeLu2: Lund Medd. Astron. Obs. Ser. II (periodical code)

MELV: Medium Expendable Launch Vehicle • Medium-light Expendable Launch Vehicle • Medium-lite Expendable Launch Vehicle

Mem: Member • Memento • Memoir • Memorandum • Memorial • Memory

Memb: Member

Meml: Memorial

MEM: Main Engine Manoeuvre • Main Engine Mode • Mars Excursion Module • Maximum Entropy Method • Membach • Memory • Memphis (USA-TN) (location identifier) • Meteoroid Exposure Module • Microelectromechanical • Microerosion Meter • Micrometeorite Exposure Module • Middeck Electronics Module • Ministry for Energy and Mines • Module Exchange Mechanism • Multimode Equivalent Network

MEMA: Microelectronic Modular Assembly

MeMaT: Mechanism and Machine Theory (periodical code)

MEME: Multitasking Extensible Messaging Environment

MEMEX: Memory Extended

MEMIC: Medical Microbiology Interdisciplinary Committee

Memo: Memorandum

MEMS: Microbial Ecological Monitoring System • Micro Electrical-Mechanical System • Micro-Electronics Mechanical System • Microelectromechanical Scale • Microelectromechanical System (*)

MEMSAC: Mega Earth Mobile Seismic Array Consortium

MEMU: Manned Extravehicular Manipulating Unit

Men: Mensa-Mensae (constellation)

MEN: Mende Brenoux (France) (location identifier) • Menu • Ministère de l'Éducation Nationale • Multiple-Event Network

MENA: Medium-Energy Neutral Atom • Mission Element Need Analysis

MENESR: Ministère de l'Éducation Nationale, de l'Enseignement Supérieur et de la Recherche (France)

MENESTRIP: Ministère de l'Éducation Nationale, de l'Enseignement Supérieur, de la Recherche et de l'Insertion Professionnelle (France)

MEng: Master of Engineering

MENJS: Ministère de l'Éducation Nationale, de la Jeunesse et des Sports

Mens: Mensile • Mensual • Mensuel

MENS: Mission Element Needs Statement

Mensur: Mensuration

Ment: Mention • Mentioned

Mentd: Mentioned

MENTOR: Meteor Echoes - No Transmitter - Only Receivers

MEO: Maintenance Engineering Order • Major Engine Overhaul • Manned Earth Orbit • Manned Extravehicular Operation • Manteo - Dare County (USA-NC) (location identifier) • Manufacturing Engineering Order • Mass in Earth Orbit • Medium Earth Orbit • Mid-Earth Orbit • Multiple-Entry Object • Municipal Environment Officer

MEOM: Manned Earth Orbit Mission

MEOP: Maximum Expected Operating Pressure

MEOS: Multidisciplinary Earth Observation Satellite

MEOSS: Monocular Electro-Optical Stereo Scanner

MEOTBF: Mean Engine Operating Time Between Failures

MEOW: Mono-Extraction Orthophographic Workstation

MEP: Main Engine Propellant • Main Entry Point • Main Equipment Platform • Maintenance and Enhancement Plan • Management Engineering Program • Management Evaluation Program • Manipulator Engineering Panel • Manufacturing Extension Partnership • Mars Entry Probe • Mars Explorer Program • Master of Engineering Physics • Mean Effective Pressure • Membre of the European Parliament • Mersing (Malaysia) (location identifier) • Meteorological Package • Meteosat Exploitation Project • Microfilm Enlarger Printer • Minimum Entry Point • Mission Effects Projector • Mission Equipment Package

MEPA: Master of Engineering and Public Administration • Meteorological and Environmental Protection Administration • Meteorology and Environmental Protection Administration (Saudi Arabia)

MEPC: Maritime Environment Protection Committee

MEPCO: Meeting of European Planetary and Cometary Observers (*) • Meeting of European Planetary and Comet Observers

MEPED: Medium Energy Proton and Electron Detector

MEPHISTO: Matériel pour l'Étude des Phénomènes Intéressant la Solidification sur Terre et en Orbite

MEPI: Medium-Energy Particle Instrument

MEPP: Management Efficiency Pilot Program • Maximum Entropy Production Principle • Middle East Peace Process

MEPS: Medium-Energy Particle Spectrometer • Modular Electrical Power Station

meq: Milli Equivalent

MEQ: Meulaboh (Indonesia) (location identifier) • Modelo de Engenharia-Qualificação

Mer: Mercantile • Merchandise • Mercoledi • Mercredi • Mercure • Mercury • Meridian • Meridional • Merionethshire (UK-Wales)

MER: Main Equipment Room • Maintenance Engineering Report • Manned Earth Reconnaissance • Mass Energy Relationship • Mechanical Equipment Room • Medium-Energy Resolution • Merced Castle (USA-CA) (location identifier) • Meridian • Metal Evaporated Resistor • Minimum Energy Requirement • Mission Evaluation Room • Multielement Radar • Multiple-Ejector Rack

MERASEX: Mesoscale Rain and Snowfall Experiment

MeRAu: Mesures Régulation Automatisme (periodical code)

Merc: Mercado • Mercantile • Mercoledi • Mercredi • Mercure • Mercurial • Mercury

MERC: Media Equipment Resource Center • Middle East Regional Cooperation

MERCATOR: Miniaturized European Carrier for Applications in Telescience in Orbit

MERCI: Montage d'Étude du Refroidissement par Convection Interne

MERCOSUR: Mercado Comun del Cono Sur

Mercu: Mercury (periodical code)

MeReC: Mechanics Research Communications (periodical code)

MERES: Mineral and Energy Resources Exploration Satellite

MERF: Monthly Estimated Rainfall

MERGE: Model for Evaluating Regional and Global Effects

MERI: Media Equipment Resource Index • Moon-Earth Radio Interferometer

MERID: Meteosat Real-time Image Dissemination

MERIS: Medium-Resolution Imaging Spectrometer

MERIT: Measure Earth Rotation and Intercompare Techniques • Monitor Earth Rotation and Intercompare Techniques • Monitoring Earth Rotation and Intercomparison of Techniques • Monitoring of Earth Rotation and Intercomparison of Techniques (*)

MERL: Materials Engineering Research Laboratory • Materials Equipment Requirement List • Mitsubishi Electric Research Laboratories

MERLIN: Multielement Radio-Linked Interferometer Network

MERMAIDS: Mediterranean Eddy Resolving Modelling and Interdisciplinary Studies

MERP: Marine Ecosystem Response Project

MERS: Message and Error Reporting System • Most Economic Route Selection

MERSAT: Meteorology and Earth Observation Satellite

MERTU: Medical Entomology Research and Training Unit

MERU: Milliearth Rate Unit

MES: Main-Engine Start • Main Equipment Supplier • Manchester Echelle Spectrograph • Manchester Echelle Spectrometer (*) • Manufacturing Execution System • Mated Elements Simulator • Mated Events Simulator • Medan Polonia (Indonesia) (location identifier) • Member and Education Services • Message Exchange System • Microsoft Education Services • Military Engineering Services • Mission Events Sequence • Mobile Earth Station • Molecular Emission Spectrometry • Mössbauer Emission Spectroscopy

MESA: Manned Environmental Systems Assessment • Mathematics - Engineering - Science Achievement • Maximum-Entropy Spectrum Analysis • Microsystems and Engineering Science Applications (USA-NM) • Miniature Electrostatic Accelerometer • Modular Equipment Stowage Area • Modularized Equipment Storage Area • Modularized Equipment Storage Assembly • Modularized Equipment Stowage Area • Modularized Equipment Stowage Assembly • Molecular Engineering, Sensors and Actuators • Multirole Electronically Scanned Array

MESAR: Multifunction Electronically-Scanned Adaptive Radar

MESc: Master of Engineering Science

MESC: Master Event Sequence Controller • Member of the Economic and Social Committee • Ministry of Education, Science and Culture • Mission Events Sequence Controller

MESCAM: Microwave Emission and Scattering Model

MESCO: Moteur d'Étage Supérieur Cryogénique Optimalisé

MeSCT: Measurement Science and Technology (periodical code)

MESE: Maximum Entropy Spectral Emission

MESEEC: Methodologies to Estimate Social, Environmental, and Economic Consequences

MESFET: Metal-Semiconductor Field Effect Transistor

MESG: Microelectrostatic Gyro

MESGA: Microelectrostatic Gyro-Accelerometer

MeSH: Medical Subject Headings

MESH: Management, Environment, Safety and Health

MESI: Modified-Exclusive-Shared-Invalid

MeSJJ: Journal of the Meteorological Society of Japan (periodical code)

MESL: Museum Educational Site Licensing

MESO: Mesoscale

MESR: Ministère de l'Enseignement Supérieur et de la Recherche

MeSRM: Mémoires Scientifiques de la Revue de Métallurgie (periodical code)

MESS: Market Expert Surveillance System • Misalignment Estimation Software System

MESSAGE: Modular Electronic Solid-State Aerospace Ground Equipment

MESSENGER: Mercury Surface, Space Environment, Geochemistry and Ranging (mission) (NASA)

MESSR: Multispectral Electronic Self-Scanning Radiometer (*) • Multispectrum Electronic Self-Scanning Radiometer

Messrs: Messieurs

MEST: Maintenance Engineering Support Team • Mobile Earth Station

MESUR: Mars Environmental Survey • Mars Environment Survey

MESZ: Mitteleuropäische Sommerszeit

Met: Metal • Metallurgical • Metallurgy • Metaphor • Metaphysics • Meteor • Meteoritic • Meteoritics • Meteorological • Meteorology • Metropolitan

Mét: Métal • Métallurgie • Métallurgique • Météorologie • Météorologique

MET: Main Entry Title • Management Entity Task • Marshall Engineering Thermosphere • Master Events Timer • Mean Elapsed Time • Mécanisme Européen de Technologie • Memory Enhancement Technology • Metafile • Meteorological • Microscope Électronique à Transmission • Midden Europese Tijd • Minimum Energy Trajectory • Minimum Exposure Time • Mission Elapsed Time • Mission Entry Time • Mission Environment Tape • Mission Events Timer • Mobile Earth Terminal • Modular Equipment Transporter • Moreton (Australia-QLD) (location identifier) • Multielement Test

MetA: Meteorological Aid Service

META: Megachannel Extraterrestrial Array • Megachannel Extraterrestrial Assay (*) • Methodology for Extracting Text Automatically • Million-channel Extraterrestrial Assay (*)

Metab: Metabolic • Metabolism

Métab: Métabolique • Métabolisme

METAC: Meteorological Advisory Committee

MetAP: Meteorology and Atmospheric Physics (periodical code)

METCAM: Metric Camera

Metal: Metallurgical • Metallurgy • Metalurgia

Metall: Metallurgical • Metallurgy

METAP: Mediterranean Environmental Technical Assistance Programme

Metaph: Metaphysical • Metaphysics

Metaphys: Metaphysics

Metapl: Metaplasmo

Metát: Metátesis

METDAS: Meteorological Data Analysis System

METDRTHIP: Mimas-Enceladus-Tethys-Dione-Rhea-Titan-Hyperion-Iapetus-Phoebe • Miriam's Enchiladas Taste Divine Recently. Tell Her I'm Proud

MetE: Metallurgical Engineer

Meteo: La Météorologie (periodical code)

Meteor: Meteorología • Meteorological • Meteorology

Météor: Météorologie • Météorologique

Meteorol: Meteorological • Meteorology

Météorol: Météorologie • Météorologique

METEOR: Multiple Experiment Transporter to Earth Orbit and Return

METEOSAT: Meteorological Satellite

Meth: Methodist

METI: Meteorological Information

Metic: Meteoritics (periodical code)

Metik: Meteoritika (periodical code)

Metit: Meteorites (periodical code)

M-et-L: Maine-et-Loire (France)

M-et-M: Meurthe-et-Moselle (France)

MetMa: Meteorological Magazine (periodical code)

m. et n.: Mane et Nocte

METOC: Météo-Océanographie

METOP: Meteorological Operational Platform • Meteorological Operational Polar (satellite) • Meteorological Operations

Métr: Métrica

Metr: Metrología

METRAN: Managed European Transmission Network

Metro: Metrologia (periodical code)

Metrol: Metrological • Metrology

Metrop: Metropolitan

Metropol: Metropolitan

MetRu: Meteorologische Rundschau (periodical code)

METS: Microwave Energy Transmission System • Multienvironment Test System

METSAT: Meteorological Satellite

METSIS: Meteorological Satellite Information System

METU: Middle East Technical University (Turkey)

METV: Main Engine Thrust Vector

METVC: Main Engine Thrust Vector Control

MetZe: Meteorologische Zeitschrift (periodical code)

MEU: Main Electronics Unit • Medium-Enriched Uranium • Message Encoder Unit • Monte Dourado (Brazil-PA) (location identifier) • Multiplexer Encoder Unit

MeUpp: Medd. Uppsala Astronomical Observatory (periodical code)

Mev: Mevrouw

MeV: Mega Electronvolts • Million Electronvolts

MEV: Manned Entry Vehicle • Mass Excursion Vehicle • Minden - Douglas County (USA-NV) (location identifier)

Mevr: Mevrouw

MEVTV: Mars: Evolution of Volcanism, Tectonics and Volatiles

MEW: Mweka (Zaire) (location identifier)

MEWG: Maintenance Engineering Working Group

MEWS: Mono-Extraction Workstation

MEWT: Matrix Electrostatic Writing Technique

Mex: Mexican • Mexico

MEX: Estado de México (Mexico) • Mercury Express (mission) • Mexico (international vehicle registration code) • Mexico (ISO country code) • México City - Benito Juarez (Mexico) (location identifier) • Mexique (ISO country code) • Modèle d'Expérimentation • Mondays Excepted

MExE: Mobile Execution Environment

Mex. Sp.: Mexican Spanish

MEY: Maximum Economic Yield • Meghauli (Nepal) (location identifier)

Mez: Mezzo • Mezza • Mezzanine

MEZ: Messina (South Africa) (location identifier) • Mitteleuropäische Zeit

MEZT: Midden Europese Zomer Tijd

m/f: Mi Favor

mf: Medium Frequency • Microfilm • Minor Frame

mF: Millifarad • mittlerer Fehler

Mf: Massif

M/F: Maintenance Factor • Male or Female

M&F: Manufacturing & Facilities • Materials & Facilities

MF: Magnetic Field • Magnetic Focus • Main Feed • Main Frame • Maintenance Factor • Major Frame • Major Function • Mandatory Frequency • Mass Function • Master File • Master Frame • Mate and Ferry • Mated Flight • Medium Frequency • Mégafranc • Membrane Filter • Metal Factor • Metallic Film • Microfiche • Middeck Forward • Middle French • Middle Frequency • Midfuselage • Millions de Francs • Modulated Frequency • Modulation de Fréquence • Modulazione de Frequenza • More Fragments • Motor Field • Move Fragment • Moyenne Fréquence • Multifield • Multifiltration • Multifractal • Multifrequency • Multiplicity Factor • Multiplying Factor

MFA: Mafia (Tanzania) (location identifier) • Manned Flight Awareness • Marconi-Franklin Antenna • Master of Fine Arts • Mean Field Annealing • Mean Field Approximation • Monofluoroacetate • Moveable Finite Automata • Multifunction Antenna • Multifurnace Assembly • Museum of Fine Arts

MFAASAR: Multifrequency Antenna Array Synthetic Aperture Radar

MFAP: Manned Flight Awareness Program

MFAS: Multifibre Area Spectrograph

MFB: Monfort (Colombia) (location identifier)

MFBM: Modified Fuzzy Bog Model

MFBP: Manufacturing Flow and Building Plan • Manufacturing Flow and Build Plan

MFC: Mafeteng (Lesotho) (location identifier) • Main Fuel Control • Manual Flight Control • Manual Frequency Control • Map Feature Coding • Master File Copy •

Maximum Fault Coverage • Microfilm Frame Card • Microgravity Facilities for Columbus • Microsoft Foundation Class • Motor Fuel Consumption • Multiple-Flight Computer • Multiple-Flight Controller

MFCO: Manual Fuel Cutoff

MFCS: Magnetic-Field Calibration System • Manual Flight Control System • Model Following Control System

Mfd: Manufactured

MFD: Magnetic Frequency Detector • Main Feed • Malfunction Detection • Malfunction Display • Manipulator Flight Demonstration (*) • Manipulator Flight Development • Mansfield Lahm (USA-OH) (location identifier) • Master File Directory • Maximum Frequency Difference • Medium-Frequency Domain • Metal Friction Device • Minimum Fatal Dose • Moving-Finite Difference • Multifunctional Display • Multifunction Display

MFDP: Mesoscale Frontal Dynamics Project

MFDSUL: Multifunction Dataset Utility Language

MFDT: Magnetic Field Directional Turning • Mean Fractional Dead Time

MFE: Magnetic-Field Energy • Magnetic-Field Experiment • Magnetic-Field Explorer • Magnetic Fusion Energy • McAllen (USA-TX) (location identifier) • Midfrequency Executive • Mission (USA-TX) (location identifier) • Moving-Finite Element

MFECC: Magnetic Fusion Energy Computer Center

MFED: Manned Flight Engineering Division

MfEE: Ministry for Education and Employment (UK)

MFENet: Magnetic-Fusion Energy Network

MFENET: Magnetic-Fusion Energy Network

MFEX: Microrover Flight Experiment

MFF: Magnetic Field Function • Magnetic Flip-Flop • Millions de Francs Français • Millions of French Francs • Moanda (Gabon) (location identifier)

MFFE: Multifrequency Frontend

MFFS: Microsoft Flash File System

MFFT: Multifunction Focal Tester

Mfg: Manufacturing

MFG: Major Functional Group • Manufacturing • Math Font Group • Muzaffarabad (Pakistan) (location identifier)

MFH: Mesquite (USA-NV) (location identifier) • Multifrequency Heterodyne

MFI: Magnetic-Field Indicator • Magnetic-Field Intensity • Magnetic-Field Investigation • Marshfield (USA-WI) (location identifier) • Medium/Format Identifier • Método del Flujo Infrarojo

MFIE: Magnetic-Field Integral Equation

MFIMR: Multi-Frequency Imaging Microwave Radiometer

MFIR: Moderate Field-of-view Imaging Radiometer

MFJ: Moala (Fiji) (location identifier) • Modified Final Judgement

MFKP: Multifrequency Key Pulsing

MFL: Magnetic-Field Line • Mount Full Stop (Australia-QLD) (location identifier)

MFLOpS: Million of Floating Point Operations per Second

MFLOPS: MegaFLOPS • Million of Floating Point Operations Per Second (*)

MFM: Magnetic-Field Modelling • Magnetic-Field Monitor • Magnetic Force Microscope • Magnetic Force Microscopy • Master File Maintenance • Modified Frequency Modulated • Modified Frequency Modulation • Multilevel Flow Modelling

MFMA: Mean Field Mean Annealing • Monolithic Ferrite Memory Array

MFMMIC: Multifunction MMIC

MFMR: Multifrequency Microwave Radiometer

MFMT: Microwave Frequency Modulation Transmitter

MFN: Milford Sound (New Zealand) (location identifier) • Most Favored Nation

MFO: Major Function Overlay • Master Frequency Oscillator • Meteorologie en Fysische Oceanografie • Mixed-Function Oxidase • Ras Nasrani (Egypt) (location identifier)

MFOD: Manned Flight Operations Directive

MFOV: Multifield Of View • Multiple Field Of View

MFP: Magnetic-Field Perturbation • Main Free Path • Management Framework Plan • Manners Creek (Australia-NT) (location identifier) • Mass Flow Parameter • Master File Program • Matched Filter Performance • Maximum Freezing Point • Mean Free Path • Middle Free Path • Mid-Frequency Preference • Mission Flight Plan • Molecular Force Probe • Molecular Free Path • Monofluorophosphate • Mosaic Focal Plane • Multifunction Peripheral • Mutualité de la Fonction Publique (France)

MFPA: Mosaic Focal Plane Array

MFPE: Mission From Planet Earth

MFPT: Machine Failure Prevention Technology

MFQ: Maradi (Niger) (location identifier) • Mouvement Français pour la Qualité (France)

Mfr: Manufacture • Manufacturer

MFr: Middle French

MFR: Major Frame Ratio • Manipulator Foot Restraint • Maximum Flight Rate • Medford (USA-OR) (location identifier) • Medium Frequency Radar • Medium Frequency Range • Multifilter Radiometer • Multifrequency Receiver • Multifunctional Receiver • Multifunctional Review • Multilink Frame Relay

m/fra: Mi factura

Mfre: Manufacture

MFRM: Mobile Foot Restraint Manipulator

Mfrs: Manufacturers

MFRS: Major Frame Ratio Signal

MFRSR: Multifilter Rotating Shadowband Radiometer

MFS: Macintosh File System • Magnetic-Field Strength • Magnetic-tape Field Search • Master Flexible System • Master of Foreign Service • Master of Foreign Study • Memory-based File System • Miraflores (Colombia) (location identifier) • Model Formulation System • Modified Filing System • Multichannel Flare Spectrograph • Multifunction Sensor • Multifunction System • Multiple-Frequency Synthetizer

MFSK: Multifrequency Shift Keying

MFSP: Macrofunction Signal Processor

MFT: Machu Pichu (Peru) (location identifier) • Master File Table • Master of Foreign Trade • Mean Flight Time • Metal Film Resistor • Metallic Facility Terminal • Meter Fix Time • Module Fixture Transporter • Multiprogramming Fixed Task • Multiprogramming with a Fixed number of Tasks

MFTAD: Master Flight Test Assignment Document

MFTIO: Minimum Flyout Time for Orbital Interception

MFTP: Multicast File Transfer Protocol

MFTL: My Favorite Toy Language

MFTS: Multilateral File Transfer System

MFU: Mfuwe (Zambia) (location identifier)

MFUA: Medical Follow-Up Agency

MFV: Magnetic-Field Vector • Main Fuel Valve • Mars Flyby Vehicle • Melfa - Accomack County (USA-VA) (location identifier) • Microfilm Viewer • Motor Fleet Vessel

MFVT: Mixed-Flow Vectored Thrust

Mfy: Manufactory

MFY: Mayfa'ah (Yemen) (location identifier)

mg: Malagasy (ISO language code) • Milligram

mG: Milligauss

Mg: Magnesium • Magnésium • Megagram

MG: Machien Gun • Madagascar (ISO country code) • Main Gauche • Major General • Malgasia (ISO country code) • Marchandises Générales • Médecin Généraliste • Megagauss • Membership Grade • Meridional Gradient • Meteorological Group • Micrograyity • Middle Gimbal • Military Government • Minas Gerais (Brazil) • Mission Generation • MIT-Greenbank • Mobile Generator • Monguor • Motor-Generated • Motor Generator • Multigrid

Mga: Maga

MGA: Managua - Augusto Cesar Sandino (Nicaragua) (location identifier) • Matrix Graphics Architecture • Media Graphics Architecture • Medium-Gain Antenna • Middle Gimbal Angle • Monochrome Graphics Adapter • Multigraph Architecture • Multimedia Graphics Architecture • Multiple Group Analysis

MGAA: Medium-Gain Autotrack Antenna

MgAL: MgII Absorption Line

MGAS: Middle Georgia Astronomical Society (USA-GA)

MGAWD: Make Good All Works Disturbed

MGB: Metropolitan Green Belt • Mount Gambier (Australia-SA) (location identifier)

MGBX: Middeck Glovebox

MGC: Main Giant Clump • Manual Gain Control • Memoryless Gaussian Channel • Michigan City (USA-IN) (location identifier) • Middle Georgia College (USA-GA) • Morphological Catalogue of Galaxies

MGCD: Microgravity Compatible Drive

MGCM: Mars General Circulation Model • Mars Global Circulation Model

MGCO: Mars Geoscience/Climatology Orbiter

MGCP: Media Gateway Control Protocol • Multiple Gateway Control Protocol

MGCS: Meteosat Ground Computer System

MGD: Magdalena (Bolivia) (location identifier) • Maximally Generalized Description • Mercury Germanium Detector • Military Geographic Documentation • Millions of Gallons per Day

MGDB: Microgravity Database

Mge: Message

MGE: Maintenance Ground Equipment • Marietta Dobbins (USA-GA) (location identifier) • Modular GIS Environment • Multi-Gaussian Expansion

MGEN: Mutuelle Générale de l'Éducation Nationale (France)

MGES: Maintenance Ground Equipment Section

Mgf: Magnify

Mgfd: Magnified

MgF: Magnesium Fluoride

MGF: Franc (Madagascar) (ISO currency code) • Magnetic Field • Maringá (Brazil-PR) (location identifier)

MGFLD: Multigroup Flux-Limited Diffusion

MGG: Margarima (Papua New Guinea) (location identifier) • Marine Geology and Geophysics • Memory Gate Generator

MGGE: Mongolian Geophysical and Geological Exploration

MGGF: Monitoring Global Geophysical Fluids

MGGI: Morphic Generator Grammatical Inference

MGH: Margate (South Africa) (location identifier)

MG/I: Major Gateway/Internet

MGI: Management Group Identifier • Matagorda Island (USA-TX) (location identifier)

MGID: Military Geographic Information and Documentation

MGIM: Micro-Gravity Isolation Mount

MGIO: Mount Graham International Observatory (USA-AZ) • Mt. Graham International Observatory (USA-AZ)

MGIS: Micro Geographic Information System • Military Geographic Information System

MGk: Medieval Greek

MGK: Mong Ton (Myanmar) (location identifier) • Multigraph Kernel

Mgl: Mitglied • Mogul

MGL: Matrix Generator Language

MGLH: Multivariate General Linear Hypothesis

Mglia: Mongolia

MGM: Master Group Multiplexer • Mechanics of Granular Materials • Metro-Goldwyn-Mayer • Modified Gaussian Model • Montgomery Dannelly (USA-AL) (location identifier)

Mgmt: Management

MGMT: Management

Mgn: Magneto • Margin

MGN: Magangue Baracoa (Colombia) (location identifier) • Magellan

Mgna: Montagna

Mgne: Montagne

Mgng: Managing

MGO: Main Geophysical Observatory (Russia) • Manega (Gabon) (location identifier)

mgp: Mètre Géopotentiel

MGP: Magnetic Granulometry Plot • Manga (Papua New Guinea) (location identifier) • Manufactured-Gas Plant • Multiannual Guidance Programme

MGpD: Millions of Gallons per Day

MGQ: Maximum Guaranteed Quantity • Mogadishu (Somalia) (location identifier)

Mgr: Manager • Monseigneur • Monsignor

MGr: Medieval Greek

MGR: Manager • Modular Gas-cooled Reactor • Moultrie Thomasville (USA-GA) (location identifier)

MGRA: Master Geographical Reference Area

Mgrs: Managers • Monseigneurs • Monsignors

MGRS: Military Grid Reference System

MGRW: Matrix Generator and Report Writer

mgs: Meter-Gram-Second

MGS: Mangaia Island (Cook Islands) (location identifier) • Marine Geophysical Survey • Mars Geoscience Surveyor • Mars Global Surveyor (*) • Microcomputer Graphic System • Mobile Ground Station • Mobile Ground System • Moment Gyro System • Multilayer Grating Spectrometer

MGSE: Maintenance Ground Support Equipment • Mechanical Ground Support Equipment

MGSO: Multimission Ground Systems Office

Mgt: Management

MGT: Major Ground Test • Millingimbi (Australia-NT) (location identifier) • Mobile Ground Terminal

MGTP: Model Generation Theorem Prover

MGU: Manaung (Myanmar) (location identifier)

MGV: Margaret River (Australia-WA) (location identifier)

MGVI: MERIS Global Vegetation Index

MGVT: Mated Ground Vibration Test

MGW: Maximum Gross Weight • Morgantown (USA-WV) (location identifier)

MGWR: Morgan-Good Wolf-Rayet

MGX: Moabi (Gabon) (location identifier)

MGY: Dayton - Montgomery County (USA-OH) (location identifier)

MGZ: Mergui (Myanmar) (location identifier) • Myeik (Myanmar) (location identifier)

mH: Millihenry • Millihour

MH: Îles Marshall (ISO country code) • Magnetic Head • Magnetic Heading • Magnetic Hole • Mail Header • Main Hatch • Man-Hour • Marital History • Marshall Islands (ISO country code) • Medal of Honor • Meierhof • Mental Health • Message Handling • Metal Halide • Mexican Hat • Mobile Host • Modified Huffman • Mount Wilson H-alpha • Mt. Wilson H-alpha

Mha: Million Hectares

MHA: Mahdia (Guyana) (location identifier) • Merrill H-Alpha • Minimum Hold Altitude • Minimum Holding Altitude • Mission Hydrographique de l'Atlantique • Modular Hypermedia Application • Multiple Headset Adapter

MHAA: Mid-Hudson Astronomy Association (USA-NY)

MHB: Auckland - Mechanics Bay (New Zealand) (location identifier) • Maintenance Handbook

MHC: Macmahon - Camp 4 (Australia-SA) (location identifier) • Manipulator Hand Controller • Most Helpful Cycle • Mount Holyoke College

MHCC: Mount Hook Community College (USA-OR)

Mhd: Masthead

MHD: Magnetohydrodynamics • Magnétohydrodynamique • Mashad (Iran) (location identifier) • Met Hartelijke Dank • Michigan Henry Draper (catalogue) • Mihalas-Hummer-Däppen • Minimum Hamming Distance • Multihead Disk

MHDR: MNG Header

MHDS: Mosaic Data Handling System

MHE: Material Handling Equipment • Mitchell (USA-SD) (location identifier)

MHEG: Multimedia and Hypermedia Experts Group

MHEO: Multiregional HEO

MHF: Massive Hydraulic Fracture • Massive Hydraulic Fracturing • Medium High Frequency • Mexican Hat Function • Ministerium für Hoch- und Fachschulwesen • Mixed Hydrazine Fuel • Morichal (Colombia) (location identifier)

MHG: Mannheim (Germany) (location identifier) • Middle High German

MHH: Marsh Harbour (Bahamas) (location identifier)

MHI: Mitsubishi Heavy Industries • Musha (Djibouti) (location identifier)

MHK: Manhattan (USA-KS) (location identifier)

MHL: Îles Marshall (ISO country code) • Marshall (USA-MO) (location identifier) • Marshall Islands (ISO country code) • Metastable Helium Level • Microprocessor Host Loader • Minimum Helium Loss

MHM: Minchumina (USA-AK) (location identifier)

MHMT: Multihead-Multitail

MHN: Mullen (USA-NE) (location identifier)

MHO: Massive Halo Object • Millstone Hill Observatory • Mount Hopkins Observatory (USA-AZ) • Mt. Hopkins Observatory (USA-AZ) • Mount House (Australia-WA) (location identifier)

MHOA: Mount Hook Observatory Association (USA-OR)

M. Hon.: Most Honourable

MHP: Move and Handling Procedure

MHPCC: Maui High-Performance Computing Center (USA-HI)

MHQ: Mariehamn (Finland) (location identifier)

MHR: Member of the House of Representatives • Microwave Hologram Radar • Mission Hydrographique de Recherche • Sacramento Mather (USA-CA) (location identifier)

MHRS: Message Handling and Retrieval System • Modified Hazard Ranking System

MHS: Magnetohydrostatistics • Mail Handling System • Many Hilbert Spaces • Medical History Sheet • Message Handling Service • Message Handling System (*) • Microwave Humidity Sounder • Mount Shasta (USA-CA) (location identifier) • Multi-aircraft Horizontally Separated

MHSA: Mars Horizon Sensor Assembly (MO)

MHSC: Manipulator Handset Controller

MHT: Manchester (USA-NH) (location identifier)

MHTC: Most Helpful Total Cut

MHTCC: Most Helpful Total Cut Cancelling

MHTGR: Modular High-Temperature Gas-cooled Reactor • Modular High Temperature Gas Reactor

MHU: Mount Hotham (Australia-VIC) (location identifier)

MHV: Miniature Homing Vehicle • Mojave - Kern County (USA-CA) (location identifier)

MHW: Mean High Water • Multihundred-Watt

MHX: Manihiki Island (Cook Islands) (location identifier)

MHY: Moorehead (Papua New Guinea) (location identifier)

mHz: Millihertz

MHz: Megahertz

MHZ: Mildenhall (UK-England) (location identifier)

mi: Maori (ISO language code) • Mile • Mill (currency)

Mi: Mile • Minor • Mittwochs

M/I: Minimum Impulse

M&I: Management & Integration • Municipal & Industrial

MI: Magnetoimpedance • Main Impulse • Maintenance Instruction • Major Issue • Malleable Iron • Management and Integration • Management Information • Management Interface • Management of Information • Manual Input • Mare Imbrium (Moon) • Marshall Islands (international aircraft registration code) • Master Index • Master Item • Medical Imaging • Medium Intensity • Michigan (USA) • Micro-Image • Mijns Inziens • Military Intelligence • Minimum Impulse • Ministry of Information • Minority Institution • Misiones (Argentina) • Mission Independent • Mode Indicate • Modèle d'Identification • Moment of Inertia • Monitoring Information • Monotically Increasing • Municipal and Industrial • Mutual Inductance • Mutual Interference • Phobos (satellite of Mars)

MIA: Master of International Affairs • Meeting of International Authorities • Miami (USA-FL) (location identifier) • Michigan Array • Minimum IFR Altitude • Missing In Action • Mission-Independent Area • Multiple-Interface Adapter • Multiplexer Interface Adapter • Multiplex Interface Adapter • Multivariate Image Analysis • Multivendor Integration Architecture

MIAC: Meteorites and Impacts Advisory Committee (Canada)

MIADS: Map Information Assembly and Display System

MIAeE: Member of the Institute of Aeronautical Engineers

MIALS: Medium Intensity Approach Light System

MIAM: Member of the Institute of Administrative Management

MIAOU: Mathématiques et Informatique de l'Automatique et de l'Optimisation pour l'Utilisateur

MIAP: Member of the Institution of Analysts and Programmers

MIAPR: Medical Image Analysis and Pattern Recognition

MIARG: Management Information Analysis and Report Generator

MiARI: Mitteilungen Astronomisches Rechen Institut (periodical code)

MIARS: Maintenance Information Automated Retrieval System

MIAS: Management Information and Accounting System • Marine Information Advisory Service • Marine Information and Advisory Service • Multipoint Interactive Audiovisual System • Multispectral Image Analysis System

MI&B: Ministry of Information & Broadcasting

MIB: Management Information Base • Management Information Block • Manual Input Buffer • Master Instruction Book • Master Interconnect Board • Message Information Base • Minimum-Impulse Bit • Minot (USA-ND) (location identifier) • Mission Information Base • Multilayer Interconnection Board

MIBK: Methyl Isobutyl Ketone

MiBud: Mitt. Sternw. Ungar. Akad. (periodical code)

Mic: Micrometer • Microphone • Microscopium-Microscopii (constellation)

MIC: Macro Interpretive Command • Management Information Center • Management Information Class • Management Integration Consortium • Mathematics Information Center • Media Interface Connector • Medium Interface Connector • Memory In Cassette • Memory Interface Connection • Memory Interface Connector • Message Identification Course • Message Integrity Check • Message Integrity Code • Metastable Interstitial Composite • Meteorological Information Committee • Microsoft Internet Chat • Microwave Integrated Circuit • Microwave Integrated Circuitry • Microwave Interference Coordination • Minimum-Impulse Controller • Minneapolis Crystal (USA-MN) (location identifier) • Mission Catalogue • Model-Integrated Computing • Modulation d'Impulsions Codées • Modulation par Impulsions Codées (*) • Module Interface Card • Monitoring, Identification and Correlation • Monolithic Integrated Circuit • Multichannel Image Camera

MICA: Macroinstruction Computer Assembler • Mirror Coronagraph for Argentina • Modem ISDN Channel Aggregation • Multiyear Interactive Computer Almanac

MICADO: Michelson Interferometer for Coordinated Auroral Doppler Observations

MICAP: Monte Carlo Ionization Chamber Analysis Package

MICAS: Miniature Integrated Camera Spectrometer

MICB: Mission Integration Change Board

MICCG: Modified ICCG

MICE: Map Image Congruency Evaluation • Member of the Institution of Civil Engineers • Modular Integrated Communications Environment

MicEc: Microbial Ecology (periodical code)

MICEI: Member of the Institution of Civil Engineers of Ireland

MICG: Mercury Iodide Crystal Growth

MicgQ: Microgravity Quarterly (periodical code)

Mich: Michigan (USA) • Michoacan (Mexico)

MICH: Michoacan (Mexico)

MICIS: Material Inventory Control and Information System • Mechanized Information Control Inventory System • Mechanized Inventory Control Information System

MICNS: Modular Integrated Communication and Navigation System

MICOS: Multifunctional Infrared Coherent Optical Sensor

Micr: Microscope • Microscopy

MICR: Magnetic Ink Character Recognition

MICRA: Motion, Imaging Contouring and Roughness Amplitude

MICREX: Microgravity Research Experiment

Micro: Micronesia • Microscope • Microscopist • Microscopy

MICRO: Multiple Indexing with Console Retrieval Operations

Microbiol: Microbiología

Micros: Microscope • Microscopist • Microscopy

MICROSAT: Microwave Satellite

MICRS: Magnetic Ink Character Recognition System • Main Instrument Console and Readout Station

MICS: Macro Interpretive Commands • Magnetospheric Ion Composition Spectrometer • Management Information and Control System • Management Integrated Control System • Mathematical, Information and Computational Science • Mid-Infrared Camera and Spectrometer

MICSIS: Micro Spatial Information System

MicWa: Microwaves (periodical code)

Mid: Middle • Midlands • Midnight • Midshipman

MID: Magnetically Insulated Diode • Management Information Division • Master of Industrial Design • Mérida (Mexico) (location identifier) • Message Input Device • Middle • Multiple-Ion Detection

MIDAC: Management Information for Decision And Control

MIDAN: Microprocessor Data Analyzer

MIDAP: Multichannel Image Display and Analysis Program

MIDAR: Microwave Detection And Ranging

MIDAS: Maintenance Information Data Automation System • Maklumat Industri dan Sains (Malaysia) • Management Integrated Data Accumulating System • Management Issues in a Distributed ADA Support • Map Information and Data Analysis System • Measurement Information and Data Analysis System • Memory-Implemented Data Acquisition System • Metrology Information And Documentation System • Microcomputer-Interfaced Data Acquisition System • Microimaged Data Addition System • Micro Imaging Dust Analysis System • Miniature Data Acquisition System • Mission Integrated Design Analysis Simulator • Modified Integration Digital Analog Simulator • Modular Interactive Data Acquisition System • MOTIF Integrated Data Applications System • Multidiscipline Interactive Display and Analysis System • Multimedia Industry Development for Academic Software • Multimedia Information Demonstration And Support • Multiple Index Direct Access System • Multiple-Input Data Acquisition System • Multiple-Input Digital Autocorrelation Spectrometer • Multiproject Investigation During the Alignment of Spacecraft • Multisensor Image Database System • Multitask Ice Data Analysis System • Multivariate Interactive Digital Analysis System • München Image Data Analysis System (ESO) • Munich Image Data Analysis System (ESO)

MIDDLE: Microprogram Design Description Language

MIDDS: Meteorological Interactive Data Display System

Middx: Middlesex (UK)

MidEx: Medium-class Explorer (*) • Medium Explorer

MIDEX: Medium-class Explorer (*) • Medium Explorer

MIDF: Major Item Data File • Multiple Input Describing Function

MIDILAH: Mail Distribution Instructions and Labelling Handbook

MIDIST: Mission Interministérielle de l'Information Scientifique et Technique (France)

Midl: Midlands

MIDLA: Media Independent Data Link Architecture

MIDMS: Machine-Independent Data Management System

Midn: Midshipman

MIDN: Midnight

MIDO: Manufacturing Inspection District Office

MIDOT: Multiple Interferometer Determination Of Trajectories

MIDP: Major Item Distribution Plan • Mars Instrument Development Program

MIDPM: Member of the Institute of Data Processing Management

MIDR: Mosaicked Image Data Record

MIDS: Management Information and Data System • Management Information Display System • Matrix Information and Directory Service • Matrix Information Directory Service • Mechanized Information and Documentation System • Multifunctional Information Distribution System • Multifunction Information Distribution System (*) • Multimode Information Distribution System

MIDSIM: Midcourse Simulation

M&IE: Meals & Incidental Expenses

MIE: Master of Industrial Engineering • Meals and Incidental Expenses • Meteor Ionizing Efficiency • Mission-Independent Equipment • Model Institution for Excellence • Most Intense Earthquake • Muncie (USA-IN) (location identifier)

MIEC: Meteorological Information Extraction Centre

MIEE: Member of the Institution of Electrical Engineers

MIEEE: Member of the Institute of Electrical and Electronics Engineers

MIEG: Mission à Intérêt Économique Général

MiElC: Military Electronics Countermeasures (periodical code)

MIEP: Multiplicative Inverse Eigenvalue Problem

MIEWF: Master Integrated Engineering Work File

MIF: Main Image File • Maker Interchange Format • Management Information File • Management Information Format • Master Index File • Master Item File • Mechanical Integration Frame • Mode Indication Function • Module Integration Facility • Module Interconnect Framework • Module Interface Formalism • Monahans - Roy Hurd (USA-TX) (location identifier) • Monopulse Interference Filter • Multiscale Image Fusion

MIFF: Management Information Format File

MIFR: Master International Frequency Register • Monitored International Frequency Register

MIFS: Multiplex Interferometric Fourier Spectroscopy

MIG: MACH Interface Generator • Magnetized Ionized Gas • Metal-Inert Gas • Metal-In-Gap • Miniature Integrating Gyroscope • München Neubiberg (Germany) (location identifier)

MIGA: Message-based Interface Gate Array

MIGMAS: Microgravity Mass Spectrometer

MIGRACION: Milieu Interstellaire Galactique, Rayons Cosmiques et Ionisation dans les Nuages

MIH: Major Industrial Hazard • Mitchell Plateau (Australia-WA) (location identifier)

MiHam: Mitteilungen Hamburger Sternwarte (periodical code)

MII: Deimos (satellite of Mars) • Marília - Dr. Gastão Vidigal (Brazil-SP) (location identifier) • Master Inspection Item • Mathematics In Industry • Microsoft-IBM-Intel • Ministry of Information Industry (China-PRC)

MIICS: Master Item Identification Control System

MIIE: Member of the Institution of Industrial Engineers

MIIF: Master Item Intelligence File

MIIS: Mid-Infrared Imager Spectrometer • Miscellaneous Inputs Information System

Mij: Maatschappij

MIJ: Mili Island (Marshall Islands) (location identifier)

MIJA: Multiple Impact Jet Apparatus

MiJo: Microwave Journal (periodical code)

MIK: Mikkeli (Finland) (location identifier)

MIKE: Mass-analyzed Ion Kinetic Energy

MIKES: Mass-analyzed Ion Kinetic Energy Spectrometry

MIKSAM: Multiple Indexed Keyed Sequential Access Method

Mil: Milano (Italy) • Mileage • Milicia • Military • Militia • Million

MIL: Machine Interface Layer • Magnetic Indicator Loop • Malfunction Investigation Laboratory • Man In Loop • Master Index List • Master Instrumentation List • Micro-Implementation Language • Milano (Italy) (location identifier) • Militaire • Military • One Million

MILA: Merritt Island Launch Area (NASA)

MILC: Modified Intermediate Load Cycle

Milit: Military

Mill: Million

MILLIPOL: Millimeter Polarimeter

MILNET: Military Network

MILO: Magnetically Insulated Line Oscillator

MILOC: Military Oceanography

MILP: Mixed-Integer Linear Program • Mixed-Integer Linear Programming

MILS: Microwave Instrument Landing System • Minicomputer Library System

Milw: Milwaukee (USA-WI)

MIM: Magnetic Interaction Mechanism • Maintenance Instructions Manual • Maquette d'Interrogation Multilingue • Market Information Machine • Master in International Management • Merimbula (Australia-NSW) (location identifier) • Metal Injection Moulding • Metal-Insulator-Metal • Microgravity Isolation Mount • Minicomputer Interface Module • Minimum Information Method • Misappropriation-Interference-Misrepresentation • Modem Interface Module • Modified Index Method • Multiple Instrument Mount

MIMA: Mighty Image Analyzer

MIMC: Multiple Indicator - Multiple Cause

MIMD: Multiple Instruction - Multiple Data

MIME: Multimedia Internet Mail Extension • Multipurpose Internet Mail Extension (*)

MIMI: Magnetospheric Imaging Instrument (Cassini) • Millimetric Ice Mapping Instrument

MIMIC: Micro-Moulding In Capillaries • Microwave Monolithic Integrated Circuit • Millimeter-wave Monolithic Integrated Circuit

MIMLT: Magnetospheric, Ionospheric, Mesospheric and Lower Thermospheric

MIMO: Man In - Machine Out • Multiple Input - Multiple Output

MIMOSA: Mission Modes and Space Analysis

MIMR: Magnetic Ink Mark Recognition • Microwave Imaging Radiometer • Monthly Integration Management Review • Multiband Imaging Microwave Radiometer • Multifrequency Imaging Microwave Radiometer

MIMS: Major Item Management System • Material Information Management System • Medical Information Management System • Meteosat Infrared and Microwave Sounder

MIMSE: Microwave Infrared Mesopheric Stratospheric Experiment

MIMU: Miniature Inertial Measurement Unit

MiMue: Mitteilungen Sternwarte Universität Münster (periodical code)

MIMUSA: Matrix Iteration Method of Unfolding Spectra

Min: Mineral • Minéral • Mineralogical • Mineralogy • Minéralogie • Minería • Minim • Minimum • Mining • Minister • Ministerial • Ministre • Ministry • Minor • Minority • Minuscule • Minute

MIN: Minimum • Minnipa (Australia-SA) (location identifier) • Minute • Mobile Identification Number • Mobile Identity Number

MINC: Minimum Curvature

MIND: Magnetic Integration Neutron Duplicator

MIndE: Master of Industrial Engineering

Mineral: Mineralogía • Mineralogical • Mineralogy

Minéral: Minéralogie • Minéralogique

MINIS: Multipurpose Interactive NASA Information System

Minist: Ministerial • Ministry

MINK: Missouri-Iowa-Nebraska-Kansas

MINLP: Mixed-Integer Non-Linear Program • Mixed-Integer Non-Linear Programming

MinM: Mineralogical Magazine (periodical code)

Minn: Minnesota (USA)

MINOS: Main Injector Neutrino Oscillation Search • Marks Information Optically Stored • Minimum Norm Solution

MINQE: Minimum Invariant Quadratic Estimate

Min. Res.: Minister Resident • Minister Residentiary

MINS: Minor In Need of Supervision

MINT: Magnetic Information Technology • Managing the Integration of New Technology • Marée Interne

Mintr: Miniature

MINUET: Minnesota Internet Users Essential Tool

MINW: Master Interface Network

Min. Wt.: Minimum Weight

MINX: Multimedia Information Network Exchange

Mio: Million • Miocene

MIO: Maidanak International Observatory (Uzbekistan) • Management Information Office • Management Integration Office • Map Information Office • Meteoritic Impact Origin • Metric Information Office • Miami (USA-OK) (location identifier) • Mobile Ionospheric Observatory • Modular Input/Output • Multipurpose Input/Output

MIONP: Micriwave-Induced Optical Nuclear Polarization

MIOP: Mission Instruction and Operation Plan • Multiplexing Input-Output Processor • Musée de l'Instrumentation Optique (France)

MIOS: Multispectral Imaging Optical Sensor

MiOTL: Microwave and Optical Technology Letters (periodical code)

MIP: Machine Instruction Processor • Mandatory Inspection Point • Man Instrument Panel • Maintenance Improvement Program • Maintenance Index Page • Management Improvement Plan • Management Improvement Program • Management Incentive Program • Mandatory Inspection Point • Manual Input Processing • Manual Input Program • Mars In-situ Propellant • Matrix Inversion Program • Maximum Investment Plan • Mean Indicated Pressure • Measurement Integration Panel • Mercury Intrusion Porosimetry • Message Input Processing • Methods Improvement Program • Microwave-Improved Plasma • Microwave-Induced Plasma • Microwave Interference Protection • Minimum Impulse Pulse • Minimum Ionising Particle • Mission Implementation Plan • Mission Integration Panel • Mitspeh Ramon (Israel) (location identifier) • Mixed Integer Programming • Model Implementation Plan • Modification Instruction Package • Monthly Investment Plan • Most Important Person • Multiband Imaging Photometer

MIPAS: Management Information Planning and Accountancy Service • Michelson Interferometer for Passive Atmospheric Sounding

MIPC: Metropolitan Information Processing Conference • Multimission Image Processing Laboratory

MIPE: Modular Information Processing Equipment

MIPG: Master Index Pulse Generator

MIPI: Mission Interministérielle pour la Promotion Informatique (France)

MIPL: Multimission Image Processing Laboratory (*) • Multimission Image Processing Library

MIPO: Multiple-Item Purchase Order

MIPR: Monthly Interim Progress Report

MIpS: Mega Instructions per Second • Million Instructions per Second

MIPS: Management Information Processing System • Map and Image Processing System • Mega Instructions Per Second (*) • Microcomputer Image Processing System • Million Instructions Per Second • Mini-Image Processing System • Mir Interface to Payload Systems • Mission and Information and Planning System • Mission and Information Planning System • Mission Information and Planning System (*) • Mission Information Planning System • MSS Image Processing System • Multiband Imaging Photometer for SIRTF • Multi-Image Processing System • Multimedia Information Presentation System • Multimedia Information Processing System • Multimission Image Processing System • Multipurpose Image Processing System

MIQ: Machine Intelligence Quotient • Omaha Millard (USA-NE) (location identifier)

MIr: Middle Irish

MIR: Magnetic Ink Reader • Malfunction Investigation Report • Management Information Report • Management Information Retrieval • Master Inventory Record • Memory Information Register • Memory Input Register • Merged Interaction Region • Micro-Instruction Register • Microwave Imaging Radiometer • Middle-Infrared • Mid-Infrared • Millimeter-wave Imaging Radiometer • Mirror • Mission Integration Review • Monastir (Tunisia) (location identifier) • Montgolfière Infrarouge • Multilingual Information Retrieval • Multimedia Information Retrieval • Multiple Internal Reflection • Multiple Isomorphous Replacement • Music Information Retrieval

MIRA: Medium-wave Infrared Array • Monterey Institute for Research in Astronomy (USA-CA) • Multifunctional Inertial Reference Assembly

MIRABU: Mid-Infrared Array for Boston University

MIRAC: Management Information Research Assistance Center • Mid-Infrared Array Camera

MIRACL: Mid-Infrared Advanced Chemical Laser • Missouri Research Consortium of Libraries (USA-MO)

MIRACLE: Magnetometer Ionospheric Radars All-sky Camera Large Experiment • Mid-Infrared Array Camera of Least Effort • Multidisciplinary Integrated Research Activities in Complex Laboratory Environments

MIRAGE: Multimedia Information Retrieval and Gathering Environment

MIRAM: Multiple Indexed Random Access Method

MIRAS: Mid-Infrared Array Spectrometer • Mid-Infrared Array Spectropolarimeter • Mortgage Interest Relief At Source

MIRC: Manufacturing Information Resource Center • Mosaic Infrared Camera

MIRCEN: Microbial Resource Center

MIRD: Medical Internal Radiation Dose • Mission Implementation Requirements Document

MIRDC: Metals Industry Research and Development Center

MiRe: Microelectronics Reliability (periodical code)

MIRF: Multiple Instantaneous Response File

MIRFAC: Mathematics In Recognizable Form Automatically Compiled

MIRFI: Mid-Infrared Fabry-Pérot Interferometer

MIRI: Mid-Infrared Imager • Multichannel Infrared Interferometer

MIRIAD: Multichannel Image Reconstruction, Image Analysis and Display (*) • Multichannel Image Reconstruction with Interactive Analysis and Display

MIRIS: Michigan Inventory and resource Information System • Modified Infrared Interferometer Spectrometer

MIRO: Microwave Instrument for the Rosetta Orbiter

MIROS: Microwave Remote Sensor for the Ocean Surface • Modulation Inducing Retrodirective Optical System

MIRP: Manipulated Information Rate Processor • Minnesota Infrared Polarimeter

MIRPS: Multiple-Information Retrieval by Parallel Selection

MIRROR: Management Information Reporting and Review of Operational Resources System

MIRROS: Modulation Inducing Reactive Retrodirective Optical System

MIRS: Maintenance Information Retrieval System • Management Information and Reporting System • Mid-Infrared Spectrometer • MOTS Information Retrieval System • Multiple Internal Reflection Spectroscopy

MIRSE: Multipurpose Imaging Radiometer Spectrometer Equipment

MIRSL: Microwave Remote Sensing Laboratory

MIRSP: Mobile Infrared Scene Projector

MIRST: Multiple Infrared Scattered Light Recorder

MIRT: Mission Integration Readiness Test

MIRTL: Model of Information Relevant to Terminological Logic • Multimedia Information Retrieval Terminological Logic

MIRV: Multiple Independently-targeted Reentry Vehicle • Multiple Independent Reentry Vehicle

Mis: Mississippian

MIS: Magnetic Isolation System • Managed Internet Service • Management Information Science • Management Information Server • Management Information Service • Management Information Specialist • Management Information System • Management of Information Systems • Man In Space • Manufacturing Information System • Map Information System • Master Index Schedule • Master Integrated Schedule • Master of Information Systems • Master of Science in Information Systems • Mathematik Informationssystem • Matière Interstellaire • Metal-Insulated Semiconductor • Metal-Insulator-Semiconductor • Meteorological Impact Statement • Meteorological Information System • Metrology Information Service • Microcomputer Information Service • Microencapsulation In Space • Milieu Interstellaire • Misima Island (Papua New Guinea) (location identifier) • Mission Information System • Moyens Informatiques et Services • Multimedia Information Source • Multispectral Imaging Spectrometer • Multispectral Imaging System

MISAM: Multiple Index Sequential Access Method

MISAR: Miniature Information Storage and Retrieval

Misc: Miscellaneous • Miscellany

MISC: Micro Image Scanner • Miscellaneous

MISEP: Mutual Information System of Employment Policies

MISER: Management Information System for Expenditure Reporting

MISETA: Multi-Instrumented Studies of Equatorial Thermospheric Aeronomy

MISCF: Miscellaneous Frame

MISCHA: Magnetic fields in Interplanetary Space during Comet Halley's Approach

MISD: Multiple Instruction - Single Data

MISDE: Mir Structural Dynamics Experiment

MISFET: Metal-Insulator-Semiconductor Field-Effect Transistor

MISG: Mathematics-in-Industry Study Group • Missing

MISHA: Magnetic fields in Interplanetary Space during Halley's Approach

MISI: Mathematics and Information Sciences for Industry • Modular Interactive Simulation

MISL: Malfunction Investigation Support Laboratory • Management Information System Laboratory

MISM: Metal-Insulator-Semiconductor Metal

MISMA: Micro Structures Magnetic Atmospheres (project)

MISMAC: Matrix-Isolation Spectroscopy of Mass-selected Clusters

MISO: Management Information Systems Office • Manufacturing Inspection Satellite Office • Multiple Input - Single Output

MISOE: Management Information System for Occupational Education

MiSon: Mitteilungen der Sternwarte Sonneberg (periodical code)

MISP: Management Information System Plan • Minnesota Infrared Spectropolarimeter

MISQ: Management Information Systems Quarterly

MISr: Muy Ilustre Señor

MISR: Major Item Status Report • Mars ISRU Sample Return • Multiangle Imaging Spectroradiometer (*) • Multidirectional Imaging Spectroradiometer

Miss: Mission • Missionary • Mississippi (USA)

MISS: Man In Space Simulator • Man In Space Soonest • Meteorological Interactive Software System • Minicomputer Interfacing Support System • Multiband Image Scanning System

MISSIL: Management Information System Symbolic Interpretive Language

MisSp: Missiles and Spacecraft (periodical code)

mist: Mistura

Míst: Mística

MiST: Microgravity Science and Technology (periodical code)

MIST: Maintenance International Standardization System • Meteorological Information Self-briefing Terminal • Microburst and Severe Thunderstorm • Mission de l'Information Scientifique et Technique • Multi-Input Standard Tape • Multispectral Image Storage Tape Format

MISTE: Miniaturized Integrated Satellite Terminal Equipment

MISTER: Model-based Intelligent Star Target Exploitation and Recognition

Mistrans: Mistranslation

MISVE: Management Information System for Vocational Education

Mit: Mitología • Mittwoch

MIT: Management Information Tree • Massachusetts Institute of Technology (USA-MA) • Master Instruction Tape • Master of International Trade • Memorandum for Information Technology • Mercury Ion Transfer • Miles In Trail • Minimum Individual Training • Mission Integration Team • Most Important Test • Most Intense Tornado • Schaffer - Kern County (USA-CA) (location identifier)

MITA: Microcomputer Industry Trade Association • Modified Iterative Transform Algorithm

MitAe: Max-Planck-Institut für Aeronomie - Mitteilungen (periodical code)

MitAG: Mitteilungen der Astronomischen Gesellschaft (periodical code)

MITAS: Martin Marietta Interactive Thermal Analysis System

MITC: Microfilm and Information Technology Center • Mortgage Investment Tax Credit

MITE: Magnetic Insulation Test Experiment • Master Instrumentation Timing Equipment • Megapixel Imaging Technology Experiment • Microelectronic Integrated Test Equipment • Microelectronics Test and Evaluation • Multiple Input Terminal Equipment

MITEC: Microwave Technology Centre

MITES: Miniature Interferometer Terminal for Earth Survey • Model-driven Iterative Texture Segmentation

Mitgl: Mitglied

MITH: Milli-Kelvin Thermostat

MITI: Ministry of International Trade and Industry

MITL: Magnetically-Insulated Transmission Line • Matsushita Information Technology Laboratory (USA-NJ)

MITLL: Massachusetts Institute of Technology Lincoln Laboratory (USA-MA)

MITM: Magnetospheric-Ionospheric-Thermospheric-Mesospheric

MITN: Michigan Interferometer Technology Network

MITO: Millimetre and Infrared Testagrigia Observatory (Italy)

MITOC: Multiple Intercommunications Technical Operations Communications

MITRE: Miniature Individual Transmitter Receiver Equipment

MITS: Management Information and nText System • Microcomputer Interactive Test System • Microfiche Image Transmission System

Mitt: Mitte • Mitteilung

Mittelfr: Mittelfranken (Germany)

Mittelholst: Mittelholstein (Germany)

MITTS: Minimalist Tutorial and Tools for SmallTalk

MITTSE: Mobile Integrated Tracking Telescope System

Mittw: Mittwochs

MitVS: Mitt. Verand. Sterne (periodical code)

MIU: Maiduguri (Nigeria) (location identifier) • Malfunction Insertion Unit • Message Interception Unit • Multiplex Interface Unit • Multistation Interface Unit

MIUS: Modular Integrated Video System • Modulated Integrated Video System

miv: Met Ingang Van

MIV: Main Instrumentation Van • Millville (USA-NJ) (location identifier) • Mobile Instrumentation Van

MIVIS: Multispectral Infrared and Visible Imaging Spectrometer

MIVP: Mission Interface Verification Plan

MIW: Marshalltown (USA-IA) (location identifier)

MiWie: Mitteilungen Univ. Sternw. Wien (periodical code)

Mix: Mixture

MIX: Magnetic Ionization Experiment • Member Information Exchange • Metropolitan Information Exchange • Miriti (Colombia) (location identifier)

Mixt: Mixture

MIY: Mittiebah (Australia-NT) (location identifier)

MIZ: Mainoru (Australia-NT) (location identifier) • Management-Informationszentrum • Marginal Ice Zone

MIZEX: Marginal Ice Zone Experiment

mJ: Millijoule

MJ: Mari • Master of Journalism • Mechanical Joint • Megajoule • Modular Jack

MJA: Manja (Madagascar) (location identifier)

MJB: Mejit Island (Marshall Islands) (location identifier)

MJC: Maison des Jeunes et de la Culture • Man (Ivory Coast) (location identifier)

MJCA: Mid-body Jettison Control Assembly

MJD: Management Job Description • Modified Julian Date • Modified Julian Day (*) • Mohenjodaro (Pakistan) (location identifier)

MJE: Majkin (Marshall Islands) (location identifier)

MJF: Mosjøen (Norway) (location identifier)

MJG: Mayajigua (Cuba) (location identifier)

MJH: Majma (Saudi Arabia) (location identifier)

MJI: Maji (Ethiopia) (location identifier)

MJJ: May-June-July • Moki (Papua New Guinea) (location identifier)

MJK: Monkey Mia (Australia-WA) (location identifier)

MJL: Mouila (Gabon) (location identifier)

MJM: Mbuji-Mayi (Zaire) (location identifier)

MJN: Majunga (Madagascar) (location identifier)

MJO: Madden-Julian Oscillation • Mariner Jupiter Orbit • Mount Etjo Lodge (Namibia) (location identifier) • Mt. Etjo Lodge (Namibia) (location identifier)

MJP: Manjimup (Australia-WA) (location identifier)

MJQ: Jackson (USA-MN) (location identifier)

MJR: Maintenance Job Request • Management Job Review • Mariner Jupiter-Saturn • Miramar (Argentina-BA) (location identifier)

MJS: Maganja da Costa (Mozambique) (location identifier) • Manipulator Jettison System • Member of the Japan Society

MJSD: March-June-September-December

MJT: Mytilene (Greece) (location identifier)

MJU: Mamuju (Indonesia) (location identifier) • Mariner Jupiter-Uranus • Multijunction Unit

MJUO: Mount John University Observatory (New Zealand) • Mt. John University Observatory (New Zealand)

MJV: Mercredi-Jeudi-Vendredi • Murcia (Spain) (location identifier)

MJX: Toms River - Robert J. Miller (USA-NJ) (location identifier)

mJy: Millijansky

MJY: Mangunjaya (Indonesia) (location identifier)

MJZ: Mirnyj (Russia) (location identifier)

mk: Macedonian (ISO language code) • Mark (currency) • Markka (currency)

MK: Macédoine (ISO country code) • Macedonia (ISO country code) • Malawi Kwatcha (currency) • Mark • Metaknowledge • Milton Keynes (UK postal code) • Modification Kit • Morgan-Keenan

MKA: Marianske Lazne (Czech Rep.) (location identifier)

MKB: Mekambo (Gabon) (location identifier) • Methods Knowledge Base

MKC: Kansas City (USA-MO) (location identifier)

Mkd: Marked

MKD: Chagni (Ethiopia) (location identifier) • Macédoine (ISO country code) • Macedonia (ISO country code) • Mehrkörperdynamik • Metekel (Ethiopia) (location identifier)

mkdir: Make Directory

MKE: Milwaukee - General Mitchell (USA-WI) (location identifier) • Molecular Kinetic Energy

MKF: Columbus McKenna (USA-GA) (location identifier)

MKG: Muskegon (USA-MI) (location identifier)

MKH: Mokhotlong (Lesotho) (location identifier)

MKI: M'Boki (Central African Rep.) (location identifier)

MKJ: Makoua (Congo) (location identifier)

MKK: Hoolehua (USA-HI) (location identifier) • Morgan-Keenan-Kellman

MKL: Jackson (USA-TN) (location identifier)

MKM: Mukah (Malaysia) (location identifier)

MKMP: Modular Key Management Protocol

MKN: Malekolon (Papua New Guinea) (location identifier)

MKO: Mauna Kea Observatory (USA-HI) • Muskogee Davis (USA-OK) (location identifier)

MKP: Makemo (Tuamotu Islands) (location identifier)

MKQ: Merauke (Indonesia) (location identifier)

Mkr: Marker

MKR: Meekatharra (Australia-WA) (location identifier)

mks: Marks • Metre-Kilogram-Second

MKS: Mehrkörpersysteme • Mekane Selam (Ethiopia) (location identifier) • Metre-Kilogram-Second • Microwave Keying Switch

mksA: Metre-Kilogram-Second-Ampere

MKSP: Mount Kobau Star Party (Canada-BC)

MKSS: Microwave Keying Switch Station

MKST: Multidimensional k-ary Search Tree

Mkt: Market • Markt

MKT: Mankato (USA-MN) (location identifier)

Mktg: Marketing

MKTG: Marketing

MKTI: Mission Kit Technical Instruction

MKU: Makokou (Gabon) (location identifier)

MKV: Master Key Vector • Mount Cavenagh (Australia-NT) (location identifier)

MKW: Manokwari (Indonesia) (location identifier)

MKX: Mukalla (Yemen) (location identifier)

MKY: MacKay (Australia-QLD) (location identifier)

MKZ: Malacca (Malaysia) (location identifier)

ml: Mail • Malayalam (ISO language code) • Milliliter

mL: Millilambert

M/L: Maintenance Loop • Manual/Local • Mass/Light • Mass/Luminosity

M&L: Maintenance & Logistics

ML: Machine Language • Machine Learning • Macro Language • Magnetisches Loch • Maintenance and Logistics • Major Lobe • Mali (ISO country code) • Management List • Master Library • Master of Law • Master of Letters • Maximum Likelihood • Mean Level • Medieval Latin • Medium Level • Memory Load • Mesa Laboratory (NCAR) • Message Line • Metalanguage • Microprogramming Language • Middeck Left • Mill • Mini Laboratory • Mirroir Liquide • Mission Life • Mission Load • Mixed Layer • Mobile Launcher • Mold Line • Monolayer • Motherwell (UK postal code) • Motor Launch • Multilayer • Multiple Lobe • Multiplexed Local

MLA: Magnetic Lens Assembly • Malta (location identifier) • MDM Launch Aft • Microprocessor Language Assembler • Microwave Linear Accelerator • Modern Language Association • Monochrome Lens Assembly • Multiline Addressing • Multilinear Array • Multiplex Line Adapter • Multispectral Linear Array

MLAP: Multilingual Action Plan

MLAPI: Multilingual Application Programming Interface

MLAT: Multilevel Adaptive Technique

MLB: Mass-Loading Boundary • Melbourne (USA-FL) (location identifier) • Multiband Layer • Multilayer Board

MLBR: Medium Low Bit Rate

MLBSR: Multilingual Bank Statistics Reporting

MLC: Machine Level Control • Maison des Loisirs et de la Culture • Manoeuvre Load Control • Maximum Likelihood Classification • McAlester (USA-OK) (location identifier) • Member of the Legislative Council • Memory Load Command • Micellar Liquid Chromatography • Microelectric Logic Circuit • Midland Lutheran College (USA-NE) • Minimum Level Cataloguing • Mobile Launcher Computer • MOL Launch Complex • Multilayer Circuit • Multilens Camera • Multilevel Cell • Multiline Control

MLCB: Multilayer Circuit Board

MLCC: Multilook Cross-Correlation

MLCD: Maximum-Likelihood Convolutional Decoder

MLCS: Multicolor Lightcurve Shape (method)

Mld: Mould • Moulded

MLD: Malad City (USA-ID) (location identifier) • Maximum-Likelihood Detection • Median Lethal Dose • Mild • Minimal Lethal Dose • Mixed Layer Depth • Multilevel Decoding

MLDAS: Meteorological and Lighting Data Acquisition System

MLDC: Museum Digital Licensing Collective

Mldg: Moulding

MLDI: Meter List Display Interval

MLDR: Maximum Likelihood Decision Rule

MLE: Male (Maldives) (location identifier) • Manned Lunar Exploration • Maximum-Likelihood Estimate • Maximum-Likelihood Estimation • Maximum-Likelihood Estimator • Maximum Likely Earthquake • Mesoscale Lightning Experiment • Microprocessor Language Editor

MLEP: Manned Lunar Exploration Program

MLF: Milford (USA-UT) (location identifier) • Mobile Launcher Facility • MOL Launch Facility

MLFA: Machine-Learned Fragment Analyis

MLFAT: MOL Launch Facility Acceptance Team

MLFG: Multiplicative Lagged-Bibonacci Generator

MLFM: Multilayer Feedforward Network

MLFN: Multiple-Layer Feedforward Network

Mlg: Malagasy

MLG: Main Landing Gear • Malang (Indonesia) (location identifier) • Maximum-Likelihood Gaussian • Middle Low German • Modular Logic Grammar

MLGD: Main Landing Gear Door

MLGS: Microwave Landing Guidance System

MLH: Bâle Mulhouse (Switzerland) (location identifier) • Basel Mulhouse (Switzerland) (location identifier) • Basle Mulhouse (Switzerland) (location identifier) • Mulhouse Bâle (France) (location identifier) • Mulhouse Basel (France) (location identifier) • Mulhouse Basle (France) (location identifier)

MLHR: Maximum-Likelihood Ratio • Maximum-Likelihood Reduction

MLI: Machine Language Instruction • Machine Learning and Inference • Magnetic Level Indicator • Mali (ISO country code) • Master Listing Index • Matrix Lightcurve Inversion • Mixing-Length Theory • Moline (USA-IL) (location identifier) • Multilayered Insulation • Multilayered Insulator • Multilayer Insulation • Multilayer Insulator • Multiple Link Interface

MLib: Master of Library Science

MLibSc: Master of Library Science

MLID: Monitor Location Identifier • Multilink Interface Driver

MLIR: Maximum-Likelihood Image Restoration • Monolingual Information Retrieval • Multilingual Information Retrieval

M&LIS: Mapping & Land Information Service

MLIS: Molecular Laser Isotope Separation • Multilevel Information System • Multilingual Information Society • Multiple-Level Indexing Scheme

MLISP: Meta LISP

MLJ: Milledgeville - Baldwin County (USA-GA) (location identifier)

MLK: Malta (USA-MT) (location identifier) • Maschinenlochkarte

MLL: Manned Lunar Landing • Marshall (USA-AK) (location identifier) • Mars Logistics Lander • Multilayer Logic

Mlle: Mademoiselle

Mlles: Mademoiselles

MLLP: Manned Lunar Landing Program

MLM: Manufacturing-Logistics-Management • Maximum-Likelihood Method • Mid-Level Manager • Mixed Level Matrix • Morelia (Mexico) (location identifier)

MLN: Melilla (Spain) (location identifier) • Metropolitan Library Network

MLO: Manned Lunar Orbiter • Master Local Oscillator • Mauna Loa Observatory (USA-HI) • Milos (Greece) (location identifier)

MLOC: Million Lines Of Code

MLOF: Mount Lemmon Observing Facility (USA-AZ) • Mt. Lemmon Observing Facility (USA-AZ)

MLOPEX: Mauna Loa Observatory Photochemistry Experiment

MLP: Machine Language Program • Malabang (Philippines) (location identifier) • Maximum-Likelihood Poisson • Maximum-Likelihood Programme • Microsoft Licence Pack • Mid-Level Processor • Mobile Launcher Platform • Mobile Launch Platform • Modern Laplacian Theory • Moon Literature Project • Multilayer Perception • Multilayer Perceptron (*) • Multilevel Programmer • Multilink Procedure • Multiple Line Printer

MLPC: Multilayer Printed Circuit

MLPF: Miniature Low Pass Filter

MLPPP: Multilink Point-to-Point Protocol

MLQ: Malalaua (Papua New Guinea) (location identifier)

M/LR: Maintenance Loop Recorder

MLR: Maintenance Loop Recorder • Mass-Luminosity Relation • Memory Lockout Register • Millicent (Australia-SA) (location identifier) • Minimum Lending Rate • Monodisperse Latex Reactor • Monotone Likelihood Ratio • Multilevel Relational • Multiple Linear Regression

MLRA: Major Land Resource Area

MLRO: Matera Laser Ranging Observatory (Italy)

MLRS: McDonald Laser Ranging Station (USA-AZ) (*) • McDonald Laser Ranging System • Multilingual Reference System • Multiple-Launch Rocket System

MLRT: Multilevel Regression Testing

MLS: Machine Literature Searching • Manned Lunar Surface • Master of Library Science • Master of Library Service • Microwave Landing System • Microwave Limb Scanner • Microwave Limb Sounder • Mid-Latitude Summer • Miles City (USA-MT) (location identifier) • Mills • Mixed Language System • MOL Launch Site • Multilanguage System • Multilevel Security • Multiline Selection • Multimedia Learning Station • Multiple Listings Service

MLSc: Master of Library Science

MLSE: Mechanical Launch Support Equipment

MLSO: Mauna Loa Solar Observatory (USA-HI)

MLSP: Multiple-Link Satellite Program

MLSPH: Moving Least Squares Particle Hydrodynamics

MLT: Machine Learning Toolbox • Magnetic Local Time • Malta (ISO country code) • Malte (ISO country code) • Manned Lunar Test • Manufacturing Lead Time • Master Library Tape • Mean Life Time • Mechanized Loop Testing • Mesosphere - Lower Thermosphere • Millinocket (USA-ME) (location identifier) • Mixing-Length Theory • Mobile Launch Tower • Modulated Lapped Transform • Multiple Logical Terminal

MLTA: Multiple Line Terminal Adapter

MLTCS: Mesosphere Lower-Thermosphere Coupling Study

MLTI: Mesosphere-Lower Thermosphere-Ionosphere

MLU: Mean Length of Utterance • Memory Lock-Up • Memory Logic Unit • Mid-Life Update • Module Lighting Unit • Monroe (USA-LA) (location identifier) • Multiplicity Logic Unit

MLV: Medium Launch Vehicle • Medium-Lift Vehicle • Merluna (Australia-QLD) (location identifier)

MLVP: Manned Lunar Vehicle Program

MLVQ: Modified Learning Vector Quantization

MLW: Maximum Landing Weight • Mean Low Water • Mid-Latitude Winter • Monrovia Sprigg-Payne (Liberia) (location identifier)

MLWFA: Multilingual Weather Forecast Assistant System

MLX: Malatya (Turkey) (location identifier)

MLY: Manley Hot Springs (USA-AK) (location identifier)

MLZ: Melo (Uruguay) (location identifier)

m/m: Même Mois • Moi-Même

mm: Millia • Millimeter • Millimetric • Moi-Même • Mutatis Mutandis

mM: Millimolare • Millimole

M/M: Maximum/Minimum • Multimission

M&M: Materials & Maintenance

MM: Magnitude Method • Main Memory • Main Module • Maintenance Manual • Maintenance Mission • Major Mode • Management Manual • Management Meeting • Man-Month • Manufacturing Manual • Mariner Mark • Mass Memory • Master Monitor • Material Manual • Mathematical Model • Mathematical Morphology • Math Model • Mechanics and Mathematics • Memory Module • Memory Multiplexer • Mesoscale Model • Messieurs • Mickey Mouse • Middle Marker • Minutes • Mirror Model • Mirror Module • Mission Manager • Mixed Morphology • Modified Mercalli • Molecular Magnet • Month • Motor Module • Mount Model • Multimedia • Mutatis Mutandis • Myanmar (ISO country code) • Two Thousand

MMA: Main Mission Antenna • Malmö Metropolitan (Sweden) (location identifier) • Masse Maximum Aurotisée • Memory-to-Memory Adapter • Metropolitan Museum of Art • Michigan Muon Array (USA-MI) • Microcomputer Managers Association • Microgravity Measurement Assembly • Microgravity Measuring Assembly • Micro Mirror Array • Millimeter Array • Moving Mechanical Assembly • Multimedia Assistant • Multimeter Array • Multiple-Module Access • Multiplexed Matrix Array

MMAC: Multimedia Mathematics Across the Curriculum

MMACE: Millimeter-wave Advanced Computer Environment

MMACS: Maintenance Management And Control System

MMAD: Mass Median Aerodynamic Diameter

MmARB: Mémoires de l'Académie Royale de Belgique (periodical code)

MMAS: Martin Marietta Astro Space • Material Management Accountability System • Material Management Accounting System • Material Management and Accounting System • Mechanized Material Accountability System

MMath: Master of Mathematics

MMB: Memanbetsu (Japan) (location identifier) • Middle Mouse Button • Mirror Maintenance Building

MMBH: Modest-Mass Black Hole

MMC: Ciudad Mante (Mexico) (location identifier) • Magnesium Methyl Carbonate • Maintenance Monitoring Console • Man-Machine Communication • Manual Mode Command • Martin Marietta Corporation • Master Microprocessor Control • Matched Memory Cycle • Meet-Me Conference • Metal Matrix Composite • Microcomputer Marketing Council • Microsoft Management Console • Midcourse Measurement Correction • Midmotor Controller • Minicomputer Maintenance Center • Mission Management Center • Modular Mission Computer • Monopolies and Mergers Commission

MMCA: Mid-body Motor Control Assembly • Midmotor Controller Assembly

MMCBT: Multimedia Computer-Based Training

MMCC: Mission Management and Control Centre • Mission Management Control Centre • Modèle de Markov Caché Continu • Multimedia Conference Control

MMCCB: Multimission Change Control Board

MMCD: Multimedia Compact Disk

MMCF: Multimedia Communications Forum

MMCL: Master Measurement and Control List

MMCM: Mars Mesoscale Circulation Model

MMCR: Millimeter Cloud Radar

MMCS: Multiple-Mission Command System

MMCX: Multimedia Communications Exchange

MMD: Masse Maximum au Décollage • Mass Median Diameter • Master Monitor Display • Material, Maintenance and Distribution • Mean Mass Density • Mean Mission Duration • Memory Module Directory • Minami Daito (Japan) (location identifier) • Minimization of Maximum Deviation • Mobile Maintenance Depot • Modified Metropolis Dynamics • Multimode Display

MMDA: Money Market Deposit Account

MMDB: Main-Memory Database • Mass Memory Database • Master Measurement Database • Mirror Metrology Database

MMDD: Matra Marconi - DASA - Dornier

MMDF: Mission Model Data File • Multichannel Memorandum Distribution Facility

MMDS: Maintenance Management Data System • Multichannel Multipoint Distribution Service • Multichannel Multipoint Distribution System • Multimegabit Data Service • Multipoint Microwave Distribution System • Multipoint Multichannel Distribution System

Mme: Madame

MME: Master of Mechanical Engineering • Master of Mining Engineering • Micrometeoritic Erosion • Microsoft Multimedia Extension • Mobile Meßeinheit • Tees-Side (UK-England) (location identifier)

MMEB: Martin Marietta Enhanced Black (coating)

MMED: Mariner Mars Spectral Editor

MMEL: Master Minimum Equipment List

Mmes: Mesdames

MMES: Martin Marietta Energy Systems

MMF: Magnetomotive Force • Mamfe (Cameroon) (location identifier) • Mean Magnetic Field • Missing Minor Frame • Mission Management Facility • Mobile Magnetic Field • Module Maintenance Facility • Moving Magnetic Feature

MMFS: Manufacturing Message Format Standard • Mean Magnetic Field of the Sun

MMG: Mount Magnet (Australia-WA) (location identifier)

MMgtE: Master of Management Engineering

MMH: Maintenance Man-Hour • Mammoth Lakes (USA-CA) (location identifier) • Monomethyl Hydrazine

mmHg: Millimeter of Mercury

MMHH: Mijne Heren

MMI: Athens - McMinn County (USA-TN) (location identifier) • Maintenance Micro-Informatique • Management and Maintenance Inspection • Man-Machine Interface • Manpower Management Information • Mutual Mortgage Insurance • 2001

MMIA: Methods and Models for Integrated Assessment • Monolithic Microwave Integrated Amplifier

MMIC: Millimeter-wave Monolithic Integrated Circuit • Monolithic Microwave Integrated Circuit

MMICS: Maintenance Management Information and Control System • Maintenance Management Information and Control System

MMIDS: Man-Machine Interface Development Station

MMII: Mariner Mark II

MMIMU: Micromechanical Inertial Measuring Unit

MMIO: Multimedia Input/Output

MMIP: Mass Memory Integration Plan

MMIPS: Man/Machine Interactive Processing System

MMIR: Multimedia Information Retrieval

M-MIS: Multimissions Integrating System

MMIS: Materials Manager Information System • Minimum Mean-square Image Spread • Multimedia Information System • Multimissions Integrating System • Multimissions Integration System • Multimodel Information System

MMIT: Man-Machine Integration Technique

MMJ: Mardi-Mercredi-Jeudi • Matsumoto (Japan) (location identifier) • Modified Modular Jack

MMK: Kyat (Myanmar) (ISO currency code) • Mensch-Maschine-Kommunikation • Murmansk (Russia) (location identifier)

MML: Man-Machine Language • Marker Mark Language • Marshall (USA-MN) (location identifier) • Master Measurement List • Multimedia Laboratory

MMLE: Modified Maximum Likelihood Estimate

MMLP: Microsoft Multilicenses Pack

MMLS: Martin Marietta Launch Services • Model-Modes-Loads-Stresses

MMM: Magnetospheric Mapping Mission • Maintenance Management Manual • Manned Mars Mission • Mars Mesoscale Model • Mars Mission Module • Measuring Monitoring Module • Mesoscale and Microscale Meteorology • Middlemount (Australia-QLD) (location identifier) • Mobile Media Mode • Multihead Multitail Mosaic • Multimission Module

MMMIS: Maintenance and Material Management Information System

MMMOS: Mobile Micrometeorological Observation System

MMMS: Maintenance and Material Management System

MmMtS: Memoirs of the Mount Stromlo Observatory (periodical code)

MMN: Mismatch Negativity • Stow - Minute Man (USA-MA) (location identifier)

MMNA: Mathematical Modelling and Numerical Analysis (periodical code)

MMO: Main Meteorological Office • Maio (Cape Verde) (location identifier) • Maria Mitchell Observatory (USA-MA) • Middelbaar Middenstandonderwijs • Mission Management Organization

MMOC: Minicomputer Maintenance Operations Center

mmol: Millimole

MMOS: Multimode Optical Sensor

MMP: Maintenance Management Plan • Maintenance Monitor Panel • Microphysical Measurement Package • Millimeter-wave Polarimeter • Mission Management Plan • Mixture Melting Point • Modular Multiband Radiometer • Momentum Management Program • Mompos (Colombia) (location identifier) • Monthly Maintenance Plan • Multiplexed Message Processor

MMPA: Marine Mammal Protection Act

MMPI: Minnesota Multiphasic Personality Inventory

MMPM: Multimedia Presentation Manager

MMPO: Mission Management and Planning Office

MMPP: Markov Modulated Poisson Process

MMPS: Multimicroprocessing System

MMPSE: Multiuse Mission Payload Support Equipment

MMPT: Man-Machine Partnership Translation

MMQ: Mbala (Zambia) (location identifier) • Minimum Manufacturing Quantity

MMR: Austin Maybry (USA-TX) (location identifier) • Magnetic Memory Record • Maintenance Management Review • Mass Memory Ready • Maximal Marginal Relevance • Mean Motion Resonance • Milton Multiband Radiometer • Modified Modified Read • Modular Multiband Radiometer • Multimode Radiometer • Multimode Receiver • Myanmar (ISO country code)

MmRAS: Memoirs of the Royal Astronomical Society (periodical code)

MMRC: Mars Mission Research Center

MMRD: Maximum Magnitude versus Rate of Decline

MMRS: Multimukrorechnersystem • Multinational Meeting on Remote Sensing

Mms: Manuscritos

MMS: Magnetospheric Multiscale (mission) • Main Memory Status • Maintenance Mode System • Maintenance Monitoring System • Malaysian Meteorological Service • Man-Machine System • Manned Maneuvering System • Manufacturing Message Specification • Mardi Mercredi Samedi • Marine Meteorological Service • Maritime Mobile Service • Marks Selfs (USA-MS) (location identifier) • Mars Mai Septembre • Mass Memory System • Mathematical Modelling of Systems • Matin Midi Soir • Matra Marconi Space • Media Management System • Memory Management System • Mes Meilleurs Souvenirs • Meteorological Measurement System • Meteorological Measuring System • Michigan Multispectral Scanner • Microfiche Management System • Micrometeorological Measurement System • Middelbare Meisjesschool • Minerals Management Service • Mission Modular Spacecraft • Modern Men of Science (book) • Modular Measurement System • Modular Multiband Scanner • Modular Multispectral Scanner • Modulation Measuring System • Module Management System • Monolithic Miniature Spectrometer • Multimedia System • Multimission Modular Spacecraft • Multimission Spacecraft • Multimission Spectrum • Multimission System • Multiplex Modulation System

MmSAI: Memorie della Societá Astronomica Italiana (periodical code)

MMSc: Master of Medical Science

MMSD: Madison Metropolitan School District (USA-WI) • Metric Mapping Support Data

MMSE: Minimum Mean Squared Error • Minimum Mean Square Error • Minimum Mean Square Estimation • Mission Module Simulation Equipment • Multimission Support Equipment • Multiuse Mission Support Equipment

MMSI: MMS Interface

MMSL: Microgravity Materials Science Laboratory

MMSR: Multimission Support Recording

MMSS: Manned Maneuverable Space System • Manual Mode Space Simulator • Marine Meteorological Services System • Maritime Mobile Satellite Service • Martin Marietta Space Systems • Multimode Space Station

MMST: Microelectronics Manufacturing Science and Technology

MMSW: Multimedia Software

MMT: Columbia McEntire (USA-SC) (location identifier) • Main Mantle Thrust • Mass Memory Test • Methylcyclopentadienyl Manganase Tricarbonyl • Mission Management Team • Monolithic Mirror Telescope • Monthly Mean Temperature • Multiple-Mirror Telescope (USA-AZ) • Multiple-Mission Telemetry • Multiresolution Median Transform

MMTO: Multiple-Mirror Telescope Observatory (USA-AZ)

MM&TS: Master Manufacturing & Test Schedule

MMTS: Master Manufacturing and Test Schedule • Maximum-Minimum Temperature System • Multiple-Mission Telemetry System

MMU: Manned Maneuvering Unit • Mass Memory Unit • Memory Management Unit • Memory Mass Unit • Midcourse Maneuvering Unit • Midcourse Measurement Unit • Million Monetary Units • Mobile Meteorological Unit • Mobile Monitoring Unit • Modular Maneuvering Unit • Morristown (USA-NJ) (location identifier)

MMUI: Multimedia User Interface

MMUS: Mass Memory Unit Simulator

mmv: Met Medewerking Van

MMV: Mal (Papua New Guinea) (location identifier)

MMVP: Model-Map-View-Praxis

MMW: Major Midwinter Warming • Millimeter Wave • Millimeter Wavelength • Moma (Mozambique) (location identifier) • Multimega Watt

MMX: Malmö Sturup (Sweden) (location identifier) • Matrix Manipulation Extension • Memory Multiplexer • Multimedia Extension

MMY: Miyako Jima (Japan) (location identifier)

MMZ: Maimana (Afghanistan) (location identifier)

mm3: Cubic Millimeter

m/n: Moneda Nacional

mn: Met Name • Moneda Nacional • Mongolian (ISO language code) • Mutato Nomine

Mn: Main • Mangan • Manganese • Manganèse

MN: Magnetic North • Main • Management Network • Mande • Manitoba (Canada) • Mantle Nerve • Météorologie Nationale • Midnight • Mine • Minnesota (USA) • Moneda Nacional • Mongolia (ISO country code) • Mongolie (ISO country code) • Month • Monthly Notices

MNA: Melangguane (Indonesia) (location identifier) • Methylnitroaniline • Modèle Numérique d'Altitude

MNAS: Mathematical Methods in the Applied Sciences (periodical code) • Member of the National Academy of Sciences

MNASSA: Monthly Notes of the Astronomical Society of Southern Africa

MNB: Moanda (Zaire) (location identifier)

MNBA: Mono-Normal-Butylamine

MNBDF: Metanitrobenzenediazonium

MNBLE: Modified Nearly Best Linear Estimator

MNC: Multinational Centre • Multinational Company • Multinational Corporation • Nacala (Mozambique) (location identifier)

MnCAR: Minnesota Center for Automotive Research (USA-MN)

MNCIS: Management Numerical Control Information System

Mncpl: Municipal

MNCS: Multipoint Network Control System

MND: Medina (Colombia) (location identifier)

MNDP: Multinational Data Processing

MNE: Master of Nuclear Engineering • Mnemonic • Modèle Numérique d'Élévation • Multinational Enterprise • Mungeranie (Australia-SA) (location identifier)

MNEE: Mission Non-Essential Equipment

MNEF: Mutuelle Nationale des Étudiants de France

MNF: Mana Island (Fiji) (location identifier) • Minimum Noise Fraction

Mnfrm: Mainframe

MNFS: Minor Frame Synchronization

Mng: Managing

MNG: Maningrida (Australia-NT) (location identifier) • Microwave Negative Grid • Mongolia (ISO country code) • Mongolie (ISO country code) • Multiple Network Graphics

MNGE: Mission Nationale de la Gestion des Échanges (France)

Mngmnt: Management

Mngmt: Management

Mngr: Manager • Monsignor

MNH: Minneriya (Sri Lanka) (location identifier)

MNHN: Muséum National d'Histoire Naturelle

MNI: Minimum Number of Individuals • Montserrat (Leeward Islands) (location identifier)

MNIR: Mid and Near Infrared

MNISM: Mongolian National Institute for Standardisation and Metrology

MNJ: Manajary (Madagascar) (location identifier)

MNK: Maiana (Kiribati) (location identifier)

Mnl: Manila (Philippines) • Manual

MNL: Manila - Ninoy Aquino (Philippines) (location identifier)

MNLD: Mainland

MNM: Menominee (USA-MI) (location identifier) • Minimum

MNN: Marion (USA-OH) (location identifier)

MNO: Manono (Zaire) (location identifier)

MNOS: Metal Nitride Oxide Semiconductor

MNP: Îles Mariannes du Nord (ISO country code) • Magnetic North Pole • Mariana Islands (Northern_) (ISO country code) • Maron (Papua New Guinea) (location identifier) • Microcom Networking Protocol • Microcom Network Protocol (*) • Microsoft License Pack • Northern Mariana Islands (ISO country code)

MNPS: Minimum Navigation Performance Specification

MNPSA: Minimum Navigation Performance Specifications Airspace

MNPSD: Mean Normal Point System Delay

MNQ: Monto (Australia-QLD) (location identifier)

Mnr: Manor • Mijnheer

MNR: Manor • Ministry of Natural Resources • Mongu (Zambia) (location identifier)

MNRAS: Monthly Notices of the Royal Astronomical Society (periodical code)

Mnrl: Mineral

MNRS: Manors

Mns: Moins

MNS: Mansa (Zambia) (location identifier) • Mean Number of States • Meta-Nitride Semiconductor

MnSCU: Minnesota State Colleges and Universities (USA-MN)

MNSSA: Monthly Notes of the Astronomical Society of South Africa (periodical code)

Mnstb: Monostable

MNT: Minto (USA-AK) (location identifier) • Modèle Numérique de Terrain • Modern Network Theory • Mononitrotoluene • Mount • Tugrik (Mongolia) (ISO currency code)

MNTN: Mountain

MNTNS: Mountains

Mntrl: Montréal (Canada-QC)

MNu: Mare Nubium (Moon)

MNU: Mathematischer und Naturwissenschaftler Unterricht • Maulmyine (Myanmar) (location identifier) • Menu • Moulmein (Myanmar) (location identifier)

MNV: Mountain Valley (Australia-NT) (location identifier)

MNVR: Maneuver

MNW: MacDonald Downs (Australia-NT) (location identifier)

MNX: Manicoré (Brazil-AM) (location identifier)

MNY: Mono (Solomon Islands) (location identifier)

MNZ: Manassas (USA-VA) (location identifier)

m/o: Mi Orden

mo: Mail Order • Modus Operandi • Moldavian (ISO language code)

Mo: Mégaoctet • Missouri (USA) • Molybdän • Molybdène • Molybdenum • Moment • Monday • Montags • Month • Mouth

M/O: Meteorological/Oceanographic

M&O: Machinery & Optics • Maintenance & Operations • Management & Operating • Management & Organization • Manpower & Organization

MO: Macau (ISO country code) • Machine Operation • Machine-Outil • Maître Ouvrier • Modus Operandi • Magneto-Optic • Magneto-Optical • Mail Order • Major Objective • Managed Object • Management Office • Manila Observatory (Philippines) • Manned Orbiter • Manner of Operating • Manually Operated • Manual Orientation • Manual Output • Manufacturing Order • Mark Off • Mars Observer • Mars Orbiter • Massive Object • Mass Observation • Master Oscillator • Medical Officer • Member Organization • Memory Operation • Meteorological Observatory • Meteorological Office • Meteorological Officer • Meteorology Officer • Method of Operation • Middeck Overhead • Middelbaar Onderwijs • Miscellaneous Operation • Mission of Opportunity • Mission Operations • Mission Order • Mission Oriented • Missouri (USA) • Model Orientation • Mode of Operation • Modus Operandi • Molecular Orbital • Money Order • Mongolia • Monitor Output • Mordvinian • Motor Operated • Move • Municipal Officer

MoA: Memorandum of Agreement

MOA: Macho Observations in Astrophysics • Made On Assembly • Make On Arrival • Matrix Output Amplifier • Memorandum Of Agreement • Microlensing Observations in Astrophysics • Military Operations Area • Minute Of Angle • Moa (Cuba) (location identifier)

MOAD: Mother Of All Demos

MOAT: Mechanical/Optical Alignment Test

Mob: Mobile

MOB: Make Or Buy • Medisch-Opvoedkundig Bureau • Mobile • Mobile (USA-AL) (location identifier) • Memory Order Buffer • Pascagoula (USA-AL) (location identifier)

MOBIS: Management-Oriented Budget Information System • Mobile Information Station

MOBL: Macro-Oriented Business Language

MOBLAS: Mobile Laser System

MOBO: Machinenorientierte Bibliotheksorganisation

MOBS: Multiple Observation Sequence

MoC: Memorandum of Consultations

MO&C: Mission Operations & Control

MOC: Magnetic Optic Converter • Management and Operating Contractor • Manual Operations Control • Marine Operation Center • Mars Observer Camera • Mars Orbiter Camera (*) • Master Operational Controller • Master Operations Center • Master Operations Console • Master Operations Control • Mathematical Operations Computer • Memorandum Of Conditions • Memorandum Of Cooperation • Method Of Characteristics • Microfilm Output Computer • Mission Operations and Control • Mission Operations Center • Mission Operations Complex • Mission Operations Computer • Mission Operations Contractor • Mobile-Originated Call • Montes

Claros (Brazil-MG) (location identifier) • Monthly Operating Cost

MoCA: Movie Content Analysis

MOCA: Minimum Obstacle Clearance Altitude • Minimum Obstruction Clearance Altitude • Museum Of Contemporary Arts

MOCAM: Mosaic Camera

MOCC: Meteosat Operations Control Center • Mission Operations Control Center

MOCCA: Mobile Operations Center for Communications and Analysis

MOCCT: Method of Characteristics-Constrained Transport

MOCF: Mission Operations Computational Facility

MOCIC: Molecular Orbital Constraint of Interaction Coordinates

MOCR: Mission Operations Control Room

MOCS: Master Operations Control System • Multichannel Ocean Colour Sensor

MOCSW: Monitor and Operations Control Software

MOCV: Manual Oxygen Control Valve

MOCVD: Metallo-Organic Chemical Vapor Deposition (*) • Metal Organic Chemical Vapor Deposition

Mod: Model • Moderate • Moderator • Modern • Modernization • Modernize • Moderno • Modification • Modified • Modifier • Modify • Modulate • Modulated • Modulation • Modulator • Module • Modulo • Modulus

MoD: Ministry of Defense

MOD: Magnetic Optical Display • Magneto-Optical Disk • Magneto-Optical Drive • Mail-Order Delivery • Mail-Order Department • Management Of Data • Masters Of Deception • Masters Of Destruction • Microfilm-Output Device • Microwave Oscillating Diode • Milliseconds of Day • Ministry of Defence • Mission Objectives Document • Mission Operations Department • Mission Operations Director • Mission Operations Directorate • Mission Operations Division • Modem • Moderate • Modesto (USA-CA) (location identifier) • Modification • Modify • Modular • Modulation • Modulator • Module • Modulus • Moving Domain Memories

MO&DA: Mission Operations & Data Analysis

MODA: Magneto-Optical Doppler Analyzer • Mission Operations and Data Analysis

MOD/AO: Modular Adaptive Optics

MODAP: Multiple-Operational Data Acquisition Program

MODAPS: Modal Data Acquisition And Processing System

MODAPTS: Modular Arrangement of Predetermined Time Standards

MODAS: Maintenance and Operational Data Access System • Modular Ocean Data Assimilation System

MODB: Mission Operational Database • Mission Operations Database

MODCA: Mixed Object Document Content Architecture

MODCPS: Multiple-Output Direct-Current Power Supply

mod. dict.: Modo Dicto

MODE: Middeck 0-Gravity Dynamics Experiment • Mid-Ocean Dynamics Experiment

MODEM: Modulate/Demodulate • Modulation/Demodulation • Modulator/Demodulator

MODFET: Modulation-Doped Field-Effect Transistor

ModGe: Modern Geology (periodical code)

ModGk: Modern Greek

Modif: Modification

MODIS: Moderate-resolution Imaging Spectrometer • Moderate-resolution Imaging Spectroradiometer (*)

MODM: Magneto-Optical Display Memory • Multi-Objective Decision Making

MODOK: Modulare Dokumentation

mod. praes.: Modo Praescripto

MODS: Manned Orbital Development Station • Manned Orbital Development System • Mission Operations Data System • Modifications

MO&DSD: Mission Operations & Data Systems Division

MODSD: Mission Operations and Data Systems Division

MoE: Measure of Effectiveness

MOE: Mars Orbit Ejection • Measure Of Effectiveness • Mission Operations Engineer • Mission-Oriented Equipment • Modèle d'Objets d'Entreprise • Momeik (Myanmar) (location identifier)

MOEBIUS: Mobile Experimental Broadcast Interconnection Using Satellites

MOEMS: Micro-Opto-Electro-Mechanical System

MOEP: Meteorological and Oceanographic Equipment Program

MOES: Multiorder Etalon Sounder

MOF: Magneto-Optical Filter • Managed Object Format • Manned Orbital Flight • Maumere (Indonesia) (location identifier) • Maximum Observed Frequency • Metal-Oxide Film • Metaobject Facility • Michoud Operations Facility (NASA) • Mini-Orange Filter • Mission Operations Facility

MOFAT: Multiphase Flow And Transport

MOFFIT: Moon Free-Flyer Interferometer Trade-off

MoFi: Mondfinsternis

MOFS: Multiobject Fibre Spectrograph

MOFT: Manoeuvre-Oriented Flight Training

MOFWG: Magneto-Optical Filter Working Group

MOG: Minicomputer Operations Group • Mong Hsat (Myanmar) (location identifier) • Monthly Orbit Geometry

MOGA: Microwave and Optical Generation and Amplification

MOGR: Moderate or Greater

MOGS: Mission Operation Ground System

MOGUNTIA: Model of the Global Universal Tracer In the Atmosphere

moh: Meter ober Hav

MOH: Maximum Operating Hours • Meteorologisches Observatorium Hamburg (Germany) • Mohanbari (India) (location identifier)

MOHAWC: Model Of Human Actions in Work Context

Mohd: Mohamad • Mohamed • Mohammad • Mohammed

MOHLL: Machine-Oriented High-Level Language

MOHp: Meteorologisches Observatorium Hohenpeissenberg (Germany)

MOI: Magneto-Optical Imaging • Mars Orbital Insertion • Mars Orbit Insertion • Mars Orbiter Insertion • Maximum Obtainable Irradiance • Mercury Orbital Insertion • Mercury Orbit Insertion • Methods Of Instruction • Ministry Of Information • Ministry Of Interior • Ministry Of the Interior • Mitiaro Island (Cook Islands) (location identifier) • Moment Of Inertia • Moon Orbital Insertion • Moon Orbit Insertion

MOIS: Mission Operations Intercommunication System

MOJ: Moengo (Suriname) (location identifier)

MOK: Mankono (Ivory Coast) (location identifier)

MOKA: Mitaka-Okayama-Kiso Archive

MOKE: Magneto-Optic Kerr Effect

mol: Mole

Mol: Molecular • Molecule • Molokai (USA-HI)

MOL: Machine-Oriented Language • Mail-Order Library • Manned Orbital Laboratory • Manned Orbiting Laboratory • Maximum Operating Level • Maximum Output Level • Metallo-Organic Laser • Meteorologisches Observatorium Lindenberg (Germany) • Method Of Lines • Microsoft Open License • Ministry Of Labour • Molde (Norway) (location identifier) • Molecular

MOLA: Mars Observer Laser Altimeter • Mars Orbiter Laser Altimeter (*) • Multiplicative Optimally Localized Average

MOLAP: Multidimensional Online Analytical Processing

Mold: Moldavia • Moldavian

MOLD: Model Of Light Diode

MOLDS: Management Online Data System

Moldv: Moldavia • Moldavian

MOLE: Molecular Optics Laser Examiner

Molec: Molecular • Molecule

Molek: Molekulare

MOLI: Microsoft Online Institute

MOLL: Metallo-Organic Liquid Laser

MolPh: Molecular Physics (periodical code)

MOLS: Magnetic-Operated Limit Switch • Mutually Orthogonal Latin Squares

MOLTOL: Manned Orbiting Laboratory Test-Oriented Language

mol. wt.: Molecular Weight

Mom: Moment • Momentary • Momentum

MoM: Method of Moments

MOM: Manned Orbiting Mission • Message-Oriented Middleware • Metal-Oxide Metal • Microsoft Office Management • Microsoft Office Manager (*) • Middle Of Mission • Middle Of Month • Minutes Of Meeting • Mission Océanographique de la Méditerranée • Mission Operations Manager • Modular Ocean Model • Moment • Monthly Orbit Milestones • Moudjeria (Mauritania) (location identifier)

MOMA: Museum Of Modern Art

MOMBE: Metal Organic Molecular Beam Epitaxy

MOMI: Manchester Occulting Mask Imager • Museum Of Moving Images (USA-CA)

MOML: Moslem Meal

MOMO: MOS Management and Operation

MOMS: Map, Operator and Maintenance Station • Mobile Optical Measurement System • Modular Optical Multispectral Scanner • Modular Optoelectronic Multispectral Scanner • Monospectral Optical Monitoring System • Multiple Orbit - Multiple Satellite • Multispectral Optical Monitoring System

MOMV: Manned Orbital Maneuvering Vehicle

Mon: Maison • Monaco • Monasterio • Monastery • Monat • Monatlich • Monday • Monetary • Mongolian • Monitor • Monmouthshire (UK-Wales) • Monoceros-Monocerotis (constellation) • Monsignor • Monsoon • Montag • Month • Monthly • Monument

MON: Mixed Oxides of Nitrogen • Monday • Monitor • Motor Octane Number • Mount Cook (New Zealand) (location identifier) • Mt. Cook (New Zealand) (location identifier)

MONA: Manager Of Network Activity

MONA LISA: Modelling Natural Images for Synthesis and Animation

MOND: Modified Newtonian Dynamics

MONET: Meeting On the Network

MONEX: Monsoon Experiment

Mong: Mongolia • Mongolian

Mongol: Mongolia • Mongolian

MONICA: Montréal Infrared Camera

MONISAR: Monitoring with Synthetic Aperture Radar

Mono: Monophonic

Monogr: Monograph • Monographie

Mons: Monsenhor • Monseñor • Monsieur

MONS: Measuring Oscillations in Nearby Stars

MONSEE: Monitoring the Sun-Earth Environment

Mont: Montags • Montana (USA) • Montería • Montgomeryshire (UK-Wales) • Monument

MONTBLEX: Monsoon Through Boundary Layer Experiment

Montep: Montepio

Montgom: Montgomeryshire (UK-Wales)

Montr: Montréal (Canada-QC)

MONTR: Monitor

MOO: Moomba (Australia-SA) (location identifier) • MUD Object-Oriented • Multiuser Object-Oriented

MOONLIT: Motif OpenLook Intrinsics Toolkit

MoonP: Moon and Planets (periodical code)

MOOSE: Man Orbital Operations Safety Equipment • Man Out Of Space Easiest

MOOSSE: Manned Orbital Oceanographic Survey System Experiment

MOP: Maintenance Operations Protocol • Manned Orbital Platform • Manual Operations Panel • Manual Override Panel • Manuscript On Paper • Matrix Operations Programming • Maximum Operating Pressure • Measure Of Performance • Memorandum Of Policy • Message Output Processing • Meteorologisches Observatorium Potsdam (Germany) • Meteosat Operational Programme • Méthode de l'Onde de Pression • Microwave Observing Project • Mission Océanographique du Pacifique • Mission Operations Plan • Mission Operations Processor • Model Operational Plan • Modular Operating Procedure • Modulation On Pulse • Monthly Observation Plan • Monthly Observing Plan • Monthly Operating Plan • Mount Pleasant (USA-MI) (location identifier) • Multiple Online Programming • Pataca (Macau) (ISO currency code)

MOPA: Master Oscillator Power Amplifier

MOPAR: My Only Problems Are Repairs

MOPCA: May Our Paths Cross Again

MOPE: Multiple Object Parameter Estimation

MOPI: Maximum Rate Output Initiator

MOPITT: Measurements Of Pollution In The Troposphere

MOPL: Machine-Oriented Program Language

MOPR: Mission Operations Planning Review • Mission Operations Planning Room

MOPS: Machine-Oriented Programming System • Man-Operated Propulsion System • Maneuver Operations Program System • Marine Oil Pollution System • McDonald Observatory Planetary Search • Military Operations Phone System • Million Operations Per Second • Minimum Operational Performance Specifications • Minimum Operational Performance Standards • Minimum Operating Performance Standard • Mission Operations • Mission Operations Planning System

MOPSY: Multiprogramming Operating System

MOpt: Master of Optometry

MOPU: Ministerio de Obras Públicas y Urbanismo (Spain)

MOPWG: Meteosat Operational Programme Working Group

MOQ: Minimum Order Quantity • Morondava (Madagascar) (location identifier)

m. or.: Mismo Origen

Mor: Moroccan • Morocco

MOR: Magneto-Optical Rotation • Management Operating Ratio • Manufacturing Operations Record • Mars Orbital Rendezvous • Memory Output Register • Middle Of the Road • Mission Objectives and Requirements • Mission Operations Report • Mission Operations Room • Modulus Of Rupture • Morelia (Mexico) • Morristown - Moore-Murrell (USA-TN) (location identifier)

MORA: Minimum Off Route Altitude

MORACC: Modelling Radiation, Clouds, and Convection

Morad: Moradia

MORANE: Mobile Radio for Railway Networks in Europe

MORAS: Mobile Radiation Station

MORB: Mid-Ocean Rich Basalt • Mid-Ocean Ridge Basalt (*) • Mid-Ocean Rigid Basalt

MORD: Mission Operations Requirements Document

mor. dict.: More Dicto

MORDS: Manned Orbital Research and Development System

MORE: MARC Optical Recognition • Multimedia Object Retrieval Environment • Multimedia-Oriented Repository Environment

MORECS: Meteorological Office Rainfall and Evapotranspiration Calculation System

MORENet: Missouri Research and Education Network

MORF: Manned Orbital Research Facility

Morg: Morgado

MORIS: Moderate Resolution Imaging Spectrometer

MORL: Manned Orbital Research Laboratory • Manned Orbiting Research Laboratory

Morn: Morning

MORO: Moon Orbiting Observatory

MORP: Meteorite Observation and Recovery Programme • Meteorite Observation and Recovery Project

Morph: Morphological • Morphology

Morphol: Morphological • Morphology

MORPHS: Minicomputer-Operated Retrieval Partially Heuristic System

MORSE: Metal Organic Research for Semiconductor Epitaxy

mor. sol.: More Solito

Mort: Mortal • Mortality • Mortgage • Mortuary

MORT: Management Oversight and Risk Tree • Ministry Of Research and Technology

Mos: Months • Mosaic • Moscow (Russia) • Moselle

MoS: Ministry of Supply

MÖS: Mössbauer Spectrometer

MOS: Magneto-Optical System • Maintenance Operating System • Management Operating System • Manned Orbital Station • Manual Override Switch • Manufacturing Operating System • Mapping-Ordering-Searching • Marine Observation Satellite • Maritime Observation Satellite • Machineorientierte Sprache • Machineorientiertes System • Mean Opinion Score • Media Operating Software • Metal-Oxide Semiconductor • Metal-Oxide-Silicon • Mission Operations Strategy • Mission Operations System • Model Output Statistics • Modular Operating System • Modular Opto-electronic Scanner • Modular Opto-electronic System • Moses Point (USA-AK) (location identifier) • Mössbauer Spectrometer • Multibeam Optical Sensor • Multiobject Spectrograph • Multiobject Spectrometer • Multiobject Spectroscopy • Multiprogramming Operating System • Multiple Optical Switch • Multispectral Optoelectronical Scanner • Multiuser Operating System

MOSAIC: MDM-Ohio State Active Infrared Camera • Meteosat Operational System for Acquisition and Interchange • Modular Open System Architecture for Industrial Control • Mosaicking of Satellite (images) using Automated Image Correlation

MOSAICS: Multiobservation Satellite Image Correction System

MOSC: Manned Orbital Systems Concept • Mission Operations Support Center • Mixed Open Systems Consortium

MOSCA: Multi-Objekt Spectrograph für den Calar Alto

MOSES: Massive Open System Environment Standard • Mission Operations and Systems Engineering Software (*) • Mission Operations and Systems Engineering Support • Molecules in Outer Space and the Earth's Stratosphere

MOSFET: Metal-Oxide Semiconductor Field-Effect Transistor

MoSI: Museum of Science and Industry

MOSIS: MOS Implementation System

MOSLM: Magneto-Optic Spatial Light Modulator

MOSM: Mission Operations System Manager

MOSN: Magnetic and Optical Storage on Network

MOSO: Multimission Operations Systems Office

MOSP: Mission Operations Support Program

MOSPF: Multicast Open Shortest Path First

MOSPO: Mobile Satellite Photometric Observatory

MOSR: Mission and Observation Scheduling Room • Mission Operations Support Requirement

MOSS: Maintenance and Operator Subsystem • Manned Orbital Space Station • Map Overlay and Statistical System • Map Overlay and Statistics System • Meteorological Observing System for Ships • MIME Object Security Service • Miniaturized Optimized Smart Sensor • Mission Operations Support Services • Modular Star Sensor • Monitor Output Signal Strength • Moving Object Scheduling System • Moving Object Support System

MOSST: Ministry of State for Science and Technology

MOST: Management Of Social Transformation • Management Operation System Technique • Manned Operational System Test • Manned Orbital Solar Telescope • Metal-Oxide-Semiconductor Transistor • Metal-Oxide-Silicon Transistor • Microvariability and Oscillations of Stars • Microwave Observation of Stratosphere and Troposphere • Ministry Of Science and Technology • Mobile Station • Molonglo Observatory Synthesis Telescope (Australia-NSW) • Multimodal Occupation Strategy • Multiple Operating System Tool • Museum Of Science Technology

MOSTAC: Mosaic Sensor Test And Calibration

MOSTE: Ministry Of Science, Technology and Environment (Malaysia)

MOSTL: Metal-Oxide Semiconductor Transistor Logic

MOSTT: Mosaic Optical Sensor Technology Testbed

Mot: Motor • Motorized

MOT: Managed Object Toolkit • Magneto-Optical Trap • Manned Orbital Telescope • Maximum Operating Time • Mean Operating Time • Mechanical Operability Test • Member Of the Tribe • Ministry of Transport • Minot (USA-ND) (location identifier) • Molecular-Orbital Theory • Motor • Motor Operating Time

MOTAR: Modular Thermal Analyzer Routine

MOTAS: Mission-Oriented Terminal Area Simulator

MOT&E: Multiservice Operational Test & Evaluation

MOTF: Manganese-Oxide Thin Film

MOTHS: Mirror Over-The-Horizon Simulation

MOTIF: Maui Optical Tracking and Identification Facility (USA-HI)

MOTIS: Message-Oriented Text Interchange System

MOTIVE: Modular Timing Verifier • Motion Video Engine

MOTNE: Meteorological Operational Telecommunications Network Europe

MOTOR: Mobile Oriented Triangulation Of Reentry

MOTR: Multiple-Object Tracking Radar

MOTS: Minitrack Optical Tracking Station • Minitrack Optical Tracking System • Mobile Optical Tracking System • Module Test Set

MOTU: Mobile Optical Tracking Unit

MOTV: Manned Orbital Transfer Vehicle

MoU: Memorandum of Understanding

MOU: Memorandum Of Understanding • Mountain Village (USA-AK) (location identifier)

MO&UCB: Multilateral Operations & Utilization Control Board

MOUCB: Multilateral Operations and Utilization Control Board

MOUDI: Micro-Orifice Uniform Deposit Impactor

MOUSE: Minimum Orbital Unmanned Satellite of the Earth

MOUT: Message Out

Mov: Movimiento

MOV: Main Oxidizer Valve • Manned Orbiting Vehicle • Metal-Oxide Varistor • Moranbah (Australia-QLD) (location identifier) • Motor-Operated Valve • Move • Moving

MOVE: Manipulator Operation using Value Encoding • Microsoft Overlay Virtual Environment

MOVG: Moving

MOVPE: Metallo-Organic Vapor Phase Epitaxy • Metal Organic Vapor Phase Epitaxy

MOVS: Move String

Movt: Movement

MOW: Ministerie van Openbare Werken • Mission Operations Wing • Moscow (Russia) (location identifier)

MOWG: Management and Operations Working Group • Management Operations Working Group • Mission Operations Working Group

MOWS: Make Only What's Sold

MOX: Mars Oxidant Experiment • Mixed Oxide • Morris (USA-MN) (location identifier)

MOXE: Monitoring X-ray Experiment

MOY: Monterrey (Colombia) (location identifier)

MOYRDA: Month-Year-Day

Moz: Mozambique

MOZ: Mittlere Ortszeit • Moorea (Society Islands) (location identifier) • Mozambique (ISO country code)

Mozár: Mozárabe

mp: Modo Praescripto

M/P: Main Parachute • Mass Properties • Memorandum of Partnership

M&P: Materials & Processes • Moon and Planets (periodical code)

MP: Îles Mariannes du Nord (ISO country code) • Madhya Pradesh (India) • Magnetic Particle • Magnetic Point • Magnetic Pressure • Magnetopause • Main Program • Mains Propres • Maintainability Plan • Maintenance Panel • Maintenance Period • Maintenance Plan • Maintenance Point • Maintenance Print • Maintenance Procedure • Maintenance Program • Major Program • Maladie Professionnelle • Management Package • Mandat-Poste • Manifold Pressure • Manpower • Manufacturing Process • Maquette de Pilotage • Mariana Islands (Northern_) (ISO country code) • Maritime Patrol • Massively Parallel • Mass Properties • Mathematical Programming • Matsya Purāna • Measuring Point • Medium Pressure • Meeting Point • Melting Point • Melting Pot • Member of Parliament • Membre du Parlement • Mercator's Projection • Merchant Profiling • Meteorological Package • Meteorology Panel • Metropolitan Police • Mezzo Piano • Michoud Plant (NASA) • Microcode Processor • Microgravity Pressure • Microprint • Microprocessor • Middle Point • Mid-Phase • Mile Post • Military Police • Minister Plenipotentiary • Miscellaneous Paper • Miscellaneous Publication • Mission Payload • Mission Planner • Mission Planning • Mission Processing • Modèle Ponctuel • Modus Ponens • Molonglo Pulsar • Monitor Panel • Monitor Printer • Months after Payment • Mounted Police • Mouvement Perpétuel • Moyenne Pression • Multilayer Perceptron • Multilink PPP • Multiple Processors • Multiplexing Procedure • Multiplier Phototube • Multipole • Multiprocessing • Multiprocessor • Multipurpose • Northern Mariana Islands (ISO country code)

MPa: Megapascal

MPA: Magazine Publishers' Association (USA) • Magnetospheric Plasma Analyzer • Manufacturing Producibility Analysis • Maritime Patrol Aircraft • Master of Professional Accounting • Master of Public Administration • Master of Public Affairs • Max-Planck-Institut für Astrophysik (Germany) • Master of Public Administration • Microgravity Pressure - Ambient • Microprocessor Application • Microwave Power Amplifier • Miniature Pendulum Accelerometer • Mission/Payload Assessment • Mission Performance Assessment • Mission Phase Analysis • Mission Profile Analysis • Modification Proposal and Analysis • Mpacha (Namibia) (location identifier) • Multiple Parameter Analysis • Multiprecision Arithmetic • Multiuse Planning Area

MPAA: Motion Picture Association of America

MPAC: Multipurpose Application Console

MPACS: Management Planning And Control System

MPAD: Mission Planning and Analysis Division (JSC)

MPAe: Max-Planck-Institut für Aeronomie (Germany)

Mpal: Municipal

MPAM: Maritime Polar Air Mass

MPAR: Maintainability Problem and Assessment Report • Maintenance Program Analysis Report

MPASS: Modular Processing And Support System

MPB: Magnetic Particle Brake • Maintenance Parts Breakdown • Mathématiques et Physique de Base • Matrix Program Board • Miami (USA-FL) (location identifier) • Minor Planet Bulletin • Missing Persons Bureau

MPBB: Maximum Permissible Body Burden

MPBS: Main Power Bus Supply

Mpc: Megaparsec

MP&C: Maintenance Planning & Control

MPC: Machine Punch Card • Magnetic Particle Clutch • Maintenance Parts Catalog • Maintenance Procedure Chart • Manpower Planning Council • Manpower Priorities Committee • Manual Pointing Controller • Manufacturing, Planning and Control • Manufacturing-Planning-Control • Mars Polar Camera • Massively Parallel Computer • Material Procurement Cost • Materials Property Council • Mathematics-Physics-Chemistry • Mathématiques-Physique-Chimie • Maximum Permissible Concentration • Mechanical Positioning Control • Memory Protection Check • Meteorological Prediction Center • Midbody Pyro Controller • Mid-Power Controller • Minor Planet Center (USA-MA) • Minor Planet Circular • Mission Processing Center • Mitsubishi Precision Co. • Monitoring Proportional Counter • Monitor Proportional Counter • Monterey Peninsula College (USA-CA) • Most Positive Cut • Motion Picture Camera • Muko-Muko (Indonesia) (location identifier) • Multichannel Pulse Count • Multimedia Personal Computer • Multiple-Process Chart • Multiple-Purpose Communications • Multiplicateur de Puissance Crête • Multiprobe Carrier • Multiprocessor Computer • Multiprogram Control • Multipurpose Center • Multipurpose Container

MPC&A: Materials, Protection, Control & Accounting

MPCA: Materials, Protection, Control and Accounting • Mid-Power Controller Assembly

MPCC: Motion Picture Control Panel • Multiprocessor Computer Complex • Multiprotocol Communications Controller • Multipurpose Coma Corrector

MPCD: Manufacturing Process Control Document • Minimum Perceptible Color Difference

MPCG: Modified Preconditioned Conjugate Gradient

MPCH: Main Parallel Channel

MPCMP: Mass Properties Control and Management Plan

MPCRL: Massively Parallel Computing Research Laboratory

MPCS: Management Performance Control System • Mission Planning and Control Station • Multiprocessing Control System

MPCT: Modular Parametric Current Transformer

MPCU: Main Power Control Unit

MPCWG: Multimedia Personal Computer Working Group

MPD: Magneto Plasma Dynamic • Magneto Plasma Dynamics • Main Power Distributor • Maintenance Policy Document • Management Policy and Directives • Management Position Description • Materials Property Data • Maximum Permissible Dose (*) • Minimum Permissible Dose • Mini Port Driver • Mirpur Khas (Pakistan) (location identifier) • Multiple Personality Disorder • Multipurpose Display

MPDD: Meteorological Penetration Detection Development

MPDI: Microwave Polarization Difference Index

MPDM: Maintenance Planning Data Manual

MPDN: Materials Property Data Network

MPDS: Maintenance Planning Data Support • Micrometeorid/Debris Protection Shield

MPDT: Microwave Polarization Difference Temperature

MPDU: Mutiplexing Protocol Data Unit

MPE: Madison (USA-CT) (location identifier) • Map Publishing Environment • Maximum Permissible Exposure • Max-Planck-Institut für Extraterrestrische Physik (Germany) • Memory Protection Electronics • Minimum Potential Energy • Mission-Peculiar Equipment • Mission to Planet Earth • Mobile Platform Equipment • Molecular Probe Ellipsometry • Monthly Project Evaluation • Most Probable Earthquale • Multiple Phase Ejector

MPEC: Mathematical Program with Equilibrium Constraint • Minor Planet Electronic Circular • Multipurpose Experiment Canister

MPECC: Multiprocessor Experimental Computer Complex

MPEF: Meteorological Products Extraction Facility

MPEG: Motion Picture Experts Group (*) • Moving Picture Experts Group

MPEL: Maximum Permissible Exposure Level

MPESS: Mission Peculiar Equipment Support Structure • Mission Peculiar Experiment Support Structure • Multipurpose Equipment Support Structure

MPF: Mapoda (Papua New Guinea) (location identifier) • Mars Pathfinder • MasPar Fortran • Materials Processing Facility • Message Parameter File • Mindless Pop Fluff • Mission Plan File • Momentary Power Failure • Multiple Primary Feed • Multipurpose Fotometer • Multispectral Photographic Facility

MPFE: Modified Poole-Frenkel Effect

MPFM: Multipurpose Fork Mount

MPFS: Multiple Primary Feed System • Multipupil Field Spectrograph

mpg: Miles per Gallon

MPG: Magnetoplasma Generator • Makini (Papua New Guinea) (location identifier) • Matched Power Gain • Max-Planck-Gesellschaft (Germany) • Meteor Physics Group • Miles Per Gallon • Multiple Point Ground • Multipoint Grounding • Multiprogramme Groundstation

MPGHM: Mobile Payload Ground Handling Mechanism

MPGS: Multipurpose Graphic Software

mph: Miles per Hour

MPh: Master of Philosophy

MPH: Caticlan Malay (Philippines) (location identifier) • Malay (Philippines) (location identifier) • Master of Public Health • Miles Per Hour

MPHCC: Maui High Performance Computing Centre (USA-HI)

M/PHE: Material or Personnel Handling Equipment

MPHE: Material and Personnel Handling Equipment

MPHF: Minimal Perfect Hash Function

MPI: Magnetic Particle Inspection • Malaysian Press Institute • Mamitupo (Panama) (location identifier) • Martin-Puplett Interferometer • Master Patient Index • Max-Planck-Institut (Germany) • Mean Point of Impact • Message-Passing Interface • Microprecision Interferometer • Mission Payload Integration • Molecular Parameter Index • Multiphoton Ionization • Multiple Protocol Interface • Multiprecision Integer

MPIA: Max-Planck-Institut für Astronomie (Germany)

MPIAT: Mars Program Independent Assessment Team (NASA)

MPIC: Message Processing Interrupt Count

MPICD: Mission-Peculiar Interface Control Document

MPID: Multiphoton Ionization and Dissociation

MPIE: Mixed Potential Integral Equation

MPIfR: Max-Planck-Institut für Radioastronomie (Germany)

MPIIN: Modification Procurement Instrument Identification Number

MPIK: Max-Planck-Institut für Kernphysik (Germany)

MPIO: Mission and Payload Integration Office • Mission Payload Integration Office

MPIP: Max-Planck-Institut für Plasmaphysik (Germany)

MPIQO: Max-Planck Institut für Quantum Optik (Germany)

MPIR: Multispectral Pushbroom Imaging Radiometer

MPJ: Morrilton - Petit Jean (USA-AR) (location identifier)

MPK: Message Passing Kernel • Mokpo (Korea-ROK) (location identifier)

MPL: Machine-Processable Language • Maintenance Parts List • Manipulation Positioning Latch • Marine Physical Laboratory • Mars Polar Lander (*) • Mars Probe Lander • MasPar Language • Master Program Library • Master Procedure List • Material and Process List • Mathematical Programming Language • Maximum Penalized-Likelihood • Maximum Permissible Level • Maximum Potential Loss • Mechanical Parts List • Megaprogramming Language • Microprocessor Programming Language • Microprogramming Language • Micropulse Lidar • Minimum Power Level • Mission Planning Laboratory • Montpellier Fréjorgues (France) (location identifier) • Montpellier Méditerranée (France) (location identifier) • Multiple Payload Launcher • Multiple-Purpose Language • Multiprogramming Level

MPLA: Multipoint Link Aggregation

MPLIS: Multipurpose Land Information System

MPLM: Main Pressurized Logistics Module • Mini Pressurized Logistics Module • Multipurpose Logistics Module (*) • Multipurpose Pressurized Logistics Module

MPLN: Maintenance Planning

MPLS: Maximal Principle Least Squares • Multiprotocol Label Swapping • Multiprotocol Label Switching (*)

mpm: Meters per Minute

MPM: Magnetic Phase Modulator • Main Propulsion Motor • Maintenance Planning Manual • Major Project Manager • Manipulator Positioning Mechanism • Manpower Planning Model • Maputo (Mozambique) (location identifier) • Marginal Probability Maximum • Maximizer of Posterior Marginals • Meters Per Minute • Metra-Potential Method • Microphysical Model • Microwave Power Module • Millimeter-wave Propagation Model • Milwaukee Public Museum (USA-WI) • Monolithic Piezoelectric Mirror • Multiple-Processor Module

MPMD: Multiple Program Multiple Data

MPMI: Micro-Phase Measuring Interferometer

MPMP: Mass Properties Management Plan • Microwave Point-to-Multipoint

MPMS: Magnetic Property Measurement System

MPMSE: Multiuse Payload and Mission Support Equipment

MPN: Most Probable Number • Mout Pleasant (Falkland Islands) (location identifier)

MPNE: Microgravity Plant Nutrient Experiment

MPO: Magnetospheric Plasma Oscillation • Management and Personnel Office • Maximum Power Output • Message-Passing Object • Mobile Post Office • Mount Pokono (USA-PA) (location identifier) • Multiproduct Operation

MPOA: Multiprotocol Over ATM

MPOD: Multimode Proximity Operations Device

MPOR: Master Power-On Reset

MPOW: Multiple Purpose Operator Workstation

MPP: Maintainability Program Plan • Major Program Proposal • Manufacturing Process Procedure • Markov Point Process • Mars Polar Path Finder • Martens Polarization Photometer • Massively Parallel Processing • Massively Parallel Processor • Material Processing Procedure • Maximum Power Point • Max-Planck-Institut für Physik (Germany) • Mechanical Parts Production • Member of the Provincial Parliament • Message Processing Program • Microfilm Printer-Plotter • Microprogrammable Processor Part • Mission Planning Phase • Mission Preparation Processor • Most Probable Position • Mulatupo (Panama) (location identifier) • Multiphase Pinned • Multiphase Pinning • Multipinned Phase • Multipin Phase • Multiple Particle Plasma

MPPD: Multipurpose Peripheral Device

MPPL: Multipurpose Processing Language • Multipurpose Programming Language

MPPM: Mission Prediction and Performance Module

MPPS: Master Program Planning Schedule

MPPSE: Multipurpose Payload Support Equipment

MPPT: Maximum Power Point Tracking

MPQ: Maan (Jordan) (location identifier) • Max-Planck Institut für Quantenoptik (Germany)

MPR: Maintainability Problem Report • Malfunction Problem Record • Management Program Review • Mars Pressurized Rover • Mass Properties Report • McPherson (USA-KS) (location identifier) • Message Processing Region • Microburst Prediction Radar • Microwave Precipitation Radiometer • Mission Planning Room • Mockup Purchase Request • Monthly Program Review • Monthly Progress Report • Monthly Project Report • Multiband Profile Radiometer • Multipart Repeater • Multiprotocol Router • Multipurpose Recorder

MPRAS: Multipath Redundant Avionics Suite

MPRD: Modified Proportionnal Radial Displacement

MPRI: Multiphoton Resonance Ionization

MPRL: Massively Parallel Richardson-Lucy

MPRSA: Marine Protection, Research and Sanctuaries Act

mps: Meters per Second

MP&S: Manufacturing, Planning & Scheduling

MPS: Magnetic Pole Strength • Mail Preference Service • Main Power Switch • Main Propulsion System • Maintenance Problem Summary • Maintenance Processor System • Management Policy Statement • Maneuvering Propulsion System • Manpower System • Manual Phase Shifter • Manufacturer's Part Specification • Manufacturing, Planning, and Scheduling • Manufacturing Process Specification • Maschinenprogrammsprache • Maschinenprogrammsystem • Massive Passive Service • Master Production Schedule • Master Program Schedule • Material Processing in Space • Material Processing Science • Material Processing Specification • Material Processing System • Mathematical and Physical Sciences • Mathematical Processing System • Mathematical Programming Society (USA) • Mechanical Phase Shifter • Memory Processor Switch • Message Passing Service • Message Passing System • Message Processing System • Meters Per Second • Microlensing Planet Search • Microwave Pressure Sounder • Milwaukee Public Schools (USA-WI) • Mission Planning Simulator • Mission Planning System • Mission Preparation Sheet • Mission Preparation Software • Mission Preparation System • Mission Profile Simulator • Moteur à Propergol Solide • Motor Pump System • Mount Pleasant (USA-TX) (location identifier) • Moving Platform Simulator • Multiple Protective Structure • Multiprocessor Specification • Multiprogramming System • Multipurpose Presentation System

MPSA: Multiprocessor Parallel Server Architecture • Multiprocessor System Architecture

MPSCL: Mathematical Programming System Control Language

MPSE: Mexican Payload Specialist Experiment

MPSI: Magnetic Plage Strength Index

MPSL: Materials and Processes Specification List

MPSO: McMath Pierce Solar Observatory (USA)

MPSOC: Multipurpose Satellite Operations Center

MPS&P: Material Processing Specifications & Procedures

MPSP: Material Processing Specifications and Procedures • Mathematical Problem-Solving Project (NSF)

MPSR: Mission Profile Storage and Retrieval • Multipurpose Support Room

MPSS: Mission Payload System Segment • Multipurpose Sampling System

MPST: Multipurpose Support Team

MPSX: Mathematical Programming System Extended

M. Pt.: Melting Point

MPT: Main Propulsion Test • Male Pipe Thread • Maliana (Indonesia) (location identifier) • Maneuver Planning Table • Mean Pulse Time • Mechanical Power Transmission • Microprogramming Technique • Minimum Process Time • Ministry of Post and Telecommunications • Ministry of Posts and Telecommunications • Mission Planning Table • Mission Planning Team • Mission Planning Terminal • Multipurpose Telescope

MPTA: Main Propulsion Test Article

MPTE: Mission Planning and Training Environment • Multipurpose Test Equipment

MPTF: Main Propulsion Test Facility • Mission Planning Task Force

MPTI: Moscow Physical-Technical Institute (Russia)

MPTN: Multiprotocol Transport Networking

MPTP: Main Propulsion Test Program

MPTS: Mobile Photographic Tracking Station • Multipurpose Tool Set • Multipurpose Tracking System

MPTWT: Medium Power Travelling Wave Tube

MPU: Main Processing Unit • Main Propulsion Unit • Mapua (Papua New Guinea) (location identifier) • Microprocessor Unit • Mission Planning Unit • Monitor Printing

Unit • Motorola Processor Unit • Motor Pressurization Unit

MPV: Barre (USA-VT) (location identifier) • Magnetic Polarization Vector • Man-Powered Vehicle • Mars Piloted Vehicle • Methane-Powered Vehicle • Montpelier (USA-VT) (location identifier) • Multipurpose Vehicle

MPVA: Main Propellant Valve Actuator

MPW: Macintosh Programmer's Workshop • Mac Path Way • Mariupol (Ukraine) (location identifier) • Maximum Permitted Weight • Microsoft Programmer's Workshop • Multiproject Wafer

MPX: Miyanmin (Papua New Guinea) (location identifier) • Mutiplex • Mutiplexer • Multiprocessor Extension • Multiprogramming Executive

MPY: Maripasoula (French Guiana) (location identifier)

MPZ: Mount Pleasant (USA-IA) (location identifier)

Mq: Mosque

MQ: Marketing Questionnaire • Martinique (ISO country code) • Merit Quotient • Message Queuing • Metro Quadrato • Modèle de Qualification

MQA: Mandora (Australia-WA) (location identifier)

MQAM: Message Queue Access Method

MQB: Macomb (USA-IL) (location identifier)

MQBO: Mesospheric Quasibiennial Oscillation

MQC: Many-Quark Cluster • Miquelon (Saint Pierre & Miquelon) (location identifier)

MQD: Maquinchao (Argentina-RN) (location identifier) • Monolithic Quad Device

MQDP: Méthodologie et Qualité de Développement de Projets

MQDT: Multichannel Quantum Defect Theory

Mqe: Martinique

MQE: Marqua (Australia-NT) (location identifier)

MQF: Magnitogorsk (Russia) (location identifier) • Mobile Quarantine Facility

MQH: Minacu (Brazil) (location identifier)

MQI: Message Queue Interface • Message Queuing Interface • Quincy (USA-MA) (location identifier)

MQIL: Miniature Quartz Incandescent Lamp

MQL: Mildura (Australia-VIC) (location identifier) • Miniature Quartz Lamp

MQN: Magnetic Quantum Number • Mo I Rana (Norway) (location identifier)

MQP: Motor Qualification Program

MQQ: Moundou (Chad) (location identifier) • Multiple Quick Query

MQR: Modified Quick Rank • Mosquera (Colombia) (location identifier) • Multiply Quotient Register

MQS: Modified Q-Switching • Mustique (Grenadine Islands) (location identifier)

M-QSPR: Multivariate Quantitative Structure-Property Relationship

MQSPR: Multivariate Quantitative Structure-Property Relationship

MQT: Macroscopic Quantum Tunnelling • Marquette County (USA-MI) (location identifier) • Model Qualification Test • Motor Qualification Test

MQU: Mariquita (Colombia) (location identifier) • Multiplier Quotient Unit

MQW: McRae Telfair-Wheeler (USA-GA) (location identifier) • Multiple Quantum Well • Multiquantum Well

MQX: Makale (Ethiopia) (location identifier)

MQY: Smyrna (USA-TN) (location identifier)

m/r: Mi Remesa

mr: Marathi (ISO language code) • Milliroentgen

mR: Milliroentgen

Mr: Meester • Mister • Monsieur

M/R: Mixture Ratio

M&R: Maintenance & Refurbishment • Maintenance & Repair

MR: Macchia Rossa • Magnetic Reconnection • Magnetic Recorder • Magnetic Resonance • Magnetic Rotation • Magneto-Resistance • Magneto-Resistive • Main Resonance • Main Ring • Maintenance and Refurbishment • Maintenance Report • Maintenance Review • Malfunction Report • Management Requirement • Management Reserve • Management Review • Manpower Requirement • Many Rows • Map Reference • March • Mars • Master Reset • Material Request • Material Review • Matériau de Référence • Mauritania (ISO country code) • Mauritanie (ISO country code) • Medium Range • Medium Resolution • Memorandum for Record • Memorandum Receipt • Memorandum Report • Memory Recall • Memory Register • Message Register • Message Repeat • Message Router • Metabolic Rate • Methane Recovery • Microwave Radiometer • Minister Residentiary • Minnesota Pulsar • Minor Repair • Miscellaneous Report • Mission Reliability • Mission Report • Mixed Reality • Mixture Ratio • Modem Ready • Modification Request • Modification Requirement • Molecular Replacement • Monitor Recorder • Monthly Report • Monthly Review • Morocco • Motivational Research • Motivation Research • Mover Ring

MRA: Macintosh Reference Architecture • Main Ring Assembly • Management and Reconfiguration Assembly • Maximum Rendezvous Altitude • Mechanical Readiness Assessment • Minimum Reception Altitude • Mission Requirements Analysis • Mission Requirements and Allocation • Misurata (Libya) (location identifier) • Mobile Robotic Actor • Multimode Radar Altimeter • Multiresolution Analysis • Multistaged Radial Axial • Mutual Recognition Agreement

mrad: Nilliradian

MRAD: Mass Random Access Disk

MRADS: Mass Random Access Data Storage

MRAM: Magneto-resistive Random Access Memory

MRAO: Mullard Radio Astronomy Observatory (UK)

MRAS: Member of the Royal Astronomical Society

MRASS: Modular Radar Analysis Software System

MRB: Magnetospheric Radio Burst • Maintenance Review Board • Malfunction Review Board • Martinsburg (USA-WV) (location identifier) • Material Review Board • Mission Reconfiguration Board • Mission Review Board • Modification Requirements Board • Modification Review Board • Multinational Review Board • Multireference Bit

MRBAB: Musées Royaux des Beaux-Arts de Belgique (Belgium)

MRBB: Master Reference Blackbody

MRBSR: Material Review Board Standard Repair

MRC: Columbia - Maury County (USA-TN) (location identifier) • Machine-Readable Character • Machine-Readable Code • Magnetic Research Corporation • Maintenance and Repair Cycle • Maintenance Requirement Card • Maître de Recherche Contractuel • Major Regional Conflict • Major Regional Contingency • Matériau de Référence Certifié • Mathematics Research Center • Measurement Requirements Committee • Medical Research Council (UK) • Meteorological Research Committee • Mineralogy Research Centre • Minimum Resolvable Contrast • Mission Requirements Change • Molonglo Reference Catalogue • Monthly Recurring Charge • Multiple Register Counter • Multiple Regression/Correlation

MRCAT: Materials Research Collaborative Access Team

MRCF: Microsoft Real-time Compression Format

MRCI: Microsoft Real-time Compression Interface

MRCR: Measurement Requirement Change Request

mrd: Millirutherford

Mrd: Milliarde

MRD: Machine-Readable Data • Machine-Readable Dictionary • Material Review Disposition • Memory Raster Display • Memory Reduction Technology • Mérida (Venezuela) (location identifier) • Mission Requirements Definition • Mission Requirements Document

MRDA: Marketing Regional Development Assistant • Mission Requirements Definition and Analysis

MRDB: Mapping Requirements Database • Mission Requirements Database • Multiple Representative Database

MRDE: Multireference Double Excitation

MRDF: Machine-Readable Data File

MRDOS: Mapped Real-time Disk Operating System

MRDR: Material Review Disposition Record

MRe: Mauritian Rupee (currency)

MRE: Mara Lodges (Kenya) (location identifier) • Material Review Engineering • Meal Ready-to-Eat • Mean Radial Error • Microrocket Engine • Mid-Range Estimate • Mineral Resources Engineering • Ministère de la Recherche et de l'Espace • Multiple Representation Environment

M/ref: Mi Referencia

mrem: Millirem

MREN: Metropolitan Research and Education Network • Mission des Relations avec les Entreprises

MRES: Ministère de la Recherche et de l'Enseignement Supérieur

M&RF: Maintenance and Refurbishing Facility

MRF: Magnetic Resonance Facility • Maintenance and Refurbishment Facility • Maintenance Repair Facility • Maintenance Replacement Factor • Maintenance Reset Function • Maintenance Responsibility File • Marfa (USA-TX) (location identifier) • Marine Research Facilities • Markov Random Field • Maximum Retarding Force • Measurements Request Form • Medium-Range Forecast • Meteorological Research Facility • Meteorological Rocket Facility • Metering Research Facility • Multipath Reduction Factor

MRFM: Magnetic Resonance Force Microscopy

MRFPI: Moderate-Resolution Fabry-Pérot Imaging

MRG: Magnetic Radiation Generator • Mareeba (Australia-QLD) (location identifier) • Mission Rules Guideline

MRGL: Marginal

MRGS: Member of the Royal Geographical Society

MRH: Magnetic Recording Head • May River (Papua New Guinea) (location identifier) • Mechanical Recording Head • Mission-Related Hardware • Module de Ressources Hermès

MRI: Anchorage Merrill (USA-AK) (location identifier) • Magnetic Resonance Image • Magnetic Resonance Imager • Magnetic Resonance Imaging • Magnetic Resonance Institute • Magneto-Rotational Instability • Major Research Instrumentation • Mathematical Research Institute • Measurement Requirements and Interface • Mehta Research Institute (India) • Metadata Reference Information • Meteorological Research Institute • Molecular Research Institute

MRINDO: Modified Rydberg Intermediate Neglect of Differential Overlap

MRIR: Medium-Resolution Imaging Radiometer • Medium-Resolution Infrared Radiometer

MRIS: Machine-Readable Identification System • Moderate Resolution Imaging Spectrometer

MRJ: Marcala (Honduras) (location identifier)

Mrk: Markarian

MRK: Marco Island (USA-FL) (location identifier)

Mrkr: Marker

Mrktg: Marketing

MRL: Machine Representational Language • Maintenance Repair Level • Maintenance Requirements List • Manipulator Retention Latch • Manipulator Retention Lock • Master Report List • Material Requirements List • Maximum Recording Level • Maximum-Residual Likelihood • Maximum Residue Limit • Measurement Requirements List • Media Response Lines • Miners Lake (Australia-QLD) (location identifier) • Multiple Rocket Launcher • Multiple Ruby Laser

MRLU: Map Reproduction and Liaison Unit

MRM: Management Responsibility Matrix • Manare (Papua New Guinea) (location identifier) • Manufacturing Resource Management • Mid-term Review Meeting • Mission Reconfiguration Master • Mission Requirements Matrix • Motif Resource Manager

MRMS: Maritime Resource Management Service • Mission Reconfiguration Management System • Mobile Remote Manipulator System

MRN: Magnetic Reynolds Number • Maritime Radionavigation • Meteorological Rocket Network • Minimum Rejection Number • Morganton Lenoir (USA-NC) (location identifier)

mRNA: Messenger Ribonucleic Acid

Mrng: Morning

MRNG: Morning

MRNS: Maritime Radionavigation Service

MRNSS: Maritime Radionavigation Satellite Service

MRO: Machine-Readable Output • Magdalena Ridge Observatory (USA-NM) • Maintenance-Repair-Operation • Maintenance-Repair-Overhaul • Management Review Officer • Manastash Ridge Observatory (USA-WA) • Masterton (New Zealand) (location identifier) • Material Release Order • Multiregion Operation • Ouguiya (Mauritania) (ISO currency code)

MROA: Magnetic Raman Optical Activity

MROC: Miniature Ruggedized Optical Correlator

MROF: Maintenance, Repair and Operation of Facility

MROL: Minimum Resolvable Object Length

MRP: Manned Reusable Payload • Manufacturers' Recommended Price • Manufacturing Requirements Planning • Manufacturing Resource Planning • Marla (Australia-SA) (location identifier) • Master in Regional Planning • Master of Regional Planning • Material Requirement Planning • Material Review Process • Mathematics Resources Project (NSF) • Maximum Resolving Power • Medium-Range Plan • Medium-Range Planning • Microfiche Reader/Printer • Microwave Radiometric Profiler • Mid-Range Plan • Mission Reference Plane • Monthly Report of Progress • Multisecond Radio Pulsar

MRPA: Modified Random Phase Approximation

MRPC: Multicost Remote Control Procedure • Multigap Resistive Plate Chamber

MRPL: Main Ring Path Length

MRQ: Marinduque (Philippines) (location identifier) • Maximum Release Quantity

MRR: Macara (Ecuador) (location identifier) • Manufacturing Readiness Review • Mars Return Rate • Mass Return Rate (*) • Master Release Record • Material Rejection Report • Material Reliability Report • Microfilm Reader Recorder • Milestone Readiness Review • Minimum Reporting Requirement • Mission Readiness Review • Mission Reconfiguration Request • Mission Requirements Review • Molecular Rotational Resonance • Monthly Review Report

MRRL: Midlands Regional Research Laboratory (UK)

MRRT: Maintenance Requirements Review Team

Mrs: Messieurs • Mistress • Missus

MRS: Magnetic Resonance Spectroscopy • Magnetic Resonance Spectrum • Maintenance, Repair and Service • Management and Reporting System • Management Review System • Marseille - Marignane (France) (location identifier) • Marseille - Provence (France) (location identifier) • Master Repair Schedule • Material Request Summary • Material Requirement Summary • Material Routing Slip • Materials Research Society (USA-PA) • Mean Root Square • Media Recognition System • Medium-Range Sensor • Medium-Resolution Spectrograph • Medium-Resolution Spectrometer • Memo Routing Slip • Microwave Radiometer Spacecraft • Microwave Ranging System • Mission Rehearsal System • Mission-Related Software • Mobile Remote Servicer • Mobility Requirement Study • Modification Record Sheet • Monitored Retrievable Storage • Multidiscipline Refurbishable Satellite • Multiresonance Spectrum • Multiresonance Structure • Multispectral Resource Sampler

MRSA: Major River Systems Analysis • Microwave Radiometer/Scatterometer and Altimeter

MRSC: Manitoba Remote Sensing Centre (Canada-MT) • Maritime Remote Sensing Committee • Member of the Royal Society of Chemistry • Mississippi Remote Sensing Center (USA-MS)

MRSD: Maximum Rated Standard Deviation • Mission Requirements on System Design

MRSE: Microwave Remote Sensing Experiment

MRSEC: Materials Research Science and Engineering Center

MRSH: Marsh

MRSP: Microwave Remote Sensing Program • Münster Redshift Survey Project

MRSR: Mars Rover and Sample Return (*) • Mars Rover Sample Return

MRSRM: Mars Rover and Sample Return Mission (*) • Mars Rover Sample Return Mission

MRSV: Maneuverable Recoverable Space Vehicle

MRT: Machine-Readable Tape • Magnetic Resonance Tomography • Magnetresonanztomograph • Management Review Team • Mass Rapid Transit • Mass Rapid Transport • Mauritania (ISO country code) • Mauritanie (ISO country code) • Mauritius Radio Telescope • Mean Radiative Transfer • Mean Repair Time • Mean Residence Time • Mean Response Time • Miniature Receive Terminal • Minimum Resolvable Temperature • Ministère de la Recherche et de la Technologie • Ministry of Research and Technology • Mission Readiness Test • Mission Recovery Team • Morehead Radio Telescope • Moroak (Australia-NT) (location identifier)

MRTA: Mechanical Response Tissue Analyzer

MRTC: Multiple Real-Time Command

MRTD: Minimum Resolvable Temperature Difference

MRTFB: Major Range and Test Facility Base

MRTM: Maritime

MRTS: Material Requirements Tracking System • Meteorological Real-Time System • Modification and Repair Tracking System

MRTT: Multirole Tanker Transport

MRTU: Multiplex Remote Terminal Unit

MRU: Machine Record Unit • Main Regulator Unit • Management and Regulation Unit • Manpower Research Unit • Material Recovery Unit • Mauritius (location identifier) • Message Retransmission Unit • Meteorological Research Unit • Microwave Relay Unit • Minimum Replacement Unit • Mobile Repair Unit • Monitoring and Reconfiguration Unit • Most Recently Used • Multicampus Research Unit • Port Louis (Mauritius) (location identifier)

MRV: Mineralnye Vody (Russia) (location identifier) • Mitte-Rand-Variation • Multiple Reentry Vehicle • Multiple Rotational Variant

MRVI: Medium Resolution Visible Imager

MRVS: Meet-, Registratie- en Verwerkingssysteem

MRW: Maribo (Denmark) (location identifier) • Maximum Ramp Weight

MRWC: Multiple Reading and Writing with Computing

MRWG: Mission Requirements Working Group

MRWS: Monoscopic Revision Workstation

MRX: Magnetic Reconnection Experiment • Magneto-Resistive Extended

MRY: Carmel (USA-CA) (location identifier) • Monterey Peninsula (USA-CA) (location identifier)

MRZ: Meteorologisches Rechenzentrum • Moree (Australia-NSW) (location identifier)

m/s: Meter per Second

ms: Malay (ISO language code) • Manuscript • Manuscrit • Millisecond

M/s: Meter per Second

Ms: Miss • Mistress

M&S: Maintenance & Security • Maintenance & Supply • Mapping & Surveying • Modelling & Simulation

M/S: Main Stage • Measurement Stimulus • Months after Sight • Motorship

MS: Machine Screw • Machine Steel • Magellanic Storm • Magellanic Stream • Magnetic South • Magnetic Storage • Magnetic Susceptibility • Mail Steamer • Main Sequence • Main Stage • Main Store • Main Structure • Main Switch • Maintenance and Service • Maintenance Software • Maintenance Standard • Maintenance State • Maîtrise Spécialisée • Management Science • Manned System • Mano Sinistra • Manufacturing Specification • Manufacturing Standard • Manufacturing Status • Manufacturing Support • Manufacturing System • Manuscript • Manuscrit • Manuskript • Mare Serenitatis (Moon) • Margin of Safety • Marine Stratocumulus • Mark Start • Massenspektrometrie • Massive Star • Mass Spectrometer • Mass Spectrometry • Mass Spectroscopy • Mass Spectrum • Mastère Spécialisé • Master of Science • Master Schedule • Master Scheduler • Master Station • Master Switch • Material Specification • Material Standard • Material Support • Mating Sequence • Mato Grosso do Sul (Brazil) • Mauritius (international vehicle registration code) • Maximum Stress • Mean Seasonal • Mean Square • Measuring Set • Measuring System • Mechanical System • Medical Staff • Medium Shot • Memoriae Sacrum • Memory System • Message Store • Meteoritical Society • Meteoroid Shield • Metric System • Microsecond • Microsoft Corporation • Microstrip • Milchstraße • Milestone • Military Standard • Millisecond • Minisymposium • Minus • Mission Schedule • Mission Scientist • Mission Sequence • Mission Simulator • Mission Specialist • Mission Station • Mississippi (USA) • Mobile Station • Modèle Standard • Modular SmallTalk • Molecular Sieve • Monitor Station • Monochromatic Scanner • Montserrat (ISO country code) • Moonshine • Mössbauer Spectrometer • Mössbauer Spectroscopy • Motor Ship • Multiple Sclerose

MSA: Main Solar Array • Major Subcontract Administrator • Major Statistical Area • Malta Standardization Authority (formerly MBS) • Management System Analysis • Master Schedule Authorization • Material Service Area • Meteorological Support Activity • Methane Sulfonic Acid • Metropolitan Statistical Area • Metropolitan Survey Area • Microgravity Science and Applications • Millennium Star Atlas • Mineralogical Society pf America • Minimum Safe Altitude • Minimum Sector Altitude • Minimum Surface Area • Mission Support Area • Mobile Service • Mobile Station • Monitor and Switching Assembly • Multiple Statistical Analysis • Muskrat Dam (Canada-ON) (location identifier)

MSAA: Microsoft Active Accessibility

MSAAS: Microgravity Science and Application Advisory Subcommittee

MSAC: Multiservice Access Concentrator

MSACM: Microsoft Audio Compression Manager

MSACP: Microgravity Science and Applications Cooperative Program

MSACSS: Mount Stromlo Abell Cluster Supernova Search

MSAD: Materials Summary Acceptance Document • Microgravity Science and Applications Division • Multisatellite Attitude Determination • Multisensor Algorithm Development

MSAE: Master of Science in Aeronautical Engineering

MSAI: Memorie della Società Astronomica Italiana

MSAM: Master of Science in Applied Mechanics • Medium-Scale Anisotropy Measurement • Multisequential Access Method

MSAP: Multisatellite Attitude Prediction

MSAR: Microwave Spectrometer And Radiometer • Miniature Synthetic Aperture Radar

MSART: Multifunction Synthetic Aperture Radar Technology

MSAS: Mansfield and Sutton Astronomical Society (UK) • Mexborough and Swinton Astronomical Society (UK) • Mountain Skies Astronomical Society (USA-CA) • MSAT Satellite-based Augmentation System

MSAT: Minnesota Scholastic Aptitude Test • Mobile Satellite

MSATT: Mars' Surface and Atmosphere Through Time • Martian Surface and Atmosphere Through Time

MSAU: Multistation Access Unit

MSAV: Microsoft Antivirus

MSAW: Minimum Safe Altitude Warning

MSAWS: Minimum Safe Altitude Warning System

msb: Minor Significant Breakthrough

MSB: Magnetschwebebahn • Maintenance Standard Book • Major Significant Breakthrough • Manufacturing Studies Board • Material Support Branch • Mission Simulator Building • Mission Status Board • Mitteilungen der Sternfreunde der VHS Buchloe (Germany) • Most Significant Bit • Most Significant Byte • Saint-Martin Marigot (Guadeloupe) (location identifier)

MSBA: Master of Science in Business Administration

MSBC: Master of Science in Building Construction

MSBF: Mean of Swaps Between Failures

MSBLS: Microwave Scanning Beam Land Station • Microwave Scanning Beam Landing System

MSBM: Maîtrise en Sciences Biologiques et Médicales

MSBN: Mobile Satellite Business Network

MSBP: Multiseed Backpropagation

Msc: Miscellaneous

MSc: Marine Stratocumulus • Master of Science

MSC: Macro Selection Compiler • Magnetic Surface Current • Main Storage Control • Maintenance Support Center • Manned Spacecraft Center • Mapping Science Committee • Marine Science Council • Maritime Safety Committee (IMO) • Mass Spin Coupling • Mass Storage Control • Master Schedule Chart • Master Sequence Controller • Materials Service Center • Memory Storage Control • Mesa Falcon (USA-AZ) (location identifier) • Message Sequence Chart • Message Switching Center • Meteorological Satellite Center • Meteorological Synthesising Center • Michigan Spectral Catalogue • Military Staff Committee • Minimal Sufficient Condition • Ministerio de Sanidad y Consumo (Spain) • Minnesota Supercomputer Center (USA-MN) • Minnesota Supercomputing Center • Mirror Sign Convention • Miscellaneous • Mission Support Calibration • Mission Support Center • Mission Support Component • Mobile Service Carrier • Mobile Service Center • Mobile Servicing Centre • Mobile Switching Centre • Modal Spiral Case • Model Selection Criterion • Mode Selector Controller • Mode Sequencing and Control • Moding-Sequencing-Control • Most Significant Character • Moved-Seconded-Carried • Moving Switching Centre • Multiple Scan Correlator • Multiple Star Catalog • Multiple Systems Coupling • Multiplicative Signal Correction

MSCADC: Miniature Standard Central Air Data Computer

MSCB: Manufacturing Status Control Board

MSCC: Manned Space Laboratories Control Centre

MScD: Master of Medical Science

MSCD: Mirror Scan Correction Data

MSCDEX: Microsoft CD-ROM Extension • Microsoft Compact Disk Extension

MSCE: Master of Science in Civil Engineering

MSChE: Master of Science in Chemical Engineering

M/SCI: Mission/Safety Critical Item

MSCI: Minnesota Supercomputer Center Inc. (USA-MN) • Mission Safety Critical Item

MSCL: Measurement Standards and Calibration Laboratory

MScMed: Master of Medical Science

MSCP: Mass Storage Control Processor • Mass Storage Control Protocol

MSCR: Measurements/Stimuli Change Request

MSCS: Microsoft Cluster Server • Mobile Service Center System

MSCTN: Meteorological Satellite Center Technical Note (periodical code)

MSCU: Modulated Shot and Continuum Uniform • Multisytem Communications Unit

MSD: Doctor of Medical Science • Magnetic Storage Drum • Major Seismic Disturbance • Mapping Support Data • Mass Storage Device • Material Support Data • Materials and Structures Division • Mean Solar Day • Mean Squared Deviation • Mean Squared Difference • Mean Square Deviation • Mean Square Difference • Mean Square Displacement • Message Sequence Diagram • Method of Steepest Descent • Microsoft Diagnostics • Microsoft System Diagnostics • Military Survey Department • Minimization of the Sum of Deviations • Minimum Safe Distance • Mission Support Directorate • Mission Systems Data • Most Significant Digit • Mount Pleasant (USA-UT) (location identifier) • Multiband Selection Device • Multiple Site Damage • Multisensor Display • Multispark Discharge • Museum of Scientific Discovery (USA-PA)

MSDB: Main Storage Database

MSDC: Mass Spectrometry Data Centre (RSC) • Midwest Space Development Corporation (USA)

MSDL: Microsoft Download Service

MSDN: Microsoft Developer Network • Microsoft Developer's Network

MSDO: Mapping Support District Office

MS-DOS: Microsoft Disk Operating System

MSDOS: Mass Storage Digital Operating System • Microsoft Disk Operating System

MSDP: Mission System Distribution Panel • Multichannel Subtractive Double Pass (spectrograph)

MSDR: Maintenance Signal Data Recorder • Master Sensor Data Record • Multiplexed Streaming Data Request

MSDRS: Master Signal Data Recording Set

MSDS: Magnetic Storage Drum System • Materials Safety Data Sheet • Microsoft Developer Support • Multispectral Scanner and Data System

MSDT: Multiple Shuttle Data Sheet

MS&E: Materials Science & Engineering

MSE: Maintenance Support Equipment • Manned-Spaceflight Engineer • Manston (UK) (location identifier) • Mass Storage Editor • Master of Science in Education • Master of Science in Engineering • Mean Squared Error • Mean Square Error • Measuring and Stimuli Equipment • Mechanical Support Equipment • Medical Support Equipment • Microelectronics Systems Education • Milton S. Eisenhower • Mission Staff Engineer • Mission Support Equipment • Mobile Subscriber Equipment • Motional Stark Effect • Multiple Site Event

MSEA: Member of the Society of Engineering Associates

MSEB: Mathematical Sciences Education Board

m/sec: Meters per Second

msec: Millisecond

MSEC: Master Separation Events Controller

MSECV: Mean Square Error of Cross Validation

MSEd: Master of Science in Education

MSEDS: Maryland Students for the Exploration and Development of Space (USA-MD)

MSEE: Master of Science in Electrical Engineering • Mean Square Error Efficiency • Mean Square Error of Estimation

MSEI: Mean Square Error Inefficiency

MSEIF: Mean Square Error Improvement Factor

MSELC: Mean Square Error of Leverage Correction

MSEM: Master of Science in Engineering Mechanics • Mission Status and Evaluation Module

MSEng: Materials Science Engineering (periodical code)

MSEO: Modified Simulated Evolutionary Optimization

MSEP: Mean Square Error of Prediction

MSEPS: Modular Space Electrical Power Station

MSER: Management System Evaluation Review

MSET: Macvicar School of Education and Technology • Maintenance Standardization and Evaluation Team • Mathematics, Science, Engineering, and Technology

MSF: Maintenance Source File • Manned Space Flight • Manufacturing, Science and Finance • Massive Star Formation • Master Security File • Master Source File • Médecins Sans Frontières • Medium Standard Frequency • Message Scenario File • Mission Simulator Facility • Month-Second-Foot • Mount Swan (Australia-NT) (location identifier)

MSFC: Marshall Space Flight Center (NASA)

MSFDPS: Manned Space Flight Data Processing System

MSFEB: Manned Space Flight Experiments Board

MSFM: Magnetospheric Specification and Forecast Model

MSFN: Manned Space Flight Network

MSFNOC: Manned Space Flight Network Operations Center

MSFO: Manned Space Flight Operations

MSFP: Manned Space Flight Program

MSFSG: Manned Space Flight Support Group

MSFSRD: Manned Space Flight Support Requirements Documentation

Msg: Message

MSG: Maintenance Steering Group • Maintenance Study Group • Maîtrise des Sciences de Gestion • Management Supervisory Group • Marine Security Guard • Matsaile (Lesotho) (location identifier) • Mechanical System Group • Message • Meteosat Second Generation • Microgravity Science Glovebox • Miscellaneous Simulation Generator • Mission Support Group • Monosodium Glutamate

MSGC: Maryland Space Grant Consortium (USA-MD)

MSGCM: Mars Simple General Circulation Model

MSGM: Master of Science in Government Management

MSGMgt: Master of Science in Government Management

Msgr: Messenger • Monseigneur • Monsignor

MSH: Masirah (Oman) (location identifier) • Melanocyte-Stimulating Hormone

MSHA: Mine Safety and Health Administration • Mine Safety and Health Agency

MSHE: Master of Science in Home Economics

Mshl: Marshal

MSI: Magnetic Shearing Instability • Maintenance Significant Item • Manned Satellite Inspector • Masalembo (Indonesia) (location identifier) • Mass Sharing Index • Mathematical Science Institute • Medina Software Inc. (USA-FL) • Medium-Scale Integration • Minnesota Supercomputer Institute (USA-MN) • Moisture Stress Index • Moon Sphere of Influence • Multisensor Imagery • Multisensor Integration • Multispectral Imager • Multispectral Imagery • Museum of Science and Industry

MSIA: Multispectral Image Analyzer

MSIAS: Multispectral Image Analysis System

MSIC: Medical Support and Integration Contract

MSID: Mass Spectrometric Isotope Detection • Master Stimulus Identification Document • Measurement and Stimulation Identification • Measurement Stimulation Identification • Measurement Stimulus Identification • Medium-Scale Integration Device • Minimization of the Sum of Interior Distances • Mobile Station Identification Number

MSIE: Master of Science in Industrial Engineering • Microsoft Internet Explorer

MSIF: Multisystem Integration Facility

MSIFO: Marine Stratocumulus Intensive Field Observation

MSII: Matra-SEP Image Informatique

MSIM: Mirror Simulator

MSIMD: Mutiple Single Instruction - Multiple Data

MSIS: Management Science and Information Systems • Manned Satellite Inspection System • Man-Systems Integration Standards • Mass Spectrometer and Incoherent Scatter • Mass Spectrometer Incoherent Scatter • Mass Spectrometer Incoherent Scanner • Microgravity Science Information System • Model State Information System

MSISE: Mass Spectrometer and Incoherent Scatter Extended

MSJ: Magnetic Society of Japan • Master of Science in Journalism • Misawa (Japan) (location identifier) • Multimedia Systems Journal (ACM)

MSK: Manual-Selection Keyboard • Mastic Point (Bahamas) (location identifier) • Minimum Shift Keying • Mission Support Kit

Mskg: Masking

Mskr: Manuskript

MSL: Florence (USA-AL) (location identifier) • Machine Specification Language • Main Sea Level • Map Specification Library • Marvel Strategy Language • Master of Science in Linguistics • Master System Library • Materials Science Laboratory • Maximum Segment Lifetime • Mean Sea Level • Measurement Standards Laboratory • Mechanical Systems Laboratory • Meteorological Satellite Laboratory • Microgravity Science Laboratory • Microsoft Software Library • Mode Select Logic • Multistrategy Learning • Muscle Shoals (USA-AL) (location identifier) • Sheffield (USA-AL) (location identifier)

MSLAP: Modular Spectral Line Analysis Program

MSLCC: Manned Space Laboratory Control Centre

MSLD: Mass Spectrometer Leak Detector

MSLIP: Multistage Stochastic Linear Programming System

MSLS: Maneuverable Satellite Landing System • Master of Science in Library Science • Multiservice Launch System

MSLT: Multiple Sleep Latency Test

Msly: Mostly

MSM: Magnetosphere Specification Model • Magnetospheric Specification Model • Manned Spaceflight and Microgravity • Manned Support Module • Manufacturing Standards Manual • Mars Surface Module • Masi Manimba (Zaire) (location identifier) • Materials Science and Mechanics • Mechanics Structures Machines (periodical code) • Meritorious Service Medal • Mission Simulation Model • Mode Selection Mechanism • Mode Select Mechanism • Multilayer Switch Module • Multiple Shooting Method • Multispectral Mapper

MSMC: Master Schedule and Milestone Chart

MSME: Master of Science in Mechanical Engineering

MSMIS: Master of Science in Management Information Systems

MSMP: Manned Support Module Program • Multispectral Measurements Program

MSMQ: Microsoft Message Queuing System

MSMR: Multispectral Multiresolution

MSMS: Meteorological Systems Management Section

MSMU: Moscow State Mining University (Russia)

MSMW: Magnetically Suspended Momentum Wheel

Msn: Mission

MSN: Madison (USA-WI) (location identifier) • Manned Space Network • Material Supply Notice • Message Switching Network • Microsoft Network • Mission

MSNF: Multisystem Network Facility

Msngr: The Messenger (periodical code)

Msnry: Masonry

MSO: Main Scientific Organizer • Maintenance Standard Order • Management Science Office • Manned Solar Observatory • Manned Spacecraft Operation • Manufacturing Safety Office • Manufacturing Sequence Outline • Mars Surface Operation • Mass Spectrometer Outgasing • Mees Solar Observatory (USA-HI) • Missoula (USA-MT) (location identifier) • Model for Spares Optimization • Mountain Skies Observatory (USA-WY) • Multiple-Systems Operator

MSOB: Manned Spacecraft Operations Building

MSOCC: Multisatellite Operations Control Center

MSOE: Milwaukee School Of Engineering (USA-WI) • Multiband Spectral Observation Equipment

MSOS: Mass Storage Operating System

msp: Manuscript Page

MSP: Magnetic Scalar Potential • Magnetic South Pole • Maintenance Support Plan • Maintenance Support Program • Manuscript Page • Master Simulator Program • Medium-Speed Printer • Message Security Protocol • Message System Primitive • Microgravity Sensor Package • Microsoft Paint (format) • Mid-level Service Provider • Military Space Program • Millisecond Pulsar • Minimum Sustaining Power • Minneapolis (USA-MN) (location identifier) • Mission Security Panel • Mission Sequence Plan • Mission Support Plan • Modular System Program • Motion-Sensing Platform • Multiband Sensor Package • Multiscale Profiler • Multispectral Photography • Multispectral Sensor Package • Saint Paul (USA-MN) (location identifier)

MSPB: Merit Systems Protection Board

MSPC: Microstrip Proportional Counter

MSPD: Million Subscriber-Page-Days

MSPF: Multispectral Photographic Facility

MSPH: Master of Science in Public Health

MSPhar: Master of Science in Pharmacy

MSPHE: Master of Science in Public Health Engineering

MSPLT: Master Source Program Library Tape

MSPO: Middelbaar Sociaal-Pedagogische Opleiding

MSPRB: Meteorological Satellite Program Review Board

MSPS: Modular Space Power Station

MSQ: Managing Service Quality • Minsk (Belarus) (location identifier)

MSR: Machine-Specific Register • Machine State Register • Magnetic Shift Register • Magnetic Storage Ring • Main Supply Route • Manufacturing Specification Request • Mark Sheet Reader • Mars Sample Return • Material Status Report • Mean Spring Rise • Measurement Stimuli Request • Measurement Support Request • Mechanized Storage and Retrieval • Meteorological Sounding Rocket • Microwave Scanning Radiometer • Minimum Security Requirement • Minimum Spatial Representation • Mission Support Recording • Mission Support Room • Modification Status Report • Modified Stepwise Regression • Module Support Rack • Monthly Status Report • Monthly Status Review • Montserrat (ISO country code) • Mus (Turkey) (location identifier)

MSRC: Major Shared Resource Center • Marine Sciences Research Center (USA-NY)

MSRE: Moon Signal Rejection Equipment

MSRF: Microwave Space Research Facility

MSRI: Mathematical Sciences Research Institute (USA-CA)

MSRM: Mars Sample Return Mission • Mass Storage Reference Model

MSRP: Meteorological Sounding Rocket Program • Mission Support Real Property • Multichannel Solar Radio Polarimeter

MSRR: Mission and System Requirements Review

MSRS: Meteoroid Shield Release System

MSRSL: Mémoires de la Société Royale des Sciences de Liège (periodical code)

MSRT: Meter-wave aperture Synthesis Radiotelescope • Miyun Synthesis Radiotelescope (China-PRC)

MSRTP: Micelle-Stabilized Room temperature Phosphorescence

mss: Manuscripts • Manuscrits

Mss: Manuskripte

MsS: Magnetostrictive Sensor

MSS: Magnetic Sensing System • Magnetic Storm Satellite • Magnetostrictive Sensor • Main Stellar Spectrograph • Main Support Structure • Maintenance Status System • Management Support System • Manned Space Station • Manual Safety Switch • Manufacturers Standardization Society • Manuscripts • Manuscrits • Marconi Space Systems • Maritime Surveillance System • Massena (USA-NY) (location identifier) • Mass Storage System • Master of Social Sciences • Master of Social Service • Master of Space Studies • Maximum Segment Size • MBS Base System • Mechanical Support System • Medium Satellite Servicing • Meeting Support System • Message Source Selector • Message Switching Station • Meteorological Satellite Section • Meteorological Service Singapore • Michigan Spectral Survey • Microwave Switching Station • Midlands Spaceflight Society (UK) • Mission Simulation System • Mission Specialist Station • Mission Status Summary • Mission Support Software • Mobile Satellite Service • Mobile Service Structure • Mobile Servicing System • Mobile Support Station • Modular Space Station • Multichannel Spectrum Scanner • Multishaker System • Multispectral Scanner • Multispectral Scanner System • Multisystem Software

MSSAR: Multispectral Synthetic Aperture Radar

MSSc: Master of Social Sciences

MSSC: Mass Storage System Communications • Maui Space Surveillance Complex (USA-HI)

MSSCB: Mass Storage and Scientific Computing Branch

MSSCC: Multicolour Spin-Scan Cloud Camera

MSSCS: Manned Space Station Communications System

MSSL: Mullard Space Science Laboratory (UK)

MSSM: Mars Spinning Support Module • Minimal Supersymmetric Standard Model

MSSN: Mission

MSSP: Mechanical Systems and Signal Processing (periodical code)

MSSR: Mars Soil Sample Return • Mars Surface Sample Return • Mission and Science Scheduling Room • Monopulse Secondary Surveillance Radar

MSSRV: Mars Surface-Sample Return Vehicle

MSSS: Malin Space Science Systems • Manned Space Station Simulator • Manned Static Space Simulator • Mass Spectral Search System • Maui Space Surveillance Site (USA-HI) • Multiple-Start Systematic Sampling • Multisatellite Support System

MSSSO: Mount Stromlo and Siding Spring Observatories (Australia-ACT) • Mt. Stromlo and Siding Spring Observatories (Australia-ACT)

MSST: Mean Sea Surface Temperature • Ministry of State for Science and Technology

MSSTA: Multispectral Solar Telescope Array

MST: Maastricht (Netherlands) (location identifier) • Magnetospheric-Stratospheric-Tropospheric • Main Spacecraft Thruster • Maîtrise des Sciences et des Techniques • Management Survey Team • Mass Spectrometer Tube • Master of Science in Teaching • MCCC Telemetry System • Mean Sidereal Time • Mean Solar Time • Mean Survival Time • Measurement Status Table • Mechanical Support Team • Mesosphere-Stratosphere-Troposphere • Minimal Spanning Tree • Ministry of Science and Technology • Mission Scheduling Test • Mission Science Team • Mission Selection Team • Mission Sequence Test • Mission Support Team • Mobile Service Tower • Module Service Tool • Mountain Standard Time • Multimission Software Transmission

M-STAR: Merced Society for Telescope and Astronomy Recreation (USA-CA)

MSTAR: Merced Society for Telescope and Astronomy Recreation (USA-CA)

MSTB: Mission Simulator and Training Building

MSTC: Manned Spacecraft Test Center

MSTE: Mechanical Special Test Equipment

MSTG: Mass Storage Task Group

MSTI: Miniature Seeker Technology Initiative • Miniature Seeker Technology Integration • Miniature Sensor Technology Integration (*)

MSTLY: Mostly

MSTO: Main-Sequence Turnoff

MSTP: Manned Space Transportation Programme

Mstr: Master • Moisture

MSTR: Master • Moisture

MSTS: Multisource Tactical System • Multisubscriber Time-Sharing

MSTV: Microsoft Television

MSU: Main Storage Unit • Management System Unit • Mankato State University (USA-MN) • Maseru Moshoeshoe (Lesotho) (location identifier) • Mass Storage Unit • Mathematical Study Unit • Measuring Stimuli Unit • Memory Service Unit • Memphis State University (USA-TN) • Message Switching Unit • Michigan State University (USA-MI) • Microwave Sounder Unit • Microwave Sounding Unit (*) • Million of Service Units • Minot State University (USA-ND) • Mississippi State University (USA-MS) • Montana State University (USA-MT) • Montclair State University (USA-NJ) • Moorhead State University (USA-MN) • Morgan State University (USA-MD) • Moscow State University (Russia) • Murray State University (USA-KY)

mSv: Millisievert

MSV: Maintenance Support Vessel • Major Software Vendor • Manned Space Vehicle • Martian Surface Vehicle • Mean Square Velocity • Mechanisms of Solar Variability • Mission Support Vehicle • Model and Scenario Visualization • Monitored Sine Vibration • Monticello (USA-NY) (location identifier) • Multispectral Video

MSVC: Mass Storage Volume Control

MSVIAS: Multispectral Video Imaging and Analysis System

MSVP: Master Shuttle Verification Plan

MSW: Machine Status Word • Magnetic Surface Wave • Mammals Species of the World • Massawa (Ethiopia) (location identifier) • Master in Social Work • Master of Social Welfare • Master of Social Work • Microswitch • Mikheyev-Smirnov-Wolfenstein • Multiple Shock Wave • Municipal Solid Waste

MSWG: Multilateral Science Working Group

MSWM: Microsoft Windows Manager

MSX: Midcourse Space Experiment • Mossendjo (Congo) (location identifier)

MSY: Maximum Sustainable Yield • New Orleans (USA-LA) (location identifier)

MSZ: Namibe (Angola) (location identifier)

MSZH: Hungarian Office for Standardization

MS2I: Matra SEP Imagerie et Informatique

mt: Maltese (ISO language code) • Metric Ton

Mt: Measurement • Megatonne • Metical (currency) • Mont • Mount • Mountain

M/T: Empty • Mail Transfer

MT: Empty • Machine Translation • Maddalena Thaddeus • Mean Time • Magnetic Tape • Magnetic Torquer • Magnetic Tube • Magnetotelluric • Mail Transfer • Main Tank • Main Telescope • Maintenance Technician • Maintenance Time • Malta (ISO country code) • Malte (ISO country code) • Mandated Territory • Man-Tended • Manual Test • Manufacturing Technology • Many Thanks • Mare Tranquilitatis (Moon) • Maritime Tropic • Mass Transport • Master Timer • Master Tool • Mato Grosso (Brazil) • Maximum Torque • Mean Tide • Mean Time • Measured Time • Mechanical Technician • Mechanical Translation • Mechanical Transport • Media Table • MediaTheque • Medio-Temporal • Megaton • Member Traceability • Même Temps • Metatheory • Metric Ton • Microptic Theodolite • Microwave Transducer • Mismo Tiempo • Mission Time • Mission Timer • Mission Trainer • Mission Trajectory • Mittagshöhe • Mobile Transfer • Mobile Transporter • Modus Tolens • Monitor Telescope • Monotype • Monotype Times • Montana (USA) • More Than • Mount • Mountain • Mountain Time • Moving Target • Multiple Transfer • Multithreading

Mta: Meseta

MTA: Macintosh Telephony Architecture • Mail Transfer Agent • Mail Transport Agent • Maintenance Task Analysis • Major Test Article • Major Trading Area • Mass Thermal Analysis • Master Time Assembly • Materials Transfer Agreement • Matamata (New Zealand) (location identifier) • Mathematics Teaching Assistant • Message Transfer Agent • Message Transport Agent • Metering Truss Assembly • Metropolitan Transit Authority • Micro-Thermal Array • Military Training Area • Minimum Terms Agreement • Mojave Test Area (USA-CA) • Motion-Time Analysis • Multiple-Terminal Access • Multithread Architecture • Multithreaded Architecture

MTACCS: Marine Tactical Command and Control System

MTAG: Moving Target Advisory Group

Mtagne: Montagne

MTAM: Maritime Tropical Air Mass

MTAP: Multitask Analysis Program

MTAS: Middle Tennessee Astronomical Society (USA-TN)

MTAT: Mean Turn-Around Time

MTB: Magnetic-Torquer Bar • Main Terminal Board • Mass Transfer Binary • Materials Testing Branch • Monte Libano (Colombia) (location identifier)

MTBB: Mean Time Between Breakdowns

MTBCF: Mean Time Between Critical Failures

MTBD: Mean Time Between Demands • Mean Time Between Discrepancies

MTBE: Mean Time Between Events

MTBF: Mean Time Before Failure • Mean Time Between Failures (*)

MTBFA: Mean Time Between False Alarms

MTBFF: Mean Time Between First Failures

MTBFL: Mean Time Between Function Losses

MTBFRO: Mean Time Between Failures Requiring Overhaul

MTBI: Mean Time Between Incidents

MTBJ: Mean Time Between Jams

MTBM: Mean Time Between Maintenances • Mean Time Between Modifications

MTBMA: Mean Time Between Maintenance Actions

MTBO: Mean Time Between Overhauls

MTBR: Mean Time Between Removals • Mean Time Between Repairs • Mean Time Between Replacements

MTBS: Mean Time Between Services

MTBSD: Mean Time Between Supply Demands

MTBSF: Mean Time Between System Failures

MTBUR: Mean Time Between Unscheduled Removals • Mean Time Between Unscheduled Replacements

MTC: Magnetic Tape Control • Maintenance Task Cycle • Man-Tended Capability • Manual Thrust Controller • Manual Traffic Control • Manufacturing Technology Center • Massachusetts Technology Collaborative • Master Tape Control • Master Thrust Control • Master Thrust Controller • Mathematics Throughout the Curriculum • Maximum Track Capacity • Measurement Technology Center • Members-Team-Committee • Memory Test Computer • Meteorological Training Center • Midlands Technical College (USA-SC) • Mission and Test Computer • Mission and Traffic Control • Mitsui Trading Co. (Japan) • Mount Clemens - Selfridge (USA-MI) (location identifier)

MTCA: Minimum Terrain Clearance Altitude • Monitor and Test Control Area • Multiple-Terminal Communication Adapter

MTCAC: Midlands Technical College Astronomy Club (USA-SC)

MTCC: Master Timing and Control Circuit

MTCE: Million Tonnes of Carbon Equivalent

MTCF: Mean Time to Catastrophic Failure

Mtchd: Matched

MTCP: Modern Technology Cockpit Program

MTCR: Million Tonnes of Coal Equivalent • Mission Technology Control Regime

MTCS: Meteor Trail Communications System • Minimal Terminal Communications System • Modular Terminal Communications System

MTCU: Magnetic Tape Control Unit

Mtd: Mounted

MTD: Machine-Traceable Dictionary • Master Tape Data • Master Time Display • Master Tracking Data • Material Technology Division • Maximum Tolerated Dose • Mean Temperature Difference • Microelectronics Technology Division • Mounted • Mount Sandford (Australia-NT) (location identifier) • Multimedia Transport Protocol

MTDA: Mesocyclone Tornado Detection Algorithm

MTDC: Modified Total Direct Cost

MTDE: Multimedia Title Development Environment

MTDF: Master Tracking Data File

mtDNA: mitochondrial DNA

MTDS: Masuda Tracking and Data Station (Japan)

MTDSK: Magnetic Tape Disk

Mte: Monte

MTE: Magnetic-Tape Emulator • Magnetic-Torquer Electronics • Magnetosphere Thermosphere Explorer • Maintenance Test Equipment • Maintenance Training Equipment • Maximum Tracking Error • Mean Transverse Energy • Modular Toolkit Environment • Monte Alegre (Brazil-PA) (location identifier) • Multisystem Test Equipment

MTEC: Maintenance Test Equipment Catalog

MTEE: Mission Time Extreme Environment

MTEM: Maintenance Test Equipment Module

MTEP: Maritime Technology Enhancement Program

Mtes: Montes

MTF: Master Test Frame • Materials Test Facility • Mean Time to Failure • Message Text Format • Microsoft Tape Format • Mid-Tandem Fan • Military Treatment Facility • Mizan Teferi (Ethiopia) (location identifier) • Modulated Transfer Function • Modulation Transfer Frequency • Modulation Transfer Function (*) • Move To Front

MTFD: Minimum Tracking Flux Density

MTFE: Mid-Thrust Family Engine • Multifrequency Front End

MTFF: Man-Tended Free Flyer • Mean Time to First Failure

MTFFCC: MTFF Control Center

MTFO: Modular Training Field Option

mTFS: Multichannel Fourier Transform Spectrometer

MTFS: Magnetic Tape Field Search

Mtg: Meeting • Mortgage • Mounting

MTG: Master Timing Generator • Mato Grosso (Brazil-MT) (location identifier) • Mounting

MTGCM: Mars Thermospheric General Circulation Model

Mtge: Mortgage

Mtgee: Mortgagee

Mtgne: Montagne

Mtgor: Mortgagor

MTGP: Monitor Tables Generator Program

MTGW: Maximum Total Gross Weight

Mth: Month

MTH: Magnetic Tape Handler • Marathon (USA-FL) (location identifier)

Mthbd: Motherboard

Mthd: Method

MTHM: Metric Ton of Heavy Metal

Mths: Months

Mti: Monti

MTI: Méthode et Traitement de l'Information • Morton Thiokol Inc. (USA) • Mosteiros (Cape Verde) (location identifier) • Moving-Target Identification • Moving-Target Identificator • Moving-Target Imaging • Moving-Target Indication • Moving-Target Indicator • Multilayer Thermal Insulation

MTIGCM: Mesosphere/Thermosphere/Ionosphere General Circulation Model

MTIE: Microthrust Ion Engine

MTIM: Mass Transfer Instability Model

MTIS: Maintenance Task Information System

MTJ: Magnetic Tunnel Junction • Montrose (USA-CO) (location identifier)

MTK: Makin Island (Kiribati) (location identifier) • Mechanical Time Keeping • Mechanized Time Keeping

Mtl: Material • Monatlich

MTL: Lira (Malta) (ISO currency code) • Main Transfer Line • Maitland (Australia-NSW) (location identifier) • Master Tape Loading • Master Timeline • Material • Mean Tide Level • Merged-Transistor Logic • Minimum Time Limit • Mission Target List • Mission Time Life • Mission Time Line • Multiconductor Transmission Line

MTLP: Master Tape Loading Program • Monitor Table Listing Program

MTLRS: Modular Transportable Laser Ranging System

MTM: Maintenance Test Module • Mean Time Measurement • Mechanical Test Model • Method of Task Measurement • Methods-Time Measurement • Metlakatla (USA-AK) (location identifier) • Midnight Temperature Mission • Modification Transmittal Memorandum • Modified Trimmed Mean • Monsieur Tout le Monde • Multitaper Method • Multitarget Model • Multiterminal Monitor

MTMC: Military Traffic Management Command

MTMM: Multitrait Multimethod • Multitrait Multimethod Model

MTMO: Mated Test Management Office

MTMS: Methyl Tri-Methoxy Silane

Mtn: Motion • Mountain

MTN: Baltimore - Glenn L. Martin (USA-MD) (location identifier) • Méthodes et Techniques Nouvelles • Mobile Telephone Network • Mountain • Multilateral Trade Negotiations • Multilevel Tree Network

MTNS: Mathematical Theory of Networks and Systems • Mountains

Mto: Monumento

MTO: Made To Order • Magnetic Tape Operator • Master Timing Oscillator • Mattoon (USA-IL) (location identifier) • Maximum Timeout • Mission-Task-Objectives • Mississippi Test Operations • Mobile Technical Office • Modification Task Outline • Mount Tamborine Observatory (Australia-QLD) • Mt. Tamborine Observatory (Australia-QLD) • Multilateral Trade Organization

MTOB: Manned Test Operations Board

MTOE: Million Tonnes of Oil Equivalent

MTOF: Mass Time Of Flight

MTOGW: Maximum Takeoff Gross Weight

MTOpS: Million Theoretical Operations per Second

MTOPS: Million Theoretical Operations Per Second (*)

MTOS: Magnetic Tape Operations System

MTOW: Maximum Takeoff Weight

MTP: Mail Transfer Protocol • Maintenance Test Package • Manufacturing Technical Procedure • Master Test Plan • Master Test Procedure • Master Timing Pulse • Mathematical Tables Project • Mediterranean Targeted Project • Medium-Term Perspective • Medium-Term Plan • Medium-Term Prediction • Message Transfer Part • Meteosat Transitional Programme • Meteosat Transition Programme • Microwave Temperature Profile • Mission Test Plan • Montauk (USA-NY) (location identifier) • Multiple-Task Performance

MTPE: Mission To Planet Earth

MTPN: Multiprotocol Transport Network

MTPS: Magnetic Tape Plotting System • Multitransport Protocol Service

MTQ: Martinique (ISO country code) • Mitchell (Australia-QLD) (location identifier)

Mtr: Meter

MTR: Magnetic Tape Reader • Magnetic Tape Recorder • Magnetic Tape Recording • Maximum Tracking Range • Mean Time to Removal • Mean Time to Repair • Mean Time to Replacement • Mean Time to Restore • Mechanized Time Reporting • Medium-Term Research • Meter • Minimum Time Rate • Minitemperature Recorder • Monteria (Colombia) (location identifier) • Motor • Multiple Tracking Range

MTRC: Mobile Teacher Resource Center

MTRD: Master Test Requirement Document

MTRE: Magnetic-Tape Recorder End

Mtrg: Metering

MTRS: Magnetic-Tape Recorder Start • Magnetic-Tape Reformatting System

Mts: Monts • Mountains • Mounts

MTS: Magnetic-Tape Station • Magnetic-Tape System • Magnetic Torquing System • Maintenance Training Set • Main Thruster System • Manzini (Swaziland) (location identifier) • Master Test Station • Master Timing System • Material Test Specification • Medium-Term Schedule • Medium-Term Scheduler • Medium-Term Scheduling • Memory Test System • Message Telecommunications Service • Message Telephone Service • Message Transfer Service • Message Transfer System • Meteoroid Technology Satellite • Metering Truss Structure • Meter-Ton-Second • Metric Time System • Microsoft Transaction Server • Microwave Temperature Sounder • Microwave Tracking System • Middelbare Technische School • Millimeter-wave Temperature Sounder • Mission Telemetry System • Mobile Telephone Service • Mobile Tracking Station • Modular Training System • Modulation Transfer Service • Modulation Transfer System • Motor-operated Transfer Switch • Motor Turbine Ship • Mountains • Moving Time Series • MSX Tracking Station • Multichannel Television Sound • Multiple Time Scale • Multithreaded Server

MTSA: Multitaper Spectral Analysis

MTSAT: Multifunction Transport Satellite

MTSE: Magnetic-Tape Splicing Equipment

MTSL: Microelectronics Technology Support Laboratory

MTSO: Mobile Telephone Switching Office

MtSOM: Mount Stromlo Observatory Mimeographs (periodical code)

MTSP: Multiple Travelling Salesmen Problem

MTSR: Magnetic Tape Selective Recorder • Mid-Term Status Report

MTSS: Magnetic-Tape Storage System

MTSU: Magnetic-Tape Search Unit • Middle Tennessee State University (USA-TN)

MTT: Magnetic-Tape Terminal • Maintenance Training Team • Maximum Touch Temperature • Mean Transit Time • Minatitlan (Mexico) (location identifier) • Mobile Training Team • Multiple Telescope Telescope • Multitelescope Telescope

MTTA: Mean Time To Accomplish

MTTB: Mean Time To Bench

MTTD: Mean Time To Defect • Mean Time To Diagnose • Mean Time To Diagnosis

MTTE: Magnetic-Tape Terminal Equipment • Mean Time To Exchange

MTTF: Mean Time To Failure

MTTFF: Mean Time To First Failure

MTTM: Mean Time To Maintain

MTTR: Mean Time To Removal • Mean Time To Repair • Mean Time To Replacement • Mean Time To Restore • Medium-Term Technology Review

MTTRF: Mean Time To Restore Function

MTTRS: Mean Time To Restore Service

MTTS: Microwave Theory and Technical Society

MTTU: Modular Timing Terminal Unit

MTU: Magnetic-Tape Unit • Maintenance Termination Unit • Master Time Unit • Master Timing Unit • Maximum Transfer Unit • Maximum Transmission Unit • Michigan Technological University (USA-MI) • Mission Timer Unit • Mobile Training Unit • Montepuez (Mozambique) (location identifier) • Motoren- und Turbinen-Union • Multiplexer and Terminal Unit

MTV: Mars Transfer Vehicle • Mean Transformed Value • MEA Test Vehicle • Mostly Traditional Value • Mota Lava (Vanuatu) (location identifier) • Motoren und Turbinen Verband

MTVC: Manual Thrust Vector Control

MTVS: Mission Test and Video System

MTW: Manitowoc (USA-WI) (location identifier) • Maximum Taxi Weight • Mechanical and Thermal Works • Monday-Tuesday-Wednesday • Mountain Waves

MTWA: Maximum Total Weight Authorized

MTWY: Motorway

Mtx: Matrix

MTX: Fairbanks Metro (USA-AK) (location identifier) • Mobile Telephone Exchange

MTY: Empty • Monterrey - General Mariano Escobedo (Mexico) (location identifier)

MTZ: Massada (Israel) (location identifier)

m/u: Make Up

M/U: Mockup

MU: Île Maurice (ISO country code) • Machine Unit • Macquarie University (Australia) • Mail Unit • Maintenance Unit • Maneuvering Unit • Master Unit • Math Unit • Mauritania • Mauritius (ISO country code) • Measurement Unit • Memory Unit • Message Unit • Microwave Unit • Middle and Upper • Minority University • Mobile Unit • Mockup • Model Updating • Monetary Unit • Multiple Unit • Multiplexing Unit

MUA: Mail User Agent • Materials Usage Agreement • Maximum Usable Altitude • Memorandum of Understanding and Agreement • Munda (Solomon Islands) (location identifier) • Mutual Usage Agreement

MUB: Maun (Botswana) (location identifier)

MUBAR: Messed Up Beyond All Recognition

MUC: Mesoscale Uncinus Complex • Message Understanding Conference • Million d'Unités de Compte • Mount Union College (USA-OH) • München - Franz-Josef Strauss (Germany) (location identifier) • Munich - Franz-Josef Strauss (Germany) (location identifier)

MUCELS: Multiple Communications Emitter Locating System

MUCS: Modulated Uniform and Continuum Shot • Multiuser Compliant Scheduling

MUD: Memory Unit Drum • Miscellaneous User Documents • Mueda (Mozambique) (location identifier) • Multiple User Dialogue • Multiple User Dimension • Multiple User Dungeon • Multiuser Dimension • Multiuser Domain • Multiuser Dungeon

MUDAS: Modular Universal Data Acquisition System

MUDSS: Mobile Underwater Debris Survey System

MUE: Kamuela (USA-HI) (location identifier) • Mission Unique Electronics

MUET: Mission Universitaire et Éducative de Tunisie (Tunisia)

MUF: Material Unaccounted For • Maximum Urgency First • Maximum Usable Frequency • Muting (Indonesia) (location identifier)

MUFFIN: Multichannel Finite-impulse-response Filter and Integrator

MUFON: Mutual Unidentified Flying Object Network

MUG: Maximum Usable Gain • McIDAS Users Group • Microelectronics Users Group • Mulege (Mexico) (location identifier)

MUH: Mersa Matruh (Egypt) (location identifier)

MUI: Fort Indiantown - Muit (USA-PA) (location identifier) • Multimedia User Interface

MUID: Map Unit Identification Data

MUIR: Map Unit Interpretation Record

MUJ: Mui (Ethiopia) (location identifier)

MUK: Mauke Island (Cook Islands) (location identifier)

MUKLIMO: Mikroskaliges Urbanes Klima-Modell

MUL: Material Usage List • Mission-Unique List • Moultrie Spence (USA-GA) (location identifier) • Multiply

MULC: Monthly Usage License Charge

MULDEM: Multiplexer-Demultiplexer

MULE: Manned-Unmanned Lunar Explorer • Modular Universal Laser Equipment

MULSAM: Multispectral Auger Microscope • Multispectral Auger Microscopy

MULSORANNS: Multiple Linked Self-Organized Artificial Neural Networks

Mult: Multiple • Multiplexer • Multiplication

MULT: Multiple • Multiply

MULTDOP: Multiple Doppler Radar Analysis Program

MULTICS: Multiplexed Information and Computing Service • Multiplex Information Computer Service

MULTIFADS: Multispectral Fast Area Digitizing Scanner

Multr: Multiplier

MUM: Maximum Useful Magnification • Mesoscale Unified Model • Method of Unweighted Means • Multiuse Manuscript • Multiuser Minicomputer • Mumias (Kenya) (location identifier)

Mun: Municipal • Municipality

MUN: Maturin (Venezuela) (location identifier) • Memorial University of Newfoundland (Canada-NF)

Muni: Municipal

Munic: Municipal • Municipality

MUO: Mountain Home (USA-ID) (location identifier)

MUP: Master of Urban Planning • Mulga Park (Australia-NT) (location identifier)

MUPPET: Maryland University Project in Physics and Educational Technology

MUPPS: Mission-Unique Planning and Processing Schedule

MUPS: Momentum Unloading Propulsion System • Multiple-Utility Peripheral System

MUQ: Muccan (Australia-WA) (location identifier)

MUR: Manpower Utilization Report • Marudi (Malaysia) (location identifier) • Rupee (Mauritius) (ISO currency code)

MURA: Midwestern Universities Research Association (USA)

Murc: Murcia (Spain)

MURC: Minority University Research Center (USA)

MURG: Macquarie University Research Grant

MURI: Multidisciplinary University Research Initiative (USA)

MURS: Management Update and Retrieval System

MURST: Ministero dell'Università, della Ricerca Scientifica e Tecnologica

Mús: Música

Mus: Musca-Muscae (constellation) • Museum • Music • Musical • Musician • Muslim

MUS: Île Maurice (ISO country code) • Marcus Island (Japan) (location identifier) • Mauritius (ISO country code) • Mission-Unique Software • Multilateral Utilisation Study • Multiutility System

MUSA: Multiple-Unit Steerable Antenna

MUSAP: Multisatellite Augmentation Program

MUSC: Microgravity User Support Center

MUSCLE: Multiple Scattering Lidar Experiment

MUSE: Mapping Utility Software Environment • Monitor of Ultraviolet Solar Energy • Multielement Ultraviolet Spectroscopic Experiment • Multimedia-based System Engineering • Multiple Sub-Nyquist Encoding • Multiuser Science Environment • Multiuser Shared Environment • Multiuser Simulated Environment

MUSES: Mu Space Engineering Satellite

MUSH: Multiuser Shared Hallucination

MUSIC: Management - User interface - System interface - Information - Communication • Metrics for Usability Standards In Computing • Multifunctional Stereo Imaging Camera • Multiple Signal Classification • Multispectral Infrared Camera

MUSICOS: Multisite Continuous Spectroscopy

MUSIP: Multisensor Image Processing • Multisensor Image Processor

MUSS: Module Utility Support Structure

MUSSL: Marine Upwelled Spectral Signature Laboratory

MUSSWAS: Multisensor Surface Wind Acquisition System

MUSTARD: Multiunit Space Transport And Recovery Device

MUSTIP: Multilateral Strategic and Tactical Investigation Process

Mut: Mutilated • Mutual

MUT: Mean Up Time • Measurements Unit Table • Model Updating Technique • Module Under Test • Muscatine (USA-IA) (location identifier)

MUTTS: Multiple-Unit Terminal Test Set

MUU: Mount Union (USA-PA) (location identifier)

muv: Met Uitzondering Van

MUV: Middle Ultraviolet • Mid-Ultraviolet • Philadelphia Mustin (USA-PA) (location identifier)

MUVE: Multiuser Virtual Environment

MUW: Mascara (Algeria) (location identifier)

MUX: Multan (Pakistan) (location identifier) • Multiplex • Multiplexer

MUY: Mouyondzi (Congo) (location identifier)

MUZ: Musoma (Tanzania) (location identifier)

mv: Visual Magnitude

mV: Millivolt

Mv: Absolute Visual Magnitude • Mendelevium • Move

M/V: Motor Vessel

MV: Main Verb • Maldives (ISO country code) • Manned Vehicle • Manual Valve • Manufacturing Verification • Mare Vaporum (Moon) • Market Value • Mean Value • Mean Variation • Measured Value • Medium Voltage • Meervoud • Megavolt • Mercury Vapor • Mitgliederversammlung • Mixed Valence • Mobile VLBI • Motor Vessel • Multivibrator

MVA: Main Valve Actuator • Manufacturing Value Added • Mean Vertical Acceleration • Megavolt-Ampere • Migration Velocity Analysis • Minimum Vectoring Altitude • Motion Video Accelerator • Multidomain Vertical Alignment • Multivendor Architecture • Mývatn Reykiahlid (Iceland) (location identifier)

MVAAC: Martha's Vineyard Amateur Astronomy Club (USA-MA)

MVACS: Mars Volatiles And Climate Surveyor

MVAS: Magic Valley Astronomical Society (USA-ID) • Mahoning Valley Astronomical Society (USA-OH) • Miami Valley Astronomical Society (USA-OH) • Mohawk Valley Astronomical Society (USA-NY) • Multipurpose Ventricular Actuating System

MVB: Franceville (Gabon) (location identifier) • Minimum Variance Bound • Multimedia Viewer Book

Mvbl: Movable

MVC: Manual Volume Control • Master Volume Control • Maximum Value Composite • Medium-Velocity Cloud • Model View Controller • Monroeville - Monroe County (USA-AL) (location identifier) • Multimedia Viewer Compiler • Multivolume Cartridge

MVCU: Multivariable Control Unit

MVD: Map and Visual Display • Mission Verification Data • Montevideo Carrasco (Uruguay) (location identifier)

MVDA: Multivariate Discriminant Analysis

MVDM: Multiple Virtual DOS Machine

MVDS: Microwave Video Distribution System

MVE: Minimum-Volume Ellipsoid • Montevideo (USA-MN) (location identifier) • Multivalued Dependency • Multivariate Exponential (distribution)

MVEd: Master of Vocational Education

MVEMJSUNP: Mein Vater erklärt mir jeden Sonntag unsere niedlichen Planeten • My Very Educated Mother Just Showed Us Nine Planets • Mercury-Venus-Earth-Mars-Jupiter-Saturn-Uranus-Neptune-Pluto (*)

MVF: Manned Vertical Flight • Mossoró - Dixsept Rosado (Brazil-RN) (location identifier)

MVFR: Marginal Visual Flight Rules

MVFV: Manned Venus Flyby Vehicle

Mvg: Moving

MVG: Mevang (Gabon) (location identifier)

MVGA: Monochrome Video Graphics Array

MVGVT: Mated Vertical Ground Vibration Test

MVH: Macksville (Australia-NSW) (location identifier)

MVI: Medium Value Item • Menatai (Papua New Guinea) (location identifier) • Microgravity Vestibular Investigation • Multizonal Visual Information

MVIEP: Multivariate Inverse Eigenvalue Problem

MVIP: Multivendor Integration Protocol

MVJ: Mandeville Marlboro (Jamaica) (location identifier)

MVK: Mulka (Australia-SA) (location identifier)

MVL: Main Vertical List • Morrisville Stowe (USA-VT) (location identifier)

MVLP: Microsoft Variable Licence Pack

MVLT: Multi-Valued Logic Tree

MVLU: Minimum Variance Linear Unbiased

MVM: Kayenta (USA-AZ) (location identifier) • Mariner Venus/Mercury • Medium-Voltage Mode • Metal-Vacuum-Metal • Mission and Vehicle Management • Multiple Virtual Machine • Multiscale Vision Model

Mvmt: Movement

MVN: Mount Vernon (USA-IL) (location identifier) • Mt. Vernon (USA-IL) (location identifier) • Multivariate Normal

MVO: Mongo (Chad) (location identifier)

MVOS: Mobile Video Optical System

MVP: Magnetic Vector Potential • Maintenance Verification Plan • Master Verification Plan • Mechanical Vacuum Pump • Member Value Plus (ACM) • Mitu (Colombia) (location identifier) • Modular Voice Processor • Most Valuable Player • MOS Verification Programme • Multiline Variety Package • Multimedia Value Point • Multimedia Video Processor • Multimission Vicar Planner

MVPG: Move Page

MVPS: Mechanical Vacuum Pump System

MVQ: Mogilev (Belarus) (location identifier)

MVR: Rufiyaa (Maldives) (ISO currency code) • Maneuver • Maroua (Cameroon) (location identifier) • Measurement Velocity Radial • Mechanical Vapour Recompression • Mesure de Vitesse Radiale

MVS: Middle-Value Select • Mid-Value Selected • Minimum Variance Simulation • Mission Video System • Mixed Variable Symplectic • Modularized Vehicle Simulation • Monthly Visits Statistics • Mucuri (Brazil-BA) (location identifier) • Multi-aircraft Vertically Stacked • Multiple Virtual Storage • Multiple Virtual System

Mvt: Movement

MVT: Mataiva (Tuamotu Islands) (location identifier) • Multiprogramming Variable Tasks

MVU: Minimum Variance Unbiased • Musgrave (Australia-QLD) (location identifier)

MVUE: Minimum Variance Unbiased Estimate • Minimum Variance Unbiased Estimator

MVULE: Minimum Variance Unbiased Linear Estimator

MVV: Mean Vertical Velocity • Mégève (France) (location identifier)

MVW: Mount Vernon - Skagit (USA-WA) (location identifier) • Mt. Vernon - Skagit (USA-WA) (location identifier)

MVWT: Mated Vehicle Wind Tunnel

MVX: Minvoul (Gabon) (location identifier)

MVY: Martha's Vineyard (USA-MA) (location identifier)

MVZ: Masvingo (Zimbabwe) (location identifier)

mW: meines Wissens • Milliwatt

Mw: Megaword • Mevrouw

M/W: Make Work • Microwave

MW: Magellansche Wolken • Malawi (international vehicle registration code) • Malawi (ISO country code) • Man-Week • Medium Wavelength • Medium Wave • Megawatt • Message Waiting • Microwave • Milky Way • Mittelwasser • Mixed Waste • Module Width • Molecular Weight

MWA: Marion (USA-IL) (location identifier) • Midwestern Astronomers (USA-KY) • Momentum Wheel Assembly

MWAP: Marshall W. Alworth Planetarium (USA-MN)

MWB: Master Workbook • Maxwell-Wien Bridge • Microwave Background • Monitoring Workbench • Morawa (Australia-WA) (location identifier)

MWC: Mary Washington College (USA-VA) • Milwaukee - Lawrence J. Timmerman (USA-WI) (location identifier) • Mount Wilson Catalog

MWCS: Mini-World Coordinate System

MWD: Magnetic White Dwarf • Material Withdrawal Document • Measurement While Drilling • Mianwali (Pakistan) (location identifier) • Microwell Detector • Millimeter Wave Device • Molecular Weight Distribution

MWDI: Master Water Data Index

MWE: Merowe (Sudan) (location identifier) • Multiword Expression

MWF: Maewo (Vanuatu) (location identifier) • Ministerium für Wissenschaft und Forschung

MWG: Magnetospheric Working Group • Maintainability Working Group • Marawaka (Papua New Guinea) (location identifier) • Meteorological Working Group • Milky Way Galaxy

MWh: Megawatt Hour

MWH: Moses Lake - Grant County (USA-WA) (location identifier)

MWI: Malawi (ISO country code) • Many-Worlds Interpretation • Maramuni (Papua New Guinea) (location identifier) • Message Waiting Indicator • Mount Wilson Institute (USA-CA)

MWIR: Medium-Wave Infrared • Medium-Wavelength Infrared • Mid-Wave Infrared

MWIS: Millemeter-Wave Imaging System

MWJ: Matthews Ridge (Guyana) (location identifier)

MWK: Kwacha (Malawi) (ISO currency code) • Matak (Indonesia) (location identifier) • Ministerium für Wissenschaft und Kunst (Germany)

mWL: Milliwatt Logic

MWL: Mean Water Level • Mineral Wells (USA-TX) (location identifier)

MWLCO: Mt. Wilson and Las Campanas Observatories

MWLS: Microwave Line Scanner

MWM: Maxwell-Wagner Mechanism • Motif Window Manager • Windon (USA-MN) (location identifier)

MWN: Message Waiting Notification • More Work Needed • Mwadui (Tanzania) (location identifier)

MWO: Maintenance Work Order • Manufacturing Work Order • Meteorological Watch Office • Meyer-Womble Observatory (USA-CO) • Middletown Hook (USA-OH) (location identifier) • Millimeter Wavelength Oscillator • Millimeter Wave Observatory • Modification Work Order • Mt. Wilson Observatory (USA-CA)

MWOA: Mt. Wilson Observatory Association (USA-CA)

MWOR: Maintenance Work Order Request

MWP: Maneuvering Work Platform • Maximum Working Pressure • Meteorologist Weather Processor • Millimeter Wave Propagation

MWPC: Multiwire Proportional Counter

MWPR: Monthly Work Package Report

MWPS: Millimeter Wave Power Source

MWQ: Magwe (Myanmar) (location identifier)

MWR: Maintenance Work Request • Maintenance Work Requirement • Mean Width Ratio • Meteor Wind Radar • Method of Weighted Residuals • Microwave Radiometer • Mikrowellen-Wasserdampf-Radiometrie • Motswari (South Africa) (location identifier) • Multiwavelength Radiometer

MWRv: Monthly Weather Review (periodical code)

MWS: Major Weapon System • Management Work Station • Maximum Wind Speed • Microwave Station • Microwave Wind Scatterometer • Millimeter-Wave Sounder • Minimum Wheel Speed • Mount Wilson (USA-CA) (location identifier) • Mt. Wilson (USA-CA) (location identifier) • Multimedia Wireless System • Multimedia Workgroup System

MWSC: Microwave Scatterometer

MWSI: Mount Wilson Sunspot Index

MWSR: Magnetic Wire Shift Register • Microwave Scanning Radiometer

MwSt: Mehrwertsteuer

MWT: Millimeter Wave Tube • Moolawatana (Australia-SA) (location identifier)

MWU: Mussau (Papua New Guinea) (location identifier)

MWV: Maximum Working Voltage

MWY: Miranda Downs (Australia-QLD) (location identifier)

MWZ: Mwanza (Tanzania) (location identifier)

mx: Mexico

Mx: Maxwell • Middlesex (UK-England)

MX: Mail Exchange • Mail Exchanger • Massachusetts X-ray • Merged Extracted (image) • Mexico (ISO country code) • Mexique (ISO country code) • Mixed • Multiplex • Multiplexer

MXA: Manila (USA-AR) (location identifier)

MXB: Masamba (Indonesia) (location identifier) • Massachusetts X-ray Burster

MXC: Monticello - San Juan County (USA-UT) (location identifier)

Mxd: Mixed

MXD: Marion Downs (Australia-QLD) (location identifier) • Mixed

MXE: Maxton (USA-NC) (location identifier)

MXF: Message Transfer Facility • Montgomery Maxwell (USA-AL) (location identifier)

Mxg: Mixing

MXG: Marlborough (USA-MA) (location identifier)

MXH: Moro (Papua New Guinea) (location identifier)

MXI: Mati (Philippines) (location identifier) • Multisystem Extension Interface

MXJ: Minna (Nigeria) (location identifier)

MXK: Mindik (Papua New Guinea) (location identifier)

MXL: Mexicali (Mexico) (location identifier)

MXLO: Mixed Extracted Low-dispersion (IUE image)

MXM: Morombe (Madagascar) (location identifier)

MXN: Morlaix (France) (location identifier)

MXO: Monticello (USA-IA) (location identifier)

MXP: Milano Malpensa (Italy) (location identifier) • Peso (Mexico) (ISO currency code)

MXQ: Mitchell River (Australia-QLD) (location identifier)

Mxr: Mixer

MXR: Mirgorod (Ukraine) (location identifier) • Multiplexer

MXRB: Massive X-Ray Binary

MXS: Maota - Savail Island (Samoa) (location identifier)

Mxt: Mixture

MXT: Maintirano (Madagascar) (location identifier)

MXU: Mullewa (Australia-WA) (location identifier) • Multiplexer Unit

MXV: Moron (Mongolia) (location identifier)

MXW: Mandalgobi (Mongolia) (location identifier)

MXX: Mora (Sweden) (location identifier)

MXY: McCarthy (USA-AK) (location identifier)

MXZ: Meixian (China-PRC) (location identifier)

my: Burmese (ISO language code) • Malaysia

MY: Malaysia (ISO country code) • Malaysie (ISO country code) • Man-Year • Mayan • Million Years • Minyong • Multiyear

MYA: Moruya (Australia-NSW) (location identifier)

MYB: Mayoumba (Gabon) (location identifier)

MYC: Maracay (Venezuela) (location identifier)

MYD: Malindi (Kenya) (location identifier)

MYDL: Middle Years Digital Library (project)

MYE: Man-Year Equivalent • Miyake Jima (Japan) (location identifier)

MYF: San Diego - Montgomery (USA-CA) (location identifier)

MYG: Mayaguana (Bahamas) (location identifier)

MYH: Marble Canyon (USA-AZ) (location identifier)

MYI: Murray Island (Australia-QLD) (location identifier)

MYJ: Matsuyama (Japan) (location identifier)

MYK: May Creek (USA-AK) (location identifier)

MYL: McCall (USA-ID) (location identifier)

mym: Myriameter

MYM: Managing Your Money • Monkey Mountain (Guyana) (location identifier)

MYN: Mareb (Yemen) (location identifier)

MYO: Myroodah (Australia-WA) (location identifier)

MYOB: Mind Your Own Business

MYP: Mary (Turkmenia) (location identifier)

MYQ: Mysore (India) (location identifier)

Myr: Megayear • Millions of Years

MYR: Ringgit (Malaysia) (ISO currency code) • Myrtle Beach (USA-SC) (location identifier)

MYRA: Multiyear Rescheduling Agreement

MYS: Malaysia (ISO country code) • Malaysie (ISO country code) • Moyale (Ethiopia) (location identifier)

Myst: Mystery

MYT: Mayotte (ISO country code) • Myitkyina (Myanmar) (location identifier)

Myth: Mythological • Mythology

Mythol: Mythological • Mythology

MYU: Mekoryuk (USA-AK) (location identifier)

MYV: Marysville Yuba (USA-CA) (location identifier)

MYW: Mtwara (Tanzania) (location identifier)

MYX: Menyamya (Papua New Guinea) (location identifier)

MYY: Miri (Malaysia) (location identifier)

MYZ: Monkey Bay (Malawi) (location identifier)

mZ: mangels Zahlung

Mz: Mehrzahl • Mesozoic

MZ: Mach-Zehnder • Mozambique (ISO country code)

MZA: Muzaffarnagar (India) (location identifier)

MZB: Mocimboa Praia (Mozambique) (location identifier)

MZC: Mitaka Zenith Catalogue • Mitzic (Gabon) (location identifier)

MZD: Mendez (Ecuador) (location identifier)

MZE: Manatee (Belize) (location identifier)

MZF: Manganese Zinc Ferrite • Mzamba (South Africa) (location identifier)

MZFW: Maximum Zero Fuel Weight

MZG: Makung (China-ROC) (location identifier)

MZH: Merzifon (Turkey) (location identifier)

MZI: Mopti (Mali) (location identifier)

MZJ: Marana (USA-AZ) (location identifier)

MZK: Marakei (Kiribati) (location identifier)

MZL: Manizales (Colombia) (location identifier)

MZM: Metical (Mozambique) (ISO currency code) • Metz (France) (location identifier)

MZN: Minj (Papua New Guinea) (location identifier)

MZO: Manzanillo (Cuba) (location identifier)

MZP: Motueka (New Zealand) (location identifier)

MZQ: Mkuze (South Africa) (location identifier)

MZR: Mazar-I-Sharif (Afghanistan) (location identifier)

MZS: Mostyn (Malaysia) (location identifier)

MZT: Mazatlán (Mexico) (location identifier) • Mercury Zinc Telluride

MZU: Muzaffarpur (India) (location identifier)

MZV: Mulu (Malaysia) (location identifier)

MZX: Mena (Ethiopia) (location identifier)

MZY: Mossel Bay (South Africa) (location identifier)

MZZ: Marion (USA-IN) (location identifier)

M0DE: Middeck O-gravity Dynamics Experiment

M1: Primary Mirror

M2: Medium Two

M2A: Multimedia Asia (satellite)

M2P2: Multimission Production Plan

M2S: Modular Multiband Scanner

M4: Milky-Way Magnetic-field Mapping Mission

N

n/: Near • Nuestro

n: Amount of Substance • And • Electron Number Density • Indefinite Number • Index of Refraction • Load Factor • Mean Daily Angular Planet Motion • Molecule Number Density • Nach • Nacional • Name • Nano • Natus • Né • Near • Negative • Nephew • Net • Neuter • Neutro • Neutron • Neutron Number Density • New • Night • Nominative • Noon • Nordlich • Normal • Nota • Notre • Noun • Nous • Nuestro • Nuevo • Number • Number Density • Number Density of Molecules • Number of Degrees of Freedom • Number of Stages • Order of Reflexion • Particle Density • Principal Quantum Number • Refractive Index • Rotational Frequency

N: And • Aperture • Asparagine • Avogadro Number • Azote • London (North_) (UK postal code) • Loudness • Molecule Number • Naira (currency) • National • Navy • Nebula • Neptune • Neutral • Neutron Number • New • Newton • Ngultrum (currency) • Night • Ninety (Roman numeral) • Nitrogen • No • Noon • Nord • Norden • Nordlich • Normal • Norte • North • Northern • North London (UK postal code) • Norway (international vehicle registration code) • Norway (postal country code) • Note • Notice • Nova • November (international alphabet) • Nuclear • Nucleon • Number of Molecules • Number of Objects per Square Degree • Number of Particles • Number of Particles per Cell • Number of Turns • Nutation Constant • Quantum Number of Total Angular Momentum • Particule Number • Precession in Declination • Spectral Type • Stickstoff • Telescope Aperture • Turn Number • United States of America (international aircraft registration code) • USA (international aircraft registration code) • 90 (Roman numeral)

n/a: No Account • Not Applicable • Not Available

na: Nauru (ISO language code)

nA: Nanoampere

Na: Fast Nova • Natrium • Nuestra • Sodium

N/A: New Account • New Address • Next Assembly • No Action • No Advice • Non Acceptance • Non Applicable • Not Applicable • Not Authorized • Not Available

NA: Namibia (ISO country code) • Namibie (ISO country code) • Nanai • Narrow Angle • National Academician • National Archives • National Atlas • Nautical Almanac • Nébuleuse Annulaire • Netherlands Antilles (international vehicle registration code) • Network Analysis • Network Archivist • Neutral Atom • Neutral Axis • New Account • New Address • Next Action • Next Address • Night Assistant • No Account • Nodal Architect • Non-African • North America • North American • Not Add • Not Applicable • Not Authorized • Not Available • Now Available • Numerical Analysis • Numerical Aperture • Numérotation Abrégée

NAA: Naperville Astronomical Association (USA-IL) • Narrabri (Australia-NSW) (location identifier) • Narrow-Angle Astrometry • National Aeronautic Association (USA) • National Airports Authority (USA) • National Airworthiness Authority • Network Analysis Area • Neutron Activation Analysis • North American Aircraft • North American Aviation • Northants Amateur Astronomers (UK) (formerly KDAAS) • Nürnberger Astronomische Arbeitsgemeinschaft (Germany)

NAAC: National Aviation Associations Coalition (USA)

NAACL: North American Association of Computational Linguistics

NAACP: National Association for the Advancement of Colored People (USA)

NAAG: National Association of Attorneys General (USA)

NaAIJ: Nanjing Aeronautical Institute Journal (periodical code)

NAAIL: NSSDC Application of Artificial Intelligence Laboratory

NAAL: Network of Alabama Academic Libraries (USA-AL) • North American Aerodynamic Laboratory

NAAP: Nederlands Antartisch Onderzoek Programma

NAAQ: National Ambient Air Quality

NAAQS: National Ambient Air Quality Standards

NAARS: National Automated Accounting Research Service (USA)

NAAS: Northeast Arkansas Astronomical Society (USA-KS)

NAB: Albany (USA-GA) (location identifier) • National Association of Broadcasters (USA) • Nordic Advisory Board • Northern Arctic Band

NABD: Normenausschuß Bibliotheks- und Dokumentationswesen

NABE: National Association of Book Editors (USA) • North Atlantic Bloom Experiment

NABI: NASA Astrobiology Institute

NABP: National Association of Book Publishers (USA)

NABST: National Advisory Board on Science and Technology (Canada)

NABTS: North American Broadcast Teletext Standard • North American Broadcast Television System

NABUCO: Nouvelle Approche Budgétaire et Fiscale

Nac: Nacelle • Nacional

NAC: Nacelle • Naracoorte (Australia-SA) (location identifier) • Narrow-Angle Camera • NASA Advisory Council • National Accelerator Centre (South Africa) • National Aerospace Centre • National Advisory Council • National Aerospace Centre • National Archives Council • Naval Avionics Center • Near Ambient Chamber • Nederlandse Astronomenclub • Nepean Astronomy Centre (Australia-NSW) • Network Adapter Card • Network Administration Center • Network Advisory Committee • North American Craton

NACA: National Advisory Committee for Aeronautics (USA) (now NASA)

NACAA: National Australian Convention of Amateur Astronomers

NACC: Necessary Ad-hoc Coordinating Committee • North American Calibration Cooperation • North Atlantic Cooperation Council

NACCB: National Accreditation Council for Certification Bodies (USA)

NACCIRN: North American Coordinating Committee for Intercontinental Research Networking

NACD: National Association of Computer Dealers (USA) • North Atlantic Climatological Dataset

NACE: North American Commission on the Environment

NacEl: Nachrichten Elektronik (periodical code)

NACEPT: National Advisory Committee for Environmental Policy and Technology

NACES: Navy Aircrew Escape System

Nachf: Nachfolger

Nachm: Nachmittags

Nachr: Nachricht

NACIS: North American Cartographic Information Society

NACK: Negative Acknowledgement

Nacl: Nacional

NaCl: Sodium Chloride

NACL: National Advisory Committee on Libraries (USA)

NACLIS: National Advisory Commission on Libraries and Information Science (USA) • National Advisory Council for Library and Information Science (South Africa)

NACM: New Area Calculation Method

NACME: National Action Council for Minorities in Engineering (USA)

NACo: National Association of Counties (USA)

NACO: Name Authority Cooperative Project

NACS: National Advisory Committee on Semiconductors (USA) • National Association of College Stores (USA) • National Association of Computer Stores (USA) • National Association of County Surveyors (USA) • NetWare Asynchronous Communications Service • Network-Assisted Coordinated Science • Neutral Atmosphere Composition Spectrometer

NACSEM: National Communications Security Emanations Memorandum

NACSIS: National Centre for Science Information System (Japan)

NACSM: National Advisory Committee on Surveys and Mapping (Canada)

NACU: National Association of Colleges and Universities (USA)

NacZe: Nachrichtentechnische Zeitschrift (periodical code)

NAD: Macanal (Colombia) (location identifier) • Nansen Arctic Drilling • Network Access Device • Network Adaptor Device • Nikotinamid-Adenin-Dinukleotid • No Apparent Defect • No Appreciable Difference • Noise Amplitude Distribution • North American Datum • Not on Active Duty

NADAMS: National Agricultural Drought Assessment and Monitoring System (India)

NADAR: Nadir Altimeter for Rain

NADB: National Air Data Branch

NADC: National Alumni Data Center • Naval Air Development Center (USA) • North American Digital Cellular • North American Dual-mode Cellular

NADEP: Naval Aviation Depot

NADF: North American Directory Forum

NADIN: National Airspace Data Interchange Network

NADIS: Nachrichtendienstliches Informationssystem • National Aerometric Data Information System (USA)

NADP: National Atmospheric Deposition Program • Nicotinamide Adenine Dinucleotide Phosphate

NADUC: Nimbus/ATS Data Utilization Center

NADW: North Atlantic Deep Water

NAE: National Academy of Engineering (USA) • National Aeronautical Establishment (USA) • Natitingou (Benin) (location identifier) • Not Above or Equal • Nuclear Astrophysics Explorer • Núcleo de Assuntos Europeus

NAEB: National Association of Educational Broadcasters (USA)

NAEC: National Atomic Energy Commission (USA) • Naval Air Engineering Center

NAEP: National Assessment of Educational Progress (USA)

NAf: Netherlands Antillean Guilder (international vehicle registration code)

NAF: Banaina (Indonesia) (location identifier) • National Analytical Facility • Negation As Failure • Nernst Approximation Formula • Network Access Facility • Norsk Astronautisk Forening (Norway)

NAFC: National Average Fuel Consumption

NAFD: Non-Associated Forbush Decrease

NAFI: National Association of Flight Instructors (USA)

NAFIPS: North American Fuzzy Information Processing Society

NAFS: Name Authority File Service • Novell Advanced File System

NAFSMA: National Association of Flood and Stormwater Management Agencies (USA)

NAFT: Network for Analysis of Fireball Trajectories

NAFTA: New Zealand and Australia Free Trade Agreement • North American Free Trade Agreement (*) • North American Free Trade Area

Nag: Nagasaki (Japan)

NAG: Nagpur (India) (location identifier) • Net Annual Gain • Netherlands Aerospace Group • Network Advisory Group • Network Architecture Group • Noyau Actif de Galaxie • Numerical Algorithms Group

NAGAP: National Acid Precipitation Assessment Program

NAGARA: National Association of Government Archives and Records Administration (USA)

NAGC: National Association of Government Communicators (USA)

NaGe: National Geographic (periodical code)

NAGE: National Association of Government Employees (USA)

NAGPS: National Association of Graduate-Professional Students (USA)

NAGT: National Association of Geoscience Teachers (USA)

NAGUA: Numerical Algorithms Group Users Association

NAH: Naha (Indonesia) (location identifier) • Not At Home

NAI: Annai (Guyana) (location identifier) • NASA Astrobiology Institute (USA) • National Archives of Ireland • Neutral Atom Imager • No Action Indicated • No-Address Instruction • Non-Accidental Injury

NAIADE: Normalized Automatic Interpreter in ADA for Data Exchange

NAIC: National Astronomy and Ionosphere Center (USA) • Network Applications and Information Center (USA-CA) • Newly Agro-Industrialized Country

NAICC: National Association of Independent Computer Companies (USA)

NAICS: North American Industry Classification System

NAICU: National Association of Independent Colleges and Universities (USA)

NAID: Note Added In Proof

NAIF: Nanjing Astronomical Instruments Factory (China-PRC) (now NAIRC) • Navigation Ancillary Information Facility

NAIIU: Not Authorized If Issued Under

NAILS: National Airspace Integrated Logistics Support • NCAR Airborne Infrared Lidar System

NAIR: Narrow Absorption Infrared • Network Action Item Report

NAIRC: Nanjing Astronomical Instruments Research Facility (China-PRC) (formerly NAIF)

NAIS: NASA Acquisition Internet Service • National Atlas Information Service (Canada)

NAISS: New Abstracting and Indexing Support System

NAIST: Nara Institute of Science and Technology (Japan)

NAJ: National Association of Journalists (UK)

NAK: Nakhon Ratchasima (Thailand) (location identifier) • Negative Acknowledge • Negative Acknowledgement • Normenausschuss Klassifikation • Norrköpings Astronomiska Klubb (Sweden)

NAL: Nalchik (Russia) (location identifier) • Narrow Absorption Line • National Acoustics Laboratory • National

Aeronautical Laboratory • National Aerospace Laboratory • NetWare Application Launcher • Novell Application Launcher • Numerical Analysis Laboratory

NALIN: National Library Information Network

NALM: Non-linear Amplifying Loop Mirror

NALMS: North American Lake Management Society

NALPS: North-American Presentation-Level Protocol Syntax

N. Am.: North America • North American

NAM: Name and Address Module • Namibia (ISO country code) • Namibie (ISO country code) • Namlea (Indonesia) (location identifier) • National Account Manager • National Association of Manufacturers (USA) • Nautical Air Mile • Navy Aerosol Model • Nederlandse Aardolie Maatschappij • Network Access Machine • Network Access Method • Network Analysis Model • Non-Aligned Movement • Normalized Axial Moment • Nuclear Assessment Methodology • Number Assignment Module

NAMAC: North American Market Awareness Committee

NAMAS: National Measurement Accreditation Service

NAMEDA: National Association of Minority Program Administrations (USA)

NAMG: Narrow-Angle Mars Gate

NAML: Naval Aircraft Materials Laboratory

NAMN: North American Meteor Network (USA-SC)

NAMP: National Association of Magazine Publishers (USA)

NAMPS: Narrow-band Advanced Mobile Phone Service

NAMRIA: National Mapping and Resources Information Authority (Philippines)

NaN: Not a Number

NAN: Nadi (Fiji) (location identifier) • No Action Necessary • Numerical Assignment Number

NANBA: North American National Broadcasters Association

NAND: Not And

NA-NET: Numerical Analysis Network

NANET: Numerical Analysis Network

NANOFET: Nanometer-scale Field Effect Transistor

NANOG: North American Network Operators Group

NANP: North American Numbering Plan

NANU: Notice Advisory to NAVSTAR Users

NAO: Nanchong (China-PRC) (location identifier) • National Astronomical Observatory • National Audit Office • Nautical Alamanac Office (USNO) • Navigation Assistée par Ordinateur • Nikolaev Astronomical Observatory (Ukraine) • North Atlantic Oscillation • Notation Assistée par Ordinateur

NAOJ: National Astronomical Observatory of Japan

NAOM: National Astronomical Observatory - Misuzawa (Japan)

NAOMZ: National Astronomical Observatory - Misuzawa (Japan)

NAOPIA: New Architectures for Optical Processing in Industrial Applications

NAOS: NASA Aircrew Oxygen System • Nasmyth Adaptive Optics System • North Atlantic Ocean Stations

NAOSMM: National Association Of Scientific Material Managers (USA)

NAOT: National Astronomical Observatory - Tokyo (Japan)

Nap: Napoli (Italy)

NAP: Naples Capodochino (Italy) (location identifier) • Napoli Capodochino (Italy) (location identifier) • National Academy Press (USA-DC) • National Action Plan • National Action Programme • National Association of Publishers (USA) • National Awareness Partner • Navigation Analysis Program • Network Access Point • Network Applications Platform • Noise Abatement Procedure • Northern Affairs Program • Not A Priori • Not At Present • Nuclear-Active Particle • Nuclear Auxiliary Power

NAPA: National Academy of Public Administration (USA)

NAPAG: National Academies Policy Advisory Group

NAPAMA: National Association of Performing Artists Managers and Agents (USA)

NAPAP: National Acidic Precipitation Assessment Program • National Acid Precipitation Assessment Program (*)

NAPCAE: National Association for Public and Continuing Education (USA)

NAPE: National Association for Primary Education (UK)

NAPF: North Atlantic Polar Front

Naph: Naphtha

NAPIS: National Agricultural Pest Information System

NAPL: National Airphoto Library (Canada)

NAPLPS: North American Presentation Level Protocol Syntax

NAPM: National Association of Photographic Manufacturers (USA)

NAPNOC: No Acceptable Price - No Contract

NAPO: National Association of Planetary Observers (USA)

NAPP: National Aerial Photography Program (USA)

NAPRS: National Airspace Performance Reporting System

NAPS: Numerical Analysis and Prediction System

NAPSS: Numerical Analysis Problem Solving System

NAPU: Nuclear Auxiliary Power Unit

NAPUS: Nuclear Auxiliary Power Unit System

NAPWPT: National Association of Professional Word Processing Technicians (USA)

NAQ: National Agency Questionnaire • Qanaq (Greenland) (location identifier)

Nar: Narrow

NAR: Nare (Colombia) (location identifier) • Net Assimilation Rate • No Action Required • No Answer Required • Noise-Adding Radiometer • Non-Advocate Review • Nonlinear Autoregression • North American Route • North Atlantic Route • Not According to Routine • Nuclear Acoustic Resonance • Numerical Analysis Research

NARA: National Archives and Records Administration (USA)

NARC: National Association of Regional Councils (USA) • National Atmospheric Radiation Centre (USA)

NARCOM: North Atlantic Relay Communication

NARE: North Atlantic Regional Experiment

NARESA: Natural Resources, Energy and Sciences Authority (Sri Lanka)

NARFA: National Allied Radio Frequency Agency

NARL: National Arctic Research Laboratory • National Air and Radiation Laboratory • Naval Arctic Research Laboratory

NARMA: Non-linear Auto-Regressive Moving Average

NARPA: National Advanced Research Projects Agency

NARS: National Archives and Records Services • Netwerk voor Autonoom Registrerende Seismografen • Network of Autonomously Recording Stations

NARSA: National Advanced Remote Sensing Application

NARSIM: NLR ATC Research Simulator

NARSTO: North American Research Strategy for Tropospheric Ozone

NARTS: National Association of Rocketry Technical Services (USA)

Nas: Nasmyth

NAS: Name and Addressing Service • Nassau (Bahamas) (location identifier) • National Academy of Sciences •

National Aerospace System • National Aircraft Standard • National Airspace System • Naval Air Station • Netware Access Server • Network Application Support • Network Attached Storage • Newcastle Astronomical Society (Australia-NSW) • Newcastle-upon-Tyne Astronomical Society (UK) • Nigerian Academy of Sciences • Noise Abatement System • Non-Avionics Simulator • Non-linear Antenna System • Norwich Astronomical Society (UK) • Nottingham Astronomical Society (UK) • Numerical Aerodynamic Simulation • Numerical Aerodynamic Simulator • Numerical Analysis Subroutine • Numerical And Atmospheric Sciences

NASA: National Aeronautics and Space Act (USA) • National Aeronautics and Space Administration (USA)

NASAP: Network Analysis for Systems Application Program

NASAR: National Aeronautics and Space Administration Reference Publication (periodical code)

NASC: National Aeronautics and Space Council (USA) • National Aircraft Standards Committee • North American Stratigraphic Code

NASCA: Nagoya Air Shower Counter Array (Japan)

NASCAP: NASA/Air Force Spacecraft Charging Analyzer Program

NASCOM: NASA Communications (network)

NASCOP: NASA Communications Operating Procedures

NASD: National Association of Securities Dealers (USA)

NASDA: National Space Development Agency (Japan)

NASDAQ: National Association of Securities Dealers Automatic Quotation

NASENI: National Agency for Science and Engineering Infrastructure (Nigeria)

Nash: Nashville (USA-TN)

NASI: NetWare Asynchronous Interface • NetWare Asynchronous Services Interface

NASIG: North American Serials Interest Group

NASIP: NASA Airframe Structural Integrity Program

NASIRC: NASA Automated Systems Incident Response Capability (*) • NASA Automated Systems Internet Response Capability

NASIRE: National Association of State Information Resources Executives (USA)

NASIS: NASA Aerospace Safety Information System • Northeast Australian Satellite Imagery System

NASL: National Association of State Libraries (USA)

NASM: National Air and Space Museum (USA-DC) • National Association for School Magazines (USA)

NASO: North American Space Operations

NASP: National Aero-Space Plane • National Airspace System Plan • National Atmospheric Sciences Program

NASPAC: National Airspace System Performance Analysis Capability

NASPR: NASA Procurement Regulation

NASREC: National Soil Reference Collection

Nass: Nassau (Bahamas)

NASS: National Agricultural Statistics Service • National Air and Space Society (USA-DC) • North American Sundial Society (USA-CT)

NASSCOM: National Association of Software and Service Companies (USA)

NASSE: National Association of Space Simulating Educators (USA-OH)

NASSL: National Association of Spanish Speaking Librarians (USA)

NAST: National Academy of Science and Technology (USA)

NASTeC: North-American Small Telescope Cooperative (now ISTeC)

NASU: National Association of State Universities (USA)

NASW: National Association of Science Writers (USA)

NASWSC: North American Society for Water and Soil Conservation

Nat: Natal (South Africa) • National • Nationalist • Nationality • Native • Natur • Natural • Nature • Naturel • Natus

NAT: Narrow Angle Tail • Narrow Angle Tailed • Narrow Angle Tracker • Natal - Augusto Severo (Brazil-RN) (location identifier) • Natural Unit of Information • Network Address Translation (*) • Network Address Translator • Network Analysis Team • New Alias Table • Nitric Acid Trihydrate • No Action Taken • Non-conformance Action Team • Normal Allowed Time • Normenausschuss Terminologie • North Atlantic Traffic

NATA: National Air Transportation Association (USA)

NATAS: NOAA AVHRR Transcription and Archive System

NATCO: Northwest Aerospace Training Corporation

NATF: Naval Air Test Facility

NATFB: National Archives Trust Fund Board(USA)

NatH: Natural History (periodical code)

NATICH: National Air Toxics Information Clearinghouse (USA)

Natl: National • Nationality

NATL: National

NATMA: Nonlinear Analysis Theory Methods and Applications (periodical code)

NATMAP: National Mapping

NATO: Narrow-Angle Target of Opportunity • North Atlantic Treaty Organization

NATOTS: North Atlantic Organized Track System

NATS: National Air Traffic Service • Naval Air Transport service • North American Terrestrial System

Nat.Sc.: Natural Science

Nat.Sci.: Natural Science

NATT: Northern Australian Tropical Transect

Natur: Nature (periodical code)

Naturwiss: Naturwissenschaft • Naturwissenschaftlich

Natuurkd: Natuurkunde

NAU: Napuka Island (Tuamotu Islands) (location identifier) • Network Access Unit • Network Addressable Unit • Network Address Unit • Northern Arizona University (USA-AZ)

NAUCA: Nomenclatura Arancelaria Uniforme Centroamericana

NAUI: National Association for Underwater Instructors (USA)

NAUN: Nearest Active Upstream Neighbor

Náut: Náutica

Naut: Nautic • Nautical • Nautique

nav: Naar Aanleiding van

Nav: Naval • Navarra (Spain) • Navigable • Navigate • Navigation • Navigational • Navigator

NAV: Navigate • Navigation • Net Asset Value • Norton Anti-Virus

NAVA: National Audiovisual Association (UK)

NAVAID: Navigational Aid

NAVAIDS: Navigational Airborne Integrated Data Systems

NavAu: Navigation Australia (periodical code)

NAVD: North American Vertical Datum

NAVDAD: Navigational Derived Air Data • Navigationally-Derived Air Data

NAVEX: Navigation Expert • NAVNET Experimental

Navig: Navigation • Navigation (periodical code) • Navigator

NAVMN: Navigation Monitor and Control

NAVNET: Navy VLBI Network

NAVO: Noordatlantische Verdragsorganisatie

NavPa: Navigation Paris (periodical code)

NAVSAT: Navigational Satellite (*) • Navigation Satellite

NAVSOC: Naval Satellite Operations Center

NAVSPASUR: Naval Space Surveillance Radar

NAVSSI: Navigation Sensor System Interface

NAVSTAR: Navigation Satellite Timing And Ranging (*) • Navigation System with Time And Ranging

NAVT: NASA Altimeter Verification Telemetry

NAW: Narathiwat (Thailand) (location identifier) • National Astronomy Week

NAWAU: National Aviation Weather Advisory Unit

NAWDEX: National Water Data Exchange

NAWPF: National Aviation Weather Processing Facility

NAWPS: National Association of Word Processing Specialists (USA)

NAWQUA: National Water Quality Assessment

NAWT: Negri-Adler-Wetzel Technique

NAX: Barbers Point (USA-HI) (location identifier)

NAY: Beijing Nanyuan (China-PRC) (location identifier) • Nayarit (Mexico)

Naz: Nazionale

NAZ: Normal Analytical Zone

Nazl: Nazionale

Nb: Niob • Niobium • Slow Nova

N/B: Narrow Band

NB: Naine Blanche • Naive Bayes (method) • Narrow Band • Narrow Beam • Nationalbibliothek • National Body • Navigation Base • Negative Binomial • Neutral Beam • New Brunswick (Canada) • Nitrogen Base • No Bias • No Bid • No Bidder • Normal Branch • Northbound • Nota Bene • Nouveau Brunswick (Canada)

NBA: Nambaiyufa (Papua New Guinea) (location identifier) • Narrow-Band Adapter • Narrow-Band Amplifier • Narrow-Band Analyzer • National Basketball Association (USA) • National Book Award

NBAA: National Business Aircraft Association (USA)

NBAANB: New Brunswick Astronomy - Astronomie Nouveau Brunswick Inc. (Canada-NB)

NBB: Barrancominas (Colombia) (location identifier) • Number of Bytes of Binary

NBBI: Nederlands Bureau voor Bibliotheekwezen en Informatieverzorging (Netherlands)

NBC: Naberevnye Chelnye (Russia) (location identifier) • National Broadcasting Corporation • Noise Balancing Circuit • Nuclear-Bacteriological-Chemical • Nuclear-Biological-Chemical • Number Base Conversion

NBCD: Natural Binary-Coded Decimal

NBD: Narrow-Band Data • Narrow-Band Detector • National Book Day • Nederlands Bibliotheek Dienst (Netherlands) • Negative Binomial Distribution • Neutral Beam Divider

NBDL: Narrow-Band Data Line

NBDM: Non-Baryonic Dark Matter

NbE: North by East

NBE: Not Below or Equal

NBER: National Bureau of Economic Research

Nbf: Nebenbahnhof

NBF: Narrow-Band Filter • Neutral Buoyancy Facility

NBFCP: NetBIOS Frames Control Protocol

NBFM: Narrow-Band Frequency Modulation

NBG: New Orleans - Alvin Callendar (USA-LA) (location identifier)

NBH: Nambucca Heads (Australia-NSW) (location identifier)

NBI: Neutral-Beam Injection • Niels Bohr Institute (Denmark) • Nothing But Initials • Nuclear Burst Indicator

NBII: National Biological Information Infrastructure

NBIOME: Northern Biosphere Observation and Modelling Experiment

NBIOS: Network Basic Input Output System

NBIS: Neutral Beam Injection System • Northern Biosphere Information System

NBIT: New Bedford Institute of Technology (USA-MA)

NBL: Neutral Beam Line • Neutral Buoyancy Laboratory (USA-TX) • San Blas (Panama) (location identifier)

NBLC: Nederlands Bibliotheek en Lektuur Centrum (Netherlands)

NBM: National Book Month • National Bureau of Metrology • Network Business Management • Nothing By Mouth

NBMA: Non-Broadcast Multiaccess

NBN: Narrow-Band Network • Narrow-Band Noise • National Bibliography Number • National Book Number

NBNC: Noted But Not Corrected

NBND: Northbound

NBO: Nairobi - Jomo Kenyatta (Kenya) (location identifier) • Network Build Out • Networked Business Object • Normal Branch Oscillation

NBOSS: New-Business Operations Support System

NBP: Name Binding Protocol • Network Bright Point • New York - Battery Park City (USA-NY) (location identifier) • Normal Boiling Point

NBpC: Number of Bits per Column

NBPC: Neutral Beam Power Conversion • Number of Bits Per Column

NBPM: Narrow-Band Phase Modulation

NBpP: Number of Bits per Pixel

NBPP: Navy/Berkeley Pulsar Processor • Number of Bits Per Pixel

NBpR: Number of Bits per Row

NBPR: Number of Bits Per Row

n. Br.: nördliche Breite

Nbr: Nombreux • Number

NBR: Nambour (Australia-QLD) (location identifier) • Norges Byggstandardiseringsråd (Norway)

Nbre: Noviembre

NBRF: National Biomedical Research Foundation • Neutral Buoyancy Research Facility

NBRI: National Botanical Research Institute

Nbrx: Nombreux

NBS: National Biological Survey (USA) • National Board of Survey • National Bureau of Standards (USA) (now NIST) • Neutral Buoyancy Simulation • Neutral Buoyancy Simulator • New British Standard • Notice Board System • Numeric Backspace

NBSAMRF: National Bureau of Standards Advanced Manufacturing Research Facility

NBSM: National Bureau of Surveying and Mapping

NBSS&LUP: National Bureau of Soil Survey & Land Use Planning

NBSSP: National Bureau of Standards Special Publication (periodical code)

NBST: National Board for Science and Technology (Indonesia)

NBSTN: National Bureau of Standards Technical Note (periodical code)

NBT: National Book Trust (India) • Neutral Buoyancy Trainer • Next Big Telescope

NBTR: Narrow-Band Tape Recorder

NBU: Glenview (USA-IL) (location identifier) • National Bibliographical Unit • Neil Brown Unit

NBV: Cana Brava (Brazil-MG) (location identifier) • Net Book Value

NBVCXO: Narrow-Band Voltage-Controlled Crystal Oscillator

NBVM: Narrow Band Voice Modulation

NBVS: Neutral Beam Vacuum System

NbW: North by West

NBW: Guantanamo (Cuba) (location identifier) • National Book Week • Noise Bandwidth

NBX: Nabire (Indonesia) (location identifier)

NBY: National Book Year

NBZ: Neutral Buoyancy Zone

n/c: No Charge • Nuestra Casa • Nuestra Cuenta • Nuestro Cargo

Nc: Very Slow Nova

N-C: Nitrogen-Carbon

N/C: No Change • No Charge • No Connect • No Coverage • Normally Closed • Nose Cone • Not Critical • Numerical Control • Numerically Controlled

NC: National Certificate • National Coarse • National Committee • National Conference • National Congress • National Coordinator • National Council • National Course • National Curriculum • Navigation Computer • Navigation Console • Neanderthal Conservative • Network Computer • Network Connect • Network Controller • Network Countdown • Neutral Current • Neutralizing Capacitance • Neutralizing Coil • New Caledonia (ISO country code) • Newtonian Collimator • Nickel-Cadmium • Nitrocellulosis • Nitrogen Compound • No Carrier • No Carry • No Change • No Charge • No Circuit • No Comment • No Connection • No Contact • No Correction • No Cost • Noise Criterion • Nomenclature Combinée • Nominal Correction • Nominating Committee • Non Communiqué • Non Coupé • Nordic Council • Normal Corrective • Normally Closed • North Carolina (USA) • Nose Cone • Not Carried • Not Changing • Not Coded • Not Collected • Not Connected • Not Critical • Notification Control • Nouvelle-Calédonie (ISO country code) • Number of Columns • Numerical Control • Numerical Controller • Numerically Controlled

NCA: National Capital Astronomers (USA-DC) • National Command Authority • National Committee for Astronomy • National Committee of Astronomy • National Computer Association • National Cooperative Account • Nederlandse Comité Astronomie • Network Communications Adapter • Network Computing Architecture • Network Control Analysis • Nickel Copper Alloy • No Control Access • North Caicos (Turks & Caicos Islands) (location identifier) • Northern Control Area

NCAA: National Collegiate Athletic Association (USA)

NCAC: National Copyright Administration of China (China-PRC) • Nicolaus Copernicus Astronomy Center (Poland)

NCAER: National Council of Applied Economic Research (USA)

NCAHF: National Council Against Health Fraud (USA)

NCAI: National Committee on Atomic Information (USA)

NCAICM: National Center for Advanced Information Components Manufacturing (USA)

NCAIR: National Center for Automated Information Retrieval (USA)

NCAL: New Caledonia • Nouvelle-Calédonie

NCAM: National Center for Accessible Media (USA) • Network Communications Access Method

NCAP: Nature Center And Planetarium • Northwest Coalition for Alternatives to Pesticides (USA)

NCAPI: Netscape Client Application Programming Interface

NCAR: National Center for Atmospheric Research (USA-CO) • National Conference on the Advancement of Research (USA) • Non-conformance and Corrective Action Reporting

NCARAI: Navy Center for Applied Research in Artificial Intelligence (USA)

NCARC: National Civil Aviation Review Commission (USA)

NCAS: Network and Customer Administration System • Northern Colorado Astronomical Society (USA-CO)

NCASI: National Council on Air and Stream Improvement (USA)

NCASR: National Conformity Assessment System Recognition (USA)

NCAT: National Center for Advanced Technologies (USA)

NCATE: National Council for Accreditation of Teacher Education (USA)

NCB: Near-Contact Binary • NetBIOS Control Block • Network Control Block • Nickel-Cadmium Battery

NCC: NASA Class Code • NASA Communications Control • National Climate Center • National Climatic Center (NOAA) • National Computer Center • National Computer Conference • National Computing Centre • National Consultative Committee • National Consultative Council • National Consumer Council • National Coordination Centre • National Coordination Committee • Navigation Computer Control • Navigation Control Console • NAVSTAR Control Center • Nederlandse Centrale Catalogus • Nederlandse Classificatie Commissie (Netherlands) • Netware Console Commander • Network Control Center • Network Coordination Center • Neural Correlate of Consciousness • New Common Carrier • New Computational Challenge • Noise Control Committee • Nominal Corrective Combination • Non-Commutative Collision • Non-Commuting Collision • Normal Corrective Combination • North Cusp Cap • Nouveau Centre de Calcul • Numerical Control Code

NCCA: National Council for Curriculum and Assessment

NCCAS: National Center of Communication Arts and Sciences (USA)

NCCB: National Configuration Control Board • NITF Configuration Control Board

NCCC: North Cusp Cap Collar

NCCDS: Network Control Center Data System

NCCE: Novell Collaborative Computing Environment

NCCF: Network Communications Control Facility • NOAA Central Computer Facility

NCCN: National Committee on Computer Networks

NCCP: Navigation Control Console Panel

NCCS: NASA Center for Computational Sciences • Navigation Command and Control System

NCCUSL: National Conference of Commissioners on Uniform State Laws (USA)

NCD: National Capability Demonstrator • Network Computing Devices Inc. (USA-CA) • No Can Do • Normalized Cumulative Deviation

NCDB: National Cartographic Database

NCDC: National Climate Data Center • National Climatic Data Center (USA-NC) (*)

NCDCDS: National Committee on Digital Cartographic Data Standards (USA)

NCDD: No Change in the Due Date

NCDM: Numerically-Controlled Drafting Machine

NCDNHC: Non-Commercial Domain Name Holders Constituency

NCDP: Navigation Control/Display Panel

NCDS: NASA Climate Data System • National Cartographic Data Standard (USA) • National Centre for Development Studies (USA)

NCDSB: National Cartographic Data Standard Base (USA)

NCDU: Navigation Control and Display Unit

NCDW: Network and Control Development Network

NCE: National Committee for the Environment • Net Carbon Emission • Network Centre of Excellence • Network Control Engine • New Chemical Entity • Nice - Côte d'Azur (France) (location identifier) • No Change in Estimates • Noise Control Engineering (periodical code) • Nomadic Computing Environment • Nouvelle Compagnie Européenne

NCEAS: National Center for Ecological Analysis and Synthesis (USA-CA)

NCEC: National Committee Exchange Controller

NCEP: National Center for Environmental Prediction (USA)

NCER: National Center for Earthquake Research (USGS)

NCERT: National Council for Educational Research and Training (India)

NCES: National Center for Education Statistics (USA)

NCET: National Council for Educational Technology (UK)

NCF: National Computing Facility • National Control Facility • Nationale Computerfaciliteiten (Netherlands) • Near-Caustic Field • NetWare Command File • Newton-Cotes Formula • Non-Cold Front • NSSDC Computer Facility • Nucleonics Calibration Facility

NCFC: National Computer Facility of China

NCG: National Control Group • Non-Condensible Gas • Nuclear Criteria Group • Nuevas Casas Grandes (Mexico) (location identifier) • Numerical Control Graphics

NCGA: National Computer Graphics Association (USA)

NCGIA: National Center for Geographic Information and Analysis (USA)

n/cgo: Nuestro Cargo

NC&GS: National Charting & Geodetic Survey (USA)

NCH: Nachingwea (Tanzania) (location identifier) • National Clearinghouse

NCHARAN: New Characteristic Analysis

Nchf: Nachfolger

NCHP: Nickel-Chromium Honeycomb Panel

NCHPC: National Consortium for High-Performance Computing (USA)

n. Chr.: Nach Christus

NCHS: National Center for High-performance Supercomputing (China-ROC)

nCi: Nanocurie

NCI: National Cancer Institute (USA) • National Computing Industry • Navigation Control Indicator • Necocli (Colombia) (location identifier) • Negative Chemical Ionization • Network Computing Infrastructure • Network Control Interface • New Community Instrument • No Common Interest • No-Cost Item • Nomenclature Color Index • Nuclear Cities Initiative

NCIC: National Career Information Center (USA) • National Cartographic Information Center (USGS) • National Crime Information Center (USA) • Non-Confidential Information Center

NCIIA: National Collegiate Inventors and Innovators Alliance (USA)

NCIMS: Negative Chemical Ionization Mass Spectra • Numerical Control Information Management System

NCIR: National Conference on Industrial Research (USA)

NCIS: Nadir Climate Interferometer Spectrometer • National Chemical Information System (USA)

NCIU: Network Common Interface Unit • NEXRAD Communications Interface Unit

NCK: Neck

NCL: National Central Library • National Chemical Laboratory • National Consumers League • Net Capacity Load • Network Control Language • New Caledonia (ISO country code) • Newcastle (UK-England) (location identifier) • Nouvelle-Calédonie (ISO country code)

NCLIS: National Commission on Libraries and Information Science (USA)

NCLP: Numerically Controlled Line Plotter

NCM: Network Control Manager • Non-Corrosive Metal • Non-Crew Member • Nozzle Closure Mechanism • Núcleo de Cooperação Multilateral • Numerical Controlled Machine

NCMC: National Critical Materials Council • NORAD Cheyenne Mountain Complex

NCMLS: North Carolina Museum of Life and Science (USA-NC)

NCMN: National Crustal Motion Network

NCMP: Nonconstituency Member of Parliament

NCMR: Non-Conforming Material Report

NCMRWF: National Centre for Medium-Range Weather Forecasting

NCMS: National Center for Manufacturing Sciences (USA) • National Classification Management Society (USA)

NCMT: Numerical Control for Machine Tools

NCN: New Chenega (USA-AK) (location identifier)

NCO: National Coordination Office • Non-Commissioned Officer • Northern Cross Observatory (USA-MI) • Number-Controlled Oscillator • Numerically-Controlled Oscillator • Quonset Point (USA-RI) (location identifier)

NCOS: National Commission On Space

NCP: Luzon Island - Cubi Point (Philippines) (location identifier) • National Climate Program • National Committee President • National Computer Policy • National Contact Point • National Contingency Plan • Natural Current Profile • Net Control Procedure • NetWare Core Protocol • Network Control Point • Network Control Processor • Network Control Program • Network Control Protocol • Network Core Protocol • No-Copy Paper • Noctilucent Cloud Particle • Non-Consultative Party • Non-linear Complementarity Problem • Normal Circular Pitch • North Celestial Pole • Not Copy Protected • Nouveau Code Pénal (France)

NCPAS: National Computer Program Abstract Service

NCPCO: National Climate Program Coordinating Office

NCPDI: National Coastal Pollutant Discharge Inventory

NCPI: National Computer Program Index (UK)

NCPO: National Climate Program Office (NOAA)

NCPS: NetWare Cross-Platform Services • Non-Contributory Pension Scheme

NCPTWA: National Clearinghouse for Periodical Title Word Abbreviations (USA)

NCQ: Marietta Atlanta (USA-GA) (location identifier)

NCR: National Capital Region • National Cash Register Co. • National Council for Research • Nippon Cataloguing Rules • Network Change Request • No Calibration Required • No Carbon Required • Non-Conformance Record • Non-Conformance Report • Notification of Change Report • San Carlos (Nicaragua) (location identifier)

NCRA: National Center for Radio Astrophysics (India) • National Cooperative Research Act

NCRC: Nickel-Cadmium Rechargeable Cell

NCRD: National Council for Research and Development

NCREMP: North Central Region Educational Materials Project

NCREN: North Carolina Research and Education Network

NCRP: National Climatic Research Program • National Council on Radiation Protection • Non-Compulsory Reporting Point

NCRS: Non-Conformance Report Supplement

NCRT: National Research Council of Thailand

NCS: National Cartographic Standards • National Communications System • Natural Color System • Navigational Computer Set • Navigation Control Simulator • Negative Central Shear • Net Control Station • NetWare Connect Services • Network Computing System • Network Control Series • Network Control Station • Network Coordination Station • Neural Computer Science • Neutron Compton Scattering • Newcastle (South Africa) (location identifier) • New Classification Society (UK) • NICMOS Cooling System • Numerical Category Scaling • Nutation Control System

NCSA: National Center for Supercomputing Applications (USA-IL) • National Computer Security Agency • National Computer Security Association

NCSC: National Computer Security Center (USA) • North Carolina State College (USA-NC)

NCSD: Natrona County School District (USA-WY)

NCSE: National Center for Science Education (USA-CA)

NCSES: Nuclear Cross-Section Evaluation System

NCSF: Northern Cross Science Foundation (USA-WI)

NCSG: National Center Sponsor Group

NCSI: National Centre for Science Information (India) • Network Communications Services Interface

NCSL: National Conference of Standards Laboratories • National Conference of State Legislatures (USA)

NCSM: National Council of Science Museums (India)

NCSR: National Centre for Scientific Research • National Committee for Space Research • National Committee on Space Research • National Council for Scientific Research

NCSSM: North Carolina School of Science and Mathematics (USA-NC)

NCS&T: National Council for Science & Technology

NCST: National Center for Science and Technology (Saudi Arabia) (formerly SANCST) • National Committee on Science and Technology • National Council for Science and Technology • Naval Center for Space Technology (USA)

NCSTC: National Council for Science and Technology Communications

NCSTRL: Networked Computer Science Technical Reports Library

NCSU: North Carolina State University (USA-NC)

NCT: Nanosatellite Constellation Trailblazer • Nicoya Guanacaste (Costa Rica) (location identifier)

n/cta: Nuestra Cuenta

NCTA: National Cable Television Association (USA)

NCTAMS: Naval Computer and Telecommunications Area Master Station

NCTD: National Committee for Technological Development • Non-Consecutive Triple Difference

NCTE: National Council of Teachers of English • Network Channel-Terminating Equipment

NCTL: National Computer and Telecommunications Laboratory

NCTM: National Council of Teachers of Mathematics (USA)

NCTRS: Network Command and Telemetry Router System

NCTS: Network Configuration and Test Section • Network Configuration and Test Station • Not Clear To Send

NCTU: National Chiao Tung University (China-ROC)

NCU: National Central University (China-ROC) • Navigation Computer Unit • Network Configuration Unit • Network Control Unit • Nitrogen Control Unit • Nozzle Control Unit • Nukus (Uzbekistan) (location identifier)

NCUF: Natinal Computer Users Forum

NCUP: No Commission Until Paid

NCUR: National Conference on Undergraduate Research

NCURA: National Council of University Research Administrators (USA)

NCURE: National Center for Ultra-Reliability Engineering

NCV: Net Calorific Value • Net Contract Value • No Commercial Value • Normalized Critical View

NCVQ: National Council for Vocational Qualification (UK)

NCWM: National Conference on Weights and Measures

NCWTD: Nationaal Centrum voor Wetenschappelijke en Technische Dokumentatie (Belgium)

NCX: Non Charge Exchange

NCXC: National Committee Exchange Controller

NCY: Annecy (France) (location identifier)

n/d: No Date

nD: n-Dimensional • n-Directional

Nd: Néodyme • Neodymium

N-D: N-Dimensional

N/D: Need Date

ND: NASA Document • National Debt • Navigation Display • N-Dimensional • Need Date • Negative Declaration • Neutral Density • Neutron Detector • Neutron Diffraction • Next Day • No Data • No Date • No Decision • No Deed • No Delay • No Demand • No Detect • No Drawing • Non Delayed • Nondelivery • Nondimensional • Normalized Difference • North Dakota (USA) • Not Dated • Not Deeded • Not Detected • Not Determined • Not Done • Not Drawn • Nothing Doing • Notre Dame

NDA: Bandanaira (Indonesia) (location identifier) • Nançay Decametric Array (France) • Neutral Detector Assembly • No Detectable Activity • Non-Destructive Analysis • Non-Destructive Assay • Non-Dimensional Analysis • Non-Disclosure Agreement

NDAB: Numerical Data Advisory Board

NDAC: Northern Data Analysis Consortium (Hipparcos) • Not Data Accepted

NDADS: NASA Data Archive and Distribution Service (*) • NASA Data Archive and Distribution System • NSSDC Data Archive and Distribution Service • NSSDC Data Archive and Distribution System

NDAG: Notre Dame Astrophysics Group (USA-IN)

N. Dak.: North Dakota (USA)

NDAP: Neptune Data Analysis Program

NDAR: National Directory of Australian Resources

NDAS: North Devon Astronomical Society (UK-England)

NDATS: Neutral Density And Temperature Spectrometer

NDB: N-Dibutylamine • Non-Directional Beacon • Nouadhibou (Mauritania) (location identifier) • Numerical Database • Numeric Database

NDBC: National Data Buoy Center

NDBO: NOAA Data Buoy Office

NDC: Nanded (India) (location identifier) • National Data Center • National Data Communications • National Documentation Center • Network Data Center • Nippon Decimal Classification • Non-Dairy Cattle • Normalized Device Coordinates • Notre Dame College (USA-IN) • NRSA Data Centre

NDCDB: National Digital Cartographic Database (USGS)

NDCC: National Disaster Coordinating Council (Philippines) • Network Data Collection Center

NDCS: National Digital Cartographic Standard

NDD: Norton Disk Doctor • Sumbe (Angola) (location identifier)

NDDK: Network Device Development Kit

NDDL: Neural Data Definition Language

NDDO: New Drug Development Office

NDDS: Nuclear Detonation Detection System

NdE: Note de l'Éditeur

NDE: Mandera (Kenya) (location identifier) • Near-Death Experience • News Development Environment • No Delay Expected • Non-Destructive Evaluation • Non-Destructive Examination • Non-Linear Differential Equation

NDEF: Not to be Defined

NDF: Narrative Data File • N-dimensional Data Format • Neutral Density Filter

NdFeB: Neodymium-Irion-Boron

NDG: No Date Given • Qiqihar (China-PRC) (location identifier)

NDGL: Neodymium-Doped Glass Laser

NDGPS: Nationwide Differential Global Positioning System

NDH: No Damage History

NDI: Namudi (Papua New Guinea) (location identifier) • Non-Destructive Inspection • Non-Developmental Item • Non-Development Item

NDIC: National Defence Industries Council (UK) • Nuclear Data Information Center (USA)

NDIR: Non-Dispersive Infrared • Non-Dispersive Infrared Radiation

NDIS: Network Device Interface Specification • Network Device Interface System • Network Driver Interface Specification • Non-Dimensional Impact Stress

NDIU: Network Data Interface Unit

NDJ: Ndjamena (Chad) (location identifier) • November-December-January

NDK: Namorik Island (Marshall Islands) (location identifier)

NDL: National Digital Library • Ndele (Central African Rep.) (location identifier) • Network Database Language • Network Definition Language • Numerical Drawing List

NdlR: Note de la Rédaction

NDLR: Note de la Rédaction

NDLTD: National Digital Library of Theses and Dissertations • Networked Digital Library of Theses and Dissertations (*)

NDM: Mendi (Ethiopia) (location identifier) • Negative Dispersion Mirror • Network Data Mover • Neutron Dose Monitor • No Dark Matter

NDMP: Network Data Management Protocol

NDMS: NetWare Distributed Management Services

NDN: Nadunumu (Papua New Guinea) (location identifier) • Nuclear Data Network

NDO: Network Development Office

NDOS: New Disk Operating System

NDP: National Defense Plan • Net Domestic Product • Neutron Depth Profiling • Normal Diametric Pitch • Numeric Data Package • Numeric Data Processor • Pensacola Ellyson (USA-FL) (location identifier)

NDPC: National Data Processing Centre

NDPD: National Data Processing Division • Numeric Data Processing Division

NDPF: NASA Data Processing Facility • National Data Processing Facility

NDPS: National Data Processing Service • Novell Distributed Print Services

NDQ: NASA Delta Quotation

Ndr: Nieder

NdR: Note du Rédacteur

NDR: Nador (Morocco) (location identifier) • Negative Differential Resistance • Network Data Reduction • New Document Reference • Non-Destructive Read • Norddeutscher Rundfunk (Germany)

NDRC: Natural Resources Defense Council (USA-DC)

NDRE: Norwegian Defence Research Establishment (Norway)

NDRO: Non-Destructive Readout

NDRW: Non-Destructive Read/Write

NDS: Nationalverband Deutscher Schriftsteller (Germany) • Natural Decision System • Navigation Display System • NetWare Directory Services • Network Distribution Service • Non-parametric Detection Scheme • Novell Directory Services • Nuclear Data Section • Nuclear Detection System • Nuclear Detonation System • Sandstone (Australia-WA) (location identifier)

NDSC: Network for the Detection of Stratospheric Change

NDSE: Non-Deliverable Support Equipment

NDSG: Network for the Detection of Stratospheric Change

NDSI: Normalized Difference Snow Index

NDSIC: National Documentation and Scientific Information Centre

NDSL: National Direct Student Loan • Non-Domestic Substances List

NDSOS: National Deep-Sky Observers Society (USA-KY)

NDST: Non-Dimensional Special Tool

NDSU: North Dakota State University (USA-ND)

NdT: Note du Traducteur

NDT: Net Data Throughout • Newfoundland Daylight Time • Non-Destructive Test • Non-Destructive Testing • Non-deterministic Decision Table • Non-Distributive Trade

NDTA: Night Driving Training Aid

NDTF: Non-Destructive Test Facility

NDTI: Non-Destructive Testing and Inspection

NDTIB: Non-Destructive Test and Inspection Building

NDTL: Non-Destructive Test Laboratory

NDU: National Defense University • Navigation Data Unit • Nuclear Data Unit • Rundu (Namibia) (location identifier)

NDV: Anacostia (USA-DC) (location identifier) • NASP Derived Vehicle

NDVI: Normalized Data Vegetation Index • Normalized Difference Vegetation Index (*)

NDW: Norton Desktop for Windows

NDX: Index

NDY: Sanday (UK) (location identifier)

NDZ: Nordholz-Spieka (Germany) (location identifier)

n/e: New Edition • No Effects

ne: Nepali (ISO language code) • Nueva Época

Ne: Neon • Néon • Nepal • Nepalese • Netherlands

N/E: New Edition • Not Entered

NE: Named Entity • National Emergency • National Engineer • National Estate • National Executive • Naval Engineer • Near Encounter • Nebraska (USA) • Neghidhian • Nepal • Neuguen (Argentina) • Neutral Excitation • Newcastle upon Tyne (UK postal code) • New Edition • New England • Niger (ISO country code) • No Effects • No Evolution • Non-Emptiness • Non Essential • Nord-Est • Northeast • Northeastern • Not Equal • Not Essential • Not Exceeding • Notice of Exception • Nuclear Energy • Nuclear Explosion

NEA: Glynco (USA-GA) (location identifier) • National Education Association (USA) • National Energy Authority (Iceland) • Near-Earth Asteroid • Negative Electron Affinity • Network Equivalent Analysis • Noise Equivalent Angle • Non-Explosive Actuator • Northeastern African • Nuclear Energy Agency • Null Error Amplifier

NEAB: Northern Examinations and Assessment Board (UK)

NEAD: Negative Electron Affinity Detector

NEAF: Northeast Astronomy Forum (USA)

NEAM: Non-volatile Electrically Alterable Memory

NEAP: National Environmental Action Plan (USA) • Near-Earth Asteroid Prospector

NEAR: National Electronic Accounting and Reporting • Near-Earth Asteroid Rendezvous

NEARnet: New England Academic and Research Network

NEARS: Near Earth Asteroid Returned Sample

NEARSS: Northeast Area Remote Sensing System

NEASC: New England Association of Schools and Colleges (USA)

NEAT: Natural-source Electromagnetic Array Technique • Near-Earth Asteroid Tracking • Novell Easy Administration Tool

NEATM: Near-Earth Asteroid Thermal Model

NEATT: Northeast Association of Technical Teachers (USA)

neb: Nebulosus

Neb: Nebraska (USA) • Nebulous

NEB: National Environmental Board • Net Environmental Benefit • Noise-Equivalent Bandwidth • North Equatorial Belt

NEBA: NASA Employees Benefit Association • North East Bolivian Airways

NEBD: North Equatorial Band Disturbance

NEbE: Northeast by East

NEbN: Northeast by North

NEBn: North Equatorial Belt - Northern Component

Nebr: Nebraska (USA)

NEBs: North Equatorial Belt - Southern Component

NEBS: New Equipment Building System

NEBz: North Equatorial Belt - Internal Zone

Nec: Necessary • Necessity

NEC: National Economic Council • National Electrical Code (*) • National Electric Code • National Engineering Centre • National Executive Committee • Necochea (Argentina-BA) (location identifier) • New Eruptive Cosmogony • Nippon Electric Co. • Noise Electron Count • North Equatorial Current • Not Elsewhere Classified • Null Energy Condition

NECAP: NASA Energy-Cost Analysis Program • Navigation Equipment Capability Analysis

Necr: Necrosis

Necrol: Necrology

NECS: National Electrical Code Standard

NECSTS: Network of European Centers in Science and Technology Studies

NECTAR: Network of European Communication and Transport Activities Research • Network of European Communications and Transport Activity Research (*)

NECUSE: New England Consortium for Undergraduate Science Education

NECZ: Near-Equatorial Convergence Zone

NED: NASA Extragalactic Database • Navigation Error Data • Neutron Detector • New English Dictionary • No Evidence of Desease • No Expiration Date • Normal Equivalent Deviation • North-East-Down

NEDA: National Economic and Development Agency • National Economic Development Authority • National Electronics Distributors Association (USA)

NEDC: National Economic Development Council

NEDR: Noise Equivalent Dalta Reflectivity

NEDRES: National Environmental Data Referential Service • National Environmental Data Referral System

NEDS: National Emission Data System

NEE: Noise-Equivalent Energy • Non-linear Evolution Equation

NEED: Non-Maxwellian Electron Energy Distribution

NEEDS: NASA End-to-End Data System • National Engineering Education Delivery System • Nikkei Economic Electronic Data System

NEEP: Nuclear Engineering and Engineering Physics

Néérl: Néérlandais

Neerl: Neerlandés

NEESDI: Netherlands Environmental Earth Systems Dynamics Initiative

NEF: National Extra Fine • Neftekamsk (Russia) (location identifier) • Noise-Equivalent Flux • Noise Exposure Forecast • Nuit des Étoiles Filantes

NEFAS: Northeast Florida Astronomical Society (USA-FL)

NEFD: Noise-Equivalent Flux Density

Neg: Negate • Negación • Negation • Negative • Negatively • Neglect • Negligence • Negligible • Negotiate

NEG: Negate • Negative • Neglect • Negligible • Negril (Jamaica) (location identifier) • Non-Evaporable Getter • Numerical Experimentation Group

Negat: Negativo

NEH: National Endowment for the Humanities (USA)

NEHA: National Environmental Health Association (USA)

NEHRP: National Earthquake Hazards Reducation Program

NEI: National Eye Institute (USA) • Noise-Equivalent Input • Noise-Equivalent Intensity • Noise-Equivalent Irradiance • Non-Equilibrium Ionization • Non Est Inventus • Non-Explosive Initiation • Northern Engineering Industries • Not Elsewhere Included • Not Elsewhere Indicated • Nouvel État Indépendant

NEIA: National Energy Information Administration (USA)

NEIC: National Earthquake Information Center (USA) • National Energy Information Center (USA) • National Enforcement Investigations Center (USA)

NEIF: Near-Earth Instrumentation Facility

NEIR: Narrative End Item Report

NEIS: National Earthquake Information Service (USGS) • National Energy Information System (USA) • National Environmental Information System

NEJ: Nejjo (Ethiopia) (location identifier)

NEK: Nekemt (Ethiopia) (location identifier) • Norsk Elektroteknisk Komite (Norway)

NEKAAL: Northeast Kansas Amateur Astronomers League (USA-KS)

NEL: Lakehurst Naec (USA-NJ) (location identifier) • Narrow Emission Line • National Exploitation Laboratory

NELAS: North East London Astronomical Society (UK)

NELG: Narrow-Emission-Line Galaxy

NELH: Natural-Energy Laboratory of Hawaii (USA-HI)

NELLCO: New England Law Library Consortium (USA)

NELPAG: New England Light Pollution Advisory Group (USA-MA)

NELS: Northern European LORAN System • Nuclear Effects Link Simulator

NELY: Norteasterly

NEM: Negative Energy Mode • Noise-Equivalent Magnitude • Non-Electronic Maintenance • Not Elsewhere Mentioned

NEMA: National Electrical Manufacturers Association (USA)

NEMAC: Normal Error Model Analysis Chart

NEMAG: Negative Effective Mass Amplifiers and Generators

nem. con.: Nemine Contradicente

nem. dis.: Nemine Dissentiente

NEMO: Neutral Extinction of Massive Objects • Neutrinoless Experiment on Molybdenum • Non-Empirical Molecular Orbital • Not Everybody Must Observe • Nouvelle Expérience Majeure de l'Observatoire

NEMP: Nuclear Electromagnetic Pulse

NEMPS: National Environmental Monitoring and Prediction System

NEMS: NASA Equipment Management System • National Energy Modelling System • Near-Earth Magnetospheric Satellite • New England Meteoritical Service (USA-MA) • Nimbus E Microwave Spectrometer

NEN: Whitehouse (USA-FL) (location identifier)

N. Eng.: New England • Northern England

NEO: Near-Earth Object • Near-Earth Orbit • Network Enterprise Objects • Northeastern Operations Office

NEOC: National Earth Observations Center (USA)

NEODyS: Near-Earth Object Dynamics Site

Neol: Neologism • Neologisme • Neologismo

NEOME: New Electroactive Organic Materials for Electronics

NEON: Network of European Observatories in the North

NEONET: Netherlands Earth Observation Network

NEOS: National Earth Orientation Service (USA-DC)

NEOSC: Near-Earth Object Search Committee

Nep: Nepal • Nepalese • Neptune

NEP: National Extension Program • Near-Earth Phase • Nearest Equivalent Product • Negative Equally Probable • Net Ecosystem Production • New Economic Policy • New Edition Pending • Noise Equal to Power • Noise-Equivalent Power • Nominal Entry Point • Non-Electronic Part • Non-Elutable Polar • Normal Entry Point • North Ecliptic Pole • Nuclear Electric Propulsion • Nuclear Energy Power

NEPA: National Electric Power Authority • National Energy Policy Plan • National Environmental Policy Act (USA) • National Environmental Protection Agency • National Environment Protection Authority

NEPAE: Núcleo de Estudos e Pesquisas Aeroespaciais (Brazil-RS)

NEPAL: Nouvelles Expériences de Physique de l'Accélérateur du LAL

NEPC: National Environment Protection Council

NEPD: Noise-Equivalent Power Density

NEPM: National Environment Protection Measure

NEPN: Near-Earth Phase Network

NEPP: National Environment Policy Plan • National Environment Protection Plan

NEPR: North Ecliptic Pole Region

NEPRF: Naval Environmental Prediction and Research Facility

Nept: Neptune

NEPTUNE: New European Programme for Technology Utilization in Education

NEQ: Não Equivalente • Non-Equivalence • Non-Équivalence • Normal Equation

NER: Neryungri (Russia) (location identifier) • Never Exceed Redline • Niger (ISO country code) • Noise-Equivalent Radiance • Non-conformance Event Record

NERA: Northern (extremity of) Earth's Rotation Axis

NERAL: Northeast Region of the Astronomical League (USA)

NERC: National Environmental Research Centre • Natural Environment Research Council (UK)

NERCS: National Energy Research Supercomputer Center

NERDAS: NASA Earth Resources Data Annotation System

NERF: Naval Electromagnetic Radiation Facility

NERI: National Environmental Research Institute

NERL: Northeast Research Libraries

NERO: Near-Earth Rescue and Operations

NEROC: Northeast Radio Observatory Corporation (USA-MA)

NERC: National Environmental Research Policy

NERN: Northeastern

NERS: Natural Environmental Radioactivity Survey

NERSC: National Energy Research Scientific Center (USA-CA)

NERVA: Nuclear Energy for Rocket Vehicle Application • Nuclear Engine for Rocket Vehicle Application (*)

NES: National Education Supercomputer • National Election Studies • National Energy Strategy • National Environment Secretariat • Network Encryption System • New York - East 34th Street (USA-NY) (location identifier) • Non-Erasable Storage • Not Elsewhere Specified • Nuclear Elastic Scattering

NESB: Non-English-Speaking Background

NESC: National Electrical Safety Code • National Energy Software Center • National Energy Software Corporation • National Environmental Satellite Center • National Environmental Supercomputing Center • New England Science Center (USA-MA)

NeSDaC: Netherlands SAX Data Centre

NESDD: NOAA Earth System Data Directory

NESDIS: National Environmental Satellite Data and Information Service (USA-NC) (*) • National Environmental Satellite Data Information Service

NESDP: National Economic and Social Development Plan

NESE: New and Renewable Source of Energy

NESHAP: National Emission Standards for Hazardous Air Pollutants

NESLI: National Electronic Site License Initiative

NESPUT: National Educational Simulations Project Utilizing Telecommunications

NESR: Natural Environment Support Room • NGST External Review Board • Noise Equivalent Spectral Radiance

NESS: National Earth Satellite Service (NOAA) (now merged with EDIS into NESDIS) • National Environmental Satellite Service (NOAA) • New England Skeptical Society (USA)

NEST: Novell Embedded Systems Technology • Nuclear Emergency Search Team

NESTA: National Earth Science Teachers Association (USA) • National Endowment for Science, Technology and Arts (UK)

NESTOR: Netherlands Educational and Scientific Titles for Online Retrieval • Numérisation et Sériation pour Traitement des Informations Radar

NET: National Educational Television • Net Energy Thrust • Net Equivalent Temperature • Network • Network of Expanded Telecommunications • Newton Emission Theory • Next European Torus • Nimbus Experiment Team

• Noise-Equivalent Target • Noise-Equivalent Temperature • Norme Européenne des Télécommunications • Not Earlier Than

NETA: No Electron Theft Act (USA)

NetBEUI: NetBIOS Extended User Interface

NETBEUI: NetBIOS Extended User Interface

NetBIOS: Network Basic Input/Output System

NETBIOS: Network Basic Input/Output System

NetBlT: Network Block Transfer

NETBlT: Network Block Transfer

NETCOM: Network Communications

NETD: Noise Equivalent Temperature Difference

NetDDE: Network Dynamic Data Exchange

NETDDE: Network Dynamic Data Exchange

NETDS: Near-Earth Tracking and Data System

NETFD: Non-Equilibrium Thermo-Field Dynamics

Neth: Netherlands

Neth. Ant.: Netherlands Antilles

NETID: Network Identification

NETSET: Network Synthesis and Evaluation Technique

NETSIM: Network Simulation

NETT: Network for Environmental Technology Transfer • Network for Environment Technology Transfer

NETV: National Educational Television

NEU: Northeastern University (USA-MA) • Sam Neua (Laos) (location identifier)

Neurol: Neurological • Neurology

Neut: Neuter • Neutral • Neutralize • Neutralized

Nev: Nevada (USA)

NEV: Nevis (Saint Kitts & Nevis) (location identifier)

NEVADA: Net Energy Verification And Determination Analyzer

NEVEC: NOVA-ESO VLTI Expertise Centre

NEVS: NASA Equipment Visibility System

New: Newton • Newtonian

NEW: Net Economic Welfare • New Orleans - Lakefront (USA-LA) (location identifier)

NEWEST: NASA Educational Workshop for Elementary School Teachers

Newf: Newfoundland (Canada)

New M: New Mexico (USA)

NEWMAST: NASA Educational Workshop for Math, Science and Technology

NEWPIMS: Neutral Environment With Plasma Interactions Monitoring System

NeWS: Network Extensible Windows System

NEWS: National Early Warning System • NetWare Early Warning System • Network Extensible Windows System • New Engineering Work Schedule • New Engineering Work System • Non-linear Elastic Wave Spectroscopy • Novell Electronic Webcasting Service

NewSc: New Scientist (periodical code)

NEWSIPS: New Spectral Image Processing System

Newsl: Newsletter

NEWSTAR: Northeast Wisconsin Star Gazers (USA-WI)

NEXAFS: Near-Edge Absorption Fine Structure

NExT: New Experiences in Teaching

NEXT: Near-End Crosstalk • NLR Engineering Expert Toolkit

n/f: No Funds • Nuestro Favor

N/F: No Funds

NF: Île Norfolk (ISO country code) • Nanofarad • National Fine • National Formulary • Natural Flood • Navigation Filter • Near Face • Neue Folge • Neuro-Fuzzy • Neutral Fraction • Newfoundland (Canada) • No Fool • No Form • No Funds • Noise Factor • Noise Figure • Noise Frequency • Non-Ferrous • Non-Fiction • Non-Filtered • Non-Flickering • Non-Fundable • Norfolk Island (ISO country code) • Norges Forskningsråd (Norway) • Normal Formula • Norman French • Norme Française (France) • Number of Fields

NFA: Nitrogen Filling Assembly • No Further Action • Null Filling Antenna

NFAIS: National Federation of Abstracting and Indexing Services (*) • National Federation of Abstracting and Information Services • National Federation of American Information Services

NFAM: Network File Access Method

NFB: Mount Clemens - Detroit (USA-MI) (location identifier) • National Film Board • Negative Feedback • Nordisk Forskningsbiblioteken

NFBC: National Film Board of Canada

NFC: Negative Factor Counting • Negative Feedback Circuit • No Further Consequence • Not Favorably Considered

NFCT: Non-Federal Control Tower

Nfd: Newfoundland (Canada)

NFD: Net Filter Discrimination • Neutron Flux Density • Newfoundland (Canada) • No Fixed Date • Nürnberger Flugdienst

NFDC: National Flight Data Center

NFE: Normalized Fuzzy Entropy

NFER: National Foundation for Educational Research (UK)

NFF: No Fault Found • Non-Formula Funded

NFFE: National Federation of Federal Employees (USA)

NFG: Nefteyugansk (Russia) (location identifier) • Non-Functional Goodie

NFGS: Nearby Field Galaxy Survey

NFI: Narrow-Field Instrument • Narrow-FOV Instrument • Nationale Faciliteit Informatica (Netherlands) • Neutron Flux Integrator • No-Frame Inconditional • Noise Figure Indicator • Nouvelles Formations d'Ingénieurs

NFIC: National Fraud Information Center

NFIP: National Flood Insurance Program

NFK: Île Norfolk (ISO country code) • Norfolk Island (ISO country code)

NFL: Fallon (USA-NV) (location identifier) • National Film Library • Nerve Fiber Layer

Nfld: Newfoundland (Canada)

NFLD: Newfoundland (Canada)

NFLNA: Noise Figure Low-Noise Amplifier

NFM: Narrow-band Frequency Modulation • Network Facilities Manual • Noise Figure Meter • Non-Ferrous Metal

NFMA: National Forest Management Act (USA)

NFMS: Nitrogen Flow Measuring System • Noise Figure Meter System

NFN: No Form Necessary • No Further Need

NFO: Niuafo'Ou (Tonga) (location identifier)

NFOM: Near-Field Optical Microscope

NFOV: Narrow Field Of View (*) • Near Field Of View

NFOVR: Narrow Field Of View Radiometer

NFP: National Focal Point • National Forschungsprogramme • National Fusion Program (Canada) • Normal Failure Period

NFPA: National Fire Protection Association (USA)

NFPM: Nuclear Flight Propulsion Module

NFPO: Non For Profit Organization

NFPS: Nuclear Flight Propulsion System

NFR: Naturvetenskapliga Forskningsrådet (Sweden) • Net Flux Radiometer • Network Flight Recorder • No Further Requirements • Non-Functional Requirement • Nuclear Fission Reactor

n/fra: Nuestra Factura

NFRA: Netherlands Foundation for Radio Astronomy • Netherlands Foundation for Research in Astronomy (now ASTRON)

NFS: NASA FAR Supplement • National and Foreign Stations • Network File Server • Network File Service • Network File Structure • Network File System (*) • Neutron Flux Spectrum • Nitrogen Flow System • Noise Frequency Spectrum • Not For Sale • Not on Flying Status • Nozzle Flow Sensor

NFSI: National Foundation for Science and Industry (USA)

NFSS: New Font Selection Scheme

NFSU: Non-Flying Support Unit

NFT: National Fly-by-wire Testbed • Network File Transfer • Non-oscillatory Forward-in-Time • Normal-Field Telescope

NFTC: National Flight Test Center

N/FTS: Near-/Far-Term System

NFU: Not For Us • Not Frequently Used

NFW: Non-Fuel-Wasting

NFWO: Nationaal Fonds voor Wetenschappelijk Onderzoek (Belgium)

NFZ: Nuclear Free Zone

NF2: Non-First Normal Form

N&G: Navigation & Guidance

ng: Nanogram

Ng: Norwegian

N&G: Navigation & Guidance

NG: Narrow Gauge • National Gallery • National Geographic (magazine) • Natural Gas • Navigation and Guidance • Nebulizer Gas • New Growth • New Guinea • Next Generation • Niger • Nigeria (ISO country code) • Nitrogen Gauge • Nitroglycerine • Nitroglycerol • No Go • No Good • Non-Graduate • Nose Gear • Not Given • Not Good • Nottingham (UK postal code)

NGA: National Governors' Association • Nigeria (ISO country code) • Young (Australia-NSW) (location identifier)

NGAO: North Georgia Astronomical Observatory (USA-GA)

NGASS: Next Generation of Australian Scientists Society

NGAT: New Generation Astronomical Telescope

NGB: Next Generation Browse • Ningbo (China-PRC) (location identifier)

NGBN: National Gravity Base Net

NGC: Grand Canyon - North Rim (USA-AZ) (location identifier) • National Geophysical Committee • National Grid Co. plc (UK) • Navigation-Guidance-Control • New General Catalogue • Next Generation Computer • North Galactic Cap • North Georgia College (USA-GA) • Nozzle Gap Control

NGCIA: National Centre for Geographic Information and Analysis

NGCP: Navigation, Guidance and Control Processor

NGCR: Next Generation Computer Resource

NGCSU: North Georgia College and State University (USA-GA)

NGD: Anegada (British Virgin Island) (location identifier)

NGDB: National Geo-Database • National Geodetic Database

NGDC: National Geophysical Data Center (USA-CO)

NGE: Navigation Guidance Equipment • N'Gaoundere (Cameroon) (location identifier) • Not Greater or Equal

NGF: Nerve Growth Factor • Nervous Growth Factor • Next Generation Focuser

NGfL: National Grid for Learning (UK)

NGG: Negative Grid Generator

NGH: NASA Grant Handbook

NGI: Nationaal Geografisch Instituut (Belgium) • National Geographic Institute • National Geographic Interactive • Nederlands Genootschap voor Informatica • Next Generation Internet • Ngau Island (Fiji) (location identifier) • Non-Governmental Institution • Norwegian Geotechnical Institute (Norway)

NGIMS: Neutral Gas and Ion Mass Spectrometer

NGIO: Next Generation Input/Output

NGIS: National Geographic Information System • Next Generation Information System

NGISE: National Geographic Information System on Environment

NGITS: Next Generation Information Technologies and Systems

NGk: New Greek

NGL: Natural-Gas Liquid • Neodymium Glass Laser • Neon Glow Lamp • Ngala (South Africa) (location identifier)

NGld: Nederlandse Gulden

NGLR: Neodymium Glass Laser Rod

NGLT: Next Generation Lowell Telescope

NGM: Guam Agana (Guam) (location identifier) • Nested-Grid Model

NGN: Naira (Nigeria) (ISO currency code) • Nargana (Panama) (location identifier) • National Geodetic Network

NGNP: Nominal Gross National Product

NGO: Nagoya (Japan) (location identifier) • National Gemini Office • Non-Governmental Organization

NGP: Corpus Christi (USA-TX) (location identifier) • Nearest Grid Point • North Galactic Pole

ngr: Nanogram

NGR: New General Relativity • Ningerum (Papua New Guinea) (location identifier)

NGRI: National Geophysical Research Institute (India)

NGRS: National Geodetic Reference System (USA)

N&GS: Navigation & Guidance System

NGS: Nagasaki (Japan) (location identifier) • National Geodetic Society • National Geodetic Survey • National Geographic Society (USA-DC) • Natural Guide Star • NSSDC Graphics System • Nuclear Generating Station

NGSDC: National Geophysical and STP Data Center

NGSIC: National Geodetic Survey Information Center (NOAA)

NGSO: Non-Geostationary Satellite Orbit • Non-Geosynchronous Orbit

NGSOC: National Geodetic Survey Operations Center (NOAA)

NGSP: National Geodetic Satellite Program

NGSS: Nearby Galaxies Supernova Search • Next-Generation Sky Survey

NGST: New-Generation Space Telescope • Next-Generation Space Telescope (*)

NGSTDC: National Geophysical and Solar-Terrestrial Data Center (NOAA)

NGT: NASA Ground Terminal • New-Generation Telescope • Next-Generation Telescope • Night • Noise Generator Tube

NGTS: New-Generation Training System

NGU: Norfolk Chambers (USA-VA) (location identifier)

NGV: Natural-Gas for Vehicles • Natural-Gas Vehicle • Navire à Grande Vitesse • Ngiva (Angola) (location identifier)

NGVCC: Natural Gas Vehicle Composite Cylinders

NGVD: National Geodetic Vertical Datum

NGW: Corpus Christi - Cabaniss (USA-TX) (location identifier)

NGWIC: National Ground Water Information Center

NGX: Manang (Nepal) (location identifier)

NGXO: Next Generation X-ray Observatory

NGZ: Alameda (USA-CA) (location identifier)

NH: National Heritage (UK) • National Highway • National Hot • New Hampshire (USA) • Nitrogen Hydride • Nominal Height • Normal Height • Northern Hemisphere • Nuclear Head

NHA: Next Higher Assembly • Next Higher Authority • Nha Trang (Viet Nam) (location identifier) • Non-Hydrogen Atom

NHAB: Nürnberger Hobbyastronomenbund (Germany)

NHAO: Nishi-Harima Astronomical Observatory (Japan)

NHAP: National High-altitude Aerial Photography

NHAPP: National High-altitude Aerial Photographic Program

NHar: Não Harmonizada

NHB: NASA Handbook • Nickel-Hydrogen Battery

NHC: National Hurricane Center (NOAA)

NHD: Minhad Ab (UAE) (location identifier)

NHDR: Network Header

NHE: National Heat Exchange • No High Energy • Normal Hydrogen Electrode

N. Heb.: New Hebrides

N. Héb.: Nouvelles Hébrides

NHEML: National Hurricane Experimental Meteorological Laboratory

NHF: New Halfa (Sudan) (location identifier)

NHG: New High German

NHGDS: Non-Hydrogenous Gas Delivery System

NHI: National Health Insurance • Nortwestern Hawaiian Islands (USA-HI)

NHK: Patuxent River (USA-MD) (location identifier)

NHL: National Hockey League (USA) • Normal Hydrogen Line

NHLBI: National Heart, Lung and Blood Institute

NHM: No Hot Metal • Nuclear Hyperfine Magnetic

NHMFL: National High-Magnetic-Field Laboratory (USA-FL)

NH&MRC: National Health & Medical Research Council

NHMRC: National Health and Medical Research Council

NHMTF: Natural Hazards Mitigation Trust Fund

NHP: Nitrogen High Pressure • Nominal Horsepower

NHPN: National Highway Planning Network (USA)

NHQ: Nuclear Hyperfine Quadrupolar

NHR: Non-Hierarchical Routing

NHRC: National Hydrology Research Centre

NHRE: National Hail Research Experiment

NHRI: National Health and Retirement Insurance • National Hydrology Research Institute

NHRP: Next-Hop Resolution Protocol • Next-Hop Routing Protocol

NH&S: Nuclear Hardening & Survivability

NHS: National Health Service • National Highway System (USA) • Next-Hop Server • Nuclear Hardening and Survivability • Nushki (Pakistan) (location identifier)

NHSE: National HPCC Software Exchange

NHT: Northolt (UK-England) (location identifier)

NHTA: Numerical Heat Transfer Applications (periodical code)

NHTSA: National Highway Traffic Safety Administration (USA)

NHV: Nuku Hiva (Marquesas Islands) (location identifier)

NHX: Foley Barin (USA-AL) (location identifier)

NHZ: Brunswick (USA-ME) (location identifier) • Non-Habitable Zone

NH3: Amonia

NH4: Hydrazine

ni: Nicaragua

Ni: Nickel

N/I: Non-Interlaced

NI: National Instruments • Network Interface • New Impression • Nicaragua (ISO country code) • Nigeria • No Information • Noise Index • Non-Inductive • Non-Interlaced • Non-Intervention • Norfolk Island • Northern Ireland • North Island (New Zealand) • Not Interested • Not Issued • Numerical Index • Triton (satellite of Neptune)

NIA: National Information Account • National Institute on Aging (USA) • National Intelligence Authority • National Irrigation Administration • Network Interoperability Alliance • Newspaper Institute of America • Newspaper Institute on Aging • Nickel-Iron Alloy • Nimba (Liberia) (location identifier) • No Information Available • No Input Acknowledge

NIAAA: National Institute on Alcohol Abuse and Alcoholism

NIAC: NASA Industrial Application Center • NASA Institute for Advanced Concepts • National Information Accounts Clearinghouse

NIAE: National Institute of Adult Education (UK)

NIAG: NATO Industrial Advisory Group • Netherlands Institute of Applied Geoscience • Northern Indiana Astronomical Group (USA-IN)

NIAID: National Institute of Allergies and Infectious Diseases

NIAM: Nijssen Information Analysis Methodology • Nijssen Information Analysis Model

NIAMS: National Institute of Arthritis and Musculoskeletal and Skin Diseases

NIAR: National Institute of Atmospheric Research

NIAS: National Institute of Advanced Studies • National Institute of Aeronautical Sciences

NIB: National Information Bureau • Negative Impedance Booster • Negative Ion Beam • Negative Ion Blemish • Nikolai (USA-AK) (location identifier) • Non-Interference Basis • Nordic Investment Bank

NIBS: Nuffield Interactive Book System

Nic: Nicaragua • Nicaraguan

NIC: Córdoba (Nicaragua) (ISO currency code) • National Industrial Council • National Informatics Centre (India) • National Information Centre • National Insurance Contribution • Near-Infrared Camera • Negative Impedance Converter • Negative Ion Chamber • Networked Improvement Community • Networked Information Community • Network Imaging Corporation • Network Information Center • Network Information Community • Network Interface Card • Network Interface Controller • Newly Industrialized Country (*) • Newly Industrializing Country • Nicaragua (international vehicle registration code) • Nicaragua (ISO country code) • Nicosia (Cyprus) (location identifier) • Not In Contact • Not In Contract • Nouvel Instrument Communautaire • Numeric Intensive Computing

Nica: Nicaragua • Nicaraguan

NICA: Non-Interactive Computer Application

NiCad: Nickel-Cadmium

NICAD: Nickel-Cadmium

NICAM: Near-Instantaneous Companded Audio Multiplex

NICAP: National Investigation Committee on Aerial Phenomena (USA)

Nicar: Nicaragua • Nicaraguan

NICC: National Inventory Control Center

Ni/Cd: Nickel-Cadmium

NiCd: Nickel-Cadmium

NICE: Narrative, Immersive, Constructionist/Collaborative Environment • NASA Science Internet Interactive Computer Environment

NICHD: National Institute on Child Health and Human Development

NICI: Near-Infrared Coronograph/Imager • Nijmeegs Instituut voor Cognitieonderzoek en Informatietechnologie (Netherlands)

NICK: Nature's Initial Cosmic Kickstart

NICMOS: Near-Infrared Camera and Multiobject Spectrometer • Near-Infrared Camera Multiobject Spectrograph (*)

NICNET: National Informatics Centre Network

NICOLAS: Network Information Center Online Aid System

NICS: National Inter-facility Communications System • Novell Intenet Caching System

NID: Network In-Dial • Network Interactive Display • New Interactive Display • Next Identification • Non-equilibrium Ionospheric Disturbance

NIDC: National Information and Documentation Center

NIDCD: National Institute of Deafness and Other Communications Disorders

NIDR: Networked Information Discovery and Retrieval

NIDRR: National Institute on Disability and Rehabilitation Research • National Institute on Disability Rehabilitation and Research

NIDS: National Institute for Discovery Science (USA-NV) • Network Interface Data System

NIE: National Institute for the Environment • National Institute of Education • National Intelligence Estimate • Negative Ion Erosion • Networking Infrastructure for Education • Networking Infrastructure in Education • Neutron Ionization Effect • Newly Industrializing Economy • Niblack (USA-AK) (location identifier) • Not Included Elsewhere

NIED: National Institute for Earth and Disasters

Niederbay: Niederbayern (Germany)

Niedersachs: Niedersachsen (Germany)

NIEHS: National Institute of Environmental Health Sciences

NIEI: NASA-Industry Education Initiative

NIEM: Network for Industrial Environmental Management

NIEO: New International Economic Order

NIES: National Institute for Environmental Studies • National Institute of Environmental Science

NIF: National Ignition Facility (USA) • National Ignition Furnace • Neighbourhood Information Frame • Network Information File • Nordisk Industrifond • Normalized Impact Factor • Note Issuance Facility • Not In File • Nuclear Information File • Número de Identificación Fiscal (Spain)

NIFO: Next In - First Out

NIFOE: Naval Observatory - Indiana University Fiber Optics Echelle

NIFTP: Network Independent File Transfer Protocol

NIFTR: Near Isothermal Flowing Test Rig

NIFS: National Institute for Fusion Science (Japan)

Nig: Nigeria • Nigerian

NIG: Nikunau (Kiribati) (location identifier)

NIGEC: National Institute for Global Environmental Change

NIH: National Institute of Health (USA) • Not Invented Here

NIHL: Noise-Induced Hearing Loss

NII: National Information Infrastructure (USA) • Nereide (satellite of Neptune) • Nuclear Installation Inspectorate (UK)

NIIAC: National Information Infrastructure Advisory Council

NIICP: No Increase In Contract Price

NIII: Naiad (satellite of Neptune)

NIIIP: National Industrial Information Infrastructure Protocol

NIIM: National Integrated Information Management

NIIN: National Item Identification Number

NIIR: Non-Imaging Infrared

NIIT: National Information Infrastructure Testbed • National Institute for Information Technology • National Institute of Instructional Technology (USA)

NIK: Niokolo Koba (Senegal) (location identifier)

NIKHEF: Nationaal Instituut voor Kernfysica en Hoge-Energiefysica (Netherlands)

NILU: Norwegian Institutt for Luftforskning (Norway)

NIM: National Information Market • National Institute of Metrology • Near Infrared Magnetograph • Network Install Manager • Network Interface Module • Niamey (Niger) (location identifier) • Non-Interrupt Mode • Normal Integration Mode • Nothing In Mind • Nuclear Instrumentation Module

NIMA: National Imagery and Mapping Agency (USA)

NIMBY: Not In My Backyard

NIMD: National Institute of Management and Development

NiMH: Nickel-Metal-Hydride

NIMH: National Institute of Mental Health

NIMHANS: National Institute of Mental Health And Neuro Sciences

NIMIS: NRSA Integrated Management Information System

n. imp.: New Impression

NIMPH: Network Interface Message Processing Host

NIMPHE: Nuclear Isotope Monopropellant Hydrazine Engine

NIMPOL: N-band Imaging Polarimeter

NIMPR: Nuclear Instruments and Methods in Physics Research (periodical code)

NIMROD: Northern Illinois Meteorological Research on Downbursts

NIMS: Near-Infrared Mapping Spectrometer (Galileo) • NSSDC Information Management System

NIMT: National Institute for Management Technology

NIN: Name Index Number • National Information Network • Ninilchik (USA-AK) (location identifier)

NINCH: National Initiative for a Networked Cultural Heritage (USA)

NINDS: National Institute on Neurological Disease and Stroke • National Institute on Neurological Disorders and Stroke

NINO: No Inflow - No Outflow • No Inspector - No Operator • Nothing In - Nothing Out

NIO: National Institute of Oceanography • Nioki (Zaire) (location identifier) • Noeud d'Interconnexion d'Opérateurs • No Inflow - Outflow

NIOBE: Numerical Integration Of the Boltzmann Equation

NIOD: Network In-Out Dial

NIOSH: National Institute of Occupational Safety and Health (USA)

NIOZ: Nederlands Instituut voor Onderzoek der Zee (Netherlands)

Nip: Nipple • Nippon • Nipponese

NIP: Jacksonville (USA-FL) (location identifier) • National Indicative Programme • National Information Point • National Information Policy • National Inventory Programme • Negative Ion Plasma • Network Input Processor • Network Integrated Processing • Network Interface Processor • New Investigator Program • Normal Incidence Pyrheliometer • Nuclear Initialization Program • Numeric Indicator Performance

NIPALS: Non-linear Iterative Partial Least Squares

NIPC: National Infrastructure Protection Center (USA)

NIPD: Not In the Public Domain

NIPDWS: National Interim Primary Drinking Water Standard

NIPO: Nederlands Instituut voor de Publieke Opinie • Negative Input - Positive Output

NIPR: National Institute of Polar Research (Japan)

ni. pri.: Nisi Prius

NIpS: Network I/Os per Second

NIPS: National Information Processing System • Network I/Os Per Second • Neural Information Processing System • News in Physiological Sciences • Nottingham Image Processing System

NIPSS: Near-Infrared Photographic Sky Survey

N. Ir.: Northern Ireland

NIR: Beeville Chase (USA-TX) (location identifier) • Near Infrared • Networked Information Resource • Networked Information Retrieval • Network Information Retrieval • Night-time Infrared • Non-Ionizing Radiation

NIRA: National Industrial Recovery Act • Near Infrared Analysis • Near Infrared Reflectance Analysis

NIRC: Near Infrared Camera • Negative Ion Recombination Chamber

N. Ire.: Northern Ireland

NIRI: Near-Infrared Imager

NIRIM: Near-Infrared Imager

NIRMOS: Near-Infrared Multiobject Spectrograph

NIRP: National Institute for Radiation Protection

NIRRL: Northen Ireland Regional Research Laboratory

NIRS: Near-Infrared Spectrometer • Northern International Reference Stars • Nuclear Information and Resource Service

NIRST: Near Infrared Solar Telescope

NIS: Israeli Shekel (currency) • Name Information Service • Naming Information Service • National Information Service • National Information System • National Institute for Standards • National Institute of Science • Near-Infrared Spectrograph • Near-Infrared Spectrometer • Negative Ion Source • Network Information Service • Network Information System • Neutron Inelastic Scattering • New Independent State • Newly Independent State • Nickel-Iron System • Night Illumination System • Noise Information System • Non-proliferation and International Security • Normal Incidence Spectrometer • Not In Stock • Not In System

NISARC: National Information Storage And Retrieval Center

NISC: NAS Implementation Support Contract • National Industrial Safety Committee • National Industrial Space Committee • National Information System for Chemistry

NISE: National Institute for Science Education (USA) • Naval In-Service Engineering • Normalized Integral Squared Error

NISM: National Information System for Mathematics

NISN: NASA Integrated Services Network

NISO: National Information Standards Organization (USA-MD) • Netherlands Industrial Space Organisation

NISP: National Industrial Security Program • National Information System for Physics • Networked Information Services Project

NISS: National Information on Software and Services • National Information Services and Systems (UK) • National Integrated Scheduling System • Nuclear Information Sharing System

NISSBB: National Information on Software and Services Bulletin Board

NISSPAC: National Information on Software and Services Public Access Collections

NIST: National Information System for Science and Technology • National Institute of Science and Technology • National Institute of Standards and Technology (USA-CO)

NISTADS: National Institute of Science Technology and Development Studies

NISTED: National Institute of Science and Technology Policy (Japan)

NISTJ: National Institute of Standards and Technology Journal (periodical code)

NIT: National Institute of Technology • Negative Income Tax • Niort (France) (location identifier)

NITC: National Information Technology Center • National Information Transfer Center • Northern Ireland Technology Centre

NITEC: NASA Information and Technology Exchange Council • National Imaging Technology in Education Conference (USA)

NITF: National Imagery Transmission Format

NITFS: National Imagery Transmission Format Standard

NITFTP: National Imagery Transmission Format Transfer Protocol

NITG: Nederlands Instituut voor Toegepaste Geowetenschappen (Netherlands)

NITI: National Institute of Technical Information

NITR: National Institute for Telecommunications Research

NITS: Network Independent Transport Service

NITSC: National Information Technology Steering Committee

NIU: NASCOM Interface Unit • National Information Unit • Network Interface Unit • New Internet User • Niue (ISO country code) • Northern Illinois University (USA-IL)

NIUA: National Institute of Urban Affairs (India)

NIUF: North-American ISDN Users' Forum

NIUNA: Napoli Istituto Universitario Navale - Annali (periodical code)

NIV: Thalassa (satellite of Neptune)

NIVR: Nederlands Instituut voor Vliegtuigontwikkeling en Ruimtevaart (Netherlands)

NIWF: Nationaal Instituut voor de Wetenschappelijke Film (Belgium)

NIX: NASA Image Exchange • Nioro (Mali) (location identifier)

NIXT: Normal Incidence X-ray Telescope

NIZ: Nationales Informationszentrum

nJ: nächsten Jahres • Nanojoule

NJ: Nächsten Jahres • New Jersey (USA)

NJA: Atsugi (Japan) (location identifier)

NJAA: New Jersey Astronomical Association (USA-NJ)

NJAC: National Joint Advisory Council

NJAG: North Jersey Astronomical Group (USA-NJ)

NJC: National Joint Council • Nizhnevartovsk (Russia) (location identifier)

NJCC: National Joint Computer Committee • National Joint Consultative Committee

NJCL: Network Job Control Language

NJE: Network Job Entry

NJFA: NATO Joint Frequency Agreement

NJIT: New Jersey Institute of Technology (USA-NJ)

NJK: El Centro (USA-CA) (location identifier)

NJNC: National Joint Negotiating Committee

NJP: Network Job Processing • New Journal of Physics

NK: Not Known

NKA: Nkan (Gabon) (location identifier)

NKB: Noonkanbah (Australia-WA) (location identifier)

NKC: Nouakchott (Mauritania) (location identifier)

NKG: Nanjing (China-PRC) (location identifier) • Nishimura-Kamata-Greisen (function) • Non-linear Klein-Gordon

NKI: Naukiti (USA-AK) (location identifier)

Nkl: Nickel

NKL: Nkolo (Zaire) (location identifier)

NKN: Nankina (Papua New Guinea) (location identifier)

NKO: Nederlandse Kalibratie Organisatie (Netherlands)

nKOM: Narrow-band Kilometric Radiation

NKOS: Network Knowledge Organization Source • Network Knowledge Organization System

NKr: Norwegian Krone (currency)

NKR: Norwegian Krone (currency)

NKRE: Non-Key Recovery Encryption

NKS: Nkongsamba (Cameroon) (location identifier)

NKU: Nkaus (Lesotho) (location identifier) • Northern Kentucky University (USA-KY)

NKV: Nichen Cove (USA-AK) (location identifier)

NKX: San Diego - Miramar (USA-CA) (location identifier)

NKY: Nkayi (Congo) (location identifier)

NKz: New Kwanza (currency)

n/l: Nuestra Letra

nl: Dutch (ISO language code) • nachleuchtend • Namelijk • Netherlands • New Line • Non Licet • Non Liquet

Nl: Namelijk • National • Nova-Like

N/L: No Limit

NL: National League • National Library • Natural Language • Negative Large • Neon Lamp • Netherlands (international vehicle registration code) • Netherlands (ISO country code) • Netherlands (postal country code) • New Latin • New Line • Newsletter • Niederlausitz (Germany) • Night Letter • No Label • No Liability • No Limit • Non-Linear • North Latitude • Not Listed • Not Located • Nouvelle Lune • Nova-Like • Nuevo León (Mexico) • Number of Layers • Pays-Bas (ISO country code)

NLA: National Land Agency (Japan) • National Libraries Authority (UK) • National Library of Australia • Ndola (Zambia) (location identifier) • New Large Aircraft • Next-Level Aggregator • Next Lower Assembly • Nine-Letter Acronym • No Longer Affiliated • No Longer an Aircraft • Non-Linear Amplifier • Numerical Linear Algebra

NLAB: National Library Advisory Board

NLAC: National Library Advisory Council

NLANR: National Laboratory for Applied Network Research (USA)

NLAPS: National Landsat Archive Production System (USGS)

N. Lat.: North Latitude

NLB: Network Load Balancing • Number of Lines of Binary

NLC: Lemoore Reeves (USA-CA) (location identifier) • Narrow-Line Component • National Laboratory Consortium • National Libraries Committee (UK) • National Library of Canada • New Line Character • Next Linear Collider • Noctilucent Cloud

NLD: NASA Launch Director • Netherlands (ISO country code) • Network Logistics Depot • Nuevo Laredo (Mexico) (location identifier) • Pays-Bas (ISO country code)

NLDA: National Library Development Act

NLDB: Natural Language Database

NLDC: National Legal Data Center

NLDM: Network Logical Data Manager

NLDP: National Launch Development Program

NLDS: National Leadership Development Seminar

NLE: National Library for the Environment • Niles - Jerry Tyler (USA-MI) (location identifier) • Non-Linear Element • Not Less or Equal

NLEA: Nutrition Labelling and Education Act (USA)

NLF: Darnley Island (Australia-QLD) (location identifier) • National Laser Facility • Natural Laminar Flow • Nearest Landing Field

NLFF: Non-Linear Force-Free

NLFM: Noise-Level Frequency Monitor

NLFSR: Non-Linear Feedback Shift Register

NLG: Guilder (Netherlands) (ISO currency code) • Natural-Language Generation • Nelson Lagoon (USA-AK) (location identifier) • Nose Landing Gear • Null Line Gap

NLGD: Nose Landing Gear Door

NLI: National Library of Ireland • Natural Language Inc. • Neodymium Laser Illuminator • Noise Limit Indicator

NLII: National Learning Infrastructure Initiative

NLIN: Non-Linear

NLIS: National Land Information Service (UK)

NLK: Norfolk Island (location identifier)

NLKF: Non-Linear Kalman Filter

NLL: Negative Logic Level • Nullagine (Australia-WA) (location identifier)

NLLR: Non-Linear Lorentz Resonance

NLLST: National Lending Library for Science and Technology

NLLU: National Lending Library Unit (UK)

NLM: National Library of Medicine (USA) • Nautical Mile • NetWare Loadable Module • Non-Linear Mapping • Non-Linear Material

NLMC: National Library of Medicine Classification

NLML: No Lactose Meal

NLN: National Learning Network • National Library Network (USA) • No Longer Needed

NLnr: Non-Linear

NLO: Kinshasa N'Dolo (Zaire) (location identifier) • National Liaison Officer • Next-to-Leading Order • Non-Linear Optical • Norman Lockyer Observatory (UK-England)

NLOS: Natural Language Operating System • Non-Line Of Sight

NLP: Name Looking Protocol • Natural Language Processing • Neglected Language Program • Nelspruit (South Africa) (location identifier) • Neurolinguistic Programming • Non-Linear Potential • Non-Linear Programming • Normal Lead Point

NLPIP: National Lighting Product Information Program

NLQ: Near-Letter Quality

NLQP: Natural Language Query Processor

NLR: Narrow-Line Region • Nationaal Lucht- en Ruimtevaartlaboratorium (Netherlands)

NLRB: National Labor Relations Board • National Level Bibliographic Record

NLRG: Narrow-Line Radio Galaxy

NLRGC: Nationaal Lucht- en Ruimtevaartgeneeskundig Centrum (Netherlands)

NLRO: North Liberty Radio Observatory (USA-NM)

NLRRN: Non-Linear Random Resistor Network

NLRS: National Laser Ranging System • NATMAP Laser Ranging System

NLS: On-Line System • Narrow-Line Seyfert (galaxy) • National Land Survey • National Language Support • National Launch System • National Library Service • Natural Language System • Negative Lens System • Neodymium Laser System • Nestor Learning System • Network License Server • Nicholson (Australia-WA) (location identifier) • Niet-Lineaire Systemen • No-Line-of-Sight • Non-Linear Smoothing • Non-Linear System • Northern Loop System • Online System

NLSE: Non-Linear Schrödinger Equation

NLSFUNC: National Language Support Function

NLSM: Non-Linear Sigma Model

NLSP: NetWare Link Services Protocol • NetWare Link State Protocol • Network Layer Security Protocol

NLST: Name List

NLSy: Narrow-Line Seyfert (galaxy)

NLT: Natural Language Technology • Not Later Than • Not Less Than

NLTE: Non-Local Thermal Equilibrium • Non-Local Thermodynamic Equilibrium (*)

NLTS: Near Launch Tracking System

NLTSG: Natural Language Translation Specialist Group

NLTT: Natural Language Theory and Technology • New Luyten Two-Tenth (catalogue)

NLU: Mexico City - Santa Lucia (Mexico) (location identifier) • National Louis University (USA-IL) • Natural Language Understanding

NLUUG: Netherlands Local Unix systems User Group

NLV: Nikolaev (Ukraine) (location identifier)

NLVD: Natural Language Voice Document

NLW: National Library Week

NLXG: Narrow(-emission)-Line X-ray Galaxy

Nly: Northerly

nm: Nachmittags • Namiddag • Nanometer • Nocte et Mane • Nutmeg

nM: Nächsten Monats • Nanomole

N/m: No Mark

Nm: Nachmittag • Next Matter

NM: Nächsten Monats • Namibia • Nautical Mile • Nedler-Mead • Network Management • Network Manager • Network Module • Neutron Monitor • New Mexico (USA) • New Moon • Newton-Meter • Nieuwe Maan • Night and Morning • Night Message • Nissan Motors (Japan) • No Mark • No Message • Non-Metallic • Not Marked • Not Measured • Noun Modifier • Nuclear Magnetic • Nuclear Magnetron

NMA: Namangan (Uzbekistan) (location identifier) • National Management Association • Network Management Architecture • Nobeyama Millimeter Array (Japan) • Normal Method of Acquisition

NMAB: National Materials Advisory Board (USA)

NMAC: Near Mid-Air Collision

NMag: Non-Magnetic

NMAH: National Museum of American History (USA-DC)

NMAS: National Map Accuracy Standard • Novell Modular Authentication Services

NMAT: Non-Mathematical Astronomical Text

NMB: Daman (India) (location identifier) • National Mapping Bureau • National Mediation Board • National Metric Board • Nucleaire Magnetische Resonantie

NMBMMR: New Mexico Bureau of Mines and Mineral Resources (USA-NM)

NMBPC: Neural Model Based Predictive Control

NMBR: Number

NMBS: Nationale Maatschappij van de Belgische Spoorwegen (Belgium)

NMC: NASA's Master Catalog • Natal Midlands Centre (ASSA) • National Map Collection • National Mathematical Centre • National Meteorological Center • National Metrological Center • Network Management Center • Network Management Console • New Media Center • No More Credit • Non-Modifying Construction • Norman's Cay (Bahamas) (location identifier) • Northwestern Michigan College (USA-MI) • Not Mission Capable • Nozzle Closure Mechanism • NSSDC Master Catalog • Numerical Meteorological Centre

NMCE: Network Monitoring and Control Equipment

NMCS: Network Monitoring and Control System • Not Mission Capable Supply

NMCV: Non-Magnetic Cataclysmic Variable

NMD: NASA Management Delegation • NASA Master Directory • National Mapping Division • Network Measurements Division

NMDG: Network Management Development Group

NMDR: Nuclear Magnetic Double Resonance

NME: Nightmute (USA-AK) (location identifier) • Noise-Measuring Equipment • Non-linear Mesoscopic Elastic • Non-linear Mesoscopic Elasticity

NMEE: Numerische Mathematik Electronic Edition

NMERI: New Mexico Engineering Research Institute (USA-NM)

N. Mex.: New Mexico (USA)

NMF: Nación Más Favorizada • Network Management Forum

NMGECC: National Magnet Fusion Energy Computer Center

NMFS: National Marine Fisheries Service (USA)

NMG: San Miguel (Panama) (location identifier)

NMH: Nickel-Metal-Hydride

NMHC: Non-methane Hydrocarbon

NMHS: National Meteorological and Hydrological Service

nmi: Nautical Mile

NMi: Nederlands Meetinstituut

NMI: NASA Management Instruction • National Metrology Institute • National Multimedia Institute • Netherlands Measurement Institute • No Middle Initial • Non-Maskable Interrupt • Non-Masking Interrupt

NMIMT: New Mexico Institute of Mining and Technology (USA-NM)

NMIS: Nuclear Materials Information System

NMKW: Nichtmethankohlenwasserstoffe

Nml: Normal

NML: National Measurement Laboratory • Network Management Layer

NMLA: New Mexico Library Alliance

Nmlz: Normalize

NMM: NASA Management Manual • NetWare Management Map • Neutron Magnetic Moment

NMN: Niedersächsischer Monographiennachweis • Nomane (Papua New Guinea) (location identifier) • No Middle Name

NMNH: National Museum of Natural History

Nmns: Néanmoins

NMO: National Member Organization • Network Management Option • Normal Manual Operation • Normal Mode Operation

NMOC: National Meteorological Operations Centre

NMOS: Negative-channel Metal-Oxide Semiconductor • Negative-type Metal-Oxide Semiconductor • Negatively-dopes Metal-Oxide Semiconductor

NMP: National Mapping Program • New Millennium Program • New Moon (Australia-QLD) (location identifier) • Nominated Member of Parliament

NMPDN: National Materials Property Data Network

NMPF: Normal Magnitude Probability Function

NMPH: Nautical Miles Per Hour

NMPP: Nouvelles Messageries de la Presse Parisienne (France)

NMpS: Nautical Miles per Second

NMPS: Nautical Miles Per Second

Nmr: Namur (Belgium)

NMR: Nappa Merry (Australia-QLD) (location identifier) • Nautical Mile Radius • No Maintenance Required • No Maintenance Requirement • Normal Mode Rejection • Nuclear Magnetic Relaxation • Nuclear Magnetic Resonance (*)

NMRAS: Nuclear Material Report and Analysis System

NMRF: National MST Radar Facility

NMRI: Nuclear Magnetic Resonance Imaging

NMRON: Nuclear Magnetic Resonance of Oriented Nuclei

NMRP: Nuclear Magnetic Resonance Program

NMRQ: Non-Master Request

NMRS: Numbers

NMRT: Nimbus Meteorological Radiation Tape

NMS: Namsang (Myanmar) (location identifier) • National Master Specification • National Meteorological Service • National Metrology Service • Naval Meteorological Service • Navigational Management System • Navigation Management System • Navire Météorologique Stationnaire • Network Management Station • Network Management System • Network Measurement System • Network Monitoring Station • Neutral Mass Spectrometer • Nippon Meteor Society (Japan) • Numerical Methods Symposium

NMSA: National Meteorological Services Agency

NMSAE: Netherlands Ministry for Social Affairs and Employment

NMSD: Next Most Significant Digit

NMSE: Normalized Mean Squared Error

NMSI: National Museum of Science and Industry (UK)

NMSL: NetWare Mirrored Server Link

NMSS: Non-Minimum State Space

NMSU: New Mexico State University (USA-NM)

NMT: Namtu (Myanmar) (location identifier) • Nordic Mobile Telephone • Nordic Mobile Telephony • Notification of Master Tool • Not More Than

NMTA: National Metal Trades Association (USA)

NMTBA: National machine Tool Builders' Association (USA)

NMTRO: New Mexico Tech Robotic Observatory (USA-NM)

NMTS: Next Manned Transportation System • Noise Measurement Test Set

NMTU: Net Manufacturing Time per Unit

NMU: Namu (Marshall Islands) (location identifier) • Northern Michigan University (USA-MI)

NMV: Nitrogen Manual Valve

NMVT: Network Management Vector Table

NMW: National Minimum Wage

NMWC: New Mexico Western College (USA-NM)

NMZ: Nationales Metrologisches Zentrum

nn: Names • Notes

N/N: Not to be Noted

NN: Far North • Far Northern • NASA Notice • Nearest Neighbour • Network Node • Neural Network • Neutralization Number • Nomen Nescio • No Name • No News • Normalnull • Northampton (UK postal code) • Nucleon-Nucleon

NNA: Kenitra (Morocco) (location identifier) • Neural Network Analysis

NNAC: Northeast Nebraska Astronomy Club (USA-NE)

NNB: Santa Ana Island (Solomon Islands) (location identifier)

NNC: Neural Network Classifier • Neural Network Council • Normes Nationales du Canada

NND: Nangade (Mozambique) (location identifier)

NNDC: National Nuclear Data Center

NNE: Non-Nuclear Energy • Nord-Nord-Est • North-Northeast • North-Northeastern

NNEO: Natural Near-Earth Object

NNERN: North-Northeastern

NNG: Nanning (China-PRC) (location identifier) • National Numbering Group

NNHS: Northamptonshire Natural History Society (UK-England)

NNI: Namutoni (Namibia) (location identifier) • Nederlands Normalisatie-Instituut (Netherlands) • Network-Network Interface • Network Node Interface • Network-to-Network Interface • Neural Network Interpolation • Neural Network Interpolator • Noise and Number Index • Noise Nuisance Index • Nucleon-Nucleon Interaction

NNK: Naknek (USA-AK) (location identifier) • Non-Nuclear Kill

NNL: Neural-Network Language • Nondalton (USA-AK) (location identifier)

NNM: Naryan-Mar (Russia) (location identifier) • Network Node Manager • Non-linear Normal Mode

NNMF: Nearest Neighbour Median Filter

NNMI: Nonlinear Normal Mode Initialization

NNMRTP: Northern New Mexico Rural Telemedicine Project

NNNDO: Neglect of Non-Neighbor Differential Overlap

NNNTB: North North North Temperate Belt

NNO: Noordnoordoost • Nord-Nord-Ouest

NNP: Negative Node Point • Net National Product

NNR: National Nature Reserve • No-Net-Rotation • Spiddal (Ireland) (location identifier)

NNRIC: National Natural Resources Information Centre

NNRMS: National Natural Resources Management System

NNRP: Network News Reader Protocol

NNRT: Not-so-Near Real Time

NNS: Navy Navigation Satellite • Nearest-Neighbor Spacing • Nearest-Neighbors Statistics • Neural Network Simulator • Nucleon-Nucleon Scattering

NNSA: National Nuclear Safety Administration (USA)

NNSC: NSF Network Service Center

NNSF: National Natural Science Foundation

NNSFC: National Natural Science Foundation of China (PRC)

NNSG: NASCOM Network Scheduling Group

NNSS: National Navigational Satellite System • Naval Navigational Satellite System • Navy Navigational Satellite System (*)

NNT: Nan (Thailand) (location identifier) • Not Necessarily True • Nouvelle Numérotation Téléphonique

NNTB: North North Temperate Belt

NNTP: National News Transfer Protocol • Net News Transfer Protocol • Network News Transfer Protocol (*) • Network News Transport Protocol • News Network Transfer Protocol

NNTT: National New Technology Telescope

NNTZ: North North Temperate Zone

NNU: Nanuque (Brazil-MG) (location identifier)

NNV: Nederlandse Natuurkundige Vereniging (Netherlands)

NNW: Noordnoordwest • North-Northwest • North-Northwestern

NNWRN: North-Northwestern

NNWS: Non-Nuclear Weapon State

NNWWD: North-Northwestward

NNX: Network Numbering Exchange • Nunukan (Indonesia) (location identifier)

NNY: Nanyang (China-PRC) (location identifier)

n/o: Notre Ordre • Nuestra Orden

no: Norway • Norwegian (ISO language code) • Número

No: Nobelium • Nobélium • North • Northern • Norway • Norwegian • Nuestro • Number • Numero • Numéro

N/O: No Orders • Normally Open

NO: Neptune Orbiter • New Orleans (USA-LA) • Nitric Oxide • Nitrous Oxide • Nitrogen Monoxide • Non-Official • Non-Original • Noordoost • Nordosten • Nord-Ouest • Normally Open • Norvège (ISO country code) • Norway (ISO country code) • Not Operational • Not Or • Not Out • November • Novembre • Nuclear Orientation • Nuffield Observatory (UK-England)

N-O-A: Not-Or-And

NOA: National Observatory of Athens (Greece) • National Optometric Association • New Obligation Authority • Not-Or-And • Not Otherwise Authorized • Nowra (Australia-NSW) (location identifier)

NOAA: National Oceanic and Atmospheric Administration (USA-DC)

NOADN: National Oceanic and Atmospheric Data Network

NOAEL: No Observed Adverse Effect Level

NOAO: National Oceanographic and Atmospheric Organization • National Optical Astronomy Observatories (USA-AZ)

NOARL: Naval Oceanographic and Atmospheric Research Laboratory

Nob: Nobis • Noble

NOB: Nosara Beach (Costa Rica) (location identifier)

NOC: Connaught (Ireland) (location identifier) • National Organizing Committee • Nederlands Olympisch Comité • Network Operation Control • Network Operations Center • Notation Of Content • Notice Of Change • Notice Of Completion • Notice Of Contents • Not Otherwise Classified • Not Otherwise Coded • Number Of Children • Numerical Optimisation Centre

NOCC: Network Operations Control Center

NOCLIMP: Nordic Climate Modelling Project

No.Co.: Northern Countries

NOCS: Network Operations Center System

Noct: Nocte

NOD: Network Operations Directive • News On Demand • Night Observation Device • Norden (Germany) (location identifier) • Notice Of Discrepancy

NODC: National Oceanographic Data Center (USA-DC)

NODDS: NOAA Oceanographic Data Distribution System

NODEID: Node Identification

NODIS: NASA Online Data and Information Service • NASA Online Directory and Information System • NSSDC Online Data and Information Services (*)

NODO: NASA Orbital Debris Observatory

NODS: NASA Ocean Data System

NoE: Number of Employees

NOE: Nap Of the Earth • Norddeich (Germany) (location identifier) • Not Otherwise Enumerated • Notice Of Exception • Notice Of Execution • Nuclear Overhauser Enhancement • Number Of Employees

NOEL: No-Observed-Effect Level

NOES: National Operational Environmental Satellite

NOESS: National Operational Environmental Satellite System

NOF: National Optical Font • Network Operations and Facilities • Not On File

NOFS: Naval Observatory Flagstaff Station (USA-AZ)

NOG: Nogales (Mexico) (location identifier)

NOGAPS: Navy Operational Global Atmospheric Prediction System

NOH: Chicago (USA-IL) (location identifier)

NOHP: Not Otherwise Herein Provided

NOI: National Optics Institute (Canada) • Neptune Orbital Insertion • Neptune Orbit Insertion • Notice Of Intent • Not Otherwise Identified • Not Otherwise Indexed • Novorossijsk (Russia) (location identifier)

NOIBN: Not Otherwise Indexed By Name

NOIP: National Office of Industrial Property

NOJ: Nojabrxsk (Russia) (location identifier)

NOK: Krone (Norway) (ISO currency code) • Next Of Kin

NOL: Naval Ordnance Laboratory (USN)

nol. con.: Nolo Contendere

nol. pros.: Nolo Prosequi

Nom: Nomenclature • Nominal • Nominate • Nominated • Nomination • Nominative

NOM: Network Operations Manager • Network Output Multiplexer • Nomad River (Papua New Guinea) (location identifier) • Normas Oficiales de México

NOMAD: Navy Oceanographic and Meteorological Device • Neutrino Oscillation Magnetic Detector • Non-Mass Dependent

Nom. Cap.: Nominal Capital

Nomen: Nomenclature

Nomin: Nominatif • Nomination • Nominative

Nominat: Nominativo

nom. nov.: Nomen Novum

NOMS: Network Operations Management System

NOMSS: National Operational Meteorological Satellite System

Non: Nonas

NON: Nonouti (Kiribati) (location identifier)

NONA: Notice Of Non-Availability • NASA Science Internet Online Network Aid

Noncom.: Non-Commissioned

Noncon: Non-Conformist

non cul.: Non Culpabilis

non obs.: Non Obstante

non pros.: Non Prosequitur

non rep.: Non Repetatur

non repetat.: Non Repetatur

non seq.: Non Sequitur

Nonstand: Non-Standard

Nonstd: Non-Standard

Nonsyn: Non-Synchronous

NOO: Naoro (Papua New Guinea) (location identifier) • National Oceanographic Office • Naval Oceanographic Office (USA)

no-op.: No Operation

NOOS: Nuclear Orbit-to-Orbit Shuttle

NOOU: Not One Of Us

NOP: Mactan Island - Nab (Philippines) (location identifier) • National Opinion Poll • Navigation Operating Procedure • Network Operations Procedure • No Operation • Normal Operating Procedure • Not Otherwise Provided • Not Our Publication

NOPA: National Office Products Association • Network Operations Performance Analysis

NOPAC: North Pacific

NOPAR: Non-Orbiter Parts Authorization Request

NOPC: NOAA Ocean Products Center

NOPEX: Northern Hemisphere Processes Experiment

NOPPA: National Ocean Pollution Protection Act (USA)

NoPT: No Procedure Turn

Nor: Norma-Normae (constellation) • Normal • Norman • North • Northern • Norway • Norwegian • Norwich (UK-England)

NOR: Nordfjördur (Iceland) (location identifier) • Norvège (ISO country code) • Norway (ISO country code) • Notice Of Revision • Not Or

NORAD: North American Aerospace Defence Command (now USSC) • North American Air Defense Command

NORADS: Naval Operations Regional Atmospheric Prediction System (model)

NORC: National Opinion Research Center (USA-IL)

NORCSEX: Norwegian Continental Shelf Experiment

Nórd: Nórdico

Nord: Nordic

NORDA: Naval Ocean Research and Development Activity

Nordb: Nordbahn

NORDITA: Nordic Institute for Theoretical Physics (Denmark) • Nordisk Institut for Teoretisk Atomfysik (Denmark)

NORDUNET: Nordic University Network

NOREK: Nordiske Elektrotekniske Komiteer

NORESS: Norwegian Regional Seismic Array System (Norway)

Norf: Norfolk

NORGEN: Network Operations Report Generator

Norm: Normal • Normalize • Normalized • Normalizer • Norman

NORM: Non-Operational Ready Maintenance • Not Operationally Ready Maintenance

NORPAX: North Pacific Experiment

NORS: Not Operationally Ready Supplies

NORSE: Nuclear Optical and Radar Systems Effects

NORSEX: Norwegian Remote Sensing Experiment

NORT: Network of Oriental Robotic Telescopes

Northants: Northamptonshire (UK-England)

Northd: Northumberland (UK-England)

Northld: Northumberland (UK-England)

Northumb: Northumberland (UK-England)

NORTIM: Nordiska Föeningen För Tillämpad och Industriell Mathematik

Norw: Norway • Norwegian

Nos: Numbers • Numéros

NOS: National Occupational Standards (UK) • National Oceanic Service • National Ocean Service • National Ocean Survey • Network Operating System • Non-Oriented Satellite • Nossi-Be (Madagascar) (location identifier) • Not Otherwise Specified • Not Otherwise Stated

NOSB: National Organic Standards Board (USA)

NOSC: Naval Ocean Systems Center (USA-CA)

NOSes: Network Operating Systems

NOSFERATU: New Object-oriented System For Effective Retrieval From Texts with Uncertainty

NOSHDB: National Ocean Service Hydrographic Database

NOSIE: NODC SPAN Information Exchange

NOSIG: No Significant Change

NOSL: Night/Day Optical Survey of Lightning

NOSP: Near-Object Scanning Probe • Network Operating Support Plan • Network Operations Support Plan • Network Operation Support Program

NOSS: National Oceanic Satellite System • National Orbiting Space Station

NOST: NASA/OSSA Office of Standards and Technology

Not: Notación • Notary • Notationes • Note • Notes • Notice • Notices • Notizi

NOT: Nordic Optical Telescope • Nottingham (Starlink internal code) • Novato (USA-CA) (location identifier)

NOTA: None Of The Above

NOTAM: Notice to Airmen

NOTAS: Naval Observatory Time Alternate Station

NOTIPG: Nordic Optical Telescope Image/Photo Gallery

NOTIS: Network Operator Trouble Information System

NOTS: Naval Ordnance Test Station

NOTSA: Nordic Optical Telescope Scientific Association

Nott: Nottingham (UK-England)

Nottm: Nottingham (UK-England)

Notts: Nottinghamshire (UK-England)

Notwg: Notwithstanding

NOU: Nouméa Tontouta (New Caledonia) (location identifier)

Nov: Novel • Novelist • November • Novembre • Novembro • Noviembre • Novice

NOV: Huambo (Angola) (location identifier) • November • Novembre • Novembro

NOVA: Nederlandse Onderzoekschool Voor Astronomie

NOVAC: Northern Virginia Astronomy Club (USA-VA)

NOVAM: Naval Oceanic Vertical Aerosol Model

NOVAS: Naval Observatory Vector Astrometry Software (*) • Naval Observatory Vector Astrometry Subroutines

NOVIB: Nederlandse Organisatie voor Internationale Bijstand

Nouv: Nouveau

NOW: Negotiable Order of Withdrawal • Network Of Workstations • NOAA Ocean Wave

NOWESP: Northwest European Shelf Programme

NOx: Nitrogen Oxide

NOX: Nitrogen-Oxygen

NOY: Not Out Yet

NOYAA: National Outstanding Young Astronomer Award

Noz: Nozzle

NOZ: Novokuznetsk (Russia) (location identifier)

NOZE: National Ozone Expedition

NO2: Nitrogen Dioxide

np: Nisi Prius • Nombre Proprio

Np: Neper • Neptunium

N/P: Not Provided

NP: National Park • National Pipe • Natural Park • Navigation Payload • Near Point • Nébuleuse Planétaire • Nebulosa Planetaria • Negative Polarization • Nepal (ISO country code) • Net Proceeds • Network Planning • Network Program • Neutral Porosity • Newman-Penrose • New Paragraph • New Pattern • New Pence (currency) • Newport (Gwent) (UK postal code) • New Providence (Bahamas) • Nitronium Perchlorate • Nobel Prize • Nombre Proprio • Non-deterministic Polynomial • Non-Participating • Non-Polar • Non-Polynomial • Non-Print • No Pagination • No Parking • No Printer • No Protest • No Publisher • Normal Pitch • Normal Point • Normal Pressure • Norma Portuguesa • North Pole • Notary Public • Not Paginated • Not Provided • Noun Phrase • Nouvelle Physique

NpA: Neutrons per Absorption

NPA: Network Performance Analyzer • Network Principal Astronomer • Network Printer Alliance • Neutrons Per Absorption • Non-Principal Axis • No Power Alarm • Nordic Planetarium Association (Sweden) • Notice of Proposed Amendment • Nulle Part Ailleurs • Numbering Plan Area • Number Plan Area • Pensacola (USA-FL) (location identifier)

NPAC: National Program for Acquisitions and Cataloging • National Provisioning Administration Center • Network Principal Astronomer's Computer • Number Portability Administration Center

NPACI: National Partnership for Advanced Computational Infrastructure • National Partnership for Advanced Computational Initiative (*) • National Partnership for Advanced Computing Initiative

NPAI: N'habite Pas à l'Adresse Indiquée

NPAP: Network Printing Alliance Protocol

NPB: Neutral Particle Beam

NPBG: Nébuleuse Planétaire du Bulbe Galactique

NPBS: Non-Primary Basic Subject

NPC: NASA Procurement Circular • NASA Publication Control • National Patent Council (USA) • National Photocolor Corporation (USA-NY) • National Point of Contact • National Project Coordinator • Needle Punch Card • Network Parameter Control • Non-Printing Character • Normal Plane Change • North Polar Cap

NPCI: National Programme of Community Interest

NPCR: National Point of Contact Representative • No Periodic Calibration Required • No Programmed Calibration Required

NPCSD: Normal Point Calibration Standard Deviation

NPCV: Non-Polar Cataclysmic Variable

NPD: Named Protection Domain • NASA Policy Directive • New Product Development • Nitrogen-Phosphorous Detector • Nominal Percent Defective • No Place or Date • North Polar Distance

NPDA: Network Problem Determination Aid • Network Problem Determination Application

NPDES: National Pollutant Discharge Elimination System • National Pollution Discharge Elimination System

NPDF: N-dimensional Probability Density Function • Normal Probability Distribution Function

NPDN: Nordic Public Data Network

NPDS: Nuclear Particle Detection System

NPDSA: National Public Domain Software Archive

NPDWR: National Primary Drinking Water Regulation

NPE: Napier Hastings (New Zealand) (location identifier) • New Preliminary Evaluation • Non-Potential Energy • Nuclear Photographic Emulsion

NPEA: National Printing Equipment Association (USA)

NPEF: New Product Evaluation Form

NPER: National Professional Engineers Register

NpF: Neutrons per Fission

NPF: National Park Foundation (USA) • Neutrons Per Fission • Newtonian Potential Function • North Polar Feature • Not Provided For • Nuclear Power Facility

NPFF: Normal Probability Frequency Function

NPFR: Northern Polarity Flare Ribbon

NPG: NASA Procedures and Guidelines • Naval Proving Ground • Nipa (Papua New Guinea) (location identifier) • Nuclear Planning Group

NPH: Nephi (USA-UT) (location identifier) • Non-Proton Hydrate • Normal Paraffin Hydrocarbon

NPI: Navy Prototype Interferometer • Network Printer Interface • Neutral Particle Imager • Non-Precision Instrument • Non-static Presentation Information • Nouveau Pays Industrialisé • Nuclear Physics Institute (Russia)

NPIC: National Photographic Interpretation Center • National Photo Interpretation Center (*)

NPIRS: National Pesticide Information Retrieval System

NPIS: National Physics Information System (USA)

NPISH: Non-Profit Institution Serving Households

NPK: Nitrogen-Phosphorus-Potassium

Npl: Nameplate

NPL: National Physical Laboratory • National Priority List • National Project Leader • Nepal (ISO country code) • New Plymouth (New Zealand) (location identifier) • New Product Line • New Programming Language (now PL/1) • Noise Pollution Level • Non-Personal Liability • Non-Procedural Language • Normal Power Level • Nuclear Physics Laboratory

NPLI: National Physical Laboratory of India

Np/M: Neper per Meter

NPM: Network Performance Monitor • Non-Print Media • Northern Proper Motion

NPMAS: Neville Public Museum Astronomical Society (USA-WI)

NPMG: Nordic Planetarium Management Group

NPMLE: Non-Parametric Maximum Likelihood Estimate • Non-Parametric Maximum Likelihood Estimator

N-P-N: Negative-Positive-Negative

NPN: NASA Part Number • Negative-Positive-Negative • New Public network

NP/ND: Not Published - No Date

NPND: Not Published - No Date

npo: Ne per Oris

NPO: Nangapinoh (Indonesia) (location identifier) • National Program Office • National Project Office • Non-Profit Organization • Non-Purchase Offer • Nuclear Propulsion Office

NPOC: National Point Of Contact • Naval Polar Oceanography Center

NPOE: Numerical Predictions of Orbital Elements

NPOESS: National Polar-orbiting Operational Environmental Satellite System • National Polar-orbiting Operational Environment Satellite System (*)

NPOI: Navy Prototype Optical Interferometer

NPOP: NASA Polar Orbiting Platform

NPOST: Non-Perturbative Open-Shell Theory

NPP: Napperby (Australia-NT) (location identifier) • Net Primary Production • Net Primary Productivity • New Policy Proposal • No Passed Proof • No Prior Publication • Nuclear Power Plant • Numéro de Patient Permanent

NPPO: NASA Parts Project Office

NPR: National Partnership for Reinventing (government) (USA) • National Performance Review • National Public Radio (USA) • Noise Power Ratio • Noise Prediction and Reduction • Non-Processor Request • North Polar Region • Nuclear Pulse Rocket • Rupee (Nepal) (ISO currency code)

NPRL: National Physical Research Laboratory

NPRM: Notice of Proposed Rule-Making

Nprn: Neoprene

NPRS: NASA Procurement Regulation Supplement • National Programme for Remote Sensing • Non-Persistent

NPRT: National Performance Review Team

NPRV: Nitrogen Pressure Relief Valve

NPS: NASA Planning Study • National Park Service (USA) • National Periodicals System (USA) • Naval Postgraduate School (USA) • Netware Protocol Stack • Non-Point Source • Non-Propulsive Stage • No Prior Service • North Polar Sequence • North Polar Spot • North Polar Spur • Nuclear Power Source • Nuclear Power Station • Nuclear Power System • Numerical Plotting System

NPSG: Nuclear Physics Support Group

NPSH: Net Positive Suction Head

NPSO: Non-Paired Spatial Orbital

NPSP: Net Positive Static Pressure • Net Positive Suction Pressure

NPSS: NASA Packet Switched System • NASA Packet Switching System

NPT: National Pipe Thread • Navy Pointer and Tracker • New Planetary Telescope • Newport (USA-RI) (location identifier) • Noise Producing Test • Non-Proliferation Treaty • Normal Pressure and Temperature

NPTF: National Pipe Thread - Female

NPTM: National Pipe Thread - Male

NPTN: National Pesticide Telecommunications Network • National Public Telecomputing Network

npu: Ne Plus Ultra

NPU: Natural Processing Unit • Nitrogen Purge Unit • San Pedro Uraba (Colombia) (location identifier)

NPUJ: Northwestern Polytechnical University Journal (periodical code)

NPV: Net Present Value • Nitrogen Pressure Valve • No Par Value

NPX: Numeric Processor Extension

NP0: Negative Power Zero

NQ: Neural Quantum

NQA: Memphis (USA-TN) (location identifier) • National Quality Assurance • Niveau de Qualité Acceptable • No Questions Asked

NQAM: Normalized Quadratic Axial Moment

NQAS: Not Quite A Smile

NQC: NASA Quality Control

NQD: Notice of Quality Discrepancy

NQE: Network Querying Environment

NQF: National Qualifications Framework (UK)

NQI: Kingsville (USA-TX) (location identifier) • Nuclear Quadrupole Interaction

NQL: Niquelândia (Brazil-GO) (location identifier)

NQN: Neuquen (Argentina-NE) (location identifier)

NQOC: Not Quite Our Class

NQR: Nuclear Quadrupole Resonance

NQS: Network Queue Scheduler • Network Queue System • Network Queuing System

NQU: Nuqui (Colombia) (location identifier)

NQX: Key West (USA-FL) (location identifier)

NQY: Newquay (UK-England) (location identifier)

n/r: Nuestra Remesa

Nr: Near • Numero • Numéro • Nummer • Recurrent Nova

N/R: Not Required

N-R: Newton-Raphson

NR: National Record • Natural Resources • Nauru (ISO country code) • Negative Resistance • Newton-Raphson • Nitrat-Reduktase • Noise Rating • Noise Ratio • Non-Rated • Non-Reflective • Non-Registered • No Ranking • No Release • No Report • No Requirement • No Risk • Normalisatieraad (Norway) • Normal Range • Norwich (UK postal code) • Not Rated • Not Reported • Not Required • Nuclear Reactor • Number of Records • Number of Reports • Number of Rows

NRA: Narrandera (Australia-NSW) (location identifier) • NASA Research Announcement • National Recovery Administration • National Recreation Area • National Rifle Association (USA) • National Rivers Authority (UK) • Natural Resources Authority • Network Resolution Area • Never Refuse Anything • Nuclear-Reaction Analysis

NRA&A: Nouvelle Revue d'Aéronautique et d'Astronautique (periodical code)

NRAC: National Radio Astrophysics Centre • National Resources Analysis Cener

NRAL: National Radio Astronomy Laboratory • Nuffield Radio Astronomy Laboratories (UK)

NRAO: National Radio Astronomy Observatory (USA-VA)

NRB: Mayport (USA-FL) (location identifier)

NRC: Crows Landing (USA-CA) (location identifier) • National Regulatory Commission (USA) • National Research Council • National Response Center • Neutron Radiation Capture • Non-Recurring Charge • Non-Recurring Cost • Nuclear Regulatory Commission (USA)

NRCAN: National Resources of Canada

NRCC: National Research Council of Canada

NRCP: National Research Council of the Philippines

NRCS: National Radio Communications System • Natural Resources Conservation Service • Normalized Radar Cross Section

NRCSE: National Resource Centre for Satellites in Education

NRCT: National Research Council of Thailand

NR/D: Non-Relevant Document

NRD: National Range Division • Negative Resistance Diode • Noeud Régional Distribué • Non-Relevant Document • Norderney (Germany) (location identifier) • Nuclear Radiation Detector

NRDC: National Research and Development Corporation • Natural Resources Defense Council

NRDD: National Resources Data Directory • Natural Resources Data Directory

NRDF: Nuclear Reaction Data File

NRDS: Nuclear Rocket Development Station

NRE: Namrole (Indonesia) (location identifier) • Negative Resistance Effect • Non-Recurring Engineering • Non-Relativistic Electron • Non-Rotating Earth • Nuclear Radiation Effect • Nuclear Rocket Engine

NREC: National Robotics Engineering Continuum

NRED: Natural Resources and Energy Division

n/ref: Nuestra Referencia

NREL: National Renewable Energy Laboratory (USA)

NREM: Non-Rapid Eye Motion • Non-Rapid Eye Movement

NREN: National Research and Education Network

NRER: Non-Rejected Earth Radiance

NRF: Nouvelle Revue Française

NRFA: Navigation Radio Frequency Assembly • Netherlands Foundation for Radio Astronomy

NRFD: Not Ready For Data

NRFI: Not Ready For Issue

NRG: Énergie • Energy • Narrogin (Australia-WA) (location identifier)

NRH: Nançay Radioheliograph (France) • Nobeyama Radio Heliograph (Japan)

NRI: Natural Resources Institute • Non-Recurrent Investment • Shangri-La (USA-OK) (location identifier)

NRIA: National Research Investment Act (USA)

NRIAG: National Research Institute for Astrophysics and Geophysics (Egypt)

NRIC: National Resource Information Centre • Network Reliability and Interoperability Council (USA-DC)

NRIHVE: National Research Institute for Higher and Vocational Education

NRIM: National Research Institute for Metals • National Research Institute for Materials (Japan) (*) • National Research Institute of Metals (Japan)

NRIS: Natural Resource Information System

NRK: Norrköping (Sweden) (location identifier)

NRL: National Reference Library • Natural Rubber Latex • Naval Research Laboratory (USA-DC) • Normal Response Level • North Ronaldsay (UK) (location identifier)

NRLM: National Research Laboratory of Metrology

NRM: Nara (Mali) (location identifier) • National Resources Management • Natural Remanent Magnetism • Natural Remanent Magnetization • Non-Recurring Maintenance • Non-Redundant Masking • Normal Response Mode

NRMC: National Resources Management Centre

NRMDP: Natural Resource Management and Development Project

Nrml: Normal

NRMP: Natural Remanent Polarization

NRMS: National Resource Management System

NRN: Neutron-Rich Nucleus • No Reply Necessary • Northern • Novell Remote Network

NRO: National Reconnaissance Office (USA) • National Reconnaissance Organization • Negative Resistance Oscillator • Nichtregierungsorganisation • Nobeyama Radio Observatory (Japan)

NROMM: Netherlands Register Of Microform masters

NROp: Nouvelle Revue d'Optique (periodical code)

NROSS: Navy Remote Ocean Sensing System

NRP: National Research Program • National Route Program • NEA Review Panel (IAU) • Non-Radial Pulsation • Non-Registered Publication • Normal Rated Power • Notice of Research Project • Nuclear Reprocessing Plant

NRPB: National Radiological Protection Board (UK) • National Resources Planning Board

NRPGC: National Research Program for Global Change

NRPIF: Nordic Regional Planetary Image Facility (Finland)

NRQ: N'Riquinha (Angola) (location identifier)

NRQCB: Non-Relativistic Quantum Chromodynamics

NRQZ: National Radio Quiet Zone

NRR: Net Reproduction Rate • Noise Rejection Ratio • Nuclear Rocket Reactor • Roosevelt Roads (Puerto Rico) (location identifier)

NRRv: Naval Research Reviews (periodical code)

NRs: Nepalese Rupee (currency)

NRS: Imperial Beach (USA-CA) (location identifier) • Name Registration Scheme • Name Repository Server • National Reference System • Nitrogen Recharge Station • Non-conformance Reporting System • Nuclear Radiation Shield • Nuclear Rocket Shuttle

NRSA: National Remote Sensing Agency (India) • National Remote Service Award (USA)

NRSC: Nansen Remote Sensing Centre (Norway) • National Remote Sensing Centre (UK)

NRSF: National Remote Sensing Facility (Zimbabwe)

NRSP: National Remote Sensing Programme

NRSQA: National Registration Scheme for Quality Assessors

NRT: Near Real Time • Net Registered Tonnage • Network Readiness Test • News Retrieval Tool • Non-Real Time • Tokyo Narita (Japan) (location identifier)

NRTA: Near Real-Time Analysis

NRtn: Non-Return

NRTS: National Reliable Transfer Server • Near Real-Time System • Network Resources and Training Site • NITF Reliable Transfer Server • Non Repairable at This Station

NRTU: Non-Real-Time Unit

nrtVFR: Non-Real-Time Variable Frame Rate

NRU: Nauru (ISO country code)

NRUS: Nonlinear Resonant Ultrasound Spectroscopy

NRV: Guam - Uscg Shore (Guam) (location identifier) • Natural Resource Valuation • Net Realizable Value • Non-Return Valve

NRvsbl: Non-Reversible

Nrw: Norwegian

NRW: Narrow • Nebular Rossby Wave

NRWS: Narrows

NRY: Newry (Australia-NT) (location identifier)

NRZ: Non-Return to Zero

NRZI: Non-Return to Zero IBM • Non-Return to Zero Indicated • Non-Return to Zero Inverse (*) • Non-Return to Zero Inverted

NRZL: Non-Return to Zero Level

NRZM: Non-Return to Zero Mark

NRZS: Non-Return to Zero Space

n/s: Non-Smoker • Not Sufficient

ns: Nanosecond

Ns: Nimbostratus

N-S: Navier-Stokes

N/S: Nord-Sud • North-South

NS: Nachschrift • Nanosecond • Nassi-Schneidermann • National Service • National Society • Natural Science • Natural State • Navier-Stokes • Near Side • Near Space • Nederlandse Spoorwegen (Netherlands) • Negative Small • NetScape • Network Service • Network Supervisor • Neutron Spectrometer • Neutron Star • New Series • New South Wales (Australia) • New Style • New System • Next System • Nickel Steel • Nimbostratus • Nitrogen Sulfide Radical • Nitrogen Supply • Nitrogen System • Non-Scheduled • Non-Serviceable • Non-Smoker • Non-Specified • Non-Standard • Non-Stop • Non-Sufficient • Nord-Sud • Norse • Norsk Standard (Norway) • North-South • No Smoking • No Sort • Nossa • No Swap • Not Satisfactory • Not Significant • Not Specified • Not Sufficient • Not Suppressed • Not Switchable • Nova Scotia (Canada) • Nuclear Science • Nuclear Ship • Nuclear Shuttle • Nuclear System

NSA: National Security Agency (USA) • National Seismological Archive (UK) • National Shipping Authority • National Sound Archive • National Standards Association • National Student Association • Network Supported Account • Noise Suppressor Assembly • Non-Standard Analysis • Non-Sterling Area • Noosa (Australia-QLD) (location identifier) • Northwest Suburban Astronomers

(USA-IL) • Nosewheel Steering Amplifier • No Suitable Applicant • Not Seasonally Adjusted

NSAC: NRSA Scientific Advisory Council • Nuclear Safety Information Center (USA) • Nuclear Science Advisory Committee

NSAI: National Standards Authority of Ireland

NSAP: National Service Assurance Plan • Network Service Access Point

NSAPI: Netscape Server Application Programming Interface

NSAS: Northern Skies Astronomical Society (USA-ND) • Northern Sydney Astronomical Society (Australia-NSW)

NSAU: National Space Agency of Ukraine

NSB: Bimini North (Bahamas) (location identifier) • National Science Board • National Standards Board • Naval Studies Board • Neutron Star Binary • Night Sky Brightness

NSBD: Narrow Spectral Band Detection

NSBE: National Society of Black Engineers

NSBF: National Scientific Balloon Facility (USA-NM)

NSBRI: National Space Biomedical Research Institute (USA)

NSC: National Safety Council • National Science Council • National Security Council • National Seismic Center • National Simulation Capability • National Standards of Canada • National Steering Committee • Navigation and Sensor Computer • Navigational Star Catalogue • Navigation Star Catalogue • Network Service Center • Network Support Committee • Network Switching Center • Network Systems Corporation • Nordic Satellite Co. • Norwegian Space Centre (Norway) • No Significant Cloud • Nuclear Science Centre (India) • Numerical Sequence Code

NSCAT: NASA Scatterometer

NSCCA: Nuclear Safety Cross-Check Analysis

NSCD: National Spatial Cadastral Database

NSCORT: NASA Specialized Center Of Research and Training

NSCP: National Scalable Cluster Project

NSCS: Network Service Center System

NSC1: Network Systems Cypher One

NSD: Network Statistics Database • Network Status Display • Night-Side Darker

NSDC: National Serials Data Centre (UK) • Natural Satellites Data Center

NSDD: National Security Decision Directive • Nuclear Structure and Decay Data

NSDF: NSSDC Supplementary Data File

NSDI: National Spatial Data Infrastructure • National Spatial Data Initiative

NSDS: Neutron Spectrometer Digital System

NSDSSO: NASA Science Data Systems Standards Office

NSE: Milton Whiting (USA-FL) (location identifier) • National Software Exchange • National Space Enterprise • National Stock Exchange • Natural Space Environment • Navier-Stokes Equation • Navigation Support Equipment • Network Software Environment • Noordwijk Space Expo (Netherlands) • Muclear Science Engineering (periodical code) • Nuclear Statistical Equilibrium

nsec: Nanosecond

NSEC: Network Switching Engineering Center

NSERC: Natural Sciences and Engineering Research Council (Canada)

NSESCC: NASA Space and Earth Sciences Computing Center

N/S/F: Not Sufficient Funds

NSF: Camp Springs - Andrews (USA-MD) (location identifier) • Nationale Simulatie Faciliteit (Netherlands) • National Science Foundation (USA-DC) • National Simulation Facility • Non-Spin-Flip • Norges Standardiseringsforbund (Norway) • Norwegian Standards Association (Norway) • Not Sufficient Funds • Nuclear Structure Facility

NSFC: National Science Foundation of China

NSFL: National Science Film Library (Canada)

NSFLEX: Navier-Stokes solver using characteristic Flux Extrapolation

NSFNET: NSF Network

NSFS: Newcastle Space Frontier Society (Australia)

NSG: Naturschutzgebiet • Nippon Silica Glass • Non-Statutory Guidelines • Not So Good • Nuclear Supplier Group

NSGT: Non-Self-Governing Territory

NSH: Now Shahr (Iran) (location identifier)

NSHC: North Sea Hydrographic Commission

NSHS: Nosewheel Steering Hydraulic System

NSI: NASA Science Internet • NASA Standard Initiator • National Security Information • National Soil Institution • National Space Institute • Network and Systems Integration • Network Support Item • Nitrogen Solubility Index • Noise Source Instrumentation • Non-Standard Item • Nouveaux Systèmes Informatiques • Yaounde Nsimalen (Cameroon) (location identifier)

NSIC: National Storage Industry Consortium • Nuclear Safety Information Center (ORNL)

NSIDC: National Snow and Ice Data Center (USA)

NSIF: Near-Space Instrumentation Facility

NSINIC: NASA Science Internet Network Information Center

NSIP: NASA Student Involvment Program

NSIPO: NASA Science Internet Project Office

NSIUWG: NASA Science Internet Users Working Group

NSJI: North-South Joint Implementation

NSK: Non-Stop Kernel • Noril'sk (Russia) (location identifier)

NSKS: Navier-Stokes-Kuramoto-Sivashinsky

NSL: National Science Laboratories • National Science Library • National Storage Laboratory • Non-Standard Logics

NSLA: National Science Library of Australia

NSLI: National Service Life Insurance

NSLR: Nuclear Spin-Lattice Relaxation

NSLRSDA: National Satellite Land Remote Sensing Data Archives

NSLS: National Synchrotron Light Source (BNL)

NSM: Network Status Monitor • Network Systems Management • Neutron Star Magnetosphere • Neutron Star Merger • Niet-Stationaire Meetmethode • Norseman (Australia-WA) (location identifier)

NSMGC: Neutron Star Merger Grand Challenge

NSN: NASA Science Network • National Satellite Network • National Stock Number • Nelson (New Zealand) (location identifier) • No Stock Number

NSO: NASA Support Operation • National Solar Observatory (USA-AZ) • Network Support Office • Scone (Australia-NSW) (location identifier)

NSOM: Near-field Scanning Optical Microscope

NSOR: Nederlandse Stichting voor Operations Research (Netherlands)

NSP: NASA Support Plan • National Service Provider • National Space Program • Native Signal Processing • Nebraska Star Party (USA-NE) • Network Service Provider

• Network Services Processor • Network Services Protocol • Network Signal Processing • Network Signal Processor • Network Support Plan • Non-volatile Serially Programmable • Normalized Stokes Parameter • Stanley Point (Philippines) (location identifier)

NSPAC: Nova Scotia Planetarium Advisory Committee (Canada-NS)

NSPAR: Non-Standard Parts Approval Request

NSpC: National Space Committee • National Space Council

NSPD: National Space Policy Directive

NSPE: National Society of Professional Engineers (USA)

NSPF: Not Specifically Provided For

NSPMP: Network Switching Performance Evaluation Plan

NSPO: National Space Program Office

NSPR: Netscape Portable Runtine

NSPS: National Society of Professional Surveyors (USA) • New Source Performance Standard • Nutrient Supply Predicting Submodel

NSPSR: New Source Performance Standard Review

NSR: National Scientific Register (USA) • Network Service Request • Network Surveillance and Reconfiguration • Neutron Source Reactor • Night Sky Radiation • NISN Service Request • Nominal Slow Rate • Non-Shared Resource • Normal Slow Rate • Nosso Senhor • Nuclear Spin Relaxation • Nutrient Supply Response

NSRA: Nossa Senhora

NSRC: National Science Research Council • National Science Resources Center (USA) • Network Startup Resource Center

NSRDC: National Standard Reference Data Center (USA)

NSRDP: National Standard Reference Data Program (USA)

NSRDS: National Standard Reference Data Series • National Standard Reference Data System (USA)

NSRF: Nutrient Supply Response Function

NSRM: New Solid Rocket Motor

NSRS: National Spatial Reference System

NSS: National Seismic Station • National Space Society (USA-DC) • National Stream Survey • Navigational Safety System • NERC Scientific Services • Network Simulation System • Neutron Spectrometer System • Nitrogen Supply System • Nodal Switching System • Non-Sea Salt • Non-Sequential Surface • Nordic Statistical Secretariat • Normal Saline Solution • North Star Systems (USA-NY) • NSSDC Storage System

NSSA: National Space Society of Australia

NSSB: National Skills Standard Board (USA)

NSSC: NASA Safety Standards Committee • NASA Standard Spacecraft Computer • National Space Science Centre (UK) • Natural Science Societies of Canada

NSSDC: NASA Space Science Data Center • National Space Science Data Center (USA-MD) (*)

NSSFC: National Severe Storms Forecast Center (USA)

NSSK: North-South Station Keeping

NSSL: National Severe Storms Laboratory (USA)

NSSO: National Solar Space Observatory

NSSS: National Space Surveillance System

NSSTS: Near-Surface Salinity and Temperature System

NST: Nakhon Si Thammarat (Thailand) (location identifier) • Network Support Team • Newfoundland Standard Time • Not Sooner Than

NSTA: National Safe Transit Association • National Science Teachers Association (USA)

NSTAC: National security Telecommunications Advisory Committee

NSTAG: National Science and Technology Analysis Group

NSTAR: NASA SEP Technology Application Readiness

NSTB: National Science and Technology Board (Singapore)

NSTC: National Science and Technology Centre • National Science and Technology Council • Nova Scotia Technical College (Canada-NS)

Nstd: Non-Standard

NSTDA: National Science and Technological Development Agency

NSTDC: National Scientific and Technical Documentation Center

NSTIC: National Scientific and Technical Information Center

NSTL: National Software Testing Laboratory • National Space Technology Laboratory

NSTR: National Solar-Terrestrial Research • Nordic Sea Test Range • Nothing Significant To Report

NSTS: National Space Transportation System

NSTSPO: National Space Transportation System Project Office

NSTSS: National Space Transportation and Support Study • National Space Transportation System

NSTX: National Spherical Tokamak Experiment

NSU: Network Synchronisation Unit • Nitrogen Supply Unit • Node Switching Unit • Norfolk State University

NSV: Negative Supply Voltage • New Suspected Variable • Non-Spinning Vehicle • Noosaville (Australia-QLD) (location identifier)

NSW: Nebenschlagwort • New South Wales (Australia) • No Significant Weather • NSP Status Word

NSWC: Naval Surface Weapons Center (USA)

NSWR: Nuclear Salt Water Rocket

NSWS: National Surface Water Survey

NSWTI: Nationales System für die Wissenschaftliche und Technische Information

NSX: North Sound (British Virgin Islands) (location identifier)

NSY: Sigonella (Italy) (location identifier)

nT: Nanotesla

Nt: Net • Newton • Nit • Niton

NT: Narrower Term • National Treatment • Neighbour Table • Net Terms • Net Tonnage • Network Termination • Network Terminator • Neutral Zone between Iraq and Saudi Arabia (ISO country code) • Newfoundland Time • New Technology • New Testament • Northern Territory (Australia) • Northwest Territories (Canada) • No Test • Not Tested • Not Titled • Nouvelle Technologie

Nta: Nuestra

NTA: Natadola (Fiji) (location identifier) • National Technology Agency • National Telecommunications Administration • News Transfer Agent • Nippon Total Access Communication System • Nitrogen Tank Assembly

NTAC: National Time Allocation Committee (Gemini)

NTAD: National Transportation Atlas Data

NTAM: Notice To Airmen

NTAP: Notice To Airmen Publication

NTAS: New Technology Advanced Server

NTB: National Technical Board • National Testbed • NITF Technical Board • Non Tariff Barrier • Non-Thermal Bremsstrahlung • North Temperate Belt • Notodden (Norway) (location identifier)

NTBR: Not To Be Resuscitated

Ntc: Notice

NTC: National Technological Capability • National Telecommunications Conference • Negative Temperature Coefficient • Negative Torque Control • Network and Training

Center • Noshiro Testing Center (Japan) • Santa Carolina (Mozambique) (location identifier)

NTCDA: Naphthalenetetracarboxylic Dianhydride

NTCo: Non-destructive Testing Communications (periodical code)

NTCS: National Television Committee System

NTD: NASA Test Director • National Topographic Database • Neutron Transmutation-Doped (*) • Neutron Transmuted-Doped • Nova Technology Development Pte. Ltd. (Singapore) • Port Hueneme - Point Mugu (USA-CA) (location identifier)

NTDB: National Topographic Database • National Trade Databank

Nte: Norte

NTE: Nantes Atlantique (France) (location identifier) • NASA Technology Experiment • Navigation Telemetry Encoder • Neutron Transient Effect • Not To Exceed • Nouvelles Technologies d'Enseignement

NTEC: Network Technical Equipment Center

NTER: National Engineering Technologists Register • Normalized Transmission Energy Requirement

NTEU: National Treasury Employees Union (Australia)

NTESS: NASA Technical Experience for Select Students

NTeZ: North Temperate Zone

NTF: National Test Facility • National Transfer Format • Non-Thermal Filament • No Trouble Found • Nuclear Test Facility

NTFS: New Technology File System

Ntfy: Notify

NTG: Nantong (China-PRC) (location identifier) • Narrower Term - Generic • NGST Task Group • Nitroglycerin • Nordic Telescope Group

NTGDS: Non-Tritium Gas Delivery System

Nth: North

NTH: Norges Tekniske Hogskole (Norway)

Nthn: Northern

NTHU: National Tsing-Hua University (China-ROC)

NTI: Bintuni (Indonesia) (location identifier) • Nanyang Technology Institute (Singapore) • National Technology Initiative • Noise Transmission Impairment • Nouvelles Technologies de l'Information • Nouvelle Technologie de l'Information • New Technology Inc.

NTIA: National Telecommunications and Information Administration (*) • National Telecommunications and Information Agency

NTIC: Non-destructive Testing Information Center (USA) • Nouvelles Technologies de l'Information et de la Communication

NTIM: Not That It Matters

NTIMM: Not That It Matters Much

NTIPS: National Total Image Processing System

NTIS: National Technical Information Service (USA-VA)

NTISSC: National Telecommunications and Information System Security

NTJ: Manti Ephraim (USA-UT) (location identifier)

NTK: Need To Know • Net Tonne Kilometer

NTL: National • Natural Thermoluminescence • Newcastle Williamtown (Australia-NSW) (location identifier) • Non-Threshold Logic • No Time Lost • Nuclear Thermionics Laboratory

NTM: Miracema do Norte (Brazil-TO) (location identifier) • Nombre de Transparents par Minute • Non-Tariff Measure • Notice To Mariners • Number of Transparencies per Minute

NTMS: National Traffic Management System

NTN: National Trends Network • Neutralized Twisted Nematic • Neutral Tree Network • New Technology News (periodical code) • Normanton (Australia-QLD) (location identifier)

NTNU: National Taiwan Normal University (China-ROC)

Nto: Nuestro

NTO: Network Terminal Option • Nitrogen Tetroxide • Not Taken Out • Nuevas Tecnologías Observacionales (Spain) • Santo Antao (Cape Verde) (location identifier)

NTOT: New Technology Orbiting Telescope

NTP: Narrower Term - Partitive • National Technology Programme • National Transportation Policy • Network Test Panel • Network Time Protocol • New Technology Prototype • Normal Temperature and Pressure • Notice To Proceed • No Title Page • Nuclear Test Plant • Nuclear Thermal Propulsion • Numerical Tape Punch

NTpM: Number of Transparencies per Minute

NTPM: Number of Transparencies Per Minute

NTPS: National Transportation Planning System

NTR: Monterrey Norte (Mexico) (location identifier) • National Tape Repository • Nernst-Thomson Rule • Neutron Test Reactor • New Technology Report • Nightly Total Rate • Nothing To Report • Nuclear Test Reactor • Nuclear Thermal Rocket • Nuclear Thermal Rocketry

Ntra: Nuestra

NTRAS: New Technology Remote Access Services

NTrB: North Tropical Belt

NTRI: NCP/Token Ring Interconnection

Ntro: Nuestro

NTRS: NASA Technical Report Server

NTrZ: North Tropical Zone

NTS: NASA Test Support • National Topographic System (USA) • Navigation Technology Satellite • Near-Term Schedule • Negative Torque Sensing • Negative Torque Signal • Negative Torque System • Network Task Scheduler • Network Technical Support • Network Test System • Network Time System • Nevada Test Site (USA-NV) • New Typesetting System • Night-Targeting System • Non-avionics Test Station • Non-Technical Skill • Not To Scale

NTSA: NetWare Telephony Services Architecture

NTSB: National Transportation Safety Board (USA)

NTSC: National Television Standards Committee • National Television System Committee • Never The Same Color

NTSI: Non-linear Thin Shell Instability

NTSO: NASA Test Support Office

NTSS: Nitrogen Transfer and Storage System

NTT: Naked T Tauri • New Technology Telescope • Nippon Telegraph and Telephone (Japan) • Niuatoputapu (Tonga) (location identifier) • Numbered Test Trunk

NTTC: National Technology Transfer Center (USA-WV)

NTTF: Networking and Telecommunications Task Force • Network Test and Training Facility

NTTS: Naked T Tauri Star

NTTUP: New Technology Telescope Upgrade Plan

NTU: National Technological University • Network Terminating Unit • Network Termination Unit • Normalized Turbidity Unit • Northern Territory University (Australia-NT) • Oceana (USA-VA) (location identifier)

NTUA: National Technical University of Athens (Greece)

NTV: Nader Te Verdelen • NASA Television • Nippon Television • Number To View

Ntwk: Network

nt. wt.: Net Weight

NTX: Natuna Ranai (Indonesia) (location identifier)

NTY: Sun City (South Africa) (location identifier)

NTZ: North Temperate Zone • North Tropical Zone • No Transgression Zone

Nu: Ngultrum (currency) • Nusselt Number

NTZ: Neutral Zone between Iraq and Saudi Arabia (ISO country code)

NU: Naciones Unidas • Name Unknown • National Union • Nations Unies • Natural Uranium • Nebraska University (USA-NE) • Niue (ISO country code) • Northern Union • Northwestern University (USA-IL) • Number Uniformity • Number Unobtainable

NUA: Network User Address • News User Agent

NUAAJ: Nanjing University Aeronautics and Astronautics Journal (periodical code)

NUAC: Nuage Acide

NUB: Nederlandse Uitgeversbond (Netherlands) • Numbulwar (Australia-NT) (location identifier)

NUBIA: Nubus Interface Adaptor

Nuc: Nuclear

NUC: National Union Catalogue

NUCATE: Núcleo de Ciência, Aplicações e Tecnologia Espaciais (Brazil)

NucFS: Nuclear Fusion Supplement (periodical code)

NucF: Nuclear Fusion (periodical code)

NucIM: Muclear Instruments and Methods (periodical code)

Nucl: Nuclear • Nucléaire • Núcleo • Nucleus

NucPh: Nuclear Physics (periodical code)

NucTe: Nuclear Technology (periodical code)

NucTF: Nuclear Technology Fusion (periodical code)

NUD: En Nahud (Sudan) (location identifier)

NUDET: Nuclear Detonation

NUDETS: Nuclear Detection System • Nuclear Detonation System

NUDIAS: National University Digital Image Analysis System

NUDTJ: National University of Defense Technology Journal (periodical code)

NUE: Nuremberg (Germany) (location identifier) • Nürnberg (Germany) (location identifier)

NuEnD: Nuclear Engineering and Design (periodical code)

NUFEM: Nagoya University Faculty of Engineering Memoirs (periodical code)

NUFFIC: Netherlands Universities Foundation For International Cooperation

NUG: Nuguria (Papua New Guinea) (location identifier)

NUGRO: Nucleation and Growth

NUH: Nunchia (Colombia) (location identifier)

NUI: National University of Ireland • Network User Identification • Network User Identifier • Network User Interface • Notebook User Interface • Nuiqsut (USA-AK) (location identifier) • Numéro Unique d'Identification

NUK: Nukutavake (Tuamotu Islands) (location identifier)

Nukl: Nuklear

NUL: Non-Utilization List • Nulato (USA-AK) (location identifier) • Null

NULO: NASA Unmanned Launch Operations

Núm: Número

Num: Number • Numeral • Numeric • Numerical • Numérique

NUM: Northern University of Malaysia • Número

NUMA: Non-Uniform Memory Access (*) • Non-Uniform Memory Architecture

NuMat: Numerische Mathematik (periodical code)

Numb: Number • Numbered

NumHT: Numerical Heat Transfer (periodical code)

NuMI: Neutrinos at the Main Injector

NUMI: Number of Images

Núms: Números

NUN: Pensacola Saufley (USA-FL) (location identifier)

NUP: Nunapitchuk (USA-AK) (location identifier)

NUPACU: Nuclear Powered Air Conditioning Unit

NuPECC: Nuclear Physics European Collaboration Committee

NUQ: Mountain View - Moffett (USA-CA) (location identifier)

NUR: Nullarbor (Australia-SA) (location identifier)

NURBS: Non-Uniform Rational B-Spline

NURE: National Uranium Resource Evaluation (USA)

NUREC: Network on Urban Research in th European Community

NURO: National Undergraduate Research Observatory (USA)

NURWEC: Netherlands UK Radar Wavebuoy Experimental Comparison

NUS: National Union of Students (UK) • National University of Singapore • New Upper Stage • Norsup (Vanuatu) (location identifier)

NUSAC: Nuclear Science Advisory Committee

NUSC: Naval Underwater Systems Center

NUSEM: Nagoya University School of Engineering Memoirs (periodical code)

NUSP: National User Support Programme

NUSS: Nuclear Safety Standards

NUSSAG: Nuclear Safety Standards Advisory Group

N-u-T: Newcastle-upon-Tyne (UK-England)

NuT: Newcastle-upon-Tyne (UK-England)

NUT: National Union of Teachers • Nutuve (Papua New Guinea) (location identifier)

NUTEC: Norwegian Underwater Technology Centre (Norway)

NuTeV: Neutrinos at the Tevatron

NUTI: NASCOM User Traffic Interface

NUTIS: NERC Unit for Thematic Information Systems • Numerical and Textual Information System

Nutr: Nutrition

NUTS: Nomenclature des Unités Territoriales Statistiques

NutzEV: Nutzungsentgeltverordnung

NUU: Nakuru (Kenya) (location identifier) • New University of Ulster

NuUB: National- und Universitätsbibliothek

NUV: Near Ultraviolet

NUVI: Near Ultraviolet Imager

NUVIEWS: Narrow-band Ultraviolet Imaging Experiment for Wide-field Surveys

NUVO: Near Ultraviolet/Optical

NUW: Whidbey Island - Ault (USA-WA) (location identifier)

NUX: Novyj Urengoj (Russia) (location identifier)

N/V: No Value

NV: Despina (satellite of Neptune) • Naamloze Vennootschap • Naked Vision • Nationale Vereinigung • Navajo • Near Vertical • Nevada (USA) • New Version • Node Visit • No Overflow • Nuclear Vehicle

NVA: Negative Vorticity Advection • Neiva (Colombia) (location identifier) • Norske Vitenskapakademi (Norway)

NVAFB: North Vandenberg Air Force Base

NVB: Navigational Base

NVC: Non-Verbal Communication • Non-Verbal Conversation

NVCN: National Vertical Control Network (USA)

NVD: Nevada (USA-MO) (location identifier)

NVE: Network Visable Entry • Nonvisual Eyepiece • Non-Volatile Electronics

NVEOL: Night Vision Electro-Optics Laboratory

NVFEL: National Vehicle and Fuel Emissions Laboratory

NVFRAM: Non-Volatile Ferroelectric Random Access Memory

NVG: Night Vision Goggles • Nueva Guinea (Nicaragua) (location identifier)

NVI: Galatea (satellite of Neptune) • Non-Value Indicator • Normalized Vegetation Index

NVII: Larissa (satellite of Neptune)

NVIII: Proteus (satellite of Neptune)

NVIS: Night Vision Imaging Spectrometer

NVK: Narvík (Norway) (location identifier)

NVLAP: National Voluntary Laboratory Accreditation Program

NVLP: Nederlandse Vereniging van Lucht- en Ruimtevaart Publicisten (Netherlands)

NVLSPM: Night Vision Laboratories Static Performance Model

NVM: Non-Volatile Matter • Non-Volatile Memory

NVO: National Virtual Observatory (USA)

NVOD: Near Video On Demand

NVP: Network Voice Protocol • Nominal Velocity of Propagation • Novo Aripuanã (Brazil-AM) (location identifier)

NVPOWG: NASA/VAFB Payload Operations Working Group

NVQ: National Vocational Qualification

NVR: Nederlandse Vereniging voor Ruimtevaart (Netherlands) • Non-Volatile Residue • No Verification Required • Novgorod (Russia) (location identifier) • No Voltage Release

NVRAM: Non-Volatile Random Access Memory

NVRP: Northern Virginia Recreational Parks (USA-VA)

NVS: Nevers (France) (location identifier) • Non-Verbal Signal • Non-Volatile Storage

nvSRAM: Non-Volatile Static Random Access Memory

NVSS: NRAO VLA Sky Survey

NVT: Navegantes (Brazil-SC) (location identifier) • Netware Virtual Terminal • Network Virtual Terminal • Novell Virtual Terminal

NVTM: Northern Virginia Telescope Meet (USA-VA)

NVTS: Network Virtual Terminal Service

NVWS: Nederlandse Vereniging voor Weer- en Sterrenkunde (Netherlands)

NVY: Neyveli (India) (location identifier)

nw: nordwestlich

nW: Nanowatt

N/W: Network

NW: London (Northwest_) (UK postal code) • Nachwort • NASA Waiver • Naturwissenschaften (periodical code) • Net Weight • Network • Net Worth • Niedrigwasser • Nominal Width • Noordwest • Nordrhein-Westfalen (Germany) • Nordwest • North Wales • Northwest • Northwestern • Northwest London (UK postal code) • Nose Wheel • No Wind

NWA: Moheli (Comoros) (location identifier) • National Water Alliance • National Weather Association • Neighbourhood Watch Area

NWACC: Northwest Academic Computing Consortium (USA)

NWAG: North West Astronomy Group (USA-OR)

NWbW: Northwest by West

NWC: NetWare Connect

NWCS: NetWare Workstation Compatible Services

NWDA: Northwest Deformation Array

NWDB: National Wasteland Development Board

nwe: Nieuwe

NWE: Nuclear Weapon Effect

NWF: Numerical Weather Facility

Nwfld: Newfoundland (Canada)

NWFS: Netware File System

NWG: Natural Work Group • Network Working Group

NWGAS: North West Group of Astronomical Societies (UK)

NWH: Newport Parlin (USA-NH) (location identifier) • Normal Working Hours

NWI: Norwich (UK-England) (location identifier)

NWL: Natural Wavelength

NWLY: Northwesterly

NWMP: National Water Management Programme • National Weapons Monitor Plan

NWN: Non-White Noise

NWO: NASA Washington Office • Nederlandse Organisatie voor Wetenschappelijk Onderzoek (Netherlands)

NWP: Argentia (Canada-NF) (location identifier) • National Weather Prediction • North-Western Province (India) • Northwest Passage • Numerical Weather Prediction

NWPA: Nuclear Waste Policy Act (USA)

NWPM: Numerical Weather Prediction Model

NWR: Next Work Request

NWRC: National Weather Records Center

NWRI: National Water Research Institute

NWRN: Northwestern

NWS: National Weather Service (USA) • New York - Pier 11 - Wall Street (USA-NY) (location identifier) • Nimbus Weather Satellite • Normal Water Surface • Nose Wheel Steering • Novell Web Server • Nuclear Weapon State

NWSA: Nose Wheel Steering Amplifier

NWSC: National Weather Service Center

NWSMC: National Weather Service Modernization Committee (USA)

NWSRFS: National Weather Service River Forecast System

NWSS: National Weather Satellite System

n. wt.: Net Weight

NWT: Non-Watertight • Northwest Territories (Canada) • Nowata (Papua New Guinea) (location identifier)

NWTC: National Wind Tunnel Complex

NWU: Bermuda (location identifier) • Nebraska Wesleyan University (USA-NE) • Northwestern University (USA-IL) • North-West-Up

NWV: Numerische Wettervorhersage

NWW: Nose Wheel Well

NWWA: National Water Well Association

NWWS: NOAA Weather Wire Service

NX: Normal Crossover • Normal to X-axis • Not Exceeding • Not Expendable

NXB: Non-X-ray Background

nXD: Non X-ray Dominant

NXLG: Narrow-emission-line X-ray-Luminous Galaxy

NXX: Willow Grove (USA-PA) (location identifier)

NY: New Year • New York (USA) • Normal to Y-axis • Number of Years

NYA: National Youth Administration • Not Yet Answered

NYAA: National Young Astronomer Award (AL) • North York Astronomical Association (Canada-ON)

NYAS: New York Academy of Sciences (USA-NY)

NYASA: New York Academy of Sciences Annals (periodical code)

NYC: New York (USA-NY) (location identifier) • New York City (USA-NY)

NYCRL: New York Comprehensive Research Libraries (USA-NY)

NYD: Not Yet Diagnosed

NYE: Nyeri (Kenya) (location identifier)

NYG: Quantico (USA-VA) (location identifier)

NYHoS: New York Hall of Science (USA-NY)

NYHS: New York Hall of Science (USA-NY)

NYI: Sunyani (Ghana) (location identifier)

NYK: Nanyuki (Kenya) (location identifier)

Nyl: Nylon

NYM: Nadym (Russia) (location identifier)

NYN: Nyngan (Australia-NSW) (location identifier)

NYNEX: New York - New England Exchange

NYO: Nyköping (Sweden) (location identifier)

NYP: Not Yet Published

NYPL: New York Public Library

NYPTC: New York Patent, Trademark and Copyright Inc. (USA-NY)

NYR: Not Yet Reported • Not Yet Required • Not Yet Returned

NYS: New York State (USA)

NYSDOL: New York State Department Of Labor

NYSE: New York Stock Exchange (USA-NY)

NYSERNet: New York State Educational and Research Network

NYSTEC: New York State Technology Enterprise Corporation (USA-NY)

NYT: New York Times

NYU: New York University (USA-NY) • Nyaung-u (Myanmar) (location identifier)

nz: New Zealand

NZ: Neutrality Zone • Neutral Zone • New Zealand (international vehicle registration code) • New Zealand (ISO country code) • Normal Acceleration • Northern Zodiacal • Nouvelle-Zélande (ISO country code)

NZA: Niobium Zinc Alloy

NZAP: New Zealand Antarctic Programme • New Zealand Associated Press

NZB: Non Zero Binary

NZBC: New Zealand Broadcasting Corporation

NZBN: New Zealand Bibliographic Network

NZC: Jacksonville Cecil (USA-FL) (location identifier)

NZCSICOP: New Zealand Committee for the Scientific Investigation of Claims of the Paranormal

NZD: Dollar (New Zealand) (ISO currency code)

NZDL: New Zealand Digital Library

NZDS: New Zealand Daylight Saving (time)

NZDSIR: New Zealand Department of Scientific and Industrial Research

NZDT: New Zealand Daylight Time

NZE: Nzerekore (Guinea) (location identifier)

N. Zeal.: New Zealand

NZF: Near Zero Field

NZG: Near Zero Gravity

NZHRM: Noord- en Zuidhollandse Reddingsmaatschappij (Netherlands)

NZIIR: New Zealand Institute for Industrial Research

NZIP: New Zealand Institute of Physics

NZJS: New Zealand Journal of Science (periodical code)

NZL: New Zealand (ISO country code) • Nouvelle-Zélande (ISO country code)

NZMG: New Zealand Map Grid

NZMS: New Zealand Meteorological Service

NZN: Niedersächsischer Zeitschriftennachweis

NZO: Nzoia (Kenya) (location identifier)

NZOI: New Zealand Oceanographic Institute

NZP: National Zoological Park

NZPA: New Zealand Press Agency • New Zealand Publishers' Association

NZSA: New Zealand Spaceflight Association

NZSI: New Zealand Standards Institute (now SANZ)

NZSLO: New Zealand Scientific Liaison Office

NZST: New Zealand Standard Time

NZW: South Weymouth (USA-MA) (location identifier)

NZY: San Diego - North Island (USA-CA) (location identifier)

NZZ: Neue Zürcher Zeitung

N1NF: Non-First Normal Form

N2: Nitrogen

N2HO4: Nitrogen Peroxide

N2H2: Hydrazine

N2H4: Hydrazine

N2O: Nitrogen Oxide

N2O4: Nitrogen Tetroxide

N30: Normal System Epoch 1930

N°: Numéro

O

o/: Orden

o: Obra • Occasional • Octarius • Octavo • Off • Old • Only • Optimus • Orden • Order • Ortho • ostlich • Over • Overcast

O/: Orden

O: Eleven (Roman numeral) • Oberösterreich (Austria) • Object • Observer • Ocean • Octarius • Octavo • October • Oculus • Oeste • Office • Officer • Officina • Ohio (USA) • Old • Ontario (Canada) • Oost • Open • Operator • Orden • Order • Order of Magnitude • Ordinary • Oregon (USA) • Organization • Oscar (international alphabet) • Ost • Osten • Ostlich • Ostrov • Ottava • Ouest • Output • Ouvert • Over • Ovest • Owner • Oxydizer • Oxygen • Quantum Number of Orbital Angular Momentum due to Molecular Rotation • Sauerstoff • Spectral Type • Zero • 11 (Roman numeral)

o/a: On Account • On or About

oa: Onder Andere • Onder Anderen • Overall

O/a: On Account of

O/A: On or About • Overall

OA: Objective Analysis • Observatoire de l'Aigle (Canada-QC) • Off-Axis • Office Address • Office Automation • Office of Administration • Office of Aeronautics • Office of Applications • Office of the Administrator • Officier d'Académie • Old Account • Omnidirectional Antenna • Onboard Acquisition • Onder Andere • Open Account • Operating Agency • Operating Authorization • Operational Aft • Operational Analysis • Operationally Available • Operations Adviser • Operations Analysis • Operations Area • Operations Astronomer • Optique Adaptative • Opto-Acoustic • Orbit Adjust • Orbital Assembly • Orbiter Assembly • Other Articles • Output Amplitude • Output Axis • Overall

ØAA: Østjyske Amateur Astronomers (Denmark)

OAA: Observatório Astronômico Antares (Brazil-BA) • Oglethorpe Astronomical Association (USA-GA) • Ontario Association of Archivists (Canada-ON) • Open Agent Architecture • Optical Acquisition Aid • Orbiter Access Arm • Orbiter Alternate Airfield • Organisation pour l'Alimentation et l'Agriculture • Orient Airlines Association • Osservatorio Astronomico di Arcetri (Italy)

ÖAB: Öffentliche Allgemeinbildende Bibliothek

OAB: Ocean Affairs Board • One-to-All Broadcast • Operational Advisory Broadcast • Ordnance and Assembly Building • Osservatorio Astronomico di Brera (Italy)

OABNN: Orthogonal Activation Based Neural Network

oac: Obra Antes Citada

OAC: Oakland Astronomy Club (USA-MI) • Observatoire de la Côte d'Azur (France) • Observatorio Astronómico de Córdoba (Argentina-CO) • Observatory Advisory Committee • Office of Academic Computing • Online Analysis Computer • On-orbit Activation and Calibration • Ontario Academic Credit • Operations Analysis Center • Optimal Automatic Control • Osservatorio Astrofisico di Catania (Italy) • Osservatorio Astronomico di Cagliari (Italy) • Osservatorio Astronomico di Capodimonte (Italy) • Outer Arm Cloud

OACCB: Office Automation Computer Coordination Board • Office Automation Configuration Control Board

OACD: Observatoire de l'Association Culturelle de Dax (France)

OACH: Office of the Assistant Chief Hydrologist

OACI: Organisation de l'Aviation Civile Internationale • Organización de Aviación Civil Internacional

OACL: Osservatorio Astronomico Colle Leone (Italy)

OACT: Office of Advanced Concepts and Technology • Organisation Africaine de Cartographie et Télédétection • Osservatorio Astronomico di Colluriana-Teramo (Italy)

OACUL: Ontario Association of College and University Libraries (Canada-ON)

OAD: Orbit and Attitude Division • Orbit Attitude Division • Orbiter Atmospheric Drag • Overall Depth

OADCES: Online Aerotriangulation Data Collection and Editing System

OADR: Originating Agency's Determination Required

ÖAdW: Österreichische Akademie der Wissenschaften (Austria)

OAE: Optical Alignment Equipment • Organisatie van Afrikaanse Eenheid

OAEM: Observatorio Astronômico da Escuela de Minas (Brazil-MG)

OAET: Office of Aeronautics, Exploration and Technology

OAF: Optical Alignment Facility • Optimum Array Filter

OAFA: Observatorio Astronómico Felix Aguilar (Argentina-SJ)

OAFD: Orbiter Aft Flight Deck

OAFTO: Orbiter Atmospheric Flight Test Office

OAG: Open Applications Group • Orange Springhill (Australia-NSW) (location identifier) • Osservatorio Astronomico di Genova (Italy)

OAGH: Observatorio Astrofísico Guillermo Haro (Mexico)

OAGM: Off-Axis Generating Machine

OAH: Organization of American Historians • Overhall Height

OAHE: Observatório Astronômico Herschel-Einstein (Brazil)

OAHTS: Orbiter Avionics Hardware Tracking System

OAI: Ohio Aerospace Institute (USA-OH) • OK As Is • Open Applications Interface

OAIA: Organisation des Agences d'Information d'Asie

OAII: Ocean-Atmosphere-Ice Interaction

OAIS: Open Archival Information System

OAJ: Jacksonville (USA-NC) (location identifier)

OAK: Oakland (USA-CA) (location identifier) • Object Application Kernel

OAL: Cacoal (Brazil-TO) (location identifier) • Observatoire Astronomique de Lyon (France) • Observatório Astronômico de Lisboa (Portugal) • ODF Access Layer • Office of Arts and Libraries (UK) • Open Access Layer • Overall Level

OALA: Ohio Academic Libraries Association (USA-OH)

OALC: Ogden Air Logistics Center

OALD: Oxford Advanced Learner's Dictionary

OALM: Observatorio Astronomico "Los Molinos" (Uruguay)

OALP: Observatorio Astronómico de La Plata (Argentina-BA)

OALT: Ontario Association of Library Technicians • Operational Acceptable Level of Traffic • Operational Acceptance Level of Traffic

o. alt. hor.: Omnibus Alternis Horis

OA&M: Operation, Administration & Maintenance

OAM: Oamaru (New Zealand) (location identifier) • Observatoire des Alpes Maritimes (France) • Observatorio Astronómico de Mallorca (Baleares) • Observatorio Astronómico Móbil • Observatório Astronômico Monoceros

(Brazil-MG) • Office of Administration and Management • Office of Aerospace Medicine • Office of Alternative Medicine (USA) • Office of Automation and Manpower • Operation-Administration-Maintenance • Operation And Maintenance • Operations-Administration-Maintenance • Operations-Administration-Management • Operations And Maintenance • Oracle Access Manager • Orbit Adjust Module • Orbital Assembly Module

OAMCE: Optical Alignment, Monitoring and Calibration Equipment

OAME: Orbital Attitude and Maneuvering Electronics

OAMEX: Ocean-Atmosphere Materials Exchange

OAMF: Observatorio Astronómico Municipal de Funes (Argentina)

OAMM: Observatoire Astronomique du Mont Mégantic (Canada-QC) • Observatorio Astronómico Municipal de Mercedes (Argentina) • Office of Acquisition and Material Management

OAMP: Optical Airborne Measurements Program • Optical Analog Matrix Processing

OAMR: Observatorio Astronómico Municipal de Rosario (Argentina-SF)

OAMS: Onboard Alignment Measurement System • Orbital Attitude and Maneuvering System • Organic and Atmospheric Mass Spectrometer

OAN: Observatório Astronômico Nacional • Observatorio Astronómico Nacional • Olanchito (Honduras) (location identifier)

OANA: Organization of Asian News Agencies

OAO: Off And On • Office of Aircraft Operations (NOAA) • One And Only • Orbital Astronomical Observatory • Orbited Assembly Operation • Orbiting Astronomical Observatory

OAOG: Ottawa Valley Astronomy and Observers Group (Canada-ON)

OAOI: On And Off Instrument

OAP: Observatoire Astronomique des Pises (France) • Observatorio Astronómico de Patacamaya (Bolivia) • Observatório Astronômico de Piedade (Brazil-MG) • Odessa Astronomical Publications • Old-Age Pension • Old-Age Pensioner • Orbit Activity Plan • Osservatorio Astronomico di Padova (Italy) • Osservatorio Astronomico di Palermo (Italy) • Oxygen at Atmospheric Pressure

OAPEC: Organization of Arab Petroleum-Exporting Countries

OAPI: Organisation Africaine de la Propriété Intellectuelle

OAPM: Optimal Amplitude and Phase Modulation

OAPS: Orbit Adjust Propulsion System

OAQ: Observatorio Astronómico de Quito (Ecuador)

OAQPS: Office of Air Quality Planning and Standards

OAQPSTTN: Office of Air Quality Planning and Standards Technology Transfer Network

OAR: Monterey Fritzsche (USA-CA) (location identifier) • Oceanic and Atmospheric Research • Off-Axis Rejection • Office of Aerospace Research • Office of Air and Radiation • Office of Analysis and Review • Office of Atmospheric Research • Open Architecture for Reasoning • Operations Analysis Report • Optical Automatic Ranging • Osservatorio Astronomico di Roma (Italy) • Overtime Authorization Request

OARE: Orbital Acceleration Research Experiment (*) • Orbiter Acceleration Research Experiment

OARL: Ontario Academic Research Libraries (Canada-ON)

OARM: Office of Administration and Resources Management

OARME: Oxygen Atom Reaction Materials Experiment

OARNET: Ohio Academic Resources Network (USA-OH)

OARS: Online Automated Reference Service • Ontario Association for Remote Sensing • Orbiter Avionics Radiation Specification

OARSA: Optical Aberration Reduction and System Analysis

ÖAS: Östergötlands Astronomiska Sällskap (Sweden)

OAS: Observatoire Astronomique de Strasbourg (France) • Office Automation System • Ogden Astronomical Society (USA-UT) • Olympic Astronomical Society (USA-WA) • Omaha Astronomical Society (USA-NE) • On Active Service • One-to-All Scatter • Open Accounting System • Optical Absorption Spectroscopy • Optical Alignment Sight • Optical Assembly Simulator • Orbit Adjustment System • Orbiter Aeroflight Simulator • Orbiter Atmospheric Simulator • Orbiter Atmospheric Sounder • Orbiter Avionics Simulator • Orbiter Avionics System • Organisation de l'Armée Secrète • Organization of American States • Orpington Astronomical Society (UK)

OASCB: Orbiter Avionics Software Control Board

OASD: Office of the Assistant Secretary of Defense

OASES: Outpost for Analysis and Spectroscopy of Extrasolar Systems

OASF: Orbiting Astronomical Support Facility

OASI: Orwell Astronomical Society Ipswich (UK) • Osservatorio Antarctica Submillimetrico Infrarosso

OASIS: Oceanic Analysis and Surveying Information System • Oceanic and Atmospheric Satellite Imaging System • Oceanic and Atmospheric Scientific Information System • OEX Autonomous Supporting Instrumentation System • Open Access Strategic Information System • Open Access Symbiotic Information System • Open Architecture Scientific Information System • Operational Airport Safety Information System • Operational Automatic Scheduling Information System • Operations And Science Instrument Support • Operations And Science Instrument System • Optically Adaptive System for Imaging Spectroscopy • Orbit Avionics Software Integration Study • Orbiter Autonomous Supporting Instrumentation System • Organization for the Advancement of Space Industrialization and Settlement • Outil d'Aide à la Spécification des Informations-Systèmes

OASM: Office of Aerospace Medicine

OASPL: Overall Sound Pressure Level

OASSIS: Operations And Support for Science Instrument Systems

OASSO: Operational Applications of Satellite Snowcover Operations

OAST: Office of Aeronautics and Space Technology

OASV: Orbital Assembly Support Vehicle

OASYS: Obstacle Avoidance System • Office Automation System

OAT: Obligation Assimilable du Trésor • Operating Acceptance Test • Operating Ambient Temperature • Operational Acceptance Test • Optical Adaptive Technique • Original Alias Table • Osservatorio Astronomico di Trieste (Italy) • Outside Air Temperature • Overall Test

OATA: Optical Acquisition and Tracking Aid

OATM: Orbiter Antenna Test Model

OATo: Osservatorio Astronomico di Torino (Italy)

OATP: Operational Acceptance Test Procedure

OATS: Office Automation Technology and Services • Orbit Adjust Thrust System • Orbiter Avionics Tracking System • Organización Andina de Telecomunicaciones por Satélite • Original Article Tear Sheet • Original Article Text Service • Overall Test Set

OATUU: Organization of African Trade Union Unity

OAU: Observatorio Astronômico de Uberlândia (Brazil) • Optical Alignment Unit • Organization of African Unity

OAV: Operational Aerospace Vehicle

ÖAW: Österreichische Akademie der Wissenschaften (Austria)

OAW: Off-Airway Weather

OAWP: Operations Analysis Working Paper

OAX: Oaxaca (Mexico) • Oaxaca Xoxocotlan (Mexico) (location identifier)

OAXTC: Ocean-Atmosphere Exchange of Trace Compounds

ob: ober • Obiit • Obiter • Obra • Observation

oB: Ohne Befund

Ob: Ober • Obligatory

O/B: Octal to Binary • On Board • Operational Base • Ordered Back

OB: Oberbegriff • Observation Block • Observed Bearing • Off Broadway • Official Board • Onboard • Openbare Bibliothek • Opening of Books • Operating Budget • Operational Base • Optical Bench • Ordered Back • Ordinary Building • Ordinary Business • Outboard • Outside Broadcast • Outside Broadcasting • Outside Broadcasting • Outward Bound • Peru (international aircraft registration code)

OBA: Oban (Australia-QLD) (location identifier) • Object Behavior Analysis • Olimpiada Brazileira de Astronomia (Brazil) • Omnibase Administrator • Optical Base Assembly • Optical Bench Assembly • Outer-Belt Asteroid • Outer Buffer Assembly

OBAFGKMRNS: O Be A Fine Girl, Kiss Me Right Now Sweatheart • Oh Backward Astronomer, Forget Geocentricity: Kepler's Motions Reveal Nature's Simplicity

obb: Obbligato

ÖBB: Österreichische Bundesbahnen (Austria)

OBC: Obock (Djibouti) (location identifier) • Onboard Checkout • Onboard Computer

OBCE: Onboard Checkout Equipment

OBCCS: Onboard Computer Calibration Source

OBCM: Object-Based Conceptual Modelling

OBCO: Onboard Checkout

OBCP: Onboard Computer Program

OBCS: Onboard Calibration Source • Onboard Checkout System

OBCTE: Onboard Computer Test Equipment

OBD: Obano (Indonesia) (location identifier) • Optimal Brain Damage

OBDD: Ordered Binary Decision Diagram • Ordered Boolean Decision Design

OBDH: Onboard Data Handler • Onboard Data Handling

OBDHTE: Onboard Data Handling Test Equipment

OBDI: Onboard Decoder Interface

OBDM: Onboard Data Management

OBDS: Onboard Decoder Simulator

Obdt: Obedient

OBDU: Onboard Diagnostic Unit

OBE: Observation Equipment • Okeechobee County (USA-FL) (location identifier) • Onboard Equipment • Optical Bench Enclosure • Order of the British Empire • Out-of-Body Experience • Overcome By Events • Overtaken By Events

OBECO: Outboard Engine Cutoff

Oberbay: Oberbayern (Germany)

Oberfr: Oberfranken (Germany)

Oberpf: Oberpfalz (Germany)

OBES: Orthonormal Basis of an Error Space

OBETC: Optical Bench Enclosure Thermal Control

OBEX: Object Exchange

OBF: Oberpfaffenhofen (Germany) (location identifier)

OBI: Obidos (Brazil-PA) (location identifier) • Observation Interval

OBIC: Optical Beam Induced Current

OBIGGS: On-Board Inert-Gas-Generating System

OBIGS: On-Board Inert-Gas-generating System

OBIPS: Optical Bench Imager and Photometer System

OBIS: Object-Based Information System

Obit: Obituary

Obj: Object • Objection • Objective • Objekt

OBJ: Object • Objective

Objv: Objective

OBK: Northbrook (USA-IL) (location identifier)

Obl: Obligation • Oblige • Oblique • Oblong

OBM: Morobe (Papua New Guinea) (location identifier) • Object Manager

OBMC: Outbound Midcourse Correction

OBMM: Onboard Mission Manager

OBMS: Onboard Management System

OBN: Oban Connel (UK-Scotland) (location identifier) • Open Bibliothek Netwerk (Netherlands)

OBNI: Objet Brilliant Non Identifié

OBO: Obihiro (Japan) (location identifier) • O'Brien Observatory (USA) • Output Back-Off

OBOGS: Onboard Oxygen Generating System

OBP: Odorant Binding Protein • Onboard Processing • Onboard Processor

OBPC: Optical Bar Panoramic Camera

OBRA: Omnibus Budget Reconciliation Act

Obre: Octobre

OBRG: Onboard Range Compression

OBRM: Observatorio del Roque de los Muchachos (Canary Islands)

OBRP: Object Representation

Obs: Obscure • Observações • Observateur • Observation • Observatoire • Observatorio • Observatório • Observatory • Observatory (periodical code) • Observe • Observed • Observer • Observing • Obsolete • The Observatory (periodical code)

OBS: Aubenas Vals-Lanas (France) (location identifier) • Observatorium Bad Salzschlirf (Germany) • Observatory Scientist • Observe • Ocean Bottom Seismometer • Omni Bearing Selector • On-Board Software • Operational Bioinstrumentation System • Operational Biomedical System • Optimal Brain Surgery • Organic Brain Syndrome

Obsc: Obscure • Obscured

OBSC: Obscure • Obscured

OBSCD: Obscured

Observ: Observateur • Observatoire • Observatorio • Observatório • Observatory • Observe • Observed • Observer • Observing

OBSET: Observations Set

OBSM: On-Board Software Maintenance

Obsol: Obsolete

obsp: Obiit Sine Prole

Obst: Obstacle • Obstruction

Obstn: Obstruction

Obsv: Observation

OBSW: On-Board Software

Obt: Obedient • Obtained

OBT: Oakland-Coliseum (USA-MD) (location identifier) • Observation de la Terre • Onboard Time • Onboard Training • Organically-Bound Tritium

Obtd: Obtained

OBU: Kobuk (USA-AK) (location identifier) • Offshore Banking Unit • Oklahoma Baptist University (USA-OK)

Obv: Observe • Observed

OBV: Oxidizer Bleed Valve

OBX: Obo (Papua New Guinea) (location identifier)

OBY: Scoresbysund (Greenland) (location identifier)

o/c: Overcharge

oc: Obra Citada • Occitan (ISO language code) • Ocean • On Center • Opere Citato

Oc: Ocean

O/C: Operations/Critical • Overcharge

O-C: Observed-Computed

O&C: Operations & Checkout • Operations & Control

OC: Object Calculus • Object Calibration • Object Class • Object Classification • Observational Catalogue • Observatory Council • Oceania • October • Octobre • Of Course • Office Copy • Office of the Comptroller • Official Classification • On Center • On Condition • On Course • Ondes Courtes • Open Circuit • Open Collector • Open Connect • Open Contract • Operating Characteristic • Operating Curve • Operational Capability • Operational Check • Operations Control • Operations Coordinator • Operator Centralization • Operator Centre • Opere Citato • Optical Carrier • Oral Contribution • Orbital Check • Ordinary Chondrite • Original Cover • Outer Cluster • Outflow Connected • Output Computer • Outside Circumference • Over Center • Overcurrent

OCA: Observatoire de la Côte d'Azur (France) • Observatorio Cerro Armazones (Chile) • Obstacle Clearance Altitude • Ocean and Coastal Areas • Ocean Reef (USA-FL) (location identifier) • Operational Control Authority • Optical Corporation of America • Orbiter Capability Analysis • Orbiter Capability Assessment • Oxygen Control Assembly

OCAC: Orbiter Cabin Air Cleaner

OCAM: Optimal Control Applications and Methods (periodical code) • Organisation de la Communauté Africaine et Malgache

OCAN: Observatoire de la Côte d'Azur, Nice (France)

OCARINA: Ocean, Atmosphere, Research and Investigation with Acoustics

OCAS: Organization of Central American States

OCAST: Oklahoma Center for the Advancement of Science and Technology (USA-OK)

OCB: Office Central des Bibliothèques(France) • Operational Configuration Baseline • Output Current Booster • Override Control Bits

Occ: Occasion • Occasional • Occasionally • Occident • Occidental • Occidente • Occupation • Occupé • Occupied • Occupy • Occur • Occurred • Occurrence

OCC: Coca (Ecuador) (location identifier) • Observation Control Center • Office of Contract Compliance • Operational Computer Complex • Operations and Control Centre • Operations Control Centre • Operations Control Console • Operator Control Command • Optically Coupled Common • Organic Carbon Cycle • Other Common Carrier

OCCAM: Ocean Circulation Climate Advanced Modelling

Occas: Occasion • Occasional • Occasionally

OCCB: Operational Configuration Control Board

OCCC: Operations and Communications Control Center

OCCE: Office Central de Coopération à l'École

OCCG: Operations Center Coordinator Group

OCCI: Optical Coincidence Coordinate Indexing

Occit: Occitan • Occitano

OCCL: Online Computer Library Center

Occn: Occasion

OCCP: Orbit Command Computer Program

OCCS: Operator Control Console System

OCCULT: Orser Complete Conversational User-Language Translator

OCD: Obsessive Compulsive Disorder • Ocean Chemistry Division • Octal Communications Device • Online Communications Driver • Operational Capability Date • Operational Capability Development • Operational Concept Document • Operations Concept Document • Out-of-Cell Delineation

OCDE: Organisation de Coopération et de Développement Économiques (*) • Organisation pour la Coopération et le Développement Économiques • Organización para la Cooperación y el Desarollo Económico

OCDM: Open Cold Dark Matter

OCDR: Orbiter Critical Design Review

OCDU: Optics Coupling Data Unit

OCE: Ocean • Ocean City (USA-MD) (location identifier) • Ocean Color Experiment • Office of the Chief Engineer • Open Collaboration Environment • Open Collaborative Environment (*) • Open Computing Environment • Optical Control Electronics • Oscillating Current Element • Other Common Equipment

OCEAN: Ocean Colour Environment Archive Network • Ocean Colour European Archive Network • Open Corrosion Expertise Accesses Network

OCEANIC: Ocean Network Information Center

Oceanogr: Oceanografía

OcEng: Ocean Engineering (periodical code)

OCenW: Onderwijs-Cultuur-Wetenschappen

OCEP: Office Central d'Édition et de Publicité (France)

OCEPA: Office of Communications, Education and Public Affairs

OCETA: Ontario Centre for Technology Advancement (Canada-ON)

OCF: Object Component Framework • Object Correlation Frame • Observer Comments File • Ocala Taylor (USA-FL) (location identifier) • Onboard Computational Facility • Open Channel Flow • Orbiter Computational Facilities

OCFNT: Occluded Front

OCG: Optimal Code Generation

ÖCG: Österreichischen Computergesellschaft (Austria)

OCGA: Office of Congressional and Government Affairs

OCH: Nacogdoches (USA-TX) (location identifier) • Obstacle Clearance Height • Office of the Chief Hydrologist • Orbiter Common Hardware

OCHA: Office of Coordination for the Humanitarian Assistance (UNO)

OCI: Object Code Insertion • Ocean Color Imager • Oceanic (USA-AK) (location identifier) • Office of Computer Information • Oficina Central de Información • Organizational Conflict of Interest • Organisation du Congrès Islamique • Ouvrages-Cartes-Instruments

OCIA: Onizuka Center for International Astronomy (USA-HI)

OCIAM: Oxford Centre for Industrial and Applied Mathematics (UK-England)

OCIF: Out Card In File

OCIMF: Oil Companies International Maritime Forum

OCIP: Ottawa Carleton Institute of physics (Canada-ON)

OCIR: Optical Coincidence Information Retrieval

OCIS: Organized Crime Information System

OCIW: Observatories of the Carnegie Institution of Washington

OCJ: Ocho Rios (Jamaica) (location identifier)

OCl: Open Cluster

OCL: Object Connecting Language • Object-Constraint Language • Obstacle Clearance Limit • Occlude • Open Clusters • Operational Check List • Operational Control Level • Operation Control Language • Operator Control Language

OCLC: Ohio College Library Center (USA-OH) • Online Computer Library Center

OCld: Oil-Cooled

OCLD: Occlude

OCLI: Optical Coating Laboratories Inc. • Optical Coating Laboratory Inc. (USA-CA)

OCLIP: Optical Cellular Logic Image Processor

OCLN: Occlusion

OCLSC: Ontario Centre for Large-Scale Computations (Canada-ON)

OCM: Object Class Model • Ocean Color Monitor • Ocean Color Monitoring • On-Channel Map • Optical Countermeasure • Orbiter Celestial Mechanics • Organic Content Monitor • Organisation Civile et Militaire • Organisation Commune des Marchés • Organisation du Commerce Mondial

OCMR: Ontario Centre for Materials Research (Canada-ON)

OCMW: Openbaar Centrum voor Maatschappelijk Welzijn

OCN: Occasion • Oceanside (USA-FL) (location identifier) • Operation Completion Notice • Operation Control Number • Order Change Number • Order Control Number • Organization Change Notice

OCNL: Occasional • Occasionally

OCNLY: Occasionally

OCO: Object Code Only • Open-Close-Open • Operational Checkout • Operations Control Office • Operations Control Operator • Optimiseur de Code Objet

OCOE: Overall Checkout Equipment

OCOG: Offset Centre Of Gravity

OCP: Office of Commercial Programs • One-Component Plasma • Open-Cluster Pattern • Operational Control Panel • Operations Checkout Procedure • Operations Control Plan • Optical Continuum Peak • Orbital Combustion Process • Orbital Correction Program • Organic Conducting Polymer • Output Control Pulse

OCPC: Office of Compensation and Position Classification

OCPI: Office Central de la Presse Illustrée (France)

OCPP: Orbital Cloud Photopolarimeter (*) • Orbiter Cloud Photopolarimeter

OCPREP: Object Correlation Preparation

OCPS: Orbiter Camera Payload System • Oxygen Cabin Pressurization Section

OCPSF: Organic Chemicals, Plastics, and Synthetic Fibers

OCR: Observatorio Cristo Rey (Argentina) • Occur • Ocean Colour Radiometer • Office for Civil Rights • Operational Capability Readiness • Operational Concept Review • Optical Character Reader • Optical Character Recognition (*) • Optical Character Recognizer • Optimal Character Recognition • Optimum Charge Regulator • Organization Change Request • Overhaul Component Requirement

OCRA: One-Centimeter Receiver Array • Overseas Companies Registration Agents

OCRE: Optical Character Recognition Element

OCRIT: Optical Character Recognizing Intelligent Terminal

OCRM: Orbiter Crash and Rescue Manual

OCRS: Ontario Centre for Remote Sensing • Optical Character Recognition System

OCRUA: Optical Character Recognition Users Association

OCS: Carbonyl Sulfide • Observatory Control System • Ocean Colour Scanner • Office of Cadastral Survey • Office of the Chief Scientist • Officer Candidate School • Onboard Checkout System • One Chip System • Operational Control Segment • Operational Control System • Operations Control System • Optical Character Scanner • Optical Character Scanning • Optical Control System • Outer Continental Shelf

OCSE: Organizzazione di Cooperazione e Sviluppo Economico

OCSG: Open Computing Security Group

Ocsnl: Occasional

Ocst: Overcast

OCST: Office of Commercial Space Transportation

Oct: Octagon • Octans-Octantis (constellation) • Octavo • October • Octobre • Octubre

OC&T: Office of Certification & Training

OCT: Observation Control Table • Observatório das Ciências e das Tecnologias (Portugal) • Octal • Octave • October • Octobre • Office of Certification and Training • Operations Control Team • Optimal Control Theory • Orbital Circularization Technique • Overseas Countries and Territories

Octl: Octal

OCTL: Open-Circuited Transmission Line

Octn: Octane

OCTOPUS: Ocean Colour Techniques for Observation, Processing and • Utilization Systems

OCTS: Ocean Color and Temperature Scanner

OCTV: Open-Circuit Television

OCTW: Optical Communication Thru the Shuttle Window

OCU: Office Channel Unit • Oklahoma City University (USA-OK) • Operational Control Unit • Operational Conversion Unit • Optical Conversion Unit • Ordnance Control Unit • Output Control Unit

OCUFA: Ontario Confederation of University Faculty Associations (Canada-ON)

OCUL: Ontario Council of University Libraries

OCV: Ocaña (Colombia) (location identifier) • Open-Circuit Voltage

OCVD: Open-Circuit Voltage Decay

OCW: Operation Command Word • Washington Warren (USA-NC) (location identifier)

OCWG: Ocean Colour Working Group

OCX: OLE Control • OLE Custom Control

OCXO: Oven-Controlled Crystal Oscillator

OCYT: Oficina de Ciencia y Tecnología

o/d: On Demand • Overdose • Overdraft

od: Oculus Dexter

O/d: On Demand • Overdraft

Od: Oder

O&D: Origin & Destination

O/D: On Demand • On Dock • Outside Diameter • Overdraft

OD: Absorbance • Doctor of Optometry • Lebanon (international aircraft registration code) • Observable Difference • Observational Data • Observatory Director • Observed Drift • Oculus Dexter • Officer of the Day • Old Dutch • On Demand • On Duty • Operational Downlink • Operations Directive • Operations Director • Operations Division • Optical Density • Optical Disk • Orbit Determination • Ordinateur Domestique • Ordnance Department • Organization Development • Original Design • Original Dirac • Other Denominations • Outer Detector • Outer Diameter • Outside Diameter • Outside Dimensions • Overdose • Overdraft • Overdrawn • Oxygen Drain

ODA: Octal Debugging Aide • Office D'Annonces • Office Document Architecture • Official Development Assistance • Omnidirectional Antenna • Open Document Architecture • Operational Data Analysis • Optical Data Associates • Optical Diagnostics Assembly • Optical Diffraction Analysis • Ouadda (Central African Rep.) (location identifier) • Overseas Development Administration • Overseas Development Agency (UK) • Oxydianiline

ODAC: ODA Consortium • Operations Distribution Administration Center

ODAL: Omnidirectional Approach Lighting System

ODAPI: Open Database Application Programming Interface

ODAPS: Oceanic Display And Planning System • Oceanic Display And Processing Station

ODAR: Omnidirectional Airborne Radar

ODAS: Ocean Data Acquisition System • Ocean Data Station • Oceanic Data Assimilation System • Optical Disk Archive System

OdB: Oeil-de-Boeuf

ODB: Córdoba (Spain) (location identifier) • Operational Database • Operational Data Book • Operations Database • Orbit Determination Beacon • Output to Display Buffer

ODBC: Open Database Connectivity

ODBMS: Object-oriented Database Management System

ODBS: Optical Disk-Based System

ODC: Open Design Center • Operational Data Center • Operational Document Control • Optical Disk Cartridge • Orbital Data Collector • ORFEUS Data Centre • Oscilloscope Digital Control • Other Data Centre • Other Direct Costs

ÖDD: Österreichischer Dokumentationsdienst (Austria)

ODD: Oodnadatta (Australia-SA) (location identifier) • Optical Data Digitizer • Optical Data Disk • Optical Digital Disk • Orbiting Debris Device • Ordered Decision Design • Overdetermined Dual-Doppler

ODDL: Onboard Digital Data Load

ODDP: Object-Dependent Data Processor

ODDR&E: Office of the Director of Defense Research & Engineering

ODDS: Operational Data Delivery Services

ODDSC: Optical Disk Data Storage Capability

Ode: Oude

OD&E: Operational Demonstration & Evaluation

ODE: Odense (Denmark) (location identifier) • Office de Documentation Économique • Office Developer Edition • Online Data Entry • Open Development Environment • Operational Demonstration and Evaluation • Ordinary Differential Equation

ODECA: Organización de Estados Centroamericanos

Odenw: Odenwald (Germany)

ODERACS: Orbital Debris Radar Calibration Spheres

ODES: Ocean Data Evaluation System • Optical Discrimination Evaluation Study

ODESSA: Oceanographic Database and Environmental Satellite System • Ordinary Differential Equation Solver with Sensitivity Analysis

ODEX: Optical Dynamics Experiment

ODF: Observation Data File • One-Dimensional Flow • Optimal Decision Function • Orbit Determination Facility • Orientation Distribution Function • Original Data File • Output Data Formatter

OdG: Ordine del Giorno

ODG: Office of the Director General • Operational Data Group • Operational Design Group • Orbit Data Generator

ODGSE: Operational Development Ground Support Equipment

ODH: Odiham (UK-England) (location identifier)

ODI: Office de Documentation Industrielle • Office Document Index • On Demand Information • Open Datalink Interface • Open Device Interconnect • Optical Digital Image • Optical-Disk Image

ODID: Operational Display and Input Development

ODIF: Office Document Interchange Format • Open Document Interchange Format

ODIN: Oosterhouts Digitaal Informatie Netwerk (Netherlands) • Orbital Design Integration (system) • Outsourcing Desktop Initiative for NASA

ODISS: Optical Digital Image Storage System

OdJ: Ordre du Jour

ODJ: Ouanda Djalle (Central African Rep.) (location identifier)

ODL: Cordillo Downs (Australia-SA) (location identifier) • Object Definition Language • Observation Data Log • Oceanic Data Link • Office Document Language • Open and Distance Learning • Optical Delay Line

ODM: Oakland (USA-MD) (location identifier) • Object Data Manager • Office of Defense Mobilization (USA) • Oil Debris Monitoring • One-Day Mission • Open Drakmatter Model • Operating Data Manual • Operational Data Management • Operational Data Message • Operational Development Memorandum • Operation Data Management • Operation Data Message • Operation Development Memorandum • Optical Demonstration Model • Optical Display Memory • Orbital Determination Module • Orbit Deflection Maneuver • Orbiter Deflection Maneuver • Orbiter Disconnect Mechanism

ODMA: Open Document Management API

ODMG: Object Database Management Group • Object Data Management Group

ODMR: Optical Detection of Magnetic Resonance • Optically Detected Magnetic Resonance

ODN: Long Seridan (Malaysia) (location identifier) • Open Data Network • Outdial Notification

ODNAO: Orbital Dynamics of Natural and Artificial Objects

ODO: Orbital Debris Object

ODP: Ocean Drilling Program (*) • Ocean Drilling Project • Official Development Planning • Open Data Path • Open Distributed Processing • Open Distributing Processing • Open-Door Policy • Operational Development Plan • Operational Development Program • Operational Display Procedure • Optical Data Processing • Orbit Determination Program • Original Document Processing • Overall Development Planning • Ozone-Depleting Potential

ODPCS: Oceanographic Data Processing and Control System

ODPI: Office of the Director of Public Information

ODPS: Optical Data Processing System

ODR: Orbital Data Record • Ord River (Australia-WA) (location identifier) • Original Data Record • Output Data Redundancy

ODRAM: Optocal Dynamic Random Access Memory

ODRAN: Operational Drawing Revision Advance Notice

ODRC: Orbiter Data Reduction Center

ODRN: Orbiting Data Relay Network

ODRS: Orbiting Data Relay System

ODRSS: Orbiting Data Relay Satellite System

ODS: Object Description Skeleton • Observation Datafile System • Observation Data System • Odessa (Ukraine) (location identifier) • Open Data Services • Operational Data Server • Operational Data Store • Operational Data

Summary • Operator and Display System • Optical Data System • Optical Detector System • Optical Disk Storage • Optical Docking System • Orbital Disconnect Support • Orbit Determination System • Orbiter Docking System • Orbiter Dynamic Simulator • Output Data Strobe • Overhead Data Stream • Oxide Dispersion Strengthening • Oxygen Depletion Sensor • Ozone Depleting Substance

ODSC: Optical Disk Storage Capability

ODSI: Open Directory Services Interface

ODT: Omnidirection Transmission • Open Desktop • Operational Demonstration Test • Optical Data Transmission • Optical Detector Team • Order-Disorder Transition

ODTF: Operational Development Test Facility

ODTM: Orbiter Dynamic Test Model

ODTS: Operational Development Test Site • Optical Data Transmission System

ODU: Old Dominion University (USA-VA) • Optical Display Unit • Output Display Unit

ODV: Omni-Directional Vehicle

ODVAR: Orbit Determination and Vehicle Attitude Reference

Odw: Odenwald (Germany)

ODW: Oak Harbor (USA-WA) (location identifier)

ODWE: Oblique Detonation Wave Engine

ODY: Oudomxay (Laos) (location identifier)

oe: Omissions Excepted

Oe: Oersted

O/E: Optical to Electrical

ÖE: Öleinheit

OE: Austria (international aircraft registration code) • Observatoire de l'Énergie • Office of Enforcement • Old English • Omissions Excepted • Open End • Operating Engineer • Operational Environment • Operations Engineer • Operations Engineering • Order Entry • Original Error • Outboard Engine • Output Enable • Ouvrier d'État • Oxygen-Enriched Air

OEA: Observatoire Européen de l'Audiovisuel • Office for External Activities • Operator Error Analysis • Organización de los Estados Americanos • Organisation des États Américains • Ouvrier d'Entretien et d'Accueil • Vincennes Oneal (USA-IN) (location identifier)

OEAS: Orbital Emergency Arresting System

OEB: Office Européen des Brevets

OEC: Ocussi (Indonesia) (location identifier) • Office of Energy Conservation • Office of Environmental Compliance • Orbital Electron Capture • Orbiting Experimental Capsule • Other Early Capability • Oxygen-Enriched Combustion • Oxygen-Evolving Complex

OECD: Organization for Economic Cooperation and Development (formerly OEEC)

OECE: Organisation Européenne de Coopération Économique • Organizzazione Europea per la Cooperazione Economica

OECO: Outboard Engine Cutoff

OECS: Organization of Eastern Caribbean States

OED: Online Event Display • Oxford English Dictionary (OUP)

OEDIP: Outil Expert de Diagnostic Interactif de Panne

OEE: Observation of Earth and its Environment • Open Enterprise Environment • Overall Equipment Effectiveness

OEEC: Organization for European Economic Cooperation (now OECD)

OEEPE: Organisation Européenne d'Études Photogrammétriques Expériment ales

OEF: Olimpiada Española de Física (Spain) • Optical Evaluation Facility

OEFD: Orbiter Electric Field Detector

OEG: Office of Environmental Guidance

OeGA: Österreichische Gesellschaft für Akustik (Austria)

OeGAI: Österreichische Gesellschaft für Artificial Intelligence (Austria)

OEH: Order Entry Handling

OEI: Oficina de Educación Ibero-americana (Spain) • One-Engine Inoperative • Optical-Electronic Interface • Organisation Européenne d'Information • Organización de Estados Iberoamericanos • Overall Efficiency Index

OEIC: Open-Ended Investment Company • Opto-Electronic Integrated Circuit

OEID: Origin and Evolution of Interplanetary Dust

OEIS: Orbiter Electrical Interface Simulator

OEL: Occupational Exposure Limit • Orel (Russia) (location identifier) • Organic Electroluminescence

OEM: Object Exchange Model • Office of Emergency Management • Office of Executive Management • Onde Électromagnétique • Optimal Estimation Method • Original Equipment Manufacturer • Own Equipment Material

OEMA: Organización Europea de Medio Ambiente

OEMCP: Optical Effects Module Electronic Controller and Processor

OEMI: Office Equipment Manufacturer's Institute (USA)

OEMM: Operational Experiment on Mesoscale Meteorology

OEN: Operational Evaluation Network

O. en W.: Onderwijs en Wetenschappen

OEO: Office of Economic Opportunity • Office of Equal Opportunity • Osceola (USA-WI) (location identifier)

OEOB: Old Executive Office Building

OEP: Office of Economic Planning • Office of Economic Preparedness • Office of Environmental Policy • Operand Execution Pipeline

OEPM: Oficina Española de Patentes y Marcas (Spain)

OEPS: Operationally Efficient Propulsion System

OEQ: Organisation Européenne pour la Qualité

OER: Odd-Even Rule • Office of Energy Research • Office of Environmental Restoration • Operational Equipment Requirement • Operations Engineering Report • Organization for European Research • Original Equipment Request • Örnsköldsvik (Sweden) (location identifier) • Ornskoldsvik (Sweden) (location identifier) • Overhead Expenditure Request • Oxygen Enhancement Ratio

OERC: Optimum Earth Reentry Corridor

OERD: Office of Energy Research and Development

OERI: Office of Educational Research and Improvement (USA)

OERR: Office of Emergency and Remedial Response

OERS: Operational Earth Resources System

OES: Ocean Energy System • Office of Economic Stabilization • Officer Evaluation System • Optical Emission Spectrometer • Optical Emission Spectroscopy • Optische und Elektronische System GmbH (Germany) • Orbital-Escape System • Orbital Extravehicular Suit • Orbiter Emergency Site • Organization of European States • San Antonio Oeste (Argentina-RN) (location identifier)

OESA: Office of Earth Science Applications

OESD: Organic Electronic Spectral Data

OESO: Organisatie voor Economische Samenwerking en Ontwikkeling

OESS: Orbiter/ET Separation System

OESZ: Ostereuropäische Sommerszeit

OET: Object Engineering Technique • Objective End Time • Orbiter Engineering and Test • Orbiter External Tank

OETF: Observing Efficiency Task Force

OETP: Orbiter Electron Temperature Probe

OEW: Operating Empty Weight

OEX: Orbiter Experiment

OEXP: Office of Exploration

of: Oficio

Of: Oficina

O/F: Oxidizer/Fuel

OF: Old Face • Old French • Open File • Operational Forward • Optical Frequency • Optional Form • Orbital Flight • Orbit File • Order Form • Out File • Output Format • Output Formatter • Outside Face • Overflow • Overflow Flag • Oxidizer-to-Fuel (ratio) • Oxidizing Flame • Oxygen Fill

OFA: Object-Free Area • Observatoire Fédéral d'Astrophysique (Canada-BC) • Office Fédéral de l'Air (Switzerland) • Off-Line Archive • Organisme de Formation Agréé

OFAC: Office Fédéral de l'Aviation Civile (Switzerland)

OFAD: Optical Field Angle Distortion

OFAP: Observational Facilities Advisory Panel

OFAT: One Factor At a Time

OFB: Operational Facilities Branch

Ofc: Office

OFC: Office • Online File Corrector • Open Financial Exchange • Operational Flight Control • Optical Frequency Conversion • Oxygen-Free Copper

OFCC: Office of Federal Contract Compliance (USA)

OFCCP: Office of Federal Contract Compliance Programs (USA)

Ofcl: Official

Ofcr: Officer

OFCS: Operational Flight Control System

OFD: One-Function Diagram

OFDA: Office of Foreign Disaster Assistance

OFDD: Ordered Functional Decision Design

OFDM: Orthogonal Frequency Division Multiplex • Orthogonal Frequency Division Multiplexing

OFDPS: Offshore Flight Data Processing System

OFDS: Orbiter Flight Dynamics Simulator • Oxygen Fluid Distribution System

OFE: Office Français d'Édition (France)

OFEFP: Office Fédéral de l'Environnement, des Forêts et des Paysages (Switzerland)

OFELIA: Optical Fibres for Electrical Industry Applications

OFES: Office Fédéral de l'Éducation et de la Science (Switzerland)

Off: Offer • Offered • Office • Office • Officer • Official • Officiate • Officiel • Officinal

OFF: Object Feasibility Filtering • Omaha Offutt (USA-NE) (location identifier)

OFFCDT: Officer Cadet

Offcl: Official • Officinal

OFFM: Office of Federal Financial Management

Offn: Offnung

Offr: Officer

OFG: Optical Frequency Generator

OFHC: Oxygen-Free High-conductive Copper • Oxygen Free High Conductivity

OFI: Operational Flight Instrumentation • Ouango Fitini (Ivory Coast) (location identifier)

OFIA: Office of Federal and International Affairs (USA)

OFID: Optical Free Induction Decay

OFIL: Observatoire Français des Industries de la Langue (France)

OFJ: Ólafsfjördur (Iceland) (location identifier)

OFK: Official Flight Kit • Optical Flight Kit • Norfolk (USA-NE) (location identifier)

Ofl: Offline

OFLS: Off-Line System

OFltr: Oil Filter

OFM: Office of Fiscal Management • Office of Flight Missions • Operational Forecast Model

OFMET: Office Fédéral de Métrologie (Switzerland)

OFML: Open Forms Markup Language

OFMS: Overseas Financial Management System

OFMT: Output Format

OFNPS: Outstate Facility Network Planning System

OFO: Office of Flight Operations • Orbiting Frog Otolith

OFP: Occluded Frontal Passage • On-the-Fly Processing • Operating Force Plan • Operational Flight Profile • Operational Flight Program • Operational Format Program • Orbiter Flight Program • Original Flight Path • Ozonforschungsprogramm

OFPP: Office of Federal Procurement Policy

OFPS: Office of Field Project Support

OFQJ: Office Franco-Québecquois de la Jeunesse

OFr: Old French

OFR: Observatoire Fédéral de Radioastrophysique (Canada-BC) • Off Frequency Rejection • Open Failure Report • Open File Report • Organization Form Registry • Original Format Recognition

OFRATEME: Office Français des Techniques Modernes d'Éducation (France)

OFRP: On-the-Fly Reprocessing

O&FS: Operations & Flight Support

OFS: Object File System • Office of the Foreign Secretary • Operations and Flight Support • Optics-Free Spectrometer • Orange Free State (South Africa) • Orbital Flight System • Orbiter Flight Simulator • Orbiter Flight System • Orbiter Functional Simulator • Output Field Separator

OFSHR: Offshore

OFSS: Old Font Selection Scheme • Open-File Services Section

Oft: Outfit

OFT: Office of Fair Trading • Operational Flight Test • Operational Flight Trainer • Optical Fiber Thermometry • Optical Fourier Transform • Orbital Flight Test • Orbital Flight Training • Outer Fix Time

OFTA: Office For Telecommunications Authority • On-The-Fly Adder

OFTDA: Office of Flight Tracking and Data Acquisition

OFTDS: Orbital Flight Test Data System

OFTE: Oscillatory Thermocapillary Flow Experiment

OFTM: On-orbit Flight Technique Meeting

OFTR: Orbital Flight Test Requirement

OFU: Observation and Fixation Unit • Ofu (American Samoa) (location identifier)

OFVR: Out-of-Field-of-View Rejection

OFX: Open Financial Exchange

OFWG: OGIP FITS Working Group

OFXB: Observatoire François-Xavier Bagnoud (Switzerland)

OFZ: Obstacle-Free Zone

ÖFZS: Österreichisches Forschungszentrum Seibersdorf (Austria)

og: Ogee

OG: Ogee • Olympic Games • Onde de Gravité • Open Group • Organizational Genetics • Or Gate • Original Gum • Ostrogoth • Outer Gimbal • Outer Gradient • Outgoing • Overground • Oxygen Gauge

OGA: Ogallala Searle (USA-NE) (location identifier) • Open GIS Architecture • Opérateur de Gestion Aéronautique • Outer Gimbal Angle • Outer Gimbal Axis

OGAR: On-Ground Attitude Reconstruction

OGAW: Organismus für gemeinsame Anlagen in Wertpapieren

OGB: Olbers Gesellschaft Bremen eV (Germany) • Orangeburg (USA-SC) (location identifier)

OGBD: Orbiter Gamma Burst Detector

OGC: Office of General Counsel

OGCM: Ocean General Circulation Model

OGD: Ogden (USA-UT) (location identifier)

OGDC: Office of Geographic Data Coordination

OGE: Office of Government Ethics • Ogeranang (Papua New Guinea) (location identifier) • Open GIS Environment • Operating Ground Equipment • Operational Ground Equipment • Optical Gyro Encoder • Out of Ground Effect

OGF: Offspring Generation Function • Operational Green's Function

OGG: Kahului - Maui Island (USA-HI) (location identifier)

OGI: Oculogyral Illusion • Oregon Graduate Institute (USA-OR) • Output Graphic Interface

OGIP: Office of Guest Investigator Programs

OGIRS: Oklahoma Geographic Information Retrieval System

OGIS: Open Goodata Interoperability Specification

OGIST: Oregon Graduate Institute of Science and Technology (USA-OR)

OGL: Ogle (Guyana) (location identifier) • Open General Licence • Overlay Generation Language

OGLE: Optical Gravitational Lensing Experiment

OGM: Optimum Gradient Method • Ordinary General Meeting • Organisme Génétiquement Modifié

OGMS: Open Gradient Magnetic Separation

OGMT: Orbiter Greenwich Mean Time

OGN: Yonaguni Jima (Japan) (location identifier)

OGO: Abengourou (Ivory Coast) (location identifier) • Orbiting Geophysical Observatory

OGOT: Oficina General de Operaciones Técnicas

OGP: Office of Global Programme

OGR: Bongor (Chad) (location identifier) • Outgoing Repeater

OGRC: On-Ground Range Compression

OGRIP: Ohio Geographically Referenced Information Program

O&GS: Operations & Ground Systems

OGS: Objective Grating Spectrometer • Offset Guidance Sensor • Ogdensburg (USA-NY) (location identifier) • Operation Ground Segment • Optical Ground Station • Optical Guidance System • Optical Guidance Systems (USA-PA) • Outer Glideslope • Oxide Glass-like Semiconductor • Oxide Glassy Semiconductor • Oxygen Generation System

OGSE: Operational Ground Support Equipment • Optical Ground Support Equipment

OGSEL: Operational Ground Support Equipment List

OGT: Observatoire Géodésique de Tahiti • Outgoing Trunk

OGV: Outlet Guide Vane • Oxygen Gauge Valve

OGW: Office of Ground Water

OGWP: Office of Ground Water Protection

OGX: Ouargla (Algeria) (location identifier)

OGY: Ott-Grebogi-York

OGZ: Vladikavkaz (Russia) (location identifier)

oh: Omni Hora

OH: Finland (international aircraft registration code) • Hydroxyl Radical • Off Hook • Office Hours • Ohio (USA) • Ohmic Heating • On Hand • Open House • Overhaul • Overhead

OHA: Ohakea (New Zealand) (location identifier) • Operating Hazard Analysis • Operational Hazard Analysis • Orbital Height Adjustment • Oscillator Housing Assembly

OHC: Northeast Cape (USA-AK) (location identifier) • Onboard Hard Copier • Optics Hand Controller • Overhead Camshaft • Oxygen Hole Center

OHD: Ohrid (Macedonia) (location identifier)

OHEA: Office of Health and Environmental Assessment

OHER: Office of Health and Environmental Research

OHF: Occupational Health Facility • Online Help Facility

OHG: Offene Handelsgesellschaft • Old High German • Operators Harmonization Group

OHGVT: Orbital Horizontal Ground Vibration Test

OHI: Organisation Hydrographique Internationale • Oshakati (Namibia) (location identifier)

OHIM: Office for Harmonization in the Internal Market (EU)

OHMS: Oil and Hazardous Materials • On Her Majesty's Service • His Majesty's Service

OHO: Ohotsk (Russia) (location identifier)

OHP: Oban (UK-Scotland) (location identifier) • Observatoire de Haute-Provence (France) • Open Hypermedia Protocol • Operational Hydrological Program • Operational Hydrology Programme • Overhead Projector

OHR: Office of Human Resources • Operational Hydrology Report • Wyk auf Föhr (Germany) (location identifier)

OH&S: Occupational Health & Safety

OHS: Hydroxyl Radical Suppressor • Occupational Health Services • Open Hyperdocument System • Open Hypermedia System (*) • Open Hypertext System

OHSU: Oregon Health Sciences University (USA-OR)

OHSWG: Open Hypermedia System Working Group

OHT: Kohat (Pakistan) (location identifier) • Orbital History Table

OHTEX: Ocean Heat Transport Experiment

OHV: Off-Highway Vehicle

OHW: Office of Hazardous Wastes

OHWM: Ordinary High Water Mark

oi: Onzes Inziens

OI: Office of Information • Office Instruction • Onzes Inziens • Operating Instruction • Operational Increment • Operational Instrumentation • Operations Interface • Optimal Interpolation • Optimum Interpolation • Orbital Insertion • Orbit Insertion • Orbiter Insertion • Orbiter Instrumentation • Ordinateur Individuel • Other Investigator • Output Impedance • Overstemming Index

OIA: Office of Internal Affairs • Office of International Activities • Office of International Administration • Office of International Affairs • Operator Information Area • Orbiter Interface Adapter

OIAC: Organisation Internationale de l'Aviation Civile

OIB: Oceanic Island Basalt • Ocean Island Basalt • Office International de Bibliographie • Official Information Base

• Operating Improvement Board • Operations Integration Branch • Orbiter Interface Box

OIBD: Office International de Bibliographie et de Documentation

OIbF: Olimpiada Iberoamericana de Física

OIC: Norwich Eaton (USA-NY) (location identifier) • Office of Information Control • Office of Instructional Consultation • Officer In Charge • Oh I See • Operations Instrumentation Coordinator • Operator's Instruction Chart • Orbital Integrated Checkout • Organisation Internationale du Commerce • Organization of the Islamic Conference

OICC: Operations Interface Control Chart

OICETS: Optical Inter-orbit Communications Engineering Test Satellite

OICO: Office of Integration and Checkout

OICP: Office of International Communications Policy

OICV: Organisation Internationale de la Commission de Valeurs

oid: Of Iets Dergelijks

OID: Object Identification • Object Identifier • Object Interaction Diagram • Octal Identifier • Oficina de Información Diplomática • Optical Image Database • Organisation Intégrée de Développement

OIDD: Orbiter Internal Density Distribution

OIDL: Object Interface Definition Language

OIE: Office International de l'Eau • Orbiter Integration Engineering • Organisation Internationale des Employeurs

OIEA: Office of Integrated Environmental Analysis • Organización Internacional para la Energía Atómica

OIEC: Organization for International Economic Cooperation

OIES: Office for Interdisciplinary Earth Studies

OIF: Old IRAF Format • Optical Interface • Optimum Index Factor • Original IRAF Format (*)

OIG: Office of Inspector General • Office of International Geology • Office of the Inspector General • Optical Imager • Organisation Intergouvernementale

OIH: Office of International Hydrology

OII: Office of Invention and Innovation (USA) • Open Information Interchange • Operations Integration Instruction

OIJ: Organisation Internationale des Journalistes

OIL: Oil City - Splane (USA-PA) (location identifier) • Optical Imaging Instrument • Orbital International Laboratory

OILR: Outer Inner Lindblad Resonance

OILSAR: Ocean-Ice-Land Synthetic Aperture Radar

OIM: Optical Interference Microscope • Orbit Insertion Maneuver • Oshima (Japan) (location identifier) • Oxygen Interaction with Materials

OIML: Organisation Internationale de Métrologie Légale

OIMS: Orbiter Ion Mass Spectrometer

OIN: Office International de Normalisation

OING: Organisation Internationale Non-Gouvernementale

OIO: Onderzoeker-In-Opleiding • Onderzoek-In-Opleiding • Operations Integration Officer

OIP: Operational Improvement Plan • Operational Improvement Program • Operations Interface Procedure • Operator Interface Process • Operator Interface Program • Optical Image Processor • Optronics Instruments and Products • Orbital Improvement Program

OIPN: Office International pour la Protection de la Nature

OIPP: Office of Intellectual Property Policy

OIPS: Optical Image Processing System

OIr: Old Irish

OIR: Office of Industrial Relations • Okushiri (Japan) (location identifier) • Open Item Report • Open Item Review • Operations Integration Review • Optical and Infrared • Optical Infrared • Orbiter Infrared Radiometer • Organisation Internationale de Radiodiffusion

OIRA: Office of Information and Regulatory Affairs

OIRE: Optical Infrared Equipment

OIRM: Office of Information Resources Management

OIRT: Organisation Internationale de Radio et Télévision

OIS: Observation Implementation Scientist • Office Information System • Operating Information System • Operational Information Service • Operational Instrumentation System • Operational Intercommunication System • Optical Imager Study • Optical Image Sensor • Orbiter Insertion System • Orbiter Instrumentation System • Orbit Injection System • Orbit Inserting System • Orbit Insertion System • Other Information Source

OISAR: Optical Inverse Synthetic Aperture Radar

OISCA: Organization for Industrial, Spiritual and Cultural Advancement

OISD: Outsourced Information Sciences Development

OISR: Open Item Status Report

OISS: Open Item Status System

OISSP: Office of Interim Space Station Program

OISTP: Oxygen-Iodine Supersonic Technology Program

OIt: Old Italian

OIT: Object Identification Test • Oita (Japan) (location identifier) • Operations-Information-Technology • Orbiter Integrated Test • Orbiter Integration Test • Orbiter Interface Test • Organisation Internationale du Travail

OITDA: Optoelectronic Industry and Technology Development Association

OITP: Office for Information Technology Policy

OIU: Observer Interface Unit • Office Interface Unit • Operator Interface Unit

OIV: Oxidizer Isolation Valve

OIVS: Orbiter Interface Verification Set

OIW: OSE Implementation Workshop

OIWFS: On-board Instrument Wavefront Sensor • On-Instrument Wavefront Sensor (*)

oJ: ohne Jahr

OJ: Official Journal • Orange Juice • Ordre du Jour • Outer Join

OJAE: Online Journal of Astronomy Education

OJC: Kansas City - Johnson Executive (USA-MO) (location identifier)

OJCCT: Online Journal of Current Clinical Trends

OJD: Official Journal Date

OJE: On-the-Job Education • On-the-Job Experience

OJEC: Official Journal of the European Communities

OJL: Optoelectronics Joint Laboratory

OJPS: Online Journal Publishing Service

OJS: Online Journal Service

OJSA: Online Journals Subscription Agreement

OJT: Office de Justification des Tirages (France) • On-the-Job Training

oK: ohne Kosten

OK: All Correct • All Right • Czech Republic (international aircraft registration code) • Oklahoma (USA) • Oklahoma-Kansas • Operatiekamer • Oxidation Kinetics • Slovakia (international aircraft registration code)

OKA: Okinawa Naha (Japan) (location identifier) • Otherwise Known As

OKAPI: Open Kernel for Access to Protected Interoperable (services)

OKB: Orchid Beach - Fraser Island (Australia-QLD) (location identifier)

OKC: Oklahoma City - Will Rogers (USA-OK) (location identifier)

OKCAC: Oklahoma City Astronomy Club (USA-OK)

OKD: Sapporo Okadama (Japan) (location identifier)

OKE: Okino Erabu (Japan) (location identifier) • Optical Kerr Effect

OKF: Okaukuejo (Namibia) (location identifier)

OKFDD: Ordered Kronecker Functional Decision Diagram

OKG: Okoyo (Congo) (location identifier)

ÖKG: Österreichische Klimatechnik Gesellschaft (Austria)

OKH: Oakham Cottesmor (UK-England) (location identifier)

OKI: Oki Island (Japan) (location identifier)

OKJ: Okayama (Japan) (location identifier)

OKK: Kokomo (USA-IN) (location identifier)

OKL: Oksibil (Indonesia) (location identifier)

Okla: Oklahoma (USA)

OKM: Okmulgee (USA-OK) (location identifier)

OKN: Okondja (Gabon) (location identifier) • Optokinetic Nystagmus

OKO: Tokyo Yokota (Japan) (location identifier)

OKP: Oksapmin (Papua New Guinea) (location identifier)

OKQ: Okaba (Indonesia) (location identifier)

OKR: Yorke Island (Australia-SA) (location identifier)

OKS: Oshkosh (USA-NE) (location identifier)

Okt: Oktober

OKT: Oktiabrskij (Russia) (location identifier)

OKU: Mokuti Lodge (Namibia) (location identifier)

OKY: Oakey (Australia-QLD) (location identifier)

ol: Oculus Laevus

Ol: Oil • Olympiade

O/L: Open Loop • Overload

OL: Oberlausitz (Germany) • Occupational Level • Oculus Laevus • Offered Load • Oldham (UK postal code) • Old Latin • Oosterlengte • Open Loop • Operating Licence • Operating Location • Operating Log • Operational Lifetime • Optimal Level • Orbital Launch • Ordered List • Organization List • Or Less • Orthogonal Linear • Overflow Level • Overlap • Overlay • Overload

ÖL: Östliche Länge

OLA: Off-Line Analysis • Optimal Localized Average • Optimally Localized Average • Orbital Lock Assembly • Orland (Norway) (location identifier)

OLAA: Open Learning Agency of Australia

OLAC: Off-Line Adaptive Computer

OLADE: Organización Latinoamericana de Energía

OLAE: Opération Locale Agri-Environnement

OLAF: Office de Lutte Anti-Fraude (UE) • On-Line Archive Facility

OLAFS: Orbiting and Launch Approach Flight Simulator

OLAP: Online Analytical Processing • Online Analytical Processor

OLAS: On-Line Archive System

OLB: Olbia (Italy) (location identifier)

OLBIN: Optical Long Baseline Interferometry Newsletter

OLBW: Open-Loop Bandwidth

OLC: Office of Library Coordination (Canada) • Online Computer

OLCA: Online Circuit Analysis

OLCP: Online Complex Processing

OLD: Off-Load Device • Old Town (USA-ME) (location identifier)

Oldb: Oldenburg (Germany)

OLDB: Online Data Bank • Online Database

OLDC: Online Data Collection

OLDECC: Ordinary Linear Differential Equation with Constant Coefficients

OLDI: Online Data Interchange

OLDP: Online Data Processing

OLDS: Online Display System • Online Log Dataset

OLDT: Ordered selection strategy with Linear resolution for Definite classes with Tabling

OLE: Object Linking and Embedding • Olean (USA-NY) (location identifier) • Opto and Laser Europe (UK) • Organizational Learning in Enterprises

OLEB: Origins of Life and Evolution of the Biosphere (periodical code)

OLED: Organic Light-Emitting Diode

OLERD: Orbiter Landing Emergency Rescue Data

OLEW: Open Learning Experimental Workshop

OLF: Orbital Launch Facility • Orbiter Landing Facility • Orbiter Lifting Frame • Outlying Field • Wolf Point (USA-MT) (location identifier)

OLG: Observatory Lustbühel Graz (Austria) • Office of Local Government • Old Low German • Open-Loop Gain

OLGA: Organic Land Grid Array

OLH: Old Harbor (USA-AK) (location identifier)

Oli: Oligocene

OLI: Ólafsvík (Iceland) (location identifier) • Optical Line Interface

OLIC: Official Language Implementation Committee

OLIF: Orbiter Landing Instrumentation Facility

OLIP: Online Instrument Package

OLIPaC: Office of Land Information Policy and Coordination

OLIS: Oxford Library Integrated System

OLJ: Olpoi (Vanuatu) (location identifier)

OLL: Organic Liquid Laser • Output Logic Level

OLLRC: Ontario Laser and Lightware Research Centre (Canada-ON)

OLM: Olympia (USA-WA) (location identifier) • One-Leg Method • Online Monitor

OLMC: Output Logic Macrocell

OLMOS: Onboard Life Monitoring System

OLMSA: Office of Life and Microgravity Sciences and Applications

OLN: Colonia Sarmiento (Argentina-CB) (location identifier)

OLO: Olomouc (Czech Rep.) (location identifier) • Online Operation • Operations Launch Order • Orbital Launch Operation

OLOM: Orbiter Lift-Off Mass

OLOS: Online Order Status • Operational Land Observation System

OLOW: Orbiter Lift-Off Weight

OLP: Off-Line Processing • Olympic Dam (Australia-SA) (location identifier) • Online Programming

OLPAC: Ohio Light Pollution Advisory Council (USA-OH)

OLPARS: Online Pattern Analysis and Recognition System

OLPR: Office for Library Personnel Resources

OLPS: Online Programming System

OLQ: Olsobip (Papua New Guinea) (location identifier)

OLR: Off-Line Reader • On-Load Reactor • Open-Loop Receiver • Open-Loop Receiver • Outer Lindblad Resonance • Outgoing Longwave Radiation

OLRT: Online Real Time

OLS: Nogales (USA-AZ) (location identifier) • Oceanographic Lidar System • Omnilink System • Online Library System • Online Search • Online Service • Online System • Operational Launch Station • Operational License Stage • Operational Line Scanner • Operational Linescan System • Optical Landing System • Optical Line Scanner • Optical Line System • Orbiting Lunar Station • Ordinary Least Squares • Over Larger Scale

OLSAP: Online Support Application Processor

OLSC: Online Scientific Computer • Online Scientific Computing

OLSCA: Orientation Linkage for a Solar Cell Array

OLSE: Ordinary Least Squares Estimators

OLSP: Operational Logistics Support Plan • Orbiter Logistics Support Plan

OLSS: Off-Line Systems List

OLT: Observatory Level Test • Online Test

OLTE: Organizational Level Test Equipment

OLTEP: Online Test Executive Program

OLTM: Optical Line Terminating Multiplexer

OLTP: Online Transaction Processing

OLTS: Online Test System • Online Time Share

OLU: Columbus (USA-NE) (location identifier)

OLUIT: Object-oriented Librarian User Interface Tool

OLUQ: Optimum Learning Vector Quantization

OLUS: Online Users Service

OLUT: Output Lookup Table

OLV: Olive Branch (USA-MS) (location identifier) • Onze Lieve Vrouw • Orbital Launch Vehicle

OLVG: Open-Loop Voltage Gain

OLVM: Open Look Window Manager

OLWM: Ordinary Low Water Mark

OLY: Olney Noble (USA-IL) (location identifier)

om: Omni Mane • Oromo (ISO language code)

Om: Ostmark (currency) (now obsolete)

O&M: Operations & Maintenance • Organization & Management • Organization & Members • Organization & Methods • Overhaul & Modification

Ö/M: Öffentlichkeitsarbeit/Marketing

OM: Object Manager • Observatoire de Marseille (France) • Office Management • Office Manager • Old Measurement • Oman (ISO country code) • Openbare Ministerie • Open Market • Operate Mode • Operating Memorandum • Operating Memory • Operations and Maintenance • Operations Maintenance • Operations Manager • Operator's Manual • Optical Marker • Optical Master • Optical Microscope • Optical Microscopy • Optical Model • Optical Module • Optical Monitor • Optimus Maximus • Orden Ministerial • Ordinary Member • Ordre de Mission • Organic Matter • Organisation and Management • Organizational Maintenance • Organizational Memory • Our Message • Outer Marker • Output Magnitude

OMA: Object Management Architecture • Observatório Municipal de Americana (Brazil-SP) • Omaha Eppley (USA-NE) (location identifier) • Operational Maintenance Activity • Operations Maintenance Area • Operations Management Application • Optical-Mechanical Assembly • Optical Multichannel Analyzer • Optical Multichannel Array • Optomechanical Assembly • Orbiter Maintenance Area • Orderly Marketing Agreement • Organizational Maintenance Activity

OMAP: Operations, Maintenance, and Administration Part

OMAPI: Office Malgache de la Propriété Industrielle (Madagascar)

OMAS: Operations Management Application Software

OMAT: Optimal Maturity Parameter

OMB: Office of Management and Budget • Omboue (Gabon) (location identifier)

OMBUU: Orbiter Mid-Body Umbilical Unit

OMC: Object Modelling Card • Observatório Municipal de Campinas (Brazil-SP) • Organisation Mondiale du Commerce • Operating and Maintenance Cost • Operation and Maintenance Center • Operation and Maintenance Cost • Operations Monitoring Computer • Optical Memory Card • Optical Monitor Camera • Optical Monitoring Camera • Orbiter Maintenance and Checkout • Organic Molecular Crystal • Organometallic Compound • Orion Molecular Cloud • Ormoc (Philippines) (location identifier)

OMCF: Operations and Maintenance Control File • Orbiter Maintenance and Checkout Facility

OMCS: Orbiter and Maintenance Communications System • Ozone Monitor Comparison System

OMCVD: Organometallic Chemical Vapor Epitaxy

OMD: Operations and Maintenance Documentation • Oranjemund (Namibia) (location identifier) • Orbiter Mating Device

OMDM: Optomechanical Display Module

OMDP: Orbiter Maintenance Down Period

OMDR: Operations and Maintainability Data Record • Operations and Maintenance Data Record • Optical Memory Disk Recorder

OME: Nome (USA-AK) (location identifier) • Open Messaging Environment • Opérateur Mathématique Électronique • Operational Mission Environment • Orbital Maneuvering Engine • Orbiter Main Engine

OMED: Ocean Minerals and Energy Division

OMEGA: Orbiting Medium Explorer for Gravitational Astrophysics

OMET: Orbiter Mission Elapsed Time

OMEWG: Orbiter Maintenance Engineering Working Group

OMEX: Ocean Margin Exchange • Ocean Margin Experiment

OMF: Mafraq - King Hussein (Jordan) (location identifier) • Object Management Facility • Object Module Format • Oil Mist Filter • Open Message Format • Operation and Maintenance of Facilities • Optical Matched Filter • Oscillatory Magnetic Field

OMFT: Optical Matched Filter Technique

OMG: Object Management Group • Omega (Namibia) (location identifier)

OMGA: Operations Management Ground Application

OMGS: Orbite, Manoeuvre et Gestion du Système

OMH: Officier Marinier Hydrographe • Orszagos Méréstügyi Hivatal (Hungary) • Urmieh (Iran) (location identifier)

OMI: Observatoire du Marché de l'Information • Office des Migrations Internationales • Open Messaging Interface • Open Microprocessor Initiative • Operational Maintenance Instruction • Operation, Maintenance and Instruction • Operations and Maintenance Instruction • Optical Mapping Instrument • Optical Media International • Organisation Maritime Internationale • Organisation Météorologique Internationale (now OMM) • Ozone Monitoring Instrument

OMIJC: Operations and Maintenance Instruction Job Card

OMIM: Online Mendelian Inheritance in Man

OMIMO: Online Multimedia Information for Maintenance and Operation

OMIPE: Office Mondial d'Information sur les Problèmes d'Environnement

OMIS: Operational Maintenance Information System • Operational Management Information System (*) • Operations Management and Information Systems • Operations and Management Information Systems • Organisation Mondiale de l'Image et du Son

OMISS: Operation and Maintenance Instruction Summary Sheet

OMJ: Omura (Japan) (location identifier)

OMK: Omak (USA-WA) (location identifier) • Orbiter Mission Kit

OML: Omkalai (Papua New Guinea) (location identifier) • Orbiter Mold Line • Organizational Maintenance Level • Outer Mold Line • Outside Mold Line

OMLA: Organizational Maintenance Level Activity

OMM: Marmul (Oman) (location identifier) • Observatoire du Mont-Mégantic (Canada-QC) • Operation and Maintenance Manual • Organisation Météorologique Mondiale • Output Message Manual

OMMH: Optical Multi-Mark Hypercube • Orbiter Maintenance Man-Hour

OMMSQA: Office of Modelling, Monitoring Systems and Quality Assurance

OMN: Oman (ISO country code) • Osmanabad (India) (location identifier)

OMNI: Organising Medical Networked Information

OMNS: Open Network Management System

OMO: Mostar (Bosnia-Hercegovina) (location identifier) • Occupational Medicine Office • One-Man-Operated • One-Man-Operation • One-Man-Operator • Other Mapping Organization

OMP: Observatoire des Matières Premières (France) • Observatoire Magnétique Planétaire • Observatoire Midi-Pyrénées (France) • Observatório Municipal de Piracicaba (Brazil-SP) • Ocean Margins Program • Operating Maintenance Plan • Operations and Maintenance Plan

OMPA: Octamethyl Pyrophosphoramide

OMPF: Operation and Maintenance Processor Frame

OMPI: Organisation Mondiale de la Propriété Intellectuelle

OMPP: Organisation Mondiale de la Presse Périodique

OMPR: Operational Maintainability Problem Reporting • Optical Mark Page Reader • Optimal Multiple Point Reassignment

OMPRA: One-Man Propulsion Research Apparatus

OMPS: Orbit Maneuvering Propulsion System

OMPT: Observed Man Point Trajectory • Observed Mass Point Trajectory (*)

OMPTE: Office of Mission To Planet Earth

OMR: Operation Management Room • Operations and Maintenance Requirement • Operations Management Room • Operations Manager's Report • Optical Mark Reader • Optical Mark Reading • Optical Mark Recognition • Oradea (Romania) (location identifier) • Orbiter Management Review • Organic Magnetic Resonance • Rial (Oman) (ISO currency code)

OMRB: Object Modelling Resource Base • Operating Material Review Board

OMRC: Operational Maintenance Requirements Catalog

OMRF: Orbiter Maintenance and Refurbishment Facility • Orbiter Modification and Refurbishment Facility

OMRP: Operations and Maintenance Requirements Plan

OMRS: Operations and Maintenance Requirements and Specifications • Operations and Maintenance Requirements Specification • Operations and Maintenance Requirements System

OMRSCB: Operations and Maintenance Requirements and Specifications Control Board

OMRSD: Operations and Maintenance Requirements and Specifications Document • Operations and Maintenance Requirements Specifications Documentation

OMRV: Operational Maneuvering Reentry Vehicle

OMS: Object Management System • Observatory Monitoring System • Omsk (Russia) (location identifier) • Onboard Maintenance System • Opcode's MIDI System • Operational Meteorological Satellite • Operations Management System • Optical Maser System • Optical Modulation System • Optimization Methods and Software • Orbital Manoeuvering System • Orbital Mapping System • Orbital Multifunction Satellite • Orbiter Manoeuvering System • Organic Mass Spectroscopy • Organisation Mondiale de la Santé • Origin of the Modern State • Output per Man Shift

OMSF: Office of Manned Space Flights

OMSI: Oregon Museum of Science and Industry (USA-OR)

OMSP: Operational Maintenance Support Plan

OMT: Object-Modelling Technique • Object Model Template • Optical Modelling Tool • Orthomode Transducer

OMTPE: Office of Mission To Planet Earth

OMU: Operational Measuring Unit • Operational Mock-Up • Optical Measuring Unit • Other Minority University (USA)

OMV: Orbital Manoeuvering Vehicle • Orbiter Manoeuvering Vehicle • Oxygen Manual Valve

OMVA: Optimum Magnified Visual Angle

OMW: Object Management Workbench

OMY: Oddor Meanche (Cambodia) (location identifier)

on: Omni Nocte

ÖN: Österreichisches Normungsinstitut (Austria)

ON: Observatório Nacional • Observatorio Nacional • Occultation Newsletter • Octane Number • Ontario (Canada) • Operation Notice • Optical Nerve • Original Negative

ONA: Office of National Assessments • Online Archive • Winona (USA-MN) (location identifier)

ONAC: Operations Network Administration Center

ONAL: Off Network Access Line

ONAP: Orbit Navigation Analysis Program

ONB: Ononge (Papua New Guinea) (location identifier) • Orthonormal Basis

ONBA: Observatorio Naval de Buenos Aires (Argentina-BA)

ONBD: On Board

ONBI: Office National des Brevets d'Inventions (France)

ONBT: Orbiter Neutral Buoyancy Trainer

ONC: Open Network Computing • Open Networking Computing • Operational Navigational Chart • Operational Navigation Chart • Ordinary National Certificate • Orion Nebula Cluster

ONCE: Online Network for Continuing Education

OND: October-November-December • Ondangwa (Namibia) (location identifier) • Operator Need Date • Ordinary National Diploma

ONDCP: Office of National Drug Control Policy

ONDINES: Ondes Internes

ONE: Onepusu (Solomon Islands) (location identifier) • OPAC Network in Europe • Open Network Environment • Organization Number of Employees

ONEB: Office of National Environmental Board

ONER: Oceanic Navigational Error Report

ONERA: Office National d'Études et de Recherches Aérospatiales (France)

ONERN: Oficina Nacional de Evaluación de Recursos Naturales (Peru)

ONF: Office National des Forêts (France) • Old North French

ONFCSM: Observatorio Nacional de Física Cosmica - San Miguel (Argentina)

ONFr: Old North French

ONG: Mornington Island (Australia-QLD) (location identifier) • Organisation Non-Gouvernementale • Organización No-Gubernamental

ONGC: Oil and Natural Gas Commission

ONH: Oneonta (USA-NY) (location identifier)

ONI: Moanamani (Indonesia) (location identifier) • Oficina Nacional de Información • Operator Number Identification • Optical Network Interface • Organismo Nacional de Información • OSI Network Interface

ONIITEM: Oficina Nacional de Invenciones, Información Técnica y Marcas (Cuba)

ONIM: Oh No It's Monday

ONISEP: Office National d'Information sur les Enseignements et les Professions (France)

ONIST: Office National d'Information Scientifique et Technique

ONL: O'Neill (USA-NE) (location identifier) • Online • Optique Non Linéaire

ONLS: Online System

ONM: Office National Météorologique • Socorro (USA-NM) (location identifier)

ONMS: Open Network Management System • Orbiter Neutral Mass Spectrometer

ONN: Onion Bay (USA-AK) (location identifier)

ONNUM: Oficina Nacional de Normas y Unidades de Medida (Costa Rica)

ONO: Ontario (USA-OR) (location identifier) • Or Nearest Offer • Ouest-Nord-Ouest

Onomat: Onomatopée • Onomatopeya

ONP: Newport (USA-OR) (location identifier) • Open Network Provision

ONPI: Office National de la Propriété Industrielle (Algeria) • Office of News and Public Information

ONPT: Office National des Postes et Télécommunications

ONR: Monkira (Australia-QLD) (location identifier) • Office of Naval Research (USA)

ONRJ: Observatório Nacional de Rio do Janeiro (Brazil-RJ)

ONRS: Organisme National de la Recherche Scientifique (Algeria)

ONS: Online News Service • Onslow (Australia-WA) (location identifier) • Oversampled Noise-Shaping

ONSHR: Onshore

Ont: Ontario (Canada)

ONT: Ontario (USA-CA) (location identifier)

ONTC: Optical Networks Technology Consortium

ONU: Ohio Northern University (USA-OH) • Olivet Nazarene University (USA-IL) • Ono i Lau (Fiji) (location identifier) • Optical Networking Unit • Optical Network Unit • Organisation des Nations Unies

ONUDI: Organisations des Nations Unies pour le Développement Industriel

ONWI: Office of Nuclear Waste Insulation

ONX: Colón (Panama) (location identifier)

ONY: Olney (USA-TX) (location identifier)

Onz: Onzijdig

o/o: On Order • Order of

oo: On Order • Order of

oO: ohne Ort

O/o: Offers Over

O/O: On/off • Optical to Optical

OO: Belgium (international aircraft registration code) • Object-Oriented • Once Over • On Order • Operation Order • Organisateur d'Ordre

OOA: Object-Oriented Analysis • Optimum Orbital Altitude • Oskaloosa (USA-IA) (location identifier) • Out Of Action • Out Of Area

OOA&D: Object-Oriented Analysis & Design

OOAD: Object-Oriented Analysis and Design • Object-Oriented Application Design

OOB: Object-Oriented Bibliography • Out Of Band

OOBE: Object-Oriented Business Engineering • Out-Of-Box Experience

OOBI: Objet Orbitant Bien Identifié

OOC: On-Off Control

OOD: Object-Oriented Design • Object-Oriented Development • Object-Oriented Directory • Officer Of the Day • Officer Of the Deck • Operational Oceanography Center • Orbiter On Dock

OODA: Observation-Orientation-Décision-Action • Observe-Orient-Decide-Act

OODB: Object-Oriented Database

OODBMS: Object-Oriented Database Management System

OODCE: Object-Oriented Distributed Computing Environment

OODP: Out-Of-Detent Pitch

OODR: Out-Of-Detent Roll

OODV: Orbit-On-Demand Vehicle

OOE: Object-Oriented Environment • Out-Of-Ecliptic (later ISPM, now Ulysses)

OOERM: Object-Oriented Entity-Relationship Model

OOFDM: Object-Oriented Functional Data Model

OOFF: Object-Oriented Feature Finder

OOFS: Object-Oriented File System

OOG: Object-Oriented Graphics

OOGL: Object-Oriented Graphics Language

OOGP: Object-Oriented Genetic Programming

OOH: Orbital Operations Handbook

OOHDM: Object-Oriented Hypermedia Design Method • Object-Oriented Hypermedia Design Model (*)

OOI: Other Online Issue

OOIS: Object-Oriented Information System

OOK: Toksook Bay (USA-AK) (location identifier)

OOL: Gold Coast - Coolangatta (Australia-QLD) (location identifier) • Object-Oriented Language • Object-Oriented Library • Object-Oriented Listing • Operator-Oriented Language • Optimized Optical Link • Out Of Limits • Out Of Lock

OOM: Cooma (Australia-NSW) (location identifier) • On-Orbit Maintenance

OOMM: Organizational Operations and Maintenance Manual

OOMP: Object-Oriented Management Planning

OOMPS: Open Ocean Monitoring Payload Studies

OOMR: Oscillating Oblique Magnetic Rotator

OOMS: Open Ocean Monitoring Studies

OON: Object-Oriented Numerics

OONI: Objet Orbitant Non Identifié

OOO: Out Of Order • Ozone

OOOI: Out-Off-On-In

OOOS: Object-Oriented Operating System

OOP: Object-Oriented Pattern • Object-Oriented Programming • Office for Official Publications • Out Of Plane • Out Of Position • Out Of Print

OOPEC: Office for Official Publications of the European Communities

OOPL: Object-Oriented Programming Language

OOPS: Object-Oriented Programming and Systems • Object-Oriented Programming System • Okayama Optical Polarimetry and Spectroscopy • Operational Ozone Processing System • Operational Ozone Product System

OOPSLA: Object-Oriented Programming Systems Language and Applications

OOR: Mooraberree (Australia-QLD) (location identifier) • On-Orbit Review

OORA: Object-Oriented Requirement Analysis

OORALD: Object-Oriented Requirements Analysis and Logical Design

OOS: Object-Oriented Software • Object-Oriented System • Office for Outer Space • Off-line Operating Simulator • On-Orbit Station • On-Orbit Support • Optical Oscillator Strength • Orbit-to-Orbit Shuttle • Orbit-to-Orbit Stage • Out Of Service

OOSA: Object-Oriented Systems Analysis • Office for Outer Space Affairs (UNO)

OOSD: Object-Oriented Software Development • On-Orbit Support Document • Orbital Operations Support Document

OOSDP: On-Orbit Station Distribution Panel

OOSE: Object-Oriented Software Engineering

OOSS: On-Orbit Support System

OOT: Object-Oriented Technology • Onotoa (Kiribati) (location identifier) • Out Of Tolerance • Out Of Town

OOTM: Object-Oriented Testing and Maintenance

OOTW: Operations Other Than War

OOUI: Object-Oriented User Interface

oOuJ: ohne Ort und Jahr

OOV: Object-Oriented View • Out Of Vocabulary

OOZE: Object-Oriented Z Environment

o/p: Out of Print

op: Opacus • Opaque • Opera • Operator • Opinion • Opposite • Optical • Optimus • Opus

oP: Ordentlicher Professor

O/p: Orden de Pago

Op: Operation • Operational • Operator • Optical • Opus • Output

O/P: Out of Print • Output

O&P: Operations & Procedures

OP: Observation Post • Observatoire de Paris (France) • Office of Personnel • Office of Policy • Oficina Principal • Old Persian • Ooty Pulsar • Opacity Project • Open • Open Policy • Operate • Operating Plan • Operating Procedure • Operation • Operational • Operational Phase • Operational Priority • Operational Procedure • Operations Plan • Operator • Optical • Optical Probe • Orbital Period • Orbital Probe • Ordering Possibility • Ordinateur Personnel • Orthogonal Polarization • Osmotic Pressure • Other People • Other Procurement • Outer Panel • Out of Print • Output • Output Primary • Outside Plant • Ouvrier Professionnel • Overproof • Oxide Precipitate • Oxygen Purge

Opa: Opaque

OPA: Kópasker (Iceland) (location identifier) • Office of Price Administration • Office of Public Affairs • Offre Publique d'Achat • Online Privacy Alliance (USA) • Opaque • Operational Amplifier • Operations Planning Analysis • Optical Parametric Amplification • Orbiter Plasma Analyzer • Output Plate Assembly • Overseas Publishers Association

OPAAL: Optimized Portable Algorithms and Application Libraries

OPAC: Office Public d'Aménagement et de Construction (France) • Online Payment and Collection System • Online Public Access Catalog (*) • Open Public Access Catalog • Operations Planning Advisory Committee • Organisme Paritaire Collecteur Agréé (France)

OPAEP: Organisation des Pays Arabes Exportateurs de Pétrole

OPAL: Omni-Purpose Apparatus for LEP • Operational Performance Analysis Language • Optical Platform Alignment Linkage • Optimized Portable Application Library

OPANAL: Organismo para la Proscripción de las Armas Nucleares en la America Latina

OPAQUE: Optical Atmospheric Quantities in Europe

OPAS: Operational Public Address System • Overpass

OPAX: Opération Pilote Aristote X500

OPB: Observation Planning Branch • Open Bay (Papua New Guinea) (location identifier) • Oxidizer Preburner

OPBOV: Oxidizer Preburner Oxidizer Valve

OPBS: Orbiter Payload Bay Simulator

OPC: Observing Programme Committee • Ocean Products Center • Office Products Center • OLE for Process Control • Online Public Catalogue • OpenGL Performance Characterization • Operations Planning Center • Optical Particle Counter • Optical Phase Conjugation • Optical Photoconductor • Optical Proximity Correction • Organic Photoconducting Cartridge • Organization Postcode • Originating Point Codes

OPCA: Organisme Collecteur Paritaire Agréé (France)

OPCGA: Observable Protein Crystal Growth Apparatus

op. cit.: Opera Citata • Opere Citato

OPCODE: Operational Code

OPCS: Office of Population Censuses and Surveys (UK)

OPCT: Operations Planning and Control Team

OPCVM: Organisme de Placements Collectifs en Valeurs Mobilières (France)

OPD: Observatorio do Pico dos Dias (Brazil) • Office of Program Development • Operand • Operations and Processing Division • Operations Department • Optical Particle Detector • Optical Path Delay • Optical Path Difference • Optical Proximity Detector • Orbiting Propellant Depot

OPDAR: Optical Detection And Ranging

OpE: Once per Exposure

OPE: Offre Publique d'Échange • Once Per Exposure • Operational Planning Estimate • Operational Processing Equipment • Operations Project Engineer • Operator Product Expansion • Optical Pointing Error • Optical Probe Experiment • Orbiting Primate Experiment • Other Project Element • Outer Planets Explorer

OPEC: Organization of Petroleum-Exporting Countries

OPECST: Office Parlementaire d'Évaluation des Choix Scientifiques et Technologiques

OPEEPE: Organisation Européenne d'Études Photogrammétriques Expériment ales

OPEF: Overall Plume Enhancement Factor

OPEI: Office of Public Education and Information

OPEIU: Office and Professional Employees International Union

OPEN: Optical Pan European Network • Origin of Plasma in Earth's Neighborhood

OPEOS: Outside Plant Engineering and Operations System

OPEP: Orbital-Plane Experiment Package • Organisation des Pays Exportateurs de Pétrole

Oper: Operate • Operating

OPER: Operate • Operations • Operator

OPERA: Onde di Plasma e Radiazioni Aurorali • Operational Programme for the Exchange of (weather) Radar (information) • Operations for ALIS

OPEST: Oil Protection of Emission System Test

OPET: Organization for the Promotion of Energy Technologies

OPEX: Olympus Propagation Experiment • Olympus Propagation Experimenter • Operating Expenses

OPF: Miami - Opa Locka (USA-FL) (location identifier) • Optical Propagation Facility • Orbiter Processing Facility

OPFC: Orbiter Pre-Flight Checklist

OPFL: Observer Personal Feed Line

ÖPG: Österreichische Physikalische Gesellschaft (Austria)

OPG: Optical Parametric Generation • Optical Precipitation Gauge

OPGC: Observatoire de Physique du Globe de Clermont-Ferrand (France)

OPGT: Outer Planets Grand Tour

OPGUID: Optimum Guidance

Oph: Ophiucus-Ophiuchi (constellation)

OPH: Observatoire Planétarium de la Hague (France) • Operational Propellant Handling

OPHF: Orbital Polarized Hartree-Fock

OPHLM: Office Public d'Habitations à Loyer Modéré (France)

OPHS: Operational Propellant Handling System

Ophtal: Ophtalmologist • Ophtalmology

OP&I: Office of Patents & Inventions

OPI: Oenpelli (Australia-NT) (location identifier) • Office de la Propriété Industrielle • Office of Primary Interest • Office of Public Information • Open Prepress Interface • Operations Procedure Instruction • Operator Interface • Opposite Polarity Intrusion • Orbital Position Indicator • Orbiter Payload Interrogator • Organismo Público de Investigación

OPIDF: Operational Planning Identification File

OPIE: One-time Password In Everything

OPIM: Order Processing and Inventory Monitoring

OPIS: Orbiter Prime Item Specification • Organismo Público de Investigación (Spain)

OPIT: Opération Pilote Interministérielle de Télédétection (France) • Oxide Powder in Tube

OPJ: Officier de Police Judiciaire

Opl: Operational

OPL: Observatoire Populaire de Laval (France) • Online Programming Language • Opelousas - Saint Landry Parish (USA-LA) (location identifier) • Open Problem List • Operational • Optical Path Length • Orbiter Parts List • Organiser Programming Language

OPLA: Optical Programmable Logic Array

op. laud.: Opere Laudato

OPLE: Omega Position Location Experiment • Optical Path Length Equalizer

OPLF: Orbiter Processing and Landing Facility

OPLIN: Ohio Public Library Information Network

OPLSS: Optimized Portable Life-Support System

OPLT: Operational Probable Level of Traffic

OpM: Operations per Minute • Output per Man

OPM: Object Protocol Model • Observatoire de Paris-Meudon (France) • Office of Personnel Management • Office of Prime Minister • Office of Program Management • Operations Message • Operations Per Minute • Operations Procedure Manual • Operator Programming Method • Optical Power Meter • Optical Properties Monitor • Orbital Propulsion Module • Other People's Money • Outer Planet Mission • Output Per Man • Outside Plant Module

OpMS: Output per Man Shift

OPMS: Output Per Man Shift

OPMT: Observatoires du Pic du Midi et de Toulouse (France) (now OMP)

Opn: Operation • Opinion • Option

OPN: Ondes dans les Plasmas Naturels

Opnr: Opener

Opns: Operations

OPO: Office of Public Outreach • Offre à Prix Ouvert • One-Person Operation • One-Person Operator • Operations Order • Operations Project Office • Oporto - Dr. Francisco Sa Carneiro (Portugal) (location identifier) • Optical Parametric Oscillator • Orbiter Project Office • Orbiting Planetary Observatory • Oscillation Paramétrique Optique • Porto - Dr. Francisco Sa Carneiro (Portugal) (location identifier)

OPOCE: Office des Publications Officielles des Communautés Européennes

OPOL: Optimization-Oriented Language

OPOV: Oxidizer Preburner Oxidizer Valve

opp: Opera • Opuses

Opp: Opera • Opportunity • Oppose • Opposed • Opposite • Opposition • Opuses

OPP: Octal Print Punch • Office of Pesticide Programs • Office of Polar Programs • Oficina de Planeamiento y Presupuesto (Uruguay) • Operational Positional Procedure • Opportunity • Opposite • Optical Precipitate Profiles • Other Personal Property • Other Physical Principles • Other Programme Participation • Outer Planet Project • Out of Print at Present • Oxygen Partial Pressure

OPPAR: Orbiter Project Parts Authorization Request

OPPE: Office of Policy, Planning, and Evaluation • Operations Planning Project Engineer

OPPL: Orbiter Project Parts List

OPPS: Operations Planning and Project Support

OPPT: Office of Pollution Prevention and Toxics

OPPTS: Office of Prevention, Pesticides and Toxic Substances

OPPTY: Opportunity

Oppy: Opportunity

OPQ: Occupational Personality Questionnaire

Opr: Operate • Operator

OP&R: Operations Planning & Requirements

OPR: Office of Planning and Research • Office of Primary Responsibility • Officer Performance Report • Operational Project Requirement • Operations Planning Review • Operations Problem Report • Operator • Optical Page Reader • Optical Page Reading • Optical Pattern Recognition • Orbiter Payload Recorder • Organisme Public de Recherche • Ortho/Para Ratio • Output Product Request • Overall Pressure Ratio • Oxygen Pressure Regulator

OPRA: Occupational Pension Regulatory Authority

OPRD: Organisme Public de Recherche et Développement

Oprg: Operating

OPRI: Office de Protection contre les Rayonnements Ionisants (France)

OPROM: Optical Programmable Read-Only Memory

OPRV: Oxygen Pressure Relief Valve

Ops: Operates • Operations • Operations per Second

OpS: Operations per Second

OPS: Office of Price Stabilization • Off-Premises Station • Open Profile Standard • Open Profiling Standard • Operational Paging System • Operational Power Supply • Operational Protection System • Operational Sequence • Operations • Operations Per Second • Operations Sequence • Opération Ponctuelle de Symétrie • Optical Processing

System • Optical Propagation Study • Optical Sensor • Oracle Parallel Server • Orbiter Project Schedule • Orbiting Primate Spacecraft • Oxygen Purge System • Sinop (Brazil-MT) (location identifier)

OPSA: Origin and evolution of Planetary and Satellite Atmospheres

OPSB: Orbiter Processing Support Building • Output Product Status Block

OpSCN: Optical Sciences Center Newsletter (periodical code)

OPScr: Opération Ponctuelle de Symétrie Cristallographique

OPSDB: Operations Database

OPSDBA: Operations Database Administrator

OPSEC: Open Platform for Secure Enterprise Connectivity

OPSF: Orbital Propellant Storage Facility

OPSR: Operations Supervisor

OPSS: Office of Public Service and Science • Orbital Propellant Storage System

OPSN: Output Product Status Number

OPSR: Output Product Status Request

OPSW: Operational Switch

OPSWG: Outer Planets Science Working Group

Ópt: Óptica

Opt: Optical • Optics • Optimal • Optimum • Option • Optional • Optique

OPT: Observatory-Planetarium-Theater • Open Protocol Technology • Operational Pressure Transducer • Operations Planning Team • Optical Point Transfer • Optics • Optimum • Option • Orbiting Planetary Telescope • Orientation Phase Transition • Overhead Projection Transparency

OPTB: Operational Program Time Base

OptCo: Optics Communications (periodical code)

OptD: Doctor of Optometry

OPTE: Organisme pour la Promotion des Technologies Énergétiques

OptEn: Optical Engineering (periodical code)

OPTIC: Optical Procedural Task Instruction Compiler

OPTICAL: Optics and Topography Involving the Cornea And Lens

OPTID: Optical Identification

Optl: Optional

OptL: Optics Letters (periodical code)

OptLE: Optics and Lasers in Engineering (periodical code)

OptLT: Optics and Laser Technology (periodical code)

OptN: Optics News (periodical code)

OPTOL: Optimized Test-Oriented Language

OPTOMA: Ocean Prediction Through Observation, Modelling and Analysis

OptQE: Optical Quantum Electronics (periodical code)

OPTR: Optical Punched-Tape Reader

OptSp: Optical Spectra (periodical code)

Opty: Opportunity

OPU: Balimo (Papua New Guinea) (location identifier) • Open University (Starlink internal code) • Operations Priority Unit • Overlay Processor Unit

OPUS: Observation/Post-observation Unified System • Octal Program Updating System • Office of Public Understanding of Science • Olympus Payload Utilisation Secretariat • Offshore Permanent Upwelling Structure • Organization of Persistant Upwelling Structure • OSS and PODPS Unified Systems • Outstation Production Unified System

OpV: Ohms per Volt

OPV: Offre Publique de Vente • Ohms Per Volt

OPW: Office of Public Works • Opuwa (Namibia) (location identifier)

OPWHEBPTYD: Other People Who Have Even Bigger Problems Than You Do

OPWS: Orbital Payload Work Station

OPX: Off-Premises Exchange • Off-Premises Extension

OQ: Optical Quality

OQB: Oscillation Quasi-Biennale

OQC: Office of Quality Control

OQD: Optical Quantum Detector

OQL: Object Query Language • Observed Quality Level • Online Query Language

OQP: Optimum Qualification Procedure

OQPSK: Offset Quadratic Phase Shift Keying

OQQ: Optically Quiet QSO

OQS: Object Query Service

or: Oriya (ISO language code)

Or: Orient • Oriental • Oriente • Origen • Original • Other

O/R: On Request • Outside Radius • Outside Relief • Owner's Risks • Oxygen Relief

O&R: Overhaul & Repair

OR: Object Relationship • Observatoire de Rouen (France) • Occurrence Report • Odds Ratio • Official Receiver • Old Roman • Olympic Record • Ondernemingsraad • On Return • Operating Resources • Operating Room • Operationally Risky • Operational Readiness • Operational Report • Operational Requirement • Operational Research • Operational Right • Operation Record • Operations Requirement • Operations Research • Operations Review • Operations Room • Optical Reader • Orderly Room • Oregon (USA) • Originating Register • Outer Roll • Out of Range • Output Register • Output Relevance • Outside Radius • Owner's Risk • Own Risk • Oxygen Relief

OR&A: Operational Research & Analysis • Operations Research & Analysis

ORA: Observatoire Régional de l'Air (France) • Occultation Radiometer • Operational Readiness Assessment • Operational Requirements Analysis • Optical Reference Axis • Optical Research Associates • Optimum Repair Analysis • Orán (Argentina-SA) (location identifier) • O'Reilly and Associates • Output Register Address

ORAC: Observatory Reduction and Acquisition Control • Oxygen Radical Absorbance Capacity

ORACLE: Oak Ridge Automatic Computer and Logical Engine • Online Retrieval and Computational Language for Economists • Optional Reception of Announcements by Coded Line Electronics

ORACT: Operational Readiness and Confidence Test

ORAM: Output Random Access Memory

ORAN: Organisation Régionale Africaine de Normalisation

ORAP: Organisation Associative Parallèle

ORAS: Oil Region Astronomical Society (USA-PA)

ORASA: Operational Research And Systems Analysis

Orat: Oration • Orator • Oratorical • Oratorically • Oratorio • Oratory

ORATOS: Orbit and Attitude Operating System • Orbit and Attitude Operations System • Orbit Attitude Operations System

ORAU: Oak Ridge Associated Universities (USA-TN)

Orb: Orbit • Orbital • Orbiter • Orbiting

ORB: Object Request Broker • Observatoire Royal de Belgique (Belgium) • Omnidirectional Radio Beacon • Operations Review Board • Optimum Radiation Baffle • Örebro (Sweden) (location identifier) • Orbit • Orbital • Orbiter • Orebro (Sweden) (location identifier) • Orthogonal Recursive Bisection • Outer Radiation Belt

ORBIS: Orbiting Radio Beacon Ionospheric Satellite

ORBS: Orbital Rendezvous Base System • Oregon Rule-Based System

ORC: Operational Readiness Check • Operations Request Card • Operations Research Center • Orbital Research Centrifuge • Organic Ranking Cycle • Orocue (Colombia) (location identifier)

ORCH: Orchard

ORCHIS: Oak Ridge Computerized Hierarchical Information System

ORCHRD: Orchard

ORCLAC: Oficina Regional para la Cultura en Latina America y el Caribe

ORCUS: Orbital Recording Capability and Update System

ORCYT: Oficina Regional de Ciencia Y Tecnologia

Ord: Orden • Order • Ordinaire • Ordinary • Ordinal • Ordinance • Ordinary • Ordnance

ORD: Chicago O'Hare (USA-IL) (location identifier) • Object Relation Diagram • Office of Research and Development • Operational Rate Distortion • Operational Readiness Data • Operational Readiness Date • Operational Readiness Demonstration • Operational Ready Data • Operational Ready Date • Operational Requirements Document • Operations Requirements Document • Optical Reference Device • Optical Rotary Dispersion • Orbital Requirements Document • Ordnance

ORDB: Object Relational Database

ORDBMS: Object Relational Database Management System

ORDC: Orbiter Data Reduction Center

ORDEAL: Orbit Rate Display - Earth And Lunar

ORDIS: Optical Reading Direct Input System

Ordn: Ordnance

Ordza: Ordenanza

Ore: Oregon (USA)

ORE: Occupational Radiation Exposure • On-orbit Repair Experiment • Operational Readiness Evaluation • Operational Readiness Exercise • Orléans (France) (location identifier) • Output Request Element

Oreg: Oregon (USA)

ORELA: Oak Ridge Electron Linear Accelerator

OREO: Orbiting Radio Emitting Observatory

OREX: Orbital Reentry Experiment • Orbiting Reentry Experiment

Orf: Orifice

OR/F: Orbiter Refrigerator/Freezer

ORF: Norfolk (USA-VA) (location identifier) • Open Reading Frame • Orbiter Refrigerator/Freezer • Orifice • Portsmouth (USA-VA) (location identifier) • Virginia Beach (USA-VA) (location identifier)

ORFA: Oxygen Resonance Fluorescence and Absorption

Orfebr: Orfebrería

ORFEUS: Orbiting and Retrievable Far and Extreme Ultraviolet Spectrometer (*) • Orbiting Far and Extreme Ultraviolet Spectrometer • Orbiting Retrievable Far and Extreme Ultraviolet Spectrometer

ORFI: Optical-RF Interface

Org: Organic • Organism • Organization • Organizational • Organize • Organized • Organizer • Origin • Original

ORG: Optical Rain Gauge • Organization • Paramaribo - Zorg en Hoop (Suriname) (location identifier)

Organ: Organic • Organization

OrGeo: Organic Geochemistry (periodical code)

ORGPHC: Orographic

Orgzn: Organization

ORH: Office of the Regional Hydrologist • Operations Requirement Handbook • Worcester (USA-MA) (location identifier)

Ori: Origin • Original • Orion-Orionis (constellation)

ORI: Occurrence of Reinforcing Information • Ocean Research Institute • Office of Research and Information • Office of Research Integrity • Ongoing-Research Information • Operational Readiness Inspection • Operations Research Inc. • Orbital Replacement Instrument (*) • Orbit Replaceable Instrument • Original • Port Lions (USA-AK) (location identifier)

ORID: Orthorectified Image Data

ORIDE: Override

ORIEN: Orientation

Orig: Origin • Original • Originally • Originate • Originated

ORIG: Origin

ORION: Oak Ridge Isochronous Observation Network (USA-TN) • Optically Recorded Information Online Network

ORISE: Oak Ridge Institute for Science Education (USA-TN)

ORJ: Orinduik (Guyana) (location identifier)

Ork: Orkney Islands

ORK: Cork (Ireland) (location identifier) • Optimizable Recursive Keyword

Orkn: Orkney Islands

ORL: Optical Return-Loss • Orbital Research Laboratory • Orlando (USA-FL) (location identifier)

ORLA: Optimum Repair Level Analysis

OrLi: Origins of Life (periodical code)

ORM: Northampton (UK-England) (location identifier) • Object Role Modelling • Observatorio del Roque de los Muchachos (Canary Islands) • Ole Rømer Museet (Denmark) • Operations Reference Model • Optical Reference Manual • Optical Reference Measurement • Optical Remote Module

ORMS: Operating Resource Management System

ORMSS: Ohio River Main Stem Study

ORMU: Orbital Remote Manoeuvering Unit

Orn: Orange • Ornitología

ORN: Oran - Es Senia (Algeria) (location identifier)

OrNav: Ortung und Navigation (periodical code)

ORNL: Oak Ridge National Laboratory (USA-TN)

ORO: Office of Regional Operations • Old Royal Observatory (UK-England) • Orbiter Radio Occultation • Yoro (Honduras) (location identifier)

OROM: Optical Read-Only Memory

OROS: Optical Read-Only Storage

ORP: Optical Reflecting Projector • Optional Retirement Program • Orapa (Botswana) (location identifier) • Orbital Rendezvous Procedure • Oxidation-Reduction Potential

ORPA: Office of Research and Project Administration • Orbiter Retarding Potential Analyzer

ORPC: Object Remote Call Procedure

ORPICS: Orbital Rendezvous Positioning, Indexing and Coupling System

ORQ: Norwalk (USA-CT) (location identifier)

ORR: Operational Readiness Review • Operational Ready Rate • Operations Requirements Review • Orbital Readiness Review • Orbital Rendezvous Radar • Orbit Readiness Review • Yorketown (Australia-SA) (location identifier)

ORRH: Office of the Regional Research Hydrologist

ORRT: Operational Readiness and Reliability Test

Ors: Others

ORS: Old Red Sandstone • Online Retrieval Service • Online Retrieval System • Operational Research Society (UK) • Optical Redshift Survey • Orbital Refueling System • Orbital Reservicing System • Orbital Resupply System • Orbiter Refueling System • Orbiter Relay Simulator • Orbiting Research Satellite • Organization Rating Scale • Orpheus Island Resort (Australia-QLD) (location identifier) • Oscillographic Recording System • Output Record Separator • Overseas Research Student

ÖRS: Österreichische Raumfahrt- und Systemtechnik GesmbH (Austria)

ORSA: Oceanic Remote Sensing Assembly • Ogive Recovery System Assembly • Operations Research Society of America (now merged with TIMS into INFORMS) • Organizzazione Ricerche e Studi d'Astronomia (Italy)

ORSAC: Oak Ridge Systems Analysis Code

ORSDI: Oak Ridge Selective Dissemination of Information (system)

ORSER: Office for Remote Sensing of Earth Resources • Office of Remote Sensing for Earth Resources

ORSO: Ottawa River Solar Observatory (Canada-ON)

ORSTOM: Office de la Recherche Scientifique et Technologique d'Outre-Mer (France) • Office de Recherche Scientifique et Technique d'Outre-Mer (France)

ORT: Northway (USA-AK) (location identifier) • Ooty Radiotelescope (India) • Operational Readiness Test • Optical Ranging Telescope • Optimum Resolution Technique • Orbital Rendezvous Technique • Orbiting Radio Telescope • Orbit Readiness Test

ORTF: Office de Radiodiffusion-Télévision Française (France) (now Radio France)

Orth: Orthodox

ORTHOG: Orthogonal

ORTN: Oriental Robotic Telescope Network

Ortogr: Ortografía

ORU: Operational Readiness Unit • Optical Reference Unit • Orbital Replaceable Unit • Orbital Replacement Unit (*) • Orbit Replaceable Unit • Orbit Replacement Unit • Organized Research Unit • Oruro (Bolivia) (location identifier)

ORV: Noorvik (USA-AK) (location identifier) • Off-Road Vehicle • Orbital Rescue Vehicle • Orbital Research Vehicle

ORW: Ormara (Pakistan) (location identifier)

ORX: Oriximiná (Brazil-PA) (location identifier)

ORY: Paris Orly (France) (location identifier)

ORZ: Orange Walk (Belize) (location identifier) • Outer Radiation Zone

o/s: Old Style • On Sale • Out of Service • Out of Stock • Outsize • Outstanding

os: Oculus Sinister • Old Series • On Station • Outside

O/s: Out of Stock • Outstanding

Os: Osmium

O&S: Operations & Support

O/S: Old Style • Operating System

OS: Object Space • Observation Software • Observation Summary • Observatory Scientist • Observing Station • Observing System • Oculus Sinister • Odd Symmetric • Old Saxon • Old Series • Old Style • On Sale • On Sample • On Schedule • On Side • Open System • Operating Schedule • Operating Software • Operating System • Operational Sequence • Operational Sheet • Operator Service • Opposite Sense • Opposition Spike • Optical Scanning • Optogalvanic Signal • Optospinography • Ordinary Seaman • Ordnance Survey • Organizational Source • Oscan • Ostyak • Other Sources • Otherwise Specified • Out of Stock • Output Secondary • Outsize • Outstate • Ouvrier Spécialisé • Sulfur Monoxide

OSA: Object System Adapter • Old Style Antique • Open Systems Adapter • Open Systems Architecture • Operational Support Area • Operations Support Area • Optical Society of America (USA-DC) • Optimal Scheduling Algorithm • Osaka (Japan) (location identifier) • Outer Soundings Array • Overseas Sterling Area

OSAC: Oxford System of Automated Cartography

OSAD: Outer Space Affairs Division

OSAJ: Optical Society of America Journal (periodical code)

OSAM: Overflow Sequential Access Method

OSAR: Optical Storage And Retrieval

OSAS: Observation du Soleil Au Sol • Oxford Space and Astronomical Society (UK-England) (see now OUSAS)

OSAT: Office of Space Access and Technology • Optical Sensor And Tracker

OSATES: Ocean Space Advanced Technologies European Show

OSB: Ocean Studies Board • Operations Software Branch • Operations Support Building • Optimal Search Behavior • Osage Beach (USA-MO) (location identifier)

Osc: Oscillate • Oscillation • Oscillator

OSC: Oblate Spherical Coordinates • Observatorio San Calixto (Bolivia) • Ocean Science Committee • Office of Scientific Computing • Office of Special Counsel • Off-line Satellite Calibration • Ohio Supercomputer Center (USA-OH) • On-Scene Coordinator • Ontario Science Centre (Canada-ON) • Open Star Cluster • Operational Simulator Console • Operational Summary Console • Operational Switching Cabinet • Operations Sequence Chart • Operation Subcommittee • Operations Support Center • Operations Support Contract • Optical Sciences Center • Optical Signatures Code • Orbital Sciences Corporation (USA-VA) • Orbit-Spin Coupling • Organic Suflur Compound • Oscillator • Oscoda Wurtsmith (USA-MI) (location identifier) • Outer Space Committee • Outsourcing Service Centre

OSCA: Optical Sensors Collaborative Association

OSCAR: Online Satellite Catalog And Request (system) • On-Screen Configuration and Activity Reporting • Operating Sequence Control Array • Operations, Scheduling, Control And Reporting • Optically Scanned Character Automatic Reader • Optical Systems Characterization and Analysis Research • Orbiting Satellite Carrying Amateur Radio

OSCE: Office Statistique des Communautés Européennes

OSCF: Operations Support Computing Facility

OSCL: Operating System Control Language

OSCN: Orbiter Specification Change Notice

OSCR: Ocean Surface Current Radar • Orbiter Software Change Request

OSCRL: Operating System Command and Response Language • Operating System Command Response Language

OSCRS: Orbital Spacecraft Consumables Resupply System

ÖSD: Österreichische Staatsdrückerei (Austria)

OSD: Object State Diagram • Office of the Secretary of Defense (USA) • On-Screen Display • Open Software Description • Open Software Distribution • Operational Sequence Diagram • Optical Scanning Device • Organization Subject Detail • Östersund (Sweden) (location identifier) • Ostersund (Sweden) (location identifier) • Overseas Surveys Directorate (UK)

OSDA: Optical Shallow-Defect Analyzer

OSDH: Orbiter System Definition Handbook

OSDI: Operating Systems Design and Implementation

OSDIT: Office of Software Development and Information Technology

OSDP: Onboard Signal and Data Processing • Onboard Signal and Data Processor • On-Site Data Processing • On-Site Data Processor • Operational System Development Program

OSDPD: Office of Satellite Data Processing and Distribution

OSDS: Oceanic System Development and Support • Operating System for Distributed Switching

OSDU: Output Signal Distribution Unit

OS/E: Operating System/Environment

OS&E: Optical Science & Engineering

OSE: Observing System Experiment • Omora (Papua New Guinea) (location identifier) • Open Systems Environment • Operating Support Equipment • Operational Status Equipment • Operational Support Equipment • Orbital Support Equipment • Orbiter Support Equipment • Other Support Equipment

OSEC: Operations Security

OSEDA: Office of Social and Economic Data Analysis

OSEM: Opto-electronic Sensor Electromagnetic Motor

OSEP: Office of Scientific and Engineering Personnel

OSES: Operation Support Equipment Software

OSETI: Optical Search for Extraterrestrial Intelligence

OSF: Observation Summary Form • Office of Space Flights • Oldenburger Sternfreunde eV (Germany) • Open Software Foundation • Optronique Secteur Frontal • Orbiter Simulation Facility • Order Statistic Filter • Ordnance Storage Facility

OSFM: Office of Spacecraft and Flight Missions

OSG: Observatory for Satellite Geodesy • Observatory Support Group • Open Systems Group • Operand Select Gate • Ossima (Papua New Guinea) (location identifier)

OSGB: Ordinance Survey of Great Britain

OSGC: Ohio Space Grant Consortium (USA-OH)

OSH: Occupational Safety and Health • Online Shorthand • Oshkosh (USA-WI) (location identifier)

OSHA: Occupational Safety and Health Act (USA) • Occupational Safety and Health Administration • Operating and Support Hazard Analysis

OSI: Office of Science Information • Office of Scientific Information • Office of Scientific Integrity • Office of Special Investigations • On-Site Inspection • Open Systems Interconnect • Open Systems Interconnection • Operational Status Indicator • Optical Stellar Interferometer • Optical Storage International • Orbiting Stellar Interferometer • Organisation du Système d'Information • Osijek (Croatia) (location identifier)

OSIA: On-Site Inspection Agency

OSID: Operating System Interface Definition

OSIE: Operational Support Integration Engineering

OSIP: Operational and Safety Improvement Program • Operational Satellite Improvement Program

OSIR: Office of Scientific Investigation and Research

OSIRIS: Ohio State Infrared Imager and Spectrometer • Ohio State Infrared Imaging Spectrograph • Ohio State Infrared Imaging System • Optical Simulation for Imaging Reconnaissance and Intelligence Sensors • Oxide Superalloy Improved by Resolidification In Space

OSIS: Office of Science Information Services • Open Shop for Information Services • Optionally Stabilized Imager and Spectrometer

OSJ: Objet Stellaire Jeune • Observatorio San José (Argentina)

OSJaJ: Oceanographical Society of Japan Journal (periodical code)

OSK: Oskarshamn (Sweden) (location identifier)

Osl: Oslo (Norway)

OSL: Old Style Latin • On-Site Laboratory • Operating System Language • Optically Stimulated Luminescence • Optical Science Laboratory • Orbiting Solar Laboratory • Orbiting Space Laboratory • Oslo (Norway) (location identifier)

OSLM: Operations Shop or Lab Manager

OSLR: Ocean Sciences and Living Resources

OSM: Mosul (Iraq) (location identifier) • Office of Special Mining • Off-line Services Manager • Off-Screen Model • On Station Mode • Operating System Manual • Operating System Module • Operational Status Monitoring • Operations Status Monitoring • Operator's Service Manual • Orbital Service Module • Orbiter Service Module • Oscillating Secondary Mirror

OSME: Open Systems Message Exchange

OSMP: Operational Support Maintenance Plan

OSMV: One-Shot Multivibrator

OSN: Osan (Korea-ROK) (location identifier)

OSNI: Ordnance Survey of Northern Ireland

OSO: Ocean Systems Operations • Office of Space Operations • Onsala Space Observatory (Sweden) • Orbiting Solar Observatory • Originating Signaling Office • Ouest-Sud-Ouest

OSOA: Optimal Statistical Objective Analysis

OSOD: On-Site Orbit Determination

OSOP: Orbiter Systems Operating Procedure

osp: Obiit Sine Prole

OSp: Old Spanish

OSP: Observer Selection Program • Office of Special Projects • Office of State Planning • Office of Strategic Phenomena (NRL) • Online Service Provider • Operating System Principle • Operations Support Plan • Optical Storage Processor • Optimal Scanning Path • Orbital Support Plan • Outer Space Place • Outside Plant • Slupsk (Poland) (location identifier)

OSPA: Open Signal Coprocessing Architecture

OSPAT: Operations and Spacecraft Performance Assessment Team

OSPF: Open Shortest Path First

OSPFIGP: Open Shortest-Path First Internet Gateway Protocol

OSPI: Office of the Superintendent of Public Instruction

OSPREY: Open Structure for Protecting Rights Electronically

OSQL: Object Structured Query Language

OSR: Occupational Survey Report • OEM Service Release • OEM System Release • Office of Science and Research • Office of Scientific Research • Operational Scanning Recognition • Operational Support Readiness • Operations Safety Requirement • Operations Support Room • Optical Scanning Recognition • Optical Solar Reflector • Optical Surface Reflector • Organization for Scientific Research • Ostrava (Czech Rep.) (location identifier) • Outgoing Shortwave Radiation

OSRA: Observatorium für Solare Radioastronomie (Germany)

OSRBD: Operational Software Reliability Benchmark Development

OSRD: Office of Scientific Research and Development

OSRMS: Ocean Surface Roughness Measurement System

OSRS: Oceanic Satellites and Remote Sensing • Operational Status Recording System

OSRT: Ooty Synthesis Radio Telescope (India)

Oss: Osservatorio

OSS: Observation Support Software • Observation Support System • Observatoire du Sahara et du Sahel • Observatory Support System • Observer Support Software •

Ocean Surveillance Ship • Office of Space Science • Office of Space Sciences • Office of Space Station • Office of Strategic Services (now CIA) (USA) • Office Support System • Operational Support System • Operations Support System • Optical Subsystem • Optical Support Structure • Optics Subsystem • Optics Support Structure • Orbital Space Station • Orbital Stabilization System • Orbiting Space Station • Osh (Kyrgyzstan) (location identifier) • Outer Solar System • Oxigen Storage System

OSSA: Office of Space Science and Applications (CSIRO)

OSSE: Observing-System Simulation Experiment • Office of Space Science Education • Oriented Scintillation Spectrometer Experiment (*) • Oriented Scintillation Spectroscopy Experiment

OSSI: Optronic Sensor for Stereographic Imagery

OSSIU: Orbiter-Space Station Interface Unit

OSSL: Operating System Simulation Language

OSSO: Observation Science Support Office

OSSP: Outer Solar System Probe

OSSRH: Orbiter Subsystem Requirements Handbook

OSSS: Optical Space Surveillance System • Orbital Space Station Study • Orbital Space Station System

OSSWG: Operating Systems Standards Working Group

OST: Objectif Scientifique et Technique • Objective Start Time • Objectives, Strategy, and Tactics • Observational Sequence Table • Observation Sequence Table • Observation Summary Table • Observatoire des Sciences et des Techniques (France) • Ocean Surface Temperature • Office of Science and Technology • Office of the Secretary of Transportation (USA) • Operational System Test • Operationnelle Systemtest • Operations Support Team • Optical Star Tracker • Optic Support Table • Orbiter Support Trolley • Orbit Stay Time • Organisation Scientifique du Travail • Oostende (Belgium) (location identifier)

OSTA: Office of Space and Terrestrial Applications • Optical Storage Technology Association

OSTC: Optical Science and Technology Center

OSTDS: Office of Space Tracking and Data Systems

Osterzgeb: Osterzgebirge (Germany)

OSTF: Off-Site Test Facility • Ordnance Survey Transfer Format

OSTI: Office of Science and Technology Infrastructure • Office of Scientific and Technical Information • Organization for Social and Technological Innovation

OSTO: Office of Space Transportation Operations

OSTP: Office of Science and Technology Policy (USA) • On-Site Test Procedure • Orbiting System Test Plan

OSTS: Office of Space Transportation Systems

OSU: Columbus OSU (USA-OH) (location identifier) • Observatoire des Sciences de l'Univers • Ohio State University (USA-OH) • Oklahoma State University (USA-OK) • Operational Switching Unit • Optical Simulation Unit • Oregon State University (USA-OR)

OSUMMPMRSFC: Ohio State University Multimode Prism Monochromator, Reflectomator and Spectral Flux Calibrator

OSURO: Ohio State University Radio Observatory (USA-OH)

OsUTR: Osaka University Technology Reports (periodical code)

OSV: Ocean Station Vessel • Orbital Support Vehicle • Oriented Space Vehicle

OSVS: Orbiter Space Vision Systems

OSW: Office of Solid Waste • Office of Surface Water • Orsk (Russia) (location identifier)

OSWER: Office of Solid Waste and Emergency Response

OSWS: Operating System Workstation

OSX: Kosciusko - Attala County (USA-MS) (location identifier)

OSY: Namsos (Norway) (location identifier)

OSZ: Koszalin (Poland) (location identifier)

OSZE: Organisation für Sicherheit und Zusammenarbeit in Europa

OS/2: Operating System Two

O/T: Other Times

OT: Oberer Tubus • Object Technology • Object Theory • Observatorio del Teide (Canary Islands) • Observing Tool • Occupational Therapist • Occupational Therapy • Offensive Threat • Off Time • Old Testament • Ontario (Canada) • On Time • Open Technology • Open Transport • Operating Temperature • Operating Time • Operational Task • Operational Test • Operational Testing • Operational Trajectory • Opposite Term • Optical Tracker • Optical Tracking • Optical Transient • Optimality Theory • Optimal Trajectory • Original Title • Other Than • Other Time • Ouest • Output Term • Out Temperature • Overtime • Overtone

OTA: Mota (Ethiopia) (location identifier) • Observatory for Time-domain Astronomy • Office of Technical Assistance • Office of Technology Assessment (USA-DC) (now defunct) • Operational Trajectory Analysis • Optical Telescope Assembly • Optical Tracking Aid • Optical Tube Assembly • Orbital Thruster Assembly • Organic Trace Analysis • Over The Air

OTAA: Ohio Turnpike Astronomers Association (USA-OH)

OTAC: Observatory Time Allocation Committee • Observing Time Allocation Committee

OTADA: Office of Tracking And Data Acquisition

OTAF: Office of Technology Assessment and Forecats

OTAN: Organisation du Traité de l'Atlantique Nord

OTASE: Organisation du Traité de l'Asie du Sud-Est

OTASIM: Optical Telescope Assembly Simulator

OTAU: Optical Test Access Unit

OTB: Object Transaction Broker • Orbiting Tanker Base • Oxide Titanium Bronze

Otbd: Outboard

OTBD: Outboard

OTC: Bol (Chad) (location identifier) • Objective, Time and Cost • Operational Test Center • Optical Test Chamber • Optical Transient Camera • Orbiter Test Conductor • Orbiting Trajectory Computation • Organization for Trade Cooperation • Over The Counter

OTCC: Operator Test Control Console

OTCCD: Orthogonal-Transfer Charged-Couple Device

OTCM: Orbiter Thermal Control Model

OTCS: Operational Teletype Communications System

OTD: Contadora (Panama) (location identifier) • Office of Technology Development • Offset Tilted Dipole • Operational Technical Documentation • Optical Tracking Device • Optical Transient Detector • Orbital Transfer Device • Orbiter Test Director

OTDA: Office of Tracking and Data Acquisition

OTDM: Optical Time-Division Multiplexing

OTDR: Optical Test Data Receiver • Optical Time Domain Reflectometer • Optical Time Domain Reflectometry

Ote: Oriente

OT&E: Operational Test & Evaluation • Optical Test & Evaluation

OTE: Open Toolkit Environment • Operational Test and Evaluation • Operational Test Environment • Operational Test Equipment • Optical Test and Evaluation • Optical Tracking Electronics • Optically Transparent Electrode • Orbital Transfer Equipment • Oriente

OTEC: Ocean Thermal-Energy Conversion • Operational Test and Evaluation Center

OTECI: Office Technique d'Études et de Coopération Internationale

OTEMP: Overtemperature

OTES: Orbiter Thermal Effects Simulator

OTF: Observation Test File • Off The Film • On-Target Flag • On-The-Fly • Open Token Foundation • Operational Time Fraction • Optical Transfer Function • Orbital Test Flight

OTFC: On-The-Fly-Calibration

OTG: Oracle Transport Gateway • Worthington (USA-MN) (location identifier)

OTH: North Bend (USA-OR) (location identifier) • Over the Horizon

OTHB: Over the Horizon Backscatter

OTHQ: Ouvrier Très Hautement Qualifié

OTHT: Over the Horizon Targeting

OTI: Morotai Island (Indonesia) (location identifier) • Office of Technical Information • Open Topology Interface • Ordered Tree Inclusion • Organización de la Televisión Iberoamericana

OTIA: Office of Technical Information Agency

OTIS: Occupational Training Information System • Oklahoma Telecommunications Interlibrary System (USA-OK)

OTIU: Overseas Technical Information Unit

OTJ: Off-The-Job • On-The-Job

OTK: Ocean Toolkit • Oostelik Transvaalse Kooperasie (South Africa) • Oxidizer Tank

OTL: Boutilimit (Mauritania) (location identifier) • Ooty Lunar Occultations • Operating Temperature Limit • Operating Time Log • Ordnance Test Laboratory • Out To Lunch

OTLAMR: Ocean Temperature Large Antenna Microwave Radiometer

OTLC: Orbiter Timeline Constraint

OTLK: Outlook

OTLT: One Time, Last Time

OTM: Observatory Technique Magazine • On-Time Marker • Orbit Transfer Manoeuvre • Orbit Trim Manoeuvre • Ottumwa (USA-IA) (location identifier) • Output Transformation Matrix

OTMS: Operational Technical Managerial System

OTN: Oaktown Green (USA-IN) (location identifier) • Octal Track Number • Operational Teletype Network

OTO: One-Time Only • Optical Tracker Operator • Otto Vor (USA-NM) (location identifier)

OTOH: On The Other Hand

OTOS: Orbit-To-Orbit Stage

OTP: Bucharest Otopeni (Romania) (location identifier) • Bucureşti Otopeni (Romania) (location identifier) • Office of Telecommunications Policy • On-Time Password • On-Time Programmable • On Top • Operational Test Procedure • Operations Test Plan • Operations Turnaround Plan

OTPC: Oficina Técnica para la Protección del Cielo (IAC)

OTR: Coto 47 (Costa Rica) (location identifier) • Observed Temperature Rise • Oceanic Transition Route • Ooty Radio Telescope (India) • Operating Temperature Range • Operating Time Record • Operational Time Record • Optical Transition Radiation • Other • Outer • Outer Tracking • Output Routine

OTRC: Optoelectronics Technology Research Corporation

OTRI: Oficina de Transferencia de Resultados de Investigación (IAC)

OTS: Anacortes (USA-WA) (location identifier) • Object Transaction Service • Office of Technical Services • Office of Thrift Supervision • Office of Toxic Substances • Off-The-Shelf • Open Trading System • Operational Test Site • Operational Time Synchronization • Opportunities To See • Optical Technology Satellite • Optical Tracking System • Optical Transient Source • Optimum Telemetry System • Orbital Test Satellite • Orbital Test System • Orbital Transfer System • Orbital Transport Services • Orbital Transport System • Orbiter Transfer System • Organized Track System

OTSDF: Optimal Trade-off Spatial Distance Filter

OTSI: Office du Tourisme et Syndicat d'Initiative • Oscillating Two-Stream Instability

OTSS: Operational Test Support System

Ott: Ottava • Ottawa (Canada-ON) • Ottobre • Ottoman

OTT: Oficina de Transferencia de Tecnología • One-Time Tape • Open Tower Telescope (now DOT) • Operational Training Test • Optional Team Targeting • Over The Top

OTTE: Operational Testing, Training and Evaluation

OTTER: Oregon Transect Ecosystem Research

OTTLE: Optically Transparent Thin-Layer Electrode

OTTOMH: Off The Top Of My Head

OTU: One-Time Use • Operational Taxonomic Unit • Operational Test Unit • Operational Training Unit • Organisme pour le Tourisme Universitaire • Otu (Colombia) (location identifier)

OTV: Operational Television • Orbital Transfer Vehicle • Orbiter Transfer Vehicle • Orbit Test Vehicle

OTVS: Onboard TV System

OTW: Off The Wall

OTY: Oria (Papua New Guinea) (location identifier)

OTZ: Kotzebue (USA-AK) (location identifier)

ou: Oculi Uterque • Oculus Uterque

OU: Oculi Uterque • Oculus Uterque • Open Under • Open University (UK) • Organizational Unit • Organization Unit • Oxford University (UK-England)

OUA: Organisation de l'Unité Africaine • Ouagadougou (Burkina Faso) (location identifier)

OUAS: Oxford University Astronomical Society (UK-England)

OUAT: Olsztyn University of Agriculture and Technology (Poland)

OUD: Operational Use Data • Oujda - Les Angads (Morocco) (location identifier)

OUE: Operational Utility Evaluation • Orbital Uncertainty Estimate • Ouesso (Congo) (location identifier)

OUG: Ouahigouya (Burkina Faso) (location identifier)

OUH: Oudtshoorn (South Africa) (location identifier)

OUI: Ban Houei (Laos) (location identifier)

OUK: Outer Skerries (UK) (location identifier)

OUL: Oulu (Finland) (location identifier)

OULIPO: Ouvroir de Littérature Potentielle

OUM: Oum Hadjer (Chad) (location identifier)

OUN: Norman - Max Westheimer (USA-OK) (location identifier)

ouO: ohne unser Obligo

OUO: Official Use Only

OUP: Oxford University Press

OUR: Batouri (Cameroon) (location identifier)

OURS: Open Users Recommended Solution • Orbiting Unification Ring Satellite

OUS: Oscillateur Ultra-Stable • Ourinhos (Brazil-SP) (location identifier) • Outer User System

OUSAS: Oxford University Space and Astronomical Society (UK-England)

OUST: Office of Underground Storage Tanks

Out: Outlet • Output

OUT: Bousso (Chad) (location identifier) • Orbiter Utility Tray • Outlet • Outline • Output • Outubro

Outbd: Outboard

OUTBD: Outboard

Outg: Outgoing

OUTR: Outeiro

OUTS: Output String

OUTWATS: Outward Wide Area Telephone Service

OUU: Ouanga (Gabon) (location identifier)

OUVS: Orbiter Ultraviolet Spectrometer

OUZ: Zouerate (Mauritania) (location identifier)

Ov: Over

O/V: Overvoltage

OV: Object-Verb • Oklahoma Variables • Optical Verification • Orbital Vehicle • Orbital Verification • Orbiter Vehicle • Output Voltage • Overflow • Oxygen Vent

Ova: Ostrova

OVA: Bekily (Madagascar) (location identifier) • Optical Visual Analysis

OVAL: Object-based Virtual Application Language • Object-based Visual Application Language

OVAM: Orbital Vehicle Assembly Mode

OVAOG: Ottawa Valley Astronomy and Observers Group (Canada-ON)

OVAS: Ocular Vergence and Accommodation Sensor • Ohio Valley Astronomical Society (USA-KY) • Ohio Valley Astronomical Society (USA-WV)

OVB: Novosibirsk (Russia) (location identifier)

Ovbd: Overboard

OVBD: Overboard • Overboard Dumping Assembly

OVC: Ocean Variability and Climates • Optical Velocities Catalogue • Other Valuable Consideration • Outer Vacuum Chamber • Overcast

OVCO: Operational Voice Communication Office

OVCP: Orbiting Vehicle Checkout Procedure

OVCS: Operational Voice Communication System

OVD: Asturias (Spain) (location identifier) • Optical Video Disk • Oviedo (Spain) (location identifier)

OVDLM: Open View Data Line Monitor

OVDS: Oxidizer Vent Drain System

ÖVE: Österreichischer Verband für Elektrotechnik (Austria)

OVE: Oroville (USA-CA) (location identifier)

OVEI: Orbiter Vehicle End Item

OVF: Organe (de calcul) en Virgule Flottante • Overfill • Overflow

Ovfl: Overflow

OVFL: Overflow

OVH: Overhead

Ovhd: Overhead

OVHD: Overhead

Ovhl: Overhaul

OVHT: Overheat

OVI: Operational Validation Inspection • Oxygen Ventilation Network

OVIP: Orbital Verification Implementation Plan

OVIS: Optical Visual Information System

OVIT: Orbital Verification Integrated Timeline

OVIWG: Orbital Verification Implementation Working Group

OVL: Oval • Ovalle (Chile) (location identifier) • Overlay

OVLA: Optical Very Large Array

OVLBI: Orbiting Very Large Baseline Interferometry

Ovld: Overload

OVLD: Overload

Ovlp: Overlap

OVMA: Owens Valley Millimeter Array

OVN: Optical - Visible - Near-infrared

OVNI: Objet Volant Non-Identifié

OVNO: Or Very Near Offer

OVOP: Orbital Verification Operating Procedure

OVOT: Orbital Verification Operating Timeline

OvP: Oval Point

OVP: Over Voltage Protection

Ovpd: Overpaid

OVPT: Orbital Verification Performance Test

OVR: Orbiting Vehicle Requirement • Outer Vertical Resonance • Over

Ovrd: Override

OVRD: Override

Ovrn: Overrun

OVRNG: Overrunning

OVRO: Owens Valley Radio Observatory (USA-CA)

OVS: Operational Voice System • Orbiting Vehicle System

OVSB: Oberbayerische Volkssternwarte Berg (Germany) • Orbiting Vehicle Support Building

ÖVSV: Österreichischer Versuchssenderverband (Austria)

OVSW: Orbital Verification Software

Ovsz: Oversize

OVT: Object au Voisinage de la Terre • Operational Verification Test

Ovtr: Overtravel

OVTR: Operational Video Tape Recorder

OVU: Optical Visualization Unit

Ovv: Overvoltage

OVV: Optically Violently Variable • Optically Violent Variable • Optically Violent Variation • Optical Violent Variable (*) • Optical Violent Variation

O&W: Onderwijs & Wetenschappen • Onderzoek & Wetenschappen

OW: Office of Water • Oil Well • One Way • Openbare Werken • Open Workstation • Optical Window • Order Wire • Orientation Workshop • Over-Write

OWA: Open-World Assumption • Ordered Weighting Averaging • Owatonna (USA-MN) (location identifier)

OWB: Owensboro (USA-KY) (location identifier)

OWCP: Office of Workers Compensation Program

OWD: Norwood (USA-MA) (location identifier) • One-Way Doppler

OWDC: Office of Water Data Coordination

OWDE: One-Way Doppler Extraction

OWE: Operating Weight Empty • Optimum Working Efficiency • Owendo (Gabon) (location identifier)

OWF: Optimum Working Frequency

OWI: Open Widget Interface

OWK: Norridgewock - Central Maine (USA-ME) (location identifier)

OWL: Object Windows Library • Object Wrapper Layer • Office Workstation Ltd. • Open Windows Library • Orbiting (array of) Wide-angle Light-collectors • Overwhelmingly Large (telescope)

OWLT: One-Way Light Time

OWM: Office of Weights and Measures

OWPE: Office of Waste Programs Enforcement

OWRR: Office of Water Resources and Research

OWRT: Office of Water Research and Technology

OWS: Ocean Weather Service • Ocean Weather Ship • Ocean Weather Station • Odyssey Worldwide Services • Operational Weather Support • Orbital Workshop • Oxford WFC Survey

OWSE: Operational WWW System Evaluation

OWU: Ohio Wesleyan University (USA-OH)

OWWS: Operational Windshear Warning System

Ox: Oxford (UK-England) • Oxidator Mass

OX: Order Crossover • Oxford (UK postal code)

OXB: Bissau (Guinea Bissau) (location identifier)

OXC: Waterbury Oxford (USA-CT) (location identifier)

Oxd: Oxidized

OXD: Oxford - Miami University (USA-OH) (location identifier) • Oxide • Oxidizer

OXDZR: Oxidizer

Oxf: Oxford (UK-England) • Oxfordshire (UK-England)

OXF: Oxford (Starlink internal code)

OXID: Oxidizer

OXO: Orbiting X-ray Observatory • Orientos (Australia-QLD) (location identifier)

Oxon: Oxoniensis • Oxfordshire (UK-England)

OXR: Oxnard Ventura (USA-CA) (location identifier)

OXS: Objective Crystal Spectrometer • Objective X-ray Spectrometer

OXVAD: Oxford (Starlink internal code)

OXY: Morney (Australia-QLD) (location identifier) • Oxygen

OY: Denmark (international aircraft registration code)

OYA: Goya (Argentina-CR) (location identifier)

OYE: Oyem (Gabon) (location identifier)

OYG: Moyo (Uganda) (location identifier)

OYK: Oiapoque (Brazil-AP) (location identifier)

OYL: Moyale (Kenya) (location identifier)

OYN: Ouyen (Australia-VIC) (location identifier)

OYO: Tres Arroyos (Argentina-BA) (location identifier)

OYP: Saint-Georges-de-l'Oyapock (French Guiana) (location identifier)

OYS: Yosemite National Park (USA-CA) (location identifier)

oz: Ounce

Oz: Ozero

OZA: Ozona (USA-TX) (location identifier)

oz. ap.: Ounce (Apothecaries)

oz. av.: Ounce (Avoirdupois)

OZC: Ozamis City - Labo (Philippines) (location identifier)

OZE: Outer Zone Electron

OZEP: Outer Zone Electron Precipitation

OZH: Zaporozhye (Ukraine) (location identifier)

OZP: Morón (Spain) (location identifier)

OZR: Ozark Cairns (USA-AL) (location identifier)

ozs: Ounces

oz. t.: Ounce (Troy)

OZZ: Ouarzazate (Morocco) (location identifier)

O2: Object Oriented • Oxygen

O2A: Object-Oriented Analysis

O2AD: Object-Oriented Analysis and Design

O2B: Object-Oriented Bibliography

O2C: Organisation de la Culture Commerciale

O2D: Object-Oriented Design • Object-Oriented Development • Object-Oriented Directory

O2DB: Object-Oriented Database

O2DBMS: Object-Oriented Database Management System

O2DCE: Object-Oriented Distributed Computing Environment

O2E: Object-Oriented Environment

O2ERM: Object-Oriented Entity-Relationship Model

O2FS: Object-Oriented File System

O2G: Object-Oriented Graphics

O2GL: Object-Oriented Graphics Language

O2L: Object-Oriented Language • Object-Oriented Library • Object-Oriented Listing

O2MP: Object-Oriented Management Planning

O2N: Object-Oriented Numerics

O2P: Object-Oriented Pattern • Object-Oriented Programming

O2PL: Object-Oriented Programming Language

O2PS: Object-Oriented Programming and System • Object-Oriented Programming System

O2RA: Object-Oriented Requirement Analysis

O2RALD: Object-Oriented Requirements Analysis and Logical Design

O2S: Object-Oriented Software • Object-Oriented System

O2SA: Object-Oriented Systems Analysis

O2SE: Object-Oriented Software Engineering

O2T: Object-Oriented Technology

O2UI: Object-Oriented User Interface

O2V: Object-Oriented View

O2ZE: Object-Oriented Z Environment

O3: Ozone

P

p: Electric Dipole Moment • Electron Spectral Index • Generalized Momentum • Hole Number Density • Molecular Momentum • Momentum • Page • Página • Pamphlet • Para • Paragraph • Parallax • Part • Participio • Particle • Partim • Pass • Passed • Past • Pater • Peak • Peculiar • Pence • Penny • Per • Perch • Permanent Dipole Moment of a Molecule • Person • Peseta (currency) • Peso (currency) • Phosphate • Piano • Piastre (currency) • Pico • Pied • Pint • Pipe • Pius • Plasma Pressure • Polar • Pole • Pole Pairs Number • Population • Por • Port • Post • Pouce • Pour • President • Pressure • Primus • Pro • Probability Density Function • Production • Professional • Proton • Resonance Escape Probability • Shower • Sound Pressure • Static Pressure

P/: Periodic Comet

P: Dielectric Polarization • DPRK (international aircraft registration code) • Electric Polarization • Four Hundred (Roman numeral) • Gauss Vector • Korea (DPRK) (international aircraft registration code) • Papa (international alphabet) • Paragraph • Parental • Parity • Park • Parking • Parkplatz • Paseo • Pass • Passing • Pater • Pedestrian Crossing • Pence • Penny • Père • Perimeter • Period • Permian • Personnel • Peseta (currency) • Peso (currency) • Peta • Pharmacy • Phase • Phosphate • Phosphor • Phosphore • Phosphorus • Pic • Pico • Piso • Pitch • Plate • Pluto • Poids • Poise • Polar • Polynomial • Poor • Populus • Portugal (international vehicle registration code) • Portugal (postal country code) • Position Angle • Positive • Post • Postage • Poster • Posterior • Post Office • Power • Precision • Predicate • Pregunta • Preliminary • President • Pressure • Priest • Prince • Probability • Probability Density • Probability Distribution • Progressive • Prohibited • Proposed • Proprietary • Proton • Public • Pula (currency) • Pulau • Pulsar • Punta • Pursuit • Quantum Number of Component of Angular Momentum Along Symmetry Axis • Radiant Flux • Radiant Power • Sidereal Period • Sound Energy Flux • Sound Power • Space Inversion • Weight • 400 (Roman numeral)

p/a: Personal Account

pa: Panama • Per Adres • Per Annum • Por Ausencia • Punjabi (ISO language code)

pA: per Adresse • Picoampere

Pa: Pascal • Pennsylvania (USA) • Protactinium

P/A: Particular Average • Preamplifier • Pressure-Actuated • Private Account • Problem Analysis

P&A: Physics & Astronomy • Planning & Assessment

PA: Pad Abort • Paging and Area • Paisley (UK postal code) • Panama (international vehicle registration code) • Panama (ISO country code) • Pará (Brazil) • Paraguay • Parametric Amplifier • Par Avion • Parents' Association • Parent Atom • Participial Adjective • Particular Average • Passenger Agent • Pattern Analysis • Paying Agent • Pedagogische Academie • Pending Availability • Pennsylvania (USA) • Per Adres • Per Annum • Performance Analysis • Performance Assessment • Performance Assurance • Permanent Address • Permanent Appointment • Personal Accident • Personal Account • Personal Allowance • Personal Assistant • Petites Annonces • Photoacoustic • Photo Area • Photodiode Amplifier • Pitch Angle • Pitch Axis • Plasma Analyzer • Plasma Armature • Plateau Age • Polar Angle • Popular Astronomy (periodical code) • Por Autorización • Port A • Position Angle • Posta Aerea • Power Alarm • Power Amplifier • Power of Attorney • Preamplifier • Prearbitrated • Precision Angle • Predictive Analyzer • Predictive Avoidance • Preliminary Acceptance • Preliminary Assessment • Presidential Appointment • Press Agent • Press Association • Press Attaché • Pressure Altitude • Primary Amplifier • Principal Astronomer • Principal Axes • Private Account • Probability of Acceptance • Problem Analysis • Procedimiento Administrativo • Process Algebra • Processing Application • Product Analysis • Product Assurance • Professional Association • Program Address • Program Analysis • Program Assessment • Program Assistant • Program Authorization • Protected Area • Provider Agent • Public Accountant • Public Address • Public Affairs • Publicity Agent • Publishers Association (UK) • Pulsating Arc • Pulse Amplifier • Pulsed Annealing • Purchasing Agent

PAA: Pa-an (Myanmar) (location identifier) • Phase Array Antenna • Photon Activation Analysis • Piedmont Amateur Astronomers (USA-SC) • Planar Array Antenna • Ports Application Analysis • Power Amplifier Assembly • Priority Abatement Area • Problem Assessment Area • Program Associate Administrator • Prueba de Aptitud Académica

PAABS: Panamerican Association of Biochemical Societies

PAAC: Program Analysis Adaptable Control

PAACE: Pennsylvania Association for Adult Continuing Education (USA-PA)

PAAD: Program Acronyms, Abbreviations and Definitions • Program Assistance Approval Document

PAARL: Philippines Association of Academic and Research Libraries

PAAS: Pakistan Association for the Advancement of Science • Phased Array Antenna System • Publications of the American Astronomical Society (periodical code)

PAASCU: Philippine Accrediting Association of Schools, Colleges and Universities (now AACCUP)

Pab: Pabellón

PAB: Balboa (Panama) (ISO currency code) • Bilaspur (India) (location identifier) • Panamean Balboa (ISO currency code) • Periodic Area Balancing • Personal Address Book • Policy Advisory Board • Program Advisory Board • Public Affairs Bureau

PABD: Precise Access Block Diagram

PABF: Precision Air Bearing Facility

PABLA: Problem Analysis By Logical Approach

PABO: Pedagogische Academie voor het Basisonderwijs

PABX: Private Automated Branch Exchange • Private Automatic Branch Exchange

Pac: Pacific

PAC: Pacific • Pacific Accredition Cooperation • Package Attitude Control • Packaged Assembly Circuit • Packet • Panama City - Paitilla (Panama) (location identifier) • Parametric Amplifier Converter • Parent-Adult-Child • Passive Air Cycle • Performance Analysis and Control • Performance Appraisal Committee • Personal Analog Computer • Pereturbed Angular Correlation • Photo Annotation/Classification • Physics Action Council • Planetarium Association of Canada • Planetary Astronomy Committee • Pneumatic Analog Computer • Pneumatic Auxiliary Console • Polar Atmospheric Chemistry • Policy Advisory Committee • Political Action Committee • Politique Agricole Commune • Popular Astronomy Club Inc. (USA-IL) • Positional Astronomy Centre (India) • Post-Attenuation Correction • Post-Award Conference • Powdered Activated Carbon • PPP Access Concentrator • Prairie Astronomy Club (USA-NE) • Preservation And Conservation • Pressure Alpha Center • Primary Address Code • Prime Associate Contractor • Principal Associate

Contractor • Principal Astronomer's Computer • Privilege Attribute Certificate • Probe Aerodynamic Center • Problem Action Center • Productibility Assurance Center • Program Activity Center • Program Advisory Committee • Program Analyzer Console • Promotion et Assistance Commerciale • Public Access Catalog • Publications Advisory Committee • Pure and Applied Chemistry

PACA: Payload Accommodation Configuration Assessment • Physics And Chemistry of the Atmosphere • Provence - Alpes - Côte d'Azur (France)

PACAS: Personnel Access Control Accountability System

PACB: Programme Atmosphère et Cycle Biogéochimique

PACC: PERT-Associated Cost Control • Problem Action Control Center

PACCS: Post-Attack Command Control System

PACE: Pacific-Arizona Crustal Experiment • Parallel Automated Coding Expert • Passive Attitude Control Experimental • Performance And Cost Evaluation • Physics And Chemistry Experiments • Pilot and Aircraft Courtesy Evaluation • Plasma-Assisted Chemical Etching • Plymouth Atmospheric Corrosion Experiment • Post-Appointment and Continuing Education • Power And Control Electronics • Precision Analog Computing Equipment • Pre-flight Acceptance Checkout Equipment • Prelaunch Automatic Checkout Equipment • Priority Access Control Enabled • Producible Alternative to Cadmium-Telluride Epitaxy • Product Acceptance Configuration Element • Programme for Advanced Continuing Education • Property Advisor to Common Estate

PACEM: Propagation en Air Clair et Météorologie

PACER: Part And Component Evaluation Report • Process Assembly Case Evaluator Routine • Program for the Acceleration of Commercial Energy Research • Public Access to Count Electronic Records

PACES: Pan American Center for Earth and Environmental Studies • Political Action Committee for Engineers and Scientists

PACI: Partnership for Advanced Computational Infrastructure • Payload and Attitude Control Interface

Pacif: Pacific

PACIMS: Passive Chemical Ionization Mass Spectrometry

PACIR: Propulsion, Aerodynamics, Control Integration Research

PACIS: Pulsed Arc Cluster Ion Source

PACK: Payload Active Cooling Kit

PACLING: Pacific Association for Computational Linguistics

PACM: Passive-Active Collective Memory • Pulse Amplitude Code Modulation

PACMIG: Pressurized Air-Cooled Metal Inert Gas

PACNET: Pacific Network

PACO: Parts Control • Prelaunch Automatic Checkout

PACP: Propulsion Auxiliary Control Panel

PACS: Pacific Area Communications System • Personal Access Communications System • Photoconductor Array Camera and Spectrometer • Physical Abstract Classification Scheme • Physics and Astronomy Classification Scheme • Picture Archiving and Communication System • Planet Asteroid and Comet Survey • Pointing and Attitude Control System • Programming And Computation Service • Public Access Computer System

PACT: Parchmarki Array of Čerenkov Telescopes (India) • Palo Alto Collaborative Testbed • Pay Actual Computer Time • PCTE Added Common Tools • Power-Augmented Catalytic Thruster • Practical Application of Constraint Technology • Prefix Access Code Translator • Primary Access to Chemical Titles • Program for Automatic Coding Techniques • Programmed Analysis Computer Transfer • Programmed Automatic Circuit Tester • Project for the Advancement of Coding Techniques

PACU: Philippines Association of Colleges and Universities

PACX: Private Automatic Computer Exchange

PAD: Packet Assembler/Disassembler • Packet Assembly/Disassembly • Paderborn (Germany) (location identifier) • Parts Approval Document • Payable After Death • Pitch Angle Distribution • Pitch Axis Definition • Plan d'Action Discrétion • Polar and Auroral Dynamics • Polyaperture Device • Poor Acquisition Data • Portable Attachment Device • Power Amplifier Device • Power Amplifier Driver • Precise Access Diagram • Preferred Arrival Date • Preliminary Advisory Data • Prêt À Diffuser • Program Action Directive • Program Analysis for Documentation • Program Approval Document • Programmation Automatique des Données • Project Approval Document • Propellant Acquisition Device • Proposal Administrative Data • Protoplanetary Accretion Disk • Provisional Acceptance Date • Pulse Amplitude Detector • Pulse Amplitude Discriminator

PADAR: Photoacoustic Detection And Ranging

PADEM: Projet Amélioration des Enquêtes de Ménage

PADI: Preserving Access to Digital Information

PADISNet: Pan African Development Information System Network

PADL: Part and Assembly Description Language

p. Adr.: Per Adresse

PADR: Preferential Arrival and Departure Route

PADRE: Portable Automatic Data Recording Equipment

p. ae.: Partes Aequales

PADS: Parallel And Distributed Simulation • Passive-Active Data Simulation • Pen Application Development System • Performance Analysis and Design Synthesis • Performing Arts Data Service • Problem Action Data System • Program Access to Dataset

PADSIM: Precision Attitude Determination Simulation

PA&E: Program Analysis & Evaluation

PAE: Everett - Snohomish County (USA-WA) (location identifier) • Parc d'Activités Économiques • Performance Assurance Engineer • Phase Angle Error • Pilote Automatique Embarqué • Presses Académiques Européennes (Belgium) • Preventive Action Engineer • Problem Assessment Engineer • Problem Assessment Engineering • Projet d'Action Éducative

PAEC: Planned and Archival Exposures Catalog (HST)

PAET: Planetary Atmosphere Experimental Test

PAEW: Personnel And Equipment Working

PAF: Pakuba (Uganda) (location identifier) • Particle and Field Package • Payload Attach Fitting • Payload Attachment Fitting • Paysage Audiovisuel Français • Peak Annual Funding • Peripheral Address Field • Plan Académique de Formation (France) • Police de l'Air et des Frontières (France) • Postcode Address File • Processing and Archiving Facility • Processing Archiving Facility • Puissance Au Frein

PAFB: Patrick Air Force Base (USA)

Pág: Página

PAG: Pagadian (Philippines) (location identifier) • Phenylacetyl Glutamine • Plena Analiza Gramatiko • Precision Alignment Gyrocompass • Precursor Active Galaxy • Preliminary Analysis Group

PAGASA: Philippine Atmospheric, Geophysical and Astronomical Service Administration

PAGB: Post-Asymptotic Giant Branch

PAGE: Polyacrylamide Gel Electrophoresis • Program for the Advancement of Geoscience Education • Programmable Aerospace Ground Equipment

PAGEOS: Passive Geodetic Earth Orbiting Satellite

PAGES: Past Global Changes

PAGIS: Pastoral and Agricultural Geographic Information System • Planning And Geographic Information System • Programmable Agents for Group Interaction Systems

PAGODA: Profile Alignment Group on ODA

Págs: Páginas

PAH: Paducah (USA-KY) (location identifier) • Payload Accommodations Handbook • Polycyclic Aromatic Hydrocarbon • Push And Hold

PAHO: Pan American Health Organization

PAI: Pailin (Cambodia) (location identifier) • Paysage Astronomique International • Point d'Accueil International • Portable Auroral Imager

PAIB: Publications of the Astronomy Institute of Bonn (periodical code)

PAIC: Publications of the Astronomical Institutes of Czechoslovakia

PAIDAI: Planetary Astronomy InSb Detector Array Infrastructure

PAIGH: Pan American Institute of Geography and History

PAIH: Public-Access Internet Host

PAIO: Permanence d'Accueil, d'Information et d'Orientation

PAIP: Performance Assurance Implementation Plan

PAIR: Partnership Award for the Integration of Research • Polarization-Agile Instrumentation Radar

PAIS: Prestigious Applications of Intelligent Systems (conference) • Product Assurance Information System • Project Analysis Information System • Public Access Internet Site • Public Affairs Information Service

PAIV: Particle-Acceleration-Induced Viscosity

PAJ: Para Chinar (Pakistan) (location identifier)

Pak: Pakistan • Pakistani

PAK: Hanapepe - Port Allen (USA-HI) (location identifier) • Packed • Pakistan (international vehicle registration code) • Pakistan (ISO country code) • Program Attention Key

PAKD: Practical Application of Knowledge Discovery

PAKISTAN: Pendjab-Afghania-Kashmir-Iran-Sind-Turkmenistan

PAKM: Practical Aspects of Knowledge Management

Pal: Palace • Palacio • Palais • Palencia (Spain) • Paleocene • Paleography • Paleontology • Palestine • Palomar

PAL: Palácio • Palanquero (Colombia) (location identifier) • Pallet • Paradox Application Language • Parallel Aligned Liquid-crystal • Parc d'Activités Logistiques • Parcel Airlift • Parts Authorization List • Performance Assessment Logic • Peripheral Availability List • Phase-Alternate Line • Phase-Alternating Line • Phase-Alternation Line • Present Atmospheric Level • Price and Availability List • Prime Agricultural Land • Privileged-Architecture Library • Process Assembler Language • Process Asset Library • Process Automation Language • Programmable Array Logic • Programmed Array Logic • Programmed Application Library • Programmer Assistance and Liaison • Programming Assembly Language • Propulseur Auxiliaire Liquide • Propulseur d'Appoint Liquide • Propulsion Auxiliaire Liquide • Public Ada Library • Push And Latch

PALAO: Palomar Adaptive Optics

PALC: Plasma-Addressed Liquid Crystal • Point Arguello Launch Complex (USA-CA) • Professional-Amateur Liaison Committee

PALCD: Plasma-Addressed Liquid Crystal Display

PALCI: Pennsylvania Academic Library Connection Initiative (USA-PA)

PALE: Paleoclimates of Arctic Lakes and Estuaries

Paleo: Paleocene • Paleography • Paleontology

Paleog: Paleography

Paleogr: Paleografía

Paleon: Paleontology

Paleont: Paleontología

Paleontol: Paleontology

Palest: Palestine • Palestinian

PALILA: Photon Acquisition at Low Intensity Level for Astronomy

PALIS: Polarized Airborne Laser Imaging Sensor • Ponds And Lakes Information System

PALL: Pallet

PAllO: Publications of the Allegheny Observatory (periodical code)

PALME: Procédure d'Acquisition de Logiciels Microsoft pour l'Enseignement

PALMS: Partnership Advancing the Learning of Math and Science

PALOP: Países de Língua Oficial Portuguesa

PALS: Permissive Action Link System • Photo Area and Location System • Point Arguello Launch Site (USA-CA) • Precision Approach and Landing System • Principle of the Alphabet Literacy System • Project for Automated Library System • Public Access Library System

Pam: Pamphlet

PAM: Panama City - Tyndall (USA-FL) (location identifier) • Parallel Associative Memory • Parameter Adjusting Mechanism • Parameter Assignment Map • Payload Assist Module • Periodic Anderson Model • Peripheral Adapter Module • Phased Array Module • Physics-Astronomy-Mathematics • Plataforma Antiruido de Mejorada • Point Access Method • Portable Automated Mesonet • Preliminary Aerosol Monitor • Programme Alimentaire Mondial • Propulsion Avionics Module • Pulse-Address Modem • Pulse-Amplification Modulation • Pulse-Amplifier Modulation • Pulse-Amplitude Modulation (*) • Pupil Adjust Mechanism • Pupil Alignment Mechanism (*)

PAMA: Pre-Assigned Multiple Access • Professional Aviation Maintenance Association • Pulse-Address Multiple Access

PAM-D: Payload Assist Module - Delta class

PAMD: Price And Management Data • Public Access Machine-Readable Document

PAMELA: Phased Array Mirror Extensible Large Aperture

PAMIPS: Parallel Algorithm for Mixed Integer Programming in Scheduling

PAMIR: Passive and Active Microwave and Infrared Radiometer

PAMIRASAT: Passive Microwave Radiometer Satellite

PAMIS: Processing And Manufacturing In Space

PAMOS: Programme Atmosphère Météorologique et Océan Superficiel

PAMOY: Programme Atmosphère Moyenne

PAMPL: Program Approved Materials and Processes Selection List

PAMRI: Peripheral Adaptor Module Replacement Item

PAMRS: Parameter Adaptive Model Reference System

PAMS: Pad Abort Measuring System • Partitioned Access Method System • Passive Aerodynamically-stabilized Magnetically-damped Satellite • Pentax Analytical Measurement System • Photogrammetric Analytical Measurement System • Portable Automated Mesonet Station

PAMSCO: Program Article Master Schedule Commitment

pan: Pannus

Pan: Panamá • Panchromatic • Panoramic

PAN: Panama (ISO country code) • Passive/Active Neutron • Pattani (Thailand) (location identifier) • Peroxyacetylnitrat • Peroxyacetyl Nitrate • Personal Account Number • Personal Area Network • Personennamen der

Antike • Plan D'Aménagement National • Polska Akademia Nauk (Poland) • Polyacrylonitrile • Probabilistic Analysis Network • Public Access Network • Project voor Automatische Namenverwerting

PANAFTEL: Pan-African Telecommunications Network

PANASH: Paleoclimates of the Northern And Southern Hemispheres

Pan. Can.: Panama Canal

PANDA: Pacific Neotectonic Deformation Array • Planetary Atmospheres Node Data Analysis

PANES: Program for Analysis of Non-linear Equilibrium and Stability

PANH: Polycyclic Aromatic Nitrogen Heterocyclic

PANI: Phénomène Aérospatial Non Identifié

PANIC: Potential And Needs - Investments - Capabilities • Prototype Adaptive-optics Near-Infrared Camera • PtSi Array Near-Infrared Camera • PtSi Astronomical Near-Infrared Camera

PANS: Pretty Advanced New Stuff • Procedures for Air Navigation Services

PAO: Palo Alto (USA-CA) (location identifier) • Physical Abstract Object • Polissage Assisté par Ordinateur • Polyalpha Olefin • Post-Academiaal Onderwijs • Public Affairs Office • Public Affairs Officer • Production Assistée par Ordinateur • Programmation Assistée par Ordinateur • Publication Assistée par Ordinateur • Pyongyang Astronomical Observatory (Korea-DPRK)

PAONET: Public Astronomical Observatory Network

PAOS: Program in Atmospheric and Oceanic Sciences (Univ. Colorado)

PAOT: Persons At One Time

Pap: Papel • Paper • Papier • Papua • Papuan

PAP: Password Authentication Protocol • Payload Accommodation Plan • Payload Adapter Plate • Peak Advertised Performance • Performance Assurance Plan • Personnel Assurance Program • Physics and Astronomy Program • Plug And Play • Port au Prince (Haiti) (location identifier) • Practical Application of PROLOG • Prêt À Poster • Printer Access Protocol • Product Assurance Plan • Propulseur d'Appoint à Poudre • Pulmonary Arterial Desease

PAPA: Photon Analog Precision Address • Precision Analog Photon Address

PAPCO: Panel Plot Composer

PApGe: Pure and Applied Geophysics (periodical code)

PAPI: Precision Approach Path Indicator

PAPL: Program Approved Parts List • Project Approved Parts List

PAPS: Pooled Analytical Plotter System

PAQ: Palmer (USA-AK) (location identifier) • Plan d'Amélioration de la Qualité • Point Association Question • Position Analysis Questionnaire • Process Average Quality • Processus d'Amélioration de la Qualité

Pár: Párrafo

Par: Parágrafo • Paragraph • Paragraphe • Paraguay • Parallax • Parallel • Parallèle • Parameter • Paramètre • Parenthèse • Parenthesis • Parish

P/AR: Peak-to-Average Ratio

PAR: Pacific Antarctic Ridge • Parallel • Parametrized Action Representation • Paris Metropolitan (France) (location identifier) • Partial Acceptance Review • Participatory Action Research • Payload Adapter Ring • Peak-to-Average Ratio • Performance Analysis and Review • Performance Appraisal Report • PERT Analysis Report • Phased-Array Radar • Photosynthetically Active Radiation • Photosynthetically Available Radiation • Planned All Around • Planning Action Request • Positive Acknowledgment and Retransmission • Precision Approach Radar • Preferential Arrival Route • Priority Action Report • Priority Action Request • Probabilistic Analysis of Risk • Problem Accountability Record • Problem Action Record • Problem Action Request • Problem Analysis Report • Problem Analysis and Resolution • Product Acceptance Review • Production Analysis Report • Program-Aid Routine • Program Analysis Review • Program Assessment Report • Program Assessment Review • Programmable Access Router • Project Appraisal Report • Project Assurance Requirements • Project Authorization Request • Projet d'Acte Règlementaire • Provisional Acceptance Review • Pulse Acquisition Radar • Purchasing Approval Request • Push And Release

Para: Paragraph • Paraguay

PARA: Paragraph

Parab: Parabola

PARABOLA: Portable Apparatus for Rapid Acquisition of Bidirectional Observations of Land and Atmosphere

PARADISE: Piloting A Researchers Directory Service for Europe • Piloting A Researchers Directory Service in Europe

PARAGON: Processing and Archiving of Radar And Gauge data Off-line and in Near-real-time

Paragr: Paragraph • Paragraphe

PARAM: Parameter

PARB: Problem Assessment Review Board

Parc: Parcella

ParC: Parallel Computing (periodical code)

PARC: Palo Alto Research Center (XEROX) • Pennsylvania Astronomical Research Consortium (USA-PA) • Polarimetric Active Radar Calibrator

PARCA: Pan American Railway Congress Association • Parallel Channel Adapter

PARCC: Precision-Accuracy-Representativeness-Completeness-Correctness

Parch: Parchemin • Parchment

PARD: Periodic And Random Deviation • Precision Annotated Retrieval Display

PARE: Physiological and Anatomical Rodent Experiment

Paren: Parenthesis

Parens: Parentheses

PARES: Preprocessing of Airborne Remote Sensing

PARG: Perth Astronomy Research Group (Australia-WA)

PARI: Pisgah Astronomical Research Institute (USA-NC)

PARINFO: Programme d'Aide à la Recherche en Informatique

PARIS: Parallel Instruction Set • Passive Reflectometry and Interferometry System • Pulse Analysis-Recording Information System

Parl: Parliament • Parliamentary • Parralel

PARM: Program Analysis for Resource Management

PARMA: Program for Analysis, Reporting and Maintenance

PARNet: Palermo Academic and Research Network

PARR: Procurement Authorization and Receiving Report • Program Analysis and Resources Review

PARRSSIS: Program for Advanced Research into the Regulation, Steering and Stability of the International System

PARS: Property Accountability Record System • Property Accounting and Reporting System

Parsec: Parallax-Second • Parser and Extensible Compiler

PARSECS: Program for Astronomical Research and Scientific Experiments Concerning Space

PARSIFAL: Programming And Real-time System Interface For ALIS

PARSYN: Parametric Synthesis

Part: Parte • Partial • Partícula • Particular • Partie • Participate • Participial • Particle • Partícula • Particular • Particule • Partition • Partner • Partnership • Party

PART: Particular • Planning for Activities with Resources and Time Constraints • Planning of Activities, Resources and Technology

part. aeq.: Partes Aequales

PARTNER: Proof of Analog Results Through a Numerical Equivalent Routine

PARTS: Parts Assembly and Reuse Tool Set

part. vic.: Partibus Vicibus

PARU: Photographic And Reproduction Unit

Parv: Parvis

Pas: Pasarela • Passage • Passive

PA&S: Product Assurance & Safety

PAS: Pacific Aerial Survey • Pan American Satellite • Parameter Adaptive System • Parametric Amplifier System • Paros (Greece) (location identifier) • PASCAL • Payload Accommodations Study • Peninsula Astronomical Society (USA-CA) • Peoria Astronomical Society (USA-IL) • Perigee-Apogee Satellite • Perigee-Apogee Stage • Perigee-Apogee System • Phase Address System • Phase Alteration Standard • Phase Array System • Philippine Astronomical Society (Philippines) • Phoenix Astronomical Society (USA-AZ) • Photo-Acoustic Spectroscopy • Physics Academic Software (USA-NC) • Pilot Alert System • Plasma-Activated Source • Planning and Analysis Section • Plasma Analyzer Spectrometer • Plymouth Astronomical Society (UK-England) • Point Antisolaire • Pontchartrain Astronomy Society (USA-LA) • Positron Annihilated Spectroscopy • Power-Assisted Steering • Precision Acquisition System • Primary Alerting System • Primary Ascent System • Principal Assistant Secretary • Principal Axis System • Probabilistic Argumentation System • Problem Assessment System • Processed Array Signal • Product Assurance and Safety • Productivity-Affordability-Supportability • Programme d'Ajustement Structurel • Programme d'Application SPOT • Program of Advanced Studies • Public Access System • Public Address System • Public Announcement Service • Publicly Available Specification • Publicly Available Standard • Publicly Available Submitter

PASA: Participating Agency Service Agreement • Publications of the Astronomical Society of Australia

PASAu: Proceedings of the Astronomical Society of Australia (periodical code)

PASC: Pacific Area Standards Congress • Pan-American Standards Commission • Planning Advisory Subcommittee • Polar Atmospheric and Snow Chemistry

PASCAL: Philips Automatic Sequence Calculator

PASDE: Photogrammetric Appendage Structural Dynamics Experiment

PASE: Polar Air-Snow Experiment

PASEG: Parliamentary Astronomy and Space Environment Group (UK)

PASH: Polynuclear Aromatic Sulphur Heterocyclic

PASI: Pragmatic Approach to Subject Indexing

PASIPS: Parallel Signal and Image Processing System

PASJ: Publications of the Astronomical Society of Japan (periodical code)

PASK: Publications of the Academy of Sciences of Kazakhstan (periodical code)

PASKA: Polarized Atomic Beam Source Karlsruhe (Germany)

PASO: Particle Acceleration Solar Orbiter

PASP: Photovoltaic Array Space Power • Publications of the Astronomical Society of the Pacific (periodical code)

PASR: Problem Analysis Status Report

pass: Passim

Pass: Passage • Passenger • Passim • Passive

PASS: Passegem • Passager • Passenger • Payload Assist Subsystem • Personal Access Satellite System • Photogrammetric Archival Storage System • Physics and Astronomy Student Society (Canada-BC) • Planar Array of Superheated Superconductors • Planetarium Activities for Student Success • Planning And Scheduling System • POCC Application Software Support (*) • POCC Application Support Software • Ports Application Software System • Precision Autocollimating Solar Sensor • Prediction of Activity Spectra for Substances • Primary Avionics Software System • Prince Albert Satellite Station (Canada) • Professional Airways Systems Specialist • Program Alternative Simulation System • Prometheus All Sky Survey

PASSCAL: Program for Array Seismic Studies of the Continental Lithosphere

PASSIM: Presidential Advisory Staff on Scientific Information Management (USA)

PASSION: Parallel And Scalable Software for Input-Output • Program for Algebraic Sequences Specifically of Input-Output Nature

passwd: Password

PASSWORD: Piloting Authentication and Security Services Within OSI for the Research and Development

PASSYS: Passive Solar System

Past: Pasteur • Pastoor

PAST: Professeur Associé Temporaire

PASTEC: Passager Technologique

PASTEL: Passager SPOT de Télécommunications par Laser • Passeport pour Téléphoner

PASTIC: Pakistan Science and Technology Information Centre

PASTIS: Projet d'Analyse de Scènes et de Traitement d'Images Symboliques

Pat: Patent • Patente • Patented • Pattern

P. At.: Peso Atomico • Poids Atomique

PAT: Patch • Patent • Patna (India) (location identifier) • Pattern Analysis Test • Physics Achievement Test • Point After Touchdown • Polygon Attribute Table • Polyvalent Alias Table • Positron Annihilation Technique • Post Acoustic Test • Power Alarm Test • Preauthorized Automatic Transfer • Prediction Analysis Technique • Preliminary Acceptance Testing • Primary Access Terminal • Priority Action Table • Problem Action Team • Process Action Team • Process Analysis Team • Production Acceptance Test • Production Assessment Test • Program Analysis Team • Programme in Advanced Technologies • Pseudo-Adder Tree • Pseudo-Automatic Translation

Pata: Patagonia

PATAC: Prévision Améliorée - Techniques d'Affinement de la Climatologie

PATC: Peninsula Advanced Technology Center (USA-VA)

Patd: Patented

PATE: Performance Analysis, Test, and Evaluation • Programmed Automatic Test Equipment

PatG: Patentgesetz

PATG: Productivity Assessment Task Group

Path: Pathological • Pathology

PATH: Performance Analysis and Test History • Post-flight Attitude and Trajectory History

Pathol: Pathological • Pathology

PATHS: Precursor Above-The-Horizon Sensor

PATMOS: Power And Timing Modelling System

Pat. Off.: Patent Office

PATOM: Programme Atmosphère et Océan à Moyenne échelle

PATOME: Programme Atmosphère et Océan à Moyenne Échelle

PATP: Parallel Application Technology Program • Preliminary Authority To Proceed

Pat. Pend.: Patent Pending

PATRACK: Package for Tracking Anomalous Changes of Knowledge

PATRICIA: Practical Algorithm To Retrieve Information Coded In Alphanumeric • Proving And Testing for Reliability Improvement of Complex Integrated Architectures

PATS: Payload Avionics Test Station • Predicasts Abstract Terminal System • Program for Analysis of Time Series • Propulsion Analysis Trajectory Simulation

PATSY: Parametric Test Synthesis

Patt: Pattern

PATT: Panel for the Allocation of Telescope Time • Partial Automatic Translation Technique • Project for the Analysis of Technology Transfer

PATU: Pan-African Telecommunications Union

PATWAS: Pilots Automatic Telephone Weather Answering Service

PAU: Pan American Union • Pattern Articulation Unit • Pauk (Myanmar) (location identifier) • Present Address Unknown • Probe Aerodynamic Upper • Programs Analysis Unit • Pupil Alignment Unit

PAUSD: Palo Alto Unified School District (USA-CA)

Pav: Pavillon • Pavo-Pavonis (constellation)

PAV: Parallel Access Volume • Paulo Afonso (Brazil-BA) (location identifier) • Pavilhão • Phase Angle Voltmeter • Pool Adjacent Violators (method) • Position And Velocity • Pressure-Actuated Valve

PAVM: Phase Angle Voltmeter

PAVT: Position And Velocity Tracking

PAW: Pambwa (Papua New Guinea) (location identifier) • Peachtree Accounting for Windows • Physics Analysis Workstation • Plasma Arc Welding • Powered All the Way

PAWE: Program for Analysis of the World Ecosystem

PAWOS: Portable Automatic Weather Observing Station

PAWS: Parallel Application Workstation • Physics Analysis Workstation • Polar Automatic Weather Station • Portable Automatic Weather Station • Protection Against Wrapped Sequence • Provision Analyst Workstation

PAX: Passenger • Payload Accelerometer Experiment (Olympus) • Portable Archive Exchange • Port de Paix (Haiti) (location identifier) • Private Automatic Exchange • PSDE Accelerometer Experiment (Olympus)

PAY: Pamol (Malaysia) (location identifier)

PAYDIRT: Processing Architecture Yielding Deductions In Real Time

PAYE: Pay As You Earn • Pay As You Enter

Paymr: Paymaster

Payt: Payment

PAYV: Pay As You View

PAZ: Partial Annealing Zone • Poza Rica - Tajin (Mexico) (location identifier)

PAZh: Pis'ma Astronomicheskij Zhurnal (periodical code)

p/b: Paperbound

Pb: Blei • Lead • Plomb • Problem • Problème

P/B: Piggyback • Playback • Preburner • Pushbutton

P-B: Pays-Bas

PB: Palomar-Berger • Pañcavimśa Brāhmana • Paraíba (Brazil) • Particle Beam • Peripheral Buffer • Personal Best • Petabyte • Pharmacopoeia Britannica • Phonetically Balanced • Photovoltaic Battery • Plage Brightness • Plasma Boundary • Playback • Port B • Prayer Book • Preburner • Precision Bus • Pressure-Balanced • Pressure Breathing • Pressure Broadening • Primary Buffer • Printer Background • Product Bulletin • Program Board • Prominence Brightness • Publications Board • Publications Branch • Pulsating Band • Pulse Beacon • Pushbutton

PBA: Barrow (USA-AK) (location identifier) • Pencil Beam Antenna • Pill Box Antenna • Point Barrow (USA-AK) (location identifier) • Pool Boiling Apparatus • Programming By Analogy • Public Buildings Administration

PBAB: Please Bring A Bottle

PBAC: Program Budget Advisory Committee

PBAN: Polybutadiene-Acrylonitrile

PBAO: Publications of the Beijing Astronomical Society (periodical code)

PBB: Parallel By Bit • Paranaíba (Brazil-MS) (location identifier) • Polybrominated Biphenyl

PBC: Parallel By Character • Performance-Based Contract • Performance-Based Contracting • Period Batch Control • Peripheral Bus Computer • Pressure Beta Center • Processor Bus Controller • Puebla (Mexico) (location identifier)

PBCS: Post Boost Control System

PBCT: Proposed Boundary Crossing Time

PBD: Payload Bay Door • Photothermal Beam Deflection • Playback Drive • Porbandar (India) (location identifier) • Precise Block Diagram • Pressboard • Program Budget Directive • Programmer Brain Damage • Programming By Demonstration

PBDF: Payload Bay Door Forward

PBDI: Position Bearing and Distance Indicator

PBDM: Payload Bay Door Mechanism

Pbdo: Poblado

PBDS: Photothermal Beam Deflection Spectroscopy

PBE: Paschen-Back Effect • Piggyback Experiment • Pool Boiling Experiment • Programming By Example • Prompt By Example • Protective Breathing Equipment • Proton Balance Equation • Proton Binding Energy • Puerto Berrio (Colombia) (location identifier)

PBeiO: Publications of the Beijing Astronomical Society (periodical code)

PBEO: Programme Board Earth Observation

Pbf: Personenbahnhof

PBF: Pine Bluff - Grider (USA-AR) (location identifier)

PBFA: Particle Beam Fusion Accelerator

PBG: Plattsburgh (USA-NY) (location identifier) • Program and Budget Guidance

PBH: Paro (Bhutan) (location identifier) • Physikzentrum Bad Honnef (Germany) • Primordial Black Hole

PBHT: Polybutadiène Hydroxy-Téléchélique

PBI: Pacific Bell Internet • Polybenimidezole • Post-Burst Increase • Primeval Baryon Isocurvature • Producto Bruto Interno • Pushbutton Indicator • West Palm Beach (USA-FL) (location identifier)

PBIC: Programmable Buffer Interface Card

PBIM: Programmable Buffer Interface Module

PBIMR: Push Broom Imaging Microwave Radiometer

PBIO: Portable Binary Input/Output

PBIS: Peachtree Business Internet Suite

PBJ: Paama (Vanuatu) (location identifier)

Pbk: Paperback

PBK: Pack Creek (USA-AK) (location identifier) • Payload Bay Kit

PBL: Payload Bay Liner • Planetary Boundary Layer • Probable • Puerto Cabello (Venezuela) (location identifier)

Pbld: Published

Pblg: Publishing

Pblr: Publisher

PBM: Paramaribo - John Adolf Pengel (Suriname) (location identifier) • Paramaribo Zanderli (Suriname) (location identifier) • Peribacteroid Membrane • Permanent Bench Mark • Physically-Based Modelling • Portable Bitmap • Probability-Based Matching • Program Budget Manager • Program Business Management

PBMR: Push Broom Microwave Radiometer

PBMS: Pacific Bell Mobile Services

PBN: Porto Amboim (Angola) (location identifier) • Pyrolitic Boron Nitride

PBNI: Pacific Bell Network Integration

PBO: Paraburdoo (Australia-WA) (location identifier) • Performance-Based Organization • Photon Bubble Oscillation • Pine Bluff Observatory (USA-WI)

PBP: Pellin-Broca Prism • Performance-Based Pay • Point By Point • Preburner Pump • Program and Budget Planning • Program Board Panel

PBPF: Propeller Blade Passing Frequency

PBPS: Post-Boost Propulsion System

PBq: Petabecquerel

PBQ: Pimenta Bueno (Brazil-RO) (location identifier)

PBQS: Palomar Bright Quasar Survey

PBR: Packed Bed Reactor • Payment By Results • Puerto Barrios (Guatemala) (location identifier)

PBRF: Parallel Blind Relevance Feedback • Pilot Briefing

PbS: Lead Sulfide

PBS: Payload Berthing System • Personal Bibliographic System • Polarized Beam Structure • Polarizing Beam Splitter • Portable Base Station • Power Bus Simulator • Pressure Balance Structure • Primary Basic Subject • Profit Before Tax • Program Breakdown Structure • Programming By Selection • Protective Breathing System • Public Buildings Service • Public Broadcasting Service • Public Broadcasting System (*)

PBSRAM: Pipelined Burst Static Random Access Memory

PBT: Payback Time • Permeable Base Transistor • Piggyback Tape

PBTS: Payload Bay Transfer System

PBU: Peribacteroid Unit • Putao (Myanmar) (location identifier)

PBV: Post-Boost Vehicle

PBW: Palibelo (Indonesia) (location identifier) • Personal Bandwidth • Power-By-Wire • Proportional Bandwidth

PBWA: Plasma Beat-Wave Accelerator

PBWU: Projektgruppe Bayern zur Erforschung der Wirkungen von Umweltschadstoffen

PBX: Playback Transmitted • Princeton Beta Experiment • Private Branch Exchange

PBX-M: Princeton Beta Experiment - Modified

PBXM: Princeton Beta Experiment - Modified

PBZ: Plettenberg Bay (South Africa) (location identifier)

p/c: Pour Compte • Price Current

pc: Parsec • Percent • Percentage • Pica • Picocurie • Piece • Post Cibos • Pour Cent • Pour Condoléance • Price

pC: Post Christo

P/C: Pay Code • Petty Cash • Physico-Chemical • Pitch Control • Price Current

P-C: Performance-Cost • Period-Colour • Polar to Cartesian • Price Current • Primitive-Composite

P&C: Procurement & Contracts • Purchasing & Contracting

PC: Package Control • Package Controller • Pad Character • Pad Coordinator • Palomar Confirmed • Panama Canal • Panel Coordinator • Panoramic Camera • Paper Chromatography • Paper Copy • Parliamentary Commissioner • Partially Cloudy • Parts Catalogue • Patent Country • Peace Corps • Peripheral Control • Personal Computer • Personal Correction • Personality Conflict • Petty Cash • Phase Coherent • Phase Conjugation • Photocell • Photoconductive • Photoconductor • Photon Counter • Physico-Chemical • Pitch Circle • Pitch Control • Plane Change • Planetary Camera • Plenipotentiary Conference • Polar Cap • Police Constable • Political Correctness • Politically Correct • Polycarbonate • Port C • Position Control • Postal Card • Postcard • Post Cibum • Post Commander • Post Computer • Poste de Commandement • Potentially Correct • Power Control • Praça • Predictor-Corrector • Preparatory Commission • Preparatory Committee • Press Club • Press Conference • Pressure Chamber • Pressure Controller • Price Current • Primary Center • Primary Circuit • Primary Control • Prime Contractor • Prime Cost • Prince Consort • Principal Component • Printed Circuit • Printer Control • Privy Council • Process Control • Process Controller • Processor Control • Processor Controller • Production Control • Production Cost • Professional Corporation • Program Change • Program Commission • Program Control • Program Coordinator • Program Counter • Programmable Controller • Programma Commissie • Programmatic Compression • Programmed Check • Programming Committee • Project Control • Project Coordinator • Proportional Counter • Proposal Coordinator • Proposed Change • Propositional Calculus • Propulsive Coefficient • Propulsive Consumable • Propylene Carbonate • Provisional Cost • Pseudocode • Public Convenience • Pulsating Current • Pulse Command • Pulse Counter • Pulverized Coal • Punched Card • Purchasing and Contracting

Pça: Praça

PCA: Parallel Channel Adapter • Particle-Cluster Agglomeration • Parts Control Area • Patient Care Assistant • Performance and Coverage Analyzer • Performance Coverage Analyzer • Permanent Court of Arbitration • Personal Channel Agent • Perturbative Convolution Approximation • Photon-Counting Array • Physical Configuration Assessment • Physical Configuration Audit • Pitch Control Assembly • Plane Circular Aperture • Plasma-Covered Antenna • Pneumatic Control Assembly • Point of Closest Approach • Polar-Cap Absorption • Polar-Cap Anomaly • Policy Certification Authority • Portage Creek (USA-AK) (location identifier) • Positive Control Area • Positively-Controlled Airspace • Power Conditioning Assembly • Power Control Assembly • Power Controller Assembly • Precontractual Authorization • Pressure Control Assembly • Prime Contractor Address • Principal-Component Analysis • Production Configuration Audit • Program Change Analysis • Program Commitment Agreement • Proportional Counter Array • Propulsion-Controlled Aircraft • Pyrotechnic Control Assembly

PCAC: Porterville College Astronomy Club (USA-CA)

PCAD: Pointing Control and Aspect Determination

PCAL: Position Calibration

PCAM: Protein Crystallization Apparatus for Microgravity • Punched Card Accounting Machine

PCAS: Planet-Crossing Asteroid Survey • Punched Card Accounting System

PCASP: Passive Cavity Aerosol Spectrometer

PCASS: Parts Control Automated Support System • Program Compliance Assurance and Status System

PCAST: President's Council of Advisors on Science and Technology

PCASTRD: Philippine Council for Advanced Science and Technology Research and Development

PCB: Parts Control Board • Petty Cash Book • Physique-

Chimie-Biologie • Polychlorinated Biphenyl • Pondok Cabe (Indonesia) (location identifier) • Power Circuit Breaker • Power Control Box • Printed-Circuit Board • Process Control Block • Processor Control Block • Product Configuration Baseline • Program Change Board • Program Control Block • Project Change Board • Project Coordination Board • Propulsion Control Box • Protocol Control Box

PCBA: Printed Circuit Board Assembly

PCBS: Programme on Capacity Building in Science

PCC: Pad Control Center • Panama Canal Commission • Panel on Climate Change • Pasadena City College (USA-CA) • Payload Control and Checkout • Per Copia Conforme • Permanent Cool Container • Personal Communications Computer • Photon Counting Camera • Physics and Chemistry of Comets • Plastic Chip Carrier • Polarity Coincidence Correlator • Policy Coordinating Committee • Portable C Compiler • Post-Core Collapse • Poste de Commande Centralisé • Pour Copie Conforme • Power Control Console • Process Control Computer • Production Control Center • Professional Competency Credit • Program Controlled Computer • Program Coordinating Centre • Program for Cooperative Cataloging • Proto Cluster Cloud • Protoglobular Cluster Cloud • Puerto Rico (Colombia) (location identifier)

PCCA: Physical Configuration Communications Area • Portable Computing and Communication Association

PCCB: Program Change Control Board • Program Configuration Control Board

PCCD: Portable Charged Couple Device

PCCE: Payload Common Communication Equipment

PCCIP: President's Commission on Critical Infrastructure Protection (USA)

PCCM: Program Change Control Management • Program Change Control Manager

PCCN: Provisioning Contract Control Number

PCCP: Preliminary Contract Change Proposal

PCCS: Photographic Camera Control System • Program Change Control System

PCCU: Power Conditioning and Control Unit • Punched Card Control Unit

PCD: Payload Correction Data • Photo Compact Disk • Photon Counting Detector • Pitch Circle Diameter • Plataforma de Coleta de Dados • Polycrystalline Diamond • Power Control and Distribution • Prairie du Chien (USA-WI) (location identifier) • Precision Course Direction • Prime Contractor Department • Procedural Change Directive • Procurement Control Document • Procurements and Contracts Division • Product Confidence Data • Program Control Document

PCDA: Program Controlled Data Acquisition

PCDMA: Personal Computer Direct Marketers Association • Personal Computer Direct Marketing Association

PCDN: Plan Communal de Développement de la Nature (Belgium)

PCDP: Punched Card Data Processing

PCDS: Pilot Climate Data System (now NCDS) • Power Converter and Distribution System • Power Conversion and Distribution System

PCDU: Payload Command Decoder Unit • Power Control and Distribution Unit

Pce: Princesse

PCE: Painter Creek (USA-AK) (location identifier) • Peripheral Control Element • Physics and Chemistry of the Earth (journal) • Power Conditioning Electronics • Power Conditioning Equipment • Pressure to Clutch Engage • Process-Centered Environment • Program Cost Estimate • Punch Card Equipment

PCEA: Pointing Control Electronics Assembly

PCEB: Post-Common Envelope Binary

PCEI: Prime Contract End Item

PCEM: Propulsion Contamination Effects Module

PCES: Physics - Chemistry - Earth Sciences

PCETF: Power Conversion Equipment Test Facility

pcf: Pounds per Cubic Foot

PCF: Packet Control Facility • Payload Control Facility • Point Coordination Function • Portable Compiled Format • Portable Complied Format • Possible Cooling Flow • Potentially Critical Failure • Primary Calibration File • Probability of Consequence Factor • Processed Citation File • Program Change Factor • Programmable Command Format • Protein Crystallization Facility

PCFC: Polytechnics and Colleges Funding Council (UK)

PCFE: Prime Contractor Furnished Equipment

PCFG: Probabilistic Context-Free Grammar

PCFOV: Partially Coded Field Of View

PCFP: Predicted Comparative Failure Probability

PCFS: Personal Computer Field System

PCG: Parts Control Group • Paso Caballos (Guatemala) (location identifier) • Phonocardiogram • Planetário Calouste Gulbenkian (Portugal) • Planning and Control Guide • Power Conditioning Group • Preconditioned Conjugate Gradient • Printed Circuit Generator • Production Control Group • Protein Crystal Growth

PCGC: Pulverized Coal Gasification and Combustion

PCGD: Pollution Control Guidance Document • Post-, Cheque- en Girodienst

PCGN: Permanent Committee on Geographic Names

PCH: Pacific Coastal Highway • Palacios (Honduras) (location identifier) • Parallel Channel • Port C High • Professional Competency Hour

Pchs: Purchase

Pchsg: Purchasing

Pcht: Parchment

pci: Pounds per Cubic Inch

P/CI: Payload/Cargo Integration

PCI: Patent Citation Index • Pattern Correspondence Index • Payload/Cargo Integration • Peripheral Command Indicator • Peripheral Component Interconnect • Peripheral Component Interface • Peripheral Controller Interface • Poste Cartographique Interactif • Probability of Chance Improvement • Procedure Change Instruction • Product Configuration Identification • Program Check Interruption • Program Control Input • Program-Controlled Input • Programmable Communication Interface • Protocol Control Information

PCIA: Personal Communications Industry Association • Principal Component Image Analysis

PCIE: President's Council on Integrity and Efficiency

PCIERD: Philippine Council for Industry and Energy Research and Development

PCIL: Pilot-Controlled Instrument Landing

PCIN: Program Change Identification Number • Program Change Integration

PCinO: Publications of the Cincinnati Observatory (periodical code)

PCIPI: Permanent Committee on Industrial Property Information

PCIPS: Personal Computer Image Processing System

PC-I/O: Program Controlled Input/Output

PCIO: Program Controlled Input/Output

PCIS: Personal Computer Information Service

PCK: Payload Command Kit • Porcupine Creek (USA-AK) (location identifier)

PCKB: Printed Circuit Keyboard

Pcl: Parcel • Pencil

PCL: Pallet Coolant Loop • Parallel Communications Link • Parcel • Pilot-Controlled Lighting • Portable Common Loop • Port C Low • Power Control List • Primary Coolant Line • Primary Coolant Loop • Printer Command Language • Printer Control Language • Procedural Control Language • Process Control Language • Programming Checklist • Pucallpa (Peru) (location identifier)

PCLA: Power Control Linkage Assembly

PCLP: PC LAN Program

PCLS: Portable Chopped Light Source

pcm: Per Calendar Month

PCM: Package Command Manager • Parabolic Collimator Mirror • Parallel Climate Model • Parity Check Matrix • Parts Control Management • Personal Computer Manufacturer • Phase Change Material • Phase Code Modulation • Phase Comparison Monopulse • Phase Contrast Microscope • Phase Contrast Microscopy • Photo-Chemical Machinery • Photoelectric Cell Module • Physical Connection Management • Physics and Chemistry of Minerals (periodical code) • Pitch Control Motor • Plane Change Mechanism • Playa del Carmen (Mexico) (location identifier) • Plug-Compatible Mainframe • Plug-Compatible Manufacturer • Polyatomic Ceramic Material • Power Control Mission • Profiling Current Meter • Project Cycle Management • Propose-Critique-Modify • Protein-Calorie Malnutrition • Pulse Coded Modulation • Pulse Code Modulated • Pulse Code Modulation (*) • Pulse Code Modulator • Pulse Counting Mode • Punch Card Machine

PCMC: Precision Coherent Monopulse C-band

PCMCIA: Personal Computer Memory Cards Industries Association • Personal Computer Memory Cards International Association (*)

PCMD: Pulse Code Modulation Digital

PCMDI: Program for Climate Model Diagnosis and Intercomparison

PCME: Pulse-Code Modulation Event

PCMGI: Physico-Chimie des Molécules et des Grains Interstellaires • Physico-Chimie des Molécules et Grains Interstellaires

PCMI: Parks City Mathematics Institute (USA-NJ) • Photochromic Microimage • Physico-Chimie des Molécules Interstellaires • Physique et Chimie du Milieu Interstellaire (*) • President's Council on Management Improvement

PCMIM: Personal Computer Media Interface Module

PCMM: People Capability Maturity Model

PCMMU: Pulse Code Modulation Master Unit

PCMOS: Positive Channel Metal Oxide Semiconductor

PCMP: Parts Control Management Program

PCMR: Probability of Correct Message Receipt

PCMS: Process and Configuration Management Software • Process Configuration Management Software • Product Configuration Management System • Pulse Code Modulation Shared • Punched Card Machine System

PCMTE: Pulse Code Modulation and Timing Equipment

PCMTEA: Pulse Code Modulation and Timing Electronics Assembly

PCMU: Programmable Control and Monitor Unit

PCN: Package Control Number • Page Change Notice • Part Control Number • Pavement Classification Number • Personal Communications Network • Personal Computer Network • Picton (New Zealand) (location identifier) • Pitcairn (ISO country code) • Planning Change Notice • Polyester-Coated Nylon • Procedure Change Notice • Processing Control Number • Procurement Control Number • Product Control Number • Program Change Notice • Program Control Number • Project Control Number • Proposal Control Number • Publication Change Notice

PCNB: Pentachloronitrobenzene

PCNFS: Personal Computer Network File System

PCNS: Polar Coordinates Navigation System

Pco: Pico

PCO: Polar Cap Observatory • Postcheckout • Postcheckout Operations • Principal Contracting Officer • Procuring Contracting Officer • Professional Conference Organizer • Program Change Order • Program Contracting Officer • Program-Controlled Output • Program Control Officer • Program Coordination Office • Publications Coordination Office • Punta Colorada (Mexico) (location identifier) • Purchase Change Order

PCOS: Proportional Chamber Operating System

PCOT: Payload Center Operations Team

PCP: Parts Control Program • Perfect Cosmological Principle • Peripheral Control Program • Peripheral Control Pulse • Planning Command Pool • Poisson Cluster Process • Power Control Panel • Preliminary Cost Proposal • Pressurization Control Panel • Primary Control Program • Prime Commercial Paper • Principe Island (location identifier) • Prise de Culot Pneumatique • Private Carrier Paging • Processor Control Program • Program Change Proposal • Project Change Proposal • Project Cost Plan

PCPA: Polar Cap Phase Anomaly

PCPC: Personal Computers Peripheral Corporation

PCPD: Project Control and Planning Division

PCPG: Primary Clock Pulse Generator

PCPGA: Printed-Circuit Pin-Grid Array

PCPI: Permanent Committee on Patent Information

PCPL: Proposed Change Point Line

PC&PM: Program Control & Performance Management

PCPM: PERT Cost Performance Measurement • Program Control and Performance Management

PCPN: Precipitation

PCPP: Parallel Computing using the Prefix Problem

PCPS: Pulse-Coded Processing System

PCR: Payload Changeout Room • Payload Checkout Room • Payload Clean Room • Peak Cell Rate • Pen Character Recognition • Phase-Conjugate Reflectivity • Pitch Circle Radius • Phase Change Rewritable • Planning Change Request • Planning Check Request • Polar-Cap Radar • Polymerase Chain Reaction • Power Change Request • Primary Cosmic Radiation • Primary Cosmic Ray • Prime Contractor Region • Principal-Component Regression • Procedure Change Request • Production Control Record • Program Change Request • Program Control Register • Program Control Report • Project Control Room • Project Cost Record • Project Cost Report • Publication Change Request • Puerto Carreno (Colombia) (location identifier) • Punched Card Reader • Punched Card Requisition

PCRB: Program Change Review Board • Project Change Review Board

PCRD: Programme Cadre de Recherche et de Développement (France)

PCRS: Payload Cover Removal System • Performance and Cosr Report System

Pcs: Pieces • Prices

PCS: Panoramic Camera System • Particle Counting System • Partition Communication System • Patchable Control Store • Patent Classification Service • Pattern Code Security • Payload Calibration System • Payload Checkout System • Payload Control Supervisor • Payload Control System • Payload Correction System • Periodical Control System • Permanent Change of Station • Performance Control System • Permit Compliance System

• Personal Communications Service • Personal Communications System • Personal Conferencing Specification • Photo Correlation Spectroscopy • Photo Correlation System • Photon Correlation Spectroscopy (*) • Photon Counting Spectrometer • Photon Counting System • Physical Control System • Picos (Brazil-PI) (location identifier) • Pitch Control System • Planning and Control System • Planning Control Sheet • Plasma Current Switch • Pointing and Control System • Pointing Control System • Portable Computer System • Position-Course-Speed • Power Conversion System • Precision Clock System • Precision Correction System • Preliminary Component Specification • Pressure Control System • Primary Calibration System • Primary Control Station • Primary Coolant System • Principal Component Substitution • Principal Coordinating Scientist • Print Contrast Signal • PRISMA Computer System • Print Contrast System • Probability of Crew Safety • Procedure Completion Sheet • Process Control System • Product Control Service • Program Control System • Project Control System • Project Coordination Staff • Program Counter Store • Project Control System • Proxy Cache Server • Punched Card System

PCSA: Personal Computer System Architecture

PCSC: Payload and Crew Support Computer • Power Conditioning, Switching and Control

PCSD: Philippine Council for Sustainable Development

PCSIU: Pointing Control System Interface Unit

PCSP: Polar Continental Shelf Project

PCSR: Planck Calibrated Sky Radiometer

Pct: Percent

PCT: Pacific Crest Trail (USA) • Parallel Change Task • Pass Control • Patent Cooperation Treaty • Percent • Performance Correlation Technique • Photometric Calibration Target • Photometric Calibration Test • Picture • Picture Contour Table • Plan à Court Terme • Polychlorinated Triphenyl • Praceta • Prime Contract Termination • Princeton (USA-NJ) (location identifier) • Principal Component Transform • Private Communication Technology • Process Control Table • Processor Control Table • Product Consistency Test • Program Coordinator Team • Pulse Compression Test

PCTC: Payload Crew Training Complex

PCTE: Portable Commercial Test Equipment • Portable Common Tool Environment

PCTF: Passive Cooler Test Facility

PCTJ: Parallel Connected Twin Junction

PCtM: Pulse Count Modulation

PCTO: Payload Cost Tradeoff Optimization

PCTR: Pad Connection Terminal Room • Pipeline Control • Prominence-Corona Transition Region

PCTS: POSIX Conformance Test Suite

PCU: Packet Control Unit • Passenger Car Unit • Passenger Control Unit • Payload Checkout Unit • Payload Control Unit • Picayune (USA-MS) (location identifier) • Plant Control Unit • Platinum Cobalt Unit • Portable Control Unit • Power Conditioning Unit • Power Control Unit • Power Conversion Unit • Power Convertor Unit • Pressure Control Unit • Pressurization Control Unit • Primary Conditioning Unit • Print Control Unit • Process Control Unit • Progress Control Unit • Propeller Control Unit • Punched Card Utility

PCV: À Percevoir • Passenger-Controlled Vehicle • Performance, Calibration and Verification • Polar Cataclysmic Variable (star) • Pollution-Controlled Valve • Positive Crankcase Ventilation • Potential Cataclysmic Variable • Precheck Verification • Premeasurement-Calibration-Verification • Pressure Containment Vessel • Punta Chivato (Mexico) (location identifier) • Purge Control Valve

PCVB: Pyro Continuity Verification Box • Pyrotechnic Continuity Verification Box

PCVL: Pilot-Controlled Visual Landing

PCW: Personal Computer World • Previously Complied With

PCWG: Personal Conferencing Work Group • Personal Conferencing Working Group

PCWQM: Process Control Water Quality Mirror

PCX: Picture Image • Plasma Confinement Experiment

PCZ: Panama Canal Zone • Physical Control Zone • Positive Control Zone

PCZA: Pseudo-Circumzenithal Arc

PCZST: Panama Canal Zone Standard Time

p/d: Post-Dated

pd: Per Diem • Postage Due • Post-Dated

P/d: Porte Debido

Pd: Paid • Palladium • Passed

P&D: Panels & Displays

P/D: Price-Divident

PD: Doctor of Pharmacy • Packet Decimal • Parameter Data • Participatory Design • Per Diem • Performance Demonstration • Peripheral Decoder • Peripheral Device • Phase Discriminator • Photodiode • Photoelectron Diffraction • Photon Drag • Phrase Dictionary • Physics Department • Pitch Diameter • Pitch Down • Planning Directive • Planning Document • Plasma Desorption • Polar Distance • Police Department • Policy Deployment • Policy Directive • Policy Division • Port Dû • Port Dues • Posdata • Position Description • Positive Displacement • Postage Due • Postage District • Postal District • Potential Difference • Power Distribution • Precision Digitizing • Preliminary Design • Presidential Directive • Price-Divident • Prime Driver • Printer Driver • Priority Designator • Priority Directive • Prism Diopter • Privatdozent • Probability of Detection • Problem Description • Processed Data • Procurement Data • Procurement Directive • Procurement Division • Procurement Document • Procurement Drawing • Production Department • Program Development • Program Directive • Program Director • Project Directive • Project Director • Project Document • Propellant Dispersal • Property Damage • Proportional Differential • Protective Device • Public Domain • Pulse Duration • Punch Driver

PDA: Packet Data Analysis • Packet Data Analyzer • Parallel Data Adapter • Parametric Design Analysis • Partial Discriminant Analysis • Payload Disconnect Assembly • Percent Defect Allowed • Per Diem Allowance • Personal Digital Assistant • Phased Doppler Anemometer • Photodiode Array • Photon Detector Assembly • Polydiacethylene • Portable Data Access • Post-Design Analysis • Pour Dire Adieu • Power Distribution Assembly • Precision Drive Axis • Predelivery Acceptance • Predicted Drift Angle • Preliminary Design Approval • Probability Distribution Analyzer • Processor and Distribution Assembly • Production Data Analysis • Propellant Drain Area • Public Display of Affection • Public Domain Algorithm • Puerto Inirida (Colombia) (location identifier) • Pump Drive Assembly

PDAC: Paris Data Analysis Center (France)

PDAD: Proposed Draft Addendum

PDAF: Probabilistic Data Acquisition Filter

PDAO: Publications of the Dominion Astrophysical Observatory (periodical code)

PDAP: Portable Data Acquisition Package

PDAR: Preferential Departure and Arrival Route • Program Description And Requirements

PDAS: Photo Data Analysis System

PDAUC: Publications of the Department of Astronomy of the University of Chile (periodic al code)

PdB: Bachelor of Pedagogy • Plateau de Bure (France)

PDB: Parachute Database • Pedro Bay (USA-AK) (location identifier) • Performance Data Book • Personal Database • Planar-Doped Barrier • Power Distribution Box • Primary Database • Project Database • Project Development Brochure • Protein Databank

PdBI: Plateau de Bure Interferometer (IRAM)

PDBS: Personalized Database System

P&DC: Printing & Distribution Center

PDC: Mueo (New Caledonia) (location identifier) • Parallel Data Controller • Pattern Description Compiler • Personal Digital Cellular • Personnel Dispatch Centre • Personnel Dispersal Centre • Power Disk Cartridge • Power Distribution and Control • Power Distribution Control • Pre-Defined Command • Pre-Departure Clearance • Pre-Deployment Checkout • Predefined Command • Premission Documentation Change • Primary Domain Controller • Printing and Distribution Center • Problem Domain Component • Procurement Document Change • Program Data Coordinator • Program Delivery Control • Program Designator Code • Program Development Center • Project Data Card • Project Data Control • Pulse-Discharge Cleaning

PDCCM: Pas De Ça Chez Moi

PDCM: Product Data Control Model

PDCR: Project Data Compliance Report • Proprietary Data Control Record

PDCS: Palomar Distant Cluster Survey • Parallel Digital Computing System • Performance Data Computer System • Power Distribution and Control System • Predictably Dependable Computing System • Processing Distribution and Control System

PDCU: Plotting Display Control Unit • Power Distribution and Control Unit

PdD: Doctor of Pedagogy • Palladium Deuteride

PDD: Past Due Date • Payload Description Document • Physical Device Driver • Portable Delivery Device • Post Dial Delay • Preliminary Design and Development • Presidential Decision Directive • Priority Delivery Date • Probability Density Distribution • Probability Distribution Density • Program Description Document • Program Definition Document • Program Design Data • Project Disclosure Document • Prospective Decision Date • Provisioning Description Data • Pulsating Delayed Detonation

PDDF: Preliminary Data Description File

PDDI: Product Data Definition Interface • Product Definition Data Interface

PDDO: Publications of David Dunlap Observatory (periodical code)

PDDR: Programme de Développement des Données Radar • Projet de Développement de Données Radar

PDDS: Program Definition Data Sheet

PDDU: Power Distribution and Drive Unit

PDE: Pandie Pandie (Australia-SA) (location identifier) • Partial Differential Equation • Process Directory Entry • Programa de Divulgación Escolar • Promenade • Pulsed Detonation Engine • Pulse Detonation Engine (*) • Pure Density Evolution

P-de-C: Pas-de-Calais (France)

P-de-D: Puy-de-Dôme (France)

PDEF: Protean Data Exchange Format

PDEL: Partial Differential Equation Language

PDES: Product Data Exchange Specification • Product Data Exchange Standard • Product Data Exchange using STEP • Product Description Exchange Standard

Pdf: Densely Fracture Plain

PDF: Parameter Definition File • Particle Distribution Function • Parton Distribution Function • Planar Deformation Feature • Plan Départemental de Formation (France) • Platform Data File • Portable Document File • Portable Document Format • Postdoctoral Fellow • Power Density Function • Power Distribution Frame • Prado (Brazil-BA) (location identifier) • Printer Description File • Probability Density Function • Probability Distribution Function • Procedure Definition File • Processor-Defined Function • Program Data File • Program Development Facility • Programmable Data Formatter • Pulsar Data File

PDFRR: Program Directors Flight Readiness Review

PDG: Padang (Indonesia) (location identifier) • Particle Data Group • Patent Documentation Group • Payload Data Generator • Paymaster Director-General • Polarization-Dependent Gain • Président Directeur Général • Program Development Group • Proposal Development Group

PDGF: Power and Data Grapple Fixture (*) • Power Data Grapple Fixture

PDH: Plesiochronous Digital Hierarchy • Professional Development Hour

PDHID: Pulsed Discharge Helium Ionization Detector

PDI: Palmer Drought Index • Partial Differential Inequality • Patent Documentation and Information • Payload Data Interleaver • Persistent Document Identifier • Pindiu (Papua New Guinea) (location identifier) • Power and Data Interface • Powered Descent Initiation • Predelivery Inspection • Project Data Index

PDIAL: Public Dialup Internet Access List

PDIEP: Partially Described Inverse Eigenvalue Problem

PDII: Pusat Dokumentasi Ilmiah dan Informasi (Indonesia)

PDIS: Parallel and Distributed Information System

PDIU: Propulsion Discrete Interface Unit

PDK: Atlanta - De Kalb - Peachtree (USA-GA) (location identifier) • Planare Driftkammer • Publishing Development Kit

pdl: Poundal

P&DL: Program & Data Library

PDL: Page Description Language • Pattern Description Language • Picture Description Language • Polarization-Dependent Loss • Polarization Diversity Lidar • Polarized Diversity Lidar • Ponta Delgada (Azores) (location identifier) • Poverty Datum Line • Procedure Definition Language • Procedure Distribution List • Procurement Data List • Product Development Laboratory • Program Definition Language • Program Description Language • Program Design Language • Programmed Digital Logic • Project Document List • Pulsed Dye Laser • Pushed Down List

PDLAC: Preliminary Dynamic Loads Analysis Cycle

PDLC: Preliminary Design Load Cycle

PdM: Master of Pedagogy

PDM: Partial Descriptive Method • Payload Deployment Module • Percent Deviation from the Median • Phase Dispersion Minimization • Physical Data Model • Physical Device Manager • Physical Distribution Management • Precedence Diagramming Method • Precision Doppler Measurement • Primary Deployment Mechanism • Primary Drive Mechanism • Primary Drive Motor • Probabilistic Dilution Model • Probe Data Model • Processor Data Monitor • Product Data Management • Program Data Manager • Program Decision Memorandum • Programmed Depot-level Maintenance • Project Design Memo • Property Driven Model • Pulse-Duration Modulation • Push Down Memory

P-DME: Precision Distance Measuring Equipment

PDME: Precision Distance Measuring Equipment

PDMF: Present-Day Mass Function

PDMFM: Pulse-Duration Modulation Frequency Modulation

PDML: Procedural Data Manipulation Language

PDMM: Push Down Memory Modem

PDMP: Project Data Management Plan

PDMS: Plasma Desorption Mass Spectroscopy • Program Definition and Management System

PDMT: Predominant

PDMU: Passive Data Memory Unit

Pdn: Production

PDN: Parndana (Australia-SA) (location identifier) • Position-Dependent Normalization • Power Dividing Network • Public Data Network

Pdo: Pasado

PDO: Pendopo (Indonesia) (location identifier) • Physical Data Object • Portable Distributed Object • Pseudo-Differential Operator

PDOP: Positional Dilution Of Precision • Position Dilution Of Precision

Pdor: Parador

PDOS: Parallel and Distributed Operating Systems • Partial Density Of States

PDP: Parallel Distributed Processing • Passive Dosimeter Packet • Payload Data Pack • Phase Doppler Pneumometer • Piecewise Deterministic Process • Plasma Diagnostics Package • Plasma Display Panel • Postinsertion Deorbit Preparation • Power Distribution Panel • Preliminary Definition Plan • Preliminary Design Phase • Preliminary Design Proposal • Procurement Data Package • Professional Development Plan • Professional Development Program • Program Definition Phase • Program Development Plan • Programmable Data Processor • Programmed Data Processor • Programmed Digital Processor • Project Data Package • Project Definition Phase • Project Development Plan • Punta del Este (Uruguay) (location identifier)

PDPF: Project Data Processing Facility

PDPS: Program Data Processing Section

PDQ: Parallel Data Query • Peachtree Data Query • Pertinent Data Quest • PODPS Data Quality • Position Description Questionnaire • Pre-route Delay Qualifier • Price and Delivery Quotation • Procedural Data Quality • Protocol Data Query

PDQS: Portable Document Query System

Pdr: Powder

PDR: People's Democratic Republic • Phase Data Recorder • Phase Delay Rectifier • Photodissociated Region • Photodissociation Region (*) • Photon-Dominated Region • Pitch-to-Diameter Ratio • Preferential Departure Route • Preliminary Data Report • Preliminary Data Requirements • Preliminary Design Review • Preliminary Document Review • Previous Document Reference • Priority Data Reduction • Processed Data Recorder • Procurement Data Reference • Program Design Review • Program Director's Review • Program Document Requirement • Program Drum Recording • Project Definition Report • Project Design Review • Publications Data Request

PDRA: Postdoctoral Research Assistant

PDRB: Preliminary Design Review Baseline

PDRD: Procurement Data Requirements Description • Procurement Data Requirements Document • Program Data Requirements Document • Program Definition and Requirements Document

PDRF: Post-Doctoral Research Fellow

PDRL: Preliminary Design Requirements List • Procurement Data Requirements List

PDRM: Payload Deployment and Retrieval Mechanism

PDRP: Program Data Requirement Plan

PDRR: Payload Design Requirements Review • Preliminary Design Requirements Review • Program Definition and Risk Reducation

PD&RS: Payload Deployment & Retrieval System

PDRS: Payload Data and Retrieval System • Payload Deployment and Retrieval System • Primary Data Receiving Station • Probe Data Relay System

PDRSS: Payload Deployment and Retrieval System Simulation • Payload Deployment and Retrieval System Simulator

PDRSTA: Payload Deployment and Retrieval System Test Article

PDRW: Post-Doctoral Research Worker

PDS: Package Data System • Packet Driver Specification • Page Data Set • Page Display System • Parachute Design System • Paradynamic Sketch • Partitioned Data Set • Pathfinder Demonstration Satellite • Payload Data Segment • Personal Data Server • Personal Dictation System • Personal Display System • Phase Differential Scattering • Phased Development Shuttle • Phoswich (crystal) Detector System • Photodischarge Spectroscopy • Photometric Data Systems • Photothermal Deflection Spectroscopy • Piedras Negras (Mexico) (location identifier) • Pilot Data System • Pilot Distributed System • Planetary Data System • Planning Data Sheet • Pneumatic Distribution System • Polydimethylsiloxane • Portable Data System • Posture Dorsale de Sauvegarde • Power Density Spectrum • Power Distribution System • Premises Distribution System • Problem Data System • Processed Data Stream • Processor Direct Slot • Procurement Data Sheet • Product Design Standard • Product Distribution System • Production Data Sheet • Professional Development Seminar • Program Data Source • Program Development Section • Program Distribution System • Programming Documentation Standards • Project Definition Statement • Pyrotechnic Device Simulator

PDSA: Plan-Do-Study-Act

PDSAR: Public Document Status of Assessment Report

PDSF: Physics Detector Simulation Facility

PdSi: Palladium Silicide

PDSI: Palmer Drought Severity Index • Phase-Diverse Speckle Imaging

PDSL: Public Domain Software Library

PDSP: Peripheral Data Storage Processor • Position Determination on Schmidt Plates

PDSR: Principal Director of Scientific Research

PDSS: Payload Data Services System • PIMOS Development Support System • Post Development and Software Support

PDSSBN: Planetary Data System Small Bodies Node

PDST: Pacific Daylight Saving Time

PDSU: Power Distribution and Switching Unit

PDT: Pacific Daylight Time • Parallel Data Transmission • Payload Data Tape • Pendleton (USA-OR) (location identifier) • Performance Demonstration Test • Personal Development Technology • Photodynamic Therapy • Plasma Display Terminal • Precision Digitizing Table • Predictor Display Technique • Printer Definition Table • Product Development Team • Programmable Drive Table

PDTA: Power and Data Transfer Assembly

Pdte: Presidente

PDTS: Program Development Tracking System

PDU: Paysandú (Uruguay) (location identifier) • Pilot Display Unit • Plan de Déplacements Urbains • Plug Distribution Unit • Position Display Unit • Power Distribution Unit • Power Drive Unit • Pressure Distribution Unit • Process Development Unit • Professional Development Unit • Programme de Développement axé sur les Utilisateurs • Protocol Data Unit • Pulse Detection Unit

PDUS: Primary Data User Station (*) • Principal Data User Station

PdV: Pressure-driven Volume

PDV: Personaldatenverarbeitung • Plovdiv (Bulgaria) (location identifier) • Pressure Disconnect Valve

PDVS: Patentdatenverarbeitungssystem

PDW: Priority Delayed Weather

PDWE: Pulsed Detonation Wave Engine

PDWS: Passive Doppler Wind Sensor • Precision Digitizing Workstation

PDX: Paradox • Phone Data Exchange • Poloidal Divertor Experiment • Portland (USA-OR) (location identifier) • Princeton Divertor Experiment

PDZ: Pedernales (Venezuela) (location identifier)

PDZAMF: Present-Day Zero-Age Mass Function

Pdzo: Pasadizo

p/e: Por Ejemplar • Price Earnings Ratio

pe: Par Exemple • Per Esempio • Perlucidus • Peru • Photoelectric • Photoelectron • Por Ejemplo • Probable Error

Pe: Péclet Number

P/E: Port of Embarkation • Precipitation/Evaporation • Price-to-Earnings • Profit-to-Earnings

P-E: Perkin-Elmer Corporation

P&E: Plant & Equipment • Pricing & Estimating • Propellant & Explosives

PE: Padre • Parity Error • Parity Even • Parlamento Europeo • Parlement Européen • Payload Element • Pellets • Peripheral Equipment • Perkin-Elmer • Permissible Error • Pernambuco (Brazil) • Pérou (ISO country code) • Persian • Personal Effects • Peru (international vehicle registration code) • Peru (ISO country code) • Peterborough (UK postal code) • Phase Encode • Phase Encoded • Phase Encoder • Phase Encoding • Photoelectric • Photoelectron • Photogrammetric Engineer • Physical Education • Planetary Explorer • Planning Estimate • Plant Equipment • Plasma Effect • Plasma Emission • Pluto Express • Pocket Edition • Polyethylene • Port Elizabeth - H.F. Verwoerd (South Africa) • Port of Embarkation • Position Error • Post Encounter • Potential Energy • Potential Evapotranspiration • Power Enrichment • Precipitation/Evaporation • Precision Electronics • Prediction Efficiency • Pre-Equilibrium • Preliminary Estimator • Preliminary Evaluation • Presiding Elder • Prime Equipment • Primitive Equation • Prince Edward Island (Canada) • Principal Engineer • Printer's Error • Probable Error • Processeur Élémentaire • Processing Element • Processor Element • Procurement Executive • Professional Engineer • Program Element • Program Exception • Project Engineer • Project Engineering • Protect Enable • Purchased Equipment

PEA: Pattern Error Analysis • Payload Electronics Assembly • Payload Enclosure Assembly • Paysage Électronique Astronomique • Penneshaw (Australia-SA) (location identifier) • Phillips Exeter Academy (USA-NH) • Pitch Error Amplifier • Platform Electronics Assembly • Pocket Ethernet Adapter • Preliminary Environmental Analysis • Proposal Expansion Award

PEACE: Plasma Electron And Current Experiment (CLUSTER) • Plasma Energy Angle and Composition Experiment • Protection of the Environment for Assuring a Cleaner Earth

PEAGB: Post-Early-Asymptotic Giant Branch

PEAM: Portable Electronic Aid for Maintenance

PEAN: Pan European ATM Network

PEAP: Pad Emergency Air Pack • Personal Egress Air Pack

PEAR: Polyether Amide Resin • Program Explanation Analysis Required

PEARLS: Physics Electronic Academic Reference Library for Simulation

PEARLSS: Particulate Effects Analysis on Response Levels of Space-borne Sensors

PEAT: Programme Élargi d'Assistance Technique

PEB: Payload Equipment Bay • PCM Expansion Bus • Pebane (Mozambique) (location identifier) • Performance Evaluation Board • Prêt Entre Bibliothèques • Program Element Breakdown

PEBB: Power Electronic Building Block

PEbd: Polyéthylène Basse Densité

PEBS: Pulsed Electron Beam Source

Pec: Peculiar

PEC: Peak Electrode Current • Pelican (USA-AK) (location identifier) • Periodic-Error Correction • Perkin-Elmer Corporation • Photoelectric Cell • Photoelectrochemical Cell • Physics-Engineering-Chemistry • Planetary Entry Capsule • Pre-Existing Condition • Prêt à Être Changé • Previous Element Coding • Primary Environment Care • Program Element Code • Program Environment Control • Program Evaluation Committee • Programme Européen de Coopération • Projet Européen Commun • Proposal Evaluation Criteria • Propulsion Environmental Chamber • Pulsed Eddy Current • Pulse-height to Energy Conversion

PECASE: Presidential Early Career Awards for Scientists and Engineers

PECC: Product Engineering Control Center

PECD: Photoelectric Conversion Device

PECE: Proposed Engineering Change Estimate

PECF: Profile of the Effective Cloud Fraction

PECH: Poly Epicholorohydrine

PECI: Preliminary Equipment Component Index • Preliminary Equipment Component Inventory

PECO: Pays d'Europe Centrale et Orientale

PECOS: Perspectives on Cooperative Systems • Picture Enhancement Computer Operating System • Program Environment Checkout System

PECP: Preliminary Engineering Change Proposal

PECR: Program Error Correction Report

PECS: Portable Environmental Control System • Powered Elevator Control System

PECU: Powered Elevator Control Unit

PECVD: Plasma-Enhanced Chemical-Vapor Deposition

Ped: Pedal • Pedestal • Pedestrian

PEd: Proposal Editor

PED: Pays En Développement • Pedal • Pedestal • Pedestrian • Period End Date • Personal Electronic Device • Personnel Equipment Data • Photoemission Diode • Photon-Emitting Defect • Plasma Emission Detector • Pôle Européen de Développement • Portable Electronic Device • Procurement Early Development • Production Eligibility Date • Program Editor • Program Element Directive • Program Execution Directive • Proposal Editor • Proton-Enhanced Diffusion • Pulsed Electrochemical Detection

Pedag: Pedagogía

PEDAS: Panel on Environmentally Detrimental Space Activities • Potentially Environmentally Detrimental Activities in Space

PEDB: Payload Engineering Database

PedD: Doctor of Pedagogy

PEDN: Planned Event Discrepancy Notification

PEDS: Pedestrians

PEE: Perm (Russia) (location identifier) • Photoelectron Emission • Photoemission Effect • Polychromatic Encircled Energy • Poste d'Expansion Économique • Program Estimating Equation

PEEK: Polyether Ether Ketone

PEEL: Programmable Electrically Erasable Logic

PEEM: Panel of Experts on Environmental Management • Photoelectron Emission Spectroscopy

PEEP: Planetary Event and Eclipse Predictor • Porous Electrode Electrostatic Precipitation

PEERS: Physics Encyclopaedia of Email Records

PEF: Personal Effects Floater • Physical Electronics Facility • Polyethylene Foam • Prediction Error Filter • Program Estimating Factor • Proposal Evaluation Form • Pulsed Electric Field

PEFO: Payload Effects Follow-On

PEFOS: Payload Effects Follow-On Study

PEFPN: Physica Energiae Fortis et Physica Nuclearis (periodical code)

PEFU: Payload Effects Follow-Up

Peg: Pegasus-Pegasi (constellation)

PEG: Performance Evaluation Group • Perugia - Sant Egidio (Italy) (location identifier) • Polyethylene Glycol • Powered Explicit Guidance

PEGA: Program in Extragalactic Astronomy

PEGASUS: Preliminary Expert Ground Analysis Scheduling User System

PEGC: Professeur d'Enseignement Général des Collèges

PEGI: Programme Environnement de la Géosphère Intertropicale

PeGRS: Penetrator Gamma-Ray Spectrometer

PEH: Pehuajo (Argentina-BA) (location identifier)

PEhd: Polyéthylène Haute Densité

PEHD: Polyéthylène Haute Densité

PEHG: Payload Ethernet Hub Gateway

PEI: Pereira (Colombia) (location identifier) • Phase-Encoded Information • Preliminary Engineering Inspection • Prince Edward Island (Canada) • Professional Engineering Institution

PEIC: Periodic Error Integrating Controller

PEIDA: Photométrie et Étude d'Images Destinées à l'Astrophysique

PEIR: Project Equipment Inspection Record

PEIS: Programmatic Environmental Impact Statement

p. ej.: Por Ejemplo

Pek: Peking (China-PRC)

PEK: Beijing Capital (China-PRC) (location identifier) • Peking Capital (China-PRC) (location identifier)

Pel: Picture Element

PEL: Pelaneng (Lesotho) (location identifier) • Permissible Exposure Level • Permissible Exposure Limit • Photoelectron Layer • Physics and Engineering Laboratory • Physics Express Letters • Picture Element • Precision Elastic Limit

PElec: Photoelectric

PEM: Pacific Exploratory Mission • Parametric Earth Model • Particle Environment Monitor • Payload Ejection Mechanism • Payload Electronics Module • Pelletier Effect Module • Photoelastic Modulator • Photoelectromagnetic • Plant Engineering and Maintenance • Plastic-Encapsulated Microcircuit • Plastic-Encapsulated Microelectronics • Positive Energy Mode • Postal Electronic Mail • Preemptive Multitasking • Privacy-Enhanced Mail (*) • Privacy-Enhanced Messaging • Probable Error of Measurement • Processing Element Memory • Product Effectiveness Manual • Product Error Message • Program Element Monitor • Programmer's Environment Manager • Project Engineering Memorandum • Proton-Exchange Membrane • Puerto Maldonado (Peru) (location identifier)

Pembs: Pembrokeshire (UK)

PEMD: Personalized Educational Materials Development

PEMF: Pulsating Electromagnetic Field

PEMO: Production Engineering and Manufacturing Organization

PEMS: Portable Environmental Measuring System

Pen: Penal • Penetration • Península • Peninsular • Peninsule • Penisola • Pennsylvanian

PEN: Nuevo Sol (Peru) (ISO currency code) • Particle Entity Notation • Penang (Malaysia) (location identifier) • Peninsula • Poets-Essayists-Novelists • Program Element Number • Program Error Note • Public Education Network • Public Electronic Network

Penn: Pennsylvania (USA)

Penna: Pennsylvania (USA)

Pens: Pensamiento

Pent: Pentagon • Pentecost

PEO: Polyethene • Polythylene • Post Emergency Organisation • Professional Employment Organization • Program Evaluation Office • Program Executive Office • Program Executive Officer • Project Engineering Office

PEP: Packet Exchange Protocol • Packetized Ensemble Protocol • Paperless Environment Project • Partitioned Emulation Program • Pauli Exclusion Principle • Payload Exploitation Plan • Peak Energy Product • Peak Envelope Power • Peer Evaluation Program • Peppimenarti (Australia-NT) (location identifier) • Performance Evaluation Plan • Peripheral Event Processor • Perkin-Elmer Processor • Personal Equity Plan • Phase-Encoded Part • Photoelectric Potential • Photoemission Polarimeter • Physics and Engineering Physics • Piper External Power • Planetary Ephemeris Program • Planetary Exploration Plan • Plant Engineering Procedure • Plasma Exhaust Process • Platform Electronic Package • Platform Evaluation Program • Polar Energetic Particle • Pole-Equator-Pole • Political and Economic Planning • Positron-Electron-Project • Positron-Electron-Proton • POSIX Environment Platform • Power Evaluation Program • Power Extension Package • Precipitation Enhancement Program • Pre-Ejection Period • Productivity Enhancement Project • Program Evaluation Procedure • Proposal Entry Processing • Proposal Entry Processor (*) • Proposal Entry Proposer • Proposal Evaluation Panel • Protocol Extension Protocol • Proton-Electron-Proton

PEPAT: Protocol on Environment Protection of the Antarctic Treaty

PEPC: Polynomial Error Protection Code

PEPDB: Proposal Entry Processing Database

PEPE: Parallel Element Processing Ensemble • Persistant Elevated Pollution Episode • Plasma Experiment for Planetary Exploration

PEPG: Piezoelectric Power Generation

PEPGB: Plasma Exhaust Pumping Glovebox

PEPI: Physics of the Earth and Planetary Interiors (periodical code)

PEPL: Photo-Electron-enhanced Plasma Line

PEPP: Planetary Entry Parachute Program

PEPR: Precision Encoding and Pattern Recognition

PEPS: Preliminary Evaluation Programme for SPOT • Programme d'Étude Probatoire de SPOT • Programme d'Évaluation Préliminaire SPOT

PEPSI: Proposal Entry Processor System

PEPSOC: Portable ESOC Package for Synchronous Orbit Control

PEPSY: Precision Earth-Pointing System • Proposal Entry Processor System

PEQ: Pecos City (USA-TX) (location identifier)

PEQUOD: Pacific Equatorial Ocean Dynamics

Per: Percentile • Period • Periodic • Periodical • Perigee • Perseus-Persei (constellation) • Persia • Persian • Person

Pér: Période • Périodique

PER: Packet Error Rate • Parity Error Rate • Performance Evaluation Report • Perigee • Period • érou (ISO country code) • PERT Event Report • Perth (Australia-WA) (location identifier) • Peru (ISO country code) • Physics Education Research • Pre-Environmental Review • Preliminary Engineering Report • Price Earning Rate • Price-Earnings Ratio • Professional Employment Register • Program Event Recording • Program for Ecological Research • Public Environment Report

PERA: Planning and Engineering for Repair Activity

per an: Per Annum

Perc: Percussion

PERC: Portable Executive for Reliable Control

per con: Per Contra

PERD: Panel on Energy Research and Development

Perf: Perfect • Perfection • Perfecto • Perforated • Perforation • Performance • Performed

PERF: Perform • Performance

PerFECT: Peroxide in Fuel Estimation and Concentration Test

PERGS: Portable Earth Resources Ground Station

Perh: Perhaps

Peri: Perigee

PERIDOT: Prévision à Échéance Rapprochée Intégrant des Données Observées et Télédétectées

Périph: Périphérique

Period: Periodic • Periodical

PERL: Pathologically Eclectic Rubbish Lister • Practical Extraction and Reporting Language • Practical Extraction and Report Language (*) • Practical Extraction Reporting Language

Perm: Permanent • Permanently • Permission

PERM: Permanent

PERMANENT: Projects to Employ Resources of the Moon and Asteroids Near Earth in the Near Term

Permb: Permeability

PERMIT: Polar Orbiter Effective Rainfall Monitoring Integrative Technique

PerMS: Perceptual Motor Skills (periodical code)

Perp: Perpendicular • Perpetual

PERP: Parts and Equipment Reliability Program

per pro: Per Procurationem

per proc: Per Procurationem

Pers: Persia • Persian • Person • Persona • Personal • Personalize • Personalized • Personally • Personen • Personnel • Perspective

PE&RS: Photogrammetric Engineering & Remote Sensing (journal)

Persp: Perspectiva • Perspective

Pert: Pertaining

PERT: Performance Evaluation and Review Technique • Program Evaluation and Research Task • Program Evaluation and Review Technique (*) • Project Evaluation and Review Technique

PERTCO: Program Evaluation and Review Technique with Cost

PERTEC: Pôle Européen de Recherches en Turbulence et Convection

Peruv: Peruvian

PES: Nuevo Sol (Peru) (ISO currency code) • Parallel Enterprise Server • Patent Examining System • Personal Earth Station • Petrozavodsk (Russia) (location identifier) • Photoelectric Scanner • Photoelectron Spectroscopy • Planning and Evaluation Service • Pointing Error Sensor • Polyethersulfone • Positioning Error Signal • Post-Enumeration Survey • Potential-Energy Surface • Processor Enhancement Socket • Production Engineering Specification • Programmable Electronic System • Public Enquiry System

PESA: Particle Elastic Scattering Analysis

PESC: Physical and Engineering Sciences • Physical and Engineering Sciences Committee • Politique Étrangère et de Sécurité Commune • Politique Extérieure et de Sécurité Commune • Public Expenditure Survey Committee

PESD: Program Execution Subdirective

PESDS: Program Element Summary Data Sheet

Pesh: Peshawar (Pakistan)

PESM: Photoelectron Spectromicroscope • Photoelectron Spectromicroscopy

PESR: Planning Element System Report

PEST: Parameter Estimation by Sequential Testing • Political-Environmental-Social-Technological

PESTO: Public Engagement and Science/Technology Options

Pet: Petroleum • Petrological • Petrologist • Petrology

PET: Particle Electrostatic Thruster • Pelotas (Brazil-RS) (location identifier) • Performance/Endurance Tradeoff • Performance Enhancement Team • Performance Evaluation Test • Periodic Environmental Test • Periodic Evaluation Test • Peripheral Equipment Tester • Personal Electronic Transactor • Phase Elapsed Time • Photoemission Tube • Phototropic Energy Transfer • Photovoltaic Engineering Testbed • Physical Equipment Table • Polyéthylène Téréphtatale • Polyethylene Terephtalate • Portable Earth Terminal • Portable Electronic Telephone • Portable Electronic Translator • Portable Executive Telephone • Position-Event-Time • Positron Emission Tomography • Potentially Exempt Transfer • Preliminary Evaluation Team • Preliminary Examination Team • Print Enhancement Technology • Probe Ephemeris Tape • Production Environmental Test • Production Evaluation Test • Program Elapsed Time • Program Evaluator and Tester • Proton/Electron Telescope

PETA: Performance Evaluation and Trend Analysis • PetaFLOPS Enabling Technologies and Applications

PetE: Petroleum Engineer

PETFE: Polyethylene Tetrafluoroethylene

Petn: Petition

PETN: Pentaerythritoltetranitrate

Petro: Petroleum

Petrog: Petrography

Petrol: Petrology

PETS: Payload Environmental Transportation System • Performance Endurance Tradeoff Study • POCC Experiments Timeline System • Posting and Enquiry Terminal System • Programmed Extended Time Sharing

PETSC: Portable Extensible Toolkit for Scientific Computation

P. et T.: Postes et Télécommunications

PEU: Peripheral Electronics Unit • Pod Electronics Unit • Puerto Lempira (Honduras) (location identifier)

PEUL: Probabiliste Eulérien Lagrangien

PeV: Petaelectronvolt

PEW: Peshawar (Pakistan) (location identifier)

p. ex.: Par Exemple • Por Extensión • Por Extenso

PEX: Pechora (Russia) (location identifier) • PHIGS Extensions to X • Pipeline Executive

PEY: Penong (Australia-SA) (location identifier)

PEZ: Penza (Russia) (location identifier)

PE3: Protocol Engineering Environment in Estelle

pf: Per Favore • Perfect • Pfarrer • Pfennig • Power Factor • Preferred • Pro Forma • Proof

pF: Picofarad

Pf: Pfeiler • Pfennig • Postfach

P/F: Poll/Final

PF: French Polynesia (ISO country code) • Packet Format • Page Footing • Page Formatter • Panchromatic Film • Parachute Facility • Parallel Fold • Parallel Fortran • Payload Function • Payload Forward • Perforated Layer • Performance Factor • Permeability Factor • Pérot-Fabry • Pluto Flyby • Poloidal Field • Polynesia (French_) (ISO country code) • Polynésie Française (ISO country code) • Powered Flight • Power Factor • Prefetch • Pre-Flight • Primary Feed • Prime Focus • Prime Function • Priming Fluid • Probability of Failure • Production Function • Program Function • Programmable Format • Programmed Function • Project File • Pseudo Format • Public Funding • Pulse Frequency • Pulverized Fuel

PFA: Paf Warren (USA-AK) (location identifier) • Parallel Finite Automaton • Peachtree First Accounting • Planet Finder Array • Post-Flight Analysis • Prime Factor Algorithm • Printer Font ASCII • Probabilité de Fausse Alarme • Probability of False Alarm • Pulverized Fuel Ash

PFAM: Programmed Frequency Amplitude Modulation

PFAR: Power Fail Automatic Restart • Preliminary Failure Analysis Report

PFB: Passo Fundo (Brazil-RS) (location identifier) • Payload Feedback • Payload Forward Bus • Position Feedback • PostScript Font Binary (*) • Pressure-Fed Booster • Pressurized Fluidized Bed • Printer Font Binary

PFBC: Pressurized Fluidized Bed Combustion

PFBR: Prototype Fast Breeder Reactor

PFC: Pacific City (USA-OR) (location identifier) • Passenger Facility Charge • Perfluorocarbon Compound • Performance Flight Certification • Plasma Fusion Center • Porous Friction Course • Portable Four Channel • Post-Flight Checklist • Power Factor Corrector • Pre-Flight Console • Preliminary Flight Certification • Preliminary Flight Configuration • Primary Flight Control • Primary Focus Camera • Prime-Focus Corrector • Priority Foreign Country • Private First Class

PFCD: Primary Flight Control Display

Pfce: Performance

PFCF: Payload Flight Control Facility

PFCS: Primary Flight Control System

Pfd: Pfund • Preferred

PFD: Particle Flux Density • Peak Flux Density • Personal Floatation Device • Port Frederick (USA-AK) (location identifier) • Position-Fixing Device • Power Flux Density • Preliminary Functional Description • Primary Flight Display • Primary Function Display • Proton Flux Density

PFDA: Post-Flight Data Analysis • Precision Frequency Distribution Amplifier

Pfd. Sp.: Preferred Spelling

PFE: Photometric Front End • Poole-Frenkel Effect • Post-Flight Evaluation • Professional Flight Engineer

PFEP: Programmable Front-End Processor

PFES: Portable Field Emission Spectrometer

PFEU: Phase and Frequency Estimation Unit

PFF: Page Fault Frequency • Physical Fitness Facility • Pluto Fast Flyby • Post-Factory Functional • Presidential Faculty Fellow • Primary Focus Feed

Pfg: Pfennig

PFG: Primary Frequency Generator

PFGS: Potential Fields Graphics System

PFHE: Plate-Fin Heat Exchanger

PfHSN: Protocol for High-Speed Network

PFI: Pack File Indexing • Port-Fuel Injection • Power Failure Indicator • Pricing For Information • Printing For Information • Private Finance Initiative • Provisional Full Internet

PFIAB: President's Foreign Intelligence Advisory Board

PFIP: Prime Focus Instrument Package • Prime Focus Instrument Platform

PFIRCAM: Prime Focus Infrared Camera

PFIS: Property Fabric Information System

PFJ: Patreksfjördur (Iceland) (location identifier) • Proposed Final Judgment

PFK: Payload Function Key • Programmed Function Keyboard

PFL: Post-Flare Loop • Primary Freon Loop • Propulsion Field Laboratory

PFLP: Point-Feature Label Placement

PFM: Peak Fitting Module • Plasma Facing Material • Platform • Potential Field Model • Predictor Frame Memory • Pre-Flight Module • Pressure Flow Meter • Printer Font Metrics • Protoflight Model • Psion Freeware Movement • Pulse-Frequency Modulation

PFN: Panama City - Bay County (USA-FL) (location identifier) • Pulse Forming Network

PFNA: Pulsed Fast Neutron Analysis

PFO: Paphos (Cyprus) (location identifier) • Particulate Fallout • Piñon Flat Observatory (USA-CA)

Pfor: Professor

PFOV: Pixel Field Of View

PFP: Personal Financial Planning • Powered Flight Program • Probability of Performance Failure • Program Financial Plan • Programmable Function Panel

PFPD: Pulse-Flame Photometric Detector

PFPU: Processor Frame Power Unit

PFQ: Pre-Flight Qualification

Pfr: Fractured and Ridged Plain

PFR: Ilebo (Zaire) (location identifier) • Portable Foot Restraint • Power-Fail Restart • Präzisions-Filterradiometer • Precision Filter Radiometer • Pre-Flight Review • Prime Focal Ratio • Problem/Failure Report • Prototype Fast Reactor

PFRA: Prairie Farm Rehabilitation Administration

PFRC: Pacific Forest Research Centre

Pfrs: Parfois

PFRS: Portable Field Recording System • Portable Field Reflectance Spectrometer • Power Failure Recovery System

PFRT: Preliminary Flight Rating Test • Preliminary Flight Readiness Test

P&FS: Particles & Fields Satellite

PFS: Parallel File System • Particles and Fields Satellite • Percent Full Scale • Perfect-Forward Secrecy • Performance Funding System • Picture Frustration Study • Planetary Fourier Spectrometer • Precision Frequency Source • Primary Flight Software • Primary Flight System • Programmable Frequency Standard • Propellant Feed System

PFSA: Pour Faire Ses Adieux

PFSMETE: Postdoctoral Fellowship in Science, Mathematics, Engineering and Technology Education

PFSS: Potential Field Source Surface

PFT: Parallel Fourier Transform • Perfluorocarbon Tracer • Pimply Faced Teenager • Post-Flight Testing • Pressed For Time • Production Flight Test • Projective Field Theory • Pulse Fourier Transform

PFTA: Payload Flight Test Article

PFTB: Pre-Flight Test Bus

PFTI: Plateforme Technique pour l'Indemnisation

PFTM: Preliminary Flight Test Memo

PFTR: Preliminary Flight Test Report

PFU: Peoples Friendship University (Russia) • Please Follow Up • Prime Focus Unit • Protoflight Unit

PFV: Pour Faire Visite • Probability of Failure of Vehicle

PFW: Progressive Free Wave

Pfx: Prefix

pg: Page • Página • Persona Grata • Picogram • Photographic

Pg: Page • Palaeogene • Petagram • Photogrammetria (periodical code) • Portugal • Portuguese

P&G: Penny & Giles

PG: Page • Palomar-Green (survey) • Papouasie Nouvelle-Guinée (ISO country code) • Papua New Guinea (ISO country code) • Parental Guidance • Pasquill-Gifford • Past Grand • Paying Guest • Penumbral Grain • Persona Grata • Planetary Geophysics • Planning Group • Plasmagrenze • Portugal • Postgraduate • Power Gain • Power Generation • Pregnant • Pressure Gauge • Primeval Galaxy • Procurator General • Program Guidance • Proving Ground • Pulse Generator • Pyrotechnic Gyro

PG/A: Programmable Gain Amplifier

PGA: Page (USA-AZ) (location identifier) • Parallel Genetic Algorithm • Pendulous Gyro Accelerometer • Pin Grid Array • Power Gain Antenna • Power Generating Assembly • Pressure Garment Assembly • Professional Graphics Adapter • Programmable Gain Amplifier • Programmable Gate Array • Proposal Generation Aid

PGADS: Personnel Guided Aerial Delivery System

PGB: Pangoa (Papua New Guinea) (location identifier) • Pico Gravity Box

pgc: Per Gyrocompass

PGC: Pacific Geosciences Centre (Canada) • Petersburg - Grant County (USA-WV) (location identifier) • Plant Growth Chamber • Policía Guardia Civil • Predicted Group Crossing • Prediction of Group Crossings • Pressure Gamma Center • Principal Galaxies Catalogue • Produit de Grande Consommation • Programmed Gain Control • Propeller Gear Case • Proto-Galactic Clump

PGCD: Plus Grand Commun Dénominateur • Plus Grand Commun Diviseur

PGCE: Postgraduate Certificate of Education

PGCS: Photogrammetric Graphics Compiling System

PGD: Planetary Geosciences Division • Pulse Generator Display • Punta Gorda - Charlotte County (USA-FL) (location identifier)

PGD&CS: Power Generation, Distribution, & Control System

PGDCS: Power Generation, Distribution, and Control System

PGDF: Product Generation and Dissemination Facility

PgDn: Page Down

PGDN: Page Down

PGDS: Pioneer Ground Data System • Pulse Generator Display System

PgE: Photogrammetric Engineering (periodical code)

PGE: Platinum Group Element • Product Generation Engine • Purge • Yegepa (Papua New Guinea) (location identifier)

PGen: Publications de l'Observatoire de Genève (periodical code)

PgERS: Photogrammetric Engineering and Remote Sensing (periodical code)

PGF: Perpignan Rivesaltes (France) (location identifier) • Plant Growth Facility • Presentation Graphics Facility • Pressure Gradient Force • Probability Generating Function • Protein Growth Facility

PGG: Project Galactic Guide

PGGA: Permanent Geodetic GPS Array • Permanent GPS Geodetic Array

PGH: Pantnagar (India) (location identifier) • Programme du Génome Humain

PGHM: Payload Ground Handling Mechanism

PGI: Chitato (Angola) (location identifier) • Plan de Gestion Informatique • Progiciel de Gestion Intégré • Programme Général d'Information

PGID: Process Group Identification • Process Group Identificator • Process Group Identifier

PGK: Kina (Papua New Guinea) (ISO currency code) • Pangkalpinang (Indonesia) (location identifier)

PGL: Pascagoula - Jackson County (USA-MS) (location identifier) • Peer-Group Leader • Planetary and Geology Laboratory • Portable Gas Laser • Professional Graphics Language • Pulsed Gas Laser

Pgm: Program

PGM: Permanent Ground Monument • Portable Greymap • Port Graham (USA-AK) (location identifier) • Pragmatic Group Multicast • Precision-Guided Munition • Privacy Graphic Mail • Product Generation Module • Program

PGML: Precision Graphics Markup Language

Pgmt: Pigment

PGN: Pangia (Papua New Guinea) (location identifier) • Pôle Galactique Nord • Positive Gauss-Newton

PGNCS: Primary Guidance, Navigation, and Conrol System

PGO: Pettinger-Guilay Observatory (USA-WA) • Principal Guest Observer • Productivity of Global Ocean

PGOC: Payload Ground Operations Contract • Payload Ground Operations Contractor

PGOR: Payload Ground Operation Requirements

PGORS: Payload Ground Operation Requirements Study

PGP: Physique des Gaz et des Plasmas • Porto Alegre (Portugal) (location identifier) • Pretty Good Privacy • Primary Geodetic Parameter • Program Parameter

PGPS: Parallel Genetic Programming System

PGR: Paragould (USA-AR) (location identifier) • Parental Guidance Recommended • Planning and Ground Rule • Planning Ground Rule • Population Growth Rate • Precision Graphic Recorder

PGRo: Publications of the Kapteyn Astronomical Observatory in Groningen (periodical co de)

PGRWG: Payload Ground Requirements Working Group

PGS: Passive Geodetic Satellite • Passive Gravity Stabilization • Peach Springs (USA-AZ) (location identifier) • Plane Grating Spectrograph • Plasma Generator System • Pôle Galactique Sud • Portable Ground Station • Power Generation System • Precision Gamma-ray Spectrometer • Primary Ground Station • Product Generation Segment • Product Generation System • Program Generation System • Proposal Generation System

PGSC: Payload and General Support Computer • Payload General Support Computer

PGSE: Payload Ground Support Equipment

PGT: Peace and Global Transformation • Per Gross Ton

PGU: Plant Growth Unit • Power Generating Unit

PgUp: Page Up

PGUP: Page Up

PGV: Greenville (USA-NC) (location identifier)

PGVF: Plant Genome Venture Fund

PGWG: Particles and Gases Working Group

PGX: Périgueux (France) (location identifier)

PGZ: Ponta Grossa - Sant'Ana (Brazil-PR) (location identifier)

ph: Philippines

pH: Potential of Hydrogen Ions

Ph: Phase • Phenyl • Philosophy • Phone • Phosphorus • Phot

PH: Netherlands (international aircraft registration code) • Packet Handler • Pädagogische Hochschule • Page Heading • Paralaje Horizontal • Parallaxe Horizontale • Parity High • Penthouse • Personal Hygiene • Perth (UK postal code) • Petroleum Hydrocarbon • Phase • Philippines (ISO country code) • Phrase Hypothesis • Pinhole • Power House • Precipitation Hardening • Proposal Handling • Proton Hydrate • Provincial Highway • Public Health • Pulse Height

PHA: Phan Rang (Viet Nam) (location identifier) • Photoelectric Astrolabe • Potentially Hazardous Asteroid • Preliminary Hazard Analysis • Public Health Act • Public Housing Administration • Public Housing Authority • Pulse Height Amplitude • Pulse Height Analysis • Pulse Height Analyzer

PHAK: Pilot's Handbook of Aeronautical Knowledge

PHAMOS: Premote Hemodynamics and Metabolism in an Orbiting Satellite

Phar: Pharmaceutical • Pharmacology • Pharmacopoeia • Pharmacy

PHARAO: Projet d'Horloge Atomique à Refroidissement d'Atomes en Orbite

PharB: Bachelor of Pharmacy

PharD: Doctor of Pharmacy

Pharm: Pharmaceutical • Pharmacology • Pharmacopoeia • Pharmacy

PharM: Master of Pharmacy

PharmB: Bachelor of Pharmacy

PharmD: Doctor of Pharmacy

PharmM: Master of Pharmacy

PHAROS: Proposal for a High Angular Resolution Optical Survey

PHARS: Phased Array SAR

PHARUS: Phased Array Universal SAR

Phas: Phase

PhASp: Physics and Astrophysics from Spacelab (periodical code)

PHAST: Physics-Astronomy

PHAZ: Poleward-moving High-latitude Activity Zone

PhB: Bachelor of Philosophy • Physics Bulletin (periodical code)

PHB: Parnaíba - Santos Dumont (Brazil-PI) (location identifier) • Per-Hop Behavior

PhC: Pharmaceutical Chemist

PHC: Pacific Hurricane Center • Port Harcourt (Nigeria) (location identifier) • Primary Health Care • Probable Hydrologic Consequence • Probe Heater Controller • Public Health Committee

PHCA: Proyecto Hidrometeorológico Centroamerico

PHCF: Pituitary-growth Hormone Cell Function

PhChE: Physics and Chemistry of the Earth (periodical code)

PhChH: Physico-Chemical Hydrodynamics (periodical code)

PhD: Doctor of Philosophy • Philosophiae Doctor

PHD: New Philadelphia - Harry Clever (USA-OH) (location identifier) • Payload Handling Device • Phase Shift Driver • Photohydrodynamic • Project History Document • Pulse Height Discrimination • Pulse Height Distribution

PhDEd: Doctor of Philosophy in Education

PHDR: Preliminary Hardware Design Review

PHDS: Parallel Hypertext Data Structure

PhDT: PhD Thesis

Phe: Phoenix-Phoenicis (constellation) • Photoelectric

PHE: Port Hedland (Australia-WA) (location identifier) • Pre-flight Heat Exchanger • Public Health Engineer

PHEDRE: Partenariat Hermes de Développement Régional

PHEL: Patton Hawksley Education Ltd. (UK)

PHEMT: Pseudo-morphic High Electron Mobility Transistor

PHEMU: Phénomènes Mutuels

Phen: Phenolic

PHEN: Phenomenon

PHENIX: Pioneering High-Energy New Ion Experiment

PHF: Hampton (USA-VA) (location identifier) • Newport News (USA-VA) (location identifier) • Parts History File • Payload Handling Fixture • Perfect Hash Function • Personal History Form • Personal Hygiene Facility • Williamsburg (USA-VA) (location identifier)

PhFl: Physics of Fluids (periodical code)

PhG: Graduate in Pharmacy

PHH: Phan Thiet (Viet Nam) (location identifier)

PHHC: Programmable Hand-Held Calculator

PHI: Permanent Health Insurance • Pinheiro (Brazil-MA) (location identifier) • Procedural Hierarchy of Issues • Public Health Inspector

PHIAT: Passive Hybrid Imaging on the AMOS Telescope

PHIGS: Programmer's Hierarchical Interactive Graphics Standard • Programmer's Hierarchical Interactive Graphics System (*) • Programmer's Hierarchical Interface to Graphics Standard

PhiJR: Philips Journal Research (periodical code)

Phil: Philadelphia (USA-PA) • Philippines • Philosopher • Philosophical • Philosophy

PHIL: Personal Hot Interest List

Phila: Philadelphia (USA-PA)

Phil. I.: Philippines Islands

Philipp: Philippines

Phil. Is.: Philippines Islands

Philol: Philological • Philology

Philos: Philosopher • Philosophical • Philosophy

Phil. Trans.: Philosophical Transactions

PhInt: Photo Interpretation (periodical code)

PHIP: Photo-Injecteur Prototype

PhiRR: Philips Research Reports (periodical code)

PHIS: Population and Housing Information System

PhiTR: Philips Technical Review (periodical code)

PHIVOLCS: Philippines Institute of Volcanology and Seismology

PHJ: Port Hunter (Australia-NS) (location identifier)

PHK: Pahokee - Palm Beach County (USA-FL) (location identifier) • Personal Hygiene Kit

PhL: Licentiate in Philosophy • Licentiate of Philosophy • Physics Letters (periodical code)

PHL: Palomar-Haro-Luyten (survey) • Peculiar Hydrogen Line • Philadelphia (USA-PA) (location identifier) • Philippines (ISO country code) • Presse Hebdomadaire Locale

PhLB: Physics Letters B (periodical code)

PhLC: Physics Letters C (periodical code)

PhLF: Physics Laser Fusion (periodical code)

PHLIS: Public Health Laboratory Information System

PHLS: Public Health Laboratory Service (UK)

PhM: Master of Philosophy

PHM: Patterson-Harker Method • Per Hundred Million • Phase Modulation • Power Hybrid Microcircuit • Prognostics and Health Management • Pump Head Module • Pyrheliometer

PHMC: Probe Heater Motor Controller

PHN: Port Huron - Saint Claire County (USA-MI) (location identifier) • Presse Hebdomadaire Nationale

PHO: Phone • Point Hope (USA-AK) (location identifier) • Potentially Harzardous Object

PHOKA: Photosphärenkamera

Phon: Phonetics

Phonet: Phonetics

Phono: Phonograph

Phonol: Phonology

Phos: Phosphate

PHOSI: Preliminary Handbook of Operations and Service Instructions

PhoSp: Photonics Spectra (periodical code)

Phot: Photograph • Photographer • Photographic • Photographical • Photographie • Photographique • Photography • Photometric

Photog: Photograph • Photographer • Photographic • Photographical • Photographie • Photographique • Photography

Photogr: Photograph • Photographer • Photographic • Photographical • Photographie • Photographique • Photography

Photom: Photometric • Photometrical • Photométrie • Photométrique • Photometry

php: Pounds per Horsepower • Pump Horsepower

PHP: Payload Handling Panel • Personal Handy Phone • Peso (Philippines) (ISO currency code) • Philip (USA-SD) (location identifier) • Planetary Horizon Platform • Planetary Horizontal Platform • Preferred Hiring Program

PHPA: Professional Helicopter Pilots Association

PHPL: Parallel Hardware Processing Language

Phr: Phrase

PhR: Physical Reports (periodical code) • Physical Reviews

PHR: Pacific Harbour (Fiji) (location identifier) • Payload Hazardous Report • Power-to-Heat Ratio • Presse Hebdomadaire Régionale

PHRI: Public Health Research Institute

PHRIR: Pointable High Resolution Imaging Radiometer

PHRK: Power and Heat Rejection Kit

Phrm: Pharmacy

PHRR: Payload High-Rate Recorder

PHRS: Portable Heat Rejection System • Proposal Handling and Reporting System

PhRv: Physical Review (periodical code)

PhRvA: Physical Review A (periodical code)

PhRvB: Physical Review B (periodical code)

PhRvC: Physical Review C (periodical code)

PhRvD: Physical Review D (periodical code)

PhRvE: Physical Review E (periodical code)

PhRvL: Physical Review Letters (periodical code)

PHS: Payload Handling Station • Payload Handling System • Personal Handyphone Service • Personal Handyphone System • Personal Hygiene System • Phitsanulok (Thailand) (location identifier) • Photohydrography System • Proposal Handling System • Public Health Service

PHSA: Public Health Service Act (USA)

PHSF: Payload Hazardous Servicing Facility

PhSFK: Phase-Shift Keying

PHS&T: Packaging, Handling, Storage, & Transportation

PHST: Packaging, Handling, Storage, and Transportation

PhT: Physics Today (periodical code)

PH&T: Packaging, Handling & Transportation

PHT: Paris - Henry County (USA-TN) (location identifier) • Plötzlicher Herztod

PhTec: Physics Technology (periodical code)

PHTR: Packaging, Handling, and Transportation Record

PHU: Phu Vinh (Viet Nam) (location identifier) • Primary Header Unit

PHVPS: Primary High-Voltage Power Supply

PHW: Phalaborwa (South Africa) (location identifier)

PHWR: Pressurized Heavy-Water Reactor

Phx: Phoenix (USA-AZ)

PHX: Phoenix (USA-AZ) (location identifier)

PhyEs: Physics Essays (periodical code)

PhyL: Physics Letters (periodical code)

Phys: Physica (periodical code) • Physical • Physically • Physician • Physicist • Physics • Physiological • Physiologist • Physiology • Physique

PhyS: Physica Scripta (periodical code)

PhysB: Physica B (periodical code)

PhySB: Physica Scripta B (periodical code)

PhysC: Physica C (periodical code)

PhysD: Physica D (periodical code)

Physiol: Physiological • Physiologist • Physiology • Physique

PHYSnet: High-Energy Physics Network

Phys. Rev.: Physical Review

PhysS: Physica Section (periodical code)

Phys. Sc.: Physical Science

PHz: Petahertz

PHZ: Phi Phi Island (Thailand) (location identifier)

pi: Par Intérim

Pi: Piaster (currency) • Piastre (currency)

P&I: Performance & Interface

PI: Charon (satellite of Pluto) • Parallel Input • Par Intérim • Patent Information • Path Integral • Payload Interrogator • Pen and Ink • Performance and Interface • Performance Improvement • Performance Index • Periodic Inspection • Peripheral Interface • Personal Identification • Personal Injury • Philippine Islands • Photographic Interpretation • Photographic Interpreter • Photointerpretation • Photointerpreter • Physik Instrumente (Germany) • Piauí (Brazil) • Point of Intersection • Politically Incorrect • Polynomial Interpolation • Position Indicator • Position Invariant • Power Injection • Power Input • Precipitation Index • Precision Instrument • Preliminary Impulse • Preliminary Inspection • Preliminary Investigation • Preliminary Issue • Pressure Indicator • Préstamo Interbibliotecario • Primary Impulse • Primary Imput • Principal Investigator • Priority Interrupt • Pristine Ice • Private Investigator • Process Intermittent • Process of Induction • Procurement Item • Program Instruction • Program Interrupter • Program Interruption • Program Introduction • Programmed Information • Programmed Instruction • Programmed Introduction • Program of Instrumentation • Protocol Interpreter • Pseudo-Intelligence • Pubblica Istruzione • Public Information • Pulse Invariant

PI&A: Personnel Identification & Authentication

PIA: Particle Impact Analyser • Payload Interface Adapter • Peoria (USA-IL) (location identifier) • Peripheral Interface Adapter • Personnel Identification and Authentication • Personal Investment Authority • Photometers for Imaging the Aurora • PHT Interactive Analysis • Plug-In

Administrator • Plug-In Amplifier • Portable Image Analizer • Positive Ion Accelerator • Pre -Inspection Acceptance • Pre -Installation Acceptance • Processor Interface Adapter • Project Impact Analysis • Public Information Access • Public Information Act

PIAB: Particle In A Box

PIACH: Proactive Interactive Adaptive Computer Help

PIAF: Placement Interactif et Automatique de Formes

PIAMA: Photo-Induced Activation of mm-wave Absorption

PIANC: Permanent International Association of Navigation Congresses

PIAR: Project Impact Analysis Report

PIARC: Permanent International Association of Road Congresses

Pias: Piaster (currency)

PIB: Laurel Pine Belt (USA-MS) (location identifier) • Particle In a Box • Payload Integration Bay • Polar Ionospheric Beacon • Preliminary Instruction Book • Producto Interior Bruto • Produit Intérieur Brut • Propellant Inspection Building • Pyrotechnic Installation Building

PIBE: Process Integration in Biochemical Engineering

PIBOR: Paris Interbank Offered Rate

PIBS: Polar Ionospheric Beacon Satellite

Pic: Pictor-Pictoris (constellation) • Pictorial • Picture

P-I-C: Particle-In-Cell

PIC: Particle-In-Cell • Particulate Inorganic Carbon • PASSCAL Instrument Center • Payload Integration Center • Payload Integration Committee • Payload Integration Contractor • Performance Indicator Chart • Personal Information Carrier • Personal Intelligent Communicator • Personal Interest Checklist • Photonic Integrated Circuit • Pilot In Command • Pine Cay (Turks and Caicos Islands) (location identifier) • Plastic Insulated Cable • Polyethylene Insulated Cable • Position Independent Code • Positive Ion Chamber • Precision Industrial Components Corporation (USA-CT) • Preferred Interexchange Carrier • Pre-Installation Calibration • Pre-Installation Checkout • Pressure Indicator Controller • Primary Independent Carrier • Primary Interexchange Carrier • Primordial Isocurvature • Principal Interexchange Carrier • Prior Informed Consent • Priority Interrupt Controller • Product Information Catalog • Product of Incomplete Combustion • Professional Interest Committee • Program Identification Code • Program In Computing • Program Information Center • Program Interrupt Control • Program Interrupt Controller • Programmable Interrupt Controller • Programmable Interval Clock • Programme Interuniversitaire de Coopération • Proton-Initiated Cascade • Public Information Center • Pulsed Ionization Chamber • Pyro Ignition Control • Pyro Initiator Capacitor • Pyro Initiator Controller • Pyrotechnic Ignition Control • Pyrotechnic Initiator Capacitor • Pyrotechnic Initiator Circuit • Pyrotechnic Initiator Controller

PICA: Performance-enhanced I/O and CPU Architecture • Picosecond Imaging Circuit Analysis • Platform-Independent Cryptography API • Pre-Integrated Columbus Assembly • Pre-Integrated Columbus ATM • Press Independence and Critical Ability • Project for Integrated Catalogue Automation

PICAO: Provisional International Civil Aviation Organization (now ICAO)

PICAPI: Platform-Independent Cryptography Application Programming Interface

Picc: Piccolo

PICC: Payload Interface Control Computer

PICCA: Positive Ion Cluster Composition Analyzer

PICE: Programmable Intergrated Control Equipment

PICF: Payload Integration and Checkout Facility

PICG: Programme International de Corrélation Géologique

PICMA: Program for International Cooperation in Mathematics and its Applications

PiCO: Portable Interactive Computing Object

PICO: Pic du Midi Coronagraph

PICP: Program Interface Control Plan

PICRS: Program Information Control and Retrieval System • Program Information Coordination and Review Service

PICS: Parts Interpretive Catalog System • Perpetual Inventory Control System • Personnel Information Communication System • Personnel Information Control System • Photographic Integrated Control System • Photo Index and Cataloging System • Photo Indexing and Cataloging System • Pioneer Image Converter System • Planetary Image Cartography System • Planetary Integrated Camera-Spectrometer • Platform for Internet Content Selection • Plug-in Inventory Control System • Pre-defined Input Control Sequence • Procurement Information Control System • Production Information and Control System • Production Inventory Control System • Programme International de Coopération Scientifique • Protocol Implementation Conformance Statement

Pict: Picture

PICT: Picture • Programme in Information and Communication Technologies

PICU: Pediatric Intensive Care Unit • Priority Interrupt Control Unit

PiD: Proportionnel Intégral Dérivé

P&ID: Process & Instrumentation Diagram

PID: Nassau - Paradise Island (Bahamas) (location identifier) • Patentinformationsdienst • Payload Insertion Device • Permanent Station Identifier • Personal Identification Device • Photoionization Detector • Pictorial Information Digitizer • Prestation Invalidité Décès • Prime Item Development • Privater Informationsdienst • Process Identification • Process Identifier • Program Information Document • Program Introduction Document • Project Initiation Document • Proportional-Integral-Derivative • Proportional-Integral-Differential • Proportional-Integral-Differentiated • Proportional Integrated-Derivative • Prototype International Directory • Public Information Division

PIDA: Payload Installation and Deployment Aid

PIDAS: Portable Instant Display and Analysis Spectrometer • Portable Intelligent Display and Analysis Spectrometer

PIDDP: Planetary Instrument Definition and Development Program

PIDP: Portable Image Display System • Product Initialization Data Pack • Programmable Indicator Data Processor

PIDS: Parameter Inventory Display System • Pattern Identification and Detection System • Polarization Intensity Differential Scattering • Portable Image Display System • Prime Item Development Specification

PIDT: Photometric Instrument Dedicated Team

PIE: Clearwater (USA-FL) (location identifier) • Parallel Instruction Execution • Payload Integration Equipment • Payload Interface Engineer • Paysage Internet Européen • Personal Integrated Electronics • Personal Interactive Electronics • Physics Information Exchange • Plug-In Electronics • Polar Ice Extent • Positive-Ion Emission • Post-Irradiation Examination • Process Improvement Experiment • Process-Integrated Environment • Product Integrity Engineering • Program Interrupt Entry • Proto-Indo-European • Saint Petersburg (USA-FL) (location identifier)

PIECP: Preliminary Impact Engineering Change Proposal

PIES: Planning-Implementation-Evaluation System • Program Information and Evaluation System

PIF: Paid In Full • Payload Integration Facility • Paysage Internet Français • Picture Interchange Format • Pingtung (Taiwan-ROC) (location identifier) • Process Information File • Program Information File

PIFOV: Planet In Field Of View

PIFS: Palomar Integral Field Spectrograph • Process Interchange Format Standard

PIFT: Package for Information Formatting and Transformation

PIG: Partially Ionized Globule • Pendulous Integrating Gyro • Pitinga (Brazil-AM) (location identifier) • Post-burst Infrared Galaxy • Primary Image Guider • Process Ink Gamut • Product Information Guide • Projet d'Intérêt Général

PIGA: Pendulous Integrating Gyro Accelerometer

PIGB: Programme International Geosphère-Biosphère

PIGE: Proton-Induced Gamma-ray Emission

PIGMI: Position Indicating General Measuring Instrument

Pigmt: Pigment

PIGS: Portable Inertial Guidance System

PIGU: Pendulus Integrating Gyro Unit

PIH: Pocatello (USA-ID) (location identifier) • Public and Indian Housing

PII: Fairbanks Phillips (USA-AK) (location identifier) • Paysage Internet International • Persatuan Insinyur Indonesia • Procurement Instrument Identification • Programa Intergubernamental de Informatica • Publisher Item Identification • Publisher Item Identifier (*)

PIIC: Politique de l'Information, de l'Informatique et des Communications

PIIN: Procurement Instrument Identification Number

PIIT: Programa Interamericano de Información Popular

PIK: Glasgow Prestwick (UK-Scotland) (location identifier) • Payment In Kind • Potsdamer Institut für Klimaforschung (Germany)

pil: Pileus

PIL: Payment In Lieu • Personal Interest List • Pilar (Paraguay) (location identifier) • Pitt Interpretive Language • Processing Information List

PILL: Programmed Instruction Language Learning

PILOT: Program Integration and Loading Tool • Programmed Inquiry, Learning Or Teaching

PILP: Parametric Integer Linear Program

PILS: Payload Integration Library System • Payload Interactive Library System

PIM: Parallel Inference Machine • Parametrized Ionosphere Model • Payload Integration Manager • Paysage Internet Mondial • Personal Information Manager • Pine Mountain - Garden Harris County (USA-GA) (location identifier) • Planning Interchange Meeting • Plug-In Module • Precision Indicator of the Meridian • Precision Instrument Mount • Presentation and Interaction Module • Prestation Interministérielle • Processor-In-Memory • Program Integration Manual • Programme Intégré Méditerranéen • Protocol-Independent Multicast • Pulse Intensity Modulation • Pulse Interval Modulation

PIMA: Photographic and Imaging Manufacturers Association • Plug-In Module Assembly

PIMOS: Parallel Inference Machine Operating System • Personal Information Manager Operating System

PIMMS: Portable Interactive Multimission Simulator • Project to Increase Mastery of Mathematics and Science

PIMR: Polar Ice Mapping Radiometer • Program Integration Management Review • Pushbroom Imaging Microwave Radiometer

PIMS: Pacific Institute for the Mathematical Sciences • Payload Information Management System • Personnel Information Management Service • Photoionization Mass Spectroscopy • Profit Impact of Market Strategy • Project Information Management System

PIMSL: Personal Information Manager System Language

PIN: Pacific Island Network • Page and Item Number • Parcel Identification Number • Parintins (Brazil-AM) (location identifier) • Patent Information Network • Personal Identification Number • Pesticide Information Network • Piece Identification Number • Plan Identification Number • Positive Intrinsic Negative • Primary Interactive Network • Process Identification Number • Processor-Independent NetWare • Produit Intérieur Net • Program Identification Number • Programme Indicatif National • Proposal Identification Number

PINC: Printer's Industry of Northern California (USA-CA) • Programme Indicatif Nucléaire pour la Communauté

PINCH: Payload Integration Nominal Cost Hardware

PIND: Particle Impact and Noise Detection • Particle Impact Noise Detection

PINE: Pine Is Not Elm

PINET: Physics Information Network

PING: Packet Internet Groper

PINO: Positive Input - Negative Output

PINR: Peak Intensity Response

PINS: Person In Need of Supervision • Proliferation Information Network System

Pint: Pintura

PIO: Parallel Input/Output • Pilot-Induced Oscillation • Pisco (Peru) (location identifier) • Position Iterative Operation • Processor Input/Output • Production Intégrée par Ordinateur • Program Integration Office • Programmable Input/Output • Programmed Input/Output • Project Implementation Order • Provisioned Item Order • Public Information Office • Public Information Officer • Public Inquiries Office

PIOCS: Physical Input-Output Control System

PIOL: Periodic Input/Output Link

PION: Physics In Our Neighborhood

PIONIER: Persoonsgerichte Impuls voor Onderzoekgroepen met Nieuwe Ideeën voor Excellente Research

PIOP: Payload Integration Operations Plan

PIOS: Planning Information Overlay System

PIOUS: Parallel Input/Output System

PIP: Packet Interface Port • Packet Interface Processor • Parallel Image Processor • Payload Integration Plan • Payload Interface Plan • Performance Improvement Program • Peripheral Interchange Program • Persistent Internal Polarization • Personal Incentive Program • Personal Identification Project • Personal Interview Program • Photo Interpretive Program • Picture In Picture • Pilot Point (USA-AK) (location identifier) • Plant Instrumentation Program • Policy Improvement Program • Portable Instrument Pedestal • Power Input Panel • Precipitation Intercomparison Project • Preliminary Information Pamphlet • Presence Information Protocol • Pretty Important Person • Priority Information Programme • Probabilistic Information Processing • Probability Information Processing • Problem Input Preparation • Problem Isolation Procedure • Procedural Information Pamphlet • Production Instrumentation Package • Program Integration Plan • Program In Process • Programmable Interconnect Point • Programme d'Information Prioritaire • Project Implementation Plan • Probability Information Processing • ProjecInformation Package! Initiation Period • Proof In Print • Proportional-Integral-Plus • Proposal Implementation Plan • Proposal Information Package • Proposal Instruction Package • Public Information Package • Pyro Interception Plug

PIPA: Pulse-Integrating Pendulum Accelerometer • Pulse-Integrating Pendulum Assembly

PIPAC: Philippines Institute for Pure and Applied Chemistry

PIPE: Parallel Image processing Engine • Pulsed Irradiance of a Particle Element

PIPO: Parallel In - Parallel Out

PIPOR: Programme for International Polar-Ocean Research

PIPPS: Publication Information Processing and Printing Standard

PIPPY: Person Inheriting Parent's Property

PIPS: Passivated Implanted Planar Silicon • Passively Ion-Planted Silicon (detector) • Personal Image Processing System • Postinjection Propulsion System

PIPSS: PPARC Industrial Programme Support Scheme

PIQ: Pipillipai (Guyana) (location identifier)

PIR: Page Information Request • Peak Intensity Ratio • Peak Irradiance Response • Phase Increment Register • Pierre (USA-SD) (location identifier) • Precision Infrared Radiometer • Pressure Ignition Rocket • Programme Interdisciplinaire de Recherche • Programming Information Requirement • Project Independence Report • Protein Identification Resource • Publications and Information Resources

PIRAMHYD: Passive Infrared Atmospheric Measurement of Hydroxyl

PIRAQ: PC Integrated Radar Acquisition

PIRCS: Payload Interactive Requirements Collection System

PIRD: Program Instrumentation Requirements Document • Project Instrumentation Requirements Document

PIRE: Puissance Isotrope Rayonnée Équivalente

PIREN: Programme Interdisciplinaire de Recherche sur l'Environnement

PIREP: Pilot Report

PIRG: Public Interest Research Group

PIRL: Patent Information Retrieval Language • Planetary Image Research Laboratory

PIRMAT: Programme Interdisciplinaire de Recherche sur les Matériaux

PIRN: Preliminary Interface Revision Notice • Proposed Interface Revision Notice

PIROCEA: Programme Interdisciplinaire de Recherche Océanographique

PIROG: Pointed Infrared Observing Gondola

PIRR: Parts Installation and Removal Record • Problem Investigation and Repair Record

PIRS: Personal Information Retrieval System • Photo-Image Retrieval System • Program Information Retrieval System • Project Information Retrieval System

PIRSEM: Programme Interdisciplinaire de Recherche sur les Sciences pour l'Énergie et les Matières Premières

PIRMAT: Programme Interdisciplinaire de Recherche sur les Matériaux

PIRT: Precision Infrared Tracking • Precision Infrared Triangulation • Projet d'Inventaire des Ressources Terrestres

PIRTTEM: Programme Interdisciplinaire de Recherche sur la Technologie, le Travail, l'Emploi et les Modes

PIRU: Public Information Reference Unit

PIS: Patent Information Service • Personal Information System • Personalinformationssystem • Pointable Imaging Spectrometer • Poitiers Briard (France) (location identifier) • Positive Ion Source • Process Instrument Sheet • Product Information Specialist • Prospero Interface Server • Provisional International Standard • Pulse Integration System • Purchase Information System

PISA: Position, Intensity and Shape Analysis • Prototype Image Spectrum Analyzer

PISCO: Polarimeter with Instrumental and Sky Compensation • Polarization with Instrumental and Sky Compensation

PISH: Program Instrumentation Summary Handbook

PISN: Private Integrated Services Network

PISO: Parallel In - Serial Out

PISP: Polar Ice Sheet Program

PIST: Plasma Initiation Separated from the Projectile

PISW: Process Interrupt Status Word

PIT: Page Information Transfer • Passive Intermediate Tracker • Paul Ion Trap • Performance Improvement Test • Peripheral Input Tape • Personal Income Tax • Phase Inversion Temperature • Photographic Interpretation Technique • Photointerpretation Tool • Pittsburgh (USA-PA) (location identifier) • Post-Installation Test • Pre-Installation Test • Pre-Integrated Truss • Principal-Interests-Taxes • Processor Interface Table • Programmable Interval Timer • Programmed Instruction Test • Projected Inactive Time • Project Interface Team

PITAC: President's Information Technology Advisory Committee (USA)

PITCOM: Parliamentary Information Technology Committee

PITG: Payload Integration Task Group

PITMAS: Paul Ion Trap Mass Spectrometer

PITN: Pain In The Neck

PITS: Payload Integration Test Set • Payload Interface Test System

Pitts: Pittsburgh (USA-PA)

PIU: Path Information Unit • Photographic Interpretation Unit • Piura (Peru) (location identifier) • Plasma Interface Unit • Plug-In Unit • Power Interface Unit • Process Input Unit • Project Implementation Unit • Pyro Initiator Unit • Pyrotechnic Initiator Unit

PIUP: Planetary Instrument Update Program

PIUS: Process-Inherent Ultimate Safety

PIV: Partial Image Velocimeter • Partial Image Velocimetry • Particle Image Velocimeter • Particle Image Velocimetry • Payload Integration and Verification • Payload Interface Verification • Peak Inverse Voltage • Personal Independent Vehicle • Philippines Institute of Volcanology • Planet In View • Plena Ilustrita Vortaro

PIVR: Payload Interface Verification Requirement

PIVS: Payload Interface Verification Summary • Payload Integration and Verification Simulator

PIW: Pikwitonei (Canada-MB) (location identifier)

PIWG: Propulsion Instrumentation Working Group

PIX: Pico Island (Azores) (location identifier)

PIXE: Particle-Induced X-ray Emission (*) • Proton-Induced X-ray Emission

PIXEL: Picture Element

PIXIE: Polar Ionospheric X-ray Imaging Experiment

PIXIT: Protocol Implementation Extra Information for Testing

PIZ: Partially Ionized Zone • Point Lay (USA-AK) (location identifier)

PJ: Netherlands Antilles (international aircraft registration code) • Pajama • Perijove • Petajoule • Pièce Jointe • Police Judiciaire • Provincia Jujuy (Argentina) • Pyjama

PJB: Payson (USA-AZ) (location identifier)

PJC: Pedro Juan Caballero (Paraguay) (location identifier) • Pensacola Junior College (USA-FL)

Pje: Pasaje

PJF: Photogrammetric Journal of Finland

PJG: Panjgur (Pakistan) (location identifier)

PJM: Puerto Jimenez (Costa Rica) (location identifier)

PJNAO: Publications of the Japan National Astronomical Observatory (periodical code)

PJO: Pioneer Jupiter Orbit

PJOP: Preliminary Joint Operation Procedure

PJS: Port San Juan (USA-AK) (location identifier)

Pjtr: Projector

PJZ: Puerto Juarez (Mexico) (location identifier)

Pk: Pack • Package • Park • Peak • Peck

P/K: Pluto/Kuiper

PK: Indonesia (international aircraft registration code) • Pakistan (international vehicle registration code) • Pakistan (ISO country code) • Park • Peak • Perek-Kohoutek • Personal Knowledge • Psychokinesis

PKA: Napaskiak (USA-AK) (location identifier)

PKB: Marietta (USA-OH) (location identifier) • Parkersburg Marietta (USA-WV) (location identifier) • Portable Keyboard

PKC: Petropavlovsk-Kamchatsky (Russia) (location identifier) • Public Key Cryptography

PKCS: Public Key Cryptographic Standard • Public Key Cryptography Standard (*)

PKD: Parallel k-d Tree • Park Rapids (USA-MN) (location identifier) • Programmable Keyboard and Display

PKE: Parkes (Australia-NSW) (location identifier) • Private Key Encryption • Public Key Encryption

PKF: Park Falls (USA-WI) (location identifier)

Pkg: Package • Packing

PKG: Package • Pangkor (Malaysia) (location identifier)

Pkge: Package

PKH: Porto Kheli - Alexion (Greece) (location identifier)

PKI: Philips Kommunikations Industrie AG (Germany) • Public Key Infrastructure

PKIS: Public Key Infrastructure Service

PKK: Pakokku (Myanmar) (location identifier)

PKL: Pakatoa Island (New Zealand) (location identifier)

PKM: Perigee Kick Motor • Port Kaituma (Guyana) (location identifier)

PKN: Pangkalanbun (Indonesia) (location identifier) • Polski Komitet Normalizacyjny (Poland)

PKNM: Polski Komitet Normalizacji Miar

PKO: Parakou (Benin) (location identifier)

PKP: Public Key Partners • Puka Puka (Tuamotu Islands) (location identifier)

PKR: Pokhara (Nepal) (location identifier) • Rupee (Pakistan) (ISO currency code)

Pks: Parks • Peaks

PKS: Paksane (Laos) (location identifier) • Parkes Catalogue • Peaks

Pkt: Packet • Paket • Pocket • Punkt

PKT: Phase Keying Technique • Pocket • Port Keats (Australia-NT) (location identifier)

PKU: Pekanbaru (Indonesia) (location identifier)

PKV: Pskov (Russia) (location identifier)

PKW: Personenkraftwagen • Selebi-Phikwe (Botswana) (location identifier)

Pkwy: Parkway

PKWY: Parkway

PKWYS: Parkways

PKY: Palangkaraya (Indonesia) (location identifier) • Parkway

PKZ: Pakse (Laos) (location identifier)

pl: Poland • Polish (ISO language code)

Pl: Plaats • Place • Plain • Planche • Plate • Platz • Plaza • Plug • Plural • Pole

P/L: Partial Loss • Parts List • Payload • Postlanding • Profit and Loss • Purchased Labor

P&L: Profit & Loss

P-L: Periode-Leuchtkraft • Period-Luminosity

PL: Packing List • Parity Low • Partial Loss • Parting Line • Parts List • Passenger Liability • Payload • Periode-Leuchtkraft • Period-Luminosity • Personal Librarian • Phillips Laboratory • Photoluminescence • Piecewise Linear • Place • Plain Language • Plate • Pleine Lune • Plug • Plus • Plymouth (UK postal code) • Poids Lourd • Point de Lagrange • Poland (international vehicle registration code) • Poland (ISO country code) • Poland (postal country code) • Pologne (ISO country code) • Positive Large • Post-Landing • Power Level • Pre-Launch • Price List • Private Line • Production Language • Product Liability • Product Licence • Programmed Learning • Programming Language • Project Leader • Propellant Loading • Property List • Public Law • Public Library • Purchased Labor

Pla: Península • Plaza

PLA: Panchromatic Linear Array • Parachute Location Aid • Perceptron Learning Algorithm • Perturbative Least Action • Phase Locked Array • Planadas (Colombia) (location identifier) • Planned Landing Area • Plan of Launch Azimuth • Power Lever Angle • Practice Low Approach • Programmable Logic Array • Programmed Logic Array • Proton Linear Accelerator • Public Library Association

PLAC: Post-Launch Analysis of Compliance

PLACE: Position Location Aircraft Communications Equipment • Position Location And Communications Experiment • Post-Landsat Advanced Concept Evaluation • Programming Language for Automatic Checkout Equipment

PLACID: Payload Aboard - Caution In Descent

PLADS: Pulsed Light Airborne Depth Sounder

PLAID: Precision Location And Identification

PLAIL: Public Libraries And Independent Learners

PLAIT: Pupils' Learning and Access to Information Technology

PLAM: Price-Level Adjusted Mortgage

Plan: Planet • Planetarium

PLAN: Payload Activity Network • Public Library Access Network • Public Library Automation Network

Planet: Planetary

PLANET: Packetized Lightwave Architecture Network • Personal Laconic Access Network/Ethereal Transmission • Probing Lensing Anomalies Network

PLANS: Programming Language for Allocation and Network Scheduling

PLAO: Poste de Lecture Assistée par Ordinateur

PLAR: Private-Line Automatic Ring-down

Plas: Plaster • Plastic

PLAS: Professional Library Automation System

PLASMAE: Plasmas Magnétisés de l'Environnement

PlaSpaSci: Planetary and Space Sciences

PLAST: Propellant Loading and All Systems Test

Plat: Plateau • Platform • Platinum

PLAT: Plateau • Platform • Pre-Launch Acceptance Test

Platf: Platform

PLATF: Plain Language Terminal Forecast

PLATO: Planetary Analysis Tools • Programmed Learning for Automated Teaching Operations • Programmed Logic for Automatic Teaching Operations (*) • Programmed Logic for Automatic Training Operations

PLATT: Page Level Availability Time Test

PLB: Payload Bay • Perioden-Leuchtkraft-Beziehung • Personal Locator Beacon • Picture Level Benchmark • Platform Bay • Plattsburgh (USA-NY) (location identifier)

PLBD: Payload Bay Door

PLBK: Playback

plc: Public Ltd. Co.

P-L-C: Period-Luminosity-Colour

PLC: Pacific Lumber Co. (USA-CA) • Payload Carrier • Payload Commander • Period-Luminosity-Colour • Planeta Rica (Colombia) (location identifier) • Power-Line Carrier • Preliminary Load Cycle • Pressurized Logistics Carrier • Product Life Cycle • Programmable Logic Control • Programmable Logic Controller • Programmation Logique avec Contraintes • Programming Language Committee • Propellant Loading Console • Public Limited Company

PLCA: Product Life Cycle Assessment

PLCB: Pseudoline Control Block

PLCC: Payload Launch Control Center • Plastic Leaded Chip Carrier (*) • Plastic Leadless Chip Carrier

PLCI: Pressurized Logistics Carrier Interface

PLCM: Propellant Loading Control Monitor

PLCOP: Prelaunch Checkout Plan

PLCP: Physical Layer Convergence Protocol • Planning Command Pool

PLCR: Period-Luminosity-Colour Relation

PLCS: Propellant Loading Control System

PLCU: Propellant Level Control Unit

Plcy: policy

Pld: Payload

PLD: Playa Samara (Costa Rica) (location identifier) • Polar Layered Deposit • Potentially Lethal Damage • Program Listing Document • Programmable Logic Device • Public Libraries Division • Pulsed Laser Deposition

PLDC: Preliminary List of Design Changes

PLDF: Payload Data File

PLDI: Payload Data Interleaver

PLDIS: Public Libraries Development Incentive Scheme

PLDM: Payload Management

PLDMI: Precise Laser Distance Measuring Instrument

PLDP: Photon-noise Limited Doppler Precision

PLDS: Payload Support • Pilot Land Data System

PLDTS: Propellant Loading Data Transmission System

ple: Pleistocene

P.le: Piazzale

PLE: Paiela (Papua New Guinea) (location identifier) • Partial Lunar Eclipse • Phased Loading Entry • Phase Locked Epitaxy • Photoluminescence Excitation • Preliminary Logistics Evaluation • Programmable Logic Element • Public Local Exchange • Pulsed Laser Experiment • Pure Luminosity Evolution

PLED: Polymeric Light-Emitting Diode

PLEGPAY: Plasma Electron Gun Payload

Plen: Plenipotentiary

PLEVAL: Payload Evaluation

plf: Pounds per Linear Foot

Plf: Plaintiff

PLF: Pala (Chad) (location identifier) • Phase Lock Frequency • Point Location File

Plff: Plaintiff

PLG: La Plagne (France) (location identifier) • Public Libraries Group • Pulsed Light Generator

PLGR: Precision Lightweight GPS Receiver • Precision Low-cost GPS Receiver

PLGSS: Payload Ground Support System

PLH: Payload Handling • Plymouth (UK-England) (location identifier)

PLHX: Payload Heat Exchanger

Pli: Pliocene

PLI: Palm Island (Saint Vincent & The Grenadines) (location identifier) • Payload Interrogator • Power Level Indicator • Programming Language Interface • Programming Language One • Propellant-Level Indicator • Pulsed Laser Interferometry

PLIC: Payload Intelligent Controller • Public Libraries and Information Council

PLicO: Publications of Lick Observatory (periodical code)

PLIF: Planar Laser-Induced Fluorescence

PLILP: Programming Languages - Implementations - Logics - Programs

PLIM: Payload Installation Mechanism • Post-Launch Information Message • Propellant Linear Insulation Matrix

PLIMHR: Payload Integration Management Hardware Review

PLIMR: Payload Integration Management Review

PLINS: Payload Integration Review

PLIS: Plannungs-Informationssystem

PLISN: Provisioning List Item Sequence Number

PLJ: Placencia (Belize) (location identifier)

PLK: Branson (USA-MO) (location identifier) • Poincaré-Lighthill-Kuo • Point Lookout - M. Graham Clark (USA-MO) (location identifier)

Pll: Pallet

PLL: Phased Lock Loop • Phase-Locked Loop • Phase Lock Loop • Ponta Pelada (Brazil-AM) (location identifier) • Positive Logic Level • Pre-Linked Library

PLLR: Phase Lock Loop Receiver

plm: Por lo Mejor

PL/M: Programming Language for Microcomputers • Programming Language for Micros

PLM: Palembang (Indonesia) (location identifier) • Paris-Lyon-Méditerranée • Passive Lunar Marker • Payload Management • Payload Module • Payload Monitoring • Polarized Light Microscopy • Power Line Modulation • Prelaunch Monitor • Programming Language for Microcomputers • Programming Language for Micros • Programming Logic Manual • Pulse-Length Modulation

Plmb: Plumber • Plumbing

PLMD: Payload Mating Dolly

Plmg: Plumbing

PLMN: Public Land Mobile Network

PLMP: Program Logistics Management Plan

Plmr: Plumber

PLMR: Post-Launch Memorandum Report

PLMS: Program Logistics Master Schedule

Pln: Plane • Plein

PLN: Pellston (USA-MI) (location identifier) • Perusahaan Listrik Negara (Indonesia) • Plain • Plan • Program Line Number • Program Logic Network

PLNAG: Public Libraries Networking Advisory Group

PLNASG: Public Libraries Networking Awereness Steering Group

Plng: Planning

Plnm: Plenum

Plnr: Planar

PLNS: Plains

Plnt: Planet

Plntrm: Planetarium

Plnty: Planetary

PLO: Pacific Launch Operation • Perform Locked Operations • Phase-Locked Oscillator • Please Leave On • Port Lincoln (Australia-SA) (location identifier) • Proton Liberation Organization • Public Liaison Officer • Pulsed Laser Oscillator • Pulsed Locked Oscillator

PLOD: Periodic List of Data

PLOO: Pacific Launch Operations Office

PLOP: Planetary Landing Observation Package

PLOS: Primary Line of Sight • Probe Line of Sight

PLOT: Porous Layer Open Tubular • Probability of Launch On Time

PLP: La Palma (Panama) (location identifier) • Packet Level Procedure • Packet Level Protocol • Parallel Logic Programming • Post-Launch Phase • Power Law Process • Pulsed Laser Photolysis

PlPh: Plasma Physics (periodical code)

PLPla: Publications of La Plata Observatory (periodical code)

PLPS: Propellant Loading and Pressurization System

Plq: Plaque

PLQ: Palanga (Lithuania) (location identifier)

Plr: Pillar • Plier • Puller

PLR: Pell City - Saint Claire County (USA-AL) (location identifier) • Period-Luminosity Relation • Planar Lightwave Circuit • Power Line Radiation • Pre-Layout Review • Public Lending Right

PLRD: Payload Requirements Document

PLRG: Public Libraries Research Group

P&LRP: Productivity & Long-Range Planning

PlRp: Planetary Report (periodical code)

PLRP: Productivity and Long-Range Planning

Plrs: Pelorus

Plrty: Polarity

PLRV: Payload Launch Readiness Verification

Pls: Plates • Please • Plus • Pulse

PLS: Palomar-Leiden Survey • Partial Least Squares • Payload Simulator • Payload System • Personal Launch System • Personal Librarian System • Personal Library Software • Personal Library System • Personnel Launch System • Phone Lattice Scanning • Plasma • Plasma Light Source • Plasma Science • Post Landing and Safing • Preliminary Landing Site • Primary Landing Site • Primary Link Station • Probability of Launch Success • Professional Land Surveyor • Propellant Loading Sequencer • Propellant Loading System • Providenciales (Turks & Caicos Islands) (location identifier) • Pulsed Laser System • Pulsed Light Source

PLSA: Partial Length Shield Assembly • Probabilistic Latent Semantic Analysis • Public Library Services Act (USA)

PLSC: Plasma Science Committee

PLSI: Probabilistic Latent Semantic Indexing

PLSL: Propellants and Life Support Laboratory

PLSP: Payload Signal Processor • Prelaunch Survival Probability

PLSR: Partial Least Squares Regression • Post-Launch Service Review

PLSS: Personal Life Support System • Portable Life Support System • Post-Landing Survival System • Prelaunch Status Simulator • Primary Life Support System • Public Land Survey System (USA)

Plstc: Plastic

PLSV: Polar Launch Satellite Vehicle

PLSVP: Preliminary SV Planning

Plt: Pilot • Plant

PLT: Pallet • Pilot • Plan à Long Terme • Plato (Colombia) (location identifier) • Precision Live Timing • Princeton Large Tokamak • Princeton Large Torus • Production Lead Time • Program Lock Token • Pulsed Light Theodolite

Plta: Plazoleta

Pltc: Political

Pltf: Plaintiff • Platform

Pltg: Planting • Plating

PLTS: Precision Laser Tracking System

Plu: Plural • Pluto • Pluton

PLU: Belo Horizonte - Pamphula (Brazil-MG) (location identifier) • People Like Us • Plage de Lumière Uniforme • Position Location Uncertainty • Price Lookup • Primary Logical Unit • Propellant Loading and Utilization

PLUM: Payload Umbilical Mast • Planning Land Use Model • Polymerization with Light Under Microgravity

PLUNA: Primeras Lineas Uruguayas de Navegación Aérea

Plur: Plural • Plurality

PLUS: Platform Utilisation Study • Pragmatic Language Understanding System • Program Library Update System • Programmed Learning Under Supervision • Programming Language for UNIVAC Systems • Project Literary United States (USA)

Plut: Pluto • Pluton

PLUTO: Pipe Line Under the Ocean

PLV: Payload Launch Vehicle • Payload Vehicule • Poltava (Ukraine) (location identifier) • Postlanding Valve • Postlanding Vent • Postlanding Ventilation • Powered-Lift Vehicle • Production Level Video • Pseudo-Local Vibration • Publicité sur le Lieu de Vente

PLVC: Post-Landing Vent Control

PLW: Palau (ISO country code) • Palu (Indonesia) (location identifier) • Plowed • Plowed

PLX: Parallax • Propellant Loading Exercise • Semipalatinsk (Kazakhstan) (location identifier)

Ply: Plymouth (UK)

PLY: Play • Plymouth (USA-IN) (location identifier)

Plywd: Plywood

PL/Z: Programming Language Zilog

Plz: Plaza • Please

PLZ: Plaza • Port Elizabeth (South Africa) (location identifier) • Postleitzahl • Programming Language for the Zilog • Złoty (Poland) (ISO currency code)

Plza: Plaza

Plzd: Polarized

PL/1: Programming Language One

pm: Picometer • Post Meridiem

Pm: Premium • Promethium

P&M: Processes & Materials • Production & Maintenance

PM: Miquelon (Saint Pierre et_) (ISO country code) • Pad Mechanic • Panama • Parabolic Mirror • Partial Match • Particle Mesh • Particulate Matter • Past Master • Payload Management • Payload Midbody • Payload Module • Paymaster • Per Month • Performance Management • Performance Monitor • Periodic Maintenance • Peripheral Module • Permanent Magnet • Permanent Motion • Permanent Mouvement • Permittivity Meter • Personality Module • Perturbation Method • Phase Modulated • Phase Modulation • Phase Monitor • Photomultiplier • Photosynthetic Measurement • Planetary Mission • Plan de Matière • Plant Management • Plus Minus • Point Mass • Polarization Modulation • Polar Motion • Police Magistrate • Policy Manual • Polymer Morphology • Portable Magnetometer • Postmaster • Post Meridiem • Post Mortem • Potentiometer • Pour Mémoire • Power Module • Predictive Maintenance • Premier Ministre •

Presentation Manager • Pressurized Module • Preventive Maintenance • Primary Memory • Primary Mirror • Prime Minister • Principle of Multiplying • Problem Mode • Processes and Materials • Process Manager • Process Manual • Procurement Manager • Product Manager • Program Management • Program Manager • Program Memo • Program Milestone • Program Monitoring • Project Management • Project Manager • Propellant Management • Proper Motion • Propulsion Module • Pulse Modulation • Saint Pierre et Miquelon (ISO country code)

PMA: Parts Manufacturer Approval • Parts Manufacturing Approval • Pays Moins Avancé • Pemba (Tanzania) (location identifier) • Performance Monitor Annunciator • Permanent Magnet Alternator • Permanent Magnet Array • Permanent Magnet Assembly • Permanent Mailing Address • Personal Mower Assistant • Personennamen des Mittelalters • Policy Management Architecture • Portable Maintenance Aid • Poste Médical Avancé • Post-Mission Archive • Preliminary Mission Analysis • Pressurized Mating Adapter • Primary Mirror Assembly • Procréation Médicalement Assistée

PMAC: Parallel Memory Address Counter • Peripheral Module Access Controller • Portable Multipurpose Applications Console • Programmable Multiple-Axis Controller

PMAD: Performance Monitor Annunciation Driver • Performance Monitor Annunciator Driver • Power Management And Distribution • Professional Map Analysis Package

PMAG: Photographic Magnitude

PMALS: Preliminary Miniature Air-Launched System

PMAP: Performance Monitor Annunciation Panel • Port Mapper • Professional Map Analysis Package

PMAR: Preliminary Maintenance Analysis Report

PMAS: Phase-coupled Multiaperture Spectrograph • Potsdamer Multi-Apertur Spektrophotometer • Potsdam Multiaperture Spectrophotometer

PMAT: Page Map Address Table • Portable Maintenance Access Terminal

PMATS: Performance Monitor And Tracking System

PMB: Pembina (USA-ND) (location identifier) • Performance Management Baseline • Performance Measurement Baseline • Produit Matériel Brut • Programme Management and Budgeting

PMBX: Private Manual Branch Exchange

PM&C: Plant Maintenance & Construction

PMC: Payload Module Computer • Payload Monitoring and Control • Payload Monitoring Control • Permanently Manned Capability (*) • Permanently Manned Configuration • Permanent Manned Capability • Permanent Manned Configuration • Photoelectric Meridian Circle • Pitman's Measure of Closeness • Plan de Mesure et de Commande • Planning Ministers Council • Plant Maintenance and Construction • Polar Mesospheric Cloud • Pollen Mother Cell • Post-Maintenance Check • Post-Manufacturing Checkout • Post-Ministerial Conference • President's Management Council • Pressure-Modulator Cell • Private Medical Communication • Procurement Method Code • Programmable Machine Controller • Program Management Center • Program Management Committee • Program Management Council • Program Modification Card • Programme Management Council • Project Management Committee • Propellant Monitor and Control • Pseudo Machine Code • Puerto Montt (Chile) (location identifier)

PMCC: Platforms Mission Control Center

PM&CE: Plant Maintenance & Construction Engineer

PMCE: Plant Maintenance and Construction Engineer

PMCM: Programmable Mobile Communications Modem

PMCS: Primary Mirror Control System

P&MD: Productivity & Manufacturing Development

PMD: Pakistan Meteorological Department • Palmdale (USA-CA) (location identifier) • Physical-layer Medium-Dependent • Physical-Media Dependent • Physical-Medium Dependent • Polarization Measurement Device • Polarization Mode Dispersion • Power Management and Distribution • Productivity and Manufacturing Development • Program for Management Development • Program Management Department • Programmed Multiple Development • Program Monitoring and Diagnosis • Propellant Management Device

PMDAR: Promedios Móviles Determinísticos Autoregresivos

PMDB: Photographic Meteor Database (IMO) • Proposal Management Database

PMDF: PASCAL Memo Distribution Facility • Project Master Data File

PMDM: Payload Multiplexer/Demultiplexer

PMDR: Parametric Monotone Decreasing Ratio

PME: Paramagnetic Meissner Effect • Payload Mechanical Engineer • Performance Monitoring Equipment • Petites et Moyennes Entreprises • Photomagnetoelectric • Porte-Monnaie Électronique • Portsmouth (UK-England) (location identifier) • Precision Measuring Equipment • Protective Multiple Earthing • Pyramid Maximum Entropy

PMEE: Prime Mission Electronic Equipment

PMEL: Pacific Marine Environmental Laboratory • Pacific Marine Environment Laboratory • Precision Measurement Equipment Laboratory

PMES: Physical, Mathematical, and Engineering Sciences

PMEST: Personality-Matter-Energy-Space-Time (formula)

PMetR: Papers in Meteorological Research (periodical code)

PMF: Parameter Management Frame • Parma (Italy) (location identifier) • Performance Monitor Function • Performance Monitoring Function • Perigee Motor Firing • Permanent Magnetic Field • Planetary Mass Function • Probability Mass Function • Probable Maximum Flood • Program Management Facility • Pulsating Magnetic Field

PMFA: Permanent Magnet Focus Assembly • Permanent Magnetic-Focus Assembly

PMFBI: Pardon Me For Butting In

PMFJI: Pardon Me For Jumping In

Pmflt: Pamphlet

PMFS: Pulsed Magnetic Field System

PM&G: Planetary Materials & Geochemistry

PMG: Papers on Meteorology and Geophysics (periodical code) • Parts Management Group • Paymaster General • Permanent Magnet Generator • Phase Modulation Generator • Plasma Motor Generator • Ponta Porã (Brazil-MS) (location identifier) • Postmaster General

pmh: Per Man-Hour

PMH: Portsmouth (USA-OH) (location identifier) • Previous Medical History

PMHEMT: Pseudo-Morphic High Electron Mobility Transistor

PMHL: Preferred Measurement Hardware List

pmi: Photographic Microimage

PMI: Palma de Mallorca (Baleares) (location identifier) • Passive Microwave Imager • Personnel Management Information • Petites et Moyennes Industries • Physical Model Interface • Point of Maximum Intensity • Primary Maintenance Inspection • Preventive Maintenance Inspection • Preventive Maintenance Instruction • Primary Measurement Instrument • Principal Maintenance Inspector • Private Medical Insurance • Programmable Machine Interface • Programmable Memory Interface • Programmable Multispectral Imager • Program Management Instruction • Programme du Marché Intérieur

PMIA: Parallel Multiplexer Interface Adapter

PMIC: Personnel Management Information Center

PMIG: Programmer's Minimal Interface to Graphics

PMIP: Paleoclimate Modelling Intercomparison Project

PMIR: Program Manager's Integration Review

PMIRR: Pressure Modulated Infrared Radiometer • Pressure Modulation Infrared Radiometer (*) • Pressure Modulator Infrared Radiometer

PMIS: Passive Microwave Imaging System • Personnel Management Information System • Printing Management Information System • Production Management Information System • Project Management Information System

PMJI: Pardon My Jumping In

Pmk: Postmark

PMK: Palm Island (Australia-QLD) (location identifier)

PML: Perfectly Matched Layer • Port Moller (USA-AK) (location identifier) • Preliminary Materials List • Prins Maurits Laboratorium (TNO) • Probable Maximum Loss • Process Modelling Language

PMLCD: Passive-Matrix LCD

PMM: Phanom Sarakham (Thailand) (location identifier) • Planetary Mission Module • Precision Measuring Machine • Precision Measuring Microdensitometer • Precision Metrology Mount • Probability Matching Method • Property Management Manual • Pulse-Mode Multiplex

PMMA: Polymethylmethacrylate • Polyméthylméthacrylate

PMMD: Programmable Magnetic Momentum Dump

PMMEL: Proposed Master Minimum Equipment List

PMMP: Preventive Maintenance Management Program

PMMR: Project Manager Monthly Review

PMMRS: Passive Microwave Mapping and Retrieval Simulator

PMMU: Paged Memory Management Unit

PMMW: Passive Millimeter Wavelength (technology)

PMN: Parkes-MIT-NRAO • Premanufacture Notification • Presse Mensuelle Nationale • Produit Matériel Net • Program Management Network • Pumani (Papua New Guinea) (location identifier)

PMO: Palermo - Punta Raisi (Italy) (location identifier) • Perturbation Molecular Orbital • Pine Mountain Observatory (USA-GA) • Prime Minister's Office • Principal Meteorological Office • Program Management Office • Project Management Office • Project Management Organization • Purple Mountain Observatory (China-PRC)

PMOC: Pioneer Mission Operations Center (NASA)

PMOD: Physikalisch-Meteorologisches Observatorium Davos (Switzerland)

P. Mol.: Peso Moleculare • Poids Moléculaire

PMOM: Performance Management Operating Manual • Performance Management Operations Manual

PMON: Performance Management Operations Network • Process Monitor

PMOS: Positively-doped Metal Oxide Semiconductor

PM&P: Parts, Materials & Processes

PMP: Parallel Microprogrammed Processor • Parts, Materials and Packaging • Parts, Materials and Processes • Passive Measurement Program • Performance Management Package • Pimaga (Papua New Guinea) (location identifier) • Point to Multipoint • Polymethylpentene • Pontryagin Maximum Principle • Pre-launch Monitor Program • Preliminary Mission Plan • Preliminary Mission Profile • Premodulation Processor • Preventive Maintenance Plan • Prism-Mirror-Prism • Probable Maximum Precipitation • Program Management Plan • Program Monitor Panel • Project Management Plan • Pump

PMPCB: Parts, Materials and Processes Change Board • Parts, Materials and Processes Control Board

PMPE: Program Materials and Processes Engineer

PMPFR: Program Manager's Pre-Flight Review

PMPL: Preferred Mechanical Parts List

PMPS: Program Management Performance System

PMQ: Perito Moreno (Argentina-SC) (location identifier)

Pmr: Paymaster

PMR: Palmerston North (New Zealand) (location identifier) • Parabolic Microwave Reflector • Parts Material Request • Passive Microwave Radiometer • Payload Manifest Review • Payload Mass Ratio • Payload Mixing Review • Performance Measurement Report • Performance Monitoring Receiver • Portable Microfiche Reader • Portable Microfilm Reader • Power Monitor Relay • Precision Microwave Radiometer • Preliminary Materials Review • Presse Mensuelle Régionale • Pressure Modulated Radiometer • Pressure Modulation Radiometer • Private Mobile Radio • Program Management Review • Program Manager's Review • Project Management Report • Project Management Review • Proton Magnetic Resonance • Provisioning Master Record • Pyrradiometer

PMRB: Primary Material Review Board

PMRC: Programme Mondial de Recherche sur le Climat

PMRO: Producibility-Maintainability-Reliability-Operations

PMRR: Pre-Manufacturing Readiness Review • Pre-Mate Readiness Review

PMRS: Performance Management and Recognition System

PMS: Palmyra (Syria) (location identifier) • Pantone Matching System • Payload Monitoring System • Payload Mounting Structure • People Moving System • Performance Management System • Performance Measurement System • Performance Monitoring System • Periodical Management System • Personal Mail Service • Physiological Monitoring System • Planar Magnetic Structure • Policy Management System • Pollution Monitoring Satellite • Power Management Software • Power Management System • Precision Metrology Station • Pre-Main Sequence • Premenstrual Syndrome • Preventive Maintenance System • Probability of Mission Success • Processor-Memory-Switch • Production Management Segment • Product Monitoring System • Program Management Support • Program Management System • Program Master Schedule • Project Management System • Propulsion Module System • Protection Management System • Psycho-Médico-Social

PMSA: Primary Metropolitan Statistical Area

PMSC: Programme Management Support and Coordination

PMSE: Percentage Mean Squared Error • Polar Mesosphere Summer Echo • Polar Mesospheric Summer Echo (*)

PMSEIC: Prime Minister's Science, Engineering and Innovation Council (Australia)

PMSFN: Planetary Manned Space Flight Network

PMSP: Photon-counting Microspectrophotometer

PMSS: Physique Mathématique - Modélisation - Simulation

Pmt: Payment

PMT: Paramakotoi (Guyana) (location identifier) • Phase-Modulated Transmission • Photoelectric Magnetograph Telescope • Photomechanical Transfer • Photomultiplier Tube • Plan à Moyen Terme • Planning/Management Team • Portable Magnetic Tape • Positive Media Transfer • Premenstrual Tension • Prepare Master Tape • Preventive Maintenance Time • Production Monitoring Test • Program Master Tape • Project Management Team • Prominent • Pyramidal Median Transform

PMTP: Pay Me Transfer Protocol

PMTS: Predetermined Motion Time System

PMU: Paimiut (USA-AK) (location identifier) • Payload Monitoring Unit • Performance Monitor Unit • Physical Mock-Up • Plaster Mock-Up • Portable Memory Unit • Precision Measurement Unit • Pressure Measuring Unit • Pulse Modulation Unit

PMV: Panneau à Message Variable • Permanently-Manned Vehicle • Porlamar del Caribe (Venezuela) (location identifier) • Predictive Mean Vote

PMW: Palmas (Brazil-TO) (location identifier) • Passive Microwave • Pitch Maneuver Wheel

PMWO: Preventive Maintenance Work Order

PMX: Packet Multiplexer • Palmer (USA-MA) (location identifier) • Partially Matched Crossover • Private Manual Exchange

PMY: Puerto Madryn - El Tehuelche (Argentina-CB) (location identifier)

PMZ: Palmar (Costa Rica) (location identifier)

PMZF: Programmable Multizone Furnace

Pn: Prochain

P-N: Positive-Negative

P/N: Part Number • Promissory Note

PN: Parc National • Parque Nacional • Part Number • Performance Number • Peso Neto • Phosphorus Nitride • Physical Network • Physique Nucléaire • Pitcairn (ISO country code) • Planetarischer Nebel • Planetary Nebula • Please Note • Poids Net • Pôle Nord • Positive-Negative • Post-Newtonian • Previous Node • Primordial Nucleosynthesis • Prior Notice • Processing Node • Programme National • Program Note • Project Number • Promissory Note • Proton Number • Pseudo Noise • Pulse Network

Pna: Panama

PNA: Pacific - North American • Pamplona (Spain) (location identifier) • Peptide Nucleic Acid

PNAA: Prompt-Neutron Activation Analysis

PNAF: Potential Network Access Facility

PNAOJ: Publications of the National Astronomical Observatory of Japan (periodical code)

PNAS: Proceedings of the National Academy of Sciences (periodical code)

PNB: Pôrto Nacional (Brazil-TO) (location identifier) • Producto Nacional Bruto • Produit National Brut

PNC: Parity Non Conserving • Personnel Navigant de Cabine • Physique Nucléaire et Corpusculaire • Ponca City (USA-OK) (location identifier) • Programmed Numerical Control • Programme National de Cosmologie (France)

PNCA: Programme National de Chimie Atmosphérique (France)

PNCGC: Programme National de Cartographie Géoscientifique du Canada

PNCI: Plan Nacional de Calidad Industrial (Spain)

Pnd: Pond

PND: Passive Nutation Damper • Personennamendatei • Present Next Digit • Presse-Nachrichtendienst (Austria) • Probabilité de Non-Détection • Punta Gorda (Belize) (location identifier)

PNdB: Perceived Noise Decibel

Pndg: Pending

PNDR: Programme National sur le Déterminisme du Recrutement (France)

PNDS: Ponds

PNe: Planetary Nebulae

PNE: North Philadelphia (USA-PA) (location identifier) • Pacific National Exhibition • Peaceful Nuclear Explosion • Pine

PNEA: Programma Nazionale di Ricerche in Antàrtide

PNEAT: Programme National d'Effluorescences Algales Toxiques (France)

PNEDC: Programme National d'Étude de la Dynamique du Climat (France)

PNES: Pines

PNETS: Public Non-Exclusive Telecom Service

Pneu: Pneumatic

Pneum: Pneumatic

PNF: Peterson's Point (USA-AK) (location identifier) • Plan National de Formation (France)

PNG: Papouasie Nouvelle-Guinée (ISO country code) • Papua New Guinea (international vehicle registration code) • Papua New Guinea (ISO country code) • Paranaguá (Brazil-PR) (location identifier) • Persona Non Grata • Portable Network Graphics

PNGV: Partnership for a New Generation of Vehicles

PNH: Phnom-Penh (Cambodia) (location identifier)

PNHRA: Programme National de Haute Résolution Angulaire

PNHRAA: Programme National de Haute Résolution Angulaire en Astronomie • (France)

PNI: Pharmaceutical News Index • Pictorial Navigation Indicator • Pohnpei (Caroline Islands) (location identifier) • Ponape (Caroline Islands) (location identifier) • Prix Nobel Ignoré

PNIC: Programme National d'Intérêt Communautaire • Prototype Near-Infrared Camera

PNK: Pontianak (Indonesia) (location identifier) • Proper Noun Keyword

Pnl: Panel

PNL: Pacific Northwest Laboratory (USA) • Panel • Pantelleria (Italy) (location identifier) • Programmable Netlist Library • Programmation Neuro-Linguistique

PNLA: Pseudo Next-Level Aggregator

Pnld: Paneled

PNLF: Planetary Nebulae Luminosity Function

Pnlg: Paneling

PNM: Petit Nuage de Magellan • Pulse Number Modulation

PNMS: Permanent Noise Monitoring System

PNN: Planetary Nebula Nucleus • Princeton (USA-ME) (location identifier) • Probabilistic Neural Network • Producto Nacional Neto • Profesor No Numerario

PNNI: Private Network-Network Interface • Private Network-to-Network Interface (*)

PNNL: Pacific Northwest National Laboratory (USA-WA)

Pnnt: Pennant

PNNV: Planetary-Nebula Nucleus Variable

PNO: Pinotepa (Mexico) (location identifier) • Public Network Operator

PNOC: Programme National d'Océanographie Côtière (France) • Proposed Notice Of Change

PnP: Plug and Play

PNP: Personnel Non Permanent • Plug and Play • Popondetta (Papua New Guinea) (location identifier) • Positive-Negative-Positive • Pre-negociation Position • Private Non-Profit (sector) • Programme National de Planétologie (France)

PNPI: Petersburg Nuclear Research Institute (Russia)

PNPN: Positive-Negative-Positive-Negative

PNQ: Poona (India) (location identifier)

PNQR: Pure Nuclear Quadrupole Resource

Pnr: Pioneer

PNR: Passenger Name Record • Pointe Noire (Congo) (location identifier) • Point of No Return • Prior Notice Required • Programme National de Recherche • Prompt Nuclear Reaction

PNRA: Programma Nazionale di Ricerche in Antartide (Italy)

PNRCO: Programme National des Récifs Coraliens (France)

PNRH: Programme National de Recherche en Hydrologie (France)

PNRI: Philippines Nuclear Research Institute

PNRN: Programme National sur les Risques Naturels (France)

PNRP: Pollution and Natural Resources Programme

PNRS: Preliminary Natural Resources Survey

PNS: Parabolized Navier-Stokes • Pensacola (USA-FL) (location identifier) • Peripheral Nervous System • Philippine National Standard • Pilot Navigation System • Projeto de Novos Satélites

PNSC: Palomar-Norris Sky Catalog • Palomar Northern Sky Catalog

PNSS: Philippine National Science Society

Pnt: Paint

PNT: Puerto Natales - Teniente J. Gallardo (Chile) (location identifier)

Pntgn: Pentagon

PNTPM: Physique Nucléaire Théorique et Physique Mathématique

PNTS: Programme Nationale de Télédétection Spatiale (France)

PNU: Panguitch (USA-UT) (location identifier) • Philippine Normal University (Philippines)

PNUD: Programme des Nations-Unies pour le Développement (ONU)

PNUE: Programme des Nations-Unies pour l'Environnement (ONU)

PNVG: Panoramic Night Vision Goggle

PNW: Pacific Northwest • Personal NetWare • Program Notes and Waivers

PNX: Sherman-Denison - Grayson County (USA-TX) (location identifier)

pnxt: Pinxit

PNY: Pondicherry (India) (location identifier)

PNZ: Petrolina (Brazil-PE) (location identifier)

p/o: Part of

po: Par Ordre • Per Ommegaande • Per Order • Per Os • Por Orden

Po: Paseo • Paso • Polonium

P&O: Peninsular & Oriental Steam Navigation Co. • Plans & Operations

P/O: Provisioning Order

PO: Parallel Output • Parity Odd • Parking Orbit • Par Ordre • Partitioned Organization • Passport Office • Patent Office • Personnel Office • Personnel Officer • Petites Ondes • Phoenician • Photogrammetric Operations • Planetary Orbit • Planning Objective • Planning Office • Plans and Operations • Point d'Observation • Poland • Polar Orbiter • Polynesia • Por Orden • Portsmouth (UK postal code) • Postal Order • Poste d'Observation • Post Office • Post Orbit • Power On • Power-Operated • Power Oscillator • Power Output • Preoperational • Previous Order • Primary Output • Printout • Prisme-Objectif • Privately Owned • Programmation Objects • Programmed Oscillator • Programme Opérationnel • Program Objective • Program Office • Program Originator • Project Office • Province of Ontario (Canada) • Publication Officielle • Public Office • Pulkovo Observatory (Russia) • Purchase Order • Pyrénées-Orientales (France)

POA: Persistent Object Adapter • Plan d'Option sur Actions • Plan Of Action • Polaris Observatory Association (USA-CA) • Portable Object Adapter • Porto Alegre - Salgado Filho (Brazil-RS) (location identifier) • Port Of Arrival • Post Office Agent • Post-Oxidation Annealing • Power Open Association • Primary Optical Area

POAA: Planetary Operations Analysis Area

POAM: Polar Ozone and Aerosol Measurement

POB: Fayetteville Pope (USA-NC) (location identifier) • Peripheral Order Buffer • Place Of Birth • Point Of Beginning • Post Office Box

POBeo: Publi. Obs. Astron. Beograd (periodical code)

Pobl: Población

POBol: Pubblicazione dell'Osservatorio Astronomico di Bologna (periodical code)

POC: La Verne - Brackett (USA-CA) (location identifier) • Parallel Optical Computer • Particulate Organic Carbon • Payload Operations Center • Physical Optics Corporation • Plan d'Opérations Combinées • Point Of Contact • Point Of Control • Policy Oversight Committee • Portable Occultation CCD • Port Of Call • Precision Oscillator Crystal • Probe Operations Centre • Process Operator Console • Product Of Combustion • Program Organizing Committee • Proof Of Concept • Public Call Office • Purchase Order Closeout • Purchase Order Contract

PoCa: Polar Camera

POCC: Payload Operations Control Center • Platform Operations Control Center • Project Operations Control Center

POCCNET: Payload Operations Control Center Network

POCE: Pantone Open Color Environment • Proof-Of-Concept Experiment

POCIBO: Polar Circling Balloon Observatory

POCM: Parallel Ocean-Climate Model

POCN: Purchase Order Change Notice

POCO: Physiology Of Chimpanzees in Orbit • Power On - Clock On • Purchase Order Change Order

PoCom: Polymer Composites (periodical code)

POCP: Program Objectives Change Proposal

POCR: Program Objectives Change Request

POCS: Passice Ocean Colour System • Patent Office Classification System (USA) • Projection Onto Convex Sets • Projection Over Convex Sets

pocul: Poculum

POD: Parameter Override Display • Payload Operations Director • Payload Operations Division • Payment On Delivery • Pay On Death • Pay On Delivery • Perl Online Documentation • Piece Of Data • Place Of Discharge • Planning and Observational Data • Plan Of the Day • Podor (Senegal) (location identifier) • Port Of Delivery • Port Of Disembarkation • Post Office Department • Post Office Directory • Precise Orbit Determination • Precision Orbit Determination • Preliminary Orbit Determination • Probability Of Detection • Production Operations Directive • Program Operation Description • Program to Output Data • Project Operations Director • Project Organization Demographic • Protective Oceanic Device • Pulse Omission Detector • Purchase Order Deviation

PODAAC: Physical Oceanography Distributed Active Archive Center

PODAF: Post Operation Data Analysis Facility

PODAPS: Portable Data Processing System

PODAS: Portable Data Acquisition System

PODC: Principles Of Distributed Computing

PODIUM: Presentation Overlay Display for Interactive Users of Media

PODM: Preliminary Orbit Determination Method

PODP: Principles Of Document Processing

PODPS: Post-Observation Data Processing System

PODS: Passive Orbital Disconnect Strut • Pilot Ocean Data System • Post-Observation Data Processing System • Precise Orbit Determination System • Precision Orbit Determination System

PODTG: Precision Orbit Determination Task Group

POE: Fort Polk (USA-LA) (location identifier) • Pilot Operational Equipment • Port Of Embarkation • Port Of Entry • Precise Orbit Ephemeris • Print Out Effect • Projected Operational Environment • Property Owning Expenses • Proximity Optical Device

POEAS: Planetary Orbiter Error Analysis Study

POED: Program Organization for Evaluation and Decision

POEM: Photonics and Optoelectronic Material • Physical Oceanography of the Eastern Mediterranean • Polar Orbit Earth-observing Mission • Polar Orbiting Earth Mission • Polar Orbiting Earth Monitoring • Polar Orbiting Environmental Mission

POEMS: Positron Electron Magnetic Spectrometer • Positron Electron Magnet Spectrometer (*)

POES: Polar Operational Environmental Satellite • Polar Orbiting Environmental Satellite

Poet: Poetic • Poetical • Poetry

P/OF: Pinhole/Occulter Facility

POF: Payload Operations Facility • Pinhole/Occulter Facility • Planned Observation File • Plastic Optical Fibre • Poplar Bluff - Earl (USA-MO) (location identifier) • Programmable Operator Facility

POFS: Private Operational Fixed Service

POG: Passion-Orange-Guava • Peace-Order-Government • Port Gentil (Gabon) (location identifier) • Pressure Omega Center • Proposers' Observatory Guide • Propulsion Operating Guide

POGASIS: Planetary Observation Geometry And Science Instrument Sequence

POGO: Polar Orbiting Geophysical Observatory

POGS: Polar Orbiting Geomagnetic Survey

POGSI: Policy Group on Scientific Information

POH: Pilot's Operating Handbook • Pocahontas (USA-IA) (location identifier) • Power On Hours

POHM: Page-Oriented Holographic Memory

POHP: Publications de l'Observatoire de Haute-Provence (periodical code)

POI: Parking Orbit Injection • Période d'Observation Intensive • Plan d'Opérations Intercalées • Plan Of Instruction • Pluto Orbital Insertion • Pluto Orbit Insertion • Point Of Interest • Potosí (Bolivia) (location identifier) • Power-On/Off Interrupt • Principal Operations Inspector • Probability Of Intercept • Product Of Inertia • Program Of Instruction • Purchase Order Item

POIC: Payload Operations and Integration Center • Payload Operations Integration Center

POID: Power-On/Off Interrupt Delay

POINTER: Particle Orientation Interferometer

POINTS: Precision Optical Interferometer in Space (*) • Precision Optical Interferometry in Space

POIROT: Passively-cooled Orbiting Infrared Observatory Telescope

POIS: Prototype Online Instrument System • Purchase Order Information System

POISE: Panel On In-flight Scientific Experiments • Photosynthetic Oxygenation Illuminated by Solar Energy

Pol: Poland • Polar • Polarity • Polarization • Polarize • Polarized • Polarizer • Police • Policía • Policy • Polish • Political • Politics

POL: Pemba (Mozambique) (location identifier) • Petroleum-Oil-Lubricants • Physics Online Library • Poland (ISO country code) • Pologne (ISO country code) • Port Of Loading • Problem-Oriented Language • Procedure-Oriented Language • Program-Oriented Language • Proudman Oceanographic Laboratory (UK)

POLA: Pointable Optical Linear Array

POLAR: Potentials Of Large objects in the Auroral Region • Production Order Location And Record • Production Order Location And Reporting

POLARIS: Polar motion Analysis by Radio Interferometric Studies

POLARSCAT: Polarimetric Scatterometer

PolAt: Pollution Atmosphérique (periodical code)

POLC: Pennsylvania Outdoor Lighting Council (USA-PA)

POLDER: Polarization and Directionality of Earth Radiation • Polarization and Directionality of Earth Reflectances (*)

POLDIRAD: Polarimetric Diversity Radar

POLDPS: Pioneer Off-Line Data Processing System

POLE: Polar Ozone Lidar Experiment

POLEX: Polar Experiment

Polg: Poligono

POLGEN: Problem-Oriented Language Generator

POLID: Personnes-Organismes-Langues-Implantations-Dates

POLIS: Parliamentary Online Information System • Polarimetric Littrow Spectrometer • Promoting Operational Links with Integrated Services

Polít: Política

Polit: Political • Politician

POLITE: Plasmaspheric Observations of Light Ions in the Topside Exosphere

Poll: Pollution

POLO: Problem-Oriented Language Organizer

PolRe: Polar Research Memoirs (periodical code)

Polstr: Polystyrene

Polthn: Polyethylene

Poly: Polynesia • Polynesian • Polytechnic

Polyn: Polynesia • Polynesian

POLyo: Publications de l'Observatoire de Lyon (periodical code)

PoM: Plan of Management

POM: Pallet-Only Mode • Particulate Organic Matter • Période Opérationnelle Moyenne • Persistent Object Manager • Petite Opération Millimétrique • Phase Of the Moon • Pilot's Operating Manual • Polycyclic Organic Matter • Polyoxymethylene • Port Moresby (Papua New Guinea) (location identifier) • Prefix-Omission Method • Printer Output Microfilm • Program Objective Memorandum • Program Operation Mode • Project Office Memorandum • Proximity Operation Mode

POMA: Pointable Optical Mosaic Array

POME: Programs-Objectives-Methods-Evaluation

POMI: Preliminary Operating and Maintenance Instruction

POMic: Publications of Michigan Observatory (periodical code)

POMin: Publications of the Astronomical Observatory of the University of Minnesota (periodical code)

POMOLA: Poor Man's Optical Landing System

POMS: Panel on Operational Meteorological Satellites • Parallel Observation Matching System • Polar Orbiting Meteorological Satellite • Post-Main-Sequence • Probability Of Mission Success

POMSEE: Performance, Operating and Maintenance Standards for Electronic Equipment

POMT: Planning Operations Management Team

PON: Particulate Organic Nitrogen • Passive Optical Network • Poptun (Guatemala) (location identifier)

PONAM: Polar North Atlantic Margin

PONS: Picture of NAG Splinefit

PONZ: Precise Object Near Zenith

POO: Poços de Caldas (Brazil-MG) (location identifier) • Post Office Order • Programmation Orientée Objet

POOF: Peripheral Online Oriented Function

POOL: Parallel Object-Oriented Language • Probabilistic Object-Oriented Logic

POOZ: Permanently Open Ocean Zone

Pop: Popular • Popularly • Populate • Populated • Population

PoP: Point of Presence

POP: Parallel Observing Program • Parallel Ocean Program • Payload Operations Plan • Payload Optimized Program • Period Of Performance • Perpendicular-to-Orbit Plane • Persistent Organic Pollutant • Physics Outreach Program • Picture Out of Picture • Plug-in Optronic Payload • Point Of Presence • Point Of Purchase • Polar Orbiting Platform • Polar Outflow Probe • Post Office Preferred • Post Office Protocol • Preburner Oxidizer Pump • Pre-flight Operations Procedure • Pre-launch Operations Plan • Principal Oscillation Pattern • Printing-Out Paper • Printout Paper • Programmed Operators and Primitives • Programme d'Orientation Pluriannuel • Program Operating Plan • Project Objective Plan • Project Operating Plan • Proof-Of Principle • Proof Of Purchase • Puerto Plata (Dominican Rep.) (location identifier)

POPA: Panel On Public Affairs • Payload Ordnance Processing Area • Programme d'Orientation Pluriannuel

POPad: Pubblicazione dell Osservatorio di Padova (periodical code)

POPAE: Protons On Protons And Electrons

POPB: Palais Omnisports de Paris-Bercy (France)

PoPC: Pile of Personal Computers

POPE: Polar Orbiting Platform Element

POPIN: Population Information Network

POPL: Principles Of Programming Languages • Problem-Oriented Programming Language

POPO: Push-On - Pull-Off

POPot: Publ. Astrophys. Obs. Potsdam (periodical code)

POPs: Principal Oscillation Patterns

POPS: Paperless Order Processing System • Participatory Orientated Planetarium in the Schools • Participatory-Oriented Planetarium Show • People-Oriented Parsing Service • Positioning Orbital Propulsion System • Principal Oscillation Pattern System • Program for Operator Scheduling • Pyrotechnic Optical Plume Simulator

POPSAT: Precise Orbit Positioning Satellite

POPSI: Postulate-based Permuted Subject Indexing

POPU: Push Over/Pull Up

POQ: Polk Inlet (USA-AK) (location identifier)

Por: Porosity • Porous • Portion • Portuguese

POR: Pacific Ocean Region • Parking Orbit Rendezvous • Pay On Receipt • Pay On Return • Payable On Receipt • Periodic Operation Report • Point Of Reference • Point Of Resolution • Pori (Finland) (location identifier) • Port Of Refuge • Power On Reset • Price On Request • Purchase Order Request

PORB: Production Operations Review Board

PORCN: Production Order Records Change Notice

POrCo: Progress in Organic Coatings (periodical code)

PORD: Performance and Operations Requirements Document

PORE: Procurement-Oriented Requirements Engineering

PORM: Plus Or Minus

PORR: Preliminary Operations Requirements Review

PORS: Periodiek Overleg Remote Sensing

PORSEC: Pacific Ocean Remote Sensing Conference

Port: Portable • Portrait • Portuario • Portugal • Portugués • Portuguese

PORT: Photo-Optical Recorder Tracker • Portable

Porta: Portada

PORTAL: Polar Orbiter Remapping and Transformation Application Library • Process-Oriented Real-Time Algorithmic Language

PORTS: POCC Real-Time Support • Preliminary Operations Requirements and Test Support

PORV: Pilot-Operated Relief Valve • Pressure-Operated Relief Valve

Pos: Position • Positional • Positive • Possession • Possessive

PoS: Point of Sale

POS: Pacific Ocean Ship • Packet Over Sonet • Part Of Speech • Payload Operations Simulation • Period Of Service • Photo Optic System • Physics On Stage (project) • Plane Of the Sky • Point Of Scale • Point Of Sale • Portable Oxygen System • Port of Spain - Piarco (Trinidad & Tobago) (location identifier) • Position • Positive • Preferred Observation Schedule • Preferred Observation Sequence • Primary Oxygen System • Probability Of Sufficiency • Probably Old System • Professional Operating System • Programmable Object Select • Programmable Option Select • Program Order Sequence • Purchase Order Supplement • Trinidad (Trinidad & Tobago) (location identifier)

POSANG: Position Angle

POSD: Pointed Observation by Survey Detectors • Program for Optical System Design • Project Operation Support Division

POSDCORB: Planning, Organizing, Staffing, Directing, Coordinating, Reporting and Budgeting

POSE: Picture-Oriented Software Engineering

POSEIDON: Positioning Ocean Solid Earth Ice Dynamics Orbiting Navigator • Projet d'Observatoire de Surveillance et d'Études Intégrées de la Dynamique des Océans

POSH: Parents for Objective Science and History (USA)

POSIX: Portable Operating System Interface for Unix

Posn: Position

POSO: People Over Sixty Online

Poss: Possession • Possessive • Possibility • Possible • Possibly

POSS: Palomar Observatory Sky Survey • Passive Optical Satellite Surveillance • Prototype Optical Satellite Surveillance

Post: Posterior • Postulación

POST: Parliamentary Office of Science and Technology • Part-Of-Speech Tagger • Passive Optical Scan Tracker • Payload Operations Support Team • Piezoelectric-Oscillator Self-Tuned • Point Of Sale Terminal • Polar Observatory Stratospheric Telescope • Polar Stratospheric Telescope • Positive • Power-On Self-Test • Predevelopment Operational System Test • Production-Oriented Scheduling Technique • Programmer Operating Standards Technique • Program to Optimize Shuttle Trajectories • Program to Optimize Simulated Trajectories

Postf: Postfach

PostGebO: Postgebührenordnung

Posth: Posthume • Posthumous • Posthumously

Posthum: Posthumous • Posthumously

PoSTI: Policies for Sustainable Technological Innovation

POSTNET: Postal Numeric Encoding Technique

Postop: Postoperative • Postoperatively

POStr: Publications de l'Observatoire de Strasbourg (periodical code)

Pot: Potassium • Potential • Potentiometer

POT: Picture Object Table • Pitch-Orthogonal Thrust • Point Of Termination • Port Antonio (Jamaica) (location identifier) • Potable • Potentiometer • Preplanned Observation Tape • Proof Of Technology • Proximity Operations Trainer

POTP: Power On Time Pad

POTS: Pan-Only Triangulation System • Plain Old Telephone Service • Plain Old Telephone System • Precision Optical Tracking System

POTUS: President Of The United States

POTV: Personnel Orbit Transfer Vehicle

POTW: Publicly Operated Treatment Work • Publicly Owned Treatment Work

PoU: Point of Use

POU: Point Of Use • Poughkeepsie (USA-NY) (location identifier)

POUS: Pousada

P-ov: Poluostrov

POV: Peak Operating Voltage • Persistence Of Vision • Pneumatically-Operated Valve • Pneumatic-Operated Valve • Point Of View • Póvoa • Presov (Slovakia) (location identifier) • Privately Owned Vehicle • Program-Oriented View • Proximity Operations Vehicle

POVRO: Publications of the Owens Valley Radio Observatory (periodical code)

POW: Please Oblige With • Portoroz (Slovenia) (location identifier) • Prisoner Of War

Powd: Powder

POWER: Performance Optimization With Enhanced RISC • Performance Optimized With Enhanced RISC • Portable Workstation for Education in Europe

POWG: Payload Operations Working Group • Payload Orbital Working Group

POWR: Predictive Optical Word Recognition

POWS: Project Operating Work Statement

POX: Point Of Exit • Pontoise (France) (location identifier) • Prism Objective X

POY: Powell (USA-WY) (location identifier)

POZ: Poznań (Poland) (location identifier)

P°: Paseo • Paso

pp: Pages • Páginas • Par Procuration • Past Participle • Per Person • Perpetuo • Per Procura • Per Procurationem • Postpaid • Prepaid

Pp: Pappband • Porte Pagado

P-P: Peak-to-Peak • Pitch-Pitch • Probability-Probability • Push-Pull

P&P: Policies & Procedures • Postage & Packing

P/P: Patch Panel • Prelaunch/Postlanding • Printer/Plotter

PP: Brazil (international aircraft registration code) • Pacco Postale • Pacific Plate • Page Printer • Parcel Post • Parish Priest • Parliamentary Paper • Partial Pressure • Partial Program • Participatory Program • Particle-Particle • Partitioned Parallelism • Past Participle • Patch Panel • Pay Period • Paysan Pingre • Peak-to-Peak • Perfect Program • Peripheral Processor • Permanent Pass • Personal Property • Piecewise Polynomial • Pipeline Processing • Pisces-Perseus • Pitch-Pitch • Planetary Program • Planning Package • Planning Purpose • Plasma Processing • Play or Pay • Polypropylene • Port Payé • Posa Piano • Positive Polarization • Possibly Present • Postage Paid • Postpaid • Post Pay • Postponed • Post-Process • Pouvoirs Publics • Power Package • Power Plant • Power Pole • Pre-modulation Processor • Prepaid • Preparatory Programme • Prepositional Phrase • Preprinted • Preprocessor • Preproduction • Present Position • Primary Point • Principal Point • Printer Page • Printer-Plotter • Print Punch • Prior Permission • Privately Printed • Probability-Probability • Process Planning • Procurement Plan • Program Performance • Programming Plan • Program Performance • Program Plan • Projection Pursuit • Project Proposal • Proof Plot • Proof Plotter • Propeller Pitch • Propulsion Power • Proton-Proton • Pruppacher-Pitter • Public Property • Pulse Polarography • Pushchino Pulsar • Push-Pull

PPA: Pacific Planetarium Association (USA-CA) • Pampa - Perry Lefors (USA-TX) (location identifier) • Parallel Performance Analysis • Parallel Plate Analyzer • Parité du Pouvoir d'Achat • Periodical Publishers' Association (UK) • Pilots Personal Assistant • Pitch Precession Amplifier • Pollution Prevention Act • Power Pitch-Around • Precision Pulsar Astrophysics • Professional Programmers Association • Program Problem Area • Proposal Preparation Assistant • Pulsed Power Amplifier • Pulse Plasma Accelerator • Pupil Plane Assembly

PPAC: Parallel Plate Avalanche Counter

PPAR: Program/Project Approval Review

PPARC: Particle Physics and Astronomy Research Council (UK)

Ppart: Plupart

PPAS: Parallel Poll Active State • Photometric Periods of Artificial Satellites

ppb: Parts per Billion

PPB: Paper-Printing-Binding • Parts Per Billion • Planning-Programming-Budgeting • Polar Patrol Balloon • Positive Pressure Breathing • Presidente Prudente A. de Barros (Brazil-SP) (location identifier) • Primary Propulsion Branch • Private Posting Box • Program Performance Baseline • Provisioning Parts Breakdown

PPBAS: Planning-Programming-Budgeting-Accounting System • Program Planning-Budgeting-Accounting System

PPBES: Planning-Programming-Budgeting-Evaluation System • Program Planning-Budgeting-Evaluation System

PPBG: Preliminary Program and Budget Guidance

ppbm: Parts per Billion by Mass

PPBM: Pulse Polarization Binary Modulation

PPBS: Planning, Programming and Budgeting System • Positive Pressure Breathing System

ppbv: Parts per Billion per Volume

PP&C: Program Planning & Control • Project Planning & Control

PPC: Parallel Poll Configure • Payload Preparation Complex • Phased Provisioning Code • Platform Position Computer • Positive Peer Control • Pour Prendre Congé • Power Pack Charger • Power Plant Change • Pre-Processing Center • Primary Power Control • Production Planning and Control • Programmable Protocol Converter • Program Planning and Control • Program-to-Program Communication • Project Planning and Control • Prospect Creek (USA-AK) (location identifier)

PPCC: Peaks Pike Community College (USA-CO)

PpCF: Pounds per Cubic Foot

PPCF: Plasma Physics and Controlled Fusion (periodical code) • Pounds Per Cubic Foot

PpCI: Pounds per Cubic Inch

PPCI: Pounds Per Cubic Inch

PPCM: Plus Petit Commun Multiple

ppCO2: Partial Pressure of Carbon Dioxide

PPCP: Post-Processing Computer Program

PPCS: Precision Pointing Control System

PPCU: Partial Payload Checkout Unit

Ppd: Postpaid • Postponed • Prepaid

PPD: Humacao - Palmas del Mar Hotel (Puerto Rico) (location identifier) • Partial Packet Discard • Particle Physics Department • Particle Physics Division • Payload Position Data • Peripheral Pulse Distributor • Plan-Position Display • Point Position Data • PostScript Printer Description • Postponed • Power Pitch-Down • Pre-defined Process Diagram • Primary Point Digitization • Printer Page Description • Program Planning Directive • Program Planning Document • Project Planning Directive • Propulsion and Power Division • Provinciale Planologische Dienst • Provisioning Procurement Data

PPDB: Point Positioning Database

PPDD: Preliminary Project Design Description

PPDF: Poisson Probability Distribution Function

p. pdo.: Próximo Pasado

PPDO: Per Person - Double Occupancy

PPDP: Preliminary Project Development Plan • Preprogram Definition Phase

PPDS: Particle Physics Data System • Personal Printer Data Stream • Physical Property Data Service • Pilot Planetary Data System

PPDU: Payload Power Distribution Unit

PPE: Personal Protection Equipment • Personal Protective Equipment (*) • Personnel Protection Equipment • Phase Partitioning Experiment • Philosophy-Politics-Economics • Platform Position Equipment • Portable Purge Equipment • Predicted Period of Effect • Pre-modulation Processing Equipment • Pre-production Evaluation • Program Planning and Evaluation • Prototype Production Evaluation • Puerto Penasco (Mexico) (location identifier) • Purchasing Power Equivalent

PPEC: Permanent Periodic Error Correction

PPECO: Pre-Programmed EVA Controlled Operations

PPEDC: Printing and Publishing Economic Development Committee

PPEE: Periodica Polytechnica on Electrical Engineering (periodical code)

PPEP: Pen Plotter Emulation Program • Plasma Physics and Environment Perturbation

PPEUC: Particle Physics and Early Universe Conference

PPF: Parsons (USA-KS) (location identifier) • Payload Processing Facility • Personal Property Floater • Phase Pushing Factor • Polar Platform • Preprints in Particles and Fields (SLAC) • Primary Processing Facility

PPFA: Page Printer Formatting Aid

PPFD: Photosynthetic Photon Flux Density

PPFF: Poisson Probability Frequency Function

PPFS: Portable Parallel File System

PPG: Pago Pago (American Samoa) (location identifier) • Piezoelectric Power Generation • Plasma Power Generator • Program Planning Group • Propulsion and Power Generation

pph: Parts per Hundred

Pph: Pamphlet

PpH: Pounds per Hour • Pulses per Hour

PPH: Paid Personal Holidays • Parts Per Hundred • Peraitepuy (Venezuela) (location identifier) • Pounds Per Hour • Primeval Particle Horizon • Pulses Per Hour

PPHA: Peak Pulse Height Analysis

pphm: Parts per Hundred Million

PPHM: Parts Per Hundred Million

PpHP: Pounds per Horsepower

PPHP: Pounds Per Horsepower

PpI: Particle Per Inch • Pixel per Inch

PPI: Packing, Postage and Insurance • Parcel Post Insured • Particle Per Inch • Patient Package Insert • Pay Per Incident • Pixel Per Inch • Planetary Plasma Interaction • Plan Particulier d'Intervention • Plan Position Indicator • Policy Proof of Interest • Port Pirie (Australia-SA) (location identifier) • Producer Price Index • Programmable Peripheral Interface • Program Package for Improvement • Project Procurement Instruction • Proposal Preparation Instruction • Public-Private Interface • Pulse Position Indicator

PPIC: Pollution Prevention Information Clearinghouse

PPID: Parent Process Identification • Parent Process Identificator • Parent Process Identifier

PPIE: Pollution Prevention Information Exchange

PPIF: Photo Processing Interpretation Facility

PPIL: Priced Provisioned Item List

PPIMS: Personal Property Information Management System

PPIP: Physics Post-doctoral Information Pool

PPITB: Printing and Publishing Industry Training Board (UK)

PPJ: Pulau Panjang (Indonesia) (location identifier)

PPK: Personal Preference Kit • Petropavlovsk (Kazakhstan) (location identifier)

PPKDD: Principles and Practice of Knowledge Discovery in Databases (conference)

Ppl: Participle

P/PL: Primary Payload

PPL: Packet-by-Packet Layer • Phaplu (Nepal) (location identifier) • Planning Parts List • Plasma Physics Laboratory • Plasma Propulsion Laboratory • Polymorphic Programming Language • Populated Place • Predictive Period Laser • Preferred Parts List • Preliminary Parts List • Priced Parts List • Primary Payload • Process Programming Language • Production Payload • Program Parts List • Program Production Library • Provisional Parts List • Provisioning Parts List • Pyrotechnically-Pumped Laser

Pple: Participle

PpLF: Pounds per Linear Foot

PPLF: Pounds Per Linear Foot

PPLL: Programmable Phase-Locked Loop

PP&LS: Production Planning & Logistics Support

PPLS: Production Planning and Logistics Support • Propellant and Pressurant Loading System

PPLUS: Polar Platform Utilisation Study

ppm: Pages per Minute • Parts per Million • Part per Minute • Pulses per Minute

PPM: Parametric Potential Method • Payload Project Manager • Peak Power Meter • Peak Programme Meter • Perfect Prog Method • Periodic Permanent Magnet • Piecewise Parabolic Method • Pompano Beach (USA-FL) (location identifier) • Portable Pixmap • Positions and Proper Motions • Pounds Per Minute • Programmable Pointing Mode • Project Profile Manual • Pulse-Position Modulation • Punch Press Machine

PPMA: Printing Paper Manufacturers Association

PPME: Pacific Plate Motion Experiment • Periodica Polytechnica on Mechanical Engineering (periodical code)

PPMG: Professional Publishers Marketing Group

PPMS: Photogrammetry Production Monitoring System • Programme and Project Management System • Project Performance Measurement System

PPMtO: Publications of the Purple Mountain Observatory (periodical code)

ppmv: Parts per Millions per Volume

PPN: Parameterized-Post-Newtonian • Popayan (Colombia) (location identifier) • Pre-Planetary Nebula • Project Programmer Number • Proto-Planetary Nebula (*) • Public Packet Network

PPNe: Pre-Planetary Nebulae • Proto-Planetary Nebulae (*)

PPNL: Pacific Northwest National Laboratory

PPO: Petit Prisme-Objectif (OHP) • Photographic Program Office • Pikes Peak Observatory (USA-CO) • Planetary Protection Officer • Powell Point (Bahamas) (location identifier) • Power Plant Operating • Preferred-Provider Organization • Prior Permission Only • Procurement Planning Officer • Projected Program Objective • Push-Pull Output

PPOV: Pilot Point Of View

ppO2: Partial Pressure of Oxygen

ppp: Pixels par Pouce • Points par Pouce

PPP: Parallel Pattern Processor • Payload Patch Panel • Peak Pulse Power • Permanent Power Pack • Personal Pension Plan • Personne Physique Payée • Phased Project Planning • Planning Purpose Proposal • Plasma Physics Package • Points Par Pouce • Point-to-Point Protocol • Polluter Pays Principle • Pople-Parr-Pariser • Preliminary-Phase Pilot • Preliminary Program Plan • Pretty Poor Planning • Proserpine (Australia-QLD) (location identifier) • Psychology-Philosophy-Physiology • Purchasing Power Parity

PPPL: Princeton Plasma Physics Laboratory (USA-NJ) • Printed Planning Parts List

PPPM: Particle-Particle-Particle-Mesh

PPPP: Proposed Partial Package Program

PPPS: Post-Post-Post-Scriptum

P&PQ: Productivity & Product Quality

PPQ: Paraparaumu (New Zealand) (location identifier) • Possible Parliamentary Question • Productivity and Product Quality

P&PQI: Productivity & Product Quality Improvement

PPQI: Productivity and Product Quality Improvement

P&PQIC: Productivity & Product Quality Improvement Council

PPQIC: Productivity and Product Quality Improvement Council

Ppr: Paper • Proper

PPR: Pasir Pangarayan (Indonesia) (location identifier) • Payload Preparation Room • Peak Power Response • Periodic Personnel Report • Permanent Pay Record • Photopolarimeter Radiometer • Present Participle • Printed Paper Rate • Prior Permission Required • Programme Prioritaire • Program Planning Review • Program Progress Review • Projection Pursuit Regression • Proprietary Procurement Request • Purchase Parts Request

PPRB: Preliminary Proposal Review Board

PPRI: Plasma Physics Research Institute (USA-CA)

PPRN: Preliminary Publication Revision Notice • Purchased Parts Requirement Notice

PPRNG: Parallel Pseudo-Random Number Generator

p. pro.: Per Procurationem

pps: Packets per Second • Post Postscriptum • Pulses per Second

PpS: Packets per Second • Pounds per Second • Pulses per Second

PPS: Additional Postscript • Packets Per Second • Page Printing System • Parallel Processing System • Paraphenylene Sulfide • Partial-Pressure Suit • Patchboard Programming System • Payload Pointing System • Payload Positioning System • Payload Power Switch • Peripheral Processing System • Photographic Processing System • Photopolarimeter Spectrometer • Photopolarimeter System • Photovoltaic Power Supply • Piece Part Specification • Pipeline Processing System • Plasma Power Supply • Pneumatic Power System • Post-Postscriptum • Pounds Per Second • Power Personal Systems • Precise Positioning Service • Precise Positioning System • Precision Power Supply • Precision Processing System • Primary Power Standard • Primary Power Supply • Primary Propulsion System • Princeton Planetary Society (USA-NJ) • Printer/Plotter System • Product Performance Survey • Product Performance System • Programmable Patch System • Programmable Power Supply • Program-Provided Summary • Projektplannungsgesellschaft • Provisioning Performance Schedule • Puerto Princesa (Philippines) (location identifier) • Pulses Per Second • Purchasing Power Standard • Second Postcript

PPSC: Programming Practices, Standards and Conventions

PpSF: Pounds per Square Foot

PPSF: Pounds Per Square Foot

PpSI: Pounds per Square Inch

PPSI: Pounds Per Square Inch • Prompt Photometric Sunspot Index

PPSL: Program Parts Selection List

PPSN: Parallel Problem Solving from Nature • Public Packet Switching Network

PPSS: Precise Positioning Service Security

ppt: Parts per Thousand • Parts per Trillion

Ppt: Plupart • Precipitate

PPT: Papeete (Tahiti) (location identifier) • Parts Per Thousand • Parts Per Trillion • Peak Power Transfer • Periodic Programs Termination • Petite Propriété Terrienne • Picture Page Table • Picture Planet Table • Pitch Precision Torquer • Plasma Phase Transition • Pre-launch Package Test • Pulsed-Plasma Thruster • Pulse Pair Timing • Pulse Plasma Thruster • Punched Paper Tape

Pptd: Precipitated

PPTD: Products/Process Technology Division

PPTE: Periodica Polytechnica on Transportation Engineering (periodical code) • Polyphenyleneterephthalamide

PPtF: Passivated Platinum Film

Pptg: Precipitating

Pptn: Precipitation

PPTP: Point-to-Point Tunneling Protocol

PPTR: Punched Paper Tape Reader

pptv: Parts per Trillion per Volume

Ppty: Property

PPU: Papun (Myanmar) (location identifier) • Parallel Poll Unconfigure • Peripheral Processing Unit • Power Processor Unit • Program-Produced Underlining

PpV: Pay per View

PPV: Pay Per View • Plus Proche Voisin • Polyphényle de Vinyle • Port Protection (USA-AK) (location identifier) • Post-Production Verification

PPW: Papa Westray (UK) (location identifier) • Plane-Polarized Wave

PP&WC: Process Planning & Work Control

PPWC: Process Planning and Work Control

PPX: Param (Papua New Guinea) (location identifier)

PPY: Pouso Alegre (Brazil-MG) (location identifier)

PPZ: Proton Polar Zone • Puerto Paez (Venezuela) (location identifier)

P&Q: Productivity & Quality

PQ: Parliamentary Question • Parque • Personality Quotient • Planetary Quarantine • Preceding Question • Premier Quartier • Previous Question • Primo Quarto • Province de Québec (Canada) • Québec (Canada)

PQA: President's Quality Award • Procurement Quality Assurance • Protected Queue Area

PQAP: Product Quality Assurance Plan

PQC: Phu Quoc - Duong Dang (Viet Nam) (location identifier) • Precision Quartz Crystal • Procurement Quality Control • Production Quality Control • Product Quality Council

PQD: Percentage Quartile Deviation • Prequalification Document • Pyroelectric Quad Detector

PQDMB: Percentage Quartile Deviation Median Bias

Pqe: Parque

PQE: Procurement Quality Engineering • Progress in Quantum Electronics (periodical code)

PQET: Print Quality Enhancement Technology

PQFP: Plastic Quad Flat Pack • Plastic Quad Flat Package

PQG: Plasma of Quarks and Gluons

PQGS: Propellant Quantity Gauging System

PQI: Presque Isle (USA-ME) (location identifier) • Professional Qualification Index • Propellant Quantity Indicator

PQIC: Productivity and Quality Improvement Council • Product Quality Improvement Council

PQL: Picture Query Language • Presse Quotidienne Locale

PQLI: Physical Quality of Life Index

PQM: Palenque (Mexico) (location identifier) • Pulse Quaternary Modulation

PQMD: Propellant Quantity Measuring Device

PQN: Presse Quotidienne Nationale • Principal Quantum Number

PQP: Planetary Quarantine Plan • Prequalification Prototype

PQPF: Probabilistic Quantitative Precipitation Forecasting

PQQ: Port Macquarie - Wauchope (Australia-NSW) (location identifier)

PQR: Presse Quotidienne Régionale • Project Quality Responsible

PQRVU: Programmazione Quadratica Ricorsiva con Vincoli di Uguaglianza

PQS: Parallel Query Server • Pilot Station (USA-AK) (location identifier)

PQT: Preliminary Qualification Test • Professional Quick Take

pr: Per

Pr: Pair • Paper • Power • Praça • Prandtl Number • Praseodym • Praséodyme • Praseodymium • Preference • Preferred • Present • Pressure • Price • Priest • Priester • Prime • Prince • Prins • Print • Printed • Printer • Printing • Private • Professeur • Pronoun • Proper • Prove • Proved • Provençal • Provence (France) • Provincial

P-R: Period-Radius • Pitch-Roll • Precision-Recall

P/R: Payroll • Planner/Replanner • Productivity/Respiration

P&R: Performance & Resources

PR: Pair • Parallax and Refraction • Paraná (Brazil) • Parcel Receipt • Parliamentary Report • Partially Reusable • Partial Response • Pattern Recognition • Payroll • Pearson-Readhead • Peer Review • People's Republic • Percentage Rate • Percentile Rank • Performance and Resources • Performance Ratio • Performance Report • Performance Requirement • Period-Radius • Personal Representative • Photographic Reconnaissance • Photoreflectance • Photo-Refractive • Photo Revision • Phrase Recognition • Physical Review • Pitch Ratio • Pitch-Roll • Planetary Radar • Planning Reference • Please Return • Polarization Ratio • Populus Romanus • Portaria • Porto Rico (ISO country code) • Position Record • Postal Regulations • Poste Restante • Potential Requirement • Power Reactor • Poynting-Robertson • Prairie • Precipitation Radar • Precipitation Rate • Precision-Recall • Préfecture • Preliminary Report • Press Release • Press Representative • Pressure • Pressure Ratio • Pressure Regulator • Preston (UK postal code) • Primary • Priority Regulation • Prize Ring • Probabilistic Reasoning • Problem Report • Procurement Regulation • Procurement Request • Procurement Requisition • Production Rate • Profit Rate • Program Register • Program Requirement • Progress Report • Project Report • Proportional Representation • Proposed Regulation • Proposed Request • Pro Rata • Public Relations • Puerto Rican • Puerto Rico (ISO country code) • Pulse Rate • Purchase Request • Pyrolitic Release

pra: Praecipitatio

PRA: Paperwork Reduction Act • Parabolic Reflector Antenna • Parana (Argentina-ER) (location identifier) • Partly Resolved Array • Physical Review A • Physical Review Abstracts • Planetary Radio Astronomy • Precession Axis • Precision Radar Approach • Print Alphanumerically • Probabilistic Relational Algebra • Probabilistic Risk Analysis • Probabilistic Risk Assessment • Programme de Recherches Avancées • Program Reader Assembly • Public Relations Adviser • Public Roads Administration

PrAA: Progress in Astronautics and Aeronautics (periodical code)

PRAC: Public Relations Advisory Committee

PRACA: Problem Reporting, Analysis and Corrective Action • Problem Reporting And Corrective Action

PRACDA: Polarimetric Radar Control and Data Acquisition

PRACL: Page-Replacement Algorithm and Control Logic

PRACSA: Public Remote Access Computer Standards Association

Prad: Pradesh

PRAD: Pitch Ratio Adjust Device

PrAeS: Progress in Aerospace Sciences (periodical code)

Prag: Pragmatic • Pragmatism

Pral: Principal

PRAM: Parallel Random Access Machine • Parallel Random Access Memory • Parameter Random Access Memory • Programmable Random Access Memory • Program Requirements Analysis Method

Prama: Pramana (periodical code)

PRAME: Pôle de Recherches Aquitain sur les Matériaux de l'Espace (France)

prand: Prandium

PRAO: Présentation Assistée par Ordinateur • Pushchino Radioastronomy Observatory (Russia)

PRARE: Positioning Range And Rate Experiment • Precise Range And Range-rate Equipment (*) • Precise Range And Range-rate Experiment • Precise Range And Rate Equipment

PRAREE: Precise Range And Range-rate Equipment Extended

PRARS: Pitch, Roll, Azimuth Reference System

PRAS: Passenger Revenue Accounting System

PRAT: Prediction Resolution Analysis Tool

PRB: Panel Review Board • Parachute Refurbishment Building • Paso Robles (USA-CA) (location identifier) • Physical Review B • Polar Research Board • Preliminary Review Board • Procurement Review Board • Program Request Block • Program Requirement Baseline • Program Review Board • Project Review Board

Prblc: Parabolic

PRBLTY: Probability

PRBS: Pseudo-Random Binary Sequence • Pseudo-Random Bit Sequence

PRBSG: Pseudo-Random Binary Sequence Generator

PRC: People's Republic of China • Physical Review C • Planning Research Corporation (USA) • Postal Rate Commission • Post Roman Conditam • Power Reflection Coefficient • Precipitation • Prescott (USA-AZ) (location identifier) • Price Regulation Committee • Procedure Review Committee • Programme de Recherches Coordonnées • Program Review Committee • Project Review Committee

PRCA: Pitch and Roll Channel Assembly • Problem Reporting and Corrective Action • Professional Rodeo Cowboys Association

PRCB: Program Requirements Change Board • Program Requirements Control Board • Program Review Control Board

PRCBD: Program Requirements Control Board Directive • Program Review Control Board Directive

Prchte: Parachute

PRCP: Power Remote Control Panel

Prcs: Process

PRCS: Primary Reaction Control System • Propulsion and Reaction Control System

Prcsr: Processor

Prcst: Precast

PRCU: Power Regulating and Control Unit

PRD: Pardoo (Australia-WA) (location identifier) • Partial Redistribution • Payload Requirements Document • Period • Personal Radiation Dosimeter • Physical Review D • Piezoelectric Resonating Device • Power Requirement Data • Preliminary Requirements Document • Presidential Review Directive • Printer Driver • Procurement Regulation Directive • Procurement Requirements Document • Program Requirements Data • Program Requirements Document • Projected Rotation Date • Project Requirements Data • Project Requirements Document

PRDA: Parada • Program Research and Development Announcement

PRDDO: Partial Retention of Diatomic Differential Overlap

PRDMD: Private Directory Management Domain

Prdn: Production

Prdr: Producer

PRE: Partial Reflection Experiment • Personal Rescue Enclosure • Pore (Colombia) (location identifier) • Power Regulation Equipment • Preamp/Readout Electronics • Preliminary Research Equipment • Preston (Starlink internal code) • Pulse Radiation Effect

PREAMP: Preamplifier

PréAO: Présentation Assistée par Ordinateur

Prec: Precede • Preceded • Precedence • Preceding • Precision

Préc: Précédent • Précéder • Précision

Precdg: Preceding

Precip: Precipitation

PRECIS: Preserved Context Index System

PRECISE: Precision Estimation and Control In Scientific and Engineering (computing)

Precp: Precipitation

PrECS: Progress in Energy and Combustion Science (periodical code)

Pred: Predicate • Predicative • Predicted • Predicting • Prediction

Préd: Prédicat • Prédiction • Prédire • Prédit

Predic: Predicate • Predicative

Préf: Préface • Préfecture • Préférence • Préférer • Préfixe

Pref: Preface • Prefaced • Prefatory • Prefazione • Prefecture • Prefer • Preferably • Preference • Preferred • Prefijo • Prefix • Prefixed

PREF: Preference • Propulsion Research Environmental Facility

Prefab: Prefabricated

Prefect: Prefecture

PREFLT: Pre-flight

Prej: Prejudice

Prel: Preliminary

PREL: Preliminary

Prél: Préliminaire

Prelim: Preliminary

Prélim: Préliminaire

Prem: Premium

PrEM: Probabilistic Engineering Mechanics (periodical code)

PREM: Preliminary Reference Earth Manual

prEN: Project of European Norm

Preop: Preoperative • Preoperatively

PREOS: Performance and Range of Electro-Optical Systems

Prep: Preparation • Preparatory • Prepare • Prepared • Preparing • Preposición • Preposicional • Preposition

Prép: Préparation • Préparer • Préposition

PReP: PowerPC Reference Platform

PREP: Preparation • Prepare

Prepd: Prepared

Prepg: Preparing

PREPHA: Programme de Recherche sur la Propulsion Hypersonique Avancée

Prepn: Preparation

PREPNet: Pennsylvania Research and Economic Partnership Network (USA-PA)

Prepr: Preprint

Pres: Presbyterian • Presence • Present • Presente • Presentation • Preserve • Preserved • Presidency • President • Presidential • Pressure • Presumed • Presumptive

Prés: Présence • Présent • Préservé • Préserver • Présidence • Président

PRES: Present

PRESAC: Photographic Reconnaissance System Analysis by Computer

Presb: Presbyterian

Presbyt: Presbyterian

Prescr: Prescription

PRSFR: Pressure Falling Rapidly

PRSRR: Pressure Rising Rapidly

Press: Pression • Pressure • Pressuriser • Pressurize

PRESS: Pressurant • Pressure

PREST: Policy Research in Engineering, Science and Technology • Politique de la Recherche Scientifique et Technique

PRESTO: Precursory Research for Embryonic Science and Technology • Program for Rapid Earth-to-Space Trajectory Optimization • Programmable Etalon Spectrometer for Twilight Observations • Program Reporting and Evaluation System for Total Operation • Project to Reengineer Space Telescope Observing

Presv: Preservation • Preserve

Pret: Preterit

Prév: Préventif • Prévention

Prev: Prevention • Preventive • Previous • Previously

PREV: Previous

PREVLV: Prevalve

PrEx: Propellants and Explosives (periodical code)

PrExP: Propellants and Explosives Pyrotechnics (periodical code)

Prf: Proof

PRF: Parachute Refurbishment Facility • Partial Recursive Function • Personal Resource Form • Pixel Response Function • Point Response Function • Port Johnson (USA-AK) (location identifier) • Potential Requirements File • Preference • Processing and Reference Facility • Promotion Recommendation Form • Pulse Rate Frequency • Pulse Recurrence Frequency • Pulse Repetition Frequency

PRFC: Plastique Renforcé de Fibres de Carbone

PRFE: Polar Reflection Faraday Effect

Prfm: Performance

PRFM: Pseudo-Random Frequency Modulated

Prfnl: Professional

Prfrd: Proofread

Prfrdg: Proofreading

Prfrdr: Proofreader

Prfs: Parfois

PRFVT: Plastiques Renforcés et Fibres de Verre Textile (periodical code)

Prfx: Prefix

Prg: Pergamino

PRG: Peculiar Red Giant • Polar-Ring Galaxy • Powerful Radio Galaxy • Prague Ruzyne (Czech Rep.) (location identifier) • Praha Ruzyne (Czech Rep.) (location identifier) • Program

Prgm: Program

Prgmg: Programming

Prgmr: Programmer

PRH: Phrae (Thailand) (location identifier) • Probability Ranking Hypothesis • Programme de Recherche en Hydrologie • Program Requirements Handbook

Prhps: Perhaps

Pri: Primary • Primer • Priority • Private

PRI: Porto Rico (ISO country code) • Praslin Island (Seychelles) (location identifier) • Primary • Primary-Rate Interface • Princeton Research Instruments (USA-NJ) • Public Radio International • Puerto Rico (ISO country code) • Pulse Rate Indicator • Pulse Recurrence Interval • Pulse Repetition Interval • Pupil Research Initiative

PRIA: Publicación en Revista Internacional con Arbitros • Public Relations Institute of Australia

PRIAM: Plasmas Réactifs en Interaction Avec les Matériaux

Prign: Prignitz (Germany)

Prim: Primaire • Primary • Primer • Primitive

PRIM: Primary

PRIMA: Phase-Referenced Imaging and Microarcsecond Astrometry (*) • Phase-Referencing Imaging and Microarcsecond Astrometry

PRIMAS: Precipitation Radar Image Management System

PRIMATE: Personal Retrieval of Information by Microcomputer And Terminal Ensemble

PRIME: Precision Recovery Including Maneuvering Entry • Program Independence, Modularity, Economy • Program of Research in International Management and Economy • Promoting Reform In Mathematics Education

Prin: Principal • Principally • Principle

PRINC: Principe

PRINCE: Parts Reliability Information Center

PRIND: Present Indications

Print: Printing

PRIOR: Program for In-Orbital Rendezvous

PRIP: Pattern Recognition and Image Processing

PRIS: Payload Remote Integration System • Power Reaction Information System • Power Reactor Information System (*) • Program Resource Information System

PRISE: Page Reader Input System with Editing

Prism: Prismatic

PRISM: Perkins Reimaging System • Power Reactor Inherently Safe Module • Programmed Integrated System Maintenance • Program Reliability Information System for Management

PRISMA: Parallel Inference and Storage Machine • Probing Rotation and Interior of Stars: Microvariability and Activity

PRISME: Programmation des Robots Intelligents et des Systèmes Manipulateurs Évolués

PrISMS: Program Information Systems Mission Services

PRISMS: Plan of Research for Integrated Soil Moisture Studies

PRISTE: Programme de Recherches Interdisciplinaires et Techniques de l'Environnement (CNRS)

Priv: Private • Privative • Privativo • Privé • Privilege • Privilège • Privilegio

Priv. Comm.: Private Communication

PRJ: Capri (Italy) (location identifier) • Paper in a Refereed Journal

PRJMP: Pressure Jump

PRK: Corée du Nord (ISO country code) • DPRK (ISO country code) • Korea (DPRK) (ISO country code) • Park • Prieska (South Africa) (location identifier)

Prl: Parallel

PRL: Page Revision Log • Pattern Recognition Logic • Payload Retention Latch • Photonics Research Laboratory • Physical Research Laboratory (India) • Physical Review Letters • Plasma Research Laboratory • Port Oceanic (USA-AK) (location identifier) • Priority Rate Limiting • Program Requirements Logic • Project Research Laboratory • Propulsion Research Laboratory

PRLA: Payload Retention Latch Actuator • Payload Retention Latch Assembly

PRLP: Planetary Rocket Launcher Platform

PRLS: Pulsed Ruby Laser System

Prm: Premium

PRM: Partially Redundant Mask • Partially Reflecting Mirror • Partially Regulated Module • Partial Response Method • Payload Retention Mechanism • Performance and Reliability Module • Perijove Raise Maneuver • Personal Radiation Monitor • Pocket Radiation Meter • Portimão (Portugal) (location identifier) • Posigrade Rocket Motor • Precision Runway Monitor • Preliminary Requirements Model • Programme Review Meeting • Programming and Resources Management • Pulse Rate Modulation

PRMC: Partial-Response Maximum Likelihood

PrMD: Private Management Domain

PRMD: Private Management Domain

PRML: low Purin Meal • Partial-Response Maximum Likelihood

PRMO: Productibility-Reliability-Maintainability-Operability

PRMP: Programme de Restitution des Mots de Passe

PrMS: Progress in Materials Science (periodical code)

Prmtr: Parameter

prn: Pro Re Nata

PRN: Pôle de Recherche National • Presse Quotidienne Régionale • Printer • Pristina (Yugoslavia-Serbia) (location identifier) • Program Release Notice • Pseudo-Random Noise • Pseudo-Random Number • Pulse Range Navigation • Pulse Ranging Navigation • Pulse Ranging Network

PRNG: Pseudo-Random Number Generator

Prnl: Pronominal

Prntd: Printed

Prntg: Printing

Pro: Professional • Project

Pr. O.: Press Officer

PRO: Perry (USA-IA) (location identifier) • Planning Resident Order • PLL Reference Oscillator • Precision RISC Organization • Proceed • Profile • Programme • Public Records Office (UK) • Public Records Officer • Public Relations Office • Public Relations Officer • RAL (Astrophysics cluster) (Starlink internal code)

Prob: Probabilité • Probability • Probable • Probably • Probate • Probation • Problem

PROB: Probabilities • Problem

PROBAR: Programmable Band Radiometer

PROBE: Pilot Radiation Observation Experiment • Portable Remote Observation of the Environment

Probl: Problem

PROBS: Prototype RVD Onboard Software

Proc: Procedure • Procédure • Proceed • Proceedings • Process • Processing • Processor • Proclamation • Proctor • Procure • Profession • Professional

PROC: Procedure • Process • Processor • Procurement

PROCEDE: Process Engineering Design Environment

PROCOM: Projection Compositor

Prod: Prodigy • Produce • Produced • Producer • Product • Production • Productor • Produire • Produit

PRODAC: Programmed Digital Automatic Control

PRODEX: Programme for the Development of Experiments (ESA)

PROE: Publications of the Royal Observatory of Edinburgh (periodical code)

Prof: Profesor • Professeur • Profession • Professional • Professionnel • Professor • Proficiency • Profile

PROFDA: Processed Frame Database

PROFIT: Programmed Reviewing, Ordering and Forecasting Inventory Technique • Propulsion Flight Control Integration Technology

PROFOEG: Programa de Fotometría y Estructura Galáctica

Profrship: Professorship

PROFS: Professional Office System (IBM) • Program for Regional Observing and Forecasting Services • Prototype for Regional Observing and Forecasting Services

Prog: Prognosis • Program • Programmable • Programmer • Programming • Progress • Progressive

PROG: Prognosis • Prognostic • Program • Programmer • Progress

PROGERT: Projet de Recherche pour l'Observation et la Gestion des Ressources Terrestres

Progr: Program • Programmable • Programmer • Programming • Progress • Progressive

PROI: Public Records Office of Ireland

Proj: Project • Projected • Projectile • Projection • Projector

PROJ: Project

PROJTA: Projectada

PROJTO: Projectado

Pról: Prólogo

Prol: Prologue

PROLA: Physical Review Online Archive

PROLOG: Programming in Logic

Prolong: Prolongación • Prolongamento

Prolyd: Proto-Planetary Disk

Prom: Prominent • Promontory • Promote • Promoter • Promotion • Promulgate

PROM: Pockels Readout Optical Modulator • Programmable Read-Only Memory • Prominence Research: Observation and Modeling

PROMATS: Programmable Magnetic Tape System

PROMENADES: Projet de Mesures en Astrométrie de Distances d'Étoiles avec un Schmidt

PROMIS: Polar Regions and Outer Magnetosphere International Study • Polar Regions Outer Magnetosphere International Study • Program for an Operational Meteorological Information System • Project-Oriented Management Information System

PROMISE: Process-Oriented Machine Independent Simulation Environment

PROMPT: Production, Reviewing, Organizing and Monitoring of Performance Techniques • Project Management and Production Team • Project Reporting Organization and Management Planning Technique

PROMYS: Program in Mathematics for Young Scientists

Pron: Pronombre • Pronominal • Pronoun • Pronounce • Pronounceable • Pronounced • Pronouncement • Pronouncing • Pronunciation

PRONAOS: Projet National d'Astronomie Submillimétrique (France) • Programme National d'Astronomie Submillimétrique (France) • Programme National d'Observation Submillimétrique (France) (*)

PRONTO: Program for Numeric Tool Operation

Pronun: Pronunciación

Pronunc: Pronunciation

Prop: Propagate • Propagation • Propellant • Propeller • Propelling • Proper • Properly • Property • Proportion • Proportional • Proposal • Propose • Proposed • Proposer • Proposition • Proprietary • Proprietor • Propulse • Propulsion

PROP: Planetary Rocket Ocean Platform • Propellant • Proportioning • Propulsion

Propag: Propagate • Propagation

Propl: Proportional

PROPLT: Propellant

PROPLYDS: Protoplanetary Disks

Propn: Proportion

Propr: Proprietor

PRORISC: Programme for Research on Integrated Systems and Circuits

ProRR: Production Readiness Review

Pros: Prosodia • Prospectus

PROS: Post-Reduction Off-line Software • Prevention, Repair and Operational Services

PROSE: Problem Solution Engineering

ProSEDS: Propulsive Small Expandable Deployer System

Prot: Protective • Protectorate

PROTEL: Procedure-Oriented Type Enforcing Language

pro tem: Pro Tempore

PROTEUS: Profile Telemetry of Upper Ocean Currents • Prototype Real-time Operational Test, Evaluation and User Simulation

Prov: Provençal • Provence (France) • Proverb • Proverbial • Proverbially • Provide • Provided • Provider • Province • Provincia • Provincial • Provinz • Provision • Provisional • Provost

Provis: Provision • Provisional

prox: Proximo

Prox: Proximité

PROX: Proximity

prox. acc.: Proximo Accessit

prox. luc.: Proxima Luce

Prp: Purpose

PRP: Peak Radiated Power • Performance-Related Pay • Personnel Reliability Program • Pipelined Resampling Processor • Potentially Responsible Party • Precision Registration Processor • Probability Ranking Principle • Profit-Related Pay • Program Redefinition Proposal • Proposal Review Panel • Propriano (France) (location identifier) • Pseudo-Random Pulse • Pulse Recurrent Period • Pulse Repetition Period

PRPDC: Pacific Regional Planetary Data Center

Prphl: Peripheral

pr. pr.: Praeter Propter

PRPS: Programming Requirements Process Specification

Prpsl: Proposal

PRQ: Prerequisite • Presidente Roque Saenz Peña (Argentina-CH) (location identifier)

PRQA: Plan Régional pour la Qualité de l'Air (France)

PR&R: Project Report & Review

PRR: Paruima (Guyana) (location identifier) • Parts Replacement Request • Parts Rework or Replacement • Peak Radiance Response • Pre-flight Readiness Review • Preliminary Requirements Review • Production Readiness Review • Program Requirements Review • Program Revision Report • Project Requirements Review • Projet de Recherche en Réseau • Proton Relaxation Rate • Pulse Recurrence Rate • Pulse Repetition Rate

PRRB: Payload Requirements Review Board

PRRC: Petroleum Recovery Research Station (USA-NM)

PRRD: Proprietary Rights to Registration Data

PRRDA: Public Relations Regional Development Assistant

PRRM: Program Review and Resources Management

PRRS: Positioning Reporting Recording System

Prs: Pairs

PRs: Pakistan Rupee (currency)

PRS: Pacific Rocket Society (USA-CA) • Parasi (Solomon Islands) (location identifier) • Passive Ranging Subsystem • Pattern Recognition Society (USA) • Pattern Recognition System • Payload Retention System • Personal Recording System • Personal Rescue System • Personal Response System • Personnel Rescue Service • Personnel Rescue System • Pointing Reference System • Position Reference System • Power Reactant System • Power Relay Satellite • Precision Ranging System • Preliminary Requirements Specification • Primary Recovery Ship • Primary Recovery Site • Primary Reference Source • Primary Rescue Site • Process Radiation Sampler • Program Requirements Summary • Provisioning Requirements Statement • Pseudo-Random Sequence • Public Relations Section • Public Research System

PRSA: Power Reactant Storage Assembly

Prsd: Pressed

PRSD: Power Reactant Storage and Distribution • Power Reactant Supply and Distribution

PRSDS: Power Reactant Storage and Distribution System

PRSN: Provisional Relative Sunspot Number

PRSNT: Present

Prsrz: Pressurize

PrST: Prairie Standard Time

PRST: Persist

Prt: Print

P&RT: Physical & Recreational Training

PRT: Pattern Recognition Technique • Personal Rapid Transit • Poincaré Recurrence Theorem • Point Retreat (USA-AK) (location identifier) • Port • Portable Remote Terminal • Porteiro • Portugal (ISO country code) • Precision Radiation Thermometer • Preliminary Reference Trajectory • Prestation Technologique Réseau • Primary Ranging Terminal • Printer • Procurement Review Team • Program Reference Table • Pulse Recurrence Time • Pulse Repetition Time • Pulsed Radar Transmitter

Prtg: Printing

Prtl: Partial

PRTLS: Powered Return to Launch Site

PRTM: Printing Response-Time Monitor

Prtng: Printing

PRTOT: Prototype Real-Time Optical Tracker

Prtr: Printer

PRTS: Ports

PRTSC: Print Screen

PRTU: Packet Remote Terminal Unit

PRTV: Peterlee Relational Test Vehicle • Production Representative Test vehicle

PRU: Pattern Recognition Unit • Primary Replacement Unit • Program Research Unit • Prome (Myanmar) (location identifier) • Pyro Regulation Unit • Pyrotechnic Regulation Unit

Prus: Prussia • Prussian

Pruss: Prussia • Prussian

PRV: Peak Reverse Voltage • Perfect Rotational Variant • Pour Rendre Visite • Prerov (Czech Rep.) (location identifier) • Pressure-Reducing Valve • Pressure Reduction Valve • Pseudo-Relative Velocity

PrVP: Proceedings of Vibration Science (periodical code)

Prvw: Preview

PRW: Peachtree Report Writer • Planetary Rossby Wave • Prentice (USA-WI) (location identifier)

PRWD: Priority Regular World Day

Prx: Prefix

PRX: Paris (USA-TX) (location identifier)

PRY: Paraguay (ISO country code) • Pretoria Wonderboom (South Africa) (location identifier)

PRZ: Prineville (USA-OR) (location identifier)

PRZM: Pesticide Root Zone Model

ps: Pashto (ISO language code) • Picosecond • Pieces • Pseudonym • Pushto (ISO language code)

Ps: Positronium

P&S: Planning & Scheduling • Purchase & Sales

P/S: Parallel to Serial • Payload Specialist • Power Supply

PS: Palomar Schmidt • Parachute System • Parity Switch • Parity Symmetry • Parliamentary Secretary • Participating Scientist • Particle Size • Passenger Steamer • Payload Specialist • Payload Station • Payload Support • Performance Standard • Permanent Secretary • Personal Station • Personal System • Pferdestärke • Phase Shift • Philippine Standard • Photographic Service • Photometric System • Phrase Structure • Physical Sequential • Physics Specification • Picosecond • Planetary Society • Planetary System • Planning and Scheduling • Planning Study • Plasma Switch • Plotting System • Plus • Pointing System • Point of Symmetry • Point Sémaphore • Point Source • Polar Stereographic • Pôle Sud • Polystyrene • Polystyrène • Positive Small • Poste Supplémentaire • PostScript • Post Scriptum • Postskript • Post-sleep Activity • Power Section • Power Source • Power Spectrum • Power Steering • Power Supply • Preliminary Study • Pre-sleep Activity • Press Secretary • Pressure Switch • Pressurized Section • Primary Structure • Prime Select • Private Secretary • Privy Seal • Probability of Success • Probability of Survival • Problem State • Process Specification • Process Status • Procurement Specification • Production Sheet • Production System • Product Standard • Programmed Symbol • Programming System • Program Specification • Program Start • Program Store • Program Summary • Project Scientist • Propellant Supply • Propellant System • Proportional Spacing • Proton Synchrotron • Public School • Public Service • Publication Standard • Publishing Service • Pull Switch • Pulsating Surface

Psa: Presa

PsA: Piscis Australis - Piscis Australis (constellation) • Piscis Austrinus - Piscis Austrini (constellation)

PSA: Pacific Science Association • Pacific-South American • Pacific Southwest Airlines • Panoramic Spectral Analyzer • Particle Size Analyzer • Particle Surface Area • Path of Steepest Ascent • Payload Service Area • Payload Support Avionics • Peacekeeping Services Agreement •

Personal Satellite Assistant (NASA) • Phase-Shift Analysis • Philosophy of Science Association • Photographic Society of America • Pisa (Italy) (location identifier) • Plataforma Solar de Almería (Spain) • Post-Sleep Activity • Power and Servo Assembly • Power Servo Amplifier • Power Servo Assembly • Power Spectral Analysis • Power Supply Assembly • Power Switching Assembly • Preferred Storage Area • Prefixed Save Area • Prefixed Storage Area • Pre-Sleep Activity • Pressure Suit Assembly • Pressure Swing Adsorption • Pressure Switch Assembly • Prices Surveillance Authority • Probabilistic Safety Analysis • Probabilistic Safety Assessment • Problem Statement Analyzer • Processor Save Area • Productibility-Supportability-Affordability • Program Study Authorization • Prostatic Specific Agent • Provisional Site Acceptance • Provisions Stowage Assembly • Public Service Announcement • Pulse Shape Analysis

PSAC: Parent Science Advisory Council • Passive Satellite Attitude Control • Presidential Scientific Advisory Committee (USA) • President's Science Advisory Committee (USA)

PSAD: Predicted Site Acquisition Data • Prediction-Simulation-Adaptation-Decision

PSAL: Programming System Activity Log

PSAM: Partitioned Sequence Access Method

PSAP: Public Safety Answering Point

PSAR: Preliminary Safety Analysis Report

PSAS: Phenytoin for Space Adaptation Syndrome • Physical-Space Assimilation Analysis System • Physical-space Statistical Analysis System

PSAT: Predicted Site Acquisition Table • Preliminary Scholastic Aptitude Test

PSB: Bellefonte (USA-PA) (location identifier) • Clearfield (USA-PA) (location identifier) • Panel on Scientific Ballooning • Philipsburg Mid-State (USA-PA) (location identifier) • Post-Starburst • Prepress Service Bureau • Productivity and Standards Board (Singapore) • Program Specification Block

PSBL: Planetary Sheet Boundary Layer • Possible

PSBLS: Permanent Space-Based Logistics System

PSBW: Program Specification Block Workpool

Psc: Pisces-Piscium (constellation)

PSC: Pacific Science Center • Pacific Science Council • Parallel Subspace Correction • Paralle Switch Control • Parallel to Serial Converter • Pasco (USA-WA) (location identifier) • Payload Support Contractor • Peachtree Support Center • Pembroke State College (USA) • Performance Seeking Control • Periodic Surveillance Committee • Personal Service Contractor • Personnel Status Change • Phase-Sensitive Converter • Physical Sciences Center • Physical Sciences Committee • Pittsburgh Supercomputer Center (USA-PA) • Pittsburgh Supercomputing Center (USA-PA) • Planetary Spacecraft • Planning-Scheduling-Control • Platform Support Center • Point Source Catalogue • Polar Stratospheric Cloud • Post-Storage Checkout • Potomac State College (USA-MD) • Power Supply Calibrator • Pressure Suit Circuit • Pressure System Control • Prime Service Contractor • Print Server Command • Problem-Solving Context • Processing and Spectral Control • Processor Support Circuit • Procurement Source Code • Product Service Center • Professional Support Centre • Profile Steering Committee • Program Schedule Chart • Program Science Center • Program Status Chart • Program Structure Code • Program Support Center • Program Support Communication • Program Support Contract • Program Support Contractor • Progress in Scientific Culture • Project Systems Control • Provide Station Control • Public Service Commission • Public Service Company

PSCA: PASS SMS Composition Aids • Pressure Suit Conditioning Assembly

PSCB: Padded Sample Collection Bag • Protected Step Control Block

PSCDS: Publications Spéciales du CDS Strasbourg (periodical code)

PSci: Pour la Science (periodical code)

PSCL: Programmed Sequential Control Language • Propellants System Cleaning Laboratory • Propellants System Components Laboratory

PSCM: Problem Space Computational Model • Progression-Southerly-Cyclonicity-Meridionality

PSCN: Permanent System Control Number • Preliminary Specification Change Notice • Program Support and Communications Network • Program Support Communications Network • Project Support Communications Network • Proposed Specification Change Notice

PSCNET: Pittsburgh Supercomputing Center Network

PSCNi: Program Support Communications Network Internet

PSCRD: Program Support Communications Requirement Document

PSCS: Pacific Space Centre Society (Canada-BC) • Program Support Control System

PSCU: Power Supply Control Unit • Propellant Storage and Control Unit

PSD: Panel on Satellite Dynamics • Particle Size Distribution • Patent Search Documentation • Pattern Standard Deviation • Phase-Sensitive Demodulator • Phase-Sensitive Detector • Philadelphia School District (USA-PA) • Photometric Sunspot Deficit • Photon-Sensitive Device • Photon Stimulated Desorption • Pitch Servo Drive • Platform-Specific Driver • Port Said (Egypt) (location identifier) • Position-Sensitive Detector • Positive Semi-Definite • Post Source Decay • Potsdamer Spektraldurchmusterung • Power Spectral Density • Power Spectrum Density • Power Systems Division • Prestation Spécifique Dépendance • Prevention of Significant Deterioration • Procedural Support Data • Processing Status Display • Program Status Document • Program Support Document • Projects and Systems Division • Propellant Storage Depot • Proportional Scintillation Detector • Pseudo-Single Domain • Pulse Shape Discrimination • Pulse Shape Discriminator

PSDC: Public Switched Digital Capability

PSDD: Preliminary System Design Description

PSDE: Payload and Spacecraft Demonstration and Experiment • Payload and Spacecraft Development and Experimentation

PSDF: Power Spectral Density Function • Printer Snap Definition File • Propulsion System Development Facility

PSDN: Packet Switched Data Network • Public Switched Data Network

PSDP: Payload Station Distribution Panel • Payload System Distribution Panel

PSDPC: Polar Satellite Data Processing Centre

PSDR: Planning and Scheduling Document Record • Public Sector Debt Requirement

PSDS: Public Switched Data Service • Public Switched Digital Service

PSDU: Power and Signal Distribution Unit • Power Switching Distribution Unit

PSE: Pacific Stock Exchange • Packet-Switch Exchange • Packet-Switching Exchange (*) • Partial Solar Eclipse • Passive Seismic Experiment • Payload Service Equipment • Payload Servicing Equipment • Payload Support Engineer • Payload Support Equipment • Peculiar Support Equipment • Periodic System of Elements • Physiological Systems Experiment • Pidgin Sign English • Pitch Steering Error • Planned Special Exposure • Polystyrène Expansible • Ponce Mercedita (Puerto Rico) (location identifier) • Probe Support Equipment • Problem-Solving Environment • Product Subsidy Equivalent • Professional Server Expert • Programming Support Environment • Project Support Environment • Psychological Stress Evaluator

PSEA: Pointing and Safemode Electronics Assembly • Pointing/Safing Electronics Assembly

psec: Picosecond

PSEP: Passive Seismic Experiments Package • Professional Student Exchange Program

PSERVER: Print Server

Pseud: Pseudonym • Pseudonymous

psf: Pounds per Square Foot

PSF: Pakistan Science Foundation • Passive Solar Foundation • Payload Structure Fuel • Permanent Swap File • Pittsfield (USA-MA) (location identifier) • Planetary Studies Foundation (USA-IL) • Planning Skeleton File • Point Spread Function • Print Services Facility • Probability of Stress Failure • Processing and Staging Facility • Processing and Storage Facility • Program for the Study of the Future • Programme Star File • Progressive Space Forum • Projection Specification Field • Propellant Servicing Facility • Punktbild-Streufunktion

PSFAM: Parameter Sensitive Frequency Assignment Method

PSFB: Plan Spécialisé de Fortes Précipitations

PSFC: Permanent Service on the Fluctuation of Glaciers • Power Supply Fuel Cell

PSFG: Permanent Service on Fluctuation of Glaciers

PSG: Passage • Passing • Petersburg (USA-AK) (location identifier) • Phrase-Structure Grammar • Power Subsystem Group • Programmable Sound Generator • Program Security Guard • Program Steering Group • Project Science Group

PSGD: Position Sensitive Germanium Detector

PSGE: Passage

Psgr: Passenger

PSGS: Palomar Scan Grism Survey

PSH: Program Support Handbook • Push • Sankt Peter (Germany) (location identifier)

PSHS: Philippine Science High School

psi: Pounds per Square Inch

PSI: Packetnet System Interface • Packet-Switched Interface • Pakistan Standards Institution • Pasni (Pakistan) (location identifier) • Paul Scherrer Institute (Switzerland) • Perceptive Solutions Inc. (USA) • Performance System International Inc. (USA) • Personalized System of Instruction • Photometric Sunspot Index • PI-class Science Instrument (SOFIA) • Planetary Science Institute (USA-AZ) • Plasma Source Instrument • Pollutant Standards Index • Power per Square Inch • Power Static Inverter • Present Serviceability Index • Pre-Shipment Inspection • Princeton Synergetics Inc. (USA-NJ) • Process System Index • Program Status Information • Project Starlight International

psia: Pounds per Square Inch - Absolute

Psic: Psicólogo

PSICP: Program Support Inventory and Control Point

psid: Pounds per Square Inch - Differential

PSID: PostScript Image Data

PSIF: Packet and Science Instrument File

psig: Pounds per Square Inch - Gauge

PSIG: Propulsion Systems Integration Group

PSII: Plasma Source Ion Implementation

psis: Pounds per Square Inch - Scaled

PSIS: Programming Systems Information System

PSIU: Packet Switch Interface Unit • Power and Signal Interface Unit

PSIWA: Penn State In-service Workshops in Astronomy (USA-PA)

PSJ: Physical Society of Japan • Poso (Indonesia) (location identifier)

PSJaJ: Physical Society of Japan Journal (periodical code)

PSK: Dublin - New River Valley (USA-VA) (location identifier) • Phase Shift Key • Phase Shift Keyed • Phase Shift Keying (*) • Power Supply Kit • Program Selection Key

PSKM: Phase-Shift Keying Modem

PSKMP: Photuris Session Key Management Protocol

PSL: Personal Launch System • Perth (UK-Scotland) (location identifier) • Photographic Science Laboratory • Physical Sciences Laboratory • Pipeline Scripting Language • Polystyrene Latex • Portable Standard LISP • Potential Source List • Power Source Logic • Pressure Seal • Pre-Stabilized Laser • Primary Standards Laboratory • Princeton Science Library • Private Sector Liquidity • Problem Specification Language • Problem Statement Language • Programming Systems Laboratory • Program Statement Language • Program Support Library • Propulsion Systems Laboratory • Protocol Specification Language • Public Sector Loan

PSLI: Pilot Site Licence Initiative

PSLIC: Pseudo-Stray Light Isochronal Contour

PSLLS: Pulse Solid-State Laser Light Source

PSLR: Portable Satellite Laser Ranging

PSLS: Programming Support Library System

PSLT: Programme Spatial à Long Terme

PSLV: Polar Satellite Launch Vehicle • Polar Space Launch Vehicle

Psm: Prism

P&SM: Procurement & Subcontract Management

PSM: Packet Service Module • Peak Selector Memory • Phase-Sensitive Modulation • Phase-Sensitive Modulator • Phase Shifter Module • Please See Me • Point Scattering Model • Point-Source Method • Portsmouth Pease (USA-NH) (location identifier) • Position Switching Module • Power Storage Module • Power System Module • Problem-Space-level Model • Procurement and Subcontract Management • Product Sales Manager • Product Support Manual • Program Support Management • Project Support Manager • Propellant Storage Module • Pyro Substitute Monitor • Pyrotechnic Substitute Monitor

PSMA: Prototype Secondary Mirror Assembly

PSME: Philippine Society for Mechanical Engineers

PSMG: Parallel Superconvergent Multigrid

PSML: Processor System Modelling Language

PSMN: Pôle Scientifique de Modélisation Numérique

PSMOS: Planetary Scale Mesopause Observing System

PSMSL: Permanent Service for Mean Sea Level

PSN: Packet Switched Network • Packet Switching Network • Palestine (USA-TX) (location identifier) • Pasifik Satelit Nusantara (Indonesia) • Piano Spaziale Nazionale (Italy) • Position • Processor Serial Number • Process Serial Number • Program Schedule Network • Program Summary Network • Public Switched Network

Psnl: Personal

PSNR: Power Signal-to-Noise Ratio

Pso: Paso

PSO: Pad Safety Officer • Paranal Science Operations (ESO) • Passeio • Pasto (Colombia) (location identifier) • Pauli Spin Operator • Pending Service Order • Permanent Solar Orientation • Persistent SQL Object • Planet Sensor Output • Point Symmetry Operation • Polar Stratospheric Ozone • Primary Standardization Office • Principal Scientific Officer • Project Science Office

PSOC: Preliminary System Operational Concept

PSOLA: Pitch-Synchronous Overlap and Add

PSOP: Payload Systems Operating Procedure • Peak Side On Pressure • Power System Optimization Program

P/SOS: Payload/Spacecraft Operations Simulation

PSP: Packet Switch Process • Palm Springs (USA-CA) (location identifier) • Payload Signal Processor • Payload Specialist Panel • Payload Support Plan • Peak Sideband Power • Peinture Sensible à la Pression • Performance Standards Program • Personal Software Products • Personal Success Program • Pierced Steel Planking (*) • Pierced Steel Plate • Planetary Scan Platform • Planetary Scan Probe • Planet Scan Platform • Planned Standard Programming • Platform Sensor Package • Positive Static Pressure • Postage Stamp Pipeline • Power System Planning • Precision Solar Pyranometer • Precision Spectral Pyranometer • Pré-Séquence Principale • Pressure Sensitive Paint • Private Sector Partnership • Programmable Signal Processor • Program Segment Prefix • Program Support Plan • Project Schedule Plan

PSPA: Pressure Static Probe Assembly

PSPC: Permanent South Pacific Commission • Position Sensitive Proportional Counter

PSPDC: Polar satellite Precipitation Data Centre

PSPDN: Packet Switched Public Data Network

PSPF: Potential Single-Point Failure

PSPFLI: Pulsed Single-Photon Fluorescence Lifetime Instrumentation

PSPL: Picosecond Pulse Labs Inc. (USA-CO) • Priced Spare Parts List

PSPM: Penn State Pulsar Machine

PSPMT: Position-Sensitive Photomultiplier Tube

PSPS: Passive Surface Pressure Sounder • Program Support Plan Summary

PSPT: Planar Silicon Power Transistor • Position Sensitive Phototube • Precision Solar Photometric Telescope

PSQ: Personnel Security Questionnaire • Philadelphia (USA-PA) (location identifier)

PSQA: Pageable System Queue Area

PSR: Packed Snow on Runway • Pad Safety Report • Payload Safety Review • Performance Status Report • Performance Summary Report • Pescara (Italy) (location identifier) • Phrase-Structure Rule • Polarimeter Sensitive Reflector • Precision Segmented Reflector • Pre-Ship Review • Problem Status Report • Processor State Register • Procurement Status Report • Production Safety Review • Programming Status Report • Programming Support Representative • Program Status Report • Program Status Review • Program Study Request • Program Support Requirement • Progress Summary Report • Project Safety Review • Project Scan Record • Project Status Report • Project Status Review • Project Summary Report • Project Support Room • Proton Storage Ring • Public Sector Research • Pulsar

P-SRAM: Pseudo-Static Random Access Memory

PSRAM: Pseudo-Static Random Access Memory

PSRB: President's Scientific Research Board (USA)

PSRC: Physical Sciences Resource Center

PSRD: Planetary Science Research Discovery • Program Support Requirements Document

PSRG: Planetary Science Research Group • Privacy and Security Research Group

PSRI: Planetary Sciences Research Institute

PSRO: Professional Standards Review Organization

PSRR: Preliminary Systems Requirements Review • Primary-Shadow RTO Replication

PSRS: Plant Service Recall System • Portable Seismic Recording System

PSRT: Passive Satellite Research Terminal

P&SS: Planetary and Space Science (periodical code) • Planning & Scheduling System

PSS: Packet-Switched Service • Packet Switching Service • Packet Switching Stream • Packet Switching System • Packet Switch Stream • Pad Safety Supervisor • Palomar Sky Survey • Panel Support Scientist • Pauli Spin Susceptibility • Payload Safing Sequence • Payload Specialist Station • Payload Status System • Payload Support System • Performance Summary System • Performance Support System • Photostationary State • Pipeline Subsystem • Planetary and Space Science • Planetary Scan System • Planning and Scheduling System • Portable Satellite Simulator • Portable Survival Shelter • Posadas (Argentina-MI) (location identifier) • Postscripts • Power Subsystem • Power Supply Section • Power System Simulator • Power System Stabilizer • Printing and Stationery Service • Procedures, Specifications and Standards • Professional Server Specialist • Professional Services Section • Programme des Sciences Spatiales • Propellant Supply System • Proprietary Software System • Propulsion Subsystem • Propulsion System Structure • Propulsion Support System

PSSC: Physical Science Study Committee

PSSG: POSIX Standards Style Guide • Protocol Standards Steering Group

PSSI: Pacific/Sulfur/Stratus Investigation

PSSM: Photographic Systems Simulation Model

PSSO: Permanent Solar-Stellar Orientation

PSSP: Payload Specialist Station Panel

PSSR: Performance Status Summary Report • Performance Status Summary Review • Problem Status and Summary Report

PSSS: Packet Scan Status Stack

PSSST: Project on the Social Studies of Science and Technology

PSST: Periodic Significant Scheduled Task

PSSU: Patch Survey and Switching Unit

PSSWG: Planetary System Science Working Group

Pst: Paste

P&ST: Physics & Space Technology

PST: Pacific Standard Time • Pacific Summer Time • Paired Selected Ternary • Parallel/Serial Technology • Paulucci Space Theatre (USA-MN) • Performance Specification Tree • Periodic Self-Test • Photometric Solar Telescope • Photometrisches Sonnenteleskop • Phototronics Solartechnik GmbH (Germany) • Physics and Space Technology • Point Source Transmission • Point Source Transmittance • Polar Star Tracker • Polar Stratospheric Telescope • Preliminary Systems Test • Pre-service School Teacher • Preston (Cuba) (location identifier) • Primary School Teacher • Product Support Team • Professional, Scientific and Technical • Project Science Team • Provincial Sales Tax

PSTA: Packaging Science and Technology Abstracts

PSTC: Public Service Training Council

PSTJ: Provence Sciences Techniques Jeunesse (France)

Pstl: Pistol • Postal

PSTL: Pressure Statis and Transient Launch

Pstn: Piston

PSTN: Public Switched Telephone Network

Psto: Puesto

PSTOL: PORTS System Test and Operations Language • Pseudo STOL

PSU: Passenger Service Unit • Penn State University (USA-PA) • Pennsylvania State University (USA-PA) • Power Supply Unit • Power Switching Unit • Primary Sampling Unit • Processor Speed Up • Process Support Unit • Program Storage Unit • Publications Services Unit • Public Service Union • Putussibau (Indonesia) (location identifier) • Pyrotechnic Switching Unit

PSV: Papa Stour (UK) (location identifier) • Photographic Spatial Volume • Pilotage Sans Visibilité • Planetary Space Vehicle • Public Service Vehicle

PSVD: Partial Singular Value Decomposition

Psvtn: Preservation

PSW: Passos (Brazil-MG) (location identifier) • Processus Status Word • Program Status Word

PSWO: Picture and Sound World Organization

PSX: Palacios (USA-TX) (location identifier)

PSY: Port Stanley (Falkland Islands) (location identifier)

Psych: Psychic • Psychical • Psychological • Psychologist • Psychology

Psychol: Psychological • Psychologist • Psychology

PSZ: Partially Stabilized Zirconia • Puerto Suarez (Bolivia) (location identifier)

p/t: Part-Time

pt: Pint • Portugal • Portuguese (ISO language code) • Pro Tempore

pT: Picotesla

Pt: Part • Patient • Payment • Petit • Platin • Platine • Platinum • Point • Pont • Poort • Port • President • Président • Puerto • Punt

P/T: Part-Time • Power Turbine • Prototype

P&T: Packaging & Transportation

PT: Brazil (international aircraft registration code) • Pacific Time • Package Table • Packaging and Transportation • Page Table • Paper Tape • Part Task • Part Time • Payload Type • Perte Totale • Perturbation Theory • Phase Transition • Photothermal • Physical Therapy • Physical Training • Physics Today (periodical) • Pint • Pitch Trim • Plein Traitement • Pneumatic Tube • Point • Point of Tangency • Point of Turn • Point of Turning • Porta • Portugal (ISO country code) • Postal Telegraph • Post Town • Power Transfer • Preferential Treatment • Pressure Transducer • Previously Treated • Primary Target • Primary Term • Printer Terminal • Procedure Turn • Productivity Team • Product Team • Proficiency Testing • Program Tab • Program Timer • Project Team • Propulsion Test • Protected Text • Protocol Translation • Protocol Translator • Publication Type • Punched Tape • Pupil Teacher • Purchase Tax

Pta: Peseta (currency) • Ponta • Pretoria (South Africa) • Puerta • Punta

PTA: Parent-Teacher Association • Personnel Telecommunications Assistant • Peseta (currency) • Phototransistor Amplifier • Pointing and Tracking Assembly • Port Alsworth (USA-AK) (location identifier) • Posts and Telecommunications Administration • Post-Test Analysis • Potential Toxic Area • Preferential Trade Area • Prepaid Ticket Advice • Program Time Analyzer • Property Transfer Authorization • Propulsion Test Article • Pulse Torquing Assembly

PTAC: Poids Total Autorisé en Charge

PTarO: Publications of Tartu Observatory (periodical code)

Ptas: Pesetas

PTAS: Performance Trend Analysis System

PTB: Pass-Through Bus • Payload Timing Buffer • Petersburg (USA-VA) (location identifier) • Physikalisch-Technische Bundesanstalt (Germany) • Pounds per Thousand Barrels • Producto Territorial Bruto • Project Testbed

Ptbl: Portable • Potable

PTBR: Punched Tape Block Reader

PTC: Pacific Telecommunications Council • Packet Transmission Controller • Parent-Teacher Consultation • Passive Thermal Conditioning • Passive Thermal Control • Personnel Transfer Capsule • Photographic Type Composition • Portable Temperature Controller • Port Alice (USA-AK) (location identifier) • Positive Temperature Coefficient • Positive Train Control • Power Transfer Coefficient • Presentation Technique Component • Pretaxi Clearance • Primary Training Centre • Programmable Test Control • Proof Test Capsule • Propellant Tanking Console • Propulsion Test Complex • Pulse Time Code

PTCC: Propulsion Test Control Center

PTCDA: Perylenetetracarboxylic Dianhydride

PTCE: Pilot Tropical Cirrus Experiment

PTCHY: Patchy

PTCR: Pad Terminal Connection Room

PTCS: Passive Thermal Control Section • Passive Thermal Control System • Payload Test and Checkout System • Planning, Training, and Checkout System • Pointing, Tracking, and Control System • Propellant Tanking Computer System

Ptd: Painted • Printed

PTD: Parafoil Technology Demonstration • Parallel Transfer Disk • Physical Theory of Diffraction • Port Alexander (USA-AK) (location identifier) • Preliminary Technical Directive • Programmable Threshold Detector • Provisioning Technical Documentation

PTDD: Parallel Transfer Disk Drive

PTDLDP: Pas Tombé De La Dernière Pluie

PTDOM: Pays, Territoires et Départements d'Outre-Mer

Pte: Plate • Pointe • Poniente • Ponte • Porte • Private • Puente

P&TE: Packaging & Transport Engineering

PTE: Escudo (Portugal) (ISO currency code) • Packaging and Transport Engineering • Page Table Entry • Peculiar Test Equipment • Personal Tax Edge • Poniente • Ponte • Portable Test Equipment • Port Stephens (Australia-NSW) (location identifier) • Portuguese Escudo (ISO currency code) • Post-Test Examination • Preliminary Test Estimator

PTER: Processed Term Error Rate

PTF: Malololailai (Fiji) (location identifier) • Payload Test Facility • Phase Transfer Function • Pilot Training Facility • Productividad Total de los Factores • Program Temporary Fix • Program Trouble Fix • Propellant Tank Flow

PTFE: Polytetrafluorethylene

Ptg: Pleistogene • Portugal • Portuguese • Printing

PTG: Personal Travel Guide • Pietersburg (South Africa) (location identifier)

PTGS: Palomar Transit Grism Survey • Portable Telemetry Ground Station

PTH: Plated-Through Hole • Port Heiden (USA-AK) (location identifier)

PThPh: Progress in Theoretical Physics (periodical code)

PThPS: Progress in Theoretical Physics Supplement (periodical code)

PTI: Palomar Testbed Interferometer • Parkes-Tidbindilla Interferometer (Australia) • Path Table Information • Payload Type Indicator • Personal TEX Inc. (USA-CA) • Physical Trait Identification • Physico-Technical Institute • Port Douglas (Australia-QLD) (location identifier) • Preliminary Test Information • Press Trust of India • Programmed Test Input • Public Tool Interface

PTIC: Program Track Interface Computer

PTIS: Programmed Test Input System • Program Test Input System • Propulsion Test Instrumentation System

PTIT: Planetary Target Implementation Team

PTJ: Portland (Australia-VIC) (location identifier)

PTK: Passport To Knowledge • Pontiac (USA-MI) (location identifier)

Ptl: Patrol

PTL: Payload Test Laboratory • Phase Tracking Loop • Photographic Technology Laboratory • Plasma Turbulence Layer • Port Armstrong (USA-AK) (location identifier) • Potential Termination Liability • Probabilistic Terminological Logic • Process and Test Language

PTLA: Pseudo Top-Level Aggregator

PTLS: Partial Total Least Squares

Ptly: Partly

PTLY: Partly

PTM: Packet Transfer Mode • Palmarito (Venezuela) (location identifier) • Pattern Transformation Memory • Pays Tiers Méditerranéens • Phase Time Modulation • Pneumatic Telescope Mast • Program Timing and Miscellaneous • Proof Test Model • Pulse Time Modulation • Pulse Time Multiplex • Pulse Transmission Mode

PTMLE: Preliminary Test Maximum-Likelihood Estimator

PTMP: Point-To-Multipoint

PTMS: Pattern Transformation Memory System • Precision Torque Measuring System

PTMU: Power and Temperature Monitor Unit

Ptn: Partition • Pattern • Portion

PTN: Morgan City (USA-LA) (location identifier) • Partly • Patterson Williams (USA-LA) (location identifier) • Physique et Techniques Nucléaires • Private Telecommunications Network • Procedure Turn • Public Telephone Network • Public Transportation Network

Ptnr: Partner

PTNX: Private Telecommunications Network Exchange

Pto: Portillo • Porto • Puerto • Puesto

PTO: Parent-Teacher Organization • Participating Test Organization • Part-Time Operation • Patent and Trademark Office (USA) • Pátio • Pato Branco (Brazil-PR) (location identifier) • Please Turn Over • Post and Telephone Operator • Power Takeoff • Power Test Operation • Proof Test Orbiter • Propellant Transfer Operation • Public Telecommunications Operator • Public Telephone Operator • Pulse Tube Refrigerator

PTOM: Pays et Territoires d'Outre-Mer

PTOP: Program Test and Operations Plan

PTOS: Papertape-Oriented Operating System

PTP: Packet Translator Process • Paper Tape Perforator • Paper Tape Printer • Paper Tape Punch • Peak-To-Peak • Piezoelectric Tilting Platform • Pointe-à-Pitre (Guadeloupe) (location identifier) • Point-To-Point • Postgraduate Training Partnership • Preliminary Task Plan • Preoperational Test Procedure • Primary Target Point • Programmable Text Processor • Program Task Planning

PTPS: Package Test Power Supply • Propellant Transfer Pressurization System

pt/pt: Point to Point

PTPU: Program Tape Preparation Unit

PTQ: Pôrto de Moz (Brazil-PA) (location identifier)

Ptr: Printer

PTR: Pacific Test Range • Paper Tape Reader • Perforated-Tape Reader • Photoelectric Tape Reader • Pleasant Harbor (USA-AK) (location identifier) • Pole Tip Recession • Portable Tape Recorder • Precision Transmitter Receiver • Preliminary Technical Report • Preliminary Technical Review • Preliminary Test Report • Printer • Processor Tape Read • Program Test Review • Program Trouble Report • Protocol de Terminal Réseau • Punched-Tape Reader

PTRA: Physics Teaching Resource Agent

PTRM: Partial Thermoremanent Magnetization

PTRSL: Philosophical Transactions of the Royal Society of London (periodical code)

Pts: Parts • Payments • Pesetas (currency) • Pints • Points • Ports • Portsmouth (UK-England)

PTS: Parallel Transaction Server • Payload Test Set • Payload Transportation System • Photothermal Spectroscopy • Phototypesetting • Pittsburgh (USA-KS) (location identifier) • Plasma Technology Section • Plasmaturbulenz-Schicht • Point de Transfert Sémaphore • Points • Polar Track Structure • Polished Thin Section • Portas • Positive Torque Signal • Power Transfer Switch • Practical Test Standard • Precision Timing System • Pressurized Thermal Shock • Principal Test Stand • Propellant Transfer System

PTSD: Post-Traumatic Stress Disorder

PTSE: PNNI Topology Sate Element

PTSF: Payload Trunnion Support Fixture

pts/hr: Parts per Hour

PtSi: Platinium Silicide

PTSI: Politique des Télécommunications et des Services de l'Information

Ptsmth: Portsmouth (UK-England)

PTSP: PNNI Topology Sate Packet

PTSR: Preliminary Technical System Review

PTSS: Precision Time Synchronization Service

Ptt: Partout • Petit • Pourtant

P&TT: Pointing & Target Tracking

PTT: Pacific Telephone and Telegraph Co. • Part Task Trainer • Performances and Phonograms Treaty • Platform Terminal Transmitter • Pointing and Target Tracking • Posterijen-Telegrafie-Telefonie • Postes-Télécommunications-Télédiffusion • Postes-Télégraphes-Téléphones • Post-Telegraph-Telephone • Pratt (USA-KS) (location identifier) • Projet de Transfert Technologique • Push To Talk

PTTC: Papertape Transmission Code

PTTI: Precise Time and Time Interval

pt-tm: Part Time

pttm: Part Time

PTTX: Phase Transfer Transmitter

PTU: Parallel Transmission Unit • Plan and Tilt Unit • Platinum (USA-AK) (location identifier) • Power Terminal Unit • Pression-Température-Humidité • Probe/Transducer Unit

PTUV: Pression-Température-Humidité-Vent

PTV: Passenger Transport Vehicle • Pathfinder Test Vehicle • Peak To Valley • Penetration Test Vehicle • Pitch Thrust Vector • Porterville (USA-CA) (location identifier) • Programmierte Textverarbeitung • Projection Television • Propulsion Test Vehicle • Prototype Test Vehicle • Public Television • Punched Tape Verifier

PTVA: Propulsion Test Vehicle Assembly

PTVC: Pitch Thrust Vector Control

PTVE: Propulsion Test Vehicle Engineering

PTW: Per Thousand Words • Playing To Win • Pottstown Limerick (USA-PA) (location identifier) • Primary Translation Word

PTWC: Pacific Tsunami Warning Center (USA-HI)

PTX: Pitalito (Colombia) (location identifier) • Pressure Transducer

Pty: Party • Principality • Property • Proprietary

PTY: Panama City - Balboa - Omar Torrijos (Panama) (location identifier) • Panama City - Tocumen (Panama) (location identifier)

PTZ: Pastaza (Ecuador) (location identifier)

p/u: Pick up

Pu: Plutonium • Purple

PU: Paid Up • Peripheral Unit • Physical Unit • Picked Up • Pick Up • Plant Unit • Polyurethane • Potchefstroom University (South Africa) • Potins d'Uranie • Power Unit

• Pressure Unit • Princeton University (USA-NJ) • Processing Unit • Production Unit • Propellant Unit • Propellant Utilization • Propulsion Unit • Prototype Unit • Public Utility • Pueblo

PUA: Profiling User Agent • Puas (Papua New Guinea) (location identifier)

PUAE: Peaceful Uses of Atomic Energy

Pub: Public • Publication • Publish • Published • Publisher • Publishing

PUB: Physical Unit Block • Public • Publication • Publish • Published • Publisher • Publishing • Pueblo (USA-CO) (location identifier)

Pubbl: Pubblicazione

Pubd: Published

Pub. Doc.: Public Document

Publ: Public • Publication • Publicity • Published • Publisher • Publishers • Publishing

PUBLICA: Public Libraries Concerted Action

Pubn: Publication

Pubr: Publisher

Pub. Wks.: Public Works

PUC: Pacific Union College (USA-CA) • Paper Under Consideration • Peripheral Unit Controller • Permanent Unit Code • Pick Up Car • Presse Universitaire Canadienne (Canada) • Price (USA-UT) (location identifier) • Processing Unit Cabinet • Public Utilities Commission • Public Utility Commission (USA)

PUCE: Pattern-Unit-Component-Evaluation • Pour l'Utilisation de Composants Électroniques

PUCS: Propellant Utilization Control System

PUCU: Propellant Utilization Control Unit

PUD: Pick Up and Delivery • Planned Unit Development • Puerto Deseado (Argentina-SC) (location identifier)

PUDT: Propellant Utilization Data Translator

Pue: Puebla (Mexico)

PUE: Puebla (Mexico) • Puerto Obaldia (Panama) (location identifier) • Puesta

PUEO: Probing the Universe with Enhanced Optics

PUF: Pau Pyrénées (France) (location identifier) • Presses Universitaires de France

PUFS: Programmer's Utility Filing System

PUG: Pascal Users' Group • Port Augusta (Australia-SA) (location identifier) • Prime Users Group • PROS Users' Guide

PUGS: Propellant Utilization and Gauging System

PUH: Pochutla (Mexico) (location identifier)

PUI: Pen User Interface • Perceptual User Interface • Pick-Up Ion • Pureni (Papua New Guinea) (location identifier)

PUJ: Punta Cana (Dominica) (location identifier)

PUK: Personal Unlocking Key • Pukarua (Tuamotu Islands) (location identifier)

PUL: Poulsbo (USA-WA) (location identifier) • Program Update Library • Propellant Utilization and Loading

PULS: Persoonsgericht Universitair Postdoc Stimulering • Propellant Utilization and Loading System

Pulsar: Pulsating Star

PULSAUR: Pulsating Aurora

pulv: Pulvis

PUM: Pomala (Indonesia) (location identifier) • Presses Universiatires de Montréal (Canada-QC)

PUMA: Pour l'Utilisation de Matériaux Avancés • Punching Machine

PUMAS: Practical Uses of Maths And Science

PUMS: Physical Unit Management Services • Public Use Microdata Samples

PUN: Physical Unit Number • Punia (Zaire) (location identifier)

Punc: Punctuation

Punj: Punjab (India)

PUO: Prudhoe Bay (USA-AK) (location identifier)

PUOW: Proposed Unit Of Work

Pup: Puppis (constellation)

PUP: PARC Universal Packet • Peripheral Unit Processor • Po (Burkina Faso) (location identifier) • Polytechnic University of the Philippines • Power Upgrade Program • Princeton University Press • Principal User Processor

PUPO: Pull Up - Push Over

PUQ: Punta Arenas - Presidente Ibañez (Chile) (location identifier)

Pur: Purchase • Purchaser • Purple

PUR: Payload Under Review • Pileup Rejection • Program of University Research • Puerto Rico (Bolivia) (location identifier)

Purch: Purchase • Purchaser

PUREX: Plutonium and Uranium Extraction

PURL: Persistent Uniform Resource Locator

Purp: Purple • Purpose

PUS: Packet Utilisation Standard • Parliamentary Undersecretary • Performance Upgrade Socket • Permanently Unfit for Service • Permanent Undersecretary • Processor Upgrade Pocket • Propellant Utilization System • Public Understanding of Science • Pusan (Korea-ROK) (location identifier)

PUSNO: Publications of the US Naval Observatory - Second Series (periodical code)

PUSPIPTEK: Pusat Penelitian Ilmu Pengetahuan dan Teknologi (Indonesia)

PUSS: Pallet Utility Support Structure

PUSSYCATS: Pedestrian Urban Safety System and Comfort at Traffic Signal

PUST: Public Understanding of Science and Technology

PUSV: Packet Utilisation Standard Validation

PUT: Programmable Unijunction Transistor • Puttaparthi (India) (location identifier)

PUTT: Propellant Utilization Time Trace

PUU: Puerto Asis (Colombia) (location identifier)

PUV: Poum (New Caledonia) (location identifier) • Propellant Utilization Valve

PUW: Planetarium und Uraniasternwarte Wien (Austria) • Pullman Moscow (USA-WA) (location identifier)

PUY: Pula (Croatia) (location identifier)

PUZ: Puerto Cabezas (Nicaragua) (location identifier)

pv: Cauchy Principal Value • Per Value

P/V: Photovoltaic • Portable/Vehicle

PV: Parallel Vector • Parity Violation • Peak to Valley • Peculiar Velocity • Pequeña Velocidad • Performance Verification • Petite Vitesse • Photovoltaic • Piccola Velocità • Pixel Value • Planetary Vehicle • Plan View • Plena Vortaro • Potential Velocity • Potential Vorticity • Power Voltage • Pressure Vessel • Pressurization Valve • Prevalve • Printed Version • Procès-Verbal • Program Visualization • Pyro Valve • Pyrotechnic Valve

PVA: Perigee Velocity Augmentation • Pivalic Acid • Polyvinyl Acetate • Polyvinyl Alcohol • Population Viability Analysis • Portable Voice Amplifier • Positive Vorticity Advection • Preburner Valve Actuator • Propellant Valve Actuator • Providencia (Colombia) (location identifier)

PVAA: Pomona Valley Amateur Astronomers (USA-CA)

PVASI: Pulsating Visual Approach Slope Indicator

PVB: Polyvinyl Butyral

PV&C: Performance Verification & Calibration

PVC: Performance Verification and Calibration • Permanent Virtual Circuit • Permanent Virtual Connection • Pigment Volume Concentration • Polychlorure de Vinyle • Polyvinylchloride • Position and Velocity Computer • Potential Volume Change • Pressure Vacuum Chamber • Pressure Vacuum Compensator • Pressure-Volume Compensated • Provincetown (USA-MA) (location identifier)

PVCS: Popular Version Control System

PV&D: Purge, Vent & Drain

PVD: Pairwise Velocity Dispersion • Paquet avec Valeur Déclarée • Passive Vibration Damping • Pays en Voie de Développement • Physical Vapour Deposition • Plan View Display • Plasma Vapor Deposition • Primary Volume Descriptor • Prise de Vues Directe • Providence (USA-RI) (location identifier) • Pulsed Voltammetric Detection • Purge-Vent-Drain

PVDF: Polyvinyl Difluoride • Polyvinylidene Fluoride

PVDS: Propellant Vent and Drain System

P&VE: Propulsion & Vehicle Engineering

PVE: El Porvenir (Panama) (location identifier) • Photovoltaic Effect • Point de Vente Électronique • Propulsion and Vehicle Engineering

PVED: Propulsion and Vehicle Engineering Division

PVF: Peripheral Visual Field • Placerville (USA-CA) (location identifier) • Polyvinyl Fluoride

PVFS: Parallel Virtual File System

PVGC: Pioneer Venus Gas Chromatograph

PVH: Pôrto Velho - Belmonte (Brazil-RO) (location identifier) • Premier Vol Habité

PVI: Paranavaí (Brazil-PR) (location identifier) • Perpendicular Vegetation Index • Point of Vertical Intersection

PVK: Polyvinylcarbazole • Preveza Lefkas (Greece) (location identifier)

PVL: Parameter Value Language • Prevail

PVLA: Purchase Versus Lease Analysis

PVLAN: Protocol Virtual Local Area Network

PVM: Parallel Virtual Machine • Particle Volume Monitor • Pass-through Virtual Machine • Potentiometric Voltmeter • Projection Video Monitor

PVN: Pleven (Bulgaria) (location identifier) • Private Virtual Network

Pvnt: Prevent

Pvntv: Preventive

PVO: Pioneer Venus Orbiter • Portoviejo (Ecuador) (location identifier) • Private Voluntary Organization

PVOC: Polar Volatile Organic Compound

PVOUVS: Pioneer Venus Orbiter Ultraviolet Spectrometer

PVP: Packet Video Processor • Parallel Vector Processing • Parallel Vector Processor • Plasma Vaporization Process • Precio de Venta al Público • Prix de Vente Public

PVPP: Polyvinylpolypyrrolidon

PV/PS: Pressure Vessel and Pressure System

PVPS: Pressure Vessel and Pressure System

PVR: Photovoltaic Radiator • Precision Voltage Reference • Procedure Validation Report • Puerto Vallarta (Mexico) (location identifier)

PVRD: Purge-Vent-Repressurize-Drain

PVS: Parameter Variation Space • Performance Verification System • Performance Visualization System • Personal Vision System • Physical Vapor Synthesis • Physical Volume Repository • Pressure Vacuum System • Program Validation Service • Project Visualization System • Propellant Venting System • Provideniya (Russia) (location identifier)

PVSG: Penobscot Valley Star Gazers (USA-ME)

PVSM: Periodic Volume Stimulus Method

Pvt: Private

PVT: Page View Terminal • Par Voie Télégraphique • Passive Visible Tracker • Performance Verification Test • Physical Vapor Transport • Point of Vertical Tangent • Polyvinyltoluene • Position-Velocity-Time • Position-Vitesse-Temps • Pre-flight Verification Test • Pressure-Volume-Temperature • Previous Value Table • Private • Probe Velocity Transducer • Projet de Validation Technologique • Pyrotechnic Verification Test

PVTCS: Photovoltaic Thermal Control System

PVTOS: Physical Vapor Transport of Organic Solids

PVTR: Portable Video Tape Recorder

PVU: Potential Vorticity Unit • Provo (USA-UT) (location identifier)

PVV: Pressure, Vent and Vacuum

PVW: Plainview - Hale County (USA-TX) (location identifier)

PVWA: Planned Value of Work Accomplished

PVWS: Planned Value of Work Scheduled

PVX: Provedenia (Russia) (location identifier)

PVY: Pope Vanoy (USA-AK) (location identifier)

PVZ: Painesville Casement (USA-OH) (location identifier)

pW: Picowatt

P&W: Pratt & Whitney

P/W: Password

PW: Palau (ISO country code) • Password • Pawnee • Per Week • Plain Washer • Planetary Wave • Plane Wave • Plan Washer • Positionswinkel • Precipitable Water • Precipitation Water • Printed Wiring • Public Works • Publieke Werken • Pulse Width

PWA: Oklahoma City - Wiley Post (USA-OK) (location identifier) • Permittivity and Wave Analyzer • Pratt and Whitney Aircraft • Printed Wire Assembly • Printed Wiring Assembly • Private Write Area • Probabilistic Web Algebra • Product Work Authorization • Project Work Authorization • Public Works Administration

PWB: Printed Wire Board • Printed Wired Board • Printed Wiring Board • Programmer's Workbench

PWBA: Printed Wiring Board Assembly

PWBS: Program Work Breakdown Structure

PWC: Pacific Weather Centre • Physical Work Capacity • Proportional Wire Chamber • Proportional Wired Chamber • Pulse-Width Coded

pwd: Password

PWD: Participatory Work Design • Plentywood Sherwood (USA-MT) (location identifier) • Pre-White Dwarf • Print Working Directory • Proceed With Design • Public Works Department

PWDS: Protected Wire Distribution System

PWE: Pauli-Weisskopf Equation • Pevek (Russia) (location identifier) • Pulse-Width Encoder

PWEA: Printed Wiring and Electronic Assemblies

PWEAR: Preliminary WBS Element Audit Review

PWEN: Physics World Electronic News

PWF: Photoelectric Work Function • Propellant Weight Fraction • Pulse Wave Form

PWFS: Peripheral Wavefront Sensor

PWG: Parallels Working Group • Payload Working Group • Permanent Working Group

PWI: Pawi (Ethiopia) (location identifier) • Plasma Waves Instrument • Plasma Waves Investigation • Public Windows Interface

PWIA: Plane Wave Impulse Approximation

PWK: Chicago Palwaukee (USA-IL) (location identifier)

PWL: Password List • Purwokerto (Indonesia) (location identifier)

PWLA: Portable Work Light Assembly

PWLL: Profile Word Logic List

PWM: Portland (USA-ME) (location identifier) • Probability-Weighted Moment • Pulse-Width Modulated • Pulse-Width Modulation • Pulse-Width Modulator

PWMD: Printed Wiring Master Drawing

PWMR: Pulse Width Modulated Regulator

PWN: Pitts Town (Bahamas) (location identifier)

PWO: Pweto (Zaire) (location identifier)

PWP: Planning Work Package • Portable Work Platform

PWQ: Pavlodar (Kazakhstan) (location identifier)

Pwr: Plain with Wrinkle Ridges • Power

PWR: Port Walter (USA-AK) (location identifier) • Power • Pressurized-Water Reactor • Program Work Requirement • Publication Work Request

PWRI: Public Works Research Institute

Pwr. Sup.: Power Supply

PWS: Parallel Working System • Peer Web Services • Plane-Wave Spectrum • Plasma and Wave System • Plasma Wave Subsystem • Possible World Semantics • Predictive Wind Shear • Preliminary Work Statement • Pricing Work Statement • Programmable Workstation • Program Work Statement • Public Water System

PWSCS: Programmable Workstation Communication Services

PW&SO: Publications of the Warner & Swasey Observatory (periodical code)

pwt: Pennyweight

PWT: Bremerton (USA-WA) (location identifier) • Progressive Wave Tube • Propulsion Wind Tunnel

Pwtr: Pewter

PWV: Precipitable Water Vapor • Pretoria-Witwatersrand Vereeniging (South Africa) • Pulse Wave Velocity

PWW: Project West Wing

PX: Physical Examination • Pixel • Please Exchange • Post Exchange • Primary Index • Private Exchange

PXL: Polacca (USA-AZ) (location identifier) • Pulsed Xenon Laser

PXLS: Pulsed Xenon Light Source

PXLSS: Pulsed Xenon Light Source System

PXM: Puerto Escondido (Mexico) (location identifier)

PXO: Porto Santo (Madeira Islands) (location identifier)

PXS: Pulsed Xenon System

PXSS: Pulsed Xenon Solar Simulator

pxt: Pinxit

PXU: Pleiku (Viet Nam) (location identifier) • Portable X-ray Unit

py: Paraguay • Pyridine

P-Y: Pitch-Yaw

P/Y: Pitch/Yaw

PY: Paraguay (international vehicle registration code) • Paraguay (ISO country code) • Pygmy • Pitch-Yaw • Prior Year • Program Year • Publication Year

PYA: Puerto Boyaca (Colombia) (location identifier)

PYB: Jeypore (India) (location identifier)

PYC: Playon Chico (Panama) (location identifier)

PYE: Penrhyn Island (Cook Islands) (location identifier)

PYerO: Publications of Yerkes Observatory (periodical code)

PYF: French Polynesia (ISO country code) • Polynesia (French_) (ISO country code) • Polynésie Française (ISO country code)

PYG: Guarani (Paraguay) (ISO currency code)

PYH: Puerto Ayacucho (Venezuela) (location identifier)

PYJ: Polyarnyj (Russia) (location identifier)

PYL: Perry Island (USA-AK) (location identifier)

PYM: Plymouth (USA-MA) (location identifier)

PYME: Pequeña Y Mediana Empresa

Pymt: Payment

PYN: Payan (Colombia) (location identifier)

PYO: Pick Your Own • Putumayo (Ecuador) (location identifier)

PYP: Photoactive Yellow Protein

Pyr: Pyramid

P-Y-R: Pitch-Yaw-Roll

PYR: Pitch-Yaw-Roll • Prior Year Report • Pyrgos (Greece) (location identifier) • Pyro Release • Pyrotechnic Release

PYREX: Pyrénées Experiment

Pyro: Pyrotechnics

Pyrotech: Pyrotechnics

PYS: Probably Young System

PYTC: Parent-Year Tutor Consultation

PYunO: Publications of Yunnan Observatory (periodical code)

PYV: Yaviza (Panama) (location identifier)

Pyx: Pyxis-Pyxidis (constellation)

PYX: Pattaya (Thailand) (location identifier)

Pz: Palaeozoic

PZ: Phase Zero • Perem. Zvezdy (periodical code) • Suriname (international aircraft registration code)

Pza: Piazza • Plaza

PZA: Paz de Ariporo (Colombia) (location identifier) • Postzeitungsamt

PZAA: Polarized Zeeman Atomic Absorption

PZB: Pädagogische Zentralbibliothek • Pietermaritzburg (South Africa) (location identifier)

PZC: Point of Zero Charge

PZE: Penzance (UK-England) (location identifier)

PZH: Zhob (Pakistan) (location identifier)

PZJA: Protected Zone Joint Authority

PZL: Phinda (South Africa) (location identifier)

Pzla: Plazuela

PZO: Puerto Ordaz (Venezuela) (location identifier)

PZP: Phase Zero Program

PZT: Lead Zirconate Titanate • Photographic Zenithal Telescope • Photographic Zenithal Tube • Photographic Zenith Telescope • Photographic Zenith Tube • Piezo Translator

PZU: Port Sudan (Sudan) (location identifier)

PZY: Piestany (Slovakia) (location identifier)

P2: Papua New Guinea (international aircraft registration code)

P2C2G: Process Too Complicated To Explain

P2I: Preplanned Productivity Increase

P2N: Physique Neutronique et Nucléaire

P2PP: Phase Two Proposal Preparation

P3: Perception and Performance Prototyper • Platform for Privacy Preferences

P3I: Preplanned Product Improvement • Preplanned Production Improvement

P3M: Particle-Particle/Particle-Mesh

P3P: Platform for Privacy Preferences Project

P4: Aruba (international aircraft registration code)

Q

q: Charge of Particle • Circular Wave Vector • Constant of Proportionality • Deceleration Constant • Density of Heat Flow Rate • Dynamic Pressure • Electric Charge • Generalized Coordinate • Heat Current Density • Heat Flow Rate • Heat Flux • Local Speed • Nodal Displacement • Partition Function • Perihelion Distance • Pitch Angular Velocity • Propagation Vector • Quadrans • Quadrat • Quaestio • Quark • Quart • Quarter • Quarterly • Quarto • Quasi • Que • Query • Question • Quick • Quintal • Quire • Quit • Relative Quantum Efficiency • Slowing Down Density • Solar Forcing Ratio • Specific Humidity • Squall

Q: Aphelion Distance • Desintegration Energy • Electric Charge • Electricity Quantity • Ensemble des Nombres Rationnels • Five Hundred (Roman numeral) • Gauss Vector • Heat • Heat Addition • Heating Rate • Heat Quantity • Injection Rate of Relativistic Electrons • Integral Heat Load • Light Quantity • Nuclear Quadrupole Moment • Partition Function • Pitch Rate • Pulsational Constant • Quadrupole Moment • Qualification • Qualified • Qualität • Qualité • Quality • Quality Factor • Quantität • Quantité • Quantity • Quantity of Electricity • Quantity of Heat • Quantity of Light • Quart • Quartal • Quarterly • Quartier • Quarto • Quarz • Quaternary • Québec (international alphabet) • Québec (Canada) • Queen • Queensland (Australia) • Query • Question • Quetzal (currency) • Queue • Quintal • Quintus • Quit • Quitter • Radiant Energy • Reaction Energy • Reactive Power • Set of Rational Numbers • Solution Vector • Squall • Throughput • 500 (Roman numeral)

qa: Quick Assembly

Q&A: Qualification & Acceptance • Question & Answer

Q/A: Qualification Approval • Question/Answer • Questioning/Answering

QA: Qatar (ISO country code) • Qualification Approval • Quality Assurance • Quality Audit • Question and Answer • Queue Arbitrated • Quiescent Aerial • Quiescent Antenna

QAA/P: Quality Assurance Assessment and Plan

QAAP: Quality Assurance Assessment and Plan

QAC: Qatar (ISO country code) • Quality Assurance Chart

QACAD: Quality Assurance Corrective Action Document

QACC: Quality Assurance Consultative Committee

QAD: Quality Assurance Division

QADS: Quality-Assurance Data System

QAF: Query Absolute Frequency

QAFM: Quality Assurance Forms Manual

QAI: Quality Assurance Institute

QALC: Quasar Absorption Line Cloud

QAM: Quadratic Amplitude Modulation • Quadrature Amplitude Modulation (*) • Quality Assurance Manual • Queued Access Method

QAMS: Quality Assurance Management Staff

QANGO: Quasi-Autonomous Non-Government Organisation

QANTAS: Queensland And Northern Territory Airline System

QAOP: Quality Assurance Operating Procedure

QAP: Qualification Acceptance Program • Qualification and Acceptance Program • Quality Assurance Procedure • Quality Assurance Product

QApMa: Quarterly of Applied Mathematics (periodical code)

QAQC: Quality Assurance - Quality Control

QA/R: Quality Assurance/Reliability

QAR: Quality Assurance/Reliability • Quality Assurance Report • Quality Assurance Responsible • Quasi-Adiabatic Representation • Quick-Access Record • Quick-Access Recorder • Quick-Access Recording • Rial (Qatar) (ISO currency code)

QAS: Quaternary Ammonium Sulphanilamide • Question Answering System

QASAR: Quality and Safety Achievement Recognition (award)

QASL: Quebec Association of School Librarians (Canada-QC)

QA&T: Quality Assurance & Test

QAT: Qatar (ISO country code) • Quality Assurance and Test • Query Alias Table

QATS: Quality Assurance Test Suite

QATT: Qualification for Acceptance Thermal Testing

QAU: Quality Assurance Unit

QAVT: Qualification Acceptance Vibration Test • Qualification for Acceptance Vibration Testing

QAW: Anniston - Fort McLellan (USA-AL) (location identifier)

QB: Quality Board • Quick Basic

Qbc: Québec (Canada)

QBC: Bella Coola (Canada-BC) (location identifier) • Query By Committee

QBD: Query By Diagram

QBE: Query By Example

QBF: Query By Form

QBFJOTF: Quick Brown Fox Jumped Over The Fence

QBFJOTLD: Quick Brown Fox Jumped Over The Lazy Dog

QBE: Query By Example

QBF: Brack (Libya) (location identifier) • Query By Form

QBIC: Query By Image Content

QBO: Quasi-Biennial Oscillation

QBS: Quality-Based Selection • Quick Blue Survey

QBSA: Quarterly Bulletin on Solar Activity

QBT: Quadrax Biaxial Tape • Query By Table

QBX: Quick Basic Extended

Q/C: Quality Control • Quality Controller

QC: Qualification Course • Quality Circle • Quality Control • Quality Controller • Quantum-Cascade • Quasi-Continuous • Québec (Canada) • Query Case • Query Coordinator • Quick Change • Quit Claim

QCA: Qualification and Curriculum Authority (UK) • Quality Control Analysis

QCAC: Queen's College Astronomers' Club (Hong Kong/PRC)

QCAM: Queued Communications Access Method

QCAS: Quad Cities Astronomical Society (USA-IA) • Quality Control and Applied Statistics

QCB: Quality Control Board • Queue Control Block

QCC: Quality Control Committee • Quantum Computing and Communications • Quenched Carbonaceous Composite

QCCOD: QMW-Cambridge-CalTech-Oxford-Durham

QCD: Quality Control Data • Quantenchromodynamik • Quantum Chromodynamics

QCDP: Quality Color Dithering Process

QCDR: Quality Control Deficiency Report

QCE: Quality Control and Evaluation • Quality Control Engineering • Quality Control Expert

QCF: Quasi-Continuous Fragmentation

QCGD: Quality Circle Gesellschaft Deutschland

QCI: Quality Control Information

QCIF: Quarter Common Interchange Format • Quarter Common Intermediate Format

QC Is: Queen Charlotte Islands (Canada)

QCL: Quality Control Level

qcm: Quadratzentimeter

QCM: Quality Control Manager • Quality Control Manual • Quark Cluster Model • Quartz Crystal Microbalance • Quartz Crystal Monitor • Question à Choix Multiple • Questionnaire à Choix Multiple

QCO: Quality Control Officer • Quality Control Organization • Quartz Crystal Oscillator

Qconque: Quiconque

QCOP: Quality Control Operating Procedure

QCP: Quality Control Procedure

QCQP: Quadratically Constrained Quadratic Program

QCR: Quality Control Reliability • Quality Control Report • Quality Control Review • Quality Control Room

QCRT: Quick Change in Real Time • Quick Change Real Time

QCS: Qualified Calibration Source • Quality Control Standard • Quality Control System

QCT: Quality Control Technology • Quantitative Computed Tomography • Quasi-Classical Trajectory

QCTR: Quality Control Test Report

QCU: Quartz Crystal Unit

qd: Quaque Die • Quasi Dicat • Quasi Dictum

Qd: Quand

Q&D: Quick & Dirty

QD: Quantum Dot • Quaque Die • Query Decomposition • Quick and Dirty • Quick Disconnect

QDA: Quadratic Discriminant Analysis • Quantitative Descriptive Analysis • Quantity Discount Agreement

QDBMS: Query for Database Management System

qdbv: Quod Deus Bene Vertat

QDC: Charge to Digital Converter

QDE: Quantitative Differential Equation • Quantum Detection Efficiency

QDHT: Quasi-Discrete Hankel Transform

Qdisc: Quick Disconnect

QDLL: Quick Data Definition Language

qdm: Quadratdezimeter

QDM: Dammam (Saudi Arabia) (location identifier)

QDMBPT: Quasi-Degenerate Many-Body Perturbation Theory

QDN: Query Direct Number • Quick Disconnect Nipple

QDO: Quantum Defect orbital

QDOS: Quick and Dirty Operating System

QDOT: Queen Mary and Westfield - Durham - Oxford - Toronto

QDP: Qualified Data Package • Quick and Dandy Plotter

QDR: Quadrennial Defense Review • Qualification Design Review

QDRI: Qualitative Development Requirement Information

Qdrnt: Quadrant

Qdrtr: Quadrature

QDS: Quality Data System • Quick and Dirty Spectrograph

QDTA: Quantitative Differential Thermal Analysis

QDU: Düsseldorf Hauptbahnhof (Germany) (location identifier) • Düsseldorf Hbf. (Germany) (location identifier) • Dusseldorf Main Station(Germany) (location identifier) • Quantized Distortion Unit

QD3D: Quickdraw Tridimensional

qe: Quod Est

QE: Quadranteinheit • Quadrant Elevation • Quality Engineer • Quality Engineering • Quality Evaluation • Quantum Efficiency • Quantum Electronics • Query Expansion

QEC: Quick Engine Change

qed: Quod Erat Demonstrandum

QED: Quality Education Data • Quantenelectrodynamik • Quantum Electrodynamics • Quantum Electronics Division • Quasar Energy Distribution • Query-Explore-Demonstrate • Quick Easy Disk • Quick Epicenter Determination • Quick-Erection Dome

QEDR: Quick-look Experiment Data Record

qef: Quod Erat Faciendum

QEH: Quantum Efficiency Hysteresis

qei: Quod Erat Inveniendum

QEII: Queen Elizabeth II

QEL: Quality Evaluation Laboratory • Queue Element

QELS: Quality Evaluation of Literature System • Quantum Electronics and Laser Science

QEM: Quality Education for Minorities

QEMM: Quarterdeck Expanded Memory Manager • Quarterdeck Extended Memory Manager

QEP: Query Execution Plan

QEPCAD: Quantifier Elimination by Partial Cylindrical Algebraic Decomposition

QERD: Quick-look Experiment Data Record

QEST: Quality Evaluation System Test

QET: Quantum Effect Transistor

QF: Qualitätsfaktor • Quality Factor • Quality Form • Quick Fix

QFA: Quick File Access

QFAA: Quartz Furnace Atomic Absorption

QFC: Quantum Flow Control

QFD: Quality Functional Deployment • Quality Function Deployment • Quality Function Development • Quantum Flavourdynamics

QFF: Quadrupole Flip-Flop

QFH: Frederikshavn (Denmark) (location identifier)

QFM: Qualification Flight Model

QFP: Quad Flat Pack • Quad Flat Package • Question Fréquemment Posée

QFT: Quantum Field Theory

QG: Qualitätsgruppe • Quantum Gravity • Quartier Général • Quasigeostrophic

QGP: Quark-Gluon Plasma

QGS: Quantity Gauging System

QGSM: Quark-Gluon String Model

QGV: Frankfurt - Neu-Isenburg (Germany) (location identifier)

qh: Quaque Hora

QH: Quadruple Hydrant

QHE: Quantum Hall Effect

QHJ: Hjørring (Denmark) (location identifier)

QHO: Quantum Harmonic Oscillator

QHR: Quantized Hall Resistance

QI: Quality Improvement • Quality Index • Quality Information • Quantity Index • Quartz Iodine • Quasi-Inertial • Quotient d'Intelligence • Quotient Intellectuel

QIC: Quality Information Control • Quality Information using Cycle • Quarter-Inch Cartridge

QICC: Quarterly Index to Current Contents

qid: Quater in Die

QIEC: Quasi-Isoenergetic Complex

QIF: Quicken Interchange File • Quicken Interchange Format

QINMS: Quadrupole Ion Neutral Mass Spectrometer

QIO: Queue Input/Output

QIP: Quad Inline Package • Quartz Iodine Polarized • Quiescat In Pace

QIRA: Quantitative Infrared Analysis

QIS: Quellenorientiertes Informationssystem

QISAM: Queued Indexed Sequential Access Method

QISM: Quantum Inverse Scattering Method

QIT: Quadrupole Ion Trap • Quality Improvement Team • Quantum Information Theory • Queensland Institute of Technology (Australia-QLD)

QIV: Qualité Image Végétation

QIZ: Quellenorientiertes Informationszentrum

QJ: Quarterly Journal

QJLC: Quarterly Journal of the Library of Congress

QJMAM: Quarterly Journal of Mechanics and Applied Mathematics (periodical code)

QJRAS: Quarterly Journal of the Royal Astronomical Society (periodical code)

QJRMS: Quarterly Journal of the Royal Meteorological Society (periodical code)

Qk: Quick

QKL: Cologne Main Station (Germany) (location identifier) • Köln Hauptbahnhof (Germany) (location identifier) • Köln Hbf. (Germany) (location identifier)

qkm: Quadratkilometer

QKU: Cologne-Deutz (Germany) (location identifier) • Köln-Deutz (Germany) (location identifier)

ql: Quantum Libet • Quintal

Q/L: Quick Look

QL: Quantitative Literacy • Queensland (Australia) • Query Language • Quick Look

QLA: Quebec Library Association (Canada-QC) • Quick-Look Analysis • Quick-Look Assessment

QLAP: Quick-Look Analysis Program

QLB: Quasi-Lower Bound

Qlconque: Quelconque

Qld: Queensland (Australia)

QLD: Queensland (Australia)

QLDA: Quick-Look Data Analysis

QLDAC: Quick-Look Data Analysis Center

QLDS: Quick-Look Data Station

QLF: Quasar Luminosity Function • Quick-Look Facility

QLI: Quality-of-Life Index • Query Language Interpreter • Quick-Look Image

QLIRS: Quick-Look Image Recording System

QLIS: Quality of Life in Information Society

QLLC: Qualified Logical Logical Link

QLM: Qualified Left Monotonicity • Quasi Littrow Mode

QLN: Quantitative Links Network

QLOPR: Quick-Look Product

QLPS: Quick-Look Processing System

QLS: Quick-Look Station

QLT: Quasi-Linear Theory

Qlty: Quality

QLX: Lauterach (Austria) (location identifier)

Qly: Quarterly

qm: Quadratmeter • Quaque Mane • Quomodo • Quoque Matutino

Qm: Quintal Métrico

QM: Qualification Model • Qualification Motor • Quality Management • Quality Memorandum • Quality Monitoring • Quantenmechanik • Quantum Mechanics • Quark Model • Query Manager

QMC: Quantum Monte Carlo • Queen Mary College (UK-England)

QMCMV: Queen Mary and Westfield College (Starlink internal code)

QMCS: Quantitative Methods and Computer Science

QMD: Quantum Molecular Dynamics

QMF: Quadrature Mirror Filter • Quadrupole Mass Filter • Query Management Facility

QMG: Qualité Maximale Garantie

QMI: Quality Management Institute • Quotidien des Marchés de l'Informatique

QMIPS: Quantitative Methods In Parallel Systems

QML: Qualified Manufacturing Listing

qmm: Quadratmillimeter

QMM: Marina di Massa (Italy) (location identifier)

QMP: Quality Measurement Plan

QMQ: Murzuq (Libya) (location identifier)

QMR: Qualitative Material Requirement • Quasi-Minimal Residual

QMRS: Quasi-Minimal Residual Squarred

QMS: Quadrupole Mass Spectroscopy • Qualification Model Spacecraft • Quality Management System • Queue Mess System

QMV: Montvale (USA-NY) (location identifier)

QMW: Queen Mary and Westfield • Queen Mary and Westfield College (Starlink internal code)

QMWC: Queen Mary and Westfield College (Starlink internal code)

qn: Quaque Nocte

Qn: Queen • Quelqu'un • Question • Quotation

QN: Quantum Number • Queuing Network • Quiescent Nova • Quinoline

QNA: Qatar News Agency • Quarterly National Accounts

QND: Quantum Non-Demolition

QNM: Queuing Network Model

QNS: Quantity Not Sufficient

Qnsld: Queensland (Australia)

Qnty: Quantity

QNY: New York - Marine (USA-NY) (location identifier)

QOF: Kiev Borispol (Ukraine) (location identifier)

QOH: Kiev Hotel Rus (Ukraine) (location identifier)

QOR: Qualitative Operational Requirement • Queue Owning Region

QORGS: Quasi-Optimal Rendezvous Guidance System

QoS: Quality of Service

QOS: Quality Of Service • Query Optimization System

QOT: Qualification Of Tooling • Quasi-Optical Technique

qp: Quantum Placet

QP: Quadratic Programming • Qualification Pay • Qualification Proposal • Quasi-Periodic • Query Processing •

Query Processor • Quick Propagation • Quiescent Prominence

QPAF: Quarter Panel Alignment Fixture

QPD: Quality Planning Document

QPE: Query by Pictorial Example

QPG: Singapore - Paya Lebar (Singapore) (location identifier)

QPIG: Quadrupole Ion Guide

q. pl.: Quantum Placet

QPL: Qualified Parts List • Qualified Products List

QPLT: Quiet Propulsion Lift Technology

QPM: Quality Program Manager • Quick Panel Manager

QPO: Quasi-Periodic Oscillation

QPOE: Quick Position-Ordered Event (*) • Quick Position-Oriented Event

QPP: Quality Program Plan

QPR: Quality Project Requirement • Quarterly Progress Report

QPRD: Quality Planning Requirements Document • Quality Program Requirements Document

QPRH: Quality Planning Requirements Handbook

QPS: Quality Planning Specification • Quantity Planning Specification • Quantitative Physical Science

QPSD: Quality and Productivity Systems Division

QPSK: Quadrature-Phase Shift Key • Quadrature-Phase Shift Keying (*) • Quadri-Phase Shift Keying • Quaternary-Phase Shift Key • Quaternary-Phase Shift Keying

QPT: Quantum Phase Transition

qq: Questions

Qq: Quartos • Quelques

Q-Q: Quantile-Quantile

QQ: Quantile-Quantile • Quick Query

Qqch: Quelque Chose

Qqe: Quelque

Qqes: Quelques

Qqf: Quelquefois

Qqfois: Quelquefois

Qqfs: Quelquefois

QQO: Quasi-Quadrennial Oscillation

QQP: Quick Query Program

QQPRI: Qualitative, Quantitative Personnel Requirements Information

Qqprt: Quelque Part

Qque: Quelque

Qques: Quelques

Qqu'un: Quelqu'un

qq. v.: Quae Vide • Quod Vide

qr: Quadrans • Quarter • Quarterly • Quire

Qr: Quarter

Q&R: Quality & Reliability

QR: Qatar Riyal (currency) • Quadrupole Resonance • Quality Review • Quarterly Review • Quick Rank • Quick Response

QRA: Johannesburg Randgermiston (South Africa) (location identifier) • Quick Reaction Alert

QRAL: Quality and Reliability Assurance Laboratory

QRC: Quantum Research Corporation (USA-MD) • Quick Reaction Communications

QRCG: Quasi-Random Code Generator

QRE: Quick Reaction Estimate • Quick Reaction Event

QREI: Quality Reliability Engineering International (periodical code)

QRF: Query Relative Frequency

QRFL: Queries-R-Focused Link

QRH: Quick Reference Handbook • Rotterdam - Central Station (Netherlands) (location identifier)

QRI: Quick-Reaction Integration

QRIA: Quick-Reaction Integration Activity

QRL: Queries-R-Linked

QRM: Qualified Right Monotonicity

QRMP: Quick-Response Multicolor Printer

QRO: Querétaro (Mexico)

QROO: Quintana Roo (Mexico)

QRP: Quick-Reaction Program • Quick-Reference Page • Quick-Response Program

QRPA: Quasi-particle Random-Phase Approximation

QRR: Qualification Results Review

Qrs: Quarters

QRS: Qualification Review Sheet • Quantum Readout System • Quartz Rate Sensor • Query Response System • Quick-Reaction Sortie

QRSIN: Queensland Remote Sensing Information Note

QRSL: Quick Reaction Space Laboratory

QRSS: Quasi Random Signal Source

QRT: Quarto

Qry: Quarry

qs: Quantum Satis • Quantum Sufficit

QS: Quadraphonic-Stereophonic • Quadratic Splines • Qualitative State • Quantity Surveyor • Quarter Section • Quasi-Stellar • Question Statement • Quick Sort • Quiet Sun

QSA: Qualification Site Approval

QSAM: Quasi-Static Acceleration Measurement • Quasi-Steady Acceleration Measurement • Queued Sequential Access Method

QSAR: Quality System Assessment Recognition • Quantitative Structure-Activity Relation • Quantitative Structure-Activity Relationship

QSB: Quarter Swing Baffle

QSC: Quadrilaterized Spherical Cube

QSCB: Quality System Certification Body

QSCV: Quasi-Stationary Convection Vortex

QSD: Quantum Spin Dynamics • Quantum Structural Diagram • Quasi-Sudden Discharge

QSDAM: Queued Sequential Data Access Method

QSDC: Quantitative Structural Design Criteria

QSDG: Quality of Service Development Group

QSE: Qualified Scientist and Engineer • Qualitätssicherungseinheit

QSF: Quasi-Stationary Floculus

QSG: Quasi-Steady Glide • Quasi-Stellar Galaxy

QSGVT: Quarter Scale Ground Vibration Test

QShAR: Quantitative Structure-Shape Relation

QSI: Quality Service Indicator • Quality Step Increase

QSIF: Quarter Source Input Format

QSIM: Queuing Simulation

QSJ: Que Sais-Je?

QSL: Q-Switch Laser • Qualification Status List • Qualified Source List • Quarter Scan Line

QSM: Quad Surface Mount • Quantum-Statistical Model • Quarter Scale Model

QSMVT: Quarter Scale Model Vibration Testing

QSO: Quasi-Stellar Object

QSOALS: Quasi-Stellar Object Absorption Line System

QSPR: Quantitative Structure-Property Relationship

QSQ: Query/Subquery

QSR: Quasar • Quasi-Stellar Radiosource

QSS: Quadrant Star Sensor • Quality-Securing System • Quality Surveillance System • Quasi-Steady State • Quasi-Stellar Source • Quick Start Service

QSSC: Quasi-Steady-State Cosmology

QSSS: Quasi-Stationary Spiral Structure

Qstn: Question

QSTNRY: Quasistationary

QSTOL: Quiet Short Takeoff and Landing

QSY: Quiet Sun Year

Qt: Quant • Quantity • Quart • Quiet

QT: Quad Tree • Qualification Test • Quantum Theory • Quarteirão • Quasi-Transverse • Queuing Theory • Quick Time • Quiet • Quiet Trader

Qta: Quinta

QTA: Quinta

QTAI: Quick Turn-Around Interconnect

QTAL: Quintal

QTAM: Queued Telecommunications Access Method • Queued Teleprocessing Access Method • Queued Transmission Access Method

QTAS: Quintas

QTC: Charge-to-Time Converter • Quick Transfer Centre

QTD: Quasi-Two-Dimensional

QTDAM: Queued Telecommunications Data Access Method

QT&E: Qualification Test & Evaluation

QTEL: Quartel

QTGG: Quantum Theories and Gravitation Group

QTH: Quartz Tungsten Halogen

QTIG: Queensland Telecommunications Industry Group

QTL: Quantitative Trait Locus

Qtly: Quarterly

QTM: Quadratic Texture Map • Qualification Test Model • Quaternary Triangular Mesh • Quick-Time Movie • Quick To Market

QTNHA: Quintinha

Qto: Quarto

QTO: Quasi-Triennial Oscillation

QTOL: Quiet Takeoff and Landing

QTP: Qualification Test Plan • Qualification Test Procedure • Qualification Test Program • Quality Test Plan • Quality Through Partnership • Quantum Test Programme

Qtr: Quarter • Quarterly

QTR: Qualification Test Report • Quarter

QTRS: Quarters

Qts: Quarts

QTU: Qualification Test Unit

QTV: Qualification Test Vehicle

QTW: Quick-Time for Windows

Qty: Quantity

qu: Quasi • Quechua (ISO language code)

Qu: Quart • Quartal • Quarter • Quarterly • Quelle • Query • Question

QU: Québec (Canada)

Quad: Quadrangle • Quadrant • Quadratic • Quadrilateral • Quadruple

QUAD: Quadrangle • Quadrant • Quadratic • Quadrilateral • Quadruple

quadr: quadratisch

Quadr: Quadragesima • Quadrant • Quadratic • Quadratur • Quadrilateral • Quadrille • Quadrupel • Quadruple

QUADS: Quadrants • Quality Achievement Data System

qual: qualifizieren • Qualifiziert • Qualitativ

Qual: Qualification • Qualified • Qualifikation • Qualify • Qualitative • Quality

QUAL: Qualification • Qualified • Qualifikation • Qualify • Qualitative • Quality

Qualn: Qualification

QUANGO: Quasi-Autonomous Non-Governmental Organization

Quant: Quantifier • Quantität • Quantitative • Quantity

Quar: Quartal • Quarter • Quarterly

QUARG: Queen's University Astronomy Research Group (Canada-ON)

Quart: Quarter • Quarterly

QUASAR: Quasi-Stellar Radiosource

QUASAT: Quasar Satellite

Quat: Quattuor

QUAT: Quaternion

QUB: Queen's University of Belfast (Northern Ireland) • Ubari (Libya) (location identifier)

QudV: Qualität durch Verantwortung

Que: Québec (Canada)

QUE: Québec (Canada)

QUEEN: Quaternary Environment for the Eurasian North

QUEL: Quelha • Query Language

QUELD: Queen's University Experiment in Liquid Diffusion

QUELH: Quelhas

Ques: Question

Quest: Quest (periodical code)

QUEST: Quasar Equatorial Survey Team

QUESTA: Querying Systems: Theory and Applications

QUESTER: Quick and Effective System to Enhance Retrieval

Questn: Questionnaire

QUESTOL: Quiet Experimental Short Takeoff and Landing

Quí: Química

QUIC: Quadrupole Ioffe Configuration • Quality Information and Control • Quick Infrared Camera

QUIDS: Quick Interactive Documentation System

QUILT: Query User Interface with Light Translations

Quím: Química

Quin: Quintuple

Quinq: Quinque

Quint: Quintuple • Quintuplicate • Quintus

QUIP: Query Interactive Processor • Questionnaire Interpreter Programme

QUIRC: Quick Infrared Camera

QUIS: Queen's University Information System • Questionnaire for User Interaction Satisfaction

Quitt: Quittung

QUK: Queen's University Kingston (Canada-ON)

QUL: Ulm Bus Station (Germany) (location identifier)

QUJM: Quincy University Journal of Mathematics (periodical code)

QUnJ: Qinghua University Journal (periodical code)

Quor: Quorum

Quot: Quotation • Quote • Quotient

QUOTA: Query Optimization Transformation Approach
QUOTE: Quality and User Oriented Typesetting Editor
Quotid: Quotidie • Quotidien
QUP: Quality Unit Pack
QUT: Queensland University of Technology (Australia-QLD)
QUVAD: Belfast (Starlink internal code)
qv: Quantum Vis • Quantum Vult • Quod Vide
QV: Quick Visual
QVANTOS: Quantum-Optical Spectrometer
QVLBI: Quasi-Very-Long-Baseline Interferometry
QVP: Quality Verification Plan
QVR: Quality Verification Report
QVT: Qualified Verification Testing
QVVT: Qualified Verification Vibration Testing
QW: Quality of Water • Quantum Well
QWA: Quarter-Wave Antenna
QWD: Quarterly World Day
QWG: Charlotte Wilgrove (USA-NC) (location identifier)
QWICPIC: QWIP Wide-field Imaging Color Prime Focus Camera
QWIKTRAN: Quick Fortran
QWIP: Quantum Well Infrared Photodetector
QWIPS: Quantum Well Infrared Photodetectors
QWL: Quality of Working Life • Quality of Work Life • Quick Weight Loss
Qy: Query
Qz: Quartz
QZ: Qualitätszirkel
QZG: Qualitätszielgruppe
Q9: Interbrasil Star

R

r: Distance • Distance from the Sun • Distance to Earth Center of Gravity • Heliocentric Distance of Perihelion • Internal Resistance • Mass Fraction of Individual Stage • Mixture Ratio • Molecular Position • Mole Ratio of Solution • Particle Position Vector • Polar Coordinate • Position • Radius • Railroad • Railway • Rain • Range • Rare • Rate • Ratio • Ray • Rayon • Read • Received • Rechts • Recipe • Recto • Red • Reflection Coefficient • Replacing • Reply • Reserve • Residence • Resistance • Response • Retirado • Retired • Revolution • Right • Río • River • Rivière • Road • Rod • Roentgen • Rouble (currency) • Royal • Rubber • Ruble (currency) • Rue • Ruled • Run • Rupee (currency)

R: Cosmic Scale Factor • Distance from the Galactic Center • Eighty (Roman numeral) • Gas Constant • Gauss Vector • Lattice Vector • Linear Range • Mass Ratio • Molar Gas Constant • Multiple Correlation Coefficient • Nuclear Radius • Particle Position Vector • Radar • Radial • Radical • Radiancy • Radiometer • Radiosonde • Radius • Railroad • Railway • Rain • Ram • Rand (currency) • Range • Rankine • Rapido • Rate • Ratio • Rayleigh • Reaction Force • Reaction Rate • Read • Real • Rear • Réaumur • Receive • Recht • Recipe • Recommandation • Recommandé • Rector • Red • Redactor • Redistribution • Redundant • Reflectance • Refraction • Regina • Registered • Registro • Regular • Regulate • Reine • Relative • Release • Reliability • Reluctance • Replace • Reply • Réponse • Report • Republic • Republican • Reserve • Resistance • Resolution • Resonance Effect • Response • Respuesta • Restricted • Return • Reverse • Revolution • Rex • Ria • Rial • Right • Río • River • Rivière • Road • Rod • Roentgen • Roi • Roll • Roman • Romania (international vehicle registration code) • Romania (postal country code) • Romeo • Rotated • Rouble (currency) • Rough • Royal • Rua • Ruble (currency) • Rúbrica • Rue • Ruin • Run • Rupee (currency) • Rydberg Constant • Set of Real Numbers • Sound Reduction Index • Spectral Type • Thermal Resistance • Total Mass Ratio • Wolf Number • 80 (Roman numeral)

ra: Radiatus

Ra: Radium • Range • Rayleight Number

R/A: Radar Altimeter • Return to Author

R&A: Report & Analysis • Research & Analysis • Responsibility & Action • Review & Analysis • Review & Approval

RA: Argentina (international vehicle registration code) • Radar Altimeter • Radar Altimetry • Radio Antenna • Radio Astronomy • Radiocommunication Assembly • Radiocommunications Agency • Range Area • Real Audio • Real-time Analysis • Rain • Random Access • Raster • Reduction of Area • Regional Association • Registration Act • Registration Authority • Relational Algebra • Relative Address • Release Authorization • Remedial Action • Repeat to Address • República Argentina • Research Assistant • Resident Astronomer • Resolution Advisory • Return Authorization • Ricerche Astron. Specola Vaticana (periodical code) • Right Aft • Right Ascension • Roll Axis • Royal Academician • Royal Academy • Russia (international aircraft registration code)

RAA: Rainwater Astronomical Association (USA-MS) • Rakanda (Papua New Guinea) (location identifier) • Rapid Access Archive • Regional Airline Association • Rockford Amateur Astronomers (USA-IL) • Root Addressable Area

RAAB: Remote Amplifier and Adaptation Box • Remote Amplifier and Advisory Box

RAAC: Remote Active Archive Center

RAAF: Royal Australian Air Force

RAAGE: Regionale Astronomische Arbeitsgemeinschaft Euskirchen (Germany)

RAAN: Right Ascension of Ascending Node

RAAO: Remote Access Automatic Observatory

RAAP: Remote Access Astronomy Project

RAAT: Remote Access Automatic Telescope

Rab: Rabat (Morroco)

RAB: Rabaul (Papua New Guinea) (location identifier) • Radiometer Box • Rapid Application Builder

RABAL: Radiosonde Balloon

RABAR: Radiosonde Balloon Release

RABATS: Rapid Analytical Block Aerial Triangulation System

RABF: Radioastronomy Bandstop Filter

RABIN: Raad van Advies voor Bibliothekwezen en Informatieverzorging (Netherlands)

RABMN: Remote Area Business and Messaging Network • Remote Area Business Management Network • Remote Area Business Message Network

RAC: Racine Horlick (USA-WI) (location identifier) • Radio Astronomy Centre • Radio Astronomy Controller • Radiometric Area Correlator • Raleigh Astronomy Club (USA-NC) • Read Address Counter • Receiver-Absorber-Converter • Reflective Array Compressor • Reliability Action Center • Reliability Analysis Center • Remote Access Computer • Remote Access Concentrator • Remote Analysis Computer • Repair, Alignment and Calibration • Remote Assessment Commission • Robotic Arm Camera • Rockland Astronomy Club (USA-NY)

RACAR: Remote Access Computer-Assisted Retrieval

RACCS: Remote Automatic Checkout and Calibration System

RACE: Random Access Card Equipment • Random Access Computer Equipment • Rapid Automatic Checkout Equipment • R&D in Advanced Communications for Europe • Research in Advanced Communications Environment • Research on Advanced Communications in Europe • Research on Automatic Computation Electronics

RACEP: Random Access and Correlation for Extended Performance

RACES: Remote Acoustic Electronic System

RACF: Resource Access and Control Facility • Resources Access Control Facility

RACI: Royal Australian Chemical Institute

RACIP: Research And Communications In Physics

RACMO: Regional Atmospheric Climate Model

RACS: Random Access Communications System • Redundant Attitude Control System • Remote-Access Computing System • Remote Automatic Calibration System • Remote Automatic Control System • Rotating Antenna C-band Scatterometer

RACT: Reasonably Available Control Technology

RACU: Remote Acquisition and Control Unit • Russian-to-American Converter Unit

rad: Radian

Rad: Radian • Radiatif • Radiation • Radiative • Radical • Radio • Radium • Radius • Radnorshire (UK-Wales)

R&AD: Research & Advanced Development

RAD: Radar • Radial • Radian • Radiation Absorbed Dose • Radiator • Radical • Radio • Radius • Radix • Random Access Data • Random Access Device • Rapid Access Disk

• Rapid Access Drive • Rapid Application Design • Rapid Application Development • Rapid Automatic Drill • Remote Antenna Driver • Research Alert Direct • Research and Advanced Development • Review and Approval Document • Roentgen Absorbed Dosage • Tortola (British Virgin Islands) (location identifier)

RADAC: Rapid Digital Automatic Computing

RADAR: Radio Detection and Ranging

RADARSAT: Radar Satellite

RADAS: Real-time And Distributed ADA Services

RADC: Reliability Analysis Data Center

RADCC: Radiation Control Center

RadEf: Radiation Effects (periodical code)

RADEM: Rocket Ascent Demonstrator Mission

RADEO: Rapid Design Exploration and Optimization

Radiad: Radiatif • Radiation • Radiative

RADIG: Radar Investigation Group

RADIR: Random Access Document Indexing and Retrieval

RaDIUS: Research and Development in the United States

RADIUS: Remote Authentication Dial-In User Server • Remote Authentication Dial-In User Service • Research And Development for Image-Understanding Systems • Research Association for the Development of the Industrial Use of Space • Residue Arithmetic Digital Image Understanding System

RADL: Radial

RADM: Regional Acid Deposition Model

Radn: Radiation

RADN: Radiation

RADOME: Réseau d'Acquisition de Données d'Observations Météorologiques Étendu

RADOT: Real-time Automatic Digital Optical Tracker

RADS: Risk Analysis Decision Support

RADSL: Rate-adaptive Asymmetrical Digital Subscriber Line • Rate-Adaptive Digital Subscriber Line (*) • Rate-adopted Asymmetrical Digital Subscriber Line

RADT: RPC-ASO Data Transfer

RADVS: Radar Altimeter and Doppler Velocity Sensor

RADWASS: Radioactive Waste Safety Standards

RAE: Arar (Saudi Arabia) (location identifier) • Radio Astronomy Explorer • Range-Azimuth-Elevation • Real Academia Española (Spain) • Research Aircraft Experiment • Research Assessment Exercise • RGS Analogue Electronics • Right Ascension Encoder • Royal Aerospace Establishment (UK) • Royal Aircraft Establishment

RaEE: Radio and Electronic Engineer (periodical code)

RAeS: Royal Aeronautical Society

RAES: Remote Access Editing System

RAET: Range-Azimuth-Elevation-Time

RAF: Ras an Naqb (Egypt) (location identifier) • Requirements Analysis Form • Research Aviation Facility • Resource Allocation Frame • Royal Air Force (UK)

RAFC: Regional Area Forecast Centre

RAFI: Rural Advancement Foundation International

RAFP: Repartição de Aministração Financeira e Patrimonial

RAFS: Regional Analysis and Forecast System

RAG: Radio Astronomy Group • Radiocommunications Advisory Group • Raglan (New Zealand) (location identifier) • Rama Asintótica de Gigantes • Readiness Analysis Group • Responsable Assurance Qualité • Reusable Agenda • Row Address Generator

RAH: Rafha (Saudi Arabia) (location identifier)

RAHF: Research Animal Holding Facility

RAI: Praia (Cape Verde) (location identifier) • Radar Altimeter Indicator • Radioactive Interference • Radio Audizioni Italia • Rainfall Anomaly Index • Review Action Item • Roll Attitude Indicator • Runway Alignment Indicator

RAID: Recherche-Assistance-Intervention-Dissuasion • Recherche Automatique de l'Information Disponible • Redundant Array of Independent Disks • Redundant Array of Independent Drives • Redundant Array of Inexpensive Disks (*) • Research Amplifying Imaging Detector

RAIDS: Rapid Information Dissemination System • Remote Atmospheric and Ionospheric Detection System

RAIE: Random Access Index Edit

RAIL: Runway Alignment Indicator Lights

RAILE: Retro-Assisted Imaging Laser Experiment

RAIM: Receiver Autonomous Integrity Monitoring

RAIN: Relation Algebraic Interpreter • Réseau d'Accès à Internet • Rochester Astronomical Information Network

RAIR: Random Access Information Retrieval • Remote-Access Immediate Response

RAIRO: Revue d'Automatique, Informatique et Recherche Opérationnelle

RAIS: Range Automated Information System

RAISE: Rigorous Approach to Industrial Software Engineering

RAIST: Réseau Africain d'Institutions Scientifiques et Techniques

RAIUB: Radioastronomisches Institut der Universität Bonn (Germany)

Raj: Rajasthan (India)

RAJ: Rajkot (India) (location identifier)

RAJPO: Range Application Joint Program Office

RAK: Marrakech (Morocco) (location identifier) • Regelwerk für die Alphabetische Katalogisierung

RAL: Radio Astronomy Laboratory • RAL (Astrophysics cluster) (Starlink internal code) • Random Assortive Law • Rapid Access Loop • Responsibility Assignment List • Revue d'Aptitude au Lancement • Ringer Approximate Loading • Riverside (USA-CA) (location identifier) • Rutherford Appleton Laboratory (UK-England)

RALI: Radar Lidar

RALPH: Reduction and Acquisition of Lunar Pulse Heights

RALS: Random Access Logical Subfile • Remote Augmented Lift System

RALSA: Random Access Line Scanned Array

RALSDC: RAL Space Data Centre

RALT: Radar Altimeter

RALU: Register Arithmetic Logic Unit

RAM: Radar-Absorbent Material • Radar-Absorbing Material • Radar-Absorption Material • Ramal • Ramingining (Australia-NT) (location identifier) • Random Access Machine • Random Access Measurement • Random Access Memory (*) • Random-Axis Model • Rank Annihilation Method • Read Active Memory • Rechargeable Alkaline Manganesedioxide • Reliability Analysis Methodology • Reliability Assessment Model • Remote-Access Modem • Remote-Access Monitor • Requirements Allocation Matrix • Research Allocation Methodology • Research and Applications Module • Resident Access Method • Resource Allocation Manager • Resource Allocation Model • Responsibility Assignment Matrix • Reverse Annuity Mortgage • Rocket-Assisted Motor

RAMA: Remote Access to Museum Archives

RAMAC: Random Access Method of Accounting and Control

RAMB: Random Access Memory Buffer

RAMC: Regional Administrative Management Center • Regional Assessment Management Committee

RAMCS: Remote Automatic Measuring and Calibration System

RAMD: Random Access Memory Device

RAMDAC: Random Access Memory Digital-to-Analog Converter

RAMEAU: Répertoire d'Autorité-Matières Encyclopédique et Alphabétique Unifié

RAMIS: Rapid Access Management Information System

RAMM: Random Access Memory Module • Rapid Agile Metrology for Manufacturing • Regional And Mesoscale Meteorology

RAMMS: Regional Atmospheric Mesoscale Modelling System

RAMOS: Radiation Monitoring in Space • Remote Automatic Meteorological Observation System • Remote Automatic Meteorological Observing Station

RAMP: Rapid Acquisition of Manufactures Parts • Rating Maintenance Phase • Remote Access Maintenance Protocol • Remote Atmospheric Measurements Program • Remote Atmospheric Monitoring Project • Replenishment Activity Management Plan • Results, Applications and Modelling Phase • Rural Abandoned Mine Program

RAMPS: Resource Allocation and Multiproject Scheduling • Resource Analysis and Mission Planning System

RAMS: Random Access Measurement System • Regional Agricultural Management System • Regional Atmospheric Modelling System • Reliability-Availability-Maintainability-Safety • Remote Atmospheric Measurements System • Research Aircraft Measurement System • Results Analysis and Management System

RAMSES: Radar Aéroporté Multi-Spectral d'Étude des Signatures • Recherche Appliquée aux Méthodes de Séparation en Électrophorèse Spatiale • Recherche Appliquée sur les Méthodes de Séparation en Électrophorèse Spatiale (*)

RAMSQ: Reliability, Availability, Maintainability, Safety and Quality

RAMUS: Remote Access Multiuser System

RAN: Ravenna - La Spreta (Italy) (location identifier) • Remote Access Node • Royal Australian Navy

RANC: Radar Absorption Noise and Clutter

RANCID: Real And Not Corrected Input Data

RAND: Radio Array Neutrino Detection • Random • Request Activity and Name Directory • Research And Development • Rural Area Network Design

RANDAM: Random Access Nondestructive Advanced Memory

RANDID: Rapid Alphanumeric Digital Indicating Device

RangKoM: Rangkaian Komputer Malaysia

RANLP: Recent Advances in Natural Language Processing (conference)

RANN: Research Applied to National Needs

RANS: Reynolds-Averaged Navier-Stokes

RAO: Radio Astronomical Observatory • Recherche Assistée par Ordinateur • Regional Acquisition Office • Research Announcement of Opportunity • Revenue Accounting Office • Ribeirão Preto - Leite Lopes (Brazil-SP) (location identifier) • Rothney Astronomical Observatory (Canada-AB)

RAOB: Radio Atmospheric Observation Balloon

RAOOS: Radio Astronomy Observatory Operating System

RAORE: RPC-ASO Oneshot Reply

RAORQ: RPC-ASO Oneshot Request

Rap: Rapids

RAP: Radio Astronomy Preprint • Ready-Assembled Price • Random Access Program • Rapid Application Prototyping • Rapid Assessment Procedure • Rapid City (USA-SD) (location identifier) • Real Application Performance • Reconnaissance Automatique de la Parole • Regional Access Point • Regional Application Program • Regression Analysis Program • Relative Ability of Prediction • Reliability Assurance Program • Remedial Action Programme • Research Applications Program • Research Assessment Panel • Research Associateship Program • Reserve Auxiliary Payload • Responsibility Accounting Program • Right-Angle Program • Root Anchor Point • RSVP Admission Policy

RAPAT: Radiation Protection Advisory Team

RAPCO: Radar Approach Control

RAPCOE: Random Access Programming and Checkout Equipment

RAPCON: Radar Approach and Control • Radar Approach Control

RAPD: Response Amplitude Probability Data

RAPI: Royal Australian Planning Institute

RAPID: Random-Access Personnel Information Disseminator • Random-Access Photographic Index and Display • Relative Address Programming Implementation Device • Remote-Access Planning for Institutional Development • Remote Atmospheric Probing Information Display • Research in Automatic Photocomposition and Information Dissemination • Research with Adaptive Particle Imaging Detectors • Retrieval through Automated Publication and Information Digest • Retrieval And Processing Information for Display • Riparian Aerial Photographic Inventory of Disturbance

RAPIDS: Random Access Personnel Information Dissemination System • Real-time Acquisition and Processing Integrated Data System

RAPIV: Radiographic Particle Image Velocimeter

Rapp: Rapport

RAPP: Resource And Planning Program

RAPS: Radio Planning System • Remote Area Power System • Retrieval Analysis and Presentation System • Right Aft Propulsion System

RAPSE: Report on Active and Planned Spacecraft and Experiments

RAQ: Raha (Indonesia) (location identifier)

RAR: Radar Arrival Route • Rarotonga (Cook Islands) (location identifier) • Real-Aperture Radar • Regulatory Affairs Review • Resource Accommodation Report • Return Address Register

RARA: Retractable Advanced Rigid Array

RARC: Regional Administrative Radio Conference • Ritter Astrophysical Research Center (USA-OH)

RaRe: Radiation Research (periodical code)

RARE: Réseaux Associés pour la Recherche Européenne

RaRLJ: Radio Research Laboratories Journal (periodical code)

RaRLR: Radio Research Laboratories Review (periodical code)

RARP: Reverse Address Resolution Protocol

Ras: Raspado • Raspadura

RAS: Racine Astronomical Society (USA-WI) • Radar Altimeter System • Radar à Antenne Synthétique • Radio Astronomy Service • Random-Access Storage • Rapid-Access Storage • Rasht (Iran) (location identifier) • Raw Address Strobe • Reacquisition Sensor • Reader Admission System • Reading Astronomical Society (UK-England) • Rectified Airspeed • Redémarrage Automatique du Serveur • Refractive Aberrated Simulator • Région Administrative Spéciale • Relay Antenna System • Reliability-Availability-Serviceability • Remote Access Server • Remote Access Service • Remote Access Set • Remote Analysis Station • Remote Analysis System • Requirements Accounting System • Requirements Allocation Sheet • Reuse All Systems • Richland Astronomical Society (USA-OH) • Richmond Astronomical Society (USA-VA) • Rien À Signaler • Riverside Astronomical Society (USA-CA) • Rotating Anode Source • Row Address Select • Row Address Strobe • Royal Academy of Sciences (UK) • Royal Astronomical Society (UK) • Runway Alert System • Rural Adjustment Scheme • Russian Academy of Sciences • Ryerson Astronomical Society (USA-IL)

RASAPI: Remote Access Service Application Programming Interface

RaSc: Radio Science (periodical code)

RASC: Radio Atmospheric Science Center • Radio Atmospheric Study Centre • Residence Account Service Center • Royal Army Service Corps • Royal Astronomical Society of Canada

RASCAL: Rotorcraft Aircrew Systems Concepts Airborne Laboratory

RASCAS: Recherches sur l'Aérodynamique, la Structure et la Commande de l'Avion Souple

RASCIBA: Radar-SCIDAR-Ballon

RASCOM: Regional African Satellite Communication System

RaScS: Radio Science Supplement (periodical code)

RASD: Radio and Atmospheric Science Division

RASER: Radio-frequency Amplification by Stimulated Emission of Radiation

RASH: Rain Shower

RASM: Remote Analog Submultiplexer • Romanian Astronomical Society for Meteors (Romania)

RASN: Rain and Snow

RASNZ: Royal Astronomical Society of New Zealand

RASO: Rear Airfield Supply Organisation

RASP: Remote Antenna Signal Processor • Retrieval And Sort Processor

RASS: Radar Altimeter Simulation System • Radar Altimeter System Simulator • Radar Analysis and Support System • Radar and Acoustic Sensing System • Radio Acoustic Sounding System • ROSAT All-Sky Survey

RAST: Reliability and System Test

RASTAC: Random Access Storage And Control

RASTAD: Random Access Storage And Display

Rat: Rateable • Rating • Ratio

R&AT: Research & Advanced Technology

RAT: Radiation Air Thermometer • Radioactive Tracer • Raduzhnyi (Russia) (location identifier) • Ram Air Turbine • Regional Assessment Team • Register Alias Table • Remote Access Terminal • Riverside Astronomical Telescope (USA-CA) • Robust Audio Tool

RATAN: Radar and Television Aid to Navigation

RATCC: Radar Air Traffic Control Center

RATCF: Radar Air Traffic Control Facility

RATE: Remote Automatic Telemetry Equipment

RATER: Response Analysis Tester

RATO: Rocket-Assisted Takeoff

RATP: Régie Autonome des Transports Parisiens (France) (*)

RATS: Radio Amateur Telecommunications Society • Random-Accessible Text System

RATTS: Runaway T Tauri Star

RAU: Rand Afrikaans University (South Africa) • Rangpur (Bangladesh) (location identifier) • Red Académica Uruguaya • Remote Access Unit • Remote Acquisition Unit • République Arabe Unie

RAUIS: Remote Acquisition Unit Interconnecting Station

RAV: Cravo Norte (Colombia) (location identifier) • Risk-Adjusted Value

RAVE: Random-Acess Video Editing • Relevance Assessment Vehicle • Rendering Acceleration Virtual Engine

RAVEN: Radar Analysis and Visualization Environment

RAVI: Représentation des Applications audioVisuelles Interactives

RAVU: Radiosonde Analysis and Verification Unit

RAW: Arawa (Papua New Guinea) (location identifier) • Read After Write • Read And Write

RAWB: Raad van Advies voor het Wetenschapsbeleid (Netherlands)

RAWD: Read-After-Write Dependency

RAWG: Regional Association Working Group

RAWS: Radar Wind Sounder • Remote Automatic Weather Station

RAWT: Remote-Access Wind Tunnel

RAX: Oram (Papua New Guinea) (location identifier) • Remote Access • Rural Automatic Exchange

RAY: Rothesay (UK-Scotland) (location identifier)

RAZ: Rawala Kot (Pakistan) (location identifier) • Remise À Zéro

RAZEL: Range-Azimuth-Elevation

Rb: Rubidium

R&B: Rehabilitation & Betterment

RB: Botswana (international vehicle registration code) • Radar Beacon • Radiation Belt • Radio Beacon • Radiocommunication Bureau • Rapid Burster • Read Buffer • Record Byte • Reduction Block • Reference Burst • Regular Budget • Rehabilitation and Betterment • Relative Bearing • Representative Body • Request Block • Requirements Board • Response Bus • Review Body • Ridge Belt • Right Back • Right Button • Rigid Body • República Boliviana • Rubber Band • Rubber Base • Rule Based

RBA: Rabat (Morocco) (location identifier) • Radar Beacon Antenna • Rapid Biodiversity Assessment • Recovery Beacon Antenna • Relative Byte Address • Rigid Bottom Approximation

RBAC: Role-Based Access Control

RBB: Borba (Brazil-AM) (location identifier) • Residential Broadband

RBBS: Remote Bulletin Board System

RBC: Radial Bay Camera • Radio Beam Communications • Red Blood Cell (*) • Red Blood Count • Redundant Battery Charger • Remote Balance Control • Return Beam Camera • Robinvale (Australia-VIC) (location identifier) • Rotating Beam Ceilometer

RBCC: Rocket-Based Combined Cycle

RbCP: Retrieval by Joint Probability

RBD: Dallas Redbird (USA-TX) (location identifier) • Randomized Block Design • Relative Band Depth

RBDE: Radar Bright Display Equipment

RBDNRQ: Received But Did Not Returned Questionnaire

RBDP: Rocket Booster Development Program

RBDPE: Radar Beacon Data Processing Equipment

RBDS: Radio Broadcast Data System

RBE: Radiation Biological Effectiveness • Ratanankiri (Cambodia) (location identifier) • Red Brown Earth • Relative Biological Effectiveness • Replacement Battery Equipment • Résultat Brut d'Exploitation • Rule-Based Engine

RBF: Big Bear (USA-CA) (location identifier) • Radial Basis Function • Rockefeller Brothers Fund

RBG: Roseburg (USA-OR) (location identifier)

RbGLI: Retrieval by General Logic Imaging

RBH: Brooks Lodge (USA-AK) (location identifier)

RBHC: Regional Bell Holding Co.

Rbí: Recibí

RBI: Rabi (Fiji) (location identifier) • Remote Bus Interface • Require Better Information • Resource-Based Industry • Response Bus Interface • River Basin Investigation • Runs Batted In

RBJ: Rebun (Japan) (location identifier)

RbJP: Retrieval by Joint Probability

RBK: Right Bank

Rbl: Rouble (currency)

RBL: Radio BL Lacertae (object) • Red Bluff (USA-CA) (location identifier) • Right Buttock Line

Rbla: Rambla

RBM: Radiation Burst Monitor • Running-Box Mean • Rutherford Backscattering Method • Straubing (Germany) (location identifier)

RBMS: Resource Base Monitoring System

RBMT: Retrospective Bibliographies on Magnetic Tape

RBN: Radio Beacon • Ribbon

RBO: Robore (Bolivia) (location identifier)

RBOC: Regional Bell Operating Co.

RBOR: Request Basic Output Report

RBP: Rabaraba (Papua New Guinea) (location identifier) • Rat Bundesdeutscher Planetarien (Germany)

RBPE: Red de Bibliotecas Públicas del Estado (Spain)

RBQ: Rurrenabaque (Bolivia) (location identifier)

Rbr: Rubber

RBR: Relevance-Based Ranking • Reserve Book Room • Rio Branco - Presidente Medici (Brazil-AC) (location identifier)

RBrFi: Revista Brasileira de Fisica (periodical code)

RBS: Orbost (Australia-VIC) (location identifier) • Ranked Bubble Sort • Rare Book Section • Real Big Scope • Reduction Block Scheduler • Reliable Broadcast System • Robbed-Bit Signaling • Rule-Based System • Rutherford Backscattering • Rutherford Backscattering Spectroscopy

RBSE: Repository-Based Software Engineering (*) • Repository-Base for Software Engineering • Research-Based Science Education

RBSS: Recoverable Booster Support System

Rbt: Roundabout

RBT: Marsabit (Kenya) (location identifier) • Random Boundary Tolerance • Random Breadth Testing • Remote Batch Terminal

RBU: Response Bus Unit • Roebourne (Australia-WA) (location identifier) • Répertoire Bibliographique Universel

RBV: Return Beam Vidicon

RBW: Resolution Bandwidth • Walterboro (USA-SC) (location identifier)

Rbx: Roubaix (France)

RBY: Ruby (USA-AK) (location identifier)

R-C: Radiative-Convective

R/C: Radio Command • Radio Control • Rate of Climb • Recredited • Remote Control

RC: China (international vehicle registration code) • PRC (international vehicle registration code) • Radiative-Convective • Radio Code • Radio Command • Radio Control • Radio Controlled • Range Clearance • Range Command • Range Contractor • Range Control • Range Correction • Rapid Change • Rate of Change • Rate Command • Rate of Climb • Reaction Control • Read Cursor • Reader Code • Recognition Context • Recovery Controller • Recurring Cost • Red Clump • Red Cross • Reference Catalogue • Regional Center • Regional Council • Reinforced Concrete • Relay Computer • Release Clause • Remote Control • Reproductive Capacity • Rés-do-Chão • Research Centre • Reverse Course • Research Corporation • Resin Coated • Resistance-Capacitance (*) • Resistance-Conductance • Resistor-Capacitor • Resonance Circuit • Ressources Centrales • Reticular Complex • Return Code • Reverse Current • Revised Catalog • Right Centre • Riser Cable • Ritchey-Chrétien • Rotary Combustion • Rotational Control • Rotation Control • Taiwan (China-ROC) (international vehicle registration code)

Rca: Rocca

RCA: Central African Republic (international vehicle registration code) • Radar-Controlled Approach • Radiative-Convective Atmospheric • Radio Corporation of America • Radiological Control Area • Rapid City - Ellsworth (USA-SD) (location identifier) • Rate Change Authorization • Reaction Control Assembly • Receiver Collision Avoidance • Regional Cooperative Agreement • Remote Control Amplifier • République Centrafricaine • Revealed Comparative Advantage • Right Coronary Artery • Rose City Astronomers (USA-OR) • Royal Canadian Academy

RCAC: Remote Computer Access Communications

RCAF: Royal Canadian Air Force

RCAG: Remote Center Air/Ground • Remote Communications Air/Ground • Remote Control Air/Ground

RCAS: Reserve Component Automation System

RCB: Congo (international vehicle registration code) • Radiation Control Board • Randomized Complete Block • R Coronae Borealis • Region Control Block • Remote Circuit Breaker • Richards Bay (South Africa) (location identifier) • Right Centre Back

RCBG: Reference Catalogue of Bright Galaxies

RC/C: Reinforced Carbon-Carbon

R&CC: Recorder & Communication Control

RCC: Radio Common Carrier • Range Control Center • Recorder and Communication Control • Recovery Control Centre • Redundancy Connection Control • Regional Census Center • Regional Commonwealth of Communications • Regional Control Center • Reinforced Carbon-Carbon • Remote Center Compliance • Remote Cluster Controller • Remote Communications Console • Rescue Coordination Centre • Reset Control Circuit • Reusable Carbon-Carbon • Reverse Command Channel • Ritchey-Chrétien-Coudé • Riverside Community College (USA-CA) • Rough Combustion Cutoff • Routing Control Center

RCCA: Rough Combustion Cutoff Assembly

RCCB: Remote Central Circuit Breaker • Remote Control Circuit Breaker

RCCC: Regional Communications Control Center

RCCL: Robot C Control Language • Royal Caribbean Cruise Line (now RCI)

RCCP: Recorder and Communications Control Panel

RCCS: Rate Command Control System • Rotary Cell Culture System

Rcd: Received

RCD: Radar Cloud Detection • Record • Residual Current Device • Revue Critique de Définition

RCDC: Radiation Chemistry Data Center

RCDD: Remote Colour Display Driver

RCDI: Rate of Climb and Descent Indicator

RCDP: Record Parallel

Rcdr: Recorder

RCDR: Recorder

RCE: Reaction Control Equipment • Remote-Control Equipment • Réseau Climatologique d'État • Réseau et Communication d'Entreprise • Restricted Coulomb Energy • Roche Harbor (USA-WA) (location identifier) • Rotary Combustion Engine

RCEC: Repayment of Capital-Excess Compensation

RCEM: Research Center for Energetic Materials (USA-NM)

RCF: Relative Centrifugal Force • Remote Call Forwarding • Remote Communications Facility

RCFT: Randomized Controlled Field Trial • Rational Conformal Field Theory

RCfTA: Research Centre for Theoretical Astrophysics

RCG: Radiation Concentration Guide • Regulatory Coordinating Group • Relativity and Cosmology Group • Rose Compact Group

RCGS: Royal Canadian Geographic Society

RCH: Chile (international vehicle registration code) • Reach • Riohacha (Colombia) (location identifier)

RCHCS: Regenerable CO and Humidity Control System

RCHG: Reaching

RCI: Radar Coverage Indicator • Range Communications Instructions • Registre de Contrôle Individuel • Relative Citation Impact • Relativistic Configuration Interaction • Remote Control Interface • Request for Contract Investigation • Rodent Cage Interface • Royal Canadian Institute • Royal Caribbean International (formerly RCCL)

RCIC: Regional Coastal Information Center

RCIP: Replacement Cost Integration Program

RCIS: Resource Center Information System

RCIU: Remote Control and Interface Unit

RCJ: Reaction Control Jet • Royal Court of Justice

RCK: Radial Conformal Killing • Rockdale Coffield (USA-TX) (location identifier)

RCKY: Rocky

Rcl: Recall

RCL: Radio Communications Link • Redcliffe (Vanuatu) (location identifier) • Reliability Control Level • Robotics Control Library • Rotate Carry Left • Ruling Case Law • Runway Centerline

RCLF: Robust Control Lyapunov Function

RCLIS: Research Center for Library and Information Science

Rclm: Reclaim

RCLM: Runway Centerline Marking

RCLS: Runway Centerline Light System

RCM: Radiative Convective Model (*) • Radioactive Convective Model • Random-Choice Method • Range Cell Migration • Recent Crustal Movement • Reliability-Centered Maintenance • Remote Carrier Module • Reverse Cuthill-McKee • Rice Convection Model • Richmond (Australia-QLD) (location identifier)

RCMP: Royal Canadian Mounted Police

Rcn: Recreation

RCN: American River (Australia-SA) (location identifier) • Reactor Centrum Nederland (Netherlands) • Receipt of Change Notice • Record Control Number • Recovery Communications Network • Report Change Notice • Report Control Number • Requirements Change Notice • Royal Canadian Navy

Rcndt: Recondition

RCO: Range Control Office • Remote Communications Outlet • Remote-Control Office • Rendezvous Compatible Orbit • Rochefort Saint-Agnan (France) (location identifier)

RCP: Radiation Constraints Panel • Radon Contractor Proficiency • Real Continuous Pairwise • Recherche Coopérative sur Programme • Recording Control Panel • Rectangular Coordinate Plotter • Recurring Content Pattern • Regional Cooperation Project • Registration Control Point • Reliability Critical Problem • Remote Communications Processor • Remote-Control Panel • Remote Copy • Remote Copy Protocol • Restore Cursor Position • Right Circularly-Polarized • Right Circular Polarization • Right Circular Polarizer

RCPA: Ratio of Cited to Published Articles

RCPL: Right Circularly Polarized Light

RCPSP: Resource Constrained Project Scheduling Project

RCPSPDCF: Resource Constrained Project Scheduling Project with Discounted Cash Flows

Rcpt: Receipt

RCPT: Receipt • Receptacle • Recipient

Rcptn: Reception

RCQ: Reconquista (Argentina-SF) (location identifier)

RCR: Radio Communications Report • Requirements Change Request • Retrofit Configuration Report • Reusable Component Recovery • Reverse Current Relay • Rochester - Fulton County (USA-IN) (location identifier) • Rotate Carry Right • Runway Condition Reading

RCRA: Resource Conservation and Recovery Act (USA) (*)

RCRE: Required Collaborative Research Experience

RCRS: Regenerative Carbon-dioxide Removal System

RCRU: Radio Communications Research Unit

RCS: Radar Cross-Section • Radio Source Coordinates • Rate Control System • Reaction Control System • Readiness Control System • Reconnecting Current Sheet • Record Creation System • Reentry Control System • Reference Color Space • Remote Computing Service • Remote Connection Service • Remote Control System • Remote Core Sampler • Requirements Collection System • Research Camera System • Retrofit Configuration System • Revision Control System • Rochester (UK-England) (location identifier) • Rochester Connectionnist Simulator • Rudder Control System • Rural Counselling Service

RCSC: Reaction Control System Controller • Remote Spooling Communications System

RCSDE: Reaction Control System Drive Electronics

RCSS: Random Communications Satellite System • Records Communications Switching System

RCSSMRS: Regional Centre for Services in Surveying, Mapping and Remote Sensing

Rct: Receipt

RCT: Randomized Clinical Trial • Randomized Control Trial • Reaction Control Thruster • Reed City - Miller (USA-MI) (location identifier) • Region Control Table • Region Control Task • Remote Control Transmitter • Resource Control Table

RCTC: Reversible Cooke Transit Circle

RCTL: Resistor-Capacitor-Transistor Logic

Rcto: Recinto

RCTS: Remote Commercial Television Service

RCTT: Regional Center for Technology Center

RCU: Radio Channel Unit • Recovery and Control Unit • Reference Control Unit • Relay Control Unit • Remote-Control Unit • Rio Cuarto (Argentina-CD) (location identifier)

RCUA: Responsable de la Charge Utile d'Ariane

Rcv: Receive

RCV: Receive

Rcvd: Received

Rcvg: Receiving

Rcvr: Receiver

RCVR: Receiver

Rcvs: Receives

RCVS: Remote Control Video Switch • Remote-extension of Concurrent Versions System

RCVY: Recovery

RCW: Rapidly Changing World • Reaction Control Wheel • Rodgers-Campbell-Whiteoak

RCWA: Rigorous Coupled Wave Analysis

RCWP: Rapidly Changing World Policy

RCY: Recovery • Remaining Cycles • Rum Cay (Bahamas) (location identifier)

RCZ: Radiation Control Zone

RC2: Second Reference Catalogue • Second Revised Catalogue

RC3: Third Reference Catalogue • Third Revised Catalogue

Rd: Radium (now Ra) • Read • Rendered • Road • Rod • Rond • Round • Rund • Rutherford

R&D: Read & Destroy • Recherche & Développement • Research & Development

R/D: Rate of Descent • Refer to Drawer • Relevant Document • Research and Development • Resolver to Digital

RD: Radiation Detection • Radiation-Dominated • Radiation Dose • Radiation-Driven • Radio Determination • Random Drift • Range Development • Raster Device • Rate of Detection • Read • Read Data • Readiness Data • Readiness Date • Real Decreto • Real Drift • Rear Door • Receive Data • Received Data • Reference Designator • Reference Directory • Reference Document • Regional Director • Relative Density • Released Drawing • Relevance Description • Remove Directory • Remove Duplicate • República Dominicana • Requirements Document • Research and Development • Research Department • Responsibility/Distribution • Restricted Data • Rive Droite • Road • Root Diameter • Round • Rotation Dominated • Rotational Discontinuity • Route Départementale • Running Days • Rural Delivery

Rda: Redonda • Ronda

RDA: Read Data Available • Receive Data • Recommended Daily Allowance • Recommended Dietary Allowance • Regional Development Assistant • Remote Data Access • Remote Database Access • Remote Dial-in Analysis • Representational Difference Analysis • République Démocratique Allemande (now merged into RFA) • Resident Data Area • Rockhampton Downs (Australia-NT) (location identifier) • ROSAT Data Archive • Rotary Drive Actuator

RDAAC: Remote Data Analysis and Archive Center

RDAP: Remote Data Access and Protocol

RDAT: Rotary-Head Digital Audio Tape

RDAU: Remote Data Acquisition Unit

RDAF: Regional Data Analysis Facility

RDB: Receive Data Buffer • Red Dog (USA-AK) (location identifier) • Relational Database • Requirements Databank • Research and Development Board

RDBI: Relational Database Interface

RDBM: Relational Database Management

RDBMS: Relational Database Management System

RDBS: Relational Database System

RdC: Rez-de-Chaussée

RDC: Real-time Data Cell • Redenção (Brazil-PA) (location identifier) • Reference Designator Code • Regional Data Center • Regional Dissemination Center • Regional Documentation Center • Remote Detector Chamber • Request for Design Change • Request for Document Change • Request for Drawing Changes • Research and Documentation Center • Running Down Clause • Rural District Council

RDCLK: Received Clock

RDCP: Remote Display Control Panel

RDCS: Reconfiguration Data Collection System • Regional Data Collecting System • ROSAT Deep Cluster Survey

RD&D: Research, Development & Demonstration

RDD: Radar Detector Detector • Random Digital Dialing • Redding (USA-CA) (location identifier) • Replaceable Database Driver • Requirements Definition Document • Required Delivery Date • Research-Development- Demonstration • Residual Dark-current Distribution • Residual Dark Distribution • Responsibility-Driven Design

RDDP: Radar Data Development Programme

RDDS: Recoil-Distance Doppler-Shift

RDDT&E: Research, Development, Demonstration, Testing & Evaluation

RDDTE: Research-Development-Demonstration-Testing-Evaluation

RD&E: Research, Development & Engineering

RDE: Merdey (Indonesia) (location identifier) • Radiation Detection Experiment • Radiation-Dominated Era • Remote Data Entry • Research and Development Establishment • RGS Digital Electronics • Rotating-Disc Electrode

RdeC: Rez-de-Chaussée

R.-de-Ch.: Rez-de-Chaussée

RdeCh: Rez-de-Chaussée

RDEMS: Regional Disaster Emergency Management System

RDES: Remote Data Entry System

RDF: Radio Direction Finder • Radio Direction Finding • Rapid Deployment Force • Rationalized Data Format • Real Data Fine • Record Definition Field • Regional Development Fund • Request Denied Frame • Resource Data File • Resource Description Framework

RDFR: Restricted Direct Fuzzy Reasoning

RDFT: Real Discrete Fourier Transform

RDG: Reading (USA-PA) (location identifier) • Ridge

RDGE: Ridge

RDGS: Reconfiguration Display Generation System • Ridges

RDI: Range-Doppler Interferometer • Range-Doppler Interferometry • Released Data Index • Remote Data Input • Remote Defect Indication • Research, Development and Industrialization

RDIS: Research and Development Information System

RDJFCB: Read Job File Control Block

RDL: Reconfigurable Design Language • Remote-Display Link

RD&M: Research, Development & Marketing

RDM: Bend (USA-OR) (location identifier) • Real-time Data Manager • Recoil Distance Method • Redmond (USA-OR) (location identifier) • Relational Data Model • Relational Document Manager • Report Display Manager • Research, Development and Marketing • Rotterdamsche Droogdok Mij. BV

RDMD: Regional Dark Mantle Deposit

RDMS: Relational Database Management System

RDMSR: Read Machine Specific Register

RDMTR: Radiometer

RDN: Relative Distinguished Name • Resources Discovery Network

rDNA: Recombinant DNA

RDNH: Reverse Domain Name Hijacking

Rdo: Radio • Recuerdo

R&DO: Research & Development Operations

RDO: Radio • Relational Database Operator • Remote Data Object • Research and Development Operations • Resource Definition Online

RDOC: Reference Designation Overflow Code

RDOP: Relative Dilution Of Precision

RDOS: Real-time Disk Operating System

RDP: Radar Data Processing • Radar Development Project • Rat Deutscher Planetarien (Germany) • Receiver and Data Processor • Reconstruction and Development Programme • Reliable Datagram Protocol • Remote Data Processor • Remote Desktop Protocol • Report Documentation Page • Requirements Data Plan • Requirements Development Plan • Research Data Program • Ribosomal Data Project

RdPt: Rond Point • Rond Punt

RDPL: Laos (international aircraft registration code)

Rdr: Radar

RDR: Radar Departure Route • Raw Data Recorder • Red River - Grand Forks (USA-ND) (location identifier) • Reduced Data Record • Requirements Definition Review • Research and Development Report • Rudder

RDRAM: Rambus Dynamic Random Access Memory

RDRO: Robinson/Downs Rooftop Observatories (USA-CA)

RDS: Radio Data System • Radio Determination Service • Radio Digital System • Rat Deutscher Sternwarten (Germany) (formerly RWS) • Real Data Smooth • Relational Data System • Remboursement de la Dette Sociale • Remote Data Service • Replaceable Disk Store • Report Distribution Software • Rincón de los Sauces (Argentina-NE) (location identifier) • Roads • ROSAT Deep Survey • Rossiter Double Star

RDSI: Red Digital de Servicios Integrados

RDSR: Receiver Data Service Request

RDSS: Radio Determination Satellite Service • Research and Data Support System

RDT: Radio Digital Terminal • Recherche et Développement Technologique • Reference Data Table • Remote Data Transmission • Remote Data Transmitter • Remote Destination Table • Réseau de Diffusion Technologique • Richard Toll (Senegal) (location identifier) • Rise Time Discriminator • Rudimentary Debugging Tool

RDT&E: Research, Development, Technology & Evaluation • Research, Development, Test & Evaluation • Research, Development, Testing & Engineering

RDTE: Research, Development, Technology and Evaluation • Research, Development, Test and Evaluation • Research, Development, Testing and Engineering

RDTR: Radiator • Research Division Technical Report

RDTS: Research and Development Tracking System

RDU: Durham (USA-NC) (location identifier) • Raleigh (USA-NC) (location identifier) • Resource Development Unit

RDV: Red Devil (USA-AK) (location identifier) • Relevance Description Vector • Rendezvous

RDVU: Rendezvous

RDW: Record Descriptor Word • Response Data Word • Return Data Word

RDX: Cyclotrimethylene Trinitramine • Realistic Display Mixer

RDY: Ready

RDZ: Radiation Danger Zone • Rodez (France) (location identifier)

Rdzvs: Rendez-vous

re: in Regards to • rechter

Re: Reference • Regarding • Reinsurance • Reply • Respecting • Reynolds Number • Rhenium • Rhénium • Rupee (currency)

R&E: Research & Engineering • Research & Experimentation

R/E: Reentry • Right-hand Engine

RE: Île de la Réunion (ISO country code) • Rare Earth • Rate of Exchange • Real Estate • Record End • Recovery-Enabled • Regular Exposure • Renewable Energy • Répondeur-Enregistreur • Resolution Element • Resonant Excitation • Responsible Engineer • Retirado • Réunion Island (ISO country code) • Reverse Engineering • Revised Edition • Riccati Equation • Richards Equation • Right End • Right Excellent • Robust Estimation

REA: Raad van Economische Aangelegenheden • Range Error Average • Reao (Tuamotu Islands) (location identifier) • Resource-Event-Agent • Rocket Engine Assembly • Rural Electrification Administration (USA)

ReAA: Recherches Aéronautiques et Aérospatiales (periodical code)

Reac: Reactor

REACH: Remote Access to Circular History • Remote External Access to Circular History

REACQ: Reacquisition

REACT: Rain Echo Attenuation Compensation Technique • Research and Education Automatically Controlled Telescope

READ: Reader for Event Analysis and Detection • Realtime Electronic Access and Display

READI: Rights for Electronic Access to Delivery of Information (*) • Rights to Electronic Access and to Delivery of Information

READS: Re-Entry Air Data System

ReAer: La Recherche Aérospatiale (periodical code)

REAFAMS: Rapid Easy Accurate Finderscope Alignment Mounting System

REAM: Rocket Engine Assembly Module

REAP: Regional Environment Assessment Program • Resource Evaluation Aerial Photography • Rural Education Access Program

REARS: Redundant Attitude Reference System

REAUMUR: Réseau Aquitain des Utilisateurs des Milieux Universitaires et de Recherche (France)

REB: Relativistic Electron Beam

REBA: Radiometer Electronics Box Assembly

REBIUN: Red de Bibliotecas Universitarias (Spain)

REBS: Radiation Energy Balance System

Réc: Réception

Rec: Receipt • Receive • Recensement • Recent • Recepción • Reception • Recipe • Recommend • Recommendation • Recommended • Record • Recorded • Recorder • Recording • Recreation • Recreational • Rectification

REC: Receive • Recife Guararapes (Brazil-PE) (location identifier) • Record • Recorder • Recuado • Regional Environment Center

Recap: Recapitulate

RECAP: Reliability Evaluation Corrective Action Program

Recd: Received

Recep: Reception

Recept: Receptacle

RECERT: Recertification

RECFM: Record Format

Rech: La Recherche (periodical code) • Recherche

RECHG: Recharge

Recíp: Recíproco

Recip: Reciprocal • Reciprocating • Reciprocity

RECIP: Reciprocate

Recirc: Recirculate

RECITE: Regions and Cities for Europe

Reclam: Reclamation

Recm: Recommend

Recmd: Recommended

RECODIS: Recherche Collaborative à Distance

Recog: Recognition • Recognize

RECOG: Recognition

Recogd: Recognized

RECON: Reconnaissance • Remote Console • Rendezvous with a Comet Nucleus • Retrospective Conversion

RECONFIG: Reconfiguration

RECONN: Reconnaissance

RECONS: Research Consortium on Nearby Stars

RECORD: Russia-Europe Cooperation on Rocket engine Demonstration • Russian-European Cooperation on Rocket Demonstration (*)

RECOV: Recovery

RECP: Receptacle • Record Engineering Change Proposal • Request for Engineering Change Proposal

Recpt: Receipt

RECRC: Recirculate

RECS: Representative Environmental Control System • Reseller Electronic Communication System

Rect: Receipt • Rectangle • Rectangular • Rectify

RECT: Rectaguarda • Rectifier • Rectify

RECTAS: Regional Centre for Training in Aerospace Surveys

RECV: Receive

RECVR: Receiver

Red: Redeemable • Reduce • Reduced • Reduction • Redundant

RE&D: Research, Engineering & Development

RED: Lewiston - Mifflin County (USA-PA) (location identifier) • Random Early Detect • Random Early Detection • Range Error Detector • Reduction • Reedsville - Mifflin County (USA-PA) (location identifier) • Relais Électroniques Deutsch • Research-Engineering-Development • Revision Effective Date

REDAR: Radiation and Environmental Data Acquisition and Recording

REDARS: Reference Engineering Data Automated Retrieval System

REDCAP: Real-time Digitally Controlled Analyzer and Processor

REDG: Réseau Européen de Documentation Géologique

REDI: Relational Database for Images

REDOSI: Répertoire de la Documentation Scientifique et Informelle

REDSTAR: Resource Data Storage And Retrieval

REDUN: Redundancy

REDVLPG: Redeveloping

REE: Lubbock Reese (USA-TX) (location identifier) • Rare-Earth Element • Régimen Especial de Estudiantes • Remote Exploration and Experimentation (NASA) • Responsible Equipment Engineer

REEP: Regional Environment Employment Program

Réf: Référence

Ref: Refer • Referee • Reference • Referencia • Referred • Refining • Reform • Reformed • Refrán • Refugio • Refund • Refunding

REF: Range Error Function • Reference • Reference File • Refurbish • Refurbishment • Relative Efficiency Factor • Répartition Électronique de Freinage

Refd: Referred • Refund

Refl: Reflect • Reflection • Reflective • Reflex • Reflexive

REFLEX: Remote Fluvial Experiment • ROSAT ESO Flux Limited X-ray

Refrig: Refrigerate • Refrigeration • Refrigerator

REFRIG: Refrigerate

REFRMG: Reforming

REFSPEC: Reflectance Spectrometer

REFURB: Refurbish

Reg: Region • Register • Registered • Registration • Registre • Registro • Registry • Regla • Regulación • Regular • Regularly • Regulated • Regulation • Regulator • Regulatory

Rég: Région

REG: Reggio Calabria (Italy) (location identifier) • Register • Regourd Aviation • Regulate • Regulator

REGAL: Range and Elevation Guidance for Approach and Landing • Rigid Epoxy Glass Acrylic Laminate

Reg.-Bez.: Regierungsbezirk

Reg.Bez.: Regierungsbezirk

Regd: Registered

REGEN: Regenerate

REGIS: Regional Geographical Information System • Remote Graphics Instruction Set

Regr: Registrar

Regres: Regresivo

Regt: Regent • Regiment

REH: Rehoboth Beach (USA-DE) (location identifier)

REHE: Relativistic Effects in Heavy Elements

REI: Range from Entry Interface • Reactor Experiments Inc. (USA-CA) • Réduction pour Embauche et Investissement • Regina (French Guiana) (location identifier) • Rigid External Insulation • Runway End Identification

REIC: Rare Earth Information Center (USA-IA)

REIL: Runway End Identification Lights

Reimp: Reimpresión

Reinf: Reinforce

REINF: Reinforce

Reinfd: Reinforced

Reinfmt: Reinforcement

REINIT: Reinitialize

REIO: Regional Economic Integration Organization

Reit: Reiteration

REIT: Real-Estate Investment Trust

REJ: Reject

REK: Reykjavík (Iceland) (location identifier)

REKLIP: Regio-Klima-Projekt

Rej: Reject

REJ: Reject

REJASE: Reusing Junk As Something Else

Rel: Relate • Related • Relating • Relation • Relative • Relatively • Release • Released • Relié • Religión • Religion • Religious

REL: Recommended Exposure Limit • Relative • Renaissance Experimental Laboratory • Restricted Energy Loss • Ripper Excavator Loader • Trelew (Argentina-CB) (location identifier)

RELAB: Reflectance Experiment Laboratory

RelComm: Reliable Communications

Relctd: Relocated

Relig: Religion • Religious

RELISH: Responses Evaluated by Listing Information in SEED Headers

RELMAT: Relative Matrix

Reloj: Relojería

RELOS: Regional Locator System • Relative radiometer for Oscillations

RELSAT: Reliability for Satellite

RELSECT: Relative Sector

Rem: Remainder • Remaining • Remark • Remarque • Remit • Remittance • Remote • Remove

REM: Radiation Environment Monitor • Random Entry Memory • Rapid Eye Movement • Rare-Earth Metal • Rat Enclosure Module • Reaction Engine Module • Recognition-Extended Memory • Reflection Electron Microscope • Relabeling Exchange Method • Release Engage Mechanism • Release Engagement Monitor • Release Engine Module • Release Escape Mechanism • Remaining • Remark • Remote • Remote Equipment Module • Remote Ethernet Management • Resource and Environmental Management • Ring Error Monitor • Rocket Engine Module • Röntgen Equivalent in Man

REMAP: Regional Environmental Mapping and Analysis Process

REMAPP: Remote Sensing Array Processing Procedures

Remb: Remboursable • Remboursé • Remboursement

REMD: Red Medium Dispersion

REMIDAS: Resource Evaluation and Monitoring Integrated Data Analysis System

Remitt: Remittance

REML: Radiological Environmental Monitoring Laboratory • Residual Maximum Likelihood Estimation

REMOB: Remote Observation

REMOBS: Remote Observation System

REMPEC: Regional Maritime Pollution Emergency Centre

REMPI: Resonantly-Enhanced Multiphoton Ionization (*) • Resonance Enhanced Multiphoton Ionization

REN: Orenburg (Russia) (location identifier) • Reduced Exterior Noise • Remote Enable • Rename • Ring Equivalence Number

RENASS: Réseau National de Surveillance Sismique (France)

RENATER: Réseau National de Télécommunications pour la Technologie, l'Enseignement et la Recherche (France) (*)

RENC: Regional Electronic Navigation Center • Regional Electronic Navigation Chart

REND: Rendezvous

ReNEv: Research in Nondestructive Evaluation (periodical code)

Renf: Renfrewshire (UK-Scotland)

RENFE: Red Nacional de Ferrocarriles Españoles (Spain)

Renor: Renormalization (periodical code)

REO: Rare-Earth Oxide • Removable-Erasable-Optical • Rome State (USA-OR) (location identifier)

REOSC: Recherches et Études d'Optique et de Sciences Connexes (France)

REOT: Right-End-Of-Tape

rep: Repetatur

Rep: Repair • Repeat • Repetition • Report • Reported • Reporter • Represent • Representative • Representing • Reproducción • Reprint • Republic • República • Republican • Republik

Rép: Réponse • Républicain • République

REP: Réacteur à Eau sous Pression • Regional Environmental Plan • Regional Exchange Point • Remote Experimental Package • Repeat • Reporting Point • Reproducing Program • Research Initiation Program • Roentgen Equivalent Physical • Siem Reap (Cambodia) (location identifier)

REPART: Repartição

REPE: Repeat while Equal

Repl: Replace • Replaced • Replacement

REPL: Replace

REPLEX: Redundancy Planning Expert

REPLI: Radar-coupled Emitter/Passive Location and Identification

REPNE: Repeat while Not Equal

REPNZ: Repeat while Not Zero

Repr: Représentant • Represent • Representation • Représentation • Representative • Represented • Représenter • Representing • Reprint • Reprinted • Reproduce • Reproductif • Reproduction • Reproductive • Reproduire

REPR: Repressurization

Repres: Represent • Representation • Representative

Représ: Représentant • Représentation • Représenter

Repro: Reproduce • Reproduced • Reproduction

REPRO: Reproduction

Rept: Receipt • Report

Repub: Republic • Republican

REPZ: Repeat while Zero

Req: Request • Require • Required • Requisition

REQ: Request • Require

Reqd: Required

REQD: Required

REQMTS: Requirements

Reqt: Requirement

RER: Renewable Energy Resource • Réseau Express Régional • Reusable-Expendable-Reusable • Rough Endoplasmic Reticulum

RERC: Regional Educational Resource Center

REREX: Remote Readout Experiment

RERF: Radiation Effects Research Foundation

RERIC: Renewable Energy Resources Information Center

RERV: Radar Estimated Rain Volume

Res: Rescue • Research • Researcher • Reservation • Reserve • Reserved • Reservoir • Residence • Resident • Residue • Resigned • Resistance • Resistor • Resolution • Resource • Resume

Rés: Réserve • Réservé • Réserver • Résidence

RES: Rachat d'Entreprise par les Salariés • Rachat d'Entreprise par ses Salariés • Radiation-Enhanced Sublimation • Radio Equipment and Systems • Record Evaluation System • Regional Economic Study • Regional Evaporation Study • Remote Execution Service • Renewable Energy Source • Renewable Energy Systems Ltd. • Reserve • Reservoir • Reset • Resistencia (Argentina-CH) (location identifier) • Resolution • Resource

RESAC: Regional Earth Science Applications Center

RESACENT: Remote Sensing Applications Centre

Resc: Rescind

RESCEU: Research Center for the Early Universe (Japan)

RESCU: Remote Emergency Satellite Cellular Unit

Resd: Residência • Residêncial

RESDA: Remote Sensing Development Agency

RESEDA: Remote Sensor Data Analysis

Resgnd: Resigned

Resid: Residual

RESID: Residencial • Residual

Resig: Resignation

RESINS: Redundant Strap-down Inertial Navigation System

RESIS: Remote Sensing Information Sensing • Resistance

Resln: Resolution

ReSLT: Reusable Space Launch Technology

RESMO: Respiratory Monitoring

Resn: Resonant

RESORS: Remote Sensing Online Retrieval Service • Remote Sensing Online Retrieval System (*)

Resp: Respectif • Respective • Respectively • Respectivement • Respektive • Respondent • Response • Respublika

ReSp: La Recherche Spatiale (periodical code)

RESP: Responsibility

RESPAS: Remote Sensing Preprocessing and Archiving System • Remote Sensing Processing and Archiving System

RESPECT: Rapid Electromagnetic Scattering Predictor for Extremely Complex Targets

Ress: Ressource

RESSAC: Radar pour l'Étude du Spectre des Surfaces par Analyse Circulaire

RESSYS: Regional Ecosystem Simulation System

Rest: Restaurado • Restauré • Restoration • Restored • Restrict • Restricted • Restriction

REST: Reader Evaluation and Selection Technique

RESTEC: Remote Sensing Technology Center (Japan)

Restr: Restaurant • Restorer • Restrict • Restricted • Restriction

RESVR: Reservoir

Ret: Retail • Retain • Retained • Reticulum-Reticuli (constellation) • Retired • Retórica • Return • Returnable • Returned

RE&T: Research Engineering & Test

RET: Reliability Evaluation Test • Renewable Energy Technology • Resolution Enhancement Technology • Retract • Return • Røst (Norway) (location identifier)

RETA: Regional Environmental Technical Assistance • Retrieval of Enriched Textual Abstracts • Revised Estimated Time of Arrival

RETAIN: Remote Technical Assistance and Information Network

Retd: Retained • Retired • Returned

RETD: Red Especial de Transmisión de Datos (Spain)

RETE: Research on Electrodynamic Tether Effects

RETEN: Retention

RETIM: Réseau Télé-Informatique de la Météorologie

RETINA: Réseau Téléinformatique National (France)

RETM: Rare-Earth Transition Metal

Retn: Retain

Retnr: Retainer

RETOOLE: Remote Tools for EUVE

RETP: Research Experience in Teacher Preparation

Retr: Retract • Retractable • Retrieval • Retrieve • Retrieved

RETR: Retrieval

Retro: Retroactive • Retrograde

RETRO: Retrofire

RETRV: Retrieve

RETS: Reconfigurable Electrical Test Stand

RETV: Retrieval

REU: Île de la Réunion (ISO country code) • Remote Electronics Unit • Research Experience for Undergraduates • Research Evaluation Unit • Réunion Island (ISO country code) • Reus (Spain) (location identifier)

REUN: Réseau Européen Unifié de Nivellement

REUNIR: Réseaux des Universités et de la Recherche

Rév: Révérend • Révision • Révolution

Rev: Revelation • Revenue • Reverend • Reverse • Reversed • Review • Reviewed • Revisado • Revise • Revised • Revision • Revisión • Revolution • Revolve • Revolving • Revue

REV: Reentry Vehicle • Reverendo • Reverse • Review • Revision • Revolution

REVAD: Royal Observatory Edinburgh (Starlink internal code)

REVCON: Revolutionary Concept

REVP: Research Expansion Verification Program • Research Experimental Verification Program

REVS: Requirements Engineering and Validation System

REVTOLC: Retrieval Experiment - Virginia Tech Online Catalogue

Rew: Reward • Rewind

REW: Rewa (India) (location identifier) • Rewind

REX: Radiation Experiment • Relocatable Executable • Reynosa (Mexico) (location identifier) • RIU Exerciser

REXEC: Remote Executable • Remote Execution

REXX: Restructured Extended Executor

REY: Reyes (Bolivia) (location identifier)

Rf: Reef • Rufiyaa (currency) • Rutherfordium

RF: Radio Facility • Radio Fréquence • Radio Frequency • Range Finder • Raumfahrtforschung (periodical code) • Read Forward • Reception Fair • Redistribution Function • Reducing Flame • Reef • Relative Flow • Relevance Feedback • Representative Fraction • Reproduction Fidelity • Reprographic Facilities • République Française • Research Foundation • Retention Factor • Return Flux • Reusable Function • Right Forward • Risk Factor • Roll Film • Rwanda Franc (currency)

RFA: Rafai (Central African Rep.) (location identifier) • Rate of False Alarms • République Fédérale d'Allemagne • Request For Action • Request For Algorithm

RFAD: Relative Frequency in All Documents

RFB: Ready For Baseline • Recording For the Blind • Request For Bid

RFB&D: Recording For the Blind & Dislexic

RFBD: Recording For the Blind and Dislexic

RFBR: Russian Foundation for Basic Research

RFC: Radio Frequency Chart • Radio Frequency Choke • Radio Frequency Compatibility • Regenerative Fuel Cell • Remote Function Call • Request For Change • Request For Comment • RGS Focal Camera • Royal Flying Corps

RFCA: Radiator and Flow Control Assembly • Radiator Flow Control Assembly

RFCM: Radio Frequency Countermeasure

RFCP: Request For Computer Program

RFCT: Radio-Frequency Compatibility Test

RFD: Radio-Frequency Device • Range Focused Doppler • Reporting For Duty • Request For Deviation • Request For Discussion • Requirement Formulation Document • Revue Finale de Définition • Rockford - Greater Rockford (USA-IL) (location identifier) • Rural Free Delivery

RFDA: Radio Frequency Deployed Assembly

RFDD: Range Focused Doppler Display

RFDL: Radio Frequency Data Link

RFDP: Rotating Full Disk Photometer

RFDS: Royal Doctor Flying Service (Australia)

RFDU: Radio Frequency and Digital Unit • Radio Frequency Distribution Unit

RFE: Regional Finite Element • Request For Enhancement • Request For Estimate • Requisition For Expenditure

RFECP: Request For Engineering Change Proposal

RFEEI: Radio Frequency Engineers Education Initiative (UK)

RFEI: Request For Engineering Information

RFF: Request For File • Resources For the Future

RFG: Refugio Rooke (USA-TX) (location identifier)

Rfgt: Refrigerant

RFHA: Reusable Flight Hardware Authorization

RFI: Radio Frequency Interface • Radio Frequency Interference • RCRA Facility Investigation • Relative Fluorescence Intensity • Remote Facility Inquiry • Remote File Inquiry • Request For Information

RFID: Radio-Frequency Identification

RFIS: Radio Frequency Instrument System • Remote Frequency Indicator Selector

RFK: Anguilla Rollang (USA-MS) (location identifier)

RFL: Reflect

Rflg: Refueling

RFLP: Restriction Fragment Length Polymorphism

Rflx: Reflex

RFM: Refocusing Mechanism

RFMS: Remote File Management System

RFN: Radio Frequency Noise • Raufarhöfn (Iceland) (location identifier) • Reflection Nebula

RFNA: Red Fuming Nitric Acid

RFNM: Request For Next Message

RFO: Request For Offer • Request For Order

RFOI: RF-Optical Interface

RFOV: Resolution Field Of View

RFP: Raiatea Island (Society Islands) (location identifier) • Raumflugplanetarium (Halle, Germany) • Request For Program • Request For Proposal • Requirements and Formulation Phase • Requirements Formulation Phase • Reversed-Field Pinch • Rutgers Fabry-Pérot

RFPA: Request For Proposal Authorization

RFPC: Radio-Frequency Pulse Compression

RFPU: Request For Proposal Update

RFQ: Radio-Frequency Quadrupole • Request For Quotation • Request For Quote

RFR: Rate of False Resolutions • Refrigerator Freezer Rack • Réseau Formation-Recherche • Retract Feed Rate • Rio Frio (Costa Rica) (location identifier)

Rfrd: Referred

RFrM: Revue Française de Mécanique (periodical code)

RFS: Radio Frequency Sensor • Radio Frequency System • Remote File Service • Remote File Sharing • Remote File System (*) • Renormalized Forward Scattering • Request For Service • Research Flight Simulator • Response Feedback System

RFSC: Rubberized Friction Seal Coat

RFSE: Radio Frequency Shielded Enclosure

RFSH: Refresh

RFT: Ready For Training • Real Fourier Transform • Request For Technology • Revisable Form Text • Richest-Field Telescope • Rich-Field Telescope

RFTC: Radio Frequency Test Console

RFTCP: Request For Test Change Proposal

RFTF: Radio Frequency Test Facility

RFTPCE: Reduced-Fill Tank Pressure Control Experiment

RFTS: Radio Frequency Transmission System • Remote Fiber Test System

RFU: Radio Frequency Unit • Ready For Use • Reserved for Future Use

RFV: Radiometro de Frequencia Variavel

RFW: Request For Waiver

RG: Radio Galaxy • Radio Guide • Rate Gyro • Reading (UK postal code) • Recursive Grid • Red Giant • Régimen General • Regular Grammar • Renormalization Group • Response Generator • Retractable Gear • Rio Group • Rive Gauche

RGA: Rate Gyro Assembly • Reduction Gas Analyzer • Reflection Grating Array • Relative Growth Rate • Remote Geological Analysis • Residual-Gas Analyser • Residual-Gas Analysis • Return Goods Authorization • RGS Grating Assembly • Río Grande (Argentina-TF) (location identifier)

RGB: Red-Giant Branch • Red-Green-Blue • ROSAT - Green Bank

RGBBL: ROSAT - Green Bank BL Lacertae (object)

Rgbf: Rangierbahnhof

RGC: Reference Great Circle • Remote Global Contact

RGCPA: Rapporteur Group on Coordination of Profile Activities

RGCT: Rapporteur Group on Conformance Testing • Reflective Gross Cloud Test

RGCV: Randomized Generalized Cross-Validation

Rgd: Registered • Rigid

RGD: Ragged • Raleigh-Gans-Debye • Rarefied Gas Dynamics • Regular Geophysical Day • Rijksgebouwendienst • Rijks Geneeskundige Dienst • Rijks Geologische Dienst (Netherlands)

Rge: Range • Ridge

RGE: Porgera (Papua New Guinea) (location identifier) • Range

RGF: Regional Growth Forecast

RGH: Balurghat (India) (location identifier)

RGI: Rangiroa Island (Tuamotu Islands) (location identifier) • Regional-Global Interaction

RGIC: Rare Gas Ionization Cell

RGIS: Regional Geographic Information System

RGISM: Raster Geographic Information System for Mapping

RGL: Rate Gyroscope Limit • Rio Gallegos (Argentina-SC) (location identifier)

Rglr: Regular

Rglt: Regulate

Rgltd: Regulated

Rgltr: Regulator

RGM: Redundant Gyro Monitor • Reflection Grating Module • Remote Geophysical Monitor

Rgn: Rangoon (Myanmar) • Region

RGN: Rangoon Mingaladon (Myanmar) (location identifier) • Region • Yangon Mingaladon (Myanmar) (location identifier)

RGNP: Real Gross National Product

RGO: Royal Greenwich Observatory (UK-England) (now closed)

RGOB: Royal Greenwich Observatory Bulletin (periodical code)

RGP: Raster Graphics Processor • Rate Gyro Package • Reduced Game Property

RGPS: Radarsat Geophysical Processing System • ROSAT Galactic Plane Survey

RGR: Ranger (USA-TX) (location identifier) • Recursive Grid Refinement • Regueirãa • Relative Growth Rate • Repairable Generator Rate

RGRF: Repair Generation Rate Forecast

RGRMA: Rate Gyro Redundancy Management Algorithm

RGRP: Regional Geochemical Reconnaissance Programme

RGS: Radio Guidance System • Rate Gyro System • Reflection Grating Spectrometer • Reflection Grating System • Remote Ground Switching • Rocket Guidance System • Royal Geographical Society (UK)

RGT: Relativistic Gravitational Theory • Rengat (Indonesia) (location identifier) • Renormalization Group Theory • Réseau de Gestion des Télécommunications

Rgtr: Register

RGU: Rate Gyroscope Unit • Robert Gordon University (UK-Scotland) • Rot-Gelb-Ultraviolet

Rh: Rhodium

RH: Haiti (international vehicle registration code) • Radio-Héliographe • Razdow Heliograph • Redhill (UK postal code) • Relative Humidity • Remote Handling • Ressources Humaines • Riemann-Hilbert • Right Half • Right Hand • Right-Handed

RHA: Rama Horizontal Azul • Reykholar (Iceland) (location identifier)

RHAPIT: Rhine/Main Area Project for Integrated Traffic

RHB: Red Horizontal Branch

RHBNC: Royal Holloway and Bedford New College (UK-England)

RHC: Rao-Hartley-Cochran (method) • Remote Hand Controller • Right-Hand Circular • Rotational Hand Controller • Rotation Hand Controller

RHCM: Relative Humidity Control/Monitor

RHCP: Right-Hand Circularly-Polarized • Right-Hand Circular Polarization

RHCTO: Rotational Hand Controller Takeover

Rhd: Railhead

RHD: Right-Hand Drive • Rio Hondo (Argentina-SE) (location identifier)

RHDP: Right Hand Decimal Point

RHE: Radiation Hazards Effect • Reims (France) (location identifier)

RHEB: Right-Hand Equipment Bay

RHEC: Radiation-Hardened Emulating Computer

RHEED: Reflected High-Energy Electron Diffraction • Reflection High-Energy Electron Diffraction

Rheinhess: Rheinhessen (Germany)

Rheinl: Rheinland (Germany)

Rheo: Rheostat

RHEO: Rheostat

Rhet: Rhetoric • Rhetorical

RHFEB: Right-Hand Forward Equipment Bay

RHG: Ruhengeri (Rwanda) (location identifier)

RHI: Radar Height Indicator • Range-Height Indicator • Relative Humidity Indicator • Rhinelander (USA-WI) (location identifier)

RHIC: Relativistic Heavy-Ion Collider (*) • Relativistic Heavy-Ion Collision

RHIP: Rank Has Its Privileges

RHISE: Remote Hitchhiker Investigation Support Environment

RHKO: Royal Hong Kong Observatory (now HKO)

RHL: Ranked Half-Life • Residual Hazards List • Roy Hill (Australia-WA) (location identifier)

RHM: Relative Humidity Meter • Relative Humidity Monitor • Rodent Habitat Module

Rho: Rhodesia

RHO: Rhodes (Greece) (location identifier) • Rosemary Hill Observatory (USA-FL)

Rhod: Rhodesia

RHP: Ramechhap (Nepal) (location identifier) • Revised Harvard Photometry • Riemann-Hilbert Problem • Right-Hand Panel

RHRI: ROSAT High-Resolution Imager

RHRZ: Regionales Hochschulrechenzentrum

RHS: Right-Hand Side • Riverview High School (USA-FL) • Rocketdyne Hybrid Simulator • Round-Headed Screw

RHSC: Right-Hand Side Console

RHT: Radiant Heat Temperature • Reynolds Hydrodynamic Theory

RHU: Radiactive Heater Unit • Radio Heating Unit • Radioisotope Heater Unit

RHV: San Jose - Reid-Hillview (USA-CA) (location identifier)

RHW: Response Header Word • Router Header Word

RHZ: Root Hair Zone

Ri: Richardson Number

R&I: Receiving & Inspection • Regina & Imperatrix • Rex & Imperator

R/I: Receiving Inspection

RI: Indonesia (international vehicle registration code) • Radiation Intensity • Radio Interference • Radioisotope • Range Instrumentation • Raw Image • Read-In • Real Image • Receiving Inspection • Recherche d'Information • Recomendaçao Internacional • Referential Integrity • Reflective Insulation • Refractive Index • Regina et Imperatrix • Register Immediate • Relational Indexing • Relative Intensity • Reliability Index • Remedial Investigation • Remote Inhibit • Report of Investigation • Représentation Intermédiaire • Reverberation Index • Rex et Imperator • Rhode Island (USA) • Right Intermediate • Ring Indicator • Rockwell International Corporation • Routing Identifier • Routing Indicator • Royal Institution

RIA: Robotic Industries Association • Robot Institute of America • Royal Irish Academy • Santa Maria (Brazil-RS) (location identifier)

RIAA: Recording Industry Association of America (USA)

RIACA: Research Institute for Applications in Computer Algebra

RIACS: Research Institute for Advanced Computer Science (USA-CA)

RIAGG: Regionaal Instituut voor de Ambulante Gestelijke Gezonheidszorg

RIAO: Recherche d'Information Assistée par Ordinateur

RIAS: Radar à Impulsion et Antenne Synthétiques • Radio In the American Sector

RIASS: ROSAT-IUE All Sky Survey

RIB: Radioactive Ion Beam • Recoverable Item Breakdown • Relevé d'Identité Bancaire • Research Internet Backbone • Réseau Informatique des Bibliothèques (France) • Riberalta (Bolivia) (location identifier) • Right Inboard

RIBE: Reactive Ion Beam Etching

Ric: Ricerca

RIC: Radiation-Induced Change • Radiation-Induced Conductivity • Regional Information Center • Relative In Centrality • Remote Indicating Compass • Remote Information Center • Request for Instrument Clarification • Research and Information Centre • Resistance-Inductance-Capacitance • Resource Information Center • Richmond (USA-VA) (location identifier) • Rockwell International Corporation • Routing Identification Code

RICA: Radar Image Classification Aid • Regole Italiane di Catalogazione per Autori

RICC: Remote Intercomputer Communications

RICE: Radio Ice Čerenkov Experiment • Relevé d'Identité de Caisse d'Épargne • Remote Inertial and Control Equipment

RICERCAR: Reading Initials Clearly Exhibits RACRECIR's Concealed Auto-Reference • Rigid Internal Codes Exclusively Rule Computers and Robots

RICH: Ring-Imaging Čerenkov

RICIS: Research Institute for Computing and Information System

RICL: Roditi International Corporation Ltd. (UK)

RICP: Real-time Interactive Computer Program

RICS: Range Instrumentation Control System • Remote IMAX Camera System

RICST: Real-time Interferometer Control Software Testbed

RID: Rack Insertion Device • Radial Immunodiffusion • Radiation-Induced Defect • Radiation-Induced Diffusion • Range Instruments Development • Remote Isolation Device • Report Identifikationsnummer • Reset Inhibit Drive • Review Item Discrepancy (*) • Review Item Disposition • Richmond (USA-IN) (location identifier) • Row Identifier

RIDA: Raster Image Device Accelerator • Reverse Isotope Dilution Assay

RIDAC: Range Interference Detection And Control

RIDDLE: Rapid Information Display and Dissemination in Library Environnements

RIDE: Research Issues in Data Engineering

RIDF: Random Input Describing Function

RIDGE: Ridge Interdisciplinary Global Experiment

RIDI: Receiving Inspection Detailed Instruction • Receiving Inspection Detail Instruction

RIDS: Radar Image Display System • Radio Information Distribution System • Range Information Display System • Receiving Inspection Data Status

RIDT: Raster Imaging and Digital Typography

RIE: Range of Incentive Effectiveness • Reactive Ion Etching • Rice Lake (USA-WI) (location identifier) • Right Inboard Elevon

Rif: Rifugio

RIF: Radio Interference Filter • Rate Input Form • Recherche dans l'Intérêt des Familles • Reduction In Force • Relative Importance Factor • Reliability Improvement Factor • Remote Interface • Richfield (USA-UT) (location identifier) • Routing Information Field

RIFA: Retry, Ignore, Fail, Abort?

RIFF: Raster Image File Format • Resource Interchange File Format

RIFI: Radio Interference Field Intensity • Radio-Interference-Free Instrument

RIFIM: Radio Interference Field Intensity Meter

RIFL: Random Item File Locator

RIFMA: Roentgen-Isotope Fluorescent Method of Analysis

RIFS: Radioisotope Field Support

RIFT: Reactor-In-Flight Test

RIFTS: Reactor-In-Flight Test System

RIG: Radio Inertial Guidance • Rate-Integrating Gyro • Rate Integration Gyro • Regional Information Group • Remote Imaging Group • Reuse Interoperability Group • Rio Grande (Brazil-RS) (location identifier) • Rivista Italiana di Geofisica (periodical code)

RIGA: Rate-Integrating Gyro Assembly

RIGES: Renewable Intensive Global Energy System

RIGGS: Ross Ice Shelf Geophysical and Glaciological Survey

RIGI: Receiving Inspection General Instruction

RIGIS: Relational Image-based Geographic Information System

RIGS: Radio Inertial Guidance System

RIGSA: Rivista Italiana di Geofisica e Scienze Affini (periodical code)

RIGTC: Research Institute of Geodesy, Topography and Cartography

RIHED: Regional Institute of Higher Education and Development

RII: Receiving Inspection Instructions • Required Inspection Item

RIIA: Royal Institute of International Affairs

RIJ: Rioja (Peru) (location identifier)

RIK: Carillo (Costa Rica) (location identifier) • Replacement In Kind

RIKES: Raman-Induced Kerr Effect Scattering

RIKZ: Rijksinstituut voor Kust- en Zeeonderzoek (Netherlands)

RIL: Radio Influence Level • Radio Interference Level • Recherche et Ingénierie Linguistiques • Recoverable Items List • Red Indicator Light • Repairable Items List • Representation Independent Language • Rifle - Garfield County (USA-CO) (location identifier)

RILA: Répertoire International de la Littérature sur l'Art

RILB: Radio Interférence à Longue Base

RILEM: Réunion Internationale des Laboratoires d'Essais sur les Matériaux

RILIS: Resonance Ionization Laser Ion Source

RILO: Research and Industrial Liaison Officer

RILS: Regularized Iterative Least Squares

R/IM: Refrigerator/Incubator Module

RIM: Mauritania (international vehicle registration code) • Read-In Mode • Read Interrupt Mask • Regional Ice Model • Rekaleszenz In Matrix • Reliability and Integrity of Media • Remote Installation and Maintenance • Resource Inventory Management • Rodriguez de Men (Peru) (location identifier)

RIME: Real-time Information Management Environment • Relaynet International Message Exchange

RIMF: Remote Integrated Medical Facility

RIMI: Research Improvement in Minority Institutions

RIMP: Remote Input Message Processor

RIMS: Radiant Intensity Measuring System • Radio Interference Measuring System • Research Institute for Mathematical Sciences • Resonance Ionization Mass Spectrometry • Resource Information Management System • Retarding Ion Mass Spectrometer

RIMU: Redundant Inertial Measurement Unit

RIN: Rapporteur Group for Internationalization • Record Identification Number • Reference Indicator Number • Relative Intensity Noise • Research Internet • Ringi Cove (Solomon Islands) (location identifier) • Royal Institute of Navigation (UK)

RINET: Reinsurance Network

RINEX: Receiver Independent Exchange (format)

RING: Relational Information Grouping

RINT: Réseau International de Néologie et de Terminologie

RIO: Radar Intercept Officer • Relative Ionospheric Opacity • Reporting In and Out • Réseau Intertropical d'Ordinateurs • Rio de Janeiro (Brazil-RJ) (location identifier) • Russian Interface Officer

RIOA: Real Instituto y Observatorio de la Armada (Spain)

Rioj: Riojano

RIOS: Radio, Infared and Optical Sensors • Rotating Image Optical Scanner

RIOT: Real-Time Input-Output Transducer • Real-Time Input-Output Translator • Remote Input-Output Terminal • Retrieval of Information by Online Terminal

RIP: Radar Image Processor • Raster Image Processing • Raster Image Processor (*) • Raster Input Processor • Readiness Improvement Program • Receiving Inspection Plan • Receiving Inspection Procedure • Register Indicator Panel • Relevé d'Identité Postal • Reliability Improvement Program • Remain In Place • Remote Imaging Protocol • Report on Individual Personnel • Requiescat In Pace • Respiratory Impedance Plethysmograph • Rest

In Peace • Router Information Protocol • Routine Information Protocol • Routing Information Protocol (*) • Routing Interchange Protocol

RIPAS: Regional Integrated Protected Areas System

RIPC: Remote Interprocess Communication

RIPE: RACE Integrity Primitive Evaluation • Range Instrumentation Performance Evaluation • Réseaux IP Européens

RIPFCOMTF: Rapid Item Processor to Facilitate Complex Operations on Magnetic-Tape Files

RIPL: Representation Independent Programming Language

RIPM: Répertoire International de la Presse Musicale

RIPP: Regional Industrial Property Programme

RIPS: Radar Impact Prediction System • Radioisotope Power Supply • Radioisotope Power System • Remote Image Processing System • Remote Information Processing System • Road Ice Prediction System

RIPSO: Revised INTERNET Protocol Security Option

RIPT: Réseau d'Informations Pédagogiques et Techniques

RIQ: Radio Intermediate QSO

RIQS: Remote Information Query System

RIR: Receiving Inspection Report • Relative Index Register • Reportable Items Report • Riverside Fla-Bob (USA-CA) (location identifier)

RIRIG: Reduced-Excitation Inertial Reference Integrating Gyro

RIRMS: Remote Information Retrieval and Management System

RIRS: Reliability Information Retrieval System

RIS: Radiocommunication Information System • Random Input Sampling • Rapid Imaging System • Receiving Inspection Segment • Record Input Subroutine • Reflector In Space • Regional Information Service • Regional Intensive Session • Registered Image Sequence • Regulatory Information Service • Regulatory Information System • Remote Information System • Report Identification Symbol • Reporting Identification Symbol • Research Information Service • Research Information System • Resonance Ionization Spectroscopy • Retroreflector In Space • Rishiri (Japan) (location identifier) • Rotating Image Scanner • Rural Information System

RISAR: Rotation-Invariant Simultaneous Autoregressive

RISC: RadioAstron International Science Committee (Russia) • Radioastronomy International Science Council • Radioastronomy International Scientific Council (*) • Reduced Instruction Set Computer • Reduced Instruction Set Computing • Remote Information Systems Center • Rockwell International Science Center (USA-CA)

RISD: Richardson Independent School District (USA-TX)

RISDE: Russian Institute of Space Device Engineering

RISE: Radiative Input from the Sun to Earth • Radiative Input of the Sun to Earth (*) • Readiness Improvement Status Evaluation • Readiness Improvement Summary Evaluation • Regional Initiative in Science Education • Register for International Service in Education • Relatively Invariant under Strategic Equivalence • Reliability Improvement Selected Equipment • Research Information Service for Education • Resources for Involving Scientists in Education (project)

RISLU: Remote Integrated Services Line Unit

RISOP: Red Integrated Strategic Operations Plan

RISP: Robot d'Intervention en Site Planétaire • Robot d'Intervention sur Site Planétaire • Ross Ice Shelf Project

RISRJ: Report Ionosphere Space Research Japan (periodical code)

R&ISS: R&I SIC&DH Simulator

RISS: Real-time IR/EO Scene Simulation • Refractive Interstellar Scintillation • Relational Inquiry and Storage System

RISSAC: Research Institute for Soil Science and Agricultural Chemistry

Rist: Ristorante

RIST: Radioactive Ion Source Testbed

RIT: Radio-Immunotherapy • Radio Information Test • Rate of Information Throughput • Red Interamericana de Telecomunicaciones • Relative Ignition Temperature • Request for Interface Tool • Request for Interface Tooling • Rio Tigre (Panama) (location identifier) • Rochester Institute of Technology (USA-NY)

RITA: Radio-frequency Ionisation Thruster Assembly • Radio-frequency Ion Thruster Assembly • Recognition of Information Technology Achievement (UK) • Recoverable Interplanetary Transport Approach • Reinvented Triple Axis • Reusable Interplanetary Transport Approach

RITC: Regional Information Technology Coordinator

RITE: Rapid Information Technique for Evaluation

RITS: Radiatively Important Trace Substance

RITSEC: Regional Information Technology and Software Engineering Center

RITSI: Radiative Ignition and Transition to Spread Investigation

RITS: Rapid Information Transmission System • Remote Input Terminal System

RITSS: Raytheon Information Technology and Scientific Services (*) • Raytheon Information Technology and Systems Services

RITZ: Regulatory and Investigative Treatment Zone

RIU: Remote Interface Unit

Riv: River

RIV: Recirculation Isolation Valve • Retrospective Informationsverbreitung • River • Riverside March (USA-CA) (location identifier) • Riverview College Observatory (Australia-NSW)

RiVas: Ridges and Valleys

RIVM: Rijksinstituut Voor Milieuhinder • Rijksinstituut voor Volksgezondheid en Milieuhygiëne (Netherlands)

RIVR: River

RIW: Recht der Internationalen Wirtschaft • Riverton (USA-WY) (location identifier)

RIX: Riga Spilve (Latvia) (location identifier) • Russia Internet Exchange

RIXOS: ROSAT International X-ray Optical Survey

RIY: Relative Ion Yield • Riyan Mukalla (Yemen) (location identifier)

RIZ: Rio Alzucar (Panama) (location identifier)

RIZIV: Rijksdienst voor Ziekte- en Invaliditeitsverzekering • Rijksinstituut voor Ziekte- en Invaliditeitsverzekering (Belgium)

RJ: Ramjet • Rayleigh-Jeans • Reaction Jet • Rio de Janeiro (Brazil) • Road Junction

RJA: Rajahmundry (India) (location identifier)

RJB: Rajbiraj (Nepal) (location identifier)

RJC: Reaction Jet Control

RJD: Reaction Jet Device • Reaction Jet Driver

RJDA: Reaction Jet Driver Aft

RJDF: Reaction Jet Driver Forward

RJE: Rayleigh-Jeans Equation • Remote Job Entry

RJEC: Reaction Jet Engine Control

RJEHM: Regional Journal of Energy, Heat and Mass Transfer (periodical code)

RJET: Russian Journal of Engineering Thermophysics (periodical code)

RJETS: Remote Job Entry Terminal System

RJFC: Revised Jagiellonian Field Catalogue

RJGC: Revised Jagiellonian General Catalogue

RJH: Rajshahi (Bangladesh) (location identifier) • Rigal-Jung-Heck

RJI: Rajouri (India) (location identifier)

RJK: Rijeka (Croatia) (location identifier)

RJNP: Robert J. Novins Planetarium (USA-NJ)

RJOD: Reaction Jet OMS Driver

RJP: Remote Job Processing • Rocket Jet Plume

RJS: Remote Job System

RJTAM: Russian Journal of Theoretical and Applied Mechanics (periodical code)

RK: Runge-Kutta (method)

RKAS: Richmond and Kew Astronomical Society (UK)

RKC: Yreka (USA-CA) (location identifier)

RKD: Rockland (USA-ME) (location identifier)

RKE: København Roskilde (Denmark) (location identifier)

RKH: Rock Hill (USA-SC) (location identifier)

RKI: Rokot (Indonesia) (location identifier)

RKM: Runge-Kutta Method

RKN: Runge-Kutta-Nystroem

RKO: Sipora (Indonesia) (location identifier)

RKP: Rockport - Aransas County (USA-TX) (location identifier)

RKR: Poteau - Robert S. Kerr (USA-OK) (location identifier)

RKS: Relational Knowledge Structure • Robotics Kernel System • Rock Springs (USA-WY) (location identifier)

Rkt: Rocket

RKT: Ras Al Khaimah (UAE) (location identifier)

RKU: Yule Island (Papua New Guinea) (location identifier)

RKV: Reykjavík (Iceland) (location identifier)

RKW: Rockwood (USA-TN) (location identifier)

Rky: Rocky

RKY: Rokeby (Australia-QLD) (location identifier)

R-L: Richardson-Lucy

R/L: Rate Limited • Remote/Local

RL: Lebanon (international vehicle registration code) • Radiation Laboratory • Radio-Loud • Radio Location • Radioluminescent • Raman Laser • Random Logic • Reading List • Real Life • Record Length • Reduced Level • Reference Library • Reinforcement Learning • Research Laboratory • Residual Layer • Resistance-Inductance • Return Link • Return Loss • Richardson-Lucy • Right Line • Rocket Launcher • Roll Lift • Runway Light

RLA: Remote Link Adapter • Repair-Level Analysis • Repair-Line Agreement • Rolla (USA-MO) (location identifier)

RLAM: Relocatable Limited Area Model

RLAN: Radio Local Area Network

RLAS: Rocket Lunar Attitude System

RLC: Regional Library Bureau (UK) • Remote Load Controller • Resistance-Inductance-Capacitance • Resistor-Inductor-Capacitor • Run Length Coding

RLCE: Request Level Change En route

RLCM: Remote Line Concentrating Module

RLCS: Radio Launch Control System

Rlct: Reluctance

Rld: Rolled

RLD: Reciprocal Linear Dispersion • Recodable Locking Device • Remote Link Director • Richland (USA-WA) (location identifier) • Rijksluchtvaartdienst (Netherlands)

Rle: Ruelle

RLE: Request Loading Entry • Research Laboratory for Electronics • Research Laboratory of Electronics • Réseau Local d'Entreprises • Responsible Laboratory Engineer • Run Length Encoded • Run Length Encoding

RLEMR: Research Laboratory Engeneering Materials Report (periodical code)

RLEO: Request for Liaison Engineering Order

Rlf: Relief

RLF: Radio Luminosity Function • Relief • Robust Luminosity Function • Russian Long-term Flight

RLFOC: Regional Laser and Fiber Optics Center

RLG: Research Libraries Group • Ring Laser Gyro • Ring Laser Gyroscope • Rostock-Laage (Germany) (location identifier)

RLI: Anniston Reilly (USA-AL) (location identifier) • Radiation Level Indicator • Right/Left Indicator

RLIN: Research Libraries Information Network

RLL: Radio in the Local Loop • Radio Local Loop • Run Length Limited • Rocket Launcher Locator

RLLO: Regional Library Liaison Officer

RLLS: Runway Lead-in Lighting System

RLM: Richardson-Lucy Method • Robert Lee Moore

RLN: Remote LAN Node

RLNZ: Rate-Limited Non-Zero

RLO: Research Liaison Officer • Returned Letter Office

Rloc: Relocate

RLOF: Roche Lobe Overflow

RLOGIN: Remote Login

RLP: Resource Location Protocol • Rosella Plains (Australia-QLD) (location identifier) • Rotating Linear Polarization

RLPA: Rotating Log Periodic Antenna

RLPME: Research Laboratory on Precision Machinery Electronics (periodical code)

RLQ: Radio-Loud Quasar

Rlr: Roller

Rls: Release • Rial (currency)

RLS: Radio Location Service • Radius of Landing Site • Raman Lidar System • Record-Level Sharing • Recursive Least Squares • Regional Library System • Regularized Least Squares • Release • Restless Legs Syndrome • Restricted Least Squares • Reweighted Least Squares • Rocket Launching System • Roll Limit Switch

RLSAC: RAL (Astrophysics cluster) (Starlink internal code)

RLSD: Received Line Signal Detected

Rlse: Release

RLSMC: Reactive Local Search for Maximum Clique

RLSS: Regenerative Life Support System

Rlt: Relate

RLT: Arlit (Niger) (location identifier) • Remote Line Test • Return Line Tether

Rltn: Relation

Rltv: Relative

RLU: Bornite Upper (USA-AK) (location identifier)

Rlv: Relieve

RLV: Reusable Launch Vehicle • Roving Lunar Vehicle

RLVAD: RAL (Project cluster) (Starlink internal code)

RLVS: Recoverable Launch Vehicle Structure

RLW: Regie der Luchtwegen • Regularized Long Wave

Rlxn: Relaxation

Rly: Railway • Relay

RLY: Relay

rm: Raummeter • Remove • Rhaeto-Romance (ISO language code)

Rm: Magnetic Reynolds Number • Ream • Remarks • Room

R&M: Reliability & Maintainability • Reliability & Marketing • Reports & Memoranda

R/M: Reliability/Maintainability

RM: Madagascar (international vehicle registration code) • Radar Mapper • Radial Marking • Radiation Measurement • Radiation Monitor • Radiation Monitoring • Rate Management • Rate Monotonic • Raw Material • Readout Matrix • Receiving Memo • Recovered Memory • Redundancy Management • Reed-Muller • Reference Material • Reference Method • Reference Mission • Reference Model • Reference Monument • Regional Manager • Regional Meeting • Registre des Métiers • Reichsmark (currency) • Relational Model • Relationship Management • Relative Mobility • Relay Mirror • Remedial Maintenance • Remote Manipulator • Remote Module • Remote Monitoring • Rendezvous Maneuver • Rescue Module • Research Memorandum • Reservation Management • Reset Mode • Residual Map • Résonnance Magnétique • Resource Management • Resource Manager • Resource Module • Resource Monitor • Resupply Module • Retrieval Mode • Richtmeyer-Meshkov • Right Mid • River Mile • Roman • Romford (UK postal code) • Room • Rotational Modulation • Rotation Measure • Rotation Module • Routine Maintenance • Royal Mail

RM&A: Reliability, Maintainability & Analysis • Reliability, Maintainability & Availability

RMA: Radio Manufacturers Association • Rate Monotonic Analysis • Realizable Multiplex Advantage • Relaxation Map Analysis • Reliability-Maintainability-Availability • Remote Manipulation Activity • Remote Manipulator Arm • Return Material Authorization • Return Merchandise Authorization • Return to Manufacturer Authorization • Revolution in Military Affairs • Right Mental Attitude • Roma (Australia-QLD) (location identifier) • Roper Mountain Astronomers (USA-SC) • Royal Military Academy

RMAA: Revista Mexicana de Astronomía y Astrofísica

RM&AE: Reliability, Maintainability & Availability Evolution

RMAE: Reliability, Maintainability and Availability Evolution

RMAS: Remote Memory Administration System

RMAT: Rhein-Main-Astrotreff (Germany)

RMATS: Remote Maintenance Administration and Traffic System

Rmb: Renminbi (currency)

RMB: Buraimi (Oman) (location identifier) • Renminbi (currency) • Right Mouse Button

RMC: Radcliffe Magellanic Clouds • Radiative Muon Capture • Randolph-Macon College (USA-VA) • Redundancy Management Control • Référentiel Momentanément Comobile • Regional Meteorological Centre • Rocket Motor Case • Rockford Machesney (USA-IL) (location identifier) • Rocky Mountain College (USA-MT) • Rod Memory Computer • Rosette Molecular Cloud • Rotary Mirror Camera • Rotating Modulation Collimator • Rotation Modulation Collimator • Royal Meteorological Centre • Royal Military College

RMCC: Remote Monitor Control Center

RMCF: Remote Monitor Control Facility

RMCU: Responsable des Moyens Charges Utiles

Rmd: Remedy

RMD: Ramagundam (India) (location identifier) • Relationship Management Data • Relationship Management Design • Repair and Modification Directive

RMDI: Radio-Magnetic-Deviation Indicator

rmdir: Remove Directory

RMDM: Relationship Management Data Model

Rmdr: Remainder

RMDR: Remainder

RME: Radiation Monitor Experiment • Radiation Monitoring Equipment • Relay Mirror Experiment • Request Monitor Entry • Resident Maintenance Engineer • Rocket Mission Evaluator • Rome Griffiss (USA-NY) (location identifier)

RMEA: Retrieval Mode Electronics Assembly

RMES: Relay Mirror Experimental Satellite

RMetA: Rivista di Meteorologia Aeronautica (periodical code)

RMF: Redistribution Matrix Field • Redistribution Matrix File • Relativistic Mean Field • Resource Management Facility • Response Matrix File

RMFT: Relativistic Mean-Field Theory

RMG: Regularized Multi-Gaussian • Right Main Gear • Rome - Richard B. Russell (USA-GA) (location identifier)

RMGA: Retrieval Mode Gyro Assembly

R-MHD: Radiation-Magnetohydrodynamics

RMHD: Radiation-Magnetohydrodynamics

RMI: Radio Magnetic Indicator • Remote Method Invocation • Resource Manager Interface • Revenu Minimum d'Insertion • Rimini (Italy) (location identifier) • Rocket Motor Igniter

RMIP: Reentry Measurements Instrumentation Package

R&MIS: Resource & Management Information System

RMIS: Resource and Management Information System • Resource Management Information System

RMIT: Royal Melbourne Institute of Technology (Australia-VIC)

Rmk: Remark

RMK: Remark • Renmark (Australia-SA) (location identifier) • Rotations-Modulations-Kollimator

RMKS: Remarks

RML: Colombo Ratmalana (Sri Lanka) (location identifier) • Radar Microwave Link • Radio Microwave Link • Relational Machine Language • Requirement Modelling Language • Rotating Mirror Laser

RMLR: Recueil Méthodique de Lois et Règlements

RMM: Mali (international vehicle registration code) • Read-Mostly Memory • Read-Mostly Mode • Relationship Management Methodology • Relative Molecular Mass • Remote Maintenance Monitor • Remote Maintenance Monitoring

RMMC: Residual Mean Meridional Circulation

RMMS: Remote Maintenance Monitoring System

RMN: Remain • Résonnance Magnétique Nucléaire • Réunion des Musées Nationaux (France) • Rumginae (Papua New Guinea) (location identifier)

RMNO: Raad voor Milieu- en Natuuronderzoek (Netherlands)

RMNT: Recherche en Micro- et Nano-Technologies

RMO: Resources Management Office • Retrieval Mode Operations • Risk Management Office

RMODP: Reference Model of Open Distributed Processing

RMON: Remote Monitor • Remote Monitoring

RMOP: Resource Management Operation Plan

RMOSLM: Reflected Magneto-Optic Spatial Light Modulator

RM&P: Reliability, Maintainability & Producibility

RMP: Rampa • Rampart (USA-AK) (location identifier) • Rate Measuring Package • Rauch Memorial Planetarium

(USA-KY) • Reentry Measurement Program • Reliability, Maintainability and Producibility • Reliable Multicast Protocol • Remote Maintenance Processor • Reviews of Modern Physics • Risk Management Plan • Rocket Motor Plume • Rocket Motor Propellant • Royalty Management Program

RMPA: Replacement Maritime Patrol Aircraft • Rocky Mountain Planetarium Association (USA)

RMPT: Reliable Multicast Transfer Protocol

RMQ&A: Reliability, Maintainability, & Quality Assurance

RMQA: Reliability, Maintainability, and Quality Assurance

RMR: Rain Mapping Radiometer • Remote Message Register

RMRK: Remark

RMRS: Recoverable Maintenance Recall System • Repeatable Maintenance and Recall System • Repeatable Maintenance Recall System

RM/S: Resources Monitor/Scheduler

RM&S: Reliability, Maintainability & Supportability

RMS: Radiation and Meteoroid Satellite • Radiation Monitoring System • Railway Mail Service • Ramstein (Germany) (location identifier) • Random Mass Storage • Random Motion Simulator • Rate Monotonic Scheduling • Record Management System • Receivables Management System • Reconnaissance Management System • Record Management Service • Record Management System • Recovery Management Support • Redundancy Management System • Reentry Measurement System • Relationship Management System • Reliable Multicast Software • Remote Maneuvering System • Remote Manipulating System • Remote Manipulator System • Remote Measurement System • Remote Measurement Systems (USA-WA) • Remote Monitoring System • Republic Management Systems Inc. • Resonant Macrosonic Synthesis • Resource Management Science • Respiratory Monitoring System • Revised Management Scheme • Root Mean Square • Root Mean Squared • Royal Mail Service • Royal Mail Ship • Royal Mail Steamship • Royal Meteorological Society (UK)

RMSAFE: Radiation Monitoring System Applicable to Fusion Experiments

RMSC: Remote Monitoring System Concentrator • Rochester Museum and Science Center (USA-NY) • Roper Mountain Science Center (USA-SC)

RMSD: Root-Mean-Square Difference

RMSDE: Root-Mean-Square Distance Error

RMSE: Reduced Mean Squarred Error • Root Mean Square Error

RMSECV: Root Mean Square Error of Cross-Validation

RMSEE: Root Mean Square Error of Estimation

RMSELC: Root Mean Square Error of Leverage-Correction

RMSEP: Root Mean Square Error of Prediction

RMSS: Range Meteorological Sounding System • Rocky Mountain Star Stare (USA)

RMST: Republic Management System Technologies Inc.

RMSVP: Remote Manipulator System Verification Plan

Rmt: Remote

RMT: Radiometric Moon Tracer • Random Matrix Theory • Rapidly Moving Telescope • Rapid Modelling Technique • Recrutement à Mi-Temps • Regional Management Team • Remote • Remote Magnetic Tape • Ring Management

RMTC: Regional Meteorological Training Centre • Russian Meteorological and Training Centre

RMTG: ROSAT Mission Timeline Generator

RMTP: Reliable Multicast Transport Protocol

RMU: Radio Maintenance Unit • Radio Management Unit • Radiomessagerie Unidirectionelle • Reference Measurement Unit • Remote Maneuvering Unit • Remote Multiplexer Unit

Rmv: Remove

RMV: Reentry Measurement Vehicle • Registry of Motor Vehicles • Regularly Monitored Variable (star)

Rmvbl: Removable

Rmvl: Removal

RMW: Read-Modify-Write • Roll Moneuver Wheel

RMWC: Randolph-Macon Woman's College (USA-VA)

RMxAA: Revista Mexicana de Astronomía y Astrofísica (periodical code)

RMZ: Rijksdienst voor Maatschappelijke Zekerheid

rn: Kiruni (ISO language code)

Rn: Radon • Rain

RN: Niger (international vehicle registration code) • Radio Navigation • Radio Networking • Rational Negation • Raumfahrtnutzung • Read News • Red Noise • Reference Noise • Reference Number • Registered Nurse • Registry Number • Reissner-Nordstrom • Research Note • Resource Node • Revenu National • Reynold's Number • Rio Grande do Norte (Brazil) • Rio Negro (Argentina) • Route Nationale • Royal Navy • Ruta Nacional

RNA: Remote Network Access • Réseau de Neurones Artificiels • Ribonucleic Acid

RNAAC: Regional Network Associated Analysis Center

RNAO: Results of National Argentine Observatory (periodical code)

RNB: Ronneby (Sweden) (location identifier)

RNC: McMinnville - Warren County (USA-TX) (location identifier) • Reflective Null Corrector • Resource Network Committee

RNCH: Ranch

Rnd: Round

RND: Random • San Antonio Randolph (USA-TX) (location identifier)

Rndm: Random

RNDZ: Rendezvous

RNE: Réseau National d'Essais • Roanne (France) (location identifier)

R-Nets: Requirements Networks

RNets: Requirements Networks

Rnf: Renfrewshire (UK-Scotland)

RNF: Registry Nomenclature File

RNFC: Reference Number Format Code

RNFL: Rainfall

Rng: Range

RNG: Random Number Generator • Range • Rangely (USA-CO) (location identifier) • Renormalization Group

RNGC: Revised New General Catalogue

Rnge: Range

Rngg: Ringing

Rngr: Ringer

RNH: New Richmond (USA-WI) (location identifier)

RNI: Corn Island (Nicaragua) (location identifier) • Réseau National d'Interconnexion

RNIPP: Répertoire National de l'Identité des Personnes Physiques

RNIS: Réseau Numérique à Intégration de Services

RNJ: Yoronjima (Japan) (location identifier)

Rnl: Renewal

RNL: Rennell (Solomon Islands) (location identifier)

RNM: Réseau Neuro-Mimétique

RNMG: Rosetta Nucleus Modelling Group

RNMI: Royal Netherlands Meteorological Institute

RNN: Bornholm (Denmark) (location identifier) • Reciprocal Nearest Neighbours • Rønne (Denmark) (location identifier)

RNO: Red Nebulous Object • Reno (USA-NV) (location identifier) • Tahoe (USA-NV) (location identifier)

RNOA: Rede Nacional de Observação Astronómica (Portugal)

RNOC: Regional Network Operations Center

RNODC: Responsible National Oceanographic Data Center

RNP: Rede Nacional de Pesquisas • Remote Network Processor • Required Navigation Performance • Rongelap Island (Marshall Islands) (location identifier) • Rotation-Nutation-Precession

RNPC: Required Navigation Performance Capability

RNPIC: Residual North Pole Ice Cap

RNPP: Répertoire National des Personnes Physiques

RNQ: Réseau National de la Qualité (France)

RnR: Registration and Resolution

RNR: Receiver Not Ready • Robinson River (Papua New Guinea) (location identifier)

RNRT: Réseau National de Recherche en Télécommunications (France)

Rns: Runs

RNS: Radio Navigation Service • Reduced Navier-Stokes • Remote Note Service • Rennes Saint-Jacques (France) (location identifier) • Residue Number System • Reusable Nuclear Shuttle • Reusable Nuclear Stage

RNSC: Reference Number Status Code

RNSS: Radio Navigation Satellite Service

RNT: Random Numerical Terminal • Renton (USA-WA) (location identifier)

Rntl: Rental

RNTL: Réseau National en Technologies Logicielles (France)

RNTS: Regenerative Non-venting Thermal System

RNU: Ranau (Malaysia) (location identifier)

RNV: Random Noise Voltmeter • Reusable Nuclear Vehicle

RNVA: Réserve Naturelle Volontaire Agréée

Rnwbl: Renewable

Rnwy: Runway

RNWY: Runway

Rny: Rainy

RNZ: Rensselaer (USA-IN) (location identifier)

RNZAF: Royal New Zealand Air Force

RNZN: Royal New Zealand Navy

ro: Romania • Romanian (ISO language code)

Ro: Recto • Río

R/O: Read Only • Readout • Roll Out • Room Only

R-O: Read Only

RO: Radar Operator • Radiation Office • Radio Operator • Range Only • Read Only • Readout • Reality Orientation • Real Orden • Receive Only • Receiving Office • Receiving Officer • Record Office • Recovery Operations • Recruiting Officer • Registered Office • Regulated Output • Remote Observatory • Remote Observe • Remote Observer • Remote Observing • Remote Operation • Reserved Occupation • Reverse Osmosis • Rial Omani (currency) • Rollover • Romania (international vehicle registration code) • Romania (ISO country code) • Rondônia (Brazil) • Room Only • Roumanie (ISO country code) • Royal Observatory • Run On • Runout • Runover

RoA: Record of Achievement

ROA: Raman Optical Activity • Real Instituto y Observatorio de la Armada (Spain) • Record Of Achievement • Reference Optical Alignment • Regional Office for Africa • Return On Assets • Roanoke (USA-VA) (location identifier)

ROADS: Real-time Optical Alignment and Diagnostic System • Resource and Operations Analysis Display System • Resource Organisation And Discovery Service

ROAI: Robinet Obturateur Automatique Intégré

RoAJ: Romanian Astronomical Journal (periodical code)

ROAM: Return On Assets Managed

ROAn: Royal Observatory Annals (periodical code)

roAp: Rapidly Oscillating Ap

ROAR: Rights Of Admission Reserved • Royal Optimizing Assembly Routine

ROB: Monrovia Roberts (Liberia) (location identifier) • Recorder Buffer • Recovery Operations Branch • Refocusing Opportunity Bulletin • Remote Order Buffer • Reorder Buffer • Right Outboard • Rigid Optical bench • Royal Observatory of Belgium

ROBEPS: Radar Operating Below Prescribed Standard

ROBIN: Rotterdam Bibliotheek en Informatienetwerk (Netherlands)

Robot: Robotica (periodical code)

ROBOT: Robotic Observatories Basic Operating Tasks

ROBS: Rapid Optical Beam-steering System

ROC: Radius of Curvature • Rate Of Change • Rate Of Climb • Receiver Operating Characteristics • Reconocimiento Óptico de Caracteres • Reconnaissance Optique de Caractères • Record Of Comments • Regional Operating Co. • Relative Out Centrality • Remote Operations Center • Remote Operations Controller • Republic of China (Taiwan) • Request Of Change • Research Organisation Cooperation • Return On Capital • Reusable Orbital Carrier • Rochester (USA-NY) (location identifier)

ROCC: Range Operational Control Center • Range Operations Control Center • Region Operational Control Center

ROCE: Return On Capital Employed

ROCES: Ranger Orbiter Crew Escape System

ROCR: Remote Optical Character Recognition

RoD: Record of Decision

ROD: Radiometric and Orbit Data • Record Of Decision • Rate Of Descent • Record Of Discussion • Regional Operational Data • Remotely-Operated Door • Required Operational Date • Robertson (South Africa) (location identifier) • ROSAT Observational Dataset

RODC: Regional Oceanographic Data Center

RODEO: Rotating Delivery of Excitation Off-resonance

RODS: Record Of Decision System

ROE: Readout Electronics • Référentiel d'Objets d'Entreprise • Reflector Orbital Equipment • Report of Expenditures • Return On Equity • Round Off Error • Royal Observatory Edinburgh (Scotland) • Royal Observatory Edinburgh (Starlink internal code)

ROEN: Rádio Observátorio Espacial do Nordeste (Brazil)

ROER: Remote Operation Error

ROEU: Remotely-Operated Electrical Umbilical

ROF: Réseau Optique Flexible

ROFF: Run-Off

ROFL: Rolling On the Floor Laughing

ROFT: Rapid Optics Fabrication Technology

ROG: Receipt Of Goods • Rogers (USA-AR) (location identifier)

ROGUE: Radio, Optical, Gamma-ray Undergraduate Experiment • Rapid Off-axis Guider Experiment

ROH: Receiver Off Hook • Robinhood (Australia-QLD) (location identifier)

ROHK: Royal Observatory Hong Kong (now HKO)

ROI: Rádio Observátorio do Itapetinga (Brazil) • Radio-Optical-Inertial • Range Operations Instruction • Rectangle Of Interest • Region Of Interest • Rendezvous Orbit Insertion • Report Of Investigation • Return On Investment • Rotating Optical Interferometer

ROIV: Remote Operation Invoke

ROK: Republic of Korea (international vehicle registration code) • Rockhampton (Australia-QLD) (location identifier) • South Korea (international vehicle registration code)

ROL: Leu (Romania) (ISO currency code) • Régie On-Line • Remote Operating Location • Roosevelt (USA-UT) (location identifier) • Rotate Left

ROLAP: Relational On-line Analytical Processing

ROLO: Robotic Lunar Observatory

ROLS: Recoverable Orbital Launch System

ROLTP: Remote On-Line Transaction Processing

ROLun: Reports of the Lund Observatory (periodical code)

Rom: Roman • Romania • Romanian

ROM: Range Of Motion • Read-Only Memory • Remote Opacity Monitor • Roma (Italy) (location identifier) • Romania (ISO country code) • Rome (Italy) (location identifier) • Rough Order of Magnitude • Roumanie (ISO country code) • Royal Ontario Museum (Canada-ON)

ROMARIN: Read-Only Memory of Madrid Actualized Registry Information

ROMBUS: Reusable Orbital Module Booster and Utility Shuttle

ROME: Résolution Océanique à Moyenne Échelle

ROMM: Read-Only Memory Module

ROMPS: Robotic Materials Processing in Space • Robotic Materials Processing System • Robot-Operated Materials Processing System

RON: Readout Noise • Remaining Overnight • Research Octane Number • Rondón (Colombia) (location identifier)

RONA: Return On Net Assets

RONALD: Rocketborne Optical Neutral-gas Analyzer with Laser Diodes

RONAST: Royal Nepal Academy of Science and Technology

RONSARD: Recherche sur les Orages et les Nuages par un Système à Radars Doppler

ROO: Radio Optical Observatory • Rondonôpolis (Brazil-MT) (location identifier)

ROOM: Real-time Object-Oriented Modelling

ROOST: Reusable One-stage Orbital Space Truck

ROP: Raster Operation • Raster Operations Processor • Read-Only Printer • Receive-Only Printer • Reference Observing Proposal • Republic of the Philippines • RISC Operation • Rota (Mariana Islands) (location identifier) • Run Of Paper

ROPA: Realising Our Potentials Award

ROPE: Research on Orbital Plasma Electrodynamics • Rip it Out and Put in Ethernet

ROPME: Regional Organization for the Protection of the Marine Environment

ROPP: Receive-Only Page Printer

ROPS: Roll Over Protective Structure

ROR: Koror (Palau Islands) (location identifier) • Rate Of Read • Rate Of Return • Released on Own Recognizance • Robust Orthogonal Regression • ROSAT Observation Request • Rotate Right

RORD: Rapid Onset - Rapid Decay

RORF: Radio Optical Reference Frame

RORJ: Remote Operation Reject

RO-RO: Roll-On - Roll-Off

RORO: Roll-On - Roll-Off

RORS: Remote Operation Result

RORSAT: Radar Ocean Reconnaissance Satellite

ROS: Radar à Ouverture Synthétique • Radarsat Optical Sensor • Rapport d'Ondes Stationnaires • Read-Only Storage • Real-time Operating System • Regulated Oxygen Supply • Remote Optical System • Removable Overhead Structure • Research Optical Sensor • Retail Operation System • Rosario Fisherton (Argentina-SF) (location identifier) • Rotating Optical Scanner

ROSA: Report Of Service Activity • Romanian Space Agency

ROSAR: Rotating Synthetic Aperture Radar

ROSAT: Radar Ocean Surveillance Satellite • Röntgen Satellite

ROSDR: Read-Only Storage Data Register

ROSE: RATS Open Systems Environment • Remotely Operated Special Equipment • Remote-Operations Service Element • Remote-Operations Service Entity • Remote Optical Sensing of Emission • Reporting On the State of the Environment • Request-Oriented Scheduling Engine • Research Open Systems for Europe • Research Open Systems in Europe • Retrieval by Online Search • Rising Observational Sounding Equipment • Rocket and Scatter Experiment • Rural Oxidant in the Southern Environment

ROSI: Rome Observatory Schmidt Imager

ROSIG: ROSAT Science Interest Group

ROSIS: Reflective Optics System Imaging Spectrometer

ROSO: Review Of Starlink Operations

ROSQ: Return On Software Quality

ROSS: Reconstruction Of Serial Sections • Remote Orbital Servicing System

ROSSA: Robotics Servicing and Space Assembly

ROSSWG: Radarsat Optical Sensor Science Working Group

ROSTA: Regional Office for Science and Technology, Africa (UNESCO)

ROSTAS: Regional Office for Science and Technology, Arab States (UNESCO)

ROSTLAC: Regional Office for Science and Technology, Latin America and Caribbean Islands (UNESCO)

ROSTSCA: Regional Office for Science and Technology for South Central Asia (UNESCO)

ROSTSEA: Regional Office for Science and Technology for Southeast Asia (UNESCO)

rot: Curl

Rót: Rótula

Rot: Rota • Rotary • Rotating • Rotation • Rotor • Rotule

ROT: Rate Of Turn • Refurbish-Operate-Transfer • Remaining Operating Time • Remotely Operated Telescope • Remote Observation Telescope • Reusable Orbital Transport • Rotate • Rotation • Rotorua (New Zealand) (location identifier) • Rotterdam (Netherlands) • Rotunda • Rule Of Thumb • Runway Occupancy Time

ROTAM: Request-Oriented Telecomuunications Access Method

ROTAT: Robot Observatory Theoretische Astrophysik Tübingen (Germany)

ROTC: Reserve Officers' Training Corps

ROTEX: Robotic Technology Experiment • Robot Technology Experiment

ROTFL: Rolling On The Floor Laughing

Rotg: Rotating

ROTHR: Relocatable Over-The-Horizon Radar

ROTI: Recording Optical Tracking Instrument

ROTL: Remote Office Test Line

Rotr: Rotator

ROTSCA: Regional Office for Science and Technology in South Central Asia (UNESCO)

ROTSE: Robotic Optical Transient Search Experiment

ROTV: Returnable Orbital Transfer Vehicle • Reusable Orbital Transport Vehicle (*)

ROU: Ruse (Bulgaria) (location identifier) • Uruguay (international vehicle registration code)

Roum: Roumanian • Roumanie

ROUS: Rodent Of Unusual Size

Rout: Routine

ROUT: Roll Out

ROV: Remotely-Operated Vehicle • Remote Operating Vehicle • Remote Optical Viewing • Report Of Visit • Rostov (Russia) (location identifier)

ROVE: Radar Observation of Vegetation

ROVS: Remote Optical Viewing System

ROW: Rest Of World • Right Of Way • Roswell (USA-NM) (location identifier)

ROWG: RadioAstron Operations Working Group (Russia)

ROWS: Radar for Observing the Wave Spectrum • Radar Ocean Wave Spectrometer • Remote Ocean Wave Spectrometer • Remote Orbital Wave Spectrometer

ROX: ROSAT X-ray Optical (survey) • Roseau (USA-MN) (location identifier) • Roxburghshire (UK-Scotland)

Roy: Royal • Royaume

ROY: Rio Mayo (Argentina-CB) (location identifier)

ROYGBIV: Read Out Your Good Book In Verse • Red-Orange-Yellow-Green-Blue-Indigo-Violet • Richard Of York Gave Battle In Vain

R°: Río

Rp: Rappen • Réponse Payée • Rupiah (currency)

R&P: Reserve & Process

R-P: Roll-Pitch

RP: Philippines (international aircraft registration code) • Philippines (international vehicle registration code) • Radiation Pressure • Radiation Protection • Reachable Path • Reaction Product • Reader Printer • Readiness Potential • Read Printer • Real Part • Received Pronunciation • Reception Poor • Recette Principale • Recommended Practice • Recommended Price • Recovery Phase • Recreation Park • Redox Potential • Redundancy Payment • Reference Paper • Reference Pattern • Reference Point • Reference Publication • Reference Pulse • Reformed Presbyterian • Regius Professor • Reinforced Plastic • Relative Pressure • Remote Printer • Repair Period • Reply Paid • Réponse Payée • Représentation Proportionnelle • República Portuguesa • Research Paper • Reserve and Process • Resolving Power • Restoration Priority • Retournement de Pupille • Return Premium • Révérend Père • Reverendus Pater • Revision Proposal • Rocket Projectile • Rocket Propellant • Roll-Pitch • Ruddlesden-Popper • Rules of Procedure • Runtime Process • Ruta Provincial

RPA: Radiation Protection Adviser • Random Phase Approximation • Rapid Prototyping Array • Reactive Plate Assembly • Record and Playback Assembly • Record of Personal Achievement • Rème Plasma Analyser • Remotely Piloted Aircraft • Remote Passphrase Authentication • Reply All • Request for Procurement Action • Request for Procurement Authorization • Resting Point of Accommodation • Retarded Potential Analyzer • Retarding Potential Analyzer • Rolpa (Nepal) (location identifier)

RPAA: Rotating Phase Array Antenna

RPAE: Retarding Potential Analyzer Experiment

RPB: Recognized Professional Body • Reports and Publications Branch • Result Producing Branch • Roper Bar (Australia-NT) (location identifier)

RPBF: Reference Pattern-Based Feature

RPBL: Reduced Parameter Bilinear Model

RPBR: Result Producing Branch Response

RPC: Real Procedure Call • Records Processing Center • Regional Program Committee • Regional Pyrheliometer Comparison • Regras Portuguesas de Catalogação (Portugal) • Relations Publiques et Communication • Remote-Position Control • Remote Power Control • Remote Power Controller • Remote Procedural Call • Remote Procedure Call (*) • Request the Pleasure of your Company • Report Generator • Resistive Plate Chamber • Resistive Plate Counter • Restoration Priority Code • RGS Power Converter • Ring Plane Crossing • Rosetta Plasma Consortium

RPCL: Remote Procedure Call Language

RPCVE: Remote Plasma-enhanced Chemical Vapor Epitaxy

RPD: Radiation Pressure Dominated • Rapid • Reconfiguration Production Department • Relative Power Density

RPDA: Robust Partial Discriminant Analysis • ROSAT Public Data Archive

RPDLY: Rapidly

RPDS: Rapids

RPE: Regional Political Ecology • Relative Pointing Error • Reliability Project Engineer • Resource Planning and Evaluation

RPECS: Rapid Prototyping Electronic Control System

RPEG: Repartição de Pessoal e Expediente Geral

RPF: Rapid Prototyping Facility • ROSAT Proposal Form

RPFITS: Radiophysics Flexible Image Transport System

RPG: Radar Processing Group • Report Program Generator • Research Planning Grant • Risk Purchasing Group • Rôle-Playing Game

RPGL: Report Program Generator Language

rph: Revolutions per Hour

RPI: Radio Plasma Imager • Relay Position Indicator • Remote Process Innovation • Remote Process Invocation • Rensselaer Polytechnic Institute (USA-NY) • Repetitive-Motion Injury • Retail Price Index • Retrieval with Probabilistic Indexing • Roll Position Indicator • Rotating Polarizer Interferometer • Runway Point of Intercept • Runway Point of Interception

RPIE: Real-Property Installed Equipment • Replacement Photographic Image Equipment

RPIF: Regional Planetary Image Facility

RPIO: Registered Publication Issuing Office

RPK: Revenue Passenger Kilometer

RPL: Radio-Photo Luminescent • Rated Power Level • Relational Production Language • Remote Procedure Load • Replaceable Parts List • Requested Privilege Level • Resident Programming Language • Rocket Propulsion Laboratory

RPLC: Replace • Reversed-Phase Liquid Chromatography

Rplr: Repeller

Rplsn: Repulsion

Rplt: Repellent

RPLV: Reentry Payload Launch Vehicle

rpm: Revolutions per Minute (*) • Rotations per Minute • Runs per Minute

RpM: Rate per Minute

R&PM: Research & Program Management

RPM: Ngukurr (Australia-NT) (location identifier) • Radiation Polarization Measurement • Random Positioning Machine • Rate Per Minute • Read Program Memory • Recursive Projection Method • Reduced Proper Motion • Register of Preservation Microfilms • Relaxation Potential Model • Reliability Performance Measure • Remote Payload Monitoring • Remote Performance Monitoring •

Resale Price Maintenance • Research and Program Management • Research Program Management • Resource Planning and Management • Resupply Provisions Module • Retro-Propulsion Mode • Retro-Propulsion Module • Revenue Passenger Miles • Roll Position Mechanism

RPMI: Radiant Power Measuring Instrument • Revolutions-Per-Minute Indicator

RPMT: Radiant Power Measuring Instrument

RPN: Recherche en Prévision Numérique • Reverse Polish Notation • Risk Priority Number • Rosh Pina (Israel) (location identifier)

RpO: Revolution per Orbit

RPO: Radiation Protection Officer • Railway Post Office • Recovery Point Objective • Regional Publishing Office • Research Programs Office • Revolution Per Orbit

RPOC: Remote Payload Operations Center

RPOP: Rendezvous and Proximity Operations Program

RPOS: Replanned Preferred Observation Schedule • Replanned Preferred Observation Sequence

RPP: Random Phase Plate • Reinforced Pyrolytic Plastic • Reliability Program Plan • Rendezvous Point Position • Représentation des Phénomènes de Physique • Request Parallel Poll • Reverse Pulse Polarography • Rural Policy Plan

RPPE: Research, Program, Planning and Evaluation

RPPh: Reports of Progress in Physics (periodical code)

RPPROM: Reprogrammable PROM

RPQ: Request for Price Quotation

Rpr: Repair

RPR: Raipur (India) (location identifier) • Rassemblement pour la République (France) • Read Printer • Rear Projection Readout • Rectangular Parallelepiped Resonance • Remote Process Redesign • Roll-Pitch Resolver

RPRINTER: Remote Printer

Rprt: Report

RPRV: Remotely Piloted Research Vehicle

rps: Revolutions per Second • Runs per Second

RpS: Reductions per Second

RPS: Rapid Processing System • Real-time Photogrammetry System • Real-time Processing System • Record and Playback System • Reductions Per Second • Redundant Power System • Regulated Power Supply • Remote Printing System • Remote Processing Service • Remote Proposal Software • Remote Proposal Submission • Remote Proposal System • Responsable à la Préparation du Satellite • Rijkspostspaarbank • Roadway Package System • ROSAT Proposal Submission • Royal Photographic Society

RPSEC: Ruth Patrick Science Education Center (USA-SC)

RPSF: Rotation Processing and Surge Facility

RPSM: Resources Planning and Scheduling Method

RPSS: Relative Partial Sum of Squares • Remote Proposal Submission System

Rpt: Repeat • Report • Reprint

RPT: Reference Point Tracking • Regular Public Transport • Repeat • Resident Provisioning Team • Rudder Pedal Transducer

RPTA: Rudder Pedal Transducer Assembly

RPTEC: Rapid Prototyping, Tooling and Evaluation Center

Rptn: Repetition

Rptr: Repeater

RPU: Receiver/Processor Unit • Rectifier Power Unit • Registered Publication Unit • Remote Positioning Unit • Remote Processing Unit

RPV: Reactor Pressure Vessel • Reactor Pressurized Vessel • Remotely Piloted Vehicle • Réseau Privé Virtuel • Roper Valley (Australia-NT) (location identifier)

RPWS: Radio and Plasma Wave Science

RPX: Ring Plane Crossing • Roundup (USA-MT) (location identifier)

RPY: Roll-Pitch-Yaw

RPZ: Runway Protection Zone

RQ: Radio-Quiet • Repeat-Request • Request for Quotation • Respiratory Quotient

R&QA: Reliability & Quality Assurance

RQA: Reliability and Quality Assurance • River Quality Assessment

RQBE: Relational Query By Example

RQC: Reliability and Quality Control

R&QE: Radiophysics and Quantum Electronics (periodical code)

RQE: Relative Quantum Efficiency • Responsive Quantum Efficiency

RQFT: Relativistic Quantum Field Theory

RQL: Reference Quality Level • Rich-Quick-Lean

Rqmt: Requirement

RQMT: Requirement

RQMTS: Requirements

RQN: Radial Quantum Number

RQQ: Radio-Quiet Quasar

Rqr: Require • Requirement

Rqrd: Required

RQS: Rate/Quote System

R-R: Roll-Roll • Rolls-Royce

R&R: Remove & Replace • Rendezvous & Recovery • Rescue & Resuscitation • Rest & Recreation • Rest & Rehabilitation

R/R: Remove/Replace

RR: Bororo • Radiation Resistance • Radiative Recombination • Radio Range • Radio Regulation • Railroad • Rainfall Rate • Range Rate • Readiness Review • Readout and Relay • Real Reality • Receiver Ready • Receiving Report • Recurrence Rate • Reduced Resolution • Redundancy Reduction • Reentry Range • Register to Register • Relative Rate • Relative Ratio • Relative Relevance • Remington Rand • Removal-Replacement • Remove and Replace • Rendezvous Radar • Rendezvous and Recovery • Respiration Rate • Repetition Rate • Requirements Review • Research Report • Resource Records • Respiration Rate • Return Rate • Ridge Regression • Rights Reserved • Rising Rapidly • Roll-Roll • Rolls-Royce • Roraima (Brazil) • Route Relay • Routine Relay • Rural Route • Rush Release

R&RA: Research & Related Activities

RRA: Realistic Risk Assessment • Research and Related Activities • Resident Research Associate • ROSAT Result Archive

RRB: Radio Regulation Board • Red Rectangle Band • Requirements Review Board • Resource Review Board

RRC: Radio Relay Center • Reentry Rate Command • Regional Research Centre • Report Review Committee • Rocket Research Co. • Roll Ratio Controller

RRCA: Radar Reference Coverage Area

RRCT: Reflectance Ratio Cloud Test

RRD: Receive, Record and Display • Reformatted Raw Data • Reliability Requirements Directive

RRDA: Range-Resolved Differential Absorption • Rendezvous Radar Deployment Actuator • Rendezvous Retrieval, Docking and Assembly

RRDCT: Risk Reduction Data Collection Task

RRDE: Rotating Ring-Disk Electrode

RRDS: Reduced Resolution Data Set

RRE: Marree (Australia-SA) (location identifier) • Range Rate Error

RREA: Rendezvous Radar Electronics Assembly

RREL: Risk Reduction Engineering Laboratory

RREU: Rendezvous Radar Electronics Unit

RRF: Rank Revealing Factorization • Relation-Revealing Formula • Retirement Register File

RRFID: Radio Reference Frame Image Database

RRG: Rodrigues Island (Mauritius) (location identifier) • Roll Reference Gyro • Very Red Giant

RRH: Relay Radio Hardware • Remote Reading Hygrometer

RRHT: Remote Reading Hygrothermometer

RRI: Barora (Solomon Islands) (location identifier) • Radio Refractive Index • Radio Research Instrument Inc. (USA-CT) • Raman Research Institute (India) • Range Rate Indicator • Rectangular Region of Interest • Rendezvous Radar Indicator • Riverside Research Institute (USA-CA) • Rocket Research Institute

RRIF: Reading Research Impact Factor

RRK: Rourkela (India) (location identifier)

RRL: Merrill (USA-WI) (location identifier) • Radio Recombination Line • Radio Relay Link • Radio Research Laboratory • Rayleigh Radiation Law • Reusable Rocket Launcher • Rocket Research Laboratory • Rudder Reference Line • Runway Remaining Lighting

RRLA: Reunión Regional Latinoamericana de Astronomía

RRLD: Recoverable Rocket Launcher Demonstrator

RRM: Marromeu (Mozambique) (location identifier) • Rayleigh-Ritz-Method • Renegociable-Rate Mortgage

RRMD: Real-time Radiation Monitoring Device

RRMPA: Revue Roumaine de Mathématique Pures et Appliquées (periodical code)

RRN: Rapid Response Network • Regional Research Network • Relative Record Number • Serra Norte (Brazil-PA) (location identifier)

RRO: Rate and Route Operator • Reproduction Rights Organisation • Sorrento (Italy) (location identifier)

RRP: Recommended Retail Price • Response Retrieval Project • Rover Robotic Payload • Rudder Reference Plane

RRPA: Relativistic Random Phase Approximation

RRPF: Réseau de Référence de Polynésie Française

RRPEV: Reusable Rocket Powered Experimental Vehicle

R&RR: Range & Rate Rate

RRR: Reconnaissance Rapide des Radiances • Reflectivity and Roughness Radar • Return Receipt Requested • Rhône-Poulenc-Rohrer

RRS: Radio Receiver Set • Radio Recording Spectrophotometer • Radio Relay Station • Radio Relay System • RAL Research Services • Reaction Research Society (USA-CA) • Regional Research Site • Relay Radio System • Remote Recovery Support • Research Referral Service • Restraint Release System • Retrograde Rocket System • Retrorocket System • River and Rainfall Station • Roll Rate Sensor • Røros (Norway) (location identifier) • Rudolf-Römer-Sternwarte (Germany)

RRSF: RACF Remote Sharing Facility • Regional Remote Sensing Facility

RRSP: Regional Remote Sensing Programme

RRSSC: Regional Remote Sensing Service Centre

RRT: Rail Rapid Transit • Rapid Rail Transit • Raster Ray Tracer • Reentry Reference Time • Remote Ranging Terminal • Rendezvous Radar Transducer • Rendezvous Radar Transfer • Rendezvous Radar Transponder • Requirements Review Team

RRTM: Rapid Radiative Transfer Model

RRTS: Range-Rate Tracking System

RRU: Remote Readout Unit

RRV: Remote Reconnaissance Vehicle • Robinson River (Australia-NT) (location identifier)

RRWDS: Radar Remote Weather Display System

RRZ: Radar Regulation Zone • Restricted Release Zone

RRZK: Regionales Rechenzentrum Köln (Germany)

rs: Rain and Snow • Sleet

Rs: Reis (currency) • Rupees (currency)

R/S: Range Safety • Redundant Set • Remote Sensing

R&S: Reliability & Safety

RS: Radio Science • Radio Shack • Radio Simulator • Radiosondage • Radiosounding • Radio Star • Radio Station • Raman Scattering • Random Search • Range Safety • Rawinsonde • Reactive Search • Reader Service • Reader Stop • Reading of Standard • Ready Service • Received Standard • Receiver Station • Reception Station • Recommended Specification • Recommended Standard • Reconnaissance Satellite • Recording Secretary • Record Separator • Record Start • Record Status • Recruiting Service • Red Straggler • Redundancy Status • Redundant Set • Reed-Solomon • Reentry System • Reference Side • Reference Standard • Reformed Spelling • Refurbishment Spare • Register and Storage • Register to Storage • Regulating Station • Reliability and Safety • Reliability Summary • Relocation Site • Remote Sensing • Remote Site • Remote Station • Representative Subset • Request to Send • Requirement Submission • Research Scientist • Research Station • Research Summary • Research System • Reset-Set • Resource Scheduler • Respiratory System • Retrieval Strategy • Retrieval System • Retrospective Search • Revised Status • Revised Statutes • Right Side • Rio Grande do Sul (Brazil) • Rocket System • Route Selector • Royal Society (UK) • Russian Segment • Rymdstyrelsen

RSA: Radar with Synthetized Aperture • Range Safety Approval • Rapid Simulation Aid • Rate Sensor Assembly • Rate System Analyst • Relative Specific Activity • Relative Standard Accuracy • Remote Sensing Application • Remote Standardization and Automation • Remote Storage Activity • Republic of South Africa • République Sud-Africaine • Revised Shapley-Ames • Revue de Statistique Appliquée • Rheometrics Solids Analyzer • Rivest-Shamir-Adleman • Runway Safety Area • Rural Services Area • Russian Space Agency • Santa Rosa (Argentina-LP) (location identifier)

RSAA: Remote Sensing Association of Australia • Research School of Astronomy and Astrophysics

RSAC: Recreational Software Advisory Council • Remote Sensing Applications Centre • Remote Sensing Applications Committee

RSADU: Remote Sensing Applications Development Unit

RSAI: R.S. Automation Industrie (France) • Royal Society of Antiquaries of Ireland

RSAL: Remote Sensing Applications Laboratory

RSAM: Remote Sensing Applications Mission

RSAW: Resistive Shear Alfvén Wave

RSB: Range Safety Beacon • Rapport Signal sur Bruit • Reference Standards Book • Research Support Branch • Roseberth (Australia-QLD) (location identifier) • Rudder/Speed Brake

RSBE: Repository-Based Software Engineering

RSBMS: Rotating Shadow Band Multispectral Radiometer

RSBR: Réseau Synoptique de Base Régional

RSBS: Rotating Shadow Band Spectrometer

Rsc: Rescue

RSC: Radar Scan Converter • Radio Source Coordinates • Range Safety Command • Range Safety Control • Reader

Service Card • Record Status Code • Relevance Score Contribution • Remote Sensing Centre • Remote Sensing Committee • Remote Sensing Course • Remote Switching Center • Request System Control • Research School of Chemistry • Richard Stockton College (USA-NJ) • Riga Skulte (Latvia) (location identifier) • RISC Single Chip • Royal Society of Canada • Royal Society of Chemistry (UK) • Runway Surface Condition

RSCAT: Radar Scatterometer

RSCC: Remote-Site Command Computer • Remote-Site Computer Complex • Russian Segment Cargo Crane

RSCE: Rocket Space Company Energia (Russia)

RSCG: Radio Set Control Group

Rsch: Research

RScI: Review of Scientific Instruments (periodical code)

RSCIE: Remote-Station Communication Interface Equipment

RSCL: Radioactive Sodium Chemistry Loop • Reactor Simulation and Control Laboratory

RSCN: Registered State Change Notice

RSCR: Range Safety Command Receiver

RSCS: Range Safety Command System • Rate Stabilization and Control System • Remote Source Control Center • Remote Spooling and Connection System • Remote Spooling Communications System

RSCSS: Range Safety Command Shutdown System

RSCU: Regional Soil Conservation Unit

RSD: Radiance Spectral Distribution • Radio Science Division • Rainbow Schlieren Deflectometry • Range Support Document • Relative Standard Deviation • Remote Sensing Data • Remote Sensing Division • Requirements and Specifications Document • Residual Standard Deviation • Rock Sound (Bahamas) (location identifier)

RSDB: Remote Sensing Database

RSDC: ROSAT Science Data Centre (*) • ROSAT Scientific Data Center

RSDIS: Remote Sensing Data and Information Services

RSDO: Rapid Spacecraft Development Office (NASA)

RSDP: Remote-Site Data Processor • ROSAT Science Data Processing • Routine Science Data Processing • Routine Scientific Data Processing

RSDT: Remote-Station Data Terminal

Rse: Remise

R&SE: Research & Systems Engineering

RSE: Radio Science Experiment • Received Standard English • Receiving Site Equipment • Rees-Sciama Effect • Reference Standards Equipment • Request Select Entry • Revised Standard Edition • Royal Society of Edinburgh (UK-Scotland) • Sydney - Au-Rose Bay (Australia-NSW) (location identifier)

RSEC: Representative Shuttle Environmental Control

RSECS: Representative Shuttle Environmental Control System

RSED: Research and Scientific Exchanges Division

RSEF: Real Sociedad Española de Física (Spain)

RSEMS: Remote Sensing Electro Magnetic Spectrum (periodical code)

RSEnv: Remote Sensing of Environment (periodical code)

RSER: Remote Sensing of Earth Resources

RSES: Research School of Earth Sciences

RSETI: Radio Search for Extraterrestrial Intelligence

RSEXEC: Resource Sharing Executive

RSF: Receiving-Safing Facility • Refurbish and Subassemblies Facility • Refurbishment and Subassembly Facility • Region of Star Formation • Remote Support Facility

RSFF: Resonant Spin-Flavor Precession

RSFQ: Rapid Single Flux Quantum

RSG: Range Safety Group • Rate Signal Generator • Rate Support Grant • Rate Switching Gyro • Recirculating Steam Generator • Red Supergiant • Reference Signal Generator • Regional Sponsor Group • Remote Sensing Group • Research Study Group • Rising • Serra Pelada (Brazil-PA) (location identifier)

RSH: Remote Shell • Restricted Shell • Russian Mission (USA-AK) (location identifier)

RSI: Radiation Systems Inc. • Radio Science Investigation • Rate Stability Indicator • Record Status Indicator • Red Spot Hollow (Jupiter) • Remotely Sensed Imagery • Remote Sensing Institute • Repetitive Strain Injury • Repetitive Stress Injury • Research Studies Institute • Research Systems Inc. • Responsable du Système d'Information • Retrospective Searching Information • Reusable Surface Insulation • Rio Sidra (Panama) (location identifier) • Roll Stability Indicator • Rotational Shearing Interferometer • Rules, Standards and Instructions • Rural Systemic Initiative

RSIAL: Remote Sensing and Image Analysis Laboratory

RSIC: Radiation Shielding Information Center • Research Institute for Social Development

RSID: Radar Simulator Instrument-on-a-Disk

RSIM: Retrospective Single-Ion Monitoring • ROSAT Simulator

RSIMS: Remote Sensing Information Management System

RSIOP: Red Single Integrated Operations Plan

RSIP: Remote Sensing and Image Processing

RSITA: Règlement du Service International des Télécommunications de l'Aéronautique

RSJ: Rosario (USA-WA) (location identifier)

RSK: Ransiki (Indonesia) (location identifier)

RSL: Radio Standards Laboratory • RAISE Specification Language • Reference Standards Laboratory • Remote-and-Status Link • Remote Sensing Laboratory • Request-and-Status Link • Requirement Specification Language • Resource Support List • Reusable Software Library • Runtime System Library • Russell (USA-KS) (location identifier)

RSLA: Range Safety Launch Approval

RSLPT: Royal Society of London Philosophical Transactions (periodical code)

RSLS: Redundant Set Launch Sequence • Runway Status Light System

RSLV: Resolve

Rslvr: Resolver

RSM: Radiation Signature Measurement • Radiation Survey Meter • Rapid Search Machine • Reference Sky Map • Regional Sales Manager • Remote Switching Module • Rendezvous Safety Monitoring • Replication Server Manager • Reppublicca di San Marino • Response Surface Methodology • Resume • Royal Society of Medicine (UK) • San Marino (international vehicle registration code) • San Marino (postal country code)

RSMA: Remote Sensing of the Middle Atmosphere

RSMC: Regional Specialized Meteorological Centre

RSMIP: Remote Sensing Management Information Program

RSMM: Redundant System Monitor Model

RSMSW: RVD Safety Monitoring Software

RSMU: Remote Satellite Measurement Unit

Rsn: Reason

RSN: Radio Supernova • Real Soon Now • Relative Sunspot Number • Ruston (USA-LA) (location identifier)

RSNC: Réseau Synoptique National Complémentaire

RSNe: Radio Supernovae

RSNs: Remnantes de Supernova

RSNZ: Royal Society of New Zealand

RSO: Radiological Safety Office • Radiosonde Observation • Radio-Source Object • Range Safety Officer • Range Support Operation • Recruiting Staff Officer • Red Stellar Object • Research in the Sociology of Organizations • Resident Space Object • Rural Suboffice

RSOC: Rosetta Science Operations Centre

RSOG: RadioAstron Science Operations Group (Russia)

RSOI: Radius of Sphere Of Influence

RSOP: Range Safety Operational Plan

RSOR: Range Safety Operations Requirements

RSP: Radar Signal Processor • Radio Switch Panel • Range Solar Panel • Range Sorting Program • Range Support Plan • Rapid Site Preparation • Raspberry Strait (USA-AK) (location identifier) • Rate Sensing Package • Record Select Program • Reinforced Structural Plastic • Remote Sensing Project • Rendezvous Station Panel • Request Serial Poll • Required Space • Research and Survey Programme • Resources Survey Project • Responder Beacon • Roll Stabilization Platform

RSPA: Research and Special Programs Administration

RSPAC: Remote Sensing Public Access Center

RSPC: Relevé Systématique du Plan Galactique

RSPCA: Royal Society for the Prevention of the Cruelty to Animals (UK)

Rspd: Respond

RSPIC: Residual South Pole Ice Cap

RSPL: Recommended Spare Parts List

RSPP: Remote Sensing Preparatory Programme

Rsps: Response

RSPS: Range Solar Panel Substrate • Research School of Pacific Studies

Rspsb: Responsible

RSPT: Real Storage Page Table

Rspv: Respective • Responsive

Rsq: Rescue

RSQ: Remote Sensing Quarterly (periodical code)

RSQC: Reliability, Safety and Quality Control

RSR: Rain Scattering Radiometer • Rain Scatterometer and Radiometer • Range Safety Report • Release Schedule Record • Request for Scientific Research • Resources Status Report • Resources Status Review • Rohini Sounding Rocket • Rotating Shadowband Radiometer

RSRB: Redesigned Solid Rocket Booster

Rsrc: Resource

RSRE: Royal Signals Research Establishment (UK)

RSRF: Relative Spectral Response Flux • Relative Spectral Response Function (*)

RSRM: Redesigned Solid Rocket Motor • Reusable Solid Rocket Motor

RSRMP: Regional Solar Radiation Monitoring Project

RSRP: Remote Sensing Research Program

RSRS: Range Safety Receiving Station

RSRT: Resort

RS&S: Receiving, Shipping & Storage

RSS: Radar Subsystem • Radiometric Sextant System • Radio Science System • Range Safety Switch • Range Safety System • Rapid Scanning of Spectra • Rapid Structure Search • Reactants Supply System • Receiving, Shipping and Storage • Reference Sound Source • Regional Support System • Remote Sensing Science • Remote Sensing Society • Remote Speaking System • Remote Switching System • Requirements Status System • Réseau Santé Social (France) • Resident Set Size • Residual Sum of Squares • Resource Sharing System • Resource Survey Satellite • Restart Selectover System • Return Signal Simulator • Rib Structure Station • Rigid Space Structure • Root-Sum-Square • Root-Sum-Squared • Roseires (Sudan) (location identifier) • Rotating Service Structure • Rotating Shadowband Spectrometer • Routing and Switching System • Royal Statistical Society (UK) • RTG Support Structure

RSSC: Remote Sensing Science Council • Remote-Site Simulator Console

RSSI: Responsable de la Sécurité des Services Informatiques • Russian Space Science Internet

RSSN: Research Space Surveillance Network

RSSPO: Resident Space Shuttle Project Office

RSSS: Reusable Space Shuttle System

RSSU: Remote-Site Simulation Unit

Rst: Restore

RST: Radiometric Sun Tracer • Range Search and Track • Read Symbol Table • Readibility-Strength-Tone • Reentry System Technology • Relation Soleil-Terre • Representative Style Table • Research Study Team • Reset • Reset-Set Trigger • Resolution Syntax Tree • Resonance Scattering Theory • Resonant Sphere Technique • Rest • Restart • Revue Scientifique et Technique (periodical code) • Rochester (USA-MN) (location identifier) • Rough Set Theory • Rough-Surface Tester • Routine Sequence Table

RSTC: Remote-Site Telemetry Computer

RSTER: Radar Surveillance Technology Experimental Radar

RSTF: Remote Sensing Task Force

Rstg: Roasting

RSTM: Relay Signal Transmitting Monitor

RSTN: Radio Solar Telescope Network • Railroad Station • Regional Seismic Test Network

Rstnt: Restaurant

RSTP: Remote-Site Telemetry Processor

Rstr: Raster • Restrain • Restrict • Restricted

RSTRT: Restart

RSTS: Resource System Time-Sharing

RSTS-E: Resource System Time-Sharing - Enhanced

RSTSE: Resource System Time-Sharing - Enhanced

RSU: Rate Sensing Unit • Rate Sensor Unit • Recommended Service Upgrade • Relay Storage Unit • Relay Switching Unit • Remote Sensing Unit • Remote Service Unit • Remote Switching Unit • Rescue Support Umbilical • Ridiculously Small Unit • Rivers State University (Nigeria) • Rostov State University (Russia) • Yosu (Korea-ROK) (location identifier)

Rsv: Reserve

RSV: Radio-Sondage Vent • Ready Storage Vessel • Red Supergiant Variable • Request Service • Reserve • Reservoir • Retrieval Status Value • Revised Standard Version

RSVA: Réseaux et Services à Valeur Ajoutée

RSVD: Rapid Sequential Visual Display

Rsvn: Reservation

RSVP: Rapid Serial Visual Presentation • Répondez S'il Vous Plaît • Reservation Protocol • Resource Reservation Protocol

Rsvr: Reservoir

RSVR: Reservoir

RSW: Fort Myers - SW Florida (USA-FL) (location identifier)

RSWC: Ride Side up With Care

RSWG: Response Strategies Working Group

RSWGSC: Response Strategies Working Group Steering Committee

RSWK: Regeln für den Schlagwortkatalog

RSWP: Remote Sensing Working Party

RSX: Realistic Sound Experience • Real-time Resource-sharing Executive • Rouses Point (USA-NY) (location identifier)

RSYS: Responsible System

Rt: Rate • Right

R/T: Radio-Telegraph • Radio-Telegraphy • Real Time • Receiver/Transmitter • Receive/Transmit • Reference Trajectory • Reflection/Transmission • Room Temperature

R&T: Recherche & Technologie • Research & Technology

R-T: Resistance Test

RT: Radar Tracking • Radiative Transfer • Radio Technician • Radio-Telegraph • Radio-Telegraphy • Radio-Telephone • Radio-Telescope • Radio-Télescope • Radioteleskoop • Radio Tower • Radio Tracking • Radio Transmitter • Range Timing • Range to Target • Range Tracking • Rate Transmitter • Ratio Transformer • Rayleigh-Taylor • Reaction Time • Reading Test • Real Time • Received Text • Receiver-Transmitter • Record Terminator • Record Transfer • Reference Trajectory • Related Term • Relativitätstheorie • Relay Tester • Relay Transmitter • Released Time • Relevance Threshold • Remote Terminal • Remote Transmitter • éseau de Télécommunications • Resistance Test • Resource Team • Response Time • Retention Time • Retro Table • Return Ticket • Right • Right Turn • RISC Technology • Rise Time • Room Temperature • Root Task • Round Table • Round Trip • Run Time • Ryle Telescope

RTA: Rapid Thermal Annealing • Real-Time Access • Real-Time Accumulator • Real-Time Activity • Real-Time Analyser • Real-Time Analysis • Real-Time Assessment • Reciprocal Trade Agreement • Reliability Test Assembly • Remote Trunk Arrangement • Rewriting Techniques and Applications • Roads and Traffic Authority (Australia) • Rotuma Island (Fiji) (location identifier)

RTAC: Real-Time Atmospheric Compensation • Regional Technical Assistance Center • Research and Technology Advisory Committee • Research and Technology Advisory Council

RTACF: Real-Time Auxiliary Computing Facility

RTAD: Real-Time Attitude Determination • Remote Tower Alphanumerics Display

RTAM: Remote Telecommunications Access Method • Remote-Terminal Access Method • Resident Terminal Access Method

RTAP: Real-Time Application Platform

RTB: Real-Time Base • Resistance Temperature Bulb • Response/Throughput Bias • Return To Base • Return To Brightest • Roatan (Honduras) (location identifier) • Robot Tour Problem

RTBA: Rate To Be Agreed

RTBF: Radio Télévision Belge de la Communauté Française (Belgium) • Radio Télévision Belge Francophone (Belgium)

RTBP: Restricted Three-Body Problem

RTC: Radio Transmission Control • Ratnagiri (India) (location identifier) • Real-Time Clock • Real-Time Command • Real-Time Compression • Real-Time Computer • Real-Time Control • Real-Time Counter • Red Telefónica Comutada • Relative-Time Clock • Removable Top Closure • Room Temperature Cure • Research and Training Centre • Réseau Téléphonique Commuté • Resolution Trust Corporation • Return To Control • Round Table Conference

RTCA: Radio-Technical Commission for Aeronautics

RTCAC: Regional Tropical Cyclone Advisory Centre

RTCC: Real-Time Command Controller • Real-Time Computer Center • Real-Time Computer Command • Real-Time Computer Complex

RTCE: Rotation/Translation Control Electronics

RTCF: Real-Time Computer Facility

Rtcl: Reticle

RTCL: Reticle

RTCM: Radio-Technical Commission for Maritime services

RTCP: Real-Time Communications Processor • Real-time Transport Control Protocol • Relative Time Command Processor

RTCPS: Real-Time Computer Program System

RTCS: Real-Time Command Sequence • Real-Time Communications System • Real-Time Computer System • Real-Time Computing System

RTCU: Real-Time Control Unit

Rtd: Returned

RTD: Range Time Decoder • Rate Damping • Read Tape Decimal • Real-Time Data • Real-Time Decision • Real-Time Disk • Real-Time Display • Real-Time Drive • Recherche et Développement Technologique • Reliability Technical Directive • Remote Temperature Detector • Research and Technological Development • Research and Technology Development • Resistance Temperature Detector • Resistance Temperature Device • Resistance Thermal Device • Resonant Tunnelling Device • Resonant Tunnelling Diode • Rise Time Discriminator • Rotunda (USA-FL) (location identifier) • Round-Trip Delay

Rtda: Rotunda

RTDAS: Real-Time Data Automation System

RTDBS: Real-Time Database System

RTDC: Real Time Data Compressor

RTD&D: Research, Technological Development & Demonstration

RTDD: Real-Time Data Distribution • Release To Design Development • Remote Timing and Data Distribution • Research, Technological Development and Demonstration

RTDDC: Real-Time Digital Data Conversion

RTD&E: Research, Development, Test & Evaluation

RTDE: Range Time Data Editor

RTDF: Residence Times Density Function

RTDGDS: Real-Time Differential Global Positioning System

RTDHS: Real-Time Data Handling System

RTDM: Real-Time Data Migration • Réseau de Transmission et de Données de la Marine

RTDMP: Real-Time Deformation Matrix Production

RTDR: Reliability Test Data Report

RTDS: Real-Time Data System

RTDT: Real-Time Data Translator

Rte: Remite • Remitente • Route

RT&E: Research, Test & Evaluation

RTE: Marguerite Bay (USA-AK) (location identifier) • Radiative Transfer Equation • Real-Time Execution • Real-Time Executive • Receiver Test Equipment • Recovery Techniques Evaluation • Refractory Trace Element • Reliability Test Evaluation • Remote Tele-Excavation • Remote-Terminal Emulator • Research, Test and Evaluation • Responsible Test Engineer • Return To Earth • Route

RTEAM: Real-Time Environmental Agriculture Monitor

RTECS: Registry of Toxic Effects of Chemical Substances

RTED: Return-To-Earth Digital

RTEE: Real-Time Engineering Environment

RTEL: Reverse Telnet

RTEP: Real-Time Event Processor

RTES: Real-Time Executive System

RTF: Radiodiffusion Télévision Française (France) • Radiotelephone • Real-Time Fortran • Regional Task Force

• Resistance Transfer Factor • Return To Flight • Revision Task Force • Rich-Text File • Rich-Text Format (*) • Robotic Training Facility • Rocket Test Facility • Rotational Test Facility

RTFF: Read The FAQ First

RTFSS: Real-Time Flight Safety Software

Rtg: Rating

RTG: Radioisotope Thermal Generator • Radioisotope Thermoelectric Generator • Radiotelegraph • Range To Ground • Relativistic Theory of Gravitation • Requirements Tape Generator • Ruteng (Indonesia) (location identifier)

RTGC: Real-Time Gaze Control

RTH: Real-Time Muon Histogram • Regional Telecommunication Hub

RTHC: Rotational-Translational Hand Controller • Rotation-Translation Hand Controller

RTHS: Real-Time Hybrid System

RTI: Radiation Transfer Index • Range-Time Indicator • Range Time Intensity • Real-Time Image • Real-Time Interface • Real-Time Interrupt • Referred-To-Input • Relational Technology Inc. • Research Triangle Institute • Rise-Time Indicator • Road Transport Informatics • Roti (Indonesia) (location identifier) • Run Time Infrastructure • Run Time Instrumentation

RTIC: Real-Time Information to the Cockpit • Real-Time Into the Cockpit • Regional Technical Information Center

RTIF: Real-Time Interface

RTIO: Real-Time Input/Output • Remote Terminal Input/Output

RTIP: Radar Technology Insertion Program • Remote-Terminal Interactive Processor

RTIRS: Real-Time Information Retrieval System

RTIS: Raytheon Texas Instryments Systems

RTK: Real-Time Kinematic • Recognition Toolkit • Revenue Ton Kilometer • Right To Know

RTL: Radio Télévision Luxembourg • Raster Transfer Language • Real-Time Language • Register Transfer Language • Resistor-Transistor Logic • Right To Left • Run Time Library • Resistor-Transistor Logic • Spirit Lake (USA-IA) (location identifier)

RTLS: Return To Launch Site

RTLT: Round Trip Light Time

RTM: Radiation Test Model • Radiative Transfer Model • Read The Manual • Real-Time Metric • Real-Time Monitor • Receiver-Transmitter-Modulator • Recovery Termination Management • Regional Telecommunications Management • Register Transfer Module • Release To Manufacturing • Remote Test Module • Research Technical Memorandum • Resin Transfer Molding • Resin-Transfer Moulding • Response Time Monitor • Robot Task Measurement • Rolling Target Mount • Rotterdam (Netherlands) (location identifier) • Running Time Meter • Run-Time Manager

RTMB: Routing Table Maintenance Protocol

RTMC: Riverside Telescope Makers Conference (USA-CA)

RTMD: Real-Time Multiplexer Display

RTMM: Reduced Thermal Mathematical Model

RTMOS: Real-Time Multiprogramming Operating System

RTMP: Real-Time Media Protocol • Routine Table Maintenance Protocol • Routing Table Maintenance Protocol (*)

RTMS: Real-Time Multiprogramming System • Rocket Thrust Measuring System

Rtn: Retain • Return

RTN: Radial-Tangential-Normal • Radio Telescope Network • Raton Crews (USA-NM) (location identifier) • Recursive Transition Network • Registered Trade Name • Remote-Telescope Network • Remote-Terminal Network • Remote-Tracking Network • Report Test Number • Resistor Terminating Network • Return • Rymdteckniknätverk

Rtng: Retaining • Returning

Rtnr: Retainer

RTO: Colton Morrow (USA-CA) (location identifier) • Rapid Turn-On • Real-Time Object • Real-Time Observation • Real-Time Operation • Real-Time Option • Recovery Time Objective • Referred-To-Output • Reflecting Telescope Optics • Rejected Takeoff • Research and Technology Organization • Responsible Test Organization • Retransmit Time Out

RTOC: Real-Time Out of the Cockpit

RTOK: Retest OK

RTOL: Reduced Takeoff and Landing

RTOP: Research and Technology Objectives and Plans • Research and Technology Operating Plan • Research and Technology Operations Plan

RTOPS: Real-Time Operations

RTOS: Real-Time Operating System

RTP: Rapid Thermal Processing • Rapid Transport Protocol • Rated Thermal Power • Real-Time Photogrammetry • Real-Time Position • Real-Time Positioning • Real-Time Processor • Real-time Transport Protocol • Remote-Transfer Point • Request To Purchase • Research and Training Programme • Research Triangle Park (USA-NC) • Reservation Transfer Protocol • Rijnhiuzen Tokamak Project • Room Temperature and Pressure • Room Temperature Phosphorescence • Rutland Plains (Australia-QLD) (location identifier)

RTPE: Real-Time Parameter Estimation

RTpH: Round Trips per Hour

RTPH: Round Trips Per Hour

RTPL: Real-Time Procedural Language

RTPM: Real-Time Program Management

RTPMS: Real-Time Pressure Monitoring System

RTQ: Response To Query

RTQC: Real-Time Quality Control

Rtr: Rotor

RTR: Ready To Run • Reel-To-Reel • Reliability Test Requirement • Remote Transmitter/Receiver • Rotor

RTRC: Radiotelemetry and Remote Control • Regional Teacher Resource Center

RTRD: Remote Tower Radar Display • Retard

RTRN: Return

RTRO: Real-Time Readout

RTRSOC: Real-Time Reporting System in Oceanic Conditions

Rtrv: Retrieve

Rtry: Rotary

Rts: Rights

R&TS: Research & Technical Standards

RTS: Range Time Signal • Range Timing System • Range Tracking Station • Rapid Targeting System • Reactive Tabu Search • Ready To Send • Real-Time Simulation • Real-Time Supply • Real-Time System • Recorded Time Signal • Relative Time Sequence • Relay Telemetry System • Relay Test System • Reliability Test Specimen • Remote Takeover System • Remote Terminal Simulator • Remote Testing System • Remote Tracking Station • Remote Triaxial Sensor • Request To Send • Research Test Site • Residual Time Stamp • Return To Search • Return To Sender • Return To Service • Rottnest Island (Australia-WA) (location identifier) • Run-Time System

RTSC: Reuse Technology Support Center

RTSD: Resources and Technical Services Division

RTSDS: Real-Time Scheduling Display System

RTSE: Reliable Transfer Service Element

RTSEE: Remote Triaxial Sensor Electronics Enclosure

RTSF: Real-Time Simulation Facility

RTSI: Real-Time System Integration

RTSMP: Real-Time Symmetric Multiprocessor

RTSP: Real-Time Signal Processor • Real-Time Streaming Protocol

RTSQL: Real-Time Structured Query Language

RTSS: Real-Time Scientific System • Real-Time Simulation System

RTSW: Real-Time Solar Wind

RTSX: Ranger Telerobotic Shuttle Experiment

RTT: Radioteletype • Réduction du Temps de Travail • Régie des Télégraphes et des Téléphones • Regie van Telegraaf en Telefoon • Regie voor Telefoon en Telegrafie • Road Transport Telematics • Round-Trip Time • Rural Telegraph Terminal

RTTB: Radio To The Building

RTTC: Regional Technology Transfer Center

RTTD: Real-Time Telemetry Data

RTTDS: Real-Time Telemetry Data System

RTTF: Red Taurus Tunable Filter

RTTI: Run Time Type Information

RTTM: Retransmission Time Measurement

RTTS: Real-Time Telemetry System • Real-Time Transmission System

RTTV: Real-Time Television

Rtty: Radioteletype

RTTY: Radioteletype • Radio Teletypewriter

RTU: Real-Time Unit • Recovery Task Unit • Remote Telemetry Unit • Remote Terminal Unit • Remote Test Unit • Response Test Unit • Right To Use

RTUTE: Remote-Terminal Unit Test Equipment

RTV: Raster To Vector • Real-Time Video • Recoverable Test Vehicle • Recovery Test Vehicle • Returned To Vendor • Return To Vendor • Rocket Test Vehicle • Room-Temperature Vulcanized • Room-Temperature Vulcanizer • Room-Temperature Vulcanizing

rTVFR: Real-Time Variable Frame Rate

RTVM: Requirements and Test Verification Matrix

RTVP: Real-Time Video Processing

RTW: Ready To Wear • Round The World • Saratov (Russia) (location identifier)

RTWUS: Research and Technology Work Unit Summary

RTX: Real-Time Transmitted

RTY: Merty (Australia-SA) (location identifier)

RTZ: Return To Zero

ru: Russia • Russian (ISO language code)

Ru: Ruthenium • Ruthénium

RU: Are You? • Burundi (international vehicle registration code) • Range Unit • Receive Unit • Registered User • Reino Unido • Remote Unit • Reprocessed Uranium • Reproducing Unit • Request Unit • Response Unit • Restaurant Universitaire • Rijksuniversiteit • Royaume-Uni • Russia (ISO country code) • Russie (ISO country code)

RUA: Arua (Uganda) (location identifier)

RUB: Ruhr-Universität Bochum (Germany)

RUC: Reporting Unit Code • Roskilde University Centre (Denmark)

RUCA: Rijksuniversitair Campus Antwerpen (Belgium) • Rijksuniversitair Centrum Antwerpen (Belgium)

RuCCS: Rutgers Center for Cognitive Science (USA-NJ)

Rud: Rudder

RUD: Rudder

RUDI: Resource for Urban design Information

RUE: Rational Use of Energy

RUF: Rough

RUFHQ: Regroupement des Universités de la Francophonie Hors-Québec

RUFIS: Rôle of Universities in the Future Information Society

RUG: Regional User Group • Remote User Gateway • Remote User Group • Restricted Users Group • Rijksuniversiteit Gent (Belgium) • Rijksuniversiteit Groningen (Netherlands) • Rugao (China-PRC) (location identifier)

RUH: Riyadh - King Khalid (Saudi Arabia) (location identifier) • ROSAT Users' Handbook

RUI: Remote User Interface • Research at Undergraduate Institutions • Research in Undergraduate Institutions • Royal University of Ireland • Ruidoso (USA-NM) (location identifier)

Ruimt: Ruimtevaart (periodical code)

RUK: Rukumkot (Nepal) (location identifier)

RUL: Rijksuniversiteit Leiden (Netherlands) • Rijksuniversiteit Limburg (Netherlands)

RULER: Remaining Useful Life Evaluation Routine

Rum: Rumanian

RUM: Recinto Universitario de Mayagüez (Puerto Rico) • Remote Underwater Manipulator • Remote-Unit Monitor • Remote User Multiplex • Rotating Unbalanced Mass • Rumjatar (Nepal) (location identifier)

RUMBLE: Requests Users Make By Listings Events

RUMS: Resource Use Management Subgroup

RUN: Saint-Denis de la Réunion (Reunion Island) (location identifier)

RUNCIBLE: Revised Unified New Compiler with Its Basic Language Extended

RUP: Rupsi (India) (location identifier)

RUPT: Rupture

Rur: Rural

RUR: Ruble (Russia) (ISO currency code) • Rurutu (Tubuai Islands) (location identifier)

RURE: Regional and Urban Restructuring in Europe

RURPIS: Resource Utilization and Regional Planning Information System

Rus: Russia • Russian

RUS: Marau Sound (Solomon Islands) (location identifier) • Rechenzentrum Universität Stuttgart (Germany) • Remote Update System • Resonant Ultrasound Spectroscopy • Russia (international vehicle registration code) • Russia (ISO country code) • Russie (ISO country code)

Russ: Russia • Russian

RUSSWO: Revised Uniform Summary of Surveyed Weather Observations

Rúst: Rústico

RUSTI: Research Unit for Surfaces, Transforms and Interfaces

Rut: Rutland

RUT: Reflectance Uniformity Test • Resource Utilization Time • Ruble (Tajikistan) (ISO currency code) • Rutland (USA-VT) (location identifier)

RUTCOR: Rutgers Center for Operational Research (USA-NJ)

Rutd: Rutland

RUU: Rijksuniversiteit Utrecht (Netherlands) • Ruti (Papua New Guinea) (location identifier)

RUV: Rubelsanto (Guatemala) (location identifier)

RUWS: Remote Underwater Work System • Remote User Workstation

RUY: Copan (Honduras) (location identifier)

R/V: Research Vessel • Reality/Virtuality

RV: Radar Vector • Radial Velocity • Radio-Vent • Random Variable • Rated Voltage • Réalité Virtuelle • Reality/Virtuality • Rear View • Recirculation Valve • Recovery Vehicle • Recovery Vessel • Recreational Vehicle • Reentry Vehicle • Reference Voltage • Regionalverband • Relief Valve • Rendezvous • Rendez-Vous • Rendezvous Vehicle • Rescue Vessel • Research Vehicle • Research Vessel • Residual Volume • Retrieval Vessel • Revised Version • Rigveda • Rotovibrational • Roving Vehicle

RvA: Raad van Arbeid

RVA: Farafangana (Madagascar) (location identifier) • RAMAC Virtual Array • Reactive Volt-Ampere • Régie des Voies Aériennes • Relative Virtual Address • Reliability Variation Analysis • Remote Voltage Adjustment • Réseau à Valeur Ajoutée • Rijksdienst voor Arbeidsvoorziening (Netherlands)

RvAc: Revue d'Acoustique (periodical code)

RVAM: Reactive Volt-Ampere Meter

RVAO: Reentry Vehicle Associated Object

RvAqS: Reviews in Aquatic Sciences (periodical code)

RVAS: Requirements Verification and Accounting System • Roanoke Valley Astronomical Society (USA-VA)

RVB: Rendezvous and Berthing • Rouge-Vert-Bleu

RvC: Raad voor de Certificatie (Netherlands)

RVC: Random Vibration Control • Relative Velocity Computer • Rivercess (Liberia) (location identifier)

RvCF: Revista Colombiana de Física (periodical code)

RVCF: Remote Vehicle Checkout Facility

RVCN: Requirement Verification Completion Notice

RVD: Remote Vehicle Docking • Remote Virtual Disk • Rendezvous and Docking • Rijksvoorlichtingsdienst

RVDISC: Rendezvous Discrimination

Rvdo: Reverendo

RVDT: Rotary Variable Differential Transducer • Rotary Variable Differential Transformer

RVE: Representative Volume Element • Saravena (Colombia) (location identifier)

RvFE: Revue Française de l'Électricité (periodical code)

RVFL: Random Vector Functional Link

RVG: Reference Voltage Generator

RvGeo: Reviews of Geophysics (periodical code)

RvGSP: Reviews of Geophysics and Space Physics (periodical code)

RVH: Rzhevka (Russia) (location identifier)

RVI: Ratio Vegetation Index • Reverse Interrupt • Rotation Variant of an Image

RVK: Roervík (Norway) (location identifier)

Rvlg: Revolving

Rvlv: Revolve

RVM: Radial-Velocity Measurement • Reactive Voltmeter • Reentry Vehicle Module • Répertoire de Vedettes-Matière • Requirements Verification Network

RvMin: Reviews of Mineralogy (periodical code)

RvMP: Reviews of Modern Physics (periodical code)

RVN: Republic of Viet Nam • Requirements Verification Network • Rovaniemi (Finland) (location identifier)

RVO: Reichsversicherungsordnung • Reivilo (South Africa) (location identifier)

RvOp: Reviews of Optics (periodical code)

RVP: Radar Verification Programme • Raster-to-Vector Processor • Regional Vice President • Reid Vapor Pressure • Remote Voice Protocol • Rotary Vacuum Pump

RvPA: Revue de Physique Appliquée (periodical code)

RvPD: Revue du Palais de la Découverte (periodical code)

RVQ: Recursive Vector Quantization

RVR: Green River (USA-UT) (location identifier) • Requirement Verification Revision • River • Runway Visual Range

Rvrb: Reverberation

RVRM: Runway Visual Range Midpoint

RvRP: Revue Roumaine de Physique (periodical code)

RVRR: Runway Visual Range Rollout

RVRS: Requirements Verification Review Sheet

RvRST: Revue Roumaine des Sciences Techniques (periodical code)

RVRT: Runway Visual Range Touchdown

Rvs: Reverse • Revise

RVS: Radial Velocity Scanner • Radial Velocity Spectrometer • Radius Vector Subroutine • Reentry Vehicle Separation • Reentry Vehicle Simulator • Remote Viewing System • Rendezvous Sensor • Reverse • Tulsa - R. Lloyd Jones (USA-OK) (location identifier)

Rvsbl: Reversible

RVSM: Reduced Vertical Separation Minimum

Rvsn: Revision

RVSVP: Répondez Vite, S'il Vous Plaît

RvT: Revue Technique Thomson (periodical code)

RVT: Remote Virtual Terminal • RV Tauri

RvTS: Rivista Tecnica Selenia (periodical code)

RVU: Radio Volksuniversiteit • Research Vessel Unit

RVV: Runway Visual Value

Rvw: Review

RVY: Rivera (Uruguay) (location identifier)

rw: Kinyarwanda (ISO language code)

Rw: Rwanda

R/W: Read/Write • Right of Way • Runway

RW: Random Wander • Reaction Wheel • Read and Write • Real World • Relative Wind • Response Word • Right Wing • Rijksweg • Robertson-Walker • Runway • Rwanda (ISO country code)

RWA: Reaction Wheel Assembly • Real-World Application • Rwanda (international vehicle registration code) • Rwanda (ISO country code)

RWAB: Regionale Wissenschaftliche Allgemeinbibliothek

RWB: Responsive Workbench • Rowan Bay (USA-AK) (location identifier)

RWC: Read-Write-Compute • Read-Write-Continue • Regional Warning Centre • Relative Water Content • Remote Work Center

RWCC: Recycled Water Coordination Committee • Regional Workshop Coordinating Committee

RWCP: Real World Computing Partnership (*) • Real World Computing Project

RWD: Radioactive-Waste Disposal • Regular World Day • Right Wing Down

RWE: Radio Wave Experiment • Read/Write Electronics

RWF: Franc (Rwanda) (ISO currency code) • Redwood Falls (USA-MN) (location identifier)

RwFr: Rwanda Franc (currency)

RWG: Rechtzentrale Württembergischer Genossenschaften (Germany) • Reliability Working Group

RWI: Read-Write-Initialize • Regular Word Interval • Respective World Interval • Rocky Mount (USA-NC) (location identifier)

RWL: Rawlins (USA-WY) (location identifier) • Reaction Wheel

R/WM: Read/Write Memory

RWM: Radioactive-Waste Management • Read-Write Memory

RWMC: Reactive Waste Management Committee

RWN: Rovno (Ukraine) (location identifier)

RWOD: ROSAT WFC Observation Dataset

RWP: Radiation Work Permit • Radio Wave Propagation • Rainwater Pipe • Real-time Weather Processor • Real World Processing • Rechts- und Wirtschaftspraxis

RWPG: Real World Problem Generation

RWPV: Rotating-Wall Perfused Vessel

RWPWZ: Rostock Wilhelm Pieck Universität Wissenschaftliche Zeitschrift (periodica l code)

RWR: Radar Warning Receiver

RWS: Franc (Rwanda) (ISO currency code) • Rat Westdeutscher Sternwarten (Germany) (now RDS) • Reactive Wind Shear • Regional Weather Service • Rijkswaterstaat (Netherlands) • Robotic Workstation • Rossby Wave Source

RWT: Read-Write Tape

RWU: Robertson-Walker Universe

RWV: Read-Write-Verify • Remote Work Vehicle • Rotary-Wing Vehicle • Rotating Wall Vessel

RWVD: Real World Visual Display

RWW: Rijksgroepsregeling Werkloze Werknemers

Rwy: Railway • Runway

RWY: Runway

RX: Receive • Receiver • Remote Exchange

RXA: Raudha (Yemen) (location identifier)

RxD: Receive Data

RXD: Received Data

RXO: Redundant Crystal Oscillator

RxR: Railroad

RXRC: Rank Xerox Research Center

RXS: Roxas City (Philippines) (location identifier)

RXTE: (Bruno B.) Rossi X-ray Timing Explorer (formerly XTE)

ry: Rydberg

Ry: Railway • Runway

R-Y: Roll-Yaw

RY: Roll-Yaw

RYB: Rybinsk (Russia) (location identifier)

RYD: Real Year Dollars

RYDMAR: Reaction-Yield-Detected Magnetic Resonance

RYK: Rahim Yar Khan (Pakistan) (location identifier)

RYN: Royan (France) (location identifier)

RYO: Rio Turbio (Argentina-SC) (location identifier)

R&Z: Range & Zero

RZ: Range and Zero • Rechenzentrum • Return to Zero

RZA: Santa Cruz (Argentina-SC) (location identifier)

RZE: Rzeszow (Poland) (location identifier)

RZG: Rechenzentrum Garching (Germany)

RZH: Lancaster - Quartz Hill (USA-CA) (location identifier)

RZM: Stichting voor Radiostraling van Zon en Melkweg (Netherlands)

RZN: Ryazan (Russia) (location identifier)

RZR: Ramsar (Iran) (location identifier)

RZSM: Root Zone Soil Moisture

Rzvs: Rendez-vous

RZZ: Roanoke Rapids - Halifax County (USA-NC) (location identifier)

R2PI: Resonant Two-Photon Ionization

R3: Rotary Reciprocating Refrigerator

R3S: Rapid Records Retrieval System

S

s/: Su

s: Distance • Path • Path Length • Laplace Variable • Length of Arc • Long-Range Order Parameter • Response Curve • Saint • Satisfactory • Save • School • Scientific • Sculpsit • Sea • Second • Secondary • Section • Sedimentation Coefficient • Segundo • See • Semi • Sensitivity Curve • Series • Set • Shilling • Siècle • Siège • Siehe • Siglo • Sign • Signature • Signed • Silver • Sin • Sine • Singlet • Singular • Sinister • Sinistra • Sister • Small • Snow • Society • Soft • Solidus • Solo • Son • Southerly • Specific Entropy • Spectral Transmission • Spherical • Spin Quantum Number • Steamer • Steel • Stem • Stere • Stock • Stoichiometric Oxidant-to-Fuel Ratio • Strange • Stratus • Strong Absorption • Su • Südlich • Sun • Sunny • sur • Sustantivo • Symmetrical • Symmetry Number

S/: Su

S: Apparent Power • Area • Characteristic Function of Hamilton • Entropy • Flux Density • Linear Stopping Power • Magnetic Solar Daily Variation • Poynting Vector • Probability Current Density • Quantum Number of Electronic Spin • Saint • Salar • Salina • San • Sankt • Santa • Santo • São • Satisfactory • Saturday • Saturn • Save • Saxon • Schilling (currency) • School • Schwefel • Scientific • Scotland • Scottish • Scout • Sea • Second • Secondary • Secret • Secretary • Sector • See • Segundo • Seite • Senate • Señor • Señora • Sentence • September • Series • Seven • Seven (Roman numeral) • Seventy • Seventy (Roman numeral) • Sheffield (UK postal code) • Shilling • Siècle • Siemens • Sierra (international alphabet) • Siglo • Signature • Signor • Signora • Single • Sinister • Slow • Small • Snow • Society • Socius • Soft • Sol • Sol (currency) • Solar • Solar Constant • Solar Mass • Soufre • South • Southern • Spectral Response • Spectral Type • Spin Quantum Number • Spiral Galaxy • Staff • State • Statute • Stere • Stimulus • Store • Strangeness Quantum Number • Stress • Subject • Submarine • Substitution Reaction • Sucre (currency) • Sud • Süd • Suden • Südlich • Sulfur • Sulphur • Summer • Sun • Sunday • Supplement • Sur • Sweden (international vehicle registration code) • Sweden (postal country code) • Synodic Period • Total Neutron Source Density • 7 (Roman numeral) • 70 (Roman numeral)

s/a: Sin Año

sa: Sanskrit (ISO language code) • Saudi Arabia • Semiannual • siehe auch • Sin Año • Sine Anno • Symmetrical

Sa: Samarium (now Sm) • Samstag • Saturday • Serra • Serrania • Sierra • Summa

S&A: Safe & Arm • Sampling & Analysis • Science & Application

S/A: Safe and Arm • Scheduled/Actual • Site Activation • Solar Array • Spacecraft Adapter • Subaccount • Subassembly • Subject to Acceptance • Such As • Supplemental Agreement • Swing Arm

SA: Arabia (Saudi_) (ISO country code) • Arabie Saoudite (ISO country code) • Safe and Arm • Safety Analysis • Safing Area • Sales Analysis • Salta (Argentina) • Samoa • Sandstorm • Sanskrit • Saskatchewan (Canada) • Saudi Arabia (ISO country code) • Saudi Arabian • Scheduled/Actual • Scientific Area • Scintillation Adjust • Secrétaire Administratif • See Also • Select Address • Selected Area • Selected Availability • Selective Acquisition • Selective Availability • Semiannual • Sensitivity Analysis • Separated Atom • Sequential Automated • Service Assistant • Service d'Aéronomie (France) • Set Attribute • Shaft Angle • Shaping Amplifier • Shapley-Ames • Shop Aid • Sierra • Signal Accuracy • Signal Analyzer • Signature Analysis • Simulated Annealing • Sine Anno • Single Access • Single Aircraft • Single Aisle • Site Activation • Situational Awareness • Sky Atlas • Small Array • Sobolev Approximation • Sociedad Anónima • Sociedade Anónima • Società Anonima • Société Anonyme • Solar Activity • Solar Array • Source Acquisition • South Africa • South African • South America • South American • South Atlantic • South Australia • South Australian • Spacecraft Adapter • Special Agent • Spectral Albedo • Spectrograph Assembly • Spectrum Analysis • Spectrum Analyzer • Spin Axis • Splitting Amplifier • Station Address • Station Automatique • Storage Alteration • Storage Area • Store Address • Strongly Agree • Structural Analyzer • Structured Analysis • Su Alteza • Subaccount • Subassembly • Subject to Approval • Subsistence Allowance • Supplemental Agreement • Support Area • Surface Area • Swansea (UK postal code) • Swath Angle • Swing Arm • System Address • System Analysis • System Architect

SAA: Safety Assurance Analysis • Saratoga (USA-WY) (location identifier) • Satellite Attitude Acquisition • Seminario de Astronomía y Astrofísica • Service Aspects and Applications • Servo-Actuated Assembly • Shift And Add • Shift-and-Add Algorithm • Single-Access Antenna • Single Article Announcement • Slot Array Antenna • Society of American Archivists (USA) • Solar Aspect Angle • South-Atlantic Anomaly • Space Association of Australia • Speech Association of America • Standards Association of Australia • Stellar Astronomy and Astrophysics • Step Adjustable Antenna • Systems Application Architecture

SAAA: San Antonio Astronomical Association (USA-TX) • Shoreline Amateur Astronomical Association (USA-MI)

SAAAA: San Angelo Amateur Astronomy Association (USA-TX)

SAAAS: South African Association of the Advancement of Science

SAAB: Svensk Aeroplan Aktiebolag

SAAC: Schedule Allocation And Control • Seismic Array Analysis Center • Space Applications Advisory Committee

SAAD: Sociedade de Astronomia e Astrofísica de Diadema (Brazil) • South Atlantic Anomaly Detector

SAAL: Signaling ATM Adaptation Layer

SAALC: San Antonio Air Logistics Center

SAAM: Simulation Analysis And Modelling

SAAMEX: Surface and Atmospheric Airborne Microwave Experiment

SAAO: South African Astronomical Observatory

SAAOC: South African Astronomical Observatory Circulars (periodical code)

SAAP: South-Atlantic Anomaly Probe • Subsonic Aircraft Assessment Program

SAAR: Seasonally Adjusted at Annual Rates

SAARC: South Asian Association for Regional Cooperation

SAARU: Secondary Air and Attitude Reference Unit • Secondary Attitude and Air Reference Unit

SAAS: Science Achievement Award for Students • Shuttle Aerospace Actuator Simulator • Shuttle Aerosurface Actuator Simulation

SAASIN: Sociedad Astronómica Amateur de Sinaloa (Mexico)

SAASS: South African Airborne Scanner Survey

SAASTEC: South African Association of Science and Technology Centers

SAAT: Satellite Attitude Acquisition Technique

SAATN: Shuttle/Atlas Advanced TIROS Satellite

Sab: Sábado • Sabato • Sabbath

SAB: Saba Island (Netherlands Antilles) (location identifier) • Same As Basic • Satellite Assembly Building • Science Advisory Board • Shuttle Assembly Building • Shuttle Avionics Breadboard • Simulation of Adaptive Behavior • Site Activation Board • Sociedad de Astronomía Balear (Baleares) • Sociedade Astronômica Brasileira (Brazil) • Société Astronomique de Bordeaux (France) • Société Astronomique de Bourgogne (France) • Solar-Array Battery • Source Approval Bulletin • Spacecraft Assembly Building • Standards Activities Board • Storage and Assembly Building • Synoptic Analysis Branch

Sabat: Sabatique • Sabbatical

SABC: South African Broadcasting Corporation

SABCA: Société Anonyme Belge de Constructions Aéronautiques (Belgium)

SABENA: Société Anonyme Belge d'Exploitation de la Navigation Aérienne (Belgium)

SABER: Situational Awareness Beacon with Reply

SABEW: Society of American Business and Economic Writers (USA)

SABINET: South African Bibliographic and Information Network

SABIR: Semiautomatic Bibliographic Information Retrieval • Space-Based Interceptor

SABIRS: Semiautomatic Bibliographic Information Retrieval System

SABIT: Special American Business Internship Training

SABLE: Simulator Atmospheric Boundary Layer Environment

SABOR: Same As Basic Or

SABRE: Semiautomatic Business Research Environment • Sweden And Britain Radar Experiment

SABREX: Satellite-Borne Radiometer Experiment

SABRO: South American Biological Research Organization

SABS: Shared-Aperture Boresight Sensor • South African Bureau of Standards • Stanford Automated Bibliographic System • Statistical Analysis of Bibliographical Structure

SABV: Suid-Afrikaanse Biblioteekvereniging (South Africa)

S/AC: Stabilization/Attitude Control

SAC: Sacramento (USA-CA) (location identifier) • Saguaro Astronomy Club (USA-CA) • Satélite de Aplicaciones Científicas • Satellite Applications Centre • Science Abstracting Committee • Science Activity Centre • Scientific Advisory Committee • Scientific Advisory Council • Scientific Analysis Component • Secondary Address Code • Sectional Aeronautical Chart • Semiautomatic Coding • Senate Appropriations Committee • Sequencing and Application Controller • Service Area Code • Service Area Computer • Shannonside Astronomy Club (Ireland) • Shapley Ames Catalog • Short-run Average Cost • Single-Address Code • Slovak Accredition Council • Solar Activity Cycle • Solar Array Carrier • South Atlantic Coast • Space Activities Commission • Space Advisory Committee • Space and Aviation Control • Space Applications Centre • Space Applications Corporation • Special Analysis Centre • Special Area Code • Special Area of Conservation • Spherical Aberration Corrector • Stabilization/Attitude Control • State Affairs Committee • Stateful Access Control • Store And Clear • Strategic Air Command • Structural Adjustment Committee • Sub-Array Center • Subject Analysis Committee • Support Action Center • Symposium on Applied Computing • Système d'Annonce des Crues • Systems Advisory Committee

SACA: Société d'Astronomie de Cannes (France)

SACAC: South African Council for Automation and Computation

SACAP: Selection, Acquisition, Cataloguing, And Processing

SACC: Satellite Archive Coordination Center • Science Action Coordinating Committee • Science Archive Coordination Center • Spatial Analysis Coordinating Committee

SACCS: SAC Automated Command Control System

SACDA: System Analysis, Control and Design Activity

SACEM: Société des Auteurs, Compositeurs et Éditeurs de Musique

SACEP: South Asia Cooperative Environment Program

SACEPO: Standing Advisory Committee before the European Patent Office

Sachs: Sachsen (Germany)

Sächs: Sächsische (Germany)

Sachs-Anh: Sachsen-Anhalt (Germany)

SACI: Secondary Address Code Indicator

SACK: Selective Acknowledgement

SACL: Standards And Calibration Laboratory

SACM: Smart Acoustic Current Meter • Sociedad de Autores y Compositores de México (Mexico)

SACMA: Suppliers of Advanced Composite Materials Association

SACN: Secrétaire Administratif de Classe Normale

Sacr: Sacramento (USA-CA)

SACRD: Safety Analysis Computerized Reactor Data

SACRS: South Australian Centre for Remote Sensing

SACS: Satellite Attitude Control Simulator • Software Avionics Command Support • Solar Attitude Control System • Synchronous Attitude Communications Satellite

SACSIR: South African Council for Scientific and Industrial Research

SACSIRO: South African Commonwealth Scientific and Industrial Research Organization

SACTO: Sacramento Test Operations

SACU: South African Customs Union

SACV: Spacecraft Autonomy Concept Validation

SACZ: South Atlantic Convergence Zone

S&AD: Science & Applications Directorate

SAD: Safety Analysis Document • Safety Assessment Document • Safford (USA-AZ) (location identifier) • Seasonal Affective Disorder • Secondary Angular Resolution • Shuttle Authorized Document • Simple, Average or Difficult • Single Administrative Document • Situation Attention Display • Software Architecture Document • Solar-Array Drive • Spacecraft Attitude Display • Spaced Antenna Drift • Special Adapter Device • Standard Abundance Distribution • System Allocation Document • System Analysis Drawing

SADA: Solar-Array Drive Assembly

SADAP: Space Astrophysics Data Analysis Program

SADAPTA: Solar-Array Drive And Power Transfer Assembly

SADC: Sequential Analog-Digital Computer • Southern Africa Development Community

SADCC: Southern Africa Development Coordination Conference

SAdeCV: Sociedad Anónima de Capital Variable

SAdeCVyRL: Sociedad Anónima de Capital Variable y Responsabilidad Limitada

SADE: Signal Analysis and Dissemination Equipment • Solar-Array Deploy Electronics • Solar-Array Drive Electronics • Spark-Assisted Diesel Engine

SADEC: Spin Axis Declination

SADEYA: Sociedad Astronómica de España y America (Spain)

SADF: Semi-Automatic Document Feed • Statistical Analysis of Documentation Files

Sadha: Sadhana (periodical code)

SADI: Secretariat Automated Document Index

SADIE: San Diego Imaging Echelle • Scanning Analog-to-Digital Input Equipment • Secure Automatic Data Information Exchange • Silicon Anode Detector with Integrated Electronics

SADIM: Stanford Art Directors' Invitational MacWorkshop

SADIO: Sociedad Argentina de Investigación Operativa

SADIS: Satellite Distribution Service • Southern African Development Information System • Southern African Documentation and Information System

SADL: Sterilization Assembly Development Laboratory • Synchronous Data Link Control

SADM: Solar-Array Drive Mechanism • Solar-Array Drive Motor

SADP: Synthetic Array Data Processor

SADPT: Solar Array Drive and Power Transfer

SADR: Spatial Address • System Audit and Design Review

SADRS: Simplified Autonomous Data Receiving Station

SADS: Solar-Array Drive System

SADT: Structured Analysis and Design Technique • Systems Analysis and Design Technique

SADV: Semiannual Density Variation

SAE: Sangir (Indonesia) (location identifier) • Self-Addressed Envelope • Shaft Angle Encoded • Simple Arithmetic Expression • Società fra gli Autori e gli Editori (Italy) • Society for the Advancement of Education (USA) • Society of Automotive Engineers • Stamped Addressed Envelope • Standard Average European • Steering Angle Error • Stop At Expiration

SAED: Selected Area Electron Diffraction

SAEF: Spacecraft Assembly and Encapsulation Facility (USA-FL)

SAENET: South Australia Educational Network

SAERIS: South Australian Education Resources Information System

SAES: Service d'Allocations d'Études Supérieures • Stand Alone Engine Simulator

SAESAC: Space And Earth Sciences Advisory Committee

SAETA: Sociedad Anónima Española de Traductores y Autores (Spain) • Sociedad Ecuatoriana de Transportes Aéreos Ltda. (Ecuador)

Saf: Safety

SAF: Santa Fe (USA-NM) (location identifier) • Science Archive Facility • Search And Find • Sección de Astrofotografía • Secured Access Facility • See Also From • Service Access Facility • Service Aérien Français • Société Astronomique de France • Space Agency Forum • Spacecraft Assembly Facility • Special-Access Facility • Standard Associative Form • Symmetry-Adapted Function • System Authorization Facility

SAfA: Sydney Association for Astrophysics (Australia-NSW)

SAFA: Safety Assessment of Foreign Aircraft • Solar-Array Failure Analysis

SAFARI: South-African Fire-Atmosphere Research Initiative • Swedish And French Auroral Radar Investigation • Système Automatisé pour les Fichiers Administratifs et les Répertoir es d'Individus (France)

SAFE: Braathens South America and Far East Air Transport SA • Safing and Failure Detection Expert • San Andreas Fault Experiment • Security And Freedom through Encryption • Solar Array Flight Experiment • South Africa - Far East • Stamped Addressed Foolscap Envelope

S. Afr.: South Africa • South African

SAFER: Simplified Aid For EVA Rescue • Société d'Aménagement Foncier et d'Établissement Rural • Switch Architecture for Extreme Resiliency

SAFGA: Société Astronomique de France - Groupe d'Alsace

SAFI: Semi-Automatic Flight Inspection

SAFIR: Sociedad Astronómica de la Faculdad de Ingenería (Mexico) • Système d'Alerte Foudre par Interféromètre Radioélectrique • Système d'Alerte Foudre par Interférométrie Radioélectrique (*) • Systèmes Algébriques Formels pour l'Industrie et la Recherche

SAFIRE: Spectroscopy of the Atmosphere using Far-Infrared Emission • Submillimeter And Far Infrared Experiment

SAFISY: Space Agencies Forum for the International Space Year

SAFM: Stuck-At Fault Model

SAFR: Schweizerische Arbeitsgemeinschaft für Raumfahrt (Switzerland)

SAFRAS: Self-Adaptive Format Retrieval and Storage System

SAFRS: Saudi Arabian Facility for Remote Sensing

SAFSA: Study Abroad and Foreign Student Adviser

SAG: Sagwon (USA-AK) (location identifier) • Schweizerische Astronomische Gesellschaft (Switzerland) • Scottish Astronomers Group (UK-Scotland) • Sociedad Astronómica Granadina (Spain) • Space Astrophysics Group • SQL Access Group • Starlink Advisory Group • Street Address Guide • Systems Analysis Group

SAGA: Short Arc Geodetic Adjustment • Solar-Array Gain Augmentation • South Atlantic Geomagnetic Anomaly • Soviet-American Gases and Aerosols (experiment)

SAGAS: Southern Area Group of Astronomical Societies (UK)

SAGE: Schéma d'Aménagement et de Gestion de l'Eau • Semi-Automatic Ground Environment • Solar-Assisted Gas Energy • Soviet-American Gallium Experiment • Sterilization Aerospace Ground Experiment • Strategic Aerosol and Gas Experiment • Stratosphere Aerosol and Gas Experiment (*) • Stratospheric Aerosol and Gas Experiment • Stratospheric Aerosol Gas Experiment • Structuring and Analysis of Gamma-ray Events • Systems Administrators Guild

SAGEECE: Schéma d'Aménagement et de Gestion Écologique des Cours d'Eau

SAGEM: Société d'Applications Générales d'Électricité et de Mécanique (France)

SAGENAP: Scientific Assessment Group for Experiments in Non-Accelerator Physics

SAGEP: Système Automatisé pour la Gestion Électronique de Périodiques

SAGET: Satellite-Geodynamical Traverse

SAGGE: Synchronous Altitude Gravity Gradient Experiment

SAGS: Semiactive Gravity Gradient System

SAGSTRAM: Standing Advisory Group on the Safe Transport of Radioactive Material

SAGW: Schweizerische Akademie der Geistes- und Sozialwissenschaften (Switzerland)

SAH: Sana'a - Al'Rahaba (Yemen) (location identifier)

SAHC: South Asia Hydrological Cycle

SAHL: Société d'Astronomie du Haut-Léman (Switzerland)

SAHM: Société Astronomique de Haute-Marne (France)

SAHR: Société Astronomique du Haut-Rhin (France)

SAHSA: Servicios Aéreos de Honduras SA

SA&I: Systems Analysis & Integration

SAI: San Marino (location identifier) • Scan Auroral Imager • Scanning Auroral Imager • Serving Area Interface

• Società Astronomica Italiana (Italy) • Space Application Institute • Spin-scan Auroral Imager • Sternberg Astronomical Institute (Russia) • Su Alteza Imperial • Sudden Auroral Intensity • Symbolic Automatic Integrator • Synoptic Atmospheric Imager • Système d'Analyse des Images • Systems Analysis and Integration

SAIC: Science Applications International Corporation (USA-VA) • State Administration for Industry and Commerce

SAICYT: Sistema Automatizado de Información Cientifica y Técnica

SAID: Safety Action Item Database • Safety Analysis Input Data • Semiautomatic Integrated Documentation • Shuttle Authorized Item Document • System Analysis Index for Diagnosis • System for Automated Information Dissemination

SAIDAS: Sistema de Adquisición e Integración de Datos de Satelite

SAIDET: Single-Axis Inertial Drift Erection Test

SAIE: Special Acceptance Inspection Equipment

SAIF: South African Industrial Federation • Spatial Archive Interchange Format

SAIL: Scattering from Arbitrarily Inclined Leaves • Serial ASCII Instrumentation Loop • Shuttle Avionics Integration Laboratory • Space and Atmospheric Instrumentation Laboratory • Spacecraft Artificial Intelligence Laboratory • Stanford Artificial Intelligence Laboratory • Structural Analysis Input Language

SAILIS: South African Institute of Librarianship and Information Science

SAILS: Space Applications of Industrial Laser Systems • Submillimeter Astronomy Investigations of Line Spectra

SAIM: Satellite Autonomous Integrity Monitoring • Systems Analysis and Integration Model

SAIMS: Selected Acquisition Information and Management System

SAINT: Satellite Array for International and National Telecommunications • Symbolic Automatic Integrator

SAIO: Service Académique d'Information et d'Orientation (France)

SAIP: Semi-Automated IMINT Processing • South African Institute of Physics • Spares Acquisition Integrated with Production

SAIR: Synthetic Aperture Interferometric Radiometer

SAIRE: Scalable Agent-based Information Retrieval Engine

SAIS: School of Advanced International Studies • Science and Applications Information System • Scientific Apparatus Information Service • Selector Automático de Imágenes Solares • Sistema de Adquisición de Imágenes Solares • Standard Aeronautical Indexing System • Staring Array Imaging Spectrometer

SAIt: Società Astronomica Italiana (Italy)

SAIT: Samsung Advanced Institute of Technology • Service d'Analyse de l'Information Technologique • South Australian Institute of Technology

SAJ: Sirajganj (Bangladesh) (location identifier) • Station astronomique Jansky (France) • Subassembly Jig

SAJAC: Satellite Japan Corporation (now merged with JC SAT into JSS)

SAJPh: South African Journal of Physics (periodical code)

SAJSc: South African Journal of Science (periodical code)

SAK: Saudarkrókur (Iceland) (location identifier)

Sal: Salaire • Salamanca (Spain) • Salarié • Salary • Salizzada

SAL: Safeguards Analytical Laboratory • Salida • San Salvador (El Salvador) (location identifier) • Scientific Airlock • Shift Arithmetic Left • Shuttle Avionics Laboratory • Simple Author Language • Société Astronomique de Liège (Belgium) • Société Astronomique de Lyon (France) • Solar Arc Lamp • Solar-Array Leaf • Space Astronomy Laboratory • Space Astrophysics Laboratory • Spread Across Liquids • SQL Application Language • Strategic Arms Limitation • Structured Assembly Language • Subject Authority List • Supersonic Aerophysics Laboratory • Symbolic Assembly Language • Synthetic Aperture Lidar • Systems Assembly Language

SALA: Solar Arc Lamp Assembly

SALAT: System for Automatic Language Analysis and Translation

SALC: Sacramento Air Logistics Center

SALCW: Space Astrophysics Laboratory Compound Widget

SALE: Simple Algebraic Language for Engineers

SALEN: Students Across Europe Language Network

SALI: Standalone Automatic Location Identification • Surface Analysis by Laser Ionization

SALINET: Satellite Library Information Network

SaLIS: Surveying and Land Information Systems (journal)

SALIS: Set-Aside Land Information System

SALM: Society of Air Line Meteorologists

SALP: Special Adult Learning Programme

SALR: Saturated Adiabatic Lapse Rate • Synthetic Aperture Laser Radar

SALRO: Saudi Arabian Laser Ranging Observatory

SALS: Short Approach Lighting System • Small Angle Light Scattering • Statistical Analysis with Least Squares (fitting)

SALT: Society for Applied Learning Technology • South African Large Telescope • Standard Analysis Language Translator • Strategic Arms Limitation Talks • Symbolic Algebraic Language Translator

SALTE: Semiautomatic Line Test Equipment

Salv: Salvador • Salvage

SALWA: Semiarid Lowlands of West Africa

Sam: Samedi • Samoa • Samstag

S. Am.: South America • South American

SAM: Safety Analysis Methodology • Salamo (Papua New Guinea) (location identifier) • Scan Analysis Module • Scanning Auger Microprobe • Scanning Auger Microscope • Scanning Auger Microscopy • School of Aerospace Medicine (USAF) • Script Applier Mechanism • Secure Application Module • Security Accounts Manager • Security Application Module • Selective Automatic Monitoring • Self-Assembled Monolayer • Semantic Analysis Machine • Semantic Analyzing Machine • Semianalytical Model • Sensing with Active Microwave • Sequential-Access Memory • Sequential-Access Method • Serial Access Memory • Shared Appreciation Mortgage • Short-Axis Mode • Shuttle Activation Monitor • Shuttle Attachment Manipulator • Shuttle Attachment Mechanism • Signal Analyzing Monitor • Simulated Assignment Model • Simulation of Analogue Methods • Single Aviation Market • Small Angle Manoeuver • Smart Acronym Macro • Sociedad Astronómica de México (Mexico) • Société Astronomique de Montpellier (France) • Société d'Astronomie de Montréal (Canada-QC) • Software Archive Meeting • Sort And Merge • Sourcebook in Applied Mathematics • Space Available Mail • Spatial Access Method • Special Advisory Message • Special Air Mission • Spectrum Analyzer Module • Speech Assessment Methodology • SQL Application Message • Squarewave Amplitude Modulation • Stage Assembly and Maintenance • Standard Additive Model • Standard Assembly Module • Static Address Mapping • Stratospheric Aerosol Measurement • Stratospheric Aerosol Monitor • Structural Assembly Model • Subtraction-Addition-Multiplication • Sun Acquisition Mode • Supportability Assessment Model • Synchronous Amplitude Modulation • System Accuracy Model • System Activation and Monitoring • System Activity Monitor • System

Administration Manager • System Administration Menu • System Analysis Machine

SAMA: Scientific Apparatus Manufacturers Association • Sociedade Astronômica Maranhense de Amadores (Brazil) • Step-by-step Automatic Message Accounting

SAMAS: Semi-Automated Mesoscale Analysis System

SAMBA: Satellite for the Measurement of Background Anisotropies • Satellite to Measure Background Anisotropies • Satellite to measure the Microwave Background Anisotropies

SAMBO: Simultaneous Auroral Multiballoons Observations

SAMEC: Science And Mathematics Education Center

S. Amer.: South America • South American

SAMEER: Society for Applied Microwave Electronics Engineering and Research (India)

SAMER: Système Automatique de Mesure de l'Évaporation Régionale

SAMEX: Shuttle Active Microwave Experiment

SAMICS: Solar-Array Manufacturing Industry Costing Standards

SAMIR: Satellite Microwave Radiometer

SAMIS: Structural Analysis and Matrix Interpretation System

SAML: Send And Mail

SAMMIE: Scheduling Analysis Method for Mission Integrated Experiments • System for Aiding Man-Machine Interaction Evaluation

Samml: Sammlung

SAMMS: Siemens Advanced Motor Master System

SAMO: School voor Atmosferisch en Marien Onderzoek (Netherlands)

SAMOS: Semi-Automatic Meteorological Observing System • Structure Adaptée à la Modélisation Opérationnelle des Systè mes

Samp: Sample

SAMP: Shuttle Automated Mass Properties

SAMPD: Science Analysis and Mission Planning Directorate

SAMPE: Society for the Advancement of Material Process Engineers

SAMPEX: Solar And Magnetospheric Particle Explorer (*) • Solar Anomalous and Magnetic Particle Explorer • Solar Anomalous and Magnetospheric Particle Explorer • Solar Anomalous Magnetospheric Particle Explorer

SAMPIE: Solar-Array Module Plasma Interaction Experiment

SAMPJ: SAMPE Journal (periodical code)

SAMPLE: Single-Assignment Mathematical Programming Language • Survey Analysis Macro Programming Language

SAMPQ: SAMPE Quarterly (periodical code)

SAMRO: Satellite Militaire de Reconnaissance Optique

SAMS: Sampling Analog Memory System • Satellite Automatic Monitoring System • Segment Alignment and Maintenance System • Shuttle Acceleration Measurement System • Shuttle Altitude Measuring System • Shuttle Attached Manipulator System • Shuttle Attachment Manipulator System • Space Acceleration Measurement System • Spatial Analysis and Modelling System • Stratospheric And Mesospheric Sounder

SAMSHA: Substance Abuse and Mental Health Services Administration

SAMSON: System Analysis of Manned Space Operations

SAMU: Service d'Aide Médicale d'Urgence

SAMW: Schweizerische Akademie der Medizinischen Wissenschaften (Switzerland)

San: Sanitary

SAN: San Diego - Lindbergh (USA-CA) (location identifier) • Serveur d'Archivage Numérique • Severe Acoustic Noise • Société d'Astronomie de Nantes (France) • Space Age News • Standard Access Number • Standard Account Number • Standard Address Number • Standard Address Numbering • Standard Article Number • Storage-Area Network • Styrene Acrylonitrile • System Area Network • Système d'Archivage Numérique

SANAGA: Système d'Acquisition Numérique pour l'Analyse des Grains Africains

SANAMET: Servicio Autónomo Nacional de Metrología (Venezuela)

SAnAp: Suppléments aux Annales d'Astrophysique (periodical code)

Sanat: Sanatorium

SANCST: Saudi National Center for Science and Technology (Saudi Arabia) (now NCST)

Sand: Sandwich

SAND: Site Activation Need Data (*) • Site Activation Need Date • Sorting and Assembly of New Data • Space-Assisted Network against Desertification • Spectro-Angular Normalized Difference

SANDS: Structural Analysis Numerical Design System

SANE: Severe Acoustic Noise Environment • Standard Apple Numeric Environment

SANET: South African Network

SANIL: Système d'Analyse des Images Landsat

Sanit: Sanitary • Sanitation

SANOS: Stabilization Airborne Night Observation System

SANOVA: Simultaneous Analysis Of Variance

SANR: Subject to Approval - No Risk

SAN REMO: Système Analytique de Répartition de Moyens

Sans: Sanserif • Sanskrit

SANS: Small-Angle Neutron Scattering • Stellar Autonomous Navigation System • Systems And Network Security

SANSA: Servicios Aéreos Nacionales SA

Sánscr: Sánscrito

SANTA: Science Analysis Near-Term Activity

SANW: Schweizerische Akademie der Naturwissenschaften (Switzerland)

SANZ: Standards Association of New Zealand (now SNZ)

SAO: São Paulo (Brazil-SP) (location identifier) • Science Assessment Observation • Select Address and Operate • Semi-Annual Oscillation • Shaanxi Astronomical Observatory (China-PRC) • Shanghai Astronomical Observatory (China-PRC) • Smithsonian Astrophysical Observatory (USA-MA) • Sobaeksan Astronomy Observatory (Korea-ROK) (see now SOAO) • Special Adaptive Optics • Special Astrophysical Observatory (Russia) • Spectrographe Acousto-Optique • Spectromètre Acousto-Optique • Surface Airways Observation • Systems Analysis Office

SAOC: Smithsonian Astrophysical Observatory Catalog • Space and Astronautics Orientation Course

SAOS: Select Address Output Signal

SAOSR: Smithsonian Astrophysical Special Report (periodical code)

SAOZ: Système d'Analyse par Observations Zénithales

SAp: Service d'Astrophysique

SAP: Safety Analysis Program • San Pedro Sula - Villeda Morales (Honduras) (location identifier) • Sans Aucun Problème • Scanned Absolute Power • Science Archive Process • Science Assessment Panel • Scientific Advisory Panel • Second Audio Program • Sensor AVE Package • Service Access Point • Service Access Provider • Service Advertisement Protocol • Service Advertising Protocol (*) • Share Assembly Program • Sociedad Astrofísica de Paraguay • Société d'Astronomie Populaire (France) •

Software Automation Panel • Soon As Possible • Speaker-Adaptation Process • State Assessment Panel • Strain Arrestor Plate • Strong Anthropic Principle • Structural Adjustment Program • Structural Analysis Program • Subject Access Project • Superadditivity Property • Symbolic Address Program • Symbolic Assembly Program • System Access Protocol • System Alignment Procedure • Systolic Arterial Pressure

SAPA: South African Publishers' Association • System for Analysis of Pyramid Algorithms

SAPAM: Society of African Physicists and Mathematicians

SAPC: Société Astronomique Populaire du Centre (France)

SAPFU: Surpassing All Previous Foul-Ups

SAPI: Service Access Point Identifier • Speech Application Programming Interface

SAPIR: System of Automatic Processing and Indexing of Reports

SAPL: Société d'Astronomie Populaire de Limoges (France)

SAPLA: Standing Advisory Panel on Library Automation

SAPM: Société d'Astronomie du Planétarium de Montréal (Canada-QC)

SAPMD: Stand-Alone Pressure Measurement Device

SApMM: North Holland Series on Applied Mathematics and Mechanics (periodical code)

SAPP: Société d'Astronomie Populaire Poitevine (France)

SAPR: Semiannual Progress Report

SAPRR: Société des Autoroutes Paris-Rhin-Rhône

SAQ: San Andros (Bahamas) (location identifier) • Sociedad Astronómica de Quebradillas (Puerto Rico)

SAQC: Statistical Analysis and Quality Control

Sar: Sarawak • Sardinia • Sardinian

SAR: Riyal (Saudi Arabia) (ISO currency code) • Safety Analysis Report • Safety Assessment Report • Scanned Absolute Response • Schedule Allocation Requirements • Science Applications Research • Science Assessment Report • Search And Release • Search And Rescue • Second Assessment Report • Secrétaire d'Administration de la Recherche • Segment And Reassemble • Segmentation And Reassembly • Selected Acquisition Report • Selling, Administrative and Research • Semiannual Report • Shift Arithmetic Right • Simultaneous Autoregressive • Single-Axis Reference • Single-Bit Alternation Recording • Site Acceptance Review • Site Action Request • Slower-than-Average Rotation • Sociedade Astronômica do Recife (Brazil) • Sodium Adsorption Ratio • Software Analysis Room • Son Altesse Royale • South African Republic • Spacecraft Acceptance Review • Sparta (USA-IL) (location identifier) • Special Access Required • Special Action Request • Special Action Requirement • Special Administrative Region • Special Analysis Report • Specific Absorption Rate • Specific Activity Report • Specification Approval Report • Stable Auroral Red • Stock Appreciation Right • Storage Address Register • Store Address Register • Su Alteza Real • Subauroral Red • Subcommittee on Atmospheric Research • Subsequent Applications Review • Successive Approximation Register • Summary Analysis Report • Support Action Request • Surveillance Assessment Report • Symbol Acquisition Routine • Synthetic Aperture Radar • Syrian Arab Republic • System Acceptance Review • System Analysis Report • System Assessment Review

SARA: Sampled Aperture Receiving Array • Satellite Amateur de Radioastronomie • Satellite pour Astronomie Radio Amateur • Single Aliquot/Regeneration and Added • Society of Amateur Radio Astronomers (USA-NY) • Southeastern Association for Research in Astronomy • Superfund Amendments and Reauthorization Act • Synthetic Aperture Radar Antenna • Systems Analysis and Resource Accounting

SARAH: Search And Range Homing

SARAS: Synthetic Aperture Radar Advanced Simulator

SARB: Sélecteur Automatique de Références Bibliographqiues • Surface and Atmospheric Radiation Budget

SARC: Satellite Archive Coordination Center • Science Archive Research Center • Scientific and Applied Research Centre (Qatar) • Space and Astronomy Research Center • Symantec Antivirus Research Center

SARD: Solar-Array Release and Deployment • Swedish Agency for Research and Development

SARDAPF: Synthetic Aperture Radar Data Acquisition and Processing Facility

SARDPF: Synthetic Aperture Radar Data Processing Facility

SARE: Science Archive Research Environment • Système Automatique de Radiosondage Embarqué

SAREC: Swedish Agency for Research Cooperation

SAREX: Search And Rescue Exercise • Shuttle Amateur Radio Experiment • South American Radar Experiment

SARF: Secrétaire d'Administration de Recherche et de Formation

SARG: Sargento • Space Assets Review Group

SARI: Sistema de Almacenamiento y Recuperación de la Información • Subduction Accelerated Research Institute • Synthetic Aperture Radar Industrial

SARIS: Scattering And Recoiling Imaging Spectroscopy

Sarl: Société à Responsabilité Limitée

SARL: Société À Responsabilité Limitée

SARO: Servicios Aéreos Rutas Oriente

SAROAD: Storage And Retrieval Of Aerometric Data

SARP: Safety Analysis Report for Packaging • Satellite Acquisition Request Programming • Schedule, Analysis and Review Procedure • Schedule And Resources Procedure • Schedule And Reporting Procedure • Schedule And Review Procedure • Shuttle Astronaut Recruitment Program • Small Autonomous Research Package • Space Allocation Requirement Procedure • Space Allocation Reservation Program • Standards And Recommended Practices

SARS: Single-Axis Reference System • Solar-Array Reorientation System • Stellar Attitude Reference Study

SARSAT: Search And Rescue Satellite • Search And Rescue Satellite-Aided Tracking (*)

SARSDM: State Agency Remote Sensing Data Management

SARSEX: Synthetic Aperture Radar Signature Experiment

SART: Search And Rescue Transponder • Stimuli Analog Refresh Table

SARTS: Switched Access Remote Test System

SARUMAN: Secure Access and Restricted Use of Images on the Net

SARUS: Search And Rescue Using Satellites

SARV: Small-Amplitude Red Variable (star)

SAS: Sales Accounting System • Salford Astronomical Society (UK) • Salisbury Astronomical Society (UK) • Salton City (USA-CA) (location identifier) • Sample Acquisition System • Sample Array System • Science Analysis Software • Science Analysis System • Science and Arts Society • Science Archive System • Secure Authentication Service • Self-Adaptive System • Service d'Analyse Stratégique • Sheboygan Astronomical Society (USA-WI) • Shenandoah Astronomical Society (USA-VA) • Sheffield Astronomical Society (UK) • Shropshire Astronomical Society (UK) • Signal Analysis System • Single-Attached Station • Single-Attach Station • Single Audio System • Slovak Astronomical Society • Small-Angle Scattering • Small Applications Satellite • Small Astronomical Satellite • Small Astronomy Satellite • Società in Accomandita Semplice • Société Astronomique de Suisse (Switzerland) • Society for Amateur Scientists (USA-CA) • Society for Applied Spectroscopy (USA-AZ) • Software Approval Sheet • Software Automation Subpanel • Solar-Array Simulator • Solar-Array Structure • Solar-Array System • Solar-Aspect Sensor • Solid-state Array Spectroradiometer • Son Altesse Sérénissime • Source Acquisition Segment • Source Assessment System • Southampton Astronomical Society (UK) • South Astronomical Society (UK) • Southend Astronomical Society (New Zealand) • Southern Astronomical Society (Australia) • Space Activity Suit • Space Adaptation Syndrome • Space Applications Section • Spacecraft Antenna System • Spokane Astronomical Society (USA-WA) • Stability Augmentation System • Stability Augmented System • Station Automatique Secondaire • Statistical Analysis System • Statistically Assigned Sockets • Stockton Astronomy Society (USA-CA) • Structurally Adaptive System • Su Alteza Serenísima • Sub-Arctic Summer • Sun Acquisition Sensor • Surveillance Architecture Study • Swiss Accreditation Service • Syracuse Astronomical Society (USA-NY) • System Analysis Study • System Application Software • System of Aeronomical Satellites

SASC: Science Analysis Support Centre

SASCB: Shuttle Avionics Software Control Board

SASCO: Sudanese Aeronautical Services Co. Ltd.

SASCOE: Solar-Array Special Checkout Equipment

SASDI: South African Selective Dissemination of Information

SASE: Self-Addressed Stamped Envelope • Society for the Advancement of Socio-Economics • Standards Analysis, Synthesis and Expression

SASER: Sound Laser

SASHR: Senior Assistant Secretary on Human Resources

SASI: S-Asteroid Spectral Interpreter • Shugart Associates System Interface • South African Standards Institution • Statistical Analysis Systems Institute (USA-NC) • Sutherland Astronomical Society Inc. (Australia)

SASIDS: Stochastic Adaptive Sequential Information Dissemination System

SASIFY: Space Agency Forum for the International Space Year

Sask: Saskatchewan (Canada)

SASK: Saskatchewan (Canada)

SASKR: Semiactive Seeker

SASL: Simple Authentication Security Layer • Société d'Astronomie de Saône et Loire (France)

SASLO: South African Scientific Liaison Office

SASM: Smithsonian Air and Space Museum (USA-DC) • Society for Automation in the Sciences and Mathematics (USA) • Synthetic Aperture Strip Map

SASMO: Syrian Arab Organization for Standardization and Metrology

SASN: Sedentary-Actor Subnetwork

SASO: Saudi Arabian Standards Organization • Senior Air Staff Officer

SASP: Science and Application Space Platform • Stand Alone Support Program

SASPRSC: South African Society for Photogrammetry, Remote Sensing and Cartography (South Africa)

SAS&R: Satellite-Aided Search & Rescue

SASR: Satellite-Aided Search and Rescue • Shuttle Avionics Status Review • Shuttle Avionics System Review

SASRS: Satellite-Aided Search and Rescue System

SASS: San Antonio Space Society (USA-TX) • Saturn Automatic Software System • Schedules And Status Summary • Scientific Analysis Software System • Seasat Active Scatterometer System • Shanghai Academy of Social Sciences (China-PRC) • Spectral Agile Sensor System • Standard Analysis and Software System • Standard Analysis Software System (*)

SASSE: Synchronous Altitude Spin-Stabilized Experiment

SASSIF: Self-Adjusting System of Scientific Information Flow

SASSY: Selected Areas Starcounts Survey

SAST: Science Analysis Software Team • Scientific Assessment and Strategy Team • Society for the Advancement of Space Travel • Statistical Analysis of Sparse Times • Strategic Analysis in Science and Technology • Strategic Assumption Surfacing Technique

SASTAR: Saint Andrews (Starlink internal code)

SASTP: Stand-Alone Self-Test Program

SASU: Secrétaire d'Administration Scolaire et Universitaire (France)

SASY: Software Architects' Synthesis

Sat: Satellite • Saturate • Saturated • Saturday • Saturn

S. At.: South Atlantic

SAT: Sampler Address Translator • San Antonio (USA-TX) (location identifier) • Satellite • Saturate • Saturated • Saturday • Scholastic Aptitude Test • Science Assessment Team • Self-Avoiding Trail • Service Autonome de Télédétection • Servicing Aid Tool • Simulator Acceptance Test • Site Assignment Time • Small Area Triangulation • Société Anonyme de Télécommunications • South Australian Time • Space Available Travel • Special Access Termination • Special Alias Table • Standard Assessment Task • Static Air Temperature • Stepped Atomic Time • Stochastic Aptitude Test • Strömgren Automatic Telescope • Support Analysis Test • Surveillance-Acquisition-Tracking • Symmetric Axis Transform • Synthetic Aperture Telescope • System Access Technique • System Alignment Test • Systems Approach to Training

SATA: Servicio Açoreana de Transportes Aéreos • Servicios Auxiliares de Transportes Aéreos

SATAF: Shuttle Activation Task Force • Site Activation Task Force

SATAN: Satellite Automatic Tracking Antenna • Security Administrator Tool for Analyzing Networks

SATANS: Static and Transient Analysis, Non-linear, Shells

SATAR: Satellite for Aerospace Research

SATB: Soprano-Alto-Tenor-Bass

SatCo: Satellite Communications (periodical code)

SATCO: Signal Automatic Air-Traffic Control

SATCOM: Satellite Communications

SATCON: Satellite Connectivity

SATDAS: Satellite Data Acquisition System

SATE: Semiautomatic Test Equipment

Satell: Satellite

SATEX: Saturdays Excepted • Semiautomatic Telegraph Exchange

SATH: Society for the Advancement of Travel for the Handicapped (USA-NY)

SATIF: Scientific And Technical Information Facility

SATIRE: Semiautomatic Technical Information Retrieval

SATIS: Science And Technology In Society • Scientific And Technical Information Service

SATKA: Surveillance, Acquisition, Tracking, and Kill Assessment

SATMOS: Système d'Archivage et de Traitement Météorologique des Observations Spatiales

Satn: Saturation

SATNAV: Satellite Navigation

SATNET: Satellite Network

SATO: Shuttle-Attached Teleoperator • Supply And Transportation Operations • Système d'Analyse de Textes par Ordinateur

SATODP: Satellite Tracking Orbit Determination Program

SATOR: Stratospheric And Tropospheric Ozone Research

SATORI: Systematic Air Traffic Operations Research Initiative

SATP: Stabilization, Acquisition, Tracking and Pointing

SATRA: South African Telecommunications Regulatory Authority • South African Telephone Regulatory Authority

SATRAC: Satellite Automatic Terminal Rendezvous And Coupling

SATRO: Science And Technology Regional Organization

SATS: Satellite Antenna Test System • Shuttle Avionics Test Set • Shuttle Avionics Test Station • Shuttle Avionics Test System • Single-Array Test System • Small Applications Technology Satellite • Spacecraft Automated Test System • Station Accommodation Test Site • Statistical Analysis of Sparse Times • Supplier Assessment Tracking System • Supply And Transportation Support

SATSI: Segmented Aperture Tilted Shearing Interferometer

SATSS: SEASAT Antenna Temperature Software System

SATSU: Science And Technology Studies Unit (UK)

SATT: Semiautomatic Transistor Tester

SATW: Schweizerische Akademie der Technischen Wissenschaften (Switzerland)

SAU: Arabia (Saudi_) (ISO country code) • Arabie Saoudite (ISO country code) • Saint Ambrose University (USA-IA) • Saudi Arabia (ISO country code) • Sawu (Indonesia) (location identifier) • Serial Acquisition Unit • Signal Acquisition Unit • Società Astronomica Urania (Italy) • Spectrum Analysis Unit • Standard Acquisition Unit • Statistical Analysis Unit • Strap-Around Unit

SAUCERS: Space And Unexplained Celestial Events Research Society

S. Aus.: South Australia • South Australian

S. Austral.: South Australia • South Australian

SAV: Sale At Valuation • Savannah (USA-GA) (location identifier) • Saved • Sector Access Velocity • Service Après-Vente • Societö Astronomica Versiliese (Italy) • Start of Active Video • Stock At Valuation

SAVAR: Société d'Astronomie du Valais Romand (Switzerland)

SAVASI: Simplified Abbreviated Visual Approach Slope Indicator

SAVB: Scientific Applications and Visualization Branch

SAVDM: Single Application Virtual DOS Machine

SAVE: Satellite Validation Experiment • South Atlantic Ventilation Experiment • Special Action for Vigorous Efficiency

SAVER: Shuttle Avionics Verification

SAVI: Soil-Adjusted Vegetation Index • Space Active Vibration Isolation

SAVS: Space Analysis and Visualization System • Status And Verification System

SAW: Marquette - Sawyer (USA-MI) (location identifier) • Satellite Attack Warning • Scottish Astronomy Weekend • Self-Avoiding Walk • Software Analysis Workstation • Software Audio Workshop • Solar-Array Wing • Sound Acoustic Wave • Space At Will • Spill Analysis Workstation • Sub-Arctic Winter • Surface Acoustic Wave

SAWB: South African Weather Bureau

SAWBOC: System Analysis With Blockout Considerations

SAWD: Solid Amine Water Desorption • Solid Amine Watersteam Desorption

SAWEGS: System for Automatic Wind Extraction from Geostationary Satellite

SAWG: Special Advisory Working Group

SAWRS: Supplemental Aviation Weather Reporting System • Supplementary Aviation Weather Reporting System

SAWS: Solar-Array Wing Simulator • Synoptic Automatic Weather Station

SAW/V: Satellite Attack Warning/Verification

SAWV: Satellite Attack Warning/Verification

Sax: Saxon • Saxony

SAX: Sambu (Panama) (location identifier) • Satellite per Astronomia in raggi X • Simple API for XML

SAXA: Slotted Array X-band Antenna

SAXON: SAR X-band and Ocean Non-linearity

SAXS: Small-Angle X-ray Scattering

SAY: Siena (Italy) (location identifier)

SAYE: Save As You Earn

SAZ: Sasstown (Liberia) (location identifier) • Standards Association of Zimbabwe

SAZU: Slovenian Academy of Sciences and Arts

sb: Stilb

Sb: Antimoine • Antimon • Antimony

S/B: Serial Block • Signal/Bruit • Standby

SB: Bachelor of Science • Barred Spiral Galaxy • Îles Salomon (ISO country code) • Sales Book • Satapatha Brāhmana • Selbstbedienung • Selection Board • Selective Bibliography • Serial Binary • Serial Block • Service Bulletin • Short Bill • Sideband • Signal Book • Simultaneous Broadcast • Simultaneous Broadcasting • Slettebak-Brundage • Small Business • Sodium Borate • Software Bus • Solomon Islands (ISO country code) • Sound Board • South Britain • Space Base • Space-Based • Special Bibliography • Special Branch • Spectroscopic Binary • Speed Brake • Spiral Galaxy - Barred • Staatsbibliothek • Standby • StarBits • StarBriefs • StarBurst • State Beach (USA) • Statsbibliotek • Statute Book • Storage Battery • Subbranch • Sub-Negotiation • Subsidiary Body • Successful Branch • Superbubble • Surface Brightness • Synchronization Base • Synchronization Bit

SBA: Santa Barbara (USA-CA) (location identifier) • Scene Balance Algorithm • School of Business Administration • Seat-Back Assembly • Secondary Baffle Assembly • Set Buffer Address • Shaped Beam Antenna • Singapore Broadcast Authority • Small Business Administration • Spreader Bar Assembly • Sandard Beam Approach • Structure-Borne Acoustic • Switchable Beam Antenna • Synchro Booster Amplifier • Synchronous Bandwidth Allocator

SBAA: Sociedade Brasileira dos Amigos da Astronomia (Brazil)

SBAC: Santa Barbara Astronomy Club (USA-CA) • Society of British Aerospace Companies (UK)

SBAS: S-Band Antenna Switch • Shreveport-Bossier Astronomical Society (USA-LA) • South Bay Astronomical Society (USA-CA)

SBB: Santa Barbara (Venezuela) (location identifier) • Schweizerische Bundesbahnen • Stadt- und Bezirksbibliothek

SBBN: Standard Big Bang Nucleosynthesis • Standard Black Body Nucleosynthesis

SBC: Sanriku Balloon Center (Japan) • Selbang (Papua New Guinea) (location identifier) • Simple Bayesian Classifier • Single-Board Computer • Solar Blind Camera (*) • Solar Blind Channel • Space Biotechnical Complex • Spaceborne Computer • Spectral Brightness Coefficient • Speed Brake Command • Standard Boundary Condition • String Bottom Controller • Sum of Base Cations

SBCA: Satellite Broadcasting and Communications Association (USA-VA)

SBCDP: Small Business Competitiveness Demonstration Program

SBCR: Stock Balance and Consumption Report

SBCS: Single-Byte Character Set

SBD: Dollar (Solomon Islands) (ISO currency code) • San Bernardino - Norton (USA-CA) (location identifier) • S-Band • Scene Block Diagram • Schematic Block Diagram • Schottky Barrier Detector • Schottky Barrier Diode • Simulated Based Design • Simulation-Based Design • Smart Battery Data • Standard Bibliographic Description • Surface Barrier Diode

SBDEW: Space-Based Directed-Energy Weapon

SBDO: Space Business Development Operation

SBDS: Solid-Based Density Standards • Structure-Based Deduction System

SbE: South by East

SBE: Scene Balance Algorithm • Simple Boolean Expression • Solar Beam Experiment • Suabi (Papua New Guinea) (location identifier) • Sub-Bit Encoder

SBEA: Sociedade Brasileira para o Encino da Astronomia (Brazil)

SBED: Serial Bit Error Detector

SBENTO: Standard Bento Entity for Natural Transport of Objects

SBER: Subbit Error Rate

sbf: Salvo Buen Fin • Sauf Bonne Fin

SBF: Sans Bureau Fixe • Sardeh Band (Afghanistan) (location identifier) • Scientific Balloon Facility • Société des Bourses Françaises (France) • Surface Brightness Fluctuation

SBFS: Space Business Financial Services

SBG: Sabang (Indonesia) (location identifier) • Starless Bok Globule

SBH: Saint Barthelemy (Leeward Islands) (location identifier)

SBHC: Speed Brake Hand Controller

SBI: Koundara Sambailo (Guinea) (location identifier) • Single Burning Item • Solar Bolometric Imager • Space-Based Interceptor • Subsidiary Body for Implementation

SBIC: Small Business Investment Co.

SBIG: Santa Barbara Instrument Group (USA-CA) • Short Baseline Interferometer • Short Baseline Interferometry

SBIR: Small Business Innovative Research • Space-Based Infrared

SBIRE: Space-Based Infrared Experiment

SBIRS: Space-Based Infrared System

SBIS: Sustaining Base Information Services

SBITP: Santa Barbara Institute of Theoretical Physics (USA-CA)

SBJ: São Mateus (Brazil-ES) (location identifier)

SBK: Saint-Brieuc (France) (location identifier)

SBKEW: Space-Based Kinetic-Energy Weapon

SBKEWS: Space-Based Kinetic-Energy Weapon System

SBKKV: Space-Based Kinetic Kill Vehicle

SBKKVWS: Space-Based Kinetic Kill Vehicle Weapon System

SBL: Santa Ana (Bolivia) (location identifier) • Sealed Beam Lamp • Space-Based Laser • Stable Boundary Layer • Super Basic Language • Surface Boundary Layer

SBLC: S-Band Linear Collider

SBLI: Savings Bank Life Insurance

SBLIP: Space-Based Laser Intercept Program

SBLS: Spaceborne Laser Ranging

SB/M: Switchable Bandpass/Mirror

SBM: Shared Bandwidth Manager • Sheboygan (USA-WI) (location identifier) • Single Buyer Model • Switchable Bandpass/Mirror • System Balance Measure

SBMAC: Sociedade Brasileira de Matématica Aplicada e Computacional

SBMB: Supervisory Bus Monitor Buffer

SBMS: Spaceborne Management Satellite • Store Business Management System

SBN: Servizio Bibliotecario Nazionale • Small Bodies Node • South Bend (USA-IN) (location identifier) • Standard Bok Nummer • Standard Book Number (now ISBN) • Standard Book Numbering

SBNC: Small Bodies Names Committee

Sbnd: Southbound

SBNPB: Space-Based Neutral Particle Beam

SBO: Salina (USA-UT) (location identifier) • Science-Based Organization • Sommers-Bausch Observatory (USA-CO)

S'board: Starboard

Sboard: Starboard

SBOF: Standard Book Order Form

SBOS: Sound Board Operating System

SBP: San Luis Obispo (USA-CA) (location identifier) • Sedimentary Basins Project • Société Belge de Physique (Belgium) • Sonic Boom Panel • Space-Based Radar • Spaceborne Programmer • Strategic Business Plan • Surface Brightness Profile • Systolic Blood Pressure

SBPA: S-Band Power Amplifier • Señal Binaria Pseudo-Aleatoria • Singapore Book Publishers' Association

SBPM: Société Belge des Professeurs de Mathématiques (Belgium)

SBPTC: Société Belge de Photogrammétrie, de Télédétection et de Cartographie (Belgium)

SBQ: Sibi (Pakistan) (location identifier)

s. Br.: südliche Breite

SBR: Saibai Island (Australia-QLD) (location identifier) • Size-Based Ranking • Space-Based Radar • Space-Borne Radar • Styrene Butadiene Rubber • Surface-Based Radiometer

SBRC: Santa Barbara Research Center (USA-CA) • Sensor-Based Robot Control

Sbre: Septiembre

SBRTN: Satellite-Based Rural Telegraph Network

SBS: Satellite Business Systems • Satellitenbodenstation • Scientific Business Satellite • Secondary Basic Subject • Second Byurakan Survey • Serially Balanced Sequence • Silicon Bilateral Switch • Simulated Brilloin Scattering • Small Bomb System • Sortie Banalisée Synchrone • Space Bio Separation • Special Broadcasting Service • Steamboat Springs (USA-CO) (location identifier) • Stimulated Brilloin Scattering • Système Binaire Serré

SBSC: Supplement to the Bright Star Catalog

SBSD: Subside

SBSE: Science-Based System Engineering

SBSG: Second Byurakan Survey of Galaxies

SBSL: Single Bubble-Sonoluminescence

SBSS: Second Byurakan Survey of Stars • Space-Based Space Surveillance • Space-Based Surveillance System

Sbstr: Substrate

SBSTA: Subsidiary Body for Scientific and Technological Advice

SBSTTA: Subsidiary Body on Scientific, Technical and Technological Advice

SBT: San Bernardino - Tri-City (USA-CA) (location identifier) • Scanbeam Table • Space-Based Tug • Spanning Binomial Tree • Statens Bibliotekstjenste

SBTC: Speed Brake/Thrust Controller

SBTO: Speed Brake Takeover

SBts: StarBits

SBTS: Shore-Based Tracking System • State Bureau of Technical Supervision

SBU: Springbok (South Africa) (location identifier)

SBUE: Switch-Backup Entry

SBURCS: Six-Bit Universal Random Character Set

SBUV: Solar and Backscattered Ultraviolet • Solar Backscatter Ultraviolet

SBV: Sabah (Papua New Guinea) (location identifier) • Space-Based Visible • Space Biospheres Ventures (USA-AZ)

SBVAA: San Bernardino Valley Amateur Astronomers (USA-CA)

SbW: South by West

SBW: Sibu (Malaysia) (location identifier) • Spectral Bandwidth

SBWR: Simplified Boiling Water Reactor

SBX: S-Band Transponder • Shelby (USA-MT) (location identifier) • Student Book Exchange

Sby: Standby

SBY: Salisbury (USA-MD) (location identifier)

SBZ: Sibiu (Romania) (location identifier)

s/c: Self-Contained • Son Compte • Sous le Couvert • Su Cargo • Su Casa • Su Cuenta

sc: Scientific • Scilicet • Scruple • Sculpsit • Small Capitals • Supercalendered

Sc: Scale • Scandinavia • Scandinavian • Scandium • Scène • Scene • Schmidt Number • Science • Scientific • Scientifique • Scotch • Scotland • Scots • Scottish • Screw • Scruple • Sculpsit • Sculpsit Carved • Stratocumulus

S/C: Santa Cruz • Security Code • Sensor/Controller • Signal Conditioner • Software Contractor • Spacecraft • Splitter/Combiner • Stabilization/Control • Stripchart • Subcontractor

S-C: Serbo-Croat

S&C: Stabilization & Control

SC: Safe Custody • Same Case • Same-sense Circular • Santa Catarina (Brazil) • Santa Cruz (Argentina) • Satellite Cell • Satellite Communication • Scale • Scaling Record • Scan Cursor • Scanner Controller • Scheduled Caste • Schmidt-Cassegrain • School Certificate • Science Case • Science Center • Science Coordinator • Sciences Chimiques • Scientific Committee • Scotland • Script • Sculpsit Carved • Scythia • Secondary Channel • Security Code • Security Council • See Copy • Select Committee • Self-Check • Self-Contained • Semiconductor • Senatus Consultum • Sensor/Controller • Separate Cover • Separately Compiled • Separation and Control • Sequence Counter • Service Certificate • Service Charge • Service Component • Service Corporation • Seychelles (ISO country code) • Short Camera • Short Circuit • Short Course • Signal Comparator • Signal Conditioner • Significant Characteristics • Simulation Coordinator • Single Case • Single Chip • Single Column • Single Contact • Single Crystal • Slope Code • Small Capitals • Small Coronagraph • Small Craft • Snow Cover • Société Commune de Satellites • Software Change • Software Contractor • Solar Cell • Solar Constant • Source Code • South Carolina (USA) • Southern Clusters • Spacecraft • Spain-California • Spatial Coherence • Special Case • Special Command • Special Commission • Specialized Corpus • Spin Coupling • Splitter/Combiner • Spread Correlation • Spreading Coefficient • Stabilization and Control • Standards and Coordination • Standing Committee • Standing Conference • Starlink Cookbook • Statement of Capability • Statement of Compatibility • Statistical Control • Statutory Committee • Steering Committee • Stellar Camera • Storage Capacity • Stored Command • Strategic Command • Stratocumulus • Stress Corrosion • String Controller • Strip Chart • Strongly Connected • Structural Change • Structural Computing • Structure Chart • Subcommittee • Subcontractor • Successive Correction • Sudden Commencement • Supercalendered • Supercluster • Supercomputer • Superconductivity • Superimposed Coding • Superimposed Current • Supervisor's Console • Support Contractor • Support Controller • Support Coordinator • Supreme Court • Surface brightness - Colour • Surface Command • Surface Contamination • Symbolic Code • Synchro-Cyclotron • System Characterization • System Concept • System Controller

SCA: Santa Catalina (USA-CO) (location identifier) • Schedule Change Authorization • Scientific Computing Associates Corporation • SDI Communications Architecture • Sensor Chip Array • Sensor Chip Assembly • Sequence Control Area • Sequence Control Assembly • Sequencer Control Assembly • Serveur de Centre d'Appel • Service Contract Act • Shared Communications Area • Shuttle Carrier Aircraft • Signal Conditioning Assembly • Simulation Control Area • Simulation Conversion Assembly • Single-Channel Analyzer • Single-Connector Attachment • Sneak Circuit Analysis • Snow-Covered Area • Software Command Analysis • Software Control Authorization • Solar-Cell Assembly • Spacecraft Adapter • Subcarrier Adapter • Subchannel Adapter • Subcontract Authorization • Subcritical Assembly • Subsidiary Communications Authority • Summary Cost Account • Switch Control Assembly • System Comparison Analysis • System Controller Assembly • Système de Contrôle d'Altitude

SCAA: School Curriculum and Assessment Authority (UK)

SCAAMP: Scientific Applications on Arrays of Multiple Processors

SCAC: Santa Cruz Astronomy Club (USA-CA) • Science Computing Advisory Committee

SCAD: Scan Converter And Display • Small Current Amplifying Device

SCADA: Supervisory Control And Data Acquisition

SCADC: Standard Central Air Data Computer

SCADS: Scanning Celestial Attitude Determination System • Superfund Chemical Analysis Data System

SCAEPA: Society for Computer Applications in Engineering, Planning and Architecture

SCAFEDS: Space Construction Automated Fabrication Experiment Definition Study

SCAFI: Studies of Computer Algebra for Industry

SCALE: Scattering Layer Experiment • Shuttle Coherent Atmospheric Lidar Experiment • Space Checkout And Launch Equipment

SCALIR: Symbolic and Connectionnist Approach to Legal Information Retrieval

SCALOP: Standing Committee on Antarctic Logistics and Operations

SCALP: Solar Calibration Package

SCAM: SCSI Configured Auto Magically • Stanford Copy Analysis Mechanism • Station Control And Monitoring

SCAMA: Skewed Circular Arc Method of Analysis • Station Conferencing And Monitoring Agreement • Station Conferencing And Monitoring Arrangement • Switching, Conferencing And Monitoring Arrangement

SCAMP: Standard Configuration And Modification Program • Supplemental Camera And Maneuvering Platform

SCAMS: Scanning Microwave Spectrometer

Scan: Scandinavia • Scandinavian

SCAN: Scientific Content Analysis • Selected Current Aerospace Notice • Selected Current Awareness Notification • Self-Correcting Automatic Navigator • Shuttle Connector Analysis Network • Spoken Content-based Audio Navigation • Switched Circuit Automatic Network

Scand: Scandinavia • Scandinavian

SCANIQ: Système Conjoint d'Analyse Numérique d'Images du Québec (Canada-QC)

SCANS: Scheduling and Control by Automated Network System • System Checkout Automatic Network Simulator

SCAO: Système de Contrôle d'Attitude et d'Orbite

SCAP: Scanning Photometer And Polarimeter • Shuttle Configuration Analysis Program • Standing Committee of Astronomy Professors • Standing Conference of Astronomy Professors (UK) • States Cooperative Assistance Program

SCAPE: Self-Contained Air Pressure Environment • Self-Contained Atmosphere Protective Ensemble • Self-Contained Atmospheric Protective Ensembly • Simulating Composition of Atmospheric Particles at Equilibrium • Slantwise Convective Available Potential Energy

S. Caps.: Small Capitals

SCAR: Safety Change Assessment Report • Scientific Committee on Antarctic Research (*) • Satellite Capture And Retrieval • Spacecraft Analysis Room • Spacecraft Assessment Report • Special Committee on Antarctic Research • Starlink Catalogue Access and Reporting • Starlink Catalogue Access and Retrieval • Supplier Corrective Action Request

SCARA: Selective Compliance Assembly Robot Arm

ScaRaB: Scanner for Radiation Budget

SCARAB: Scanner for Radiation Budget

ScaRaBE: Scanner for Radiation Budget Experiment

SCARDE: Study Committee on Analysis of Research, Development and Engineering

SCARI: Self-Compensating All-ReflectivonInterferometer (*) • Self-Compensating All-Reflective Interferometric

SCARP: Skylab Concentrated Atmospheric Radiation Project

SCARS: Scientific Computing And Research Support • Serial Control And Recort System • Serialized Control And Reporting System • Sneak-Circuit Analysis Report Summary

SCARSD: Science Computing And Research Support Division

SCAS: Scan String • Seven Color Asteroid Survey • Soil, Crop, and Atmospheric Sciences (Cornell) • Sonoma County Astronomical Society (USA-CA) • Southern Cross Astronomical Society (USA-FL) • Subsystem Computer Application Software

SCASIS: Seeing Cell Aperture Synthesis Imaging Spectrometer

SCAST: Social Change And Sustainable Transport

SCASU: Secrétaire en Chef d'Administration Scolaire et Universitaire

Scat: Scatter

SCAT: Secondary Combustor Augmented Thruster • Share Compiler-Assembler Translator • Space Communications And Tracking • Storage, Checkout And Transport • Storage, Checkout And Transportation • Supersonic Commercial Air Transport • System Commonality Analysis Tool

SCATE: Self-Checking Automatic Testing Equipment

SCATHA: Spacecraft Charging At High Altitude

SCATS: Self-Contained Automatic Test System • Simulation, Checkout And Tracking System • Simulation, Checkout And Training System (*) • Simulation Control And Training System

Scatt: Scatterometer

SCATT: Scatterometer • Scientific Communication And Technology Transfer • Shared Catalog Accessed Through Terminals

SCAUL: Standing Conference of African University Librarians

ScB: Bachelor of Science

SCB: Sample Collection Bag • Schedule Change Board • Schedule Control Board • Scheduled Change Board • Scribner (USA-NE) (location identifier) • Selection Control Board • Selector Control Box • Session Control Block • Software Control Board • Space Control Block • Specification Control Board • Steering Control Box • Subsystem Control Block

SCBA: Self-Contained Breathing Apparatus

ScBC: Bachelor of Science in Chemistry

SCBE: Seller's Configuration Baseline Document

ScBE: Bachelor of Science in Engineering

SCBE: Simcoe County Board of Education

scc: Single Column Centimeter

SCC: Prudhoe Bay - Deadhorse (USA-AK) (location identifier) • Safety Control Center • Sagamihara Control Center (Japan) • Satellite Communications Control • Satellite Communications Controller • Satellite Control Centre • Satellite-Controlled Clock • Scholarly Communication Center • Scientific Computing Center • Secondary Control Code • Security Control Center • Security Coordination Committee • Selectional Classification Code • Seminole Community College (USA-FL) • Sequence Control Chart • Sequential Control Counter • Serial Communications Controller • Serial Controller Chip • Signal Conditioning Component • Simplified Computer Code • Simulation Configuration Control • Simulation Control Center • Single Column Centimeter • Slidell Computer Complex (NASA) • South Cusp Cap • Space Commerce Corporation • Space Communications Corporation • Space Components Coordination • Space Control Centre • Spacecraft Clock • Spacecraft Control Centre • Spacecraft Coordinates • Specialized Common Carrier • Specific Clauses and Conditions • Standard Cubic Centimeter • Standards Coordinating Committee • Standards Council of Canada • Standing Consultative Commission • State Consultative Council • Station Control Center • Stress Corrosion Cracking • String Center Controller • Structured Conceptual Clustering • Super Cloud Cluster • Supervisor Control Console • Supply Control Center • Switching Control Center • Synchronous Channel Check • System Command Center • Systems Control Center

SCCA: Statistical Consulting Center for Astronomy • Subcontract Change Authorization

SCCAC: Society for Conceptual and Content Analysis by Computer

SCCB: Safety Change Control Board • Site Configuration Control Board • Software Change Control Board • Software Configuration Control Board

SCCC: Satellite Communications Control Centre • Scottish Consultative Committee on the Curriculum (UK-Scotland) • Seattle Central Community College (USA-WA) • Seward County Community College (USA-KS) • Sociedad Chilena de Ciencia de la Computación (Chile) • South Cusp Cap Collar • Spacecraft Command and Control Center

SCCCB: Science Computing Configuration Control Board

SCCE: Scientific Computing in Chemical Engineering • Signal Conditioning Control Electronics

SCCER: System Command and Control Equipment Rack

SCCF: Satellite Communication Control Facility

SCCG: Space Components Coordination Group • Station Communications Control Group

SCCH: Standard Cubic Centimeters per Hour

SCCI: Source Code Control Interface

SCCM: Standard Cubic Centimeters per Minute

SCCO: Soret Coefficient in Crude Oil

SCCP: Signaling Connection Control Part • Standing Committee of Physics Professors

SCCPG: Satellite Communications Contingency Planning Group

SCCR: Subcontractor Change Request

SCCS: Source Code Control System • Specialized Common Carrier Service • Standard Cubic Centimeters per Second

• Switching Control Center System

SCCT: Steering Committee on Conformance and Testing

SCCU: Spacecraft Command Control Unit

Scd: Scheduled

Sc. D.: Doctor of Science

ScD: Doctor of Science

SCD: Satélite de Coleta de Dados • Satellite Control Department • Satellite de Coleta Dados • Science Communications Division • Scientific Capabilities Document • Scientific Computing Division • Scrolling Display • Service Commun de Documentation • Software Conceptual Design • Source Control Document • Source Control Drawing • Space Control Document • SPARC Compliance Definition • Specification Change Drawing • Specification Control Document • Specification Control Drawing • Speed Control Device • Standard Color Display • Subcontract Deviation • Subject Cataloging Division • Sudden Cardiac Death • Sulaco (Honduras) (location identifier) • Synthetic Clock Drive • System Conception Definition • System Correction Data

SCDA: Safing, Cool-down and Decontamination Area

SCDC: Source Coding and Data Collection

SCDDB: Satellite Climate Diagnostics Database

SCDI: Scottish Council for Development and Industry

SCDL: Saturated Current Demand Logic • Stabilized Carbon Dioxide Laser • Structural Dynamics and Control Laboratory • Surveillance and Control Data Link

SCDM: Standard Cold Dark Matter

SCDMA: Synchronous Code-Division Multiple Access

SCDMS: Soil-Climate Database Management System

SCDP: Simulation Control Data Package • Specialized Centre for Data Processing

Scdr: Screwdriver

SCDR: Seller Critical Design Review • Shuttle Critical Design Review • Software Critical Design Review • Special Committee for Disaster Research • Subcontractor Critical Design Review

SCDS: Scan Converter Display System • Secure Central Data System • Sensor Communication and Display System • Stock Control and Distribution System

SCDU: Satellite Control Display Unit • Signal Conditioning and Display Unit

Sce: Scenario • Source

SCE: Schedule Compliance Evaluation • Scottish Certificate of Education • Selection Control Element • Service Central d'Exploitation • Service Creation Environment • Shuttle Cargo Element • Signal Conditioning Equipment • Signal Conversion Equipment • Single Corporate Entity • Situationally Caused Error • Small Current Element • Société Coopérative Européenne • Solar Corona Experiment • Sont Computer Entertainment • Sortie Cargo Element • Space Cabin Environment • Spacecraft Command Encoder • Special Conditioning Equipment • Stabilization Control Electronics • Standard Calomel Electrode • State College (USA-PA) (location identifier) • Structure Chart Editor • Superior Council of Education

SCEA: Signal Conditioning Electronics Assembly

SCEAL: Standing Conference of East African Librarians

SCEC: Southern California Earthquake Center (USA-CA)

SCED: Stations and Communications Engineering Division

SCEES: Service Central des Enquêtes et Études Statistiques (France)

SCEI: Special Committee on Environmental Information

SCEM: Service Central d'Exploitation de la Météorologie (*) • Service Central d'Exploitation Météorologique

SCEP: Simple Certificate Enrollment Protocol

SCEPS: Solar-Cell Electric Power System

SCEPTRE: System for Constant Elevation Precipitation Transmission and Recording

SCERT: Systems and Computers Evaluation and Review Technique

SCES: Satellite Control Earth Station

SCET: Scottish Council for Educational Technology • Spacecraft Event Time

scf: Standard Cubic Foot

SCF: Phoenix - Scottsdale (USA-AZ) (location identifier) • Sampled Channel Filter • Satellite Control Facility • Science Computer Facility • Scientific Committee on Food • Scientific Computing Facility • Selective Call Forwarding • Self-Consistent Field • Sequence Compatibility Firing • Sequenced Compatibility Firing • Simulator Command File • Single Catastrophic Failure • Single Cost Factor • Small Cooling Flow • Sociedad Colombiana de Física (Colombia) • Spacecraft Checkout Facility • Spacecraft Control Facility • Specialized Computing Facility • Standard Charge Factor • Star Catalogue Facility • Statistical Collection File • Steam Condensation Front • Stock and Checkout Facility • Supercritical Fluid • System Control Facility

SCFCS: Standing Committee on the Free Circulation of Scientists • Standing Committee on the Freedom in the Conduct of Science (*)

SCFD: Scientific and Cultural Facilities District

SCFE: Special Cassette Flow Experiment

scfh: Standard Cubic Feet per Hour

scfm: Standard Cubic Feet per Minute

scfph: Standard Cubic Feet per Hour

scfpm: Standard Cubic Feet per Minute

SCFN: Seed Capital Fund Network • Spherical Convergent Flap Nozzle

SCFS: Standard Cubic Feet per Second

Scg: Scoring

SCG: Security Classification Guide • Sequential Control Guidance • Single Cell Gel • Space Charge Grid • Spring Creek (Australia-QLD) (location identifier)

SCGD: Specification Control Group Directive

SCGI: Secrétariat Général du Conseil Interministériel

s/cgo: Su Cargo

SCGSS: Supercritical Gas Storage System

SCGUI: Steering Committee for Graphical User Interface

Sch: Schedule • Scheduled • Scheduling • Schema • Schematic • Schilling (currency) • Schmidt • Scholar • Scholarship • Scholastic • School • Schule

SCh: Sciences Chimiques

SCH: Schenectady County (USA-NY) (location identifier) • School • Software Clearinghouse • Space Center Houston (USA-TX) • Spherical Cap Harmonic • Statistics Collection Host

SCHA: Spherical Cap Harmonic Analysis

SCHD: Schedule

SchDm: Schematic Diagram

SCHe: Supercritical Helium

Sched: Schedule

SCHED: Scheduled • Scheduling

Schem: Schematic • Schematics

SCHEM: Schematic • Schematics

SCHG: Supercharge

Schill: Schilling

SCHINA: Schaltanlagen-Informations-Anlagen

Schl: Schloß

SCHM: State Committee on Hydrometeorology

SCHMM: Semicontinuous Hidden Markov Model

Schol: Scholar • Scholarship • Scholastic

Schr: Schrift

Schurw: Schurwald (Germany)

Schw: Schwester

Schwäb: Schwäbisch

Schwab: Schwaben (Germany)

Schwarzw: Schwarzwald (Germany)

Sci: Science • Science (periodical code) • Scientific • Scientifique • Scientist

sci: Single Column Inch

ScI: Science Institute

SCI: San Cristobal (Venezuela) (location identifier) • Scalable Coherent Interface • Science • Science Citation Index • Scientific Computing and Imaging • Selected Configuration Item • Seller Configuration Inspection • Sensitive Compartmented Information • Sensitive Compartmented Intelligence • Sensor Chip Imager • Serial Communications Interface • Servicio de Calibración Industrial • Société de Chimie Industrielle • Software Customer Inspection • Space Commerce International Ltd. • Space Craft International (USA-CA) • Strategic Computing Initiative • Stratospheric Circulation Index • Switch Closure In • Systemics-Cybernetics-Information

SCIA: Smart Card Industry Association

SciAm: Scientific American (periodical code)

SCIAMACHY: Scanning Imaging Absorption Spectrometer for Atmospheric Cartography

SCIBP: Scientific Committee for the International Biological Programme

SCIC: Satellite Cloud Information Chart • Semiconductor Integrated Circuit

SCICEX: Submarine Science Ice Experiment

SCIDAR: Scintillation Detection And Ranging

SCIDF: System-Independent Data Format

SciDi: Scientific Dimension (periodical code)

SCIDNDR: Special Committee for the International Decade on Natural Disaster Reduction

SCIE: School on Information Extraction

SCIENCE: Simulation des Coopérations Internationales et des Échanges Nécessaires aux Chercheurs Européens

SCIES: Science Catalog Information Exchange System

ScIF: Science Institute Facility

SCIF: Sensitive Compartmented Information Facility • Single Console Image Facility

sci. fa.: Scire Facias

SCIFER: Sounding the Cleft Ion Fountain Energization Region

Sci. Fi.: Science Fiction

SCIGBP: Special Committee for the International Geosphere-Biosphere Program

SCIGN: Southern California Integrated GPS Network (USA-CA)

SCIGY: Special Committee for the International Geophysical Year

scil: Scilicet

SCIL: Simulation Command and Input Language

SCILS: School of Communication, Information and Library Studies

SCIM: Shuttle Cargo Integration Manager • Standard Cubic Inches per Minute

SciN: Science News (periodical code)

SCINATT: Spectra, Composition, and Interactions of Nuclei Avove Ten TeV

SCINS: Small Common Inertial Navigation System

SCIP: Self-Contained Instrument Package • Shuttle Cargo Integration Plan • Society of Competitive Intelligence Professionals

Scis: Sciences (periodical code)

SCIS: Satellite Coverage Information System • Science Curriculum Improvement Study • Spacecraft Interface Specification • Standard Cubic Inches per Second

SciSim: Science Simulator

SciSn: Scientia Sinica (periodical code)

SCISRG: Semicontinuous Inferred Stochastic Regular Grammar

SCIT: Standard Change Integration and Tracking • System Concept and Integrated Technology

SCIU: Selector Control Interface Unit • Sortie Control Interface Unit • Spacecraft Interface Unit

SCJ: Smith Cove (USA-AK) (location identifier)

SCK: Stockton (USA-CA) (location identifier) • Studiecentrum voor Kernenergie

Scl: Scluptor-Sculptoris (constellation)

SCL: Safety Checklist • Santiago - Comodoro Arturo Merino Benitez (Chile) (location identifier) • Secondary Coolant Line • Secondary Coolant Loop • Serial Clock • Service Control Level • Sociedad Cooperativa Libre • SOGS Command Language • Spacecraft Command Language • Specification Change Log • Standard Classification List • Standards and Calibration Laboratory • String Control Language • Symbolic Correction Loader • Systems Command Language • Systems Component List • Systems Control Language

SCLC: Space-Charge-Limited Current

SCLERA: Santa Catalina Laboratory for Experimental Relativity by Astrometry (USA-CA)

SCLK: Spacecraft Clock Kernel

SCLM: Software Configuration and Library Manager • Stability, Control, and Load Maneuvers

Sclr: Scaler

ScM: Master of Science • Scripta Metallurgica (periodical code)

SCM: Scammon Bay (USA-AK) (location identifier) • Scanning Capacitance Microscope • Scanning Capacitance Microscopy • Schedule/Configuration Management • Service Command Module • Service Control Manager • Signal Conditioning Module • Single-Column Model • Single-Crystal Meteorite • Site Configuration Message • Small-Core Memory • Sociedad de Crédito Mercantil • Société de Cautionnement Mutuel • Society of Computer Medicine • Software Configuration Management • Software Configuration Manager • Solar-Cell Module • Solar-Constant Monitor • Space Cargo Modification • Specification Change Memorandum • Spinal Change in Microgravity • Standard Cosmological Model • Standard Cubic Meter • Station Class Mark • Streamline Curvature Method • Subsidies and Countervailing Measures • Supply Chain Management • Suppressed-Carrier Modulation • System Configuration Management • System Configuration Monitoring • System Control Module

SCMA: Société Canadienne des Mathématiques Appliquées (Canada)

SCMAI: Société Canadienne de Mathématiques Appliquées et Industrielles (Canada)

SCMB: SOGS Configuration Management Bulletin • System Configuration Management Board

SCMFT: Self-Consistent Mean-Field Theory

SCMGGI: Semicontinuous Morphic Generator Grammatical Inference

ScMM: Scripta Metallurgica et Materiala (periodical code)

SCMM: Single-Connection - Multiple-Measurement

SCMO: Software Configuration Management Office

SCMP: Software Configuration Management Plan • System Contractor Management Plan

SCMR: Surface Composition Mapping Radiometer

SCMS: Serial Copy Management System • Signal Command Management System • Small Cumulus Microphysical Study • Spacecraft Control and Monitoring System • STScI Code Management System

SCMSS: Standing Committee on Membership, Structure and Statutes

SCMT: Security Crisis Management Training • Southern Columbia Millimeter Telescope

SCMU: Satellite Control and Monitor Unit

SCN: Saarbrücken (Germany) (location identifier) • Satellite Control Network • Schedule Change Notice • Schematic Change Notice • Seattle Community Network (USA-WA) • Self-Checking Number • Self-Contained Navigation • Sensitive Command Network • Show Cause Notice • Space Communications Network • Special Change Notice • Specification Change Notice • Specific Control Number • Succinonitrile • Suprachiasmatic Nucleus • System Change Notice • System Control Number

SCNA: Sudden Cosmic-Noise Absorption

Scng: Scanning

SCNP: Specification Change Notice Proposal

Scnr: Scanner

SCNR: Scanner • Sequence Control Number Register • Solid-Core Nuclear Rocket

SCNS: Self-Contained Navigation System

Sco: Scenario • Scorpius-Scorpii (constellation)

S/CO: Spacecraft Observer

SCO: Sabino Canyon Observatory (USA-AZ) • Santa Cruz Operation Inc. • Shevchenko (Kazakhstan) (location identifier) • Single Copy Order • Slovak Central Observatory (Slovakia) • Software Change Order • Spacecraft Observer • Spacecraft Operations • Spacecraft Operator • Start Checkout • Statistical Control Office • Statistical Control Officer • Subcarrier Oscillator • Supercritical Oxygen • Switch Closure Out

SCOCP: State-Constrained Optimal Control Problem

SCODS: Standing Committee on Ocean Data Stations

SCOE: Special Checkout Equipment • Specific Checkout Equipment

SCOL: Simulation Control Level

SCOM: Science Operations Manager

SCOMO: Satellite Collection Of Meteorological Observations

SCOMP: Secure Communications Protocol

SCONUL: Standing Conference of National and University Libraries (UK)

SCOOPS: Scheme Object-Oriented Programming System

SCOP: Single-Copy Order Plan • Société Canadienne d'Optique et de Photonique (Canada)

SCOPE: Scientific Committee On Problems of the Environment • Setting-Content-Objectives-Program-Evaluation • Simple Communications Programming Environment • Software Certification On Programme in Europe • Standing Committee on Professional Education • System for Control of Operations of Payload for EuroMir

SCOPL: Simple Checkout-Oriented Program Language

SCOR: Scientific Committee on Oceanic Research

SCORE: Satellite Computer-Operated Readiness Equipment • Signal Communication by Orbiting Relay Equipment • SIRTF-Cornell Echelle • System Cost and Operational Resource Evaluation

SCORT: Specialized Center Of Research and Training

SCORUS: Standing Committee on Regional and Urban Statistics

SCOS: Spacecraft Control and Operations System • Spacecraft Control Operations System • Spacecraft Operation System • Subsystem Computer Operating System

SCOST: Special Committee on Space Technology

SCOSTEP: Scientific Committee on Solar-Terrestrial Physics

Scot: Scotland (UK) • Scottish

SCOT: Satellite Communication Terminal • Satellite Communications Overseas Terminal • Small Communication Terminal • Standby Compatible On-Tape

SCOTS: Surveillance and Control Of Transmission Systems • System Checkout Test Set

SCOTT: Synchronous Continuous Orbital Three-Dimensional Tracking

SCOTUS: Supreme Court of the United States

SCOUT: Serial Checkout Unit

SCOWAR: Scientific Committee on Water Research

SCP: Saint-Crépin (France) (location identifier) • Satellite Cloud Photograph • Satellite Commissioning Phase • Saunders College Publishing (USA-PA) • Save Cursor Position • Scanner Control Power • Scanning Phased Array • Schematic Change Proposal • Sección de Cielo Profundo • Security Classification Procedure • Service Central des Prêts • Server Control Protocol • Service Connection Point • Service Control Point • Session Control Protocol (*) • Simulator Control Panel • Simulator Converter Program • Single-Color Photometry • Single Control Point • Site Configuration Package • Société Civile Professionnelle • Solar-Cell Panel • South Celestial Pole • Spacecraft Platform • Specific Candle Power • Standard Corporate Protocol • Station Communications Processor • Stored Command Processor • Subsystem Control Port • Supernova Cosmology Project • Symbolic Conversion Program • System Control Program • System Control Programming

SCPA: Semi-Conductor Protection Act • Solar-Cell Panel Assembly • Spacecraft Payload Adapter

SCpB: Single Channel per Burst

SCPB: Single Channel Per Burst

SCpC: Single Channel per Carrier

SCPC: Single Channel Per Carrier • Single Channel Per Connection

SCPD: Stanford Center for Professional Development (USA-CA)

SCPI: Standard Command for Programmable Instrumentation • Standard Command for Programmable Instruments (*) • Standard Command for Programmable Interface

SCPP: Seasonal Climate Prediction Program • Sierra Cooperative Pilot Project

ScPr: Science Progress (periodical code)

SCPR: System Concept and Programmatic Review

SCPRI: Service Central de Protection contre les Rayonnements Ionisants

SCPS: Servo-Controlled Positioning System • Support Computer Programming System

SCPT: Self-Consistent Perturbation Theory • Single Channel Per Transponder

SCPWG: Solar Corona and Particle Working Group

SCQ: Santiago de Compostela (Spain) (location identifier)

Scr: Screw • Scrip • Scruple

SCR: Scanning Control Register • Schedule Change Request • Schedule Confirmation Request • Science Certification Review • Scranton (USA-PA) (location identifier) • Script • Selective Catalytic Reaction • Selective Catalytic Reactor • Selective Chopper Radiometer • Semiconductor Controlled Rectifier • Sequence Checking Routine • Seychelles Rupee (ISO currency code) • Signal Conditioner • Silicon-Controlled Rectifier • Single-Channel Reception •

Sneak Circuit Report • Software Change Request • Software Cost Reduction • Solar Cosmic Radiation • Space-Constructible Radiator • Special Certification Review • Speed Change Rate • Stirling Cryogenics and Refrigeration • Strip Chart Recorder • Summary Control Report • Support Control Room • Surface Contour Radar • Surveillance Cost Report • Sustainable Cell Rate • Sustained Cell Rate • System Change Request • System Concept Review • System Control Routine

SCRAM: Several Compilers Reworked And Modified • Space Capsule Regulator And Monitor • Spares Control, Release And Monitoring • Station Crew Return Alternative Module • Support Center for Regulatory Air Models

SCRAP: Series Computation of Reliability And Probability

SCRB: Software Configuration Review Board

SCRE: Scottish Council for Research in Education

SCREMS: Scientific Computing Environments for the Mathematical Sciences

SCRES: Standing Committee on Responsability and Ethics in Science

SCREWS: Solar Cosmic Ray Early Warning System

SCRG: Stationary Cosmic Ray Gas • System Change Review Group

SCRI: Serviço de Cooperação e Relações Internacionais • Supercomputer Computations Research Institute

SCRIA: Supply Chain Relationships In Aerospace

SCRIBE: Stratospheric Cryogenic Interferometer Balloon Experiment

Script: Scriptural • Scripture

SCRIPT: Scientific and Commercial Routine Interpreter and Program Translator • Support for Creative Independent Production Talent

SCRL: Station Configuration Requirement List

SCRM: Safety Control and Risk Management

Scrn: Screen

SCRN: Screen

SCROLL: String and Character Recording Oriented Logogrammatic Language

SCRP: Secure Communications Requirements Panel

SCRS: Serialized Control and Reporting System • Strip Chart Recorder System • Strip Chart Recording System

SCRSY: Serialized Control and Reporting System

Scrt: Security

SCRT: Sealed Cathode Ray Tube • Silicon Compton Recoil Telescope

SCS: Safety Control Switch • Sales/Customer Service • Satellite Control Section • Satellite Control System • Scan Converter System • Scan Counter System • Science Commanding System • Scientific Control System • Secondary Control System • Secondary Coolant System • Secret Control Station • Secret Cover Sheet • Sequence Control System • Serialization Card System • Service and Checkout System • Shetland Islands - Scatsta (UK) (location identifier) • Signal Conditioning System • Signature Control System • Silicon-Controlled Switch • Simple Classifier System • Simulation Control Subsystem • Singapore Computer Society • Single Channel Simplex • Single Control Support • Sira Certification Service • Slaving Control System • Slow Code Scanner • Société de Commercialisation de Services • Society for Computer Simulation (USA) • Soil Conservation Service • Space Command Station • Space Communications System • Spacecraft Computer System • Spacecraft Control System • Stabilization and Control System • Standard Coordinate System • State Conservation Strategy • Statistical Control System • Stimulated Compton Scattering • Suit Communication System • Supervisory Control System • Switch Control Server • Syrian Cosmological Society • System Control Station

SCSA: Signal Computing System Architecture • Soil Conservation Society of America • Standard Consolidated Statistical Area

SCSBR: Southern California Space Business Roundtable (USA-CA)

SCSC: Scientific and Commercial Systems Corporation • Service Culturel Scientifique et de Coopération • South Carolina State College (USA-SC) • Southern Connecticut State College (USA-CT) • Standing Committee on Soil Conservation • Starin Gauge Signal Conditioner • Subcommittee on Standards and Conformance

SCSDI: Service Central de Statistique et de Documentation Industrielles

SCSE: State Commission for Space Exploration

SCSG: Silicon-Chip Strain Gauge • Single-Channel Signal Generator

SCSI: Small Computer System Interface • Steering Committee for Statistical Information

SCSL: Standing Lenticular Stratocumulus

SC/SM: Spacecraft Systems Monitor

SCSM: Spacecraft Systems Monitor

SCSN: Space Centre for Satellite Navigation (QUT) • Standard Computer Software Number

SCSNT: Space Centre for Satellite Navigation Technology

SCSO: Space Communications Station Operation • Superconducting-Cavity Stabilized Oscillator

SCSP: Standing Committee on Science Policy

SCSPS: Standing Committee on the Safeguard of the Pursuit of Science

SCSR: Scientific Council on Space Research

SCSS: Satellite Communications System Control • Satellite Control Simulation System • Self-Contained Starting System • Shoe Cove Satellite Station (Canada-NF) • Specialist Computer Systems and Software • Standing Committee for Social Sciences • Standing Committee on Structure and Statutes

SCSSI: Service Central de la Sécurité des Systèmes Informatiques (France)

SCST: State Commission for Science and Technology

SCSU: Saint Cloud State University (USA-MN) • South Carolina State University (USA-SC) • Southern Connecticut State University (USA-CT)

sct: Scattered • Sculpsit

Sct: Scutum-Scuti (constellation)

ScT: Spacecraft Time

SCt: Supreme Court

SCT: Scanning Telescope • Scattered • Scheduled Completion Time • Schmidt-Cassegrain Telescope • Second Coding Theorem • Sentence Completion test • Sequence Checking Tape • Short Collision Time • Sine-Cosine Transformer • Single-Channel Transponder • Société Canadienne de Télédétection (Canada) • Socotra (Yemen) (location identifier) • Spacecraft Team • Spectral Control Technique • Stratus-to-Cumulus Transition • System Capability Test • System Circuit Test • System Compatibility Test

s/cta: Su Cuenta

SC/TC: Spacecraft Test Conductor

SCTC: Spacecraft Test Conductor

Sctd: Scattered

SCTD: Scattered • Systems Consultancy Training Development

Sc&Te: Sciences & Techniques (periodical code)

SCTE: S-band Communication and Tracking Equipment • Society of Cable Telecommunications Engineers • Solar-Cell Technology Experiment • Spacecraft Central Timing Equipment

ScTEn: Science of the Total Environment (periodical code)

SCTI: Service Central des Télécommunications et de l'Informatique • Special Committee for Technical Information

SCTL: Surface Compiler for Temporal Logic

SCTOC: Satellite Communications Test Operations Center

SCTP: Screened Twisted Pair • Straight Channel Tape Print

SCTPP: Straight Channel Tape Print Program

SCTR: Sector • Standing Conference on Telecommunications Research

Sctrd: Scattered

SCTS: SFOF Communications Terminal System

SCU: Santiago (Cuba) (location identifier) • Scanner Control Unit • Secondary Control Unit • Selector Control Unit • Sensor Control Unit • Sequence Control Unit • Sequential Control Unit • Service and Cooling Umbilical • Service Command Unit • Shuttle Configuration Unit • Signal Conditioner Unit • Signal Conditioning Unit • Signal Control Unit • Southern Cross University (Australia-NSW) • Special Care Unit • Storage Control Unit • Synchronous Controller Unit • System Control Unit

SCUBA: Self-Contained Underwater Breathing Apparatus • Submillimeter Common-User Bolometer Array

SCUD: Structuring Knowledge from Unstructured Data

SCUDS: Simplification, Clarification, Unification, Decimalization, Standardization

SCUIO: Services Communs Universitaires d'Information et d'Orientation (France)

Sculp: Sculpsit • Sculptor • Sculptural • Sculpture

Sculpt: Sculptor • Sculptural • Sculpture

SCUM: Small Cumulus Microphysics

SCUMM: Script Creation Utility for Maniac Mansion

SCV: Stato della Città del Vaticano (Vatican) • Suceava (Romania) (location identifier) • System Component Verification • Vatican (international vehicle registration code)

SCVDR: Surface Characteristics Vector Data Record

SCVE: Spacecraft Vicinity Equipment

SCVTR: Scan Converting Video Tape Recorder

SCVTS: Switched Compressed Video Telecommunications Service

SCW: Search-Complete Word • Slowly Changing World • State College of Washington (USA-WA) • Syktyvkar (Russia) (location identifier)

SCWC: Special Commission on Weather Modification

SCWO: Supercritical Water Oxidation

SCWP: Satellite Communications Working Party • Slowly Changing World Policy

SCX: Salina Cruz (Mexico) (location identifier) • Single-Charge Exchange • Solar Coronal X-ray • Specialized Communications Exchange

SCXI: Signal Conditioning Extension for Instrumentation

SCZ: Santa Cruz Island (Solomon Islands) (location identifier) • Solar Convection Zone • Solar Convective Zone

sd: Semidetached • siehe dies! • Sindhi (ISO language code) • Sine Die • Subdwarf

Sd: Said • Side • Signed • Sound • Sund

S&D: Storage & Distribution

S/D: School District • Sight Draft • Synchro to Digital

SD: Doctor of Science • Safe Deposit • Salutem Dicit • Same Date • Sample Data • Sample Delay • Sans Date • Scaling and Display • Scandinavian • Schematic Diagram • Science Data • Scientific Design • Screening Device • Sea-Damaged • Self-Dual • Semidetached • Semidiameter • Senatus Decreto • Send Data • Send Direct • Senior Deacon • Sequence Date • Several Dates • Shared Disk • Short Delay • Short Delivery • Short Duration • Shut Down • Sicherheitsdienst • Sight Draft • Signal Digitizer • Sine Die • Single Degenerate • Single Density • Single Determination • Single Domain • Sit Down • Situation Display • Size Distribution • Slowdown • Smoke Detector • Software Development • Software Distributor • Solar Dynamic • Soudan (ISO country code) • South Dakota (USA) • Space Division • Spare Disposition • Speaker Dependent • Special Delivery • Special Document • Special Duty • Specification Document • Specification for Design • Spectral Distribution • Spectroscopy Detector • Standard Deviation • Standardization Data • Start Delimiter • State Department • Station Director • Statistical Division • Stellar Dynamics • Strongly Disagree • Structural Description • Structural Determinism • Structured Design • Structured Document • Student Director • Study Director • Sudan (ISO country code) • Südliche Durchmusterung • Superdeformed • Super Density • Superdirectory • Supply Depot • Support Directive • Support Document • Supporting Document • Swaziland (international vehicle registration code) • Symmetry Determination • System Demonstration • System Description • System Development

Sda: Senda

SDA: Baghdad Saddam (Iraq) (location identifier) • Schutzverband Deutscher Autoren (Germany) • Schweizerische Depeschen Agentur (Switzerland) • Science Data Analysis • Science Data Analyst • Science Data Archive • Scientific Data Analysis • Scientific Data Analyst • Sélection Directe à l'Arrivée • Sequential Degradation Analysis • Service Delivery Arrangement • SGML Document Access • Shaft Drive Axis • Significant Digit Arithmetic • Simple Doublet Antenna • Sleeve Dipole Antenna • Slowdown Area • Software Disk Array • Solar Depression Angle • Source Data Acquisition • Source Data Automation • Spaceborn Distributed Aperture • Space Digest Australia • Special Duty Assignment • Standard Descriptive Abstract • Statistical Distribution Analyzer • Step Down Amplifier • Stepwise Discriminant Analysis • Supporting Data Analysis • Swappable Data Area • Symbolic Disk Address • Symbols-Digits-Alphabetics • System Display Architecture

SDAA: San Diego Astronomy Association (USA-CA)

SDAC: Satellite Data Analysis Center • Scientific Data Analysis Coordinator • Seismic Data Analysis Center (USA) • Solar Data Analysis Center • Space Data Analysis Center • Surveillance Data Analysis Center

Sdad: Sociedad

SDAD: Satellite Digital and Analog Display

SDADS: Satellite Digital And Display System

SDAE: Source Data Automation Equipment

SDAF: Science Data Analysis Facility

SDAG: Science Data Analysis Group • Space Debris Advisory Group • Strategic Defense Advisory Group

SDAI: Standard Data Access Interface

SDAIR: Symposium on Document Analysis and Information Retrieval

S. Dak.: South Dakota (USA)

SDAP: SAR Data Acquisition Plan • Shuttle Descent Analysis Program • System Development And Performance • System Development and Acquisition Plan • Systems Development Analysis Program

SDAS: Science Data Analysis Software • Science Data Analysis System • Scientific Data Automation System • Source Data Automation System • South Dows Astronomical Society (UK) • Stafford and District Astronomical Society (UK) • Systems Data Analysis Section

S-DAT: Stationary Digital Audio Tape

SDAT: Sector Design Analysis Tool • Spacecraft Data Analysis Team • Stanford Diagnostic Arithmetic Test • Stationary Digital Audio Tape

SDB: Saldanha Bay - Langebaanweg (South Africa) (location identifier) • Secondary Data Block • Shallow Draft Barge • Shallow Draft Board • Small and Disadvantaged Business • Small Disadvantaged Business • Soil Database • Solnechnye Dannye Byulleten • Spatial Database • Spectrum Database • Stellar Database • Survey Database • Symbolic Debugger • System-Determined Blocksize • Système de Données Bibliographiques • Système de Documentation Bibliographique

SDBC: Software Development Baseline Control

SDBS: Scientific Database System • Sodium-Dodecil-Benzene-Sulphonate • Spatial Database System

SDC: Sample Data Collection • Sandcreek (Guyana) (location identifier) • SAX Data Centre • Scandinavian Data Centre • Science Data Catalogue • Science Data Centre • Science Data Coordinator • Scientific Data Center • Scientific Documentation Center • Secondary Distribution Center • Seismological Data Center • Selenoidal Detector Collaboration • Serial Digital Commands • Service Definition Coefficient • Signal Data Converter • Single Data Control • Software Design Definition • Software Design Document • Software Development Computer • Solenoidal Detector Collaboration • Source Debugger • Space Data Centre • Space Data Corporation • Space Development Corporation • Spares Disposition Code • Spatial Distortion Correction • Stabilization Data Computer • Support Design Change • Swiss Development Cooperation • System Debugger

SDCD: Space Data and Computing Division

SDCE: Scientific Data Collection Exercise

SDCF: Sample Data Channel Filter

SDCI: Single-Dish Calibration and Imaging

SDCL: Supplier Documentation Checklist • Structural Dynamics and Control Laboratory

SDCMO: Structural Dynamics and Control MATLAB Operator

SDCN: Space Data Communication Network

SDCP: Self-Designing Controller Program • Serial Data Control Protocol • Summary Development Cost Plan

SDCR: Source Data Communication Retrieval

SDCS: SAIL Data Communications System • SAR Data Catalog System • Sample Data Control System • Science Data Conditioning System • Simulation Data Conversion System • Single Differential Cross Section

SD&D: Scientific Development & Design

SDD: Lubango (Angola) (location identifier) • Selective Dissemination of Documentation • Semi-Discrete Decomposition • Shuttle Design Directive • Silicon Drift Detector • Single Data Disable • Site Description Database • Software Description Database • Software Description Document • Software Design Document • Stored Data Description • Subscriber Direct Dialing • Synthetized Directivity Diagram • System Definition Directive • System Design Description • System Design Document • System for Distributed Databases • Systems Development Device

SDDAS: SwRI's Data Display and Archival System

SDDC: Science Data Decoder Chassis

SDDD: Software Detailed Design Document

SDDF: Self-Defining Data Format

SDDL: Software Design and Document Language • Stored-Data Definition Language

SDDPT: Science Digital Data Preservation Task

SDDS: Satellite Data Distribution System • Scientific Document Delivery System • Secondary Data Display System • Solid-Discharge Data System

SDDVD: Super Density Digital Video Disc

SDE: Santiago del Estero (Argentina-SE) (location identifier) • Security Data Exchange • Shadow Dose Equivalent • Software Development Environment • Space Division Evaluator • Stellar Density Enhancement • Stochastic Differential Equation

SDEB: Science Data Evaluation Board

SDECE: Service de Documentation Extérieure et de Contre-Espionnage

SDEF: Service Development Evaluation Facility

SDES: Spatial Data Exchange Standard

SDESG: Strapdown Electrically Suspended Gyro

SdF: Sûreté de Fonctionnement

SDF: Louisville Standiford (USA-KY) (location identifier) • Safing and Deservicing Facility • Science Data File • Science Data Formatter • Screen Definition Facility • Screen Definition File • Shared Data Facility • Shuttle Development Facility • SIG Discretionary Fund (ACM) • Simplified Directional Facility • Simplified Direction Finding • Simulation Data File • Single Degree of Freedom • Slew Data File • Software Description File • Software Development Facility • Software Development File • Space Delimited File • Space Delimited Format • Spatial Distance Filter • Special Dispersion Focuser • Spectral Density Function • Standard Data Format • Standard Delimited Format • Standard Distribution Format • Stored Data File • Strap-Down Gyro • Sudden Disappearing Filament • Supplementary Data File • Sûreté De Fonctionnement • Synthetic Discriminant Function • System Data Format (*) • System Development Facility

SDFEM: Streamline Diffusion Finite Element Method

SD/FS: Smoke Detector/Fire Suppression

SD&FS: Smoke Detector & Fire Suppression

SDFS: Smoke Detector and Fire Suppression • Standard Disk File System

SDFU: Standard Format Data Unit

Sdg: Sending • Siding

S/DG: Strap-Down Gyro

SDG: Sample Delivery Group • Sanandaj (Iran) (location identifier) • Scan Display Generator • Simulated Data Generator • Situation Display Generator • Snow Depth Gauge • Software Development Group • Soli Deo Gloria • Strap-Down Gyro • Supermassive Disk Galaxy • Supplier Documentation Group • System Development Group

SDGE: Situation Display Generator Element

SDH: Santa Rosa - Copan (Honduras) (location identifier) • Science Data Header • Select Drive Head • Software Development Handbook • Spatial Data Handling • Synchronous Digital Hierarchy • Synchrotron Digital Handling • System Development Handbook • System Development Hardware

SDHE: Spacecraft Data Handling Equipment

SDHS: Satellite Data Handling System

SD&I: System Development & Integration

SDI: Saidor (Papua New Guinea) (location identifier) • Science Discrepancy Item • Selected Descriptive Item • Selected Dissemination of Information • Selective Dissemination of Information (*) • Serial Digital Interface • Service de Documentation et d'Information • Service de Documentation Industrielle • Service de Documentation Interministérielle • Service de Développement Informatique • Serviço de Documentação e Informação • Single-Document Interface • Société de Développement Informatique • Software Development Interface • Source Data Information • Spatial Data Infrastructure • Standard Data Interface • Standard Disk Interconnect • Strategic Defense Initiative • Support Directive Instruction • System Development and Integration

SDIF: SGML Document Interchange Format • Software Development and Integration Facility

S-DIMM: Solar Dual Image Motion Monitor

SDIMM: Solar Dual Image Motion Monitor

SDIO: Strategic Defense Initiative Office • Strategic Defense Initiative Organization (*)

SDIP: Service de Documentation et d'Information Pédagogiques (France) • Strategic Defense Initiative Program

SDIPC: Software Digital Image Processing Calculator

SDIRK: Singly Diagonally Implicit Runge-Kutta

SDIS: Satellite Data Input System • Switched Digital Integrated Service

SDIT: Service de Documentation et d'Information Technique

SDJ: Sendai (Japan) (location identifier)

SDK: Sandakan (Malaysia) (location identifier) • Software Developer's Kit • Software Development Kit

Sdl: Saddle

SDL: Samedi-Dimanche-Lundi • Satellite Data Laboratory • Scenario Description Language • Semiconductor Diode Laser • Serpentine Delay Line • Shielded Data Link • Software Design Language • Software Development Laboratory • Software Development Language • Software Development Library • Space Dynamics Laboratory • Special Duties List • Specification and Description Language • Standard Distribution List • Sundsvall (Sweden) (location identifier) • Supporting Document List • Surface Display Library • System Descriptive Language • System Design Language • System Development Language • System Directory List • Systematic Design Language

SDLC: Software Development Life Cycle • Synchronous Data-Link Control • System Data-Link Control • Systems Development Life Cycle

SDLV: Shuttle-Derived Launch Vehicle

SDM: Samsonov Density Meter • San Diego - Brown (USA-CA) (location identifier) • Scientific Data Management • Secondary Deployment Mechanism • Selective Dissemination of Microfiche • Semantic Database Model • Semantic Data Model • Service de Détermination des Manoeuvres • Shape Deposition Manufacturing • Shuttle Data Management • Shuttle Definition Manual • SOC Development Manager • Space Disaster Mitigation • Space Division Multiplexing • State-Dependent Model • Statistical-Dynamical Model • Subrate Data Multiplexer • Subtractive Double Monochromator • Su Divina Majestad • System Data Mover • System Definition Manual • System Design Manual

SDMA: Shared Direct Memory Access

SDMB: Shuttle Discrepancy Management Board

SDME: Synchronous Data Modem Equipment

SDMI: Secure Digital Music Initiative

SDMIS: Standardization Data Management Information System

SDMJ: September-December-March-June

SDMS: Shuttle Data Management System • Spatial Data Management System • Supplier Data Management System

SDMTG: Space Data Management Task Group

Sdn: Sedan • Sendirian

SdN: Société des Nations

SDN: Safety Discrepancy Notice • Sandane (Norway) (location identifier) • Schedule Deletion Notification • Simulator Data Name • Société Des Nations • Software-Defined Network • Software Delivery Notice • Software Development Note • Soudan (ISO country code) • Space Data Network • Sudan (ISO country code) • Sustainable Development Networking • System Development Network

SDNE: Services in Distributed and Networked Environments

SDNS: Secure Data Network Service • Secure Data Network System

SDO: Ryotsu Sado Island (Japan) (location identifier) • Semi-Definite Optimization • Senior Duty Officer • Shared Data Object • Shielded Diatomic Orbital • Spatial Data Option • Special Development Order • Standards-Developing Organization • Standards Development Organization • Station Duty Officer • Synthetic Display Output

S. Doc.: Senate Document

SDOC: Science Data Operations Center • Specific Direct Operating Cost

SDOF: Single Degree Of Freedom

SDOH: SOC Data Operations Handbook

SDOI: Spatial Data Organization Information

SDP: Pound (Sudan) (ISO currency code) • Safety Data Package • Sand Point (USA-AK) (location identifier) • Science Data Packet • Science Data Processing • Semidefinite Programming • Sensor Data Processing • Service Delivery Point • Servicio de Documentación Periodista • Shower Detector Plate • Shuttle Data Processor • Signal Data Processor • Silicon Diode Pellet • Site Data Processor • Small Distribution Phenomena • Social-Domestic-Pleasure • Software Development Plan • Solar Drum Positioner • Source Data Processing • Spectral Diagnostics Package • Standard Data Processing • Standard Data Processor • Station Data Processing • Steyr-Daimler-Puch • Strategic Defense Program • Supplier Data Package • Supply Distribution Point • Surface Deformation Pattern • Survey Data Processing • System Development Package • System Development Plan

SDPC: Science Data Processing Computer • Shuttle Data Processing Complex

SDPF: Satellite Data Processing Facility • Science Data Processing Facility

SDPLV: Shuttle-Derived-Propellant Launch Vehicle

SDPR: System Design and Performance Requirements

SDPS: SAR Data Processing System • Science Data Processing Segment • Signal Data Processing System • Spatial Data Production System • Standard Data Processing System

SDPT: South Downs Planetarium Trust (UK-England)

SDQ: Santo Domingo - Las Americas (Dominican Rep.) (location identifier)

SDQA: Space Division Quality Assurance

SDQL: Structured Document Query Language

Sdr: Sender

SDR: Same Day Return • Santander (Spain) (location identifier) • S-band Data Relay • Science Data Reader • Scientific Data Recorder • Search Decision Rule • Seismic Detection and Ranging • Sensor Data Record • Sensor Data Recorder • Serial Data Receiver • Series Dissipating Regulator • Service Detail Record • Signal Data Record • Signal Data Recorder • Signal Distribution Room • Signal-to-Distortion Ratio • Software Design Requirement • Software Design Review • Space Division Regulation • Spacelab Disposition Record • Spatial Dynamic Range • Special Drawing Right • Spectrum Data Record • Spoken Document Retrieval • Statistical Data Recorder • Storage Data Register • Streaming Data Request • Subjective-Distance Retrieval • Survey of Doctoral Recipients • System Data Record • System Definition Requirement • System Definition Review • System Design Report • System Design Review • System Discrepancy Report

SDRA: Shaped Dual Reflector Antenna

SDRAM: Synchronous Dynamic Random Access Memory

SDRB: Software Design Review Board • Supplier Documentation Review Board

SDRC: Structural Dynamics Research Corporation

SDRD: Supplier Data Requirements Description • Supplier Documentation Review Data

SDR&I: Space Data Resources & Information

SDRL: Subcontract Data Requirements List • Subcontractor Data Requirements List • Supplier Data Requirements List

SDRP: Simulated Data Reduction Program

SDRPC: Satellite Data Receiving and Processing Center

SDRPS: Satellite Data Receiving and Processing System

SDRS: Satellite Data Reference System • Shuttle Display Reconfiguration System • Signal Data Recording Set

SDRSI: South Dakota Remote Sensing Institute (USA-SD)

SDRT: Science Data Restoration Task

SDRW: Société de Développement Régional pour la Wallonie (Belgium)

SdS: Surveillance du Satellite

SDS: Sado Shima (Japan) (location identifier) • Safety Disconnect Switch • Satellite Data System • Science Data Store • Scientific Data Set • Scientific Data Store • Scientific Data System • Self-defining Data System • Service de Documentation Spatiale • Servo Drive System • Shuttle Dynamic Simulation • Shuttle Dynamic Simulator • Signal Distribution System • Sign-Digit Subtractor • Simulation Data System • Site Data System • Smoke Detection System • Sodium Dodecyl Sulphate • Software Design Specification • Solar Diameter Sensor • Solar Disk Sextant • Solar Disk Simulator • Spacecraft Data Simulator • Spacecraft Design Specification • Space Documentation Service • Special Docking Simulator • Spectrometer Digital System • Spelda Dedicated Satellite • Steering and Damping System • Steering Damping System • Stored Data Smooth • Streaming Data Strobe • Switched Data Service • Switched Digital Service • Synchronous Data Set • SYSOPs Distribution System • System Data Synthesizer • System Definition Study • System Design Specification

SDSB: Satellite Data Services Branch

SDSC: San Diego State College (USA-CA) • San Diego Supercomputer Center (USA-CA) • SAR Data Catalog System • Synchronous Data Set Controller

SDSD: Satellite Data Services Division • Single Disk Storage Device

SDSF: Spool Display and Search Facility • System Display and Search Facility

SDSI: Shared Dataset Integrity • Short Duration and Specific Date Impact • Simple Distributed Security Information

SDSL: Single-line Digital Subscriber Line • Symmetric Digital Subscriber Line (*)

SDSM&T: South Dakota School of Mines & Technology (USA-SD)

SDSS: Self-Deploying Space Station • Sloan Digital Sky Survey • Space Division Shuttle Simulator • Spatial Decision Support System

SDSSE: Science Data System Support Equipment

SDST: Science Data Support Team • Small Deep Space Transponder

SDSU: San Diego State University (USA-CA)

SdT: Service de Traduction

SDT: Saidu Sharif (Pakistan) (location identifier) • Scaling and Display Task • Science Data Team • Science Definition Team • Scientific Distribution Technique • Serial Data Transmission • Shuttle Data Tape • Shuttle Data Type • Signal Detection Theory • Simulated Data Tape • Source Distribution Technique • Specification-Driven Tool • Spin-Dependent Tunneling • Structural Dynamic Test • Structured Data Transfer

SDTA: Structural Dynamic Test Article

SDTI: Selective Dissemination of Technical Information • Serial Digital Transmission Interface

SDTP: Spatial Data Transfer Processor • Startover Data Transfer and Processing

SDTS: Satellite Data Transmission System • Spatial Data Transfer Standard

SDTT: Similarly Dependence Textural Transformation

SDTV: Standard-Definition Television • Structural Dynamics Test Vehicle

SDTX: Science Data Transmitter

SdU: Sciences de l'Univers

SDU: Rio de Janeiro - Santos Dumont (Brazil-RJ) (location identifier) • Satellite Data Unit • Sciences De l'Univers • Service Data Unit • Service Delivery Unit • Signal Distribution Unit • Spectrum Display Unit

SDUC: Secondary Data Utilisation Centre

SDUK: Society for the Diffusion of Useful Knowledge

SDUS: Satellite Data Users Station • Secondary Data User Station • Small-scale Data Utilization Station

SDV: Sciences De la Vie • Shuttle-Derived Vehicle • Solar Daily Variation • Switched Digital Video • Tel Aviv - Yafo Sde-Dov (Israel) (location identifier)

SDVF: Software Development and Verification Facility

SDW: Sandwip (Bangladesh) (location identifier) • Science Data Writer • Serial Digital Word • Spiral Density Wave

SDWA: Safe Drinking Water Act

SDX: Satellite Data Exchange • Sedona (USA-AZ) (location identifier)

SDY: Sidney (USA-MT) (location identifier)

SDZ: Shetland Islands (UK) (location identifier)

s/e: Sin Editorial

se: Sweden

Se: Selen • Selenium • Sélénium

S&E: Salaries & Expenses • Science & Engineering • Scientific & Engineering • Sensors & Effectors • Services & Equipment • Signature Error • Single-Excited

S-E: Survivability-Endurance

S/E: Service Equipment • Stock Exchange

SE: London (Southeast_) (UK postal code) • Safety Evaluation • Santiago del Estero (Argentina) • Saponification Equivalent • Schrödinger Equation • Scientific Editor • Secretaría de Estado • Secrétariat d'État • Sensors and Effectors • Sequence of Events • Seltene Erden • Semitic • Seneca • September • Septembre • Sergipe (Brazil) • Service Element • Service Engineer • Service Equipment • Session Environment • Shared Everything • Single End • Single-Ended • Single Engine • Single Entry • Software Engineering • Solar Eclipse • Solar Ecliptic • Solar Explorer • Son Éminence • Son Excellence • Southeast • Southeastern • Southeast London (UK postal code) • Space Exploration • Special Edition • Special Equipment • Specialized Engineering • Spectroscopic Ellipsometer • Spherical Equivalent • Spring Equinox • Stable Element • Staff Engineer • Stamped Envelope • Standard Engine • Standard English • Standard Error • Stock Exchange • Straight Edge • Strategy Evaluation • Stress Engineer • Student Experiment • Subset Entry • Sud-Est • Suède (ISO country code) • Su Excelencia • Support Element • Support Equipment • Supporter Equipment • Surface Element • Survivability-Endurance • Sweden (international aircraft registration code) • Sweden (ISO country code) • Synthetic Environment • Système Expert • System Effectiveness • System Element • System Engineer • System Engineering • System Evaluator

SE&A: Safety Engineering & Analysis

SEA: Safety Engineering Analysis • Scanning Electrostatic Analysis • Scientific Exchange Agreement • Scientist's Expert Assistant • SeaTac (USA-WA) (location identifier) • Seattle (USA-WA) (location identifier) • Security Extension Architecture • Self-Extracting Archive • Sensor Electronics Assembly • Service d'Étude des Accélérateurs • Silicon Elastimeter Ablator • Small Earth Approacher • Small Earth-Approaching • Sociedad Española de Astronomía (Spain) • Society of Engineering Associates (Australia) • Southeast Asia • Spherical Electrostatic Analyzer • Splitting Equilibration Algorithm • Standard Electronic Assembly • Standard Extended Architecture • Statistical

Energy Analysis • Strategic Environmental Assessment • Sudden Enhancement of Atmospherics • System Engineering Analysis • System Error Analysis • System for Environmental Analysis • Tacoma (USA-WA) (location identifier)

SEAB: Secretary of Energy Advisory Board (USA)

SeAc: Sensors and Actuators (periodical code)

SEAC: School Examination and Assessment Council • Société Européenne pour l'Astronomie dans la Culture • Standard Eastern Automatic Computer • Standard Eastern Automating Computing

SEACF: Support Equipment Assembly and Checkout Facility

SEACOF: Support Equipment Assembly and Checkout Facility

SEAD: Suppression of Enemy Air Defenses

SEADEX: Shoreline Environment Atmospheric Dispersion Experiment

SEADS: Shuttle Entry Air Data Sensor

SEAES: Spacecraft Environmental Anomalies Expert System

SEAGA: Selective Employment of Airborne and Ground Alert

SEAI: Secrétaire Exécutif aux Affaires Industrielles

SEAIC: Southeast Asia Information Center

SEAID: Support Equipment Abbreviated Items Description

SEAIG: Southeast Asia Information Group

SEAL: Sea-Air-Land • Secure Electronic Authorization Laboratory • Segmentation and Assembly Layer • Simple and Efficient Adaptation Layer • Standard Electronic Accounting Language

SEALION: Sea Level Instrumentation and Observation Network

SEALITE: Systematic Evaluation and Analysis of a Laser in a Test Environment

SEAM: Saber Enterprise Applications Manager • Spectral-fitting Expanding Atmosphere Method • System Engineering and Analysis Management

SEAMEC: Southeast Asian Ministers of Education Council

SEAMEO: Southeast Asian Ministers of Education Organization

SEAMES: Southeast Asian Ministers of Education Secretariat

SEAMS: Southeast Asian Mathematical Society

SEAN: Scientific Event Alert Network (now GVN)

SEAOP: Sociedade de Estudos Astronômicos de Ouro Preto (Brazil-MG)

SEAP: Science and Engineering Apprenticeship Program

SEAPG: Support Equipment Acquisition Planning Group

SEAQ: Stock Exchange Automated Quotation

SEAR: System Engineering Analysis Report

SEARCH: Study of Earth Rapid Changes

SEARNG: Southeast Asian Region Network for Geosciences

SEAS: School of Engineering and Applied Science • Shipboard Environmental Acquisition System • Shipboard Environmental Automated System • Study of the European Arctic Shelf • Synthetic Economy for Analysis and Simulation (project)

SEASAR: Sea Synthetic Aperture Radar

SEASAT: Sea Satellite

SEASTARS: Satellite Engineering Assisted Shipboard Test and Repair System

SeaTac: Seattle-Tacoma (USA-WA)

SEATO: South-East Asia Treaty Organization

SEB: Scientific Equipment Bay • Sebha (Libya) (location identifier) • Source Evaluation Board • South Equatorial Belt • State Electricity Board • Strip Electron Beam • Support Equipment Building • Systems Engineering Branch

SEBA: Support Electronics Bay Assembly

SEBC: Système Européen de Banques Centrales

SEBD: Software Engineering Bibliographic Database • South Equatorial Band Disturbance

SEbE: Southeast by East

SEBEX: Sahelian Energy Balance Experiment

SEBn: South Equatorial Belt - Northern Component

SEbS: Southeast by South

SEBs: South Equatorial Belt - Northern Component

SEBV: Soil Environmental Background Value

SEBz: South Equatorial Belt - Internal Zone

sec: Secant • Secolo

Sec: Sección • Second • Secondaire • Secondary • Seconded • Secrétaire • Secrétariat • Secret • Secrétaire • Secretaría • Secretary • Section • Sector • Secular • Secundum • Security

SEC: Science Executive Committee • Scientific and Engineering Computation • Second • Secondary • Secondary Electron Conduction • Secondary Emission Conductivity • Secondary Evaluation Committee • Section • Securities and Exchange Commission (USA) • Security Exchange Commission • Sequential Events Controller • Serre-Chevalier (France) (location identifier) • Single-Edge Contact (*) • Single-Edge Contract • Single Error Correction • Size-Exclusion Chromatography • Société Européenne de Culture • Society for Educative Communication • Solar Electric Company • Solar Energy Collector • Solar Energy Concentrator • Solid Electrolyte Capacitor • Source Evaluation Committee • Space Education Council • Space Environmental Chamber • Space Environment Center • Split Expander Cycle • Sponsor Executive Committee • Standard Error of Calibration • Sun-Earth Connection

SECAM: Séquence de Couleurs Avec Mémoire • Sequential Color And Memory

SECAP: Systems Experiment Correlation and Analysis Program

SECAS: Ship Equipment Configuration Accounting System

Secc: Sección

SECDEC: Single Error Correction - Double Error Correction

SECDED: Single Error Correction - Double Error Detection

SECDS: Sun-Earth Connection Data System

SECEF: Sun-Earth Connection Education Forum

SECEM: Support Equipment Cost Effectiveness Model

SECFO: System Engineering and Consensus Formation Office

SECFT: Sociedad Española de Cartografía, Fotogrametría y Teledetecció n (Spain)

Sec. Gen.: Secretary General

sech: Hyperbolic Secant

SECHO: Supervised Extraction and Classification of Homogeneous Objects

SECI: Southeast European Cooperation Initiative

SECIS: Solar Eclipse Coronal Imaging System

sec. leg.: Secundum Legem

SECM: Science and Engineering of Composite Materials (periodical code)

SECME: Southeastern Consortium for Minorities in Engineering

sec. nat.: Secundum Naturam

SECO: SEC Orthicon • Sun-Earth Connection Observatory • Sustainer Engine Cutoff

SECOIA: Systèmes Experts et Conceptions d'Outils pour l'Intelligence Artificielle

Secr: Secrétaire • Secretariat • Secrétariat • Secretary

sec. reg.: Secundum Regulam

Secs: Seconds • Sections

SECS: Sequential Events Control System • Shuttle Events Control System • Simulation and Evaluation of Chemical Synthesis • Simulation/Experiment Control System • Small Experimental Communications Satellite • Solar Electric Communication Satellite • Space Environmental Control System

Sect: Section • Sector

Secy: Secretary

Secy.-Gen.: Secretary-General

SEC/YR: Seconds per Year

SECYT: Secretaría de Ciencia Y Tecnologia

sed: Stream Editor

Sed: Sediment

SED: Saisie Électronique de Documents • Saturn Electrostatic Discharge • Science Education Department • Secondary Energy Distribution • Secure Encryption Device • Sedom Min'hat Hashnayim (Israel) (location identifier) • Sensor Evolutionary Development • Service d'Étude des Détecteurs • Service d'Études et de Documentation • Smoke-Emitting Diode • Solar Energy Density • Space Electronics Division • Space Engineering Document • Space Environment Division • Space Experiments Division • Special Engineering Detachment • Spectral Energy Distribution • Spheroidal Equidensity • Stream Editor • Surface-conduction Emitter Display • Système d'Équations Différentielles • Système d'Exploitation de Disquettes • System Effectiveness Demonstration

SEDAC: Socio-Economic Data and Applications Center

SEDAR: Submerged Electrode Detection and Ranging

SEDAS: Satellite Environmental Data Acquisition System • Spurious Emission Detection Acquisition System

SEDB: Specifically Extragalactic Database

SEDF: Système d'Équations Différentielles Finies

SEDFB: Surface-Emitting Distributed Feedback

SEDI: Study of the Earth's Deep Interior

Sedim: Sedimentology (periodical code)

SEDIX: Selective Dissemination of Indices

SEDL: Software Engineering Design Language

SEDN: Système Européen de Documentation Nucléaire

SEDO: Sociedad Española De Optica (Spain)

SEDPC: Scientific and Engineering Data Processing Center (USA)

SEDR: Science Experiment Data Record • Supplemental Experiment Data Record • Supplementary Experiment Data Record • Supplementary Experimenter Data Record • System Effective Data Rate • System Engineering Department Report

SEDS: Small Expendable Deployable System • Small Expendable Deployer System • Small Expendable Deployment System • Small Expendable-tether Deployer System • Space Electronics Detection System • Students for the Exploration and Development of Space • Support Equipment Data System

Sedt: Sediment

Sedtn: Sedimentation

SEE: San Diego - Gillespie (USA-CA) (location identifier) • Satellite Energy Exchange • Secondary Electron Emission • Senior Electrical Engineer • Service de l'Énergie de l'État (Luxemburg) • Signing Essential English • Single Event Effect • Société des Électriciens et Électroniciens (France) • Society of Environment Engineers • Software Engineering Economics • Software Engineering Environment • Software Engineering Exercise • Software Equality Evaluation • Solar Extreme-ultraviolet Experiment • Solar EUV Experiment • Spectral Energy Emission • Standard End Effector • Standard Error of Estimation • Stimulated Electromagnetic Emission • Sun-Earth Explorer • System Effectiveness Evaluation

SEEA: Southeast European Airlines

SEEC: Scottish Environmental Education Centre

SEECM: Shuttle Environmental Effects on Coated Mirrors

SEED: Science Education for Early Development • Self Electro-optic Effect Device • Standard for the Exchange of Earthquake Data • Support for Eastern European Democracy

SEEDS: Spacelab End-to-End Data System

SEEJ: Society of Environmental Engineers Journal (periodical code)

SEEK: System Evaluation and Exchange of Knowledge

SEEP: Shelf-Edge Exchange Processes • Students to Explore and Experience Physics

SEER: Science Education for Equity Reform • Sensor Experimental Evaluation and Review • Social and Economic Environmental Research • Surveillance, Epidemiology, and End Results • System Engineering, Evaluation and Research • System for Electronic Evaluation and Retrieval

SEES: Students to Experience Engineering and Science

SEET: Science End-to-End Test

SEF: Sebring (USA-FL) (location identifier) • Software Entrepreneurs' Group • Solar Energy Flux • Spacecraft Event File • Space Environmental Facility • Space Experiment Facility • Stability Enhancement Function • System Engineering Facility

SEFD: Solar Energy Flux Density

SEFI: Service-Formation-Information

SEFIP: Statistical Estimation Fault Isolation Procedure

SEFOR: Serviço de Estatísticas e Fomento de Recursos

seg: Segue • Segundo

Seg: Segment • Segmented • Segovia (Spain) • Segregate • Segregated • Seguente

SEG: Scientific Expert Group • Segment • Selinsgrove - Penn Valley (USA-PA) (location identifier) • Sequence of Events Generator • Simulation Engineering Group • Society of Economic Geologists (USA) • Society of Exploration Geophysicists (USA) • Software Engineering Group • Solid-Earth Geophysics • Standardization Evaluation Group • System Evaluation Group

SEGF: Surface Embedded Green Function

SEGPA: Section d'Enseignement Général et Professionnel Adapté

SEGS: Sequence of Events Generation System

SEGway: Science Education Gateway

SEH: Senggeh (Indonesia) (location identifier) • Shuttle Electronic Hardware • Solar Equivalent Hour • Solar EUVE Hitchhicker • Structured Exception Handling

SE&I: Systems Engineering & Integration

SEI: Salzburger Elektronik Industrie (Austria) • Science Education Institute • Seller End Item • Senhor do Bonfim (Brazil-BA) (location identifier) • Service d'Électronique et d'Informatique • Service des Échanges Internationaux • Service Européen d'Information • Sobolev with Exact Integration (method) • Software Engineering Institute • Space Exploration Initiative • Special Engineering Investigation • Special Equipment Item • Specific Emitter Identification • Statistical Engineering Institute • Stockholm Environment Institute (Sweden) • Support Equipment Installation • Systems Engineering and Integration • System/Equipment Inventory

SEIA: Solar-Energy Industries Association • Solar-Energy Institute of America (*)

SEIAC: Southeastern Iowa Astronomy Club (USA-IA)

SEIC: Solar-Energy Information Center • State Educational Information Center (USA) • System Engineering and Integration Contract

SEICO: Support Equipment Installation and Checkout

SEICT: Sistema Estadual de Informação Científica e Tecnológica (Brazil)

SEID: Systems Engineering and Integration Division

SEIDAM: System of Experts for Intelligent Data Management

SEII: Société d'Échanges Industriels Internationaux (France)

SEIR: Software and Electronic Information Resources • Southeast Indian Ridge

SEIS: Shuttle Experiments Integration Support • Single Element Imaging Spectrograph • Solar-Energy Information Service • Spacelab Experiments Integration Support • Standard End Item Specification • Submarine Escape Immersion Suit

Seismol: Seismological • Seismology

SEIST: System Engineering, Integration, and Support Team

SEIT: Satellite Educational and Informational Television • System Engineering, Integration and Test • System Evaluation, Integration and Test

SEIU: Service Employees International Union

SEJ: Seydisfjördur (Iceland) (location identifier)

Sek: Sekunde

SEK: Crown (Sweden) (ISO currency code) • Krona (Sweden) (ISO currency code) • Ksar es Souk (Morocco) (location identifier) • Svenska Elektriska Kommissionen

SEKAS: South East Kent Astronomical Society (UK-England)

Sel: Select • Selected • Selecting • Selection • Selectivity • Selector • Selig

Sél: Sélection • Sélectionné • Sélectionner

SEL: Satellite Experiment Laboratory • Select • Selected Equipment List • Selection • Selector • Self-Extensible Language • Seoul Kimpo (Korea-ROK) (location identifier) • Single-Event Latch-up • Software Engineering Laboratory • Space Environment Laboratory • Spontaneously Emitted Light • Stabilized Electromagnetic Levitator • Standard Elektrik Lorenz • Système d'Échange Local • Système d'Équations Linéaires • Système Électrique et Logiciel • Systems Electronics Laboratory

SeLA: Seven-Letter Acronym

SELA: Sistema Económico Latino-Americano • Système Économique Latinoaméricain

SELDADS: Space Environment Laboratory Data Acquisition and Display System

SELENE: Selenological and Engineering Explorer

SELF: Self-Eject Launch Facility • Solar Electric Light Fund

Selk: Selkirk

SELPER: Sociedad de Especialistas Latino-americanos en Percepción Remota

SELRAS: Space Environment Laboratory Retrieval and Analysis System

SELT: Self-Eject Launch Technique

SELV: Separated Extra-Low Voltage • Small Expendable Launch Vehicle

SELVS: Small Expendable Launch Vehicle Services

SELY: Southeasterly

Sem: Semaine • Semana • Semanal • Semester • Semicolon • Seminar • Seminary • Standard Error of the Mean

Sém: Séminaire

SEM: Safe Emergency Mode • Scanning Electron Microscope • Scanning Electron Microscopy • Science, Engineering and Mathematics • Secondary Electron Model • Secondary Electron Mirror • Secondary Emission Monitor • Seismometer • Seller's Engineering Memorandum • Selma Craig (USA-AL) (location identifier) • Semi-Empirical Model • Service d'Échange des Médias • Shared Equity Mortgage • Shuttle Experiment Module • Sociedad Espacial Mexicana (Mexico) • Société d'Économie Mixte • Solar Environmental Monitor • Solar EUV Monitor • Space Environment Monitor • Space Environment Monitoring • Space Experiment Module • Spectral Evolutionary Model • Standard Electronic Module • Standard Error of the Mean • Stray Energy Monitor • String Environmental Module • Structural Econometric Model • Structural Equation Modelling • Structured Entry Model • Subarray Electronics Module • Subcontractor Engineering Memorandum • Sun-Earth-Moon • Système d'Entretien Mobile • System Engineering Management • System Environment Monitor • System Error Message

SEMA: Sociedad Española de Matematica Aplicada (Spain) • Société d'Études de Mathématiques Appliquées (France)

SEMAA: Science, Engineering, Mathematics and Aerospace Academy (USA)

SEMANE: Sortie d'Eaux Méditerranéennes en Atlantique Nord-Est

SEMD: Stray Energy Monitor Device

SEMECCEL: Société d'Économie Mixte d'Exploitation de Centres Culturels, Éducatifs et de Loisirs (France)

SEMECO: Sección de Meteoros y Cometas

SEMER: Schéma d'Ensemble pour la Météorologie Routière

SEMI: Semiconductor Equipment and Materials Institute • Special Electromagnetic Interference

Semicnd: Semiconductor

semih: Semilhora

SEMIR: Support, Exploitation, Maintenance, Informatique et Réseaux

SEMIRA: System of Electronic Marks' Interrogation, Registration and Administration

SEMIS: Solar Energy Monitor In Space

SEMM: Solar Electric Multiple Mission

SEMMS: Solar Electric Multiple-Mission Spacecraft

semp: Sempre

SEMP: System Engineering Management Plan

SEMPA: Scanning Electron Microscopy with Polarization Analysis

SEMS: Space Environment Monitor System • Stray Energy Monitor System • System Engineering Management Standard

SEMSQ: Société des Éditeurs de Manuels Scolaires du Québec (Canada-QC)

Sen: Senate • Senator • Senior • Senza

SEN: Senegal (ISO country code) • Sensor • Southend (UK-England) (location identifier) • Special Educational Needs • System Engineering Notice

SENAMHI: Servicio Nacional de Meteorología e Hidrología (Bolivia)

SENAMI: Servicio Nacional de Meteorología e Hidrología (Peru)

SenAR: Sendai Astronomiaj Raportoj (periodical code)

SEND: Shared Equipment Need Date

SENEAM: Servicios a la Navegación en el Espacio Aéreo Mexicano

Senr: Senior

Sens: Sensitive • Sensitivity • Sensor

SENS: Sensitive

Sensi: Sensible

Sent: Sentido • Sentier

Sen. VP: Senior Vice-President

SEO: Satellite for Earth Observation • Sauf Erreur ou Omission • Seguela (Ivory Coast) (location identifier) • Senior Executive Officer • Senior Experimental Officer • Société Européenne d'Optique • Society of Education Officers • Socio-Economic Objective • Special Engineering Order • Special Equipment Operator • Synchronous Equatorial Orbit • Synchronous Equatorial Orbiter

SEOCS: Sun Earth Observatory and Climatology Satellite

SEON: Solar Electronic Observing Network • Solar Electro-Optical Network

SEOO: Sauf Erreur Ou Omission

S/EOS: Standard Earth Observation Satellite

SEOS: Standard Earth Observation Satellite • Stationary Earth Orbit Satellite • Synchronous Earth Observation Satellite • Synchronous Earth Observatory Satellite

Sep: Separable • Separate • Separated • Separation • September • Septembre

SEP: Scientific and Engineering Personnel • Sclérose En Plaques • Separation • September • Septembre • Serial Entry Printer • Service de l'Électronique Physique • Shuttle Electronics Package • Société Européenne de Propulsion • Solar Electric Power • Solar Electric Propulsion • Solar Energetic Particle • Solar Energetic Proton • Source Evaluation Panel • South Ecliptic Pole • Space Electronics Package • Special Embassy Programme • Spherical Error Probable • Spherical Error Probability • Spheroidal Equipotential • Standard Electronic Package • Standard Engineering Practice • Standard Error of Prediction • Stephenville Clark (USA-TX) (location identifier) • Supervisor Executive Program • Support Equipment Package • Surface d'Énergie Potentielle • System Enhancement Plan • Systems Effectiveness Plan • Systems Engineering Plan • Systems Engineering Process

SEPA: Scottish Environment Protection Agency (UK-Scotland) • Site Engineering Project Authorization • Southeastern Planetarium Association (USA-FL) • State Environmental Protection Agency

SEPAC: Space Experiment with Particle Accelerators

SEPAP: Shuttle Electrical Power Analysis Program

SEPAR: Shuttle Electrical Power Analysis Report

Separat: Separativo

SEPC: Space Exploration Program Council

SEPD: Systems Engineering and Programmatics Department

SEPG: Software Engineering Process Group

SEPIA: Structured Elicitation and Processing of Ideas for Authoring

SEPICA: Solar Energetic Particle Charge Analyzer

SEPIL: Selective Excitation of Probe Ion Luminescence

SEPL: Set-oriented Picture Programming Language

Sepn: Separation

SEPO: Space Electric Power Office

SEPOR: Service des Programmes des Organismes de Recherche

SEPP: Secure Electronic Payment Protocol • System Effectiveness Program Plan

SEPR: South Ecliptic Pole Region

SEPRA: Service d'Expertise des Phénomènes de Rentrées Atmosphériques

SEPS: Solar Electric-Propelled Spacecraft • Solar Electric Propulsion System • Solar Electronic Propulsion System

SEPST: Solar Electric Propulsion System Technology

SEPSU: Science and Engineering Policy Studies Unit

Sept: September • Septembre • Septiembre

SEPT: Service d'Étude de la Poste et des Télécommunications • Société Européenne de Programmes de Télévision

SEPTR: Solar Energetic Particle Tracer

SEPUP: Science Education for Public Understanding Program

seq: Sequens • Sequente • Sequitur

Seq: Sequel • Sequence • Sequential

SEQ: Semantically Equivalent Query • Sequence • Sequencer • Sungai Pakning (Indonesia) (location identifier)

SEQAS: South East Queensland Astronomical Society (Australia-QLD)

seqq: Sequentia • Sequentibus

SEQT: Sequential Threshold

SEQUAL: Seasonal Equatorial Atlantic Experiment

SEQUEL: Structured English Query Language

Ser: Serbian • Serial • Serie • Series • Sermon • Serpens-Serpentis (constellation) • Service

Sér: Série

SER: Safety Evaluation Report • Scanning Emission Radiometer • Secrétariat d'État à la Recherche • Section Efficace Radar • Sequential Events Recorder • Serial • Service Employment Redevelopment • Seymour Freeman (USA-IN) (location identifier) • Site Evaluation Report • Solar Extinction Radiometer • Special Environmental Report • State of the Environment Support • Stop End of Runway • Support Equipment Requirement • Surface Electrical Resistivity • System Environment Recording • Systems Engineering Report

SERA: Southern (extremity of the) Earth's Rotation Axis

SERAL: Southeastern Region of the Astronomical League (USA)

SERAPHIM: Segmented Rail Phased Induction Motor

Serb: Serbia • Serbian

SERB: Shuttle Engineering Review Board • Southeast Regional Bureau • Strategic Education Research Board • Study of the Enhanced Radiation Belt • Support Equipment Review Board • Systems Engineering Review Board

SERBS: Surface Energy and Radiation Balance System

SERC: Satellite Education Resource Consortium • Scientific and Engineering Research Council • Smithsonian Environmental Research Center • Software Engineering Research Centre • Space Engineering Research Centre • State Emergency Response Commission

SERCnet: SERC Network

SERD: Socio-Économie de la Recherche et du Développement • Stored Energy Rotary Drive • Support Equipment Recommendation Data

SERDP: Strategic Environmental Research and Development Program

SERE: Spatial Effective Resolution Element

SERENDIP: Search for Extraterrestrial Radio Emission from Nearby Developed Intelligent Populations

SEREP: System Environment Recording, Editing and Printing

SERF: Solar Electromagnetic Radiation Flux • Space Environment Research Facility • Study of Energy Release in Flares

Serg: Sergeant • Sergent

SERG: Saint Empire Romain Germanique

SERI: Solar Energy Research Institute • System for Earth Resource Information • Systems Engineering Research Institute

SERIES: Satellite Emission Radio Interferometric Earth Surveying • Satellite Emission Range Infrared Earth Surveying

SERIN: Synthetic Environments Research Initiative

SERIS: State of the Environment Reporting Information System

SERIWA: Solar Energy Research Institute of Western Australia

SERME: Sign Error Root Modulus Error

SERN: Single-Expansion Ramp-Nozzle • Southeastern

SERNG: Saint Empire Romain de la Nation Germanique

serno: Serial Number

SERO: System Engineering Release Order

SERPA: Société d'Enquête et de Recherche sur les Phénomènes Aérospatiaux (France)

SERPIA: Service d'Études et de Réalisation des Produits d'Information Avancés

SERS: Shuttle Enhanced Reconfiguration System • Shuttle Equipment Record System • Support Equipment Requirements Sheet • Surface-Enhanced Raman Spectroscopy

SERT: Service d'Études et de Réalisations Techniques • Society of Electronic and Radio Technicians • Space Electrical Rocket Test • Space Electronic Rocket Test

SERTIT: Service Régional de Traitement d'Image et de Télédétection

SERTS: Solar EUV Rocket Telescope and Spectrograph

SERURE: Service des Énergies Renouvelables et de l'Utilisation Rationnelle de l'Énergie (France)

Serv: Service

SERV: Service • Space Emergency Reentry Vehicle

SES: Satellite Earth Station • Sciences Économiques et Sociales • Scientific Exploration Society • SCSI Enclosure Services • Secondary Electron Scattering • Second Epoch Survey • Section d'Éducation Spécialisée • Section d'Études Spécialisées • Selma Selfield (USA-AL) (location identifier) • Senior Executive Service • Service Evaluation System • Severely Errored Seconds • Ship Earth Station • Shuttle Engineering Simulation • Shuttle Engineering Simulator • Signal Enhancement Seismograph • Simulation Environment System • Société Européenne des Satellites • Socioeconomic Status • Software Evaluation Subpanel • Solar Eclipse Sensor • Solar Energy Society • Solar Environment Simulator • Space Environment Simulator • Space Erectable Structure • Special-Emphasis Study • Spinning Earth Sensor • Stand-alone Edit Station • Structure of Earnings Statistics • Sun-Earth-Satellite • Super-Eddington Source • Suprathermal Electron Spectrometer • Surface Effect Ship • Systems Engineering Simulator

SESA: Solar-Energy Society of America • Special Equipment Stowage Assembly

SESAC: Space and Earth Sciences Advisory Committee

SESAM: Secours dans un Établissement Scolaire face à un Accident Majeur • Surface Effective Sample Monitor • Surface Effects Sample Monitor (*) • System for Electronic Support of Academic Material

SESAME: Satellite Experiments Simultaneous with Antarctic Measurements • Search for Excellence in Science and Mathematics Education • Second European Stratospheric Arctic and Mid-latitude Experiment • Service, Sort And Merge • Severe Environmental Storms And Mesoscale Experiment • Solid Earth Science and Application Missions for Europe • Structure and Energetics of the Stratosphere and Mesosphere

SESC: Shuttle Events Sequential Control • Société d'Éditions Scientifiques et Culturelles (France) • Société d'Étude de la Solidification et de la Cristallogénès e (France) • Software Engineering Standards Committee • Space Environment Services Center

SESD: Science and Engineering Systems Division • Software Engineering Systems Division

SESE: Space Electronics Support Equipment

SESL: Self-Erecting Space Laboratory • Space Environmental Simulation Laboratory • Space Environment Simulation Laboratory

SESMI: Systems Engineering Support and Management Integration

SESNET: Space and Earth Science Network

SESP: Space Environmental Satellite Program • Space Experiment Support Program

SESPD: Space and Earth Science Programs Directorate

SESQUINET: Sesquicentennial Network

Sess: Session

SESS: Space Environmental Support System • Space Environment Sensor Suite

SEST: School of Engineering, Systems and Technologies • Swedish-ESO Submillimeter Telescope

Set: Setembro • Settembre • Settlement

SET: San Estebán (Honduras) (location identifier) • Satellite Environmental Terminal • Science Engineering and Technology • Science-Engineering-Technology • Scientists-Engineers-Technicians • Secured Encrypted Transaction • Secure Electronic Transaction (*) • Setembro • Simplified Engineering Technique • Simultaneous Engineering Technology • Simulated Emergency Test • Single-Electron Transistor • Single-Electron Transport • Single-Electron Tunnelling • Single Escape Tower • Software Engineering Technology • Solar Energy Thermionic • Spacecraft Elapsed Time • Space Education Trust • Space Electronics and Telemetry • Standard d'Échange et de Transfert • Stress-Energy Tensor • Système d'Échange et de Transfert • Systems Engineering Test

SETA: Satellite Electrostatic Triaxial Accelerometer • Systems Engineering and Technical Assistance

SETAC: Society of Environmental Toxicology and Chemistry • System Engineering and Technical Assistance Contract • System Engineering and Technical Assistance Contractor • System Engineering Technical Advancement Center

SETAR: Serial Event Timer And Recorder

SETAS: Space Environments and Technology Archive System

SETC: Space and Environment Technology Center

SE&TD: System Engineering & Technical Direction

SETD: Space Environment Test Division • System Engineering and Technical Direction

SETE: Support and Electronic Test Equipment

SETEXT: Structure Enhanced Text

SETF: Small-Engine Test Facility

Setg: Setting

SETHI: Système Expérimental de Télédétection Hyperfréquence Imageur

SETI: Search for Extraterrestrial Intelligence

SETIM: Service des Équipements et des Techniques Instrumentales de la Météorologie

S-et-L: Saône-et-Loire (France)

SETL: Science Experiment Test Laboratory

Setlg: Settling

S-et-M: Seine-et-Marne (France)

SETN: Science and Engineering Television Network

S-et-O: Seine-et-Oise (France)

SETP: System Engineering Test Phase

SETS: Set Equation Transformation System • Shuttle Electrodynamic Tether System • Solar Electric Test Satellite

Sett: Settembre

SEU: Scan Electronics Unit • Seronera (Tanzania) (location identifier) • Service des Échanges Universitaires • Single-Event Upset • Smallest Executable Unit • Soft Error Upset • Solar Energy Update (periodical) • Source Entry Utility • Structure and Evolution of the Universe

SEUL: Support Equipment Utilization List

seuo: Salvo Error u Omisión

Sev: Several • Sevilla (Spain)

SEV: Sección de Estrellas Variables • Sensor Equivalent Visibility • Severe • Severodoneck (Ukraine) (location identifier) • Status-Error-Validity

SEVA: Surface Extravehicular Activity

SEVIRI: Spinning Enhanced Visible and Infrared Imager

Sevl: Several

SEVP: Status-Error-Validity-Parity

SEVPCD: Syndicats des Entreprises de Vente par Correspondance et à Distance (France)

SEVS: Simulation and Evaluation System

SEVUPS: South Eastern Virginia Urban Plume Study

SEW: Siwa (Egypt) (location identifier) • Software Engineering Workshop

SEWB: System Engineering Workbench

SEWG: Scheduling Efficiency Working Group • Sustaining Engineering Working Group • System Effectiveness Working Group

SEWP: Scientific and Engineering Workstation Procurement

SEWS: Satellite Early Warning System • Sun-End Work Station

Sex: Sextans-Sextantis (constellation)

SEX: Sundays Excepted • Sembach (Germany) (location identifier) • Software Exchange • Système d'Exploitation

Sext: Sextant

SEY: Secondary Electron Yield • Selibaby (Mauritania) (location identifier)

Seyf: Seyfert

SEYM: Secondary Electron Yield Measurement

SEYMS: Secondary Electron Yield Measurement System

SEYS: Secondary Electron Yield System

SEZ: Mahe Island (Seychelles) (location identifier) • Special Economic Zone

s/f: Sin Fecha

sf: Sin Fecha • Sin Foliar • Square Foot • Sub Finem

Sf: Suriname Guilder (currency)

S/F: Safety Factor • Scale Factor • Single Flow • Static Firing

SF: Finland (international vehicle registration code) • Finland (postal country code) (now FIN) • Safe • Safety Factor • San Francisco (USA-CA) • Sans Frais • Santa Fe (Argentina) • Scale Factor • Science Fiction • Science File • Screening Factor • Secondary Failure • Selection Filter • Senior Fellow • Separation Factor • Shift Forward • Short-term Fluctuation • Shrink Factor • Side Frequency • Signal Frequency • Signaling Frequency • Sign Flag • Simulation Facility • Sin Foliar • Single Flow • Single Frequency • Skip Flag • Sliding Filter • Small Field • Snow Flurries • Software Factory • Solar Flare • Source Function • Special Facility • Spontaneous Fission • Spot Face • Square Foot • Standard Form • Standard Frequency • Star Field • Star Formation • Star*s Family • Start Field • Static Firing • Strong Field • Structured Field • Structure Function • Subcontractor-Furnished • Sub Finem • Successful Flight • Superfast • Superfund • Surge Filament • Swiss Franc

S&FA: Shipping & Forwarding Agent

SFA: Sales Force Automation • Scientific Film Association (UK) • Securities and Futures Authority (UK) • Segmented Flow Analysis • Selling Forces Automation • Sfax (Tunisia) (location identifier) • Single Failure Analysis • Single-Frequency Approach • Small Fine Arm • Spaceport Florida Authority (USA-FL) • Spatial Frequency Analyzer • Star Formation Activity • Step Frequency Analyser • Stephen F. Austin • Stereo Field Analyzer • Sun Finder Assembly • Supplementary Failure Analysis • Surface Force and Adhesion • Surface Force Assembly • Surface Forces Adhesion • Survey Feedback Action • Sweet Fanny Adams

SFAA: San Francisco Amateur Astronomers (USA-CA)

SFAAA: South Florida Amateur Astronomers Association (USA-FL)

SFACI: Software Flight Article Configuration Inspection

SFAPS: Space Flight Acceleration Profile Simulator

SFAR: Special Federal Aviation Regulation (USA) • Star*s Family of Astronomy Resources (now SFARR)

SFARR: Star*s Family of Astronomy and Related Resources (formerly SFAR)

SFAS: Statement of Financial Accounting Standards

SFASU: Stephen F. Austin State University (USA-TX)

SFB: Sanford (USA-FL) (location identifier) • Société Française de Biométrie (France) • Sonderforschungsbereich • Spezialforschungsbereich • Sternfreunde Breisgau (Germany) • Super Fast Build

SFBA: Société Française de Bibliométrie Appliquée (France)

SFBI: Share Frame Buffer Interconnect

Sfc: Surface

SFC: Saint-François (Guadeloupe) (location identifier) • Secondary Focus Camera • Selection Filter Control • Shopfloor Control • Société Française de Chimie (France) • Société Francophone de Classification • Space Flight Center • Specific Fuel Consumption • Standing Finance Committee • Star Field Camera • Subject Field Code • Subject Field Coder • Supercritical Fluid Chromatography • Surface • Synchronized Framing Camera

SFCC: Santa Fe Community College (USA-MN)

SFCG: Space Frequency Coordination Group

SFCS: Survival Flight Control System

SFD: San Fernando de Apure (Venezuela) (location identifier) • Solar Flux Density • Source Follower per Detector • Start of Frame Delimiter • Sudden Frequency Deviation • Switching Field Distribution • System Functional Diagram • System Functional Drawing

SFDF: Satellite Field Distribution Facility

SFDI: Solar Facility Design Integration • Standard Format for Digital Images

SFDK: Stepped-Frequency Delta-K

SFDR: Supplementary Format Data Record • Supplementary Formatted Data Record

SFdS: Société Française de Statistique (France)

SFDS: Secondary Flight Display System • Société Française De Statistique (France) • System Functional Design Specification

SFDT: Site Format Dump Tape

SFDU: Standard Format Data Unit • Standard Formatted Data Unit

SFE: San Fernando (Philippines) (location identifier) • Scale Factor Error • Schottky Field Effect • Seller-Furnished Equipment • Société Française d'Électrophorèse (France) • Solar-Flare Effect • Solid-Fuel Engine • Star Formation Efficiency • Start Field Extended • Static Feed Electrolysis • Stellar Formation Efficiency • Supercritical Fluid Extraction • System Functional Element

SFEF: Standard Format Exchange File

SFEP: Society of Freelance Editors and Proofreaders (UK)

SFEX: Solar Flare X-ray

SFF: Small Form Factor • Solar Forecast Facility • Solid Freeform Fabrication • Spokane Felts (USA-WA) (location identifier) • Standard File Format • Static Firing Facility • Sternfreunde Franken eV (Germany)

SFFF: Sedimentation Field Flow Fractionation

SFFS: Satellite Freeze Forecast System • Satellite Frost Forecast System

SFFT: Short Form Functional Test • Superconducting Flux Flow Transistor

SFG: Saint-Martin Espérance (French Antilles) (location identifier) • Star Formation Group

SFGC: Space Frequency Coordination Group

Sfgd: Safeguard

SFGEP: Space Flight Ground Environment Panel

SFGI: Syndicat Français des Grossistes Informatiques

SFH: San Felipe (Mexico) (location identifier) • Slow Frequency Hopping • Star Formation History

SFHA: Special Flood Hazard Area

SFHT: Superfluid Helium Tanker

SFI: Safi (Morocco) (location identifier) • Santa Fe Institute (USA-NM) • Schweizerisches Forschungsinstitut (Switzerland) • Société Fédérale d'Investissement • Space Flight Instrumentation • Step Function Input • Synthetic Final Index

SFIA: San Francisco International Airport (USA-CA)

SFINX: SRON Fabry-Pérot Interferometer Experiment

SFIR: Sequential Filter Imaging Radiometer

SFJ: Sondre Stromfjord (Greenland) (location identifier)

SFJTI: Société Franco-Japonaise des Techniques Industrielles

SFK: SmallTalk Frame Kit • Special Function Key

SFL: São Felipe (Cape Verde) (location identifier) • Secondary Freon Loop • Sequenced Flashing Lights • Société Française du Livre (France) • Structured Function Language • Symbolic Flowchart Language

SFLA: Stupid Four-Letter Acronym

SFLC: Software First Life Cycle

sfm: Surface Feet per Minute

SFM: Sanford (USA-ME) (location identifier) • Scanning Force Microscope • Scanning Force Microscopy • Senior Functional Manager • Sky Flux Map • Solar Flux Monitor • Spectrographe à Fentes Multiples • Spool File Manager • Steering Front Mechanism • Structure Framework Model • Sysplex Failure Management

SFMC: San Fernando Meridian Circle • Satellite Facility Management Center

SFMD: Storable Fluid Management Demonstration

SFME: Storable Fluid Management Experiment

SFMOMA: San Francisco Museum Of Modern Art (USA-CA)

SFMR: Stepped Frequency Microwave Radiometer

SFN: Santa Fe (Argentina-SF) (location identifier) • Single Frequency Network • Star Formation Newsletter

SFO: San Fernando Observatory (USA-CA) • San Francisco (USA-CA) (location identifier) • Sensor de Frente Onda • Serious Fraud Office • Simulated Flameout • Single-Frequency Outlet • Société Française d'Optique (France) • Space Flight Operations

SFOC: Shuttle Flight Operations Contractor • Space Flight Operations Center • Space Flight Operations Complex • Space Flight Operations Contract

SFOD: Space Flight Operations Director

SFOF: Space Flight Operations Facility

SFOM: Shuttle Flight Operations Manual • SOC Facility Operations Manual • Space Flight Operations Manual • Space Flight Operations Memorandum

SFOP: Space Flight Operating Procedure • Space Flight Operations Plan • Space Flight Operations Procedure

SFOT: Solar Flare Observation Technique

SFOV: Sounding Field Of View

SFP: Scanning Fabry-Pérot • Single Failure Point • Société Française de Physique (France) • Solar Flare Proton • Space Flight Participant • Standard Format Part • Standard Form of Payment • Standard Front Page • Summary Flight Plan • Surfers Paradise (Australia-QLD) (location identifier) • Sustainer Fuel Pressure

SFPA: Simplified Floating-Point Addition • Single Failure Point Analysis

SFPI: Scanning Fabry-Pérot Interferometer

SFpM: Surface Feet per Minute

SFPM: Surface Feet Per Minute

SFPPL: Short Form Provisioning Parts List

SFPS: Scanning Fabry-Pérot Spectrograph • Secure Fast Packet Switching • Single Failure Point Summary

SFPT: Société Française de Photogrammétrie et Télédétectio n (France)

SFPTB: Bulletin de la Société Française de Photogrammétrie et Tél 'edétection (periodical code)

SFQ: Sanliurfa (Turkey) (location identifier)

SFQL: Structured Full-text Query Language

SFr: Swiss Franc (currency)

SFR: San Fernando (USA-CA) (location identifier) • Selective File Retrieval • Shop Floor Reporting • Société Française de Radioprotection (France) • Société Française de Radiotéléphone (France) • Solar Flare Radiation • Star Formation Rate • Star Forming Region • Swiss Franc (currency)

s/fra: Su Factura

SFRS: Service du Film de Recherche Scientifique (France)

SFS: Finnish Standards Association (Finland) • Sequential Forward Selection • Shared File System • Shuttle Flight Status • Société Française de Sociologie (France) • Solar Flux Spectrometer • Special Flights Section • Star Field Sensor • Sufficient Feature Set • Suomen Standardisoimisliito (Finland) • Synchronous Fluorescence Spectroscopy • System Failure Summary

SFSA: Société Française des Spécialistes d'Astronomie (France) (now SF2A)

SFSM: South Florida Science Museum (USA-FL)

SFSS: Satellite Field Service Site • Satellite Field Service Station

SFSU: San Francisco State University (USA-CA)

Sft: Shaft

SFT: Shaft • Short Functional Test • Simulated Flight Test • Skellefteå (Sweden) (location identifier) • Special Flight Test • Static Firing Test • Structural Fatigue Test • Supercritical Fluid Technology • System Fault Tolerance • System Fault Tolerant • System File Table

SFTA: Structural Fatigue Test Article

SFTF: Static Firing Test Facility

SFTP: Shielded Foil Twisted Pair • Société Française de Photogrammétrie et de Télédétec tion (France)

SFTS: Space Flight Test System

SFTSS: Standard Frequency and Time Signal Satellite • Standard Frequency and Time Signal Service

SFTSSS: Standard Frequency and Time Signal Satellite Service

Sftw: Software

SFTWE: Software

SFU: Safia (Papua New Guinea) (location identifier) • Sample Fixation Unit • Secure Frame Unit • Simon Fraser University (Canada-BC) • Solar Flux Unit • Space Flyer

Unit • Specimen Fixation Unit • Standard Firing Unit • Suitable For Upgrade

SFUG: Security Features User's Guide

SFUR: Self-Filtering Unstable Resonator

SFV: Safety of Fishing Vessels • Santa Fé do Sul (Brazil-SP) (location identifier)

SFVN: Secure Fast Virtual Networking

SFW: Santa Fe (Panama) (location identifier) • Software

SFWD: San Francisco Water District (USA-CA)

SFWE: Static Feed Water Electrolysis

SFWK: Spatial Framework

SFWP: San Francisco Web Professionals (USA-CA)

Sfx: Suffix

SFX: San Felix (Venezuela) (location identifier) • Sound Effect

SFY: Springfield (USA-MA) (location identifier)

SFZ: Smithfield North Central (USA-RI) (location identifier) • Sonnenfleckenzyklus

SF2A: Société Française d'Astronomie et d'Astrophysique (France) (formerly SFSA)

s/g: Su Giro

sg: Sangho (ISO language code) • Singapore • Singular

s/G: Sin Gastos

Sg: Sungei

S/G: Salary Grade • Strain Gauge

SG: Sa Grâce • Sa Grandeur • Salary Grade • Scenario Generation • Secrétaire Général • Secrétariat Général • Secretary General • Senior Grade • Seyfert Galaxy • Ship and Goods • Signal Generator • Signal Ground • Singapore (ISO country code) • Singapour (ISO country code) • Small Gain • Snow Grains • Sollicitor General • Specific Gravity • Spinglass • StarGuides • Starlink Guide • Stern-Gerlach (force) • Stevenage (UK postal code) • Strain Gauge • Subgroup • Superconducting Gravimeter • Super Guppy • Supergroup • Surgeon General • Symbol Generator • Systems Generation

SG&A: Selling, General & Administrative

SGA: Shared Global Area • Sheghnan (Afghanistan) (location identifier) • Shocked Graphite Aggregate • Simple Genetic Algorithm • Slave Gyro Assembly • Société de Géologie Appliquée • Spectrometric Gas Analysis • Suppress Go Ahead • Système de Gestion des Abonnés • System Global Area

SGAA: Schweizerische Gesellschaft für Astrophysik und Astronomie (Switzerland)

SGAE: Sociedad General de Autores de España (Spain) • Sociedad General de Autores y Editores (Spain)

SGAF: Serviço de Gestão Administrativa e Financeira

SGAP: Secrétariat Général pour l'Administration de la Police (France)

SGAPO: Space Geodesy and Altimetry Project Office

SGASAA: Standard Generic Approach to Spacecraft Autonomy and Automation

SGATC: Second-Generation Advisory Technical Committee

SGB: Service Géologique de Belgique (Belgium) • Singaua (Papua New Guinea) (location identifier) • Subgiant Branch • Supergiant Branch • Système de Gestion Bord

SGBD: Système de Gestion de Base de Données

SGBDOO: Système de Gestion de Base de Données Orienté Objet

SGBDR: Système de Gestion de Base de Données Relationnelle

SGBDT: Système de Gestion de Base de Données Techniques

SGC: Solar Great Circle • Southern Galaxy Catalog • South Galactic Cap • Space Geodesy Centre • Stabilized Ground Cloud • Stabilizer Gyro Circuit • Superior Geocentric Conjunction • Surgut (Russia) (location identifier)

SGCN: Secrétariat Général du Comité National

SGCP: Simple Gateway Control Protocol

SGCR: Subcommittee on Global Change Research

SGCS: Slave Gyro Control System

Sgd: Signed

SGD: Dollar (Singapore) (ISO currency code) • Solar Geophysical Data (journal) • Sønderborg (Denmark) (location identifier) • Symbolized Graphic Data

SGDE: Sustainable Gross Domestic Expenditure

SGDG: Sans Garantie Du Gouvernement

SGDN: Secrétariat Général à la Défense Nationale (France)

SGDP: Sustainable Gross Domestic Product

SGDT: Store Global Descriptor Table • Système de Gestion des Données Techniques

Sge: Sagitta-Sagittae (constellation)

SGE: Secondary Grid Emission • Simulation Graphics Environment • Système de Gestion de l'Environnement • Système de Gestion Environnementale

SGED: Système de Gestion Électronique de Documents

SGEMP: Space-Generated Electromagnetic Pulse • System-Generated Electromagnetic Pulse

SGEN: Signal Generator • Syndicat Général de l'Éducation Nationale • System Generator

SGEOS: Second Generation Earth Observation Satellite

SGEP: Signaux Géophysiques et Environnements Planétaires

SGEPSS: Society of Geomagnetism and Earth, Planetary and Space Sciences (Japan)

SGER: Small Grant for Exploratory Research

SGES: Specialist Group on Expert Systems (BCS)

SGET: Spacecraft Ground Elapsed Time

SGF: Solution Growth Facility • Space Generation Foundation (USA-CA) • Springfield (USA-MO) (location identifier) • Système de Gestion de Fichiers

SGFGAS: Société de Gestion des Fonds de Garantie de l'Accession Sociale

SGFI: Schweizerische Gesellschaft zur Förderung der Informatik (Switzerland)

SGFS: Special Group on Functional Standardization

SGG: Simanggang (Malaysia) (location identifier) • Sozialgerichtsbuch

SGGA: Species-Generating Generic Algorithm

SGGM: Superconducting Gravity Gradiometer

SGGS: Satellite Geodesy and Geodynamics Systems (JPL)

SGH: Simpson Gumpertz & Heyer Inc. • Springfield (USA-OH) (location identifier)

SGHL: Schweizerische Gesellschaft für Hydrologie und Limnologie (Switzerland)

SGI: Sargodha (Pakistan) (location identifier) • Science Group Identifier • Silicon Graphics Inc. • Specific Gravity Indicator • Sun-Glint Imager

SGIM: Shape Globe Invariance Map

SGITS: Spacecraft/GOSS Interface Test System

SGJ: Sagarai (Papua New Guinea) (location identifier)

SGJP: Satellite Graphic Job Processor

SGK: Sangapi (Papua New Guinea) (location identifier) • Schweizerische Geologische Kommission (Switzerland) • Small-Generating-Kernel

Sgl: Single

SGL: Manila - Sangley Point (Philippines) (location identifier) • Schweizerische Gesellschaft für Limnologie (Switzerland) • Service Géologique du Luxembourg • Surface Gridding Library

SGLDM: Spatial Grey Level Dependence Matrix

SGLS: Satellite-Ground Link System • Space-Ground Link Station • Space-Ground Link System • Space-to-Ground Link System

SGLSY: Space Ground Link System

SGM: San Ignacio (Mexico) (location identifier) • Schweizerische Gesellschaft für Meteorologie (Switzerland) • Second-Generation Meteosat • Second-Generation Model • Shape Group Method • Special General Meeting

SGMC: Supergiant Molecular Cloud

SGML: Standard Generalized Markup Language • Standard General Markup Language

SGMP: Simple Gateway Management Protocol

SGMS: Satellite Geodetic Measuring Site • Space Geodetic Measurement Site

SGMT: Simulated Greenwich Mean Time

SGN: Ho Chi Minh City - Tan Son Nhat (Viet Nam) (location identifier)

SGO: Saint George (Australia-QLD) (location identifier) • Satellite Geodetic Observatory • Sodankylä Geophysical Observatory (Finland) • Sugar Grove Observatory (USA-IL)

SGOAA: Space Generic Open Avionics Architecture

SGOEM: Schweizerische Gesellschaft für Optik und Elektronenmikroskopie (Switzerland)

SGOMSEC: Scientific Group on Methodologies for the Safety Evaluation of Chemicals

SGOS: Shuttle Ground Operations Simulation • Shuttle Ground Operations Simulator

SGP: Serviços Geológicos de Portugal • Shay Gap (Australia-WA) (location identifier) • Singapore (international vehicle registration code) • Singapore (ISO country code) • Singapour (ISO country code) • Single Gauss Point • Single Ground Point • Southern Great Plains (USA) • South Galactic Pole • Space Geodesy Program • Space Geodesy Project • Standard Genetic Programming • Standard Guidance Package • Starlink General paper • Static Ground Point • Structure Ground Point

SGPI: Service de Gestion des Programmes et d'Ingénierie

SGpK: Schweizerische Geophysikalische Kommission (Switzerland)

SGPPV: Système de Gestion du Pay-Per-View

Sgpr: Singapore

SGPS: South Galactic Plane Survey

SGPW: Schweizerische Gesellschaft für Pflanzenbauwissenschfaten (Switzerland)

SGQ: Sanggata (Indonesia) (location identifier)

Sgr: Sagittarius-Sagittarii (constellation)

SGR: Houston Hull (USA-TX) (location identifier) • Science and Government Reports • Seismic Group Recorder • Set Graphics Rendition • Soft Gamma-ray Repeater • Soft Gamma Repeater • Specific Growth Rate • Sugar Land - Hull (USA-TX) (location identifier)

SGRAM: Synchronous Graphics Random Access Memory

SGRBE: Swift Gamma-Ray Burst Explorer

SGRS: Spaceborne Geodynamics Ranging System

SGS: Georgia and South Sandwich Islands (South_) (ISO country code) • Sandwich Islands (South Georgia and Sandwich_) (ISO country code) • Sanga Sanga (Philippines) (location identifier) • Satellite Ground Station • Science Ground Segment • Segmented Gamma Scanner • Sequence Generation System • Sequence Generator System • Simple Graphics System • Sixth Generation Systems • Société Générale de Surveillance • Solar Generator System • South Georgia and South Sandwich Islands (ISO country code) • South Sandwich Islands (South Georgia and_) (ISO country code) • Space Gateway Support (USA-VA) • Statistics Gathering System • Steep Glide Slope • Subgrid Scale • Supergiant Shell • Symbol Generation and Storage • Symbol Generator and Storage

SGSC: Strain Gauge Signal Conditioner

SGSE: Standard Ground Support Equipment

SGSM: Subgrid Scale Modelling

SGSN: Serving GPRS Support Node

SGSO: Space Ground Support Operations

SGSP: Subgroup on Standard Procedures

SGSTS: Space-to-Ground System Test Set

Sgt: Sergeant

SGT: Society of Glass Technology • Software-Guided Telescope • Stuttgart (USA-AR) (location identifier)

SGTE: Scientific Group Thermodata Europe

SGtK: Schweizerische Geotechnische Kommission (Switzerland)

Sgto: Sargento

SGts: StarGates

SGU: Saint George (USA-UT) (location identifier)

SGV: Sierra Grande (Argentina-RN) (location identifier)

SGW: Saginaw Bay (USA-AK) (location identifier)

SGX: Songea (Tanzania) (location identifier) • Spacecraft Glow Experiment

SGY: Skagway (USA-AK) (location identifier)

SGZ: Songkhla (Thailand) (location identifier)

s/h: Shipping/Handling • Shorthand

sh: Hyperbolic Sine • Serbo-Croatian (ISO language code) • Shall • Share • Sheet • Shilling (currency)

Sh: Shall • Share • Sheet • Shilling • Shower

S/H: Sample/Hold

S&H: Sample & Hold • Shipping & Handling

SH: Île Sainte-Hélène (ISO country code) • Saint Helena (ISO country code) • Sample and Hold • School House • Second Hand • Sequence History • Shack-Hartmann • Shoshoni • Showers • Southern Hemisphere • Speckle Holography • StarHeads • State Highway • Subset Header • Sulfhydril • Système Harmonisé

SHA: Secure Hash Algorithm • Secure Hashing Algorithm • Semantic Hypermedia Architecture • Shanghai (China-PRC) (location identifier) • Sidereal Hour Angle • Spherical Harmonic Analysis • System Hazard Analysis

SHAB: Schweizerisches Handelsamtsblatt

SHAD: Stellar Horizon Atmospheric Dispersion • Stellar Horizon Attitude Determination

SHADE: Share Dependency Engineering

SHAEF: Supreme Headquarters Allied Expeditionary Force

SHAG: Simplified High-Accuracy Guidance

Shak: William Shakespeare (1564-1616)

Shaks: William Shakespeare (1564-1616)

SHAO: Shanghai Astrononomical Observatory (China-PRC)

SHAPE: Second-generation Hypermedia Application Programming Environment • Second-generation Hypermedia Application Project Environment • Supreme Headquarters of Allied Powers in Europe • Systematic Horizontal Adjustment of Positional Errors

SHAR: Shell Archive • Shell Archiver • Stellar Horizon Atmospheric Refraction

SHARC: Serendipitous High-redshift Archival ROSAT Cluster • Submillimiter High-Angular Resolution Camera

SHARE: Safety and Hazard Assessment Research in Europe • Satellites in Health And Rural Education • Software

Help for Applications, Research, and Education • Space-station Heat-pipe Advanced Radiator Element

SHARED: Satellite Health Access for Remote Environment Demonstrator

SHARES: Sharing of Experience in Space

SHARK: Standard HRTP Archiving and Reprocessing Kernel • Standard HRTP Archiving and Retrieving Kernel

SHARMOR: Spectrographe Haute-Résolution Meudon-Orsay

SHARP: Slender Hypervelocity Aerothermodynamic Research Probe • Spacecraft Health Automated Reasoning Prototype • Standard Hardware Acquisition and Reliability Program • Standard HRTP Archiving and Request Product • Stationary High-Altitude Relay Platform • Summer High-school Apprenticeship Research Program • Super High Altitude Research Project • System for High-Angular-Resolution Pictures

SHASA: Servicios Aéreos de Honduras SA • Small High-Altitude Science Aircraft

SHAVE: Shared Virtual Environment

SHB: Nakashibetsu (Japan) (location identifier) • Second-Harmonic Band

SHBD: Spectral Hole Burning Device

SHC: Shire Indaselassie (Ethiopia) (location identifier) • Specific Heat Capacity • Spherical Harmonic Coefficient • Spontaneous Human Combustion

Shd: Should

SHD: Shadow • Staunton (USA-VA) (location identifier) • Strahlungshydrodynamik • Super-High Definition

SHDF: Sliding Hinge Double Fold

SHDM: Seeded Hot Dark Matter

SHDN: Shutdown

S-HDSL: Single-line High-bit-rate Digital Subscriber Line

SHDSL: Single-line High-bit-rate Digital Subscriber Line • Symmetric High-bit-rate Digital Subscriber Line

s/he: She or He

SHe: Supercritical Helium

S/HE: Sundays and Holidays Excepted

SHE: Sheffield (Starlink internal code) • Shenyang (China-PRC) (location identifier) • Signal Handling Equipment • Special Handling Equipment • Standard Hydrogen Electrode • Supercritical Helium

SHEAL: Shuttle High-Energy Astrophysics Laboratory

SHEBA: Surface Heat Budget of the Arctic (ocean)

SHED: Segmented Hypergraphics Editor

Shef: Sheffield (UK)

SHEF: Standard Hydrologic Exchange Format

SHEFC: Scottish Higher Education Funding Council

Sheff: Sheffield (UK)

SHER: Spectroscopic High-Energy Resolution

SHERB: Sandia Human Error Rate Bank • Spectroscopic High-Energy Resolution Burst

SHERPA: Sources de Haute Énergie Relativistes et Processus d'Accrétion-éject ion Structurée

Shet: Shetland Islands

SHET: Soleil, Héliosphère et Environnement Terrestre

Shetl: Shetland Islands

SHEVE: Southern Hemisphere VLBI Experiment

SHEX: Sundays and Holidays Excepted

SHF: Shanhaiguan (China-PRC) (location identifier) • Short History File • Short-term History File • Superhigh Frequency

SHFE: Space Human Factors Engineering

SHFP: Space Human Factors Program

SHFS: Superhyperfine Structure

SHFT: Shift

SHG: Second-Harmonic Generation • Segmented Hypergraphics • Shungnak (USA-AK) (location identifier)

SHGN: Service Hydrologique et Géologique National

SHGS: Service Hydrologique et Géologique Suisse (Switzerland)

SHH: Shishmaref (USA-AK) (location identifier)

SHI: Shimojishima (Japan) (location identifier)

SHIC: Social History and Industrial Classification

SHIE: Spacecraft Health Insurance Engine

SHIEP: Schur-Horn Inverse Eigenvalue Problem

SHIMS: System Hazard Information Management System

SHINE: Solar, Heliospheric, and Interplanetary Environment

SHIP: Standard Hardware Interface Program

Shipt: Shipment

SHIRL: Serial Holdings in Irish Libraries

SHIVA: SDSS Human Interface and Visualization

SHJ: Sharjah (UAE) (location identifier)

SHK: Schweizerische Hochschulkonferenz (Switzerland) • Sehonghong (Lesotho) (location identifier) • Sternwarte Heiligkreuz

SHL: Science Header Line • Semihard Lander • Shift Logical Left • Shillong (India) (location identifier) • Shoal

SHLB: Simulation Hardware Load Box

Shld: Shield

SHLD: Shield

Shldr: Shoulder

SHLDR: Shoulder

SHLS: Shoals

Shltr: Shelter

SHLW: Shallow

SHM: Nanki Shirahama (Japan) (location identifier) • Shutter Mechanism • Simple Harmonic Motion • Space Hydrogen Maser • Spectromètre Hétérodyne Submillimétrique • Stage Handling Manual

SHME: Single High Magnification Event

SHN: Île Sainte-Hélène (ISO country code) • Saint Helena (ISO country code) • Shelton Sanderson (USA-WA) (location identifier)

SHO: Scheduled Order • Service Hydrographique et Océanographique • Simosato Hydrographic Observatory (Japan) • Sokch'o (Korea-ROK) (location identifier)

SHOALS: Scanning Hydrographic Operational Airborne Lidar Survey

SHOE: Shadow-Hiding Opposition Effect

SHOFM: Système Hydrologique Opérationnel à Fins Multiples

SHOL: Shoal

SHOLAS: Simosato Hydrographic Observatory Laser Ranging System

SHOM: Service Hydrographique et Océanographique de la Marine (France)

SHOOT: Shuttle Helium On-Orbit Transfer • Superfluid Helium On-Orbit Transfer (*)

SHORAN: Short-Range Navigation

SHOS: Segmentation Hiérarchique basée sur l'Optimisation Séquentielle

SHOT: Society for History Of Technology • Space Hardware Optimization Technology

shp: Shaft Horsepower

SHP: Pound (Saint Helena) (ISO currency code) • Qinhuangdao (China-PRC) (location identifier) • Science Header Packet • Selective Hydrogenation Process • Shaft Horsepower • Standard Header Packet • Sustainer Hydraulic Pressure

SHPE: Society of Hispanic Professional Engineers

SHPF: Standard Header Packet File

Shpg: Shipping

Shpng: Shipping

SHPO: State Historic Preservation Officer

Shpt: Shipment

SHQ: Southport (Australia-QLD) (location identifier)

Shr: Share

SHR: Sheridan (USA-WY) (location identifier) • Shift Logical Right • Shore • Solar-Heat Reflecting

SHRD: Shroud • Supplemental Heat Rejection Device

SHRI: Spectrometer for Heavy Recoil Ions • Stereo High Resolution Imager

SHRIF: System for Heuristic Retrieval of Information and Facts

SHRIMP: Shallow-water Real-time Imaging Polarimeter

SHRIP: Stereo High-Resolution Imager Panchromatic

SHRM: Society for Human Resource Management

Shrops: Shropshire (UK-England)

SHRS: Shores

SHRTLY: Shortly

SHS: Sciences de l'Homme et de la Société • Sciences Humaines et de la Société • Sciences Humaines et Sociales • Secure Hash Standard • Shashi (China-PRC) (location identifier) • Simulation Hardware System • Spatial Heterodyne Spectrometer • Spatial Heterodyne Spectroscopy • Special Hardware System

SHSG: Spherical Harmonic Spatial Grid

SHSU: Sam Houston Sate University (USA-TX)

Sht: Sheet

SHT: Shepparton (Australia-VIC) (location identifier) • Solar Horizontal Telescope • Space Hand Tool

ShTC: Short Time Constant

Shtdn: Shutdown

Shtg: Shortage

SHTR: Shutter

SHTTP: Secure Hypertext Transfer Protocol (*) • Secure Hypertext Transport Protocol

SHU: Seton Hall University (USA-NJ) • Smith Point (Australia-NT) (location identifier)

SHUD: Smart Head-Up Display

SHUTDN: Shutdown

shv: Sub Hac Voce • Sub Hoc Verbo

Shv: Sheave

SHV: Safe High Voltage • Safety-Health-Welfare • Shreveport (USA-LA) (location identifier)

SHW: Sharurah (Saudi Arabia) (location identifier) • System High Workstation

ShWav: Shock Waves (periodical code)

Shwr: Shower

SHWR: Shower

SHX: Shageluk (USA-AK) (location identifier)

SHY: Shinyanga (Zimbabwe) (location identifier)

SHZ: Seshutes (Lesotho) (location identifier)

SH2: Supercritical Hydrogen

si: Sauf Imprévu • Singhalese (ISO language code) • Sin Impresión • Slovenia

Si: Silicium • Silicon

S/I: System Integration

SI: Mimas (satellite of Saturn) • Sample Implementation • Sample Interval • Sandwich Islands • Science Instrument • Scientific Information • Scientific Instrument • Search Information • Selected Item • Semiinsulating • Sensitive Information • Sensitivity Index • Senter for Industriforskning • Seriously Ill • Service d'Information • Shetland Islands • Shift-In • Shipping Instructions • Shuttle Integration • Siberia • Signal Interface • Signature Invoke • Site Inspection • Slovenia (ISO country code) • Slovènie (ISO country code) • Smithsonian Institution (USA-DC) • Social Information • Societatis Iesu • Society Island • Society of Indexers (UK) • Solar Inertial • Solar Influence • Solar Intertial • Solubility Index • Source Impedance • Source Index • South Island (New Zealand) • Space Invariant • Spark-Ignited • Spark Injection • Sparse Image • Speaker Independent • Special Instruction • Special Interest • Speckle Interferometer • Speckle Interferometry • Spectrum Index • Square Inch • Staff Inspector • Staten Island (USA-NY) • Station Identification • Status Indicator • Statutory Instrument • Stewart Island (New Zealand) • Storage Immediate • Straight In • Sudden Impulse • Sum Insured • Sustainable Indicator • Sympletic Integrator • Syndicat d'Initiative • Système d'Information • Système International • System Information • System Integration

SIA: Satellite Industry Association (USA-VA) • Science Image Access • Science Information Association • Self-Interstitial Atom • Semiconductor Industry Association • Sensor Interface Assembly • Serial Input Adapter • Service d'Information Aéronautique • Shuttle Induced Atmosphere • Social Impact Assessment • Société des Ingénieurs de l'Automobile • Software Impact Assessment • Solar Inertial Attitude • Speaker Intercom Assembly • Structural Interface Adapter • System Information Agent • System Integration Area • Xi'An Xiguan (China-PRC) (location identifier)

SIAB: Société Industrielle d'Automatisation Bancaire (France)

SiAc: Silicon Accelerometer

SIAC: Science Instrument Aperture Coordinates • Southeastern Iowa Astronomy Club Inc. (USA-IA) • Spatial Information and Analysis Consortium

SIAD: Système Interactif d'Aide à la Décision (*) • Système Interactif d'Aide et de Décision

SIADIS: Sistema de Interpretación Automática De Imagenes de Satelite

SIAE: Salon International de l'Aéronautique et de l'Espace (France)

SIAF: Science Instrument Aperture File

SIAG: SIAM Activity Group

SIAM: Separate Index Access Method • Shipboard Ice Alert and Monitoring • Society for Industrial and Applied Mathematics (USA-PA) • Système d'Information et d'Aide pour les Mutations (France) • System Integrated Access Method

SIAMR: Society for Industrial and Applied Mathematics Review (periodical code)

SIAP: Standard Instrument Approach Procedure • Statistical Institute for Asia and the Pacific • System Integrity Assurance Program

SIAPP: System Integrity Assurance Program Plan

SIAR: Scientific Instrument Analysis Room

SIAS: Scientific Instrument Aperture System

SIASP: Society for Italian-American Scientists and Physicians (USA)

SIASRNR: Scientific and Industrial Association for Space Research of Natural Resources

SIAT: Scene Image Annotation Tape • Shuttle Independent Assessment Team (NASA) • Special Image Annotation Tape

Sib: Siberia • Siberian

SIB: Satellite Integrated Buoy • Satellite Ionospheric Beacon • Science Instrument Branch • Sibiti (Congo) (location identifier) • Simulation Interface Buffer • Small Innovative Business • Sociedade Interplanetaria Brasileira (Brazil) • Software Information Base • Solar Indices Bulletin • Special Investigation Branch • Systems Integration Board • Systems Interface Bus

SIBAN: Sistema Informativo Biblioteche Ateneo Napoletano

SIBEX: Submillimeter Infrared Balloon Experiment

SibIZMIR: Siberian Institute of Terrestrial Magnetism, Ionosphere and Radio Wave Propagation (Russia)

SIBO: Single Board Organiser

sic: Siccus

Sic: Sicilian • Sicily

SiC: Silicon Carbide

SIC: Science Information Council • Scientific Information Center • Second In Command • Semantic Integrity Constraint • Send Interface Clear • Service de l'Information et de la Communication • Service Initiation Charge • Set Initial Condition • Sextant Installation Correction • Silicon Integrated Circuit • Simulated Interface Calibration • Sinop (Turkey) (location identifier) • Software Integration Contractor • Solar Information Center • Spacecraft Identification Code • Special Interest Committee • Specialized Information Center • Standard Industrial Classification • Standard Industrial Code • Standard Instrument Controller • Standards Information Center • Subject Index Classification • Subject Index Code • Subject Information Center • Survey Information Center • Synchronization and Imaging Converter • Synchronization Image Channel • Système d'Informations Commerciales

SICA: Structural Interface Compatibility Assessment

SICAR: Sea Ice Chart Archival

SICAS: SSR Improvement and Collision Avoidance System

SICC: Surveying Instrumentation and Coordinate Computation

SICD: System Interface Control Document

SI/C&DH: Science Instrument Control & Data Handling

SICI: Serial Item and Contribution Identifier

SICL: Standard Instrument Control Library

SICLOPS: Scanner Imagery Correction and Line-by-line Orientation Program

SICM: Scientific Instrument Central Module

SICMT: Space Industry Centre for Microwave Technology (Australia)

SICP: Selected Ion Current Profile

SICS: Scientific Instrument Coordinate System • Scientific Instrument Corrected System • Swedish Institute for Computer Sciences (Sweden) • Swedish Institute of Computer Science (Sweden) (*)

SICSP: Space Industry Centre for Signal Processing

SICT: Serviço de Informação Científica e Técnica

SICWG: SOHO Intercalibration Working Group

Sid: Sidérurgie

SID: Sal (Cape Verde) (location identifier) • Scanned Image Data • Search In Depth • Secondary Imagery Dissemination • Sequence Information Data • Serial Input Data • Service d'Information et de Diffusion • Service d'Information et de Documentation • Session Identification • Session Identificator • Session Identifier • Shuttle Integration Device • Simulator Interface Device • Society for Information Display (USA) • Society for International Development • Software Implementation Directive • Sound Interface Device • Source Identification • Source Identifier • Source Image Distortion • Specification Interpretation Documentation • Standard Instrument Departure • Standard Interface Document • Standard Items Database • Station Identification • Station Identifier • Stored Ion Device • Structure Identifier • Sudden Ionospheric Disturbance • Surface-Induced Disorder • Surface-Induced Dissociation • Symbolic Interactive Debugger • Système d'Information Distribué • Système d'Information Douanier • System for Information on Demand • System Identification • System Identifier • System Interactive Debugger • System Interface Document • System of Information and Documentation

SIDA: Swedish Industrial Development Agency (Sweden) • Swedish International Development Agency (Sweden) (*) • Système Informatisé de Documentation Automatique

SIDAR: Selective Information Dissemination And Retrieval

SIDC: Space Industry Development Centre • Sunspot Index Data Centre

SIDD: Scientific Information and Documentation Division

SIDES: Source Input Data Edit System

SIDF: System-Independent Data Format

SIDL: System Identification List

SIDPC: Service Interministériel de Défense et de Protection Civile

SIDS: Satellite Image Display System • Satellite Imagery Dissemination System • Scientific Instrument Detector System • Screening Information Data Set • Secondary Imagery Dissemination System • Space Investigations Documentation System • Stellar Inertial Doppler System • Support Integrated Data System

SIDT: Store Interrupt Descriptor Table

SIE: Science Information Exchange • Secondary Ion Emission • Selected Information Exchange • Shuttle Interface Equipment • Simple Imaging Extension • Sines (Portugal) (location identifier) • Single Instruction Execute • Society of Industrial Engineers • Sounding Interactive Evaluation • Surface Ionization Engine • System Integration Engineer • System Integration Equipment • System Investigation Equipment

SIEA: Sensor Interface Electronics Assembly

SIEB: Security Impact Evaluation Board

SIECA: Summer Institute for Engineering and Computer Applications

Siegerl: Siegerland (Germany)

SIEP: Screening Inspection for Electronic Parts • Service d'Information sur les Études et les Professions • Structured Inverse Eigenvalue Problem

SIES: Supervision-Inspection-Engineering-Services

SIESTA: Simulation of Interferometric Detector Sensitive to Gravitational Waves

SIF: Satellite Integration Facility • Science Information Facility • Science Instrument Formatter • Scientific Instrument Formatter • Selective Identification Feature • Simra (Nepal) (location identifier) • Società Italiana di Fisica (Italy) • SOGS Integration Facility • Source Input Format • Spannungsintensitätsfaktor • Standard Input Format • Standard Interchange Format • Status Information Frame • Stress-Intensity Factor • Symbol Information File • System Integration Facility

SIFDT: Selected Ion Flow Drift Tube

SIFE: Second ISLSCP Field Experiment • Stage d'Insertion et de Formation à l'Emploi

SIFES: SAR Image Feature Extraction System

SiFoE: Siemens Forschungs und Entwicklungsberichte (periodical code)

SIFT: Selected Ion Flow Tube • Software Implemented Fault Tolerance • Stanford Information Filtering Tool

sig: Signa • Signetur • Siguiente • Siguientes

Sig: Signal • Signal (periodical code) • Signature • Signification • Signifies • Signor • Signore

SIG: San Juan - Isla Grande (Puerto Rico) (location identifier) • SERIN Infrastructure Group • Service d'Information du Gouvernement • Service d'Instrumentation Générale • Signal • Sistema de Informacões Geograficas • Special Interest Group • Strasbourg-Illkirch-Graffenstaden (France) • Système d'Informations de Gestion • Système d'Informations Géographiques • Système d'Informations Géophysiques • Système Intégré de Gestion

SIGACT: Special Interest Group on Automata and Computability Theory (ACM)

SIGADA: Special Interest Group on ADA (ACM)

SIGAL: Système Intégré de Gestion Automatisée du Livre (Belgium)

SIGAPL: Special Interest Group on APL (ACM)

SIGAPP: Special Interest Group on Applied Computing (ACM)

SIGARCH: Special Interest Group on Computer Architecture (ACM)

SIGART: Special Interest Group on Artificial Intelligence (ACM)

SIGB: Standard Interface Glovebox

SIGBIO: Special Interest Group on Biomedical Computing (ACM)

SIGBIT: Special Interest Group on Business Information Technology (ACM)

SIGCAPH: Special Interest Group on Computers and the Physically Handicapped-Print Edition (ACM)

SIGCAS: Special Interest Group on Computers and Society (ACM)

SIGCAT: Special Interest Group on CD-ROM Applications and Technology (ACM)

SIGCG: Special Interest Group on Computer Graphics (ACM)

SIGCHI: Special Interest Group on Computer and Human Interaction (ACM)

SIGCOMM: Special Interest Group on Data Communications (ACM)

SIGCPR: Special Interest Group on Computer Personnel Research (ACM)

SIGCSE: Special Interest Group on Computer Science Education (ACM)

SIGCUE: Special Interest Group on Computer Uses in Education (ACM)

SIGDA: Special Interest Group on Design Automation (ACM)

SIGDOC: Special Interest Group on Documentation (ACM)

SIGFORTH: Special Interest Group on Forth (ACM)

SIGGRAPH: Special Interest Group on Computer Graphics (ACM)

SIGH: Système d'Information Géographique Hydro-océanique

SIGHT: Scintillation Imaging Gas-filled Hard X-ray Telescope

SIGICE: Special Interest Group on Individual Computing Environments (ACM) (formerly SIGSMALL/PC)

sigill: Sigillum

SIGINT: Signal Intelligence

SIGIR: Special Interest Group on Information Retrieval (ACM)

SIGLE: System for Information on Grey Literature in Europe

SIGLINK: Special Interest Group on Hypertext/Hypermedia (ACM)

SIGLIT: Special Interest Group on Library Information Technology (USA)

SIGMA: Science In General management • Scientist's Intelligent Graphical Modelling Assistant • Stochastic-Integration Global Minimization Algorithm • Système d'Imagerie Gamma à Masque Aléatoire • Système d'Information Géographique Maritime • Système Informatique de Gestion du Matériel

SIGMET: Significant Meteorological Phenomenon

SIGMETRICS: Special Interest Group on Measurement and Evaluation (ACM)

SIGMICRO: Special Interest Group on Microprogramming (ACM)

SIGMOD: Special Interest Group on Management of Data (ACM)

Sign: Signature

SIGN: Special Interest Group Network (ACM)

Signa: Signorina

Signal: Signalétique

sig. n. pro.: Signa Nomine Proprio

SIGNUM: Special Interest Group on Numerical Mathematics (ACM)

SIGOIS: Special Interest Group on Office Information Systems (ACM)

SIGOPS: Special Interest Group on Operating Systems (ACM)

SIGP: Scientific Instrument Gas Purge

SIGPLAN: Special Interest Group on Programming Languages (ACM)

SigPr: Signal Processing (periodical code)

Sigra: Signora

SIGRAV: Società Italiana di Relatività Generale e Fisica della Gravitazione (Italy)

SIGS: Simplified Inertial Guidance System • Stellar Inertial Guidance System • Système d'Information Géographique du SHOM

SIGSAC: Special Interest Group on Security, Audit and Control (ACM)

SIGSAM: Special Interest Group on Symbolic and Algebraic Manipulation (ACM)

SIGSIM: Special Interest Group on Simulation (ACM)

SIGSMALL/PC: Special Interest Group on Small and Personal Computing Systems and Applications (ACM) (now SIGICE)

SIGSOFT: Special Interest Group on Software Engineering (ACM)

SIGTRANS: Signal Transmissions

SIGUCCS: Special Interest Group on University and College Computing Services (ACM)

SIGWEX: Significant Weather

SIG3C: Special Interest Group for Computing at Community Colleges • Special Interest Group on Computing in Community Colleges (*)

SIH: Silgadi Doti (Nepal) (location identifier) • Système Informatique Hospitalier

SIHI: Sampled Images - High dispersion (IUE)

SIHRS: Système d'Information Hydrogéologique à Référence Spatiale

SIHT: Space Impact Hand Tool

SII: Enceladus (satellite of Saturn) • Science Information Infrastructure • Sidi Ifni (Morocco) (location identifier) • Space Industries Inc. (USA-TX) • Standards Institution of Israel • Static Invocation Interface • System Internal Interface

SIIA: Software and Information Industry Association (USA-DC)

SIIB: Sistema Integrado de Informação Bibliográfica da Região Centro (Portugal)

SIICT: Sistema Internacional de Información Científica y Ténica

SSID: Système Interactif d'Interrogation à Distance

SIIG: Service Interuniversitaire d'Informatique et de Gestion

SIII: Tethys (satellite of Saturn)

SIIMOD: Scalable Intelligent Image Module

SIIP: Standard Information Interchange Process

SIIR: Smithsonian Institution Information Retrieval (system)

SIIRS: Smithsonian Institution Information Retrieval System

SIIS: Sea Ice Information System

SIIT: Stateless Ip/ICMP Translation • Stateless Ip/ICMP Translator

SIJ: Siglufjördur (Iceland) (location identifier)

SIK: Sikeston (USA-MO) (location identifier)

Sil: Silence

SIL: Sample In Liquid • Satellites International Ltd. • Scanner Input Language • Selected Item List • Sila (Papua New Guinea) (location identifier) • Silver • Single In Line • Societas Internationalis Limnologiae • Sound Interference Level • Speech Interference Level • Sudden Iron Layer • System Implementation Language • Systems Integration Laboratory

SiLA: Six-Letter Acronym

SILEX: Satellite Interlink Experiment • Satellite Interorbit Laser Experiment • Semiconductor Intersatellite Laser Experiment • Semiconductor (laser) Intersatellite Link Experiment

SILMA: Structur Interne Lancement Multiple Ariane

SILO: Sampled Images - Low dispersion (IUE)

SILOON: Scripting Interface Language for Object-Oriented Numerics

SILS: Satellite Interference Location System

SILTS: Shuttle Infrared Leeside Temperature Sensing

sim: Simile

Sim: Similar • Similarly • Simulate • Simulated • Simulation • Simulator

SIM: Satellite Interpretation Message • Scaling Index Method • Science Instrument Module (*) • Scientific Instrument Module • Scientific Instrumentation Module • Secondary Ion Microscope • Secondary Ion Microscopy • Seismometer • Selected Ion Monitoring • Serial Interface Module • Servicio de Información Militar • Simbai (Papua New Guinea) (location identifier) • Simulation • Simulator • Sistema Interamericano de Metrología • Society for Information Management • Solar Interplanetary Model • Solar Irradiance Measurement • Solar Irradiance Monitor • Space Interferometry Mission • Special Interaction Model • Stellar Image Monitor • Subscriber Identification Module • Subscriber Identify Module • Subscriber Identity Module (*) • Survey Information on Microfilm • Systems Integration Model

SIMA: Secondary-mirror Incremental Motor Actuator • Small Incremental Motion Actuator

SIMAI: Società Italiana di Matematica Applicata e Industriale (Italy)

SIMAN: System for Interactive Multispectral Analysis

SIMAP: Système d'Information pour les Marchés Publics

SIMAS: Shuttle Information Management Accountability System

Símb: Símbolo

SIMBA: SEST Imaging Bolometer Array • Solar Irradiance Monitoring from Balloons

SIMBAD: Set of Identifications, Measurements and Bibliography for Astronomical Data • Sistema Municipal de Bases de Datos

SIMC: Surveillance Infrastructure Management Concept

SimCom: Simulation Complex

SIMD: Salon International du Marketing Direct • Single Instruction - Multiple Data

SIMECA: Simulations pour Étoiles Chaudes Actives

SIMEX: Simulation and Extrapolation

SIMFAC: Simulation Facility

SIMKO: Société Immobilière de Kourou (French Guyana)

SIMM: Single In-line Memory Module • SME Internal Market Monitor • Standard In-line Memory Module • Symbolic Integrated Maintenance Manual • System-Integrated Memory Module

SIMO: Stage d'Initiation aux Méthodes d'Observation

SIMOX: Separation for Implantation of Oxygen

SIMP: Satellite Information Message Protocol

SIMPL: Scientific, Industrial, and Medical Photographic Laboratories • Synoptic Interplanetary Monitor Platform

SIMPLE: Semi-Implicit Method for Pressure-Linked Equation • Spatial and Inductive Modelling Package for Land Evaluation

SIMPLEX: Shuttle Ionospheric Modification with Pulsed Localized Exhaust • Standard Incubator Module for Portable Late-access Experiment

SIMPO: Simulation Promotion Office

SIMPR: Structured Information Management, Processing and Retrieval

SIMR: Systems Integration Management Review

SIMS: Sea Ice Monitoring Site • Secondary Ion Mass Spectrometer • Secondary Ion Mass Spectrometry • Secondary Ion Mass Spectroscopy • Selective Interference Modulation Spectrometer • Shuttle Imaging Microwave System • Shuttle Inventory Management System • Stellar Inertial Measurement System • Symbolic Integrated Maintenance System

SIMSAT: Simulation Infrastructure for Modelling Satellites

SIMSWG: Space Interferometry Mission Science Working Group

SIMU: Simulated Inertial Measurement Unit

Simul: Simulation (periodical code)

SIMULA: Simulation

SIMURIS: Solar-system Interferometric Mission for Ultrahigh-Resolution Imaging and Spectroscopy • Stellar Interferometric Mission for Ultrahigh-Resolution Imaging and Spectroscopy

sin: Sine • Sinecure • Sinestra • Sinus

SiN: Silicon Nitride

SIN: Sensitive Information Network • Sinaloa (Mexico) • Sine • Singapore Changi (Singapore) (location identifier) • Support Information Network • Swedish Ionosonde Network

SINAP: Satellite Input to Numerical Analysis and Prediction

SINAPS: Système Intégré d'Acquisition par Satellite • System of Integrated Acquisition Procedures for Satellite

SINAS: Sir Isaac Newton Astronomical Society (USA-PA)

SINC: Site of Important Nature Conservation

SINCHARS: Single-Channel Radio System

SIND: Satellite Inertial Navigation Determination

SINDA: Systems Improved Numerical Differences Analyzer • Systems Improved Numerical Differencing Analyzer • Systems Integrated Numerical Differencing Analyzer

SINE: Spacecraft-Induced and Natural Environment • Summer In Northern Europe

SINET: Schlumberger Information Network

SINEX: Software-Independent Exchange (format) • Solution Independent Exchange

Sing: Singapore • Singular • Singulorum

sinh: Hyperbolic Sine • Sinus Hyperbolic

SINS: Ships' Inertial Navigation System • Sphère Isotherme Non-Singulière • Supernova Intensive Study

SINSS: Shipboard Ice Navigation Support System

SINTEF: Stiftelsen for Industriell og Teknisk Forskning (Norway)

SiO: Silicon Monoxide

SIO: Scalable Input/Output • Scripps Institution of Oceanography (USA-CA) • Serial Input/Output • Service Informatique de l'Observatoire • Smithton (Australia-TAS) (location identifier) • Staged In Orbit • Systems Integration Office

SIOE: Service Information-Orientation-Emploi

SION: Stichting Informatica Onderzoek in Nederland (Netherlands)

SIOP: Single Integrated Operational Plan • Single Integrated Operations Plan

SIP: Satellite Information Processor • Scanner Interface Processor • Scientific Instrument Package • Semi-Infinite Programming • Separation Instrumentation Package • Separation Instrument Package • Service Identification Packet • Session Initiation Protocol • Short Implementation Plan • Simferopol (Ukraine) (location identifier) • Simulated Input Processor • Simulated Input Profile • Single-Image Pattern • Single Inhibition Process • Single In-line Package • Single-Investigator Project • Sky Image Processor • SMDS Interface Protocol • SOC Implementation Plan • Software Implementation Plan • SOGS Invocable Procedure • Solar and Interplanetary Program • Solar Influence Program • Solar Instrument Probe • Solar Irradiance Photometer • Standard Inspection Procedure • Standard Interest Profile • Standard Interface Panel • Standard Interface Plan • Strain Isolation Pad • Strain Isolator Pad • Strategic Information Plan • Supplemental Income Plan • Symbolic Input Program • System Improvement Program • System Interface Processing

SIPAD: Système d'Information, de Présentation et d'Accès aux Données

SIPB: Student Information Processing Board • Scientific Information Policy Board (CERN)

SIPC: Simply Interactive Personal Computer

SIPE: Science Instrument Payload Enclosure • Science Instrument Payload Envelope • Science Instrument Protective Enclosure • Science Instrument Protective Envelope • Scientific Instrument Payload Enclosure • Scientific Instrument Payload Envelope • Scientific Instrument Protective Enclosure • Scientific Instrument Protective Envelope

SIPES: Society of Independent Professional Earth Scientists

SIPHES: Sistema de Predicción de Heladas por Satelites

SIPO: Serial In - Parallel Out • Spacecraft Integration Project Office

SIPOP: Satellite Information Processor Operational Program

SIPP: Single In-line Pinned Package • Single In-line Pin Package • System Information Processing Program

SIPPEREC: Syndicat Intercommunal de la Périphérie de Paris pour l'Électricité et les Réseaux de Communication (France)

SIPRI: Stockholm International Peace Research Institute (Sweden)

SIPS: Satellite Image Processing System • Satellite Imagery Processing System • Satellite Input Processing System • Small Instrument Pointing System • SOGS Invokable Procedures • Spatial Information Processing System • Spectral Image Processing System • State Implementation Plan System

SIPSF: Space-Invariant Point Spread Function

SIPT: Société Internationale de Photogrammétrie et de Télé dé tection

SIPWR: System Integrated Pressurized Water Reactor

SIQ: Singkep (Indonesia) (location identifier)

SIQA: Sistema Integrado da Qualidade do Ambiente (Portugal)

SIR: Safe Integral Reactor • Scientific Information and Retrieval • Scientific Information Retrieval • Selective Item Reporting • Selective Information Retrieval • Semantic Information Retrieval • Serial Infrared • Service d'Informations Rapides • Service Informatique Régional • Shuttle Imaging Radar • Signal-to-Interference Ratio • Signature Identification Record • Sion (Switzerland) (location identifier) • Small Income Relief • Software Initiated Review • Spaceborne Imaging Radar • Space Imaging Radar • Special Information Retrieval • Special Inspection Requirement • Special Investigative Requirement • Specialized Information Retrieval • Spectral Irradiance Response • Statistical Information Retrieval • Subsurface Interface Radar • Surveys, Investigations, and Research (USGS) • Sustainable Information Rate • Sustained Information Rate • Symbolic Input Routine • Système Informatisé de Réservation • System Initialization Routine • System Interface Requirement • Systems Integration Review

SIRA: Scientific Instrument Research Association • Stable Isotope Radio Analysis

SIRC: Science Information Research Center

SIRCA: Simulador Reactivo de Conducción de Automoviles • Stockholm Infrared Camera

SIRD: SOC Implementation Requirements Document • Support Instrumentation Requirements Document

SIRE: Satellite Infrared Experiment • Space Infrared Experiment • Stichting Ideële Reclame

SIREV: Sector Imaging Radar for Enhanced Vision • SIAM Review

SIRGAS: Sistema de Referencia Geocéntrico para América del Sur

SIRH: Système d'Information des Ressources Humaines

SIRI: Service Intégré de Ressources d'Information

SIRIEL: Service Informatique de Renseignements Internationaux Électroniques

SIRIM: Synthetic Infrared Imaging Model • Standards and Industrial Research Institute of Malaysia

SIRIS: Stratospheric Infrared Interferometric Spectrometer

SIRM: Scanning Infrared Microscopy

SIRMO: Senior Information Resource Management Officer

SIRN: Società Italiana di Reti Neuroniche (Italy)

SIROG: Situation-Related Operational Guidance

SIROS: Solar and Infrared Radiation Observation Station

SIRQ: Session Initiation Request

SIRR: Software Integration Readiness Review

SIRS: Satellite Infrared Sounder • Satellite Infrared Spectrometer • Selective Information Retrieval System • Session Initiation Response • Soils Information Retrieval System • Southern International Reference Stars • Système d'Information à Référence Spatiale • System for Information Retrieval and Storage

SIRT: Simultaneous Iterative Reconstruction Technique • Synchronous Intelligent Remote Terminal

SIRTF: Shuttle Infrared Telescope Facility • Space Infrared Telescope Facility (*)

SIRUS: Système Intégré de Réception et d'Utilisation des Satellites

Sis: Sister

SiS: Silicon Sulfide

SIS: Safety Interlock System • Scanning Image Spectrometer • Scanning Imaging Spectrometer • Schengen Information System • School of Information Science • Schwer

Ionen Synchrotron • Science Information Service • Scientific Information System • Scientific Instrument System • Semantic Index System • Semiconducteur-Isolant-Semiconducteur • Semiconductor-Insulator-Semiconductor • Serial Input Special • Shimano Index System • Shock Isolation System • Shuttle Imaging Spectrometer • Shuttle Information Service • Shuttle Information System • Shuttle Integration Schedule • Shuttle Interface Simulator • Simulation Interface System • Simulator Interface System • Singular Isothermal Sphere • Sishen (South Africa) (location identifier) • Society for Interdisciplinary Studies (UK) • Software Implementation Specification • Software Integrated Schedule • Solar Irradiance Spectrometer • Solar Isotope System • Solid-state Imaging Spectrometer • Space and Information System • Spatial Information Services • Spatial Information System • Speaker Intercom System • Special Information System • Specialized Information System • Specification Information System • Sphère Isotherme Singulière • SPOT Imaging Services Pty. Ltd. (Australia-NSW) • Standard Indexing System • Standard Interface Specification • Standardiseringskommissionen i Sverige (Sweden) • Steam Inerting System • Strategic Information System • Strategy Implementation Study • Subarcsecond Imaging Spectrograph • Superconducteur-Isolant-Superconducteur • Superconducting-Isolating-Superconducting • Superconductor-Insulator-Superconductor • Super Index System • Superleiter-Isolator-Superleiter • Swedish Standards Institution (Sweden) • Système d'Information Schengen • Système Informatique de Santé • Systems Integration Schedule • Systems Integration Support

SISAC: Serials Industry Advisory Committee

SISAM: Spectrometer with Interference Selective Amplitude Modulation

SISB: Scientific Instrument Support Branch

SISC: Science Information System Center

SISD: Scientific Information Systems Department • Simulation Interface and Standards Document • Single Instruction - Single Data • Software and Information Systems Division

SISE: Science Instrument Systems Engineer • Système d'Information de Suivi des Étudiants

SISEX: Shuttle Imaging Spectrometer Experiment

SISIC: Science Information Systems Interoperability Conference

SISIR: Singapore Institute of Standards and Industrial Research

SISO: Serial In - Serial Out • Shift In - Shift Out • Single Input - Single Output

SISP: Sudden Increase of Solar Particles • Surface Imaging and Sounding Package

SISS: Scientific Instrument Support Structure • System Integration Support Service

SISSA: Scuola Internazionale Superiore di Studi Avanzati (Italy)

SIST: Space Interferometry Study Team • Stellar Interferometry Study Team

SISTER: Special Institution for Scientific and Technological Education and Research

SISTERS: Space Interferometer for the Search for Terrestrial Exoplanets by Rotation Shearing

SISTO: Software and Intelligent Systems Technology Office

SISWG: Space Interferometry Science Working Group • STS Integrated Schedule Working Group

Sit: Sitting • Situation

SIT: Satellite Interactive Terminal • Scientific Imaging Technology • Sección de Instrumentación y Taller • Self-Ignition Temperature • Self-Induced Transparency • Sequential Interval Timer • Service d'Information Technique • Shuttle Integrated Test • Shuttle Interface Test • Silicon-Intensified Target • Silicon Intensifier Target • Simulation Input Tape • Single Instrument Test • Site Investigation Team • Sitka (USA-AK) (location identifier) • Situation • Small-to-Intermediate-size Telescope • Small-to-Intermediate Telescope • Society of Industrial Technology • Software Integrated Test • Space Impact Tool • Special Information Tone • Spectro-Imageur Transportable • Spontaneous Ignition Temperature • SSV Integrated Test • Stopping In Transit • Storing In Transit • Strategic Implementation Team • Structural Integrity Test • Structured Inventive Thinking • Stuff It • Syndicat des Industries de Télécommunication • Système Interbancaire de Télécompensation (France) • System Initialization Table • System Integration Test • System Integrator Toolkit • Systems Interface Test • Tolar (Slovenia) (ISO currency code)

SITA: Société Internationale de Télécommunications Aéronautiques

SITB: System Integration and Testing Branch

SITC: Standard International Trade Classification

SITE: Satellite Instructional Television Experiment • School In The Exploratorium • Search Information Tape Equipment • Sensor and Instrument Test and Evaluation • Single Ion Track Electronics • Society for Information Technology and Teacher Education • Spacecraft Instrumentation Test Equipment • Superfund Innovative Technology Education

SITEC: Scientifique et Technique

SITEF: Salon International des Technologies et des Énergies du Futur (France)

SITEL: Stage Interdisciplinaire de Télédétection

SITI: Schweizerisches Institut für Technische Information (Switzerland) • Search for Intra-Terrestrial Intelligence

SITN: Stanford Instructional Television Network (USA-CA)

SITP: System Integration Test Plan • System Integration Test Program

SITPB: System Integration Test Program Board

SITS: System Integration Test Service

Sitt: Sitting

Sit. Vac.: Situation Vacant

SITVC: Secondary Injection Thrust Vector Control

SIU: Sequence Initiate Update • Serial Interface Unit • Signal Interface Unit • Siuna (Nicaragua) (location identifier) • SOGS Interface Unit • Southern Illinois University (USA-IL) • Standard Interface Unit

SIUAPS: Service Interuniversitaire des Activités Physiques et Sportives

SIUP: Southern Illinois University Press (USA-IL)

Siv: Sieve

SIV: Dione (satellite of Saturn) • Sample In Vacuum • Selektive Informationsverbreitung • Solar and Interplanetary Variability • Sullivan County (USA-IN) (location identifier)

SIVD: Speaker Independent and Vocabulary Dependent

SIVE: Shuttle Interface Verification Equipment • Spacelab Interface Verification Equipment

SIVIT: Siemens Virtual Touchscreen

SIW: Sektion Informationswissenschaft • Sibisa (Indonesia) (location identifier) • Station-In-Wolke

SIWG: Simulation Interoperability Working Group

SIX: Phoebe (satellite of Saturn) • Singleton (Australia-NSW) (location identifier)

SIY: Montague - Siskiyou County (USA-CA) (location identifier)

SIZ: Seasonal Ice Zone • Sissano (Papua New Guinea) (location identifier)

SIZEX: Seasonal Ice Zone Experiment

sj: Sub Judice

SJ: Jan Mayen Islands (Svalbard and_) (ISO country code) • San Jose (USA-CA) • San Juan (Argentina) • Slip Joint • Societas Jesu • Societatis Jesu • Society of Jesus • Supersonic Jet • Svalbard and Jan Mayen Islands (ISO country code)

SJA: San Juan (Peru) (location identifier) • Société Jurassienne d'Astronomie (Switzerland)

SJAA: San Jose Astronomical Association (USA-CA)

SJAC: Society of Japanese Aerospace Companies • South Jersey Astronomy Club (USA-NJ)

SJAM: SIAM Journal of Applied Mathematics (periodical code)

SJB: San Joaquin (Bolivia) (location identifier)

SJC: San Jose (USA-CA) (location identifier) • Silicon Valley (USA-CA) (location identifier) • Slit Jaw Camera • Spacecraft Jettison Controller • Standing Joint Committee • Supreme Judicial Court

SJCO: SIAM Journal of Control Optimization (periodical code)

SJD: Doctor of Juridical Science • San José del Cabo (Mexico) (location identifier)

SJE: San José del Guaviare (Colombia) (location identifier)

SJF: Saint John Island (US Virgin Islands) (location identifier)

SJG: San Pedro - Jagua (Colombia) (location identifier)

SJH: San Juan del Cesar (Colombia) (location identifier)

SJI: San Jose (Philippines) (location identifier) • San Juan Institute (USA-AZ)

SJJ: Sarajevo (Bosnia-Hercegovina) (location identifier)

SJK: São José dos Campos (Brazil-SP) (location identifier)

SJL: São Gabriel da Cachoeira (Brazil-AM) (location identifier)

SJM: Jan Mayen Islands (Svalbard and_) (ISO country code) • San Juan de Maria (Dominican Rep.) (location identifier) • Special Joint Meeting • Svalbard and Jan Mayen Islands (ISO country code) • System Junction Module

SJMA: SIAM Journal of Mathematical Analysis (periodical code)

SJMN: San Jose Mercury News (USA-CA)

SJN: Saint Johns (USA-AZ) (location identifier)

SJNA: SIAM Journal of Numerical Analysis (periodical code)

SJO: San José - Juan Santamaria (Costa Rica) (location identifier)

SJP: São José do Rio Preto (Brazil-SP) (location identifier)

SJPMEA: Si Je Puis M'Exprimer Ainsi

SJQ: Sesheke (Zambia) (location identifier)

SJR: San Juan (Colombia) (location identifier) • Software Justification Report

SJS: San José (Bolivia) (location identifier)

SJSA: Sun-Jupiter-Saturn-Asteroid

SJSU: San Jose State University (USA-CA)

SJT: San Angelo - Mathis (USA-TX) (location identifier)

SJTI: Service Juridique et Technique de l'Information

SJTU: Shanghai Jiao Tong University (China-PRC)

SJU: San Juan - Luis Muñoz Marin (Puerto Rico) (location identifier)

SJUSD: San Jose Unified School Districy (USA-CA)

SJV: San Javier (Bolivia) (location identifier)

SJW: Shijiazhuang (China-PRC) (location identifier)

SJX: Sartaneja (Belize) (location identifier)

SJY: Seinäjoki Ilmajoki (Finland) (location identifier)

SJZ: São Jorge Island (Azores) (location identifier)

sk: Slovak (ISO language code) • Slovakia

Sk: Sack • Sanduleak • Sick • Sink • Sketch

SK: Sachkatalog • Saskatchewan (Canada) • Service Kit • Slovakia (ISO country code) • Slovaquie (ISO country code) • Station Keeping • Stockport (UK postal code) • Stransky-Krastonov

SKA: Spokane Fairchild (USA-WA) (location identifier) • Square Kilometer Array

SKAI: Square Kilometer Array Interferometer

SKAMP: Station Keeping And Mobile Platform

SKAT: Science, Knowledge, And Technology

SKB: Nevis (Leeward Islands) (location identifier) • Saint Kitts (Leeward Islands) (location identifier) • Stadt- und Kreisbibliothek

SKBO: Scattered Kuiper Belt Objects

SKC: Sky Clear • Suki (Papua New Guinea) (location identifier)

SKD: Samarkand (Uzbekistan) (location identifier)

SKDR: S/Ka-band Data Relay

SKE: Skien (Norway) (location identifier) • Software Key Escrow

SKED: Schedule

SKEME: Secure Key Exchange Mechanism

S. Ken.: South Kensington (UK-England)

SKEPP: Studiekring voor Kritische Evaluatie van Pseudowetenschap en het Paranormale (Belgium)

SKF: San Antonio - Kelly (USA-TX) (location identifier)

SKG: Thessaloniki (Greece) (location identifier)

SKH: Surkhet (Nepal) (location identifier)

SKI: Skikda (Algeria) (location identifier)

SkICAT: Sky Image Cataloging and Analysis Tool

SKICAT: Sky Image Cataloging and Analysis Tool

SKIE: Second-Kind Integral Equation

SKIP: Simple Key for Internet Protocol

SKIRT: Shuttle Kinetic Infrared Test • Spacecraft Kinetic Infrared Test

SKJ: Sitkinak Island (USA-AK) (location identifier)

SKK: Koruna (Slovakia) • Shaktoolik (USA-AK) (location identifier)

SKL: Isle of Skye (UK) (location identifier)

Sklt: Skylight

Skm: Stockholm (Sweden)

SKM: Skeldon (Guyana) (location identifier)

SKN: Stokmarknes (Norway) (location identifier)

SKO: Sokoto (Nigeria) (location identifier)

SKP: Skopje (Macedonia) (location identifier)

SKPT: Skeleton Cursor Table

SKQ: Sekakes (Lesotho) (location identifier)

Skr: Sanskrit

SKr: Swedish Krona (currency)

SKR: Saturn Kilometric Radiation • Shakiso (Ethiopia) (location identifier)

Skrt: Sanskrit

SKRT: Square Kilometer Radio Telescope

SKS: Skrydstrup (Denmark) (location identifier) • Vojens (Denmark) (location identifier)

Skt: Sankt • Sanskrit • Skirt • Socket

SKT: Sialkot (Pakistan) (location identifier)

Sktd: Skirted

SKU: Seamless Knowledge Universe • Skiros (Greece) (location identifier) • Sri Krishnadevaraya University (India) • Stock Keeping Unit

SKUB: Schweizerische Kommission für Umweltbeobachtung (Switzerland)

SKV: Santa Katerina (Egypt) (location identifier)

SKW: Skwentna (USA-AK) (location identifier) • Structured Keyword

SKWOC: Structured Keyword Out of Context

SKWY: Skyway

SKX: Saransk (Russia) (location identifier)

SKY: Sandusky Griffing (USA-OH) (location identifier)

SKYLAB: Sky Laboratory

SKZ: Sukkur (Pakistan) (location identifier)

s/l: Su Letra

sl: Secundum Legem • Sine Loco • Sin Lugar • Slovenian (ISO language code) • Sus Labores

Sl: Sleet • Sliding • Slight • Slightly • Slip • Slovakia • Slovakian

S/L: Sea Level • Shop/Laboratory • Space Laboratory

S&L: Savings & Loans

SL: Safety Limit • Sample Laboratory • Sampling Line • Sandqvist-Lindroos • San Luis (Argentina) • Sans Lieu • Savings and Loans • Scanning Slit • Scripting Language • Seal • Sea Level • Secundum Legem • Self-Linearized • Service Layer • Service Lecteurs • Shelf Life • Shoemaker-Levy • Shop Laboratory • Sick Leave • Sierra Leone (ISO country code) • Signal Level • Simulation Language • Sine Loco • Single Line • Single Lobe • Slavic • Slavonic • Slough (UK postal code) • Slovenia (international aircraft registration code) • Sociedad Limitada • Sound Level • Source Language • South Latitude • Spacelab • Space Laboratory • Space Launch • Specialized Lexicon • Special Linear • Standard of Living • StarLabels • Star Line • Start Line • Statistical List • Storage Location • Straight Line • Streamline • Strong Line • Superlattice • Support Line • System Language

Sla: Slovak

SLA: Salta - General Belgrano (Argentina-SA) (location identifier) • Satellite Linear Accelerometer • Scanning Laser Altimeter • Scottish Library Association • Second-Language Acquisition • Semiconductor Laser Amplifier • Service Level Agreement • Shuttle Laser Altimeter • Société Lorraine d'Astronomie (France) • Special Landscape Area • Special Libraries Association (USA-DC) • Square Loop Antenna • Statistical Local Area • Straight Line Approximation • Superlight Ablative • Support and Logistics Area • Synchronous Line Adapter

SLAC: Solar-Lunar Almanac Core • Stanford Linear Accelerator Center (USA-CA)

SLAD: System Logic and Algorithm Development

SLAHTS: Stowage List And Hardware Tracking System

SLAK: Spacelab Late Accessory Kit

SLALOM: Space Laser Low-Orbit Mission

SLAM: Scanning Laser Acoustic Microscope • Scanning Laser Acoustic Microscopy • Scanning Low Acquisition Manoeuvre • Side Load Arrest Mechanism • Simulation Language for Alternative Modelling • Station Logistics And Maintainability

SLAMAP: Space Laser Mission Analysis Program

SLAMMR: Side-Looking Airborne Multimission Radar

slan: Sine Loco, Anno, vel Nomine

SLAP: Soils Landscape Analysis Project • Solar Lunar Almanac Program • Space Link Access Protocol • Supplemental Low Altitude Photography • SUPREMUM Linear Algebra Package • Symbolic Language Assembly Program

SLAR: Side-Looking Airborne Radar

SLART: Simultaneous Left And Right Truncation

SLAS: Saint Louis Astronomical Society (USA-MO) • Salt Lake Astronomical Society (USA-UT) • Society for Latin American Studies

SLASO: Space Licensing And Safety Office (Australia)

S. Lat.: South Latitude

SLAT: Spacecraft Location/Attitude Tape • Supersonic Low-Altitude Target

SLATE: Second Line Automatic Test Equipment • Structural Linguistic Analysis and Text Evaluation

SLATEC: Sanida Los Alamos Technical Exchange Committee

Slav: Slavic • Slavonia • Slavonian • Slavonic

SLB: Îles Salomon (ISO country code) • Schweizerische Landesbibliothek (Switzerland) • Second Level Buffer • Server Load Balancing • Solomon Islands (ISO country code) • Storm Lake (USA-IA) (location identifier) • Strong Light Bridge

SLBD: Sea-Light Beam Director

Slbl: Soluble

SLC: Salt Lake City (USA-UT) (location identifier) • Scan Line Corrector • Sea Level Change • Single Launch Contractor • Single-Layer Ceramic • Single Line Control • SLAC Linear Collider • Space Launch Complex • Spacelab Control • Standard Launch Complex • Stanford Linear Collider • Statute Law Committee • Subscriber Line Charge • Subscriber Loop Carrier

SLCA: Spacelab Contract Administration

SLCL: Spacelab Configuration Layout

SLCM: Spacelab Configuration Management • Spacelab Contract Management

SLCS: Software Library Control System

SLCSAT: Submarine Laser Communications Satellite

Sld: Sailed • Sealed • Sold • Solid

SLD: Sea Level Datum • Self-Locking Device • Side-Looking Detector • Simulated Launch Demonstration • SLAC Large Detector • SLC Large Detector • Slope Line Detection • Solid • Stanford Large Detector • Stiff-Leg Derrick • Supercooled Large Droplet • Superluminescent Diode

SLDC: Synchronous Data Link Protocol

SLDI: Sector List Drop Interval

SLDMS: SAR/Landsat Data Mapping System • SAR/Landsat Data Merging System

SLDP: Spacelab Double Pallet

SLDPF: Spacelab Data Processing Facility

Sldr: Solder

SLDS: Scanning Laser Doppler System • Swedish Land Information System

SLDT: Store Local Descriptor Table

SLDV: Scanning Laser Doppler Velocimeter

SLE: Salem (USA-OR) (location identifier) • Screening Line Editor • Serveur Logique d'Édition • Sierra Leone (ISO country code) • Small Lattice Experiment • Spacelab Engineering • Station Liaison Engineer

SLED: SBV Language for Experiment Design • Single Large Expensive Disk • Software Life Cycle Empirical Database • Space Laser Experiment Definition • Stable Large E-mail Database • State Libraries Electronic Doorway

SLEE: Service Logic Execution Environment

SLEL: Service Logic Execution Language

SLEM: Solution of Linearized Equations of Motion

SLEMSA: Soil Loss Estimation Modern for Southern Africa

SLEMU: Spacelab Engineering Model Unit

SLEP: Service Life Extension Programme

SLF: Scientific Laboratory Facility • Service de la Législation Fiscale • Shuttle Landing Facility • Site Listing Filter • Standard Linear Format • Straight Line Frequency • Sulayel (South Africa) (location identifier) • System Library File

SLFC: Selectable Low-Frequency Clock • Supersonic Laminar Flow Control

SLFCS: Survivable Low-Frequency Communication System

SLG: Sloam Springs - Smith (USA-AR) (location identifier) • Symbolized Line Graph • Synthetic Line Generation

SLGT: Slight

SLGTLY: Slightly

SLH: Sola (Vanuatu) (location identifier)

SLHS: Sum of Left-Hand Side

SL&I: System Load & Initialization

SLI: Solwezi (Zambia) (location identifier) • Spacelab Integration

SLIAKI: Számitástechnikai és Automatizálási Kutató Intézete (Hungary)

SLIC: Simple Low-cost Innovative Concept • Spectral Line Imaging Camera • State Land Information Council • Subscriber-Line Interface Card • System-Level Integrated Circuit • System Level Interface Compatibility • System Licensed Internal Code

SLIES: Stratospheric Limb Infrared Emission Spectrometer

SLIM: Scammed Longwave Infrared Module • Semiempirical Low-latitude Ionospheric Model • Simpler Legislation for the Internal Market

SLIP: Serial Line Interface Protocol • Serial Line Internet Protocol (*) • Serviceability Level Indication Processing • Simulation Language for Information Processing • Software for Landsat Image Processing • Symmetric List Interpretive Program • Synchronous Link Internet Protocol

SLIPAR: Short Light Pulse Alerting Receiver

SLIPS: Scattered Light Intensity Profile Sensor

SLIS: School of Library and Information Science • Shared Laboratory Information System

SLISIR: Sri Lanka Institute of Scientific and Industrial Research

SLISR: Stanford-Lockheed Institute for Solar Research (USA-CA)

SLJ: Chandler Stellar (USA-AZ) (location identifier) • Sobre Loja

Slk: Slovak

SLK: Saranac Lake (USA-NY) (location identifier) • Sichtlochkarte

SLKT: Survivability, Lethality, and Key Technologies

SLL: Leone (Sierra Leone) (ISO currency code) • Salalah (Oman) (location identifier) • Shelf Life Limit

SLLN: Strong Law of Large Numbers

SLLW: Solid Low-Level Waste

SLM: Salamanca Matacán (Spain) (location identifier) • Senior Level Management • Service Level Management • Service Level Monitor • Servicio Local Medido • Simulated Laboratory Module • Sound Level Meter • Space Laboratory Module • Spatial Light Modulator • Structure Line Model • Sul Livello del Mare • System Level Manager

SLMC: Source Locking Microwave Counter

Slmnt: Seulement

SLMP: Stray Light Modelling Program

SLMR: Silly Little Modem Reader

Sln: Slovene

SLN: Salina (USA-KS) (location identifier) • Science Learning Network

slnd: Sine Loco Nec Data

SLND: Sans Lieu Ni Date

SLNDF: Skewed Log Normal Distribution Function

Slo: Slovenian

SLO: Salem Leckrone (USA-IL) (location identifier) • Saturn-Like Object • Senior Liaison Officer • Society of Low-energy Observers (USA-TN) • Spacelab Only • Spacelab Operations

SLOC: Software Lines of Code • Source Lines of Code

SLODB: Spacelab Operational Databook

SLOPE: Study of Lunar Orbiter Photographic Evaluation

SLOS: Star Line Of Sight • Sun Line Of Sight

SLOSH: Sea, Lake and Overland Surge from Hurricane

SLOT: Solar Luminosity Oscillations Telescope

SLOTH: Strange Loop Or Tangled Hierarchy

Slovak: Slovakia • Slovakian

slp: Sine Legitime Prole

Slp: Slip • Slope

SLP: San Luis Potosí (Mexico) (location identifier) • Sea Level Pressure • Seismic Line Program • Service de Lecture Publique • Service Location Protocol • Service Logic Programme • Significant Lexical Pattern • Silicon Light Pulser • Slope • Slopes • Source Language Processor • Spacelab Pallet • Spacelab Payload • Spacelab Program • Sturm-Liouville Problem • Superlong Play

SLPB: Spacelab Program Board

SLPO: Spacelab Program Office

SLPOCC: Spacelab Payload Operations Control Center

SLQ: Sleetmute (USA-AK) (location identifier)

SLR: Satellite Laser Ranger • Satellite Laser Ranging (*) • Scanning Laser Radar • Sea Level Rise • Side-Looking Radar • Single-Lens Reflex • Solar Laser Ranging • Spacelift Range • Spin-Lattice-Relaxation • State and Local Relations • Sulphur Springs (USA-TX) (location identifier) • Symbolic Logic Retrieval

SL Rs: Sri Lanka Rupee (currency)

SLRS: Satellite Laser Ranging System

SLRU: Sequential Least Recently Used

SLRV: Shuttle Launch Research Vehicle • Surveyor Lunar Roving Vehicle

SLS: Satellite Landing System • Saturn Longitude System • School of Library Science • Secondary Landing Site • Sequential Least Squares • Side Lobe Suppression • Silicon Light Source • Silistra (Bulgaria) (location identifier) • Sodium Lauryl Sulphate • Software Loadable System • Solid-state Linear Sensor • Sortie Laboratory Simulator • Source Library System • Southern Loop System • Southern Luminous Stars • Spacelab Life Sciences • Spacelab Simulator • Space Launch System • Space Life Sciences • Specific Living Space • Statement Level Simulator • Stereo Line Scanner • Strained Layer Superlattice • Structure of the Lower Stratosphere • Submillimeter Limb Sounder • Surface Laboratory System

SLSA: Shuttle Logistics Support Aircraft

SLSC: Saint Louis Science Center (USA-MO)

SLSG: Scanning Laser Slope Gauge • Suit Loop Stimuli Generator

SLSI: Sri Lanka Standards Institution • Super Large-Scale Integration

SLSMS: Spacelab Support Module Simulator

SLSS: Secondary Life Support System • Super Large Scale Structure • Systems Library Subscription Service

SLSTP: Space Life Sciences Training Program

Slt: Sleet

SLT: Salida (USA-CO) (location identifier) • Second Level Trigger • Self-Loading Tape • Shuttle Loop Transit • Simulated Launch Test • Sleet • Solid Logic Technology • Spacelab Technology • Special Launch Trajectory • Specific Launch Trajectory • Superlight Telescope • Système de Localisation et de Trajectographie

Sltd: Slotted

SLTDP: Special Laser Technology Development Program

SLTE: System Level Test Equipment

SLTV: System Level Thermal Vacuum

SLU: Saint Lawrence University • Saint Louis University (USA-MO) • Saint Lucia - Vigie (Windward Islands) (location identifier) • Secondary Logical Unit • Southeastern Louisiana University (USA-LA)

SLUG: Starlink Local Users Group • Superconducting Low-inductance Undulating Galvanometer

SLUGS: SCUBA Local Universe and Galaxy Survey

SLUK: Spacelab Utility Kit

SLUP: Spacelab Utilisation Programme

SLURP: Simple Linear Unit Reservoir Program

Slv: Sleeve

SLV: El Salvador (ISO country code) • Salvador (El_) (ISO country code) • Satellite Launching Vehicle • Saturn Launch Vehicle • Simla (India) (location identifier) • Simulated Launch Vehicle • Soft Landing Vehicle • Southern Launch Vehicle • Space Launch Vehicle • Sport-Luxury Vehicle • Standard Launch Vehicle

Slvg: Salvage

SLVG: Special Launch Vehicle Group

Slvt: Solvent

SLW: Saltillo (Mexico) (location identifier) • Slow • Solid Low-level Waste • Spectral Line Width • Stichting Levenwetenschappen (Netherlands) • Store Logical Word • Straight-Line Wavelength • Supercooled Liquid Water

Slwstr: Solwaster (Belgium)

SLWT: Super Lightweight Tank

SLX: Salt Cay (Turks & Caicos Islands) (location identifier)

Sly: Slowly • Southerly

SLY: Salehard (Russia) (location identifier)

SLZ: São Luís Tirrical (Brazil-MA) (location identifier)

SL9: Shoemaker-Levy 9

sm: Samoan (ISO language code) • Seemeile • Small

Sm: Small • Samarium (formerly Sa)

S&M: Stock & Machinery • Structural & Mechanical

S/M: Scheduled Maintenance • Service/Maintenance • Service Module • Simulation • Structural/Mechanical • Systems Management

SM: Master of Science • Safe Mode • Saint Marin (ISO country code) • Sales Manager • Sa Majesté • Samoyed • Sanctae Memoriae • San Marco • San Marino (ISO country code) • Scan Mechanism • Scheduled Maintenance • Scientific Management • Scientific Memorandum • Secondary Memory • Secondary Mirror • Seine Majestät • Seismometer • Semimonthly • Senior Magistrate • Sequence Monitor • Serial Magnitude • Service Maintenance • Service Manual • Service Mark • Service Mode • Service Module • Servicing Mission • Session Monitor • Set Mode • Shared Memory • Shipment Memorandum • Shopping Mall • Short Module • Shuttle Management • Silver Medal • Simulation Model • Sinus Medii (Moon) • Software Manual • Solar Magnetic • Solar Magnetospheric • Special Memorandum • Specification Memorandum • Stabilized Member • Stable Member • Staff Manager • Staff Memorandum • Standard Model • Station Manager • Status Mile • Statute Mile (*) • Storage Manager • Structural Mechanics • Structural Model • Sua Majestà • Subcontract Management • Successful Mission • Su Majestad • Summary Memorandum • Super Mongo • Supply Manual • Support Module • Surface Mount • Survey Manual • Suspended Matter • Sutton (UK postal code) • Synchrotron Maser • System Management • System Manager • System Mechanics • System Memory • System Monitoring • Systems Management

SMA: Safety and Mission Assurance • Santa Maria (Azores) (location identifier) • Schweizerische Meteorologische Anstalt (Switzerland) • Science Monitoring Area • Secondary Mirror Actuator • Secondary Mirror Assembly • Semimajor Axis • Service Management Architecture • Shape Memory Alloy • Simulation Model Adapter • Site Maintenance Area • Slave Manipulator Arm • Small-Angle Maneuver • Sociedad Malagueña de Astronomía (Spain) • Solar-Maximum Analysis • State Meteorological Administration • Submillimeter Array • Super Modified Achromat • Support Management Area • Surface Movement Advisor • Survey and Mapping Alliance • Système Mécanique Automatisé • System Management Architecture

SMAAC: Southern Michigan Amateur Astronomy Conference

SMAB: Solid Motor Assembly Building

SMAC: Scene Matching by Area Correlation • Sequential Multiple Analysis Computer • Serial Memory Address Counter • Simplified Method for Atmospheric Correction • Simultaneous Multiframe Analytical Calibration • Spacecraft Maximum Allowable Concentration • Space Maximum Allowable Concentration • State Mapping Advisory Committee • State Mapping Advisory Council • Stellar Mass Catalogue • Streaming Motions of Abell Clusters • System Monitor And Control

SMACK: Streaming Motions of Abell Cluster Kernels

SMAG: Star Magnitude

SMAGLE: Stationary Magnetic Levitation Experiment

SMAI: Société de Mathématiques Appliquées et Industrielles (France)

SMAL: Structural Microassembly Language

SMAP: Short and Medium-term Action Programme

SMARA: Safety and Mission Assurance Risk Assessment • Schenectady Museum Amateur Radio Association (USA-NY) • Servicio de Meteorología de la Armada de la Republica Argentina • Surface Mining and Reclamation Act (USA)

SMArc: Solid Mechanics Archives (periodical code)

SMARF: Solid Motor Assembly and Readiness Facility

SMART: Science and Mathematics Achievement Required for Tomorrow • Science, Mathematics, Aerospace, Research and Technology Inc. • Science, Mathematics And Relevant Technology • Self-Monitoring, Analysis, and Reporting Technology • Sequential Mechanism for Automatic Recording and Testing • Simulator for Multi-Radar Analysis for Realistic Traffic • Simultaneous Multispectral Absolute Radiometer and Transmissometer • Small Ada Run Time • Small Mission for Advanced Research in Technology • Smooth Multiple Additive Regression Technique • Sobolev Modified Approximate Radiative Transfer • Software Metering And Resource Tracking • Space Maintenance And Repair Technique • Storage Modification and Retrieval Transaction • Stereo Mapping with Radar Techniques • Stretched Mesh Attached to Rope Trusses • Structural Maintenance And Repair Team • Submillimeter Array of Taiwan (China-ROC) • System for Management and Allocation of Resources Technique • System for Manipulation And Retrieval of Text • System for Mechanical Analysis and Retrieval of Text • Systems Management Analysis, Research and Testing

SMARTS: Shared Memory Asynchronous Runtime System • Simulation, Monitoring, Analysis, Reducation, and Test System • Status Memory And Real-Time System

SMAS: Smoky Mountain Astronomical Society (USA-TN) • Southern Maryland Astronomical Society (USA-MD) • Southwest Montana Astronomical Society (USA-MT)

SMASS: Small Main-belt Asteroid Spectroscopic Survey

SMASSIR: Small Main-belt Asteroid Spectroscopic Survey in the Infrared

SMATH: Satellite Materials And Target Hardening

SMATV: Satellite Master Antenna Television • Small Master Antenna Television

SMB: Cerro Sombrero (Chile) (location identifier) • Secondary Mirror Baffle • Server Message Block • Service Message Block • Session Message Branch • Society for

Mathematical Biology • Space Meteorology Branch • Submillimeter Background • Surveys and Mapping Branch • System Management Bus • Systems Management Branch

Smbdy: Somebody

SMBH: Supermassive Black Hole

Smthg: Something

SMBus: System Management Bus

SMC: Safe-Mode Command • Safe-Mode Computer • Santa Maria (Colombia) (location identifier) • Scientific Manpower Commission • Sequential Machine Controller • Serial Magnitude Command • SFOC Monitor and Control • Sheet Molding Compound • Small Magellanic Cloud • Special Mapping Center • Speech and Music Compression • Spin Myon Collaboration • Standard Mean Chord • Standard Microsystems Corporation • Stepping Motor Controller • Stichting Matematisch Centrum (Netherlands) • Strategic Manufacturing Center • Super-Multi-Coating • Surface Material Class • System Management Center • System Monitor and Control • Systems, Man and Cybernetics

SMCAT: Strong Motion Catalog

SMCC: Shuttle Mission Control Center • Simulation Monitor and Control Console • Sun Microsystems Computer Corporation

SMCD: Simple Code Multiple Data

SmCES: Smithsonian Contributions to Earth Sciences (periodical code)

SMCH: Standard Mixed Cargo Harness

SMCL: Secondary Maximum Contaminant Level

SMCM: Small Module Core Memory

SMCRA: Surface Mining Control and Reclamation Act (USA)

SMCS: Secondary Mirror Control System • Separation Monitor and Control System

SMCU: Separation Monitor Control Unit • Separation Monitoring Control Unit

SMCV: Side-Mount Cargo Vehicle

S&MD: Science & Mission Design

SMD: Fort Wayne - Smith (USA-IN) (location identifier) • Scanning Microdensitometer • Scheduling Management Display • Science Mission Division • Scientific Missions Division • Shop Manufacturing Development • Silicon Microstrip Detector • Singular Multinomial Distribution • Soil Moisture Deficit • Sonntag-Montag-Dienstag • Spacelab Mission Development • Special Measuring Device • Standard Magnet Dewar • Storage Module Device • Structures and Mechanics Division • Surface Mount Design • Surface Mount Device • Surface-Mounted Device • Surveys and Mapping Division • System Management Device • System Management Directive • System Measuring Device • Systems and Mission Division • Systems Monitor Display

SMDB: Summary Database

SMDC: Shielded Mild Detonation Cord • Superconductive Materials Data Center

SMDE: Static Mercury Drop Electrode

SMDF: Soil Moisture Deficit Function

SMDL: Standard Musical Description Language • Standard Music Description Language

SMDM: Small Multiplexer/Demultiplexer

SMDR: Station Message Detail Recording

SMDS: Switched Multimegabit Data Service

SMDSM: Système Mondial de Détresse et de Sécurité en Mer

SME: Sakharov's Metric Elasticity • School of Mechanical Engineering • Seismic Margin Earthquake • Sistema Monetario Europeo • Small and Medium Enterprises • Small and Medium-sized Enterprises • Society of Mining Engineers • Solar Mesosphere Explorer • Somerset - Pulaski County (USA-KY) (location identifier) • Stellar Mass Ejection • Subject Matter Expert • Super-Maximum Entropy • Suriname (international vehicle registration code) • Système de Management de l'Environnement • Système de Management Environnemental • Système Monétaire Européen • System Management Environment

SMEA: Safe Mode Electronics Assembly

SMEGMA: Sophisticated Modern Editor with Gloriously Magnificent Abilities

SMEI: Solar Mass Ejection Imager

SMES: Shuttle Mission Engineering Simulator • Shuttle Mission Evaluation Simulation • Superconducting Magnetic Energy Storage

SMET: Science, Mathematics, Engineering and Technology • Shuttle Mission Engineering Simulator • Shuttle Mission Evaluation Simulator • Simulated Mission Endurance Test • Superconducting Magnetic Energy Storage

SMETE: Science, Mathematics and Technology Education

SmEX: Small Explorer

SMEX: Small Explorer

SMF: Sacramento (USA-CA) (location identifier) • Sociedad Mexicana de Física (Mexico) • Société Météorologique de France • Solar Magnetic Field • Space Manufacturing Facility • Special Modifying Factor • Standardized Monitoring Framework • Standard Message Format • Static Magnetic Field • Strong Magnetic Field • System Management Facility • System Management Framework • System Manager Facility

SMG: Santa Maria (Peru) (location identifier) • Spaceflight Meteorology Group • Special Mobile Group • Super Master Group

SMGC: Surface Movement Guidance and Control

SMGCS: Surface Movement Guidance and Control System

SMH: Sapmanga (Papua New Guinea) (location identifier) • Service de Métrologie Habilité • Spectromètre à Modulation Hétérodyne • Spectromètre Millimétrique Hétérodyne

SMHI: Swedish Meteorological and Hydrological Institute

SMHN: Servicio de Meteorología e Hidrología Nacional (El Salvador)

SMHP: Summer Microwave Hurricane Program

SMI: Sa Majesté Impériale • Samos (Greece) (location identifier) • Servo Moteur Intégré • Sistema Monetario Internacional • Slovak Metrology Inspectorate • Sortie Mission Integrator • Standard Measuring Instrument • Static Memory Interface • Structure of Management Information • Su Majestad Imperial • Svensk Materialinformation (Sweden) • Swiss Meteorological Institute • System Management Interrupt

SMIA: Single Multiplexer Interface Assembly

SMIC: Salaire Minimum Interprofessionnel de Croissance • Sortie Mission Integration Contractor • Study of Man's Impact on Climate • Surveying and Mapping Industry Council

SMIF: Standard Mechanical Interface

SMIFTS: Spatially-Modulated Imaging Fourier Transform Spectrometer

SMIIS: Solar Microwave Interferometer Imaging System

SMIL: Statistics and Market Information Library • Synchronized Multimedia Integration Language

SMILE: Spherical Micro-Integrated Lens

SMILI: Superconducting Magnetic Instrument for Light Isotopes • Superconducting Magnet Instrument for Light Isotopes • Synthetic Modal Interferometric Laser Imaging

SMILS: Submillimeter IR Line Survey

SMIM: Submillimeter Intermediate Mission

SMIPS: Small Interactive Image Processing System

SMIR: Service Météorologique Interrégional • Shuttle Multispectral Infrared Radiometer

SMIRAG: Service Météorologique Interrégional Antilles-Guyane (France)

SMIRO: Service Météorologique Interrégional Ouest (France)

SMIRR: Shuttle Multispectral Infrared Radiometer

S-MIS: Single-Mission Integrating System

SMIS: Single-Mission Integrating System • Single-Mission Integration System • Standards and Metrology Institute of Slovenia • STDN Management Information System

SMIT: Systems Management Interface Tool

SMJ: Sim (Papua New Guinea) (location identifier)

Smk: Smoke

SMK: Saint Michael (USA-AK) (location identifier) • Smoke • Software Migration Kit

Sml: Simulate • Simulation • Simulator • Small

SML: Science Museum Library • Search Mode Logic • Serials Master List • Simple Macro Language • Small • Standard Markup Language • Stella Maris (Bahamas) (location identifier) • Structure Mold Line • Support Material List • Surface Mixed Layer • Symbolic Machine Language

SMLC: Suivi Médical Longitudinal Contrôlé

SMLE: Suboptimal Maximum-Likelihood Estimator

SMLI: Stateful Multilayer Inspection

SMLR: Stepwise Multiple Linear Regression • Stepwise Multivariate Linear Regression

SMLTM: Spectral Mesosphere/Lower Thermosphere Model

S/MM: Shuttle MIR Mission

SMM: Scattering Matrix Method • Science Mission Management • Science Museum of Minnesota (USA-MN) • Semporna (Malaysia) (location identifier) • Server Management Module • Server Manager Module • Shuttle MIR Mission • Société de Mécanique Magnétique (France) • Solar Maximum Mission • Spectral Matrix Method • Spectro-Mirror Mechanism • Start of Manual Message • Subcontract Management Manual • Submillimeter Mission • Subsystem Management Manual • Surface Micromachining • System and Mission Management • System Maintenance Management • System Maintenance Manual • System Management Mode • System Management Module • System Measurement Management • System Mission Management

SMMART: Society for Mass Media and Resources Technology (Australia)

SmMaS: Smart Materials and Structures (periodical code)

SMMD: Specimen Mass Measurement Device

SMME: Small-, Medium- and Micro-Enterprise

SMMI: Structured Multimedia Information

SMMM: Submillimeter Moderate Mission

SMMR: Scanning Multichannel Microwave Radiometer • Scanning Multifrequency Microwave Radiometer • Scanning Multispectral Microwave Radiometer

SMMT: Society of Motor Manufacturers and Traders • Submillimeter Telescope

SMN: Salmon (USA-ID) (location identifier) • Sécurité Multiniveau • Servicio Meteorológico Nacional • Software Modification Notice

SMO: Sampple Management Office • Santa Monica (USA-CA) (location identifier) • Senior Meteorological Officer • Sequential Minimal Optimization • Service Module Oxidizer • Servicing Mission Office • Single-Market Observatory (EU) • Small Magnetospheric Observatory • Small Missions Opportunities • So Much Of • Stabilized Master Oscillator • Supplementary Meteorological Office • Support Materials Organization (USA-CA) • Systems Management Officer

SMOC: Science and Mission Operations Center • Simulation Mission Operation Computer

SMOG: Smoke and Fog • Structural Modelling Oriented Graphics • Structured Model Generator • Surface Meteorology and Ozone Generation

SMOH: Since Major Overhaul

SMOKE: Surface Magneto-Optical Kerr Effect

SMON: System Monitor

SMOP: Satellite Measurement of Oceanographic Parameters • Simple Matter Of Programming • So Much Of Paragraph

SMORS: Service Mode Optimised Reduction Scheme

SMOS: Surface Meteorological Observation System

SMOV: Servicing Mission Observatory Verification • Servicing Mission Orbital Verification

SMOW: Standard Mean Ocean Water

smp: Sine Mascula Prole

SMP: Safety Management Plan • Sample • Scanning and Measuring Projector • Scheduled Maintenance Program • See Me Please • Sensitized Material Print • Servo Meter Panel • Shared Memory Processor • Signaling Mediation Point • Simple Management Protocol • Software Management Plan • Software Management Practice • Soil Management Program • Special Manufacturing Procedure • Spectral Mapping Theorem • Standard Maintenance Procedure • Standards, Methods and Planning • Statutory Maternity Pay • Stockholm (Papua New Guinea) (location identifier) • Structure Mechanism Policy • Summary Maneuver Plan • Symbolic Manipulation Program • Symmetrical Multiprocessing • Symmetrical Multiprocessor • Symmetric Multiprocessing • Symmetric Multiprocessor • System Maintenance Procedure • System Monitoring Panel

SMPA: Solid-Motor Processing Area

SMPC: Shared Memory Parallel Computer

SMPG: Standardization Management Policy Group

SM/PM: System Management/Performance Monitor

SMPM: Structural Materials Property Manual • System Management Performance Monitor

SMPQ: Software Management Practice Questionnaire

SMPS: Sector Maintenance Processor System • Semiconductor Materials Processing System • Service Module Propulsion Stage • Switching Mode Power Supply

SMPTE: Society of Motion Picture and Television Engineers

SMPTRB: Shuttle Main Propulsion Test Requirement Board

SMPZ: Système Mondial de Prévisions de Zone

SMQ: Sampit (Indonesia) (location identifier) • Structure Module Qualification

SMQT: Structure Module Qualification Test

SMR: Saint Marin (ISO country code) • Sa Majesté Royale • San Marco Range • San Marino (ISO country code) • Santa Marta (Colombia) (location identifier) • Scanning Microwave Radiometer • Sites and Monuments Records • Software Management Report • Software Modification Record • Software Modification Report • Solid Moderated Reactor • Source-Maintenance-Recoverability • Spaceborne Meteorological Radar • Specialized Mobile Radio • Special Mobile Radio • Standardized Mortality Rate • Status Monitoring Routine • Super-Metal-Rich • Supply Management Report • System Malfunction Report

SMRA: Simultaneous Multicomponent Rank Annihilation

SMRAM: System Management Random Access Memory

SMRB: Senior Material Review Board

SMRD: Science Modelling Requirement Document • Spin Motor Rotational Detector • Spin Motor Rotation Detector • Spin Motor Rotation Device • Spin Motor Run Detector • Spin Motor Run Direction • Spin Motor Run Discrete • Spin Motor Running Detector • Spin Motor Running Director

SMRM: Solar Maximum Repair Mission

SMRSS: Surveys, Mapping and Remote-Sensing Sector

SMRT: Single-Message Rate Timing

SMRWG: Science and Mission Requirements Working Group

SM&S: Systems Management & Sequencing

SMS: Safety Management System • Sainte-Marie (Madagascar) (location identifier) • Sales Management System • Satellite Motion Simulator • Scanning Mirror System • Schema Management System • School of Mathematical Sciences • School of Mathematics and Statistics • Science Mission Schedule • Science Mission Specification • Scientific Mission Support • Semesterwochenstunden • Semiconductor-Metal-Semiconductor • Separation Mechanism System • Sequence Milestone System • Service Management System • Service Manipulator System • Short Message Service • Short Messaging Service • Shuttle Mission Simulator • Single-Molecule Spectroscopy • Site-Managed Software • Small Magnetospheric Satellite • Société Mathématique Suisse (Switzerland) • Software Maintenance Specialist • Software Modification Sheet • Solar-Mass Sensor • Solar-Maximum Satellite • Sonne-Mond-Sterne • Space Motion Sickness • Spares Management System • Spin Motor Supply • Standard Modular System • Standard Molecular System • Stationary Manipulator System • Stationary Meteorological Satellite • Storage Management Services • Storage Management System • Stores Management System • Subject Matter Specialist • Suprathermal Mass Spectrometer • Sveriges Mekanstandisering (Sweden) • Switching and Maintenance Set • Synchronous Meteorological Satellite • Synoptic Meteorological Sounding • System Maintenance Sector • System Maintenance Service • System Managed Storage • System Management Server • Systems Management and Sequencing

SMSA: Secondary Mirror Subassembly • Standard Metropolitan Statistical Area • Standard Metropolitan Statistical Average

SMSC: Semi-Magnetic Semiconductor • Service Module Sequence Controller • Shared-Memory Supercomputer • Short Message Service Center • Software Moding, Sequencing, and Control • Standard Modular System Card

SMSCC: Shuttle Mission Simulator Computer Complex

SMSG: Shuttle Maintenance Steering Group

SMSGEN: Science Mission Specification Generation

SMSI: Standard Manned Space Flight Initiator

SMSK: Serial Minimum-Shift Keying

SMSP: Sodium Metasilicate Peutahydrate

SMSR: Side-Mode Suppression Ratio

SMSS: School of Management and Social Sciences • Shuttle Modular Scanning Spectroradiometer

SMSST: Satellite-Measured Skin Surface Temperature

SMSSV: Space Mission Scenario Simulation and Visualization

SMST: Science Mission Specification Timeline

SMSU: Southwest Missouri State University (USA-MO)

SMSW: Store Machine Status Word

SMt: Station Management

SMT: Selective Message Transaction • Service Module Technician • Ship's Mean Time • Software Maintenance Team • Standard Measurement Technique • Station Management • Steiner Minimum Tree • Submillimeter Telescope • Summit • Sunday-Monday-Tuesday • Sun Moon Lake (China-ROC) (location identifier) • Surface-Mounted Technology • Surface-Mount Technology • Système Modulaire Thermique • System Maintenance Test • Système Mondial de Télécommunications

SMTAS: Shuttle Model Test and Analysis System

SMTF: Shuttle Mission Training Facility • Spacecraft Magnetic-Test Facility

SMTH: Smooth

SMTO: Submillimeter Telescope Observatory

SMTP: Simple Mail Transfer Protocol

SMTS: Synchronous Meteorological Test Satellite

SMU: Saint Mary's University (Canada) • Sheep Mountain (USA-AK) (location identifier) • Single-Motor Unit • Software Management Utility • Software Mockup • Source Management Utility • Southeastern Massachusetts University (USA-MA) • Southern Methodist University (USA-TX) • Speaker/Microphone Unit • Spectrum Monitoring Unit • System Management Utility

SMUR: Service Mobile d'Urgence et de Réanimation

SMURFS: Southern Michigan Unorganized Regional Festival of Stargazers (USA-MI)

SMV: Saint-Moritz Samedan (Switzerland) (location identifier) • Satellite Mutual Visibility • Science Museum of Virginia (USA-VA) • Symbolic Model verifier

SMVP: Shuttle Master Verification Plan

SMVRD: Shuttle Master Verification Requirements Document

SMW: Smara (Morocco) (location identifier) • Standard Materials Worksheet • Standard Metal Window

SMWHT: Somewhat

SMWT: Self-Managing Work Team

SMWV: Science Museum of Western Virginia (USA-VA)

SMX: Santa Maria (USA-CA) (location identifier)

SMXP: Simple MIME Exchange Protocol

SMY: Scientist-Man Year • Simenti (Senegal) (location identifier) • Solar-Maximum Year

SMYA: Solar-Maximum Year Analysis

SMZ: Stoelmans Eiland (Suriname) (location identifier)

SM2: Second Servicing Mission

s/n: Sin Número

sn: Sans Nom • Secundum Naturam • Shona (ISO language code) • Sine Nomine • Sin Numerar • Sub Nomine

Sn: Étain • Snow • Tin • Zinn

S-N: Stree Number

S/N: Serial Number • Service Note • Shipping Note • Signal to Noise • Stress Number

SN: Sample Name • Scientific Note • Scope Note • Senegal (international vehicle registration code) • Senegal (ISO country code) • Sequence Number • Serial Number • Series Number • Service Number • Shared Nothing • Shipping Note • Shipping Number • Shot Noise • Side Note • Snow • Space Network • Strategic Navigation • Student Number • Supernova • Swindon (UK postal code)

SNA: Laguna Beach - John Wayne (USA-CA) (location identifier) • Orange County - John Wayne (USA-CA) (location identifier) • Santa Ana - John Wayne (USA-CA) (location identifier) • Seltsames Nichtchaotisches Attraktor • Service National de l'Adresse (France) • Société Neuchâteloise d'Astronomie (Switzerland) • Strange Nonchaotic Attractor • Synchronous Network Architecture • System of National Accounts • Systems Network Architecture

SNACP: Systems Network Architecture Control Protocol

SNACRP: Société Nouvelle des Ateliers et Chantiers de La Rochelle - La Pallice (France)

SNACS: System Network Architecture Character Stream • System Network Architecture Character String

SNADS: Systems Network Architecture Distribution Service

SNAFU: Situation Normal, All Fouled Up • Southern Nevada Fantasy and Fiction Union (USA-NV)

SNAIR: Smart Navigators to Access and Integrate Resources

SNAP: Selective Niobium Anodization Process • Semantic Network Array Processor • SIMD Numerical Array Processor • Simplified Numerical Automatic Programmer • Six Node Averaging Program • Small Nuclear Auxiliary Power • Space Nuclear Auxiliary Power • Special Needs Access Program • Standard Network Access Protocol • Sterile Nitrogen Atmosphere Processing • Structural Network Analysis Program • Student Needs Assessment Project • Stylus Not A Pen • Subnet Access Protocol • Subnetwork Access Protocol • Supernova Acceleration Probe • Supersonic Non-equilibrium Analysis Program • Switching Network Analysis Program • System for Nuclear Assistant Power • System for Nuclear Auxiliary Power

SNAS: Slovak National Accreditation System • Strategic National Alternatives Study

SNB: Snake Bay (Australia-NT) (location identifier)

SNBNK: Snowbank

SNBO: Supernova Neutrino Burst Observatory

SNC: Salinas (Ecuador) (location identifier) • Shergottite-Nakhlite-Chassignite • Shergotty-Nakhla-Chassigny • Société en Nom Collectif • Special Nominating Committee • System Network Controller

SNCB: Société Nationale des Chemins de Fer Belges (Belgium)

SCND: Syndicat National de la Communication Directe (France)

SNCF: Société Nationale des Chemins de Fer Français (France)

SNCI: Service National des Champs Intenses (France)

SNCL: Serial Number Configuration List

SNCs: Shergottites-Nakhlites-Chassignites

SNCS: Syndicat National des Chercheurs Scientifiques

Snd: Sand • Sanded • Sound • Sounding

SND: Seno (Laos) (location identifier) • Signal Network Distributor • Sound • Standardized Normal Distribution • Street Name Database

SNDE: Society for Nonlinear Dynamics and Econometrics

SNDP: Sustainable National Domestic Product

SNe: Supernovae

SNE: São Nicolau (Cape Verde) (location identifier) • Severe Noise Environment • Syndicat National de l'Édition (France)

SNEA: Société Nouvelle Europe Aéroservice

SNEAP: Symposium of Northeastern Accelerator Personnel (USA)

SNECMA: Société Nationale d'Études et de Construction de Moteurs d'Avions (France)

SNEPT: Space Nuclear Electric Propulsion Test

SNEWS: Supernova Neutrino Early Warning System

SNF: San Felipe (Venezuela) (location identifier) • Schweizerischer Nationalfonds (Switzerland) • Second Normal Form • Secret - No Foreigners • Server Natural Format • Server Normal Format • Schweizerischen Nationalfonds zur Förderung der wissenschaftlichen Forschung (Switzerland) • Spent Nuclear Fuel • System Noise Figure

SNFL: Superficial Nerve Fiber Layer

SNFLK: Snowflake

Sng: Singapore

SNG: San Ignacio de Velasco (Bolivia) (location identifier) • Satellite News Gathering • Schweizerische Naturforschende Gesellschaft (Switzerland) • Substitute Natural Gas • Synthetic Natural Gas

Sngl: Single

SNGR: Sin Nuestra Garantia ni Responsabilidad

SNH: Stanthorpe (Australia-QLD) (location identifier) • Sternwarte Neanderhöhe Hochdahl eV (Germany)

SNI: Selected Notification of Information • Sequence Number Indicator • Siemens Nixdorf Informationsysteme AB • Sinoe - R.E. Murray (Liberia) (location identifier) • Sistema Nacional de Información • Source Non-Identifiée • Subscribers Network Interface • System Network Interconnect

SNIA: Storage Network Industry Association (USA-CA)

SNIAS: Société Nationale Industries Aérospatiales (France)

SNICT: Sistema Nacional de Informação Cientifica e Tecnológica (Brazil) • Sistema Nacional de Información Científica y Técnica (Uruguay) • Sistema Nacional de Información Científica y Técnológica (Mex ico)

SNID: Start Note Identification

SNIFTIRS: Subtractively Normalized Interfacial Fourier Transform Infrared Spectroscopy

SNIG: Sistema Nacional de Informação Geografica (Portugal)

SNIMA: Service de Normalisation Industrielle Marocaine (Morocco)

SNIP: Space Network Interoperability Panel

SNIRD: Supposedly Noiseless Infrared Detector

SNIRS: Syndicat National Indépendant de la Recherche Scientifique

SNJ: San Julián (Cuba) (location identifier)

SNK: Snyder Winston (USA-TX) (location identifier)

SNL: Sample Noise Level • Sandia National Laboratories (USA-NM) • Shawnee (USA-OK) (location identifier) • Standard Nomenclature List

SNLE: Sous-marin Nucléaire Lanceur d'Engins

SNLR: Services No longer Required

SNM: San Ignacio (Bolivia) (location identifier) • Satellite Navigation Map • Sobre Nivel del Mar • Society of Nuclear Medicine • Special Nuclear Material • Square Nautical Mile • SunNet Manager

SNMC: Système National de Mesure du Canada

SNMP: Simple Network Management Protocol

SNMS: Sonet Network Management System • Sputtered Neutral Mass Spectrometry

SNN: Shannon (Ireland) (location identifier) • Shared Nearest Neighbor • Structure-Nomenclature Notation

SNNE: Sustainable Net National Expenditure

SNNS: Stuttgart Neural Network Simulator

Sno: Snow

SNO: Sakon Nakhon (Thailand) (location identifier) • Snow • Sudbury Neutrino Observatory (Canada-ON)

SNOBOL: String-Oriented Symbolic Language

SNODAQ: Sudbury Neutrino Observatory Data Acquisition

SNOE: Student Nitric Oxide Explorer

SNOM: Scanning Near-field Optical Microscope • Scanning Near-field Optical Microscopy

SNOMED: Standard Nomenclature of Medicine

SNOPT: Sparse Non-linear Optimizer

SNP: Saint Paul Island (USA-AK) (location identifier) • Science Network Planning • Secure Network Programming • Serial Number/Password • Silicon Nanoparticle • Single Nucleotide Polymorphism • Space Nuclear Propulsion • Suspected, Not Proved • Synchronous Network Processor • Système Net Protégé • System Network Processor

SNPE: Société Nationale de Poudres et Explosifs (France)

SNPO: Space Nuclear Propulsion Office

SNQ: San Quitín (Mexico) (location identifier)

Snr: Senior

SNR: Saint-Nazaire (France) (location identifier) • Science Networking Representative • Signal-to-Noise Ratio • Supernova Remnant (*) • Supernova Reste

SNRI: Short-Notice Random Inspection

SNRS: Sunrise

SNS: Salinas (USA-CA) (location identifier) • Shuttle Network System • Simple Name Service • Simulated Network Simulation • Space Nuclear System • Spallation Neutron Source

SNSO: Space Nuclear Systems Office

SNSB: Swedish National Space Board (Sweden)

SNSH: Snow Showers

SNSM: Société Nationale de Sauvetage en Mer (France)

SNSO: Space Nuclear Systems Office

Snsr: Sensor

SNSR: Sensor

SNST: Sunset

SNT: Sabana de Torres (Colombia) (location identifier) • Schmidt Newtonian Telescope

SNTP: Space Nuclear Thermal Propulsion

Sntzd: Sensitized

SNU: Santa Clara (Cuba) (location identifier) • Seoul National University (Korea-ROK) • Solar Neutrino Unit

SNUS: Swedish Network Users' Society

SNV: Santa Elena (Venezuela) (location identifier) • Schweizerische Normen-Vereinigung (Switzerland) • Single Network View • Statens Naturvårdsverk (Sweden) • Swiss Association for Standardisation

SNW: Sandoway (Myanmar) (location identifier) • Snow • Thandwe (Myanmar) (location identifier)

SNWFL: Snowfall

SNX: Sabana de Mar (Dominican Rep.) (location identifier)

Sny: Sunny

SNY: Sidney (USA-NE) (location identifier)

SNZ: Santa Cruz (Brazil) (location identifier) • Standards New Zealand (formerly SANZ)

s/o: Shipping Order • Son of • Su Orden

so: Seller's Option • Siehe Oben! • Solución • Somali (ISO language code) • South • Southern

So: Sonntag • South • Southern

S-O: Shutoff

S/O: Shutoff • Switchover

SO: Saturn Orbiter • Science Operations • Scientific Officer • Security Office • Senior Officer • Serial Output • Shift Out • Shop Order • Shuttle Operations • Shut Off • Siehe Oben! • Significant Other • Slow Operate • Slow Operating • Small Outline • Somalia (ISO country code) • Somalie (ISO country code) • Sorting Office • Southampton (UK postal code) • Southern Oscillation • Soviet • Special Order • Spherical Overdensity • Stable Oscillator • Staff Officer • Standing Order • Stationery Office • Statistical Office • Steward Observatory (USA-AZ) • Sud-Ouest • Sulfur Monoxide • Supply Office • Supply Officer • Support Operations • Suroeste • Surveillance Officer • Switching Oscilloscope • Switchover • System Override

SOA: Self-Optimizing and Adaptive • Shuttle Orbital Application • Sky Observers' Association (Hong Kong) • Soc Trang (Viet Nam) (location identifier) • Source Of Assignment • Special Operating Area • State Of the Art • Switch Off Assembly

SOAC: Satellite Ozone Analysis Center

SOAL: Search Optical Augmentation Laser

SOAO: Sobaeksan Optical Astronomy Observatory (Korea-ROK)

SOAP: Satellite Orbit Analyzer Predictor • Seal Of Approval • Selective Optical Atmospheric Probe • Simplify Obscure ALGOL Programs • Single Operator Automation System • Spectrometric Oil Analysis Program • Subjective-Objective-Analysis-Plan • Symbolic Optimizing Assembler Program • Symbolic Optimum Assembly Program • Système Opérationnel d'Analyse et de Prévision • Systems Operational Analysis Plan

SOAR: Shuttle Orbital Applications and Requirements • Software for Optical Archival and Retrieval • Southern Observatory for Astronomical Research (*) • Southern Observatory for Astrophysical Research • State-Of-the-Art Report • Stratospheric Observatory for Astronomical Research (*)

SOARD: Steward Observatory Asteroid Relational Database

SOARS: Shuttle Operations Automated Reporting System • Significant Opportunities in Atmospheric Research and Science

SOATS: Shuttle Operations Automated Training System

SOB: Saarmelleek Balaton (Hungary) (location identifier) • Science Observatory Branch • Science Operations Branch • Senate Office Building • Soul On Board • State Office Building

SOBC: SSM Onboard Computer

SOBEPS: Société Belge d'Études des Phénomènes Spatiaux (Belgium)

SOBLI: Study of Optical Baseline Instrumentation

SOBOL: Second-Order Boundary Layer

SoByu: Soob. Byurakan Obs. (periodical code)

Soc: Social • Sociedad • Società • Société • Society • Sociología • Sociologie • Sociology

SOC: Satellite Operations Center • Satellite Orbit Control • Science Observations Center • Science Operations Center • Science Operations Coordinator • Scientific Operations Center • Scientific Organizing Committee • Scientific Oversight Committee • Security Operations Center • Self-Organized Critical • Self-Organizing Criticality • Service Oversight Center • Shuttle Operations Center • Simulated Operational Computer • Simulation Operations Center • Simulation Operations Computer • Single-Orbit Computation • Soil Organic Carbon • Solo City (Indonesia) (location identifier) • Space Operations Center • Space Operations Controller • Specialized Oceanographic Center • Specific Optimal Control • Start Of Climb • State Of Charge • Stationary Overturning Convection • Surveillance Operations Center • Synthetic Organic Chemical • System On a Chip • System Operational Concept • System Option Controller

SOCA: Second-Order Correlation Approximation

So-Cal: Southern California (USA)

SOCAR: Shuttle Operational Capability Assessment Report • Systems Operational Compatibility Assessment Review

SOCC: Satellite Oceanic Control Center • Satellite Operations Command and Control • Satellite Operations Control Center • Shuttle Operations Computing Center

SOCCB: Secure Operations Configuration Control Board

SOCCER: Sample Of Comet Coma Earth Return

SoCe: Solar Cells (periodical code)

SOCEX: Southern Ocean Cloud Experiment

SOCF: Spacecraft Operations and Checkout Facility

SOCH: Spacelab Orbiter Common Hardware

Sociol: Sociología • Sociological • Sociology

SOCKO: Systems Operational Checkout

SOCKS: Socket Secure

SOCO: Switched Out for Checkout

SOCOM: Special Operations Command

SOCP: Space Oblique Conformal Projection

SOCR: Synchronous Orbit Communications Relay

SOCRATES: Strategy for Organisations Concerned in Rural Advanced Telecommunications Experiments • System Of Cellular Radio for Traffic Efficiency and Safety

SOCS: Spacecraft Orientation Control System • System Operation and Checkout System

Socy: Society

Sod: Sodium

SOD: Science Operations Database • Science Operations Director • Second Order Design • Serial Output Data • Service d'Orbitographie DORIS (CNES) • Site Occupation Designator • Small Object Detector • Sorocaba (Brazil-SP) (location identifier) • Spectral Optical Depth • Staff Operations Division • Standing Order for Data • Super Oxyde Dismutase • Sustained Operational Date • System Operational Description • System Operational Design

SODA: Science Operations and Data Analysis • Small-Object Database • Solar Dynamo Amplitude • Source-Oriented Data Acquisition • Source-Oriented Data Automation • Symposium On Discrete Algorithms

SODALS: Simplified Omnidirectional Approach Lighting System

SODAR: Sound Detecting And Ranging

SODART: Soviet-Danish Röntgen Telescope

SODAS: Synoptic Oceanographic Data Assimilation System

SODB: Shuttle Operational Data Book • Spacecraft Operational Data Book • Start Of Data Block

SODIM: Standard Operational Data Item Manual

SODRS: Synchronous Orbit Data Relay Satellite

SODS: Saturn Operational Display System • Shuttle Operational Data Submittal • Shuttle Operational Data System • Skylab Orbit-Deorbit System • Subordinate Operations Data System

SODSIM: Strategic Offense-Defense Simulation

SODSSWG: Satellite Ocean Data System Science Working Group

SoE: State of the Environment

SOE: Science Operations Element • Sequence Of Events • Server Operating Environment • Skylab Operational Environment • Slater Orbital Exponent • Sonnenobservatorium Einsteinturm (Germany) • Souanke (Congo) (location identifier) • Special Operation Executive • Standard Operating Environment • State Of the Environment • State-Owned Enterprise • Super Orbit Entry

SOED: Shorter Oxford English Dictionary

SoEn: Solar Energy (periodical code)

SoEnM: Solar Energy Materials (periodical code)

SOEP: Solar-Oriented Experimental Package • Sozio-Ökonomische Panel

SOEST: School of Ocean and Earth Science and Technology

SOF: Safety Of Flight • Science Operations Facility • Scintillating Optical Fiber • Sofia (Bulgaria) (location identifier) • Soluble Organic Fraction • Sound On a Film • Special Operation Forces • Special Operations Force

SOFA: Selectable Optical Filter Assembly • Sevon-0-Five Astronomers (USA-CA) • Standard Objects For Astronomy • Standards Of Fundamental Astronomy • Statement Of Financial Activities

SOFAR: Sound Fixing And Ranging

SOFAS: Survivable Optical Forward Acquisition System

S/Ofc: Site Office

SOFC: Saturn Operational Flight Control • Solid Oxide Fuel Cell

SOFCAL: Scintillating Optical Fiber Calorimeter

SOFCS: Self-Organizing Flight Control System

S/OFF: Sign Off

SoFi: Sonnenfinsternis

SOFI: Son of ISAAC • Spray-On Foam Insulation • Système d'Ordinateurs pour le Frêt International

SOFIA: Stratospheric Observatory For Infrared Astronomy • Surface de l'Océan - Flux et Interactions avec l'Atmosphère • Surface of the l'Ocean Fluxes and Interaction with the l'Atmosphere

SOFIE: Scintillating Optical Fiber Isotope Experiment

SOFIN: Soviet-Finnish (spectrograph)

SOFM: Self-Organizing Feature Map

SofR: Society of Rheology

SOFRADIR: Société Française de Détecteurs Infrarouges (France)

SOFRES: Société Française d'Enquêtes et de Sondages (France)

SofS: Secretary of State

SOFSEM: Software Seminar

SOFT: Solar Flare Telescope • Space Operations and Flight Techniques • Suborbit Flight Test

SOFTE: Software Estimator

SOFTEKS: Software Engineering and Knowledge-base System

Sog: Sogenannt

SOG: Same Output Gate • Scientific Operations Group • Second-Order Gradient • Sogndal (Norway) (location identifier) • Stuck On the Ground

SOGA: Schedule Optimizing Genetic Algorithm • SNA Open Gateway Architecture

SOGS: Science Operations Ground System

SOH: Since Overhaul • Skylab Operations Handbook • Solita (Colombia) (location identifier) • Start Of Header • Start Of Heading • State Of Health

SOHF: Sense Of Humor Failure

SOHMA: Servicio de Oceanografía, Hidrografía y Meteorología de la Armada (Uruguay)

SoHo: South of Houston (Street) (New York)

SOHO: Small Office - Home Office • Solar and Heliospheric Observatory (*) • Solar Heliospheric Observatory

SOHR: Solar Hydrogen Rocket Engine

SOI: Saturn Orbital Insertion • Saturn Orbit Insertion • Scan Ozone Imager • Scientific and Optical Instruments • Silicon On Insulator • Solar Oscillation Imager • Solar Oscillation Investigation • Southern Oscillation Index • South Molle Island (Australia-QLD) (location identifier) • Space Object Identification • Specific Operating Instruction • Sphere Of Influence • Standard Operating Instruction • Statement Of Intent • Survey Of India

SOIC: Small Outline Integrated Circuit

SOIF: Structure Object Interchange Format • Summary Object Interchange Format

SOILIMS: Soil Laboratory Information and Management System

SOIN: Solar Interferometer

SOIR: Simultaneous Operations on Intersecting Runways

SOIS: Space Object Identification System • Spacelab/Orbiter Interface Simulator

SOIWR: Simultaneous Operations on Intersecting Wet Runways

SOJ: Sorkjosen (Norway) (location identifier)

SOJAAM: Société des Jeunes Astronomes Amateurs de Montréal (Canada-QC)

SOK: Semongkong (Lesotho) (location identifier)

Sol: Solaire • Solar • Solenoid • Solicitor • Solid • Soluble • Solution

SOL: Science On Line • Scrape-Off Layer • Short On Landing • Simulation-Oriented Language • Small Outline • Solar • Solenoid • Solomon (USA-AK) (location identifier) • Space-On-Line • Strickly Out of Luck • Structure-Oriented Language • System-Oriented Language

SOLA: Subtractive Optimally Localized Average

SOLAIR: Solomon Islands Airways Ltd.

SOLAR: Solar On-Line Activity Resources • Solid Application Resource

SOLAS: Safety Of Life At Sea

SOLB: Start Of Line Block

SOLCON: Solar Constant

SOLD: Simulation Of Logic Design • Source Online Debugger

SOLE: Star-Obvious Low-level Elongated (PPN)

SOLEM: Solar Experiment Module

SOLERS: Solar Electromagnetic Radiation Study

SolFDI: Solar Flare Dynamics Imager

SOLFEGE: Service d'Observation Localisé Favorisant l'Exploitation des Paramètres Géophysiques

SOLIS: Sydney Outdoor Lighting Improvement Society (Australia-NSW) • Synoptic Optical Long-term Investigations of the Sun

SOLMC: Surface-Oriented Locally Monolithic Coordinates

Soln: Solution

SOLN: Solution

Sol. Phys.: Solar Physics

Solr: Solicitor

SOLRAD: Solar Radiation

SOLSE: Shuttle Ozone Limb Sounding Experiment

SOLSPEC: Solar Spectrum

SOLSTICE: Satellite Online Searching Training Interactive Conferencing Experiment • Solar-Stellar Irradiance Comparison Experiment

SOLT: Short-Open-Load-Through • Short-Open-Load-Thru

SOLTIP: Solar Transient Interplanetary Phenomenon

Solv: Solvent

SOLV: Solenoid Valve

SOLVE: SAGE III Ozone Loss and Validation Experiment • Solar Variability Enterprise

Soly: Solubility

Som: Somali • Somerset (UK-England)

SoM: School of Management

SOM: San Tomé (Venezuela) (location identifier) • Science Operations Manager • Scripting Object Model • Section of Orbital Mechanics • See Our Message • Self-Organized Map • Self-Organizing Machine • Self-Organizing Map • Ship Operations Manager • Simulation Object Model • Society Of Mind • Somalia (ISO country code) • Somalie (ISO country code) • Spacecraft Operations Manager • Spacecraft Operations Manual • Space Oblique Mercator (projection) • Spares Optimization Model • Stage Operating Manual • Standard Operating Manual • Start Of Message • Structured Object Method • Suborbital Mission • System Object Model

SOMA: Semantic Object Modelling Approach • Support Opto-Mechanical Assembly

SOMALAIR: Somali Airlines

SOMEI: Sociedad Mexicana de Estudios Interplanetarios (Mexico)

SOMF: Start Of Minor Frame

SOML: Steward Observatory Mirror Laboratory (USA-AZ)

SOMM: Software Operations and Maintenance Manual

SOMMA: Single-Operator Multiparameter Metabolic Analyzer

SOMO: Space Operations and Management Office

SOMOWG: Science Operations Management Operations Working Group

SOMP: State Optical and Mechanical Plant (USSR)

SOMS: Shuttle Orbiter Medical System • Synchronous Operational Meteorological Satellite

Som. Sh.: Somali Shilling (currency)

SOMYCE: Sociedad de Observadores de Meteoros y Cometas de España (Spain)

S/ON: Sign On

SON: Espíritu Santo (Vanuatu) (location identifier) • September-October-November • Son Of NAIF • Sonneberg Variable • Sonora (Mexico) • Standards Organization of Nigeria • Statement Of Need • Stichting voor Scheikundig Onderzoek in Nederland (Netherlands)

SONA: Society of Neuroscientists of Africa

SONAAS: Stratford-on-Avon Astronomical Society (UK)

SONAR: Sound Navigation And Ranging (*) • Sound Navigation Ranging

Sonderdr: Sonderdruck

SONET: Synchronous Optical Network

SONEX: Subsonic assessment of Ozone and Nitrogen Experiment

SONG: Space Oceanography, Navigation and Geodynamics • Stellar Oscillations Network Group • Survey Of Nearby Galaxies

SONIA: Southeast Oxidant and Nitrogen Intensive Analysis

SONIC: SAO Near-Infrared Camera • Simultaneously Operating Numerical Integration Computer • System Online Network for Information Control

SONITA: Société Nigérienne de Transports Aériens

SONNET: System Of Neighboring Networks in Ohio (USA-OH)

SOO: Soderhamn (Sweden) (location identifier) • Söderhamn (Sweden) (location identifier)

SOON: Solar Observing Optical Network

SOOS: Stacked OSCAR On Scout

Sop: Soprano

SOP: Safety and Operations Plan • Safety Operating Plan • Safety Oversight Program • Saturn Orbiter Probe • Secondary Oxygen Pack • Secure Operations Preboard • Semester Observation Plan • Semester Observing Plan • Ship of Opportunity Programme • Simulated Output Program • Simulation Operations Plan • Small Outline Package • Southern Pines (USA-NC) (location identifier) • Spacecraft Operations Plan • Spacelab Opportunity Payload • Special Observation Period • Special Observing Period • Standard Operating Procedure (*) • Standing Operating Procedure • Structural Optimization Procedure • Sum Of Products • Surveillance Opportunities File • Symbolic Optimum Program • Systems Operating Plan • Systems Operating Procedure • Systems Operating Program • Systems Operation Plan

SOPAC: Scripps Orbit and Permanent Array Center • South Pacific • South Pacific Applied Geosciences Commission • South Pacific Countries

SOPC: Science Operations and Planning Computer • Science Operations Planning Computer • Shuttle Operations and Planning Center • Shuttle Operations Planning Complex

SOPE: Simulated Off-the-Pad Ejection • Small ORU Protective Envelope

SOPG: Science Operation Planning Group

SoPh: Solar Physics (periodical code)

SOPH: SOC Operation Procedures Handbook

SOPHIA: SFOC Operational Pilot Hands-on Interactive Activity

SOPITA: South Pacific Isotopic and Thermal Anomaly

SOPM: Standard Orbital Parameter Message

SOPRI: Space Object Photometry, Radiometry and Identification

SOPS: Satellite Output Processing System • Science and Observation Planning System • Southern Objective Prism Survey • Spacecraft Operations Planning Section

SOPSA: Shuttle Orbit-injection Propulsion System Analysis

SOPSO: Space Operations Support Office

SOPT: Science Operations Planning Team

SOQ: Sorong (Indonesia) (location identifier) • System Optical Quality

SOQM: STSF Requirements Qualification Matrix

Sor: Soria (Spain)

SoR: Sale or Return

SOR: Al Thaurah (Syria) (location identifier) • Sale Or Return • Sampling Oscilloscope Recorder • Science Operations Room • Simultaneous Over-Relaxation • Specification Operational Requirement • Specific Operational Requirement • Stable Orbit Rendezvous • Standard Operating Report • Standard Operating Rule • Starfire Optical Range (USA-NM) • Start Of Record • Successive Overrelaxations • Symposium on Operations Research • System Operational Requirement

SORD: Submerged-Object Recovery Device

SOREL: Sun-Orbiting Relativity

SORIAP: Satellite Ocean-Related Imagery Applications Program

SORL: Space Optics Research Labs (USA-MA)

SORM: System of Efficient Research Measures

SORPTR: South Repeater

SORS: Same Origin Ray Set • Solar Radiospectroscopy • Spacecraft Oscillograph Recording System

SORSA: Spatially-Oriented Referencing Systems Association

SORT: Solar Orbit Relativity Test • Structures for Orbiting Radio Telescope

SORTI: Solar-Radiance-Transmission Interferometer • Star-Oriented Real-time Teaching Instrument

SORTIE: Superorbital Reentry Test Integrated Environment

SORV: Surface Operations Research Vehicle

sos: Si Opus Sit

SoS: Secretary of State

SOS: Satellite Observation System • Satellite Operational Services • Save Our Science • Save Our Souls • Science Operations Specialist • Sección de Observación Solar • Secretary Of State • Section d'Optique et de Spectroscopie • Self-Optimized Sector • Self-Organizing System • Serial Output Special • Services Of Supply • Share Operating System • Shilling (Somalia) (ISO currency code) • Short Orbit Spectrometer • Silicon on Sapphire • Sophisticated Operating System • Source Of Supply • Southern Oxidants Study • Spacecraft Operations Simulation • Space Operation Service • Space Option Society (Switzerland) • Space Ordnance System • Special Ordered Set • Spectroscopy and Optics Section • Speed Of Sound • SPOCC Operational Status • Stabilized Optical Sight • Standards and Open Systems • Start Of Significance • Suborbital Sequence • Sum Of Squares • Supervising Observation Software • Supplemental Oxygen System • Surface Of Section • Symbolic Operating System • Symmetry, Orbitals and Spectra • Synchronous-Orbit Satellite • System Operational Safety

SOSA: Sustained Operations Support Area • System-Of-Systems Architecture

SOSC: Second-Order Sufficient Condition

SOSD: Spatial Operational Sequence Diagram

SOSE: Science Operations Support Equipment

SoSh: Somali Shilling (currency)

SOSI: Service de l'Organisation et du Système d'Information (CNRS) • Space Operations Scientific Investigation

SOSIG: Social Science Information Gateway

SOSL: Saturn Orbiter Satellite Lander

SOSO: Safety and Operating Systems Office • Solar Oriented Space Observatory • Synchronous Orbiting Solar Observatory

SOSP: Solar Spectrum

SoSR: Software Specification Review

SOSRM: Strap-On Solid Rocket Motor

SOSS: Satellite Optical Surveillance Station • Structurally Oriented Simulation System

SOT: Science Observing Team • Science Operations Team • Sodankylä (Finland) (location identifier) • Solar Optical Telescope • Sótão • Start Of Tape • Strap-On Tank

SOTA: State Of The Art

SOTD: Steward Observatory Technical Department (USA-AZ)

SOTE: Standard Optical Test Equipment

SOTM: Science On The Move

Soton: Southampton (UK-England)

SOTON: Southampton (Starlink internal code)

SOTS: Some Old Tried Stuff • Suborbital Tank Separation • Synchronous Orbiting Tracking Station

SOTUS: Sequentially-Operated Teletypewriter Universal Selector

SOTV: Solar Orbit Transfer Vehicle

Sou: South • Southampton (UK-England) • Southern

SOU: Southampton (Starlink internal code) • Southampton Eastleigh (UK-England) (location identifier)

SOUP: Solar Optical Universal Polarimeter

SOUSY: Sounding System

Sov: Sovereign • Soviet

SOV: Science Opportunity Visualizer • Seldovia (USA-AK) (location identifier) • Shut-Off Valve • Solenoid-Operated Valve • Subject-Object-Verb

SOVA: Solar Variability (experiment) • Solar Variations

SOVAFA: Sociedad Venezolana de Aficionados a la Astronomía (Venezuela)

SOVIM: Solar Variability and Irradiate Monitor

SOVT: System Operations Validation Test

SOW: Show Low (USA-AZ) (location identifier) • Skylab Orbital Workshop • Start Of Word • Statement Of Work • Subdivision Of Work

SOWETO: Southwestern Township

SOWG: Science Operations Working Group

SOWICOMS: Solar-Wind Composition Mass Spectrometer

SOWP: Science Operations Working Group

SOx: Sulphur Oxide

SOX: Sogamoso (Colombia) (location identifier) • Sound Exchange • Supercritical Oxygen

SOY: Stronsay (UK) (location identifier)

SOZ: Solenzara (France) (location identifier) • Stichting Onderzoek der Zee (Netherlands)

SO2: Sulphur Dioxide

sp: Sine Prole

Sp: Space • Spain • Spaniard • Spanish • Spare • Spatial • Special • Species • Specific • Specimen • Speed • Spelling • Spiral • Spirit • Spitze • Sport

S/P: Serial to Parallel • Signal Processor

SP: Poland (international aircraft registration code) • Sacramento Peak (USA-NM) • Safety Plan • Salisbury (UK postal code) • São Paulo (Brazil) • Satellite Processor • Scalable Parallel • Scaled Point • Schwerpunkt • Scientific Paper • Scientific Processor • Scientific Product • Scientific Publication • Secteur Postal • Security Protocol • Self-Propelled • Sensor Processor • Séquence Principale • Sequence Programmer • Serial to Parallel • Service Pack • Service Provider • Servicio Público • Short Period • Short Play • Shoulder Pitch • Shuttle Project • Signal Processor • Single Phase • Single Pole • Single Precision • Single Purpose • Small Photometer • Software Process • Solar Panel • Solar Physics • Solid Propellant • Sortie Payload • Sous-Préfecture • South Pole • SPACE • Spain • Spare • Sparking Plug • Spark Plug • Special • Specialist • Special Performance • Special Programme • Special Publication • Special Purpose • Spectrometer Processor • Spin • Split Phase • Spontaneous Potential • Stable Platform • Stack Pointer • Standardized Profile • Standard or Peculiar • Standard Practice • Standard Procedure • Standard Program • Standing Procedure • StarPages • Starting Point • Starting Price • State Park (USA) • Static Pressure • Station Pressure • Stochastic Programming • Stop Payment • Stop Press • Stream Protocol • Structured Programming • Summus Pontifex • Sun Point • Super-Plössl • Supplement • Suppply Point • Supra Protest • Surge Prominence • Surveillance Payload • Swimming Pool • Switch Panel • Symbol Programmer • System Parameter • System Product • Systems and Procedures • Systems Procedure

Spa: Spanish

SpA: Società per Azioni

S&PA: Safety & Product Assurance

SPA: Safety and Product Assurance • S-band Power Amplifier • Scatter Propagation Antenna • Self-Phasing Array • Service Provider Application • Servopower Amplifier • Servo Power Assembly • Shared Peripheral Area • Shortest Path Algorithm • Signal Pre-Amplifier • Signal Processor Assembly • Single-Parameter Analysis • Small-Payload Accommodation • Smooth Potential Approximation • Society for Popular Astronomy (UK) • Software Performance Analyzer • Software Process Assessment • Software Product Assurance • Software Producers Association (UK) • Software Publishers Association • Solar-Panel Assembly • Solar-Power Array • South Pole-Aitken • Space Physics and Aeronomy • Space Processing Application • Spartanburg (USA-SC) (location identifier) • Specially Protected Area • Special Project Agreement • Spectral Processing Analysis • Spectrograph Assembly • Spin Phase Angle • Standard de Pouvoir d'Achat • Statistical Processing and Analysis • Steering Position Amplifier • Storage Planning and Allocation • Subprogramme Area • Sudden Phase Anomaly

SPAA: Sociedad Paraguya de Astronomía Amateur • Symposium on Parallel Algorithms and Architectures

SPAC: Saint Petersburg Astronomy Club (USA-FL) • Scientific Program Advisory Committee • South Plains Astronomy Club (USA-TX) • Space Program Advisory Council • Spacecraft Performance Analysis and Command • Spectrum Planning and Advisory Committee

SP'ACE: Space Advanced Communication Engine

SPACE: Saint Petersburg Astronomy Club Examiner • Satellite Precipitation And Cloud Experiment • Self-Programming Automatic Circuit Evaluator • Seminario Permanente de Astronomia y Ciencias Espaciales • Sidereal Polar Axis Celestial Equipment • Single Point of Access to Citizens of Europe • Space Advanced Communication Engine • Spacecraft Prelaunch Automatic Checkout Equipment • Space, Physics and the Advancement of Cosmic Exploration • Space, Planetary, and Astronomical Cyberexperience • Space Program American Citizen's Effort • Spacecraft Prelaunch Automatic Checkout Equipment • Strategic Preparedness in Advancing Careers in Engineering • Student Programs for Achievements in Careers and Education • Symbolic Programming Anyone Can Enjoy

SPACECOM: Space Communications Corporation

SPACELAB: Space Laboratory

SPACEWARN: World Warning Agency for Satellites

SPACON: Spacecraft Controller

SPAD: Sensor Processing And Display • Shuttle Payload Accommodation Document • Signal Passed At Danger • Single Photon Avalanche Diode • Société pour l'Aviation et ses Dérivés • Space Principles, Applications and Doctrines • Système Portable pour l'Analyse de Données

SPADE: Signal Processing And Display Equipment • Spare Parts Analysis, Documentation and Evaluation • Stratospheric Photochemistry, Aerosols, and Dynamics Experiment

SPADOC: Space Defence Operation Center

SPADS: Satellite Position And Display System • Satellite Processing And Display System • Shuttle Problem Analysis Data System • Small Parcel Explosive Detection System

SPADVOS: Spaceborne Direct Viewing Optical System

SPAF: Simulation Processor And Formatter

SPAG: Space Psychology Advisory Group • Standards and Practices Action Group • Standards Promotion and Application Group

SPAH: Spacelab Payload Accommodation Handbook

SPALNA: Soil and Plant Analytical Laboratories Network of Africa

Sp. Am.: Spanish America • Spanish American

SPAM: Satellite Processor Access Method • Spectral Analysis Manager • Spectroscopy and Polarimetry At Mees • Stewart Platform Augmented Manipulator • System for Photointerpretation of Airports using Maps

Span: Spaniard • Spanish

SPAN: Solar-Particle Alert Network • SONAR Program Analyzer • Spacecraft Analysis • Space Physics Analysis Network • Spiral Anode • Statistical Processing and Analysis • Stored Program Alphanumerics • Successive, Proportionate, Additive Numeration • System Planning and Analysis

SPAND: Solar Proton Albedo Neutron Decay

SPANL: Spectral Analysis Library

SPANS: Simple Protocol for ATM Network Signalling • Spatial Analysis System

SPAP: System for Publications Acquisition and Processing

Spar: Separata

Sp. Ar.: Spanish Arabic

SPAR: Shuttle Pointed Autonomous Research • Space Processing Applications Rocket • Special Prelaunch Analysis Request • Standard Payload Assurance Requirements • Superprecision Approach Radar • Supportability Problem Assessment Report • Symbolic Program Assembly Routine

SpARC: Space Automation and Robotics Center

SPARC: Scalable Performance Architecture • Scalable Processor Architecture • Scaled Processor Architecture • Scholarly Publishing and Academic Resources Coalition • Space Applications and Research Centre • Space Automation and Robotics Center • Spectral Analyzer and Recognition Computer • Standards, Planning And Requirements Committee • Strategic Planning And Research Coordination • Stratospheric Processes And their Role in Climate

SPARCENT: Space and Atmospheric Research Center

SPARCLE: Spaceborne Aerosols and Cloud Lidar Earthprobe

SPARCS: Solar Pointing Aerobee Rocket Control System • Solar Pointing Attitude Rocket Control System

SPARK: Student Program for Aeronautics Resource and Knowledge

SPARMO: Solar Particles And Radiations Monitoring Organization

SPARO: Submillimeter Polarimeter for Antarctic Remote Observations

SPARQ: Serving with Pride And Reaching for Quality

SPARRSO: Space Research and Remote Sensing Organization (Bangladesh)

SPARS: Space and Atmospheric Research Station • Space Precision Altitude Reference System • Space Precision Attitude Reference System • Spatial Analysis and Remote Sensing

SPARTAN: Shuttle Pointed Autonomous Research Tool for Astronomy • System for Personnel Automated Reports, Transactions and Notices

SPARTECA: South Pacific Regional Trade and Economic Cooperation Agreement

SPAS: Safety Performance Analysis System • Shuttle Pallet Application System • Shuttle Pallet Satellite • Solar Proton Alpha Spectrometer

SPASA: Servicios Politécnicos Aéreos SA • South Pole Air Shower Array

SpaSci Rev: Space Science Reviews

SPASE: South Pole Air Shower Experiment

SPaSM: Scalable Parallel Short-range Molecular

SPASM: Smithsonian Package for Algebra and Symbolic Mathematics • Space Propulsion Automated Synthesis Modelling

Spat: Spatial

SPAT: Satellite Performance Analysis Team • Silicon Precision Alloy Transistor • Small-Payload Accommodation Terminal

SPATE: Stress Pattern Analysis by Thermal Emission

SPAW: Specially Protected Areas and Wildlife

SPB: Saint Thomas Island (US Virgin Islands) (location identifier) • Science Planning Branch • Shergottite Parent Body • Solar-Particle Beam • Space Planning Bureau

SPC: Santa Cruz de la Palma (Canary Islands) (location identifier) • Satellite Processing Center • Scholastic Programming Contest • Science Policy Committee • Science Programme Committee • Scientific Policy Committee • Scientific Programme Committee • Scientific Programmes Committee • Scintillation Proportional Counter • Security Processing Center • Semi-Permanent Connection • Shipping and Packing Cost • Short-Period Comet • Short Periodic Comet • Shuttle Processing Contract • Signal Processing Center • Single-Probe Carrier • Sistema Panamericano de Clasificación • Sky Publishing Corporation (USA-MA) • Signal Processing Center • Small Peripheral Controller • Software Productivity Consortium • Software Publishing Corporation • Solar Pointing Control • South Pacific Commission • South Polar Cap • Space • Space Projects Center • Spare Parts Catalog • Special Political Committee • Special Purpose Committee • Special Purpose Computer • Spectroscopic and Polarimetric Coronagraph • Starting-Point Code • State Plane Coordinates • Statistical Process Control • Stored Program Command • Stored Program Control • Strategic Planning Council • Supplementary Protection Certificate • Synoptic Properties Code • System Parameter Control

SPCA: Secondary Power Control Assembly • Short Posterior Ciliary Artery

SPCBT: Self-Paced Computer-Based Training

SPCC: Spare Parts Control Center • Stored Program CAMAC Channel

SPCD: Space Data and Computing Division

Spchg: Supercharge

Spcl: Special

SPCM: Single Photon Counter Module

SPCN: Sciences Physiques, Chimiques, Naturelles

SpCoB: Space Communication Broadcasting (periodical code)

SpCom: Space Communications (periodical code)

Spcr: Spacer

SPCR: Software Problem Change Request

SPCS: Satellite Personal Communication Services • Schedule Planning and Control System • Selective Paging Communications System • Service and Performance Checkout System • State Plane Coordinates System • Static Power Conversion System

SPCU: Simulation Process Control Unit • Skylab Process Control Unit

SPCZ: South Pacific Convergence Zone

Spd: Speed

SPD: Safety Program Directive • Saidpur (Bangladesh) (location identifier) • Science Programs Division • Scientific Passenger Pod • Serviço de Processamento de Dados • Skylab Program Directive • Solar Physics Division • South Polar Distance • Space Physics Division • Spectral Power Distribution • Standard Practice Directive • Standard Processed Data • Stored Program Decoder • Stored Program Decommutation • Subjective Probability Distribution • Subject to Permission to Deal • Symmetric Positive Definite • System Preliminary Design • System Program Documentation

SPDA: Single-Premium Deferred Annuity

SPDB: System Power Distribution Box

SPDC: Special Programme for Developing Countries

SPDE: Solar Plasma Diagnostic Experiment

Sp. Del.: Special Delivery

SPDF: Sensor Data Processing Facility • Space Physics Data Facility

SPDIF: Sony-Philips Digital Interface • Sony-Philips Digital Interface Format

Spdl: Spindle

SPDL: Standard Page Description Language

SPDM: Special-Purpose Dexterous Manipulator

SPDP: Society of Professional Data Processors (USA) • Stored Program Data Processor

SPDR: Scheduled Planning Drawing Release • Short-Pulse Doppler Radar • Software Preliminary Design Review

SPDS: Scalable Performance Data Server • Sequential Payload Delivery System • Space Physics Data System • Stabilized Payload Deployment System

SPDT: Single-Pole - Double-Throw

SPE: Satellite Precipitation Estimate • Senior Project Engineer • Sepulot (Malaysia) (location identifier) • Shrinkage Preliminary Estimator • Single Photoelectron • Single Programming Document • Software Performance Engineering • Software Production Environment • Software Process Engineering • Solar Particle Event • Solar Proton Event • Solid-Phase Epitaxy • Solid-Phase Extraction • Solid Polymer Electrolyzer • Space Processing Equipment • Special-Purpose Equipment • Spherical Probable Error • Standard Problem Exercise • Static Phase Error • Station Project Engineer • Stored Program Element • Stork Product Engineering BV (Netherlands) • Sun-Planet-Earth • Symbolic Programming Environment • System Performance Evaluation • Systems Performance Effectiveness

SPEAR: Small Payload Ejection And Recovery • Source Performance Evaluation And Reporting • Spaceborne Earth Applications Ranging System • Space Power Experiments Aboard Rockets • Stanford Positron Electron Accelerating Ring • Support Programme for the Evaluation Activities in Research

SPEARNET: South Pacific Education And Research Network

SPEARS: Satellite Photoelectric Analog Rectification System • Screener Proficiency Evaluation And Reporting System • Supplier Performance Evaluation And Rating System

Spec: Special • Specially • Specialist • Specialization • Specialize • Specialized • Specific • Specifically • Specification • Specified • Specify • Specimen • Spectrum • Speculation

Spéc: Spécial • Spécialisé • Spécialiser • Spécialiste • Spécification • Spécifier • Spécifique

SPEC: South Pacific Bureau for Economic Cooperation • Stored Program Educational Computer • Support Programme for Employment Creation • System Performance Evaluation Cooperative • System Performance Evaluation Corporation

Special: Specialized

Specif: Specific • Specifically • Specification

SpecL: Spectroscopy Letters (periodical code)

SPECMAP: Spectral Mapping

SpECS: Special Education of Computational Science

SPECS: Solar Physics Element on Columbus/Space Station • Specification and Programming Environment for Communication Software • Submillimeter Probe for the Evolution of Cosmic Structure

SPECT: Single Photon Emission Computed Tomography (*) • Single Photon Emission Computerized Tomography

SPECTA: Structure-Preserved Error-Correction Tree Automaton

Spectr: Spectral • Spectre • Spectrograph • Spectrographe • Spectrometer • Spectromètre • Spectrum

SPECTRE: Spectral Radiance Experiment • Student-run Project for Exo-atmopsheric Colliding Technologies and Rocket Experiment

SPECTRUM: Special Experiment Concerning Typhoon Recurvature and Unusual Movement

Sp. Ed.: Specialist in Education

SpEd: Space Education (periodical code)

SPED: Supersonic Planetary Entry Decelerator

SPEDE: System for Processing Educational Data Electronically

SPEDO: Solar-Powered Extended Duration Orbiter

SPEDTAC: Stored Program Educational Transistorized Automatic Computer

SPEE: Special-Purpose End Effector

SPEEA: Society of Professional Engineering Employees in Aerospace (USA)

SPEED: Self-Programmed Electronic Equation Delineator • Systematic Plotting and Evaluation of Enumerated Data • Systems Planning and Effectiveness Evaluation Device

SPEEDUP: Simulation Program for the Economic Evaluation and Design of Unsteady-state Processes

SPEFO: Spectrofotometry

SPELDA: Structure Porteuse Externe pour Lancement Double Ariane

SPELTRA: Structure Porteuse Externe pour Lancement Triple Ariane

SpEnvIS: Space Environment Information System

SPEO: Space Physics Educational Outreach

SPEPD: Space Power and Electric Propulsion Division

SPER: Syndicat Professionnel d'Électronique et de Radiocommunications • Syndicat Professionnel Électronique et Radioélectrique

SPERAD: Spectrally-scanning Radiometer

SPERT: Schedule Performance Evaluation and Review Technique • Simplified Program Evaluation and Review Technique • Special Program Evaluation and Review Technique

SPES: Spacecraft Performance and Evaluation Service • Spacecraft Performance Evaluation System • Stimulation Plan for Economic Science

SPET: Solid-Propellant Electric Thruster

SPEVAL: Spacecraft Evaluation

SPEX: Satellite Potential Experiment • Spectral X-ray • Stellar and Planetary Explorer

SPF: Science Policy Foundation • Ship Point Formation • Shortest Path First • Single-Point Failure • Sociedade Portuguesa de Física (Portugal) • Software Production Facility • South Pacific Forum • South Polar Feature • Space Power Facility • Spacelab Processing Facility • Spearfish (USA-SD) (location identifier) • Specific Pathogen Free • Spectrofluorometer • Structured Programming Facility • Sun Protection Factor • Super-Plastically Formed • Super Plastic Forming • Survival Probability Function • System Productivity Facility • System Programming Facility

SPFA: Single-Point Failure Analysis

SPFD: Spectral Power Flux Density

SpFl: Spaceflight (periodical code)

SPFL: Single-Point Failure List

SPFM: Single-Point Failure Mode

SPFP: Single-Pass Fit Program • Single-Point Failure Potential

SPFPAD: Spacecraft Performance and Flight Path Analysis Directorate

SPFR: Southern Polarity Flare Ribbon • Spectral Flux Radiometer

SPFS: Small Payload Flight System

SPFT: Strong Permittivity Fluctuation Theory

Spg: Spring

SPG: Saint Petersburg - Whitted (USA-FL) (location identifier) • Schweizerische Physikalische Gesellschaft (Switzerland) • Short-Pulse Generator • Single-Point Ground • Sort Program Generator • Source Power Gain • Special Projects Group • Spring • Système de Préférences Généralisées

SPGAI: Surface Position Gimbal Angle Indicator

Sp. Gr.: Specific Gravity

SPGS: Springs

SPH: Smoothed Particle Hydrodynamics (*) • Smooth Particle Hydrodynamics • Soft-Particle Hydrodynamics • Sopu (Papua New Guinea) (location identifier)

Spher: Spherical

SPHINX: Space Plasma High-voltage Interaction Experiment

SPhN: Service de Physique Nucléaire

SPHN: Société de Physique et d'Histoire Naturelle (Switzerland)

SPhT: Service de Physique Thérique

Sp. Ht.: Specific Heat

spi: Spissatus

SPI: Schedule Performance Index • Sciences Physiques pour l'Ingénieur • Sciences Pour l'Ingénieur • Security Parameters Index • Security Protocol Identifier • Segmented Pupil Imager • Selected Period Investment • Selective Profile Indexing • Serial Peripheral Interface • Serial to Parallel Interface • Service de la Propagande et de l'Information • Service Provider Interface • Single-Process Initiative (*) • Single-Program Initiation • Software Patent Institute • Software Process Improvement • Space Power Institute (USA-AL) • Spectrographic Particle Imager • Spectrometer Integral • Springfield (USA-IL) (location identifier) • Surface Position Indicator • Surveillance Principal Investigator • Systèmes et Périphériques Informatiques • System Performance Indicator • System Provider Interface

SPICA: Special Instrument for Cosmic-ray Astrophysics • Support Program for Instructional Competency in Astronomy

SPICE: Shuttle Payload Integration Cargo Evaluation • Simulator Program for Integrated Circuit Emulator • Software Process Improvement and Capability • Solar-Particle Intensity Composition Experiment • Spacecraft - Planet - Instrument - C-matrix - Events • Space Integrated Control Experiment • Space Particle Impact Capture Experiment • Space Power Internal Combustion Engine • Spacelab Payload Integration and Coordination in Europe • Suprathermal Plasma Investigation of Cometary Environments • Survey Probe Infrared Celestial Experiment • Survey Program of Infrared Celestial Experiments

SPID: Service Profile Identification • Service Profile Identifier • Speckle Imaging by Deconvolution

SPIDAB: Space Industrial Database

SPIDASL: Space Plasma Interactive Data Analysis and Simulation Laboratory

SPIDER: Solar-Powered Ion Driven Eclipsing Rover

SPIDF: Support-Planning Identification File

SPIDPO: Shuttle Payload Integration and Development Program Office

SPIDR: Space Physics Interactive Data Resource

SPIDS: Stabilized Payload Interface Deployment System

SPIE: Shuttle Plume Impingement Experiment • Simulated Problem Input Evaluation • Society of Photo-optical Instrumentation Engineers

SPIEL: Signal Processing Interactive Extensible Language

SPIF: Shuttle Payload Integration Facility • Shuttle-POCC Interface Facility • Shuttle Processing Interface Facility • Small Photometric Imaging Facility • Spacecraft Planetary Imaging Facility • Standard Payload Interface

SPIFEX: Shuttle Plume Impingement Flight Experiment

SPIFI: South Pole Imaging Fabry-Pérot Interferometer

SPIFR: Single-Pilot Instrument Flight Rules

SPIHT: Set Partitioning In Hierarchical Trees

SPII: Shuttle Program Implementation Instruction

SPIKE: Science Planning Interactive Knowledge Environment • Science Planning using Intelligent Knowledge-based Environment • Scientific Planning Interactive Knowledge Expert

SPIM: Spectrographic Imager

SPIMS: Shuttle Program Information Management System

SpIN: Space Industry News (CSIRO)

SPIN: Science Policy Information News • Searchable Physics Information Notices • Software Process Improvement Network • Space Interdisciplinary Network • Space Physics Information Network • Special-Purpose Inexpensive Satellite Program

SPINE: Space Informatics Network Experiment

SPINES: Science Policy Information Exchange System

SPINR: Spectrograph for Photometric Imaging with Numeric Reconstruction

SPINS: Stabilized Platform Inertial Navigation System

SPIPR: Sistema de Procesamiento de Imagenes de Perceptión Remota

SPIR: Spectropolarimeter for Infrared

SPIRAC: South Pole Infrared Array Camera

SPIRAL: Surface Photometry Interactive Reduction and Analysis Library

SPIRE: Sensor Performance in a Radiation Environment • Spatial Inertial Reference Equipment • Spatial Paradigm for Information Retrieval and Exploration • Spectral and Photometric Imaging Receiver • Spectral Infrared Rocket Experiment • Stop-Protect-Insulate-Rewarm

SPIRES: Stanford Public Information and Retrieval System (USA-CA) • Stanford Public Information Retrieval System (USA-CA)

SPIREX: South Polar Infrared Explorer • South Pole Infrared Explorer (*)

SPIRIT: Space Infrared Imaging Telescope • Space Infrared Interferometry Trailblazer • Spatial Infrared Imaging Telescope • Spatial Infrared Rocket-borne Interferometric Telescope

SPIRITS: Spectral and In-band Radiometric Imaging of Targets and Scenes

SPIRS: SuperJANET Project on Information Resources

SPISS: Shuttle Polar Ice Sheet Sounder

SPIT: Selective Printing of Items from Tapes • Starburst Pixel Interleaving Technique • Strategic Plan for Information Technology

SPJ: Service Pages Jaunes • Sparta (Greece) (location identifier)

SPK: Sapporo (Japan) (location identifier) • Spacecraft-Planet Kernel

SPKAC: Signed Public Key And Challenge

SPKI: Simple Public Key Infrastructure

Spkl: Sprinkler

SPKL: Sprinkle

Spkr: Speaker

SPKR: Sprinkler

spl: Sine Prole Legitima

SPL: Scratch Pad Line • Sección Planetaria y Lunar • Secondary Payload • Serialized Parts List • Set Priority Level • Signal Processing Language • Simple Programming Language • Sonic Pressure Level • Sound Pressure Level • Space Photography Laboratory • Space Physics Laboratory • Space Programming Language • Spare Parts List • Special-Purpose Language • Spellchecker • Split-Phase Level • Spooler • Sun-Pumped Laser • Supplies-Procurement-Logistics • System Programming Language

SPLAASH: Spacecraft Protective Landing Area for the Advancement of Sciences and Humanities

SPLE: Singular Power Law Ellipsoid

SPLIT: Space Program Language Implementation Tool • Spectro-Polarimetric Lightweight Telescope

SPLL: Self-Propelled Launcher Loader

SPLM: Space Programming Language Machine

Splmt: Simplement

SPLNS: South Plains (USA)

SPLOM: Scatter Plot Matrix

Splt: Spotlight

Spltr: Splitter

Sply: Supply

spm: Sine Prole Mascula

SPM: Miquelon (Saint Pierre et_) (ISO country code) • Saint Pierre et Miquelon (ISO country code) • San Pedro Mártir (Mexico) • Scanning Probe Microscope • Scanning Probe Microscopy • Scanning Proton Microscope • Scanning Proton Microscopy • Sciences Physiques et Mathématiques • Scratch Pad Memory • Security Plug-in Module • Seeker and Processor Model • Sequential Processing Machine • Simple Plasma Monitor • Single Plasma Monitor • Small-Powers Minimization • Smoothed Particle Method • Software Performance Monitor • Software Programmer's Manual • Software Project Management • Solar Photometer • Solar Polar Mission • Solar Probe Mission • Solar Proton Monitor • Sonnephotometer • Source Program Maintenance • Southern Proper Motion • South

Proper Motion • Spangdahlem (Germany) (location identifier) • Spectromagnetograph • Spectrophosphorimeter • Spectrophotomètre Multibandes • Standard Parts Manual • Standard Payload Module • Standard Practice Manual • Standard Practice Memo • Standard Procedure Manual • Standard Process Manual • Subproject Manager • Sue Proprie Mani • Summary for Policy Makers • Sun Pointing Mode • Sun Point Mode • Sun Probe Mars • Superparamagnetic • Support Program Management • Suspended Particulate Matter • Symposium • System Performance Monitor • Systems Program Manager

SpMar: Space Markets (periodical code)

SPMD: Single Program/Multiple Data • Single Program/Multiple Data-stream

SPME: Solid-Phase Micro Extraction

SPMF: Solar Photosphere and Magnetic Fields

Spmkt: Supermarket

SPMM: Schweizerisches Patent, Muster und Markenblatt

SpMME: Space Medicine Medical Engineering (periodical code)

SPMO: San Pedro Mártir Observatory (Mexico)

SPMP: Software Project Management Plan

SPMS: Software Process Modelling System • Solar-Particle Monitoring System • Special-Purpose Manipulator System

SPN: Saipan (Mariana Islands) (location identifier) • Servicio de Parques Nacionales • Shuttle Project Notice • Special Program Number • Spectropyranometer • Stop Press News

SPND: Self-Powered Neutron Detector

SPNG: Spring

SPNGS: Springs

SPNFZ: South Pacific Nuclear Free Zone

SPNFZT: South Pacific Nuclear Free Zone Treaty

Spnr: Spanner

SPO: Sacramento Peak Observatory (USA-NM) • San Pablo (Spain) (location identifier) • Schmidt Prisme-Objectif • Section des Publications Officielles • Senior Press Officer • Short-Period Oscillation • Shuttle Program Office • Shuttle Project Office • South Pole Observatory • Spacecraft Program Office • Spacelab Program Office • Spare Parts Order • Special Projects Office • State Planning Office • System Program Office • System Project Office

SPOA: Single Point Of Administration

SPOC: Satellite Payload Operations Centre • Search Point Of Contact • Shuttle Payload Operations Contract • Shuttle Portable Onboard Computer • Single Point Of Control • Single-Point Orbit Calculator • Singularly Perturbed Optimal Control • Spacecraft Payload Operations Centre • SPAS Payload Operations Center

SPOCC: SBV Processing, Operations, and Control Center • South Pacific Organizations Coordinating Committee

SPOCT: SBV Processing, Operations, and Control Team

SPODP: Single-Precision Orbit Determination Program

SPOE: Standard Payload Outfitting Equipment

SPOF: Science Planning and Operations Facility (*) • Science Planning Operational Facility

SPONG: Secrétariat Permanent des Organisations Non-Gouvernementales

SPOOL: Simultaneous Peripheral Operations Online

S'pore: Singapore

SPORTS: Second-generation Portable Remote Telecommunications System • Simplified Payload/Orbiter Thermal Simulator

SPOS: Solid-Phase Organic Synthesis

SPOT: Satellite Pour l'Observation de la Terre • Satellite Probatoire d'Observation de la Terre • Smithsonian Precision Optical Tracking • South Pole Optical Telescope • Système Probatoire d'Observation de la Terre • Système Probatoire d'Observation Terrestre

Spp: Species

SPP: Menongue (Angola) (location identifier) • Scalable Parallel Protocol • Schwerpunktsprogramm • Science and Public Policy • Science Power Platform • Secular Periodic Perturbation • Security Practices and Procedures • Sequenced Packet Protocol • Service de Physique des Particules • Serviço de Programas e Projectos • Signal Preprocessor • Simulation Planning Panel • Single Process Purge • Solar Photometry Probe • Solar Physics Payload • Solar Power Platform • Solid Performance Program • Space Plasma Physics • Space Policy Project • Standard Parallel Port • Standard Practice Procedure • Subset Preprocessor • Swiss Priority Programme • Switched Poisson Process

SPPD: Space Propulsion and Power Division

SPPIL: Shuttle Preferred Pyrotechnic Items List

SPPL: Spare Parts Provisioning List

SPPM: System Performance Prediction Model

SPPO: Spacelab Payload Project Office

SpPol: Space Policy (periodical code)

SpPow: Space Power (periodical code)

SPPPI: Secrétariat Permanent pour la Prévention des Pollutions Industrielles (CSG)

SPPS: Service de Programmation de la Politique Scientifique • Subsystem Program Preparation Support • System Plot Package Simulator

SPQ: San Pedro (USA-CA) (location identifier) • Sistema Portugês da Qualidade • Sociedade Portuguesa de Química (Portugal)

SPQL: Space Physics Query Language

SPQR: Senatus Populusque Romani • Small Profit and Quick Return

Spr: Spring • Sprinkle

SpR: Seconds per Revolution • Seconds per Rotation

SPR: San Pedro (Belize) (location identifier) • Satellite Parametric Reduction • Seconds Per Revolution • Seconds Per Rotation • Sequential Probability Ratio • Single-Point Resistance • Smoothed Particle Rendering • Software Performance Report • Software Problem Report • Solid-Propellant Rocket • South Polar Region • Special Program Requirement • Special Project Report • Special-Purpose Register • Subcontractor Performance Review • Supplementary Progress Report • Supply Performance Report • Surface Plasmon Resonance • System Performance Rating • System Program Review

SPRAG: STS Payload Requirements and Analysis Group

SPRD: Spread • System Performance Requirements Document

SPRDG: Spreading

Sprdr: Spreader

SPRE: Solid-Propellant Rocket Engine • Spartan Packet Radio Experiment

SPREC: South Pacific Regional Environment Convention

SPREE: Shuttle Potential and Return Electron Experiment

Spreew: Spreewald (Germany)

SPREP: South Pacific Regional Environmental Programme • South Pacific Regional Environmental Protection • South Pacific Region Environmental Programme • South Pacific Region Environmental Programme

SPREX: Spring Runoff Experiment

SPRF: Space Propulsion Research Facility • Special-Purpose Receiving Facility

SPRI: Scott Polar Research Institute (UK) • Space Physics Research Institute (South Africa)

SPRINTEL: Speedy Retrieval of Information on the Telephone

SPRITE: Solid-Propellant Rocket Ignition Test and Evaluation • Storage, Processing and Retrieval of Information in a Technical Environment

Sprl: Spiral

SPRL: Société de Personnes à Responsabilité Limitée • Space Physics Research Laboratory • Spiral

SPRMD: Space, Power Resources, and Manufacturing Development (periodical code)

SPRNG: Spring

SPRNGS: Springs

SPROB: Solid Propellant Space Booster Plant

Sprt: Support

SPRT: Scientific Programme Review Team • Sequential Probability Ratio Test • Single-Pole Single-Throw

SPRU: Science Policy Research Unit • Spacecraft Power Regulator Unit • Standard Power Regulator Unit

Sprvsr: Supervisor

sps: Sine Prole Supersite

SpS: Samples per Second • Symbols per Second

SP&S: Special Processes & Sequencing

SPS: Samples Per Second • Sanitaire et Phytosanitaire • Sanitary and Phytosanitary • Satellite Power System • Scientific Power Switching • Secondary Power System • Secondary Propulsion System • Self-Provided Summary • Service Propulsion System • Short Prototype String • Shuttle Procedures Simulator • Shuttle Processing Site • Signal Processing System • Single-Pole Switch • Society of Physics Students • Soft-Particle Spectrometer • Software Productivity Solution • Software Product Specification • Solar-Panel Substrate • Solar-Physics Section • Solar-Power Satellite • Solar-Power System • Solar-Probe Spacecraft • Source Preparation Segment • South Polar Spot • South Pole Station • Space Peace Sculpture • Space Power System • Spacecraft Propulsion System • Special Processes and Sequencing • Spectral Planning Subcommittee • Standard Positioning Service • Standard Process Specification • Standby Power Supply • Standby Power System • Statement of Prior Submission • Static Power System • Statistical Problem Solving • Stellar Photometer Software • Stellar Photometry Software • Stellar Population Synthesis • Stereo Photographic System • Stored Program Simulator • Super Proton Synchrotron • Swiss Physical Society (Switzerland) • Symbolic Programming System • Symbols Per Second • Syndiotactic Polystyrene • Wichita Falls - Sheppard (USA-TX) (location identifier)

SPSA: Signal-Processor Subassembly • Simple Problem Statement Analyzer

SPSC: Signal Processing and Spectral Control

SPSD: Space Power Systems Division • Systems and Project Support Department

SPSE: Spacecraft Power System Electronics

SPSF: Spectral Point Spread Function

SPSI: Society for the Promotion of Scientific Industry (UK)

SPSL: Society for the Protection of Science and Learning • Spare Parts Selection List

SPSME: Spacelab Payload Standard Modular Electronics

SPSN: Submitted Package Sequence Number

SPSO: Science Processing Support Office • Science Programs Selection Office • Senior Principal Scientific Officer

SPSP: Spare Parts Support Package

SPSS: Science Planning and Scheduling System • Statistical Package for the Social Sciences

SPST: Science Planning and Scheduling Team • Single Pole - Single Throw

SPSTD: Spectrophotometric Standard

Sp. Surf.: Specific Surface

Spt: Seaport • Support

SpT: Sectors per Track

SPT: Scaled-Particle Theory • Sectors Per Track • Signal Processing Technique • Silicon Planar Transistor • Silicon-Powered Transistor • Sipitang (Malaysia) (location identifier) • Small-Perturbation Theory • Software Porting Tool • Space Power Tool • Special Performance Test • Special-Purpose Test • Stationary Plasma Thruster • Steering Position Transducer

SPTC: Specified Period of Time Contract

SPTD: Signal Processor Techniques Department • Supplemental Provisioning Technical Documentation

SPTDP: South Pacific Telecommunications Development Programme

SPTF: Signal Processing Test Facility

SPTI: Structural Program Test Input

SPTLB: Short-Period Tidally-Locked Binary

SPTMLE: Shrinkage Preliminary Test Maximum-Likelihood Estimator

SPTN: SPARTAN

SPTP: Space Physics Theory Program • Space Power Test Project • Special-Purpose Test Program

SPU: Scientific Programs Unit • Self-Provided Underlining • Service Propulsion Unit • Signal Processing Unit • Sky Protection Unit • Spacecraft Processing Unit • Split (Croatia) (location identifier) • System Power Unit • System Processor Unit

SPUC: Serial Peripheral Unit Controller

SPUDS: Single Platform Uniform Diagnostic System

SPUIAGG: Secção Portuguesa das Uniões Internacionais Astronômica e Geodesica e Geofisica (Portugal)

SPUP: Space Platform Users Project

SPUR: Separate Parts Usage Record

SPV: Sepik Plains (Papua New Guinea) (location identifier) • Slow Phase Velocity • System Processing Unit

SPVC: Soft Permanent Virtual Circuit

SPVM: Saddle-Point Variational Method

Spvn: Supervision

SPVPF: Shuttle Payload Vertical Processing Facility

Spvw: Spaceview (periodical code)

SPW: Sheath-Plasma Wave • Signal Processing Worksystem • Spencer (USA-IA) (location identifier)

SpWd: Space World (periodical code)

SPWG: Science Planning Working Group • Scientific Priorities Working Group • Spare Parts Working Group

SPX: Houston Spaceland (USA-TX) (location identifier) • Sequenced Packet Exchange

SPY: San Pedro (Ivory Coast) (location identifier) • Spray

SPyCAT: Space Physics Catalog

SPYDER: System to Provide You Data from Earthquakes Rapidly

SPZ: Sovetskaya Peremenna ya Zvezda • Springdale (USA-AR) (location identifier)

sq: Albanian (ISO language code) • Sequens • Sequentia

Sq: Sequence • Squadron • Square • Squawk

SQ: Safety Quotient • Shocked Quartz • Significant Quantity • Social Question • Squall • Square • Squeezed • Standard Quality • Stephan's Quintet • Stereophonic-Quadraphonic • Survival Quotient

SQA: Shocked Quartz Aggregate • Software Quality Assurance • Subcontractor Quality Assurance • System Queue Area

SQAL: Squall

SQAP: Software Quality Assurance Plan

SQB: Santa Ana (Colombia) (location identifier) • Space-Qualified Booster

SQBC: Space-Qualified Booster Charger

SQC: Southern Cross (Australia-WA) (location identifier) • Statistical Quality Control

sq. cm.: Square Centimeter

SQCP: Statistical Quality Control Procedure

SQD: Signal Quality Detector • Standards and Quality Division

Sqdn: Squadron

SQE: San Luis de Pale (Colombia) (location identifier) • Signal Quality Error • Software Quality Evaluation • Supplier Quality Evaluation

SQEP: Software Quality Evaluation Plan

SQF: Software Quality Forum • Solano (Colombia) (location identifier) • Solidification Quenching Furnace

SQFD: Software Quality Function Deployment

sq. ft.: Square Foot

SQG: Sintang (Indonesia) (location identifier) • Small Quantity Generator

SQH: Supplier Quality History

SQI: Rock Falls - Whiteside County (USA-IL) (location identifier) • Software Quality Investment • Sterling (USA-IL) (location identifier)

SQID: Simultaneous Quad Infrared Detector • Super Quantum Interface Device

SQIDS: Satellite Quick-look Interactive Display System

SQIID: Simultaneous Quad-color Infrared Imaging Device • Simultaneous Quad Infrared Imaging Device

sq. in.: Square Inch

SQK: Sidi Barani (Egypt) (location identifier)

sq. km.: Square Kilometer

SQL: San Carlos (USA-CA) (location identifier) • Space-Qualified Laser • Standard Quantum Limit • Standard Query Language • Structural Query Language • Structured Query Language (*) • System Query Language

SQLC: Structured Query Language Constructor

SQLDA: Structured Query Language Descriptor Area

SQLN: Squall Line

SQLNS: Squall Lines

SQLPT: Structured Query Language Parser and Translator

sq. m.: Square Meter

SQM: São Miguel do Araguaia (Brazil-GO) (location identifier) • Software Quality Maintenance • Square Meter • Stochastic Quantization Method • Strange Quark Matter

sq. mi.: Square Mile

sq. mm.: Square Millimeter

SQN: Sanana (Indonesia) (location identifier) • Spin Quantum Number

SQO: Storuman (Sweden) (location identifier)

SQP: Sequential Quadratic Programming • Starcke (Australia-QLD) (location identifier)

SQPI: Software Quality Profitability Index

SQPN: Staggered Quadriphase Pseudo-random Noise

SQPSK: Staggered Quadrature Phase Shift Keying

sqq: Sequentia • Sequentes

sq. r.: Square Rod

SQR: Software Quality Revenues • Soroako (Indonesia) (location identifier) • Square

sq. rd.: Square Rod

SQRE: Square

sq. rt.: Square Root

SQRT: Square Root

SQS: San Ignacia (Belize) (location identifier) • Securing Quality System • Squares • Stochastic Queuing System

SQT: Sphere Quadtree • Student Quick Take (ACM) • System Qualification Test

SQU: Square

SQUARE: Specifying Queries As Relational Expressions

SQUEL: Spatial Query Language

SQUID: Simultaneous Quad Imaging Device • Superconducting Quantum Interface Detector • Superconducting Quantum Interface Device • Superconducting Quantum Interference Detector • Superconducting Quantum Interference Device (*)

SQV: Sequim (USA-WA) (location identifier)

SQW: Square Wave

sq. yd.: Square Yard

SQZ: Scampton (UK) (location identifier)

s/r: Su Remesa

sr: Serbian (ISO language code) • Steradian • Sunrise

Sr: Senhor • Senior • Señor • Sir • Sister • Strontium • Strouhal Number

S/R: Sale or Return • Send and Receive • Shift Register • Stimulus/Response

S&R: Storage & Retrieval

S-R: Stimulus-Response

SR: Safety Recommendation • Safety Requirement • Safety Review • Sample Rate • Saturable Reactor • Saudi Riyal (currency) • Scanning Radiometer • Scan Ratio • Schedule Result • Scientific Report • Scientific Research • Search and Retrieval • Search and Retrieve • Section Report • Selenium Rectifier • Self-Raising • Semantic Reaction • Semiregular • Senate Resolution • Send and Receive • Senhor • Server Ring • Service Record • Service Report • Seychelles Rupee (currency) • Shift Register • Shipping Receipt • Shop Release • Short Range • Short Rate • Shutdown Request • Signature Result • Simulation Report • Size Resolving • Slow Release • Software Requirement • Solar Radiation • Solar Reference • Solid Rocket • Source Route • Spares Requirement • Special Relativity • Special Report • Specific Range • Specification Requirement • Squareness Ratio • Sri Lanka • Staff Report • Standard Rate • Standard Repair • Standard Requirement • Standardization Report • Star Route • Statement of Requirements • Station Regulation • Status Register • Status Report • Status Review • Steradian • Stimulus-Response • Stochastic Resonance • Stoichiometric Ratio • Storage Room • Strehl Ratio • Subroutine • Summary Report • Sunderland (UK postal code) • Sunrise • Supplemental Report • Supplementary Regulation • Supporting Research • Support Request • Support Requirement • Support Room • Suriname (ISO country code) • Surveillance Requirement • Sveriges Radio • Synchrotron Radiation • Synthetic Rubber • System Requirement

Sra: Senhora • Señora

SR&A: Supporting Research & Analysis

SRA: Santa Rosa (Brazil-RS) (location identifier) • Satanic Ritual Abuse • Scanning Radar Altimeter • Science Research Associate • Selective Routing Arrangement • Senhora • Society of Risks Analysis • Special Rules Area • Spin Reference Angle • Spin Reference Axis • Stability Robustness Analysis • Staff Research Associate • Standard Radiation Atmosphere • Standard Range Analyzer • Standards of Readiness and Availability • Stanford Radiation Atmosphere • State Recreation Area (USA) • Static

Robotic Actor • Stochastic Relaxation with Annealing • Sulforicinoleic Acid • Supporting Research and Analysis • Support Requirements Analysis • Surveillance Radar Approach • System Reaction Analysis • System Replacement Assembly • System Requirements Analysis • Systems Research Aircraft

SRAC: Société Royale d'Astronomie du Canada

SRAD: Shuttle Radar Assembly Demonstration (*) • Shuttle Radiator Assembly Demonstration • Space Radiator Assembly Demonstration • Spectral Radiometer

SRAE: Solar Radio Astronomy Experiment

SRAEC: Scene Radiation and Atmospheric Effects Characterization

SRAG: Space Radiation Analysis Group

SRAM: Scientific Review of Alternative Medicine • Sequential Random Access Memory • Shadow Random Access Memory • Static Random Access Memory (*)

SRAP: Système de Reconnaissance Automatique de la Parole

SRAT: Short-Range Advanced Technology

SRATS: Solar Radiation And Thermospheric Structure

SRB: Safety Review Board • Santa Rosa (Bolivia) (location identifier) • Service Request Block • Software Review Board • Solar Radiation Budget • Solar Reflectory Beacon • Solid-fuel Rocket Booster • Solid Rocket Booster • Source-Route Bridge • Source-Route Bridging • Source Routing Bridging • Specialist Regional Body • Storage Resource Loader • Surface Radiation Budget • System Requirement Baseline • System Review Board

SRBA: Société Royale Belge d'Astronomie, de Météorologie et de Physique du Globe (Belgium)

SRBAB: Solid-Rocket Booster Assembly Building

SRBDB: Solid-Rocket Booster Disassembly Building

SRBL: Solar Radio Burst Locator

SRBP: Synthetic Resin-Bonded Paper

SRBSEP: Solid Rocket Booster Separation

SRC: Safety Review Committee • Sample Recovery Container • Sample Return Canister • Sample Return Container • Sample Rock Container • Saskatchewan Research Council (Canada-SK) • Science Research Council • Science Review Committee • Searcy (USA-AR) (location identifier) • Semiconductor Research Corporation • Send Register Control • Se Ruega Contestación • Sociedad Regular Colectiva • Société de Recherche sous Contrat • Solvent-Refined Coal • Source • Southern Redshift Catalogue • Southwest Research Corporation • Space Regatta Consortium • Space Research Council • Spares Receiving Checklist • Standard Reference Code • Standard Reference Compound • Station Reliability Coordinator • Stray Radiation Chamber • Support Review Code • Synchrotron Radiation Center

SRCB: Simulation Requirements Control Board • Software Requirements Change Board

SRCBD: Software Requirements Change Board Directive

SRCD: Set-Reset Clocked Data

srCDM: Standard Reionized Cold Dark Matter (model)

Srch: Search

SRCM: Slow Rotating Centrifuging Microscope

SRCT: Standard Recovery Completion Time

SRD: San Ramón (Bolivia) (location identifier) • Scheduled Release Date • Schellreferatdienst • Science Requirements Document • Screen Reader • Screen-Reader Device • Selective Radiation Detector • Selective Retrieval of Documentation • Short-Range Device • Shuttle Requirements Definition • Shuttle Requirements Document • Simultaneous Ranging Difference • Small Rigid Dome • Software Requirements Document • Spacecraft Research Division • Standard Reference Data • Station de Réception Directe • Step Recovery Diode • System-derived Requirements Document • Système de Recherche Documentaire • Systems Requirements Document

SRDC: Space Research in Developing Countries • Standard Reference Data Center • Surface Reference Data Centre

SRDCS: Simulation Reconfiguration Data Collection System

SRDH: Systems Requirements Definition Handbook

SRD&I: Space Data Resources & Information

SRDP: Standard Reference Data Program

SRDS: Single Requirements Determination System • Standard Reference Data System

SRE: Secondary Reflection Effect • Send Remote Enable • Shrinkage Restricted Estimator • Software Reliability Engineering • Sounding Rocket Experiment • Stellar Research and Education • Sucre (Bolivia) (location identifier) • Surveillance Radar Element • System Research Environment

SR&ED: Scientific Research & Experimental Development

s/ref: Su Referencia

SREL: Space Radiation Effects Laboratory

SREM: Software Requirements Evaluation Methodology

SRES: Special Report on Emission Scenarios

SRF: San Rafael - Hamilton (USA-CA) (location identifier) • Shuttle Refurbish Facility • Shuttle Refurbishment Facility • Solar Radiation Facility • Solar Radiation Flux • Solar Radiation Function • Spectral Response Function • Stable Radio Frequency • State Revolving Fund • Status Re/45tpFrame • Systematic Reconnaissance Flight • System Requirements Handbook

SRFF: Set-Reset Flip-Flop

SRG: Guilder (Suriname) (ISO currency code) • Scientific Review Group • Security Rapporteur Group • Semarang (Indonesia) (location identifier) • Shift-Register Generator • Small Research Grant • Special Research Group • Spectrum Röntgen Gamma • Spinning Rotor Gauge • SRON Groningen (Netherlands)

SRGP: Simple Raster Graphics Package

Srgto: Sargento

SRH: Sarh (Chad) (location identifier) • Search • Searching • Systems Requirements Handbook

SRHE: Society for Research into Higher Education

SRHIT: Small Radar Homing Interceptor Technology

SRHP: Space Radiation Health Program

SRHS: Sum of Right-Hand Side

SRI: Sacrum Romanum Imperium • Samarinda (Indonesia) (location identifier) • Service des Relations Internationales • Soil Research Institute • Space Research Institute • Spatial Reference Information • Stanford Research Institute (USA-CA) • Statistical Reference Index • Système de Recherche d'Informations • Système de Référence Inertielle

SRIC: Short-Rotation Intensive Culture

SRIF: Square Root Information Filter

SRIM: Selected Research In Microfiche • Simulated Radar Image Model

SRIP: Security Requirements Implementation Plan

SRIS: Science and Reference Information Service • Science Reference Information Service • Station de Réception des Images Spatiales

SRIT: Système de Rectification d'Images en Télédétection

SRIV: Station de Réception d'Images pour la Végétation

SRJ: San Borja (Bolivia) (location identifier)

SRK: Sierra Leone (location identifier)

Srl: Società a Responsabilità Limitata

SRl: Saudi Riyal (currency)

SRL: Santa Rosalia (Mexico) (location identifier) • Scientific Research Laboratory • Shuttle Research Laboratory • Single-Rack Laboratory • Space Radar Laboratory • Space Radiation Laboratory • Speed of Relative Movement • SRON Leiden (Netherlands) • Standard Reference Library • Study Reference List • Systems Requirements Letter

SRM: Sandringham (Australia-QLD) (location identifier) • Scalable Reliable Multicast • Scanning Radar Microscope • Security Reference Monitor • Shift Register Memory • Solid-fuel Rocket Motor • Solid Rocket Motor • Spares Requirement Model • Specifications Requirement Manual • Specified Risk Material • Square Root Mode • Stability Robustness Measure • Standard Reference Material • Standard Reference Module • Strength Ratio Method • Structural Risk Minimization • Su Real Majestad • System for Resources Management • System Resources Manager

SRMC: Stimulus-Response Measurements Catalog

SRMD: Slant Range Measuring Device

SRML: Solar Radiation Monitoring Laboratory

SRMLE: Shrinkage-Restricted Test Maximum-Likelihood Estimator

SRM&QA: Safety Reliability, Maintainability & Quality Assurance

SRMQA: Safety Reliability, Maintainability, and Quality Assurance

SRMS: Shuttle Remote Manipulator System

SRMU: Solid Rocket Motor Upgrade

SRN: Serial Reference Number • Serpens Reflection Nebula • Simulation Reference Number • Software Release Note • Software Release Notice • Specification Revision Notice • Star in Reflection Nebula • Strahan (Australia-TAS) (location identifier) • Stretched Receptor Neurone

SRNU: Service du Réacteur Nucléaire Universitaire

SRO: Santana Ramos (Colombia) (location identifier) • Self-Regulatory Organization • Senior Resource Official • Sensitive Reconnaissance Operation • Sharable and Read Only • Short-Range Order • Single-Room Occupancy • Site Recovery Option • Solar Radio Observatory • Specification Release Order • Standing Room Only • Statutory Rules and Orders • Stony Ridge Observatory (USA-CA) • Superintendent of Range Operations • Supervisor of Range Operations

SROC: Space Remote Operations Center

SROD: Shared Read-Only Data

SROM: Spectro-Radiometer for Ocean Monitoring

SRON: Space Research Organization Netherlands • Stichting Ruimte Onderzoek Nederland (Netherlands)

SROSS: Stretched Rohini Satellite Series

SRP: Sale-Recommended Price • Science Review Panel • Scientific Research Proposal • Seat Reference Plane • Seat Reference Point • Short-Range Plan • Short-Range Planning • Solar Radiation Pressure • Soluble Reactive Phosphorus • Spatial Reuse Protocol • Special Research Programme • Stabilizer Reference Plane • Stabilizer Reference Plane • Standard Repair Procedure • Stord (Norway) (location identifier) • Stratospheric Research Program • Suggested Retail Price • Surplused Research Professor • Sustained Readiness Program • Switch Route Processor • Système de Retenue Programmée

SR&PA: Safety, Reliability & Product Assurance

SRPA: Safety, Reliability, and Product Assurance • Small Reclamation Projects Act • Spherical Retarding Potential Analyzer

SRPI: Server-Requester Programming Interface

SRPS: Secure Record and Playback System • Sensor-Referenced Positioning System • Subrecursive Programming System

SR&Q: Safety, Reliability & Quality

SRQ: Bradenton (USA-FL) (location identifier) • Sarasota (USA-FL) (location identifier) • Service Request • Service Request Queue

SR&QA: Safety, Reliability & Quality Assurance

SRQA: Safety, Reliability, and Quality Assurance

SRR: Scanned Relative Response • Serially Reusable Resource • Short-Range Radio • Shuttle Requirements Review • Site Readiness Review • Slow Rotation Room • Software Requirements Review • Solar Radiation Response • Specification Readiness Review • Stradbroke Island - Dunwich (Australia-QLD) (location identifier) • Subcontractor Readiness Review • System Readiness Review • System Requirements Review

SRRC: Synchrotron Radiation Research Center

SRRF: Solar Radiation Response Function

SRRID: Site Readiness Review Item Document

SRS: Fellow of the Royal Society • Safety and Reliability Society • Same Runway Separation • San Marcos (Colombia) (location identifier) • Satellite Receiving Station • Satellite Remote Sensing • Schweizerische Radiosonde • Science Requirements Strategy • Seismic Recording System • Selective Record Service • Shared Registry System • Silicon Readout Sequence • Simple Random Sample • Simulated Raman Scattering • Simulated Remote Site • Simulated Remote Station • Simulation Reconfiguration System • Simulation Recording System • Single Resonance Structure • Small Research Satellite • Smog Regulation System • Societatis Regiae Sodalis • Software Requirements Specification • Solar Radiation Simulator • Sortie Réservée Synchrone • Sounding Rocket System • Sound Retrieval System • Southern Reference Stars • Southern Reference System • South Research Station • Space Recovery System • Space Research Service • Spatial Reference System • Specification Requirement Sheet • Specification Review Sheet • Specification Revision Sheet • Speed Reference System • Stanford Research Systems (USA-CA) • Statistical Reporting Service • Stellar Radiation Sensor • Stimulated Raman Scattering • Superluminal Radio Source • Supplemental Restraint System • Support Requirement System • Synchro Retransmission System • Synchrotron Radiation Source

SRSA: Scientific Research Society of America

SRSC: Satellite Remote Sensing Center • Space Remote Sensing Center

SRSCC: Simulated Remote Station Control Center • Simulated Remote Station Control Console

SRSG: Science Review Steering Group

SRSI: Simulated Reusable Surface Insulation

SRSL: Société Royale des Sciences de Liège (Belgium)

SRSR: Schedule and Resources Status Report

SRSS: Shuttle Range Safety System • Solar Radiation Simulator System • Square-Root Sum of Squares

SRSU: Sul Ross State University (USA-TX)

SR&T: Science Research & Technology • Support Research & Technology • Supported Research & Technology • Supporting Research & Technology

SRT: Safety Review Team • Science Recommendation Team • Science Research and Technology • Shuttle Requirements Traceability • Small Radiotelescope • Solar Radiation Test • Solar Radio Telescope • Soroti (Uganda) (location identifier) • Source Route Transparent • Source Routing Transparent • Spacecraft Received Time • Special Rated Thrust • Specification Requirements Table • Station Readiness Test • Step Recovery Transistor • Student Radio Telescope • Sun Reference Sensor • Supporting Research and Technology • Sweeney-Robertson-Tocher • Synthese Radio Teleskoop • System Reaction Time • System Readiness Test • System Reinvention Team • System Reliability Test • System Requirements Test

Srta: Senhorita • Señorita

SRTB: Source Route Transparent Bridging

SRTC: Special Real-Time Command

SRTE: Scalar Radiative Transfer Equation

SRTF: Subregional Task Force

SRTG: Source-Route Translation Gateway

SRTM: Shuttle Radar Terrain Mapping • Shuttle Radar Topography Mission (*)

SRTN: Solar Radio Telescope Network

SRTS: Synchronous Residual Time Stamp

SRU: Santa Cruz (USA-CA) (location identifier) • Shop Repairable Unit • Shop Replaceable Unit • Signal Responder Unit • Simulation Resource Utilization • Space Replaceable Unit • SRON Utrecht (Netherlands) • Support Resource Unit • System Register Unit • System Replaceable Unit

Srv: Service

SRV: Satellite Reentry Vehicle • Scientific Remote Viewing • Semiregular Variable • Southern Republic of Vietnam • Space Rescue Vehicle • Stony River (USA-AK) (location identifier) • Surface Roving Vehicle • System Readiness Verification

SRVIFS: Service Installable File System

Srvln: Surveillance

SRVT: Suivi des Ressources Végétales par Télédétection

SRW: Salisbury - Rowan County (USA-NC) (location identifier) • Standard Reaction Wheel

SRWCP: Short Rotation Woody Crop Program

SRX: Sert (Libya) (location identifier)

SRY: Sary - Dashte Naz (Iran) (location identifier)

SRZ: Santa Cruz - El Trompillo (Bolivia) (location identifier)

SRZM: Stichting Radiostraling van Zon en Melkweg (Netherlands)

s/s: Samples per Second • Symbols per Second

ss: Scilicet • Sections • Seguro Servidor • Semis • Sensu Stricto • Sentencias • Siglos • Siguientes • Siswati (ISO language code) • Sunset • Supra Scriptum

S/S: Same Size • Samples per Second • Sector Suite • Service Station • Signal Strength • Single Sideband • Sous-Sol • Space Shuttle • Station-Service • Steamship • Subsystem

SS: Far South • Far Southern • Sacra Scriptura • Safing System • Saints • Saint-Siège • Same Sense • Same Size • Sancti • Sanctissimus • Sandstorm • Santa Sede • Santi • Santissimi • Sa Sainteté • Satellite-Switched • Satellite System • Schutzstaffel • Science Service • Screw Steamer • Secondary School • Seconds • Second Structure • Secretary of State • Secret Service • Sector Suite • Sécurité Sociale • Security Service • Segment Specification • Seguridad Social • Semana Santa • Sequential Signature • Set Screw • Seyfert Sextet • Shell Sort • Shore Station • Short Sleeve • Shuttle System • Side by Side • Signaling System • Simultaneous Sampling • Single Sideband • Single-Sided • Single Signal • Single Star • Single String • Singapore Standard • Sky Survey • Slow Steady • Social Security • Sodium Sulphate • Solar System • Solid State • Source of Supply • Sous-Sol • Southend-on-Sea (UK postal code) • Space Segment • Space Shuttle • Space Sphere • Space Station • Special Series • Special Session • Special Study • Speech Sequence • Speech Sequencer • Speech Sequencing • Spider Squad • Spin-Stabilized • Sporadic Server • Stabilization System • Stack Segment • Stainless Steel • Standard Stage • Standard Size • Star Selector • StarSets • Station Set • Statistical Standard • Steady State • Steamship • Steering System • Storage to Storage • Strada Statale • Strain Sensor • Stricto Sensu • Strict Sense • Structure System • Study Scientist • Subsection • Subsolar • Substructure • Subsystem • Summer Solstice • Sun Sensor • Sun Simulator • Sun Synchronous • Supersonic • Support Software • Support System • Surveying Ship • Survey Scientist • Su Santidad • Su Señoría • Svensk Standard (Sweden) • Symbiotic Star • Système Solaire • System Segment • System Sensitivity • System Simulator • System Software • System Specification • System Summary

SS&A: Space Systems & Applications

SSA: Salvador - Dois de Julho (Brazil-BA) (location identifier) • S-band Single Access • School of Sciences and Applications • Segment Search Argument • Seismological Society of America (USA) • Serial Storage Architecture • Shuttle Simulation Aircraft • Simple Shift-and-Add • Single-Service Access • Singular Spectrum Analysis • Slitless Spectra Analysis • Small Science Aperture • Small Scientific Aperture • Social Security Act • Social Security Administration • Software Safety Analysis • Software Support Agency • Solid-State Amplifier • Space Suit Assembly • Stable Solar Analyzer • Staging and Support Area • Standard Spending Assessment • Star Selector Assembly • Sub-Saharan African • Superconducting Six-axis Accelerometer • Symbol Synchronizer Assembly • Synchro Signal Amplifier • System Support Analysis

SSAA: Société Suisse d'Astrophysique et d'Astronomie (Switzerland) • Space Science Analysis Area • Sus Altezas

SSAAC: Space Science and Applications Advisory Committee

SSAAT: Sun Sensor Attitude Angle Transducer

SSAC: Selective Self-Adaptive Crossover • Space Science Advisory Committee • Space Science Analysis and Command

SSAD: Software System Analysis Document • Space Systems Analysis and Design • System Summary Acceptance Document

SSADM: Structured Systems Analysis and Design Method

SSAE: Stamped Self-Addressed Envelope

SSAIAF: Space Systems Automated Integration and Assembly Facility

SSALR: Simplified Short Approach Lighting System

SSALT: Solid-State Altimeter

SSAM: Simple Sequential Access Method

SSAP: Source Service Access Point • Statement of Standard Accounting Practice

SSAR: Solid-State Adaptive Rotor

SSARA: Scanning Synthetic Aperture Radar Altimeter

SSARR: Streamflow Synthesis And Reservoir Regulation

SSASC: Space Science and Applications Steering Committee

SSASP: Small-Sized Aerospace Plane

SSAT: S-band Single Access Transmitter • Shuttle Service and Access Tower • SIRTF Science Activities Timeline • Space Shuttle Access Tower • System Segment Acceptance Test

SSB: Saint Croix Island (US Virgin Islands) (location identifier) • Science Software Branch • Scientific Software Branch • Sea State Bias • Selective Service Board • Silicon Surface Barrier • Single Sideband • Social Security Board • Société Spécialisée de Biotechnologie • Solar Sector Boundary • Solar-System Barycenter • Source Selection Board • Space Science Board • Space Studies Board • Spontaneous Symmetry Breaking • State Seismological Bureau

SSBAM: Single Sideband Amplitude Modulation

SSBC: Summary Sheet Bar Chart

SSBF: Single Sideband Filter

SSBG: Single Sideband Generator

SSBM: Single Sideband Modulation

SSBP: Self-Supervised Back Propagation

SSBS: Space-Based Surveillance System

SSBUV: Shuttle Solar Backscatter Ultraviolet

SSBW: Spread-Spectrum Bandwidth

S&SC: Sized & Supercalendered

SS&C: Scope, Sequence & Coordination

SSC: Satellite Situation Center • Scanning Shadow Telescope • Science Steering Committee • Science Support Center • Science Survey Consortium • Scientific Steering Committee • Secondary School Certificate • Sensor Signal Conditioner • Sequential System Controller • Serendipitous Survey Catalog • Serial Shift Counter • Set of Station Coordinates • Shapley Supercluster • Shapley Superconcentration • Short Service Commission • Shuttle System Contractor • Simulated Spacecraft • Single-Stage Command • SIRTF Science Center • Site Selection Criterion • Small-Saver Certificate • Small Source Catalog • Solar Stabilization Computer • Solid-Solution Cermet • Solid-State Circuit • Spacecraft System Console • Space Science Center • Space Surveillance Center • Space Systems Center • Spin Synchronous Clock • Staff Selection Committee • Standardization Status Code • Standard Spacecraft Cryocooler • Standards Steering Committee • Starlink Software Change • Starlink Software Collection • Static Standby Computer • Station Selection Code • Station Support Computer • Statistical Society of Canada • STEIS Steering Committee • Stennis Space Center (USA-MS) • Stepping Switch Counter • Stockholm Spectral Catalogue • Storm Sudden Commencement • Subsystem Computer • Subsystem Controller • Subsystem Sequence Controller • Successive Subspace Correction • Sudden Storm Commencement • Sumter Shaw (USA-SC) (location identifier) • Sunspot Count • Superconducting Supercollider • Super Star Cluster • Survey Science Center • Survey Scientist Consortium • Suspended Sediment Concentration • Swedish Space Corporation • Synchrotron Self-Compton • System Sequence Controller • Systems Science and Cybernetics • Systems Support Center

SSCA: Statistical Consulting Center for Astronomy (USA-PA) • Surface Sampler Control Assembly

SSCB: Space Station Control Board

SSCC: SATCOM System Control Center • Second-Stage Conduit Container • Space Section Control Center • Space Station Control Center • Space Surveillance Control Center • Spin-Scan Cloud Camera • Support Services Control Center

SSCD: Suspended Solids Concentration Determination

SSCE: Solid-Surface Combustion Experiment • Sun-Spacecraft-Earth

SSCF: Space Station Control Facility • Space System Control Facility

SSCHF: Space Shuttle Cargo Handling System

SSCI: Social Sciences Citation Index • Société de Services et de Conseil en Informatique

SSCL: Shuttle System Commonality List • Sunspot Cycle Length • Supercomputing Supercollider Laboratory

SSCM: Segment Sol de Contrôle et de Mission • Space Station Cost Model • Sun-Spacecraft-Mars

SSCMP: Sequenced Synchronized Clock Multicast Protocol

SSCom: Solid State Communications (periodical code)

SSCOP: Service Specified Convergence Protocol

SSCP: Small Self-Contained Payload • Springer Series on Chemical Physics (periodical code) • Sum of Squares and Cross-Products • System Services Control Point

SSCR: Support Software Change Request

SSCS: Shipboard Satellite Communications System • Side-Stick Control System • Single-Sideband Communications System • Solar-System Creation Sequence • Space Suit Communications System • Stand Satellite Control System • Synchronous Satellite Communications System

SSCSP: Space Shuttle Crew Safety Panel

SScT: Speculations in Science and Technology (periodical code)

SSCT: Science User Resource Expert • Solid-State Celestial Tracker

SSCU: Spacecraft Systems Controller Unit • Special Signal Conditioning Unit • Store Station Control Unit

SSCV: Solar Sail Cargo Vessel

SSCVR: Solid-State Cockpit Voice Recorder

SSCVT: Subsystem Communications Vector Table

SS.D: Sanctissimus Dominus

SSD: Sacrae Scripturae Doctor • Safe Separation Device • Sanctissimus Dominus • San Felipe (Colombia) (location identifier) • Satellite Superhighway Demonstrator • Scaling-Steepest Descent • Science Support Division • Seasat Decimal • Semiconductor Silicon Detector • Sequence Switch Driver • Software Specification Document • Software System Design • Solid-State Detector • Solid-State Device • Solid-State Disk • Solid-state Storage Device • Spacecraft Software Division • Space-Scale Decomposition • Space Science Department • Space Science Division • Space Sciences Division • Space Shuttle Display • Space Systems Division • Split-Screen Display • Start of Shell Delimiter • Statistical Subdivision • Sun Shadow Device • Surface Sampler Device • Surveillance Situation Display • Survival Situation Display • Synthetic State Determination • System Sciences Division • System Status Display • System Summary Display

SSDA: Sequential Similarity Detection Algorithm • Space Systems Development Agreement

SSDAC: Space Station Data Acquisition and Control

SSDB: Scientific and Statistical Database • Specifically Stellar Database • Spin System Distribution Box

SSDBMS: Statistical and Scientific Database Management System

SSDC: Space Science Data Center

SSDD: Single-Side Double-Density • Software System Design Document

SSDF: Space Science Development Facility

SSDH: Subsystem Data Handbook

SSDI: Support System Design Integration

SSDM: Shielding Standard Design Method

SSDO: Steady-State Design and Optimization

SSDOO: Space Science Data Operations Office (NASA)

SSDPS: Solar System Data Processing System

SSDR: Satellite Situation Display Room • Subsystem Development Requirement

SSDS: Space Sciences Data Service • Space Sciences Data System • Space Shuttle Display and Simulation • Space Station Data System

SSDSAS: Space Station Data System Architecture Study

SSE: Safety System Engineering • Second Starlink Environment • Self-Sustained Emission • Sholapur (India) (location identifier) • Signal Security Element • Single Sideband Exciter • Single Site Event • Société Suisse des Écrivains (Switzerland) • Society for Scientific Exploration (USA-VA) • Software Support Environment • Software Support Equipment • Software System Environment • Solar-System Exploration • Solid-State Electronics • Sortie Support Equipment • South-Southeast • South-Southeastern • Spacecraft Simulation Equipment • Spacecraft Support Equipment • Space Science Enterprise • Space Shuttle Engine • Space Support Equipment • Special Support Equipment • Squared Sum of Errors • Stage Systems Engineer • Star Selector Encoder • Subsystem Element • Sud-Sud-Est • Sum of Squared Errors • Sum-Squared Error • Sun-Spacecraft-Earth • Support System Evaluation • Surface Solar Energy • Surface Support Equipment • Système Statistique Européen • System Safety Engineering • System Security Engineering • System Support Engineering • System Support Equipment

S'sea: Swansea (UK-Wales)

SSEA: Separate Sampling and Excitation Analysis • System Safety Engineering Analysis

SSEAF: Space Science Experimenters Analysis Facility

SSEB: Source Selection Evaluation Board

SSEC: Selective Sequence Electronic Calculator • Solar-System Exploration Committee • Solid-State Electronic Chronograph • Space Science and Engineering Center

SSECO: Second-Stage Engine Cutoff

SSECS: Space Shuttle Environmental Control System

S&SED: Satellite & Space Electronics Division

SSED: Solar-System Exploration Division

SSEDF: Software Support Environment Development Facility

SSEED: Symmetric Self-Electro-optic Effect Device

SSEL: Solid-State Electronics Laboratory • Space Science and Engineering Laboratory

SSEle: Solid State Electronics (periodical code)

SSEM: Space System Effectiveness Model • System Security Engineering Manager

SSEMC: Solar-System Exploration Management Council

SSEOO: Space Shuttle Earth Observation Office • Space Shuttle Earth Observation Optics

SSEOP: Space Shuttle Earth Observations Photography • Space Shuttle Earth Observations Photos • Space Shuttle Earth Observations Project (*)

SSEOS: Space Shuttle Engineering and Operations Support • Space Station External Operations Simulator

SSEp: Springer Series on Electrophysics (periodical code)

SSEP: Solar-System Exploration Project • Space Systems for Economic Purposes • System Safety Engineering Plan • System Security Engineering Phase • System Security Engineering Plan

SSERN: South-Southeastern

SSERC: Scottish Schools Equipment Research Centre (UK-Scotland)

SSES: Software System Environment System • Solar-System Exploration Subcommittee

SSEU: System Selector Extension Unit

SSEWD: South-Southeastward

SSF: San Antonio - Stinson (USA-TX) (location identifier) • Service Storage Facility • Single-Sideband Filter • Single Solar Flare • Société Statistique de France • Space Simulation Facility • Space Station Freedom • Space Systems Finland Ltd. • Special Security Facility • Split-Step Fourier • Stiff Shipping Frame • Stiftelsen för Strategisk Forskning (Sweden) • Super Skyflux

SSFCS: Space Shuttle Flight Control System

SSFD: Solid-State Floppy Disk

SSFDC: Solid-State Floppy Disk Card

SSFDR: Solid-State Flight Data Recorder

SSFF: Space Station Furnace Facility • Space Station Future Fighters (USA-TX)

SSFGSS: Space Shuttle Flight and Ground System Specification

SSFL: Steady-State Fermi Level

SSFLC: Surface-Stabilized Ferroelectric Liquid Crystal

SSFM: Single-Sideband Frequency Modulation

SSFN: Space Station Flight Network

SSFP: Space Station Freedom Program

SSFPO: Space Station Freedom Project Office

SSFR: Solar Spectral Flux Radiometer

SSFS: Space Shuttle Functional Simulation • Space Shuttle Functional Simulator • Sydney Space Frontier Society (Australia)

SSFVT: Subsystems Functional Verification Test

SSG: Malabo (Equatorial Guinea) (location identifier) • Safety Study Group • Science Software Group • Science Steering Group • Scientific Steering Group • Single-Sideband Generator • Software Strategy Group • Specialization Subgraph • Special Study Group • Special Support Group • Stonehenge Study Group • Supply Spectrum Generator

SSGA: System Support Gate Array

SSGC: Short System Ground Check

SSGD: Superheated Superconducting Granule Detector

SSGDAC: Space Shuttle Guidance, Dynamics And Control

SSGS: Solid-State Gamma Switch • Standard Space Guidance System

SSGSR: Springer Series on Geophysics and Space Research (periodical code)

SSGT: Subsystem Ground Test

SSH: Schwartz-Slawsky-Herzfeld • Sea Surface Height • Second-Stage Hydraulics • Secure Shell • Sharm El Sheik (Egypt) (location identifier) • Simulation Support Hardware • Spatial-Spectral Holography • Starkenburg-Sternwarte Heppenheim eV (Germany)

SSHA: Shuttle System Hazard Analysis • Subsystem Hazard Analysis

SSHB: Station Set Handbook

SSHD: Single-Side High-Density

SSHL: Société Suisse d'Hydrologie et de Limnologie (Switzerland)

SSHRC: Social Sciences and Humanities Research Council

SSHT: Sea Surface Height

SSI: Brunswick (USA-GA) (location identifier) • San Salvador Island • Schedule Support Impact • Second-Stage Ignition • Sector Scan Indicator • Sécurité des Systèmes d'Information • Serial Synchronous Interface • Server-Side Includes • Shore Station Interface • Significant Structural Item • Single System Image • Site of Scientific Interest • Sky Survey Instrument • Small-Scale Integration • Social-Security Information • Solid-State Imager • Solid-State Imaging • Solid-State Inverter • Spacecraft System Integration • Spaceport System International • Space Science Institute (USA-CO) • Space Science Instrumentation (periodical code) • Space Segment Integration • Space Services Inc. (USA-TX) • Space Software Italia • Space Studies Institute (USA-NJ) • Special-purpose-class Science Instrument (SOFIA) • Spectral Statistical Interpolation • Spectrum Sciences Inc. (USA-CA) • Statewide System Initiative • Structural Significant Item • Supplemental Security Income • Surface Stereo Imager • System Software Interface

SSIBD: Shuttle System Interface Block Diagram

SSIC: Small-Scale Integrated Circuit • Standard Subject Identification Code

SSID: Survey Science Interface Document

SSIE: Smithsonian Science Information Exchange

SSIES: Special Sensor for Ions, Electrons, and Scintillations

SSII: Société de Service et d'Ingénierie Informatique • Société de Services en Ingénierie Informatique • Solid-State Image Intensifier

SSIM: Solar Spectral Irradiance Monitor • Statistical, Sampling Inventory Method

SSIP: Shuttle Student Involvement Program • Space-science Student Involvement Program • Space Surveillance Interface Processor • Standard Systems Improvement Program • Student Science Investigation Project • Subsystem Integration Plan • Subsystem Integration Program • System Setup Indicator Panel • System Software Interface Processing

SSIR: Sun Shade Interface Ring

SSIS: Solid-State Imaging System • Spacecraft System Integration Support • Space Station Information System

SSISS: Spacecraft System Integration Support Service

SSITP: Shuttle System Integrated Test Plan

SSIU: Subsystem Interface Unit • Systems Simulator Interface Unit

SSIV: Segment Sol Image Végétation

SSIZ: Seasonal Sea Ice Zone

SSJ: Sandnessjoen (Norway) (location identifier)

SSK: Strahlenschutz-Kommission • Sturt Creek (Australia-WA) (location identifier)

SS/L: Space Systems/Loral

SSL: Santa Rosalia (Colombia) (location identifier) • Secure Socket Layer • Selected Source List • Shore Station Laboratory • Site Support List • Small Space Laboratory • Software Specification Language • Software Support Laboratory • Solid-State Lamp • Solid-State Laser • Space Science Laboratory • Space Sciences Laboratory • Space Simulation Laboratory • Space Systems Laboratory • Storage Structure Layout • Sudden Sodium Layer • Super-Scanning Locator • Support Status List • System Software Loader • System Specification Language

SSLDEF: Space Shuttle Long-Duration Exposure Facility

SSLO: Solid-State Local Oscillator

SSLOR: Solar Stellar Limb Occultation Radiometer

SSLP: Secure Socket Layer Protocol

SSLS: Solid-State Laser System • Standard Space Launch System

SSLT: Solid-State Logic Timer

SSLV: Standard Small Launch Vehicle • Standard Space Launch Vehicle

SSM: Samstag-Sonntag-Montag • Satellite Survivability Model • Saturday-Sunday-Monday • Sault Sainte Marie (USA-MI) (location identifier) • Second-Stage Motor • Second Surface Mirror • Signal Strength Monitor • Single-Sideband Modulation • Site Status Message • Société de Secours Mutuel • Soft Systems Methodology • Solar Standard Model • Solar Stereoscopic Mission • Solid-State Material • Spacecraft Support Module • Spacecraft Systems Monitor • Space Station and Microgravity • Space Station Module • Special Sensor for Microwave • Special Sensor Microwave • Spread Spectrum Modulation • Springfield Science Museum (USA-MA) • Standard Solar Model • Station Spatiale et Microgravité • Steady-State Model • Subsystem Manager • Subsystem Model • Supply Support Management • Supply Support Manager • Support Systems Module • System Safety Manager • System Support Management • System Support Manager • System Support Module

SSMA: Solid-State Microaccelerometer • Solid-State Microwave Amplifier • Spread-Spectrum Multiple Access

SSMB: Space Shuttle Maintenance Baseline • Space Station Manned Base

SSMC: Second-Stage Motor Container • Space Shuttle Mission Control

SSME: Satellite System Monitoring Equipment • Space Shuttle Main Engine

SSMEC: Space Shuttle Main Engine Controller

SSMECA: Space Shuttle Main Engine Controller Assembly

SSMF: Science Studies Methods Forum • Space Station Millimeter Facility

SSM/I: Special Sensor Microwave/Imager

SSMI: Special Sensor Microwave Instrument

SSM/IS: Special Sensor Microwave Imager/Sounder

SSMIS: Special Sensor Microwave Imager/Sounder

SSMM: Space Station Mathematical Model • Sus Majestades

SSMO: Southern Skies Mobile Observatory (Australia) • Summary of Synoptic Meteorological Observations

SSMOC: SOFIA Science and Mission Operations Center

SSMP: Stockpile Stewardship and Management Program

SSMS: Solid-State Mass Spectrometer • Spark Source Mass Spectroscopy • Support Systems Module Simulator

SSM/T: Special Sensor Microwave/Temperature

SSMTF: Space Station Mockup and Trainer Facility

SSN: Seoul (Korea-ROK) (location identifier) • Severely Subnormal • Social-Security Number • Space Surveillance Network • Specification Serial Number • Standard Serial Number • Standard Star Newsletter • Starlink System Note • Standard Serial Number • Station Serial Number • Sunspot Number • Switched Service Network

SSND: Solid-State Neutral Dosimeter

SSNS: Standard Study Numbering System

SSO: Safety Signification Operation • Sahara and Sahel Observatory • Sales Support Officer • São Lourenço (Brazil-MG) (location identifier) • Satellite Systems Operation • Satellite Switching Office • Sayan Solar Observatory (Russia) • Senior Scientific Officer • Single Sign-On • Solar-System Object • Solid-State Oscillator • Source Selection Official • Space Shuttle Orbiter • Special Studies Office • Spurious Shut-Off • Staff Security Officer • Steady-State Oscillation • Structure, Sequence and Organization • Substellar Object • Subsystem Operation • Sud-Sud-Ouest • Sun Synchronous Orbit • Swedish Solar Observatory • System Security Officer • System Software Office

SSOB: Senior Scientist Onboard

SSOC: Satellite System Operational Center

SSOCR: Superscale Optical Character Reader

SSOD: Solid-State Optical Detector

SSODB: Space Station Operational Databook

SSODC: Steady-State Off-Design Code

SSODCM: Space Systems Operational Design Criteria Manual

SSOG: Satellite Systems Operations Guide

SSOM: Solid-State Optical Maser

SSONET: Structured Semantic Object Network

SSOP: Space Systems Operating Procedure

SSOR: Symmetric Successive Over-Relaxation

SSOS: Severe Storms Observation Satellite • Springer Series on Optical Sciences (periodical code)

SSOSC: Strong Second-Order Sufficient Condition

SSP: Scanning Spectral Polarimeter • Scientific Subroutine Package • Security System Plan • Seismic Section Profiler • Sensor Select Panel • Service des Soins Palliatifs • Service Switching Point • Silver Plains (Australia-QLD) (location identifier) • Simple Solar Population • Simple Stellar Population (*) • Simulation Support Processor • Single-generation Stellar Population • Small Sortie Payload • Société Statistique de Paris (France) • Société Suisse de Physique (Switzerland) • Society of rScholarly Publishing (USA) • Soil Survey of Pakistan • Solid-State Photodiode • Solid-State Preamplifier • Source Selection Plan • Space Science Program • Space Shuttle Program • Space Station Program • Spread Spectrum • Standard Shop Practice • Standard Stability Prediction • Standard Switch Panel • Standby Status Panel • Star/Sky Photometry • Statutory Sick Pay • Stellar Synthesis Population • Subsatellite Point • Subsolar Point • Summer Science Program • Summer Session Program • Summer Star Party • Super Sans Plomb • Superstable Platform • Surface Science Package • System Safety Plan • System Security Plan • System Support Plan • System Support Program • System Status Panel

SSPA: Solid-State Power Amplifier

SSPC: Solid-State Power Controller • Spacelab Stored Program Command

SSPD: Shuttle System Payload Data • Shuttle System Payload Definition • Shuttle System Payload Description

SSPDA: Shuttle System Payload Data Activity

SSPDB: Subsystem Power Distribution Box

SSPDS: Shuttle System Payload Data Study • Space Shuttle Payload Data Study

SSPE: Space Station Program Element • Support System Project Engineer

SSPF: Sortie Software Production Facility • Space Shuttle Programming Facility • Space Station Processing Facility (USA-FL)

SSPFA: Space Shuttle Program Flight Assignment

SSPI: Security Service Provider Interface • Society of Satellite Professionals International (USA-VA)

SSPGSE: Space Shuttle Program Ground Support Equipment

SSPII: Space Shuttle Program Implementing Instruction

SSPM: Software Standards and Procedures Manual • Solid-State Photomultiplier • Space Shuttle Program Manager • System Safety Program Manager

SSPO: Space Shuttle Program Office • Space Shuttle Projects Office

SSPP: Shuttle Small Payloads Project • Solar Sea Power Plant • Space Station Program Participant • System Safety Program Plan

SSPPSG: Space Shuttle Payload Planning Steering Group

SSPRO: Space Shuttle Program Resident Office

SSPRv: Space and Solar Power Review (periodical code)

SSPS: Satellite Solar Power Station • Satellite Solar Power System • Sky Survey Prototype System • Solar Satellite Power Station • Spacecraft Support Planning Section • Space Shuttle Program Schedule

SSPSF: Stochastic Self-Propagating Star Formation

SSPTF: Santa Susana Propulsion Test Facility

SSQ: La Sarre (Canada-QC) (location identifier) • Simple Sinusoidal Quantity • Steep-Spectrum Quasar

SSQM: Scanning Superconducting Quantum Microscope

SSR: Sara (Vanuatu) (location identifier) • Satélite de Sensoriamento Remoto • Satellite Situation Report • Scanning Spectral Radiometer • Secondary Surveillance Radar • Selenium Stack Rectifier • Shop Support Request • Site Suitability Report • Software Specification Review • Solid-State Reaction • Solid-State Recorder • Solid-State Relay • Soviet Socialist Republic (now obsolete) • Special Scientific Report • Spin-Stabilized Rocket • Staff Support Room • Station Set Requirement • Statistical Summary Report • Stock Status Report • Structured Semantic Relationship • Subsurface Radar • Summer Solstice Sunrise • Sum of the Squared Residuals • Supersynthesis Radiometer • Surface Search Radar • System Safety Review • System Specification Review • System Status Review • System Study Requirement • System Support Record

SSRA: Subsurface Sounding Radar/Altimeter

SSRAAO: Space Station Remote Access Automatic Observatory

SSRAAT: Space Station Remote Access Automatic Telescope

SSRAM: Synchronous Static Random Access Memory

SSRB: Subsystem Review Board

SSRC: Social Sciences Research Council (UK) • Strict Source and Record Route

SSRD: Specific Science Requirements Document • Station Set Requirements Document

SSRE: Society for Social Responsibility in Engineering

SSRF: Screen Scraper Read File

SSRG: Simple Shift Register Generator

SSRI: Système de Stockage et de Recherche de l'Information

SSRL: Stanford Synchrotron Radiation Laboratory (USA-CA)

SSRM: Scanning Spreading Resistance Microscopy • Second-Stage Rocket Motor

SSRMS: Space Station Remote Manipulating System • Space Station Remote Manipulator System (*)

SSRN: System Software Reference Number • System Software Requirement Number

SSROP: Space Station Resource Optimization Program

SSRP: Severe Storms Research Program • Simple Server Redundancy Protocol • Spectral Surface Radiation Program • System Safety Review Panel

SSRPI: Selectable Spectral Resolution Pointing Imager

SSRQ: Steep-Spectrum Radio Quasar

SSRR: Station Set Requirement Review

SSRS: Society for Social Responsibility in Science • Southern Sky Redshift Survey • Start-Stop-Restart System

SSRT: Siberian Solar Radio Telescope • Single-Stage Rocket Technology • Subsystem Readiness Test

SSRTF: Solid Surface Room Temperature Fluorescence

SSRTP: Solid Surface Room Temperature Phosphorescence

SSRU: Standard Speech Reproducing Unit

SSRv: Space Science Reviews (periodical code)

SSRV: Solar-Sail Race Vehicle

SSRvE: Soviet Scientific Reviews Section E (periodical code)

SSS: Safety Status System • Satellite Servicer System • Scanning Star Simulator • Scatterometer System Simulator • Scientific Subroutine System • Sea Surface Salinity • Selective Service System • Sensor Support System • Siassi (Papua New Guinea) (location identifier) • Signal Switching System • Silicon-Symmetrical Switch • Simulation Study Series • Single-Screw Ship • SIPE Support Structure • Sira Safety Services • Small-Scale Structure • Small Scientific Satellite • Small Solar Satellite • Small Starlight Scope • Solar and Skylight Spectrophotometer • Solar-Stellar Spectrograph • Solar-Stellar Spectrophotometry • Solid-State Spectrometer • Solid-State Switching • Solid-State System • Sortie Support System • Sound Suppression System • Spacecraft Software System • Spacecraft System Support • Space Shuttle Segment • Space Shuttle Simulation • Space Shuttle System • Space Station Simulator • Space Surveillance System • Spinning Space Station • Spinning Sun Sensor • Spin-Stabilized Spacecraft • Stage Separation System • Standard Software System • Star Selection Servo • Static Spherically Symmetric • Station Set Specification • Steep-Spectrum Source • Strategic Satellite System • Subject Specialization Scheme • Subsystem Segment • Summer Solstice Sunset • Sun Screen Server • Supersoft Source • Support Software System • Su Seguro Servidor • Switching Subsystem • Symbolic Shorthand System • System and Segment Specification • System Security Standard • System Segment Specification

SSSA: Soil Science Society of America (USA)

SSSAAS: Space Station Science and Applications Advisory Subcommittee

SSSAJ: Soil Science Society of America Journal (periodical code)

SSSC: SAX Scientific Steering Committee • Solid State Sciences Committee • Space Science Steering Committee • Space Science Support Center • Sydney Southern Star Catalogue

SSSD: Second-Stage Separator Device • Single-Side Single-Density • Solid-State Solenoid Driver • Space Shuttle Simulation Display • State Soil Survey Database

SSSF: Surface and Sounding Systems Facility

SSSI: Science of Soft Surfaces and Interfaces • Site of Special Scientific Interest

SSSP: Space Shuttle Synthesis Program

SSSR: Summer Solstice Sunrise

SSSS: Society for Social Studies of Science (USA-MA) • Space Shuttle System Specification • Spherically Symmetric Self-Similar

SSSSR: Space Shuttle System Safety Review

SSSSS: Simulation of the Satellite Signal in the Solar Spectrum • Springer Series on Solid State Sciences (periodical code)

SSST: Space Station System Trainer

SSSTB: South South South Temperate Belt

SSSyn: Springer Series on Synergetics (periodical code)

SST: Sample Sound Technology • Santa Teresita (Argentina-BA) (location identifier) • Santé et Sécurité du Travail • Satellite-to-Satellite Tracking • Scalable Systems Technology • Seamless Steel Tubing • Search Strategy and Terms • Sea Surface Temperature • Sea Surface Topography • Shuttle System Test • Simple Software Tool • Simulated Structural Test • Simulation and Systems Test • Single-Sideband Transmission • Single System Trainer • Society of Surveying Technicians • Software Support Team • Solid-State Telescope • Spacecraft Systems Test • Space Solar Telescope • Space Surveillance and Tracking • Spectroscopic Survey Telescope • Spread-Spectrum Technology • Stably Stratified Turbulence • Structural Static Test • Student Science Training • Subsystem Terminal • Subsystem Test • Supersnelle Trein • Supersone Snelheids Tunnel • Supersonic Transport • Superstring Theory

SSTA: Support System Task Analysis

SSTAC: Sea Surface Temperature Analysis and Composite

SSTB: South South Temperate Belt

SSTC: Solid-State Timer-Controller • Spacecraft System Test Console • Space Shuttle Test Conductor • Specialized System Test Contractor • State Science and Technology Commission

SSTCLS: Space Shuttle Test, Checkout, and Launch Support

SSTD: Space Science and Technology Department (RAL)

SSTDMA: Satellite-Switched Time-Division Multiple Access

SSTec: Solid State Technology (periodical code)

SSTEP: System Support Test Evaluation Program

SSTF: Space Shuttle Task Force • Space Shuttle Training Facility • Space Simulation Test Facility • Space Station Task Force • Space Station Training Facility

SSTG: Space Station Task Group • Space Systems and Technology Group

SSTI: Small Satellite Technology Initiative • Small Spacecraft Technology Initiative

SSTIMR: Sea Surface Temperature Imaging Microwave Radiometer

SSTL: Solar System Target List • Surrey Satellite Technology Ltd. (UK)

SSTNET: Sociology of Science and Technology Network

SSTO: Single-Stage-to-Orbit

SSTP: Small Satellite Technology Program • Subsystem Test Procedure

SSTR: Safing System Tape Recorder

SSTS: Satellite Surveillance and Tracking Satellite • Space-based Surveillance and Tracking Satellite • Space Surveillance and Tracking Satellite • Space Surveillance and Tracking System

SSTV: Slow-Scan Television

SSTWG: SETI Science and Technology Working Group

SSTZ: South South Temperate Zone

SSU: Sample Storage Unit • Secondary Sampling Unit • Self-Service Unit • Semiconductor Storage Unit • Sensor Simulator Unit • Solar-Sail Union • Sonoma State University (USA-CA) • Southwest State University (USA-MN) • Spacecraft Support Unit • Space Science Update • Star Sensor Unit • Station Supervisory Unit • Strategic Services Unit • Stratospheric Sounding Unit • White Sulphur Springs - Greenbrier (USA-WV) (location identifier)

SSUAS: Space Station Utilization Advisory Subcommittee

SSUP: Space Station User Panel • Space Station Utilization Preparedness

SSUS: Spin-Stabilized Upper Stage • Spinning Solid Upper Stage

SSUSI: Special Sensor for Ultraviolet Spectrographic Imaging

SSUSP: Spinning Solid Upper Stage Project

SSUT: Sous-Système Utilisateur Téléphonique

SSV: Satellite Servicing Vehicle • Set of Stations Velocities • Siasi (Philippines) (location identifier) • Solar System Visualization • Space Shuttle Vehicle • Space Sortie Vehicle • Standard Service Volume • Supersatellite Vehicle

SSVN: Structure Searching Via Nomenclature

SSVS: Subsystem Verification Specification

SSVT: Swedish Solar Vacuum Telescope

SSW: Schwäbische Sternwarte eV (Germany) • South-Southwest • South-Southwestern • Stuart Island (USA-WA) (location identifier) • Sudden Stratospheric Warming

SSWC: Set of Significant Wavelet Coefficients

SSWG: Servicing Science Working Group • Site Selection Working Group • Shuttle Security Selection Working Group • Solar-System Working Group • Space Science Working Group

SSWP: Space Station Work Package • Springer Series on Wave Phenomena (periodical code)

SSWRD: Systems Software Requirements Document

SSWRN: South-Southwestern

SSWTI: Staatliches System der Wissenschaftlich-Technischen Information

SSWWD: South-Southwestward

SSX: Samsun (Turkey) (location identifier) • Space Ship Experimental

SSXB: Supersoft X-ray Binary

SSXTF: Solar Soft X-ray Telescope Facility

SSY: M'Banza Congo (Angola) (location identifier)

S/Sys: Subsystem

SSZ: Santos (Brazil-SP) (location identifier)

st: Sesotho (ISO language code) • Stere • Stet

St: Saint • Sanct • Sankt • Sant • Sint • Sitio • Stanton Number • Stanza • State • Statement • Station • Statute • Stem • Stitch • Stokes • Stone • Strait • Straße • Stratus • Street • Stück • Stunde

S&T: Science & Technology • Scientific & Technical • Sky & Telescope (periodical code) • Supply & Transport

ST: Principe (Sao Tome &_) (ISO country code) • Saint • Sample Tray • Sao Tome & Principe (ISO country code) • Scalar Totalizer • Scenario Template • Scheduled Tribe • Schmidt Telescope • Schmitt Trigger • Science Team • Search and Track • Select Time • Self-Test • Sequence Timer • Sequential Timer • Serial Traceability • Service des Transmission • Shipping Ticket • Shock Tunnel • Short Ton • Sidereal Time • Silicon Target • Simplification Task • Simulator Training • Single Throw • Singularity Theorem • Slow Technology • Solar-Terrestrial • Spacelab Technology • Space Telescope • Space-Time • Spacelab Technology • Special Test • Special Tooling • Spécification Technique • Spring Tide • Standard Thickness • Standard Time • Start • Star Tracker • Static Test • Static Thrust • Steam Turbine • Stoke-on-Trent (UK postal code) • Störmer Predictor • Stratus • Street • Submarine Tender • Substitution Theorem • Subtask •

Sudan (international aircraft registration code) • Summer Time • Superconducting Tunnel • Superior Technology • Survival Time • Synthesis Telescope • System Test

Sta: Santa • Station • Stationary

StA: Staatsarchiv

STA: Saint Andrews (Starlink internal code) • Santa • Satellite Tracking Annex • S-band Test Antenna • Science and Technology Agency • Shuttle Training Aircraft • SIGIR Technology Affiliate • Space Technology Application • Space Transportation Association (USA-VA) • Spanning-Tree Algorithm • Special Temporary Authorization • State Transit Authority • Static Test Article • Station • Stauning (Denmark) (location identifier) • Strategic Transportation Analysis • Strategy and Technical Assessment • Structural Test Article • Student Travel Agency

STAARTE: Scientific Training and Access to Aircraft for Atmospheric Research Throughout Europe

STAAC: Space and Terrestrial Applications Advisory Committee • Systems Technology and Advanced Concepts

STAAR: Submillimeter Telescope for Astronomy and Atmospheric Research

Stab: Stabilization • Stabilizer • Stable

StAB: Staatliche Allgemeinbibliothek

STAC: Science and Technical Advisory Committee • Science and Technology Advisory Committee • Space Telescope Advisory Committee (*) • Space Telescope Advisory Council

STACC: Small Telescope Array with CCD Cameras • Standard Telemetry and Command Component • Standard Telemetry and Communications Component

STACE: STScI Advisory Council on Education

STACEE: Solar Tower Atmospheric Čerenkov Effect Experiment (*) • Solar Two Atmospheric Čerenkov Experiment

STACS: Serials Tracking And Control System • Subtropical Atlantic Climate Studies

STAD: Société de Traitement Automatique des Données (France)

STADAC: Station Data Acquisition and Control

STADAD: RAL (Starlink internal code) • Satellite Tracking And Data Acquisition Department

STADAN: Satellite Tracking And Data Acquisition Network • Space Tracking And Data Acquisition Network

STADAT: Starlink Database

STADU: System Termination And Dipslay Unit

STAE: Specific Task Asynchronous Exit

STAEDT: Simulation Tools for Analysis and Evaluation of Data Transmission

STAF: Science Team Analysis Facility • Service de Transfert et d'Archivage de Fichiers

STAFF: Spatio-Temporal Analysis of Field Fluctuations • Stellar-Acquisition Flight Feasibility

Staffs: Staffordshire (UK-England)

STAFT: Steerable Antenna Focusing Technique

STAG: Secure Text And Graphics • Shuttle Turnaround Analysis Group • Space Telescope Advisory Group

STAGN: Stagnation

STAIR: Structural Analysis Interpretive Routine

STAIRS: Storage and Information Retrieval System (IBM)

STAIS: Student Advisory Information Service

STALAS: Stationary Laser Site • Stationary Laser Station

STALO: Stable Local Oscillator

StAM: Studies in Applied Mathematics (periodical code)

STAMOS: Sortie Turn Around Maintenance Operations Simulation

STAMP: Satellite Telecommunications Analysis and Modelling Program • Space Technology Analysis and Mission Planning • System Testability And Maintainability Program

STAN: Space Telescope Analysis Newsletter

Stand: Standard

STANet: Science and Technology Awareness Network

STANTEK: Standardiseringstekniska Föreningen (Sweden)

STAP: Schmidt Telescope Advisory Panel • Scientific and Technical Advisory Panel • Scientific and Technical Analysis and Program • SmallTalk Apprentice Program • Special Technical Assistance Program

STAPP: Simulation Tape Print Program

STAPS: Sciences et Techniques des Activités Physiques et Sportives

STAR: Satellite Telecommunications Automatic Routing • Satellites in Telematics Applications and Research (ESA) • Satellite Tracking using RF • Science Teaching Achievement Recognition • Science Teaching through its Astronomical Roots • Science Technology And Research • Scientific and Technical Aerospace Reports (NASA) • Scientific Toolkit for Astrophysical Research • Sea And Terrain Assessment Radar • Self-Test Automatic Readout • Self-Testing And Repairing • Shuttle Turnaround Analysis Report • Significant Technological Achievement in Research • Significant Technology Achievement through Research • SOFIA Terahertz Array Receiver • Software Tools And Requirements • Solar, Terrestrial and Astrophysical Research • Solenoidal Tracker at RHIC • Source Transmission And Response • South Pacific Tectonics and Resources • Space Technology and Advanced Research • Space Telescope Anomaly Room • Space Telescope Audit Report • Space Telescope Axial Replacement • Space Three-Axis Accelerometer for Research • Space Transportation Automated Reconfiguration • Spectral Technology and Applied Research • Spectrograph/Telescope for Astronomical Research • Special Telecommunication Action of Regional Development • Standard Telecommunications Automatic Recognizer • Standard Terminal Arrival • Standard Terminal Arrival Route • Statistical Table Assembly and Retrieval • Stellar Attitude Reference • Stockholms Amatörastronomer (Sweden) • Streamer Tubes Analog Readout • String Array • Surface Target Attenuation Radar • Système Terrestre Automatique de Radiosondage

STARAN: Stellar Attitude Reference And Navigation

STARCAL: Star Calibration

STARCAT: Space Telescope Archive and Catalog • Space Telescope Archive Catalog

STARCH: Stressed Arch

STARE: Scandinavian Twin Auroral Radar Experiment • Southern Tropical Atlantic Region Experiment • Steerable Telemetry Antenna Receiving Equipment

STARFIRE: System To Accumulate and Retrieve Financial Information with Random Extraction

STARLAB: Space Technology Applications and Research Laboratory

Starlink: Satellite Telemetry And Return Link

STARR: Schedule, Technical And Resources Report

STaRS: Science Teacher and Research Scientist

STARS: Satellite Telemetry Automatic Reduction System • Satellite Tracking And Ranging Station • Satellite TV Airborne Receiving System • Science Teachers Active in Real Science • Seismic Telescope for Astrophysical Research from Space • Software Technology for Adaptable and Reliable Systems • Space Transportation Automated Reconfiguration System • Stabilized Twin-gyro Attitude Reference System • Standard Terminal Automation Replacement System • Status Analysis And Reporting System • Stellar Tracking Attitude Reference System • Strategic Targeting Activities Reporting System • Structural Target Analysis and Recognition System • Students Taking

Astronomy-Related Subjects (USA-CA) • Subaru Telescope Archive System • Support Tracking Analysis Reporting System • Surveillance and Target Attack Radar System • System Test and Astronaut Requirement Simulation

START: Selection To Activate Random Testing • Space Test And Reentry Technology • Space Transport And Reentry Test • Strategic Arms Reduction Talks • Strategic Arms Reduction Treaty • System for Analysis, Research and Training

STARTED: Strategies, Tools, and Resources for Team-based Early Design

STAS: Space Transportation Architecture Study

stat: Statim

Stat: Statement • Static • Station • Stationary • Statistical • Statistics • Statuary • Statue • Status • Statute

STAT: Space Telescope Advisory Team • Space Telescope Astrometry Team • System Testability and Analysis Tool

STATE: Structure and Atmospheric Turbulence Environment

STATN: Station

STATS: Star Tracker Alignment Test Set

STATSGO: State Soil Geographic Database

STAU: Standard Acquisition Unit

StAV: Steirischer Astronomen Verein (Austria)

STAVF: Space Telescope Assembly and Verification Facility

STAWA: Science Teachers of Western Australia

Stb: Steinbruch

StB: Stadtbibliothek

STB: Santa Barbara - Zulia (Venezuela) (location identifier) • Save The Bits • Set-Top Box • SIM Testbed • Solid Target Boronization • South Temperate Belt • South Tropical Belt • Strobe • System Test Bed

STBA: Selective Top-to-Bottom Algorithm

Stbd: Starboard

STBE: Space Transportation Booster Engine

StBGebV: Steuerberatergebührenverordnung

STBL: Stable • Stable Boundary Layer • Stratus-topped Boundary Layer

Stbln: Stabilization

STBP: Sir Thomas Brisbane Planetarium (Australia-QLD)

Stbt: Steamboat

Stby: Standby

STBY: Standby

STC: Saint Cloud (USA-MN) (location identifier) • Satellite Test Center • Satellite Tracking Center • Science and Technical Committee • Science and Technology Center • Scientific and Technical Committee • Scientific Technical Committee • Semivariogram Textual Classifier • Sensivity Time Control • Services Techniques Centraux • Set Carry • Short-Time Constant • Short-Title Catalog • Shuttle Test Conductor • Simulation Tape Conversion • Society for Technical Communication (USA-VA) • Spacecraft Test Conductor • Space Technology Center • Space Test Center • Space-Time Continuum • Spécification Technique du Besoin • Specific Transport Cost • Spectral Transfer Coefficient • Standard Test Condition • Standard Test Configuration • Standard Transmission Code • Started Task Console • Station Test and Calibration • Storage Technology Corporation • Stored Time Command • Subtechnical Committee • Suffix Tree Clustering • Supplemental Type Certificate • Supplemental Type Certification • Switching Technical Center • Système de Trajectographie Captive • System Test Complex • System Test Configuration • System Test Console • System Timing Controller • System Transfer Constant

STCB: Survey Technician Certification Board

STCC: Spacecraft Technical Control Center

STCDHS: Spacecraft Telemetry Command Data Handling System

STCDS: System Test Complex Data System

STCE: System Test Complex Equipment

STCK: Store Clock

STCM: Service Technique de Cryogénie et de Magnétisme

STCN: Short Title Catalogue Netherlands

STCP: Short-Term Cost Plan

STCS: Special Telemetry Conversion System

STCT: Small Transportable Communications Terminal

STCU: Science and Technology Center in Ukraine

STCW: Standards of Training, Certification and Watchkeeping • System Time Code Word

Std: Standard • Started • Stunde

STD: Dobra (Sao Tome & Principe) (ISO currency code) • Salinity-Temperature-Depth • Santo Domingo (Venezuela) (location identifier) • SATCOM Technologies Division • Save The Data • Science and Technology Development • Scientific and Technical Data • Sensor and Test Driver • Servo Tape Display • Set Direction • Shuttle Test Detector • Software Test Description • Software Transfer Document • Solar-Terrestrial Dispatch • Spacecraft Technology Division • Special Test Device • Standard • Standard Document • Standard Test Device • State Transition Diagram • Subscriber Toll Dialing • Subscriber Trunk Dialing • Supporting Technology Development

STDA: StreetTalk Directory Assistance

STDADS: Space Telescope Data Archive and Distribution System

STDAUX: Standard Auxiliary

STDB: Space Telescope Database • Standard Database

STDBM: Spatio-Temporal Database Management (workshop)

STDC: Space Technique Development Center

STDCE: Surface Tension Drive Convection Experiment • Surface Tension Driven Convection Experiment (*)

Std. Dev.: Standard Deviation

Stde: Stunde

STDERR: Standard Error

STDIN: Standard Input

STDIOH: Standard Input/Output Header

stdl: stündlich

STDL: Standard Distribution List

STDM: Stadium • Statistical Time-Division Multiplexer • Statistical Time-Division Multiplexing (*) • Synchronous Time-Division Multiplexing

STDMA: Self-organizing Time Division Multiple Access

STDMS: Space Telescope Data Management System

STDN: Spacecraft Tracking and Data Network (*) • Spaceflight Tracking and Data Network

STDOUT: Standard Output

STDPRN: Standard Printer

STDSD: Solar-Terrestrial Data Services Division

STDT: Science Technology Definition Team

Stdy: Steady

STDY: Steady

Stdzn: Standardization

Ste: Sainte • Suite

Sté: Société

ST&E: Security Test & Evaluation

STE: Security Test and Evaluation • Segment Table Entry • Shrinkage Test Estimator • Simulation and Test Equipment • Spacecraft Test Engineering • Space-Time Editor • Special Test Equipment • Standard Test Environment • Star Tracker Electronics • Station Test Equipment • Stevens Point (USA-WI) (location identifier) • Suitability Test Evaluation • Suite • Support Test Equipment • System Test Engineer

STEA: System Test, Evaluation and Assembly

STEAP: Simulated Trajectories Error Analysis Program

ST-ECF: Space Telescope - European Coordinating Facility

STECF: Space Telescope - European Coordinating Facility

STED: Science, Technology and Economic Development

STEDI: Student Explorer Demonstration Initiative

STEEP: STX Terrain Elevation Extraction Package

STEI: Scientific, Technical, and Economic Information

STEIN: Short-Term Experimental Information Network

STEIS: Space Telescope Electronic Information System • STScI Electronic Information Service

STel: Stanford Telecommunications

STEL: Short-Term Exposure Level • Short-Term Exposure Limit • Solar-Terrestrial Environment Laboratory

STELAB: Solar-Terrestrial Environment Laboratory

STELAR: Study of Electronic Literature for Astronomical Research

STELLA: Satellite Transmission Experiment Linking Laboratories • System To Estimate Latitude and Longitude Astronomically

Stellv: Stellvertretende • Stellvertretender • Stellvertreter • Stellvertreterin

STEM: Scanning Transmission Electron Microscope • Scanning Transmission Electron Microscopy • Science, Technology, Engineering and Mathematics • Solar-Terrestrial Environment Model • Space-Time-Energy-Matter • Stay Time Extension Module • Structural-Thermal-Engineering Model • System Test Equipment Mission

STEMPRA: Science, Technology, Engineering, and Medicine Public Relations • Association (UK)

Sten: Stencil • Stenographer • Stenographic • Stenography

Steno: Stenographer • Stenographic • Stenography

Stenog: Stenographer • Stenographic • Stenography

STeP: Stanford Temporal Prover

STEP: Satellite Test of the Equivalence Principle • Science and Technology for Environmental Protection • Science, Technology and Economic Policy • Scientific and Technical Exploitation Program • Scientists and Teachers in Educational Partnership • Secure Tunnel Establishment Protocol • Sensitivity Temperature Error Program • Simple Transition to Electronic Processing • Society of Trust and Estate Practitioners • Software Technology and Engineering Practice • Solar-Terrestrial Energy Program • Space Technology Experiments Program • Space Telescope Exhibited Preprint • Space Terminal Evaluation Program • Space Test Experimental Program • Space Test Experiment Platform • Space Test Experiment Program • Special Temporary Employment Programme • Standard for the Exchange of Product • Standard Tape Executive Package • Standard Tape Executive Program • Standard Terminal Program • Standard Test Equipment Procedure • Standard To Exchange Product • Station de Transfert d'Énergie par Pompage • Statistical Trajectory Estimation Program • Stratosphere-Troposphere Exchange Project • Supervisory Tape Executive Program • Surveillance and Tracking Experiment Program

STEPAN: Science and Technology Policy Asian Network

STEPS: Science, Technology, the Economy, Politics, Society • Single Telescope Extrasolar Planet Survey • Solar Thermionic Electrical Power System • Stellar Planetary Survey • Stored Thermal Energy Propulsion System

Ster: Stereophonic • Stereotype • Sterling

STER: Space Techniques for Environmental Risks • Stichting Ether Reclame

STERAO: Stratospheric-Tropospheric Experiments: Radiation, Aerosols and Ozone

Stereo: Stereophonic • Stereotype

STEREO: Solar-Terrestrial Relations Observatory

STERJ: Solar Terrestrial and Environmental Research in Japan (periodical code)

Stern: Die Sterne (periodical code)

Sternw: Sternwarte

STES: Single-locker Thermal Enclosure System

STESA: Solar-Terrestrial Environment Sensing Assembly

STESOP: Space Telescope Elementary Outreach Program

STETF: Solar Total Energy Test Facility

STETS: Solar-Terrestrial Energy Transfer Studies

STEVE: Space Tool for Extravehicular Emergencies

St. Ex.: Stock Exchange

STEX: Sensor Technology Experiment • Space Technology Experiment • Static Test and Evaluation Complex

STeZ: South Temperate Zone

STF: Safety Test Facility • Satellite Test Facility • Satellite Tracking Facility • Save The Frames • S-band Transmit Filter • Shaping The Future • Short Title File • Software Test Facility • Space Telescope Fix • Space Telescope Format • Space Track Facility • Speckle Transfer Function • Spin Test Facility • Source Test File • Spin Test Facility • Standard Task File • Static Test Facility • Stephen Island (Australia-QLD) (location identifier) • Stratiform • Structured File • Summary Tape File • System Test Facility • System Transfer Function

STFRM: Stratiform

STFS: Space Telescope Flight Software

STFSDB: Space Telescope Flight Software Database

STFSDS: Standard Time and Frequency Signal Dissemination Service

STFT: Short-Time Fourier Transform

STFVS: Space Telescope Functional Verification Simulation

Stg: Stage • Sterling

STG: Saint George Island (USA-AK) (location identifier) • Satellite Terminal Guidance • Scene Transition Graph • Schiffbautechnische Gesellschaft (Germany) • Scientific and Technical Group • Shuttle Test Group • Space Task Group • Space Telescope Guidance • Special Technology Group • Steering Task Group • Strong

Stge: Storage

StGG: Studia Geophysica et Geodaetica (periodical code)

STGG: Superconducting Tensor Gravity Gradiometer

STGP: Strongly Typed Genetic Programming

Stgt: Suttgart (Germany)

STGT: Second TDRS Ground Terminal (*) • Second TDRSS Ground Terminal

Sth: Something • South

STH: Strathmore (Australia-QLD) (location identifier)

Sthn: Southern

STHSPAAL: Space Telescope High-Speed Photometer Acronym and Abbreviation List

STI: Santiago (Dominican Rep.) (location identifier) • Save The Images • Science and Technology Information • Science-Technology-Innovation • Scientific and Technical Information • Scientific and Technological Information • Scientific Technologies Inc. (USA-CA) • Screw Thread Insert • Service Technique Informatique • Set Interrupt • Signal Technology Inc. • Software Testing International •

Standard Tape Interconnect • Supplemental Technical Instruction • Système de Traitement d'Images • System of Topographic Information • Systolic Time Interval

STIB: Scientific and Technical Information Branch • Stratosphere-Troposphere Interactions and the Biosphere

STIC: Science and Technology Information Center • Science, Technology and Industry Collection • Scintillation Tile Calorimeter • Space Technical Information Control • Space Telescope Institute Council • Space Test Integration Contract

STICA: Space Technology Industrial and Commercial Applications (periodical code)

STICS: Suprathermal Ion Composition Spectrometer

STID: Science and Technology Information and Documentation (EU) • Scientific and Technical Information Dissemination • Scientific and Technical Information Division • Statistiques et Traitement Informatique des Données

STIF: Science and Technology Information Facility • Scientific and Technical Information Facility • Semantically Transparent Interfacing • Software Test and Integration Facility • Spectral Transmission Interference Filter

STII: Science and Technology Information Institute • Software Test and Integration Instruction

STIK: STIS Interface Kit

STIKSCAT: Stick Scatterometer

STIL: Software Test and Integration Laboratory • Statistical Interpretive Language • Streak Tube Imaging Lidar

STIM: Scanning Transmission Ion Microscope • Scanning Transmission Ion Microscopy • Science-Technology-Information-Medicine • Solar Total Irradiance Monitor

STIMS: Scientific and Technical Information Modular System • Shuttle Thermal Infrared Multispectral Scanner

STINGS: Stellar Inertial Guidance System

STINT: Standard Interface

STIO: Scientific and Technical Information Office (NASA)

Stip: Stipend • Stipendiary • Stipulation

STIP: Scientific and Technical Information Processing • Smart Thermal Interface Plate • Study of Travelling Interplanetary Phenomena

Stir: Stirlingshire (UK-Scotland)

STIR: Surplus To Immediate Requirements

STIS: Science and Technology Information System • Scientific and Technical Information Services • Scientific and Technical Information System • Scientific and Technical Intelligence System • Space Telescope Imaging Spectrograph (*) • Space Telescope Imaging Spectrometer • Stabilized Thermal Imaging System • Standards and Technology Information System

STISS: Scientific and Technical Information Systems and Services

s/tít: Sin Título

STITE: Scientific and Technical Information Transfer for Education

STIVE: Standard Interface Verification Equipment

STJ: Saint Joseph - Rosecrans (USA-MO) (location identifier) • Superconducting Tunnel Junction • Superconducting Tunnelling Junction

Stk: Stock

STK: Satellite Tool Kit • Standard Test Key • Sterling Crosson (USA-CO) (location identifier)

Stl: Steel

STL: Saint Louis - Lambert (USA-MO) (location identifier) • Save The Logs • Science and Technology Laboratory • Science Trailer Line • Short-term Timeline • Simulated Tape Load • Small Text List • Space Technology Laboratory • Space Tissue Loss • Standardiseringsorganens Ledningsgrupp (Sweden) • Standard Telecommunications Laboratories • Standard Template Library • Studio-to-Transmitter Link • Studio-Transmitter Link • Synchronous Transistor Logic • Systems Techniques Laboratory • System Test Laboratory

STLC: Small Time Local Controllability • Small Time Local Controllable

STLCC: Small Time Local Configuration Controllable

STLDD: Software Top Level Design Document

Stlg: Sterling

STLM: Spatial-Temporal Light Modulator

STLO: Scientific Technical Liaison Office • Scientific Technical Liaison Officer

STLP: Stochastic-excited Linear Prediction

STLV: Slow Turning Lateral Vessel

Stm: Steam • Storm

STM: Santarém - Eduardo Gomes (Brazil-PS) (location identifier) • Scanning-Tunnelling Microscope • Scanning-Tunnelling Microscopy • Science and Technology Month (Malaysia) • Scientific and Technical Manuscript • Scientific-Technical-Medical • Send Test Message • Service Test Model • Short-Term Memory • Signal Termination Module • Specification Test Material • Spin Tuned Magnetron • Springfield Telescope Makers (USA-VT) • Standard Thermal Model • Static Test Model • Store Multiple • Storm • Stream • Structural Test Model • Structural and Thermal Mode • Structural and Thermal Model • Structural Test Model • Structure and Thermal Model • Structure Thermal Model • Synchronous Transfer Mode • System Termination Model • System Test Mode • System Training Mission

STME: Space Transportation Main Engine

STMEM: Structural/Thermal Model and Engineering Model

StMLU: Staatministerium für Landesentwicklung und Umweltfragen

STMPLP: Stochastic-Multipulse-excited Linear Predictive

STMS: Short-Term Monetary Support

Stmt: Statement

STMU: Special Test and Maintenance Unit

STMUX: Statistical Data Multiplexer

STMV: Satellite Tobacco Mosaic Virus

STMW: Subtropical Mode Water

Stmy: Stormy

Stn: Station • Stazione

STN: London Stansted (UK-England) (location identifier) • Satellite Tracking Network • Scientific and Technical Network • Software Trouble Note • Solar Telescope Network • Station • Supertwisted Nematic • Supertwist Nematic • Switched Telephone Network • Systemetech Nord GmbH

STNA: Service Technique de la Navigation Aérienne

Stng: Sustaining

STNLCD: Supertwist Nematic LCD

STNMTE: Service Technique de la Navigation Maritime et des Transmissions de l'Équipement

STNR: Stationary

STNRY: Stationary

StNws: Standardization News (periodical code)

Sto: Santo

STO: Santo • Save The Optics • Senior Technical Officer • Service du Travail Obligatoire (WWII) • Short-Term Objective • Slater-Type Orbital • Solar-Terrestrial Observatory • Standing Order • Stockholm (Sweden) (location identifier) • Sub-Task Officer • System Test Objective

STOA: Scientific and Technical Options Assessment • Scientific and Technological Options Assessment

StoAn: Stockholm Observatory Annals (periodical code)

STOAT: Space Telescope Observatory Assessment Team

STOCC: Space Telescope Operations Control Center

STOF: Slosh Test Orbital Facility

STOG: Science and Technology Objectives Guide

STOH: Since Top Overhaul

STOIC: Stratospheric Ozone Intercomparison

STOL: Short Takeoff and Landing • Systems Test and Operations Language

STOLVCD: Short Takeoff and Landing - Vertical Climb and Descent

STOM: System Test and Operations Manual

STOMS: Space Telescope Observatory Management System

S'ton: Southampton (UK-England)

Ston: Southampton (UK-England)

STOP: Selected Test Optimization Program • Short-Term Operation Plan • Single-Title Order Plan

STOPAT: Space Telescope Observatory Performance Assessment Team

Stor: Storage

STOR: Segment Table Origin Register • System for Table-Oriented Reports • System Test and Operations Report

STORE: Shuttle Test Of the Relativity Experiment • Storage Technology for Operational Readiness

STORM: Stormscale Operational and Research Meteorology • Storm Tracking and Observation for Rainfall-runoff Monitoring

STORMS: Standardized Operation Research Management System

StoS: Ship to Shore • Station to Station

STOS: Space Test Operations Section • Store String

STOSC: Space Telescope Operations Support Cluster

STOVL: Short Take Off and Vertical Landing

Stp: Stamp

STP: Minneapolis Saint Paul (USA-MN) (location identifier) • Principe (Sao Tome &_) (ISO country code) • Sao Tome & Principe (ISO country code) • Save The Plots • Satellite Technology Program • Satellite Test Program • Satellite Tracking Program • Scanning Tunnelling Potentiometry • Schmidt Telescope Panel • School of Theoretical Physics • Science and Technology Project • Scientifically Treated Petroleum • Secure Transaction Protocol • Selective Tape Print • Self-Test Program • Seminar Traineeship Program • Service Transfer Point • Shielded Twisted Pair • Short-Term Plan • Short-Term Prediction • Shuttle Technology Panel • Signaling Transfer Point • Signal Transfer Point • Simultaneous Test Procedure • Simultaneous Track Processor • Software Technology Park • Software Technology Programme • Software Test Plan • Software Test Procedure • Software Through Pictures • Soil Temperature Probe • Solar-Terrestrial Physics • Solar-Terrestrial Probe • Space Technology Payload • Space Telescope Project • Space Test Payload • Space Test Program • Spanning Tree Protocol • Special Technical Publication • Spectrum of Time Project • Standardized Test Program • Standard Processor • Standard Temperature and Pressure • Standard Test Procedure • Static Pressure • Static Test Program • Stop • Subsurface Transport Program • Supporting Technology Programme • System Technology Program • System Test Plan

STPB: Scientific and Technical Publications Branch

STPC: Society of Technical Publications Contractors

STPCOE: Special-To-Project Checkout Equipment

Stpd: Stripped

STPDF: Solar-Terrestrial Physics Data Facility

Stpg: Stepping

STPL: Standard Test Processing Language

STPM: Structural/Thermal/Pyro Model

STPMET: Solar-Terrestrial Physics - Meteorology

StPO: Strafprozeßordnung

Stpr: Stepper

STPRC: Space Telescope Proposal Review Committee

STPS: Summary Task Planning Sheet

STQ: Saint Marys (USA-AK) (location identifier)

str: Stratiformis

Str: Seater • Steamer • Straat • Straight • Strait • Straße • Streamer • Street • Strength • String • Strong • Structural • Structure

STR: Satellite Transceiver • Science Tape Recorder • Scientific and Technological Research • Scientific Technical Report • Service Test Review • Service Trouble Report • Short-Term Research • Software Tape Recorder • Software Test Report • Software Turnover Record • Solar-Terrestrial Relation • Special Theory of Relativity • Star Tracker • Store Task Register • Street • Structure • Stuttgart Echterdingen (Germany) (location identifier) • Sub-THz Receiver • Summary Technical Report • Surplus To Requirements • Synchronous Transmitter Receiver • System Test Review

STRA: Stravenue

STRAC: Solar-Terrestrial Relations Advisory Committee

STRAD: Signal Transmitting, Receiving, and Distribution

STRAM: Synchronous Transmit Receive Access Method

STRAP: Star Tracking Rocket Attitude Positioning • Stellar Tracking Rocket Attitude Positioning

Strat: Strategic

STRAT: Stratospheric Tracers of Atmospheric Transport

Stratig: Stratigraphy

Strato: Stratosphere

STRAV: Stravenue

STRAVE: Stravenue

STRAVEN: Stravenue

STRAVN: Stravenue

STRB: Strobe

STRC: Science and Technology Research Center • Scientific, Technical and Research Commission

STRCH: Strectch

Strd: Stranded

STREAM: Stratosphere-Troposphere Exchange by Aircraft Measurements

STREP: Space Trajectory Radiation Exposure Procedure • Standard Repair

STRESS: Sediment Transport Events on Shelves and Slopes • Structural Engineering Systems Solver

STRETCH: Space Technology Requirements Engineering Test of Component Hardware

STREX: Storm Transfer and Response Experiment

STRF: SLR Terrestrial Reference Frame

STRFLD: Star Field

Strg: Steering • Strong

STRG: Steering • String

STRÍ: Icelandic Council for Standardization

STRI: Smithsonian Tropical Research Institute

STRICOM: Simulation, Training and Instrumentation Command

STRIDE: Science and Technology for Regional Innovation and Development in Europe • Science and Technology Reporting Information Distribution Enhancement

STRING: Soil and Terrain Resources Information Network Generation

STRIP: String Processing Language

STRIVE: Science Teachers Research Involvment for Vital Education

Strk: Stroke

Strl: Structural

Strln: Streamline

Strm: Storeroom • Stream

STRM: Stream

Strn: Strainer

STRN: Standard Technical Report Number

STROBE: Satellite Tracking Of Balloons and Emergencies

STROGANOFF: Structured Representation On Genetic Algorithm for Non-linear Function Fitting

StrOp: Structural Optimization (periodical code)

STRQ: Sensor Termination Request • Session Termination Request

STRS: Sensor Termination Response • Space-Time Resolved Spectrograph

STRT: Strait • Street

STRUC: Structure

STRUCT: Structure

STRUDL: Structural Design Language

STRV: Space Technology Research Vehicle

STRVN: Stravenue

STRVNUE: Stravenue

STrZ: South Tropical Zone

Struct: Structural • Structure

sts: Status

Sts: Saints • Streets

STS: Santa Rosa (USA-CA) (location identifier) • Satellite To Satellite • Satellite Tracking Station • Satellite Transmission Systems Inc. • Save The Spectra • S-band Transmitter System • Science and Technology Section • Science Talent Search • Science-Technologie-Société • Science-Technology-Society • Scientific Terminal System • Section des Techniciens Supérieurs • Selective Tape Service • Sensor Test Set • Servo Test System • Short-Term Schedule • Short-Term Scheduler • Short-Term Scheduling • Shuttle Test Station • Shuttle Transportation System • Shuttle Transport System • Skylab Terminal System • Slicing Tree Structure • Socio-Technical System • Solar Tracking System • Spacecraft Telecommunications System • Spacecraft Tracking Station • Space Transportation System • Space Transportation System Ltd. • Stabilized Telescope System • Standard Technical Specification • Static Test Stand • Station To Station • Status Transfer Switch • Streets • Structural Transition Section • Supercooled Ternary Solution • Synchronous Transport Signal • Système de Télécommunication Sol • System Test Set • System Test Station • System Trouble Shooting

STSA: Space Telescope Solar Array • State Technical Service Act (USA)

STSAM: Space Transportation System Analysis Model

STSB: Scientific and Technical Services Branch

STSCHG: Status Change

STScI: Space Telescope Science Institute (USA-MD)

STScIF: STScI Facility

STSCM: Space Transportation System Cost Model

STSDAS: Space Telescope Science Data Analysis Software • Space Telescope Science Data Analysis System (*) • Space Telescope Scientific Data Analysis System

STSF: Science and Technology through Science Fiction

StSky: Star Sky (periodical code)

STSM: Space Transportation System Management

STSO: Senior Technical Staff Officer

STSOC: Space Telescope Science Operations Center • Space Transportation System Operations Contract

STSOF: Space Telescope Science Operations Facility

STSOP: Space Telescope Science and Operations Project

STSOPO: Space Transportation System Operations Program Office

STSP: Same Time - Same Place • Solar-Terrestrial Science Program • Symmetric Travelling Salesman Problem

STSR: System Test Summary Report

STSS: Sensitive Thrust Stand System • Space Telescope Safing System

STSWG: Space Telescope Science Working Group

STT: Saint Thomas - Cyril E. King (US Virgin Islands) (location identifier) • Scientific Tools and Techniques • Secure Transaction Technology • Site Testing Telescope • Spacecraft Terminal Thrust • Spacelab Transfer Tunnel • Spontaneous Trait Transference • Star Tracker

STTE: Special Tools and Test Equipment • Spectroscopic Time-Tagged Event

STTF: Space Telescope Test Facility

Sttg: Starting

STTG: Solar-Terrestrial Theory Group

STTP: SGML Tree Transformation Process • Space Technology Training Program

Sttr: Starter • Stator

STTR: Short-Term Technology Review • Small-business Technology Transfer

STTT: Software Tools for Technology Transfer • Space Telescope Task Team

StTur: Studies in Turbulence (periodical code)

STU: Santa Cruz (Belize) (location identifier) • Schmidt Telescope Unit • Secure Telephone Unit • Servo Test Unit • Signal Transfer Unit • Space-Time Unit • Special Test Unit • Special Training Unit • Star Tracker Unit • Static Test Unit • Streaming Tape Unit • Systems Test Unit • System Timing Unit

StUB: Stadt- und Universitätsbibliothek

STUB: Stadt- und Universitätsbibliothek

STUC: Space Telescope Users Committee

Stud: Student

STUTEL: Spécifications Techniques d'Utilisation du Téléchange de Logiciel

STV: Satellite Television • Scottish Television • Separation Test Vehicle • Single Transferable Vote • Solar Thermal Vacuum • Space Test Vehicle • Special Test Vehicle • Standard Test Vehicle • Stichting Technology Vlanderen (Belgium) • Structural Test Vehicle • Subscription Television • Surat (India) (location identifier)

STVNS: Spherical Tokamak Volume Neutron Source

StVO: Straßenverkehrsordnung

STVS: Secure Television System

STVSS: Space Telescope Verification Software System

STVSW: Space Telescope Verification Software

STVT: Space Telescope Verification Team

STVTL: Space Telescope Verification Team Leader

StVZO: Straßenverkehrszulassungsordnung

Stw: Steenweg

STW: Save The Waves • Standard Typical Waste • Stichting voor de Technische Wetenschappen (Netherlands) • Stravopol (Russia) (location identifier)

STWE: Society of Technical Writers and Editors

STWG: Scientific and Technical Working Group • Space Telescope Working Group

STWP: Society of Technical Writers and Publishers (USA)

Stwy: Stairway

STX: Saint Croix - Alexander Hamilton (US Virgin Islands) (location identifier) • Start of Text

STY: Salto (Uruguay) (location identifier) • Style

STZ: Santa Terezinha Confresa (Brazil-MT) (location identifier) • South Temperate Zone • South Tropical Zone

su: siehe unten! • Sudanese (ISO language code) • Superuser

Su: Sudan • Sudanese • Sunday

S/U: Setup

SU: Egypt (international aircraft registration code) • Scheduling Unit • Setup • Siehe Unten! • Signaling Unit • Single User • Soviet Union (now obsolete) • Space Unit • Special Unitary • Standard Update • Stand Up • Stanford University (USA-CA) • Station Unit • Storage Unit • Strontium Unit • Student Union • Sudan • Sumerian • Superuser • Support Unit • Switching Unit • Syracuse University (USA-NY)

SUA: Satellite Unfurlable Antenna • State Universities Association • Stuart Witham (USA-FL) (location identifier) • Suppression of Unlawful Acts

SUAR: Start Unload Address Register

SUAS: System for Upper Atmosphere Sounding

Sub: Subida • Subito • Subject • Subjective • Subjectively • Submissive • Subordinated • Subscription • Subsidiary • Subsidy • Subsistence • Substitute • Suburb • Suburban • Subvention • Subway

SUB: Staats- und Universitätsbibliothek • Subject • Surabaya Juanda (Indonesia) (location identifier) • Subroutine • Substitute • Subtract • Supplemental Unemployment Benefit

Subd: Subdivision

Subj: Subject • Subjective • Subjectively • Subjugate • Subjugated • Subjunctive • Subjuntivo

SUBJ: Subject

Subl: Sublimation

SUBOP: Submersible Operations

Subq: Subsequent

SUBR: Southern University - Baton Rouge (USA-LA)

Subs: Subsequent • Subsequently • Subsidiary • Subsistence

SUBS: Single-bus Unidirectional Broadcast System

Subsc: Subscription

Subsec: Subsecretaría • Subsection

Subseq: Subsequent • Subsequently

Subsp: Subspecie

Subspp: Subspecies

Subst: Substantive • Substantively • Substitute

Substand: Substandard

Subte: Subterráneo

Subtr: Subtract

Suc: Succeed • Succeeding • Success • Successor • Suction • Sucursal

SUC: Southern Union College • Starlink Users Committee • Sundance Schloredt (USA-WY) (location identifier)

Succ: Succeed • Succeeded • Succeeding • Success • Successful • Succession • Successor • Succursale

SUCCESS: Subsonic Aircraft: Contrail and Cloud Effects Special Study

SUCE: Spatial Understanding Critical Examination

SUCOSIP: Subcommittee on Site Properties

Suct: Suction

SUD: Stroud (USA-OK) (location identifier) • Subjective Unit of Distress

SUDS: Software Update and Distribution System • Subjective Unit of Discomfort Scale

SUDSS: Spatial Understanding and Decision Supporting System

SUE: Seismic Underwater Explorer • Strontium Unit Equivalent • Stupid User Error • Sturgeon Bay (USA-WI) (location identifier) • Systematic Usability Evaluation

SUEA: Scan Unit Electronics Assembly

Suf: Sufficient • Suffix • Sufijo

SUF: Lamezia Terme (Italy) (location identifier) • Scientific Update Facility • Service Update Facility • Software Update Facility

Suff: Sufficient • Suffix • Suffolk (UK-England)

Sug: Suggest • Suggestion

SUG: Science Users Group • Starlink User Guide • SUN Users Group • Surigao (Philippines) (location identifier)

SUH: Sur (Oman) (location identifier)

SUI: Sensory User Interface • Sound User Interface • Speech User Interface • Standard User Interface • State Unemployment Insurance • Sukhumi (Georgia) (location identifier)

SUIM: Station Unit Interface Module

SUIT: Simple User Interface Toolkit

Suiv: Suivant

SUJ: Satu Mare (Romania) (location identifier)

SUK: Samchok (Korea-ROK) (location identifier) • Someraj Universitataj Kursoj

SUL: Software Utilization List • Sui (Pakistan) (location identifier)

Sult: Sultan

sum: Sumat • Sumendum

Sum: Summary • Summer

SUM: Software User Manual • Statistical Utility Module • Summary • Sumter (USA-SC) (location identifier)

SUMAS: Submillimeter Atmospheric Sounder

SUMC: Satellite Ultrareliable Modular Computer • Space Ultrareliable Modular Computer

SUMER: Solar Ultraviolet Measurement of Emitted Radiation

Summ: Summary • Summer

SUMMIT: Sydney University Multiple Mirror Telescope

SUMO: Supermirror Observatory

SUMP: Submicron Photometer

SUMS: Shuttle Upper-Atmosphere Mass Spectrometer • Specialized Unit Maintenance Support

SUMSS: Sydney University Molonglo Sky Survey

SUMT: Sequential Unconstrained Minimization Technique

Sun: Sunday

SUN: Hailey - Sun Valley (USA-ID) (location identifier) • Single User Network • Solar Ultraviolet Network • Standard Units and Nomenclature • Stanford University Networks (USA-CA) • Starlink User Note • Students Understanding Nature • Sunday • Sun Valley (USA-ID) (location identifier) • Symbols-Units-Nomenclature

SUNA: Sociedad Universitaria de Astronomía (Venezuela)

SUNAMCO: Symbols, Units, Nomenclature, Atomic Masses and Fundamental Constants

SUNBEAMS: Students United with NASA Becoming Enthusiastic About Math and Science

Sund: Sunday

SUNDIAL: Speech Understanding and Dialogue

SUNET: Swedish University Network

SUNIST: Serveur Universitaire National pour l'Information Scientifique et Technique

SUNRS: Sunrise

Sunwo: Sunworld (periodical code)

SUNY: State University of New York (USA-NY)

SUNYA: State University of New York at Albany (USA-NY)

SUNYSB: State University of New York at Stony Brook (USA-NY)

SUO: Sun River (USA-OR) (location identifier)

SUOA: Super-Unit Of Assessment

SUOC: Science User Operation Centre

sup: Supra • Supremum

Sup: Superficial • Superfine • Supérieur • Superior • Superlative • Superlativo • Supplement • Supplementary • Supply • Supreme • Suprême

SUP: Single Unit Parameter • Soluble Unreactive Phosphorus • Special Utility Program • Statistical Utility Program • Sumenep Trunojoyo (Indonesia) (location identifier) • Supplemental • Supply • Support • Suspected Unapproved Part • System Utilization Procedure

SUPARCO: Space and Upper Atmosphere Research Commission (Pakistan)

Super: Superficial • Superfine • Superintendent • Superior • Supernumerary

Superg: Supergiant

Superl: Superlative • Superlativo

SUPIDEN: Support Identification

SUPIM: Sheffield University Plasmasphere and Ionosphere Model

Supp: Supplement • Supplementary

Suppl: Supplement • Supplément • Supplementary • Supplémentaire • Supplied • Supply • Support • Supported • Supporting • Suppress • Suppressed • Suppressing • Suppression

SUPPORT: Study to Understand Prognoses and Preferences for Outcomes and Risks of Treatment

SUPPR: Suppression

SUPPS: Supplemental Procedures

Supr: Superior • Supervisor • Suppress • Supreme

SUPRA: Service Universitaire Pour la Recherche et les Applications en Supraconductivité

Supt: Superintendent

SUPT: Support

Supv: Supervise

SUPV: Supervisor

Supvr: Supervisor

SUQ: Sucua (Ecuador) (location identifier)

Sur: Surface • Surplus • Surrey (UK) • Surveyor

SUR: Rouble (Russia) (ISO currency code) • Shared University Research • Summer Beaver (Canada-ON) (location identifier) • Suriname (ISO country code)

SURA: Southeastern Universities Research Association (USA)

SURANET: Southeastern Universities Research Association Network

SURAP: Summer Undergraduate Research Assistantship Program

SURBAT: Simultaneous Unlimited Rigorous Block Analytical Triangulation

SURC: Space Utilization Research Center

SURE: Shuttle Users Review and Evaluation • Simplicity-Useability-Reliability-Economy

Surf: Surface

SURF: SCUBA User Reduction Facility • Stichting Samenwerkende Universitaire Rekenfaciliteiten (Netherlands) • Summer Undergraduate Research Fellowship • Surface • Synchrotron Ultraviolet Radiation Facility

SURFNET: Dutch University Network (Netherlands)

SURFSAT: Summer Undergraduate Research Fellowship Satellite

Surg: Surgeon • Surgery • Surgical

SURGE: Sorting-Updating-Report-Generating-Etc.

SURPC: Scottish Universities Research Policy Consortium

SURPIC: Surface Picture

Surr: Surrey (UK)

SURRVS: Standard Umbilical Retraction Retention Vent System

SURS: Standard Umbilical Retraction System

SurSc: Surface Science (periodical code)

SURTEX: Shuttle-Released Target Experiment • Shuttle-Released Technology Experiment

Surv: Surveillance • Survey • Surveying • Surveyor • Survival • Survive • Surviving

SURV: Standard Underwater Research Vessel

SURVIAC: Survivability/Vulnerability Information Analysis Center

Sus: Sussex (UK-England)

SUS: Single Unix Specification • Speech-Understanding System • Spirit of Saint Louis (USA-MO) (location identifier) • Suit Umbilical System • Sussex (Starlink internal code)

SUSI: Stanford University Speckle Interferometry • Superb Seeing Imager • Superseeing Imager • Sydney University Stellar Interferometer

SUSIE: Satellites for the Understanding of Space Instrumentation and Experimentation

SUSIM: Solar Ultraviolet Spectral Irradiance Monitor

Susp: Suspected • Suspend • Suspended • Suspending • Suspension

SUSP: Suspend

Suss: Sussex (UK-England)

SUSS: Spatial Understanding Support System

SUST: Shanghai University of Sciences and Technology (China-PRC)

SUSTAR: Sussex (Starlink internal code)

SuStB: Staats- und Stadtbibliothek (Germany)

SUSY: Supersymmetry

SUT: Satellite Under Test • Science University of Tokyo (Japan) • Set-Up Time • Software Under Test • Sumbawanga (Tanzania) (location identifier) • Suranaree University of Technology (Thailand) • Swinburne University of Technology (Australia) • System Under Test

SUTAGS: Shuttle Uplink Text And Graphics Scanner

Suth: Sutherland (UK-Scotland)

SUU: Fairfield Travis (USA-CA) (location identifier)

SuUB: Staats- und Universitätsbibliothek (Germany)

SUV: Solar Ultraviolet • Sport-Utility Vehicle • Suva (Fiji) (location identifier)

SUVDDP: Space Ultraviolet/Visible Detector Development Program

SUVE: Solar Ultraviolet Experiment

SuW: Sterne und Weltraum

SUW: Superior - Richard I. Bong (USA-WI) (location identifier)

SUX: Sioux City (USA-IA) (location identifier)

SUY: Sudureyri (Iceland) (location identifier)

SUZ: Suria (Papua New Guinea) (location identifier)

SuZIE: Sunyaev-Zel'dovich Infrared Experiment

s/v: Sub Voce

sv: Sub Verbo • Sub Voce • Swedish (ISO language code)

Sv: Sievert

S/V: Space Vehicle • Supply Valve • Surface/Volume • Survivability/Vulnerability

SV: El Salvador (ISO country code) • Rhea (satellite of Saturn) • Safety Valve • Sailing Vessel • Salvador (El_) (ISO country code) • Sancta Virgo • Satellite Vehicle • Satellite Virus • Science Verification • Sea Vessel • Secondary Verification • Secular Variation • Security Violation • Selenoid Valve • Self-Ventilated • Self-Verification • Short Vector • Shuttle Vehicle • Side Valve • Side View • Significant Vector • Small Volume • Software Visualization • Solenoid Valve • Space Variant • Space Vehicle • Spatially-Variant • Speaker Verification • State Vector • State Violation • Stiftverband • Stroke Volume • Sub Verbo • Sub Voce • Supply Valve • Surrender Value • System Verification

SvA: Soviet Astronomy (now obsolete) (periodical code)

SVA: Savoonga (USA-AK) (location identifier) • Sectionalized Vertical Antenna • Service à Valeur Ajoutée • Shared Virtual Area • Skokie Valley Astronomers (USA-IL) • Société Vaudoise d'Astronomie (Switzerland) • Spatially Varying Apodization • Speech-Vision Association • Speech-Vision Associator • Statistical Vibration Analysis • Stator Vane Actuator

SVAA: Susquehanna Valley Amateur Astronomers (USA-PA)

SVAB: Shuttle Vehicle Assembly Building

SVA&C: Shuttle Vehicle Assembly & Checkout

SVAC: Shuttle Vehicle Assembly and Checkout

SVACO: Shuttle Vehicle Assembly and Checkout

SVAFB: South Vandenberg Air Force Base

SvAL: Soviet Astronomy Letters (now obsolete) (periodical code)

S/VAR: Survivability/Vulnerability Analysis Report

SVAS: Sacramento Valley Astronomical Society (USA-CA)

SVAT: Soil-Vegetation-Atmosphere Transfer

SVATS: Soil-Vegetation-Atmosphere Transfer Scheme

SvB: Stichting voor Biofysica (Netherlands)

SVB: Sambava (Madagascar) (location identifier) • Shuttle Vehicle Booster • Space Vehicle Booster • Stichting Voor Biofysica (Netherlands)

SVBT: Space Vehicle Booster Test

Svc: Service

SVC: Colón (El Salvador) (ISO currency code) • Science VAX Cluster • Serviço • Silver City (USA-NM) (location identifier) • Single Version Charging • Spacecraft Vehicle Contractor • Space Vehicle Code • Steele Visitor Center (Lowell Obs.) • Store-Value Card • Supervisor Call • Switched Virtual Circuit (*) • Switched Virtual Connection • Switch Virtual Circuit

SVCDS: Space Vehicle Computer Design System • Space Vehicle Control/Display Simulation

Svce: Service

SVCMU: Space Vehicle Control and Monitor Unit

SVD: Saint Vincent (location identifier) • Seismic Velocity Discontinuity • Shock Vibration Digest (periodical code) • Simultaneous Voice and Data • Simultaneous Voice/Data • Singular Value Decomposition • Small-Volume Detector • Space Vehicles Division • Supplementary Volume Descriptor

SVDP: Skylab Video Documentation Project

SVDS: Space Vehicle Dynamic Simulation • Space Vehicle Dynamic Simulator

SVDT: Space Vehicle Development Team

SVE: Space Vehicle Equipment • Susanville (USA-CA) (location identifier)

SVEB: Schweizerische Vereinigung für Erwachsenenbildung (Switzerland)

SVEC: Space Vacuum Epitaxy Center (USA-TX)

SVF: Save (Benin) (location identifier) • Simple Vector Format • Software Validation Facility • State Variable Feedback • State Variable Filter

SVFR: Special Visual Flight Rule

SVG: Scalable Vector Graphics • Silicon Valley Group (USA-CT) • Stavanger (Norway) (location identifier)

SVGA: Super Video Graphics Adapter • Super Video Graphics Array (*)

Svgs: Savings

SVH: Statesville (USA-NC) (location identifier)

SVHS: Super Video Home System

SVI: San Vincente del Caguan (Colombia) (location identifier) • Serveur Vocal Interactif • Titan (satellite of Saturn)

SVIC: Shock and Vibration Information Center • State Variable Inequality Constraint

SVICB: Shock Vibration Information Center Bulletin (periodical code)

SVICD: Shock Vibration Information Center Digest (periodical code)

SVID: System V Interface Definition

SVII: Hyperion (satellite of Saturn)

SVIII: Japetus (satellite of Saturn)

SVIR: Scanning Visible and Infrared Radiometer

SVIRI: Spinning Visible and Infrared Imager

SVISSR: Strechted Visible and Infrared Spin Scan Radiometer

SVIU: Space Vehicle Interface Unit

SVJ: Svolvaer (Norway) (location identifier)

SvJPP: Soviet Journal of Plasma Physics (periodical code)

SVJPS: Springer-Verlag Journal Preview Service

SvJQE: Soviet Journal of Quantum Electronics (periodical code)

SVK: Silver Creek (Belize) (location identifier) • Slovakia (ISO country code) • Slovaquie (ISO country code)

SVL: Savonlinna (Finland) (location identifier)

SVM: Saint Paul's Mission (Australia-QLD) (location identifier) • Scaling Vector Mothod • Science Verification Manager • Service Module • Storage Vault Management • Storage Vault Manager • Strong Vertical Mixing • Support Vector Machine • System Validation Model • System Virtual Machine

SVMA: Space Vehicle Mission Analysis

SVMG: Solar Vector Magnetograph

SVN: Savannah Hunter (USA-GA) (location identifier) • Slovenia (ISO country code) • Slovénie (ISO country code) • Space Vehicle Number

SVO: Moscow Sheremetyevo (Russia) (location identifier) • Space Vehicle Operations • Space Volcanic Observatory • Subject-Verb-Object

SVOD: Space Vehicle Operations Director

SVOSA: Solar Variability and the Outer Solar Atmosphere

SVP: Kuito (Angola) (location identifier) • Saturated Vapour Pressure • Science Verification Phase • S'il Vous Plaît • Software Verification Procedure • Space Vehicle Pressure • Star-Vaporizing Pulsar

SVPCS: Springer Series on Physics Chemistry Space (periodical code)

SvPhA: Soviet Physics-Astronomy (periodical code)

SvPhD: Soviet Physics-Doklady (periodical code)

SvPhU: Soviet Physics-Uspekhi (periodical code)

SVPN: Secure Virtual Private Network

SVPSF: Space-Variant Point Spread Function

SVQ: Sevilla (Spain) (location identifier)

Svr: Severe

SVR: Server • Severe • Smart Video Recorder • Space Vehicle Requirement • Spiritus Vini Rectificatus • Svay Rieng (Cambodia) (location identifier) • Synthetic Variable Ratio

SVRB: Supervisor Request Block

SVRL: Several

SVRR: Software Verification Readiness Review

SVS: Scientific Visualization Studio • Service Vision System • Shuttle Vehicle Simulator • Single Virtual Storage • Soviet Variable Stars • Space Vehicle Simulator • Space Vehicle System • Space Vision System • Spinning Vehicle Simulator • Stevens Village (USA-AK) (location identifier) • Still-camera Video System • Suit Ventilation System • Synthetic Vision System

SVSM: Semantic Vector Space Model

SVST: Swedish Vacuum Solar Telescope

Svt: Souvent

SVT: Savuti (Botswana) (location identifier) • Sciences de la Vie et de la Terre • Software Validation Test • Solar Vacuum Telescope • Space Vehicle Test • System Validation Test

SVTN: Secure Voice Telecommunications Network • Secure Voice Teleconference Network

SVTP: Sound-Velocity-Temperature-Pressure

SVU: Savusavu (Fiji) (location identifier) • Sri Venkateswara University (India)

svv: Sub Verbis

SVV: San Salvador (Venezuela) (location identifier)

SVVP: Software Verification and Validation Plan

SVVS: System V Verification Suite

svw: soviel wie

SVW: Sparrevohn (USA-AK) (location identifier)

SVWG: Science Verification Working Group

SVX: Ekaterinburg (Russia) (location identifier) • Silicon Vertex Detector

Svy: Survey

SVY: Savo (Solomon Islands) (location identifier)

SVZ: San Antonio (Venezuela) (location identifier)

s/w: Sea Water

Sw: Swahili (ISO language code) • Sweden • Swedish • Swiss • Switch • Switzerland

S/W: Software

S&W: Sterne und Weltraum (periodical code)

SW: London (Southwest_) (UK postal code) • Salt Water • Schwassmann-Wachmann • Science Watch (journal) • Sea Water • Shipper's Weight • Shock Wave • Shortwave • Short Wavelength • Sidewall • Single Weight • Single Wheel • Software • Solar Wind (*) • Solar Wing • South Wales (UK) • Southwest • Southwestern • Southwest London (UK postal code) • Space Wing • Spares Warehouse • Specific Weight • Spot Weld • Standard Weight • StarWorlds • Statistical Weight • Status Word • Stellar Wind • Stemming Weight • Subset Word • Superworld • Surface Water • Swap • Sweden • Switch • Switchboard Wound • Switcher

SWA: Namibia (international vehicle registration code) • Seewetteramt • Shantou (China-PRC) (location identifier) • Slow Wave Activity • Southwest Africa • Superwide Angle • Support Work Authorization • Swiss World Airways

Swab: Swabia • Swabian

SWACOM: Standard Water Column Model

SWACS: Space Warning and Control System

SWAD: Shop Work Authorization Document • Subaccount Work Authorization Document • Subdivision of Work Authorization Document

SWADE: Surface Wave Dynamic Experiment

SWADS: Scheduler Work Area Data Set

SWAG: Simulated Waste Access to Ground

SWAIS: Simple Wide-Area Information Server

SWAK: Sealed With A Kiss

SWAL: Southwest Airlines Inc. (Japan)

SWALK: Sealed With A Loving Kiss

S/WAN: Secure Wide-Area Network

SWAN: Secure Wide-Area Network • Solar Wind Anisotropies (SOHO)

SWANA: Solid Waste Association of North America

SWAP: Severe Weather Avoidance Plan • Southwestern Association of Planetariums (USA-TX)

S/WARE: Software

SWARE: Software

SWARM: Solar Warning And Real-time Monitor • Space Weather Aeronautical Responses Model

SWAS: Submillimeter Wave Astronomy Satellite

SWAT: Scandinavian Workshop on Algorithm Theory • Short Wavelength Adaptive Technology • Software Assessment Team • Software Assurance and Test • Software Assurance Team • Space Warfare ATM Timing • Special Weapons And Tactics • Stress Wave Analysis Technique • Subjective Workload Assessment Technique • Swift Action Team

SWATEC: Short-Wavelength Thermoelectrically Cooled

SWATH: Space Weather And Terrestrial Hazards

Swazil: Swaziland

SWB: Searle-Wilkinson-Bagnuolo (catalogue) • Shaw River (Australia-WA) (location identifier) • Solder Waiver Board • Sternwarte Bremen (Germany) • Subtract With Borrow

SWBB: Standard Wind - Blown Bubble

Swbd: Switchboard

SWbS: Southwest by South

SWbW: Southwest by West

SWC: Solar-Wind Compensation • Solar-Wind Compensator • Space Warfare Center • Stawell (Australia-VIC) (location identifier)

SWCA: Soil and Water Conservation Act (USA)

SWCC: Second World Climate Conference • Southwestern Computing Center

SWCDR: Software Critical Design Review

SWCG: Shane-Wirtanen Clusters of Galaxies

SWCI: Software Configuration Inspection • Software Configuration Item

SWD: Schlagwortnormdatei • Seward (USA-AK) (location identifier) • Synchronous Wave Device

SWDA: Solid Waste Disposal Act (USA) • Stepwise Discriminant Analysis

SWD&I: Software Development & Integration

SWDI: Software Development and Integration

SWds: StarWords

SWDS: Software Development System

SWDVS: Software Development and Verification System

Swe: Swedish

SWE: Scalar Wave Equation • Shallow-Water Equation • Siwea (Papua New Guinea) (location identifier) • Snow Water Equivalent • Solar-Wind Experiment • Suède (ISO country code) • Sweden (ISO country code)

SWEAP: SOTER Water Erosion Assessment programme

SWECC: Software Engineering Coordinating Committee

Swed: Sweden • Swedish

SWEDARP: Swedish Antarctic Research Programme

SWEE: Software Engineering Environment

SWEEP: Structures With Error Expurgation Program

SWEN: Space Weather Euro News

SWERVE: Sandia Winged Energetic Reentry Vehicle Experiment

SWET: Signal Weighted Extraction Technique • Slit-Weighted Extraction Technique

SwF: Swiss Franc (currency)

SWF: Newburgh Stewart (USA-NY) (location identifier) • Poughkeepsie Stewart (USA-NY) (location identifier) • Shortwave Fade • Shortwave Fading • Shortwave Fadout • Social Welfare Function

SWFAS: South West Florida Astronomical Society (USA-FL)

Swg: Sewage

SWG: Satwag (Papua New Guinea) (location identifier) • Schedule Working Group • Science Working Group • Security Working Group • Shock Wave Generator • Shuttle Working Group • Sine Wave Generator • Software Working Group • Special Working Group • Squarewave Generator • Standards Working Group • Standard Wire Gauge

Swgr: Switchgear

SWH: Significant Wave Height • Soft Warehouse (USA-HI) • Solar Water Heating • Sternwarte Höfingen (Germany) • Swan Hill (Australia-VIC) (location identifier)

SWHAS: South West Herts Astronomical Society (UK)

SWHE: Soft Warehouse Europe

Swi: Swiss • Switzerland

SWI: Schlagwortindex • Sine Wave Inverter • Software Interrupt • Sternwarte Ilsede (Germany) • Swindon (UK-England) (location identifier)

SWICS: Solar-Wind Ion Composition Spectrometer

SWIFET: Shortwave Infrared Field Effect Transistor • Switched Infrared Field Effect Transistor

SWIFT: Selected Words In Full Title • Significant Words In Full Title • Society for the Worldwide Interbank Financial Telecommunications

SWIM: Solar Wind Interplanetary Measurements • Spherical-Wave Interferometry Module

SWING: Stage, Work placement and Internship Networking Group

SWIP: Secret Work In Process

SWIR: Shortwave Infrared • Shortwave Infrared Radiometer • Shortwavelength Infrared

SWIRLS: Stratospheric Wind Infrared Limb Sounder

SWIRNS: Short Wave Infrared Nadir Spectrometer

SWIRT: Supernova Watchdogging Infrared Telescope

SWIS: Satellite Weather Information System • State-Wide Information System • Straßenzustands- und Wetterinformationssystem

SWISH: Simple Web Indexing System for Humans

SWISSAR: Swiss Interdisciplinary Study with Synthetic Aperture Radar

SWISTO: Swingby-Stopover Optimization

Swit: Switzerland

Switz: Switzerland

SWJ: South West Bay (Vanuatu) (location identifier)

SWK: Milano Segrate (Italy) (location identifier) • Schlagwortkatalog

SWL: Safe Working Load • Spanish Wells (Bahamas) (location identifier) • Sternwarte Langwedel (Germany)

SWLA: Short Wavelength Large Aperture

SWlds: StarWorlds

SWLG: Swelling

SWLIB: Software Library

SWLY: Southwesterly

SWM: Space Weather Month • Suía-Missu (Brazil-MT) (location identifier)

SWMDP: Software Management and Development Plan

SWMM: Storm Water Management Model

SWMS: Single Wafer Multiple Sensor

SWN: Sahiwal (Pakistan) (location identifier) • Scientist Workstation Network

SWNT: Single-Walled Nanotube

SWO: Space Weather Operations • Squarewave Oscillator • Standards-Writing Organization • Stillwater Searcy (USA-OK) (location identifier)

SWOB: Salaries-Wages-Overhead-Benefits

SWOC: Subject-Word of Context

SWOOPS: Solar-Wind Observations Over the Poles of the Sun

SWOP: Second Window Of Protection • Specifications Web Offset Publication • Stereo Wave Observation Project

SWOT: Strengths-Weaknesses-Opportunities-Threats

Swp: Sweep

SWP: Safe Working Pressure • Science Working Panel • Short-Wavelength Primary • Short-Wavelength Prime (camera) • Southwest Pacific • Space-Weight-Power • Swakopmund (Namibia) (location identifier) • Swap

SWPA: Shuttle Wing Panel Analysis • Southwest Pacific Area

SWPDR: Software Preliminary Design Review

SWPWR: Switched Power

SWQ: Sumbawa (Indonesia) (location identifier)

SWR: SAR Wave Mode • Schweizerischer Wissenschaftsrat (Switzerland) • Sea Water Return • Short-Wavelength Radar • Short-Wavelength Radiation • Short-Wavelength Redundant (camera) • Silur (Papua New Guinea) (location identifier) • Sine Wave Response • Standing Wave Ratio • Stepwise Regression

SWRAL: Southwest Region of the Astronomical League (USA)

SwRI: Southwest Research Institute (USA-TX)

SWRIMS: Southwest Regional Institute for Mathematical Sciences (USA)

SWRR: Software Readiness Review • Software Requirements Review

SWRU: Software Replaceable Unit

SWS: SAS Web Server • Semesterwochenstunden • Short-Wavelength Spectrometer • Software System • Solar Wind Spectrometer • Static Water Supply • Swansea Fairwood (UK-Wales) (location identifier) • Switch Scan

SWSA: Short Wavelength Small Aperture • Solid Waste Storage Area • Space West Science Activities Inc. (Canada-SK)

SWSL: Supplemental Weather Service Location

SWSR: Sum of the Weighted Squares of the Residuals

SWT: Science Working Team • Shortwave Transmitter • Southwest Texas (USA) • Sternwarte Trier (Germany) • Strezhevoy (Russia) (location identifier) • Strzhewoi (Russia) (location identifier) • Supersonic Wind Tunnel

Swtr: Sea Water

SWTR: Surface Water Treatment Rule

SWTSU: Southwest Texas State University (USA-TX)

Swtz: Switzerland

SWU: Su Won City (Korea-ROK) (location identifier)

SWUIS: Southwest Ultraviolet Imaging System

SWW: Space Weather Week • Sweetwater (USA-TX) (location identifier)

SWWAS: Southwest Washington Astronomical Society (USA-WA)

SWWG: Software Working Group

SWX: Shakawe (Botswana) (location identifier)

Swy: Stopway

SWY: Sitiawan (Malaysia) (location identifier)

SWys: StarWays

SWZ: Swaziland (ISO country code)

Sx: Sussex (UK-England)

SX: Essex (UK-England) • Greece (international aircraft registration code) • Janus (satellite of Saturn) • Saturday Excepted • Sioux • Soft X-ray • Stability Index • Sundays Excepted

SXA: Sialum (Papua New Guinea) (location identifier) • Stored Index to Address

SXAO: Shaanxi Astronomical Observatory (China-PRC)

SXAPS: Soft X-ray Appearance Potential Spectrometer • Soft X-ray Appearance Potential Spectroscopy

SXB: Strasbourg Entzheim (France) (location identifier)

SXC: Catalina Island (USA-CA) (location identifier)

SXD: Sophia Antipolis (France) (location identifier)

SXE: Sale (Australia-VIC) (location identifier) • Soft X-ray Experiment

SXES: Soft X-ray Electron Spectroscopy

SXF: Berlin Schönefeld (Germany) (location identifier) • Solid Exchange Format

SXG: Senanga (Zambia) (location identifier) • Spectrum X-Gamma

SXG-CF: Spectrum X-Gamma Coordinating Facility

SXGCF: Spectrum X-Gamma Coordinating Facility

SXH: Sehulea (Papua New Guinea) (location identifier)

SXI: Epimetheus (satellite of Saturn) • Sirri Island (Iran) (location identifier) • Solar X-ray Imager

SXII: Helene (satellite of Saturn)

SXIII: Telesto (satellite of Saturn)

SXIV: Calypso (satellite of Saturn)

SXJ: Shanshan (China-PRC) (location identifier)

SXK: Saumlaki (Indonesia) (location identifier)

SXL: Sligo (Ireland) (location identifier)

SXLS: Superconductivity X-ray Lithography Synchrotron

SXM: Sint Maarten - Juliana (Netherlands Antilles) (location identifier)

SXN: Section • Suapan (Botswana) (location identifier)

SXNS: Sections

SXO: São Félix do Araguaia (Brazil-MT) (location identifier)

SXP: Sheldons Point (USA-AK) (location identifier)

SXPS: Soft X-ray Photoemission Study

SXQ: Soldotna (USA-AK) (location identifier)

SXR: Soft X-Ray • Soft X-ray Region • Srinagar (India) (location identifier)

SXRB: Soft X-Ray Background

SXRP: Stellar X-Ray Polarimeter

SXRT: Soft X-Ray Transient

SXS: Sahabat 16 (Malaysia) (location identifier) • Stellar X-ray Spectrum

SXT: Sextant • Soft X-ray Telescope • Soft X-ray Transient • Spectroscopy X-ray Telescope • Stable X-ray Transmitter • Taman Negara (Malaysia) (location identifier)

SXTF: Satellite X-ray Test Facility

SXU: Soddu (Ethiopia) (location identifier)

SXV: Atlas (satellite of Saturn)

SXVI: Prometheus (satellite of Saturn)

SXVII: Pandora (satellite of Saturn)

SXVIII: Pan (satellite of Saturn)

SXW: Sauren (Papua New Guinea) (location identifier)

SXX: São Félix do Xingu (Brazil-PA) (location identifier)

SXY: Sidney (USA-NY) (location identifier)

SXZ: Siirt (Turkey) (location identifier)

Sy: Seychelles • Supply • Surrey (UK) • Syria

SY: Seasonal Year • Seychelles (international vehicle registration code) • Shoulder Yaw • Shrewsbury (UK postal code) • Square Yard • Syria (ISO country code) • Syrie (ISO country code) • System

SYA: Shemya (USA-AK) (location identifier)

SYB: Seal Bay (USA-AK) (location identifier)

SYC: Seychelles (ISO country code) • Shiringayoc (Peru) (location identifier)

Syd: Sydney (Australia-NSW)

S. Yd.: Scotland Yard

SYD: Sum of the Year's Digits • Sydney - Kingsford Smith Mascot (Australia-NSW) (location identifier)

SYE: Sa Dah (Yemen) (location identifier)

SYH: Syangboche (Nepal) (location identifier)

SYI: Sediment Yield Index • Shelbyville Bomar (USA-TN) (location identifier) • Synchronization In

SyK: Systematischer Katalog

SYK: Stykkishólmur (Iceland) (location identifier)

Sy'kat: Syarikat

Syl: Syllable • Syllabus

SYL: San Miguel - Roberts (USA-CA) (location identifier)

SYLK: Symbolic Link

Syll: Syllable • Syllabus

Sym: Symbol • Symbolic • Symmetrical • Symmetry • Symphony • Symptom

SYM: Simao (China-PRC) (location identifier) • Supersymmetric Yang-Mills • Symbolic • Symbol

Symb: Symbiotic

Symm: Symmetrical

SYMM: Symmetrical

Symp: Symposium

Symph: Symphonic • Symphony

Syn: Synchronize • Synonym • Synonymous • Synonymy • Synthetic

SYN: Stanton Carleton (USA-MN) (location identifier) • Synchronous • Synonym • Synthetic

Sync: Synchronization • Synchronize • Synchronizous

SYNC: Synchronize • Synchronous

SYNCOM: Synchronous Communications

Synd: Syndicate • Syndicated

SYNMOD: Synoptic Monitoring of Orbital Debris

Synon: Synonymous

Synop: Synopsis

SYNOP: Synoptic

SYNS: Synopsis

Synth: Synthèse • Synthesis • Synthetic • Synthétique

SYO: Sakata-shi (Japan) (location identifier) • Shonai (Japan) (location identifier) • Synchronization Out

SYP: Pound (Syria) (ISO currency code) • Society of Young Publishers (UK)

Syr: Syria • Syrian • Syrup

SYR: Syracuse (USA-NY) (location identifier) • Syria (international vehicle registration code) • Syria (ISO country code) • Syrie (ISO country code)

SYRIANAIR: Syrian Arab Airlines

Sys: System

SYS: Sunchon Yosu (Korea-ROK) (location identifier) • System

SYSADMIN: System Administrator

SysCoP: System for Copyright Protection

SYSEX: System Exclusive

SYSGEN: System Generator

SYSIN: System Initialization

SYSLOG: System Log

SYSMOD: System Modification

SYSOP: System Operator

SYSOUT: System Output

SYSREQ: System Request

Syst: System • Systematic • Systématique • Système

SYT: Saint-Yan - Charolais Bourgogne (France) (location identifier)

SyTec: Systems Technology (periodical code)

SYU: Sue Island - Warraber Island (Australia-QLD) (location identifier) • Warraber Island (Australia-QLD) (location identifier)

SYV: Sylvester (USA-GA) (location identifier)

SYX: Sanya (China-PRC) (location identifier)

SYY: Stornoway (UK-Scotland) (location identifier)

SYZ: Shiraz (Iran) (location identifier)

sZ: seinerzeit

SZ: Size • Sonnenzirkel • Southern Zodiacal • Sternzeit • Sunyaev-Zel'dovich • Swaziland (ISO country code) • Switzerland

SZA: Solar Zenith Angle • Soyo (Angola) (location identifier) • Sun Zenith Angle

SZB: Silver-Zinc Battery • Stadt- und Zentralbibliothek

SZC: Santa Cruz - Guanacaste (Costa Rica) (location identifier)

SZE: Sunyaev-Zel'dovich Effect

SZG: Salzburg (Austria) (location identifier)

SZH: Senipah (Indonesia) (location identifier)

SZK: Skukuza (South Africa) (location identifier)

SZL: Lilangeni (Swaziland) (ISO currency code) • Warrensburg Whiteman (USA-MO) (location identifier)

SZM: Shielded Zone of the Moon • Stereo Zoom Microscope • Stereo Zoom Microscopy

SZN: Santa Barbara - Santa Cruz Island (USA-CA) (location identifier)

SZP: Santa Paula (USA-CA) (location identifier)

SZQ: Saenz Peña (Argentina-BA) (location identifier)

SZR: Stara Zagora (Bulgaria) (location identifier)

SZS: Saverni Zavod za Standardizaciju (Yugoslavia-Serbia) • Stewart Island (New Zealand) (location identifier) • Yugoslav Institute of Standardization

SZTS: Stereo Zoom Transferscope

SZU: Segou (Mali) (location identifier)

SZV: Suzhou (China-PRC) (location identifier)

SZX: Shenzhen (China-PRC) (location identifier)

SZZ: Szczecin (Poland) (location identifier)

S2: Bangladesh (international aircraft registration code) • Diatomic Sulfur

S3: Sortie Support System • Spaceborne Sensor-to-Shooter

S3C: Service Central de la Communication et de la Commercialisation

S3OS2: Small Solar System Objects Spectroscopic Survey

S3R: Sequential Switching Shunt Regulator

S5: Slovenia (international aircraft registration code)

S7: Seychelles (international aircraft registration code)

S9: Sao Tome and Principe (international aircraft registration code)

T

t: Age • Celsius Temperature • Cosmic Time • Duration • Hour Angle • Metric Ton • Student's Distribution • Table • Tabulated • Täglich • Taken from • Tare • Tariflich • Technisch • Temperature • Tempo • Temporal • Tempore • Tense • Terminación • Terminal • Territorial • Territory • Tertiary • Thunder • Tief • Tiempo • Time • Time Interval • Time Scale • Tome • Tomo • Tomus • Ton • Tonelada • Tonne • Tonneau • Total • Tour • Town • Township • Transit • Transitive • Transport Number • Troy • Tun • Turn • Tutti

T: Absolute Temperature • Absolute Time • Couple Moment • Hundred Sixty (Roman numeral) • Internal Transmittance • Kinetic Energy • Kinetic Temperature of Free Particles • Meridian Angle • Moment of a Couple • Orbital Period • Period • Periodic Time • Reactor Time Constant • Reverberation Time • Static Temperature • Surface Tension • Table • Tableau • Tactical • Tactics • Tactique • Tag • Takt • Tal • Tall • Tangage • Tango (international alphabet) • Tank • Tankstelle • Target • Tarif • Tasse • Tat • Tau • Taufe • Tausch • Tausend • Tautomeric Effect • Taxe • Té • Teacher • Technicien • Technik • Teil • Teilung • Telefon • Telegraf • Telegraph • Telegraphic • Telephone • Temperatur • Température • Temperature • Tempo • Temporary • Temps • Tenor • Tension • Tensor • Tera • Termin • Terminal • Terrain • Terre • Terrestre • Territorial • Territory • Tertiaire • Tesla • Test • Testament • Tête • Text • Thailand (international vehicle registration code) • Thermodynamic Temperature • Thermometer • Threonine • Thrust • Thunderstorm • Thursday • Thymine • Tief • Tiefdruck • Tiefe • Tiefgang • Time • Time Reversal • Tirant • Tireur • Tirol (Austria) • Titel • Titus • Tochter • Toise • Tolérance • Tome • Tomus • Tonikum • Tor • Torpedo • Torque • Torre • Total Light Transmission • Tour • Tourbillon • Touren • Tourist • Touristenklass • Township • Trace • Traceur • Tracté • Tracteur • Traction • Train • Traînée • Tranchée • Transaction • Transducteur • Transformateur • Transistor • Transition • Translation • Transmetteur • Transmissibility • Transmission • Transmittance • Transport • Transportation • Transverse • Treasury • Tribord • Triebwagenzug • Tritium • Tropical • Troupe • True • Tube • Tubulaire • Tuesday • Tür • Turbine • Turboprop • Turbopulseur • Turbulent • Turkish • Turm • Turnen • Turning • Turnus • Tuyère • Twist • Twisting Number • Typ • Type • Typus • 160 (Roman numeral)

ta: Tamil (ISO language code)

Ta: Tank • Tanne • Tantal • Tantale • Tantalum • Tanz • Tara • Tarif • Tasche • Tasse • Taste • Température Ambiante • Température d'Arrêt

T/A: Temporary Assistant • Training As • Turnaround

TA: Tableau d'Avancement • Tableau des Alarmes • Tables Astronomiques • Tanzania • Tape Adapter • Tape Advance • Target Acquisition • Tarifangestellter • Taschenausgabe • Taschenbuchausgabe • Task Analysis • Task Assessment • Task Assignment • Task Authorization • Taunton (UK postal code) • Taux d'Amortissement • Teaching Assistant • Teal Amber • Technical Advisor • Technical Analysis • Technical Assessor • Technical Assistance • Technical Assistant • Technicien d'Atelier • Technische Abteilung • Technische Akademie • Technische Anleitung • Technische Anweisung • Technischer Angestellter • Technischer Assistent • Technischer Ausschuß • Technology Assessment • Teilamt • Teilnehmerausweis • Telefon-Anrufbeantworter • Telegrammadresse • Telegraphenamt • Telegraphic Address • Telemetry Address • Téléphone Automatique • Telescope Assembly • Température Absolue • Temperature Anisotropy • Température d'Arrêt • Temporary Admission • Tension Artérielle • Tension Auxiliaire • Tension d'Asservissement • Terme Associé • Terminal Adapter • Terminal Adaptor • Test Accessory • Test Announcer • Test Article • Textausgabe • Thermal Analysis • Timbre Amende • Time Actual • Tonabnehmer • Tonne d'Affrètement • Torpille Aérienne • Total Abstinence • Toutes Armes • Trace Analytical • Trachée-Artère • Tracteur Automobile • Traction Avant • Traduction Automatique • Traffic Advisory • Traffic Authority • Train Auxiliaire • Train Avant • Train d'Atterrissage • Traînée Aérodynamique • Training Adviser • Training Agency • Traité d'Alliance • Transactional Analysis • Transaktionsanalyse • Transfer Aisle • Transfer Allowed • Transformateur d'Alimentation • Transit Authority • Transition Altitude • Translation Ambiguity • Transmetteur Automatique • Transport Aérien • Transportation Authority • Transverse Acoustic • Travel Allowance • Travel Authorization • Travelling Allowance • Trip Authorization • Trou d'Air • Troupes Auxiliaires • True Altitude • True Anomaly • Trunnion Angle • Turbine Aérienne • Turnaround • Type Approval

TAA: Tarapaina (Solomon Islands) (location identifier) • Technical Assistance Administration • Technical Assistance Agreement • Temporary Access Authorization • Three-Axis Accelerometer • Trade Agreement Act (USA) • Trans-Atlantic Agreement • Trieste Aurélie Archive

TAAA: Tucson Amateur Astronomy Association (USA-AZ)

TAAATS: The Australian Advanced Air Traffic System

TAAC: Technology Assessment Advisory Council • Triad Action Astronomy Club (USA-NC)

TAAF: Terres Australes et Antarctiques Françaises

TAAP: Tethered Aerostat Antenna Program • Three-Axis Antenna Positioner • Transient Analysis Array Program

TAAS: Telescope Assembly Alignment Simulator • Terminal Advance Automation System • Thames Amateur Astronomical Society (USA-CT) • The Albuquerque Astronomical Society (USA-NM) • Triple Azimuth Angle Scan

TAAU: Telescope Assembly Alignment Unit

tab: Tabella

Tab: Tabelle • Tabla • Table • Tabular • Tabulate • Tabulated • Tabulation • Tabulator

TAB: Tabasco (Mexico) • Tab-Automated Bonded • Tabulator • Tape Automated Bonding • Targets And Background • Technical Advisory Board • Technical Assistance Board • Technology Assessment Board • Telecommunications Advisory Board • Temporarily Able Body • Time After Burst • Tobago (Trinidad & Tobago) (location identifier) • Training Aid Booklet • Transportes Aéreos Bolivianos • Transports Aériens du Bénin

TABA: Transportes Aéreos de Bacia Amazonica

TABD: Transatlantic Business Dialogue

TABI: Tailorable Advanced Blanket Insulation

Tabl: Tablett • Tablette

TABL: Tableau

TABLE: Tsukuba Atmospheric Boundary Layer Experiment

TABS: Terminal Access to Batch Service • Time Analysis and Billing System

Tabu: Taschenbuch

Tac: Tactical

TAC: Tacloban (Philippines) (location identifier) • Tactical Air Command • Taiwan Aerospace Corporation • Target Allocation Committee • Technical Advisory Committee • Technical Advisory Council • Technical Assistance Center • Technical Assistance Committee • Technical Assistance Contract • Technical Assistance Council • Technicien de l'Aviation Civile • Technology Advisory Council • Technology Application Center • Telecommunications Advisory Committee • Telemetry And Command • Telescope Allocation Committee • Tendeur Automatique de Câbles • Terminal Access Circuit • Terminal Access Controller • Terminal Area Chart • Terrain Analysis Center • Test Advisory Committee • Test And Configuration • Tester Analyzer Controller • The Astronomy Club (USA-CA) • Theoretical Astrophysics Center • Tilted Aligned Coupling • Time Allocation Committee • Time Assignment Committee • Time At Completion • Time-to-Amplitude Converter • Timing and Address Counter • Tomografía Axial Computerizada • Total Accumulated Cycles • Total Allowable Catch • Total Average Cost • Traceability And Configuration • Tracking Accuracy Control • Transformer Analog Computer • Transistorized Automatic Control • Transition Assurance Center • Transportes Aéreos Coyhaique • Trouble Analysis Chart • Turbulence en Air Clair • Twin Astrograph Catalog

TACA: Transportes Aéreos Centroamericanos

TACACS: Terminal Access Controller Access Control System

TACAMO: Take Charge And Move Out

TACAN: Tactical Air Control And Navigation • Tactical Air Navigation

TACANS: Tactical Air Control And Navigation System

TACAS: Tools and Algorithms for the Construction and Analysis of Systems

TACC: Tactical Air Control Center • Technical Assistance Coordination Committee • Tracking And Control Center

TACCS: Time, Age, and Cycle Control System

Tach: Tachometer

TACH: Tachometer • Tachymètre

TACHIR: Tool for Automatic Construction of Hypertexts for Information Retrieval

Tacho: Tachometer

TACIS: Technical Assistance to the Commonwealth of Independent States

TACL: Time And Cycle Log

TACO: Test And Checkout • Test And Checkout Operation

TACOS: Time Allocation Committee Operations Support

TACP: Tactical Air Control Party • Training And Certification Plan

TACR: Time And Cycle Record

TACS: Technical Assignment Control System • Test Assembly Conditioning Station • Theoretical Aspects of Computer Software • Thruster Attitude Control System • Total Access Communications System

TACT: Transactional Analysis in Customer Treatment • Truc À Côté de la Télé

TACTIC: TeV Atmospheric Čerenkov Telescope with Imaging Camera

TACV: Tracked Air-Cushion Vehicle • Transportes Aéreos de Cabo Verde

TAD: Target Acquisition Data • Task Assignment Directive • Task Assignment Drawing • Teaching for Astronomy Development (*) • Teaching in Astronomy Development • Technical Acceptance Date • Technical Analysis Division • Technical Assistance Data • Telephone Answering Device • Temperature And Dewpoint • Test And Development • Thermal Analysis Data • Throw Away Detector • Timbre À Date • Time Array Detection • Time Available for Delivery • Top Assembly Drawing • Traitement Automatique de Documents • Traitement Automatique de Données • Trajectoire d'Alignement de Descente • Transport Automatique de Documents • Travelling Around Drunk • Travelling Atmospheric Disturbance • Trinidad - Las Animas (USA-CO) (location identifier) • Type Abstrait de Données

TADAT: Tauern-Dachstein Testite

TADC: Tunable Atmospheric Dispersion Compensator

TADIC: Telemetry Analog-to-Digital Information Converter

TADIL: Tactical Digital Information Link

TADP: Terminaison Abonné à Débit Primaire

TADS: Target Acquisition and Designation System • Tracking And Display System • Trade and Analysis Data System • Translation Assistant Dictionary System

TAE: Taegu (Korea-ROK) (location identifier) • Tasa Anual Equivalente (Spain) • Techniques d'Applications Énergétiques • Technische Akademie Esslingen (Germany) • Toroidal Alfvén Eigenmodes • Toroidicity-induced Alfvén Eigenmodes • Trait d'Atterrissage Escamotable • Transportable Applications Environment • Transportable Applications Executive

TAEM: Terminal Area Energy Management

TAER: Time-Azimuth-Elevation-Range

TAES: Técnicas Aéreas de Estudios y Servicios SA • Transportes Aéreos de El Salvador SA de CV

TAESA: Transportes Aéreos Ejecutivos SA de CV

Taf: Tafel

TAF: Oran Tafaraoui (Algeria) (location identifier) • Task Analysis Form • Terminal Aerodrome Forecast • Terminal Area Forecast • Test-Analyze-Fix • Trait d'Atterrissage Fixe • Truss Assembly Fixture

TAFE: Technical And Further Education

TAFG: Two-Axis Free Gyro

TAFIM: Technical Architecture For Information Management

TAFM: Tunneling Atomic Force Microscopy

TAFTA: Trans-Atlantic Free-Trade Agreement • Trans-Atlantic Free-Trade Area

TAFTS: Tropospheric Airborne Fourier Transform Spectrometer

TAG: Tabular And Graphic • Tagbilaran (Philippines) (location identifier) • Technical Advisory Group • Technical Air-to-Ground • Technical Assistance Group • Techniques d'Avant-Garde • Telescope Allocation Group • Term Association Graph • Test Analysis Guide • Time Assignment Group • Transient Analysis Generator • Travelling Astronomer's Guide • Tree-Adjoining Grammar • Troupes Auxiliaires du Génie • Two-Axis Gimbal

TAGA: Trace Atmospheric Gas Analyser

TAGB: Transportes Aéreos da Guiné Bissau

TAGD: Technologie de Groupe Assisté par Ordinateur

TAGIU: Tracking And Ground Instrumentation Unit

Tägl: Täglich

TAGS: Technique for Automated Generation of Systems • Technology for Automatic Generation of Systems • Text And Graphics System • Thermal Analysis Graphics System • Total Absorption Gamma-ray Spectrometer • Training And Guided Study

TAH: Tanna (Vanuatu) (location identifier)

TAHD: Tunnel Aérohydrodynamique

Tai: Taiwan (China-ROC)

TAI: Taiz (Yemen) (location identifier) • Temps Atomique International • Tiempo Atómico Internacional • Time to Autoignition • Traitement Automatique d'Information • Transmission Automatique Informatique • Transverse Acceleration Instability

TAIDET: Triple-Axis Inertial Drift Erection Test

TAIR: Test And Inspection Record • Test Assembly Inspection Record

TAJ: Aitapa Tadji (Papua New Guinea) (location identifier)

TAK: Takamatsu (Japan) (location identifier) • Tierps Astronomiska Klubb (Sweden)

TAKO: Timeline Assembler, Keyword-Oriented

tal: Talis

TAL: Tanana (USA-AK) (location identifier) • Tandem Accelerator Laboratory • Terminal Application Language • Trait d'Atterrissage Largable • Transatlantic Abort Landing • Transatlantic Landing • Transoceanic Abort Landing • Turbulence en Air Limpide

TalAE: Taligent Application Environment

TALAR: Tactical Approach and Landing Radar

TALC: Très À La Carte

TALES: The Association of Library Equipment Suppliers (UK)

TALIF: Two-photon Atom Light Induced Fluorescence

TALK: Titles Alphabetically Listed by Keyword

TALN: Traitement Automatique du Langage Naturel

TALO: Time After Lift-Off

TALON: Texas - Arkansas - Louisiana - New Mexico (USA)

TALOS: Tropical Atmospheric Lidar Observations from Space

Tal. qual.: Talis Qualis

Talsp: Talsperre

Tam: Tamil

TAM: Tampico (Mexico) (location identifier) • Tangent Approximating Manifold • Technical Account Manager • Technology Acceptance Model • Telecommunications Access Method • Telephone Answering Machine • Temps d'Arrêt Moteur • Thermal Analytical Model • Throw-Away Maintenance • Tir-Armement-Munition • Trainer-Akademie München (Germany) • Trajectory Application Method • Transmission Automatique de Messages • Transportación Aérea Mexicana • Transport Aérien Militaire • Triangle Amplitude Modulation • Tropical Asian Monsoon • Type-Approval Model • Typed Access Matrix

TAMA: Technical Assistance and Manufacturing Agreement

TAMAC: Three-Axis Manual Attitude Controller

Tamb: Tambour

TAMB: Température Ambiente

TAME: Transportes Aéreos Militares Ecuatorianos

TAMER: Testing, Analysis, and Measurement Environment for Robustness

TAMEX: Taiwan Area Mesoscale Experiment

TAMPA: Transportes Aéreos Mercantiles Panamericanos SA

TAMPER: Tables for Approximation of Midpoints for Exponential Regression

TAMPS: Tamaulipas (Mexico)

TAMS: Teachers Academy for Mathematics and Science (USA-IL) • Technical and Acquisition Management Support • Telenet Access Management System • Terminal-Age Main Sequence • Total Airport Management System • Trade and Analysis Management System • Transient Amplitude Monitor System

TAMSAT: Tropical Agricultural Meteorology using Satellite

TAMU: Texas A&M University (USA-TX)

TAMUCC: Texas A&M University at Corpus Christi (USA-TX)

tan: Tangent

TAN: Tangalooma (Australia-QLD) (location identifier) • Total Area Networking • Traité de l'Atlantique Nord • Twilight All Night

TANet: Taiwan Academic Network

Tang: Tanganika • Tangente • Tangentiel • Tanger (Morocco)

tanh: Hyperbolic Tangent

TANS: Terminal-Area Navigation System • Trimble Advanced Navigation Sensor

TANSTAAFL: There Are No Such Things As A Free Lunch

TAO: Qingdao (China-PRC) (location identifier) • Technical Assistance Operation • Technology Applications Office • Télémanipulation Assistée par Ordinateur • Telephony Application Object • Test Analysis Outline • Tokyo Astronomical Observatory (now NAOJ) • Tomographie Acoustique Océanique • Toyama Astronomical Observatory (Japan) • Tradeable Absorption Obligation • Traduction Assistée par Ordinateur • Transport in the Atmospheres and the Oceans • Tropical Atmosphere Ocean • TUVA Astronomy Organization (USA-OK)

TAOF: Tunable Acousto-Optical Filter

TAOS: Taiwan-American Occultation Survey • Technology for Autonomous Operational Survivability • Thrust-Assisted Orbiter Shuttle

TàP: Tirage-à-Part • Tiré-à-Part

TAP: Table of Authorized Personnel • Tapachula (Mexico) (location identifier) • Target Array Payload • Task Area Plan • Technical Achievement Plan • Technical Advancement Program • Technical Advisory Panel • Technical Assistance Plan • Technical Assistance Program • Technical Assistance Project • Technological Assistance Program • Technology Application Program • Telelocator Alphanumeric Protocol • Telematics Applications Programme • Telemetry Acceptance Pattern • Telemetry Antenna Pedestal • Telemetry Archive Process • Ter Aangehaalder Plaatse • Terminal Area Productivity • Terrestrial Auxiliary Power • Test Administration Plan • Thermal Analysis Program • Thermal Analyzer Program • Thermal-Array Probe • Thermoviscoelastic Analysis Program • Three-Axis Package • Timeslot Access Point • Total Air Pressure • Traitement Automatique de la Parole • Trajectory Analysis Program • Trans-Alaska Pipeline • Trans-Arabian Pipeline • Transboundary Air Pollution • Transit Access Point • Transports Aériens Privés • Trend Analysis Program • Troupes Aéroportées

TAPAC: Tape Automatic Positioning And Control

TAPAT: Tape Programmed Automatic Tester

TAPC: Terrestrial Aeronautical Public Correspondence • Traitement Automatique de la Parole Continue

TAPCIS: The Access Program for CompuServe Information Service

TAPE: Tape Automatic Preparation Equipment • Teaching Aids in Physics Education • Telematics Automated Proposal Editor • Tentative Annual Planning Estimate • Totally Automated Programming Equipment

TAPI: Technology Application and Promotion Institute • Telephony Applications Program Interface • Telephony Applications Programming Interface

TAPIR: Theoretical Astrophysics Including Relativity

TAPS: Time Analysis of Program Status • Trans-Alaska Pipeline System • Two-Axis Pointing System

TAQ: Tarcoola (Australia-SA) (location identifier)

TAQL: Task Acquisition Language

Tar: Tara • Tariff

TAR: Tape Address Register • Tape Archive • Tape Archiver • Taranto (Italy) (location identifier) • Technical-Analysis Request • Technical-Assistance Request • Terminal Area Radar • Terrain-Avoidance Radar • Test Action Requirement • Test Agency Report • Test Analysis Report • Test Authorization Request • Threshold Autoregression • Thrust-Augmented Rocket • Track Address Register • Train Arrière • Trajectory Adjustment Request • Trajectory Analysis Room • Tribunal Administratif Régional

TARA: Tools for ACIS Real-time Analysis • Train d'Atterrissage à Roue Avant

TARAN: Test And Repair As Necessary • Test And Replace As Necessary

TARC: Through Axis Rotational Control • Transmutation by Adiabatic Resonance Crossing

TARD: Towed Active Radar Decoy

TARDIS: Targeted Delivery of Information and Services • Titles Automated Register and Document Information System

TARE: Telegraphic Automatic Relay Equipment • Telegraphic Automatic Routing Equipment • Telemetry Automatic Reduction Equipment

TARFOX: Tropospheric Aerosol Radiative Forcing Observational Experiment

TARFU: Things Are Really Fouled Up

TARGA: Truevision Advanced Raster Graphics Adapter

TARGET: Training And Rehearsal Generation Toolkit • Trans-European Automated Real-time Gross-settlement Express Transfer

TARM: TIROS Atmospheric Radiance Module

TarOT: Taru Astrof. Obs. Teated (periodical code)

TARP: Test And Repair Processor

TARPS: Tactical Air Reconnaissance Pod System

TARS: Terminal Automated Radar Service • Theater Airborne Reconnaissance System • Three-Axis Reference System

Tas: Tasmania (Australia) • Tasmanian

TAS: Table d'Atmosphère Standard • Tacoma Astronomical Society (USA-WA) • Tallahassee Astronomical Society (USA-FL) • Tape Accounting System • Tape Alteration Subroutine • Taree Astronomical Society (Australia) • Target Acquisition System • Tashkent (Uzbekistan) (location identifier) • Tasmania (Australia) • Technology Applications and Science • Telecommunication Analysis System • Telecommunications Authority of Singapore • Telemetry Antenna System • Telephone Access Server • Telephone Answering Service • Telephone Answering System • Teleprocessing-Auskunftssystem • Tenure Administration System • Terminal Address Selector • Test And Set • Test Article Specification • Texas Astronomical Society (USA-TX) • Thai Astronomical Society (Thailand) • Three-Axis Stabilization • Total Analysis System • Tracking Antenna System • Transporteur Aérospatial • Triangulum Astronomical Society (USA-VA) • Tritium Analysis System • True Airspeed • Tullamore Astronomical Society (Ireland)

TASA: Telecomunicações Aeronáuticas SA • Telecommunication Alarm Sequence Analyzer • Test Area Support Assembly

TASC: Technical Assistance Service Center • Telecommunications Alarm Surveillance and Control • The Analytical Sciences Corporation • Total Absorption Shower Counter • The Analytical Science Corporation (USA)

TASCC: Tandem Accelerator Superconducting Cyclotron

TASE: Test Automation Support Environment • Traité de l'Asie du Sud-Est

TASI: Technical Advisory Source for Images • Time Assignment Speech Interpolation

TASK: Technologien für Aktive Strukturen und Komponenten

Tasm: Tasmania (Australia) • Tasmanian

TASM: Turbo Assembler

TASOS: The Astronomical Society Of Singapore

TASP: Telemetry Analysis and Simulation Program • Tropical Area Special Product

TASPAC: Total Analysis System for Production, Accounting and Control

TASPR: Technical And Schedule Performance Report

TASR: Terminal Area Surveillance Radar

TASS: Technische Arbeitsgemeinschaft für Schrift und Sprache (Austria) • Telegrafnoye Aguentstvo Sovietskogo Soyuza (USSR) • Teleprinter Automatic Switching System • The Amateur Sky Survey • Trinity Astronomy and Space Society (Ireland) • Two-Axis Sun Sensor

TASSEL: Three-Astronaut Space System Experimental Laboratory

TASSF: The Atlantic Space Sciences Foundation (Canada-NS)

TAT: Tatry Poprad (Slovakia) (location identifier) • Technical Acceptance Team • Technical Approval Team • Technical Area Task • Technical Assistance Team • Technical Assistance and Training • Technology Application Team • Température d'Air Totale • Thematic Apperception Test • Thrust-Augmented Thor • Total Air Temperature • Total Alias Table • Toxine-Antitoxine • Trans-Atlantic Trunk • Transparent Assembly Test • Turnaround Time • Twin Astrographic Telescope • Two-Axis Tracking

TATCA: Terminal Air Traffic Control Automation

TATDL: Tabulated Assembly Technical Data List

TATI: Total Air Temperature Indicator

TATP: Two-Axis Tracking Pedestal

TATRP: Texas Advanced Technology Research Program (USA-TX)

tats: tatsächlich

Tats: Tatsache

TATS: Test And Training Satellite

TATT: Total Average Talk Time

Tau: Taurus-Tauri (constellation)

TAU: Tauramena (Colombia) (location identifier) • Tel Aviv University (Israel) • Thesaurus Alphabetical Update • Thousand Astronomical Units

TAUMOK: Tautenburger Multi-Objekt Kamera

TAUN: Technical Assistance of the United Nations

TAUP: Topics in Astroparticle and Underground Physics

Taut: Tautological • Tautology

TAUVEX: Tel Aviv Ultraviolet Experiment

t.-à-v.: Tout-à-Vous

TAV: Target Acquisition and Verification • Tau (American Samoa) (location identifier) • Ten Aanzien Van • Test And Validation • Test And Verification • Train Avant

TAVERNS: Test And Verification Environment for Remotely Networked Systems

TAVT: Terminal Airspace Visualization Tool

TAW: Tacuarembo (Uruguay) (location identifier) • Thrust-Augmented Wing • Twice A Week

TAWRS: Temperature And Wind Rocket Spectrometer

TAWS: Terrain Avoidance Warning System • Theater Airborne Warning System

Tax: Taxation

TAX: Taliabu (Indonesia) (location identifier)

TAXATA: Things Are Exactly As They Appear

TAXI: Transparent Asynchronous Transceiver Interface

Taxn: Taxation

TAY: Tartu (Estonia) (location identifier)

TAZ: Tagesanzeiger • Tagesauszug • Tashauz (Turkmenia) (location identifier) • Traffic Analysis Zone

t&b: Top & Bottom

Tb: Tabelle • Tabulator • Taschenbuch • Tatbestand • Taube • Terabit • Terbium • Tube • Tuberculosis • Turbulence

T&B: Turn & Bank

T/B: Talkback • Thermal Balance • Thermal Barrier • Title Block • Top/Bottom

TB: Tableau de Bord • Tagungsbüro • Taittirīya Brāhmana • Talk Back • Talking Book • Tarifbestimmung • Tätigkeitsbericht • Technical Board • Technical Bulletin • Technische Bestimmungen • Technische Bibliothek • Technischer Beamter • Technischer Bericht • Technisches Büro • Téléphone de Bord • Température de Bruit • Température du Boîtier • Tension aux Bornes • Terabyte • Terminal Base • Terminal Board • Text Begin • Thermal Bake • Thermal Balance • Thermal Barrier • Tibet • Time Base • Time of Beginning • Time of Burn • Title Block • Toggle Buffer • Tonnage Brut • Train Baladeur • Training Board • Tranquility Base (Moon) • Translation Buffer • Transfer Building • Transformateur Biphasé • Transient Brightening • Transparent Bridging • Transverse Boom • Treasury Bill • Trial Balance • Tribord • Tropical Band • True Bearing • Tuberculosis • Turbopropulseur • Turn and Bank

TBA: Tabibuga (Papua New Guinea) (location identifier) • Teaching By Analogy • To Be Activated • To Be Added • To Be Advised • To Be Agreed • To Be Announced • To Be Assigned • To Be Avoided

TBAI: Temporary Base Activation Instruction

TBAMS: Test-Bed Aircraft Multispectral Scanner

TBAX: Tube Axial

TBB: Tuy Hoa (Viet Nam) (location identifier)

Tbc: Tuberculosis

TBC: Tape Block Channel • Terminal Buffer Controller • The Boeing Company • Thermal Barrier Coating • Time-Base Corrector • Time-Based Corrector • To Be Confirmed • Tuba City (USA-AZ) (location identifier)

TBCC: Thallium-Barium-Calcium-Copper

TBCCO: Thallium-Barium-Calcium-Copper Oxide

TBCF: To Be Called For

TBCM: Target Background Characterization Model

TBCO: To Be Carried Over

TBD: Teaching By Demonstration • Temps Barycentrique Dynamique • Timbiqui (Colombia) (location identifier) • To Be Decided • To Be Defined • To Be Determined • To Be Developed • To Be Done • Trade Development Board • Troubleshooting Block Diagram

TBDF: Transborder Dataflow

TBE: Teaching By Example • Teledyne Brown Engineering (USA-AL) • Timbunke (Papua New Guinea) (location identifier) • Time Base Error • To Be Evaluated • To Be Expanded • Transmitter Buffer Empty

TBF: Tabiteuea North (Kiribati) (location identifier) • Test de Bon Fonctionnement • Très Basse Fréquence

TBG: Tabubil (Papua New Guinea) (location identifier)

TBH: Tablas (Philippines) (location identifier)

TBI: Test Bench Installation • The Bight (Bahamas) (location identifier) • Throttle-Body Injection • Through-Bulkhead Initiator • Thru-Bulkhead Initiator • Time Between Inspection • To Be Inactivated • Total Body Irradiation • Traumatic Brain Injury • Turn and Bank Indicator

TBIG: Technicien en Bureautique et Informatique de Gestion

TBJ: Tabarka (Tunisia) (location identifier)

TBK: Timber Creek (Australia-NT) (location identifier) • Toolbook

Tbl: Tabelle • Tablette • Tabulator • Titelblatt

TBL: Tableland (Australia-WA) (location identifier) • Télescope Bernard Lyot • The British Library • Timothy Berners-Lee • Turbulent Boundary Layer

Tblr: Tumbler

Tblsht: Troubleshoot

TBM: Table Buffer Management • Temporary Benchmark • Terabit Memory • Tumbang Samba (Indonesia) (location identifier) • Typenbegleitmannschaft

TBMS: Thirteen Band Multispectral Scanner

TBN: Fort Leonard Wood (USA-MO) (location identifier) • To Be Negotiated • To Be Nominated

TBO: Tabora (Tanzania) (location identifier) • Time Before Overhaul • Time Between Overhauls (*) • Total Blackout • Tycho Brahe Observatory (Sweden)

TBP: Technology Business Process • To Be Provided • True Boiling Point • Tumbes (Peru) (location identifier) • Tyauterie de Basse Pression • Two-Body Problem

TBPS: Terabytes Per Second

TBQ: Tarabo (Papua New Guinea) (location identifier)

TBR: Statesboro (USA-GA) (location identifier) • Ter Beschikking van de Regering • To Be Received • To Be Resolved • To Be Revised • Toolbar

tb&s: Top, Bottom & Sides

TBS: Talk Between Ships • Tape and Buffer System • Task Breakdown Structure • Tbilisi (Georgia) (location identifier) • Test Bench Set • Tethered Balloon System • To Be Selected • To Be Specified • To Be Superseded • To Be Supplied • Turner Broadcasting System

Tbsp: Tablespoon

TBSV: Time Between Scheduled Visits

TBT: Tabatinga (Brazil-AM) (location identifier) • Technical Barrier to Trade • Time Ball Tower • Tributyl Tin

TBTD: That'll Be The Day • Thermal Bus Technology Demonstrator

TBU: Nuku Alofa - Fua'Amotu (Tonga) (location identifier) • Time Base Unit • Tongatapu (Tonga) (location identifier)

TBV: Tabal (Marshall Islands) (location identifier) • Ten Behoeve Van • Ten Bate Van • To Be Verified

TBW: Tambov (Russia) (location identifier) • To Be Withheld • To Be Written • Total Bandwidth

TBX: Taabo (Ivory Coast) (location identifier)

TBY: Tsabong (Botswana) (location identifier)

TBYB: Try Before You Buy

Tbyte: Terabyte

TBZ: Tabriz (Iran) (location identifier)

Tc: Critical Temperature • Technetium • Technétium

T/C: Telecommunications • Temperature Compensating • Temperature/Conductivity • Termination Check • Test Conductor • Thermal Capacitor • Thermocouple • Thrust Chamber • Training and Certification

T&C: Telemetry & Command • Telemetry & Control • Terms & Conditions • Test & Checkout • Tracking & Communication

TC: Caicos Islands (Turks &_) (ISO country code) • Îles Turks & Caicos (ISO country code) • Tableau de Commande • Table de Composition • Tagungscenter • Tariff Commission • Taurid Complex • Taux de Charge • Taux de Compression • Taux de Confiance • Technical Characteristics • Technical Circular • Technical College • Technical Committee • Technical Communication • Technical Control • Technical Council • Technology Center • Telecommand • Télécommande • Telecommunications • Telecontrol • Telemetry and Command • Telemetry and Control • Telemetry Calibration • Télévision en Couleurs • Temperature Coefficient • Temperature Compensated • Temperature Compensating • Température Constante • Temperature Control • Température Critique • Temporary Clerk • Temporary Correction • Temps Clair • Tendeur de Câble • Term Combination • Term Coordination • Terminal Computer • Terminal Controller • Termination Check • Terms and Conditions • Test and Checkout • Test Cell • Test Change • Test Condition • Test Conductor • Test Console • Test Container • Test Control • Test Controller • Test Coordinator • Testing Center • Text Categorization • Text Companion • Thermal Capacitor • Thermal Conductivity • Thermal Control

• Thermocontact • Thermocouple • Thermomètre Centigrade • Thiokol Corporation • Thrust Chamber • Time Check • Time Compensation • Time Constant • Time to Circular • Timing Channel • Timing Counter • Tissu Cellulaire • Toll Center • Top Center • Total Capacity • Total Cost • Tour de Contrôle • Town Council • Town Councillor • Traceability Code • Tracking and Communication • Tracking Camera • Tracking Console • Training and Certification • Training Center • Transaction Capability • Transfer Control • Transformateur de Courant • Transit Circle • Transit Corridor • Transition Control • Translation Controller • Transmission Control • Transmission Controller • Transmission Convergence • Travail à la Chaîne • Travel Club • Traveller's Check • Trinity College • Trip Completed • Tronc Commun • Tropical Cyclone • True Course • Trusteeship Council • Tube Capillaire • Tube Cathodique • Tube de Choc • Turbocharger • Turbocompresseur • Turkey (international aircraft registration code) • Turks & Caicos Islands (ISO country code) • Type Certificate • Type Certification

TCA: Caicos Islands (Turks &_) (ISO country code) • Îles Turks & Caicos (ISO country code) • Tachyon Cross Assembler • Technical Change Analysis • Technical Contract Administrator • Technology Cooperative Agreement • Télécommande Automatique • Télécommunication Automatique • Telemetering Control Assembly • Temperature Control Amplifier • Temperature Control Assembly • Tennant Creek (Australia-NT) (location identifier) • Terminal Control Area • Thrust Chamber Assembly • Thrust Control Assembly • Time of Closest Approach • To Come Again • Tour de Contrôle d'Aérodrome • Tower Control Airport • Trace Contamination Analysis • Traffic Control Airport • Translation Control Assembly • Translation Controller Assembly • Trichloroacetic Acid • Turks & Caicos Islands (ISO country code)

TCAA: Taiwan Civil Aeronautics Administration • Twin City Amateur Astronomers (USA-IL)

TCAB: Temperature of Cabin

TCAC: Tri-City Astronomy Club (USA-WA)

TCACCIS: Transportation Coordinator Automated Command and Control Information System

TCAD: Technology Computer-Aided Design

TCAL: Thermal Calibration

TCAM: Telecommunications Access Method

TCAMS: Technical Control, Automation, Maintenance and Support

TCAO: Tradeable Carbon Absorption Obligation

T/CAP: Thermal Capacitor

TCAP: Thermal Capacitor • Transaction Capabilities Applications Port

TCAS: Terminal Collision Avoidance System • Traffic-alert and Collision Avoidance System • Traffic-alert Collision Avoidance System • Traffic and Collision Avoidance System (*)

TCAT: Tape-Controlled Automatic Testing

TCB: Take Care of Business • Taking Care of Business • Task Control Block • Technical Coordination Bulletin • Temps-Coordonnée Barycentrique • Test Cell Basement • Test Certification Board • Test Connector Buffer • Time Correlation Buffer • Treasure Cay (Bahamas) (location identifier) • Trouble Came Back • Trusted Computing Base

TCBL: Tradewind Cumulus Boundary Layer

TCC: Technical Control Center • Technical Coordination Committee • Technological Change Committee • Temperature Coefficient of Capacitance • Test Classification Code • Test Conductor's Console • Test Control Center • Thermal Control Coating • Thrust Control Command • Time Compression Coding • Timing-Clock-Control • Total Change Control • Trace Contaminant Control • Tracking and Control Center • Traffic Control Centre • Transport and Communications Commission • Transportation Computer Center • Transportation Control Center • Trinity College Cambridge (UK-England) • Tropical Cyclone Centre • Tucumcari (USA-NM) (location identifier) • Turbulence par Ciel Clair • Tuyère à Corps Central

TCCA: Trace Contaminant Control Assembly

TCCC: Technical Committee on Computer Communications • Tower Control Computer Complex

TCCIS: Texas Center for Computational and Information Sciences (USA-TX)

TCCM: Temperature-Controlled Crystal Microbalance • Three-Corner Coding Method

TCCS: Timed Calculus of Communicating Systems

T&CD: Timing & Countdown

T/CD: Time/Command Distributor

TCD: Chad (ISO country code) • Tarapaca (Colombia) (location identifier) • Task Completion Date • Tchad (ISO country code) • Teaching Company Directorate • Technical Cooperation for Development • Telemetry and Command Data • Test Completion Date • Test Control Document • Thermal Conductivity Detector • Time/Command Distributor • Time Correlation Data • Timing and Countdown • Timing Control Distributor • Tour Completion Date • Transmitted Current Detector • Transport de Chalands de Débarquement • Trinity College Dublin (Ireland) • Two-Colour Diagramme

TCDC: Technical Cooperation among Developing Countries (*) • Technical Cooperation between Developing Countries

TCDD: Tetrachlorodibenzodioxin

TCDM: Tilted Cold Dark Matter

TCDS: Type Certification Data Sheet

TCE: TC Electronics • Technology Center of Excellence • Telemetry Checkout Equipment • Temperature Coefficient of Expansion • Terrace • Thermal Canister Experiment • Thermal Control Electronics • Thomson Electronics Consumer • Thread-based Communication Environment • Tone Control Equipment • Total Composite Error • Tout Corps d'État • Transaction Cost Economics • Trichloroethylene • Tulcea (Romania) (location identifier)

TCEE: Temperature-Controlled Emission Enclosure

TCEF: Tableau des Consommations d'Énergie en France

tcf: Trillion Cubic Feet

TCF: Tank Checkout Facility • Technical Control Facility • Télévision en Circuit Fermé • Time-Correction Factor • Tocoa (Honduras) (location identifier) • Touring-Club de France

TCFH: Télescope Canada-France-Hawaii

TCFM: Temperature Control Flux Monitor

TCFP: Thrust Chamber Fuel Purge

TCG: Technical-Coordination Group • Telecommunications Group • Teleport Communication Group • Temps Coordonné Géocentrique • Test Control Group • Time Code Generator • Time Control Gain • Time-Controlled Gain • Transponder Control Group • Tune-Controlled Gain

TCGF: T-Cell Growth Factor

TCGS: Tennessee Conference of Graduate Students (USA-TN)

TCH: Tchibanga (Gabon) (location identifier) • Threshold Crossing Height

Tchg: Teaching

Tchr: Teacher

TCI: Tableau de Composition Illustré • Tape-to-Computer Interface • Technical Critical Item • Telecommunications

Inc. • Telnet Command Interpreter • Temperature Control Instrument • Tenerife (Canary Islands) (location identifier) • Terrain-Clearance Indicator • Test Control Instruction • Through Check-In • Time Change Item • Touring-Club Italiano

TCIC: Technical Committee on INdustrial Classification

TCID: Terminal Computer Identification • Test Configuration Identifier

TCIF: Telecommunications Industry Forum

TCJ: Torembi (Papua New Guinea) (location identifier)

TCK: Tinboli (Papua New Guinea) (location identifier)

TCL: TAE Command Language • Terminal Command Language • Terminal Control Language • Time and Cycle Log • Tool Command Language • Tool Control Language • Transformation Control Language • Transistor Coupled Logic • Transportable Calibration Laboratory • Trichloroethylene • Triple Communication Line • Tuscaloosa (USA-AL) (location identifier)

TCLC: Tri-state College Library Cooperation

TCLE: Thermal Coefficient of Linear Expansion

TCLP: Toxicity Characteristic Leaching Procedure

TCLT: Tentative Calculated Landing Time

TCLZ: Texas Learning and Computation Center (USA-TX)

TCM: Tableau Climatologique Mensuel • Tacoma McChord (USA-WA) (location identifier) • Target Cost per Mission • Technical Control Management • Technical Coordinating Meeting • Technologie de Couches Minces • Télécommande • Télécommunication Manuelle • Telemetry Code Modulation • Telemetry Control and Monitoring • Temperature Control Model • Terminal-to-Computer Multiplexer • Tetrachloromerkurat • Thermal Conduction Module • Thermal Control Model • Time Compression Multiplexer • Time Compression Multiplexing • Traditional Chinese Medicine • Trajectory Control Maneuver • Trajectory Correction Maneuver • Trellis-Coded Modulation • Trellis Code Modulation • Tropical Cyclone Motion

TCMD: Transportation Control and Movement Document

TCMO: Travail Collectif Médiatisé par Ordinateur

TCMP: Trunk Control Message Protocol

TCMS: Telecommunications System • Test Control and Monitoring System

TCN: Tehuacan (Mexico) (location identifier) • Test Change Notice • Tracing Change Notice • Training Course Number • Transportation Control Number

TCNE: Tetracyanoethylene

TCNJ: The College of New Jersey (USA-NJ)

TCNQ: Tetracyanoquinodimethane

TCNRDI: Transmission Corrected Normalized Reflectance Difference Index

TCO: Tableau de Contrôle Optique • Taken Care Of • Telecommunications Certification Officer • Telemetry and Control • Terminal Control Office • Test and Checkout • Test Control Office • Test Control Officer • Thomson CSF Optronique (France) • Three College Observatory (USA-NC) • Total Cost of Ownership • Trinity College Oxford (UK-England) • Tumaco (Colombia) (location identifier)

TCOM: Terminal Communications

TCOP: Test and Checkout Plan

TCOS: Technical Committee on Operating Systems

T&CP: Test & Checkout Procedure

TCP: Tape Carrier Packaging • Tape Conversion Program • Task Change Proposal • Task Control Package • Task Control Packet • Technical Change Proposal • Technical Cooperation Development • Technical Copyright Protection • Technical Cost Proposal • Temporary Change Procedure • Test and Checkout Procedure • Test Case Procedure • Test Checkout Procedure • Test Compare Results • Test Conductor • Test Constraints Review • Test Control Package • Tester Converter Program • TEX Code Page • Thermal Concept Review • Thrust Chamber Pressure • Time-Cost-Performance • Timing and Control Panel • Transmission Control Protocol (*) • Transport Connection Protocol • Transport Control Protocol • Très Courte Portée • Tricalcium Phosphate • Trichlorophenoxyacetic • Tromsø CCD Photometer • Tropical Cyclone Programme

TCPA: Telephone Consumer Protection Act

TCPC: Technical Committee for Physics Computing

TCPE: Total Cargo Project Engineer

TCPI: Technology Consumer Price Index

TCPS: Technical Copyright Protection System

TCQ: Tacna (Peru) (location identifier)

TCQAM: Trellis-Coded Quadrature Amplitude Modulation

TCQO: Trellis-Coded Quasi-Orthogonal

TCR: Tape Cassette Recorder • T-Cell Receptor • Technical Change Request • Technical Characteristics Review • Telemetry Command and Ranging • Temperature Coefficient of Resistance • Test Compare Result • Test Constraint Review • Thermal Concept Review • Tonnes of Coal Replacement • Total Contractual Requirements • Transfer Control Register • Transient Call Record • Traveling Compression Region

TCRB: Test Change Review Board • Test Criteria Review Board

T/CRD: Telemetry/Command Requirements Document

TCRD: Test and Checkout Requirements Document

TCRS: Tennessee Consolidated Retirement System (USA-TN) • Training and Certification Records System

TCRSD: Test and Checkout Requirements and Specifications Document • Test and Checkout Requirements Specification Document

TCS: Teaching Company Scheme • Technical Change Summary • Technical Concurrence Sheet • Telecommunications Control System • Teleconference System • Telescope Control Software • Telescope Control System • Telescope Coordination Software • Telescopio Carlos Sánchez • Television Camera Set • Temperature Control System • Terminal Computer System • Terminal Control System • Test Control Sequence • Test Control Supervisor • Test Control System • Theoretical Computer Science • Thermal Conditioning System • Thermal Control System • TNG Control System • Toroidal Confinement System • Torus Cleanup System • Total Cross Section • Touting-Club der Schweiz • Touting-Club de Suisse • Tower Communications System • Traffic Control Station • Traitement Chimique de Surface • Trajectory Control Sensor • Transfer Carry Subtract • Truth or Consequences (USA-NM) (location identifier) • Two Channel Switch

TCSC: Trainer Control and Simulation Computer

TCSEC: Trusted Computer System Evaluation Criteria

TCSPC: Time-Correlated Single Photon Counting Technique

TCSSS: Thermal Control Subsystem Segment

TCT: Takotna (USA-AK) (location identifier) • Tarski-Church-Turing • Test Coordination Team • Tilted-Component Telescope • Time Code Translator • Trans-Continental Traverse • Transonic Cryogenic Tunnel • Tribunal Central de Trabajo (Spain)

TCTA: Transcontinental Control Area

TCTD: Technical Countdown Demonstration Unit

TCTE: Telecommand Test Equipment

TCTI: Time Compliance Technical Instruction

TCTO: Time-Compliance Technical Order

TCTR: Time-Compliance Technical Requirement

TCTS: Trans-Canada Telephone System

TCTT: Tarski-Church-Turing Theorem

TCTUA: Terminal Control Table User Area

TCu: Tribunal de Cuentas (Spain)

TCU: Tape Control Unit • Teletype Communications Unit • Teletypewriter Control Unit • Temperature Control Unit • Terminal Computer Unit • Test Computer Unit • Test Control Unit • Texas Christian University (USA-TX) • Thaba Nchu (South Africa) (location identifier) • Thermal Control Unit • Threshold Control Unit • Thrust Control Unit • Time Change Unit • Time Code Update • Towering Cumulus • Transmission Control Unit

TCV: Technischer Cartellverband • Temperature-Controlled Voltage • Temperature Control Voltage • Tete (Mozambique) (location identifier) • Thrust Chamber Valve • Time Code Validity • Tokamak with Variable Configuration • Travelling Convection Vortex

TCW: Time Code Word • Tocumwal (Australia-NSW) (location identifier) • Track Confirmation Word

TCXO: Temperature-Compensated Crystal Oscillator

td: Ter in Die

Td: Tausend • Tedeum • Tender • Trend

T/D: Time Delay • Time Deposit • Touchdown

T&D: Training & Development • Transportation & Docking

TD: Chad (ISO country code) • Galashiel's (UK postal code) • Table Data • Tabular Data • Tagesdosis • Tankdampfer • Task Description • Task Design • Task Directive • Taux de Débit • Taux de Dilution • Tchad (ISO country code) • Teaching Diploma • Technical Data • Technical Demonstration • Technical Design • Technical Development • Technical Direction • Technical Directive • Technical Director • Technical Directorate • Technical Division • Technical Drawing • Technischer Dienst • Technischer Direktor • Technological Determinist • Technology Data • Technology Demonstration • Technology Development • Technology Document • Télédétection • Telemetry Data • Température Diurne • Temporal Difference • Temporary Duty • Temps de Descente • Tension Disruptive • Terminal Display • Terminal Distributor • Terminating Descriptor • Terminator Deformation • Test Data • Test Directive • Test Director • Test Drawing • Testing Device • Thematic Data • Thor-Delta • Threshold Densitometry • Threshold Detection • Tiefendosis • Time Delay • Time Difference • Time of Departure • Topic-Directed • Topographic Division • Total Depth • Total Drift • Touchdown • Track Data • Track Display • Traffic Director • Training Division • Transmission de Données • Transmit Data • Transportation and Docking • Transportdienst • Transverse Direction • Travail Dirigé • Travaux Dirigés • Treasury Department • Tribord • Tridirectional • Triple Difference • Tropical Depression • Trust Deed • Tube à Décharge • Tunisian Dinar (currency) • Tunnel Diode • Turbodynamo

Tda: Tienda

TDA: Tactical Decision Aid • Target Docking Adapter • Technical Direction Agent • Technology Development Approach • Telemetric Data Analyzer • Text-based Discriminant Analysis • Thermodifferential Analysis • Time Delay Amplifier • Time Division Arrival • Today • Tracking and Data Acquisition • Tracking Data Analysis • Transit Detection Algorithm • Trinidad (Colombia) (location identifier) • Trunnion Drive Axis • Tuning Device Assembly • Turkheim Detection Algorithm • Type de Données Abstrait

TDAAS: Two-dimensional Data Assessment and Applications Study

TDAB: Terrestrial Digital Audio Broadcasting

TDAC: Tycho Data Analysis Consortium

TDAF: Tracking and Data Analysis Facility

TDAS: Technical Data Automation System • Tracking and Data Acquisition System • Traffic Data Administration System • Tunnel-Diode Amplifier System

TDB: Target Deployment Bus • Task Database • Technical Directive Board • Temps Dynamique Barycentrique • Terminology Database • Terrestrial Dust Belt • Tetabedi (Papua New Guinea) (location identifier) • Tiempo Dinámico Baricéntrico • Time Duration of Burn • Total Disability Benefit • Trade and Development Board • Tranet Doppler Beacon

TDBU: Top-Down Bottom-Up

TDC: Tape Data Controller • Tarif Douanier Commun • Technical Data Center • Technical Development Center • Technical Directive Compliance • Technical Document Center • Telemetry Data Center • Terrestrial Data Circuit • Test Director Console • Text Database • Thermal Diffusion Chamber • Thermal Diffusion Coefficient • Time Data Card • Time-Delay Closing • Time-to-Digital Converter • Top Dead Centre • Total Distributed Control • Trade Development Council • Tropical Data Centre • True Digital Control • Two-Dimensional Cepstrum

TDCC: Transportation Data Coordinating Committee

TDCD: Transmitter Duty Cycle Distortion

TDCM: Transistor Driver Core Memory

TDCN: Technical Data Change Notice

TDCR: Technical Data Change Request

TDD: Task Description Document • Techniques Development Department • Technology Development Division • Telecommunications Device for the Deaf • Telemetry Data Digitizer • Telephone Device for the Deaf • Test Definition Document • Test Design Description • Test Development Director • Time Data Dependency • Time Division Duplex • Trinidad (Bolivia) (location identifier)

TDDB: Thermodynamic Design Data Block

TDDL: Time-Division Data Link

TDDP: Trajectory Design Data Package

TDDR: Technical Data Department Report • Total Direct-Diffuse Radiometer

TDE: Technical Data Engineer • Technical Data Evaluation • Terminal Display Editor • Total Differential Equation • Total Dose Experiment • Transition Diagram Editor • Two-Dimensional Equilibrium

TDEC: Technical Development Evaluation Center

TDEP: Tracking Data Editing Program

TDF: Take Data Flag • Télédiffusion De France • Telemetry Data File • Telemetry Data Format • Telemetry Distribution File • Telemetry Distribution Format • Test Data Folder • Trace Definition File • Transport Data File • Two Degrees of Freedom • Typeface Definition File

TDFC: Transfert des Données Fiscales et Comptables

TDFCHB: Telemetry Data Format Control Handbook

TDFD: Telemetry Downlink File Description

TDG: Tandag (Philippines) (location identifier) • Telemetry Data Generation • Test Data Generator • Test Display Generator • Test Documentation Group • Tidal Dwarf Galaxy • Time Delay Generator • Twist Drill Gauge

TDH: Tracking Data Handling

TDHS: Tape Data Handling System

TDI: Task Description Item • Telecommunications Data Interface • Temporary Disability Insurance • Test Data Interpolation • Time Delay and Integrate • Thompson Digital Image • Time Delay and Integration • Time-Delayed Integration • Time Delay Integrating • Time Delay Integration • Traitement Digital des Informations • Transport Device Interface • Transport Driver Interface • Trusted Database Interpretation • Trusted DBMS Interpretation

TDIF: TOPEX/Poseidon Data Information Facility

TDIO: Timing Data Input-Output

TDIS: Technical Data Impact Summary • Technical Documentation Information System • Terminal Data Input System

TDJ: Tadjoura (Djibouti) (location identifier) • Transfer Delay Jitter

TDK: Taldy-Kurgan (Kazakhstan) (location identifier)

TDL: Tandil (Argentina-BA) (location identifier) • Task-Directed Learning • Technical Data Laboratory • Technical Document List • Techniques Development Laboratory • Temporal-Difference Learning • Threshold Detection Level • Translation Definition Language • Translator Definition Language • Tunable Diode Laser • Tunnel-Diode Logic

TDLAS: Tunable Diode Laser Absorption Spectroscopy

TDLHS: Tunable Diode Laser Hetrodyne Spectrometer

TDLS: Tower Data-Link Service • Tunable Diode Laser Spectroscopy

Tdm: Tandem

TDM: Tableau de Définitions et de Mesures • Task Description Memo • Tausend Deustche Mark • Technical Data Management • Technical Division Manager • Technology Development Mission • Telemetric Data Monitor • Test Data Memorandum • Thermal Diffusion Method • Thermodynamic Molding • Time-Division Multiplex • Time-Division Multiplexer • Time-Division Multiplexing • Time Duration Modulation • Trouble Detection and Monitoring • Tunnel-Diode Mixer

TDMA: Tandem Differential Mobility Analyzer • Time-Division Multiple Access (*) • Time-Domain Multiple Access

TDMC: Technical Data Management Center

TDMD: Time Division Multiplex Device

TDMF: Transparent Data Migration Facility

TDMG: Telegraph and Data Message Generator

TDMS: Technical Data Management System • Telegraph Distortion Measuring System • Telemetry Data Monitor Set • Terminal Data Management System • Terminal Display Management System • Time Division Multiplex System • Transmission Distortion Measuring Set

TDMUX: Time-division Data Multiplexer

TDN: Time-Dependent Normalization • Topografische Dienst Nederland • Totally Digestible Nutrient

TDNN: Time-Delay Neural Network

TDNO: Thermal Detection of Nuclear Orientation

TDO: Technical Development Objective • Technical Divisions Office • Technology Development Office • Time Delay Opening • Toledo Winlock (USA-WA) (location identifier) • Tornado

TDOA: Time Difference Of Arrival

TDOP: Time Dilution Of Precision

TDOS: Tape Disk Operating System

TdP: Turbine de Puissance

TDP: Tag Distribution Protocol • Technical Data Package • Technical Development Plan • Technical Dissemination Project • Technological Demonstration Programme • Technology Demonstration Plan • Technology Demonstration Programme • Teledata Processing • Telelocator Data Protocol • Temperature Density Plotter • Temperature and Dewpoint • Tidal Deformation Polarization • Total Development Plan • Tracking Data Processor

TDPAC: Time Differential Perturbed Angular Correlation

TDPAD: Time Differential Perturbed Angular Distribution

TDPD: Technical Data Program Document

TDPL: Technical Data Package List

TDPSK: Time Differential Phase Shift Keying

TDQAS: Training Development Quality Assurance System

TDQM: Total Data Quality Management

Tdr: Tender

TDR: Tape Data Register • Technical Data Report • Technical Data Request • Technical Deficiency Report • Technical Design Report • Technical Design Review • Technical Development Requirement • Technical Dispatch Reliability • Technical Documentary Report • Technical Documentation Report • Test Data Report • Test Discrepancy Report • Theodore (Australia-QLD) (location identifier) • Time Division Reflectometer • Time-Domain Reflectometer • Time-Domain Reflectometry • Tous Droits Réservés • Tracking and Data Relay • Transistorized Digital Readout • Transmission de Données par Radio • Transmit Data Register • Treasury Deposit Request

TDRB: Tool Design Review Board

T&DRE: Tracking & Data Relay Equipment

TDRE: Tracking and Data Relay Equipment

TDRL: Terminal Distinguishable Regular Language

TDRR: Test Data Recording and Retrieval

TDRRB: Technical Data Requirement Review Board

TDRS: Telemetry Downlink Receiving Site • Text Data Retrieval System • Tracking and Data Relay Satellite

TDRSS: Tracking and Data Relay Satellite System

TDRSSGT: Tracking and Data Relay Satellite System Ground Terminal

TDRST: Tracking and Data Relay Satellite Terminal

tds: Ter Die Sumendum

TDS: Tabular Data Stream • Tape Data Selector • Technical Data System • Technical Directive System • Technology Demonstration Satellite • Test Data Sheet • Test Data System • TEX Directory Structure • Thermal Degradation Sample • Thermal Design System • Thermal Desorption Spectroscopy • Thermal Diffuse Scattering • Time Delay Switch • Time-Distance-Speed • Total Dissolved Solids • Track Data Simulator • Track Data Storage • Tracking and Data System • Transfert des Données Sociales • Transputer Development System

TD&SA: Telephone, Data & Special Audio

TDSA: Technical Data Status Accounting • Technical and Data Signals Analyzer • Telephone, Data and Special Audio • Tribal-Designated Statistical Area

TDSCC: Tidbindilla Deep Space Communications Complex

TDSE: Task Data Space Element

TDSI: Traitement du Signal et de l'Image

TDSP: Time-Dependent Signal Processor

TDSR: Transmitter Data Service Request

TDSS: Technical Data Storage System • Texas Deep Sky Survey

TdT: Traitement de Texte

TDT: Tanda Tula (South Africa) (location identifier) • Target Docking Trainer • Telemetry Data Tape • Temps Dynamique Terrestre • Terrestrial Dynamical Time • Terrestrial Dynamic Time • Test Data Team • Test Direction Team • Test Dwell Time • Tiempo Dinámico Terrestre • Topic Detection and Tracking • Tower Disconnect Technician • Transonic Dynamic Tunnel • Tunnel-Diode Transducer

TDTL: Test Data Team Leader • Tunnel-Diode Transistor Logic

TDU: Third Dredge-Up • Time Display Unit • Time Distribution Unit • Tracking Display Unit

TDUM: Tape Dump and Utility Monitor

TDUP: Technical Data Usage Program

TDV: Tanandava (Madagascar) (location identifier) • Test Data Van • Tool Design Variance • Touchdown Velocity

TDW: Amarillo Tradewind (USA-TX) (location identifier) • Twisted Double Shielded

TDWO: Test and Development Work Order

TDWR: Terminal Doppler Weather Radar

TDWT: Transonic Dynamic Wind Tunnel

TDy: Temporary Duty

TDY: Temporary Duty

TDZ: Toledo (USA-OH) (location identifier) • Touchdown Zone

TDZE: Touchdown Zone Elevation

TDZL: Touchdown Zone Lights

te: Tegulu (ISO language code)

Te: Tellur • Tellure • Tellurium • Température d'Équilibre • Tetanus

T/E: Table of Equipment • Transporter/Erector

T&E: Test & Evaluation • Testing & Evaluation • Training & Education • Travel & Entertainment • Trial & Error

TE: Task Element • Taux d'Erreurs • Technical Engineer • Technical Exchange • Technicien d'Études • Technicien d'Exploitation • Technische Einheit • Tecklenburg (Germany) • Teileinheit • Teilentladung • Telecommunications Engineering • Template Element • Temps d'Érection • Temps d'Escale • Temps d'Établissement • Temps d'-Exploitation • Temps Estimé • Tension à l'Entrée • Tension d'Essai • Tension Électrique • Terminal Emulator • Terminal Equipment • Terrain d'Envol • Terrestrial Ecosystem • Test and Engineering • Test and Evaluation • Test Equipment • Tête d'Enregistrement • Text End • Thermal Efficiency • Thermal Element • Thermoelectric • Thermomètre Enregistreur • Tiempo de las Efemérides • Time Earliest • Time Expected • Time of End • Timing Electronics • Tirant d'Eau • Topographic Engineer • Total Effect • Total Energy • Trace Element • Traction Électrique • Trading Expert • Trägerfrequenzerzeuger • Trailing Edge • Training and Education • Training Equipment • Trajectory Engineer • Trans-Earth • Transequatorial • Transformateur Électrique • Transfusioneinheit • Transmission Error • Transporter/Erector • Transverse Electric • Transverse Électrique • Transverse Electrostatic • Trennungsentschädigung • Trial and Error • Turbine à Engrenages • Turbine Engine • Tuyère d'Échappement • Twin Engine

TEA: Task Equipment Analysis • Technical Exchange Agreement • Tela (Honduras) (location identifier) • Temperate East Asia • Temps d'Entretien Actif • Test Equipment Accessory • Test Equipment Analysis • Test and Evaluation Agency • Thermal Energy Analysis • Thermal Energy Analyzer • Torque Equilibrium Attitude • Track Error Angle • Training Effectiveness Analysis • Transferred Electron Amplifier • Translation Exception Address • Transverse Electric Atmospheric • Transversely Excited Atmospheric • Triethylamonium

TEAC: Technical Educational Advisory Council

Teach: Teacher • Teaching

TEAL: Transversely Excited Atmospheric Laser

TEAM: Technique for Evaluation and Analysis of Maintainability • Top European Advertising Media • Total Environment Analysis and Management

TEAMS: Technology Experiments for Advancing Missions in Space • Terrestrial Ecosystem Analysis and Modelling System • Test Evaluation And Monitoring System • Time-of-flight Energy Angle Mass Spectrometer

TEAP: Terminating Extended-Area Descriptor • Trajectory Error Analysis Program • Transversely Excited Atmospheric Pressure

TEARRR: Times, Elevations, Azimuths, Ranges and Range Rates

TEB: Tender Evaluation Board • Teterboro (USA-NJ) (location identifier) • Thread Environment Block • Tone-Encoded Burst • Tropical Experiment Board

TEBALDI: Télécommande Ballon Longue Distance

TEC: Teaching Fxcellence Center • Technology Exchange Co. • Technology-Employment-Competitiveness • Temperature Electric Cooler • Test Equipment Center • Test Equipment Committee • Thermal Electric Cooler • Thermal Expansion Coefficient • Thermionic Energy Conversion • Thermo-Electric Cooled • Thermo-Electric Cooler • Tokyo Electronics Corporation • Tonne d'Équivalent Charbon • Topographic Engineering Center • Total Electric Content • Total Electron Content (*) • Total Environment Centre • Total Estimated Cost • Tower En route Control • Training Evaluation and Control • Trans-Earth Coast • Transistor à Effet de Champ • Transports Européens Combinés

TECE: Teleprinter Error Correction Equipment

Tech: Technical • Technician • Technics • Technique • Technische • Technological • Technologie • Technologique • Technology • Technology (periodical code)

TECH: Team Encouraging Career Horizons • Technical • Technician

Techn: Technical • Technician • Technics • Technique • Technische • Technological • Technologie • Technologique • Technology

Technol: Technological • Technologie • Technologique • Technology

TECL: Test Equipment Configuration Log

TECM: Test Equipment Commodity Manager

Tecn: Tecnicismo

TecN: Technical News (periodical code)

Tecnol: Tecnología

TECO: Tape Editor and Corrector • Text Editor and Corrector • Turbine Engine Checkout

TECOMAC: Technical Conference on Observation and Measurements of Atmospheric Contaminants

TECR: Test Equipment Change Requirement

TECRO: Taiwan's Economic and Cultural Representative Office

TecRv: Technology Review (periodical code)

TECS: Tether Elevator/Crawler System

Tecto: Tectonophysics (periodical code)

TED: Tableau Effectif-Dotation • Technological and Economic Dynamics • Tenders Electronic Daily • Test, Evaluation and Development • Test and Evaluation Division • Test Evaluation Directorate • Thermoelectric Device • Thermoionic Emission Detector • Thisted (Denmark) (location identifier) • Threat Environment Description • Tiny Editor • Total Energy Detector • Tracking Error Detector • Trailing Edge Down • Traitement Électronique des Données • Translation Error Detection • Translation Error Detector

TEDI: Terminology Editor

TEDIS: Trade Electronic Data Interchange System

TEE: Tape Editing Equipment • Tbessa (Algeria) (location identifier) • Test Equipment Engineering • Trans Europ Express • Triaxial Earth Ellipsoid

TEEM: Techno-Economico-Environmental Model • Trans Europ Express Marchandises • Two-phase Extended Evaluation in Microgravity

TEES: Texas Engineering Experiment Station (USA-TX)

TEF: Technology Experiment Facility • Technology Exposure Facility • Telfer (Australia-WA) (location identifier) • Toxicity Equivalence Factor

TEFF: Turbine Engine Fatigue Facility

TEFL: Teaching English as a Foreign Language

TEFLON: Polytetrafluoroethylene

TEG: Taux Effectif Global • Technical Evaluation Group • Tenkodogo (Burkina Faso) (location identifier) • Thermoelectric Generator • Top Edges Gilt • Tuyère d'Éjection des Gaz

TEGa: Triethyl Gallium

TEGA: Thermal and Evolved Gas Analyzer

TEGD: Technical Enforcement Guidance Document

Teh: Tehran (Iran)

TEH: Tetlin (USA-AK) (location identifier)

TEHM: Terrestrial Ecosystem Hydrology Model

TEHR: Tension Elliptical Hollow Rod

TEI: Technical and Economic Information • Technical Engineering Item • Text Electronic Interchange • Text Encoding Initiative • Tezu (India) (location identifier) • The Education Initiative • Thermoelectric Incubator • Thermostat Électronique Intelligent • Traitement Électronique de l'Information • Trans-Earth Injection • Tube à Entretien d'Images • Tusas Engine Industries

teilw: teilweise

Teilz: Teilzahlung

TEIn: Triethyl Indium

TEISC: Test and Evaluation Information Support Capability

TEK: Tatitlek (USA-AK) (location identifier)

Tél: Télécommunications • Télémètre • Télémétrie • Téléphone • Télescope • Télescopique

Tel: Telecommunications • Telegram • Telegraph • Telegraphic • Telemeter • Telemetry • Telephone • Telescope • Telescopic • Telescopium-Telescopii (constellation)

TEL: Technische Einsatzleitung • Telecommunications • Telupid (Malaysia) (location identifier) • Transporter-Erector-Launcher

TELB: Test Engineering Logbook

TelCo: Telephone Co.

Telec: Telecommunicación

Télécom: Télécommunication • Télécommunications

Telecom: Telecommunication • Telecommunications

TELECON: Telephone Conference • Telephone Conversation

Télég: Télégramme • Télégraphe • Télégraphie • Télégraphique

Teleg: Telegram • Telegraph • Telegraphic • Telegraphy

Télém: Télémètre • Télémétrie

Telem: Telemeter • Telemetry

Téléph: Téléphone • Téléphonie

Teleph: Telephone • Telephony

TELERJ: Telecomunicações do Rio de Janeiro (Brazil)

Télesc: Télescope • Télescopique

Telesc: Telescope • Telescopic

TELESAT: Telecommunications Satellite

Télév: Télévision

Telev: Television

TELEX: Teleprinter Exchange • Teletypewriter Exchange

TELH: Telheiro

TELIAS: Telidon Image Analysis System

TelIn: Telematics and Informatics (periodical code)

Tell: Tellus (periodical code)

TELMS: Telecommunications Management System

TELNET: Telecommunications Network • Teletype Network

Tel. No.: Telephone Number

TelOp: Teleconference Operator

TELOPS: Telemetry On-line Processing System

TELOS: The Electronic Library Of Science

TELSAT: Teledetection by Satellite

TEM: Tableau Effectif-Matériel • Target Engine Module • Temora (Australia-NSW) (location identifier) • Terrestrial Ecosystem Model • Tête Électromécanique • Thermoelectric Module • Tot En Met • Transformed Eulerian Mean • Transform Engine Motor • Transition Electron Microscope • Transition Electron Microscopy • Transmission Electron Microscope (*) • Transmission Electron Microscopy (*) • Transverse Electric Mode • Transverse Electromagnetic • Travail Électrique des Métaux

TEMA: Training, Educational and Mutual Assistance • Training, Education and Mutual Assistance

TEMAC: Trailing Edge Mean Aerodynamic Chord

TEMIC: Telecommunications Executive Management Institute of Canada

TEML: Turbo Editor Macro Language

Temp: Temperate • Temperature • Température • Tempered • Template • Tempo • Temporal • Temporary • Tempore

TEMP: Test and Evaluation Master Plan • Test Evaluation Master Plan • Thermal Energy Management Process

TEMPEST: Transient Electromagnetic Emanations Standard

TEMPO: Technical Electronic Management Planning Organization • Time and Earth Motion Precision Observation • Tyrrhenian Eddy Multiplatform Observation

TEMPUS: Tiegelfreies Elektro-Magnetisches Prozessieren Unter Schwerelosigkeit • Trans European Mobility Programme for University Studies

TEMS: Test Equipment Maintenance Set • Transport Environment Monitoring System

TEMSI: Temps Significatif

Ten: Tenant • Tenement

TEN: Tenente • Tongren (China-PRC) (location identifier) • Trans-European Network • Trans-Europ-Nacht • Trans-Europ-Nuit

TENCAP: Tactical Exploitation of National Capabilities

Tency: Tenancy

Tend: Tendenz • Tender • Tendeur

Tenn: Tennessee (USA)

Tens: Tension

TENS: Transcutaneous Electrical Nerve Stimulation • Transcutaneous Electrical Nerve Stimulator • Trans European Networks

TEO: Technology Enhancement Office • Terapo (Papua New Guinea) (location identifier) • Test Equipment Operator • Transferred Electron Oscillator

TEOC: Trømsø Earth Observation Center

Teol: Teología

TEOM: Tapered Element Oscillating Microbalance

TEOS: Tetraethoxysilane

tep: Top Edges Gilt

TEP: Teaching Electronic Publishing • Technical Evaluation Panel • Technology Enhancement Program • Temps d'Entretien Préventif • Temps Écoulé Prévu • Teptep (Papua New Guinea) (location identifier) • Terminal Edge Pixel • Test and Evaluation Plan • Test Evaluation Plan • Thermoelectric Power • Thermostat Électronique Programmable • Tonne Équilavent Pétrole • Traded Endowment Policy • Trans-Atlantic Economic Partnership • Transformable Edge Pair • Transit of Extrasolar Planet

TEPA: Taiwan Environmental Protection Agency

TEPC: Tissue Equivalent Proportional Counter

TEPCO: Tokyo Electronic Power Co.

TEPP: Tetraethyl Pyrophosphate

TEQB: Thermal Equilibration

Ter: Terrace • Terrain • Territorial • Territory • Tertiary • Teruel (Spain)

TER: Terceira Island - Lajes (Azores) (location identifier) • Term Error Rate • Terrace • Test Effectiveness Ratio • Test Equipment Readiness • Test Evaluation Report • Thermal Eclipse Reading • Time Estimating Relationship • Total Expense Ratio • Travel Expense Report

TERA: Terminal Effects Research and Analysis

TERAPIX: Traitement Élémentaire de Réduction et Analyse de Pixels

Terat: Teratology

TERC: Technical Education Research Center • Turbine Engine Research Center

TERCOM: Terrain Contour Matching

TERENA: Trans-European Research and Education Networking Association

TEREZA: Température Rotationnelle en Zone Aurorale

TERI: Tata Energy Research Institute (India) • Terrestrial Ecosystem Research Initiative

TERL: Test Equipment Readiness List

TERLS: Thumba Equatorial Rocket Launching Station (India)

Term: Terminación • Terminal • Terminate • Terminated • Terminateur • Terminating • Termination • Terminator • Terminological • Terminologie • Terminologique • Terminology

TERM: Tackling Environmental Resource Management • Temporary Equipment Recovery Mission • Terminal • Terminate

TERMPWR: Terminator Power

TERPS: Terminal Enquiry/Response Programming System

Terr: Terrace • Terrain • Terrestre • Terrestrial • Territoire • Territorial • Territory

TERR: Terrace • Terreiro

TERRA: Tecniche di Elaborazione e Rilevamento delle Risorse Ambientali • Terrestrial Ecosystems Regional Research and Analysis

TERRIERS: Tomographic Experiment using Radiative Recombinative Ionospheric EUV and Radio Sources

Territ: Territorial

TERS: Tropical Earth Resources Satellite • Tropical Environments Remote Sensing

TERSE: Tunable Etalon Remote Sounder for the Earth

TERSS: Tasmanian Earth Resources Satellite Station (Australia-TAS)

TERSSE: Total Earth Resources System for the Shuttle Era

TERSWG: Tropical Environments Remote Sensing Working Group

Tert: Tertiaire • Tertiary

Terz: Terrazzo

TES: Telemetry Evaluation Station • Telephony Earth Station • Terminal Encounter System • Terrain Edit System • Tessenei (Ethiopia) (location identifier) • Text Editing System • Thermal Emission Spectrometer • Thermal Enclosure System • Thermal Energy Storage • Thermophysikalische Eigenschaften Schmelzen • Tilting Etalon Spectrometer • Time-Enriched Statechart • Training of European Statisticians • Transition Edge Sensor • Travel Expert System • Tropospheric Emission Spectrometer

TESA: Table Équatoriale Solaire Automatisée

TESCAS: Tropospheric Environmental Studies by Laser Soundings

TESH: Technical Shop

TESL: Teaching English as a Second Language • Teaching of English as a Second Language

TESLA: TeV Electron Superconducting Linear Accelerator • TeV Energy Superconducting Linear Accelerator

TESLAS: Tropospheric Environmental Studies by Laser Sounding

TESOL: Teaching English to Speakers of Other Languages

TESOS: Telecentric Solar Spectrograph • Telecentric Solar Spectrometer (*)

TESR: Test Equipment Status Report • Torsion Elliptical Solid Rod

TESRE: Technologie E Studio delle Radiazioni Extraterrestri (CNR)

TESRP: Test and Evaluation Support Resource Plan

TESS: Tactical Environmental Support System

TESSA: Testbed for Evaluating Spatial Similarity Algorithms • Texas Severe Storm Association

Test: Testament • Testator • Testatrix • Testigo • Testimonial • Testimonio • Testimony

TEST: Thesaurus of Engineering and Scientific Terms • Time-Efficient Sequencer of Tests • Transferable European Space Technology

Testa: Testament • Testamento

TESTA: Trans-European Services for Telematics between Administrations

Testis: Testigos

TESTS: Technical European Support for Testing Standards

TESZ: Trans-European Suture Zone

Tet: Tetrahedron

TET: Technicien d'Études et de Travaux • Test Environment Toolkit • Test Equipment Team • Test Equipment Tool • Test Evaluation Team • Tete Matunda (Mozambique) (location identifier) • Theoretical Evolutionary Track • Thermionic Emission Technique • Thermometric Enthalpimetric Titration • Total Elapsed Team • Trajectory Events Timeline • Transistor Evaluation Test • Turbine Entry Temperature • Turbine Exhaust Temperature

TETM: Thermal Effects Test Model

Tetr: Tetragonal

TETRA: Terrestrial Trunked Radio • Trans-European Trunked Radio • Trans-European Trunk Radio

TEU: Te Anau (New Zealand) (location identifier) • Télé-Enseignement Universitaire • Telemetry Equipment Unit • Treaty on European Union • Twenty-foot Equivalent Unit

Teut: Teuton • Teutonic

Teutón: Teutónico

TeV: Teraelectronvolt

TEV: Target Engine Vehicle • Taux d'Entrée en Virage • Thermoelectric Voltage

TEW: Total Equivalent Weight

TEWS: Tactical Electronic Warfare System

Tex: Texan • Texas (USA)

TEx: Transition Experiment

TEX: Telluride (USA-CO) (location identifier)

TEXES: Texas Echelon-Cross-Echelle Spectrograph

TEXnet: Texas Network

TEXT: Texas Experimental Tokamak

TEXUS: Technologische Experiment Unter Schwerelosigkeit

TEY: Thingeyri (Iceland) (location identifier)

TEZ: Tezpur (India) (location identifier)

t/f: True/False

tf: Tonne-Force

T/F: True/False • Turbofan

T&F: Time & Frequency

TF: French Southern Territories (ISO country code) • Iceland (international aircraft registration code) • Southern Territories (French_) (ISO country code) • Tabulating Form • Task Force • Taux de Fiabilité • Tax Free • Telefax • Telemetry Format • Telford (UK postal code) • Temperature Factor • Temperature Function • Temps de Fonctionnement • Tenue en Fatigue • Term Frequency • Terrain Following • Territoire Français Austral (ISO country code) • Test Facility • Test Fixture • Test Flight • Text Flow • Thermomètre Fahrenheit • Thin Film • Throughfall • Tidal Field • Tierra del Fuego (Argentina) • Time and Frequency • Time Frame • To Fill • To Follow • Traînée de Frottement • Train Fixe • Transfer Function • Transformateur • Transformée de Fourier • Très Fort • Très Forte • Triple Frequency • Trust Framework • Trust Funds • Tube à Flamme • Tully-Fisher • Turbo-Fusée • Tuyau Flexible

TFA: Target Factor Analysis • Taux de Fausse Alarme • Tilfalmin (Papua New Guinea) (location identifier) • Transfer Function Analyzer • Transistor Feedback Analyzer • Transparent File Access • Trifluoroacetic Acid

TFAC: Taux de Fausse Alarme Constant • To Facility

TFAP: Tropical Forestry Action Plan

Tfc: Traffic

TFC: Telemetry Format and Control • Time From Cutoff • Total Fixed Cost • Transfer Function Computer • Transistorized Frequency Converter • Trigonometric Function Computer • Tunable Frequency Converter

TFCS: Triplex Flight Control System

TFD: Test Flow Diagram • Thermo Field Dynamics • Thin-Film Detector • Thin-Film Diode • Time Frequency Digitizer • Total Frequency Deviation • Transcription Factor Database • Transformée de Fourier Discrète

TFDD: Text File Device Driver

TFDM: Technical Feasibility Demonstration Model

TFE: Teflon • Tetrafluoroethylene • Thermionic Fuel Element • Time From Event

TFEL: Thin-Film Electroluminescent

TFF: Tefé (Brazil-AM) (location identifier) • Time of Free Fall

TFFF: Thermal Field Flow Fractionation

TFGS: Time and Frequency Generation System

TFH: Technische Fachhohschule (Germany) • Timing Failure Handler

TFHR: Task Force Human Resources

TFI: Technology Flight Instrumentation • Temporal Fractional Integral • Thin-Film Inductive • Time From Ignition • Tufi (Papua New Guinea) (location identifier)

TFIDF: Term Frequency times Inverse Document Frequency

TFK: Turbulent Flare Kernel

TFL: Telemetry Format Load • Teófilo Otoni (Brazil-MG) (location identifier) • Time From Launch

TFLOPS: TeraFLOPS

TFM: Tagged Font Metrics • Telefomin (Papua New Guinea) (location identifier) • TEX Font Metric • Traffic Flow Management • Transmit Frame Memory • Transmitter Frequency Multiplier • Trusted Facility Manual

TFMRP: Transfer Function Mode Reduction Package

TFMS: Text and File Management System • Time-of-Flight Mass Spectrometer

TFN: Tenerife Norte (Canary Islands) (location identifier) • Till Further Notice

Tfno: Teléfono

TFO: Technology Flight Opportunity • Transferring From Original

TFODM: Task Force on Observations and Data Management

TFOL: Tape File Octal Load • Thermo Field Dynamics

TFOV: Target Field Of View • Telescope Field Of View • Total Field Of View

TFP: Test Facility Program • Thermal Front Parameter • TOPS Filing Protocol • Tunable Fabry-Pérot

TFPC: Thin-Film Polarizing Cube • Toll Fraud Prevention Committee

TFPF: Table Formats and Parameters File

TFQMR: Transpose-Free Quasi-Minimal Residual

Tfr: Transfer

TFR: Temporary Flight Restriction • Terrain-Following Radar • Tightly Folded Resonator • Tokamak de Fontenay-aux-Roses (France) • Total Final Report • Transfer Function Response • Transformación de Fourier Rápida • Transformée de Fourier Rapide • Traveller Failure Report • Trouble and Failure Report • Tully-Fisher Relation • Tunable Frequency Range

TFS: Table File System • Taux de Formation Stellaire • Telemetry Format Selection • Tenerife Sur - Reina Sofia (Canary Islands) (location identifier) • Thermal and Fluid Systems • Thursday-Friday-Saturday • Time and Frequency Standard • Translucent File System • Tunable Frequency Source

TFSS: Technical Facilities Subsystem

TFT: Taftan (Pakistan) (location identifier) • Technology For Technology • Tensor Field Theory • Thermal Fatigue Test • Thin-Film Transistor • Tit For Tat

TFTP: Trivial File Transfer on PCs • Trivial File Transfer Protocol (*)

TFTR: Tokamak Fusion Test Reactor

TFTS: Terrestrial Flight Telecommunications System • Terrestrial Flight Telephone System (*)

TFTT: Thanks For The Thoughts

TFU: Test Facility Utilization • Theoretical First Unit

TFxIDF: Term Frequency and Inverse Document Frequency

TFY: Tarfaya (Morocco) (location identifier)

TFYT: Thanks For Your Time

TFZ: Tropospheric Frontal Zone

tg: Tajik (ISO language code) • Tangent • Tangente

Tg: Tanjong • Tanjung

TG: Guatemala (international aircraft registration code) • Tail Gear • Task Group • Technical Guidance • Technical Guide • Technology Gap • Telegiornale • Telegramme • Téléguidage • Terme Générique • Terminal Guidance • Test Group • Test Guideline • Test Guaranteed • Thermogravimetry • Timing Gate • Togo (international vehicle registration code) • Togo (ISO country code) • Trace Gas • Tracking and Guidance • Training Group • Transformational Generative • Transformational Grammar • Triglyceride • Tube de Guidage • Turbine à Gaz • Turbine Générateur • Turbogénérateur

TGA: Tandy Graphics Adapter • Task Group on AXAF • Télégestion des Abonnés • Thermal Gravimetric Analysis • Thermogravimeter Analyzer • Thermogravimetric Analysis • Trace Gas Analyzer • Trusted Global Advisor • Turbogénérateur d'Air

TGAO: Technologie de Groupe Assistée par Ordinateur

TGB: Tagbita (Philippines) (location identifier) • Très Grande Bibliothèque

TGBNTOO: Task Group on the BMDO New Technology Orbital Observatory

TGC: Transmit Gain Control

TGCM: Thermospheric General Circulation Model

TGCP: Test Ground Control Point

TGCR: Thermal Gross Cloud Restoral

TGCT: Thermal Gross Cloud Test

TGD: Titograd (Yugoslavia-Montenegro) (location identifier) • Trajectory and Guidance Data

TGDDIS: Trace Gas Dynamics Data Information System

TGDF: Turbulent Gasjet Diffusion Flame

TGE: Transverse Gravimeter Experiment • Très Grand Équipement • Turbogénérateur Externe • Tuskegee - Sharpe Field (USA-AL) (location identifier)

TGF: Tignes (France) (location identifier) • Transforming Growth Factor

TGG: Kuala Terengganu (Malaysia) (location identifier) • Third-Generation Gyro • Transformational-Generative Grammar • Turbogénérateur de Gaz

TGGPB: Task Group on Gravity Probe B

TGH: Tongoa (Vanuatu) (location identifier)

TGI: Target Group Identifier • Technique Graphique Industrielle • Temperature Growth Index • Tingo Maria (Peru) (location identifier) • Tribunal de Grande Instance

TGID: Trunk Group Identification Number

TGIR: Test Generation with Infrared Reasoning

TGIRS: Two Grating Infrared Spectrometer

TGJ: Tiga (Loyalty Islands) (location identifier)

Tgl: Toggle

TGL: Tagula (Papua New Guinea) (location identifier) • Task Group Leader • Touch and Go Landing

TGM: Tirgu Mures (Romania) (location identifier)

TGME: Total Generalized M-Estimation

TGMT: Temperature Gradient Mass Transport

tgn: Tangent

TGN: Traralgon (Australia-VIC) (location identifier)

TGO: Time To Go • Togo (ISO country code) • Tongliao (China-PRC) (location identifier)

Tgov: Tegenover

TGOWG: Teleoperator Ground Operations Working Group

TGP: Tidal Generating Potential • Très Gros Plan

TGR: Timing Gap Record • Touggourt (Algeria) (location identifier)

TGRA: Très Grand Rayon d'Action

TGRAP: Task Group on Research and Analysis Programs

TGRD: Turbine à Gaz à Réaction Directe

TGRS: Transient Gamma-Ray Spectrometer

TGS: Chokwe (Mozambique) (location identifier) • Telemetry Ground Station • Telemetry Ground System • Terminal Guidance Sensor • Terminal Guidance System • Text and Graphics Server • Thermogravimetric System • Thermogravimetry System • Ticket-Granting Server • TOPEX Ground System • Transmission Grating Spectrometer • Transportable Ground Station • Triglycine Sulfate

TGSAA: Task Group on Space Astronomy and Astrophysics

TGSE: Telemetry Ground Support Equipment • Test Ground Support Equipment

TgSF: Télégraphie Sans Fil

TGSS: Terminal Guidance Sensor System

Tgt: Target

TGT: Tanga (Tanzania) (location identifier) • Target • Temperature Gradient Technique • Ticket-Granting Ticket • TDRSS Ground Terminal • Turbine Gas Temperature

TGU: Target Generation Unit • Technical Guidance Unit • Tegucigalpa Toncontin (Honduras) (location identifier) • Théorie de Grande Unification

TGV: Targovishte (Bulgaria) (location identifier) • Ten Gevolge Van • Train à Grande Vitesse (*) • Tren de Gran Velocidad • Très Grande Vigilance • Très Grande Vitesse

TGW: Tropospheric Gravity Wave

TGX: Tingrela (Ivory Coast) (location identifier)

TGZ: Tuxtla Gutierrez - Llano San Juan (Mexico) (location identifier)

th: Hyperbolic Tangent • Tangente Hyperbolique • Thai (ISO language code) • Thailand

Th: Thai • Thermal • Thorium • Thursday

T&H: Test & Handling

T/H: Track/Hold

TH: Table Header • Technische Hochschule • Technische Hogeschool • Temporary Hold • Territory of Hawaii • Thailand (ISO country code) • Thailande (ISO country code) • Thesaurus • Third Harmonic • Threshold • Totalisateur Horaire • Town Hall • Track and Hold • Train Haut • Triple Hydrant • True Heading • Turbine à Hélice • Turbine Hydraulique • Turbo-Hélice

THA: Thailand (ISO country code) • Thailande (ISO country code) • Thermal Hydraulic Analyzer • Tullahoma (USA-TN) (location identifier)

THAAD: Theater High-Altitude Area Defense

ThAFM: Theoretical and Applied Fracture Mechanics (periodical code)

Thai: Thailand

THAP: Thermal Hydraulic Analyzer Program

THB: Bath (Thailand) (ISO currency code) • Technische-Hochschule-Bibliothek • Thaba-Tseka (Lesotho) (location identifier) • Third-Harmonic Band

THC: Tchien (Liberia) (location identifier) • Temperature and Humidity Control • Thermohaline Circulation • Through Hole Connection • Thrust Hand Controller • Total Hydrocarbon • Translational Hand Control • Translation Hand Controller

ThCFD: Theoretical and Computational Fluid Dynamics (periodical code)

THCS: Temperature and Humidity Control System

Thd: Thread

ThD: Doctor of Theology

THD: Tag Hierarchy Diagram • Technische Hochschule Darmstadt (Germany) (now TUD) • Technische Hogeschool Delft (Netherlands) • Thread • Thunderhead • Total Harmonic Distortion • Tunnel Hydrodynamique

THDA: Temperature of Camera Head Amplifier

THDB: Technology History Database

THD+N: Total Harmonic Distortion plus Noise

THDN: Total Harmonic Distortion plus Noise

THDR: Thunder • Transport Layer Header

THDS: Time Homogeneous Data Set

THE: Tape-Handling Equipment • Technische Hogeschool Eindhoven (Netherlands) • Teresina (Brazil-PI) (location identifier)

THEA: Thousand Element Array

Theat: Theater • Theatrical

THEL: Tactical High-Energy Laser

THEMA: Thermal Mapper

THEMIS: Télescope Héliographique pour l'Étude du Magnétisme et des Instabilités Solaires

THEMISTOCLE: Tracking High-Energy Muons In Showers Triggered On Čerenkov Light Emission

THENet: Texas Higher Education Network (USA-TX)

THEO: Theoretical

Theol: Theologian • Theological • Theology

Theor: Theorem • Theoretical • Theory

Théor: Théorique

Theoret: Theoretical

Theos: Theosophical • Theosophy

Therm: Thermal • Thermique • Thermometer • Thermometry • Thermostat • Thermostatic

THERM: Thermal

Thermochem: Thermochemistry

Thermodyn: Thermodynamics

Thermodynam: Thermodynamics

Thermom: Thermometer • Thermometry

Thes: Thesis

THESEO: Third European Stratospheric Experiment on Ozone

THEX: Thursdays Excepted

THF: Berlin Tempelhof (Germany) (location identifier) • Très Haute Fréquence

THG: Thangool Biloela (Australia-QLD) (location identifier) • Third Harmonic Generation

THH: Taharoa (New Zealand) (location identifier)

ThI: Thallium Iodide

THI: Temperature-Humidity Index • Tichitt (Mauritania) (location identifier)

THIA: Toledo Heavy Ion Accelerator

THIR: Temperature/Humidity Infrared Radiometer • Thermal Infrared

THIS: Telescope Human Interface System • Total Health Information System • Transportable Hemispherical Illumination System

THISTLE: Tracking Heavy Isotope Spectrometer Telescope for Low-Energies • Tracking Heavy Isotope Spectrometer Telescope for Low-Energy (cosmic rays) (*)

Thk: Thick

THK: Thakhek (Laos) (location identifier) • Thick

ThkF: Thick Film

Thkns: Thickness

Thl: Thermal

THL: Tachilek (Myanmar) (location identifier) • Trans-Hybrid Loss

ThLA: Three-Letter Acronym

Thld: Threshold

THLF: Thermal Hydraulic Load Flow

ThM: Master of Theology

THM: Thomsonfield (USA-MO) (location identifier) • Travelling Heater Method • Tricode Hexaphase Modulation • Trihalomethane

Thml: Thermal

Thmom: Thermometer

Thms: Thermistor

THN: Thin • Trollhättan (Sweden) (location identifier) • Trollhattan (Sweden) (location identifier)

ThnF: Thin Film

Thnr: Thinner

THO: Thórshöfn (Iceland) (location identifier)

THOLD: Threshold

THOR: Tandy High-intensity Optical Recording • Tandy High-performance Optical Recording • Tape-Handling Option Routine • Technology by Highly Oriented Research

THORP: Thermal Oxide Reprocessing Plant

Thou: Thousand

THOUS: Thousand

Thp: Thrust Horsepower

THP: Thermopolis - Hot Springs (USA-WY) (location identifier) • Thrust Horsepower • Turbine à Haute Pression

Thr: Their • Threshold • Through • Thrust

THR: Tehran Mehrabad (Iran) (location identifier) • Terahertz Heterodyne Receiver • Thruster • THz Receiver • Transmit Holding Register

THRFTR: Thereafter

Throt: Throtle

THROT: Throtle

Thrt: Throtle

THRU: Through

THRUT: Throughout

THS: Thesaurus • Trimmable Horizontal Stabilizer

THSD: Thousand

Thstm: Thunderstorm

THT: Tamchakett (Mauritania) (location identifier) • Token Holding Time • Token Hold Time • Très Haute Tension

THTN: Threaten

THTR: Thorium High-Temperature Reactor

Thu: Thursday

THU: Thule (Greenland) (location identifier) • Thursday

THUDD: Thermal Uplink Data Display

Thur: Thursday

Thür: Thüringen (Germany)

Thüringer W: Thüringer Wald (Germany)

Thurs: Thursday

THV: York (USA-PA) (location identifier)

THW: Technisches Hilfswerk

Thwy: Throughway

THY: Thohoyandou (South Africa) (location identifier)

Thyr: Thyristor

THz: Terahertz

THZ: Tahoua (Niger) (location identifier)

ti: Tigrinya (ISO language code)

Ti: Titan • Titan • Titanium

TI: Costa Rica (international aircraft registration code) • Tape Inverter • Target Indicator • Taux d'Incidents • Technical Information • Technical Inspection • Technical Institute • Technical Instruction • Technical Instructor • Technical Integration • Technical Interchange • Technical Investigation • Technicien d'Instruments • Technologies de l'Information • Temperature Indication • Temperature Indicator • Temporary Installation • Temps d'Indisponibilité • Terminal Initiation • Terminator Irregularity • Test Instruction • Test Instrumentation • Texas Instruments Inc. • Thermal Imagery • Thermal Imaging • Tiempo Internacional • Timing Interface • Title Information • Topographic Instruction • Total Immersion • Trademark Information • Traditional Instruction • Traînée d'Interaction • Traînée Induite • Training Integrator • Traitement de l'Information • Transmetteur d'Impulsion • Transfer Initiate • Translation Invariant

TIA: Task Item Authorization • Telecommunications Industries Association • Telecommunications Industry Association • Territorial and International Affairs • Test Interface Assembly • Thanks In Advance • The Internet Adapter • Time Interval Analyzer • Tirana (Albania) (location identifier) • Trans-Impedance Amplifier

TIAC: Technical Information Analysis Center • Texas Instruments Automatic Computer

TIACA: The International Air Cargo Association

TIADD: Texas Instruments Automatic Data Display System

TIALD: Thermal Imaging Airborne Laser Designator

TIATA: Technical Information and Training Authoring

Tib: Tibet • Tibetan

TIB: Technische Informations-Bibliothek • Telescope and Instruments Branch • Telescope Instrument Branch • Technical Information Branch • Technical Information Bulletin • Technical Information Bureau • Technsiche Informationsbibliothek • Thread Information Block • Tibu (Colombia) (location identifier) • Traitement Indiciaire Brut

TIBAQ: Text-and-Reference-Based Answering of Questions

TIBL: Thermal Internal Boundary Layer

TIBM: Time-Integration Block Marching

TIBS: Tactical Information Broadcast Service • Tactical Information Broadcast System • Through-Ice Bathymetry System • Transcribed Information Briefing Service • Trends in Biochemical Sciences

TIC: Taken Into Consideration • Tape Identification Card • Tape Index Count • Technical Implementation Committee • Technical Information Capability • Technical Information Center • Technical Interface Committee • Technologie de l'Information et de la Communication • Temperature Indicator Controller • TermInfo Compiler • Thermal Imaging Camera • Thin Ionization Calorimeter • Timed Calculus • Time Interval Counter • Timing Interface Controller • Tinak Island (Marshall Islands) (location identifier) • Token-ring Interface Coupler • Total Inorganic Carbon • Total Item Change • Tourist Information Centre • Trade Information Committee • Translation Information Center • Trigger Interface Controller • Tuned Integrated Circuit • Turbulence Instrument Cluster

TICA: Technical Information Center Administration

TICCE: Time-band Capture Cell Experiment

TICCIT: Time-Shared Interactive Computer-Controlled Information Television

TICE: Technologies de l'Information et de la Communication à l'Enseignement • Time Integral Cost Effectiveness

TICES: Type-In Coding and Editing System

TICHS: Thermal Ion Capped Hemisphere Spectrometer

TICL: Transport Interface Compatibility Layer

TICLER: Technical Input Checklist/Evaluation Report

TICM: Test Interface and Control Module

TICR: Tycho Input Catalogue Revision

TICT: Twisted Intramolecular Charge Transfer

tid: Ter in Die

TID: Target Identification • Technical Information Division • Ter In Die • Thread Identifier • Tiaret (Algeria) (location identifier) • Time Integration Delay • Traitement Informatique des Données • Traitement Intégré des Données • Transaction Identification • Travelling Ionospheric Disturbance

TIDAS: Totally Integrated Data System

TIDB: Technical Information Dissemination Bureau

TIDDB: Thermal Interface Design Data Block

TIDE: Tachyon Integrated Development Environment • Telematics Initiative for the Disabled and Elderly • Thermal Infrared Differential Emission • Thermal Ion Dynamics Experiment

TIDER: Technical-Information Deficiency Report

TIDMA: Tape Interface Direct Memory Access

TIDP: Telemetry and Image Data Processing

TIDS: Technical Information Distribution Service

TIDU: Technical Information and Documents Unit

TIDY: Teletypewriter Integrated Display

TIE: Targeted Image Extraction • Taux d'Intérêt Équivalent • Technical Idea Exchange • Technical Information Exchange • Technology Information Exchange • Technology Integration Experiment • Teleform Interpretation Engine • Telescopes In Education • Thermosphere-Ionosphere Electrodynamics • Tippi (Ethiopia) (location identifier) • Tool for Image Encryption • Translated-Image Environment

TIE-GCM: Thermosphere/Ionosphere/Electrodynamic General Circulation Model

TIEGCM: Thermosphere/Ionosphere/Electrodynamic General Circulation Model

TIERCE: Transport des Ions Et des Rayonnements dans les Cibles Épaisses

TIES: Technological Information Exchange System • Telecom Information Exchange Service • The Interactive Encyclopaedia System (now HyperTIES) • Time-Independent Escape Sequence • Transmission and Information Exchange System

TIEU: Technical Information and Educational Unit

TIF: Tagged Image File • Taif (Saudi Arabia) (location identifier) • Technical Information File • Telephone Influence Factor • Telephone Interference Factor

TIFDL: TOPEX Inter-Facility Data Link

TIFF: Tagged Image File Format • Tag Image File Format

TIFR: Tata Institute of Fundamental Research (India)

TIFS: Total In-Flight Simulator

TIG: Tingwon (Papua New Guinea) (location identifier) • Travail d'Intérêt Général • Tungsten Inert Gas

TIGA: Texas Instruments Graphics Architecture

TIGCM: Thermosphere/Ionosphere General Circulation Model

TIGER: Tasman International Geospace Environment Radar • Terrestrial Initiative in Global Environment Research (*) • Terrestrial Initiative on Global Environment Research • Thermal Infrared Ground Emission Radiometer • Topologically Integrated and Geophysically Encoded Reference • Topologically Integrated Geographic Encoding and Referencing • Traitement Intégral des Galaxies par l'Étude de leurs Raies • Trans-Iron Galactic Element Recorder

TIGO: Transportable Integrated Geodetic Observatory (*) • Transportable Integrated Geophysical Observatory

TIGR: The Institute for Genomic Research (USA-MD) • TOVS Initial Guess Retrieval

TIGRE: Tracking and Imaging Gamma-Ray Experiment

TIGRIS: Televised Images of Gaseous Region in Interplanetary Space • Topologically Integrated Geographically Referenced Information System • Topologically Integrated Geographic Resource and Information System

TIH: Tikehau Atoll (Tuamotu Islands) (location identifier)

TIHS: Torino Impact Hazard Scale

TII: Technology Independent Interface • Texas Instruments Inc. • Tirinkot (Afghanistan) (location identifier) • Tooling Inspection Instrumentation

TIIAP: Telecommunications and Information Infrastructure Assistance Program (now TOP)

TIIS: Thermal Infrared Imaging Spectrometer

TIJ: Tijuana (Mexico) (location identifier) • Tribunal Internacional de Justicia

TIK: Oklahoma City - Tinker (USA-OK) (location identifier)

TIL: Inverlake (USA-CA) (location identifier) • Targeted Institutional Link • Technology Insertion Laboratory • Until

TILL: Total Initial Lamp Lumens

TILO: Technical Industrial Liaison Office • Technical Information Liaison Office

TILS: Technical Information and Library Service

TILT: Teaching and Learning Technology Programme • Technique d'Images Laser Tridimensionnelle

TIM: Technical Information Manual • Technical Information on Microfilm • Technical Interchange Meeting • Technical Interface Meeting • Telescope and Image Modelling • Telescope Image Model • Telescope Image Modelling • Telescope Imaging Model • Telescope Instrument Model • Telescope Instrument Modelling • Tembagapura (Indonesia) (location identifier) • Temperature Indicator Monitor • Ternary Ionic Mixture • Test Item Malfunction • Time Interval Measurement • Time Interval Meter • Time Interval Monitor • Time Is Money • Topic Indexing Matrix • Total Information Management • Tracking Information

Memorandum • Tracking Instruction Manual • Tracking Instrument Mount

TIMA: Techniques de l'Informatique et de la Microélectronique pour l'Architecture (d'ordinateurs)

TIMC: Techniques de l'Imagerie, de la Modélisation et de la Cognition

TIME: Technique for Information Management and Employment • Thermosphere-Ionosphere-Mesosphere Electrodynamics • Total Inventory Management in Engineering

TIMED: Thermosphere-Ionosphere-Mesosphere Electrodynamics • Thermosphere-Ionosphere-Mesosphere Energetics and Dynamics (*)

TIME-GCM: Thermosphere/Ionosphere/Mesosphere/Electrodynamic General Circulation Model

TIMEGCM: Thermosphere/Ionosphere/Mesosphere/Electrodynamic General Circulation Model

TIMES: Thermoelectric Integrated Membrane Evaporation System

TIMEX: Time Multiplexer

TIMI: Technical Information Maintenance Instruction

TIMIS: Totally Integrated Management Information System

TIMM: Thermally-Integrated Micromodule • Toshiba Integrated Multimedia Monitor

TIMMI: Thermal Infrared Multimode Imager • Thermal Infrared Multimode Instrument

TIMO: Temporary Installation Manufacturing Order

TIMS: Technical Information Management System • Telecommunications Information Management System • Thaad Information Management System • The Institute of Management Sciences (now merged with ORSA into INFORMS) • Thermal Imaging Spectrometer • Thermal Infrared Mapping Spectrometer • Thermal Infrared Multispectral Scanner • Thermal Ionization Mass Spectrometry • Total Information Management System • Traditional Interactive Multiuser System • Training Integration Management System • Transmission Impairment Measurement Set

TIMSS: Third International Mathematics and Science Study

TIM3: Techniques de l'Informatique, des Mathématiques, de la Microélectronique et de la Microscopie

TIN: Task Implementation Notice • Taxpayer Identification Number • Temporary Identification Number • Temporary Instruction Notice • Tindouf (Algeria) (location identifier) • Transaction Identification Number • Triangulated Irregular Network

TINA: Telecommunication Information Networking Architecture • There Is No Alternative

TINS: Transinertial Navigation System

TINTIN: Terahertz Integrated Technology Initiative

TIO: Target Insertion Orbit • Task Input/Output • Technical Information Office • Technical Information Officer • Tilin (Myanmar) (location identifier) • Transistorized Image Orthicon

TIOM: Terminal Input/Output Module

TIOP: Taux Interbancaire Offert á Paris

TIOT: Task Input/Output Table

TIP: Target Instrument Payload • Task Initiation and Prediction • Technical Improvement Plan • Technical Improvement Panel • Technical Improvement Program • Technical Improvement Project • Technical Information Panel • Technical Information Program • Technical Integration Panel • Technical Interest Profile • Technical Interest Profiling • Teletype Input Processing • Template-controlled Image Processor • Tenerife Infrared Polarimeter (*) • Tenerife Island Polarimeter • Terminal Interface Package • Terminal Interface Processor • Terrestrial Interface Port • Textual Information Processing • The Industrial Physicist (periodical) • Thesis In Progress • Thrust Inlet Pressure • Tilted Payload • TIROS Information Processor • Titre Interbancaire de Paiement • To Insure Promptitude • Total Information Processing • Tracking and Impact Prediction • Tracking Impact Prediction • Transaction Internet Protocol • Transit Improvement Program • Tripoli (Libya) (location identifier) • Truly Important Person • Turbine Inlet Pressure

TIPA: TEX International Phonetic Alphabet

TIPACS: Texas Instruments Planning And Control System

TIPE: Travail d'Initiative Personnel Encadré

TIPEX: Test-In-Progress Executive

TIPL: Teaching Information Processing Language

TIPMEX: Tibetan Plateau Meteorology Experiment

TIPP: Taxe Intérieure sur les Produits Pétroliers (France)

TIPPS: Total In-house Publication Production System

TiPS: Tether Physics and Survivability

TIPS: Teaching Information Processing System • Technical Information Processing Service • Technical Information Processing System • Technological Information Pilot System • Terminal Information Processing System • Tether Physics and Survivability (experiment) • Text Information Processing Service • Text Information Processing System • Thematic Mapper Image Processing System • Theory of Inventive Problem Solving • Thruster Instability Protection System • Tile Information Processing System • Traitement d'Images Perceptifs Spécifiques • Truevision Image Processing Software • Tutorial Image Processing System

TIPSH: Translation Invariant Poisson Smoothing using Haar (wavelets)

TIPSY: Theoretical Image Processing System

TIPT: Technology Integrated Product Team

TIPTOP: Tape Input - Tape Output • The Internet Pilot To Physics

TIQ: Tinian Island (Mariana Islands) (location identifier)

Tir: Tirage

TIR: Technical Information Release • Technical Information Retrieval • Telescopio Infrarosso • Temporary Installation Record • Terminal Imaging Radar • Thermal Infrared • Thermal Infrared Radiometer • Time In Rate • Tirupati (India) (location identifier) • Total Indicator Reading • Total Internal Reflection • Total Item Record • Transport International Routier

TIRACS: Telecommanded Inertially Referenced Attitude Control System

TIRAD: Tool for Infrared Algorithm Development

TIRAS: Technical Information Retrieval and Analysis System

TIRC: Technical Information Retrieval Center • Technical Information Retrieval Committee

TIREC: TIROS Ice Reconnaissance

TIREX: TIROS Ice Experiment

TIRGO: Telescopio Infrarosso del Gornergrat

TIRL: Telematique International Research Laboratories

TIRLS: Thermal Infrared Linescan Imagery

TIRM: Training for Information Resources Management • Transparent Infrared Material

TIROS: Television and Infrared Observation Satellite (*) • Television and Infrared Observing Satellite • Television Infrared Observation Satellite

TIRP: Tenerife Infrared Polarimeter • Total Internal Reflection Prism

TIRPC: Transport-Independent Remote Procedure Call

TIRV: Tuned Inter-Range Vector

TIRS: Text Information Retrieval System • The Integrated Reasoning Shell • Thermal Infrared Scanner

TIRSS: Technical Information Retrieval and Storage System

TIS: Teacher In Space • Technical Information Service • Technical Information System • Technical Inspection and Safety • Technical Interface Specification • Telemetry Input System • Temperature Indicating Switch • Test Information Sheet • Test Instruction Set • Test Instrumentation System • Test Interface Summary • Test Interface System • The Information Society • Thermal Imaging Scanner • Thermal Imaging System • Thermal Insulation System • Thematic Information System • Thursday Island (Australia-QLD) (location identifier) • Time In Service • Total Information System • Tracking and Injection Station • Tracking Instrumentation System

TISA: Technical Information Support Activity • Technique for Interactive Systems Analysis

TISAB: Total Ionic Strength Adjustment Buffer

TISAP: Technical Information Support Activity Project

TiSAT: Time Series Analysis Tool

TISB: Technical Information Services Branch

TISE: Technical Information Service Extension

TISI: Thai Industrial Standards Institute (Thailand) • Thermal Infrared Spectral Index

TISN: Tokyo International Science Network

TISO: Technical Information Services Office • Terrestrial Impact of Solar Output • Tropical Intraseasonal Oscillation

TISP: Teacher In Space Program

TISS: Terrestrial Impacts and Spherules Symposium

TISSEC: Transactions on Information and Systems Security (ACM)

TISTR: Thailand Institute for Scientific and Technical Research

Tít: Título

Tit: Titel • Title • Titre • Titular • Titulaire

TIT: Tecnologias da Informação e Telecomunicações • Thermal Inactivation Time • Thermoinsulation Technique • Total Insertion Time • Turbine Inlet Temperature

Titr: Titration

TIU: Tape Identification Unit • Technical Information Unit • Terminal Interface Unit • Timaru (New Zealand) (location identifier) • Time Interval Unit • Time Isolation Unit • Time In View

TIV: Tableau Indicateur de Vitesse • Torus Interface Valve

TIVA: Transmission d'Images par Voie Acoustique

TIW: Tacoma (USA-WA) (location identifier) • Topic Index Word

TIWE: Tropical Instability Wave Experiment

TIWG: Technical Interface Working Group • Test Integration Working Group

TIX: Titusville Space Center (USA-FL) (location identifier)

TIY: Tidjikja (Mauritania) (location identifier)

TIZ: Tari (Papua New Guinea) (location identifier)

T/j: Tonne par Jour

T/J: Turbojet

TJ: Cameroon (international aircraft registration code) • Tajik • Tajikistan (ISO country code) • Technical Journal • Telephone Jack • Temperature Junction • Terajoule • Test Jack • Thermal Junction • Tower Jettison • Trajectoire • Transmetteur de Jaugeur • Turbojet

TJA: Tarija (Bolivia) (location identifier)

TJAu: Telecommunication Journal of Australia (periodical code)

TJB: Tanjung Balai (Indonesia) (location identifier)

TJC: Tyler Junior College (USA-TX)

TJCM: Télescope James Clerk Maxwell

TJG: Tanjung Warukin (Indonesia) (location identifier)

TJH: Toyooka (Japan) (location identifier)

TJI: Tonneau de Jauge International • Trujillo Capiro (Honduras) (location identifier)

TJJPT: Tuijin Jishu Journal of Propulsion Technology (periodical code)

TJK: Tajikistan (ISO country code)

TJM: Tower Jettison Motor • Tyumen (Russia) (location identifier)

TJNAF: Thomas Jefferson National Accelerator Facility (USA-VA) (formerly CEBAF)

TJQ: Tanjung Pandan (Indonesia) (location identifier)

Tjrs: Toujours

TJS: Tanjung Selor (Indonesia) (location identifier)

TJSA: Tribal Jurisdiction Statistical Area

TJV: Thanjavur (India) (location identifier)

tk: Turkmen (ISO language code)

Tk: Taka (currency) • Teluk • Telukan • Track

TK: Tank • Teknisk Kommitté • Tokelau (ISO country code) • Tool Kit • Tonne-Kilomètre • Track • Turkmenistan

TKA: Talkeetna (USA-AK) (location identifier)

TKB: Tekadu (Papua New Guinea) (location identifier)

TKBS: Token Bus Controller

TKC: Tiko (Cameroon) (location identifier)

TKD: Takoradi (Ghana) (location identifier)

TKE: Tenakee Springs (USA-AK) (location identifier) • Terminology and Knowledge Engineering • Thermodynamic Kinetic Equilibrium • Track Error • Turbulence Kinetic Energy • Turbulent Kinetic Energy (*)

TKF: Take-off • Truckee (USA-CA) (location identifier)

TKG: Bandar Lampung (Indonesia) (location identifier)

TKH: Takhli (Thailand) (location identifier)

TKI: Tokeen (USA-AK) (location identifier) • Trail Kit Installation

TKJ: Tok (USA-AK) (location identifier)

TKK: Truk (Caroline Islands) (location identifier)

TKL: Taku Lodge (USA-AK) (location identifier) • Tokelau (ISO country code)

TKM: Tikal - El Peten (Guatemala) (location identifier) • Turkmenistan (ISO country code)

TKN: Tokunoshima (Japan) (location identifier)

TKO: Technical Knockout • Tlokoeng (Lesotho) (location identifier)

Tkof: Take-off

TKP: Takapoto (Tuamotu Islands) (location identifier)

TKQ: Kigoma (Tanzania) (location identifier)

TKR: Terrestrial Kilometric Radiation • Thakurgaon (Bangladesh) (location identifier)

Tks: Thanks

TKS: Tokushima (Japan) (location identifier)

TKSC: Tsukuba Space Center (Japan)

Tkt: Ticket

TKT: Tak (Thailand) (location identifier)

TKTS: Thermodynamic Kelvin Temperature Scale

TKU: Turku (Finland) (location identifier)

TKV: Tatatoko (Tuamotu Islands) (location identifier)

TKW: Tekin (Papua New Guinea) (location identifier)

TKX: Takaroa (Tuamotu Islands) (location identifier)

TKY: Turkey Creek (Australia-WA) (location identifier)

TKZ: Tokoroa (New Zealand) (location identifier)

tl: Tagalog (ISO language code) • Thunderstorm

Tl: Teil • Thallium

T/L: Talk/Listen • Timeline • Time Loan • Total Loss • Trace by Lot • Transformer Load

TL: Central African Rep. (international aircraft registration code) • Tableau Lumineux • Table de Lancement • Table de Logarithmes • Talk and Listen • Talk-Listen • Tape Library • Target Language • Task Leader • Task List • Taxe Locale • Taxi Lane • Team Leader • Technical Letter • Technical Library • Telecommunication Laboratories • Temps Local • Ten Laatste • Terminological Logic • Terminological Logics • Test Laboratory • Test Link • Test Load • Tête de Lecture • Thermoluminescence • Thermoluminescent • Thrust Level • Thrust Line • Thrust Link • Time Latest • Time Length • Time Limit • Timeline • Tired Light • Title List • Total Length • Total Load • Total Loss • Total Luminescence • Tour de Lancement • Trace by Lot • Trade List • Transformée de Laplace • Transformer Load • Transient Load • Transition Level • Transmission Level • Transmission Line • Transmittal Letter • Treuil de Levage • Tube de Lancement • Turbine Libre • Turkish Lira (currency) • Until

TLA: Tableau Lumineux des Alarmes • Teller (USA-AK) (location identifier) • Thin-Layer Activation • Three-Letter Abbreviation • Three-Letter Acronym • Time Line Analysis • Top-Level Aggregator • Transmission Line Adapter • Transmission Line Assembly

TLAP: Token Ring Link Access Protocol

TLAX: Tlaxcala (Mexico)

TLB: Tarbela (Pakistan) (location identifier) • Translation Lookaside Buffer (*) • Transmission Lookaside Buffer

TLC: Talker-Listener-Controller • Tangent Latitude Computer • Teacher-Librarian Certificate • Telecommand • Télécommande • Telemetry Limit Check • Thin-Layer Chromatography • Time Line Controller • Toluca (Mexico) (location identifier) • Translunar Coast

TLCA: Tangent Latitude Computer Amplifier

TLCE: Transmission Line Conditioning Equipment

TLCF: Team Leader Computing Facility

TLCN: Test Logic and Constraint Network

TLCSC: Top-Level Computer Software Component

TLD: Target Level Debugger • Thermal Luminescence Detection • Thermal Luminescence Detector • Thermoluminescence Dosimeter • Thermoluminescent Device • Thermoluminescent Dosimeter • Thermoluminescent Dosimetry • Top Level Domain • Tuli Lodge (Botswana) (location identifier)

Tle: Teile

TLE: Technical Liaison Engineer • Temperature-Limited Emission • Temporal Lobe Epilepsy • Thin-Layer Electrode • Thin-Layer Electrophoresis • Thin Leading Edge • Top Loading Electrical • Total Lunar Eclipse • Tower Lighting Equipment • Tracking Light Electronics • Transfert Linéaire d'Énergie • Treaty-Limited Equipment • Tulear (Madagascar) (location identifier) • Two-Line Element

TLF: Telida (USA-AK) (location identifier) • Terminal Launch Facility • Time Line Form • Trésors de la Langue Française

Tlg: Telegraph

TLG: Tail Landing Gear • Tulagi Island (Solomon Islands) (location identifier)

TLH: Tallahassee (USA-FL) (location identifier)

TLI: Telephone Line Interface • Terrain Line Interpolation • Thermionics Laboratory Inc. (USA-CA) • Time-Life International • Tolitoli (Indonesia) (location identifier) • Translunar Injection • Transport Layer Interface (*) • Transport Level Interface

TLIEF: Thin-Layer Isoelectric Focusing

TLJ: Tatalina (USA-AK) (location identifier)

TLK: Talknafjördur (Iceland) (location identifier) • Terrace-Ledge-Kink

TLL: Tallinn Ulemiste (Estonia) (location identifier) • Telelink • Television Laser Link

TlLd: Total Load

TLM: Technical Liaison Memo • Telemeter • Télémesure • Télémètre • Telemetry • Télémétrie • Tlemcen (Algeria) (location identifier) • Top Loading into Mixture • Topographic Line Map • Transmission Line Modelling • Trouble Locating Manual

TLMS: Tape Library Management System

Tlmy: Telemetry

TLN: Toulon Hyères (France) (location identifier) • Toulon - Le Palyvestre (France) (location identifier) • Trunk Line Network

TLO: Technical Liaison Officer • Terminal Learning Objective • The Last One • Tol (Papua New Guinea) (location identifier) • Total Loss Only

TLOTOS: Timed Language Of Temporal Ordering Specification

TLP: Technology Leadership Programme • Telescope Local Processor • Transient Lunar Phenomenon • Transmission Level Point • Transparent Language Processing • Transport Layer Connection • Triangular Libration Point • Trouble Location Problem • Tumolbil (Papua New Guinea) (location identifier)

TLR: Tape Loop Recorder • Trailer • Tulare (USA-CA) (location identifier) • Twin Lens Reflex

TLRP: Track Last Reference Position

TLRS: Transportable Laser Ranging System

TLS: Technical Library Service • Terminal Landing System • Thailand Landsat Station • Thüringer Landessternwarte (Germany) • Time Limited Signal • Time Line Sheet • Top Left Side • Total Least Squares • Total Luminescence Spectroscopy • Toulouse Blagnac (France) (location identifier) • Training Launch Station • Transponder Landing System • Transport Layer Security • Two-Level System

TLSA: Torso Limb Suit Assembly

Tlscp: Telescope

TLSCP: Telescope

TLSP: Transport Layer Security Protocol

TLSS: Technical Library Services Section • Texas Large-scale Structure Survey

TLT: Trademark Law Treaty • Tuluksak (USA-AK) (location identifier)

TLTP: Teaching and Learning Technology Program • Télétype

Tltr: Translator

TLTSN: Teaching and Learning Technology Support Network (UK)

TLU: Table Lookup • Texas Lutheran University (USA-TX) • Threshold Logic Unit • Time of Last Update • Tolu (Colombia) (location identifier) • Transportable Laser Unit

TLv: Transition Level

TLV: Target Launch Vehicle • Tel Aviv Yafo - Ben Gurion (Israel) (location identifier) • Threshold Limit Value • Transporter-Loader Vehicle

TLVO: Type Length Value Object

tlw: teilweise

TLW: Talasea (Papua New Guinea) (location identifier)

TLX: Talca (Chile) (location identifier) • Telex

TLZ: Catalão (Brazil-GO) (location identifier) • Titanium-Lead-Zinc • Transfer on Less than Zero

t/m: Tot en Met

tm: Telemetry • Time • Trademark

Tm: Thulium • Tonelada Métrica

T&M: Time & Materials

T/M: Trace by Member

TM: Table de Mortalité • Table de Multiplication • Table Maintenance • Talking Machine • Tambour Magnétique • Tangent Mechanism • Tape Mark • Tape Module • Task Management • Task Master • Taux de Modulation • Team Member • Technical Management • Technical Manager • Technical Manual • Technical Memo • Technical Memorandum • Technical Minutes • Technical Monograph • Technicien de la Météorologie • Technology Mission • Telemeter • Telemetered • Télémètre • Télémétrie • Telemetry • Télémétrie • Telescope Maker • Telescope Model • Teleskopischer Meteor • Temperature Meter • Temperature Monitor • Température Moyenne • Template Matching • Temps de Maintien • Temps de Mission • Temps de Montée • Temps Moyen • Terminal Man • Terminological Model • Terminological Module • Territory Manager • Test Maintenance • Test Manual • Test Message • Test Model • Tête Motrice • Thematic Mapper • Thermal Mapper • Time and Materials • Time Management • Time Mission • Time Monitor • Time of Maximum • Tone Modulation • Tonne Métrique • Top Management • Torque Motor • Tot en Met • Tour de Montage • Trace by Member • Tracer Model • Trade Mark • Traffic Manager • Traffic Model • Training Manual • Transaction Manager • Transcendental Meditation • Transfer Memorandum • Transition Metal • Transmission and Multiplexing • Transverse Magnetic • True Mean • Turing Machine • Turkmenistan (ISO country code)

TMA: Target Motion Analysis • Technology Mirror Assembly • Telecommunications Management Association (USA) • Telescope Makers Association • Tense Mood Aspect • Thermomechanical Analysis • Thermomechanical Analyzer • Three Mirror Anastigmat • Tierce Maintenance Applicative • Tifton - Henry Tift Myers (USA-GA) (location identifier) • Time-Modulated Antenna • Traffic Management Advisor • Tycho Magnetic Anomaly

TMAC: Technical Mapping Advisory Council (USA)

TMAG: Travel More Advantageous to the Government

TMB: Miami - Tamiami (USA-FL) (location identifier) • Task Maintenance Burden • Technical Management Board • Transient Microwave Brightening • Transportation Management Bulletin

TMBF: Temps Moyen de Bon Fonctionnement

TMBPS: Thematic Mapper Bulk Processing System

Tmbr: Timber

TMBU: Table Maintenance Block Update

TMC: Tambolaka (Indonesia) (location identifier) • Tape Management Catalog • Task Management Component • Taurus Molecular Cloud • Technical Management Committee • Technology Management Consultants • Test Mass Chamber • Test Monitoring Console • The Map Collector (journal) • Thinking Machines Corporation • Third Mediterranean Country • Thrust Magnitude Control • Titanium Matrix Composite • Total Manufacturing Cost • Traffic Management Center • Traffic Management Coordinator • Traffic Message Channel • Triple Magnetic Configuration • Triple Molecular Collision

TMCA: Thrust Management Control Analysis

TMCC: Traffic Management Computer Complex

Tmd: Timed

TMD: Thematic Mapper Derivative • Theoretical Maximum Density • Timbedra (Morocco) (location identifier)

TMDE: Test - Measurement - Diagnostic Equipment

TMDF: Trunk Main Distribution Frame

TMDT: Total Mean Downtime

Tme: Time

TME: Tame (Colombia) (location identifier) • Test and Measurement Equipment • Test Maintenance Equipment • Tivoli Management Environment • Tokyo-Mitaka catalogue of Equatorial stars • Tropical Mountain Ecosystem

TMEC: Tempo Medio dell'Europa Centrale • Temps Moyen de l'Europe Centrale

TMECO: Time of Main Engine Cuttoff

TMED: Temps Moyen Entre Défaillances

TMEP: Temps Moyen Entre Pannes

TMER: Temps Moyen Entre Révisions

TMF: Telemetry Module Facility • Time Marker Frequency • Traffic Management Facility • Training Mission Facility • Transporter Maintenance Facility

TMFG: Telemetry Format Generator

TmFl: Time of Flight

Tmg: Timing

TMG: Temps Moyen de Greenwich • Thermal Meteoroid Garment • Thermal Model Generator • Time Mark Generator • Tomanggong (Malaysia) (location identifier) • Traffic Management

TMGE: Thermomagnetic-Galvanic Effect

T/MGS: Transportable Mobile Ground Station

TMGS: Terrestrial Magnetic Guidance System • Transportable Mobile Ground Station • Two Micron Galactic Survey

TMH: Tanahmerah (Indonesia) (location identifier)

TMI: Technical Management Item • Technical Measurements Inc. • Technical Meetings Information • Telesat Mobile Inc. • Test Method Instruction • Three Mile Island (USA) • Trans-Mars Injection • TRMM Microwave Imager • Tumling Tar (Nepal) (location identifier)

TMIC: Technical Meetings Information Center

TMIS: Technical Management Information System • Technical Meetings Information Service • Telecommunications Management Information System • Total Management Information System

TMJ: Temporomandibular Joint • Termez (Uzbekistan) (location identifier)

TMK: Tamky (Viet Nam) (location identifier)

TMKF: Technische Mitteilungen Krupp Forschungsberichte (periodical code)

Tmkpr: Timekeeper

Tml: Terminal

TML: Tamale (Ghana) (location identifier) • Television Microwave Link • Temps Moyen Local • Terminal • Tetramethyllead • TEX Macro Language • Three-Mile Limit • Total Mass Loss

TMLI: Television Microwave Link Indicator

TMLR: Television Microwave Link Repeater

TMLT: Television Microwave Link Terminal

TMM: Manat (Turkmenistan) (ISO currency code) • Tamatave (Madagascar) (location identifier) • Tape Mount Management • Télescope du Mont Mégantic (Canada-QC) • Test Message Monitor • Theory of Moon's Motion • Thermal Mathematical Model • Thermal Math Model • Total Mass Loss

TMMP: Technical Manual Management Program

TMMS: Telephone Message Management System

TMN: Tamana Island (Kiribati) (location identifier) • Tatebayashi-Matsuzaki-Newman • Telecommunications Management Network • Time Management Networking • Tokyo Meteor Network (Japan) • Transmission

TM&O: Telecommunications Management & Operations

TMO: Table Mountain Observatory (USA-CA) • Tabor Mountain Observatory (Canada-BC) • Target Mount • Technique Magnéto-Optique • Telecommunications Management and Operations • Telegraphic Money Order • Télémesure Opérationnelle • Telephone Money Order • Température Maximale Opérationnelle • Test Management Order • Test Manufacturing Order • Tool Manufacturing Order • Tumeremo (Venezuela) (location identifier)

TMOD: Telecommunications and Mission Operations Directorate

TMOP: Transitional Meteosat Operational Programme

TMP: East Timor (ISO country code) • Tampere (Finland) (location identifier) • Taux Moyen Pondéré • Technical Management Plan • Temperature • Temporary • Terminal Monitor Program • Test Maintenance Panel • Test Methods and Procedures • Theodolite Measuring Point • Thermal Modelling Program • Thermomechanical Processing • Thermomechanical Pump • Time Management Processor • Timor (East_) (ISO country code) • Timor Orientale (ISO country code) • Traffic Management Processor • Transistor Mounting Pad • Transportable Measurement Package • Transversely Magnetized Plasma • Turbomolecular Pump

TMPA: Traffic Management Program Alert

Tmpl: Template

TMPL: Temps Moyen de Préparation au Lancement

Tmpry: Temporarily • Temporary

TMPRY: Temporary

TMPS: Temperature Monitoring Power Supply • Test Maintenance Panel Subassembly

TMPSU: Telescope Module Power Supply Unit

TMPV: Torque Motor Pilot Valve

TMQ: Tambao (Burkina Faso) (location identifier)

Tmr: Timer

TMR: Tamanrasset (Algeria) (location identifier) • Technical Memorandum Report • Télémesure Rapide • Temperature Minimum Region • Time Meter Reading • Timer • TOPEX Microwave Radiometer • Training and Mobility of Researchers • Transient Memory Record • Triple Modular Redundancy

TMRAO: Table Mountain Radio Astronomy Observatory (USA-CA)

TMRF: Technical Meeting Request Form

TMRS: Traffic Measurement and Recording System • Traffic Metering Remote System

TMS: Sao Tome (location identifier) • Tanzanian Meteorological Society (Tanzania) • Tape Management System • Target Monitoring System • Task Management System • Technical Management System • Telecommunications and Microgravity Section • Telemarketing Management System • Telemetry Modulation System • Telemetry Multiplexer System • Telemobiloskop • Teleoperator Maneuvering System • Telescope Management System • Temperature Management Station • Test and Measurement Station • Test and Monitoring Station • Test and Monitor Station • Test Monitor System • Tetramethylsilane • Thematic Mapper Simulator • Thermal Maneuvering System • Thermomechanical System • Thrust Measuring System • Time-Multiplexed Switch • Topographic Mapping Service • Tracing Management System • Traffic Management Specialist • Traffic Management System • Traffic Monitoring System • Transcranial Magnetic Simulation • Trend Monitoring System • Truth Maintenance System • Turbulence Measuring System • Twist Monitoring System • Two-Micron Survey

TMSC: Talcott Mountain Science Center (USA-CT)

TMSI: Temporary Mobile Subscriber Identity

TMSL: Test and Measurement Systems Language

TMSP: Table Mountain Star Party

TMSPA: Table Mountain Star Party Association (USA-WA)

TMSS: Technical Manuals Specifications and Standards • Two-Micron Sky Survey

TMT: Technique Moderne de Transformation • Télémesure Technologique • Terminal Monitor Program • Thermal Measurement Treatment • Trombetas (Brazil-PA) (location identifier) • Turbo Motor Train

TMTE: Telemetry Test Equipment

TMTR: Thermistor • Transmetteur

TMTS: Thematic Mapper Transcription System

TMTSF: Tetramethyltetraselenofulvalene

TMU: Tambor (Costa Rica) (location identifier) • Temperature Measurement Unit • Test Maintenance Unit • Thermal-Mechanical Unit • Time Measurement Unit • Tokyo Metropolitan University (Japan) • Traffic Management Unit • Transmission Message Unit

TMUNER: Tokyo Metropolitan University Network for Education and Research (Japan)

TMV: Temps Moyen de Vol • True Mean Value • Tube à Modulation de Vitesse

TMW: Tamworth (Australia-NSW) (location identifier) • To-morrow • Transverse Magnetic Wave

TMX: Timimoun (Algeria) (location identifier)

TMXRT: Three-Mirror X-Ray Telescope

TMY: Tiom (Indonesia) (location identifier)

TMZ: Thames (New Zealand) (location identifier)

tn: Setswana (ISO language code) • Tunisia

Tn: Thoron • Ton • Tonne • Town • Train • Transportation • Ton • Tonne • Town • Train • Transportation

TN: Congo (international aircraft registration code) • Technical Note • Telephone Number • Telnet • Tempérament Nerveux • Température Nocturne • Température Normale • Tennessee (USA) • Tension Normale • Test Number • Tête Nucléaire • Thermonuclear • Tonbridge (UK postal code) • Tonnage Net • Trade Name • Traineeship Number • Transaction Number • Transfer on Negative • Translator's Note • Trou Noir • True Negative • True North • Tunisia (international vehicle registration code) • Tunisia (ISO country code) • Tunisie (ISO country code) • Twisted Nematic

TNA: Jinan (China-PRC) (location identifier) • Telematics Network between Administrations • Télémètre de Nuages Analogique • Thermal Neutron Activation • Thermal Neutron Analysis

TNAU: Tamil Nadu Agricultural University (India)

TNB: Tanah Grogot (Indonesia) (location identifier) • Terra Nova Bay • Trinitrobenzene

TNBS: Taiwan National Bureau of Standards (China-ROC)

TNC: Tekniska Nomenklaturcentralen • Terminal Node Controller • Tin City (USA-AK) (location identifier) • Too Numerous to Count • Total Numerical Control • Trade Negotiating Committee • Transnational Corporation

TND: Dinar (Tunisia) (ISO currency code) • Telecommunications Network for the Deaf • The Next Definition • The Next Dimension • Trinidad (Cuba) (location identifier)

TNDCY: Tendency

TNDS: Total Network Data System

TNDT: Thermal Non-Destructive Testing

TNE: Tanega Shima (Japan) (location identifier)

TNEF: Transport Neutral Encapsulation Format

TNF: Third Normal Form • Timing Negative Film • Toussus-le-Noble (France) (location identifier)

Tng: Tongue • Training • Turning

TNG: Tanger Boukhalef (Morocco) (location identifier) • Telescopio Nazionale Galileo (Italy) • The Next Generation

TNGT: Tonight

TNH: Temperate Northern Hemisphere • Tonghua (China-PRC) (location identifier)

TNI: Satna (India) (location identifier) • Tennelec/Nucleus Inc. • Trusted Network Interpretation

TNJ: Tanjung Pinang (Indonesia) (location identifier)

Tnk: Trunk

TNK: Tununak (USA-AK) (location identifier)

Tnl: Tunnel

TNL: Technological Research Programme • Télémètre de Nuages à Laser • Terminaison Numérique de Ligne • Terminal Net Loss • Ternopol (Ukraine) (location identifier) • Traitement en Langage Naturel • Tunnel

TNM: Teniente R. Marsh (Antarctica) (location identifier) • Tetranitromethane

TNMR: Tritium Nuclear Magnetic Resonance

TNN: Tainan (China-ROC) (location identifier) • Transport Network Node • Trunk Network Number

TNO: Toegepast-Natuurwetenschappelijk Onderzoek (Netherlands) • Tamarindo (Costa Rica) (location identifier) • Transnational Operation • Trans-Neptunian Object

TNOP: Total Network Operation Plan

TNP: Traité de Non-Prolifération • Transportation Network Profile • Twentynine Palms (USA-CA) (location identifier)

TNPC: Taiwan New PC Consortium (China-ROC) • Traffic Network Planning Center

Tnpk: Turnpike

TNR: Antananarivo (Madagascar) (location identifier) • Terminaison Numérique de Réseau • Thermonuclear Runaway • Transit Nuclear Radiation

TNRE: Transit Nuclear Radiation Effect

TNRIS: Texas Natural Resources Information System

TNRSP: Thailand National Remote Sensing Programme

TNS: The Next Step • Tungsten (Canada-NT) (location identifier)

TNSC: Tanegashima National Space Center (Japan) • Trailside Nature and Science Center (USA-NJ)

TNSE: Total Normalized Stress Error

Tnsn: Tension

Tnt: Torrent

TNT: Miami - Dade Collier (USA-FL) (location identifier) • Theoretical Nose Tip • Tightened-Normal-Tightened • Trinitrotoluene • Typographical Number Theory

TNTC: Too Numerous To Count

TNTCAM: Ten and Twenty Micron Camera

TNTDL: Tabulated Numerical Technical Data List

Tntv: Tentative

TNU: Newton (USA-IA) (location identifier)

TNV: Ten name Van • Total Net Value • Total Network Visibility

TNX: Stung Treng (Cambodia) (location identifier) • Thanks

t&o: Taken & Offered

to: Tonga (ISO language code)

To: Togo

T/O: Table of Organization • Takeoff • Takeover • Target of Opportunity • Turnover

T&O: Test & Operations

TO: Table of Organization • Takeoff • Takeover • Target of Opportunity • Task Order • Technical Observer • Technical Officer • Technical Operation • Technical Order • Technical Office • Technical Officer • Technical Order • Telecommunication Operator • Telegraphic Order • Télégraphie Optique • Telegraph Office • Téléobjectif • Telephone Office • Telephone Order • Telescope Operation • Telescope Operator • Température Optimale • Teraoctet • Terminal Ordinaire • Term-Ordered • Test and Operations • Thermo-Optic • Tiempo Official • Timeout • Tocantins (Brazil) • Togo • Tonga (ISO country code) • Tonnage Officiel • Tool Order • Torsional Oscillation • Train d'Ondes • Trained Operator • Transfer Optics • Transfer Order • Transmitter Oscillator • Transporteur Orbital • Transport Officer • Transverse-Optical • Transverse Optics • Travel Order • Turnoff • Turnout • Turnover • Tuyère Orientable

TOA: Tarifordnung für Angestellte • Time Of Arrival • Top Of the Atmosphere • Torrance (USA-CA) (location identifier) • Tropical Ocean-Atmosphere

TOAD: Tremendous Outburst Amplitude Dwarf

TOADS: Transportable Optical Atmospheric Data System

TOAE: Terre-Océan-Air-Espace • Terre-Océan-Atmosphère-Espace (*)

TOAR: Top Of Atmosphere Radiation

TOAST: The Orpington Astronomical Society Times

TOASTE: Transport of Ozone and Stratosphere/Troposphere Exchange

TOB: Telemetry Output Buffer • Temporary Office Building • Tobruk (Libya) (location identifier)

ToC: Table of Contents • Time of Coincidence

TOC: Table Of Contents • Tactical Operations Center • Technical Order Compliance • Telescope Operations Control • Telescope Operations Crew • Test Operations Center • Test Operations Change • Time Of Coincidence • Toccoa (USA-GA) (location identifier) • Total Organic Carbon • Total Organic-carbon Content • Total Ozone Content • Training Observation Center • Transfer Of Control • Trouble Obsessionnel Convulsif • Trousse d'Outils de Communication

TOCC: TDRS Operations Control Center • Telescope Operations Control Center • Test Operations Control Center

TOCE: Tonnes Of Carbon Equivalent

TOCHI: Transactions On Computer Human Interaction (ACM)

TOCL: Transactions On Computational Logic (ACM)

TOCOTOX: Too Complicated To Explain

TOCS: Terminal Operations Control System • Terminal-Oriented Computer System • Term-Oriented Classification System • Transactions On Computer Systems (ACM)

TOD: Technical Operations Department • Technical Order Distribution • Test Operations Directorate • Time Of Day • Time Of Delivery • Time Of Departure • Time-Ordered Data • Tioman (Malaysia) (location identifier) • TIROS Oceanographic Drifter • Top Of Descent • Total Oxygen Demand • Tour Of Duty • Triangle Orientation Dissemination • Turnover Device

TODA: Takeoff Distance Available

TODAES: Transactions on Design Automation of Electronic Systems

TODARS: Terminal-Oriented Data Analysis and Retrieval System

TODCAR: Two-Dimensional Correlation

TODS: Transaction On Database Systems (ACM)

TOE: Theory Of Everything • Ton Oil Equivalent • Tozeur (Tunisia) (location identifier)

TOEFL: Testing Of English as a Foreign Language • Test Of English as a Foreign Language

TOEIC: Test Of English for International Communication

TOET: Test Of Elementary Training

TOF: Test Operations Facility • Theoretical Orthogonal Function • Time Of Flight • Tomsk (Russia) (location identifier) • To Order From • Top Of Form

TOFMS: Time-Of-Flight Mass Spectrometer • Time-Of-Flight Mass Spectrometry • Time-Of-Flight Mass Spectroscopy

TOFS: Time-Of-Flight Spectrometer • Time-Of-Flight Spectrometry

ToFU: To FITS Utilities

Tog: Together

TOG: Toggle • Togiak Village (USA-AK) (location identifier) • Transactions On Graphics (ACM)

TOGA: Takeoff/Go-Around • Tropical Ocean Global Atmosphere • Tropical Oceans and the Global Atmosphere (*) • Tropical Oceans - Global Atmosphere

Togr: Together

TOGW: Take Off Gross Weight

TOH: Torres (Vanuatu) (location identifier)

TOI: Technical Operation Instruction • Transfert Orbit Insertion • Troy (USA-AL) (location identifier)

TOIS: Transactions On Information Systems (ACM)

TOJ: Madrid Torrejón (Spain) (location identifier)

Tok: Tokyo (Japan)

TOK: Torokina (Papua New Guinea) (location identifier)

TokAB: Tokyo Astronomical Bulletin (periodical code)

Tol: Toledo (Spain) • Tolerance

TOL: Temporary Occupation Licence • Test-Oriented Language • Toledo Express (USA-OH) (location identifier) • Tolerance

TOLIP: Trajectory Optimization and Linearized Pitch

TOLIS: Total Online Information System

Tom: Tomus

TOM: Target Object Map • Task Ordering Management • Territoire d'Outre-Mer (France) • Thermal Optical Modelling • Tilt-Over Mode • Time of Orbital Midnight • Tombouctou (Mali) (location identifier) • Total Organic Matter • Trunk Operations Monitor

TOMA: Telescope Optics Manufacturers' Association

TOMACS: Transactions On Modelling And Computer Graphics (ACM)

TOMCAT: Telemetry Online Monitoring Compression And Transmission • Toulouse Off-line Model of Chemistry And Transport

TOMIS: Task-Oriented Multipurpose Information System

TOMS: Total Ozone Mapping Spectrometer (*) • Total Ozone Mapping System • Transactions On Mathematical Software (ACM)

Ton: Tonelaje • Tonne

TON: Taiwan Oscillations Network • Telescope Operations News • Test Operation Notice • Tonga (ISO country code) • Tonu (Papua New Guinea) (location identifier) • Total Organic Nitrogen • Transactions on Networking (ACM)

Tonn: Tonnage

TONS: TDRSS Onboard Navigation System

ToO: Target of Opportunity

TOO: Target Of Opportunity • Target of Opportunity Observation • Test Operations Order • Thousand Oaks Optical (USA-MI) • Threshold Of Odor • Time Of Origin • To Order Only

TOOIS: Transactions On Office Information Systems (ACM)

TOOL: Test-Oriented Operated Language

TOOLS: Technology for Object-Oriented Linking and Sharing

Top: Topographic • Topological • Topology

TOP: Pa'anga (Tonga) (ISO currency code) • Tactical Operations Plan • Take Or Pay • Technical and Office Protocol • Technical Office Protocol • Technical Operating Procedure • Technical Operations Procedure • Technology Opportunities Program (formerly TIIAP) • Telecoms Operating Protocol • Temporarily Out of Print • Terrestrial Observation Panel • Tertiary Operation • Test and Operations Plan • Test Operational Plan • Thunderstorm Overflight Program • Time-averaged Orbiting Potential • Topeka (USA-KS) (location identifier) • Tracker Optical Package • Transfer Orbit Phase • Tube à Ondes Progressives • Turbine Outlet Pressure

TOPAS: Terra Observation Pattern Analysis System • Tropospheric Optical Absorption Spectroscopy

TOPEX: Topographic Experiment • (ocean) Topography Experiment • Topography Explorer • Typhoon Operational Experiment

TOPIAS: Topographic Information and Archiving Software

TOPIC: Tactical Operations and Increment Centre • Teletext Output Price Information Computer • Time-Ordered Programmer Integrated Circuit

TOPLAS: Transactions On Programming Languages And Systems (ACM)

TOPM: Takeoff Performance Monitor

TOPMS: Takeoff Performance Monitoring System

Topog: Topographer • Topographical • Topography

Topogr: Topografía

Topol: Topological • Topology

TOPP: Terminal-Operated Production Program • Torino Observatory Parallax Program

TOPS: Technology Opportunities Showcase • Telecommunications Orbiting and Pricing System • Telemetry Online Processing System • Terminal Operating System • Text-Oriented Processing System • Thermodynamic Ocean Prediction System • The Operational PERT System • Total Operations Processing System • Towards Observation of Planetary Systems • Towards Other Planetary Systems (*) • Transcendental Operating System • Transfer of Optical Processing to Systems • Transfer Orbit Perigee Stage • Transistorized Operational Phone System • Transition of Optical Processing to Systems • Transition of Optical Processors into Systems • Trends in Optics and Photonics (OSA)

TOPSAR: Topographic Synthetic Aperture Radar

TOPSIM: Torino Politecnico Simulator

TOPSS: Tenerife Observatory for Planetary and Stellar Science (Canary Islands)

TOQ: Table d'Observations Quotidiennes • Tocopilla Barriles (Chile) (location identifier)

Tor: Toronto (Canada-ON)

TOR: Technical Operating Report • Technical Operations Research • Technical Override • Tentative Operational Requirement • Terminal Owning Region • Terms Of Reference • Test Operation Report • Test Operating Report • Time Of Receipt • Time Of Reception • Torrington (USA-WY) (location identifier) • Torsion • Torsional • Tout Ou Rien • Tropospheric Ozone Research • Tube à Onde Régressive

TORA: Take-Off Run Available

TORES: Text Oriented Editing System

TORF: Time of Retrofire

Torn: Tornado

TORPEDO: The Optical Retrieval Project: Electronic Documents Online

TORS: Technical Order Status Report • Time-Ordered Reporting System • Torsion

TOS: Tachyon Operating System • Tape Operating System • Taux d'Ondes Stationnaires • Technical and Operational Support (*) • Technical Operational Support • Teleoperator System • Temporarily Out of Service • Temporarily Out of Stock • Terms Of Service • Test Operating System • Text Organizing System • The Oceanography Society • Time-Ordered System • TIROS Operational Satellite • Towarzystwo Obserwatorów Slonca (Poland) • Traffic Orientation System • Transfer Orbital Stage • Transfer Orbit Stage (*) • Tromsø (Norway) (location identifier) • Type Of Service

TOSCA: Text and Office Systems Color Architecture • Text-Oriented System Controller for ALIS

TOSCANE: Travaux d'Océanographie Spatiale dans l'Atlantique Nord-Est

TOSEM: Transactions On Software Engineering and Methodology (ACM)

TOSP: Taux d'Ondes Stationnaires en Puissance

TosRv: Toshiba Review (periodical code)

TOSS: Test Operation Support Segment • TIROS Operational Satellite System

TOST: Taux d'Ondes Stationnaires en Tension

Tot: Total

TOT: Telescope Optical Throughput • Time Of Transmission • Time On Target • Time Over Target • Total • Total Operating Cost • Totness Coronie (Suriname) (location identifier) • Transfer Of Technology • Triumph Over Tragedy • Turbine Outlet Temperature • Turn-On Time

TOTO: Totable Tornado Observatory

TOTOVS: Total Ozone from TOVS

TOTS: Transportable Orbital Tracking Station

Tou: Toulon (France)

TOU: Touho (New Caledonia) (location identifier)

Tour: Tourism • Tourist

TOUS: Test On Understanding Science • Transmission Oscillator Ultrasonic Spectrometer

TOV: Telemetering Oscillator Voltage • Ten Opzichte Van • Ten Overstaan Van • Tolbert-Oppenheimer-Volkoff • Tortola West End (British Virgin Islands) (location identifier)

TOVS: TIROS Operational Vertical Sounder

TOW: Tank and Orbiter Weight • Time Of Week • Toledo (Brazil-PR) (location identifier)

TOWARD: Tower Ocean Wave and Radar Dependence

Tox: Toxicology

TOX: Tobolsk (Russia) (location identifier)

TOY: Time Of Year • Toyama (Japan) (location identifier)

TOYal: Trans. Astron. Obs. Yale Univ. (periodical code)

TOZ: Touba (Ivory Coast) (location identifier)

t&p: Theft & Pilferage

Tp: Telephone • Township • Troop • Type

T/P: TOPEX/Poseidon

T&P: Tracking & Processing

TP: East Timor (ISO country code) • Tape Punch • Target Practice • Taux de Pannes • Taux de Pression • Taxe Professionnelle • Taxpayer • Taylor Polynomial • Teaching Practice • Technical Pamphlet • Technical Paper • Technical Performance • Technical Preprint • Technical Problem • Technical Proposal • Technical Publication • Technographic Publication • Technology Parameter • Telemetry Processor • Telephone • Teleprinter • Teleprocessing • Telescope Project • Temperature Pod • Temperature Probe • Temperature Profile • Tempo Primo • Temps de Partage • Temps de Préparation • Temps de Propagation • Temps Partiel • Temps Passé • Tension de Polarisation • Test Panel • Test Plan • Test Point • Test Port • Test Position • Test Pressure • Test Procedure • Text Processing • Theorem Proving • Thermally Pulsing • Thermal Pulse • Third Party • Time Pulse • Time to Perigee • Timor (East_) (ISO country code) • Timor Orientale (ISO country code) • Title Page • Tool Permit • To Pay • Top Priority • Totally Positive • Total Pressure • Tout Payé • Tracking and Processing • Tracking Program • Traînée de Pression • Traînée de Profil • Training Parameter • Training Plan • Traité de Paix • Traitement Préventif • Trajectographie Passive • Transaction Processing • Transformation Procedure • Transition Period • Translation Process • Transpondeur • Transport Protocol • Transputer • Travail Pratique • Travaux Pratiques • Travaux Publics • Trésor Public (France) • Trigonometric Point • Triple Pole • Tropical Pacific • True Position • True Positive • Turbopompe • Turbopropulseur • Turbo Pump • Turning Point • Twisted Pair • Type

TPA: Clearwater (USA-FL) (location identifier) • Saint Petersburg (USA-FL) (location identifier) • Tampa (USA-FL) (location identifier) • Tape Pulse Amplifier • Technical Publications Announcement • Telemetry Power Amplifier • Telemetry Processing Assembly • Temporary Program Area • Test Planning Analysis • Test Preparation Area • Thermal Plasma Analyzer • Third-Party Access • Third-Party Application • Tissue Plasminogen Activator • Traffic Pattern Altitude • Transistor Power Amplifier • Trans-Polar Arc • Tunable Parametric Amplifier • Turbopump Assembly • Two-Photon Absorption

TPAC: Telescope Precision Angle Counter

TPAD: Trunnion Pin Attachment Device

TPAGB: Thermally Pulsating Asymtotic Giant Branch (star)

TPAM: Teleprocessing Access Method

TPAS: Technical Publication Automated System • The Phoenix Astronomical Society (New Zealand)

TPAT: Test Point Algorithm Technique

TPAW: Tiger Parallel Architecture Workbench

TPB: Tape Playback Bit • Tetraphenyl-Butadiene • Theory of Planned Behavior

TPBVP: Two-Point Boundary Value Problem

TP&C: Thermal Protection & Control

TPC: Tactical Pilotage Chart • Tarapoa (Ecuador) (location identifier) • Technical Performance Council • Technical Performance Criterion • Technical Prime Contractor • Technical Processing Council • Telemetry Preprocessor Computer • The Peace Corps • Thermal Protection and Control • Time Polarity Control • Time Projection Chamber • Total Program Cost • Trade Practices Commission • Tropical Prediction Center

TPCA: Test Procedure Change Authorization

TPCC: Trade Promotion Coordinating Committee

TPCE: Tank Pressure Control Experiment

TPCF: Two-Point Correlation Function

TPCN: Task Plan Change Notice • Technical Program Change Negotiation • Test Plan Change Notice • Test Procedure Change Notice

TPCR: Task Plan Change Request

tpd: Tons per Day

Tpd: Tapped

TpD: Tons per Day

TPD: Technical Products Division • Technisch Physische Dienst • Temperature Programmed Desorption • Terminal Protection Device • Test Point Data • Test Procedure Drawing • Third-Party Developer • Time Pulse Distributor • Tons Per Day • Trailing Packet Discard • Train Principal Droit

TPDA: Tapada

TPDAD: Time Differential Perturbed Angular Distribution

TPDC: Trade Point Development Centre

TPDK: Teleform Programmer's Developer Kit

TPDS: Test Procedures Development System

TPDU: Transport Protocol Data Unit

TPE: Escudo (East Timor) (ISO currency code) • Tableau de Protection Électrique • Tableau Périodique des Éléments • Taipei - Chiang Kai Shek (China-ROC) (location identifier) • Teacher Preparation and Enhancement • Terminal de Paiement Électronique • Test Planning and Evaluation • Test Project Engineer • Threshold Photoelectron • Total Potential Energy • Tracking and Pointing Experiment • Travail Personnel Encadré • Très Petite Entreprise

TPED: Tasking-Processing-Evaluation-Dissemination

TPEG: Technical Planning and Evaluation Group

TPEP: Trusted Product Evaluation Program

TPES: Threshold Photoelectronic Spectrum

TPESP: Technical Panel on the Earth Satellite Program

TPF: Tampa - Peter O'Knight (USA-FL) (location identifier) • Technology Performance Financing • Terminal Phase Final • Terminal Phase Finalization • Terminal Phase Finish • Terrestrial Planet Finder • Thermal Phase

Finish • Tibetan Plateau Field Experiment • Transfer Phase Final • Tug Processing Facility • Two-Phase Flow

TPFE: Two-Phase Flow Experiment

TPFLEX: Two-Phase Fluid Loop Experiment

TPFO: TOPEX-Poseidon Follow-On

Tpg: Tapping

TPG: Taiping (Malaysia) (location identifier) • Test Pulse Generator • Thermal Plasma Generator • Thermionic Power Generator • Timing Pulse Generator • Topping • Train Principal Gauche • Trigger Primitive Generator

TPGID: Terminal Process Group Identification • Terminal Process Group Identificator • Terminal Process Group Identifier

tph: Tons per Hour

TPH: Tonopah (USA-NV) (location identifier) • Transmetteur de Pression Hydraulique

tpi: Teeth per Inch • Tracks per Inch • Turns per Inch

TPI: Tape Position Indicator • Tapini (Papua New Guinea) (location identifier) • Tax and Price Index • Technical Proposal Instructions • Tennessee Polytechnic Institute (USA-TN) • Terminal Phase Ignition • Terminal Phase Initiate • Terminal Phase Initiation • Terminal Phase Insertion • Test Process Improvement • Thermal Protection Investigation • Threads Per Inch • Title Page and Index • Tons Per Inch • Tracks Per Inch • Tribunal de Première Instance • Tribunal Pénal International (ONU) • Tritium Pellet Injector • Turns Per Inch

TPITS: Two-Phase Integrated Thermal System (*) • Two-Phase Integration Thermal System

TPJ: Taplejung (Nepal) (location identifier)

Tpk: Turnpike

TPK: Tapaktuan (Indonesia) (location identifier) • Turnpike

TPKE: Turnpike

Tpl: Triple

TPL: Table Producing Language • Teacher Programming Language • Technical Publications Library • Temple (USA-TX) (location identifier) • Terminal Processing Language • Test Point Logic • Trans-Pacific Landing • Tunable Pulsed Laser

TPLS: Terminal Position Location System • Tunable Pulsed Laser System

tpm: Tons per Minute • Tours par Minute • Transactions per Minute

TPM: Tape Preventive Maintenance • Technical Performance Measurement • Technical Performance Module • Telemetry Processor Module • Terminal Phase Maneuver • Terminal Phase Midcourse • Test Planning Manager • Third-Party Maintenance • Tide-Predicting Machine • Token Passing Model • Tons Per Minute • Tons Per Month • Torsion Photoelectric Magnetometer • Total Productive Maintenance • Tours Par Minute • Transactions Per Minute • Transfer Phase Midcourse • Transmission and Processing Model • Tree-Particle Mesh (code)

TPMP: Total Performance Measurement Plan

TPMS: Technical Performance Measurement System

TPN: Température et Pression Normales • Tiputini (Ecuador) (location identifier) • Titulaire de Physique Nucléaire

TPNS: Teleprocessing Network Simulator

TPO: Tanalian Point (USA-AK) (location identifier) • Tank Pressurizing Orifice • Technical Planning Office • Test Program Outline • Travelling Post Office • Tube à Propagation d'Onde

TPOCC: Transportable Payload Operations Control Center

TPOCP: Turbopump Oxidizer Cavity Purge

TPOD: Test Plan Of the Day

TPOM: Tube à Propagation d'Onde Magnétique

TPON: Telephony on a Passive Optical Network

TPoP: Traditional Point of Presence

TPOP: Traditional Point Of Presence

TPP: Tarapoto (Peru) (location identifier) • Technical Performance Parameter • Telecommunications Preparatory Programme • Terminal Procedures Publication • Tetraacrylporphyrin • Thermal Power Plant • Thermal Protection Panel • Total Program Planning • Total Program Procurement • Trans-Pluto Probe • Two-Phase Principle • Two-Photon Propagation

TPPC: Transaction Processing Performance Council

TPPD: Twisted Pair - Physical-Media Dependent

TPPS: Tape Post-Processing System

TPQ: Tepic (Mexico) (location identifier) • Terms Per Query

Tpr: Taper • Teleprinter

TPR: Tape Programmed Row • Technical Progress Report • Technical Progress Review • Teleprinter • Telescopic Photographic Recorder • Temperature Profile Recorder • Température-Pouls-Respiration • Temperature-Pulse-Respiration • Test Performance Recorder • Test Phase Report • Test Problem Report • Test Procedure Report • Tom Price (Australia-WA) (location identifier) • Training Personnel Requirement • Trappe de Positionnement de Réception

TPRL: Transport Properties Research Laboratory

tps: Tours par Seconde • Transactions per Second

TPS: Tape Plotting System • Technical Publishing Society • Technical Publishing Software • Telemetry Processing System • Telephone Preference Service • Teleprocessing Services • Television Par Satellite • Terminal Performance Specification • Test Plotting System • Test Preparation Sheet • Test Procedure Specification • Test Program Set • The Planetary Society • Thermal Protection System • Thin-Plate Splines • Total Parameter Space • Training Planning System • Transaction Processing System • Transactions Per Second • Transfer Propulsion System • Translunar Propulsion Stage • Trapani (Italy) (location identifier) • Tritium Purification System • Turbine-Powered Simulation • Twisted Pair Shielded

TPSA: Telekommunikacji Polskiej SA (Poland)

TPSB: Telemetry Processing System Buffer

TPSE: Thermal Protection System Experiment

TPSO: TOPEX-Poseidon Follow-On

TPSS: Test Preparation Sheet Summary

Tpt: Transport

TPT: Tapeta (Liberia) (location identifier) • Technicien Principal Nucléaire • Time Priority Table • Transport

TPTA: Transient Pressure Test Article

TPTB: The Powers That Be

TPTG: Tuned-Plate Tuned-Grid

TPU: Tape Preparation Unit • Taux de Panne en Utilisation • Text Processing Utility • Tikapur (Nepal) (location identifier) • Time Processing Unit • Tomsk Polytechnic University (Russia)

TPUN: Test Procedure Update Notice

TPV: Terminal au Point de Vente

TPVS: Transaction Program Verb Set

TpW: Tons per Week

TPW: Tons Per Week • True Polar Wander • True Polar Wandering • Turbo Pascal for Windows

TPX: Tokamak Physics Experiment • Two-Phase Experiment

TpY: Tons per Year

TPY: Tons Per Year

TP2: Transmission Protocol Class 2

TP3: Transmission Protocol Class 3

TP4: Transmission Protocol Class 4

TQ: Tel Quel • Topographic Quadrangle • Torquay (UK postal code) • Total Quality

T&QA: Test & Quality Assurance

TQA: Test and Quality Assurance • Transformational Question Answering

TQC: Théorie Quantique des Champs • Time-Quality-Cost • Total Quality Commitment • Total Quality Control • Three-Quark Cluster

TQCM: Temperature-controlled Quartz Crystal Microbalance

TQDM: Total Quality Data Management

TQE: Timer Queue Element

TQF: Threshold Quality Factor

TQFP: Thin Quad Flat Pack

TQI: Total Quality Involvment

TQL: Total Quality Leadership

TQM: Total Quality Management • Total Quality Movement

TQMS: Total Quality Management System • Triple-Quadrupole Mass Spectrometer

TQMSC: Total Quality Management Steering Committee

TQN: Taluqan (Afghanistan) (location identifier)

TQP: Total Quality Program • Transistor Qualification Program

TQS: Term Question Specificity • Total Quality of Service • Total Quality System • Tres Esquinas (Colombia) (location identifier)

TQT: Transistor Qualification Test

TQTP: Transistor Qualification Test Program

tr: Translucidus • Turkey • Turkish (ISO language code)

Tr: Tare • Trace • Track • Train • Tranche • Transaction • Transitif • Transitive • Transitivo • Translate • Translated • Translation • Translator • Transpose • Transposition • Treasurer • Trench • Triassic • Troop • Through • Trust • Trustee

T-R: Transmit-Receive

T/R: Tape Recorder • Technical Report • Transformer/Rectifier • Transmit/Receive • Transmitter/Receiver • Transportation Request • Travel Request • Turnaround Requirements

TR: Gabon (international aircraft registration code) • Table Row • Tape Reader • Tape Recorder • Tape Register • Tariff Reform • Taux des Retards • Technical Reference • Technical Regulation • Technical Report • Technical Representative • Technical Requirement • Technical Review • Teknologirådet • Telecommunications Request • Teleoperator Retrieval • Temperature Range • Temperature Recorder • Tempore Regis • Temps de Ravitaillement • Temps de Réaction • Temps de Roulage • Temps Réel • Terminal Ready • Terms of Reference • Terres Rares • Test Register • Test Report • Test Request • Test Response • Test Retest • Test Rocket • Test Routine • Test Run • Tête de Rotor • Text Retrieval • Theodore Roosevelt (1858-1919) • Thermomètre Réaumur • Thermorégulation • Thrust Reverser • Time Record • Time for Retrofire • Time to Retrofire • Time to Retrograde • Tons Registered • Tool Request • Top Return • Total Regulation • Total Revenue • Tracking Radar • Trail • Train Radio • Transfer • Transfer Register • Transformation Ratio • Transformer-Rectifier • Transient Response • Transition Radiation • Transition Region • Transmetteur-Récepteur • Transmission-Réception • Transmit-Receive • Transmitter-Receiver • Transportation Request • Trip Report • Trouble Report • Truro (UK postal code) • Trust Receipt • Turboréacteur • Turkey (international vehicle registration code) • Turkey (ISO country code) • Turquie (ISO country code) • Turkey (postal country code) • Turnaround Requirements

TrA: Triangulum Australe - Trianguli Australis (constellation)

TRA: Tape Recorder Amplifier • Taramajima (Japan) (location identifier) • Teacher Resource Agent • Teledyne Ryan Aeronautical • Temporary Reserved Airspace • Temporary Restricted Area • Test Requirements Analysis • Theory of Reasoned Action • Token Ring Adapter • Traffic Routing Administration • Training Requirements Analysis • Triaxial Recording Accelerometer • Turnaround Requirements Analysis

TRAA: Tule River Amateur Astronomers (USA-CA)

Trac: Tractatus • Tracto

TRAC: Targeting and Reflective Alignment Concept • Teaching, Reflection And Collaboration • Technical Recommendations Application Committee • Text Reckoning And Compiling • Total Record Access Control • Transaction Reporting and Control System • Transient Radiation Analysis by Computer

TRACALS: Traffic Control And Landing System

TRACE: Task Reporting And Current Evaluation • Technical Report Analysis, Condensation, Evaluation • Test Equipment for Rapid Automatic Checkout Evaluation • Total Risk Assessment Cost Estimate • Trace Remote Atmospheric Chemical Evaluation • Transitional Region And Coronal Explorer • Transition Region And Coronal Explorer (*) • Transport and Atmospheric Chemistry near the Equator • Transport And Chemistry near the Equator • Trans-Regional Academic mobility and Credential Evaluation

TRACER: Tropospheric Radiometer for Atmospheric Chemistry and Environmental Research

TRACON: Terminal Radar Approach Control

TRACS: Tool-Record Accountability System • Trouble Adviser for Customer Services

TRACT: Transportable Remote-Area Communication Terminal

Trad: Tradition • Traditional • Traducción • Traducteur • Traduction • Traductor • Traduit • Traduttore • Traduzione

TRADE: Transatlantic Research and Development Environment

TRADIC: Transistorized Airborne Digital Computer

TRADOC: Traitement de Données Océanographiques

Trafo: Transformator

TRAGB: Tritium Receiving and Analytical Glove Box

TraGe: Travaux Géophysiques (periodical code)

TRAI: Telecom Regulatory Authority of India

TRAIPR: Trade-Related Aspects of Intellectual-Property Rights

Traj: Trajectory

TRAJ: Trajectory

TRAK: Track

TRAM: Transputer Module

TRAMAS: Transportable Microwave Active Spectrometer

TRAMP: Temperature Regulation And Monitor Panel • Testing, Reporting, and Maintenance Program • Test Retrieval And Memory Print • Three-dimensional Radiation-hydrodynamical Modelling Project

TRANET: Tracking Network

Tranfd: Transferred

TRANID: Transaction Identification

Trans: Transaction • Transfer • Transferred • Transform • Transformation • Transformed • Transformer • Transit • Transition • Transitive • Transitory • Translate • Translated • Translation • Translator • Transmission • Transmit • Transmittance • Transmitted • Transparence • Transparent • Transport • Transportation • Transpose • Transposition • Transvaal (South Africa) • Transverse

TRANS: Transition • Transitory • Translation
Transa: Transaction
TRANSAC: Transistorized Automatic Computer
Transcr: Transcribed • Transcription
Transf: Transferred
Transfd: Transferred
Transl: Translate • Translated • Translation • Translator
TRANSL: Translation
Transp: Transparent • Transport • Transportation
TRANSV: Transversal
Trany: Transparency
TRAO: Taeduk Radio Astronomy Observatory (Korea-ROK) (formerly DRAO) • Torun Radio Astronomy Observatory (Poland)
TRAP: Tactical Recovery of Aircraft and Personnel • Transient Response Analysis Program • Transmettre-Ranger-Agir-Poubelle
TRASYS: Thermal Radiation Analysis System • Thermal Radiation Analyzer System
TRAU: Transcoding Rate Adaption Unit
Trav: Travail • Travaux • Travel • Traveller
TRAWL: Tape Read And Write Library
Trb: Trebe
TRB: Technical Review Board • Test Requirements Bulletin • Test Review Board • Transportation Research Board • Turbo (Colombia) (location identifier)
TRBL: Troubleshooting
TRC: Tape Reader Calibrator • Tape Reader Control • Tape Reader Coordinator • Teacher Resource Center (NASA/JSC) • Technical Reference Center • Technical Repair Center • Technical Reports Center • Technical Research Center • Technical Resources Center • Technical Review Committee • Technology Resource Center • Test Readiness Certificate • Thermodynamics Research Center (USA-TX) • Token Ring Connection • Torreón (Mexico) (location identifier) • Transition Region Camera • Tsukuba Research Consortium • Tube à Rayons Cathodiques • Tycho Reference Catalogue
TRCE: Thermionic Reactor Critical Experiment • Trace
Tr. Co.: Trust Co.
TRCP: Tape Recorder Control Panel
TRCR: Traceur
Trd: Tread • Trinidad
TR&D: Technology Research & Development
TRD: Technical Requirements Data • Technical Requirements Document • Technology Research and Development • Test Requirements Document • Transition Radiation Detector • Trondheim (Norway) (location identifier)
TRDA: Technological Research and Development Authority
TRDI: Technical Research and Development Institute
TRDPS: Topographic Retrieval and Data Presentation System
TRDS: Text Retrieval Data System
TRE: Timing Read Error • Tiree (UK) (location identifier) • Transient Radiation Effect • True Radiation Emittance
Treas: Treasurer • Treasury
Treasr: Treasurer
TREAT: Trouble Report Evaluation and Analysis Tool
TREC: Text Retrieval Conference • Tracking Radar Echoes by Correlation
TRECAMS: Tritium Remote Control And Monitoring System
T-RECS: Thermal-Region Camera System
TRECS: Thermal-Region Camera System
Tree: Trustee
TREE: Trans-European Employment
TREES: Tropical Ecosystem Environment Satellite
TREF: Tripartite Employment Conditions Forum (CERN)
TREIS: Tropical Radar Environmental Information System
TREM: Tropical Rainfall Explorer Mission
TREMI: Tracking Radar Echoes by Moment Invariants
TREMP: Trapped Radiation Environment Modelling Program
TREND: Technical Requirements Engineering Network Database • Terrestrial Radiation Environment Development
Trés: Trésor • Trésorier
TRES: Time-Resolved Emission Spectrum
TRESR: Time-Resolved Electron Spin Resonance
TRESTLE: Text Retrieval, Extraction and Summarisation Technologies for Large Enterprises
TRET: Telecommunications and Radio Engineering Telecommunications
Trf: Tariff • Transfer
TRF: Sandefjord (Norway) (location identifier) • Technical Reference File • Term Relevance Feedback • Terrestrial Reference Frame • Transfer • Tribunal Regional Federal • Tube à Refroidissement Forcé • Tuned Radio Frequency • Turbine Research Facility • Two-Ribbon Flare
Trfc: Traffic
TRFY: Trafficway
Trg: Training • Triangle
TRG: Tailed Radio Galaxy • Tauranga (New Zealand) (location identifier) • Teaching, Research and Development • Trigger
TRGB: Tip of the Red Giant Branch
TRGD: Technical Requirements and Goals Document
Trgf: Traffähigkeit
TrGRC: Transactions of the Geothermal Resources Council (periodical code)
TRGS: Technische Regelung Gefahrtstoff
TRH: Trona (USA-CA) (location identifier)
Tri: Triangulum-Trianguli (constellation)
TRI: Bristol (USA-VA) (location identifier) • Johnson City (USA-TN) (location identifier) • Kingsport (USA-TN) (location identifier) • Tape Recorder Interface • Technical Report Instruction • Time Reversal Invariance • Toxic Release Inventory • Tri-Cities (USA-TN) (location identifier)
TrIA: Trainable Information Assistant
TRIAD: Timesharing Research Image Analysis and Display • Tucson Revised Index of Asteroid Data
TRIAL: Technique to Retrieve Information from Abstracts of Literature
Trib: Tributary
TRIB: Transfer Rate of Information Bits • Tributary
TRIBECA: Triangle Below Canal (Street) (New York)
TRIC: Transition Radiation and Ion Calorimeter
TRICS: Trajectory Incremental Correction System
trid: Triduum
TRIDAC: Three-Dimensional Analog Computer
TRIFAR: Three Frequency Airborne Radar
Trig: Trigger • Trigonometría • Trigonometric • Trigonometrical • Trigonometry
TRIG: Trigger
Trigon: Trigonometría • Trigonometric • Trigonometrical • Trigonometry
Trim: Trimester

TRIM: Terrain Resource Information Management • Testing for Reliability by Incremented Maximization • Time Resolution Imaging Mode • Time Resolved Imaging • Time Resolved Imaging Mode • Transport of Ions in Matter

TRIMS: Time-Resolved Ion Momentum Spectrometry

Trin: Trinidad

Trip: Triple • Triplicate

TRIP: Thunderstorm Research International Program • Thunderstorm Research International Project • Trade-Related aspects of Intellectual Property • Trade-Related Intellectual Property

Tripl: Triplet • Triplicate

TRIPS: Trade-Related Aspects of Intellectual Property Rights • Trillion Instructions Per Second

TRIS: Test and Reliability Information System • Trapped Ions in Space • Tropical Region Imaging Spectrometer

TRIUMF: Tri-University Meson Facility (Canada-BC)

TRIZ: Theory of Inventive Problem Solving (Russian)

TRJ: Tarakbits (Papua New Guinea) (location identifier)

Trk: Track • Truck

TRK: Tarakan (Indonesia) (location identifier) • Track

TrKaz: Trudy Kazan Gorod. Astron. Obs. (periodical code)

Trkg: Tracking

Trkr: Tracker

TRKR: Tracker

TKS: Tracks

Trl: Trail

TRL: Lira (Turkey) (ISO currency code) • Tarasko-Richardson-Lucy • Teacher Resource Laboratory • Technical Report Literature • Technology Readiness Level • Terrell (USA-TX) (location identifier) • Test Readiness List • Thermodynamics Research Laboratory • Through-Reflect Line • Thru-Reflect Line • Trail • Transistor Resistor Logic

TRLF: Time-Resolved Lidar Fluorosensor

Trlg: Trailing

Trlr: Trailer

TRLR: Trailer

TLS: Trails

Trly: Trolley

TRM: Terminal • Test Request Message • Test Requirements Manual • Thermal (USA-CA) (location identifier) • Thermal Remanent Magnetisation • Thermal Resistance Measurement • Thermoremanent Magnetism • Thermoremanent Magnetization • Time Ratio Modulation • Time-Reversal Mirror • Totally Reflective Mirror • Track Recovery Module • Trademark • Training and Research Mobility • Training Manual

Trml: Terminal

TRML: Terminal

TRMM: Tropical Rainfall Mapping Mission • Tropical Rainfall Measurement Mission • Tropical Rainfall Measuring Mission (*)

TrMO: Tabor Mountain Observatory (Canada-BC)

Trmr: Trimmer

TRMS: Technical Requirements Management System • Test Resource Management System

TRMTR: Transmitter

Trn: Train

TRN: Task Requirements Notice • Technical Research Note • Temporary Record Number • Terrain-Referenced Navigation • Threaded Read News • Torino (Italy) (location identifier) • Trunnion • Turin (Italy) (location identifier)

tRNA: Transfer Ribonucleic Acid

Trnbkl: Turnbuckle

Trnd: Turned

Trng: Training

TRNG: Training

Trngl: Triangle

TRNL: Triangle Research Libraries Network

TRNPK: Turnpike

Trnr: Trainer

Trnsp: Transportation

Trnspn: Transportation

TRO: Taree (Australia-NSW) (location identifier) • Technical Reviewing Office • Temporary Restraining Order • Test Requirements Outline • Thermal Relaxation Oscillation

TROFF: Text Run Off

Trol: Troland

TROLL: Transmission and Receiving Of Laser Light

TROMEX: Tropical Meteorology Experiment

TRON: The Real-time Operating-system Nucleus • Time Rationale Organizational Notation

Trop: Tropic • Tropical

Trop. Can.: Tropic of Cancer

Trop. Cap.: Tropic of Capricorn

TROP: Tactical Resource and Operations Plan • Tropopause

TROPIC: Token-Ring Protocol Interface Controller

TROPICS: Transparent Object-oriented Parallel Information Computing System

Tropo: Troposphere

TROPOZ: Tropospheric Ozone

TROS: Televisie Radio Omroep Stichting

TROV: Telepresence Remotely Operated Vehicle

Trp: Troop

TRP: Technical Review Panel • Technological Research Program • Technology Reinvestment Program • Technology Reinvestment Project • Technology Research Plan • Technology Research Program • Terminal Rendezvous Phase • Three Point (USA-AK) (location identifier) • Thunderstorm Research Project • Tiepoint Registration Processor • Time Reduction Processor • Transformation-Ring-Projection • Trip

TRPC: Transaction Remote Procedure Call

TRPCL: Tropical

TRPD: Test Requirements and Plan Documentation

TRPK: Turnpike

TRPS: Temperature Regulating Power Supply

TrPul: Trudy Glav. Astron. Obs. Pulkovo Ser. 2 (periodical code)

Trq: Torque

TRQ: Tarauacá (Brazil-AC) (location identifier) • Tariff Rate Quota • Torque

TRR: Tape Recorder Reproducer • Technical Requirements Review • Test Readiness Review • Test Requirements Review • Theoretical Research Report • Time-Resolved Reflectivity • Torre • Trincomalee - China Bay (Sri Lanka) (location identifier) • Tropical Report Request • Tropical Report Review

TRRB: Test Readiness Review Board

TRRN: Terrain

TRRU: Tape Recorder Reproducer Unit

Trs: Transfer • Transpose • Trustees

TRS: Tandy Radio Shack • Tape Recorder System • Technical Repair Standard • Technical Report Service • Technical Requirements Specification • Technical Research Support • Teleoperator Retrieval System • Telerobotics Servicer • Term-Rewriting System • Test Requirements Sheet • Test Requirements Specification • Test Requirements Summary • Test Research Service • Test Research Station • Thermal Radiation Source • Time Reference Signal • Time Reference System • Time-Resolved Spectrometry • Time-Resolved Spectroscopy • Tone-Ranging System • Top Right Side • Total Routhian Surface • Township Range System • Traceability and Reporting System • Tracking and Reporting System • Training Record System • Trieste (Italy) (location identifier) • Tritium Receiving System • Tropical Revolving Storm • Troubleshooting Record Sheet • Tug Rotational System

TRSA: Terminal Radar Service Area

TRSB: Time Reference Scanning Beam

Trsbr: Transcriber

TRSC: Thailand Remote Sensing Centre

Trscb: Transcribe

Trsd: Transferred • Transposed

TRSD: Test Requirements and Specification Document • Test Requirements/Specification Document • Total Radiance Spectral Distribution

TRSE: Token Ring Switching Engine

Trsfr: Transfer

Trsfrt: Transfert

TrSht: Trudy Gos. Astron. Inst. Shternberga (periodical code)

Trsn: Transition

TRSN: Torsion

Trsp: Transport

TRSP: Total Radiance Spectral Polarization

TRSR: Taxi and Runway Surveillance Radar

TRSU: Tropical Remote Sensing Unit

TRSV: Time-Resolved Surface Viscometer

Trt: Turret

TRT: Test Reference Tool • Token Rotation Time • Tremonton (USA-UT) (location identifier) • Triply Reflecting Telescope • Triply Reflection Telescope

Trtd: Treated

Trtmt: Treatment

TRU: Test Replaceable Unit • Time Release Unit • Transformer-Rectifier Unit • Transmit-Receive Unit • Transuranic • Transuranium • Trujillo (Peru) (location identifier)

TRUCE: Tropical Urban Climate Experiment

TRUFO: True Unidentified Flying Object

TRUMP: Teaching Resources Unit for Modern Physics • Technical Review Updated Manuals and Publications • Total Revision and Upgrading of Maintenance Procedures

Trv: Travesia

TRV: Trivandrum (India) (location identifier)

TRW: Tarawa (Kiribati) (location identifier) • Thompson Raine Woolridge Inc.

TRWY: Throughway

Trx: Triplex

TRX: Trenton (USA-MO) (location identifier)

Try: Truly

TRY: Tororo (Uruguay) (location identifier)

TRZ: Tiruchirapally (India) (location identifier)

t&s: Toilet & Shower

ts: Tsonga (ISO language code)

Ts: Tosyl Group • Tous

T&S: Transport & Supply

T/S: Temperature/Salinity • Thermal/Structural • Time Sharing • Traceability by Serial • Troubleshoot • Troubleshooting • Turboshaft

TS: Cleveland (UK postal code) • Taittirīya Samhitā • Targeted Search • Target Suppression • Target System • Tasmania (Australia) • Team Supervisor • Technical Secretariat • Technical Service • Technical Specification • Technical Support • Technicien Supérieur • Telecommunications System • Telescope Scientist • Telescope Section • Temperature Switch • Temporary Storage • Tensile Strength • Tentative Specification • Terme Spécifique • Terminatorschatten • Terminator Shadow • Term Specificity • Test Set • Test Site • Test Specification • Test Stand • Test Station • Test Summary • Text Split • Thermal Spray • Thunderstorm • Tiempo Sideral • Time Scheduled • Time Series • Time Sharing • Time Switch • Timing Selector • Timing System • Tool Steel • Top Secret • Traceability by Serial • Tracking Scope • Tracking Station • Tracking System • Transfer Set • Transient Source • Transient State • Transient Steady • Transition State • Transit Storage • Transmission Set • Transparent Substrate • Transport Stream • Très Secret • Tropical Storm • Troubleshoot • Troubleshooting • Tunisia (international aircraft registration code) • Twin Screw • Type Specification • Typescript

TSA: Taipei - Sung Shan (China-ROC) (location identifier) • Target Service Agent • Taxiway Safety Area • Technical Support Alliance • Technologie de Systèmes Automatisés • Test and Simulation Assembly • Test Site Activation • Test Start Approval • Time Scale Analysis • Time Series Analysis • Time-Shared Amplifier • Time Study Analysis • Toronto Sidewalk Astronomers (Canada-ON) • Total Surface Area • Two-Step Antenna

TSAC: Time of Spectrum Acquisition • Title, Subtitle And Caption • Total System Analysis and Cost • Tracking System Analytical Calibration

TSACC: Telecommunications Standards Advisory Council of Canada

TSAG: Telecommunications Standardization Advisory Group

TSAP: Time Series Analysis Package • Transport Service Access Point

TSAPI: Telephone Services Application Program Interface • Telephone Services Application Programming Interface • Telephony Services Application Program Interface • Telephony Services Application Programming Interface

TSAPTS: The Southern Automatic Patrol Telescope Service

TSAR: Télécommande de Sauvegarde Ariane • Time-series Storage And Retrieval

TSAVI: Transformed Soil Adjusted Vegetation Index

TSB: Technical Sector Board • Technical Service Bulletin • Telecommunications Standards Bureau • Telematics Supervisory Board • Temporary Stowage Bag • Terminal Status Block • Transportation Safety Board • Tsumeb (Namibia) (location identifier) • Twin Sideband

TSBF: Tropical Soil Biology and Fertility

TSC: Taisha (Ecuador) (location identifier) • Tangential Spherical Cube • Taxes and Social Contributions • Technical Standards Committee • Technical Subcommittee • Technical Support Center • Teledyne Systems Corporation • Television Scan Converter • Tension Sous Charge • Terminal Sterilization Chamber • Terrestrial Science Chamber • Test Set Computer • Test Set Connection • Test Setup Complete • Test Support Complex • Test Support Controller • Test Support Coordinator • Thermally-Stimulated Current • Time-Sharing Control • Tooling Supplemental Control • Transmission System Component • Transportation Systems Center • Triangular-Shape Cloud • Trimble System Controller • Twin-assigned Synchronous Composite • Two-Stage Command

TSCA: Toxic Substances Control Act (USA)

TSCC: Test Support Control Center • Tilt-Scan CCD Camera

TSCD: Test Specification and Criteria Document • Test Specification Control Document

TSCLT: Transportable Satellite Communications Link Terminal

TSCM: Taylor Series Correction Method • Technical Survey Countermeasure • Test Station Configuration Model

TSCO: Test Support Coordination Office • Test Support Coordinator

TSCP: Training Simulator Control Panel

TSCPF: Time Switch and Call Processor Frame • Time Switch and Central Processor Frame

TSCS: Test Support Computer System

TSCT: Transportable Satellite Communications Terminal

Tsd: Tausend

TSD: Technical Services Division • Technical Support Document • Test Start Date • Text-Speech-Dialogue • Thermionic Specific Detector • Total Spectral Density • Total System Design • Transient Signal Detector • Treatment-Storage-Disposal • Tshipise (South Africa) (location identifier)

TSDA: Theory of Signal Detection Analysis

TSDCS: Tritium Storage and Delivery Cleanup System

TSDF: Telemetry Subset Definition File

TSDGB: Tritium Storage and Delivery Glovebox

TSDIS: TRMM Science Data and Information System

TSDMO: Tellus Series on Dynamic Meteorology and Oceanography (periodical code)

TSDMS: Time-Shared Data Management System

TSDN: Transfer Syntax Description Notation

TSDS: Tritium Storage and Delivery System

TSDU: Transport Service Data Unit

TSE: Team Supportability Element • Technical Support Effort • Technicien Supérieur d'Exploitation • Tender Support Equipment • Terminal Server Edition • Test of Spoken English • Test Scoring Equipment • Test Set Electrical • Test Support Equipment • Thermoelastic Stress Analysis • Total Solar Eclipse • Transmissible Spongiform Encephalopathy • Transmission Secondary Emission • Transportation Support Equipment • Tselinograd (Kazakhstan) (location identifier) • Turkish Standards Institution (Turkey) • Türk Standardlari Enstitüsü (Turkey)

TSEC: Translinear Scanning Electro-optical Camera

TSEE: Test Support Equipment Evaluation

TSEF: Technicien Supérieur d'Étude et de Fabrication

TSEM: Transmission Secondary Emission Multiplier

TSER: Targeted Socio-Economic Research

TSES: Technical Simulation and Evaluation System • Transportable Satellite Earth Station

TS&ET: Test Support & Evaluation Team

TSET: Test Support and Evaluation Team

TSF: Télégraphie Sans Fil • Telemetry Scalar File • The Space Foundation • Thin Solid Films (periodical code) • Topographic Shell Fabrication • Typical Shape Function • Venice Treviso (Italy) (location identifier)

TSFAR: The Star*s Family of Astronomy Resources

TSFC: Thrust Specific Fuel Consumption

Tsfr: Transfer

TSG: Table de Situation Générale • Tanacross (USA-AK) (location identifier) • Technical Specialty Group • Technical Study Group • Technical Subgroup • Technology Support Group • Température de Sortie des Gaz • Term Support Graph • Test Signal Generator • Threat Scenario Generator • Time Signal Generator • Timing Systems Group

TSGP: Test Sequence Generator Program

TSGS: Time Series Generation System

TSh: Tanzanian Shilling (currency)

TSH: Telephony Signal Handler • Télescope Spatial Hubble • Temperate Southern Hemisphere • Tshikapa (Zaire) (location identifier)

TsHUJ: Tsin Hua University Journal (periodical code)

TSHWR: Thundershower

tsi: Tons per Square Inch

T/SI: Turn and Slip Indicator

TSI: Technicien Supérieur en Instruments • Technique et Science Informatiques • Teinte-Saturation-Intensité • Temporal Status Interval • Test Support Instruction • Time Slot Interchanger • Time Status Interval • Traitement du Signal et des Images • Tsili Tsili (Papua New Guinea) (location identifier) • Turn and Slip Indicator

TSIG: Trusted Systems Interoperability Group

TSIM: Total Solar Irradiance Mission

TSIO: Time-Shared Input/Output

TSIPS: Teachers and Students Investigating Plants in Space

TSIR: Time-Shared Information Retrieval

TSJ: Tsushima (Japan) (location identifier)

TSK: Tagaki-Sugeno-Kang • Taskul (Papua New Guinea) (location identifier) • Time Shift Keying

TSL: Tamuin (Mexico) (location identifier) • Technical Scripting Language • Technical Services Laboratory • Temporal Surface Language • Temps Sidéral Local • Test Set Logic • Test Source Library • Test Support Language • The Software Link (USA-GA) • Three Searching Language

TSLC: Taijuan Satellite Launch Center (China-PRC)

TSLD: Troubleshooting Logic Diagram

TSLS: Two-Stage Least Squares

TSM: Tail Service Mast • Taos (USA-NM) (location identifier) • Tax System Modernization • Technology-Science-Mathematics • Temps Sidéral à Minuit • Test Standards Module • Test Support Manager • Test Support Methodology • Textron Specialty Materials • Time Scheduled Maintenance • Time Sharing Monitor • TIROS Stratospheric Mapper • Total Suspended Matter • Toyama Science Museum (Japan) • Trade Study Management

TSMD: Time-Stress Measurement Device

TSMG: Tiempo Sideral Medio en Greenwich

TSMT: Technical Support Management Team

TSN: Tape Serial Number • Term Semantic Network • Tianjin (China-PRC) (location identifier)

Tsnt: Transient

TSO: Isles of Scilly - Tresco (UK) (location identifier) • Table Structure Overview • Technical Service Order • Technical Standard Order • Technical Standing Order • Technical Support Office • Technisch Secundair Onderwijs • Terminator Sensor Output • Test Site Office • Test Support Operation • Time-Sharing Operations • Time-Sharing Option • Time Since Overhaul • Turtle Star Observatory (Germany)

TSO/E: Time-Sharing Option Extended

TSOE: Time-Sharing Option Extended

TSOP: Technical Standard Operating Procedure

TSOR: Tentative Specific Operational Requirement

TSORT: Transmission System Optimum Relief Tool

TSOSC: Test Set Operational Signal Converter

Tsp: Teaspoon

TSP: Technical Support Package • Tehachapi - Kern County (USA-CA) (location identifier) • Telecommunications Service Priority • Telecommunications Service Provider • Telemetry Simulation Package • Teleprocessing Services

Program • Temperature-Sensitive Parameter • Temperature-Sensitive Period • Temporary Standard Practice • Temporary Standard Procedure • Test Site Position • Test Software Program • Test Status Panel • Test Supervisor • Test Support Package • Test Support Position • Test Support Program • Texas Star Party (USA-TX) • Thermospray • Time Series & Polarimetry • Time Series Polarimetry • Time Sorting Program • Time Synchronization Protocol • Total Soluble Phosphorus • Total Suspended Particles • Traffic Service Position • Trajectory Shaping Program • Trajectory Synthesis Program • Travelling Salesman Problem • Triple Super Phosphate • Trisodium Phosphate • Twisted Shielded Pair

TSPC: Time Stored Program Command

TSPD: Technology and Space Program Development

tspi: Tons per Square Inch

TSPI: Time-Space-Position-Information

TsPK: Tsentr Podgotovska Kosmonavti (Russia)

TSPM: Total Suspended Particulate Matter

TSPS: Time-Sharing Programming System • Traffic Service Position System

TSPU: Test, Synchronisation and Pulsargate-Unit • Test, Synchronisation and Pulsar-gating Unit

TSQLS: Thundersqualls

TSR: Technical Safety Requirement • Technical Status Review • Technical Study Report • Technical Summary Report • Telecommunications Service Request • Temps Significatif Routier • Terminal and Stay Resident • Terminate and Stay Resident (*) • Terminate Stay Resident • Test Schedule Report • Test Schedule Request • Test Status Report • Test Storage Ring • Test Support Requirement • Time Status Register • Timişoara (Romania) (location identifier) • Torsion Solid Rod • Trade Study Report • T-shirts over Suits Ratio

TSRA: Total System Requirements Analysis

TSRM: Third-Stage Rocket Motor

TSRO: Temps Significatif Routier Observé

TSRP: Temps Significatif Routier Prévu

TSRPC: Transport-Specific Remote Procedure Call

TSRV: Transport Systems Research Vehicle

TSS: New York - East 34th Street (USA-NY) (location identifier) • Tank Subsystem • Tape Search System • Targeted Search System • Task State Segment • Technical Specification Sheet • Technical Support Service • Telecommunications Standards Sector • Telecommunication Switching System • Télémanipulateur de la Station Spatiale • Telemetry Simulation System • Telescope Software System • Terminal Security System • Terminal Support System • Test Set Simulator • Test Software Support • Test Support System • Tethered Satellite System • Time-Sharing System • Time-Slot Shifter • Top Side Sounder • Toroidal Space Station • Total Surveying System • Total Suspended Solids • TPS Summary Sheet • Trading Support System • Traffic Separation Scheme • Trajectory Subsystem • Trapezoid Self-Scheduling • Tromsø Satellite Station (Norway) • Tropospheric Scatter System • Tug Structural Support • Turbine Steamship • Twin-Screw Steamer

TSSA: Test Scorer and Statistical Analyzer • Test Site Support Activity

TSSC: Technical Services Support Contract

TSSD: Typesetting System for Scientific Documents

TSSF: Tape Staging and Storage Facility

TSSG: Top-Seeded Solution Growth

TSSM: Thruster Subsystem Module

TSSMS: Time Sharing Service Management System

TSSO: Time Sharing System Option

TSSSO: Time Sharing Subsystem Option

TSST: Time-Space-Space-Time

TSSU: Test Signal Switching Unit

Tst: Test

TST: Telefunken Systems Technik (Germany) • Test • Test Status Tracker • Time-Sharing Terminal • Time-Space-Time • Trang (Thailand) (location identifier)

TSTA: Tritium System Test Assembly

TSTE: Training System Test and Evaluation

Tstg: Testing

TSTM: Thunderstorm • Tropical Science Team Member

TSTN: Triple Super-Twisted Nematic

TSTO: Two-Stage-To-Orbit

TSTP: Test of Selected Topics in Physics • Transaction Switching and Transport Protocol

Tstr: Tester

TSTR: Transistor

TSTRF: Terrestrial Space-Time Reference Frame

Tstrz: Transistorize

TSTS: Third Stage Test Set • Time-Space-Time-Space • Tracking System Test Set

TSTV: Two-Stage Transport Vehicle

TSU: Tabiteuea South (Kiribati) (location identifier) • Tajik State University (Tajikistan) • Tape Search Unit • Technical Service Unit • Technical Support Unit • Tennessee State University (USA-TN) • Texas Southern University (USA-TX) • Thermal Structural Unit • This Side Up • Time-Sharing User • Time Standard Unit • Towson State University (USA-MD) • Transfer Switch Unit

TSUM: Troy State University Montgomery (USA-AL)

TSV: Tab Separated Values • Townsville (Australia-QLD) (location identifier)

TSVD: Truncated Singular Value Decomposition

TSVP: Tournez S'il Vous Plaît

TSVQ: Tree-Structured Vector Quantizer

TSVS: Time-Sharing Virtual System

TSW: Test Software • Test Switch • Tsewi (Papua New Guinea) (location identifier)

TSX: Tanjung Santan (Indonesia) (location identifier)

TSY: Tasikmalaya (Indonesia) (location identifier)

TSZ: Tsetserleg (Mongolia) (location identifier)

T&T: Travel & Tourism • Trinidad & Tobago

tt: Tatar (ISO language code) • Teletypewriter • Tomos

T/T: Terminal Timing • Timing/Telemetry

TT: Chad (international aircraft registration code) • Tartar • Task Team • Technical Task • Technical Test • Technical Training • Technical Translation • Technicien Titulaire • Technology Transfer • Telegraphic Transfer • Telescope Team • Teletype • Teletypewriter • Temps Terrestre • Terminal Timing • Terrestrial Time • Tertiary • Therapeutic Touch • Think Tank • Thrust Termination • Tiefe Temperatur • Time Trial • Timing/Telemetry • Tobago (Trinidad &_) (ISO country code) • Top Term • Total Temperature • Total Time • Tracking Telescope • Transit Temporaire • Travel and Tourism • Travelling Telescope • Tremblement de Terre • Trinidad & Tobago (international vehicle registration code) • Trinidad & Tobago (ISO country code) • True Type

TTA: Tan Tan (Morocco) (location identifier) • Technological/Technical Assistance • Technology Teachers Association • Technology Trend Assessment • Telemetry and Telecommand Assembly • Thermomechanical Test Area • Time To Apogee • Transmission Temporelle Asynchrone • Transport-Triggered Architecture • Travel Time Authorized • Triplet-Triplet Annihilation

TTAE: Total Time Aircraft Engine

TTAF: Total Time Aircraft Frame

TTAPS: Turco-Toon-Ackerman-Pollack-Sagan

TTB: Technology Testbed • Telescience Testbed • Tetragonal Tungsten Bronze • Tortoli Arbatax (Italy) (location identifier)

TTBS: Trinidad and Tobago Bureau of Standards

TT&C: Telemetry, Tracking & Command • Telemetry, Tracking & Communications • Telemetry, Tracking & Control • Tracking, Telemetry & Command • Tracking, Telemetry & Control

TTC: Taltal (Chile) (location identifier) • Training Training Centre • Technology Transfer Center • Telecommunications Techniques Corporation • Telemetry and Telecommand • Telemetry, Telecommand and Control • Telemetry, Tracking and Command • Telemetry, Tracking and Control • Thompson Techniques de Communication • Time To Circularize • Time To Control • Time To Customer • Timing-Trigger-Control • Toutes Taxes Comprises • Transportation Technology Center • Tunnel Thermal Control

T/TCA: Thrust Translation Controller Assembly

TTCA: Thrust Translation Controller Assembly

TTC&M: Telemetry, Tracking, Control & Monitoring

TTCN: Tree and Tabular Combined Notation

TTCV: Tracking Telemetry, Command, and Voice

TTD: Dollar (Trinidad & Tobago) (ISO currency code) • Tactical Terrain Data • Technology Transfer Division • Test Technique Demonstrator • Troutdale (USA-OR) (location identifier)

TTDL: Terminal-Transparent Display Language

TTDR: Tracking Telemetry Data Receiver

Tte: Teniente

TT&E: Test, Teardown & Evaluation

TTE: Technology Transfer Institute • Telephone Terminal Equipment • Temporary Test Equipment • Ternate (Indonesia) (location identifier) • Test, Teardown and Evaluation • Time-Tagged Event • Time To Event

TTEL: Tool and Test Equipment List

Ttes: Toutes

TTESP: Test Time Event Sequencer Plan

TTF: Taurus Tunable Filter • TESLA Test Facility • Tetrathiofulvalene • Transition Task Force • TrueType Font

TTFF: Time To First Fix

TTFN: Tah Tah, For Now • Ta-Ta For Now

Ttfs: Toutefois

TTG: Tartagal (Argentina-SA) (location identifier) • Time To Go • Transition Task Group

TTH: Tactical Transport Helicopter • Thumrait (Oman) (location identifier)

TTI: TechTrans International Inc. (USA-TX) • Télécommunications et Traitement de l'Information • Telecommunication Training Institute • Tetiaroa Island (French Polynesia) (location identifier) • Theme-Topic Interface • Transitional Technology Inc.

TTIC: Technology Transfer In Cryogenics

TTIS: Translation and TEchnical Information Service

TTJ: Tottori (Japan) (location identifier)

TTK: Tottenham - Hale Station (UK-England) (location identifier)

Ttl: Total

TTL: Telescope Technologies Ltd. (UK-England) • Thermal Test Laboratory • Through The Lens • Time To Live • Total Time to Launch • To Take Leave • Tranfer Tube - Low-loss • Transistor-to-Transistor Logic • Transistor-Transistor Logic • Turtle Island (Fiji) (location identifier)

TTLM: Through-The-Lens Metering

TTM: Tablón de Tamara (Colombia) (location identifier) • Temperature Test Model • Temps de Traversée Moyenne • Thermal Test Model • Time To Market • Two-Tone Modulation

TTMM: True Tape Motion Monitor

TTMP: Transit Time Magnetic Pumping

TTN: Technology Transfer Network • Technology Transfer Node • TEX and TUG News • Transfer Technology Node • Transfer Tune - Normal-loss • Trenton (USA-NJ) (location identifier)

Tto: Toronto (Canada-ON)

TTO: Britton (USA-SD) (location identifier) • Tailored Technical Order • Tobago (Trinidad &_) (ISO country code) • Total Toxic Organic (compound) • Transient Tracers in the Ocean • Trinidad & Tobago (ISO country code)

TTOC: Tape Table Of Contents

TTOF: Thermally Tunable Optical Filter

TTP: Tape-To-Print • Technical Task Plan • Tecnología de Telescopios Pequeños • Telescience Testbed Project • Thermal-Transfer Printing • Thermistor Test Program • Time To Perigee • Trusted Third Party

TTPP: Telescience Testbed Pilot Program

TTQ: Tortuquero (Costa Rica) (location identifier)

TTR: Tana Toraja (Indonesia) (location identifier) • Time-Temperature Recorder • Time Token Rotation • Transient Thermal Radiation

TTRB: Telescope Time Review Board

TTRT: Target Token Rotation Time

TTRTO: Time-Triggered Real-Time Object

TTS: Teachers Touch the Sky (project) • Telemetry Transmission System • Telemetry Tracking System • Teletypesetter • Teletypesetting • Temperature Test Set • Temporary Threshold Shift • Test and Training Satellite • Text-To-Speech • Time To Spill • Time To Station • Transaction-Tracking System • Transponder Test Set • Trouble Ticketing System • True To Scale • Tsaratanana (Madagascar) (location identifier) • T Tauri Star

TTSF: Time To Subsequent Fix

TTSI: Tandem Telecommunications Systems Inc.

TTSN: Taiwan Telemetered Seismographic Network (China-ROC) • Total Time Since New

TTT: Taitung (China-ROC) (location identifier) • Team Time Trial • Through The Thickness • Trunk-to-Trunk Transfer

TTTS: Tanker/Transport Training System

TTU: Tape Transport Unit • Technology Transfer Unit • Tennessee Technological University (USA-TN) • Texas Tech University (USA-TX) • Terminal Timing Unit • Tetuan (Morocco) (location identifier) • Time Tag Unit • Time Terminal Unit • Timing Terminal Unit

TTUC: Technical Teachers Upgrading Centre

TTV: Tartu Variable • Termination, Test, and Verification • Time To View

TTVC: Test Termination and Verification Contractor

TTX: Teletext

TTY: Teletype • Teletypewriter

TTYL: Talk To You Later

ttz: 't is te zeggen

Tu: Thulium (now Tm) • Tuesday • Turkish

TU: Côte d'Ivoire (international aircraft registration code) • Ivory Coast (international aircraft registration code) • Tape Unit • Task Unit • Technical University • Technical Utilization • Technische Universität • Technology Utilization • Temps d'Utilisation • Temps Universel • Terminal Unit • Teutonic • Thermal Unit • Tiempo Universal • Timing Unit • Toamotu • Trade Union • Traffic Unit • Training Unit • Transfer Unit • Transmission Unit • Transport Unit • Tucuman (Argentina) • Tungu • Turkey

TUA: Tulcan (Ecuador) (location identifier)

tub: Tuba

Tub: Tubular

TUB: Technische Universität Berlin (Germany) • Tubuai Island (location identifier)

TUBSAT: Technical University of Berlin Satellite

Tuc: Tucana-Tucanae (constellation)

TUC: Temps Universel Coordonné • Tiempo Universal Coordinado • Time of Useful Consciousness • Trade Union Congress (UK) • Travail d'Utilité Collective • Tucuman - Benjamin Matienzo (Argentina-TU) (location identifier) • Type Unit Code

Tud: Tudmanyos

TUD: Tambacounda (Senegal) (location identifier) • Technische Universität Darmstadt (Germany) (formerly THD) • Technische Universiteit Delft (Netherlands) • Technology University of Denmark

Tue: Tuesday

TUE: Technische Universiteit Eindhoven (Netherlands) • Tratado de la Unión Europea • Tuesday • Tupile (Panama) (location identifier)

Tues: Tuesday

TUEX: Thuesdays Excepted

TUF: Technology Utilization Foundation • Tours (France) (location identifier)

TUFI: Toughened Unipiece Fibrous Insulation

TUG: Technische Universität Graz (Austria) • TEX Users Group • TRANSAC Users Group • TUBITAK Ulusal Gozlemevi (Turkey) • Tuguegarao (Philippines) (location identifier) • Turbo User Group

TUH: Tullahoma Arnold (USA-TN) (location identifier)

TUI: Text-based User Interface • Turaif (Saudi Arabia) (location identifier)

TUJ: Tum (Ethiopia) (location identifier)

TUK: Turbat (Pakistan) (location identifier)

TUL: Tulsa (USA-OK) (location identifier)

TULE: Transistorized Universal Logic Element

TULIP: The University Licensing Programme

TULIPS: Terrain Understanding and Land Information Processing System

TUM: Table Universelle des Marées • Technische Universität München (Germany) • Tumut (Australia-NSW) (location identifier)

Tun: Tuning • Tunisia • Tunnel

TUN: Tunis Carthage (Tunisia) (location identifier) • Tunisia (ISO country code) • Tunisie (ISO country code)

Tung: Tungsten

TUNIS: Toronto University System (Canada-ON)

TUNL: Tunnel

TUNS: Technical University of Nova Scotia (Canada-NS)

TUO: Taupo (New Zealand) (location identifier) • Technology Utilization Office

TUP: Teknologiske Udviklingsprogram (Denmark) • Temple University Press (USA) • Titre Universel de Paiement • Tupelo (USA-MS) (location identifier)

TUPR: Total Uplink Power Requirement

TUQ: Tougan (Burkina Faso) (location identifier)

Tur: Turret

TUR: Thermal Uniformity Restoral • Tucuruí (Brazil-PA) (location identifier) • Turkey (ISO country code) • Turquie (ISO country code)

Turb: Turbine • Turbulence

TURB: Turbine • Turbulence

TURBC: Turbulence

TURBT: Turbulent

Turk: Turkey • Turkish

Turkmen: Turkmenistan

TUS: Tucson (USA-AZ) (location identifier)

TUSA: Tactical Unmanned Surveillance Aircraft

TUT: Tauta (Papua New Guinea) (location identifier) • Telemetry User Table • Thermal Uniformity Test • Tutorial

TUU: Tabuk (Saudi Arabia) (location identifier)

TUV: Technischer Überwachungsverein • Tucupita (Venezuela) (location identifier) • Tuvalu (ISO country code)

TUVR: Total Ultraviolet Radiation

TUW: Technische Universität Wien (Austria) • Tubala (Panama) (location identifier)

TUX: Tumbler Ridge (Canada-BC) (location identifier)

TUY: Tulum (Mexico) (location identifier)

TUZ: Tucumã (Brazil-PA) (location identifier)

T/V: Testing Verification • Thermal Vacuum

TV: Target Vehicle • Téléviseur • Television (*) • Télévision • Temps de Vol • Temps Variable • Temps Vrai • Tension de Vapeur • Terminal Velocity • Terre Végétale • Test Vehicle • Textverarbeitung • Thermal Vacuum • Thrust Vector • Thrust Vectoring • Time-Varying • Tokyo Variable • Tololo-Vienna • Total Variation • Total Vision • Total Volume • Trajectoire de Vol • Transport Vehicle • Travessa • Tuva • Tuvalu (ISO country code)

TVA: Morafenobe (Madagascar) (location identifier) • Taxe à la Valeur Ajoutée • Taxe sur la Valeur Ajoutée • Temporary Variance Authority • Tennessee Valley Authority (USA) • Thrust Vector Alignment

TVAC: Taxe sur la Valeur Ajoutée Comprise

TVAR: Test Variance

TVASI: Tee Visual Approach Slope Indicator

TVB: Television Broadcast

TVBS: Television Broadcast Satellite

TVC: Television Camera • Télévision Communautaire par Câble • Télévision en Couleurs • Thermal Vacuum Chamber • Thermal Voltage Converter • Thrust Vector Control • Time Variable Constant • Time-Varying Coefficient • Trapped Vortex Combustor • Traverse City (USA-MI) (location identifier) • Troisième Vitesse Cosmique

TVCA: Thrust Vector Control Actuator • Thrust Vector Control Assembly

TVCD: Thrust Vector Control Driver

TVCS: Thrust Vector Control System • Thrust Vectoring Control System • Tritium Vault Cleanup System

TVD: Torque Velocity Dynamometer • Total Variation Diminishing

TVDPM: Total Variation Diminishing Particle Mesh

TVDS: Transportable VLBI Data System

TVE: Technology Validation Experiment • Temps de Vol des Équipements • Test and Verification Equipment • Test Vehicle Engine • Tête de Visée Électronique • Thermal Vacuum Environment • Thermo-Visco-Elasticity

TVF: Television Video File • Thief River Falls (USA-MN) (location identifier) • TOC Verbosely from File

TVFI: Television Frame International

TVFS: Thermal Video Frame Scanner

TVG: Test Vector Generator • Threshold Voltage Generator • Time Variation of Gain

TVGDHS: Television Ground Data Handling System

TVHD: Télévision à Haute Définition

TVI: Télévision Industrielle • Television Interference • Terminal de Vente Image • Thomasville (USA-GA) (location identifier) • Transformed Vegetation Index • Translation Variant of an Image • Tutored Videotape Instruction

TVIG: Television and Inertial Guidance

Tvl: Transvaal (South Africa) • Transversal • Travel

TVL: South Lake Tahoe (USA-CA) (location identifier) • Transversal-Vertical-Longitudinal

Tvlg: Travelling

Tvlr: Traveller

TVM: Television Memory • Test Verification Matrix • Tower Vector Magnetograph

TVMA: True Visual Magnitude Atlas

TVN: Test Verification Network

TVOC: Television Operations Center

TVOR: Terminal VHF Omnidirectional Range

TVP: Test Verification Program • Time-Variable Parameter • Topological Vector Profile

TVPS: Torus Vacuum Pumping System

TVQ: Taxe sur les Ventes du Québec

TVr: Temps Vrai

TVR: Télétel à Vitesse Rapide • Television Rating • Temperature Variation of Resistance

TVRO: Television Receive Only

TVS: Telecommunicatie- en Verkeersbeleidingssystemen • Television System • Temporal Variance Spectrum • Term Vector Space • Thermal Video System • The Virtual School • Thrust Vectoring System • Thrust Vector System • Topological Vector Space • Tornado Vortex Signature • Toxic Vapor Suit • Tri-Valley Stargazers (USA-CA)

TVSA: Thrust Vector Servoamplifier

TVSD: Time-Varying Spectral Display

TVSL: Thin Viscous Shock Layer

TVSO: Television Space Observatory

TVT: Thermal Vacuum Test

TVTA: Thermal Vacuum Test Article

TVU: Taming Virtual University • Taveuni (Fiji) (location identifier)

TVY: Dawe (Myanmar) (location identifier) • Tavoy (Myanmar) (location identifier)

TV3D: Télévision en Relief

tw: Taiwan (China-ROC) • Te Weten • Twi (ISO language code)

T/W: Tailwind • Thrust/Weight

TW: ROC (ISO country code) • Tactical Warning • Tail Wind • Taiwan (ISO country code) • Taxiway • Teamwork • Terawatt • Thermal Windowing • Threat Warning • Thrust-Weight • Thumbwheel • Time Words • Total Work • Travelling Wave • Turbulent Wake (region) (Jupiter) • Twickenham (UK postal code) • Typewriter

TWA: Time-Weighted Average • Transaction Work Area • Travelling Wave Amplifier • TW Hya Association • Twin Hills (USA-AK) (location identifier)

TWAIN: Technology Without An Important Name • Toolkit Without An Important Name

TWARS: Third World Association for Remote Sensing

TWAS: Third World Academy of Sciences (Italy)

TWAT: Travelling Wave Amplifier Tube

TWB: Terrestrial Wideband • Toowoomba (Australia-QLD) (location identifier) • Travelling-Wave Beam

TWC: The Weather Channel • Three-Way Catalytic • Total Water Content

TWD: New Dollar (China-ROC) (ISO currency code) • Port Townsend (USA-WA) (location identifier) • Tail Wags Dog • Total World Domination • Toward

TWDS: Twisted Double Shielded

TWE: Taylor (USA-AK) (location identifier) • Test of Written English • Threat Warning Equipment

TWEB: Transcribed Weather Broadcast • Transcribed Weather En route Briefing

TWF: Twin Falls (USA-ID) (location identifier)

TWG: Technical Working Group • Technology Working Group • Telescope Working Group • Test Working Group • TEXnical Working Group • Transfermium Working Group • Transfer Working Group • Transition Working Group

TWGMLC: Technical Working Group on Multiple Language Coordination

TWh: Terawatt Hour

TWH: Catalina Island - Two Harbors (USA-CA) (location identifier)

TWI: Trade Weighted Index • Training Within Industry

TWICE: Travelling Wave-tube Interactive Computational Environment

TWIG: Technology Watch Information Group

TWIMC: To Whom It May Concern

TWIN: Third-World Information Network

TWINS: Two Wide-Angle Neural-atom Spectrometer

TWK: Too Well Known • Typewriter Keyboard

TWL: Threat Wavelength • Total Weight Loss • Travelling-Wave Laser

TwLA: Two-Letter Acronym

TWLT: Twente Workshop on Language Technology

TWM: Tab Windows Manager • Travelling-Wave Maser • Two-Way Mirror

TWN: ROC (ISO country code) • Taiwan (China-ROC) (ISO country code) • Teleprinter Weather Network • Tewantin (Australia-QLD) (location identifier) • Third-World Network

TWNC: Two-Way Non-Coherent

TWNSO: Third World Network of Scientific Organizations

TWOM: Travelling-Wave Optical Maser

TWOV: Transit Without Visa

Twp: Township

TWP: Task Work Package • Torwood (Australia-QLD) (location identifier) • Township • Tropical Western Pacific

TWPA: Travelling-Wave Parametric Amplifier

TWPB: Total Work Package Budget

TWPO: Tropical Western Pacific Ocean

Twr: Tower

TWR: Total Weighted Ranks • Tower

TWRG: Towering

TWS: Telefonisch Waarschuwings-Systeem • Telemanipulation Workstation • Test Workstation • Thermal Wavelength Shift • Trader Workstation System

TWSTT: Two-Way Satellite Time Transfer

TWT: Tawitawi (Philippines) (location identifier) • Transonic Wind Tunnel • Travelling-Wave Tube • Trisonic Wind Tunnel • Tuesday-Wednesday-Thursday

TWTA: Travelling-Wave Tube Amplifier! (* • Travelling-Wave Tube Assembly

TWU: Tawau (Malaysia) (location identifier) • Technical Writing Unit • Transportation Workers Union • Transport Workers Union

TWV: Ten Waarde Van

TWX: Teletypewriter Exchange

Twy: Taxyway • Twenty

TWY: Tawa (Papua New Guinea) (location identifier)

TWZ: Mount Cook - Pukaki-Twizel (New Zealand) (location identifier) • Mt. Cook - Pukaki-Twizel (New Zealand) (location identifier)

Tx: Tax • Taxation • Tonneaux

TX: Telex • Texas (USA) • Transmit • Transmitter

TxD: Transmit Data

TXD: Transmitted Data

TXF: Tax Exchange Format

TXG: Taichung (China-ROC) (location identifier)

TXI: Tunable X-ray Imager

TXK: Texarkana (USA-TX) (location identifier)

TXL: Berlin Tegel (Germany) (location identifier)

TXM: Teminabuan (Indonesia) (location identifier)

TXN: Tunxi (China-PRC) (location identifier)

Txp: Transponder

Txpdr: Transponder

TXR: Tanbar (Australia-QLD) (location identifier)

TXRF: Total X-Ray Fluorescence

TXT: Text

Txtl: Textile

TXT2STF: Text To Structured File

TXU: Tabou (Ivory Coast) (location identifier)

Ty: Territory • Truly

TY: Benin (international aircraft registration code) • Thank You • Total Yield • Typhoon

TYA: Tula (Russia) (location identifier)

TYB: Tibooburra (Australia-NSW) (location identifier)

TYC: Two-Year College

TYD: Tynda (Russia) (location identifier)

TYDAC: Typical Digital Automatic Computer

TYE: Tyonek (USA-AK) (location identifier)

TYF: Torsby (Sweden) (location identifier)

TYG: Thylungra (Australia-QLD) (location identifier)

TYL: Talara (Peru) (location identifier)

TYM: Staniel Cay (Bahamas) (location identifier)

TYN: Taiyuan (China-PRC) (location identifier)

TYNIAC: Tiny Almost Automatic Computer

TYO: Tokyo (Japan) (location identifier)

Typ: Typical • Typing • Typist • Typographer • Typographic • Typographical • Typography

TYP: Tobermorey (Australia-NT) (location identifier) • Typical

Typh: Typhoon

Typo: Typographer • Typographic • Typographical • Typography

Typog: Typographer • Typographic • Typographical • Typography

Typstg: Typesetting

Typw: Typewriter • Typewriting • Typewritten

TYR: Tyler (USA-TX) (location identifier)

TYS: Knoxville (USA-TN) (location identifier)

TYT: Treinta-y-Tres (Uruguay) (location identifier)

TYVM: Thank You Very Much

TYZ: Taylor (USA-AZ) (location identifier)

TZ: Mali (international aircraft registration code) • Tanzania (ISO country code) • Tanzanie (ISO country code) • Time Zero • Truncated Zel'dovich (approximation)

TZA: Belize City (Belize) (location identifier) • Tanzania (ISO country code) • Tanzanie (ISO country code) • Truncated Zel'dovich Approximation

TZB: Technische Zentralbibliothek

TZD: True Zenith Distance

TZK: Temperatur-Zeit-Kristallisation

TZM: Titanium-Zirconium-Molybdenum • Tizimin (Mexico) (location identifier) • Travelling Zone Method

TZN: South Andros (Bahamas) (location identifier)

TZO: Thorne-Zytkow Object

TZP: Time Zero Pulse

TZS: Shilling (Tanzania) (ISO currency code)

TZT: Te Zijner Tijd

TZW: Technologiezentrum Wasser (Germany)

TZX: Trabzon (Turkey) (location identifier)

T1M: Télescope de 1m

T1T: Treburer-1m-Teleskop (Germany)

T2: Tuvalu (international aircraft registration code)

T2L2: Time Transfer by Laser Link

T2M: Télescope de 2m

T3: Kiribati (international aircraft registration code)

T7: San Marino (international aircraft registration code)

T9: Bosnia-Hercegovina (international aircraft registration code)

U

ú: Úsase

u: Atomic Mass Unit • Control Vector • Displacement of Ion • Energy Density • Instantaneous Potential Difference • Lethargy • Micro • Molecular Velocity • Object Distance • Relative Velocity • Sound Particle Velocity • Specific Internal Energy • Speed • Ugly Threatening Sky • Und • Underfloor • Ungerade • Unified Atomic Mass Unit • Uniform • Unit • Unsatisfactory • Unter • Up • Upper • Utility • Velocity Component

Ú: Úsase

U: Electric Tension • Gas Ionization Parameter • Internal Energy • Potential Difference • Rate of Heat Loss • Tension • Total Energy • Unclassified • Uncle • Und • Unidirectional • Unidirectionnel • Uniform (international alphabet) • Union • Unit • Unite • Unité • United • Universe • Université • University • Unix • Unknown • Unsatisfactory • Unterer • Untergrundbahn • Up • Upper • Uranium • Uranus • Uruguay (international vehicle registration code) • Utilisation • Velocity Component • You

ua: Ukraine • unter anderen

uä: und ähnlich

UA: Ukraine (ISO country code) • Unauthorized Absence • Und Andere • Und Anderes • Under Age • Unexplained Anomaly • Unité Administrative • Unité Associée • Unité Astronomique • Unité d'Acquisition • Unité d'Armée • Universitätsarchiv • University of Alabama (USA-AL) • University of Alaska (USA-AK) • University of Arizona (USA-AZ) • Unter Anderem • Unter Anderen • Upper Air • Urbanized Area • User Agent • User Area • Usine d'Aviation • Usque Ad

UAA: Ukrainian Astronomical Association • Universidad Autónoma de Aguascalientes (Mexico) • University Aviation Association (USA) • University of Alaska in Anchorage (USA-AK) • Upper Advisory Area • Uppsala Amatörastronomer (Sweden)

UAAARA: Unión de Asociaciones y Aficionados a la Astronomía de la República Argentina

UAB: Unione Astrofili Bresciani (Italy) • Unitary Awarding Body (UK) • Universitat Autònoma de Barcelona (Spain) • University of Alabama in Birmingham (USA-AL) • Until Advised By

UAC: San Luis - Rio Colorado (Mexico) (location identifier) • Unattended Assembly Call • Universele Artikelcodering • Universele Auteursrechtconventie • Unseen Astrometric Companion • Upper Atmosphere Composition • User Access and Control • User Access Component

UACC: Upper Area Control Center

UACPyP: Unidad Académica de los Ciclos Profesional y de Posgrado

UAD: Unit Assembly Drawing • Universal Antennas Division • User Accommodation Document

UADP: Uniform Automated Data Processing • Uniform Automatic Data Processing

UADPS: Uniform Automated Data Processing System • Uniform Automatic Data Processing System

UADS: User Attribute Data Set

UAE: Mount Aue (Papua New Guinea) (location identifier) • United Arab Emirates • Uninterruptible Application Error • Unrecoverable Application Error

UAEU: United Arab Emirates University

UAF: Unit Authorization File • University-Affiliated Facility • University of Alaska in Fairbanks (USA-AK) • Upper Atmosphere Facility • User Authorization File

UAG: Hryvna (Ukraine) (ISO currency code) • Upper-Atmosphere Geophysics • User Advisory Group

UAH: Ua Huka (Marquesas Islands) (location identifier) • University of Alabama in Huntsville (USA-AL)

UAI: Suai (Indonesia) (location identifier) • Uncertainty Avoidance Index • Union Astronomique Internationale • Union des Associations Internationales • Unione Astrofili Italiani (Italy) • Universal Access to Information • Universal Astronomy Index • Universal Azimuth Indicator • Universitair Instelling Antwerpen (Belgium) • Use As Is

UAK: Narssarssuaq (Greenland) (location identifier)

UAL: Luau (Angola) (location identifier) • Unit Area Loading • Unit Authorization List • Universal Assembly Language • Upper Acceptance Limit • User Adaptive Language • Usinage Assisté par Laser

UALI: Unit Authorization List Item

UALR: University of Arkansas at Little Rock (USA-AR)

uam: und andere mehr • und anderes mehr

UAM: Guam - Anderson (Guam) (location identifier) • Universidad Autónoma de Madrid (Spain) • Universidad Autónoma Metropolitana • University of Arkansas at Monticello (USA-AK)

UAM-I: Universidad Autónoma Metropolitana - Iztapalapa (Mexico)

UAMI: Universidad Autónoma Metropolitana - Iztapalapa (Mexico)

UAMS: Upper-Atmosphere Mass Spectrometer

UAMU: User Assistance and Marketing Unit

UAO: Unexplained Aerial Object • University of Arizona Observatories (USA-AZ) • Urumqi Astronomical Observatory (China-PRC) • Usinage Assisté par Ordinateur • Ussurijsk Astrophysical Observatory (Russia)

UAOB: Upper Air Observation

UAOS: User Alliance for Open Systems

UAP: Ua Pou (Marquesas Islands) (location identifier) • Unidentified Atmospheric Phenomenon • Union des Assurances de Paris • Universal Availability of Publications • Universidad Autónoma de Puebla (Mexico) • University-Affiliated Program • Usine d'Aviation Pilote

UAPI: Uniform Application Programming Interface

UAPT: Union Africaine des Postes et Télécommunications • Union of African Posts et Telecommunications

UAPV: Université d'Avignon et des Pays de Vaucluse (France)

UAQ: San Juan (Argentina-SJ) (location identifier)

UAR: United Arab Republic • Upper Air Route • Upper-Atmosphere Research • Use As Required • User Action Routine

UARC: Upper-Atmosphere Research Collaboration • Upper-Atmosphere Research Collaboratory

UARP: Upper-Atmosphere Research Program

UARS: Upper-Atmosphere Research Satellite • Upper-Atmosphere Research Spacecraft • Upper Atmospheric Research Satellite (*)

UART: Universal Asynchronous Receiver/Transmitter

UARTE: Unidade de Apoio à Rede Telemática Educativa (Portugal)

UAS: Samburu (Kenya) (location identifier) • Uniform Accounting System • Union of African States • Unit Approval System • United Arab States • University of Agricultural Sciences • Upper Airspace • Upper-Atmospheric Sounder • Urban-Area Study • Urgent Action Survey

UASB: Upflow Anaerobic Sludge Blanket

UASIF: Union des Associations Scientifiques et Industrielles Françaises (France)

UAST: Unité Administrative du Service de Transmission

UAT: Ultraviolet Acquisition Technique • Union Africaine des Télécommunications • Universal Access Transceiver • Universal Alias Table

UATI: Union des Associations Techniques Internationales

UATL: Unité d'Action, de Temps et de Lieu • Universal ADA Test Language

UATE: Universal Automatic Test Equipment

UATI: Union des Associations Techniques Internationales

UATM: Unité d'Acquisition de Télémétrie

UAV: Uninhabited Aerial Vehicle • Unmanned Aerial Vehicle (*) • Unmanned Aerospace Vehicle • Upper Atmosphere Vehicle

UAW: United Astronomers of the World • United Auto Workers • Upper-Atmosphere Wind

UAwg: Um Antwort wird gebeten

UAX: Uaxactun (Guatemala) (location identifier) • Unit Automatic Exchange

UAZ: Universidad Autónoma de Zacatecas (Mexico)

U/B: Usage Block

UB: Southall (UK postal code) • Ulaan Baatar (Mongolia) • Ultraviolet-Blue • Undistributed Budget • Unité de Bataille • Universitätsbibliothek • Universiteitsbibliotheek • Unterbegriff • Upper Brace • Usage Block • Utility Bridge • User Background

UBA: Uberaba (Brazil-MG) (location identifier) • Umweltbundesamt (Germany) • Unibus Adapter • Union Bank of Africa

UBB: Mabuiag Island (Australia-QLD) (location identifier)

UBC: Universal Bibliographic Control • Universal Buffer Controller • University of British Columbia (Canada-BC)

UBCIM: Universal Bibliographic Control and International MARC

UBD: User Brain Damage

UBE: Universal Bus Exercisor

UBER: Undetected Bit Error Rate

UBF: Universal Birefringent Filter

UBFFP: Universal Birefringent Filter Fabry-Pérot

UBI: Buin (Papua New Guinea) (location identifier)

UBIC: Universal Bus Interface Controller

UBIDEP: Universal Building Information and Data Exchange Protocol

UBIS: Universitätsbibliothek Informationsstelle • Users of Book Industry Standards

UBITRON: Undulating Beam Interaction Electron

UBJ: Ube (Japan) (location identifier) • Univariant Box-Jenkins

UBK: Universelle Bibliographische Kontrolle

UBM: Unité de Bande Magnétique • Unpressurized Berthing Mechanism

UBMTA: Uniform Biological Materials Transfer Agreement

UBO: Université de Bretagne Occidentale (France)

UBP: Ubon Ratchathani (Thailand) (location identifier) • Umweltbelastungspunkt

UBR: Uniform Business Rate • Unspecified Bit Rate • Upper Brazing Region

UBS: Columbus (USA-MS) (location identifier) • Université de Bretagne-Sud (France) • US Biological Survey

UBT: Ubatuba (Brazil-SP) (location identifier)

UBU: Kalumburu (Australia-WA) (location identifier)

UBV: Ultraviolet-Blue-Visual

u/c: Under Charge

uc: Upper Case

U/C: Under Current • User Code

UC: Ultracompact • Ultracompression • Ultracompressor • Under Construction • Undercover • Under Current • Unemployment Compensation • Union Catalogue • Unit Chairman • Unit Cost • Unité Centrale • Unité de Cadre • Unité de Combat • Unité de Compte • University of California (USA-CA) • University of Chicago (USA-IL) • University College • Universidad Complutense (Spain) • Unoperated Control • Unsatisfactory Condition • Unseen Companion • Uplink Command • Upper Canada • Upper Case • Usable Control • User Code • User Committee • Users Committee • Using Command • Utilization Control

UCA: Unitary Convolution Approximation • University of Central Arkansas (USA-AR) • Upper-Case Alphabet • Universal Calibration Adapter • Upper Control Area • USNO CCD Astrograph • Utica (USA-NY) (location identifier)

UCAC: USNO CCD Astrograph Catalog

UCAID: University Corporation for Advanced Internet Development

UCAR: University Corporation for Atmospheric Research • Unsatisfactory Condition Action Report

UCAS: Union of Central African States • Universities and Colleges Admissions Service (UK)

UCASS: University Computer-Assisted Search Service

UCAV: Uninhabited Combat Aerial Vehicle • Uninhabited Combat Air Vehicle (*)

UCB: Ufficio Centrale Brevetti (Italy) • Ultraviolet Cosmic Background • Union Chimique Belge (Belgium) • Unit Control Block • University of California at Berkeley (USA-CA) • Unless Caused By • Utilization Control Board

UCBL: Université Claude Bernard de Lyon (France)

UCBT: Universal Circuit Board Tester

UCC: Umbilical Checkout Cable • Unadjusted Contractual Cable • Unified Classification Code • Uniform Commercial Code • Union Carbide Corporation • Unité de Chaleur Centésimale • Unité de Conditionnement Centrale • Universal Checkout Console • Universal Copyright Convention (1952) • University College Cork (Ireland) • Upper Control Center • Utility Control Console • Yucca Flat (USA-NV) (location identifier)

UCCD: Undergraduate Course and Curriculum Development

UCCEGA: Union des Chambres de Commerce et des Établissements Gestionnaires d'Aéroports (France)

UCCIFE: Union des Chambres de Commerce et d'Industrie Françaises à l'Étranger

UCCS: Universal Camera Control System • University of Colorado at Colorado Springs (USA-CO)

UCCSN: University and Community College System of Nevada (USA-NV)

UCD: Uniform Call Distribution • Uniform Call Distributor • University College Dublin (Ireland) • University of California at Davis (USA-CA) • Unstructured Cell Data • Upper Critical Depth • Urine Collection Device • User-Centered Design

UCDL: University of California Digital Library

UCDP: Uncorrected Data Processor

UCE: Eunice (USA-LA) (location identifier) • Ultimate Computer Environment • Unió de Consumidores Españoles (Spain) • Unit Checkout Equipment • Unit Correction Entry • Unité de Compte Européenne • Unité de Contrôle Espace • Unsollicited Commercial E-mail

UCEC: University of Chicago Engineering Center (USA-IL)

UCET: Universities Council for the Education of Teachers

UCF: Union Catalogue File • Union Circulation File • University of Central Florida (USA-FL)

UCFM: Unitree Central File Manager

UCG: Unité de Calcul et de Gestion • Uppsala Catalog of Galaxies

UCGIS: University Consortium for Geographic Information Science

UCHII: Ultra-Compact HII (region)

UCHR: Ultra-Compact HII Region

UCI: Unidad de Cuidados Intensivos (Spain) • Union Cycliste Internationale • Unit Construction Index • University of California at Irvine (USA-CA) • Utility Card Input

UCIS: Uniterm Coordinate Index System • Univesity Center for International Studies (USA) • Uprange Computer Input System

UCITA: Uniform Computer Information Transactions Act (USA)

UCITS: Undertaking for Collective Investment in Transferable Securities

UCK: Lutsk (Ukraine) (location identifier)

Uckerm: Uckermark

UCL: Universal Communications Language • Université Catholique de Louvain (Belgium) • University College London (Starlink internal code) • University College London (UK-England) • Update Control List • Upper Control Limit • User Control Language

UCLA: University of California at Los Angeles (USA-CA)

UCLES: University College London Echelle Spectrograph

UCLGES: Unified Cartographic Line Graph Encoding System

UCLJ: University of California at La Jolla (USA-CA)

UCLXE: University College of London Experiment

UCM: Unité de Calcul Miniaturisée • Universal Cable Module • Universidad Complutense de Madrid (Spain)

UCMS: Unit Capability Measurement System

UCN: Buchanan (Liberia) (location identifier) • Ultra-Centrifuge Nederland BV • Uniform Control Number • Universidad Católica del Norte (Chile)

UCNW: University College of North Wales (UK-Wales)

UCO: University of California Observatories (USA-CA) • User Consultancy Office

UCOAC: University of California Observatories Advisory Council

UCOIF: University of California Observatories Instrumentation and Facilities

UCOS: Universal Checkout Station • Uprange Computer Output System

UCOSP: University of California Observatories Science Policy Committee

UCP: Unified Command Plan • Unité de Commande du Processeur • University of Chicago Press (USA-IL) • Usinage Chimique Progressif • Utility Control Program

UCPTE: Union for the Coordination of the Production and Transport of Electricity

UCPU: Urine Collection and Pretreatment Unit

UCR: Unconditioned Reflex • Under-Color Removal • Uniform Crime Report • Unit Card Reader • Unité de Circulation Routière • Unité de Commande du Réseau • Unité de Contrôle de Transmission • Universidad de Costa Rica • University of California at Riverside (USA-CA) • Unsatisfactory Condition Report • User Control Routine

UCRC: University Cooperative Research Center

UCRL: University of California Radiation Laboratory

UCRP: University Collaborative Research Programme

UCS: Unconditioned Stimulus • Uniform Chromaticity Scale • Uniform Color Scale • Union of Concerned Scientists (USA) • Unit Count System • Universal Card Scanner • Universal Character Set • Universal Classification System • Universal Command System • Universal Control System • University Computer Services • Urine Collection System • Utilities Control System

UCSB: University of California at Santa Barbara (USA-CA)

UCSC: Universal Communications Switching Device • University of California at Santa Cruz (USA-CA)

UCSD: University of California at San Diego (USA-CA)

UCSF: University of California at San Francisco (USA-CA)

UCSTRI: Unified Computer Science Technical Report Index

UCSW: University College of South Wales (UK-Wales)

UCT: Ukhta (Russia) (location identifier) • Uncorrelated Target • Units Compatibility Test • University of Cape Town (South Africa) • User Communication Tracking

UCTA: Urine Collection Transfer Assembly

UCTM: Unité de Contrôle de la Télémétrie • Unité de Contrôle et de Télémesure

UCV: Unidade Ciência Viva (Portugal) • Universidad Central de Venezuela

UCW: Ulysses Comet Watch • Unit Control Word • University College of Wales (UK-Wales)

UCWE: Upper Coupling of Wave Energy • Upward Coupling of Wave Energy

UCY: Union City - Everett-Stewart (USA-TN) (location identifier)

ud: Ut Dictum

Ud: Usted

U/D: Under Deed • Update

UD: Udmurtian • Unavoidable Delay • Unidirectional • Uniform Distribution • Uniquely Decipherable • Unit Designation • Unit Director • Unité à Disque • Unité de Documentation • Unité Dynamique • Universal Dipole • Universitair Docent • Update • Usage Data • Ut Dictum

UDA: User Data Area

UDAS: Unified Direct-Access System

UDB: Unified Database • Universal Database • Up-Data Buffer • Update Buffer • Urban Development Board

UDC: Universal Decimal Classification • Up-Down Counter • Urban Data Center • User-Defined Command • Utility Distribution Center

UDD: Palm Springs - Bermuda Dunes (USA-CA) (location identifier)

UDDF: Updrafts/Downdrafts

UDE: Uden Volkel (Netherlands) (location identifier) • Universal Data Exchange

UDEAC: Union Douanière des États de l'Afrique Centrale • Union Douanière et Économique de l'Afrique Centrale

UDEAO: Union Douanière des États d'Afrique de l'Ouest

UDES: User Operational Evaluation System

UDF: UAV Demonstration Flight • UHF Direction Finder • Unducted Fan • Unit Development Folder • Universal Data Format • Universal Disk Format • User-Defined Function • Utility and Data Flow

UdeG: Universidad de Guadalajara (Mexico)

UdG: Universitat de Girona (Spain)

u. dgl.: und dergleichen

UDHD: User Defined Data Field

UDHR: UNO Declaration on Human Rights

UDI: Uberlândia - Eduardo Gomes (Brazil-MG) (location identifier) • Uniform Driver Interface • Unique Data Item • Universal Document Identifier

UDID: Universal Data Identification • Unique Data Item Description • User Defined Image Data Field

UDJ: Uzhgorod (Ukraine) (location identifier)

UDK: Universal Decimalklassifikationssystem • User Defined Key • User Domain Knowledge

UDL: Uniform Data Language • Uniform Data Link • Unique Data Log • Up-Data Link • Update Link

UDLC: Universal Data Link Control

udM: unter dem Meeresspiegel

üdM: Über dem Meeresspiegel

UDM: Universal Docking Module

UDMH: Unsymmetrisches Dimethylhydrazine • Unsymmetrical Dimethylhydrazine

UDN: Udine (Italy) (location identifier) • User Densification Network

UDO: Udomxay (Laos) (location identifier)

UDOP: Ultrahigh Doppler

UDP: Uniform Data Processing • Unitary Development Plan • Unit Data Processing • Unix Datagram Protocol • Upper Deck Pressure • User Datagram Protocol (*) • User Data Packet • User Data Protocol • User Development Program

UDR: Udaipur (India) (location identifier) • Ultra-Dipole Radiation • Universal Digital Readout • Universal Document Reader • Unprocessed Data Record • Urgent Data Request • Usage Data Report • Utility Data Reduction

UDRC: Utility Data Retrieval Control

UDRO: Utility Data Retrieval Output

Uds: Ustedes

UDS: Ulysses Data System • Unified Data System • Unité de Détection au Sol • Universal Digital Switch • Universal Documentation System • Unresolved Double Star • US Differential System • User Database Searching • User Data System • Utility Distribution System

UDSC: Usuda Deep Space Center (Japan)

UDSL: Universal Digital Subscriber Line

UDSPL: Ultrasonics and Digital Signal Processing Laboratory

UdSSR: Union der Sozialistischen Sowjetrepubliken

udT: unter dem Titel

UDT: Uniform Data Transfer • Universal Dataflow and Telecommunications • User Display Terminal

Udtd: Updated

UDTS: Universal Data Transfer Service • Universal Data Transmission System

UDU: Underwater Dive Unit

UDW: University of Durban-Westville (South Africa)

uE: unseres Erachtens

UE: Unión Europea • Union Européenne (formerly CEE) • Unité Électronique • Unit Entry • Unit Equipment • Unit Establishment • Univers Émotionnel • University Entrance • University Extension • University of the East (Philippines) • Unrestricted Estimator • Until Exhausted • Useful Elevation • User Enterprise • User Environment • User Equipment

UEA: Unattended Equipment Area • University of East Anglia (UK-England) • University of East Asia

UEAC: Union des États de l'Afrique Centrale

UEB: Upper Equipment Bay

UEBL: Union Économique Belgo-Luxembourgeoise

UEBS: Union des Entreprises de la Base Spatiale

UEC: Unité d'Échange et de Couplage • User Environment Component • Users Executive Council

UECBu: University Electro Communications Bulletin (periodical code)

UED: University Education Diploma

UEE: Queenstown (Australia-TAS) (location identifier) • Unit Essential Equipment

UEF: Uniform Electric Field • Universal Extra Fine

UEG: Union Européenne des Géosciences

UEGN: Unified European Gravity Network

UEI: Union of Educational Institutions (UK) • United Electronic Industries

UEIP: Ural Electrochemical Integrated Plant

UEL: Quelimane (Mozambique) (location identifier)

UELF: Union des Éditeurs de Langue Française (France)

UELV: Ultralight Expandable Launch Vehicle • Ultra-lite Expandable Launch Vehicle

UEM: Unión Económica y Monetaria • Union Économique et Monétaire • Union Européenne Monétaire • Unité Électromagnétique • Universal Electron Microscope • Universal Electron Microscopy

UEMMA: Unsollicited Electronic Mail with Multiaddressing

UEN: Universität Erlangen-Nürnberg (Germany)

UEO: Kume Jima (Japan) (location identifier) • Union de l'Europe Occidentale (*) • Unione dell'Europa Occidentale • Union Européenne Occidentale

UEP: Uniform External Pressure • Union Européenne des Paiements • Unplanned Event Pickup

UER: Union Européenne de Radiodiffusion • Unique Equipment Register • Unité d'Enseignement et de Recherche • University Entrance Requirements • Unplanned Event Record • Unsatisfactory Equipment Report

UERE: User Equivalent Range Error

UERL: Unplanned Event Record Log

UES: Uniform Emission Standard • Union Européenne et Suisse • Unité Électrostatique • Utrecht Echelle Spectrograph • Waukesha (USA-WI) (location identifier)

UESAC: Uppsala-ESO Asteroid and Comet Survey

UET: Quetta (Pakistan) (location identifier) • User Earth Terminal

UETA: Uniform Electronic Transactions Act (USA)

UEX: Underexposed

U/F: Unique File

UF: Ultrafilter • Ultrafiltration • Ultrafine • Ultrasonic Frequency • Unique File • Unité de Force • Universal Format • University of Florida (USA-FL) • Used For • User Foreground • Usine Frigorifique • Utilization Factor • Utilization Flight

UFA: Ufa (Russia) (location identifier) • University Film Association (USA) • Until Further Advised • Upper (tropospheric) Formation of Aerosols • Use Frequency Analysis

UFACNC: Ultrafine Aerosol Condensation Nucleus Counter

UFAG: Union Française des Associations Graphiques (France)

UFAM: Universal File Access Method

UFAS: Unified Flight Analysis System

UFBE: Union Francophone des Belges à l'Étranger

UFC: Unified Flight Classification • Universal Flight Computer • Universities Funding Council

UFCV: Union Française des Centres de Vacances (France)

UFD: User File Directory

UFE: Undergraduate Faculty Enhancement

UFES: Universidade Federal de Espíritu Santo (Brazil-ES)

Uff: Ufficiale

UF-F: Universal Flip-Flop

UFF: Universal Flip-Flop • Universality of Free Fall • Upper Forward Fuselage

UFG: Universidade Federal de Goiás (Brazil-GO)

UFI: Uniform File Identifier • Union des Foires Internationales • Universal Fermi Interaction • User-Friendly Index • User-Friendly Interface

UFIRS: Universal Far-Infrared Sensor

UFJF: Universidade Federal de Juiz de Fora (Brazil-MG)

UFL: Upper Flammability Limit

UFMG: Universidade Federal de Minas Gerais (Brazil-MG)

UFN: Until Further Notice

UFO: UHF Follow-On • Unbekanntes Flugobjekt • Unidentifiable Format Object • Unidentified Flying Object • Universal Fiber Optic

UFOD: Union Française des Organismes de Documentation (France)

UFOE: Upgraded Fiber Optic Echelle

UFOIRC: Unidentified Flying Object Information Retrieval Center

UFOP: Universidade Federal de Ouro Preto (Brazil-MG)

UFP: Ultra-Fine Powder • Université Française du Pacifique • Unix File Transfer

UFPR: Universidade Federal de Parana (Brazil)

UFR: Ultra-Fast Rotator • Unité de Formation et de Recherche (France)

UFRGN: Universidade Federal do Rio Grande do Norte (Brazil-RN)

UFRGS: Universidade Federal do Rio Grande do Sul (Brazil-RS)

UFRJ: Universidade Federal de Rio de Janeiro (Brazil-RJ)

UFRO: University of Florida Radio Observatory (USA-FL)

UFS: Umweltforschungssatellit • Unconnected Fine Structure • United Feeder Service • Unix File System • Unresolved Fine Structure

UFSA: Union Française des Sociétés d'Auteurs (France)

UFSC: Universidade Federal de Santa Catarina (Brazil-SC)

UFSM: Universidade Federal de Santa Maria (Brazil-RS)

UFSS: Unified Flexible Spacecraft Simulation

UFT: Ultrasonic Frequency Transformer • Ultrasonic Frequency Translator • Unified Field Theory • United Federation of Teachers

UFTI: UKIRT Fast Track Imager

UFTP: Upper Feedthru Plate

UFVA: University Film and Video Association (USA)

UFZ: Umweltforschungszentrum

u/g: Underground

Ug: Uganda

UG: Uganda (ISO country code) • Underground • Unité de Gestion • Universal Generalization • Universal Grammar • User Group

UGA: Bulgan (Mongolia) (location identifier) • Uganda (ISO country code) • Unity Gain Amplifier • University of Georgia at Athens (USA-GA) • User Global Area

UGAMP: Universities Global Atmospheric Modelling Programme (*) • Universities Global Atmospheric Modelling Project

UGB: Pilot Point - Ugashik Bay (USA-AK) (location identifier) • Urban Growth Boundary

UGBW: Unity Gain Bandwidth

UGC: University Grants Commission • University Grants Committee • Uppsala General Catalogue • Urgench (Uzbekistan) (location identifier)

UGCG: Uppsala General Catalogue of Galaxies

UGGI: Union Géodésique et Géophysique Internationale

UGI: Uganik (USA-AK) (location identifier) • Union Géographique Internationale

UGID: Universal Geographic Identity

UGLI: Universal Gate for Logic Implementation

UGM: Universitas Gadjah Mada (Indonesia) • Unusual Ground Marking • User Group Maintenance

UGN: Waukegan Memorial (USA-IL) (location identifier)

Ugnd: Underground

UGO: Uige (Angola) (location identifier) • User-Group-Other

UGS: Undergraduate Student • Uniaxial Gyrostabilizer

UGSJR: US Geological Survey Journal Research (periodical code)

UGV: Unmanned Ground Vehicle • Usinage à Grande Vitesse

UGX: Shilling (Uganda) (ISO currency code)

UH: Ultra Heavy • Unit Head • University of Hawaii (USA-HI) • University of Houston (USA-TX) • Upper Half • Usine Hydraulique

UHA: Université de Haute-Alsace (France) • Upper Half Assembly • Uppsala H-Alpha

UHAO: University of Hawaii Adaptive Optics

UHB: Ultra-High Bypass • Urban Haute Bourgeoisie • User Home Base

UHC: Ultrahigh Contrast • Unburnt Hydrocarbon

UHCA: Ultra-High Capacity Aircraft

UHCI: Universal Host Controller Interface

UHCL: University of Houston - Clear Lake (USA-TX)

UHCR: Ultra-High-energy Cosmic-Ray

UHCRE: Ultra-High Cosmic-Ray Experiment

UHCS: Ultrahigh Capacity Storage

UHD: Universitair Hoofdocent

UHE: Uherske Hradiste (Czech Rep.) (location identifier) • Ultrahigh Energy

UHECR: Ultra-High-Energy Cosmic Ray

UHF: Ultra-Haute Fréquence • Ultrahigh Frequency • Unrestricted Hartree-Fock • Upper Heyford (UK) (location identifier)

UHFF: Ultrahigh Frequency Filter

UHFG: Ultrahigh Frequency Generator

UHFO: Ultrahigh Frequency Oscillator

UHFR: Ultrahigh Frequency Receiver

UHI: Urban Heat Island

UHM: Ultra-High Modulus • Universal Horizontal Metroscope

UHMS: Urban Hydrology Monitoring System

UHMW: Ultrahigh Molecular Weight

UHO: University of Hertfordshire Observatory (UK-England)

UHP: Ultrahigh Power

UHPTRS: Ultra-High Precision Time Reference Station

UHR: Ultra-Harmonic Resonance • Ultra-High Reduction • Ultra-High Resolution • Upper Hybrid Resonance

UHRF: Ultra-High Resolution Facility

UHRXS: Ultrahigh Resolution XUV Spectroheliograph

UHS: Ultrahigh Speed • Ultramedia Host Support

UHT: Ultra-Haute Température • Ultra-Heat Treated • Ultrahigh Temperature • Umbilical Handling Technician • Universal Hand Tool

UHV: Ultrahigh Vacuum • Ultrahigh Voltage

UHVC: Ultrahigh Vacuum Chamber

UHVCVD: Ultrahigh Vacuum Chemical Vapor Deposition

UHVS: Ultrahigh Vacuum System

u/i: Under Instruction

ui: Ut Infra

U/I: Unidentified

U-I: Union-Intersection

UI: Ariel (satellite of Uranus) • Understemming Index • Unemployment Insurance • Union-Intersection • Unité Internationale • Unit of Issue • Universal Instantiation • University of Ibadan (Nigeria) • University of Illinois (USA-IL) • Unix International • Unnumbered Information • Upgren Intermediate • Upper Intercept • User Institution • User Interaction • User Interface

UIA: Ukraine International Airlines • Union Internationale des Architectes • Union of International Associations • Universidad Iberoamericana • Universitaire Instelling Antwerpen (Belgium) • Usable Inside Area

UIAA: Union Internationale des Astronomes Amateurs

UIAS: University of Illinois Astronomical Society (USA-IL)

UIB: Quibdo (Colombia) (location identifier) • Unidentified Infrared Band • Unidentifiziert Infrarotband • Unité d'Interface Bus • Universidad de las Islas Baleares (Spain)

UIC: Underground Injection Control • Union des Industries Chimiques (France) • Union Internationale des Chemins de fer • University of Illinois at Chicago (USA-IL) • Urban Information Center • User Identification Code

UICA: Union of Independent Colleges of Art (USA)

UICC: Union Internationale Contre le Cancer

UICN: Union Internationale pour la Conservation de la Nature

UICPA: Union Internationale de Chimie Pure et Appliquée

UID: Unique Identifier • User Identification • User Identificator • User Identifier

UIDE: User Interface Design Environment

UIDS: User Interface Design System

UIE: UNESCO Institute for Education • Union Internationale d'Électrothermie • Union Internationale des Éditeurs • Union Internationale des Étudiants

UIEO: Union of International Engineering Organizations (now UITA)

UIF: User Interface

UIFN: Universal International Freephone Number

Uig: Uighur

UIG: User Instruction Group

UIH: Qui Nhon (Viet Nam) (location identifier)

UIHPS: Union Internationale d'Histoire et de Philosophie des Sciences

UII: Umbriel (satellite of Uranus) • Utila (Honduras) (location identifier)

UIIG: Union Internationale de l'Industrie du Gaz

UIII: Titania (satellite of Uranus)

UIIP: User Integration Implementation Plan

UIJ: Union Internationale des Journalistes

UIJPLF: Union Internationale des Journalistes de la Presse de Langue Française

UIK: Ust-Ilimsk (Russia) (location identifier)

UIL: Quillayute State (USA-WA) (location identifier) • Undergraduate Instructional Laboratory • Undergraduate Instruction Language • User Interface Language

UIM: Unit Information Module • User Interface Module

UIMP: Universidad Internacional Menéndez Pelayo (Canary Islands)

UIMS: User Interface Management System

UIN: Quincy (USA-IL) (location identifier)

UIO: Quito - Mariscal Sucre (Ecuador) (location identifier) • Unique Input/Output • Utility Iterative Operation

UIOD: User Input/Output Device

U/IP: User/International Partner

UIP: Quimper Cornouaille (France) (location identifier) • Quimper Pluguffan (France) (location identifier) • Union Internationale de Physique • Union Interparlementaire • Universal International Pictures • Université Interdisciplinaire de Paris (France) • User Interaction Protocol • User Interface Processor • User/International Partner • Utility Interface Panel

UIPE: United Institute for Physics of the Earth (Russia)

UIPPA: Union Internationale de Physique Pure et Appliquée

UIPPI: Union Internationale pour la Protection de la Propriété Industrielle

UIPRE: Union Internationale de la Presse Radiotechnique et Électronique

UIPT: Ukrainian Institute of Physics and Technology • Utilization Integrated Product Team

UIQ: Quine Hill (Vanuatu) (location identifier)

UIR: Quirindi (Australia-NSW) (location identifier) • Unidentified Infrared • University Industrial Research • Upwelling Infrared

UIRB: Unidentified Infrared Band

UIS: Ubiquitous Information System • University of Illinois at Springfield (USA-IL) • Urban Information System • User Information Service • User Interface Service • User Interface System • User Interface Specification

UISB: Union Internationale des Sciences Biologiques

UISM: User Interface Standards Manual

UIST: User Interface Software and Technology

UIT: Jaluit Island (Marshall Island) (location identifier) • Ultraviolet Imaging Telescope • Union Internationale des Télécommunications • Unit Investment Trust

UITA: Union of International Technical Associations

UITF: Union des Ingénieurs et Techniciens utilisant la langue Française

UITP: Union Internationale des Transports Publics

UIUC: University of Illinois at Urbana-Champaign (USA-IL)

UIV: Oberon (satellite of Uranus)

UIW: University of the Incarnate Word (USA-TX)

UIWG: User Interface Working Group

UIX: Cressida (satellite of Uranus)

UIW: Usable Inside Width

UIZ: Utica Berz-Macomb (USA-MI) (location identifier)

UJ: Union Jack • Universal Joint

UJCL: Universal Job Control Language

UJE: Ujae Island (Marshall Island) (location identifier)

UJF: Université Joseph Fourier (France)

UJJEF: Union des Journaux et Journalistes d'Entreprises de France

UJT: Unijunction Transistor

uk: Ukrainian (ISO language code) • United Kingdom

UK: Ukraine • Underkommitté • United Kingdom • University of Kansas (USA-KS) • Uzbekistan (international aircraft registration code)

UKAC: United Kingdom Automation Council

UKAEA: United Kingdom Atomic Energy Authority

UKAIS: United Kingdom Academy for Information Systems

UKAPE: United Kingdom Association of Professional Engineers

UKATC: United Kingdom Astronomy Technology Centre

UKB: Kobe (Japan) (location identifier) • Universal Keyboard

UKCIS: United Kingdom Chemical Information Service

UKCOSA: United Kingdom Council for Overseas Student's Affairs

UKCSU: University of Kent at Canterbury Students' Union (UK-England)

UKDC: United Kingdom Data Centre

UKERNA: United Kingdom Education and Research Networking Association

UKESRU: United Kingdom Earth Satellite Research Unit

UKFCC: United Kingdom Federation for Culture Collections

UKgal: United Kingdom Gallon

UKGSG: United Kingdom Gemini Support Group

UKHG: Ultrakompakte H-II Gebiete

UKI: Ukiah (USA-CA) (location identifier)

UKIRT: United Kingdom Infrared Telescope

UKIS: Universal Knowledge-based Imaging System

UKISC: United Kingdom Industrial Space Committee

UKK: Ust-Kamenogorsk (Kazakhstan) (location identifier)

UKL: Pound Sterling (currency)

UKLT: United Kingdom Large Telescope

UKM: Universiti Kebangsaan Malaysia

UKMO: United Kingdom Meteorological Office

Ukn: Unknown

UKN: Unknown • Waukon (USA-IA) (location identifier)

UKNCPRS: United Kingdom National Committee for Photogrammetry and Remote Sensing

UKOL: United Kingdom On Line

UKOLN: United Kingdom Office for Library Networking

UKOLUG: United Kingdom Online User Group

UKOP: United Kingdom Official Publications

Ukr: Ukraine • Ukrainian

UKR: Mukeiras (Yemen) (location identifier) • Ukraine (ISO country code)

UKRDAC: United Kingdom ROSAT Data Archive Centre

UKS: United Kingdom Satellite • United Kingdom Schmidt

UKSC: United Kingdom Survey Centre

UKSEDS: United Kingdom Students for the Exploration and Development of Space

UKST: United Kingdom Schmidt Telescope

UKSTU: United Kingdom Schmidt Telescope Unit

UKT: Quakertown - Upper Bucks (USA-PA) (location identifier)

UKTS: United Kingdom Treaty Series

UKU: Nuku (Papua New Guinea) (location identifier)

UKW: Ultrakurzwellen

UKX: Ust-Kut (Russia) (location identifier)

UKY: Kyoto (Japan) (location identifier)

U/L: Uplink

UL: Ulci • Ultraléger • Ultralinear • Ultraluminous • Underload • Underwriters Laboratories • Unit of Length • Universiteit Leiden (Netherlands) • University Libraries • University Library • Unordered List • Upgren Late • Uplink • Upload • Upper Left • Upper Level • Upper Limb • Upper Limit • Upper Lobby • User Language

ULA: San Julian (Argentina-SC) (location identifier) • Uncommitted Logic Array • Universidad de Los Andes (Venezuela)

ULANA: Unified Local-Area Network Architecture

ULAS: Union List of Astronomy Serials

ULB: Ulei (Vanuatu) (location identifier) • Universitäts- und Landesbibliothek (Germany) • Université Libre de Bruxelles (Belgium)

u&lc: Upper & Lower Case

ULC: Santiago - Los Cerrillos (Chile) (location identifier) • Underwriters' Laboratories of Canada • Unitary Launch Concept • Unitary Launch Control • Unité Logique de Contrôle • Universal Logic Circuit • Unpressurized Logistics Carrier

ULCAS: Unpressurized Logistics Carrier Attach System

ULCC: Ultra Large Crude Carrier • University of London Computer Centre • Utility Location and Coordination Council

ULCI: Unpressurized Logistics Carrier Interface

ULD: Ulundi (South Africa) (location identifier) • Unit Load Device • Unit Logic Device • Upper Level Discriminator

ULDA: Uniform Low-Dispersion Archive

ULDB: Ultra-Long-Duration Balloon

UL/DL: Uplink/Downlink

ULDL: Uplink-Downlink

ULDMI: Ultraprecise Laser Distance Measuring Instrument

ULE: Sule (Papua New Guinea) (location identifier) • Ultra-Low Expansion • Unit Location Equipment

ULEIS: Ultra-Low-Energy Isotope Spectrometer

ULET: Ultra-Low Energy Telescope

ULEV: Ultra-Low Emission Vehicle

ULF: Ultralow Field • Ultralow Frequency • Upper Limiting Frequency

ULg: Université de Liège (Belgium)

ULG: Ulgit (Mongolia) (location identifier)

ULI: Ulithi (Caroline Islands) (location identifier) • Universal Laboratory Interface • Universal Language Interface • Universal Logic Implementer

ULIBG: Ultra-Luminous Infrared Bright Galaxy

ULICS: University of London Institute of Computer Science

ULIG: Ultraluminous Infrared Galaxy

ULIPC: Upper Layer Interprocess Communication

ULIRG: Ultra-Luminous Infrared Galaxy • Ultra-Luminous IRAS Galaxy

ULL: Mull (UK) (location identifier) • Unit Local Loading

ULLV: Unmanned Lunar Logistics Vehicle

ULM: New Ulm (USA-MN) (location identifier) • Ultra-Léger Motorisé • Ultrasonic Light Modulator • Universal Logic Module

ULN: Ulaanbataar (Mongolia) (location identifier) • Unité de Liaison Numérique • Universal Link Negotiation

ULO: Ulaangom (Mongolia) (location identifier) • University of London Observatory (UK-England) • Unmanned Launch Operations • Unmanned Lunar Orbiter

ULP: Quilpie (Australia-QLD) (location identifier) • Unfair Labor Practice • Université Louis Pasteur (France) • University of London Press (UK-England) • Upper-Layer Protocol • Upper-Level Protocol

ULQ: Tulua Farfán (Colombia) (location identifier)

ULS: Mulatos (Colombia) (location identifier) • Union List of Serials • Union List System • Upload Station • Upward-Looking Sonar • User Location Service

ULSB: Ultra-Low Surface Brightness

ULSD: Ultra-Low-Sulphur Diesel

ULSI: Ultra-Large-Scale Integration • United Launch Systems International (Australia)

ULST: Uniform Low Stratus Test

ULSV: Unmanned Launch Space Vehicle

Últ: Último

Ult: Ultimate • Ultimately • Ultimo

ULT: Ultimate

ULTD: Unit Level Training Device

UltIm: Ultrasonic Imaging (periodical code)

Ultmi: Ultramicroscopy (periodical code)

Ulto: Ultimo

Ultra: Ultrasonics (periodical code)

ULU: Gulu (Uganda) (location identifier)

ULUT: UCB Lookup Table

ULV: Ultralow Volume • Unmanned Launch Vehicle

ULX: Ulusaba (South Africa) (location identifier)

ULY: Ulyanovsk (Russia) (location identifier)

ULZ: Uliastai (Mongolia) (location identifier)

u/m: Undermentioned

uM: unter dem Meeresspiegel

üM: Über dem Meeresspiegel

U/M: Unit of Measure • Unmanned • Unscheduled Maintenance

UM: Mauritanian Ouguiya (currency) • Minor Outlying Islands (United States_) (ISO country code) • Umbilical Mast • Under Mentioned • Unité de Mesure • United States Minor Outlying Islands (ISO country code) • Unité Marine • Unité Motrice • Unit of Measure • Universal Monitor • Universiteit van Maastricht (Netherlands) • University of Maryland (USA-MD) • University of Massachusetts (USA-MA) • University of Miami (USA-FL) • University of Michigan (USA-MI) • University of Minnesota (USA-MN) • Unmanned • Unscheduled Maintenance • User Model • User's Machine • Usine Mécanique

UMa: Ursa Major - Ursae Majoris (constellation)

uma: Unité de Masse Atomique

UMA: Punta de Maisi (Cuba) (location identifier) • Unified Memory Architecture • Union du Maghreb Arabe • Unité de Masse Atomique • Universal Mail Agent • Universal Measurement Assembly • Universal Measurement Amplifier • Universal Multiple Access • Upper Memory Area • Uppsala Model Atmosphere

UMAB: University of Maryland at Baltimore (USA-MD)

UMaG: Ursa Major Group

UMASS: Unlimited Machine Access from Scattered Sites

Umb: Umbilical

UMB: Umbilical • Umnak Island (USA-AK) (location identifier) • Université Marc Bloch (France) • Upper Memory Block

Umbc: Umbilical

UMBC: Umbilical Cord • University of Maryland - Baltimore County (USA-MD)

Umbl: Umbilical

UMC: Umba (Papua New Guinea) (location identifier) • United Microelectronics Corporation • Universal Multizone Crystallizator • Universidade de Mogi das Cruzes (Brazil) • University of Missouri at Columbia (USA-MO) • Upper Mantle Committee

UMCP: University of Maryland at College Park (USA-MD)

UMCS: University Marie Curie-Sklodowska (Poland)

UMD: University of Minnesota at Duluth (USA-MN) • Uummannaq (Greenland) (location identifier)

UMDL: University of Michigan Digital Library (USA-MI)

UME: Ulusai Metroloji Enstitüsü (Turkey) • Umeå (Sweden) (location identifier) • Undergraduate Mathematics Education • Union Monétaire Européenne • Unit Mission Equipment

UMF: Ultra-Microfiche • Uniform Magnetic Field • User Master File

UMH: Université de Mons-Hainaut (Belgium)

UMi: Ursa Minor - Ursae Minoris (constellation)

UMI: Minor Outlying Islands (United States_) (ISO country code) • Quincemil (Peru) (location identifier) • UMIST (Starlink internal code) • United Microelectronics Inc. • United States Minor Outlying Islands (ISO country code) • University Microfilms International • University of Manchester Institute of Science and Technology (Starlink internal code)

UMIACS: University of Maryland's Institute for Advanced Computer Studies (USA-MD)

UMIF: Upper Mechanical Integration Frame

UMIS: Urban Management Information System

UMIST: University of Manchester Institute of Science and Technology (UK-England)

UMK: Universal Meßkamer

UML: Unified Modelling Language

UMLE: Unrestricted Maximum-Likelihood Estimator

UMLS: Unified Medical Language System

UMM: Summit (USA-AK) (location identifier) • Universal Measuring Machine

UMO: Unmanned Orbital

UMOC: Ugliest Man On Campus

UMOL: Unmanned Orbital Laboratory

UMP: Unité Modulaire de Pilotage • Unité Modulaire Programmable • Upper Mantle Programme

UMPA: Unité de Mathématiques Pures et Appliquées

UMPS: Universal Multiprogramming System • Up Mission Processing Start

UMR: Unité Mixte de Recherche • Woomera (Australia-SA) (location identifier)

UMRAO: University of Michigan Radio Astronomy Observatory (USA-MI)

UMS: Ultrasound Monitoring Experiment • UNESCO Member State • Unité Mixte de Service • University of Massachusetts System • Unmanned Multifunction Satellite • Unresolved Multiple Star • Urine Monitoring System • User Management System

UMSA: Universidad Mayor de San Andrés (Bolivia)

UMSI: University of Minnesota Supercomputer Institute (USA-MN)

UMSL: University of Missouri in Saint Louis (USA-MO)

UMT: Umiat (USA-AK) (location identifier) • Unachieved Merge Total • Underground Muon Telescope

UMTA: Urban Mass Transportation Administration

UMTF: User Migration Task Force

UMTRAP: Uranium Mill Tailings Remedial Action Program

UMTRCA: Uranium Mill Tailings Radiation Control Act (USA)

UMTS: Universal Mobile Telecommunications System (*) • Universal Mobile Telephone System

UMU: Umuarama - Ernesto Geisel (Brazil-PR) (location identifier)

UMVF: Université Médicale Virtuelle Francophone • Unmanned Vertical Flight

UMW: Ultramicrowave • University of Wisconsin-Milwaukee (USA-WI)

UMWD: Ultra-Massive White Dwarf

UMY: Sumy (Ukraine) (location identifier)

un: Undulatus • Unit • Unité

Un: Unified • Union • United • Unsatisfactory

UN: Union • United Nations

UNA: United Nations Association • Universitäts-Netz • University of North Alabama (USA-AL)

Unab: Unabridged

UNAC: United North American Conference

Unacc: Unaccompanied

UNACC: United Nations Administrative Committee and Coordination

Unaccomp: Unaccompanied

UNADS: UNIVAC Automated Documentation System

UNAH: Universidad Nacional Autónoma de Honduras

UNALM: Universidad Nacional Agraria La Molina (Peru)

UNAM: Universidad Nacional Autónoma de México

Unan: Unanimous

UNAP: Universidad Nacional de la Amazonia Peruviana (Peru)

Unasgd: Unassigned

Unatnd: Unattended

Unatt: Unattached

Unattrib: Unattributed

Unauth: Unauthorized

Unauthd: Unauthorized

Unavbl: Unavailable

UNAVCO: University NAVSTAR Consortium (USA)

UNAVF: Union Nationale des Accueils des Villes Françaises (France)

Unb: Unbound

UNB: Universal Navigation Beacon • University of New Brunswick (Canada-NB)

UNBAL: Unbalance

Unbd: Unbound

UNBIS: United Nations Bibliographic Information System

unc: Uncinus

Unc: Uncertain • Uncirculated

UNC: Uncoated • Unencoded Netnews Collator • Unguia (Colombia) (location identifier) • Unified Coarse • Universal Naming Convention • Universal Navigation Computer • Universidad Nacional de Cuyo (Argentina) • University of North Carolina (USA-NC) • University of Northern Colorado (USA-CO) • Uranyl Nitrate Concentrate • Uuencoded Netnews Collator

UNCA: University of North Carolina at Asheville (USA-NC)

UNCAST: United Nations Conference on the Applications of Science and Technology

UNCC: United Nations Cartographic Commission

UNCCC: United Nations Convention on Climate Change

UNCDF: United Nations Capital Development Fund

UNCED: United Nations Conference on Environmental Development • United Nations Conference on Environment and Development (*)

UNCG: University of North Carolina at Greensboro (USA-NC)

UNCHE: United Nations Conference on Human Environment

UNCHS: United Nations Centre for Human Settlements

UNCIO: United Nations Conference on International Organizations

UNCITRAL: United Nations Commission on International Trade Law

UNCJIN: United Nations Crime and Justice Information Network (*) • United Nations Criminal Justice Information Network

UNCLA: Unclassified

Unclas: Unclassified

Unclass: Unclassified

UNCLE: United Nations Command for Law Enforcement • Unsteady Computation of Linear Equations

UNCLOS: United Nations Conference on the Law of the Sea

UNCOD: United Nations Conference On Desertification

UNCOL: Universal Computer-Oriented Language

Uncond: Unconditional

Unconst: Unconstitutional

UNCOPUOS: United Nations Committee On the Peaceful Uses of the Outer Space

Uncor: Uncorrected

UNCRD: United Nations Centre for Regional Development

UNCSTD: United Nations Centre for Science and Technology for Development • United Nations Conference on Science and Technology for Development

UNCTAD: United Nations Conference on Trade And Development

UNCTC: United Nations Centre for Transnational Corporations

Und: Under

UND: Kunduz (Afghanistan) (location identifier) • University of Natal at Durban (South Africa) • University of North Dakota (USA-ND) • University of Notre Dame (USA-IN) • User Need Date

Undc: Undercurrent

UNDC: United Nations Disarmament Commission

UNDCP: United Nations Drug Control Programme

Undef: Undefined

Undetm: Undetermined

Undf: Underfrequency

UNDHA: United Nations Department of Humanitarian Affairs

UNDK: Undock

Undld: Underload

UNDP: United Nations Development Program

UNDRO: United Nations Disaster Relief Organization

Undsgd: Undersigned

UNDTCD: United Nations Department of Technical Cooperation for Development

Undtkr: Undertaker

Undv: Undervoltage

UNDV: Undervoltage

UNE: Qachas Nek (Lesotho) (location identifier) • Underground Nuclear Explosion • University of New England (Australia-NSW)

UNEC: United Nations Education Conference

UNECA: United Nations Economic Commission for Asia

UNECE: United Nations Economic Commission for Europe

UNECLAC: United Nations Economic Commission for Latin America and the Caribbean

UnECR: University Electro Communications Reports (periodical code)

UNED: Universidad Nacional de Educación a Distancia (Spain)

UNEDA: United Nations Economic Development Administration

UNEF: Unified Extra-Fine • Union Nationale des Étudiants de France

UNEP: United Nations Environmental Program • United Nations Environment Program (*)

UNESCAP: United Nations Economic and Social Commission for Asia and the Pacific

UNESCO: United Nations Educational, Scientific and Cultural Organization

UNESIS: United Nations Economic and Social Information System

UNESP: Universidade Estadual Paulista (Brazil-RJ)

UnESS: University Earth System Science (project)

UNETPSA: United Nations Educational and Training Program for Southern Africa

UNEX: University-class Explorer (*) • University Explorer

Unexpl: Unexplained • Unexplored

UNF: Unified Fine • University of North Florida (USA-FL)

UNFAO: United Nations Food and Agriculture Organization

UNFB: United Nations Film Board

UNFCC: United Nations Framework for Climate Change

UNFCCC: United Nations Framework Convention on Climate Change

Unfin: Unfinished

UNFPA: United Nations Fund for Population Activities

UNFSTD: United Nations Fund for Science and Technology Development

ung: Unguentum

UNG: Kiunga (Papua New Guinea) (location identifier)

UNGA: United Nations General Assembly

UNGASS: United Nations General Assembly Special Session

UNGCRP: United Nations Global Change Research Program

UNGEGN: United Nations Group of Experts on Geographical Names

ungt: Unguentum

Unh: Unnilhexium

UNH: University of New Hampshire (USA-NH) • University of New Haven (USA-CT) • Uranyl Nitrate Hexahydrate

UNHCR: United Nations High Commission for Refugees

UNHQ: United Nations Headquarters

Uni: Universität

UNI: Ente Nazionale Italiano di Unificazione (Italy) • Union Island (Grenadine Islands) (location identifier) • Unité de Navigation Inertielle • University of Northern Iowa (USA-IA) • User-Network Interface • User-to-Network Interface (*)

UNIC: United Nations Information Centre

UNICC: United Nations International Computing Center

UNICE: Union des Industries de la Communauté Européenne • Union of Industrial and Employers' Confederation of Europe

UNICEF: United Nations International Children's Emergency Fund

UNICOM: Universal Integrated Communication

UNIDO: United Nations Industrial Development Organization

UNIDRV: Universal Driver

Unif: Uniform

UNIFET: Unipolar Field-Effect Transistor

UNII: Unlicensed National Information Infrastructure

UNINETT: Nordic University Network

UNIPEDE: Union Internationale des Producteurs et Distributeurs d'Électricité

UNIPOL: Universal Problem-Oriented Language • Universal Procedure-Oriented Language

UNIPS: Unité d'Indicateurs de Politique Scientifique

UNIS: Unified Information System

UNISA: University of South Africa

UNISIST: Universal System for Information in Science and Technology

UNISTAR: United Nations Institute for Scientific Training And Research • UNIVAC Storage And Retrieval • User Network for Information Storage, Transfer Acquisition and Retrieval

UNIT: Instituto Uruguayo de Normas Técnicas (Uruguay) • Universal Numerical Interchange Terminal

UNITAR: United Nations Institute for Training and Research

Univ: Universal • Universally • Universe • Universel • Universidad • Universidade • Università • Universität • Université • Universiteit • Universiti • University

UNIV: Universal

UNIVAC: Universal Automatic Computer

UNJSPF: United Nations Joint Staff Pension Fund

Unk: Unknown

UNK: Unalakleet (USA-AK) (location identifier) • Unknown

Unkn: Unknown

Unl: Unloading

UNL: Universidad Nacional del Litoral (Argentina) • University of Nebraska at Lincoln (USA-NE) • Unlimited

UNLCH: Unlatch

Unlgtd: Unlighted

Unlim: Unlimited

UNLK: Unlock

Unlkg: Unlocking

UNLP: Universidad Nacional de La Plata (Argentina-BA)

UNLV: University of Nevada at Las Vegas (USA-NV)

Unm: Unmarried

UNM: University of New Mexico (USA-NM)

UNMA: Unified Network Management Architecture

UNMC: United Nations Mediterranean Commission

Unmkd: Unmarked

Unmon: Unmonitored

Unrelbl: Unreliable

UNMS: Slovak Office of Standards, Metrology and Testing

UNMSM: Universidad Nacional Mayor de San Marcos (Peru)

Unmtd: Unmounted

UNN: Ranong (Thailand) (location identifier) • University of Nigeria at Nsukka

UNNE: Universidad Nacional del Nordeste (Argentina)

UNO: University of Nebraska at Omaha (USA-NE)

UNOASD: United Nations Outer Space Affairs Division

UNOG: United Nations Office in Geneva (Switzerland)

Unop: Unopposed

UNOV: United Nations Office in Vienna (Austria)

Unp: Unnilpentium • Unpaged • Unpaid

Unpd: Unpaid

UNP: Uluru National Park (Australia-NT) • University of Natal at Pietermaritzburg (South Africa)

UNPA: United Nations Postal Administration

Unpub: Unpublished

Unpubl: Unpublished

Unq: Unnilquadium

UNR: Underkhaan (Mongolia) (location identifier) • Université Nationale du Rwanda • University of Nevada in Reno (USA-NV)

UNRFNRE: United Nations Revolving Fund for Natural Resources Exploration

Unrgltd: Unregulated

UNRISD: United Nations Research Institute for Social Development

UNRRA: United Nations Relief and Rehabilitation Administration

UNRSTD: Unrestricted

UNRWA: United Nations Relief and Works Agency

Uns: Unserviceable

UNS: Umnak Island (USA-AK) (location identifier) • Unions • Universidad Nacional del Sur (Argentina)

UNSA: Unconsolidated Sedimentary Aquifer • Universidad Nacional San Agustín (Peru)

UNSAC: United Nations Scientific Advisory Committee

Unsat: Unsatisfactory • Unsaturated

UNSBL: Unseasonable

UNSC: United Nations Security Council

UNSCC: United Nations Standards Coordinating Committee

UNSCCUR: United Nations Scientific Conference on the Conservation and Utilization of Resources

UNSCEAR: United Nations Scientific Committee on Effects of Atomic Radiations

UNSD: United Nations Statistical Division

UNSE: Union Nationale des Séjours à l'Étranger (France)

UNSG: United Nations Secretary-General

UNSIS: United Nations Statistical Information System

UNSL: Universidad Nacional de San Luis (Argentina)

UNSM: United States National Museum (USA)

UNSO: United Nations Statistical Office

UNSR: United Nations Space Registry

UNSTBL: Unstable

UNSTDY: Unsteady

UNSTL: Unstettle

UNSTLD: Unstettled

Unstpd: Unstamped

UNSW: University of New South Wales (Australia-NSW)

UNSWIRF: University of New South Wales Infrared Fabry-Pérot

UNSWPWR: Unswitched Power

Unsym: Unsymmetrical

Unt: Unter

UNT: Universidad Nacional de Tucumán (Argentina) • University of Newcastle-on-Tyne (UK) • University of North Texas (USA-TX) • Unst - Shetland Island - Baltasound (UK) (location identifier)

UNTAA: United Nations Technical Assistance Administration

UNTAB: United Nations Technical Assistance Board

UNTAM: United Nations Technical Assistance Mission

UNTAP: United Nations Technical Assistance Programme

UNTCD: United Nations Technical Cooperation for Development

Unterfr: Unterfranken (Germany)

UNTPDC: United Nations Trade Point Development Centre

Untrtd: Untreated

UNTSFA: United Nations Trust Fund for Southern Africa

UNTSO: United Nations Truce Supervision Organization

UNU: Juneau - Dodge County (USA-WI) (location identifier) • United Nations University

UNUIST: United Nations University Institute for Software Technology

UNUIIST: United Nations University International Institute for Software Technology

Unusbl: Unusable

UNV: United Nations Volunteers

UNX: Underground Nuclear Explosion

uö: und öfter

u/o: Used on

U&O: Use & Occupancy

U/O: Used On

UO: University of Oregon (USA-OR) • Used On

UoA: University of Arizona (USA-AZ)

UOA: Unit Of Assessment

UOC: Ultimate Operating Capability • Ultimate Operational Configuration • Unequilibrated Ordinary Chondrite • Universal Output Computer • Universitat Oberta de Catalunya (Spain) • Unusual Occurrence Control • Uranium Ore Concentrate

UOD: Ultimate Oxygen Demand

UODDL: User-Oriented Data Display Language

UOE: Unit Of Error

UOES: User Operational Evaluation System

UOF: User Operations Facility

UofA: University of Alaska (USA-AK)

UofNC: University of North Carolina (USA-NC)

UofS: University of Orange Free State (South Africa) • University of Saskatchewan (Canada-SK)

UofT: University of Toronto (Canada-ON)

UofW: University of Washington

UOG: Universitätsorganisationsgesetz (Austria) • University Of Guam

UOI: Uranus Orbital Insertion • Uranus Orbit Insertion • User Online Interaction

UOL: Buol (Indonesia) (location identifier) • Utility Octal Load • Utility-Oriented Language

UOLI: User-Oriented Language Interface

UoM: University of Mauritius

UOMS: Unmanned Orbital Multifunction Satellite

UON: Muong Sai (Laos) (location identifier) • Uniform Object Notation

UOP: UCAR Office of Programs

UOR: Unusual Occurrence Report

UOS: Sewanee - Franklin County (USA-TN) (location identifier) • Ultraviolet Ozone Spectrometer • Unless Otherwise Specified • Unmanned Orbital Satellite

UoSAT: University of Surrey Satellite

UOTS: Untouched Off-The-Shelf

UOV: Unit Of Variance

UoW: University of Wollongong (Australia)

UOX: Oxford University (USA-MS) (location identifier)

Up: Upper

UP: Ultra-Puissant • Uncertainty Principle • Under Proof • United Press • Unit Pack • Unit Price • Universidade do Porto (Portugal) • Universidad Panamericana • Université de Paris (France) • University of Pennsylvania (USA-PA) • University of Pittsburgh (USA-PA) • University of the Philippines • University Press • Unknown Precipitation • Unpublished • Unsaturated Polyester • Unsollicited Proposal • Unusual Programme • Upper • Upper Peninsula • Urban Planning • Uttar Pradesh (India)

UPA: Punta Alegre (Cuba) (location identifier) • Ultrafine Particle Analyzer • Ultra Port Architecture • Uncooled Parametric Amplifier • Union Postale Arabe • Union Professionnelle Artisanale (France) • Unitary Pole Approximation

UPADI: Unión Panamericana de Asociaciones De Ingenieros

UPAJ: Union Panafricaine des Journalistes

UPAR: Uninstantiated Parametrized Action Representation

UPAS: Underpass • Universal Pilot Application Service

UPB: Uniform Pixon Basis

UPC: Puerto la Cruz (Spain) (location identifier) • UKIRT Programme Committee • Uniform Price Code • Unit Processing Code • Universal Postal Convention • Universal Product Code • Universitat Politecnica de Catalunya (Spain) • Uplink Power Control • Usage Parameter Control • User Parameter Control

UPCA: Uniform Phase Evolution and Chaotic Amplitude

UPCC: Uniform Product Code Council (USA)

UPCE: University of the Philippines College of Engineering

UPCHUK: University Program for the Comprehensive Handling and Utilization of Knowledge

Upd: Unpaid

UPD: Update

UPDFTS: Updrafts

Updt: Update

UPE: Unitary Pole Expansion • Université Populaire Européenne • University of Port Elizabeth (South Africa)

UPEPI: Union des Practiciens Européens en Propriété Industrielle

UPF: Pforzheim (Germany) (location identifier) • Universal Patch Format • Universitat Pompeu Fabra (Spain)

UPG: Ujung Pandang (Indonesia) (location identifier) • Usine de Propergol de Guyane

UPGMA: Unweighted Pair Group Method using arithmetic Averages

UPGMC: Unweighted Pair Group Method using Centroids

UPGWA: United Plant Guard Workers of America

Uphd: Uphold

UPI: Uncertain Prior Information • United Press International • Universal Presentation Interface

UPJV: Université de Picardie Jules Verne (France)

UPL: Unapproved Parts List • Universal Programming Language • Upala (Costa Rica) (location identifier) • User Program Language

UPLIFT: Utrecht Project: Linguistic Information Free Text

UPLK: Uplink

UPM: Universidad Politécnica de Madrid (Spain) • University Pertanian Malaysia • Unix Programmer's Manual • Urban Pollution Management • User Profile Management

UPMAIL: Uppsala Programming Methodology and Artificial Intelligence Laboratory (Sweden)

UPN: Unique Program Number • Unique Project Number • Universidad Pedagógica Nacional (Mexico) • Uruapan (Mexico) (location identifier)

UPNE: University Press of New England (USA-NH)

UPO: Unstable Periodic Orbit

UPOCC: Ulysses Payload Operations Control Centre

Upp: Upper

UPP: Universal Payment Preamble • Universal Procedure Pointer • University of Pennsylvania Press (USA-PA) • Upolu Point (USA-HI) (location identifier) • User Parameter Processing • User Partyicipation Program • User Preference Profiling • Utility Print Punch

UppAn: Uppsala Astronomical Observatory Annals (periodical code)

UPPB: Union de la Presse Périodique Belge (Belgium)

UppOR: Uppsala Astronomical Observatory Reports (periodical code)

UPPR: Upper

UPPS: Universal Projection Plotting System

Upr: Upper

UPR: Unité Propre de Recherche • Universidad de Puerto Rico • University of Puerto Rico • Upiara (Papua New Guinea) (location identifier) • Upper • User Preferred Routing

UPRA: University Postgraduate Research Award

UPRM: University of Puerto Rico at Mayaguez (Puerto Rico)

UPRS: Unique Property Reference Number

Uprt: Upright

UPS: Ultraviolet Photoelectron Spectroscopy • Ultraviolet Photoemission Spectroscopy • Uninterrupted Power Supply • Uninterruptible Power Supply (*) • Uninterruptible Power System • Unité Propre de Service • Universal Polar Stereographic (projection) • Università Popolare Sestrese (Italy) • Université de Paris-Sud (France) • Université Paul Sabatier (France) • University of Puget Sound (USA-WA) • Upright Perigee Stage

UPSI: User Program Switch Indicator

UPSLP: Upslope

UPSO: Uttar Pradesh State Observatory (India)

UPSTAIRS: Universal Processing System for Treasuring Up and Analyzing Images in Remote Sensing

UPSTM: Upstream

UPT: Universal Personal Telecommunications • User Profile Table

UPTCAGP: University of the Philippines Training Center for Applied Geodesy and Photogrammetry

UPTLM: Uplink Telemetry

UPTP: Universal Package Test Panel

UPU: Union Postale Universelle • Unión Postal Universal • Universal Postal Union

UPV: Upavon (UK-England) (location identifier)

uPVC: Unplasticized Polyvinyl Chloride

UPWARDS: Unified Procedures for World-wide Access to Research Databases

Upwd: Upward

UPWD: Upward

UPWT: Unitary Plan Wind Tunnel

UQ: Ultimo Quarto • University of Queensland (Australia-QLD) • Upper Quartile

UQAC: Université du Québec à Chicoutimi (Canada-QC)

UQAH: Université du Québec à Hull (Canada-QC)

UQAM: Université du Québec à Montréal (Canada-QC)

UQAR: Université du Québec à Rimouski (Canada-QC)

UQAT: Université du Québec en Abitibi-Témiscamingue (Canada-QC)

UQCP: Uniform Quality Control Program

UQE: Queen (USA-AK) (location identifier) • Uniform Query Engine

UQEI: Uniform Query Engine Interface

UQF: Uniform Query Format

UQG: Uniform Query Gateway

UQGT: Unified Quantum Gravitational Theory

UQPSK: Unbalanced Quadrature Phase-Shift Keying

UQTR: Université du Québec à Trois-Rivières (Canada-QC)

UQTSM: Unit Quality and Technical Support Manager

ur: Urdu (ISO language code)

Ur: Urdu • Uruguay

U/R: Urban/Rural

UR: Ukraine (international aircraft registration code) • Ultra-Rapide • Unconditioned Reflex • Uniform Regulations • Unité de Recherche • Unité de Reconnaissance • Unité de Réserve • Unit Record • Universal Rectifier • Universal Relation • University of Rochester (USA-NY) • Unsatisfactory Report • Unspecified Reception • Upper Right • Uruguay • Usage Restreint • User Requirement • Utility Room

Ura: Uranus

URA: Uniformly Redundant Array • Unité de Raccordement d'Abonnés • Unité de Recherche Associée • Universities Research Associates • Universities Research Association Inc. (USA) (*) • Ural'sk (Kazakhstan) (location identifier) • Urine Receptable Assembly • User Radar Altimeter • User Requirement Analysis

URACA: Union Rhône-Alpes des Clubs d'Astronomie (France)

Uran: Uranus

URANIA: Universal Research Archive of Networked Information in Astronomy

URAP: Ulysses Radio And Plasma • Unified Radio And Plasma-wave (*)

URAV: Uninhabited Reconnaissance Aerial Vehicle

Urb: Urban • Urbanism • Urbanisme • Urbanismo • Urbanización

URB: Universal Radio Background • Urbanizaçao • Urubupunga - Ernesto Pochler (Brazil-SP) (location identifier)

UR&C: Usual, Reasonable & Customary

URC: Uniform Resource Characteristic • Uniform Resource Citation • Unit Record Card • Urumqi (China-PRC) (location identifier) • Usual, Reasonable and Customary

URCN: Unité de Raccordement de Circuits Numériques

URD: User Requirements Document

URDU: Urban Resource Development Unit

URE: Undergraduate Record Examination • User Range Error • Utilisation Rationnelle de l'Énergie

UREC: Unité Réseaux du CNRS

UREF: Universités des Réseaux d'Expression Française

UREP: Unix RSCS Emulation Program • Unix RSCS Emulation Protocol

URF: Unité de Responsabilité Fonctionnelle

URFIST: Unité Régionale de Formation pour l'Information Scientifique et Technique (France)

URFPIST: Unité Régionale de Formation et de Promotion pour l'Information Scientifique et Technique

Urg: Urgence • Urgent

URG: Urgent • Uruguaiana - Ruben Berta (Brazil-RS) (location identifier)

URGENT: Upper Rhine Graben Environmental Tectonics

URI: Uniform Resource Identifier • Universal Resource Identifier • Universal Resource Indicator • University of Rhode Island (USA-RI) • University Research Initiative • Uribe (Colombia) (location identifier)

URIF: Universal Research Incentive Fund

URIFT: University of Regina Image File Translator

URJ: Uraj (Russia) (location identifier)

URL: Uniform Record Locator • Uniform Resource Locator (*) • Unité Remplaçable en Ligne • Universal Resource Language • Universal Resource Locator • Universidad Ramón Llull (Spain) • Upper Range Limit • User Requirements Language

URM: Ulysses Reference Mission • Unlimited Register Machine • Urban-to-Regional Multiscale • Uriman (Venezuela) (location identifier)

UrMad: Urania Madrid (periodical code)

URMS: Universal Reproducing Matrix System

URM3: Universitá di Roma 3 (Italy)

URN: Uniform Random Numerator • Uniform Resource Name (*) • Uniform Resource Naming • Uniform Resource Number • Universal Resource Number • Urgoon (Afghanistan) (location identifier) • Urine

URO: Rouen (France) (location identifier)

URP: Unité Remplaçable sur Piste

URQ: Unsatisfactory Report Questionnaire

URR: Ultra-Rapid Reader • Urrao (Colombia) (location identifier) • User Requirement Review

URS: Kursk (Russia) (location identifier) • Universal Reference System • Université Robert Schuman (France) • Unrestricted Random Sample

URSI: Union Radio Scientifique Internationale

Urspr: Ursprünglich

URSS: Union des Républiques Socialistes Soviétiques

URSSAF: Union de Recouvrement pour la Sécurité Sociale et les Allocations Familiales (France)

URT: Surat Thani (Thailand) (location identifier) • University Research and Training • Upright

URTNA: Union des Radiodiffusions et Télévisions Nationales d'Afrique

Uru: Uruguay

Urug: Uruguay

URUPABOL: Uruguay-Paraguay-Bolivia

URV: Unité à Réponse Vocale

URY: Gurayat (Saudi Arabia) (location identifier) • Uruguay (ISO country code)

URZ: Universitätsrechenzentrum • Uruzgan (Afghanistan) (location identifier)

u/s: Unserviceable • Useless

us: Ubi Supra • United States of America • Ut Supra

Us: Usuel

U/S: Unserviceable • Useless

US: États-Unis d'Amérique (ISO country code) • Ufficio Stampa • Ultrasonic • Ultrason • Ultrasound • Ultrasonic Scanning • Ultrasound Scanning • Uncle Sam • Unconditioned Stimulus • Uniform System • United Service • United States • United States of America (ISO country code) • Unités et Symboles • Unit Separator • University of Singapore • Untreated Sample • Update Server • Upper Stage • USA (ISO country code) • User Segment

USA: États-Unis d'Amérique (ISO country code) • Unconventional Stellar Aspect • Unified Solutions Algorithm • Union of South Africa • United Space Alliance • United States of America (international vehicle registration code) • United States of America (ISO country code) • USA (international vehicle registration code)

USAA: United Services Automobile Association

USAAF: United States Army Air Force

USAATS: United States of America Air Traffic Service

Usáb: Usábase

USABC: United States of America Battery Consortium

USACA: United States Advanced Ceramics Association (USA)

USACE: United States Army Corps of Engineers (USA)

USACOM: United States Atlantic Command (USA)

USAE: Unconventional Stellar Aspect Experiment

USAEC: United States Atomic Energy Commission (USA)

USAETL: United States Army Engineering Topographic Laboratories (USA)

USAF: United States Air Force (USA)

USAFA: United States Air Force Academy (USA-CO)

USAFE: United States Air Force in Europe

USAFETAL: United States Air Force Environmental Technical Applications Center

USAID: United States Agency for International Development (USA)

USAFE: United States Air Forces in Europe

USAFR: United States Air Force Reserve (USA)

USAM: Unique Sequential Access Method

USAMO: United States of America Mathematical Olympiad

USAN: University Satellite Network

USAP: United States Antarctic Program

USARP: United States Antarctic Research Program

USART: Universal Synchronous/Asynchronous Receiver/Transmitter

USASCSOCR: United States of America Standard Character Set for Optical Character Recognition

USASDC: United States Army Strategic Defense Command (USA)

USASDG: Unites States Army Space Defense Group

USASI: United States of America Standards Institute

USAT: Ultra-Small Aperture Terminal

USATS: United States Air Traffic Service (USA)

U-SB: Unified S-Band

USB: Unified S-Band • Universal Serial Bus • Universidad Simon Bolívar (Mexico + Venezuela) • Universitäts-Sternwarte Bonn (Germany) • Unused Serial Bus • Upper Sideband • User Support Branch

USBC: United States Bureau of Census (USA) • Universal Standard Book Code

USBCF: United States Bureau of Commercial Fisheries (USA)

USBE: Unified S-Band Equipment • United States Book Exchange (USA) • Universal Serials and Book Exchange

USBI: United Space Boosters Inc.

USBLS: United States Bureau of Labor Statistics (USA)

USBM: United States Bureau of Mines (USA)

USBR: United States Bureau of Reclamation (USA)

USBS: Unified S-Band System • United States Bureau of Standards (USA)

USBUCEN: United States Bureau of Census (USA)

USC: Ultrasonic Storage Cell • Under Separate Cover • Unified Soil Classification • United States Code (USA) • United States Congress (USA) • University of San Carlos (Philippines) • University of Santa Clara (USA-CA) • University of South Carolina (USA-SC) • University of Southern California (USA-CA) • Upper Stage Contractor • User Security Component • User Spacecraft

USCAR: United States Council for Automotive Research (USA)

USCC: United States Chamber of Commerce (USA) • User Support Coordinating Committee

USCF: United States Chess Federation (USA)

USCG: United States Coast Guard (USA)

USC&GS: United States Coast & Geodetic Survey (USA)

USCGS: United States Coast and Geodetic Survey (USA)

USCM: United States Conference of Mayors (USA)

USCMI: United States Commission on Mathematical Instruction (USA)

USCO: United States and Canada Only

USCS: United States Civil Service (USA) • United States Customary System (USA) • Urine Sampling and Collection System

USCW: Understanding Space, Climate and Weather

USD: Dollar (USA) (ISO currency code) • Unified School District • United States Dollar (currency) • University of South Dakota (USA-SD)

USDA: United States Department of Agriculture (USA)

USDC: United States Department of Commerce (USA)

USDD: United States Department of Defense (USA)

USDE: United States Department of Education (USA) • United States Department of Energy (USA)

USDHEW: United States Department of Health Education and Welfare (USA)

USDHUD: United States Department of Housing and Urban Development (USA)

USDI: United States Department of the Interior (USA)

USDJ: United States Department of Justice (USA)

USDL: Ultra-Stable Diode Laser • United States Department of Labor (USA)

USDMA: United States Defense Mapping Agency (USA)

USDOC: United States Department Of Commerce (USA)

USDOD: United States Department Of Defense (USA)

USDOE: United States Department Of Energy (USA)

USDOI: United States Department Of the Interior (USA)

USDT: United States Department of Transportation (USA)

USDW: Underground Source of Drinking Water • Usable Supply of Drinking Water

Use: Usine

USE: Undersea Scientific Expedition • Unified Stochastic Engine • Unique Support Equipment • Unit Support Equipment • UNIVAC Scientific Exchange • University Space Experiment • Unmanned Surveillance Equipment • Unobtrusive Sensors and Effectors • User Support Environment

USEIT: User Station Evaluation Integration and Test

USEMAP: Urban Survey Experimental Method for Analysing Photo-data

USENET: Users' Network

USEPA: United States Environment Protection Agency (USA)

USER: User Systems Ergonomics Research (IBM)

USERID: User Identification

USERP: United Scientists for Environmental Responsibility and Protection

USES: United States Employment Service (USA)

usf: und so fort

USF: Universal Service Fund

USFFT: Unequally Spaced Fast Fourier Transform

USFS: United States Forest Service (USA) • United States Frequency Standard (USA)

USFWS: United States Fish and Wildlife Service (USA)

USG: Ultrasonic Space Grating • Unité Secondaire de Gestion • Universitäts-Sternwarte Göttingen (Germany) • User Support Group

USgal: United States Gallon

USGCA: United States Government Contract Awards (USA)

USGCRP: United States Global Change Research Program (USA)

USGCRPO: United States Global Change Research Program Office (USA)

USGLI: United States Government Life Insurance

USGP: United States Gemini Program

USGPO: United States Gemini Project Office (USA) • United States Government Printing Office (USA)

USGS: United States Geological Service (USA) • United States Geological Survey (USA) (*)

USh: Uganda Shilling (currency)

USH: User Shell • Ushuaia (Argentina-TF) (location identifier)

USHA: United States Housing Authority (USA)

USHDI: United States Historical Documents Institute (USA)

US-HEPNET: United States High-Energy Physics Network

USHS: Université des Sciences Humaines de Strasbourg (France)

USI: Mabaruma (Guyana) (location identifier) • Ultrasound Imaging • Ultraviolet Spectroheliographic Instrument • Update Software Identity • Urban Systemic Initiative • User System Interface • Usine de Séparation Isotopique

USIA: United States Information Agency (USA)

USIBA: United States International Book Association (USA)

USIC: United States Information Center (USA)

USICA: United States International Communication Agency (USA)

USIS: Ultraviolet Stratospheric Imaging Spectrometer • United Society In Space (USA-CO) • United States Imagery System (USA) • United States Information Service (USA)

USIT: University System for Information Technology

USITC: United States International Trade Commission (USA)

USJ: Universitäts-Sternwarte Jena (Germany)

USJGOFS: United States Joint Global Ocean Flux Study

USJPRS: United States Joint Publications Research Service (USA)

USK: Usinsk (Russia) (location identifier)

USL: Universal Sign Language • University of Southern Louisiana (USA-LA) • Unix Systems Laboratories Inc. • Useless Loop (Australia-WA) (location identifier)

USLC: United States Liaison Committee

USLE: Universal Soil Loss Equation

USM: Koh Samui (Thailand) (location identifier) • Ultimate Safe Mode • Universitäts-Sternwarte München (Germany) • Universiti Sains Malaysia • United States Mail (USA) • United States Mint

USMA: United States Military Academy (USA)

USMAPS: United States Map Analysis Package System

USMARC: Unites States Machine Readable Cataloging Standard

USMC: United States Marine Corps (USA) • United States Maritime Commission (USA)

USMES: Unified Science and Mathematics for Elementary Schools

USMI: Universal Software Market Identifier

USML: United States Microgravity Laboratory (USA)

USMP: United States Microgravity Payload

USMTF: United States Message Text Format

USN: Ulsan (Korea-ROK) (location identifier) • United States Navy (USA) • USA Dollar (next day) (ISO currency code) • User Subnetwork

USNA: United States Nationl Army (USA) • United States Naval Academy (USA-MD)

USNARS: United States National Archives and Record Service (USA)

USNBS: United States National Biological Survey (USA)

USNC: United States National Committee (USA)

USNCHC: United States National Committee on Scientific Hydrology (USA)

USNDC: United States Nuclear Data Committee (USA)

USNMAS: United States National Map Accuracy Standard

USNO: United States Naval Observatory (USA-DC)

USNOFS: United States Naval Observatory Flagstaff Station (USA-AZ)

USNOS: United States National Ocean Service (USA) • United States National Ocean Survey (USA)

USNPS: United States National Parks Service (USA)

USNR: United States Navy Reserve (USA)

USNRC: United States National Research Council (USA)

USNRL: United States Naval Research Laboratory (USA-DC)

USNS: United States Navy Ship

USO: Udaipur Solar Observatory (India) • Ultrastable Oscillator • Unidentified Superconducting Object • Unidentified Supraconducting Object • United Service Organization • United States Only • Universal Service Order • Unix Software Operation • Unmanned Seismological Observatory • User Support Office • User Support Organization • Usino (Papua New Guinea) (location identifier)

USOC: Unified System for Orbit Computation • Uniform Service Order Code • Universal Service Order Code • User Support and Operations Centre

USOE: United States Office of Education (USA)

USofA: United States of America

USP: Uniform Specification Program • Unique Selling Proposition • United States Patent (USA) • United States Pharmacopeia (USA) • United States Property (USA) • Universal Sampling Plan • Universal Signal Processor • Universidade de São Paulo (Brazil-SP) • University of the South Pacific • Unusual Stellar Pattern • Urgent Special Report • User-Sensitive Pricing • Utility Summary Program

USPat: United States Patent

Uspd: Underspeed

USPDI: United States Professional Development Institute (USA)

USPHS: United States Public Health Service (USA)

USPLSS: United States Public Land Survey System (USA)

USPO: United States Patent Office (USA) • United States Post Office (USA)

USPS: United States Postal Service (USA)

USPTO: United States Patent and Trademark Office (USA)

USQ: Universal Structured Query Language • University of Southern Queensland (Australia-QLD) • Unsqueezed

USQBR: United States Quarterly Book Review

Usr: User

USR: Unité de Service et de Recherche • Unité de Surveillance et de Reconfiguration • Universal Space Rectangular (projection) • User • User Service Request • User Service Routine

USRA: Universities Space Research Association

USRDA: United States Recommended Daily Allowance

USRLVC: United States Reusable Launch Vehicle Corporation (USA)

USRN: Unité de Synchronisation de Réseau Numérique

USRSDC: United States ROSAT Science Data Center (USA)

USRT: Universal Synchronous Receiver/Transmitter

USS: Sancti Spiritus (Cuba) (location identifier) • Ultrasteep Spectrum • Umbilical Separation System • Understanding Support System • Unique Support Structure • United States Senate (USA) • United States Service • United States Ship • United States Standard • United States Steamer • United States Steamship • Unit for Space Sciences • Unmanned Space Satellite • Unstructured Software System • Upper Stage System • USA Dollar (same day) (ISO currency code) • Utility Support Structure

USSA: Uniaxial Split Sphere Apparatus • Unit for Space Sciences and Astrophysics • User Supported Software Association

USSB: United States Satellite Broadcasting • United States Shipping Board (USA)

USSC: United Space Space Command (USA) (formerly NORAD) • United States Space Command (USA) (formerly NORAD) • Unix Starlink Software Collection

USSCS: United States Soil Conservation Service (USA)

USSCt: United States Supreme Court (USA)

USSD: Unstructured Supplementary Services Data

USSDTS: United States Spatial Data Transfer Specification

USSF: United States Space Foundation (USA)

USSI: United States Standards Institution (USA)

USSIE: Ultraviolet Solar Spectral Irradiance Experiment

USSP: ULDA Software Support Package • User Support Software Package

USSPA: United States Student Press Association (USA)

USSPACECOM: United States Space Command (USA)

USSR: Union of Soviet Socialist Republics (now obsolete)

USSS: United States Steamship • Unmanned Sensing Satellite System • User Services and Systems Support

USSSP: United States Science Support Program (USA)

UST: Saint Augustine (USA-FL) (location identifier) • Ultra Surface Treatment • Underground Storage Tank • Undersea Technology • Uniform Specification Tree • Universal Servicing Tool • Universal Standard Time • University of Science and Technology (Ghana)

USTA: United States Telephone Association (USA) • United States Trademark Association (USA) (now INTA)

UStB: Universitäts- und Stadtbibliothek

UStÄR: Umsatzsteuer-Änderungsrichtlinien

USTC: United States Tariff Commission (USA) • University of Science and Technology of China (PRC)

UStDV: Umsatzsteuer-Durchführungsverordnung

USTIS: Ubiquitous Scientific and Technical Information System

USTL: Université des Sciences et Techniques du Languedoc (France)

UStR: Umsatzsteuer-Richtlinien

USTR: United States Trade Representative

USTS: United States Travel Service (USA)

USTTA: United States Travel and Tourism Administration (USA) • United States Travel and Tourism Association (USA)

Usu: Usual • Usually

USU: Busuanga (Philippines) (location identifier) • Ural State University (Russia) • Utah State University (USA-UT)

USUHS: Uniformed Services University of the Health Sciences

USVI: United States Virgin Islands

USVMS: Urine Sample Volume Measuring System

usw: und so weiter

USW: Ultrashort Wave • Unterschlagwort

USWA: United Steelworkers of America

USWB: United States Weather Bureau (USA)

USWG: User Services Working Group

USWI: United States West Indies

USWRP: United States Weather Research Program (USA)

USX: Ultra-Soft X-ray

Usz: Undersize

Ut: Utah (USA) • Utility

U/T: Under Trust

UT: Umbilical Tower • Uncontrolled Term • Unité de Traitement • Unité de Transport • United Technologies • Unité Territoriale • Unit Telescope • Unit Terminator • Unit Test • Unit Tester • Universal Time • Universiteit Twente (Netherlands) • University of Tasmania (Australia-TAS) • University of Tennessee (USA-TN) • University of Texas (USA-TX) • University of Tulsa (USA-OK) • Unorganized Territory • Unspecified Temperature • Unterer Tubus • Upper Troposphere • User Terminal • User Test • Usine Thermique • Utah (USA)

UTA: Mutare (Zimbabwe) (location identifier) • Undergraduate Teaching Assistant • Unité de Transport Automobile • University of Texas at Arlington (USA-TX) • Unresolved Transition Array • Upper Terminal Area

UTB: Muttaburra (Australia-QLD) (location identifier)

UTBM: Université de Technologie de Belfort-Montbéliard (France)

UTC: Coordinated Universal Time • Corrected Universal Time • Undergraduate Teaching Committee • Unité de Traitement Centrale • United Technologies Corporation • United Technology Centre • Unit Test Case • Unit Time Coding • Unit Type Code • Universal Test Console • Universal Time Code • Universal Time Compensated • Universal Time Constant • Universal Time Coordinated (*) • Université de Technologie de Compiègne (France) • University of Tennessee at Chattanooga (USA-TN) • User Time Clock • Utrecht Soesterberg (Netherlands) (location identifier)

UTCQ: Universal Trellis-Coded Quantization

UTCSR: University of Texas Center for Space Research (USA-TX)

Utd: United

UTD: Nutwood Downs (Australia-NT) (location identifier) • Unit Training Device • University of Texas at Dallas (USA-TX) • User Terminal and Display

UTDF: Universal Tracking Data Format

ut dict: Ut Dictum

UTE: Butterworth (South Africa) (location identifier) • Union Technique de l'Électricité • Unité Technologique Élémentaire • Unit Test Equipment • Universal Test Equipment

UTEC: Universal Test Equipment Compiler

utend: Utendum

UTEP: University of Texas at El Paso (USA-TX)

UTF: Underwater Test Facility

UTG: Quthing (Lesotho) (location identifier)

UT&GS: Uplink Test & Graphics System

UTGS: Uplink Test and Graphics System

UTH: Udon Thani (Thailand) (location identifier) • Unité de Travail Homme • Upper Tropospheric Humidity

UTI: Kouvola Utti (Finland) (location identifier) • Unité de Traitement d'Images • Universal Text Interchange • Universal Text Interface • Unordered Tree Inclusion

UTIAS: University of Toronto Institute for Aerospace Studies (Canada-ON)

Util: Utilisateur • Utilisation • Utility • Utilization

UTIL: Utility

ut inf: Ut Infra

UTK: University of Tennessee at Knoxville (USA-TN) • Utirik Island (Marshall Islands) (location identifier)

UTL: Torremolinos (Spain) (location identifier)

UTLAS: University of Toronto Library Automation Systems

UTLB: Unified Translation Look-aside Buffer

UTLM: Up Telemetry

UTM: Unified Transport Model • Unified Type Model • Universal Testing Machine • Universal Test Message • Universal Transverse Mercator (projection)

UtMat: Utilitas Mathematica (periodical code)

UTMC: United Technologies Microelectronics Center

UTMOSST: University of Texas McDonald Observatory Supernovae Search Team

UTMS: User Total Management System

UTN: Upington (South Africa) (location identifier)

UTO: Unité de Transformation Opérationnelle • United Towns Organization • Utopia Creek - Indian Mountain (USA-AK) (location identifier)

UTOL: Universal Translator-Oriented Language

UTOPIA: University of Tsukuba Online Processing of Information Archives

UTOPIO: University of Texas Orbit Processor

UTP: Unified Test Plan • Unit Test Plan • Universal Tape Processor • Unscreened Twisted Pair • Unshielded Twisted Pair (*) • Untwisted Pair • Utapao (Thailand) (location identifier) • Utility Tape Processor

UTR: Unité de Turboréacteur • Unprogrammed Transfer Register • Uttaradit (Thailand) (location identifier)

UTRAO: University of Texas Radioastronomical Observatory (USA-TX)

UTRC: United Technologies Research Center (USA-CT)

UTRS: Universal Trouble Reporting System

UTS: Umbilical Test Set • Unified Transfer System • Unité de Travail Séparé • Universal Test Set • Universal Test Station • Universal Thrust Stand • Universal Time-sharing System • Universal Time Standard • University of Technology in Sydney (Australia-NSW) • Unmanned Teleoperator Spacecraft • Urine Transfer System

UTSI: University of Tennessee Space Institute (USA-TN)

UTSO: University of Toronto Southern Observatory (Chile)

ut sup: Ut Supra

UTT: Umtata (South Africa) (location identifier) • Université de Technologie de Troyes (France) • University of Tennessee at Tullahoma (USA-TN)

UTTC: Universal Tape-to-Tape Converter

UTTS: Universal Target Tracking Station

UTU: Ustupo (Panama) (location identifier)

UTW: Queenstown (South Africa) (location identifier)

uU: unter Umständen

UU: Universiteit Utrecht (Netherlands) • University of Ulster (Northern Ireland) • Usenet University

UUA: Bugul'ma (Russia) (location identifier) • UNIVAC Users Association

UUB: Uppsala Universitetsbibliothek (Sweden)

UUC: UKIRT Users' Committee • Unix-to-Unix Copy

UUCICO: Unix-to-Unix Copy Incoming Copy Outgoing

UUCP: Unix-to-Unix Copy Program

UUD: Ulan-Ude (Russia) (location identifier) • Unix-to-Unix Decoding

UUE: Unix-to-Unix Encoding • Use Until Exhausted

UUG: Unix User Group

UUI: Uniform User Interface • User-to-User Information

UUID: Universally Unique Identification • Universally Unique Identifier

UUK: Kuparuk (USA-AK) (location identifier)

UUN: Baruun-Urt (Mongolia) (location identifier)

UUS: Yuzhno-Sakhalinsk (Russia) (location identifier)

UuStB: Universitäts- und Stadtbibliothek

UUT: Unit Under Test

UUU: Manumu (Papua New Guinea) (location identifier)

uüV: unter üblichem Vorbehalt

UUV: Unmanned Underwater Vehicle

U/V: Under Voltage

UV: Miranda (satellite of Uranus) • Ultraviolet • Under Voltage • Unité de Valeur • Unit Value • Universitäts-Verlag

uva: und viele andere • und vieles andere

UvA: Universiteit van Amsterdam (Netherlands)

UVa: University of Virginia (USA-VA)

UVA: Ultraviolet A • Unidad Vecinal de Absorción • Uvalde Garner (USA-TX) (location identifier)

UVAI: Ultraviolet Auroral Imager

UVAL: Ultraviolet Argon Laser

UVAS: Ultraviolet Astronomical Satellite

UVASER: Ultraviolet Amplification by Stimulated Emission of Radiation

UVAT: Ultraviolet Automatic Tracker

UVB: Ultraviolet B

UVBS: Ultraviolet Bright Star

UVC: Ultrahigh Vacuum Chamber • Ultraviolet Camera • Ultraviolet Converter • Ultraviolet Coronagraph • Ultraviolet Court • Ultraviolet to Visible Converter

UVCS: Ultraviolet Coronal Spectrometer • Ultraviolet Coronagraph and Spectrometer (*) • Ultraviolet Coronagraph Spectrometer

UVD: Ultraviolet Detector • Under Voltage Device

UVDD: Ultraviolet Detector Development

UVDDP: Ultraviolet Detector Development Program

UVE: Ouvea (Loyalty Islands) (location identifier)

UVES: Ultraviolet Echelle Spectrograph • Ultraviolet/Visible Echelle Spectrograph • Ultraviolet/Visual Echelle Spectrograph (*)

UVF: San Lucia - Hewanorra (Windward Islands) (location identifier) • Ultraviolet Filter • Unmanned Vertical Flight

UVGA: Ultraviolet, Visible, and Gravitational Astrophysics

UVI: Cordelia (satellite of Uranus) • Ultraviolet Irradiation • Unidad de Vigilancia Intensiva (Spain)

UVII: Ophelia (satellite of Uranus)

UVIII: Bianca (satellite of Uranus)

UVIL: Ultraviolet Ion Laser

UVIS: Ultraviolet Imaging Spectrograph • Ultraviolet Imaging Spectrometer

UVISI: Ultraviolet Visible Imager and Spectrographic Imager

UVL: New Valley (Egypt) (location identifier) • Ultraviolet Lamp • Ultraviolet Laser • Ultraviolet Light • Ultraviolet Long

UVLIM: Ultraviolet Limb Imaging

UVLMB: Ultraviolet Limb

UVLS: Ultraviolet Light Stabilizer • Ultraviolet Limb Spectrometer

UVM: Ultraviolet Moyen • Universal Vendor Marking

UVMCS: Ultraviolet Multiple Channel Spectrometer

UVMS: Urine Volume Measuring System

UVNO: Ultraviolet Nitric Oxide Spectrometer

UVO: Uvol (Papua New Guinea) (location identifier)

UVOIR: Ultraviolet-Optical-Infrared

UVOT: Ultraviolet/Optical Telescope

UVOWG: Ultraviolet-Optical Working Group

UVP: Ultrahigh Vacuum Pump • Ultraviolet Photometry

UVPE: UV/Video Processing Electronics

UVPI: Ultraviolet Plume Imager

UVPS: Ultrahigh Vacuum Pumping Station

UVR: Ultraviolet Radiation • Ultraviolet Radiometer • Ultraviolet Receiver • Ultraviolet Rocket • Under Voltage Relay

UVS: Ultraviolet Sensor • Ultraviolet Spectrometer • Under Voltage Sensing

UVSC: Ultraviolet Solar Constant • Utah Valley State College (USA-UT)

UVSE: Ultraviolet Small Explorer • Ultraviolet Survey Explorer

UVSP: Ultraviolet Spectral Photometer • Ultraviolet Spectrometer and Polarimeter • Ultraviolet Spectro-Polarimeter

UVSQ: Université de Versailles/Saint-Quentin-en-Yvelines (France)

UVSTAR: Ultraviolet Spectrograph Telescope for Astronomical Research

UVT: Ultraviolet Transmission • Ultraviolet Tube • Universal Voltage Tester

UVV: Upward Vertical Velocity

UVVIS: Ultraviolet Visible

UVVS: Ultraviolet Visible Spectrometer

u/w: Used With

U/W: Under Will • Underwriter • Used With

UW: Unique Word • University of Wales (UK-Wales) • University of Washington (USA-WA) • University of Wisconsin (USA-WI) • Upstream Wave • Usable Width • Used With

UWa: University of Washington (USA-WA)

UWA: Ultra-Wide Angle • University of Western Australia (Australia-WA) • Ware (USA-MA) (location identifier)

UWB: Ultrawide Band

UWC: Universal Wireless Communications • University of Western Cape (South Africa)

UWCC: University of Wales - College of Cardiff (UK-Wales) • University of Wisconsin Computer Center (USA-WI)

UWCG: University of Wisconsin Controls Group

UWE: Wiesbaden (Germany) (location identifier)

UWF: University of West Florida (USA-FL)

UWG: User Working Group

UWGB: University of Wisconsin at Green Bay (USA-WI)

UWi: University of Wisconsin (USA-WI)

ÜWI: Überwachungsinstrument

UWI: Uniform Window Interface • University of West Indies (Jamaica)

UWIST: University of Wales Institute of Science and Technology (UK-Wales)

UWL: Underwater Launch

UWM: University of Wisconsin at Madison (USA-WI) • University of Wisconsin at Milwaukee (USA-WI)

UWNDS: Upper Winds

UWO: University of Western Ontario (Canada-ON)

UWPC: United World Press Cooperative (USA)

UWS: University of Western Sydney (Australia-NSW) • Unmanned Weather Station • User Workstation

UWSP: University of Wisconsin at Stevens Point (USA-WI)

UWT: User's Working Team

Uwtr: Underwater

UWY: Upper Airway

ux: Uxor

UX: Desdemona (satellite of Uranus) • Unité X

UXI: Juliet (satellite of Uranus)

UXII: Portia (satellite of Uranus)

UXIII: Rosalind (satellite of Uranus)

UXIV: Belinda (satellite of Uranus)

UXOR: UX Orionis (type)

UXV: Puck (satellite of Uranus)

uy: Uruguay

UY: Uruguay (ISO country code)

UYL: Nyala (Sudan) (location identifier)

UYN: Yulin (China-PRC) (location identifier)

UYP: Peso (Uruguay) (ISO currency code)

uz: Uzbek (ISO language code)

Uz: Uzbekistan

UZ: Uzbekistan (ISO country code)

UZB: Uzbekistan (ISO country code)

Uzbek: Uzbekistan

UZC: Updated Zwicky Catalog

UZGOST: Uzbek State Centre for Standardization, Metrology and Certification (Uzbekistan)

UZH: Unayzah (Saudi Arabia) (location identifier)

UZS: Sum (Uzbekistan) (ISO currency code)

UZU: Curuzu Cuatia (Argentina-CR) (location identifier)

u. zw.: und zwar

U3A: Université du Troisième Age • University of the Third Age

U3M: Université du Troisième Millénaire (France)

U3P: Union pour la Promotion de la Propulsion Photonique (France)

V

v/: Vista • Visto

v: Flight Velocity • Image Distance • Instantaneous Potential Difference • Instantaneous Voltage • Quantum Number of Vibrational Mode • Specific Volume • Speed • Vacuum • Vale • Valley • Valve • Van • Variable • Variable Absorption • Véase • Vedi • Vel • Velocity Component • Velocity Vector • Verb • Verbal • Vers • Version • Verso • Versus • Vertical • Very • Via • Vibrational Quantum Number • Vice • Vicinal • Vide • Village • Visibility • Vision • Voice • Voir • Volcano • Volt • Voltage • Volume • Volumen • vom • von • vor • Votre • Voyez • Vuelta • Vuestra • Vuestro

V: Dwarf • Electric Potential • Five (Roman numeral) • Luminous Efficiency • Macroscopic Speed • Main-Sequence • Potential Difference • Potential Energy • Potential of Conservative Forces • Speed • Stokes Parameter • Tension • Usted • Val • Validated • Valle • Valley • Valve • Vanadin • Vanadium • Variable • Vatican (international vehicle registration code) • Vatican (postal country code) • Véase • Vector • Vehicle • Velocity Component • Venerable • Ventana • Venus • Verb • Vers • Verse • Version • Versus • Vertical • Very • Via • Vice • Victor (international alphabet) • Victoria (Australia) • Vide • Ville • Visa • Vision • Visual Acuity • Vitesse • Vocative • Voice • Voir • Volcán • Volt • Voltmeter • Volume • Vortex • Vortrag • Vowel • Voyez • 5 (Roman numeral)

Va: Vanadium • Virginia (USA)

V-A: Vibroacoustic • Volt-Ampere

V/A: Video/Analog • Visualization/Analysis • Voucher Attached

VA: Valeur Absolue • Valeur Ajoutée • Value Added • Value Analysis • Variable Aléatoire • Vatican City State (ISO country code) • Véhicule Aérien • Velocity at Apogee • Veterans Affairs • Vibroacoustic • Video Amplifier • Vie Artificielle • Virginia (USA) • Virtual Address • Visiting Astronomer • Vistas in Astronomy (periodical code) • Visual Accumulation • Visual Acuity • Visualization-Analysis • Vitesse Absolue • Vitesse Accélérée • Vitesse Acquise • Vitesse Angulaire • Vitesse d'Atterrissage • Voice of America • Voie Aérienne • Volcanic Ash • Volt-Ampere • Vor Allem • Vostra Altezza • Votre Altesse • Vuestra Alteza

VAA: Vaasa (Finland) (location identifier) • Vehicle Assembly Area • Viersener Astronomischer Arbeitskreis (Germany) • Viewpont Adapter Assembly • Visible Acquisition Aid • Vorarlberger Amateur Astronomen (Austria)

VAAC: Vectored-thrust Aircraft Advanced Control • Volcanic Ash Advisory Centre

VAACP: Visible Acquisition Aid Control Panel

VAAE: Vereinigte Amateur-Astronomen Eschwege 1975 eV (Germany)

VAAS: Virginia Association of Astronomical Societies (USA-VA)

VAB: Van Allen Belts • Vehicle Assembly Building • Vertical Assembly Building • Yavarate (Colombia) (location identifier)

VABD: Van Allen Belt Dositometer

VABI: Variable Area Bypass Injection

VABM: Vertical Angle Benchmark

VABR: Vehicle Assembly Building Repeater

Vac: Vacancy • Vacant • Vacation • Vacuum • Vacuum (periodical code)

VAC: Vacuum • Value-Added Carrier • Value-Added Center • Variance At Completion • Vector Analog Computer • Vehicle Assembly and Checkout • Versatile Advection Code • Video Access Center • Vidicon Alignment Coil • Visiting Advisory Committee • Visual-Aid Console • Voies de l'Aviation Civile (periodical code) • Voltage/Alternate Current • Volts Alternating Current

VACA: Volatile Aerosol Component Analysis • Volatile Aerosol Composition Analysis

Vac.Dist.: Vacuum Distilled

VACM: Vector Averaging Current Meter

VACO: Vehicle Assembly and Checkout

VACP: Virtual Active Control Point

Vac.Pmp.: Vacuum Pump

VACTL: Vertical Assembly Component Test Laboratory

VAD: Valdosta Moody (USA-GA) (location identifier) • Value-Added Dealer • Value-Added Detailer • Vapor-phase Axial Deposition • Velocity-Azimuth Display • Voice Activity Detection

VADAS: Voice-Activated Domestic Appliance System

VADC: Venting Analysis Digital Computer

VADD: Value-Added Disk Driver

VADE: Versatile Automatic Data Exchange

VADIS: Value-Added Distributed-Information System

VADS: Verdix ADA Development System • Visual Assist Drive System

VAE: Visible Airglow Experiment

VAEP: Verify And Enter Program

VAF: Valence Chabeuil (France) (location identifier)

VAFB: Vandenberg Air Force Base (USA-CA)

VAFC: VESA Advanced Feature Connector

VAFTAD: Volcanic Ash Forecast Transport And Dispersion

VAG: Varginha - Maj. Brig. Trompowsky (Brazil-MG) (location identifier) • VRML Architecture Group

VAGI: VESA Advanced Graphics Interface

VAGO: Vssessojusnoje Astronomo-Geodesycheskoje Obtshestvo (Russia)

VAH: Vallegrande (Bolivia) (location identifier) • Vanne d'Alimentation en Hydrogène

VAI: Vanimo (Papua New Guinea) (location identifier) • Video-Assisted Instruction • Vol Aux Instruments • Vorticity Area Index

VAIB: Vehicle Assembly and Integration Building

VAIO: Video Audio Input/Output • Video Audio Integration Operation

VAIP: Vietnamese Association for Information Processing (Viet Nam)

VAIS: Value Added Information Service

Vaiss: Vaisseau

VAK: Chevak (USA-AK) (location identifier) • Vertical Assembly Kit

Val: Valencia (Spain) • Valeur • Valid • Validate • Validated • Validating • Validation • Validity • Valley • Valuation • Value • Valued • Valuta • Wert

VAL: Valença (Brazil-BA) (location identifier) • Valid • Validity • Value • Véhicule Automatique Léger • Voice Application Language

VALD: Vienna Atomic Line Database

Valdtn: Validation

Valid: Validate • Validation

VALID: Validation

Vall: Valladolid (Spain)

Vallad: Valladolid (Spain)

VALM: Vice-Almirante

Valn: Valuation

Valpo: Valparaiso (Chile)

VALT: Variable-stress Accelerated Life-testing Trial

VALUE: Valorisation et Utilisation pour l'Europe

Vam: Voltammeter

VAM: Variational Analysis Method • Virtual-Access Method • Vogel's Approximation Method

VAMAS: Versailles Project on Advanced Materials and Standards

VAMOS: Velocity And Magnetic Observations of the Sun

VAMP: Validation of Model Prediction • Valise Automatique Modulaire Programmable • Vibroacoustic Modelling and Prediction

Van: Vancouver (Canada-BC)

VAN: Value-Added Network • Van (Turkey) (location identifier) • Vehicle Area Network

Vanc: Vancouver (Canada-BC)

VANS: Value-Added Network Service

VANT: Virginia Network for Technology (USA-VA)

VAO: Suavanao (Solomon Islands) (location identifier) • Vie Assistée par Ordinateur

Vap: Vapeur

VAP: Validation des Acquis Professionnels • Valparaiso (Chile) (location identifier) • Value-Added Process • Value-Added Product • Value-Added Publishing • Vapor • Vector Analysis Power • Vehicle Antenna Position • Véhicule Automatique Planétaire • Velocity Analysis Program • Vendor Assistance Phase • Verification Acceptance Procedures • Verification and Acceptance Program • Verlagsabgabepreis • Visitor Academic Programme • Voluntary Assistance Programme (now VCP)

VAPC: Variable Absolute Pressure Controller • Vector Adaptive Predictive Coding

VAPCS: Verification and Acceptance Program Computer System

VAPEPS: Vibroacoustic Payload Environment Prediction System

VAPGS: Verification and Acceptance Program Ground System

VAPI: Visual Approach Path Indicator

Vapor: Vaporization

Vapp: Vitesse d'Approche

Vap. Prf.: Vapour Proof

VAPS: Virtual Applications Prototyping System • Virtual Avionics Prototyping System • Volume-Article-Paragraph-Sentence

Var: Variable • Variant • Variante • Variation • Variety • Variometer • Various

VAR: Value-Added Reseller • Value-Added Retailer • Variable • Variable Angle Reflectance • Variance • Variance Analysis Report • Variation • Varna (Bulgaria) (location identifier) • Verification Analysis Report • Verification Analysis Review • Vertical Air Rocket • Visual Aural Range • Volt-Ampere Reactive • Votre Altesse Royale • Vuestra Alteza Real

VARIG: Viação Aérea Rio-Grandense

VARIS: Virginia Resource Information System (USA-VA)

VARS: Vertical Azimuth Reference System

VARZ: Várzea

VAS: Sivas (Turkey) (location identifier) • Value-Added Service • Véhicule Aérospatial • Verein der Amateurastronomen des Saarlandes (Germany) • Vermont Astronomical Society (USA-VT) • Vibration Analysis System • Villanova Astronomical Society (USA-PA) • Visible Acquisition Sensor • Visible Atmospheric Sounder • Visiting Astronomers Section • VISSR Atmospheric Sounder • Visual Analogue Scale • Vlašim Astronomical Society (Czech Rep.)

Vasc: Vascuense

VASC: Virginia Air and Space Center (USA-VA)

VASCAR: Visual Average Speed Computer And Recorder

VASE: Variable Angle Spectroscopic Ellipsometer • Variable Angle Spectroscopic Ellipsometry

VASI: Visual Approach Slope Indicator

VASIMIR: Variable Specific Impulse Magnetoplasma Rocket

VASP: Value-Added Service Provider • Variable Automatic Synthesis Program • Vicação Aérea São Paulo SA

VAST: Versatile Avionics System Tester • Virtual Advanced Switch Test • VLAN Advanced Switch Test

Vat: Vatican

VAT: Value-Added Tax • Vatican (ISO country code) • Vatomandry (Madagascar) (location identifier) • Vibration Acceptance Test • Vibroacoustic Test • Video Alias Table • Virtual-Address Translation • Visual Audio Tool • Voice-Activated Typewriter • Volontaire d'Aide Technique

VATA: Vertical Assembly and Test Area • Vibroacoustic Test Article

VATE: Value-Added Tax Excluded • Versatile Automatic Test Equipment

VATF: Vibration and Acoustic Test Facility

VATI: Value-Added Tax Included

VatOP: Vatican Observatory Publication (periodical code)

VATP: Vector Adaptive Transform Processing

VATPOL: Vatican Polarimeter

VatRA: Specola Astronomica Vaticana - Ricerche Astronomiche (periodical code)

VATS: Vertical Assembly Test Stand • Video-Augmented Tracking System

VATT: Vatican Advanced Technology Telescope

VAU: Vatukoula (Fiji) (location identifier)

VAUS: Value-Added Utilisation System

VAV: Variable Air Volume • Vava'u Lupepau'u (Tonga) (location identifier)

VAW: Vardø (Norway) (location identifier)

VAWP: Voice-Activated Word Processor

VAWR: Vector Averaging Wind Recorder

VAWT: Vertical-Axis Wind Turbine

Vax: Vitesse Axiale

VAX: Virtual Address Executive • Virtual-Address Extended • Virtual-Address Extension • Virtual Array Executive

VAXBI: VAX Bus Interconnect

VAZ: Val d'Isère (France) (location identifier)

vb: Volledige Betrekking

Vb: Verb • Verbal

VB: Vacío de Boyero • Variable Block • Variables Bologna • Verb • Virtual Basic • Visual Basic • Vitesse Balistique • Vitesse de Balayage

VBA: Veterans Benefits Administration • Vibrating Beam Accelerometer • Visual Basic Application • Visual Basic for Applications (*)

VBAS: Von Braun Astronomical Society (USA-AL)

VBB: Verkehrsgemeinschaft Berlin-Brandenburg (Germany) • Very Broad Band

VBC: Video Bandwidth Compressor

VBD: Voice Band Data

VBE: VESA BIOS Extension

Vbf: Verschriebebahnhof

VBG: Lompoc Vandenberg (USA-CA) (location identifier)

VBH: Volkshochschule Bad Homburg (Germany)

VBI: Vertical Blanking Interrupt • Vertical Blanking Interval (*) • Virtual Binary Interface

Vbl: Verbal

VBL: Voyager Biological Laboratory

VBLR: Very Broad Line Region

vBNS: Very-high-performance Backbone Network Service

VBNS: Very-high-speed Broadband Network Services • VHS Broadband Network Services

VBO: Vainu Bappu Observatory (India) • Verbond van Belgische Ondernemingen (Belgium) • Views Brothers Observatory (USA-CO)

VBPS: Variable Bandpass Filter

VBR: Variable Bit Rate

VBS: Very Broad Structure • Voice Broadcast System

VBSMCVC: Variable Block Size Motion Compensated Video Color

VBSP: Video Baseband Signal Processor

VBT: Vainu Bappu Telescope (India) • Virtual Browsing Tool

Vbtm: Verbatim

VBV: Vanuabalavu (Fiji) (location identifier)

VBW: Video Bandwidth

VBWL: Variable Block-Word-Length

VBX: Virtual Basic Extension • Visual Basic Extension

VBY: Visby (Sweden) (location identifier)

v/c: Votre Compte

V/C: Vector Control • Velocity Counter

VC: Grenadine Islands (Saint Vincent and_) (ISO country code) • Saint Vincent and Grenadine Islands (ISO country code) • Vaisseau Capillaire • Vapour Compression • Variable Cataclísmica • Variable Cataclysmique • Vatican City • Vector Character • Vector Control • Vehicular Communication • Velocity Counter • Venture Capital • Venture Capitalist • Version Control • Vice-Chairman • Vice-Chamberlain • Vice-Chancellor • Vice-Consul • Vicinity • Video Camera • Video Centroid • Video Conferencing • Video Correlator • Vinyl Chloride • Virtual Channel • Virtual Circuit • Virtual Container • Visiting Committee • Visual Communication • Vital Capacity • Vitesse Calibrée • Vitesse Corrigée • Volt-Coulomb • Voltage Comparator • Voltage Compensator • Volume Control • Votre Compte

VCA: Can Tho (Viet Nam) (location identifier) • Vehicle Checkout Area • Véhicule à Coussin d'Air • Very Compact Array • Vinyl Carbonate • Voice Communication Assembly • Voltage Control of Amplification

VCAP: Vehicle Charging and Potential Experiment

VCAS: VCOS Application Server • Ventura County Astronomical Society (USA-CA)

VCATS: Visually Coupled Acquisition and Targeting System

VCB: Vertical of Center of Buoyancy • Video Converter Board • View Cove (USA-AK) (location identifier) • Visual Control Board

VCC: Variable Ceramic Capacitor • Vehicle Command Count • Vehicle Crew Chief • Velocity Code Counter • Verification Code Counter • Very Close Call • Very Cold Cloud • Victoria (Cameroon) (location identifier) • Video Coaxial Connector • Video Compact Cassette • Video Conferencing Communication • Video Conferencing Connection • Virgo Cluster Catalog • Virtual Channel Connection • Virtual Circuit Connection • Virtual Concurrent Copy • Vitrociset-CRI Consortium • Voice Control Center • Voltage Change Controller • Voltage Coefficient of Capacitance • Voltage-Controlled Capacitor

VCCA: Video Connect Card Adapter

VCCR: Vienna Convention on Consular Relations

VCCS: Voltage-Controlled Current Source

VCD: Vapor Compression and Distillation • Vapor Compression Distillation • Variable-Capacitance Diode • Verification Control Document • Vernier Engine Cutoff • Vibrational Circular Dichroism • Victoria River Downs (Australia-NT) (location identifier) • Virtual Communications Driver • Virtual COM-port Driver • Voltage Crossing Detector

Vce: Venice (Italy)

VCE: Vanne à Commande Électrique • Variable-Cycle Engine • Venezia - Marco Polo (Italy) (location identifier) • Venice - Marco Polo (Italy) (location identifier) • Virginia Center of Excellence (USA-VA) • Voice

VCF: Valcheta (Argentina-RN) (location identifier) • Variable Crystal Filter

VCG: Vapor Crystal Growth • Vector Cardiogram • Vertical at Centre of Gravity • Vice-Consul-General • Vice-Consul-Général

VCGL: Versatec Color Graphics Language

VCGS: Vapor Crystal Growth System

VCH: Vichadero (Uruguay) (location identifier) • Vinyl Cyclohexene

VCHAD: VGA Change Display

VCHP: Variable Conductance Heat Pipe

v. Chr.: Vor Christus

VCI: Vegetation Condition Index • Velocity Change Indicator • Verband der Chemischen Industrie (Germany) • Virtual Channel Identifier • Virtual Circuit Identifier • Volatile Corrosion Inhibitor

VCITE: Vertical Cargo Integration Test Equipment

Vcl: Vehicle

VCL: Vegetation Canopy Lidar • Vehicle Centerline • Vertical Component of Lift • Virtual Channel Link • Visual Component Library

VCLC: Virtual Channel Link Control

VCM: Variable-Curvature Mirror • Vector Correlation Map • Vehicle Control Module • Ventilation Control Module • Vibrating Coil Magnetometer • Video Control Module • Vinyl Chloride Monomer • Voice Coil Motor • Volatile Condensable Material • Volatile Condensed Materials • Volatile Condensible Materials • Voronezh Crystal Massif

VCMC: Vocational Curriculum Management Center (USA)

VCN: Vehicle Configuration Number • Verification Change Notice • Verification Completion Notice

VCNO: Verkenningscommissie Natuurkundig Onderzoek (Netherlands)

VCNR: Voltage-Controlled Negative Resistance

Vcnty: Vicinity

VCO: Voltage-Controlled Oscillator • Voltage Control Operator • Voltage Control Oscillator • Voluntary Conservation Organisation

VCON: Voltage-Controlled Oscillator-model Neutron

VCOS: Visible Catching Operating System • Visual Catching Operating System

VCP: São Paulo - Viracopos (Brazil-SP) (location identifier) • Vandenberg Contract Report • Vehicle Control Processor • Velocity Control Program • Velocity Control Programmer • Vertical Complementarity Problem • VideoRISC Compression Processor • Virtual Channel Processor • Visual Context Processing • Visual Context Processor • Voluntary Cooperation Programme (formerly VAP)

VCPI: Virtual Control Program Interface

VCPS: Velocity Control Propulsion System • Verification Completion Planning Sheet

Vcr: Vancouver (Canada-BC)

VCR: Carora (Venezuela) (location identifier) • Video Cassette Recorder (*) • Video Cassette Recording • Visual Control Room

VCRI: Verification Cross Reference Index

VCRT: Variable Contrast Resolution Test

Vcs: Voices

VCS: Vacuum Control Switch • Vapor Coating System • Vapor Control System • Vapor-Cooled Shield • Vapor Cooling System • Variable Correlation Synchronization • Vehicle Cargo Ship • Vehicle Cargo System • Vehicle Computer Simulation • Vehicular Communications System • Velocity Cutoff System • Ventilation Cutoff System • Verification Control Sheet • Vernier Control System • Video Communications System • Video Computer System • Vidicon Camera System • VILSPA Computer Services (ESA) • Virtual-Circuit System • Virtual Content Store • VLT Control Software • VLT Control System • Voice Command System • Voice Controlled System • Voice Control Switch • Voltage Calibration Set

VCSEL: Vertical-Cavity Surface-Emitting Laser (*) • Vertical-Cavity Surface-Emitting Laser-diode

VCSU: Valley City State University (USA-ND)

VCT: Grenadine Islands (Saint Vincent and_) (ISO country code) • Saint Vincent and the Grenadines (ISO country code) • Victoria (USA-TX) (location identifier) • Voice Code Translation • Voice Congestion Threshold • Voltage Control Transfer

Vctr: Vector

VCTR: Vector

VCU: Video Control Unit • Virginia Commonwealth University (USA-VA) • Voltage Control Unit

VCV: Variance-Covariance

VCVS: Voltage-Controlled Voltage Source

VCW: Vertical Component of Weight

VCXO: Voltage-Controlled Crystal Oscillator

V. Cz.: Vera Cruz (Mexico)

VC3: Victoria Computational Cosmology Conference (Canada-BC)

vd: van de • vor dem • vor der

Vd: Usted • Vend • Void

V/D: Voice/Data

VD: Democratic Viet-Nam (ISO country code) • Vapor Density • Various Dates • Vector Displacement • Venereal Disease • Vent Debout • Video Disk • Viet-Nam (Democratic_) (ISO country code) • Virtual Destination • Voltage Detector • Voltage Drop

Vda: Viuda

V&DA: Video & Data Acquisition • Video & Data Assembly

VDA: Ovda (Israel) (location identifier) • Valve Driver Assembly • Vapor-Deposited Aluminium • Vapor Diffusion Apparatus • Variable Data Area • Verband Deutscher Autoren (Germany) • Verein für das Deutschtum im Ausland • Video and Data Acquisition • Video and Data Assembly • Video Data Assembly • Video Distribution Amplifier • Visual Discriminatory Acuity

VDADC: Virtual Domain Application Data Center

VDAL: Variable Datalength Assembly Language

VDAM: Virtual Data Access Method

VDAS: Vibration Data Acquisition System

VDASD: Virtual Direct Access Storage Device

VDB: Fagernes (Norway) (location identifier) • Verification Database • VISSR Database

VDBG: Van den Bergh Galaxies

VDBMS: Video Database Management System

VDC: Virgo Data Center • Virtual Device Coordinates • Virtual Display Coordinates • Vitória da Conquista (Brazil-BA) (location identifier) • Voice Data Control • Voltage Direct Current • Voltage-to-Digital Converter • Volts Direct Current

VDCP: Video Data Collection Program

VDCT: Viaduct

VDD: Verification Description Document • Version Description Document • Vienna Danubepier (Austria) (location identifier) • Virtual Device Driver • Virtual Display Driver • Visual-Display Data • Voice Digital Display • Wien Danubepier (Austria) (location identifier)

VDDI: Voyager Data Detailed Index

VDDL: Voyager Data Detailed List

VDDM: Virtual Device Driver Manager

VDDP: Video Digital Data Processing

VDDS: Voyager Data Description Standard

VDE: Valverde (Canary Islands) (location identifier) • Verband Deutscher Elektroniker (Germany) • Verband Deutscher Elektrotechniker (Germany) • Verein Deutscher Elektroniker (Germany) • Video Display Editor • Visual Development Environment • Visual Display Equipment

VDEh: Verein Deutscher Eisenhüttenleute (Germany)

VDEM: Velocity Differential Emission Measure

VDES: Voice Data Entry System

VDETS: Voice Data Entry Terminal System

VDEW: Vereinigung Deutscher Elektrizitätswerke (Germany)

VDF: Validation Data File • VHF Direction Finder

VDFM: Virtual Disk File Manager

VdG: Van de Graaf

VDG: Vertical and Direction Gyro

VDI: Vector Display Indicator • Vendor Documentation Inventory • Verein Deutscher Ingenieure (Germany) • Vertical Display Indicator • Vidalia (USA-GA) (location identifier) • Video Device Interface • Video Display Interface • Virtual Data Interchange • Virtual Device Interface • Voice-Data-Image • Voix-Données-Image

VDIG: Visualization and Digital Information Group

VDISK: Virtual Disk

VDJ: Verein der Deutschen Journalisten (Germany)

Vdkhr: Vodokhranilishche

VDL: Van Diemen's Land (Australia-TAS) • VHF Data Link • Vienna Definition Language • Voice Direct Line

VDLF: Variable Depth Launch Facility

VDM: VAS Digital Multiplexer • Vector Dominance Model • Vector Drawn Map • Viedma (Argentina-RN) (location identifier) • Vienna Development Method • Virtual Device Machine • Virtual Device Metafile • Virtual Display Manager • Virtual Document Management • Virtual DOS Machine

VDMA: Verband Deutscher Maschinen und Anlagenbau eV (Germany)

VDMAT: Virtual Direct Memory Access Driver

VDME: VOR with Distance Measuring Equipment

VDMS: Vocal Data Management System

VDO: Video

VDOP: Vertical Dilution Of Precision

VDP: Valle de Pascua (Venezuela) (location identifier) • Variable-Delivery Pump • Ventura Database Publisher • Verband der Deutschen Presse (Germany) • Vertical Data Processing • Vibration Diagnostic Program • Vibration Dissociation Process • Video Datagram Protocol • Video Data Processor

VDPI: Voyager Data Processing Instruction

VDPM: Variable-Probability Detection Model

VDPS: Voice Data Processor System

VDR: Variable-Diameter Rotor • VHF Data Radio • Video-Disc Recorder • Video-Disc Recording • Video-Disk Recorder • Video-Disk Recording • Villa Dolores (Argentina-CD) (location identifier) • Voltage-Dependent Resistor

VDRZ: Verband Deutscher Rechenzentren (Germany)

Vds: Ustedes

VdS: Vereinigung der Sternfreunde (Germany)

VDS: Vadsø (Norway) (location identifier) • Variable-Depth SONAR • Vehicle Dynamics Simulator • Verband Deutscher Schriftsteller (Germany) • Versatec Data Standard • Vertex Detector System • VHDL Development System • Video Display System • Video Distribution System • Viewfinder Detector System • Virtual DMA Service • Virtual Dome System • Visual Display System

VDSÄ: Verein Deutscher Strahlenschutzärzte (Germany)

VDSL: Very-high-data-rate Digital Subscriber Line • Very-high-speed Digital Subscriber Line

VDSS: Vehicle Dynamic Simulator System

VDT: Vehicle Data Table • Video Data Terminal • Video Dial Tone • Video Display Terminal • Visual Database Tool • Visual-Display Terminal

VDU: Refugio - Tom O'Connor (USA-TX) (location identifier) • Video Display Unit • Visual Display Unit

VDUC: VAS Data Utilization Console

vdW: van der Waals

VDW: Venus Departure Window • Verband Deutscher Wissenschaftler (Germany)

VDWQT: Vertical Deferred Write Queue Threshold

VDZ: Valdez (USA-AK) (location identifier) • Verband Deutscher Zeitschriftenverleger (Germany)

ve: van een • Venezuela • Vertebratus

Ve: Veuve

VE: Valence Electron • Valeur Énergétique • Vehicle Engineering • Velocity Error • Venezuela (ISO country code) • Vent Effectif • Vertex Element • Victory in Europe • Virtual Enterprise • Virtual Environment • Visual Efficiency • Visualisation Électronique • Vitesse d'Écoulement • Vitesse d'Éjection • Vitesse d'Entrée • Vitesse d'Évasion • Vitesse Équivalente • Vocational Education • Vol d'Essai • Voltage Efficiency • Vostra Eccellenza • Votre Éminence • Votre Excellence • Vuestra Excelencia

VeARI: Veröff. Astron. Rechen-Inst. Heidelberg (periodical code)

VEB: Bolivar (Venezuela) (ISO currency code) • Vehicle Equipment Bay • Venus Entry Body • Volkseigener Betrieb • Volkssternwarte Erich Bär (Germany)

VeBoc: Veröfftlichungen Astronomisches Institut Ruhr-Universität Bochum (periodical code)

VeBon: Veröfftlichungen Astronomisches Institut Bonn (periodical code)

Vec: Vector

VECAM: Veille Européenne et Citoyenne sur les Autoroutes de l'Information et le Multimédia

VECI: Virtual Environment Control Interface

VECIB: Vehicle Engineering Change Implementation Board

VECM: Vocational Education Curriculum Material

VECO: Vernier Engine Cutoff

VECOS: Vehicle Checkout Set

Ved: Vedova

VED: Virtual Environment Display

VEDA: Validation of European Design Advances

VEDS: Vehicle Emergency Detection System

VEE: Venetie (USA-AK) (location identifier)

VEEGA: Venus-Earth-Earth Gravity-Assist • Venus-Earth-Earth Gravity-Assisted

VEEI: Vehicle Electrical Engine Interface

VEF: Variable Eddington Factor • Variable Electronic Filter • Virtual Event File

VEFCO: Vertical Functional Checkout

VEFI: Vector Electric Field Instrument

VEFR: Vehicle Frame

Veg: Vegetable • Vegetarian • Vegetation

VEG: Maikwak (Guyana) (location identifier) • Virtual Environment Group • Vitesse d'Éjection des Gaz

VEGA: Validation Expérimentale en Grille Annulaire • Vector-Evaluated Genetic Algorithm • Venus-Earth Gravity-Assist • Video Enhanced Graphics Adapter

VEGF: Vascular Endothelial Growth Factor

Véh: Véhicule

Veh: Vehicle • Vehicular

VEH: Vehicle

VeHei: Veröfftlichungen des Sternwarte Heidelberg-Königstuhl (periodical code)

VEI: Vehicle End Item • Vocabulário Electrotécnico Internacional • Volcano Explosivity Index

VEIN: Vocational Education Information Network

VEIS: Vocational Education Information System

VEJ: Vejle (Denmark) (location identifier)

vel: Velum

Vel: Vela-Velorum (constellation) • Vellum • Velocity • Velvet

VEL: Velocity • Vernal (USA-UT) (location identifier) • Virtual Electronic Library

VELCRO: Velours et Crochets

VEM: Viscoelastic Material • Volume Emission Measure • VSCS Electronic Module

VEMM: Virtual Expanded Memory Manager

VEMMI: Videotex Enhanced Multimedia Interface

VEMR: Virtual Electronic Medical Record

VeMun: Veröffentlichungen München (periodical code)

Ven: Vendredi • Venerable • Venerdi • Venezuela • Venezia (Italy) • Ventral • Venus

VEN: Venezuela (ISO country code)

Venez: Venezuela

Vent: Ventilate • Ventilateur • Ventilation • Ventilator • Venture

VENUS: Vehicle for Education of the Network of Universities in Space

VEO: Voluntary Environmental Organisation

VEOS: Versatile Electro-Optical System • Virtual-Environment Operating System

VEP: Vehicle Evaluation Payload • Violation of Equivalence Principle

VEPIS: Vocational Education Program Information System

VEQ: Vertically-loaded External Quench

Ver: Verein • Vereinigung • Verification • Verified • Verify • Verse • Version • Vertical

VER: Vera Cruz (Mexico) • Vera Cruz - La Bajadas (Mexico) (location identifier) • Verified • Verify • Version • Vertical

VERA: Versatile Reactor Assembly • Vision Electronic Recording Apparatus

verb: verbessert

Verb. et Lit.: Verbatim et Literatim

Verf: Verfasser • Verfassung

VerglO: Vergleichsordnung

verh: verheiratet

Verif: Verification • Verified • Verify

Vérif: Vérification • Vérifié • Vérifier

VERI: Virtual Environments Research Institute (USA-TX)

VERIF: Verification

VERITAS: Very Energetic Radiation Imaging Telescopes Array System

Verl: Verlag

Vern: Vernacular

VERN: Vernier

Veröff: Veröffentlichung

VERONICA: Very Easy Rodent-Oriented Net-wide Index to Computerized Archives

VERR: Verify Read • Verrière

Verrou: Verrouillage

Vers: Version

Versch: Verschieden

Verst: Versatile

Vert: Vertebra • Vertebrate • Vertica (periodical code) • Vertical

VERT: Vertical

Vertal: Vertical

VERTEX: Vertical Transport Experiment

Vertf: Vertiflite (periodical code)

VERTIS: Vehicle, Road and Traffic Intelligence Society

VERVIS: Vertical Visibility

Verw: Verwaltung • Verwitwet

VERW: Verify Write

Ves: Vespere • Vessel

VES: Vacuum Exhaust System • Vatican Emission Stars • Vertical Equivalent Size

VESA: Video Electronics Standards Association • Visualization and Earth Science Applications

VESAT: Venus Environmental Satellite

VeSon: Veröffentlichungen des Sternwarte Sonneberg (periodical code)

VESPER: Venus Sounder for Planetary Exploration

VESS: Visual Environment Simulation System

VEST: Vehicle Electrical System Test

Vet: Veteran

VET: Video Editing Terminal • Virtual Environment for Training • Vocational Education and Training

VETA: Verification Engineering Test Article • Virginia Educational Technology Alliance

VeUSB: Veröffentlichungen Universität Sternwarte Bonn (periodical code)

VEV: Barakoma (Solomon Islands) (location identifier) • Vacuum Expectation Valve • Vernier Engine Vibration

VEVI: Virtual Environment Vehicle Interface

VEW: Vereinigte Edelstahlwerke

VEX: Tioga (USA-ND) (location identifier)

VEXUVIO: Visible Explorer and Ultraviolet Infrared Observer

VEY: Vestmannaeyjar (Iceland) (location identifier)

Vf: Verfasser • Veuf

VF: Vacuum Fluorescent • Vanne de Flux • Variable Frequency • Variation de Force • Vector Feature • Vector Field • Velocity Failure • Ventricular Fibrillation • Version Française • Vertical Flight • Very Fair • Very Fine • Video Frequency • Virtual Floppy • Visiting Fellowship • Visual Field • Voice Frequency • Voie Ferrée

VFA: Variation Flow Analysis • Victoria Falls (Zimbabwe) (location identifier) • Video Frequency Amplifier

VFAT: Virtual File Allocation Table (*) • Virtual File Attribute Table

VFC: Variable Frequency Control • Vector Function Chainer • V Fast Class • Video Feature Connector • Video Frequency Carrier • Video Frequency Channel • Voice Frequency Carrier • Voice Frequency Channel • Voltage-to-Frequency Converter

VFCS: Vehicle Flight Control System • Vienna Fortran Compilation System

VFCT: Variable-Frequency Carrier Tone

VFD: Vacuum Fluorescent Display • Variable Field Data • Verified Free Distribution

VFDR: Variable Flow Ducted Rocket

VFE: Vendor-Furnished Equipment

VFF: Voice Frequency Filter

VFG: Video Frame Grabber

Vfg. v.: Verfügung von

VFI: Verification Flight Instrumentation

VFL: Variable Focal Length

VFLFS: Very Fast and Large File System

VFM: Value For Money • Viewfinder Monitor

VFN: Verification Change Notice • Virtual File Name

VFO: Variable-Frequency Oscillator

VFOAR: Vandenberg Field Office of Aerospace Research

VFP: Variable-Factor Programming • Visiting Fellows Program

VFPA: Visible Focal-Plane Assembly

VFR: Vector Relational Format • Very Far Radiation • Visual Flight Rules

VFS: Vertical Full Scale • Virtual File Store • Virtual File System • Visual Flight Rule • Volume and File Structure

VFSC: Very Faint Source Catalogue

VFSS: VAP Flight Simulation Software

VFST: Vacuum Fine Structure Telescope (China-PRC)

VFT: Verification Flight Test • Vertical Flight Test • Very Fast Train • Very Forward Tracker • Visual Function Test • Visual Function Tester • Voice Frequency Terminal

VFU: Vertical Format Unit

VFV: Venus Flyby Vehicle

VFW: Variable/Fixed Wavelength • Veterans of Foreign Wars (USA) • Video For Windows

Vfy: Verify

VFY: Verify

vg: Verbi Gratia

VG: British Virgin Islands (ISO country code) • Îles Vierges Britanniques (ISO country code) • Variable Geometry • Variable Gradient • Velocity Generator • Velocity/Gravity • Verlagsrechtsgesetz • Vertical Gyro • Verwertungsgesellschaft • Very Good • Voie de Garage • Voltage Gain • Votre Grâce • Votre Grandeur • Virgin Islands (British_) (ISO country code) • Vraie Grandeur

VGA: Variable Gain Amplifier • Vertical Gyro Alignment • Very General Algorithm • Video Graphics Adapter • Video Graphics Array • Vijayawada (India) (location identifier)

VGB: British Virgin Islands (ISO country code) • Îles Vierges Britanniques (ISO country code) • Virgin Islands (British_) (ISO country code)

VGC: Very Good Condition • Video Graphics Controller

VGCS: Voice Group Call System

VGD: Vologda (Russia) (location identifier)

VGE: Visual Gross Error • Voluntary Group of Experts

VGF: Vapor Growth Facility • Voice Grade Facility

VGG: Vangrieng (Laos) (location identifier)

VGH: Velocity - Ground range - Height • Very Good Health

VGI: Vertical Gyro Indicator

Vgl: Vergelijk • Vergleiche • Vergleichung

VGLIS: Video Guidance, Landing and Imaging System

VGML: Vegetarian Meal

VGNA: Video Graphics Network Adapter

VGO: Vigo (Spain) (location identifier)

VGOR: Vandenberg Ground Operations Requirement • Vehicle Ground Operations Requirement

VGP: Vehicle Ground Point • Virtual Geomagnetic Pole

v. gr.: Verbi Gratia

VGRF: Variable-Gravity Research Facility

VGS: General Villegas (Argentina-BA) (location identifier) • Vereinigung Gandersheimer Sternfreunde (Germany) • Versatec Graphics Software

VGSR: Vacuum Gas-Solid Reaction

VGSW: VAP Ground Software

VGT: La Vegas North (USA-NV) (location identifier) • Vacuum Gregory Telescope • Vegetation • Vehicle Ground Test

VGTM: Variable Geometry-Truss Manipulator

vgu: vorgelesen-genehmigt-unterschrieben

VGV: Variable Guide Vane

VGVI: Vereniging voor Geografische Informatie en Vastgoedinformatie

VGVT: Vertical Ground Vibration Test

VGVTA: Vertical Ground Vibration Test Article

VGW: Veiligheid-Gezondheid-Welzijn

VGZ: Mocoa (Colombia) (location identifier) • Villagarzón (Colombia) (location identifier)

vh: van het

vH: vom Hundert

Vh: Véhicule

V/H: Velocity to Height

V&H: Vertical & Horizontal

VH: Australia (international aircraft registration code) • Vertical Height • Very Heavy • Very High • Visibilité Horizontale • Vitesse Horaire • Vitesse Horizontale • Vitesse Hyperbolique • Vol Horizontal

VHA: Very High Altitude

VHAA: Very-High-Altitude Abort

VHARC: Very High-Angular-Resolution Camera • Visible High-Angular-Resolution Camera

VHC: Saurimo (Angola) (location identifier) • Very Highly Commended • Virgo-Hydra-Centaurus • Volatile Halocarbon Compound

VHD: Video High Density

VHDL: VHSIC Hardware Description Language

VHE: Very High Energy

VHEGRA: Very High Energy Gamma-Ray Astronomy

VHF: Very High Fidelity • Very High Frequency

VHFF: Very-High-Frequency Filter

VHFG: Very-High-Frequency Generator

VHFO: Very-High-Frequency Oscillator

VHFR: Very-High-Frequency Receiver

VHLL: Very High Level Language

VHLW: Vitrified High-Level Waste

VHM: Vector Helium Magnetometer • Vehicle Health Monitoring • Vilhelmina (Sweden) (location identifier)

VHN: Van Horn - Culberson County (USA-TX) (location identifier) • Vickers Hardness Number

VHO: Very High Output

VHOL: Very High Order Language

VHP: Very High Performance • Very High Pressure • Video-to-Hardcopy Recorder

VHR: Very High Reduction • Very High Resolution

VHRA: Very-High-Resolution Apodizer

VHRFC: Very-High-Resolution Framing Camera

VHRR: Very-High-Resolution Radiometer

Vhs: Véhicules

VHS: Variable Hard Sphere • Very High Speed • Very High State • Video Home System • Virtual Host Storage • Vitesse Hypersonique • Volkshochschule

VHSA: Variable Hydrologic Source Area

VHSIC: Very-High-Speed Integrated Circuit

VHST: Very-High-Speed Transit

VHT: Very High Temperature

VHU: Véhicule Hors d'Usage

VHV: Very High Vacuum

VHY: Vichy (France) (location identifier)

VHZ: Vahitahi (Tuamotu Islands) (location identifier)

vi: Vietnamese (ISO language code) • Visual Editor

Vi: Vitesse

VI: Îles Vierges Américaines (ISO country code) • Subdwarf • US Virgin Islands (ISO country code) • Vancouver Island (Canada-BC) • Variable Interval • Vegetation Index • Vertical Interval • Victoria (Australia) • Vide Infra • Video Input • Video Integrator • Virgin Islands (US_) (ISO country code) • Virtual Impactor • Virtual Instrument • Virtual Interface • Viscosity Index • Visual Impairment • Visual Inspection • Visual Interactive • Vitesse d'Impact • Vitesse d'Injection • Vitesse Indiquée • Vitesse Initiale • Vitesse Intermédiaire • Vocabulary Independent • Volume Indicator

Via: Viaduct

VIA: Vendors ISDN Association • Versatile Integrated Avionics • Viaduct • Virtual Interface Architecture • Vistas In Astronomy

VIABLE: Vertical Installation Automation Baseline

VIAEA: Vienna International Atomic Energy Agency (periodical code)

VIASA: Venezolana de Aviación SA

Vib: Vibrant • Vibrate • Vibrating • Vibration • Vibrer

VIB: Vehicle Integration Building • Vertical Integration Building • Vibrate • Vibration • Villa Constitución (Mexico) (location identifier)

VIBE: Visual Information Browsing Environment

VIBES: Vibration Isolation Box Experiment System • Vibration Isolation in a Box Experiment System

Vic: Vicinity • Victoria (Australia)

VIC: Variable Instruction Computer • Very Important Customer • Vehicle Identification Code • Vicenza (Italy) (location identifier) • Victoria (Australia) • Vienna International Centre (Austria) • Visitor Information Center • Vitesse Indiquée Corrigée • VLT Instrument Consortium • Vortex In Cell • VXI Interactive Control

VICAR: Video Image Communication And Retrieval (*) • Video Information Communication And Retrieval

Vice-Prés: Vice-Président

Vice-Pres: Vice-President

VICI: Velocity Indicating Coherent Integrator • Video Input Code Identifier

VICOS: Verification, Integration and Checkout Software

VICS: Vehicle Information and Communications System

Vict: Victoire • Victoria • Victorian • Victory

Vid: Vidange • Vide • Video • Videtur • Vidua

VID: Video • Vidin (Bulgaria) (location identifier) • Virtual Image Display • Visual Information Design • Visual Interactive Debugger • Visual Interactive Debugging

VIDAC: Visual-Information Display And Control

VIDAP: Vibration Data Accuracy Program • Video Data Processor

VIDAR: Velocity Integration, Detection And Ranging

VIDC: Video Controller

VIDD: Vertical Interval Data Detector

Vide: Videtur

VIDEO: Visual-Data Entry Online

VidF: Video Frequency

VIDI: Visual-Input Detection Instrumentation

VIDR: Visual Information Display and Retrieval

VIDS: VICAR Interactive Display System • Visual Integrated Display System

VIE: Vienna Schwechat (Austria) (location identifier) • Virtual Information Environment • Wien Schwechat (Austria) (location identifier)

VIEF: Volume Integration Equation Function

VIEL: Viela

Vien: Vienna (Austria)

VIES: VAT Information Exchange System

VIEW: Video for Exploring the World • Virtual Environment Workstation

VIF: Variable Import Fee • Variance Inflation Factor • Virtual Interface • Virtual Interrupt Flag

VIFC: Vertical Interface Frame Code

VIFIS: Variable Interference Filter Imaging Spectrometer

Vig: Vignette

VIG: El Vigia (Venezuela) (location identifier) • Video Image Generator

Vign: Vignette

VIGRE: Vertical Integration of Research and Education • Vertically Integrated Grants for Research and Education (*)

VIGS: Visual Image Generating System

VIH: Vichy Rolla (USA-MO) (location identifier)

VII: White Dwarf

VIJ: Virgin Gorda (British Virgin Islands) (location identifier)

VIK: Kavik (USA-AK) (location identifier) • Value-In-Kind

Vil: Village

VIL: Dakhla (Morocco) (location identifier) • Vehicle Integration and Launch • Vertical-Injection Logic • Very Important Lady • Very Important Launch • Volume Imaging Lidar

VilCo: Villanova University Observatory Contributions (periodical code)

Vill: Villa • Village

VILSPA: Villafranca del Castillo Satellite Tracking Station (ESA)

VILT: Visible Light Tomography

VIM: Variability-Induced Mover • Vendor-Independent Mail • Vendor-Independent Messaging • Video Interface Module • Vision Immersion Module • Voyager Interstellar Mission

VIMA: VLT Interferometer Main Array (*) • VLT Interferometric Main Array • VLT Interferometric mode using the Medium Array

VIMAD: Voice Interactive Maintenance Aiding Device

VIMS: Visible Imaging Spectrometer • Visual and Infrared Mapping Spectrometer

VIMSARS: Video Image Map Storage And Retrieval System

Vin: Vinum

VIN: Vehicle Identification Number • Vinnica (Ukraine) (location identifier)

VINCI: VLT Interferometer Commissioning Instrument

VINES: Virtual Network Software

VIO: Véhicule Interorbital • Video Input/Output • Virtual Input/Output

VIOC: Variable Input-Output Code

VIOS: Voltage Imaging Optical System

VIP: Vacuum Image Pipeline • Variable Information Processing • Variable Input Phototypesetter • Versatile Information Processor • Very Important Paper • Very Important Passenger • Very Important Person • Video Information Provider • Video Intertial Pointing • Video Integrator and Processor • Video Integrator Processor • Virtual Interface Processor • Virtual Internet Protocol • Virtual Interrupt Pending • Visual Information Processing • Visual Information Processor • Visual Interactive Preview • Vocabulary Improvement Project • Volcanisme Intraplaque

VIPC: Vegetation Image Processing Centre

VIPER: Verifiable Integrated Processor for Enhanced Reliability

VIPOR: Visual Investigation Program on Orbiter Operations

VIPS: Video Ice Particle Sampler • Video Image Processing System • VIZIR Image Processing System • Voice Interruption Priority System

VIQ: Variation In Quantity • Viqueque (Indonesia) (location identifier)

vir: Virga

Vir: Virage • Virgo-Virginis (constellation) • Viridis

VIR: Durban Virginia (South Africa) (location identifier) • Îles Vierges Américaines (ISO country code) • US Virgin Islands (ISO country code) • Victoria Imperatrix Regina • Virgin Islands (US_) (ISO country code) • Visible and Infrared Radiometer • Visual Information Retrieval

VIRA: Venus International Reference Atmosphere

VIRAC: Ventspils International Radioastronomy Centre (Latvia) (formerly VIRC)

VIRAF: Visible Infrared Reflectance Absorbance Fluorescence

VIRC: Ventspils International Radioastronomy Centre (Latvia) (now VIRAC)

VIRD: Verification and Integration Requirements Document

Virg: Virginia (USA) • Vírgula

VIRGO: Variability in Irradiance and Gravity Oscillations • Variability of Irradiance and Gravity Oscillations • Variability of solar Irradiance and Gravity Oscillations

VIRGS: VISSR Interactive Registration and Gridding System

VIRI: Visible/Infrared Imager • Visual Information Retrieval Interface

VIRIS: Visible-Infrared Intelligent Spectrometer

VIRL: Visible and Infrared Lidar

VIRMOS: Visible Infrared Multi-Object Spectrograph

VIRR: Visible and Infrared Radiometer

VIRS: Vegetation Image Receiving Station • Visible and Infrared Scanner • Visible Infrared Scanner

VIRSR: Visible and Infrared Scanning Radiometer

VIRT: Visible to Infrared Transducer

VIRTIS: Visible and Infrared Thermal Imaging Spectrometer

Vis: Verbis • Viscosity • Viseur • Visibilité • Visibility • Visible • Visit • Visite • Visité • Visited • Visiteur • Visiting • Visitor • Vista • Visual • Visuel

VIS: VAT Information System • Vector Instruction Set • Véhicule d'Intervention et de Sauvetage • Verification Information System • Video Information System • Virtual Information Storage • Visalia (USA-CA) (location identifier) • Visibility • Visible • Visible Imaging System • Visite d'Intégrité de Structure • Vista • Visual Imagery System • Visual Imaging System • Visual Information Seeking • Visual Information Storage • Visual Information System • Visual Instruction Set • Visual Instrumentation System • Vital Information System • Voice Information System

VISA: VLT Interferometric mode using the Small Array (*) • VLT Interferometer Subarray (*) • VLT Interferometric Subarray

VISAM: Virtual Index Sequential Access Method

VISAR: Velocity Interferometer System for Any Reflector

Visc: Viscosity • Viscount

ViSC: Visualization in Scientific Computing

VISC: Visconde

ViSCA: Video System Control Architecture

VISCSS: Viscondessa

VISG: Visual Information Services Group

Visigót: Visigótico

VISIR: VLT Imager/Spectrometer in the Infrared • VLT Imaging and Spectrometry in the mid-Infrared

VISITT: Vendor Information System for Innovative Treatment Technologies

VISN: Virtual Integrated Sky Network

VISR: Virginia Institute for Scientific Research (USA-VA)

VISS: VHS Index Search System

VISSR: Visible and Infrared Spin Scan Radiometer • Visible Infrared Spin Scan Radiometer • Visual Infrared Spin Scan Radiometer

VIST: Vista

VISTA: Variable In-flight Stability Test Aircraft • Verbal Information Storage and Text Analysis • Visible and Infrared Survey Telescope for Astronomy • Visible Interferometer with Separate Telescope Assemblies • Visual Information for Satellite Telemetry Analysis • Visual Inspection Software for Technical Applications • Visual Interface for Space and Terrestrial Analysis • Volunteers In service to America

VisUV: Visible-Ultraviolet

VIS-UV: Visible-Ultraviolet

Vit: Vitesse • Vitreous

VIT: Vehicle Integration Test • Very Important Traveller • Vitoria (Spain) (location identifier)

VITA: Virginia Institute for Theoretical Astronomy (USA-VA) • Volunteer In Technical Assistance

VITAC: Voice Input To Automatic Computer

VITAL: Variably Initialized Translator for Algorithmic Languages • Virtually Integrated Technical Architecture Lifecycle

VITB: Vocational Industrial Training Board

VITC: Vertical Interval Time Code

Vitr: Vitreous • Vitreum

VITS: Videoconferencing Interaction for Teaching Science

Vit.Stat.: Vital Statistics

VITT: Vehicle Integration Test Team

VIU: Vehicle Interface Unit • Venice International University (Italy) • Video Interface Unit • Viru (Solomon Islands) (location identifier) • Voiceband Interface Unit

VIV: Vivigani (Papua New Guinea) (location identifier)

VIVA: Viajes Internacionales de Vacaciones SA • Virtual Library of Virginia

VIX: Vitória - Eurico Sales (Brazil-ES) (location identifier)

viz: Namely • Videlicet

Viz: Vizcaya (Spain)

VIZ: Vizag (India) (location identifier)

Vizc: Vizcaya (Spain)

vJ: vorigen Jahres

VJ: Vacuum-Jacketed • Vedāñga Jyotisha • Victory over Japan • Visite Journalière • Vitesse de Jet

VJB: Xai Xai (Mozambique) (location identifier)

VJQ: Gurue (Mozambique) (location identifier)

VJS: Virginia Journal of Science (periodical code)

VK: Verbundkatalog • Voyageur-Kilomètre

VKD: Virtual Keyboard Device • Virtual Keyboard Driver

VKE: Von Karman Equation

VKF: Von Karman Facility

VKG: Rach Gia (Viet Nam) (location identifier)

VKH: Vitesse en Kilomètres par Heure

VKI: Von Kármán Institute

VKIFD: Von Karman Institute for Fluid Dynamics

VKO: Moscow Vnukovo (Russia) (location identifier)

VKS: Vicksburg (USA-MS) (location identifier) • Visuelles Kommunikationssystem

VKT: Vorkuta (Russia) (location identifier)

VKW: West Kavik (USA-AK) (location identifier)

vl: Varia Lectio

VL: Véhicule Léger • Vent Latéral • VESA Local • Viking Lander • Vila • Ville • Virtual Life • Virtual Liquidity • Visible Light • Vitesse de la Lumière • Vitesse de Lancement • Vitesse de Libération • Vitesse Linéaire • Voie Lactée • Vorlesung • Vulgar Latin

VLA: Vandalia (USA-IL) (location identifier) • Verification Loads Analysis • Vertical Line Array • Very Large Aircraft • Very Large Array • Very Low Altitude • Video Logarithmic Amplifier • Volume Licensing Agreement

VLACC: Vlaamse Geautomatiseerde Centrale Catalogus

Vlad: Vladivostok (Russia)

VLAIS: Very Large Array Information System

VLAM: Variable Level Access Method

vLAN: Virtual Local Area Network

VLAT: Very Large Array Telescope

VLB: Very Long Baseline • Verzeichnis Liefbarer Bücher • VESA Local Bus • VESA Logical Bus • Visual Laser Beam

VLBA: Very Long Baseline Array

VLB: Vertical-Lift Bridge

VLBC: Very Large Bulk Carrier

VLBI: Very Long Baseline Interferometry • Viking Lander Biological Instrument

VLBP: Very Long Baseline Polarimetry

VLC: Valencia Manises (Spain) (location identifier) • Variable-Length Coder • Verification Load Cycle • Very Large Corpus • Viking Lander Capsule • Volcano • Vulpecula-Lyra-Cygnus

VLCC: Very Large Cargo Carrier • Very Large Crude Carrier

VLCS: Voltage-Logic-Current-Switching

VLCT: Very Large Capacity Transport • Very Large Commercial Transport

VLCTY: Velocity

VLD: Vacuum Leak Detector • Valdosta (USA-GA) (location identifier) • Variable-Length Decoder

VLDB: Very Large Database

VLDC: Variable Length Don't Care

VLDL: Very-Low-Density Lipoprotein

Vle: Viale

VLE: Valeur Limite d'Exposition • Valle (USA-AZ) (location identifier) • Vapor-Liquid Equilibrium • Very-Low Excitation • Visible Light Emission

VLF: Variation Linéaire de Fréquence • Vertical Launch Facility • Very Low Frequency • Virtual Lookaside Facility

VLFA: Very Low-Frequency Array • Very Low-Frequency Astronomy

VLFN: Very-Low-Frequency Noise

VLFR: Very-Low-Frequency Receiver

Vlg: Volgend

VLG: Village • Villa Gesell (Argentina-BA) (location identifier)

VLGS: Villages

VLGT: Very Large Gravitational Telescope

VLHC: Very Large Hadron Collider

VLI: Port Vila (Vanuatu) (location identifier) • Violation of Lorentz Invariance

VLIS: Viking Lander Imaging System

VLISM: Very Local Interstellar medium

VLIW: Very Long Instruction Word

VLK: Volgodonsk (Russia) (location identifier)

VLL: Valladolid (Spain) (location identifier) • Video Layout Terminal

VLLRD: Virtual Law Library Reference Desk

VLLW: Very Low-Level Waste

VLLY: Valley

VLM: Variable Length Multiply • Very Low Mass • Villamontes (Bolivia) (location identifier) • Virginia Living Museum (USA-VA) • Virtual Loadable Module (*) • Virtual Load Module • Vortex Lattice Method

VLML: Lacto-ovo Vegetarian Meal

VLMS: Very Low Mass Star

Vlmtrc: Volumetric

VLN: Valencia (Venezuela) (location identifier)

vlnr: van links naar rechts • von links nach rechts

VLNT: Violent

VLO: Vallejo Stolport (USA-CA) (location identifier) • Vernieuwd Lager Onderwijs • Very Low Observable

VLOT: Very Large Optical Telescope

VLP: Vertical Long Period • Very Long Period • Very Long Play • Very Long Player • Vila Rica (Brazil-MT) (location identifier) • Visiting Lecturer Programme

VLPC: Visible Light Photon Source

VLPE: Very-Long-Period Experiment

VLPRT: Very-Low-Power Radio Transmitter

VLPS: Vandenberg Launch Processing System

VLR: Vallenar (Chile) (location identifier) • Very Long Range • Very Low Range • Visiting Location Register • Visitor Location Register

VLS: Valesdir (Vanuatu) (location identifier) • Vandenberg Launch Site • Vapour-Liquid-Solid • Véhicule Logistique Spatial • Veículo Lançador de Satélites • Velocity Lowest Selectable • Vertical Launching System • Vertical Launch System • Very Low Speed • Viking Lander System • Visible-Light Sensor

VLSAL: Very-Large-Scale Application of Logic

VLSI: Very-Large-Scale Integrated • Very-Large-Scale Integration (*)

VLSIC: Very-Large-Scale Integration Circuit

VLSIPS: Very Large Scale Immobilized Polymer Synthesis

VLSM: Variable-Length Subnet Mask

VLSR: Velocity of Local Standard of Rest

VLSST: Vandenberg Launch Site Support Team

VLST: Vertical Launch/Landing Space Transport

vlt: Vliet

VLT: Variable List Table • Very Large Telescope • Video Layout Terminal • Video Lookup Table

VLTC: Very Long Time Coming

Vltg: Voltage

VLTI: Very Large Telescope Interferometer • Very Large Telescope Interferometry

VLTICS: Very Large Telescope Interferometer Control Software

VLU: Velikiye Luki (Russia) (location identifier) • Video Logic Unit

VLV: Valera (Venezuela) (location identifier) • Value • Valve

VLVS: Voltage-Logic-Voltage-Switching

VLY: Valley

VLYS: Valleys

vm: Vormittags

vM: vorigen Monats

V/m: Volts per Meter

V/M: Velocity Meter

VM: Velocity Meter • Velocity Modulation • Vertical Magnet • Video Machine • View Manager • Virtual Machine • Virtual Memory • Visitor Mode • Vitesse de Montée • Vitesse Maximum • Vitesse Moyenne • Vlasov-Maxwell • Voice Modulation • Volatile Matter • Voltmeter • Von Neumann-Morgenstern • Voormiddag • Votre Majesté • Voyager Mars • Vuestra Majestad

VMA: Valid Memory Address • Vehicle Maintenance Area • Virtual Memory Address

VMAG: Visual Magnitude

VMB: Virtual Machine Boot • Voice Mail Box

VMBF: Ventilation Mécanique Bi-Flux

VMC: Velocity Modulating Camera • Ventilation Mécanique Continue • Ventilation Mécaniquement Contrôlée (*) • Vertical Microcode • VESA Media Channel • Video-Mediated Communication • Visual Meteorological Conditions • Vitesse Maximum de Contrôle • VLT Master Catalogue

VMCF: Virtual Machine Communications Facility

VMCM: Vector Measuring Current Meter

VMD: Virtual Manufacturing Device • Virtual Mouse Driver

VMDB: Visual Meteor Database

VMDK: VCOS Multimedia Development Kit

VME: Valeur Moyenne d'Exposition • Versabus-Module-European • Versa Modular Eurocard • Versa Module Eurocard • Versa Module Europe • Versa Multibus Eurostandard • Vertical Memory Expansion • Villa Mercedes (Argentina-SL) (location identifier) • Virtual Machine Environment • Virtual Memory Environment

VMF: Vacuum Melting Furnace • Variable Message Format • Vertical Maintenance Facility • Voice Message Format

VMG: Vector Magnetograph • Video Magnetograph

VMH: Vitesse Moyenne à l'Heure

VMI: Variable Moment of Inertia • Vegetation Monitoring Instrument • Vibration Measurement Integrator • Virginia Military Institute (USA-VA)

VMIR: Vandenberg Management Integration Review

VMIS: VAS Model Impact Study

VML: Vector Markup Language • Virtual Machine Language

VMM: Veille Météorologique Mondiale • Virtual Machine Management • Virtual Machine Manager • Virtual Machine Monitor • Virtual Memory Management • Virtual Memory Manager • Virtual Memory Monitor • Vlaamse Milieumaatschappij (Belgium)

VMMM: Virtual Machine Memory Manager

VMMPS: Vehicle Management and Mission Planning System

VMO: Velocity-Modulated Oscillator • Very Massive Object • Vitesse de Mise en Orbite • Vitesse Maximale Opérationnelle

VMOA: Valley of the Moon Observatory Association (USA-CA)

VMOS: Vertical Metal Oxide Semiconductor

VMP: Valve de Mise en Pression • Vol de Mise au Point

VMPM: Venus Multiprobe Mission

VMQ: Vote à la Majorité Qualifiée

VMR: Variance-to-Mean Rate • Vela Molecular Ridge • Video Mail Retrieval • Virtual Meeting Room • Voltmeter Reverse • Volume Mixing Ratio

VMRR: Vendor Material Review Report

VMRS: Voice Message Relay System

VMS: Variable Mass System • Variable Message Sign • VAX Management System • Vehicle Management System • Velocity Measurement System • Velocity Measuring System • Vertical Motion Simulator • Very Massive Star • Vessel Monitoring System • Vibration Measuring System • Video Modulation System • Virtual Machine System • Virtual Memory System • Visual Management System • Visual Motion Simulator • Voice Mail System • Voice Management System • Voice Messaging System • Voyage Management System

VMSG: Vening Meinesz School for Geodynamics

VMT: Validate Master Tape • Variable Mass Theory • Very Many Thanks • Virtual Memory Technique • Volcano Mitigation Team

VMTF: Vertical Magnet Test Facility

VMU: Baimuru (Papua New Guinea) (location identifier) • Volume Maintenance Utility

VMV: Vision-Mission-Value • Vitesse Maximum de Vol

VMVM: Vector Measuring Current Meter

vn: Viet Nam

Vn: Volcán

VN: Vacancy Notice • Valeur Nominale • Verbal Noun • Vereinte Nationen • Verenigde Naties • Version Numérique • Viet Nam (international vehicle registration code) • Viet Nam (ISO country code) • Virtual Network • Vitesse Normale • Voie Navigable • Volmètre Numérique • Von Neumann

Vna: Vienna (Austria)

VNA: Saravane (Laos) (location identifier) • Video Network Adapter • Video Noise Analyzer • Virtual Network Architecture

VNAP: Vertical Noise Abatement Procedure

VNAV: Vertical Navigation

VNB: Very Narrow Band

VNC: Valeur Nette Comptable • Venice (USA-FL) (location identifier)

VND: Dong (Viet Nam) (ISO currency code)

VNE: Vannes (France) (location identifier)

VNESSA: Voyager Neptune Encounter Science Support Activity

VNET: Virtual Network

VNF: Virtual Network Feature

VNG: Viengxay (Laos) (location identifier)

VNIIFTRI: Institute for Physical Technical and Radio-Technical Measurements (Russia)

VNIP: Virtual Network Interactive Protocol

VNIR: Very Near Infrared • Visible and Near Infrared • Visible and Near Infrared Radiometer

VNIRP: Visible and Near Infrared Polarimeter

vnl: Voornamelijk

VNL: Variable Name Length

VNM: Viet Nam (ISO country code) • Vitesse au Niveau de la Mer

VNO: Value Not Obtained • Verbond van Nederlandse Ondernemingen • Vilnius (Lithuania) (location identifier) • Vitesse Normale Opérationnelle

VNR: Van Nostrand Reinhold • Vanrook (Australia-QLD) (location identifier) • Video News Release • Vol Non Régulier

VNS: Varanasi (India) (location identifier) • Virtual Network Switching • Virtual Notebook System

VNSU: Vereniging van Samenwerkende Nederlandse Universiteiten (Netherlands)

VNT: Vacuum Newton Telescope • Vakuum Newton Teleskop

vnw: Voornaamwoord

VNX: Vilanculos (Mozambique) (location identifier)

VNY: Los Angeles - Van Nuys (USA-CA) (location identifier)

v/o: Votre Compte

vo: Volapuk (ISO language code) • von oben

Vo: Verbo • Verso

VO: Vatican Observatory • Vehicle Operations • Verbal Order • Verb Object • Verordnung • Versión Original • Version Originale • Very Old • Video Output • Viking Orbiter • Virtual Observatory • Visualization Option • Visual Object • Vitesse Orbitale • Voice • Votre Ordre

VÖAA: Vereinigung der Österreichischen Amateurastronomen (Austria)

VOA: Voice Of America

VOB: Vacuum Optical Bench • Verdingungsordnung für Bauleistungen • Versioned Object Base • Vertical Optical Bench

Vobf: Vorbahnhof

V°B°: Visto Bueno

Voc: Vocal • Vocation

VOC: Verenigde Oostindische Compagnie • Video Output Card • Video Output Control • Volatile Organic Carbon • Volatile Organic Chemical • Volatile Organic Compound (*)

VOCA: Voice Communications Assembly

Vocab: Vocabulary

Vocat: Vocativo

VOCC: Venous Occlusion Cuff and Controller • Voltage-Operated Calcium Channel

VOCP: Vehicle Out-of-Commission Part

Voctl: Vocational

VoD: Video on Demand • Voice over Data

VOD: Video On Demand • Virtual Optical Disk • Voice Over Data

VODAS: Voice Over Data Access Station

VODAT: Voice-Operated Device for Automatic Transmission

VOE: Venus Orbit Ejection

VOF: Vatican Observatory Foundation

VoFR: Voice over Frame Relay

VOFR: Voice Over Frame Relay

VOG: Video-Oculographic • Video-Oculography • Video-Okulographie • Video-Okulographisch • Voice Over Frame • Volgograd (Russia) (location identifier)

VOGAA: Voice-Operated Gain-Adjusting Amplifier

VOGAD: Voice-Operated Gain-Adjusting Device

Vogtl: Vogtland (Germany)

VOH: Vohemar (Madagascar) (location identifier)

VOI: Venus Orbital Insertion • Venus Orbit Insertion • Voinjama (Liberia) (location identifier) • Vol Opérationnel aux Instruments

VOICE: Vertical Observations Involving Convective Exchange Experiment

Voil: Voilure

VOILA: Variable Object Identification, Location and Acquisition

VoIP: Voice over Internet Protocol

VOIR: Veille Omnidirectionnelle Infrarouge • Venus Orbiting Imaging Radar (now Magellan) • Visualization Of Information Retrieval

VOIS: Visual-Observation Instrumentation System • Visual-Observation Integration System • Voice-Operated Information System

VOK: Camp Douglas - Volk (USA-WI) (location identifier)

Vol: Volans-Volantis (constellation) • Volant • Volatería • Volatile • Volcán • Volcanic • Volcánico • Volcano • Volet • Voltige • Volume • Volumen

VOL: Variable Orientation Launcher • Volos - Nea Anchialos (Greece) (location identifier) • Volume

Volc: Volcanic • Volcánico • Volcano

Vols: Volúmenes • Volumes

VOLT: Visual Observations Layout Tool

Volum: Volumetric

VOM: Vacuum Oscillation Mechanism • Versioned Object Manager • Vitesse Optimale de Montée • Volt-Ohm-Meter

VOMD: Vandenberg Operations and Maintenance Documentation

VOMP: Vandenberg Operational Management Plan

VONA: Vehicle Of the New Age

voorpl: Voorplein

VOP: Vertical Ozone Profile

VOR: Vestibular-Ocular Reflex • VHF Omnidirectional Radiorange • VHF Omnidirectional Range • VHF Omnirange • Voice-Operated Relay

VORAD: Vehicle On-board Radar

VORG: Vatican Observatory Research Group (USA-AZ)

vorm: vormals • vormittags

Vors: Vorsitzender

VORTAC: VHF Omnirange TACAN

VORTEX: Verification of the Origins of Rotation in Toradoes Experiment

VOS: Verbal Operating System • Viking Orbiter System • Virtual Operating System • Voice-Operated Switch • Voluntary Observing Ship • Volunteer Observer Ship

VOSS: Vatican Observatory Summer School

Vot: Voto

VOT: VHF Omnidirectional Test • Voice-Onset-Time • VOR Test • Votuporanga (Brazil-SP) (location identifier)

Vou: Voucher

VOU: Virtual Online University

VOV: Video Output Voltage

VOWB: Vlaams Overlegorgaan voor Wetenschappelijk Bibliotheekwerk

VOX: Voice Operated Transmit • Voice-Operated Transmitter

VOXEL: Volume Picture Element

VOZ: Voronezh (Russia) (location identifier)

Vp: Vitesse de Périgée

VP: Falkland Islands (international aircraft registration code) • Islas Malvinas (international aircraft registration code) • Vacuum Pump • Vaglia Postale • Valve à Pointeau • Vanishing Point • Vanne Pneumatique • Vapour Phase • Vapour Pressure • Variable Physique • Variable Pitch • Variable Procedure • Variation Périodique • Vāyu Purāna • Vector Process • Vector Processing • Vector Processor • Vector Program • Véhicule Porteur • Verb Phrase • Verification Polarization • Vertical Pipe • Vertical Parallelism • Vertical Polarization • Very Poor • Vía Preferente • Vice President • Vice-Président • Vice-Principal • Video Processor • Viewing Period • View Point • Virgin Islands (international aircraft registration code) • Virtual Path • Virtual Pool • Virtual Processor • Vision Processing • Visite Périodique • Vitesse de Pointe • Vitesse Parabolique • Vitesse Périphérique • Vitesse Propre • Vlasov-Poisson • Volume Partiel • Vuestra Paternidad

VPA: Vanne de Prélèvement d'Air • Variable Profile Antenna • Virtual Personal Assistant • Virtual Population Analysis

VPAM: Virtual Partitioned Access Method

VPB: Variable Procrustean Bed

VPBA: Varipolarization Beacon Antenna

VPC: Vacuum Pump Chamber • Vapour-Phase Chromotography • Vente Par Catalogue • Vente Par Correspondance • Video Processor Control • Viewport Coordinate • Virgo Photometric Catalogue • Virtual Path Connection

VPCA: Video Prelaunch Command Amplifier

VPCAR: Vapor Phase Catalytic Ammonia Removal

VpD: Vehicles per Day

VPD: Vapor Pression Deficit • Variable Power Divider • Vector-Point Diagram • Vehicles Per Day • Vertically Polarized Dipole • Virtual Printer Device • Volume Partition Descriptor

VPDN: Virtual Private Data Network

VPDS: Virtual Private Data Service

VPE: Ongiva (Angola) (location identifier) • Vapour-Phase Epitaxy • Video Processing Equipment • Virtual Prototyping Environment • Visualization for Planetary Exploration

VP-F: Falkland Islands (international aircraft registration code)

VPF: Falkland Islands (international aircraft registration code) • Vacuum Pump Filter • Variable Phase Filter • Vector Product Format • Vertical Processing Facility • Void Probability Function

VPFG: Variable Phase Function Generator

VpH: Vehicles per Hour

VPH: Vehicles Per Hour • Vertical Photography • Volume-Phase Holographic

VPHD: Vertical Payload Handling Device

VPI: Valve Position Indicator • Vapour-Phase Inhibitor • Versatec Parallel Interface • Vibration Pattern Imager • Virginia Polytechnic Institute (USA-VA) • Virtual Path Identifier • Virtual Private Intranet

VPICD: Virtual Programmable Interrupt Controller Driver

VPIM: Virtual Personal Information Manager • Voice Profile for Internet Mail

VPI&SU: Virginia Polytechnic Institute & State University (USA-VA)

VPK: Volts Peak

VPL: Virtual Private Line • Visual Process Language • Volkssternwarte und Planetarium Laupheim eV (Germany)

VP-LA: Anguilla (international aircraft registration code)

VPLCC: Vehicle Propellant Loading Control Center

VP-LMA: Montserrat (international aircraft registration code)

VP-LV: Virgin Islands (international aircraft registration code)

VpM: Vehicles per Mile • Vibrations per Minute • Volts per Meter

VPM: Vacuum Pumping Module • Variability/Proper Motion • Vehicle Project Manager • Vehicles Per Mile • Vertical Polarization Mode • Vibrations Per Minute • Virtual Parallel Machine • Volts Per Meter

VPN: Virtual Page Number • Virtual Private Network • Virtual Private Networking • Vopnafjördur (Iceland) (location identifier)

Vpo: Valparaiso (Chile)

VPO: Vernonia Peak Observatory (USA)

vPoP: Virtual Point of Presence

VPP: Value Payable Post

VPPAW: Variable-Polarity Plasma Arc Welding

VPPSEQT: Virtual Pool Parallel Sequential Threshold

V. Pres.: Vice-President

VPR: Video Production Recorder

VpS: Vibrations per Second

VPS: Destin (USA-FL) (location identifier) • Fort Walton Beach (USA-FL) (location identifier) • Vakuum-Plasma-Spritzen • Valparaiso (USA-FL) (location identifier) • Variable Pump System • Variable Parameter System • Variable Phase Shifter • Vernier Propulsion System • Versatec Plot Server • Vibrations Per Second • Vibrator Power Supply • Video Program System

VPSA: Vehicle Processing Storage Area

VPSEQT: Virtual Pool Sequential Threshold

VPT: Vente Par Téléphone • Virtual Printer Technology

VPU: Vacuum Penetration Unit • Video Processing Unit • Vision Processing Unit • VME Pattern Unit

VPUS: Vacation Programme for University Students

VPW: Vente Par WWW • Vertically Polarized Wave

VPZ: Valparaiso - Porter County (USA-IN) (location identifier)

VQ: Saint Helena and Ascension (international aircraft registration code) • Turks and Caicos (international aircraft registration code) • Vector Quantization • Vector Quantizer • Ventilateur de Queue

VQC: Vector Quantization Coding

VQL: Visual Query Language

VQM: Valeur Quadratique Moyenne

VQS: Vieques (Puerto Rico) (location identifier)

VQ-T: Turks and Caicos Islands (international aircraft registration code)

VQT: Turks and Caicos Islands (international aircraft registration code)

v/r: Valor Recibido

vr: Vrouwelijk

V-R: Vibrational-Rotational

VR: Bermuda (international aircraft registration code) • Cayman Islands (international aircraft registration code) • Gibraltar (international aircraft registration code) • Hong Kong (international aircraft registration code) (obsolete) • Valeur Résiduelle • Valve de Remplissage • Vanne de Régulation • Variable Reluctance • Variable Response • Vedi Retro • Veer • Véhicule de Reconnaissance • Velocity Ratio • Vent Relatif • Verlagsrecht • Via Rápida • Video Recorder • Virtual Reality • Visual Reference • Vitesse de Rafale • Vitesse de Rapprochement • Vitesse de Rentrée • Vitesse de Rotation • Vitesse de Roulage • Vitesse Radiale • Vitesse Relative • Voie Routière • Voiture Radio • Vol de Réception • Vol Rectiligne • Vol Régulier • Voltage Reference • Voltage Regulator • Voltage Relay • Vorstandsrat

VRA: Varadero (Cuba) (location identifier) • Vertical Reference Altitude • Vertical Resolution Advisory • Viking Radar Altimeter • Voltage Reference Amplifier

VRAM: Variable Random Access Memory • Video Random Access Memory • Virtual Random Access Machine • Virtual Random Access Memory

VRAP: Visiting Research Assistant Professorship

VRAS: Verification Requirements Analysis Sheet

VRASP: Virtual Reality Alliance of Students and Professionals

VR-B: Bermuda (international aircraft registration code)

VRB: Variable • Vero Beach (USA-FL) (location identifier) • VHF Recovery Beacon

VRBL: Variable

VR-C: Cayman Islands (international aircraft registration code)

VRC: Cayman Islands (international aircraft registration code) • Vertical Redundancy Character • Virac (Philippines) (location identifier)

VRCS: Vernier Reaction Control System

VRD: Vanne Régulatrice de Débit • Virtual Retinal Display • Vitesse de Rotation au Décollage

VRDA: Vereda

VRDI: Virtual Raster Display Interface

VRE: Visual Recognition Engine • Vredendal (South Africa) (location identifier)

VRF: Versatec Random Format • Vertical Removal Fixture • Vestibular Research Facility • Visual-Reproduction Facility

VRFD: Virtual Reality Flight Deck

Vrfy: Verify

VRFY: Verify

VR-G: Gibraltar (international aircraft registration code)

VRG: Gibraltar (international aircraft registration code) • Variable Report Generation

VR-H: Hong Kong (international aircraft registration code) (obsolete)

VRH: Hong Kong (international aircraft registration code) (obsolete) • Variable Range Hopping

VRI: Vanguard Research Inc. (USA-CA) (formerly JSE) • Virtual Radio Interferometer • Visual Rule Instrument

VRIFS: Vector Recurrent Iterated Function System

VRISL: Vancouver Island (Canada-BC)

VRK: Varkaus (Finland) (location identifier) • Video Recorder Kit

VRL: Vertical Recovery Line • Vertical Reference Line • Vila Real (Portugal) (location identifier)

VRLA: Valve-Regulated Lead-Acid (battery)

VRLI: Virtual Reality Laboratories Inc. (USA-CA)

VRLSA: Virtual-Reality Large-Scale Application

VRM: Variable-Rate Mortgage • VCOS Resource Manager • Venus Radar Mapper (formerly VOIR - now Magellan) • Video Record Machine • Virtual Register Machine • Virtual Resource Manager

VRML: Virtual Reality Markup Language (*) • Virtual Reality Modelling Language

VRMP: Voltage Reference Maintenance Program

VRMS: Voltage Root Mean Square

VRN: Vernier • Verona (Italy) (location identifier)

vrnl: van rechts naar links • von rechts nach links

VRNR: Vernier

Vro: Ventisquero

VRO: Vehicle Registration Office • Vermilion River Observatory

VROM: Vocabulary Read-Only Memory • Volkshuisvesting Ruimtelijke Ordening en Milieubeheer (Netherlands)

VROOMM: Virtual Real-time Object-Oriented Memory Manager

VRP: Variability Reduction Program • Vector-to-Raster Processor

VRPM: Visible Radiation Polarization Monitor

VRPS: Voltage-Regulated Power Supply

VRR: Verification Readiness Review

VRRP: Virtual Router Redundancy Protocol

vrs: Versus

VRS: Vacuum Resources System • Velocity Response Shape • Vendor Rating System • Verkehr Raumfahrt und Systemtechnik GmbH (Germany) • Versailles (USA-MO) (location identifier) • Video Reception System • Video Recognition System • Video Reference Service • Video Relay System • Video Response System • Virtual Reality and Simulation • Virtual Route Switching • Visual Reference System • Voice Recording System • Voice Response System

VRSD: Verification Requirements and Specifications Document

VRSS: Virginia Remote Sensing Study

VRT: Video Recording Tape • Voltage Reduction Technology • Voltage Reference Tube • Voltage Regulation Technology (*) • Voltage Regulator Tube

VRTM: Vacuum Resin Transfer Molding

VRTMOTN: Vertical Motion

VRU: Velocity Reference Unit • Vertical Reference Unit • Voice Response Unit • Voltage Readout Unit • Vryburg (South Africa) (location identifier)

VRUI: Virtual-Reality User Interface

VRY: Vaerøy (Norway) (location identifier)

vs: Verse • Versus • Vide Supra

V/S: Vertical Speed

VS: Vaisseau Sanguin • Vanne de Sécurité • Vanne de Surpression • Vapeur Saturée • Variable Speed • Vehicle Station • Véhicule de Sauvetage • Véhicule de Service • Véhicule Spatial • Velocity Search • Ventilateur de Sustentation • Ventilation System • Vent Solaire • Verbal Scale • Verenigde Staten • Vertical Speed • Vertical Stereoscopic • Vertical Structure • Vertical Sync • Vide Spatial • Vide Supra • Vie de Stockage • Vie Scolaire • Vieux Style • Virtual Simulation • Virtual Source • Virtual Storage • Viseur Stabilisé • Visite de Sécurité • Visite Systématique • Visual Storage • Vitesse à la Sortie • Vitesse de Sécurité • Vitesse de Sortie • Vitesse du Son • Vitesse-Sol • Voile Solaire • Volatile Solid • Vol Spatial • Vol Stationnaire • Volumetric Solution • Vostra Santità • Votre Sainteté • Vuestra Señoría

VSA: Vacuum Society of Australia • Variable-Stability Aircraft • Velocity Sensor Antenna • Verenigde Staten van Amerika • Verification Site Approval • Vernier Solo Accumulator • Vertical Sensor Assembly • Very Small Array • Villahermosa (Mexico) (location identifier) • Very Small Array • Volkssterrenwacht Amsterdam (Netherlands) • Voltage-Sensitive Amplifier • Volume Stimulus Actuator

VSAM: Virtual Sequential Access Method • Virtual Storage Access Method • Virtual System Access Method

VSAR: Versatile Synthetic Aperture Radar

VSASSY: Very Small Array Sky Survey

VSAT: Very Small Angle Tagger • Very Small Aperture Terminal

VSB: Vereinigten Staatsbibliotheken (Germany) • Vertical Sideband • Vestigial Sideband (*) • Visible • Volkssternwarte Bonn (Germany) • Volkssterrewacht Bussloo (Netherlands)

Vsbl: Visible

VSBR: Visual Surface Brightness Ratio

Vsby: Visibility

VSBY: Visibility

VSC: Vehicle Support Contractor • Vibration Safety Cutoff • Video Standards Council • Virtual Science Centre • Volatile Sulphur Compound • Voltage-Saturated Capacitor • Volumes-Seats-Circulation

VSCE: Variable Stream Control Engine

VSCF: Variable Speed Constant Frequency

VSCS: Vertical Stabilizer Control System • Voice Switching and Control System

VSD: Variable Spark Duration • Variable-Speed Drive • Vendor's Shipping Document • Vendredi-Samedi-Dimanche • Vertical Situation Display • Visualisation-Saisie-Déport • Volkssternwarte Darmstadt eV (Germany)

VSE: Vector Symbol Editor • Velocity-Silence-Efficiency • Virtual Storage Extended • Virtual Storage Extension • Viseu (Portugal) (location identifier) • Visite Selon État • Vol Sans Escale

VSEL: Vickers Shipbuilding & Engineering Ltd.

VSEP: Visiting Students Enrichment Program • Visiting Summer Enrichment Program

VSEPR: Valency Shell Electron Pair Repulsion

VSF: Springfield (USA-VT) (location identifier) • Variable Star Field • Vertical Scanning Frequency

VSG: Vernier Step Gauge • Very Small Gain • Very Small Grain • Lugansk (Ukraine) (location identifier)

VSGC: Virginia Space Grant Consortium (USA-VA)

VSH: Vaisseau Spatial Habité

VSHPS: Vernier Solo Hydraulic Power System

VSI: Vents Stellaires en Interaction • Vertical Speed Indicator • Video Simulation Interface • Volcanological Survey of Indonesia • Voluntary Severance Incentive • Vuestra Señoría Ilustrísima

VSIO: Virtual Serial Input/Output

VSIP: Valence State Ionization Potential

VSIPL: Vector/Signal/Image Processing Laboratory (DARPA)

VSK: Volkssternwarte Köln (Germany)

VSL: Van Swinden Laboratorium (Netherlands) • Very Strong Line • Very Strong-Lined • Virtual Shareware Library • Viscous Shock Layer

VSLAN: Verdix Secure Local Area Network

VSM: Vector Space Model • Vector Spectromagnetograph • Véhicule Sous-Marin • Verein Schweizerischer Machinen-Industrieller (Switzerland) • Vibrating Sample Magnetometer • Video Switching Matrix • Virtual Shared Memory • Virtual Storage Management • Virtual Storage Manager • Virtual Storage Memory • Volkssternwarte Mainz (Germany) • Vol Sans Moteur

VSMF: Vendor Specification Microfile

VSN: Video Switching Network • Virtual Station Number • Vision • Volume Serial Number

VSNET: Variable Star Network

VSNL: Videsh Sanchar Nigam Ltd. (India)

VSNS: Virtual School of Natural Sciences

VSNU: Vereniging van Samenwerkende Nederlandse Universiteiten (Netherlands)

VSO: Phuoclong (Viet Nam) (location identifier) • Verb-Subject-Object • Vernieuwd Secundair Onderwijs • Very Stable Oscillator • Very Superior Old • Virtual Storage Option

VSOLJ: Variable Star Observing League of Japan

VSOP: Very Superior Old Pale • VLBI from Space Orbiting Platform • VLBI Science Observatory Program • VLBI Space Observatory Program (now HALCA) (*)

VSOS: Virtual Storage Operating System

VSP: Vacation Students Programme • Vehicle Scheduling Program • Vehicle Synthesis Program • Vertical Seismic Profile • Vertical Speed • Very Short Period • Video Signal Processor • Vienna Software Publishing • Visible Sensor Package • Visiting Scientist Program • Vol Spatial Piloté

VSPC: Virtual-Storage Personal Computing

VSPN: Virtual Secure Private Network

VSPS: Vernier Solo Power Supply

VSPX: Vehicle Scheduling Program Extended

VSR: Vaisseau Spatial Réutilisable • Very Short Range • Very Special Reserve • Vibration Sensitive Relay • Voice Storage and Retrieval • Voltage-Sensing Relay

VSRA: V/STOL System Research Aircraft

VSRF: Very-Short-Range Forecasting

VSRR: Verein Sternwarte Rotgrueb Rümlang (Switzerland)

vss: Versions

VSS: Variable Stability System • Variable Stars Section • Velocity Sensor System • Video Signal Simulator • Video Storage System • Virtual Simulation Software • Virtual Storage System • Visual Scene System • Vitesse Supersonique • Vocabulary Switching System • Voice Storage System • Vol Supersonique • Voltage-Sensing Switch • Voyager Spacecraft System

VSSC: Vikram Sarabhai Space Centre (India)

VSSHG: Video Spectra-Spectroheliogram

VSSM: Video Scanner Switch System

VSST: Vacuum Solar Spectrum Telescope (China-PRC)

VST: Västerås (Sweden) (location identifier) • Velocity-to-Space Transformation • Vista • VLT Survey Telescope

VSTAG: Vandenberg Shuttle Turnaround Analysis Group

VSTAR: Vandenberg Shuttle Turnaround Assessment Report

V/STOL: Vertical/Short Takeoff and Landing

VSTOL: Vertical/Short Takeoff and Landing

VSTP: Visual Satellite Tracking Program

VStR: Vermögensteuer-Richtlinien

VSU: Valdosta State University (USA-GA)

VSV: Variable Stator Vane • Véhicule à Stabilité Variable • Vol Sans Visibilité

VSW: VAP Software • Very Shallow Water

VSWR: Visual Standing Wave Ratio • Voltage Standing Wave Ratio (*)

VSYNC: Vertical Synchronization

VSZ: Verzeichnis Schweizerischer Zeitschriften

vT: vom Tausend

Vt: Vermont (USA) • Vietnamese

V&T: Verification & Test

V/T: Voltage/Temperature

VT: India (international aircraft registration code) • Vacuum Technology • Vacuum Tube • Vaisseau de Transport • Vanne de Transfert • Variable Temperature • Variable Thrust • Variable Time • Variable Transmission • Variation Thermique • Vatu (currency) • Véhicule de Transport • Vent • Vermont (USA) • Vertical Tab • Vertical Tabulation • Vibration Testing • Videotape • Videotext • Viet Nam • Vie Totale • Virginia Tech (USA-VA) • Virial Theorem • Virtual Terminal • Virtual Tributary • Vitesse de Tangage • Vocational Training • Voice Tube • Voie de Transmission • Vol de Transition • Vol en Translation • Vol Télécommandé • Volume Total

Vta: Vuelta

VTA: Vacuum Test Article • Vacuum Tube Amplifier • Valley Transportation Authority (USA-CA) • Variable Transfer Address • Vertex Time of Arrival • Victoria (Honduras) (location identifier) • Voie du Train d'Atterrissage

VTAM: Virtual Telecommunications Access Method (*) • Virtual Teleprocessing Access Method • Virtual Terminal Access Method

VTAR: Verification Test/Analysis Report

VTAS: Visual Target Acquisition System

VTB: Vitebsk (Belarus) (location identifier) • Voltage Time to Breakdown

VTBC: Vertical Time Bar Chart

VTC: Video Tape Camera • Video Time Code • Vidicon Television Camera • Virtual Tape Control • Virtual Training Centre • Vol Télécommandé • Voting Trust Certificate

VTCS: Virtual Tape Control System

VTCW: Vehicle Time Code Word

VTD: Variable Time Delay • Variation Totale Décroissante • Vertical Tape Display • Virtual Tape Drive • Virtual Timer Device • Virtual Timer Driver • Voting District

VTDA: Verification Team Data Analyst

Vte: Variante

VTE: Vapor Transport Equilibrated • Variable Thrust Engine • Véhicule Transporteur-Érecteur • Vibration Test Equipment • Vientiane (Laos) (location identifier) • Virtual Transmit Engine • Viscous Transonic Equation

VTELN: Vapor Transport Equilibrated Lithium Niobate

VTES: Variable Thrust Engine System • Vinyltriethoxysilane

VTF: Vacuum Test Furnace • Variable Temperature Facility • Vatulele (Fiji) (location identifier) • Vertical Test Facility • Vertical Test Flight • Vibration Test Facility • Voltage Transfer Function

Vtg: Voting

VTG: Video Timing Generator

VTGCM: Venus Thermospheric General Circulation Model

VTH: Virtual Town Hall

VTHL: Vertical Takeoff and Horizontal Landing

VTI: Variable Temperature Insert • Vidéotransmission Interactive • Virtual Terminal Interface

VTIR: Visible and Thermal Infrared • Visible and Thermal Infrared Radiometer (*)

VTJ: Vitro Technical Journal (periodical code)

VTL: Vacuum-Tube Launcher • Variable Threshold Logic • Verification Test Laboratory • Video Tape Lecture • Vittel (France) (location identifier)

VTM: Vacuum-Tube Module • Verification Test Matrix • Versatile Tracking Mount • Vibration Test Module • Voltage Tunable Magnetron

VTMIS: Vessel Traffic Management and Information Service

VTML: Versioned Text Markup Language

VTMO: Voltage Tunable Microwave Oscillator

VTMS: Vinyltrimethysilane

VTN: Valentine Miller (USA-NE) (location identifier) • Verification Test Network • Video Tape Network

Vto: Vencimiento • Vuelto

VTO: Véhicule de Transfert Orbital • Vertical Takeoff • Voltage Tunable Oscillator

VTOA: Voice and Telephony Over ATM

VTOC: Visual Table Of Contents • Volume Table Of Contents (*)

VTOGW: Vertical Takeoff Gross Weight

VTOHL: Vertical Takeoff - Horizontal Landing

VTOI: Verification Team Operating Instruction

VTOL: Vertical Takeoff and Landing

VTOVL: Vertical Takeoff - Vertical Landing

VTP: Vehicle Test Plan • Verification Test Plan • Verification Test Program • Virtual Tape Protocol • Virtual Terminal Protocol • Voronoi Tessellation and Percolation

VTPR: Vertical Temperature Profile Radiometer

VTPS: Vibration Test Plotting System • Vibroacoustic Test Plan Summary

VTR: Verification Test Report • Vertical Test Range • Video Tape Recorder • Video Tape Recording • Voltage Transformation Ratio

VTRE: Vented Tank Resupply Experiment

VTRF: VLBI Terrestrial Reference Frame

VTRS: Vandenberg Tracking Station (USA-CA) • Videotape Recording System • Videotape Response System

VTS: Vandenberg Tracking Station • Variable Time Step • Vereinigung Technischer Schriftsteller (Austria) • Verification Test Station • Versatile Training System • Vertical Test Site • Vertical Test Stand • Vertical Test System • Vessel Traffic Service • Vessel Traffic System • Vibration Test Specification • Vibration Test System • Vibroacoustic Test Summary • Video Teleconferencing System • Video Test Set • Video Timesharing Service • Viewfinder Tracking System • Virtual Tape Server • Virtual Tape Solution • Virtual Tape System • Visual Telemetry System • Vitesse Transsonique • Voice Telecommunications System

VTSS: Values, Technology, Science and Society

VTT: Vacuum Thermal Testing • Vacuum Tower Telescope • Vacuum-Tube Transmitter • Vakuum Turm-Teleskop • Validation Test Suite • Valtion Teknillinen Tutkimuskeskus (Finland) • Variable Threshold Transmitter • Véhicule Tout Terrain • Vol Tous Temps

VTU: Las Tunas (Cuba) (location identifier) • Virtual Time Unit • Vital Telemetry Unit

VTV: Verband für Textverarbeitung (Germany) • Virtual Tape Volume

VTVL: Vertical Takeoff and Vertical Landing

VTVM: Vacuum-Tube Voltmeter

VTW: Variable Transmission Window • Vehicle Time Word

VTX: Vertex

VTZ: Vishakhapatnam (India) (location identifier)

vu: von unten

V/U: Vehicle Utility

VU: Vanuatu (ISO country code) • Vehicle Unit • Vehicle Utility • Véhicule Utilitaire • Vertical Unit • Very Urgent • Virtual Universe • Virtual University • Vitesse Uniforme • Voice Unit • Voie Unique • Volume Unit • Vrije Universiteit

VUA: Virtual Unit Address • Vrije Universiteit Amsterdam (Netherlands)

VUB: Vrije Universiteit Brussel (Belgium)

VUBIS: Vrije Universiteit Brussel Information System

VUE: Visible Ultraviolet Experiment • Visual User Environment

VUI: Video User Interface • Visual User Interface

Vul: Vulpecula-Vulpeculae (constellation)

Vulg: Vulgaire • Vulgar

VUP: Valledupar (Colombia) (location identifier) • VAX Unit of Performance

VUS: Velikij Ustyug (Russia) (location identifier) • Versatile Upper Stage

VUSW: Vehicle Utility Software

VUT: Vanuatu (ISO country code) • Vervroegde Uitreding

VUTS: Verification Unit Test Set

VUV: Vacuum Ultraviolet • Vatu (Vanuatu) (ISO currency code) • Visible-Ultraviolet

VUVES: Vacuum Ultraviolet Echelle Spectrograph

VUW: Victoria University of Wellington (New Zealand)

v/v: Volume in Volume

vv: Varios • Vice Versa • Volúmenes

V/V: Validation & Verification • Vent Valve • Vertical Velocity

V&V: Validation & Verification • Verification & Validation

V-V: Vibrational-Vibrational

VV: Ustedes • Validation and Verification • Vanne de Ventilation • Varios • Vatican Variables • Velocity Vector • Vent Valve • Verification and Validation • Vertical Velocity • Vertical-Vertical • Vertical Visibility • Vibrational-Vibrational • Vice Versa • Visibilité Verticale • Visibility Value • Vitesse de Virage • Vitesse de Vol • Vitesse du Vent • Vitesse Verticale • Vitesse Vraie • Vivenda • Vol à Vue • Vorontsov-Vel'yaminov

VVA: Verein für Volkstümliche Astronomie • Verstoring Van Aardsystemen

VVB: Mahanoro (Madagascar) (location identifier)

VVC: Variable Vacuum Capacitor • Vertical-Velocity Console • Victor Valley College (USA-CA) • Villavicencio (Colombia) (location identifier) • Voltage Versatile Capacitor

VVCCC: Version Virtuelle du Catalogue Collectif Canadien (Canada)

VVCD: Voltage Variable Capacitance Diode

VVCS: Vernier Velocity Correction System

VVD: Voltage Variable Diode

Vve: Veuve

VVH: Very Very Heavy

VVI: Santa Cruz - Viru Viru (Bolivia) (location identifier) • Vertical Velocity Indicator • Vitesse Verticale Instantanée • Voltage Variation Indicator

VVK: Västervik (Sweden) (location identifier) • Verwaltung Vermessungs- und Kartenwesen

VVM: Vector Voltmeter • Velocity Vector Measurement

VV.MM.: Vos Majestés

VVMS: Velocity Vector Measurement System

VVO: Van Vleck Observatory (USA-CT) • Very Very Old • Vladivostok (Russia) (location identifier)

VVP: Vitesse de Variation de Pression

VVR: Variable Voltage Rectifier

VVS: Vereniging voor Sterrenkunde, Meteorologie, Geophysica en Aanverwante Wetenschappen (Belgium) • Very Very Superior • Vol à la Vitesse du Son

VVSA: Velocity Vector Sensor Assembly

VVSR: Vegetation Visible Spectral Reflectance

VVTP: Verification and Validation Test Plan

VVV: Vereneging voor Vreemdelingen Verkeer

VVZ: Illizi (Algeria) (location identifier)

V&W: Verkeer & Waterstaat

VW: Vector Warning • Verkeer en Waterstaat • Very Weak • View • Virtually Worthless • Visual Worth

VWA: Vacuum Window Assembly

VWB: Visual Workbench

VWFC: Very Wide Field Camera

VWG: Verification Working Group

VwGO: Verwaltungsgerichtsordnung

VWL: Variable Word Length

VWO: Voorbereidend Wetenschappelijk Onderwijs

VWP: Variable Width Pulse • Visible Web Presence

VWRS: Variable Wire Rate Sensor

VWS: Vertical Wind Shear • Views

VWSI: Vegetation Water Stress Index

Vx: Vertex • Vieux

VXC: Lichinga (Mozambique) (location identifier)

VxD: Virtual Extended Driver

VXD: Virtual X Device

VXDS: VETA X-ray Detector System

VXE: São Vicente (Cape Verde) (location identifier)

VXI: Virtual Extension for Instrumentation • Virtual Extension Interface • VME Extension for Instrumentation

VXO: Variable Crystal Oscillator • Växjö (Sweden) (location identifier) • Vaxjo (Sweden) (location identifier)

Vy: Various Years • Very

VYD: Vryheid (South Africa) (location identifier)

VYS: Peru (USA-IL) (location identifier)

Vz: Voorzitter

VZ: Virtual Zero

vzm: Voor Zover Mogelijk

vZM: von Zeipel Method

VZM: Vereniging Zonder Winstoogmerk

V2: Antigua and Barbuda (international aircraft registration code)

V3: Belize (international aircraft registration code)

V4: Saint Kitts and Nevis (international aircraft registration code)

V5: Namibia (international aircraft registration code)

V6: Micronesia (international aircraft registration code)

V7: Marshall Islands (international aircraft registration code)

V8: Brunei (international aircraft registration code)

W

w/: With

w: Axial Velocity • Coefficient of Concordance • Concordance • Dew • Electromagnetic Energy Density • Mass Fraction • Radiant Energy Density • Sound Energy Density • Velocity Component • Warden • Warehouse • Water • Weak Absorption • Walk • Warm • Water • Weather • Week • Weight • West • Western • Westlich • Wet • White • Wide • Width • Wife • Wind • Wire • With • Withdrawn • Withdrew • Withheld • Won • Word • Work • Wrong

W/: With

W: Energy • Load • London (West_) (UK postal code) • Mechanical Work • Radiant Energy • Radiant Flux • Tungsten • Tungstène • Velocity Component • Wagon • Wald • Wales (UK) • Warden • Warehouse • Warm • Warning • Washington • Water • Watt • Wednesday • Weight • Well • Welsh • Wesleyan • West • Westen • Western • Westerhout • Westlich • West London (UK postal code) • Whisky (international alphabet) • White • Width • Wide • Widow • Widowed • Width • Wien Constant • Window • Windows • With • Withdrawn • Withdrew • Withheld • Wolfram • Won (currency) • Work

wa: With Answers

W/A: Work Around

WA: Wales • Warrington (UK postal code) • Washington (USA) • Watt Angle • Weather Advisory • West Africa • Western Arctic • Western Australia • Wetteramt • Wide Angle • With Average • Withholding Agent • Work Around • Work Authorization

WAA: Wales (USA-AK) (location identifier) • Washington Area Astronomers (USA-DC) • Westchester Amateur Astronomers (USA-NY) • Wide-Angle Astrograph • Wide-Aperture Array • Wiener Arbeitsgemeinschaft für Astronomie (Austria) • With Abject Apologies

WAAC: Wollongong Amateur Astronomy Club (Australia-NSW)

WAAE: World Association for Adult Education

WAAL: Wisconsin Association of Academic Librarians (USA-WI)

WAAS: Wide-Area Augmentation System • World Academy of Art and Science

WAAT: Wide-Area Application Testbed

WAB: Wabag (Papua New Guinea) (location identifier) • When Authorized By • Wissenschaftliche Allgemeinbibliothek

WABA: Western Australia Booksellers' Association (Australia-WA)

WABI: Windows Application Binary Interface

WABS: Wide-Angle Broadband Spectrometer

WAC: Waca (Ethiopia) (location identifier) • Wide-Angle Camera • World Aeronautical Chart • World Aeronautical Corporation • Wyoming Academic Consortium (USA-WY)

WACC: Work Activities and Coordination and Collaboration

WACCC: Worldwide Air Cargo Commodity Classification

WACK: Wait before Acknowledge

WACL: Worcester Area Cooperating Libraries (USA-MA)

WACO: Whatcom Association of Celestial Observers (USA-WA)

WACS: Wire Automated Check System

WACT: Wide-Angle Čerenkov Telescope

Wad: Wadham

WAD: Andriamena (Madagascar) (location identifier) • Wide-Area Development • Work Authorization Document • World Academic Database • World Astronomy Day • World Aviation Directory

WADB: West African Development Bank

WADD: Wright Air Development Division

WADE: World Association of Document Examiners

WADEX: Word and Author Index

WADGPS: Wide-Area Differential Global Positioning System

WADS: Wide-Angle Display System • Wide-Area Data Service • Wide-Area Differential System • Workshop on Algorithms and Data Structures • Write-Ahead Data Set

WAE: Wadi ad Dawasir (Saudi Arabia) (location identifier) • When Actually Employed

WAEA: World Airline Entertainment Association

WAF: Wana (Pakistan) (location identifier) • Width Across Flats • Word Address Format • Work Approval Form

WAFC: World Area Forecast Centre

WAFL: Write Anywhere File Layout

W. Afr.: West Africa • West African

WAFS: World Area Forecast System

WAG: Gambia (international vehicle registration code) • Wanganui (New Zealand) (location identifier) • Waste Area Grouping

WAGS: Worldwide Acoustic Gravity-wave Study

WAH: Wahpeton (USA-ND) (location identifier) • Wide-Area Hypertext

WAHA: Wide-Angle High Aperture

WAI: Antsohihy (Madagascar) (location identifier) • Web Accessibility Initiative • Wörterbuchentwicklung für Automatisches Indexing

WAIF: Weighted Average Impact Factor

WAIM: Web-Age Information Management (conference)

WAIOP: Will Accept, If Offered, the Position

WAIPA: World Association of Investment Promotion Agencies

WAIS: Wechsler Adult Intelligence Scale • West Antarctic Ice Sheet • Western Antarctic Ice Sheet • Wide-Area Information Server • Wide-Area Information Service • Wide-Area Information System

WAIT: Western Australia Institute of Technology

WAITIO: World Association of Industry and Trade Information Offices

WAITRO: World Association of Industrial and Technological Research Organizations

WAITS: Wide-Area Information Transfer System

WAJ: Wawoi Falls (Papua New Guinea) (location identifier)

WAK: Ankazoabo (Madagascar) (location identifier)

WAKS: Wiener Allgemeines Kategorienschema

Wal: Wallon • Wallonie (Belgium)

WAL: Chincoteaque Wallops (USA-VA) (location identifier) • Sierra Leone (international vehicle registration code) • Wide-Angle Lens

WALA: West African Library Association

WALC: Washington Accelerated Learning Center (USA-DC)

WahlO: Wahl Ordnung

WALIS: Western Australia Land Information System

WALL: Western Association of Law Libraries (USA-CA)

WALN: Western Australia Library Network (Australia-CA)

WALP: Wide-Array Linear Planar

WALS: Wide-Angle Light Scattering

WAM: Ambatondrazaka (Madagascar) (location identifier) • Warren Abstract Machine • Work Analysis and Measurement • Wraparound Mortgage

WAMA: Weight After Mars Arrival

WAMAP: Waste Management Advisory Programme

WAMDII: Wide-Angle Michelson Doppler Imaging Interferometer

WAMEX: West African Monsoon Experiment

WAMI: Washington-Alaska-Montana-Idaho • World Association for Medical Informatics

WAML: Western Association of Map Libraries

WaMot: Wave Motion (periodical code)

WAMP: Weight And Mass Property

WAMS: Wide-Area Master Station

WAN: Nigeria (international vehicle registration code) • Waverney (Australia-QLD) (location identifier) • Wide-Area Network • Work Authorization Number

WANO: World Association of Nuclear Operators

WAO: Wabo (Papua New Guinea) (location identifier) • Wallace Astrophysical Observatory (USA-MA) • Wet op de Arbeidsongeschiktheidsverzekering • Wissenschaftliche Arbeitsorganisation • Work Assignment Order

WAOS: Wide-Angle Optical System

WAOSS: Wide-Angle Optical Stereo Scanner • Wide-Angle Optoelectronic Stereo Scanner (*)

WAP: Alto Palena (Chile) (location identifier) • Weak Anthropic Principle • Wireless Application Protocol • Wissenschaftlerarbeitsplätze • Work Analysis Program • Work Assignment Plan • Work Assignment Procedure

WAPI: World Aerial Photographic Index • World Air Photography Index

WAPOR: World Association for Public Opinion Research

WAQ: Antsalova (Madagascar) (location identifier)

War: Warrant • Warsaw Okecie (Poland) • Warszawa Okecie (Poland) • Warwickshire (UK-England)

WAR: Waris (Indonesia) (location identifier) • Wide Area Routing • Work Authorization Report • Work Authorization Request • Write After Read

WARAL: World Amateur Radio Astronomers League

WARC: World Administrative Radio Conference (*) • World Allocations Radio Conference

WARDA: West Africa Rice Development Association

WARF: Wide-Aperture Research Facility

Warks: Warwickshire (UK-England)

WARLA: Wide-Aperture Radio Location Array

Warn: Warning

WARN: Warning

WARP: Weather And Radar Processor • Work And Recreation Party

WARPS: Wide-Angle ROSAT Pointed Survey

Warr: Warranty

WARR: Wide-Angle Reflection and Refraction

Warrty: Warranty

WARS: Wedge Absorption Remote Sensor • Wide-Area Reference Station • Wide-Area Remote Sensor

WARSIDEC: Western Australia Remote Sensing Industry Development and Education

Warwicks: Warwickshire (UK-England)

WAS: Wadhurst Astronomical Society (UK-England) • Walsall Astronomical Society (UK-England) • Warren Astronomical Society (USA-MI) • Warsaw Astronomical Society (USA-IN) • Washington (USA-DC) (location identifier) • Wehr Astronomical Society (USA-WI) • Wellington Astronomical Society (New Zealand) • Weltastronomie Studienreisen (Germany) • Wessex Astronomical Society (UK-England) • Westminster Astronomical Society (USA-MO) • Westport Astronomical Society (USA-CT) • Wide-Angle Sensor • Wideband Antenna System • Worthing Astronomical Society (UK) • Wycombe Astronomical Society (UK)

WASA: West African Science Association • Wide Angle Star Acquisition

WASAR: Wide-Angle Synthetic Aperture Radar

WASC: Washington Analytical Service Center (EG&G) • Western Association of Schools and Colleges (USA)

Wash: Washington (USA)

WASL: Windows Aspect Script Language

WAsP: Wind Atlas Analysis and Application Programme

WASP: Water Analysis Simulation Program • Water And Steam Program • Weather-Atmospheric Sounding Projectile • Weightless Analysis Sounding Probe • Wideband Autocorrelation Spectrometer • Wien Automatic System Planning • Window Atmosphere Sounding Projectile • Workshop Analysis and Scheduling Programming • Wyoming-Arizona Search for Planets

WASS: Wide-Area Surveillance System

WAST: Western Australia Standard Time

WASTAC: Western Australian Satellite Technology Application Consortium

Wat: Waterford (Ireland)

WAT: Waterford (Ireland) (location identifier) • Wavelet Alias Table • Web Action Time • Weight-Altitude-Temperature • Weighted-Average Toxicity • Wide-Angle Tail • Wide-Angle Tailed • Word Association Test

WATA: World Association of Travel Agencies

WATCH: Wide-Angle Telescope for Cosmic Hard (X-rays)

WATERS: Wide-Area Technical Report Server • Wide-Area Technical Report Service

WATIS: Work and Training Information System

WATM: Wide-area Asynchronous Transfer Mode

WATOX: Western Atlantic Ocean Experiment

WATS: Website Activity Tracking Statistics • Wide-Area Telecommunications Service • Wide-Area Telephone Service • Wind And Temperature Spectrometer • World Air Transport Statistics

WAU: Wauchope (Australia-NSW) (location identifier) • Wavelet Audio Unit

WAUBA: West African University Booksellers Association

W. Aus.: Western Australia

WAV: Waveform • Wave Hill - Kalkgurung (Australia-NT) (location identifier)

WAVE: Water Vapour Experiment • Web Analysis and Visualization Environment • Wide Area Voice Exchange • World Association of Video-Makers and Editions

WAVECAL: Wavelength Calibration

WavEl: Wave Electronics (periodical code)

WAW: Warszawa (Poland) (location identifier) • Weaving the Astronomy Web • Write After Write

WAWD: Write-After-Write Dependency

WAWG: Where Are We Going?

WAX: Warroad (USA-MN) (location identifier)

WAXD: Wide-Angle X-ray Diffraction

WAXS: Wide-Angle X-ray Scattering • Wide-Area X-ray Survey

WAY: Waynesburg - Green County (USA-PA) (location identifier)

WAZ: Warwick (Australia-QLD) (location identifier)

W/B: Waybill • Weight & Balance • Wideband

W&B: Weight & Balance

WB: Warehouse Book • Water Boiler • Waveband • Waybill • Weather Bureau • Wechsler-Bellevue • Weight and Balance • West Berlin (ISO country code) (now obsolete) • Westbound • Wet Bulb • Wet Bulk • Wetenschappelijke Bibliotheek • Wheel Base • Wideband • Will Be • Wissenschaftliche Bibliothek • Word-Based • Word Before • Workbench • World Bank • Write Back • Write Buffer

WBA: Welsh Border Astronomers (UK-Wales) • Wideband Amplifier • Worn By Astronaut

WBAFC: Weather Bureau Area Forecast Center

WBB: Stebbins (USA-AK) (location identifier) • Wiener Brückenbau (Austria) • Wind-Blown Bubble

WBC: Western Boundary Current • White Blood Cell • Wideband Channel • Wideband Coupler • Wien Bridge Circuit • Wire Bridge Circuit

WBCT: Wideband Current Transformer

WBD: Befandriana (Madagascar) (location identifier) • Wideband Data • Wire-Boom Drive

WBDA: Wideband Data Assembly

WBDCS: Wideband Data Collection System

WBDI: Wideband Data Interleaver

WBDTS: Wideband Data Transmission System

WBDX: Wideband Data Switch

WBE: Bealanana (Madagascar) (location identifier) • Wideband Electronics • Wire-Boom Electronics

WBEM: Web-Based Enterprise Management

WBF: Wegener-Bergeron-Findeisen

WBFDM: Wideband Frequency Division Multiplexer

WBFM: Wideband Frequency Modulation

Wbg: Webbing

WBGT: Wet-Bulb Globe Temperature

WBGU: Wissenschaftlicher Beirat der Globale Umweltveränderungen

WBI: Boulder - Broker Inn (USA-CO) (location identifier) • Whole-Body Irradiation • Will Be Issued

WBIZ: Wissenschaftliches Bibliotheks- und Informationszentrum

WBL: Wideband Laser

WBM: Wapenamanda (Papua New Guinea) (location identifier)

WBMC: Weight Before Mars Capture

WBMCR: Wideband Multichannel Receiver

WBMOD: Waideband Model

WbN: West by North

WBN: Woburn - Cummings Park (USA-MA) (location identifier)

WBND: Westbound

WBNL: Westbound Noise Limiting

WBNV: Westbound Noise Voltage

WBO: Beroroha (Madagascar) (location identifier) • Wideband Oscilloscope • Wideband Overlap • Wien Bridge Oscillator

WBQ: Beaver (USA-AK) (location identifier)

WBR: Big Rapids (USA-MI) (location identifier) • Whole Body Radiation • Wideband Radiometer • Wideband Receiver • Word Buffer Register • Workbench Rack

WBRR: Weather Bureau Radar Remote

WbS: West by South

WBS: Walking Beam Suspension • Welsh Bibliographical Society (UK-Wales) • Whole-Body Scan • Wideband Spectrometer • Wideband Spectrum • Wideband System • Without Benefit of Salvage • Work Breakdown Schedule • Work Breakdown Structure • Work Breakdown System

WBSA: Wideband Spectrum Analyzer

WBSC: Wideband Signal Conditioner

WBSO: Wet Bevordering Speur- en Ontwikkelingswerk (Netherlands)

WBSS: Wide-Binary Supersoft Source

WBT: Wet-Bulb Temperature • Wideband Terminal • Wideband Transformer • Wideband Transmitter

WBTCH: Water Boiler Thermal Control Hydraulics

WBTS: Wideband Transmission System

WBTV: Weather Briefing Television

WBU: Boulder (USA-CO) (location identifier)

WBV: Wideband Voltage

WBVCO: Wideband Voltage-Controlled Oscillator

WBVCXO: Wideband Voltage-Controlled Crystal Oscillator

WBVT: Wideband Video Tape

WBVTR: Wideband Video Tape Recorder

WBW: Wilkes-Barre - Wyoming Valley (USA-PA) (location identifier) • Word By Word

wc: Without Charge • Word Count

W&C: Wire & Cable

W/C: Workmen's Compensation

WC: Carbon Wolf-Rayet (star) • London (West Central_) (UK postal code) • Tungsten Carbide • Wagon-Citerne • Watch Committee • Water Clock • Water Closet • Web Club • West Central London (UK postal code) • Wire and Cable • Without Charge • Word Contribution • Word Count • Work Center • Work Component • Work Control • Working Capital • Workmen's Compensation • World Coordinates • Write and Compute

WCA: Castro Gamboa (Chile) (location identifier) • Wideband Cassegrain Antenna • Wind Correction Angle • Worst Case Analysis

WCAC: Western Colorado Astronomy Club (USA-CO) • Wilderness Center Astronomy Club (USA-OH)

WCAFS: Wideband Cassegrain Antenna Feed System

WCAP: World Climate Applications Programme

WCASP: World Climate Applications and Services Programme

WCB: World Council for the Biosphere • WWW Catalogue Browser

WCC: Washington Comprehensive Catalog • Western Computing Center • Wilson Cloud Chamber • Wire Chamber Conference • Work Control Center • World Computer Congress

WCCE: World Computer Conference on Education

WCCP: Web Cache Control Protocol

WCCS: Window Contamination Control System

WCCU: Wireless Crew Communication Unit

WCD: Wind-Compressed Disk • World Commission on Dams

WCDA: Western Canada Deformation Array

WCDB: Work Control Database

WCDDT: Wet Countdown Demonstration Test

WCDMA: Wideband Code Division Multiple Access

WCDMP: World Climate Data and Monitoring Programme

WCDP: World Climate Data Programme

WCE: Early-type Carbon Wolf-Rayet (star) • Wavefront Control Experiment

WCED: World Commission on Environment and Development

WCEFA: World Conference on Education For All

WCET: Worst-Case Execution Time

WCF: Working Capital Fund

WCGM: Writable Character Generation Module

WCGR: Waterloo Center for Groundwater Research (Canada-ON) (now CRESTech)

WCH: Chaiten (Chile) (location identifier)

WCHR: Wheelchair

WCI: Wildlife Conservation International

WCIP: World Climate Impacts Programme

WCIRP: World Climate Impact and Response Programme

WCISP: World Climate Impact Studies Program

WCL: Water Coolant Line • Water Coolant Loop • World Confederation of Labor

WCLS: Water Coolant Loop System

WCM: Wired-Core Matrix • Wired-Core Memory • Word Combine and Multiplexer • Writable Control Memory

WCMC: World Conservation Monitoring Centre

WCNDR: World Conference on Natural Disaster Reduction (UN)

WCNN: World Conference on Neural Networks

WCOT: Wall-Coated Open Tubular

WCOTP: World Confederation of Organisations of the Teaching Professions

WCP: Weapon Control Platform • Weather Communications Processor • Weight Control Program • Wing Chord Plane • Work Control Plan • Working Capital Productivity • World Climate Programme

WCPT: West Coast Precision Traverse

WCR: Chandalar (USA-AK) (location identifier) • Water-Cooled Reactor • Wire Contact Relay • Word Control Register

WCRA: Weather Control Research Association

WCRP: World Climate Research Programme

WCS: Waste Collection System • Waste Collector System • Wavelength Coincidence Statistics • Wing Center Section • Wireless Communication Service • Work Control System • World Communications System • World Conference of Science • World Conservation Strategy • World Coordinate System • Writable Control Storage • Writeable Control Store

WCSAR: Wisconsin Center for Space Automation and Robotics (USA-WI)

WCSC: Walker County Science Center (USA-GA) • West Coast Switching Center • Western Commercial Space Center (USA-CA)

WCSI: World Centre for Scientific Information

WCSMP: World Climate System Monitoring Programme

WCSU: Western Connecticut State University (USA-CT)

WCT: World Class Test

WCU: West Chester University (USA-PA) • World Customs Union

WCY: World Communications Year

WCZ: Wind-Compressed Zone

w/d: Warranted • Well Developed • Withdrawn

wd: Ward • When Distributed • White Dwarf • Width • Wind • Wood • Word

Wd: Warranted • Wind • Wood • Word • Would

W/D: Weight/Drag • Withdrawal

WD: Dominica (international vehicle registration code) • Wage Determination • Walking Distance • Warnier Diagram • Watford (UK postal code) • Wave Drag • Western Digital • White Dwarf • Wilson-Devinney • Wind Direction • Wired Discrete • Wiring Diagram • Withdrawn • Work Day • Working Day • Working Distance • Working Document

WDA: Wadi Ain (Yemen) (location identifier) • Wheel Drive Assembly

Wdb: Woordenboek

WDB: Deep Bay (USA-AK) (location identifier) • World Databank

WDC: World Data Center

WDCGG: World Data Center for Greenhouse Gases

WDCM: World Data Center on Microorganisms

WDCO: World Data Center for Oceanography

WDCRS: World Data Center for Rockets and Satellites

WDCSA: World Data Center for Solar Activity

WDD: World Data Depository

WDDES: World Digital Database for Environmental Sciences

WDDHRS: Wideband Digital Decimeter High-Resolution Spectroscope

WDF: Wave Distribution Function • Weather Data Facility

Wdg: Winding

WDG: Enid Woodring (USA-OK) (location identifier)

WDH: Windhoek - J.G. Strijdom (Namibia) (location identifier)

WDI: Wind Direction Indicator • Wondai (Australia-QLD) (location identifier)

WDL: Weak Double Layer • Western Development Laboratory • Wien Displacement Law • Windows Description Language • Windows Driver Library • Wireless Data Link

WDLF: White Dwarf Luminosity Function

Wdlg: Widerlager

WDLY: Widely

WDM: Wave Division Multiplexer • Wave Division Multiplexing • Wavelength Division Multiplex • Wavelength Division Multiplexing • Weight after Departure from Mars • Win32 Driver Model

WDMF: Weak Disordered Magnetic Field

WDMO: Weight before Departure from Mars Orbit

WDMS: Watershed Data Management System

WDN: Waldron Island (USA-WA) (location identifier)

Wdo: Window

WDP: Wavelength-Dependent Polarization • Work Distribution Policy

WDPC: Western Data Processing Center (USA-CA)

WDR: Winder (USA-GA) (location identifier)

WDRC: World Data Referral Centre

WDS: Wake Deluge Supply • Washington Double Star (catalogue) • Wavelength-Dispersive Spectrometer • Wetterdienstschule • Wire Data Service

WDSD: Water Data Sources Directory

WDSH: Work- and Data-Sharing

WDSPR: Widespread

WDSPRD: Widespread

WDSH-PTc: Work- and Data-Sharing Parallel Tree-code

WDSHPTc: Work- and Data-Sharing Parallel Tree-code

WDT: Watch Dog Timer

Wdth: Width

WDV: Wissenschaftliche Datenverarbeitung • Written-Down Value

WDVL: Web Developers Virtual Library

WDVI: Weighted Difference Vegetation Index

WDY: Kodiak (USA-AK) (location identifier)

w/e: Week End • Week Ending

We: Weber Number

W/E: Wing Elevon • Write/Erase

WE: Westinghouse Electric • Wing Elevon • Write Enable • Write-Erase • Writing Environment • Wyld Equation

Wea: Weather

WEA: Weather • Weatherford - Parker County (USA-TX) (location identifier)

WEAAP: Western European Association of Aviation Psychologists

WEAG: Western European Armament Group

WEAP: World Environment Action Plan

WEAR: WBS Element Audit Review

WEB: Whole Earth Blazar

WEBB: Water, Energy, and Biogeochemical Budget

WEBS: Woven Electronic Book System

WEBT: Whole Earth Blazar Telescope

WEC: Wave Energy Converter • Wave Experiment Consortium • Weak Energy Condition • Wind Energy Converter • World Energy Conference • World Energy Council • World Environment Centre

WECC: Western European Calibration Cooperation (*) • West European Calibration Cooperation

WECo: Western Electric Co.

WECOM: Western Electric Satellite Communications

Wed: Wednesday

WED: Wedau (Papua New Guinea) (location identifier) • Wednesday • World Environment Day

WEDC: West Engineering for Developing Countries

WEDOS: World Environment Disaster Observation System

Weds: Wednesday

WEDSS: Whole Earth Decision Support System

WEE: What Everyone Eats • Wind Erosion Equation

WEED: Windows Extracted from Earthquake Data

WEEG: Wednesday Evening Evaluation Group

WEENIAC: Weeny Almost Automatic Computer

WEF: Water Environment Federation • Weifang (China-PRC) (location identifier) • With Effect From • World Economic Forum

WEFAX: Weather Facsimile

WEFT: Wings-Engine-Fuselage-Tail

WEG: Washington Energy Group

WEGA: Wind Energy Groß-Anlagen

WEGENER: Working group of European Geoscientists for the Establishment of Networks for Earthquake Research

WEGS: Western European Geological Survey

WEH: Weihai (China-PRC) (location identifier)

WEHI: Walter and Elisa Hall Institute

WEI: Weipa (Australia-QLD) (location identifier)

WeiEn: Weight Engineering (periodical code)

WEIZAC: Weizmann Automatic Computer

WEJ: Westward Electrojet

Wel: Welsh

WEL: Welkom (South Africa) (location identifier)

WELAC: Western European Laboratory Accreditation Cooperation

Weld: Welding

WeldJ: Welding Journal (periodical code)

Weled: Weledele

Weledel: Weledelgeboren

Well: Wellington (New Zealand)

WELL: Whole Earth 'Lectronic Link

WELMEC: Western European Legal Metrology Cooperation

WEM: West Malling (UK-England) (location identifier) • World Event Management

WEMC: Western European Metrology Club

WeMFAS: West Midlands Federation of Astronomical Societies (UK)

WEN: Write Enable

WEND: Worldwide Electronic Navigational Database

WENDS: World Energy Data System

WENET: World Energy Network

WENSS: Westerbork Northern Sky Survey

WEO: Western European Organization

WEP: Weak Equivalence Principle • Weam (Papua New Guinea) (location identifier) • Wisconsin Experiment Package

WEPT: Week on Electronic Publishing and Typography

WER: Weight Estimating Relationship • Word Error Rate

WERT: Wide Energy Range Telescope

WERTID: West European Round Table on Information and Documentation

Wes: Wesleyan

WES: Waterways Experiment Station • Weasua (Liberia) (location identifier) • What Everyone Sees • World Economic Survey

West: Western

WEST: West European Science and Technology • Wideband European Satellite Telecommunication (*) • Wideband European Space Telecommunication

WESTAR: Westinghouse Space Transportation And Recovery

Westerw: Westerwald (Germany)

Westf: Westfalen (Germany)

Westm: Westminster

WestPac: Western Pacific

WESTPAC: Western Pacific

WET: Wagethe (Indonesia) (location identifier) • Waste Extraction Test • Western European Time • West European Time • Wetland Evaluation Technique • Whole Earth Telescope

Wetb: Wetboek

WETF: Weightless Environment Training Facility

WETN: Whole Earth Telescope Network

WEU: Western European Union (*) • West European Union

WEV: Water-Energy-Vegetation

WEVS: Wetland Ecosystem Vulnerability Study

WEW: Wee Waa (Australia-NSW) (location identifier) • What Everyone Wants

WEX: Wednesdays Excepted • Wexford Castlebridge (Ireland) (location identifier)

WEZ: Westeuropäische Zeit

wf: Wrong Font

wF: wahrscheinlicher Fehler

WF: Futuna Islands (Wallis &_) (ISO country code) • Wakefield (UK postal code) • Wallis & Futuna Islands (ISO country code) • Wave Function • Wide Field • Wind Force • Write Forward

WFA: Wavefront Analyzer • Weak-Field Approximation • Weighted Factor Analysis • Wide-Frequency Antenna

WFB: Ketchikan Waterfront (USA-AK) (location identifier)

WFC: Wallops Flight Center (NASA) • Washington Fundamental Catalog • Wave Front Curvature • Wide-Field

Camera • Wide-Field Channel • Windows Foundation Class • World Food Council

WFCC: World Federation for Culture Collections

WFCCI: Wide Field Camera Consortium Institutes (ROSAT)

WFE: Wavefront Error

WFEA: World Federation of Educational Associations

WFEO: World Federation of Engineering Organizations

WFF: Wallops Flight Facility (NASA) • Well-Formed Formula

WFG: Waveform Generator

WFI: Fianarantsoa (Madagascar) (location identifier) • Wide-Field Imager • Wide-Field Imaging • Wide-Field Instrument • Wide-FOV Instrument

WFIR: Wide-Field Imaging in the Red • Wide-Field Imaging Radiometer

WFIS: Wide-Field Imaging Spectrograph • Wide-Field Imaging Spectrometer

WFK: Frenchville (USA-ME) (location identifier)

WFL: Work Flow Language

WFM: Work Flow Management

WFMP: Work Force Motivation Plan

WFMS: Work Flow Management System

WFMU: Weather Fixed Map Unit

WFNA: White Fuming Nitric Acid

WFO: Weather Forecast Office • Wide-Field Optics • Work For Others

WFOF: Wide-Field Optical Filter

WFOV: Wide Field Of View

WFP: Warm Frontal Passage • Warm Front Passage • World Food Programme

WFPA: Wide-Field Plate Archive

WF/PC: Wide-Field/Planetary Camera

WFPC: Wide-Field and Planetary Camera

WFPC2: Second Wide-Field and Planetary Camera

WFPDB: Wide-Field Plate Database

WFQ: Weighted Fair Queuing

wfr: Wafer

WfS: Workgroup for Spectrography

WFS: Wavefront Sensor • Wilhelm-Foerster-Sternwarte (Germany)

WFSW: World Federation of Scientific Workers

WFT: Windowed Fourier Transform

WFTA: Winograd Fourier Transform Algorithm

WFTU: World Federation of Trade Unions

WfS: Workgroup for Spectrography

WFS: Wave Front Sensor • Wilhelm-Foerster-Sternwarte (Germany) • WORM File System

WFSW: World Federation of Scientific Workers

WFU: Wake Forest University (USA-NC)

WfW: Word for Windows

WFW: Windows For Workgroups • World Forest Watch

WFWG: Windows For Workgroups

WFXT: Wide-Field X-ray Telescope

Wg: Wing

W-G: Water-Glycol

W/G: Water/Glycol

WG: Grenada (international vehicle registration code) • Water Gauge • Water-Glycol • Waveguide • Weight Guaranteed • West Germany • Wigner Distribution • Windgeschwindigkeit • Wired Glass • Wire Gauge • Working Group

WGA: Wagga Wagga - Forrest Hill (Australia-NSW) (location identifier) • Waveguide Assembly • Weekly Government Abstracts • Writer's Guild of America

WGADBRA: Working Group on Archiving and Databases for Radioastronomy

WGAE: Working Group on Astronomy Education

WGAI: Working Group Agenda Item

WGAIT: Working Group on Astronomical Imaging Technology

WGAR: Working Group on Astronomical Refraction

WGAS: Working Group on Astronomical Software

WGB: Bahawalnagar (Pakistan) (location identifier)

WGC: Warrangal (India) (location identifier)

WGCV: Working Group on Calibration and Validation

WGD: Working Group on Data

WGDA: Working Group for the Development of Astronomy

WGDAA: Working Group for Development of Antarctic Astronomy

WGDC: Waveguide Directional Coupler

WGDL: Waveguide Delay Line

WGDM: Working Group on Data Management

WGE: Walgett (Australia-NSW) (location identifier)

WGEN: Weather Generator

WGEP: Working Group on Energy Policy

W. Ger.: West Germany

WGES: World's Greatest Environment Statement

WGF: Waveguide Filter

WGGC: Working Group on Global Change

WGGGI: Working Group on Geodesy and Geographic Information

WGI: World Geophysical Interval • World Glacier Inventory

WGII: Working Group on Internal Instrumentation

WGISS: Working Group on Information Systems and Services

W. Gl.: Wired Glass

WGL: Wavefront Generation Language • Waveguide Load • Wissenschaftsgemeinschaft Gottfried (Wilhelm) Leibniz (Germany)

WGLAM: Working Group on Limited Area Modelling

WGM: Waveguide Meter • Working Group Meeting

WGMAM: Working Group on Modern Astronomical Methodology

WGmc: West Germanic

WGMS: World Glacier Monitoring Service

WGN: Waitangi (New Zealand) (location identifier) • Working Group News

WGNE: Working Group on Numerical Experimentation

WGNEO: Working Group on Natural Near-Earth Objects

WGO: Winchester (USA-VA) (location identifier)

WGP: Waingapu (Indonesia) (location identifier) • Working Group on Policy

WGPAC: Working Group on Professional-Amateur Collaboration (AAS)

WGPSN: Working Group on Planetary System Nomenclature

WGR: Waste Generation Rate • Working Group Report

WGRA: Working Group on Resource Assessment

WGRF: Working Group on Radiation Fluxes • Working Group on Reference Frames (IAU)

WGRS: Working Group on Reference Systems

WGS: Work Group Server • Work Group System • Working Group on Software • World Geodetic Spheroid • World Geodetic Survey • World Geodetic System (*)

WGSIM: Working Group on Satellite Ionospheric Measurements

WGSPR: Working Group for Space Physics Research

WGSS: Working Group on Standard Stars (IAU)

WGSU: Water-Glycol Service Unit

Wgt: Weight

WGT: Wangaratta (Australia-VIC) (location identifier)

WGTC: Working Group on Tracking and Computation

WGU: Wagau (Papua New Guinea) (location identifier) • Western Governors University (USA)

WGWFI: Working Group on Wide-Field Imaging

WGWDA: Working Group for the Worldwide Development of Astronomy

WGWWDA: Working Group for the Worldwide Development of Astronomy

WGY: Wagny (Gabon) (location identifier)

w/h: Withholding

W/h: Watt/Hour

Wh: Watt-Hour • White

W/H: Wire Harness

WH: Warehouse • Warhead • Water Heater • White House • Widrow-Hoff • Wire Harness • Wirtshaus • Withholding • Word Hypothesis • World Heritage

WHA: Waste Handling Area • World Heritage Area

WHA/FH: Waste Handling Area Fume Hood

WHAFH: Waste Handling Area Fume Hood

WHAM: Waveform Hold And Modify • Wisconsin H-Alpha Mapper • Wisconsin Hydrogen-Alpha Mapper

W'hampton: Wolverhampton (UK)

WHAT: Wind Height And Temperature

WHAT A CAT: West Hellenic Arc Tectonics And Calabrian Arc Tectonics

WHC: Western Hemisphere Cooperation

WHD: Hyder (USA-AK) (location identifier)

WHDTV: Wide High-Definition Television

WHEPP: Workshop on High-Energy Physics Phenomenology

Whf: Wharf

WHF: Wadi Halfa (Sudan) (location identifier)

WHG: Walter-Horn-Gesellschaft (Germany)

WHH: Boulder - Hiltons Har (USA-CO) (location identifier)

WHI: Hwange (Zimbabwe) (location identifier) • Werner-Heisenberg-Institut (Germany)

WHIP: Wormhole Induction Propulsion

WHIRCAM: William Herschel Infrared Camera

WHIRL: Wide-angle High-Resolution Line

WHISP: Westerbork HI Spiral-galaxy Project

WHISPER: Waves of High frequency and Sounder for Probing of density by Relaxation

WHK: Whakatane (New Zealand) (location identifier)

Whl: Wheel

WHL: Welshpool (Australia-VIC) (location identifier)

WHO: Franz Josef (New Zealand) (location identifier) • White House Office • Wilfred Hall Observatory (UK-England) • World Health Organization • Worth Hill Observatory (UK-England)

WHOI: Woods Hole Oceanographic Institute

Whol: Wholesale

WHOSIS: World Health Organization Statistical Information System

WHP: Los Angeles - Whiteman (USA-CA) (location identifier) • WOCE Hydrographic Programme

WHPT: While-light and H-alpha Patrol Telescope

Whr: Watt-Hour • Whether

WHR: Vail Avon Stolport (USA-CO) (location identifier) • Watt Hour

Whs: Warehouse

WHS: Whalsay (UK) (location identifier) • White Sands (USA-NM) • World Heritage Site

WHSC: White House Science Council (USA)

Whse: Warehouse

Whsle: Wholesale

Wht: White

WHT: Wharton (USA-TX) (location identifier) • William Herschel Telescope

WHU: Wuhu (China-PRC) (location identifier)

WHYCOS: World Hydrological Cycle Observing System

w/i: Within

W&I: Weighting & Inspection

WI: Wallops Island (USA-VA) • Weizmann Institute (Israel) • West Indian • West Indies • When Issued • Windward Islands • Wisconsin (USA) • Workstation Interface • Wrought Iron

WIA: World Institute of Achievement

WIB: When Interrupt Block • Wissenschaftliche Internationale Bibliographie

WIBNI: Wouldn't It Be Nice If

WIC: Warning Information Correlation • Westindische Compagnie • Whole Internet Catalog • Wick (UK-Scotland) (location identifier)

WICAT: World Institute of Computer-Aided Teaching

WICEM: World Industry Conference on Environment Management

WICHE: Western Interstate Commission for Higher Education (USA)

Wid: Widow • Widower

WID: Wildenrath (Germany) (location identifier)

WIDA: Weather Impact Division Aids

WIDER: World Institute for Development Economics Research

WIDL: Web Interface Definition Language

WIE: Wiesbaden (Germany) (location identifier) • With Immediate Effect

WIEBS: Winter Ice Experiment in Beaufort Sea

WIF: Water Immersion Facility

WiFS: Wide Field Sensor

WIFS: Wide Field Sensor

WIG: Wing In Ground

WIHG: Wadia Institute of Himalayan Geology (India)

WII: Worldwide Information Infrastructure

WIK: Surfdale (New Zealand) (location identifier)

WIL: Nairobi Wilson (Kenya) (location identifier)

WILCO: Will Comply

WILMA: Wire List Management and Accounting

Wilts: Wiltshire (UK-England)

WIM: WAP Identity Module • Warm Interstellar Medium • Warm Ionized Medium

WIMC: Whom It May Concern

WIMP: Weakly Interactive Massive Particle • Window-Icon-Mouse-Pointer • Window - Icon - Menu - Pop up •

Windows-Icons-Menus-Pointer • Windows - Icons - Mouse - Pull-down

WIMS: Weather Information Management System

WIN: Walden Information Network • Wallonie Internet • Weather Information Network • Welfare Information Network • Western Information Network • Westlaw Is Natural • Winton (Australia-QLD) (location identifier) • Wisconsin-Indiana-NOAO • Wissenschaftsnetz (Germany) • Work Incentive

WINC: Wavelength Insensitive Coupler

Winch: Winchester (UK-England)

W. Ind.: West Indian • West Indies

WIND: Weather Information Display • Weather Information Network and Display

WINDEX: Window Experiment

Wind.I.: Windward Islands

WINDII: Wind Imaging Interferometer

WINDO: Wide Information Network for Data Online

WINDS: Weather Information Network and Display System • Weather Information Network Display System

WINDII: Wind Imaging Interferometer

WINDSAT: Wind Satellite

WINE: Windows Emulator • Winter In Northern Europe

WinEn: Wind Engineering (periodical code)

WING: Western Pacific Integrated Network of GPS • Work and Internship Networking Group

WinHEC: Windows Hardware Engineering Conference

Winn: Winnipeg (Canada-MB)

WINS: Weather Information Network Service • Windows Internet Name Service • Windows Internet Naming Service (*) • Wiskunde-Informatica-Natuurkunde-Sterrenkunde

WINSOCK: Windows Socket

Wint: Winter

WINT: Winter

WINTER: Web Internationalization

WIO: Wilcannia (Australia-NSW) (location identifier)

WIP: Waste Incineration Plant • Wissenschaftler-Integrations-Programm • Wissenschaftler-Integrations-Projekt • Work In Place • Work In Process • Work In Progress

WIPERS: Weakly Ionized Powerful Extragalactic Radio Source

WIPG: West Information Publishing Group

WIPICS: Work-In-Proces Inventory Control System

WIPIS: Whi Is Publishing In Science

WIPLA: World Intellectual Private Law Agency (France)

WIPO: World Intellectual Property Organization

WIPP: Waste Isolation Pilot Plant

WIPS: Weather Image Processing System

WIR: Wet op de Investeringsrekening • Wide Irradiance Response • Woomera Instrument Range

WIRDS: Weather Information Reporting and Display System

WIRE: Web/Internet Resource Executive • Wide-Field Infrared Explorer

WIRESCRIPT: Web Interactive Review of Scientific Culture, Research, Innovation Policy and Technology

WIRL: West India Reference Library

WIRO: Wyoming Infrared Observatory (USA-WY)

WIRP: Watershed Integrated Research Program

Wis: Wisconsin (USA)

WIS: Wallops Island Station (USA-VA) • Weather Information System • Web-based Information System • Wedge Imaging Spectrometer • Work Information System • WWMCCS Information System

WISARD: Web Interface for Searching Archival Research Data • Wideband System for Acquiring and Recording Data

WisBT: Wissenschaftliche Berichte (periodical code)

Wisc: Wisconsin (USA)

WISC: Wechsler Intelligence Scale for Children • Writable Instruction Set Computer

WISE: Washington Information Systems for Education • Whole Information System Expert • WordPerfect Information System Environment • Workshop on Information Systems and Economics • World Information Service for Energy • World Inventory of Soil Emissions • Worldwide Information System for Engineering

WISI: World Index of Space Imagery

WISIWYS: What I See Is What You See

WISLAD: World Information System for Library Automation Development

WISP: Waves In Space Plasma • Wide-field Imaging Survey Polarimeter • Wide-range Imaging Spectrometer • Winter Icing and Storms Project

Wiss: Wissenschaft • Wissenschaftlich

WISS: World Institute on Space and Society

Wissensch: Wissenschaft • Wissenschaftlich

WisZe: Wissenschaftliche Zeitschrift (periodical code)

Wit: Witness

WIT: Washington Interactive Television • Wittenoom (Australia-WA) (location identifier)

Withdrl: Withdrawal

Witht: Without

WITID: Wissenschafliche und Technische Information und Dokumentation

WITS: Windows Integrated Test Suite • Workshop on Internet Technology and Systems • World Ionosphere Thermosphere Study • Worldwide Information and Trade System

WIU: Western Illinois University (USA-IL) • Witu (Papua New Guinea) (location identifier)

WIX: Wait for Index

WIYN: Wisconsin-Indiana-Yale-NOAO

WJA: Woja (Marshall Islands) (location identifier)

WJEC: Welsh Joint Education Committee (UK-Wales)

WJF: Lancaster - William J. Fox (USA-CA) (location identifier)

WJR: Wajir (Kenya) (location identifier)

Wk: Walk • Weak • Week • Work

WK: Weak • Well Known • Wissenschaftskolleg

WKA: Wanaka (New Zealand) (location identifier)

WKB: Warracknabeal (Australia-VIC) (location identifier) • Wentzel-Kramers-Brillouin • Workbook

WKBJ: Wentzel-Kramers-Brillouin-Jeffreys

Wkdays: Weekdays

Wkds: Weekdays

Wkends: Weekends

WKF: Well-Known Factor • Worksheet

Wkg: Working

WKJ: Wakkanai (Japan) (location identifier)

WKK: Aleknagik (USA-AK) (location identifier)

WKL: Waikoloa (USA-HI) (location identifier)

Wkly: Weekly

WKM: Wechselkursmechanismus • Hwange National Park (Zimbabwe) (location identifier)

WKN: Wakunai (Papua New Guinea) (location identifier) • Weaken • Weakening

WKQ: Spreadsheet

Wkr: Worker

WKR: Walkers Cay (Bahamas) (location identifier)

Wks: Weeks • Works • Workshop

WKS: Well-Known Service • Werner-Keller-Schinke • Worksheet

WKU: Western Kentucky University (USA-KY)

WKZ: Compressed Spreadsheet

wL: westliche Länge

WL: Saint Lucia (international vehicle registration code) • Wagon-Lit • Waiting List • Water Line • Waterloopkundig Laboratorium • Wavelength • Well • Werkstofflabor • Western Laboratories • White Light • Whiteshell Laboratories • Will • Wing Left • Wing Load • Wire List

WLA: Wallal (Australia-WA) (location identifier)

WLAN: Wireless Local Area Network

Wlb: Wallboard

WLB: Labouchere Bay (USA-AK) (location identifier)

WLC: Walcha (Australia-NSW) (location identifier) • White-Light Channel • White-Light Cooling • White-Light Coronagraph

Wld: Welded • Would

WLD: Wärmeleitfähigkeitsdetektor • West-Longitude Date • Winfield - Arkansas City (USA-KS) (location identifier)

WLDC: Wall-Less Drift Chamber

Wldr: Welder

WLE: Weak Localization Effect

WLF: Futuna Islands (Wallis &_) (ISO country code) • Wallis & Futuna Islands (ISO country code) • White-Light Flare

WLG: Wellington (New Zealand) (location identifier)

WLH: Walaha (Vanuatu) (location identifier)

WLI: Workload Index

WLIF: Wireless LAN Interoperability Forum

Wlk: Walk

WLK: Selawik (USA-AK) (location identifier)

WLL: Wireless Local Loop • Wollogorang (Australia-QLD) (location identifier)

WLLN: Weak Law of Large Numbers

WLM: Waltham (USA-MA) (location identifier) • Wolf-Lundmark-Melotte • Workload Manager

Wln: Wellington (New Zealand)

WLN: Little Naukati (USA-AK) (location identifier) • Washington Library Network • Western Library Network • Wiswesser Line Notation

WLO: Waterloo (Australia-NT) (location identifier) • Windows Library for OS/2

W. Long.: West Longitude

WLOS: Window Line Of Sight

WLR: Loring (USA-AK) (location identifier) • Wind-Luminosity Relationship • Wrong Length Record

WLRG: Weak Line Radio Galaxy

WLRS: Wettzels Laser Ranging System

WLS: Wallis Island (location identifier) • Wavelength Scanning • Wavelength Shifter • Weighted Least Squares • Wells • Western Launch Site • Wide Latitude Storm • World Library Service • World Listening Service

WLSP: World List of Scientific Periodicals

WLT: White-Light Telescope

WLU: Wilfrid Laurier University (Canada-ON)

WLW: Willows - Glenn County (USA-CA) (location identifier)

Wly: Westerly

WLY: Westerly

Wm: William

W/m: Words/Minute

W/M: Weight/Measurement

WM: Wagon de Marchandises • Waste Management • Wattmeter • Webmaster • Window Manager • Work Manager

WMA: Mandritsara (Madagascar) (location identifier)

WMAC: Windows Media Access Control

WMB: Warrnambool (Australia-VIC) (location identifier)

WMC: Waste Management Compartment • Ways and Means Committee • Western Maryland College (USA-MD) • Winnemucca (USA-NV) (location identifier) • World Manufacturing Congress • World Meteorological Centre

WMD: Mandabe (Madagascar) (location identifier) • Water Management District • Weapon of Mass Destruction

WMDB: Waste Management Database

WME: Mount Keith (Australia-WA) (location identifier) • Window Meteoroid Experiment

WMES: Workload Management Expert System

WMF: Windows Metafile • Window Metafile Format • World Monuments Fund

WMH: Mountain Home (USA-AR) (location identifier)

WMI: World Metal Index • World Meteorological Interval

Wmk: Watermark

WMK: Meyers Chuck (USA-AK) (location identifier)

WMKO: W.M. Keck Observatory (USA-HI)

WML: Malaimbandy (Madagascar) (location identifier) • Wireless Markup Language

WMLR: Wind-Momentum-Luminosity Relation

WMN: Maroantsetra (Madagascar) (location identifier)

WMO: White Mountain (USA-AK) (location identifier) • World Meteorological Organization

WMP: Waste Management Plan • Weather Modification Program • WindowsMaker Professional • With Much Pleasure

WMR: Mananara (Madagascar) (location identifier)

WMRA: Write Many Read Always

WMS: Waste Management System • Watchdog Monitoring System • Wind Measuring System • Work Management System • World Magnetic Survey • World Mapping System

WMSC: Weather Message Switching Center

WMSCR: Weather Message Switching Center Replacement

WMT: Weighting More Than • West-Meridian Time • Wrongly Merge Total

WMU: Western Michigan University (USA-MI) • World Maritime University

WMV: Madirovalo (Madagascar) (location identifier)

WMX: Wamena (Indonesia) (location identifier)

WMY: World Mathematical Year

WN: Nitrogen Wolf-Rayet (star) • Wigan (UK postal code)

WNA: Napakiak (USA-AK) (location identifier)

WNB: Will Not Be

WNC: Nichin Cove (USA-AK) (location identifier) • Tuxekan Island - Nichen Cove (USA-AK) (location identifier)

Wnd: Waarnemend • Wound

WND: Wind • Windarra (Australia-WA) (location identifier)

WNDP: With No Down Payment

Wndr: Winder

WNE: Wora na Ye (Gabon) (location identifier)

WNIC: Wide-area Network Interface Coprocessor

WNILS: WHOIS and Network Information Lookup service

WNIM: Wide-area Network Interface Module

WNL: Late-type Nitrogen Wolf-Rayet (star) • Within Normal Limits

WNM: Warm Nuclear Medium

WNN: Wunnummin Lake (Canada-ON) (location identifier)

WNP: Naga (Philippines) (location identifier) • Will Not Proceed • Will Not Process • World Numeracy Program

Wnr: Weeknummer

WNr: Week Number

WNR: Windorah (Australia-QLD) (location identifier)

WNRC: Washington National Records Center (USA-DC)

WNS: Nawabshah (Pakistan) (location identifier) • World News Service

WNSL: Wright Nuclear Structure Laboratory (USA-CT)

WNU: Wanuma (Papua New Guinea) (location identifier)

WNV: World New Vistas

WNW: West-Northwest • West-Northwestern • What's New on the Web

WNWRN: West-Northwestern

WNWWD: West-Northwestward

WNY: Wynyard (Australia-TAS) (location identifier)

WNZ: Wenzhou (China-PRC) (location identifier)

w/o: Without • Written Off

wo: Waaronder • wie oben • Wolof (ISO language code)

W/O: Without

WO: Oxygen Wolf-Rayet (star) • Wait Order • Western Observatorium (USA-CA) • Wireless Operator • Work Order • Write Once • Write Only • Write Out • Written Order

WOA: Without Answers • Wonerara (Papua New Guinea) (location identifier)

WOB: Suttonheath Woodbridge (UK-England) (location identifier)

WOC: Weekly Operating Cost • Without Compensation • Work Order Control

WOCAT: World Overview of Conservation Activities and Technologies

WOCE: World Ocean Circulation Experiment

WOCS: WIYN Open Cluster Survey • Workshop on Optics and Computer Science

WOD: Wood River (USA-AK) (location identifier) • Word Of Day

WODC: World Ozone Data Center

WODD: World Oceanographic Data Display

WODDIN: Worldwide Online Data and Documents Intelligence System

WOE: Weight Of Evidence • Without Enclosure • Without Equipment

WOFOST: World Food Study

WOFS: Write-Once File System

WOG: Water-Oil-Gas • Weekly Orbit Geometry (file) • With Other Goods • Woodgreen (Australia-NT) (location identifier)

WOGS: World Organization for Cybernetics and Systems

WoGV: Wohngeldverordnung

WOI: Wind Outflow Inflow • Wologissi (Liberia) (location identifier)

WOIEP: World Office of Information on Environmental Problems

WOK: Wonken (Venezuela) (location identifier)

WOL: Wollongong (Australia-NSW) (location identifier)

Woll: Wollongong (Australia-NSW)

WOM: Web Object Management • Word Of Month • Write-Only Memory

WOMBAT: Waste Of Money, Brains, And Time

WON: Wondoola (Australia-QLD) (location identifier) • Work Order Number

WONG: Weight On Nose Gear

WoO: Werk ohne Opuszahl

WOO: Western Operations Office • Woodchopper (USA-AK) (location identifier) • World Oceanographic Organization

WOOL: Words Out of the Ordinary Language

WOP: Web Offensive Presence • Weekly Observing Plan • Weekly Operating Plan • Wing Outer Panel • With Other Property • Without Passport • Without Payment • Without Personnel • Without Preference • Without Priority

WOPE: Without Personnel and Equipment • Without Personnel or Equipment

WOR: Without Our Responsibility • Work Order Release

WORAM: Word-Oriented Random Access Memory

WORCC: World Optical depth Research and Calibration Center

Worcs: Worcestershire (UK-England)

WORF: Window Observational Research Facility • WWW Observing Remote Facility

WORM: Write Once - Read Many

WORMTOMS: WORM To Mass Storage System

WOROM: Write-Only Read-Only Memory

WOS: White Oval Spot (Jupiter) • Workplace Operating System • Workstation Operating System

WOSA: Windows Open Services Architecture • Windows Open Systems Architecture

WOSAC: Worldwide Synchronization of Atomic Clocks

WOSE: Weather Officer in Space Experiment

WOT: Wide Open Throttle • Wonan (China-ROC) (location identifier)

WOTRO: Wetenschappelijk Onderzoek van de Tropen in Ontwikkelingslanden

WOTS: Worldwide Origin Thread System

WOVO: World Organization of Volcano Observatories

WOW: Waiting On Weather • Web-Oracle-Web • Weight On Wheels • Willow (USA-AK) (location identifier) • Wisconsin Observer's Weekend • World Of Wonders • Worst On Worst

WOZ: Wahre Ortszeit

W/P: Waypoint

WP: Wagon Particulier • Wagon Planeur • Wagon-Poste • Warsaw Pact • Waste Paper • Water Plane • Water Processor • Water Pump • Waterproof • Way Point • Weapons Platform • Weather Permitting • Weatherproof • Wegener-Panda • Western Province (South Africa) • White Paper • Will Proceed • Wind Profile • Wind Profiles • Wire Payment • Without Prejudice • WordPerfect • Word Processing • Word Processor • Working Package • Working Paper • Working Party • Working Pressure • Working Protocol • Work Package • Work Program • Work Progress • Wrist Pitch • Write Protect • Write Protected

WPA: Puerto Aisen (Chile) (location identifier) • Water Pump Assembly • With Particular Average • Work Package Address • Work Package Agreement • Works Progress Administration • Works Projects Administration (*) • World Patents Abstracts • World Press Archives

WPAFB: Wright-Patterson Air Force Base

WPANL: Wright-Patterson Aeronautical Laboratory

WPB: Port Berge (Madagascar) (location identifier) • Waste Paper Basket • Write Printer Binary

Wpc: Watts per Candle

WPC: Pincher Creek (Canada-AB) (location identifier) • Washington Photoelectric Catalogue • Wired Program Computer • Wood-Plastic Composite • Word Processing Center • Work Package Concept • World Power Conference

WPCF: Water Pollution Control Federation

WPD: WordPerfect under DOS • Work Package Description • Work Program Directive • Write Printer Decimal

WPDL: Workflow Process Definition Language

WPDN: Wind Profile Demonstration Network

WPESS: Within Pulse Electronic Sector Scanning

WPF: Wissenschaftliche Projektführung • Work Process Flow

Wpg: Waterproofing

WPG: WordPerfect Graphics • Work Package Grouping

WPGMA: Weighted Pair Group Method using arithmetic Averages

WPGMC: Weighted Pair Group Method using Centroids

WPHD: Write-Protect Hard Disk

WPI: Wholesale Price Index • Worcester Polytechnic Institute (USA-MA) • Work Process Indicator • Work Progress Indicator • World Patent Index • World Patent Information • World Press Institute

WPK: WordPerfect Keyboard • Wrotham Park (Australia-QLD) (location identifier)

WPL: Powell Lake (Canada-BC) (location identifier) • Warning-Point Level • Wave Propagation Laboratory • Wesenheit-Period-Luminosity • Windows Personal Librarian

WPLS: Western Pacific Laser Satellite

WPLTN: Western Pacific Laser Tracking Network

WPLTO: Western Plateau

wpm: Words per Minute

WPM: Wipim (Papua New Guinea) (location identifier) • WordPerfect Macro • Words Per Minute • Work Package Management • Work Package Manager • Write Program Memory

WPMCP: Work Package Manpower and Cost Plan

WPMPR: Work Package Milestone Progress Report

Wpn: Weapon

WPO: Paonia - North Fork Valley (USA-CO) (location identifier) • Wirtschaftsprüfungsordnung • Woomera Project Office • World Packaging Organization

WPOM: Word Processing Output Microform

WPOS: Workplace Operating System

WPP: Water Pump Package • Weibull Probability Paper • White Pages Pilot • Work Package Plan

WPPS: Work Package Planning Sheet

WPPSI: Wechsler Preschool and Primary Scale of Intelligence

WPR: Porvenir (Chile) (location identifier)

wps: Words per Second

WpS: Watts per Steradian • Words per Second

WPS: Wallops Island Tracking Station (NASA) • Watts Per Steradian • Weather Protection System • Windows Printing System • With Prior Service • Word Processing Specialist • Word Processing System • Words Per Second • Work Package Structure • Workplace Shell • World Patent Service

WPSF: Waste Processing and Storage Facility

Wpt: Waypoint

WPT: Wireless Power Transmission • World Press Trends

WPU: Puerto Williams (Chile) (location identifier) • William Penn University (USA-IA)

WPW: WordPerfect under Windows

WPWOD: Will Proceed Without Delay

WQAS: Water Quality Analysis System

WQC: Waveform Quality Center

WQCD: Water Quality Criteria Document

WQM: Water Quality Monitor

WQP: Water Quality Parameter

WQS: Water Quality Standard

Wr: Washroom • Writer

W/R: Water/Rock

WR: Wagon-Réservoir • Wagon-Restaurant • Waiver Request • Warehouse Receipt • Water Resources • Weber Variables • Western Range • Western Region • Wet Runway • Whole Rock • Windrichtung • Wing Right • Wire Recorder • Wire Reply • Wissenschaftsrat • Wolf-Rayet • Worcester (UK postal code) • Word Recognition • Word Restoration • Work Request • World Record • Wrinkle Ridge • Wrist Roll

WRA: Warder (Ethiopia) (location identifier) • Web Research Alliance • Whiteshell Research Area

WRAM: Window Random Access Memory

WRAP: Water Resources Assessment Program • Western Region Applications Program • Winter Rainfall Acceptance Proposal

Wrb: Wardrobe

WRB: Macon Robins (USA-CA) (location identifier) • World Reference Base

WRBP: W. Russell Blake Planetarium (USA-MA)

WRC: Water Research Commission • Water Resources Council • Weather Relay Center • World Radiation Centre • World Radiocommunication Conference

WRCB: Water Resources Control Board

WRD: Water Resources Department • Water Resources Division

WRDC: Wright Research and Development Center

WRDD: Water Resources Development Department

WRE: Whangarei (New Zealand) (location identifier)

WREBUS: Water Reactor Extended Burnup Study

WRED: Weighted Random Early Discord

W. Ref.: With Reference to

WRG: Wrangell (USA-AK) (location identifier)

WRGP: Weak Reduced Game Property • Western Regional Graduate Program (USA)

WRH: World Radio Handbook

WRI: Fort Dix - McGuire (USA-NJ) (location identifier) • Web Report Interface • World Resources Institute (USA-DC) • Write

WRIST: Weather Radar Identification of Severe Thunderstorms

Wrk: Work

WRK: Window Resource Kit

Wrkg: Working

WRL: Willow Run Laboratory (NASA) • Wing Reference Line • Worland (USA-WY) (location identifier)

WRLC: Washington Research Library Consortium

WRM: Warm • Water Recovery and Management

WRMFNT: Warm Front

WRMS: Weighted Root Mean Square

WRMSR: Write Machine Specific Register

Wrn: Warning • Western

WRNG: Warning

Wrnt: Warrant

WRO: Warren Rupp Observatory (USA-OH) • Work Release Order • Wroclaw (Poland) (location identifier)

WRP: Water Resource Planning • Weight Reduction Proposal • Wiener Random Process • Wing Reference Plane

WRQ: Walker-Richer-Quinn

WRR: Water Resources Research (periodical code) • Weighted Round Robin • World Radiometric Reference • World Resources Report

WRRC: Water Resources Research Center

WRS: Warning and Report System • Waveguide Raman Spectroscopy • Wave Radiometer System • Wide-Range Sensor • Word Recognition System • World Reference System • Worldwide Reference System

WRSIC: Water Resources Scientific Information Center

WRSP: World Register of Scientific Periodicals

Wrt: Wrought

WRT: With Regards To • With Request To • With Respect To • Write

Wrtr: Writer

WRU: Who Are You?

WRW: Warrawagine (Australia-WA) (location identifier)

WRY: Westray (UK-Scotland) (location identifier)

W/S: Workstation

W&S: Wärme und Stoffübertragung (periodical code) • Whisky & Soda

WS: Samoa Occidentale (ISO country code) • Samoa (Western_) (international vehicle registration code) • Samoa (Western_) (ISO country code) • Wait State • Walsall (UK postal code) • Water Servicer • Water Soluble • Water Supply • Water Surface • Water System • Weather Service • Weather Ship • Weather Station • Weather Stripping • Western Samoa (international vehicle registration code) • Western Samoa (ISO country code) • Windshear • Wind Shield • Wind Speed • Wind Sonde • Wind Speed • Winter Solstice • Withholding Statement • Word Spotting • WordStar • Working Storage • Work Statement • Workstation

WSA: Wasua (Papua New Guinea) (location identifier)

WSAA: Waveguide Slot Array Antenna

WSAAG: Western Sydney Amateur Astronomy Group (Australia-NSW)

W. Sam.: Western Samoa

WSAPI: Web Site Application Programming Interface

WSB: Steamboat Bay (USA-AK) (location identifier) • Water Spray Boiler

WSBS: Wide-Sweep Backscatter Sounder

WSc: Wind Scatterometer

WSC: Water Survey of Canada • Wayne State College (USA-NE) • Westfield State College (USA-MA) • Whole Sun Catalog • Wideband Signal Conditioner • Wind Scatterometer • World Space Congress

WSCC: Work Station Control Center

WSCG: West Superclusters of Galaxies

WSCMO: Weather Service Contract Meteorological Observatory

WSD: Wet Server Director • White Sands - Condron (USA-NM) (location identifier) • Wieland System Design • Wind Speed Detector • Working Stress Design

WSDA: World Storage, Documentation and Abstracting

WSDB: Working Survey Database

WSDDM: Worldwide Solutions Design and Delivery Methods

WSDL: Web Structure Definition Language

WSDRN: Wide Satellite Data Relay Network

WSDS: Working Space Dataset

WSE: Workstation Environment • Workstation Support Environment

WSEN: Western Space Education Network (Canada-SK)

WSF: Sarichef (USA-AK) (location identifier) • Wake Shield Facility • Water Soluble Fraction • World Space Foundation

WSFO: Weather Service Forecast Office

WSFS: World Science Fiction Society

WSG: Washington County (USA-PA) (location identifier) • Weltstandardgruppe • World Standard Group

WSGC: Wisconsin Space Grant Consortium (USA-WI)

WSGR: White Sands Ground Relay

WSGS: White Sands Ground Station (USA-NM)

WSGT: White Sands Ground Terminal

w/sh: Workshop

Wsh: Workshop

WSH: Shirley Brookhaven (USA-NY) (location identifier) • Windows Scripting Host

WSHFT: Wind Shift

Wshg: Washing

Wshld: Windshield

Wshr: Washer

WSI: Wafer-Scale Integration • Whole-Sky Imager • Wind Speed Indicator • World Synoptic Interval

WSIR: Wide Swath Imaging Radar

WSIS: World Space Industry Survey

WSJ: San Juan (USA-AK) (location identifier) • Wall Street Journal

WSK: Webmaster Starter Kit

WSL: Westerbork Source List

WSM: Samoa (Western) (ISO country code) • Samoa Occidentale (ISO country code) • Watershed Model • Western Samoa (ISO country code) • Wet Satellite Model • Whole Sun Month • Wigner-Seitz Method • Wiseman (USA-AK) (location identifier)

WSMO: Weather Service Meteorological Observatory

WSN: South Naknek (USA-AK) (location identifier) • Washington State Network (USA-WA) • Water-Soluble Nitrogen

WSO: Washabo (Suriname) (location identifier) • Water Servicer Operator • Weapon System Officer • Weather Service Office • Western Support Office (NASA) • Wilcox Solar Observatory (USA-CA) • World Space Observatory

WSOF: Wavelength-Shifting Optical Fibre

WSOM: Weather Service Operations Manual

WSP: Wagon-Salon Pullman • Waspam (Nicaragua) (location identifier) • Water Supply Paper • Water Supply Point • Wideband Signal Processor • Winter Star Party • Work Sequence Plan • World Scientific Publishing

WSPC: World Scientific Publishing Co.

WSPR: Weapon System Problem Report

WSQJ: Webb Society Quarterly Journal

WSR: Wasior (Indonesia) (location identifier) • Weak Signal Reception • Weather Service Radar • Weather Surveillance Radar • Weekly Summary Report • Weltstrahlungsreferenz • Winter Solstice Sunrise

WSRH: Weather Service Regional Headquarters

WSRT: Westerbork Synthesis Radio Telescope (Netherlands)

WSS: Wall Shear Stress • Winter Solstice Sunset • Workstation Software System • Work Summarization System

WSSI: World Seismic Safety Initiative

WSSP: White Sands Star Party (USA-NM)

WSSR: Winter Solstice Sunrise

WSSU: Weather Service Support Unit

Wst: Wetterstation

WST: Tala (Samoa) (ISO currency code) • Weapon System Trainer • Westerly (USA-RI) (location identifier) • Wholesale Sales Tax • Write Symbol Table

WSTA: Weapon System Testability Analyzer

WSTAWG: Weapon Systems Technical Architecture Working Group

WSTB: Water Science and Technology Board • Westerbork (Netherlands)

WStDV: Wechselsteuer-Durchführungsverordnung

WSTF: White Sands Test Facility (NASA)

WSTN: World Service Television News

WSTV: World Service Television • W-Sensor Test Vehicle

WSU: Washington State University (USA-WA) • Wasu (Papua New Guinea) (location identifier) • Water Servicing Unit • Wayne State University (USA-MI) • Weber State University (USA-UT) • Wichita State University (USA-KS) • Wright State University (USA-OH)

WSUIPA: Washington State University International Phonetic Alphabet

WSV: Weapon System Vehicle

WSW: West-Southwest • West-Southwestern

WSWRN: West-Southwestern

WSWWD: West-Southwestward

WSX: Westsound (USA-WA) (location identifier)

WSY: Airlie Beach - Whitsunday (Australia-QLD) (location identifier)

WSZ: Westport (New Zealand) (location identifier)

Wt: Warrant • Weight • Without

W&T: Wear & Tear

W/T: Walkie-Talkie • Wind Tunnel

WT: Waist Tether • Walkie-Talkie • Watchdog Timer • Watch Time • Water Tank • Watertight • Waveguide Transmission • Wavelet Transform • Weather Trend • Weight • Wind Tunnel • Winter Time • Wire Ticket • Wireless Telegraphy • Wireless Telephony • Wireless Transmitter • With Tape • Write Through

WTA: Tambohorano (Madagascar) (location identifier) • West Texas Astronomers (USA-TX) • Willingness To Avoid • Winner Takes All • Wire Traceability and Accountability

WTB: Wissenschaftliche Taschenbücher

WTC: New York - World Trade Center (USA-NY) (location identifier) • World Trade Center

Wtd: Wanted • Warranted

WTD: West End (Bahamas) (location identifier) • Wind Tunnel Data

WTDC: World Telecommunication Development Conference

WTE: World Tapes for Education • Wotje Island (Marshall Islands) (location identifier)

Wtf: Waterford

WTF: Wednesday-Thursday-Friday

Wtg: Waiting • Weighting

WTG: Wind Turbine Generator

WTH: Whole-Tree Harvest

Wthr: Weather • Weather (periodical code)

WTHR: Weather

WTI: Wetenschappelijke en Technische Informatie • Wissenschaftliche und Technische Information • World Translation Index

WTIC: World Trade Information Center

WTID: Wissenschaftlich-Technische Information und Dokumentation

WTIE: Wastewater Treatment and Information Exchange

WTIS: World Trade Information Service • World Trade Information System

WTK: Noatak (USA-AK) (location identifier)

WTL: Tuntutuliak (USA-AK) (location identifier)

WTM: Wind Tunnel Model

WTN: Waddington (UK) (location identifier) • Worldwide Television Network • Worldwide Television News

WTO: World Tourism Organization • World Trade Organization (formerly GATT) • Wotho Island (Marshall Islands) (location identifier) • Write To Operator

WTOR: Write To Operator with Reply

WTP: Willingness To Pay • Wireless Transaction Protocol • Woitape (Papua New Guinea) (location identifier)

WTPF: World Telecommunication Policy Forum

Wtr: Water • Winter • Writer

WTR: Water • Weighted Total Rank • Western Test Range • White River (USA-AZ) (location identifier) • Work Transfer Record • Work Transfer Request

WTROD: Western Test Range Operations Division (NASA)

Wtrprf: Waterproof

WTRS: Waters

Wtrtt: Watertight

WTS: Tsiroanomandidy (Madagascar) (location identifier) • Wavelet Transform Space • Westward Travelling Surge • Windows Terminal Server • Wind Test Site • Wind Tunnel Study • Word Terminal Synchronous

WTSC: Wet Tantalum Slug Capacitor • World Telecommunication Standardization Conference

WTSPT: Waterspout

WTT: Wantoat (Papua New Guinea) (location identifier) • Weak-emission T Tauri • Weak-lined T Tauri • Weapons and Tactics Trainer • Wind Tunnel Test

WTTC: World Travel and Tourism Council

WTTM: WIYN Top-Tilt Module

WTTS: Weak-lined T Tauri Star (*) • Weak-line T Tauri Star • Weak T Tauri Star • Weak-wind T Tauri Star

WTU: Wuhan Technical University (China-PRC)

WTUC: World Trade Union Conference

WTUSM: Wuhan Technical University of Surveying and Mapping (China-PRC)

WTW: Worldwide Transaction Web

WTWIWYH: What They Want Is What You Have

WTWS: Windshear and Turbulence Warning System

WTZ: Whitianga (New Zealand) (location identifier) • Wissenschaftlich-Technisches Zentrum

Wu: Puissance Utile

WU: Washington University (USA-MI) • Wesleyan University (USA-CT) • Western Union • Work Unit

WUA: Weighted Usable Area

WUC: Work Unit Code • Writers Union of Canada

WUCF: Work Unit Code File

WUD: Wudinna (Australia-SA) (location identifier)

WUE: Water Use Efficiency • Western Undergraduate Exchange (USA)

WUG: Wau (Papua New Guinea) (location identifier) • Webmasters User Group

WUH: Wuhan (China-PRC) (location identifier)

WUIS: Work Unit Information System

WUM: Work Unit Manager

WUN: Wiluna (Australia-WA) (location identifier)

WUPL: Where-Used Parts List

WUPPE: Wisconsin Ultraviolet Photo-Polarimeter Experiment

WUR: Wageningen Universiteit en Researchcentrum (Netherlands)

Württ: Württemberg (Germany)

WUS: World University Service

WUSCI: Western Union Space Communications Inc.

WUT: Warm Up Time • Warsaw University of Technology (Poland) • Washburn University of Topeka (USA-KS)

WUTC: Washington Utilities and Transportation Commission (USA-DC)

WUU: Wau (Sudan) (location identifier)

WUUA: World Union for a Universal Alphabet

WUV: Wuvulu Island (Papua New Guinea) (location identifier)

WUX: Wuxi (China-PRC) (location identifier)

WUZ: Wuzhou Changzhoudao (China-PRC) (location identifier)

w/v: Weight in Volume

W/V: Wind Vector

WV: Saint Vincent and the Grenadines (international vehicle registration code) • Wagon de Voyageurs • Water Valve • Water Vapor • Wave • Westdeutscher Verlag • West Virginia (USA) • Wind Velocity • Wolverhampton (UK postal code) • Working Voltage

W. Va.: West Virginia (USA)

WVA: Warenverkehrsanmeldung

WVAS: Wabash Valley Astronomical Society (USA-IN) • Wake Vortex Avoidance System

WVB: Walvis Bay (South Africa) (location identifier)

WVC: Water Vapor Column • Water Vapor Content • West Valley College (USA-CA)

Wvd: Waived

WVE: Water Vapor Electrolysis • Wave

Wvfm: Waveform

WVI: Watsonville (USA-CA) (location identifier) • Work Volume Indicator

WVIS: Work Volume Indicator System

WVK: Manakara (Madagascar) (location identifier)

WVL: Waterville (USA-ME) (location identifier)

WVN: Wilhelmshaven (Germany) (location identifier)

WVNET: West Virginia Network for Educational Telecomputing (USA-WV)

WVO: Wet Verontreiniging Oppervlaktewateren (Netherlands)

WVR: Water Vapor Radiometer

WvS: Wetboek van Strafrecht

WVS: Weekly Visits Statistics • World Vector Shoreline

WVSC: West Virginia State College (USA-WV)

WVSGC: West Virginia Space Grant Consortium (USA-WV)

WVT: Water Vapour Transfer • Water Vapour Transmission

WVTM: Wharton Visual Test Market

WVU: West Virginia University (USA-WV)

WVWV: Water-Vapour Wind Vector

w/w: Weight for Weight • Weight in Weight

W/W: Wall to Wall • Weather • Wheel Well • Wheel Wing

WW: Wall to Wall • Warehouse Warrant • Water Waste • Werkloosheidswet • Wheel Wing • Who's Who • Wire Wound • Wire Wrap • With Warrants • World War • Worldwide

WWA: Wasilla (USA-AK) (location identifier) • Weighted Wavelet Amplitude • Western Writers of America • World Warning Agency

WWAN: Wireless Wide Area Network

WWAR: World-Wide Arts Resources

WWAS: World Warning Agency for Satellites

WWB: Wavelet Workbench • Westerly Wind Burst

WWC: World Water Council • World-Wide Computer

WWD: Weather Working Day • Westward • Wildwood (USA-NJ) (location identifier)

WWDGPS: Worldwide Differential Global Positioning System

WWDMS: World-Wide Data Management System

Wwe: Witwe

WWF: World Wildlife Fund

WWFPP: Weit-Winkel-Flächen-Photometer-Polarimeter

WWI: Woodie Woodie (Australia) (location identifier) • World War I • World Watch Institute

WWII: World War II

WWIS: World-Wide Information Services • World-Wide Information System

WWK: Wewak (Papua New Guinea) (location identifier)

WWLC: Worldwide Launching Campaign

WWM: Weizsäcker-Williams Method • Welded Wire Matrix

WWMCCS: Worldwide Military Command and Control System

W/WMS: Water and Waste Management System

WWMS: Waste Water Management System • Water and Waste Management System

ww/o: With or Without

WWO: Waste Water Outflow

WWP: Whale Pass (USA-AK) (location identifier) • World Weather Program • Write Without Program

WWR: West Woodward (USA-OK) (location identifier)

WWRS: Wash-Water Recovery System

WWS: Water and Waste System • Wildwood (USA-AK) (location identifier) • World Weather System

WWSE: Who's Who in Science and Engineering

WWSN: Worldwide Seismology Net

WWSSN: Worldwide Standardized Seismograph Network • Worldwide Standard Seismograph Network

WWST: Who's Who in Science and Technology

WWT: Newtok (USA-AK) (location identifier) • Waste Water Treatment

WWTP: Waste Water Treatment Plan

WWU: Western Washington University (USA-WA) • Westfälische Wilhelms-Universität (Germany) • Wirtschafts- und Währungsunion

WWW: Warm Wet Weather • Web-Weit Welt • Who's Who in the World • Who Was Who • World Weather Watch • World-Wide Waste (of time) • World-Wide Wealth (of data) • World-Wide Web (*)

WWWC: World-Wide Web Consortium

WWWI: First World-Wide Web Conference

WWWO: World-Wide Web Organization

WWWS: World Who's Who in Science

WWWT: Worldwide Windows on TeX

WWWVL: World-Wide Web Virtual Library

WWWW: Who-Where-When-Why • World-Wide Web Worm

WWW1: First World-Wide Web Conference

WWW2: Second World-Wide Web Conference

WWW3: Third World-Wide Web Conference

WWY: West Wyalong (Australia-NSW) (location identifier)
WWZ: Weighted Wavelet Z-transform
WX: Weather
WXF: Braintree Wether (UK-England) (location identifier)
WXM: Wide-field X-ray Monitor
WXN: Wanxian (China-PRC) (location identifier)
Wy: Wyoming (USA)
WY: Way • Working Year • Work Year • Wyoming (USA)
WYA: Whyalla (Australia-SA) (location identifier)
WYAIO: Will You Accept, If Offered
WYAS: West Yorkshire Astronomical Society (UK)
WYB: Yes Bay (USA-AK) (location identifier)
WYE: Yengema (Sierra Leone) (location identifier)
WYFFOS: Wide-Field Fibre Optics Spectrograph
WYN: Wyndham (Australia-WA) (location identifier)
Wyo: Wyoming (USA)
WYS: West Yellowstone (USA-MT) (location identifier)
WYSBYGI: What You See Before You Get It
WYSIAWYG: What You See Is Almost What You Get
WYSIAWYH: What You See Is (Almost) What You Hear
WYSIAYG: What You See Is All You'll Get
WYSIWES: What You See Is What Everyone Sees
WYSIWIS: What You See Is What I See
WYSIWYG: What You See Is What You Get
WYSIWYP: What You See Is What You Print
WYWIWYG: What You Want Is What You Get
Wz: Warenzeichen
WZ: Weißer Zwerg • Weltzeit
WZN: Weltzentrum für Niederschlagsklimatologie
WZW: Wess-Zumino-Witten
WZY: Nassau (Bahamas) (location identifier)
W3: World-Wide Web
W3A: World-Wide Web Applet
W3C: World-Wide Web Consortium
W3DS: World-Wide Web Data System
W3J: World-Wide Web Journal
W3O: World-Wide Web Organization
W4: What Works With What • World-Wide Web Wanderer
W4G: World-Wide Web Working Group
W4WG: Windows For Workgroups

X

x: Algebraic Variable • Cartesian Abcissa • Cartesian Coordinate • Choice • Coordinate Axis • Cross • Crossed with • Degree of Ionization • Distance • Ex • Excess • Experimental • Extra • Hoar Frost • Kiss • Magnification • Mole Fraction • Multiplication • Optimization Parameter • Out of • Path Length • Space Coordinate • State Vector • Times • Travel Distance • Without

X: Air Mass • Anonyme • Anonymous • Cloud Depth • Cloud Diameter • Cross • Crossing • École Polytechnique (France) • Error • Exchange • Experimental • Explosive • Exposure • Exposure Dose • Extension • Extra • Extraordinary • Fractional Mass Abundance of Hydrogen • Halogen • Inconnu • Index • Location • Mole Fraction • Multiplication • North Component of the Magnetic Vector • Reactance • Ten (Roman numeral) • Times • Unknown • Unspecified • Xenon (now Xe) • X-ray (international alphabet) • Xylonite • 10 (Roman numeral)

xa: Ex All

XA: Extended Architecture • Extended Attribute • Mexico (international aircraft registration code)

Xaa: Unknown Amino Acid

XAA: Extended Attribute Area

XADC: Extended Atlantic Data Coverage

XAF: Extended Attribute Field • Franc (CFA/Banque des États d'Afrique Centrale) (ISO currency code)

XAFS: X-ray Absorption Fine Structure

XAL: Alamos (Mexico) (location identifier)

Xals: Crystals

XALS: Extended Application Layer Structure

Xan: Xanthene • Xanthic

XANADU: X-ray Analysis and Data Utilization

XANES: X-ray Absorption Near-Edge Structure

XAO: Ingénierie Assistée par Ordinateur

XAP: Chapecó (Brazil-SC) (location identifier)

XAPIA: X.400 Application Program Interface Association

XAR: Aribinda (Burkina Faso) (location identifier) • Extended Attribute Record

Xarm: Crossarm

XAS: X-ray Absorption Spectroscopy • X-ray Analysis System

XAT: Extended Alias Table

XAU: Saul (French Guyana) (location identifier)

XAY: Xayabury (Laos) (location identifier)

xb: Ex Bonus

XB: Mexico (international aircraft registration code) • X-ray Burst • X-ray Burster

Xbar: Crossbar

XBB: Blubber Bay (Canada-BC) (location identifier)

XBE: Bearskin Lake (Canada-ON) (location identifier)

Xber: December

XBG: Bogande (Burkina Faso) (location identifier) • X-ray Background

XBJ: Birjand (Iran) (location identifier)

XBL: Buno Bedelle (Ethiopia) (location identifier) • X-ray BL Lacertae (object)

XBM: X Bitmap

XBN: Biniguni (Papua New Guinea) (location identifier)

XBO: Boulsa (Burkina Faso) (location identifier) • X-ray Blow Out

XBP: X Bright Point • X-ray Bright Point

XBR: Brockville (Canada-ON) (location identifier)

Xbra: Crossbracing

Xbre: Décembre

xbt: Exhibit

XBT: Crossbar Tandem • Expandable Bathythermograph

XBW: Port Burwell (Canada-QC) (location identifier)

xc: Ex Capitalization • Ex Coupon

XC: Cross Connect • Cross Correlation • Cross Country • Mexico (international aircraft registration code)

XCAL: Cross-Calibration

XCD: Dollar (East Caribbean) (ISO currency code)

XCES: XML Corpus Encoding Standard

XCF: Cross Communication Facility • Cross Coupling Facility • Cross-system Coupling Facility • Experimental Computing Facility

XCH: Christmas Island (location identifier)

XCHG: Exchange

XCL: Cluff Lake (Canada-SK) (location identifier) • Excess Current Liabilities

XCLF: X-ray Cluster Luminosity Function

XCM: Chatham (Canada-ON) (location identifier) • Extended Computer Modern

XCMD: External Command

XCN: Coron (Philippines) (location identifier)

XCO: Colac (Australia-VIC) (location identifier)

Xconn: Crossconnection

XCOPY: Extended Copy

xcp: Ex Coupon

Xcp: Except

Xcpt: Except

XCS: Xerox Computre Services

XCT: X-ray Computed Tomography

XCTD: Extended Conductivity-Temperature-Depth

XCVR: Transceiver

XCY: Cross Country

X'd: Executed

xd: Ex Dividend

XD: Exothermic Dispersion • X-ray Diffraction • X-ray Dominant

XDA: X-ray Detector Aperture

XDACS: X-ray Data Acquisition and Control System

XDB: Extended Database

XDCR: Transducer

XDE: Diebougou (Burkina Faso) (location identifier)

XDF: Exchange Data Format • Extended Density Format • XTE Data Finder

xdiv: Ex Dividend

XDJ: Djibo (Burkina Faso) (location identifier)

XDL: Crossed Delay Line

XDLB: Extended Digital Line Backend

XDM: X Display Manager • Xerox Dry Microfilm

XDMCP: X Display Manager Control Protocol

XDR: Extended Dynamic Range • Extended Data Representation • External Data Representation • X-ray Dominated Region

XDS: Dollar (Saint Christopher) (ISO currency code) • X Display Station • Xerox Data Systems • X/Open Directory Service • X-ray Detection System • X-ray Detector System

XDUP: Extended Disk Utilities Program

Xe: Xenon • Xénon

XEN: Xingcheng (China-PRC) (location identifier)

Xer: Xerox

Xerogr: Xerographie

Xerok: Xerokopie

XES: Lake Geneva Municipal (USA-WI) (location identifier) • Xerox Escape Sequence • X-ray Emission Spectrum • X-ray Emission Spectrometry • X-ray Emission Spectroscopy

XEU: European Currency Unit (ISO currency code) • Xerox Encryption Unit

XEUS: X-ray Early Universe Spectroscopy • X-ray for Evolving Universe Spectroscopy • X-ray (mission) for Evolving Universe Spectroscopy (*)

XF: Extra Fine

XFA: X-ray Fluorescence Absorption • X-ray Fluorescence Analysis

XFC: X-band Frequency Converter

XFCN: External Function

XFD: X-ray Flow Detection

XFDB: XTE FITS Database

XFDS: XMM Flight Dynamics System

Xfer: Transfer

XFER: Transfer

XFF: XTE FITS Format (*) • XTE FITS Formatter

XFM: X-ray Flux Monitor

Xfmr: Transformer

XFN: Xiangfan (China-PRC) (location identifier)

Xformer: Transformer

Xfr: Transfer

XFR: Transfer

XFW: Finkenwerder (Germany) (location identifier)

XFX: Crosstalk Network Simulator

Xg: Crossing

XGA: Extended Graphics Array • Gaoua (Burkina Faso) (location identifier)

XGG: Gorom-Gorom (Burkina Faso) (location identifier)

XGL: Granville Lake (Canada-MB) (location identifier)

XGN: Xangongo (Angola) (location identifier)

XGP: Xerox Graphic Printer

XGR: Kangiqsualujjuaq (Canada-QC) (location identifier)

XGRS: X and Gamma-Ray Spectrometer

xh: Xhosa (ISO language code)

XHD: Crosshead • Extended Header Data

XHF: Extrahigh Frequency

XHR: Extrahigh Reliability

Xhst: Exhaust

Xhvy: Extra Heavy

xi: Ex Interest

XI: X-ray Imaging

XIC: Xichang (China-PRC) (location identifier)

XICS: Xerox Integrated Composition System

XID: Cross Identify • Cross Identity • Exchange Identifier • X-ray Identification • X-window Identification • X-window Identifier

XIDT: X-ray Identification Team

XIE: Xienglom (Laos) (location identifier) • X Image Extension • X-ray Imaging Experiment

XIG: Xinguara (Brazil) (location identifier)

XIL: Xilinhot (China-PRC) (location identifier)

xin: Ex Interest

XIN: Xingning (China-PRC) (location identifier)

XING: Crossing

XIOS: Extended Input/Output System

XIP: Execute In Place • Execution In Place

XIPC: X Interprocess Communication

XIPS: Xenon-Ion Propulsion System

XIR: X Irradiance Response at Wavelength X

XIS: X-ray Imaging Spectrometer

XITE: X-ray Imaging Telescope

XIWG: Cross Industry Working Group

XIWT: Cross Industry Working Team

XIY: Xi'An - Xianyang (China-PRC) (location identifier)

XKA: Kantchari (Burkina Faso) (location identifier)

XKH: Xieng Khouang (Laos) (location identifier)

XKO: Kemano (Canada-BC) (location identifier)

XKS: Kasabonika (Canada-ON) (location identifier)

XKY: Kaya (Burkina Faso) (location identifier)

XL: Extended Language • Extra Large • Extra Long

XLAT: Translate

XLB: Lac Brochet (Canada-MB) (location identifier)

XLD: X-ray Luminosity Distribution

XLF: Leaf Bay (Canada-QC) (location identifier) • X-ray Luminosity Function

XLFD: X Logical Font Description

XLL: Extensible Linking Language

XLM: Excel Macro Language

Xlnt: Excellent

XLO: Long Xuyen (Viet nam) (location identifier)

XLPE: Cross-Linked Polyethylene

XLS: Saint-Louis (Senegal) (location identifier)

XLT: Extremely Large Telescope

XLU: Leo (Burkina Faso) (location identifier)

XLW: Lemwerder (Germany) (location identifier)

XM: Extended Memory

XMA: Extended Middleware Architecture • Maramag (Philippines) (location identifier) • X-ray Mirror Assembly • XRT Mirror Assembly

XMAP: X-ray (instrument for) Mineralogical Analysis of Planetary (material)

Xmas: Christmas

XMB: M'Bahiakro (Ivory Coast) (location identifier)

XMC: Mallacoota (Australia-VIC) (location identifier)

XMCS: XMM Mission Conreol System

XMD: Madison (USA-SD) (location identifier)

XMG: Mahendranagar (Nepal) (location identifier)

XMH: Manihi (Tuamotu Islands) (location identifier)

XMI: Exchange Marketing Information • Masasi (Tanzania) (location identifier)

Xmit: Transmit

Xmits: Transmits

X-ml: Ex Mill

Xml: Ex Mill

XML: Extensible Markup Language • Minlaton (Australia-SA) (location identifier)

X-mll: Ex Mill

Xmll: Ex Mill

XMM: Extended Memory Manager • X-ray Multimirror • X-ray Multimirror Mission • X-ray Multiple Mirror

XMN: Xiamen (China-PRC) (location identifier)

XMP: McMillan Pass (Canada-YT) (location identifier) • Xanthosine Monophosphate • X-ray Microprobe

XMS: Extended Memory Services • Extended Memory Specification (*) • Extended Memory System • Macas (Ecuador) (location identifier)

Xmsn: Transmission

Xmt: Transmit

XMT: Transmit

Xmtd: Transmitted

Xmtg: Transmitting

Xmtr: Transmitter

XMTR: Transmitter

XMY: Yam Island (Australia) (location identifier)

xn: Ex New

XNG: Quang Ngai (Viet Nam) (location identifier)

XNN: Xining (China-PRC) (location identifier)

XNS: Crosstalk Network Simulator • Xerox Network Services • Xerox Network System

X/nt: Excellent

XNT: Xingtai (China-PRC) (location identifier)

Xnty: Christianity

XNU: Nouna (Burkina Faso) (location identifier)

XO: Crystal Oscillator • Exactly One • Executive Officer • Xylenol Orange

XOC: Xanadu Operating Co.

XOF: Franc (CFA/Banque Centrale des États de l'Afrique de l'Ouest) (ISO currency code)

XOFF: Transmitter Off

XOL: X Open Look

XOM: External Output Management

XON: Transmitter On

XOR: Exclusive Or

XOS: Xerox Operating System

XP: Cross Polar • Exprès Payé • Express • Express Paid

XPA: Pama (Burkina Faso) (location identifier) • X Public Access • X Public Address

XPARS: External Research Publication and Retrieval System

XPC: Expect

Xpct: Expect

XPD: Cross-Polar Discrimination

Xpdr: Transponder

XPE: Cross Platform Extension • X-ray Polarimeter Experiment

XPES: X-ray Photoelectron Spectroscopy

XPF: Franc (CFP) (ISO currency code)

XPG: Xerox Publishing Group • X/Open Portability Guide

XPI: Xanadu Parameter Interface

XPK: Pukatawagan (Canada-MB) (location identifier)

Xpl: Explosive

XPL: Comayagua Palmerola (Honduras) (location identifier)

Xplt: Exploit

XPM: X11 Print Manager

Xpn: Expansion

XPN: X-ray Photoionized Nebula

Xpnd: Transponder

Xpndr: Transponder

XPORT: Transport

XPP: Express Pallet with Payloads • Poplar River (Canada-MB) (location identifier)

XPr: Without Privileges

XPR: Pine Ridge (USA-SD) (location identifier) • Power Response at Wavelength X

XPRS: Experimental Precision Resupply System

Xprt: Transport

XPS: X-ray Photoelectron Spectroscopy • X-ray Photoemission Spectroscopy

Xprtn: Transportation

XPt: Crosspoint

XPU: West Kuparuk (USA-AK) (location identifier)

XQ: Cross Question

XQC: X-ray Quantum Calorimeter

XQP: Quepos (Costa Rica) (location identifier)

XQSO: X-ray Quasar • X-ray Quasi-Stellar Object

XQU: Qualicum (Canada-BC) (location identifier)

xr: Ex Rights

Xr: Examiner

XR: X-Ray

XRASE: X-Ray Spectroscopic Explorer

XRB: X-Ray Background • X-Ray Beacon • X-Ray Buster

XRCF: X-Ray Calibration Facility

XRD: Crossroad • X-Ray Diffraction

Xrds: Crossroads

XRE: X-Ray Emission

XRef: Cross Reference

XRF: Experimental Reproduction Film • X-Ray Fluorescence

XRFA: X-Ray Fluorescence Analysis

XRFS: X-Ray Fluorescence Spectrometry

XRL: X-Ray Laser

XRM: X-Ray Monitor

XRMA: X-Ray Microprobe Analysis

XRP: X-Ray Polychromator • X-Ray Pulsar

XRPS: XMM Remote Proposal Submission

XRR: Ross River (Canada-YT) (location identifier) • X-Ray Reflectivity

XRS: X-Ray Sensor • X-Ray Spectrometer

XRT: Extension for Real-Time • X-Ray Telescope • X-Ray Topography

xrts: Ex Rights

XRY: Jerez de la Frontera (Spain) (location identifier)

Xs: Expenses

XS: Excess • Exit Shadow • Extra Small • Multispectral

XSA: Sinoe (Liberia) (location identifier) • X-band Satellite Antenna

X-SAR: X-band Synthetic Aperture Radar

XSAR: X-band Synthetic Aperture Radar

XSAT: Exceptional Satellite

XSBL: X-ray Selected BL Lac (object)

XSC: Extended Source Catalog • South Caicos (Turks & Caicos Islands) (location identifier)

XSCS: XMM Science Control System

XSD: Halogen-Selective Detector • Tonopah (USA-NV) (location identifier)

XSDC: XTE Science Data Center

XSE: Sebba (Burkina Faso) (location identifier)

XSECT: Cross Section
XSF: X-ray Scattering Facility
X-ship: Ex Ship
Xship: Ex Ship
X-shp: Ex Ship
Xshp: Ex Ship
XSI: South Indian Lake (Canada-MB) (location identifier)
XSL: Experimental Space Laboratory • Extensible Style Language • Extensible Stylesheet Language
XSLC: Xichang Satellite Launch Center (China-PRC)
XSLT: Extensible Stylesheet Language Transform
XSM: Saint Marys (USA-MD) (location identifier) • X/Open System Management
XSMD: Extended Storage Module Drive
XSO: Siocon (Philippines) (location identifier)
XSP: Singapore Selatar (Singapore) (location identifier)
XSPEC: X-ray Spectral Analysis Package
XSS: X-ray Source System
XSTA: X-band Satellite Tracking Antenna
Xstg: Extra Strong
Xstr: Transistor
X-stre: Ex Store
Xstre: Ex Store
xSU: Former Soviet Union
XT: Burkina Faso (international aircraft registration code) • Extended • Extended Technology • X-ray Telescope
XTA: X-band Tracking Antenna
XTAL: Crystal
Xtalk: Crosstalk
XTCLK: External Transmit Clock
XTD: Cross Track Distance
XTE: Cross Track Error • X-ray Timing Explorer (now RXTE) (*) • X-ray Transient Explorer
XTEN: Xerox Telecommunications Network
XTG: Thargomindah (Australia-QLD) (location identifier)
X-tgd: Cross-Tongued
Xtgd: Cross-Tongued
XTI: X/open Transport Interface
Xtian: Christian
XTK: Cross-Track
XTL: Tadoule Lake (Canada-MB) (location identifier)
XTLO: Crystal Oscillator
XTM: Experimental Test Model
XTND: Extend
XTO: Taroom (Australia-QLD) (location identifier)
XTP: Xpress Transfer Protocol
XTR: Tara (Australia-QLD) (location identifier)
Xtra: Extra
XTRAN: Extended Fortran
XTRM: Extreme
Xtry: Extraordinary
XTV: X Teleconferencing and Viewing
Xty: Christianity
XU: Cambodia (international aircraft registration code)
XUI: X-windows User Interface
XUL: Xavier University of Louisiana (USA-LA)
XUV: Extreme Ultraviolet • X-ray and Ultraviolet
XUVI: Extreme Ultraviolet Imager
XUVT: Extreme Ultraviolet Telescope
XUZ: Xuzhou (China-PRC) (location identifier)
XV: Viet Nam (international aircraft registration code)
XVA: X-ray Vidicon Analysis
XVL: Vinh Long (Viet Nam) (location identifier)
XVMX: Xerox Voice Message Exchange
Xvs: Transverse
XVSAT: Switched Very Small Aperture Terminal
xw: Ex Warrants
X/W: Cross Wind
XWA: Extended Wide Area • Wakeham Bay (Canada-QC) (location identifier)
XWD: X-Window Dump
X-whse: Ex Warehouse
Xwhse: Ex Warehouse
XWIB: X-Windows Interface Builder
X-wks: Ex Works
Xwks: Ex Works
XWS: Extended Web Site
XX: Heavy
XXL: Extra Extra Large
XXS: Extra Extra Small
XY: Myanmar (international aircraft registration code)
XYA: Yandina (Solomon Islands) (location identifier)
XYE: Ye (Myanmar) (location identifier)
Xyl: Xylographie • Xylometer • Xylophon
XYR: Yellow River (Papua New Guinea) (location identifier)
XYT: Toulouse-Montaudran (France) (location identifier)
XZ: Myanmar (international aircraft registration code)
XZA: Zabre (Burkina Faso) (location identifier)
X2B: Hexadecimal to Binary
X2C: Hexadecimal to Character
X2D: Hexadecimal to Decimal

Y

y: Algebraic Variable • Altitude • Cartesian Coordinate • Cartesian Ordinate • Coordinate Axis • Dry Air • Lateral Axis • Space Coordinate • Yard • Year • Yellow • Yen (currency) • Yes • Young • Youngest

Y: Admittance • East Component of the Magnetic Vector • Fractional Mass Abundance of Helium • Hundred Fifty (Roman numeral) • Hypercharge • Planck Function • Time Period of Orbit • Yankee (international alphabet) • Yaw • Yen (currency) • Yes • Yttrium • Yuan (currency) • 150 (Roman numeral)

YA: Afghanistan (international aircraft registration code) • Yao • Yaw Axis • Yet Another • Young Adult

YAA: Anahim Lake (Canada-BC) (location identifier)

YAB: Arctic Bay (Canada-NT) (location identifier)

Yac: Yacimiento

YAC: Cat Lake (Canada-ON) (location identifier)

YACC: Yet Another Compiler-Compiler

YAD: Moose Lake (Canada-MB) (location identifier)

YAE: Alta Lake (Canada-BC) (location identifier)

YAF: Asbestos Hill (Canada-QC) (location identifier)

YAFIYGI: You Asked For It, You Got It

YAG: Fort Frances (Canada-ON) (location identifier) • Yttrium-Aluminium Garnet

YAH: Lagrande 4 (Canada-QC) (location identifier)

YAHOO: Yet Another Hierarchically Officious Oracle

YAI: Chillan (Chile) (location identifier)

YAJ: Lyall Harbour (Canada-BC) (location identifier)

YAK: Yakutat (USA-AK) (location identifier) • Youth And Knowledge

YAL: Alert Bay (Canada-BC) (location identifier)

YALO: Yale-AURA-Lisbon-Ohio (consortium)

YAM: Sault Sainte Marie (Canada-ON) (location identifier) • Yaw Acquisition Mode • Yet Another Modem • Young Astronomers Meeting

YAMM: Yet Another Menu Manager

YAN: Yangambi (Zaire) (location identifier)

YANA: Yet Another Bloated Application

YAO: Yaounde (Cameroon) (location identifier) • Yunnan Astronomical Observatory (China-PRC)

YAP: Yap (Caroline Islands) (location identifier) • Yield Analysis Pattern • Young Aspiring Professional • Yttrium-Aluminium-Perowskit

YAQ: Maple Bay (Canada-BC) (location identifier)

YAR: Lagrande 3 (Canada-QC) (location identifier) • York-Antwerp Rules

YAS: Yasawa Island (Fiji) (location identifier) • Yaw Attitude Sensor • York Astronomical Society (UK)

YAT: Attawapiskat (Canada-ON) (location identifier) • Your Alias Table

YAUN: Yet Another Unix Nerd

YAV: Miner's Bay (Canada-BC) (location identifier)

YAVIS: Young-Attractive-Verbal-Intelligent-Successful

YAW: Halifax-Shearwater (Canada-NS) (location identifier)

YAWYSYA: You Are Who You Say You Are

YAX: Angling Lake (Canada-ON) (location identifier)

YAY: Saint Anthony (Canada-NF) (location identifier)

YAZ: Tofino - Metropolitan Area (Canada-BC) (location identifier)

Yb: Ytterbium

YB: Yearbook

YBA: Banff (Canada-AB) (location identifier)

YBB: Pelly Bay - Townsite (Canada-NT) (location identifier)

YBC: Baie Comeau (Canada-QC) (location identifier)

YBCO: Yttrium-Barium-Copper Oxide

YBE: Uranium City (Canada-SK) (location identifier) • Yang-Baxter Equation

YBF: Bamfield (Canada-BC) (location identifier)

YBG: Bagotville (Canada-QC) (location identifier)

YBH: Bull Harbour (Canada-BC) (location identifier)

YBI: Black Tickle (Canada-NF) (location identifier)

YBJ: Baie Johan Beetz (Canada-QC) (location identifier)

YBK: Baker Lake (Canada-NT) (location identifier)

YBL: Campbell River (Canada-BC) (location identifier)

YBM: Bronson Creek (Canada-BC) (location identifier)

YBN: Borden (Canada-ON) (location identifier)

YBO: Bobquinn Lake (Canada) (location identifier)

YBP: Yibin (China-PRC) (location identifier)

YBQ: Telegraph Harbour (Canada-BC) (location identifier)

YBR: Brandon (Canada-MB) (location identifier)

YBS: Yale Bibliographic System

YBT: Brochet (Canada-MB) (location identifier)

YBV: Berens River (Canada-MB) (location identifier)

YBW: Bedwell Harbour (Canada-BC) (location identifier)

YBX: Blanc Sablon (Canada-QC) (location identifier)

YBY: Bonnyville (Canada-AB) (location identifier)

YBZ: Toronto Downtown (Canada-ON) (location identifier)

Y/C: Luminance/Chrominance

Y-C: Yale-Columbia

YC: Yachtclub • Yale Catalogue • Yale College • Yale-Columbia

YCA: Courtenay (Canada-BC) (location identifier)

YCAS: York County Astronomical Society (USA-PA)

YCB: Cambridge Bay (Canada-NT) (location identifier)

YCbCr: Brightness-Chrominance/Blue-Chrominance/Red

YCC: Cornwall (Canada-ON) (location identifier) • Lightness-Colour-Chromatics • Youth Computing Club

YCD: Nanaimo Cassidy (Canada-BC) (location identifier)

YCE: Centralia (Canada-ON) (location identifier)

YCF: Cortes Bay (Canada-BC) (location identifier)

YCG: Castlegar (Canada-BC) (location identifier)

YCH: Chatham (Canada-NB) (location identifier)

YCI: Caribou Island (Canada-ON) (location identifier)

YCJ: Cape Saint James (Canada-BC) (location identifier)

YCK: Colville Lake (Canada-NT) (location identifier)

YCL: Charlo (Canada-NB) (location identifier)

YCM: Saint Catharines (Canada-ON) (location identifier) • Yellow-Cyan-Magenta

YCMOU: Yashwantrao Chavan Maharashtra Open University (India)

YCN: Cochrane (Canada-ON) (location identifier)

YCO: Coppermine (Canada-NT) (location identifier)

YCP: Co-op Point (Canada-MB) (location identifier)

YCPS: Yale Center for Parallel Supercomputing (USA-CT)

YCQ: Chetwynd (Canada-BC) (location identifier)

YCR: Cross Lake (Canada-MB) (location identifier)

YCS: Chesterfield Inlet (Canada-NT) (location identifier)

YCSP: Yale Catalogue of Stellar Parallaxes

YCT: Coronation (Canada-AB) (location identifier)

YCTSP: Yale Catalogue of Trigonometric Stellar Parallaxes

YCU: Cullaton Lake (Canada-NT) (location identifier)

YCV: Cartierville (Canada-QC) (location identifier)

YCW: Chilliwack (Canada-BC) (location identifier) • You Can't Win

YCX: Gagetown (Canada-NB) (location identifier)

YCY: Clyde River (Canada-NT) (location identifier)

YCZ: Creston (Canada-BC) (location identifier)

yd: Yard

YD: Democratic Yemen (ISO country code) • Yaw Deviation • Yemen (Democratic_) (ISO country code) • Yemeni Dinar (currency)

YDA: Dawson City (Canada-YT) (location identifier)

YDAC: Yohkoh Data Archive Center

Y'day: Yesterday

YDB: Burwash Landings (Canada-YT) (location identifier)

YDC: Drayton Valley (Canada-AB) (location identifier)

YDD: Dinar (Yemen) (ISO currency code)

YDE: Paradise River (Canada-NF) (location identifier)

YDF: Deer Lake (Canada-NF) (location identifier)

ydg: Yarding

YDG: Digby (Canada-NS) (location identifier)

YDH: Daniels Harbour (Canada-NF) (location identifier)

YDI: Davis Inlet (Canada-NF) (location identifier)

YDK: Main Duck Island (Canada-ON) (location identifier)

YDL: Dease Lake (Canada-BC) (location identifier)

YDN: Dauphin (Canada-MB) (location identifier)

YDO: Dolbeau - Sainte Méthode (Canada-QC) (location identifier)

YDP: Nain (Canada-NF) (location identifier)

YDQ: Dawson Creek (Canada-BC) (location identifier)

YDR: Broadview (Canada-SK) (location identifier)

yds: Yards

YDS: Desolation Sound (Canada-BC) (location identifier)

YDT: Vancouver Boundary Bay (Canada-BC) (location identifier)

YDX: Doc Creek (Canada-BC) (location identifier)

YE: Yemen (ISO country code) • Young Engineer • Your Excellency

YEA: Edmonton Metropolitan (Canada-AB) (location identifier) • Yaw Error Amplifier • Year-End Appraisal Report

Yearb: Yearbook

YEC: Yechon (Korea-ROK) (location identifier) • Young Engineer Club

YED: Edmonton Namao (Canada-AB) (location identifier)

YEG: Edmonton (Canada-AB) (location identifier)

YEI: Ennadai Lake (Canada-NT) (location identifier)

YEK: Arviat (Canada-NT) (location identifier) • Eskimo Point (Canada-NT) (location identifier)

Yel: Yellow

YEL: Elliot Lake (Canada-ON) (location identifier)

YEM: Yemen (ISO country code)

YEN: Estevan (Canada-SK) (location identifier)

YEO: Yeovilton (UK-England) (location identifier) • Your Eyes Only

YEP: Estevan Point (Canada-BC) (location identifier)

YEQ: Yenkis (Papua New Guinea) (location identifier)

YER: Fort Severn (Canada-ON) (location identifier) • Rial (Yemen) (ISO currency code) • Yearly Effective Rate

YERAC: Young European Radio Astronomers Conference

YES: Young Engineers and Scientists • Young Engineers' Satellite • Youth Extension Service

YESS: Yaw Earth Sensor System

Yesty: Yesterday

YET: Edson (Canada-AB) (location identifier)

YETA: Yet Another

YEU: Eureka (Canada-NT) (location identifier)

YEV: Inuvik (Canada-NT) (location identifier)

YEY: Amos (Canada-QC) (location identifier)

YFA: Fort Albany (Canada-ON) (location identifier)

YFB: Iqaluit (Canada-NT) (location identifier)

YFC: Fredericton (Canada-NB) (location identifier)

YFE: Forestville (Canada-QC) (location identifier)

YFG: Fontanges (Canada-QC) (location identifier) • Yttrium-Ferrite Garnet

YFH: Fort Hope (Canada-ON) (location identifier)

YFL: Fort Reliance (Canada-NT) (location identifier)

YFO: Flin Flon (Canada-MB) (location identifier)

YFR: Fort Resolution (Canada-NT) (location identifier)

YFS: Fort Simpson (Canada-NT) (location identifier) • Photopic Flux Sensitivity

YFX: Fox Harbour - Saint Lewis (Canada-NF) (location identifier)

YGA: Gagnon (Canada-QC) (location identifier)

YGB: Gillies Bay (Canada-BC) (location identifier)

YGC: Grande Cache (Canada-AB) (location identifier)

YGE: Gorge Harbour (Canada-BC) (location identifier)

YGG: Ganges Harbour (Canada-BC) (location identifier)

YGH: Fort Good Hope (Canada-NT) (location identifier)

YGJ: Yonago (Japan) (location identifier)

YGK: Kingston (Canada-ON) (location identifier)

YGL: La Grande (Canada-QC) (location identifier)

YGM: Gimli (Canada-MB) (location identifier)

YGN: Greenway Sound (Canada-BC) (location identifier)

YGO: Gods Narrows (Canada-MB) (location identifier)

YGP: Gaspé (Canada-QC) (location identifier)

YGQ: Geraldton (Canada-ON) (location identifier)

YGR: Îles de la Madeleine (Canada-QC) (location identifier)

YGS: Germansen (Canada-BC) (location identifier)

YGT: Igloolik (Canada-NT) (location identifier) • Young Graduate Trainee

YGV: Havre Saint-Pierre (Canada-QC) (location identifier)

YGW: Kuujjuarapik (Canada-QC) (location identifier)

YGX: Gillam (Canada-MB) (location identifier)

YGY: Déception (Canada-QC) (location identifier)

YGZ: Grise Fiord (Canada-NT) (location identifier)

YHA: Port Hope Simpson (Canada-NF) (location identifier)

YHB: Hudson Bay (Canada-SK) (location identifier)

YHC: Hakai Pass (Canada-BC) (location identifier) • Young Harris College (USA-GA)

YHD: Dryden (Canada-ON) (location identifier)

YHE: Hope (Canada-BC) (location identifier)

YHF: Hearst (Canada-ON) (location identifier)
YHG: Charlottetown (Canada-NF) (location identifier)
YHH: Campbell River Harbour (Canada-BC) (location identifier)
YHI: Holman Island (Canada-NT) (location identifier)
YHK: Gjoa Haven (Canada-NT) (location identifier)
YHM: Hamilton (Canada-ON) (location identifier)
YHN: Hornepayne (Canada-ON) (location identifier)
YHO: Hopedale (Canada-NF) (location identifier)
YHP: Poplar Hill (Canada-ON) (location identifier)
YHR: Chevery (Canada-QC) (location identifier) • Harrington Harbour (Canada-QC) (location identifier)
YHS: Sechelt (Canada-BC) (location identifier)
YHT: Haines Junction (Canada-YT) (location identifier)
YHU: Montréal Saint-Hubert (Canada-QC) (location identifier)
YHY: Hay River (Canada-NT) (location identifier)
YHZ: Halifax (Canada-NS) (location identifier)
YI: Iraq (international aircraft registration code) • Young Investigator
YIB: Atikokan (Canada-ON) (location identifier)
YIC: Isachsen (Canada-NT) (location identifier)
YIF: Pakuashipi (Canada-QC) (location identifier)
YIG: Big Bay - Marina (Canada-BC) (location identifier) • Yttrium-Iron Garnet
YIH: Yichang (China-PRC) (location identifier)
YIK: Ivujivik (Canada-QC) (location identifier)
YIN: Standard Lamp Intensity • Yining (China-PRC) (location identifier)
YINS: Photopic Intensity Sensitivity
YIO: Pond Inlet (Canada-NT) (location identifier)
YIP: Detroit - Willow Run (USA-MI) (location identifier)
YIQ: Intensity-Inphase-Quadrature
YIS: Photometric Illuminance Sensitivity
YIV: Island Lake - Garden Hill (Canada-MB) (location identifier)
YIW: Yiwu (China-PRC) (location identifier)
YIX: Port Radium (Canada-NT) (location identifier)
YJ: Vanuatu (international aircraft registration code)
YJA: Jasper Hinton (Canada-AB) (location identifier)
YJD: Young Jong Do (Korea-ROK) (location identifier)
YJF: Fort Liard (Canada-NT) (location identifier)
YJN: Saint Jean (Canada-QC) (location identifier)
YJO: Johnny Mountain (Canada-BC) (location identifier)
YJT: Stephenville (Canada-NF) (location identifier)
YK: Syria (international aircraft registration code) • Yakutian
YKA: Kamloops (Canada-BC) (location identifier)
YKC: Collins Bay (Canada-SK) (location identifier)
YKD: Kincardine (Canada-ON) (location identifier)
YKE: Knee Lake (Canada-MB) (location identifier)
YKF: Kitchener - Waterloo-Wellington (Canada-ON) (location identifier)
YKG: Kangirsuk (Canada-QC) (location identifier)
YKI: Kennosao Lake (Canada-MB) (location identifier)
YKJ: Key Lake (Canada-SK) (location identifier)
YKK: Kitkatla (Canada-BC) (location identifier)
YKL: Schefferville (Canada-QC) (location identifier)
YKM: Yakima (USA-WA) (location identifier)
YKN: Yankton (USA-SD) (location identifier) • Yukon (Canada)
YKQ: Waskaganish (Canada-QC) (location identifier)
YKS: Yakutsk (Russia) (location identifier)
YKT: Klemtu (Canada-BC) (location identifier)
YKU: Chisasibi (Canada-QC) (location identifier)
YKX: Kirkland Lake (Canada-ON) (location identifier)
YKY: Kindersley (Canada-SK) (location identifier)
YKZ: Toronto Buttonville (Canada-ON) (location identifier)
YL: Latvia (international aircraft registration code) • Yield Limit
YLA: Langara (Canada-BC) (location identifier)
YLB: Lac Biche (Canada-AB) (location identifier)
YLC: Lake Harbour (Canada-NT) (location identifier)
YLD: Chapleau (Canada-ON) (location identifier)
YLE: Lac La Martre (Canada-NT) (location identifier)
YLF: LaForges (Canada-QC) (location identifier)
YLG: Yalgoo (Australia-WA) (location identifier) • Youth Library Group
YLH: Lansdowne House (Canada-ON) (location identifier)
YLI: Ylivieska (Finland) (location identifier)
YLJ: Meadow Lake (Canada-SK) (location identifier)
YLL: Lloydminster (Canada-AB) (location identifier)
YLM: Clinton Creek (Canada-YT) (location identifier)
YLN: Yilan (China-PRC) (location identifier)
YLP: Mingan (Canada-QC) (location identifier)
YLQ: La Tuque (Canada-QC) (location identifier)
YLR: Leaf Rapids (Canada-MB) (location identifier)
YLS: Lebel-sur-Quevillon (Canada-QC) (location identifier) • Photometric Luminance Sensitivity
YLSTN: Yellowstone (USA)
YLT: Alert (Canada-NT) (location identifier)
YLW: Kelowna (Canada-BC) (location identifier)
YLX: Long Point (Canada-ON) (location identifier)
YLY: Lytton (Canada-BC) (location identifier)
YM: Yang-Mills • Yield Management • Your Message
YMA: Mayo (Canada-YT) (location identifier)
YMB: Merritt (Canada-BC) (location identifier)
YMC: Maricourt (Canada-QC) (location identifier) • Young Massive Cluster
YMD: Mould Bay (Canada-NT) (location identifier) • Year-Month-Day
YME: Matane (Canada-QC) (location identifier)
YMF: Montagne Harbour (Canada-BC) (location identifier)
YMG: Manitouwadge (Canada-ON) (location identifier)
YMH: Mary's Harbour (Canada-NF) (location identifier) • Yang-Mills-Higgs
YMI: Minaki (Canada-ON) (location identifier) • Why Am I ...
YMJ: Moose Jaw (Canada-SK) (location identifier)
YML: Murray Bay - Charlevoix (Canada-QC) (location identifier)
YMM: Fort McMurray (Canada-AB) (location identifier)
YMMV: Your Mileage May Vary
YMN: Makkovik (Canada-NF) (location identifier)
YMO: Moosonee (Canada-ON) (location identifier)
YMP: Port McNeil (Canada-BC) (location identifier)
YMQ: Montréal Metropolitan (Canada-QC) (location identifier)
YMR: Merry Island (Canada-BC) (location identifier)
YMS: Young Micro System • Yurimaguas (Peru) (location identifier)

YMT: Chibougamau (Canada-QC) (location identifier)

YMU: Yamanashi Medical University (Japan)

YMW: Maniwaki (Canada-QC) (location identifier) • Yaw Maneuver Wheel

YMX: Montréal Mirabel (Canada-QC) (location identifier)

YMY: Montréal - Victoria Stol (Canada-QC) (location identifier)

Y/N: Yes/No

YN: Nicaragua (international aircraft registration code)

YNA: Natashquan (Canada-QC) (location identifier)

YNB: Yanbu (Saudi Arabia) (location identifier)

YNC: Wemindji (Canada-QC) (location identifier)

YND: Gatineau (Canada-QC) (location identifier)

YNE: Norway House (Canada-MB) (location identifier)

YNG: Youngstown (USA-OH) (location identifier)

YNH: Hudson Hope (Canada-BC) (location identifier)

YNI: Nichequon (Canada) (location identifier)

YNJ: Yanji (China-PRC) (location identifier)

YNK: Nootka Sound (Canada-BC) (location identifier)

YNL: Points North Landing (Canada-SK) (location identifier)

YNM: Matagami (Canada-QC) (location identifier)

YNO: North Spirit Lake (Canada-ON) (location identifier)

YNS: Nemiscau (Canada-QC) (location identifier)

YNT: Yantai (China-PRC) (location identifier)

YNZ: Yancheng (China-PRC) (location identifier)

yo: Yoruba (ISO language code)

YO: Years Old • York (UK postal code) • Yunnan Observatory (China-PRC)

YOB: Year Of Birth • Your Office in the Box

YOC: Old Crow (Canada-YT) (location identifier)

YOD: Cold Lake (Canada-AB) (location identifier) • Year Of Death

YOE: Falher (Canada-AB) (location identifier)

YOG: Ogoki (Canada-ON) (location identifier)

YOH: Oxford House (Canada-MB) (location identifier)

YOJ: High Level (Canada-AB) (location identifier)

YOK: Yokohama (Japan) (location identifier)

YOL: Yola (Nigeria) (location identifier)

YOLT: You Only Live Twice

YOM: Year Of Marriage

YOO: Oshawa (Canada-ON) (location identifier)

YOP: Rainbow Lake (Canada-AB) (location identifier)

Yorks: Yorkshire (UK-England)

YOTO: Year Of The Ocean

YOW: Ottawa McDonald Cartier (Canada-ON) (location identifier)

YOY: Valcartier (Canada-QC) (location identifier)

Y-P: Yaw-Pitch

YP: Year's Purchase • Yellow Pages • Yield Point • Young Person

YPA: Prince Albert (Canada-SK) (location identifier) • Yaw Precession Amplifier • Yearbook Printers Association

YPB: Port Alberni (Canada-BC) (location identifier)

YPC: Paulatuk (Canada-NT) (location identifier) • Yale Parallax Catalog

YPD: Parry Sound (Canada-ON) (location identifier) • Yaw Phase Detector

YPE: Peace River (Canada-AB) (location identifier)

YPF: Esquimalt (Canada-BC) (location identifier)

YPG: Portage - La Prairie (Canada-MB) (location identifier)

YPH: Inukjuak (Canada-QC) (location identifier)

YPI: Port Simpson (Canada-BC) (location identifier)

YPJ: Aupaluk (Canada-QC) (location identifier)

YPL: Pickle Lake (Canada-ON) (location identifier)

YPM: Pikangikum (Canada-ON) (location identifier)

YPN: Port Menier (Canada-QC) (location identifier)

YPO: Peawanuck (Canada-ON) (location identifier)

YPOP: Yohkoh Public Outreach Project

YPP: Pine Point (Canada-NT) (location identifier)

YPQ: Peterborough (Canada-ON) (location identifier)

YPR: Prince Rupert - Digby Island (Canada-BC) (location identifier)

yps: Yards per Second

YPS: Port Hawkesbury (Canada-NS) (location identifier) • Yellow-Page Service

YPT: Pender Harbour (Canada-BC) (location identifier)

YPW: Powell River (Canada-BC) (location identifier)

YPX: Povungnituk (Canada-QC) (location identifier)

YPY: Fort Chipewyan (Canada-AB) (location identifier)

YPZ: Burns Lake (Canada-BC) (location identifier)

YQA: Muskoka (Canada-ON) (location identifier)

YQB: Québec (Canada-QC) (location identifier)

YQC: Quaqtaq (Canada-QC) (location identifier)

YQD: The Pas (Canada-MB) (location identifier)

YQE: Kimberley (Canada-BC) (location identifier)

YQF: Red Deer (Canada-AB) (location identifier)

YQG: Windsor (Canada-ON) (location identifier)

YQH: Watson Lake (Canada-YT) (location identifier)

YQI: Yarmouth (Canada-NS) (location identifier)

YQK: Kenora (Canada-ON) (location identifier)

YQL: Lethbridge (Canada-AB) (location identifier)

YQM: Moncton (Canada-NB) (location identifier)

YQN: Nakina (Canada-ON) (location identifier)

YQQ: Comox (Canada-BC) (location identifier)

YQR: Regina (Canada-SK) (location identifier)

YQS: Saint Thomas - Pembroke (Canada-ON) (location identifier)

YQT: Thunder Bay (Canada-ON) (location identifier)

YQU: Grande Prairie (Canada-AB) (location identifier)

YQV: Yorkton (Canada-SK) (location identifier)

YQW: North Battleford (Canada-SK) (location identifier)

YQX: Gander (Canada-NF) (location identifier)

YQY: Sydney (Canada-NS) (location identifier)

YQZ: Quesnel (Canada-BC) (location identifier)

Yr: Year • Younger • Your

Y-R: Yaw-Roll

YR: Romania (international aircraft registration code) • Yaw-Roll

YRA: Rae Lakes (Canada-NT) (location identifier)

YRB: Resolute (Canada-NT) (location identifier)

Yrbk: Yearbook

YRD: Dean River (Canada-BC) (location identifier)

YRDST: Year-Round Daylight Saving Time

YRE: Resolution Island (Canada-NT) (location identifier)

YREC: Yale Rotating Evolution Code

YRF: Cartwright (Canada-NF) (location identifier)

YRG: Rigolet (Canada-NF) (location identifier)

YRI: Rivière du Loup (Canada-QC) (location identifier)

YRIF: Yuri Rubinsky Insight Foundation

YRJ: Roberval (Canada-QC) (location identifier)

YRl: Yemen Ryal (currency)

YRL: Red Lake (Canada-ON) (location identifier)

Yrly: Yearly

YRM: Rocky Mountain House (Canada-AB) (location identifier)

YRMODA: Year-Month-Day

YRN: River's Inlet (Canada-BC) (location identifier)

YRO: Ottawa - Rockcliffe Street (Canada-ON) (location identifier)

YRQ: Trois-Rivières (Canada-QC) (location identifier)

YRR: Stuart Island (Canada-BC) (location identifier)

Yrs: Years • Yours

YRS: Red Sucker Lake (Canada-MB) (location identifier) • Ysgarth Rules System

YRT: Rankin Inlet (Canada-NT) (location identifier)

YRV: Revelstoke (Canada-BC) (location identifier)

YS: El Salvador (international aircraft registration code) • Yield Strength • Young Scholar

YSA: Sable Island (Canada-NS) (location identifier)

YSAG: Yttrium-Scandium-Aluminium Garnet

YSB: Sudbury (Canada-ON) (location identifier)

YSC: Sherbrooke (Canada-QC) (location identifier) • Yearly Spares Cost • Yokohama Science Center (Japan) • Young Stellar Cluster

YSD: Suffield (Canada-AB) (location identifier)

YSE: Swan River (Canada-MB) (location identifier) • Yaw Steering Error

YSF: Stony Rapids (Canada-SK) (location identifier) • Youth Science Foundation

YSG: Snowdrift (Canada-NT) (location identifier)

YSH: Smith Falls (Canada-ON) (location identifier)

YSI: Sans Souci (Canada-ON) (location identifier) • Young Springs Instruments (USA)

YSJ: Saint John (Canada-NB) (location identifier)

YSK: Sanikiluaq (Canada-NT) (location identifier)

YSL: Saint Leonard (Canada-NB) (location identifier)

YSM: Fort Smith (Canada-NT) (location identifier) • Yale Survey Machine

YSN: Salmon Arm (Canada-BC) (location identifier) • Young Scientists' Network

YSO: Postville (Canada-NF) (location identifier) • Yale Southern Observatory (Argentina-SJ) • Young Stellar Object

YSP: Marathon (Canada-ON) (location identifier)

YSQ: Spring Island (Canada-BC) (location identifier)

YSR: Nanisivik (Canada-NT) (location identifier)

YSS: State Island (Canada) (location identifier)

YSSP: Young Scientists' Summer Program

Yst: Youngest

YST: Sainte Therese Point (Canada-MB) (location identifier) • Yukon Standard Time (Canada)

YSU: Summerside (Canada-PE) (location identifier) • Yerevan State University (Armenia) • Youngstown State University (USA-OH)

YSV: Saglek (Canada-NF) (location identifier)

YSW: Salluit (Canada-QC) (location identifier)

YSX: Shearwater (Canada-BC) (location identifier)

YSY: Sachs Harbour (Canada-NT) (location identifier)

YSZ: Squirrel Cove (Canada-BC) (location identifier) • Yttrium Stabilized Zirconium

Y-T: Yam-Trim

YT: Mayotte (ISO country code) • Yaw Trim • Yukon Territory (Canada) • Yukon Time

YTA: Pembroke (Canada-ON) (location identifier) • Yaw Trim Angle

YTB: Hartley Bay (Canada-BC) (location identifier)

YTC: Sturdee (Canada-BC) (location identifier)

YTD: Thicket Portage (Canada-MB) (location identifier) • Year To Date

YTE: Cape Dorset (Canada-NT) (location identifier)

YTF: Alma (Canada-QC) (location identifier)

YTG: Sullivan Bay (Canada-BC) (location identifier) • Yukon Territorial Government

YTH: Thompson (Canada-MB) (location identifier)

YTI: Triple Island (Canada-BC) (location identifier)

YTJ: Terrace Bay (Canada-ON) (location identifier)

YTK: Tulugak (Canada-QC) (location identifier)

YTL: Big Trout Lake (Canada-ON) (location identifier)

YTN: Rivière au Tonnerre (Canada-QC) (location identifier)

YTO: Toronto Metropolitan (Canada-ON) (location identifier)

YTP: Tofino (Canada-BC) (location identifier)

YTQ: Tasiujuaq (Canada-QC) (location identifier)

YTR: Trenton (Canada-ON) (location identifier)

YTS: Timmins (Canada-ON) (location identifier)

YTT: Tisdale (Canada-SK) (location identifier)

YTU: Tasu (Canada-BC) (location identifier)

YTV: Yaw Thrust Vector

YTX: Telegraph Creek (Canada-BC) (location identifier)

YTZ: Toronto Island (Canada-ON) (location identifier)

YU: Yale University • York University (UK-England) • Yugoslavia (international vehicle registration code) • Yugoslavia (ISO country code) • Yugoslavia (postal country code) • Yugoslavie (ISO country code)

YUA: Yuanmou (China-PRC) (location identifier)

YUAC: York University Astronomy Club (Canada-ON)

YUB: Tuktoyaktuk (Canada-NT) (location identifier)

YUC: Yucatan (Mexico)

YUCL: Young and Upcoming Leaders

YUD: Umiujaq (Canada-QC) (location identifier)

YUE: Yuendumu (Australia-NT) (location identifier)

YUF: Pelly Bay - Dewline Site (Canada-NT) (location identifier)

YUG: Yugoslavia (ISO country code) • Yugoslavie (ISO country code)

Yugo: Yugoslavia

YUJ: Lady Franklin (Canada-NT) (location identifier)

Yuk: Yukon

YUL: Montréal Dorval (Canada-QC) (location identifier) • Yale University Library

YUM: Yuma (USA-AZ) (location identifier)

YUN: New Dinar (Yugoslavia) (ISO currency code) • Johnson Point (Canada-NT) (location identifier)

YUP: Yale University Press

Yuppie: Young Upwardly-mobile Professional People • Young Urban Professional

YUT: Repulse Bay (Canada-NT) (location identifier)

YUX: Hall Beach (Canada-NT) (location identifier)

YUY: Rouyn (Canada-QC) (location identifier)

YV: Venezuela (international aircraft registration code) • Venezuela (international vehicle registration code)

YVA: Moroni (Comoros) (location identifier)

YVAA: Yellowstone Valley Amateur Astronomers (USA-MT)

YVAC: Yakima Valley Astronomy Club (USA-WA)

YVB: Bonaventure (Canada-QC) (location identifier)

YVC: La Ronge (Canada-SK) (location identifier)

YVD: Yeva (Papua New Guinea) (location identifier)

YVE: Vernon (Canada-BC) (location identifier)

YVG: Vermilion (Canada-AB) (location identifier)

YVM: Broughton Island (Canada-NT) (location identifier)

YVN: Cape Dyer (Canada-NT) (location identifier)

YVO: Val d'Or (Canada-QC) (location identifier)

YVP: Kuujjuaq (Canada-QC) (location identifier)

YVQ: Norman Wells (Canada-NT) (location identifier)

YVR: Vancouver (Canada-BC) (location identifier)

YVS: Yearly Visits Statistics

YVT: Buffalo Narrows (Canada-SK) (location identifier)

YVV: Wiarton (Canada-ON) (location identifier)

YVZ: Deer Lake (Canada-ON) (location identifier)

YWA: Petawawa (Canada-ON) (location identifier)

YWB: Kanjiqsujuaq (Canada-QC) (location identifier)

YWF: Halifax - Downtown Waterfront (Canada-NS) (location identifier)

YWG: Winnipeg (Canada-MB) (location identifier)

YWH: Victoria - Inner Harbour (Canada-BC) (location identifier)

YWIA: You're Welcome In Advance

YWJ: Deline (Canada-NT) (location identifier) • Fort Franklin (Canada-NT) (location identifier)

YWK: Wabush (Canada-NF) (location identifier)

YWL: Williams Lake (Canada-BC) (location identifier)

YWM: Williams Harbour (Canada-NF) (location identifier)

YWN: Winisk (Canada-ON) (location identifier)

YWP: Webequie (Canada-ON) (location identifier)

YWR: White River (Canada-ON) (location identifier)

YWS: Whistler (Canada-BC) (location identifier)

YWY: Wrigley (Canada-NT) (location identifier)

YXC: Cranbrook (Canada-BC) (location identifier)

YXD: Edmonton (Canada-AB) (location identifier)

YXE: Saskatoon (Canada-SK) (location identifier)

YXF: Snake River (Canada-YT) (location identifier)

YXH: Medecine Hat (Canada-AB) (location identifier)

YXI: Killaloe (Canada-ON) (location identifier)

YXJ: Fort Saint John (Canada-BC) (location identifier)

YXK: Rimouski (Canada-QC) (location identifier)

YXL: Sioux Lookout (Canada-ON) (location identifier)

YXN: Whale Cove (Canada-NT) (location identifier)

YXP: Pangnirtung (Canada-NT) (location identifier)

YXQ: Beaver Creek (Canada-YT) (location identifier)

YXR: Earlton (Canada-ON) (location identifier)

YXS: Prince George (Canada-BC) (location identifier)

YXT: Terrace (Canada-BC) (location identifier)

YXU: London (Canada-ON) (location identifier)

YXX: Abbotsford (Canada-BC) (location identifier)

YXY: Whitehorse (Canada-YT) (location identifier)

YXZ: Wawa (Canada-ON) (location identifier)

Y-Y: Yaw-Yaw

YY: Yaw-Yaw

YYA: Big Bay - Yacht Club (Canada-BC) (location identifier)

YYB: North Bay (Canada-ON) (location identifier)

YYC: Banff (Canada-AB) (location identifier) • Calgary (Canada-AB) (location identifier)

YYD: Smithers (Canada-BC) (location identifier)

YYE: Fort Nelson (Canada-BC) (location identifier)

YYF: Penticton (Canada-BC) (location identifier)

YYG: Charlottetown (Canada-PE) (location identifier)

YYH: Spence Bay (Canada-NT) (location identifier) • Taloyak (Canada-NT) (location identifier)

YYI: Rivers (Canada-MB) (location identifier)

YYJ: Victoria (Canada-BC) (location identifier)

YYL: Lynn Lake (Canada-MB) (location identifier)

YYM: Cowley (Canada-AB) (location identifier)

YYN: Swift Current (Canada-SK) (location identifier)

YYQ: Churchill (Canada-MB) (location identifier)

YYR: Goose Bay (Canada-NF) (location identifier)

YYT: Saint John's (Canada-NF) (location identifier)

YYU: Kapuskasing (Canada-ON) (location identifier)

YYW: Armstrong (Canada-ON) (location identifier)

YYY: Mont Joli (Canada-QC) (location identifier)

YYZ: Toronto - Lester B. Pearson (Canada-ON) (location identifier)

YZ: Yale Zones

YZA: Ashcroft (Canada-BC) (location identifier)

YZC: Beatton River (Canada-BC) (location identifier) • Yale Zone Catalogue

YZE: Gore Bay (Canada-ON) (location identifier)

YZF: Yellowknife (Canada-NT) (location identifier)

YZG: Salluit (Canada-QC) (location identifier)

YZH: Slave Lake (Canada-AB) (location identifier)

YZL: Liard River (Canada-BC) (location identifier)

YZM: Buchans (Canada-NF) (location identifier)

YZP: Sandspit (Canada-BC) (location identifier)

YZR: Sarnia (Canada-ON) (location identifier)

YZS: Coral Harbour (Canada-NT) (location identifier)

YZT: Port Hardy (Canada-BC) (location identifier)

YZU: Whitecourt (Canada-AB) (location identifier)

YZV: Sept-Îles (Canada-QC) (location identifier)

YZW: Teslin (Canada-YT) (location identifier)

YZX: Greenwood (Canada-NS) (location identifier)

Y10K: Year 10,000

Y2K: Year 2000

Z

z: Algebraic Variable • Altitude • Cartesian Coordinate • Cartesian Third Axis • Charge Number of Ion • Coordinate Axis • Depth • Distance from the Galactic Plane • Doppler Shift • Haze • Height • Ion Charge Number • Redshift • Space Coordinate • Vertical Component of the Magnetic Vector • Vertical Distance • Zenith Angle • Zenith Distance • Zero • Zone • zu • zum • zur

Z: Altitude • Atomic Number • Azimuth Angle • Charge Number of Atoms • Coriolis Correction • Ensemble des Entiers Relatifs • Fractional Mass Abundance of Heavy Elements • Fractional Mass Abundance of Metals • Greenwich Mean Time • Impedance • Mechanical Impedance • Normal Acceleration • Partition Function • Proton Number • Radar Reflectivity Factor • Two Thousand (Roman numeral) • Zahl • Zahlung • Zaïre (currency) • Zambia (international vehicle registration code) • Zeichen • Zeile • Zeit • Zeitschrift • Zeitung • Zenith • Zenith Distance • Zenith Pointing • Zentralblatt • Zentrale • Zentrifuge • Zentrum • Zero • Zéro • Zettel • Zeuge • Zeugnis • Ziffer • Zimbabwe (international aircraft registration code) • Zimmer • Zitat • Zoll • Zone • Zoulou Time • Zuchthaus • Zucker • Zug • Zulu (international alphabet) • Zurücknahme • 2000 (Roman numeral)

za: South Africa • zirka

zA: zur Ansicht • zur Anstellung

ZA: Afrique du Sud (ISO country code) • Albania (international aircraft registration code) • South Africa (international vehicle registration code) • South Africa (ISO country code) • Zahlungsabkommen • Zahlungsanweisung • Zahnarzt • Zaire • Zeitschrift für Astrophysik (periodical code) • Zel'dovich Approximation • Zenith Angle • Zentralabteilung • Zentralamt • Zentralanstalt • Zentralarkiv • Zero Age • Zollamt • Zoneamt

ZAA: Alice Arm (Canada-BC) (location identifier) • Zeeman-effect Atomic Absorption • Zentralanmeldeamt

ZAB: Zollamt am Bahnhof

ZAbfO: Zollabfertigungsordnung

ZAbfSt: Zollabfertigungsstelle

ZAC: York Landing (Canada-MB) (location identifier) • Zacatecas (Mexico) • Zone d'Aménagement Concerté (France)

ZACV: Zero-Age Cataclysmic Variable

ZAD: Zadar (Croatia) (location identifier) • Zenith Angle Distribution • Zone d'Aménagement Différé (France)

ZAED: Zentralstelle für Atomkernenergie-Dokumentation

ZAF: Afrique du Sud (ISO country code) • South Africa (ISO country code)

ZAG: Zagreb Pleso (Croatia) (location identifier) • Zeitschriftenverlag Aktien-Gesellschaft • Zentralarbeitsgemeinschaft

ZAH: Zahedan (Iran) (location identifier)

ZAHB: Zero-Age Horizontal Branch

Zahl: Zahlung

Zahlm: Zahlmeister

zahlr: Zahlreich

Zahlst: Zahlstelle

ZAI: Zero Address Instruction • Zero Administrative Initiative

ZAJ: Zaranj (Afghanistan) (location identifier)

ZAK: Zeitschriftenaufsatzkatalog • Zentralarbeitskreis • Zero Administration Kit

ZAL: Financial South African Rand (ISO currency code) • Valdivia Pichoy (Chile) (location identifier)

Zam: Zamora (Spain)

ZAM: Zamboanga (Philippines) (location identifier) • Z-Axis Modulation • Zentralinstitut für angewandte Mathematik (Germany) • Zentrum für angewandte Mikroelektronik

ZAMG: Zentralanstalt für Meteorologie und Geodynamik (Austria)

ZaMM: Zeitschrift für angewandte Mathematik und Mechanik (periodical code)

ZAMO: Zero Angular Momentum Observer

ZaMP: Zeitschrift für angewandte Mathematik und Phyik (periodical code)

ZAMS: Zero-Age Main Sequence

z. Anst.: zur Anstellung

ZAnst: Zahlungsanweisung

ZAnwO: Zollanweisungsordnung

ZAO: Cahors Laberandie (France) (location identifier)

ZAPA: Zambia Press Agency

ZAPE: Zonal Available Potential Energy

ZAPO: Zulassungs-, Ausbildungs- und Prüfungsordnung

Zar: Zaragoza (Spain)

ZAR: Rand (South Africa) (ISO currency code) • Zaire (ISO country code) • Zaria (Nigeria) (location identifier) • Zentral-Archiv Retrieval

ZARM: Zentrum für Angewandte Raumfahrttechnologie und Mikrogravitation (Germany)

ZAS: Zaire Aero Service • Zentrum für Auslandsstudien • Zero Access Storage

ZASt: Zentrale Auskunftstelle • Zollaufsichtstelle

ZAT: Zero Alias Table • Zhaotong (China-PRC) (location identifier)

ZAW: Zero Administration for Windows • Zero Administration Workstation

ZAZ: Zaragoza (Spain) (location identifier)

zB: zum Beispiel • zur Beachtung

ZB: Zahlungsbefehl • Zahlungsbewilligung • Zentralbatterie • Zentralbibliothek • Zentralbild • Zentralblatt • Zentralbüro • Zero Beat • Zuiderbreedte • Zwischenbericht

ZBA: Zero Bias Anomaly

ZBB: Zero Base Budgeting • Zimbabwe Booksellers Association

ZBE: Zabreh - Dolni Benesov (Czech Rep.) (location identifier)

ZBF: Bathurst (Canada-NB) (location identifier) • Zentralbereich Förderung • Zugbahnfunk

ZBI: Zentralverband Berufsständischer Ingenieurvereine • Zero Baseline Interferometer

ZBK: Zabljak (Yugoslavia-Montenegro) (location identifier)

Zbl: Zentralblatt

ZBL: Biloela (Australia-QLD) (location identifier) • Ziegler-Biersack-Littmark

ZBM: Bromont (Canada-QC) (location identifier) • Zentralbibliothek der Medizin (Germany)

ZBO: Bowen (Australia-QLD) (location identifier)

ZBP: Baltimore Penn (USA-MD) (location identifier)

ZBR: Chah-Bahar (Iran) (location identifier) • Zero-Based Review • Zero-Base Review • Zone Bit Recording

zbV: zur besonderen Verwendung

ZBV: Beaver Creek (USA-CO) (location identifier)

ZBX: Branson Bus (USA-MO) (location identifier)

ZBY: Sayaboury (Laos) (location identifier)

ZC: Zero Crossing • Zodiacal Catalogue • Zone Convective • Zwicky Cluster

Z-CAV: Zone Constant Angular Velocity

ZCAV: Zone Constant Angular Velocity • Zoned Constant Angular Velocity

ZCG: Impedance Cardiogram • Zeolite Crystal Growth

Zch: Zürich (Switzerland)

ZCH: Zone Continuellement Habitable

Zchg: Zeichnung

Zchn: Zeichnen

Zchng: Zeichnung

ZCIT: Zona de Convergencia Intertropical • Zone de Convergence Intertropicale

ZCL: Zacatecas (Mexico) (location identifier)

ZCO: Temuco Maquehue (Chile) (location identifier)

ZCR: Zero-Crossing Rate

ZCY: Zirconium-Cerium-Yttrium

zD: zur Disposition

ZD: Zeitschriftendienst • Zeitungsdienst • Zener Diode • Zenithal Distance • Zenith Distance • Zentraldienst • Zentraldirektion • Zentrale Dokumentation • Zero Defects • Ziff-Davis • Zimmer mit Dusche • Zoned Decimal • Zusatz-Deskriptor

zdA: zu den Akten

ZDA: Zentrale Direktionsabteilung

ZDB: Zeitschriftendatenbank

ZDBOp: Ziff-Davis Benchmark Operation

ZDC: Zero-Degrees Calorimeter

ZDDSE: Zentrale Dokumentation für Entwicklungshilfe und Entwicklungsländer

ZDE: Zentrale Dokumentation Elektrotechnik • Zentralstelle Dokumentation Elektrotechnik

ZDEC: Two-Dimensional Electron Gas

ZDF: Zweites Deutsches Fernsehen (Germany)

Zdg: Zündung

ZDG: Zentrale Dokumentation für Geographie • Zentralen Dominierenden Galaxien • Zivildienstgesetz

Zdh: Zündhölzer

ZDI: Zeeman-Doppler Imager • Zeeman-Doppler Imaging • Zentralstelle für Dokumentation und Information

ZDIWD: Zentrale Dokumentations- und Informationsstelle für Werkstoffdaten

ZDJ: Berne Main Station (Switzerland) (location identifier) • Bern Hauptbahnhof (Switzerland) (location identifier) • Bern Hbf. (Switzerland) (location identifier)

ZDK: Zentrale Driftkammer

ZDL: Zero Delay Lockout

ZDN: Zentrum zur Dokumentation für Naturheilverfahren

ZDO: Zentrum für Datenverarbeitung und Organisation

ZDR: Zentrale Dokumentation für Raumordnung

ZDS: Zenith Data Systems

Zdw: Zündwaren

ZDW: Zeitung-Druck-Wirtschaft • Zentralamt des Deutschen Wetterdienstes

zE: zu Ehren • zum Exempel

ZE: Lerwick (UK postal code) • Zahlungseinheit • Zahlungseinstellung • Zeiteinheit • Zentraleinheit • Zugkrafteinheit

ZEB: Zenith-Enhanced Backscatter

ZEBRA: Zero-Energy Breeder-Reactor Assembly

ZEC: Secunda (South Africa) (location identifier)

ZEE: Zona Económica Exclusiva • Zone Économique Exclusive

ZEEP: Zero-Energy Experimental Pile

ZEFU: Zentrum Für Umweltforschung (Germany)

ZEG: Senggo (Indonesia) (location identifier) • Zero Economic Growth

ZeIE: Zeitschrift für elektrische Information und Energietechnik (periodical code)

ZeiPh: Zeitschrift für Physik (periodical code)

zeitgen: zeitgenössisch

zeitl: zeitlich • zeitlos

Zeitschr: Zeitschrift

zeitw: zeitweilig • zeitweise

Zeitw: Zeitwert • Zeitwort

ZEL: Bella Bella (Canada-BC) (location identifier) • Zell am Mosel (Germany)

Zem: Zemlya

ZEM: Eastmain (Canada-QC) (location identifier) • Zero-Effort Miss • Zero-Energy Mode

ZeMet: Zeitschrift für Meteorologie (periodical code)

Zen: Zenith

ZEN: Zenag (Papua New Guinea) (location identifier) • Zero-Effort Network • Zero-Effort Networking

ZENITH: Zero-Energy Nitrogen-Heated Thermal Reactor

Zent: Zentral

zentr: zentral • zentralisiert • Zentralistisch • zentriert • zentrisch

Zentr: Zentrale • Zentrum

zentrif: zentrifugal

ZEP: Zone d'Échanges Préférentiels • Zone d'Éducation Prioritaire (France)

ZER: Zero (India) (location identifier)

Zerst: Zerstörung

zerstr: zerstreut • zerstritten

ZES: Zone Économique Spéciale

ZESA: Zodiac Española SA (Spain)

ZETA: Zero-Energy Thermonuclear Assembly

ZETR: Zero-Energy Thermal Reactor

ZEU: Zentrum für Europäische Forschung

ZEV: Zero-Emission Vehicle

ZEVIS: Zentrales Verkehrsinformationssytem

zF: zu Fuß • zur Folge

Zf: Zufahrt • Zugführer

ZF: Zentrifuge • Zermelo-Fraenkel • Zero Flag • Zero Frequency • Zona Franca • Zone Franche • Zwischenfrequenz

ZFA: Faro (Canada-YT) (location identifier)

ZFB: Old Fort Bay (Canada-QC) (location identifier) • Zentralinstitut für Film und Bild

ZfBB: Zeitschrift für Bibliothekswesen und Bibliographie

ZFC: Zero-Field-Cooled

ZfD: Zeitschrift für Datenverarbeitung

ZFD: Fond du Lac (Canada-SK) (location identifier) • Zwei-Farben-Diagramm

zFdW: zur Förderung der Wissenschaften

Zfg: Zufügung • Zusammenfassung

ZfG: Zeitschrift für Geschichtswissenschaft

ZFL: South Trout Lake (Canada-ON) (location identifier)

ZFlu: Zeitschrift Flugwissenschaften (periodical code)

ZFlWe: Zeitschrift Flugwissenschaften und Weltraumforschung (periodical code)

ZFM: Fort McPherson (Canada-NT) (location identifier)

ZFN: Fort Norman (Canada-NT) (location identifier)

zfr: zinsfrei • zollfrei

ZfS: Zentralstelle für Standardisierung

ZFT: Fort Lauderdale - Railway Station (USA-FL) (location identifier)

ZfU: Zentrum für Unternehmenskommunikation (Germany)

ZFU: Zentralstelle für Fernunterricht

ZfV: Zeitschrift für Vermessungswesen

ZFV: Zentrale für Fremdenverkehr

ZFW: Fairview (Canada-AB) (location identifier) • Zeitschrift für Flugwissenschaften und Weltraumforschung • Zero Fuel Weight

ZFZ: Centre in Charge of a Flight Information Region or an Upper

Zg: Zeitung • Zeuge • Zeugung • Zug • Zugabe • Zugang

ZG: Zivilgericht • Zoologischer Garten • Zug

ZGB: Zivilgesetzbuch

ZGC: Lanzhou Zhongchuan (China-PRC) (location identifier)

ZGD: Zollgrenzdienst

ZGDV: Zentrum für Graphische Datenverarbeitung eV (Germany)

zge: ziemlich gut erhalten

ZGeo: Zeitschrift für Geophysik (periodical code)

ZGET: Zero-Gravity Expulsion Technique

Zgf: Zugfolge • Zugfürher

ZGF: Grand Forks (Canada-BC) (location identifier) • Zero-Gravity Facility

ZGG: Zero-Gravity Generator

Zgh: Zugehörigkeit

ZGH: Zero-Gravity Harmonic

ZGI: Gods River (Canada-MB) (location identifier) • Zentrales Geologisches Institut

Zgkw: Zugkraftwagen

zgl: zugleich

ZGL: South Galway (Australia-QLD) (location identifier)

ZGM: Ngoma (Zambia) (location identifier)

Zgn: Zeugen • Zeugnis

ZGP: Zeiss-Grossplanetarium Berlin (Germany)

ZGR: Little Grand Rapids (Canada-MB) (location identifier)

zgs: zusammengesetzt

ZGS: Gethsemani (Canada-QC) (location identifier) • Zero-Gradient Synchrotron • Zero-Gravity Shower • Zero-Gravity Simulator

zgst: zusammengestellt

ZGT: Zero-Gravity Trainer

ZGU: Gaua (Vanuatu) (location identifier)

ZGWBS: Zero-Gravity Whole-Body Shower

zh: Chinese (ISO language code)

zH: zu Händen • zur Hälfte

Zh: Zhurnal

ZH: Zentralheizung • Zero Hour • Zijne Heiligheid • Zijne Hoogheid

ZHA: Zhangjiang (China-PRC) (location identifier)

z. Hd.: zu Händen

z. Hdn.: zu Händen

z. Hdn. v.: zu Händen von

z. Hd. v.: zu Händen von

ZHF: Zodiacal History File

ZHK: Zusatzheilkosten

zhlr: zahlreich

ZHM: Shamshernagar (Bangladesh) (location identifier)

ZHO: Houston (Canada-BC) (location identifier)

ZHP: High Prairie (Canada-AB) (location identifier)

ZHR: Zenithal Hourly Rate • Zenith Hourly Rate

ZHSA: Zeiss Historica Society of America (USA-NY)

zHv: zu Händen von

ZHV: Zoll- und Handelsvertrag

Zi: Ziffer • Zimmer

ZI: Zentralinkasso • Zentralinstitut • Zielinformation • Zimbabwe • Zinner Variables • Zonal Index • Zone Industrielle • Zone of the Interior

ZIA: Zentralinstitut für Automation • Zentralinstitut für Automatisierung

ZIAP: Zentralinstitut für Astrophysik Potsdam, Germany

ZIB: Zentralinformationsbank • Zentralinstitut für Bibliothekswesen • Zentrum für Informationstechnik Berlin (Germany) • Zero-Impulse Bolt

ZIC: Victoria (Chile) (location identifier) • Zentrale Informationsverarbeitung Chemie (Germany) • Zone Intertropicale de Convergence

ZID: Zeitschrifteninhaltsdienst • Zentralstelle für Information und Dokumentation

ZIDIS: Zentrale Information und Dokumentation der Isotopen- und Strahlentechnik

ZIE: Zone Locale Élargie

ZiF: Zentrum für interdisziplinäre Forschung

ZIF: Zero Insertion Force

Ziff: Ziffer

Zig: Zigarre • Zigarette • Zigarillo

ZIG: Ziegenhain (Germany) • Ziguinchor (Senegal) (location identifier) • Z39.50 Implementation Group

ZIH: Ixtapa Zihuatanejo (Mexico) (location identifier)

ZIHD: Zentrale Informationsstelle für Hochschuldidaktik

ZIID: Zentralinstitut für Information und Datenverarbeitung • Zentralinstitut für Information und Dokumentation

ZIIS: Zentrum für Internationale und Interdisziplinärte Studien (Austria)

ZIL: Zentrale Informationsleitstelle

Zim: Zimmer

ZIMPOL: Zürich Infrared Polarimeter (*) • Zürich Imaging Polarimeter

ZIP: Zero-Inflated Poisson • Zigzag In-line Package • Zigzag In-line Pin • Zodiacal Infrared Probe • Zodiacal Infrared Project • Zone Improvement Plan • Zone Information Protocol • Zukunftinvestigationsprogramm

ZIPE: Zentralinstitut für Physik der Erde (Germany)

Zirk: Zirkular

ZIRST: Zone pour l'Innovation et la Recherche Scientifique et Technique

ZIS: Scotopic Illuminance Sensitivity • Seis Station (Switzerland) (location identifier) • Zentrales Informationssystem • Ziekehuis Informatie System • Zweiginformationssystem

zit: zitieren • zitiert

Zit: Zitadelle • Zitat

ZIT: Zone Information Table

zil: zivil • zivilisiert

Zil: Zivil • Zivilisation

ZIV: Inverness Scotrail (UK-Scotland) (location identifier) • Zentrale Informationsstelle für Verkehr

ZIZ: Zweigorientiertes Informationszentrum

ZJG: Jenpeg (Canada-MB) (location identifier)

ZJID: Zentralstelle für Journalistische Information und Dokumentation

ZJN: Swan River (Canada-MN) (location identifier)

zk: zurück

zK: zur Kenntnis

Zk: Zahlkarte • Zuschlagkarte

ZK: New Zealand (international aircraft registration code) • Zeitkontrolle • Zentrale Kommission • Zentralkatalog • Zentralkommission • Zentralkomitee • Zollkontrolle • Zugkraft • Zuschlagkarte • Zwischenklärung

ZKB: Kasaba Bay (Zambia) (location identifier) • Zimmer-Küche-Bad

ZKBB: Zimmer-Küche-Bad-Balkon

ZKE: Kaschechewan (Canada-ON) (location identifier) • Zonal Kinetic Energy

Zkft: Zukunft • Zusammenkunft

zKg: zur Kenntnis genommen

ZKG: Kegaska (Canada-QC) (location identifier)

ZKH: Zijne Keizerlijke Hoogheid • Zijne Koninklijke Hoogheid

ZKK: Zentrale Kontrolkommission

ZKL: Steenkool (Indonesia) (location identifier)

ZKM: Sette Cama (Gabon) (location identifier) • Zweikoordinatenmeßgerät

ZKP: Kasompe (Zambia) (location identifier)

Z. Kts.: Zur Kenntnis

Zkw: Zugkraftwagen

ZKZ: Zentralkatalog der Zeitschriften

Zł: Złoty (currency)

Zl: Zahl • Zeile • Ziel • Zugleitung

ZL: New Zealand (international aircraft registration code) • Zero Lift • Zwischenlandung

ZLA: Zambia Library Association • Zimbabwe Library Association

ZLD: Zero Level Drift • Zentralstelle für Luftfahrtdokumentation • Zero Lift Drag • Zodiacal Light Device

ZDLI: Zentralstelle für Luftfahrtdokumentation und Information

Zle: Zeile • Ziele

Zlg: Zahlung • Zulassung • Zuleitung

ZLG: El Gouera (Mauritania) (location identifier) • Zero Line Gap • Zero-lock Laser Gyro

ZLISP: Zilog List Processor

ZLKB: Zentrale Leit- und Koordinierungstelle der Bibliotheken

ZLM: Zone de Logistique et de Maintenance • Zweifrequenz-Laserwegmeßsystem

ZLO: Manzanillo (Mexico) (location identifier) • Zone de Logistique et d'Opérations

ZLR: Zentrum für Luft- und Raumfahrttechnik (Germany)

ZLS: Scotopic Luminance Sensitivity • Zentralstelle der Länder für Sicherheitstechnik

ZLSM: Zeiss Light Section Microscope

ZLT: La Tabatière (Canada-QC) (location identifier)

ZLV: Zero-Length Vector

Zm: Zahlmeister • Zeitmaß • Zeitmesser • Zeitmessung • Zugmaschine

ZM: New Zealand (international aircraft registration code) •

ZMD: Sena Madureira (Brazil-AC) (location identifier)

ZME: Metro Park - Railway Station (USA-NY) (location identifier)

ZMete: Zeitschrift für Meteorologie (periodical code)

ZMetl: Zeitschrift für Metallkunde (periodical code)

ZMH: 108 Mile Ranch (Canada-BC) (location identifier)

ZMI: Napoli - Mergellina Stazione (Italy) (location identifier)

ZMK: Kwacha (Zambia) (ISO currency code)

ZML: Milwaukee Railway Station (USA-WI) (location identifier)

ZMM: Zamora (Mexico) (location identifier)

ZMMAS: Zodiacal Microparticle Multiparameter Analysis System

ZMMD: Zürich-Mainz-München-Darmstadt

ZMMF: Zentrale Medizin-Meteorologische Forschungsstelle

ZMP: Zentralmeridianpassage

ZMS: Firenze - S.M. Novella Stazione (Italy) (location identifier)

ZMT: Masset (Canada-BC) (location identifier)

zn: zu nehmen • zu nennen

Zn: Zeichen • Zeitnahme • Zinc • Zink • Zone • Zoon • Zunahme • Zuname

ZN: Zahlungsnachweis • Zimmernachweis • Zusatz-Nichtdeskriptor

ZNA: Nanaimo Harbour (Canada-BC) (location identifier)

ZNat: Zeitschrift für Naturforschung (periodical code)

ZNC: Nyac (USA-AK) (location identifier)

ZND: Zentrale Namendatei • Zinder (Niger) (location identifier)

ZNE: Newman (Australia-WA) (location identifier)

ZNG: Negginan (Canada-MB) (location identifier)

ZNIEFF: Zone Naturelle d'Intérêt Écologique, Faunistique et Floristique (France)

ZNK: Zentrale Namenkartei

ZNS: Zentrales Nervensystem • Zentralnervensystem

ZNTH: Zenith

ZNU: Namu (Canada-BC) (location identifier)

ZNZ: Zanzibar (Tanzania) (location identifier)

ZO: Zahlungsordnung • Zero Output • Zollort • Zuidoost • Zulassungsordnung • Zuständigkeitsordnung

ZOA: Zone Of Avoidance

Zod: Zodiac

ZOD: Zero-Order Design • Zero-Order Detector

ZOE: Zone Of Exclusion

ZOF: Ocean Falls (Canada-BC) (location identifier) • Zone Of Fire

ZOHF: Zodiacal Observation History File

Zoochem: Zoochemistry

Zool: Zoología • Zoological • Zoologist • Zoology

Zoot: Zootecnia

ZOP: Zeroth-Order Prediction

ZOPI: Zero Order Polynomial Interpolator

ZOPP: Zero Order Polynomial Predictor

ZOS: Osorno - Cañal Bajo (Chile) (location identifier) • Zoom Optical System

ZPB: Sachigo Lake (Canada-ON) (location identifier)

ZPC: Pucón (Chile) (location identifier) • Zero Point of Change • Zero Point of Charge

ZPD: Zanithal Point Determination

ZPE: Zero-Point Energy

zPf: zur Pferd

ZPG: Zero Population Growth

ZPH: Zephyrhills (USA-FL) (location identifier)

ZPhy: Zeitschrift für Physik (periodical code)

ZPID: Zentralstelle für Pädagogische Information und Dokumentation

ZPM: Zone-Plate Mirror

ZPN: Impedance Pneumogram • Zenithal Polynomial Projection • Zentralstern eines Planetarischen Nebels

ZPO: Pine House (Canada-SK) (location identifier) • Zivilprozeßordnung

ZPR: Zivilprozeßrecht

ZPRSN: Zürich Provisional Relative Sunspot Number

ZPS: Zone de Protection Spéciale

ZPT: Zero Power Test

ZQN: Queenstown (New Zealand) (location identifier)

ZQS: Queen Charlotte Island (Canada-BC) (location identifier)

ZQZ: Centre in Charge of Flight Information Region or an Upper Flight

zr: zahlreich

zR: zu Recht • zur Revision • zur Ruhe

Zr: Zimmer • Zirconium • Zirkonium

ZR: Zaire (ISO country code) • Zentralrat • Zivilrecht

ZRD: Buffalo Depew (USA-NY) (location identifier)

ZRE: Zaire (international vehicle registration code) • Zero Rate Error

ZRF: Rockford - Clock Tower (USA-IL) (location identifier)

ZRG: Bratislava Bus Station (Slovakia) (location identifier)

ZRH: Zürich Kloten (Switzerland) (location identifier)

ZRI: Serui (Indonesia) (location identifier) • Zimbabwe Research Index

ZRJ: Round Lake (Canada-ON) (location identifier)

ZRK: Zentrale Regionenkonferenz

ZRL: Albany-Rensselaer Railway Station (USA-NY) (location identifier)

ZRM: Sarmi (Indonesia) (location identifier)

ZRP: Newark Penn (USA-NJ) (location identifier)

ZRR: Arctic Red River (Canada-NT) (location identifier)

ZRS: Zivilrechtssache

ZRT: Zero Reaction Tool

ZRU: Zone de Redynamisation Urbaine (France)

ZRV: Zero Relative Velocity

ZRZ: Zaire (Zaire) (ISO currency code)

zs: zusammen

zS: zur Sache • zur See

Zs: Zeitschrift • Zusage • Zusatz • Zusammensetzung

ZS: South Africa (international aircraft registration code) • Zahlungsstelle • Zentralsekretariat • Zentralstation • Zentralstelle • Zivilschutz

ZSE: Saint-Pierre de la Réunion (Reunion Island) (location identifier) • Zero Stress Everywhere

ZSEE: Zoneamento Socio-Economico-Ecologico

ZSF: Springfield Railway Station (USA-MO) (location identifier)

Zsfg: Zusammenfassung • Zusammenfügung

Zshg: Zusammenhang

ZSI: Zambia Standards Institute

ZSJ: Sandy Lake (Canada-ON) (location identifier)

ZSL: Zentrale Sendeleitung • Zero Slot LAN

zsm: Zo Spoedig Mogelijk

ZSM: Zavod za Standardizacijai Metrologija (Macedonia)

Z-SN: Zwicky Supernova

ZSN: Zwicky Supernova

ZSP: Saint Paul (Canada-QC) (location identifier)

ZSPN: Zentralstern eines Planetarischen Nebels

ZSR: Zugsteuerrechner

Zss: Zeitschriften

ZSS: Sassandra (Ivory Coast) (location identifier)

Zst: Zustand • Zusammenstellung • Zustellung

ZSt: Zentralstelle

ZST: Stewart (Canada-BC) (location identifier) • Zone Standard Time

ZStA: Zentrales Staatarchiv

Zstg: Zusammenstellung • Zustellung

ZSV: Saint Louis - Railway Station (USA-MO) (location identifier)

ZSW: Prince Rupert - Seal Cove (Canada-BC) (location identifier) • Zentrum für Sonnenenergie- und Wasserstoffforschung (Germany)

zT: zum Teil • zum Termin

Zt: Zeit • Zitat • Zuteilung

ZT: South Africa (international aircraft registration code) • Zahlungstermin • Zenith Telescope • Zone Time

ZTA: Tureia (Tuamotu Islands) (location identifier) • Zirconia Toughened Alumina

ZTB: Tête-à-la-Baleine (Canada-QC) (location identifier)

Ztg: Zeitung • Zuteilung

Ztgn: Zeitungen

ZTH: Zakynthos Island (Greece) (location identifier)

ztl: zeitlich • zeitlos

Ztm: Zeitmaß • Zeitmesser • Zeitmessung

ZTM: Shamattawa (Canada-MB) (location identifier)

Ztn: Zeitnahme • Zeitnehmer

ZTO: Zero Time Outage

Ztr: Zeitrechnung • Zentner • Zentrale • Zentrum • Zutritt

ZTR: Zhitomir (Ukraine) (location identifier)

ZTS: Tahsis (Canada-BC) (location identifier) • Zoom Transfer Scope

Ztschr: Zeitschrift

ztw: zeitweilig • zeitweise

Ztw: Zeitwert

ZtW: Zeitwert

ZTY: Philadelphia - 30th Station (USA-PA) (location identifier)

zu: Zulu (ISO language code)

ZUC: Ignace (Canada-ON) (location identifier)

ZUD: Ancud (Chile) (location identifier)

ZUE: Zuenoula (Ivory Coast) (location identifier)

zuf: zufällig • zufolge

Zuf: Zufall • Zufuhr

zug: zugängig • zugänglich

Zug: Zugabe • Zugang

zugel: zugelassen

zuget: zugeteilt

zugew: zugewandert • zugewiesen

zugl: zugleich

ZUH: Zhuhai (China-PRC) (location identifier)

ZUI: Zooming User Interface

zul: zulässig • zuletzt

ZUL: Zilfi (Saudi Arabia) (location identifier)

ZUM: Churchill Falls (Canada-NF) (location identifier)

ZUMA: Zeitung für Umfragen, Methoden und Analysen

ZÜN: Zentraler Übersetzungsnachweis

ZUN: Chicago Union (USA-IL) (location identifier)

ZUP: Zone à Urbaniser en Priorité (France)

zur: zurück

Zus: Zusammen • Zusatz • Zusätzlich

ZUVAD: University College London (Starlink internal code)

zuw: zuweilen

Zuz: Zuzüglich

zv: zur verkaufen • zu vermieten • zu vernichten

zV: zum Verkauf

ZV: Zeitungsverlag • Zentralvorstand • Zoomed Video • Zoom Video

ZVA: Miandrivazo (Madagascar) (location identifier) • Zeitungsvertriebsamt

ZVC: Zero-Velocity Curve

ZVE: Zentrale Verarbeitungseinheit

ZVEI: Zentralverband der Elektrotechnischen Industrie eV (Germany)

z. Vg.: zum Vorgang

ZVG: Springvale (Australia-WA) (location identifier)

ZVIS: Zentralverwaltung Informationssysteme

ZVK: Savannakhet (Laos) (location identifier)

ZVZV: Zeitungsverlag und Zeitschriftenverlag

zw: zwar • zwecks • zwei • zwischen

Zw: Zahlungsweise • Zahlwort • Zeitwort • Zuwachs • Zwang • Zweck • Zwicky • Zwillinge

ZW: Zimbabwe (international vehicle registration code) • Zimbabwe (ISO country code) • Zuidwest • Zwicky

ZWA: Andapa (Madagascar) (location identifier) • Zentralstelle für Wissenschaftliche Altbestände

ZwCl: Zwicky Clusters

ZwD: Zweigbibliothek

ZWD: Dollar (Zimbabwe) (ISO currency code)

ZWE: Zimbabwe (ISO country code)

ZWer: Zeitschrift für Werkstofftechnik (periodical code)

ZWI: Wilmington - Railway Station (USA-DE) (location identifier)

ZWL: Wollaston Lake (Canada-SK) (location identifier) • Zentralstelle für Wissenschaftliche Literatur

ZWO: Organisatie voor Zuiverwetenschappelijk Onderzoek (Netherlands)

ZWP: West Palm Beach - Railway Station (USA-FL) (location identifier)

ZWS: Stuttgart Hauptbahnhof (Germany) (location identifier) • Stuttgart Hbf. (Germany) (location identifier) • Stuttgart Main Station (Germany) (location identifier)

ZWTI: Zentrum für Wissenschaftlich-Technische Information

ZWTR: Zentraler Wissenschaftlich-Technischer Rat

ZWU: Washington Union (USA-DC) (location identifier)

ZXS: Buffalo - Exchange Street (USA-NY) (location identifier)

ZYI: Zunyi (China-PRC) (location identifier)

Zyl: Zylinder

ZYL: Sylhet (Bangladesh) (location identifier)

ZYP: New York - Penn (USA-NY) (location identifier)

zz: Ginger

zZ: zur Zeit

Zz: Ginger

ZZ: Zero-Zero • Zig-Zag • Zonenzeit

ZZB: Zirkumzenitalbogen

ZZBB: Zentralkatalog der Zeitungsbestände in Berliner Bibliotheken

ZZD: Zeitungs- und Zeitschriftendienst

zzgl: zuzüglich

ZZK: Zeitschriftenzentralkatalog

ZZO: Zuidzuidoost

z. Zt.: zur Zeit

ZZU: Mzuzu (Malawi) (location identifier)

ZZV: Zanesville (USA-OH) (location identifier)

ZZW: Zuidzuidwest

Special Sections

Numerical Beginnings

0: Super-Supergiant Star
0g: Null Gravity
0th: Zeroth
1: Monday
1A: First Ariel V Catalogue
1ASE: First American Science and Engineering Catalogue
1C: First Cambridge Catalogue of Radio Sources
1CG: First COS B Catalogue
1D: One-Dimensional
1dF: One-degree Field
1E: EINSTEIN Observatory Catalog • HEAO-3 Catalog
1HT: One Hectare Telescope
1M: Massachusetts OSO-7 Catalogue
1NF: First Normal Form
1OT: First Outer Track • First Overtone
1st: First
1U: First Uhuru Catalogue
1XRS: First X-ray Sources Catalogue • SAS-3 Catalogue
2: Tuesday
2a: Segunda-feira (Monday/Portuguese)
2A: Second Ariel V Catalogue
2ASE: Second American Science and Engineering Catalogue
2AT: Second Alias Table
2BMS: Two-Bed Molecular Sieve
2BSD: Second Berkeley Software Distribution
2B1Q: Two Binary One Quaternary
2C: Second Cambridge Catalogue of Radio Sources
2CG: Second COS B Catalogue
2D: Two-Dimensional • Two-Poisson
2DAG: Two-Dimensional Atom Gas
2DD: Two-Dimensional and Dynamic (method)
2DEG: Two-Dimensional Electron Gas
2dF: Two-Degree Field
2EG: Second EGRET Catalog
2GL: Second-Generation Language
2H: Double Hard
2MASS: Two-Micron All-Sky Survey
2nd: Second
2NF: Second Normal Form
2OT: Second Outer Track • Second Overtone
2P: Two-Poisson
2PC: Two-Phase Commit
2PCF: Two-Point Correlation Function
2SCS: Second Satellite Control Squadron
2SLS: Two-Stage Least Squares
2U: Second UHURU Catalogue
2WD: Two-Wheel Drive
2XRS: Second X-ray Sources Catalogue
3: Wednesday
3a: Terça-feira (Tuesday/Portuguese)
3A: Monaco (international aircraft registration code) • Third Ariel V Catalogue
3B: Mauritius (international aircraft registration code)
3C: Care-Companionship-Clothing • Equatorial Guinea (international aircraft registration code) • Third Cambridge Catalogue of Radio Sources
3CR: Revised Third Cambridge Catalogue of Radio Sources
3D: Désétatisation-Défiscalisation-Dérégulation • Swaziland (international aircraft registration code) • Three-Dimensional
3dF: Three-Degree Field
3DTV: Three-Dimensional Television
3F: Form-Fit-Function
3GL: Third-Generation Language
3H: Hot-High-Humid • Treble Hard
3I: Improved Initialization Inversion
3M: Maintenance and Material Management • Minnesota Mining and Manufacturing Co. (USA-MN)
3NF: Third Normal Form
3OT: Third Outer Track • Third Overtone
3PCF: Three-Point Correlation Function
3PN: Third Post-Newtonian
3R: Rapid Radiance Recognition • Reading-Writing-Arithmetic • Reconnaissance Rapide des Radiances • Reference and Research Resources
3rd: Third
3RD: Réseau Radioélectrique à Ressources pour Données
3RP: Réseau Radioélectrique à Ressources Partagées
3S: Satellite Surface Stress • Sols-Solides-Structures
3T: Twente Technology Transfer BV (Netherlands)
3U: Third UHURU Catalogue
3WC: Three-Way Calling
3X: Guinea (international aircraft registration code)
4: For • Thursday
4a: Quarta-feira (Wednesday/Portuguese)
4A: Association Andromède Astronomie Aveyronnaise (France) • Atlas Automatisé des Absorptions Atmosphériques • Automated Atmospheric Absorption Atlas
4BMS: Four-Bed Molecular Sieve
4C: Fourth Cambridge Catalogue of Radio Sources
4D: Four-Dimensional
4dF: Four-Degree Field
4GL: Fourth Generation Language
4H: Head-Heart-Hands-Health
4K: Azerbaijan (international aircraft registration code)
4MAPS: 4-Meter Active Primary Support
4NF: Fourth Normal Form
4OT: Fourth Overtone
4P: Prévision Par Petits Pavés
4R: Sri Lanka (international aircraft registration code)
4S: Society for the Social Studies of Science
4th: Fourth
4to: Quarto
4U: Fourth UHURU Catalogue
4WD: Four-Wheel Drive
4X: Israel (international aircraft registration code)
5: Friday
5a: Quinta-feira (Thursday/Portuguese)
5A: Libya (international aircraft registration code)

5As: Amherst Area Amateur Astronomers Association (USA-MA)

5B: Cyprus (international aircraft registration code)

5C: Fifth Cambridge Catalogue of Radio Sources

5dF: Five-Degree Field

5GL: Fifth Generation Language

5H: Tanzania (international aircraft registration code)

5J: Private Jet Expeditions Inc. (location identifier)

5N: Nigeria (international aircraft registration code)

5OT: Fifth Overtone

5R: Madagascar (international aircraft registration code)

5S: Satellite Signal Simulation in the Solar Spectrum

5T: Mauritania (international aircraft registration code) • Time-Turn-Twist-Throttle-Talk

5th: Fifth

5U: Niger (international aircraft registration code)

5V: Togo (international aircraft registration code)

5W: Samoa (international aircraft registration code)

5X: Uganda (international aircraft registration code)

5Y: Kenya (international aircraft registration code)

6: Saturday

6a: Sexta-feira (Friday/Portuguese)

6C: Sixth Cambridge Catalogue of Radio Sources

6dF: Six-degree Field

6O: Somalia (international aircraft registration code)

6OT: Sixth Overtone

6th: Sixth

6V: Senegal (international aircraft registration code)

6W: Senegal (international aircraft registration code)

7: Sunday

7O: Yemen (international aircraft registration code)

7OT: Seventh Overtone

7P: Lesotho (international aircraft registration code)

7QY: Malawi (international aircraft registration code)

7T: Algeria (international aircraft registration code)

7th: Seventh

8OT: Eigth Overtone

8P: Barbados (international aircraft registration code)

8Q: Maldives (international aircraft registration code)

8R: Guyana (international aircraft registration code)

8th: Eighth

8to: Octavo

9A: Croatia (international aircraft registration code)

9G: Ghana (international aircraft registration code)

9H: Malta (international aircraft registration code)

9J: Zambia (international aircraft registration code)

9K: Kuwait (international aircraft registration code)

9L: Sierra Leone (international aircraft registration code)

9M: Malaysia (international aircraft registration code)

9N: Nepal (international aircraft registration code)

9OT: Ninth Overtone

9Q: Zaire (international aircraft registration code)

9th: Ninth

9U: Burundi (international aircraft registration code)

9V: Singapore (international aircraft registration code)

9XR: Rwanda (international aircraft registration code)

9Y: Trinidad and Tobago (international aircraft registration code)

Greek Alphabet

α .... Acoustic Absorption Coefficient
alpha
Angle of Attack
Angle of Incidence
Angular Acceleration
Attenuation Coefficient
Cubic Expansion Coefficient
Dissociation Degree
Effective Thermal Diffusivity
Fine-Structure Constant
Internal-Conversion Coefficient
Internal-Conversion Factor
Lagrangian Multiplicator
Mach Angle
Madelung Constant
Molecular Electrical Polarizability
Payload Ratio
Plane Angle
Polarizability
Radio Spectral Index
Recombination Coefficient
Right Ascension
Thermal Accomodation Coefficient
Thermal Diffusivity
Thermal Expansion Coefficient
Townsend Electron Ionization Coefficient
α^h .... Absorptance
α_i .... Electron-Ion Recombination Coefficient
α_i .... Ion-Ion Recombination Coefficient
α_l .... Linear Expansion Coefficient
α_p .... Relative Pressure Coefficient
$\alpha_x\ \alpha_y\ \alpha_z$ ... Dirac (4×4) Matrices
α_T .... Thermal Diffusion Factor
$\alpha(A)$ .... Nullity Index of A
A .... Alpha
β .... Ballistic Coefficient
beta
Coefficient of Thermal Expansion
Dirac (4×4) Matrix
Dry Mass Ratio
Ecliptic Latitude
Magnetic Pressure Ratio
Phase Coefficient
Plane Angle
Pressure Coefficient
Ratio of Gas Pressure to Total Pressure
Townsend Ion Ionization Coefficient
$\beta(A)$ .... Deficiency Index of A
B .... Beta
γ .... Acceleration
Aspect Ratio of the Duct
Conductivity
Cubic Expansion Coefficient
Damping Constant
Electrolytic Conductivity
Flight Path Angle
gamma
Grüneisen Parameter
Gyromagnetic Ratio
Photon
Plane Angle
Polarizability
Propagation Coefficient
Propellant Fraction
Ratio of Specific Heat Capacities
Secondary Electron Emission Coefficient
Shear Strain
Slope of Characteristic Curve of Photographic Emulsion
Specific Heat Capacities Ratio
Surface Tension
γ_B .... Activity Coefficient
Γ .... Boudary of A
Gamma
Grüneisen Parameter
Level Width
Specific Gamma-Ray Constant
Van den Kerkhove Factor
δ .... Boundary Layer Thickness
Damping Coefficient
Declination
Deflection Angle
delta
Dissipation Coefficient
Flow Angle
Loss Angle
Thickness
Wedge Angle
δ_e .... Pitch Control Deflection
δ^i_j .... Kronecker Symbol
δ_{ij} .... Kronecker Symbol
δ_T .... Throttle Setting
$\delta(x)$ .... Dirac Delta Function
Δ .... Area of Triangle
Delta
Difference of
Distance to the Earth in AU
Geocentric Distance of Perihelion
Mass Excess
Superconductor Energy Gap
$\Delta\epsilon$ .... Nutation in Ecliptic Latitude
$\Delta\lambda$ .... Doppler Wavelength Shift
Wavelength Bandwidth
$\Delta\nu$ .... Frequency Bandwidth
$\Delta\varphi$ .... Nutation in Ecliptic Longitude
ΔS .... Preston's Metallicity Parameter (RR Lyrae stars)
ΔT .... Ephemeris Time Difference
Universal Time Difference
ϵ .... Emissivity
Energy of a Particle
epsilon
Linear Strain
Molecular Attraction Energy Parameter
Permittivity
Relative Elongation
ϵ_F .... Fermi Energy
ϵ_{ab} .... Seebeck Coefficient for Substances a and b
ϵ_o .... Permittivity of Vacuum
ϵ_r .... Relative Permittivity
ε .... Asymmetry Parameter
Collision Parameter (azimuth angle)
Ecliptical Obliquity
Emission Coefficient
Emissivity
Energy per Unit of Mass
epsilon

Error
Expansion Ratio
Fast Fission Factor
Inclination of the Galactic Plane on the Equatorial One
Nuclear Energy Generation
Permittivity
Sign Factor
Structural Fraction
Turbulence Dissipation
ε_P .... Potential Energy
ε_T .... Thrust Vector Angle
ε_i .... Energy of Level i
Energy of State i
ε_o .... Dielectric Constant of Vacuum
$\varepsilon(t)$ .... Unit Step Function
E .... Epsilon
ζ .... Apparent Zenithal Distance
Modal Damping Ratio
Normalized Pressure Drop
zeta
$\zeta(\lambda)$ .... Interstellar Extinction Law
Z .... Zeta
η .... Charge Density
Efficiency
Efficiency Rate
eta
Modal Coordinates
Neutron Yield per Absorption
Propulsion Efficiency
Sector-Triangle Ratio
Viscosity
η_m .... Magnetic Diffusivity
H .... Eta
θ .... Bulk Strain
Dimensionless Temperature
Scattering Angle
theta
Plane Angle
Regression Coefficient
Volume Strain
ϑ .... Angular Coordinate (galactocentric system)
Bragg Angle
Celsius Temperature
Elongation
Euler Angle
Flow Angle
Pitch Angle
Plane Angle
Polar Angle
Position Angle
Scattering Angle
theta
Volume Strain
Θ .... Characteristic Temperature
Current Linkage
Diffraction Angle
Local Sidereal Time
Sidereal Time
Thermodynamic Temperature
Theta
Θ_D .... Debye Temperature
Θ_E .... Einstein Temperature
Θ_W .... Characteristic Weiss Temperature
Θ_r .... Rotational Temperature
Θ_v .... Vibrational Temperature
ι .... iota
I .... Iota
κ .... Absorption Coefficient
Bulk Compressibility
Compressibility
Epicyclic Frequency
Isentropic Exponent
kappa
Landau-Ginzburg Parameter
Magnetic Susceptibility
Molar Absorption Coefficient
Opacity
Ratio of Specific Heat Capacities
Rosseland Mean Absorption Coefficient
Thermal Diffusivity
K .... Kappa
λ .... Decay Constant
Desintegration Constant
Ecliptic Longitude
Geographical Longitude
Lagrangian Multiplier
lambda
Mean Free Path
Mean Molecular Free Path
Payload Fraction
Planetographic Longitude
Thermal Conductivity
Wavelength
$\lambda\lambda$ .... Wavelength Range
λ_B .... Absolute Activity of Substance B
λ_C .... Compton Wavelength
λ_D .... Debye Length
λ_L .... London Penetration Depth
λ_i .... Quantum Number of Component of Electronic Orbital Angular Momentum Vector Along Symmetry Axis
λ_{ik} .... Thermal Conductivity Tensor
Λ .... Boundary Layer Shape Factor
Characteristic Diffusion Length
Collision Parameter (polar angle)
Cosmological Constant
Lambda
Logarithmic Decrement
Mean Free Path of Phonons
Non-Leakage Probability
Phonon Mean Free Path
Quantum Number of Component of Electronic Orbital Angular Momentum Vector Along Symmetry Axis
Shock Wave Thickness
μ .... Atomic Attenuation Coefficient
Chemical Potential
Dynamic Viscosity
Earth/Moon Mass Ratio
Electromagnetic Moment
Friction Coefficient
Friction Factor
Joule-Thomson Coefficient
Linear Attenuation Coefficient
Magnetic Moment of Particle
Mass Flow Ratio
Mass Fraction
Micro
Micron
Mobility
mu
Muon
Nucleus Magnetic Moment
Particle Magnetic Moment
Permeability
Poisson Ratio
Proper Motion
Reduced Mass
Surface Mass Density

Thompson Coefficient
Viscosity
μ_B .... Bohr Magneton
Chemical Potential of Substance B
μ_N .... Nuclear Magneton
μ_a .... Atomic Attenuation Coefficient
Bank Angle
μ_e .... Magnetic Moment of Electron
μ_l .... Linear Attenuation Coefficient
μ_m .... Mass Attenuation Coefficient
μ_n .... Magnetic Moment of Neutron
μ_o .... Magnetic Permeability of Vacuum
Permeability of Vacuum
μ_r .... Relative Permeability
M .... Mu
ν .... Frequency
Kinematic Viscosity
Neutrino
Neutron Yield per Fission
nu
Poisson Number
Sound Particle Velocity
Specific Volume
True Anomaly
ν_B .... Stoichiometric Number of Substance B
ν_c .... Collision Frequency
ν_{coll} .... Collision Frequency
ν_i .... Ionization Frequency
N .... Nu
ξ .... Average Logarithmic Energy Decrement
Coherence Length
Extent of Reaction
Inertial Parameter
Microturbulent Velocity
Molecular Velocity
Mole-Mass Ratio
Sound Particle Displacement
xi
Ξ Xi
o .... omicron
O .... Omicron
π .... Equatorial Horizontal Parallax
Parallax
pi (see also ϖ)
Pion
Ratio of a Circle Circumference to its Diameter
$\pi_\odot$ .... Solar Parallax
ϖ .... Longitude of Perihelion
pi (see also π)
Π .... Linear Polarization Fraction
Osmotic Pressure
Pi
Product
Π_{ab} .... Peltier Coefficient for Substances a and b
ρ .... Atmospheric Density
Charge Density
Charge Volume Density
Density
General Secular Precession in Longitude
Geocentric Distance
Linear Strain
Mass Density
Reactivity
Reflection Coefficient
Resistivity
rho
ρ^h .... Reflectance
ρ_R .... Residual Resistivity
ρ_{ik} .... Resistivity Tensor
$\rho(A)$ .... Resolvent Set of A
ϱ .... rho
P .... Rho
σ .... Charge Surface Density
Conductivity
Cross Section
Dispersion
Leakage Coefficient
Momentum Accomodation Coefficient
Normal Stress
Population Standard Deviation
Short-Range Order Parameter
sigma
Specific Fuel Consumption
Standard Deviation
Stefan-Boltzmann Constant
Stress Components
Surface Charge Density
Surface Tension
Symmetry Factor
Wave Number
Wave Vector
σ_i .... Quantum Number of Component of Electronic Spin Along Symmetry Axis
σ_{ik} .... Electric Conductivity Tensor
σ_x, σ_y σ_z .. Pauli Matrices
σ_1, σ_2 σ_3 .. Pauli Matrices
$\sigma(A)$ .... Spectrum of A
$\sigma_{as}(A)$ ..... Absolutely Continuous Spectrum of A
$\sigma_c(A)$ ...... Continuous Spectrum of A
$\sigma_d(A)$ ...... Discrete Spectrum of A
$\sigma_e(A)$ ...... Essential Spectrum of A
$\sigma_p(A)$ ...... Point Spectrum of A
$\sigma_r(A)$ ...... Residual Spectrum of A
$\sigma_{sc}(A)$ ..... Singular Continuous Spectrum of A
σ_T .... Thomson Cross Section
ς .... sigma
Σ .... Covariance Matrix
Macroscopic Cross Section
Quantum Number of Component of Electronic Spin Along Symmetry Axis
Sigma
Sum
Summation
τ .... Age
Carrier Lifetime
Containment Time Scale
Dead Time
Eigenvalue
Heat Release Parameter
Hour Angle
Interval
Mean Life
Molecular Cross Section
Opacity
Optical Depth
Reacting Time
Relaxation Time
Shear Stress
tau
Thompson Coefficient
Time Constant
Transmission Coefficient
Viscous-Stress Tensor
τ^h .... Transmittance
τ_A .... Light Time per Astronomical Unit
τ_c .... Mean Time Interval between Collisions
τ_{coll} .... Mean Time Interval between Collisions
T .... Tau
υ .... Neutron Speed
upsilon

Υ .... Upsilon
Vernal Equinox
ϕ .... Constraint Mode
Interaction Potential
Latitude
Merit Function
phi
Porosity
φ .... Absorption Profile
Argument of z
Azimuthal Angle
Bank Angle
Composite Mass Fraction
Eigenmode
Electric Potential
Euler Angle
Geographic Latitude
Heat Current Density
Neutron Flux Density
Osmotic Pressure
Parallactic Angle
Particle Flux Density
Phase (of z)
Phase Difference
Phase Displacement
phi
Plane Angle
Planetographic Latitude
Potential
Radiant Flux Density
Scattering Angle
Solution Osmotic Coefficient
Work Function
φ_B .... Volume Fraction of Substance B
φ_{ij} .... Intercation Energy between Molecules i and j
Φ .... Angular Diameter of a Source
Heat Flow Rate
Luminous Flux
Magnetic Flux
Maxwell-Boltzmann Velocity Distribution Function
Phi
Polarization Angle
Potential Energy
Radial Phase (density wave)
Radiant Flux
Work Function
Φ_o .... Magnetic Flux Quantum
χ .... chi
Potential
Susceptibility
χ_c .... Electric Susceptibility
χ_m .... Magnetic Susceptibility
X .... Chi
ψ .... Degeneracy Parameter
Energy Flux Density
Euler Angle
Heading
psi
Radiant Flux Density
Stream Function
Vertex Deviation
Ψ .... Electric Flux
Electron Wave Function
Galactic Gravitational Potential
Psi
ω .... Angular Frequency
Angular Velocity
Argument of Perigee
Argument of Perihelion
Circular Frequency
Longitude of Periastron
omega
Oscillation Frequency
Prandtl Angle
Pulsatance
Rotational Rate
Solid Angle
ω_D .... Debye Circular Frequency
ω_L .... Larmor Circular Frequency
ω_{cc} .... Electron Cyclotron Circular Frequency
ω_{ci} .... Ion Cyclotron Circular Frequency
ω_i .... Quantum Number of Total Electronic Angular Momentum Along Symmetry Axis
ω_{pc} .... Electron Plasma Circular Frequency
Ω .... Density Parameter
Electric Resistance
Longitude of the Ascending Node
Ohm
Omega
Open Set in a Topological Space
Pattern Speed (density wave)
Quantum Number of Total Electronic Angular Momentum Vector Along Symmetry Axis
Solid Angle
Volume in γ Phase Space

Mathematical Signs

alphabetical on the explanations)

$\sigma_{as}(A)$	Absolutely Continuous Spectrum of A
$\lvert a \rvert$	Absolute Value of a
A^*	Adjoint of A
$\aleph$	Aleph
$\dim X$	Algebraic Dimension of X
$\xrightarrow{**}$	Almost Sure Convergence
$\angle$	Angle
$\measuredangle$	Angle
$\sphericalangle$	Angle
U°	Annihilator of Subspace U
$[A, B]_+$	Anticommutator of A and B
$\rightarrow$	Approaches
$\approx$	Approximately Equal to
$\cong$	Approximately Equal to
$\arg z$	Argument of z
$\arg(z)$	Argument of z
$*$	Asterisk Operator
$\simeq$	Asymptotically Equal to
$\asymp$	Asymptotical to
e	Base of Natural Logarithms
$\in$	Belongs to
X''	Bidual Space of X
lb x	Binary Logarithm of x
$\log_2 x$	Binary Logarithm of x
$\binom{n}{p}$	Binomial Coefficient
a_o	Bohr Radius
Ω	Boundary of A
∂A	Boundary of A
$:=$	By Definition
$\stackrel{\text{def}}{=}$	By Definition Equal to
$\stackrel{\Delta}{=}$	By Definition Equal to
$\widehat{=}$	By Definition Equal to
$\cap$	Cap
$\aleph$	Cardinal Number
$A \times B$	Cartesian Product of A and B
pv	Cauchy Principal Value (of an integral)
U_A	Cayley Tranform of A
∇	Characteristic Boundary Layer Length
∇	Characteristic Boundary Thickness
r_o	Classical Electron Radius
$\bar{A}$	Closure of A
lg x	Common Logarithm of x
$\log_{10} x$	Common Logarithm of x
$[A, B]$	Commutator of A and B
$[A, B]_-$	Commutator of A and B
$F \subset\subset E$	Compact Injection of the Set F in E
A^c	Complement of A
$A \backslash B$	Complement of B in A
$\complement_A^B$	Complement of B in A
$\bar{z}$	Complex Conjugate of z
z^*	Complex Conjugate of z
A^*	Complex Conjugate of A
$g \circ f$	Composite Function of f and g
$f \circ g$	Composite Mapping of $f : F \rightarrow G$ and $g : E \rightarrow F^{(2)}$
f^n	Composite Mapping of Order n of $f : E \rightarrow E$
&	Conjunction
$\wedge$	Conjunction
$\supseteq$	Contains
$\ni$	Contains as Element or Member
$\supset$	Contains Properly
$\sigma_c(A)$	Continuous Spectrum of A
$\xrightarrow{d}$	Convergence in Distribution
$\xrightarrow{L^i}$	Convergence in L^i
$\xrightarrow{P}$	Convergence in Probability
$f * g$	Convolution Product of f and g
D_i	Cook's Influence Measure
$\amalg$	Coproduct
r_{xy}	Correlation Coefficient (linear) between x and y
$\Rightarrow$	Corresponds to
$\stackrel{\Delta}{=}$	Corresponds to
cosec x	Cosecant of x
$\cos x$	Cosine of x
$\cot x$	Cotangent of x
ctg x	Cotangent of x
$\cup$	Cup
$\nabla \times \mathbf{A}$	Curl of $\mathbf{A}$
curl $\mathbf{A}$	Curl of $\mathbf{A}$
rot $\mathbf{A}$	Curl of $\mathbf{A}$
$\Box f$	Dalembertian of f
$a_n \downarrow$	Decreasing Suite
$\beta(A)$	Deficiency Index of A
$\dot{f}$	Derivative of f
f'	Derivative of f
$\mathrm{d}f/\mathrm{d}x$	Derivative of f
$\frac{\mathrm{d}f}{\mathrm{d}x}$	Derivative of f
$f^{(n)}$	Derivative of Order n of f
$\det A$	Determinant of A
diam A	Diameter of A
$\backslash$	Difference
dm	Differential Mass
dt	Differential Time
$\partial/\partial\mathbf{r}$	Differential Vector Operator
dv	Differential Velocity
$\dim X$	Dimension of X
$\delta(x)$	Dirac Delta Function
δ_o	Dirac Distribution
$\delta(x)$	Dirac Distribution
$\delta_o(x)$	Dirac Distribution
$\alpha_x\ \alpha_y\ \alpha_z$	Dirac (4×4) Matrices

β .... Dirac (4×4) Matrix
$\sigma_d(A)$ .... Discrete Spectrum of A
$A \oplus B$ .... Direct Sum of Vector Spaces A and B
$\vee$ .... Disjunction
$d(A,B)$ .... Distance between A and B
$\nabla \cdot f$ .... Divergence of f
div f .... Divergence of f
$\neq$ .... Diversity Relation
$\not\approx$ .... Diversity Relation
$/$ .... Divides
$:$ .... Divides
$\div$ .... Division
$\notin$ .... Does not Belong to
$\not\ni$ .... Does not Contain
$\neq$ .... Does not Equal
$\nprec$ .... Does not Precede
$\npreceq$ .... Does not Precede or not Equal to
$\nsucc$ .... Does not Succede
$\nsucceq$ .... Does not Succede or not Equal to
$D(A)$ .... Domain of A
$\dot{x}$ .... Dot (here on x)
$\doteq$ .... Dot Equal
$\ddot{x}$ .... Double Dot (here on x)
X' .... Dual Space of X
$\mathbf{ab}$ .... Dyadic Product of $\mathbf{a}$ and $\mathbf{b}$
$a \circ b$ .... Dyadic Product of a and b
A_{ij} .... Element ij of A
$\emptyset$ .... Empty Set
$\square$ .... End of a Logical Sequence
$\Box$ .... End of a Logical Sequence
$\blacksquare$ .... End of a Logical Sequence
▮ .... End of a Logical Sequence
$\stackrel{\mathrm{def}}{=}$ .... Equal by Definition to
$\stackrel{\Delta}{=}$ .... Equal by Definition to
$\widehat{=}$ .... Equal by Definition to
$:=$ .... Equal by Definition to
$=$ .... Equal to
$\doteq$ .... Equal up to a Certain Order
$\leftrightarrow$ .... Equivalence
$\Leftrightarrow$ .... Equivalence
$=$ .... Equivalence Relation
$\approx$ .... Equivalence Relation
$\sigma_e(A)$ .... Essential Spectrum of A
$\hat{a}$ .... Estimated Value of a
$|x|$ .... Euclidean Norm of x
ψ .... Euler Angle
ϑ .... Euler Angle
φ .... Euler Angle
e .... Euler Number
$\underline{\vee}$ .... Exclusive Disjunction
$\underline{\vee}$ .... Exclusive or
$\exists$ .... Existential Quantifier
$< a >$ .... Expectation Value of a
$\bar{a}$ .... Expectation Value of a
E(a) .... Expectation Value of a
exp x .... Exponential of x
e^x .... Exponential of x
$\bigoplus_{n=1}^{\infty}$ .... Exterior Hilbert Sum of a Sequence H_n of Hilbert Spaces
$n!$ .... Factorial n
Δx^b .... Finite Increase of x
Fp .... Finite Part (of an integral)
$\forall$ .... For All
$\forall$ .... For Every
$\exists$ .... For Some
$:$ .... For which
$|$ .... For which
$\hat{f}$ .... Fourier Transform of f
Df .... Fréchet Derivative of f
$D^n f$ .... Fréchet Derivative of order n of f
∞ .... Freestream
$f(\cdot,t)$ .... Function Defined by $x \mapsto f(x,t)$ for fixed t
$f(x)$ .... Function of x
$\mathrm{f}(x)$ .... Function of x
$\Rightarrow$ .... Gives
∇f .... Gradient of f
grad f .... Gradient of f
$G(A)$ .... Graph of A
$>$ .... Greater than
$\geq$ .... Greater than or Equal to
$\geqq$ .... Greater than or Equal to
$\geqslant$ .... Greater than or Equal to
ent a .... Greatest Integer $\leq a$
$\Rightarrow$ .... Has for an Image
$A^\dagger$ .... Hermitian Conjugate of A
$\equiv$ .... Identically Equal to
I_E .... Identity Mapping in E
I .... Identity Operator
$=$ .... Identity Relation
$\approx$ .... Identity Relation
iff .... If and only if
$\iff$ .... If and only if
Im A .... Image of A
$R(A)$ .... Image of A
Im $f = f(E)$ .... Image of $f: E \to F$ in $F^{(3)}$
$\Im z$ .... Imaginary Part of z
Im z .... Imaginary Part of z
Im(z) .... Imaginary Part of z
z'' .... Imaginary Part of z
$\Rightarrow$ .... Implication
$\rightarrow$ .... Implication
$\Rightarrow$ .... Implies
$\rightarrow$ .... Implies
$\subseteq$ .... Inclusion
$\vee$ .... Inclusive Disjunction
$\vee$ .... Inclusive or
$a_n \uparrow$ .... Increasing Suite
inf A .... Infimum Bound of A
∞ .... Infinity
$\int$ .... Integral
$\mathring{A}$ .... Interior of A
$\cap$ .... Intersection
$G^s(A)$ .... Inverse Graph of A
f^{-1} .... Inverse of f
A^{-1} .... Inverse of A
$\in$ .... Is a Member of
$\in$ .... Is an Element of
$\approx$ .... Is Approximately Equal to
$\leftrightarrow$ .... Is Approximately Equal to
$\subseteq$ .... Is Contained as Subset in
$:=$ .... Is Defined to be
$=$ .... Is Equal to
$=$ .... Is Equivalent to
$\leftrightarrow$ .... Is Equivalent to
$\Leftrightarrow$ .... Is Equivalent to
$\approx$ .... Is Equivalent to
$=$ .... Is Identical with
$\approx$ .... Is Identical with
$\subseteq$ .... Is Included in
$\notin$ .... Is not a Member of
$\notin$ .... Is not an Element of
$\not\approx$ .... Is not Approximately Equal to

$\not\leftrightarrow$ .... Is not Approximately Equal to
$\not\subset$ .... Is not a Proper Subset of
$\neq$ .... Is not Equal to
$\neq$ .... Is not Equivalent to
$\not\approx$ .... Is not Equivalent to
$\not\leftrightarrow$ .... Is not Equivalent to
$\neq$ .... Is not Identical with
$\not\approx$ .... Is not Identical with
$\not\subset$ .... Is not Properly Included in
$\subset$ .... Is Properly Contained as Subset in
$\subset$ .... Is Properly Included in
$\Rightarrow$ .... Is Transformed into
$\vdash$ .... It is Asserted that
J_f .... Jacobian of f
$\cup$ .... Join
$\ker A$ .... Kernel of A
$N(A)$ .... Kernel of A
δ_{ij} .... Kronecker Symbol
δ^i_j .... Kronecker Symbol
$\nabla^2\varphi$ .... Laplacian φ
Δf .... Laplacian of f
$>$ .... Larger than
$\geq$ .... Larger than or Equal to
$\geqslant$ .... Larger than or Equal to
$\geqq$ .... Larger than or Equal to
$<$ .... Less than
$\leq$ .... Less than or Equal to
$\leqslant$ .... Less than or Equal to
$\leqq$ .... Less than or Equal to
$\lim_{x\to a} f(x)$ ...... Limit of $f(x)$
$\lim\limits_{x\to a} f(x)$ .... Limit of $f(x)$
r_{xy} .... Linear Correlation Coefficient between x and y
L .... Linear Integral Operator
$\log_a x$ .... Logarithm to the base a of x
$\cap$ .... Logical Product
$\cup$ .... Logical Sum
$\inf A$ .... Lower Bound of A
$<$ .... Lower than
$\leq$ .... Lower than or Equal to
$\leqslant$ .... Lower than or Equal to
$|a|$ .... Magnitude of a
$f{:}{}^{A\to F}_{x\mapsto f(x)}$ .... Mapping of A into F such that to the elements $x \in A$, there corresponds $f(x) \in F^{(1)}$
$x \in A \to f(x) \in F$ Mapping of A into F such that to the elements $x \in A$, there corresponds $f(x) \in F^{(1)}$
$\imath$ .... Mathematical i
$\jmath$ .... Mathematical j
ℓ .... Mathematical l
a_{ij} .... Matrix Element
(a_{ij}) .... Matrix whose ij-th Element is a_{ij}
$\hat{x}$ .... Maximal Value of x
$\bar{x}$ .... Mean Value of x
$< x >$ .... Mean Value of x
$-$ .... Minus
$\mp$ .... Minus-Plus
$|z|$ .... Modulus of z
$\gg$ .... Much Greater than
$\gg$ .... Much Larger than
$\ll$ .... Much Less than
$\ll$ .... Much Smaller than
$\times$ .... Multiplies
$\bullet$ .... Multiplies

∇ .... Nabla
$\ln x$ .... Natural Logarithm of x
$\log_e x$ .... Natural Logarithm of x
$\neg$ .... Negation
$\sim$ .... Negation
$'$ .... Negation
$|a|$ .... Norm of a
$\| A \|$ .... Norm of A
$\neg$ .... Not
$\sim$ .... Not
$'$ .... Not
$\not\asymp$ .... Not Asymptotical to
$\neq$ .... Not Equal to
$\not\equiv$ .... Not Equivalent to
$\not>$ .... Not Greater than
$\not\geq$ .... Not Greater than or not Equal to
$\not>$ .... Not Larger than
$\not\geq$ .... Not Larger than or not Equal to
$\not<$ .... Not Less than
$\not\leq$ .... Not Less than or not Equal to
$\not\sim$ .... Not of the Order of
$\ncong$ .... Not Similar or not Equal to
$\not\simeq$ .... Not Similar or not Equal to
$\not<$ .... Not Smaller than
$\not\leq$ .... Not Smaller than or not Equal to
$\not\subseteq$ .... Not Subset of or not Equal to
$\not\supset$ .... Not Superset of
$\not\supseteq$ .... Not Superset of or not Equal to
$\alpha(A)$ .... Nullity Index of A
$\emptyset$ .... Null Set
$\sim$ .... Of the Order of
$\leftrightarrow$ .... One-to-One Reciprocal Correspondence
$B(x,\alpha)$ .... Open Ball with Centre x and Radius $\alpha > 0$
Ω .... Open Set in a Topological Space
$\vee$ .... Or
$H^{\perp}$ .... Orthogonal Complement of H in a Pre-Hilbert Space
$\|$ .... Parallel
f_x .... Partial Derivative of f
$\partial_x f$ .... Partial Derivative of f
$\dfrac{\partial f}{\partial x}$ .... Partial Derivative of f
$\partial f/\partial x$ .... Partial Derivative of f
$D_k f$ .... Partial Derivative of f
$\partial_k f$ .... Partial Derivative of f
s .... Path Length
$\sigma_x,\ \sigma_y\ \sigma_z$ .... Pauli Matrices
$\sigma_1,\ \sigma_2\ \sigma_3$ .... Pauli Matrices
$\perp$ .... Perpendicular
$\arg z$ .... Phase of z
$+$ .... Plus
$\pm$ .... Plus-Minus
$\pm$ .... Plus or Minus
$\sigma_p(A)$ .... Point Spectrum of A
a^n .... Power n (or n-th power) of a
$\prec$ .... Precedes
$\preceq$ .... Precedes or Equal to
$\hat{a}$ .... Predicted Value of a
$'$ .... Prime
Π .... Product
$*$ .... Product
$\mathbf{S}\cdot\mathbf{T}$ .... Product of Tensor $\mathbf{S}$ and Vector $\mathbf{A}$
$\subset$ .... Proper Inclusion
$\subset$ .... Proper Subset of
$\propto$ .... Proportional to
$\sim$ .... Proportional to
A/B .... Quotient Space of Vector Spaces A and

rank A .... Rank of A
rk(A) .... Rank of A
$\hbar$ .... Rationalized Planck constant
π .... Ratio of a Circle Circumference to its Diameter
$\Re z$ .... Real Part of z
Re z .... Real Part of z
Re(z) .... Real Part of z
z' .... Real Part of z
$\sigma_r(A)$ .... Residual Spectrum of A
$R(\lambda, A)$ .... Resolvent Operator of A
$R(\lambda)$ .... Resolvent Operator of A
$\rho(A)$ .... Resolvent Set of A
$f\mid_A$ .... Restriction of $f : E \to F$ to the subset A of E
$\sqrt[n]{a}$ .... Root n of a
$\bar{x}$ .... Root-Mean-Square Value of x
R_∞ .... Rydberg Constant
$\bullet$ .... Scalar Product
$\mathbf{S} : \mathbf{T}$ .... Scalar Product of Tensors $\mathbf{S}$ and $\mathbf{T}$
sec x .... Secant of x
$\ddot{x}$ .... Second Derivative (here of x)
$\mathbf{A}$ .... Second-Order Tensor
$\{a_k\}$ .... Sequence a_1, a_2, ..., a_k, ...
(a_k) .... Sequence a_1, a_2, ..., a_k, ...
C .... Set of Complex Numbers
$\mathbb{C}$ .... Set of Complex Numbers
$\{a_1, a_2, ...\}$ .... Set of Elements
$\mathbb{Z}$ .... Set of Integers
$\mathbb{N}$ .... Set of Natural Numbers
$[a, b]$ .... Set of Objets x such that $a \leq x \leq b$
$[a, b)$ .... Set of Objets x such that $a \leq x < b$
$(a, b]$ .... Set of Objets x such that $a < x \leq b$
(a,b) .... Set of Objets x such that $a < x < b$
Q .... Set of Rational Numbers
$\mathbb{Q}$ .... Set of Rational Numbers
Q^+ .... Set of Rational Numbers ≥ 0
Q^{+*} .... Set of Rational Numbers > 0
R .... Set of Real Numbers
$\mathbb{R}$ .... Set of Real Numbers
$A \backslash B$ .... Set-Theoric Difference $A \cap (B^c)$
sgn a .... Signum a
$\cong$ .... Similar or Equal to
$\simeq$ .... Similar or Equal to
sin x .... Sine of x
$\sigma_{sc}(A)$ .... Singular Continuous Spectrum of A
sing supp$_a f$ .... Singular Analytic Support of f
sing supp f .... Singular Support of f
$<$ .... Smaller than
$\leq$ .... Smaller than or Equal to
$r(A)$ .... Spectral Radius of A
$S(A)$ .... Spectrum of A
$\sigma(A)$ .... Spectrum of A
$\sqrt{a}$ .... Square Root of a
$\sqrt{a}$ .... Square Root of a
$a^{1/2}$ .... Square Root of a
s_y .... Standard Deviation of y
x_n .... Subscript n on x
$\subseteq$ .... Subset of or Equal to
$\{x \in A; B\}$ .... Subset of A Consisting of the Elements with Property B
$\succ$ .... Succedes
$\succeq$ .... Succedes or Equal to
$|$ .... Such that
: .... Such that
x^n .... Superscript n on x
$\supset$ .... Superset of
$\supseteq$ .... Superset of or Equal to
Σ .... Sum
Σ .... Summation
$A + B$ .... Sum of Subsets A and B in a Vector Space
supp f .... Support of f
sup A .... Supremum of A
tan x .... Tangent of x
tg x .... Tangent of x
$\rightarrow$ .... Tends to
$f \otimes g$ .... Tensor Product of f and g
$\mathbf{S} \cdot \mathbf{T}$ .... Tensor Product of Tensors $\mathbf{S}$ and $\mathbf{T}$
$A \otimes B$ .... Tensor Product of Vector Spaces A and B
$\exists$ .... There Exists at Least One
$\exists ! v$ .... There Exists Exactly One v such that
$\exists\,|$ .... There Exists Uniquely
df .... Total Differential of f
tr A .... Trace of A
Tr A .... Trace of A
$\tilde{A}$ .... Transpose of A
${}^tA$ .... Transpose of A
M^T .... Transpose of Matrix M
a^T .... Transpose of Vector a
$\cup$ .... Union
$\exists ! v$ .... Uniqueness Quantifier
E .... Unit Matrix
I .... Unit Matrix
$\varepsilon(t)$ .... Unit Step Function
$\forall$ .... Universal Quantifier
sup A .... Upper Bound of A
δ .... Variational Operator
δf .... Variation of f
$\mathbf{a}$ .... Vector $\mathbf{a}$
$\mathbf{a} \wedge \mathbf{b}$ .... Vector Product of $\mathbf{a}$ and $\mathbf{b}$
$\mathbf{a} \times \mathbf{b}$ .... Vector Product of $\mathbf{a}$ and $\mathbf{b}$
V^* .... Vector Space Dual to Vector Space V
$\wp$ .... Weierstraß Function
$\wr$ .... Wreath Product

Other Signs

(alphabetical on the explanations)

@ .... About
$\mathfrak{R}$ .... Absorption (total-to-selective ratio)
á .... Acute Accent (here on a)
& .... Ampersand
& .... And
+ .... And
$\angle$ .... Angle
$\measuredangle$ .... Angle
$\triangleleft$ .... Angle
Å .... Ångström
' .... Apostrophe
$*$ .... Asterisk
@ .... At
\ .... Backslash
$\bar{a}$ .... Bar Accent (here on a)
$\hbar$ .... Barred h
a̠ .... Bar-Under Accent (here under a)
♣ .... Black Club Suit
♠ .... Black Spade Suit
¶ .... Blind P
$a_\circ$ .... Bohr Radius
m_{bol} ...... Bolometric Magnitude
$\perp$ .... Bottom Sign
$\bowtie$ .... Bowtie Sign
$\breve{a}$ .... Breve Accent (here on a)
• .... Bullet
ç .... Cedilla Accent (here under c)
¢ .... Cent Sign
č .... Check Accent (here on c)
✓ .... Check Mark
â .... Circumflex Accent (here on a)
$r_\circ$ .... Classical Electron Radius
♣ .... Clubsuit Sign
: .... Colon
$\mathfrak{N}$ .... Column Density
, ... Comma
^ .... Control Character
© .... Copyright
† .... Dagger
° .... Degree Sign
Ø .... Diameter
◇ .... Diamond
♢ .... Diamondsuit Sign
ä .... Dieresis (here on a)
† .... Diesis
$<|$ .... Dirac Bra Vector
$|>$ .... Dirac Ket Vector
" .... Ditto Mark
$ .... Dollar
/ .... Done (after making an e-mail remark)
$\dot{a}$ .... Dot Accent (here on a)
ạ .... Dot-Under Accent (here under a)
‡ .... Double Dagger
‡ .... Double Obelisk
// .... Double Prime
" .... Double Quote
$\downarrow$ .... Down Arrow
✓ .... Earmark
$\oplus$ .... Earth
□ .... End of a Logical Sequence
▯ .... End of a Logical Sequence
■ .... End of a Logical Sequence
▮ .... End of a Logical Sequence
ß .... Es-zet
+ .... Et
! .... Exclamation Point
D^a .... Feminine Ordinal Indicator (here after D)
♭ .... Flat Sign
' .... Foot
$\frown$ .... Frown
$\Re$ .... Gas Constant
à .... Grave Accent (here on a)
č .... Háček Accent (here on c)
.... Hash Mark
^ .... Hat
♡ .... Heartsuit Sign
? .... Help
$\hookleftarrow$ .... Hooked Left Arrow
$\hookrightarrow$ .... Hooked Right Arrow
... . Horizontal Ellipsis
ő .... Hungarian Umlaut (here on o)
" .... Inch
¡ .... Inverted Exclamation Mark
¿ .... Inverted Question Mark
$\langle$ .... Left Angle Bracket
$\leftarrow$ .... Left Arrow
{ .... Left Brace
[.......... Left Bracket
$\lceil$ .... Left Ceiling Bracket
$\lfloor$ .... Left Floor Bracket
(.......... Left Parenthesis
“ .... Left Double Quote
$\ll$ .... Left Pointing Guillemet
‘ .... Left Quote
$\leftrightarrow$ .... Left Right Arrow
□ .... Logical-Sequence End
▯ .... Logical-Sequence End
■ .... Logical-Sequence End
▮ .... Logical-Sequence End
$\mathfrak{L}$ .... Luminosity Density
$L_\odot$ .... Luminosity of the Sun
$\mathfrak{L}_\odot$ .... Luminosity of the Sun
$\bar{a}$ .... Macron Accent (here on a)
✠ .... Maltese Cross

D^o .... Masculine Ordinal Indicator (here after D)
$\mathfrak{M}$ .... Mass
$M_\oplus$ .... Mass of the Earth
$\mathfrak{M}_\oplus$ .... Mass of the Earth
$M_\odot$ .... Mass of the Sun
$\mathfrak{M}_\odot$ .... Mass of the Sun
' .... Minute of Arc
... . More to Come
♮ .... Natural Sign
.... Number
† .... Obelisk
¡ .... Open Exclamation Mark
¿ .... Open Question Mark
§ .... Paragraph
¶ .... Paragraph
% .... Percentage
. ... Period
m_{pg} .... Photographic Magnitude
m_{pv} .... Photovisual Magnitude
. ... Point
£ .... Pound Sign
′ .... Prime
? .... Question Mark
" .... Quotation Mark
$\mathfrak{L}$ .... Radiant Flux
$R_\oplus$ .... Radius of the Earth
$\mathfrak{R}_\oplus$ .... Radius of the Earth
$R_\odot$ .... Radius of the Sun
$\mathfrak{R}_\odot$ .... Radius of the Sun
$\hbar$ .... Rationalized Planck constant
® .... Registered Mark
‘ .... Reverse Apostrophe
⟩ .... Right Angle Bracket
→ .... Right Arrow
} .... Right Brace
] .... Right Bracket
⌉ .... Right Ceiling Bracket
" .... Right Double Quote
⌋ .... Right Floor Bracket
) .... Right Parenthesis
≫ .... Right Pointing Guillemet
' .... Right Quote
R_∞ .... Rydberg Constant
" .... Second of Arc
§ .... Section
; .... Semi-Colon
ß .... Sharp S
.... Sharp Sign
♯ .... Sharp Sign
/ .... Slash
č .... Slavic Háček Accent (here on c)
⌣ .... Smile
$L_\odot$ .... Solar Luminosity
$\mathfrak{L}_\odot$ .... Solar Luminosity
$M_\odot$ .... Solar Mass
$\mathfrak{M}_\odot$ .... Solar Mass
$R_\odot$ .... Solar Radius
$\mathfrak{R}_\odot$ .... Solar Radius
♠ .... Spadesuit Sign
□ .... Square
□° .... Square Degree
□ ' .... Square Feet
□ " .... Square Inches
□ ' .... Square Minute of Arc
□ " .... Square Second of Arc
˜ .... Squiggle Accent
★ .... Star
⊙ .... Sun
∇ .... Temperature Gradient
$M_\oplus$ .... Terrestrial Mass
$\mathfrak{M}_\oplus$ .... Terrestrial Mass
$R_\oplus$ .... Terrestrial Radius
$\mathfrak{R}_\oplus$ .... Terrestrial Radius
✓ .... Tick-off Sign
o͡o .... Tie-After Accent (here linking oo)
ñ .... Tilde (here on n)
× .... Times
⊤ .... Top Sign
$\mathfrak{R}$ .... Total-to-Selective Absorption Ratio
ö .... Umlaut (here on o)
+ .... Und
_ .... Underscore
↑ .... Up Arrow
⊥ .... Up Tack
m_v .... Visual Magnitude
¥ .... Yen Sign
+ .... &

E-mail Smileys

(alphabetical on the explanations)

=):-)= .... Abraham Lincoln
:-o .... Amazed
(:-& .... Angry
:-|| .... Angry
:-| .... Apathetic
(-: .... Australian
:~-(........... Bawling
:-X .... Big Kiss
(:+) .... Big Nose
!-(............. Black Eye
):-(............ Bull-Headed
}:-(............ Bull-Headed
:-X .... Closed Mouth
%-) .... Confused
%-(............ Confused
%-) .... Cross-Eyed
;(.............. Crying
;-(............. Crying
:'-(............ Crying
:-@ .... Cursing
>:-) .... Devil
]:-) .... Devil
]:-> .... Devil
%*} .... Drunk
:-(............. Frowning
(:-(............ Frowning
:-| .... Grim
<g> .... Grin
:-} .... Grin
:-) .... Happy
[x] .... Hugs to x
(:-$ Ill
:-* .... Kiss
:-x .... Kiss
:-D .... Laughing
(-: .... Left-Handed
=):-)= .... Lincoln (Abraham_)
:-# .... Lips Sealed
:-X .... Mouth Closed
:-O .... Mouth Open
:-{ .... Mustache
:-(............. Not Happy
:-Q .... Not Inhaling!
:-o .... Oh, no!
:-O .... Open Mouth
+-(:-) .... Pope
+<|:-) .... Pope
:-{ .... Pouting
:-? .... Pouting
+-:-) .... Priest
+:-) .... Priest
:-(............. Sad
:-] .... Sarcastic
:-[............. Sarcastic
:-x .... Say Nothing
:-# .... Sealed Lips
(-_-) .... Secret Smile
8-o .... Shocked
:-@ .... Shouting
(-::-) .... Siamese Twins
(-:|:-) .... Siamese Twins
:) .... Smiling
:-) .... Smiling
:-,, ... Smirking
:-] .... Smirking
:-i .... Smoking
:-Q .... Smoking
:-v .... Speaking
[:-] .... Square Head
:-p .... Striking Out Tongue
&-| .... Tearful
:-& .... Tied Tongue
:-p .... Tongue Striking Out
:-& .... Tongue Tied
=):-) .... Uncle Sam
:-(............. Unhappy
<g ... g> ...... Very Wide Grin
? .... Want to Ask a Question
! .... Want to Make a Comment
:-|) .... Wearing a Moustache
<ggg> .... Wide Grin
8-) .... Wide-Eyed
'-) .... Wink
;-) .... Wink
-=:-)\ .... Wizard
:-o .... Wow!
:O .... Yelling
:-O .... Yelling
$-) .... Yuppie